THE PRINCIPAL FUNCTIONAL GROUPS OF ORGANIC CHEMISTRY

	Example	Acceptable Name(s) of Example	Characteristic Reaction Type
Carboxylic acid derivatives			
Acyl halides	CH_3CCl (with =O)	Ethanoyl chloride or acetyl chloride	Nucleophilic acyl substitution
Acid anhydrides	CH_3COCCH_3 (with two =O)	Ethanoic anhydride or acetic anhydride	Nucleoph... ...tion
Esters	$CH_3COCH_2CH_3$ (with =O)	Ethyl ethanoate or ethyl acetate	Nucleophilic acyl substitution
Amides	CH_3CNHCH_3 (with =O)	N-Methylethanamide or N-methylacetamide	Nucleophilic acyl substitution
Nitrogen-containing organic compounds			
Amines	$CH_3CH_2NH_2$	Ethanamine or Ethylamine	Nitrogen acts as a base or as a nucleophile
Nitriles	$CH_3C \equiv N$	Ethanenitrile or acetonitrile	Nucleophilic addition to carbon-nitrogen triple bond
Nitro compounds	$C_6H_5NO_2$	Nitrobenzene	Reduction of nitro group to amine

ORGANIC CHEMISTRY

ORGANIC CHEMISTRY

FRANCIS A. CAREY

Department of Chemistry
University of Virginia

McGRAW-HILL BOOK COMPANY

New York   St. Louis   San Francisco   Auckland   Bogotá
Hamburg   London   Madrid   Mexico   Milan
Montreal   New Delhi   Panama   Paris   São Paulo
Singapore   Sydney   Tokyo   Toronto

ABOUT THE COVER

The cover depicts a new kind of molecular structure, one characterized by a spherical cluster of 60 carbon atoms. This compound, referred to as "buckminsterfullerene," has been described by Professor Richard E. Smalley and his coworkers in the Chemistry Department at Rice University. They suggest that it may be present among the products formed by high-vacuum laser vaporization of graphite. The interior of the molecule is large enough to accommodate other atoms and the + sign represents an atom of lanthanum trapped within the spherical cavity. The colored dots indicate the approximate van der Waals surface of the molecule. Theoretical calculations indicate that buckminsterfullerene and its metal complexes should be quite stable, yet further research is needed to conclusively establish the proposed structure.

In addition to Professor Smalley, I would also like to thank Professor Florante Quiocho and John C. Spurlino of the Biochemistry Department at Rice for permission to reproduce their computer graphics depiction of buckminsterfullerene.

2 3 4 5 6 7 8 9 0 DOWDOW 8 9 4 3 2 1 0 9 8 7

ISBN 0-07-009831-X

This book was set in Serif by Progressive Typographers, Inc. The editors were Karen S. Misler, Randi B. Kashan, and David A. Damstra; the production supervisor was Leroy A. Young; the designer was Rafael Hernandez. The drawings were done by J & R Services, Inc. R. R. Donnelley and Sons Company was printer and binder.

Library of Congress Cataloging-in-Publication Data

Carey, Francis A. (date)
Organic chemistry.

Bibliography: p.
Includes index.
1. Chemistry, Organic. I. Title.
QD251.2.C364 1987 547 86-10374
ISBN 0-07-009831-X

ABOUT
THE AUTHOR

Francis A. Carey is a native of Pennsylvania, educated in the public schools of Philadelphia, at Drexel University (B.S. in Chemistry), and at Penn State (Ph.D.). Following postdoctoral work at Harvard and military service, he was appointed to the chemistry faculty of the University of Virginia. With his students, he has published over forty research papers in synthetic and mechanistic organic chemistry. He is coauthor (with Richard J. Sundberg) of a two-volume advanced-level organic chemistry text.

The text is accompanied by a complete supporting package of instructional materials.

For the Instructor

1. *Test Bank to Accompany Carey: Organic Chemistry* by David G. I. Kingston of Virginia Tech. Contains over 990 multiple choice questions, most of them class-tested.

2. Four-color and two-color overhead transparencies. One hundred twenty-five useful figures from the text.

For the Student

3. *Study Guide to Accompany Carey: Organic Chemistry* by Robert Atkins of James Madison University and Francis A. Carey. Each chapter contains a key terms and concepts section, a list of important reactions, a self-test, and complete solutions to all text problems. Also included is a list of selected readings for students who desire further study.

4. *Experimental Organic Chemistry,* 2d edition, by H. DuPont Durst of the University of Puerto Rico and G. Gokel of the University of Miami. This laboratory manual uses the research, or investigative, approach and has an extensive coverage of qualitative analysis.

This book is dedicated to my family . . .
my parents, Francis W. and Julia Carey;
my children, Andrew, Robert, and William Carey;
and with special affection to my wife, Jilann B. Carey.

CONTENTS

CHAPTER 7
REACTIONS OF ALKENES.
ELECTROPHILIC ADDITION REACTIONS   204

CHAPTER 8
STEREOCHEMISTRY   252

CHAPTER 9
NUCLEOPHILIC SUBSTITUTION REACTIONS   294

CHAPTER 10
ALKYNES   329

CHAPTER 11
CONJUGATION IN ALKADIENES AND ALLYLIC SYSTEMS   369

CHAPTER 12
ARENES AND AROMATICITY   392

CHAPTER 13
REACTIONS OF ARENES: ELECTROPHILIC AROMATIC SUBSTITUTION REACTIONS

CHAPTER 14
SPECTROSCOPY

CHAPTER 15
ORGANOMETALLIC COMPOUNDS   521

CHAPTER 16
ALCOHOLS   556

CHAPTER 17
ETHERS AND EPOXIDES   601

CHAPTER 18
ALDEHYDES AND KETONES. NUCLEOPHILIC
ADDITION TO THE CARBONYL GROUP   639

CHAPTER 19
ENOLS, ENOLATES, AND ENAMINES   694

CHAPTER 27
CARBOHYDRATES   1002

CHAPTER 28
ACETATE-DERIVED NATURAL PRODUCTS   1048

CHAPTER 29
AMINO ACIDS, PEPTIDES, AND PROTEINS. NUCLEIC ACIDS

PREFACE

In one way or another, writing this book has been on my mind ever since I took my first course in organic chemistry as a college student in 1957. I had learned something that I found fascinating and felt the need to tell everyone else about it. Since then it has been my good fortune to continue studying organic chemistry, with the past 20 years of that study being carried out as a member of the faculty at the University of Virginia. During that time, I have studied something else as well—how best to help students learn organic chemistry. This latter study was, and continues to be, an informal one. Its results are not publishable in journals of science education but I believe them to be valid nevertheless; they are called "experience." I have called upon that experience repeatedly in writing this book. In the process of writing I discovered that the flame kindled 30 years ago still burned as brightly as ever. I hope that the enthusiasm I felt in writing this book is evident to the students who use it.

The text is organized according to functional groups because that is the approach which permits most students to grasp the material most readily. Students retain the material best, however, if they understand how organic reactions take place. Thus *reaction mechanisms are stressed, but within a functional group framework.*

Principles of structure are developed in the first three chapters. The first chapter reviews the elements of covalent bonding, applies them to organic molecules, and provides drill work designed to sharpen the student's ability to write correct structural formulas. Alkanes are examined from a structural perspective in Chapter 2 and their conformations analyzed in Chapter 3.

Most organic chemistry teachers agree that the first reaction which a student encounters should be (a) useful, and (b) illustrate features of chemical reactivity that are broadly applicable. The first reaction in this text, the conversion of alcohols to alkyl halides in Chapter 4, satisfies both criteria. Many reactions involve alkyl halides as starting materials and alkyl halides are almost always prepared from alcohols. In studying this reaction, *the student is exposed early to ionic processes such as proton transfer, carbocation formation, and capture of a carbocation by a nucleophile.* The reaction illustrates many principles of nucleophilic substitution, yet is not complicated by competition with elimination.

Whenever possible, which is almost all of the time, *reactions are illustrated with real examples taken from the chemical literature.* In addition to providing some feeling for how the reactions are carried out, this approach allows the student to see a variety of structural types and to practice identifying functional units in complex molecules.

Summary tables help students organize material and are an important part of this text. Most functional group chapters contain several types of summary tables. Typically, they include a table that recalls reactions from previous chapters which lead to the functional group, a table of reactions of the functional group encountered in earlier chapters, and conclude with summaries of those new preparative procedures and reactions introduced in the chapter. All tables of this type are annotated with a brief statement about the reaction, a general equation, and a specific example.

One way in which students can help themselves understand and retain organic chemistry is to do problems regularly, expressing their answers as written structural formulas and equations. *This text includes numerous problems, both within and at the end of each chapter.* Multipart problems within a chapter are always accompanied by a sample solution which describes the thought process that leads to the answer. Answers to all in-chapter problems are included in an appendix. Detailed solutions to all of the problems are given in the Study Guide.

The text is selective, rather than encyclopedic, in its coverage. It includes all of the fundamental processes of organic chemistry but avoids material that, while traditionally appearing in introductory organic books, seems better suited to advanced courses. By excising material of this type, more space is available for those features that are designed to help the student. It also becomes possible to include, for example, *a separate chapter on organometallic chemistry* (Chapter 15)—an important topic often relegated to a subsidiary position in a functional group chapter. A novel feature of this chapter is a section on synthetic planning that describes the principles of retrosynthetic analysis. After its introduction in Chapter 15, *retrosynthetic analysis is used frequently throughout the text.*

The last thing to be mentioned is the first thing that instructors and students will probably notice—the appearance of the text itself. The *liberal use of four-color illustrations* in selected chapters makes the text not only visually arresting but dramatically enhances its effectiveness as a teaching tool. By color-coding various atoms and groups, the text directs the reader's attention toward the important features of chemical structures.

I am pleased to acknowledge the generous assistance of a great many people. Their names are listed on the following page. I am equally pleased to be able to invite you, the student, to begin your adventure in organic chemistry. I hope you find the journey as enjoyable as I have.

Francis A. Carey

ACKNOWLEDGMENTS

This text has benefited immeasurably from the insightful comments offered by a large number of teachers of organic chemistry who reviewed the manuscript at various stages of its development. In particular, I wish to thank Robert C. Atkins of James Madison University, Raymond C. Fort, Jr. of the University of Maine, and Marye Anne Fox of the University of Texas, Austin. By assuming primary responsibility for the Study Guide, Professor Atkins freed me to devote full attention to the text. Professors Fort and Fox reviewed chapters on a regular and continuing basis for a period of more than five years; their advice was always both frank and constructive.

I appreciate the help of all the reviewers and wish to publicly acknowledge their assistance. They are:

R. Gerald Bass (Virginia Commonwealth University)
William H. Bunnelle (University of Missouri, Columbia)
Roy G. Garvey (North Dakota State University)
David M. Howell (Northeastern University)
Bruce B. Jarvis (University of Maryland)
Michael M. King (George Washington University)
George A. Kraus (Iowa State University)
Richard C. Larock (Iowa State University)
Jack E. Leonard (University of Texas, El Paso)
Ronald L. Marhenke (California State University, Fresno)
Lawrence K. Montgomery (Indiana University)
Michael A. Ogliaruso (Virginia Tech)
Bryan W. Roberts (University of Pennsylvania)
Melvyn D. Schiavelli (College of William and Mary)
Allen M. Schoffstall (University of Colorado at Colorado Springs)
Homer A. Smith, Jr. (Millikin University)
Walter Trahanovsky (Iowa State University)
Peter A. Wade (Drexel University)
David F. Wiemer (University of Iowa)
James F. Wolfe (Virginia Tech)

Compliments are also due to the editorial staff at McGraw-Hill, especially Sybil Golden, Randi Kashan, and David Damstra, to the copyeditor Ernestine Daniels, and to my typists Mary Critzer and William A. Critzer for their helpfulness and professionalism.

My thanks to the following for permission to reproduce their materials:

The ^{1}H nuclear magnetic resonance spectra are reproduced with permission from "The Aldrich Library of NMR Spectra," first edition, C. J. Pouchert and J. R. Campbell, the Aldrich Chemical Company, 1975.

Infrared spectra are reproduced with permission from "The Aldrich Library of Infrared Spectra," third edition, C. J. Pouchert, the Aldrich Chemical Company, 1981.

The ^{13}C nuclear magnetic resonance spectra are reproduced with permission from "Carbon-13 NMR Spectra: A Collection of Assigned, Coded, and Indexed Spectra," by LeRoy F. Johnson and William C. Jankowski, Wiley-Interscience, New York, 1972.

Mass spectra are reproduced with permission from "EPA/NIH Mass Spectral Data Base," Supplement I, S. R. Heller and G. W. A. Milne, National Bureau of Standards, 1980.

Figure 14.31 is adapted from R. Isaksson, J. Roschester, J. Sandstrom, and L.-G. Wistrand, Journal of the American Chemical Society, **1985,** *107,* 4074–4075 with permission of the American Chemical Society.

Figure 29.18 is adapted with permission from F. A. Quiocho and W. N. Lipscomb in "Advances in Protein Chemistry," Volume 25, C. B. Anfinsen, Jr., J. T. Edsall, and F. M. Richards, Editors, Academic Press, New York, NY, 1971.

Figure 29.21 is adapted with permission from Richard E. Dickerson in "The Proteins," second edition, Volume II, H. Neurath, Editor, Academic Press, New York, NY, 1964.

INTRODUCTION

$\mathbf{A}$t the root of all science is our own unquenchable curiosity about ourselves and our world. We marvel, as our ancestors did thousands of years ago, at the ability of a firefly to light up a summer evening. The colors and smells of nature bring subtle messages of infinite variety. Blindfolded, we know whether we are in a pine forest or near the seashore. We marvel. And we wonder. How does the firefly produce light? What are the substances that characterize the fragrance of the pine forest? What happens when the green leaves of summer are replaced by the red, orange, and gold of fall?

THE ORIGINS OF ORGANIC CHEMISTRY

As one of the tools that fostered an increased understanding of our world, the science of chemistry—the study of matter and the changes it undergoes—developed slowly until near the end of the eighteenth century. About that time, in connection with his studies of combustion the French nobleman Antoine-Laurent Lavoisier provided the clues that showed how chemical compositions could be determined by identifying and measuring the amounts of water, carbon dioxide, and other materials produced when various substances were burned in air. By the time of Lavoisier's studies, two branches of chemistry were becoming recognized. One branch was concerned with matter obtained from natural or living sources and was called *organic chemistry*. The other branch dealt with substances derived from nonliving matter—minerals and the like. It was called *inorganic chemistry*. Combustion analysis soon established that the compounds derived from natural sources contained carbon, and eventually a new definition of organic chemistry emerged: organic chemistry is the study of carbon compounds. This is the definition which is still used today.

BERZELIUS, WÖHLER, AND VITALISM

As the eighteenth century gave way to the nineteenth, Jöns Jacob Berzelius made his mark as one of the leading scientists of his generation. Berzelius, whose training was in medicine, had wide-ranging interests and made numerous contributions in diverse areas of chemistry. It was he who in 1807 coined the term organic chemistry for the study of compounds derived from natural sources. Berzelius, like almost everyone else at the time, subscribed to the doctrine known as *vitalism*. Vitalism held that living systems possessed a "vital force" which was absent in nonliving systems. A chemical consequence of this school of thought was that compounds derived from natural sources (organic) were fundamentally different from inorganic compounds and that while inorganic compounds could be synthesized in the laboratory, organic compounds could not—at least not from inorganic materials.

In 1823 Friedrich Wöhler, fresh from completing his medical studies in Germany, traveled to Stockholm to study under Berzelius. A year later Wöhler accepted a position teaching chemistry and conducting research in Berlin. He went on to have a distinguished career, spending most of it at the University of Göttingen, but is best remembered for a brief paper he published in 1828. Wöhler noted that when he evaporated an aqueous solution of ammonium cyanate he obtained "colorless, clear crystals often more than an inch long," which were not ammonium cyanate but were instead urea.

$$\text{Ammonium cyanate} \xrightarrow{\text{heat}} \text{Urea}$$

(An inorganic compound)   (An organic compound)

The transformation that occurred was one in which an *inorganic* salt, ammonium cyanate, was converted to urea, a known *organic* substance earlier isolated from urine. Wöhler's experiment is recognized as a scientific milestone, the first step toward overturning the philosophy of vitalism. Although Wöhler's synthesis of an organic compound in the laboratory from inorganic starting materials struck at the foundation of vitalist dogma, vitalism was not displaced overnight. Wöhler made no extravagant claims concerning the relationship of his discovery to vitalist theory, but the die was cast and over the next generation organic chemistry outgrew vitalism.

What particularly seemed to excite Wöhler and his mentor Berzelius about this experiment had very little to do with vitalism. Berzelius was interested in cases in which two clearly different materials had the same elemental composition and had invented the term *isomerism* to define it. The fact that an inorganic compound (ammonium cyanate) and an organic compound (urea) could be isomers of each other was an important discovery in this context.

THE STRUCTURAL THEORY

It is from the concept of isomerism that we can trace the origins of the *structural theory*—the idea that a precise arrangement of atoms uniquely defines a substance. Ammonium cyanate and urea are different compounds because they have different structures. To some degree the structural theory was an idea whose time had come. Three scientists stand out, however, in being credited with independently proposing the elements of the structural theory. They are August Kekulé, Archibald S. Couper, and Alexander M. Butlerov.

It is somehow fitting that August Kekulé's early training at the university in Giessen was as a student of architecture. Kekulé's contribution to chemistry lies in his description of the architecture of molecules. Two themes recur throughout Kekulé's reminiscences of his own work: critical evaluation of experimental information and a gift for visualizing molecules as particular assemblies of atoms. The essential features of Kekulé's theory, developed and presented while he taught at Heidelberg in 1858, were that carbon normally formed four bonds and had the capacity to bond to other carbons so as to form long chains. Isomers were possible because the same elemental composition (say, C_2H_6O) accommodates more than one pattern of atoms and bonds.

Shortly thereafter, but independently of Kekulé, Archibald S. Couper, a Scot working in the laboratory of Charles-Adolphe Wurtz at the École de Medicine in Paris, and Alexander Butlerov, a Russian chemist at the University of Kazan, proposed similar theories.

ELECTRONIC THEORIES OF STRUCTURE AND REACTIVITY

In the late nineteenth and early twentieth centuries, major discoveries about the nature of atoms allowed theories of molecular structure and bonding to be placed on a more secure foundation. Structural ideas progressed from simply identifying atomic connections to attempting to understand the bonding forces. In 1916 Gilbert N. Lewis of the University of California at Berkeley described covalent bonding in organic compounds in terms of shared electron pairs. Linus Pauling at the California Institute of Technology subsequently elaborated a more sophisticated bonding scheme based on Lewis's ideas and a concept called *resonance,* which he borrowed from the quantum mechanical treatments of theoretical physics.

Once chemists had a good appreciation of some fundamental principles of bonding, a logical next step became the understanding of how chemical reactions occurred. Most notable among the early workers in this area were two British organic chemists, Sir Robert Robinson and Sir Christopher Ingold. Both held a number of teaching positions, with Robinson spending most of his career at Oxford while Ingold was at University College, London.

Robinson, who was primarily interested in the chemistry of natural products, had a keen mind and a penetrating grasp of theory. He was able to take the basic elements of Lewis's structural theories and apply them to chemical transformations by suggesting that chemical change can be understood as being synonymous with changes in electron positions. In effect, Robinson analyzed organic reactions by looking at the electrons and understood that atoms moved because they were carried along by the transfer of electrons. Ingold applied the quantitative methods of physical chemistry to the study of organic reactions so as to better understand the sequence of events, the *mechanism,* by which an organic substance is converted to a product under a given set of conditions.

Our current understanding of elementary reaction mechanisms is quite good. Most of the fundamental reactions of organic chemistry have been scrutinized to the degree that we have a relatively clear picture of the intermediates that occur during the passage of starting materials to products. Extension of the principles of mechanism to reactions that occur in living systems, on the other hand, is an area in which a large number of important questions remain to be answered.

THE IMPACT OF ORGANIC CHEMISTRY

Many organic compounds were known to and used by primitive people. Almost every known human society has manufactured and used beverages containing ethyl alcohol and has observed the formation of acetic acid when wine was transformed into vinegar. Early Chinese civilizations (2500 to 3000 B.C.) extensively used natural materials for treating illnesses and prepared a drug known as Ma Huang from herbal extracts. This drug was a stimulant and elevated blood pressure. We now know that it contains ephedrine, an organic compound similar in structure and physiological activity to adrenaline, a hormone secreted by the adrenal gland. Almost all drugs prescribed for the treatment of disease are organic compounds—some are derived from natural sources, many others are the products of synthetic organic chemistry.

As early as 2500 B.C. in India, indigo was used to dye cloth a deep blue. The early Phoenicians discovered that a purple dye of great value, Tyrian purple, could be extracted from a Mediterranean sea snail. The beauty of the color and its scarcity made purple the color of royalty. The availability of dyestuffs underwent an abrupt change in 1856 when William Henry Perkin, an 18-year-old student, accidentally discovered a simple way to prepare a deep purple dye, which he called *mauveine,* from extracts of coal tar. This led to a search for other synthetic dyes and forged a permanent link between industry and chemical research.

The synthetic fiber industry as we know it began in 1928 when E. I. DuPont de Nemours & Company lured Professor Wallace H. Carothers from Harvard University to direct their research department. In a few years Carothers and his associates had produced *nylon,* the first synthetic fiber, and *neoprene,* a rubber substitute. Synthetic fibers and elastomers both represent important contemporary industries, with an economic impact far beyond anything imaginable in the middle 1920s.

CHALLENGES AND OPPORTUNITIES

A major contributor to the growth of organic chemistry during this century has been the accessibility of cheap starting materials. Petroleum and natural gas provide the building blocks for the construction of larger molecules. From petrochemicals comes a dazzling array of materials that enrich our lives: many drugs, plastics, synthetic fibers, films, and elastomers are made from the organic chemicals obtained from petroleum. As we enter an age of inadequate and shrinking supplies, the use to which we put petroleum looms large in determining the kind of society we will have. Alternative sources of energy, especially for transportation, will allow a greater fraction of the limited petroleum available to be converted to petrochemicals instead of being burned in automobile engines. At a more fundamental level, scientists in the chemical industry are trying to devise ways to use carbon dioxide as a carbon source in the production of building block molecules.

Many of the most important processes in the chemical industry are carried out in the presence of *catalysts.* Catalysts increase the rate of a particular chemical reaction but are not consumed during it. In searching for new catalysts, we can learn a great deal from *biochemistry,* the study of the chemical reactions that take place in living organisms. All these fundamental reactions are catalyzed by enzymes. Rate enhancements of several millionfold are common when one compares an enzyme-catalyzed reaction with the same reaction performed in its absence. Many diseases are the result of specific enzyme deficiencies which interfere with normal metabolism. In the final analysis, effective treatment of diseases requires an understanding of biological processes at the molecular level — what the substrate is, what the product is, and the mechanism by which substrate is transformed to product. Enormous advances have been made in understanding biological processes. Because of the complexity of living systems, however, we have only scratched the surface of this fascinating field of study.

You are studying organic chemistry at a time of its greatest impact on our daily lives, at a time when it can be considered a mature science, and at a time when the challenging questions to which this knowledge can be applied were never more important.

CHEMICAL BONDING: THE LEWIS APPROACH

Central to an understanding of the properties and reactions of organic compounds is an appreciation of the fundamental principles of molecular structure. This chapter reviews some concepts originally introduced by G. N. Lewis that provide the most widely used model of chemical bonding in organic compounds. By applying the Lewis approach and some extensions of it, you will learn to recognize bonding patterns that are more stable than others and begin to develop skills in communicating chemical information by way of structural formulas that will be used throughout your study of organic chemistry.

1.1 ATOMS AND ELECTRONS

Before discussing bonding principles, let us first review some fundamental relationships between atoms and electrons. Each element is characterized by a unique *atomic number Z,* which is equal to the number of protons in its nucleus. A neutral atom has the same number of electrons as the number of protons in the nucleus. These electrons spend 90 to 95 percent of their time near the nucleus in regions of space called *orbitals.*

Orbitals are described by specifying their size, shape, and spatial orientation. Those that are spherically symmetric are called *s* orbitals and are identified as $1s$, $2s$, $3s$, etc. according to the volume they enclose (Figure 1.1). An electron in a $1s$ orbital is likely to be found closer to the nucleus, is lower in energy, and is more strongly bound than an electron in a $2s$ orbital.

A hydrogen atom ($Z = 1$) has one electron; a helium atom ($Z = 2$) has two. These electrons occupy a $1s$ orbital in each case. We say that the electron configurations of hydrogen and helium are

$$\text{hydrogen: } 1s^1 \qquad \text{helium: } 1s^2$$

In addition to being negatively charged, electrons possess the property of *spin* and may have a spin quantum number of $+\frac{1}{2}$ or $-\frac{1}{2}$. According to the *Pauli exclusion*

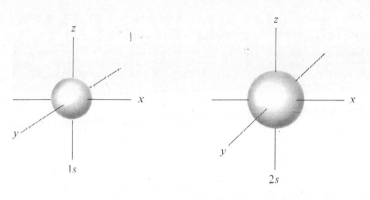

FIGURE 1.1 Representations of the boundary surfaces of a 1s orbital and a 2s orbital. The boundary surfaces enclose the volume where there is a 90 to 95 percent probability of finding the electron.

principle only two electrons may occupy the same orbital, and then only when they have opposite, or "paired" spins. Since its 1s level has a full complement of two electrons, the third electron in lithium ($Z = 3$) must occupy a higher-energy orbital. After 1s, the next higher energy orbital is 2s. Therefore, the electron configuration of lithium is

$$\text{lithium: } 1s^2 2s^1$$

The number that describes the energy level of the orbital (1, 2, 3, etc.) is termed the *principal quantum number*. It also corresponds to the *period* (or *row*) of the periodic table in which an element appears. Hydrogen and helium are first row elements and lithium is a second row element. A complete periodic table of the elements is presented on the inside back cover.

With beryllium ($Z = 4$), the 2s level becomes filled, and the next orbitals to be occupied in the remaining second-row elements are the $2p_x$, $2p_y$, and $2p_z$ orbitals. These orbitals, portrayed in Figure 1.2, are usually described as being "dumbbell-shaped." Each comprises two "lobes," i.e., slightly flattened spheres that touch each other along a surface passing through the nucleus. This flat surface is a *nodal plane* in which the probability of finding an electron is quite small. The $2p_x$, $2p_y$, and $2p_z$ orbitals are equal in energy and mutually perpendicular to each other.

The electron configurations of the first 12 elements, hydrogen through magnesium, are given in Table 1.1. In filling the 2p orbitals, notice that each is singly occupied before any one is doubly occupied. This is a general principle for orbitals of equal energy known as *Hund's rule. Of particular importance in Table 1.1 are hydrogen, carbon, nitrogen, and oxygen.* Countless organic compounds contain nitrogen, oxygen or both in addition to carbon, the essential element of organic chemistry. Most of them also contain hydrogen.

It is often convenient to speak of the *valence electrons* of an atom. These are the outermost electrons, the ones most likely to be involved in chemical reactions. For

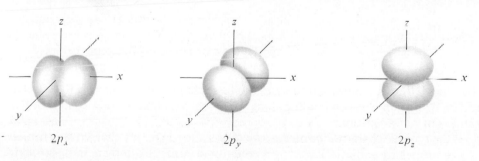

FIGURE 1.2 Representations of the boundary surfaces of the 2p orbitals.

TABLE 1.1

Electron Configurations of First Twelve Elements of the Periodic Table

Element	Atomic number, Z	Number of electrons in indicated orbital					
		1s	2s	$2p_x$	$2p_y$	$2p_z$	3s
Hydrogen	1	1					
Helium	2	2					
Lithium	3	2	1				
Beryllium	4	2	2				
Boron	5	2	2	1			
Carbon	6	2	2	1	1		
Nitrogen	7	2	2	1	1	1	
Oxygen	8	2	2	2	1	1	
Fluorine	9	2	2	2	2	1	
Neon	10	2	2	2	2	2	
Sodium	11	2	2	2	2	2	1
Magnesium	12	2	2	2	2	2	2

second row elements these are the $2s$ and $2p$ electrons. Since a total of four orbitals is involved, the maximum number of electrons in the *valence shell* of any second row element is eight. Neon, with all its $2s$ and $2p$ orbitals doubly occupied, is said to have a complete *octet* of electrons.

Once the $2s$ and $2p$ orbitals are filled, the next level is the $3s$, followed by the $3p_x$, $3p_y$, and $3p_z$ orbitals. Electrons in these orbitals are farther from the nucleus than those in $2s$ and $2p$ orbitals and are of higher energy.

PROBLEM 1.1   Refer to the periodic table as needed and write electron configurations for all the third row elements.

1.2   PROPERTIES OF ATOMS: IONIZATION ENERGY AND ELECTRON AFFINITY

A measure of how strongly an atom binds its electrons is its *ionization energy* (also called *ionization potential*). It is the energy required to remove an electron from an atom (M) in the gas phase.

$$M(g) \longrightarrow M^+(g) + e^-$$

Atom   Positive ion   Electron

The SI (Système International d'Unités) unit of energy is the *joule* (J). An older unit is the *calorie* (cal). This text follows the practice of most organic chemists by expressing energy changes that accompany chemical processes in units of kilocalories per mole (kcal/mol) rather than kilojoules per mole (kJ/mol) (1 kcal/mol = 4.184 kJ/mol).

The general trend in ionization energies can be explained by considering the effect of nuclear charge on electron binding along with what we already know of the relative energies of s and p orbitals.

As the positive charge of the nucleus increases, the negatively charged electrons in a given orbital are subject to a stronger coulombic attraction and are more strongly held. A $1s$ electron of helium ($Z = 2$) requires more energy to dislodge than does a $1s$ electron of hydrogen ($Z = 1$).

$$H(g) \longrightarrow H^+(g) + e^-$$

Hydrogen   Proton: Electron
atom: $1s^1$ $1s^0$

Ionization energy = 313 kcal/mol

Ionization energy increases from left to right, and from bottom to top

$$He(g) \longrightarrow He^+(g) + e^-$$

Helium   Helium   Electron
atom: $1s^2$ ion: $1s^1$

Ionization energy = 567 kcal/mol (largest ionization energy of any element)

The highest ionization energies in each row of the periodic table are those of the rare gases, helium in the first row, neon in the second. The rare gases are said to have a *closed-shell* electron configuration. Helium has no electrons beyond the pair that fill its $1s$ orbital. Neon has no electrons beyond the 10 that fill its $1s$, $2s$, $2p_x$, $2p_y$, and $2p_z$ orbitals.

Electrons in $2s$ orbitals are not as strongly held as those in $1s$ orbitals. Ionization of lithium ($Z = 3$) requires less energy than ionization of hydrogen.

$$Li(g) \longrightarrow Li^+ + e^-$$

Lithium   Lithium   Electron
atom: $1s^2 2s^1$ ion: $1s^2 2s^0$

Ionization energy = 124 kcal/mol

The lowest ionization energies belong to the metals in group I at the left of the periodic table. When an electron is removed from one of these atoms, an ion that has a closed-shell electron configuration identical to that of a rare gas results. Removing the $2s$ electron from lithium, for example, yields a lithium ion having the same electron configuration as a neutral helium atom.

The electron that is lost most readily from any atom is one from its highest-energy orbital. Ionization of carbon, for example, proceeds by loss of one of its $2p$ electrons rather than one of its more tightly held $2s$ electrons.

PROBLEM 1.2   What is the electron configuration of C^+?

Electron affinity is the energy liberated when an electron is added to a neutral atom.

$$X(g) + e^- \longrightarrow X^-(g)$$

Atom   Electron   Negative ion

Atoms that have low-energy unfilled orbitals have a substantial capacity to bind electrons. Fluorine is an example of an atom with a high electron affinity.

$$F(g) + e^- \longrightarrow F^-$$

Fluorine   Electron   Fluoride
atom: ion:
$1s^2 2s^2 2p^5$ $1s^2 2s^2 2p^6$

Electron affinity = 79.5 kcal/mol

Adding an electron to a fluorine atom yields a fluoride ion. Fluoride ion has a closed-shell electron configuration identical to that of neon. Adding an electron to any halogen atom generates a negatively charged ion that has a rare gas electron configuration.

PROBLEM 1.3   Which of the following ions possess a rare gas electron configuration?

(a) Na$^+$

(b) He$^+$

(c) H$^-$

(d) O$^-$

(e) Cl$^-$

(f) Ca^{2+}

SAMPLE SOLUTION   (a) Sodium is atomic number 11, so a neutral sodium atom has the electron configuration $1s^2 2s^2 2p^6 3s^1$. Sodium ion, with a charge of $+1$, will have one less electron, hence an electron configuration of $1s^2 2s^2 2p^6$. The electron configuration of Na$^+$ is the same as that of neon—it has a rare gas electron configuration.

A feeling for the comparative tendencies of atoms to release or attract electrons is quite useful in understanding organic chemistry and provides a starting point from which to explore chemical bonding.

1.3   IONIC BONDING (one loses an e⁻, the other gains)

In order to explain the ability of certain substances to conduct an electric current, the German physicist Walther Kossel proposed in 1916 that a sodium chloride molecule was made up of a positively charged sodium ion and a negatively charged chloride ion held together by the electrostatic attractive force between them. He noted that both Na$^+$ and Cl$^-$ have rare gas electron configurations and suggested that *ionic bonding*, as it is now called, is possible for compounds of the type MX when the atomic number of one of the elements is slighty greater and the atomic number of the other slightly less than that of a rare gas.

As shown in Figure 1.3, loss of a $3s$ electron from sodium gives a sodium ion, which has the same electron configuration as the rare gas neon. Addition of an

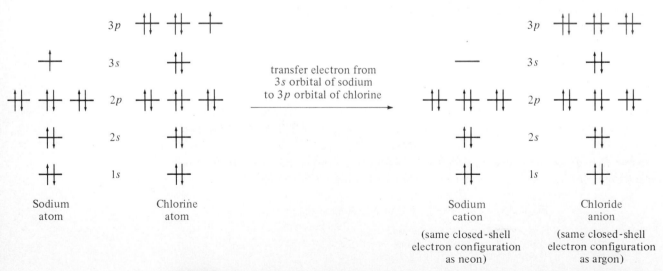

FIGURE 1.3   Transfer of an electron from sodium to chlorine produces a sodium cation (Na$^+$) and a chloride anion (Cl$^-$). Both ions have closed-shell electron configurations corresponding to that of a rare gas.

electron to chlorine fills its $3p$ level and gives a chloride ion having the same electron configuration as the rare gas argon.

The electrostatic attraction between Na^+ and Cl^- is so large that ionically bonded sodium chloride molecules are more stable than isolated sodium atoms and chlorine atoms. The overall process is written as the transfer of an electron from sodium to chlorine.

$$Na(g) \quad + \quad Cl(g) \quad \longrightarrow \quad Na^+Cl^-(g)$$

| Sodium atom | Chlorine atom | Sodium chloride molecule |

Overall, the process is exothermic. An *exothermic reaction* is one in which the products have less energy, i.e., are more stable, than the reactants. An *endothermic reaction* is just the opposite, i.e., the products have more energy, i.e., are less stable, than the reactants.

PROBLEM 1.4   Is the ionization of a sodium atom an exothermic reaction or an endothermic one? What about the capture of an electron by a chlorine atom?

It is the electrostatic attraction between oppositely charged ions that is meant by ionic bonding—not simply the fact that the ions have rare gas electron configurations. Only when they have rare gas electron configurations, however, are the ions accessible enough to permit the overall process to be exothermic.

Ionic bonding is very common in inorganic chemistry but very rare in organic chemistry. The ionization energy of carbon is too high and its electron affinity is too low for it to serve as either the positive or negative ion partner in an ionic bond. Instead, carbon compounds are characterized by a different kind of bonding, called *covalent bonding.* Covalent bonding, introduced in the following section, constitutes most of the remaining subject matter of this chapter.

1.4   COVALENT BONDING. THE SHARED ELECTRON PAIR BOND

An alternative to ionic bonding that is quite helpful in understanding the structure and properties of organic compounds is based on the *shared electron pair,* or *covalent,* bond. To illustrate the principles of covalent bonding consider first a simple example, the hydrogen molecule H_2. Here, where both atoms bonded together are the same, there is no real possibility of ionic bonding. Why should one hydrogen atom transfer an electron to another hydrogen atom to give H^+H^-? Gilbert N. Lewis of the University of California suggested in 1916 that, instead of an ionic bond, the two hydrogens were linked by virtue of sharing a pair of electrons between them.

$$\boxed{H \cdot \qquad \cdot H} \qquad\qquad H : H$$

| Two hydrogen atoms, each with a single electron | Hydrogen molecule: covalent bonding by way of a shared electron pair |

A significant contributor to the covalent bond is the increased binding force exerted on the electrons in a hydrogen molecule, where each one interacts with two protons. In a hydrogen atom an electron is associated with only one nucleus. Thus each electron is more strongly held in a hydrogen molecule than in separated hydrogen atoms. The ionization energy of H_2 is 355 kcal/mol; that of a hydrogen atom is 313 kcal/mol.

Lewis suggested a similar sharing of electrons in other diatomic molecules, as for example, in fluorine F_2.

$$:\ddot{F}\cdot \qquad \cdot\ddot{F}: \qquad\qquad :\ddot{F}:\ddot{F}:$$

| Two fluorine atoms, each with seven electrons in their valence shells | Fluorine molecule: covalent bonding by way of a shared electron pair |

At the root of the Lewis scheme is the idea that a rare gas electron configuration represents a particularly stable arrangement. A rare gas configuration might be attained, as we have seen, by the complete transfer of an electron from a sodium to a chlorine atom, but it can also be achieved by electron sharing. In H_2, the two electrons are associated with both hydrogens and give each one an electron configuration analogous to that of helium. Similarly, each fluorine in F_2 is considered to have a completed octet of electrons in its valence shell and thus has an electron configuration analogous to that of neon.

An analogous electron-sharing scheme may be applied to molecules such as hydrogen fluoride HF, in which a covalent bond connects two different atoms.

$$H\cdot \qquad \cdot\ddot{F}: \qquad\qquad H:\ddot{F}:$$

| Hydrogen atom with its single electron and a fluorine atom with its seven valence electrons | Hydrogen fluoride molecule: covalent bonding by sharing of an electron pair |

Sharing of an electron pair permits hydrogen to have its full complement of two electrons and fluorine to have eight. An important feature of the Lewis approach is that second row elements such as fluorine are limited to a total of eight electrons (shared plus unshared) in their valence shells; hydrogen is limited to two.

Lewis structures of organic substances are constructed on the basis of the two principles of shared electron pairs and attainment of rare gas electron configurations. Covalent bonding in methane and in carbon tetrafluoride, for example, follows directly from the Lewis rules.

— carbon has its 8 e-
— each Hydrogen has its 2

Combine $\cdot\overset{\cdot}{\underset{\cdot}{C}}\cdot$ and four $H\cdot$ — to write a Lewis structure for methane

$$\begin{array}{c} H \\ H:\overset{..}{\underset{..}{C}}:H \\ H \end{array}$$

Combine $\cdot\overset{\cdot}{\underset{\cdot}{C}}\cdot$ and four $\cdot\ddot{F}:$ — to write a Lewis structure for carbon tetrafluoride

$$\begin{array}{c} :\ddot{F}: \\ :\ddot{F}:\overset{..}{\underset{..}{C}}:\ddot{F}: \\ :\ddot{F}: \end{array}$$

In each of these organic molecules, carbon has eight electrons in its valence shell. By forming covalent bonds to four hydrogens, carbon has achieved a rare gas electron configuration.

PROBLEM 1.5   Write a satisfactory Lewis structure, showing all valence electrons, for each of the following:

(a) H_2O

(b) NH_3

(c) NF_3

(d) PCl_3

(e) CH_3Cl

(f) C_2H_6

SAMPLE SOLUTION   (a) Oxygen is in group VI of the periodic table and contributes six valence electrons. Each hydrogen contributes one.

Combine $\cdot \ddot{O} \cdot$ and two $H \cdot$ to write a Lewis structure for water $H : \ddot{O} : H$

In the Lewis structure shown, each hydrogen is associated with two electrons and oxygen has eight. The structure represented by $H : H : \ddot{O}$ is incorrect because its central hydrogen has too many electrons in its valence shell (four).

It is customary to represent a two-electron covalent bond by a dash (—). Thus, the Lewis structures for hydrogen fluoride, fluorine, methane, and carbon tetrafluoride become:

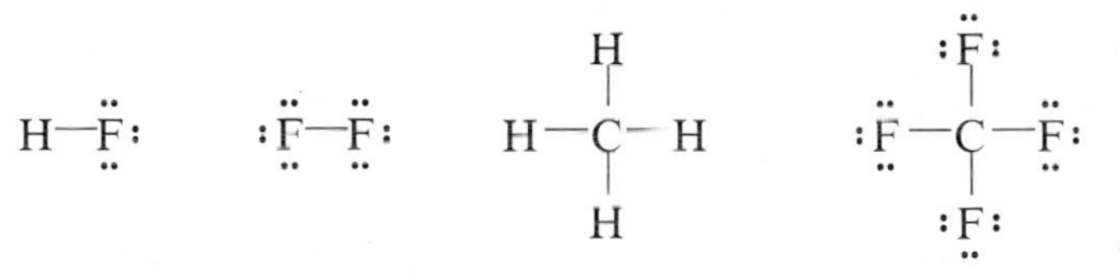

H—$\ddot{F}$:	:$\ddot{F}$—$\ddot{F}$:	H—C—H with H above and below	:$\ddot{F}$—C—$\ddot{F}$: with :$\ddot{F}$: above and below
Hydrogen fluoride	Fluorine	Methane	Carbon tetrafluoride

1.5 POLAR BONDS

Electrons in covalent bonds are not necessarily shared equally by the two atoms that they join together. If one atom has a greater tendency to attract electrons toward itself than the other, the electron distribution in the bond is said to be *polarized*, and the bond is referred to as a *polar* bond. Hydrogen fluoride, for example, has a polar covalent bond. Fluorine attracts electrons more strongly than does hydrogen. The center of negative charge in the molecule is closer to fluorine while the center of positive charge is closer to hydrogen. This polarization of electron density in the hydrogen-fluorine bond is represented in various ways.

$^{\delta+}H—F^{\delta-}$

(The symbols $\delta+$ and $\delta-$ indicate partial positive and partial negative charge, respectively)

$\overset{+\longrightarrow}{H—F}$

(The symbol $+\longrightarrow$ represents the direction of polarization of electrons in the H—F bond)

The tendency of an atom to draw the electrons in a covalent bond toward itself is referred to as its *electronegativity*. An *electronegative* element attracts electrons, an *electropositive* one donates them. In general, electronegative elements have a high electron affinity; electropositive elements have a low ionization potential. Electronegativity increases across a row in the periodic table. The most electronegative of the second row elements is fluorine; the most electropositive one is lithium. Electronegativity decreases in going down a column. Fluorine is more electronegative than chlorine. The most commonly cited electronegativity scale was devised by Linus Pauling and is presented in Table 1.2.

When centers of positive and negative charge are separated from each other, they comprise a *dipole*. The *dipole moment* (μ) of a molecule is the product of the charge (e) (either the positive charge or the negative charge since they must be equal) and the distance between them.

$$\mu = e \cdot d$$

Since the charge on an electron is 4.80×10^{-10} electrostatic units (esu) and distances within a molecule typically fall in the 10^{-8} cm range, molecular dipole moments are on the order of 10^{-18} esu·cm. The *debye* unit D (named after the Dutch chemist, Peter J. W. Debye, who pioneered the systematic study of polar molecules) is defined as equal to 1.0×10^{-18} esu·cm, a notational device that simplifies the reporting of molecular dipole moments. Thus, the experimentally determined dipole moment of hydrogen fluoride, 1.7×10^{-18} esu·cm, is stated as 1.7 D.

PROBLEM 1.6   The compounds FCl and ICl have dipole moments μ that are similar in magnitude (0.9 and 0.7 D, respectively) but opposite in direction. In one compound, chlorine is the positive end of the dipole; in the other it is the negative end. Specify the direction of the dipole moment in each compound and explain the reasoning behind your choice.

Diatomic molecules in which both atoms are the same, such as H_2, N_2, and Cl_2, have dipole moments equal to zero because the electrons are equally shared by both atoms.

The polar properties of polyatomic molecules can be considered as the aggregate of contributions from all the individual bonds. An empirically determined summary of dipole moments associated with various bond types is given in Table 1.3. In particular, notice that the polarity of a carbon-hydrogen bond is relatively low; it is substan-

TABLE 1.2

Selected Values from the Pauling Electronegativity Scale

Period	I	II	III	IV	V	VI	VII
				Group number			
1	H 2.1						
2	Li 1.0	Be 1.5	B 2.0	C 2.5	N 3.0	O 3.5	F 4.0
3	Na 0.9	Mg 1.2	Al 1.5	Si 1.8	P 2.1	S 2.5	Cl 3.0
4	K 0.8	Ca 1.0					Br 2.8
5							I 2.5

TABLE 1.3
Selected Bond Dipole Moments

Bond*	Dipole moment, D	Bond*	Dipole moment, D
H—F	1.7	C—F	1.4
H—Cl	1.1	C—Cl	1.5
H—Br	0.8	C—Br	1.4
H—I	0.4	C—I	1.2
H—C	0.4		
H—N	1.3	C—N	0.2
H—O	1.5	C—O	0.7

* The direction of the dipole moment is toward the more electronegative atom. In the above examples hydrogen and carbon are the positive ends of the dipoles. Carbon is the negative end of the dipole associated with the C—H bond.

tially less than that of carbon-oxygen and carbon-halogen bonds. As will be seen in later chapters, the kinds of reactions that an organic substance undergoes can often be related to the polar nature of the bonds it contains.

1.6 FORMAL CHARGE

Lewis structures frequently contain atoms that bear a unit positive or negative charge. If the molecule as a whole is neutral, the sum of its positive charges must equal the sum of its negative charges. An example is ammonium chloride. Ammonium chloride, like sodium chloride, is an ionic compound. The positive ion (the *cation*) is NH_4^+. The negative ion (the *anion*) is Cl^-.

$$Na^+[:\ddot{C}l:]^- \qquad [NH_4]^+[:\ddot{C}l:]^-$$
Sodium chloride   Ammonium chloride

Bonding in ammonium ion is represented by a Lewis structure in which nitrogen has an octet of electrons. Since nitrogen shares those eight electrons with four hydrogens, it can be said to have an *electron count* of four.

$$\left[\begin{array}{c} H \\ H:N:H \\ H \end{array}\right]^+$$

Ammonium ion

Electron count of nitrogen = $\frac{1}{2}(8) = 4$
Electron count of each hydrogen = $\frac{1}{2}(2) = 1$

A neutral nitrogen has five electrons in its valence shell. Nitrogen in ammonium ion has an electron count equal to four electrons and so lacks one of the electrons of a neutral nitrogen. Nitrogen in ammonium ion has a *formal charge* of +1.

Each hydrogen in ammonium ion shares two electrons with nitrogen and so has an electron count of one. Since this is exactly equal to the number of electrons of a neutral hydrogen atom, none of the hydrogen substituents in ammonium ion has a formal charge.

The total charge in ammonium ion is equal to the sum of the formal charges of all of its atoms. It is +1.

These are called formal charges because they are based on Lewis structures in which electrons are considered to be shared equally between covalently bonded atoms. In actuality, polarization of the nitrogen-hydrogen bonds leads to some transfer of positive charge from nitrogen to each of its hydrogen substituents. Charge dispersal of this kind cannot be represented in a Lewis structure.

The formal charge of chloride ion, which in this case is the same as the actual charge, is calculated in the same way. Chloride ion shares none of its electrons; it has an electron count of eight.

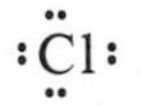

$$:\ddot{Cl}: \qquad \text{Electron count} = 8$$

A neutral chlorine atom has seven electrons. Thus, chloride ion has one electron in excess of that of a neutral chlorine. It has a charge of -1.

Look now at the species shown below. Is it neutral? Is it an anion? Is it a cation? Do any of its atoms bear formal charges? To answer these questions, determine first the formal charge on each atom. Begin by counting the electrons "owned" by each atom.

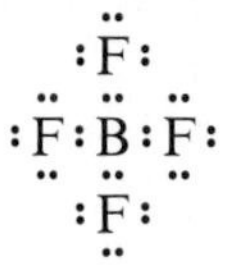

$$:\ddot{F}:$$
$$:\ddot{F}:B:\ddot{F}:$$
$$:\ddot{F}:$$

Electron count of boron $= \frac{1}{2}(8) = 4$
Electron count of each fluorine $= 6 + \frac{1}{2}(2) = 7$

unshared electrons   electrons in covalent bonds

A neutral boron atom has three valence electrons. Therefore, boron with an electron count of four has a formal charge of -1. Neutral fluorine atoms have seven valence electrons. Therefore, each fluorine substituent in this species is electrically neutral. The net charge is the sum of all the formal charges, in this case -1. This species is an anion; it occurs in salts such as $Na^+[BF_4]^-$.

PROBLEM 1.7 Determine the formal charge at all the atoms in each of the following species and the net charge on the species as a whole.

(a) H—Ö—H
        |
        H

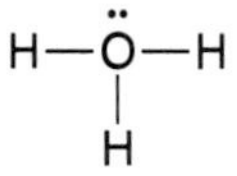

(b) H—Ċ—H
        |
        H

(c) H—Ċ—H
        |
        H

(d) H—C—H
        |
        H

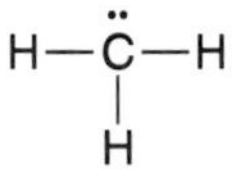

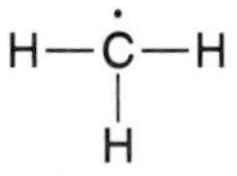

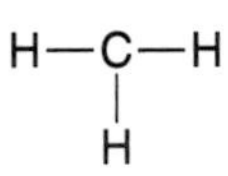

(e) H—Ö—H

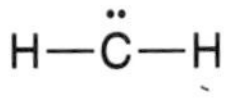

SAMPLE SOLUTION (a) Each hydrogen has a formal charge of zero, as is always the case when hydrogen is covalently bonded to one substituent by a covalent bond. Oxygen has an electron count of five.

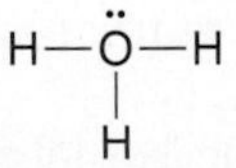

H—Ö—H   Electron count of oxygen $= 2 + \frac{1}{2}(6) = 5$
      |
      H

unshared pair   covalently bonded electrons

A neutral oxygen atom has six valence electrons; therefore, oxygen in this species has a formal charge of $+1$. The species as a whole has a unit positive charge. It is the hydronium ion H_3O^+.

Counting electrons for the purpose of computing the formal charge of an atom differs from counting electrons to see if the octet rule is satisfied. A second row element has a filled valence shell if the sum of all the electrons, shared and unshared, is eight. Electrons that connect two atoms by a covalent bond count toward filling the valence shell of both. When calculating the formal charge, however, only one-half the number of electrons in covalent bonds can be considered to be "owned" by an atom. Determination of formal charges on individual atoms of Lewis structures is an important element in good "electron bookkeeping." So much of organic chemistry can be made more understandable by keeping track of electrons that it is worth taking some time at the outset to develop a reasonable facility in the seemingly simple exercise of counting electrons.

1.7 MULTIPLE BONDING IN LEWIS STRUCTURES

A logical extension of Lewis's concept of the shared-electron-pair bond allows for four-electron *double bonds* and six-electron *triple bonds*.

$$A::B \text{ (or } A{=}B) \qquad X:::Y \text{ (or } X{\equiv}Y)$$

Four-electron
double bond

Six-electron
triple bond

Some relatively simple substances, such as molecular nitrogen, are best represented by Lewis structures involving multiple bonds. If we take two nitrogen atoms, each with five valence electrons, and connect them by a single two-electron covalent bond

Combine $:\!\overset{\cdot}{N}\!\cdot$ and $\cdot\overset{\cdot}{N}\!:$ to give $:\!\overset{\cdot}{N}\!-\!\overset{\cdot}{N}\!:$

we find that each nitrogen is associated with six electrons. Using two of the remaining four electrons on each nitrogen, two more covalent bonds can be drawn, giving a structure in which each nitrogen is associated with eight electrons.

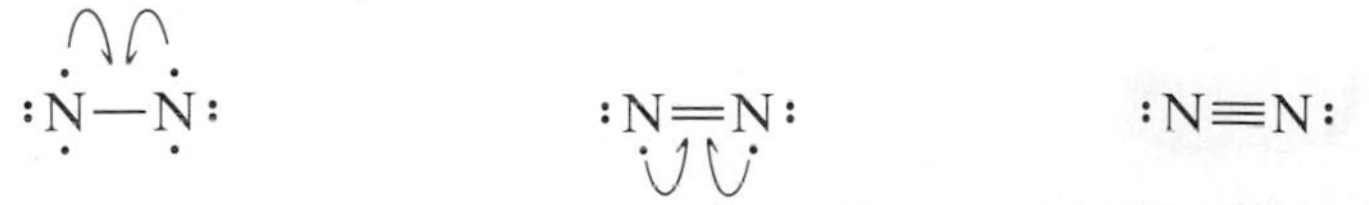

$:\!\overset{\cdot}{N}\!-\!\overset{\cdot}{N}\!:$ $:N{=}N:$ $:N{\equiv}N:$

Pair two electrons
to give double
bond

Pair two elec-
,trons to give
triple bond

Preferred (most
stable) Lewis
structure for
nitrogen

A helpful formalism that shows electron reorganization involves the use of arrows. A curved "fishhook" ($\curvearrowright$) as shown above indicates that one electron is being moved from a particular site. A curved normal arrow ($\curvearrowright$) is used to show movement of electron pairs from one site to another. Both arrows originate at the electrons and point toward the site to which they are being moved. In the above example, single electrons on adjacent nitrogens are being moved to a region between them.

In general, the most stable Lewis structure for a substance is the one that has the greatest number of covalent bonds, provided that the octet rule is not violated.

Multiple bonding is very common in organic chemistry. A singly bonded structure for ethylene (C_2H_4) has only seven electrons surrounding each carbon. Pairing the unshared electrons completes the octets of both carbons and gives a more stable Lewis structure.

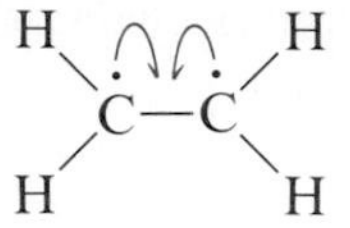

Pair two electrons
to give a double
bond

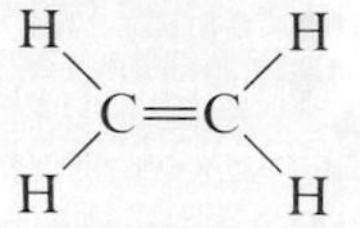

Best Lewis structure
for ethylene

A triple bond joins the carbons of acetylene (C_2H_2).

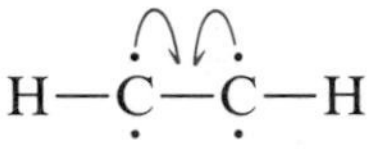

Pair two elec-
trons to give
a double bond

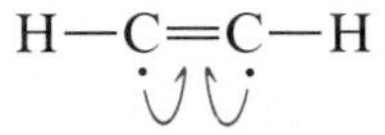

Pair remaining
two to give
a triple bond

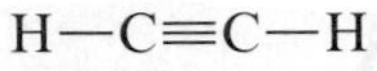

Most stable
Lewis structure
for acetylene

PROBLEM 1.8   Write the most stable Lewis structure for each of the following compounds.

(a) HCN (hydrogen is bonded to carbon)
(b) CH_2O (both hydrogens are bonded to carbon)
(c) CO_2 (only carbon-oxygen bonds are present)
(d) C_3H_4 (each carbon is bonded to the other two forming a three-membered ring)
(e) C_3HN (the bond connections are in the order NCCCH)

SAMPLE SOLUTION   (a)   Combining the atoms with the appropriate number of valence electrons (hydrogen one, carbon four, and nitrogen five) gives the singly bonded Lewis structure shown. Pairing of the remaining two electrons of carbon with two electrons of nitrogen gives a Lewis structure that has a carbon-nitrogen triple bond.

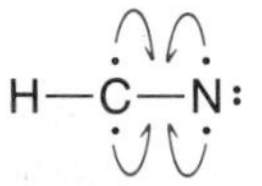

H—C—N:

Pairing of electrons as
indicated by the arrows
gives

H—C≡N:

The structure that contains the triple bond is the most stable Lewis structure for hydrogen cyanide; both carbon and nitrogen have complete octets of electrons.

Formal charges in Lewis structures containing double and triple bonds are calculated in exactly the same way as was described in Section 1.6. A double bond contributes two electrons, and a triple bond three toward the electron count of an atom.

1.8 RESONANCE

Two or more Lewis structures differing only in the arrangement of their electrons can often be written for a substance, especially when multiple bonds are present. Formaldehyde, for example, can be written as:

A B C

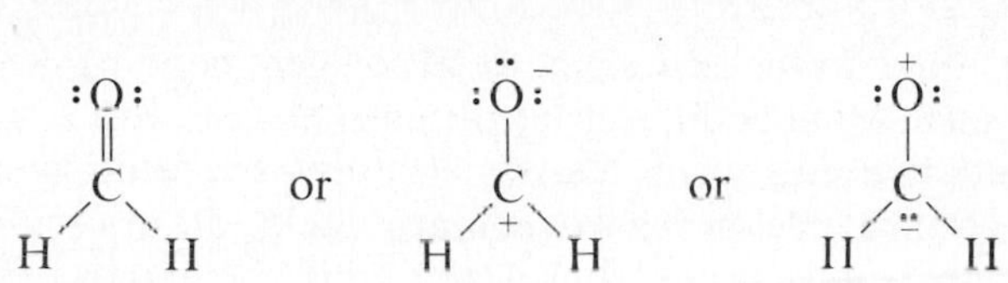

Structure A containing a carbon-oxygen double bond is far more stable than the dipolar structures B and C for several reasons. Structure A has one more covalent bond than either B or C, and carbon and oxygen both have octets of electrons in A, but not in B or C. Carbon is associated with six electrons in B, while oxygen has only six in C. Another factor that favors A is that B and C both have a separation of positive and negative charges. Charge separation is a destabilizing factor and raises the energy of Lewis structures in which it is present. The least stable Lewis structure is C. In C the more electronegative atom (oxygen) is positively charged, while the more electropositive one (carbon) bears the negative charge. This illustrates another principle to apply in differentiating between Lewis structures for a molecule. Whenever possible, negative charge should be on the most electronegative atom and positive charge on the most electropositive one.

PROBLEM 1.9 Keeping the same atomic connections and moving only electrons, write a more stable Lewis structure for each of the following. Be sure to specify formal charges, if any, in the new structure.

(*a*)

(*d*)

(*b*)

(*e*)

(*c*)

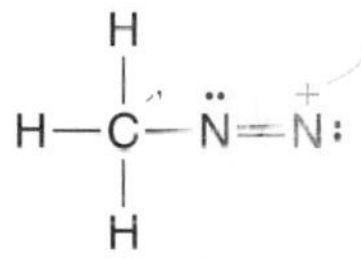

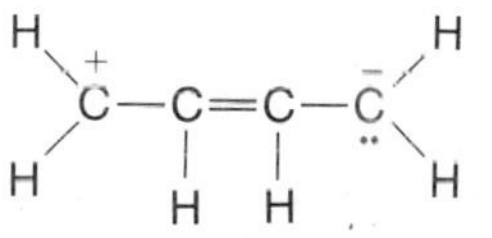

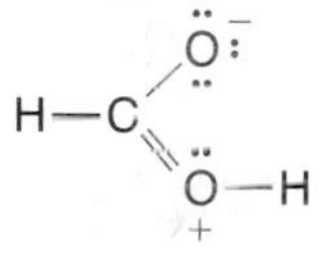

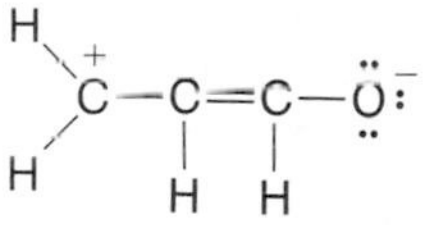

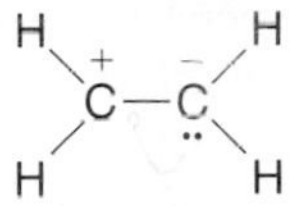

SAMPLE SOLUTION (*a*) The terminal nitrogen has only six electrons, therefore use the unshared pair of the adjacent nitrogen to form another covalent bond.

By moving both electrons of the nitrogen lone pair as shown by the arrow a structure that has octets about both nitrogen atoms is obtained

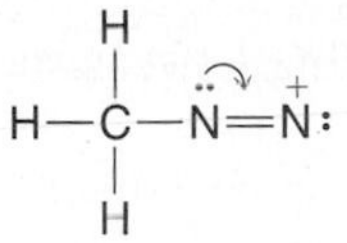

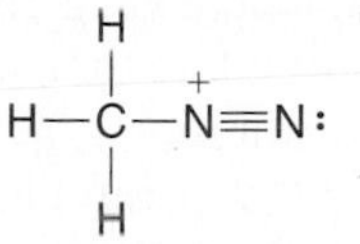

In general, move electrons from sites of high electron density toward sites of low electron density. Notice that the location of formal charge has changed, but the net charge on the species remains the same.

Writing alternative Lewis structures for the same molecule or ion has a much deeper significance than might be at first appreciated. Probably the underlying theme for all theories of chemical bonding is that of *electron delocalization*. We have already seen an example of that in the Lewis picture of covalent bonding. Rather than each electron in a covalent bond being restricted to association with a single nucleus, a more stable assembly results when the two electrons are delocalized between two nuclei. When electrons are delocalized over many nuclei, the molecule is much more stable than it would be if these electrons were localized in bonds between only two carbons. Unfortunately, individual Lewis structures do not lend themselves well to showing how electrons are distributed among various atoms in a molecule. One way to show electron delocalization is embodied in the concept of *resonance* between two or more Lewis structures.

According to the resonance theory, if a molecule can be represented by more than one Lewis structure, it is not accurately described by any single Lewis structure alone. Rather, the true structure is considered to be a hybrid of all the possible structures, possessing some of the characteristics of electron distribution of each. Returning to the case of formaldehyde, we represent this by

$$
\underset{A}{\overset{:O:}{\underset{H \quad H}{\overset{\|}{C}}}} \longleftrightarrow \underset{B}{\overset{:\ddot{O}:^-}{\underset{H \quad H}{\overset{|}{\overset{+}{C}}}}} \longleftrightarrow \underset{C}{\overset{:\ddot{O}:^+}{\underset{H \quad H}{\overset{|}{\overset{\cdot\cdot}{C}}}}}
$$

where the double-headed arrow means that we are dealing with an electron-delocalized species represented as a hybrid or composite of electron-localized Lewis structures. The most stable Lewis structure A is a better approximation of the true structure than are either of the dipolar forms B or C. The carbon-oxygen bond of formaldehyde, however, is polarized, so that there is positive character associated with carbon and negative character on oxygen. Structure B is a more important contributor than is the less stable dipolar structure C.

It is important to realize that the double-headed arrow does not represent a dynamic process; the contributing resonance forms are *not* in equilibrium with each other. The double-headed arrow means that the electron distribution in the real molecule bears a similarity to the electron distribution represented by the aggregate of all the contributing structures but is identical to no single localized structure. A number of analogies have been drawn to clarify the role of contributing Lewis structures. One of these compares the actual molecular structure to a mule. A mule is the hybrid offspring of a horse and a donkey, resembling both but being identical to neither. A more recent analogy suggests that the situation is more like the relationship between a rhinoceros, a unicorn, and a dragon. A rhinoceros is a real animal which has some characteristics of two imaginary animals, the unicorn and the dragon. A real molecule has a distinct distribution of its electrons, which is, in varying degrees, something like that seen in all the contributing Lewis structures but is exactly like none of them.

To cite an example of the way in which resonance theory is applied to structure in organic chemistry, compare the carbon-oxygen bond distances in formic acid and in formate anion.

Formic acid   Formate anion

1.22 Å   :Ö:
H—C
1.36 Å   ÖH

          :Ö:
H—C
          :Ö:⁻

each C—O bond length is 1.27 Å

[Organic chemists usually cite bond distances in angstrom units (Å). The SI unit of length is the meter; 1 Å = 10^{-10} m.] The two oxygen substituents in formic acid are very different. One is doubly bonded to carbon at a distance of 1.22 Å; the other is singly bonded to carbon at a distance of 1.36 Å. These are typical bond distances for carbon-oxygen double bonds and carbon-oxygen single bonds, respectively. In formate anion, however, the two carbon-oxygen bond lengths are exactly the same (1.27 Å), shorter than a C—O single bond but longer than a C=O double bond.

The Lewis structure for formate anion is not consistent with these bond length data. This Lewis structure implies that formate anion should have one short and one long carbon-oxygen bond. The data are correct; it is our reliance on a single Lewis structure for formate anion which is flawed.

Two identical Lewis structures for formate anion may be written. It is convenient to use curved arrows to show the electron reorganization by which one Lewis structure may be transformed to another.

The real structure of formate anion is a resonance hybrid of these two with minor contributions from other, less stable Lewis structures. Each carbon-oxygen bond is intermediate between a single and a double bond and each oxygen bears one-half of a unit negative charge. The electron distribution in a delocalized structure such as formate anion is sometimes represented by means of a modified Lewis structure in which dotted lines represent *partial bonds,* as shown.

In most cases, however, it is easier to keep track of electrons by working with one reasonable Lewis structure while remembering that it only approximates the true distribution of electrons in the molecule.

PROBLEM 1.10   Each of the following anions may be represented by at least one other Lewis structure that is equivalent to the one drawn. Use curved arrows to derive an alternative structure of each anion.

SAMPLE SOLUTION   (a)   When using curved arrows to represent the reorganization of electrons, begin at a site of high electron density, preferably an atom that is negatively charged. Move electrons continuously until a proper Lewis structure results. For nitrate ion, this can be accomplished in two ways.

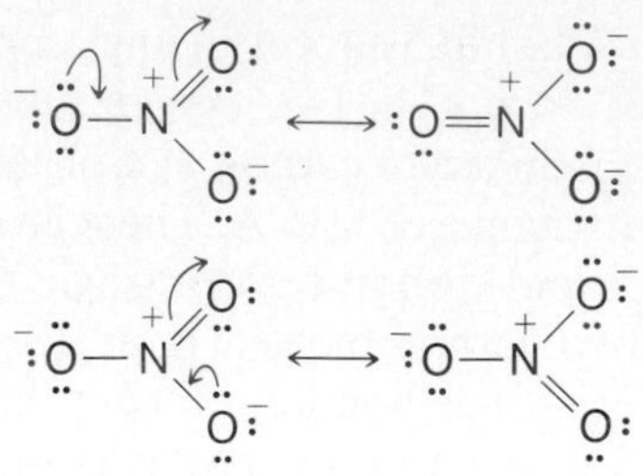

Three equivalent Lewis structures are possible for nitrate ion. The negative charge in nitrate is shared equally by all three oxygens.

In writing alternative Lewis structures for a molecule the net charge must be the same in each resonance form. Additionally, the number of unpaired electrons must be the same in each structure. Thus the resonance formulation shown for ethylene is a valid one.

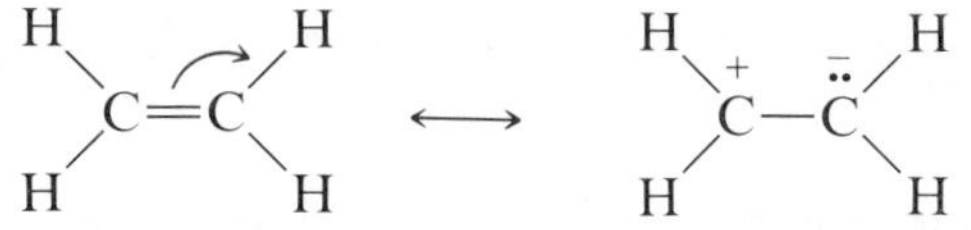

Best Lewis structure of
ethylene

Dipolar resonance form of
ethylene

The net charge is zero and all electrons are paired in both resonance forms. Another formulation is, however, incorrect:

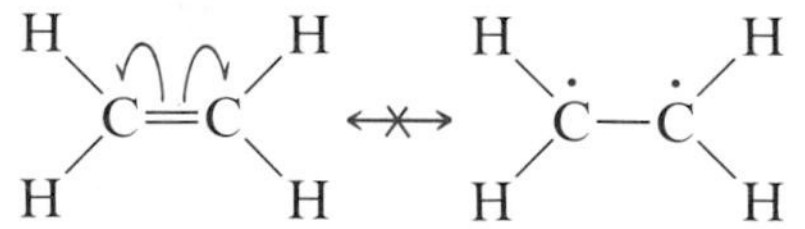

The two structures have different numbers of unpaired electrons. All the electrons are paired in the structure at the left, but two are unpaired in the structure at the right.

1.9 THE VALENCE-SHELL ELECTRON PAIR REPULSION APPROACH TO MOLECULAR GEOMETRY

Since only two electrons can occupy the same region of space and then only when their spins are paired, it follows that electron pairs in a molecule will repel other electron pairs. This reasoning provides the basis for a simple approach toward understanding and predicting the shapes of molecules. According to this *valence-shell electron pair repulsion* (VSEPR) method, a molecule such as methane CH_4 is tetrahedral because this is the geometry that permits maximum separation of its four electron pair bonds.

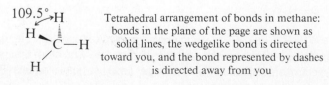

Tetrahedral arrangement of bonds in methane:
bonds in the plane of the page are shown as
solid lines, the wedgelike bond is directed
toward you, and the bond represented by dashes
is directed away from you

Carbon is at the center of the tetrahedron, and all H—C—H angles are equal to 109.5°.

In boron trifluoride, BF_3, boron shares six electrons in its valence shell with three fluorine atoms. The maximum separation of the three bonded pairs around boron is achieved in a planar geometry with 120° bond angles.

All the atoms of boron trifluoride lie in the same plane; the geometry is described as trigonal planar

Electrons in multiple bonds are treated in the same way as those in single bonds. The linear geometry in carbon dioxide allows the electrons comprising one double bond to be as far away as possible from those in the other double bond.

Linear geometry and 180° bond angles in carbon dioxide

There is a hierarchy of repulsive interactions between electron pairs.

1. The strongest repulsions are lone pair–lone pair interactions.
2. Lone pair–bonded pair repulsions come next.
3. The weakest repulsions are bonded pair–bonded pair interactions.

To illustrate the application of the VSEPR rationale to some questions of molecular geometry, consider the bond angle variation in the series methane, ammonia, and water, where the central atom in each case is surrounded by four electron pairs.

Methane   Ammonia   Water

The four bonded pairs in methane assume the normal tetrahedral value of 109.5°. In ammonia, the lone pair–bonded pair repulsions are larger than the bonded pair–bonded pair repulsions. This opens up the angle between the lone pair orbital and the bonded pair orbital and contracts the H—N—H angle to 107°—less than the tetrahedral value. In water, the lone pair–lone pair interaction is strongly repulsive. When the lone pair electrons move away from each other, the bonded pair electrons must come together, giving an H—O—H angle of 105°.

When describing the shapes of molecules, it is atomic positions that are specified. While the arrangement of electron pairs is approximately tetrahedral in ammonia and in water, these compounds are described as being pyramidal and bent, respectively.

Electrons in a double bond take up somewhat more space than those in a single bond. Thus, formaldehyde is a trigonal planar molecule with its HCO angles slightly larger than 120° and its HCH angle slightly smaller. In a species such as carbonate ion, in which resonance makes the bonds to all three oxygens equivalent, each angle is exactly 120°.

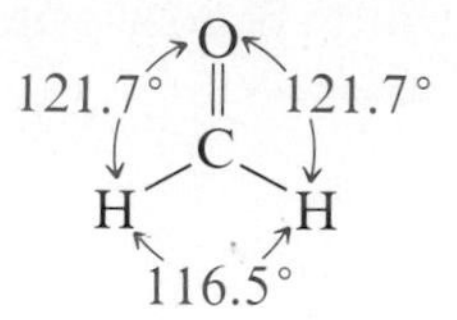

Formaldehyde
(trigonal planar)

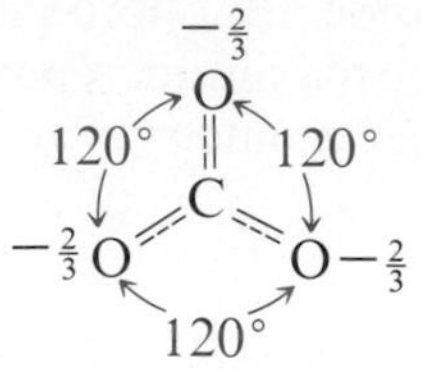

Carbonate ion
(trigonal planar)

PROBLEM 1.11 Apply the VSEPR method to deduce the shape of each of the following.

(a) NH_4^+

(b) BF_4^-

(c) H_3O^+

(d) CH_3Cl (all atoms are bonded to carbon)

(e) $COCl_2$ (all atoms are bonded to carbon)

(f) HCN (both atoms are bonded to carbon)

SAMPLE SOLUTION (a) Nitrogen in ammonium ion has four bonded pairs of electrons. A tetrahedral geometry allows for the maximum separation of these four bonded pairs.

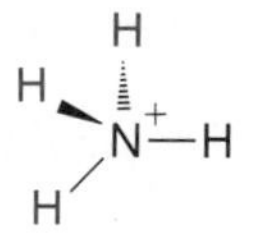

Ammonium ion: each HNH angle is 109.5°

1.10 MOLECULAR DIPOLE MOMENTS

By combining a knowledge of polar bonds with an ability to deduce molecular geometries by the VSEPR method, it is possible to make reasonable predictions about molecular dipole moments. Molecules can have polar bonds and yet because of their shape have no net dipole moment. Carbon dioxide, for example, contains polar bonds but has no dipole moment. The molecular dipole moment is the resultant, or vector sum, of all the individual bond dipole moments of a substance. In carbon dioxide, the linear arrangement of the atoms causes the two carbon-oxygen bond moments to cancel.

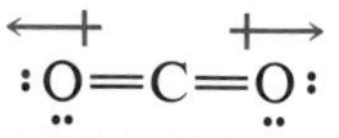

Dipole moment = 0 D

Carbon dioxide

Carbon tetrachloride, with four polar C—Cl bonds and a tetrahedral shape, has no net dipole moment because the resultant of the four bond dipoles, as shown in Figure 1.4 is zero. Dichloromethane, on the other hand, has a dipole moment of 1.62 D. The C—H bond dipoles reinforce the C—Cl bond dipoles.

PROBLEM 1.12 Which of the following compounds would you expect to have a dipole moment? If the molecule has a dipole moment, specify its direction.

(a) BF_3

(b) H_2O

(c) CH_4

(d) CH_3Cl

(e) CH_2O

(f) HCN

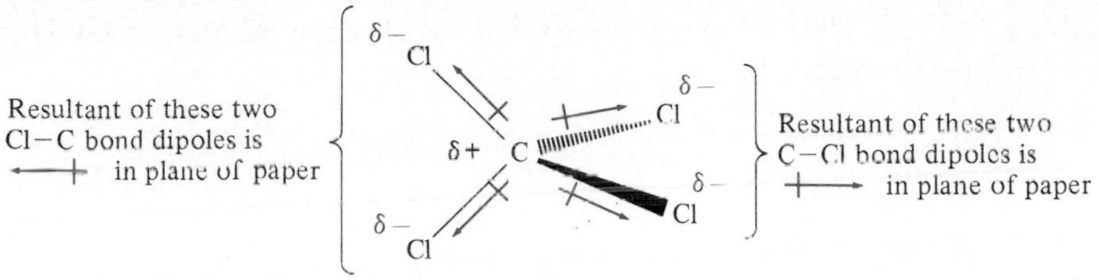

(a) There is a mutual cancellation of individual bond dipoles in carbon tetrachloride. It has no dipole moment.

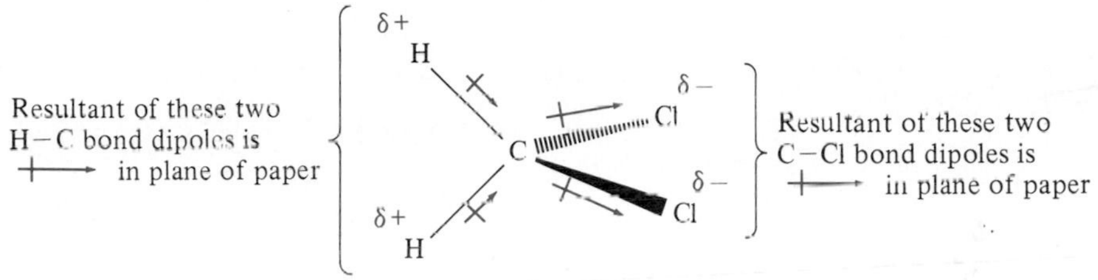

FIGURE 1.4 Contribution of individual bond dipole moments to the molecular dipole moments of carbon tetrachloride and dichloromethane.

(b) The H—C bond dipoles reinforce the C—Cl bond moment in dichloromethane. The molecule has a dipole moment of 1.62 D.

SAMPLE SOLUTION (a) Boron trifluoride is planar with 120° bond angles. Although each boron-fluorine bond is polar, their combined effects cancel and the molecule has no dipole moment.

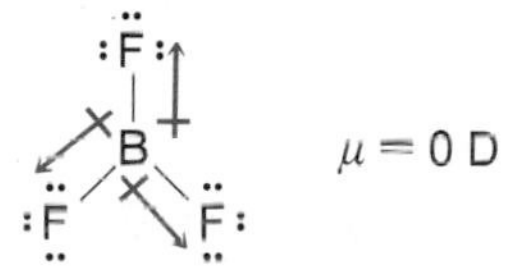

$$\mu = 0 \text{ D}$$

By using the numerical values of Table 1.2 for bond dipole moments along with molecular geometries, quantitative estimation of molecular dipole moments can be made.

1.11 WRITING STRUCTURAL FORMULAS OF ORGANIC MOLECULES

There are a number of notational shortcuts that organic chemists have developed to speed the writing of structural formulas. Representing covalent bonds by dashes is one of them. Another simplification omits unshared electron pairs from Lewis structures. This results in a less cluttered appearance with molecules such as methyl alcohol, for example.

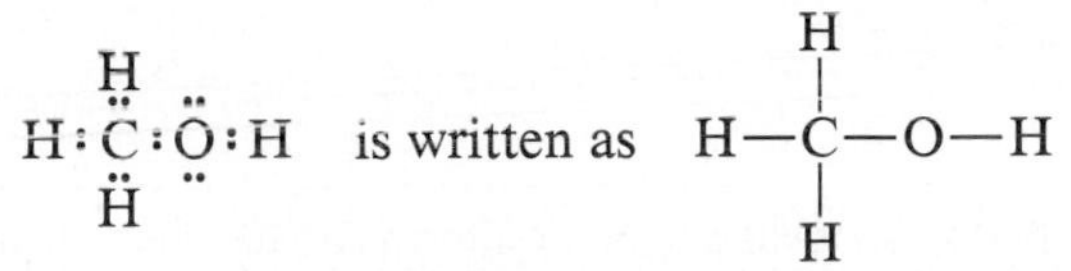

While omitting lone pairs from Lewis structures gives a cleaner presentation of structural information, it requires that you be sufficiently familiar with this practice so that you can identify the atoms that do bear unshared electron pairs.

PROBLEM 1.13   Rewrite the following Lewis structures of neutral molecules showing all the unshared pairs of electrons.

(a)
$$H-O-\underset{\underset{H}{|}}{\overset{\overset{H}{|}}{C}}-\underset{\underset{H}{|}}{\overset{\overset{H}{|}}{C}}-\underset{}{\overset{\overset{H}{|}}{N}}-H$$

(b)
$$H-\overset{\overset{H}{|}}{C}-\overset{\overset{H}{|}}{C}-H$$
with O below bridging the two C atoms

(c)
$$H-\overset{\overset{H}{|}}{C}-\overset{\overset{H}{|}}{C}-H$$
ring with $H-C$, $C-H$, $N-H$

(d)
$$H-\overset{\overset{H}{|}}{C}-H$$ above B, with two CH_3 groups attached to B

SAMPLE SOLUTION   (a)   First, count the total number of valence electrons brought to the molecule by its component atoms. Each hydrogen contributes 1, each carbon 4, nitrogen 5, and oxygen 6, for a total of 26. Since the problem specifies that the molecule is neutral, the sum of the electrons in covalent bonds and those present as unshared pairs must also be 26. There are 10 bonds shown, accounting for 20 electrons, therefore 6 must be present as unshared pairs. Add pairs of electrons to oxygen and nitrogen so that their octets are complete, two unshared pairs to oxygen and one to nitrogen.

$$H-\overset{..}{\underset{..}{O}}-\underset{\underset{H}{|}}{\overset{\overset{H}{|}}{C}}-\underset{\underset{H}{|}}{\overset{\overset{H}{|}}{C}}-\underset{..}{\overset{\overset{H}{|}}{N}}-H$$

As you develop more practice, you will find it more convenient to remember patterns of electron distribution. A neutral oxygen with two bonds has two unshared electron pairs. A neutral nitrogen with three bonds has one unshared pair.

Further simplification in presenting structural formulas can be accomplished by deleting some or all the two-electron covalent bonds and indicating the number of identical groups attached to an atom by a subscript. We call these *condensed structural formulas*. Thus, methyl alcohol is written as CH_3OH. Parentheses are used when identical assemblies of atoms are present at the same site.

$$H-\underset{\underset{H}{|}}{\overset{\overset{H}{|}}{C}}-\underset{\underset{\underset{\underset{H}{|}}{O}}{|}}{\overset{\overset{H}{|}}{C}}-\underset{\underset{H}{|}}{\overset{\overset{H}{|}}{C}}-H \quad \text{is written as} \quad (CH_3)_2CHOH$$

PROBLEM 1.14   Write condensed structural formulas for each of the following.

(a)
$$H-\underset{\underset{H}{|}}{\overset{\overset{H}{|}}{C}}-\underset{\underset{Cl}{|}}{\overset{\overset{H}{|}}{C}}-Cl$$

(b)
$$H-\underset{\underset{Cl}{|}}{\overset{\overset{H}{|}}{C}}-\underset{\underset{Cl}{|}}{\overset{\overset{H}{|}}{C}}-H$$

(c)

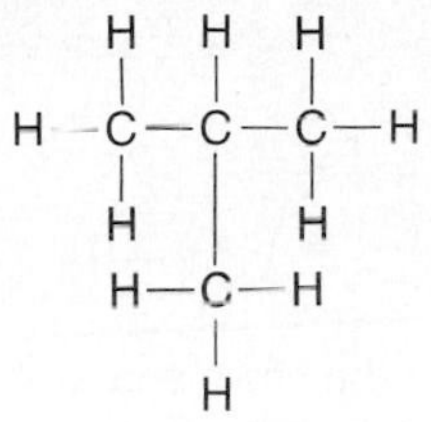

H H H
| | |
H—C—C—C—H
| | |
H | H
H—C—H
|
H

(d)

H H   H H
| | | |
H—C—C—O—C—C—H
| | | |
H H   H H

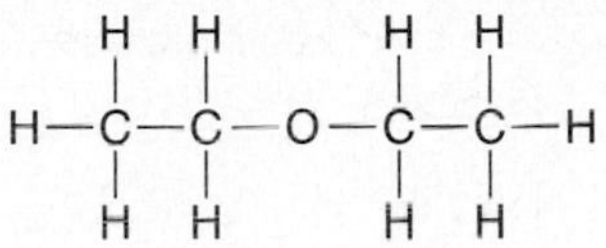

SAMPLE SOLUTION   (a)   In this compound a CH_3 group is connected to a $CHCl_2$ group.

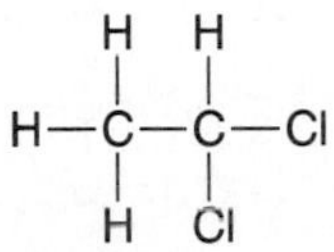

H H
| |
H—C—C—Cl   is rewritten as CH_3CHCl_2
| |
H Cl

With practice the writing of organic structures will soon become routine and some additional expedients may be employed. For example, by assuming that a carbon atom is present at both ends of a chain and at every bend in the chain, we can omit drawing individual carbons. The structures that result can be simplified even further by omitting the hydrogens attached to carbon.

$CH_3CH_2CH_2CH_3$ becomes   [structure]   simplified to   [zigzag]

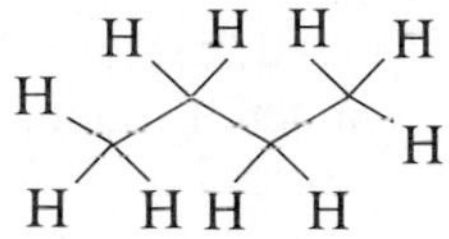

[cyclohexane structure]   becomes   [structure]   simplified to   [hexagon]

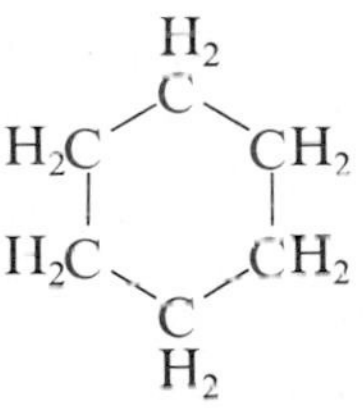

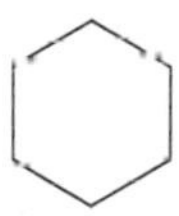

In these simplified representations, called *carbon skeleton formulas,* the only atoms specifically written in are those that are neither carbon nor hydrogen bound to carbon.

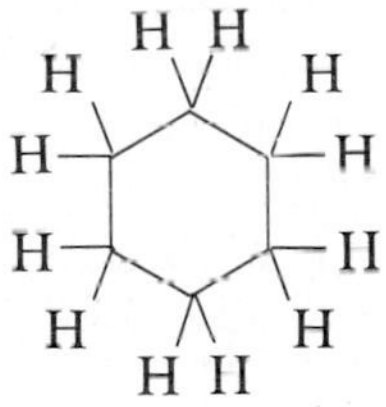

$CH_3CH_2CH_2CH_2OH$ becomes   [zigzag]OH

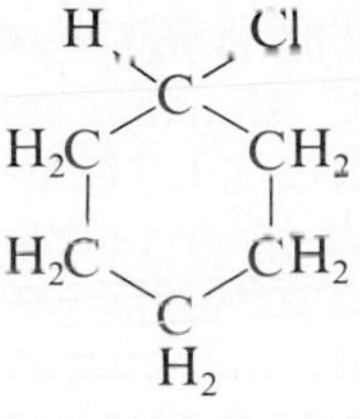

[chlorocyclohexane structure]   becomes   [hexagon with Cl]

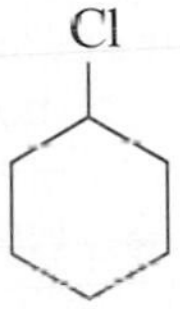

PROBLEM 1.15   Expand the following structural representations to show all the atoms, including carbon and hydrogen.

(a)  [zigzag structure]   (b) 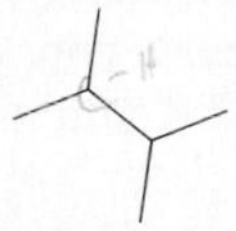 [branched structure]

(c)

(d)

SAMPLE SOLUTION (a) There is a carbon at each bend in the chain and at the ends of the chain. Each of the 10 carbon atoms bears the appropriate number of hydrogen substituents so that it has four bonds.

$$\equiv H-C-C-C-C-C-C-C-C-C-C-H$$

Alternatively, the structure could be written as $CH_3CH_2CH_2CH_2CH_2CH_2CH_2CH_2CH_2CH_3$ or in condensed form as $CH_3(CH_2)_8CH_3$.

1.12 CONSTITUTIONAL ISOMERS

In the introductory chapter it was noted that both Berzelius and Wöhler were fascinated by the fact that two clearly different compounds, ammonium cyanate and urea, possessed the same molecular formula, CH_4N_2O. Berzelius had studied examples of this phenomenon earlier and invented a word to describe it. He said that *isomers* are different compounds that have the same molecular formula. Ethyl alcohol and dimethyl ether are examples of two compounds that are isomers; both have the molecular formula C_2H_6O but are different in ways that are readily apparent. Ethyl alcohol is a liquid with a boiling point of 78.3°C. Dimethyl ether is a gas at room temperature; its boiling point is −24.9°C.

Ethyl alcohol and dimethyl ether are different compounds because they have different structures.

Ethyl alcohol (CH_3CH_2OH)

Dimethyl ether (CH_3OCH_3)

Structurally, the two compounds differ in respect to their atomic connections. Ethyl alcohol contains an O—H bond and a C—C bond; dimethyl ether has neither. Dimethyl ether has a C—O—C unit; ethyl alcohol does not. The pattern of atomic connections in a molecule is referred to as its *constitution*. Ethyl alcohol and dimethyl ether have different constitutions and are classified as *constitutional isomers*.

Opportunities for constitutional isomerism increase dramatically as the number of atoms increase. Ethyl alcohol and dimethyl ether are the only compounds possible that have the molecular formula C_2H_6O. There are seven constitutional isomers having the molecular formula $C_4H_{10}O$, however. These are shown in Table 1.4. Each compound in the table is a constitutional isomer of each of the other six. Constitutional isomers can, as the data in the table show, have different physical properties.

TABLE 1.4

Constitutionally Isomeric Molecules of Molecular Formula $C_4H_{10}O$

Structural formula	bp, °C	mp, °C	Solubility in water, g per 100 g H_2O
$CH_3CH_2CH_2CH_2OH$	118	−90	9
$(CH_3)_2CHCH_2OH$	108	−108	10

$CH_3CHCH_2CH_3$
    ｜
    OH
	99.5	−115	12.5

$(CH_3)_3COH$	83	26	∞
$CH_3CH_2OCH_2CH_3$	35	−116	7.5
$CH_3OCH_2CH_2CH_3$	39	*	3.0
$CH_3OCH(CH_3)_2$	32.5	*	*

* Data not available.

They can also have very different chemical properties, a fact which will become evident as we proceed through the course.

PROBLEM 1.16 Write structural formulas for all the constitutionally isomeric compounds having the molecular formula given.

(a) C_3H_9N
(b) C_4H_{10}
(c) C_5H_{12}

(d) $C_2H_4Cl_2$
(e) C_4H_9Br
(f) C_3H_8O

SAMPLE SOLUTION (a) In order to generate constitutionally isomeric structures having the molecular formula C_3H_9N, we need to consider the various ways in which three carbon atoms and one nitrogen can be bonded together. These are

$$C-C-C-N \quad C-C-N-C \quad C-N\!\!\begin{array}{c}C\\ \\ C\end{array} \quad C-C-N\!\!\begin{array}{c}\\ \\ C\end{array} \quad \text{and} \quad \begin{array}{c}C-N\\ |\quad\ |\\ C-C\end{array}$$

Filling in the appropriate hydrogen substituents we obtain for the first four structures

$$CH_3CH_2CH_2NH_2 \quad CH_3CH_2NHCH_3 \quad CH_3-N\!\!\begin{array}{c}CH_3\\ \\ CH_3\end{array} \quad \text{and} \quad CH_3CHNH_2\\ \qquad\qquad\qquad\qquad\qquad\qquad\qquad\qquad\qquad\qquad\qquad\qquad\qquad\qquad\qquad\ |\\ \qquad\qquad\qquad\qquad\qquad\qquad\qquad\qquad\qquad\qquad\qquad\qquad\qquad\qquad\qquad CH_3$$

When the three carbons and the nitrogen are arranged in a ring, the molecular formula based on such a structure is C_3H_7N, not C_3H_9N as required.

$$\begin{array}{c}C-C\\ |\quad\ |\\ C-N\end{array} \quad \text{gives} \quad \begin{array}{c}CH_2-CH_2\\ |\qquad\ |\\ CH_2-NH\end{array} \quad \text{(not an isomer)}$$

Later, another type of isomerism, called *stereoisomerism*, will be presented. Stereoisomers have the same constitution but differ in the way the atoms are arranged in space.

1.13 SUMMARY

Chemical bonds are classified as ionic or covalent. An *ionic bond* is the electrostatic attraction between two oppositely charged ions and occurs in substances such as sodium chloride where the energy required to transfer an electron from one element to the other is not prohibitively high. Since carbon is neither strongly electronegative nor strongly electropositive, it does not normally participate in ionic bonding, and covalent bonding is observed instead. A *covalent bond* arises, according to the model proposed by G. N. Lewis, by the sharing of a pair of electrons between two atoms. Double bonds correspond to the sharing of four electrons and triple bonds to the sharing of six. The most stable Lewis structures are characterized by rare gas electron configurations for all its component atoms. Second row elements are said to possess a rare gas electron configuration when the total number of electrons, shared plus unshared, in their valence shell is equal to eight. Two electrons confer a rare gas configuration upon hydrogen.

$$Na^+[:\ddot{C}l:]^- \qquad H:\underset{\underset{\textstyle H}{\displaystyle |}}{\overset{\overset{\textstyle H}{\displaystyle |}}{C}}:\ddot{C}l:$$

Ionic bonding in sodium chloride   Covalent bonding in CH_3Cl

The order of atomic connections that define a substance is called its *constitution.* Isomers are different compounds that have the same molecular formula, so constitutional isomers are compounds having the same molecular formula but differing in the order of their atomic connections.

Lewis structures for organic molecules restrict electrons to the region between nuclei when, in fact, they may be shared among many nuclei. *Resonance* is an adaptation of the Lewis approach to bonding that attempts to describe electron delocalization in terms of contributions from several Lewis structures. The actual molecule is always more stable than any of the individual structures written for it. In writing alternative resonance forms of a molecule, only the distribution of the electrons is varied — the positions of the atoms must remain the same.

Relatively unstable arrangement of electrons: only six electrons around carbon

Better Lewis structure: all atoms have octets but separation of opposite charges is present

Most stable Lewis structure: no charge separation and all atoms have octets

When the electron distribution in a covalent bond is distorted so that the centers of positive and negative charge do not coincide, the bond is termed a *polar covalent bond.* Electronegative elements attract electrons in their covalent bonds with carbon toward themselves, becoming negatively polarized, and leaving a partial positive charge on carbon.

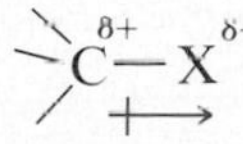

$\overset{\delta+}{\underset{\longmapsto}{\diagup\!\!\!\!C}}\!-\!\overset{\delta-}{X}$ Polarization of C—X bond; X is more
electronegative than carbon

Molecules that contain polar bonds can either have dipole moments or not depending on their geometry. The molecular dipole moment is the vector sum, or resultant, of all the individual bond dipoles. Two bond dipoles may reinforce, oppose, or even cancel each other's effect. Thus, neither CH_4 nor CCl_4 has a dipole moment, but CH_3Cl, CH_2Cl_2, and $CHCl_3$ all do.

The *valence-shell electron pair repulsion* method permits molecular geometries to be predicted on the basis of repulsive interactions between the pairs of electrons that surround a central atom. A tetrahedral arrangement provides for the maximum separation of four electron pairs; a trigonal planar geometry is optimal for three electron pairs; and a linear arrangement is best for two electron pairs.

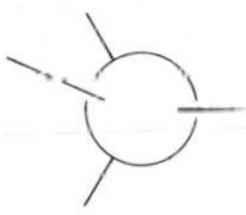

Tetrahedral arrangement of four electron pairs 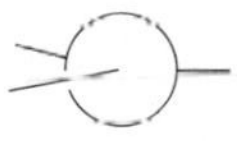   Trigonal planar arrangement of three electron pairs 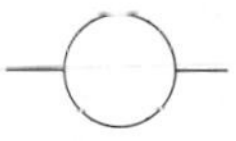   Linear arrangement of two electron pairs

Once some facility is gained in writing organic structural formulas, a number of shortcuts can be employed to simplify the task. It is essential, however, that one always be able to expand an abbreviated formula so as to locate all the constituent atoms, account for all the electrons, and assign formal charges where they are required.

PROBLEMS

1.17 Each of the following species will be encountered at some point in this text. They all have the same number of electrons binding the same number of atoms and the same arrangement of bonds, i.e., they are *isoelectronic*. Specify which atoms, if any, bear a formal charge in the Lewis structure given and the net charge for each species.

(a) :N≡N: (d) :N≡O:
(b) :C≡N: (e) :C≡O:
(c) :C≡C:

1.18 You will meet all the following isoelectronic species in this text. Repeat the previous problem for these three structures.

(a) :Ö=C=Ö: (b) :N̈=N=N̈:   (c) :Ö—N=Ö:

1.19 All the following compounds are characterized by ionic bonding between a metal cation and a tetrahedral anion. Write an appropriate Lewis structure for each anion, remembering to specify formal charges where they exist.

(a) $NaBH_4$ (c) K_2SO_4
(b) $LiAlH_4$ (d) Na_3PO_4

1.20 Consider structural formulas A, B, and C:

$$H_2\ddot{C}-N\equiv N\!: \qquad H_2C=N=\ddot{N}\!: \qquad H_2C-\ddot{N}=\ddot{N}\!:$$
$$A B C$$

(a) Are A, B, and C constitutional isomers or are they resonance forms?
(b) Which structures have a negatively charged carbon?
(c) Which structures have a positively charged carbon?
(d) Which structures have a positively charged nitrogen?
(e) Which structures have a negatively charged nitrogen?
(f) What is the net charge on each structure?
(g) Which is a more stable structure, A or B? Why?
(h) Which is a more stable structure, B or C? Why?

1.21 In each of the following pairs, determine whether the two represent resonance forms of a single species or depict different substances. If two structures are not resonance forms, explain why.

(a) $:\ddot{N}-N\equiv N\!:$ and $:N=N=N\!:$

(b) $:\ddot{N}-N\equiv N\!:$ and $:\ddot{N}-N=\ddot{N}\!:$

(c) $:\ddot{N}-N\equiv N\!:$ and $:\ddot{N}-\ddot{N}-\ddot{N}\!:$

1.22 Among the four structures given below, one of them is *not* a permissible resonance form. Identify the wrong structure. Why is it incorrect?

$$\overset{+}{C}H_2-\ddot{N}-\ddot{O}\!:^{-} \qquad CH_2=\overset{+}{N}-\ddot{O}\!:^{-} \qquad CH_2=N=\ddot{O}\!: \qquad :\overset{-}{C}H_2-\overset{+}{N}=\ddot{O}\!:$$
$$||||$$
$$CH_3 CH_3 CH_3 CH_3$$
$$D E F G$$

1.23 Write a resonance form that is more stable than the one given for each of the following:

(a) $:O=\overset{+}{C}-\ddot{O}\!:^{-}$

(b) $:O=C=\ddot{N}\!:^{-}$

(c) $:\overset{+}{O}-\ddot{O}-\ddot{O}\!:^{-}$

(d) $\underset{H}{\overset{H}{>}}\overset{+}{C}-\ddot{C}-\ddot{O}\!:^{-}$

(e) $\underset{H}{\overset{H}{>}}\overset{..}{C}-C\overset{\ddot{O}:}{<_H}$

(f) $\underset{H}{\overset{H}{>}}\overset{\ddot{}\,-}{C}-C\equiv N\!:$

(g) $H-\overset{+}{C}=\ddot{O}\!:$

(h) $\underset{H}{\overset{H}{>}}\overset{+}{C}-\ddot{O}H$

(i) $\underset{H}{\overset{H}{>}}\overset{-}{C}-\overset{+}{N}=NH_2$

1.24 What is the formal charge of oxygen in each of the following Lewis structures?

(a) $CH_3\ddot{O}\!:$

(b) $(CH_3)_2\ddot{O}\!:$

(c) $(CH_3)_3O\!:$

1.25 Write structural formulas for all the constitutional isomers of

(a) C_3H_8

(b) C_3H_6

(c) C_3H_4

1.26 Write structural formulas for all the constitutional isomers of molecular formula C_3H_6O that contain

   (a)   Only single bonds

   (b)   One double bond

1.27 Which compound in each of the following pairs would you expect to have the greater dipole moment μ? Why?

   (a)   NaCl or HCl

   (b)   HF or HCl

   (c)   HF or BF_3

   (d) $(CH_3)_3CH$ or $(CH_3)_3CCl$

   (e) $CHCl_3$ or CCl_3F

   (f) CH_3NH_2 or CH_3OH

   (g) CH_3NH_2 or CH_3NO_2

1.28 Apply the VSEPR method to deduce the geometry around carbon in each of the following species.

   (a) $:\overset{-}{C}H_3$ (b) $\overset{+}{C}H_3$ (c) $:CH_2$

1.29 Expand the following structural representations so as to more clearly show all the atoms and any unshared electron pairs.

   (a)   A component of high-octane gasoline

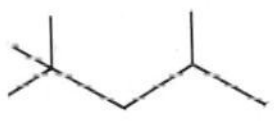

   (b)   Occurs in bay and verbena oil

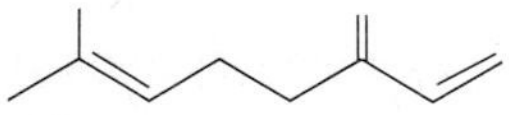

   (c)   Pleasant-smelling substance found in marjoram oil

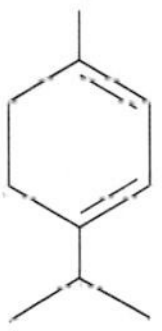

   (d)   Present in oil of cloves

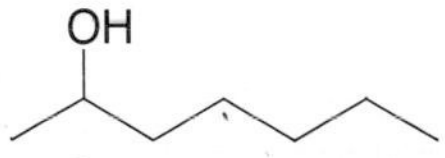

   (e)   Found in Roquefort cheese

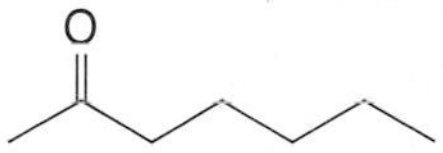

   (f)   Benzene, parent compound of a large family of organic substances

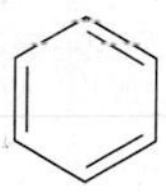

   (g)   Naphthalene: sometimes used as a moth repellent

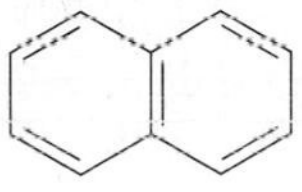

   (h)   Aspirin

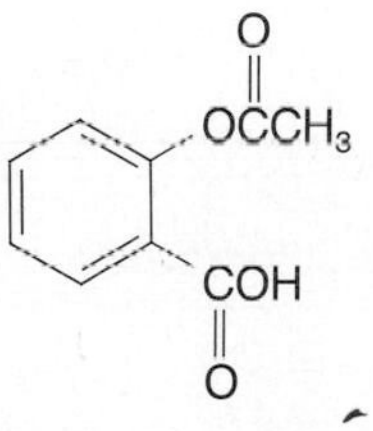

(i) Nicotine: a toxic substance present in tobacco

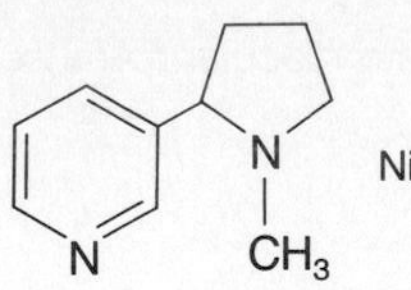

(j) Tyrian purple: a purple dye extracted from a species of Mediterranean sea snail

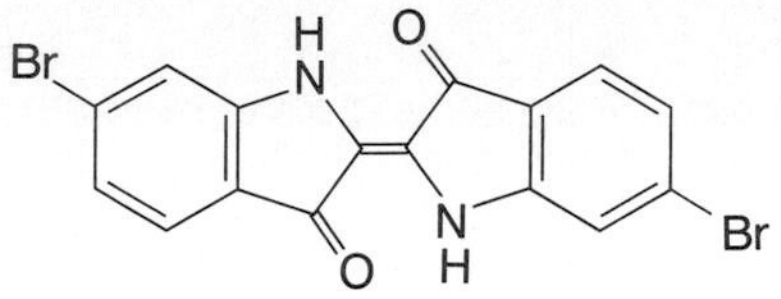

(k) Hexachlorophene: an antiseptic

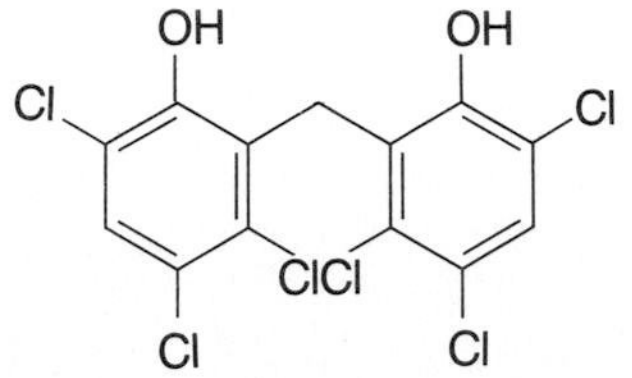

1.30 Molecular formulas of organic compounds are customarily presented in the fashion $C_2H_5BrO_2$. The number of carbon and hydrogen atoms are presented first, followed by the other atoms in alphabetical order. Give the molecular formulas corresponding to each of the compounds in the preceding problem. Are any of them isomers?

1.31 The elemental composition of organic compounds is easily determined and often cited by chemists as supporting evidence of structure and purity. An *empirical formula* is an expression of the simplest whole number ratio of atoms that satisfies the experimentally determined elemental composition. Thus an empirical formula of CH_2O can correspond to a molecular formula of CH_2O, $C_2H_4O_2$, $C_3H_6O_3$, etc. Calculate the empirical formulas corresponding to each of the elemental compositions given below. The accuracy of elemental analyses is about ±0.4 percent in absolute terms. Oxygen content is not normally determined directly but is assumed to be the difference between 100 percent and the total content of all the other elements.

(a) C 92.61%, H 7.01%
(b) C 56.28%, H 6.07%
(c) C 50.79%, H 6.52%
(d) C 55.65%, H 7.99%, N 6.47%
(e) C 45.94%, H 4.99, N 5.13% Br 27.83%

1.32 (a) Combustion analysis of glucose indicated that it contained 39.8% carbon and 6.8% hydrogen. A molecular weight determination gave a value of 175 ± 10. What is the molecular formula of glucose?
(b) Elemental analysis of the herbicide 2,4-D indicated that it contained 43.6% carbon, 2.8% hydrogen, and 32.0% chlorine. The molecular weight of 2,4-D is 225 ± 10. What is its molecular formula?

ALKANES

2

We begin our discussion of the various classes of organic compounds with *alkanes*. Alkanes are *hydrocarbons*—they contain only carbon and hydrogen. Alkane names form the foundation upon which the most widely accepted system of organic nomenclature is based. The elements of this nomenclature system, the *IUPAC rules,* are introduced in this chapter.

An examination of alkane structure permits us to extend the principles of covalent bonding discussed in the previous chapter to more modern interpretations. Molecular orbital (MO) theory is introduced in this chapter and a simplified version of it, based on the concept of *orbital hybridization,* is used to describe bonding in alkanes. We begin this discussion at a fundamental level by taking a closer look at the nature of electrons.

2.1 ELECTRONS AS WAVES

Electrons were believed to be particles until 1924, when the French physicist Louis de Broglie suggested that they have wavelike properties as well. Two years later Erwin Schrödinger expressed the energy of an electron in a hydrogen atom in terms of a *wave equation.* Wave equations have not one, but a series of solutions; each solution is called a *wave function* and is symbolized by the Greek letter ψ (psi). The wave functions that are solutions to the wave equation for a hydrogen atom are designated $\psi(1s)$, $\psi(2s)$, $\psi(2p_x)$, $\psi(2p_y)$, $\psi(2p_z)$, $\psi(3s)$, etc. A wave function can be used to calculate the energy of an electron in any of the various states that are solutions to the wave equation. According to the Heisenberg uncertainty principle, the position and momentum of an electron cannot be specified simultaneously, but it is possible to determine the probability that an electron will be found at a particular point relative to the nucleus. That probability is given by the square of the wave function (ψ^2). Electron probability distributions, represented as in Figure 2.1, are sometimes referred to as *electron clouds* because of their diffuse quality. The region of space where there is a high probability of finding an electron is known as an *orbital* (Section 1.1). Orbitals are classified as s, p, d, or f depending on their shape. The spatial properties of

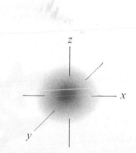

FIGURE 2.1 Probability distribution (ψ^2) for an electron in a 1s orbital.

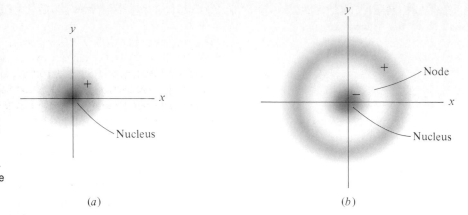

FIGURE 2.2 Cross sections of *(a)* a 1s orbital, and *(b)* a 2s orbital. The wave function has the same sign over the entire 1s orbital. It is arbitrarily shown as +, but it could just as well have been designated as −. The 2s orbital has a spherical node where the wave function changes sign.

(a)   *(b)*

d and *f* orbitals are not essential to an understanding of subsequent material in this book and will not be discussed further.

Orbitals are characterized by *nodal surfaces* where the probability of finding an electron is close to zero. The more nodes an orbital has, the higher is its energy. While all orbitals have a node an infinite distance from the nucleus, it simplifies discussion if this node is not considered explicitly when comparing nodal properties of different orbitals since it is common to them all. Thus, a 1s orbital is described as having no nodes, and a 2s orbital as having one. These two orbitals are shown in cross section in Figure 2.2. The 2s wave function changes sign on passing through the node near the nucleus, as indicated by the plus (+) and minus (−) signs in Figure 2.2. Do not confuse these algebraic signs with those that represent electric charges — they have nothing to do with electron or nuclear charge.

The 2p orbitals each have a nodal plane passing through the nucleus. The *yz* plane is a nodal surface for the $2p_x$ orbital, the *xz* plane a nodal surface for the $2p_y$ orbital, and the *xy* plane a nodal surface for the $2p_z$ orbital. As shown in the electron probability plot of Figure 2.3, the wave function has an opposite sign in each lobe of a 2p orbital; the wave associated with an electron is commonly described as having undergone a change in phase in passing from one lobe to another.

The Schrödinger equation has been solved exactly only for hydrogen, so orbital energies and shapes are not directly available for atoms such as carbon, oxygen, nitrogen, and the halogens. Through the use of certain approximations, however, wave functions may be derived for these other atoms, and hydrogenlike orbitals may be used to mimic the orbitals of those that are most commonly encountered in organic chemistry.

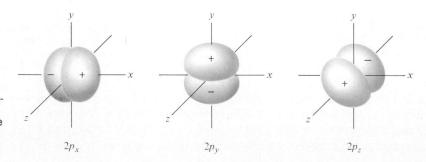

FIGURE 2.3 Probability distributions (ψ^2) for electrons in 2p orbitals. Lobes on opposite sides of the nodal plane are out of phase with each other.

$2p_x$ $2p_y$ $2p_z$

2.2 ELECTRONS IN MOLECULES. THE LCAO-MO METHOD

A fundamental principle of modern theories of structure and bonding is that electrons in molecules occupy *molecular orbitals,* just as electrons in atoms occupy atomic orbitals. A molecular orbital may be primarily associated with only one or two or a few atoms, or it may be so large as to encompass the entire molecule. Molecular orbitals, no matter how extensive, have many of the same properties that we associate with atomic orbitals. Like atomic orbitals, molecular orbitals are populated by electrons, with the lowest-energy orbitals first; an electron in a molecular orbital has a well-defined energy; each molecular orbital can contain a maximum of two electrons, and the spins of these electrons must be paired.

The system of molecular orbitals that defines the distribution of electrons in a molecule is approximated by a technique known as the *linear combination of atomic orbitals–molecular orbital (LCAO-MO)* method. According to the LCAO-MO method, molecular wave functions are expressed as combinations of atomic wave functions. Rather than showing the mathematics involved, the LCAO-MO approach will be illustrated through a qualitative presentation of some of its results.

2.3 MOLECULAR ORBITALS OF THE HYDROGEN MOLECULE

Consider first the bonding of two hydrogen atoms in a hydrogen molecule (H_2). Molecular orbitals of H_2 are generated by combining (or "mixing") the $1s$ atomic orbitals of two hydrogen atoms, as shown in Figure 2.4. The number of molecular orbitals generated must equal the number of atomic orbitals that combine to produce them. Thus, two hydrogen $1s$ atomic orbitals are combined to produce two molecular orbitals of H_2. One combination corresponds to the *in-phase* overlap of individual $1s$ orbitals, the other to an *out-of-phase* overlap. When two orbitals of the same phase overlap (Figure 2.5), their wave functions reinforce each other and increase the probability than an electron will be found in the region between the two nuclei, binding them together. We call the in-phase overlap a *bonding* interaction, and the molecular orbital that results is called a *bonding orbital.* Conversely, the out-of-phase combination of two hydrogen $1s$ orbitals leads to a node between the two nuclei and is described as *antibonding* (Figure 2.6). Both the bonding and antibonding orbitals of

bonding and anti-bonding

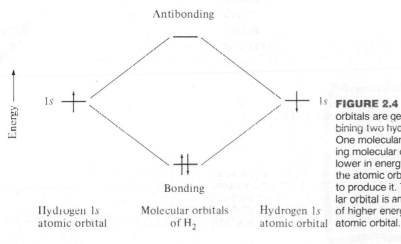

Antibonding

$1s$ $1s$

Energy

Bonding

Hydrogen $1s$ atomic orbital

Molecular orbitals of H_2

Hydrogen $1s$ atomic orbital

FIGURE 2.4 Two molecular orbitals are generated by combining two hydrogen $1s$ orbitals. One molecular orbital is a bonding molecular orbital and is lower in energy than either of the atomic orbitals that combine to produce it. The other molecular orbital is antibonding and is of higher energy than either atomic orbital.

FIGURE 2.5 When two 1s atomic orbitals overlap so that the signs of their wave functions reinforce each other, a bonding molecular orbital results.

H_2 are characterized by rotational symmetry around the internuclear axis. A cross section of the orbital taken perpendicular to the internuclear axis is a circle. Orbitals that have rotational symmetry around an internuclear axis are classified as σ (sigma) molecular orbitals. The antibonding orbital is designated by an asterisk and is referred to as a σ^* (sigma star) orbital.

The bonding molecular orbital σ is lower in energy than either of the hydrogen 1s orbitals from which it was generated, while the antibonding molecular orbital σ^* is higher in energy than the individual atomic orbitals. Since there are two electrons in a hydrogen molecule, the bonding orbital is doubly occupied and the antibonding orbital is vacant. The hydrogen molecule binds its electrons more strongly than two separated hydrogen atoms, and the two nuclei are said to be joined by a σ bond. A σ bond is one that has the symmetry characteristics of a σ orbital, i.e., an electron distribution that is symmetric with respect to rotation about the internuclear axis.

The placement of the energy levels in Figure 2.4 tells us that energy must be expended to break the σ bond, separating the hydrogen molecule into two hydrogen atoms.

$$H_2 \; + \text{energy} \longrightarrow \; 2H\cdot$$

Hydrogen molecule   Two hydrogen atoms

The amount of energy that is required is called the *bond dissociation energy.* For hydrogen the experimentally determined bond dissociation energy is 104 kcal/mol. Dissociation energies have been measured for many bond types and will be discussed in more detail in Section 4.19.

PROBLEM 2.1   The hydrogen molecule-ion H_2^+ is formed in interstellar space as a result of cosmic ray bombardment of H_2. Basing your answer on the molecular orbital diagram given for the hydrogen molecule in Figure 2-4, decide whether the dissociation of H_2^+ to a hydrogen atom (H·) and a proton (H^+) is exothermic or endothermic.

There is a limit to how closely two bonded nuclei can approach each other, and bonds between atoms have characteristic *bond lengths,* or *bond distances.* For H_2 the two nuclei are separated by 0.74 Å. At distances greater than this the bond is not as strong because the overlap of the individual 1s orbitals is not as pronounced. At distances less than 0.74 Å repulsive forces between the two positively charged nuclei increase substantially.

FIGURE 2.6 When two 1s atomic orbitals overlap so that the signs of their wave functions oppose each other, an antibonding molecular orbital results.

2.4 A MOLECULAR ORBITAL DESCRIPTION OF METHANE

When constructing a molecular orbital description of compounds that contain second row elements, the $2s$, $2p_x$, $2p_y$, and $2p_z$ orbitals of these atoms are the ones that need to be considered along with the $1s$ orbital of hydrogen. The $1s$ orbitals of second row elements can be ignored because they are too closely associated with the nucleus to overlap effectively with the orbitals of any other atom. This practice is comparable with that of focusing attention only on the valence electrons of second row elements when writing Lewis structures. For example, the molecular orbital description of methane, as shown in Figure 2.7, comprises a system of eight molecular orbitals resulting from the mixing of the $1s$ orbitals of four hydrogen atoms with the $2s$, $2p_x$, $2p_y$, and $2p_z$ orbitals of carbon. Remember, the number of molecular orbitals must equal the number of atomic orbitals that combine to produce them. Four of these orbitals are bonding, and together contain the eight valence electrons of methane. Four of them are antibonding and are unoccupied. What is important to realize about the molecular orbital description of methane is that the four bonding orbitals do not represent individual carbon-hydrogen "bonds." Rather, each molecular orbital encompasses all five atoms of methane, and the two electrons in each orbital are delocalized over all the atoms in the molecule.

PROBLEM 2.2   How many molecular orbitals are there in each of the following compounds? (For the purpose of this problem, do not include the $1s$ orbital of second row elements.) None of the molecules have orbitals that are only half-filled. How many orbitals are occupied in each case?

(a) H_2O
(b) NH_3
(c) F_2

(d) CO
(e) CO_2
(f) CH_3CH_3

SAMPLE SOLUTION   (a) Two hydrogens each contribute one $1s$ orbital. Exclusive of its own $1s$ orbital, oxygen contributes a $2s$ and three $2p$ orbitals. Thus, there are a total of

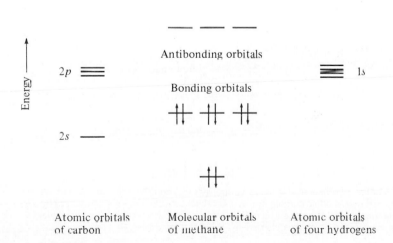

Energy →

$2p$ —

Antibonding orbitals

$1s$

Bonding orbitals

$2s$ —

Atomic orbitals of carbon

Molecular orbitals of methane

Atomic orbitals of four hydrogens

FIGURE 2.7 The energy levels of methane. The $1s$ level of carbon is not shown; it is of much lower energy than any of the other orbitals and does not interact with them to any significant extent.

six molecular orbitals. A water molecule has eight valence electrons, so four of these molecular orbitals are doubly occupied.

The pattern of the overlaps that produce the molecular orbitals of methane, as well as a molecular orbital explanation for its tetrahedral geometry, arises from an analysis that is more detailed than is necessary for an understanding of the material in this text. In the following section, an alternative approach that combines certain principles of molecular orbital theory and terminology with the more familiar picture of the shared electron pair covalent bond will be introduced.

2.5 ORBITAL HYBRIDIZATION AND THE STRUCTURE OF METHANE

The molecular orbital method plays a prominent role in theoretical treatments of molecular structure. As a practical matter, however, the properties of most organic molecules, including most of their chemical reactions, are more easily understood by considering molecules in terms of shared electron pair bonds. Organic chemists have for many years employed a bonding model that combines elements of molecular orbital theory with the Lewis model of covalent bonding. This model, based on the concept of *orbital hybridization,* was proposed by Linus Pauling in the 1930s to

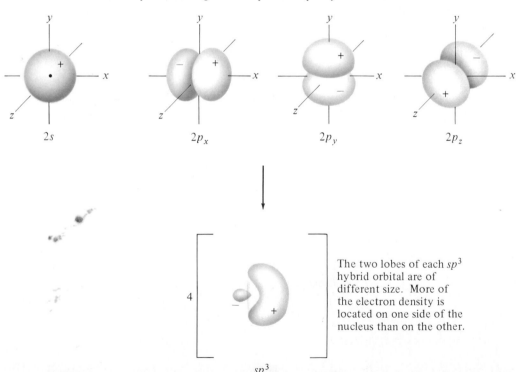

Combine one $2s$ and three $2p$ orbitals to give four equivalent sp^3-hybrid orbitals

$2s$ $2p_x$ $2p_y$ $2p_z$

4   sp^3

The two lobes of each sp^3 hybrid orbital are of different size. More of the electron density is located on one side of the nucleus than on the other.

FIGURE 2.8 Representation of orbital mixing in sp^3 hybridization. Mixing of one s orbital with three p orbitals generates four sp^3 hybrid orbitals. Each sp^3 hybrid orbital has 25 percent s character and 75 percent p character. The four sp^3 hybrid orbitals have their major lobes directed towards the corners of a tetrahedron, which has the carbon atom at its center.

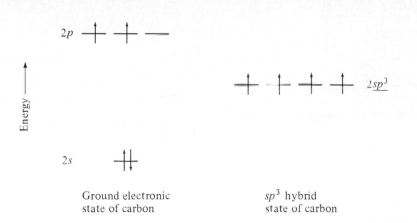

FIGURE 2.9 Electron configuration of carbon in its most stable state (left) and in its sp^3 hybridized state (right).

Ground electronic state of carbon

sp^3 hybrid state of carbon

explain the fact that carbon can form bonds to four other atoms and that these atoms arc arranged in a tetrahedral fashion around carbon. This orbital hybridization model uses terminology common to molecular orbital theory but treats atomic connections as though they were analogous to bonds in diatomic molecules. Electrons in σ bonding orbitals are localized to regions between two nuclei only. What results is a structural picture that is very similar to the Lewis shared electron pair covalent bond model but more detailed in that it includes a description of the electron distribution in the bonds. It is, in a sense, a *localized molecular orbital* treatment of bonding.

The orbital hybridization model was developed in order to explain the bonding in methane. Consider the electron configurations of the individual atoms in CH_4.

Four hydrogen atoms   each $1s^1$
One carbon atom $1s^2, 2s^2, 2p_x^1, 2p_y^1, 2p_z^0$

Since carbon has only two unpaired electrons, it would seem that it could bond to two hydrogens to form CH_2 but not to four to give CH_4. Pauling proposed that this problem could be circumvented by invoking the idea of orbital hybridization.

In orbital hybridization a new set of atomic orbitals is constructed so that carbon has four half-filled valence orbitals rather than the two that characterize its most stable electron configuration. This is accomplished by the orbital mixing scheme depicted in Figure 2.8. Combining the $2s$, $2p_x$, $2p_y$, and $2p_z$ orbitals of carbon generates a set of four new orbitals. These new orbitals are neither s nor p. Each one has 25 percent of the character of the $2s$ orbital and 75 percent p orbital character. The four new orbitals are called sp^3 hybrid orbitals. They are all of equal energy, and the four valence electrons of carbon populate them one by one, as shown in Figure 2.9. Orbital hybridization transforms the electron configuration of carbon from

$1s$ $2s$ $2p_x$ $2p_y$ $2p_z$ to $1s$ $2sp^3$ $2sp^3$ $2sp^3$ $2sp^3$

(Most stable electron configuration)   (sp^3 Hybridized state)

With four half-filled orbitals available, carbon now has the capacity to form bonds to four hydrogens to give CH_4.

Each sp^3 hybrid orbital has one node. There are two lobes of unequal size, making the electron density greater on one side of the nucleus than the other. Each of these

FIGURE 2.10 The sp^3 hybrid orbitals are arranged in a tetrahedral fashion around carbon. Each contains one electron and can form a bond with a hydrogen atom to give a tetrahedral methane molecule. (*Note:* Only the major lobe of each sp^3 orbital is shown. As indicated in Figure 2.8, each orbital contains a smaller back lobe, which has been omitted for the sake of clarity.)

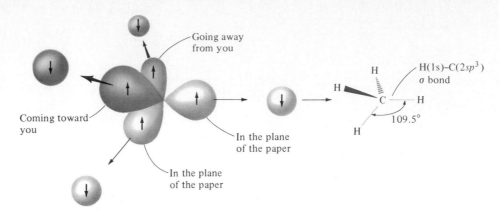

four hybrid orbitals can participate in bond formation and when it does, it is the larger lobe in each case that overlaps with the orbital of another atom. Our picture of the shape of sp^3 hybrid orbitals is derived from mathematical treatments of orbital hybridization. These mathematical approaches also give the spatial orientation of the orbitals; the axes of the four sp^3 hybrid orbitals are directed toward the corners of a tetrahedron with carbon at the center. A tetrahedral orientation of its bonds is characteristic of sp^3 hybridized carbon.

The four carbon-hydrogen bonds are formed by overlap of sp^3 hybridized orbitals of carbon with individual hydrogen $1s$ orbitals. The electron distribution in each of the bonds has rotational symmetry. Thus, we have a new kind of σ bond, one that is generated by sp^3–s overlap. This is illustrated in Figure 2.10.

There are two justifications for the orbital hybridization model of chemical bonding. First, orbital hybridization allows for the formation of more bonds. As discussed above, the sp^3 hybrid state of carbon can form four bonds while the ground electronic state can form only two. Second, the σ bonds that are formed from hybrid orbitals are stronger than the bonds from either s or p orbitals. Adding s character to p orbitals makes them more electron-attracting, enabling them to hold the bonding electrons more tightly. Adding p character to s orbitals puts more electron density along the internuclear axis, where it can be used more effectively in bond formation than it would be if it were in a spherically symmetric s orbital. The energy that is given off by forming four stronger bonds in preference to two weaker ones is more than adequate to compensate for the energy required to promote an electron from the $2s$ level to the sp^3 hybrid state.

2.6 BONDING IN ETHANE

The orbital hybridization model of covalent bonding is readily extended to compounds that contain carbon-carbon bonds—the compounds of greatest interest to organic chemists. As Figure 2.11 illustrates, ethane can be described in terms of a carbon-carbon bond joining two CH_3 (methyl) groups. Each methyl group comprises an sp^3 hybridized carbon attached to three hydrogens by sp^3–$1s$ σ bonds. Overlap of the remaining half-filled sp^3 hybridized orbital of one carbon with that of the other generates a σ bond between them. Here is yet another kind of σ bond, one that has as its basis the overlap of two sp^3 hybridized orbitals.

The remaining sp^3 hybrid orbitals on each of two methyl groups

overlap to form the two-electron σ bond between the two carbon atoms of ethane

FIGURE 2.11 Orbital overlap description of an sp^3-sp^3 σ bond between two carbon atoms.

PROBLEM 2.3 How many carbons are sp^3 hybridized in propane ($CH_3CH_2CH_3$)? How many σ bonds are there in this molecule? Identify the overlaps of atomic orbitals that give rise to each σ bond.

Carbon-carbon σ bonds are slightly weaker than carbon-hydrogen bonds but are still classified as strong bonds. The carbon-carbon bond dissociation energy of ethane, for example, is 88 kcal/mol, while the carbon-hydrogen bonds of ethane have bond dissociation energies of 98 kcal/mol.

Structural features that are characteristic of bonds involving sp^3 hybridized carbon include bond angles and bond distances. These are shown for methane and ethane in Figure 2.12. Carbon-carbon bond distances in alkanes are about 1.54 Å, and carbon-hydrogen bond distances are about 1.10 Å. Because of the symmetry of methane, all its bond angles are perfectly tetrahedral (109.5°). In ethane the H—C—C angles are slightly larger than tetrahedral, forcing the H—C—H angles to be slightly smaller.

2.7 ALKANE NOMENCLATURE. COMMON NAMES

Alkanes are represented by the general formula C_nH_{2n+2}. Methane (CH_4) is the only one-carbon alkane, ethane (CH_3CH_3) the only alkane with two carbon atoms, and propane ($CH_3CH_2CH_3$) the only one with three carbons. Beginning with the molecular formula C_4H_{10}, however, constitutional isomers appear. There are two C_4H_{10} alkanes: one is called *n-butane,* the other *isobutane.*

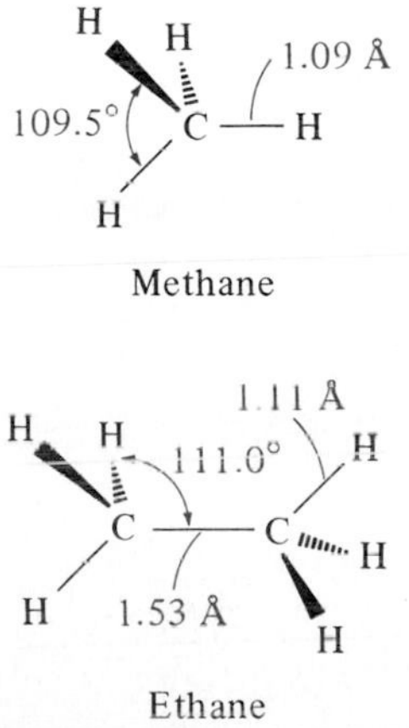

$CH_3CH_2CH_2CH_3$ CH_3CHCH_3
                                    |
                                  CH_3

n-Butane   Isobutane

The *n* in *n*-butane stands for "normal" and signifies an unbranched carbon chain. The *iso-* in isobutane indicates that the molecule is an isomer of *n*-butane in which there is a methyl branch attached to the second carbon.

There are three C_5H_{12} isomers: they are called *n*-pentane, isopentane, and neopentane.

Methane

Ethane

FIGURE 2.12 The structures of methane and ethane.

*Neo seems to be 2 branches
off the same carbon*

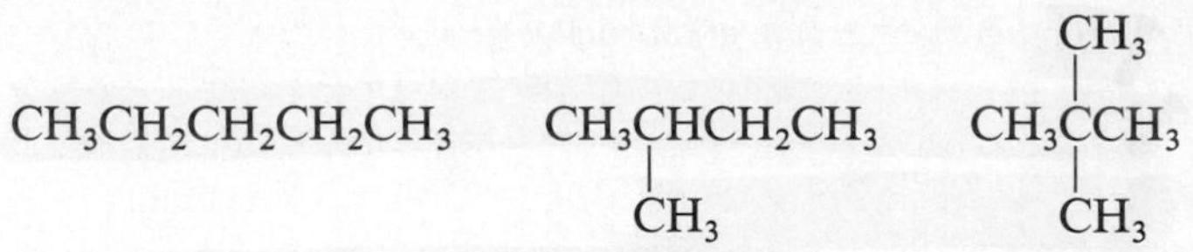

$$CH_3CH_2CH_2CH_2CH_3 \qquad CH_3CHCH_2CH_3 \qquad CH_3CCH_3$$

with CH_3 below on the isopentane, and CH_3 above and CH_3 below on neopentane.

n-Pentane	Isopentane	Neopentane

These are examples of "*common*" or *trivial* names. They are names that have a long history in organic chemistry but do not lend themselves to the development of a systematic vocabulary of chemical names based on well-defined rules. For example, while there are five C_6H_{14} isomers, only three of them have common names. We run out of unique, let alone descriptive, names before we run out of structures.

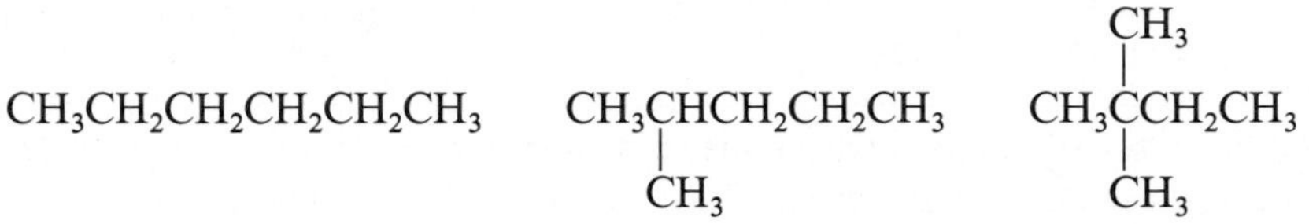

$$CH_3CH_2CH_2CH_2CH_2CH_3 \qquad CH_3CHCH_2CH_2CH_3 \qquad CH_3CCH_2CH_3$$

with CH_3 below on isohexane, and CH_3 above and CH_3 below on neohexane.

n-Hexane	Isohexane	Neohexane

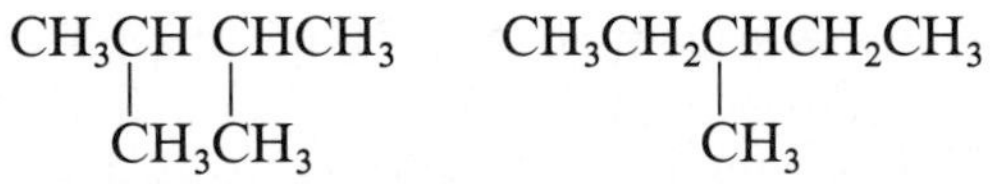

$$CH_3CH \; CHCH_3 \qquad CH_3CH_2CHCH_2CH_3$$

with CH_3CH_3 below the first, and CH_3 below the second.

(No common name)	(No common name)

PROBLEM 2.4   Write carbon skeleton formulas (Section 1.11) for all the C_6H_{14} isomers.

SAMPLE SOLUTION   *n*-Hexane has a continuous chain of six carbon atoms. It is represented in the zig-zag fashion shown below.

$$CH_3CH_2CH_2CH_2CH_2CH_3 \quad \text{may be rewritten in the form} \quad \wedge\!\!\wedge\!\!\wedge$$

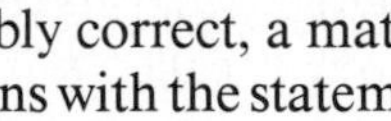

The number of constitutional isomers increases enormously with the number of carbon atoms. There are 9 isomeric C_7H_{16} alkanes; while 75 isomers are possible for $C_{10}H_{22}$, 4347 for $C_{15}H_{32}$ alkanes, 366,319 for $C_{20}H_{42}$ alkanes, and more than 62 trillion isomers for $C_{40}H_{82}$! (Lest you be tempted to try to derive the mathematical expression relating the number of possible isomers to molecular formula, it should be pointed out that while the numbers cited above are probably correct, a mathematician who attempted the problem summarized his conclusions with the statement that no simple mathematical formula exists.)

2.8 SYSTEMATIC ALKANE NOMENCLATURE. INTRODUCTION TO THE IUPAC RULES

It is axiomatic that there should be a different name for every different chemical structure. The world's organic chemists find it convenient to use certain generally accepted common names but have agreed upon a set of rules for systematic organic nomenclature in order to cope with the multitude of structural possibilities. These rules are known as the *International Union of Pure and Applied Chemistry (IUPAC) nomenclature system.* They originated in proposals made by scientists from many nations who convened for this purpose in Geneva, Switzerland in 1892. Periodically the IUPAC Commission on Nomenclature of Organic Chemistry meets to review and supplement the rules. The most recent version of the IUPAC rules was published in 1979.

In this section the IUPAC rules for alkanes are introduced. As we progress through the various classes of organic molecules, we will see how their names can be derived from the fundamental alkane names.

The IUPAC rules begin by specifying the names to be applied to the alkanes having unbranched carbon chains. Methane, ethane, propane, and butane are retained as the names for CH_4, CH_3CH_3, $CH_3CH_2CH_3$, and $CH_3CH_2CH_2CH_3$, respectively. Beginning with pentane, $CH_3CH_2CH_2CH_2CH_3$, the unbranched alkanes are named by appending the suffix *-ane* to a Latin or Greek stem that specifies the number of carbon atoms in the chain. The names of some selected alkanes are presented in Table 2.1. Notice that the prefix *n* is not a part of the IUPAC system; $CH_3(CH_2)_5CH_3$ is heptane, not *n*-heptane.

PROBLEM 2.5 Refer to Table 2.1 as needed to answer the following questions.

(a) Beeswax contains 8 to 9% hentriacontane. Write a condensed structural formula for hentriacontane.

(b) Octacosane has been found to be present in a certain fossil plant. Write a condensed structural formula for octacosane.

SAMPLE SOLUTION (a) Note in Table 2.1 that hentriacontane has 31 carbon atoms. All the alkanes in Table 2.1 have unbranched carbon chains. Hentriacontane has the condensed structural formula $CH_3(CH_2)_{29}CH_3$.

With this as a beginning, let us develop the fundamental principles of IUPAC nomenclature by returning to the question of how to name the five isomers of C_6H_{14}.

TABLE 2.1
IUPAC Names of Unbranched Alkanes

Number of carbon atoms	Name	Number of carbon atoms	Name	Number of carbon atoms	Name
1	Methane	11	Undecane	21	Henicosane
2	Ethane	12	Dodecane	22	Docosane
3	Propane	13	Tridecane	23	Tricosane
4	Butane	14	Tetradecane	24	Tetracosane
5	Pentane	15	Pentadecane	30	Triacontane
6	Hexane	16	Hexadecane	31	Hentriacontane
7	Heptane	17	Heptadecane	32	Dotriacontane
8	Octane	18	Octadecane	40	Tetracontane
9	Nonane	19	Nonadecane	50	Pentacontane
10	Decane	20	Icosane*	100	Hectane

* Spelled "eicosane" prior to 1979 version of IUPAC rules.

2.9   APPLYING THE IUPAC RULES. THE NAMES OF THE C_6H_{14} ISOMERS

The easiest of the C_6H_{14} alkanes to name is the unbranched isomer. By definition (Table 2.1) $CH_3(CH_2)_4CH_3$ is hexane.

$$CH_3CH_2CH_2CH_2CH_2CH_3$$

IUPAC name: hexane
(Common name: *n*-hexane)

Consider next the isomer represented by the structure

$$CH_3CHCH_2CH_2CH_3$$
$$|$$
$$CH_3$$

Step 1

Identify the longest continuous carbon chain and find the IUPAC name in Table 2.1 that corresponds to the unbranched alkane having that number of carbons. This is the parent alkane from which the IUPAC name is to be derived.

In this case, the longest continuous chain has *five* carbon atoms; the compound is named as a derivative of pentane. The key word here is *continuous*. It does not matter whether the carbon skeleton is drawn in an extended straight-chain form or in one with many bends and turns. All that counts is the number of carbons linked together in an uninterrupted sequence.

Step 2

Identify the substituent groups attached to the parent chain.

The parent pentane chain bears a methyl (CH_3) group as a substituent. Alkyl groups are named by dropping the *-ane* ending of the corresponding alkane and replacing it by *-yl*.

Step 3

Locate the position of the substituent groups by number. Number the longest continuous chain in the direction that gives the lowest number to the substituent groups at the first point of branching.

The numbering scheme

$$\overset{1}{C}H_3\overset{2}{C}H\overset{3}{C}H_2\overset{4}{C}H_2\overset{5}{C}H_3 \quad \text{is equivalent to} \quad \overset{2}{C}H_3\overset{3}{C}H\overset{4}{C}H_2\overset{5}{C}H_3$$
$$|\qquad\qquad\qquad\qquad\qquad\qquad\qquad |$$
$$CH_3 \qquad\qquad\qquad\qquad\qquad\qquad\quad _1CH_3$$

Both schemes count five carbon atoms in their longest continuous chain and bear a methyl group as a substituent at the second carbon. An alternative numbering sequence, one that begins at the other end of the chain, is incorrect.

$$\overset{5}{C}H_3\overset{4}{C}H\overset{3}{C}H_2\overset{2}{C}H_2\overset{1}{C}H_3 \quad \text{(methyl group attached to C-4)}$$
$$|$$
$$CH_3$$

Step 4

Write the name of the compound. The parent alkane is the last part of the name and is preceded by the names of the substituent groups and their numerical locations (locants). No punctuation is used between names, but hyphens separate the locants from the names.

$$CH_3CHCH_2CH_2CH_3$$
$$\mathclap{|}$$
$$CH_3$$

IUPAC name: 2-methylpentane
(Common name: isohexane)

Applying the same sequence of four steps leads to the IUPAC name for the isomer that has its methyl group attached to the middle carbon of the five-carbon chain.

$$CH_3CH_2CHCH_2CH_3 \qquad \text{IUPAC name: 3-methylpentane}$$
$$|$$
$$CH_3$$

Both remaining C_6H_{14} isomers have two methyl groups as substituents on a four-carbon chain. When the same group appears more than once as a substituent, the multiplying prefixes *di-, tri-, tetra-,* etc. are appended. A separate locant is used for each substituent, and the locants are separated in the IUPAC name by commas.

$$\begin{array}{c} {}_{,}CH_3 \\ | \\ CH_3CCH_2CH_3 \\ | \\ CH_3 \end{array} \qquad\qquad \begin{array}{c} CH_3 \\ | \\ CH_3CHCHCH_3 \\ | \\ CH_3 \end{array}$$

IUPAC name: 2,2-dimethylbutane   **IUPAC name: 2,3-dimethylbutane**
(Common name: neohexane)

PROBLEM 2.6   Phytane is a naturally occurring alkane produced by the alga *Spirogyra* and is a constituent of petroleum. The IUPAC name for phytane is 2,6,10,14-tetramethyl-hexadecane. Write a structural formula for phytane.

PROBLEM 2.7   Derive the IUPAC names for

(a) The isomers of C_4H_{10}
(b) The isomers of C_5H_{12}
(c) $(CH_3)_3CCH_2CH(CH_3)_2$
(d) $(CH_3)_3CC(CH_3)_3$

SAMPLE SOLUTION   (a)   There are two C_4H_{10} isomers. Butane (Table 2.1) is the IUPAC name for the isomer that has an unbranched carbon chain. The other isomer has three carbons in its longest continuous chain with a methyl branch at the central carbon; its IUPAC name is 2-methylpropane.

$$CH_3CH_2CH_2CH_3 \qquad\qquad \begin{array}{c} CH_3CHCH_3 \\ | \\ CH_3 \end{array}$$

IUPAC name: butane   **IUPAC name: 2-methylpropane**
(Common name: *n*-butane)   (Common name: isobutane)

Our discussion of the IUPAC rules for alkanes is not yet complete. Before proceeding further, however, it is necessary to shift our focus from the examination of entire molecules to the consideration of substituent groups other than methyl that may be attached to the main carbon chain.

2.10 ALKYL GROUPS

Alkyl groups are structural units that lack one of the hydrogen substituents of an alkane. Unbranched alkyl groups in which the point of attachment is at the end of the chain are named in systematic nomenclature by replacing the *-ane* endings of Table 2.1 by *-yl*. Their common names also include the prefix *n-*.

$$CH_3CH_2— \qquad CH_3(CH_2)_5CH_2— \qquad CH_3(CH_2)_{16}CH_2—$$

Ethyl group   **Heptyl** group   **Octadecyl** group
(Common name: *n*-heptyl)   (Common name: *n*-octadecyl)

The dash at the end of the chain in each of these groups represents a potential point of attachment to some other atom or group. When the carbon atom at a potential point of attachment is bonded to only one other carbon, it is referred to as a *primary* carbon. Ethyl, heptyl, and octadecyl are examples of *primary* alkyl groups.

Branched alkyl groups are named by using the longest continuous chain that begins at the point of attachment as the base name. Thus, the systematic names of the two C_3H_7 alkyl groups are propyl and 1-methylethyl. Both are better known by their common names, *n*-propyl and isopropyl, respectively.

$$CH_3CH_2CH_2— \qquad \overset{\displaystyle CH_3}{\underset{\displaystyle |}{CH_3CH—}} \quad or \quad (CH_3)_2CH—$$

Propyl group   **1-Methylethyl** group
(Common name: *n*-propyl)   (Common name: isopropyl)

An isopropyl group is a *secondary* alkyl group; its point of attachment is a carbon atom that is directly bonded to two other carbons.

The C_4H_9 alkyl groups may be derived either from the unbranched carbon skeleton of butane or from the branched carbon skeleton of isobutane. Those derived from butane are the butyl (*n*-butyl) group and the 1-methylpropyl (*sec*-butyl) group.

$$CH_3CH_2CH_2CH_2— \qquad \overset{\displaystyle CH_3}{\underset{\displaystyle |}{CH_3CH_2CH—}}$$

Butyl group   **1-Methylpropyl** group
(Common name: *n*-butyl)   (Common name: *sec*-butyl)

Those derived from isobutane are the 2-methylpropyl (isobutyl) group and the 1,1-dimethylethyl (*tert*-butyl) group.

$$\overset{\displaystyle CH_3}{\underset{\displaystyle |}{CH_3CHCH_2—}} \quad or \quad (CH_3)_2CHCH_2— \qquad \overset{\displaystyle CH_3}{\underset{\displaystyle |}{\underset{\displaystyle CH_3}{\overset{\displaystyle |}{CH_3C—}}}} \quad or \quad (CH_3)_3C—$$

2-Methylpropyl group   **1,1-Dimethylethyl** group
(Common name: isobutyl)   (Common name: *tert*-butyl)

A *tertiary* carbon, as contained within a *tert*-butyl group, is one that is directly bonded to three other carbons.

PROBLEM 2.8   Give the structures of all the C_5H_{11} alkyl groups and identify the carbon at the point of attachment as primary, secondary, or tertiary, as appropriate.

SAMPLE SOLUTION   Consider the alkyl group having the same carbon skeleton as neopentane $(CH_3)_4C$. All the hydrogens of neopentane are equivalent, so replacing any one of them is the same as replacing any of the others. The resulting alkyl group is called a neopentyl group.

$$CH_3CCH_3 \qquad CH_3CCH_2- \qquad or \qquad (CH_3)_3CCH_2-$$

Neopentane   Neopentyl group

The carbon by which the neopentyl group is attached to some other atom or group is classified as primary because it is directly bonded to only one other carbon.

In addition to methyl and ethyl groups, we shall encounter *n*-propyl, isopropyl, *n*-butyl, *sec*-butyl, isobutyl, *tert*-butyl, and neopentyl groups many times throughout this text. While these are common names, they have been integrated into the IUPAC system and are an acceptable adjunct to systematic nomenclature. You should be able to recognize these groups on sight and to produce their structures when needed.

2.11 IUPAC NAMES OF HIGHLY BRANCHED ALKANES

By combining the fundamental principles of IUPAC notation with the names of the various alkyl groups, systematic names of highly branched alkanes are readily derived. Take, for example, the alkane having the constitution

$$CH_3CHCH_3$$
$$\underset{1\quad 2\quad 3\quad 4\quad 5\quad 6\quad 7\quad 8}{CH_3CH_2CH_2CHCH_2CH_2CH_2CH_3}$$

As numbered on the structure, the longest continuous chain contains eight carbons, so the compound is named as a derivative of octane. Since it bears an isopropyl group at C-4, its IUPAC name is **4-isopropyloctane.**
     What happens to the IUPAC name when another substituent, for example, a methyl group at C-3, is added to the structure?

$$CH_3CHCH_3$$
$$CH_3CH_2CHCHCH_2CH_2CH_2CH_3$$
$$CH_3$$

The compound is named as an octane derivative that bears a C-3 methyl group and a C-4 isopropyl group. When two or more different substituents are present, they are listed in alphabetical order in the name. The IUPAC name for this compound is

4-isopropyl-3-methyloctane (because the *i* in isopropyl precedes the *m* in methyl alphabetically).

Replicating prefixes such as di-, tri-, and tetra- are ignored when substituents are arranged alphabetically. Adding a second methyl group to the original structure, at C-5, for example, converts it to **4-isopropyl-3,5-dimethyloctane.**

$$CH_3CHCH_3$$
$$|$$
$$CH_3CH_2CHCHCHCH_2CH_2CH_3$$
$$|\quad\quad|$$
$$CH_3\quad CH_3$$

Italicized prefixes such as *sec-* and *tert-* are ignored except when compared with each other. *tert*-Butyl precedes isobutyl and *sec*-butyl precedes *tert*-butyl.

PROBLEM 2.9   Give an acceptable IUPAC name for each of the following alkanes:

$$CH_3CHCH_3$$
(a) $CH_3CH_2CHCHCHCH_2CHCH_3$
$$|\quad|\quad|$$
$$CH_3\quad CH_3\quad CH_3$$

(b) $(CH_3CH_2)_2CHCH_2CH(CH_3)_2$

$$CH_3$$
$$|$$
(c) $CH_3CH_2CHCH_2CHCH_2CHCH(CH_3)_2$
$$|\quad\quad\quad|$$
$$CH_2CH_3\quad\quad CH_2CH(CH_3)_2$$

SAMPLE SOLUTION   (a)   This problem extends the preceding discussion by adding a third methyl group to 4-isopropyl-3,5-dimethyloctane, the compound just described. It is, therefore, an isopropyltrimethyloctane derivative. Notice, however, that the numbering sequence needs to be changed in order to adhere to the rule that the order must give the lowest number to the substituent at the first point of difference. The first appearing substituent in the former numbering scheme was located at C-3. When numbered from the other end, this compound has a methyl group at C-2 as its first appearing substituent.

$$CH_3CHCH_3$$
$$8\quad 7\quad 6\quad |\quad 4\quad 3\quad 2\quad 1$$
$$CH_3CH_2CHCHCHCH_2CHCH_3$$ 5-Isopropyl-2,4,6-trimethyloctane
$$|\quad 5\quad |\quad\quad |$$
$$CH_3\quad CH_3\quad CH_3$$

The IUPAC nomenclature system is inherently logical and incorporates healthy elements of common sense into its rules. Granted, some long, funny-looking, hard-to-pronounce names are generated. Once one knows the code (rules of grammar) though, it becomes a simple matter to convert that long name to a unique structural formula.

With the ever-increasing use of computer-based information retrieval systems in chemistry, attempts are being made to develop nomenclature systems better suited to such technology. The leading chemical information service, Chemical Abstracts Service, has developed its own nomenclature system to facilitate computer searching. Because it is so different from systematic IUPAC usage, it is only slowly being adopted by chemists. There seems to be little doubt, however, that computer-based information retrieval systems will be the only way to cope with the expanding chemical literature. Some notational system that facilitates this will earn acceptance but will probably never completely replace the IUPAC rules.

2.12 CYCLOALKANE NOMENCLATURE

Cyclic alkanes and their derivatives are frequently encountered in organic chemistry. These *cycloalkanes* are compounds in which the carbon skeleton forms a ring and have the molecular formula C_nH_{2n}. Some examples of cycloalkanes include:

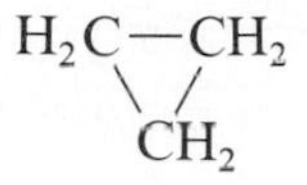

Cyclopropane 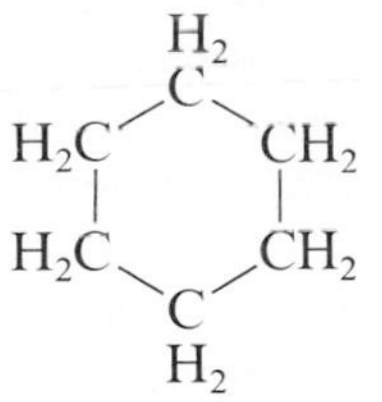   usually represented as

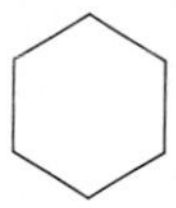

Cyclohexane   usually represented as

As you can see, cycloalkanes are named, under the IUPAC system, by adding the prefix *cyclo-* to the name of the unbranched alkane with the same number of carbons as the ring. Substituent groups are identified in the usual way. Their positions are specified by numbering the carbon atoms of the ring in the direction that gives the lowest number to the substituent groups at the first point of difference.

Ethylcyclopentane 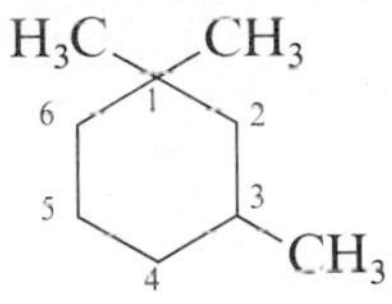   1,1,3-Trimethylcyclohexane

When the ring contains fewer carbon atoms than an alkyl group attached to it, the compound is named as an alkane and the ring is treated as a cycloalkyl substituent.

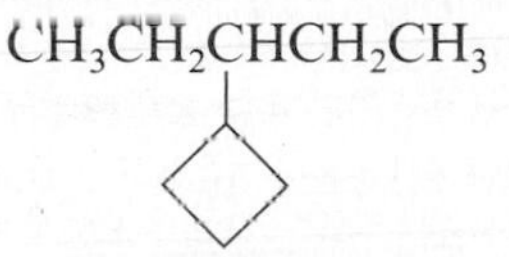

$$CH_3CH_2CHCH_2CH_3$$

3-Cyclobutylpentane

PROBLEM 2.10   Name each of the following compounds:

(a) —C(CH₃)₃   (b)

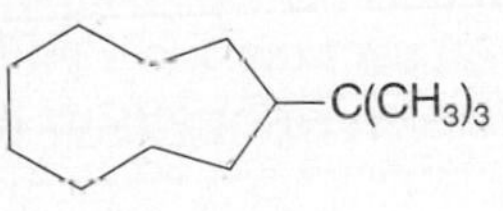

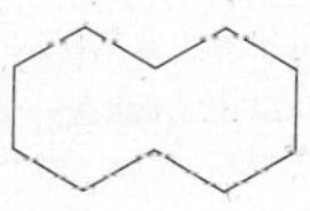

(c) $(CH_3)_2CH$ — [structure: cyclodecane ring with H₃C and CH₃ groups]

(d) [cyclopropyl]—$CHCH_2CH_2CH_3$

(e) [cyclopropyl]—$CH_2CHCH_2CH_3$ [with cyclopropyl substituent]

(f) [bicyclohexyl structure]

SAMPLE SOLUTION   (a)   The molecule has a *tert*-butyl group bonded to a nine-membered cycloalkane. It is *tert*-butylcyclononane.

2.13 SOURCES OF ALKANES

The simplest alkane, methane, comprises a significant portion of the atmospheres of Jupiter and Saturn. The atmospheres of these planets are said to be *reducing.* They are rich in hydrogen, and both nitrogen and carbon are found in their reduced forms, ammonia (NH_3) and methane (CH_4), respectively. Earth's weaker gravity permitted its molecular hydrogen to escape long ago, and its present atmosphere is considered *oxidizing.* Nevertheless, large quantities of methane are present on Earth. Marsh gas, produced by bacteria that thrive on decaying vegetable matter in the absence of oxygen, is mostly methane. Natural gas is about 75 percent methane and contains appreciable quantities of ethane, propane, and butane as well.

Natural gas is often found associated with petroleum deposits. Petroleum, from the Latin words *petra* ("rock") and *oleum* ("oil"), is a complex mixture of hydrocarbons. It can be separated into simpler mixtures by distillation. The lower-molecular-weight hydrocarbons, containing 5 to 14 carbons, are the ones most useful as engine fuels (gasoline, kerosene). Heating oil is a mixture of hydrocarbons in the 14 to 18 carbon range. Higher-molecular-weight mixtures provide lubricating oil, greases, and asphalt.

Many plants have a waxy protective coating on their leaves that contains alkanes. A prominent component of many of these waxes is hentriacontane (see Problem 2.5).

2.14 PHYSICAL PROPERTIES

Table 2.2 summarizes selected physical constants of some alkanes and cycloalkanes. Alkanes and cycloalkanes containing four or fewer carbon atoms are gases at room temperature. In general boiling points in any group of structurally related compounds increase with increasing molecular weight. Figure 2.13 shows this relationship for a series of unbranched alkanes and for their isomers that bear a methyl branch at C-2. By exploring at the molecular level the reasons for the increase in boiling point with molecular weight and the reasons for the difference in boiling points between branched and unbranched alkanes, we can begin to develop some insights into the relationship between structure and properties.

The most instructive way to consider the connection between boiling point and structure is to ask yourself why any substance, for example, pentane, is a liquid rather than a gas. Pentane is a liquid at room temperature and atmospheric pressure because

TABLE 2.2

Physical Properties of Some Alkanes and Cycloalkanes

Compound name	Molecular formula	Condensed structural formula	Melting point, °C	Boiling point, °C (1 atm)
Alkanes				
Methane	CH_4	CH_4	−182.5	−160
Ethane	C_2H_6	CH_3CH_3	−183.6	−88.7
Propane	C_3H_8	$CH_3CH_2CH_3$	−187.6	−42.2
Butane	C_4H_{10}	$CH_3CH_2CH_2CH_3$	−139.0	−0.4
Isobutane	C_4H_{10}	$(CH_3)_3CH$	−160.9	−10.2
Pentane	C_5H_{12}	$CH_3(CH_2)_3CH_3$	−129.9	36.0
Isopentane	C_5H_{12}	$(CH_3)_2CHCH_2CH_3$	−160.5	27.9
Neopentane	C_5H_{12}	$(CH_3)_4C$	−16.6	9.6
Hexane	C_6H_{14}	$CH_3(CH_2)_4CH_3$	−94.5	68.8
Isohexane	C_6H_{14}	$(CH_3)_2CHCH_2CH_2CH_3$	−154.0	60.2
3-Methylpentane	C_6H_{14}	$(CH_3CH_2)_2CHCH_3$	−118.0	63.2
2,2-Dimethylbutane	C_6H_{14}	$(CH_3)_3CCH_2CH_3$	−100.5	49.7
2,3-Dimethylbutane	C_6H_{14}	$(CH_3)_2CHCH(CH_3)_2$	−128.5	58.1
Heptane	C_7H_{16}	$CH_3(CH_2)_5CH_3$	−90.6	98.4
Octane	C_8H_{18}	$CH_3(CH_2)_6CH_3$	−56.9	125.6
Nonane	C_9H_{20}	$CH_3(CH_2)_7CH_3$	−53.6	150.7
Decane	$C_{10}H_{22}$	$CH_3(CH_2)_8CH_3$	−29.7	174.0
Dodecane	$C_{12}H_{26}$	$CH_3(CH_2)_{10}CH_3$	−9.7	216.2
Pentadecane	$C_{15}H_{32}$	$CH_3(CH_2)_{13}CH_3$	10.0	272.7
Icosane	$C_{20}H_{42}$	$CH_3(CH_2)_{18}CH_3$	36.7	205(15mm)
Hectane	$C_{100}H_{202}$	$CH_3(CH_2)_{98}CH_3$	115.1	
Cycloalkanes				
Cyclopropane	C_2H_6		−127.0	−32.9
Cyclobutane	C_4H_8			13.0
Cyclopentane	C_5H_{10}		−94.0	49.5
Cyclohexane	C_6H_{12}		6.5	80.8
Cycloheptane	C_7H_{14}		−13.0	119.0
Cyclooctane	C_8H_{16}		13.5	149.0
Cyclononane	C_9H_{18}			171
Cyclodecane	$C_{10}H_{20}$		9.6	201
Cyclopentadecane	$C_{15}H_{30}$		60.5	112.5(1mm)

there are cohesive forces between molecules (*intermolecular* forces) that are greater in the liquid state than in the vapor. In order to vaporize pentane enough energy must be added to overcome these short-range attractive forces.

Chemists recognize three kinds of cohesive forces as being of primary importance in attracting molecules to each other in the liquid phase. These forces are:

1. Hydrogen bonding
2. Dipole-dipole attractions
3. Attractive van der Waals forces (London dispersion forces)

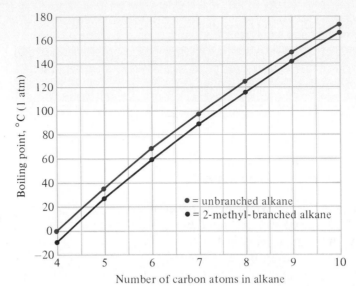

FIGURE 2.13 Boiling points of unbranched alkanes and their 2-methyl-branched isomers. [Temperatures in this text are expressed in degrees Celsius (°C). The SI unit of temperature is the kelvin (K). To convert degrees Celsius to kelvins add 273.15.]

To illustrate how these forces cause molecules to attract one another, compare boiling points among the compounds propane, ethyl fluoride, and ethyl alcohol. These three compounds have the same numbers of electrons, and their molecular weights are similar.

	$CH_3CH_2CH_3$ Propane	CH_3CH_2F Ethyl fluoride	CH_3CH_2OH Ethyl alcohol
Molecular weight	44	48	46
Boiling point	−42°C	−32°C	78°C
Dipole moment	0	1.9 D	1.7 D

Ethyl fluoride has a slightly higher boiling point than propane; thus the forces of intermolecular attraction are slightly greater in liquid ethyl fluoride than they are in liquid propane. Contributing to these forces in liquid ethyl fluoride but absent in propane are dipole-dipole attractive forces. Ethyl fluoride has a substantial dipole moment; propane does not. The positively polarized carbon of one ethyl fluoride molecule is attracted to the negatively polarized fluorine of another. An extended network of these dipole-dipole attractive forces is present in liquid ethyl fluoride, and in order for individual ethyl fluoride molecules to escape from the liquid phase and enter the gas phase, energy (heat) must be added to disrupt this network.

$$
\overset{\delta+}{---CH_2}\!-\!\overset{\delta-}{\ddot{\underset{\cdot\cdot}{F}}:}\overset{\delta+}{---CH_2}\!-\!\overset{\delta-}{\ddot{\underset{\cdot\cdot}{F}}:}\overset{\delta+}{---CH_2}\!-\!\overset{\delta-}{\ddot{\underset{\cdot\cdot}{F}}:}\overset{\delta+}{---CH_2}\!-\!\overset{\delta-}{\ddot{\underset{\cdot\cdot}{F}}:}---etc.
$$

$$
\underset{CH_3}{\quad}\quad\underset{CH_3}{\quad}\quad\underset{CH_3}{\quad}\quad\underset{CH_3}{\quad}
$$

Dipole-dipole attractive forces in ethyl fluoride

The positively polarized hydrogen of a hydroxyl (—OH) group is an especially effective participant in interactions of this type, and the term *hydrogen bonding* is used to describe dipole-dipole attractive forces involving protons bonded to electronegative elements. The considerably higher boiling point of ethyl alcohol, as compared with propane and ethyl fluoride, for example, is primarily attributable to hydrogen bonding. A network of hydrogen bonds between the positively polarized hydrogen and the negatively polarized oxygen of ethyl alcohol molecules must be broken in order to vaporize the liquid.

$$
\begin{array}{ccccccc}
\delta+ & \delta- & \delta+ & \delta- & \delta+ & \delta- & \delta+ & \delta- \\
---\text{H} & \ddot{\text{O}}:---\text{H}—\ddot{\text{O}}: & ---\text{H}—\ddot{\text{O}}:---\text{H}—\ddot{\text{O}}: & ---\text{etc.}\\
| & | & | & | \\
\text{CH}_3\text{CH}_2 & \text{CH}_3\text{CH}_2 & \text{CH}_3\text{CH}_2 & \text{CH}_3\text{CH}_2
\end{array}
$$

Hydrogen bonding in ethyl alcohol

Individually, hydrogen bond strengths are on the order of about 5 kcal/mol — not very great compared with the usual covalent bond energies of 80 to 100 kcal/mol. Nevertheless, these weak interactions can, in the aggregate, produce large effects. Hydrogen bonding can be expected in molecules that have polar covalent bonds between hydrogen and an electronegative element with unshared electron pairs — compounds with O—H and N—H bonds, especially.

PROBLEM 2.11   The molecular weight of hydrogen fluoride is the same as the atomic weight of neon (20). One of these substances has a boiling point of −246°C, the other a boiling point of 19°C. Match the substance with its boiling point and explain the reason for your choice.

Ethyl alcohol has intermolecular attractive forces (hydrogen bonding) that are not available to ethyl fluoride and is higher boiling. Ethyl fluoride has intermolecular attractive forces (dipole-dipole interactions) that are not available to propane and is higher-boiling, but not much higher. Liquid propane must have some forces that cause its molecules to "stick" together at temperatures below −42°C. What is the nature of these remaining attractive forces?

You might think that the interaction between two nonpolar molecules as they approach each other would steadily become more repulsive. Indeed, this is exactly what happens when molecules are brought too close together. At distances less than the sum of their *van der Waals radii,* two atoms that are not directly bonded resist closer contact because electron-electron repulsions between them increase dramatically. At separations slightly greater than the sum of their van der Waals radii, however, nonbonded atoms and groups tend to attract each other. Consider two nonpolar molecules A and B — molecules where the centers of positive and negative charge coincide.

The distribution of electrons in A and B is not static but fluctuates. As A and B approach each other, the electric field of A adjusts to the presence of B and vice versa. The electric field of B induces a polarization of A so that its centers of positive and negative charge no longer coincide.

A   B

Polarization in A induces a complementary polarization in B.

A   B

Molecule A and molecule B thereby experience a mutual attraction that is the result of an *induced dipole–induced dipole* interaction. Induced dipole–induced dipole interactions are also known as *attractive van der Waals interactions* and as *London dispersion forces.*

It is important to remember that these are temporary dipoles induced by the rapidly fluctuating electric fields in molecules. Their effect is to cause neighboring molecules to attract each other. Extended assemblies of induced dipole–induced dipole interactions can accumulate in large molecules, again in a complementary sense, to give rise to substantial intermolecular attractive forces.

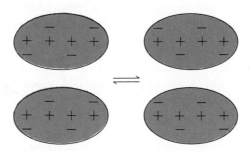

The magnitude of these London forces depends on the number of electrons in the molecules and on how easily these electrons are distorted by nearby electric fields. The reason that boiling point increases with increasing molecular weight among similar compounds, such as a series of alkanes, is simply that the higher-molecular-weight compound has more electrons and can participate in more induced dipole–induced dipole attractions.

Unbranched alkanes have higher boiling points than their branched isomers because their extended shape permits more points of contact for intermolecular associations. Compare the boiling points of pentane and its isomers.

$$CH_3CH_2CH_2CH_2CH_3 \qquad CH_3\underset{\underset{\displaystyle CH_3}{|}}{C}HCH_2CH_3 \qquad CH_3\underset{\underset{\displaystyle CH_3}{|}}{\overset{\overset{\displaystyle CH_3}{|}}{C}}CH_3$$

Pentane   Isopentane   Neopentane
(bp 36°C)   (bp 28°C)   (bp 9°C)

Neopentane has a more compact structure than either of its isomers, i.e., it has the smallest surface area over which neighboring molecules can approach each other. It engages in fewer induced dipole–induced dipole attractions than either of its isomers, is less associated in the liquid phase, and has the lowest boiling point.

London forces are very weak interactions individually, but a typical substance can participate in so many of them that they are probably the most important of all the contributors to intermolecular forces in the liquid state. They are the only forces of attraction possible between nonpolar molecules.

Solid alkanes are soft, generally low-melting materials. The forces responsible for holding the crystal together are the same induced dipole–induced dipole interactions that operate between molecules in the liquid phase. The assemblies in the solid are more highly organized than in the liquid and are arranged in a network which tends to maximize the intermolecular attractive forces. Highly symmetrical molecules usually have higher melting points than those with less symmetry because they can pack together better in a crystal lattice. *Adamantane*, for example, is more symmetrical and has a much higher melting point than its isomer *twistane*.

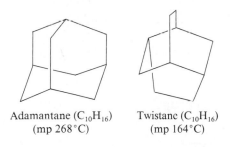

Adamantane ($C_{10}H_{16}$)   Twistane ($C_{10}H_{16}$)
(mp 268°C)   (mp 164°C)

2.15 CHEMICAL PROPERTIES. COMBUSTION OF ALKANES

An older name for alkanes is *paraffin hydrocarbons*. Paraffin is derived from the Latin words *parum affinis* ("with little affinity") and testifies to the low level of reactivity of alkanes. Like most other organic compounds, however, alkanes burn readily in air. Their combination with oxygen is known as *combustion* and is quite exothermic. All hydrocarbons yield carbon dioxide and water as the products of their combustion.

General equation for combustion of alkanes

$$C_nH_{2n+2} + \left(\frac{3n+1}{2}\right)O_2 \longrightarrow nCO_2 + (n+1)H_2O + heat$$

Alkane   Oxygen   Carbon dioxide   Water

The heat released on complete combustion of a substance is called its *heat of combustion*. The heat of combustion is equal to $-\Delta H°$ for the reaction written in the direction shown. By convention

$$\Delta H° = H°_{products} - H°_{reactants}$$

where $H°$ is the heat content, or *enthalpy*, of a compound in its standard state, i.e., the ideal gas, pure liquid, or crystalline solid at a pressure of 1 atm. In an exothermic process the heat content of the products is less than the heat content of the starting materials and $\Delta H°$ is a negative number.

$$CH_4 + 2O_2 \longrightarrow CO_2 + 2H_2O \qquad \Delta H° = -212.8 \text{ kcal/mol}$$

Methane   Oxygen   Carbon   Water
dioxide

$$(CH_3)_2CHCH_2CH_3 + 8O_2 \longrightarrow 5CO_2 + 6H_2O \qquad \Delta H° = -843.4 \text{ kcal/mol}$$

2-Methylbutane   Oxygen   Carbon   Water
(isopentane)   dioxide

Since heats of combustion correspond to $-\Delta H°$, they are positive numbers and are expressed without an algebraic sign.

The heat of combustion of methane is less than that of 2-methylbutane. Burning both alkanes converts them to carbon dioxide and water, but on a molar basis 2-methylbutane yields more of these products than does methane and gives off more heat.

Table 2.3 lists the heats of combustion of several alkanes. The unbranched alkanes constitute a *homologous series,* i.e., a group of structurally related compounds that differ among themselves by the sequential addition of methylene (CH_2) groups. There is a regular increase of about 156 kcal/mol in the heat of combustion of the unbranched alkanes for each additional methylene group that they contain. The heats of combustion of the alkanes having a methyl branch at C-2 are slightly less than those of their unbranched isomers. These methyl-branched isomers constitute a separate homologous series, and the difference between succeeding members of this series is also about 156 kcal/mol for each additional methylene group.

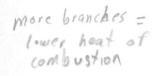
more branches = lower heat of combustion

TABLE 2.3

Heats of Combustion ($-\Delta H°$) of Representative Alkanes

Compound	Formula	$-\Delta H°$ (kcal/mol)
Unbranched alkanes (*n*-alkanes)		
Methane	CH_4	212.8
Ethane	CH_3CH_3	372.8
Propane	$CH_3CH_2CH_3$	530.6
Butane	$CH_3(CH_2)_2CH_3$	687.4
Pentane	$CH_3(CH_2)_3CH_3$	845.3
Hexane	$CH_3(CH_2)_4CH_3$	995.0
Heptane	$CH_3(CH_2)_5CH_3$	1151.3
Octane	$CH_3(CH_2)_6CH_3$	1307.5
Nonane	$CH_3(CH_2)_7CH_3$	1463.9
Decane	$CH_3(CH_2)_8CH_3$	1620.1
Undecane	$CH_3(CH_2)_9CH_3$	1776.1
Dodecane	$CH_3(CH_2)_{10}CH_3$	1932.7
Hexadecane	$CH_3(CH_2)_{14}CH_3$	2557.6
2-Methyl-branched alkanes (isoalkanes)		
2-Methylpropane	$(CH_3)_2CHCH_3$	685.4
2-Methylbutane	$(CH_3)_2CHCH_2CH_3$	843.4
2-Methylpentane	$(CH_3)_2CHCH_2CH_2CH_3$	993.6
2-Methylhexane	$(CH_3)_2CH(CH_2)_3CH_3$	1150.0
2-Methylheptane	$(CH_3)_2CH(CH_2)_4CH_3$	1306.3

PROBLEM 2.12 Using the data in Table 2.3, estimate the heats of combustion of each of the following alkanes:

(a) 2-methylnonane (b) tetradecane (c) icosane

SAMPLE SOLUTION (a) The last entry for the group of 2-methylalkanes (isoalkanes) in the table is 2-methylheptane. Its heat of combustion is 1306 kcal/mol. Since 2-methyl-nonane has two more methylene groups than 2-methylheptane, its heat of combustion is 2×156 kcal/mol higher.

Heat of combustion of 2-methylnonane $= 1306 + 2(156) = 1618$ kcal/mol

More highly branched alkanes have even lower heats of combustion, as the data obtained for selected C_8H_{18} isomers reveal.

Compound	Heat of combustion (kcal/mol)
$CH_3(CH_2)_6CH_3$ Octane	1307.5 kcal/mol
$(CH_3)_2CHCH_2CH_2CH_2CH_2CH_3$ 2-Methylheptane	1306.3 kcal/mol
$(CH_3)_3CCH_2CH_2CH_2CH_3$ 2,2-Dimethylhexane	1304.6 kcal/mol
$(CH_3)_3CC(CH_3)_3$ 2,2,3,3-Tetramethylbutane	1303.0 kcal/mol

Since these compounds are isomers, their combustion corresponds in every case to the chemical equation

$$C_8H_{18} + \tfrac{25}{2}O_2 \longrightarrow 8CO_2 + 9H_2O$$

By constructing an *energy diagram* such as that shown in Figure 2.14, it can be seen that because the final state is common to all the reactions, the energies of the starting alkanes may be compared directly. The isomer with the largest heat of combustion, octane, has the highest potential energy. The isomer with the smallest heat of combustion, 2,2,3,3-tetramethylbutane, has the lowest potential energy. *Potential energy* is comparable with enthalpy; it is the energy a molecule has, exclusive of its kinetic energy of motion. A molecule with more potential energy is less stable than one with less potential energy; thus, the most highly branched C_8H_{18} isomer has the most stable arrangement of atoms and electrons, while the unbranched isomer has the least stable arrangement.

The most reasonable explanation for the small but real difference in stability between branched and unbranched alkanes has to do with London dispersion forces. We noted earlier that the more compact shape of branched alkanes leads to decreased forces of attraction between molecules (*intermolecular* forces). For the same reason —the more compact structure of branched alkanes—the forces of attraction within a molecule (*intermolecular* forces) are greater in these compounds. Atoms or groups are sufficiently close together in highly branched alkanes that the van der Waals

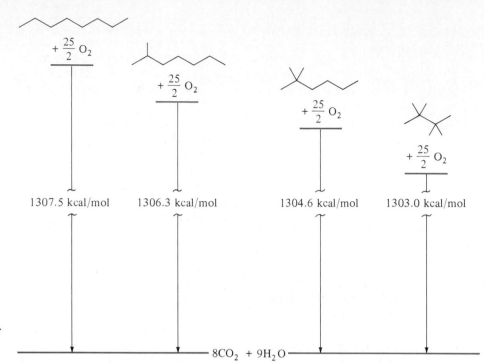

FIGURE 2.14 Energy diagram comparing heats of combustion of isomeric C_8H_{18} alkanes.

attraction between the groups confers a small increment of stabilization to the molecule compared with its less branched isomers.

In the next chapter you will see what happens when atoms and groups within a molecule approach each other too closely and how molecules adjust to van der Waals repulsive forces.

PROBLEM 2.13 Without consulting Table 2.3 arrange the following compounds in order of decreasing heat of combustion: pentane, isopentane, neopentane, hexane.

The kind of reasoning that related heats of combustion data to the relative energies of molecules and after that to an explanation based on molecular structure is typical of that used in organic chemistry. You will encounter similar thought processes frequently throughout this text.

2.16 SUMMARY

According to molecular orbital theory, electrons in molecules are not localized in individual bonds but occupy regions in space, molecular orbitals, which may be so extensive as to include all the atoms of the molecule. Molecular orbitals are approximated as combinations of atomic orbitals; overlap of two atomic orbitals generates a *bonding* molecular orbital and an *antibonding* molecular orbital. Electrons in a bonding molecular orbital are more strongly bound than in an isolated atom; electrons in an antibonding molecular orbital are less strongly bound.

Orbitals characterized by rotational symmetry about the internuclear axis are called σ orbitals. The bond in H_2 is termed a σ bond; two electrons occupy the σ orbital and bind two hydrogen atoms together.

A modified type of molecular orbital theory based on the concept of *orbital hybridization* provides a useful model of bonding in organic compounds. Mixing the

$2s$ and the three $2p$ orbitals of carbon produces a set of sp^3 hybrid orbitals. In methane each of these sp^3 hybrid orbitals can overlap with the $1s$ orbital of a hydrogen atom to form a C—H σ bond. Each of the hydrogen atoms of methane occupies the corner of a tetrahedron, with the carbon atom at the center. Carbon-carbon bonds in alkanes are viewed as σ bonds arising from overlap of an sp^3 hybrid orbital of one carbon with the sp^3 hybrid orbital of another.

Alkanes are compounds of the general formula C_nH_{2n+2}. Cycloalkanes contain a ring of carbon atoms and have the general formula C_nH_{2n}.

A single alkane may have several different names; a name may be a *common* or *trivial* name or it may be a *systematic* name developed by a well-defined set of rules. The systematic nomenclature system that is the most widely used is the *IUPAC system*. According to IUPAC nomenclature, alkanes are named as derivatives of unbranched parents. Substituents on the longest continuous chain are identified and their positions specified by number.

Alkanes and cycloalkanes are not very polar and the forces of attraction between molecules are relatively weak. These forces are primarily induced dipole–induced dipole attractions and result from the mutual and complementary polarization of the electric fields of neighboring molecules.

Alkanes and cycloalkanes burn in air to give carbon dioxide, water, and heat. The heat evolved is called the *heat of combustion.* By comparing the heats of combustion of isomeric substances, their relative energies are revealed. Branched alkanes have lower heats of combustion than their unbranched isomers because they have less potential energy. They have less potential energy because they are stabilized by intramolecular attractive forces of the induced dipole–induced dipole type.

PROBLEMS

2.14 Write the structures and give the IUPAC names for all the alkanes that have the molecular formula C_7H_{16}.

2.15 Write a structural formula for each of the following compounds:

(a) 3-Ethylhexane
(b) 2,3-Dimethyl-6-isopropylnonane
(c) 4-*tert*-Butyl-3-methylheptane
(d) 4-Isobutyl-1,1-dimethylcyclohexane
(e) *sec*-Butylcycloheptane
(f) Dicyclopropylmethane
(g) Cyclobutylcyclopentane

2.16 Which of the compounds in each of the following groups are isomers?

(a) Butane, cyclobutane, isobutane, 2-methylbutane
(b) Cyclopentane, neopentane, 2,2-dimethylpentane, 2,2,3-trimethylbutane
(c) Cyclohexane, hexane, methylcyclopentane, 1,1,2-trimethylcyclopropane
(d) Ethylcyclopropane, 1,1-dimethylcyclopropane, 1-cyclopropylpropane, cyclopentane
(e) 4-Methyltetradecane, 2,3,4,5-tetramethyldecane, pentadecane, 4-cyclobutyldecane

2.17 A certain alkane isolated from a species of blue-green alga has a molecular weight of 240 and an unbranched carbon chain. Identify this alkane.

2.18 Female tiger moths signify their presence to male moths by giving off a sex attractant. The sex attractant has been isolated and found to be a 2-methyl-branched alkane having a molecular weight of 254. What is this material?

2.19 Write a balanced chemical equation for the combustion of each of the following compounds.

    (*a*)   Decane
    (*b*)   Cyclodecane
    (*c*)   Methylcyclononane
    (*d*)   Cyclopentylcyclopentane

2.20 When used as a fuel, would methane or butane generate more heat for the same mass of gas? Which would generate more heat for the same volume of gas? (Use the data in Table 2.3 to enable you to calculate the amount of heat produced in each case.)

2.21 In each of the following groups of compounds, identify the one with the largest heat of combustion and the one with the smallest. (Try to do this problem without consulting Table 2.3.)

    (*a*)   Hexane, heptane, octane
    (*b*)   Isobutane, pentane, isopentane
    (*c*)   Isopentane, 2-methylpentane, neopentane
    (*d*)   Pentane, 3-methylpentane, 3,3-dimethylpentane
    (*e*)   Ethylcyclopentane, ethylcyclohexane, ethylcycloheptane

2.22 Account for all the electrons in each of the following species, assuming sp^3 hybridization of the second row element in each case. Which electrons are found in sp^3 hybridized orbitals? Which are found in σ bonds?

    (*a*)   Ammonia (NH_3)   (*e*)   Borohydride anion (BH_4^-)
    (*b*)   Water (H_2O)   (*f*)   Amide anion ($:\overline{N}H_2$)
    (*c*)   Hydrogen fluoride (HF)
    (*d*)   Ammonium ion (NH_4^+)   (*g*)   Methyl anion ($:\overline{C}H_3$)

2.23 Of the orbital overlaps represented below, one is bonding, one is antibonding, and the other is nonbonding (neither bonding nor antibonding). Which pattern of orbital overlap corresponds to which interaction? Why?

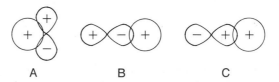

          A   B   C

2.24 A useful exercise, one that tests your spatial perception and understanding of the tetrahedral orientation of the bonds to carbon, is to imagine the carbon atom of methane as located in the center of a cube with its hydrogens at four of the corners. Two representations of a cube containing a carbon and two hydrogens are shown below; complete the drawings by adding the remaining hydrogens at the corners of the cubes so as to give the correct tetrahedral geometry of methane.

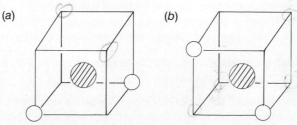

    (*a*)   (*b*)

CONFORMATIONS OF ALKANES AND CYCLOALKANES

In this chapter we expand our structural perspective by examining the *conformations* that are adopted by individual molecules. Conformations are the various spatial arrangements of a molecule that are generated by rotation about single bonds. We will focus most of our attention on three molecules: ethane, butane, and cyclohexane. Through a detailed discussion of these three substances the fundamental principles of *conformational analysis* will be developed.

The particular conformation that a molecule adopts can exert a profound influence on its physical and chemical properties. Conformational analysis is a tool used not only by organic chemists to help them understand their discipline but also by research workers in the life sciences as they attempt to develop a clearer picture of how molecules—from simple to very complex ones—interact with each other in living systems.

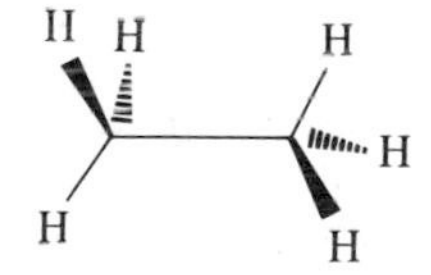

(*a*) Wedge-and-dash drawing of ethane

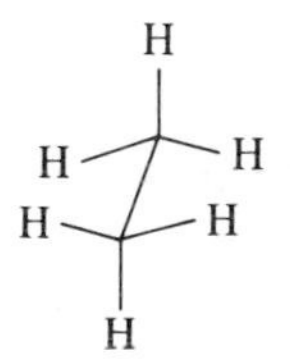

3.1 CONFORMATIONS OF ETHANE

Any discussion of conformational analysis in organic chemistry logically begins with ethane because it is the simplest hydrocarbon that possesses distinct conformations. Among the ways in which molecular conformations are depicted, the wedge-and-dash, sawhorse, and Newman projection drawings are used more often than any others. These are illustrated for ethane in Figure 3.1.

In the *wedge-and-dash* representation (Figure 3.1*a*), a wedge indicates a bond coming from the plane of the paper toward the viewer, a normal line is a bond in the plane of the paper, and a dashed line represents a bond receding from the viewer. Tetrahedral bond angles are assumed, and wedge-and-dash drawings attempt to show this.

A sawhorse drawing (Figure 3.1*b*) is rather similar in concept except that by altering the orientation of the molecule with respect to the viewer, it is possible to adequately represent the conformation without resorting to different styles of bonds. The orientation of the bonds to carbon is somewhat distorted in a sawhorse drawing, so it is important to remember that there is a tetrahedral arrangement of the four bonds to each carbon.

(*b*) Sawhorse drawing of ethane

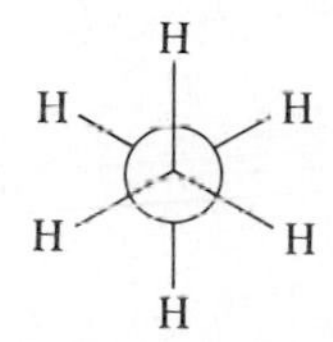

(*c*) Newman projection of ethane

FIGURE 3.1 Some commonly used representations of staggered conformations of ethane.

Newman projection formulas were devised by Professor M. S. Newman of Ohio State University. In a Newman projection formula of ethane (Figure 3.1c), we sight down the carbon-carbon bond, represent the front carbon by a point, and the back carbon by an open circle. Each carbon has three hydrogen substituents; these are placed symmetrically around each carbon.

PROBLEM 3.1 Identify the alkanes corresponding to each of the representations shown.

(a)

```
        CH₃
   H         H

   H         H
        H
```

(c)

```
        CH₃
   H         H

   H         CH₃
        CH₃
```

(b)

```
        CH₃
     H     H

  H     H
     CH₃
```

(d)

```
        CH₃
     H     CH₂CH₃

  H     H
     CH₂CH₃
```

SAMPLE SOLUTION (a) The Newman projection formula of this alkane resembles that of ethane except that one of the hydrogen substituents has been replaced by a methyl group. The drawing is a Newman projection formula of propane, $CH_3CH_2CH_3$.

The structural feature that the drawings in Figure 3.1 are designed to illustrate is the spatial relationship between substituents on adjacent carbon atoms. Each H—C—C—H unit in ethane is characterized by a *torsion angle* or *dihedral angle*. Torsion angle or dihedral angle is the angle between the H—C—C plane and the C—C—H plane of an H—C—C—H unit. It is easily seen in a Newman projection of ethane as the angle between C—H bonds of adjacent carbons.

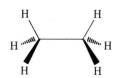

(a) Wedge-and-dash

(b) Sawhorse

(c) Newman projection

FIGURE 3.2 Some commonly used representations of eclipsed conformations of ethane.

```
   H ←60°
        H



Torsion angle = 60°
```

```
   H ←

        180°

   H ←

Torsion angle = 180°
```

When the torsion angle between two substituents is approximately 60°, we say that the spatial relationship between those substituents is *gauche*. When the torsion angle is approximately 180°, we say that the relationship between the two substituents is *anti*.

Conformations in which there are only gauche and anti relationships between substituents are said to be *staggered conformations*. All the drawings in Figure 3.1 represent staggered conformations of ethane. These are to be distinguished from the drawings in Figure 3.2, which represent *eclipsed* conformations of ethane. Conformations are said to be eclipsed when the torsion angles between adjacent substituents are 0°.

In principle there are an infinite number of conformations of ethane, differing by only tiny increments in their torsion angles. This raises the question of whether any one conformation is more stable than the others, and if so, which one is it? A related question has to do with interconversion of conformations. Rotation about the carbon-carbon bond of ethane converts a staggered conformation to an eclipsed one. How rapidly do these conformations interconvert?

3.2 INTERNAL ROTATION IN ETHANE

During the 1930s a combination of experimental measurements and theoretical calculations revealed that rotation about the carbon-carbon bond in ethane, while very rapid, is not completely "free." Figure 3.3 is a potential energy diagram that shows how the internal energy of ethane changes as one of its methyl groups rotates through an angle of 360° around the carbon-carbon bond.

Potential energy diagrams such as that of Figure 3.3, in which we plot the progress of a particular transformation as the x axis versus potential energy as the y axis, can aid the understanding of chemical and physical processes. As noted earlier in Section 2.15, potential energy is the energy a molecule possesses in excess of its kinetic energy. When comparing conformations, the one with the least potential energy is the most stable.

Figure 3.3 shows that the staggered conformations of ethane are 2.9 kcal/mol lower in energy than the eclipsed conformations. The three equivalent staggered conformations correspond to potential energy minima, while the three equivalent eclipsed conformations represent potential energy maxima.

What makes the staggered conformation of ethane more stable than the eclipsed conformation? A number of factors contribute to its greater stability, but the most important one seems to be that the staggered conformation allows for the maximum separation of bonded electron pairs. In the eclipsed conformation the electrons in the C—H bonds of neighboring carbon atoms are aligned. Electron pair repulsions are

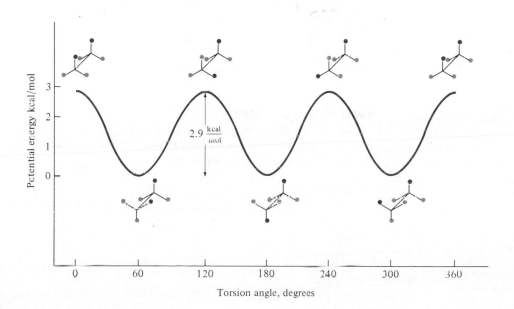

FIGURE 3.3 Potential energy diagram for internal rotation in ethane. Hydrogen substituents are represented by circles, and two of them are darkened so as to more clearly indicate the process of rotation about the carbon-carbon bond.

greater when the bonds are eclipsed than when the bonds are staggered. We call the destabilization energy associated with the eclipsing of bonds on adjacent atoms *torsional strain.*

While it might be supposed that the eclipsed form of ethane is destabilized by van der Waals repulsions between hydrogens on adjacent carbons, this is probably not very significant energetically. Hydrogens on adjacent carbons in ethane are separated by about 2.3 Å, a distance that is not very much different from the sum of the van der Waals radii of two hydrogen atoms (1.2 Å each). It is only when atoms approach each other at distances less than the sum of their van der Waals radii that repulsive interactions between them become appreciable.

The energy difference of 2.9 kcal/mol between the staggered and the eclipsed conformations is the *energy of activation* (E_{act}) for the conversion of one staggered form to another staggered form. Every dynamic process in chemistry has an activation energy associated with it. Activation energy is the energy that a molecule must possess above the energy of its ground state in order to be transformed into some other species. Molecules must become energized in order to undergo chemical reaction or as in this case, to undergo rotation around a carbon-carbon bond. Kinetic (thermal) energy is absorbed by a molecule from its surroundings and is transformed into potential energy. When the potential energy exceeds a certain minimum value (the activation energy), the unstable arrangement of atoms that exists at that instant (called the *activated complex* or the *transition state*) relaxes to a more stable arrangement, giving off its excess potential energy as kinetic energy in collisions with other molecules or with the walls of the container. The eclipsed conformation of ethane corresponds to a transition state for the conversion of one staggered conformation to another. In any reaction or process, the transition state is the point of maximum potential energy that must be traversed during the conversion of reactants to products.

An activation energy of 2.9 kcal/mol is quite small. The relationship between an activation energy of this magnitude and the amount of thermal energy available from the surroundings is such that staggered forms of ethane interconvert millions of times each second at room temperature. Internal rotation in ethane is not completely "free" but is an exceedingly rapid process. The force that resists free rotation is the torsional strain that must be overcome at the transition state, and this torsional strain results from the eclipsing of bonds on adjacent carbons in the activated complex.

3.3 CONFORMATIONAL ANALYSIS OF BUTANE

The next alkane that we will examine is butane. In particular, we will consider conformations related by internal rotation about the central carbon-carbon bond. There are two distinctly different staggered conformations of butane; one is called the *gauche* conformation, the other the *anti* conformation.

Gauche conformation
of butane

Anti conformation
of butane

In the gauche conformation the torsion angle between the two methyl groups is 60°. In the anti conformation the two methyl groups are much farther apart and define a torsion angle of 180°. The anti conformation is 0.8 kcal/mol more stable than the gauche. In butane at 25°C, approximately twice as many molecules exist in the anti conformation as in the gauche.

The gauche conformation of butane is less stable than the anti because the close approach of its two methyl groups adds an increment of *van der Waals strain* that is absent in the anti conformation. Because both conformations have a staggered arrangement of their bonds, both are free of torsional strain.

Similarly, there are two distinctly different eclipsed conformations about the C-2—C-3 bond of butane. In one each methyl group is eclipsed with a hydrogen substituent; in the second the two methyl groups eclipse each other.

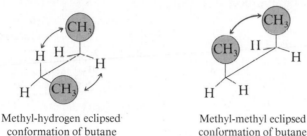

Methyl-hydrogen eclipsed conformation of butane

Methyl-methyl eclipsed conformation of butane

Both conformations have the same number of eclipsed bonds and possess about the same degree of torsional strain. The methyl-methyl eclipsed conformation, however, has significantly more van der Waals strain than the methyl-hydrogen eclipsed con-

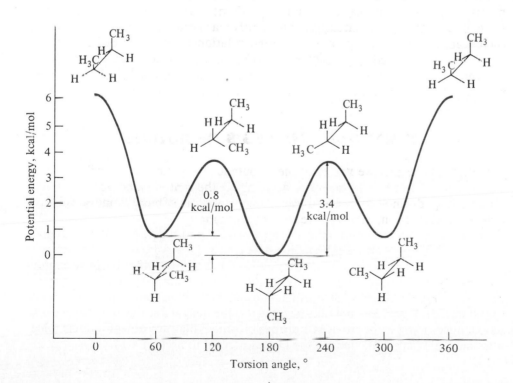

FIGURE 3.4 Potential energy diagram for internal rotation around the central carbon-carbon bond in butane.

formation and is higher in energy. Methyl groups take up more space, i.e., are larger, than are hydrogen substituents.

Figure 3.4 shows the potential energy relationships among the various conformations of butane. The staggered conformations are more stable than the eclipsed ones. As just described, however, all the staggered conformations of butane do not have the same energy, nor do all the eclipsed conformations have the same energy. The point of maximum potential energy, the methyl-methyl eclipsed conformation, is 6.1 kcal/mol higher in energy than the point of minimum potential energy, the anti conformation. This energy difference of 6.1 kcal/mol is the activation energy for rotation about the central carbon-carbon bond in butane. It is higher than the 2.9 kcal/mol energy barrier of ethane because it incorporates an increment of van der Waals strain along with the torsional strain present in the activated complex. Rotation about the C-2—C-3 bond in butane, while slower than internal rotation in ethane, is a very rapid process.

PROBLEM 3.2   Sketch a potential energy diagram for internal rotation in propane. Clearly identify each potential energy maximum and minimum with a structural formula that shows the conformation of propane at that point. Does your potential energy diagram more closely resemble that of ethane or of butane? Would you expect the activation energy for internal rotation in propane to be more than or less than that of ethane? Of butane?

3.4   CONFORMATIONS OF HIGHER ALKANES

In describing the conformations of higher alkanes it is often more helpful to look at them from the side rather than end-on as in a Newman projection. Viewed from this perspective, the most stable conformations of pentane and hexane have their carbon "backbones" arranged in a zigzag fashion. In a zigzag arrangement of this kind all the bonds are staggered and only anti arrangements of C—C—C—C units are present.

All-anti conformation of pentane   All-anti conformation of hexane

Higher alkanes having unbranched carbon chains are also most stable in their all-anti conformations.

Interconversion of conformations is rapid in the liquid and gaseous states but does not take place in crystals. The crystalline state is a rigid matrix that retards molecular motion. When a particular substance crystallizes, it usually does so in a single conformation. One of the factors that determines the conformation adopted in the crystal is the ability of molecules to pack together. Unbranched alkanes normally crystallize in the all-anti conformation, not only because it is the most stable one but also because its extended zigzag shape provides more opportunities than any other for intermolecular van der Waals attractions of the kind described in Section 2.14.

3.5 MOLECULAR MODELS

We can gain a heightened appreciation for those features which affect structure and reactivity if we can examine a molecule in three dimensions, hold it in various orientations, and measure distances between atoms. As early as the nineteenth century many chemists built scale models of molecules in order to understand them better. Figure 3.5 shows three common types of molecular models of ethane.

Ball-and-stick models (Figure 3.5*a*) are exactly what their name implies. Carbon atoms are represented by round wooden balls with four holes drilled into them at the proper tetrahedral angle. Hydrogen atoms are represented by smaller balls with a single hole; oxygen has two holes, nitrogen three, etc. Dowels inserted in these holes are the bonds that connect the atoms together.

Framework models (Figure 3.5*b*) may be of several types, but all share the common feature of showing the shape of a molecule by emphasizing its bonds. *Space-filling* models (Figure 3.5*c*), on the other hand, embody the opposite concept — they emphasize a molecule's atoms and the space that they occupy. Space-filling models are especially useful when analyzing the effects of crowding within a molecule or in seeing how two molecules can best approach each other.

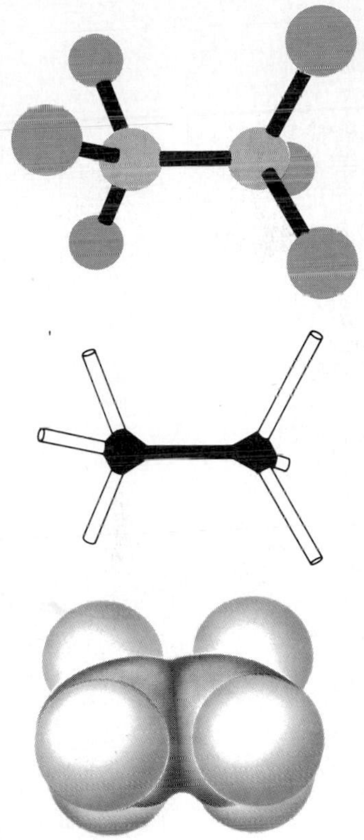

You will find that investing in even an inexpensive set of molecular models will pay large dividends. The innate ability to visualize molecules in three dimensions is a skill with which few individuals are blessed but one which almost everyone can acquire with a little practice. Working with molecular models is the best way to develop this facility. Understanding the three-dimensional nature of molecules is indispensable to the understanding of chemistry and biology. *Stereochemistry* (from Greek *stereos,* "solid") is the name given to the chemical considerations associated with the spatial arrangement of atoms.

Computers have affected many aspects of science and they will affect molecular modeling in significant ways. Already it is a routine matter to have a computer draw a three-dimensional representation of a molecule and to turn it so that it may be viewed in many different orientations. If you have access to a personal computer and a molecular modeling program, it can be a powerful complement to, but probably not a substitute for, a set of molecular models.

FIGURE 3.5 *(a)* Ball-and-stick model of ethane; *(b)* framework molecular model of ethane; *(c)* space-filling model of ethane.

3.6 THE SHAPES OF CYCLOALKANES. PLANAR OR NONPLANAR?

During the nineteenth century it was widely believed — erroneously, as we shall see — that the carbon skeletons of cycloalkanes are planar. A prominent advocate of this point of view was the German chemist Adolf von Baeyer. Noting that rings containing fewer than five or more than six carbons seemed to be less abundant among naturally occurring materials as well as less stable than those related to cyclopentane and cyclohexane, Baeyer suggested that stability was related to how closely the angles of the corresponding planar regular polygons matched the tetrahedral value of $109.5°$. Since the $60°$ bond angles required by the geometry of cyclopropane deviate by $49.5°$ from the tetrahedral value, Baeyer suggested that the *strain* in the three-membered ring of cyclopropane is responsible for its decreased stability relative to cyclopentane and cyclohexane. Similarly, each C—C—C angle of $90°$ in planar cyclobutane deviates by $19.5°$ from the ideal tetrahedral value. Cyclobutane, like cyclopropane, is said to be destabilized because of *angle strain.*

According to Baeyer, cyclopentane should be the most stable of all the cycloalkanes because the ring angles of a planar pentagon, 108°, are closer to the tetrahedral angle than those of any other cycloalkane. A prediction of the *Baeyer strain theory* is that the cycloalkanes beyond cyclopentane should become increasingly strained and correspondingly less stable. The angles of a regular hexagon are 120°, and the angles of larger polygons deviate more and more from the ideal tetrahedral angle.

PROBLEM 3.3   The angles of a regular planar polygon of *n* sides are equal to

$$\frac{n-2}{n}(180°)$$

If cyclododecane had a planar carbon skeleton, how large would its C—C—C bond angles have to be?

As was done with isomeric alkanes in Section 2-15, we can call upon heat of combustion data to probe the relative energies of cycloalkanes. By so doing, some of the inconsistencies of the Baeyer strain theory will become evident. Table 3.1 lists the experimentally measured heats of combustion for a number of cycloalkanes. The most important column in the table is the heat of combustion per methylene (CH_2) group. Cyclopropane has the highest heat of combustion per methylene group, which is consistent with the idea that its internal energy is elevated because of angle strain. Cyclobutane has less angle strain at each of its carbon atoms and a lower heat of combustion per methylene group. Cyclopentane, as expected, has a lower value still. Notice, however, that contrary to the prediction of the Baeyer strain theory, cyclohexane has a smaller heat of combustion per methylene group than does cyclopentane. Each methylene group contributes a smaller increment to the internal energy of cyclohexane than the methylene units of cyclopentane. If bond angle distortion were greater in cyclohexane than in cyclopentane, an opposite effect would have been observed.

Furthermore, the heats of combustion per methylene group of the very large rings are all about the same and similar to that of cyclohexane. Rather than rising because

TABLE 3.1
Heats of Combustion ($-\Delta H°$) of Cycloalkanes

Cycloalkane	Heat of combustion, kcal/mol	Number of CH_2 groups	Heat of combustion per CH_2 group, kcal/mol
Cyclopropane	499.8	3	166.6
Cyclobutane	650.3	4	162.7
Cyclopentane	786.6	5	157.3
Cyclohexane	936.8	6	156.1
Cycloheptane	1099.2	7	157.0
Cyclooctane	1258.8	8	157.3
Cyclononane	1418.0	9	157.5
Cyclodecane	1574.3	10	157.4
Cycloundecane	1729.8	11	157.3
Cyclododecane	1875.1	12	156.2
Cyclotetradecane	2184.2	14	156.0
Cyclohexadecane	2501.4	16	156.3

of increased angle strain in large rings, the heat of combustion per methylene group remains constant at approximately 156 kcal/mol, the value cited in Section 2.15 as the difference between successive members of a homologous series of alkanes. Therefore, it is reasonable to conclude that the bond angles of large cycloalkanes are not significantly different from the bond angles of noncyclic alkanes. The prediction of the Baeyer strain theory that angle strain increases steadily with ring size is contradicted by experimental fact.

The Baeyer strain theory is useful to us in identifying angle strain as a destabilizing structural effect. It contains a fundamental flaw, however, in its assumption that the rings of cycloalkanes are planar. With the exception of cyclopropane, cycloalkane rings are distinctly nonplanar. Six-membered rings rank as the most important ring size among cyclic organic compounds, so let us begin with cyclohexane to examine the forces that determine the shapes of cycloalkanes.

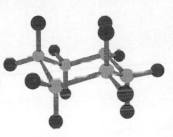

(*a*) Chair conformation

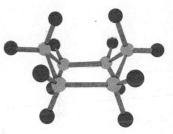

(*b*) Boat conformation

FIGURE 3.6 Conformations of cyclohexane: *(a)* the chair conformation; *(b)* the boat conformation.

3.7 CONFORMATIONS OF CYCLOHEXANE

In the preceding section we saw how the Baeyer strain theory predicted that a planar geometry of cyclohexane would be destabilized by angle strain. As long ago as 1890 chemists found that cyclohexane models constructed from carbon tetrahedra adopted nonplanar conformations with scarcely any distortion of their customary 109.5° bond angles. Two of these nonplanar conformations were given characteristic names describing their general shape; they are the *chair* (Figure 3.6*a*) and the *boat* conformation (Figure 3.6*b*).

Experimental evidence indicating that six-membered rings were nonplanar began to accumulate in the 1920s. Eventually, Odd Hassel of the University of Oslo established that the most stable conformation of cyclohexane is the chair. Later, Sir Derek Barton of Imperial College (London) showed how Hassel's structural results could be extended to an analysis of chemical reactivity. Hassel and Barton shared the Nobel prize in chemistry in 1969.

The structural features of the chair conformation of cyclohexane are summarized in Figure 3.7. With C—C—C bond angles of 111°, it is nearly free of angle strain. All its bonds are staggered and so it is free of torsional strain as well. The staggered arrangement of the bonds in cyclohexane is apparent in a Newman-style projection of the chair conformation.

Staggered arrangement of bonds in chair conformation of cyclohexane

The boat form of cyclohexane is about 6.4 kcal/mol higher in energy (less stable) than the chair. Again, because the bond angles are close to tetrahedral, there is little

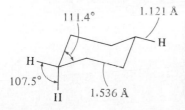

FIGURE 3.7 Bond angles and bond distances in cyclohexane. All the C—C—C angles are equal; all the H—C—H angles are equal. All the carbon-carbon bond distances are equal to each other and are approximately the same as the carbon-carbon bond distance in ethane. All the carbon-hydrogen bond distances are approximately the same as those of ethane and methane.

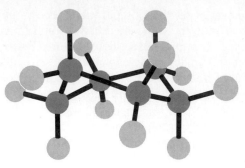

FIGURE 3.8 The skew boat conformation of cyclohexane.

angle strain. The boat conformation has eclipsed bonds on four of its carbon atoms and so has a significant amount of torsional strain.

Eclipsed bonds in boat conformation give it torsional strain

The boat conformation is further destabilized by a close contact between its "flagpole" hydrogens. These hydrogens are about 1.8 Å apart, a distance that is significantly shorter than the sum (2.4 Å) of their van der Waals radii. Thus, van der Waals repulsions between the flagpole hydrogens combine with torsional strain to make the potential energy of the boat form greater than that of the chair.

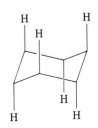

Axial C-H bonds

Van der Waals repulsion between flagpole hydrogens destabilizes boat conformation

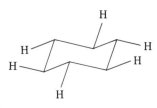

Equatorial C-H bonds

A third conformation, called the *twist* or *skew boat* (Figure 3.8), is less stable than the chair but 0.6 kcal/mol more stable than the boat. It is related to the boat by a slight twisting motion that relieves a portion of the torsional strain. At the same time the two flagpole hydrogens are moved from points of direct opposition, and the van der Waals repulsion between them decreases.

At any moment most of the molecules of cyclohexane exist in the chair conformation. The available experimental data indicate that no more than one or two molecules per thousand are in the skew boat conformation. The various conformations of cyclohexane are in rapid equilibrium with each other, and we will explore the consequences of that equilibrium in Section 3.9 after taking a closer look at some important properties of the chair conformation.

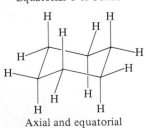

Axial and equatorial bonds together

FIGURE 3.9 Axial and equatorial bonds in cyclohexane.

3.8 AXIAL AND EQUATORIAL BONDS IN CYCLOHEXANE

One of the most significant findings to emerge from conformational studies of cyclohexane is that the spatial orientations of its 12 hydrogen atoms are not all identical but are divided into two groups, as shown in Figure 3.9. Six of the hydrogens, called *axial* hydrogens, have their bonds to the ring parallel to an axis of symmetry that

(1)   Begin with the chair conformation of cyclohexane.

(2)   Draw the axial bonds before the equatorial ones, alternating their direction on adjacent carbon atoms. Always start by placing an axial bond "up" on the uppermost carbon or "down" on the lowest carbon.

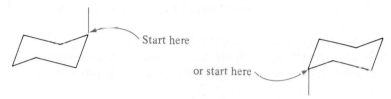

then alternate to give

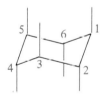

in which all the axial bonds are parallel to each other

(3)   Place the equatorial bonds so as to approximate a tetrahedral arrangement of the bonds to each carbon. The equatorial bond of each carbon should be parallel to the ring bonds of its two nearest-neighbor carbons.

Place equatorial bond at C-1 so that it is parallel to the bonds between C-2 and C-3 and between C-5 and C-6

Following this pattern gives the complete set of equatorial bonds.

(4)   Practice drawing cyclohexane chairs oriented in either direction.

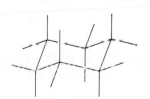 and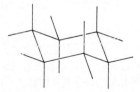

FIGURE 3.10 A guide to representing the orientations of the bonds in the chair conformation of cyclohexane.

passes through the ring's center. These axial bonds alternately are directed up and down on adjacent carbons. The second set of six hydrogens, called *equatorial* hydrogens, are located approximately along the equator of the molecule. Notice that the four bonds to each carbon are arranged tetrahedrally, as sp^3 hybridization of the ring carbons requires.

The conformational features of six-membered rings are fundamental to organic chemistry, which makes it essential that you have a clear understanding of the directional properties of axial and equatorial bonds and be able to represent them accurately. Figure 3.10 offers some guidance on the drawing of chair cyclohexane rings.

It is no accident that sections of our chair cyclohexane drawings resemble sawhorse projections of staggered conformations of alkanes. The same spatial relationships seen in alkanes carry over to substituents on a six-membered ring. In the structure

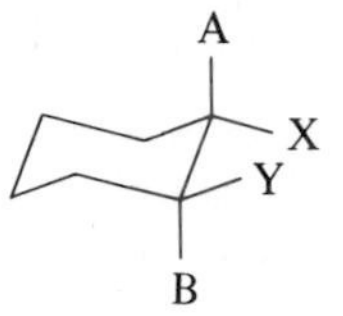

the substituted carbons have the spatial arrangement shown

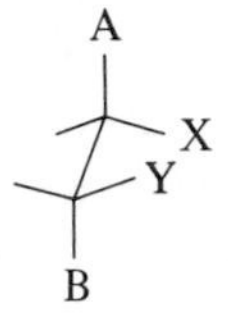

Substituents A and B are anti to each other, while the other relationships — A and Y, X and Y, and X and B — are gauche

---

PROBLEM 3.4   Given the partial structure shown below, add a substituent X to C-1 so that it satisfies the indicated stereochemical requirement.

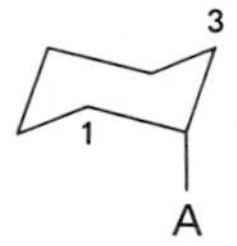

(a) Anti to A   (c) Anti to C-3

(b) Gauche to A   (d) Gauche to C-3

SAMPLE SOLUTION   (a)   In order to be anti to A, substituent X must be axial. The blue lines in the drawing show the A—C—C—X torsion angle to be 180°.

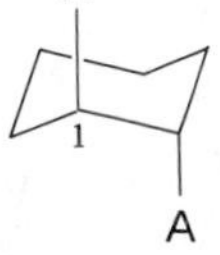

3.9 CONFORMATIONAL INVERSION (RING FLIPPING) IN CYCLOHEXANE

We have seen that alkanes are not locked into a single conformation. Rotation around the central carbon-carbon bond in butane occurs rapidly, converting the stable anti conformation back to itself via the higher-energy gauche and eclipsed conformations. Is cyclohexane a static structure or does it too possess conformational mobility?

Cyclohexane is conformationally mobile. Through a process known as *ring inversion, chair-chair interconversion,* or, more simply, *ring flipping,* one chair conformation is converted to another chair.

The process by which ring inversion takes place is illustrated in Figure 3.11.

Twisting one half of a cyclohexane chair, while keeping the other half essentially unchanged, converts the chair conformation to the skew boat. Repeating the operation on the other half of the skew boat conformation generates the inverted chair. The skew boat, then, is an intermediate structure achieved by the molecule on its passage from one chair conformation to another. Unlike a transition state, an intermediate is not a potential energy maximum but is a local minimum on the potential energy profile. The boat is a transition state for interconversion of a series of equivalent skew boat conformations.

The point of maximum potential energy in the chair-chair interconversion is a conformation called the *half-chair* or *half-twist*—it is some 10.8 kcal/mol higher in energy than the chair. This value of 10.8 kcal/mol is the activation energy for ring inversion in cyclohexane and corresponds to a very rapid process. An activation energy of this magnitude indicates that after 10^{-5} s about half of all the cyclohexane molecules in a sample have undergone ring inversion at room temperature.

An important consequence of chair-chair interconversion is that any substituent that is axial in the original chair conformation becomes equatorial in the ring-flipped

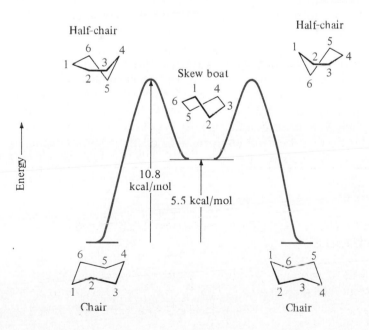

FIGURE 3.11 Energy diagram showing interconversion of various conformations of cyclohexane. In order to simplify the diagram the boat conformation has been omitted. The boat is a transition state for the interconversion of skew boat conformations.

form and vice versa. The process does *not* require any bond-breaking or bond-form-ing steps. Only ring inversion is involved.

X axial; Y equatorial   X equatorial; Y axial

3.10 CONFORMATIONAL ANALYSIS OF MONOSUBSTITUTED CYCLOHEXANES

Ring inversion in methylcyclohexane occurs at a rate comparable with that in cyclo-hexane itself. The equilibrium differs from that of cyclohexane, however, in that the two conformations are not equivalent. In one conformation the methyl group occu-pies an axial position; in the other it is equatorial. Structural studies have established that approximately 95 percent of the molecules of methylcyclohexane are in the chair conformation that has an equatorial methyl group while only 5 percent of the mole-cules have an axial methyl group at room temperature.

5%
(Less stable chair
conformation)

95%
(More stable chair
conformation)

In any equilibrium process, the species present in greatest amount is the most stable one. Therefore, we conclude that equatorial methylcyclohexane is more stable than axial methylcyclohexane. Is there a structural reason for this experimentally determined fact?

It appears that a methyl group is more stable when it occupies an equatorial site because it is less crowded there than as an axial substituent. An axial methyl group at C-1 is relatively close to the axial hydrogens at C-3 and C-5. At its closest approach a hydrogen of the methyl group is within 1.9 Å of the C-3 and C-5 axial hydrogens. This distance is significantly less than the sum of the van der Waals radii of two hydrogen atoms (2.4 Å) and indicates that the axial conformation of methylcyclohexane is destabilized by van der Waals repulsive forces.

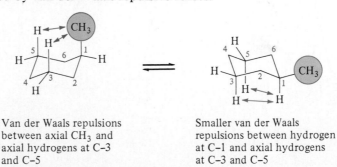

Van der Waals repulsions
between axial CH₃ and
axial hydrogens at C-3
and C-5

Smaller van der Waals
repulsions between hydrogen
at C-1 and axial hydrogens
at C-3 and C-5

The alternative chair conformation, the one with its methyl group equatorial, replaces these close methyl-hydrogen contacts with hydrogen-hydrogen contacts. The repulsive forces are much less because hydrogen is a smaller substituent than methyl; it takes up less space and crowds the axial hydrogens at C-3 and C-5 less.

PROBLEM 3.5   The following questions relate to a cyclohexane ring depicted in the chair conformation shown.

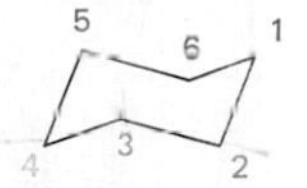

(a) Is a methyl group at C-6 which is "down" axial or equatorial?
(b) Is a methyl group which is "up" at C-1 more or less stable than a methyl group which is up at C-4?
(c) Place a methyl group at C-3 in its most stable orientation. Is it up or down?

SAMPLE SOLUTION   (a) First indicate the directional properties of the bonds to the ring carbons. A substituent is down if it is below the other substituent on the same carbon atom. Therefore, a methyl group that is down at C-6 is axial.

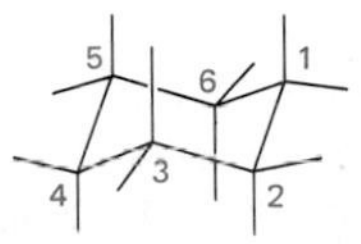

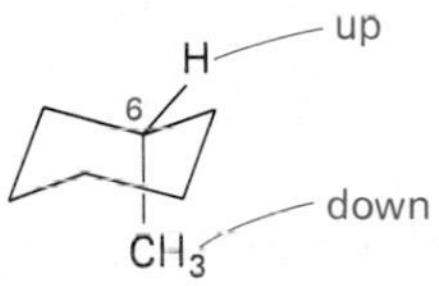

Other substituted cyclohexanes are similar to methylcyclohexane. Two chair conformations exist in rapid equilibrium, and the more stable of these two is the one in which the substituent is equatorial. The relative amounts of the two conformations depends on the effective size of the substituent. The size of a substituent, in the context of cyclohexane conformations, is related to the degree of branching at its point of connection to the ring. A single atom, such as a halogen substituent, does not take up much space and has a less pronounced preference for an equatorial orientation than does a methyl group.

40 percent   60 percent

A branched substituent such as an isopropyl group exhibits a greater preference for the equatorial orientation than a methyl group.

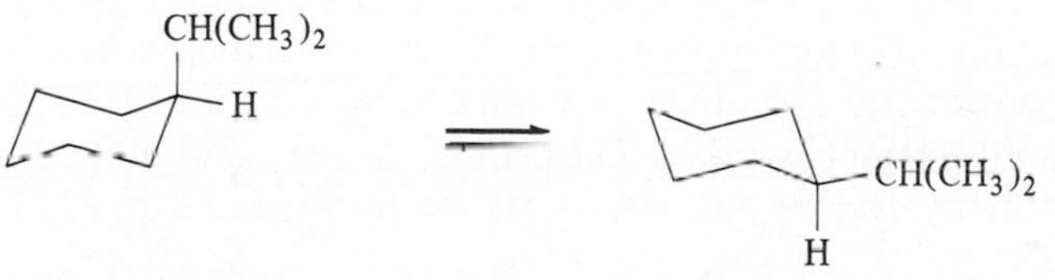

3 percent   97 percent

A *tert*-butyl group is so large that *tert*-butylcyclohexane exists almost entirely in the conformation in which the *tert*-butyl group is equatorial. The amount of axial *tert*-butylcyclohexane present is too small to measure.

Less than 0.01 percent

(Serious van der Waals repulsions involving *tert*-butyl group)

Greater than 99.99 percent

(Decreased van der Waals repulsions)

PROBLEM 3.6   Draw the most stable conformation of 1-*tert*-butyl-1-methylcyclohexane.

Effects on structure and reactivity that come about because atoms or groups occupy a particular region of space are called *steric effects*. Here we see a steric effect on relative stability that traces its origin to van der Waals repulsions between axial substituents. The process of conformational inversion permits these van der Waals repulsions to be relieved by transforming an axially oriented substituent to an equatorial one.

3.11   CONFORMATIONAL ANALYSIS OF DISUBSTITUTED CYCLOHEXANES

When a six-membered ring bears two substituents on different carbons, two methyl groups, for example, these substituents may be on the same side or on opposite sides of the ring. When the substituents are on the same side, we say they are *cis* to each other; when substituents are on opposite sides, we say they are *trans* to each other. Both terms come from the Latin, in which cis means "on this side" and trans means "across."

cis-1,2-Dimethylcyclohexane

trans-1,2-Dimethylcyclohexane

The cis and trans forms of 1,2-dimethylcyclohexane are *stereoisomers*. They have the same order of atomic connections, i.e., the same constitution, but they differ in the arrangements of their atoms in space. You learned in Section 2.15 that constitutional isomers could differ in stability. What about stereoisomers?

We can obtain a measure of the energy difference between *cis*- and *trans*-1,2-dimethylcyclohexane by comparing their heats of combustion. As illustrated in Figure 3.12, the two compounds are isomers, so the difference in their heats of combustion is a direct measure of the difference in their energies. The heat of combustion of *trans*-1,2-dimethylcyclohexane is 1.5 kcal/mol less than that of its cis stereoisomer, so *trans*-1,2-dimethylcyclohexane is 1.5 kcal/mol more stable than *cis*-1,2-dimethylcyclohexane.

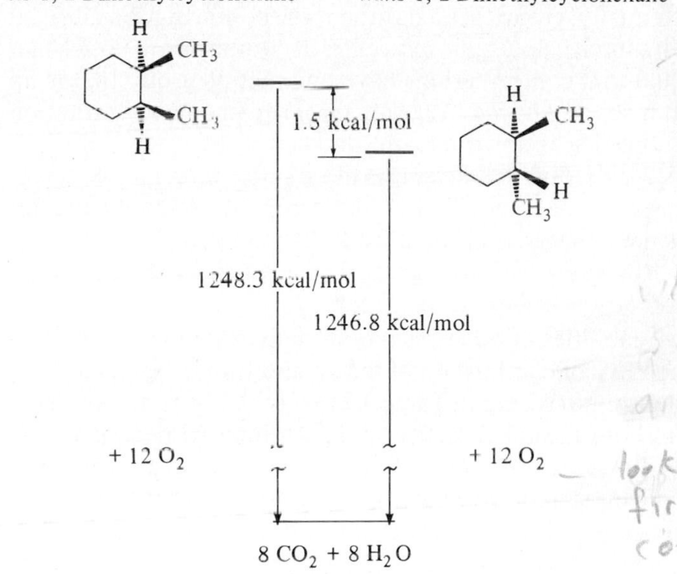

cis-1, 2-Dimethylcyclohexane   trans-1, 2-Dimethylcyclohexane

1.5 kcal/mol

1248.3 kcal/mol

1246.8 kcal/mol

+ 12 O_2 + 12 O_2

8 CO_2 + 8 H_2O

(handwritten) trans – one methyl up one methyl down

cis – both methyls are up

(handwritten) when alternating, the methyls must switch from axial to equatorial ...

(handwritten) look at pg 73 to first draw the correct structure

FIGURE 3.12 Diagram showing how the enthalpy difference between *cis*- and *trans*-1,2-dimethylcyclohexane can be determined from their heats of combustion. All energies are in kilocalories per mole (kcal/mol).

The relationship between stability and stereochemistry can be readily explained by examining these compounds in their chair conformations. *cis*-1,2-Dimethylcyclohexane can adopt either of two equivalent chair conformations, each one having one axial methyl group and one equatorial methyl group.

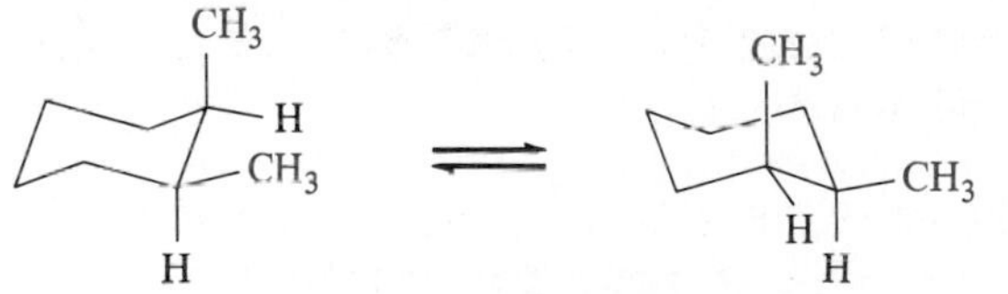

cis-1,2-Dimethylcyclohexane

These two equivalent chair conformations interconvert by a ring flipping process that is exactly analogous to that of cyclohexane and of methylcyclohexane. The axial methyl group becomes equatorial, and the equatorial methyl group becomes axial. The methyl groups are cis because both are up relative to the hydrogen substituents present at each carbon. If both methyl groups were down, they still would be cis to each other. Notice that the ring flipping process does not alter the cis relationship between the methyl groups. Nor does it alter their up versus down quality; substituents that are up in one conformation remain up in the ring-flipped form.

The two chair conformations of *trans*-1,2-dimethylcyclohexane are not equivalent to each other. In one, both methyl groups are axial; in the other, both methyl groups are equatorial.

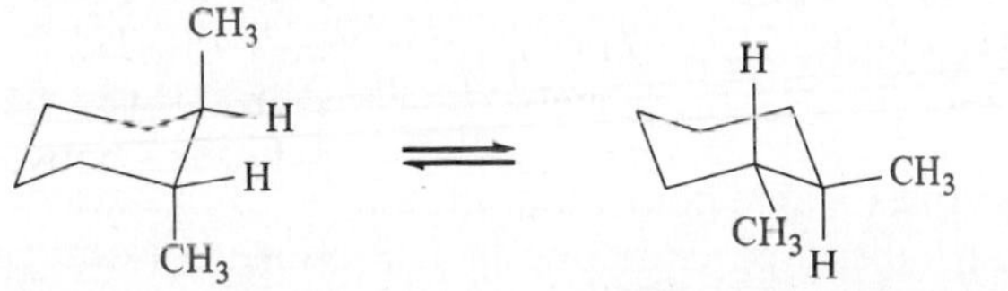

(Both methyl groups are axial; less stable chair conformation)

(Both methyl groups are equatorial; more stable chair conformation)

trans-1,2-Dimethylcyclohexane

The chair conformation that has both methyl groups equatorial is much more stable than the one in which both methyl groups are axial and is the conformation adopted by most of the *trans*-1,2-dimethylcyclohexane molecules at equilibrium. Notice that both the diaxial and diequatorial conformations have one methyl group that is up and one methyl group that is down. The methyl group that is up in one conformation remains up; the methyl group that is down remains down.

The reason why *trans*-1,2-dimethylcyclohexane is more stable than the cis stereoisomer is that the latter has one axial methyl group in its most stable conformation while both methyl groups are equatorial in the most stable conformation of *trans*-1,2-dimethylcyclohexane. Remember, it is a general rule that any substituent is more stable in an equatorial orientation than in an axial one.

What about the relative stabilities of other pairs of dimethylcyclohexane stereoisomers? Table 3.2 lists the heats of combustion of the cis and trans isomers of 1,2-, 1,3-, and 1,4-dimethylcyclohexane. We see in Table 3.2 that the trans stereoisomer is more stable than the cis in both the 1,2-dimethyl and 1,4-dimethyl derivatives. A different situation prevails, however, when *cis*- and *trans*-1,3-dimethylcyclohexane are compared. Here, it is the cis stereoisomer that is the more stable. Why?

The most stable conformation of *cis*-1,3-dimethylcyclohexane has both methyl groups equatorial.

(Both methyl groups are axial; less stable chair conformation)

(Both methyl groups are equatorial; more stable chair conformation)

(Both methyl groups are up)

cis-1,3-Dimethylcyclohexane

The two chair conformations of *trans*-1,3-dimethylcyclohexane are equivalent to each other. Both contain one axial and one equatorial methyl group.

(One methyl group is axial, the other is equatorial)

(One methyl group is axial, the other is equatorial)

(One methyl group is up, the other is down)

trans-1,3-Dimethylcyclohexane

Because the most stable conformation of *cis*-1,3-dimethylcyclohexane has both its methyl groups disposed equatorially, the cis stereoisomer is more stable than the trans, in which one methyl group must be axial in the most stable conformation of the molecule.

The situation in 1,4-dimethyl-substituted cyclohexanes resembles that of the ster-

TABLE 3.2

Heats of Combustion of Isomeric Dimethylcyclohexanes

Compound	Orientation of methyl groups in most stable conformation	Heat of combustion, kcal/mol	Difference in heat of combustion, kcal/mol	More stable stereoisomer
cis-1,2-Dimethylcyclohexane	Axial-equatorial	1248.3		
trans-1,2-Dimethylcyclohexane	Diequatorial	1246.8	1.5	trans
cis-1,3-Dimethylcyclohexane	Diequatorial	1245.7		
trans-1,3-Dimethylcyclohexane	Axial-equatorial	1247.4	1.7	cis
cis-1,4-Dimethylcyclohexane	Axial-equatorial	1247.4		
trans-1,4-Dimethylcyclohexane	Diequatorial	1245.8	1.6	trans

eoisomeric 1,2-dimethylcyclohexanes. The cis stereoisomer has two equivalent chair conformations, each containing one axial and one equatorial methyl group.

(One methyl group is axial, the other is equatorial) (One methyl group is axial, the other is equatorial)

(Both methyl groups are up)

cis-1,4-Dimethylcyclohexane

The trans stereoisomer is more stable because its most stable conformation permits both methyl groups to occupy equatorial sites.

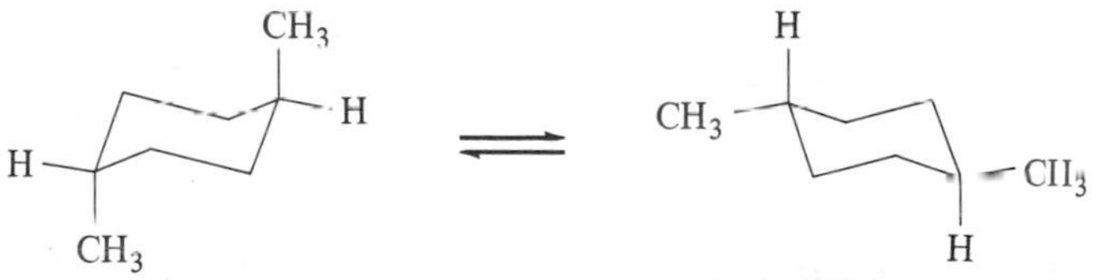

(Both methyl groups are axial: less stable chair conformation) (Both methyl groups are equatorial: more stable chair conformation)

(One methyl group is up, the other is down)

trans-1,4-Dimethylcyclohexane

The pattern established in the preceding section has continued in this one. In Section 3.10 you learned that the more stable chair conformation of a monosubstituted cyclohexane is the one in which the substituent is equatorial. In this section you have seen that the more stable of a pair of disubstituted cyclohexanes is the one that can adopt a conformation in which both substituents are equatorial.

PROBLEM 3.7 Based on what you know about disubstituted cyclohexanes, which of the two stereoisomeric 1,3,5-trimethylcyclohexanes shown below would you expect to be more stable?

cis-1,3,5-Trimethylcyclohexane   trans-1,3,5-Trimethylcyclohexane

If in a disubstituted derivative of cyclohexane the two substituents are different, then the most stable conformation will be the chair that has the larger substituent in an equatorial orientation. This is most apparent when one of the substituents is a *tert*-butyl group. A *tert*-butyl group is the largest of the C_1 to C_4 alkyl groups and has the greatest preference for an equatorial site. Thus, the most stable conformation of *cis*-1-*tert*-butyl-2-methylcyclohexane has an equatorial *tert*-butyl group and an axial methyl group.

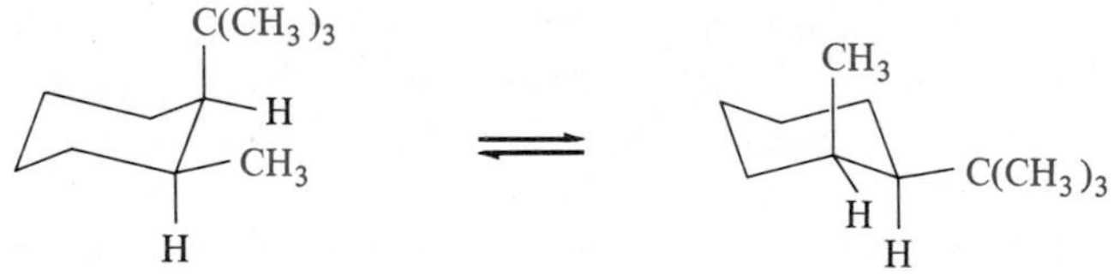

(Less stable conformation: (More stable conformation;
larger group is axial)   larger group is equatorial)

cis-1-*tert*-Butyl-2-methylcyclohexane

PROBLEM 3.8 Write structural formulas for the most stable conformation of each of the following compounds.

(a)   *trans*-1-*tert*-Butyl-3-methylcyclohexane
(b)   *cis*-1-*tert*-Butyl-3-methylcyclohexane
(c)   *trans*-1-*tert*-Butyl-4-methylcyclohexane
(d)   *cis*-1-*tert*-Butyl-4-methylcyclohexane

SAMPLE SOLUTION (a) The most stable conformation is the one that has the larger substituent, the *tert*-butyl group, equatorial. Draw a chair conformation of cyclohexane, and place an equatorial *tert*-butyl group at one of its carbons. Add a methyl group at C-3 so that it is trans to the *tert*-butyl group.

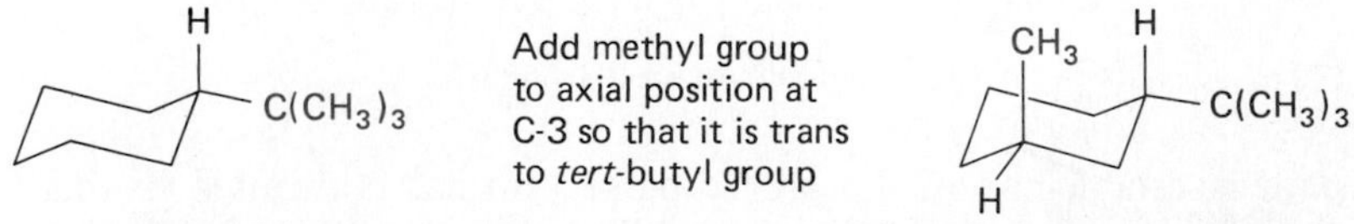

tert-Butyl group   Add methyl group   *trans*-1-*tert*-Butyl-3-
equatorial on   to axial position at   methylcyclohexane
six-membered ring   C-3 so that it is trans
                         to *tert*-butyl group

Cyclohexane rings that bear *tert*-butyl substituents are examples of conformationally biased molecules. A *tert*-butyl group has such a pronounced preference for the equatorial orientation that it will bias the equilibrium to favor such conformations. This does not mean that ring inversion does not occur, however. Ring inversion does occur, but at any instant only a tiny fraction of the molecules exist in conformations having axial *tert*-butyl groups. It is not strictly correct to say that *tert*-butylcyclohexane and its derivatives are "locked" into a single conformation; conformations related by ring flipping are in rapid equilibrium with each other, but the distribution between them strongly favors those in which the *tert*-butyl group is equatorial.

3.12 SMALL RINGS. CYCLOPROPANE AND CYCLOBUTANE

The conformational situation is far simpler in cyclopropane than in any other cycloalkane. Its three carbon atoms are, of geometric necessity, coplanar and rotation about carbon-carbon single bonds is impossible. You saw in Section 3.6 how angle strain in cyclopropane led to an abnormally large heat of combustion for a compound with only three methylene units. Let us now examine cyclopropane in more detail in order to see how our orbital hybridization bonding model may be adapted to molecules of unusual geometry.

Strong sp^3-sp^3 σ bonds are not possible for cyclopropane because the 60° bond angles of the ring do not permit the orbitals to be properly aligned for effective overlap (Figure 3-13). The less effective overlap that does occur leads to what chemists refer to as "bent bonds." The electron density in the carbon-carbon bonds of cyclopropane does not lie along the internuclear axis but is distributed along an arc between the two carbon atoms. The ring bonds of cyclopropane are weaker than other carbon-carbon σ bonds.

In addition to angle strain, cyclopropane is destabilized by torsional strain. Each C—H bond of cyclopropane is eclipsed with two others.

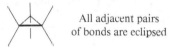
All adjacent pairs
of bonds are eclipsed

PROBLEM 3.9   The heats of combustion of *cis*- and *trans*-1,2-dimethylcyclopropane are 805.6 and 804.5 kcal/mol, respectively. Suggest an explanation for the 1.1 kcal/mol difference in energy between these two stereoisomers.

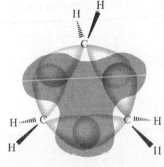

FIGURE 3.13 "Bent bonds" in cyclopropane. The orbitals involved in carbon-carbon bond formation overlap in a region that is displaced from the internuclear axis. Orbital overlap is less effective than in a normal carbon-carbon σ bond, and the carbon-carbon bond is weaker.

Cyclobutane has less angle strain than cyclopropane. Unlike cyclopropane, cyclobutane may adopt either a planar or a nonplanar conformation.

Planar conformation
of cyclobutane

Nonplanar conformations
of cyclobutane

The planar conformation, with its eclipsed bonds, has more torsional strain than the nonplanar conformations. The staggered arrangement of bonds in the nonplanar conformations makes them more stable than the planar form.

3.13 CYCLOPENTANE

The amount of angle strain present in a planar conformation of cyclopentane is, as we have seen, relatively small. The extent of torsional strain, however, is substantial since there is a set of five eclipsed bonds on the top face of the ring and another set of five eclipsed bonds on the bottom.

Eclipsing of bonds in
planar cyclopentane

Some but not all of this torsional strain is relieved in nonplanar conformations. Two nonplanar conformations of cyclopentane are of comparable energy; one is called the *envelope* conformation, the other is known as the *half-chair* or *twist* conformation.

Envelope conformation
of cyclopentane

Half-chair conformation
of cyclopentane

In the envelope conformation four of the carbon atoms are coplanar. The fifth carbon is out of the plane defined by the other four. There are three coplanar carbons in the half-chair conformation, with one carbon atom displaced above that plane and another below it. In both the envelope and the half-chair conformations, in-plane and out-of-plane carbons exchange positions rapidly. Equilibration between conformations of cyclopentane occurs at rates that are comparable with the rate of rotation about the carbon-carbon bond of ethane.

3.14 MEDIUM AND LARGE RINGS

Beginning with cycloheptane, which has four conformations of similar energy, conformational analysis of cycloalkanes becomes more complicated. The same fundamental principles apply to medium and large rings as apply to smaller ones — there are simply more atoms and more bonds to consider and more conformational possibilities. The most abundantly populated conformation of a molecule is the most stable one, and the most stable conformation is the one that is the most free of strain. Conformational depictions of two representative medium-ring cycloalkanes, cyclooctane and cyclodecane, are shown below.

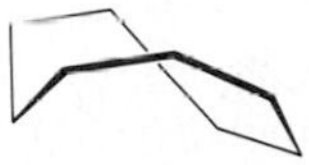

Saddle conformation
of cyclooctane

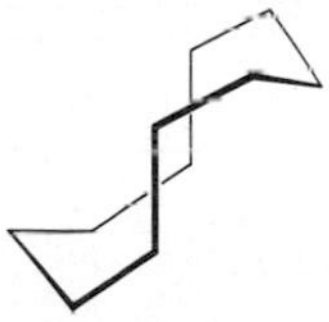

Boat chair-boat
conformation
of cyclodecane

3.15 POLYCYCLIC RING SYSTEMS

Organic molecules containing a single carbon atom that is common to two rings are called *spirocyclic* compounds. The simplest spirocyclic hydrocarbon is spiropentane, a product of laboratory synthesis. More complicated spirocyclic hydrocarbons not only have been synthesized but also have been isolated from natural sources. α-Alaskene, for example, occurs in the fragrant oil given off by the needles of the Alaskan yellow cedar; one of its carbon atoms is common to both the six-membered ring and the five-membered ring.

Spiropentane

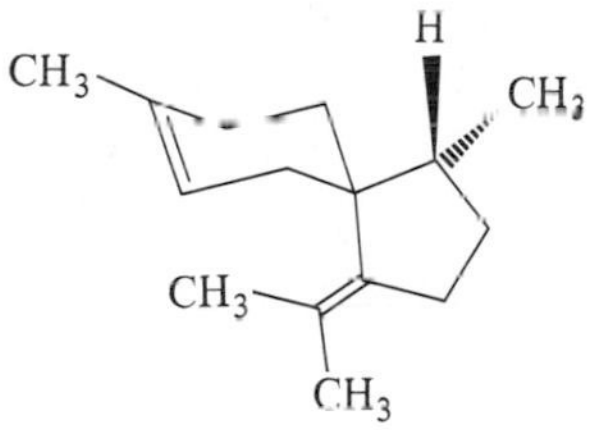

α-Alaskene

When two or more atoms are common to more than one ring, the compounds are called *polycyclic* ring systems; such compounds are subdivided into *bicyclic, tricyclic, tetracyclic,* etc. ring systems depending on the number of rings involved. Bicyclobutane is the simplest bicyclic hydrocarbon; it can be regarded as two three-membered rings that share a common side. Camphene is a naturally occurring bicyclic hydrocarbon obtained from pine oil. It is best regarded as a six-membered ring (indicated by colored bonds in the structure shown below) in which two of the carbons (designated by asterisks) are bridged by a CH_2 group.

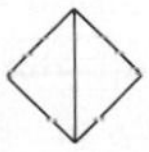

Bicyclobutane

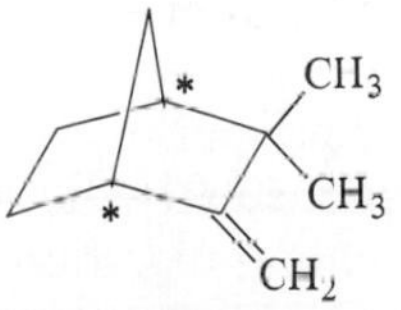

Camphene

Bicyclic compounds are named in the IUPAC system by counting the number of carbons in the ring system, assigning to the structure the base name of the unbranched alkane having the same number of carbon atoms, and attaching the prefix bicyclo-. The number of atoms in each of the bridges connecting the common atoms is then placed, in descending order, within brackets.

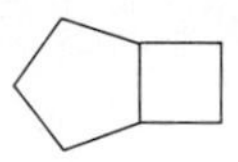

Bicyclo[3.2.0]heptane 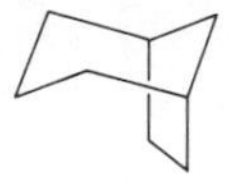   Bicyclo[3.2.1]octane

---

PROBLEM 3.10   Write structural formulas for each of the following bicyclic hydrocarbons:

(a) Bicyclo[2.2.1]heptane   (c) Bicyclo[3.1.1]heptane
(b) Bicyclo[5.2.0]nonane   (d) Bicyclo[3.3.0]octane

SAMPLE SOLUTION   (a)   The bicyclo[2.2.1]heptane ring system is one of the most frequently encountered bicyclic structural types. It contains seven carbon atoms, as indicated by the -heptane suffix. The bridging groups contain two, two, and one carbon, respectively.

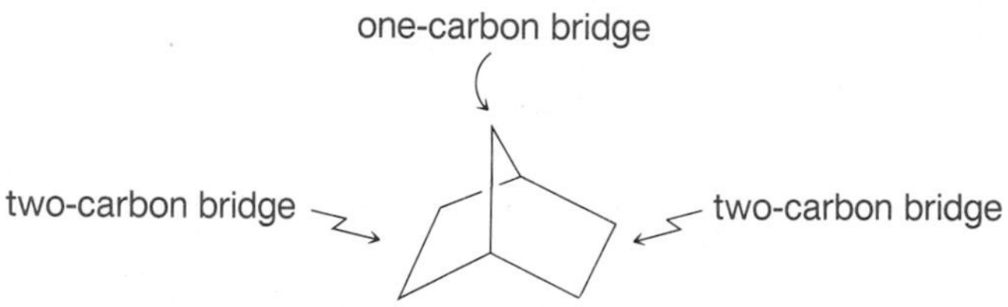

one-carbon bridge

two-carbon bridge   two-carbon bridge

Bicyclo[2.2.1]heptane

Among the most important of the bicyclic hydrocarbons are the two stereoisomeric bicyclo[4.4.0]decanes, called *cis*- and *trans*-decalin. The hydrogen substituents at the ring junction positions are on the same side in *cis*-decalin and on opposite sides in *trans*-decalin.

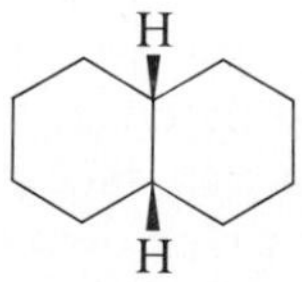

cis-Bicyclo[4.4.0]decane   *trans*-Bicyclo[4.4.0]decane
(*cis*-Decalin)   (*trans*-Decalin)

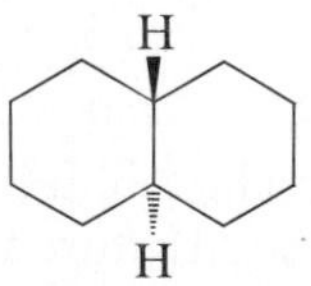

The decalin ring systems appear as structural units in a large number of naturally occurring substances, particularly the steroids. Cholic acid, for example, a steroid present in bile that promotes digestion, incorporates *cis*-decalin and *trans*-decalin units into a rather complex tetracyclic structure.

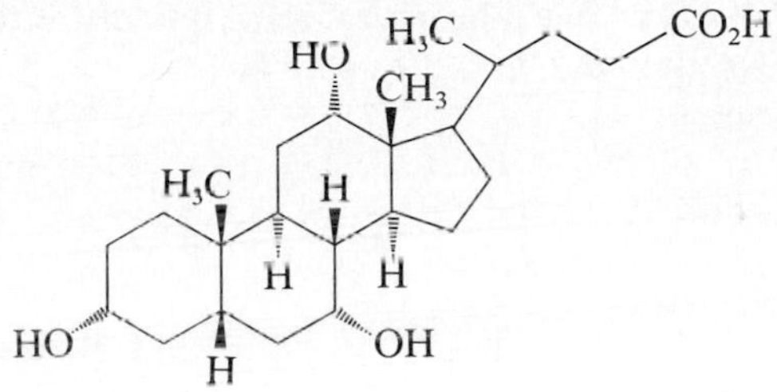

Cholic acid

3.16 SUMMARY

In this chapter we have delved more deeply into the shapes of organic molecules. Any molecule adopts the *conformation* that minimizes its total strain. The sources of internal strain in alkanes and cycloalkanes are *angle strain, torsional strain,* and *van der Waals repulsions.* Molecules are rarely frozen into a single conformation but engage in rapid equilibration among those that are energetically accessible. Interconversion of conformations occurs by rotation about single bonds. Rotation around carbon-carbon single bonds is normally very fast, occurring hundreds of thousands of times per second at room temperature. The energy of activation for rotation about the carbon-carbon bond of ethane is only about 3 kcal/mol and corresponds to the energy difference between the *staggered* and the *eclipsed* conformation. At any instant, almost all the molecules of ethane reside in the staggered conformation.

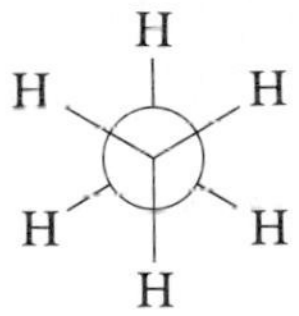

Staggered conformation
of ethane (most stable conformation)

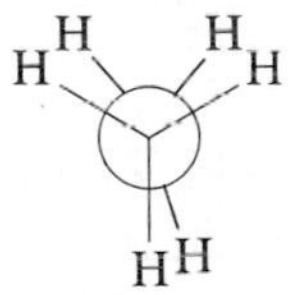

Eclipsed conformation
of ethane (least stable conformation)

The decreased stability of eclipsed conformations is given the name *torsional strain.* Torsional strain is thought to result from repulsions between the electrons in the C—H bonds of adjacent carbons.

The two staggered conformations of butane are not equivalent. The *anti* conformation is more stable than the *gauche.*

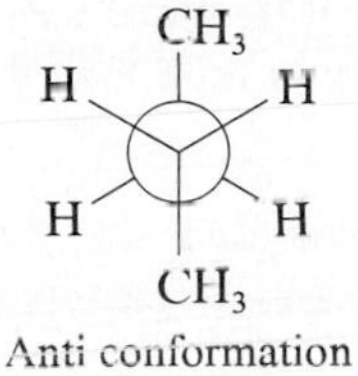

Anti conformation
of butane

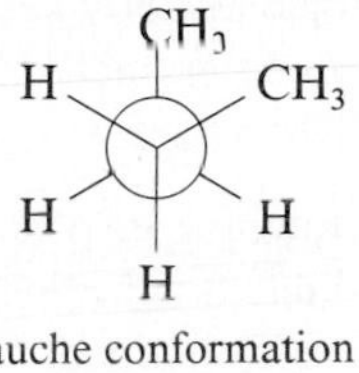

Gauche conformation
of butane

Neither conformation incorporates any torsional strain because each is a staggered form. The gauche conformation is less stable because of van der Waals repulsions between the methyl groups.

Three conformations of cyclohexane have approximately tetrahedral angles at the carbon atoms: the *chair,* the *boat,* and the *skew boat.*

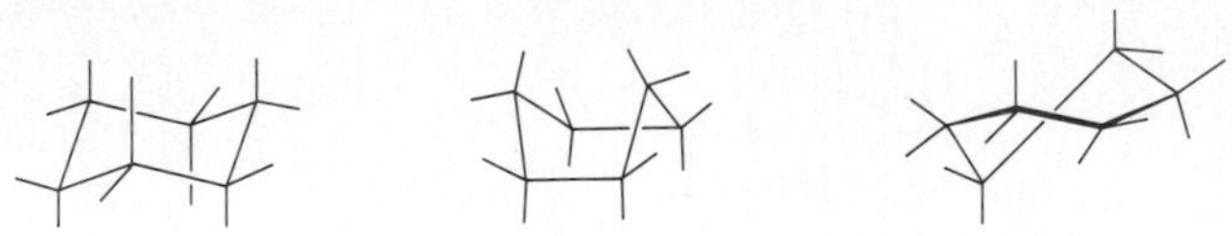

Conformations of cyclohexane

The chair is by far the most stable conformation for cyclohexane and its derivatives. The skew boat is somewhat more stable than the boat. The chair conformation is free of angle strain, torsional strain, and van der Waals strain. The C—H bonds in cyclohexane are not all equivalent but are divided into two sets of six each, called *axial* and *equatorial.* Cyclohexane undergoes a rapid conformational change referred to as *ring inversion* or *ring flipping.* The process of ring inversion causes all axial bonds to become equatorial and vice versa.

Substituents on a cyclohexane ring are more stable when they occupy equatorial sites than when they are axial. Branched substituents, especially *tert*-butyl, have a pronounced preference for the equatorial position. The relative stabilities of stereoisomeric disubstituted (and more highly substituted) cyclohexanes can be assessed by analyzing chair conformations for van der Waals repulsion involving axial substituents.

Cyclopropane is planar and strained (angle strain and torsional strain). Cyclobutane is nonplanar and less strained than cyclopropane. Cyclopentane has two nonplanar conformations which are of similar stability, the envelope and the half-chair.

PROBLEMS

3.11 Inorganic molecules also possess conformational mobility, and their various conformations can be described in ways analogous to those used to describe organic ones. Write structural representations of:

(a) Two different planar conformations of hydrogen peroxide (H_2O_2)
(b) Two different staggered conformations of hydrazine (H_2NNH_2)

3.12 Draw Newman projection formulas for the two most stable conformations of 2-methylbutane. One of these two conformations is more stable than the other. Which is the more stable one?

3.13 Even though the methyl group occupies an equatorial site, the conformation shown below is not the most stable one for methylcyclohexane. Explain why, and draw the most stable conformation of this compound.

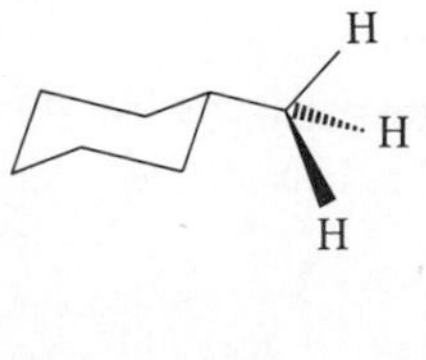

3.14 Which do you think is a more stable conformation, A or B? Why?

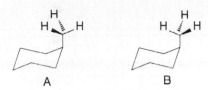

A   B

3.15 Determine whether the two structures in each of the following pairs represent constitutional isomers, stereoisomers, or different conformations of the same compound.

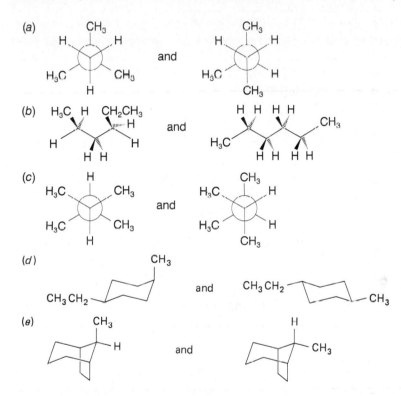

3.16 Draw structural formulas for all the compounds of molecular formula C_5H_8 that do not contain double or triple bonds.

3.17 Draw structural formulas for all the compounds of molecular formula C_6H_{10} that do not contain double or triple bonds.

3.18 In each of the following groups of compounds, identify the one with the largest heat of combustion and the one with the smallest:

(a) Cyclopropane, cyclobutane, cyclopentane
(b) Ethylcyclopropane, methylcyclobutane, cyclopentane
(c) cis-1,2-Dimethylcyclopentane, methylcyclohexane, 1,1,2,2-tetramethylcyclopropane
(d)

(e)

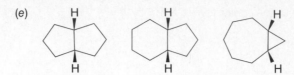

3.19 Sketch an approximate potential energy diagram similar to that shown in Figures 3.3 and 3.4 for rotation about the carbon-carbon bond in 2,2-dimethylpropane. Does the form of the potential energy curve of 2,2-dimethylpropane more closely resemble that of ethane or that of butane?

3.20 Write a structural formula for the most stable conformation of each of the following compounds.

(a) 2,2,5,5-Tetramethylhexane (Newman projection of conformation about C-3—C-4 bond)
(b) 2,2,5,5-Tetramethylhexane (zigzag conformation of entire molecule)
(c) *cis*-1-Isopropyl-3-methylcyclohexane
(d) *trans*-1-Isopropyl-3-methylcyclohexane
(e) *cis*-1-*tert*-Butyl-4-ethylcyclohexane
(f) *cis*-1,1,3,4-Tetramethylcyclohexane
(g)

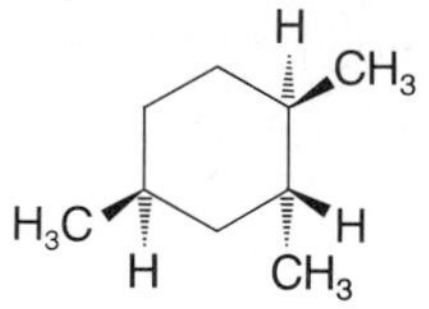

3.21 Identify the more stable stereoisomer in each of the following pairs and give the reason for your choice:

(a) *cis*- or *trans*-1-Isopropyl-2-methylcyclohexane
(b) *cis*- or *trans*-1-Isopropyl-3-methylcyclohexane
(c) *cis*- or *trans*-1-Isopropyl-4-methylcyclohexane
(d)

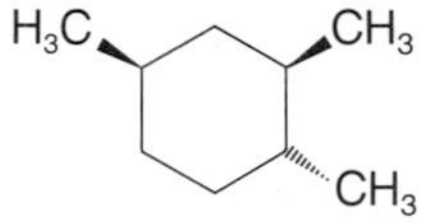

or

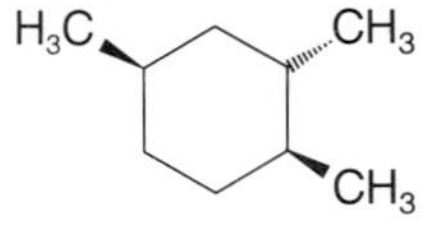

(e)

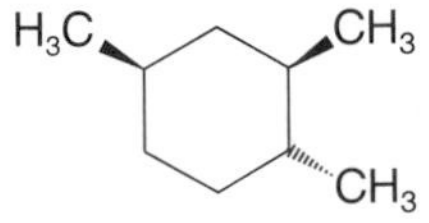

or

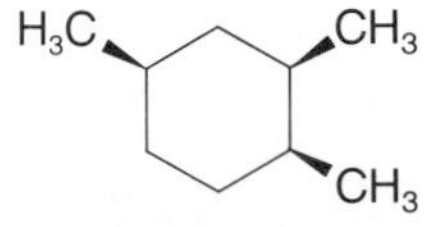

(f)

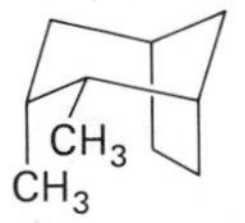

or

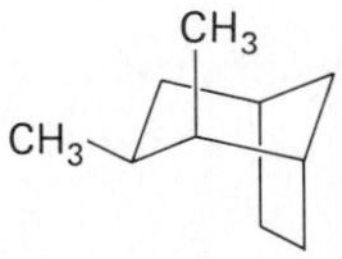

3.22 One stereoisomer of 1,1,3,5-tetramethylcyclohexane is 3.7 kcal/mol less stable than the other. Indicate which isomer is the less stable and identify the reason for its decreased stability.

3.23 One of the two tricyclooctane stereoisomers shown below is 4.9 kcal/mol less stable than the other. Indicate which isomer is the less stable and identify the reason for its decreased stability.

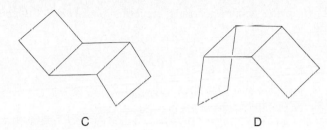

C   D

3.24 Representations of two forms of glucose are shown below. The six-membered ring is known to exist in a chair conformation in each form. Draw clear representations of the most stable conformation of each. Are they two different conformations of the same molecule or are they stereoisomers? Which substituents (if any) occupy axial sites?

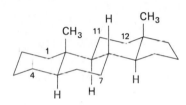

3.25 A typical steroid skeleton is shown along with the numbering scheme used for this class of compounds. Specify in each case whether the designated substituent is axial or equatorial.

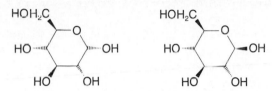

(a) Substituent at C-1 cis to the methyl groups
(b) Substituent at C-4 cis to the methyl groups
(c) Substituent at C-7 trans to the methyl groups
(d) Substituent at C-11 trans to the methyl groups
(e) Substituent at C-12 cis to the methyl groups

3.26 Repeat the preceding problem for the isomeric steroid skeleton having a cis ring fusion between the first two rings.

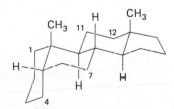

3.27 (a) Write Newman projection formulas for the gauche and anti conformations of 1,2-dichloroethane ($ClCH_2CH_2Cl$).

(b) The measured dipole moment of $ClCH_2CH_2Cl$ is 1.12 D. Which one of the following statements about 1,2-dichloroethane compound is false?

(1) It may exist entirely in the anti conformation.
(2) It may exist entirely in the gauche conformation.
(3) It may exist as a mixture of anti and gauche conformations.

3.28 (a) Sketch the planar and the nonplanar conformations of *trans*-1,3-dibromocyclobutane.

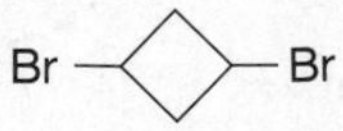

1,3-Dibromocyclobutane

(b) The measured dipole moment of *trans*-1,3-dibromocyclobutane is 1.10 D. Which one of the following statements about this compound is false? Explain.

(1) It may exist entirely in the conformation in which the ring is planar.
(2) It may exist entirely in the conformation in which the ring is nonplanar.
(3) It may exist as a mixture of conformations.

INTRODUCTION TO FUNCTIONAL GROUPS. ALCOHOLS AND ALKYL HALIDES

This chapter introduces the principles of chemical reactions of organic compounds. A particularly important category of organic reactions is concerned with the interconversion of *functional groups*. A functional group is an atom or group of atoms in a molecule that experiences chemical change under a prescribed set of reaction conditions. An example of a functional group is the hydroxyl group (OH). On treatment with hydrogen halides, alcohols — substances that have a hydroxyl group bonded to sp^3 hybridized carbon — are transformed into alkyl halides.

$$R{-}OH + HX \longrightarrow R{-}X + H_2O$$

Alcohol $\quad$ Hydrogen $\qquad$ Alkyl $\quad$ Water
$\qquad\qquad$ halide $\qquad\quad$ halide

In this general equation the symbol R designates an alkyl group, ROH is an alcohol, and RX is an alkyl halide. During the reaction the functional group OH is replaced by the functional group X, but the remainder of the molecule (R) is unchanged.

The most important of the functional groups in organic chemistry are listed in the table on the inside front cover of this text. All of them will be discussed in detail in subsequent chapters.

Even a hydrogen substituent in an alkane can be a functional group. Alkanes react with halogens to form alkyl halides. The alkane is said to undergo *halogenation*.

$$R{-}H + X_2 \longrightarrow R{-}X + HX$$

Alkane $\quad$ Halogen $\qquad$ Alkyl $\quad$ Hydrogen
$\qquad\qquad\qquad\qquad$ halide $\qquad$ halide

Both the conversion of alcohols to alkyl halides and the halogenation of alkanes will be described in this chapter, with emphasis on their applications to chemical synthesis and their *mechanism*. The mechanism of a chemical reaction is a precise description, in as much detail as experimental data permit, of the path traveled by starting materials as they are converted to the products. In this chapter you will see how the mechanisms by which alcohols and alkanes are converted to alkyl halides are strikingly different from each other.

4.1 NOMENCLATURE OF ALCOHOLS AND ALKYL HALIDES

Several alcohols are commonplace substances, well known by familiar names that reflect their origin (wood alcohol, grain alcohol) or use (rubbing alcohol). The common name of wood alcohol is *methyl alcohol;* grain alcohol is *ethyl alcohol* and rubbing alcohol is *isopropyl alcohol.*

$$CH_3OH \qquad CH_3CH_2OH \qquad \underset{\underset{OH}{|}}{CH_3CHCH_3}$$

Methyl alcohol   Ethyl alcohol   Isopropyl alcohol

The common names of alcohols are derived by naming the alkyl group to which the hydroxyl group is attached, then adding the separate word *alcohol.*

Alkyl halides are named in a similar way. After specifying the alkyl group, the halide is identified in a separate word as *fluoride, chloride, bromide,* or *iodide,* as appropriate.

$$CH_3CH_2CH_2F \qquad (CH_3)_3CCH_2Cl$$

| *n*-Propyl fluoride | Neopentyl chloride | Cyclopropyl bromide | Cyclohexyl iodide |

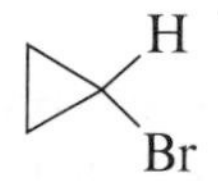

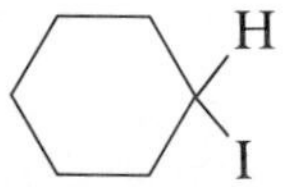

PROBLEM 4.1 Write structural formulas and give the common names of all the isomeric alkyl chlorides that have the molecular formula C_4H_9Cl.

Alcohols are given systematic IUPAC names by replacing the *-e* ending of the corresponding alkane name by *-ol.* The position of the hydroxyl group is indicated by number, choosing the sequence that assigns the lower locant to the carbon that bears the hydroxyl group.

$$\overset{3}{C}H_3\overset{2}{C}H_2\overset{1}{C}H_2OH \qquad \overset{1}{C}H_3\underset{\underset{OH}{|}}{\overset{2}{C}H}\overset{3}{C}H_2\overset{4}{C}H_2\overset{5}{C}H_3 \qquad \overset{1}{C}H_3\overset{2}{C}H_2\underset{\underset{OH}{|}}{\overset{3}{C}H}\overset{4}{C}H_2\overset{5}{C}H_3$$

1-Propanol   2-Pentanol   3-Pentanol

Hydroxyl groups take precedence over alkyl groups in determining the direction in which a carbon chain is numbered.

$$\overset{7}{C}H_3\underset{\underset{CH_3}{|}}{\overset{6}{C}H}\overset{5}{C}H_2\overset{4}{C}H_2\underset{\underset{OH}{|}}{\overset{3}{C}H}\overset{2}{C}H_2\overset{1}{C}H_3$$

6-Methyl-3-heptanol
(not 2-methyl-5-heptanol)

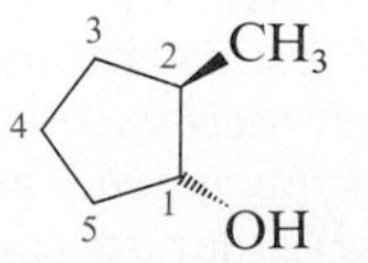

trans-2-Methylcyclopentanol

PROBLEM 4.2   Give systematic IUPAC names to all the isomeric $C_4H_{10}O$ alcohols.

Systematic nomenclature of alkyl halides treats the halogen as a substituent on an alkane chain. The carbon chain is numbered in the direction that gives the carbon bearing the halo- substituent the lower locant.

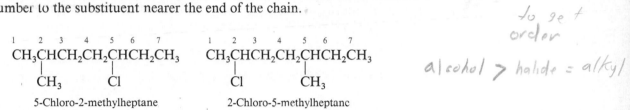

$$\overset{5}{C}H_3\overset{4}{C}H_2\overset{3}{C}H_2\overset{2}{C}H_2\overset{1}{C}H_2F$$

1-Fluoropentane

$$\overset{1}{C}H_3\overset{2}{C}HCH_2\overset{4}{C}H_2\overset{5}{C}H_3$$
$$|$$
$$Br$$

2-Bromopentane

$$\overset{1}{C}H_3\overset{2}{C}H_2\overset{3}{C}HCH_2\overset{5}{C}H_3$$
$$|$$
$$I$$

3-Iodopentane

When the carbon chain bears both a halogen and an alkyl substituent, the two substituents are considered of equal rank and the chain is numbered so as to give the lower number to the substituent nearer the end of the chain.

$$\overset{1}{C}H_3\overset{2}{C}HCH_2\overset{4}{C}H_2\overset{5}{C}HCH_2\overset{7}{C}H_3$$
$$|\qquad\qquad\quad|$$
$$CH_3\qquad\quad Cl$$

5-Chloro-2-methylheptane

$$\overset{1}{C}H_3\overset{2}{C}HCH_2\overset{4}{C}H_2\overset{5}{C}HCH_2\overset{7}{C}H_3$$
$$|\qquad\qquad\quad|$$
$$Cl\qquad\quad CH_3$$

2-Chloro-5-methylheptane

PROBLEM 4.3   Give the systematic names for all the isomeric alkyl chlorides having the molecular formula C_4H_9Cl.

Hydroxyl groups have precedence over halogen substituents in determining the direction of numbering. $FCH_2CH_2CH_2OH$, for example, is 3-fluoro-1-propanol, not 1-fluoro-3-propanol.

4.2 CLASSES OF ALCOHOLS AND ALKYL HALIDES

Alcohols and alkyl halides are classified as primary, secondary, or tertiary according to the classification of the carbon that bears the functional group (Section 2.10). In *primary alcohols* and *primary alkyl halides* the functional group is bonded to a primary carbon, that is, a carbon that bears one carbon substituent and two hydrogens.

$$\overset{\displaystyle H}{\underset{\displaystyle H}{R-\overset{|}{\underset{|}{C}}-}}$$

Primary
alkyl group

$$CH_3CH_2OH$$

Ethyl alcohol
(a primary alcohol)

$$(CH_3)_3CCH_2Br$$

Neopentyl bromide
(a primary alkyl halide)

In *secondary alcohols* and *secondary alkyl halides* the functional group is bonded to a secondary carbon atom; a secondary carbon is bonded to two other carbons and one hydrogen.

$$R-\overset{\overset{\displaystyle H}{|}}{\underset{\underset{\displaystyle R'}{|}}{C}}-$$

Secondary
alkyl group

$$CH_3CHCH_2CH_2CH_3$$
|
$$OH$$

2-Pentanol
(a secondary alcohol)

Cyclooctyl iodide
(a secondary alkyl halide)

In *tertiary alcohols* and *tertiary alkyl halides* the functional group is bonded to a tertiary carbon atom; a tertiary carbon is bonded to three other carbons.

$$R-\overset{\overset{\displaystyle R''}{|}}{\underset{\underset{\displaystyle R'}{|}}{C}}-$$

Tertiary
alkyl group

1-Methylcyclohexanol
(a tertiary alcohol)

$$CH_3\underset{\underset{\displaystyle Cl}{|}}{\overset{\overset{\displaystyle CH_3}{|}}{C}}CH_2CH_2CH_3$$

2-Chloro-2-methylpentane
(a tertiary alkyl halide)

PROBLEM 4.4   Classify the isomeric $C_4H_{10}O$ alcohols as primary, secondary, or tertiary.

Many of the properties of alcohols and alkyl halides are affected by whether their functional groups are attached to primary, secondary, or tertiary carbons. We will encounter numerous instances in which a functional group attached to a primary carbon is more reactive than if attached to a secondary or tertiary one, and numerous other instances in which the reverse is true.

4.3 BONDING IN ALCOHOLS AND ALKYL HALIDES

The carbon that bears the functional group is sp^3 hybridized in alcohols and alkyl halides. The hydroxyl group of an alcohol is attached to carbon by a σ bond generated by overlap of an sp^3 hybrid orbital of carbon with an sp^3 hybrid orbital of oxygen. Figure 4.1 illustrates this for the specific case of bonding in methanol. The bonds to carbon are approximately tetrahedral. Repulsions between the unshared electron pairs of oxygen cause the C—O—H bond angle, like the H—O—H bond angle of

FIGURE 4.1   Orbital hybridization model of bonding in methanol. (*a*) The orbitals used in bonding are the 1*s* orbitals of hydrogen, *sp*³ hybridized orbitals of carbon, and *sp*³ hybridized orbitals of oxygen. (*b*) The bond angles at carbon and oxygen are close to tetrahedral, and the carbon-oxygen σ bond distance is about 0.1 Å shorter than a carbon-carbon single bond.

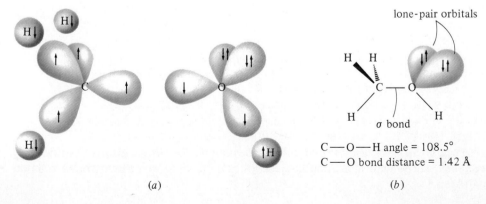

(*a*)

(*b*)

water, to be slightly less than the tetrahedral value of 109.5°. Bonding in alkyl halides is similar to that of alcohols. The halogen substituent is connected to sp^3 hybridized carbon by a σ bond.

Oxygen and halogen substituents are more electronegative than carbon, so electrons in carbon-oxygen and carbon-halogen bonds are drawn away from carbon toward the more electronegative atom. Carbon-oxygen and carbon-halogen bonds are polar bonds (Section 1.5), and alcohols and alkyl halides are polar molecules. Their dipole moments are typically on the order of 2 D and thus comparable with that of water.

the halide is not hybridized, though, it is a p-sp3 sigma bond

Water
($\mu = 1.8$ D)

Methanol
($\mu = 1.7$ D)

Chloromethane
($\mu = 1.9$ D)

PROBLEM 4.5   Bromine is less electronegative than chlorine, yet methyl bromide and methyl chloride have dipole moments that are very similar to each other. Can you think of a reason for this?

4.4 PHYSICAL PROPERTIES OF ALCOHOLS AND ALKYL HALIDES

As discussed earlier in Section 2.14, polar substances such as alkyl halides and alcohols are normally higher-boiling than alkanes of similar molecular weight. In addition to the induced dipole–induced dipole attractive forces common to all substances, alkyl halides have permanent dipole moments. This results in dipole-dipole attractive forces between molecules, which must be broken in order to cause alkyl halide molecules to pass from the liquid phase to the vapor phase. Table 4.1 lists the boiling points and densities of selected alkyl halides. For a given alkyl group the alkyl fluoride has the lowest boiling point and the lowest density of the various alkyl halides; alkyl iodides are the highest-boiling and have the highest density.

Hydrogen bonding stabilizes the liquid phase of alcohols and makes them higher-boiling than alkyl halides of comparable molecular weight. Table 4.2 presents physical data for a number of representative alcohols. Their ability to participate in intermolecular hydrogen bonding not only affects the boiling points of alcohols, but also enhances their water solubility. The lower-molecular-weight alcohols, methanol and ethanol, for example, are soluble in water in all proportions. Hydrogen-bonded networks of the type shown in Figure 4.2, in which alcohol and water molecules associate with one another, replace the alcohol-alcohol and water-water hydrogen-bonded networks present in the pure substances.

Higher alcohols become more "hydrocarbonlike" and less water-soluble. 1-Octanol, for example, dissolves to the extent of only 1 part in 2000 parts of water. London forces cause 1-octanol molecules to bind to each other through their long alkyl chains, and the energy of hydrogen bonding between water and 1-octanol molecules does not provide enough stabilization to dissociate these 1-octanol aggregates.

TABLE 4.1
Physical Properties of Some Alkyl Halides

IUPAC name	Common name	Condensed structure	Boiling point, °C (1 atm)				Density, g/ml (20°C)			
			Fluoride	Chloride	Bromide	Iodide	Fluoride	Chloride	Bromide	Iodide
Halomethane	Methyl halide	CH_3X	−78	−24	3	42				2.279
Haloethane	Ethyl halide	CH_3CH_2X	−32	12	38	72		0.903	1.460	1.933
1-Halopropane	n-Propyl halide	$CH_3CH_2CH_2X$	−3	47	71	103		0.890	1.353	1.739
2-Halopropane	Isopropyl halide	$(CH_3)_2CHX$	−11	35	59	90		0.859	1.310	1.714
1-Halobutane	n-Butyl halide	$CH_3CH_2CH_2CH_2X$		78	102	130		0.887	1.276	1.615
2-Halobutane	sec-Butyl halide	$CH_3\underset{\underset{X}{\vert}}{CH}CH_2CH_3$		68	91	120		0.873	1.261	1.597
1-Halo-2-methylpropane	Isobutyl halide	$(CH_3)_2CHCH_2X$	16	68	91	121		0.878	1.264	1.603
2-Halo-2-methylpropane	tert-Butyl halide	$(CH_3)_3CX$		51	73	99		0.847	1.220	1.570
1-Halopentane	n-Pentyl halide	$CH_3(CH_2)_3CH_2X$	65	108	129	157		0.884	1.216	1.516
1-Halohexane	n-Hexyl halide	$CH_3(CH_2)_4CH_2X$	92	134	155	180		0.879	1.175	1.439
1-Halooctane	n-Octyl halide	$CH_3(CH_2)_6CH_2X$	143	183	202	226	0.804	0.892	1.118	1.336
Halocyclopentane	Cyclopentyl halide	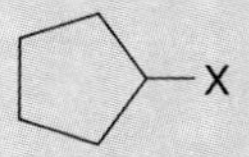		114	138	166		1.005	1.388	1.694
Halocyclohexane	Cyclohexyl halide	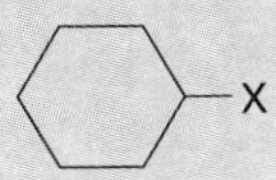		142	167	192		0.977	1.324	1.626

TABLE 4.2
Physical Properties of Some Alcohols

IUPAC name	Common name	Condensed structure	Melting point, °C	Boiling point, °C (1 atm)	Solubility, g/100 ml H_2O
Methanol	Methyl alcohol	CH_3OH	−94	65	∞
Ethanol	Ethyl alcohol	CH_3CH_2OH	−117	78	∞
1-Propanol	n-Propyl alcohol	$CH_3CH_2CH_2OH$	−127	97	∞
2-Propanol	Isopropyl alcohol	$(CH_3)_2CHOH$	−90	82	∞
1-Butanol	n-Butyl alcohol	$CH_3CH_2CH_2CH_2OH$	−90	117	9
2-Butanol	sec-Butyl alcohol	$CH_3CHCH_2CH_3$ $\quad$ OH	−115	100	12.5
2-Methyl-1-propanol	Isobutyl alcohol	$(CH_3)_2CHCH_2OH$	−108	108	10
2-methyl-2-propanol	tert-Butyl alcohol	$(CH_3)_3COH$	26	83	∞
1-Pentanol	n-Pentyl alcohol	$CH_3(CH_2)_3CH_2OH$	−79	138	
1-Hexanol	n-Hexyl alcohol	$CH_3(CH_2)_4CH_2OH$	−52	157	0.6
1-Dodecanol	n-Dodecyl alcohol	$CH_3(CH_2)_{10}CH_2OH$	26	259	insoluble
1-Octadecanol	n-Octadecyl alcohol	$CH_3(CH_2)_{16}CH_2OH$	59	332	insoluble
Cyclopentanol	Cyclopentyl alcohol	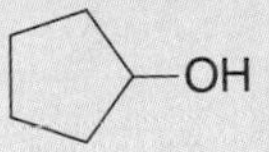 ⬠—OH		139	slightly soluble
Cyclohexanol	Cyclohexyl alcohol	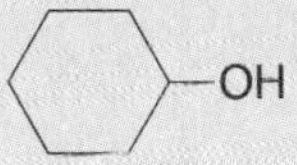 ⬡—OH	25	161	3.6

4.5 ALCOHOLS AS BRÖNSTED BASES

According to the theory proposed by Svante Arrhenius, a Swedish chemist and winner of the 1903 Nobel prize in chemistry, *acids ionize in aqueous solution to liberate protons,* and *bases ionize to liberate hydroxide ions.*

$$H\overset{\frown}{-}A \rightleftharpoons H^+ + :A^-$$

Arrhenius acid

$$M\overset{\frown}{-}\overset{..}{\underset{..}{O}}H \rightleftharpoons M^+ + {}^-:\overset{..}{\underset{..}{O}}H$$

Arrhenius base

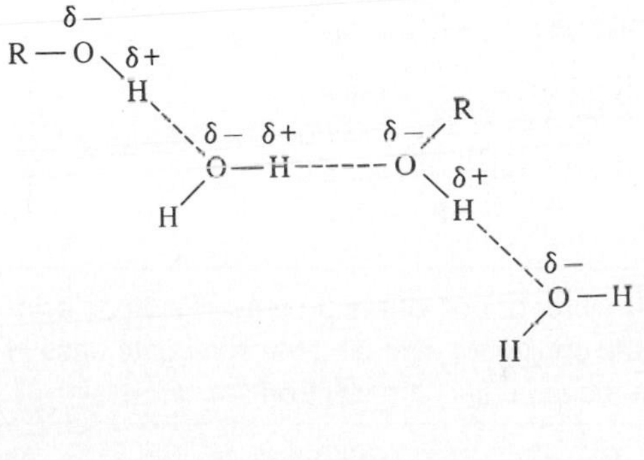

FIGURE 4.2 Hydrogen bonding between alcohol and water molecules.

Note the use of "curved arrows" here to show the nature of the ionization. A covalent bond in acid HA is broken, with both electrons in that bond becoming an unshared pair of the ion $:A^-$. Similarly, the metal-oxygen bond in the base MOH cleaves, with both electrons becoming an unshared pair of the hydroxide ion.

An alternative theory of acids and bases was devised independently by Johannes Brönsted and Thomas M. Lowry in 1923. In the Brönsted-Lowry approach, *an acid is a proton donor,* and *a base is a proton acceptor.*

$$B: + H{-}A \rightleftharpoons \overset{+}{B}{-}H + :A^-$$

Base   Acid   Conjugate acid   Conjugate base

The curved arrows in this equation show the electron pair of the base abstracting a proton from the acid. The pair of electrons in the H—A covalent bond becomes an unshared electron pair in the conjugate base $:A^-$.

In aqueous solution, an acid transfers a proton to water. Water acts as a Brönsted base.

$$\overset{H}{\underset{}{HO}}: + H{-}A \rightleftharpoons H:\overset{H}{\underset{+}{O}}:H + :A^-$$

Water (base)   Acid   Conjugate acid of water   Conjugate base

The conjugate acid of water (H_3O^+) is commonly known as *hydronium ion.* Its systematic name is *oxonium ion.*

Alcohols are structurally similar to water and can also act as Brönsted bases. The conjugate acid of an alcohol is an alkyl-substituted oxonium ion, i.e., an *alkyloxonium ion.*

$$\overset{H}{\underset{}{RO}}: + H{-}A \rightleftharpoons \overset{H}{\underset{+}{ROH}} + :A^-$$

Alcohol   Acid   Alkyloxonium ion   Conjugate base

Alcohols are comparable with water in their relative strengths as Brönsted bases. Strong acids such as sulfuric acid and hydrogen halides transfer a proton to the oxygen atom of alcohols in an acid-base reaction.

$$R\overset{..}{O}H + H{-}\overset{..}{I}: \rightleftharpoons R\overset{+}{O}H_2 + :\overset{..}{I}:^-$$

Alcohol   Hydrogen iodide   Alkyloxonium ion   Iodide ion

(base)   (acid)   (conjugate acid)   (conjugate base)

PROBLEM 4.6 Write an equation for the transfer of one of the protons of sulfuric acid to methyl alcohol. Identify the acid, the base, the conjugate acid, and the conjugate base in your equation. Using curved arrows, show the movement of the electrons.

We shall see that several important reactions of alcohols involve strong acids either as reagents or as catalysts to increase the rate of reaction. In most of these reactions the first step is formation of an alkyloxonium ion by proton transfer from the acid to the alcohol.

Proton transfers from strong acids to alcohols are exceedingly fast. Measurements of the rate of acid-base reactions of this type reveal that they take place at rates approaching the rate at which the proton donor and the proton acceptor encounter each other in solution. Proton transfers from strong acids to water or to alcohols rank among the most rapid chemical processes—faster even than the rate of rotation about the carbon-carbon bond of ethane, for example.

4.6 ALCOHOLS AS BRÖNSTED ACIDS

In the presence of strong bases, water and alcohols can act as proton donors; they are *Brönsted acids.*

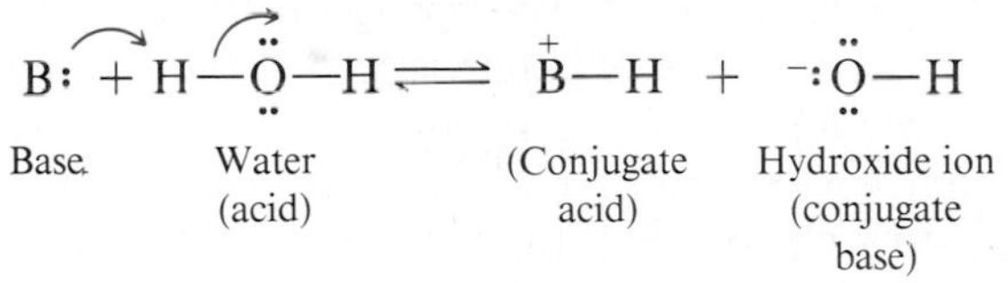

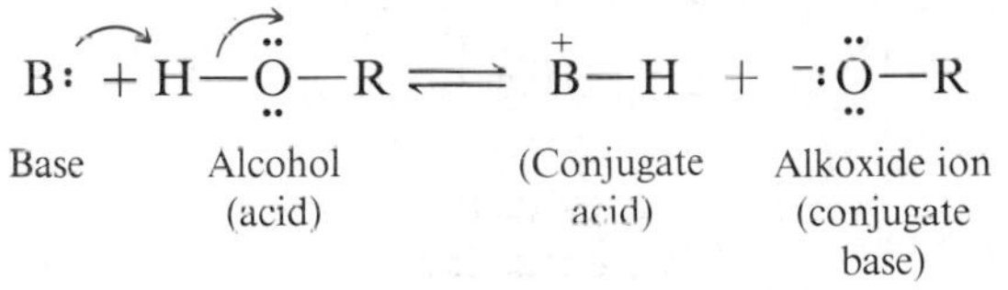

$$B: + H-\overset{..}{\underset{..}{O}}-H \rightleftharpoons \overset{+}{B}-H + \overset{..}{}\overset{-}{:}\overset{..}{O}-H$$

Base   Water   (Conjugate   Hydroxide ion
(acid)   acid)   (conjugate base)

$$B: + H-\overset{..}{\underset{..}{O}}-R \rightleftharpoons \overset{+}{B}-H + \overset{-}{:}\overset{..}{\underset{..}{O}}-R$$

Base   Alcohol   (Conjugate   Alkoxide ion
(acid)   acid)   (conjugate base)

The conjugate base of water is hydroxide ion; the conjugate base of an alcohol is an *alkoxide ion.*

Alcohols tend to be somewhat weaker acids than water; water is a better proton donor. In solution the equilibrium represented by the equation shown below usually lies to the side of hydroxide ion and the alcohol.

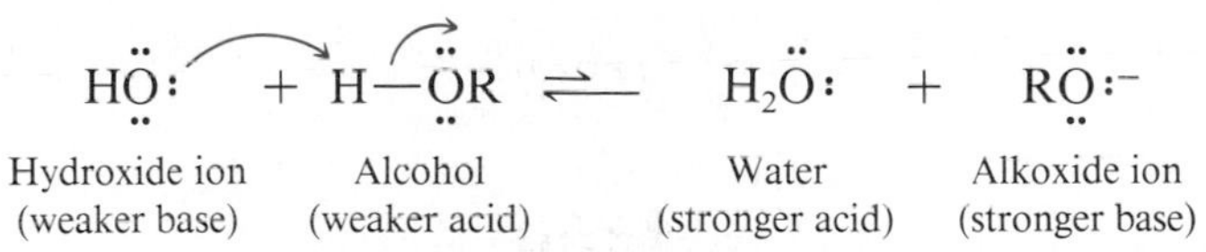

$$H\overset{..}{\underset{..}{O}}:^- + H-\overset{..}{\underset{..}{O}}R \rightleftharpoons H_2\overset{..}{O}: + R\overset{..}{\underset{..}{O}}:^-$$

Hydroxide ion   Alcohol   Water   Alkoxide ion
(weaker base)   (weaker acid)   (stronger acid)   (stronger base)

The strength of an acid in dilute aqueous solution is given by the *equilibrium constant K* for the reaction

$$HA + H_2O \overset{K}{\rightleftharpoons} H_3O^+ + A^-$$

where

$$K = \frac{[H_3O^+][A^-]}{[HA][H_2O]}$$

But, since water is the solvent and the solution is dilute, the large water concentration (55 M) does not change much. The concentration of water can thus be treated as a constant and incorporated into a new constant K_a called the *acid dissociation constant* or the *ionization constant.*

$$K_a = K[H_2O] = \frac{[H_3O^+][A^-]}{[HA]}$$

Strong acids are characterized by large values of K_a; essentially every molecule of a strong acid transfers a proton to water in dilute aqueous solution. Weak acids have small K_a values. Table 4.3 lists a number of Brönsted acids and their acid dissociation constants. Most alcohols have K_a's in the range 10^{-16} to 10^{-20}; they are extremely weak acids.

A convenient way to express acid strength is through the use of pK_a, defined as

$$pK_a = -\log_{10} K_a$$

This permits acidity to be expressed in numbers that are not exponentials. Thus, an alcohol with $K_a = 10^{-16}$ has $pK_a = 16$. The stronger the acid, the smaller the value of pK_a. The weaker the acid, the larger the value of pK_a. Table 4.3 includes pK_a as well as K_a values for acids. Both K_a and pK_a values are used extensively as measures of acid strength. Our custom will be to cite K_a but to include pK_a in parenthesis. Your instructor will, either directly or by example, inform you as to which one to emphasize. There is much to be said, however, for being conversant with both.

PROBLEM 4.7   Calculate K_a for each of the following acids, given its pK_a. Rank the compounds in order of decreasing acidity.

(a)   Aspirin: $pK_a = 3.48$
(b)   Vitamin C (ascorbic acid): $pK_a = 4.17$
(c)   Formic acid (present in sting of ants): $pK_a = 3.75$
(d)   Oxalic acid (poisonous substance found in certain berries): $pK_a = 1.19$

SAMPLE SOLUTION   (a)   This problem reviews the relationship between logarithms and exponential numbers. We need to determine K_a, given pK_a. The equation that relates the two is

$$pK_a = -\log_{10} K_a$$

Therefore

$$K_a = 10^{-pK_a}$$
$$= 10^{-3} \times 10^{-0.48} = 10^{-4} \times 10^{0.52}$$
$$= 3.3 \times 10^{-4}$$

An important corollary of the Brönsted-Lowry view of acids and bases involves the relative strengths of an acid and its conjugate base. The stronger the acid, the weaker the conjugate base; the weaker the acid, the stronger the conjugate base. Referring to Table 4.3, we see that ammonia is a very weak acid, its K_a being only 10^{-36} ($pK_a = 36$). Therefore, amide anion (NH_2^-) is a very strong base. Ethoxide ion ($CH_3CH_2O^-$) and methoxide ion (CH_3O^-) are comparable with hydroxide ion in basicity. Halide ions are very weak bases: fluoride ion is the strongest base of the halide ions but is 10^{12} times less basic than hydroxide ion based on a comparison of their K_a's.

TABLE 4.3

Acid Dissociation Constants K_a and pK_a Values for Some Brönsted Acids*

Acid	Formula†	Dissociation constant, K_a	pK_a	Conjugate base
Hydrogen iodide	HI	$\sim 10^{10}$	~ -10	I^-
Hydrogen bromide	HBr	$\sim 10^9$	~ -9	Br^-
Hydrogen chloride	HCl	$\sim 10^7$	~ -7	Cl^-
Sulfuric acid	$HOSO_2OH$	1.6×10^5	-4.8	$HOSO_2O^-$
Nitric acid	$HONO_2$	2.5×10^1	-0.6	$^-ONO_2$
Phosphoric acid	$(HO)_2P(O)(OH)$	6×10^{-3}	2.2	$(HO)_2PO_2^-$
Hydrogen fluoride	HF	3.5×10^{-4}	3.5	F^-
Acetic acid	$\overset{\overset{O}{\|\|}}{CH_3COH}$	1.8×10^{-5}	4.7	$\overset{\overset{O}{\|\|}}{CH_3CO^-}$
Water	HOH	1.8×10^{-16}	15.7	HO^-
Methanol	CH_3OH	$\sim 10^{-16}$	~ 16	CH_3O^-
Ethanol	CH_3CH_2OH	$\sim 10^{-16}$	~ 16	$CH_3CH_2O^-$
Isopropyl alcohol	$(CH_3)_2CHOH$	$\sim 10^{-17}$	~ 17	$(CH_3)_2CHO^-$
tert-Butyl alcohol	$(CH_3)_3COH$	$\sim 10^{-18}$	~ 18	$(CH_3)_3CO^-$
Ammonia	H_2NH	$\sim 10^{-36}$	~ 36	H_2N^-

* Acid strength decreases from top to bottom of the table. Strength of conjugate base increases from top to bottom of the table.

† The most acidic proton — the one that is lost on ionization — is indicated in color.

In any proton-transfer process the position of equilibrium favors formation of the weaker acid and the weaker base. Thus, when *tert*-butoxide ion is introduced into aqueous solution, the predominant base that is present is hydroxide ion.

$$(CH_3)_3CO^- + H_2O \rightleftharpoons (CH_3)_3COH + HO^-$$

tert-Butoxide ion	Water	*tert*-Butyl alcohol	Hydroxide ion
(stronger base)	(stronger acid: $K_a \cong 10^{-16}$)	(weaker acid: $K_a \cong 10^{-18}$)	(weaker base)

PROBLEM 4.8 Write an equation for the reaction between amide ion and methanol. What species are present in greatest concentration at equilibrium? Identify the various species as acids or bases and specify which are stronger and which are weaker.

4.7 REACTIONS OF ALCOHOLS WITH METALS

Alkoxide bases are among the most commonly used bases in organic chemistry. How are they prepared? Recall that metals such as sodium and potassium react vigorously —even violently—with water, liberating hydrogen and forming the corresponding metal hydroxide.

$$H_2O + 2M \longrightarrow 2MOH + H_2$$

Water	Group I metal	Group I metal hydroxide	Hydrogen

An analogous reaction takes place between metals and alcohols. The product is a metal alkoxide.

$$CH_3OH + 2Na \longrightarrow 2NaOCH_3 + H_2$$

Methyl   Sodium   Sodium methoxide   Hydrogen
alcohol

$$(CH_3)_3COH + 2K \longrightarrow 2KOC(CH_3)_3 + H_2$$

tert-Butyl   Potassium   Potassium   Hydrogen
alcohol   *tert*-butoxide

Sodium and potassium are the metals used most often in the preparation of metal alkoxides, potassium being more reactive than sodium. The order of alcohol reactivity is:

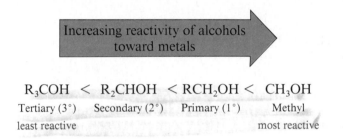

$$R_3COH < R_2CHOH < RCH_2OH < CH_3OH$$

Tertiary (3°)   Secondary (2°)   Primary (1°)   Methyl

least reactive   most reactive

When the alcohols are primary and secondary, the reactivity of sodium is sufficient for relatively concentrated solutions of the sodium alkoxide in the alcohol to be easily prepared. When the alcohol is tertiary, the more reactive metal potassium is normally used.

Like sodium and potassium hydroxide, metal alkoxides have an ionic metal-oxygen bond and are strongly basic.

4.8 PREPARATION OF ALKYL HALIDES

The remaining sections of this chapter describe methods for the preparation of alkyl halides. Chemical equations will be used extensively to illustrate the reactions that take place. In these and other examples to be cited throughout this text, the *percent yield* of isolated product is given in parentheses. The efficiency of a synthetic transformation is normally expressed as some percentage of the *theoretical yield,* which is the amount of product that could be formed if the reaction proceeded to completion and did not lead to the formation of any products other than those given in the equation. The yield data are taken from experiments reported in the chemical literature and are frequently well below 100 percent. The difference between the theoretical yield and that actually obtained results from competing side reactions that divert the reactants to other products, as well as from mechanical losses of product during its isolation and purification.

The procedures to be described use either an alkane or an alcohol as the starting material for the preparation of an alkyl halide. By knowing how alkyl halides are prepared we can better appreciate the material in succeeding chapters, where alkyl halides are featured prominently in key chemical transformations. The preparation of alkyl halides serves also as a focal point for the development of principles of reaction mechanisms. We begin with the preparation of alkyl halides from alcohols by reaction with hydrogen halides.

4.9 ALKYL HALIDES FROM ALCOHOLS AND HYDROGEN HALIDES

Alcohols react with hydrogen halides according to the general equation shown:

$$\text{ROH} + \text{HX} \longrightarrow \text{RX} + \text{H}_2\text{O}$$

| Alcohol | Hydrogen halide | Alkyl halide | Water |

As the examples given below illustrate, the alcohol may be primary, secondary, or tertiary and the hydrogen halide may be hydrogen chloride, hydrogen bromide, or hydrogen iodide.

$$(\text{CH}_3)_2\text{CHCH}_2\text{OH} + \quad \text{HI} \longrightarrow (\text{CH}_3)_2\text{CHCH}_2\text{I} + \text{H}_2\text{O}$$

2-Methyl-1-propanol (isobutyl alcohol), a primary alcohol　　Hydrogen iodide　　1-Iodo-2-methylpropane (isobutyl iodide) (64%)　　Water

$$\underset{\underset{\text{OH}}{|}}{\text{CH}_3\text{CHCH}_2\text{CH}_3} + \quad \text{HBr} \longrightarrow \underset{\underset{\text{Br}}{|}}{\text{CH}_3\text{CHCH}_2\text{CH}_3} + \text{H}_2\text{O}$$

2-Butanol (sec-butyl alcohol), a secondary alcohol　　Hydrogen bromide　　2-Bromobutane (sec-butyl bromide) (73%)　　Water

$$(\text{CH}_3)_3\text{COH} + \quad \text{HCl} \longrightarrow (\text{CH}_3)_3\text{CCl} + \text{H}_2\text{O}$$

2-Methyl-2-propanol (tert-butyl alcohol), a tertiary alcohol　　Hydrogen chloride　　2-Chloro-2-methylpropane (tert-butyl chloride) (78–88%)　　Water

---

PROBLEM 4.9 Write chemical equations expressing the reaction that takes place between each of the following pairs of reactants:

   (a) Cyclohexanol and hydrogen iodide
   (b) 3-Ethyl-3-pentanol and hydrogen chloride
   (c) 1-Tetradecanol and hydrogen bromide

SAMPLE SOLUTION (a) An alcohol and a hydrogen halide react to form an alkyl halide and water. In this case, iodocyclohexane was isolated in 79% yield.

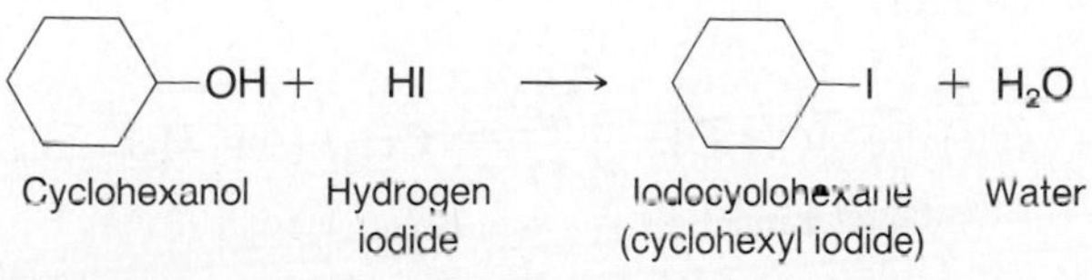

Cyclohexanol　　Hydrogen iodide　　Iodocyclohexane (cyclohexyl iodide)　　Water

---

The rate of the reaction between an alcohol and a hydrogen halide depends on the nature of both the alcohol and the hydrogen halide. Among the hydrogen halides hydrogen iodide reacts fastest and hydrogen fluoride most slowly. The relative *reactivity,* i.e., the relative rate at which the various hydrogen halides convert an alcohol to an alkyl halide, follows the same order as their acidity:

> Increasing reactivity of hydrogen halides toward alcohols

$$HF \ll HCl < HBr < HI$$

Weakest acid		Strongest acid
least reactive		most reactive

Hydrogen fluoride is the least reactive of the hydrogen halides, and the preparation of alkyl fluorides by this method is not practical.

Among classes of alcohols, tertiary alcohols are observed to be the most reactive and primary alcohols the least reactive:

> Increasing reactivity of alcohols toward hydrogen halides

$$CH_3OH < RCH_2OH < R_2CHOH < R_3COH$$

Methyl	Primary (1°)	Secondary (2°)	Tertiary (3°)
least reactive			most reactive

Thus, while tertiary alcohols are converted to alkyl chlorides in high yield within minutes at 0 to 25°C, primary alcohols react too slowly with hydrogen chloride for the reaction to be of any synthetic value. More reactive hydrogen halides such as hydrogen bromide do react with primary alcohols, but elevated temperatures are required.

$$CH_3(CH_2)_5CH_2OH + HBr \xrightarrow{120°C} CH_3(CH_2)_5CH_2Br + H_2O$$

1-Heptanol	Hydrogen bromide	1-Bromoheptane (87–90%)	Water

Alternatively, the same kind of transformation may be carried out by simply heating the alcohol with sodium bromide and sulfuric acid.

$$CH_3CH_2CH_2CH_2OH \xrightarrow[\text{heat}]{NaBr, H_2SO_4} CH_3CH_2CH_2CH_2Br$$

1-Butanol (*n*-butyl alcohol)	1-Bromobutane (70–83%) (*n*-butyl bromide)

In a similar modification alcohols are heated with potassium iodide in phosphoric acid to form alkyl iodides.

$$(CH_3)_2CHCH_2OH \xrightarrow[\text{heat}]{KI, H_3PO_4} (CH_3)_2CHCH_2I$$

Isobutyl alcohol	Isobutyl iodide (88%)

Organic chemists often find it convenient to write chemical equations in the abbreviated form shown above, in which reagents, especially inorganic ones, are not included in the body of the equation but instead are indicated over the arrow. Inorganic products —in this case water—are usually omitted. These simplifications serve to focus attention on the functional group transformation undergone by the organic reactant.

4.10 MECHANISM OF THE REACTION OF ALCOHOLS WITH HYDROGEN HALIDES

We classify the reaction of an alcohol with a hydrogen halide as a *substitution* reaction. Halogen replaces a hydroxyl group as a substituent on carbon. When, for example, *tert*-butyl alcohol reacts with hydrogen chloride, chlorine replaces a hydroxyl group and *tert*-butyl chloride is formed.

$$(CH_3)_3COH + HCl \longrightarrow (CH_3)_3CCl + H_2O$$

| *tert*-Butyl alcohol | Hydrogen chloride | *tert*-Butyl chloride | Water |

Identifying the reaction as one of substitution simply tells us the relationship between the organic reactant and its product but does not reveal how the reaction takes place.

The "how" of a chemical reaction is what we try to explain when we propose a *mechanism* for it. A reaction mechanism is a structural description of the intermediates and transition states encountered during the conversion of reactants to products. In developing a mechanistic picture for a particular reaction, some basic principles of chemical reactivity are combined with experimental observations to deduce

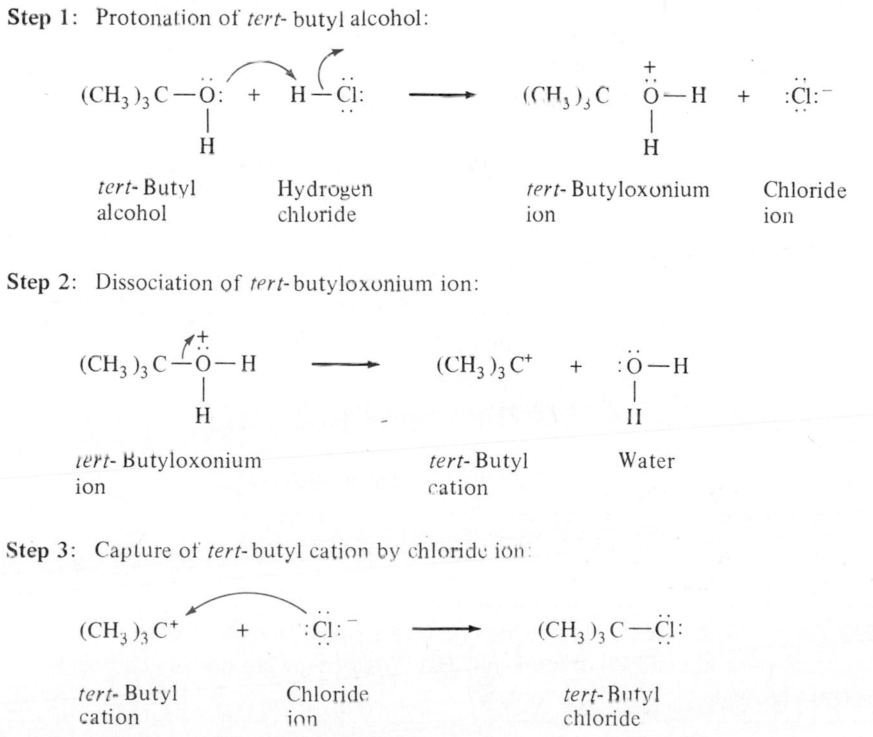

Step 1: Protonation of *tert*-butyl alcohol:

$$(CH_3)_3C-\overset{..}{\underset{|}{O}}: \quad + \quad H-\overset{..}{\underset{..}{Cl}}: \quad \longrightarrow \quad (CH_3)_3C \quad \overset{+}{\underset{|}{O}}-H \quad + \quad :\overset{..}{\underset{..}{Cl}}:^-$$
$$H H$$

| *tert*-Butyl alcohol | Hydrogen chloride | *tert*-Butyloxonium ion | Chloride ion |

Step 2: Dissociation of *tert*-butyloxonium ion:

$$(CH_3)_3C-\overset{+}{\underset{|}{O}}-H \quad \longrightarrow \quad (CH_3)_3C^+ \quad + \quad :\overset{..}{O}-H$$
$$H H$$

| *tert*-Butyloxonium ion | *tert*-Butyl cation | Water |

Step 3: Capture of *tert*-butyl cation by chloride ion:

$$(CH_3)_3C^+ \quad + \quad :\overset{..}{\underset{..}{Cl}}:^- \quad \longrightarrow \quad (CH_3)_3C-\overset{..}{\underset{..}{Cl}}:$$

| *tert*-Butyl cation | Chloride ion | *tert*-Butyl chloride |

FIGURE 4.3 Sequence of steps that describes the mechanism of formation of *tert*-butyl chloride from *tert*-butyl alcohol and hydrogen chloride.

the most likely sequence of events that occur during the functional-group transformation.

Consider a specific reaction, the one shown above in which *tert*-butyl alcohol reacts with hydrogen chloride to form *tert*-butyl chloride and water. The generally accepted mechanism for this reaction is presented as a series of equations in Figure 4.3.

We already know something about step 1 of this mechanism. Step 1 is a Brönsted acid-base reaction. Proton transfer to *tert*-butyl alcohol gives the corresponding oxonium ion. Like proton-transfer reactions between hydrogen halides and water, proton transfer from hydrogen chloride to an alcohol should be very fast. Steps 2 and 3, however, are new to us. Step 2 involves conversion of the oxonium ion to a *carbocation;* a species that contains a positively charged carbon atom. In step 3, this carbocation reacts with chloride ion to yield *tert*-butyl chloride. So that we may better understand the chemistry expressed in steps 2 and 3, we need to examine carbocations in more detail.

4.11   STRUCTURE AND BONDING IN CARBOCATIONS

Any species that incorporates a positively charged carbon is referred to as a *carbocation.* The type of carbocation that we will encounter in this text is *tricoordinate,* or *trivalent,* meaning that it has three substituents bonded to a central carbon. The central carbon bears a formal positive charge of $+1$; it has only six valence electrons, the ones involved in covalent bonds to its three substituent groups. Tricoordinate carbocations are also known as *carbonium ions* or *carbenium ions.*

Tricoordinate carbocations are classified as primary, secondary, or tertiary according to the number of carbon substituents attached to the positively charged carbon.

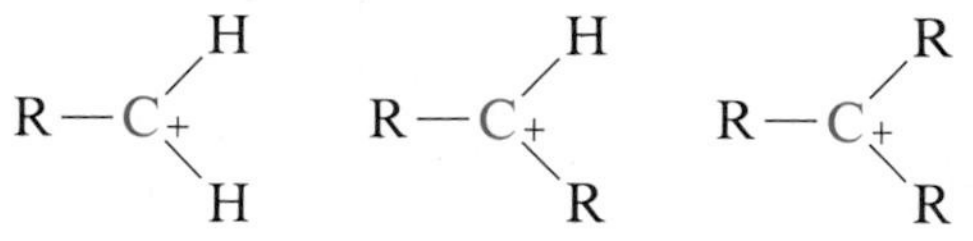

| Primary carbocation | Secondary carbocation | Tertiary carbocation |

Carbocations will be named in this text by appending "cation" as a separate word after the name of the appropriate alkyl group.

Ethyl cation
(Primary carbocation)

Isopropyl cation
(Secondary carbocation)

1-Methylcyclohexyl cation
(Tertiary carbocation)

PROBLEM 4.10   Write structural formulas and give suitable "cation" names for all the isomeric tricoordinate carbocations of formula $C_4H_9^+$. Which ones are primary carbocations? Which are secondary? Which are tertiary?

The properties of carbocations are intimately related to their structure, so let us consider the bonding in methyl cation, CH_3^+. Carbon is bonded to each of its three hydrogen substituents by a two-electron covalent bond. The valence shell of carbon contains only six electrons instead of the full complement of eight. According to the valence-shell electron pair repulsion rationale, maximum separation of three pairs of electrons is achieved by a trigonal planar geometry with bond angles of 120° at carbon, as shown in Figure 4.4.

While carbon bears a formal charge of $+1$ in methyl cation, some of that charge is shared by the hydrogen substituents. Electrons are drawn toward the positively charged carbon by polarization of the electron distribution in the C—H bonds, reducing the positive character of carbon and dispersing the charge.

Development of an orbital hybridization model of bonding in CH_3^+ begins by considering the electron configuration of C^+. A neutral carbon atom has six electrons; C^+ has five. Two of these are in the $1s$ level and are not involved in bonding. The remaining three are the valence electrons and are distributed among the $2s$ and $2p$ orbitals.

$1s$	$2s$	$2p$	$2p$	$2p$	$1s$	$2s$	$2p$	$2p$	$2p$

Electron configuration of C   Electron configuration of C^+

In order to form bonds to three hydrogens, carbon adopts an sp^2 hybridization state. The $2s$ orbital and two of the $2p$ orbitals are combined to give three new orbitals. These sp^2 hybridized orbitals have one-third s character and two-thirds p character. They are of equal energy, and the three valence electrons of C^+ occupy them one by one to give the electron configuration shown:

$1s$	$2s$	$2p$	$2p$	$2p$	$1s$	$2sp^2$	$2sp^2$	$2sp^2$	$2p$

Electron configuration of C^+ sp^2 Hybrid state of C^+

The $2p$ orbital that remains unchanged in the hybridization scheme is of higher energy than the sp^2 hybrid orbitals. It contains no electrons. Figure 4.5 gives a pictorial representation of sp^2 hybridization at carbon.

As shown in Figure 4.6, the three carbon-hydrogen bonds of methyl cation arise by overlap of sp^2 hybrid orbitals of carbon with hydrogen $1s$ orbitals. These bonds are σ bonds. *Methyl cation is planar* because of the directional properties associated with sp^2 hybrid orbitals. It can be shown mathematically that the axes of the three sp^2 hybrid orbitals are arranged in a trigonal planar geometry about carbon. The $2p$ orbital that remains unhybridized has its axis perpendicular to the plane of the ion.

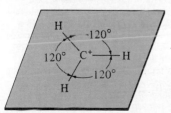

FIGURE 4.4 The electron pairs of the three C—H bonds of methyl cation are farthest apart in a trigonal planar geometry.

Carbonium ion
uses
sp^2
orbitals

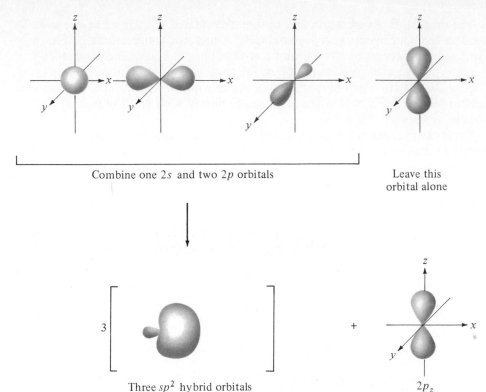

FIGURE 4.5 Representation of orbital mixing in sp^2 hybridization at carbon. Mixing of one s orbital with two p orbitals generates three sp^2 hybrid orbitals. Each sp^2 hybrid orbital has one-third s character and two-thirds p character.

Combine one $2s$ and two $2p$ orbitals

Leave this orbital alone

Three sp^2 hybrid orbitals

$2p_z$

The structural features that dominate the chemistry of carbocations are the positive charge on carbon and its vacant p orbital. These features combine to make carbocations strongly *electrophilic* ("electron-loving" or "electron-seeking"). Carbocations are *high-energy* species. They exist in the presence of some negatively charged counterion, but most carbocation salts are too reactive to isolate. The existence of

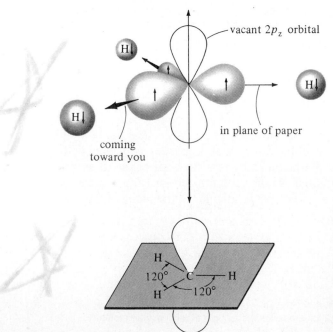

vacant $2p_z$ orbital

coming toward you

in plane of paper

FIGURE 4.6 The sp^2 hybrid orbitals of methyl cation are coplanar. Each contains one electron and can form a bond with a hydrogen atom to give a planar methyl cation. (*Note:* Only the major lobe of each sp^2 orbital is shown. As indicated in the previous figure, each sp^2 orbital contains a smaller back lobe. This has been omitted for the sake of clarity.) The axis of the $2p$ orbital is perpendicular to the plane defined by the atoms.

carbocations as intermediates in chemical reactions has been inferred from studies of reaction mechanisms, and they have been examined by various spectroscopic techniques of the type to be described in Chapter 14.

4.12 RELATIVE STABILITIES OF CARBOCATIONS

Carbocations bearing alkyl groups as substituents on the positively charged carbon are structurally related to CH_3^+. The central carbon is sp^2 hybridized, with a vacant p orbital, and its three σ bonds are coplanar. An important difference becomes apparent, however, when the stability of various classes of carbocations is compared. The order of carbocation stability is:

Increasing carbocation stability

Methyl cation	Ethyl cation (primary, 1°)	Isopropyl cation (secondary, 2°)	*tert*-Butyl cation (tertiary, 3°)
(least stable)			(most stable)

Since carbocations are electrophilic, they are stabilized by substituents that release electrons to the positively charged carbon. Alkyl groups release electrons better than do hydrogen substituents, so the more alkyl groups that are attached to the positively charged carbon, the more stable the carbocation.

PROBLEM 4.11 Of the isomeric $C_6H_{11}^+$ carbocations, which one is the most stable?

Alkyl groups release electrons through a combination of effects. One of these effects involves polarization of the electron distribution in the σ bond that connects the alkyl group to the positively charged carbon. As illustrated for ethyl cation in Figure 4.7, the electrons in the σ bond are drawn toward the positively charged carbon and away from its alkyl substituent. Electrons in a carbon-carbon σ bond tend to be more polarizable than electrons in a carbon-hydrogen σ bond, so replacing hydrogen substituents by alkyl groups stabilizes a carbocation. Structural effects that are transmitted by the polarization of electrons in σ bonds are called *inductive effects*. The inductive effect of an alkyl group is to release electrons.

A second effect of comparable importance also makes alkyl groups more electron-releasing than hydrogen substituents. This is the delocalization of electrons directly into the unfilled $2p$ orbital of the positively charged carbon. In general, any increase in the number of nuclei that an electron can interact with, i.e., its degree of delocalization, is a stabilizing effect. Figure 4.8 shows how the electrons of the CH_3 group of ethyl cation are delocalized into the vacant $2p$ orbital of the positively charged carbon. This increase in electron delocalization stabilizes the carbocation and assists in the dispersal of positive charge. In order for stabilization of this kind to occur, the substituent group must have a filled σ orbital available to overlap with the vacant $2p$ orbital. A hydrogen substituent does not meet this requirement and so cannot stabi-

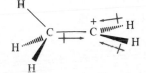

FIGURE 4.7 The charge in ethyl cation is stabilized by polarization of the electron distribution in the σ bonds between the positively charged carbon atom and its substituents. An alkyl substituent releases electrons better than does a hydrogen substituent.

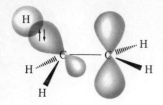

(a) Empty *p* orbital on positively charged carbon can overlap with σ orbital of methyl substituent.

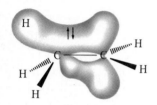

(b) Electrons in C—H σ bond are delocalized over a greater number of atoms, and the carbocation is stabilized.

FIGURE 4.8 Electrons in the C—H bonds of the methyl group are delocalized into the vacant 2*p* orbital of ethyl cation. Electron delocalization and dispersal of the positive charge stabilize the carbocation.

lize a carbocation by this mechanism in the way that alkyl groups do. The effect is cumulative: the more alkyl groups there are that are directly attached to the positively charged carbon, the greater the degree of electron delocalization, the greater the dispersal of charge, and the more stable the carbocation.

When we say that tertiary carbocations are more stable than their secondary or primary counterparts, understand that we are speaking here in a relative sense. Even tertiary carbocations are highly reactive species, normally not capable of being isolated under the conditions of their formation. In the case of the reaction of alcohols with hydrogen halides, a halide anion captures the carbocation as soon as it is formed, giving an alkyl halide. This process is represented by step 3 in our mechanism for the reaction of *tert*-butyl alcohol with hydrogen chloride and is described in more detail in the next section.

4.13 CARBOCATION-ANION COMBINATION. ELECTROPHILES AND NUCLEOPHILES

Figure 4.9 depicts the formation of an alkyl halide by the reaction of a carbocation and a halide anion. A carbocation was described in Section 4.11 as an electrophilic species; its positive charge, but especially its vacant 2*p* orbital, makes it electron-seeking. A carbocation can accept two electrons into its vacant 2*p* orbital.

Nucleophiles are species that have precisely complementary characteristics. A nucleophile is "nucleus-seeking"; it has a nonbonded pair of electrons that it can use to form a covalent bond to the positively charged carbon of a carbocation. Halide anions are examples of nucleophiles, as they are negatively charged and have unshared electron pairs.

Some time ago G. N. Lewis extended our understanding of acid-base behavior to include reactions other than proton transfers. According to Lewis, *an acid is an*

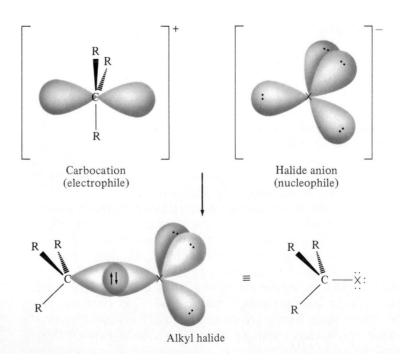

Carbocation (electrophile)

Halide anion (nucleophile)

Alkyl halide

FIGURE 4.9 Combination of a carbocation and a halide anion to give an alkyl halide.

electron pair acceptor and a base is an electron pair donor. As Figure 4.9 illustrates, carbocations are electron pair acceptors and thus are *Lewis acids.* Halide anions are electron pair donors and are *Lewis bases.* The combining of a carbocation and a halide anion is a Lewis acid – Lewis base reaction. It is generally true that electrophiles are Lewis acids and nucleophiles are Lewis bases.

Carbocations are among the most powerful electrophiles. They react rapidly under the conditions of their formation. Let us refer again to our three-step mechanism for the reaction of *tert*-butyl alcohol and hydrogen chloride:

Step 1

$$(CH_3)_3COH + HCl \underset{}{\overset{fast}{\rightleftharpoons}} (CH_3)_3C\overset{+}{O}H_2 + Cl^-$$

tert-Butyl	Hydrogen	*tert*-Butyloxonium	Chloride
alcohol	chloride	ion	ion

Step 2

$$(CH_3)_3C\overset{+}{O}H_2 \longrightarrow (CH_3)_3C^+ + H_2O$$

tert-Butyloxonium ion	*tert*-Butyl cation	Water

Step 3

$$(CH_3)_3C^+ + Cl^- \overset{fast}{\longrightarrow} (CH_3)_3CCl$$

tert-Butyl cation	Chloride	*tert*-Butyl chloride
	ion	

We note that step 1 and step 3 are rapid reactions. What about the rate of step 2? Step 2 is the slowest step in the sequence; it is the *rate-determining step.* Let us now review what we mean when we say that a particular step is rate-determining and examine the evidence suggesting that carbocation formation is the slow step in the proposed mechanism.

4.14 POTENTIAL ENERGY DIAGRAMS FOR CHEMICAL REACTIONS

Potential energy diagrams of the type used in Chapter 3 to describe conformational processes can also help us understand the mechanisms of chemical reactions. Consider the one-step transformation represented by Figure 4.10, in which reactant A is converted to product B.

$$A \longrightarrow B$$

Reactions such as this, which occur in a single step and proceed through only a single transition state (see Section 3.2), are referred to as *concerted* reactions.

As indicated in the diagram, product B is more stable than reactant A, and this particular reaction is an exothermic, or energy-yielding, one. In order for reaction to

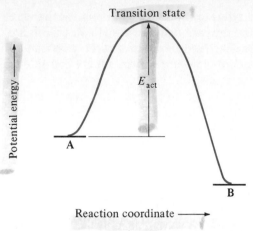

FIGURE 4.10 Energy diagram for a concerted reaction in which reactant A is converted to product B through a single transition state.

occur, however, a molecule of reactant A must possess an excess of potential energy, the *activation energy,* above that of its most stable state. Activation energy is given the symbol E_{act} or $E^{\neq}$. The superscript $\neq$ identifies the symbol as one that applies to a transition state. The abscissa in Figure 4.10 is called the *reaction coordinate.* It is an arbitrary measure of the extent to which a molecule of starting material has progressed toward product.

When large numbers of molecules of A are examined, it is seen that not all of them have the same energy. There exists a distribution of energies, such as that shown in Figure 4.11. More molecules have energies that are clustered about some median value than have energies very much greater or very much less than this. At any instant some molecules of A possess potential energy equivalent to that of E_{act} and undergo chemical change. Figure 4.11 indicates the number of molecules of A with energies greater than E_{act} as the shaded area under the curve. The precise shape of the energy distribution curve depends on the temperature. At higher temperatures the median energy is closer to the activation energy for chemical reaction, and as shown in Figure 4.11, more molecules have energies in excess of the energy of activation. Thus, more molecules of A cross the energy barrier per unit time, and an increase in temperature causes an increase in the reaction rate. The effect of temperature is quite pronounced; an increase of only 10°C in temperature typically produces a severalfold increase in the rate of most chemical reactions.

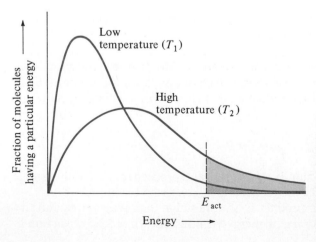

FIGURE 4.11 Distribution of molecular energies. (a) The number of molecules with energy greater than E_{act} at temperature T_1 is shown as the shaded area. (b) At some higher temperature T_2, the shape of the energy distribution curve is different and more molecules have energies in excess of E_{act}.

An example of a concerted reaction is the transfer of a proton from a Brönsted acid to a Brönsted base. In step 1 of our three-step mechanism for formation of *tert*-butyl chloride from *tert*-butyl alcohol and hydrogen chloride, the proton that is transferred is partially bonded to both chlorine and the oxygen of the alcohol at the transition state. Partial bonds are represented by dashed lines in the equation below

$$(CH_3)_3C-\overset{\cdot\cdot}{\underset{\underset{H}{|}}{O}}: + H-\overset{\cdot\cdot}{\underset{\cdot\cdot}{Cl}}: \longrightarrow \left[(CH_3)_3C-\overset{\cdot\cdot}{\underset{\underset{H}{|}}{O}}\text{---}H\text{---}\overset{\overset{\delta+}{} \cdot\cdot}{\underset{\cdot\cdot}{Cl}}:^{\delta-} \right]^{\neq} \rightsquigarrow \longrightarrow$$

tert-Butyl alcohol	Hydrogen chloride	Activated complex for proton transfer

$$(CH_3)_3C-\overset{+\cdot\cdot}{\underset{\underset{H}{|}}{O}}-H \quad + \quad :\overset{\cdot\cdot}{\underset{\cdot\cdot}{Cl}}:^-$$

tert-Butyloxonium ion	Chloride ion

Bond formation with oxygen occurs in concert with cleavage of the hydrogen-chlorine bond, although the oxygen-hydrogen bond is not necessarily developed to the same degree as the hydrogen-chlorine bond is broken at the transition state. It requires energy to break bonds, but energy is given off when bonds form. A portion of the energy required to break the hydrogen-chlorine bond is compensated for by oxygen-hydrogen bond formation, and the activation energy for proton transfer from acids to alcohols is observed to be very low. Both the alcohol and hydrogen chloride undergo a change in bonding at the transition state. We call a process in which the transition state requires an interaction between two particles a *bimolecular* reaction.

Many reactions occur in a series of steps rather than as concerted processes. Consider the conversion of reactant X to product Z via a mechanism in which X is converted to intermediate Y in a slow step, followed by rapid conversion of Y to Z.

$$X \xrightarrow{\text{slow}} Y$$

$$Y \xrightarrow{\text{fast}} Z$$

A multistep reaction can proceed no faster than its slowest step. Since the first step is slower than the second, formation of intermediate Y is rate-determining. Figure 4.12 shows a potential energy diagram for this process. As you can see, the point of maximum potential energy occurs during the formation of intermediate Y from reactant X. There is a second energy barrier, corresponding to the transition state for conversion of intermediate Y to product Z, but this is of lower energy than the transition state for the rate-determining step. The activation energy for the conversion of reactant X to product Z is equal to the activation energy for the slow step.

Now consider the same reaction but with the first step faster than the second.

$$X \xrightarrow{\text{fast}} Y$$

$$Y \xrightarrow{\text{slow}} Z$$

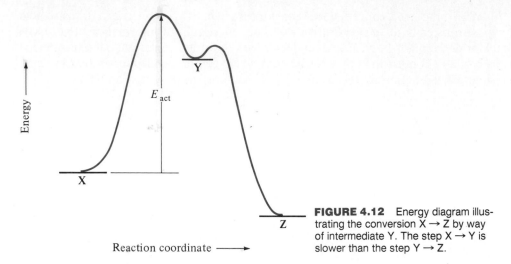

FIGURE 4.12 Energy diagram illustrating the conversion X → Z by way of intermediate Y. The step X → Y is slower than the step Y → Z.

Here, the conversion of intermediate Y to product Z is the rate-determining step. This situation is represented by Figure 4.13, which shows the point of maximum potential energy occurring during the conversion of intermediate Y to product Z. Intermediate Y can build up during the course of the reaction and, if it is sufficiently stable, can be isolated from the reaction mixture before the reaction is complete.

How does the reaction of alcohols with hydrogen halides to form alkyl halides fit into this picture? In the case of the reaction of *tert*-butyl alcohol with hydrogen chloride, as illustrated in Figure 4.3 and Figure 4.14, there are two intermediates, *tert*-butyloxonium ion and *tert*-butyl cation, and three transition states. The first transition state corresponds to transfer of a proton from hydrogen chloride to the oxygen of *tert*-butyl alcohol to form *tert*-butyloxonium ion and has already been discussed. Let us examine now the dissociation of the oxonium ion in the second step. Here, cleavage of the carbon-oxygen bond of the oxonium ion to form a carbocation is occurring, but this bond cleavage is not accompanied by bond formation.

| *tert*-Butyl oxonium ion | Activated complex for dissociation of oxonium ion | *tert*-Butyl cation | Water |

Only one particle, the oxonium ion, undergoes a change in bonding at the transition state; the second step of the mechanism is a *unimolecular* dissociation of the oxonium ion. The activation energy for this step is greater than the activation energy for formation of the oxonium ion because the energy required to break the carbon-oxygen bond is not compensated for by a simultaneous bond-forming process.

In fact, this is an oversimplification. Solvent molecules cluster about the developing carbocation and provide some transition state stabilization. This lowers the absolute value of the activation energy for carbocation formation but does not alter its ranking with respect to the other steps in the sequence.

In the third step of the reaction, the carbocation intermediate is captured by chloride ion, bond formation is the dominant process, and the energy barrier for this

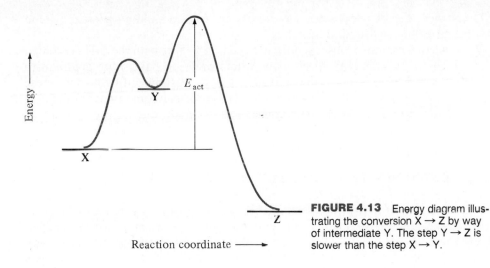

FIGURE 4.13 Energy diagram illustrating the conversion X → Z by way of intermediate Y. The step Y → Z is slower than the step X → Y.

rapid cation-anion combination is relatively low. The transition state is characterized by partial bond formation between the nucleophile (chloride anion) and the electrophile (*tert*-butyl cation).

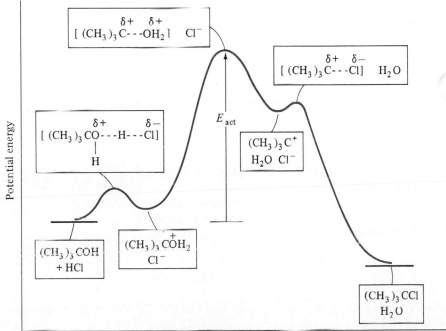

FIGURE 4.14 Energy diagram depicting the intermediates and transition states involved in the reaction of *tert*-butyl alcohol with hydrogen chloride.

Of the three steps in the reaction mechanism, the second has the highest activation energy and is the rate-determining step.

We shall analyze numerous reactions in a qualitative way with the aid of potential energy diagrams of this type. Ideally, one would prefer to have more quantitative information available, such as the activation energies associated with each of the transition states in a multistep process. Such information can sometimes be obtained, and we shall see an example of its application when we discuss the free-radical halogenation of alkanes later in this chapter.

4.15 FACTORS THAT AFFECT REACTION RATE

We have noted in the previous section that *the rate of practically every chemical reaction increases with increasing temperature.* As the experimental data cited earlier pointed out, this behavior is observed in the reaction of hydrogen halides with alcohols. Those data also established that the reactivity of the various hydrogen halides parallels their strength as acids and that tertiary alcohols are more reactive than secondary, secondary alcohols are more reactive than primary, and primary alcohols are more reactive than methyl alcohol. The three-step mechanism for conversion of alcohols to alkyl halides by hydrogen halides was developed on the basis of experimental observations such as these. Let us now see how other experimental observations are consistent with the proposed mechanism.

Let us begin with the effect of the hydrogen halide on the rate of the reaction. The hydrogen halide is involved in the first step of the reaction, a rapid equilibrium in which it transfers a proton to the alcohol to give an oxonium ion.

$$\text{ROH} + \quad \text{HX} \quad \overset{\text{fast}}{\rightleftharpoons} \quad \overset{+}{\text{ROH}_2} \; + \; \text{X}^-$$

Alcohol   Hydrogen halide   Oxonium ion   Halide ion

The equilibrium constant K for proton transfer is given by the expression:

$$K = \frac{[\text{oxonium ion}][\text{halide ion}]}{[\text{alcohol}][\text{hydrogen halide}]}$$

Rearranging this expression gives the concentration of the oxonium ion:

$$[\text{Oxonium ion}] = \frac{K[\text{alcohol}][\text{hydrogen halide}]}{[\text{halide ion}]}$$

A stronger acid has a larger value of K than a weaker acid and gives a greater concentration of the oxonium ion at equilibrium. The rate-determining step is the unimolecular dissociation of the oxonium ion to a carbocation:

$$\overset{+}{\text{ROH}_2} \quad \overset{\text{slow}}{\rightleftharpoons} \quad \text{R}^+ \quad + \text{H}_2\text{O}$$

Oxonium ion   Carbocation   Water

The rate of carbocation formation, as well as the overall rate of the reaction, is directly proportional to the concentration of the oxonium ion:

$$\text{Rate} = k \, [\text{oxonium ion}]$$

where k is a constant of proportionality called the *rate constant*. Since the concentration of the oxonium ion is directly proportional to the value of K for the proton-transfer step, an increase in the proton-donating ability of the hydrogen halide (its Brönsted acidity) increases the rate of alkyl halide formation.

The value of the rate constant k depends on, among other things, the reaction temperature: as the temperature increases, k increases. It is also related to the energy of activation for oxonium ion dissociation. The greater the value of k, the more readily the oxonium ion dissociates to a carbocation. A large value of k implies a low activation energy for carbocation formation.

Consider next what happens when the oxonium ion dissociates to a carbocation and water. The positive charge resides mainly on oxygen in the oxonium ion but is shared between oxygen and carbon as the transition state is approached.

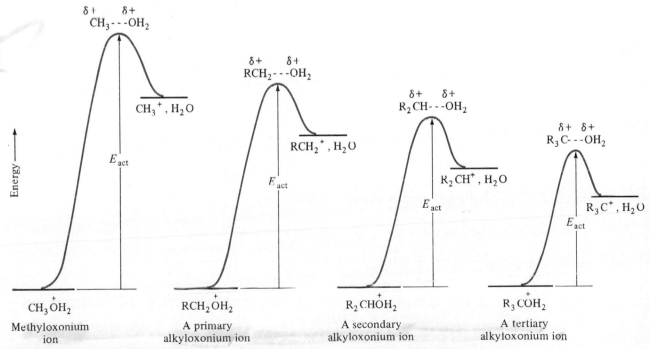

| Alkyloxonium ion | Activated complex for dissociation of alkyloxonium ion | Carbocation water | Water |

The transition state for carbocation formation begins to resemble the carbocation. If we assume that structural features that stabilize carbocations also stabilize transition states that have carbocation character, it follows that oxonium ions derived from

FIGURE 4.15 Diagrams comparing energies of activation for formation of carbocations from oxonium ions of methyl, primary, secondary, and tertiary alcohols.

tertiary alcohols have a lower energy of activation for dissociation and are converted to their corresponding carbocations faster than oxonium ions derived from secondary and primary alcohols. Figure 4.15 depicts the effect of oxonium ion structure on the activation energy for and thus the rate of carbocation formation. Once the carbocation is formed, it is rapidly captured by halide ion, so the rate of alkyl halide formation is governed by the activation energy for dissociation of the oxonium ion.

Inferring the structure of the activated complex on the basis of what is known about the molecules that lead to it or may be formed by way of it is a practice with a long history in organic chemistry. A modern justification of this practice was advanced by George S. Hammond, who reasoned that if two states, such as an activated complex and an intermediate derived from it, are similar in energy, then they are similar in structure. This rationale is known as *Hammond's postulate*. In the formation of a carbocation from an alcohol, the activated complex is closer in energy to the carbocation than it is to the alcohol, so its structure more closely resembles that of the carbocation and it responds in a similar way to the stabilizing effects of alkyl substituents.

4.16 MECHANISM OF THE REACTION OF PRIMARY ALCOHOLS WITH HYDROGEN HALIDES

Unlike tertiary and secondary carbocations, primary carbocations are of such high energy as to be inaccessible as intermediates in chemical reactions. Since primary alcohols are converted, albeit rather slowly, to alkyl halides on treatment with hydrogen halides, there must be some mechanism for the formation of primary alkyl halides that avoids carbocation intermediates. This alternative mechanism is believed to be a concerted process in which the carbon-halogen bond begins to form before the carbon-oxygen bond of the oxonium ion is completely broken.

$$:\ddot{X}:^- \ + \ RCH_2\overset{+}{\underset{}{\ddot{O}}}H_2 \longrightarrow \left[\overset{\delta-}{:\ddot{X}}\text{---}\overset{\overset{\displaystyle R}{|}}{CH_2}\text{---}\overset{\delta+}{\ddot{O}H_2} \right]^{\neq} \longrightarrow RCH_2\ddot{\ddot{X}}: \ + \ H_2\ddot{O}:$$

| Halide ion | Primary alkyloxonium ion | Activated complex | Primary alkyl halide | Water |

In effect, the halide nucleophile helps to "push off" a water molecule from the oxonium ion in a bimolecular step. Both the halide ion and the oxonium ion undergo covalency change at the transition state.

PROBLEM 4.12 1-Butanol and 2-butanol are converted to their corresponding bromides on being heated with hydrogen bromide. Write a suitable mechanism for each reaction.

4.17 OTHER METHODS FOR CONVERTING ALCOHOLS TO ALKYL HALIDES

As alternatives to reaction with hydrogen halides, a number of synthetic methods based on inorganic reagents have been developed that permit alcohols to be converted to alkyl halides. One such synthetic reagent is *thionyl chloride* ($SOCl_2$).

Thionyl chloride reacts with alcohols to give alkyl chlorides and the gaseous by-products sulfur dioxide and hydrogen chloride.

$$ROH + SOCl_2 \longrightarrow RCl + SO_2 + HCl$$

| Alcohol | Thionyl chloride | Alkyl chloride | Sulfur dioxide | Hydrogen chloride |

Since tertiary alcohols are so easily converted to chlorides with hydrogen chloride, thionyl chloride is used mainly to prepare primary and secondary alkyl chlorides. Reactions with thionyl chloride are normally carried out in the presence of potassium carbonate or the weak organic base pyridine.

$$CH_3CH(CH_2)_5CH_3 \xrightarrow[K_2CO_3]{SOCl_2} CH_3CH(CH_2)_5CH_3$$
$$\overset{|}{OH} \qquad\qquad\qquad \overset{|}{Cl}$$

2-Octanol   2-Chlorooctane (81%)

$$(CH_3CH_2)_2CHCH_2OH \xrightarrow[pyridine]{SOCl_2} (CH_3CH_2)_2CHCH_2Cl$$

2-Ethyl-1-butanol   1-Chloro-2-ethylbutane (82%)

When alkyl bromides are desired, the reaction of alcohols with *phosphorus tribromide* is frequently used.

$$3ROH + PBr_3 \longrightarrow 3RBr + P(OH)_3$$

| Alcohol | Phosphorus tribromide | Alkyl bromide | Phosphorous acid |

$$(CH_3)_2CHCH_2OH \xrightarrow{PBr_3} (CH_3)_2CHCH_2Br$$

Isobutyl alcohol   Isobutyl bromide (55–60%)

Cyclopentanol   Cyclopentyl bromide (78–84%)

Thionyl chloride and phosphorus tribromide are specialized reagents used to bring about particular functional group transformations in organic chemistry. For this reason, we have chosen not to present the mechanisms by which they convert alcohols to alkyl halides. Instead, our practice will be to emphasize those reaction mechanisms that have broad applicability and that enhance our knowledge of the fundamental principles of organic chemistry. In those instances you will find that a mechanistic understanding is of immeasurable help in organizing the large number of chemical reactions to which you will be exposed.

As it turns out, there are only a few fundamental reaction mechanisms in organic chemistry, and the countless reactions that are known are variations on these few basic themes. The key features of the reactions discussed so far are carbocation formation from alcohols in the presence of strong acids and combination of the carbocation intermediate with a nucleophile. In the remaining sections of this chapter we will describe a second type of fundamental reaction mechanism, one that proceeds by way of a different kind of reactive intermediate called a *free radical*.

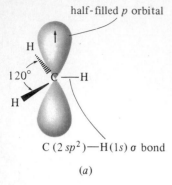

half-filled *p* orbital

120°

C (2 *sp²*)—H(1*s*) σ bond

(*a*)

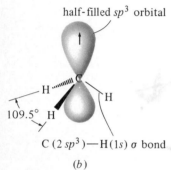

half-filled *sp³* orbital

109.5°

C (2 *sp³*)—H(1*s*) σ bond

(*b*)

FIGURE 4.16 Orbital hybridization models of bonding in methyl radical. (*a*) Carbon is *sp²* hybridized with unpaired electron in 2*p* orbital. (*b*) Carbon is *sp³* hybridized with electron in *sp³* orbital.

4.18 FREE RADICALS

Free radicals are species that contain unpaired electrons. Those of immediate concern to us are alkyl radicals, in which the unshared electron is associated with a tricoordinate carbon. Alkyl radicals are classified as primary, secondary, or tertiary according to the number of carbon substituents attached to the carbon that bears the odd electron.

Methyl radical	Primary radical	Secondary radical	Tertiary radical

Bonding in the methyl radical may be approached by comparing the radical with methyl cation. As we described in Section 4.11, methyl cation has six electrons in three σ bonds between *sp²* hybridized carbon and its hydrogen substituents. The axis of the vacant 2*p* orbital is perpendicular to the plane of the ion. Methyl radical has one more electron than methyl cation, so it is electrically neutral. This additional electron does not participate in bonds to the three hydrogens but occupies the 2*p* orbital. Thus, the 2*p* orbital is half-filled in methyl radical. Figure 4.16*a* illustrates this bonding model.

A slightly different bonding model, shown in Figure 4.16*b*, permits the unpaired electron to be more strongly bound. Since electrons in *s* orbitals are more strongly held than electrons in *p* orbitals, the unpaired electron will be more stable if it is in an orbital with some *s* character — one that is *sp³* hybridized, for example. According to the bonding model in Figure 4.16*b*, carbon is *sp³* hybridized and the unpaired electron is located in an orbital with 25 percent *s* character.

Of these two extremes, the planar *sp²* hybridization model of Figure 4.16*a* is a better bonding description than the pyramidal *sp³* hybridization model of Figure 4.16*b*. Experimental results and theoretical studies reveal that methyl radical is planar, or very nearly so. Apparently, the stronger binding of the σ electrons when carbon is *sp²* hybridized is more important than binding of the unpaired electron. More highly substituted analogs, such as *tert*-butyl radical, are flattened pyramids in which the hybridization state of the trivalent carbon is closer to *sp²* than to *sp³*.

4.19 STABILITY OF FREE RADICALS

Free radicals, like carbocations, have an unfilled 2*p* orbital and are stabilized by substituents, such as alkyl groups, that release electrons to the trivalent carbon. Consequently, the order of free-radical stability parallels that of carbocations.

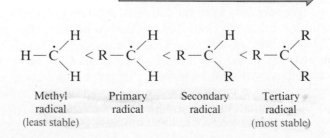

Increasing free radical stability

Methyl radical	Primary radical	Secondary radical	Tertiary radical
(least stable)			(most stable)

Included in the experimental evidence that demonstrates this order of relative stabilities are measurements of *bond dissociation energies* (BDEs). Bond dissociation energy is the energy required for *homolytic* cleavage of a bond. In a homolytic bond cleavage, each fragment retains one of the electrons in the original two-electron bond.

$$X : Y \longrightarrow X \cdot + \cdot Y$$

Homolytic bond cleavage

In contrast, in *heterolytic* bond cleavage one fragment retains both the electrons when the bond is broken.

$$X : Y \longrightarrow X^+ + : Y^-$$

Heterolytic bond cleavage

TABLE 4.4
Bond Dissociation Energies of Some Representative Compounds*

Bond	Bond dissociation energy, kcal/mol	Bond	Bond dissociation energy, kcal/mol
Diatomic molecules			
H—H	104	H—F	136
F—F	38	H—Cl	103
Cl—Cl	58	H—Br	87.5
Br—Br	46	H—I	71
I—I	36		
Alkanes			
CH_3—H	104		
CH_3CH_2—H	98	CH_3—CH_3	88
$CH_3CH_2CH_2$—H	98	CH_3CH_2—CH_3	85
$(CH_3)_2CH$—H	94.5		
$(CH_3)_3C$—H	91	$(CH_3)_2CH$—CH_3	84
		$(CH_3)_3C$—CH_3	80
Alkyl halides			
CH_3—F	108	$(CH_3)_2CH$—F	105
CH_3—Cl	83.5	$(CH_3)_2CH$—Cl	81
CH_3—Br	70	$(CH_3)_2CH$—Br	68
CH_3—I	56	$(CH_3)_3C$—Cl	79
CH_3CH_2—Cl	81	$(CH_3)_3C$—Br	63
$CH_3CH_2CH_2$—Cl	82		
Water and Alcohols			
HO—H	119	CH_3CH_2—OH	91
CH_3O—H	102	$(CH_3)_2CH$—OH	92
CH_3—OH	91	$(CH_3)_3C$—OH	91

* Bond dissociation energies refer to bond indicated in structural formula for each substance.

The bond dissociation energy is equal to the enthalpy change ($\Delta H°$) for the homolytic bond cleavage and is a positive number because energy must be added to the system to break the chemical bond. The separate fragments together have more energy than the original molecule. A list of some bond dissociation energies is given in Table 4.4.

As the table indicates, carbon-hydrogen bond energies in alkanes are approximately 90 to 105 kcal/mol. Notice that the table lists two entries for propane but only one for ethane and one for methane. This is because propane has two different kinds of carbon-hydrogen bonds, while ethane and methane have only one. It requires slightly less energy to break a methylene C—H bond in propane than it does to break a C—H bond in its methyl group.

$$CH_3CH_2CH_2—H \longrightarrow CH_3CH_2\overset{\cdot}{C}H_2 + \quad H\cdot \qquad \Delta H° = +98 \text{ kcal/mol}$$

$$\begin{array}{ccc} \text{Propane} & n\text{-Propyl} & \text{Hydrogen} \\ & \text{radical} & \text{atom} \\ & \text{(primary)} & \end{array}$$

$$\underset{\underset{\displaystyle H}{|}}{CH_3CHCH_3} \longrightarrow CH_3\overset{\cdot}{C}HCH_3 + \quad H\cdot \qquad \Delta H° = +95 \text{ kcal/mol}$$

$$\begin{array}{ccc} \text{Propane} & \text{Isopropyl} & \text{Hydrogen} \\ & \text{radical} & \text{atom} \\ & \text{(secondary)} & \end{array}$$

Since the starting material is the same in each case (propane), the heats of reaction, i.e., the bond dissociation energies, may be compared directly. Furthermore, one of the products, namely, a hydrogen atom, is the same in each case. Thus, as illustrated in Figure 4.17, the difference in the bond dissociation energies measures the difference in energy between an n-propyl radical (primary) and an isopropyl radical (secondary). The secondary radical is 3 kcal/mol more stable than the primary radical.

Similarly, comparing the bond dissociation energies of the two different types of C—H bonds in 2-methylpropane reveals a tertiary radical to be 7 kcal/mol more stable than a primary radical.

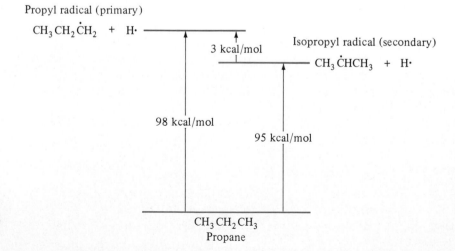

FIGURE 4.17 Diagram showing how bond dissociation energies of methylene and methyl C—H bonds in propane reveal difference in stabilities between two isomeric free radicals. The secondary radical is more stable than the primary.

$$CH_3CHCH_2{-}H \longrightarrow CH_3\overset{\cdot}{C}HCH_2 + \quad H\cdot \qquad \Delta H° = +98 \text{ kcal/mol}$$
$$\qquad | \qquad\qquad\qquad\qquad | $$
$$\qquad CH_3 \qquad\qquad\qquad CH_3$$

2-Methylpropane   Isobutyl   Hydrogen
(isobutane)   radical   atom
                        (primary)

$$\qquad H$$
$$\qquad |$$
$$CH_3\overset{}{C}CH_3 \longrightarrow CH_3\overset{\cdot}{C}CH_3 + \quad H\cdot \qquad \Delta H° = +91 \text{ kcal/mol}$$
$$\qquad | \qquad\qquad\qquad\qquad |$$
$$\qquad CH_3 \qquad\qquad\qquad CH_3$$

2-Methylpropane   *tert*-Butyl   Hydrogen
(isobutane)   radical   atom
                       (tertiary)

Methyl radical ($\cdot CH_3$) bears no stabilizing alkyl substituents. The bond dissociation energy of methane at 104 kcal/mol is higher than that of any of the alkanes in Table 4.4.

The order of carbon-hydrogen bond dissociation energies is

Increasing carbon-hydrogen bond strength ⟶

$$R_3C{-}H \quad < R_2CH{-}H < RCH_2{-}H < \quad CH_3{-}H$$

Methine   Methylene   Methyl   Methane
(tertiary)   (secondary)   (primary)

weakest C—H bond   strongest C—H bond

The greater the stability of the free radical, the less energy is required to break the carbon-hydrogen bond that produces it.

PROBLEM 4.13   Carbon-carbon bond dissociation energies have been measured for alkanes. Without referring to Table 4.4, identify the alkane in each of the following pairs that has the lower carbon-carbon bond dissociation energy and explain the reason for your choice.

(a) Ethane or propane
(b) Propane or 2-methylpropane
(c) 2-Methylpropane or 2,2-dimethylpropane

SAMPLE SOLUTION   (a)   First write the equations that describe homolytic carbon-carbon bond cleavage in each alkane.

$$CH_3{-}CH_3 \longrightarrow \quad \cdot CH_3 + \cdot CH_3$$

Ethane   Two methyl radicals

$$CH_3CH_2{-}CH_3 \longrightarrow \quad CH_3\overset{\cdot}{C}H_2 + \quad \cdot CH_3$$

Propane   Ethyl radical   Methyl radical

Cleavage of the carbon-carbon bond in ethane yields two methyl radicals, while carbon-carbon bond cleavage in propane yields an ethyl radical and one methyl radical. Ethyl radical is more stable than methyl, so less energy is required to break the carbon-carbon bond in propane than in ethane. The measured carbon-carbon bond dissociation energy in ethane is 88 kcal/mol, while that in propane is 85 kcal/mol.

When we speak of the stability of free radicals, we are comparing them with other free radicals. Simple alkyl radicals are high in energy; they are highly reactive and not stable enough to be isolated and stored. The role that alkyl radicals play as reactive intermediates is similar to that of carbocations. They are unstable species that, once formed, react rapidly with other substances present in the system. Unlike carbocations, however, alkyl radicals are electrically neutral and react with one-electron donors. Recall that carbocations react with nucleophiles, substances that are electron pair donors. In the following sections both the formation and the subsequent reactions of free-radical intermediates will be discussed.

4.20 CHLORINATION OF METHANE

Most alkyl halides are prepared in the laboratory from alcohols as the starting material. There are, however, a number of industrially important alkyl chlorides that are prepared by direct chlorination of alkanes. In the chlorination of alkanes a chlorine replaces hydrogen as a substituent on carbon.

$$RH \ + \ Cl_2 \ \longrightarrow \ RCl \ + \ HCl$$

Alkane   Chlorine   Alkyl chloride   Hydrogen chloride

The various chloro- derivatives of methane, for example, are prepared in this way.

CH_3Cl	CH_2Cl_2	$CHCl_3$	CCl_4
Chloromethane (methyl chloride)	Dichloromethane (methylene dichloride)	Trichloromethane (chloroform)	Tetrachloromethane (carbon tetrachloride)

These compounds are used as solvents and in paint removers and degreasing agents. Dichloromethane has been used to extract caffeine from coffee and as a propellant in aerosol cans, but recent concerns about its toxicity are causing it to be replaced by other compounds for these purposes.

Methane is the ultimate starting material in the preparation of each compound in the series of chloromethane derivatives. All four hydrogen atoms of methane may be replaced sequentially by reaction with chlorine at elevated temperatures.

$$CH_4 \ + \ Cl_2 \ \xrightarrow{400-440°C} \ CH_3Cl \ + \ HCl$$

Methane   Chlorine   Chloromethane   Hydrogen (bp −24°C)   chloride

$$CH_3Cl \ + \ Cl_2 \ \xrightarrow{400-440°C} \ CH_2Cl_2 \ + \ HCl$$

Chloromethane   Chlorine   Dichloromethane   Hydrogen (bp 40°C)   chloride

$$CH_2Cl_2 \ + \ Cl_2 \ \xrightarrow{400-440°C} \ CHCl_3 \ + \ HCl$$

Dichloromethane   Chlorine   Trichloromethane   Hydrogen
                                        (bp 61°C)   chloride

$$CHCl_3 \ + \ Cl_2 \ \xrightarrow{400-440°C} \ CCl_4 \ + \ HCl$$

Trichloromethane   Chlorine   Tetrachloromethane   Hydrogen
                                        (bp 77°C)   chloride

Indeed, it is difficult to control the chlorination of methane so as to introduce exactly a prescribed number of chlorine atoms. A mixture of chlorinated methane derivatives is obtained and the various components are then separated by distillation.

What about the other halogens? What about other alkanes? How do the reactions of fluorine, chlorine, bromine, and iodine with alkanes in general compare with the chlorination of methane? The scope of alkane halogenation can be explored by considering the energy changes that accompany each reaction.

4.21 THERMODYNAMICS OF METHANE HALOGENATION

To understand the difference among the halogens in their reactions with alkanes, let us examine the energy changes associated with the fluorination, chlorination, bromination, and iodination of methane. This is accomplished by using bond dissociation energies to determine whether each halogenation reaction is exothermic or endothermic. Consider first the chlorination of methane to chloromethane. The bond dissociation energies of the relevant bonds are indicated in the equation

$$CH_3{-}H \ + \ Cl{-}Cl \ \longrightarrow \ CH_3{-}Cl \ + \ H{-}Cl$$

104 kcal/mol   58 kcal/mol   83.5 kcal/mol   103 kcal/mol

In order for the reaction to be exothermic the bonds formed in the products must be stronger than the bonds broken in the reactants. As the data show, chlorination of methane is an exothermic process; bonds equivalent to 186.5 kcal/mol (83.5 kcal/mol + 103 kcal/mol) in the products are formed at the expense of 162 kcal/mol (104 kcal/mol + 58 kcal/mol) of bonds broken in the reactants. The enthalpy change for the reaction is:

$$\Delta H° = \text{BDE of bonds broken} - \text{BDE of bonds formed}$$
$$\Delta H° = 162 \text{ kcal/mol} - 186.5 \text{ kcal/mol}$$
$$\Delta H° = -24.5 \text{ kcal/mol}$$

An exothermic process usually has an equilibrium constant that favors formation of product ($K \gg 1$). More precisely, the equilibrium constant depends on $\Delta G°$, the *standard free energy change* for the process according to the equation:

$$\Delta G° = -RT \ln K$$

where T is the absolute temperature in kelvins and R is a constant equal to 1.99 cal/(mol·°C). Reactions characterized by a negative $\Delta G°$ are described as *spontaneous* in the direction written. When used in this sense, the word "spontaneous" refers

TABLE 4.5

Equilibrium Constant and Product Composition as a Function of $\Delta G°$

Free energy difference ($\Delta G°$), kcal/mol*	Equilibrium constant K†	Percent conversion to products
+1.0	0.18	15
0.0	1.00	50
−0.1	1.18	54
−0.2	1.40	58
−0.5	2.32	70
−1.0	5.41	84
−2.0	29.1	97
−3.0	158	99.4
−4.0	844	99.9
−5.0	10,000	99.99

* $G°$(product) − $G°$(reactant)

† At 25°C; $K = e^{-\Delta G°/RT}$

only to the free energy difference between initial and final states and has nothing to do with the rate at which the higher-energy state is converted to the lower-energy one. A large value of K is associated with a large negative $\Delta G°$. Table 4.5 summarizes the relationship between the standard free energy change for a reaction, its equilibrium constant, and the reaction mixture composition. As the table shows, a reaction with $\Delta G°$ more negative than about −3 kcal/mol can be considered to go to completion, meaning that starting materials are converted almost entirely to products.

Free energy is related to enthalpy by the expression

$$\Delta G° = \Delta H° - T\,\Delta S°$$

where $\Delta S°$ is the difference in *entropy* between the products and reactants. Entropy is a measure of the degree of disorder in a system; a positive $\Delta S°$ is associated with an increase in disorder or randomness in going from reactants to products. A dissociation reaction in which two neutral molecules are formed from one is an example of a process in which the degree of disorder increases. A positive value of $\Delta S°$ makes the $-T\,\Delta S°$ term negative and makes $\Delta G°$ more negative than $\Delta H°$. The equilibrium constant K in this case, then, is greater than expected solely on the basis of considering $\Delta H°$. Conversely, if $\Delta S°$ is negative, the equilibrium constant is smaller than expected on the basis of $\Delta H°$ alone. In the halogenation of alkanes, two particles, an alkane and a halogen, react to form two particles, an alkyl halide and a hydrogen halide, and $\Delta S°$ for these reactions is typically close to zero. Therefore $\Delta G°$ and $\Delta H°$ are very similar, and it is a satisfactory approximation to say that the negative $\Delta H°$ for chlorination of methane is associated with a large equilibrium constant for the reaction.

Iodination of alkanes is not a useful reaction. We can gain a greater appreciation of why chlorination is successful while iodination fails by calculating the heat of reaction for iodination of methane. Again, using the bond dissociation energy data of Table 4.4, we obtain

$$CH_3—H + I—I \longrightarrow CH_3—I + H—I$$
104 kcal/mol   36 kcal/mol   56 kcal/mol   71 kcal/mol

BDE of bonds broken = (104 kcal/mol + 36 kcal/mol) = 140 kcal/mol
BDE of bonds formed = (56 kcal/mol + 71 kcal/mol) = 127 kcal/mol
$$\Delta H° = +13 \text{ kcal/mol}$$

In contrast to the chlorination of methane, in which $\Delta H°$ is negative, in iodination the enthalpy change is positive. A positive value of $\Delta H°$ indicates an endothermic process. The reactants are more stable than the products, and $K < 1$ for the reaction in the direction written.

Fluorination of alkanes is a violent reaction and too difficult to control to be useful. One of the reasons for this behavior becomes apparent when the energetics of methyl fluoride formation are calculated.

$$CH_3—H + F—F \longrightarrow CH_3—F + H—F$$
104 kcal/mol   38 kcal/mol   108 kcal/mol   136 kcal/mol

BDE of bonds broken = (104 kcal/mol + 38 kcal/mol) = 142 kcal/mol
BDE of bonds formed = (108 kcal/mol + 136 kcal/mol) = 244 kcal/mol
$$\Delta H° = -102 \text{ kcal/mol}$$

Fluorination of methane has a very large negative value of $\Delta H°$. The reaction is quite hazardous to carry out unless provision is made to dissipate the heat evolved. Fluorination of alkanes beyond methane is also complicated by fluorine-induced carbon-carbon bond cleavage.

The remaining halogen in our series to be discussed is bromine. Just as chlorination of methane is less exothermic than fluorination, bromination is less exothermic than chlorination.

$$CH_3—H + Br—Br \longrightarrow CH_3—Br + H—Br$$
104 kcal/mol   46 kcal/mol   70 kcal/mol   87.5 kcal/mol

BDE of bonds broken = (104 kcal/mol + 46 kcal/mol) = 150 kcal/mol
BDE of bonds formed = (70 kcal/mol + 87.5 kcal/mol) = 157.5 kcal/mol
$$\Delta H° = -7.5 \text{ kcal/mol}$$

While bromination of methane is energetically feasible, economic considerations cause most of the methyl bromide prepared commercially to be made from methanol by reaction with hydrogen bromide.

4.22 MECHANISM OF CHLORINATION OF METHANE

The chlorination of methane is normally carried out at high temperature. Even though the reaction is exothermic, heat (or thermal) energy must be put into the system in order for it to occur. Alternatively, *photochemical* rather than thermal energy may be used. Photochemical reactions are reactions that take place when light

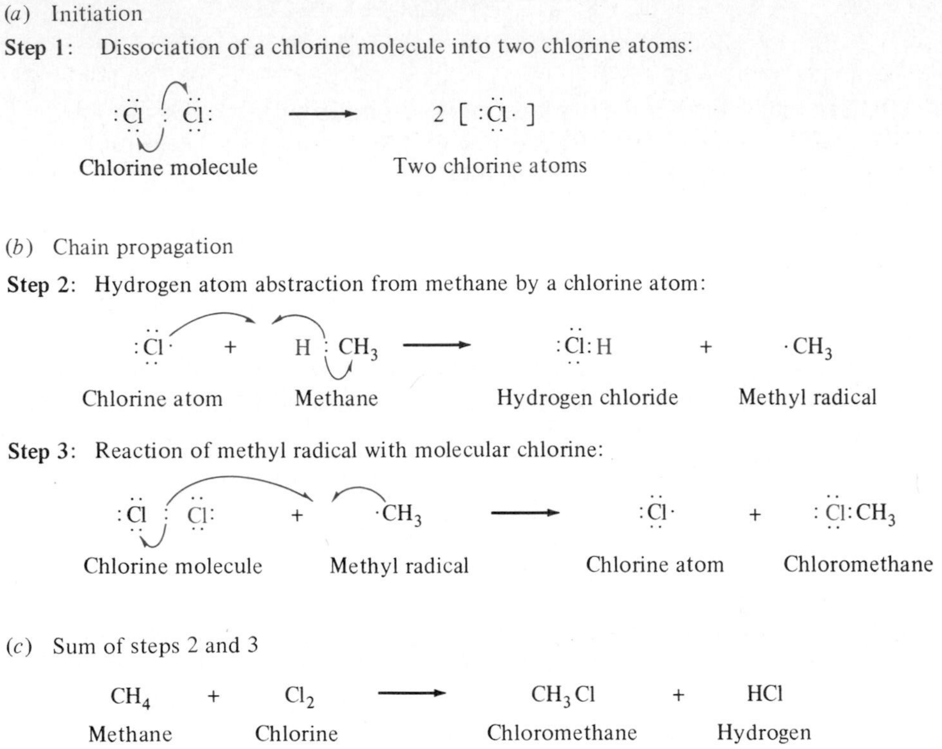

(*a*) Initiation

Step 1: Dissociation of a chlorine molecule into two chlorine atoms:

$$:\!\overset{..}{\underset{..}{Cl}}\!:\overset{..}{\underset{..}{Cl}}\!: \longrightarrow 2\,[\,:\!\overset{..}{\underset{..}{Cl}}\!\cdot\,]$$

Chlorine molecule   Two chlorine atoms

(*b*) Chain propagation

Step 2: Hydrogen atom abstraction from methane by a chlorine atom:

$$:\!\overset{..}{\underset{..}{Cl}}\!\cdot \;+\; H\!:\!CH_3 \longrightarrow :\!\overset{..}{\underset{..}{Cl}}\!:\!H \;+\; \cdot CH_3$$

Chlorine atom   Methane   Hydrogen chloride   Methyl radical

Step 3: Reaction of methyl radical with molecular chlorine:

$$:\!\overset{..}{\underset{..}{Cl}}\!:\overset{..}{\underset{..}{Cl}}\!: \;+\; \cdot CH_3 \longrightarrow :\!\overset{..}{\underset{..}{Cl}}\!\cdot \;+\; :\!\overset{..}{\underset{..}{Cl}}\!:\!CH_3$$

Chlorine molecule   Methyl radical   Chlorine atom   Chloromethane

FIGURE 4.18 Equations describing the initiation and propagation steps in the free-radical mechanism for the chlorination of methane. Together the two propagation steps give the overall equation for the reaction.

(*c*) Sum of steps 2 and 3

$$CH_4 \;+\; Cl_2 \longrightarrow CH_3Cl \;+\; HCl$$

Methane   Chlorine   Chloromethane   Hydrogen

energy is absorbed by a molecule. Photochemically induced halogenations are typically carried out by using artificial sources of illumination such as ordinary sun lamps. We indicate a source of photochemical energy in an equation by writing "light" or "*hv*" above the arrow (*hv* is the energy of a light photon; *h* is the symbol for Planck's constant and *v* is the frequency).

The generally accepted mechanism for chlorination of methane is presented in Figure 4.18. It begins with a recognition of what happens when energy, either thermal or photochemical, is absorbed by the system. The reaction is *initiated* in step 1 by a homolytic cleavage of the weakest bond in the system, the chlorine-chlorine bond.

The two chlorine atoms formed in the initiation step are very reactive species. Each has seven valence electrons and reacts by abstracting a hydrogen atom from a methane molecule, as shown in step 2. Hydrogen chloride, one of the isolated products of the overall reaction, is formed in this step, along with methyl radical. Methyl radical does not accumulate but rapidly abstracts one of the chlorine atoms from a Cl_2 molecule in step 3 to form chloromethane. This third step not only produces chloromethane but also generates a chlorine atom. This chlorine atom can cycle back to step 2, repeating the process. Step 2 and step 3, when added together, give the equation for the overall process.

Step 1 is called an *initiation step* because it gets the process going. Steps 2 and 3 together comprise the *propagation steps* of the reaction. A chlorine atom generated in the initiation step sets in motion the sequence of events: step 2 → step 3 → step 2 → step 3 → etc. Reactions that involve repetitive sequences of this type are called *chain reactions*. Since the halogenation of methane proceeds by way of a free-radical intermediate, it is a *free-radical chain reaction*.

In practice, side reactions intervene to reduce the efficiency of the propagation steps. A single chlorine atom is not sufficient to convert all the reactants to products by an unending series of propagation steps that continue until all the reactants are consumed. The chain sequence is interrupted whenever two odd-electron species combine to give an even-electron product. Reactions of this type are called *chain-terminating steps*. The important chain-terminating steps that occur during the chlorination of methane are shown in the following equations:

Combination of a methyl radical with a chlorine atom:

$$\dot{C}H_3 \quad \cdot \ddot{C}l\colon \longrightarrow CH_3\ddot{C}l\colon$$

Methyl radical   Chlorine atom   Chloromethane

Combination of two methyl radicals:

$$\dot{C}H_3 \quad \dot{C}H_3 \longrightarrow CH_3CH_3$$

Two methyl radicals   Ethane

Combination of two chlorine atoms:

$$\colon\dot{\ddot{C}}l\cdot \quad \cdot\dot{\ddot{C}}l\colon \longrightarrow Cl_2$$

Two chlorine atoms   Chlorine molecule

Termination steps are, in general, less likely to occur than the propagation steps. Each of the termination steps requires two very reactive radicals to encounter each other in a medium that contains far greater quantities of other materials (methane and chlorine molecules, for example) with which they can react. While some chloromethane undoubtedly arises via direct combination of methyl radicals with chlorine atoms, most of it is formed by the propagation sequence shown in Figure 4.18.

4.23 CHLORINATION OF HIGHER ALKANES

Higher alkanes also react with chlorine under conditions of thermal or photochemical initiation. In substances all of whose hydrogen atoms are equivalent to each other, free-radical chlorination has been used to prepare alkyl and cycloalkyl chlorides.

$$CH_3CH_3 + Cl_2 \xrightarrow{420°C} CH_3CH_2Cl + HCl$$

Ethane   Chlorine   Chloroethane (78%)   Hydrogen chloride
                                  (ethyl chloride)

$$\square + Cl_2 \xrightarrow{h\nu} \square{-}Cl + HCl$$

Cyclobutane   Chlorine   Chlorocyclobutane (73%)   Hydrogen
                                (cyclobutyl chloride)   chloride

As in the chlorination of methane, however, it is often difficult to limit the reaction to monochlorination, and significant quantities of derivatives having more than one chlorine atom are also formed.

PROBLEM 4.14 Chlorination of ethane yields, in addition to ethyl chloride, a mixture of two isomeric dichlorides. What are the structures of these two dichlorides?

A more serious problem arises in molecules in which the hydrogen atoms are not all equivalent. One reaction which has received careful study is the photochemical chlorination of butane. Two isomeric chlorides are formed, 1-chlorobutane and 2-chlorobutane.

$$CH_3CH_2CH_2CH_3 \xrightarrow[hv, \; 35°C]{Cl_2} CH_3CH_2CH_2CH_2Cl + \underset{\underset{Cl}{|}}{CH_3CHCH_2CH_3}$$

Butane   1-Chlorobutane (28%)   2-Chlorobutane (72%)
                              (*n*-butyl chloride)   (*sec*-butyl chloride)

The percentages cited in this equation reflect the composition of the monochloride fraction of the product mixture rather than the isolated yield of each component.

These two products arise because in one of the propagation steps a chlorine atom may abstract a hydrogen atom from either a methyl or a methylene group of butane.

$$CH_3CH_2CH_2CH_2{-}H + \cdot Cl \longrightarrow CH_3CH_2CH_2\overset{\cdot}{C}H_2 + HCl$$

Butane   Butyl radical

$$\underset{\underset{H}{|}}{CH_3CHCH_2CH_3} + \cdot Cl \longrightarrow CH_3\overset{\cdot}{C}HCH_2CH_3 + HCl$$

Butane   *sec*-Butyl radical

The respective free radicals react with chlorine in a succeeding propagation step to give the corresponding alkyl chlorides. Butyl radical gives only 1-chlorobutane; *sec*-butyl radical gives only 2-chlorobutane.

$$CH_3CH_2CH_2\overset{\cdot}{C}H_2 + Cl_2 \longrightarrow CH_3CH_2CH_2CH_2Cl + \cdot Cl$$

Butyl radical   1-Chlorobutane
                                       (*n*-butyl chloride)

$$CH_3\overset{\cdot}{C}HCH_2CH_3 + Cl_2 \longrightarrow \underset{\underset{Cl}{|}}{CH_3CHCH_2CH_3} + Cl \cdot$$

sec-Butyl radical   2-Chlorobutane
                                       (*sec*-butyl chloride)

If every collision of a chlorine atom with a butane molecule resulted in hydrogen abstraction, the butyl/*sec*-butyl radical ratio would be 6 : 4. There are six equivalent methyl hydrogens and four equivalent methylene hydrogens, so a chlorine atom is 1.5 times more likely to encounter a methyl hydrogen than a methylene hydrogen.

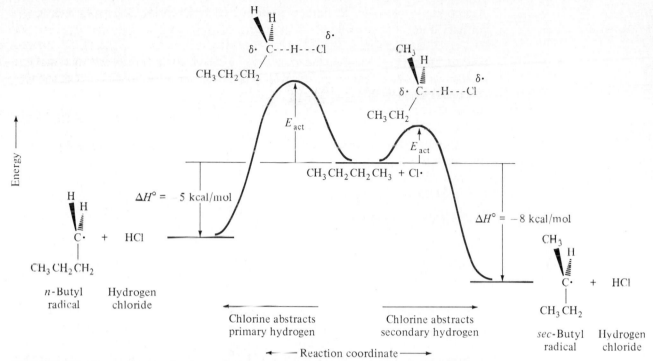

FIGURE 4.19 Energy diagram for the hydrogen atom abstraction step in the free-radical halogenation of butane. The activation energy for abstraction of a secondary hydrogen is less than that for abstraction of a primary hydrogen.

The product distribution expected on a *statistical* basis is 60 percent 1-chlorobutane to 40 percent 2-chlorobutane; however, the experimentally observed product distribution is 28 percent 1-chlorobutane to 72 percent 2-chlorobutane. Therefore, *sec*-butyl radical is formed in greater amounts and butyl radical in lesser amounts than expected statistically. The reason for this behavior stems from the greater stability of secondary compared with primary free radicals.

As the diagrams in Figure 4.19 illustrate, the carbon atom from which a hydrogen atom is removed develops free-radical character as progress is made toward the transition state. Since secondary radicals are more stable than primary, the transition state is somewhat lower in energy when radical character is developed at a secondary carbon of butane than at a primary carbon. Consequently, hydrogen atom abstraction from a methylene group has a lower activation energy than hydrogen atom abstraction from a methyl group, and a secondary hydrogen is abstracted faster than a primary hydrogen.

The relative rate of formation of isomeric free radicals reflects both the rate of hydrogen atom abstraction and the statistical factor. The rates at which secondary and primary hydrogens are removed by chlorine may be compared by adjusting the experimentally observed product distribution for the statistical bias of 6 : 4 that favors attack at the methyl groups.

$$\frac{\text{Percent 2-chlorobutane}}{\text{Percent 1-chlorobutane}} = \frac{4}{6} \times \frac{\text{rate of H abstraction from } CH_2}{\text{rate of H abstraction from } CH_3}$$

$$\frac{\text{Rate of H abstraction from } CH_2}{\text{Rate of H abstraction from } CH_3} = \frac{6}{4} \times \frac{(72)}{(28)} = 3.9$$

A secondary hydrogen in butane is abstracted by chlorine 3.9 times as fast as a primary one.

PROBLEM 4.15   Assuming the relative rate of secondary to primary hydrogen atom abstraction to be the same in the chlorination of propane as it is in that of butane, calculate the relative amounts of propyl chloride and isopropyl chloride obtained in the free-radical chlorination of propane.

A similar study of the chlorination of 2-methylpropane established that a tertiary hydrogen is removed 5.2 times as fast as a primary one.

$$
\underset{\substack{\text{2-Methylpropane} \\ \text{(isobutane)}}}{\overset{\displaystyle H}{\underset{\displaystyle CH_3}{CH_3\!-\!\overset{|}{\underset{|}{C}}\!-\!CH_3}}}
\xrightarrow[hv,\,35^\circ C]{Cl_2}
\underset{\substack{\text{1-Chloro-2-methylpropane (63\%)} \\ \text{(isobutyl chloride)}}}{\overset{\displaystyle H}{\underset{\displaystyle CH_3}{CH_3\!-\!\overset{|}{\underset{|}{C}}\!-\!CH_2Cl}}}
\quad+\quad
\underset{\substack{\text{2-Chloro-2-methylpropane (37\%)} \\ (\textit{tert}\text{-butyl chloride})}}{\overset{\displaystyle Cl}{\underset{\displaystyle CH_3}{CH_3\!-\!\overset{|}{\underset{|}{C}}\!-\!CH_3}}}
$$

An important feature to recognize in the reactions describing the chlorination of higher alkanes is that chlorination is not very *selective.* Mixtures of isomers corresponding to substitution of each of the various hydrogens in the molecule are formed. The order of reactivity per hydrogen encompasses only a factor of 5 in relative rate.

$$R_3CH \; > \; R_2CH_2 \; > \; RCH_3$$

Relative rate (chlorination)	(Tertiary)	(Secondary)	(Primary)
	5.2	3.9	1

This low level of selectivity limits chlorination of alkanes as a means of alkyl chloride preparation to alkanes and cycloalkanes in which all the hydrogen substituents are equivalent.

4.24 BROMINATION OF HIGHER ALKANES

Bromination of alkanes is much more selective than chlorination. A tertiary hydrogen of an alkane is replaced so much more readily than a secondary or primary one by bromine that photochemical bromination is a satisfactory method for preparing tertiary alkyl bromides.

$$
\underset{\substack{\text{2-Methylpentane}}}{\overset{\displaystyle H}{\underset{\displaystyle CH_3}{CH_3\!-\!\overset{|}{\underset{|}{C}}\!-\!CH_2CH_2CH_3}}}
+\;
\underset{\substack{\text{Bromine}}}{Br_2}
\;\xrightarrow[60^\circ C]{hv}\;
\underset{\substack{\text{2-Bromo-2-methylpentane} \\ \text{(76\% isolated yield)}}}{\overset{\displaystyle Br}{\underset{\displaystyle CH_3}{CH_3\!-\!\overset{|}{\underset{|}{C}}\!-\!CH_2CH_2CH_3}}}
\;+\;
\underset{\substack{\text{Hydrogen} \\ \text{bromide}}}{HBr}
$$

Free-radical bromination of alkanes is selective for the same reason that chlorination is. The transition state for the hydrogen atom abstraction step has some of the character of a free radical, and a more stable radical is formed faster than a less stable one. The spread in reactivity among primary, secondary, and tertiary hydrogens is, however, much greater for bromination than for chlorination.

$$R_3CH > R_2CH_2 > RCH_3$$

	(Tertiary)	(Secondary)	(Primary)
Relative rate (bromination)	1640	82	1

Bromination is more selective than chlorination because the transition state for hydrogen atom removal by bromine has more of the character of an alkyl radical than does the transition state for hydrogen atom removal by chlorine. The energy of the transition state is more sensitive to the stability of the incipient radical in bromination, less sensitive in chlorination. The basis for this point of view is a comparative analysis of the energetics of the hydrogen atom abstraction step in the bromination and chlorination of methane.

The heat of reaction for the hydrogen atom abstraction step in the bromination of methane can be calculated from the bond dissociation energies of Table 4.4.

$$CH_4 + \cdot Br \longrightarrow [\overset{\delta\cdot}{CH_3} --- H --- \overset{\delta\cdot}{Br}]^{\neq} \longrightarrow \cdot CH_3 + HBr$$

Activated complex
for hydrogen atom
abstraction

Bond broken $(CH_3 — H) = 104$ kcal/mol
Bond formed $(H — Br) = 87.5$ kcal/mol
$\Delta H° = 16.5$ kcal/mol

The measured activation energy for this step is 18.6 kcal/mol. Thus, the transition state is 18.6 kcal/mol higher in energy than the reactants in this step but only 2.1 kcal/mol higher in energy than the products. The transition state is closer in energy to the methyl radical and hydrogen bromide than it is to methane and a bromine atom. According to Hammond's postulate (Section 4.15), if two states are similar in energy, they are similar in structure. Therefore, the structure present at the transition state is fairly far along the way toward a methyl radical and hydrogen bromide. The carbon-hydrogen bond is significantly weakened, the hydrogen-bromine bond is well developed, and carbon has much of the character of a free radical.

For chlorination of methane, the hydrogen atom abstraction step is endothermic by only 1 kcal/mol, and the measured energy of activation is 3.8 kcal/mol.

$$CH_4 + \cdot Cl \longrightarrow [\overset{\delta\cdot}{CH_3} --- H --- \overset{\delta\cdot}{Cl}]^{\neq} \longrightarrow \cdot CH_3 + HCl$$

Activated complex
for hydrogen atom
extraction

Bond broken $(CH_3 — H) = 104$ kcal/mol
Bond formed $(H — Cl) = 103$ kcal/mol
$\Delta H° = 1$ kcal/mol

The energy separating the reactants and the transition state (3.8 kcal/mol) is not much different from that separating the transition state from the products of this propagation step (2.8 kcal/mol). Therefore, the structure at the transition state is almost as much like the reactants as it is like the products of hydrogen atom abstraction. The transition state for hydrogen abstraction by a chlorine atom does not have nearly the alkyl radical character that the transition state for hydrogen abstraction by a bromine atom does.

4.25 SUMMARY

Chemical reactivity is the unifying theme of this chapter. The reactions discussed, those that produce alkyl halides from alcohols and from alkanes, are summarized in Table 4.6. These reactions not only are important for synthetic purposes but also provide a framework in which to explore some fundamental principles of reaction mechanisms.

The reaction of secondary and tertiary alcohols with hydrogen halides proceeds by way of a *carbocation* intermediate. The carbocation is formed in the rate-determining step and then captured by a halide ion to yield an alkyl halide in the final step of a three-step sequence.

(a) $\quad\quad\quad\quad\quad\quad ROH + HX \longrightarrow R\overset{+}{O}H_2 + X^-$

$\quad\quad\quad\quad\quad\quad$ Alcohol $\quad$ Hydrogen $\quad\quad$ Oxonium $\quad$ Halide
$\quad\quad\quad\quad\quad\quad\quad\quad\quad$ halide $\quad\quad\quad\quad$ cation $\quad\quad$ anion

(b) $\quad\quad\quad\quad\quad\quad\quad\quad R\overset{+}{O}H_2 \longrightarrow R^+ + H_2O$

$\quad\quad\quad\quad\quad\quad\quad\quad$ Oxonium cation $\quad$ Carbocation $\quad$ Water

(c) $\quad\quad\quad\quad\quad\quad\quad R^+ + X^- \longrightarrow RX$

$\quad\quad\quad\quad\quad\quad\quad$ Carbocation $\quad$ Halide anion $\quad\quad$ Alkyl halide

The critical intermediate in this mechanism, the carbocation, has a positively charged carbon. The positively charged carbon is sp^2 hybridized, has a vacant $2p$ orbital, and is tricoordinate.

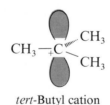

tert-Butyl cation

A different kind of intermediate, a *free radical,* is involved in the halogenation of alkanes.

(a) $\quad\quad\quad\quad X_2 \longrightarrow 2X\cdot \quad\quad\quad$ (initiation step)

$\quad\quad\quad\quad$ Halogen molecule $\quad\quad$ Two halogen atoms

(b) $\quad\quad RH + X\cdot \longrightarrow R\cdot + HX \quad\quad$ (propagation step)

$\quad\quad$ Alkane $\quad$ Halogen $\quad\quad$ Alkyl $\quad$ Hydrogen
$\quad\quad\quad\quad\quad$ atom $\quad\quad\quad$ radical $\quad$ halide

(c) $\quad\quad R\cdot + X_2 \longrightarrow RX + \cdot X \quad\quad$ (propagation step)

$\quad\quad$ Alkyl $\quad$ Halogen $\quad\quad$ Alkyl $\quad$ Halogen
$\quad\quad$ radical $\quad$ molecule $\quad\quad$ halide $\quad$ atom

TABLE 4.6

Conversions of Alcohols and Alkanes to Alkyl Halides

Reaction (section) and comments	General equation and specific example(s)
Reactions of alcohols with hydrogen halides (Section 4.9) Alcohols react with hydrogen halides to yield alkyl halides. The reaction is useful as a synthesis of alkyl halides. The reactivity of hydrogen halides decreases in the order HI > HBr > HCl > HF. Alcohol reactivity decreases in the order tertiary > secondary > primary > methyl.	$ROH + HX \longrightarrow RX + H_2O$ Alcohol / Hydrogen halide / Alkyl halide / Water 1-Methylcyclopentanol →(HCl) 1-Chloro-1-methylcyclopentane (96%)
Reaction of alcohols with thionyl chloride (Section 4.17) Thionyl chloride is a synthetic reagent used to convert alcohols to alkyl chlorides.	$ROH + SOCl_2 \longrightarrow RCl + SO_2 + HCl$ Alcohol / Thionyl chloride / Alkyl chloride / Sulfur dioxide / Hydrogen chloride $CH_3CH_2CH_2CH_2CH_2OH \xrightarrow[pyridine]{SOCl_2} CH_3CH_2CH_2CH_2CH_2Cl$ 1-Pentanol / 1-Chloropentane (80%)
Reaction of alcohols with phosphorus tribromide (Section 4.17) As an alternative to converting alcohols to alkyl bromides with hydrogen bromide, the inorganic reagent phosphorus tribromide is sometimes used.	$3ROH + PBr_3 \longrightarrow 3RBr + P(OH)_3$ Alcohol / Phosphorus tribromide / Alkyl bromide / Phosphorous acid $CH_3CHCH_2CH_2CH_3$ (OH) $\xrightarrow{PBr_3} CH_3CHCH_2CH_2CH_3$ (Br) 2-Pentanol / 2-Bromopentane (67%)
Free-radical halogenation of alkanes (Sections 4.20 through 4.24) Alkanes react with halogens by substitution of a halogen for a hydrogen on the alkane. The reactivity of the halogens decreases in the order $F_2 > Cl_2 > Br_2 > I_2$. The ease of replacing a hydrogen decreases in the order tertiary > secondary > primary > methyl. Chlorination is not very selective and so is only used when all the hydrogens of the alkane are equivalent. Bromination is highly selective, replacing tertiary hydrogens much more readily than secondary or primary ones.	$RH + X_2 \longrightarrow RX + HX$ Alkane / Halogen / Alkyl halide / Hydrogen halide Cyclodecane →(Cl₂, hν) Cyclodecyl chloride (64%) $(CH_3)_2CHC(CH_3)_3 \xrightarrow{Br_2, h\nu} (CH_3)_2CC(CH_3)_3$ (Br) 2,2,3-Trimethylbutane / 2-Bromo-2,3,3-trimethylbutane (80%)

Alkyl radicals are neutral species, characterized by the presence of an unpaired electron on a tricoordinate carbon.

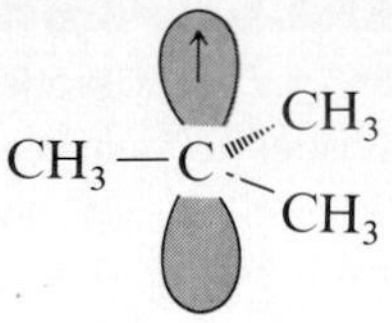

tert-Butyl radical

Multistep reaction mechanisms are commonplace in organic chemistry and will be encountered frequently as new reactions are introduced in the chapters that follow. Carbocations and free radicals are but two examples of reactive intermediates that occur during the conversion of starting materials to products. Both carbocations and free radicals are involved in reactions other than those described in this chapter and will reappear in a number of the functional group transformations that remain to be described.

PROBLEMS

4.16 Write structural formulas for each of the following alcohols and alkyl halides:

(a) Cyclobutanol

(b) *sec*-Butyl alcohol

(c) 3-Heptanol

(d) 1-Pentadecanol

(e) *trans*-2-Chlorocyclopentanol

(f) 4-Methyl-2-hexanol

(g) 2,6-Dichloro-4-methyl-4-octanol

(h) *trans*-4-*tert*-Butylcyclohexanol

(i) 1-Bromo-3-iodo-2-propanol

(j) 1-Cyclopropylethanol

(k) 2-Cyclopropylethanol

(l) Chlorotrifluoromethane

4.17 Give a systematic IUPAC name to each of the following compounds:

(a) $CH_3(CH_2)_8I$

(b) $(CH_3)_2CHCH_2CH_2CH_2Br$

(c) $(CH_3)_2CHCH_2CH_2CH_2OH$

(d) Cl_3CCH_2Br

(e) $Cl_2CHCHBr$
$\quad\quad\quad\quad|$
$\quad\quad\quad\;Cl$

(f) CF_3CH_2OH

(g)

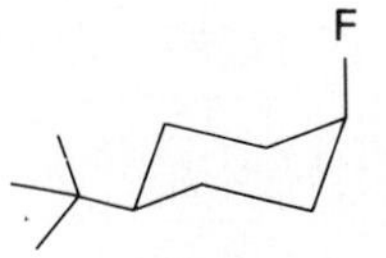

(h)

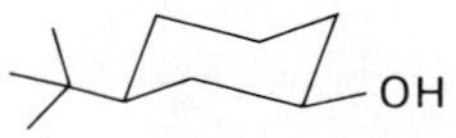

(i)

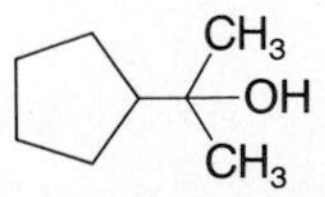

(j)

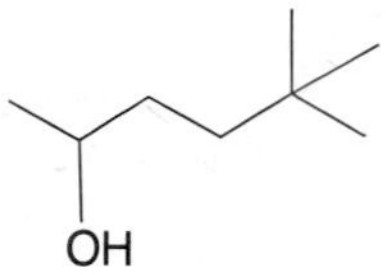

(k)

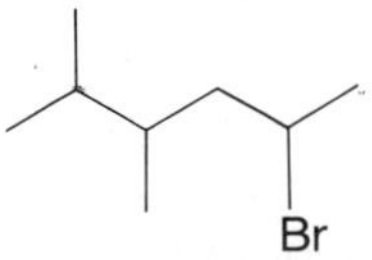

(l)

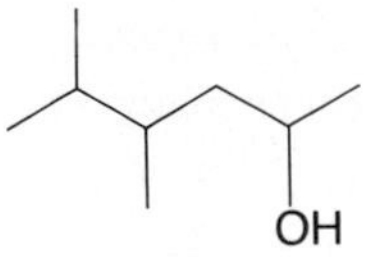

4.18 Many fluorinated hydrocarbons (known as *fluorocarbons*) were once used extensively as aerosol propellants, but evidence indicating that they accumulated in the upper atmosphere and damaged the earth's ozone layer led to their being banned for this purpose in the United States. Other uses of certain fluorocarbons are still permitted, and some of these compounds are listed below. Using the structural formula given, assign a systematic name to each one.

(a) Cl_2CF_2 (Freon 12), a refrigerant

(b) $ClCF_2CF_2Cl$ (Freon 114), a refrigerant

(c) F_2CBrCl, used in fire extinguishers

(d) $F_3CCHBrCl$ (halothane), an anesthetic

(e) 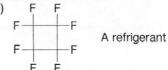 A refrigerant

4.19 Write structural formulas for all the constitutionally isomeric alcohols of molecular formula $C_5H_{12}O$. Assign a systematic IUPAC name to each one and specify whether it is a primary, secondary, or tertiary alcohol.

4.20 A hydroxyl group is a somewhat "smaller" substituent on a six-membered ring than is a methyl group. That is, the preference of a hydroxyl group for the equatorial orientation is less pronounced than that of a methyl group. Given this information, write structural formulas for all the isomeric methylcyclohexanols, showing each one in its most stable conformation. Give an acceptable IUPAC name for each isomer.

4.21 By assuming that the heat of combustion of the cis isomer was larger than the trans, structural assignments were made many years ago for the stereoisomeric 2-, 3-, and 4-methylcyclohexanols. This assumption is valid for two of the stereoisomeric pairs but is incorrect for the other. For which pair of stereoisomers is the assumption incorrect? Why?

4.22 (a) Menthol, used to flavor various foods and tobacco, is the most stable stereoisomer of 2-isopropyl-5-methylcyclohexanol. Draw its most stable conformation. Is the hydroxyl group cis or trans to the isopropyl group? To the methyl group?

(b) Neomenthol is a stereoisomer of menthol. That is, it has the same constitution but differs in the arrangement of its atoms in space. Neomenthol is the second most stable stereoisomeric form of 2-isopropyl-5-methylcyclohexanol; it is less stable than menthol but more stable than any other stereoisomer. Write the structure of neomenthol in its most stable conformation.

4.23 Each of the following pairs of compounds undergoes a Brönsted acid-base reaction for which the equilibrium constant is greater than unity. Give the products of each reaction and identify the acid, the base, the conjugate acid, and the conjugate base. Use the acid dissociation constants in Table 4.3 as a guide.

(a) $HI + HO^- \rightleftharpoons$

(b) $CH_3CH_2O^- + CH_3\overset{\displaystyle O}{\overset{\displaystyle \|}{C}}OH \rightleftharpoons$

(c) $HF + H_2N^- \rightleftharpoons$

(d) $CH_3\overset{\displaystyle O}{\overset{\displaystyle \|}{C}}O^- + HCl \rightleftharpoons$

(e) $(CH_3)_3CO^- + H_2O \rightleftharpoons$

(f) $(CH_3)_2CHOH + H_2N^- \rightleftharpoons$

(g) $F^- + H_2SO_4 \rightleftharpoons$

4.24 Write a chemical equation for the reaction of 1-butanol with each of the following:

(a) Lithium

(b) Sodium

(c) Potassium

(d) Sodium amide ($NaNH_2$)
(e) Hydrogen bromide, heat
(f) Sodium bromide, sulfuric acid, heat
(g) Phosphorus tribromide
(h) Thionyl chloride

4.25 Repeat the preceding problem for the corresponding reactions of cyclohexanol.

4.26 Each of the following reactions has been described in the chemical literature and involves an organic starting material somewhat more complex than those we have encountered so far. Nevertheless, on the basis of the topics covered in this chapter, you should be able to write the structure of the principal organic product of each reaction.

(a)

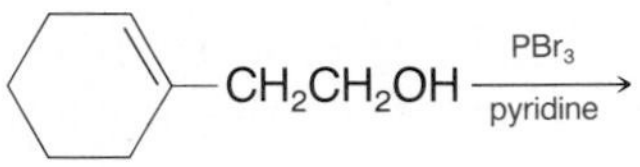

(b)

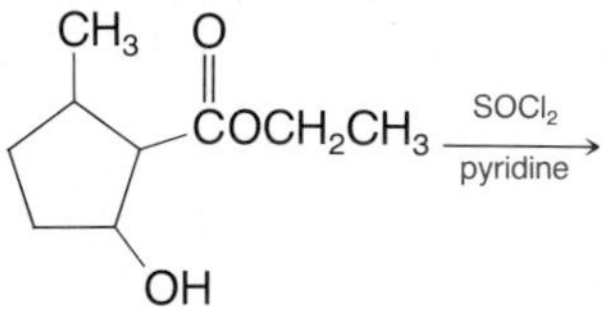

(c)

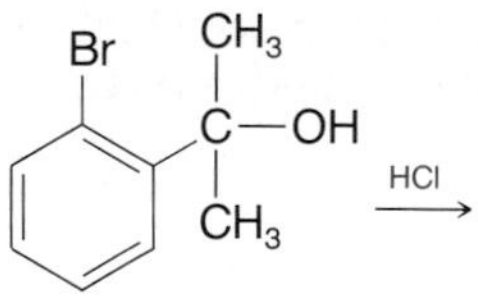

(d)

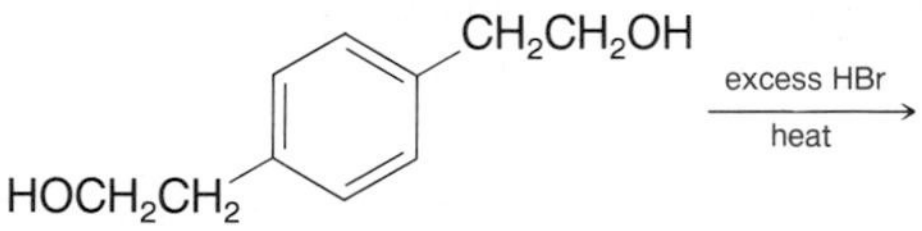

(e)

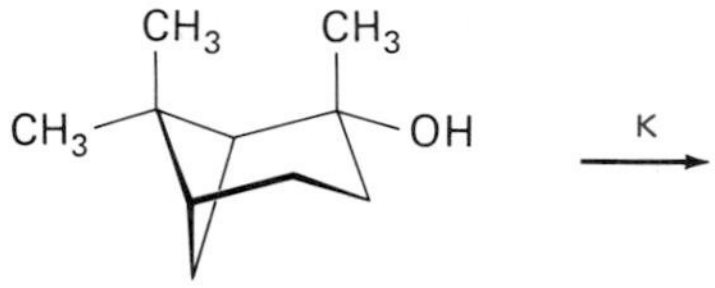

(f) 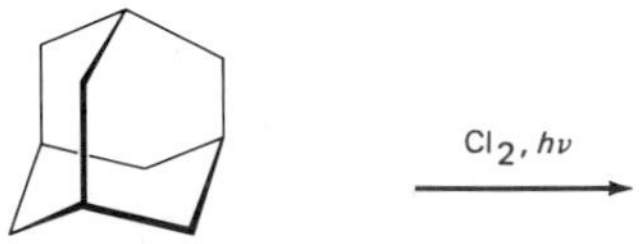   (two isomeric monochlorides)

(g) $(CH_3)_2CHCH(CH_3)_2 + 2Br_2 \xrightarrow[\text{heat}]{hv}$

4.27 Select the compound in each of the following pairs that will be converted to the corresponding alkyl bromide more rapidly on being treated with hydrogen bromide. Explain the reason for your choice.

(a) 1-Butanol or 2-butanol
(b) 2-Methyl-1-butanol or 2-butanol
(c) 2-Methyl-2-butanol or 2-butanol
(d) 2-Methylbutane or 2-butanol
(e) 1-Methylcyclopentanol or cyclohexanol
(f) 1-Methylcyclopentanol or *trans*-2-methylcyclopentanol
(g) 1-Cyclopentylethanol or 1-ethylcyclopentanol

4.28 (a) Evidence from a variety of sources indicates that cyclopropyl cation is much less stable than isopropyl cation. Why?

(b) The cation $CF_3\overset{+}{C}HCH_3$ is much less stable than isopropyl cation. Why?

4.29 The difference in energy between a secondary and a tertiary alkyl radical is about 3 to 4 kcal/mol. The difference in energy between a secondary and tertiary carbocation is about 10 to 15 kcal/mol. Can you think of a reason why carbocations are more sensitive to stabilization by substituents than free radicals are?

4.30 The acid dissociation constants for ethanol and 2,2,2-trifluoroethanol differ by a factor of 10^4. Which is the stronger acid? Why?

4.31 Basing your answers on the bond dissociation energy data in Table 4.4, calculate which of the following reactions are endothermic and which are exothermic.

(a) $(CH_3)_2CHOH + HF \longrightarrow (CH_3)_2CHF + H_2O$
(b) $(CH_3)_2CHOH + HCl \longrightarrow (CH_3)_2CHCl + H_2O$
(c) $CH_3CH_2CH_3 + HCl \longrightarrow (CH_3)_2CHCl + H_2$

4.32 By carrying out the reaction at $-78°C$ it is possible to fluorinate neopentane to yield $(CF_3)_4C$. Write a balanced chemical equation for this reaction.

4.33 In a search for fluorocarbons having anesthetic properties, 1,2-dichloro-1,1-difluoropropane was subjected to photochemical chlorination. Two isomeric products were obtained, one of which was identified as 1,2,3-trichloro-1,1-difluoropropane. What is the structure of the second compound?

4.34 Among the isomeric alkanes of molecular formula C_5H_{12}, identify the one that on photochemical chlorination yields

(a) A single monochloride
(b) Three isomeric monochlorides
(c) Four isomeric monochlorides
(d) Two isomeric dichlorides

4.35 In both of the following exercises, assume that all the methylene groups in the alkane are equally reactive as sites of free-radical chlorination.

(a) Photochemical chlorination of heptane gave a mixture of monochlorides containing 15% 1-chloroheptane. What other monochlorides are present? Estimate the percentage of each of these additional $C_7H_{15}Cl$ isomers in the monochloride fraction.

(b) Photochemical chlorination of dodecane gave a monochloride fraction containing 19% 2-chlorododecane. Estimate the percentage of 1-chlorododecane present in that fraction.

4.36 A certain compound contains only carbon and hydrogen and has a molecular weight of 72. Photochemical chlorination gave a mixture containing only one monochloride and only two dichlorides. What are the structures of the starting material and its chlorination products?

4.37 A certain substance contains only carbon and hydrogen and has a molecular weight of 70. Photochemical chlorination gave only one monochloride. Write the structure of the hydrocarbon and its monochloride derivative.

4.38 Two isomeric compounds A and B have the molecular formula C_3H_7Cl. Chlorination of A gave a mixture of two dichlorides of formula $C_3H_6Cl_2$. Chlorination of B gave three different compounds of formula $C_3H_6Cl_2$ (although these may not all be different from the dichlorides from A). What are the structural formulas of A and B and the dichlorides obtained from each?

4.39 (*a*) Write the structures of all the possible monobromides formed during the photo-chemical bromination of 2,2-dimethylbutane.

(*b*) Estimate the distribution of these monobromides in the product if the secondary/primary rate ratio for bromination is 82 : 1.

4.40 Deuterium oxide (D_2O) is water in which the protons (^1H) have been replaced by their heavier isotope deuterium (^2H). It is readily available and is used in a variety of mechanistic studies in organic chemistry and biochemistry. When D_2O is added to an alcohol (ROH), deuterium replaces the proton of the hydroxyl group.

$$ROH + D_2O \rightleftharpoons ROD + DOH$$

The reaction takes place extremely rapidly, and if D_2O is present in excess, all the alcohol is converted to ROD. This hydrogen-deuterium exchange can be catalyzed by either acids or bases. If D_3O^+ is the catalyst in acid solution and DO^- the catalyst in base, write reasonable reaction mechanisms for the conversion of ROH to ROD under conditions of (*a*) acid catalysis and (*b*) base catalysis.

ALKENES. STRUCTURE AND STABILITY

*A*lkenes are hydrocarbons that contain a carbon-carbon double bond. A carbon-carbon double bond is both an important structural unit and an important functional group in organic chemistry. The shape of an organic molecule is influenced by the presence of a carbon-carbon double bond, and the double bond is the site of most of the chemical reactions that alkenes undergo. This chapter is the first of three dealing with alkenes; it describes the principles of their structure and bonding. The following two chapters discuss the preparation of alkenes and their chemical reactions.

5.1 ALKENE NOMENCLATURE

Alkenes are named in the IUPAC system by replacing the *-ane* ending in the name of the corresponding alkane with *-ene*. Thus, the two simplest alkenes are named *ethene* and *propene*. Both these alkenes are also well known by their common names *ethylene* and *propylene*. Ethylene is an acceptable synonym for ethene in the IUPAC system. Propylene and other common names ending in *-ylene* are not acceptable IUPAC names.

$$CH_2 = CH_2 \qquad CH_3CH = CH_2$$

IUPAC name: **ethene**   IUPAC name: **propene**
Common name: ethylene   Common name: propylene

The longest continuous chain that includes the double bond forms the base name of the alkene, and the chain is numbered in the direction that gives the doubly bonded carbons their lower numbers. The locant (or numerical position) of only one of the doubly bonded carbons is specified in the name; it is understood that the other doubly bonded carbon must follow in sequence.

$$\overset{1}{C}H_2 = \overset{2}{C}H\overset{3}{C}H_2\overset{4}{C}H_3 \qquad \overset{6}{C}H_3\overset{5}{C}H_2\overset{4}{C}H_2\overset{3}{C}H = \overset{2}{C}H\overset{1}{C}H_3$$

1-Butene   2-Hexene
(not 1,2-butene)   (not 4-hexene)

Substituents are identified in the usual way, while remembering that the chain containing the double bond takes precedence in determining the parent alkene and that the numbering is dictated by the direction that gives the lower number to the doubly bonded carbons.

$$\overset{4}{C}H_3\overset{3}{C}H\overset{2}{C}H=\overset{1}{C}H_2$$
$$|$$
$$CH_3$$

3-Methyl-1-butene
(not 2-methyl-3-butene)

$$\overset{6}{C}H_3\overset{5}{C}H_2\overset{4}{C}H_2\overset{3}{C}HCH_2CH_2CH_3$$
$$\overset{2}{|}\quad\overset{1}{}$$
$$CH=CH_2$$

3-Propyl-1-hexene
(longest chain that contains
double bond is six carbons)

PROBLEM 5.1 Name the following alkenes using systematic IUPAC nomenclature.

(a) $(CH_3)_2C=C(CH_3)_2$

(b) $(CH_3)_3CCH=CH_2$

(c) $(CH_3)_2C=CHCH_2CH_2CH_3$

(d) $(CH_3)_2C=CHC(CH_3)_3$

(e) $(CH_3)_2CH$
$$\qquad\qquad\quad C=CHCH_2CH(CH_3)_2$$
$$(CH_3)_2CH$$

SAMPLE SOLUTION (a) The longest continuous chain in this alkene contains four carbon atoms. The double bond is between C-2 and C-3, so it is named as a derivative of 2-butene.

$$\overset{1}{H_3C}\qquad\overset{4}{\diagup}CH_3$$
$$\underset{2}{\diagdown}C=C\underset{3}{\diagup}$$
$$H_3C\diagup\qquad\diagdown CH_3$$

2,3-Dimethyl-2-butene

Identifying the alkene as a derivative of 2-butene leaves two methyl groups to be accounted for as substituents attached to the main chain. This alkene is 2,3-dimethyl-2-butene. (It is sometimes called tetramethylethylene, but that is a common name, not a systematic IUPAC name.)

We noted earlier in Section 2.10 that the common names of certain frequently encountered alkyl groups, such as isopropyl and *tert*-butyl, are acceptable in the IUPAC system. Three *alkenyl* groups—vinyl, allyl, and isopropenyl—are treated the same way.

$$CH_2=CH—\qquad \text{as in} \qquad CH_2=CHCl$$

Vinyl   Vinyl chloride

$$CH_2=CHCH_2—\qquad \text{as in} \qquad CH_2=CHCH_2Cl$$

Allyl   Allyl chloride

$$CH_2=C—\qquad \text{as in} \qquad CH_2=CCl$$
$$|\qquad\qquad\qquad\qquad\qquad |$$
$$CH_3\qquad\qquad\qquad\qquad\quad CH_3$$

Isopropenyl   Isopropenyl chloride

Cycloalkenes and their derivatives are named by adapting cycloalkane terminology to the principles of alkene nomenclature.

Cyclopentene   1-Methylcyclohexene   3-Chlorocycloheptene
(not 1-chloro-2-cycloheptene)

No locants are needed in the absence of substituents; it is understood that the double bond connects C-1 and C-2. Substituted cycloalkenes are numbered beginning with the double bond, proceeding through it, and continuing in sequence around the ring. The direction of numbering is chosen so as to give the lower of two possible locants to the substituent group.

PROBLEM 5.2 Write structural formulas and give the IUPAC names of all of the monochloro-substituted derivatives of cyclopentene.

5.2 SOURCES OF ALKENES

Ethylene was known to chemists in the eighteenth century and was isolated in pure form in 1795. An early name for ethylene was *gaz oléfiant* (French for "oil-forming gas"). This term was suggested by Antoine François Fourcroy, a contemporary of Lavoisier, to describe the fact that an oily liquid product is formed when two gases — ethylene and chlorine — react with each other. Fourcroy's name for ethylene survived for a long time in the general term *olefin*, formerly used for the class of compounds we now call *alkenes.*

Ethylene occurs naturally in small amounts in plants and is involved in the ripening process of many fruits. As little as 1 part per million of ethylene in the air can stimulate ripening, and the rate of ripening increases at higher ethylene concentrations. This property is used to advantage in the marketing of bananas. Bananas are picked green in the tropics, kept green by being stored with adequate ventilation to limit the concentration of ethylene, and then induced to ripen at their destination by adding ethylene.

Ethylene is the cornerstone substance of the world's mammoth petrochemical industry. Ethylene, and propene as well, is produced on an industrial scale in vast quantities. In a typical year the amount of ethylene produced in the United States is approximately the same as the combined weight of all of its people (3×10^{10} pounds). Propene production is about one-half that of ethylene. The principal use of these alkenes is as starting materials for the preparation of polyethylene and polypropylene plastics, fibers, and films. Some ethylene is converted to ethyl alcohol for solvent purposes and to ethylene glycol for use in the preparation of polyester fibers and as an antifreeze for automobile radiators.

Most of the ethylene is produced by a process called *thermal cracking* of hydrocarbons. Hydrocarbons from petroleum yield ethylene when they are heated (not burned) at high temperatures. When ethane is used, hydrogen is produced along with ethylene and the reaction is referred to as *dehydrogenation.*

$$CH_3CH_3 \xrightarrow{750°C} CH_2{=}CH_2 + H_2$$

Ethane   Ethylene   Hydrogen

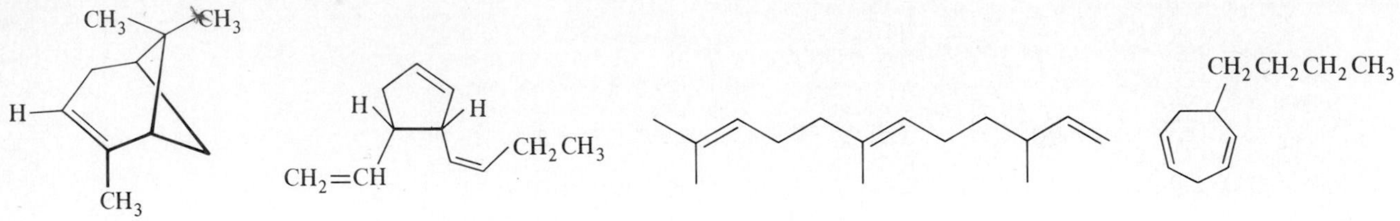

α-Pinene
(a major constituent
of turpentine)

Multifidene
(attracts male sperm cells
of a species of brown alga
to female)

Farnesene
(present in the waxy coating
found on apple skins)

Dictyopterene C′
(obtained from oil of a
species of marine alga
that grows on Hawaiian
reefs)

FIGURE 5.1 Some naturally occurring substances that contain carbon-carbon double bonds.

Similarly, thermal dehydrogenation of propane gives propene.

$$CH_3CH_2CH_3 \xrightarrow{750°C} CH_3CH{=}CH_2 + H_2$$

Propane   Propene   Hydrogen

Higher-molecular-weight alkanes also yield ethylene and propene under these reaction conditions by processes that involve the rupture of carbon-carbon bonds.

Methods that are both more versatile and better suited for laboratory-scale synthesis of alkenes will be discussed in the next chapter. The synthetic methods to be described there permit double bonds to be introduced into a wider range of structures at much lower temperatures.

Many naturally occurring organic compounds are alkenes and cycloalkenes. Figure 5.1 shows a few examples.

5.3 STRUCTURE AND BONDING IN ALKENES

The important structural features of ethylene are presented in Figure 5.2. Ethylene is a planar molecule. Its carbon-carbon double bond length is 1.34 Å, a distance significantly shorter than the carbon-carbon single bond length of 1.53 Å seen in ethane (Section 2.6). The bond angles at carbon are close to 120°.

Bonding in ethylene can be described according to the principles of orbital hybridization developed earlier. As we have seen during our discussion of bonding in methyl cation in Section 4.11, carbon is sp^2 hybridized when it is bonded to three substituents. While both carbons of ethylene have four bonds, each carbon is attached to only three substituent groups. Our bonding model for ethylene therefore is based on an sp^2 hybridization state for each of its carbons.

Electron configuration
of sp^2 hybridized carbon: $(1s)^2$, $(2sp^2)^1$, $(2sp^2)^1$, $(2sp^2)^1$, $(2p)^1$

Carbon, in its sp^2 hybridized state, has three equivalent $2sp^2$ hybridized orbitals generated by mixing the $2s$ orbital with two of the $2p$ orbitals. Each of these sp^2 hybrid orbitals contains one electron. One of the original $2p$ orbitals is not hybridized and remains as a half-filled $2p$ orbital.

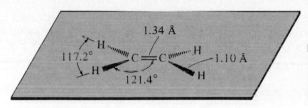

FIGURE 5.2 All the atoms of ethylene lie in the same plane. Bond angles are close to 120°, and the carbon-carbon bond distance is significantly shorter than that of ethane.

Each carbon uses two of its sp^2 hybrid orbitals to form σ bonds to two hydrogen atoms, as illustrated in the first part of Figure 5.3. The remaining sp^2 orbitals, one on each carbon, overlap to form a σ bond connecting the two carbons. As Figure 5.3 shows, each carbon atom still has, at this point, an unhybridized $2p$ orbital available for bonding. These two $2p$ orbitals overlap in a side-by-side manner to give what is called a π (pi) *molecular orbital.*

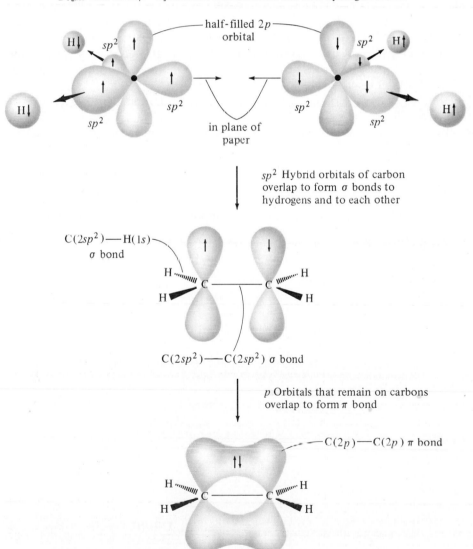

FIGURE 5.3 Orbital hybridization model of bonding in ethylene.

Each carbon contributes one electron to this π orbital, resulting in a π bond between the two. The carbon-carbon double bond in ethylene is seen, according to this analysis, to be composed of a σ component and a π component.

Before proceeding further, let us examine some features of bonding in ethylene in more detail. We learned in Section 2.3 that the number of σ molecular orbitals formed when atomic orbitals are combined is equal to the number of atomic orbitals that are mixed together. This principle applies to π orbitals as well. As Figure 5.4 illustrates, the two carbon $2p$ orbitals of ethylene overlap to give both a bonding (π) and an antibonding (π^*) combination. We shall find it convenient to depart from using plus ($+$) and minus ($-$) signs to designate the signs of p-orbital wave functions in favor of a symbolism whereby one lobe of a p orbital is shaded and the other is not. This is the practice, depicted in Figure 5.4, that is favored by most organic chemists and that avoids confusion with electronic charges.

Both the π and π^* molecular orbitals of ethylene are antisymmetric with respect to the molecular plane. The wave functions of both p orbitals and the π orbital that results from their overlap changes sign on passing through the plane of the molecule. The molecular plane corresponds to a nodal plane for the π and the π^* orbital. π Orbitals, like other atomic and molecular orbitals, are limited to containing two electrons and are filled beginning with the orbitals of lowest energy. Ethylene has two π electrons, and those electrons occupy the bonding π molecular orbital. The antibonding π^* orbital is vacant.

We shall see that much of the chemistry of alkenes is dominated by the electrons in the π orbital. Electrons in a π bond are less tightly held and are more easily polarized than electrons in a σ bond. When we discuss the chemical reactions of alkenes in Chapter 7, we shall see that most of them involve initial attack at the π electrons of the double bond.

The double bond in ethylene, while much stronger, is not twice as strong as the single bond of ethane. Estimates of the carbon-carbon double bond energy in ethylene suggest a value of about 146 kcal/mol, compared with the 83 kcal/mol bond

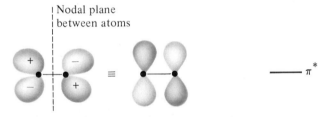

Antibonding π^* orbital of ethylene; orbital is unoccupied.

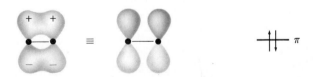

Bonding π orbital of ethylene. Orbital has no nodes; it is occupied by two electrons.

FIGURE 5.4 Representations of the bonding and antibonding π molecular orbitals of ethylene.

dissociation energy of ethane. Assuming that the σ component of the carbon-carbon double bond of ethylene is as strong as the σ bond of ethane, we conclude that the π component of the double bond is weaker that the σ component.

The structural features of propene are similar to those of ethylene. The carbon-carbon double bond length is 1.34 Å, and the bond angles involving the sp^2 hybridized carbons are close to 120°. The bond angles involving the methyl group are approximately tetrahedral.

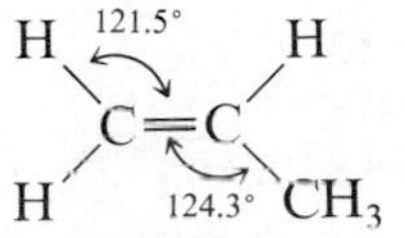

C—C bond length 1.50 Å
C=C bond length 1.34 Å

Propene

Notice that the carbon-carbon single bond length in propene is somewhat shorter than the 1.53-Å sp^3-sp^3 bond distance customarily seen in alkanes. Since one of the orbitals in the sp^2-sp^3 bond has increased s character, it will bind electrons more strongly, thereby contracting the size of the σ orbital and shortening the internuclear distance.

PROBLEM 5.3   Rank the carbon-carbon bonds of cyclobutene in order of decreasing bond length.

5.4 PHYSICAL PROPERTIES

Alkenes resemble alkanes in their physical properties. The lower-molecular-weight alkenes through C_4H_8 are gases at room temperature and atmospheric pressure. The melting points and boiling points of alkenes are comparable with those of alkanes having the same carbon skeleton. Alkenes are not very soluble in water and have very small dipole moments. Table 5.1 lists selected physical properties of some representative alkenes and cycloalkenes.

5.5 STEREOISOMERISM IN ALKENES

It has been known for a long time that there are four alkenes represented by the molecular formula C_4H_8. We can readily see that an unbranched four-carbon chain offers two different possibilities for the location of a carbon-carbon double bond—the double bond can unite C-1 and C-2 or C-2 and C-3.

$$CH_2=CHCH_2CH_3 \qquad CH_3CH=CHCH_3$$

1-Butene   2-Butene

It is further apparent that there is a branched-chain isomer.

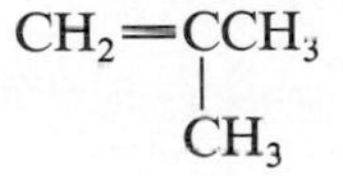

$$CH_2=CCH_3$$
$$\qquad | $$
$$\qquad CH_3$$

2-Methylpropene

TABLE 5.1

Physical Constants of Some Alkenes and Cycloalkenes

Compound name	Molecular formula	Condensed structure	Melting point, °C	Boiling point, °C (1 atm)
Alkenes				
Ethene (ethylene)	C_2H_4	$CH_2{=}CH_2$	−169.1	−103.7
Propene	C_3H_6	$CH_3CH{=}CH_2$	−185.0	−47.6
1-Butene	C_4H_8	$CH_3CH_2CH{=}CH_2$	−185	−6.1
2-Methylpropene	C_4H_8	$(CH_3)_2C{=}CH_2$	−140	−6.6
1-Pentene	C_5H_{10}	$CH_3CH_2CH_2CH{=}CH_2$	−138.0	30.2
2-Methyl-2-butene	C_5H_{10}	$(CH_3)_2C{=}CHCH_3$	−134.1	38.4
1-Hexene	C_6H_{12}	$CH_3CH_2CH_2CH_2CH{=}CH_2$	−138.0	63.5
2,3-Dimethyl-2-butene	C_6H_{12}	$(CH_3)_2C{=}C(CH_3)_2$	−74.6	73.5
1-Heptene	C_7H_{14}	$CH_3(CH_2)_4CH{=}CH_2$	−119.7	94.9
1-Octene	C_8H_{16}	$CH_3(CH_2)_5CH{=}CH_2$	−104	119.2
1-Decene	$C_{10}H_{20}$	$CH_3(CH_2)_7CH{=}CH_2$	−80.0	172.0
Cycloalkenes				
Cyclopentene	C_5H_8	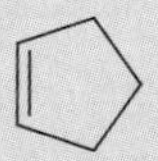	−98.3	44.1
Cyclohexene	C_6H_{10}	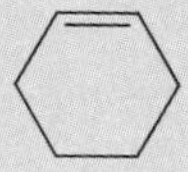	−104.0	83.1
1-Methylcyclohexene	C_7H_{12}	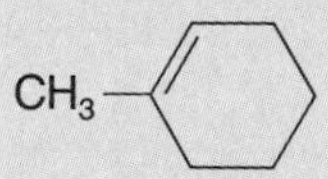 CH_3-		109.5
3-Methylcyclohexene	C_7H_{12}	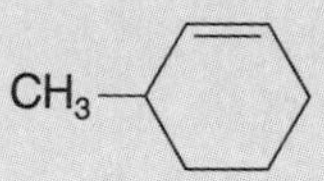 CH_3-		102.5
Cycloheptene	C_7H_{12}	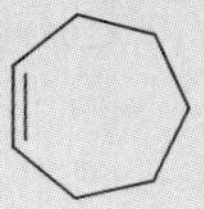		114.0

No other constitutionally isomeric C_4H_8 alkenes are possible. How then, are there four alkenes that have this molecular formula? The answer becomes apparent when the alkene named above as 2-butene is examined in more detail. There are, in fact, two alkenes represented by this constitution. The two alkenes are stereoisomers; one is known as *cis*-2-butene, the other as *trans*-2-butene.

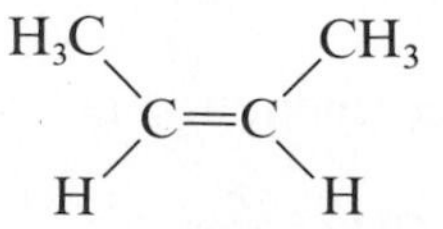

cis-2-Butene 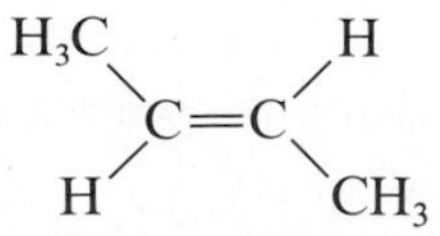   *trans*-2-Butene

The cis and trans prefixes have meanings that are similar to their meanings when we first encountered them in Section 3.11. Substituents that are on the same side of the double bond are said to be cis to each other; those on opposite sides of the double bond are trans. We specify the *configuration* of the double bond in the stereoisomeric 2-butenes as cis or trans according to whether the methyl groups (or the hydrogen substituents) are on the same or on opposite sides of the double bond.

Table 5.2 lists certain physical properties of the four C_4H_8 isomers. Notice that these properties are different for the two stereoisomers, just as they are different for a pair of constitutional isomers. Stereoisomers can have different physical properties. We shall see later that they can also have different chemical properties.

PROBLEM 5.4 How many alkenes are there of molecular formula C_5H_{10}? Write their structures and give their IUPAC names. Specify the configuration of stereoisomers as cis or trans as appropriate.

---

cis-2-Butene and *trans*-2-butene are related by rotation about their carbon-carbon double bond. Unlike rotation about carbon-carbon single bonds, which, as we have seen in Chapter 3, is normally quite fast, interconversion of stereoisomeric alkenes by rotation about a carbon-carbon double bond is so slow as not to be observable at room temperature. The activation energy for interconversion of *cis*- and *trans*-2-butene is on the order of 60 kcal/mol, or approximately 10 times that for rotation about the C-2–C-3 bond of butane.

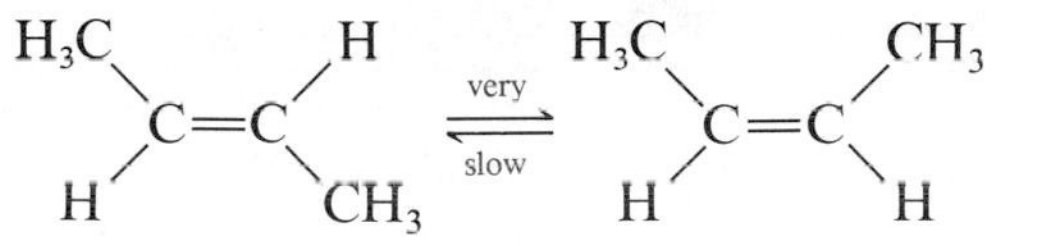

$E_{act} \sim 60$ kcal/mol

(too high an activation energy for stereoisomers to interconvert at room temperature)

Our σ-π model of bonding in alkenes provides a ready explanation for the high barrier to rotation about carbon-carbon double bonds. In order for π overlap to be effective, both p orbitals must be properly aligned. Figure 5.5 shows the proper alignment of individual p orbitals in *cis*- and *trans*-2-butene—the axis of each p orbital is perpendicular to the plane defined by the σ bonds to each sp^2 hybridized carbon. Rotation of one sp^2 hybridized carbon with respect to the other requires an activated complex in which the p orbitals are no longer properly aligned but are turned 90° with respect to

TABLE 5.2
Physical Properties of Isomeric C_4H_8 Alkenes

Compound name	Condensed structural formula	Melting point, °C	Boiling point, °C
1-Butene	$CH_2{=}CHCH_2CH_3$	−185	−6
cis-2-Butene	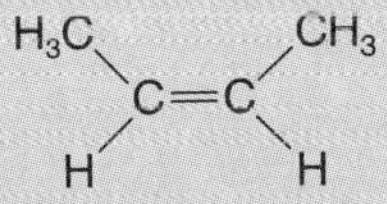	−139	4
trans-2-Butene	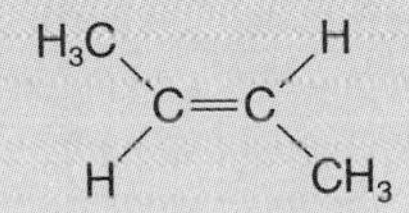	−105	1
2-Methylpropene	$CH_2{=}C(CH_3)_2$	−140	−7

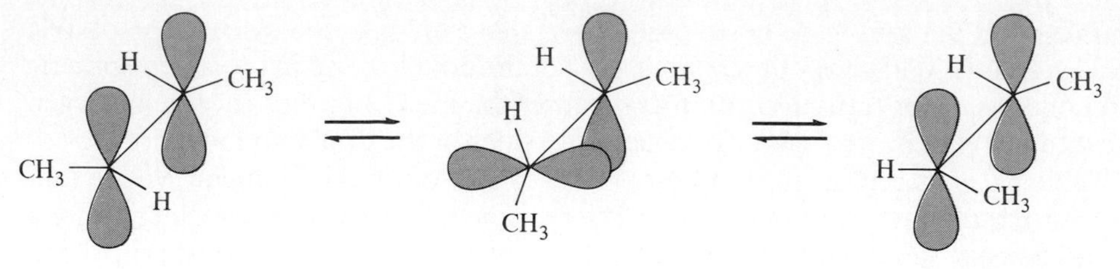

trans-2-Butene — p Orbitals aligned: optimal geometry for π bond formation

p Orbitals perpendicular: worst geometry for π bond formation

cis-2-Butene — p Orbitals aligned: optimal geometry for π bond formation

FIGURE 5.5 Interconversion of *cis*- and *trans*-2-butene proceeds by cleavage of π component of the double bond.

one another. The overlap of adjacent p orbitals is poor in the activated complex. In effect, converting *cis*-2-butene to *trans*-2-butene and vice versa involves breaking the π component of the double bond. The π bond energy of alkenes is approximately 60 kcal/mol, and this is the energy that must be expended in order to achieve the transition state for equilibration of stereoisomeric alkenes.

PROBLEM 5.5 One of the stereoisomers represented by the constitution ClCH=CHCl has a dipole moment of 1.8 D. Which stereoisomer is this? What would you expect the dipole moment of the other stereoisomer to be?

5.6 NAMING STEREOISOMERIC ALKENES BY THE *E-Z* SYSTEM

When the substituents on either end of a double bond are the same or are structurally similar to each other, it is a simple matter to apply the cis and trans descriptors to adequately specify the configuration about the double bond. Oleic acid, for example, a material that can be obtained from olive oil, has a cis double bond. Cinnamaldehyde, responsible for the characteristic odor of cinnamon, has a trans double bond.

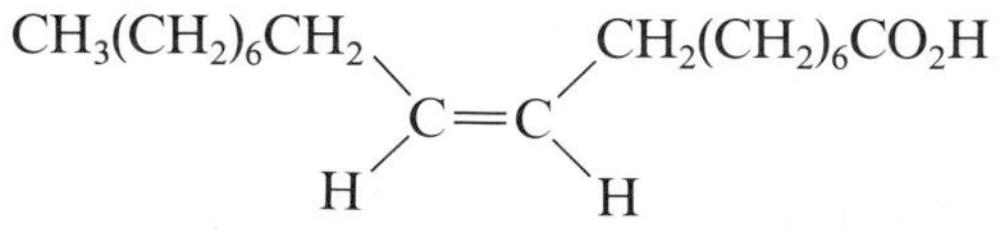

Oleic acid

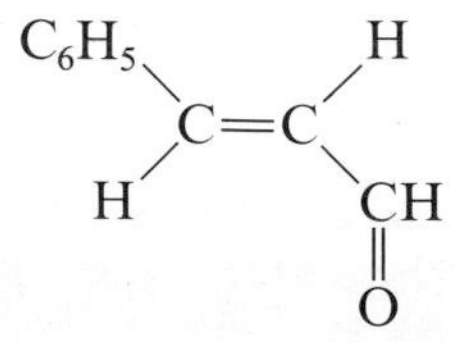

Cinnamaldehyde

PROBLEM 5.6 Female houseflies attract males by sending a chemical signal known as a *pheromone.* The substance emitted by the female housefly that attracts the male has been identified as *cis*-9-tricosene, $C_{23}H_{46}$. Write a structural formula, including stereochemistry, for this compound.

However, cis and trans are ambiguous descriptors when the similarities or differences between substituents are less clear. For example, it is not at all obvious whether

the alkene

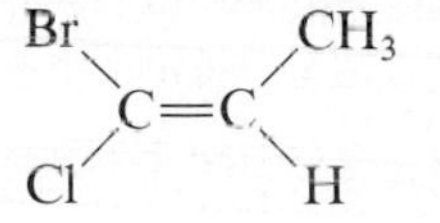

$$\underset{Cl}{\overset{Br}{\diagdown}} C = C \underset{H}{\overset{CH_3}{\diagup}}$$

should be described as cis or trans.

Fortunately, a completely unambiguous system for specifying double bond stereochemistry has been developed. Its key feature is a priority ranking of substituents, not on the subjective basis of perceived similarity, but on an absolute criterion based on atomic number.

Consider the 1-bromo-1-chloropropene stereoisomer shown above. The substituents attached to C-1 are bromine (atomic number 35) and chlorine (atomic number 17). Bromine has a higher atomic number than chlorine and is the higher-ranking substituent. At C-2 the attached atoms are hydrogen (atomic number 1) and the carbon atom of the methyl group (atomic number 6). The methyl group outranks hydrogen as a substituent because it is attached to the double bond by an atom of higher atomic number. In the stereoisomer shown, the higher-ranked substituents (bromine and methyl) are on the same side of the double bond. When higher-ranked substituents are on the same side, we say that the configuration of the double bond is Z. The Z descriptor stands for the German word *zusammen*, which means "together."

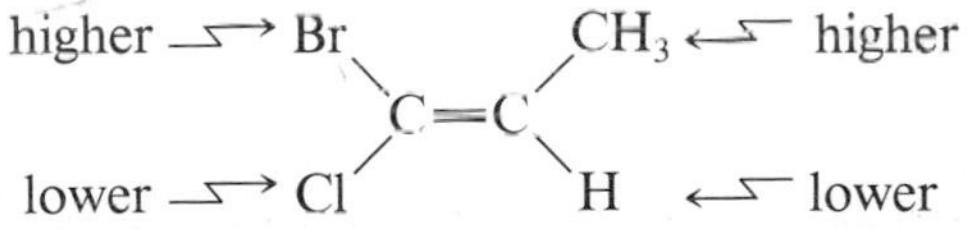

$$\text{higher} \longrightarrow \underset{\text{lower} \longrightarrow Cl}{\overset{Br}{\diagdown}} C = C \underset{H \longleftarrow \text{lower}}{\overset{CH_3 \longleftarrow \text{higher}}{\diagup}}$$

(*Z*)-1-Bromo-1-chloropropene

When higher-ranked substituents are on opposite sides of the double bond, we say the double bond has the E configuration. The symbol E stands for the German word *entgegen*, which means "opposite."

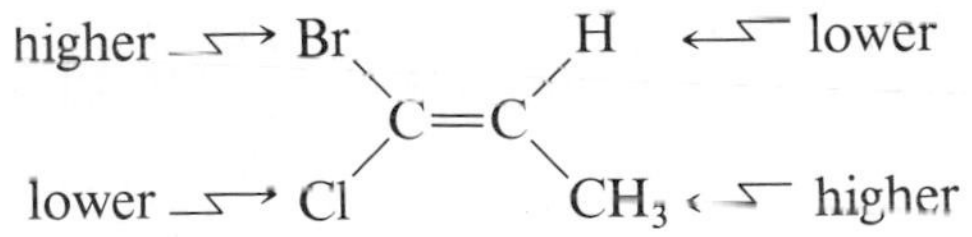

$$\text{higher} \longrightarrow \underset{\text{lower} \longrightarrow Cl}{\overset{Br}{\diagdown}} C = C \underset{CH_3 \longleftarrow \text{higher}}{\overset{H \longleftarrow \text{lower}}{\diagup}}$$

(*E*)-1-Bromo-1-chloropropene

Having described the basic principle upon which the *E-Z* notational system rests, let us turn to some more complicated examples that illustrate various subrules used to rank substituents. First, consider an alkene similar to the previous one except that this alkene has an ethyl group in place of the hydrogen substituent on the double bond.

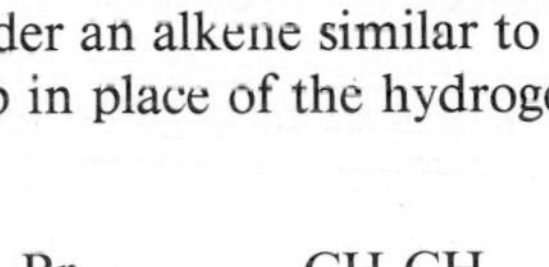

$$\underset{Cl}{\overset{Br}{\diagdown}} C = C \underset{CH_3}{\overset{CH_2CH_3}{\diagup}}$$

In deciding whether this alkene has the E or the Z configuration of its double bond, we need to determine whether ethyl or methyl is the higher-ranked alkyl group.

Subrule

When two atoms attached to the double bond are identical, examine the atoms attached to these two on the basis of their atomic numbers. Precedence is determined at the first point of difference.

The carbon that is the point of attachment of the ethyl group to the double bond has a methyl group and two hydrogens as its substituents. The substituent atoms are (C,H,H). The methyl group attached to the double bond has as its substituent atoms (H,H,H). Ethyl has precedence over methyl because the highest-ranked atom in (C,H,H) has a higher atomic number than any of the atoms in (H,H,H). Since higher-ranked substituents are on the same side of the double bond, the configuration of this alkene is Z.

higher $\longrightarrow$ Br$\diagdown$ CH_2CH_3 $\longleftarrow$ higher

$C=C$

lower $\longrightarrow$ Cl$\diagup$ CH_3 $\longleftarrow$ lower

(Z)-1-Bromo-1-chloro-2-methyl-1-butene

Among the common alkyl groups, the order of precedence is:

$$-C(CH_3)_3 > -CH(CH_3)_2 > -CH_2CH_3 > \quad -CH_3$$

| *tert*-Butyl | Isopropyl | Ethyl | Methyl |
| $-C(C,C,C)$ | $-C(C,C,H)$ | $-C(C,H,H)$ | $-C(H,H,H)$ |

Substituent atoms are always evaluated one by one, never as a group, when working outward from the point of attachment. For example, $-CH_2OH$ takes precedence over $-C(CH_3)_3$. Although *tert*-butyl has three carbon substituents on its attached atom, precedence is determined at the first point of difference and $-CH_2OH$ [$-C(O,H,H)$] outranks $-C(CH_3)_3$ [$-C(C,C,C)$]. The highest-ranked atom of (O,H,H) is oxygen, which has a higher atomic number than any of the atoms of (C,C,C).

PROBLEM 5.7   Determine the configuration of each of the following alkenes as Z or E as appropriate.

(a) $H_3C\diagdown$ CH_2OH

$C=C$

H$\diagup$ CH_3

(c) $H_3C\diagdown$ CH_2CH_2OH

$C=C$

H$\diagup$ $C(CH_3)_3$

(b) $H_3C\diagdown$ CH_2CH_2F

$C=C$

H$\diagup$ $CH_2CH_2CH_2CH_3$

(d) $\triangle\diagdown$ H

$C=C$

$CH_3CH_2\diagup$ CH_3

SAMPLE SOLUTION   (a)   There is a methyl group and a hydrogen as substituents on one of the doubly bonded carbons. Methyl is of higher rank than hydrogen. The other carbon atom of the double bond bears a methyl and a $-CH_2OH$ group. The $-CH_2OH$ group is of higher priority than methyl.

higher   (C) $\longrightarrow$ $H_3C\diagdown$ CH_2OH $\longleftarrow$ higher $-C(O,H,H)$

$C=C$

lower   (H) $\longrightarrow$ H$\diagup$ CH_3 $\longleftarrow$ lower $-C(H,H,H)$

Higher-ranked substituents are on the same side of the double bond; the configuration is Z.

Subrule

An atom that is multiply bonded to another one is considered to be replicated as a substituent on that atom.

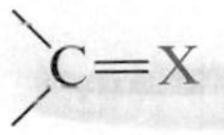

$$\text{C=X} \qquad \text{is treated as if it were} \qquad \text{C} \begin{array}{c} \text{X} \\ \text{X} \end{array}$$

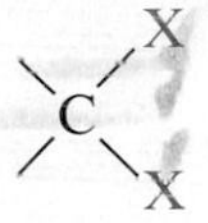

Thus, a vinyl group outranks an ethyl group.

$$-\text{CH=CH}_2 \qquad \text{is treated as} \qquad -\text{C}\overset{\text{C}}{\underset{\text{H}}{-}}\text{C}$$
  Vinyl group

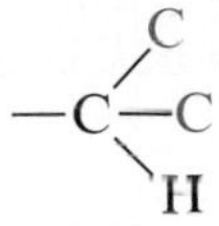

$$-\text{CH}_2\text{CH}_3 \qquad \text{is treated as} \qquad -\text{C}\overset{\text{C}}{\underset{\text{H}}{-}}\text{H}$$
  Ethyl group

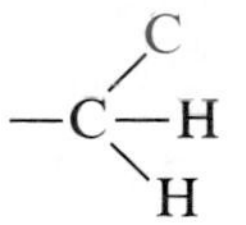

Sometimes it is necessary to evaluate the atom at the origin of the double bond in more detail in order to decide between structures of apparently equal rank. In such cases

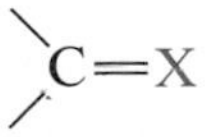

$$\text{C=X} \qquad \text{is treated as if it were} \qquad \text{C}\begin{array}{c} \text{X—C} \\ \text{X—C} \end{array}$$

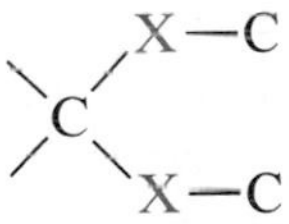

Atom X is bonded to carbon (the origin of the double bond), and this carbon atom is counted as a substituent of X. For example, a vinyl group and an isopropyl group are nominally of equal rank if only the carbons at their respective points of attachment are considered.

$$-\text{CH=CH}_2\ [-\text{C(C,C,H)}] \qquad -\text{CH(CH}_3)_2\ [-\text{C(C,C,H)}]$$
  Vinyl Isopropyl

But the situation is resolved in favor of the vinyl group once the double bond is considered in more detail. Thus

$$-\text{CH=CH}_2 \qquad \text{is treated as} \qquad -\text{C}\overset{\text{C—C}}{\underset{\text{H}}{-}}\text{C} \quad \text{C}$$
  Vinyl group

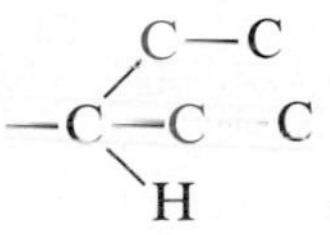

This is of higher precedence than

$$-\text{CH(CH}_3)_2 \qquad \text{is treated as} \qquad -\text{C}\overset{\text{CH}_3}{\underset{\text{H}}{-}}\text{CH}_3$$
  Isopropyl group

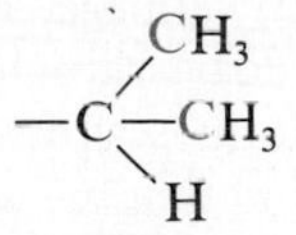

TABLE 5.3

Listing of Commonly Encountered Groups in Order of Increasing Rank in the Cahn-Ingold-Prelog System

1.	H—	14.	$CH_3O\overset{O}{\overset{\|}{C}}$—
2.	CH_3—	15.	$HSCH_2$—
3.	CH_3CH_2—	16.	H_2N—

4.	CH_3CHCH_2—
	    $\|$
	   CH_3
		17.	HO—

5.	$(CH_3)_3CCH_2$—	18.	CH_3O—
6.	$(CH_3)_2CH$—	19.	CH_3CH_2O—

7.	CH_3CH—
	    $\|$
	   CH_2CH_3
		20.	$H\overset{O}{\overset{\|}{C}}O$—

8.	$CH_2{=}CH$—	21.	$CH_3\overset{O}{\overset{\|}{C}}O$—
9.	$(CH_3)_3C$—	22.	F
10.	$HOCH_2$—	23.	HS—
11.	$H\overset{O}{\overset{\|}{C}}$—	24.	Cl
12.	$CH_3\overset{O}{\overset{\|}{C}}$—	25.	Br
13.	$HO\overset{O}{\overset{\|}{C}}$—	26.	I

Table 5.3 lists some of the more frequently encountered groups in organic chemistry in order of their precedence. The higher the number of the substituent in the table, the higher its rank.

PROBLEM 5.8 Insects use alarm pheromones to alert each other to danger. The alarm pheromone of one species of ant has the constitution shown below. The configuration of the double bond is *E*. Write a structural formula, showing stereochemistry, for this substance.

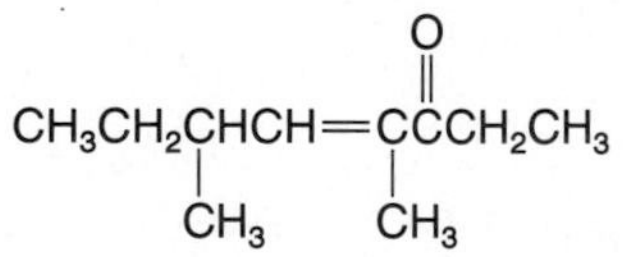

$$CH_3CH_2\underset{\underset{CH_3}{|}}{C}HCH{=}\underset{\underset{CH_3}{|}}{C}\overset{O}{\overset{\|}{C}}CH_2CH_3$$

The *E-Z* naming scheme for alkene stereochemistry was derived from a notational system developed by R. S. Cahn of the Chemical Society, London, Sir Christopher Ingold of University College, London, and Vladimir Prelog of the Swiss Federal Institute of Technology, Zürich. Their system of ranking substituents according to atomic number priority was developed to describe another aspect of organic stereochemistry of the kind that we will examine in Chapter 8. Cahn, Ingold, and Prelog have called their system of ranking substituents the *sequence rule*. Most organic chemists call it the Cahn-Ingold-Prelog convention or, more familiarly, the CIP system.

5.7 RELATIVE STABILITIES OF ALKENES

We have described on a number of earlier occasions how heats of combustion are used to evaluate relative stabilities of isomeric substances. This technique has been applied to isomeric alkenes and reveals the ways in which alkene structure affects internal energy. Alkenes have the molecular formula C_nH_{2n} and, like alkanes and cycloalkanes, burn in air to yield carbon dioxide and water.

$$C_nH_{2n} + \frac{3n}{2}\,O_2 \longrightarrow nCO_2 + nH_2O$$

Alkene   Oxygen   Carbon dioxide   Water

If two alkenes have the same molecular formula but differ in structure, the difference in their heats of combustion reveals which is the more stable isomer and by how much their energies differ from each other.

Consider, for example, the heats of combustion of the isomeric C_4H_8 alkenes. All these compounds undergo combustion according to the equation

$$C_4H_8 + 6O_2 \longrightarrow 4CO_2 + 4H_2O$$

The amount of heat evolved, however, is different for each isomer.

$CH_2 = CHCH_2CH_3$ Heat of combustion 649.3 kcal/mol   *— highest energy*
1-Butene   *— least stable*

cis-2-Butene   Heat of combustion 647.6 kcal/mol

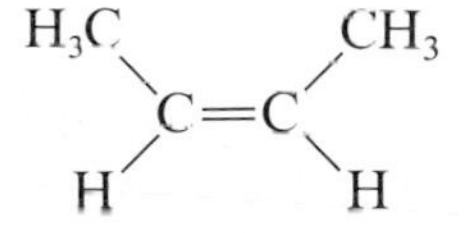

trans-2-Butene   Heat of combustion 646.9 kcal/mol

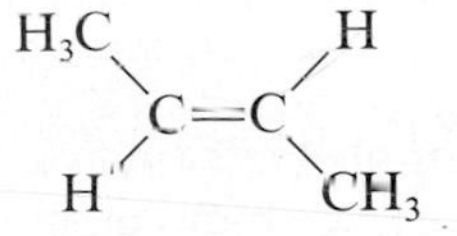

2-Methylpropene   Heat of combustion 645.2 kcal/mol   *· least energy most stable*

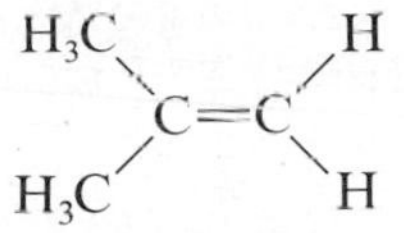

These heats of combustion are plotted on a common scale in Figure 5.6 and reveal the relative energies of the various isomers. The isomer of highest energy, i.e., the least stable one, is 1-butene. The isomer of lowest energy, the most stable one, is 2-methylpropene.

By analyzing the heats of combustion of the C_4H_8 isomers cited above, along with similar data for numerous other alkenes, some of the factors that influence the effects

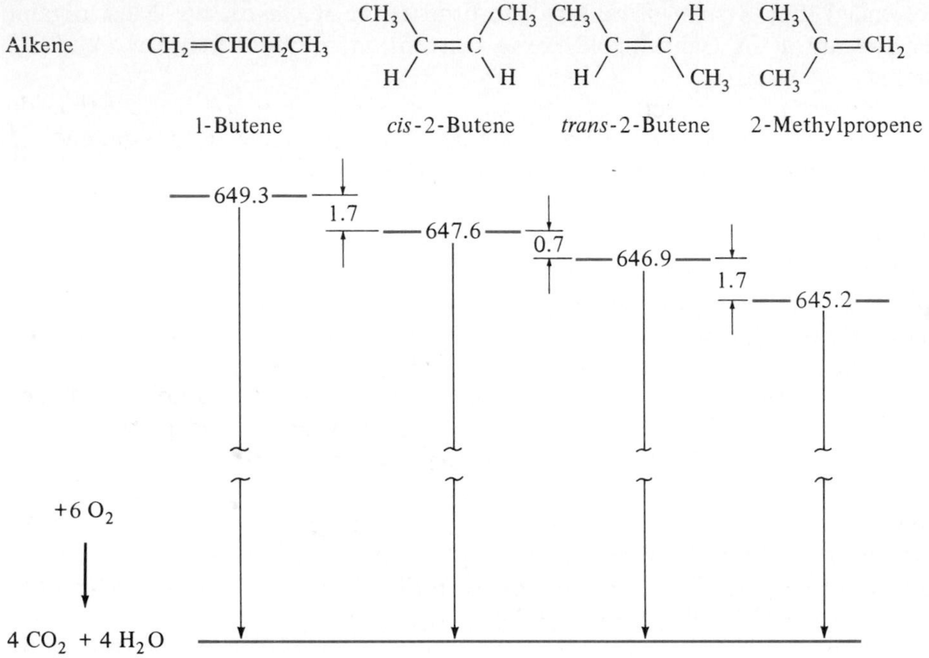

Alkene $\quad CH_2{=}CHCH_2CH_3$

1-Butene $\qquad$ *cis*-2-Butene $\qquad$ *trans*-2-Butene $\qquad$ 2-Methylpropene

FIGURE 5.6 Heats of combustion of C_4H_8 alkene isomers plotted on a common scale. All energies are in kilocalories per mole.

of substituents on the relative stabilities of alkenes can be assessed. In general, there is an observed trend toward greater stability: (1) when large substituents are trans to each other, and (2) when alkyl groups rather than hydrogen atoms are directly attached to the double bonds.

Alkenes are more stable when large substituents are trans to each other than when they are cis. As noted above and in Figure 5.6, *trans*-2-butene has a lower heat of combustion and is more stable than *cis*-2-butene.

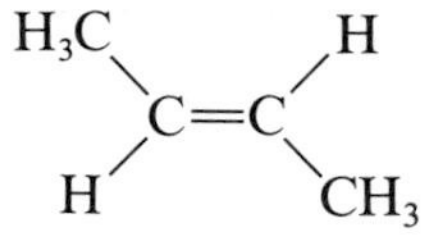

trans-2-Butene
(heat of combustion
646.9 kcal/mol)

is more stable than

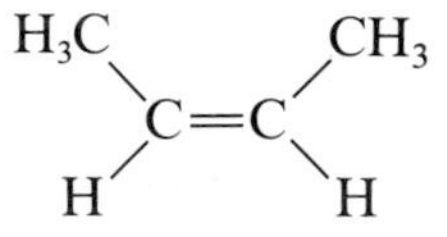

cis-2-Butene
(heat of combustion
647.6 kcal/mol)

The energy difference between the two is 0.7 kcal/mol. The principal contributor to this energy difference is the van der Waals repulsive force between methyl groups on the same side of the double bond in the cis isomer. The difference in stability between stereoisomeric alkenes is more pronounced with larger alkyl groups. A particularly dramatic example is evident on comparing (*E*)- and (*Z*)-2,2,5,5-tetramethyl-3-hexene.

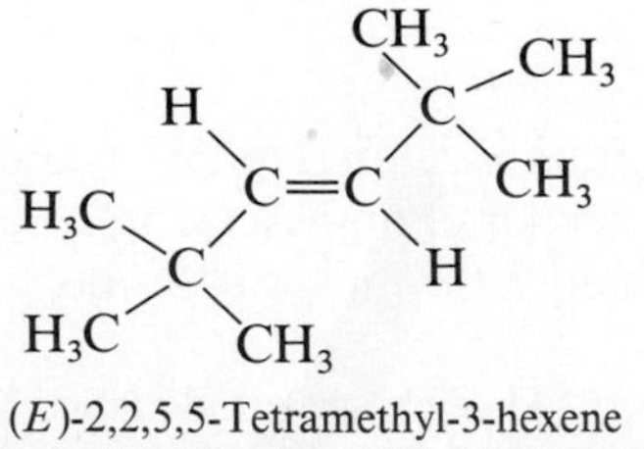

(*E*)-2,2,5,5-Tetramethyl-3-hexene
(heat of combustion 1574.1 kcal/mol)

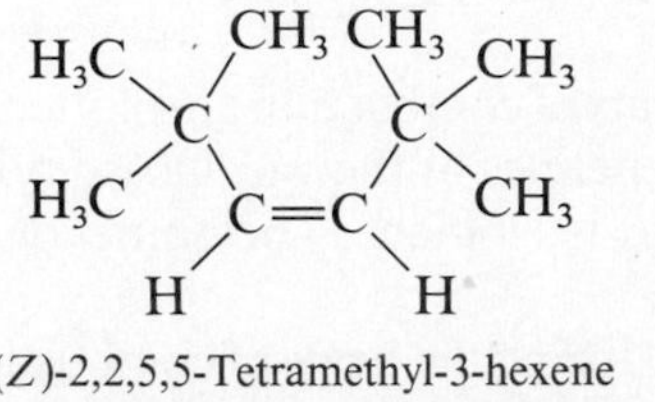

(*Z*)-2,2,5,5-Tetramethyl-3-hexene
(heat of combustion 1584.6 kcal/mol)

The Z isomer, in which the bulky *tert*-butyl groups are cis to each other, is destabilized by the large van der Waals repulsion between them, and its energy exceeds that of the trans stereoisomer by 10.5 kcal/mol.

Alkyl substituents directly attached to a double bond stabilize it more than do hydrogen substituents. For example, *cis*-2-butene is more stable than 1-butene:

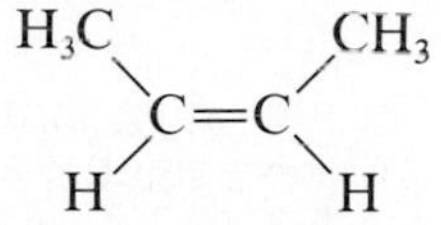

H_3C, CH_3, C=C, H, H
is more stable than
H, CH_2CH_3, C=C, H, H

cis-2-Butene
(heat of combustion 647.6 kcal/mol)

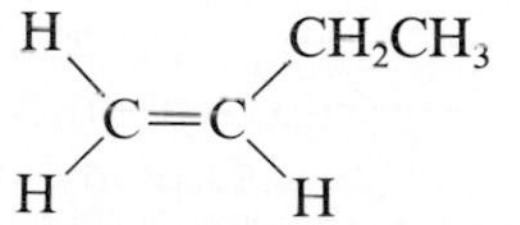

1-Butene
(heat of combustion 649.3 kcal/mol)

1-Butene is described as having a *terminal* or *monosubstituted* double bond. The double bond in 1-butene bears one alkyl group, an ethyl group, and three hydrogen substituents. The double bond in *cis*-2-butene is *disubstituted;* it bears two methyl groups.

Similarly, a *trisubstituted* double bond is normally more stable than a disubstituted one, and a *tetrasubstituted* double bond is more stable than one that is trisubstituted. Among the C_6H_{12} isomers the tetrasubstituted alkene 2,3-dimethyl-2-butene has the lowest heat of combustion and is the most stable.

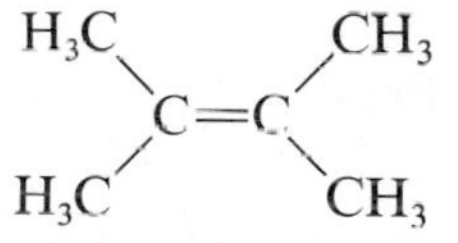

H_3C, CH_3, C=C, H_3C, CH_3
has a tetrasubstituted double bond; is the most stable of the C_6H_{12} alkenes

2,3-Dimethyl-2-butene

PROBLEM 5.9 Write structural formulas and give the IUPAC names for all of the alkenes of molecular formula C_6H_{12} that contain a trisubstituted double bond.

Alkyl groups stabilize double bonds in much the same way that they stabilize carbocations, by releasing electrons to an sp^2 hybridized carbon. The increased s character in the sp^2 hybridized carbon of a carbon-carbon double bond makes it more electron-withdrawing than the sp^3 hybridized carbon of an alkyl group to which it is attached. A shift of electron density in the σ bond that connects the alkyl substituent to the doubly bonded carbon helps to stabilize the alkene. Figure 5.7 illustrates the effect of an alkyl substituent on the stability of a carbon-carbon double bond.

PROBLEM 5.10 Arrange the following alkenes in order of decreasing stability: 1-pentene; (E)-2-pentene; (Z)-2-pentene; 2-methyl-2-butene.

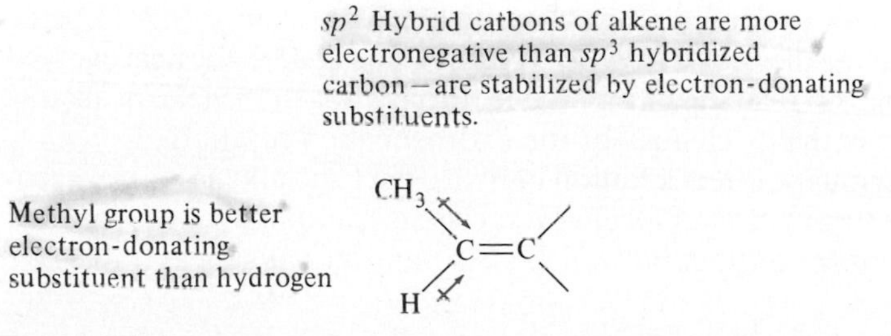

sp^2 Hybrid carbons of alkene are more electronegative than sp^3 hybridized carbon — are stabilized by electron-donating substituents.

Methyl group is better electron-donating substituent than hydrogen

CH_3, C=C, H

FIGURE 5.7 Alkyl groups stabilize carbon-carbon double bonds by donating electrons to sp^2 hybridized carbons.

In discussing the synthesis of alkenes in Chapter 6, we will see that the relative stabilities of isomeric alkenes play an important role in the selection of methods for their preparation.

5.8 STERIC EFFECTS AND ELECTRONIC EFFECTS

The relative stabilities of isomeric alkenes were analyzed in the preceding section in terms of a combination of two independent effects. In one of these van der Waals repulsions between large substituents were cited as increasing the energy of cis alkenes compared with their trans stereoisomers. The other effect concerned the greater electron-releasing power of an alkyl group compared with that of a hydrogen substituent on a double bond.

An effect that results from two or more atoms or groups being close enough in space so that a van der Waals interaction between them becomes energetically significant is called a *steric effect*. The greater stability of trans alkenes compared with their cis isomers is an example of a steric effect. An effect that results from two or more atoms or groups interacting so as to alter the electron distribution in a system is called an *electronic effect*. The increased stability of more highly substituted alkenes is an example of an electronic effect.

5.9 HYDROGENATION OF ALKENES

The characteristic reaction type exhibited by alkenes is *addition* to the double bond.

$$\diagup\!\!\!C\!\!=\!\!C\diagdown + \; X\!\!-\!\!Y \longrightarrow \quad X\!\!-\!\!\overset{|}{\underset{|}{C}}\!\!-\!\!\overset{|}{\underset{|}{C}}\!\!-\!\!Y$$

Alkene   Product of addition of X—Y
                              to the double bond

An example of this reaction is the addition of a molecule of hydrogen to the double bond of ethylene to form ethane. This process is known as *hydrogenation*.

$$CH_2\!\!=\!\!CH_2 + \quad H_2 \quad \longrightarrow CH_3CH_3$$

Ethylene   Hydrogen   Ethane

Addition of hydrogen to the double bond of an alkene leads to the formation of two new σ bonds in the product at the expense of the π component of the alkene double bond and the σ bond of H_2. Since π bonds tend to be weaker than σ bonds, the bonds in the product are stronger than the bonds in the starting state, and the enthalpy change is favorable. Hydrogenation of ethylene, like many other addition reactions of alkenes, is an exothermic reaction, its enthalpy change ($\Delta H°$) being -32.6 kcal/mol. We define the *heat of hydrogenation* of an alkene as the heat evolved (in kilocalories per mole) on reaction of the alkene with hydrogen to yield an alkane; it is the negative of the enthalpy change for the hydrogenation reaction.

In spite of its exothermicity, direct addition of hydrogen to an alkene is an exceedingly slow process. Hydrogenation of alkenes is a classic example of a reaction that has a favorable equilibrium constant but which does not occur at a measurable rate because its energy of activation is very high. The rate of hydrogenation is dramatically increased, however, by carrying out the reaction in the presence of certain finely divided metals. Platinum is the one most often used, although palladium, rhodium, and nickel are also effective. The metal acts as a hydrogenation catalyst; it does not alter the heat of hydrogenation and is not consumed in the reaction but only speeds its rate. Metal-catalyzed addition of hydrogen is normally rapid at room temperature, and the alkane is produced in high yield, usually as the only product.

$$(CH_3)_2C=CHCH_3 + \quad H_2 \quad \xrightarrow{Pt} (CH_3)_2CHCH_2CH_3$$

2-Methyl-2-butene   Hydrogen   2-Methylbutane (100%)

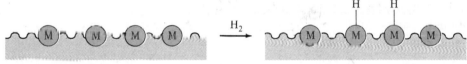

5,5-Dimethyl(methylene)cyclononane   Hydrogen   1,1,5-Trimethylcyclononane (73%)

PROBLEM 5.11   What three alkenes yield 2-methylbutane on catalytic hydrogenation?

Step 1: The catalyst activates the hydrogen by binding hydrogen atoms through metal-hydrogen bonds at reactive sites on the metal surface:

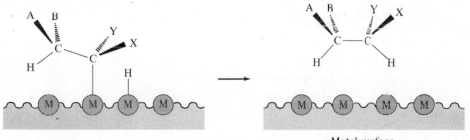

Metal surface

Step 2: A hydrogen atom is transferred from the catalyst surface to one of the carbons of the double bond—the other carbon becomes bound to the metal:

Step 3: The second hydrogen atom is transferred, forming the alkane and regenerating the catalyst:

Metal surface

FIGURE 5.8 A mechanistic representation of heterogeneous catalysis in the hydrogenation of alkenes.

not in 2 steps ↓ all at once

A mechanistic description that shows the role of the metal catalyst is presented in Figure 5.8. According to this mechanism, catalytic hydrogenation of an alkene proceeds by a series of steps. In the first step molecular hydrogen reacts with metal atoms on the catalyst surface. The hydrogen-hydrogen σ bond is broken, and two metal-hydrogen covalent bonds are formed. One of these metal-bound hydrogen atoms is then transferred to the alkene, forming a C—H σ bond. The other carbon atom of the original double bond becomes bonded to the metal. Subsequently, this metal-carbon bond is broken by transfer of a second hydrogen from the metal surface to carbon. The alkane that is produced leaves the metal surface, returning the catalyst to its original state, where it is ready to repeat the process.

5.10 HEATS OF HYDROGENATION

Heats of hydrogenation may be used to assess the relative stabilities of alkenes in much the same way as we use heats of combustion. Catalytic hydrogenation of 1-butene, *cis*-2-butene, or *trans*-2-butene yields the same product—butane. As Figure 5.9 shows, the measured heats of hydrogenation reveal that *trans*-2-butene is 1.0 kcal/mol lower in energy than *cis*-2-butene and that *cis*-2-butene is 1.7 kcal/mol lower in energy than 1-butene.

The energy differences between these isomeric alkenes as measured by their heats of hydrogenation are, within experimental error, equal to the differences in their respective heats of combustion. Both sets of data—heat of hydrogenation and heat of combustion—permit the energies of isomeric compounds to be compared by measuring the heat evolved when they are converted to a product or products common to them all. Heats of hydrogenation are also used to estimate differences in double bond stability between alkenes that are not isomers of each other. By assuming that the

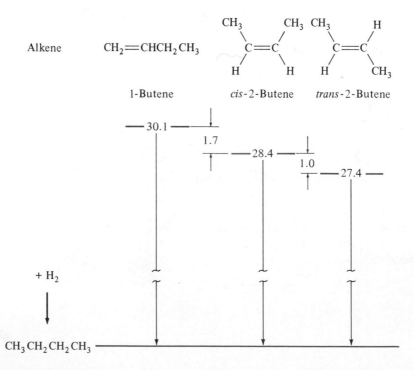

FIGURE 5.9 Heats of hydrogenation of butene isomers plotted on a common scale. All energies are in kilocalories per mole.

TABLE 5.1

Heats of Hydrogenation of Some Alkenes

Alkene	Structure	Heat of hydrogenation, kcal/mol
Ethylene	$CH_2{=}CH_2$	32.6
Monosubstituted alkenes		
Propene	$CH_2{=}CHCH_3$	29.9
1-Butene	$CH_2{=}CHCH_2CH_3$	30.1
1-Hexene	$CH_2{=}CHCH_2CH_2CH_2CH_3$	30.2
cis-Disubstituted alkenes		
cis-2-Butene	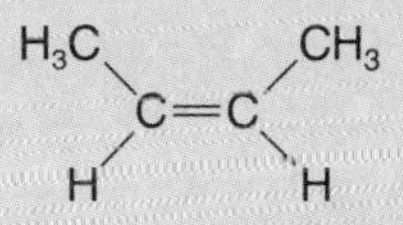 $\begin{array}{c}H_3CCH_3\\ \diagdown\diagup\\ C{=}C\\ \diagup\diagdown\\ HH\end{array}$	28.4
cis-2-Pentene	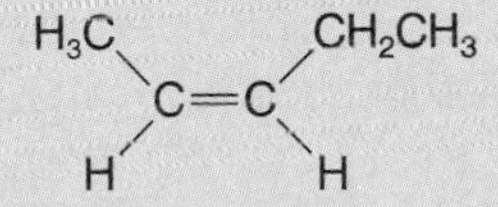 $\begin{array}{c}H_3CCH_2CH_3\\ \diagdown\diagup\\ C{=}C\\ \diagup\diagdown\\ HH\end{array}$	28.1
trans-Disubstituted alkenes		
trans-2-Butene	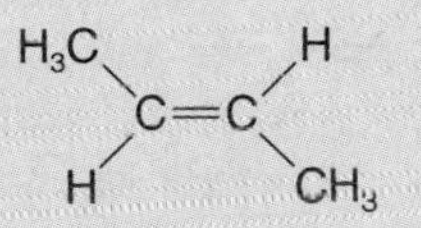 $\begin{array}{c}H_3CH\\ \diagdown\diagup\\ C{=}C\\ \diagup\diagdown\\ HCH_3\end{array}$	27.4
trans-2-Pentene	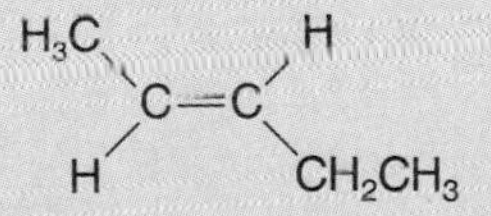 $\begin{array}{c}H_3CH\\ \diagdown\diagup\\ C{=}C\\ \diagup\diagdown\\ HCH_2CH_3\end{array}$	27.2
Terminally disubstituted alkenes		
2-Methylpropene	$CH_2{=}C(CH_3)_2$	28.1
2,3-Dimethyl-1-butene	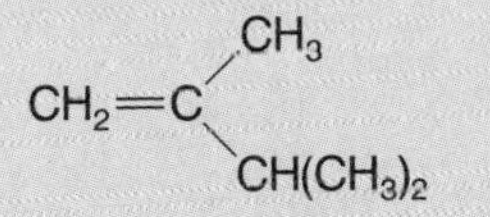 $CH_2{=}C\begin{array}{l}{}^{CH_3}\\ {}_{CH(CH_3)_2}\end{array}$	27.8
2,4,4-Trimethyl-1-pentene	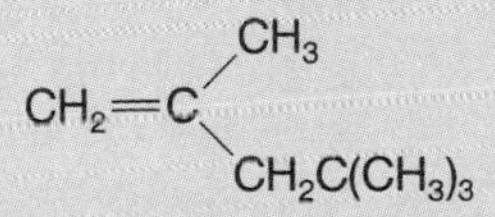 $CH_2{=}C\begin{array}{l}{}^{CH_3}\\ {}_{CH_2C(CH_3)_3}\end{array}$	27.0
Trisubstituted alkenes		
2-Methyl-2-pentene	$(CH_3)_2C{=}CHCH_2CH_3$	26.7
Tetrasubstituted alkenes		
2,3-Dimethyl-2-butene	$(CH_3)_2C{=}C(CH_3)_2$	26.4

heats of hydrogenation are little affected by the structure of the resulting alkanes, it becomes possible to compare a wider range of alkenes with each other. Table 5.4 lists the heats of hydrogenation for a representative group of alkenes.

Table 5.4 reveals a dependence of heat of hydrogenation on structure that is consistent with the generalizations derived from heat of combustion measurements.

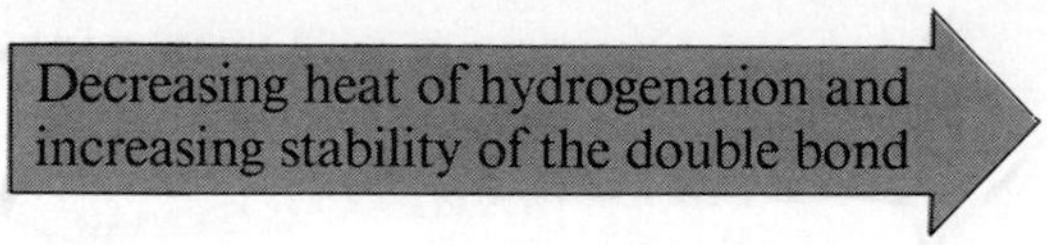

Decreasing heat of hydrogenation and increasing stability of the double bond

$CH_2{=}CH_2$	$RCH{=}CH_2$	$RCH{=}CHR$	$R_2C{=}CHR$	$R_2C{=}CR_2$
Ethylene	Monosubstituted	Disubstituted	Trisubstituted	Tetrasubstituted

Alkyl substitution lowers the heat of hydrogenation, indicating a stabilization of the double bond. Among disubstituted alkenes trans double bonds are more stable than cis, while terminally disubstituted alkenes of the type $R_2C{=}CH_2$ may have heats of hydrogenation that are larger or smaller than either cis- or trans-disubstituted alkenes, depending on the substituents.

5.11 CYCLOALKENES

Double bonds are accommodated by rings of all sizes. The simplest cycloalkene is cyclopropene. Compounds containing cyclopropene rings occur naturally; sterculic acid, for example, is a naturally occurring cyclopropene obtained from the seed oil of a number of plants.

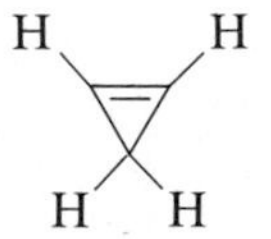

Cyclopropene
(heat of hydrogenation
54 kcal/mol)

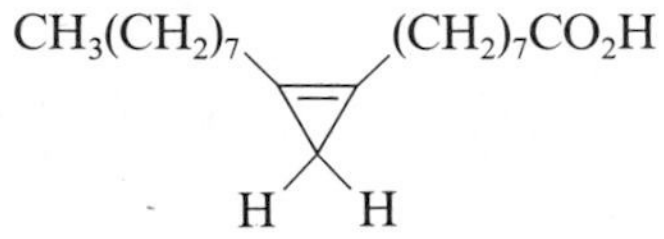

Sterculic acid

Cyclopropene is a highly strained compound, as evidenced by a heat of hydrogenation that is over 20 kcal/mol greater than that of ethylene. The angle strain that results when sp^2 hybridized carbons are incorporated into a three-membered ring is even greater than that for the sp^3 hybridized carbons of cyclopropane.

As indicated by heats of hydrogenation that are similar to those of noncyclic disubstituted alkenes, cyclopentene and cyclohexene contain double bonds that are not appreciably strained.

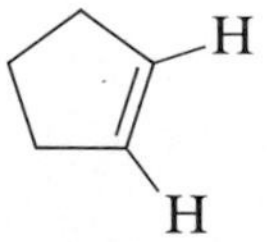

Cyclopentene
(heat of hydrogenation
26.2 kcal/mol)

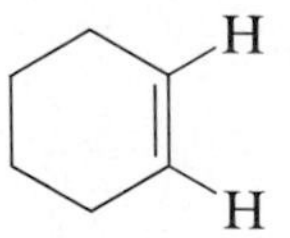

Cyclohexene
(heat of hydrogenation
26.9 kcal/mol)

So far we have shown all cycloalkenes as structures in which the double bonds are of the cis configuration. If the ring is large enough, however, a trans configuration of the double bond is also possible.

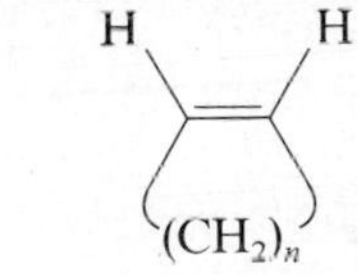

A *cis* (or Z) cycloalkene   A *trans* (or E) cycloalkene

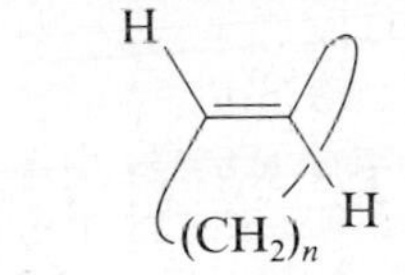

usually trans is more stable than cis, but not here, because of the big strain to bend

The smallest trans cycloalkene that is stable enough to be isolated and stored in a normal way is *trans*-cyclooctene. As measured by their heats of hydrogenation, *cis*-cyclooctene is 9.2 kcal/mol more stable than *trans*-cyclooctene.

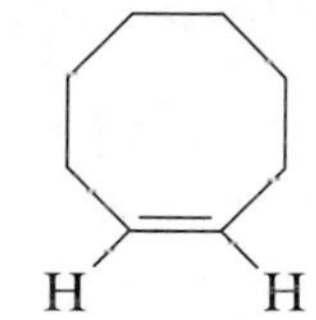

(Z)-Cyclooctene
(*cis*-cyclooctene)
(heat of hydrogenation
23.0 kcal/mol)

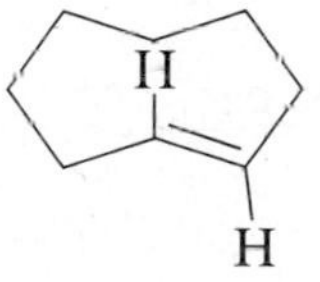

(E)-Cyclooctene
(*trans*-cyclooctene)
(heat of hydrogenation
32.3 kcal/mol)

trans-Cycloheptene has been trapped as an intermediate in certain reactions but is too reactive to be isolated and stored. Evidence has also been presented for the transitory existence of the even more strained *trans*-cyclohexene.

PROBLEM 5.12   Place a double bond in the carbon skeleton shown below so as to represent

(a) (Z)-1-Methylcyclodecene   (d) (E)-3-Methylcyclodecene
(b) (E)-1-Methylcyclodecene   (e) (Z)-5-Methylcyclodecene
(c) (Z)-3-Methylcyclodecene   (f) (E)-5-Methylcyclodecene

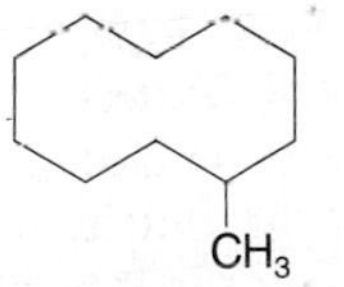

CH₃

SAMPLE SOLUTION   (a) and (b)   Since the methyl group must be at C-1, there are only two possible places to put the double bond.

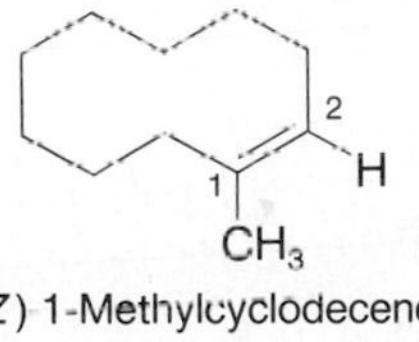

(Z)-1-Methylcyclodecene 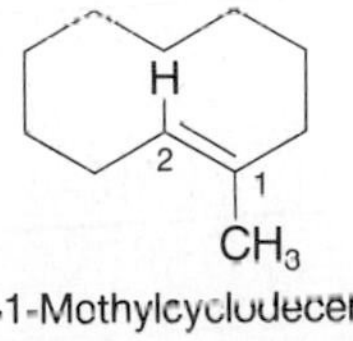   (E)-1-Methylcyclodecene

In the *Z* stereoisomer the two lower-priority substituents—the methyl group and the hydrogen—are on the same side of the double bond. In the *E* stereoisomer these substituents are on opposite sides of the double bond. The ring carbons are the higher-ranking substituents at each end of the double bond.

Because larger rings have more methylene groups with which to span the ends of a double bond, the strain associated with a trans cycloalkene decreases with increasing ring size. For example, the heats of hydrogenation of *cis*- and *trans*-cyclododecene are about the same, indicating that the two double bonds are of comparable stability.

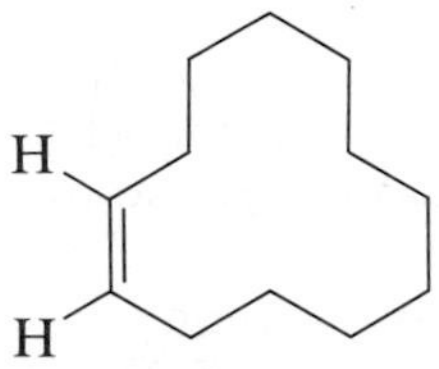

(*Z*)-Cyclododecene
(*cis*-cyclododecene)
(heat of hydrogenation
26.3 kcal/mol)

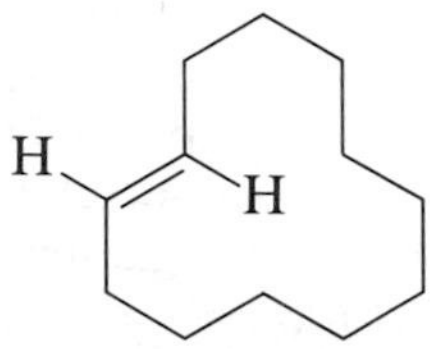

(*E*)-Cyclododecene
(*trans*-cyclododecene)
(heat of hydrogenation
26.8 kcal/mol)

When the rings are larger than 12-membered, trans cycloalkenes are more stable than cis. In these cases, the ring is large enough and flexible enough so that it is energetically similar to a noncyclic alkene. As in noncyclic cis alkenes, a van der Waals repulsion between carbons on the same side of the double bond destabilizes a cis cycloalkene.

here, the trans finally becomes more stable than cis

5.12 STEREOCHEMISTRY OF HYDROGENATION OF CYCLOALKENES

In our model for alkene hydrogenation shown in Figure 5.8, hydrogen atoms are transferred from the catalyst surface to the alkene. While the two hydrogens are not transferred simultaneously, there is a pronounced tendency for them to add to the same face of the double bond, as shown in the following example:

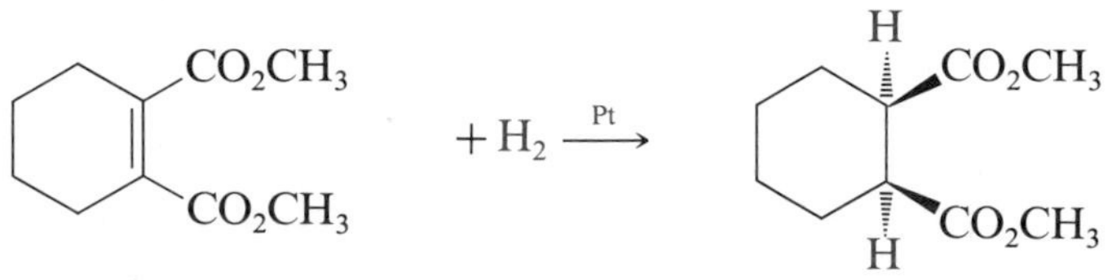

1,2-Dicarbomethoxy-1-cyclohexene

cis-1,2-Dicarbomethoxycyclohexane (100%)

When two atoms or groups add to the same face of a double bond, the process is referred to as *syn* addition.

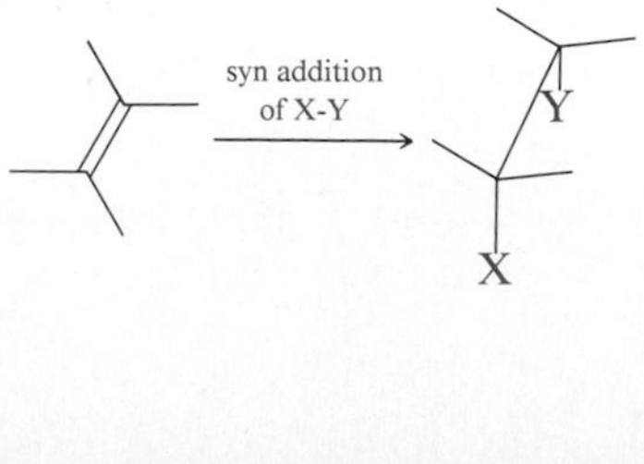

When atoms or groups add to opposite faces of the double bond, the process is referred to as *anti* addition.

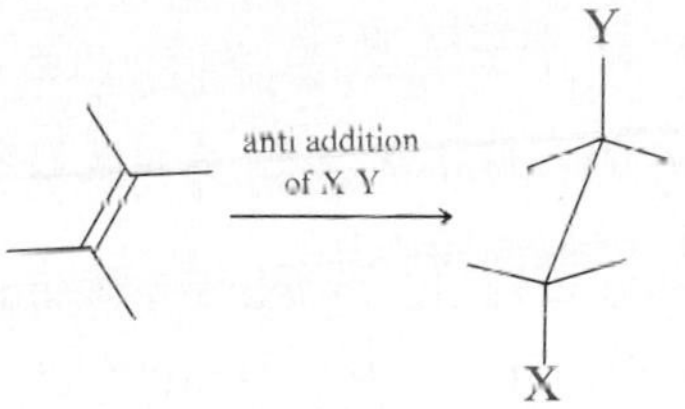

The terms *syn* and *anti* describe the stereochemical course of the addition reaction. Catalytic hydrogenation of alkenes proceeds by syn addition of hydrogen.

A second stereochemical aspect of alkene hydrogenation concerns its *stereoselectivity*. A reaction in which a single starting material can give two or more stereoisomeric products but yields one of them in greater amounts than the other (or even to the exclusion of the other) is said to be *stereoselective*. The catalytic hydrogenation of α-pinene (a constituent of turpentine) is a stereoselective reaction and can be used to illustrate this principle. Syn addition of hydrogen can in principle lead to either *cis*-pinane or *trans*-pinane depending on which face of the double bond accepts the hydrogen atoms (shown in color in the equation).

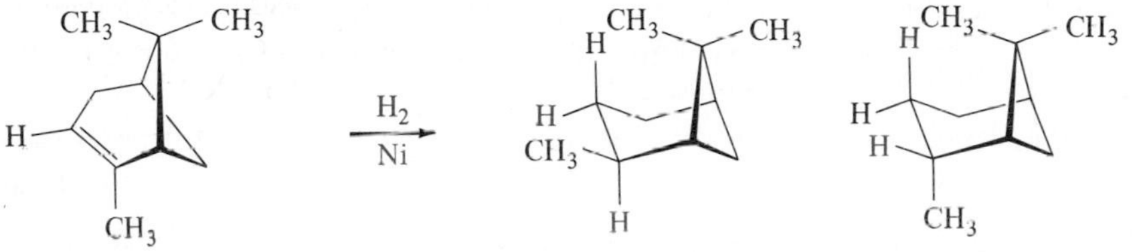

α-Pinene   *cis*-Pinane (only product)   *trans*-Pinane (not formed)

This hydrogenation is highly stereoselective, since the only product obtained is *cis*-pinane, none of the stereoisomeric *trans*-pinane being formed.

The stereoselectivity of this reaction is governed by the manner in which the alkene approaches the catalyst surface. As Figure 5.10 shows, one of the methyl groups on the bridge carbon lies directly over the double bond and blocks that face from easy access to the catalyst. The bottom face of the double bond is more exposed, and hydrogen is transferred from the catalyst to that face; this is the direction of hydrogen transfer that produces *cis*-pinane.

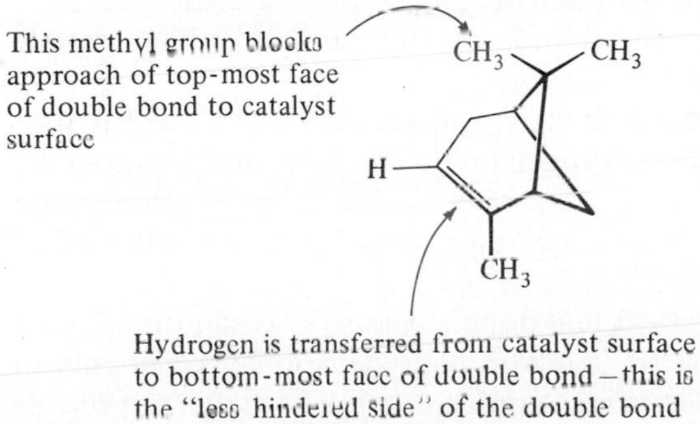

This methyl group blocks approach of top-most face of double bond to catalyst surface

Hydrogen is transferred from catalyst surface to bottom-most face of double bond — this is the "less hindered side" of the double bond

FIGURE 5.10 The methyl group that lies over the double bond of α-pinene shields one face of it, preventing a close approach to the surface of the catalyst. Hydrogenation of α-pinene occurs preferentially from the bottom face of the double bond.

Hydrogenation occurs by addition to the "less-hindered" side of an alkene. Reactions that take place at the less-hindered side of a reactant are commonplace in organic chemistry and are examples of steric effects on chemical reactivity.

5.13 MOLECULAR FORMULA AS A CLUE TO STRUCTURE

Chemists are often confronted with the problem of identifying the structure of an unknown compound. Sometimes the unknown can be shown to be a sample of a compound previously reported in the chemical literature; on other occasions, the unknown may be truly that — a compound never before encountered by anyone. An arsenal of powerful techniques is available to simplify the task of structure determination, and we will discuss some of the most important of these in Chapter 14. It should be pointed out, however, that the molecular formula of a molecule provides more information than might be apparent at first glance.

Consider, for example, a substance with the molecular formula C_7H_{16}. We know immediately that the compound is an alkane because its molecular formula corresponds to the general formula for that class of compounds, C_nH_{2n+2}, where $n = 7$.

What about a substance with the molecular formula C_7H_{14}? This compound cannot be an alkane but may be either a cycloalkane or an alkene because both these classes of hydrocarbons correspond to the general molecular formula C_nH_{2n}. Any time a ring or a double bond is present in an organic molecule, its molecular formula has two fewer hydrogen atoms than that of an alkane with the same number of carbons.

Various names have been given to this relationship between molecular formulas and classes of hydrocarbons. It is sometimes referred to as an *index of hydrogen deficiency* and sometimes as *elements of unsaturation* or *sites of unsaturation*. We will use the term *sum of double bonds and rings* because it is more descriptive than the others. In the interests of economy of space, however, the full term sum of double bonds and rings will be abbreviated as SODAR.

$$\text{Sum of double bonds and rings (SODAR)} = \tfrac{1}{2}[C_nH_{2n+2} - C_nH_x]$$

where C_nH_x is the molecular formula of the compound.

Thus, a molecule that has a molecular formula of C_7H_{14} has a SODAR of 1.

$$SODAR = \tfrac{1}{2}[C_7H_{16} - C_7H_{14}]$$
$$SODAR = \tfrac{1}{2}[2] = 1$$

Thus, the structure has one ring or one double bond. A molecule of molecular formula C_7H_{12} has four fewer hydrogens than the corresponding alkane; it has a SODAR of 2 and can have two rings, two double bonds, one ring and one double bond, or one triple bond.

A halogen substituent, like hydrogen, is monovalent, and when present in a molecular formula is treated as if it were hydrogen for the purposes of computing the number of double bonds and rings. Oxygen atoms have no effect on the relationship between molecular formulas, double bonds, and rings. They are ignored when computing the SODAR of a substance.

How does one distinguish between rings and double bonds? This additional piece of structural information is revealed by catalytic hydrogenation experiments in which the amount of hydrogen that reacts is measured exactly. Each of a molecule's double bonds consumes one molar equivalent of hydrogen, while rings do not undergo hydrogenation. A substance with a SODAR of 5 that takes up 3 moles of hydrogen must therefore have two rings.

PROBLEM 5.13   How many rings are present in each of the following compounds? Each consumes 2 moles of hydrogen on catalytic hydrogenation.

(a) $C_{10}H_{18}$ (d) C_8H_8O
(b) C_8H_8 (e) $C_8H_{10}O_2$
(c) $C_8H_8Cl_2$ (f) C_8H_9ClO

SAMPLE SOLUTION   (a)   The molecular formula $C_{10}H_{18}$ contains four fewer hydrogens than the alkane having the same number of carbon atoms ($C_{10}H_{22}$). Therefore, the SODAR of this compound is 2. Since it consumes two molar equivalents of hydrogen on catalytic hydrogenation, it must have two double bonds and no rings.

5.14 SUMMARY

Alkenes and cycloalkenes contain carbon-carbon double bonds. These double bonds unite two sp^2 hybridized carbon atoms and are made up of a σ *component* and a π *component*. The σ bond results from overlap of an sp^2 hybrid orbital on each carbon. The π bond is weaker than the σ bond and results from a side-by-side overlap of *p* orbitals.

Representations of bonding in ethylene

The sizable barrier to rotation about a carbon-carbon double bond of 60 kcal/mol corresponds to the energy required to break the π component of the double bond. Stereoisomeric alkenes, such as *cis*- and *trans*-2-butene, do not interconvert readily —they are configurationally stable under normal conditions.

The *configurations* of stereoisomeric alkenes are described according to two notational systems. The simpler system adds the prefix *cis* to the name of the alkene when similar substituents are on the same side of the double bond and the prefix *trans* when they are on opposite sides. An alternative system takes the subjectivity out of deciding when substituents are similar by ranking substituents according to a system of rules based on atomic number. The prefix *Z* is used for alkenes that have higher-priority substituents on the same side of the double bond; the prefix *E* is used when higher-priority substituents are on opposite sides.

H_3C⟍   ⟋CH_2CH_3 H_3C⟍   ⟋H

    C=C   C=C

H⟋   ⟍H   H⟋   ⟍CH_2CH_3

cis-2-Pentene   *trans*-2-Pentene
(Z)-2-Pentene   (E)-2-Pentene

Alkyl substitution tends to stabilize a double bond. The general order of alkene stability is

1. Tetrasubstituted alkenes ($R_2C=CR_2$) are the most stable.
2. Trisubstituted alkenes ($R_2C=CHR$) are next.
3. Among disubstituted alkenes, *trans*-RCH=CHR is normally more stable than *cis*-RCH=CHR. Exceptions are cycloalkenes, cis cycloalkenes being more stable than trans when the ring contains fewer than 11 carbons. Terminally disubstituted alkenes ($R_2C=CH_2$) may be slightly more or less stable than RCH=CHR, depending on their substituents.
4. Monosubstituted alkenes (RCH=CH$_2$) have a more stabilized double bond than ethylene (unsubstituted) but are less stable than disubstituted alkenes.

Alkyl groups tend to release electrons to sp^2 hybridized carbon and they stabilize double bonds by an *electronic effect*. Repulsive van der Waals interactions between cis substituents destabilize double bonds by a *steric effect*.

Catalytic *hydrogenation* converts alkenes to alkanes. Hydrogen adds to the double bond.

$$R_2C=CR_2 + \quad H_2 \quad \longrightarrow R_2CHCHR_2$$

$$\text{Alkene} \qquad \text{Hydrogen} \qquad \text{Alkane}$$

Catalysts for this reaction are finely divided metals such as platinum, palladium, nickel, and rhodium. Both hydrogen atoms add to the same face of the double bond (*syn* addition). When the two faces of the double bond are not equivalent, the reaction is stereoselective and hydrogen addition occurs at the less hindered face.

Hydrogenation of an alkene is an exothermic reaction, and the amount of heat evolved is called the *heat of hydrogenation* of the alkene. Heats of hydrogenation are used to estimate the relative stabilities of double bond types in much the same way that heats of combustion are.

PROBLEMS

5.14 Write structural formulas and give an acceptable IUPAC name for all the hydrocarbons of molecular formula C_5H_{10}.

5.15 Write structural formulas for each of the following, clearly showing the stereochemistry around the double bond where indicated:

(a) 1-Heptene
(b) 3-Ethyl-1-pentene
(c) 3-Isopropyl-2-methyl-2-hexane
(d) *cis*-3-Octene
(e) *trans*-2-Hexene
(f) (*Z*)-3-Methyl-2-hexene
(g) (*E*)-3-Chloro-2-hexene
(h) 1-Bromocyclohexene
(i) 1,3-Dibromocyclohexene
(j) 1,6-Dibromocyclohexene
(k) Vinylcycloheptane
(l) 1,1-Diallylcyclopropane
(m) *trans*-1-Isopropenyl-3-methylcyclohexane

5.16 Give the IUPAC names for each of the following compounds:

(a) $(CH_3CH_2)_2C=CHCH_3$

(b) $(CH_3CH_2)_2C=C(CH_2CH_3)_2$

(c) $(CH_3)_3CCH=CCl_2$

(d)

(e)

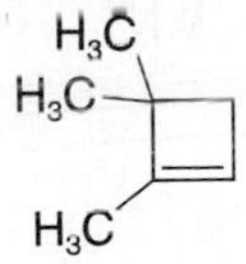

(f)

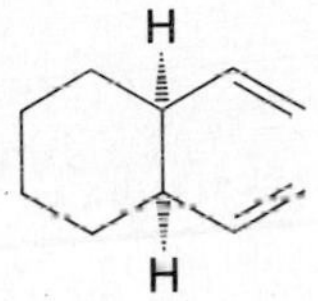

(g)

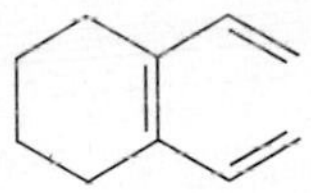

5.17 (a) A hydrocarbon isolated from fish oil and from plankton was identified as 2,6,10,14-tetramethyl-2-pentadecene. Write its structure.

(b) The sex attractant of the Mediterranean fruit fly is (E)-6-nonen-1-ol. Write a structural formula for this compound, showing the stereochemistry of the double bond.

(c) Geraniol is a naturally occurring substance present in the fragrant oil of many plants. It has a pleasing, roselike odor. Geraniol is the E isomer of

$$(CH_3)_2C=CHCH_2CH_2\underset{\underset{CH_3}{|}}{C}=CHCH_2OH$$

Write a structural formula for geraniol, showing its stereochemistry.

(d) Nerol is a naturally occurring substance that is a stereoisomer of geraniol. Write its structure.

(e) The sex attractant of the codling moth is the 2Z, 6E stereoisomer of

$$CH_3CH_2CH_2\underset{\underset{CH_3}{|}}{C}=CHCH_2CH_2\underset{\underset{CH_2CH_3}{|}}{C}=CHCH_2OH$$

Write the structure of this substance in a way that clearly shows its stereochemistry.

(f) The sex pheromone of the honeybee is the E stereoisomer of the compound shown. Write a structural formula for this compound.

$$CH_3\overset{\overset{O}{\|}}{C}(CH_2)_4CH_2CH=CHCO_2H$$

(g) A growth hormone from the cecropia moth has the structure shown. Express the stereochemistry of the double bonds according to the E-Z system.

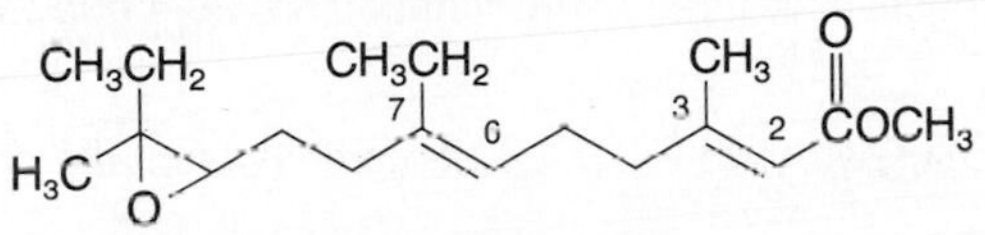

5.18 Match the following alkenes with the appropriate heats of combustion:

(a) 1-Heptene

(b) 2,4-Dimethyl-1-pentene

    (c)   2,4-Dimethyl-2-pentene
    (d)   (Z)-4,4-Dimethyl-2-pentene
    (e)   2,4,4-Trimethyl-2-pentene

Heats of combustion (kcal/mol): 1264.9; 1113.4; 1111.4; 1108.6; 1107.1.

5.19   Match the following alkenes with the appropriate heats of hydrogenation:

    (a)   1-Pentene
    (b)   (E)-4,4-Dimethyl-2-pentene
    (c)   (Z)-4-Methyl-2-pentene
    (d)   (Z)-2,2,5,5-Tetramethyl-3-hexene
    (e)   2,4-Dimethyl-2-pentene

Heats of hydrogenation (kcal/mol): 36.2; 29.3; 27.3; 26.5; 25.1.

5.20   Choose the more stable alkene in each of the following pairs. Explain your reasoning.

    (a)   1-Methylcyclohexene or 3-methylcyclohexene
    (b)   Isopropenylcyclopentane or allylcyclopentane

    (c)

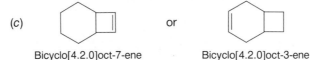

    Bicyclo[4.2.0]oct-7-ene   Bicyclo[4.2.0]oct-3-ene

    (d)   (Z)-Cyclononene or (E)-cyclononene
    (e)   (Z)-Cyclooctadecene or (E)-cyclooctadecene

5.21   (a)   You have learned that trisubstituted double bonds are normally more stable than disubstituted ones. However, as measured by their heats of combustion, the trisubstituted alkene 1-methylcyclopropene is less stable than its disubstituted isomer methylenecyclopropane.

        1-Methylcyclopropene   Methylenecyclopropane
    (heat of combustion 639.4 kcal/mol)   (heat of combustion 629.1 kcal/mol)

        Suggest a reasonable explanation for this observation.
    (b)   Based on your answer to part (a), compare the expected stability of 3-methylcyclopropene with that of 1-methylcyclopropene and that of methylenecyclopropane.

5.22   Suggest an explanation for the fact that 2,4,4-trimethyl-2-pentene, which has a trisubstituted double bond, is less stable than 2,4,4-trimethyl-1-pentene, which has a disubstituted double bond. Their respective heats of hydrogenation are 28.4 and 27.2 kcal/mol.

5.23   (a)   How many alkenes yield 2,2,3,4,4-pentamethylpentane on catalytic hydrogenation?
    (b)   How many yield 2,3-dimethylbutane?
    (c)   How many yield methylcyclobutane?
    (d)   Several alkenes undergo hydrogenation to yield a mixture of cis- and trans-1,4-dimethylcyclohexane. Only one substance, however, gives only cis-1,4-dimethylcyclohexane. What compound is this?

5.24 Compound A undergoes catalytic hydrogenation much faster than does compound B. Why?

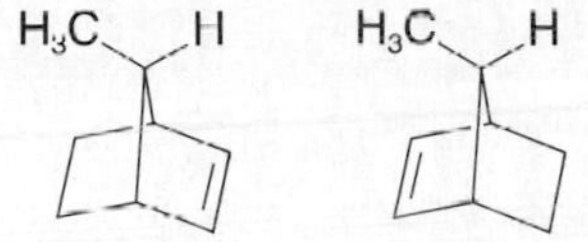

Compound A   Compound B

5.25 Catalytic hydrogenation of 1,4-dimethylcyclopentene yields a mixture of two products. Identify them. One of them is formed in much greater amounts than the other (observed ratio = 10:1). Which one is the major product?

5.26 There are two products that can be formed by syn addition of hydrogen to 2,3-dimethylbicyclo[2.2.1]-2-heptene. Write their structures.

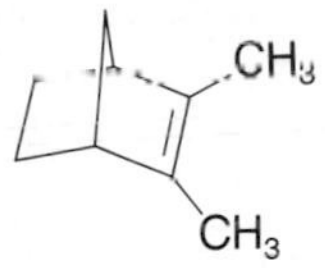

2,3-Dimethylbicyclo[2.2.1]-2-heptene

5.27 Hydrogenation of 3-carene is in principle capable of yielding two stereoisomeric products. Write their structures. Only one of them was actually obtained on catalytic hydrogenation over platinum. Which one do you think is formed? Explain your reasoning.

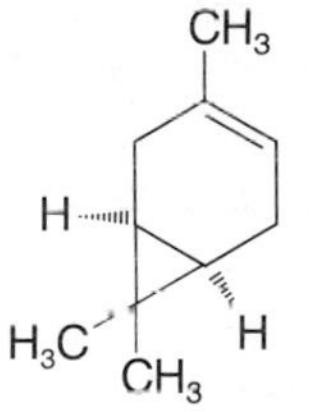

3-Carene

PREPARATION OF ALKENES. ELIMINATION REACTIONS

O ne of the distinguishing features of organic chemistry is the interrelationship that exists among many of its seemingly separate parts. Facts and concepts learned during the study of one class of compounds are used again while examining another. You were introduced to alcohols and alkyl halides in Chapter 4 and to alkenes in Chapter 5. Important elements of both these chapters meet here in Chapter 6, where you will learn how alkenes are prepared from alcohols and alkyl halides by *elimination reactions.* Another topic introduced in Chapter 4, the role of carbocations as intermediates in chemical reactions, will also make a return appearance when the mechanisms of elimination reactions are discussed.

6.1 ELIMINATION REACTIONS

Reactions of the type

$$X-\overset{\alpha}{\underset{|}{\overset{|}{C}}}-\overset{\beta}{\underset{|}{\overset{|}{C}}}-Y \longrightarrow \overset{\diagdown}{\diagup}C=C\overset{\diagup}{\diagdown} + X-Y$$

are classified as *β elimination* reactions. Substituents X and Y are abstracted from a suitable reactant to yield an alkene. Alkene formation requires that *X and Y be substituents on adjacent carbon atoms.* By arbitrarily assigning X as the reference atom and identifying the carbon attached to it as the α carbon, we see that atom Y is a substituent on the β carbon. Carbons succeedingly more remote from the reference atom are designated γ, δ, *etc.* Only β elimination reactions will be discussed in this chapter. [Beta (β) elimination reactions are also known as 1,2 eliminations.]

You are already familiar with one type of β elimination reaction, having seen in Section 5.2 that ethylene and propene are prepared on an industrial scale by *dehydrogenation* of ethane and propane. Both these reactions involve the β elimination of H_2.

$$CH_3CH_3 \xrightarrow{750°C} CH_2{=}CH_2 + H_2$$

Ethane Ethylene Hydrogen

$$CH_3CH_2CH_3 \xrightarrow{750°C} CH_3CH{=}CH_2 + H_2$$

Propane Propene Hydrogen

Alkane dehydrogenation is not a practical laboratory synthesis for the vast majority of alkenes. The principal methods by which alkenes are prepared are two other β elimination processes, the *dehydration of alcohols* and the *dehydrohalogenation of alkyl halides.* Discussions of these two methods comprise the remainder of this chapter.

6.2 DEHYDRATION OF ALCOHOLS

In the dehydration of alcohols the elements of water are eliminated from adjacent carbons. An acid catalyst is necessary.

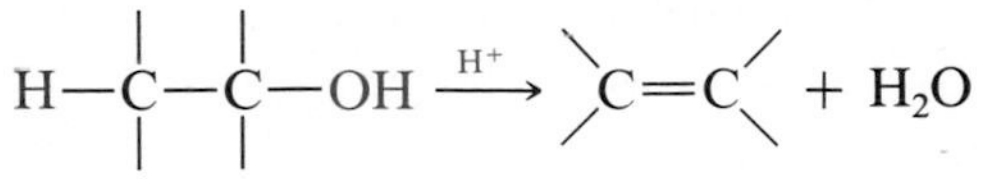

Alcohol Alkene Water

Before dehydrogenation of ethane became the dominant route, ethylene was prepared by heating ethyl alcohol with sulfuric acid.

$$CH_3CH_2OH \xrightarrow[160°C]{H_2SO_4} CH_2{=}CH_2 + H_2O$$

Ethyl alcohol Ethylene Water

Other alcohols behave similarly.

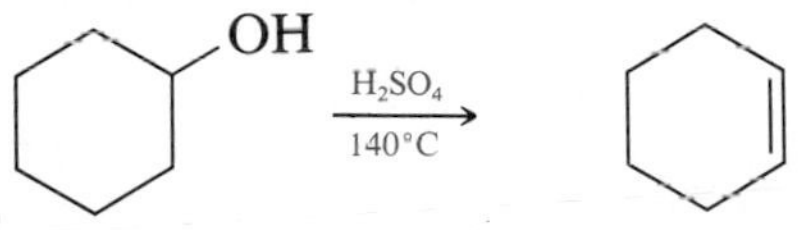

Cyclohexanol Cyclohexene (79–87%)

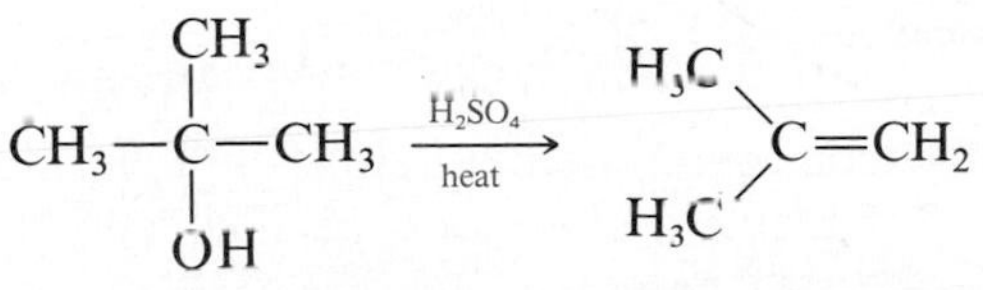

2-Methyl-2-propanol 2-Methylpropene (82%)
(*tert*-butyl alcohol) (isobutene)

Sulfuric acid (H_2SO_4) and phosphoric acid (H_3PO_4) are the acids most frequently used in alcohol dehydration reactions. Potassium hydrogen sulfate ($KHSO_4$) is also often used.

PROBLEM 6.1   Identify the alkene obtained on dehydration of each of the following alcohols:

(a) 3-Ethyl-3-pentanol
(b) 1-Propanol
(c) 2-Propanol
(d) 2,3,3-Trimethyl-2-butanol

SAMPLE SOLUTION   (a)   The hydrogen and the hydroxyl are lost from adjacent carbons in the dehydration of 3-ethyl-3-pentanol.

$$CH_3CH_2 \overset{\beta}{-} \overset{\overset{\beta}{CH_2CH_3}}{\underset{\underset{OH}{|}}{\overset{\alpha}{C}}} \overset{\beta}{-} CH_2CH_3 \xrightarrow{H^+} \overset{CH_3CH_2}{\underset{CH_3CH_2}{}}C=CHCH_3 + H_2O$$

3-Ethyl-3-pentanol   3-Ethyl-2-pentene   Water

The hydroxyl group is located on a carbon that bears three equivalent ethyl substituents in the starting alcohol. Beta elimination can occur in either of three equivalent directions to give the same alkene, 3-ethyl-2-pentene.

6.3   REGIOSELECTIVITY IN ALCOHOL DEHYDRATION REACTIONS. THE ZAITSEV RULE

In the preceding examples only a single alkene product could be formed from each alcohol by β elimination. What about elimination in alcohols such as 2-methyl-2-butanol, whose dehydration can occur in two distinctly different ways to give alkenes that are constitutional isomers? Here, a double bond can be generated between C-1 and C-2 or between C-2 and C-3. It turns out that both processes occur but not nearly to the same extent. Under the usual reaction conditions 2-methyl-2-butene is the major product and its isomer 2-methyl-1-butene is the minor product.

take this H

$$\overset{1}{CH_3} - \overset{2}{\underset{\underset{CH_3}{|}}{\overset{\overset{OH}{|}}{C}}} - \overset{3}{CH_2} \overset{4}{CH_3} \xrightarrow[80°C]{H_2SO_4} CH_2 = C \overset{CH_2CH_3}{\underset{CH_3}{}} + \overset{H_3C}{\underset{H_3C}{}}C=CHCH_3$$

take this H

2-Methyl-2-butanol   2-Methyl-1-butene   2-Methyl-2-butene
(*tert*-pentyl alcohol)   (10%)   (90%)

As a second example, consider 2-methylcyclohexanol. It undergoes dehydration to yield a mixture of 1-methylcyclohexene (major) and 3-methylcyclohexene (minor).

2-Methylcyclohexanol   1-Methylcyclohexene   3-Methylcyclohexene
                                  (84%)   (16%)

Reactions such as these, in which more than one constitutional isomer can be formed from a single reactant but where one is observed to predominate in the product, are said to be *regioselective*. Sometimes the regioselectivity is so high that one isomer is formed exclusively; in those cases the reaction is described as *regiospecific*.

In 1875 Alexander M. Zaitsev of the University of Kazan (Russia) set forth a generalization describing the regioselectivity to be expected in β elimination reactions. *Zaitsev's rule* is an empirical one and summarizes the results of numerous experiments in which alkene mixtures were produced by β elimination. In its original form Zaitsev's rule stated that *the alkene formed in greatest amount is the one that corresponds to removal of the hydrogen from the β carbon having the fewest hydrogen substituents.*

Hydrogen is lost from β carbon having the fewest hydrogen substituents

Alkene present in greatest amount in product

Zaitsev's rule as applied to the acid-catalyzed dehydration of alcohols is now more often expressed in a different way: *β elimination reactions of alcohols yield the most highly substituted alkene as the major product.* Since, as was discussed in Section 5.7, the most highly substituted alkene is also normally the most stable one, Zaitsev's rule is sometimes expressed in terms of a preference for *predominant formation of the most stable alkene that could arise by β elimination.*

PROBLEM 6.2   Each of the following alcohols has been subjected to acid-catalyzed dehydration and yields a mixture of two isomeric alkenes. Identify the two alkenes in each case and predict which one is the major product on the basis of the Zaitsev rule.

(a) $(CH_3)_2CCH(CH_3)_2$
                |
               OH

(b)

(c)

SAMPLE SOLUTION   (a)   Dehydration of 2,3-dimethyl-2-butanol can lead to either 2,3-dimethyl-1-butene by removal of a C-1 hydrogen or to 2,3-dimethyl-2-butene by removal of a C-3 hydrogen.

2,3-Dimethyl-2-butanol

2,3-Dimethyl-1-butene
(minor product)

2,3-Dimethyl-2-butene
(major product)

The major product is 2,3-dimethyl-2-butene. It has a tetrasubstituted double bond and is more stable than 2,3-dimethyl-1-butene, which has a disubstituted double bond. The major alkene product arises by loss of a hydrogen from the β carbon that has fewer hydrogen substituents (C-3) rather than from the β carbon that has the greater number of hydrogen substituents (C-1).

In addition to being regioselective, alcohol dehydration reactions are *stereoselective*. As we saw in Section 5.12, a stereoselective reaction is one in which a single substrate undergoes a reaction that can lead to two or more stereoisomeric products but exhibits a preference for forming one of them in greater amounts than any other. Alcohol dehydration reactions tend to produce the more stable stereoisomeric form of an alkene. Dehydration of 3-pentanol, for example, yields a mixture of *trans*-2-pentene and *cis*-2-pentene in which the more stable trans stereoisomer is present to the extent of 75 percent versus 25 percent for the cis.

$$CH_3CH_2CHCH_2CH_3 \xrightarrow[\text{heat}]{H_2SO_4}$$

OH

3-Pentanol

cis-2-Pentene (25%)
(minor product)

+

trans-2-Pentene (75%)
(major product)

PROBLEM 6.3   What three alkenes are formed in the acid-catalyzed dehydration of 2-pentanol?

6.4   THE MECHANISM OF ACID-CATALYZED DEHYDRATION OF ALCOHOLS

The dehydration of alcohols and the conversion of alcohols to alkyl halides by treatment with hydrogen halides are similar in two important ways:

1. Both reactions are promoted by acids.
2. The relative reactivity of alcohols decreases in the order tertiary > secondary > primary.

These common features suggest that carbocations are key intermediates in alcohol dehydration, just as they are in the conversion of alcohols to alkyl halides (Section 4.10). Figure 6.1 portrays a three-step mechanism for the sufuric acid–catalyzed dehydration of *tert*-butyl alcohol. Steps 1 and 2 describe the generation of *tert*-butyl cation by a process similar to that which led to its formation as an intermediate in the reaction of *tert*-butyl alcohol with hydrogen chloride. Step 3 in Figure 6.1, however, is new to us and is the step in which the alkene product is formed from the carbocation intermediate.

Step 3 is an acid-base reaction. In this step the carbocation acts as a Brönsted acid, transferring a proton to a Brönsted base (hydrogen sulfate ion). This is the property of carbocations that is of the most significance to elimination reactions. Carbocations are strong acids; they are the conjugate acids of alkenes and readily lose a proton to form alkenes. Even weak bases such as hydrogen sulfate ion and water molecules are sufficiently basic to abstract a proton from a carbocation.

According to the mechanism of Figure 6.1, carbocation formation is rate-determining. This is consistent with the observed reactivity of alcohols. Tertiary alcohols

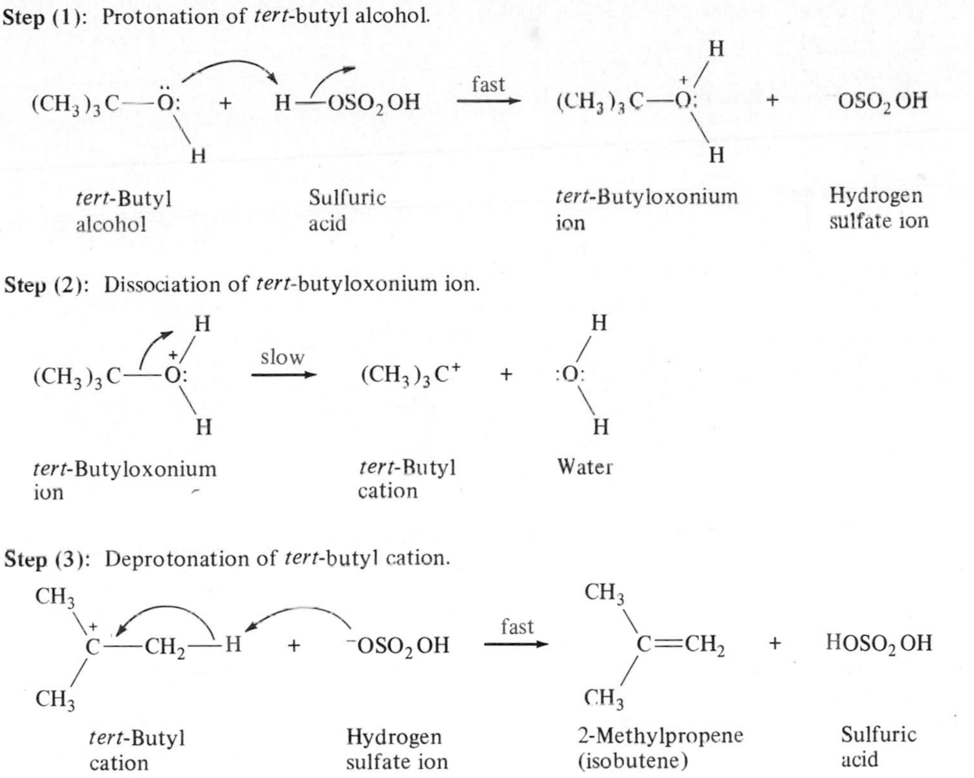

Step (1): Protonation of *tert*-butyl alcohol.

$$(CH_3)_3C\!-\!\overset{..}{\underset{H}{O}}\!: \quad + \quad H\!-\!OSO_2OH \quad \xrightarrow{\text{fast}} \quad (CH_3)_3C\!-\!\overset{H}{\underset{H}{\overset{|}{\underset{|}{O^+}}}} \quad + \quad OSO_2OH$$

| *tert*-Butyl alcohol | Sulfuric acid | *tert*-Butyloxonium ion | Hydrogen sulfate ion |

Step (2): Dissociation of *tert*-butyloxonium ion.

$$(CH_3)_3C\!-\!\overset{H}{\underset{H}{\overset{|}{\underset{|}{O^+}}}}\!: \quad \xrightarrow{\text{slow}} \quad (CH_3)_3C^+ \quad + \quad :\overset{H}{\underset{H}{\overset{|}{\underset{|}{O}}}}\!:$$

| *tert*-Butyloxonium ion | *tert*-Butyl cation | Water |

Step (3): Deprotonation of *tert*-butyl cation.

$$\overset{CH_3}{\underset{CH_3}{\overset{|}{\underset{|}{C^+}}}}\!-\!CH_2\!-\!H \quad + \quad {}^-OSO_2OH \quad \xrightarrow{\text{fast}} \quad \overset{CH_3}{\underset{CH_3}{\overset{|}{\underset{|}{C}}}}\!\!=\!\!CH_2 \quad + \quad HOSO_2OH$$

| *tert*-Butyl cation | Hydrogen sulfate ion | 2-Methylpropene (isobutene) | Sulfuric acid |

FIGURE 6.1 Sequence of steps that describes the mechanism for the acid-catalyzed dehydration of *tert*-butyl alcohol.

undergo acid-catalyzed dehydration most readily because tertiary carbocations are more stable than secondary or primary ones. The more stable the carbocation intermediate, the lower the energy of activation for its formation and the faster the rate of reaction.

The regioselectivity of the elimination process is determined in the deprotonation step. As the proton is being removed from the carbon at the transition state for step 3, a double bond begins to form between the two carbons.

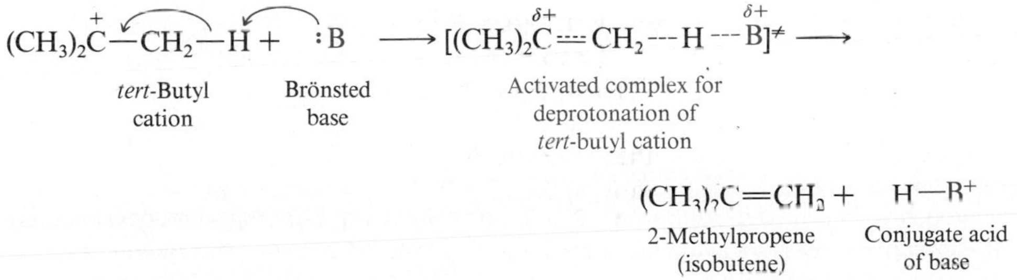

$$(CH_3)_2\overset{+}{C}\!-\!CH_2\!-\!H + \quad :B \quad \longrightarrow \quad [(CH_3)_2\overset{\delta+}{C}\!=\!=\!CH_2\text{---}H\text{---}\overset{\delta+}{B}]^{\neq} \quad \longrightarrow$$

| *tert*-Butyl cation | Brönsted base | Activated complex for deprotonation of *tert*-butyl cation |

$$(CH_3)_2C\!=\!CH_2 + \quad H\!-\!B^+$$

| 2-Methylpropene (isobutene) | Conjugate acid of base |

We say the transition state has *partial double bond character* and reason that substituents that stabilize double bonds also stabilize double bonds in the process of being formed. Thus, as illustrated in Figure 6.2, the transition state that leads to 2-methyl-2-butene from the carbocation $(CH_3)_2\overset{+}{C}CH_2CH_3$ is lower in energy than the one that leads to 2-methyl-1-butene because the greater degree of alkyl group substitution about its developing double bond stabilizes it more.

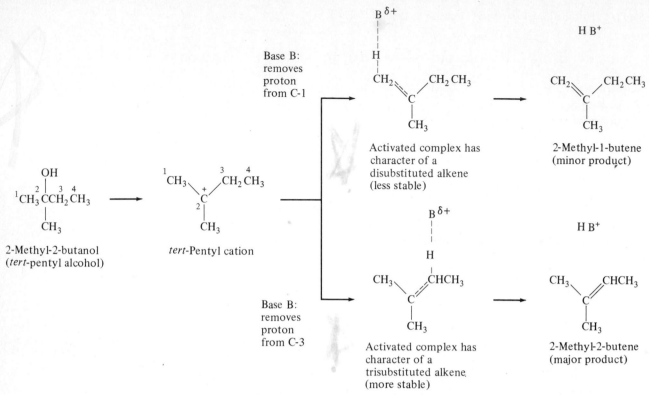

FIGURE 6.2   Diagram showing how partial double bond character developed in activation complex influences regioselectivity in dehydration of 2-methyl-2-butanol.

PROBLEM 6.4   Write a structural formula for the carbocation intermediate formed in the dehydration of each of the alcohols in Problem 6.2 (Section 6.3). Using curved arrow notation, show how each carbocation is deprotonated by hydrogen sulfate anion ($^-OSO_2OH$) to give a mixture of alkenes.

SAMPLE SOLUTION   (a)   The carbon that bears the hydroxyl group in the starting alcohol is the one that becomes positively charged in the carbocation.

$$(CH_3)_2CCH(CH_3)_2 \xrightarrow[-H_2O]{H^\pm} (CH_3)_2\overset{+}{C}CH(CH_3)_2$$
$$\quad\quad\quad | $$
$$\quad\quad\; OH$$

Hydrogen sulfate ion may remove a proton from either C-1 or C-3 of this carbocation. Loss of a proton from C-1 yields the minor product 2,3-dimethyl-1-butene. (This alkene has a disubstituted double bond.)

$$HOSO_2O^- \quad H \overset{1}{-}CH_2 \overset{2}{-} \overset{CH_3}{\underset{\underset{3}{+}}{C}} \quad \longrightarrow HOSO_2OH + CH_2 = C \overset{CH_3}{\underset{CH(CH_3)_2}{}}$$

2,3-Dimethyl-1-butene

Loss of a proton from C-3 yields the major product 2,3-dimethyl-2-butene. (This alkene has a tetrasubstituted double bond.)

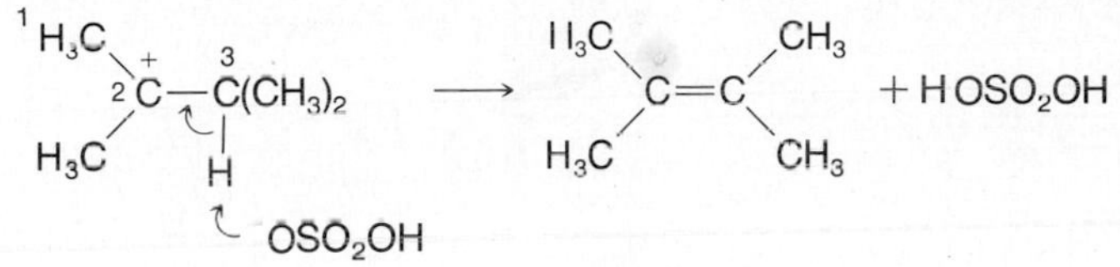

2,3-Dimethyl-2-butene

As noted earlier in Section 4.16, primary carbocations are too high in energy to be realistically invoked as intermediates in most chemical reactions. If primary alcohols do not form primary carbocations, then how do they undergo elimination? A modification of our general mechanism for alcohol dehydration offers a reasonable explanation. For primary alcohols it is believed that a proton is lost from the oxonium ion in the same step in which carbon-oxygen bond cleavage takes place. For example, the rate-determining step in the sulfuric acid–catalyzed dehydration of ethanol may be represented as:

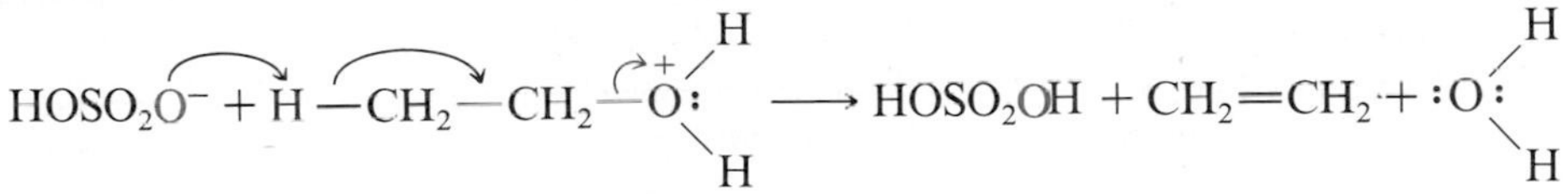

| Hydrogen sulfate ion | Ethyloxonium ion | Sulfuric acid | Ethylene | Water |

Like their tertiary alcohol counterparts, secondary alcohols normally undergo dehydration to alkenes by way of carbocation intermediates.

In Chapter 4 you learned that carbocations could be captured by halide anions to give alkyl halides. In the present chapter, a second type of carbocation reaction has been introduced—a carbocation can lose a proton to form an alkene. In the next section a third aspect of carbocation behavior will be described, the *rearrangement* of one carbocation to another.

6.5 REARRANGEMENTS IN ALCOHOL DEHYDRATION

Chemists have long recognized that some alcohols undergo dehydration to yield alkenes having carbon skeletons different from those of the starting alcohols. Not only has the functional group transformation of alcohol to alkene taken place but the arrangement of atoms in the alkene is different from that in the alcohol. A *rearrangement* is said to have occurred. An example of an alcohol dehydration that is accompanied by rearrangement is the case of 3,3-dimethyl-2-butanol. This is one of many such experiments carried out by F. C. Whitmore and his students at Pennsylvania State University in the 1930s as part of a general study of rearrangement reactions.

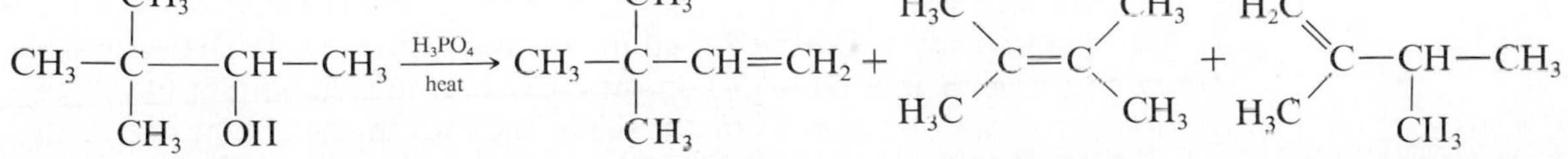

| 3,3-Dimethyl-2-butanol | 3,3-Dimethyl-1-butene (3%) | 2,3-Dimethyl-2-butene (64%) | 2,3-Dimethyl-1-butene (33%) |

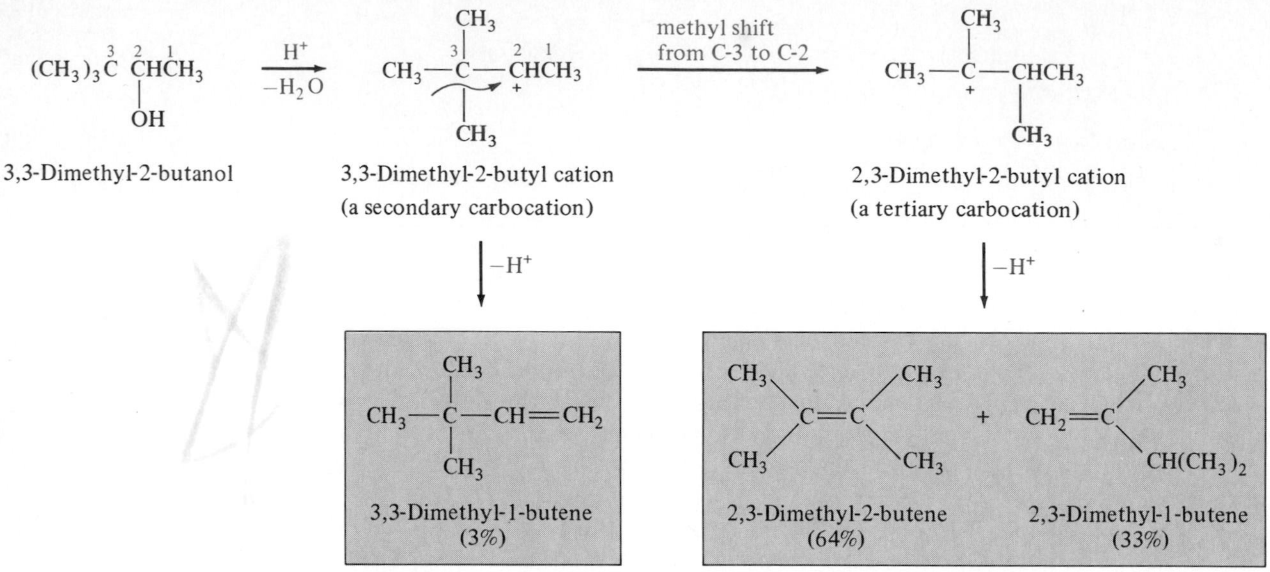

$$(CH_3)_3\overset{3}{C}\ \overset{2}{C}H\overset{1}{C}H_3 \quad \underset{-H_2O}{\overset{H^+}{\longrightarrow}} \quad CH_3-\overset{3}{\underset{CH_3}{\overset{|}{C}}}-\overset{2}{C}H\overset{1}{C}H_3$$

3,3-Dimethyl-2-butanol

3,3-Dimethyl-2-butyl cation
(a secondary carbocation)

methyl shift from C-3 to C-2

2,3-Dimethyl-2-butyl cation
(a tertiary carbocation)

$-H^+$

$-H^+$

3,3-Dimethyl-1-butene
(3%)

2,3-Dimethyl-2-butene
(64%)

2,3-Dimethyl-1-butene
(33%)

FIGURE 6.3 The first formed carbocation from 3,3-dimethyl-2-butanol is secondary and rearranges to a more stable tertiary carbocation by a methyl migration. The major portion of the alkene products is formed by way of a tertiary carbocation.

A mixture of three alkenes was obtained in 80 percent yield, having the composition shown in the equation. The alkene having the same carbon skeleton as the starting alcohol, 3,3-dimethyl-1-butene, constituted only 3 percent of the alkene mixture. The two alkenes present in greatest amount, 2,3-dimethyl-2-butene and 2,3-dimethyl-1-butene, both have carbon skeletons different from that of the starting alcohol.

Whitmore proposed that the rearrangement of the carbon skeleton occurred in a separate step following carbocation formation. Once the alcohol was converted to the corresponding carbocation, that carbocation could either lose a proton to give an alkene having the same carbon skeleton or it could rearrange to a different carbocation, as shown in Figure 6.3. The rearranged alkenes arise by loss of a proton from the rearranged carbocation.

Why do carbocations rearrange? How do carbocations rearrange? The "why" is easy to understand. Estimates of their relative energies suggest that secondary carbocations are 10 to 15 kcal/mol less stable than tertiary carbocations. Consequently, rearrangement of a secondary to a tertiary carbocation decreases its potential energy and is an exothermic process. This is exactly what happens in the dehydration of 3,3-dimethyl-2-butanol. As Figure 6.3 depicts, the first formed carbocation is secondary; the rearranged carbocation is tertiary. Almost all the alkene products are derived from the more stable tertiary carbocation.

The "how" of carbocation rearrangements can be approached by considering the nature of the structural change that takes place at the transition state. Again referring to the initial (secondary) carbocation intermediate in the dehydration of 3,3-dimethyl-2-butanol, rearrangement occurs when a methyl group shifts from C-3 to the positively charged carbon. The methyl group migrates with the pair of electrons that comprised its original σ bond to C-3. In the curved arrow notation for this methyl migration, we draw the arrow so that it shows the movement of both the methyl group and the electrons in the σ bond.

$$CH_3-\underset{\underset{4}{\overset{CH_3}{|}}}{\overset{\overset{CH_3}{|}}{C}}\overset{+}{-}\underset{1}{\overset{2}{C}}HCH_3 \longrightarrow \left[CH_3-\underset{\delta+|}{\overset{\overset{CH_3}{\cdots}}{C}}-\underset{\delta+}{C}HCH_3\right]^{\neq} \longrightarrow CH_3-\underset{\overset{|}{CH_3}}{\overset{\overset{CH_3}{|}}{\overset{+}{C}}}-CHCH_3$$

3,3-Dimethyl-2-butyl
cation (secondary, less stable)

Activated complex
for methyl migration
(dotted lines indicate
partial bonds)

2,3-Dimethyl-2-butyl
cation (tertiary, more stable)

At the transition state for rearrangement the methyl group is partially bonded both to its point of origin and to the carbon that will be its destination.

An orbital description (Figure 6.4) of the rearrangement process harmonizes nicely with what we have already learned about carbocation stabilization. The secondary carbocation is stabilized by delocalization of the electrons in the $C(3)-CH_3$ σ bond into the empty $2p$ orbital of the positively charged carbon. Some carbocations, as shown in Figure 6.4c, absorb energy from their surroundings sufficient to achieve the transition state. In a sense, the electrons involved in stabilizing the secondary carbocation are simply carrying along the methyl group. Electron delocalization is a spontaneous process, but movement of atoms and groups requires energy, so the rearrangement of one carbocation to another has an activation energy associated with it. The activation energy is modest, and carbocation rearrangements are normally quite rapid. Figure 6.4d completes the picture; the methyl group has reached its destination, and the tertiary carbocation is shown along with the relevant orbitals.

PROBLEM 6.5   The alkene mixture obtained on dehydration of 2,2-dimethylcyclohexanol contains appreciable amounts of 1,2-dimethylcyclohexene. Give a mechanistic explanation for the formation of this product.

Alkyl groups other than methyl are also capable of migrating to a positively charged carbon. Many carbocation rearrangements involve migration of a hydrogen substituent. These are called *hydride shifts.* The same requirements apply to hydride shifts as to alkyl group migrations; they proceed in the direction that leads to a more stable carbocation; the origin and destination of the migrating group are adjacent carbons, one of which must be positively charged; and the group migrates with a pair of electrons.

$$A-\underset{\underset{B}{\overset{|}{|}}}{\overset{\overset{H}{|}}{C}}\overset{+}{-}\underset{\overset{|}{Y}}{\overset{|}{C}}-X \longrightarrow A-\underset{\underset{B}{\overset{|}{|}}}{\overset{+}{C}}-\underset{\overset{|}{Y}}{\overset{\overset{H}{|}}{C}}-X \qquad \text{Hydride shift}$$

Hydride shifts often occur during the dehydration of primary alcohols. Thus, while 1-octene would be expected to be the only alkene formed on dehydration of 1-octanol, it is in fact only a minor product. The major product is a mixture of *cis-* and *trans*-2-octene.

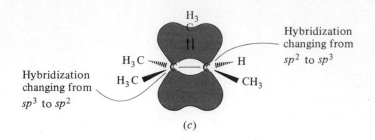

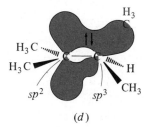

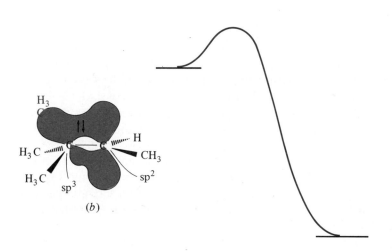

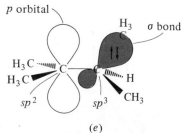

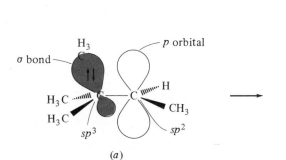

FIGURE 6.4 Representation of methyl migration in 3,3-dimethyl-2-butyl cation. Structures (a) through (e) show the orbitals that participate in the rearrangement process: (a) and (b) are equivalent representations of the secondary carbocation; (d) and (e) are equivalent representations of the tertiary carbocation. In (a) and (e) the participating orbitals are shown separately for the purpose of identifying them, while in (b) and (d) they are shown in the "orbital overlapped" presentation. Structure (c) depicts the activated complex for methyl migration; the methyl group has moved part of the way from one carbon to the other.

3,3-Dimethyl-2-butyl cation
(secondary)

2,3-Dimethyl-2-butyl cation
(tertiary)

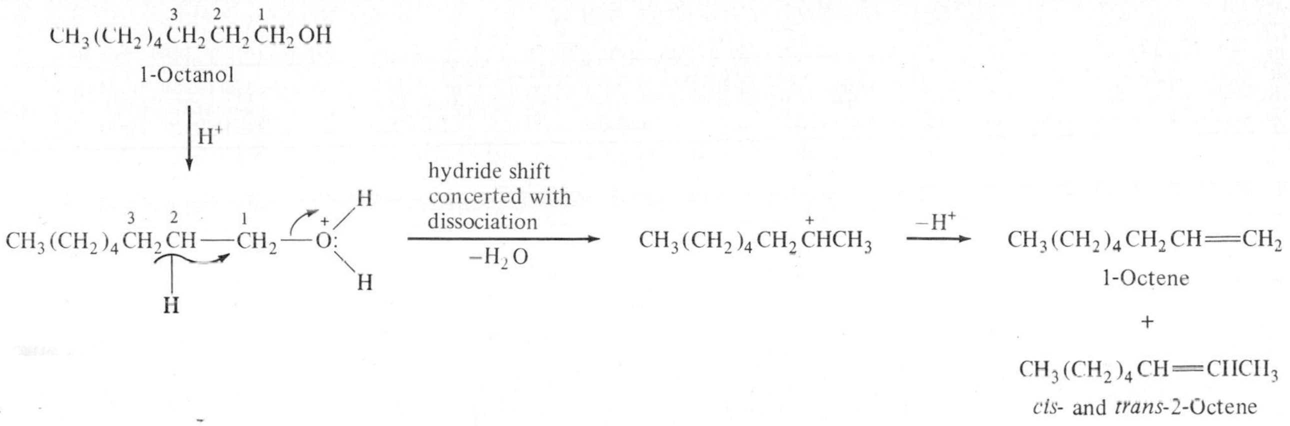

$$CH_3(CH_2)_4 \overset{3}{C}H_2 \overset{2}{C}H_2 \overset{1}{C}H_2 OH$$

1-Octanol

$$\Big\downarrow H^+$$

$$CH_3(CH_2)_4 \overset{3}{C}H_2 \overset{2}{C}H \!-\! \overset{1}{C}H_2 \!-\! \overset{+}{O}{:} \qquad \xrightarrow[\;-H_2O\;]{\substack{\text{hydride shift} \\ \text{concerted with} \\ \text{dissociation}}} \qquad CH_3(CH_2)_4 CH_2 \overset{+}{C}HCH_3 \qquad \xrightarrow{-H^+} \qquad CH_3(CH_2)_4 CH_2 CH \!=\! CH_2$$

H
H

1-Octene

+

$$CH_3(CH_2)_4 CH \!=\! CHCH_3$$

cis- and trans-2-Octene

FIGURE 6.5 Dehydration of 1-octanol is accompanied by a hydride shift from C-2 to C-1.

$$CH_3(CH_2)_4CH_2CH_2CH_2OH \xrightarrow[255°C]{H_3PO_4}$$

1-Octanol

$$CH_3(CH_2)_4CH_2CH\!=\!CH_2 + CH_3(CH_2)_4CH\!=\!CHCH_3$$

1-Octene
(minor product)

Mixture of *cis*- and *trans*-
2-octene (major product)

A mechanism consistent with the formation of these three alkenes is shown in Figure 6.5. Dissociation of the primary oxonium ion is accompanied by a shift of hydride from C-2 to C-1. This avoids the formation of a primary carbocation, leading instead to a secondary carbocation in which the positive charge is at C-2. Deprotonation of this carbocation yields the observed products. (Some 1-octene may also arise directly from the primary oxonium ion.)

This concludes our discussion of a second functional group transformation involving alcohols: the first is the conversion of alcohols to alkyl halides (Chapter 4) and the second the conversion of alcohols to alkenes. In the remaining sections of the chapter the conversion of alkyl halides to alkenes by dehydrohalogenation reactions is described.

6.6 DEHYDROHALOGENATION OF ALKYL HALIDES

By dehydrohalogenation we mean the removal of the elements of a hydrogen halide (HX) from an alkyl halide. Dehydrohalogenation of alkyl halides is one of the most useful methods for preparing alkenes by β elimination.

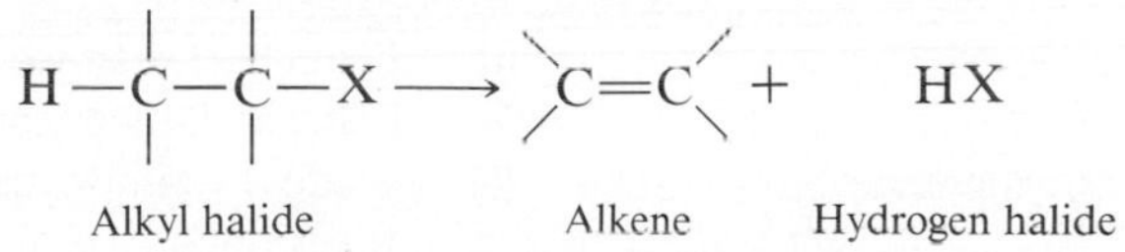

$$H \!-\! \overset{\displaystyle |}{\underset{\displaystyle |}{C}} \!-\! \overset{\displaystyle |}{\underset{\displaystyle |}{C}} \!-\! X \longrightarrow \overset{\diagdown}{\diagup}C \!=\! C\overset{\diagup}{\diagdown} + \qquad HX$$

Alkyl halide   Alkene   Hydrogen halide

When applied to the preparation of alkenes, the reaction is carried out in the presence of a strong base, such as sodium ethoxide in ethyl alcohol as solvent.

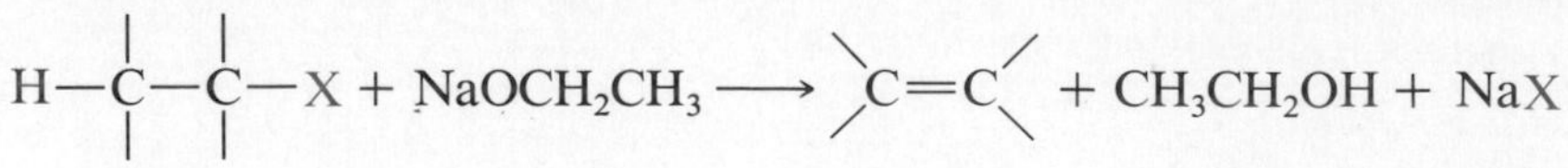

$$H-\underset{|}{\overset{|}{C}}-\underset{|}{\overset{|}{C}}-X + NaOCH_2CH_3 \longrightarrow \underset{}{\overset{}{C}}{=}\underset{}{\overset{}{C}} + CH_3CH_2OH + NaX$$

| Alkyl halide | Sodium ethoxide | Alkene | Ethyl alcohol | Sodium halide |

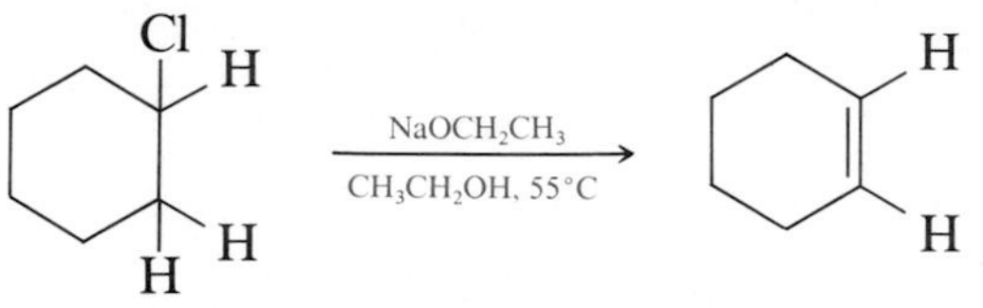

Cyclohexyl chloride   Cyclohexene (100%)

Similarly, sodium methoxide is a suitable base and is used in methyl alcohol. Potassium hydroxide in ethyl alcohol is another base-solvent combination often employed in the dehydrohalogenation of alkyl halides. Potassium *tert*-butoxide is the preferred base when the alkyl halide is primary and is used in either *tert*-butyl alcohol or dimethyl sulfoxide as solvent. Dimethyl sulfoxide has the structure $(CH_3)_2\overset{+}{S}-\overset{..}{\underset{..}{O}}:^-$ and is commonly abbreviated as DMSO.

$$CH_3(CH_2)_{15}CH_2CH_2Cl \xrightarrow[\text{DMSO, 25°C}]{\text{KOC(CH}_3)_3} CH_3(CH_2)_{15}CH{=}CH_2$$

1-Chlorooctadecane   1-Octadecene (86%)

The regioselectivity of dehydrohalogenation of alkyl halides follows the Zaitsev rule; β elimination predominates in the direction that leads to the more highly substituted alkene. The major alkene is normally the more stable one, which is formed by removing a proton from the β carbon that has the fewest hydrogen substituents.

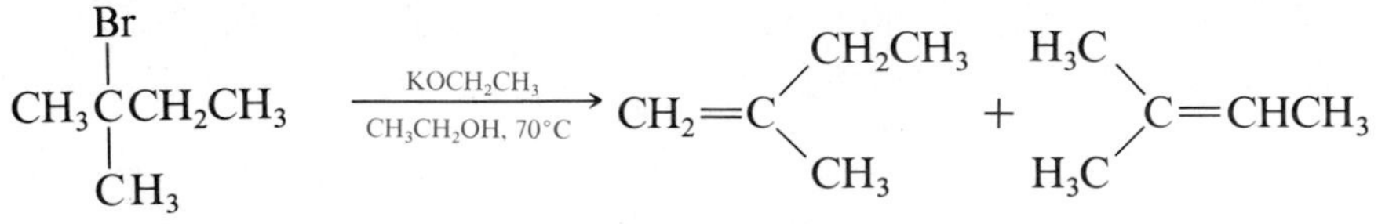

2-Bromo-2-methylbutane   2-Methyl-1-butene   2-Methyl-2-butene
(*tert*-pentyl bromide)   (29%)   (71%)

PROBLEM 6.6   Write the structures of all the alkenes capable of being produced when each of the following alkyl halides undergoes dehydrohalogenation. Apply the Zaitsev rule to predict the alkene formed in greatest amount in each case.

(a)   2-Bromo-2,3-dimethylbutane   (d)   2-Bromo-3-methylbutane
(b)   *tert*-Butyl chloride   (e)   1-Bromo-3-methylbutane
(c)   3-Bromo-3-ethylpentane   (f)   1-Iodo-1-methylcyclohexane

SAMPLE SOLUTION   (a)   First analyze the structure of 2-bromo-2,3-dimethylbutane with respect to the number of possible β elimination pathways.

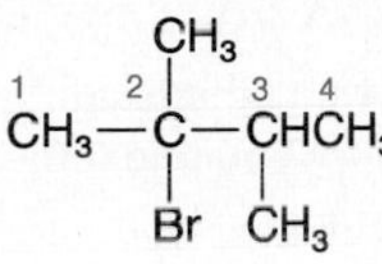

$$
\begin{array}{c}
\overset{CH_3}{\underset{1\quad 2\,|\quad 3\quad 4}{CH_3-\!\!\underset{|}{\overset{|}{C}}\!\!-\!CHCH_3}}
\end{array}
$$
Br CH$_3$

Bromine must be lost from C-2; hydrogen may be lost from C-1 or from C-3

The two possible alkenes are

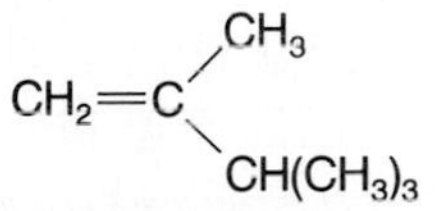

CH$_2$=C
 CH$_3$
 CH(CH$_3$)$_3$

and

H$_3$C   CH$_3$
   C=C
H$_3$C   CH$_3$

2,3-Dimethyl-1-butene
(minor product)

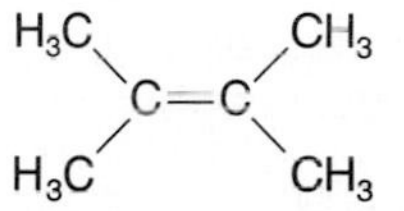

2,3-Dimethyl-2-butene
(major product)

The major product, predicted on the basis of Zaitsev's rule, is 2,3-dimethyl-2-butene. It has a tetrasubstituted double bond. The minor alkene has a disubstituted double bond.

In addition to being regioselective, dehydrohalogenation of alkyl halides is stereoselective and favors formation of the more stable stereoisomer. Usually, as in the case of 5-bromononane, the trans (or E) alkene is formed in greater amounts than its cis (or Z) stereoisomer.

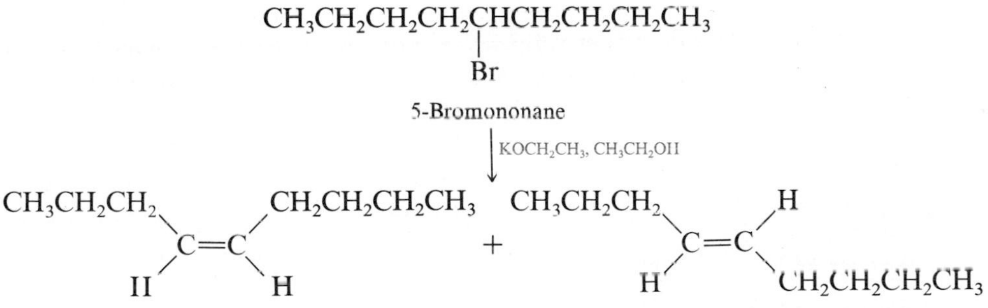

$$CH_3CH_2CH_2CH_2CHCH_2CH_2CH_2CH_3$$
|
Br

5-Bromononane

↓ KOCH$_2$CH$_3$, CH$_3$CH$_2$OH

CH$_3$CH$_2$CH$_2$ CH$_2$CH$_2$CH$_2$CH$_3$
        C=C
      H   H

+

CH$_3$CH$_2$CH$_2$ H
        C=C
      H   CH$_2$CH$_2$CH$_2$CH$_3$

cis-4-Nonene (23%)

trans-4-Nonene (77%)

PROBLEM 6.7 Write structural formulas for all the alkenes capable of being produced by dehydrohalogenation of each of the following alkyl halides:

(a) 2-Bromobutane
(b) 2-Bromohexane
(c) 2-Iodo-4-methylpentane
(d) 3-Iodo-2,2-dimethylpentane

SAMPLE SOLUTION (a) Elimination in 2-bromobutane can take place between C-1 and C-2 or between C-2 and C-3. There are three alkenes capable of being formed, namely, 1-butene, cis-2-butene, and trans-2-butene.

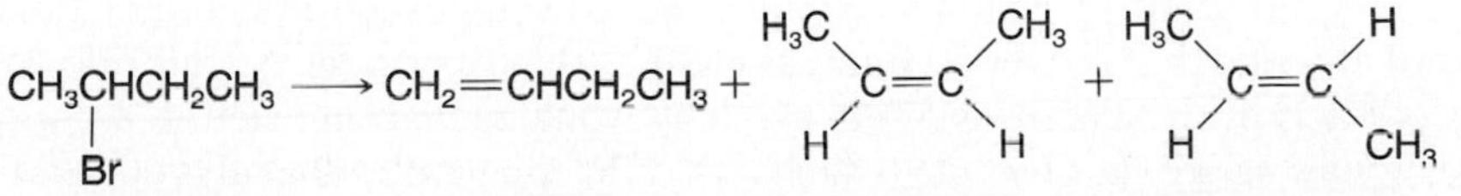

CH$_3$CHCH$_2$CH$_3$ ⟶ CH$_2$=CHCH$_2$CH$_3$ +
|
Br

H$_3$C   CH$_3$
   C=C
H   H

+

H$_3$C   H
   C=C
H   CH$_3$

2-Bromobutane   1-Butene   cis-2-Butene   trans-2-Butene

The Zaitsev rule tells us that the mixture of *cis-* and *trans-*2-butene will be present in greater amounts than 1-butene. In the experiment as actually carried out with potassium *tert*-butoxide as the base in dimethyl sulfoxide, the major product was *trans*-2-butene (60 percent); *cis*-2-butene and 1-butene each comprised 20 percent of the product.

Elimination reactions of cycloalkyl halides lead exclusively to *cis* cycloalkenes when the ring has fewer than 10 carbons. As the ring becomes larger, it can accommodate either a cis or a trans double bond, and large-ring cycloalkyl halides give mixtures of cis and trans cycloalkenes.

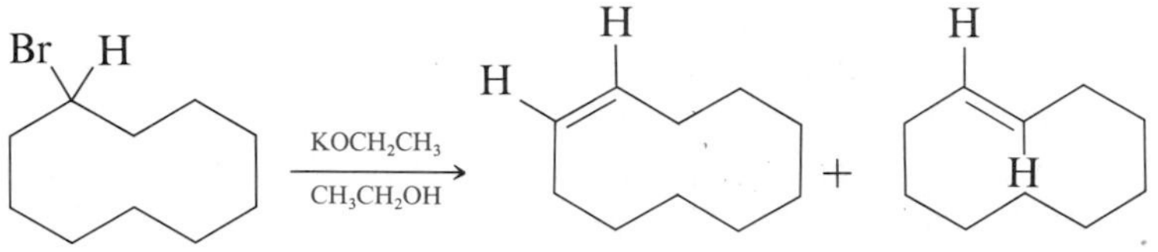

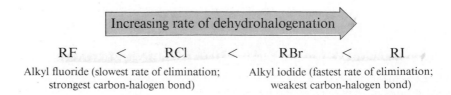

Bromocyclodecane   →   *cis*-Cyclodecene [(*Z*)-Cyclodecene] 85%   +   *trans*-Cyclodecene [(*E*)-Cyclodecene] 15%

6.7 MECHANISM OF THE DEHYDROHALOGENATION OF ALKYL HALIDES. THE E2 MECHANISM.

In the 1920s Sir Christopher Ingold proposed a mechanism for dehydrohalogenation that is still accepted as a valid description of how these reactions occur. Some of the information upon which Ingold based his mechanism included these facts:

1. The dehydrohalogenation reaction exhibits second-order kinetics; it is first-order in alkyl halide and first-order in base.
2. The rate of elimination depends on the halogen, the reactivity of alkyl halides increasing with decreasing strength of the carbon-halogen bond.

The rate equation may be written

$$\text{Rate} = k \text{ [alkyl halide][base]}$$

Doubling the concentration of either the alkyl halide or the base doubles the reaction rate. Doubling the concentration of both reactants increases the rate by a factor of 4.

The effect of the halogen may be represented as

Increasing rate of dehydrohalogenation

RF   <   RCl   <   RBr   <   RI

Alkyl fluoride (slowest rate of elimination; strongest carbon-halogen bond)   Alkyl iodide (fastest rate of elimination; weakest carbon-halogen bond)

Cyclohexyl bromide, for example, is converted to cyclohexene by sodium ethoxide in ethanol at a rate that is over 60 times as great as that of cyclohexyl chloride. We say that iodide is the best *leaving group* in a dehydrohalogenation reaction, fluoride the poorest leaving group. Fluoride is such a poor leaving group that alkyl fluorides are rarely used as starting materials in the preparation of alkenes.

What are the implications of second-order kinetics? Ingold assumed in this case that second-order kinetics suggested a bimolecular transition state for the rate-determining step—one in which both a molecule of the alkyl halide and a molecule of base contribute to the structure of the activated complex. For this reason he concluded that proton removal from the β carbon by the base occurs during the rate-determining step rather than in a separate step following the rate-determining step.

What are the implications of the effects of the various halide leaving groups? Since it is the halogen with the weakest bond to carbon that reacts fastest, Ingold concluded that carbon-halogen bond cleavage must accompany carbon-hydrogen bond cleavage in the rate-determining step. The weaker the carbon-halogen bond, the lower will be the activation energy for its cleavage.

On the basis of these observations, Ingold proposed a concerted mechanism for the dehydrohalogenation of alkyl halides in the presence of strong bases.

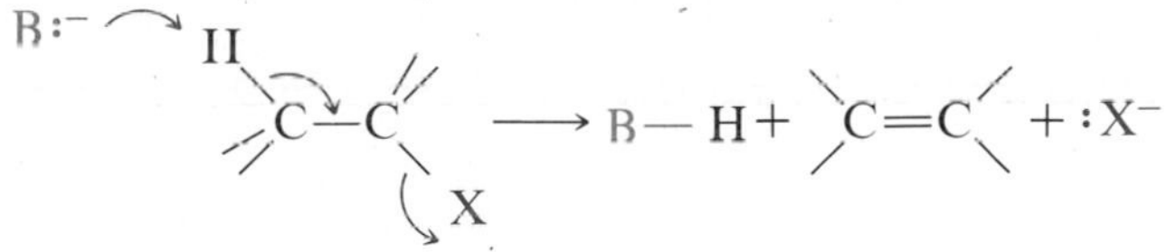

He coined the mechanistic symbol E2 to refer to this mechanism. The symbol stands for *elimination bimolecular*. In the E2 mechanism the three key elements

1. C—H bond breaking
2. C=C π bond formation
3. C—X bond breaking

all contribute to the transition state in a single-step transformation without the involvement of an intermediate. In the representation of the E2 transition state shown below, the carbon-hydrogen and carbon-halogen bonds are in the process of being broken, the base is becoming bonded to the hydrogen, a π bond is being formed, and the hybridization state of carbon is changing from sp^3 to sp^2.

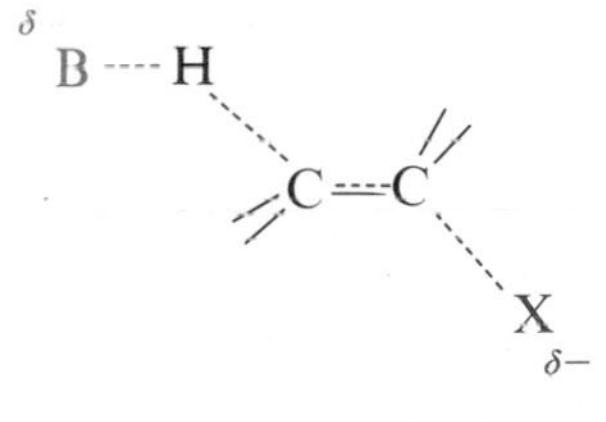

An energy diagram depicting the E2 mechanism is presented in Figure 6.6.

Certain other observations concerning elimination reactions of alkyl halides are consistent with Ingold's concerted E2 process, particularly with the idea that a partial double bond is developed at the transition state. As is the case with alcohol dehydration, the regioselectivity of the reaction corresponds to formation of the more highly substituted, more stable alkene. Again as in alcohol dehydration, we can say that because alkyl groups stabilize double bonds, they also stabilize an activated complex containing a partially formed π bond. Therefore, the more stable alkene requires a lower energy of activation for its formation than a less stable one and predominates in the product mixture because it is formed faster.

For a similar reason tertiary alkyl halides undergo E2 elimination faster than secondary alkyl halides, and secondary alkyl halides react faster than primary ones.

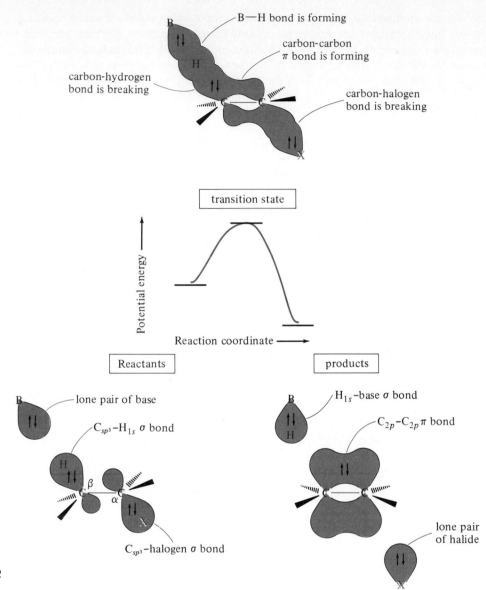

FIGURE 6.6 Potential energy diagram for concerted E2 elimination of an alkyl halide.

Increasing alkyl substitution at the site of reaction is translated into increased stabilization of the partial double bond in the activated complex. A second reason for the difference in rate of various alkyl halides is that tertiary alkyl halides have the weakest carbon-halogen bonds and primary alkyl halides the strongest.

Ingold was a pioneer in applying quantitative measurements of reaction rates to the understanding of organic reaction mechanisms. Many of the reactions to be described in this text were studied by him and his students during the period of about 1920 to 1950. The facts disclosed by Ingold's experiments have been verified many times. His interpretations, although considerably refined during the decades that followed his original reports, still serve us well as a starting point for understanding how the fundamental processes of organic chemistry take place. Elimination reactions of alkyl halides by the E2 mechanism is one of those fundamental processes.

6.8 ANTI ELIMINATION IN E2 REACTIONS. STEREOELECTRONIC EFFECTS

Further insight into the mechanism of base-promoted elimination reactions of alkyl halides has been obtained by stereochemical studies. In one such experiment the rates of elimination of the cis and trans isomers of 4-*tert*-butylcyclohexyl bromide were compared.

cis-4-*tert*-Butylcyclohexyl bromide

trans-4-*tert*-Butylcyclohexyl bromide

KOC(CH₃)₃
(CH₃)₃COH

KOC(CH₃)₃
(CH₃)₃COH

4-*tert*-Butylcyclohexene

While both bromides yield 4-*tert*-butylcyclohexene as the only alkene product, they do so at quite different rates. The cis isomer reacts over 500 times faster than the trans with potassium *tert*-butoxide in *tert*-butyl alcohol.

The disparity in reaction rate can be understood by considering bond development in the E2 transition state. In order for the activated complex to be stabilized by partial π bond formation, the p orbitals that develop as the hybridization of the two adjacent carbons changes from sp^3 to sp^2 must be suitably aligned. Since π overlap of p orbitals requires their axes to be parallel, it seems reasonable that π bond formation is best achieved when the four atoms of the H—C—C—X unit lie in the same plane at the transition state. The two conformations that permit this relationship are termed *syn periplanar* and *anti periplanar*.

Syn periplanar

Anti-periplanar

Because adjacent bonds are eclipsed when the H—C—C—X assembly is syn periplanar, a transition state derived from this conformation is less stable than one that has an anti periplanar relationship between the proton and the leaving group.

As Figure 6.7 shows, bromine is axial in the most stable conformation of *cis*-4-*tert*-butylcyclohexyl bromide, while it is equatorial in the trans stereoisomer. An axial bromine is anti periplanar with respect to the axial hydrogens at C-2 and C-6, so the proper relationship between proton and leaving group is already present in the most stable conformation of the cis bromide. This conformation undergoes E2 elimina-

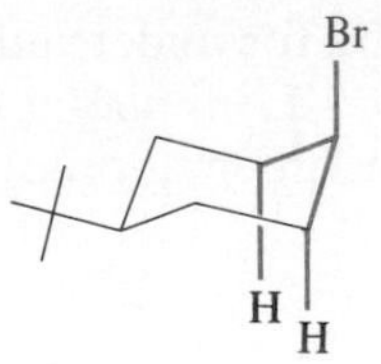

cis-4-*tert*-Butylcyclohexyl bromide

Axial halide is in proper orientation for anti elimination with respect to axial hydrogens on adjacent carbon atoms. Dehydrobromination is rapid.

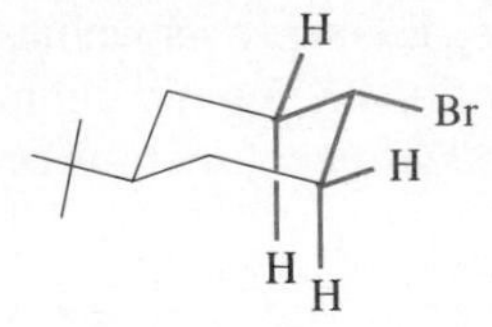

trans-4-*tert*-Butylcyclohexyl bromide

Equatorial halide is gauche to axial and equatorial hydrogens on adjacent carbon; cannot undergo anti elimination in this conformation. Dehydrobromination is slow.

FIGURE 6.7 Conformations of *cis*- and *trans*-4-*tert*-butylcyclohexyl bromide and their relationship to the preference for an anti periplanar arrangement of proton and leaving group. (Bonds indicating H—C—C—Br units are shown in color.)

tion rapidly. The less reactive stereoisomer, the trans bromide, has an equatorial bromine in its most stable conformation. An equatorial bromine is not anti periplanar with respect to any hydrogen substituents. The relationship between an equatorial leaving group and the C-2 and C-6 hydrogens is gauche in every instance. In order to achieve the geometry required for E2 elimination, the trans bromide must adopt a geometry in which the ring is distorted from its most stable conformation. Therefore, the transition state for its elimination is higher in energy, and reaction is slower.

Effects that arise because one spatial arrangement of electrons (or orbitals or bonds) is more stable than another are called *stereoelectronic effects.* We say there is a stereoelectronic preference for the anti periplanar arrangement of proton and leaving group in E2 reactions.

PROBLEM 6.8 Menthyl chloride and neomenthyl chloride have the structures shown. One of these stereoisomers undergoes elimination on treatment with sodium ethoxide in ethanol much more readily than the other. Which reacts faster, menthyl chloride or neomenthyl chloride? Why?

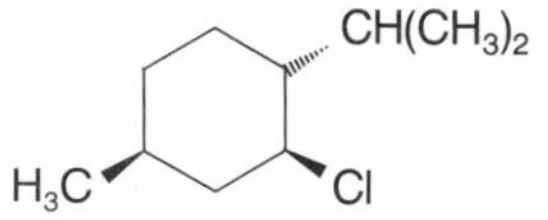

Menthyl chloride

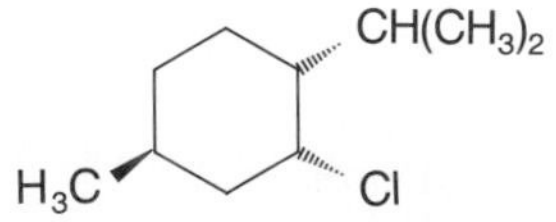

Neomenthyl chloride

6.9 STEREOSELECTIVITY IN E2 ELIMINATION REACTIONS

Some experimental results illustrating the stereoselectivity of dehydrohalogenation reactions were described earlier in Section 6.6. As noted there, alkyl halides such as *sec*-butyl bromide normally yield the more stable trans alkene as the major product on treatment with base.

$$CH_3CHCH_2CH_3 \xrightarrow[\text{CH}_3\text{CH}_2\text{OH, 70°C}]{\text{KOCH}_2\text{CH}_3} CH_2{=}CHCH_2CH_3 +$$

(structures of cis-2-Butene and trans-2-Butene shown)

2-Bromobutane   1-Butene (20%)   *cis*-2-Butene   *trans*-2-Butene
(*sec*-butyl bromide)   [(Z)-2-butene]   [(E)-2-butene]
                                                          (20%)   (60%)

We can understand the basis for this stereoselectivity with the aid of Figure 6.8. As illustrated there, methyl groups are gauche and move closer together in the activated complex leading to *cis*-2-butene. On the other hand, methyl groups are anti at the transition state leading to *trans*-2-butene. This transition state is more stable because it has less van der Waals strain, and trans-2-butene is formed faster than *cis*-2-butene.

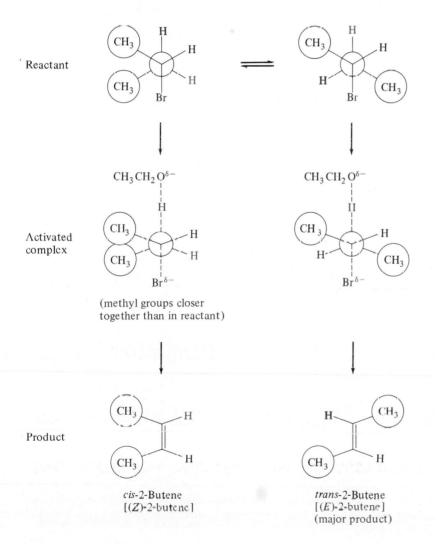

Reactant

Activated complex

(methyl groups closer together than in reactant)

Product

cis-2-Butene
[(Z)-2-butene]

trans-2-Butene
[(E)-2-butene]
(major product)

FIGURE 6.8 More *trans*-than *cis*-2-butene is formed when 2-bromobutane undergoes E2 elimination because the activated complex leading to the cis isomer is destabilized by a van der Waals repulsion between its two methyl groups.

6.10 A DIFFERENT MECHANISM FOR ALKYL HALIDE ELIMINATION. THE E1 MECHANISM

The E2 mechanism is a concerted process in which the carbon-hydrogen and carbon-halogen bonds undergo cleavage in the same elementary step. Both bonds are partially broken, though not necessarily to the same extent, at the transition state. What if these bonds broke in separate steps rather than in a concerted manner?

One possibility is the two-step mechanism of Figure 6.9, in which the carbon-halogen bond breaks first to give a carbocation intermediate, followed by deprotonation of the carbocation in a second step.

The alkyl halide, in this case 2-bromo-2-methylbutane, ionizes to a carbocation and a halide anion by a heterolytic cleavage of the carbon-halogen bond. Like the dissociation of an oxonium ion to a carbocation, this step is rate-determining. Because the rate-determining step is unimolecular — it involves only the alkyl halide and not the base — this mechanism is known by the symbol E1, standing for *elimination unimolecular*. It exhibits first-order kinetics.

$$\text{Rate} = k \text{ [alkyl halide]}$$

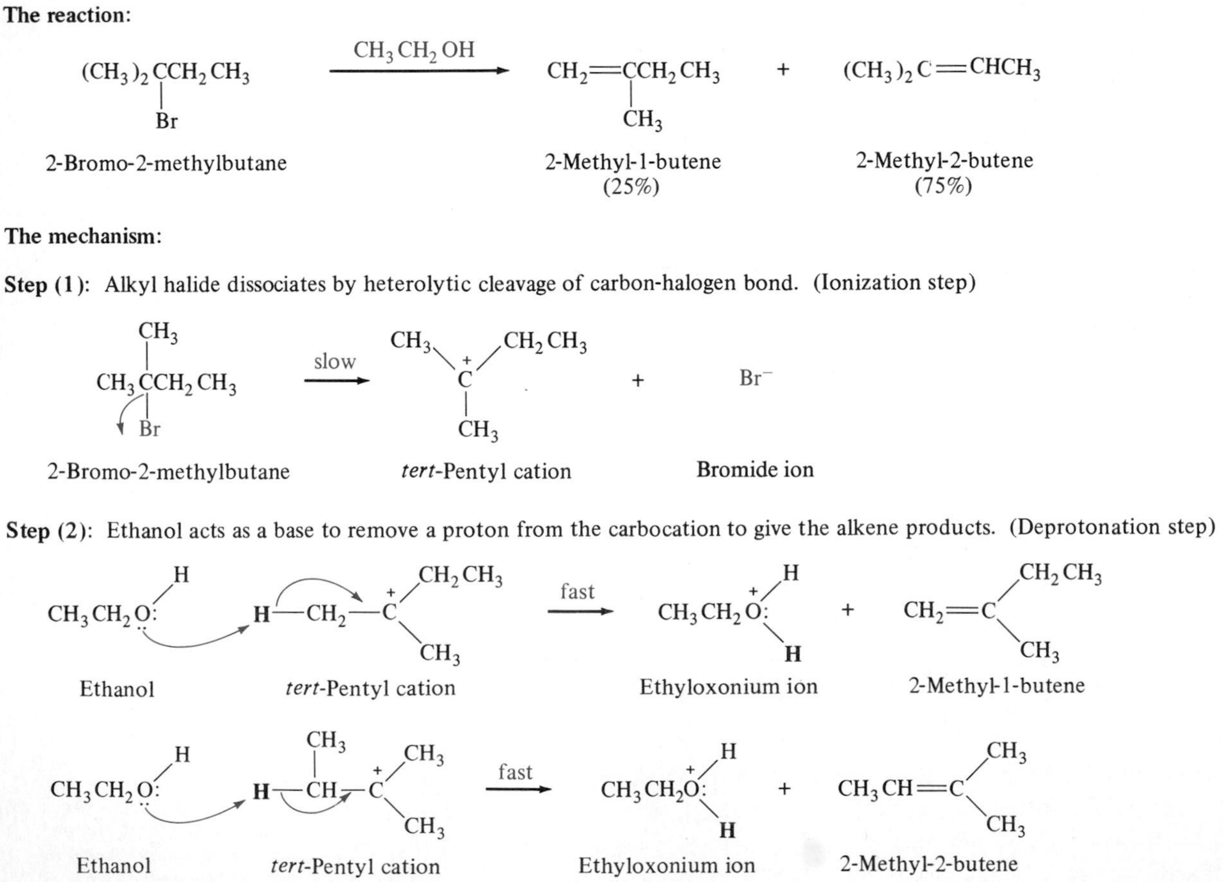

The reaction:

The mechanism:

Step (1): Alkyl halide dissociates by heterolytic cleavage of carbon-halogen bond. (Ionization step)

Step (2): Ethanol acts as a base to remove a proton from the carbocation to give the alkene products. (Deprotonation step)

FIGURE 6.9 Sequence of steps that describes the E1 mechanism for the dehydrohalogenation of 2-bromo-2-methylbutane in ethanol.

Typically, elimination by the E1 mechanism is important only for tertiary and some secondary alkyl halides, and then only when the base is weak or in low concentration. The reactivity order parallels the ease of carbocation formation.

Increasing rate of elimination by the E1 mechanism

$$RCH_2X \quad < \quad R_2CHX \quad < \quad R_3CX$$

Primary alkyl
halide: slowest
rate of E1 elimination

Tertiary alkyl
halide: fastest
rate of E1 elimination

Because the carbon-halogen bond breaks in the rate-determining step, the nature of the leaving group is important in terms of reaction rate. Alkyl iodides have the weakest carbon-halogen bond and are the most reactive; alkyl fluorides have the strongest carbon-halogen bond and are the least reactive.

The best examples of E1 eliminations are those carried out in the absence of added base. In the example cited in Figure 6.9, the base that abstracts the proton from the carbocation intermediate is a very weak one; it is a molecule of the solvent, ethyl alcohol. At even modest concentrations of strong base, elimination by the E2 mechanism is faster than E1 elimination.

There is a strong similarity between the mechanism shown in Figure 6.9 and the ones shown for alcohol dehydration in Figures 6.1 and 6.2. Indeed, we can describe the acid-catalyzed dehydration of alcohols as an E1 elimination of their conjugate acids. The principal difference between the dehydration of 2-methyl-2-butanol in Figure 6.2 and the dehydrohalogenation of 2-bromo-2-methylbutane in Figure 6.9 is the source of the carbocation. When the alcohol is the substrate, it is the corresponding oxonium ion that dissociates to form the carbocation. The alkyl halide ionizes directly to the carbocation.

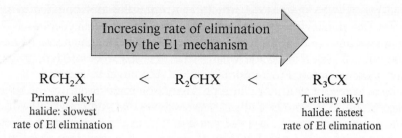

Alkyloxonium   Carbocation   Alkyl halide
ion

Like alcohol dehydrations, E1 reactions of alkyl halides can lead to alkenes with altered carbon skeletons resulting from carbocation rearrangements. Eliminations by the E2 mechanism, on the other hand, normally proceed without rearrangement. Consequently, if one wishes to prepare an alkene from an alkyl halide, conditions favorable to E2 elimination should be chosen. In practice this simply means carrying out the reaction in the presence of a strong base.

6.11 SYNTHETIC APPLICATIONS OF ELIMINATION REACTIONS

An important area of concern to chemists is synthesis, the task of preparing a particular compound in an economical way and with confidence that the synthetic method will lead precisely to the desired structure. Among the tools applied to synthetic objectives, functional group transformations are the most often encountered. In this section we will introduce the topic of synthesis, emphasizing the necessity of systematic planning in order to decide what is the best sequence of steps to convert a specified starting material to a desired product (the target molecule).

A critical feature of synthetic planning is to reason backward from the target to the starting material. A second is to always use reactions that you know will work. Let us examine a simple example. Suppose you had to prepare cyclohexane, given cyclohexanol as the starting material. We have encountered no reactions to this point that permit us to carry out the indicated conversion in a single step.

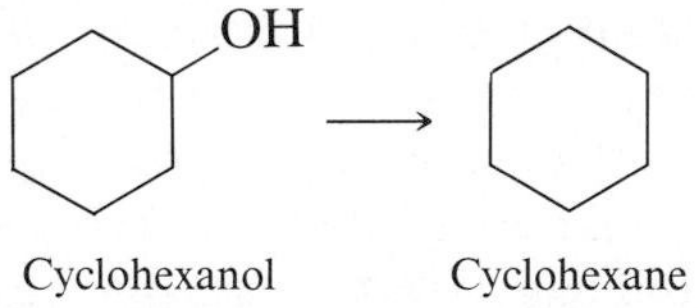

Cyclohexanol   Cyclohexane

Reasoning backward, however, we know that we can prepare cyclohexane by hydrogenation of cyclohexene. Therefore, let us use this reaction as the last step in our proposed synthesis.

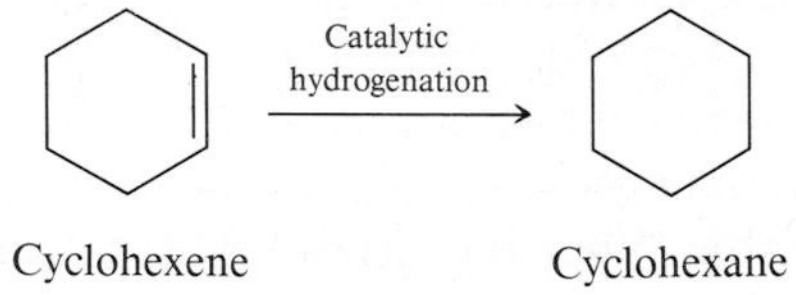

Cyclohexene   Cyclohexane

Recognizing that cyclohexene may be prepared by dehydration of cyclohexanol, a practical synthesis of cyclohexane from cyclohexanol becomes apparent.

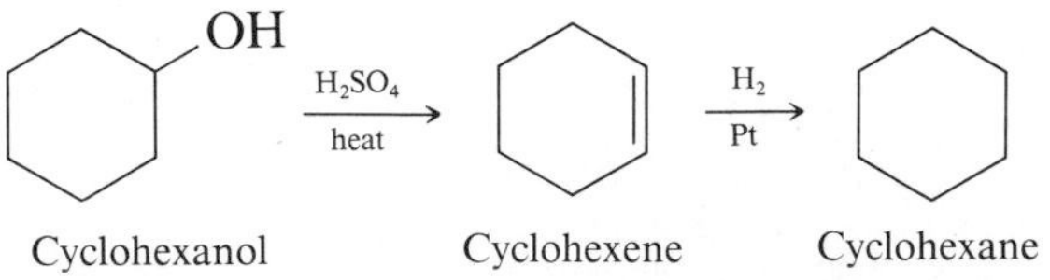

Cyclohexanol   Cyclohexene   Cyclohexane

As a second illustration of this principle, consider the preparation of cyclohexene, given cyclohexane as the starting material.

Cyclohexane   Cyclohexene

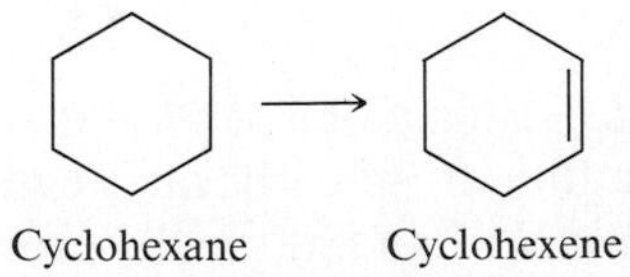

Begin by asking the question: What kind of compound is the target molecule and what methods are available for preparing that kind of compound? At this point in our coverage of organic chemistry, we know two methods for the laboratory preparation of alkenes, the dehydration of alcohols and the dehydrohalogenation of alkyl halides. Thus, a reasonable last step is the dehydrohalogenation of a cyclohexyl halide.

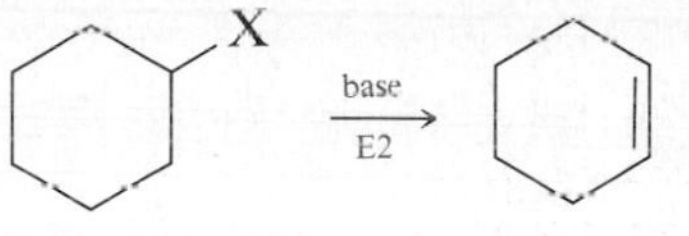

Cyclohexyl halide   Cyclohexene

We now have a new problem: Where does the necessary cyclohexyl halide come from? Alkyl halides are prepared from alcohols by reaction with hydrogen halides (Section 4.9) or from alkanes by free radical halogenation (Sections 4.24 and 4.25). Since our designated starting material is cyclohexane, we can combine free radical chlorination of cyclohexane with E2 elimination of cyclohexyl chloride to give cyclohexene.

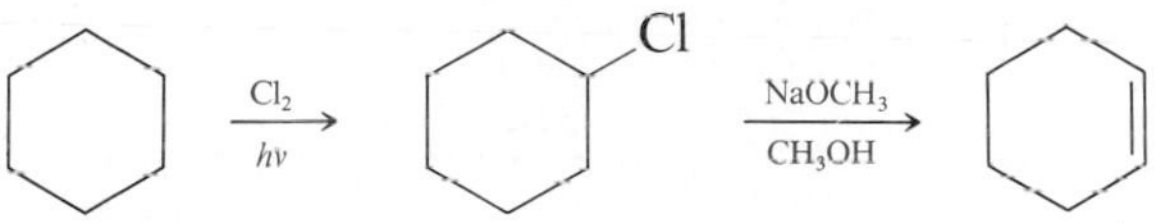

Cyclohexane   Cyclohexyl chloride   Cyclohexene

Often more than one synthetic route may be available to prepare a particular compound. Indeed, it is normal to find in the chemical literature that the same compound has been synthesized in a number of different ways. As we proceed through the text and develop a larger inventory of functional group transformations, our ability to evaluate alternative synthetic plans will increase. In most cases the best synthetic plan is the one with the fewest steps.

6.12 SUMMARY

Beta elimination reactions provide the basis for effective laboratory syntheses of alkenes. Aspects of the two elimination reactions described in this chapter, dehydration of alcohols and dehydrohalogenation of alkyl halides, are summarized in Table 6.1.

Secondary and tertiary alcohols undergo *dehydration* by way of carbocation intermediates.

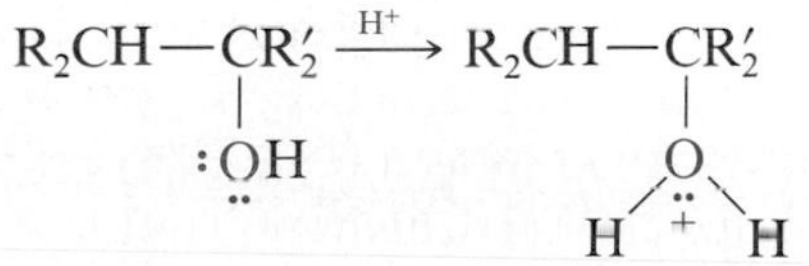

Alcohol   Alkyloxonium ion

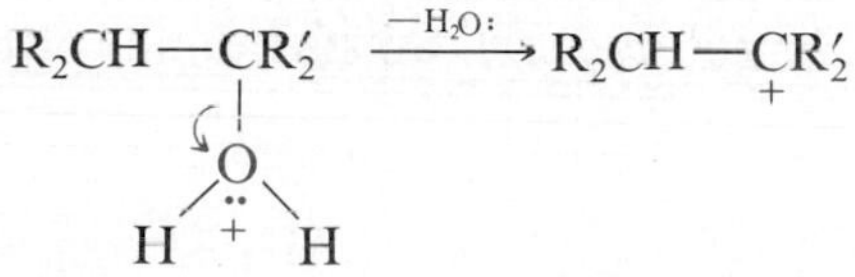

Alkyloxonium ion   Carbocation

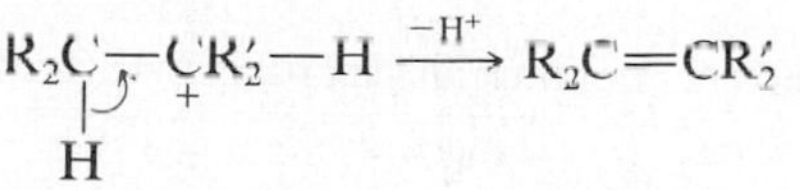

Carbocation   Alkene

TABLE 6.1

Preparation of Alkenes by Elimination Reactions of Alcohols and Alkyl Halides

Reaction (section) and comments	General equation and specific example
Dehydration of alcohols (Sections 6.2 through 6.5) Dehydration requires an acid catalyst; the order of reactivity of alcohols is tertiary > secondary > primary. Elimination is regioselective and proceeds in the direction that produces the most highly substituted double bond. When stereoisomeric alkenes are possible, the more stable one is formed in greater amounts. A carbocation intermediate is involved, and sometimes rearrangements take place during elimination.	$R_2CHCR_2' \xrightarrow{H^+} R_2C{=}CR_2' + H_2O$ OH Alcohol · · · · Alkene · · · · Water 2-Methyl-2-hexanol $\downarrow$ H_2SO_4, 80°C 2-Methyl-1-hexene (19%) · · · 2-Methyl-2-hexene (81%)
Dehydrohalogenation of alkyl halides (Sections 6.6 through 6.10) Strong bases cause a proton and a halide to be lost from adjacent carbons of an alkyl halide to yield an alkene. Regioselectivity is in accord with the Zaitsev rule. The order of halide reactivity is I > Br > Cl > F and tertiary > secondary > primary. A concerted E2 reaction pathway is followed, carbocations are not involved, and rearrangements do not normally occur. An anti periplanar arrangement of the proton being removed and the halide being lost is more stable than any other at the transition state.	$R_2CHCR_2' + :B^- \longrightarrow R_2C{=}CR_2' + H{-}B + X^-$ X Alkyl · · · Base · · · Alkene · · Conjugate · Halide halide · · · · · · · · · · · · · · · · acid of · base CH_3 Cl 1-Chloro-1-methylcyclohexane $\downarrow$ $KOCH_2CH_3$, CH_3CH_2OH, 100°C CH_2 · · · · · · · · · CH_3 Methylenecyclohexane (6%) · · · 1-Methylcyclohexene (94%)

Primary alcohols do not dehydrate as readily as secondary or tertiary alcohols, and their dehydration does not involve a carbocation. A proton is lost from the β carbon in the same step in which carbon-oxygen bond cleavage occurs.

Alkene preparation using alcohol dehydration as the synthetic method is complicated by *carbocation rearrangements*. A less stable carbocation can rearrange to a more stable one by an alkyl group migration or by a hydride shift, opening the possibility for alkene formation from two structurally distinct carbocation intermediates.

$$R{-}\underset{\underset{R}{|}}{\overset{\overset{G}{|}}{C}}{\overset{+}{-}}\underset{\underset{H}{|}}{C}{-}R \longrightarrow R{-}\underset{\underset{R}{|}}{\overset{+}{C}}{-}\underset{\underset{H}{|}}{\overset{\overset{G}{|}}{C}}{-}R$$

Secondary carbocation · · · · · · Tertiary carbocation

(G is a migrating group; it may be either a hydrogen or an alkyl group)

Dehydrohalogenation of alkyl halides by alkoxide bases is not complicated by rearrangements. The bimolecular E2 mechanism is a concerted process in which the base abstracts a proton from the β carbon atom while the bond between the halogen and the α carbon undergoes heterolytic cleavage.

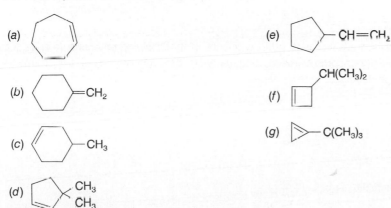

| Base | Alkyl halide | | Conjugate acid of base | Alkene | Halide ion |

In the absence of strong base, alkyl halides eliminate by the E1 mechanism. The E1 mechanism involves rate-determining ionization of the alkyl halide to a carbocation, followed by deprotonation of the carbocation.

PROBLEMS

6.9 How many alkenes would you expect to be formed from each of the following alkyl bromides under conditions of E2 elimination? Identify the alkenes in each case.

(a) 1-Bromohexane
(b) 2-Bromohexane
(c) 3-Bromohexane
(d) 2-Bromo-2-methylpentane

(e) 2-Bromo-3-methylpentane
(f) 3-Bromo-2-methylpentane
(g) 3-Bromo-3-methylpentane
(h) 2-Bromo-3,3-dimethylbutane

6.10 Write structural formulas for all the alkene products that could reasonably be formed from each of the following compounds under the indicated reaction conditions. Where more than one alkene is produced, specify the one that is the major product.

(a) 1-Bromo-3,3-dimethylbutane (potassium *tert*-butoxide, *tert*-butyl alcohol, 100°C)
(b) 1-Methylcyclopentyl chloride (sodium ethoxide, ethanol, 70°C)
(c) 3-Methyl-3-pentanol (sulfuric acid, 80°C)
(d) 2,3-Dimethyl-2-butanol (phosphoric acid, 120°C)
(e) 3-Iodo-2,4-dimethylpentane (sodium ethoxide, ethanol, 70°C)
(f) 2,4-Dimethyl-3-pentanol (sulfuric acid, 120°C)

6.11 Choose the compound of molecular formula $C_7H_{13}Br$ that gives each alkene shown as the exclusive product of E2 elimination.

(a)

(e) $-CH=CH_2$

(b) $=CH_2$

(f) $CH(CH_3)_2$

(c) $-CH_3$

(g) $-C(CH_3)_3$

(d) CH_3 CH_3

6.12   Give the structures of two different alkyl bromides both of which yield the indicated alkene as the exclusive product of E2 elimination.

(a) $CH_3CH{=}CH_2$

(c) $BrCH{=}CBr_2$

(b) $(CH_3)_2C{=}CH_2$

(d)

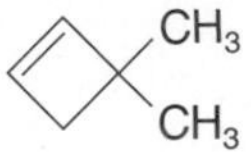

6.13   (a)   Write the structures of all the isomeric alkyl bromides having the molecular formula $C_5H_{11}Br$.
(b)   Which one undergoes E1 elimination at the fastest rate?
(c)   Which one undergoes E2 elimination at the fastest rate?
(d)   Which one is incapable of reacting by the E2 mechanism?
(e)   Which ones can only yield a single alkene on E2 elimination?
(f)   For which isomer is E2 elimination regiospecific but yet leads to the formation of two alkenes?
(g)   Which one yields the most complex mixture of alkenes on E2 elimination?

6.14   Arrange the compounds in each group in order of their decreasing rate of elimination toward potassium *tert*-butoxide in *tert*-butyl alcohol (100°C):

(a)   1-Bromopentane, 2-bromopentane, 2-bromo-2-methylbutane, 1-bromo-2,2-dimethylpropane
(b)   1-Chloropentane, 1-bromopentane, cyclopentyl bromide, cyclopentyl iodide
(c)   1-Bromo-2-methylbutane, 1-chloro-3-methylbutane, 2-bromo-2-methylbutane, 2-bromo-3-methylbutane
(d)   1-Methylcyclopentyl bromide, *cis*-2-methylcyclopentyl bromide, (bromomethyl)cyclopentane
(e)   *trans*-4-Methylcyclohexyl fluoride, *trans*-4-methylcyclohexyl bromide, *cis*-4-methylcyclohexyl bromide, 1-methylcyclohexyl bromide.

6.15   (a)   Write the structures of all the isomeric alcohols having the molecular formula $C_5H_{12}O$.
(b)   Which one will undergo acid-catalyzed dehydration most readily?
(c)   Write the structure of the most stable C_5H_{11} carbocation.
(d)   Which alkenes may be derived from the carbocation in (c)?
(e)   Which alcohols can yield the carbocation in (c) by a process involving a hydride shift?
(f)   Which alcohols can yield the carbocation in (c) by a process involving a methyl shift?

6.16   Identify the principal organic product of each of the following reactions. In spite of the structural complexity of some of the starting materials, the functional group transformations are all of the type described in this chapter.

(a)

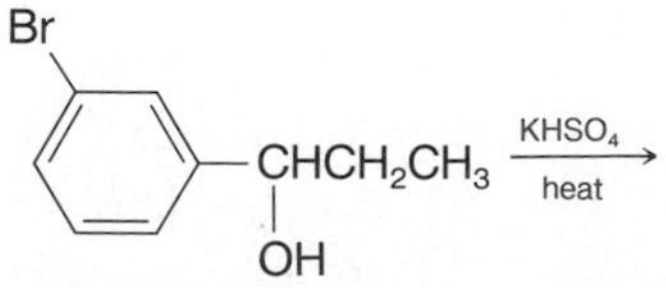

(b) $ICH_2CH(OCH_2CH_3)_2 \xrightarrow[\text{(CH}_3)_3\text{COH, heat}]{\text{KOC(CH}_3)_3}$

(c) 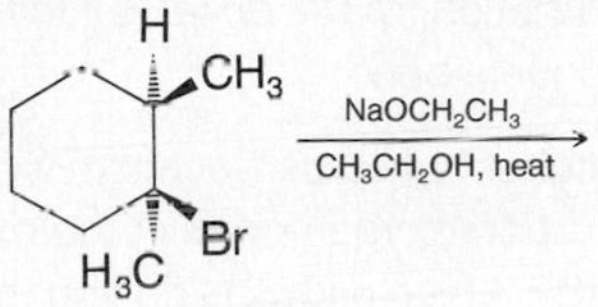
$$\xrightarrow[\text{CH}_3\text{CH}_2\text{OH, heat}]{\text{NaOCH}_2\text{CH}_3}$$

(d) 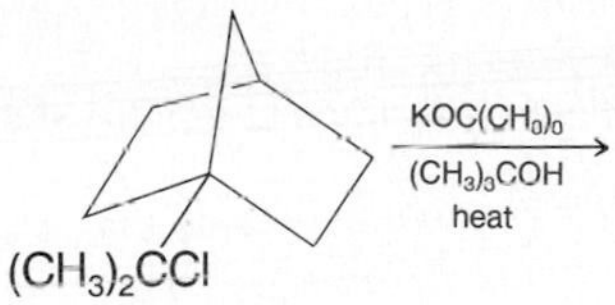
$$\xrightarrow[\substack{(\text{CH}_3)_3\text{COH} \\ \text{heat}}]{\text{KOC(CH}_3)_0}$$

(e) 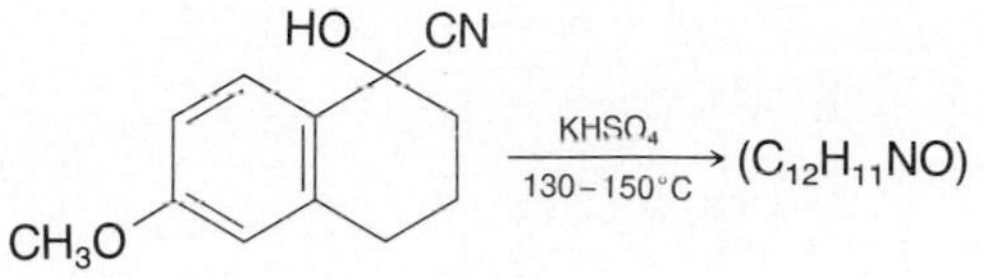
$$\xrightarrow[130-150°\text{C}]{\text{KHSO}_4} (\text{C}_{12}\text{H}_{11}\text{NO})$$

(f) 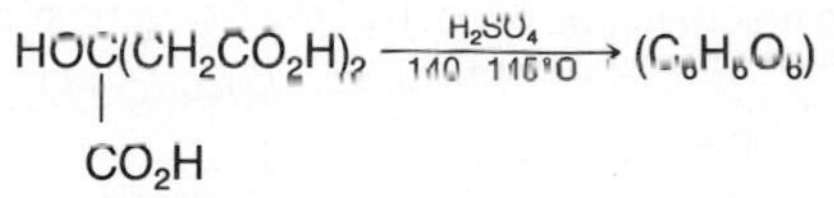 $\text{HOC(CH}_2\text{CO}_2\text{H})_2 \xrightarrow[140-145°\text{O}]{\text{H}_2\text{SO}_4} (\text{C}_6\text{H}_6\text{O}_6)$

$\quad\quad \overset{|}{\text{CO}_2\text{H}}$

Citric acid

(g) 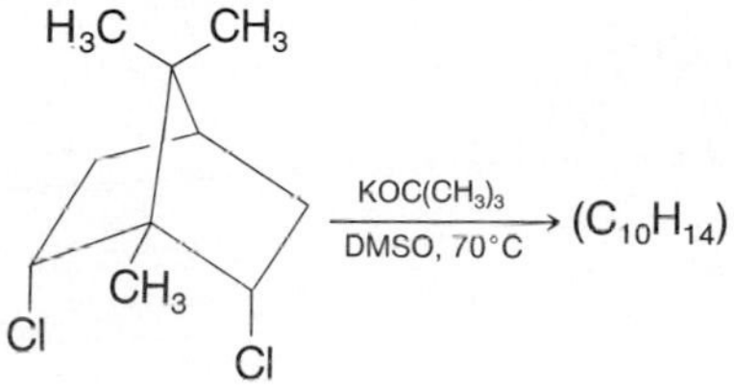
$$\xrightarrow[\text{DMSO, 70°C}]{\text{KOC(CH}_3)_3} (\text{C}_{10}\text{H}_{14})$$

(h)
$$\xrightarrow[\text{DMSO}]{\text{KOC(CH}_3)_3} (\text{C}_{14}\text{H}_{16}\text{O}_4)$$

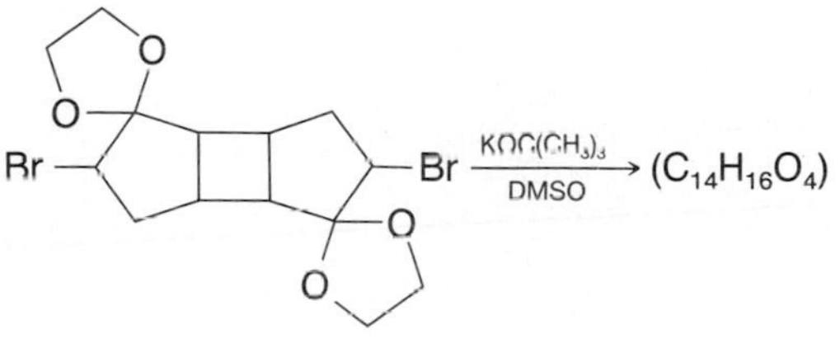

(i) 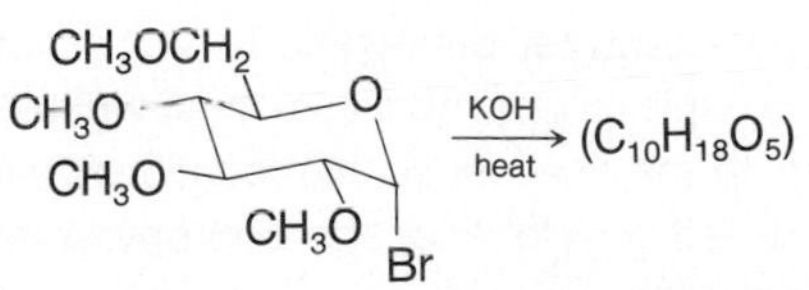
$$\xrightarrow[\text{heat}]{\text{KOH}} (\text{C}_{10}\text{H}_{18}\text{O}_5)$$

(j) 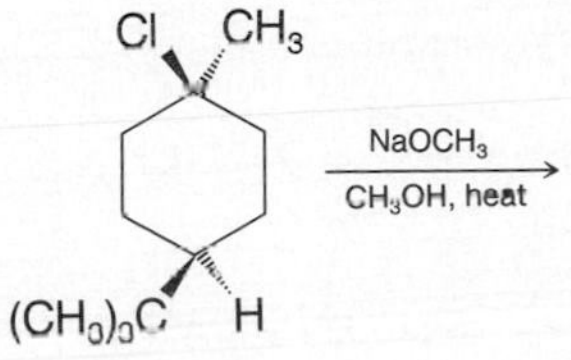
$$\xrightarrow[\text{CH}_3\text{OH, heat}]{\text{NaOCH}_3}$$

6.17 An apparent exception to the original formulation of the Zaitsev rule can be found in the elimination of 2-bromo-2,4,4-trimethylpentane, which reacts with potassium ethoxide in ethanol (70°C) to yield an alkene mixture containing 86 percent 2,4,4-trimethyl-1-pentene and 14 percent 2,4,4-trimethyl-2-pentene. Given that the heats of hydrogenation of the two alkenes are 27.2 and 28.4 kcal/mol, respectively, suggest an explanation for the observation that the major alkene is the one with the less highly substituted double bond.

6.18 In the acid-catalyzed dehydration of isobutyl alcohol, what carbocation would be formed if a hydride shift accompanied cleavage of the carbon-oxygen bond in the oxonium ion? What ion would be formed as a result of a methyl shift? Which pathway do you think will predominate, a hydride shift or a methyl shift?

6.19 Each of the following carbocations has the potential to rearrange to a more stable one. Write the structure of the rearranged carbocation.

(a) $CH_3CH_2CH_2{}^+$

(b) $(CH_3)_2CH\overset{+}{C}HCH_3$

(c) $(CH_3)_3C\overset{+}{C}HCH_3$

(d) $(CH_3CH_2)_3CCH_2{}^+$

(e)

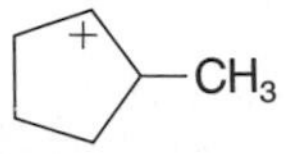

6.20 Write a sequence of steps depicting the mechanisms of each of the following reactions:

(a)

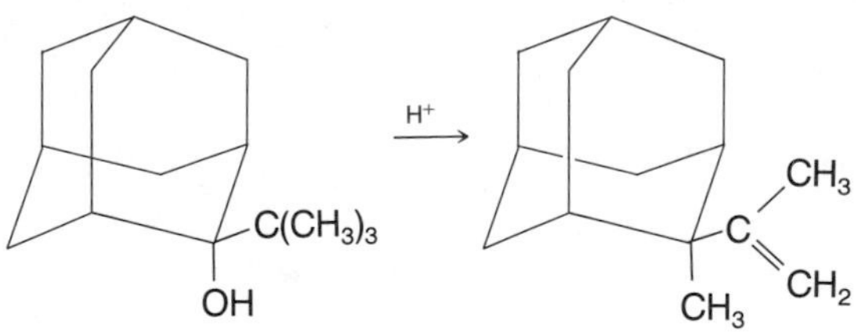

(b)

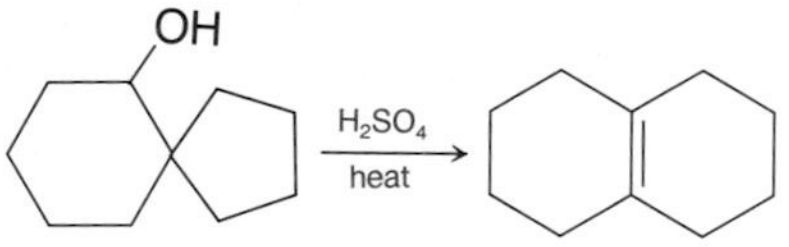

(c)

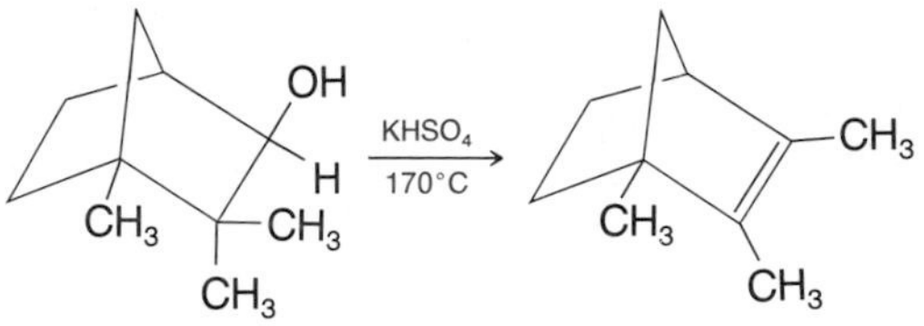

6.21 In Problem 6.5 (Section 6.5) we saw that acid-catalyzed dehydration of 2,2-dimethylcyclohexanol afforded 1,2-dimethylcyclohexene. To explain this product we must write a mechanism for the reaction in which a methyl shift transforms a secondary carbocation to a tertiary one. Another product of the dehydration of 2,2-dimethylcyclohexanol is isopropylidenecyclopentane. Write a mechanism to rationalize its formation.

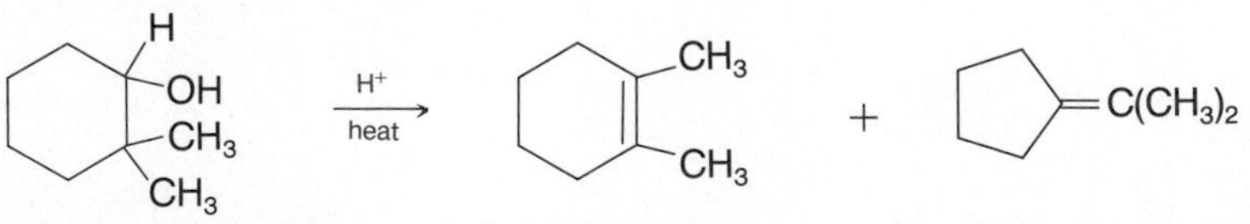

2,2-Dimethylcyclohexanol   1,2-Dimethylcyclohexene   Isopropylidenecyclopentane

6.22 On being heated with a solution of sodium ethoxide in ethanol, compound A ($C_7H_{15}Br$) yielded a mixture of two alkenes B and C, each having the molecular formula C_7H_{14}. Catalytic hydrogenation of the major isomer B or the minor isomer C gave only 3-ethylpentane. Suggest structures for compounds A, B, and C consistent with these observations.

6.23 Compound D (C_4H_{10}) gives two different monochlorides on photochemical chlorination. Treatment of either of these monochlorides with potassium *tert*-butoxide in dimethyl sulfoxide gives the same alkene E (C_4H_8) as the only product. What are the structures of compound D, the two monochlorides, and alkene E?

6.24 Compound F (C_6H_{14}) gives three different monochlorides on photochemical chlorination. One of these monochlorides is inert to E2 elimination. The other two monochlorides yield the same alkene G (C_6H_{12}) on being heated with potassium *tert*-butoxide in *tert*-butyl alcohol. Identify compound F, the three monochlorides, and alkene G.

6.25 Compound H ($C_7H_{15}Br$) is not a primary alkyl bromide. It yields a single alkene (compound I) on being heated with sodium ethoxide in ethanol. Hydrogenation of compound I yields 2,4-dimethylpentane. Identify compounds H and I.

6.26 Compounds J and K are isomers of molecular formula $C_9H_{19}Br$. Both yield the same alkene L as the exclusive product of elimination on being treated with potassium *tert*-butoxide in dimethyl sulfoxide. Hydrogenation of alkene L gives 2,3,3,4-tetramethylpentane. What are the structures of compounds J and K and alkene L?

6.27 Alcohol M ($C_{10}H_{18}O$) is converted to a mixture of alkenes N and O on being heated with potassium hydrogen sulfate ($KHSO_4$). The structure of alkene N is given below. Catalytic hydrogenation of N and O yields the same product. Assuming that dehydration of alcohol M proceeds without rearrangement, deduce the structures of alcohol M and alkene O

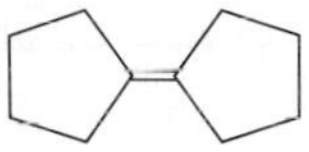

6.28 Suggest a sequence of reactions suitable for preparing each of the following compounds from the indicated starting material. You may use any necessary organic or inorganic reagents.

(a) 2-Methylhexane from 2-methyl-2-hexanol
(b) 1-Methylcyclopentene from methylcyclopentane
(c) (Z)-Cyclodecene from cyclodecane

REACTIONS OF ALKENES. ELECTROPHILIC ADDITION REACTIONS

Now that we know something of the structural nature of alkenes and have seen how they can be prepared by elimination reactions, we are ready to look at their chemical reactions. The characteristic reaction of alkenes is *addition* to the carbon-carbon double bond. You already know of one example of addition to alkenes, namely, their hydrogenation to alkanes.

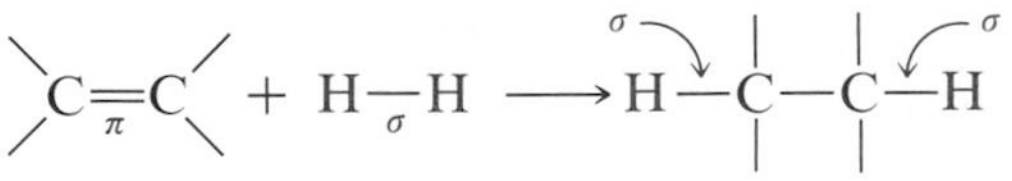

$$\underset{\text{Alkene}}{\overset{\displaystyle >\!\!C\underset{\pi}{=\!\!=}\!C\!\!<}{}} + \underset{\text{Hydrogen}}{\overset{\displaystyle H\underset{\sigma}{-\!\!}H}{}} \longrightarrow \underset{\text{Alkane}}{\overset{\displaystyle H-\overset{|}{\underset{|}{C}}-\overset{|}{\underset{|}{C}}-H}{}}$$

The elements of H_2 add to the alkene; the π component of the double bond and the σ bond of a hydrogen molecule are transformed into two σ C—H bonds of an alkane.

A large number of addition reactions of alkenes are known in which the attacking reagent, unlike H_2, is an ionic or easily polarizable molecule. If we represent the attacking reagent as $^{\delta+}E—Y^{\delta-}$, where E is an atom or group that has positive *(electrophilic)* character, the general equation for an *electrophilic addition* reaction is

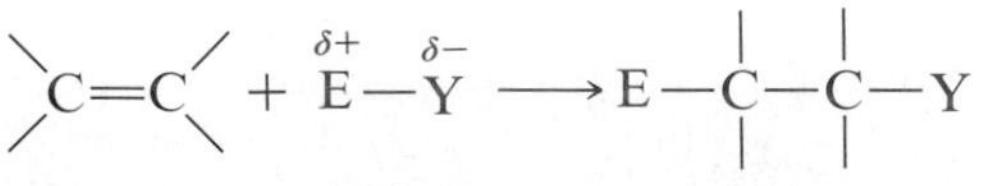

$$\underset{\text{Alkene}}{\overset{\displaystyle >\!\!C\!=\!\!C\!\!<}{}} + \underset{\substack{\text{Attacking} \\ \text{reagent}}}{\overset{\displaystyle \overset{\delta+}{E}-\overset{\delta-}{Y}}{}} \longrightarrow \underset{\substack{\text{Product of} \\ \text{electrophilic addition}}}{\overset{\displaystyle E-\overset{|}{\underset{|}{C}}-\overset{|}{\underset{|}{C}}-Y}{}}$$

The chemical equation for electrophilic addition to alkenes is similar to that for catalytic hydrogenation. The reactions are fundamentally different, however, with respect to their mechanisms.

The carbon-carbon double bond is a functional group in an alkene. The characteristic reaction of the carbon-carbon double bond is electrophilic addition, and that is the central topic of this chapter.

7.1 ADDITION OF HYDROGEN HALIDES TO ALKENES. A MODEL FOR ELECTROPHILIC ADDITION

Hydrogen halides add to alkenes to give alkyl halides.

$$\underset{\text{Alkene}}{\Large{\diagdown}\text{C}=\text{C}{\diagup}} + \underset{\text{Hydrogen halide}}{\text{HX}} \longrightarrow \underset{\text{Alkyl halide}}{\text{H}-\overset{|}{\underset{|}{\text{C}}}-\overset{|}{\underset{|}{\text{C}}}-\text{X}}$$

Ethyl chloride is prepared on an industrial scale by the addition of hydrogen chloride gas to ethylene.

$$\underset{\text{Ethylene}}{CH_2{=}CH_2} + \underset{\text{Hydrogen chloride}}{HCl} \xrightarrow[\text{AlCl}_3\text{ or FeCl}_3]{150-250°C} \underset{\text{Ethyl chloride}}{CH_3CH_2Cl}$$

Addition of hydrogen halides to alkenes also occurs rapidly in a variety of solvents, including pentane, benzene, dichloromethane, chloroform, and acetic acid.

$$\underset{cis\text{ 3-Hexene}}{\underset{H}{\overset{CH_3CH_2}{\diagdown}}\text{C}=\text{C}\underset{H}{\overset{CH_2CH_3}{\diagup}}} + \underset{\text{Hydrogen bromide}}{HBr} \xrightarrow[\text{CHCl}_3]{-30°C} \underset{\text{3-Bromohexane (76\%)}}{CH_3CH_2CH_2\underset{\underset{Br}{|}}{C}HCH_2CH_3}$$

The combination of potassium iodide and phosphoric acid (Section 4.9) is a convenient substitute for hydrogen iodide.

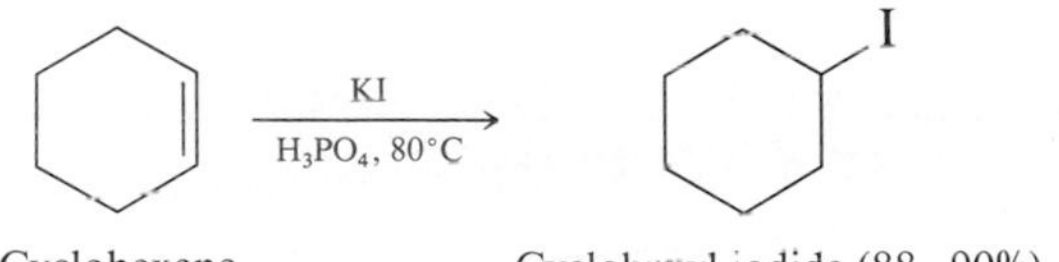

$$\underset{\text{Cyclohexene}}{\bigcirc} \xrightarrow[\text{H}_3\text{PO}_4,\ 80°C]{KI} \underset{\text{Cyclohexyl iodide (88–90\%)}}{\bigcirc\text{—I}}$$

The order of reactivity of the hydrogen halides reflects their ability to donate a proton. Hydrogen iodide is the strongest acid of the hydrogen halides and reacts with alkenes at the fastest rate.

Increasing reactivity of hydrogen halides in addition to alkenes

$$HF \ll HCl < HBr < HI$$

Slowest rate of addition;	Fastest rate of addition;
least acidic	most acidic

A general understanding of the mechanism of hydrogen halide addition to alkenes can be achieved by extending some of the principles of reaction mechanisms introduced earlier. In Section 6.4 we pointed out that carbocations are the conjugate acids of alkenes. Acid-base reactions are reversible processes. An alkene, therefore, can accept a proton from a hydrogen halide to form a carbocation.

$$R_2C=CR_2 \;+\; H-\overset{..}{\underset{..}{X}}: \;\rightleftharpoons\; R_2\overset{+}{C}-\overset{\overset{\displaystyle H}{|}}{C}R_2 \;+\; :\overset{..}{\underset{..}{X}}:^-$$

Alkene (base)   Hydrogen halide (acid)   Carbocation (conjugate acid)   Anion (conjugate base)

We have also seen (Section 4.14) that carbocations, when generated in the presence of halide anions, react with them to form alkyl halides.

$$R_2\overset{+}{C}-\overset{\overset{\displaystyle H}{|}}{C}R_2 \;+\; :\overset{..}{\underset{..}{X}}:^- \;\longrightarrow\; R_2C-\overset{\overset{\displaystyle H}{|}}{\underset{\underset{\displaystyle :\overset{..}{\underset{..}{X}}:}{|}}{C}}R_2$$

Carbocation (electrophile)   Halide ion (nucleophile)   Alkyl halide

Both steps in this general mechanism are based on precedent. It is called *electrophilic addition* because the reaction is triggered by the attack of an electrophile (a proton donor) on the π electrons of the carbon-carbon double bond. Alkenes are weak bases and the site of that basicity is the π component of the double bond. Transfer of the two π electrons to an electrophile generates a carbocation as a reactive intermediate; normally this is the rate-determining step. The general mechanism can be elaborated upon by considering how certain experimental observations bear on the details of the individual steps. Some of the most important observations are concerned with the regioselectivity of addition.

7.2 REGIOSELECTIVITY OF HYDROGEN HALIDE ADDITION. MARKOVNIKOV'S RULE

In principle a hydrogen halide can add to an unsymmetrical alkene (an alkene in which the two carbons of the double bond are not equivalently substituted) in either of two ways. In practice, addition is so highly regioselective as to be considered regiospecific.

$$RCH=CH_2 + HX \longrightarrow \underset{\underset{\displaystyle X \;\;\; H}{|\;\;\;\;\;|}}{RCH-CH_2} \quad \text{rather than} \quad \underset{\underset{\displaystyle H \;\;\; X}{|\;\;\;\;\;|}}{RCH-CH_2}$$

$$R_2C=CH_2 + HX \longrightarrow \underset{\underset{\displaystyle X \;\;\; H}{|\;\;\;\;\;|}}{R_2C-CH_2} \quad \text{rather than} \quad \underset{\underset{\displaystyle H \;\;\; X}{|\;\;\;\;\;|}}{R_2C-CH_2}$$

$$R_2C=CHR + HX \longrightarrow \underset{\underset{\displaystyle X \;\;\; H}{|\;\;\;\;\;|}}{R_2C-CHR} \quad \text{rather than} \quad \underset{\underset{\displaystyle H \;\;\; X}{|\;\;\;\;\;|}}{R_2C-CHR}$$

In 1870 Vladimir Markovnikov, a colleague of Alexander Zaitsev at the University of Kazan, noting the pattern of hydrogen halide addition to alkenes, assembled his observations into a simple statement. *Markovnikov's rule* states that when an unsymmetrically substituted alkene reacts with a hydrogen halide, the hydrogen adds to the carbon that has the greater number of hydrogen substituents, and the halogen adds to the carbon having fewer hydrogen substituents. The general equations set forth above illustrate regioselective addition in accordance with Markovnikov's rule, while the equations that follow provide some specific examples.

$$CH_3CH_2CH{=}CH_2 + \quad HBr \quad \xrightarrow{\text{acetic acid}} \quad CH_3CH_2CHCH_3$$
$$\underset{\displaystyle Br}{|}$$

| 1-Butene | Hydrogen bromide | 2-Bromobutane (*sec*-butyl bromide), 80% |

$$\underset{H_3C}{\overset{H_3C}{>}}C{=}CH_2 + \quad HBr \quad \xrightarrow{\text{acetic acid}} \quad CH_3{-}\underset{\displaystyle CH_3}{\overset{\displaystyle CH_3}{\underset{|}{\overset{|}{C}}}}{-}Br$$

| 2-Methylpropene (isobutene) | Hydrogen bromide | 2-Bromo-2-methylpropane (*tert*-butyl bromide), 90% |

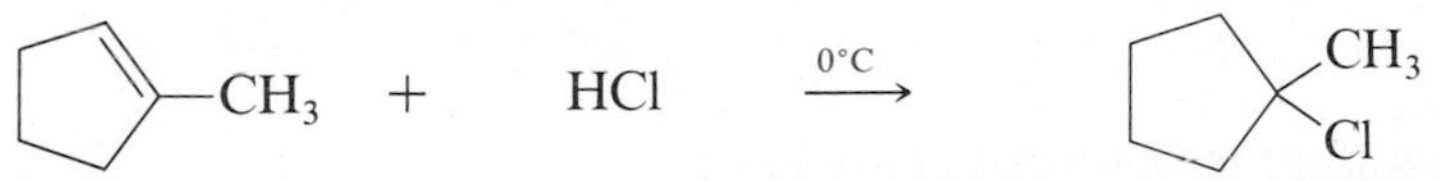

1-Methylcyclopentene + HCl $\xrightarrow{0°C}$ 1-Chloro-1-methylcyclopentane 100%

Hydrogen chloride

PROBLEM 7.1 Write the structure of the principal organic product formed in the addition of hydrogen chloride to each of the following:

(*a*) 2-Methyl-2-butene
(*b*) 2-Methyl-1-butene
(*c*) *cis*-2-Butene

(*d*) $CH_3CH{=}$ ⬡

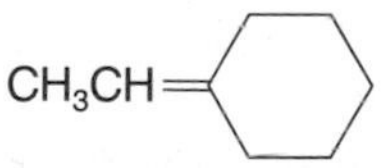

SAMPLE SOLUTION (*a*) Hydrogen chloride adds to the double bond of 2-methyl-2-butene in accordance with Markovnikov's rule. Hydrogen adds to the carbon that has one hydrogen substituent, chlorine to the carbon that has none.

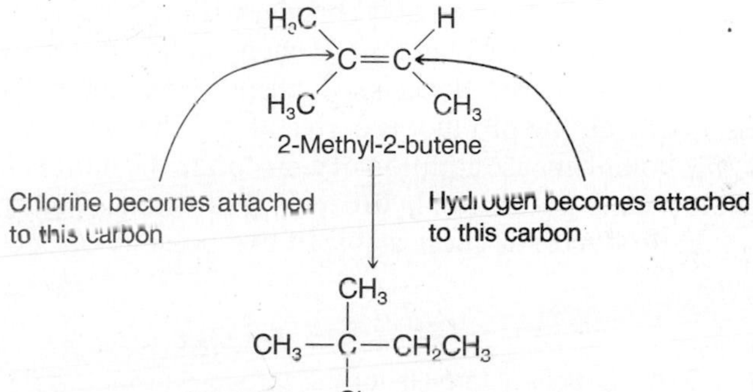

$$\underset{H_3C}{\overset{H_3C}{>}}C{=}C\underset{CH_3}{\overset{H}{<}}$$

2-Methyl-2-butene

Chlorine becomes attached to this carbon

Hydrogen becomes attached to this carbon

$$CH_3{-}\underset{\displaystyle Cl}{\overset{\displaystyle CH_3}{\underset{|}{\overset{|}{C}}}}{-}CH_2CH_3$$

2-Chloro-2-methylbutane

(Major product from Markovnikov addition of hydrogen chloride to 2-methyl-2-butene)

Markovnikov's rule, like Zaitsev's, is an empirical rule. It organizes experimental observations in a form suitable for predicting the major product of a reaction. A fuller appreciation of the reasons why it works arises from considering the mechanism of electrophilic addition in more detail.

7.3 MECHANISTIC BASIS FOR REGIOSELECTIVE ELECTROPHILIC ADDITION TO ALKENES

In the reaction of a hydrogen halide HX with an unsymmetrically substituted alkene $RCH=CH_2$, let us compare the carbocation intermediates for addition of HX according to Markovnikov's rule and contrary to Markovnikov's rule.

Addition according to Markovnikov's rule:

$$RCH=CH_2 \ \xrightarrow{\ \ H-X\ \ } \ RCH-\overset{+}{C}H_2 + \ X^- \ \longrightarrow \ RCHCH_3$$

Secondary carbocation	Halide ion	Observed product

$RCH-\overset{+}{C}H_2$ with H below; $RCHCH_3$ with X below; observed product.

Addition contrary to Markovnikov's rule:

$$RCH=CH_2 \longrightarrow RCH-\overset{+}{C}H_2 + \ X^- \longrightarrow RCH_2CH_2X$$

Primary carbocation   Halide ion   Not formed

By an extension of Hammond's postulate (Section 4.15), the transition state for proton transfer to the double bond resembles the carbocation intermediate more than it resembles the alkene, and the activation energy for formation of the more stable carbocation (secondary) is less than that for formation of the less stable (primary) one. Figure 7.1 is a potential energy diagram illustrating these two competing modes of hydrogen halide addition to an unsymmetrical alkene. Both carbocations are rapidly captured by X^- to give an alkyl halide, with the major product derived from the carbocation that is formed faster. The energy difference between a primary carbocation and a secondary carbocation is so high and their rates of formation are so different that essentially all the product is derived from the secondary carbocation. Markovnikov's rule holds because addition of a proton to the doubly bonded carbon that already has the greater number of hydrogen substituents is highly regioselective in the direction that produces the more stable of two possible carbocation intermediates.

PROBLEM 7.2 Give a structural formula for the carbocation intermediate that leads to the principal product in each of the reactions of Problem 7.1 (Section 7.2).

SAMPLE SOLUTION (a) Proton transfer to the carbon-carbon double bond of 2-

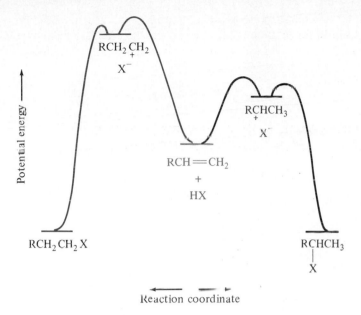

FIGURE 7.1 Energy diagram comparing addition of a hydrogen halide to an alkene in the direction corresponding to Markovnikov's rule and in the direction opposite to Markovnikov's rule. The alkene and hydrogen halide are shown in the center of the diagram. The lower-energy pathway that corresponds to Markovnikov's rule proceeds to the right and is shown in red; the higher-energy pathway proceeds to the left and is shown in blue.

methyl-2-butene can occur in a direction that leads to a tertiary carbocation or to a secondary carbocation.

The product of the reaction is derived from the more stable carbocation—in this case, it is a tertiary carbocation that is formed more rapidly than a secondary one.

Figure 7.2 depicts the electrophilic attack of a hydrogen halide on the π electrons of an unsymmetrical alkene, showing successive stages in the process of electrophilic addition. The hydrogen halide approaches the π orbital of the alkene in (*a*). As electrons flow from the π system of the alkene to the hydrogen halide in (*b*), the hydrogen-halogen bond begins to break. The π electron distribution in the alkene becomes distorted so that positive charge begins to develop at the carbon that can better support it, namely, the carbon that bears an electron-releasing alkyl group. The structure in (*c*) is that of the carbocation intermediate associated with a halide anion. Carbocation-anion combination completes the process.

Another fact that points to the intermediacy of carbocations in these reactions is the observation that rearrangements of the carbon skeleton analogous to those described in Section 6.5 sometimes occur.

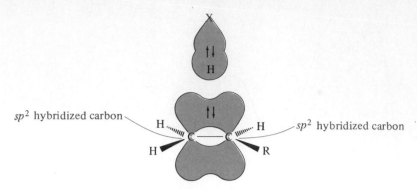

(a) Approach of a hydrogen halide to an alkene:

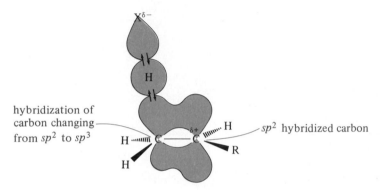

(b) Interaction of hydrogen halide with π electrons of the alkene as progress is made toward the transition state:

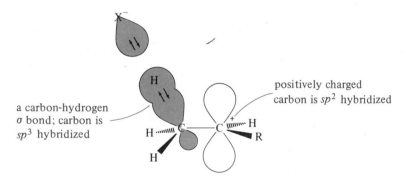

(c) A carbocation is formed. The π electrons of the alkene are used to form a carbon-hydrogen σ bond. The positive charge in the carbocation is located on the carbon that bears an electron-releasing alkyl substituent:

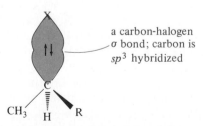

FIGURE 7.2 Description of electron flow and orbital interactions in the electrophilic addition of a hydrogen halide to an unsymmetrical alkene.

(d) Attack of the anion on the carbocation yields the alkyl halide product:

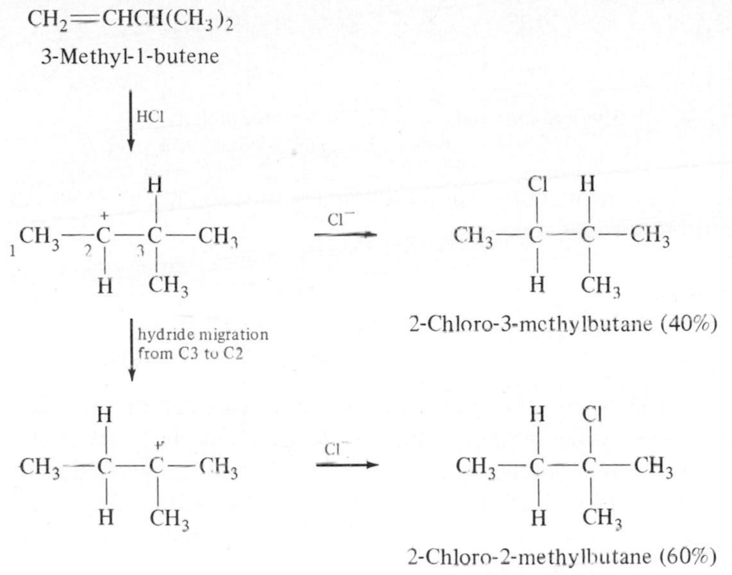

$CH_2\!=\!CHCH(CH_3)_2$

3-Methyl-1-butene

↓ HCl

$\overset{1}{C}H_3 - \overset{+}{\underset{2}{C}} - \overset{3}{C} - CH_3$ (with H above and below C2, H above and CH₃ below C3)

→ Cl⁻ →

$CH_3 - \overset{Cl}{\underset{H}{C}} - \overset{H}{\underset{CH_3}{C}} - CH_3$

2-Chloro-3-methylbutane (40%)

↓ hydride migration from C3 to C2

$CH_3 - \overset{H}{\underset{H}{C}} - \overset{+}{\underset{CH_3}{C}} - CH_3$

→ Cl⁻ →

$CH_3 - \overset{H}{\underset{H}{C}} - \overset{Cl}{\underset{CH_3}{C}} - CH_3$

2-Chloro-2-methylbutane (60%)

FIGURE 7.3 Sequence of steps describing electrophilic addition of hydrogen chloride to 3-methyl-1-butene. Addition can either occur in the manner expected to form 2-chloro-3-methylbutane or be accompanied by a hydride shift to form 2-chloro-2-methylbutane.

$CH_2\!=\!CHCH(CH_3)_2 \xrightarrow[0°C]{HCl} CH_3\underset{Cl}{CHCH(CH_3)_2} + CH_3CH_2\underset{Cl}{C(CH_3)_2}$

3-Methyl-1-butene   2-Chloro-3-methylbutane   2-Chloro-2-methylbutane
(40%)   (60%)

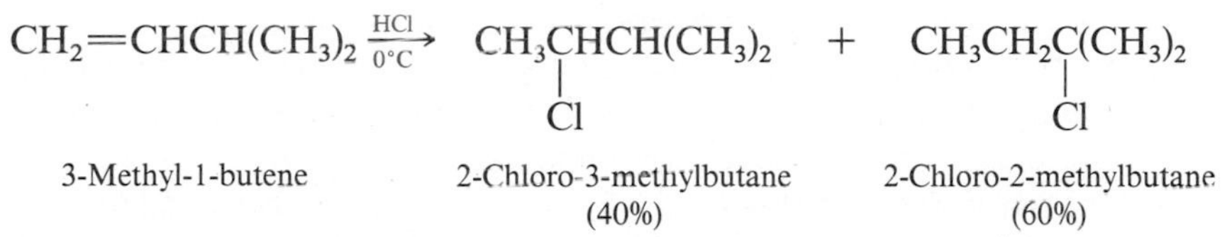

In this reaction as depicted in Figure 7.3, the initially formed secondary carbocation can either be captured by chloride ion to give 2-chloro-3-methylbutane or can rearrange to a more stable tertiary carbocation by a hydride shift. Subsequent capture of the tertiary carbocation yields 2-chloro-2-methylbutane.

In general, alkyl substituents on the double bond increase the reactivity of alkenes toward electrophilic addition of hydrogen halides. Alkyl groups are electron-releasing, and the more *electron-rich* a double bond is the better it is able to donate its π electrons to an electrophilic reagent. Along with the observed regioselectivity of addition, this supports the idea that carbocation formation, rather than carbocation capture, is rate-determining.

7.4 FREE-RADICAL ADDITION OF HYDROGEN BROMIDE TO ALKENES

For a long time the reactions of hydrogen bromide with alkenes were unpredictable in that sometimes addition occurred according to Markovnikov's rule while at other times, seemingly under the same conditions, the opposite regioselectivity *(anti-Markovnikov addition)* was observed. In 1929 Morris S. Kharasch and his students at the University of Chicago began a systematic investigation of this phenomenon. On the basis of hundreds of experiments, Kharasch discovered that anti-Markovnikov addition was promoted when peroxides, i.e., organic compounds of the type ROOR, were present in the reaction mixture. He and his colleagues found, for example, that carefully purified 1-butene reacted with hydrogen bromide to give only 2-bromobutane—the product expected on the basis of Markovnikov's rule.

$$CH_2{=}CHCH_2CH_3 \ + \qquad HBr \quad \xrightarrow[\text{peroxides}]{\text{no}} \quad CH_3\underset{\underset{\textstyle Br}{|}}{C}HCH_2CH_3$$

| 1-Butene | Hydrogen bromide | 2-Bromobutane |
| | | (only product; 90% yield) |

On the other hand when the same reaction was performed in the presence of an added peroxide, only 1-bromobutane was formed.

$$CH_2{=}CHCH_2CH_3 \ + \qquad HBr \quad \xrightarrow{\text{peroxides}} \quad BrCH_2CH_2CH_2CH_3$$

| 1-Butene | Hydrogen bromide | 1-Bromobutane |
| | | (only product, 95% yield) |

Kharasch termed this phenomenon the *peroxide effect* and reasoned that it could occur even if peroxides were not deliberately added to the reaction mixture. Unless alkenes are protected from atmospheric oxygen, they become contaminated with small amounts of alkyl hydroperoxides, compounds of the type ROOH. These alkyl hydroperoxides act in the same way as deliberately added peroxides to promote addition in the direction opposite to that predicted on the basis of Markovnikov's rule.

PROBLEM 7.3   Kharasch's earliest studies in this area were carried out in collaboration with Frank R. Mayo. Mayo performed over 400 experiments in which allyl bromide (3-bromo-1-propene) was treated with hydrogen bromide under a variety of conditions and determined the distribution of the "normal" and "abnormal" products formed during the reaction. What two products were formed? Which is the product of addition in accordance with Markovnikov's rule? Which one corresponds to addition contrary to the rule?

Kharasch proposed that hydrogen bromide can add to alkenes by two different mechanisms, both of which are, in modern terminology, regiospecific. The first mechanism is the one we discussed in the preceding section, electrophilic addition, and follows Markovnikov's rule. It is the mechanism followed when care is taken to ensure that no peroxides are present. The second mechanism is a free-radical chain process, initiated by homolytic cleavage of the weak oxygen-oxygen bond of a peroxide. It is presented in Figure 7.4.

Peroxides are *initiators* in this process; they are not incorporated into the product but act as a source of radicals necessary to get the chain started. The oxygen-oxygen bond dissociation energies of most peroxides are in the 35 to 50 kcal/mol range, and the free-radical addition of hydrogen bromide to alkenes begins when a peroxide molecule undergoes homolytic cleavage to two alkoxy radicals. This is depicted in step 1 of Figure 7.4. A bromine atom is generated in step 2 when one of these alkoxy radicals abstracts a proton from hydrogen bromide. Once a bromine atom becomes available, the propagation phase of the chain reaction begins. In the propagation phase as shown in step 3, a bromine atom adds to the alkene in the direction that produces the more stable alkyl radical.

The addition of a bromine atom to C-1 is represented by

$$\overset{4}{C}H_3\overset{3}{C}H_2\overset{2}{C}H{=}\overset{1}{C}H_2 \qquad \longrightarrow \qquad CH_3CH_2\overset{\bullet}{C}H{-}CH_2$$
$$\qquad\qquad\qquad\qquad\qquad :\overset{\bullet}{\underset{\bullet\bullet}{Br}}: \qquad\qquad\qquad\qquad :\overset{\bullet\bullet}{\underset{\bullet\bullet}{Br}}:$$

Secondary alkyl radical

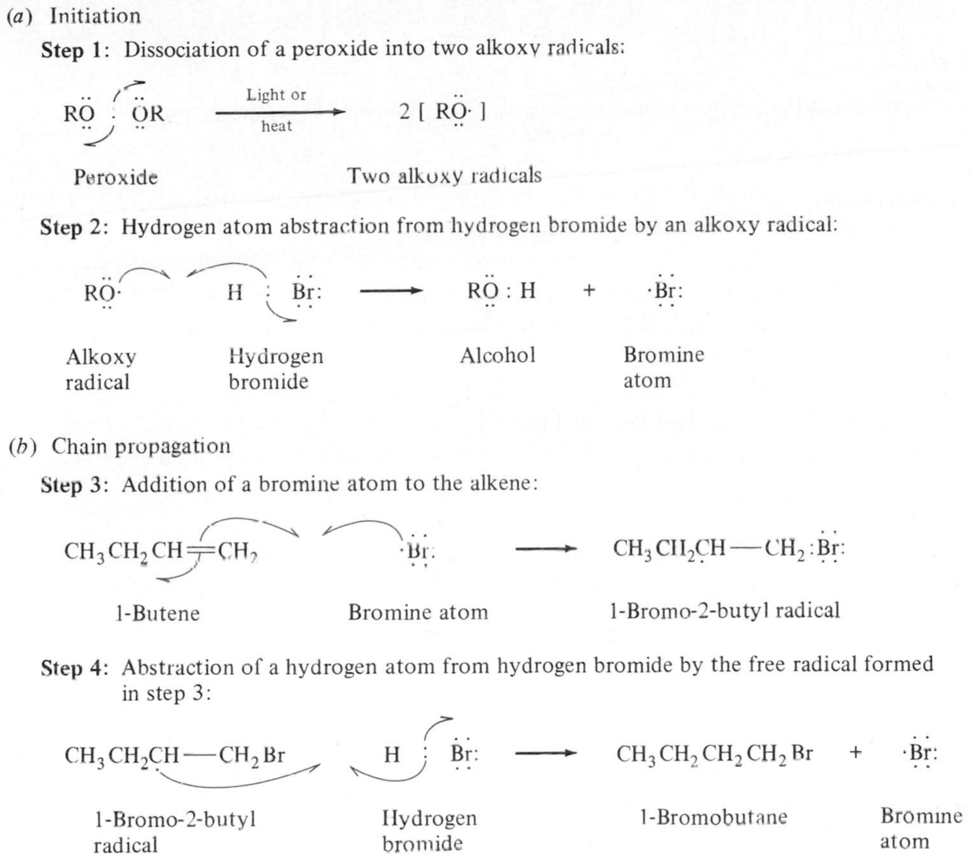

(*a*) Initiation

   Step 1: Dissociation of a peroxide into two alkoxy radicals:

   RÖ : ÖR $\xrightarrow[\text{heat}]{\text{Light or}}$ 2 [RÖ·]

   Peroxide   Two alkoxy radicals

   Step 2: Hydrogen atom abstraction from hydrogen bromide by an alkoxy radical:

   RÖ·   H : B̈r: $\longrightarrow$ RÖ : H   +   ·B̈r:

   Alkoxy   Hydrogen   Alcohol   Bromine
   radical   bromide   atom

(*b*) Chain propagation

   Step 3: Addition of a bromine atom to the alkene:

   $CH_3CH_2CH{=}CH_2$ ·B̈r: $\longrightarrow$ $CH_3CH_2CH{-}CH_2{:}$B̈r:

      1-Butene   Bromine atom   1-Bromo-2-butyl radical

   Step 4: Abstraction of a hydrogen atom from hydrogen bromide by the free radical formed
   in step 3:

   $CH_3CH_2\overset{\cdot}{C}H{-}CH_2Br$ H : B̈r: $\longrightarrow$ $CH_3CH_2CH_2CH_2Br$ +   ·B̈r:

      1-Bromo-2-butyl   Hydrogen   1-Bromobutane   Bromine
      radical   bromide   atom

FIGURE 7.4 The initiation and propagation steps in the free-radical addition of hydrogen bromide to 1-butene.

The addition of a bromine atom to C-2 is represented by

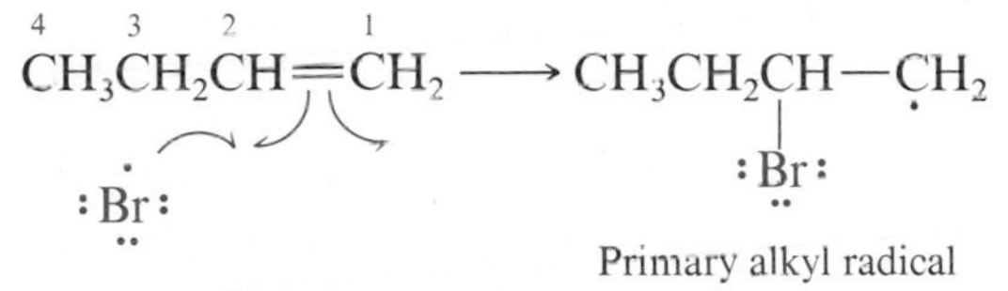

$$\overset{4}{C}H_3\overset{3}{C}H_2\overset{2}{C}H{=}\overset{1}{C}H_2 \longrightarrow CH_3CH_2CH{-}\overset{\cdot}{C}H_2$$
$$:\overset{\cdot}{B}r: \qquad\qquad\qquad :\overset{\cdot\cdot}{B}r:$$

Primary alkyl radical

A secondary alkyl radical is more stable than a primary radical. Bromine adds to the terminal carbon of 1-butene to form a secondary radical rather than adding to C-2 to form a primary radical. Once the bromine atom has added to the double bond, the regioselectivity of addition is set. The alkyl radical then abstracts a hydrogen atom from hydrogen bromide to give the alkyl bromide product as shown in step 4 of Figure 7.4.

   The regioselectivity of addition of hydrogen bromide to alkenes under normal (ionic addition) conditions is controlled by the tendency of a proton to add to the double bond so as to produce the more stable carbocation. Under free-radical conditions the regioselectivity is governed by the tendency of a bromine atom to add so as to give the more stable alkyl radical.

   Another way that the free-radical chain reaction may be initiated is photochemically, either with or without added peroxides.

$$\text{(Methylene)cyclopentane} =CH_2 \quad + \quad HBr \quad \xrightarrow{h\nu} \quad \text{(Bromomethyl)cyclopentane (60\%)}$$

| (Methylene)cyclopentane | Hydrogen bromide | (Bromomethyl)cyclopentane (60%) |

Among the hydrogen halides only hydrogen bromide reacts with alkenes rapidly by both an ionic and a free-radical mechanism. Hydrogen iodide and hydrogen chloride always add to alkenes by an ionic mechanism in accordance with Markovnikov's rule. Hydrogen bromide normally reacts by the ionic mechanism, but if peroxides are present or if the reaction is initiated photochemically, the free-radical mechanism is followed.

PROBLEM 7.4   Give the principal organic product formed when hydrogen bromide reacts with each of the alkenes in Problem 7.1 in the absence of peroxides and in their presence.

SAMPLE SOLUTION   (a)   The addition of hydrogen bromide in the absence of peroxides exhibits a regioselectivity just like that of hydrogen chloride addition; Markovnikov's rule is followed.

$$\text{2-Methyl-2-butene} \quad + \quad HBr \quad \xrightarrow{\text{no peroxides}} \quad \text{2-Bromo-2-methylbutane}$$

| 2-Methyl-2-butene | Hydrogen bromide | | 2-Bromo-2-methylbutane |

Under free-radical conditions in the presence of peroxides, addition takes place with a regioselectivity opposite to that of Markovnikov's rule.

$$\text{2-Methyl-2-butene} \quad + \quad HBr \quad \xrightarrow{\text{peroxides}} \quad \text{2-Bromo-3-methylbutane}$$

| 2-Methyl-2-butene | Hydrogen bromide | 2-Bromo-3-methylbutane |

While the possibility of having two different reaction paths available to an alkene and hydrogen bromide may seem like a complication, it can be advantageous in organic synthesis. From a single alkene one may prepare either of two different alkyl bromides, with control of regioselectivity, simply by choosing reaction conditions that favor ionic addition or free-radical addition of hydrogen bromide.

7.5   ADDITION OF SULFURIC ACID TO ALKENES

Acids other than hydrogen halides also add to the carbon-carbon bond of alkenes. Concentrated sulfuric acid, for example, reacts with certain alkenes to form alkyl hydrogen sulfates.

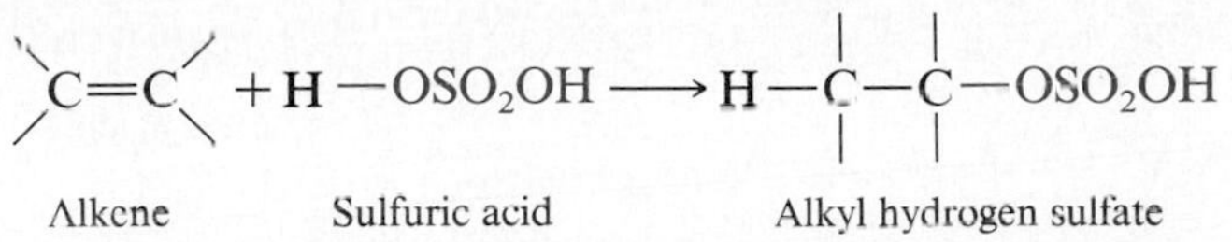

$$\text{C=C} + \text{H}-\text{OSO}_2\text{OH} \longrightarrow \text{H}-\text{C}-\text{C}-\text{OSO}_2\text{OH}$$

| Alkene | Sulfuric acid | Alkyl hydrogen sulfate |

Notice that a proton adds to the carbon that has the greater number of hydrogen substituents and the hydrogen sulfate anion ($^-\text{OSO}_2\text{OH}$) to the carbon that has the fewer hydrogen substituents.

$$\text{CH}_3\text{CH}=\text{CH}_2 + \text{HOSO}_2\text{OH} \longrightarrow \text{CH}_3\underset{\underset{\text{OSO}_2\text{OH}}{|}}{\text{CH}}\text{CH}_3$$

| Propene | Sulfuric acid | Isopropyl hydrogen sulfate |

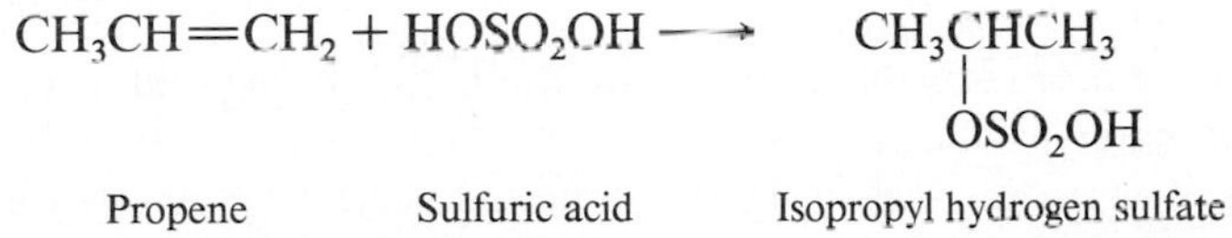

Markovnikov's rule is obeyed because the mechanism of sulfuric acid addition to alkenes, illustrated for the case of propene in Figure 7.5, is analogous to that described earlier for the ionic addition of hydrogen halides.

Alkyl hydrogen sulfates are converted to alcohols by heating them with water or steam. This is called a *hydrolysis* reaction because a bond is cleaved by reaction with water. (The suffix *-lysis* indicates cleavage.) It is the oxygen-sulfur bond that is broken when an alkyl hydrogen sulfate undergoes hydrolysis.

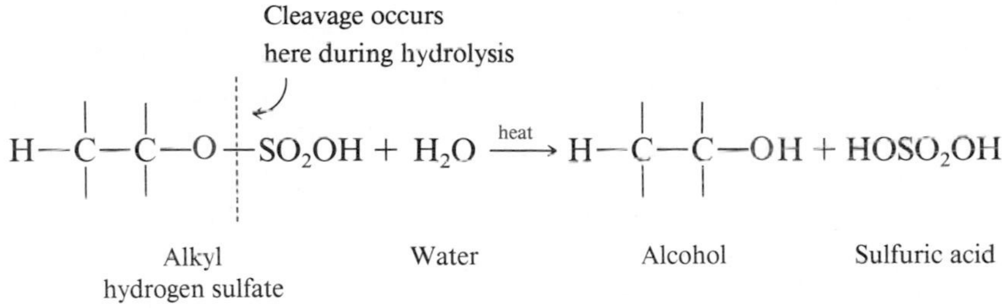

Cleavage occurs
here during hydrolysis

$$\text{H}-\text{C}-\text{C}-\text{O}+\text{SO}_2\text{OH} + \text{H}_2\text{O} \xrightarrow{\text{heat}} \text{H}-\text{C}-\text{C}-\text{OH} + \text{HOSO}_2\text{OH}$$

| Alkyl hydrogen sulfate | Water | Alcohol | Sulfuric acid |

The combination of sulfuric acid addition to alkenes with hydrolysis of the resulting alkyl hydrogen sulfate is an important industrial process for preparation of ethyl alcohol and isopropyl alcohol.

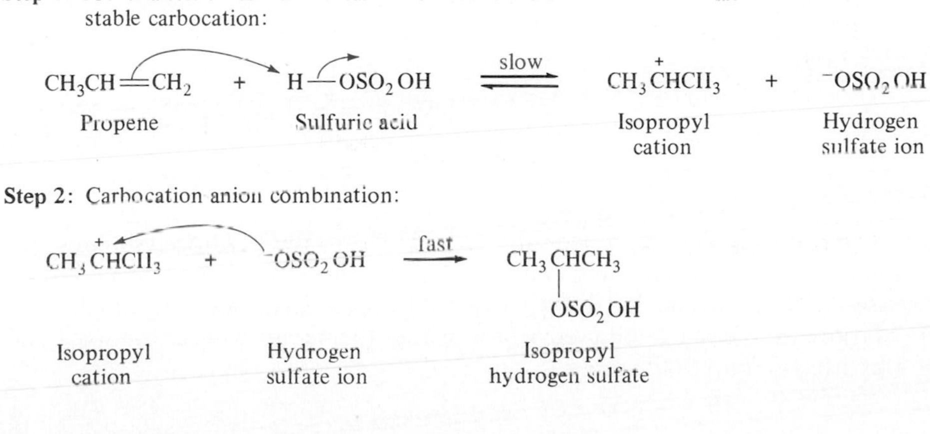

Step 1: Protonation of the carbon-carbon double bond in the direction that leads to the more stable carbocation:

$$\text{CH}_3\text{CH}=\text{CH}_2 + \text{H}-\text{OSO}_2\text{OH} \underset{\text{slow}}{\rightleftharpoons} \text{CH}_3\overset{+}{\text{CH}}\text{CH}_3 + {}^-\text{OSO}_2\text{OH}$$

| Propene | Sulfuric acid | Isopropyl cation | Hydrogen sulfate ion |

Step 2: Carbocation anion combination:

$$\text{CH}_3\overset{+}{\text{CH}}\text{CH}_3 + {}^-\text{OSO}_2\text{OH} \xrightarrow{\text{fast}} \text{CH}_3\underset{\underset{\text{OSO}_2\text{OH}}{|}}{\text{CH}}\text{CH}_3$$

| Isopropyl cation | Hydrogen sulfate ion | Isopropyl hydrogen sulfate |

FIGURE 7.5 Sequence of steps that describes the mechanism for addition of sulfuric acid to propene.

$$CH_2\!=\!CH_2 \xrightarrow{H_2SO_4} CH_3CH_2OSO_2OH \xrightarrow[\text{heat}]{H_2O} CH_3CH_2OH$$

Ethylene   Ethyl   Ethyl alcohol
hydrogen sulfate

$$CH_3CH\!=\!CH_2 \xrightarrow{H_2SO_4} \underset{\underset{OSO_2OH}{|}}{CH_3CHCH_3} \xrightarrow[\text{heat}]{H_2O} \underset{\underset{OH}{|}}{CH_3CHCH_3}$$

Propene   Isopropyl   Isopropyl
hydrogen sulfate   alcohol

We say that ethylene and propene have undergone *hydration* in this process. In effect, a molecule of water has been added across the carbon-carbon double bond by the combination of reactions shown. In the same manner, cyclohexanol has been prepared from cyclohexene.

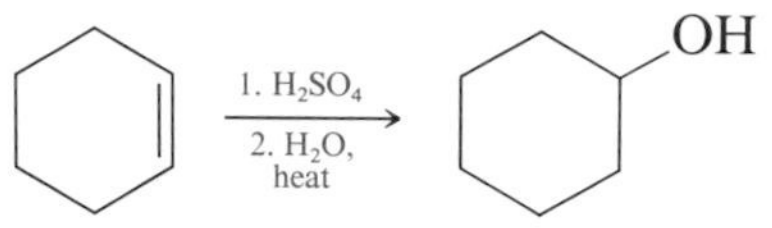

Cyclohexene   Cyclohexanol (75%)

It is convenient in synthetic transformations involving more than one step to simply list all the reagents over a single arrow. Individual synthetic steps are indicated by number. Numbering the individual steps is essential so as to avoid the implication that everything over the arrow is added to the reaction mixture at the same time.

Hydration of alkenes by this method, however, is limited to monosubstituted alkenes and disubstituted alkenes of the type $RCH\!=\!CHR$. Disubstituted alkenes of the type $R_2C\!=\!CH_2$, along with trisubstituted and tetrasubstituted alkenes, do not form alkyl hydrogen sulfates under these conditions but rather react in a more complicated way with concentrated sulfuric acid.

7.6 ACID-CATALYZED HYDRATION OF ALKENES

Another method by which alkenes may be converted to alcohols is through the addition of a molecule of water across the carbon-carbon double bond under conditions of acid catalysis.

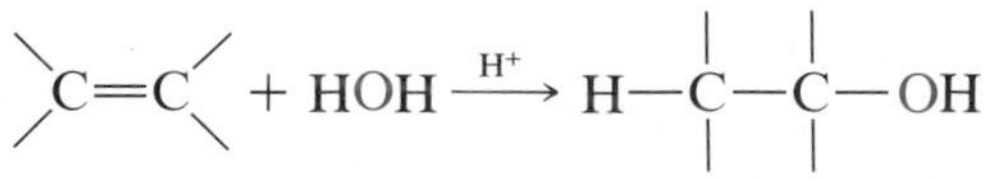

Alkene   Water   Alcohol

Unlike the addition of concentrated sulfuric acid to form alkyl hydrogen sulfates, this reaction is carried out in a dilute acid medium. A 50% water–sulfuric acid solution is often used, yielding the alcohol directly without the necessity of a separate hydrolysis step. Markovnikov's rule is followed: a proton adds to one carbon of the double bond and a hydroxyl group to the other.

Step 1: Protonation of the carbon-carbon double bond in the direction that leads to the more stable carbocation:

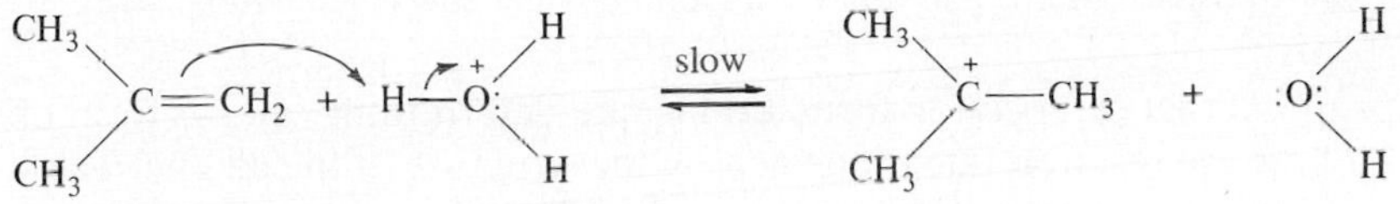

| 2-Methylpropene (isobutene) | Hydronium ion | | *tert*-Butyl cation | Water |

Step 2: Water acts as a nucleophile to capture *tert*-butyl cation:

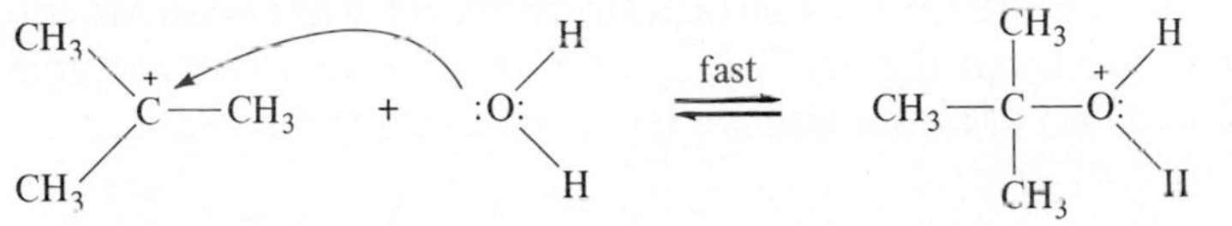

tert-Butyl cation   Water   *tert*-Butyloxonium ion

Step 3: Deprotonation of *tert*-butyloxonium ion. Water acts as a Brönsted base:

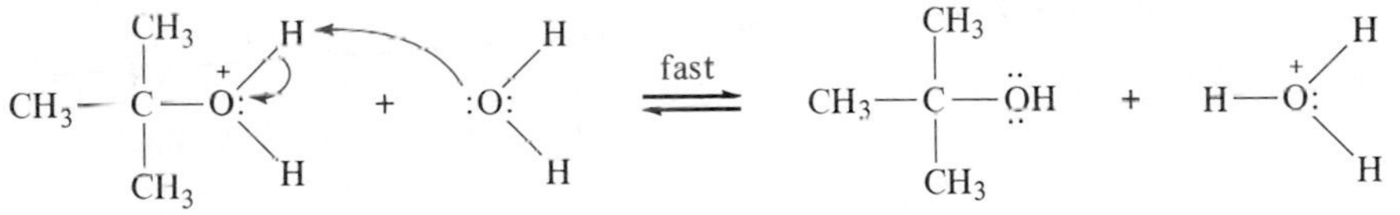

tert-Butyloxonium ion   Water   *tert*-Butyl alcohol   Hydronium ion

FIGURE 7.6 Sequence of steps describing the mechanism of acid-catalyzed hydration of 2-methylpropene.

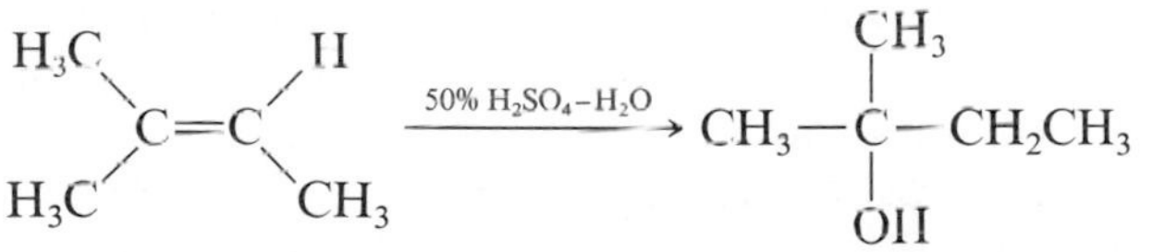

2-Methyl-2-butene   2-Methyl-2-butanol (90%)

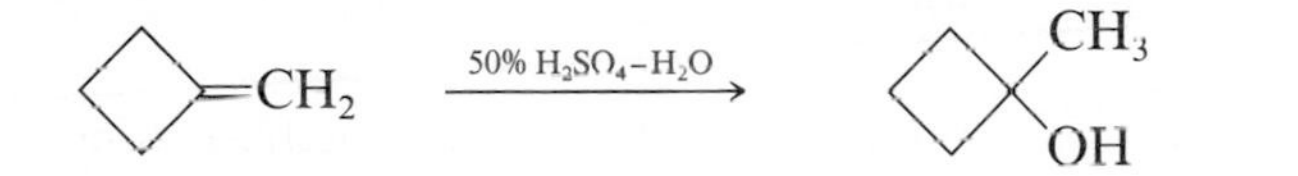

Methylenecyclobutane   1-Methylcyclobutanol (80%)

Figure 7.6 extends the general principles of electrophilic addition to alkenes to their acid-catalyzed hydration. In the specific example cited, proton transfer to 2-methylpropene (isobutene) forms *tert*-butyl cation in the first step. This is followed in step 2 by reaction of the carbocation intermediate with a molecule of water acting as a nucleophile. The product of nucleophilic capture of the carbocation by water is an oxonium ion. The oxonium ion is simply the conjugate acid of *tert*-butyl alcohol. Deprotonation of the oxonium ion in step 3 yields the alcohol product and regenerates the acid catalyst.

PROBLEM 7.5 Instead of the three-step mechanism of Figure 7.6, the following two-step mechanism might be considered:

1. $(CH_3)_2C{=}CH_2 + H_3O^+ \xrightarrow{\text{slow}} (CH_3)_3C^+ + H_2O$

2. $(CH_3)_3C^+ + HO^- \xrightarrow{\text{fast}} (CH_3)_3COH$

This mechanism cannot be correct! What is its fundamental flaw?

The proposal that carbocation formation is rate-determining follows from observations of how the reaction rate is affected by the structure of the alkene. Table 7.1 cites some data showing that alkenes that yield relatively stable carbocations react faster than those that yield less stable carbocations. Protonation of ethylene, the least reactive alkene in the table, yields a primary carbocation; protonation of 2-methyl-propene, the most reactive in the table, yields a tertiary carbocation. As we have seen on other occasions, the more stable the carbocation, the faster is its rate of formation.

You may have noticed that the acid-catalyzed hydration of an alkene and the acid-catalyzed dehydration of an alcohol are the reverse of one another.

$$\overset{\diagdown}{\diagup}C=C\overset{\diagup}{\diagdown} + H_2O \overset{H^+}{\rightleftharpoons} H-\overset{|}{\underset{|}{C}}-\overset{|}{\underset{|}{C}}-OH$$

Alkene   Water   Alcohol

According to *Le Chatelier's principle,* a system at equilibrium adjusts so as to minimize any stress applied to it. When the concentration of water is increased, the system responds by consuming water. This means that proportionally more alkene is converted to alcohol; the position of equilibrium shifts to the right. Thus, when we wish to prepare an alcohol from an alkene, we employ a reaction medium in which the molar concentration of water is high—dilute sulfuric acid, for example.

On the other hand, alkene formation is favored when the concentration of water is kept low. The system responds to the absence of water by causing more alcohol molecules to suffer dehydration, and when alcohol molecules dehydrate, they form more alkene. The amount of water in the reaction mixture is kept low by using concentrated strong acids as catalysts. Distilling the reaction mixture is an effective way of removing water as it is formed, causing the equilibrium to shift toward products. If the alkene is low-boiling, it too can be removed by distillation. This offers the additional benefit of protecting the alkene from acid-catalyzed isomerization after it is formed.

In any equilibrium process, the sequence of intermediates and transition states encountered as reactants proceed to products in one direction must also be encountered, and in precisely the reverse order, in the opposite direction. This is called the *principle of microscopic reversibility.* Just as the reaction

$$(CH_3)_2C=CH_2 + H_2O \overset{H^+}{\rightleftharpoons} (CH_3)_3COH$$

2-Methylpropene   Water   2-Methyl-2-propanol
(isobutene)   (*tert*-butyl alcohol)

is reversible with respect to reactants and products, so each tiny increment of progress along the reaction coordinate is considered reversible. Once we know the mechanism for the forward phase of a particular reaction, we also know what the intermediates and transition states must be for the reverse phase. In particular, the three-step mechanism for the acid-catalyzed hydration of 2-methylpropene in Figure 7.6 is the reverse of that for the acid-catalyzed dehydration of *tert*-butyl alcohol in Figure 6.1. The only difference is that Figure 7.6 represents the proton donor as a hydronium

TABLE 7.1
Relative Rates of Acid-Catalyzed Hydration of Some Representative Alkenes

Alkene	Structural formula	Relative rate of acid-catalyzed hydration*
Ethylene	$CH_2{=}CH_2$	1.0
Propene	$CH_3CH{=}CH_2$	1.6×10^6
2-Methylpropene	$(CH_3)_2C{=}CH_2$	2.5×10^{11}

* In water, 25°C.

ion because the hydration medium is water-rich. The dehydration mechanism in Figure 6.1 shows sulfuric acid as the proton donor in a water-poor medium.

PROBLEM 7.6   Is the electrophilic addition of hydrogen chloride to 2-methylpropene the reverse of the E1 or the E2 elimination reaction of *tert*-butyl chloride?

7.7 SYNTHESIS OF ALCOHOLS FROM ALKENES BY OXYMERCURATION-DEMERCURATION

Acid-catalyzed alkene hydration has been extensively studied from a mechanistic perspective. For the laboratory synthesis of alcohols, however, there are two alternative methods that are more effective. One of these, *oxymercuration-demercuration*, brings about the hydration of alkenes with a regioselectivity that, like acid-catalyzed hydration, obeys Markovnikov's rule. The other procedure, *hydroboration-oxidation*, allows hydration to be achieved with a regioselectivity precisely opposite to that of acid-catalyzed hydration.

Certain metal cations are sufficiently electrophilic to react with alkenes. One of these is mercury (II) in the form of salts such as mercuric chloride ($HgCl_2$) and mercuric acetate. Mercuric acetate has the formula

$$\underset{CH_3C\overset{\displaystyle O}{\overset{\|}{}}O-Hg-O\overset{\displaystyle O}{\overset{\|}{C}}CCH_3}{}$$

also written as $Hg(O_2CCH_3)_2$ or $Hg(OAc)_2$. Mercury is positively polarized in mercuric acetate and is the electrophilic atom that attacks the π electron system of an alkene. In the presence of water the reaction of mercuric acetate with alkenes leads to compounds known as *hydroxyalkylmercuric acetates*.

$$\underset{\text{Alkene}}{\overset{}{C{=}C}} + \underset{\substack{\text{Mercuric} \\ \text{acetate}}}{Hg(O_2CCH_3)_2} + \underset{\text{Water}}{H_2O} \longrightarrow \underset{\substack{\text{Hydroxyalkyl} \\ \text{mercuric acetate}}}{HO{-}\overset{|}{\underset{|}{C}}{-}\overset{|}{\underset{|}{C}}{-}HgO_2CCH_3} + \underset{\substack{\text{Acetic} \\ \text{acid}}}{CH_3CO_2H}$$

This reaction is known as *oxymercuration*. It is best carried out in a mixture of water and tetrahydrofuran as the solvent.

 Tetrahydrofuran (THF) is soluble in water in all proportions and is also a good solvent for organic compounds.

The hydroxyalkylmercuric acetate formed in the oxymercuration step is not normally isolated but is instead treated directly with sodium borohydride ($NaBH_4$). Sodium borohydride converts the hydroxyalkylmercuric acetate to an alcohol.

$$HO-\overset{|}{\underset{|}{C}}-\overset{|}{\underset{|}{C}}-HgO_2CCH_3 \xrightarrow[HO^-]{NaBH_4} HO-\overset{|}{\underset{|}{C}}-\overset{|}{\underset{|}{C}}-H + Hg^0 + CH_3CO_2^-$$

Hydroxyalkylmercuric   Alcohol   Mercury   Acetate ion
acetate

This step is called *demercuration*. The carbon-mercury bond of the hydroxyalkylmercuric acetate is replaced by a carbon-hydrogen bond.

Experimentally oxymercuration-demercuration reactions are very easy to carry out. The alkene is added to a solution of mercuric acetate in aqueous tetrahydrofuran. After about an hour the oxymercuration phase is complete. A basic solution of sodium borohydride is then added, whereupon metallic mercury separates from the reaction mixture and is removed and the alcohol is isolated.

Cyclopentene   Cyclopentanol (91%)

The oxymercuration-demercuration sequence produces alcohols with a regioselectivity identical to that of acid-catalyzed hydration. Hydrogen is introduced at the carbon that has the greater number of hydrogen substituents and hydroxyl at the carbon that has the fewer number of hydrogens.

$$CH_3CH_2CH_2CH_2CH=CH_2 \xrightarrow[\text{2. NaBH}_4, \text{ HO}^-]{\text{1. Hg(OAc)}_2, \text{ THF–H}_2\text{O}} CH_3CH_2CH_2CH_2\underset{\underset{OH}{|}}{C}HCH_3$$

1-Hexene   2-Hexanol (94%)

Rearrangements of the carbon skeleton do not occur.

$$(CH_3)_3CCH=CH_2 \xrightarrow[\text{2. NaBH}_4, \text{ HO}^-]{\text{1. Hg(OAc)}_2, \text{ THF–H}_2\text{O}} (CH_3)_3C\underset{\underset{OH}{|}}{C}HCH_3$$

3,3-Dimethyl-1-butene   3,3-Dimethyl-2-butanol (94%)

PROBLEM 7.7 Oxymercuration-demercuration of *trans*-2-pentene gives comparable amounts of two isomeric alcohols in a combined yield of 91 percent. What are these two alcohols?

PROBLEM 7.8 Each of the following alcohols may be prepared in high yield by oxymercuration-demercuration of two different alkenes. Write the structures of the two alkenes that could produce

(a) 2-Methyl-2-butanol

(b) 1-Methylcyclopentanol

(c) 3-Hexanol

SAMPLE SOLUTION (a) The alcohol is formed by addition of a hydrogen to one carbon and a hydroxyl to the other end of a double bond. Examining the structure of the given alcohol reveals a tertiary hydroxyl group at C-2. Therefore, an alkene with its double bond either between C-1 and C-2 or between C-2 and C-3 could undergo hydration according to Markovnikov's rule to give 2-methyl-2-butanol.

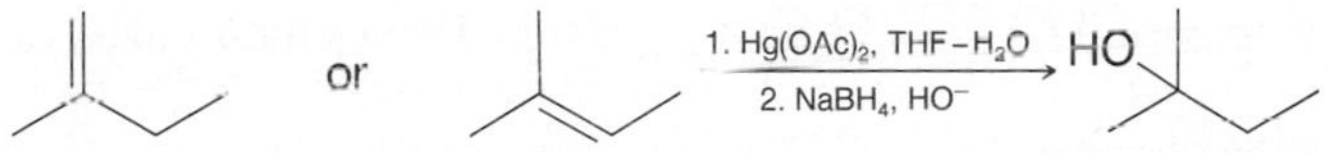

| 2-Methyl-
1-butene | or | 2-Methyl-
2-butene | | 2-Methyl-
2-butanol |

1. Hg(OAc)$_2$, THF–H$_2$O
2. NaBH$_4$, HO$^-$

Step 1: Mercuric acetate acts as an electrophile, attacking the π electrons of the carbon-carbon double bond:

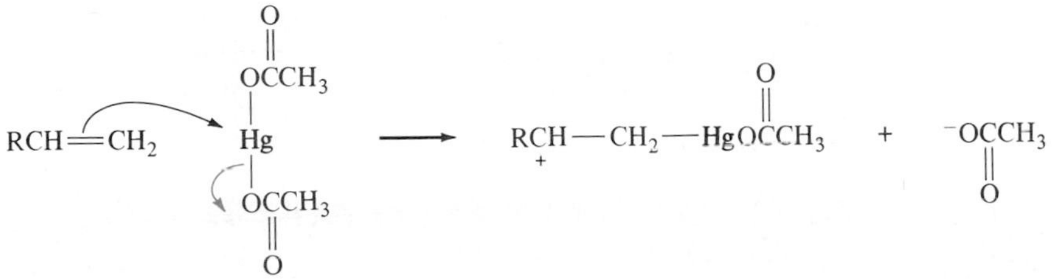

| Alkene | Mercuric
acetate | | Mercury-substituted
carbocation | Acetate
ion |

Step 2: Water acts as a nucleophile toward the mercury-substituted carbocation intermediate.

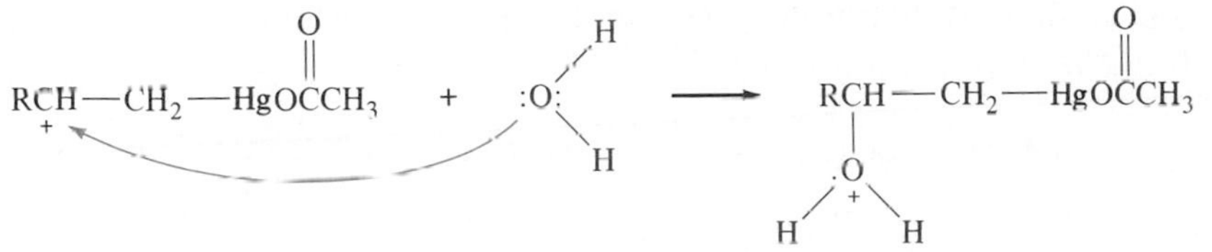

| Mercury-substituted
carbocation | Water | Mercury-substituted
oxonium ion |

Step 3: Loss of a proton from the oxonium ion yields the product of oxymercuration, a β-hydroxyalkylmercuric acetate:

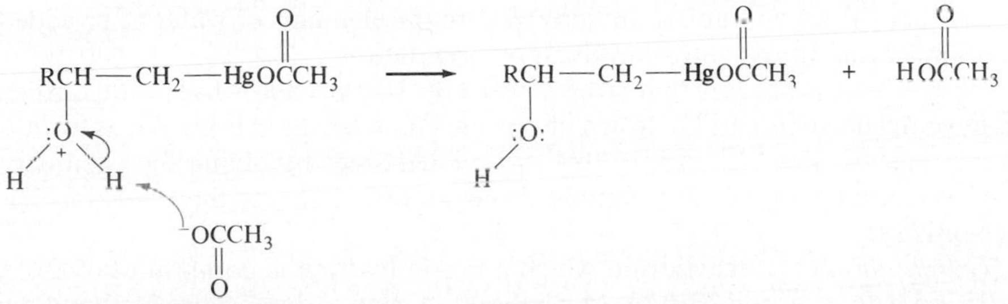

| Mercury-substituted
oxonium ion
+ acetate ion | β-Hydroxyalkylmercuric
acetate | Acetic acid |

FIGURE 7.7 A mechanistic description of the oxymercuration of an alkene.

The mechanism proposed for the oxymercuration phase of the oxymercuration-demercuration sequence is outlined in Figure 7.7. It begins in step 1 with electrophilic attack of mercuric acetate on the double bond. The pair of π electrons displaces one of the acetate groups from mercury, forming a carbon-mercury bond. The carbocation formed in this step is of a special kind, stabilized by the presence of a mercury substituent on the carbon adjacent to the positively charged one. Carbocations of this type are relatively stable, are formed readily, and do not rearrange. Like other carbocations, however, they are attacked by nucleophiles, and this is what happens in step 2, in which a molecule of water from the tetrahydrofuran-water mixture used as the solvent is the nucleophile. Loss of a proton in step 3 gives the hydroxyalkylmercuric acetate.

The second phase of the reaction, the demercuration phase, involves chemistry that is beyond the scope of an introductory course in organic chemistry and will not be discussed. The hydrogen atom that replaces the mercury substituent on carbon is derived from the sodium borohydride used in the demercuration reaction.

$$RCH{=}CH_2 \xrightarrow[\text{2. NaBH}_4, \text{ HO}^-]{\text{1. Hg(OAc)}_2, \text{ THF}-\text{H}_2\text{O}} RCH{-}CH_2{-}H$$

$$\underset{\substack{\text{From water present} \\ \text{in solvent (THF} - \text{H}_2\text{O)}}}{OH} \qquad \underset{\substack{\text{From sodium} \\ \text{borohydride (NaBH}_4)}}{}$$

7.8 SYNTHESIS OF ALCOHOLS FROM ALKENES BY HYDROBORATION-OXIDATION

Acid-catalyzed hydration and oxymercuration-demercuration add the elements of water to alkenes with Markovnikov rule regioselectivity. Frequently, however, one needs an alcohol having a structure that corresponds to hydration of an alkene with a regioselectivity opposite to that of Markovnikov's rule. Two examples illustrating the kind of transformation required are

$$CH_3(CH_2)_7CH{=}CH_2 \longrightarrow CH_3(CH_2)_7CH_2CH_2OH$$

1-Decene   1-Decanol

$$(CH_3)_2C{=}CHCH_3 \longrightarrow (CH_3)_2CHCHCH_3$$

$$OH$$

2-Methyl-2-butene   3-Methyl-2-butanol

We say that these two transformations require the elements of water to be added to the double bond in the anti-Markovnikov orientation.

The synthetic procedure that brings about anti-Markovnikov hydration of alkenes was developed by Herbert C. Brown of Purdue University and is known as *hydroboration-oxidation*. For his work in this and related areas involving the chemistry of boron-containing organic compounds, Professor Brown shared the 1979 Nobel prize in chemistry.

Hydroboration is a reaction in which a boron hydride, a compound of the type R_2BH, adds to a carbon-carbon double bond. A new carbon-hydrogen bond and a new carbon-boron bond are created.

$$\underset{\text{Alkene}}{\overset{\backslash}{/}}C=C\overset{/}{\underset{\backslash}{}} + \underset{\text{Boron hydride}}{R_2BH} \longrightarrow \underset{\text{Organoborane}}{H-\overset{|}{\underset{|}{C}}-\overset{|}{\underset{|}{C}}-BR_2}$$

The organoboranes formed by hydroboration of alkenes can be oxidized to alcohols by hydrogen peroxide in aqueous base.

$$\underset{\text{Organoborane}}{H-\overset{|}{\underset{|}{C}}-\overset{|}{\underset{|}{C}}-BR_2} + \underset{\substack{\text{Hydrogen}\\\text{peroxide}}}{3H_2O_2} + \underset{\substack{\text{Hydroxide}\\\text{ion}}}{4HO^-} \longrightarrow$$

$$\underset{\text{Alcohol}}{H-\overset{|}{\underset{|}{C}}-\overset{|}{\underset{|}{C}}-OH} + \underset{\text{Alcohol}}{2ROH} + \underset{\substack{\text{Borate}\\\text{ion}}}{B(OH)_4^-} + \underset{\text{Water}}{3H_2O}$$

The combination of these two reactions, hydroboration and subsequent oxidation, converts alkenes to alcohols in a convenient synthetic operation with a regioselectivity opposite to that of both acid-catalyzed hydration and oxymercuration-demercuration.

Before providing examples of this reaction, some introduction to boron chemistry is required. The simplest boron hydride is borane, BH_3, but it is a relatively unstable species. Individual BH_3 molecules combine in pairs to give the dimeric form of borane, diborane, B_2H_6. The equilibrium

$$\underset{\substack{\text{Borane}\\\text{(monomer)}}}{2BH_3} \rightleftharpoons \underset{\substack{\text{Diborane}\\\text{(dimer)}}}{B_2H_6}$$

lies overwhelmingly to the side of diborane under normal conditions of temperature and pressure.

Diborane is the boron hydride employed most often in the hydroboration of alkenes. A popular solvent for hydroboration reactions is called *diglyme,* an acronym for *di*ethylene *gly*col *di*methyl *e*ther. It has the structure:

$$CH_3OCH_2CH_2OCH_2CH_2OCH_3$$

All the boron-hydrogen bonds of diborane are capable of adding to alkenes. The equation describing the reaction of diborane with ethylene, for example, is

$$\underset{\text{Ethylene}}{6CH_2=CH_2} + \underset{\text{Diborane}}{B_2H_6} \xrightarrow{\text{diglyme}} \underset{\text{Triethylborane}}{2(CH_3CH_2)_3B}$$

It will be sufficient for our purposes to focus on only one of the organic groups attached to boron. Therefore, we will write equations for hydroboration reactions in the form:

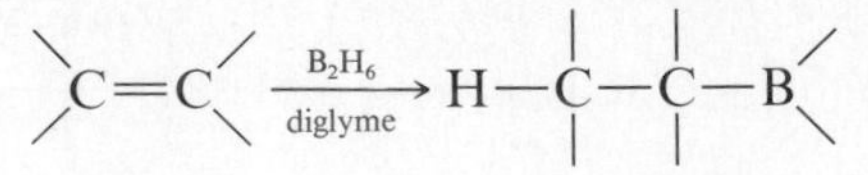

$$\begin{array}{c}\diagdown \\ {}^{/}\end{array}C=C\begin{array}{c}\diagup \\ \diagdown\end{array}\xrightarrow[\text{diglyme}]{B_2H_6}\ H-\overset{|}{\underset{|}{C}}-\overset{|}{\underset{|}{C}}-B\begin{array}{c}\diagup \\ \diagdown\end{array}$$

Alkene   Alkylborane

Returning to the synthetic question that introduced this topic, the preparation of 1-decanol from 1-decene, it is observed that hydroboration-oxidation carries out this transformation in high yield and in a regiospecific manner. 1-Decanol is formed exclusively. We can show the hydroboration-oxidation procedure in a single equation, focusing on the organic functional group transformation, by indicating the necessary reagents sequentially over the arrow.

$$CH_3(CH_2)_7CH\!=\!CH_2 \xrightarrow[\text{2. }H_2O_2,\,HO^-]{\text{1. }B_2H_6,\,\text{diglyme}} CH_3(CH_2)_7CH_2CH_2OH$$

1-Decene   1-Decanol (93%)

Another convenient hydroborating agent is the borane-tetrahydrofuran complex (BH_3—THF). It is very reactive, adding to alkenes within minutes at $0\,°C$, and is used in tetrahydrofuran as the solvent.

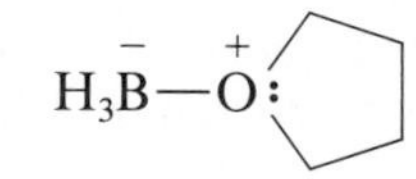

$$H_3\overset{-}{B}\!-\!\overset{+}{O}\!:\!\diagup$$

Borane-tetrahydrofuran
complex

The second of our proposed transformations noted in the introduction, the conversion of 2-methyl-2-butene to 3-methyl-2-butanol, has been accomplished by using the borane-tetrahydrofuran complex in the hydroboration step.

$$(CH_3)_2C\!=\!CHCH_3 \xrightarrow[\text{2. }H_2O_2,\,HO^-]{\text{1. }BH_3\text{–THF}} (CH_3)_2CHCHCH_3$$
$$\underset{\text{OH}}{|}$$

2-Methyl-2-butene   3-Methyl-2-butanol (98%)

Hydration of double bonds takes place without rearrangement, even in alkenes as highly branched as the one shown below.

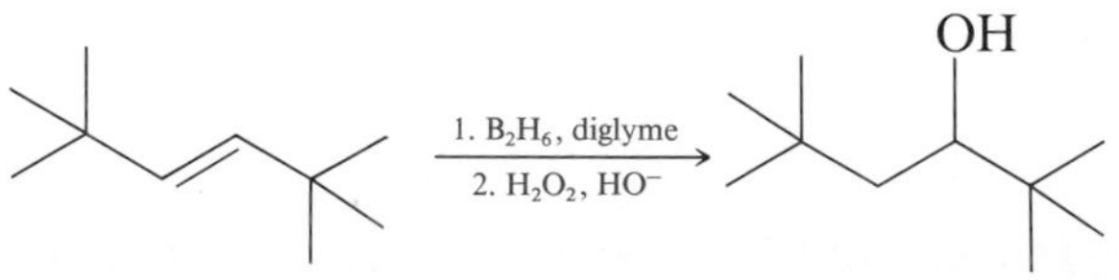

(E)-2,2,5,5-Tetramethyl-
3-hexene   2,2,5,5-Tetramethyl-
3-hexanol (82%)

PROBLEM 7.9 Write the structure of the major organic product obtained by hydroboration-oxidation of each of the following alkenes:

(a) 2-Methylpropene

(b) cis-2-Butene

(c) ⬦=CH_2

(d) Cyclopentene

(e) 3-Ethyl-2-pentene

(f) 3-Ethyl-1-pentene

SAMPLE SOLUTION   (a) In hydroboration-oxidation the elements of water (H and OH) are introduced with a regioselectivity opposite to that of Markovnikov's rule. In the case of 2-methylpropene, this leads to 2-methyl-1-propanol as the product.

$$(CH_3)_2C\!=\!CH_2 \xrightarrow[\text{2. oxidation}]{\text{1. hydroboration}} (CH_3)_2CH\!-\!CH_2OH$$

2-Methylpropene
(isobutene)

2-Methyl-1-propanol
(isobutyl alcohol)

Hydrogen becomes bonded to the carbon that has the fewer hydrogen substituents, hydroxyl to the carbon that has the greater number of hydrogen substituents. We say that hydroboration-oxidation leads to anti-Markovnikov hydration of alkenes.

7.9 STEREOCHEMISTRY OF HYDROBORATION

Hydration of alkenes by the hydroboration-oxidation procedure is stereospecific. On comparing the structure of the alkene starting material to that of the alcohol product in the hydroboration-oxidation of 1-methylcyclopentene for example, we see that the hydrogen and hydroxyl groups have been added to the same face of the double bond.

1-Methylcyclopentene

trans-2-Methylcyclopentanol
(only product, 86% yield)

Overall, the reaction leads to syn addition of the elements of water to the double bond. This fact has an important bearing on the mechanism of the process.

7.10 MECHANISM OF HYDROBORATION-OXIDATION

Among the factors that contribute to a mechanistic understanding of alkene hydration via hydroboration-oxidation are

1. Regiospecificity
2. Stereospecificity
3. Chemical nature of alkenes
4. Chemical nature of boranes

Let us begin with the last of these since it requires the most explanation and also lies at the heart of the reaction.

We can consider the hydroboration step as though it involved monomeric BH_3. It makes our mechanistic analysis easier to follow and is at variance with reality only in matters of detail. Borane is an electrophilic species; it has a vacant $2p$ orbital and can accept a pair of electrons into that orbital. The source of this electron pair is the π bond of an alkene. It is believed, as shown in Figure 7.8 for the example of the hydroboration of 1-methylcyclopentene, that the first step produces an unstable intermediate called a π *complex*. In this π complex boron and the two carbon atoms

Step 1: A molecule of borane (BH₃) attacks the alkene. Electrons flow from the π orbital of the alkene to the 2p orbital of boron. A π complex is formed.

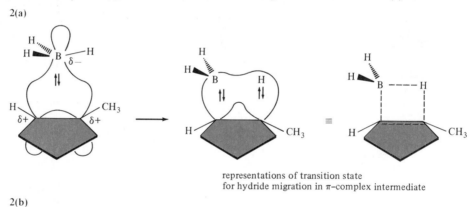

alternative representations of
π–complex intermediate

Step 2: The π complex rearranges to an organoborane. Hydrogen migrates from boron to carbon, carrying with it the two electrons in its bond to boron. Development of the transition state for this process is shown in 2(a), and its transformation to the organoborane is shown in 2(b).

2(a)

representations of transition state
for hydride migration in π–complex intermediate

2(b)

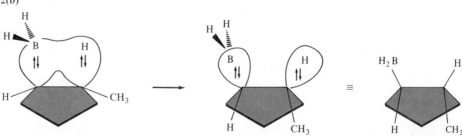

product of addition of borane (BH₃)
to 1–methylcyclopentene

FIGURE 7.8 Depiction of orbital interactions and electron redistribution in the hydroboration of 1-methylcyclopentene.

of the double bond are joined by a *three-center two-electron bond,* by which we mean that three atoms share two electrons. Three-center two-electron bonds are frequently encountered in boron chemistry. The π complex is formed by a transfer of electron density from the π orbital of the alkene to the 2p orbital of boron. This leaves each carbon of the complex with a small positive charge, while boron is slightly negative. The negative character of boron in this intermediate makes it easy for one of its hydrogen substituents to migrate with a pair of electrons (a hydride shift) from boron to carbon. The transition state for this process is shown in step 2a of Figure 7.8; completion of the migration in step 2b yields the alkylborane. According to this mechanism, the carbon-boron bond and the carbon-hydrogen bond are formed on the same side of the alkene. The hydroboration step is a syn addition process.

The regioselectivity of addition is consistent with the electron distribution in the π complex. Hydrogen is transferred with a pair of electrons to the carbon atom that can best support a positive charge, namely, the one that bears the methyl group. Steric effects are believed to be an important factor in determining the regioselectivity of

Step 1: Hydrogen peroxide is converted to its anion in basic solution:

$$H—O—O—H \quad + \quad {}^{-}OH \quad \rightleftharpoons \quad H—O—O^{-} \quad + \quad H—O—H$$

Hydrogen   Hydroxide   Hydroperoxide   Water
peroxide   ion   ion

Step 2: Anion of hydrogen peroxide acts as a nucleophile, attacking boron and forming an oxygen-boron bond:

Organoborane intermediate
from hydroboration of
1-methylcyclopentene

Step 3: Carbon migrates from boron to oxygen, displacing hydroxide ion. Carbon migrates with the pair of electrons in the carbon-boron bond; these become the electrons in the carbon-oxygen bond:

Representation of transition
state for migration of carbon
from boron to oxygen

Alkoxyborane

Step 4: Hydrolysis cleaves the boron oxygen bond, yielding the alcohol:

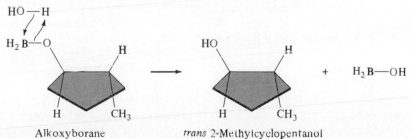

Alkoxyborane   *trans* 2-Methylcyclopentanol

FIGURE 7.9 Mechanistic description of the oxidation phase in the hydroboration-oxidation of 1-methylcyclopentene.

addition as well. According to this view, boron tends to bond to the less crowded carbon atom of the double bond—this is the one with the greater number of hydrogen substituents.

A second aspect of the electrophilic character of boron is evident when we consider the oxidation of organoboranes. In the oxidation phase of the hydroboration-oxidation sequence, as presented in Figure 7.9, the anion of hydrogen peroxide attacks boron. Hydroperoxide ion is formed in an acid-base reaction in step 1 and attacks boron in step 2. The empty $2p$ orbital of boron makes it electrophilic and permits nucleophilic reagents such as HOO^- to add to it.

The combination of a negative charge on boron and a weak oxygen-oxygen bond in its hydroperoxy substituent causes an alkyl group to migrate from boron to oxygen in step 3 of the mechanism shown in Figure 7.9. This alkyl group migration occurs with loss of hydroxide ion and is the step in which the critical carbon-oxygen bond is formed. What is especially significant about this alkyl group migration is that stereochemical orientation of the new carbon-oxygen bond is the same as that of the original carbon-boron bond. This is crucial to the overall syn stereospecificity of the hydroboration-oxidation sequence. Migration of the alkyl group from boron to oxygen is said to have occurred with *retention of configuration* at carbon. The alkoxyborane intermediate formed in step 3 undergoes subsequent base-promoted oxygen-boron bond cleavage in step 4 to give the alcohol product.

The mechanistic complexity of hydroboration-oxidation stands in contrast to the simplicity with which these reactions are carried out experimentally. Both the hydroboration and oxidation steps are extremely rapid reactions and are performed at room temperature with conventional laboratory equipment. Ease of operation, along with the fact that hydroboration-oxidation leads to syn hydration of alkenes and occurs with a regioselectivity contrary to Markovnikov's rule, makes this procedure one of great value to the synthetic chemist.

7.11 ADDITION OF HALOGENS TO ALKENES

In contrast to the free-radical substitution reaction observed when halogens react with alkanes, halogens normally react with alkenes to yield the products of electrophilic addition.

$$\underset{\text{Alkene}}{\diagdown C{=}C\diagup} + \underset{\text{Halogen}}{X_2} \longrightarrow \underset{\text{Vicinal dihalide}}{X-\overset{|}{C}-\overset{|}{\underset{|}{C}}-X}$$

The products of these reactions are called *vicinal* dihalides. Two substituents, in this case the halogen substituents, are vicinal if they are attached to adjacent carbons. The word is derived from the Latin *vicinalis,* which means "neighboring."

Bromine addition to alkenes is the most thoroughly studied of these reactions. Bromine adds rapidly to most alkenes, giving vicinal dibromides in high yield. These reactions may be performed in a variety of solvents, including acetic acid, carbon tetrachloride, chloroform, and dichloromethane.

$$\underset{\text{4-Methyl-2-pentene}}{CH_2CH{=}CHCH(CH_3)_2} + \underset{\text{Bromine}}{Br_2} \xrightarrow[0°C]{CHCl_3} \underset{\text{2,3-Dibromo-4-methylpentane (100\%)}}{CH_3CH-CHCH(CH_3)_2 \atop \underset{Br \quad\ Br}{|\quad\ |}}$$

Addition of bromine to ethylene yields 1,2-dibromoethane.

$$CH_2{=}CH_2 + Br_2 \longrightarrow BrCH_2CH_2Br$$

| Ethylene | Bromine | 1,2-Dibromoethane (ethylene dibromide) |

The common name for this substance is ethylene dibromide (EDB). Until it was banned in 1984, about 12×10^6 lb/year of EDB was produced in the United States for use as an agricultural pesticide and soil fumigant.

Bromine addition to alkenes is the basis of a qualitative test for alkenes. Solutions of bromine in carbon tetrachloride, like bromine itself, are reddish brown. When a solution of bromine in carbon tetrachloride is added dropwise to an alkene, reaction occurs practically instantaneously and the red color is discharged, giving a colorless solution. Since compounds other than alkenes may also react with bromine under conditions of this test, discharge of the bromine color suggests, but does not prove, that a carbon-carbon double bond is present. Additional evidence must then be sought to confirm that the unknown substance is an alkene.

Chlorine adds to alkenes to form vicinal dichlorides.

$$(CH_3)_3CCH{=}CH_2 + Cl_2 \xrightarrow{5°C} (CH_3)_3CCH{-}CH_2Cl$$
$$\underset{Cl}{|}$$

| 3,3-Dimethyl-1-butene | Chlorine | 1,2-Dichloro-3,3-dimethylbutane (53%) |

Fluorine addition to alkenes is too violent a reaction to be easily controlled and so is rarely used.

The equilibrium constant for formation of vicinal diiodides from alkenes and iodine is often less than unity. Vicinal diiodides have a pronounced tendency to lose I_2 and revert to alkenes. Thus, special conditions are normally required for their preparation, and their relative instability makes vicinal diiodides an infrequently encountered class of compounds.

7.12 STEREOCHEMISTRY OF HALOGEN ADDITION TO ALKENES

The additions of bromine and chlorine to alkenes are stereospecific reactions. Anti addition is observed; the two bromine atoms of Br_2 or the two chlorines of Cl_2 add to opposite faces of the carbon-carbon double bond.

| Cyclohexene | Bromine | trans-1,2-Dibromocyclohexane (73–86% yield; none of the cis isomer is formed) |

only trans is formed

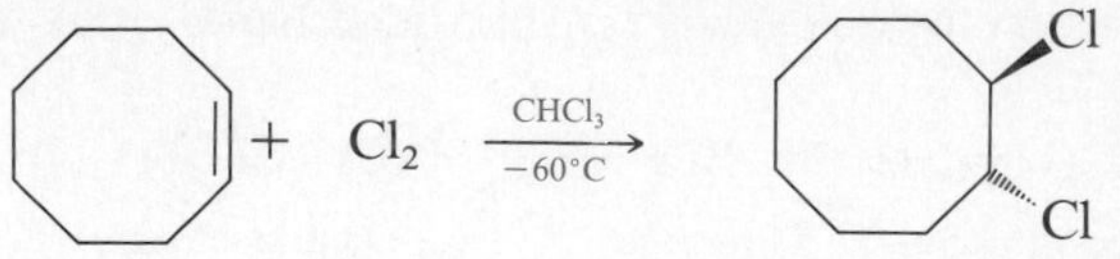

Cyclooctene   Chlorine   *trans*-1,2-Dichlorocyclooctane
(73% yield; none of the cis
isomer is formed)

These observations must be taken into account when considering the mechanism of halogen addition. They force the conclusion that a simple one-step "bond switching" process of the type shown below cannot be correct. A process of this type requires syn addition; it is *not* consistent with the anti addition actually observed.

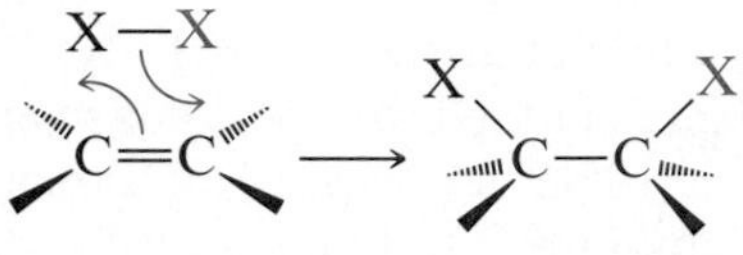

PROBLEM 7.10   The mass 82 isotope of bromine (^{82}Br) is radioactive and is used as a tracer to identify the origin and destination of individual atoms in chemical reactions and biological transformations. A sample of 1,1,2-tribromocyclohexane was prepared by adding ^{82}Br—^{82}Br to ordinary (nonradioactive) 1-bromocyclohexene. How many of the bromine atoms in the 1,1,2-tribromocyclohexane produced are radioactive? Which ones are they?

7.13 MECHANISM OF HALOGEN ADDITION TO ALKENES. HALONIUM IONS

The generally accepted mechanism for bromine and chlorine additions to alkenes begins with electrophilic attack of the halogen on the π electrons of the double bond. Bromine and chlorine are not polar compounds but they are moderately electrophilic. Nucleophilic species, such as alkenes, interact with bromine and chlorine to break the weak halogen-halogen bond. One halogen atom becomes bonded to the nucleophile, the other is lost as a halide anion. Taking bromine addition to ethylene as an example

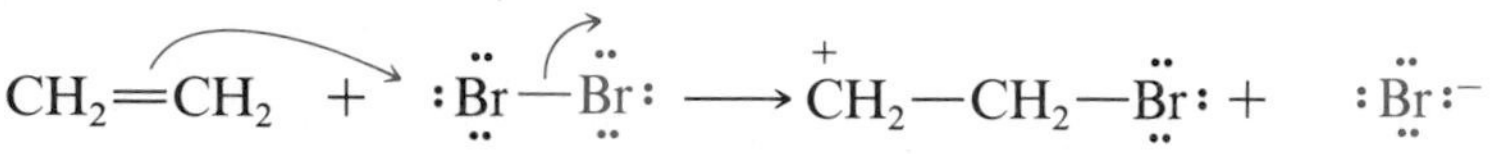

$CH_2{=}CH_2$ + $:\overset{..}{\underset{..}{Br}}{-}\overset{..}{\underset{..}{Br}}:$ $\longrightarrow$ $\overset{+}{C}H_2{-}CH_2{-}\overset{..}{\underset{..}{Br}}:$ + $:\overset{..}{\underset{..}{Br}}:^{-}$

Ethylene   Bromine   2-Bromoethyl cation   Bromide ion
(nucleophile)   (electrophile)   (leaving group)

Nominally, such a reaction leads to a carbocation. A carbocation of this type, however, contains a source of electrons (the lone pairs on the bromine substituent) in close proximity to the positively charged carbon. Carbocations of this type could, if formed, close to form a cyclic bromonium ion.

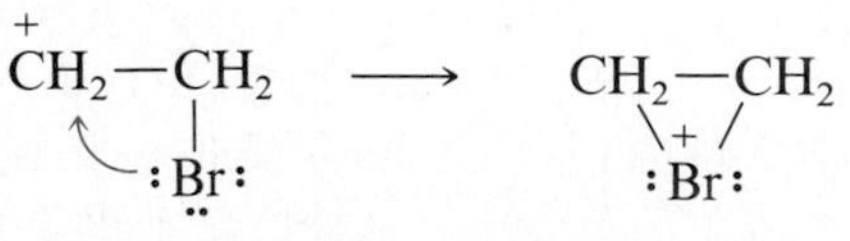

$\overset{+}{C}H_2{-}CH_2$ $\longrightarrow$ $CH_2{-}CH_2$
         | \\ + /
      $:\overset{..}{Br}:$ $:\overset{..}{Br}:$

2-Bromoethyl cation   Ethylenebromonium ion

Step 1: Reaction of ethylene and bromine to form a bromonium ion intermediate:

$$CH_2\!=\!CH_2 \quad + \quad :\overset{..}{Br}\!-\!\overset{..}{Br}: \quad \longrightarrow \quad \underset{\overset{|}{:Br:}}{CH_2\!-\!CH_2} \quad + \quad :\overset{..}{Br}:^-$$

Ethylene   Bromine   Ethylenebromonium   Bromide
                                                      ion   ion

Step 2: Nucleophilic attack of bromide anion on the bromonium ion:

$$:\overset{..}{Br}:^- \qquad \underset{\overset{|}{:Br:}^+}{CH_2\!-\!CH_2} \quad \longrightarrow \quad :\overset{..}{Br}\!-\!CH_2\!-\!CH_2\!-\!\overset{..}{Br}:$$

Bromide   Ethylenebromonium   1,2-Dibromoethane
 ion   ion

FIGURE 7.10 Mechanistic description of electrophilic addition of bromine to ethylene.

This cyclic bromonium ion is often referred to as a *bridged* bromonium ion. Bromine, not carbon, bears most of the positive charge. In spite of its three-membered ring the bromonium ion is more stable than 2-bromoethyl cation. This is because 2-bromoethyl cation has six electrons around its positively charged carbon while ethylenebromonium ion has octets of electrons around bromine and both carbons.

Since ethylenebromonium ion is appreciably more stable than bromoethyl cation, it is the intermediate formed directly from the ethylene and bromine. Step 1 in Figure 7.10 illustrates this process. Formation of the cyclic bromonium ion is followed in step 2 by its reaction with bromide ion to give the vicinal dibromide. Bromide ion acts as a nucleophile in this step. For reasons that will be explained in Chapter 9, bromide anion attacks the cyclic bromonium ion from the side opposite the carbon-bromine bond that is broken.

The mechanism in Figure 7.10 was devised to accommodate a number of experimental observations, including the observed anti stereospecificity. Thus, when bromide ion attacks the cyclic bromonium ion, the two bromine substituents must of necessity be trans to each other in the product that results from cyclic alkenes such as cyclohexene.

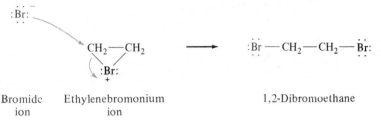

Another experimental observation consistent with the proposed mechanism is the effect of substituents on reaction rate. The data presented in Table 7.2 show that alkyl substituents on the carbon-carbon double bond increase the rate of reaction. Alkyl groups release electrons and stabilize the activated complex leading to the bromonium ion.

TABLE 7.2

Relative Rates of Reaction of Some Representative Alkenes with Bromine

Alkene	Structural formula	Relative rate of reaction with bromine*
Ethylene	$CH_2{=}CH_2$	1.0
Propene	$CH_3CH{=}CH_2$	61
2-Methylpropene	$(CH_3)_2C{=}CH_2$	5,400
2,3-Dimethyl-2-butene	$(CH_3)_2C{=}C(CH_3)_2$	920,000

* In methanol, 25°C

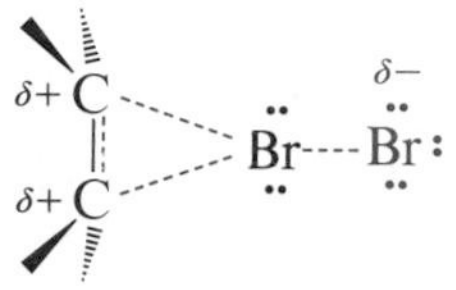

Representation of activated complex for
bromonium ion formation from an
alkene and bromine

The carbon atoms of the double bond develop carbocation character as progress is made toward the transition state. By releasing electrons, alkyl substituents lower the activation energy for bromonium ion formation and increase the rate of the reaction. This is another example of the effect of alkyl substituents in increasing the rate of electrophilic addition to alkenes by making the carbon-carbon double bond more electron-rich.

PROBLEM 7.11   Arrange the compounds 2-methyl-1-butene, 2-methyl-2-butene, and 3-methyl-1-butene in order of decreasing reactivity toward bromine.

That a cyclic bromonium ion could be an intermediate in bromine addition to alkenes was an innovative concept at the time of its proposal in 1937, but additional data accumulated since then have reinforced the conclusion that it is a correct description of the rate-determining intermediate in these reactions. Similarly, chloronium ions are believed to be involved in the addition of chlorine to alkenes. In the next section we shall see how cyclic chloronium and bromonium ions *(halonium ions)* are intermediates in another reaction, namely, the formation of halohydrins from alkenes and halogens in aqueous solution.

7.14   ADDITION OF HYPOHALOUS ACIDS TO ALKENES

In aqueous solution chlorine and bromine react with alkenes to form compounds known as *halohydrins.*

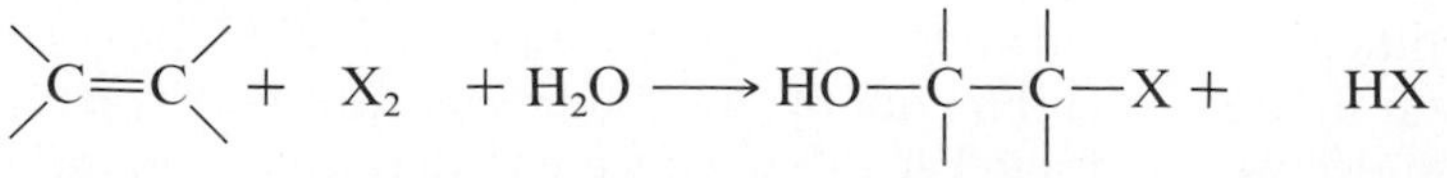

| Alkene | Halogen | Water | Halohydrin | Hydrogen halide |

When the halogen is chlorine, the product is a vicinal *chlorohydrin;* when the halogen is bromine, it is a vicinal *bromohydrin.*

$$CH_2{=}CH_2 + Br_2 \xrightarrow{H_2O} HOCH_2CH_2Br$$

Ethylene   Bromine   2-Bromoethanol (70%)
(ethylene bromohydrin)

Halohydrin formation is mechanistically related to halogen addition to alkenes. Halonium ions are formed as intermediates. In aqueous solution water, rather than a halide anion, acts as the nucleophile toward the halonium ion.

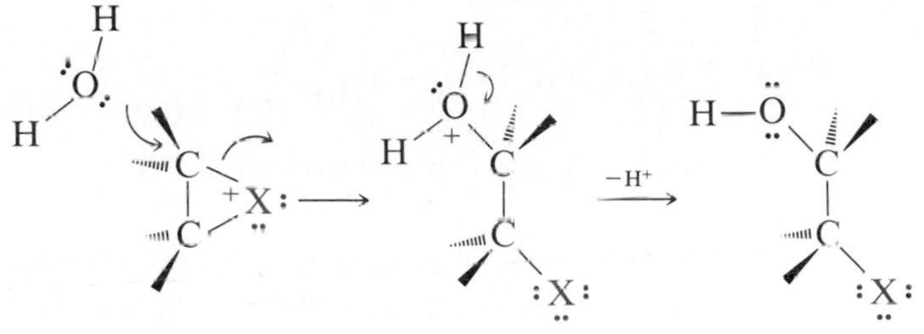

Water + halonium ion   Oxonium ion   Vicinal halohydrin

The reaction is a stereospecific anti addition; the halogen and hydroxyl groups are added to opposite faces of the double bond.

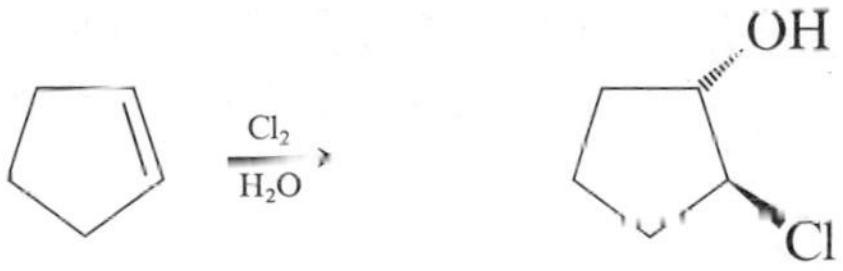

Cyclopentene   *trans*-2-Chlorocyclopentanol
(52–56% yield; cis isomer not formed)

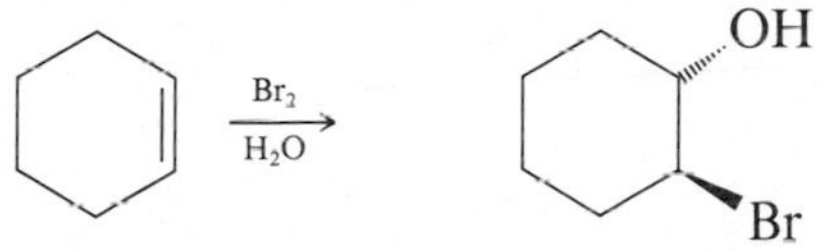

Cyclohexene   *trans*-2-Bromocyclohexanol
(79% yield; cis isomer not formed)

Nucleophilic ring opening of the halonium ion intermediate occurs by attack of water from the side opposite the carbon-halogen bond that is broken.

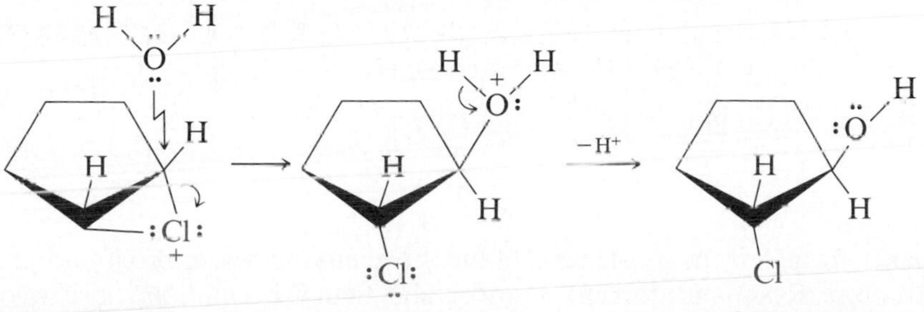

Cyclic chloronium ion   Oxonium ion   *trans*-2-Chlorocyclopentanol

In these reactions it is as if the elements of a hypohalous acid (HOX) had been added across the double bond. Both hypochlorous acid and hypobromous acid are known compounds and exist in a reversible equilibrium with the free halogens in aqueous solution.

$$Cl_2 \ + \ H_2O \longrightarrow \quad HOCl \quad + \quad HCl$$

Chlorine   Water   Hypochlorous acid   Hydrogen chloride

Oxygen is more electronegative than either chlorine or bromine, and the hypohalous acids are polarized so that the halogen bears a partial positive charge.

$$\overset{\delta-}{HO}-\overset{\delta+}{Cl} \qquad\qquad\qquad \overset{\delta-}{HO}-\overset{\delta+}{Br}$$

Polarization in hypochlorous and hypobromous acid

Since an aqueous solution of chlorine or bromine contains both the free halogen and the hypohalous acid, the halonium ion may be formed by attack of either of these species on the double bond. Hypohalous acids act as sources of electrophilic halogen.

The regioselectivity of halohydrin formation follows a slightly modified version of Markovnikov's rule such that the positive part of the addend becomes attached to the carbon atom of the double bond having the greater number of hydrogen substituents, and the negative part becomes attached to the carbon having the fewer hydrogens. The positive part of a hypohalous acid, i.e., the halogen, adds to the carbon atom of the double bond having the greater number of hydrogen substituents.

$$(CH_3)_2C{=}CH_2 \xrightarrow[H_2O]{Br_2} (CH_3)_2\underset{\underset{OH}{|}}{C}-CH_2Br$$

2-Methylpropene   1-Bromo-2-methyl-2-propanol (77%)

A mechanistic explanation for the regioselectivity observed in halohydrin formation rests on the idea that one carbon-halogen bond in the halonium ion intermediate is weaker than the other. The bridging halogen is less strongly bonded to the carbon atom that can better support a positive charge, i.e., the one with the greater number of electron-releasing alkyl substituents. In resonance terms we say the bromonium ion formed from 2-methylpropene has some of the character of the three resonance structures A, B, and C.

$$(CH_3)_2\underset{\underset{+}{\underset{:Br:}{\diagdown\diagup}}}{C}-CH_2 \longleftrightarrow (CH_3)_2\overset{+}{\underset{\underset{:Br:}{\cdot\cdot\diagup}}{C}}-CH_2 \longleftrightarrow (CH_3)_2\underset{\underset{:Br:}{\diagdown\cdot\cdot}}{C}-\overset{+}{CH_2}$$

A   B   C

Resonance form A is the most stable of the three and the best approximation of the true structure. Resonance form B is more stable than C because the positive charge resides on a tertiary carbon in B versus a primary carbon in C. Water attacks the more highly substituted carbon because this carbon atom has a greater degree of carbocation character than does its less substituted counterpart.

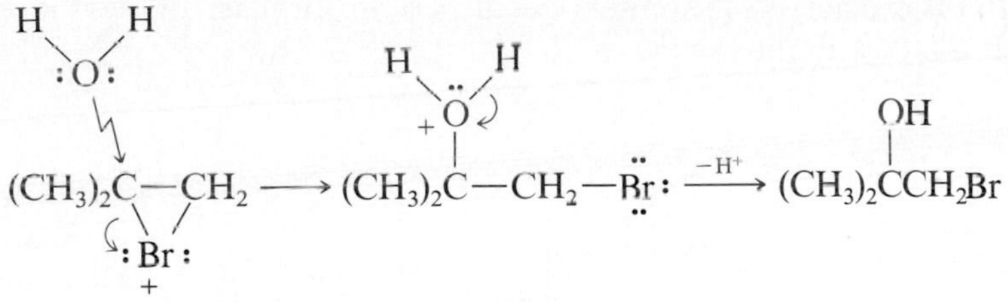

$$(CH_3)_2C—CH_2 \longrightarrow (CH_3)_2C—CH_2—\overset{..}{\underset{..}{Br}}: \xrightarrow{-H^+} (CH_3)_2\overset{OH}{C}CH_2Br$$

7.15 EPOXIDATION OF ALKENES

You have just seen that cyclic halonium ion intermediates are formed when sources of electrophilic halogen attack a double bond. Likewise, three-membered oxygen-containing rings are formed by the reaction of alkenes with sources of electrophilic oxygen.

Three-membered rings that contain oxygen are called *epoxides*. At one time, epoxides were named as oxides of alkenes. Ethylene oxide and propylene oxide, for example, are the common names of two industrially important epoxides.

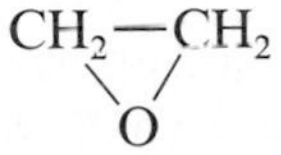

$$\underset{\text{Ethylene oxide}}{\overset{\displaystyle CH_2—CH_2}{\underset{O}{\diagdown\diagup}}} \qquad \underset{\text{Propylene oxide}}{\overset{\displaystyle CH_2—CHCH_3}{\underset{O}{\diagdown\diagup}}}$$

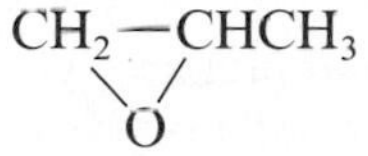

The IUPAC system permits two alternative ways to name epoxides. In one of these, epoxides are named as epoxy derivatives of alkanes. According to this system, ethylene oxide becomes epoxyethane, and propylene oxide becomes 1,2-epoxypropane. The prefix *epoxy* always immediately precedes the alkane ending; it is not listed in alphabetical order in the manner of other substituents.

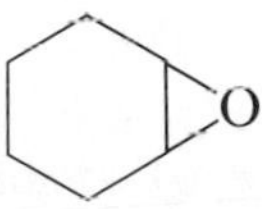

1,2-Epoxycyclohexane 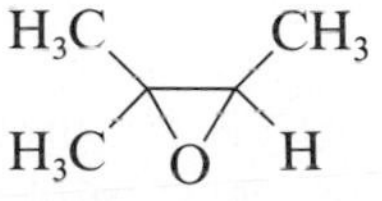   2-Methyl-2,3-epoxybutane

Another way that epoxides are named in the IUPAC system is based on *oxirane* as the parent name of the simplest epoxide, ethylene oxide. Substituents are specified in the usual way and their positions identified by number. Numbering begins at the ring oxygen and proceeds in the direction that gives the lower number to the substituent.

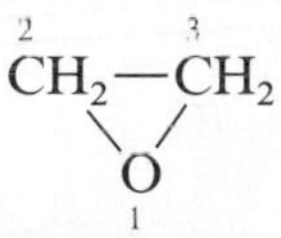

Oxirane 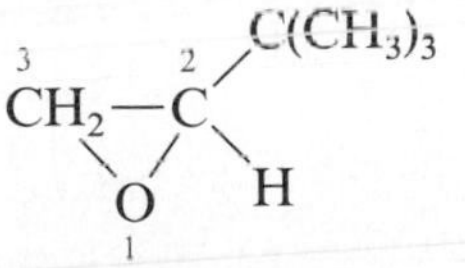   2-*tert*-Butyloxirane 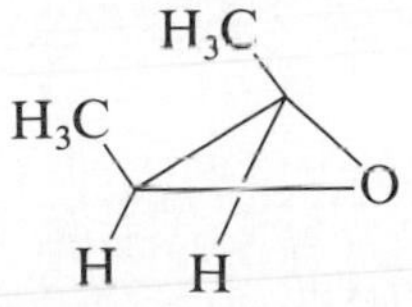   *cis*-2,3-Dimethyloxirane

Functional group transformations of epoxide groups rank among the fundamental reactions of organic chemistry, and epoxides are commonplace natural products. The female gypsy moth, for example, attracts the male by emitting an epoxide known as *disparlure.* On detecting the presence of disparlure, the male follows the scent to its origin and mates with the female.

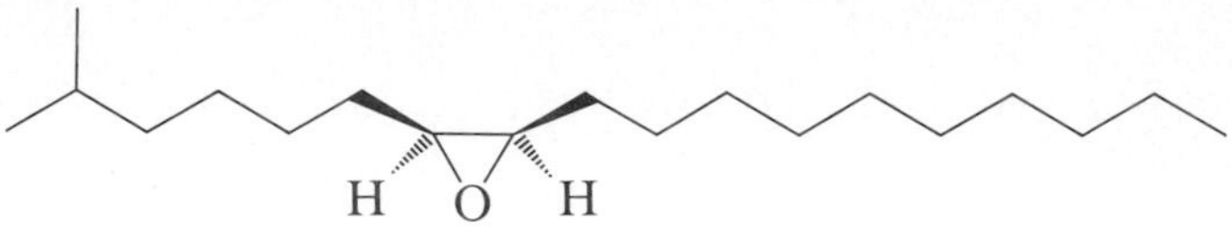

Disparlure

In one strategy designed to control the spread of the gypsy moth, infested areas are sprayed with synthetic disparlure. With the sex attractant everywhere, male gypsy moths become hopelessly confused as to the actual location of individual females. Many otherwise fertile female gypsy moths then live out their lives without producing hungry gypsy moth caterpillars.

PROBLEM 7.12   The IUPAC rules allow you to name epoxides by either the epoxy or the oxirane system. In the case of disparlure, the epoxy nomenclature is slightly easier. Give the IUPAC name, including stereochemistry, for disparlure according to the epoxy nomenclature.

Epoxides are very easy to prepare. They are the products of the reaction between an alkene and a peroxy acid. This process is known as *epoxidation.*

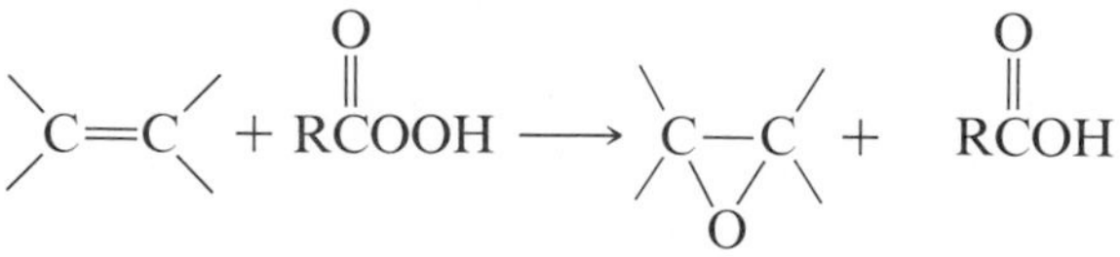

| Alkene | Peroxy acid | Epoxide | Carboxylic acid |

A commonly used peroxy acid is peroxyacetic acid, which has the formula $CH_3\overset{O}{\overset{\|}{C}}OOH$ (CH_3CO_2OH). Peroxyacetic acid is normally used in acetic acid as the solvent, but epoxidation reactions tolerate a variety of solvents and are often carried out in dichloromethane or chloroform.

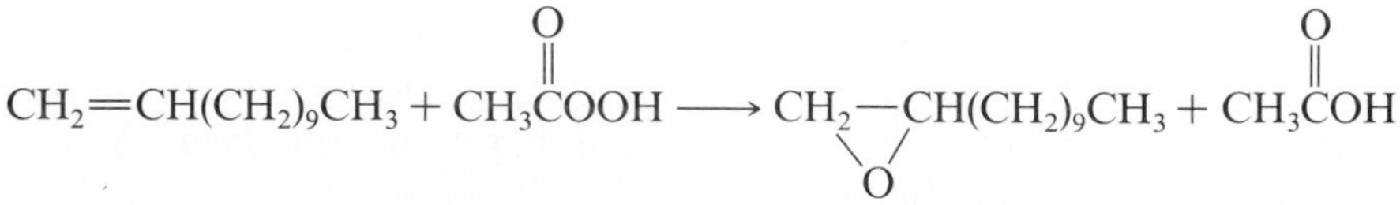

$$CH_2{=}CH(CH_2)_9CH_3 + CH_3\overset{O}{\overset{\|}{C}}OOH \longrightarrow CH_2{-}CH(CH_2)_9CH_3 + CH_3\overset{O}{\overset{\|}{C}}OH$$

| 1-Dodecene | Peroxyacetic acid | 1,2-Epoxydodecane (52%) | Acetic acid |

| Cyclooctene | Peroxyacetic acid | 1,2-Epoxycyclooctane (86%) | Acetic acid |

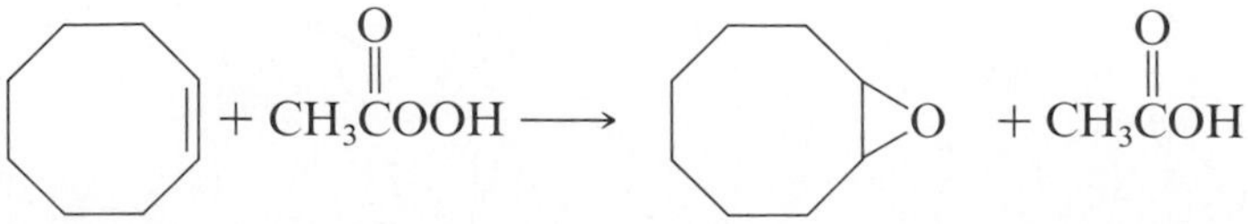

TABLE 7.3

Relative Rates of Epoxidation of Some Representative Alkenes with Peroxyacetic Acid

Alkene	Structural formula	Relative rate of epoxidation*
Ethylene	$CH_2{=}CH_2$	1.0
Propene	$CH_3CH{=}CH_2$	22
2-Methylpropene	$(CH_3)_2C{=}CH_2$	484
2-Methyl-2-butene	$(CH_3)_2C{=}CHCH_3$	6526

* In acetic acid, 26°C.

Epoxidation of alkenes with peroxy acids is a stereospecific reaction corresponding to syn addition of oxygen to the double bond. Substituents that are cis to each other in the alkene remain cis to each other in the epoxide; substituents that are trans in the alkene remain trans in the epoxide.

PROBLEM 7.13   Give the structure of the alkene, including stereochemistry, that you would choose as the starting material in a preparation of synthetic disparlure.

As shown in Table 7.3, electron-releasing alkyl groups on the double bond increase the rate of epoxidation. This suggests that the peroxy acid acts as an electrophilic reagent toward the alkene.

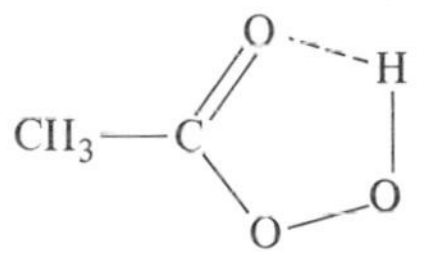

Intramolecular hydrogen bonding in peroxyacetic acid

(*a*) Peroxy acids are known to exist in a conformation in which the hydroxyl proton is intramolecularly hydrogen-bonded to the oxygen of the carbonyl group.

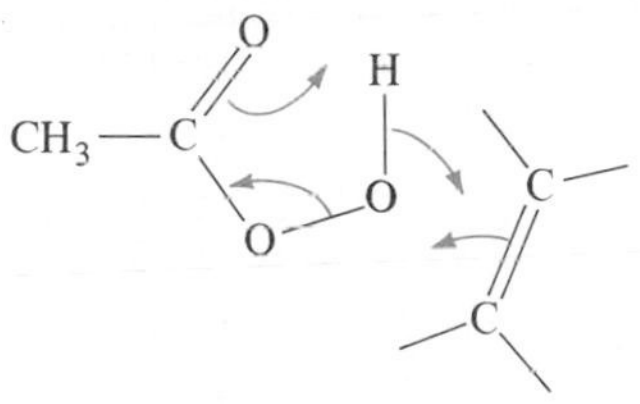

(*b*) The weak oxygen-oxygen bond of the peroxy acid breaks as the hydroxyl oxygen is transferred to the alkene.

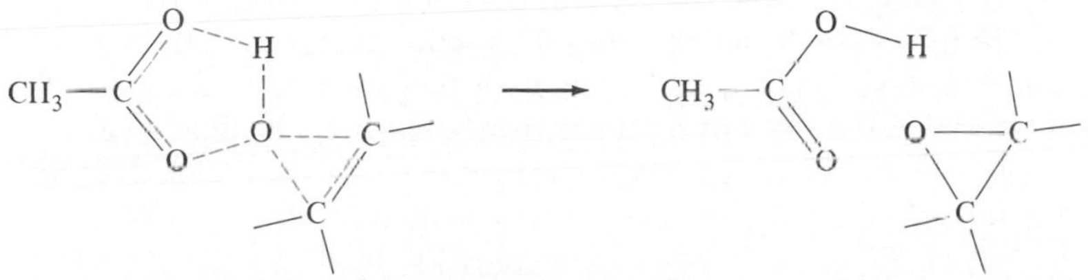

(*c*) A single transition state is believed to connect reactants and products. A representation of the activated complex is shown in which the bonds that are breaking and those that are forming are represented by dotted lines.

FIGURE 7.11   Aspects of the mechanism of alkene epoxidation. The most stable conformation of peroxyacetic acid is portrayed in (*a*). In (*b*) curved arrows are used to show the concerted transfer of oxygen from peroxyacetic acid to an alkene. A representation of the activated complex for oxygen transfer is depicted in (*c*).

The mechanism of alkene epoxidation is believed to be a concerted process, as shown in Figure 7.11.

7.16 OZONOLYSIS OF ALKENES

Ozone (O_3) is the triatomic form of oxygen. It can be represented as a combination of its two most stable Lewis structures.

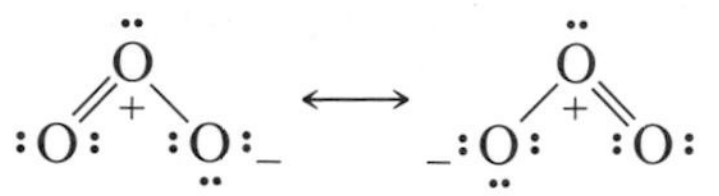

Overall, ozone is a neutral but polar molecule. The high electronegativity of oxygen makes ozone a powerful electrophile. It undergoes a remarkable reaction with alkenes in which both the σ and π components of the carbon-carbon double bond are cleaved. This reaction is known as *ozonation,* and its product is referred to as an ozonide. (The systematic IUPAC name for an ozonide is a 1,2,4-trioxolane.)

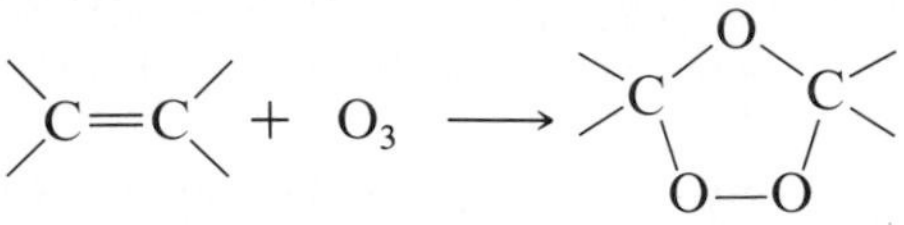

$$\ce{C=C} + O_3 \longrightarrow \text{Ozonide}$$

Alkene Ozone Ozonide

Ozonides undergo hydrolysis in water, giving carbonyl compounds.

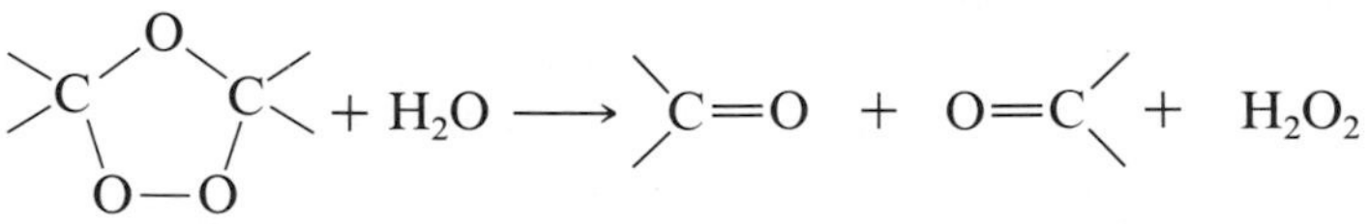

$$\text{Ozonide} + H_2O \longrightarrow \ce{C=O} + \ce{O=C} + H_2O_2$$

Ozonide Water Two carbonyl compounds Hydrogen peroxide

Two types of carbonyl compounds, aldehydes and ketones, may be formed on hydrolysis of the ozonides derived from alkenes. Aldehydes have at least one hydrogen substituent on the carbonyl group; ketones have two carbon substituents, alkyl groups, for example, on the carbonyl.

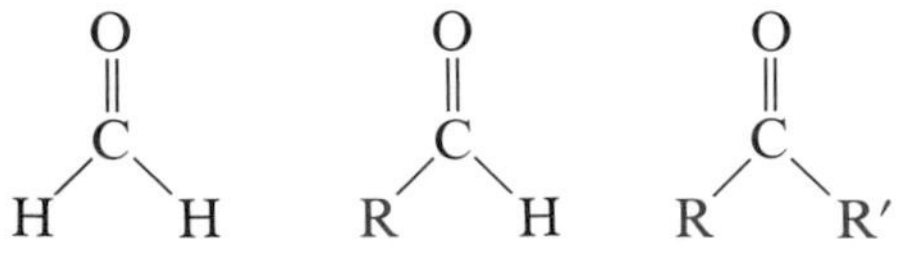

Formaldehyde Aldehyde Ketone

Aldehydes are easily oxidized to carboxylic acids under conditions of ozonide hydrolysis. When one wishes to isolate the aldehyde itself, a reducing agent such as zinc is included during the hydrolysis step. Zinc reacts with the oxidants present (excess ozone and hydrogen peroxide), preventing them from oxidizing the aldehyde.

The combination of the two reactions of ozonation and hydrolysis is called *ozonolysis.* In general, ozonolysis of an alkene may be represented by the equation

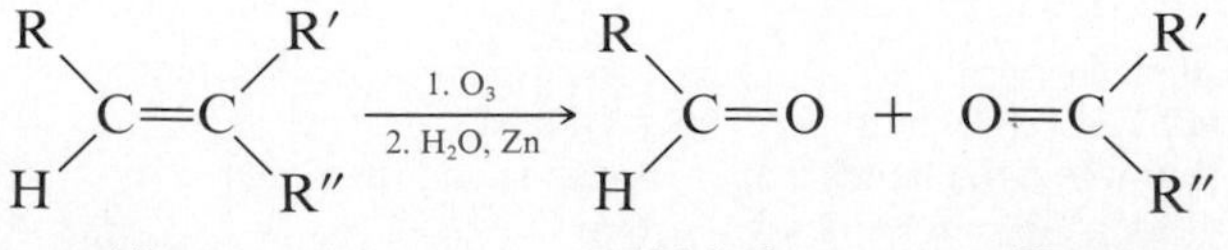

$$\ce{C=C} \xrightarrow[\text{2. } H_2O, Zn]{\text{1. } O_3} \ce{C=O} + \ce{O=C}$$

Alkene Aldehyde Ketone

Each carbon atom of the carbon-carbon double bond becomes the carbon of a carbonyl group.

Ozonation of alkenes is a complicated reaction mechanistically and has been the subject of extensive and continuing investigation. Its individual steps require a knowledge of certain types of organic reactions that we have not yet encountered. Therefore, we will not discuss the mechanism of either ozonation or ozonide hydrolysis but instead will emphasize the applications of ozonolysis to synthesis and analysis.

Ozonolysis has both synthetic and analytical applications in organic chemistry. In synthesis ozonolysis of alkenes provides a method for the preparation of aldehydes and ketones.

$$(CH_3)_2CHCH_2CH_2CH_2CH{=}CH_2 \xrightarrow[\text{2. H}_2\text{O, Zn}]{\text{1. O}_3} \underset{\text{5-Methylhexanal (62\%)}}{(CH_3)_2CHCH_2CH_2CH_2\overset{\displaystyle O}{\overset{\|}{C}}H} + \underset{\text{Formaldehyde}}{H\overset{\displaystyle O}{\overset{\|}{C}}H}$$

6-Methyl-1-heptene

$$\underset{\underset{\displaystyle CH_3}{|}}{CH_3CH_2CH_2CH_2C}{=}CH_2 \xrightarrow[\text{2. H}_2\text{O, Zn}]{\text{1. O}_3} \underset{\text{2-Hexanone (60\%)}}{CH_3CH_2CH_2CH_2\overset{\displaystyle O}{\overset{\|}{C}}CH_3} + \underset{\text{Formaldehyde}}{H\overset{\displaystyle O}{\overset{\|}{C}}H}$$

2-Methyl-1-hexene

When the objective is analytical, the products of ozonolysis are isolated and identified, thereby allowing the structure of the alkene to be deduced. In one such example, an alkene having the molecular formula C_8H_{16} was obtained from a chemical reaction and was then subjected to ozonolysis, giving acetone and 2,2-dimethylpropanal as the products.

$$\underset{\text{Acetone}}{CH_3\overset{\displaystyle O}{\overset{\|}{C}}CH_3} \qquad\qquad \underset{\text{2,2-Dimethylpropanal}}{(CH_3)_3C\overset{\displaystyle O}{\overset{\|}{C}}H}$$

Together, these two products contain all eight carbons of the starting alkene. The two carbonyl carbons correspond to those that were doubly bonded in the original alkene. Therefore, one of the doubly bonded carbons bears two methyl substituents, the other bears a hydrogen and a *tert*-butyl group. The alkene is identified as 2,4,4-trimethyl-2-pentene.

2,4,4-Trimethyl-2-pentene

Cleavage occurs here on ozonolysis; each doubly bonded carbon becomes the carbon of a C=O unit

PROBLEM 7.14 The same reaction that gave 2,4,4-trimethyl-2-pentene also yielded an isomeric alkene. This second alkene produced formaldehyde and 4,4-dimethyl-2-pentanone on ozonolysis. Identify this alkene.

$$CH_3\overset{\displaystyle O}{\overset{\displaystyle \|}{C}}CH_2C(CH_3)_3$$

4,4-Dimethyl-2-pentanone

7.17 PERMANGANATE CLEAVAGE OF ALKENES

As an alternative to ozonolysis, alkenes can be cleaved by treatment with potassium permanganate.

$$\underset{R'}{\overset{R}{>}}C=C\underset{H}{\overset{R''}{<}} \xrightarrow[\text{2. H}^+]{\text{1. KMnO}_4} \underset{R'}{\overset{R}{>}}C=O + O=C\underset{OH}{\overset{R''}{<}}$$

Alkene   Ketone   Carboxylic acid

As was the case for ozonolysis, each carbon atom of the double bond becomes the carbon atom of a carbonyl group. A hydrogen substituent on the double bond is replaced by a hydroxyl group, and a carboxylic acid is produced on permanganate oxidation in cases in which an aldehyde would result from ozonolysis.

$$CH_3(CH_2)_{10}CH=CH_2 \xrightarrow[\text{2. H}^+]{\text{1. KMnO}_4} CH_3(CH_2)_{10}\overset{\displaystyle O}{\overset{\displaystyle \|}{C}}OH + CO_2 + H_2O$$

1-Tridecene   Dodecanoic acid   Carbon   Water
                       (84%)   dioxide

Because carboxylic acids are formed in these reactions as their potassium carboxylate salts, the acidification step indicated in the above equations is necessary in order to isolate the product as the free acid. The terminal carbon of 1-tridecene bears two hydrogen substituents, so it is oxidized to carbonic acid (as its potassium salt). On acidification potassium carbonate is converted to carbonic acid, which spontaneously dissociates to carbon dioxide and water.

$$KO\overset{\displaystyle O}{\overset{\displaystyle \|}{C}}OK \xrightarrow{\text{H}^+} HO\overset{\displaystyle O}{\overset{\displaystyle \|}{C}}OH \longrightarrow CO_2 + H_2O$$

Potassium   Carbonic   Carbon   Water
carbonate   acid   dioxide

PROBLEM 7.15 Oxidation of bicyclo[2.2.1]hept-2-ene with sodium permanganate in water, followed by acidification with sulfuric acid, gave a single product having the molecular formula $C_7H_{10}O_4$. What is the structure of this product?

Bicyclo[2.2.1]hept-2-ene

7.18 REACTIONS OF ALKENES WITH ALKENES. POLYMERIZATION

While 2-methylpropene undergoes acid-catalyzed hydration in dilute sulfuric acid to form *tert*-butyl alcohol (Section 7.6 and Figure 7.6), an unusual reaction occurs in more concentrated solutions of sulfuric acid. Rather than forming the expected alkyl hydrogen sulfate (Section 7.5), 2-methylpropene is converted to a mixture of two isomeric C_8H_{16} alkenes.

$$2(CH_3)_2C{=}CH_2 \xrightarrow{\text{65\% } H_2SO_4} CH_2{=}\underset{\underset{CH_3}{|}}{C}CH_2C(CH_3)_3 + (CH_3)_2C{=}CHC(CH_3)_3$$

2-Methylpropene	2,4,4-Trimethyl-1-	2,4,4-Trimethyl-2-
(isobutene)	pentene	pentene

With molecular formulas corresponding to twice that of the starting alkene, the products of this reaction are referred to as *dimers* of 2-methylpropene. The suffix *-mer* denotes a molecule or molecular unit; in this case, 2-methylpropene is termed a *monomer*. When two monomers combine, a dimer results. Three monomeric units produce a *trimer,* four a *tetramer,* etc. A high-molecular-weight material comprising a large number of monomeric units is called a *polymer.*

PROBLEM 7.16   The two dimers of 2-methylpropene shown in the equation can be converted to 2,2,4-trimethylpentane (known by its nonsystematic name *isooctane*) for use as a gasoline additive. Can you suggest a method for this conversion?

Alkene dimers are formed by the mechanism shown in Figure 7.12 for the case of 2-methylpropene. In step 1 protonation of the double bond generates a small amount of *tert*-butyl cation in equilibrium with the alkene. The carbocation is an electrophile and attacks a second molecule of 2-methylpropene in step 2, forming a new carbon-carbon bond and generating a C_8 carbocation. This new carbocation loses a proton in step 3 to form a mixture of 2,4,4-trimethyl-1-pentene and 2,4,4-trimethyl-2-pentene.

Dimerization in concentrated sulfuric acid occurs mainly with those alkenes that form tertiary carbocations. This is because the alkene itself is nucleophilic enough to compete effectively with sulfuric acid for the carbocation intermediate. In some cases reaction conditions can be developed that favor the formation of higher-molecular-weight alkenes that are polymers of 2-methylpropene. Since these reactions proceed by way of carbocation intermediates, the process is referred to as *cationic polymerization.* It is limited to those alkenes capable of forming relatively stable carbocations.

We made special mention in Section 5.2 of the enormous volume of ethylene production in the petrochemical industry and have noted from time to time since then some of the uses of ethylene. By far, however, most of the ethylene is used to prepare *polyethylene,* a high-molecular-weight polymer of ethylene. Polyethylene cannot be prepared by cationic polymerization—ethylene does not form a carbocation and its double bond is not very nucleophilic. Ethylene can be converted to polyethylene, however, under conditions of *free-radical polymerization.*

In the free-radical polymerization of ethylene, ethylene is heated at high pressure in the presence of oxygen or a peroxide.

$$n CH_2{=}CH_2 \xrightarrow[\substack{O_2 \text{ or} \\ \text{peroxides}}]{\substack{200°C \\ 2000 \text{ atm}}} {-}CH_2{-}CH_2{-}(CH_2{-}CH_2)_{n-2}{-}CH_2{-}CH_2{-}$$

Ethylene	Polyethylene

In this reaction n can have a value of thousands.

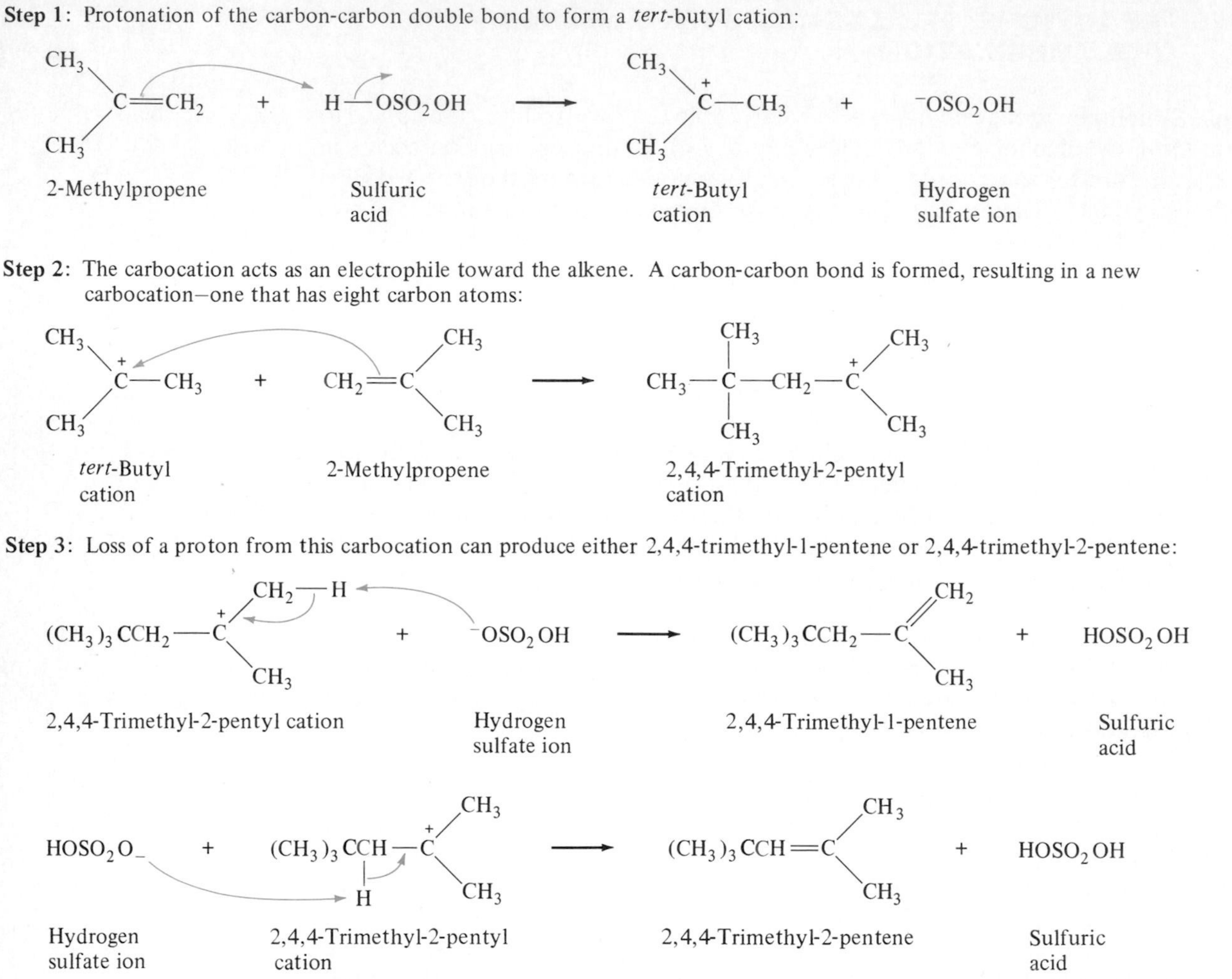

Step 1: Protonation of the carbon-carbon double bond to form a *tert*-butyl cation:

(CH₃)₂C=CH₂ + H—OSO₂OH ⟶ (CH₃)₃C⁺ + ⁻OSO₂OH

2-Methylpropene	Sulfuric acid	*tert*-Butyl cation	Hydrogen sulfate ion

Step 2: The carbocation acts as an electrophile toward the alkene. A carbon-carbon bond is formed, resulting in a new carbocation—one that has eight carbon atoms:

(CH₃)₃C⁺ + CH₂=C(CH₃)₂ ⟶ (CH₃)₃C—CH₂—C⁺(CH₃)₂

tert-Butyl cation	2-Methylpropene	2,4,4-Trimethyl-2-pentyl cation

Step 3: Loss of a proton from this carbocation can produce either 2,4,4-trimethyl-1-pentene or 2,4,4-trimethyl-2-pentene:

$(CH_3)_3CCH_2$—C⁺(CH₂—H)(CH₃) + ⁻OSO₂OH ⟶ $(CH_3)_3CCH_2$—C(=CH₂)(CH₃) + HOSO₂OH

2,4,4-Trimethyl-2-pentyl cation	Hydrogen sulfate ion	2,4,4-Trimethyl-1-pentene	Sulfuric acid

HOSO₂O⁻ + $(CH_3)_3CCH$(H)—C⁺(CH₃)₂ ⟶ $(CH_3)_3CCH$=C(CH₃)₂ + HOSO₂OH

Hydrogen sulfate ion	2,4,4-Trimethyl-2-pentyl cation	2,4,4-Trimethyl-2-pentene	Sulfuric acid

FIGURE 7.12 Sequence of steps that describes the mechanism of acid-catalyzed dimerization of 2-methylpropene.

An outline of the mechanism of free-radical polymerization of ethylene is shown in Figure 7.13. Dissociation of a peroxide initiates the process in step 1. The resulting peroxy radical adds to the carbon-carbon double bond in step 2, giving a new radical, which then adds to a second molecule of ethylene in step 3. The carbon-carbon bond-forming process in step 3 can be repeated thousands of times to give long carbon chains.

In spite of the -ene ending to its name, polyethylene is much more closely related to alkanes than to alkenes. It is simply a long chain of CH_2 groups bearing at its ends an alkoxy group (from the initiator) or a carbon-carbon double bond.

Teflon is a polymer prepared by free-radical polymerization of tetrafluoroethylene, CF_2=CF_2. It is a very inert substance used as a "nonstick" coating in cookware and as a material from which greaseless bearings and fittings are made.

Ethylene can be polymerized at low pressure and temperature by a process known as *coordination polymerization.* Coordination polymerization utilizes a mixture of titanium tetrachloride, $TiCl_4$, and triethylaluminum, $Al(CH_2CH_3)_3$, as a catalyst.

Step 1: Homolytic dissociation of a peroxide produces alkoxy radicals that serve as free-radical initiators:

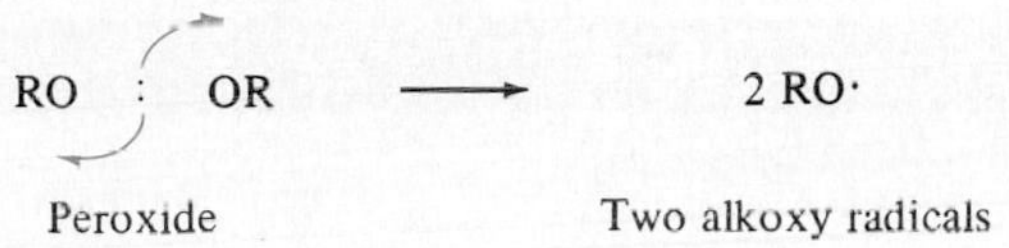

$$RO : OR \longrightarrow 2\ RO\cdot$$

Peroxide   Two alkoxy radicals

Step 2: An alkoxy radical adds to the carbon-carbon double bond:

$$RO\cdot \quad + \quad CH_2{=}CH_2 \longrightarrow RO{-}CH_2{-}\dot{C}H_2$$

Alkoxy radical   Ethylene   2-Alkoxyethyl radical

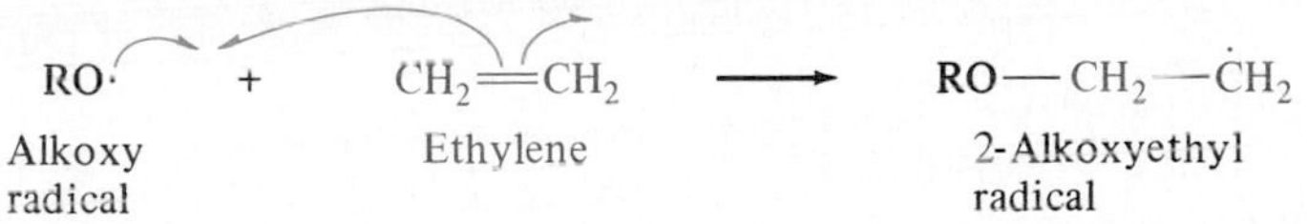

Step 3: The radical produced in step 2 adds to a second molecule of ethylene:

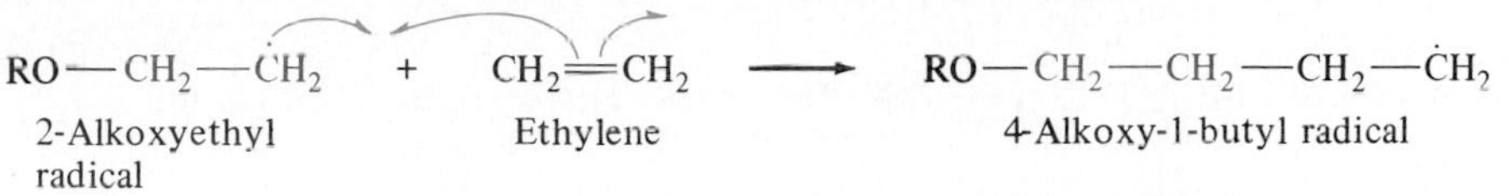

$$RO{-}CH_2{-}\dot{C}H_2 \quad + \quad CH_2{=}CH_2 \longrightarrow RO{-}CH_2{-}CH_2{-}CH_2{-}\dot{C}H_2$$

2-Alkoxyethyl radical   Ethylene   4-Alkoxy-1-butyl radical

The radical formed in step 3 then adds to a third molecule of ethylene, and the process continues, forming a long chain of methylene groups.

FIGURE 7.13 Mechanistic description of peroxide-induced free-radical polymerization of ethylene.

The polyethylene produced by coordination polymerization has a higher density than that produced by free-radical polymerization and somewhat different—in many applications, more desirable—properties. The catalyst system used in coordination polymerization was developed independently by Karl Ziegler in Germany and Giulio Natta in Italy in the early 1950s. They shared the Nobel prize in chemistry in 1963 for this work. The Ziegler-Natta catalyst system also permits polymerization of propene to be achieved in a way that gives a form of *polypropylene* suitable for plastics and fibers. When propene is polymerized under free-radical conditions, the polypropylene has physical properties (such as a low melting point) that preclude its use in plastics and fibers.

7.19 SUMMARY

The reaction path that most typifies alkenes is addition of electrophilic reagents to the carbon-carbon double bond.

$$\sideset{}{}{C}{=}C + \overset{\delta+}{E}{-}\overset{\delta-}{Y} \longrightarrow E{-}\overset{|}{\underset{|}{C}}{-}\overset{|}{\underset{|}{C}}{-}Y$$

Alkene   Electrophilic reagent   Product of addition

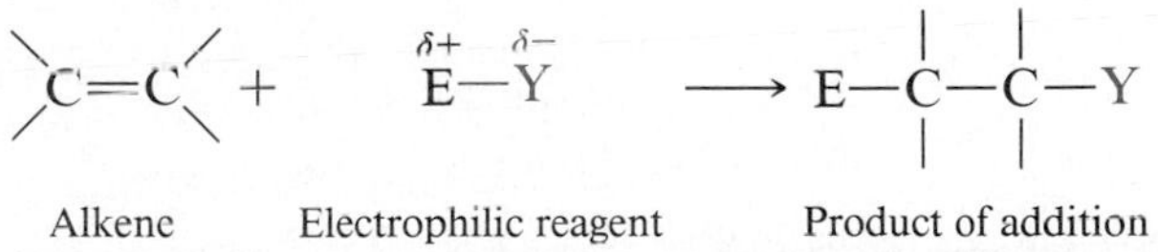

Examples of some reactions of this type are presented in Table 7.4. In all these reactions the π electrons of the alkene interact with the electrophilic reagent in the first step of the reaction. Carbocations are likely intermediates in the addition of hydrogen halides to alkenes.

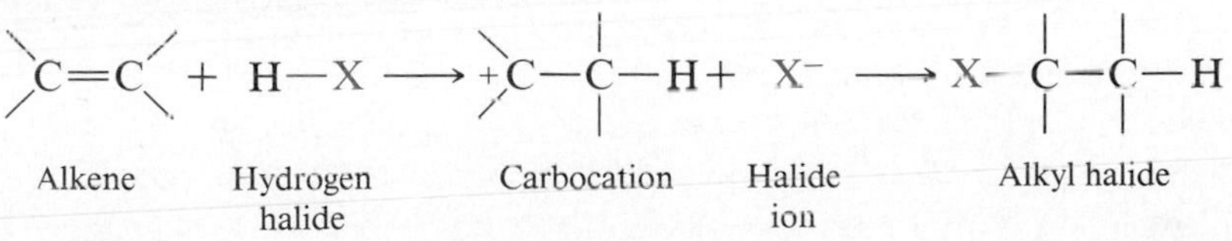

$$C{=}C + H{-}X \longrightarrow {}^+C{-}\overset{|}{\underset{|}{C}}{-}H + X^- \longrightarrow X{-}\overset{|}{\underset{|}{C}}{-}\overset{|}{\underset{|}{C}}{-}H$$

Alkene   Hydrogen halide   Carbocation   Halide ion   Alkyl halide

TABLE 7.4

Addition of Electrophilic Reagents to Alkenes

Reaction (section) and comments	General equation and specific example
Addition of hydrogen halides (Sections 7.1 through 7.3) A proton and a halogen add to the double bond of an alkene to yield an alkyl halide. Addition proceeds in accordance with Markovnikov's rule; hydrogen adds to the carbon that has the greater number of hydrogen substituents, halide to the carbon that has the fewer hydrogen substituents.	$RCH{=}CR_2' + HX \longrightarrow RCH_2{-}\underset{\underset{X}{\|}}{CR_2'}$ Alkene Hydrogen halide Alkyl halide Methylenecyclohexane + HCl $\longrightarrow$ 1-Chloro-1-methylcyclohexane (75–80%)
Addition of sulfuric acid (Section 7.5) Alkenes react with cold, concentrated sulfuric acid to form alkyl hydrogen sulfates. A proton and a hydrogen sulfate ion add across the double bond in accordance with Markovnikov's rule. Alkenes that yield tertiary carbocations on protonation tend to polymerize in concentrated sulfuric acid (Section 7.18).	$RCH{=}CR_2' + HOSO_2OH \longrightarrow RCH_2{-}\underset{\underset{OSO_2OH}{\|}}{CR_2'}$ Alkene Sulfuric acid Alkyl hydrogen sulfate $CH_2{=}CHCH_2CH_3 + HOSO_2OH \longrightarrow CH_3{-}\underset{\underset{OSO_2OH}{\|}}{CHCH_2CH_3}$ 1-Butene Sulfuric acid *sec*-Butyl hydrogen sulfate
Addition of halogens (Sections 7.11 through 7.13) Bromine and chlorine add across the carbon-carbon double bond of alkenes to form vicinal dihalides. A cyclic halonium ion is an intermediate. The reaction is stereospecific; anti addition is observed.	$R_2C{=}CR_2 + X_2 \longrightarrow X{-}\underset{\underset{R}{\|}}{\overset{\overset{R}{\|}}{C}}{-}\underset{\underset{R}{\|}}{\overset{\overset{R}{\|}}{C}}{-}X$ Alkene Halogen Vicinal dihalide $CH_2{=}CHCH_2CH_2CH_2CH_3 + Br_2 \longrightarrow BrCH_2{-}\underset{\underset{Br}{\|}}{CHCH_2CH_2CH_2CH_3}$ 1-Hexene Bromine 1,2-Dibromohexane (100%)
Halohydrin formation (Section 7.14) When treated with bromine or chlorine in aqueous solution, alkenes are converted to vicinal halohydrins. A halonium ion is an intermediate. The elements of hypobromous acid (HOBr) or hypochlorous acid (HOCl) add across the double bond in accordance with Markovnikov's rule. The positively polarized halogen adds to the carbon that has the greater number of hydrogen substituents.	$RCH{=}CR_2' + X_2 + H_2O \longrightarrow X{-}\underset{\underset{R}{\|}}{CH}{-}\underset{\underset{R'}{\|}}{\overset{\overset{R'}{\|}}{C}}{-}OH + HX$ Alkene Halogen Water Vicinal halohydrin Hydrogen halide Methylenecyclohexane $\xrightarrow[H_2O]{Br_2}$ (1-Bromomethyl)cyclohexanol (89%)

TABLE 7.4 (continued)

Reaction (section) and comments	General equation and specific example
Epoxidation (Section 7.15) Peroxy acids transfer oxygen to the double bond of alkenes to yield epoxides. The reaction is a stereospecific syn addition.	$R_2C=CR_2 + R'COOH \longrightarrow R_2C-CR_2 + R'COH$ Alkene Peroxy acid Epoxide Carboxylic acid 1-Methylcycloheptene Peroxyacetic acid 1-Methyl-1,2-epoxycycloheptane (65%) Acetic acid

Hydrogen bromide is unique among the hydrogen halides in that it can add to alkenes either by an ionic mechanism, as shown in the above equation, or by a free-radical mechanism. The ionic and free-radical additions of hydrogen bromide are opposite in their regioselectivities. Like other ionic addition reactions, hydrogen bromide normally adds to alkenes in accordance with Markovnikov's rule. Protonation of the double bond occurs in the direction that gives the more stable carbocation; bromide ion then captures the carbocation intermediate. Under photochemical conditions or in the presence of peroxides, a bromine atom is formed and then adds to the carbon-carbon double bond to give the more stable free radical. This radical then abstracts a hydrogen atom from hydrogen bromide to yield the alkyl bromide.

Methylenecycloheptane (Bromomethyl)cycloheptane (61%)

Carbocations are bypassed in certain other electrophilic addition reactions. Vicinal dihalides and halohydrins are formed by way of halonium ion intermediates, and epoxides arise by direct oxygen transfer to the alkene in a concerted process.

Several reactions of alkenes are useful in the preparation of alcohols. These are summarized in Table 7.5. All involve electrophilic addition to the double bond in the first step. Acid-catalyzed hydration proceeds by proton transfer to the double bond; mercuric acetate acts as an electrophile in oxymercuration and diborane as an electrophile in hydroboration.

Alkenes can be cleaved to carbonyl compounds by ozonolysis or by treatment with potassium permanganate. These reactions, illustrated in Table 7.6, are useful both for synthesis (preparation of aldehydes, ketones, or carboxylic acids) and analysis. When applied to analysis, the carbonyl compounds are isolated and identified, allowing the substituents attached to the double bond to be deduced.

Polymerization is the process whereby many alkene molecules react to give long hydrocarbon chains. Among the methods by which alkenes are polymerized are

TABLE 7.5

Reactions That Convert Alkenes to Alcohols

Reaction (section) and comments	General equation and specific example
Formation and hydrolysis of alkyl hydrogen sulfates (Section 7.5) Alkyl hydrogen sulfates, formed by addition of sulfuric acid to alkenes, undergo hydrolysis on heating in aqueous solution, giving alcohols.	$RCH_2CR_2' + H_2O \longrightarrow RCH_2CR' + H_2SO_4$ with OSO_2OH below left carbon and OH below right carbon. Alkyl hydrogen sulfate Water Alcohol Sulfuric acid $CH_3CH_2CHCH_3 \xrightarrow[\text{heat}]{H_2O} CH_3CH_2CHCH_3$ with OSO_2OH and OH substituents *sec*-Butyl hydrogen sulfate *sec*-Butyl alcohol
Acid-catalyzed hydration (Section 7.6) Addition of the elements of water to the double bond of an alkene takes place in aqueous acidic solution. Addition occurs according to Markovnikov's rule. A carbocation is an intermediate and is captured by a molecule of water acting as a nucleophile. For synthetic purposes this reaction ordinarily works well only for the preparation of tertiary alcohols.	$RCH{=}CR_2' + H_2O \xrightarrow{H^+} RCH_2CR_2'$ with OH substituent Alkene Water Alcohol $CH_2{=}C(CH_3)_2 \xrightarrow{50\% \ H_2SO_4-H_2O} (CH_3)_3COH$ 2-Methylpropene (isobutene) *tert*-Butyl alcohol (55–58%)
Oxymercuration-demercuration (Section 7.7) As an alternative to acid-catalyzed hydration, alkenes are treated with mercuric acetate in aqueous tetrahydrofuran, followed by sodium borohydride. The elements of water are added across the double bond with a regioselectivity corresponding to Markovnikov's rule. Rearrangements do not occur.	$RCH{=}CR_2' \xrightarrow[\text{2. NaBH}_4, \ HO^-]{\text{1. Hg(OAc)}_2, \ THF-H_2O} RCH_2CR_2'$ with OH substituent Alkene Alcohol $CH_3(CH_2)_{15}CH{=}CH_2 \xrightarrow[\text{2. NaBH}_4, \ HO^-]{\text{1. Hg(OAc)}_2, \ THF-H_2O} CH_3(CH_2)_{15}CHCH_3$ with OH substituent 1-Octadecene 2-Octadecanol (93%)
Hydroboration-oxidation (Sections 7.8 through 7.10) This two-step sequence achieves hydration of alkenes in a stereospecific syn manner, with a regioselectivity opposite to that of Markovnikov's rule. An organoborane is formed by electrophilic addition of diborane to an alkene. Oxidation of the organoborane intermediate with hydrogen peroxide completes the process. Rearrangements do not occur.	$RCH{=}CR_2' \xrightarrow[\text{2. H}_2O_2, \ HO^-]{\text{1. B}_2H_6, \ \text{diglyme}} RCHCHR_2'$ with OH substituent Alkene Alcohol $(CH_3)_2CHCH_2CH{=}CH_2 \xrightarrow[\text{2. H}_2O_2, \ HO^-]{\text{1. BH}_3-THF} (CH_3)_2CHCH_2CH_2CH_2OH$ 4-Methyl-1-pentene 4-Methyl-1-pentanol (80%)

TABLE 7.6

Reactions That Cleave Carbon-Carbon Double Bonds

Reaction (section) and comments	General equation and specific example
Ozonolysis (Section 7.16) Ozone reacts with alkenes to form compounds known as ozonides. Hydrolysis converts ozonides to two carbonyl-containing compounds. Each carbon atom of the original double bond becomes the carbon atom of a carbonyl group.	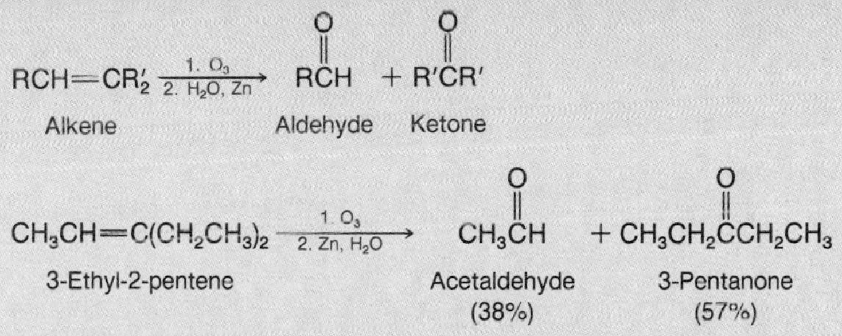 $RCH{=}CR'_2 \xrightarrow[\text{2. H}_2\text{O, Zn}]{\text{1. O}_3}$ RCH + R'CR' Alkene Aldehyde Ketone $CH_3CH{=}C(CH_2CH_3)_2 \xrightarrow[\text{2. Zn, H}_2\text{O}]{\text{1. O}_3}$ CH$_3$CH + CH$_3$CH$_2$CCH$_2$CH$_3$ 3-Ethyl-2-pentene Acetaldehyde 3-Pentanone (38%) (57%)

Permanganate Oxidation (Section 7.17) Potassium permanganate cleaves alkenes to carbonyl compounds. Unlike ozonolysis, aldehydes cannot be isolated because they are oxidized to carboxylic acids under the reaction conditions.	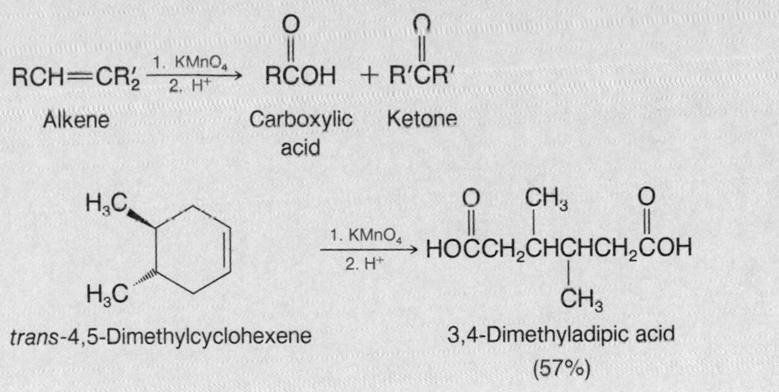 $RCH{=}CR'_2 \xrightarrow[\text{2. H}^+]{\text{1. KMnO}_4}$ RCOH + R'CR'
	 Alkene Carboxylic Ketone
	             acid
	
	 trans-4,5-Dimethylcyclohexene $\xrightarrow[\text{2. H}^+]{\text{1. KMnO}_4}$ HOCCH$_2$CHCHCH$_2$COH
	                     3,4-Dimethyladipic acid
	                       (57%)

cationic polymerization, free-radical polymerization, and coordination polymerization.

PROBLEMS

7.17 Write the structure of the principal organic product formed in the reaction of 1-pentene with each of the following:

   (a)   Hydrogen chloride
   (b)   Hydrogen bromide
   (c)   Hydrogen bromide in the presence of peroxides
   (d)   Hydrogen iodide
   (e)   Dilute sulfuric acid
   (f)   Diborane in diglyme, followed by basic hydrogen peroxide
   (g)   Mercuric acetate in aqueous tetrahydrofuran, followed by sodium borohydride
   (h)   Bromine in carbon tetrachloride
   (i)   Bromine in water
   (j)   Peroxyacetic acid
   (k)   Ozone
   (l)   Product of (k) treated with zinc and water
   (m)   Potassium permanganate, followed by acidification

7.18 Repeat Problem 7.17 for 2-methyl-2-butene.

7.19 Repeat Problem 7.17 for 1-methylcyclohexene.

7.20   Specify reagents suitable for converting 3-ethyl-2-pentene to each of the following:

(a)   2,3-Dibromo-3-ethylpentane
(b)   3-Chloro-3-ethylpentane
(c)   2-Bromo-3-ethylpentane
(d)   3-Ethyl-3-pentanol
(e)   3-Ethyl-3-pentanol (by a second route)
(f)   3-Ethyl-2-pentanol
(g)   2,2-Diethyl-3-methyloxirane
(h)   3-Ethylpentane

7.21   There are eight isomeric alcohols with the molecular formula $C_5H_{12}O$. (a) One of the primary alcohol isomers cannot be prepared from an alkene. Which one? (b) Three of the isomeric primary alcohols can be prepared from alkenes. Which ones are they? Write equations for their preparation. (c) Two of the three isomeric secondary alcohols can be prepared efficiently from alkenes. Which ones? Describe an efficient method for the preparation of these two. (d) Write the structure of the only isomer that is a tertiary alcohol. Suggest methods for its preparation from two different alkenes.

7.22   All the following reactions have been reported in the chemical literature. Give the structure of the principal organic product in each case.

(a) $CH_3CH_2CH{=}CHCH_2CH_3 + HBr \xrightarrow{\text{no peroxides}}$

(b) $(CH_3)_2C{=}C(CH_3)_2 \xrightarrow[\text{H}_3\text{PO}_4]{\text{KI}}$

(c) $CH_2{=}CHCH_2CH_2CH_2CH_3 \xrightarrow[\text{H}_3\text{PO}_4]{\text{KI}}$

(d) $(CH_3)_2CHCH_2CH_2CH_2CH{=}CH_2 \xrightarrow[\text{peroxides}]{\text{HBr}}$

(e) $(CH_3)_2C{=}CHC(CH_3)_3 \xrightarrow[\text{2. NaBH}_4,\ \text{HO}^-]{\text{1. Hg(O}_2\text{CCH}_3)_2,\ \text{THF}-\text{H}_2\text{O}}$

(f)   2-tert-Butyl-3,3-dimethyl-1-butene $\xrightarrow[\text{2. H}_2\text{O}_2,\ \text{HO}^-]{\text{1. B}_2\text{H}_6}$

(g) $\xrightarrow[\text{2. H}_2\text{O}_2,\ \text{HO}^-]{\text{1. B}_2\text{H}_6}$

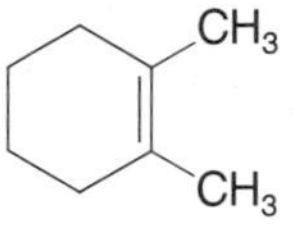

(h) $CH_2{=}\underset{\underset{CH_3}{|}}{C}CH_2CH_2CH_3 + Br_2 \xrightarrow{\text{CHCl}_3}$

(i) $(CH_3)_2C{=}CHCH_3 + Br_2 \xrightarrow{\text{H}_2\text{O}}$

(j) $(CH_3)_2C{=}C(CH_3)_2 + CH_3\overset{\overset{\text{O}}{||}}{C}OOH \longrightarrow$

(k) $\xrightarrow[\text{2. H}_2\text{O}]{\text{1. O}_3}$

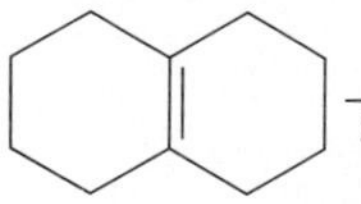

7.23   Suggest a sequence of reactions suitable for preparing each of the following compounds from the indicated starting material. You may use any necessary organic or inorganic reagents.

(a)   1-Propanol from 2-propanol
(b)   1-Bromopropane from 2-bromopropane

(c) 1,2-Dibromopropane from 2-bromopropane
(d) 1-Bromo-2-propanol from 2-propanol
(e) 1,2-Epoxypropane from 2-propanol
(f) *tert*-Butyl alcohol from isobutyl alcohol
(g) *tert*-Butyl iodide from isobutyl iodide
(h) *trans*-2-Chlorocyclohexanol from cyclohexyl chloride
(i) Cyclopentyl iodide from cyclopentane
(j) *trans*-1,2-Dichlorocyclopentane from cyclopentane

(k) $\overset{O}{\overset{\|}{HOCCH_2}}CH_2CH_2\overset{O}{\overset{\|}{COH}}$ from cyclopentanol

7.24 Two different compounds having the molecular formula $C_8H_{15}Br$ are formed when 1,6-dimethylcyclohexene reacts with hydrogen bromide in the dark and in the absence of peroxides. The same two compounds are formed from 1,2-dimethylcyclohexene. What are these two compounds?

7.25 Trimedlure is a powerful attractant for the Mediterranean fruit fly. It is a mixture of four isomers. In the synthesis of trimedlure, the step in which four isomeric products are formed involves the reaction of hydrogen chloride with compound A:

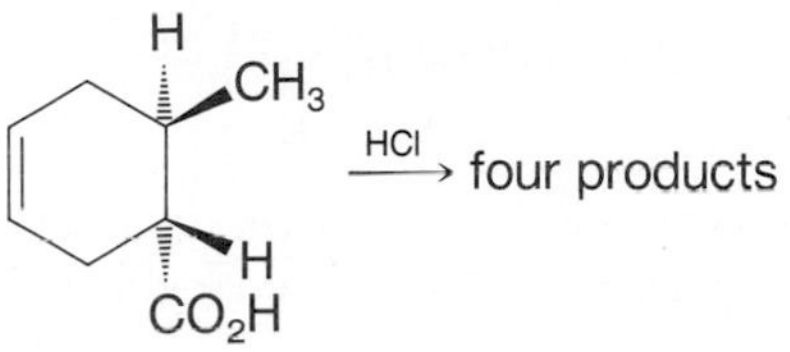

Compound A

Write the structures of four isomeric products which could reasonably be formed under these reaction conditions.

7.26 Oxymercuration-demercuration of *cis*-1,3,4-trimethylcyclopentene gives two different alcohols as products. Write their structures.

7.27 Two different vicinal dibromides are obtained when bromine is added to the alkene shown. Suggest structures for these two dibromides.

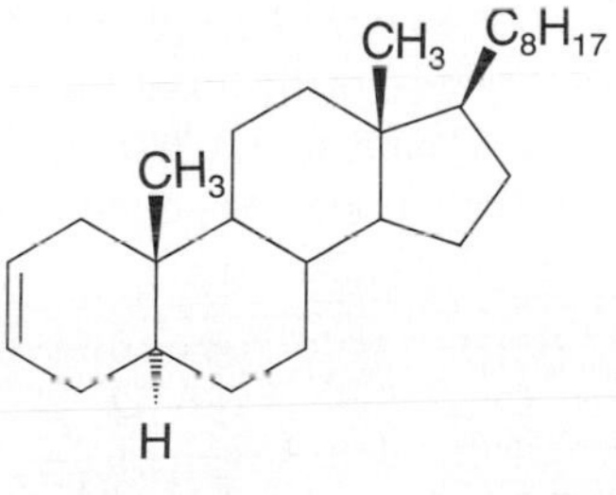

7.28 Three isomeric alkenes B, C, and D all have the molecular formula C_5H_{10} and all yield isopentane (2-methylbutane) on catalytic hydrogenation. Isomers B and C give a tertiary alcohol on oxymercuration-demercuration. Isomers C and D give different primary alcohols on hydroboration-oxidation. What are the structures of B, C, and D?

7.29 Reaction of 3,3-dimethyl-1-butene with hydrogen iodide yields two compounds E and F, each having the molecular formula $C_6H_{13}I$, in the ratio E : F = 90 : 10. Compound E, on being heated with potassium hydroxide in *n*-propyl alcohol, gives only 3,3-dimethyl-1-butene. Compound F undergoes elimination under these conditions to give 2,3-dimethyl-2-butene as the major product. Suggest structures for compounds E and F and write a reasonable mechanism for the formation of each.

7.30 Dehydration of 2,2,3,4,4-pentamethyl-3-pentanol gave two alkenes G and H. Ozonolysis of the lower boiling alkene G gave formaldehyde (CH_2=O) and 2,2,4,4-tetramethyl-3-pentanone. Ozonolysis of H gave formaldehyde and 3,3,4,4-tetramethyl-2-pentanone. Identify G and H and suggest an explanation for the formation of H in the dehydration reaction.

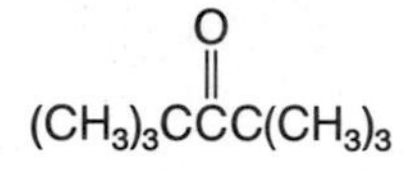

$$\begin{matrix} O \\ \parallel \\ (CH_3)_3CCC(CH_3)_3 \end{matrix} \qquad\qquad \begin{matrix} OCH_3 \\ \parallel\, | \\ CH_3CCC(CH_3)_3 \\ | \\ CH_3 \end{matrix}$$

2,2,4,4-Tetramethyl-3-pentanone 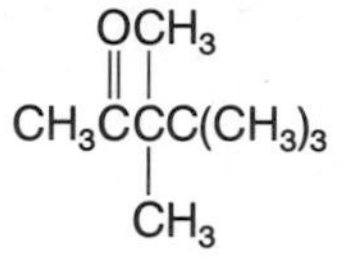   3,3,4,4-Tetramethyl-2-pentanone

7.31 Compound I ($C_7H_{13}Br$) is a tertiary bromide. On treatment with sodium ethoxide in ethanol, I is converted into J (C_7H_{12}). Ozonolysis of J gives K as the only product. Deduce the structures of I and J. What is the symbol for the reaction mechanism by which I is converted to J under the reaction conditions?

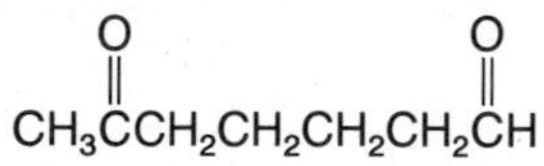

$$\begin{matrix} O & & O \\ \parallel & & \parallel \\ CH_3CCH_2CH_2CH_2CH_2CH \end{matrix}$$

Compound K

7.32 East Indian sandalwood oil contains a hydrocarbon given the name *santene* (C_9H_{14}). Ozonation of santene followed by hydrolysis gives compound L. What is the structure of santene?

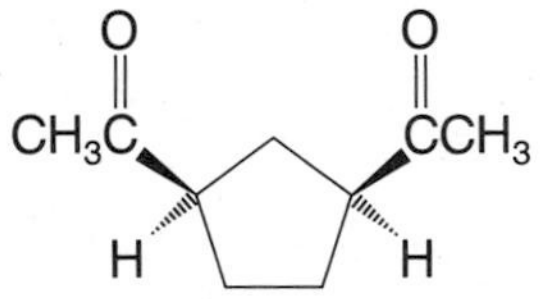

Compound L

7.33 *Sabinene, Δ^3-carene,* and *α-pinene* are isomeric natural products with the molecular formula $C_{10}H_{16}$. (*a*) Ozonolysis of sabinene followed by hydrolysis in the presence of zinc gives compound M. What is the structure of sabinene? What other compound is formed on ozonolysis? (*b*) Ozonolysis of Δ^3-carene followed by hydrolysis in the presence of zinc gives compound N. What is the structure of Δ^3-carene? (*c*) Treatment of α-pinene with potassium permanganate, followed by acidification, gives compound O. What is the structure of α-pinene?

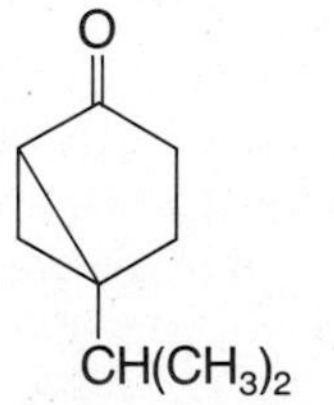

Compound M 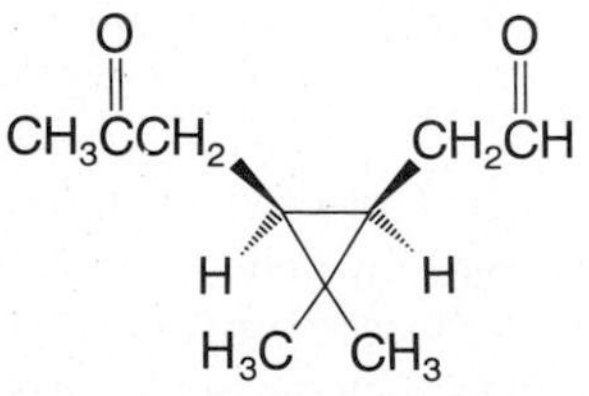   Compound N 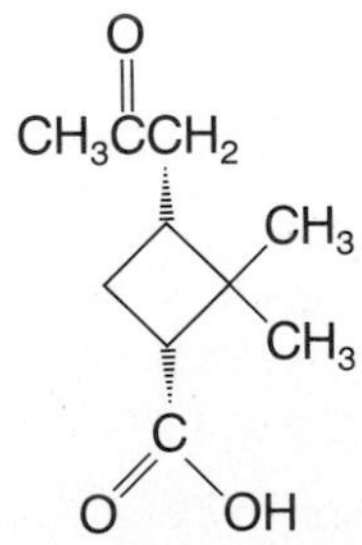   Compound O

7.34 The sex attractant by which the female housefly attracts the male has the molecular formula $C_{23}H_{46}$. Catalytic hydrogenation yields an alkane of molecular formula $C_{23}H_{48}$. Ozonation followed by hydrolysis in the presence of zinc yields $CH_3(CH_2)_7\overset{\displaystyle O}{\overset{\displaystyle \|}{C}}H$ and $CH_3(CH_2)_{12}\overset{\displaystyle O}{\overset{\displaystyle \|}{C}}H$. What is the structure of the housefly sex attractant?

7.35 A certain compound of molecular formula $C_{19}H_{38}$ was isolated from fish oil and from plankton. On hydrogenation it gave 2,6,10,14-tetramethylpentadecane. Ozonation followed by hydrolysis in the presence of zinc, gave $(CH_3)_2C{=}O$ and a 16-carbon aldehyde. What is the structure of the natural product? What is the structure of the aldehyde?

7.36 The sex attractant of the female arctiid moth contains, among other components, a compound of molecular formula $C_{21}H_{40}$ that yields $CH_3(CH_2)_{10}\overset{\displaystyle O}{\overset{\displaystyle \|}{C}}H$, $CH_3(CH_2)_4\overset{\displaystyle O}{\overset{\displaystyle \|}{C}}H$, and $H\overset{\displaystyle O}{\overset{\displaystyle \|}{C}}CH_2\overset{\displaystyle O}{\overset{\displaystyle \|}{C}}H$ on ozonolysis. What is the constitution of this material?

7.37 Cedrene is a pleasant-smelling constituent of cedar wood oil and has the molecular formula $C_{15}H_{24}$. Treatment of cedrene with potassium permanganate followed by acidification of the reaction mixture gives compound P. Deduce the structure of cedrene.

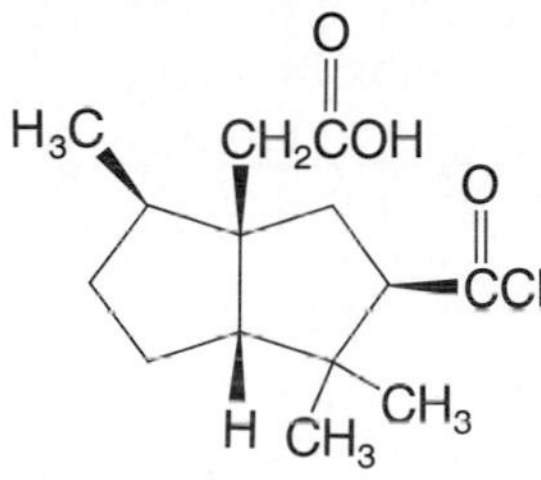

Compound P

STEREOCHEMISTRY

Stereos is a Greek word meaning "solid," and stereochemistry refers to chemistry in three dimensions. The foundations of organic stereochemistry were laid by Jacobus van't Hoff and Charles LeBel in 1874. Independently of each other, van't Hoff and LeBel proposed that the four bonds to carbon were directed toward the corners of a tetrahedron. One consequence of a tetrahedral arrangement of bonds to carbon is that two compounds may be different because the arrangement of their atoms in space is different. Isomers that have the same constitution but differ in the spatial arrangement of their atoms are called *stereoisomers.* We have already had considerable experience with certain types of stereoisomers—those dealing with cis and trans substitution patterns in alkenes and in cycloalkanes.

Our major objectives in this chapter are to develop a feeling for molecules as three-dimensional objects and to become familiar with stereochemical principles, terms, and notation. A full understanding of organic and biological chemistry requires an awareness of the spatial requirements for interactions between molecules; this chapter provides the basis for that understanding.

8.1 MOLECULAR CHIRALITY. ENANTIOMERS

Symmetry of form abounds in classical solid geometry. A sphere, a cube, a cone, and a tetrahedron are all identical to, and can be superposed point for point on, their mirror images. Mirror image superposability also exists in many objects used every day. Cups and saucers, forks and spoons, chairs and beds, are all identical to their mirror images. Many other objects, however, lack symmetry and cannot be superposed on their mirror images. Your left hand and your right hand, for example, are mirror images of each other but cannot be made to coincide point for point in three dimensions. Objects that are nonsuperposable on their mirror images are said to be *chiral.* Chiral is derived from the Greek word *cheiros,* meaning "hand." Conversely, an object that *is* superposable on its mirror image is not chiral; it is *achiral.*

Van't Hoff pointed out that an individual molecule is *asymmetric* (without symmetry) when four different groups are arranged in a tetrahedral fashion around one of its carbon atoms. An example is 2-butanol; the secondary carbon in 2-butanol (C-2) bears a hydrogen atom and hydroxyl, methyl, and ethyl groups as substituents. The two mirror image forms of 2-butanol are A and B.

$$CH_3CH_2 \overset{CH_3}{\underset{H}{\diagdown}} C—OH$$

A

$$H_3C \overset{CH_2CH_3}{\underset{H}{\diagup}} HO—C$$

B

Let us reorient B so as to see more clearly if it can be superposed on A. We do this by mentally picking up B, turning it 180°, and replacing it on the page.

$$H_3C \overset{CH_2CH_3}{\underset{H}{\diagdown}} HO—C$$

B

turn 180°

$$CH_3 CH_2 \overset{CH_3}{\underset{H}{\diagdown}} C—OH$$

B

Now compare A and B.

$$CH_3 CH_2 \overset{CH_3}{\underset{H}{\diagdown}} C—OH$$

B

$$CH_3CH_2 \overset{CH_3}{\underset{H}{\diagup}} C—OH$$

A

The two do not match. The central carbon, the hydrogen, and the hydroxyl group can be superposed, but the spatial orientations of the methyl and ethyl groups are different in A and B. Structures A and B are nonsuperposable mirror images of 2-butanol. *Since mirror image representations of 2-butanol are nonsuperposable, 2-butanol is a chiral molecule.*

The two mirror image representations of 2-butanol have the same constitution, that is, the atoms are connected to each other in the same order, but they differ in the arrangement of their atoms in space; they are *stereoisomers.* Stereoisomers that are related as an object and its nonsuperposable mirror image are classified as *enantiomers.*

The word enantiomer describes a particular relationship between two objects. One cannot look at a single molecule in isolation and ask: Is this molecule an enantiomer? any more than one can look at an individual and ask: Is that person a cousin? Furthermore, just as an object has one, and only one, mirror image, a chemical structure can have one, and only one, mirror image. When the structure is superposable on its mirror image, the two are identical. When the structure and its mirror image are not superposable, they are *enantiomeric.*

Consider next a molecule such as 2-propanol, particularly with respect to C-2 and its substituents. Two of the four groups at C-2 are the same; C-2 bears a hydrogen, a hydroxyl, and two methyl groups. A three-dimensional representation of 2-propanol is shown as C, and D is its mirror image. Are C and D superposable? Test for superposability by reorienting D so that it is arranged in the same fashion as C. This is done by mentally picking up D and turning it 180°.

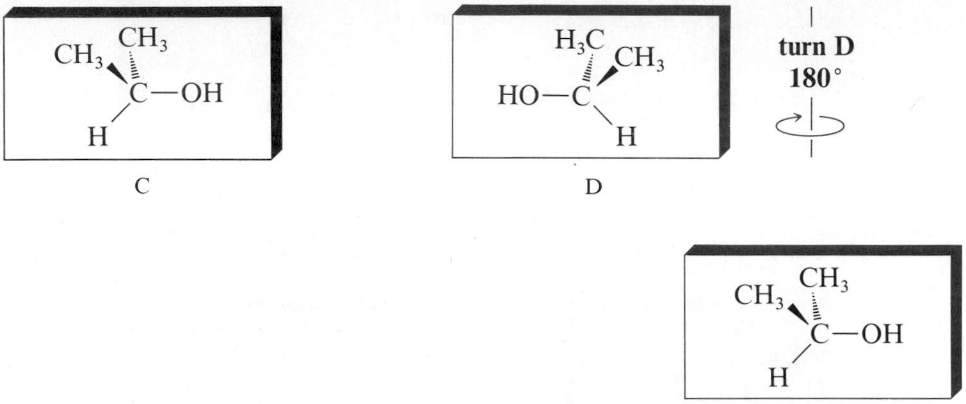

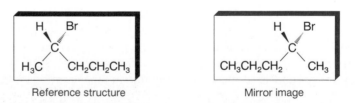

C   D   turn D 180°

                                                 Reoriented version
                                                 of D

When this is done, it is clear that C and D are superposable on each other. Structures C and D are not enantiomers: they are identical. *Since mirror image representations of 2-propanol are superposable on each other, 2-propanol is an achiral molecule.*

PROBLEM 8.1   Which of the following are chiral? Which are achiral?

(a)   2-Bromopentane
(b)   3-Bromopentane
(c)   1-Bromo-2-methylbutane
(d)   2-Bromo-2-methylbutane

SAMPLE SOLUTION   (a)   Since a chiral molecule is one which is not superposable on its mirror image, we draw the two mirror image representations of 2-bromopentane showing the position of the atoms in space as clearly as possible.

Reference structure   Mirror image

To test for superposability, mentally lift the mirror image from the page and replace it so that it is oriented in the same fashion as the reference structure.

Mirror image   turn 180°   Reoriented mirror image

Comparing the reference structure and its mirror image, we see that they cannot be superposed. The two structures are enantiomeric and 2-bromopentane is thus a chiral molecule.

The surest test for chirality is a careful examination of mirror image forms for superposability. Working with models provides the best practice in dealing with molecules as three-dimensional objects and is strongly recommended.

8.2 SYMMETRY IN ACHIRAL STRUCTURES

Certain structural features related to molecular symmetry can be applied to stereochemical analysis and often enable one to determine if a molecule is chiral or achiral by inspection.

For example, a molecule that has a *plane of symmetry* or a *center of symmetry* is superposable on its mirror image and is achiral.

A plane of symmetry bisects a molecule in such a way that one half is the mirror image of the other half. Planes of symmetry are shown in Figure 8.1 for 2-propanol and for dichloromethane. There is one plane of symmetry in 2-propanol, while dichloromethane has two.

A point in a molecule is a center of symmetry if any line drawn from that point to some element of the structure, when extended an equal distance in the opposite direction, encounters an identical element. The cyclobutane derivative in Figure 8.2 lacks a plane of symmetry, yet is superposable on its mirror image. The center of the four-membered ring is a center of symmetry. The molecule is achiral.

PROBLEM 8.2 Locate any planes of symmetry or centers of symmetry in each of the following compounds. Which of the compounds are chiral? Which are achiral?

(a) (E)-1,2-Dichloroethene
(b) (Z)-1,2-Dichloroethene
(c) cis-1,2-Dichlorocyclopropane
(d) trans-1,2-Dichlorocyclopropane

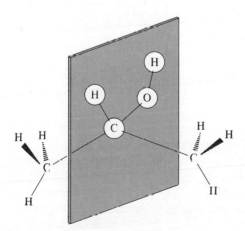

The plane of symmetry in 2-propanol is defined by the three atoms
H—(C—2)—O

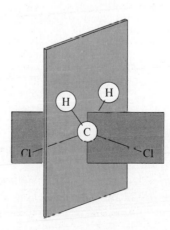

Planes of symmetry in dichloromethane: one plane is defined by the three atoms H—C—H, the other by Cl—C—Cl

FIGURE 8.1 Planes of symmetry in the achiral molecules 2-propanol and dichloromethane.

FIGURE 8.2 Center of symmetry in *trans*-1,3-dibromo-*trans*-2,4-dichlorocyclobutane. (*a*) Mirror image forms of *trans*-1,3-dibromo-*trans*-2,4-dichlorocyclobutane. (*b*) Rotate form F 180° about axis passing through center of symmetry.

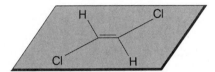

(*a*)   (*b*)

SAMPLE SOLUTION (*a*) (*E*)-1,2-Dichloroethene is a planar molecule. The molecular plane is a plane of symmetry.

Further, (*E*)-1,2-dichloroethene has a center of symmetry located at the midpoint of the carbon-carbon double bond. It is an achiral molecule.

The preceding brief discussion covers most of the situations we shall encounter that involve symmetry and allows us to identify certain, but not all, structural features that lead to achiral molecules. Other kinds of molecules may also be achiral. Thus, while any molecule with a plane of symmetry or a center of symmetry is achiral, the absence of these symmetry elements is not sufficient for a molecule to be chiral. A molecule lacking a center of symmetry or a plane of symmetry is likely to be chiral, but to be certain of this, the superposability test should be applied.

8.3 CHIRAL CENTERS IN CHIRAL MOLECULES

We have already seen in the case of 2-butanol that a molecule of the general type

$$w - \overset{\overset{\displaystyle x}{|}}{\underset{\underset{\displaystyle z}{|}}{C}} - y$$

is chiral when w, x, y, and z are different substituents. They lack all significant symmetry elements and are asymmetric. A tetrahedral carbon atom that bears four different substituents is variously referred to as a *chiral center*, a *chiral carbon atom*, an *asymmetric center*, or an *asymmetric carbon atom*. Since it is the molecule itself that is chiral rather than one of its atoms, it has been suggested that it is more strictly correct to call carbon atoms of this type *stereocenters*. While this term may eventually be adopted, the IUPAC rules for stereochemical notation use the term *chiral center*, and so will we.

Noting the presence of a chiral center in a given molecule is a simple, rapid way to determine that the molecule is chiral. For example, C-2 is a chiral center in 2-butanol; it bears a hydrogen atom and methyl, ethyl, and hydroxyl groups as its four different substituents. By way of contrast, 2-propanol is achiral; none of its carbons bear four different substituents.

$$CH_3 - \overset{\overset{\displaystyle H}{|}}{\underset{\underset{\displaystyle OH}{|}}{C}} - CH_2CH_3 \qquad\qquad CH_3 - \overset{\overset{\displaystyle H}{|}}{\underset{\underset{\displaystyle OH}{|}}{C}} - CH_3$$

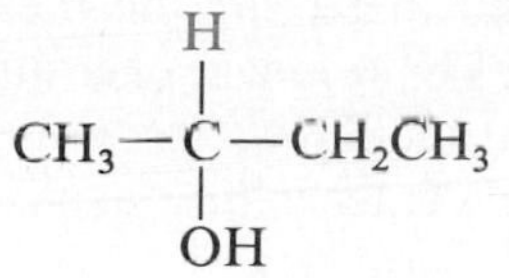

2-Butanol: chiral;
four different
substituents at C-2

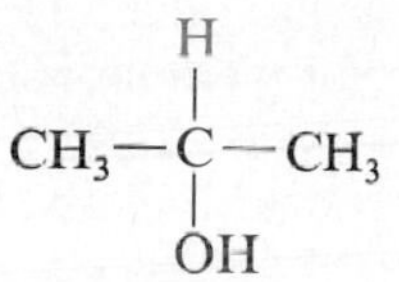

2-Propanol: achiral;
two of the substituents
at C-2 are the same

PROBLEM 8.3   Examine the following for chiral centers:

(a)   2-Bromopentane
(b)   3-Bromopentane
(c)   1-Bromo-2-methylbutane
(d)   2-Bromo-2-methylbutane

SAMPLE SOLUTION   A chiral carbon has four different substituents. (a) In 2-bromopentane, C-2 satisfies this requirement. (b) None of the carbons in 3-bromopentane have four different substituents, so none of its atoms are chiral centers.

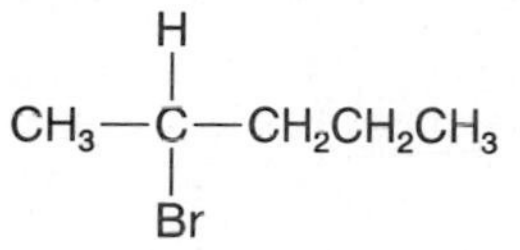

$$CH_3 - \overset{\overset{\displaystyle H}{|}}{\underset{\underset{\displaystyle Br}{|}}{C}} - CH_2CH_2CH_3 \qquad\qquad CH_3CH_2 - \overset{\overset{\displaystyle H}{|}}{\underset{\underset{\displaystyle Br}{|}}{C}} - CH_2CH_3$$

2-Bromopentane 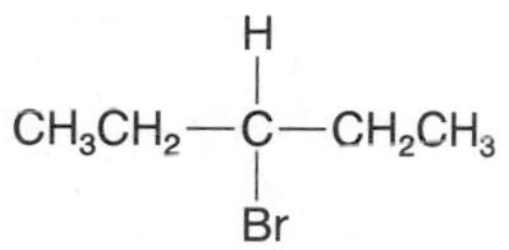   3-Bromopentane

Since chiral carbons must have four different groups attached, a carbon with a double or triple bond to another atom cannot be a chiral center. Molecules with chiral centers are very common, both as naturally occurring substances and as the products of chemical synthesis. In the examples shown below the chiral carbon is indicated by an asterisk.

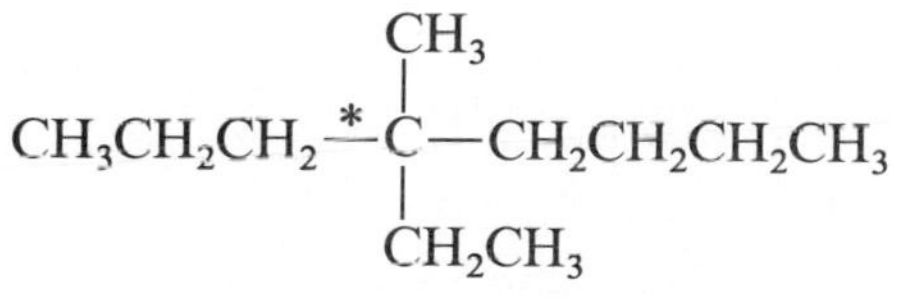

$$CH_3CH_2CH_2 \overset{*}{-} \overset{\overset{\displaystyle CH_3}{|}}{\underset{\underset{\displaystyle CH_2CH_3}{|}}{C}} - CH_2CH_2CH_2CH_3 \qquad\qquad CH_3CH_2 \overset{*}{-} \overset{\overset{\displaystyle H}{|}}{\underset{\underset{\displaystyle Br}{|}}{C}} - CH_2Br$$

4-Ethyl-4-methyloctane   1,2-Dibromobutane
(a chiral alkane)   (formed by addition
                                    of bromine to
                                    1-butene)

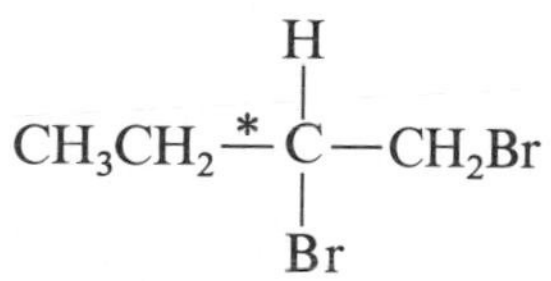

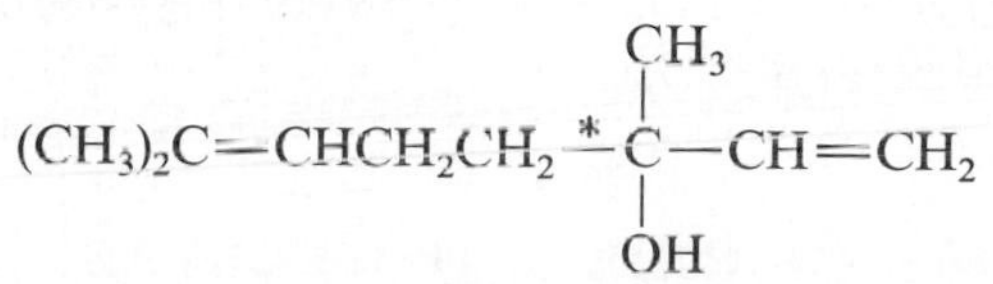

$$(CH_3)_2C = CHCH_2CH_2 \overset{*}{-} \overset{\overset{\displaystyle CH_3}{|}}{\underset{\underset{\displaystyle OH}{|}}{C}} - CH = CH_2$$

Linalool
(a pleasant-smelling oil
obtained from orange flowers)

A carbon atom in a ring can be a chiral center if it bears two different substituents and the path traced around the ring from that carbon in one direction is different from that traced in the other. The carbon atom that bears the methyl group in 1,2-epoxypropane, for example, is a chiral center. The sequence of groups is CH_2—O as one proceeds clockwise around the ring from that atom but is O—CH_2 in the anticlockwise direction. Similarly, C-4 is a chiral center in limonene.

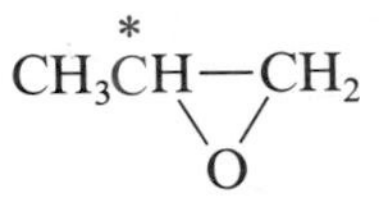

1-2-Epoxypropane (product
of epoxidation of propene)

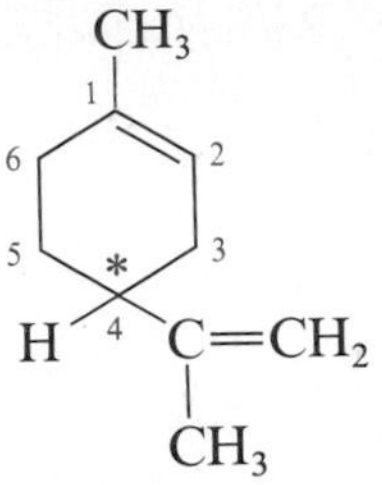

Limonene
(a constituent of lemon oil)

PROBLEM 8.4   Identify the chiral centers, if any, in

(a)   2-Cyclopenten-1-ol

(b)   3-Cyclopenten-1-ol

(c)   1,1,2-Trimethylcyclobutane

(d)   1,1,3-Trimethylcyclobutane

SAMPLE SOLUTION   (a)   The hydroxyl-bearing carbon in 2-cyclopenten-1-ol is a chiral center.

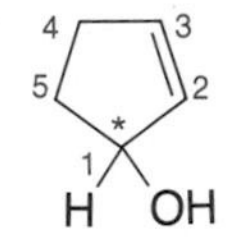

2-Cyclopenten-1-ol

(b)   There is no chiral center in 3-cyclopenten-1-ol since the sequence of atoms $1 \rightarrow 2 \rightarrow 3 \rightarrow 4 \rightarrow 5$ is equivalent regardless of whether one proceeds clockwise or anti-clockwise.

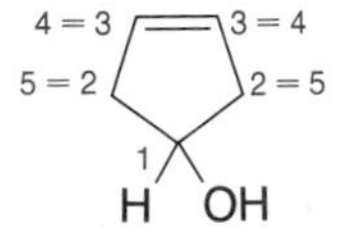

3-Cyclopenten-1-ol (does
not have a chiral carbon)

Molecules with more than one chiral center may or may not be chiral; these will be discussed in Sections 8.9 and 8.10.

8.4 PROPERTIES OF CHIRAL MOLECULES. OPTICAL ACTIVITY

The experimental facts that led van't Hoff and LeBel to propose that molecules having the same constitution could differ in the arrangement of their atoms in space concerned a property called *optical activity*. In order to understand optical activity, we first need to examine some characteristics of light.

Electromagnetic radiation, including visible light, has a wave nature and is composed of an electric field and a magnetic field. When we speak of light as a vibration, we mean that its electric and magnetic fields undergo a regular variation in their amplitudes. If we consider only the electric field, we can represent the wave property of light as shown below.

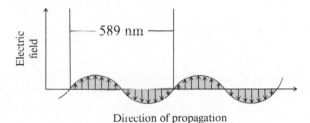

The wavelength is the distance for one complete cycle; it corresponds to 589 nm (589×10^{-9} m) for the yellow light of sodium lamps. This is called the D line of sodium and is the light used most often in measurements of optical activity.

By standing on the line of propagation and facing the light source, we would see the amplitude of the electric field vector of one wave gradually increase and then decrease with time.

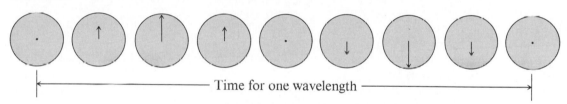

We might represent the wave simply as the aggregate of all its electric field vectors.

One wave vibrates in a single plane, but the light beam comprises many waves perpendicular to the direction of propagation.

Optical activity is measured by using an instrument called a *polarimeter*. A polarimeter contains a device called a *Nicol prism*, which transmits only those light waves having their electric field components in the same plane. This is *plane-polarized light*.

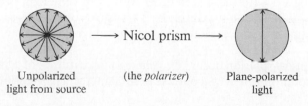

Nicol prisms are fabricated in a special way from a crystal of a transparent form of calcite (Iceland spar). They can be thought of as providing a grid through which a light beam can pass only when it is properly aligned.

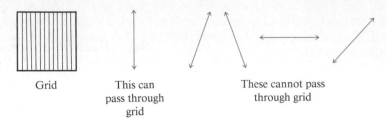

| Grid | This can pass through grid | These cannot pass through grid |

A solution containing the substance being examined is then placed in the path of the beam of plane-polarized light. When the substance is achiral or contains equal amounts of enantiomers, the plane of polarization of the emergent beam is the same as that of the incident beam.

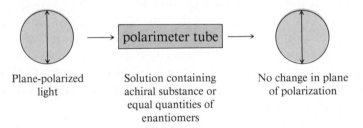

Plane-polarized light   Solution containing achiral substance or equal quantities of enantiomers   No change in plane of polarization

A substance which does not cause rotation of the plane of polarized light is said to be *optically inactive.*

When the substance is chiral and one enantiomer is present in excess of the other, then the plane of polarization is rotated through some angle α.

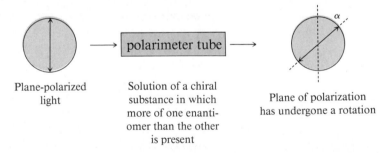

Plane-polarized light   Solution of a chiral substance in which more of one enanti-omer than the other is present   Plane of polarization has undergone a rotation

A substance which causes the plane of polarized light to undergo a rotation is said to be *optically active.* The angle of rotation α is simply referred to as the *observed rotation.*

The observed rotation α is measured by using another Nicol prism as an analyzer. Since the plane of polarization has changed, this second Nicol prism must be oriented at that same angle α with the first in order to transmit the light beam. The viewer adjusts the analyzer so as to maximize light transmission and measures the angle it makes with the polarizer. Figure 8.3 summarizes the measurement of optical rotation.

Why does optical rotation occur? The plane of polarization of a light wave undergoes a minute rotation when it encounters a chiral molecule. Enantiomeric forms of a

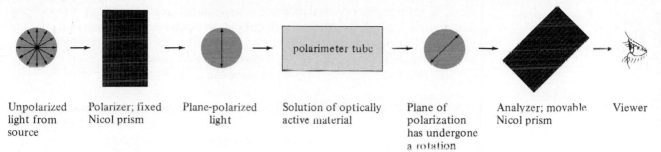

Unpolarized | Polarizer; fixed | Plane-polarized | Solution of optically | Plane of | Analyzer; movable | Viewer
light from | Nicol prism | light | active material | polarization | Nicol prism |
source | | | | has undergone | |
| | | | a rotation | |

FIGURE 8.3 Measurement of optical activity using a polarimeter.

chiral molecule cause a rotation of the plane of polarization in exactly equal amounts but in opposite directions. Therefore a solution containing equal quantities of enantiomers exhibits no net rotation because all the tiny increments of clockwise rotation produced by molecules of one "handedness" are cancelled by an equal number of increments of anticlockwise rotation produced by molecules of the opposite handedness.

Mixtures containing equal quantities of enantiomers are called *racemic mixtures.* Racemic mixtures are optically inactive.

Conversely, when one enantiomer is present in excess, a net rotation of the plane of polarization is observed. At the limit, where all the molecules are of the same handedness, we say the substance is *optically pure.* Optical purity, or *percent enantiomeric excess,* is defined as

Optical purity = percent enantiomeric excess =
      percent of one enantiomer − percent of other enantiomer

Thus, a material that is 50 percent optically pure contains 75 percent of one enantiomer and 25 percent of the other.

Rotation of the plane of polarized light in the clockwise sense is taken as positive (+), while rotation in the anticlockwise sense is taken as a negative (−) rotation. The classical terms for positive and negative rotations are *dextrorotatory* and *levorotatory,* respectively. *Dextro-* and *levo-* are Latin prefixes, meaning "to the right" and "to the left," respectively. Formerly, the symbols *d* and *l* were used to distinguish between enantiomeric forms of a substance. Thus the dextrorotatory enantiomer of 2-butanol was called *d*-2-butanol, the levorotatory form *l*-2-butanol, and a racemic mixture of the two was referred to as *dl*-2-butanol. Current custom favors using algebraic signs instead, as in (+)-2-butanol, (−)-2-butanol, and (±)-2-butanol, respectively.

The observed rotation α of an optically pure substance depends on how many molecules the light beam encounters. A filled polarimeter tube twice the length of another produces twice the observed rotation. In the same way, a solution twice as concentrated produces twice the observed rotation. In order to account for the effects of path length and concentration, chemists have defined the term *specific rotation,* given the symbol $[\alpha]$. Specific rotation is calculated from the observed rotation according to the expression

$$[\alpha] = \frac{100\alpha}{cl}$$

where c is the concentration of the sample in grams per 100 mL of solution, and l is the length of the polarimeter tube in decimeters. (One decimeter is 10 cm.)

Specific rotation is a physical property of a substance, just as melting point, boiling point, density, and solubility are. For example, the lactic acid obtained from milk is exclusively a single enantiomer. We cite its specific rotation in the form $[\alpha]_D^{25} + 3.8°$. The temperature in degrees Celsius and the wavelength of light at which the measurement was made are indicated as superscripts and subscripts, respectively.

PROBLEM 8.5 Cholesterol, when isolated from natural sources, is obtained as a single enantiomer. The observed rotation α of a 0.3-g sample of cholesterol in 15 mL of chloroform solution contained in a 10-cm polarimeter tube is $-0.78°$. Calculate the specific rotation of cholesterol.

PROBLEM 8.6 A sample of synthetic cholesterol was prepared consisting entirely of the enantiomer of natural cholesterol. A mixture of natural and synthetic cholesterol has a specific rotation $[\alpha]_D^{20}$ of $-13°$. What fraction of the mixture is natural cholesterol?

It is convenient to distinguish between enantiomers by prefixing the sign of rotation to the name of the substance. For example, we refer to one of the enantiomers of 2-butanol as (+)-2-butanol, and the other as (−)-2-butanol. Optically pure (+)-2-butanol has a specific rotation $[\alpha]_D^{27}$ of $+13.5°$; optically pure (−)-2-butanol has an exactly opposite specific rotation $[\alpha]_D^{27}$ of $-13.5°$.

8.5 ABSOLUTE AND RELATIVE CONFIGURATION

The precise arrangement of substituents at a chiral center is its *absolute configuration.* Neither the sign nor the magnitude of rotation by itself provides any information concerning the absolute configuration of a substance. Thus, one of the structures shown below is (+)-2-butanol and the other is (−)-2-butanol, but in the absence of additional information we cannot tell which is which.

While no absolute configuration was known for any substance prior to 1951, organic chemists had established the *relative configurations* of thousands of compounds on the basis of an elaborate system of chemical interconversions. To illustrate, consider (+)-3-buten-2-ol. Hydrogenation of this compound yields (+)-2-butanol.

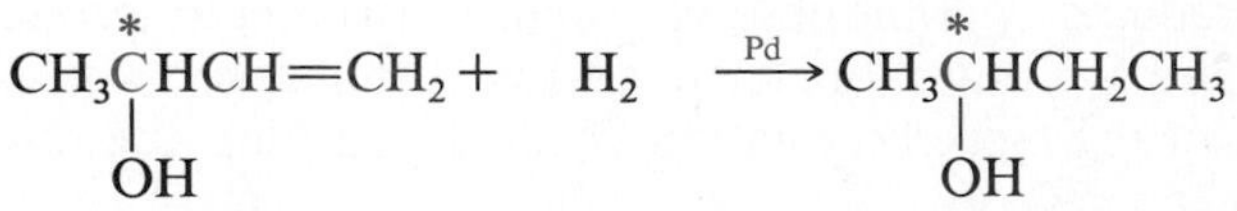

$$\overset{*}{CH_3}CHCH=CH_2 + \quad H_2 \quad \xrightarrow{\text{Pd}} \quad \overset{*}{CH_3}CHCH_2CH_3$$

$$\underset{OH}{|} \qquad\qquad\qquad\qquad\qquad \underset{OH}{|}$$

3-Buten-2-ol	Hydrogen	2-Butanol
$[\alpha]_D^{27} + 33.2°$		$[\alpha]_D^{27} + 13.5°$

Since hydrogenation of the double bond does not involve any of the bonds to the chiral center, the spatial arrangement of substituents at the chiral center in (+)-3-buten-2-ol must be the same as that of the substituents in (+)-2-butanol. The fact that these two compounds have the same sign of rotation when they have the same configuration is shown by the hydrogenation experiment; it could not be predicted in advance of the experiment.

Sometimes compounds that have the same configuration at their chiral center have optical rotations of opposite sign. For example, (−)-2-methyl-1-butanol is converted to (+)-1-bromo-2-methylbutane on treatment with hydrogen bromide.

$$\overset{*}{CH_3CH_2CHCH_2OH} + HBr \longrightarrow \overset{*}{CH_3CH_2CHCH_2Br} + H_2O$$

2-Methyl-1-butanol	Hydrogen	1-Bromo-2-methylbutane	Water
$[\alpha]_D^{25} -5.8°$	bromide	$[\alpha]_D^{25} +4.0°$	

This reaction does not involve any of the bonds to the chiral center, so both the starting alcohol and product bromide must have the same spatial arrangement of substituent groups. The two compounds have the same configuration at their chiral centers even though their signs of rotation are opposite to each other.

When, as in the above examples, we know how the configurations and signs of rotation of two compounds are related to each other, we say we know their *relative configurations*. We now know their absolute configurations as well, because in 1951 the true three-dimensional structure of one compound was determined. This compound was a salt of (+)-tartaric acid. Once its absolute configuration was determined, the absolute configurations of all other compounds whose configurations had been related to (+)-tartaric acid were also revealed.

We are now in a position not only to specify the sign of rotation of a particular enantiomer of a chiral molecule but also to describe unambiguously the spatial arrangement of substituents at the chiral center. Returning to the pair of enantiomers that introduced this section, their absolute configurations are firmly established as shown.

(+)-2-Butanol	(−)-2-Butanol

8.6 THE CAHN-INGOLD-PRELOG *R-S* NOTATIONAL SYSTEM

Just as it makes sense to have a system of nomenclature that permits structural information to be communicated without the necessity of writing a constitutional formula for each compound, so also is it reasonable to have a notational system that simply and unambiguously describes the absolute configuration of a substance. We have already had experience with this principle when we distinguished between stereoisomers in cyclic compounds and in alkenes.

cis-4-Methylcyclohexanol (*E*)-3-Isopropyl-2-hexene

In the case of alkenes, the system using the descriptors E and Z proved to be more versatile than the earlier cis and trans notation.

In the E-Z system, substituents are ranked according to sequence rule precedence. Precedence is based on an atomic number criterion, as described in Section 5.6 and summarized in Table 5.3. In fact, substituent priority based on the sequence rule was first developed in order to deal with the problem of specifying absolute configuration at chiral centers. The unambiguous designation of absolute configuration is the major application of the Cahn-Ingold-Prelog notational system. Let us see how this system deals with (+)-2-butanol as a representative example.

The absolute configuration of (+)-2-butanol is as shown:

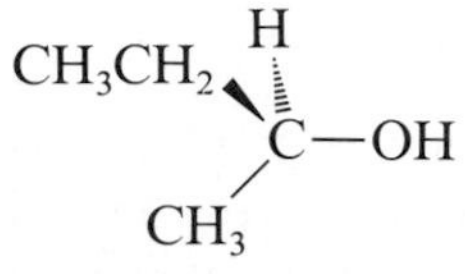

$$CH_3CH_2 \overset{\textstyle H}{\underset{\textstyle CH_3}{\diagdown\!\!\!\diagup C}}\!-\!OH$$

(+)-2-Butanol

We apply the Cahn-Ingold-Prelog system using the following rules:

1. Identify the substituents at the chiral center and rank them in order of decreasing sequence rule precedence. Sequence rule precedence is determined by atomic number, working outward from the point of attachment of the substituent to the chiral center.

   Thus, the substituents attached to the chiral center of 2-butanol are H, HO, CH_3, and CH_3CH_2. In order of decreasing sequence rule precedence (Table 5.3, Section 5.6), they are

$$HO \ > CH_3CH_2 > CH_3 > \ H$$

   Highest   Lowest

2. Orient the molecule so that the lowest-ranking substituent is pointing away from you.

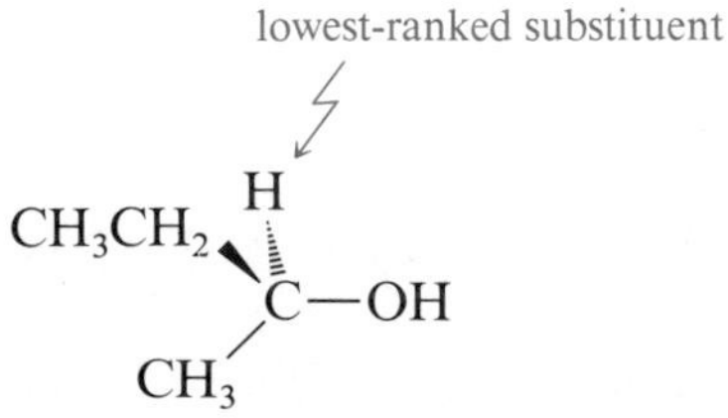

lowest-ranked substituent

$$CH_3CH_2 \overset{\textstyle H}{\underset{\textstyle CH_3}{\diagdown\!\!\!\diagup C}}\!-\!OH$$

(+)-2-Butanol

   As drawn in the perspective shown above, the molecule is already appropriately oriented. Hydrogen is the lowest-ranked substituent on the chiral center, and the molecule is oriented so that it is the substituent that is farthest away from us.

3. Depict the three highest-ranking substituents as they appear to you when the molecule is viewed from this perspective.

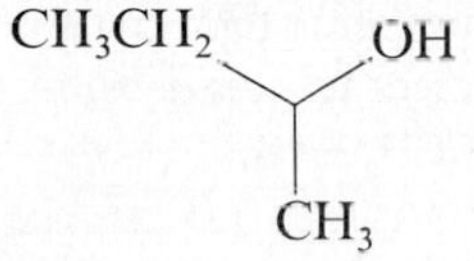

$$CH_3CH_2 \quad OH$$
$$|$$
$$CH_3$$

4. If the order of decreasing precedence of these three substituents, $HO >$ $CH_3CH_2 > CH_3$, appears in a clockwise sense, the absolute configuration is R (Latin, *rectus*, "right," "correct"). If the order of decreasing precedence of these three substituents appears in an anticlockwise sense, the absolute configuration is S (Latin, *sinister*, "left").

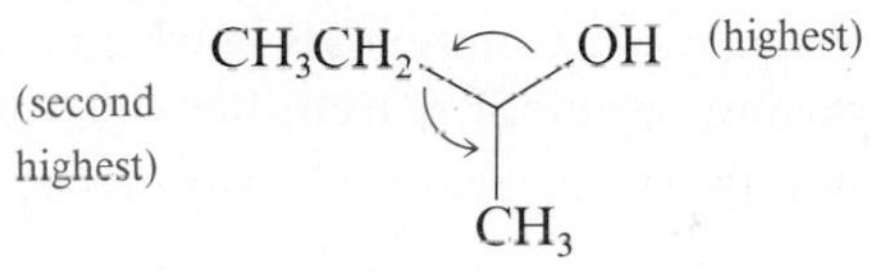

$$CH_3CH_2 \quad OH \quad \text{(highest)}$$
(second highest)
$$CH_3$$
(third highest)

Here the order of decreasing sequence rule priority is *anticlockwise*. This is the S enantiomer of 2-butanol. Its mirror image is (R)-2-butanol.

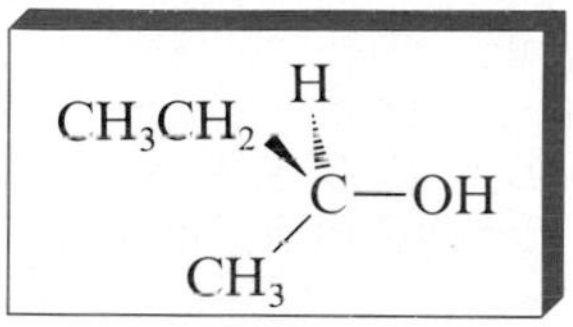

$$CH_3CH_2 \qquad H$$
$$C{-}OH$$
$$CH_3$$

and

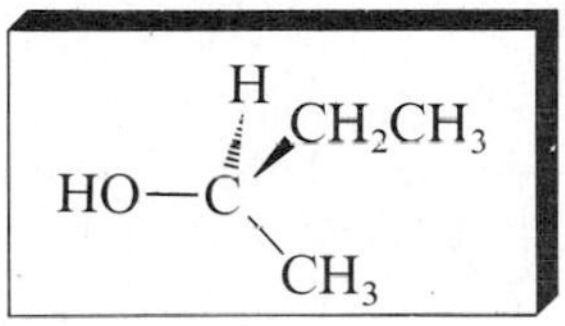

$$H \quad CH_2CH_3$$
$$HO{-}C$$
$$CH_3$$

(*S*)-2-Butanol   (*R*)-2-Butanol

Often, both the R or S descriptor of absolute configuration and the sign of rotation are incorporated into the name of the compound, as in (R)-$(-)$-2-butanol and (S)-$(+)$-2-butanol. A racemic mixture can be denoted by combining R and S, as in $(R)(S)$-2-butanol.

PROBLEM 8.7   Assign absolute configurations as *R* or *S* to each of the following compounds:

(a) 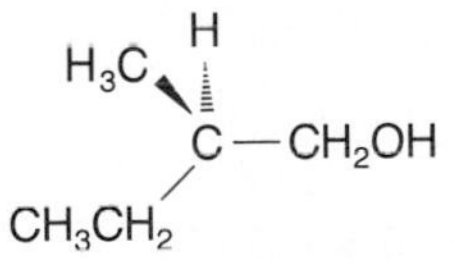
$$H_3C \quad H$$
$$C{-}CH_2OH$$
$$CH_3CH_2$$
(+)-2-Methyl-1-butanol

(c) 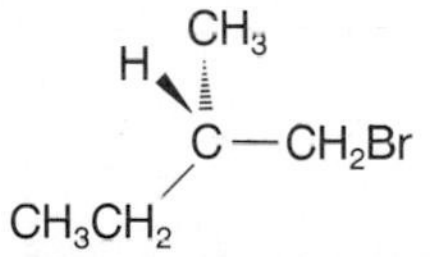
$$H \quad CH_3$$
$$C{-}CH_2Br$$
$$CH_3CH_2$$
(+)-1-Bromo-2-methylbutane

(b) 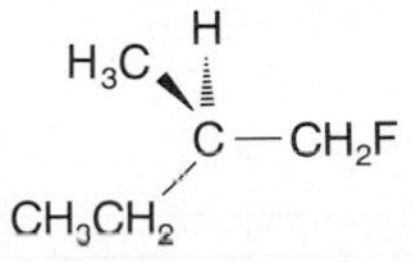
$$H_3C \quad H$$
$$C{-}CH_2F$$
$$CH_3CH_2$$
(+)-1-Fluoro-2-methylbutane

(d) 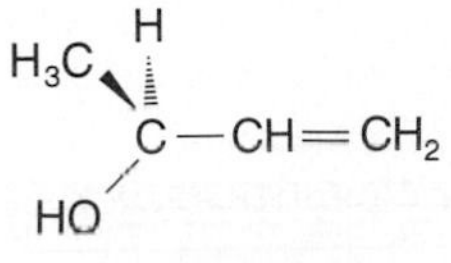
$$H_3C \quad H$$
$$C{-}CH{=}CH_2$$
$$HO$$
(+)-3-Buten-2-ol

SAMPLE SOLUTION   (a) The highest-ranking substituent at the chiral center of 2-methyl-1-butanol is CH_2OH; the lowest is H. Of the remaining two, ethyl outranks methyl.

Order of precedence: $CH_2OH > CH_3CH_2 > CH_3 > H$

The lowest-ranking substituent (hydrogen) at the chiral center points away from us, so the molecule is oriented properly as drawn. The three highest-ranking substituents trace a clockwise path from $CH_2OH \rightarrow CH_3CH_2 \rightarrow CH_3$.

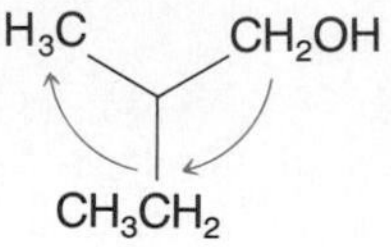

$$H_3C \qquad CH_2OH$$
$$CH_3CH_2$$

Therefore, this compound has the *R* configuration. It is (*R*)-(+)-2-methyl-1-butanol.

The system is broadly applicable. Compounds in which a chiral center is part of a ring are handled in analogous fashion. For example, (+)-4-methylcyclohexene has the absolute configuration shown:

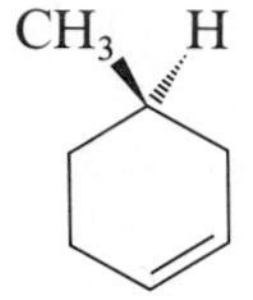

$$CH_3 \quad H$$

(+)-4-Methylcyclohexene

To determine whether this absolute configuration is *R* or *S*, treat the right- and left-hand paths around the ring as if they were independent substituents.

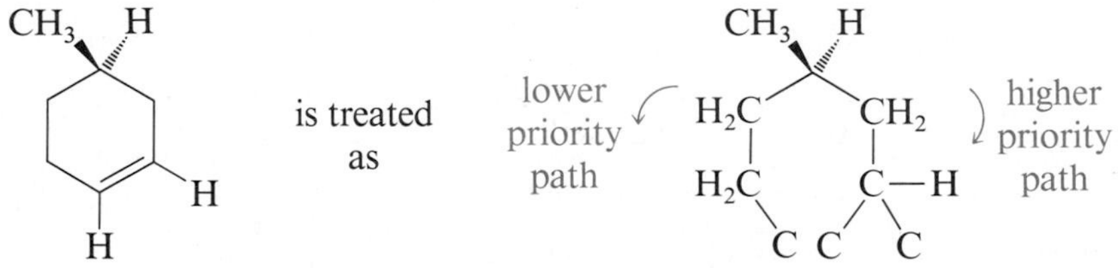

$$CH_3 \quad H \qquad \text{is treated as} \qquad \text{lower priority path} \qquad CH_3 \quad H \qquad \text{higher priority path}$$

Orienting the molecule with the hydrogen directed away from us, we see that the order of decreasing sequence rule priority is *clockwise*. The absolute configuration is *R*.

$$CH_3$$

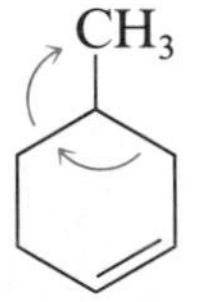

PROBLEM 8.8 Draw three-dimensional representations of

(a) the *R* enantiomer of

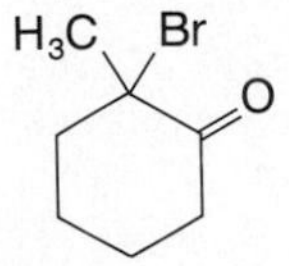

$$H_3C \quad Br$$
$$O$$

(b) the *S* enantiomer of

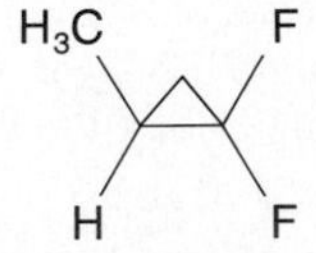

$$H_3C \qquad F$$
$$H \qquad F$$

SAMPLE SOLUTION   (a)   The chiral center is the one which bears the bromine. In order of decreasing precedence, the substituents attached to the chiral center are

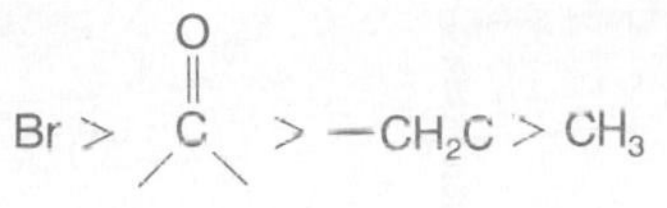

$$Br > \overset{\displaystyle O}{\underset{\displaystyle}{\overset{\|}{C}}} > -CH_2C > CH_3$$

When the lowest-priority substituent (the methyl group) is away from us, the order of decreasing sequence rule precedence of the remaining groups must appear in a clockwise sense in the *R* enantiomer.

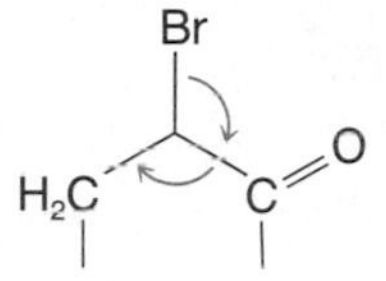

Therefore, we can represent the *R* enantiomer as

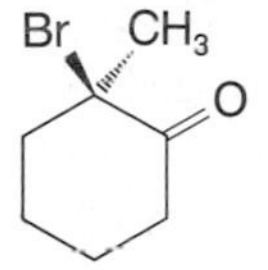

(*R*)-2-Bromo-2-methylcyclohexanone

Since its inception in 1956, the Cahn-Ingold-Prelog system has received almost universal acceptance as the preferred method of stereochemical notation.

8.7 PHYSICAL PROPERTIES OF ENANTIOMERS

The usual physical properties of density, melting point, and boiling point are identical within experimental error for both enantiomers of a chiral compound. Table 8.1 lists some selected physical properties for (*R*)- and (*S*)-2-octanol.

Enantiomers can have striking differences, however, in properties that depend on the arrangement of atoms in space. Take, for example, the enantiomeric forms of carvone. (*R*)-(−)-Carvone is the principal component of spearmint oil. Its enantiomer, (*S*)-(+)-carvone, is the principal component of caraway seed oil. The two

TABLE 8.1

Physical Properties of Enantiomeric 2-Octanols

Property	(*R*)-(−)-2-Octanol	(*S*)-(+)-2-Octanol
Specific rotation, $[\alpha]_D^{17}$	−9.9°	+9.9°
Boiling point, °C	175	175
Refractive index, n_D^{25}	1.4254	1.4258
Specific gravity, d_4^{20}	0.838	0.822

enantiomeric forms of carvone do not smell the same; each has its own characteristic odor.

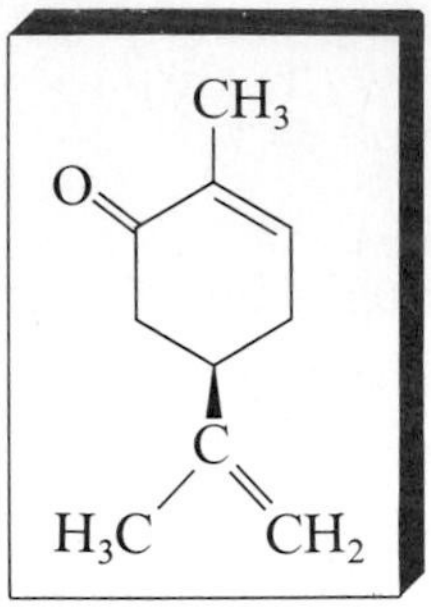

(R)-(−)-Carvone
(from spearmint oil)

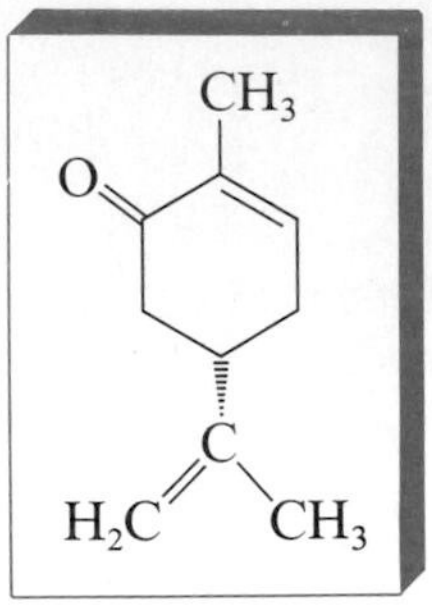

(S)-(+)-Carvone
(from caraway seed oil)

The reason for the difference in odor between (R)- and (S)-carvone results from their different behavior toward receptor sites in the nose. It is believed that volatile molecules occupy only those receptor sites that have the proper shape to accommodate them. These receptor sites are themselves chiral, so one enantiomer may fit one kind of receptor site while the other enantiomer fits a different kind of receptor. One analogy that can be drawn is to hands and gloves. Your left hand and your right hand are enantiomers. You can place your left hand into a left glove but not into a right one. The receptor site (the glove) can accommodate one enantiomer of a chiral object (your hand) but not the other.

The term *chiral recognition* has been coined to refer to the process whereby some chiral receptor or reagent interacts selectively with one of the enantiomers of a chiral molecule. Very high levels of chiral recognition are common in biological processes. (−)-Nicotine, for example, is much more toxic than (+)-nicotine, and (+)-adrenaline is more active in the constriction of blood vessels than (−)-adrenaline. (−)-Thyroxine is an amino acid of the thyroid gland, which speeds up metabolism and causes nervousness and loss of weight. Its enantiomer, (+)-thyroxine, exhibits none of these effects but is sometimes given to heart patients to lower their cholesterol levels.

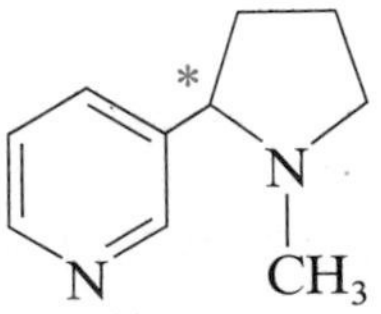

Nicotine

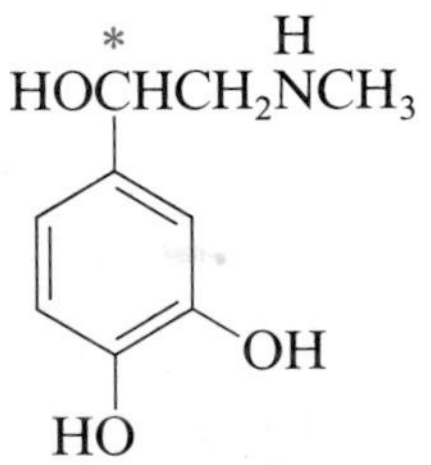

Adrenaline

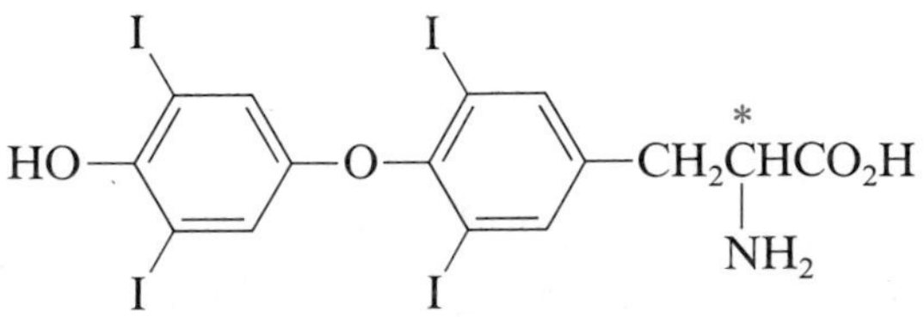

Thyroxine

(Chiral centers identified by asterisks)

8.8 STEREOCHEMISTRY OF CHEMICAL REACTIONS THAT PRODUCE CHIRAL CENTERS

Many of the reactions we have encountered in earlier chapters can produce a chiral product from an achiral starting material. A large number of the reactions of alkenes, for example, fall into this category. In the examples shown below, addition to their carbon-carbon double bonds converts alkenes to products that contain a chiral center.

$$CH_3CH=CH_2 \xrightarrow{CH_3CO_2OH} CH_3CH-CH_2$$
$$\underset{O}{\diagdown\diagup}$$

Propene   1,2-Epoxypropane
(achiral)   (chiral)

$$CH_3CH_2CH=CH_2 \xrightarrow{Br_2, H_2O} CH_3CH_2CHCH_2Br$$
$$\underset{OH}{|}$$

1-Butene   1-Bromo-2-butanol
(achiral)   (chiral)

$$CH_3CH=CHCH_3 \xrightarrow{HBr} CH_3CHCH_2CH_3$$
$$\underset{Br}{|}$$

(E)- or (Z)-2-butene   2-Bromobutane
(achiral)   (chiral)

In these and related reactions, the chiral product is formed as a *racemic mixture* and is *optically inactive.* Remember, in order for a substance to be optically active not only must it be chiral but one enantiomer must be present in excess of the other.

To understand the reason why a mixture containing equal amounts of enantiomers is produced, consider a specific case, the reaction of an achiral reagent such as peroxyacetic acid with propene to form 1,2-epoxypropane, as illustrated in Figure 8.4. The peroxy acid, as shown in the figure, is just as likely to transfer oxygen to one face of the double bond as to the other. Enantiomeric transition states have the same energy. The rates of formation of the R and S enantiomers are the same, and a racemic mixture of the two is formed.

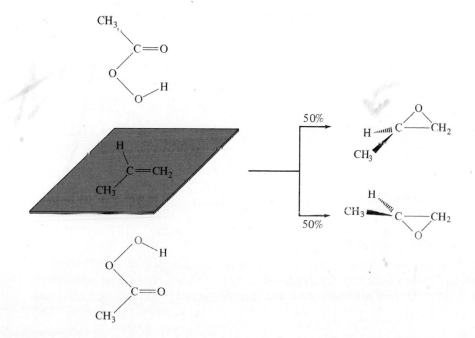

FIGURE 8.4 Epoxidation of propene produces a racemic mixture of (R)- and (S)-1,2-epoxypropane. These two epoxides are nonsuperposable mirror images, and are formed in equal amounts.

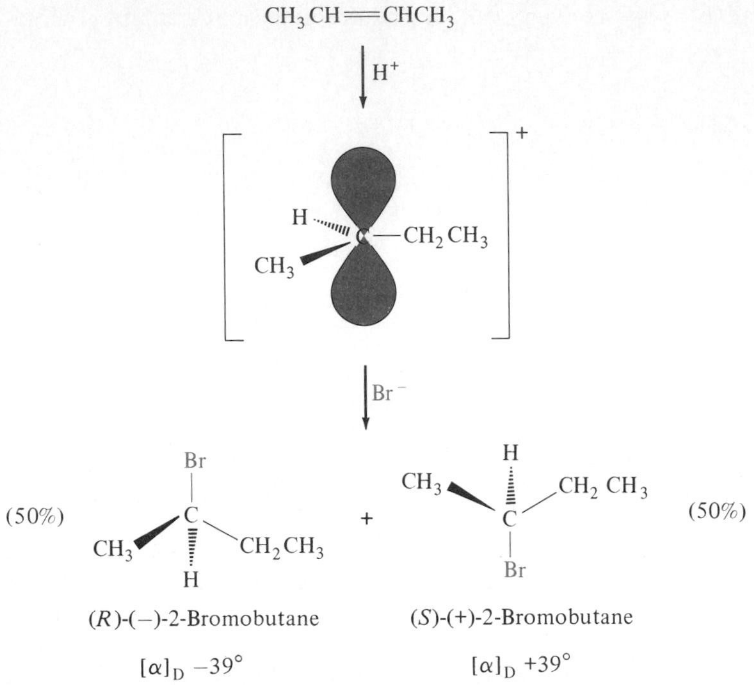

$$CH_3CH = CHCH_3$$

FIGURE 8.5 Ionic addition of hydrogen bromide to (*E*)- or (*Z*)-2-butene proceeds by way of an achiral carbocation, which leads to equal quantities of (*R*)- and (*S*)-2-bromobutane.

(50%)

(*R*)-(−)-2-Bromobutane

$[\alpha]_D$ −39°

+

(50%)

(*S*)-(+)-2-Bromobutane

$[\alpha]_D$ +39°

It is a general principle that optically active products cannot be formed when optically inactive substrates react with optically inactive reagents. This principle holds irrespective of whether the addition is syn or anti, concerted or stepwise. No matter how many steps are involved in a reaction, if the reactants are achiral, formation of one enantiomeric form is just as likely as the other and a racemic mixture results. In the stepwise ionic addition of hydrogen bromide to 2-butene, a carbocation is an intermediate. The bonds to the positively charged carbon are coplanar and define a plane of symmetry. The carbocation is achiral and may be captured by bromide ion with equal facility from the top face or the bottom face, as depicted in Figure 8.5. The chiral product, 2-bromobutane, is formed as a racemic mixture.

Many reactions convert a chiral substrate to a chiral product. If the chiral substrate is optically active, the product may or may not be optically active, depending on the kind of reaction that takes place. When, for example as described in Section 8.5, the reaction does not involve any of the bonds to the chiral center, the product must have the same optical purity as the starting material and the same arrangement of groups around the chiral center.

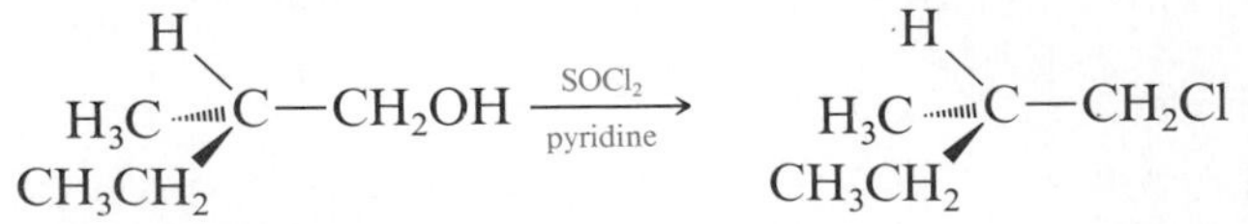

(*S*)-(−)-2-Methyl-1-butanol
$[\alpha]_D^{25}$ −5.8°

(*S*)-(+)-1-Chloro-2-methylbutane
$[\alpha]_D^{25}$ +1.7°

Substitution reactions involving replacement of one group at the chiral center by another can lead to a product that has the same configuration as the starting material, the enantiomeric configuration, or a mixture of the two. Stereochemical studies

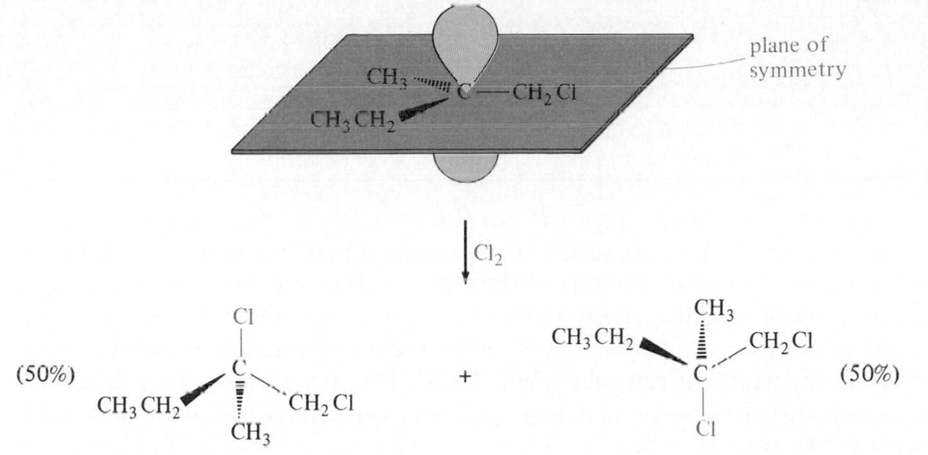

(50%)

(R)-1, 2-Dichloro-2-methylbutane

+

(50%)

(S)-1, 2-Dichloro-2-methylbutane

FIGURE 8.6 The free-radical intermediate resulting from hydrogen atom abstraction at C-2 of 1-chloro-2-methylbutane is achiral and leads to equal amounts of (R)- and (S)-1,2-dichloro-2-methylbutane.

frequently are carried out to provide information about reaction mechanisms. In one such study photochemical chlorination of optically active 1-chloro-2-methylbutane yielded, as part of a mixture of chlorination products, 1,2-dichloro-2-methylbutane that was optically inactive.

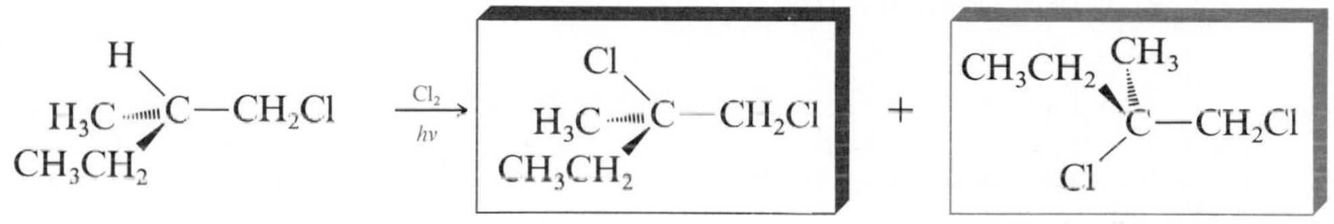

(S)-(+)-1-Chloro-2-methylbutane
$[\alpha]_D^{25} +1.7°$

R   S

Racemic mixture of (R)- and (S)-1,2-dichloro-2-methylbutane; optically inactive

Formation of a racemic mixture of (R)- and (S)-1,2-dichloro-2-methylbutanes is consistent with the existence of an achiral free-radical intermediate, as shown in Figure 8.6.

When an achiral intermediate is formed from an optically active precursor, optically inactive products result.

PROBLEM 8.9   As in alkane halogenations generally, a variety of products were obtained in the photochemical chlorination of (S)-(+)-1-chloro-2-methylbutane. Among them was 1,4-dichloro-2-methylbutane. Do you think this product was optically active? Explain your reasoning.

When a substrate is chiral but optically inactive because it is racemic, any products derived from its reactions with optically inactive reagents will be optically inactive. For example, 2-butanol is chiral and may be converted with hydrogen bromide to 2-bromobutane, which is also chiral. If racemic 2-butanol is used, each enantiomer will react at the same rate with the achiral reagent. Whatever happens to (R)-(−)-2-butanol is mirrored in a corresponding reaction of (S)-(+)-2-butanol, and a racemic, optically inactive product results.

$$(\pm)\text{-}CH_3CHCH_2CH_3 \xrightarrow{\text{HBr}} (\pm)\text{-}CH_3CHCH_2CH_3$$
$$\quad\quad\quad | \quad\quad\quad\quad\quad\quad\quad\quad\quad\quad |$$
$$\quad\quad\quad OH \quad\quad\quad\quad\quad\quad\quad\quad\quad\quad Br$$

2-Butanol   2-Bromobutane
(Chiral but racemic)   (Chiral but racemic)

Optically inactive starting materials can give optically active products if they are treated with an optically active reagent or if the reaction is catalyzed by an optically active substance. The best examples of these phenomena are found in biochemical processes. Most biochemical reactions are catalyzed by enzymes. Enzymes are chiral and enantiomerically homogeneous; they provide an asymmetric environment in which chemical reaction can take place. Ordinarily, enzyme-catalyzed reactions occur with such a high level of stereoselectivity that one enantiomer of a substance is formed exclusively even when the substrate is achiral. The enzyme *fumarase,* for example, catalyzes the hydration of fumaric acid to malic acid in apples and other fruits. Only the *S* enantiomer of malic acid is formed in this reaction.

Fumaric acid (S)-$(-)$-Malic acid

The reaction is a reversible one, and its stereochemical requirements are so pronounced that neither the cis isomer of fumaric acid (maleic acid) nor the *R* enantiomer of malic acid can serve as substrates for the fumarase-catalyzed hydration-dehydration equilibrium.

8.9 MOLECULES WITH TWO CHIRAL CENTERS

When a molecule incorporates two chiral centers, as does 2,3-dihydroxybutanoic acid, how many stereoisomers are possible?

2,3-Dihydroxybutanoic acid

The answer can be found by employing a commonsense approach. The absolute configuration at C-2 may be *R* or *S*. Likewise, C-3 may have either the *R* or the *S* configuration. There are four combinations of these two chiral centers.

(2R,3R)	(stereoisomer I)	(2S,3S)	(stereoisomer II)
(2R,3S)	(stereoisomer III)	(2S,3R)	(stereoisomer IV)

Compounds I through IV have the same constitution but differ in the arrangement of their atoms in space. They are stereoisomers. Figure 8.7 presents structural formulas for these four stereoisomers. Stereoisomers I and II are enantiomers of each other; the

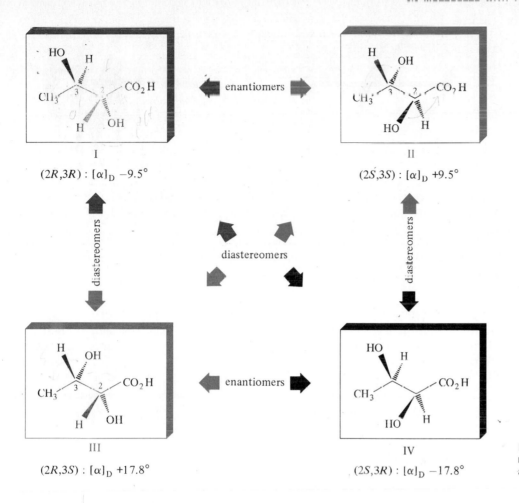

FIGURE 8.7 Stereoisomeric 2,3-dihydroxybutanoic acids.

enantiomer of (R,R) is (S,S). Likewise stereoisomers III and IV are enantiomers of each other, the enantiomer of (R,S) being (S,R).

Stereoisomer I is not a mirror image of III or IV, so it is not an enantiomer of either one. Stereoisomers that are not related as an object and its mirror image are called *diastereomers;* diastereomers are stereoisomers that are not enantiomers. Thus, stereoisomer I is a diastereomer of III and a diastereomer of IV, and similarly, II is a diastereomer of III and IV.

In order to convert a molecule with two chiral centers to its enantiomer, the configuration at both centers must be changed. Reversing the configuration at only one chiral center converts it to a diastereomeric structure.

Enantiomers must have equal and opposite specific rotations. Diastereomeric substances can have different rotations, with respect to both sign and magnitude. Thus, as Figure 8.7 shows, the $(2R,3R)$ and $(2S,3S)$ enantiomers (I and II) have specific rotations that are equal in magnitude but opposite in sign. The $(2R,3S)$ and $(2S,3R)$ enantiomers (III and IV) likewise have specific rotations that are equal to each other but opposite in sign. The magnitudes of rotation of I and II are different, however, from those of their diastereomers III and IV.

Because diastereomers are not mirror images of each other, they can have, and often do have, markedly different physical and chemical properties. For example, the

(2R,3R) stereoisomer of 3-amino-2-butanol is a liquid, while the (2R,3S) diastereomer is a crystalline solid.

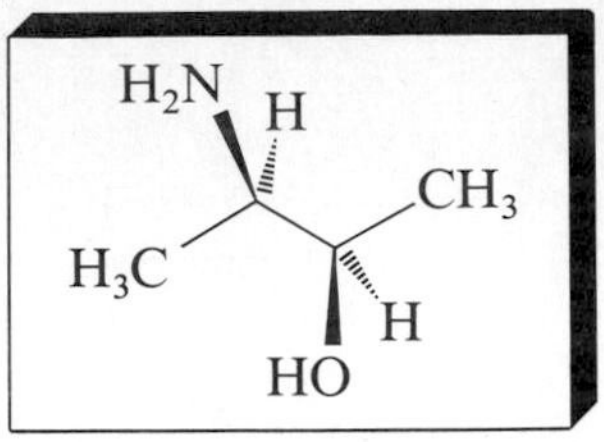

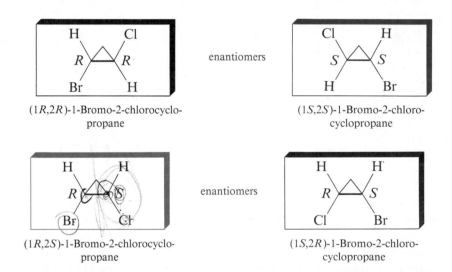

(2R,3R)-3-Amino-2-butanol
(liquid)

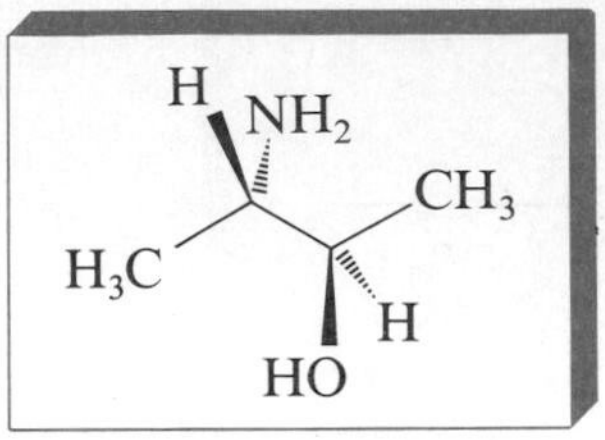

(2R,3S)-3-Amino-2-butanol
(solid, mp 49°C)

PROBLEM 8.10 One other stereoisomer of 3-amino-2-butanol is a crystalline solid. Which one?

The situation is the same when the two chiral centers are present in a ring. There are four stereoisomeric 1-bromo-2-chlorocyclopropanes, a pair of enantiomers in which the halogens are trans and a pair in which they are cis. The cis compounds are diastereomers of the trans.

(1R,2R)-1-Bromo-2-chlorocyclo-
propane

enantiomers

(1S,2S)-1-Bromo-2-chloro-
cyclopropane

(1R,2S)-1-Bromo-2-chlorocyclo-
propane

enantiomers

(1S,2R)-1-Bromo-2-chloro-
cyclopropane

Now consider what happens when a molecule has two chiral centers that are equivalently substituted, as does 2,3-butanediol.

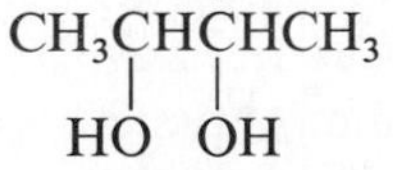

$$\overset{*}{C}H_3\overset{*}{C}H\overset{*}{C}HCH_3$$
$$HOOH$$

2,3-Butanediol

Only *three*, not four, stereoisomeric 2,3-butanediols exist. The (2R,3R) and (2S,3S) forms are enantiomers of each other and are optically active.

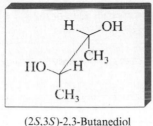

(2R,3R)-2,3-Butanediol
(liquid, $[\alpha]_D^{25} - 13°$)

(2S,3S)-2,3-Butanediol
(liquid; $[\alpha]_D^{25} + 13°$)

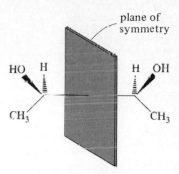

FIGURE 8.8 Plane of symmetry in eclipsed conformation of *meso*-2,3 butanediol.

A third combination of chiral centers (2R,3S) is, however, achiral and is superposable with and identical to its (2S,3R) mirror image. This stereoisomer is optically inactive.

(2R,3S) is equivalent to (2S,3R)
(solid, mp 34°C; optically inactive)

The easiest way to demonstrate that this stereoisomer is achiral is to recognize that it has a plane of symmetry that passes through and is perpendicular to the C-2–C-3 bond, as shown in Figure 8.8.

We call molecules that have chiral centers but which are themselves achiral *meso forms.* Notice that we have used an eclipsed conformation of *meso*-2,3-butanediol to identify a plane of symmetry. Even though this is an unstable conformation, it is an accessible one. If any accessible conformation has a plane of symmetry or a center of symmetry, the molecule will be achiral. In practice, planes of symmetry are easier to see than centers of symmetry and are the symmetry elements we look for first. Certain conformations of a meso compound may be chiral, but a chiral conformation of a meso compound always exists as a one-to-one mixture of enantiomers.

PROBLEM 8.11 Is there an important symmetry element present in the anti conformation of *meso*-2,3-butanediol? Explain.

Turning to cyclic compounds, we see that there are three, not four, stereoisomeric 1,2-dibromocyclopropanes. Of these, two are enantiomeric *trans*-1,2-dibromocyclopropanes. The cis diastereomer is a meso form; it has a plane of symmetry.

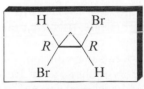

(1R,2R)-1,2-Dibromo-
cyclopropane

(1S,2S)-Dibromocyclo-
propane

meso-1,2-Dibromocyclo-
propane

PROBLEM 8.12   One of the stereoisomers of 1,3-dimethylcyclohexane is a meso form. Which one?

8.10   MOLECULES WITH MULTIPLE CHIRAL CENTERS

Many naturally occurring compounds contain several chiral centers. By an analysis similar to that described for the case of two chiral centers, it can be shown that the maximum number of stereoisomers for a particular constitution is 2^n, where n is equal to the number of chiral centers.

PROBLEM 8.13   Using R and S descriptors, write all the possible combinations for a molecule with three chiral centers.

When two or more of a molecule's chiral centers are equivalently substituted, meso forms are possible, and the number of stereoisomers is then less than 2^n. Thus, 2^n represents the maximum number of stereoisomers for a constitutional formula containing n chiral centers.

The best examples of substances with multiple chiral centers are the carbohydrates. One class of carbohydrates, called aldohexoses (we will explain this terminology in Chapter 27), has the constitution

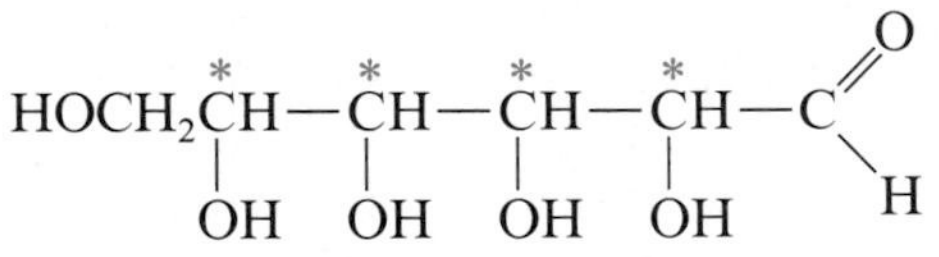

$$\text{HOCH}_2\overset{*}{\text{CH}}-\overset{*}{\text{CH}}-\overset{*}{\text{CH}}-\overset{*}{\text{CH}}-\overset{O}{\underset{H}{\text{C}}}$$
$$\phantom{\text{HOCH}_2}\underset{\text{OH}}{|}\underset{\text{OH}}{|}\underset{\text{OH}}{|}\underset{\text{OH}}{|}$$

An aldohexose

Since there are four chiral centers and no possibility of *meso* forms, there are 2^4, or 16, stereoisomeric aldohexoses. All 16 are known, having been isolated either as natural products or as the products of chemical synthesis.

PROBLEM 8.14   2-Ketohexoses are a second type of carbohydrate; they have the constitution shown below. How many stereoisomeric 2-ketohexoses are possible?

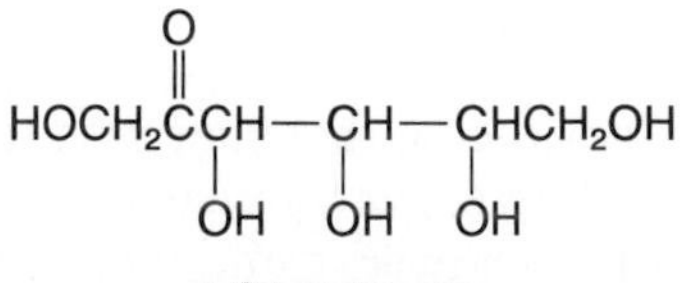

$$\text{HOCH}_2\overset{O}{\overset{||}{\text{C}}}\text{CH}-\text{CH}-\text{CHCH}_2\text{OH}$$
$$\phantom{\text{HOCH}_2\text{CC}}\underset{\text{OH}}{|}\underset{\text{OH}}{|}\underset{\text{OH}}{|}$$

A 2-ketohexose

Steroids represent another class of natural products with multiple chiral centers. One such compound is cholic acid, which can be obtained from bile. Its structural formula is given in Figure 8.9. Cholic acid has 11 chiral centers, so there are a total (including cholic acid) of 2^{11} or 2048 stereoisomers that have this constitution. Of these 2048 stereoisomers, how many of them are diastereomers of cholic acid? Remember! Diastereomers are stereoisomers that are not enantiomers, and any object can have only one mirror image. Therefore, only one of the stereoisomers is an enantiomer of cholic acid while all the rest are diastereomers. Of the 2048 stereo-

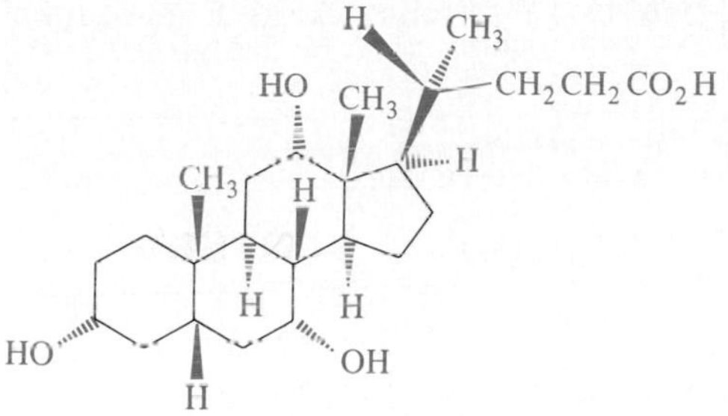

FIGURE 8.9 The structure of cholic acid. Its 11 chiral centers are those carbons at which stereochemistry is indicated in the diagram.

isomers, one is cholic acid, one is its enantiomer, and the other 2046 are diastereomers of cholic acid. Only a small fraction of these compounds are known, and (+)-cholic acid is the only one ever isolated from natural sources.

Eleven chiral centers may seem like a lot but this number is nowhere close to a world record. Palytoxin, a very poisonous polyhydroxylated substance produced by a Tahitian marine organism, has 64 chiral centers. Even this number seems modest when we note that most proteins and nucleic acids have well over 100 chiral centers.

If a molecule contains both chiral centers and double bonds, additional opportunities for stereoisomerism arise. For example, the configuration of the chiral center in 3-penten-2-ol may be either R or S, and that of the double bond may be either E or Z. Therefore, even though 3-penten-2-ol has only one chiral center, there are four stereoisomeric forms.

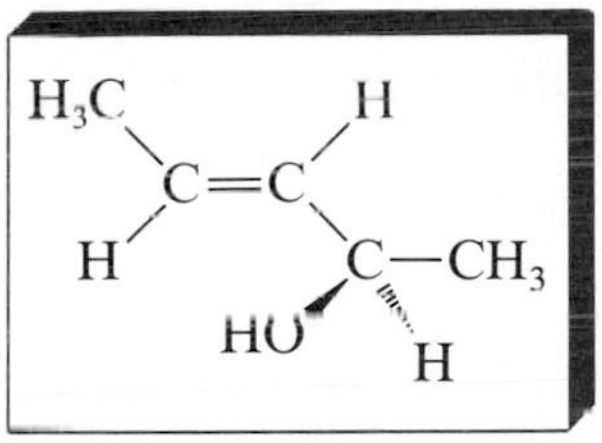

(2R,3E)-3-Penten-2-ol

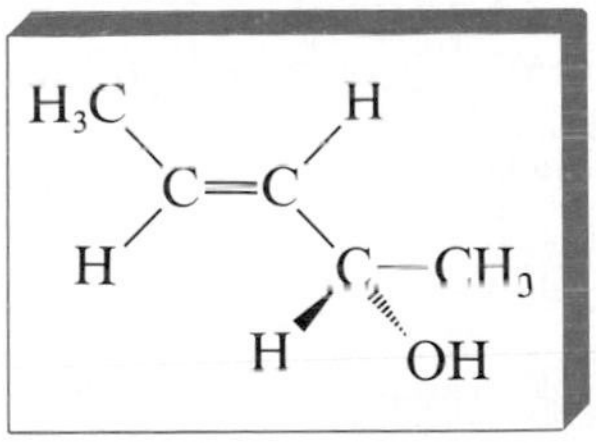

(2S,3E)-3-Penten-2-ol

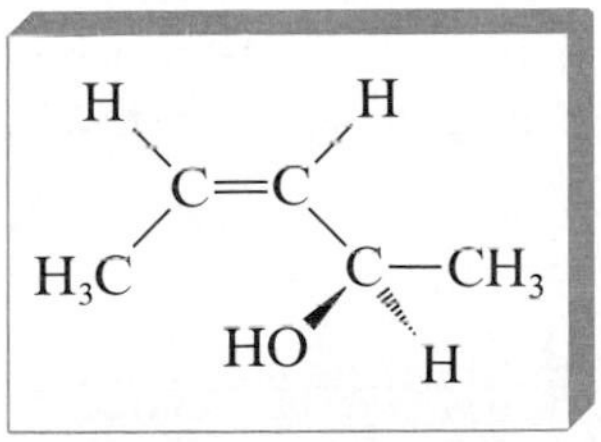

(2R,3Z)-3-Penten-2-ol

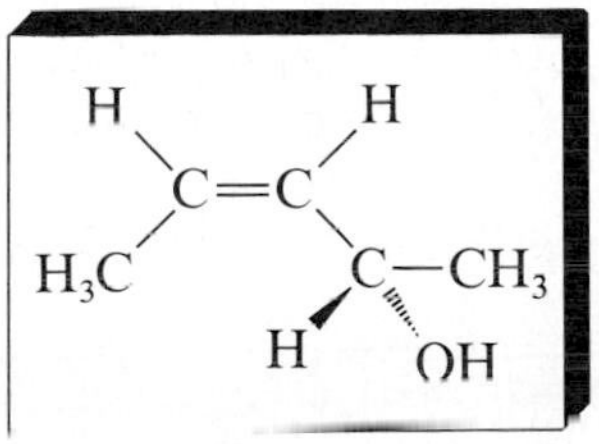

(2S,3Z)-3-Penten-2-ol

The relationship of the (2R,3E) stereoisomer to the others is that it is the enantiomer of (2S,3E)-3-penten-2-ol and is a diastereomer of the (2R,3Z) and (2S,3Z) isomers.

8.11 CHEMICAL REACTIONS THAT PRODUCE DIASTEREOMERS

Once we grasp the idea of stereoisomerism in molecules with two or more chiral centers, further details of addition reactions of alkenes can be explored.

When bromine adds to (Z)- or (E)-2-butene, the product 2,3-dibromobutane contains two chiral centers.

$$CH_3CH = CHCH_2 \xrightarrow{Br_2} CH_3\overset{*}{C}H\overset{*}{C}HCH_3$$
$$\begin{array}{cc} | & | \\ Br & Br \end{array}$$

(Z)- or (E)-2-Butene   2,3-Dibromobutane

Since the two chiral centers are equivalently substituted, three stereoisomers are possible, a pair of enantiomers and a meso form.

Which stereoisomers are actually formed in this reaction? Two factors are decisive.

1. The (Z)- or (E)-configuration of the starting alkene
2. The anti stereochemistry of the addition mechanism

Bromine adds to the cis alkene (Z)-2-butene to give a racemic mixture of (2R,3R)- and (2S,3S)-2,3-dibromobutane.

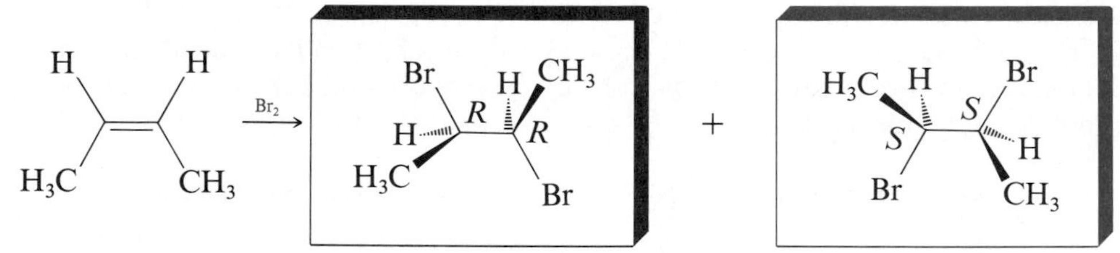

(Z)-2-Butene   (2R,3R)2,3-Dibromo-butane (50%)   (2S,3S)-2,3-Dibromo-butane (50%)

Bromine adds to the trans alkene (E)-2-butene to give only the meso form of 2,3-dibromobutane.

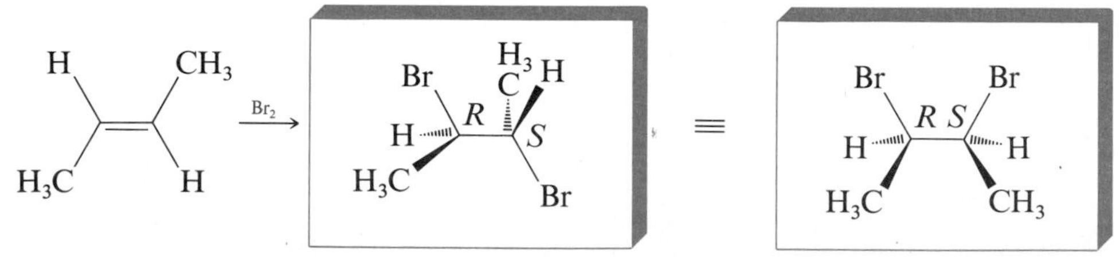

(E)-2-Butene   *meso-*   or   (2R,3S)-2,3-Dibromobutane

As shown in Figure 8.10, both reactions proceed by way of bromonium ion intermediates. In each instance, attack by bromide ion on the cyclic bromonium ion can occur equally well at either carbon. The cyclic bromonium ion from (Z)-2-butene leads to equal amounts of the (2R,3R) and (2S,3S) stereoisomers (Figure 8.10a). The optically inactive meso form results from the cyclic bromonium ion derived from (E)-2-butene (Figure 8.10b). Stereochemical results like these were instrumental in formulating the bromonium ion mechanism for addition of bromine to alkenes.

These reactions are examples of *stereospecific reactions*. A stereospecific reaction is one in which stereoisomeric starting materials yield products that are stereo-

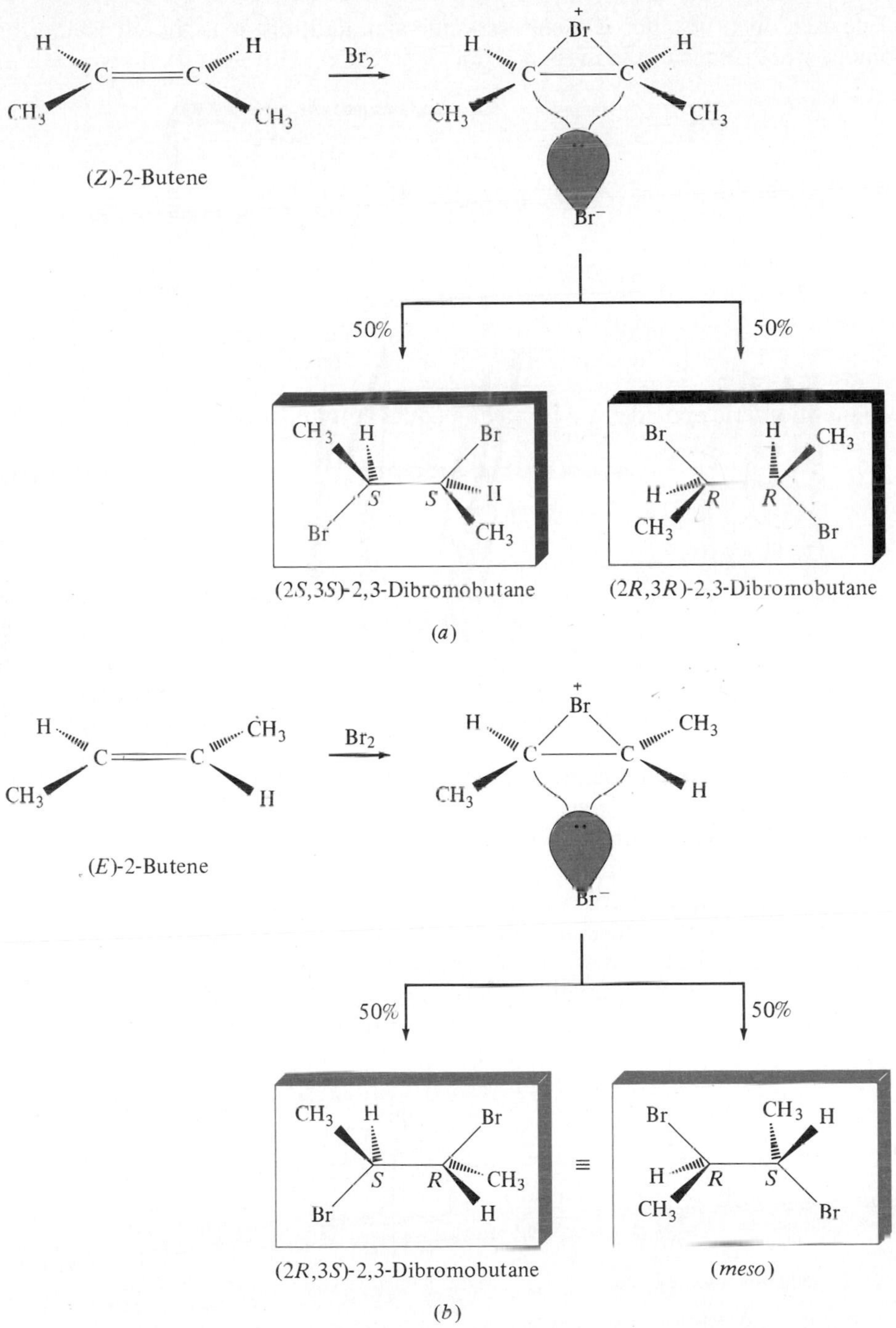

FIGURE 8.10 Stereochemical course of addition of bromine to (a) (Z)-2-butene and (b) (E)-2-butene.

isomers of each other. In this case the starting materials, in separate reactions, are the Z and E isomers of 2-butene. The products of bromination of (Z)-2-butene are stereoisomers of the product of bromination of (E)-2-butene.

PROBLEM 8.15 Suppose the reaction of bromine with alkenes were a syn addition. Which of the two stereoisomeric 2-butenes would yield *meso*-2,3-dibromobutane?

Epoxidation of alkenes is a stereospecific syn addition. It is the cis isomer of 2-butene which leads to the meso epoxide.

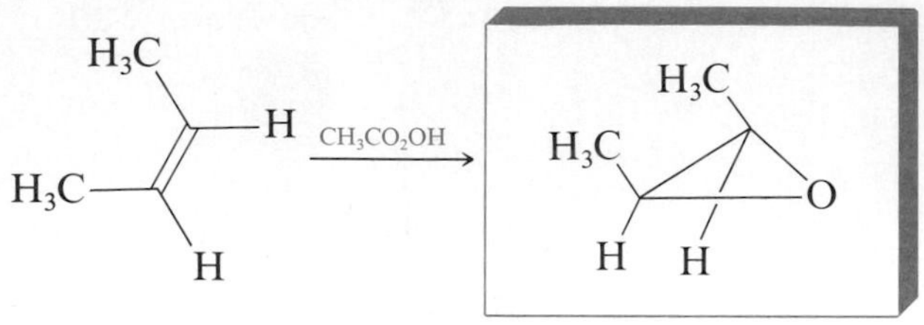

(Z)-2-Butene   cis-2,3-Epoxybutane
                                      (meso)

Two enantiomeric epoxides are formed in equal amounts from (E)-2-butene.

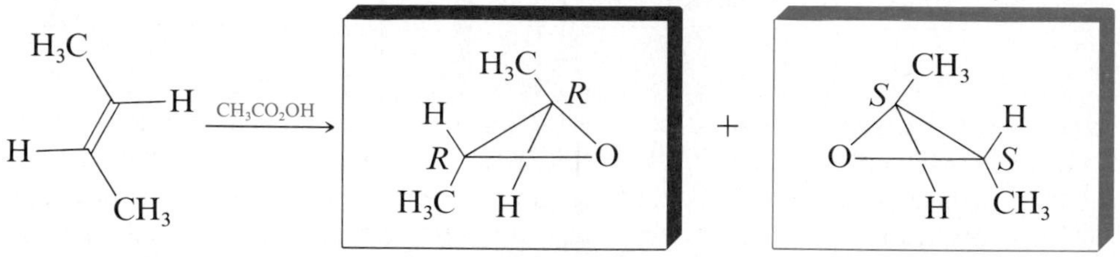

(E)-2-Butene   (2R,3R)-2,3-Dimethyl-   (2S,3S)-2,3-Dimethyl-
                       oxirane (50%)   oxirane (50%)

Alkene epoxidation is thus stereospecific: stereoisomeric products are formed from stereoisomeric starting materials.

You should notice that optically inactive (achiral) starting materials yield optically inactive products (racemic mixtures or meso structures) in these reactions in accordance with the principle described in Section 8.8.

A reaction which introduces a second chiral center into a chiral starting material need not produce diastereomeric forms of the product in equal amounts. Catalytic hydrogenation of 2-methyl(methylene)cyclohexane occurs from the less hindered face of the double bond to give more than twice as much cis- as trans-1,2-dimethylcyclohexane.

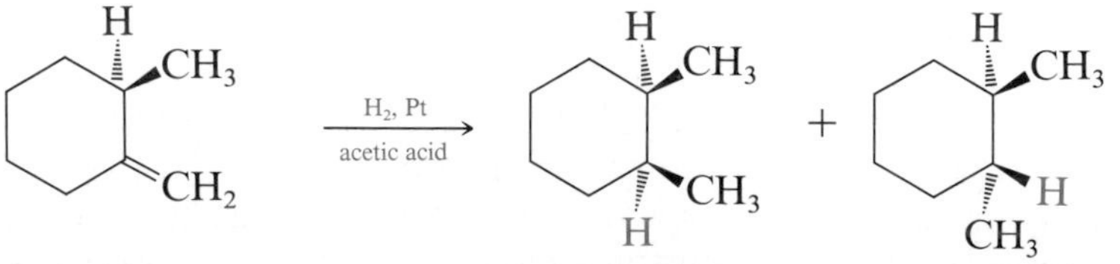

2-Methyl(methylene)cyclo-   cis-1,2-Dimethylcyclo-   trans-1,2-Dimethyl-
hexane   hexane (68%)   cyclohexane (32%)

This reaction is *stereoselective*. Stereoisomeric products are formed in unequal amounts from a single starting material.

8.12 RESOLUTION OF ENANTIOMERS

The separation of a racemic mixture into its enantiomeric components is termed *resolution*. The first resolution, that of tartaric acid, was carried out by Louis Pasteur in 1848. Tartaric acid is a by-product of wine making and almost always is found as its dextrorotatory 2*R*, 3*R* stereoisomer.

(2*R*,3*R*)-Tartaric acid
(mp 170°C,
$[\alpha]_D +12°$)

PROBLEM 8.16   There are two other stereoisomeric tartaric acids. Write their structures and specify the configuration at their chiral centers.

Occasionally, an optically inactive sample of tartaric acid was obtained. Pasteur noticed that the sodium ammonium salt of optically inactive tartaric acid was a mixture of two mirror image crystal forms. With microscope and tweezers, Pasteur carefully separated the two. He found that one kind of crystal (in aqueous solution) was dextrorotatory, while its mirror image rotated the plane of polarized light an equal amount, but was levorotatory.

Although Pasteur was not then able to provide a structural explanation — that had to wait for van't Hoff and LeBel a quarter of a century later — he correctly deduced that the enantiomeric quality of the crystals of the sodium ammonium tartrates must be a consequence of enantiomeric molecules. The rare form of tartaric acid was optically inactive because it contained equal amounts of (+)-tartaric acid and (−)-tartaric acid. It had earlier been called *racemic acid* (from Latin *racemus*, "a bunch of grapes"), a name which subsequently gave rise to our present term for a mixture of enantiomers.

PROBLEM 8.17   Could the unusual, optically inactive form of tartaric acid studied by Pasteur have been *meso*-tartaric acid?

Pasteur's technique of separating enantiomers is not only laborious but requires that the crystal habits of enantiomers be distinguishable. This happens very rarely. Consequently, alternative and more general approaches to optical resolution of enantiomers have been developed. Most of these are based on a strategy of temporarily converting the enantiomers of a racemic mixture to diastereomeric derivatives, separating these diastereomers, and then regenerating the enantiomeric starting materials.

Figure 8.11 illustrates this strategy. Let us say we have a mixture of enantiomers which, for simplicity, we label as C(+) and C(−). Further, let us say that C(+) and C(−) bear some functional group that can combine with a reagent P to yield adducts C(+)—P and C(−)—P. Now, if reagent P is chiral, and if only a single enantiomer of P — say P(+) — is added to a racemic mixture of C(+) and C(−), as shown in step 1 of Figure 8.11, then the products of the reaction are C(+)—P(+) and C(−)—P(+). These products are not mirror images; they are diastereomers. Diastereomers can have different physical properties, and this difference in physical properties can serve as a means of separating them. In step 2, these diastereomers are separated, usually by

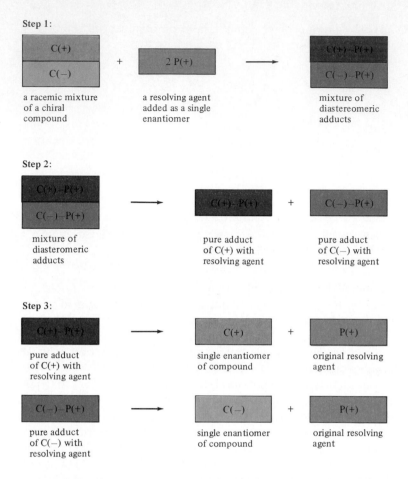

Step 1:

a racemic mixture
of a chiral
compound

a resolving agent
added as a single
enantiomer

mixture of
diastereomeric
adducts

Step 2:

mixture of
diasteromeric
adducts

pure adduct
of C(+) with
resolving agent

pure adduct
of C(−) with
resolving agent

Step 3:

pure adduct
of C(+) with
resolving agent

single enantiomer
of compound

original resolving
agent

pure adduct
of C(−) with
resolving agent

single enantiomer
of compound

original resolving
agent

FIGURE 8.11 Diagram illustrating general procedure followed in resolution of a chiral substance into its individual enantiomeric components.

recrystallization from a suitable solvent. In step 3, an appropriate chemical transformation is employed to remove the resolving agent from the separated adducts, thereby liberating the enantiomers and regenerating the resolving agent.

Whenever possible, the chemical reactions involved in the formation of diastereomers and their conversion to separate enantiomers are simple acid-base reactions. For example, naturally occurring (S)-(−)-malic acid is often used to resolve alkyl-substituted derivatives of ammonia, called *amines*. One such amine that has been resolved in this way is 1-phenylethylamine. Amines are bases, and malic acid is an acid. Proton transfer from (S)-(−)-malic acid to a racemic mixture of (R)- and (S)-1-phenylethylamine gives a mixture of diastereomeric salts.

$$C_6H_5\overset{*}{C}HNH_2 + HO_2CCH_2\overset{*}{C}HCO_2H \longrightarrow C_6H_5\overset{*}{C}H\overset{+}{N}H_3 \quad {}^-O_2CCH_2\overset{*}{C}HCO_2H$$

$$\underset{CH_3}{|} \qquad\qquad \underset{OH}{|} \qquad\qquad\qquad \underset{CH_3}{|} \qquad\qquad \underset{OH}{|}$$

1-Phenylethyl-
amine
(racemic mixture)

(S)-(−)-Malic
acid
(resolving agent)

1-Phenylethylammonium
(S)-malate
(mixture of diastereomeric salts)

The diastereomeric salts are separated and the individual enantiomers of the amine liberated by treatment with a base.

$$C_6H_5\overset{*}{C}H\overset{+}{N}H_3 \quad {}^-O_2CCH_2\overset{*}{C}HCO_2H \;+\; 2OH^- \longrightarrow$$
$$\underset{CH_3}{|} \qquad\qquad \underset{OH}{|}$$

1-Phenylethylammonium
(*S*)-malate
(a single diastereomer)

Hydroxide

$$C_6H_5\overset{*}{C}HNH_2 + {}^-O_2CCH_2\overset{*}{C}HCO_2{}^- + 2H_2O$$
$$\underset{CH_3}{|} \qquad\qquad \underset{OH}{|}$$

1-Phenylethyl-
amine
(a single
enantiomer)

(*S*)-(−)-
Malic acid
(recovered
resolving agent)

Water

PROBLEM 8.18   In the resolution of 1-phenylethylamine using (−)-malic acid, the compound obtained by recrystallization of the mixture of diastereomeric salts is (*R*)-1-phenylethylammonium (*S*)-malate. The other component of the mixture is more soluble and remains in solution in the recrystallization solvent. What is the configuration of the more soluble salt?

This method is widely used for the resolution of chiral amines and carboxylic acids. Analogous methods based on the formation and separation of diastereomers have been developed for other functional groups; the precise approach depends on the kind of chemical reactivity associated with the functional groups present in the molecule.

8.13 STEREOREGULAR POLYMERS

Prior to the development of the Ziegler-Natta catalyst systems (Section 7.18), polymerization of propene was not a reaction of much utility. The reason for this has a stereochemical basis. Consider a section of *polypropylene:*

$$\}-CH_2CHCH_2CHCH_2CHCH_2CHCH_2CHCH_2CH-\{$$
(each CH bearing a CH_3 group)

Representation of the polymer chain in an extended zigzag conformation, as shown in Figure 8.12, reveals several distinct structural possibilities differing with respect to the relative configurations of the carbons that bear the methyl groups.

One possible structure, represented in Figure 8.12*a*, has all the methyl groups oriented in the same direction with respect to the polymer chain. This stereochemical arrangement is said to be *isotactic*. Another form, shown in Figure 8.12*b*, has its methyl groups alternating in their stereochemical orientation along the chain. This arrangement is called *syndiotactic*. Both the isotactic and syndiotactic forms of polypropylene are known as *stereoregular polymers* because each is characterized by a precise stereochemistry at the carbon atom that bears the methyl group. There is a third possibility, shown in Figure 8.12*c*, which is called *atactic*. Atactic polypropylene has a random orientation of its methyl groups; it is not a stereoregular polymer.

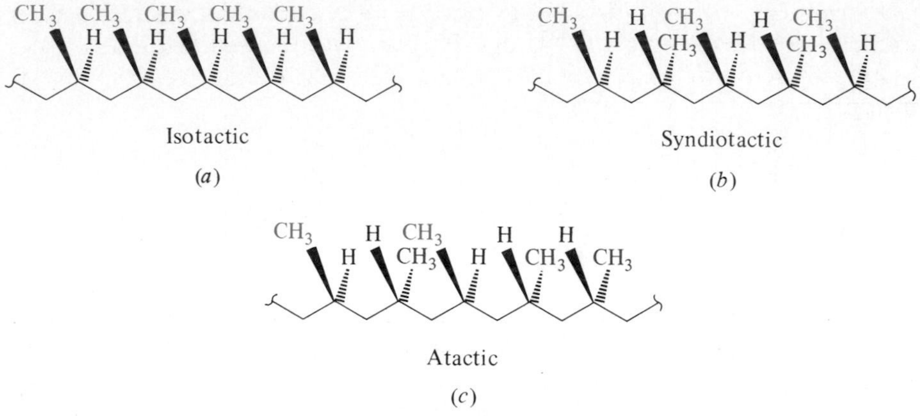

FIGURE 8.12 Polymers of propene. (a) All the methyl groups are on the same side of the carbon chain in isotactic polypropylene. (b) They alternate "front to back" in syndiotactic polypropylene. (c) Their spatial distribution is random in atactic polypropylene.

Polypropylene chains associate with each other because of attractive van der Waals forces. The extent of this association is relatively large for isotactic and syndiotactic polymers because the stereoregularity of the polymer chains permits efficient packing. Atactic polypropylene, on the other hand, does not associate as strongly. It has a lower density and lower melting point than the stereoregular forms. The physical properties of stereoregular polypropylene are more useful for most purposes than those of atactic polypropylene.

When propene is polymerized under free-radical conditions, it is the atactic form of polypropylene that results. Catalysts of the Ziegler-Natta type, however, permit the preparation of either isotactic or syndiotactic polypropylene. We see here an example of how proper choice of experimental conditions can affect the stereochemical course of a chemical reaction to the extent that entirely new materials with unique properties result.

8.14 CHIRAL CENTERS OTHER THAN CARBON

Our discussion to this point has been limited to molecules in which a carbon atom was the chiral center. Atoms other than carbon may also be chiral centers. Silicon, like carbon, is characterized by a tetrahedral arrangement of bonds when it bears four substituents. A large number of organosilicon compounds in which silicon bears four different groups have been resolved into their enantiomers.

Trigonal pyramidal molecules are chiral if the central atom bears three different groups. If one is to resolve substances of this type, however, the pyramidal inversion that interconverts enantiomers must be slow at room temperature. Such is not the case for tricoordinate nitrogen in amines. Pyramidal inversion in amines is very rapid (E_{act} = 6 to 10 kcal/mol), and attempts at resolving amines that have no chiral centers other than nitrogen have been thwarted by their immediate racemization.

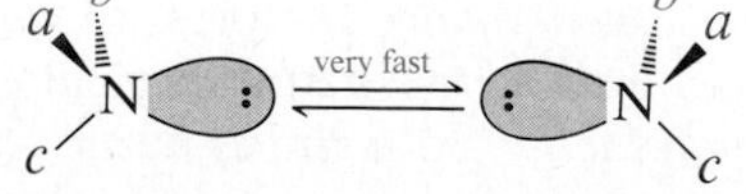

Phosphorus is in the same group of the periodic table as nitrogen, and tricoordinate phosphorus compounds (phosphines), like amines, are trigonal pyramidal. Phosphines, however, undergo pyramidal inversion much more slowly than amines (E_{act} = 30 to 35 kcal/mol), and a number of optically active phosphines have been prepared. Similarly, tricoordinate sulfur compounds are chiral when sulfur bears three different substituents because the rate of pyramidal inversion at sulfur is rather slow. Optically active sulfoxides, such as the one shown, are examples of compounds of this type.

$$CH_3CH_2CH_2CH_2 \overset{\overset{\displaystyle CH_3}{\diagdown}}{\underset{\overset{\displaystyle |}{-O}}{S^+} \hspace{-1em}\text{:}} \qquad (S)\text{-}(+)\text{-Butyl methyl sulfoxide}$$

The absolute configuration at sulfur is specified by the Cahn-Ingold-Prelog method with the provision that the unshared electron pair is considered to be the lowest-ranking substituent.

8.15 SUMMARY

Our concern in this chapter has been with the spatial arrangement of atoms and groups in molecules. Chemistry in three dimensions is known as *stereochemistry*. At its most fundamental level, stereochemistry deals with molecular structure; at another level, it is concerned with chemical reactivity. Table 8.2 summarizes some basic

TABLE 8.2
Isomers are different compounds that have the same molecular formula. Isomers may be either constitutional isomers or stereoisomers.)

Definition	Example
A. Constitutional isomers are isomers that differ in the order in which their atoms are connected.	There are three constitutionally isomeric compounds of molecular formula C_3H_8O: $\quad CH_3CH_2CH_2OH \quad CH_3\underset{\overset{\displaystyle |}{OH}}{C}HCH_3 \quad CH_3CH_2OCH_3$ 1-Propanol $\qquad$ 2-Propanol $\qquad$ Ethyl methyl ether
B. Stereoisomers are isomers that have the same constitution but differ in the arrangement of their atoms in space.	
B-1. *Enantiomers* are stereoisomers that are related as an object and its nonsuperposable mirror image.	The two enantiomeric forms of 2-chlorobutane are $(R)\text{-}(-)\text{-2-Chlorobutane}$ $\qquad$ and $\qquad$ $(S)\text{-}(+)\text{-2-Chlorobutane}$
B-2. *Diastereomers* are stereoisomers that are not enantiomers.	The cis and trans isomers of 4-methylcyclohexanol are stereoisomers, but they are not related as an object and its mirror image; they are diastereomers. *cis*-4-Methylcyclohexanol $\qquad$ *trans*-4-Methylcyclohexanol

definitions relating to molecular structure and shows where considerations of stereo-chemistry become important in classifying isomeric substances.

A molecule is *chiral* if it cannot be superposed on its mirror image. The most common kind of chiral molecule contains a carbon atom that bears four different substituents. Table 8.2 shows the nonsuperposable mirror images of 2-chlorobutane; 2-chlorobutane is a chiral molecule. The C-2 atom in 2-chlorobutane bears four different substituents; it is a *chiral center.*

Achiral molecules are superposable on their mirror images. Any structure that has a plane of symmetry or a center of symmetry is achiral. Both the cis and trans stereoisomers of 4-methylcyclohexanol shown in Table 8.2 are achiral. Each has a plane of symmetry that bisects the molecule into two mirror image halves.

The *configuration* of a molecule is a precise description of the arrangement of its atoms in space. We specify configuration by the Cahn-Ingold-Prelog notational system using the descriptors *R* and *S*. Table 8.2 identifies the enantiomers of 2-chlorobutane according to this system.

When a structure contains more than one chiral center, the maximum number of stereoisomers is 2^n, where *n* is equal to the number of structural units capable of stereochemical variation — usually this is the number of chiral centers in a molecule, but it can include *E* and *Z* double bonds as well. The number of potential stereo-isomers is reduced to less than 2^n when meso forms are possible. A meso form is an achiral molecule that contains chiral centers. Tartaric acid has two chiral centers, but only three stereoisomers are possible because one of them is a meso form.

(2*R*,3*R*)   (2*S*,3*S*)   *meso*
(achiral)

Optical activity, the capacity to rotate the plane of polarized light, is a physical property of chiral molecules. Enantiomeric forms of the same molecule rotate the plane of polarization an equal amount but in opposite directions. The enantiomer which rotates the plane of polarization in the clockwise (or positive) sense is said to be dextrorotatory; that which rotates plane-polarized light in the anticlockwise (or nega-tive) sense is said to be levorotatory. A mixture which contains equal quantities of enantiomers is optically inactive and is described as racemic. Achiral molecules are optically inactive.

The property of optical activity can serve as a probe of reaction mechanism. Optically active starting materials yield optically inactive products when the reaction proceeds by way of an achiral intermediate. Chiral products are frequently found in chemical reactions but must be formed as a racemic mixture if the reactants are optically inactive.

The use of stereochemistry in the study of reaction mechanisms will play an important role in the following chapter.

PROBLEMS

8.19 In each of the following pairs of compounds one is chiral and the other is achiral. Identify each compound as chiral or achiral as appropriate.

(a) ClCH$_2$CHCH$_2$OH   and   HOCH$_2$CHCH$_2$OH
 | |
 OH   Cl

(b) CH$_3$CH=CHCH$_2$Br   and   CH$_3$CHCH=CH$_2$
 |
 Br

(c)

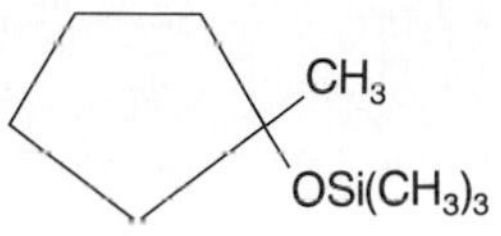

and

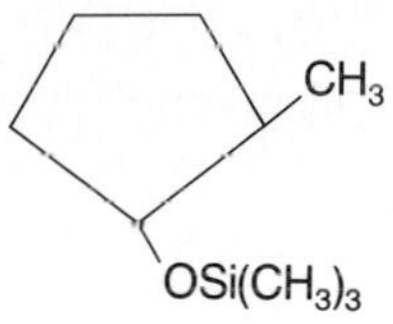

(d) H$_3$C   CH$_3$ H   H$_3$C   CH$_3$ H

and

 H$_3$C H   H   H   H   CH$_3$

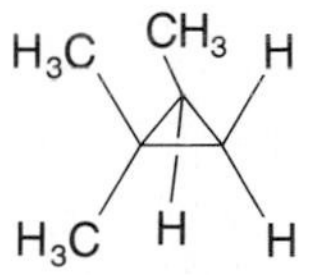

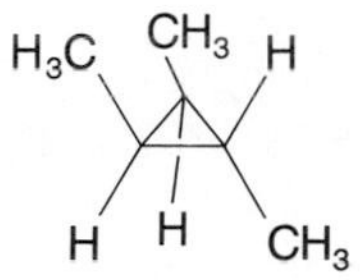

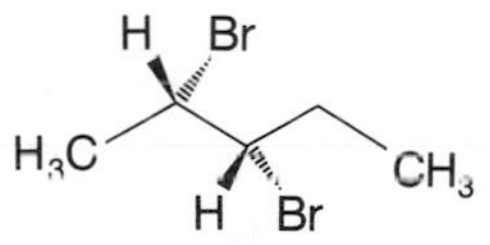

(e)

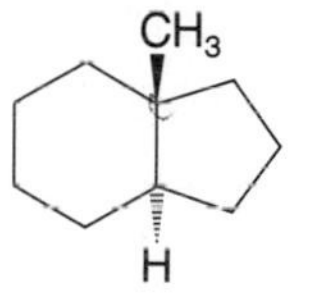

and

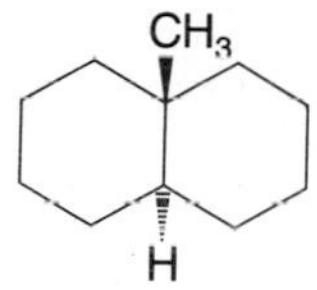

(f)   H   Br   H   Br H   Br

and

 H$_3$C   CH$_3$ H$_3$C   CH$_3$
    H   Br

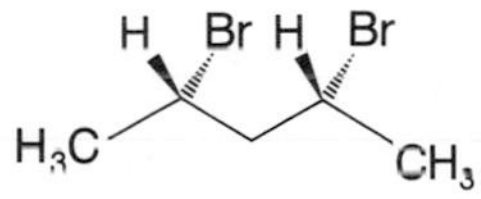

(g)

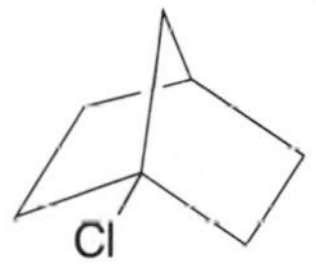

and

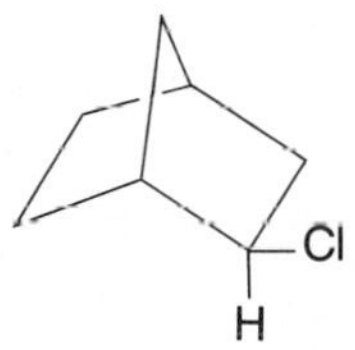

(h)

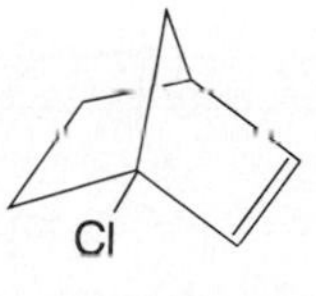

and

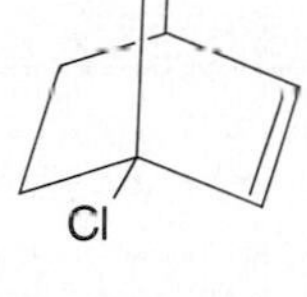

8.20 Which of the isomeric alcohols having the molecular formula C$_5$H$_{12}$O are chiral? Which are achiral?

8.21 Write the structures of all the isomers, including stereoisomers, of trichlorocyclopropane.

8.22 Compare 2,3-pentanediol and 2,4-pentanediol with respect to the number of potential stereoisomers of each. Which stereoisomers are chiral? Which are achiral?

8.23 Identify the chiral centers in each of the following:

(a)

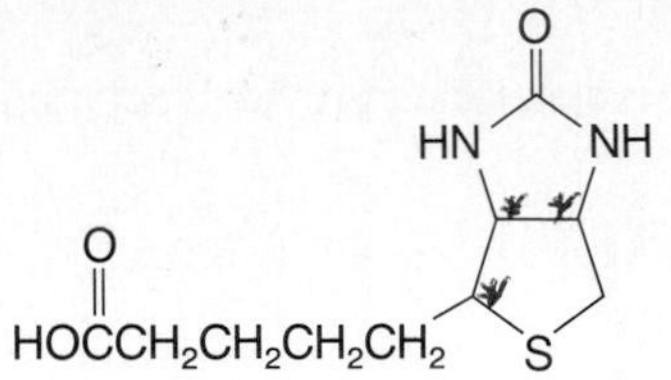

Biotin (a nutrient essential for normal growth)

(b)

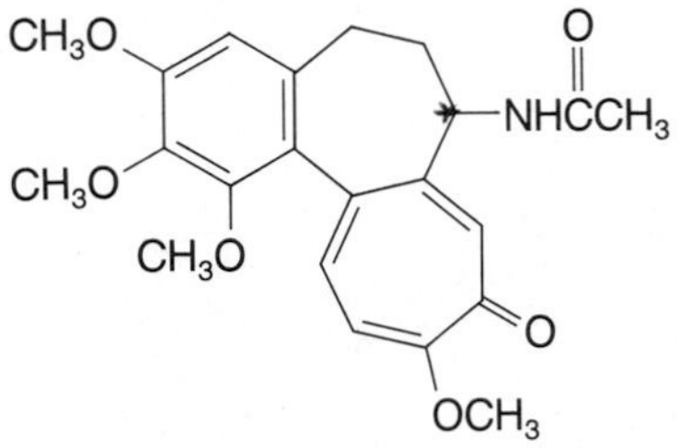

Colchicine (used in biology research to induce doubling of chromosomes)

(c)

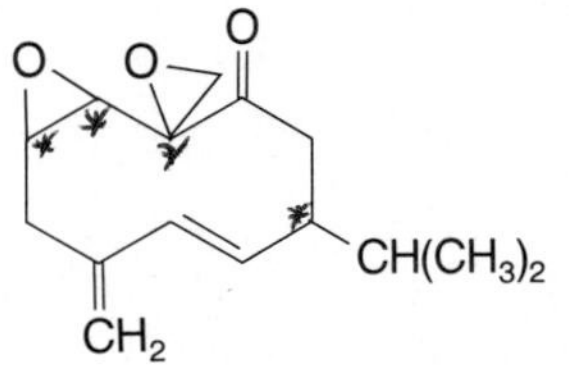

Periplanone B (sex attractant of the American cockroach)

(d)

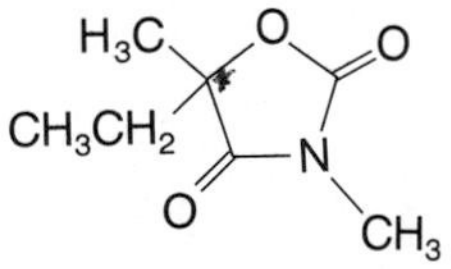

Paramethadione (an anticonvulsant drug)

(e)

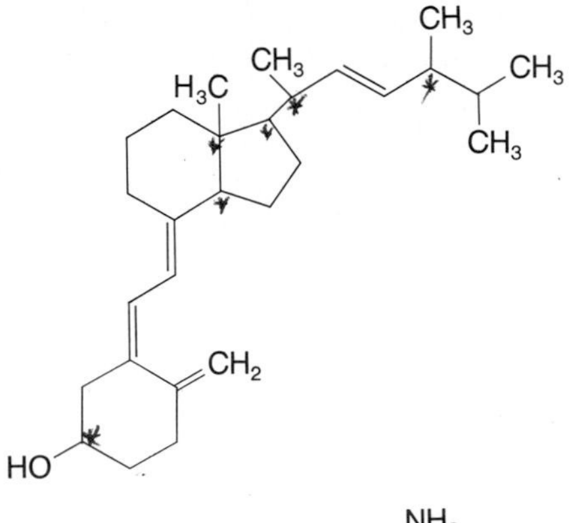

Calciferol (a hormone, also called vitamin D$_2$, which is involved in calcium deposition in bones)

(f)

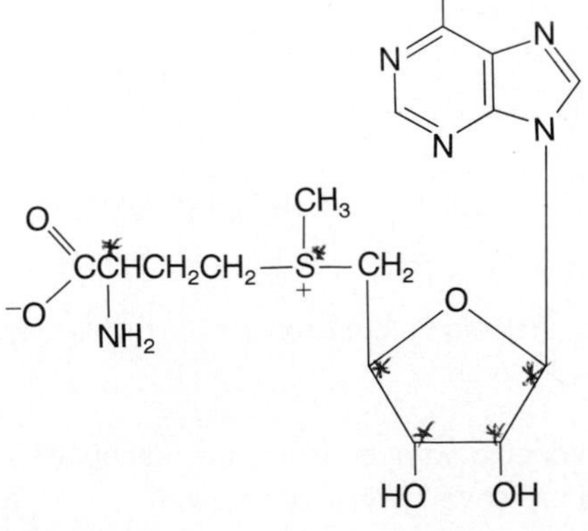

S-Adenosylmethionine (a biological methyl transfer agent)

8.24 Specify the configuration as *R* or *S* in each of the following:

(*a*) (−)-2-Octanol

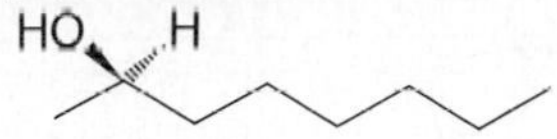

(*b*) Monosodium L-glutamate (only this stereoisomer is of any value as a flavor-enhancing agent)

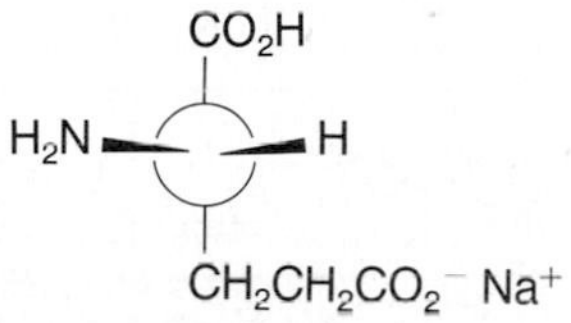

(*c*) (+)-2-Phenylbutanoic acid

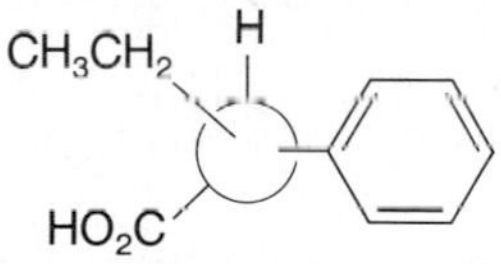

(*d*) 5-Hydroxytryptophan (an important intermediate in a complex series of chemical reactions occurring in the brain)

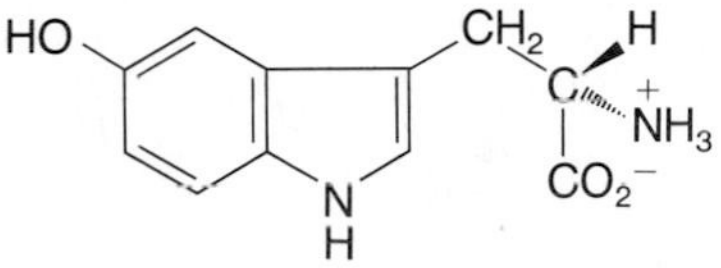

8.25 Identify the relationship in each of the following pairs. Do the drawings represent compounds which are constitutional isomers or stereoisomers, or are they identical? If they are stereoisomers, are they enantiomers or diastereomers? (Molecular models may prove useful in this problem.)

(*a*) 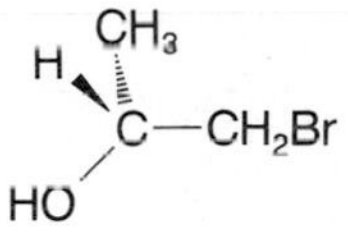 and 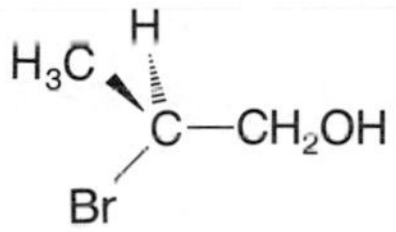

(*b*) 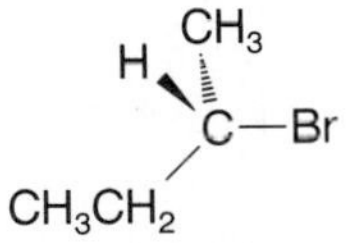 and 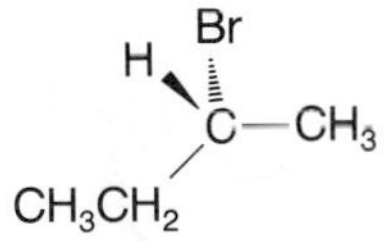

(*c*) 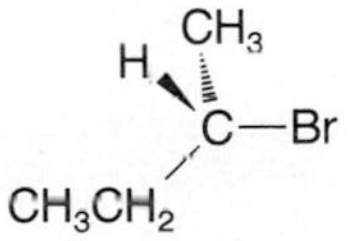 and 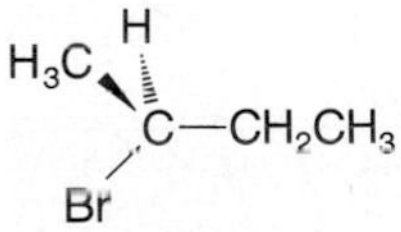

(*d*) 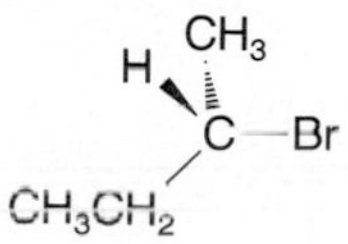 and 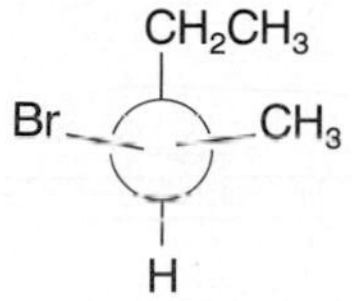

(*e*) 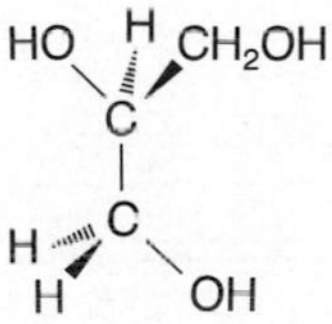 and 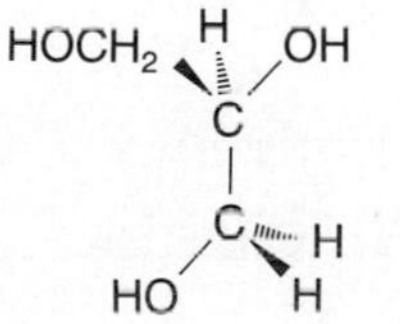

(f) and

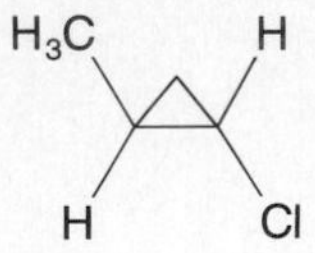

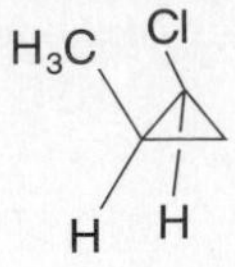

(g) 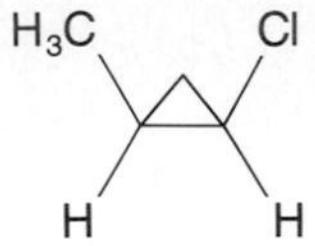 and

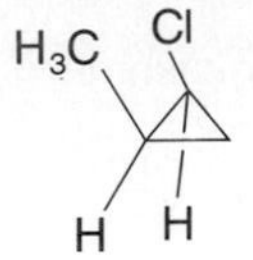

(h) 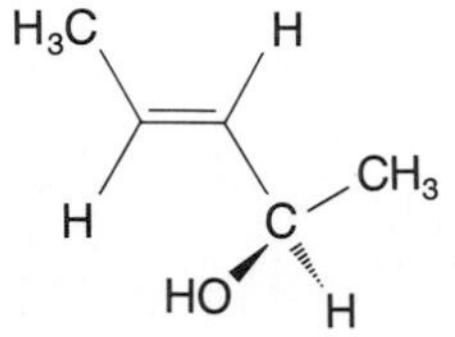 and

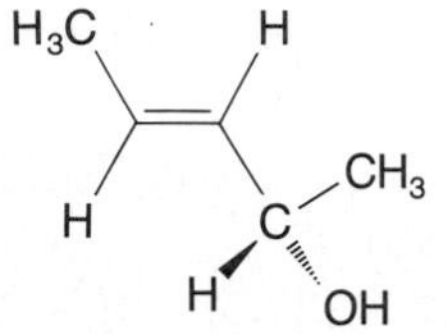

(i) 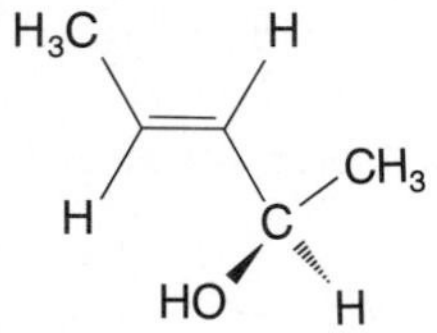 and

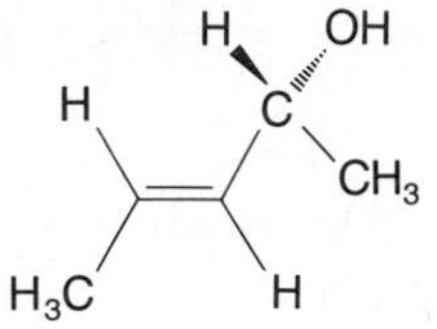

(j) 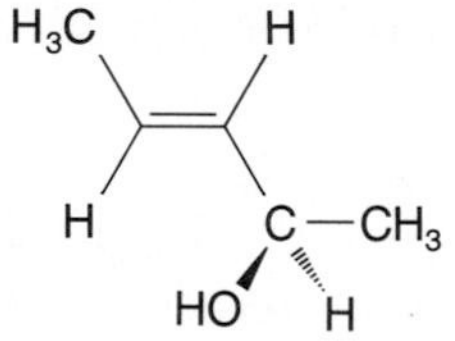 and

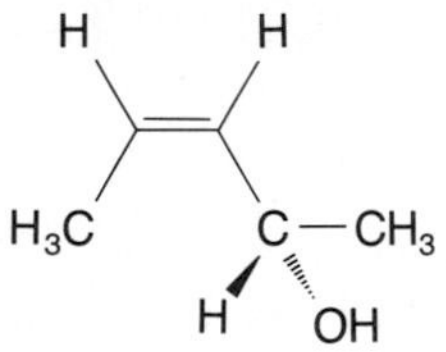

(k) 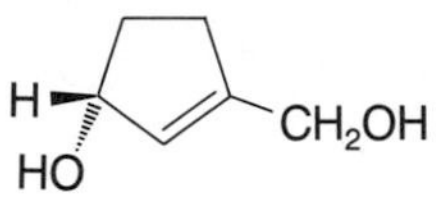 and

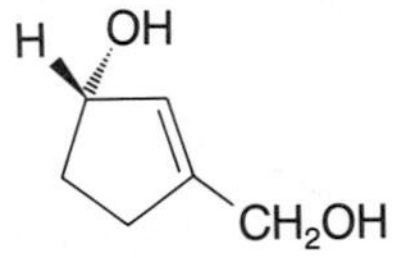

(l) 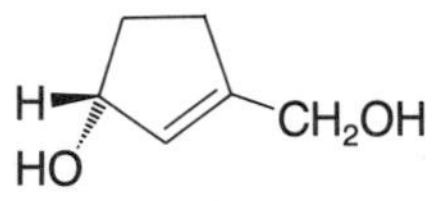 and

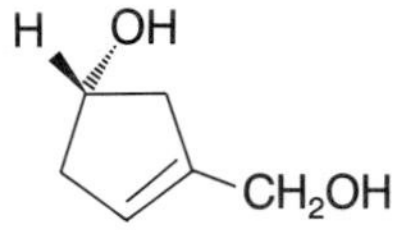

(m) 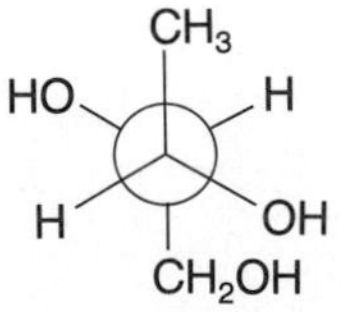 and

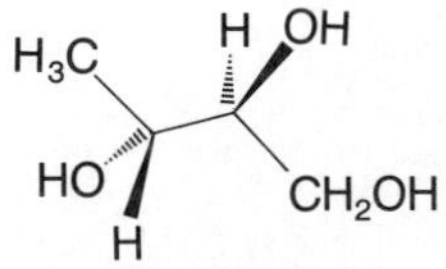

(n) 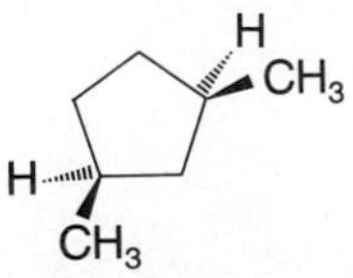 and

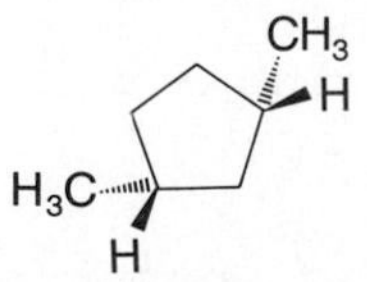

(o) 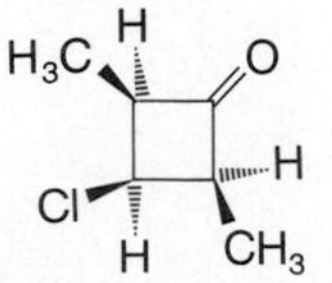 and

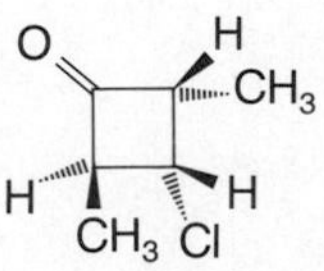

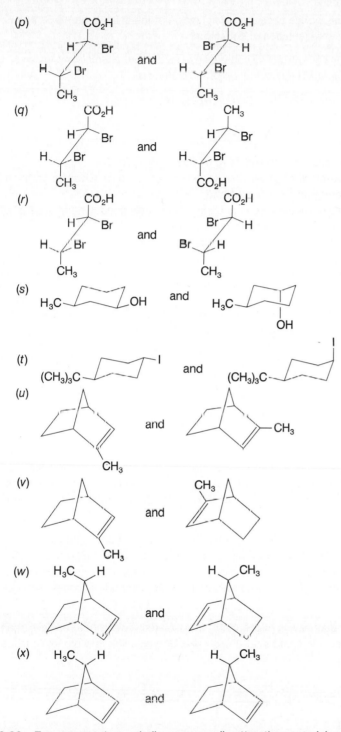

(p)

(q)

(r)

(s)

(t)

(u)

(v)

(w)

(x)

8.26 Ectocarpene is a volatile, sperm cell–attracting material released by the eggs of the seaweed *Ectocarpus siliculosus*. Its constitution is

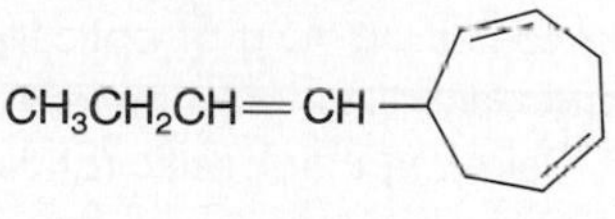

$CH_3CH_2CH=CH$

All the double bonds are cis and the absolute configuration of the chiral center is *S*. Write a stereochemically accurate representation of ectocarpene.

8.27 Multifidene is a sperm cell–attracting substance released by the female of a species of brown algae *(Cutleria multifida)*. The constitution of multifidene is

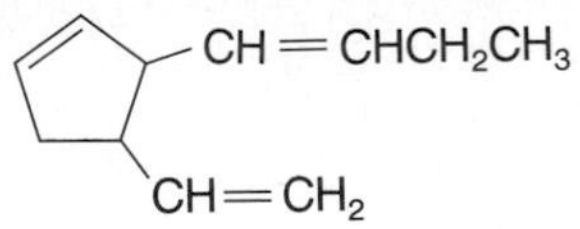

(a) How many stereoisomers are represented by this constitution?

(b) Multifidene has a cis relationship between its alkenyl substituents. Given this information, how many stereoisomers are possible?

(c) The butenyl side chain has the *Z* configuration of its double bond. On the basis of all the data, how many stereoisomers are possible?

(d) Draw stereochemically accurate representations of all the stereoisomers that satisfy the structural requirements of multifidene.

(e) How are these stereoisomeric multifidenes related (enantiomers or diastereomers)?

8.28 Streptimidone is an antibiotic and has the structure shown. How many diastereomers of streptimidone are possible? How many enantiomers? Using the *E, Z* and *R, S* descriptors, specify all essential elements of stereochemistry of streptimidone.

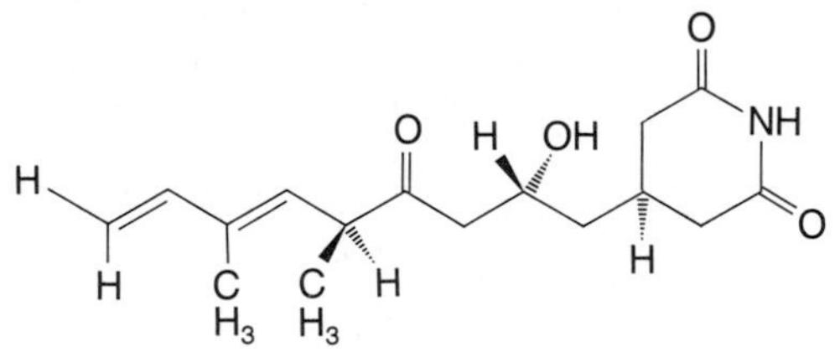

8.29 In Problem 4.22 you were asked to draw the preferred conformation of menthol based on the information that menthol is the most stable stereoisomer of 2-isopropyl-5-methylcyclohexanol. We can now completely describe (−)-menthol structurally by noting that it has the *R* configuration at the hydroxyl-substituted carbon.

(a) Write the preferred conformation of (−)-menthol in its correct configuration.

(b) (+)-Isomenthol has the same constitution as (−)-menthol. The configuration at C-1 and C-2 of (+)-isomenthol are the opposite of the corresponding chiral centers of (−)-menthol. Write the preferred conformation of (+)-isomenthol in its correct configuration.

8.30 A certain natural product having $[\alpha]_D + 40.3°$ was isolated. Two structures have been independently proposed for this compound. Which one do you think is more likely to be correct? Why?

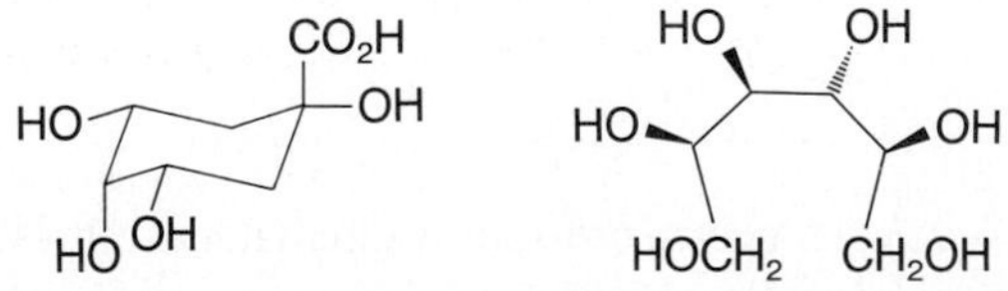

8.31 (a) An aqueous solution containing 10 g of optically pure fructose was diluted to 500 mL with water and placed in a polarimeter tube 20 cm long. The measured rotation was −5.20°. Calculate the specific rotation of fructose.

(b) If the above solution were mixed with 500 mL of a solution containing 5 g of racemic fructose, what would the specific rotation of the resulting fructose mixture be? What would be its optical purity?

8.32 Write the organic products of each of the following reactions. If two stereoisomers are formed, show both. Label all chiral centers as R or S as appropriate.

(a) 1-Butene and hydrogen iodide
(b) (E)-2-Butene and hydrogen bromide in hexane in the presence of peroxides
(c) (E)-2-Pentene and bromine in carbon tetrachloride
(d) (Z)-2-Pentene and bromine in carbon tetrachloride
(e) 1-Butene and peroxyacetic acid in dichloromethane
(f) (Z)-2-Pentene and peroxyacetic acid in dichloromethane
(g) 1,5,5-Trimethylcyclopentene and hydrogen in the presence of platinum
(h) 1,5,5-Trimethylcyclopentene and diborane in tetrahydrofuran followed by oxidation with hydrogen peroxide

8.33 The enzyme *aconitase* catalyzes the hydration of aconitic acid to two products, citric acid and isocitric acid. Isocitric acid is optically active, citric acid is not. What are the respective constitutions of citric acid and isocitric acid?

$$HO_2CCH_2 \diagdown \qquad \diagup CO_2H$$
$$C=C$$
$$HO_2C \diagup \qquad \diagdown H$$

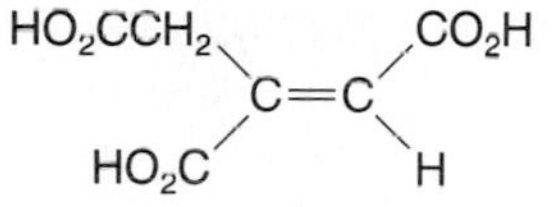

Aconitic acid

8.34 Compound A (C_6H_{10}) contains a five-membered ring. When Br_2 adds to A, two diastereomeric dibromides are formed. Suggest reasonable structures for A and the two dibromides.

8.35 When optically pure 2,3-dimethyl-2-pentanol was subjected to a dehydration reaction, a mixture of two alkenes was obtained. Hydrogenation of this alkene mixture gave 2,3-dimethylpentane, which was 50 percent optically pure. What were the two alkenes formed in the elimination reaction and what were the relative amounts of each?

8.36 When (R)-3-buten-2-ol is treated with a peroxy acid, two stereoisomeric epoxides are formed in a 60:40 ratio. The minor stereoisomer has the structure shown

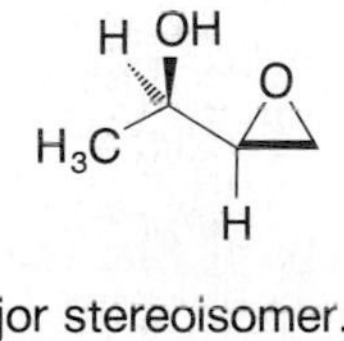

$$\begin{array}{c} H \quad OH \\ \diagdown \; \; | \\ H_3C \diagdown \; C \diagup \; \; \diagup O \\ \diagup \\ H \end{array}$$

(a) Write the structure of the major stereoisomer.
(b) What is the relationship between the two epoxides? Are they enantiomers or diastereomers?
(c) What four stereoisomeric products are formed when racemic 3-buten-2-ol is epoxidized under the same conditions? How much of each stereoisomer is formed?

NUCLEOPHILIC SUBSTITUTION REACTIONS

In our discussion of elimination reactions in Chapter 6, we learned that a Lewis base can react with an alkyl halide to form an alkene by dehydrohalogenation. In the present chapter, you shall find that the same kinds of reactants can also undergo another kind of reaction, one in which the Lewis base acts as a nucleophile to substitute for the halide substituent on carbon.

$$RX \; + \; [Y\!:]^- \; \longrightarrow \; RY \; + [:\!\overset{..}{\underset{..}{X}}\!:]^-$$

| Alkyl halide | Lewis base | Product of nucleophilic substitution | Halide anion |

We first encountered nucleophilic substitution reactions in Chapter 4, where we saw that alkyl halides could be prepared from alcohols by nucleophilic substitution. Now we will see how alkyl halides can be converted to other classes of organic compounds by nucleophilic substitution.

This chapter has a mechanistic emphasis designed to achieve a practical result. By understanding the mechanisms by which alkyl halides are converted to products in substitution reactions, intelligent decisions can be made in choosing experimental conditions best suited to carry out a particular functional group transformation. The difference between a successful reaction that leads cleanly to a desired product and one that fails is often a subtle one. Mechanistic analysis helps us to appreciate these subtleties and use them to our advantage.

9.1 FUNCTIONAL GROUP TRANSFORMATION BY NUCLEOPHILIC SUBSTITUTION

Nucleophilic substitution reactions of alkyl halides are related to elimination reactions in that the halogen acts as a leaving group on carbon, and is lost as an anion. The carbon-halogen bond of the alkyl halide is broken heterolytically; the pair of electrons in that bond are lost with the leaving group.

The carbon–halogen bond in an alkyl halide is polarized

$$\overset{\delta+}{R}-\overset{\delta-}{X} \qquad X = I, Br, Cl, F$$

and is cleaved on attack by a nucleophile so that the two electrons in the bond are retained by the halogen

$$^-Y: \overset{\curvearrowright}{R} \overset{..}{\overset{\curvearrowleft}{X}}: \longrightarrow R-Y + :\overset{..}{\underset{..}{X}}:^-$$

The most frequently encountered nucleophiles in functional group transformations are anions, which are used as their lithium, sodium, or potassium salts. If we use M to represent lithium, sodium, or potassium, some representative nucleophilic reagents are

MOR   (a metal alkoxide, a source of the nucleophilic anion $R\overset{..}{O}:^-$)
MSH   (a metal hydrogen sulfide, a source of the nucleophilic anion $H\overset{..}{S}:^-$)
MCN   (a metal cyanide, a source of the nucleophilic anion $:C\equiv N:$)
MN$_3$ (a metal azide, a source of the nucleophilic anion $:\overset{..}{N}=\overset{+}{N}=\overset{..}{\underset{..}{N}}:$)

Table 9.1 illustrates the application of each of these nucleophilic reagents to functional group transformations. The anionic portion of the salt substitutes for the halogen of an alkyl halide. The metal cation portion becomes a lithium, sodium, or potassium halide salt.

$$M^+ \quad Y: \overset{\curvearrowright}{} + \overset{\curvearrowright}{R} \overset{..}{\underset{..}{X}}: \longrightarrow \quad R-Y \ + M^+ \ :\overset{..}{\underset{..}{X}}:^-$$

| Nucleophilic reagent | Alkyl halide | Product of nucleophilic substitution | Metal halide salt |

PROBLEM 9.1 Write the structure of the principal organic product formed in the reaction of methyl bromide with each of the following compounds:

(a) NaOH (sodium hydroxide)
(b) KOC(CH$_3$)$_3$ (potassium *tert*-butoxide)
(c) LiN$_3$ (lithium azide)
(d) KCN (potassium cyanide)
(e) NaSH (sodium hydrogen sulfide)

SAMPLE SOLUTION   (a)   The nucleophile in sodium hydroxide is the negatively charged hydroxide ion. The reaction that occurs is nucleophilic substitution of bromide by hydroxide. The product is methyl alcohol.

$$HO^- \ + \ CH_3Br \longrightarrow CH_3OH \ + \ Br^-$$

| Hydroxide ion (nucleophile) | Methyl bromide (substrate) | Methyl alcohol (product) | Bromide ion (leaving group) |

In order to ensure that reaction occurs in homogeneous solution, solvents are chosen that dissolve both the alkyl halide and the ionic salt. The alkyl halide substrates are soluble in organic solvents, but the salts often are not. Inorganic salts are soluble in water, but alkyl halides are not. Mixed solvents such as ethanol-water mixtures can often dissolve enough of both the substrate and the nucleophile to give fairly concentrated solutions and thus are frequently used. Many salts, as well as most alkyl halides, possess significant solubility in dimethyl sulfoxide (DMSO), which makes this a good medium for carrying out nucleophilic substitution reactions.

TABLE 9.1

Representative Functional Group Transformations by Nucleophilic Substitution Reactions of Alkyl Halides

Nucleophile and comments	General equation and specific example
Alkoxide ion ($R\ddot{O}:^-$) The oxygen atom of a metal alkoxide acts as a nucleophile to replace the halogen of an alkyl halide. The product is an *ether*.	$R'O^- + R{-}X \longrightarrow R'OR + X^-$ Alkoxide ion Alkyl halide Ether Halide ion $(CH_3)_2CHCH_2ONa + CH_3CH_2Br \xrightarrow{\text{isobutyl alcohol}} (CH_3)_2CHCH_2OCH_2CH_3 + NaBr$ Sodium isobutoxide Ethyl bromide Ethyl isobutyl ether (66%) Sodium bromide
Hydrogen sulfide ion ($H\ddot{S}:^-$) Use of hydrogen sulfide as a nucleophile permits the conversion of alkyl halides to compounds of the type RSH. These compounds are the sulfur analogs of alcohols and are known as *thiols*.	$HS^- + R{-}X \longrightarrow RSH + X^-$ Hydrogen sulfide ion Alkyl halide Thiol Halide ion $KSH + CH_3CH(CH_2)_6CH_3 \;(\text{Br}) \xrightarrow[\text{water}]{\text{ethanol}} CH_3CH(CH_2)_6CH_3 \;(\text{SH}) + KBr$ Potassium hydrogen sulfide 2-Bromononane 2-Nonanethiol (74%) Potassium bromide
Cyanide ion ($:C{\equiv}N:$) The negatively charged carbon atom of cyanide ion is usually the site of its nucleophilic character. Use of cyanide ion as a nucleophile permits the extension of a carbon chain by carbon-carbon bond formation. The product is an *alkyl cyanide* or *nitrile*.	$:N{\equiv}C:^- + R{-}X \longrightarrow RC{\equiv}N: + X^-$ Cyanide ion Alkyl halide Alkyl cyanide Halide ion $NaCN + \text{(cyclopentyl)}{-}Cl \xrightarrow{\text{DMSO}} \text{(cyclopentyl)}{-}CN + NaCl$ Sodium cyanide Cyclopentyl chloride Cyclopentyl cyanide (70%) Sodium chloride
Azide ion ($:N{=}N{=}N:$) Sodium azide is a reagent used for carbon-nitrogen bond formation. The product is an *alkyl azide*.	$:N{=}N{=}N:^- + R{-}X \longrightarrow RN{=}N{=}N:^- + X^-$ Azide ion Alkyl halide Alkyl azide Halide ion $NaN_3 + CH_3(CH_2)_4I \xrightarrow{\text{propanol-water}} CH_3(CH_2)_4N_3 + NaI$ Sodium azide Pentyl iodide Pentyl azide (52%) Sodium iodide

With this as background, you can begin to see the important position occupied by alkyl halides in synthetic organic chemistry. Alkyl halides may be prepared from alcohols by nucleophilic substitution, from alkanes by free-radical halogenation, and from alkenes by addition of hydrogen halides. They then become available as starting materials for the preparation of other functionally substituted organic compounds by replacement of the halide leaving group with a nucleophile. The range of compounds that can be prepared by nucleophilic substitution reactions of alkyl halides is quite large; the examples shown in this section illustrate only a few of them. Numerous other examples will be added to the list in this and subsequent chapters.

9.2 SUBSTITUTION OF ONE HALOGEN BY ANOTHER

In the reactions described in the preceding section, we saw examples of nucleophilic substitution reactions involving halide leaving groups. Halide ions may also act as nucleophiles. In a reaction known as *halide-halide exchange,* one halogen displaces another from an alkyl halide.

$$:\ddot{Y}:^- \quad + \quad R - \ddot{X}: \quad \longrightarrow \quad R - \ddot{Y}: \quad + \quad :\ddot{X}:^-$$

Halide ion	Alkyl halide	Alkyl halide	Halide ion
(nucleophile)	(substrate)	(product)	(leaving group)

An equilibrium is established, and organic chemists have learned how to shift the position of equilibrium so as to make this reaction an effective one for the preparation of alkyl fluorides and alkyl iodides.

In the preparation of alkyl fluorides, an alkyl chloride, bromide, or iodide is heated with potassium fluoride in a high-boiling alcohol solvent such as ethylene glycol.

$$CH_3CH_2CH_2CH_2CH_2Br + KF \xrightarrow[120°C]{ethylene\ glycol} CH_3CH_2CH_2CH_2CH_2F + KBr$$

Pentyl bromide, bp 129°C	Potassium fluoride	Pentyl fluoride, bp 65°C (50%)	Potassium bromide

Alkyl fluorides have the lowest boiling points of all the alkyl halides and are removed from the reaction mixture by distillation as they are formed. In accordance with Le Chatelier's principle, the equilibrium responds by forming more alkyl fluoride at the expense of the original alkyl halide. Since reactions that lead to alkyl fluorides in good yield are relatively rare, this is a valuable synthetic method for these compounds.

Alkyl iodides may be prepared from the corresponding alkyl chlorides and alkyl bromides by treatment with sodium iodide in acetone as solvent.

$$CH_2{=}CHCH_2Cl + NaI \xrightarrow{acetone} CH_2{=}CHCH_2I + NaCl \downarrow$$

Allyl chloride	Sodium iodide	Allyl iodide (77%)	Sodium chloride

$$CH_3CHCH_3 + NaI \xrightarrow{acetone} CH_3CHCII_3 + NaBr \downarrow$$
$$\quad\ \ | \qquad\qquad\qquad\qquad\qquad |$$
$$\quad\ \ Br \qquad\qquad\qquad\qquad\quad\ I$$

Isopropyl bromide	Sodium iodide	Isopropyl iodide (63%)	Sodium bromide

Le Chatelier's principle is at work here as well. Sodium iodide is soluble in acetone, but sodium bromide and sodium chloride are not. In these reactions, sodium chloride and sodium bromide precipitate from the reaction mixture, causing the position of equilibrium to shift so as to favor formation of the alkyl iodides.

9.3   RELATIVE REACTIVITY OF HALIDE LEAVING GROUPS

Among alkyl halides, alkyl iodides undergo nucleophilic substitution at the fastest rate, alkyl fluorides at the slowest.

Increasing rate of substitution by nucleophiles →

$$RF \ll RCl < RBr < RI$$

Least reactive   Most reactive

As in elimination reactions of alkyl halides, iodide is the best leaving group because it has the weakest carbon-halogen bond. Alkyl iodides are several times more reactive than bromides and from 50 to 100 times more reactive than chlorides. Alkyl fluorides are rarely used as substrates in nucleophilic substitution reactions because they are many thousand times less reactive than alkyl chlorides. Fluoride is the poorest leaving group because it forms the strongest bond to carbon of any of the halogens.

PROBLEM 9.2   A single organic product was obtained when 1-bromo-3-chloropropane was allowed to react with one molar equivalent of sodium cyanide in aqueous ethanol. What was this product?

Leaving group ability is also related to basicity. A strongly basic anion is usually a poorer leaving group than a weakly basic one. Fluoride is the most basic and the poorest leaving group among the halide anions, iodide the least basic and the best leaving group.

9.4   THE BIMOLECULAR (S_N2) MECHANISM OF NUCLEOPHILIC SUBSTITUTION

The mechanisms by which nucleophilic substitution reactions take place have been the subject of much study. Extensive research by Sir Christopher Ingold and Edward D. Hughes and their associates at University College, London during the 1930s emphasized kinetic and stereochemical measurements to probe the mechanisms of these reactions.

Methyl bromide reacts with sodium hydroxide to form methyl alcohol by a nucleophilic substitution reaction.

$$CH_3Br \quad + \quad HO^- \quad \longrightarrow \quad CH_3OH \quad + \quad Br^-$$

Methyl bromide   Hydroxide ion   Methyl alcohol   Bromide ion

The rate of this reaction is directly proportional to the concentration of both methyl iodide and sodium hydroxide. It is first-order in each reactant, or second-order overall.

$$Rate = k \, [CH_3Br][HO^-]$$

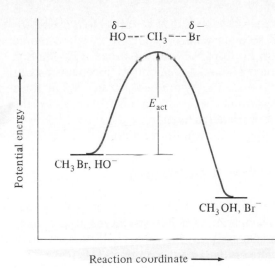

FIGURE 9.1 Energy diagram for hydrolysis of methyl bromide by the S_N2 mechanism.

Hughes and Ingold interpreted second-order kinetic behavior to mean that the rate-determining step is *bimolecular,* i.e., that both hydroxide ion and methyl bromide are involved at the transition state. The symbol given to the detailed description of the mechanism that they developed is S_N2, standing for *substitution nucleophilic bimolecular.*

Hughes and Ingold visualized the S_N2 mechanism as a concerted process, i.e., as a single-step reaction in which both the substrate and the nucleophile contribute to the activated complex. Cleavage of the bond between carbon and the leaving group is assisted by formation of a bond between carbon and the nucleophile. In effect, the nucleophile "pushes off" the leaving group from its point of attachment to carbon. For this reason, the S_N2 mechanism is sometimes referred to as a *direct displacement* process. The S_N2 mechanism for the hydrolysis of methyl bromide may be represented as

$$HO^- + CH_3Br \longrightarrow [\overset{\delta-}{HO} \text{---} CH_3 \text{---} \overset{\delta-}{Br}]^{\neq} \longrightarrow HOCH_3 + Br^-$$

| Hydroxide ion | Methyl bromide | Activated complex | Methyl alcohol | Bromide ion |

Figure 9.1 presents an energy diagram for this reaction. It is a concerted process involving only a single transition state, in which carbon in partially bonded to both the incoming nucleophile and the departing leaving group. In proceeding to the transition state, the nucleophile begins to share a pair of electrons with carbon. As the halide ion leaves, it takes with it the pair of electrons in its bond to carbon and in so doing, develops negative charge. Stabilization of this developing negative charge by hydrogen bonding of the leaving group to the solvent assists the cleavage of the carbon-halogen bond.

PROBLEM 9.3 Is the two-step sequence depicted in the following equations consistent with the second-order kinetic behavior observed for the hydrolysis of methyl bromide?

$$CH_3Br \xrightarrow{\text{slow}} CH_3^+ + Br^-$$

$$CH_3^+ + HO^- \xrightarrow{\text{fast}} CH_3OH$$

Direct displacement by the S_N2 mechanism is considered to apply to most substitutions in which simple primary and secondary alkyl halides react with anionic nucleophiles. All the examples cited in Sections 9.1 and 9.2 proceed by the S_N2 mechanism (or a mechanism very much like S_N2—remember, mechanisms can never be established with certainty but represent only our best present explanations of experimental observations). We will examine the S_N2 mechanism, particularly the structure of the activated complex, in more detail in Section 9.6 after first looking at some stereochemical studies of nucleophilic substitution reactions.

9.5 STEREOCHEMISTRY OF S_N2 REACTIONS

Assuming that the activated complex is bimolecular in reactions of primary and secondary alkyl halides with anionic nucleophiles, what is its structure? In particular, what is the spatial arrangement of the nucleophile in relation to the leaving group as reactants pass through the transition state on their way to products?

Two stereochemical possibilities present themselves. In the pathway shown in Figure 9.2a, the nucleophile simply assumes the position occupied by the leaving group. It attacks the substrate at the same face from which the leaving group departs. This is called "front-side displacement" or substitution with *retention of configuration.*

In a second possibility, illustrated in Figure 9.2b, the nucleophile attacks the substrate from the side opposite the bond to the leaving group. This is called "backside displacement" or substitution with *inversion of configuration.*

Which of these two opposite stereochemical possibilities is the correct one was determined by carrying out nucleophilic substitution reactions with optically active alkyl halides. In one such experiment, Hughes and Ingold determined that hydrolysis of 2-bromooctane in the presence of hydroxide ion gave 2-octanol having a configuration opposite to that of the starting alkyl halide.

(S)-(+)-2-Bromooctane (R)-(−)-2-Octanol

Nucleophilic substitution in this case had occurred with inversion of configuration, consistent with the transition state representation shown.

PROBLEM 9.4   Would you expect the 2-octanol formed by S_N2 hydrolysis of (−)-2-bromooctane to be optically active? If so, what will be its absolute configuration and sign of rotation? What about the 2-octanol formed by hydrolysis of racemic 2-bromooctane?

Numerous similar experiments have demonstrated the generality of this observation. Substitution reactions that involve the nucleophile in the rate-determining step

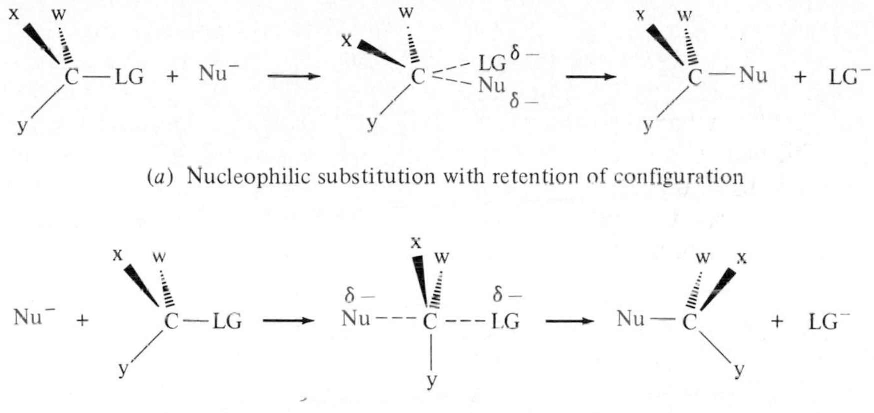

(a) Nucleophilic substitution with retention of configuration

(b) Nucleophilic substitution with inversion of configuration

FIGURE 9.2 Two contrasting stereochemical pathways for substitution of a leaving group (LG) by a nucleophile (Nu⁻). In (a) the nucleophile attacks carbon at the same side from which the leaving group departs. In (b) nucleophilic attack occurs at the side opposite the bond to the leaving group.

are stereospecific and proceed with inversion of configuration at carbon. There is a *stereoelectronic* requirement for the nucleophile to approach carbon from the side opposite the bond to the leaving group. Electrons are transferred from a lone pair orbital of the nucleophile to the antibonding component of the carbon-halogen bond. This orbital has its larger lobe at carbon on the side opposite that to which the halogen is attached.

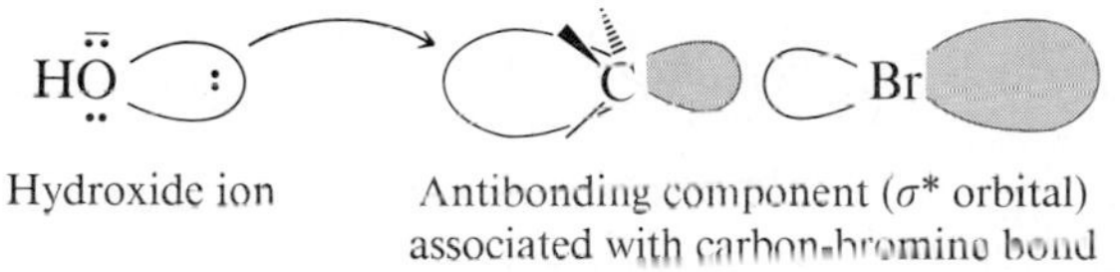

Hydroxide ion   Antibonding component (σ^* orbital) associated with carbon-bromine bond

9.6 HOW S$_N$2 REACTIONS OCCUR

Once the overall reaction stereochemistry is known and is considered along with the kinetic data, a fairly complete picture of the bonding changes that take place during S$_N$2 reactions emerges. The potential energy diagram of Figure 9.3 for the hydrolysis of (S)-(+)-2-bromooctane is one which is consistent with the experimental observations.

Hydroxide ion acts as a nucleophile, using an unshared electron pair to attack carbon from the side opposite the bond to the leaving group. The hybridization of the carbon at which substitution occurs changes from sp^3 in the alkyl halide starting material to sp^2 in the activated complex. Both the nucleophile (hydroxide) and the leaving group (bromide) are partially bonded to this carbon in the activated complex. We say the S$_N$2 transition state is *pentacoordinate;* carbon is fully bonded to three substituents and partially bonded to both the leaving group and the incoming nucleophile. The bonds to the nucleophile and the leaving group are relatively long and weak at the transition state.

Once past the transition state, the leaving group is expelled and carbon becomes tetracoordinate, its hybridization returning to sp^3.

During the passage of starting materials to products, three interdependent and synchronous changes take place. They are

1. Stretching, then breaking of the bond to the leaving group

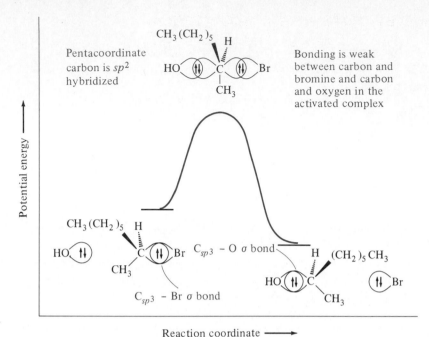

FIGURE 9.3 Hybrid orbital description of the bonding changes that take place at carbon during nucleophilic substitution by the S_N2 mechanism.

2. Formation of a bond to the nucleophile from the opposite side of the bond that is broken
3. Stereochemical inversion of the tetrahedral arrangement of bonds to the carbon at which substitution occurs

While this mechanistic picture developed from experiments involving optically active alkyl halides, it applies as well to cases not amenable to direct stereochemical study. Chemists even speak of methyl bromide as undergoing nucleophilic substitution with *inversion*. By this they mean that tetrahedral inversion of the bonds to carbon occurs as reactant proceeds to product by way of a pentacoordinated activated complex.

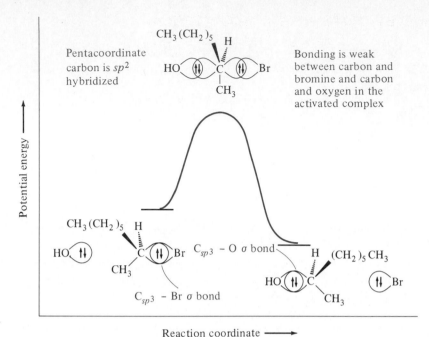

| Hydroxide ion | Methyl bromide | Activated complex | Methyl alcohol | Bromide ion |

We have noted earlier in Section 9.3 that alkyl halides differ in reactivity according to the nature of their leaving group. The bond to the leaving group is partially broken in the S_N2 transition state, and alkyl iodides react faster than other alkyl halides with nucleophilic reagents because they have the weakest carbon-halogen bonds. In the next section we will see how reactivity in nucleophilic substitution reactions is affected by another aspect of alkyl halide structure.

9.7 STERIC EFFECTS IN S$_N$2 REACTIONS

There are very large differences in reactivity among alkyl halides, which depend on the degree of substitution at the carbon that bears the leaving group. As Table 9.2 shows for the reaction

$$\text{RBr} \quad + \quad \text{LiI} \quad \xrightarrow{\text{acetone}} \quad \text{RI} \quad + \quad \text{LiBr} \downarrow$$

Alkyl bromide   Lithium iodide   Alkyl iodide   Lithium bromide

the rates of nucleophilic substitution of a series of alkyl bromides differ by a factor of over 10^6 when the most reactive member of the group (methyl bromide) and the least reactive member (*tert*-butyl bromide) are compared.

In general, nucleophilic substitution reactions that exhibit second-order kinetics reveal a substrate structure dependence indicating that reactivity is governed by *steric* factors.

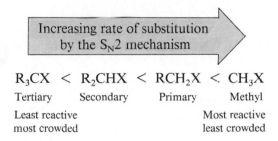

$$\text{R}_3\text{CX} \quad < \quad \text{R}_2\text{CHX} \quad < \quad \text{RCH}_2\text{X} \quad < \quad \text{CH}_3\text{X}$$

Tertiary   Secondary   Primary   Methyl

Least reactive   Most reactive
most crowded   least crowded

The activated complex in an S$_N$2 reaction is considerably more crowded than the starting material. The carbon that bears the leaving group is tetracoordinate in the substrate but becomes pentacoordinate at the transition state. Substitution occurs most readily when the substituent groups attached to this carbon are small. Therefore, as illustrated in Figure 9.4a, the best substrate for an S$_N$2 reaction is a methyl halide CH$_3$X. Approach of the nucleophile to the methyl group is relatively unhindered by the hydrogen substituents, and the activated complex is not appreciably destabilized by van der Waals repulsive forces. When, as shown in Figure 9.4b, an alkyl group is present in place of a hydrogen substituent, the activated complex is much more crowded, the activation energy is higher, and the rate of the reaction is slower. Alkyl groups sterically hinder approach of the nucleophile to the reaction site.

TABLE 9.2

Reactivity of Some Alkyl Bromides toward Substitution by the S$_N$2 Mechanism*

Alkyl bromide	Structure	Class	Relative rate†
Methyl bromide	CH$_3$Br	Unsubstituted	221,000
Ethyl bromide	CH$_3$CH$_2$Br	Primary	1,350
Isopropyl bromide	(CH$_3$)$_2$CHBr	Secondary	1
tert-Butyl bromide	(CH$_3$)$_3$CBr	Tertiary	Too small to measure

* Substitution of bromide by lithium iodide in acetone.

† Ratio of second-order rate constant k for indicated alkyl bromide to k for isopropyl bromide at 25°C.

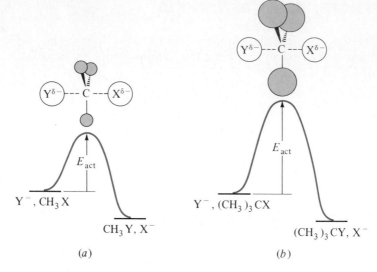

FIGURE 9.4 Steric repulsive interactions between the nucleophile and the groups attached to the reaction site strongly influence the activation energy and the rate of S_N2 reactions. (*a*) Bimolecular activated complex is least crowded when all the substituents at the reaction site are hydrogen atoms. (*b*) Bimolecular activated complex is most crowded when all the substituents at the reaction site are alkyl groups.

PROBLEM 9.5 Identify the compound in each of the following pairs that reacts with sodium iodide in acetone at the faster rate:

(*a*) 1-Chlorohexane or cyclohexyl chloride
(*b*) 1-Bromopentane or 3-bromopentane
(*c*) 2-Chloropentane or 2-fluoropentane
(*d*) 2-Bromo-2-methylhexane or 2-bromo-5-methylhexane
(*e*) 2-Bromopropane or 1-bromodecane

SAMPLE SOLUTION (*a*) Compare the structures of the two chlorides. 1-Chlorohexane is a primary alkyl chloride; cyclohexyl chloride is secondary. Primary alkyl halides are less crowded at the site of substitution than secondary ones and react faster in substitution by the S_N2 mechanism. 1-Chlorohexane is more reactive.

$$CH_3CH_2CH_2CH_2CH_2CH_2Cl$$

1-Chlorohexane
(primary, more reactive)

Cyclohexyl chloride
(secondary, less reactive)

Alkyl substituents at the carbon atom adjacent to the point of nucleophilic attack also decrease the rate of the S_N2 reaction. Compare the rates of nucleophilic substitution in the series of primary alkyl bromides shown in Table 9.3. Taking ethyl bromide as the standard and successively replacing its C-2 hydrogen substituents by methyl groups, we see that each additional methyl group decreases the rate of displacement of bromide by iodide. The effect is slightly smaller than that seen for alkyl substituents directly attached to the carbon that bears the leaving group but is still substantial. When C-2 is completely substituted by methyl groups, as it is in neopentyl bromide, we see the unusual case of a primary alkyl halide that is practically inert to substitution by the S_N2 mechanism because of steric hindrance.

TABLE 9.3
Effect of Chain Branching on Reactivity of Primary Alkyl Bromides toward Substitution under S_N2 Conditions*

Alkyl bromide	Structure	Relative rate†
Ethyl bromide	CH_3CH_2Br	1.0
Propyl bromide	$CH_3CH_2CH_2Br$	0.8
Isobutyl bromide	$(CH_3)_2CHCH_2Br$	0.036
Neopentyl bromide	$(CH_3)_3CCH_2Br$	0.00002

* Substitution of bromide by lithium iodide in acetone.
† Ratio of second order rate constant k for indicated alkyl bromide to k for ethyl bromide at 25°C.

9.8 NUCLEOPHILES AND NUCLEOPHILICITY

The Lewis base that acts as the nucleophile in a nucleophilic substitution reaction often is, but need not always be, an anion. Neutral Lewis bases can also serve as nucleophiles. Common examples of substitution processes involving neutral nucleophiles include *solvolysis* reactions, i.e., reactions in which the Lewis base is the solvent in which the reaction is carried out. Solvolysis in water converts an alkyl halide to an alcohol.

Water · Alkyl halide · Alkyloxonium halide · Alcohol · Hydrogen halide

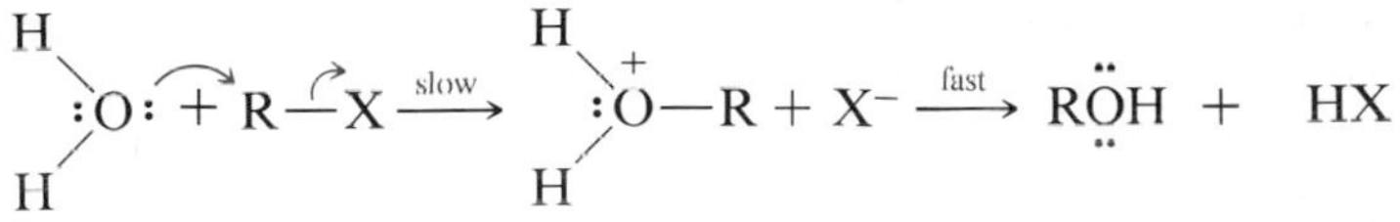

Solvolysis in methyl alcohol converts an alkyl halide to an alkyl methyl ether.

Methyl alcohol · Alkyl halide · Dialkyloxonium halide · Alkyl methyl ether · Hydrogen halide

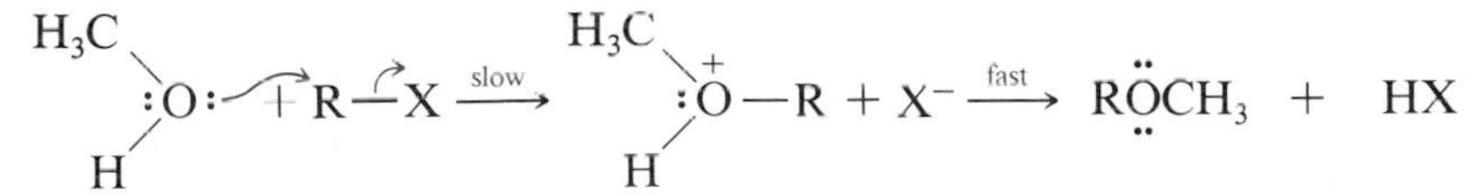

In these and related solvolysis reactions, the nucleophilic substitution step is the first one and is rate-determining. The proton transfer step that follows it is fast.

Since, as we have seen, the nucleophile interacts with the substrate in the rate-determining step of the S_N2 mechanism, it follows that the rate at which substitution occurs may vary from nucleophile to nucleophile. Just as some alkyl halides are more reactive than others, some nucleophiles are more reactive than others. Nucleophilic strength, or *nucleophilicity,* is a measure of how fast a Lewis base displaces a leaving group from a suitable substrate. By measuring the rate at which various Lewis bases react with methyl iodide in methanol, a list of their nucleophilicities relative to methanol as the standard nucleophile has been compiled. It is presented in Table 9.4.

Neutral Lewis bases such as water, alcohols, and carboxylic acids are much weaker nucleophiles than their conjugate bases. When comparing species that have the same

TABLE 9.4

Nucleophilicity of Some Common Nucleophiles

Reactivity class	Nucleophile	Relative reactivity*
Very good nucleophiles	I^-, HS^-, RS^-	$> 10^5$
Good nucleophiles	Br^-, HO^-, RO^-, CN^-, N_3^-	10^4
Fair nucleophiles	NH_3, Cl^-	10^3
	F^-, RCO_2^-	$10^1 - 10^2$
Weak nucleophiles	H_2O, ROH	1
Very weak nucleophiles	RCO_2H	10^{-2}

* Relative reactivity is k(nucleophile)/k(methanol) for typical S_N2 reactions and is approximate. Data pertain to methanol as the solvent.

nucleophilic atom, such as an alkoxide ion and an alcohol, a negatively charged nucleophile is more reactive than a neutral one.

$$RO^- \quad > \quad ROH$$

Stronger nucleophile   Weaker nucleophile

As long as the nucleophilic atom is the same, the more basic the nucleophile, the more reactive it is. An alkoxide ion is more basic and more nucleophilic than a carboxylate ion.

$$RO^- \quad > \quad R\overset{\displaystyle O}{\overset{\displaystyle \|}{C}}O^-$$

Stronger nucleophile   Weaker nucleophile
(stronger base; K_a of   (weaker base; K_a of
conjugate acid 10^{-16})   conjugate acid 10^{-5})

The connection between basicity and nucleophilicity holds when comparing atoms in the same row of the periodic table. On comparing two neutral molecules, we see that ammonia is more basic than water and is a better nucleophile. On comparing two anions, we see that hydroxide is more basic than fluoride and is more nucleophilic.

Basicity is not the sole determinant of nucleophilicity, however. This is well illustrated by comparing the nucleophilicities of halide anions. The order of halide nucleophilicity in methanol is the opposite of their basicities. Iodide, which is the least basic of the halide anions, is the strongest nucleophile; fluoride is the most basic of the halides but is the weakest nucleophile.

The factor that seems to be primarily responsible for the unusual order of nucleophilicity of the various halide anions is the different degree to which they are *solvated*. In solvents such as water and methanol, halide anions are solvated by hydrogen bonding, as illustrated in Figure 9.5. In order to react as a nucleophile, the halide must shed some of these solvent molecules. Among the halide anions, fluoride forms the strongest hydrogen bonds and iodide the weakest. Thus, the nucleophilicity of fluoride is suppressed by solvation more than that of chloride, chloride more than bromide, and bromide more than iodide.

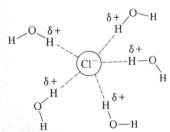

FIGURE 9.5 Solvation of a representative halide ion (chloride) by water. The negatively charged chloride ion interacts with the positively polarized hydrogens of water molecules to form hydrogen bonds.

In general, smaller anions are more highly solvated than larger ones, and nucleophilicity increases in going down the periodic table. Hydrogen sulfide anion (HS$^-$) is more nucleophilic than hydroxide ion (HO$^-$) even though it is less basic.

9.9 THE UNIMOLECULAR (S$_N$1) MECHANISM OF NUCLEOPHILIC SUBSTITUTION

Recalling from Section 9.7 that tertiary alkyl halides are practically inert to substitution by the S$_N$2 mechanism because of crowding in the transition state, we might wonder if tertiary alkyl halides undergo nucleophilic substitution at all. We shall see in this section that they do, but by a mechanism different from S$_N$2.

By studying the hydrolysis of *tert*-butyl bromide at very low concentrations of sodium hydroxide, Hughes and Ingold discovered that the reaction was characterized by a first-order rate law.

$$(CH_3)_3CBr \quad + \quad HO^- \quad \longrightarrow \quad (CH_3)_3COH \quad + \quad Br^-$$

| *tert*-Butyl bromide | Hydroxide ion | *tert*-Butyl alcohol | Bromide ion |

The rate of hydrolysis was first order in *tert*-butyl bromide but independent of the concentration of base.

$$\text{Rate} = k\,[(CH_3)_3CBr]$$

Just as second-order kinetics was interpreted as indicating a bimolecular rate-determining step, first-order kinetics was interpreted as evidence for a *unimolecular* rate-determining step.

Hughes and Ingold proposed a mechanism they called S$_N$1, standing for *substitution nucleophilic unimolecular* for this reaction. This mechanism is a two-step process in which the first step is rate-determining. It begins with unimolecular dissociation of the alkyl halide to form a carbocation intermediate.

$$(CH_3)_3C-Br \quad \overset{slow}{\rightleftharpoons} \quad (CH_3)_3C^+ \quad + \quad Br^-$$

| *tert*-Butyl bromide | *tert*-Butyl cation | Bromide ion |

Once generated, the carbocation is rapidly captured by hydroxide ion.

$$(CH_3)_3C^+ \quad + \quad {}^-OH \quad \overset{fast}{\longrightarrow} \quad (CH_3)_3C-OH$$

| *tert*-Butyl cation | Hydroxide ion | *tert*-Butyl alcohol |

In the absence of base, the rate of hydrolysis remains the same because the rate-determining step is the same. In this case the carbocation is captured by a water molecule.

$$(CH_3)_3C^+ \; + \; :\overset{\displaystyle H}{\underset{\displaystyle H}{O}} \quad \overset{fast}{\longrightarrow} \quad (CH_3)_3C-\overset{\displaystyle H}{\underset{\displaystyle H}{\overset{+}{O}}}: \quad \overset{fast}{\underset{-H^+}{\longrightarrow}} \quad (CH_3)_3C-OH$$

| *tert*-Butyl cation | Water | *tert*-Butyloxonium ion | *tert*-Butyl alcohol |

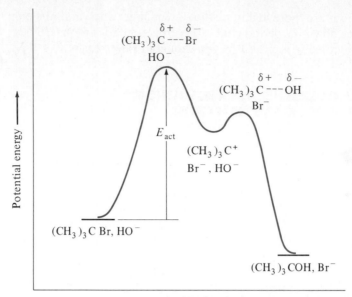

FIGURE 9.6 Energy diagram illustrating the S_N1 mechanism for hydrolysis of *tert*-butyl bromide.

An energy diagram representing hydrolysis of *tert*-butyl bromide by the S_N1 mechanism is shown in Figure 9.6. There are two transition states. The first involves stretching the carbon-halogen bond and leads to the formation of a carbocation intermediate. The second transition state is not as high in energy as the first and corresponds to capture of the carbocation by hydroxide.

The S_N1 mechanism is an ionization mechanism. The nucleophile does not participate until after the rate-determining step has taken place. Thus, the effects of nucleophile and substrate structure are expected to be different from those observed for reactions proceeding by the S_N2 pathway. How the structure of the substrate affects the rate of S_N1 reactions is the topic of the next section.

9.10 CARBOCATION STABILITY AND THE RATE OF SUBSTITUTION BY THE S_N1 MECHANISM

Tertiary alkyl halides are good candidates for reaction by the S_N1 mechanism because they are too sterically hindered to react by the S_N2 mechanism and since they form relatively stable carbocations, do not have prohibitively high activation energies for ionization. What about S_N1 reactions in other classes of alkyl halides?

In order to compare S_N1 substitution rates in a range of alkyl halides, experimental conditions are chosen in which competing substitution by the S_N2 route is very slow. One such set of conditions is solvolysis in aqueous formic acid (HCO_2H).

$$RX \quad + \ H_2O \xrightarrow{\text{formic acid}} ROH \ + \quad HX$$

Alkyl halide   Water   Alcohol   Hydrogen halide

Neither formic acid nor water is very nucleophilic, so S_N2 substitution is suppressed. The relative rates of hydrolysis of a group of alkyl bromides under these conditions are presented in Table 9.5.

TABLE 9.5

Reactivity of Some Alkyl Bromides toward Substitution by the S_N1 Mechanism*

Alkyl bromide	Structure	Class	Relative rate†
Methyl bromide	CH_3Br	Unsubstituted	1
Ethyl bromide	CH_3CH_2Br	Primary	2
Isopropyl bromide	$(CH_3)_2CHBr$	Secondary	43
tert-Butyl bromide	$(CH_3)_3CBr$	Tertiary	100,000,000

* Solvolysis in aqueous formic acid.

† Ratio of first-order rate constant k for indicated alkyl bromide to k for methyl bromide at 25°C.

The relative rate order in S_N1 reactions is exactly the opposite of that seen in S_N2 reactions.

S_N1 reactivity   methyl < primary < secondary < tertiary
S_N2 reactivity   tertiary < secondary < primary < methyl

Clearly, the steric crowding effects that influence reaction rates in S_N2 processes play no role in S_N1 reactions. The order of alkyl halide reactivity in S_N1 reactions is in the same direction as the order of carbocation stability: the more stable the carbocation, the faster an alkyl halide reacts by the S_N1 mechanism. We have seen this situation before in the reaction of alcohols with hydrogen halides (Section 4.16), in the acid-catalyzed dehydration of alcohols (Section 6.4), and in the conversion of alkyl halides to alkenes by the E1 mechanism (Section 6.10). As in these other reactions, an electronic effect, specifically the stabilization of the carbocation intermediate by alkyl substituents, is the most important factor in determining S_N1 reactivity.

PROBLEM 9.6   Identify the compound in each of the following pairs that reacts at the faster rate in an S_N1 reaction:

(a) Isopropyl bromide or isobutyl bromide
(b) Cyclopentyl iodide or 1-methylcyclopentyl iodide
(c) Cyclopentyl bromide or neopentyl bromide
(d) tert-Butyl chloride or tert-butyl iodide

SAMPLE SOLUTION   (a) Isopropyl bromide, $(CH_3)_2CHBr$, is a secondary alkyl halide, while isobutyl bromide, $(CH_3)_2CHCH_2Br$, is primary. Since the rate-determining step in an S_N1 reaction is carbocation formation and since secondary carbocations are more stable than primary carbocations, isopropyl bromide is more reactive in nucleophilic substitution by the S_N1 mechanism than is isobutyl bromide.

Primary carbocations are so high in energy that their intermediacy in nucleophilic substitution reactions is unlikely. When ethyl bromide undergoes hydrolysis in aqueous formic acid, substitution probably takes place by a direct displacement of bromide by water in an S_N2-like process.

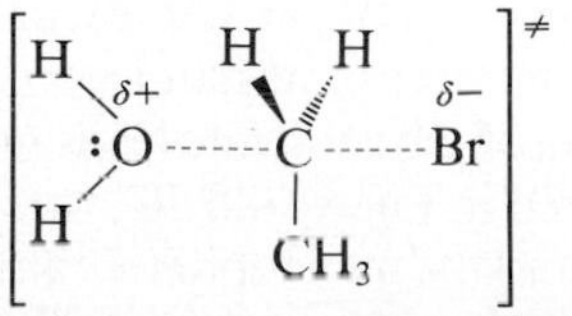

Representation of activated complex
for hydrolysis of ethyl bromide

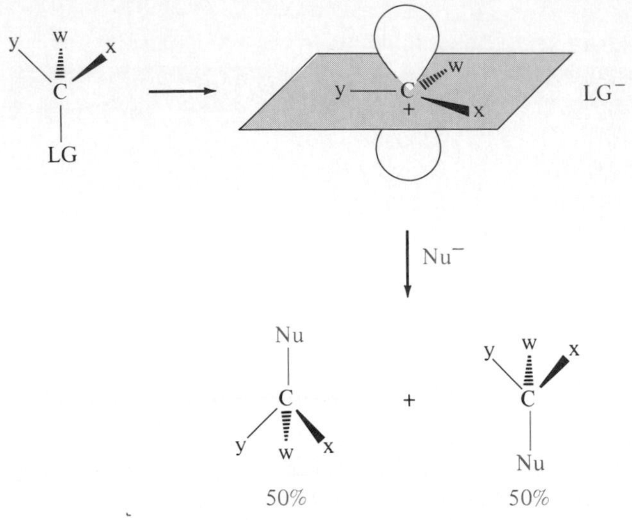

FIGURE 9.7 Formation of a racemic mixture by nucleophilic substitution via a carbocation intermediate.

9.11 STEREOCHEMISTRY OF S_N1 REACTIONS

While nucleophilic substitution reactions that exhibit second-order kinetics are stereospecific and proceed with inversion of configuration at carbon, the situation is somewhat less clear-cut for reactions that have a first-order kinetic dependence. When the leaving group is attached to the chiral center of an optically active halide, ionization leads to a carbocation intermediate that is achiral. It is achiral because the three bonds to the positively charged carbon lie in the same plane, and this plane is a plane of symmetry for the carbocation. As shown in Figure 9.7, a symmetrical carbocation should react with a nucleophile at the same rate at either of its two faces. We expect the product of substitution by the S_N1 mechanism to be formed as a racemic mixture and to be optically inactive. This outcome is rarely observed in practice. Normally, the product is formed with predominant, but not complete, inversion of configuration.

For example, the hydrolysis of 2-bromooctane in the absence of added base is a first-order reaction. When optically active substrate is used, the product 2-octanol is formed with 66 percent inversion of configuration.

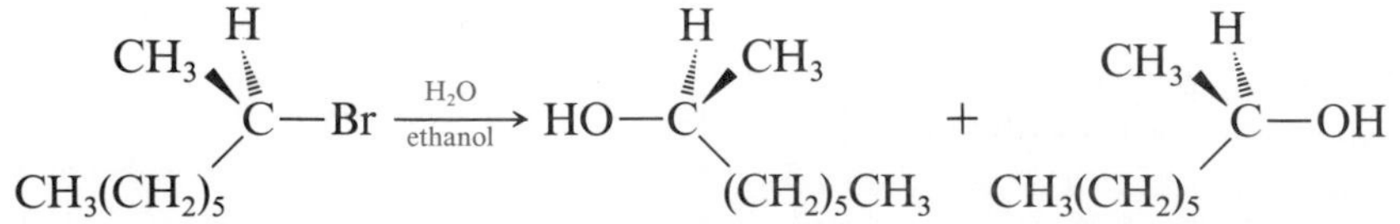

(R)-$(-)$-2-Bromooctane (S)-$(+)$-2-Octanol (R)-$(-)$-2-Octanol

66% net inversion corresponds
to 83% *S*, 17% *R*.

Partial but not complete loss of optical activity in S_N1 reactions is understood to mean that the carbocation is not free of its halide ion counterpart when it is attacked by the nucleophile. Ionization of the substrate leads not to free ions but to a carbocation-halide ion pair, as depicted in Figure 9.8. The anion of the leaving group shields one side of the carbocation, and the nucleophile captures the carbocation faster from the side opposite the leaving group. More product of inverted configuration is formed than product of retained configuration. In spite of the observation that the products of S_N1 reactions are only partially racemic, the fact that these reactions are not

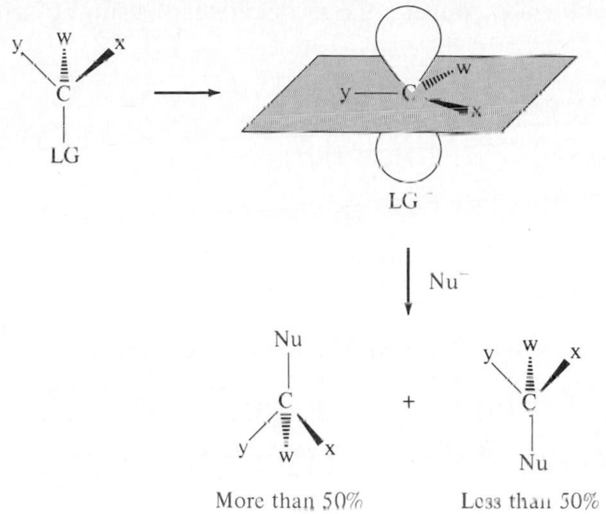

FIGURE 9.8 Inversion of configuration predominates in S$_N$1 reactions because one face of the carbocation is shielded by the leaving group.

More than 50%   Less than 50%

stereospecific is more in accord with the involvement of carbocation intermediates than with concerted displacement by way of a pentacoordinate transition state.

PROBLEM 9.7   What two stereoisomeric substitution products would you expect to isolate from the hydrolysis of cis-1,4-dimethylcyclohexyl bromide? From trans-1,4-dimethylcyclohexyl bromide?

9.12 CARBOCATION REARRANGEMENTS IN S$_N$1 REACTIONS

Further evidence for the intermediacy of carbocations in certain nucleophilic substitution reactions comes from the observation that the products of substitution must in some cases have arisen by rearrangement processes of the kind customarily associated with carbocation intermediates. For example, hydrolysis of the secondary alkyl bromide 2-bromo-3-methylbutane yields the rearranged tertiary alcohol 2-methyl-2-butanol as the exclusive product of substitution.

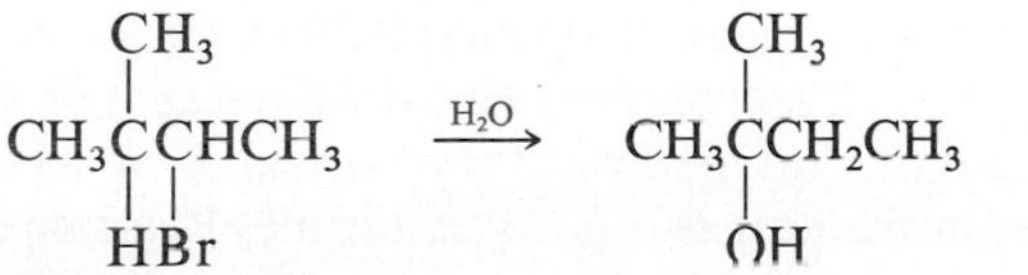

2-Bromo-3-methylbutane   2-Methyl-2-butanol (93%)

A reasonable mechanism which explains this observation assumes rate-determining ionization of the substrate as the first step.

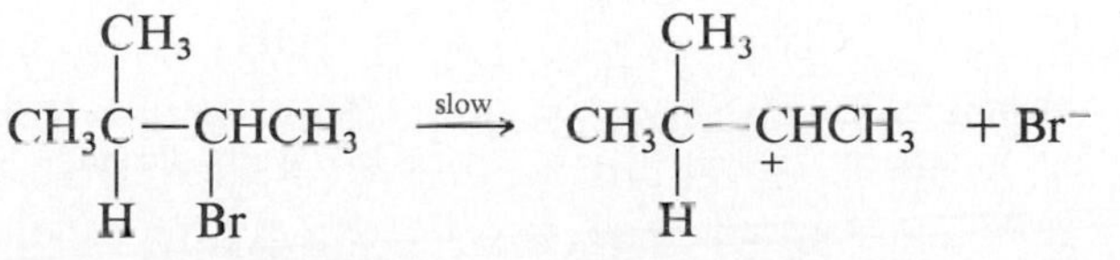

2-Bromo-3-methylbutane   1,2-Dimethylpropyl cation
                           (a secondary carbocation)

This is followed by a hydride shift which converts the secondary carbocation to a more stable tertiary one.

$$CH_3\underset{\underset{H}{|}}{\overset{\overset{CH_3}{|}}{C}}{-}\overset{+}{C}HCH_3 \quad \xrightarrow{\text{fast}} \quad CH_3\underset{\underset{H}{|}}{\overset{\overset{CH_3}{|}}{\overset{+}{C}}}CHCH_3$$

1,2-Dimethylpropyl cation   1,1-Dimethylpropyl cation
                                   (a tertiary carbocation)

The tertiary carbocation then reacts with water to yield the observed product.

$$CH_3\underset{+}{\overset{\overset{CH_3}{|}}{C}}CH_2CH_3 \xrightarrow[\text{fast}]{H_2O} CH_3\underset{\underset{+}{\overset{|}{O}H_2}}{\overset{\overset{CH_3}{|}}{C}}CH_2CH_3 \xrightarrow{\text{fast}} CH_3\underset{\underset{OH}{|}}{\overset{\overset{CH_3}{|}}{C}}CH_2CH_3$$

1,1-Dimethylpropyl cation   2-Methyl-2-butanol

PROBLEM 9.8   Why does the carbocation intermediate in the hydrolysis of 2-bromo-3-methylbutane rearrange by way of a hydride shift rather than a methyl shift?

The ionization of alkyl halides is accelerated by adding silver salts as catalysts. Silver ion is a weak Lewis acid and enhances the leaving group ability of halogen substituents by coordination to them.

$$R-\ddot{\underset{\cdot\cdot}{X}}\colon + Ag^+ \rightleftharpoons R-\overset{+}{\underset{\cdot\cdot}{X}}-Ag \longrightarrow R^+ + \colon\ddot{\underset{\cdot\cdot}{X}}-Ag$$

Alkyl   Silver ion   Carbocation   Silver
halide   halide

(Lewis base) (Lewis acid)   (Lewis acid–Lewis
                                 base complex)

This technique has been used to study solvolysis reactions of alkyl halides, such as neopentyl halides, that are characterized by a low level of reactivity in nucleophilic substitution reactions of all types. Neopentyl halides are too crowded near the reaction site to react by the S_N2 mechanism, and since they are primary, they show little tendency to react by the S_N1 pathway. Hydrolysis of neopentyl iodide has been observed, however, in the presence of silver nitrate. Rearrangement of the carbon skeleton occurs and 2-methyl-2-butanol is formed.

$$CH_3-\underset{\underset{CH_3}{|}}{\overset{\overset{CH_3}{|}}{C}}-CH_2I \xrightarrow[H_2O]{AgNO_3} CH_3\underset{\underset{OH}{|}}{\overset{\overset{CH_3}{|}}{C}}-CH_2CH_3$$

1-Iodo-2,2-dimethylpropane   2-Methyl-2-butanol (97%)
(neopentyl iodide)   (*tert*-pentyl alcohol)

The rearrangement step involves a methyl shift and is believed to occur in concert with loss of the leaving group. This leads to a tertiary carbocation intermediate instead of a far less stable primary one.

$$CH_3C \overset{\overset{\displaystyle CH_3}{|}}{\underset{\underset{\displaystyle CH_3}{|}}{}} CH_2 \overset{+}{\underset{}{}} I \text{—} Ag \longrightarrow CH_3 \overset{\overset{\displaystyle CH_3}{|}}{\underset{+}{C}} CH_2CH_3 \quad + AgI$$

Lewis acid–Lewis base
complex of neopentyl
iodide and silver ion

1,1-Dimethylpropyl cation
(*tert*-pentyl cation)

Rearrangements, when they do occur, are taken as evidence for carbocation intermediates and point to the S_N1 mechanism as the reaction pathway. Rearrangements are never observed in S_N2 reactions.

9.13 SOLVENT EFFECTS

Solvent effects are most apparent in reactions that proceed by the S_N1 mechanism. In these reactions a neutral alkyl halide molecule ionizes in the rate-determining step, forming a carbocation-halide ion pair. The activated complex for this step has polar character. A partial positive charge develops on carbon, and a partial negative charge on the halide leaving group as the reactant approaches the transition state. As shown in Figure 9.9, a polar solvent interacts more strongly with the polarized activated

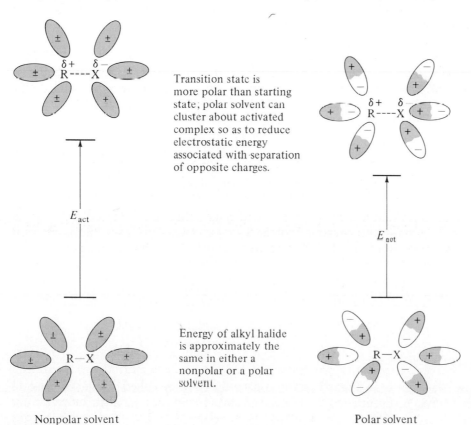

Transition state is more polar than starting state; polar solvent can cluster about activated complex so as to reduce electrostatic energy associated with separation of opposite charges.

Energy of alkyl halide is approximately the same in either a nonpolar or a polar solvent.

Nonpolar solvent

Polar solvent

FIGURE 9.9 A polar solvent stabilizes the transition state of an S_N1 reaction and increases its rate.

TABLE 9.6

Relative Rate of S_N1 Solvolysis of *tert*-Butyl Chloride as a Function of Solvent Polarity*

Solvent	Dielectric constant, ϵ	Relative rate
Acetic acid	6	1
Methanol	33	4
Formic acid	58	5,000
Water	78	150,000

* Ratio of first-order rate constant for solvolysis in indicated solvent compared with that for solvolysis in acetic acid at 25°C.

complex than with the starting alkyl halide. One way of expressing the polarity of a solvent is by its *dielectric constant* ϵ. The dielectric constant is a measure of the ability of a material to disperse the force of attraction between oppositely charged particles. Polar solvents—water, for example—have high dielectric constants and a pronounced ability to stabilize charge-separated systems. Nonpolar solvents—pentane, for example—have low dielectric constants and only a limited ability to moderate the forces of electrostatic attraction. Table 9.6 presents relative rate data for the S_N1 solvolysis of *tert*-butyl chloride in a variety of solvents arranged according to their dielectric constants. The more polar the solvent, the lower the activation energy for ionization and the faster the rate of reaction.

The effect of solvent polarity is not as dramatic in most S_N2 processes. When a reaction involves an anionic nucleophile, both the starting state ($RX + Y^-$) and the activated complex ($^{\delta-}Y\text{---}R\text{---}X^{\delta-})^{\neq}$ are negatively charged. The net change in the activation energy is small because the extent to which the transition state is stabilized by a polar solvent is offset by an equal or even larger stabilization of the reactants.

Water, alcohols, and carboxylic acids are classified as *protic solvents;* they all have hydroxyl groups that make them capable of participating in hydrogen bonds. These hydrogen bonds may be to the leaving group (Figure 9.10a) or to the nucleophile (Figure 9.10b). Hydrogen bonding to the leaving group lowers the activation energy for nucleophilic substitution by both the S_N1 and S_N2 mechanisms. Hydrogen bonding of the solvent to the nucleophile has no effect on the rate of an S_N1 reaction because the nucleophile does not become involved until after the rate-determining step. It decreases the rate of substitution by the S_N2 mechanism, because by stabilizing the nucleophile the solvent lowers the energy of the nucleophile and makes it less reactive.

A class of solvents called *polar aprotic* solvents exhibit large effects on the rate of nucleophilic substitution reactions of the S_N2 type. Two of the most commonly used polar aprotic solvents are dimethyl sulfoxide and *N,N*-dimethylformamide:

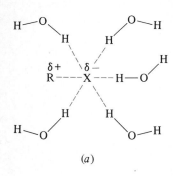

$\delta+$ $\delta-$
R----X---H—O

(a)

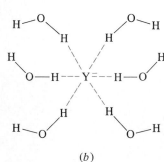

O—H---Y---H—O

(b)

FIGURE 9.10 (a) Hydrogen bonding of the solvent to the leaving group at the transition state lowers the activation energy for nucleophilic substitution by both the S_N1 and the S_N2 mechanism. (b) Hydrogen bonding of the solvent to the nucleophile stabilizes the nucleophile and makes it less reactive.

$$\overset{\displaystyle O}{\underset{\displaystyle \|}{CH_3\,S\,CH_3}}$$

Dimethyl sulfoxide (DMSO)

($\epsilon = 47$)

$$\overset{\displaystyle O}{\underset{\displaystyle \|}{(CH_3)_2NCH}}$$

N,N-Dimethylformamide (DMF)

($\epsilon = 37$)

These materials have large dielectric constants. They are called aprotic because all their hydrogen substituents are bonded to carbon and protons of this type do not form strong intermolecular hydrogen bonds. Both DMSO and DMF are quite good, however, at solvating metal ions by electrostatic interactions with their negatively polarized oxygens. Since the ability of DMSO and DMF to participate in hydrogen bonds is limited, both are relatively poor at solvating anions. Consequently, when a salt such as sodium cyanide is dissolved in a polar aprotic solvent, the metal cation is strongly solvated while the anion is not. An anion that is not encumbered by solvent molecules is a much more powerful nucleophile than a highly solvated one, and nucleophilic substitution reactions of the S_N2 type take place much faster in polar aprotic solvents than in protic ones. An example of this can be seen in the reaction of sodium cyanide with alkyl halides:

$$CH_3(CH_2)_4CH_2X + \quad NaCN \quad \longrightarrow \quad CH_3(CH_2)_4CH_2CN + \quad NaX$$

| Hexyl halide | Sodium cyanide | Hexyl cyanide | Sodium halide |

When the reaction was carried out in aqueous methanol as the solvent, hexyl bromide was converted to hexyl cyanide in 71 percent yield by heating with sodium cyanide. While this is a perfectly acceptable synthetic reaction, a period of over *20 h* was required. Changing the solvent to dimethyl sulfoxide brought about an increase in the reaction rate sufficient to allow the less reactive substrate hexyl chloride to be used instead, and was complete (91 percent yield) in only *20 min!*

The rate at which reactions occur can be important to the practice of organic chemistry in the laboratory, so understanding how solvents affect rate is of practical value in synthetic planning. As we proceed through the text, however, and see how nucleophilic substitution reactions are applied to a variety of functional group transformations, be aware that it is the nature of the substrate and the nucleophile that, more than anything else, determine what product is formed.

9.14 SUBSTITUTION AND ELIMINATION AS COMPETING REACTIONS

We have seen that an alkyl halide and a Lewis base can react together in either a substitution reaction or an elimination reaction.

Substitution can take place by the S_N1 or the S_N2 mechanism, elimination by E1 or E2.

How can we predict whether substitution or elimination will be the principal reaction observed with a particular combination of reactants? While many factors influence the competition between nucleophilic substitution and elimination, the two most important ones are the *structure of the alkyl halide* and the *basicity of the anion.*

First, recall the order of reactivity of alkyl halides in substitution and elimination reactions.

E1   tertiary > secondary > primary
E2   tertiary > secondary > primary
S_N1 tertiary > secondary > primary > methyl
S_N2 methyl > primary > secondary > tertiary

Methyl halides can only undergo substitution, and that only by the S_N2 mechanism.

The S_N2 process is very favorable for primary alkyl halides, while S_N1, E1, and E2 processes are very slow. Primary alkyl halides give predominantly the products of substitution, even with bases as strong as hydroxide and alkoxides. For example,

$$CH_3CH_2CH_2Br \xrightarrow[CH_3CH_2OH,\ 55°C]{NaOCH_2CH_3} CH_3CH_2CH_2OCH_2CH_3 + CH_3CH=CH_2$$

Propyl bromide   Ethyl propyl ether (91%)   Propene (9%)

Increasing steric hindrance at a primary site — as in isobutyl bromide, for example — decreases the rate of substitution and brings about an increase in the proportion of elimination.

$$\underset{\underset{CH_3}{|}}{CH_3CHCH_2Br} \xrightarrow[CH_3CH_2OH,\ 55°C]{NaOCH_2CH_3} \underset{\underset{CH_3}{|}}{CH_3CHCH_2OCH_2CH_3} + \underset{\underset{CH_3}{|}}{CH_3C=CH_2}$$

Isobutyl bromide   Ethyl isobutyl ether (40%)   2-Methylpropene (60%)

Crowding is more pronounced in secondary substrates, making the rate of substitution much slower. Secondary substrates also undergo E2 elimination faster than do primary ones. The result is that secondary alkyl halides yield mainly the product of elimination on reaction with alkoxide bases.

$$\underset{\underset{CH_3}{|}}{CH_3CHBr} \xrightarrow[CH_3CH_2OH,\ 55°C]{NaOCH_2CH_3} \underset{\underset{CH_3}{|}}{CH_3CHOCH_2CH_3} + CH_3CH=CH_2$$

Isopropyl bromide   Ethyl isopropyl ether (13%)   Propene (87%)

With tertiary alkyl halides, substitution by the only mechanism available (S_N1) is too slow to compete with the rapid rate of E2 elimination that takes place in the presence of strong bases.

$$\underset{\underset{CH_3}{|}}{\overset{\overset{CH_3}{|}}{CH_3CBr}} \xrightarrow[CH_3CH_2OH,\ 55°C]{NaOCH_2CH_3} \underset{\underset{CH_3}{|}}{\overset{\overset{CH_3}{|}}{CH_3COCH_2CH_3}} + CH_2=C\begin{smallmatrix}CH_3\\ \\CH_3\end{smallmatrix}$$

tert-Butyl bromide   *tert*-Butyl ethyl ether (7%)   2-Methylpropene (93%)

A good rule of thumb to take from the preceding equations is that *the principal reaction of secondary and tertiary alkyl halides with alkoxide bases is elimination rather than substitution.*

Only when the nucleophile is less basic than hydroxide will an anion react with a secondary halide by substitution. To illustrate, cyanide ion is about as nucleophilic as ethoxide according to Table 9.4 (Section 9.8). It is, however, much less basic. The conjugate acid HCN has a K_a of 7.2×10^{-10} (pK_a 9.1), while K_a for ethanol is about 10^{-16} (pK_a 16). Cyanide ion gives mainly the product of substitution with secondary alkyl halides.

$$CH_3CH(CH_2)_5CH_3 \xrightarrow[\text{DMSO}]{\text{KCN}} CH_3CH(CH_2)_5CH_3$$
$$\underset{\text{Cl}}{|} \qquad\qquad\qquad \underset{\text{CN}}{|}$$

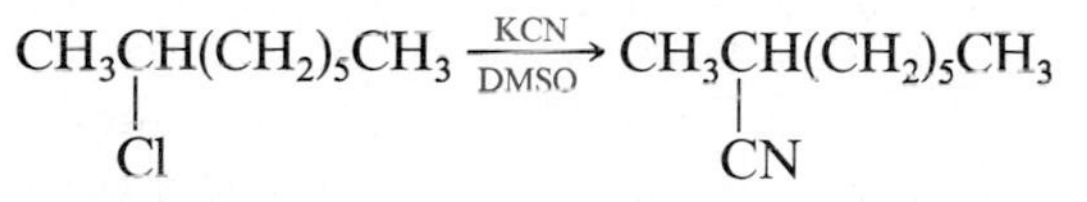

2-Chlorooctane   2-Cyanooctane (70%)

Azide ion ($:\overset{-}{N}=\overset{+}{N}=\overset{-}{N}:$) is a good nucleophile and not strongly basic; the K_a of its conjugate acid HN$_3$ is 2.6×10^{-5} (p$K_a = 4.6$). It reacts with secondary alkyl halides primarily by substitution.

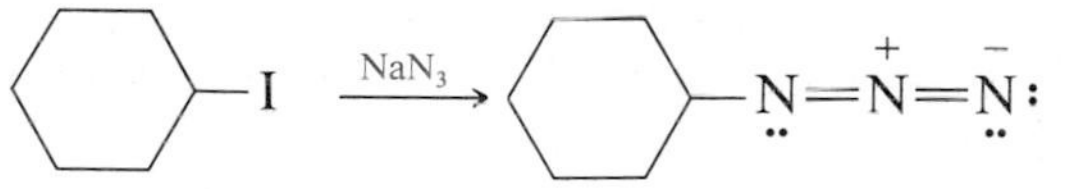

Cyclohexyl iodide   Cyclohexyl azide (75%)

PROBLEM 9.9   Predict the principal organic product of each of the following reactions:

(a) Cyclohexyl bromide and potassium ethoxide
(b) Ethyl bromide and potassium cyclohexanolate
(c) sec-Butyl bromide solvolysis in methanol
(d) sec-Butyl bromide solvolysis in methanol containing 2 M sodium methoxide

SAMPLE SOLUTION   (a) Cyclohexyl bromide is a secondary halide and reacts with alkoxide bases by elimination rather than substitution. The principal organic products are cyclohexene and ethanol.

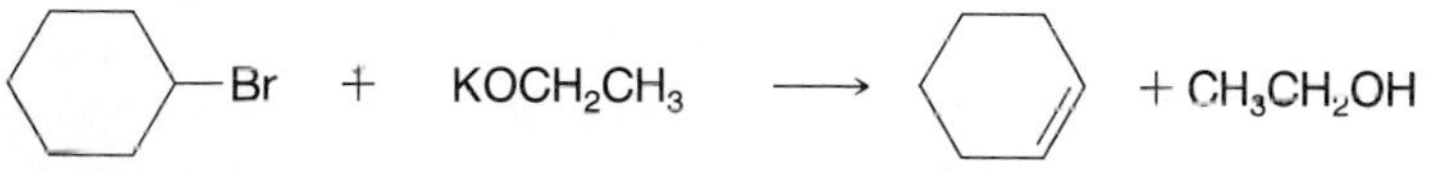

Cyclohexyl bromide   Potassium ethoxide   Cyclohexene   Ethanol

---

Hydrogen sulfide is a stronger acid than water; its K_a is 10^{-7} (pK_a 7.0), compared with a K_a of 1.8×10^{-16} (pK_a 15.7) for water. Thus, hydrogen sulfide ion HS$^-$, as well as anions of the type RS$^-$, is substantially less basic than hydroxide ion and reacts with both primary and secondary alkyl halides to yield mainly substitution products.

Tertiary alkyl halides are so sterically hindered to nucleophilic attack and so prone to elimination that the presence of any anionic Lewis base leads to elimination as the major reaction path. Usually substitution predominates over elimination in tertiary alkyl halides only under solvolysis conditions, in which the concentration of anionic Lewis bases is small. In the solvolysis of the tertiary bromide 2-bromo-2-methylbutane, for example, the ratio of substitution to elimination is 64 : 36 in pure ethanol but falls to 1 : 99 in the presence of 2 M sodium ethoxide.

$$\underset{\substack{\text{2-Bromo-2-methyl-}\\\text{butane}}}{\underset{\overset{|}{\text{Br}}}{\overset{\overset{\text{CH}_3}{|}}{\text{CH}_3\text{CCH}_2\text{CH}_3}}} \xrightarrow[\text{25°C}]{\text{ethanol}} \underset{\substack{\text{2-Ethoxy-2-}\\\text{methylbutane}}}{\underset{\overset{|}{\text{OCH}_2\text{CH}_3}}{\overset{\overset{\text{CH}_3}{|}}{\text{CH}_3\text{CCH}_2\text{CH}_3}}} + \underset{\text{2-Methyl-2-butene}}{(\text{CH}_3)_2\text{C}=\text{CHCH}_3} + \underset{\text{2-Methyl-1-butene}}{\underset{}{\overset{\overset{\text{CH}_3}{|}}{\text{CH}_2=\text{CCH}_2\text{CH}_3}}}$$

(Major product in absence of sodium ethoxide)

(Alkene mixture is major product in presence of sodium ethoxide)

Regardless of the substrate, an increase in temperature causes both the rate of substitution and the rate of elimination to increase. The rate of elimination, however, increases faster than the rate of substitution, so the proportion of elimination products increases at higher temperatures at the expense of substitution products.

As a practical matter, elimination can always be made to occur quantitatively. Strong bases, especially bulky ones such as *tert*-butoxide ion, react even with primary alkyl halides by an E2 process at elevated temperatures. The more difficult task is to find the set of conditions most conducive to substitution. In general, the best approach is to choose conditions that favor the S_N2 mechanism—an unhindered substrate, a good nucleophile that is not strongly basic, and the lowest practical temperature consistent with reasonable reaction rates.

Functional group transformations that rely on substitution by the S_N1 mechanism are not as generally applicable as those of the S_N2 type. Hindered substrates are prone to elimination by dehydrohalogenation, and the possibility of rearrangement reactions arises when carbocation intermediates are involved. Only in cases in which elimination is impossible are S_N1 reactions employed in functional group transformations of alkyl halides.

9.15 SULFONATE ESTERS AS SUBSTRATES IN NUCLEOPHILIC SUBSTITUTION REACTIONS

Two kinds of substrates have been examined in nucleophilic substitution reactions to this point. In Chapter 4 we saw how alcohols can be converted to alkyl halides by reaction with hydrogen halides, and it was pointed out that this process is a nucleophilic substitution reaction taking place on the protonated form of the alcohol, with water serving as the leaving group. In the present chapter the substrates have been alkyl halides and halide ions have been the leaving groups. There are a few other classes of organic compounds that undergo nucleophilic substitution reactions analogous to those of alkyl halides, the most important of these being alkyl esters of sulfonic acids.

Sulfonic acids are strong acids, comparable in acidity with sulfuric acid. Representative examples are methanesulfonic acid and *p*-toluenesulfonic acid.

$$\underset{\text{Methanesulfonic acid}}{\text{CH}_3-\overset{\overset{\displaystyle O}{\|}}{\underset{\underset{\displaystyle O}{\|}}{S}}-\text{OH}}$$

$$\underset{\text{\textit{p}-Toluenesulfonic acid}}{\text{CH}_3-\!\!\left\langle\!\!\bigcirc\!\!\right\rangle\!\!-\overset{\overset{\displaystyle O}{\|}}{\underset{\underset{\displaystyle O}{\|}}{S}}-\text{OH}}$$

Alkyl sulfonate esters are derivatives of sulfonic acids in which the proton of the hydroxyl group is replaced by an alkyl group. They are prepared by treating an alcohol with the appropriate sulfonyl chloride.

$$\underset{\substack{| \\ \text{H} \\ \text{Alcohol}}}{\text{R}\ddot{\text{O}}:} \;+\; \underset{\substack{|| \\ \text{O} \\ \text{Sulfonyl chloride}}}{\text{R}'\overset{\text{O}}{\overset{||}{\text{S}}}\text{Cl}} \longrightarrow \underset{\substack{|| \\ \text{O} \\ \text{Sulfonate ester}}}{\text{R}\ddot{\text{O}}\text{—}\overset{\text{O}}{\overset{||}{\text{S}}}\text{R}'} \;+\; \underset{\text{Hydrogen chloride}}{\text{HCl}}$$

These reactions are usually carried out in the presence of pyridine.

$$\underset{\text{Ethanol}}{\text{CH}_3\text{CH}_2\text{OH}} \;+\; \underset{\substack{p\text{-Toluenesulfonyl} \\ \text{chloride}}}{\text{CH}_3\text{—}\langle\!\!\!\rangle\!\!\!\text{—}\overset{\text{O}}{\underset{\text{O}}{\overset{||}{\underset{||}{\text{S}}}}}\text{Cl}} \xrightarrow{\text{pyridine}} \underset{\substack{\text{Ethyl } p\text{-toluenesulfonate} \\ (72\%)}}{\text{CH}_3\text{CH}_2\text{O}\;\overset{\text{O}}{\underset{\text{O}}{\overset{||}{\underset{||}{\text{S}}}}}\text{—}\langle\!\!\!\rangle\!\!\!\text{—}\text{CH}_3}$$

Alkyl sulfonate esters resemble alkyl halides in their ability to serve as substrates in elimination reactions and in nucleophilic substitution reactions.

$$\underset{\text{Nucleophile}}{\text{Y}:^-} \;+\; \underset{\substack{p\text{-Toluenesulfonate} \\ \text{ester}}}{\text{R}\text{—}\text{O}\overset{\text{O}}{\underset{\text{O}}{\overset{||}{\underset{||}{\text{S}}}}}\text{—}\langle\!\!\!\rangle\!\!\!\text{—}\text{CH}_3} \longrightarrow \underset{\substack{\text{Product of} \\ \text{nucleophilic} \\ \text{substitution}}}{\text{R}\text{—}\text{Y}} \;+\; \underset{\substack{p\text{-Toluenesulfonate} \\ \text{anion}}}{^-\text{O}\overset{\text{O}}{\underset{\text{O}}{\overset{||}{\underset{||}{\text{S}}}}}\text{—}\langle\!\!\!\rangle\!\!\!\text{—}\text{CH}_3}$$

The sulfonate esters used most frequently in nucleophilic substitution reactions are the *p*-toluenesulfonates. They are commonly known as *tosylates* and abbreviated ROTs.

$$\underset{\substack{\text{(3-Cyclopentenyl)methyl} \\ p\text{-toluenesulfonate}}}{\langle\!\!\!\!\!\bigcirc\!\!\!\!\!\rangle\!\!\!\!\text{—}\overset{\text{H}}{\underset{\text{CH}_2\text{OTs}}{}}} \xrightarrow[\text{ethanol-water}]{\text{KCN}} \underset{\substack{\text{4-(Cyanomethyl)cyclo-} \\ \text{pentene (86\%)}}}{\langle\!\!\!\!\!\bigcirc\!\!\!\!\!\rangle\!\!\!\!\text{—}\overset{\text{H}}{\underset{\text{CH}_2\text{CN}}{}}}$$

p-Toluenesulfonate is a very good leaving group. Alkyl *p*-toluenesulfonates react at rates comparable to those of alkyl iodides in nucleophilic substitution reactions. *p*-Toluenesulfonate can be displaced by chloride or bromide in homogeneous solution or by sodium iodide in acetone.

$$\underset{\substack{\text{sec-Butyl} \\ p\text{-toluenesulfonate}}}{\underset{\substack{| \\ \text{OTs}}}{\text{CH}_3\text{CHCH}_2\text{CH}_3}} + \underset{\substack{\text{Sodium} \\ \text{bromide}}}{\text{NaBr}} \xrightarrow{\text{DMSO}} \underset{\substack{\text{sec-Butyl} \\ \text{bromide (82\%)}}}{\underset{\substack{| \\ \text{Br}}}{\text{CH}_3\text{CHCH}_2\text{CH}_3}} + \underset{\substack{\text{Sodium} \\ p\text{-toluenesulfonate}}}{\text{NaOTs}}$$

PROBLEM 9.10   Write a chemical equation showing the preparation of octadecyl *p*-toluenesulfonate.

PROBLEM 9.11   Write equations showing the reaction of octadecyl *p*-toluenesulfonate with each of the following reagents:

(a)   Potassium acetate ($KO\overset{\displaystyle O}{\overset{\|}{C}}CH_3$)

(b)   Potassium iodide (KI)

(c)   Potassium cyanide (KCN)

(d)   Potassium hydrogen sulfide (KSH)

(e)   Sodium butanethiolate ($NaSCH_2CH_2CH_2CH_3$)

SAMPLE SOLUTION   All these reactions of octadecyl *p*-toluenesulfonate have been reported in the chemical literature, and all proceed in synthetically useful yield. You should begin by identifying the nucleophile in each of the parts to this problem. The nucleophile replaces the *p*-toluenesulfonate leaving group in an S$_N$2 reaction. In (a) the nucleophile is acetate ion, and the product of nucleophilic substitution is 1-octadecyl acetate.

$$CH_3\overset{\displaystyle O}{\overset{\|}{C}}O^- \ + \ \underset{\displaystyle (CH_2)_{16}CH_3}{CH_2-OTs} \longrightarrow CH_3\overset{\displaystyle O}{\overset{\|}{C}}OCH_2(CH_2)_{16}CH_3$$

Acetate ion   Octadecyl tosylate   Octadecyl acetate

Sulfonate esters are subject to the same limitations as alkyl halides in their use as substrates in nucleophilic substitution reactions. Competition from elimination reactions needs to be taken into consideration when planning a functional group transformation that requires an anionic nucleophile because tosylates undergo elimination reactions, just as alkyl halides do.

An advantage that sulfonate esters have over alkyl halides is that their preparation from alcohols does not involve any of the bonds to carbon. The alcohol oxygen becomes the oxygen that connects the alkyl group to the sulfonyl group. Thus, the configuration of a sulfonate ester is exactly the same as that of the alcohol from which it was prepared. If we wish to study the stereochemistry of nucleophilic substitution reactions in an optically active substrate, for example, we know that a tosylate ester will have the same configuration and the same optical purity as the chiral alcohol from which it was prepared.

(S)-(+)-2-Octanol
$[\alpha]_D^{25}+9.9°$
(optically pure)

(S)-(+)-1-Methylheptyl *p*-toluenesulfonate
$[\alpha]_D^{25}+7.9°$
(optically pure)

The same cannot be said about reactions with alkyl halides as substrates. The conversion of optically active 2-octanol to the corresponding halide *does* involve a bond to the chiral center, so the optical purity and absolute configuration of the alkyl halide need to be independently established.

The mechanisms by which sulfonate esters undergo nucleophilic substitution reactions are the same as those undergone by alkyl halides. Inversion of configuration is observed in S_N2 reactions of alkyl sulfonates and predominant inversion accompanied by racemization in S_N1 processes.

PROBLEM 9.12   The hydrolysis of sulfonate esters of 2-octanol is a stereospecific reaction and proceeds with complete inversion of configuration. Write a structural formula that shows the stereochemistry of the 2-octanol formed by hydrolysis of an optically pure sample of (S)-(+)-1-methylheptyl p-toluenesulfonate, identify the product as R or S, and deduce its specific rotation.

9.16 THE S_N1–S_N2 CONTINUUM

The view of S_N1 and S_N2 as alternative and competing mechanisms of nucleophilic substitution continues to serve organic chemists well as a means of understanding extremes of chemical behavior. In the vast range between methyl halides and *tert*-alkyl halides, it is not clear that it is correct to conclude that a fraction of the substrate reacts by the S_N1 mechanism and the rest by an S_N2 mechanism. One alternative suggestion is that there exists a continuum of levels of nucleophilic participation between the extremes associated with the S_N1 and S_N2 transition states. According to this picture, methyl halides require significant nucleophilic participation to assist the departure of the leaving group. With an increasing number of alkyl substituents, nucleophilic participation becomes more difficult for steric reasons and less necessary for electronic reasons. Consequently, interaction of the nucleophile with the carbon bearing the leaving group becomes weaker as one proceeds from methyl to primary to secondary alkyl halides and becomes quite weak with tertiary alkyl halides.

It has become increasingly apparent that *ion pairing* needs to be explicitly considered in any detailed mechanistic proposal. A carbocation and its leaving group can be associated in several different ways, as illustrated in Figure 9.11. These ions can be directly adjacent (an *intimate* ion pair), a molecule of solvent can be interposed between them (a *solvent-separated* ion pair) or they can each be heavily solvated (*dissociated* ions). A nucleophile can attack the substrate prior to its ionization, at the

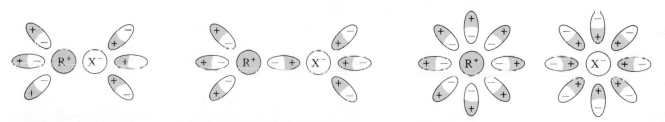

(a) Intimate ion pair   (b) Solvent-separated ion pair   (c) Solvated dissociated ions

FIGURE 9.11   An intimate ion pair (a) is formed on ionization of an alkyl halide to a carbocation and a halide anion. A solvent molecule attacks this intimate ion pair, inserting itself between the carbocation and the halide anion to give a solvent-separated ion pair (b). Further solvation of the carbocation and the anion leads to dissociated ions (c), each of which is surrounded by several molecules of the solvent.

intimate ion pair stage, at the solvent-separated ion pair stage, or after the ions have diffused apart and become completely solvated. The relative importance of the various intermediates depends on a number of factors, including the alkyl halide, the nucleophile, and the solvent.

The study of nucleophilic substitution reactions has resulted in numerous contributions to fundamental principles of organic chemistry. While research continues in this area, the $S_N1 - S_N2$ classification, although probably deficient in detail, remains a useful concept to facilitate understanding of the interplay between diverse variations in substrate, nucleophile, leaving group, and solvent.

9.17 SUMMARY

Nucleophilic substitution reactions play a prominent role in functional group transformations. Among the synthetically useful processes accomplished by nucleophilic substitution are:

1. Preparation of alkyl fluorides

$$RX + F^- \xrightarrow{\text{heat in high-boiling solvent}} RF + X^-$$

2. Preparation of alkyl iodides

$$RX + I^- \xrightarrow{\text{acetone}} RI + X^-$$

3. Preparation of alkyl cyanides

$$RX + CN^- \longrightarrow RCN + X^-$$

4. Preparation of ethers

$$RX + R'O^- \longrightarrow ROR' + X^-$$

5. Preparation of alkyl azides

$$RX + N_3^- \longrightarrow RN_3 + X^-$$

6. Preparation of thiols and thioethers

$$RX + R'S^- \longrightarrow RSR' + X^-$$

The preparation of alcohols from alkyl halides by nucleophilic substitution is feasible

$$RX + HO^- \longrightarrow ROH + X^-$$

but not often used because alkyl halides are typically prepared from alcohols (Chapter 4) rather than vice versa.

In all these just mentioned reactions the order of substrate reactivity is

$$CH_3 > \text{primary} > \text{secondary} > \text{tertiary}$$

and the order of leaving group effectiveness is

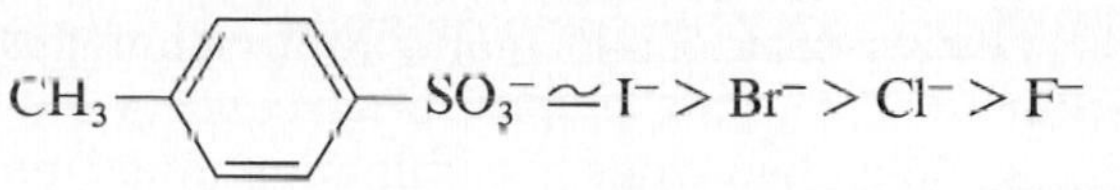

$$CH_3 \langle \text{benzene ring} \rangle SO_3^- \simeq I^- > Br^- > Cl^- > F^-$$

Nucleophilic substitution reactions can be described by two mechanisms:

1. S_N2 *(substitution nucleophilic bimolecular)*
2. S_N1 *(substitution nucleophilic unimolecular)*

The S_N2 mechanism is concerted. It is characterized by second-order kinetics, implying a bimolecular rate-determining step.

$$\text{Rate} = k \text{ [substrate][nucleophile]}$$

The nucleophile attacks carbon from the side opposite the bond to the leaving group, which leads to a pentacoordinate activated complex.

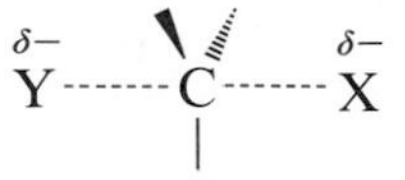

$$\overset{\delta-}{Y} \text{------} C \text{------} \overset{\delta-}{X}$$

Substitution by the S_N2 mechanism is stereospecific and takes place with *inversion of configuration*. The rate of S_N2 reactions is sensitive to steric effects. The less crowded the carbon bearing the leaving group, the faster it reacts with good nucleophiles. The S_N2 order is

$$CH_3 > \text{primary} > \text{secondary} > \text{tertiary}$$

The S_N1 mechanism is a two-step process in which the rate-determining step is carbocation formation.

$$RX \xrightarrow{\text{slow}} R^+ + X^-$$

$$R^+ + Y^- \xrightarrow{\text{fast}} RY$$

The rate of the reaction is dictated by carbocation stability. The more stable the carbocation, the faster RX will react by the S_N1 reaction. Therefore, the order of substrate reactivity in an S_N1 reaction is

$$\text{Tertiary} > \text{secondary} > \text{primary} > CH_3$$

The rate of an S_N1 reaction is independent of the concentration of the nucleophile and independent of the nature of the nucleophile.

$$\text{Rate} = k[RX]$$

The nucleophile plays no role in an S_N1 reaction until after the rate-determining step has taken place. The presence of a carbocation intermediate suggests that optically active alkyl halides should yield racemic products by this mechanism. Racemization, however, is usually incomplete and partial net inversion of configuration is normally observed.

When an alkyl halide is capable of undergoing elimination, this can compete with nucleophilic substitution. The normal reaction of a secondary alkyl halide with a base as strong as or stronger than hydroxide is elimination. Substitution predominates with primary alkyl halides. Secondary alkyl halides undergo substitution effectively with bases weaker than hydroxide. Elimination predominates when tertiary halides react with any anion. Solvolysis reactions in the absence of added anions cause nucleophilic substitution to be the major reaction of tertiary alkyl halides.

PROBLEMS

9.13   Write the structure of the principal organic product to be expected from the reaction of 1-bromopropane with each of the following.

    (a)   Sodium iodide in acetone

    (b)   Sodium acetate ($CH_3\overset{\overset{\displaystyle O}{\|}}{C}ONa$) in acetic acid
    (c)   Sodium ethoxide in ethanol
    (d)   Sodium cyanide in dimethyl sulfoxide
    (e)   Sodium azide in aqueous ethanol
    (f)   Sodium hydrogen sulfide in ethanol
    (g)   Sodium methanethiolate ($NaSCH_3$) in ethanol

9.14   Repeat the preceding problem for 2-bromopropane as the substrate.

9.15   Each of the reagents in Problem 9.13 converts 2-bromo-2-methylpropane to the same product. What is this product?

9.16   Each of the following nucleophilic substitution reactions has been reported in the chemical literature. Many of them involve reactants that are somewhat more complex than those we have dealt with to this point. Nevertheless, you should be able to predict the product based on analogy to what you know about nucleophilic substitution in simple systems.

    (a) $BrCH_2\overset{\overset{\displaystyle O}{\|}}{C}OCH_2CH_3 \xrightarrow[\text{acetone}]{\text{NaI}}$

    (b) $O_2N-\!\!\left\langle\!\!\bigcirc\!\!\right\rangle\!\!-CH_2Cl \xrightarrow[\text{acetone}]{\text{NaI}}$

    (c) $O_2N-\!\!\left\langle\!\!\bigcirc\!\!\right\rangle\!\!-CH_2Cl \xrightarrow[\text{acetic acid}]{CH_3\overset{\overset{\displaystyle O}{\|}}{C}ONa}$

    (d) $CH_3CH_2OCH_2CH_2Br \xrightarrow[\text{ethanol-water}]{\text{NaCN}}$

    (e) $NC-\!\!\left\langle\!\!\bigcirc\!\!\right\rangle\!\!-CH_2Cl \xrightarrow{H_2O,\ HO^-}$

    (f) $ClCH_2\overset{\overset{\displaystyle O}{\|}}{C}OC(CH_3)_3 \xrightarrow[\text{acetone-water}]{\text{NaN}_3}$

(g) 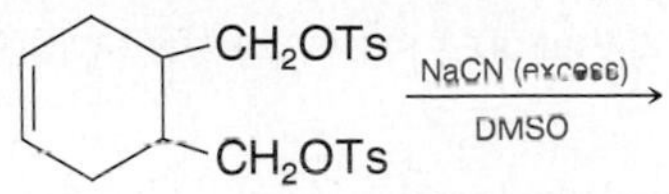 $\xrightarrow[\text{DMSO}]{\text{NaCN (excess)}}$

(h) 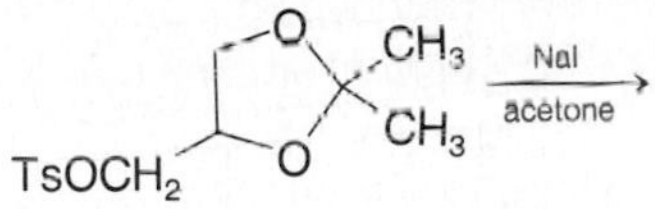 $\xrightarrow[\text{acetone}]{\text{NaI}}$

(i) 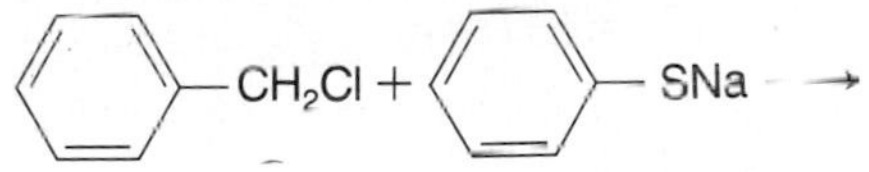 $-CH_2Cl +$ $- SNa \longrightarrow$

(j) 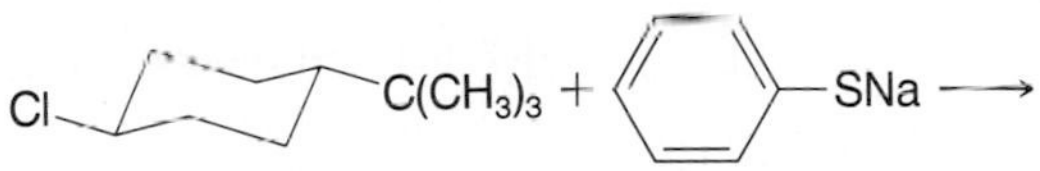 Cl $C(CH_3)_3 +$ $-SNa \longrightarrow$

(k) 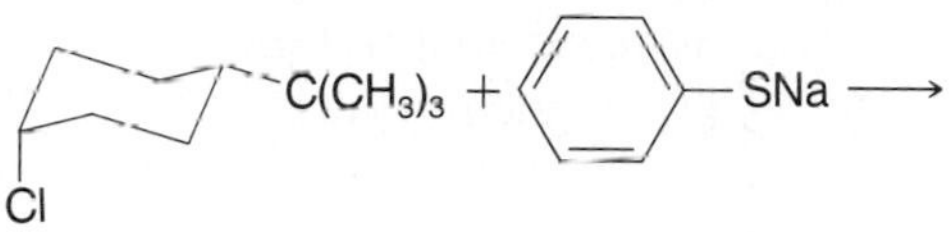 $C(CH_3)_3 +$ $-SNa \longrightarrow$

(l) 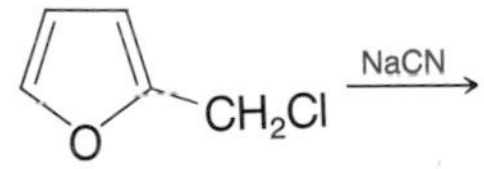 $CH_2Cl \xrightarrow{\text{NaCN}}$

(m) 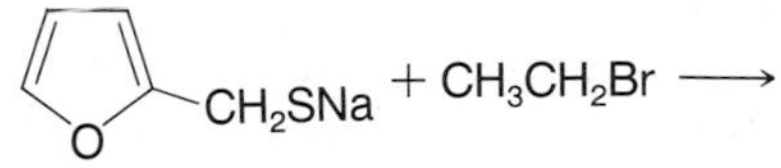 $CH_2SNa + CH_3CH_2Br \longrightarrow$

(n) 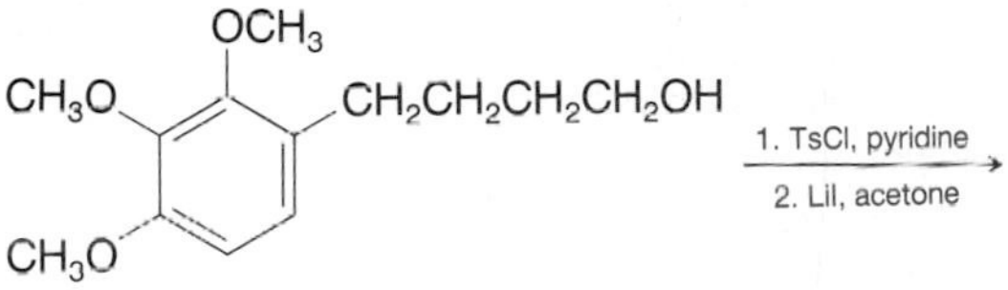 $\xrightarrow[\text{2. LiI, acetone}]{\text{1. TsCl, pyridine}}$

9.17 Arrange the isomers of molecular formula C_4H_9Cl in order of decreasing rate of reaction with sodium iodide in acetone.

9.18 In each of the following indicate which reaction will occur faster. Explain your reasoning.

(a) $CH_3CH_2CH_2CH_2Br$ or $CH_3CH_2CH_2CH_2I$ with sodium cyanide in dimethyl sulfoxide
(b) 1-Chloro-2-methylbutane or 1-chloropentane with sodium iodide in acetone
(c) *n*-Hexyl chloride or cyclohexyl chloride with sodium azide in aqueous ethanol
(d) Solvolysis of neopentyl bromide or *tert*-butyl bromide in ethanol
(e) Solvolysis of isobutyl bromide or *sec*-butyl bromide in aqueous formic acid
(f) Reaction of 1-chlorobutane with sodium acetate in acetic acid or with sodium methoxide in methanol
(g) Reaction of 1-chlorobutane with sodium azide or sodium *p*-toluenesulfonate in aqueous ethanol

9.19 Identify the product in each of the following reactions.

(a) $ClCH_2CH_2CHCH_2CH_3 \xrightarrow[\text{acetone}]{\text{NaI (1.0 equiv)}} C_5H_{10}ClI$
     $\quad\quad\quad\quad\quad |$
     $\quad\quad\quad\quad\quad Cl$
(b) $BrCH_2CH_2Br + NaSCH_2CH_2SNa \longrightarrow C_4H_8S_2$
(c) $ClCH_2CH_2CH_2CH_2Cl + Na_2S \longrightarrow C_4H_8S$

9.20 Give the mechanistic symbols (S_N1, S_N2, E1, E2) that are most consistent with each of the following statements.

(a) Methyl halides react with sodium ethoxide in ethanol only by this mechanism.

(b) Unhindered primary halides react with sodium ethoxide in ethanol mainly by this mechanism.

(c) When cyclohexyl bromide is treated with sodium ethoxide in ethanol, the major product is formed by this mechanism.

(d) The principal substitution product obtained by solvolysis of *tert*-butyl bromide in ethanol arises by this mechanism.

(e) In ethanol that contains sodium ethoxide, *tert*-butyl bromide reacts mainly by this mechanism.

(f) These reaction mechanisms represent concerted processes.

(g) Reactions proceeding by these mechanisms are stereospecific.

(h) These reaction mechanisms involve carbocation intermediates

(i) These reaction mechanisms are the ones most likely to have been involved when the products are found to have a different carbon skeleton than the substrate.

(j) Alkyl iodides react faster than alkyl bromides in reactions that proceed by these mechanisms.

9.21 Outline an efficient synthesis of each of the compounds shown below from the indicated starting material and any necessary organic or inorganic reagents.

(a) Ethyl fluoride from ethyl alcohol

(b) Cyclopentyl cyanide from cyclopentane

(c) Cyclopentyl cyanide from cyclopentene

(d) Cyclopentyl cyanide from cyclopentanol

(e) $NCCH_2CH_2CN$ from ethyl alcohol

(f) Isobutyl iodide from isobutyl chloride

(g) Isobutyl iodide from *tert*-butyl chloride

(h) Isopropyl azide from isopropyl alcohol

(i) Isopropyl azide from 1-propanol

(j) (S)-*sec*-butyl azide from (R)-*sec*-butyl alcohol

(k) (S)-$CH_3CH_2CHCH_3$ from (R)-*sec*-butyl alcohol
       |
      SH

9.22 Select the combination of alkyl bromide and potassium alkoxide that would be the most effective in the syntheses of the following ethers.

(a) $CH_3OC(CH_3)_3$

(b) ⬠—OCH_3

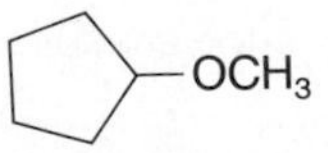

(c) $(CH_3)_3CCH_2OCH_2CH_3$

(d) (R)-$CH_3CH_2CHCH_2OCH(CH_3)_2$
              |
             CH_3

9.23 (a) Suggest a reasonable series of synthetic transformations for converting *trans*-2-methylcyclopentanol to *cis*-2-methylcyclopentyl acetate.

cis-2-Methylcyclopentyl acetate

(b) How could you prepare cis-2-methylcyclopentyl acetate from 1-methylcyclopentanol?

9.24 In one of the classic experiments of organic chemistry, Hughes studied the rate of racemization of 2-iodooctane by sodium iodide in acetone and compared it with the rate of incorporation of radioactive iodine into 2-iodooctane.

$$RI + [I^*]^- \longrightarrow RI^* + I^-$$

(I^* = radioactive iodine)

How will the rate of racemization compare with the rate of incorporation of radioactivity if

(a) Each act of exchange proceeds stereospecifically with retention of configuration?
(b) Each act of exchange proceeds stereospecifically with inversion of configuration?
(c) Each act of exchange proceeds in a stereorandom manner, i.e., retention and inversion of configuration are equally likely?

9.25 Solvolysis of 2-bromo-2-methylbutane (tert-pentyl bromide) in acetic acid containing potassium acetate gave three products. Identify them.

9.26 The ratio of elimination to substitution is exactly the same (26 percent elimination) for 2-bromo-2-methylbutane and 2-iodo-2-methylbutane in 80% ethanol/20% water at 25°C.

(a) By what mechanism does substitution most likely occur in these compounds under these conditions?
(b) By what mechanism does elimination most likely occur in these compounds under these conditions?
(c) Which substrate undergoes substitution faster?
(d) Which substrate undergoes elimination faster?
(e) What two substitution products are formed from each substrate?
(f) What two elimination products are formed from each substrate?
(g) Why do you suppose the ratio of elimination to substitution is the same for the two substrates?

9.27 Solvolysis of 1,2-dimethylpropyl p-toluenesulfonate in acetic acid (75°C) yields five different products. Three are alkenes and two are substitution products. Suggest reasonable structures for these five products.

9.28 Solution A was prepared by dissolving potassium acetate in methanol. Solution B was prepared by adding potassium methoxide to acetic acid. Reaction of methyl iodide either with solution A or with solution B gave the same major product. Why? What was this product?

9.29 If the temperature is not kept below 25°C during the reaction of primary alcohols with p-toluenesulfonyl chloride in pyridine, it is sometimes observed that the isolated product is not the desired alkyl p-toluenesulfonate but is instead the corresponding alkyl chloride. Suggest a mechanistic explanation for this observation.

9.30 The reaction of cyclopentyl bromide with sodium cyanide to give cyclopentyl cyanide

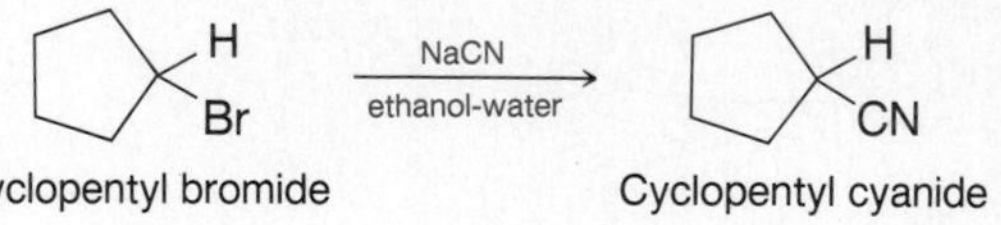

Cyclopentyl bromide → Cyclopentyl cyanide

proceeds faster if a small amount of sodium iodide is added to the reaction mixture. Can you suggest a reasonable mechanism to explain the catalytic function of sodium iodide?

ALKYNES

Hydrocarbons characterized by the presence of a carbon-carbon triple bond are called *alkynes*. Noncyclic alkynes have the molecular formula C_nH_{2n-2}. *Acetylene* (HC≡CH) is the simplest alkyne. We call compounds that have their triple bond at the end of a carbon chain (RC≡CH) *monosubstituted* or *terminal alkynes*. Disubstituted alkynes (RC≡CR′) are said to have *internal* triple bonds. You will see in this chapter that a carbon-carbon triple bond is a functional group, reacting with many of the same reagents that react with alkenes. A second, and more unique, aspect of the chemistry of acetylene and terminal alkynes is their acidity. As a class, compounds of the type RC≡CH are the most acidic of all simple hydrocarbons. The structural reasons for this property, as well as the ways in which the enhanced acidity of alkynes is used to advantage in chemical synthesis, comprise important elements of this chapter.

10.1 SOURCES OF ALKYNES

Acetylene was first characterized by the French chemist P. E. M. Berthelot in 1862 and did not command much attention until its large-scale preparation from calcium carbide in the last decade of the nineteenth century stimulated interest in industrial applications. In the first stage of that synthesis, limestone and coke are heated in an electric furnace to form calcium carbide.

$$CaO \quad + \quad 3C \quad \xrightarrow{1800-2100°C} \quad CaC_2 \quad + \quad CO$$

Calcium oxide (from limestone)	Carbon (from coke)		Calcium carbide	Carbon monoxide

Calcium carbide is the calcium salt of the doubly negative carbide ion (:C̄≡C̄:). Carbide dianion is strongly basic and reacts with water to form acetylene.

$$Ca^{2+} \begin{bmatrix} \ddot{C} \\ \parallel \\ C \\ \ddot{} \end{bmatrix}^{2-} + 2H_2O \longrightarrow Ca(OH)_2 + HC\equiv CH$$

Calcium carbide   Water   Calcium hydroxide   Acetylene

Beginning in the middle of the twentieth century, alternative methods of acetylene production became practical. One of these is based on the thermal dehydrogenation of ethylene.

$$CH_2\!=\!CH_2 \rightleftharpoons HC\equiv CH + H_2$$

Ethylene   Acetylene   Hydrogen

The position of equilibrium favors ethylene at low temperatures but shifts to favor acetylene above 1150°C. Indeed, at very high temperatures most hydrocarbons, even methane, are converted to acetylene. Higher alkynes are prepared from alkenes and from acetylene by methods to be described later in this chapter.

Natural products that contain carbon-carbon triple bonds are numerous but are encountered less frequently than those that contain double bonds. Tariric acid, from the seed fat of a Guatemalan plant, is an example of a naturally occurring alkyne.

$$CH_3(CH_2)_{10}C\equiv C(CH_2)_4\overset{\displaystyle O}{\overset{\displaystyle \parallel}{C}}OH$$

Tariric acid

10.2 NOMENCLATURE

In naming alkynes the usual IUPAC rules for hydrocarbons are followed and the *-ane* suffix is replaced by *-yne*. Both acetylene and ethyne are acceptable IUPAC names for $HC\equiv CH$. The position of the triple bond along the chain is specified by number in a manner analogous to that used in alkene nomenclature.

$HC\equiv CCH_3$	$HC\equiv CCH_2CH_3$	$CH_3C\equiv CCH_3$	$(CH_3)_3CC\equiv CCH_3$
Propyne	1-Butyne	2-Butyne	4,4-Dimethyl-2-pentyne

PROBLEM 10.1   Write structural formulas and give the IUPAC names for all the alkynes of molecular formula C_5H_8.

In cases in which the $-C\equiv CH$ group is named as a substituent, it is designated as an *ethynyl* group.

An informal nomenclature system is still encountered in the older literature. This system names alkynes as alkyl derivatives of acetylene and specifies the substituents attached to the triply bonded carbons.

$CH_3C\equiv CH$	$CH_3CH_2C\equiv CCH(CH_3)_2$
Methylacetylene	Ethylisopropylacetylene

This naming system is obsolete and will not be used in this text.

TABLE 10.1
Physical Constants of Some Alkynes

Compound name	Molecular formula	Condensed structural formula	Melting point, °C	Boiling point, °C (1 atm)
Ethyne (acetylene)	C_2H_2	HC≡CH	−81.8	−84.0
Propyne	C_3H_4	$CH_3C≡CH$	−101.5	−23.2
1-Butyne	C_4H_6	$CH_3CH_2C≡CH$	−125.9	8.1
2-Butyne	C_4H_6	$CH_3C≡CCH_3$	−32.3	27.0
1-Pentyne	C_5H_8	$CH_3CH_2CH_2C≡CH$	−106.5	40.2
2-Pentyne	C_5H_8	$CH_3CH_2C≡CCH_3$	−109.5	56.1
3-Methyl-1-butyne	C_5H_8	$(CH_3)_2CHC≡CH$		29.0
1-Hexyne	C_6H_{10}	$CH_3(CH_2)_3C≡CH$	−132.4	71.4
2-Hexyne	C_6H_{10}	$CH_3(CH_2)_2C≡CCH_3$	−89.6	84.5
3-Hexyne	C_6H_{10}	$CH_3CH_2C≡CCH_2CH_3$	−103.2	81.4
3-Methyl-1-pentyne	C_6H_{10}	$CH_3CH_2CHC≡CH$ | CH_3		57.7
4-Methyl-1-pentyne	C_6H_{10}	$(CH_3)_2CHCH_2C≡CH$	−105.3	61.3
4-Methyl-2-pentyne	C_6H_{10}	$(CH_3)_2CHC≡CCH_3$	−110.4	73.1
3,3-Dimethyl-1-butyne	C_6H_{10}	$(CH_3)_3CC≡CH$	−78.2	37.7
1-Octyne	C_8H_{14}	$CH_3(CH_2)_5C≡CH$	−79.6	126.2
1-Nonyne	C_9H_{16}	$CH_3(CH_2)_6C≡CH$	−36.0	160.6
1-Decyne	$C_{10}H_{18}$	$CH_3(CH_2)_7C≡CH$	−40.0	182.2

10.3 PHYSICAL PROPERTIES

Table 10.1 presents selected physical constants of some representative alkynes. Among the alkynes listed in the table, acetylene, propyne, and 1-butyne are gases at room temperature, the rest being liquids. Alkynes generally have slightly higher boiling points than the corresponding alkanes and alkenes.

Alkynes share with alkanes and alkenes the properties of low density and low water solubility. They are nonpolar and dissolve readily in typical organic solvents such as alkanes, diethyl ether, and chlorinated hydrocarbons.

10.4 STRUCTURE AND BONDING IN ALKYNES. *sp* HYBRIDIZATION

Acetylene is a linear molecule with a carbon-carbon bond distance of 1.20 Å and carbon-hydrogen bond distances of 1.06 Å.

$$\overset{\displaystyle 1.20\ \text{Å}}{\underset{\displaystyle 180°}{1.06\ \text{Å} \diagdown \quad \diagup 1.06\ \text{Å}} }$$

$$H—C≡C—H$$

There is a progressive shortening of the carbon-carbon bond distance in the series

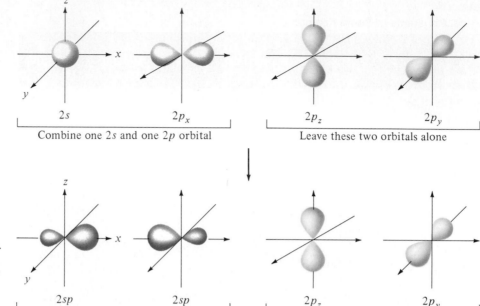

FIGURE 10.1 Representation of orbital mixing in *sp* hybridization. Mixing one 2s orbital with one of the three 2p orbitals generates two 2sp hybrid orbitals. Each *sp* hybrid orbital has 50 percent s character and 50 percent p character.

ethane (1.53 Å), ethylene (1.34 Å), and acetylene (1.20 Å). The carbon-hydrogen bond distances decrease in the series as well.

We have earlier described bonding in alkanes and alkenes on the basis of orbital hybridization models that picture carbon as sp^3 hybridized when it is bonded to four atoms or groups, as in ethane, and sp^2 hybridized when it is bonded to three, as in ethylene. Each carbon in acetylene is bonded to two atoms, and extension of the orbital hybridization model to systems of this type is based on sp hybridization of the orbitals of carbon.

Figure 10.1 depicts the orbital interactions involved in sp hybridization. According to this model the carbon 2s orbital and one of the 2p orbitals are mixed to generate a pair of two equivalent sp hybrid orbitals. Each sp hybrid orbital has 50 percent s character and 50 percent p character. These two sp orbitals share a common axis, but their major lobes are oriented in opposite directions. There is an angle of 180° between the major lobes of carbon's two $2sp$ orbitals. Two of the original 2p orbitals remain unhybridized. Their axes are perpendicular to each other and to the common axis of the pair of sp hybrid orbitals.

As portrayed in Figure 10.2, the two carbons of acetylene are connected to each other by a $2sp$–$2sp$ σ bond, and each is attached to a hydrogen substituent by a $2sp$–$1s$ σ bond. The unhybridized 2p orbitals on one carbon overlap with their counterparts on the other to form two π bonds. The carbon-carbon triple bond in acetylene, and in higher alkynes as well, is viewed as a multiple bond of the $\sigma + \pi + \pi$ type.

Table 10.2 compares some structural features in the two-carbon compounds ethane, ethylene, and acetylene. As we noted above, the carbon-carbon bonds and carbon-hydrogen bonds become shorter as one proceeds from ethane to ethylene to acetylene. As the s character of the orbitals involved in σ bonds increases from 25 percent (sp^3) to $33\frac{1}{3}$ percent (sp^2) to 50 percent (sp), the electrons in that orbital are,

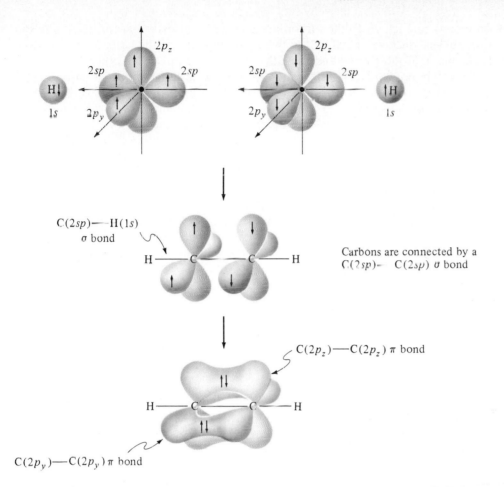

$C(2sp)$—$H(1s)$
σ bond

Carbons are connected by a
$C(2sp)$—$C(2sp)$ σ bond

$C(2p_z)$—$C(2p_z)$ π bond

$C(2p_y)$—$C(2p_y)$ π bond

FIGURE 10.2 A description of bonding in acetylene based on *sp* hybridization of carbon.

TABLE 10.2
Structural Features of Ethane, Ethylene, and Acetylene

Feature	Ethane	Ethylene	Acetylene
Systematic name	Ethane	Ethene	Ethyne
Molecular formula	C_2H_6	C_2H_4	C_2H_2
Structural formula			$H—C\equiv C—H$
C—C bond distance, Å	1.534	1.337	1.203
C—H bond distance, Å	1.112	1.103	1.060
H—C—C bond angles, °	111.0°	121.4°	180°
C—C bond dissociation energy, kcal/mol	88	146	196
C—H bond dissociation energy, kcal/mol	98	108	128
Hybridization of carbon	sp^3	sp^2	sp
s Character in C—H bonds	25%	33%	50%
Approximate acidity as measured by K_a	10^{-62}	10^{-45}	10^{-26}

on the average, closer to the carbon nucleus, and this leads to a contraction in the internuclear distance. Further, the two π components of the carbon-carbon triple bond augment the σ bond and bring the two nuclei nearer each other.

Shorter bonds tend to be stronger bonds. Table 10.2 cites bond dissociation energy data that reveal a pattern of increasing bond strength for bonds to carbon as its hybridization changes from sp^3 to sp^2 to sp. As the s character of the orbital that binds carbon to another atom increases, the pair of electrons in that orbital is more strongly held, and it requires more energy to cleave the bond in a homolytic manner.

Table 10.2 also compares the acidities of ethane, ethylene, and acetylene. Acetylene and terminal alkynes are far more acidic than alkenes, and alkenes are more acidic than alkanes—important observations related to orbital hybridization, which will be addressed in greater detail in Section 10.6.

Higher alkynes combine structural features of alkynes and alkyl groups. Propyne has a linear arrangement of its $H—C\equiv C—C$ unit, and its carbon-carbon and carbon-hydrogen bonds are shorter than those of propene, the analogous alkene.

Propyne

Propene

Not only is the $C(sp)—C(sp)$ triple bond distance shorter than that of the $C(sp^2)—C(sp^2)$ double bond distance, but the $C(sp)—C(sp^3)$ single bond distance in propyne is shorter than the $C(sp^2)—C(sp^3)$ single bond distance in propene. This is another example of the general rule that an increase in the s character of the orbitals involved leads to a shorter bond.

PROBLEM 10.2   Which has a longer carbon-methyl bond, 1-butyne or 2-butyne? Why?

10.5   CYCLOALKYNES

Among cycloalkynes, cyclononyne is the smallest one stable enough to be stored for extended periods of time at room temperature.

Cyclononyne

The triple bond limits stable cycloalkynes to those in which the ring is of sufficient size to accommodate a linear $C—C\equiv C—C$ unit. While cyclononyne satisfies this structural requirement, its smaller relatives cyclooctyne and cycloheptyne are quite strained and react rapidly with themselves to form polymers soon after they are isolated. Evidence has been presented suggesting that cyclohexyne and even cyclopentyne are formed as transitory intermediates in certain chemical reactions, but neither has been isolated as a stable compound.

10.6 ACIDITY OF ACETYLENE AND TERMINAL ALKYNES

Carbon is not very electronegative, and carbon-hydrogen bonds show little tendency to ionize. The species formed on ionization is an anion in which carbon is negatively charged. Species of this type are known as *carbanions*.

$$R_3\overset{\frown}{C}-H \longrightarrow H^+ + R_3\overset{\text{–}}{C}{:}$$

Hydrocarbon   Proton   Carbanion

The ionization constant K_a for ionization of methane is too small to be measured directly but has been estimated to be about 10^{-60} (pK_a 60).

$$CH_3\overset{\frown}{-}H \rightleftharpoons H^+ + \overset{\text{–}}{:}CH_3$$

Methane   Proton   Methide anion

Alkanes are very weak acids; carbanions are very strong bases.

Electronegativity increases in going across the periodic table. Acidity increases in the order

$$CH_4 \quad < \quad NH_3 \quad < \quad H_2O \quad < \quad HF$$

Methane	Ammonia	Water	Hydrogen fluoride
$K_a \sim 10^{-60}$	$\sim 10^{-36}$	1.8×10^{-16}	3.5×10^{-4}
$pK_a \sim 60$	~ 36	15.7	3.2
(Weakest acid)			(Strongest acid)

The basicity of the corresponding anions decreases in proceeding across the periodic table.

$$H_3\overset{\text{–}}{C}{:} \quad > H_2\overset{..}{N}{:}^{\text{–}} > \quad H\overset{..}{O}{:}^{\text{–}} \quad > \quad :\overset{..}{\underset{..}{F}}{:}^{\text{–}}$$

Methide   Amide   Hydroxide   Fluoride
(Strongest base)   (Weakest base)

As the electron-attracting power of the negatively charged atom becomes greater, the anion becomes less basic.

There are significant differences in the acidity of alkanes, alkenes, and alkynes. Acetylene is more acidic than ethylene, and ethylene is more acidic than ethane.

$$HC\equiv CH > CH_2{=}CH_2 > CH_3CH_3$$

Acetylene	Ethylene	Ethane
$K_a = 10^{-26}$	$\sim 10^{-45}$	$\sim 10^{-62}$
$pK_a = 26$	~ 45	~ 62
(Strongest acid)		(Weakest acid)

We can understand this order of acidity by examining how strongly the unshared electron pair is bound in the respective anions. The ionization of acetylene yields an anion in which the unshared electron pair occupies an *sp* hybridized orbital.

$$H-C\equiv\overset{\frown}{C}-H \rightleftharpoons H^+ + H-C\equiv C{:}^{\text{–}}$$

Acetylene   Proton   Acetylide ion

Ionization of ethylene gives an anion in which the unshared electron pair occupies an sp^2 hybridized orbital.

$$\underset{\substack{\text{Ethylene}}}{\overset{\substack{H \\ \diagdown \\ H}}{\underset{\substack{H}}{C}} = \overset{\substack{H}}{\underset{\substack{H}}{C}}} \rightleftharpoons \underset{\substack{\text{Proton}}}{H^+} + \underset{\substack{\text{Ethylenide ion} \\ \text{(vinyl anion)}}}{\overset{\substack{H}}{\underset{\substack{H}}{C}} = \overset{\substack{sp^2}}{\underset{\substack{H}}{C}}}$$

The equilibrium constant for ionization of acetylene is greater than that of ethylene because an electron pair in an orbital with 50 percent s character (sp) is more strongly bound than an electron pair in an orbital with $33\frac{1}{3}$ percent s character (sp^2). Acetylide ion holds its electron pair more strongly than ethylenide ion and is less basic. According to the same reasoning, ethane is the weakest acid of the group because the unshared electron pair in ethyl anion (CH_3CH_2 :$^-$) occupies an orbital with only 25 percent s character. An electron pair in ethyl anion is less strongly held than an electron pair in ethylenide ion. Terminal alkynes are similar to acetylene in their acidity.

$$(CH_3)_3CC{\equiv}CH \qquad K_a = 3 \times 10^{-26} \ (pK_a = 25.5)$$
$$\text{3,3-Dimethyl-1-butyne}$$

While acetylene and terminal alkynes are far stronger acids than other hydrocarbons, it must be remembered that they are, nevertheless, very weak acids—much weaker than water and alcohols, for example. Hydroxide ion is too weak a base to convert acetylene to its anion in meaningful amounts. The position of the equilibrium described by the following equation lies overwhelmingly to the left.

$$\underset{\substack{\text{Acetylene} \\ \text{(Weaker acid)} \\ K_a = 10^{-26} \\ pK_a = 26}}{H{-}C{\equiv}C{-}H} + \underset{\substack{\text{Hydroxide ion} \\ \text{(Weaker base)}}}{:\overset{..}{\underset{..}{O}}H^-} \rightleftharpoons \underset{\substack{\text{Acetylide ion} \\ \text{(Stronger base)}}}{H{-}C{\equiv}C:^-} + \underset{\substack{\text{Water} \\ \text{(Stronger acid)} \\ K_a = 1.8 \times 10^{-16} \\ pK_a = 15.7}}{H\overset{..}{\underset{..}{O}}H}$$

Because acetylene is a far weaker acid than water and alcohols, these substances are not suitable solvents for reactions involving acetylide ions. Acetylide is instantly converted to acetylene by proton transfer from compounds that contain hydroxyl groups.

Amide ion is a much stronger base than acetylide ion and converts acetylene to its conjugate base quantitatively.

$$\underset{\substack{\text{Acetylene} \\ \text{(Stronger acid)} \\ K_a = 10^{-26} \\ pK_a = 26}}{H{-}C{\equiv}C{-}H} + \underset{\substack{\text{Amide ion} \\ \text{(Stronger base)}}}{:\overset{..}{N}H_2^-} \rightleftharpoons \underset{\substack{\text{Acetylide ion} \\ \text{(Weaker base)}}}{H{-}C{\equiv}C:^-} + \underset{\substack{\text{Ammonia} \\ \text{(Weaker acid)} \\ K_a = 10^{-36} \\ pK_a = 36}}{\overset{..}{N}H_3}$$

Solutions of sodium acetylide ($HC{\equiv}CNa$) may be prepared by adding sodium amide to acetylene in liquid ammonia as the solvent. Terminal alkynes react similarly to give species of the type $RC\equiv CNa$.

PROBLEM 10.3   Complete each of the following equations to show the conjugate acid and the conjugate base formed by proton transfer between the indicated species. For each reaction, specify whether the position of equilibrium lies to the side of reactants or products.

(a) $CH_3C\equiv CH + {}^-OCH_3 \rightleftharpoons$

(b) $HC\equiv CH + CH_3CH_2^- \rightleftharpoons$

(c) $CH_2=CH_2 + H_2N^- \rightleftharpoons$

(d) $CH_3C\equiv CCH_2OH + H_2N^- \rightleftharpoons$

SAMPLE SOLUTION   (a)   The equation representing the acid-base reaction between propyne and methoxide ion is:

$$CH_3C\equiv C-H + {}^-OCH_3 \rightleftharpoons CH_3C\equiv C\overset{..}{:}{}^- + HOCH_3$$

Propyne	Methoxide ion	Propynide ion	Methanol
(Weaker acid)	(Weaker base)	(Stronger base)	(Stronger acid)

Alcohols are stronger acids than acetylene, so the position of equilibrium lies to the left. Methoxide ion is not a sufficiently strong base to abstract a proton from acetylene.

Anions of acetylene and terminal alkynes are nucleophilic species and react with unhindered alkyl halides to form carbon-carbon bonds by nucleophilic substitution. Some useful applications of this reaction will be discussed in the following section.

10.7 PREPARATION OF ALKYNES BY ALKYLATION OF ACETYLENE AND ALKYNES

Organic synthesis makes use of two major reaction types:

1. Functional group transformations
2. Carbon-carbon bond–forming reactions

Both strategies are applied to the preparation of alkynes. In this section we shall see how alkynes are prepared by combining smaller structural units to build longer carbon chains. One of these structural units may be as simple as acetylene itself. By attaching alkyl groups to this unit, more complex alkynes can be prepared.

$$H-C\equiv C-H \longrightarrow R-C\equiv C-H \longrightarrow R-C\equiv C-R'$$

Acetylene	Monosubstituted or terminal alkyne	Disubstituted derivative of acetylene

Reactions that lead to attachment of alkyl groups to molecular fragments are called *alkylation* reactions. One way in which alkynes are prepared is by alkylation of acetylene.

Alkylation of acetylene is a synthetic process comprising two separate reactions carried out in sequence. In the first stage, acetylene is converted to its conjugate base by treatment with sodium amide.

$$HC\equiv CH + NaNH_2 \longrightarrow HC\equiv CNa + NH_3$$

Acetylene   Sodium amide   Sodium acetylide   Ammonia

Next, an alkyl halide is added to the solution of sodium acetylide. Acetylide ion acts as a nucleophile, displacing halide from carbon and forming a new carbon-carbon bond. In these reactions, substitution occurs by an S_N2 mechanism.

$$HC\equiv CNa + RX \longrightarrow HC\equiv CR + NaX \qquad via \qquad HC\equiv C\colon \overset{\frown}{} R\!-\!X$$

Sodium   Alkyl   Alkyne   Sodium
acetylide   halide   halide

The synthetic sequence is usually carried out in liquid ammonia as the solvent. Alternatively, diethyl ether or tetrahydrofuran may be used.

$$HC\equiv C^- Na^+ + CH_3CH_2CH_2CH_2Br \xrightarrow{NH_3} CH_3CH_2CH_2CH_2C\equiv CH$$

Sodium acetylide   1-Bromobutane   1-Hexyne (70–77%)

An analogous sequence using terminal alkynes as starting materials yields alkynes of the type $RC\equiv CR'$.

$$(CH_3)_2CHCH_2C\equiv CH \xrightarrow[NH_3]{NaNH_2} (CH_3)_2CHCH_2C\equiv CNa \xrightarrow{CH_3Br} (CH_3)_2CHCH_2C\equiv CCH_3$$

4-Methyl-1-pentyne   5-Methyl-2-hexyne (81%)

Dialkylation of acetylene can be achieved by carrying out the sequence twice.

$$HC\equiv CH \xrightarrow[2.\ CH_3CH_2Br]{1.\ NaNH_2,\ NH_3} HC\equiv CCH_2CH_3 \xrightarrow[2.\ CH_3Br]{1.\ NaNH_2,\ NH_3} CH_3C\equiv CCH_2CH_3$$

Acetylene   1-Butyne   2-Pentyne (81%)

PROBLEM 10.4 Outline efficient syntheses of each of the following alkynes from acetylene and any necessary organic or inorganic reagents:

(*a*) 1-Heptyne   (*b*) 2-Heptyne   (*c*) 3-Heptyne

SAMPLE SOLUTION (*a*) An examination of the structural formula of 1-heptyne reveals it to have a pentyl group attached to an acetylene unit. Alkylation of acetylene, by way of its anion, with a pentyl halide is a suitable synthetic route to 1-heptyne.

$$HC\equiv CH \xrightarrow[NH_3]{NaNH_2} HC\equiv CNa \xrightarrow{CH_3CH_2CH_2CH_2CH_2Br} HC\equiv CCH_2CH_2CH_2CH_2CH_3$$

Acetylene   Sodium acetylide   1-Heptyne

The most significant limitation to this reaction is that synthetically acceptable yields are obtained only with methyl halides and primary alkyl halides. Acetylide anions are very basic, more basic than hydroxide, and react with secondary and tertiary alkyl halides by an E2 elimination pathway. The desired S_N2 substitution pathway is observed only with methyl halides and primary alkyl halides.

PROBLEM 10.5   Refer to the various alkynes of molecular formula C_5H_8 of Problem 10.1 (Section 10.2) and select those that can be prepared in good yield by alkylation or dialkylation of acetylene. Explain why the preparation of the other isomers would not be practical.

A second strategy for alkyne synthesis, involving functional group transformation reactions, is described in the following section.

10.8   PREPARATION OF ALKYNES BY ELIMINATION REACTIONS

Just as it is possible to prepare alkenes by dehydrohalogenation of alkyl halides, so may alkynes be prepared by a double dehydrohalogenation of dihaloalkanes. The dihalide may be a *geminal dihalide,* one in which both halogens are substituents on the same carbon, or it may be a *vicinal dihalide,* one in which the halogens are substituents on adjacent carbons.

Double dehydrohalogenation of a geminal dihalide

$$\underset{\text{Geminal dihalide}}{R-\overset{\overset{\displaystyle H}{|}}{\underset{\underset{\displaystyle H}{|}}{C}}-\overset{\overset{\displaystyle X}{|}}{\underset{\underset{\displaystyle X}{|}}{C}}-R'} + \underset{\text{Sodium amide}}{2NaNH_2} \longrightarrow \underset{\text{Alkyne}}{R-C\equiv C-R'} + \underset{\text{Ammonia}}{2NH_3} + \underset{\text{Sodium halide}}{2NaX}$$

Double dehydrohalogenation of a vicinal dihalide

$$\underset{\text{Vicinal dihalide}}{R-\overset{\overset{\displaystyle H}{|}}{\underset{\underset{\displaystyle X}{|}}{C}}-\overset{\overset{\displaystyle H}{|}}{\underset{\underset{\displaystyle X}{|}}{C}}-R'} + \underset{\text{Sodium amide}}{2NaNH_2} \longrightarrow \underset{\text{Alkyne}}{R-C\equiv C-R'} + \underset{\text{Ammonia}}{2NH_3} + \underset{\text{Sodium halide}}{2NaX}$$

The most frequent applications of these procedures are in the preparation of terminal alkynes. Since the terminal alkyne product is a strong enough acid to transfer a proton to amide anion, an equivalent of base in excess of the two required for double dehydrohalogenation is required. Addition of water or acid after the reaction is complete converts the sodium salt to the corresponding alkyne.

Double dehydrohalogenation of a geminal dihalide

$$\underset{\substack{\text{1,1-Dichloro-3,3-} \\ \text{dimethylbutane}}}{(CH_3)_3CCH_2CHCl_2} \xrightarrow[\text{NH}_3]{\text{3NaNH}_2} \underset{\substack{\text{Sodium salt of alkyne} \\ \text{product (not isolated)}}}{(CH_3)_3CC\equiv CNa} \xrightarrow{\text{H}_2\text{O}} \underset{\substack{\text{3,3 Dimethyl-} \\ \text{1-butyne (56–60\%)}}}{(CH_3)_3CC\equiv CH}$$

Double dehydrohalogenation of a vicinal dihalide

$$CH_3(CH_2)_7CHCH_2Br \xrightarrow[NH_3]{3NaNH_2} CH_3(CH_2)_7C\equiv CNa \xrightarrow{H_2O} CH_3(CH_2)_7C\equiv CH$$
$$\underset{Br}{|}$$

| 1,2-Dibromodecane | Sodium salt of alkyne product (not isolated) | 1-Decyne (54%) |

Double dehydrohalogenation to form terminal alkynes may also be carried out by heating geminal and vicinal dihalides with other base-solvent combinations, such as sodium amide in mineral oil and potassium hydroxide or potassium *tert*-butoxide in dimethyl sulfoxide.

PROBLEM 10.6   Give the structures of three isomeric dibromides that could be used as starting materials for the preparation of 3,3-dimethyl-1-butyne.

Since vicinal dihalides are prepared by addition of chlorine or bromine to alkenes (Section 7.11), we see that alkenes can serve as starting materials for the preparation of alkynes by carrying out the sequence of functional group transformations shown in the following equations:

$$RCH{=}CHR' + \quad X_2 \quad \longrightarrow RCH{-}CHR' \xrightarrow[2.\ H_2O]{1.\ NaNH_2,\ NH_3} RC\equiv CR'$$
$$\underset{X\quad X}{\qquad\qquad\quad |\quad\ |}$$

| Alkene | Chlorine or bromine | Vicinal dihalide | Alkyne |

$$(CH_3)_2CHCH{=}CH_2 \xrightarrow{Br_2} (CH_3)_2CHCHCH_2Br \xrightarrow[2.\ H_2O]{1.\ NaNH_2,\ NH_3} (CH_3)_2CHC\equiv CH$$
$$\underset{Br}{\qquad\qquad\qquad\qquad |}$$

| 3-Methyl-1-butene | 1,2-Dibromo-3-methylbutane | 3-Methyl-1-butyne (52%) |

We can go back one more step by recalling that alkenes may be prepared by elimination reactions of alcohols and alkyl halides. This reasoning generates the sequence

$$RCH_2CHR' \longrightarrow RCH{=}CHR' \longrightarrow RCH{-}CHR' \longrightarrow RC\equiv CR'$$
$$\underset{X}{|} \qquad\qquad\qquad\qquad\qquad \underset{X\quad X}{|\quad\ |}$$

| Alcohol or alkyl halide | Alkene | Vicinal dihalide | Alkyne |

PROBLEM 10.7   Show, by writing an appropriate series of equations, how you could prepare propyne from each of the following compounds as starting materials. You may use any necessary organic or inorganic reagents.

(a) 2-Propanol
(b) 1-Propanol
(c) Isopropyl bromide

(d) 1,1-Dichloroethane
(e) Ethyl alcohol

SAMPLE SOLUTION (a) Since we know that we can convert propene to propyne by the sequence of reactions:

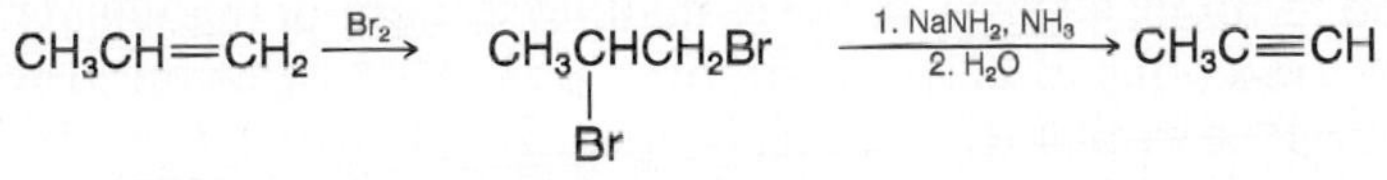

$$CH_3CH{=}CH_2 \xrightarrow{Br_2} \underset{\underset{Br}{|}}{CH_3CHCH_2Br} \xrightarrow[\text{2. } H_2O]{\text{1. } NaNH_2,\ NH_3} CH_3C{\equiv}CH$$

Propene   1,2-Dibromopropane   Propyne

all that remains in order to completely describe the synthesis is to show the preparation of propene from 2-propanol. Acid-catalyzed dehydration is suitable.

$$(CH_3)_2CHOH \xrightarrow[\text{heat}]{H^+} CH_3CH{=}CH_2$$

2-Propanol   Propene

The two dehydrohalogenation steps in the base-promoted formation of alkynes from vicinal and geminal dihalides occur sequentially. Alkenyl halides are intermediates.

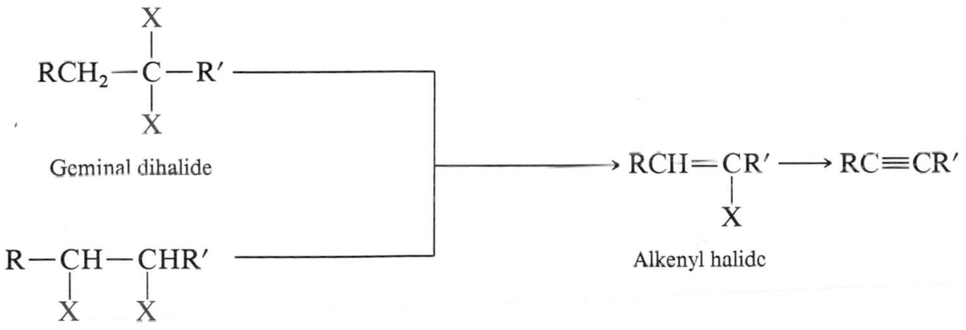

Geminal dihalide

Vicinal dihalide

Alkenyl halide

The second dehydrohalogenation step is typically more difficult than the first, so rather strongly basic conditions or high temperatures are needed to convert dihalides to alkynes. When weaker bases and lower temperatures are used, the intermediate alkenyl halide may be isolated.

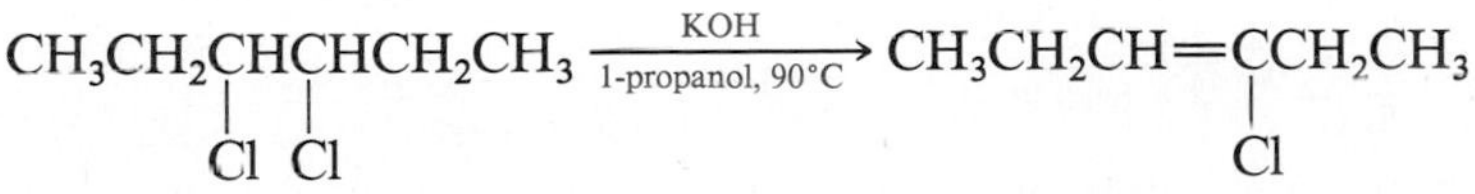

$$\underset{\underset{Cl\ \ Cl}{|\ \ \ |}}{CH_3CH_2CHCHCH_2CH_3} \xrightarrow[\text{1-propanol, 90°C}]{KOH} \underset{\underset{Cl}{|}}{CH_3CH_2CH{=}CCH_2CH_3}$$

3,4-Dichlorohexane   3-Chloro-3-hexene (90%)

Subjecting alkenyl halides to more strongly basic conditions brings about their dehydrohalogenation to alkynes.

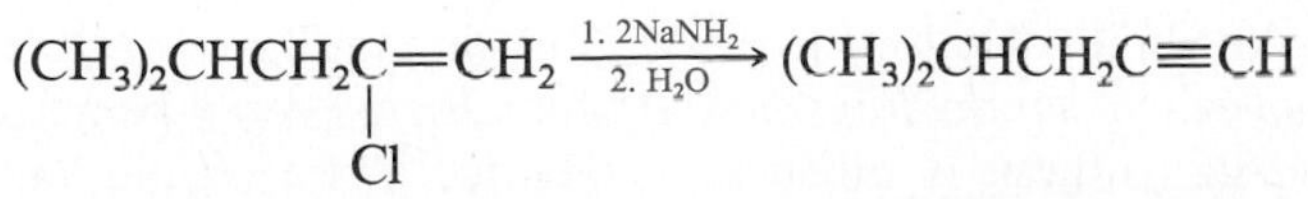

$$\underset{\underset{Cl}{|}}{(CH_3)_2CHCH_2C{=}CH_2} \xrightarrow[\text{2. } H_2O]{\text{1. } 2NaNH_2} (CH_3)_2CHCH_2C{\equiv}CH$$

2-Chloro-4-methyl-1-pentene   4-Methyl-1-pentyne (80%)

10.9 REACTIONS OF ALKYNES

We have already discussed one important chemical property of alkynes, the acidity of acetylene and terminal alkynes. In the remaining sections of this chapter additional aspects of the reactions of alkynes will be explored. Most of the reactions to be encountered will be similar to those of alkenes. Like alkenes, alkynes undergo addition reactions. We shall begin with a reaction familiar to us from our study of alkenes, namely, catalytic hydrogenation.

10.10 HYDROGENATION OF ALKYNES

The conditions for hydrogenation of alkynes are similar to those employed for alkenes. In the presence of certain finely divided metals two molar equivalents of hydrogen add to the triple bond of an alkyne to yield an alkane.

$$RC{\equiv}CR' + 2H_2 \xrightarrow[\text{catalyst}]{\text{metal}} RCH_2CH_2R'$$

| Alkyne | Hydrogen | Alkane |

$$\underset{\underset{CH_3}{|}}{CH_3CH_2CHCH_2C{\equiv}CH} + 2H_2 \xrightarrow{\text{Ni}} \underset{\underset{CH_3}{|}}{CH_3CH_2CHCH_2CH_2CH_3}$$

4-Methyl-1-hexyne   Hydrogen   3-Methylhexane (77%)

The heats of hydrogenation of alkynes are affected by alkyl substituents in the same way as are those of compounds with carbon-carbon double bonds. Alkyl groups release electrons to sp hybridized carbon, stabilizing the alkyne and decreasing the heat of hydrogenation.

	$HC{\equiv}CH$	$CH_3CH_2C{\equiv}CH$	$CH_3C{\equiv}CCH_3$
	Acetylene	1-Butyne	2-Butyne
$-\Delta H°$ (hydrogenation)	75.1 kcal/mol	69.9 kcal/mol	65.6 kcal/mol

Alkenes are intermediates in the hydrogenation of alkynes to alkanes.

$$RC{\equiv}CR' \xrightarrow[\text{catalyst}]{H_2} RCH{=}CHR' \xrightarrow[\text{catalyst}]{H_2} RCH_2CH_2R'$$

Alkyne   Alkene   Alkane

The heats of hydrogenation of alkynes are typically somewhat greater than twice the heats of hydrogenation of analogous alkenes. Thus, as shown in Figure 10.3, the energy released in the first hydrogenation step is greater than that of the second. Their carbon-carbon triple bond makes alkynes rather high in energy compared with other hydrocarbons.

Noting that alkenes are intermediates in the hydrogenation of alkynes leads us to consider the possibility of halting hydrogenation at the alkene stage. If partial hydrogenation (also called *semihydrogenation*) of an alkyne could be achieved, it would provide a useful synthesis of alkenes. In practice it is a simple matter to convert alkynes to alkenes and no further by hydrogenation in the presence of specially

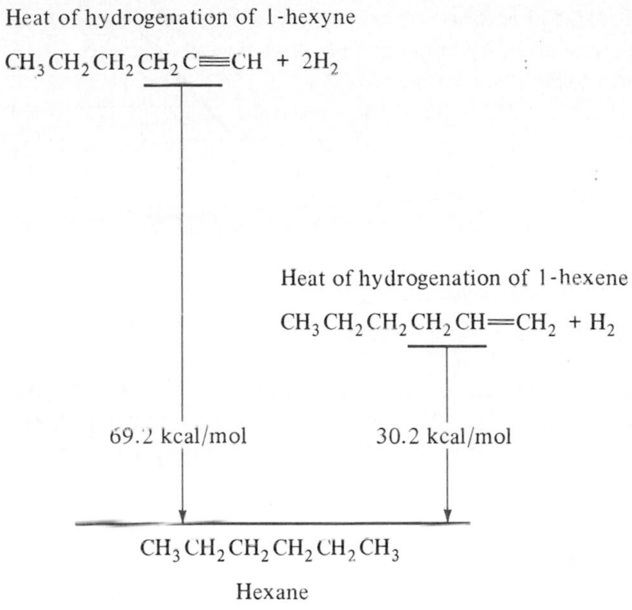
Heat of hydrogenation of 1-hexyne
$CH_3CH_2CH_2CH_2C{\equiv}CH + 2H_2$
Heat of hydrogenation of 1-hexene
$CH_3CH_2CH_2CH_2CH{=}CH_2 + H_2$
69.2 kcal/mol
30.2 kcal/mol
$CH_3CH_2CH_2CH_2CH_2CH_3$
Hexane

FIGURE 10.3 Diagram comparing heats of hydrogenation 1-hexyne and 1-hexene. By subtracting the heat of hydrogenation of 1-hexene from that of 1-hexyne, we see that addition of the first molecule of H_2 liberates 39.0 kcal/mol of heat, the second only 30.2 kcal/mol.

developed catalysts. The most frequently used catalyst for this purpose is the *Lindlar catalyst,* a palladium-on-calcium carbonate combination to which has been added lead acetate and quinoline. Lead acetate and quinoline partially deactivate (poison) the catalyst, making it a poor catalyst for alkene hydrogenation while retaining its ability to catalyze the addition of hydrogen to alkynes.

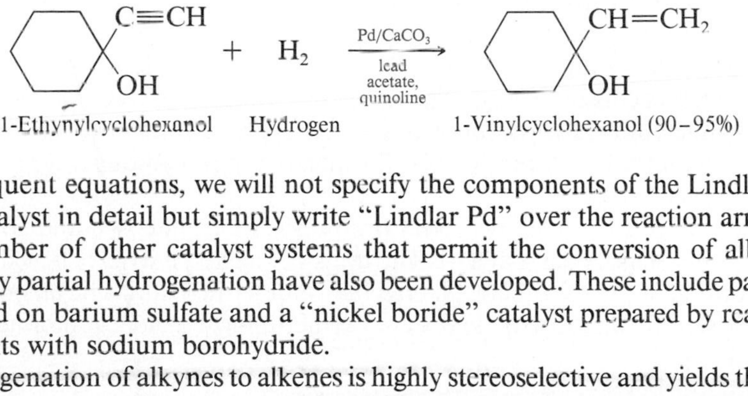
C≡CH
OH
+ H_2
$Pd/CaCO_3$
lead acetate, quinoline
$CH{=}CH_2$
OH

1-Ethynylcyclohexanol   Hydrogen   1-Vinylcyclohexanol (90–95%)

In subsequent equations, we will not specify the components of the Lindlar palladium catalyst in detail but simply write "Lindlar Pd" over the reaction arrow.

A number of other catalyst systems that permit the conversion of alkynes to alkenes by partial hydrogenation have also been developed. These include palladium supported on barium sulfate and a "nickel boride" catalyst prepared by reaction of nickel salts with sodium borohydride.

Hydrogenation of alkynes to alkenes is highly stereoselective and yields the cis (or *Z*) alkene by syn addition to the triple bond.

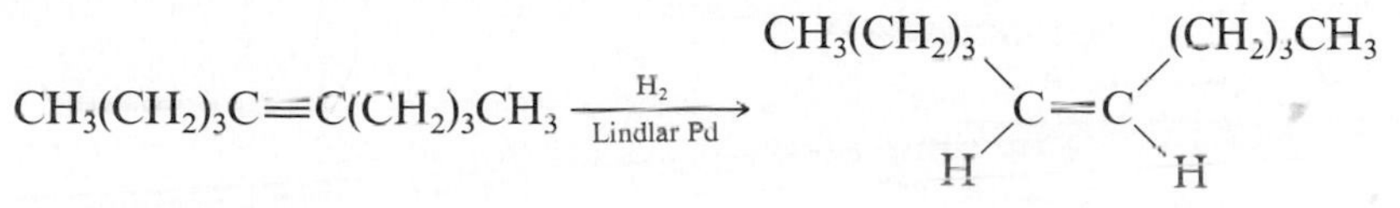
$CH_3(CH_2)_3C{\equiv}C(CH_2)_3CH_3$
H_2
Lindlar Pd
$CH_3(CH_2)_3$
$(CH_2)_3CH_3$
C=C
H
H

5-Decyne   (*Z*)-5-Decene (87%)

PROBLEM 10.8 Oleic acid and stearic acid are naturally occurring compounds, which can be isolated from various fats and oils. In the laboratory, each can be prepared by hydrogenation of a compound known as stearolic acid, which has the formula

$CH_3(CH_2)_7C{\equiv}C(CH_2)_7CO_2H$. Oleic acid is obtained by hydrogenation of stearolic acid over Lindlar palladium; stearic acid is obtained by hydrogenation over platinum. What are the structures of oleic acid and stearic acid?

10.11   METAL-AMMONIA REDUCTION OF ALKYNES

A useful alternative to catalytic partial hydrogenation as a means of converting alkynes to alkenes is reduction by a group I metal (lithium, sodium, or potassium) in liquid ammonia as the reaction medium. The unique feature of metal-ammonia reduction is that it converts alkynes to trans (or E) alkenes while catalytic hydrogenation procedures yield cis (or Z) alkenes. Thus, from the same alkyne one can prepare either a cis or a trans alkene by choosing the appropriate reaction conditions.

trans

$$CH_3CH_2C{\equiv}CCH_2CH_3 \xrightarrow[NH_3]{Na}$$

3-Hexyne

(E)-3-Hexene (82%)

PROBLEM 10.9   Sodium-ammonia reduction of stearolic acid (see Problem 10.8) yields a compound known as *elaidic acid*. What is the structure of elaidic acid?

PROBLEM 10.10   Suggest efficient syntheses of (E)- and (Z)-2-heptene from propyne and any necessary organic or inorganic reagents.

Metal-ammonia reduction of alkynes differs from catalytic hydrogenation with respect to stereochemistry because the mechanisms of the two reactions are different. The mechanism of catalytic hydrogenation of alkynes is similar to that of alkenes (Sections 5.9 and 5.12). A mechanism for metal-ammonia reduction of alkynes is outlined in Figure 10.4.

The mechanism includes two electron-transfer steps (steps 1 and 3) and two proton-transfer steps (steps 2 and 4). The trans stereoselectivity is believed to reflect the distribution of stereoisomeric alkenyl radical intermediates. The alkenyl radical formed in step 2 can exist in two stereoisomeric forms and equilibrates rapidly between them.

(Z)-Alkenyl radical
(less stable)

(E)-Alkenyl radical
(more stable)

The alkenyl radical in which the alkyl groups R and R′ are trans to each other is more stable than its stereoisomer in which they are cis and predominates at equilibrium. Both radicals are converted to alkene via steps 3 and 4, with the major product being derived from the radical present in greater concentration.

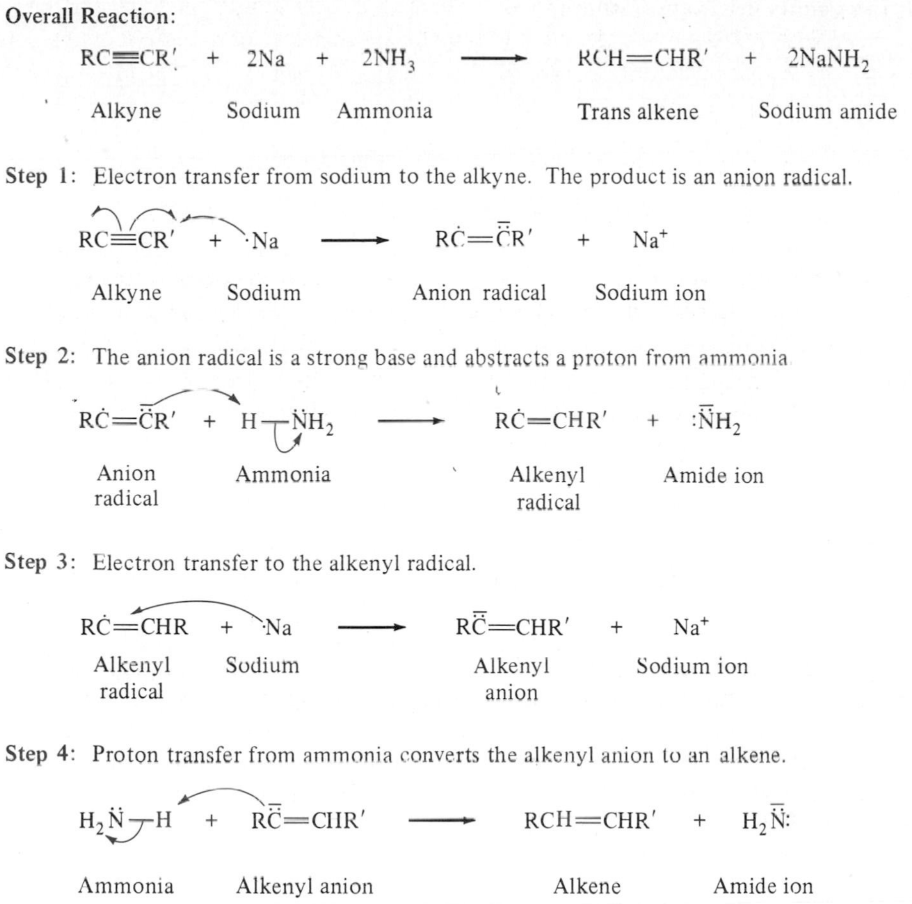

Overall Reaction:

$$RC\equiv CR' \ + \ 2Na \ + \ 2NH_3 \ \longrightarrow \ RCH\!=\!CHR' \ + \ 2NaNH_2$$

Alkyne   Sodium   Ammonia   Trans alkene   Sodium amide

Step 1: Electron transfer from sodium to the alkyne. The product is an anion radical.

$$RC\equiv CR' \ + \ \cdot Na \ \longrightarrow \ R\dot{C}\!=\!\overline{\overline{C}}R' \ + \ Na^+$$

Alkyne   Sodium   Anion radical   Sodium ion

Step 2: The anion radical is a strong base and abstracts a proton from ammonia

$$R\dot{C}\!=\!\overline{\overline{C}}R' \ + \ H\!-\!\dot{N}H_2 \ \longrightarrow \ R\dot{C}\!=\!CHR' \ + \ :\overline{N}H_2$$

Anion   Ammonia   Alkenyl   Amide ion
radical   radical

Step 3: Electron transfer to the alkenyl radical.

$$R\dot{C}\!=\!CHR \ + \ \cdot Na \ \longrightarrow \ R\overline{\overline{C}}\!=\!CHR' \ + \ Na^+$$

Alkenyl   Sodium   Alkenyl   Sodium ion
radical   anion

Step 4: Proton transfer from ammonia converts the alkenyl anion to an alkene.

$$H_2\dot{N}\!-\!H \ + \ R\overline{\overline{C}}\!=\!CHR' \ \longrightarrow \ RCH\!=\!CHR' \ + \ H_2\overline{N}:$$

Ammonia   Alkenyl anion   Alkene   Amide ion

FIGURE 10.4 Sequence of steps that describes the sodium-ammonia reduction of an alkyne.

10.12 ADDITION OF HYDROGEN HALIDES TO ALKYNES

Alkynes react with many of the same electrophilic reagents that add to the carbon-carbon double bond of alkenes. Hydrogen halides, for example, add to alkynes to form alkenyl halides.

$$RC\equiv CR' + \quad HX \quad \longrightarrow \quad \underset{\underset{X}{|}}{RCH\!=\!CR'}$$

Alkene   Hydrogen halide   Alkenyl halide

The regioselectivity of addition follows Markovnikov's rule. A proton adds to the carbon that has the greater number of hydrogen substituents, while halide adds to the carbon with the fewer hydrogen substituents.

$$CH_3CH_2CH_2CH_2C\equiv CH + \quad HI \quad \longrightarrow \quad \underset{\underset{I}{|}}{CH_3CH_2CH_2CH_2C\!=\!CH_2}$$

1-Hexyne   Hydrogen iodide   2-Iodo-1-hexene (73%)

Addition occurs in an anti fashion.

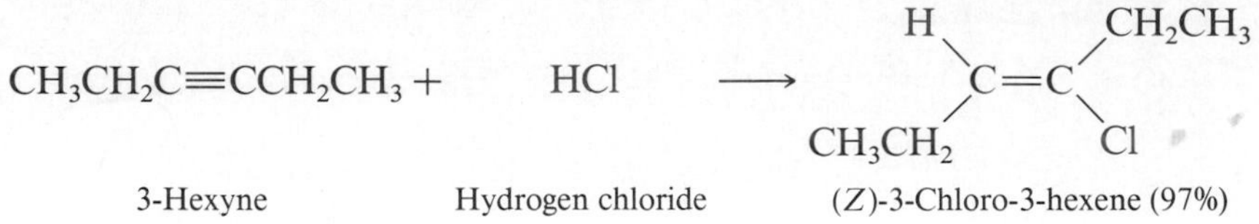

$$CH_3CH_2C\equiv CCH_2CH_3 \ + \qquad HCl \qquad \longrightarrow \qquad \underset{\underset{CH_3CH_2}{\overset{H}{\diagdown}}}{}C=C\underset{\underset{Cl}{}}{\overset{CH_2CH_3}{\diagup}}$$

3-Hexyne	Hydrogen chloride	(Z)-3-Chloro-3-hexene (97%)

PROBLEM 10.11   Write a series of chemical equations showing how you could prepare vinyl bromide (CH_2=CHBr) from ethyl bromide and any necessary inorganic reagents.

The rate of hydrogen halide addition increases with the degree of substitution of
the triple bond; acetylene is less reactive than terminal alkynes and terminal alkynes
are less reactive than disubstituted alkynes.

$$RC\equiv CR' \ > \ RC\equiv CH \ > \ HC\equiv CH$$

Disubstituted alkyne (most reactive)	Monosubstituted alkyne	Acetylene (least reactive)

The reaction mechanism of Figure 10.5 outlines a sequence of intermediates that
describe the ionic addition of hydrogen bromide to propyne. This mechanism is
analogous to the mechanism of hydrogen halide addition to alkenes. Proton transfer
to the substrate is regioselective and yields the more stable of two possible carbocation
intermediates in the rate-determining step. The carbocation intermediate is then
captured by the halide ion acting as a nucleophile to yield the alkenyl halide.

Overall Reaction:

$$CH_3C\equiv CH \quad + \quad HBr \quad \longrightarrow \quad CH_3\underset{\underset{Br}{|}}{C}=CH_2$$

Propyne	Hydrogen bromide	2-Bromopropene

Step 1: Protonation of the carbon-carbon triple bond of propyne by hydrogen bromide.
Protonation of the triple bond occurs in the direction that leads to the more stable
of two possible alkenyl cations.

$$CH_3C\equiv CH \quad + \quad H-\overset{..}{\underset{..}{Br}}: \quad \longrightarrow \quad CH_3\overset{+}{C}=CH_2 \quad + \quad :\overset{..}{\underset{..}{Br}}:^-$$

Propyne	Hydrogen bromide	2-Propenyl cation	Bromide ion

($CH_3\overset{+}{C}$=CH_2 is a more highly substituted and therefore more stable
carbocation than CH_3CH=$\overset{+}{CH}$ and is formed faster.)

Step 2: Bromide ion captures the alkenyl cation, forming 2-bromopropene.

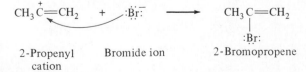

$$CH_3\overset{+}{C}=CH_2 \quad + \quad :\overset{..}{\underset{..}{Br}}:^- \quad \longrightarrow \quad CH_3\underset{\underset{:\overset{..}{\underset{..}{Br}}:}{|}}{C}=CH_2$$

2-Propenyl cation	Bromide ion	2-Bromopropene

FIGURE 10.5 Sequence of
steps that describes the
mechanism of addition of hydrogen bromide to propyne.

The carbocation formed by addition of a proton to an alkyne is an *alkenyl cation* or *vinyl cation.* The positively charged carbon of an alkenyl cation is *sp* hybridized. Because *sp* hybridized carbon is more electronegative than its sp^2 hybridized counterpart, alkenyl cations are somewhat less stable than alkyl cations.

$$\overset{+}{RCH_2-CHR'} \qquad \text{is more stable than} \qquad \overset{+}{RCH=CR'}$$

Alkyl cation
(positive charge is on
sp^2 hybridized carbon)

Alkenyl cation
(positive charge is on
sp hybridized carbon)

This relative instability of alkenyl cations contributes to the observation that the rate of hydrogen halide addition to alkynes is slightly less than the corresponding rate of addition to alkenes.

In the presence of excess hydrogen halide geminal dihalides are formed by sequential addition of two molecules of hydrogen halide to the carbon-carbon triple bond.

$$RC\equiv CR' \xrightarrow{HX} \underset{X}{RCH=CR'} \xrightarrow{HX} \underset{X}{\overset{X}{RCH_2CR'}}$$

Alkene　　　　Alkenyl halide　　　Geminal dihalide

The hydrogen halide adds to the initially formed alkenyl halide in accordance with Markovnikov's rule. Overall, both protons become bonded to the same carbon and both halogens to the adjacent carbon.

$$CH_3CH_2C\equiv CCH_2CH_3 + \quad 2HF \quad \longrightarrow CH_3CH_2CH_2\underset{F}{\overset{F}{C}}CH_2CH_3$$

3-Hexyne　　　Hydrogen fluoride　　3,3-Difluorohexane (76%)

PROBLEM 10.12　Write a series of equations showing how you could prepare 1,1-dichloroethane from

(a) Ethylene
(b) Vinyl chloride (CH_2=CHCl)
(c) Vinyl bromide (CH_2=CHBr)
(d) 1,1-Dibromoethane

SAMPLE SOLUTION　(a) Reasoning backward, we recognize 1,1-dichloroethane as the product of addition of two molecules of hydrogen chloride to acetylene. Thus, the synthesis requires converting ethylene to acetylene as a key feature. As described in Section 10.8, this may be accomplished by conversion of ethylene to a vicinal dihalide, followed by double dehydrohalogenation. A suitable synthesis based on this analysis is as shown:

$$CH_2=CH_2 \xrightarrow{Br_2} BrCH_2CH_2Br \xrightarrow[2.\ H_2O]{1.\ NaNH_2} HC\equiv CH \xrightarrow{2HCl} CH_3CHCl_2$$

Ethylene　　1,2-Dibromoethane　　Acetylene　　1,1-Dichloroethane

Overall Reaction:

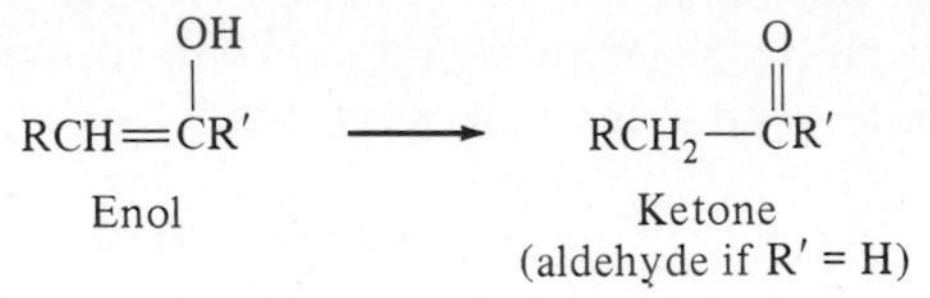

$$\underset{\text{Enol}}{RCH{=}CR'} \quad\longrightarrow\quad \underset{\substack{\text{Ketone}\\\text{(aldehyde if } R' = H)}}{RCH_2{-}CR'}$$

Step 1: The enol is formed in aqueous acidic solution. The first step of its transformation to a ketone is proton transfer to the carbon-carbon double bond.

$$\underset{\text{Hydronium ion}}{H_3O^+} \quad + \quad \underset{\text{Enol}}{RCH{=}CR'} \quad\rightleftharpoons\quad \underset{\text{Water}}{H_2O} \quad + \quad \underset{\text{Carbocation}}{RCH{-}CR'}$$

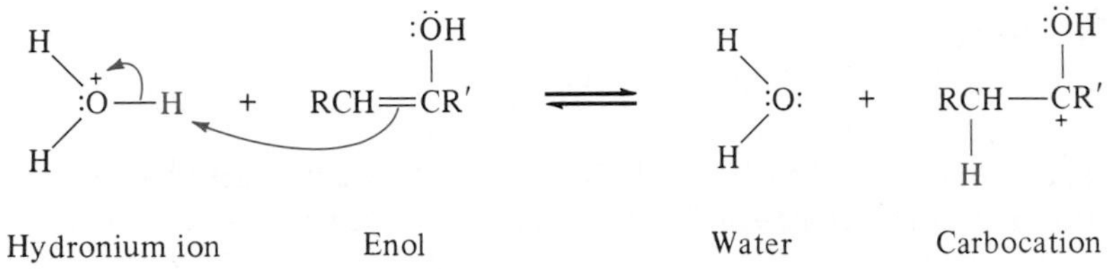

Step 2: The carbocation transfers a proton from oxygen to a water molecule, yielding a ketone.

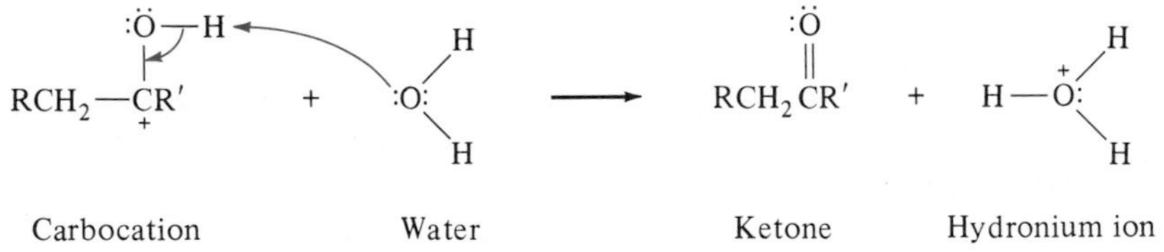

FIGURE 10.6 Conversion of an enol to a ketone takes place by way of two solvent-mediated proton transfers. A proton is transferred to carbon in the first step, then removed from oxygen in the second.

$$\underset{\text{Carbocation}}{RCH_2{-}CR'} \quad + \quad \underset{\text{Water}}{H_2O} \quad\longrightarrow\quad \underset{\text{Ketone}}{RCH_2CR'} \quad + \quad \underset{\text{Hydronium ion}}{H_3O^+}$$

10.13 HYDRATION OF ALKYNES

By analogy to the hydration of alkenes, addition of the elements of water to the triple bond is expected to yield an alcohol. The kind of alcohol produced by hydration of an alkyne, however, is of a special kind, one in which the hydroxyl group is a substituent on a carbon-carbon double bond. This type of alcohol is called an *enol* (the double bond suffix *-ene* plus the alcohol suffix *-ol*). An important property of enols is that they are rapidly converted to aldehydes or ketones under the conditions of their formation.

$$\underset{\text{Alkyne}}{RC{\equiv}CR'} + \underset{\text{Water}}{H_2O} \xrightarrow{\text{slow}} \underset{\substack{\text{Enol}\\\text{(not isolated)}}}{RCH{=}CR'} \xrightarrow{\text{fast}} \underset{\substack{R' = H,\ \text{aldehyde}\\R' = \text{alkyl, ketone}}}{RCH_2CR'}$$

The process by which enols are converted to aldehydes or ketones is called *keto-enol isomerism* (or *keto-enol tautomerism*) and proceeds by a sequence of proton transfers, as shown in Figure 10.6. Proton transfer to the double bond of an enol occurs readily because the carbocation that is produced is a very stable one. The positive charge on carbon is stabilized by electron release from oxygen. The electron delocalization responsible for this stabilization is represented in resonance terms as

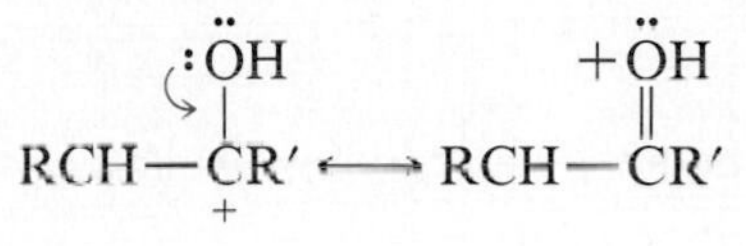

$$\underset{\text{A}}{\overset{\displaystyle :\ddot{O}H}{\underset{+}{RCH-CR'}}} \longleftrightarrow \underset{\text{B}}{\overset{\displaystyle +\ddot{O}H}{RCH-CR'}}$$

Delocalization of an oxygen lone pair stabilizes the cation. All the atoms in B have octets of electrons, making it a more stable structure than A. Only six electrons are associated with the positively charged carbon in A.

PROBLEM 10.13 Give the structure of the enol formed by hydration of 2-butyne and write a series of equations showing its conversion to its corresponding ketone isomer.

In general ketones are more stable than their enol precursors and are the products actually isolated when alkynes undergo acid-catalyzed hydration. The standard method by which alkyne hydration is carried out employs aqueous sulfuric acid as the reaction medium and mercuric sulfate or mercuric oxide as a catalyst. Because alkynes possess only limited solubility in aqueous sulfuric acid, methanol or acetic acid is often added as a cosolvent.

$$CH_3CH_2CH_2C\equiv CCH_2CH_2CH_3 + H_2O \xrightarrow{H^+,\ Hg^{2+}} CH_3CH_2CH_2CH_2\overset{\displaystyle O}{\overset{\|}{C}}CH_2CH_2CH_3$$

4-Octyne 4-Octanone (89%)

Hydration of alkynes follows Markovnikov's rule; terminal alkynes yield methyl-substituted ketones.

$$HC\equiv CCH_2CH_2CH_2CH_2CH_2CH_3 + H_2O \xrightarrow[\text{HgSO}_4]{\text{H}_2\text{SO}_4} CH_3\overset{\displaystyle O}{\overset{\|}{C}}CH_2CH_2CH_2CH_2CH_2CH_3$$

1-Octyne 2-Octanone (91%)

Because of the regioselectivity of alkyne hydration, acetylene is the only alkyne structurally capable of yielding an aldehyde under these conditions.

$$HC\equiv CH + H_2O \longrightarrow CH_2=CHOH \longrightarrow CH_3\overset{\displaystyle O}{\overset{\|}{C}}H$$

Acetylene Water Vinyl alcohol (not isolated) Acetaldehyde

At one time acetaldehyde was prepared on an industrial scale by this method. More modern methods involve direct oxidation of ethylene and are more economical.

10.14 ADDITION OF HALOGENS TO ALKYNES

Alkynes react with chlorine and bromine to yield tetrahaloalkanes. Two molecules of the halogen add to the triple bond.

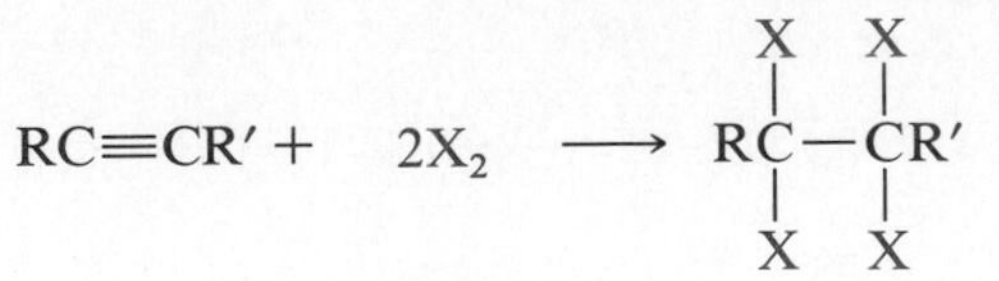

$$RC\equiv CR' + \quad 2X_2 \quad \longrightarrow \quad \underset{\underset{X}{|}}{\overset{\overset{X}{|}}{RC}}-\underset{\underset{X}{|}}{\overset{\overset{X}{|}}{CR'}}$$

Alkyne   Halogen   Tetrahaloalkane
(chlorine or
bromine)

$$CH_3C\equiv CH + \quad 2Cl_2 \quad \longrightarrow \quad \underset{\underset{Cl}{|}}{\overset{\overset{Cl}{|}}{CH_3CCCHCl_2}}$$

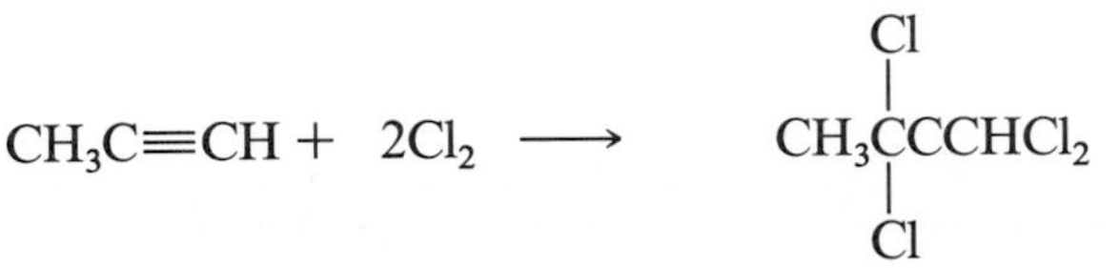

Propyne   Chlorine   1,1,2,2-Tetrachloropropane (63%)

A dihaloalkene is an intermediate and is the isolated product when the alkyne and the halogen are present in equimolar amounts. The stereochemistry of addition is anti.

(handwritten, left margin: CIS IS minor)

(handwritten: trans is major)

$$CH_3CH_2C\equiv CCH_2CH_3 + \quad Br_2 \quad \longrightarrow \quad \underset{Br}{\overset{CH_3CH_2}{}}C=C\underset{CH_2CH_3}{\overset{Br}{}}$$

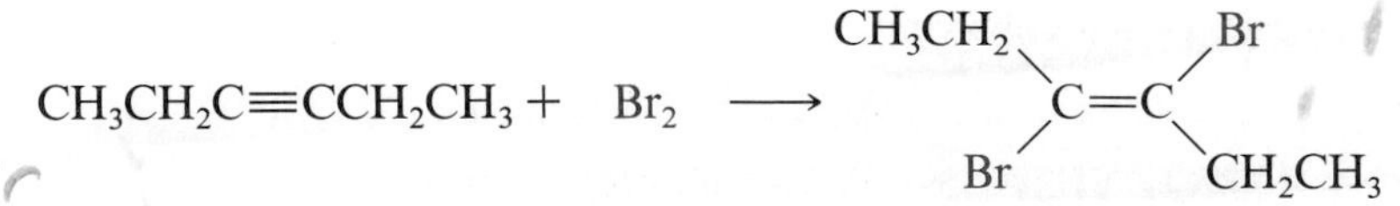

3-Hexyne   Bromine   (*E*)-3,4-Dibromo-3-hexene (90%)

In contrast to the addition of hydrogen halides to alkynes and to their acid-catalyzed hydration, reactions that take place at rates only slightly slower than the corresponding reactions of alkenes, the addition of halogens to carbon-carbon triple bonds proceeds far more slowly than halogen addition to alkenes. This is presumed to be due to the less stable nature of cyclic halonium ions formed from alkynes compared with those formed from alkenes.

Cyclic halonium ion
formed by addition
to double bond
(more stable)

Cyclic halonium ion
formed by addition
to triple bond
(more strained, less stable)

PROBLEM 10.14   There are three isomeric alkynes of molecular formula C_5H_8. One of these reacts with bromine several times faster than the other two. Which isomer is the most reactive? Why?

10.15 OXIDATION OF ALKYNES

Reagents and reaction conditions that cleave alkenes also lead to cleavage of carbon-carbon triple bonds. Carboxylic acids are produced when alkynes are subjected to ozonolysis or to oxidation by potassium permanganate.

$$RC\equiv CR' \xrightarrow[\text{cleavage}]{\text{oxidative}} R\overset{O}{\overset{\|}{C}}OH + HO\overset{O}{\overset{\|}{C}}R'$$

$$CH_3CH_2CH_2CH_2C\equiv CH \xrightarrow[\text{2. H}_2\text{O}]{\text{1. O}_3} CH_3CH_2CH_2CH_2CO_2H + HCO_2H$$

1-Hexyne   Pentanoic acid (51%)   Formic acid

$$CH_3(CH_2)_7C\equiv C(CH_2)_7CO_2H \xrightarrow[\text{2. H}^+]{\text{1. KMnO}_4, \text{HO}^-} CH_3(CH_2)_7CO_2H + HO_2C(CH_2)_7CO_2H$$

Stearolic acid   Nonanoic acid   Azelaic acid

Oxidative cleavage reactions are used primarily as a tool in structure determination. By identifying the carboxylic acids produced, it becomes a simple matter to deduce the structure of the alkyne.

PROBLEM 10.15 A certain hydrocarbon had the molecular formula $C_{16}H_{26}$ and contained two triple bonds. Ozonation followed by hydrolysis gave $CH_3(CH_2)_4CO_2H$ and $HO_2CCH_2CH_2CO_2H$ as the only products. Suggest a reasonable structure for this hydrocarbon.

10.16 FORMATION OF ACETYLIDES OF TRANSITION METALS

We have already encountered a number of metal acetylides. Calcium carbide is an ionic substance containing the dianion $[:C\equiv C:]^{2-}$. Sodium acetylide, $NaC\equiv CH$, is prepared by an acid-base reaction between acetylene and sodium amide.

A property unique to acetylene and to terminal alkynes is their reaction with salts of certain transition metal ions, most notably silver and cuprous ions, to form insoluble metal acetylides. The carbon-metal bonds are covalent and these compounds are not very basic.

$$RC\equiv CH + Ag^+ \longrightarrow RC\equiv CAg + H^+$$
Silver acetylide

$$RC\equiv CH + Cu^+ \longrightarrow RC\equiv CCu + H^+$$
Cuprous acetylide
(red precipitate)

These metal acetylides are shock-sensitive and explosive when dry. The reactions occur practically instantaneously on mixing and have been applied to the qualitative and quantitative analysis of alkynes.

PROBLEM 10.16 A certain compound C_6H_{10} gave a red precipitate on being treated with cuprous chloride in aqueous ammonia. Hydrogenation of the starting compound gave 2-methylpentane. What must be the structure of this compound?

Formation of the carbon-metal bond corresponds, at least formally, to the combining of an acetylide anion and a metal cation. It is, however, unlikely that the reaction conditions are sufficiently basic to permit the formation of a significant concentration of acetylide ions (these reactions are usually carried out in aqueous ammonia). What seems more reasonable is that the metal ion acts as an electrophile and forms a complex with the alkyne:

$$RC\equiv CH + M^+ \longrightarrow RC\!=\!\overset{+}{C}\!\!\begin{smallmatrix}M\\\\H\end{smallmatrix}$$

TABLE 10.3
Preparation of Alkynes

Reaction (section) and comments	General equation and specific example
Alkylation of acetylene and terminal alkynes (Section 10.7) The acidity of acetylene and terminal alkynes permits them to be converted to their conjugate bases on treatment with sodium amide. These anions are good nucleophiles and react with primary alkyl halides to form carbon-carbon bonds. Secondary and tertiary alkyl halides cannot be used because they yield only elimination products under these conditions.	$RC\equiv CH + NaNH_2 \longrightarrow RC\equiv CNa + NH_3$ Alkyne — Sodium amide — Sodium alkynide — Ammonia $RC\equiv CNa + R'CH_2X \longrightarrow RC\equiv CCH_2R' + NaX$ Sodium alkynide — Primary alkyl halide — Alkyne — Sodium halide $(CH_3)_3CC\equiv CH \xrightarrow[\text{2. CH}_3\text{I}]{\text{1. NaNH}_2,\ \text{NH}_3} (CH_3)_3CC\equiv CCH_3$ 3,3-Dimethyl-1-butyne — 4,4-Dimethyl-2-pentyne (96%)
Double dehydrohalogenation of geminal dihalides (Section 10.8) An E2 elimination reaction of a geminal dihalide yields an alkenyl halide. If a strong enough base is used, sodium amide, for example, a second elimination step follows the first and the alkenyl halide is converted to an alkyne.	$\underset{\text{H}\ \ \text{X}}{RC\!-\!CR'} + 2NaNH_2 \longrightarrow RC\equiv CR' + 2NaX$ (H,X above and below) Geminal dihalide — Sodium amide — Alkyne — Sodium halide $(CH_3)_3CCH_2CHCl_2 \xrightarrow[\text{2. H}_2\text{O}]{\text{1. 3NaNH}_2,\ \text{NH}_3} (CH_3)_3CC\equiv CH$ 1,1-Dichloro-3,3-dimethylbutane — 3,3-Dimethyl-1-butyne (56–60%)
Double dehydrohalogenation of vicinal dihalides (Section 10.8) Dihalides in which the halogens are on adjacent carbons undergo two elimination processes analogous to those of geminal dihalides.	$\underset{\text{X}\ \ \text{X}}{RC\!-\!CR'} + 2NaNH_2 \longrightarrow RC\equiv CR' + 2NaX$ (H,H above) Vicinal dihalide — Sodium amide — Alkyne — Sodium halide $CH_3CH_2CHCH_2Br \xrightarrow[\text{2. H}_2\text{O}]{\text{1. 3NaNH}_2,\ \text{NH}_3} CH_3CH_2C\equiv CH$ (Br below) 1,2-Dibromobutane — 1-Butyne (78–85%)

This complex then loses a proton to form the metal acetylide.

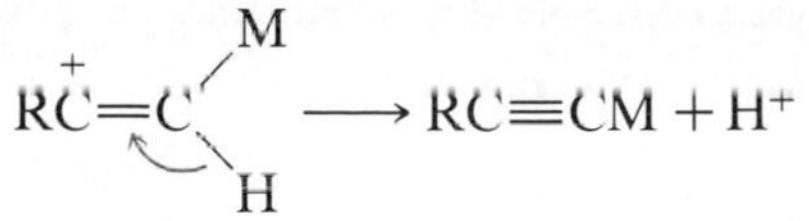

$$RC\overset{+}{=}C\underset{H}{\overset{M}{\diagdown}} \longrightarrow RC\equiv CM + H^+$$

10.17 SUMMARY

Alkynes are hydrocarbons that contain a carbon-carbon triple bond. Simple alkynes having no other functional groups or rings are represented by the general formula C_nH_{2n-2}.

The carbon-carbon triple bond in alkynes is composed of a σ and two π components. The σ component contains two electrons in an orbital generated by the overlap of sp hybridized orbitals on adjacent carbons. Each of these carbons also has two $2p$

TABLE 10.4

Conversion of Alkynes to Alkenes and Alkanes

Reaction (section) and comments	General equation and specific example
Hydrogenation of alkynes to alkanes (Section 10.10) Alkynes are completely hydrogenated, yielding alkanes, in the presence of the customary metal hydrogenation catalysts.	$RC\equiv CR' + \ 2H_2 \ \xrightarrow{\text{metal catalyst}} \ RCH_2CH_2R'$ Alkyne ⠀⠀ Hydrogen ⠀⠀⠀⠀ Alkane 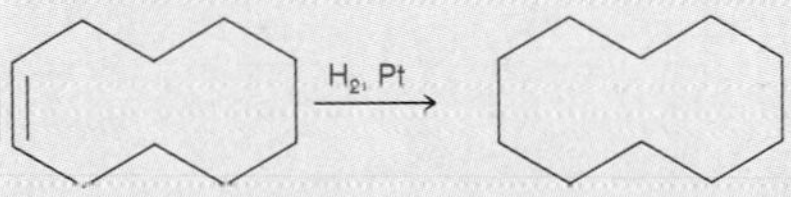 Cyclodecyne $\xrightarrow{H_2, Pt}$ Cyclodecane (71%)
Semihydrogenation of alkynes to alkenes (Section 10.10) Hydrogenation of alkynes may be halted at the alkene stage by using special catalysts. Lindlar palladium is the metal catalyst employed most often. Hydrogenation occurs with syn selectivity and yields a cis alkene.	$RC\equiv CR' + \ H_2 \ \xrightarrow{\text{Lindlar Pd}}$ 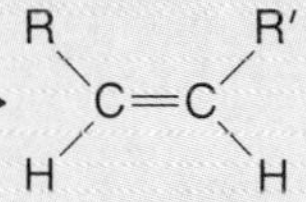 Cis alkene Alkyne ⠀⠀ Hydrogen $CH_3C\equiv CCH_2CH_2CH_2CH_3 \xrightarrow[\text{Lindlar Pd}]{H_2}$ 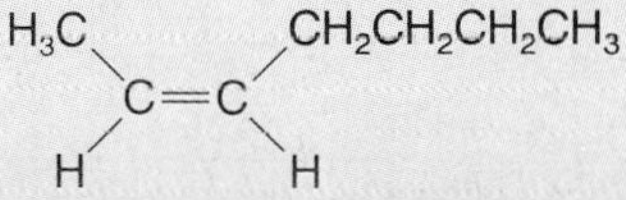 2-Heptyne ⠀⠀⠀⠀⠀⠀ cis-2-Heptene (59%)
Metal-ammonia reduction (Section 10.11) Group I metals—sodium is the one usually employed—in liquid ammonia as the solvent convert alkynes to trans alkenes. The reaction proceeds by a four-step sequence in which electron-transfer and proton-transfer steps alternate.	$RC\equiv CR' + \ 2Na \ + \ 2NH_3 \ \longrightarrow$ 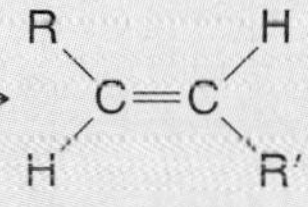 Trans alkene $ + 2NaNH_2$ Alkyne ⠀ Sodium ⠀ Ammonia ⠀⠀ Trans alkene ⠀ Sodium amide $CH_3C\equiv CCH_2CH_2CH_3 \xrightarrow[NH_3]{Na}$ 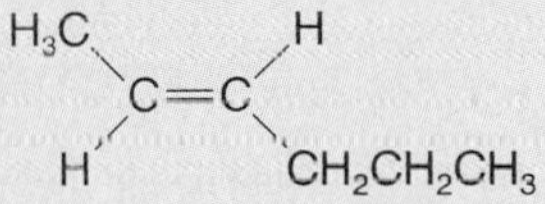 2-Hexyne ⠀⠀⠀⠀⠀⠀ trans-2-Hexene (69%)

TABLE 10.5
Electrophilic Addition to Alkynes

Reaction (section) and comments	General equation and specific example
Addition of hydrogen halides (Section 10.12) Hydrogen halides add to alkynes in accordance with Markovnikov's rule and with anti stereoselectivity to give alkenyl halides. In the presence of 2 mol hydrogen halide a second addition occurs to give a geminal dihalide.	$RC\equiv CR' \xrightarrow{HX} RCH=CR' \xrightarrow{HX} RCH_2CR'$ Alkyne Alkenyl halide Geminal dihalide $CH_3C\equiv CH + \quad 2HBr \quad \longrightarrow \quad CH_3CCH_3$ (with Br, Br) Propyne Hydrogen bromide 2,2-Dibromo-propane (100%)
Acid-catalyzed hydration (Section 10.13) Water adds to the triple bond of alkynes to yield ketones by way of an unstable enol intermediate. The enol arises by Markovnikov hydration of the alkyne, followed by rapid isomerization of the enol to a ketone.	$RC\equiv CR' + H_2O \xrightarrow[Hg^{2+}]{H_2SO_4} RCH_2CR'$ (ketone, C=O) Alkyne Water Ketone $HC\equiv CCH_2CH_2CH_2CH_3 + H_2O \xrightarrow[HgSO_4]{H_2SO_4} CH_3CCH_2CH_2CH_2CH_3$ (C=O) 1-Hexyne Water 2-Hexanone (80%)
Halogenation (Section 10.14) Addition of one equivalent of chlorine or bromine to an alkyne yields a trans dihaloalkene. A tetrahalide is formed on addition of a second mole of the halogen.	$RC\equiv CR \xrightarrow{X_2} \overset{R}{\underset{X}{}}C=C\overset{X}{\underset{R'}{}} \xrightarrow{X_2} RC-CR'$ (with X, X / X, X) Alkyne Dihaloalkene Tetrahaloalkane $CH_3C\equiv CH + 2Cl_2 \longrightarrow CH_3CCHCl_2$ (with Cl, Cl) Propyne Chlorine 1,1,2,2-Tetrachloro-propane (63%)

orbitals, which overlap in pairs so as to give two π orbitals. Carbon-carbon triple bonds are shorter and stronger than carbon-carbon double bonds. Acetylene is a linear molecule, and alkynes have a linear geometry of their C—C≡C—C units.

Acetylene and terminal alkynes are more acidic than other hydrocarbons. They have K_a's for ionization of approximately 10^{-26}, compared with about 10^{-45} for alkenes and about 10^{-60} for alkanes. An sp hybridized carbon is more electronegative than an sp^2 or sp^3 hybridized one and can better bear a negative charge.

Methods for the preparation of alkynes are summarized in Table 10.3. They include

1. Alkylation of acetylene and terminal alkynes
2. Double dehydrohalogenation of geminal dihalides
3. Double dehydrohalogenation of vicinal dihalides

Hydrogenation of alkynes can lead to a cis alkene or an alkane depending on the choice of catalyst. Triple bonds, but not double bonds, are reduced by sodium in ammonia to give trans alkenes. These reactions of alkynes are summarized in Table 10.4.

Alkynes react by electrophilic addition with many of the same reagents that add to alkenes. A summary of these reactions is presented in Table 10.5. Particularly noteworthy among these reactions is the hydration of alkynes; this reaction leads to an aldehyde or ketone by way of an unstable intermediate called an *enol*.

Alkynes are cleaved to carboxylic acids on ozonolysis or on oxidation by potassium permanganate. These reactions are often used for analytical purposes. Another analytical method that indicates the presence of the $-C\equiv CH$ functional group is the formation of insoluble metal acetylides on addition of cuprous or silver salts to a solution of an alkyne in aqueous ammonia.

PROBLEMS

10.17 Write structural formulas and give the IUPAC names for all the alkynes of molecular formula C_6H_{10}.

10.18 Provide a systematic IUPAC name for each of the following alkynes:

(a) $CH_3CH_2CH_2C\equiv CH$

(b) $ClI_3CH_2C\equiv CCH_3$

(c) $CH_3C\equiv CCHCH(CH_3)_2$
$\qquad\qquad\quad |$
$\qquad\qquad\ CH_3$

(d) ▷$-CH_2CH_2CH_2C\equiv CH$

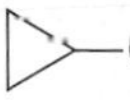

(e) [cyclic structure with $CH_2C\equiv CCH_2$]

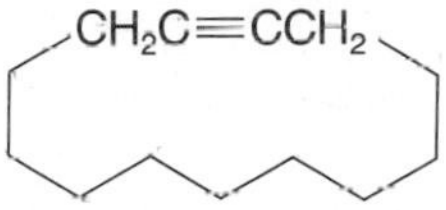

(f) $CH_3CH_2CH_2CH_2CHCH_2CH_2CH_2CH_2CH_3$
$\qquad\qquad\qquad\quad |$
$\qquad\qquad\qquad\ C\equiv CCH_3$

(g) $(CH_3)_3CC\equiv CC(CH_3)_3$

(h) $(CH_3)_3CCHC(CH_3)_3$
$\qquad\qquad |$
$\qquad\quad\ C\equiv CH$

10.19 Write a structural formula corresponding to each of the following.

(a) 1-Octyne
(b) 2-Octyne
(c) 3-Octyne
(d) 4-Octyne
(e) 2,5-Dimethyl-3-hexyne
(f) 4-Ethyl-1-hexyne
(g) Ethynylcyclohexane
(h) 3-Ethyl-3-methyl-1-pentyne

10.20 All the compounds in Problem 10.19 are isomeric except one. Which one?

10.21 Which of the compounds in Problem 10.19 will give a red precipitate with ammoniacal cuprous chloride?

10.22 An unknown acetylenic amino acid obtained from the seed of a tropical fruit has the molecular formula $C_7H_{11}NO_2$. On catalytic hydrogenation over platinum this amino acid yielded homoleucine (an amino acid of known structure) as the only product. What is the structure of the unknown amino acid?

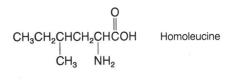

Homoleucine

10.23 Show by writing appropriate chemical equations how each of the following compounds could be converted to 1-hexyne:

(a) 1,1-Dichlorohexane
(b) 1-Hexene
(c) Acetylene
(d) 1-Iodohexane
(e) (*E*)-1-Bromohexene

10.24 Show by writing appropriate chemical equations how each of the following compounds could be converted to 3-hexyne:

(a) 1-Butene
(b) 1,1-Dichlorobutane
(c) 1-Chlorobutene
(d) Acetylene

10.25 When 1,2-dibromodecane was treated with potassium hydroxide in aqueous ethanol, it yielded a mixture of three isomeric compounds of molecular formula $C_{10}H_{19}Br$. Each of these compounds was converted to 1-decyne on reaction with sodium amide in dimethyl sulfoxide. Identify these three compounds.

10.26 Write the structure of the major organic product isolated from the reaction of 1-hexyne with:

$$CH \equiv CCH_2CH_2CH_2CH_3$$

(a) Hydrogen (2 mol), platinum
(b) Hydrogen (1 mol), Lindlar palladium
(c) Lithium in liquid ammonia
(d) Sodium amide in liquid ammonia
(e) Product in (*d*) treated with 1-bromobutane
(f) Product in (*d*) treated with *tert*-butyl bromide
(g) Hydrogen chloride (1 mol)
(h) Hydrogen chloride (2 mol)
(i) Chlorine (1 mol)
(j) Chlorine (2 mol)
(k) Aqueous sulfuric acid, mercuric sulfate
(l) Ammoniacal cuprous chloride
(m) Ammoniacal silver nitrate
(n) Ozone followed by hydrolysis

10.27 Write the structure of the major organic product isolated from the reaction of 3-hexyne with:

(a) Hydrogen (2 mol), platinum
(b) Hydrogen (1 mol), Lindlar palladium
(c) Lithium in liquid ammonia
(d) Hydrogen chloride (1 mol)
(e) Hydrogen chloride (2 mol)
(f) Chlorine (1 mol)
(g) Chlorine (2 mol)
(h) Aqueous sulfuric acid, mercuric sulfate
(i) Ammoniacal cuprous chloride
(j) Ammoniacal silver nitrate
(k) Ozone followed by hydrolysis

10.28 When 2-heptyne was treated with aqueous sulfuric acid containing mercuric sulfate, two products, each having the molecular formula $C_7H_{14}O$, were obtained in approximately equal amounts. What are these two compounds?

10.29 The alkane formed by hydrogenation of (S)-4-methyl-1-hexyne is optically active while the one formed by hydrogenation of (S)-3-methyl-1-pentyne is not. Explain. Would you expect the products of semihydrogenation of these two compounds in the presence of Lindlar palladium to be optically active?

10.30 All the following reactions have been described in the chemical literature and proceed in good yield. In some cases the reactants are more complicated than those we have so far encountered. Nevertheless, based on what you have already learned, you should be able to predict the principal product in each case.

(a) $NaC \equiv CH + ClCH_2CH_2CH_2CH_2CH_2CH_2I \longrightarrow$

(b) $BrCH_2CHCH_2CH_2CHCH_2Br \xrightarrow[\text{2. H}_2\text{O}]{\text{1. excess NaNH}_2,\ \text{NH}_3}$
         | |
        Br   Br

(c) $\xrightarrow[\text{heat}]{\text{KOC(CH}_3)_3,\ \text{DMSO}}$

(d) $-C \equiv CNa + CH_3CH_2O\overset{O}{\underset{O}{\overset{\|}{\underset{\|}{S}}}}$ $-CH_3 \longrightarrow$

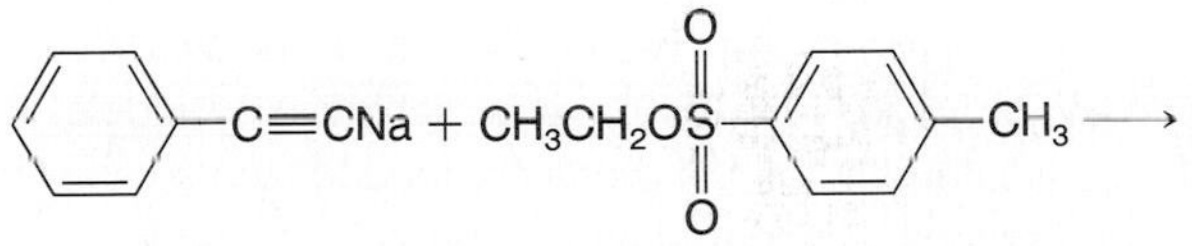

(e) Cyclodecyne $\xrightarrow[\text{2. H}_2\text{O}]{\text{1. O}_3}$

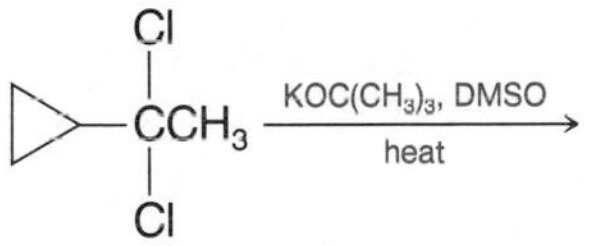

(f) $\xrightarrow[\text{2. H}_2\text{O}]{\text{1. O}_3}$

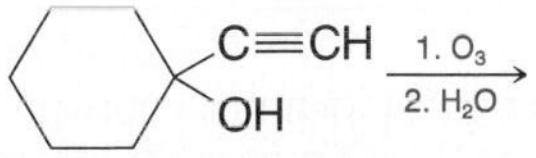

(g) $CH_3CHCH_2\overset{OH}{\underset{CH_3}{\overset{|}{\underset{|}{C}}}}C \equiv CH \xrightarrow[\text{HgO}]{\text{H}_2\text{O, H}_2\text{SO}_4}$
         |
        CH_3

(h) $(Z) - CH_3CH_2CH_2CH_2CH = CHCH_2(CH_2)_7C \equiv CCH_2CH_2OH \xrightarrow[\text{2. H}_2\text{O}]{\text{1. Na, NH}_3}$

(i) $+ NaC{\equiv}CCH_2CH_2CH_2CH_3 \longrightarrow$

(j) Product of (i) $\xrightarrow[\text{Lindlar Pd}]{H_2}$

10.31 The ketone 2-heptanone has been identified as contributing to the odor of a number of dairy products, including condensed milk and cheddar cheese. Describe a synthesis of 2-heptanone from acetylene and any necessary organic or inorganic reagents.

$$CH_3\overset{\overset{\textstyle O}{\|}}{C}CH_2CH_2CH_2CH_2CH_3$$

2-Heptanone

10.32 (Z)-9-Tricosene [(Z)—$CH_3(CH_2)_7CH{=}CH(CH_2)_{12}CH_3$] is the sex pheromone of the female housefly. Synthetic (Z)-9-tricosene is used as bait to lure male flies to traps that contain insecticide. Using acetylene and alcohols of your choice as starting materials, along with any necessary inorganic reagents, show how you could prepare (Z)-9-tricosene.

10.33 Show by writing a suitable series of equations how you could prepare each of the following compounds from the designated starting materials and any necessary organic or inorganic reagents.

(a) 2,2-Dibromopropane from 1,1-dibromopropane
(b) 2,2-Dibromopropane from 1,2-dibromopropane
(c) 2-Chloropropene from 1-bromopropene
(d) 1,1,2,2-Tetrachloropropane from 1,2-dichloropropane
(e) 2,2-Diiodobutane from acetylene and ethyl bromide
(f) 1-Hexene from 1-butene and acetylene
(g) Decane from 1-butene and acetylene
(h) Cyclopentadecyne from cyclopentadecene

(i) from and methyl bromide

10.34 Assume that you need to prepare 4-methyl-2-pentyne and discover that the only alkynes on hand are acetylene and propyne. You also have available methyl iodide, isopropyl bromide, and 1,1-dichloro-3-methylbutane. Which of these compounds would you choose in order to perform your synthesis, and how would you carry it out?

10.35 Compound A has the molecular formula $C_{14}H_{25}Br$ and was obtained by reaction of sodium acetylide with 1,12-dibromododecane. On treatment of compound A with sodium amide, it was converted to compound B ($C_{14}H_{24}$). Ozonolysis of compound B gave the diacid $HO_2C(CH_2)_{12}CO_2H$. Catalytic hydrogenation of compound B over Lindlar palladium gave compound C ($C_{14}H_{26}$), while hydrogenation over platinum gave compound D ($C_{14}H_{28}$). Sodium-ammonia reduction of compound B gave compound E ($C_{14}H_{26}$). Both C and E yielded $HO_2C(CH_2)_{12}CO_2H$ on oxidation with potassium permanganate. Compound D was inert to potassium permanganate. Assign structures to compounds A through E so as to be consistent with the observed transformations.

CONJUGATION IN ALKADIENES AND ALLYLIC SYSTEMS

Not all the properties of alkenes can be understood by focusing attention solely on its pair of doubly bonded carbons. In this chapter we will examine how the interaction of the π electrons of an alkene with some other functional unit within a molecule affects chemical reactivity.

Among the systems to be considered are two types of reactive intermediates, *allylic cations* and *allylic free radicals*.

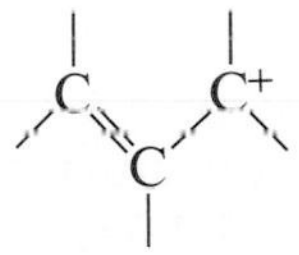

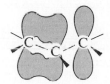

Allylic cation

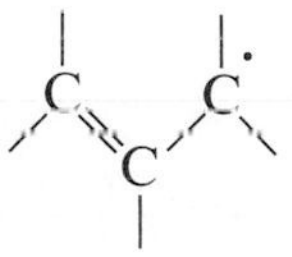

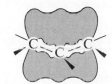

Allylic free radical

$$CH_2 = CH - CH_2$$

Allylic cations and allylic free radicals are examples of *conjugated* π electron systems, i.e., species in which electrons are delocalized over a π orbital that encompasses more than two atoms. In an allylic cation the π orbital of the double bond overlaps with the vacant $2p$ orbital of the positively charged carbon to give an extended π orbital, one that encompasses three carbon atoms. In an allylic free radical the π orbital overlaps with the half-filled $2p$ orbital of the adjacent carbon to give an extended π orbital.

π Orbital of double bond
and $2p$ orbital of carbon

Extended π orbital generated
by orbital overlap

Joining two alkene units by a single bond provides a *conjugated diene*. The two π orbitals overlap to give an extended π system encompassing four carbons.

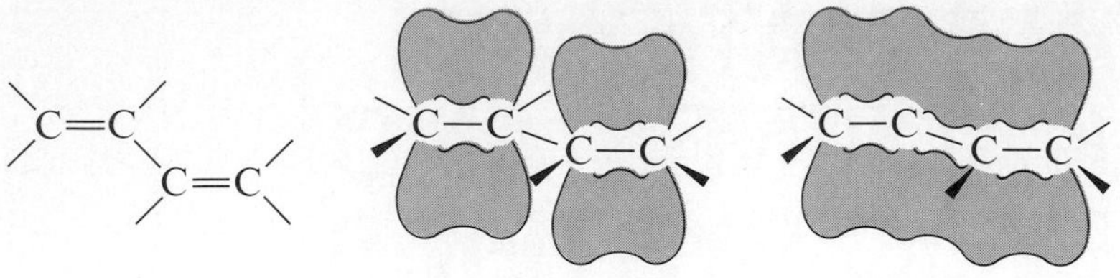

Conjugated diene	Individual π orbitals of two double bonds	Extended π orbital of conjugated system

The concept of the functional group has served us well to this point as a device to organize patterns of organic chemical reactivity. In this chapter we consider how conjugation permits two functional units within a molecule to interact with each other so as to display a kind of reactivity that is qualitatively different from that of either unit alone.

11.1 THE ALLYL GROUP

The common name for the group CH_2=$CHCH_2$— is *allyl.* The compounds shown below are much better known by their common names than by their systematic ones, and these common names are acceptable in the IUPAC system.

$$CH_2\text{=}CHCH_2OH \qquad CH_2\text{=}CHCH_2Cl \qquad CH_2\text{=}CHCH_2Br$$

Allyl alcohol (2-propen-1-ol)	Allyl chloride (3-chloro-1-propene)	Allyl bromide (3-bromo-1-propene)

The adjective *allylic* denotes the structural unit C=C—C. When we say an atom or a group is an *allylic substituent,* we mean that it is attached to the sp^3 hybridized carbon of an allylic unit, as in C=C—C—X. Thus, the following compounds represent an allylic alcohol and an allylic chloride, respectively. These compounds are named according to the IUPAC rules of systematic nomenclature.

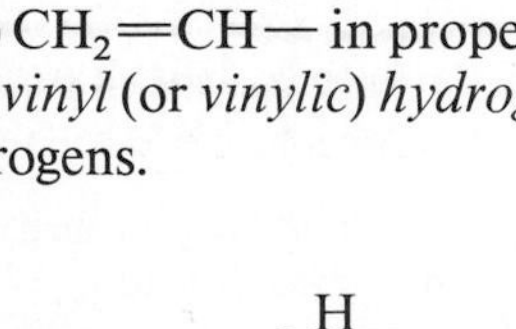

3-Methyl-2-buten-1-ol 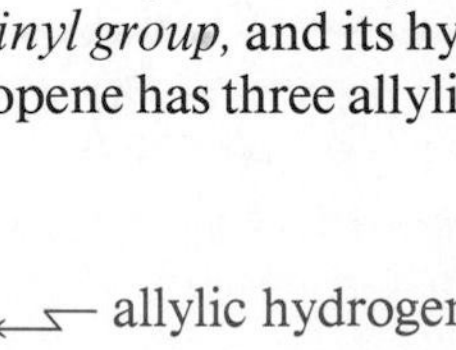   3-Chloro-3-methyl-1-butene

The hydrogens of the methyl group in propene, CH_2=$CHCH_3$, are allylic hydrogens. The group CH_2=CH— in propene is a *vinyl group,* and its hydrogen substituents are termed *vinyl* (or *vinylic*) *hydrogens.* Propene has three allylic hydrogens and three vinyl hydrogens.

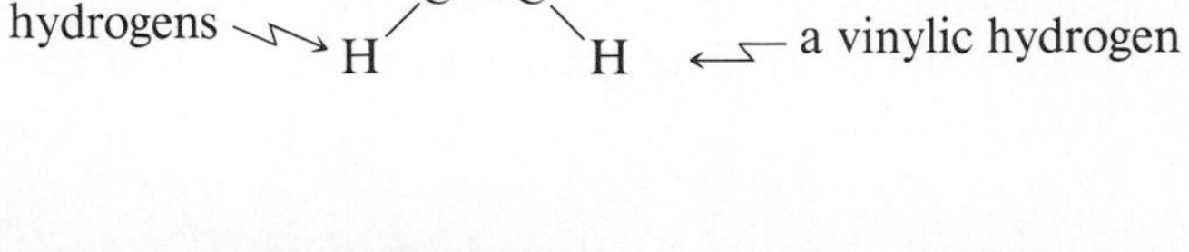

11.2   ALLYLIC CARBOCATIONS

Allylic carbocations are carbocations that have a vinyl group or substituted vinyl group as a substituent on their positively charged carbon. The allyl cation is the simplest allylic carbocation.

Representative allylic carbocations:

$$CH_2=CHCH_2^+ \qquad CH_3CH=\overset{+}{C}HCHCH_3$$

Allyl cation   1-Methyl-2-butenyl   2-Cyclopentenyl
                              cation   cation

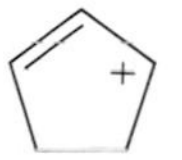

A substantial body of evidence indicates that allylic carbocations are more stable than simple alkyl cations. Compare, for example, the first-order rate constant k for the solvolysis of a typical tertiary alkyl chloride with that for solvolysis of a chloride that is both tertiary and allylic.

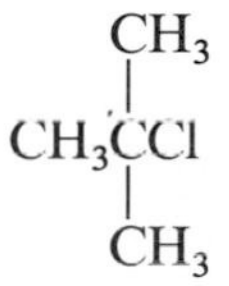

$$\underset{\underset{CH_3}{|}}{\overset{\overset{CH_3}{|}}{CH_3CCl}} \qquad\qquad \underset{\underset{CH_3}{|}}{\overset{\overset{CH_3}{|}}{CH_2=CHCCl}}$$

tert-Butyl chloride   3-Chloro-3-methyl-1-butene
Less reactive: k(rel) 1.0   More reactive: k(rel) 123

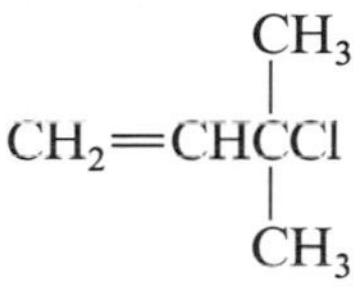

The allylic chloride 3-chloro-3-methyl-1-butene is over 100 times more reactive toward ethanolysis (at 45°C) than is *tert*-butyl chloride. Both compounds undergo ethanolysis by an S_N1 mechanism, and their relative rates reflect their activation energies for carbocation formation. Since the allylic chloride is more reactive, we infer that it ionizes more rapidly because it forms a more stable carbocation. Structurally, the two carbocations differ in that the allylic carbocation has a vinyl substituent on its positively charged carbon in place of one of the methyl groups of *tert*-butyl cation.

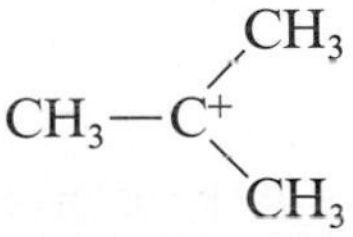

tert-Butyl cation   1,1-Dimethylallyl
(less stable)   cation
                               (more stable)

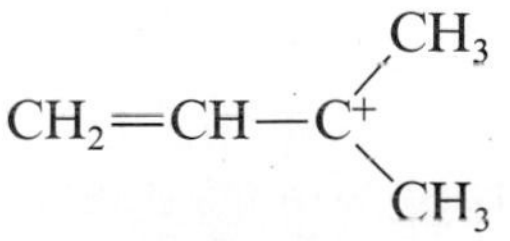

A vinyl group stabilizes a carbocation more than does a methyl group. Why?

A vinyl group is an extremely effective electron-releasing substituent. A resonance interaction of the type shown permits its π electrons to be delocalized and disperses the positive charge.

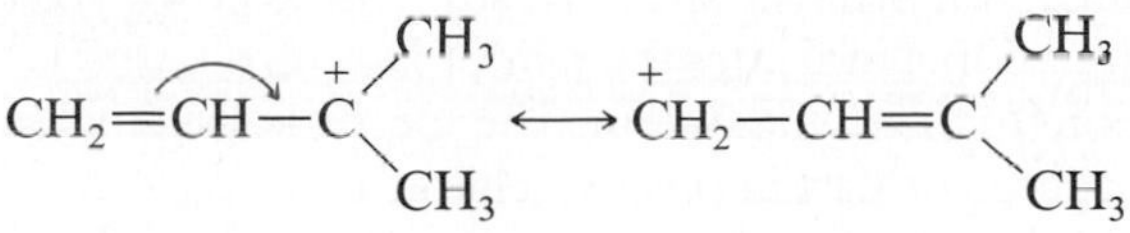

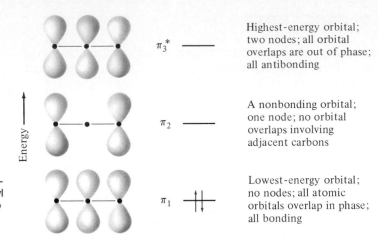

π_3^* —— Highest-energy orbital; two nodes; all orbital overlaps are out of phase; all antibonding

π_2 —— A nonbonding orbital; one node; no orbital overlaps involving adjacent carbons

π_1 ⥮ Lowest-energy orbital; no nodes; all atomic orbitals overlap in phase; all bonding

Energy →

FIGURE 11.1 The π molecular orbitals of allyl cation. Allyl cation has two π electrons, so only the lowest-energy orbital is occupied.

Because it is a resonance-stabilized species, this allylic carbocation is formed faster than *tert*-butyl cation. Allylic halides undergo ionization to form carbocations faster than do alkyl halides.

PROBLEM 11.1 Write a second resonance structure for each of the following carbocations:

(a) $CH_3CH{=}CHCH_2^+$

(b) $CH_2{=}CCH_2^+$
          |
         CH_3

(c) ⬡=$C(CH_3)_2$
       (+)

SAMPLE SOLUTION (a) When writing resonance forms of carbocations, electrons are moved in pairs from sites of high electron density toward the positively charged carbon.

$$CH_3CH{=}CH{-}\overset{+}{C}H_2 \longleftrightarrow CH_3\overset{+}{C}H{-}CH{=}CH_2$$

In orbital terms, as portrayed for allyl cation in Figure 11.1, the $2p$ orbitals of three adjacent sp^2 hybridized carbons overlap to give three π orbitals, each of which encompasses all three carbons of allyl cation. One of these orbitals is bonding, one is nonbonding, and the third is antibonding. There are two π electrons in the carbocation, so the lowest-energy orbital (bonding) is filled while the two π orbitals of higher energy are vacant.

A comparable picture of electron delocalization in allyl cation, concerned only with the lowest-energy orbital, is shown in Figure 11.2. It depicts the interaction of the π electrons of the double bond with the vacant $2p$ orbital on an adjacent positively charged carbon.

Electron delocalization in allylic species is sometimes portrayed by using a dotted-line notation. A dotted line indicates the pair of π electrons and is drawn to encompass the three carbons of the allylic system. The species is specified as a carbocation by adding a positive charge or partial positive charges.

$$\underset{\text{Allyl cation}}{\overset{\displaystyle H}{\underset{\displaystyle H}{\overset{\displaystyle \overset{\displaystyle C}{\parallel}}{\underset{\displaystyle C}{\overset{+}{\diagdown}}}}}}$$

charge divided in half, so more stable

Since the positive charge in an allylic carbocation is shared by two carbons, there are two potential sites at which it may be captured by a nucleophile. Thus, hydrolysis of 3-chloro-3-methyl-1-butene gives a mixture of two allylic alcohols.

$$(CH_3)_2\underset{\underset{Cl}{|}}{C}CH=CH_2 \xrightarrow[Na_2CO_3]{H_2O} (CH_3)_2\underset{\underset{OH}{|}}{C}CH=CH_2 + (CH_3)_2C=CHCH_2OH$$

3-Chloro-3-methyl-   2-Methyl-3-buten-   3-Methyl-2-buten-
1-butene   2-ol (85%)   1-ol (15%)

Both alcohols are formed from the same carbocation. Water may react with the carbocation so as to give either a primary alcohol or a secondary alcohol.

$$\begin{array}{c}\underset{H_3C}{\overset{H_3C}{\diagdown}}\overset{+}{C}-CH=CH_2 \\[4pt] \mathbf{A} \\[2pt] \updownarrow \\[6pt] \underset{H_3C}{\overset{H_3C}{\diagdown}}C=CH-\overset{+}{C}H_2 \\[4pt] \mathbf{B}\end{array} \qquad \xrightarrow{H_2O} (CH_3)_2\underset{\underset{OH}{|}}{C}CH=CH_2 + (CH_3)_2C=CHCH_2OH$$

2-Methyl-3-buten-2-ol   3-Methyl-2-buten-1-ol

It should be emphasized that we are not dealing with an equilibrium between two isomeric carbocations. There is only one carbocation. Its structure is not adequately represented by either of the individual resonance forms but is a hybrid having qualities of both of them. The carbocation has more of the character of A than B because tertiary carbocation A is a more stable structure than primary carbocation B. Water attacks faster at the carbon atom that bears the greater share of the positive charge. The tertiary carbon bears more of the positive charge than the primary carbon because the carbocation has more of the character of structure A than structure B.

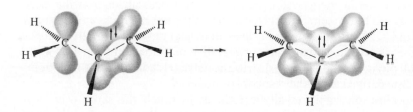

FIGURE 11.2 A representation of electron delocalization in allyl cation. The vacant 2*p* orbital of the positively charged carbon and the π orbital of the double bond overlap to give an extended π orbital that encompasses all three carbons.

The same two alcohols are formed in the hydrolysis of 1-chloro-3-methyl-2-butene.

$$(CH_3)_2C=CHCH_2Cl \xrightarrow[Na_2CO_3]{H_2O} (CH_3)_2\underset{\underset{OH}{|}}{C}CH=CH_2 + (CH_3)_2C=CHCH_2OH$$

1-Chloro-3-methyl-2-butene	2-Methyl-3-buten-2-ol (85%)	3-Methyl-2-buten-1-ol (15%)

The carbocation formed on ionization of 1-chloro-3-methyl-2-butene is the same allylic carbocation as the one formed on ionization of 3-chloro-3-methyl-1-butene. The same cation intermediate gives the same mixture of products irrespective of the starting allylic halide from which it was formed.

PROBLEM 11.2   From among the following compounds, choose the two that yield the same carbocation on ionization: 3-bromo-1-methylcyclohexene; 4-bromo-1-methylcyclohexene; 5-chloro-1-methylcyclohexene; 3-chloro-3-methylcyclohexene; 1-bromo-3-methylcyclohexene

---

Later in this chapter we will see how allylic carbocations are involved in electrophilic addition to dienes and how the principles developed in this section are applicable there as well.

11.3 ALLYLIC FREE RADICALS

Just as allyl cation is stabilized by resonance, so also is allyl radical.

$$CH_2=CH-\overset{\bullet}{C}H_2 \longleftrightarrow \overset{\bullet}{C}H_2-CH=CH_2 \quad \text{or}$$

Allyl radical

Allyl radical is a conjugated system in which a singly occupied $2p$ orbital overlaps with the π orbital of an adjacent double bond to give an extended π system. The π electrons are delocalized over all three carbons. The unpaired electron has an equal probability of being found at C-1 or C-3.

Figure 11.3 shows the π molecular orbitals of allyl radical. Everything about this diagram is the same as that of Figure 11.1 for allyl cation except for the number of electrons. There are three π electrons in allyl radical but only two in allyl cation. Two of the π electrons of allyl radical fill the lowest-energy orbital, while the third is found in the orbital next higher in energy. All these electrons are delocalized over the three carbons that make up the system, but the electron of highest energy is in an orbital that has a node at C-2. Thus, the unpaired electron in allyl radical is equally likely to be at C-1 or C-3; it has an almost zero probability of being found at C-2. The molecular orbital picture of the electron distribution in allyl radical leads to a conclusion similar to that derived from the resonance model.

Electron delocalization stabilizes allylic radicals in much the same way as it renders allylic carbocations more stable than alkyl cations. Reactions that generate

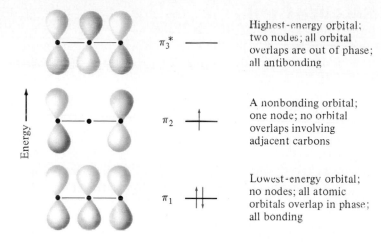

Energy →

π_3^* ——

Highest-energy orbital; two nodes; all orbital overlaps are out of phase; all antibonding

π_2 —|—

A nonbonding orbital; one node; no orbital overlaps involving adjacent carbons

π_1 —||—

Lowest-energy orbital; no nodes; all atomic orbitals overlap in phase; all bonding

FIGURE 11.3 The π molecular orbitals of allyl radical. Allyl radical has three π electrons, so π_1 is doubly occupied and π_2 is singly occupied.

allylic radicals occur more readily than those involving simple alkyl radicals. This can be seen clearly by comparing the bond dissociation energies of the primary C—H bonds of propane and propene.

$$CH_3CH_2CH_2 \!:\! H \longrightarrow CH_3CH_2\dot{C}H_2 + \cdot H \qquad \Delta H° = +98 \text{ kcal/mol}$$

| Propane | 1-Propyl radical | Hydrogen atom |

$$CH_2{=}CHCH_2 \!:\! H \longrightarrow \dot{C}H_2{=}CH\dot{C}H_2 + \cdot H \qquad \Delta H° = +88 \text{ kcal/mol}$$

| Propene | Allyl radical | Hydrogen atom |

It is easier, by 10 kcal/mol, to abstract a primary hydrogen from propene than from propane. The primary hydrogen in propene is allylic; none of the hydrogens in propane are allylic.

PROBLEM 11.3   Identify the allylic hydrogens in

(a) Cyclohexene   (c) 2,3,3-Trimethyl-1-butene
(b) 1-Methylcyclohexene   (d) 1-Octene

SAMPLE SOLUTION   (a) Allylic hydrogens are hydrogen substituents on an allylic carbon. An allylic carbon is an sp^3 hybridized carbon that is attached directly to an sp^2 hybridized carbon of an alkene. Cyclohexene has four allylic hydrogens.

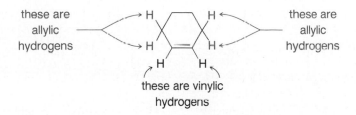

these are allylic hydrogens

these are allylic hydrogens

these are vinylic hydrogens

11.4 ALLYLIC HALOGENATION

Of the reactions that involve carbon radicals, the ones with which you are most familiar are the chlorination and bromination of alkanes (Sections 4.20 through 4.25).

$$RH + X_2 \xrightarrow{\text{heat or light}} RX + HX$$

Alkane   Halogen   Alkyl halide   Hydrogen halide

You have also seen that alkenes react with chlorine and bromine by electrophilic addition to their double bonds rather than by free-radical substitution (Sections 7.11 through 7.13).

$$R_2C{=}CR_2' + X_2 \longrightarrow R_2C{-}CR_2'$$

     | |
     X   X
Alkene   Halogen   Vicinal dihalide

At high temperatures, however, propene and other alkenes react with chlorine and bromine by substitution of their allylic hydrogens. This forms the basis of an industrial preparation of allyl chloride.

$$CH_2{=}CHCH_3 + Cl_2 \xrightarrow{500°C} CH_2{=}CHCH_2Cl + HCl$$

Propene   Chlorine   Allyl chloride (80–85%)   Hydrogen chloride

The reaction is a free-radical one, proceeding by way of the propagation steps

$$CH_2{=}CHCH_2{:}H + \cdot\ddot{C}l{:} \longrightarrow CH_2{=}CH\dot{C}H_2 + H{:}\ddot{C}l{:}$$

Propene   Chlorine atom   Allyl radical   Hydrogen chloride

$$CH_2{=}CH\dot{C}H_2 + {:}\ddot{C}l{:}\ddot{C}l{:} \longrightarrow CH_2{=}CHCH_2\ddot{C}l{:} + \cdot\ddot{C}l{:}$$

Allyl radical   Chlorine   Allyl chloride   Chlorine atom

Allylic bromination reactions are normally carried out by using one of a number of specialized reagents developed for that purpose. *N*-Bromosuccinimide (NBS) is the most frequently used of these reagents. An alkene is dissolved in carbon tetrachloride, *N*-bromosuccinimide is added, and the reaction mixture is heated, illuminated with a sun lamp, or both. The products are an allylic halide and succinimide.

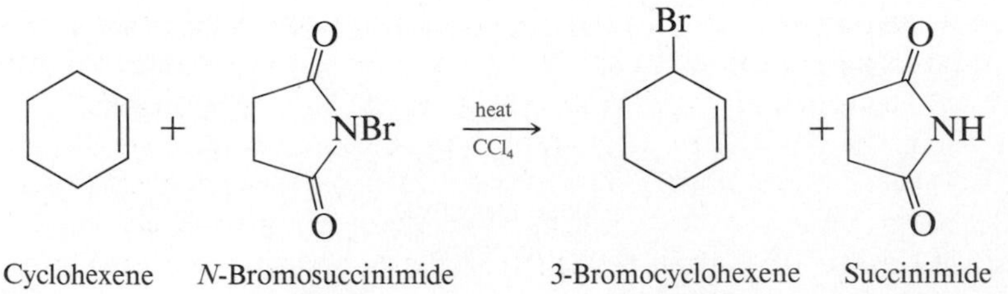

Cyclohexene   *N*-Bromosuccinimide (NBS)   3-Bromocyclohexene (82–87%)   Succinimide

N-Bromosuccinimide provides a low concentration of molecular bromine, which reacts with alkenes by a mechanism analogous to that of other free-radical halogenations. Electrophilic addition to the double bond to give a vicinal dibromide via a bromonium ion intermediate is not observed in these reactions. Because the bromonium ion intermediate is formed reversibly, at low bromine concentrations it reverts to alkene and bromine faster than it is captured by bromide ion. The free-radical intermediate in allylic halogenation, on the other hand, is formed irreversibly. While an alkene and bromine may lead more rapidly to a bromonium ion than to an allyl radical, a low concentration of bromine ensures that more of the product is derived from the radical than from the bromonium ion.

Allylic halogenation is normally used only with alkenes that yield allyl radicals in which the two resonance forms are equivalent. In the preceding example the critical intermediate is a cyclohexenyl radical. Delocalization of the unpaired electron by allylic resonance in cyclohexenyl radical leads to two equivalent structures.

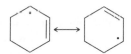

Cyclohexenyl radical

Reaction with bromine at either site yields the same product, 3-bromocyclohexene. In cases in which the resonance forms of the allylic radical are not equivalent, mixtures of allylic bromides are possible.

PROBLEM 11.4 The two alkenes 2,3,3-trimethyl-1-butene and 1-octene were each subjected to allylic halogenation with *N*-bromosuccinimide. One of these alkenes yielded a single allylic bromide, whereas the other gave a mixture of two constitutionally isomeric allylic bromides. Match the chemical behavior to the correct alkene and give the structure of the allylic bromide(s) formed from each.

In our earlier discussion of free-radical halogenation we ascribed selectivity in bromination reactions to an electronic effect. Bromine atoms tend to abstract the hydrogen atom that leaves the most stable free radical behind. Here we see another example of the same effect. Allyl radical is stabilized by resonance, and bromine atoms show a high selectivity for abstraction of allylic hydrogens. This selectivity is exploited in the preparation of allylic bromides from alkenes by using *N*-bromosuccinimide as a reagent specifically developed to carry out that transformation.

11.5 CLASSES OF DIENES

Allylic carbocations and allylic radicals are conjugated systems that are involved as reactive intermediates in chemical reactions. The third type of conjugated system that we will examine, *conjugated dienes,* consists of stable molecules.

When a molecule contains two double bonds, it is called a *diene,* and the relationship between the double bonds may be described as *isolated, conjugated,* or *cumulated. Isolated diene* units are those in which two carbon-carbon double bond units are separated from each other by one or more *sp*³ hybridized carbon atoms. 1,4-Pentadiene and 1,5-cyclooctadiene have isolated double bonds.

$$CH_2=CHCH_2CH=CH_2$$

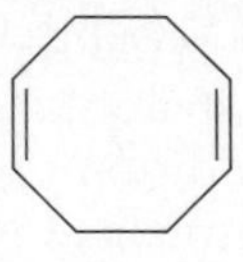

1,4-Pentadiene   1,5-Cyclooctadiene

Conjugated dienes are those in which two carbon-carbon double bond units are directly connected to each other by a single bond. 1,3-Pentadiene and 1,3-cyclooctadiene are dienes which contain conjugated double bonds.

$$CH_2=CH-CH=CHCH_3$$

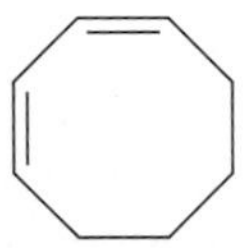

1,3-Pentadiene   1,3-Cyclooctadiene

Cumulated dienes are those in which one carbon atom is common to two carbon-carbon double bonds. The simplest cumulated diene is 1,2-propadiene, also called allene, and compounds of this class are more usually called *allenes*.

$$CH_2=C=CH_2$$

1,2-Propadiene

(Allene is an acceptable IUPAC name for 1,2-propadiene.)

PROBLEM 11.5   Many naturally occurring substances contain several carbon-carbon double bonds: some isolated, some conjugated, and some cumulated. Identify the types of carbon-carbon double bonds found in each of the following substances:

(a) β-Springene (a scent substance obtained from the dorsal gland of springboks)

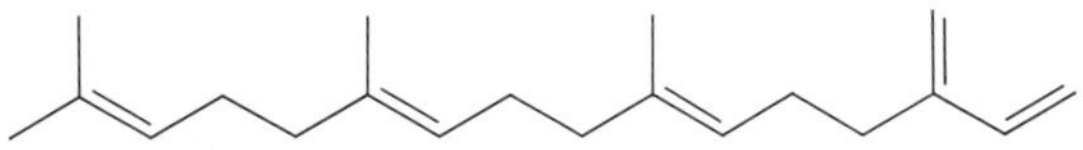

(b) Humulene (found in hops and oil of cloves)

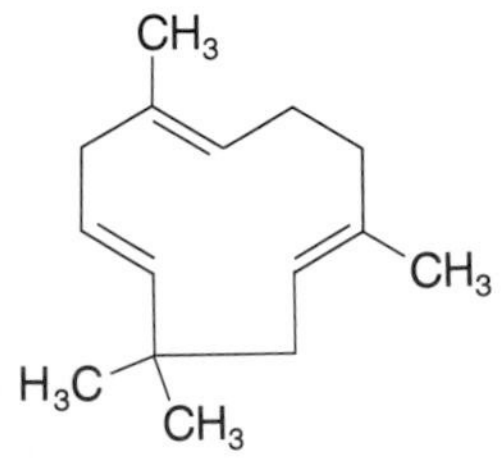

(c) Cembrene (occurs in pine resin)

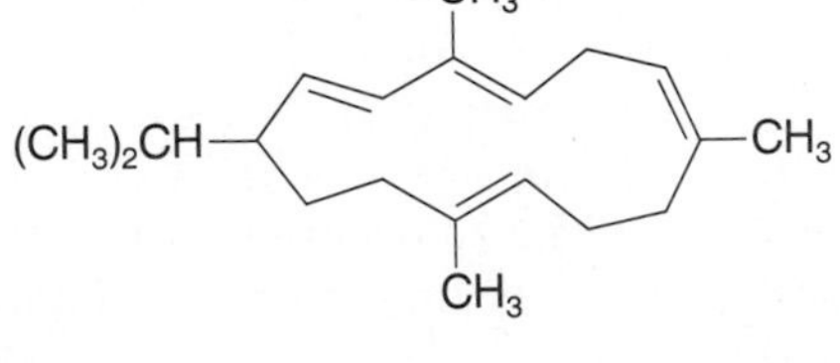

(d) The sex attractant of the male dried-bean beetle

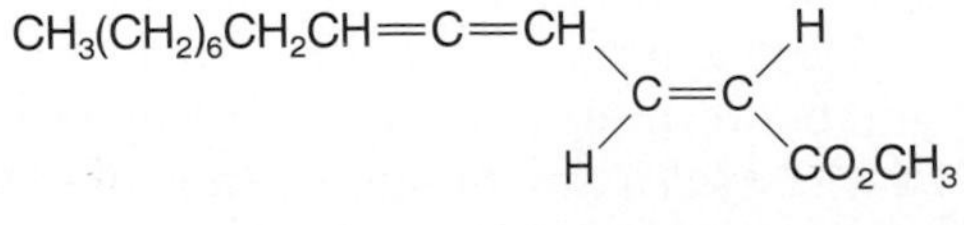

SAMPLE SOLUTION   (a)   As indicated in the structural formula shown below, β-springene has three isolated double bonds and a pair of conjugated double bonds.

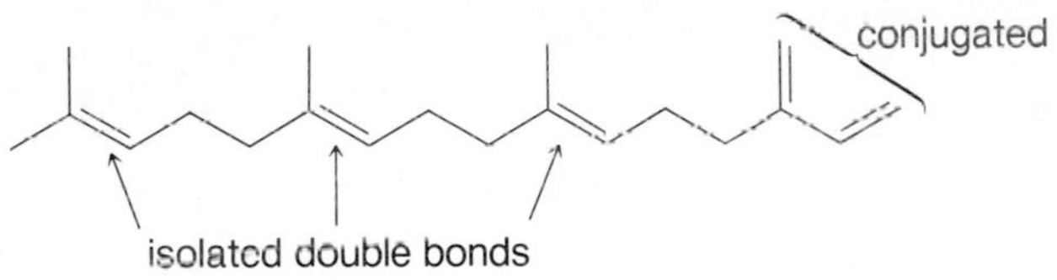

Isolated double bonds are separated from other double bonds by at least one sp^3 hybridized carbon. Conjugated double bonds are joined by a single bond.

As may be apparent from some of the examples, dienes are named according to the IUPAC convention by replacing the -*ane* ending of an alkane with -*adiene* and locating the position of each double bond by number. The IUPAC term for this general class is *alkadiene*. In an analogous manner, compounds with three carbon-carbon double bonds are called alkatrienes and named accordingly, those with four double bonds are alkatetraenes, and so on.

11.6 PREPARATION OF DIENES

Dienes with isolated double bonds are prepared by the same methods that are commonly used for alkene synthesis. These include, for example, the dehydrohalogenation of alkyl halides.

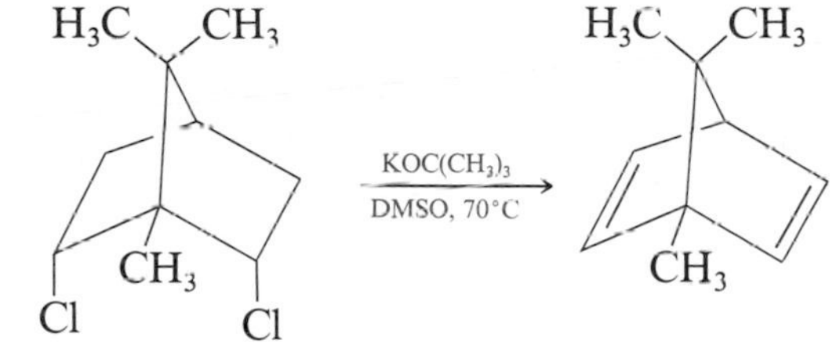

2,6-Dichlorocamphane   Bornadiene (83%)

The conjugated diene 1,3-butadiene is the starting material in the manufacture of synthetic rubber and is prepared on an industrial scale in vast quantities. Production in the United States is presently 2.5×10^9 lb/year. One industrial process is similar to that used for the preparation of ethylene: in the presence of a suitable catalyst, butane undergoes thermal dehydrogenation to yield 1,3-butadiene.

$$CH_3CH_2CH_2CH_3 \xrightarrow[\text{chromia-alumina}]{590-675^\circ C} CH_2{=}CHCH{=}CH_2 + 2H_2$$

Laboratory syntheses of conjugated dienes can be achieved by elimination reactions of alcohols and alkyl halides.

$$CH_2{=}CHCH_2\underset{\underset{OH}{|}}{\overset{\overset{CH_3}{|}}{C}}CH_2CH_3 \xrightarrow{KHSO_4,\ heat} CH_2{=}CHCH{=}\underset{}{\overset{\overset{CH_3}{|}}{C}}CH_2CH_3$$

3-Methyl-5-hexen-3-ol   4-Methyl-1,3-hexadiene (88%)

$$CH_2{=}CHCH_2\underset{\underset{Br}{|}}{\overset{\overset{CH_3}{|}}{C}}CH_2CH_3 \xrightarrow{KOH,\ heat} CH_2{=}CHCH{=}\overset{\overset{CH_3}{|}}{C}CH_2CH_3$$

4-Bromo-4-methyl-1-hexene   4-Methyl-1,3-hexadiene (78%)

$$(CH_3)_2\underset{\underset{OH}{|}}{C}{-}\underset{\underset{OH}{|}}{C}(CH_3)_2 \xrightarrow{HBr,\ heat} CH_2{=}\underset{\underset{CH_3}{|}}{C}{-}\underset{\underset{CH_3}{|}}{C}{=}CH_2$$

2,3-Dimethyl-2,3-butanediol   2,3-Dimethyl-1,3-butadiene (55–60%)

We will not discuss the preparation of cumulated dienes. The reactions used are very specialized and do not illustrate any general principles.

11.7 RELATIVE STABILITIES OF ALKADIENES

Which is the most stable arrangement of double bonds in an alkadiene—isolated, conjugated, or cumulated?

As we have seen in Chapter 5, the stabilities of alkenes may be assessed by comparing their heats of hydrogenation. Figure 11.4 depicts the heats of hydrogenation of an isolated diene (1,4-pentadiene) and a conjugated diene (1,3-pentadiene), along with the alkenes 1-pentene and (E)-2-pentene. The figure shows that an isolated pair of double bonds behaves much like two independent alkene units. The measured heat of hydrogenation of the two double bonds in 1,4-pentadiene is 60.2 kcal/mol, exactly twice the heat of hydrogenation of 1-pentene. Further, the heat evolved on hydro-

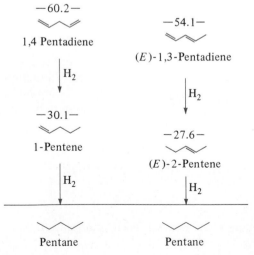

FIGURE 11.4 Heats of hydrogenation of some C_5H_{10} alkenes and C_5H_8 alkadienes. All energies are in kilocalories per mole.

genation of each double bond must be 30.1 kcal/mol, since 1-pentene is an intermediate in the hydrogenation of 1,4-pentadiene to pentane.

By the same reasoning hydrogenation of the terminal double bond in the conjugated diene (E)-1,3-pentadiene gives off only 26.5 kcal/mol when it is hydrogenated to (E)-2-pentene. Hydrogenation of the terminal double bond in the conjugated diene evolves 3.6 kcal/mol less heat than hydrogenation of a terminal double bond in the diene with isolated double bonds. A conjugated double bond is thus 3.6 kcal/mol more stable than a simple double bond. We call this increased stability due to conjugation the *delocalization energy, resonance energy,* or *conjugation energy.*

PROBLEM 11.6   Another way in which energies of isomers may be compared is by their heats of combustion. Which has the smaller heat of combustion, myrcene or *cis*-alloocimene?

Myrcene   *cis*-Alloocimene

The cumulated double bonds of an allenic system are of relatively high energy. The heat of hydrogenation of allene is more than twice that of propene.

$$CH_2{=}C{=}CH_2 + \quad 2H_2 \quad \longrightarrow CH_3CH_2CH_3 \quad \Delta H^\circ = -70.5 \text{ kcal/mol}$$

Allene   Hydrogen   Propane

$$CH_3CH{=}CH_2 + \quad H_2 \quad \longrightarrow CH_3CH_2CH_3 \quad \Delta H^\circ = -29.9 \text{ kcal/mol}$$

Propene   Hydrogen   Propane

The energy content of a cumulated alkadiene is very similar to that of an alkyne.

Thus, the order of alkadiene stability decreases in the order: conjugated diene (most stable) → isolated diene → cumulated diene (least stable). In order to understand this ranking, we need to look at structure and bonding in alkadienes in more detail.

11.8   ELECTRON DELOCALIZATION IN CONJUGATED DIENES

The factor most responsible for the increased stability of conjugated double bonds is the greater delocalization of their π electrons compared with the π electrons of isolated double bonds. As shown in Figure 11.5a, the π electrons of an isolated diene system occupy, in pairs, two noninteracting π orbitals. Each of these π orbitals encompasses two carbon atoms. An sp^3 hybridized carbon insulates the two π orbitals from each other, preventing the exchange of electrons between them. In a conjugated diene, however, mutual overlap of the two π orbitals, represented in Figure

FIGURE 11.5 (a) Isolated double bonds are separated from one another by one or more sp^3 hybridized carbons and cannot overlap to give an extended π orbital. (b) In a conjugated diene overlap of two π orbitals gives an extended π system encompassing four carbon atoms.

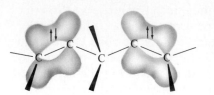

(a) Isolated double bonds   (b) Conjugated double bonds

11.5b, gives an orbital system in which each π electron is delocalized over four carbon atoms. Delocalizing of electrons lowers their energy and gives a more stable molecule.

A more detailed π molecular orbital description of bonding in 1,3-butadiene dienes is shown in Figure 11.6. The four sp^2 hybridized carbons contribute four $2p$ atomic orbitals, and their overlap leads to four π molecular orbitals. Two of these are bonding and two are antibonding; each π molecular orbital encompasses all four carbons of the diene. There are four π electrons, and these are distributed in pairs between the two orbitals of lowest energy. Both bonding orbitals are occupied; both antibonding orbitals are vacant.

At 1.46 Å the C-2—C-3 distance in 1,3-butadiene is relatively short for a carbon-carbon single bond. This is most reasonably seen as a hybridization effect. In ethane both carbons are sp^3 hybridized and are separated by a distance of 1.53 Å. The carbon-carbon single bond in propene unites sp^3 and sp^2 hybridized carbons and is shorter than that of ethane. Both C-2 and C-3 are sp^2 hybridized in 1,3-butadiene, and a decrease in bond distance between them is consistent with the tendency of carbon to attract electrons more strongly as its s character increases.

$$\overset{sp^3}{CH_3}-\overset{sp^3}{CH_3} \qquad \overset{sp^3}{CH_3}-\overset{sp^2}{CH}=CH_2 \qquad CH_2=\overset{sp^2}{CH}-\overset{sp^2}{CH}=CH_2$$

$$1.534\ \text{Å} \qquad\qquad 1.506\ \text{Å} \qquad\qquad 1.463\ \text{Å}$$

Additional evidence for electron delocalization in 1,3-butadiene can be obtained by considering its conformations. Overlap of the two π electron systems is optimal when the four carbon atoms are coplanar. There are two conformations which allow this coplanarity: they are called the *s-cis* and *s-trans* conformations.

s-Cis conformation of 1,3-butadiene   *s*-Trans conformation of 1,3-butadiene

The letter s in *s*-cis and *s*-trans refers to conformations around the *single* bond in the diene. The *s*-trans conformation of 1,3-butadiene is 2.8 kcal/mol more stable than the *s*-cis conformation; the *s*-cis conformation contains an unfavorable van der Waals interaction between the C-1 and C-4 hydrogens. These two coplanar conformations interconvert by rotation around the C-2—C-3 bond, as illustrated in Figure 11.7. The conformation at the midpoint of this rotation, the *perpendicular confor-*

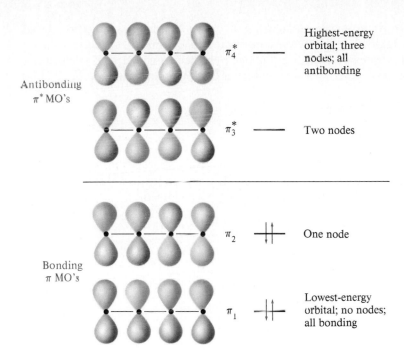

Antibonding
π^* MO's

π_4^* ——— Highest-energy orbital; three nodes; all antibonding

π_3^* ——— Two nodes

Bonding
π MO's

π_2 ⥮ One node

π_1 ⥮ Lowest-energy orbital; no nodes; all bonding

FIGURE 11.6 Electron distribution among the π molecular orbitals of 1,3-butadiene.

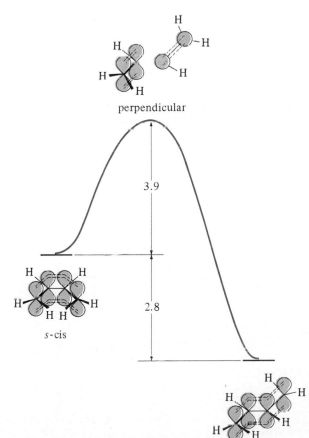

perpendicular

3.9

s-cis

2.8

s-trans

FIGURE 11.7 Conformational isomerization in 1,3-butadiene showing the alignment of 2p orbitals for maximum π electron delocalization in the s-cis and s-trans conformations. Electron delocalization is a minimum in the perpendicular conformation. Energies are in kilocalories per mole.

mation, has its 2*p* orbitals in a geometry that precludes extended conjugation. It has localized double bonds. A significant contributor to the energy of activation for rotation about the single bond in 1,3-butadiene is the decrease in electron delocalization that attends conversion of the *s*-cis or *s*-trans conformation to the perpendicular conformation.

11.9 BONDING IN ALLENES

The three carbons of allene are collinear with relatively short carbon-carbon bond distances of 1.31 Å. The central carbon, since it bears only two substituents, is *sp* hybridized. The terminal carbons of allene are *sp*^{2} hybridized.

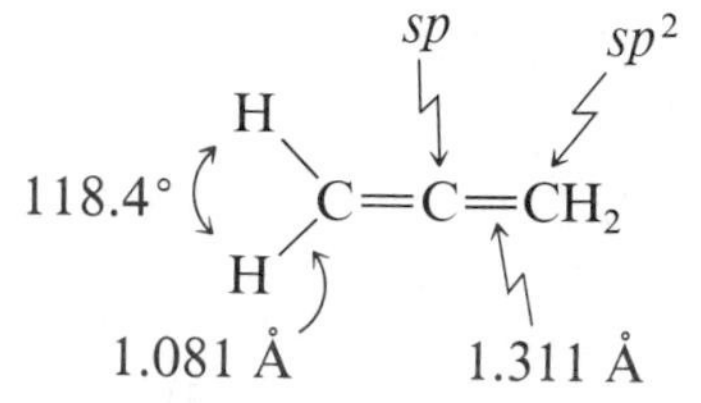

Allene

Structural studies of allene reveal it to be nonplanar. As Figure 11.8 illustrates, the plane of one HCH unit is perpendicular to the plane of the other. Figure 11.8 also portrays the reason for the molecular geometry of allene. The 2*p* orbital of each of the terminal carbons overlaps with a different 2*p* orbital of the central carbon. Since the 2*p* orbitals of the central carbon are perpendicular to each other, the perpendicular nature of the two HCH units follows naturally.

The nonplanarity of the allene unit has an interesting stereochemical consequence. 1,3-Disubstituted allenes are not superposable on their mirror images, i.e., they are chiral. Even an allene as simple as 2,3-pentadiene has been obtained in optically active form.

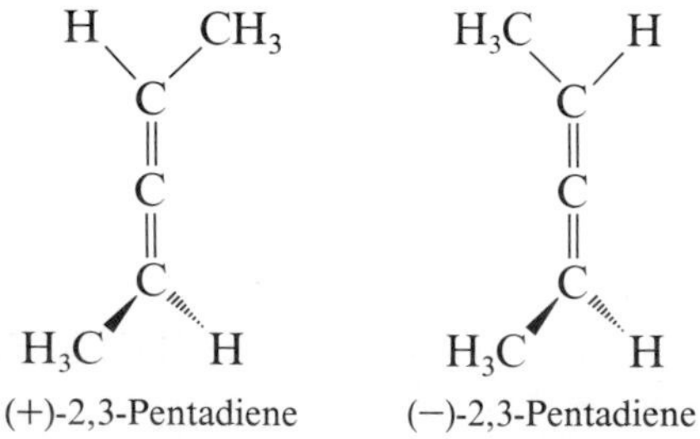

(+)-2,3-Pentadiene   (−)-2,3-Pentadiene

Because of the linear geometry required of cumulated dienes, cyclic allenes, like cycloalkynes, are strained unless the rings are fairly large. 1,2-Cyclononadiene is the smallest cyclic allene that is sufficiently stable to be isolated and stored conveniently.

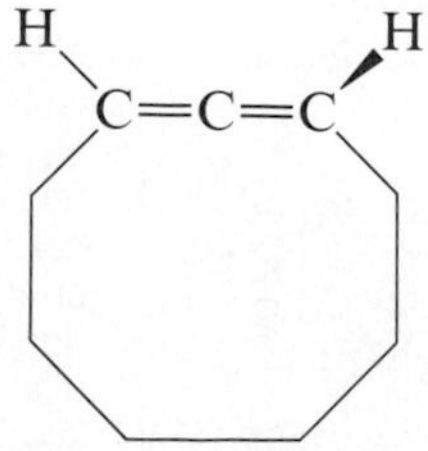

1,2-Cyclononadiene

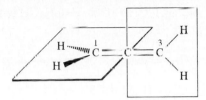

(*a*) Planes defined by H(C-1)H and H(C-3)H are mutually perpendicular

(*b*) The *p* orbital of C-1 and one of those of C-2 can overlap so as to participate in π bonding

(*c*) The *p* orbital of C-3 and one of those of C-2 can overlap to participate in a second π orbital perpendicular to the one in (*b*)

(*d*) Allene is a nonplanar molecule characterized by a linear carbon chain and two mutually perpendicular π bonds

FIGURE 11.8 Bonding and geometry in 1,2-propadiene (allene).

11.10 ADDITION OF HYDROGEN HALIDES TO CONJUGATED DIENES

Our discussion of chemical reactions of alkadienes will be limited to those of conjugated dienes. The reactions of isolated dienes are essentially the same as those of individual alkenes and demonstrate little that is new in the way of fundamental principles. The reactions of cumulated dienes are, on the other hand, so specialized that their treatment is better suited to a more advanced course in organic chemistry.

Electrophilic addition is the characteristic chemical reaction of alkenes, and conjugated dienes undergo addition reactions with the same electrophiles that react with alkenes. The reactions of 1,3-butadiene with hydrogen chloride and with hydrogen bromide are typical of electrophilic addition reactions of conjugated dienes. A mixture of the respective 3-halo-1-butene and 1-halo-2-butene isomers is formed. When the addition reaction is carried out at low temperature, the major product is the secondary allylic halide in each case.

$$CH_2\!=\!CHCH\!=\!CH_2 \xrightarrow[-80°C]{HCl} CH_3\underset{\underset{Cl}{|}}{C}HCH\!=\!CH_2 + CH_3CH\!=\!CHCH_2Cl$$

1,3-Butadiene   3-Chloro-1-butene (78%)   1-Chloro-2-butene (22%)

$$CH_2\!=\!CHCH\!=\!CH_2 \xrightarrow[\substack{-80°C \\ \text{free-radical} \\ \text{inhibitor}}]{HBr} CH_3\underset{\underset{Br}{|}}{C}HCH\!=\!CH_2 + CH_3CH\!=\!CHCH_2Br$$

1,3-Butadiene   3-Bromo-1-butene (81%)   1-Bromo-2-butene (19%)

The major product corresponds to Markovnikov addition of the elements of a hydrogen halide to one of the double bonds of the diene. This mode of reaction is referred to as *direct addition* or *1,2 addition.*

Bonding pattern in direct or 1,2 addition:

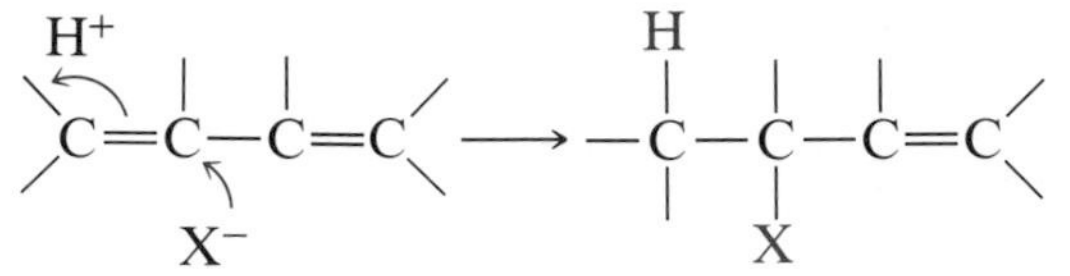

The minor product in each case, the 1-halo-2-butenes, requires addition of a proton to C-1, halide to C-4, and migration of a double bond to C-2—C-3. This mode of reaction is called *conjugate addition* or *1,4 addition.*

Bonding pattern in conjugate or 1,4 addition:

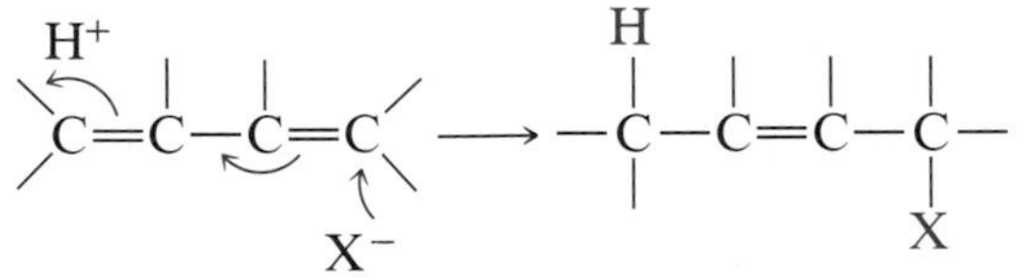

We can account for the formation of both kinds of products on the basis of the customary mechanism for hydrogen halide addition to alkenes once we recognize a unique feature of the carbocation intermediate that is formed by protonation of 1,3-butadiene.

$$H^+ + CH_2\!=\!CHCH\!=\!CH_2 \longrightarrow CH_3\overset{+}{C}HCH\!=\!CH_2$$

This carbocation is an allylic carbocation. Delocalization of its π electrons stabilizes it and causes the positive charge to be shared by two carbons. Attack by halide at one of these two carbons gives a 3-halo-1-butene, attack at the other gives a 1-halo-2-butene.

$$\boxed{\begin{array}{c} CH_3\overset{+}{C}HCH\!=\!CH_2 \\ \updownarrow \\ CH_3CH\!=\!CH\overset{+}{C}H_2 \end{array}} \xrightarrow{X^-} CH_3\underset{\underset{X}{|}}{C}HCH\!=\!CH_2 + CH_3CH\!=\!CHCH_2X$$

3-Halo-1-butene   1-Halo-2-butene
(major)   (minor)

The secondary carbon bears more of the positive charge in the allylic carbocation than does the primary carbon, and attack by the nucleophilic halide anion is faster there. Hence, the major product is the secondary halide under these conditions.

When the major product of a reaction is the one that is formed at the greatest rate, we say that the reaction is governed by *kinetic control* (or *rate control*). Most organic reactions fall into this category, and the electrophilic additions of hydrogen chloride and hydrogen bromide to 1,3-butadiene at low temperature are kinetically controlled reactions.

When, however, the ionic addition of hydrogen bromide to 1,3-butadiene is carried out at room temperature, the ratio of isomeric allylic bromides observed is different from that which is formed at $-80°C$. At room temperature, the 1,4-addition product comprises the major portion of the reaction product.

$$CH_2=CHCH=CH_2 \xrightarrow[\substack{\text{room temperature,} \\ \text{free radical} \\ \text{inhibitor}}]{\text{HBr}} CH_3\underset{\underset{Br}{|}}{CH}CH=CH_2 + CH_3CH=CHCH_2Br$$

1,3-Butadiene	3-Bromo-1-butene (44%)	1-Bromo-2-butene (56%)

Clearly, the temperature at which reaction occurs exerts a major influence on the product composition. In order to understand why, two important facts must be added. First, the 1,2- and 1,4-addition products *interconvert rapidly* at elevated temperature in the presence of hydrogen bromide. Heating the product mixture to 45°C in the presence of hydrogen bromide leads to a mixture in which the ratio of 3-bromo-1-butene to 1-bromo-2-butene is 15:85.

$$CH_3\underset{\underset{Br}{|}}{CH}CH=CH_2 \underset{}{\overset{\text{HBr}}{\rightleftharpoons}} CH_3CH=CHCH_2Br$$

3-Bromo-1-butene (15% at equilibrium)	1-Bromo-2-butene (85% at equilibrium)

Second, the product of 1,4 addition, 1-bromo-2-butene, contains an internal double bond and so is *more stable* than the product of 1,2 addition, 3-bromo-1-butene, which has a terminal double bond.

When the addition reaction is carried out under conditions such that the products may equilibrate, the composition of the reaction mixture no longer reflects the relative rates of formation of its components but tends to reflect their *relative stabilities*. Reactions of this type are said to be governed by *thermodynamic control* (or *equilibrium control*). Given sufficient time, a thermodynamically controlled reaction will give a product composition corresponding to the thermodynamic stabilities of the various products. In the case of the addition of hydrogen bromide to 1,3-butadiene, the two products equilibrate by an ionization-recombination pathway.

$$CH_3\underset{\underset{Br}{|}}{CH}CH=CH_2 \underset{\substack{\text{cation-anion} \\ \text{combination}}}{\overset{\text{ionization}}{\rightleftharpoons}} CH_3CH\overset{\overset{H}{|}}{\underset{Br^-}{\overset{C}{\cdots}}}\overset{+}{\cdots}CH_2 \underset{\text{ionization}}{\overset{\substack{\text{cation-anion} \\ \text{combination}}}{\rightleftharpoons}} CH_3CH=CHCH_2Br$$

3-Bromo-1-butene (less stable isomer)	Carbocation + bromide anion	1-Bromo-2-butene (more stable isomer)

A useful way to illustrate kinetic and thermodynamic control in this reaction is by way of the energy diagram of Figure 11.9. At low temperature addition takes place

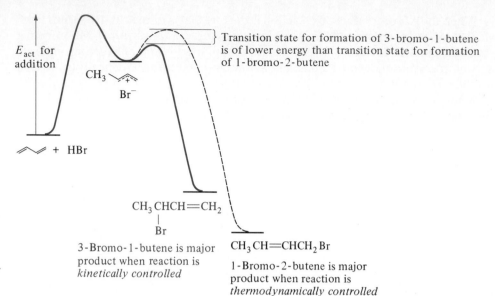

Transition state for formation of 3-bromo-1-butene is of lower energy than transition state for formation of 1-bromo-2-butene

E_{act} for addition

$CH_3 \overset{+}{\underset{Br^-}{\sim}}$

$\sim\!\!\!\sim$ + HBr

$\underset{\underset{Br}{|}}{CH_3\,CHCH}\!=\!\!CH_2$

3-Bromo-1-butene is major product when reaction is *kinetically controlled*

$CH_3\,CH\!=\!\!CHCH_2\,Br$

1-Bromo-2-butene is major product when reaction is *thermodynamically controlled*

FIGURE 11.9 Energy diagram showing relationship of kinetic control to thermodynamic control in addition of hydrogen bromide to 1,3-butadiene.

irreversibly. The activation energy for addition is relatively small and 3-bromo-1-butene is the major product. At low temperature isomerization is slow because insufficient energy is available to permit the products to pass over the relatively high barrier for their ionization. At higher temperatures both addition and isomerization are rapid and the less stable product is converted to the more stable one.

Since a carbon-chlorine bond is stronger than a carbon-bromine bond, the products of hydrogen chloride addition are much less readily isomerized by ionization-recombination than are the products of hydrogen bromide addition. Thus, even at room temperature hydrogen chloride adds to 1,3-butadiene in a kinetically controlled reaction.

11.11 HALOGEN ADDITION TO DIENES

Mixtures of 1,2- and 1,4-addition products are obtained when 1,3-butadiene is allowed to react with chlorine or bromine.

$$CH_2\!=\!\!CHCH\!=\!\!CH_2 + Br_2 \xrightarrow{CHCl_3} BrCH_2\underset{\underset{Br}{|}}{CHCH}\!=\!\!CH_2 + \overset{BrCH_2}{\underset{H}{\diagdown}}C\!=\!C\overset{H}{\underset{CH_2Br}{\diagup}}$$

| 1,3-Butadiene | Bromine | 3,4-Dibromo-1-butene (37%) | (E)-1,4-Dibromo-2-butene (63%) |

The tendency for conjugate addition is pronounced, and E double bonds are generated almost exclusively.

2,3-Dimethyl-1,3-butadiene   Bromine   (E)-1,4-Dibromo-2,3-dimethyl-2-butene (85–90%)

PROBLEM 11.7 Exclusive of stereoisomers, how many products are possible in the electrophilic addition of 1 mol of bromine to 2-methyl-1,3-butadiene?

*test 3
thru

here*

11.12 THE DIELS-ALDER REACTION

A particular kind of conjugate addition reaction earned the Nobel prize in chemistry for Otto Diels and Kurt Alder of the University of Kiel in 1950. The Diels-Alder reaction is the *conjugate addition of an alkene to a diene.* Using 1,3-butadiene as a typical diene, the Diels-Alder reaction may be represented by the general equation:

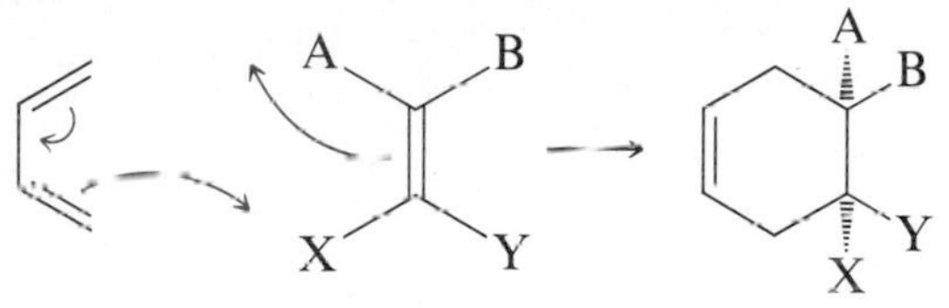

1,3-Butadiene   Dienophile   Diels-Alder adduct

The alkene that adds to the diene is called the *dienophile.* Because the Diels-Alder reaction leads to the formation of a ring, it is termed a *cycloaddition reaction.*

It is generally agreed that the Diels-Alder reaction is a concerted process. The transition state involves bond formation at both ends of the diene system. The diene must be able to achieve the *s*-cis conformation.

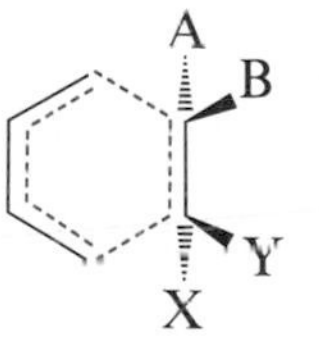

Representation of
activated complex for
Diels-Alder cycloaddition

The simplest of all Diels-Alder reactions, the addition of ethylene to 1,3-butadiene, does not proceed readily. It has a high activation energy and a low reaction rate. Certain substituents on the double bond of the dienophile, however, enhance its

reactivity. Relatively reactive dienophiles are alkenes that bear one or more $\diagup C{=}O$

or $-C{\equiv}N$ substituents on the double bond. Compounds of this type, illustrated in the following examples, react readily with dienes.

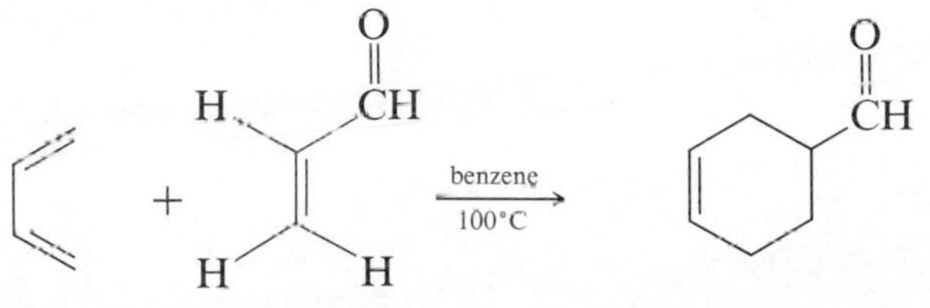

1,3-Butadiene   Acrolein   Cyclohexene-4-carboxaldehyde
(100%)

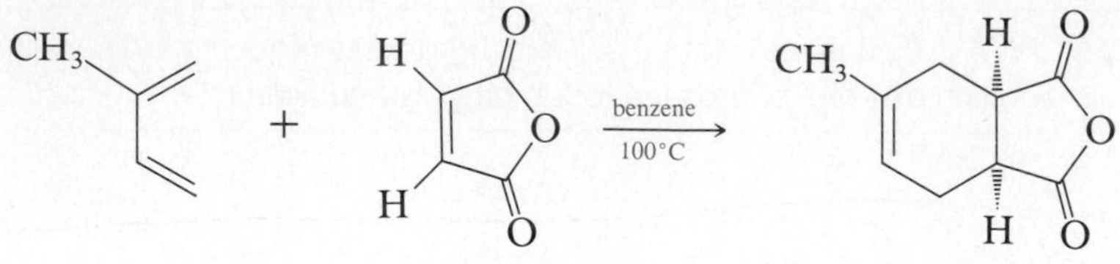

2-Methyl-1,3-butadiene Maleic anhydride 1-Methylcyclohexene-4,5-dicarboxylic anhydride (100%)

PROBLEM 11.8 Benzoquinone is a very reactive dienophile. It reacts with 2-chloro-1,3-butadiene to give a single product, $C_{10}H_9ClO_2$, in 95 percent yield. Write a structural formula for this product.

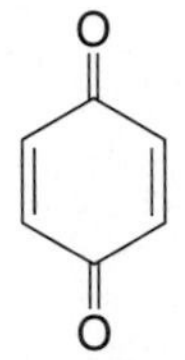

Benzoquinone

Like ethylene, acetylene is a poor dienophile but alkynes that bear $\overset{}{\underset{}{>}}C{=}O$ substituents react readily with dienes.

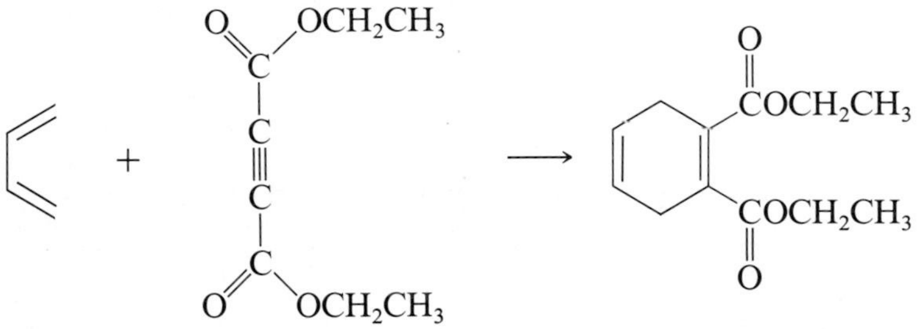

1,3-Butadiene Diethyl acetylenedicarboxylate Diethyl 1,2,4,5-cyclohexadiene-1,2-dicarboxylate (98%)

The Diels-Alder reaction is stereospecific. Substituents that are cis in the dienophile remain cis in the product; substituents that are trans in the dienophile remain trans in the product.

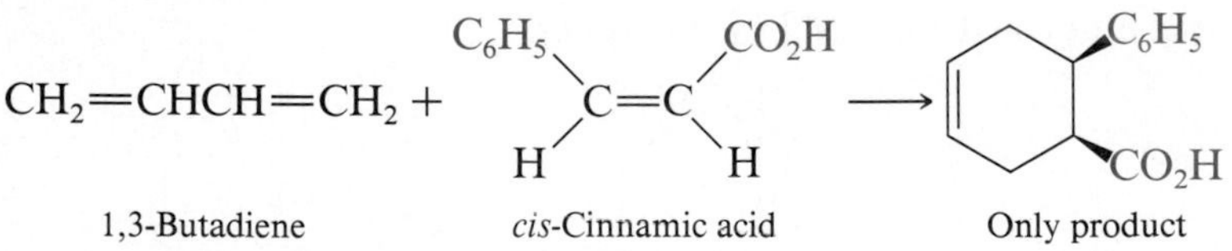

$CH_2{=}CHCH{=}CH_2 +$

1,3-Butadiene cis-Cinnamic acid Only product

$CH_2{=}CHCH{=}CH_2 +$

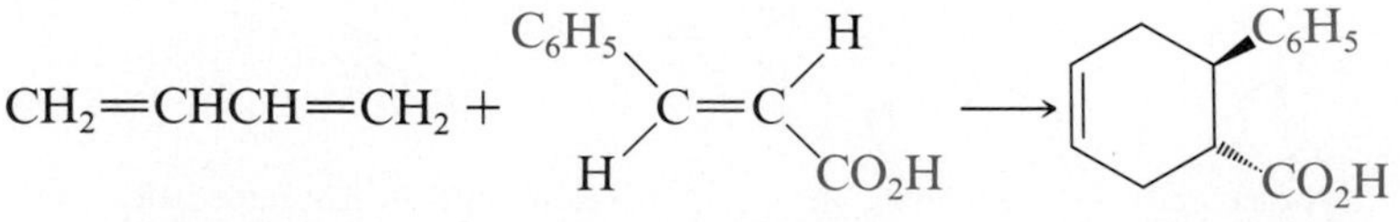

1,3-Butadiene trans-Cinnamic acid Only product

Cyclic dienes yield bicyclic Diels-Alder adducts.

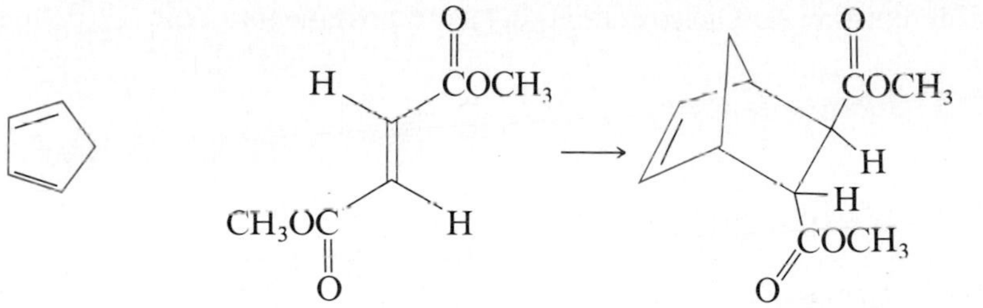

1,3-Cyclopentadiene   Dimethyl fumarate   Dimethyl-
bicyclo[2.2.1]hept-2-ene-
trans-5,6-dicarboxylate

PROBLEM 11.9   What combination of diene and dienophile would you choose in order to prepare each of the following compounds?

(a)

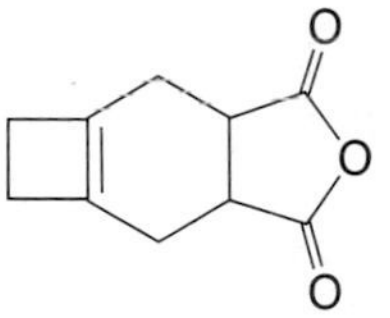

(c)

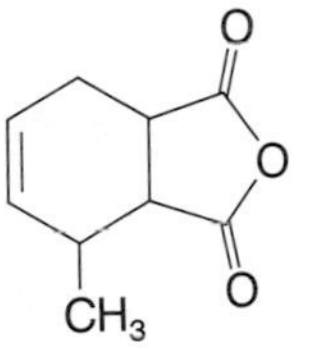

(b)

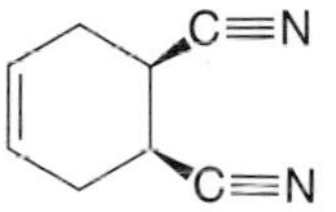

(d)

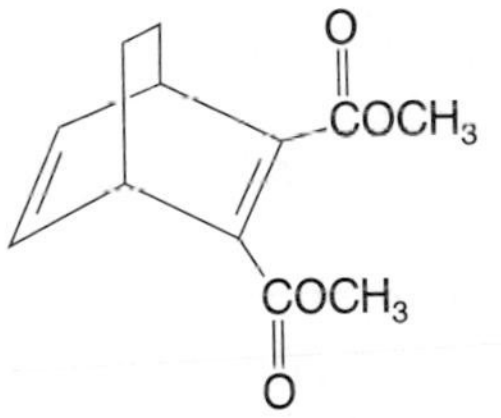

SAMPLE SOLUTION   (a)   We represent a Diels-Alder reaction according to the curved arrow formalism as

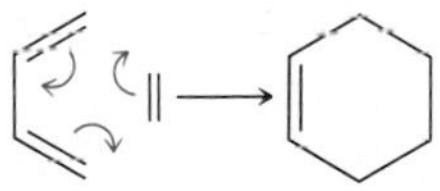

In order to deduce the identity of the diene and dienophile that leads to a particular Diels-Alder adduct, all we need to do is use curved arrows in the reverse fashion to "undo" the cyclohexene derivative. Start with the π component of the double bond, and move electrons in pairs.

Diels-Alder adduct   Diene   Dienophile

Besides being stereospecific, Diels-Alder reactions are stereoselective. The addition of methyl acrylate to 1,3-cyclopentadiene illustrates this facet of the reaction. Two stereoisomeric Diels-Alder adducts are possible and both are formed.

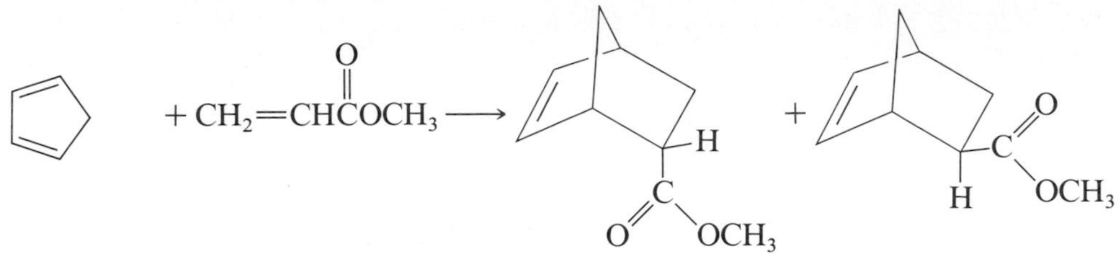

| 1,3-Cyclopentadiene | Methyl acrylate | Endo isomer (75%) | Exo isomer (25%) |

(Stereoisomeric forms of methyl bicyclo[2.2.1]hept-5-ene-2-carboxylate)

The stereoisomer formed in greater amount has its $\overset{O}{\overset{\|}{C}}OCH_3$ group syn to the CH=CH bridge and is called the *endo* isomer. The minor product has its $\overset{O}{\overset{\|}{C}}OCH_3$ group anti to the CH=CH bridge and is called the *exo* isomer. It has been shown that both isomers are of almost equal stability; therefore we are not simply observing the formation of the more stable product in an equilibrium-controlled process. There is a kinetically controlled preference for the formation of the endo product.

Observations similar to this have been made many times in Diels-Alder reactions and have led to the formulation of the following empirical rule, known as the *Alder rule* or the *rule of maximum accumulation of unsaturation:* In a Diels-Alder reaction, the major product is derived from the activated complex in which unsaturated groups in the dienophile assume a syn rather than an anti orientation with respect to the diene.

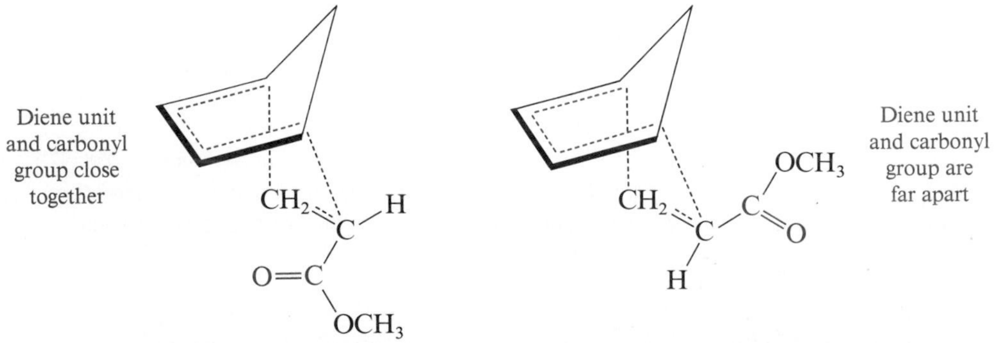

Diene unit and carbonyl group close together

More stable orientation of groups in activated complex; gives endo

Diene unit and carbonyl group are far apart

Less stable orientation of groups in activated complex; gives exo

The stereoselective bias as expressed in the Alder rule is, however, not sufficient to overcome the stereospecific requirement for syn addition. (See the example of the already illustrated reaction between 1,3-cyclopentadiene and dimethyl fumarate.

Because the $\overset{\overset{\displaystyle O}{\|}}{C}OCH_3$ groups are trans in the dienophile, they must be trans in the Diels-Alder adduct; thus, one of them is an exo substituent while the other is endo.)

The importance of the Diels-Alder reaction is in synthesis. It provides the opportunity to form two new carbon-carbon bonds in a single step in a reaction that requires no reagents, such as acids or bases, that might affect other functional groups within the molecule.

(a)

(b)

FIGURE 11.10 The frontier orbitals of 1,3-butadiene and ethylene. (a) The highest occupied molecular orbital (HOMO) of 1,3-butadiene. (b) The lowest unoccupied molecular orbital (LUMO) of ethylene.

11.13 ORBITAL SYMMETRY AND THE DIELS-ALDER REACTION

In the early stages of its development the primary emphasis of molecular orbital theory was directed toward understanding the electronic forces that influence molecular *structure*. Beginning in the 1960s, however, its application toward questions of chemical *reactivity* has received increasing attention. Cycloaddition reactions represent one area in which our grasp of reactivity has been enhanced through the use of qualitative molecular orbital theory.

In one approach to understanding the Diels-Alder reaction, attention is focused on the *frontier orbitals* of the diene and the dienophile. The frontier orbitals are those that contain the electrons of highest energy, i.e., the electrons most likely to be transferred, and the vacant orbitals of lowest energy, i.e., the orbitals that can accept electrons most readily. Chemical experience suggests that electrons flow from the diene to the dienophile in the Diels-Alder reaction. The dienophiles that are the most reactive are those that bear electron-attracting substituents. Thus, the frontier orbitals in the Diels-Alder reaction are the highest occupied molecular orbital (HOMO) of the diene and the lowest unoccupied molecular orbital (LUMO) of the dienophile. Figure 11.10 depicts these orbitals for 1,3-butadiene and ethylene.

Consider what happens as a diene and a dienophile interact with each other to produce a Diels-Alder adduct. Bond formation occurs between the ends of the diene system and the two carbons of the dienophile as electrons are transferred from the HOMO to the LUMO. As Figure 11.11 demonstrates, this interaction is accompanied by "in phase" overlap of orbitals of like symmetry in the diene and the dienophile. Cycloaddition of a diene and an alkene is said to be a *symmetry-allowed reaction* because the symmetry properties of the HOMO and LUMO permit bonding interactions between the atoms that must become joined by σ bonds in the product.

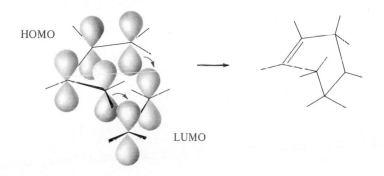

HOMO

LUMO

FIGURE 11.11 The HOMO of 1,3-butadiene and the LUMO of ethylene have the proper symmetry to allow σ bond formation to occur at both ends of the diene chain in the same transition state.

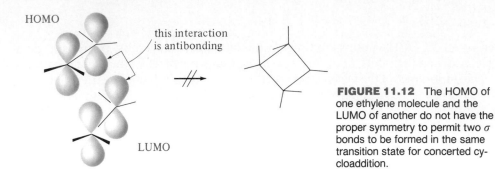

HOMO

this interaction is antibonding

LUMO

FIGURE 11.12 The HOMO of one ethylene molecule and the LUMO of another do not have the proper symmetry to permit two σ bonds to be formed in the same transition state for concerted cycloaddition.

Contrast the Diels-Alder reaction with a cycloaddition reaction that looks superficially similar, the combination of two alkenes to give a cyclobutane derivative.

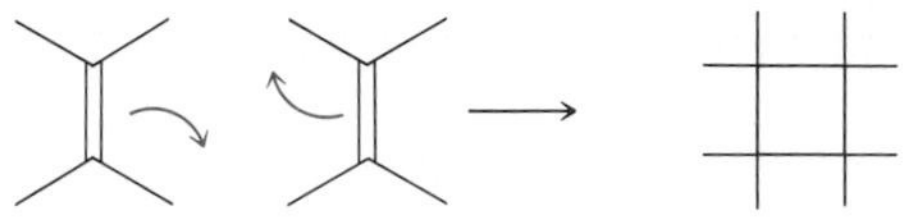

Alkene   Alkene   Cyclobutane derivative

Reactions of this type are rather rare and seem to proceed in a stepwise fashion rather than by way of a concerted mechanism involving a single transition state.

Figure 11.12 shows the interaction between the HOMO of one alkene and the LUMO of another. In particular, notice that two of the carbons that are to become σ bonded to each other in the product experience an antibonding interaction during the cycloaddition process. This raises the activation energy for cycloaddition and leads to it being classified as a *symmetry-forbidden reaction.* Reaction, if it does occur, takes place slowly and by a mechanism in which the two new σ bonds are formed in separate steps rather than by way of a concerted process involving a single transition state.

Frontier orbital analysis is a powerful technique that aids our understanding of a great number of organic reactions. Its early development is attributed to Professor Kenichi Fukui of Kyoto University, Japan. The application of frontier orbital methods to Diels-Alder reactions represents one part of what organic chemists refer

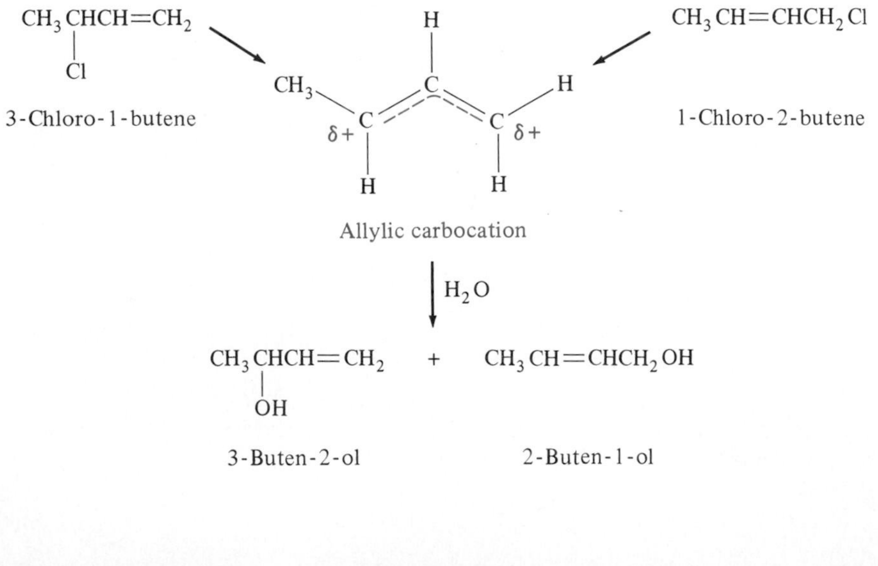

$CH_3CHCH=CH_2$
|
Cl

3-Chloro-1-butene

$CH_3CH=CHCH_2Cl$

1-Chloro-2-butene

Allylic carbocation

H_2O

$CH_3CHCH=CH_2$ + $CH_3CH=CHCH_2OH$
|
OH

3-Buten-2-ol   2-Buten-1-ol

FIGURE 11.13 The hydrolysis of 3-chloro-1-butene proceeds via the same allylic carbocation as that of 1-chloro-2-butene. This carbocation is captured by water at each of the two carbons that share the positive charge, and a mixture of allylic alcohols is formed.

to as the *Woodward-Hoffmann rules,* a beautifully simple analysis of organic reactions by Professor R. B. Woodward of Harvard University and Professor Roald Hoffmann of Cornell University. Professors Fukui and Hoffmann were corecipients of the 1981 Nobel prize in chemistry for their work.

11.14 SUMMARY

This chapter described some of the properties of three conjugated systems: allylic carbocations, allylic free radicals, and conjugated alkadienes.

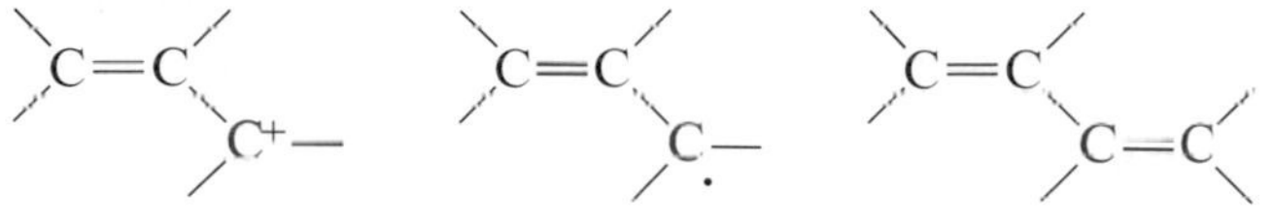

| Allylic carbocation | Allylic radical | Conjugated diene |

Each of these systems is stabilized by π electron delocalization.

Since allylic carbocations are more stable than alkyl cations, the activation energy for their formation is less, and S_N1 reactions of allylic halides proceed at faster rates than the corresponding reactions of alkyl halides. Because allylic carbocations have their positive charge distributed over two atoms, they may be captured by nucleophiles at two different sites, and mixtures of products are often isolated in solvolysis reactions of allylic halides. Figure 11.13 summarizes the solvolysis behavior of a pair of allylic halides.

Allylic radicals are more stable than structurally related alkyl radicals. Free-radical halogenation of alkenes proceeds by substitution of an allylic hydrogen. The reagent *N*-bromosuccinimide is used to carry out the allylic bromination of alkenes.

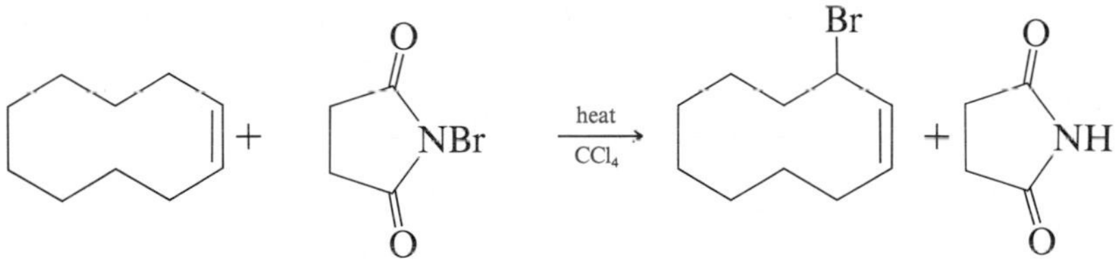

| Cyclodecene | *N*-Bromosuccinimide | 3-Bromocyclodecene (56%) | Succinimide |

Conjugated dienes are stabilized by resonance to the extent of 3 to 4 kcal/mol. Their two most stable conformations are designated as *s*-cis and *s*-trans.

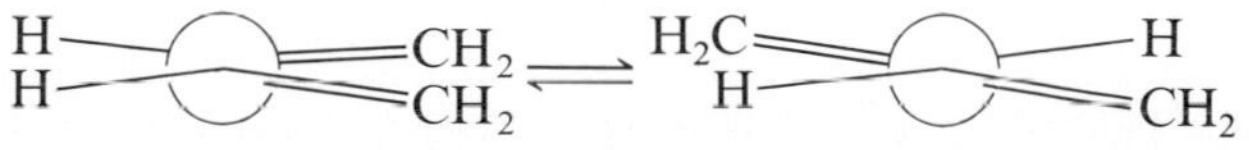

| *s*-cis | *s*-trans |

The *s*-trans conformation is normally more stable than the *s*-cis. Both these conformations are planar; this permits the four $2p$ orbitals to be aligned so as to provide the maximum degree of overlap.

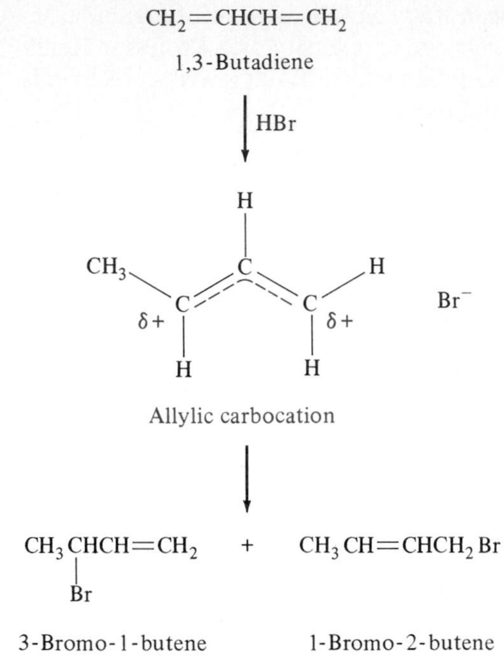

$$CH_2\!\!=\!\!CHCH\!\!=\!\!CH_2$$

1,3-Butadiene

$\downarrow$ HBr

Allylic carbocation

$CH_3CHCH\!\!=\!\!CH_2$ + $CH_3CH\!\!=\!\!CHCH_2Br$
|
Br

3-Bromo-1-butene   1-Bromo-2-butene

FIGURE 11.14 Electrophilic addition of hydrogen bromide to 1,3-butadiene proceeds by way of an allylic carbocation intermediate. The intermediate may be captured by bromide at either of the two carbons that share the positive charge.

Protonation of a conjugated diene generates an allylic carbocation. Capture of this allylic carbocation, as shown in Figure 11.14, leads to two products: one is termed the product of direct addition, the other is referred to as the product of conjugate addition. In the addition of hydrogen bromide to 1,3-butadiene, the distribution of products is temperature-dependent. At $-80°C$ the major product is 3-bromo-1-butene because it is the one that is formed faster. We say that the addition is subject to *kinetic control* at low temperature. At room temperature and above, the major product is 1-bromo-2-butene because the two allylic bromides equilibrate under these conditions. We say that the addition is subject to *thermodynamic control* at elevated temperature.

Halogens add to conjugated dienes to give mixtures of dihalides arising from both direct addition and conjugate addition.

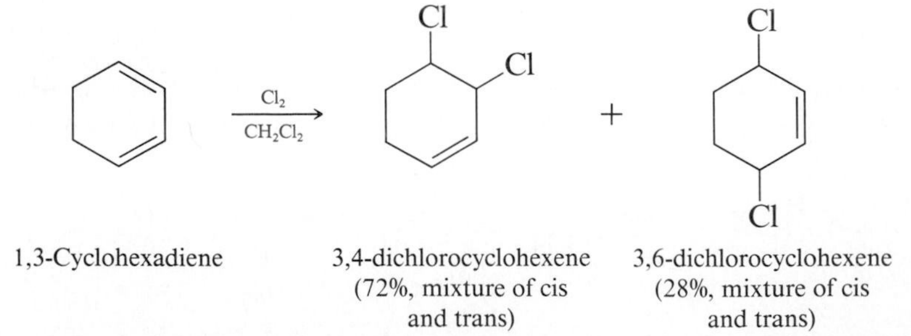

1,3-Cyclohexadiene   3,4-dichlorocyclohexene   3,6-dichlorocyclohexene
                            (72%, mixture of cis   (28%, mixture of cis
                             and trans)   and trans)

A novel kind of conjugate addition known as the *Diels-Alder reaction* takes place when a conjugated diene reacts with an alkene.

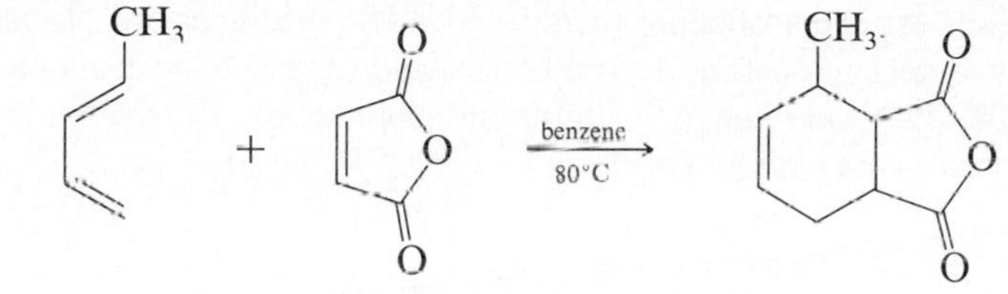

1,3-Pentadiene   Maleic anhydride   3-Methylcyclohexene-4,5-dicar-
                                                boxylic anhydride (81%)

The Diels-Alder reaction is a concerted process. It is extensively used in organic synthesis.

PROBLEMS

11.10 Write structural formulas for each of the following:

(a) 3,4-Octadiene
(b) (*E*,*E*)-3,5-Octadiene
(c) (*Z*,*Z*)-1,3-Cyclooctadiene
(d) (*Z*,*Z*)-1,4-Cyclooctadiene
(e) (*Z*,*E*)-1,5-Cyclooctadiene
(f) (2*E*,4*Z*,6*E*)-2,4,6-Octatriene
(g) 5-Allyl-1,3-cyclopentadiene
(h) *trans*-1,2-Divinylcyclopropane
(i) 2,4-Dimethyl-1,3-pentadiene

11.11 Give the IUPAC names for each of the following compounds:

(a) $CH_2{=}CH(CH_2)_6CH{=}CH_2$

(b) 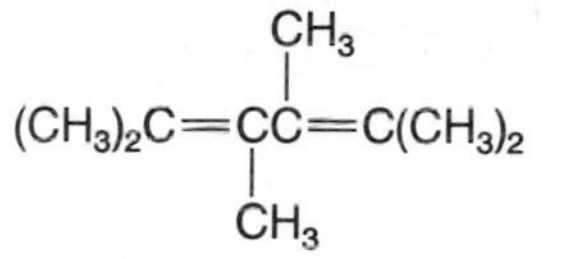 $(CH_3)_2C{=}\overset{\overset{\displaystyle CH_3}{|}}{C}C{=}C(CH_3)_2$
(with CH_3 below)

(c) $(CH_2{=}CH)_3CH$

(d)

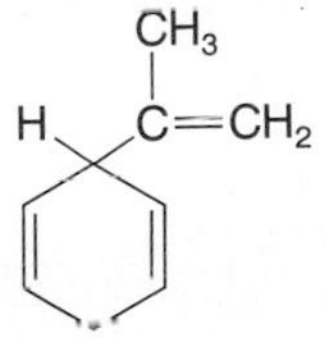

(e)

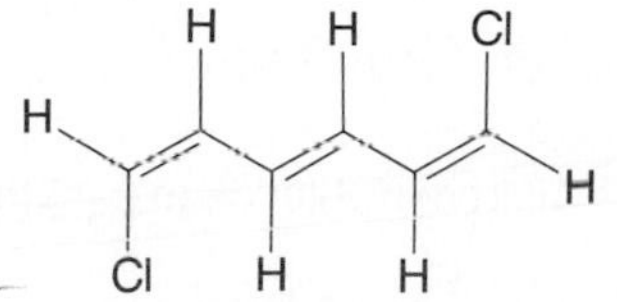

(f) $CH_2{=}C{=}CHCH{=}CHCH_3$

(g)

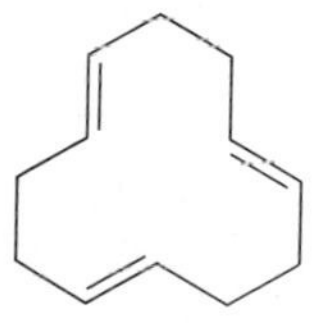

(h)

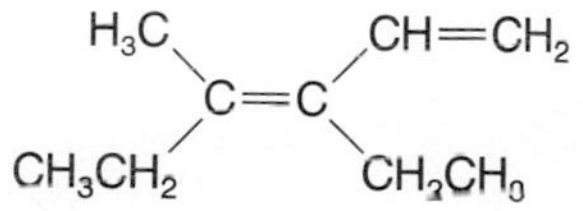

11.12 (a) What compound of molecular formula C_6H_{10} gives 2,3-dimethylbutane on catalytic hydrogenation over platinum?

(b) What two compounds of molecular formula $C_{11}H_{20}$ give 2,2,6,6-tetramethylheptane on catalytic hydrogenation over platinum?

11.13 A certain species of grasshopper secretes an allenic substance of molecular formula $C_{13}H_{20}O_3$ that acts as an ant repellent. The carbon skeleton and location of various substituents in this substance are indicated in the partial structure shown. Complete the structure, adding double bonds where appropriate.

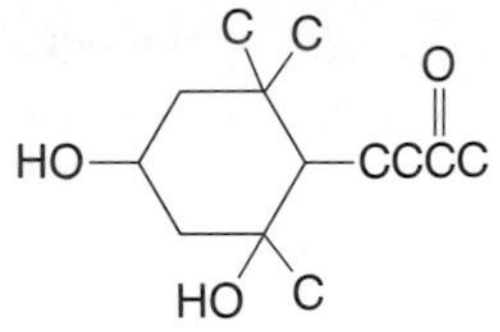

11.14 Show, by writing a suitable sequence of chemical equations, how you could prepare each of the following compounds from cyclopentene and any necessary organic or inorganic reagents.

(a) 2-Cyclopenten-1-ol
(b) 3-Iodocyclopentene
(c) 3-Cyanocyclopentene
(d) 1,3-Cyclopentadiene

(e)

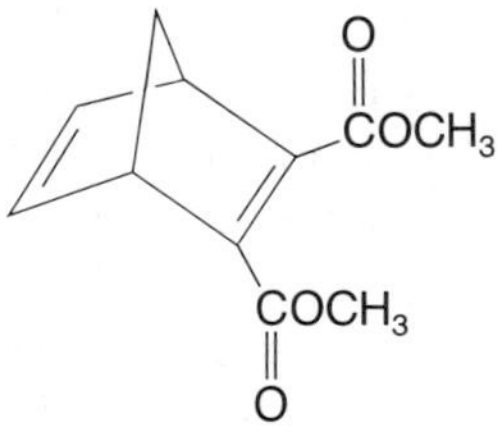

11.15 Give the structure, exclusive of stereochemistry, of the principal organic product formed on reaction of 2,3-dimethyl-1,3-butadiene with each of the following:

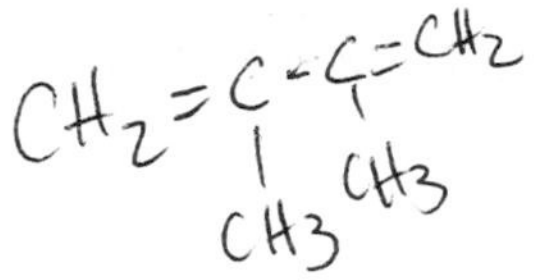

(a) 2 mol H_2, platinum catalyst
(b) 1 mol HCl (product of direct addition)
(c) 1 mol HCl (product of conjugate addition)
(d) 1 mol Br_2 (product of direct addition)
(e) 1 mol Br_2 (product of conjugate addition)
(f) 2 mol Br_2

(g)

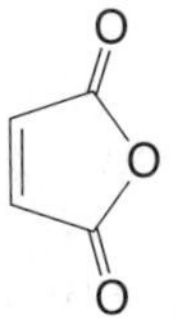

11.16 Repeat the previous problem for the reactions of 1,3-cyclohexadiene.

11.17 Predict the major product obtained on addition of hydrogen chloride to 2-methyl-1,3-butadiene under conditions of

(a) Kinetic control
(b) Thermodynamic control

11.18 Identify compound A on the basis of the following information:

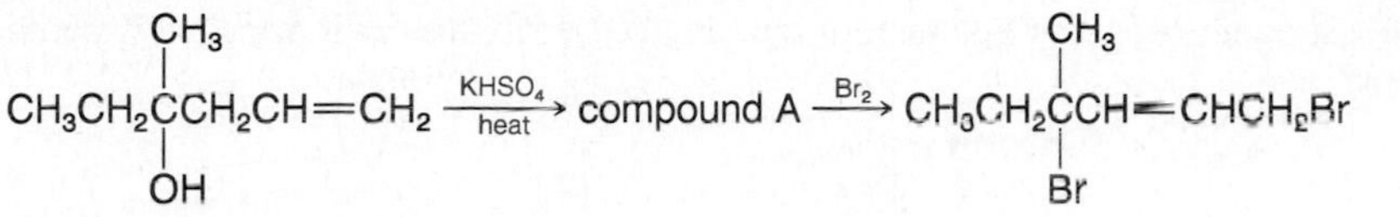

$$CH_3CH_2\underset{\underset{OH}{|}}{\overset{\overset{CH_3}{|}}{C}}CH_2CH{=}CH_2 \xrightarrow[\text{heat}]{KHSO_4} \text{compound A} \xrightarrow{Br_2} CH_3CH_2\underset{\underset{Br}{|}}{\overset{\overset{CH_3}{|}}{C}}CH{=}CHCH_2Br$$

11.19 Allene can be converted to a trimer (compound B) of molecular formula C_9H_{12}. Compound B reacts with dimethyl acetylenedicarboxylate to give compound C. Deduce the structure of compound B.

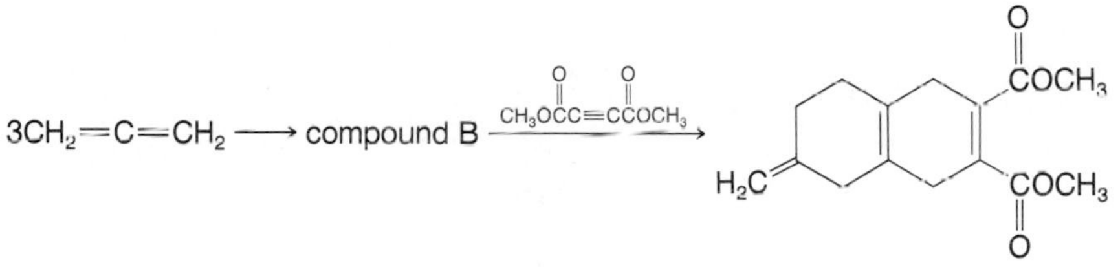

Compound C

11.20 Suggest reasonable explanations for each of the following observations:

(a) The first-order rate constant for the solvolysis of $(CH_3)_2C{=}CHCH_2Cl$ in ethanol is over 6000 times greater than that of allyl chloride (25°C).

(b) After a solution of 3-buten-2-ol in aqueous sulfuric acid had been allowed to stand for 1 week, it was found to contain both 3-buten-2-ol and 2-buten-1-ol.

(c) Treatment of $CH_3CH{=}CHCH_2OH$ with hydrogen bromide gave a mixture of 1-bromo-2-butene and 3-bromo-1-butene.

(d) Treatment of 3-buten-2-ol with hydrogen bromide gave the same mixture of bromides as in (c).

(e) The major product in (c) and (d) was 1-bromo-2-butene.

11.21 Which has the smaller heat of hydrogenation, α-phellandrene (from eucalyptus oil) or α-terpinene (from marjoram oil)?

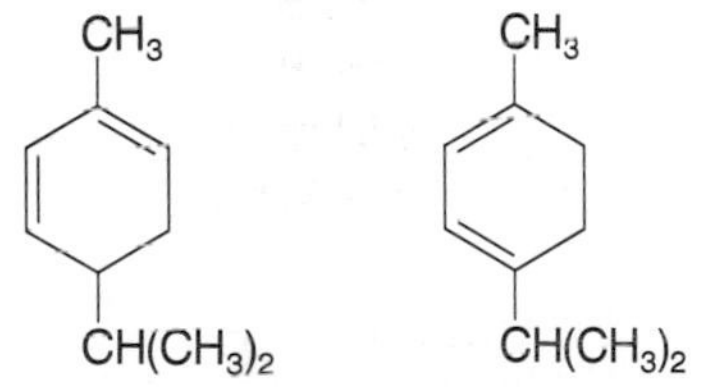

α-Phellandrene α-Terpinene

11.22 (a) Write equations expressing the s-trans $\rightleftharpoons$ s-cis conformational equilibrium for (E)-1,3-pentadiene and for (Z)-1,3-pentadiene.

(b) For which stereoisomer will the equilibrium favor the s-trans conformation more strongly? Why?

11.23 The allene 2,3-pentadiene is chiral. Which of the following are chiral?

(a) 2-Methyl-2,3-pentadiene
(b) 2-Methyl-2,3-hexadiene
(c) 4-Methyl-2,3-hexadiene
(d) 2,4-Dimethyl 2,3-pentadiene

11.24 Suggest reagents suitable for carrying out each step in the following synthetic sequence.

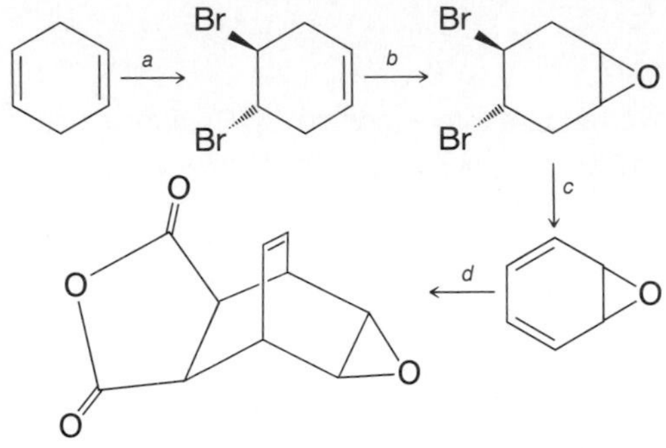

11.25 A very large number of Diels-Alder reactions are recorded in the chemical literature, many of which involve relatively complicated dienes and/or dienophiles. On the basis of your knowledge of Diels-Alder reactions, predict the constitution of the Diels-Alder adduct which you would expect to be formed from the following combination of dienes and dienophiles.

(a)   2,3-Dimethyl-1,3-butadiene $+$ C$_6$H$_5$SO$_2$CH$=$CH$_2$

(b) $+$ CH$_3$O$_2$CC$\equiv$CCO$_2$CH$_3$

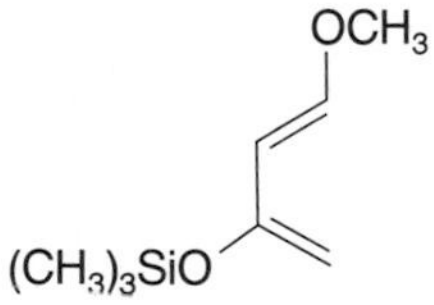

(c) $+$ CH$_2$=CHCO$_2$CH$_3$

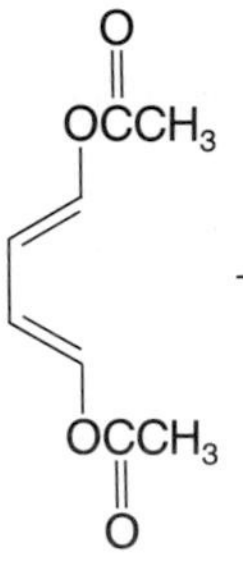

(d) $+$ CH$_3$O$_2$CC$\equiv$CCO$_2$CH$_3$

(e) $+$ CH$_2$=CHNO$_2$

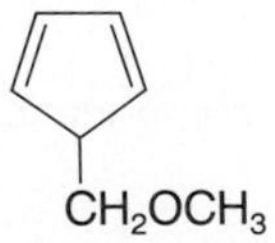

11.26 On standing, 1,3-cyclopentadiene is transformed into a new compound called *dicyclopentadiene,* having the molecular formula C$_{10}$H$_{12}$. Hydrogenation of dicyclopentadiene gives the compound shown below. Suggest a structure for dicyclopentadiene. What kind of reaction is occurring in its formation?

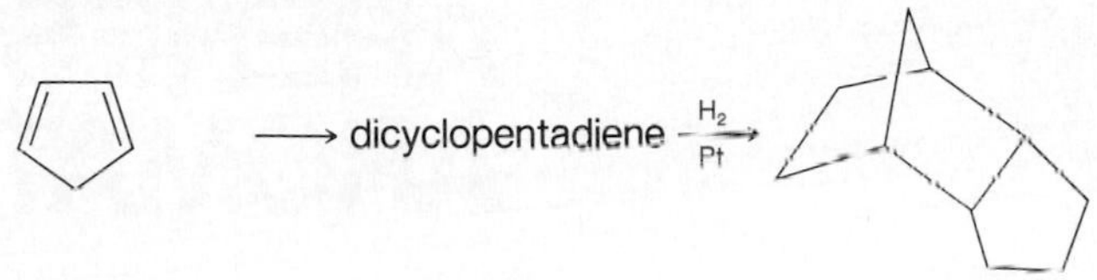

1,3-Cyclopentadiene $C_{10}H_{12}$ $C_{10}H_{16}$

11.27 Refer to the molecular orbital diagrams of allyl cation (Figure 11.1), ethylene (Figure 5.4), and 1,3-butadiene (Figure 11.6) to decide which of the reactions given below are allowed and which are forbidden according to the Woodward-Hoffmann rules.

(a)

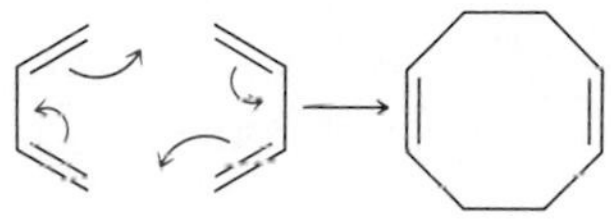

(b)

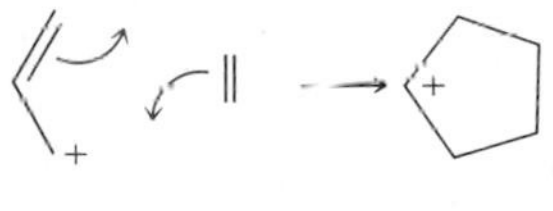

(c)

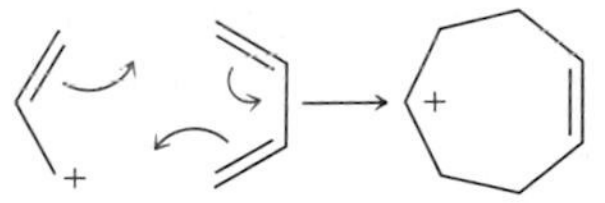

11.28 Alkenes slowly undergo a reaction in air called *autoxidation* is which allylic hydroperoxides are formed.

Cyclohexene   Oxygen   3-Hydroperoxycyclohexene

Keeping in mind that oxygen has two unpaired electrons ($\cdot \ddot{O} \! : \! \ddot{O} \cdot$), suggest a reasonable mechanism for this reaction.

12

ARENES AND AROMATICITY

In this chapter and the next we extend our coverage of conjugated systems to include *arenes*. Arenes are hydrocarbons based on the benzene ring as a structural unit. Benzene and toluene, for example, are arenes.

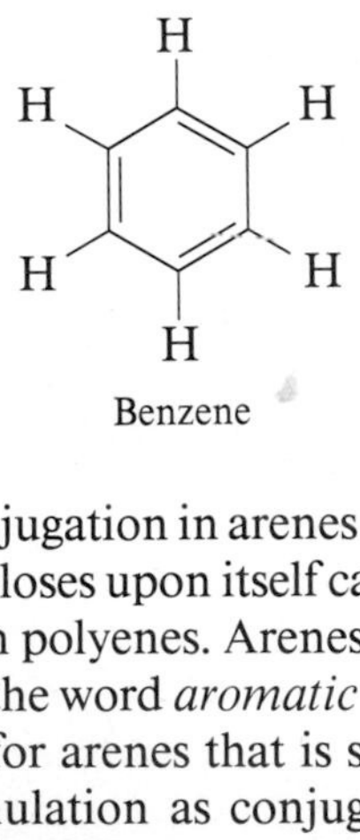

Benzene 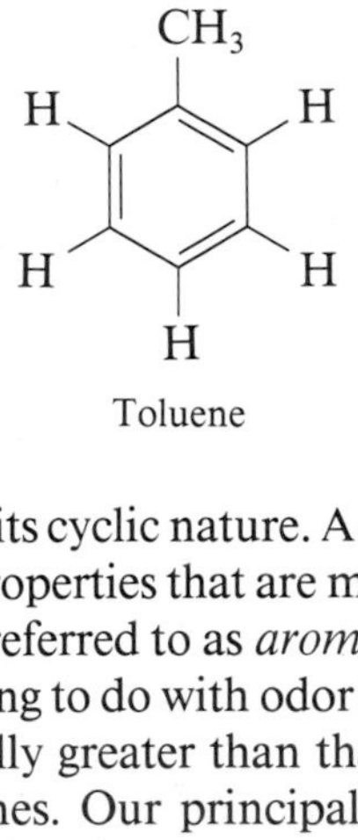   Toluene

What makes conjugation in arenes special is its cyclic nature. A conjugated system of π electrons that closes upon itself can have properties that are much different from those of open-chain polyenes. Arenes are also referred to as *aromatic hydrocarbons.* Used in this sense, the word *aromatic* has nothing to do with odor but rather refers to a level of stability for arenes that is substantially greater than that expected on the basis of their formulation as conjugated trienes. Our principal objectives in this chapter will be to develop an appreciation for the phenomenon of *aromaticity*—to see what are the properties of benzene and its derivatives that reflect this enhanced stability, and to explore the reasons for it.

Aromaticity can be profitably introduced by tracing the history of benzene, its origin, and its structure. Many of the terms we use, including *aromaticity* itself, are of historical origin. Let us begin with the discovery of benzene.

12.1 BENZENE

In 1825 Michael Faraday isolated a new hydrocarbon from illuminating gas, which he called "bicarburet of hydrogen." Nine years later this same hydrocarbon was prepared by Eilhardt Mitscherlich of the University of Berlin, who obtained it by heating benzoic acid with lime and determined its molecular formula to be C_6H_6.

$$C_6H_5CO_2H + \quad CaO \quad \xrightarrow{heat} \quad C_6H_6 + \quad CaCO_3$$

Benzoic acid   Calcium oxide   Benzene   Calcium carbonate

Eventually, because of its relationship to benzoic acid, this hydrocarbon came to be named *benzin,* then later *benzene,* the name by which it is known today in the IUPAC nomenclature system.

Benzoic acid had been known for several hundred years by the time of Mitscherlich's experiment. Many trees exude resinous materials called balsams when cuts are made in their bark. Some of these balsams are very fragrant, a property which once made them highly prized articles of commerce, especially when the trees which produced them could be found only in exotic, faraway lands. *Gum benzoin* is such a substance and is obtained from a tree indigenous to the area around Java and Sumatra. *Benzoin* is a word derived from the French equivalent *benjoin,* which in turn comes from the Arabic *luban jawi,* meaning "incense from Java." Benzoic acid is itself odorless but can be easily isolated from the mixture that makes up the material known as *gum benzoin.*

Compounds related to benzene were obtained from similar plant extracts. For example, a pleasant-smelling resin known as *tolu balsam* was obtained from a South American tree. In the 1840s it was discovered that distillation of tolu balsam gave a methyl derivative of benzene, which, not surprisingly, came to be named *toluene.*

Although benzene and toluene are not particularly fragrant compounds themselves, their origins in aromatic plant extracts led them and compounds related to them to be classified as *aromatic hydrocarbons.* Alkanes, alkenes, and alkynes belong to another class, the *aliphatic hydrocarbons.* The word *aliphatic* comes from the Greek word *aleiphar* (meaning "oil or "unguent") and arose from the observation that hydrocarbons of this class could be obtained by the chemical degradation of fats.

Benzene was prepared from coal tar by August W. von Hofmann in 1845. This remained the primary source for the industrial production of benzene for many years, until petroleum-based technologies became competitive about 1950. Although free benzene does not occur naturally, it is efficiently prepared from coal and petroleum and is thus a relatively inexpensive chemical. Current production is about 6 million tons per year in the United States. A substantial portion of this benzene is converted to styrene for use in the preparation of polystyrene plastics and films.

Toluene is also an important organic chemical. Like benzene, its early production was from coal tar, but more recently petroleum sources have come to supply most of the toluene used for industrial purposes.

12.2 BENZENE REACTIVITY

The separation of hydrocarbons into two groups designated as aliphatic and aromatic in the 1860s took place at a time when it was already apparent that there was something special about benzene, toluene, and their derivatives. The high carbon/hydrogen ratios evident in their molecular formulas (benzene is C_6H_6, toluene is C_7H_8) suggests that they, like alkenes and alkynes, should undergo addition reactions. However, under conditions in which bromine adds rapidly to alkenes and alkynes, benzene proved to be inert. When bromination was carried out in the presence of catalysts such as ferric bromide, the reaction that took place was not addition but substitution!

$$C_6H_6 \ + \ Br_2 \quad \overset{CCl_4}{\longrightarrow} \text{ no observable reaction}$$
$$\overset{FeBr_3}{\longrightarrow} \quad C_6H_5Br \ + \quad HBr$$

| Benzene | Bromine | | Bromobenzene | Hydrogen bromide |

Further, only one monobromination product of benzene was ever obtained, which implies that all the hydrogen atoms of benzene are equivalent. Substitution of one hydrogen by bromine gives the same product as substitution of any of the other hydrogens.

Benzene is not affected by oxidizing agents such as potassium permanganate under conditions that lead to cleavage of the carbon-carbon double bonds of alkenes. Toluene, a methyl derivative of benzene, is oxidized by potassium permanganate, but it is the methyl group that is oxidized.

$$C_6H_5CH_3 \xrightarrow[\text{2. H}^+]{\text{1. KMnO}_4} \overset{\displaystyle O}{\overset{\displaystyle \|}{C_6H_5C}} OH$$

| Toluene | Benzoic acid |

Organic chemists came to regard the six carbon atoms of benzene as a fundamental structural unit. Reactions could be carried out that altered its substituents, but the integrity of the benzene unit remained intact. There must be something "special" about the structure of benzene that makes it sufficiently stable to be inert to many of the reagents that react readily with alkenes and alkynes.

12.3 KEKULÉ'S FORMULATION OF THE BENZENE STRUCTURE

In 1866, only a few years after publishing his ideas concerning what we now recognize as the structural theory of organic chemistry, August Kekulé applied its principles and his own remarkable insight to the problem of a structural representation for benzene. He based his reasoning on three premises:

1. Benzene is C_6H_6.
2. All the hydrogens of benzene are equivalent.
3. The structural theory requires that there be four bonds to each carbon.

Kekulé advanced the venturesome notion that the six carbon atoms of benzene were joined together in a ring. Four bonds to each carbon could be accommodated by a system of alternating single and double bonds with one hydrogen on each carbon.

PROBLEM 12.1   All the following hydrocarbons have the molecular formula C_6H_6. Which of them, if any, satisfies the requirement of giving a single monosubstitution product on bromination?

(a)

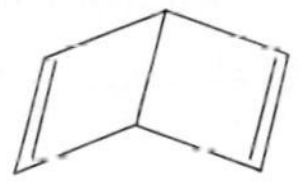

(c)

(b)

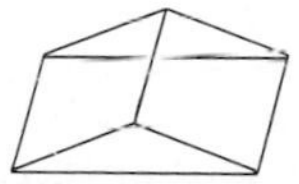

(d)

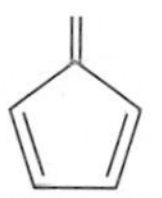

SAMPLE SOLUTION   (a)   This compound is known as "Dewar benzene." It has four equivalent vinyl hydrogens and two equivalent allylic hydrogens. Substitution at the vinylic position yields a different monobromo derivative than substitution of an allylic hydrogen.

Dewar benzene

Two isomeric monobromo derivatives of Dewar benzene

Two monobromo derivatives of Dewar benzene are possible.

A flaw in the Kekulé formula for benzene was soon pointed out. According to the Kekulé formula, 1,2-disubstitution is different from 1,6-disubstitution.

1,2-Disubstituted
derivative of benzene

1,6-Disubstituted
derivative of benzene

These two "isomers" differ in that the two carbons which bear the substituents are connected by a double bond in one structure but by a single bond in the other. Since all the available facts indicated that no such isomerism existed, Kekulé modified his benzene structure to one in which rapid bond migrations caused interconversion of the two isomers.

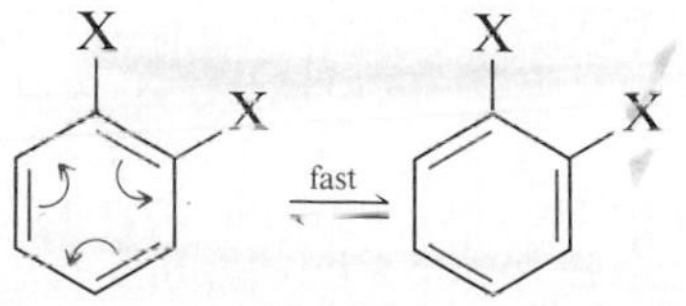

While Kekulé's formulation of the structure of benzene was a significant advance, it is important to note that the issue he addressed emphasized *structure* and left unanswered important questions concerning *reactivity*. Benzene does not behave at all the way we would expect cyclohexatriene to behave. *Benzene is not cyclohexa-*

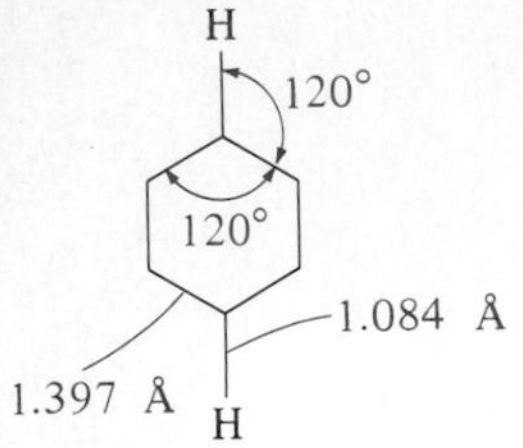

H

120°

120°

1.084 Å

1.397 Å   H

FIGURE 12.1 Bond distances and bond angles of benzene.

triene, nor is benzene a pair of rapidly equilibrating cyclohexatriene isomers. It remained for twentieth century electronic theories of organic chemistry to provide insight into why a benzene ring is such a stable structural unit. We will discuss modern interpretations of the stability of benzene shortly. First, however, let us examine what the results of a variety of experimental studies tell us about the details of its structure.

12.4   STRUCTURAL FEATURES OF BENZENE

Evidence from structural studies leaves no doubt that benzene is a planar molecule and that its carbon skeleton has the shape of a regular hexagon. There are no data to support structural formulations based on notions of alternating single and double bonds. As shown in Figure 12.1, all the carbon-carbon bonds of benzene are the same length (1.397 Å), and the 120° bond angles correspond to perfect sp^2 hybridization at carbon. Interestingly, the 1.397 Å carbon-carbon bond distances in benzene are just midway between the typical sp^2-sp^2 single-bond distance (1.46 Å) and the sp^2-sp^2 double-bond distance (1.34 Å). If bond distances are related to bond type, what kind of carbon-carbon bond is it that lies halfway between a single bond and a double bond in length?

12.5   A RESONANCE DESCRIPTION OF BONDING IN BENZENE

Twentieth century descriptions of bonding in benzene include two major approaches, resonance and molecular orbital treatments. These two approaches, especially the molecular orbital method, have given us a rather clear picture of aromaticity. We will start with the resonance description of benzene.

The two Kekulé structures for benzene have the same arrangement of atoms but they differ in the distribution of their electrons. The Kekulé structures satisfy the rules of valence by localizing electron pairs to regions between two nuclei. The Kekulé structures are not interconvertible isomers of each other but are separate *resonance forms* of the same molecule. Neither Kekulé structure correctly depicts the bonding in benzene. Benzene is a hybrid of both forms and is represented in resonance terms as

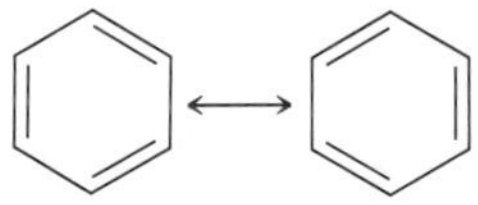

The double-headed arrow does not symbolize a process but rather the fact that the true structure of benzene can be considered to be a hybrid of the two alternative Lewis structures (the Kekulé forms).

A convenient way to represent the two Kekulé structures as resonance forms of benzene is by inscribing a circle inside a hexagon.

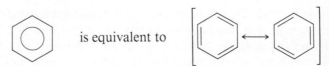

is equivalent to

The circle reminds us of the delocalized nature of the electrons. It was first suggested by the British chemist Sir Robert Robinson as a convenient symbol for the *aromatic sextet*, the six delocalized π electrons present in a benzene ring. In this text, we shall use both the Kekulé and the Robinson representations. The Robinson circle-in-a-ring symbol is a time-saving shorthand device, while Kekulé formulas are better for counting and keeping track of electrons, particularly when we study the chemical reactions of benzene and its derivatives.

PROBLEM 12.2 Write structural formulas for toluene ($C_6H_5CH_3$) and for benzoic acid ($C_6H_5CO_2H$): (*a*) as resonance hybrids of two Kekulé forms, and (*b*) using the Robinson symbol.

Since the carbons that are singly bonded in one resonance form are doubly bonded in the other, the resonance description is consistent with the observed carbon-carbon bond distances in benzene. These distances not only are all identical but are intermediate between typical single-bond and double-bond lengths.

We have come to associate electron delocalization, in conjugated dienes for example, with increased stability. On that basis alone, benzene ought to be a stabilized molecule. It differs from other conjugated systems that we have seen, however, in that its π electrons are delocalized over a cyclic conjugated system. Both Kekulé structures of benzene are of equal energy, and one of the principles of resonance theory is that stabilization is greatest when the contributing structures are of similar energy. Cyclic conjugation in benzene, then, leads to a greater stabilization than is observed in noncyclic conjugated trienes. How much greater that stabilization is can be estimated from heat of hydrogenation data.

12.6 THE STABILITY OF BENZENE

In the presence of the usual hydrogenation catalysts, benzene consumes three molar equivalents of hydrogen to give cyclohexane.

Benzene   Hydrogen   Cyclohexane (100%)
                (2–3 atm pressure)

The heat of hydrogenation of benzene to cyclohexane is 49.8 kcal/mol. To put this value into perspective, compare it with the heats of hydrogenation of cyclohexene and 1,3-cyclohexadiene, as shown in Figure 12.2.

The most striking feature of Figure 12.2 is that the heat of hydrogenation of benzene, with three "double bonds," is less than the heat of hydrogenation of the two double bonds of 1,3-cyclohexadiene. Our experience has been that some 25 to 30 kcal/mol of energy per double bond is given off as heat whenever an alkene is hydrogenated. When benzene combines with three molecules of hydrogen, the reaction is far less exothermic than we would expect it to be on the basis of a 1,3,5-cyclohexatriene structure for benzene.

How much less? Since 1,3,5-cyclohexatriene does not exist (if it did, it would instantly relax to benzene) we cannot measure its heat of hydrogenation in order to

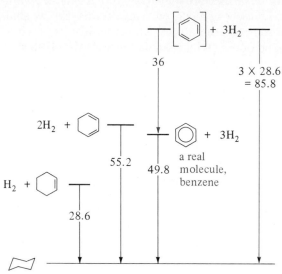

FIGURE 12.2 Heats of hydrogenation of cyclohexene, 1,3-cyclohexadiene, a hypothetical 1,3,5-cyclohexatriene, and benzene. All heats of hydrogenation are in kilocalories per mole.

allow direct comparison with benzene. We can approximate the heat of hydrogenation of our hypothetical 1,3,5-cyclohexatriene as being equal to three times the heat of hydrogenation of cyclohexene, or a total of 85.8 kcal/mol. The heat of hydrogenation, and therefore the energy, of benzene is 36 kcal/mol *less* than expected for a hypothetical 1,3,5-cyclohexatriene with noninteracting double bonds. This is the *resonance energy, stabilization energy,* or *delocalization energy* of benzene. Benzene is far more stable than predicted on the basis of its formulation as a pair of rapidly interconverting 1,3,5-cyclohexatrienes.

A similar conclusion is reached when benzene is compared with the open-chain conjugated triene (Z)-1,3,5-hexatriene. Here we are comparing two real molecules, both conjugated trienes, but one of which is cyclic and the other not. The heat of hydrogenation of (Z)-1,3,5-hexatriene is 80.5 kcal/mol, a value 30.7 kcal/mol greater than that of benzene.

$$ \text{(Z)-1,3,5-Hexatriene} + 3H_2 \longrightarrow CH_3(CH_2)_4CH_3 \qquad \Delta H^\circ = -80.5 \text{ kcal/mol} $$

(Z)-1,3,5-Hexatriene   Hydrogen   Hexane

The precise value of the resonance energy of benzene depends, as comparisons with 1,3,5-cyclohexatriene and (Z)-1,3,5-hexatriene illustrate, on the compound chosen as the reference. What is important is that the resonance energy of benzene is quite large, 6 to 10 times the resonance energy of a conjugated triene. It is this very large increment of resonance energy that places benzene and related compounds in a separate category and accords to them the description aromatic. An aromatic hydrocarbon is one that possesses an amount of resonance energy far in excess of that expected for a structure that has localized π electrons. For benzene this resonance energy is 30 to 36 kcal/mol.

12.7 AN ORBITAL HYBRIDIZATION MODEL OF BONDING IN BENZENE

Since each carbon of benzene is attached to three other atoms, since all the atoms lie in the same plane, and since all the bond angles are 120°, the framework of carbon-carbon σ bonds is reasonably described as arising from the overlap of sp^2 hybrid orbitals. Figure 12.3 illustrates the σ framework of benzene.

An unhybridized $2p$ orbital remains on each carbon. As shown in Figure 12.4, overlap of these $2p$ orbitals generates a continuous π system encompassing all the carbon atoms of the ring. The six π electrons are delocalized over all six carbons. Benzene is characterized by a cyclic conjugated π system containing six π electrons. The electron density associated with the π system of benzene is greatest in regions directly above and directly below the plane of the ring. The π electron density is lowest, approaching zero, in the plane of the ring.

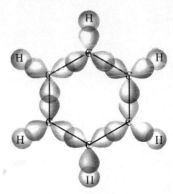

FIGURE 12.3 The assembly of σ bonds in benzene. Each carbon is sp^2 hybridized and forms σ bonds to two other carbons and to one hydrogen.

12.8 THE π MOLECULAR ORBITALS OF BENZENE

The picture portrayed in Figure 12.4 is a useful model of electron distribution in benzene and vividly depicts the delocalization of its π electrons. It is, of course, a superficial one, since six electrons cannot simultaneously occupy any one orbital, be it an atomic orbital or a molecular orbital. A more rigorous molecular orbital analysis recognizes that overlap of the six $2p$ atomic orbitals of the ring carbons generates six π molecular orbitals. These six π molecular orbitals include three which are bonding and three which are antibonding. The relative energies of these orbitals and the distribution of the π electrons among them are illustrated in Figure 12.5.

The nodal properties of the π orbitals are diagrammed in Figure 12.5. The lowest-energy molecular orbital is all bonding—it has no nodes. The highest-energy orbital is all antibonding—it has six nodes and no π bonding interactions between adjacent carbons. Benzene is said to have a *closed-shell* π-electron configuration. All the bonding orbitals are filled, and there are no electrons in antibonding orbitals.

Molecular orbital theory is capable of providing quantitative measures of orbital energies. When this approach is applied to benzene, it is determined that the total energy of the π electrons in benzene is substantially less than the π-electron energy of a localized 1,3,5-cyclohexatriene model. Electrons are more strongly bound in the six π-electron cyclic conjugated system of benzene than they are in a hypothetical model that has alternating single and double bonds.

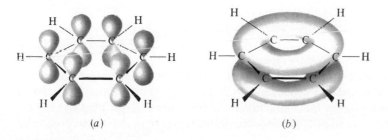

(a)   (b)

FIGURE 12.4 (a) The $2p$ orbitals of benzene carbon atoms are suitably aligned for effective π overlap. (b) Overlap of the $2p$ orbitals generates a π system encompassing the entire ring. There are regions of high π electron density above and below the plane of the ring.

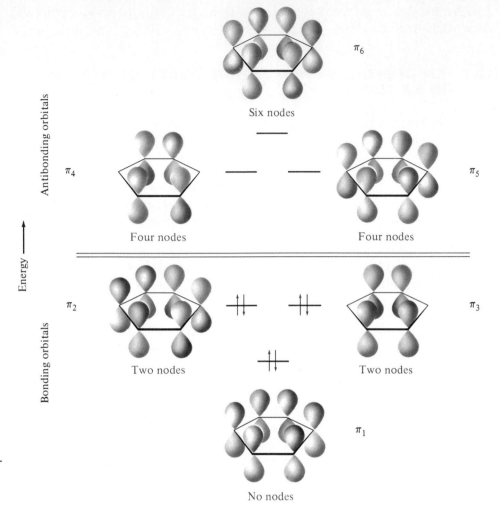

FIGURE 12.5 The π molecular orbitals of benzene arranged in order of increasing energy.

12.9 CYCLOBUTADIENE AND CYCLOOCTATETRAENE

The possibility may have already occurred to you that cyclobutadiene and cyclooctatetraene offer opportunities for stabilization through delocalization of their π electrons in a manner analogous to that in benzene.

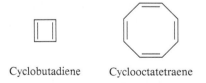

Cyclobutadiene   Cyclooctatetraene

The same thought occurred to early chemists. The complete absence of compounds based on cyclobutadiene and cyclooctatetraene contrasted starkly with the abundance of compounds based on the benzene nucleus. Attempts to synthesize cyclobutadiene and cyclooctatetraene met with failure and reinforced the growing conviction that these compounds would prove to be quite unlike benzene if, in fact, they could be isolated at all.

The first breakthrough came in 1911 when Richard Willstätter prepared cyclooctatetraene by a lengthy degradation of pseudopelleticrine, a natural product obtained from the bark of the pomegranate tree. Nowadays, cyclooctatetraene is prepared from acetylene in a reaction catalyzed by nickel cyanide.

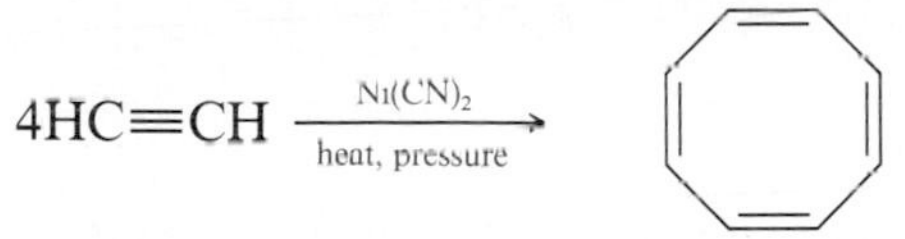

$$4HC\equiv CH \xrightarrow[\text{heat, pressure}]{\text{Ni(CN)}_2}$$

Acetylene   Cyclooctatetraene (70%)

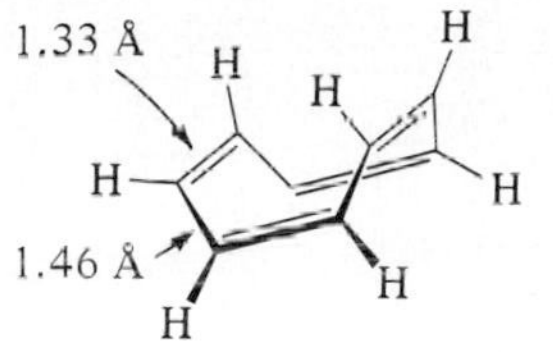

1.33 Å

1.46 Å

FIGURE 12.6 Molecular geometry of cyclooctatetraene.

Cyclooctatetraene's resonance energy, as estimated from its heat of hydrogenation, is only 5 kcal/mol, indicating a degree of stabilization far less than the aromatic stabilization of benzene.

PROBLEM 12.3   Styrene is vinylbenzene ($C_6H_5CH{=}CH_2$). Both cyclooctatetraene and styrene have the molecular formula C_8H_8 and undergo combustion according to the equation

$$C_8H_8 + 10O_2 \longrightarrow 8CO_2 + 4H_2O$$

The measured heats of combustion are 1050 and 1086 kcal/mol. Which heat of combustion belongs to which compound?

Structural studies confirm the absence of appreciable π-electron delocalization in cyclooctatetraene. Its structure is as pictured in Figure 12.6 — a *nonplanar* hydrocarbon with four short carbon-carbon bond distances and four longer carbon-carbon bond distances. Cyclooctatetraene has four noninteracting double bonds. The evidence clearly indicates that cyclooctatetraene is not at all like benzene and is more appropriately considered to be a cyclic polyene. It is satisfactorily represented by a single Lewis structure having alternating single and double bonds in a tub-shaped eight-membered ring.

Cyclobutadiene escaped chemical characterization for more than 100 years. Despite numerous attempts, all synthetic efforts met with failure. It became apparent that cyclobutadiene was not only not aromatic but that it was exceedingly unstable. Beginning in the 1950s, a variety of novel techniques succeeded in generating cyclobutadiene as a transient, reactive intermediate.

PROBLEM 12.4   One of the chemical properties which makes cyclobutadiene difficult to isolate is that it reacts readily with itself to give a dimer.

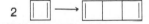

What reaction of dienes does this resemble?

Structural studies of cyclobutadiene and some of its derivatives indicate that it is best described as a diene with alternating single and double bonds and a rectangular, rather than a square, shape. Bond distances that have been determined for a stable, highly substituted derivative of cyclobutadiene clearly exhibit a pattern of alternating short and long ring bonds.

$(CH_3)_3C$ $C(CH_3)_3$

← 1.376 Å

$(CH_3)_3C$ CO_2CH_3

1.506 Å

Methyl 2,3,4-tri-*tert*-butylcyclobutadiene-1-carboxylate

All the available evidence shows that neither cyclooctatetraene nor cyclobuta-diene is aromatic. Clearly, cyclic conjugation, while necessary for aromaticity, is not sufficient for it. Some other factors must be involved that contribute to the special stability of benzene and its derivatives. To understand these factors, let us return to the molecular orbital description of benzene.

12.10   HÜCKEL'S RULE. ANNULENES

One of the early successes of molecular orbital theory was achieved by Erich Hückel, who in 1931 applied its principles to the question of why benzene was so very much different from cyclobutadiene and cyclooctatetraene. Hückel discovered an interest-ing difference in the pattern of π orbitals of benzene compared with cyclobutadiene and cyclooctatetraene. He found that cyclic systems of conjugated double bonds were described by a set of π molecular orbitals in which one orbital is lowest in energy, one is highest in energy, and the rest are distributed in pairs between them.

The arrangements of π orbitals for cyclobutadiene, benzene, and cyclooctate-traene are presented in Figure 12.7. The four π electrons of cyclobutadiene are distributed so that two are in the lowest-energy orbital and, in accordance with Hund's rule, each of the two equal-energy nonbonding orbitals is half-filled. (Re-member, Hund's rule tells us that when two orbitals have the same energy, each one is half-filled before either of them reaches its full complement of two electrons.) Ac-cording to this picture, then, cyclobutadiene should be a diradical. It should have two unpaired electrons with parallel spins. Benzene, on the other hand, has its six π electrons distributed in pairs among its three bonding orbitals. Cyclooctatetraene has six of its eight π electrons in three bonding orbitals. The remaining two electrons occupy one by one the two equal-energy nonbonding orbitals. Planar cyclooctate-traene should, like cyclobutadiene, be a diradical.

As it turns out, neither cyclobutadiene nor cyclooctatetraene is a diradical in its most stable electron configuration. The Hückel approach treats them as planar

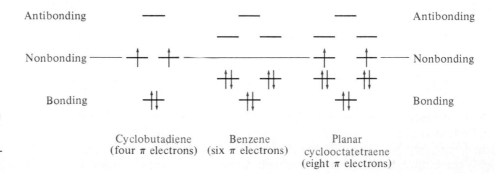

FIGURE 12.7   Distribution of π molecular orbitals and π electrons in cyclobutadiene, benzene, and planar cyclooc-tatetraene.

regular polygons. Because the electron configurations associated with these geometries are not particularly stable, cyclobutadiene and cyclooctatetraene adopt structures other than planar regular polygons. Cyclobutadiene, rather than possessing a square shape with two unpaired electron spins, is a spin-paired rectangular molecule. Cyclooctatraene is nonplanar, with all its π electrons paired in alternating single and double bonds. Only benzene of this group of three conjugated hydrocarbons has a closed-shell electron configuration.

On the basis of his analysis Hückel proposed that only certain numbers of π electrons could lead to aromatic stabilization. Only when the number of π electrons is 2, 6, 10, 14, etc. can a closed-shell electron configuration be realized. These results are summarized in *Hückel's rule: Among planar, monocyclic, fully conjugated polyenes, only those possessing (4n + 2) π electrons, where n is an integer, will have special aromatic stability.*

The general term *annulene* has been coined to apply to completely conjugated monocyclic hydrocarbons. A numerical prefix specifies the number of carbon atoms. Cyclobutadiene is [4]-annulene, benzene is [6]-annulene, and cyclooctatetraene is [8]-annulene.

PROBLEM 12.5   Represent the π-electron distribution in:

(a) [10]-Annulene
(b) [12]-Annulene

SAMPLE SOLUTION   (a) [10]-Annulene is cyclodecapentaene. It has 10 π orbitals and 10 π electrons. Like benzene, it should have a closed-shell electron configuration with all its bonding orbitals doubly occupied.

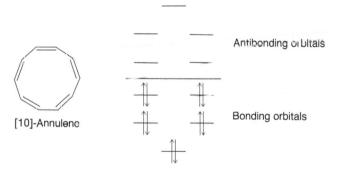

[10]-Annulene

The prospect of observing aromatic character in conjugated polyenes having 10, 14, 18, etc. π electrons spurred efforts toward the synthesis of higher annulenes. A problem immediately arises in the case of the all-cis isomer of [10]-annulene, the structure of which is shown in the preceding problem. Geometry requires a 10-sided regular polygon to have 144° bond angles; sp^2 hybridization at carbon requires 120° bond angles. Therefore, aromatic stabilization due to conjugation in all-*cis*-[10]-annulene is opposed by the destabilizing effect of 24° of angle strain at each of its carbon atoms. All-*cis*-[10]-annulene has been prepared. It is not very stable and is highly reactive.

A second isomer of [10]-annulene, the cis, trans, cis, cis, trans isomer, can have bond angles close to 120° but is destabilized by a close contact between two hydrogens directed toward the interior of the ring. In order to minimize the van der Waals repulsion between these hydrogens, the ring adopts a nonplanar geometry, which limits its ability to be stabilized by resonance. It, too, has been prepared and is not very stable.

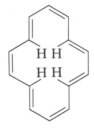

Hydrogens in interior of planar ring
are too close together.

cis,trans,cis,cis,trans-
[10]-Annulene

The next higher $(4n + 2)$ system, [14]-annulene, is also somewhat destabilized by intramolecular van der Waals repulsions between hydrogens and is nonplanar.

Intramolecular van der Waals repulsions
between indicated hydrogen atoms distort ring
from planarity.

[14]-Annulene

When the ring contains 18 carbon atoms, it is large enough to adopt a planar geometry while allowing its interior hydrogens to remain far enough apart that they do not interfere with each other. The [18]-annulene shown is planar or nearly so and has all its carbon-carbon bond distances in the range 1.37 to 1.43 Å — very much like those of benzene. Its resonance energy is about 100 kcal/mol, as estimated from its heat of combustion. The properties of [18]-annulene attest to the validity of the Hückel $(4n + 2)$ rule. [18]-Annulene is predicted to be an aromatic hydrocarbon and seems to be exactly that.

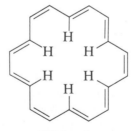

No serious repulsions between six
interior hydrogens; molecule is planar and
aromatic.

[18]-Annulene

According to Hückel's rule, annulenes with $4n$ π electrons are not aromatic. Cyclobutadiene and cyclooctatetraene are [4n]-annulenes and, as we have seen, their properties are more in accord with their classification as cyclic polyenes than as aromatic hydrocarbons. Among higher [4n]-annulenes, [16]-annulene has been prepared. [16]-Annulene is not planar and shows a pattern of alternating short (average 1.34 Å) and long (average 1.46 Å) bonds typical of a nonaromatic cyclic polyene.

[16]-Annulene

Most of the synthetic work directed toward the higher annulenes was carried out by Franz Sondheimer and his students, first at Israel's Weizmann Institute and then later at the University of London. Sondheimer's research systematically explored the chemistry of these hydrocarbons and provided experimental verification of Hückel's rule.

12.11 AROMATIC IONS

Hückel realized that his molecular orbital analysis of conjugated systems could be extended beyond the realm of neutral hydrocarbons. He pointed out that cycloheptatrienyl cation contained a π system with a closed-shell electron configuration similar to that of benzene (Figure 12.8). Cycloheptatrienyl cation has a set of seven π molecular orbitals. Three of these are bonding and are occupied by the six π electrons of the cation. These six π electrons are delocalized over seven carbon atoms, each of which contributes one $2p$ orbital to a planar, monocyclic, completely conjugated π system. Therefore, cycloheptatrienyl cation should be aromatic. It should be appreciably more stable than expected on the basis of any Lewis structure written for it.

Cycloheptatriene

Cycloheptatrienyl cation
(commonly referred to as
tropylium cation)

It is important that we recognize the difference between cycloheptatriene and cycloheptatrienyl (tropylium) cation. The carbocation, as we have just stated, is aromatic, whereas cycloheptatriene is not. Cycloheptatriene has six π electrons in a conjugated system but its π system does not close upon itself. The ends of the triene system are joined by an sp^3 hybridized carbon, which precludes continuous electron delocalization. The ends of the triene system in the carbocation are joined by an sp^2 hybridized carbon, which contributes an empty p orbital, thereby allowing continuous delocalization of the six π electrons. When we say cycloheptatriene is not aromatic but tropylium cation is, we are not comparing the stability of the two directly. Cycloheptatriene is a stable hydrocarbon but does not possess the special stability required to be called aromatic. Tropylium cation, while aromatic, is still a carbocation and reasonably reactive toward Lewis bases. Its special stability does not imply a rocklike passivity but rather a much greater ease of formation than expected on the

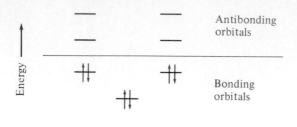

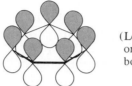

(Lowest-energy orbital; all bonding)

FIGURE 12.8 Electron distribution among the π molecular orbitals of cyclo-heptatrienyl (tropylium) cation.

basis of the Lewis structure drawn for it. To emphasize the aromatic nature of tropylium cation, it is sometimes written in the Robinson manner, including a circle to represent the aromatic sextet.

Br⁻

Tropylium bromide

Tropylium bromide was first prepared but not recognized as such in 1891. The work was repeated in 1954 and the ionic properties of tropylium bromide were demonstrated. On the basis of what we know of its stability, tropylium cation is far more stable than simple tertiary carbocations.

Cyclopentadienide anion is an *aromatic anion.* It has six π electrons delocalized over a completely conjugated planar monocyclic array of five sp^2 hybridized carbon atoms.

H H
H   H
H
Cyclopentadienide anion

≡ (−)

Figure 12.9 presents the molecular orbital description of cyclopentadienide anion. One way of demonstrating the stabilization due to resonance in cyclopentadienide anion is to measure the acidity of cyclopentadiene.

⇌ H⁺ + $K_a\ 10^{-16}$
(pK_a 16)

H H   H
Cyclopentadiene   Cyclopentadienide
anion

Energy →

Antibonding
orbitals

⧸⧹

Bonding
orbitals

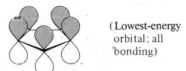

(Lowest-energy
orbital; all
bonding)

FIGURE 12.9 Electron distribution among the π molecular orbitals of cyclopentadienide anion.

Cyclopentadiene is about as strong an acid as water. The position of equilibrium for its deprotonation is more favorable than for other hydrocarbons because cyclopentadienide anion possesses the special stability of aromaticity. A striking contrast is evident when we compare this equilibrium with that for cycloheptatriene deprotonation.

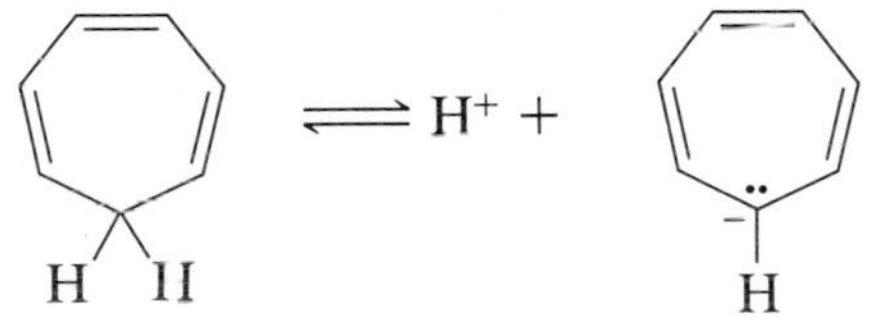

Cycloheptatriene ⇌ H⁺ + Cycloheptatrienide anion

$K_a \ 10^{-36}$
$(pK_a \ 36)$

H H

H

Cycloheptatrienide anion can have its negative charge delocalized by resonance but, because it contains *eight* π electrons, is not aromatic. The equilibrium constant for formation from their parent hydrocarbons is more favorable by 20 orders of magnitude (powers of 10) for the aromatic cyclopentadienide anion than for the nonaromatic cycloheptatrienide anion.

Hückel's rule is now taken to apply to planar, monocyclic, completely conjugated systems generally, not just to neutral hydrocarbons. *A planar, monocyclic, continuous system of p orbitals possesses aromatic stability when it contains (4n + 2) π electrons.*

Other aromatic ions include cyclopropenyl cation (2 π electrons) and cyclooctatetraene dianion (10 π electrons),

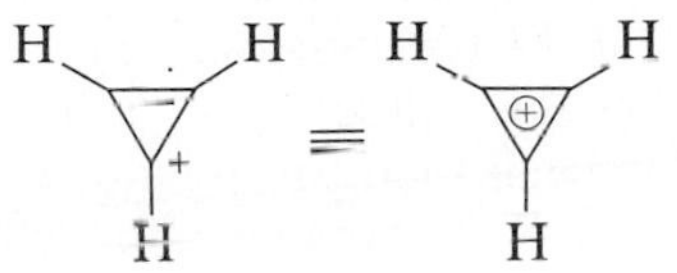

Cyclopropenyl
cation

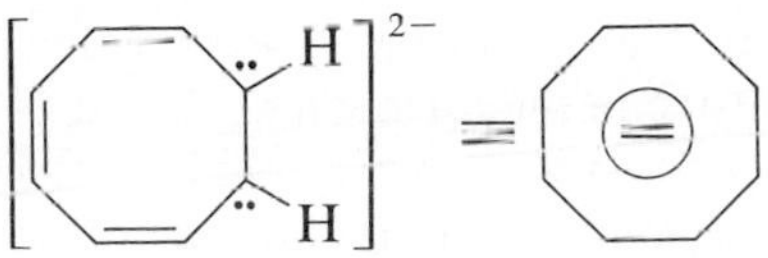

Cyclooctatetraene
dianion

Here, liberties have been taken with the Robinson symbol. Instead of restricting its use to a sextet of electrons, organic chemists have come to adopt it as an all-purpose symbol for cyclic electron delocalization.

PROBLEM 12.6   Is either of the following ions aromatic?

(a) Cyclononatetraenyl
cation

(b) Cyclononatetraenide
anion

SAMPLE SOLUTION   (a)   The crucial point is the number of π electrons. If there are (4n + 2) π electrons the ion is aromatic. Electron counting is facilitated by writing the ion as a species with localized electrons and remembering that each double bond contributes two π electrons, a negatively charged carbon contributes two, and a positively charged carbon contributes none.

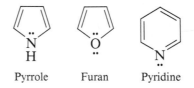

Cyclononatetraenyl cation has eight
π electrons; it is *not aromatic*.

12.12   HETEROCYCLIC AROMATIC COMPOUNDS

When oxygen and nitrogen are incorporated into a cyclic π system, they can contribute an electron pair to satisfy Hückel's rule, as in pyrrole and furan.

Pyrrole   Furan   Pyridine

Counting, as in Figure 12.10*a*, the unshared pair of electrons of nitrogen along with the four π electrons of the diene unit, pyrrole has six π electrons in a planar, conjugated, monocyclic π system so is aromatic.

Of the two unshared pairs of oxygen in furan (Figure 12.10*b*), only one pair is delocalized into the π system to produce the full complement of six π electrons. The other unshared pair occupies an *sp*² hybridized orbital perpendicular to the π system and is therefore not counted when applying Hückel's rule. Furan is aromatic.

Pyridine is aromatic because, as is seen in Figure 12.10*c*, it has six π electrons in its ring exclusive of the nitrogen unshared pair. Therefore, these electrons do not interact with the π system and remain a nitrogen lone pair in an *sp*² hybridized orbital.

Cyclic compounds that contain at least one atom other than carbon within their ring are called *heterocyclic compounds*. Depending on their structure, heterocyclic compounds may or may not be aromatic. Pyrrole, furan, and pyridine are aromatic and are therefore classified as heterocyclic aromatic compounds. We will have more to say about such compounds in later chapters. By including them here we can see how Hückel's rule applies to substances other than hydrocarbons.

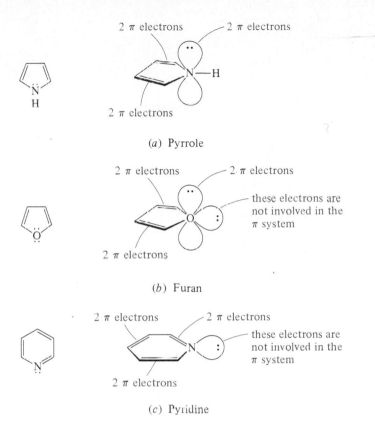

(a) Pyrrole

(b) Furan

(c) Pyridine

FIGURE 12.10 Electron distribution and aromaticity. (a) Pyrrole has six π electrons. (b) Furan has six π electrons plus an unshared pair in an oxygen sp^2 orbital, which is perpendicular to the π system and does not interact with it. (c) Pyridine has six π electrons plus an unshared pair in a nitrogen sp^2 orbital.

12.13 POLYCYCLIC AROMATIC HYDROCARBONS

Naphthalene is the simplest member of the class of arenes known as *polycyclic benzenoid aromatics*. It consists of two fused benzene rings sharing a common side.

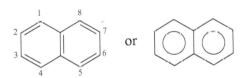

Naphthalene

Naphthalene is a white crystalline solid melting at 80°C. It sublimes readily and has a characteristic odor; it is used as a moth repellent.

Hückel's rule is irrelevant to the question of aromaticity in polycyclic benzenoid aromatics because it applies only to *monocyclic* systems. Most polycyclic benzenoid aromatic hydrocarbons have substantial resonance energies because they are essentially assemblies of benzene units. The resonance energy of naphthalene, with 10 π electrons, is 61 kcal/mol, about the same resonance energy per π electron as that of benzene itself.

Anthracene and phenanthrene are both tricyclic aromatic hydrocarbons. Anthracene has its three rings fused in a "linear" fashion while "angular" fusion occurs in phenanthrene.

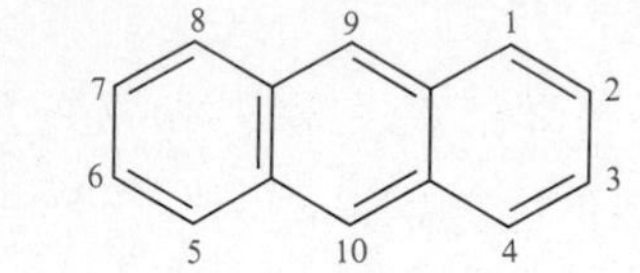

Anthracene 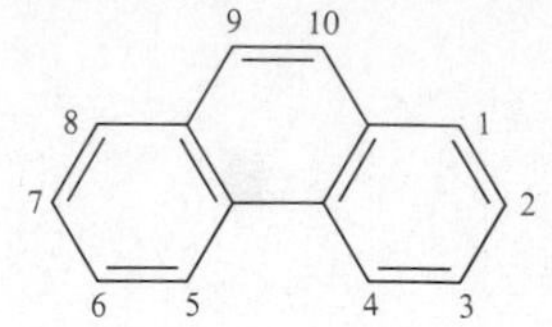   Phenanthrene

The resonance energy of anthracene is 83 kcal/mol and that of phenanthrene is 91 kcal/mol.

PROBLEM 12.7   How many monobromo derivatives are possible for:

(*a*) Anthracene   (*b*) Naphthalene   (*c*) Phenanthrene

SAMPLE SOLUTION   (*a*)   There are only three monobromo derivatives of anthracene: 1-bromoanthracene, 2-bromoanthracene, and 9-bromoanthracene. All other positions of substitution are equivalent to one of these.

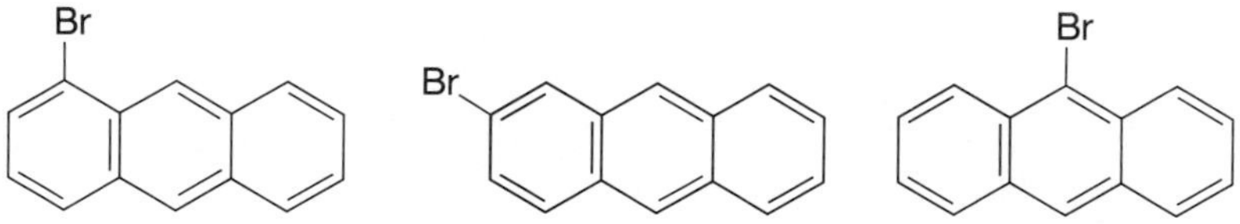

1-Bromoanthracene   2-Bromoanthracene   9-Bromoanthracene

A large number of polycyclic benzenoid aromatic hydrocarbons are known. Many have been synthesized in the laboratory, and several of the others are products of combustion. Benz[a]pyrene has been determined to be present in tobacco smoke. It contaminates food cooked on barbecue grills and collects in the soot of chimneys. Benz[a]pyrene is a *carcinogen* (a cancer-causing substance). It is converted in the liver to an epoxy diol that can induce mutations leading to the uncontrolled growth of certain cells.

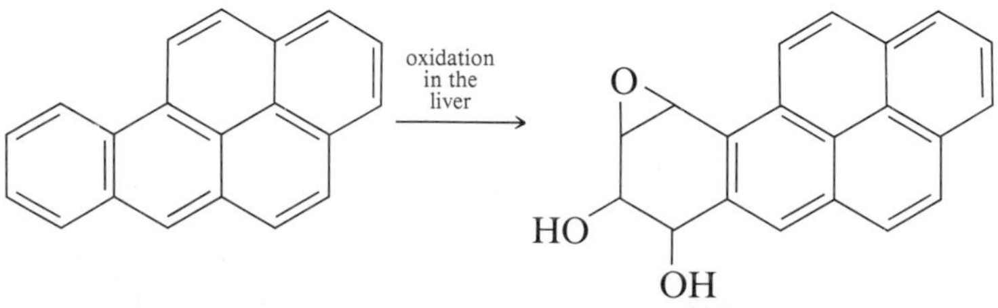

Benz[a]pyrene

7,8-Dihydroxy-9,10-epoxy-
7,8,9,10-tetrahydrobenz[a]pyrene

12.14 PHYSICAL PROPERTIES OF ARENES

Arenes resemble other hydrocarbons in their physical properties. They are nonpolar materials, insoluble in water and less dense than water. The forces of intermolecular attraction are relatively weak and in the absence of polar substituent groups are limited to van der Waals attractions of the induced dipole-induced dipole type.

Table 12.1 lists a number of aromatic hydrocarbons, along with their boiling points and melting points. Benzene is a colorless, flammable liquid with a boiling

TABLE 12.1
Physical Constants of Some Common Arenes

Compound name	Molecular formula	Structure	Melting point, °C	Boiling point, °C (1 atm)
Benzene	C_6H_6	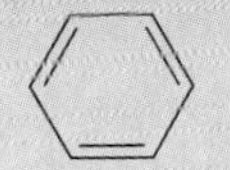	5.5	80.1
Toluene	C_7H_8	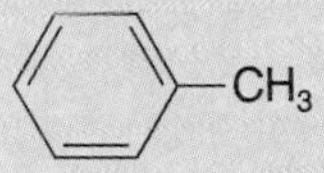 —CH_3	−95	110.6
Styrene	C_8H_8	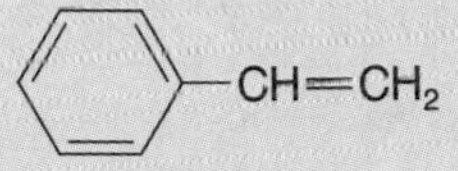 —$CH=CH_2$	−33	145
o-Xylene	C_8H_{10}	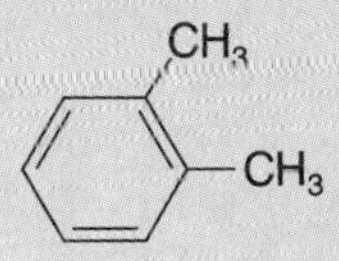 CH_3 —CH_3	−25	144
m-Xylene	C_8H_{10}	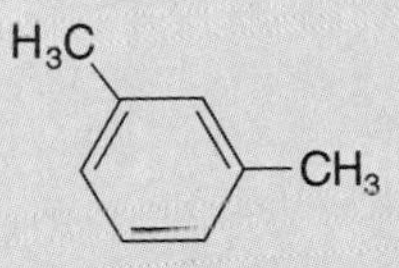 H_3C —CH_3	−47	139
p-Xylene	C_8H_{10}	 H_3C— —CH_3	13	138
Ethylbenzene	C_8H_{10}	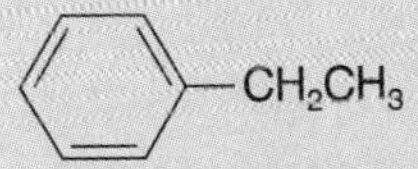 —CH_2CH_3	−94	136.2
Indene	C_9H_8	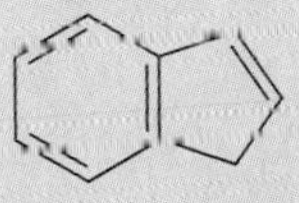	−2	182
Naphthalene	$C_{10}H_8$	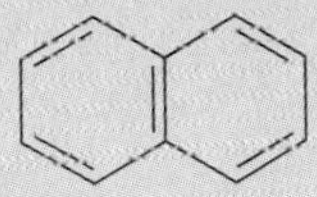	80.3	218
Diphenylmethane	$C_{13}H_{12}$	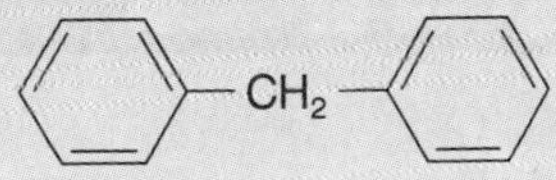 —CH_2—	26	261
Triphenylmethane	$C_{19}H_{16}$	$(C_6H_5)_3CH$	94	—

point of 80.1°C and a melting point of 5.5°C. It was at one time a popular solvent, with numerous applications. This use has been practically eliminated since it has been found that benzene is carcinogenic. Toluene has replaced benzene as an inexpensive organic solvent because it has similar solvent properties but has not been demonstrated to be carcinogenic in the cell systems at the dose levels that benzene is.

12.15 NOMENCLATURE OF SUBSTITUTED DERIVATIVES OF BENZENE

A benzene ring that has one of its hydrogens replaced by some other group is named as a substituted derivative of benzene. For example

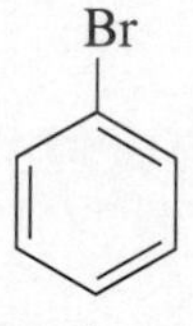

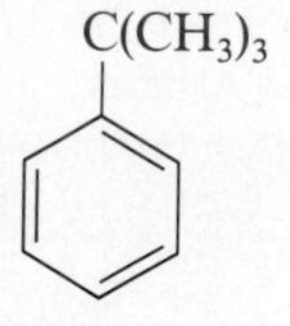

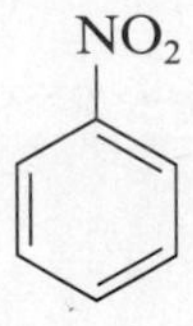

Bromobenzene   *tert*-Butylbenzene   Nitrobenzene

Many simple monosubstituted derivatives of benzene have common names of long standing that have been retained in the IUPAC system. The common names benzaldehyde and benzoic acid, for example, are used far more frequently than their systematic counterparts benzenecarbaldehyde and benzenecarboxylic acid, respectively. Table 12.2 gives the common names of some others.

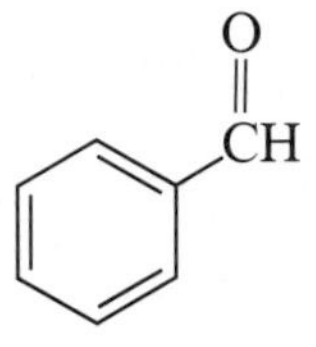

Benzenecarbaldehyde   Benzenecarboxylic acid
(benzaldehyde)   (benzoic acid)

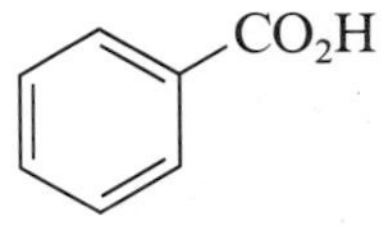

Dimethyl derivatives of benzene are called *xylenes*. There are three xylene isomers, the *ortho* (*o*)-, *meta* (*m*)-, and *para* (*p*)-substituted derivatives.

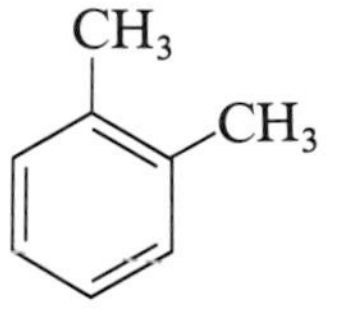

o-Xylene   *m*-Xylene   *p*-Xylene
(1,2-dimethylbenzene)   (1,3-dimethylbenzene)   (1,4-dimethylbenzene)

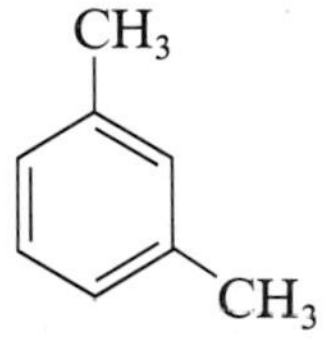

TABLE 12.2

Names of Some Common Benzene Derivatives

Structure	Name*
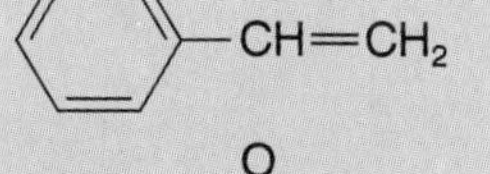 $-CH=CH_2$	Styrene
$\overset{O}{\overset{\|}{-CCH_3}}$	Acetophenone
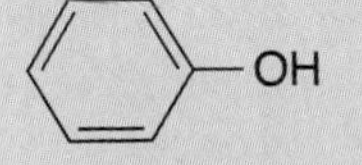 $-OH$	Phenol
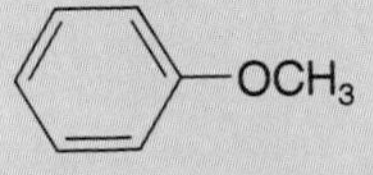 $-OCH_3$	Anisole
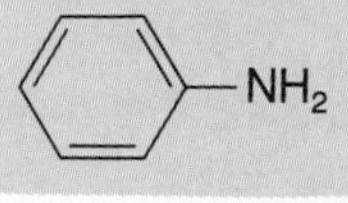 $-NH_2$	Aniline

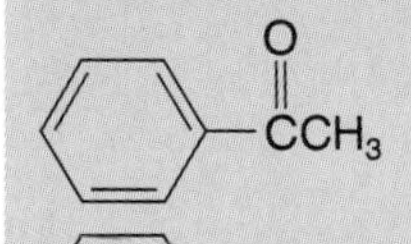

* These common names are acceptable in IUPAC nomenclature.

The prefix *ortho* is used to identify a 1,2-disubstituted benzene ring, *meta* signifies 1,3-disubstitution, and *para* signifies 1,4-disubstitution. The *o*, *m*, and *p* prefixes can be used when a substance is named as a benzene derivative or when a specific base name (such as acetophenone) is used. For example,

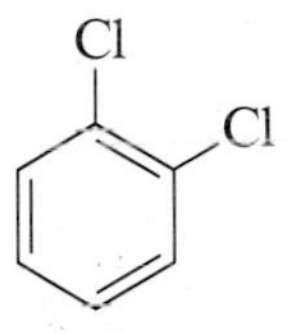

o-Dichlorobenzene
(1,2-dichlorobenzene)

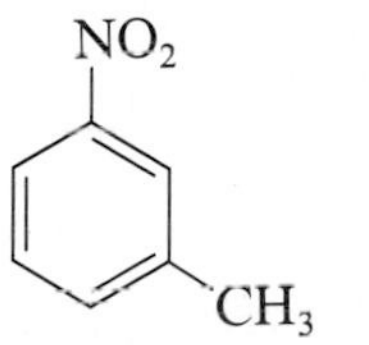

m-Nitrotoluene
(3-nitrotoluene)

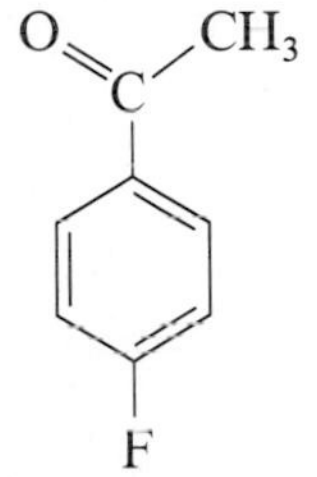

p-Fluoroacetophenone
(4-fluoroacetophenone)

PROBLEM 12.8   Write a structural formula for each of the following compounds.

(*a*) *o*-Ethylanisole   (*b*) *m*-Chlorostyrene   (*c*) *p*-Nitroaniline

SAMPLE SOLUTION   (*a*)   The parent compound in *o*-ethylanisole is anisole. Anisole, as shown in Table 12.2, has a methoxy (CH_3O—) substituent on the benzene ring. The ethyl group in *o*-ethylanisole is attached to the carbon adjacent to the one that bears the methoxy substituent.

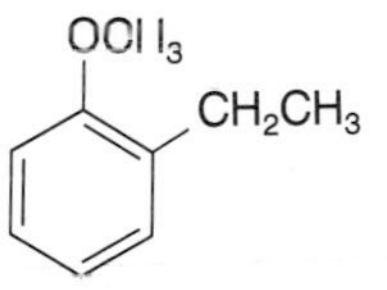

o-Ethylanisole

The *o*, *m*, and *p* prefixes are not used when three or more substituents are present on benzene. Multiple substitution is described by identifying substituent positions on the ring by number.

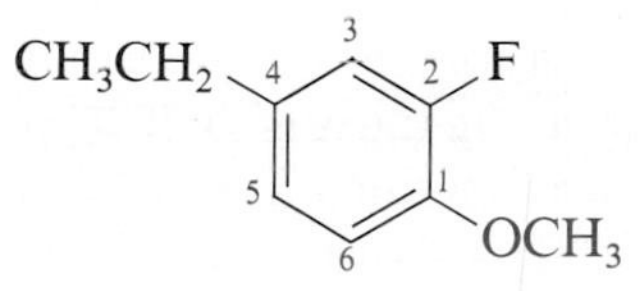

4-Ethyl-2-fluoro-
anisole

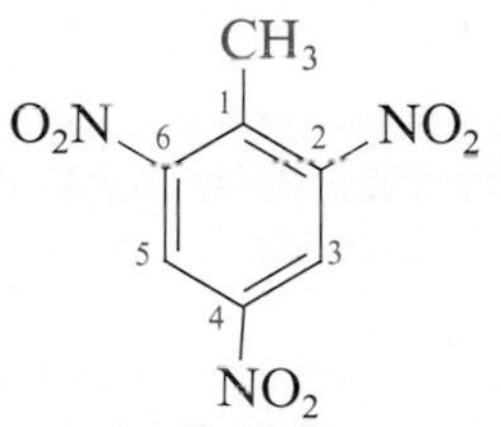

2,4,6-Trinitro-
toluene

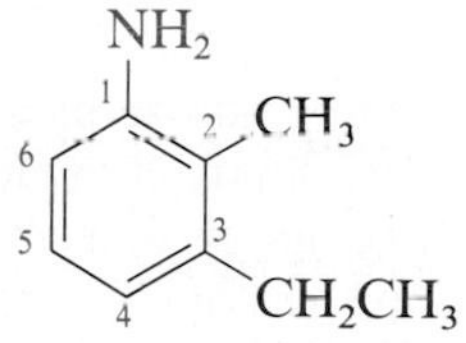

3-Ethyl-2-methyl-
aniline

In the above examples the numbering sequence is established by the base name of the benzene derivative: anisole has its methoxy group at C-1, toluene its methyl group at C-1, and aniline its amino group at C-1. The direction of numbering is chosen to give the next substituted position the lowest number irrespective of what substituent it bears. The order of appearance of substituents in the name is alphabetical. When no simple base name other than benzene is appropriate, positions are numbered so as to give the lowest number at the first point of difference. Thus, each of the following examples is named as a 1,2,4-trisubstituted derivative of benzene rather than as a 1,3,4-derivative.

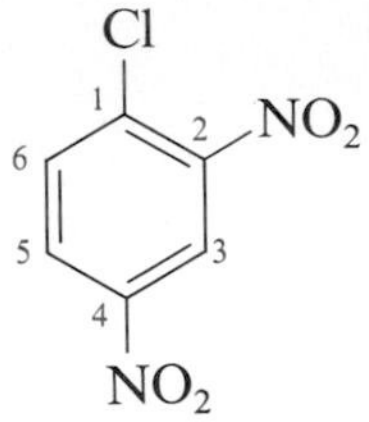

1-Chloro-2,4-dinitrobenzene

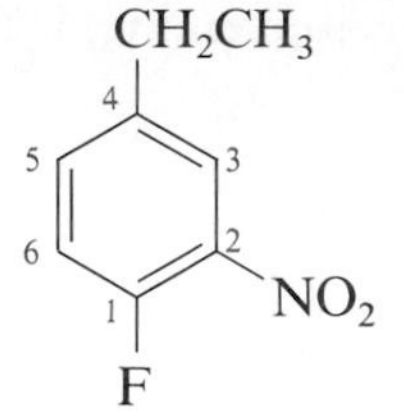

4-Ethyl-1-fluoro-2-nitrobenzene

In cases in which the benzene ring is named as a substituent, the word *phenyl* is used to stand for C_6H_5—. Similarly, an arene named as a substituent is called an *aryl* group. A *benzyl* group is $C_6H_5CH_2$—.

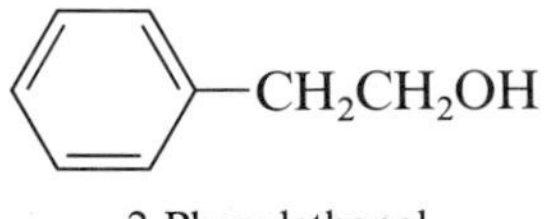

2-Phenylethanol

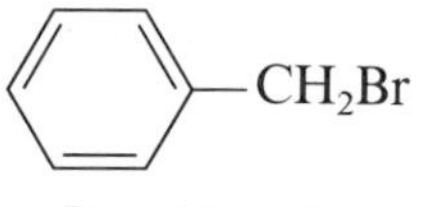

Benzyl bromide

Biphenyl is the accepted IUPAC name for the compound in which two benzene rings are connected by a single bond.

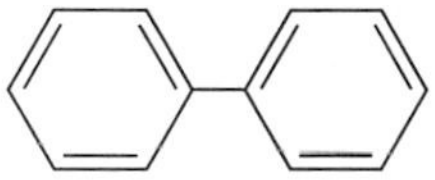

Biphenyl

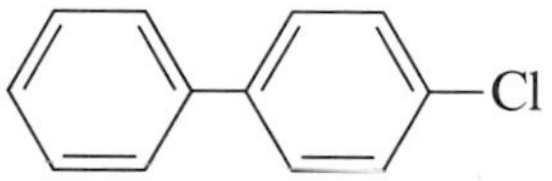

p-Chlorobiphenyl

All the substituted derivatives of benzene listed in this section, and indeed all substituted benzene derivatives, are aromatic because they contain a benzene ring.

12.16 CONJUGATION IN BENZYLIC SYSTEMS

In terms of their chemical properties we shall examine aromatic compounds from two distinct perspectives. In the first, to be described in the remainder of this chapter, an aryl group is present as a substituent and affects the reactivity of the functional unit to which it is attached. In the second, chemical reactions occur on the ring itself. This latter group of reactions, in which the aromatic ring is itself a functional group, form the basis of Chapter 13.

In Chapter 11 we looked at three types of conjugated systems: allylic carbocations, allylic radicals, and conjugated dienes. Each of these species is stabilized by delocalization of the π electrons of a vinyl group into an adjacent orbital.

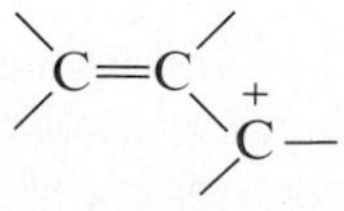

Allylic carbocation

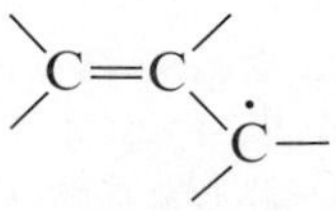

Allylic radical

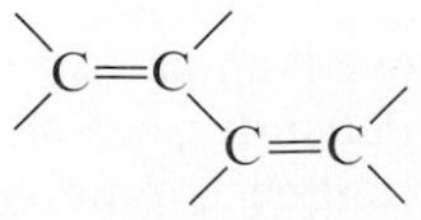

Conjugated diene

A phenyl group (C_6H_5—) is an even better conjugating substituent than a vinyl group (CH_2=CH—). Benzylic carbocations and radicals are more highly stabilized species than their allylic counterparts. The double bond of an alkenylbenzene is stabilized to about the same extent as that of a conjugated diene.

Benzylic carbocation   Benzylic radical   Alkenylbenzene

PROBLEM 12.9   Both 1,2-dihydronaphthalene and 1,4-dihydronapththalene may be selectively hydrogenated to 1,2,3,4-tetrahydronaphthalene.

1,2-Dihydronaphthalene $\xrightarrow[\text{Pt}]{H_2}$ 1,2,3,4-Tetrahydro- $\xleftarrow[\text{Pt}]{H_2}$ 1,4-Dihydronaphthalene
                                                          naphthalene

One of these isomers has a heat of hydrogenation of 24.1 kcal/mol while the heat of hydrogenation of the other is 27.1 kcal/mol. Match the heat of hydrogenation with the appropriate dihydronaphthalene.

Electron delocalization in benzylic systems may be described in resonance terms. In benzyl cation the positive charge is shared by the carbon of the methylene group and by the ring carbons that are ortho and para to it.

Most stable Lewis structure
of benzyl cation

The principal Lewis structures of benzyl radical reveal that the odd electron is shared by the methylene carbon and by the ring carbons that are ortho and para to it.

Most stable Lewis structure
of benzyl radical

In orbital terms, as represented in Figure 12.11, benzylic cations and radicals are stabilized by the delocalization of electrons throughout the extended π system formed by overlap of the p orbital of the methylene group with the π system of the ring. Orbital overlap in alkenylbenzenes involves the π system of the ring and the π component of the double bond.

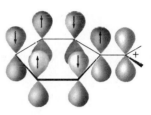

(*a*) Benzylic carbocation

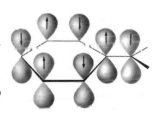

(*b*) Benzylic radical

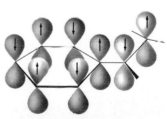

(*c*) Alkenylbenzene

FIGURE 12.11   Conjugation in benzylic cation, benzylic radical, and alkenylbenzene. (*a*) Benzylic carbocation is stabilized by overlap of its vacant p orbital with the π system of the aromatic ring. (*b*) Benzylic radical is stabilized by overlap of its half-filled p orbitals with the π system of the aromatic ring. (*c*) Alkenylbenzene is stabilized by overlap of the π bond of the alkene unit with the π system of the aromatic ring.

12.17 FREE-RADICAL HALOGENATION OF ALKYLBENZENES

Conjugation in benzylic systems affects the chemical reactivity of alkyl- and alkenyl-substituted derivatives of benzene. In general, reactions that involve the formation of carbocations or free radicals occur more readily at benzylic positions than they do at positions where stabilization due to conjugation is not possible.

Toluene reacts with chlorine at elevated temperatures or under conditions of photochemical initiation to give products resulting from free-radical substitution of the hydrogens on the methyl group.

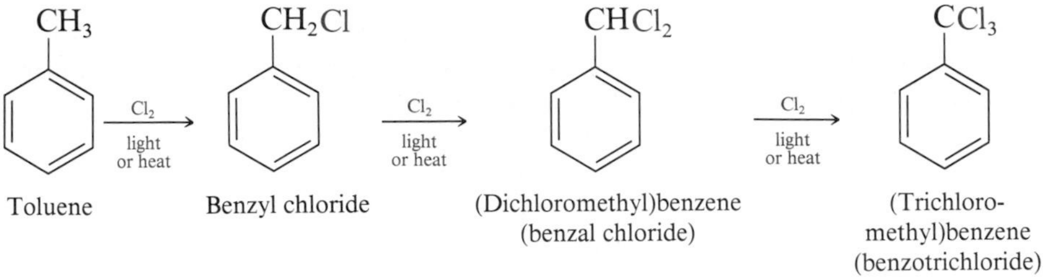

| Toluene | Benzyl chloride | (Dichloromethyl)benzene (benzal chloride) | (Trichloro-methyl)benzene (benzotrichloride) |

The propagation steps in the formation of benzyl chloride involve benzyl radical as an intermediate.

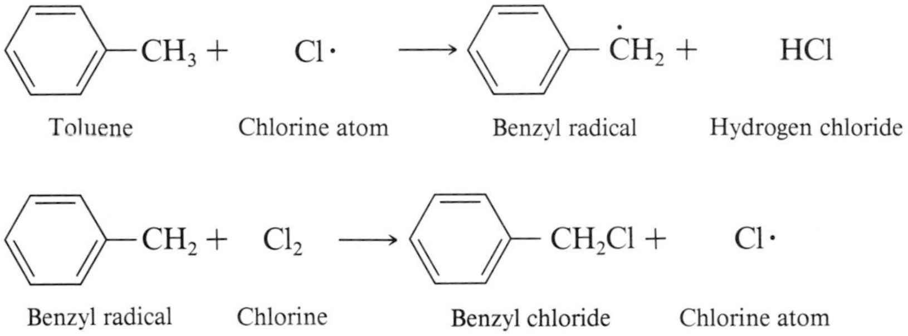

| Toluene | Chlorine atom | Benzyl radical | Hydrogen chloride |

| Benzyl radical | Chlorine | Benzyl chloride | Chlorine atom |

Benzal chloride and benzotrichloride arise by further side-chain chlorination of benzyl chloride. Chlorination of toluene is not particularly convenient to carry out in the laboratory but is a useful industrial process.

The ease with which these free-radical substitution reactions take place at benzylic positions is a direct result of the stability of benzylic radicals. Homolytic cleavage of a benzylic C—H bond in toluene requires less energy than cleavage of an allylic C—H bond in propene or a tertiary C—H bond in 2-methylpropane.

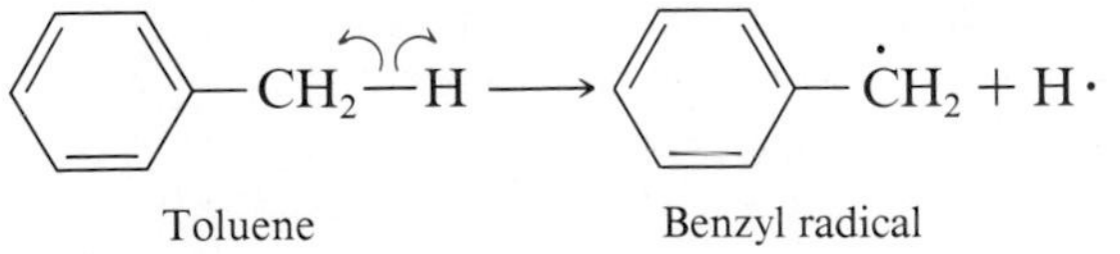

$$\Delta H° = 85 \text{ kcal/mol}$$

Toluene → Benzyl radical

$$CH_2{=}CHCH_2{-}H \longrightarrow CH_2{=}CH\dot{C}H_2 + H\cdot \qquad \Delta H° = 88 \text{ kcal/mol}$$

Propene → Allyl radical

$$(CH_3)_3C{-}H \longrightarrow (CH_3)_3C\cdot + H\cdot \qquad \Delta H° = 95 \text{ kcal/mol}$$

2-Methylpropane → *tert*-Butyl radical

Benzylic bromination is a more commonly used laboratory procedure than chlorination. Side-chain substitution by bromine at the benzylic position occurs in refluxing carbon tetrachloride under conditions of photochemical initiation.

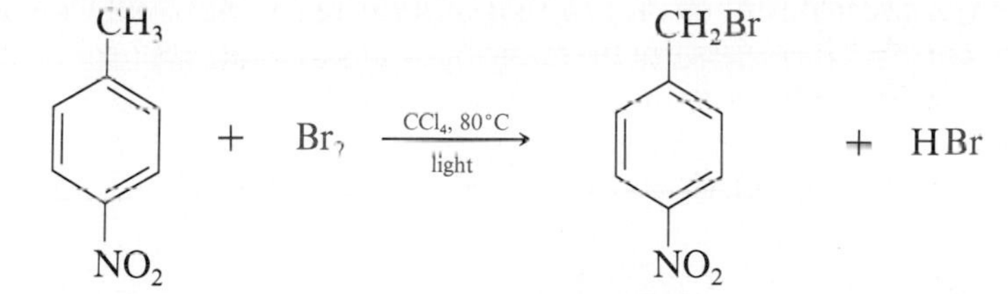

CH_3				CH_2Br		
	$+$	Br_2	$\xrightarrow[\text{light}]{CCl_4, 80°C}$		$+$	HBr
NO_2				NO_2		
p-Nitrotoluene	Bromine			p-Nitrobenzyl bromide (71%)	Hydrogen bromide	

As we saw in the discussion of allylic bromination in Section 11.4, *N*-bromosuccinimide is a convenient brominating agent. Benzylic brominations using this reagent are normally performed in carbon tetrachloride as the solvent in the presence of peroxides, which are added as initiators of free-radical reactions.

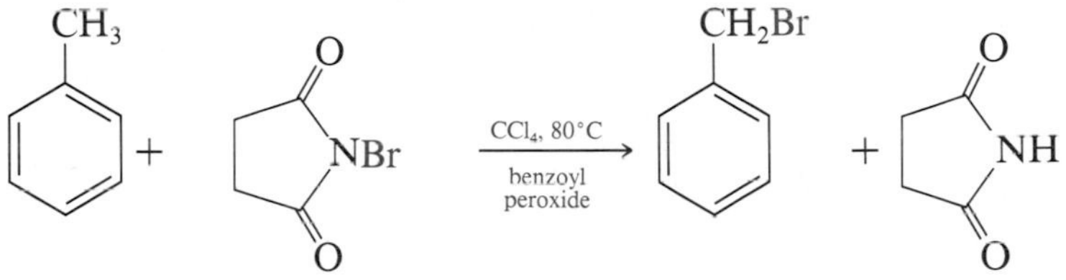

CH_3		O			CH_2Br		O
	$+$	NBr	$\xrightarrow[\text{benzoyl peroxide}]{CCl_4, 80°C}$			$+$	NH
		O					O
Toluene	N-Bromosuccinimide (NBS)				Benzyl bromide (64%)	Succinimide	

Free-radical bromination of alkylbenzenes is selective for substitution of benzylic hydrogens.

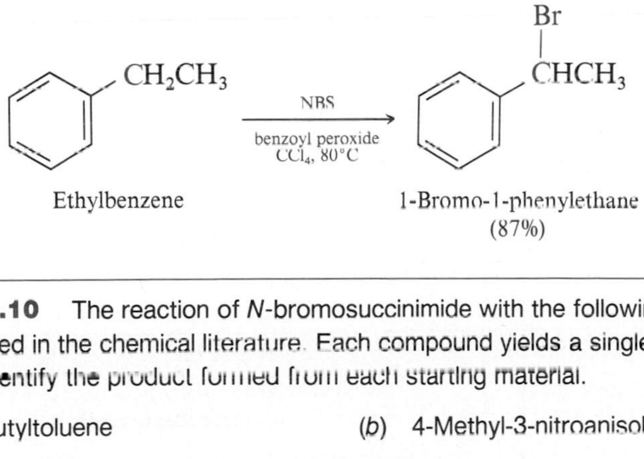

	CH_2CH_3			Br
		$\xrightarrow[\text{benzoyl peroxide} \atop CCl_4, 80°C]{NBS}$		$CHCH_3$
Ethylbenzene				1-Bromo-1-phenylethane (87%)

PROBLEM 12.10 The reaction of *N*-bromosuccinimide with the following compounds has been reported in the chemical literature. Each compound yields a single product in 95 percent yield. Identify the product formed from each starting material.

(a) *p-tert*-Butyltoluene

(b) 4-Methyl-3-nitroanisole

SAMPLE SOLUTION (a) The only benzylic hydrogens in *p-tert*-butyltoluene are those of the methyl group that is attached directly to the ring. Substitution occurs there to give *p-tert*-butylbenzyl bromide.

$(CH_3)_3C$—	—CH_3	$\xrightarrow[\text{free radical} \atop \text{initiator}]{NBS \atop CCl_4, 80°C}$	$(CH_3)_3C$—	—CH_2Br
p-tert-Butyltoluene			*p-tert*-Butylbenzyl bromide	

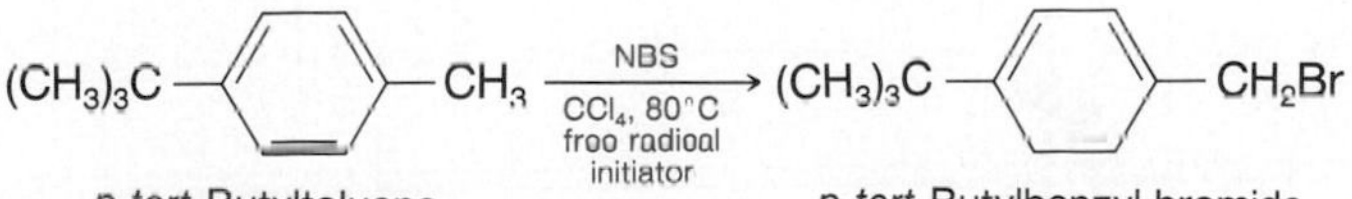

12.18 OXIDATION OF ALKYLBENZENES

A striking example of the activating effect that a benzene ring has on reactions that take place at benzylic positions may be found in the reactions of alkylbenzenes with oxidizing agents. While potassium permanganate cleaves alkenes readily, it does not react either with benzene or with alkanes.

$$RCH_2CH_2R' \xrightarrow{\text{KMnO}_4} \text{no reaction}$$

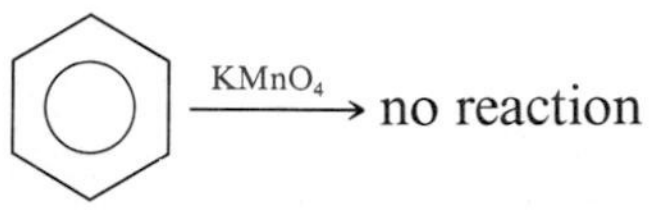

$$\text{[benzene]} \xrightarrow{\text{KMnO}_4} \text{no reaction}$$

On the other hand, an alkyl side chain on a benzene ring is oxidized on being heated with potassium permanganate. The product is benzoic acid or a substituted derivative of benzoic acid.

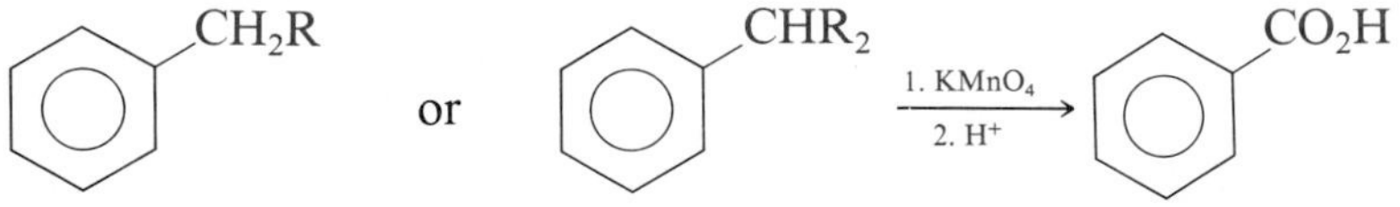

Alkylbenzene   Benzoic acid

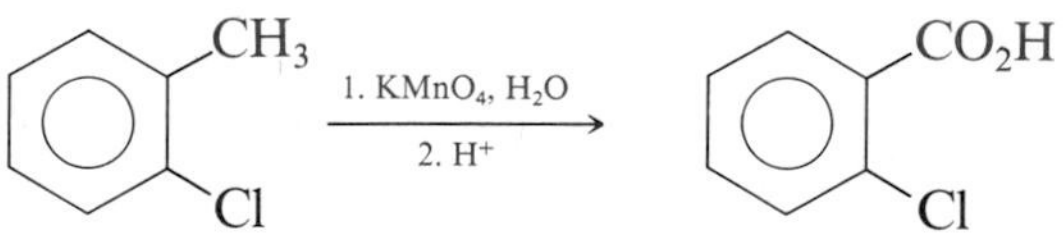

o-Chlorotoluene   *o*-Chlorobenzenecarboxylic
                                         acid (76–78%)
                                      (*o*-Chlorobenzoic acid)

Another commonly used oxidizing agent is chromic acid, prepared by adding sulfuric acid to aqueous sodium dichromate.

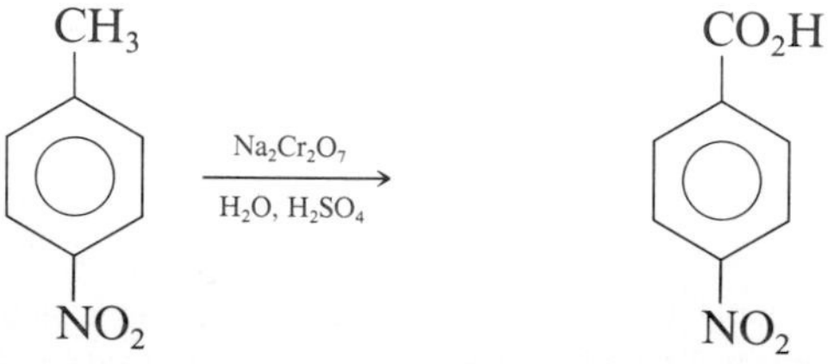

p-Nitrotoluene   *p*-Nitrobenzenecarboxylic acid (82–86%)
                                (*p*-nitrobenzoic acid)

When two alkyl groups are present on the ring, both are oxidized.

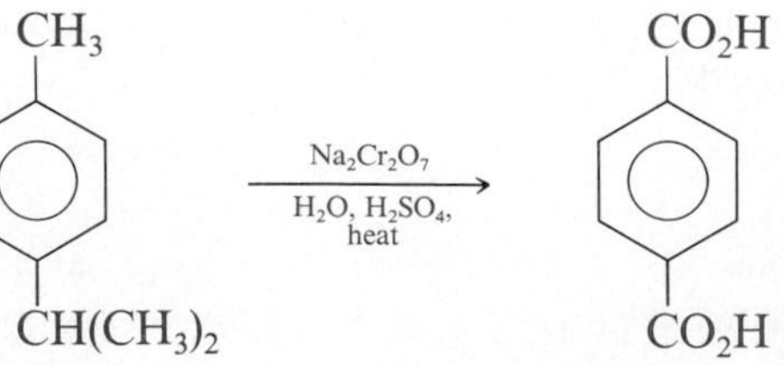

p-Cymene   *p*-Benzenedicarboxylic
(*p*-isopropyltoluene)   acid (45%)

Note that alkyl groups, regardless of their chain length, are converted to carboxyl groups ($-CO_2H$) attached directly to the ring. An exception is a *tert*-alkyl substituent. Because it lacks benzylic hydrogens, a *tert*-alkyl group is not susceptible to oxidation under these conditions.

PROBLEM 12.11 Permanganate oxidation of 1,2-dimethyl-4-*tert* butylbenzene yielded a single compound having the molecular formula $C_{12}H_{14}O_4$. What was this compound?

Side-chain oxidation of alkylbenzenes is important in certain metabolic processes. One way in which the body gets rid of foreign substances is by oxidation in the liver to compounds more easily excreted in the urine. Toluene, for example, is oxidized to benzoic acid by this process and is eliminated rather readily.

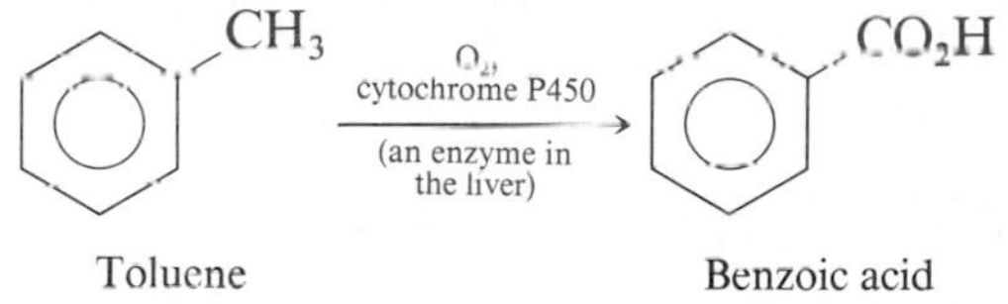

| Toluene | | Benzoic acid |

Benzene, with no alkyl side chain, undergoes a different reaction in the presence of these enzymes, which convert it to a substance capable of inducing mutations in DNA (deoxyribonucleic acid). This difference in chemical behavior seems to be responsible for the fact that benzene is carcinogenic while toluene is not.

12.19 NUCLEOPHILIC SUBSTITUTION AT BENZYLIC POSITIONS

Because an aromatic ring stabilizes a carbocation directly attached to it, benzylic halides undergo S_N1 reactions readily.

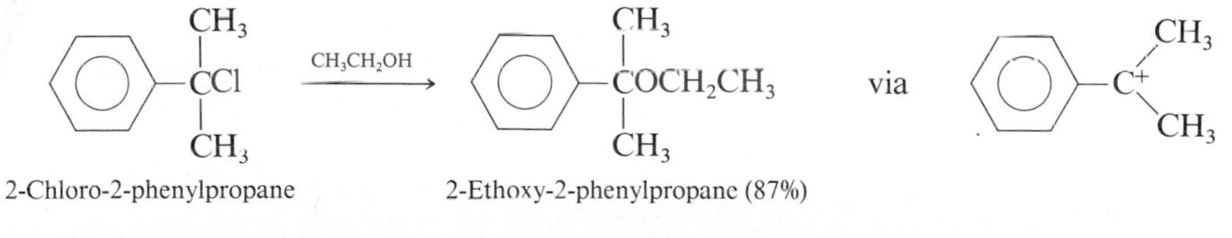

2-Chloro-2-phenylpropane   2-Ethoxy-2-phenylpropane (87%)

Benzylic halides ionize more easily than comparable alkyl halides because they form more stable carbocations. On comparing the two tertiary chlorides

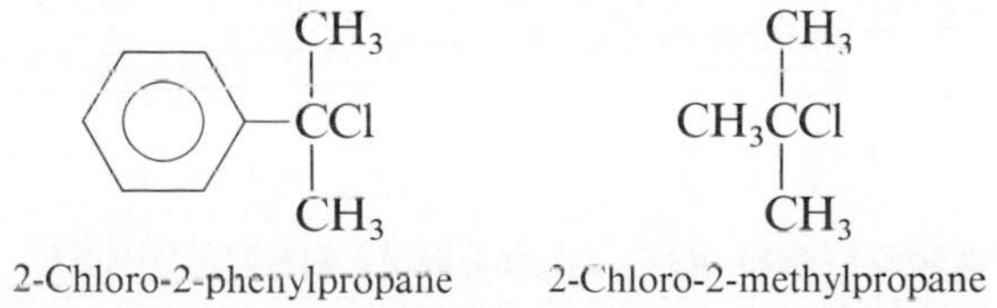

2-Chloro-2-phenylpropane   2-Chloro-2-methylpropane

we find that the benzylic chloride undergoes S_N1 hydrolysis in aqueous acetone over 600 times faster than does *tert*-butyl chloride.

Additional phenyl substituents stabilize carbocations even more. Triphenylmethyl cation is a particularly stable carbocation. Its perchlorate salt is ionic and stable enough to be isolated and stored indefinitely.

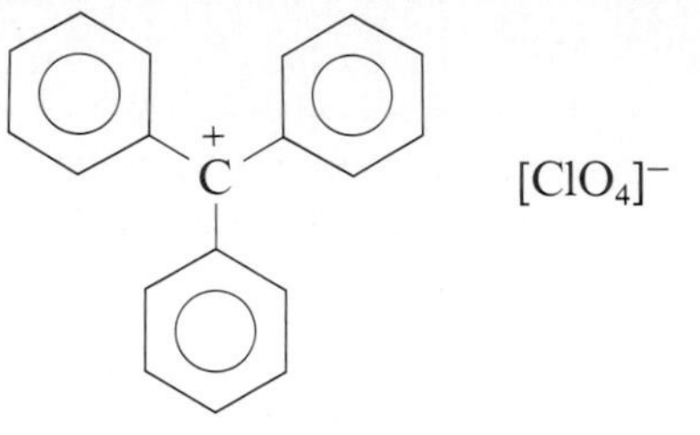

Triphenylmethyl perchlorate

Primary benzylic halides are ideal substrates for S_N2 reactions since they are very reactive toward good nucleophiles and cannot undergo competing elimination.

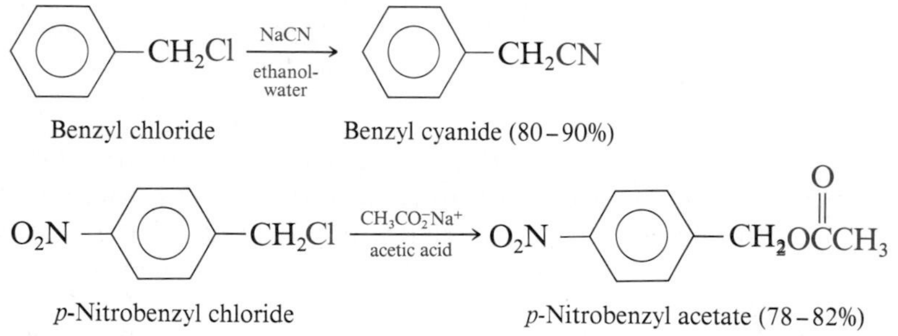

Benzyl chloride   Benzyl cyanide (80–90%)

p-Nitrobenzyl chloride   *p*-Nitrobenzyl acetate (78–82%)

PROBLEM 12.12   Give the structure of the principal organic product formed on reaction of benzyl bromide with each of the following reagents.

(*a*)   Sodium ethoxide   (*d*)   Sodium hydrogen sulfide
(*b*)   Potassium *tert*-butoxide   (*e*)   Sodium iodide (in acetone)
(*c*)   Sodium azide

SAMPLE SOLUTION   (*a*)   Benzyl bromide is a primary bromide and undergoes S_N2 reactions readily. It has no hydrogens β to the leaving group and so cannot undergo elimination. Ethoxide ion acts as a nucleophile, displacing bromide and forming benzyl ethyl ether.

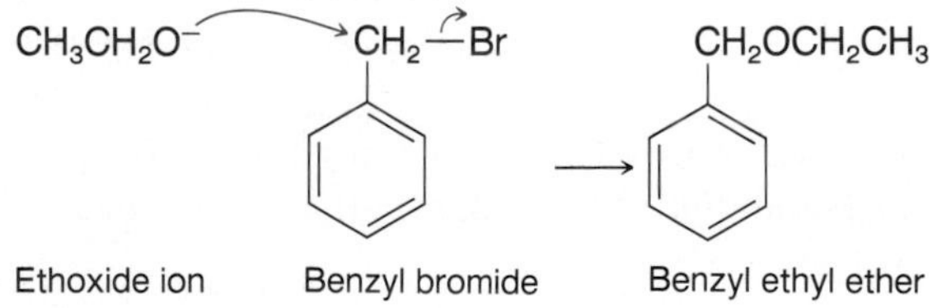

Ethoxide ion   Benzyl bromide   Benzyl ethyl ether

12.20   PREPARATION OF ALKENYLBENZENES

Alkenylbenzenes are prepared by the various methods described in Chapter 6 for the preparation of alkenes; dehydrogenation, dehydration, and dehydrohalogenation.

Dehydrogenation of alkylbenzenes is not a convenient laboratory method but is used industrially to convert ethylbenzene to styrene.

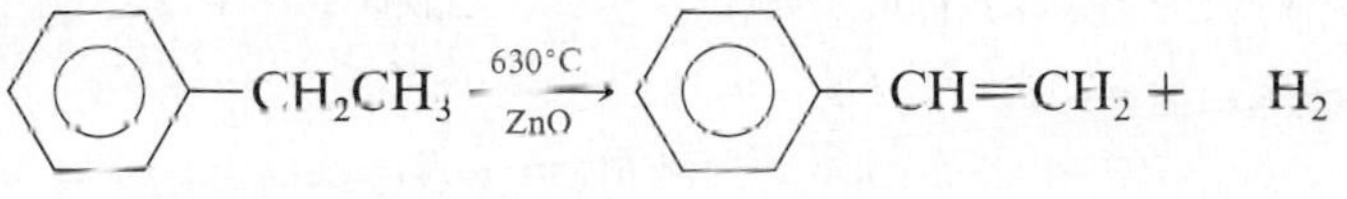

| Ethylbenzene | Styrene | Hydrogen |

Acid-catalyzed dehydration of benzylic alcohols is a useful route to alkenylbenzenes.

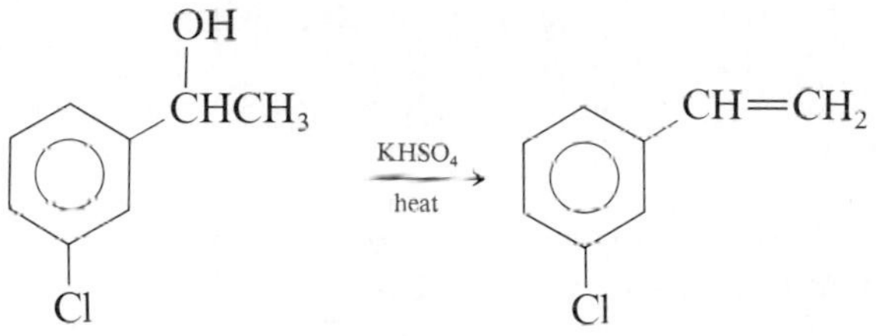

1-(*m*-Chlorophenyl)ethanol   *m*-Chlorostyrene (80–82%)

Dehydrohalogenation under E2 conditions is a reliable method.

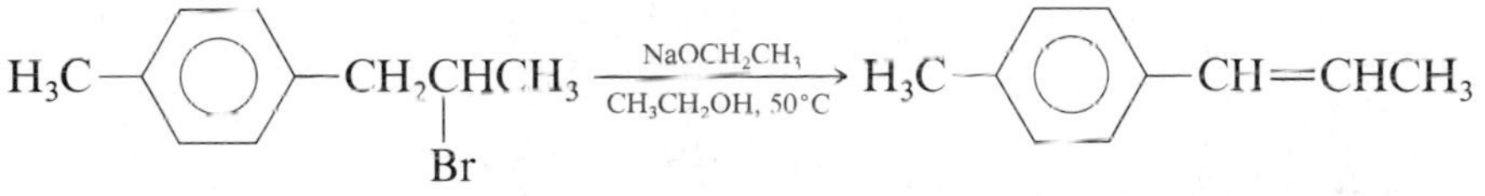

2-Bromo-1-(*p*-methylphenyl)propane   1-(*p*-Methylphenyl)propene (99%)

12.21 REACTIONS OF ALKENYLBENZENES

Most of the reactions of alkenes which were discussed in Chapter 7 find a parallel in the reactions of alkenylbenzenes.

Hydrogenation of the side-chain double bond of an alkenylbenzene is much easier than hydrogenation of the aromatic ring and can be achieved with high selectivity, leaving the ring unaffected.

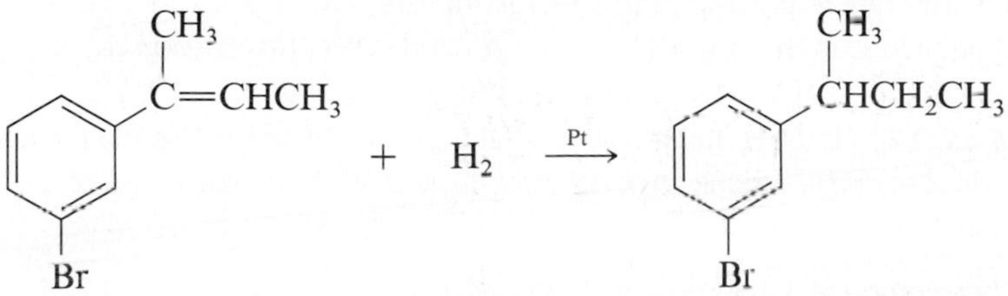

2-(*m*-Bromophenyl)-2-butene   Hydrogen   2-(*m*-Bromophenyl)butane (92%)

The double bond in the alkenyl side chain undergoes addition reactions that are typical of alkenes when treated with electrophilic reagents.

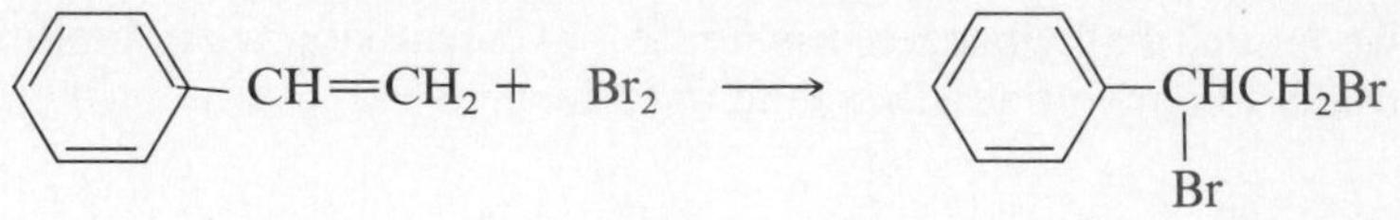

Styrene   Bromine   1,2-Dibromo-1-phenylethane (82%)

The regioselectivity of electrophilic addition is governed by the ability of an aromatic ring to stabilize an adjacent carbocation. This is clearly seen in the addition of hydrogen chloride to indene. Only a single chloride is formed.

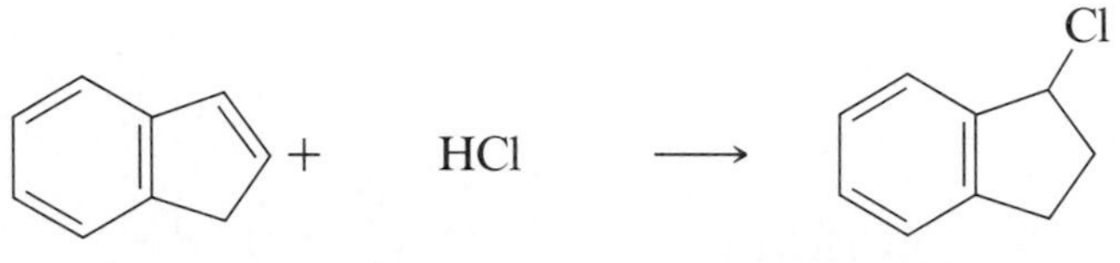

Indene   Hydrogen chloride   1-Chloroindane (75–84%)

Only the benzylic chloride is formed because protonation of the double bond occurs in the direction that gives a secondary carbocation that is benzylic.

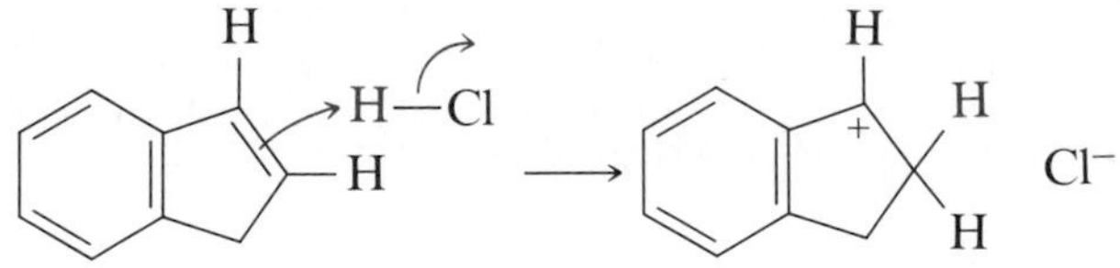

Carbocation that leads to
observed product

Protonation in the opposite direction also gives a secondary carbocation, but one that is not benzylic.

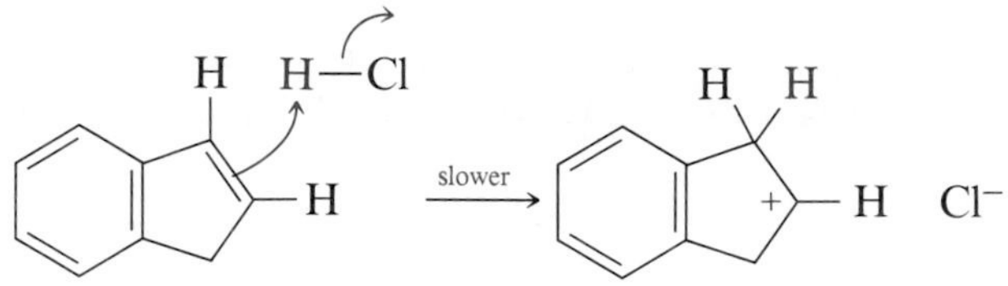

Less stable carbocation

This alternative carbocation does not receive the extra increment of stabilization that its benzylic isomer does and so is formed more slowly. The orientation of addition is controlled by the rate of carbocation formation; the more stable benzylic carbocation is formed faster and is the one that determines the reaction product.

PROBLEM 12.13   Each of the following reactions has been reported in the chemical literature and gives a single organic product in high yield. Write the structure of the product for each reaction.

(a) 2-Phenylpropene + hydrogen chloride
(b) 2-Phenylpropene treated with mercuric acetate in aqueous tetrahydrofuran followed by demercuration with sodium borohydride
(c) 2-Phenylpropene treated with diborane in tetrahydrofuran followed by oxidation with basic hydrogen peroxide

(d)   Styrene + bromine in aqueous solution

(e)   Styrene + peroxybenzoic acid (two organic products in this reaction; identify both)

SAMPLE SOLUTION   (a)   Addition of hydrogen chloride to the double bond takes place by way of a tertiary benzylic carbocation.

| 2-Phenylpropene | Hydrogen chloride | | 2-Chloro-2-phenylpropane |

12.22 SUMMARY

An *aromatic* compound is one which is substantially more stable than expected on the basis of structural formulas that restrict electron pairs to regions between two nuclei. Benzene is an aromatic hydrocarbon. Neither of the two Kekulé formulas for benzene adequately describes its structure or properties. Benzene is said to be a *resonance hybrid* of these two Kekulé forms.

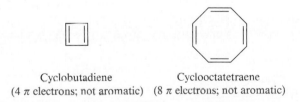

Resonance between two
Kekulé forms of benzene

alternatively
written as

Robinson circle-in-a-ring
symbol for benzene

The extent to which the energy of benzene is less than either of the Kekulé structures is its *resonance energy.* Experimental estimates of the resonance energy of benzene place it at 30 to 36 kcal/mol.

While cyclic conjugation is a necessary requirement for aromaticity, this alone is not sufficient. If it were, cyclobutadiene and cyclooctatetraene would be aromatic. They clearly are not.

Cyclobutadiene
(4π electrons; not aromatic)

Cyclooctatetraene
(8π electrons; not aromatic)

An additional requirement is that the number of π electrons in conjugated planar monocyclic species must be equal to $4n + 2$, where n is an integer. This is called *Hückel's rule.* Benzene, with six π electrons, satisfies Hückel's rule for $n = 1$. Cyclobutadiene and cyclooctetraene do not and so do not possess the closed-shell electron configuration necessary for aromatic stability.

Other than benzene, systems with six π electrons that are substantially more stable than expected include certain ionic species, such as cyclopentadienide anion and cycloheptatrienyl (tropylium) cation.

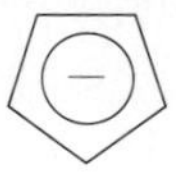

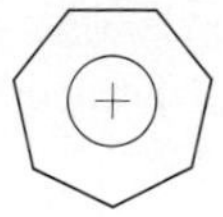

Cyclopentadienide anion

Cycloheptatrienyl cation

An aryl substituent stabilizes an adjacent radical, carbocation, or double bond by conjugation. Benzylic radicals are more stable than simple alkyl radicals and are formed more readily. This allows for efficient halogenation of alkylbenzenes at the position adjacent to the ring.

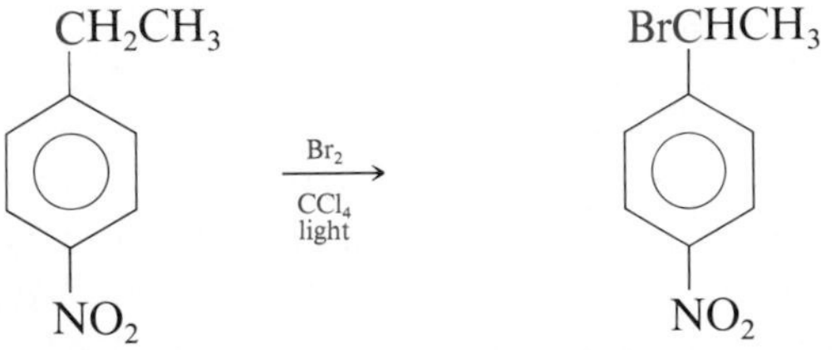

CH_2CH_3 → $BrCHCH_3$

$$\xrightarrow[\substack{CCl_4 \\ light}]{Br_2}$$

NO_2 NO_2

p-Nitroethylbenzene   1-(*p*-Nitrophenyl)ethyl bromide (77%)

Oxidation of alkyl side chains on aromatic rings leads to benzoic acid derivatives by functionalization of the benzylic carbon.

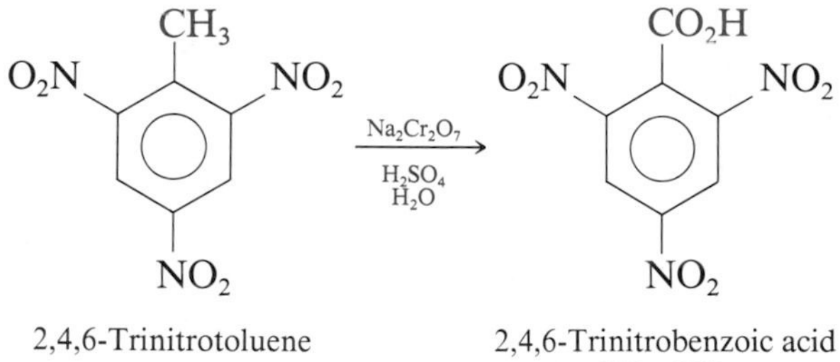

CH_3 O_2N NO_2 CO_2H O_2N NO_2

$$\xrightarrow[\substack{H_2SO_4 \\ H_2O}]{Na_2Cr_2O_7}$$

NO_2 NO_2

2,4,6-Trinitrotoluene   2,4,6-Trinitrobenzoic acid (57–69%)

Benzylic carbocations are intermediates in many of the reactions involving electrophilic addition to alkenylbenzenes. The regioselectivity of addition is strongly influenced by the stabilizing effect the aryl group exerts on an adjacent carbocation.

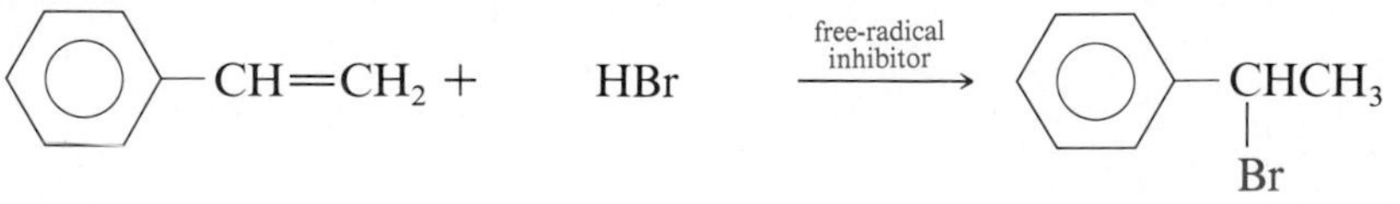

$CH{=}CH_2$ +   HBr $\xrightarrow{\substack{\text{free-radical} \\ \text{inhibitor}}}$ $CHCH_3$ | Br

Styrene   Hydrogen bromide   1-Phenylethyl bromide (85%)

Hydrogenation of aromatic rings is somewhat slower than hydrogenation of alkenes, which makes it a simple matter to convert alkenylbenzenes to alkylbenzenes by selective hydrogenation.

Br $\qquad$ Br

$$\underset{\text{CH}=\text{CHCH}_3}{} \xrightarrow[\substack{\text{Pt} \\ \text{ethanol}}]{\text{H}_2} \underset{\text{CH}_2\text{CH}_2\text{CH}_3}{}$$

1-(*m*-Bromophenyl)propene $\qquad$ *m*-Bromo-*n*-propylbenzene (85%)
1-(*m*-bromophenyl)propane

PROBLEMS

12.14 Write the structures and give the IUPAC names of all the isomers of $C_6H_5C_4H_9$ which contain a monosubstituted benzene ring.

12.15 Write the structure corresponding to each of the following:

(a) Neopentylbenzene
(b) Allylbenzene
(c) Phenylacetylene
(d) (*E*)-1-Phenyl-1-butene
(e) (*Z*)-2-Phenyl-2-butene
(f) (*R*)-1-Phenylethanol
(g) *p*-Chlorophenol
(h) 2-Nitrobenzenecarboxylic acid
(i) *p*-Diisopropylbenzene
(j) 2,4,6-Tribromoaniline
(k) *m*-Nitroacetophenone
(l) 3,5-Dichlorobenzenesulfonic acid

12.16 Write the structures and give acceptable names for all the isomeric

(a) Nitrotoluenes
(b) Dichlorobenzoic acids
(c) Tribromophenols
(d) Tetrafluorobenzenes
(e) Naphthalenesulfonic acids

12.17 Mesitylene (1,3,5-trimethylbenzene) is the most stable of the trimethylbenzene isomers. Can you think of a reason why? Which isomer do you think is the least stable?

12.18 Which one of the dichlorobenzene isomers does not have a dipole moment?

12.19 Identify the longest and the shortest carbon-carbon bonds in styrene.

12.20 (a) The molecule calicene, so-called because it resembles a chalice (*calix* is the Latin word for "cup") should be a fairly polar hydrocarbon. This prediction is based on the idea that one of the two dipolar structures A and B is expected to contribute appreciably to the resonance description of calicene. Which of these two resonance forms is more reasonable, A or B? Why?

Calicene $\qquad$ A $\qquad$ B

(*b*) Which one of the following should be stabilized by resonance to a greater extent? (*Hint:* Consider the reasonableness of dipolar resonance forms.)

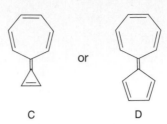

or

C   D

12.21 Each of the following may be represented by at least one alternative resonance structure in which all the six-membered rings correspond to Kekulé forms of benzene. Write such a resonance form for each.

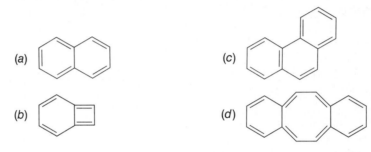

(*a*)

(*b*)

(*c*)

(*d*)

12.22 There are two different tetramethyl derivatives of cyclooctatetraene which have methyl groups on four adjacent carbon atoms. They are both completely conjugated and are not stereoisomeric. Write their structures.

12.23 From among the molecules and ions shown, all of which are based on cycloundeca-pentaene, identify those which satisfy the criteria for aromaticity as prescribed by Hückel's rule.

(*a*) Cycloundecapentaene

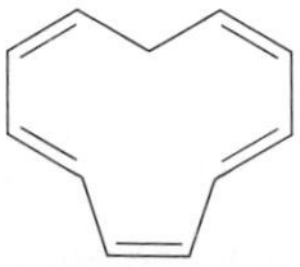

(*b*) Cycloundecapentaenyl radical

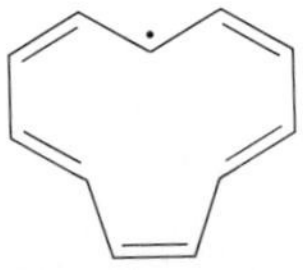

(*c*) Cycloundecapentaenyl cation

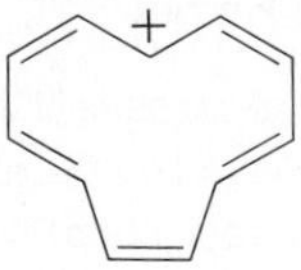

(*d*) Cycloundecapentaenide anion

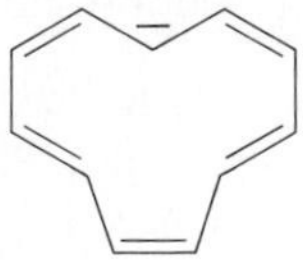

12.24 (*a*) Using the diagram in Figure 12.7 (Section 12.10) as a guide, show the electron configuration for cyclooctatetraene dianion ($C_8H_8^{2-}$).

Cyclooctatetraene dianion

(b) Suggest how one could chemically convert cyclooctatetraene to its dianion. What is the necessary chemical property of the reagent you would choose to carry out this transformation?

12.25 Classify each of the following heterocyclic molecules as aromatic or not according to Hückel's rule:

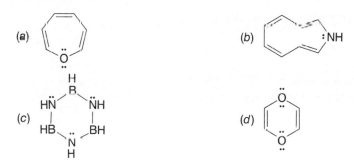

(a)

(b) :NH

(c)

(d)

12.26 Write the structure of the expected product from the reaction of styrene with

(a) Bromine in carbon tetrachloride
(b) Hydrogen chloride
(c) Hydrogen bromide in the presence of peroxides
(d) Potassium permanganate
(e) Hypochlorous acid
(f) Ozone
(g) The product of (f) subjected to hydrolysis in the presence of zinc
(h) Diborane followed by basic hydrogen peroxide
(i) m-Chloroperoxybenzoic acid
(j) One molar equivalent of hydrogen in the presence of platinum
(k) Excess hydrogen in the presence of platinum

12.27 Each of the following reactions has been described in the chemical literature and gives a single organic product in good yield. Identify the product of each reaction.

(a) C_6H_5 $\xrightarrow[\text{2. H}_2\text{O}_2, \text{ HO}^-]{\text{1. B}_2\text{H}_6, \text{ diglyme}}$

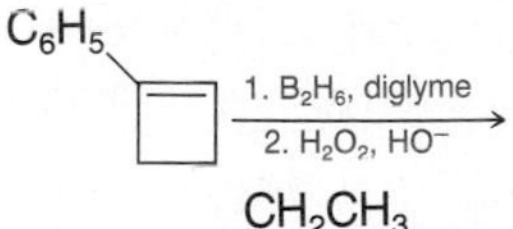

(b) CH_2CH_3 $+ H_2 \text{ (1 mol)} \xrightarrow{\text{Pt}}$

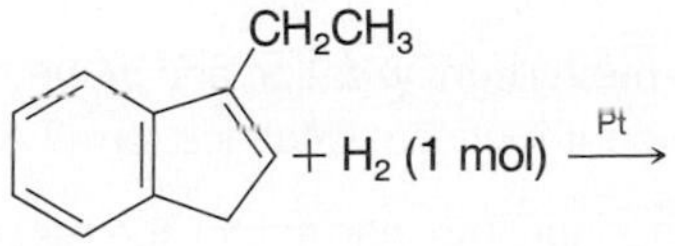

(c) $(C_6H_5)_2CH$—⟨ ⟩—CH_3 $\xrightarrow[\text{CCl}_4, \text{ light}]{\text{excess Cl}_2}$ $C_{20}H_{14}Cl_4$

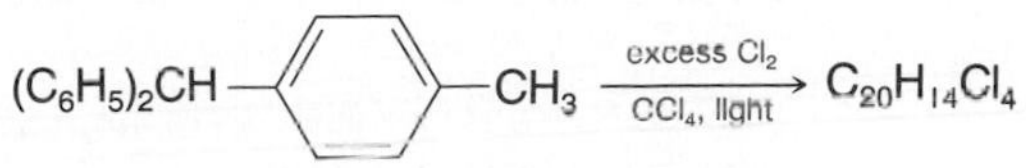

(d) (E)—C_6H_5CH=CHC_6H_5 $\xrightarrow[\text{acetic acid}]{\text{CH}_3\text{CO}_2\text{OH}}$

(e) H_3C· $\xrightarrow[\text{acetic acid}]{\text{H}_2\text{SO}_4}$

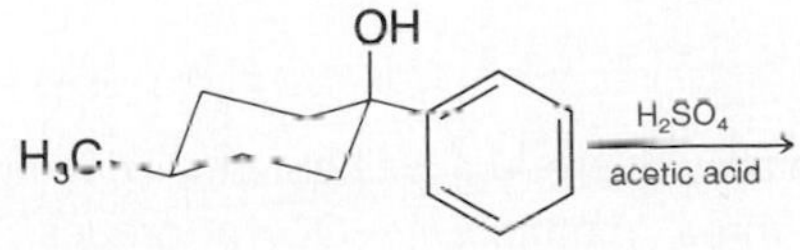

(f)

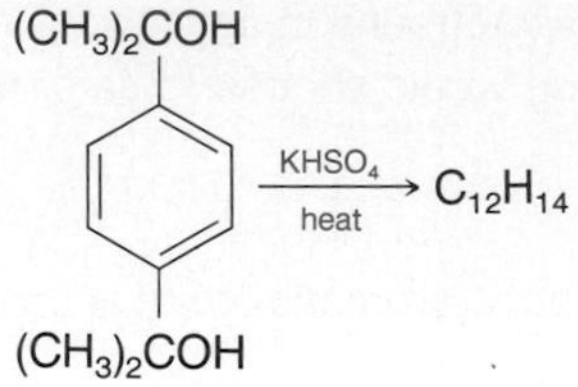

$(CH_3)_2COH$

$\xrightarrow[\text{heat}]{KHSO_4} C_{12}H_{14}$

$(CH_3)_2COH$

(g) $(Cl-\langle\rangle-)_2CHCCl_3 \xrightarrow[CH_3OH]{NaOCH_3} C_{14}H_8Cl_4$

(h)

CH_3

$\xrightarrow[CCl_4,\ \text{heat}]{N\text{-bromosuccinimide}}$

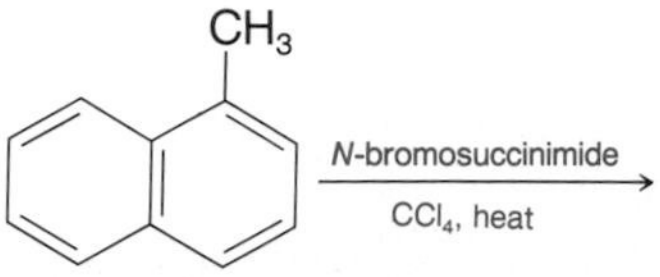

(i) 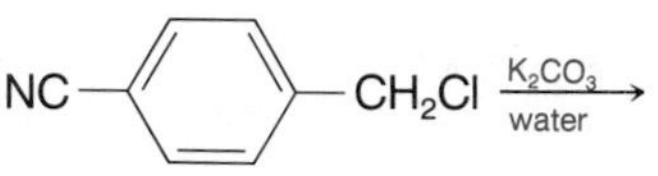 $NC-\langle\rangle-CH_2Cl \xrightarrow[\text{water}]{K_2CO_3}$

12.28 Suggest reagents suitable for carrying out each of the following conversions. In most cases more than one synthetic operation may be necessary.

(a) $C_6H_5CH_2CH_3 \longrightarrow C_6H_5\underset{\underset{Br}{|}}{C}HCH_3$

(b) $C_6H_5\underset{\underset{Br}{|}}{C}HCH_3 \longrightarrow C_6H_5\underset{\underset{Br}{|}}{C}HCH_2Br$

(c) $C_6H_5CH{=}CH_2 \longrightarrow C_6H_5C{\equiv}CH$

(d) $C_6H_5C{\equiv}CH \longrightarrow C_6H_5CH_2CH_2CH_2CH_3$

(e) $C_6H_5CH_2CH_2OH \longrightarrow C_6H_5\underset{\underset{OH}{|}}{C}HCH_3$

(f) $C_6H_5CH_2CH_2OH \longrightarrow C_6H_5CH_2CH_2C{\equiv}CH$

(g) $C_6H_5CH_2CH_2Br \longrightarrow C_6H_5\underset{\underset{OH}{|}}{C}HCH_2Br$

12.29 The relative rates of reaction of ethane, toluene, and ethylbenzene with bromine atoms have been measured. The most reactive hydrocarbon undergoes hydrogen atom abstraction 1 million times faster than does the least reactive one. Arrange these hydrocarbons in order of decreasing reactivity.

12.30 Write the principal resonance structures of *o*-methylbenzyl cation and *m*-methylbenzyl cation. Which one has a tertiary carbocation as a contributing resonance form?

12.31 Write a series of equations expressing a reasonable mechanism for the reaction

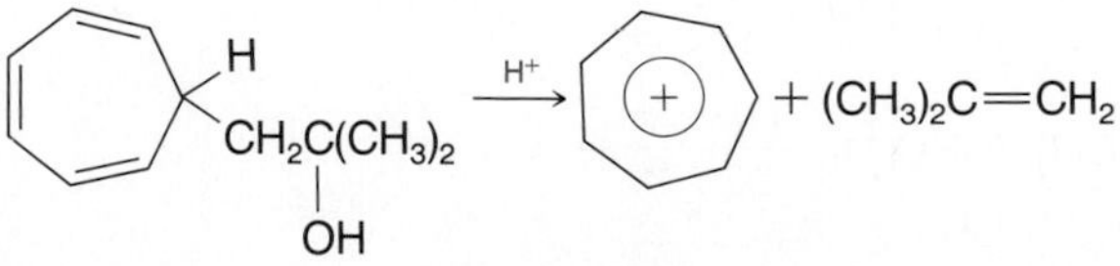

12.32 A certain compound E obtained from a natural product was determined to have the molecular formula $C_{14}H_{20}O_3$. It contained three methoxy (—OCH_3) groups and a

—CH_2CH=$C(CH_3)_2$ substituent. Oxidation with hot potassium permanganate gave 2,3,5-tri-methoxybenzoic acid. What is the structure of compound E?

12.33 Hydroboration-oxidation of (E)-2-(p-anisyl)-2-butene yielded an alcohol F, mp 60°C, in 72 percent yield. When the same reaction was performed on the (Z)-alkene, an isomeric liquid alcohol G was obtained in 77 percent yield. Suggest reasonable structures for F and G and describe the relationship between them.

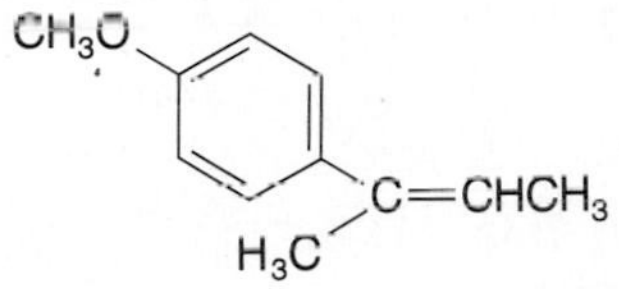

2-(p-Anisyl)-2-butene

12.34 Dehydrohalogenation of the diastereomeric forms of 1-chloro-1,2-diphenylpropane is stereospecific. One diastereomer yields (E)-1,2-diphenylpropene while the other yields the Z isomer. Which diastereomer yields which alkene? Why?

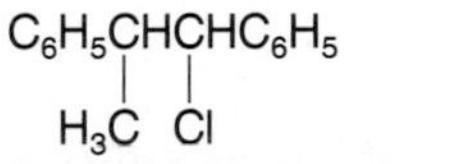

1-Chloro-1,2-diphenylpropane

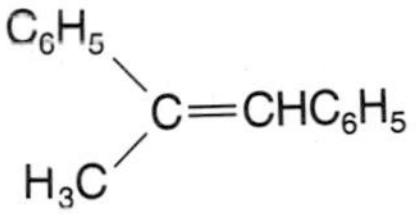

1,2-Diphenylpropene

REACTIONS OF ARENES. ELECTROPHILIC AROMATIC SUBSTITUTION REACTIONS

I n the preceding chapter the special stability of benzene was described, along with reactions in which an aromatic ring was present as a substituent. In the present chapter we move from considerations of the aromatic ring as a substituent to a study of the ring as a functional group. What kind of reactions are available to benzene and its derivatives? What sort of reagents react with arenes, and what products are formed in those reactions?

Characteristically, the reagents that react with the aromatic ring of benzene and its derivatives are *electrophiles*. We already have some experience with electrophilic reagents, particularly with respect to how they react with alkenes. Electrophilic reagents *add* to alkenes.

$$\underset{\text{Alkene}}{\overset{\displaystyle \diagup}{\underset{\diagdown}{C}}}=\underset{}{\overset{\displaystyle \diagdown}{\underset{\diagup}{C}}} + \underset{\substack{\text{Electrophilic}\\\text{reagent}}}{\overset{\delta+}{E}-\overset{\delta-}{Y}} \longrightarrow \underset{\substack{\text{Product of}\\\text{electrophilic addition}}}{E-\overset{|}{\underset{|}{C}}-\overset{|}{\underset{|}{C}}-Y}$$

A different reaction takes place when electrophiles react with arenes. *Substitution is observed instead of addition.* Representing an arene by the general formula ArH, where Ar stands for an aryl group, the electrophilic portion of the reagent replaces one of the hydrogens on the ring.

$$\underset{\text{Arene}}{Ar-H} + \underset{\substack{\text{Electrophilic}\\\text{reagent}}}{\overset{\delta+}{E}-\overset{\delta-}{Y}} \longrightarrow \underset{\substack{\text{Product of}\\\text{electrophilic aromatic}\\\text{substitution}}}{Ar-E} + H-Y$$

We call this reaction *electrophilic aromatic substitution:* an electrophile attacks an aromatic ring and substitutes for one of its hydrogens. It is one of the fundamental processes of organic chemistry.

13.1 REPRESENTATIVE ELECTROPHILIC AROMATIC SUBSTITUTION REACTIONS OF BENZENE

The scope of electrophilic aromatic substitution reactions is quite large; both the arene and the electrophilic reagent are capable of wide variation. Indeed, it is this breadth of scope that makes electrophilic aromatic substitution such an important reaction. Electrophilic aromatic substitution is the principal method by which substituted derivatives of benzene are prepared both industrially and in the laboratory. A feeling for these reactions can be developed by examining a few typical examples in which benzene is the substrate.

When benzene is warmed with a mixture of nitric acid and sulfuric acid, *nitration* of the ring occurs. A nitro group substitutes for one of the hydrogens of benzene.

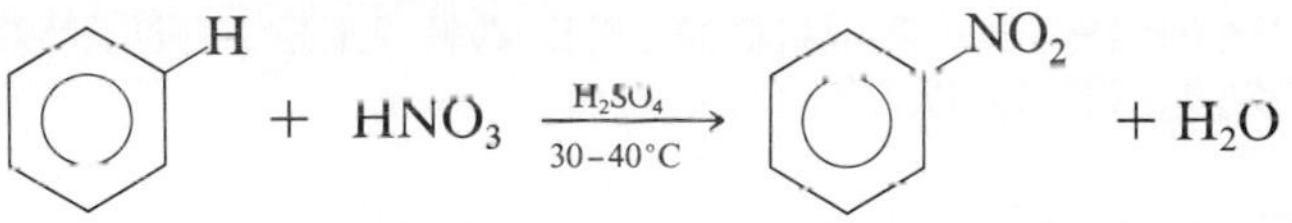

| Benzene | Nitric acid | | Nitrobenzene (95%) | Water |

Sulfonation takes place when benzene reacts with hot concentrated sulfuric acid.

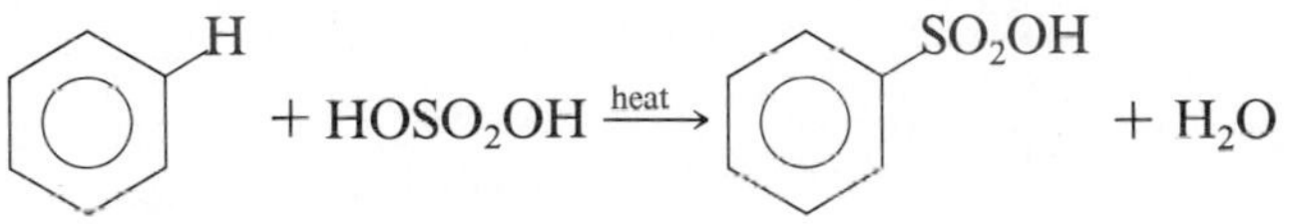

| Benzene | Sulfuric acid | | Benzenesulfonic acid (100%) | Water |

Bromination of benzene is usually carried out in the presence of a ferric bromide catalyst.

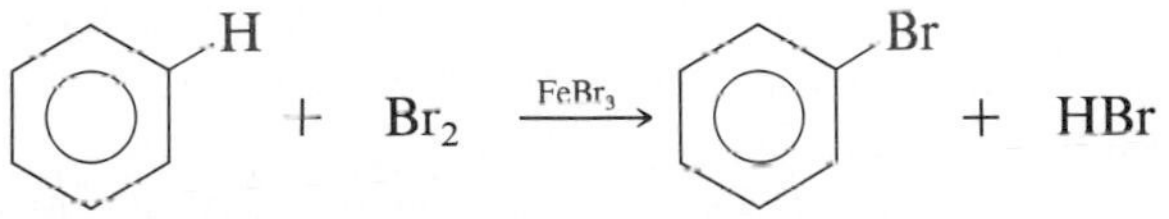

| Benzene | Bromine | | Bromobenzene (65–75%) | Hydrogen bromide |

Alkyl halides react with benzene in the presence of aluminum chloride to yield alkylbenzenes. This reaction is called *Friedel-Crafts alkylation*.

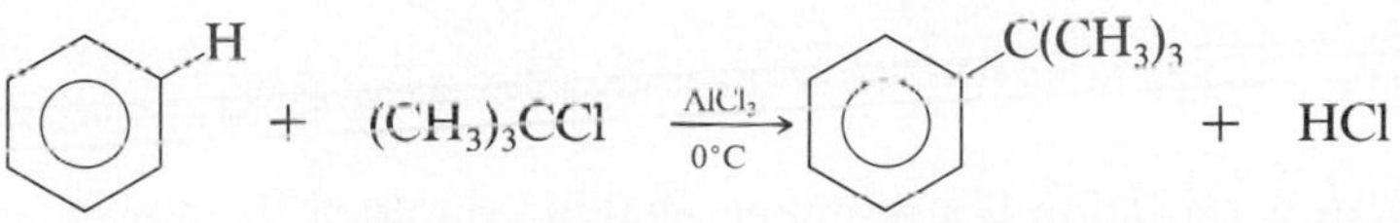

| Benzene | *tert*-Butyl chloride | | *tert*-Butylbenzene (60%) | Hydrogen chloride |

A comparable reaction called *Friedel-Crafts acylation* occurs when acyl halides are used.

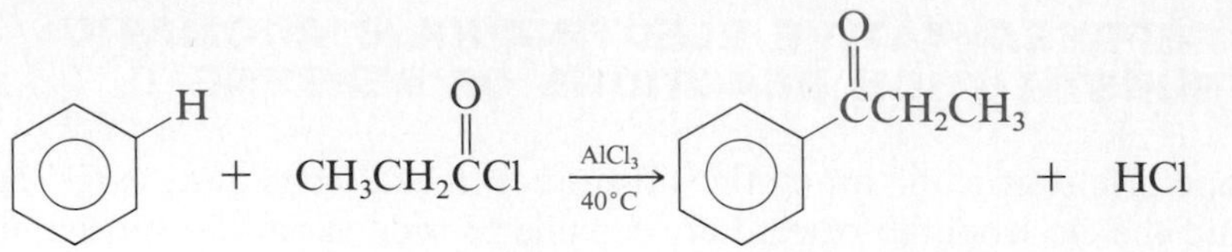

Propionyl chloride   Propiophenone (88%)   Hydrogen
(propanoyl chloride)   (1-phenyl-1-propanone)   chloride

We will shortly discuss each of these reactions in more detail. First, however, let us look at the essential mechanistic principles of electrophilic aromatic substitution.

13.2 MECHANISTIC PRINCIPLES OF ELECTROPHILIC AROMATIC SUBSTITUTION

Recall from Chapter 7 the general mechanism for electrophilic addition to alkenes.

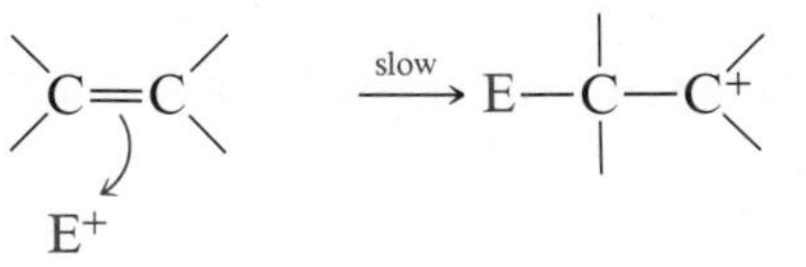

Alkene and electrophile   Carbocation

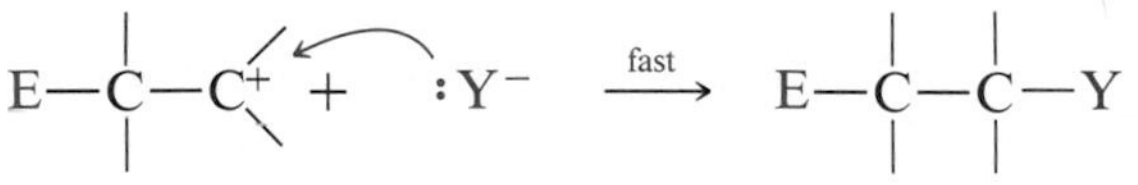

Carbocation   Nucleophile   Product of electrophilic
                                      addition

The first step is rate-determining. It is the transfer of the pair of π electrons of the alkene to the electrophile to form a high-energy intermediate, a carbocation. Following its formation, the carbocation undergoes rapid capture by some Lewis base present in the medium.

The first step in the reaction of electrophilic reagents with benzene is similar. An electrophile accepts an electron pair from the π system of benzene to form a carbocation.

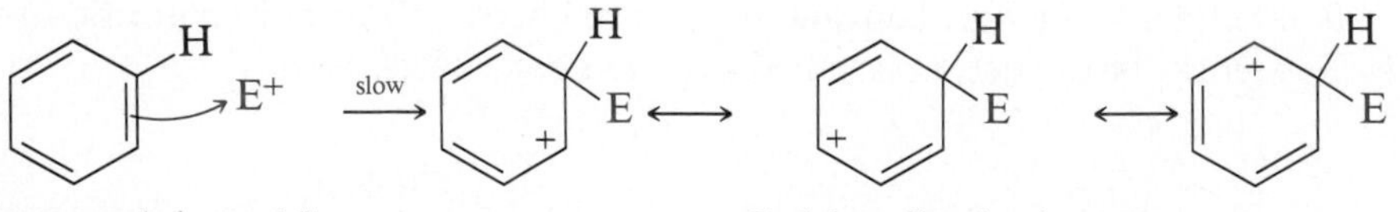

Benzene and electrophile   Cyclohexadienyl cation

This particular carbocation is a resonance-stabilized one of the allylic type. It is a *cyclohexadienyl cation* (often referred to as an *arenium ion* or as a *σ complex*).

PROBLEM 13.1 Write the three most stable resonance structures for the cyclohexadienyl cation formed by addition of a proton to benzene.

A significant portion of the 30 – 36 kcal/mol of resonance energy in benzene is lost when it is converted to the cyclohexadienyl cation intermediate. In spite of the allylic nature of a cyclohexadienyl cation, which makes it relatively stable compared with simple alkyl cations, a cyclohexadienyl cation is not aromatic and possesses only a fraction of the resonance stabilization of benzene. Once it is formed, the cyclohexadienyl cation intermediate rapidly suffers deprotonation, resulting in full restoration of the aromaticity of the ring and formation of the substitution product.

Cyclohexadienyl
cation

Product of electrophilic
aromatic substitution

If the Lewis base ($:Y^-$) acted as a nucleophile and added to carbon instead of acting as a proton acceptor, the product would be a nonaromatic cyclohexadiene derivative with only a few kilocalories per mole of resonance energy. Addition and substitution products can arise by alternative reaction paths of a cyclohexadienyl cation. Substitution occurs preferentially because there is a substantial driving force favoring rearomatization.

Cyclohexadienyl
cation

Observed product of electrophilic
aromatic substitution

Not observed — not aromatic

Figure 13.1 is a potential energy diagram describing the general mechanism of electrophilic aromatic substitution. The first step, the one in which the aromaticity of the ring is destroyed as it proceeds to the cyclohexadienyl cation, is the slow step. The second step, restoration of the aromatic stabilization of the ring by deprotonation of the cyclohexadienyl cation, is fast. In order for electrophilic aromatic substitution reactions to overcome the high activation energy that characterizes the first step, the electrophile must be a fairly reactive one. Many electrophilic reagents that react with alkenes do not react at all with benzene. Peroxy acids and diborane, for example, fall into this category. Others, such as bromine, react with benzene only in the presence of catalysts that enhance their electrophilicity. The low level of reactivity of benzene toward electrophiles stems from the substantial loss of resonance stabilization that accompanies transfer of a pair of its six π electrons to an electrophile.

With this as background, let us now examine each of the electrophilic aromatic substitution reactions presented in Section 13.1 in more detail, especially with regard to the nature of the electrophile that attacks benzene.

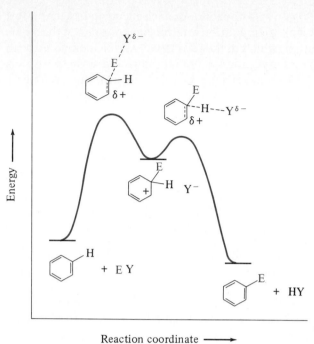

FIGURE 13.1 Energy diagram illustrating the energy changes associated with the two steps of electrophilic aromatic substitution.

13.3 NITRATION OF BENZENE

On the basis of the mechanism developed in the preceding section for the general case of an electrophile designated as E^+, we need only identify the specific electrophile in the nitration of benzene to have a fairly clear idea of how the reaction occurs.

$$\text{Benzene} + HNO_3 \xrightarrow[30-40°C]{H_2SO_4} \text{Nitrobenzene (95\%)} + H_2O$$

Benzene   Nitric acid   Nitrobenzene (95%)   Water

A combination of physical-chemical measurements and spectroscopic techniques allowed Sir Christopher Ingold to establish that nitronium cation (NO_2^+) is the electrophile that attacks benzene.

Benzene and nitronium cation $\xrightarrow{\text{slow}}$ Cyclohexadienyl cation intermediate $\xrightarrow{\text{fast}}$ Nitrobenzene + H^+

Benzene and nitronium cation   Cyclohexadienyl cation intermediate   Nitrobenzene   Proton

Nitric acid alone does not provide a high enough concentration of nitronium ion to nitrate benzene at a convenient rate. Addition of sulfuric acid to nitric acid causes the nitronium ion concentration to increase according to the following reactions:

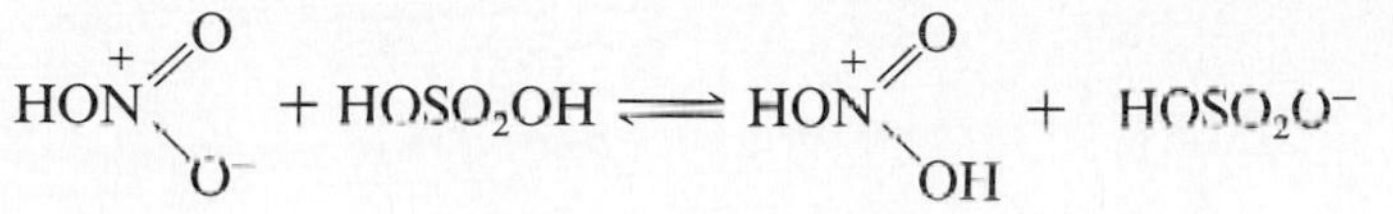

$$\underset{\substack{\text{Nitric acid}}}{\overset{\overset{+}{HON}\diagdown\overset{\displaystyle O}{\underset{O^-}{}}}{}} + HOSO_2OH \rightleftharpoons \underset{\substack{\text{Conjugate acid} \\ \text{of nitric acid}}}{\overset{\overset{+}{HON}\diagdown\overset{\displaystyle O}{\underset{OH}{}}}{}} + \underset{\substack{\text{Hydrogen sulfate} \\ \text{ion}}}{HOSO_2O^-}$$

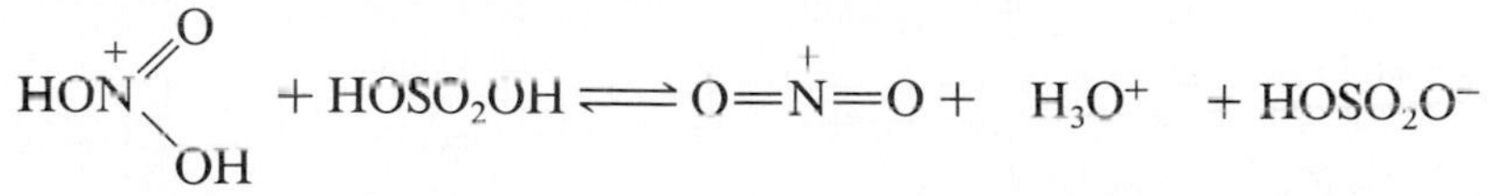

$$\underset{\substack{\text{Conjugate acid} \\ \text{of nitric acid}}}{\overset{\overset{+}{HON}\diagdown\overset{\displaystyle O}{\underset{OH}{}}}{}} + \underset{\substack{\text{Sulfuric} \\ \text{acid}}}{HOSO_2OH} \rightleftharpoons \underset{\substack{\text{Nitronium} \\ \text{ion}}}{O{=}\overset{+}{N}{=}O} + \underset{\substack{\text{Hydronium} \\ \text{ion}}}{H_3O^+} + \underset{\substack{\text{Hydrogen} \\ \text{sulfate ion}}}{HOSO_2O^-}$$

Thus, sulfuric acid promotes the nitration of benzene by converting nitric acid to nitronium ion, which is the active electrophile. Replacement of a ring hydrogen by a nitro group is a general reaction, observed when benzene and its derivatives are treated with a mixture of nitric acid and sulfuric acid.

PROBLEM 13.2 Nitration of p-xylene gives a single product having the molecular formula $C_8H_9NO_2$ in high yield. Suggest a reasonable structure for this product. (If necessary, see Section 12.15 to remind yourself of the structure of p-xylene.)

13.4 SULFONATION OF BENZENE

The reaction of benzene with sulfuric acid to produce benzenesulfonic acid

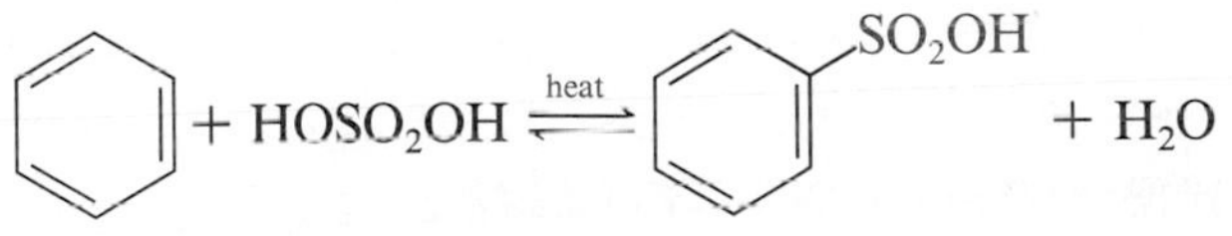

is reversible but can be driven to completion by several experimental techniques. Removing the water formed in the reaction, for example, allows benzenesulfonic acid to be obtained in virtually quantitative yield. When *oleum* (a solution of sulfur trioxide in sulfuric acid) is used as the sulfonating agent, the rate of sulfonation is much faster and the equilibrium is displaced entirely to the side of products, according to the equation:

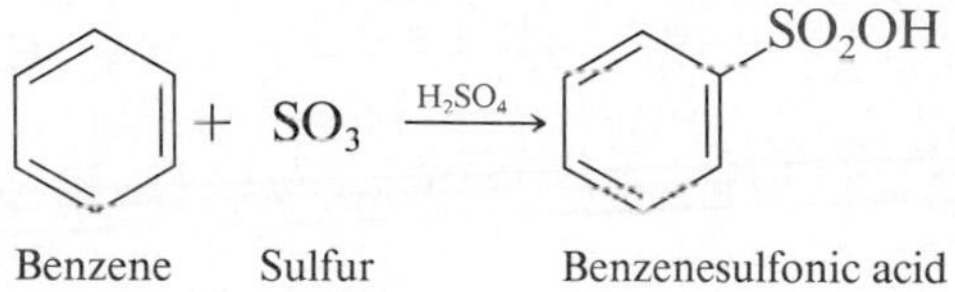

A variety of electrophilic species is present in concentrated sulfuric acid and oleum. Probably sulfur trioxide is the actual electrophile in aromatic sulfonation. We can represent the mechanism of sulfonation of benzene by sulfur trioxide by the equations:

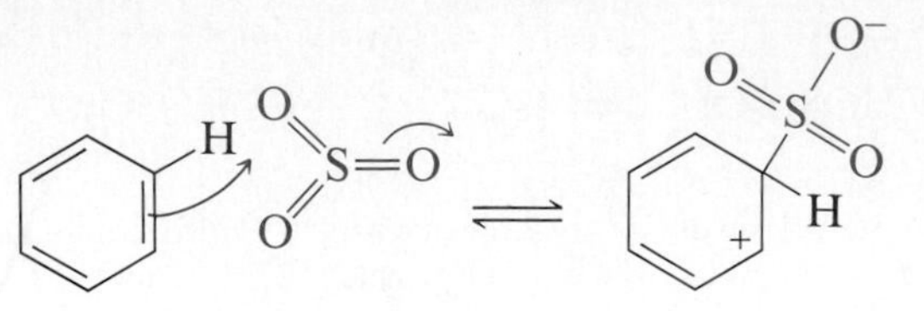

Benzene and sulfur trioxide

Cyclohexadienyl cation intermediate

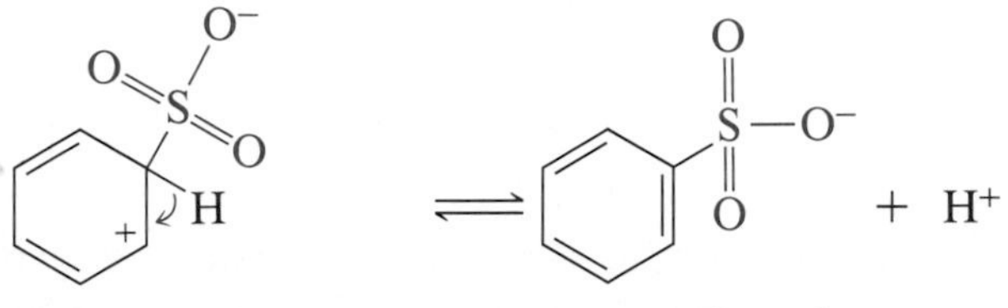

Cyclohexadienyl cation intermediate

Benzenesulfonate ion   Proton

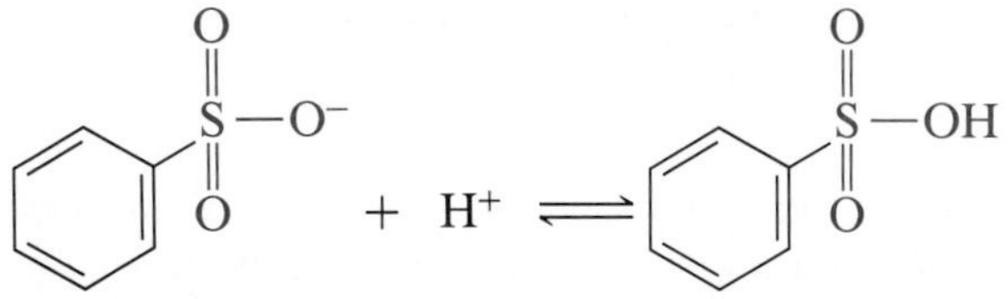

Benzenesulfonate ion   Proton   Benzenesulfonic acid

PROBLEM 13.3   On being heated with sulfur trioxide in sulfuric acid, 1,2,4,5-tetra-methylbenzene was converted to a product of molecular formula $C_{10}H_{14}O_3S$ in 94 percent yield. Suggest a reasonable structure for this product.

13.5 HALOGENATION OF BENZENE

Conversion of benzene to bromobenzene by electrophilic aromatic substitution has been a popular laboratory experiment with several generations of organic chemistry students (or at least with their teachers). According to the usual procedure, bromine is added to benzene in the presence of metallic iron (customarily a few carpet tacks) and the reaction mixture is refluxed while taking precautions to prevent the hydrogen bromide which is formed from escaping into the laboratory.

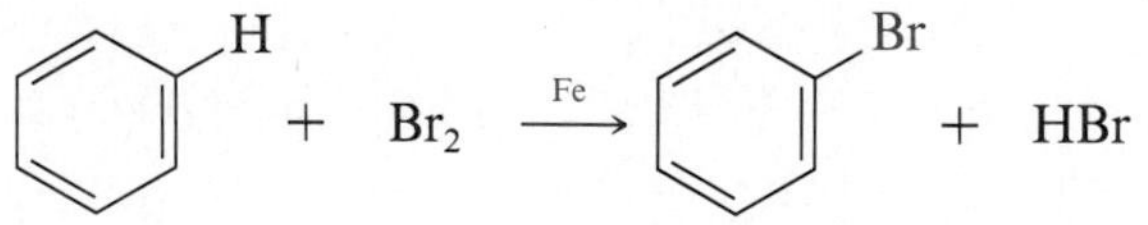

Benzene   Bromine   Bromobenzene   Hydrogen
                                      (65–75%)   bromide

Bromine, while it is a good enough electrophile to react instantaneously with alkenes, is insufficiently electrophilic to react at an appreciable rate with benzene. A catalyst must be present which lowers the activation energy for the reaction and increases its rate. The role which the catalyst plays is that of enhancing the electrophilic qualities of bromine. Somehow carpet tacks can do this. How?

The active catalyst is not iron itself but ferric bromide, formed by reaction of iron and bromine.

$$2Fe + 3Br_2 \longrightarrow 2FeBr_3$$

Iron   Bromine   Ferric bromide

Ferric bromide is a weak Lewis acid. It can coordinate to a molecule of bromine to form a Lewis acid–Lewis base complex.

:Br—Br: + FeBr$_3$ $\rightleftharpoons$:Br—Br—FeBr$_3$

Lewis base   Lewis acid   Lewis acid–Lewis base complex

The bromine-bromine bond in this complex is polarized and more easily broken in a *heterolytic* fashion than is the bond in bromine itself. Complexation of bromine with an electrophilic catalyst makes bromine more electrophilic.

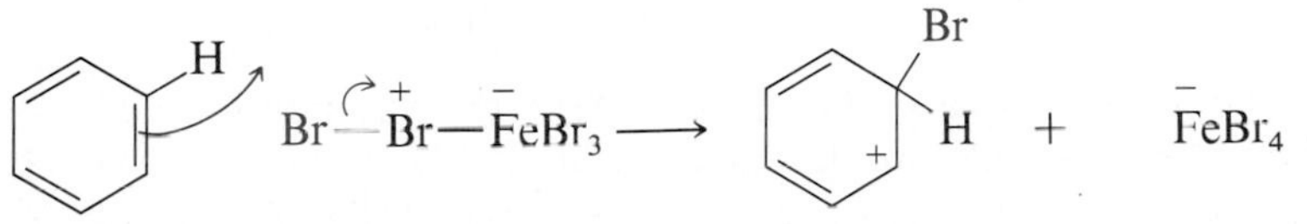

Br—Br—FeBr$_3$ $\longrightarrow$ + FeBr$_4$

Benzene and bromine–ferric bromide complex   Cyclohexadienyl cation intermediate   Tetrabromoferrate ion

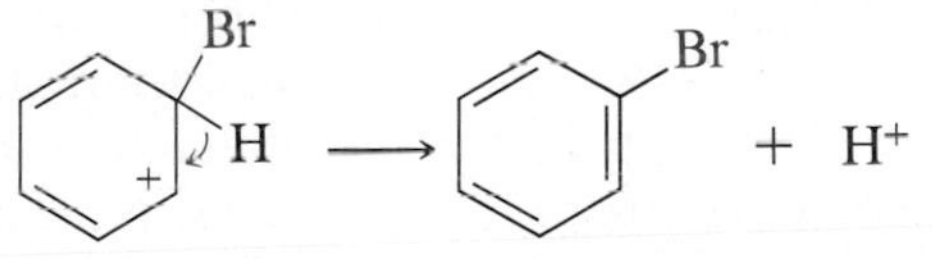

$\longrightarrow$ + H$^+$

Cyclohexadienyl cation intermediate   Bromobenzene   Proton

Only catalytic quantities of ferric bromide are required because the catalyst is regenerated by formation of hydrogen bromide.

$$H^+ + FeBr_4^- \longrightarrow HBr + FeBr_3$$

The combination Br_2—$FeBr_3$ is a source of electrophilic bromine; it acts as if it provided Br^+ to attack benzene. The active species is not Br^+ but a complex between bromine and ferric bromide. We shall see later in this chapter that some aromatic substrates are much more reactive than benzene and react rapidly with bromine even in the absence of catalysts.

Chlorination by electrophilic aromatic substitution is similar to bromination and provides a ready route to chlorobenzene and other aryl chlorides.

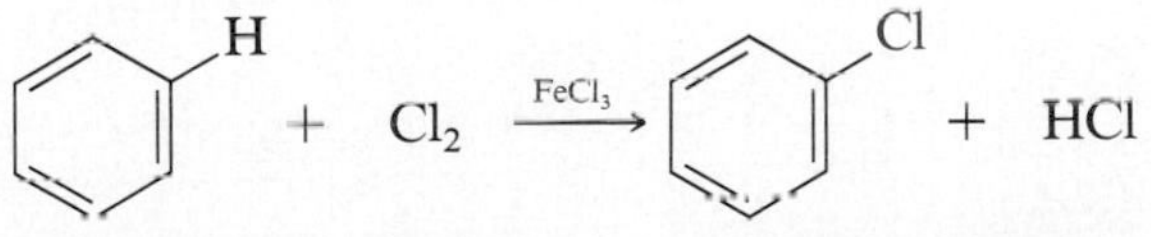

+ Cl$_2$ $\xrightarrow{FeCl_3}$ + HCl

Benzene   Chlorine   Chlorobenzene   Hydrogen chloride

Direct fluorination and iodination of benzene and other arenes are rarely used processes. Fluorine is so reactive an electrophile that its reaction with benzene is difficult to control. Iodination is both very slow and characterized by an unfavorable equilibrium constant. Syntheses of aryl fluorides and aryl iodides are normally carried out by way of functional group transformations of arylamines; these reactions will be described in Chapter 24.

13.6 FRIEDEL-CRAFTS ALKYLATION OF BENZENE

Alkyl halides react with benzene in the presence of aluminum chloride to yield alkylbenzenes.

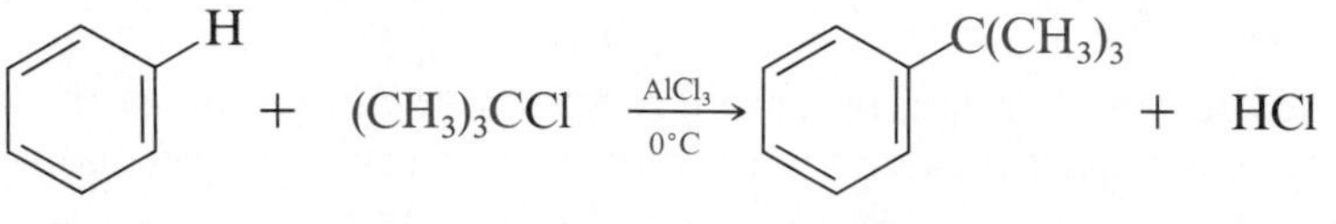

Benzene | *tert*-Butyl chloride | *tert*-Butylbenzene (60%) | Hydrogen chloride

Alkylation of benzene with alkyl halides in the presence of aluminum chloride was discovered by Charles Friedel and James M. Crafts in 1877. Crafts, who later became president of the Massachusetts Institute of Technology, collaborated with Friedel at the Sorbonne in Paris, and together they developed what we now call the *Friedel-Crafts reaction* into one of the most useful synthetic methods in organic chemistry.

Alkyl halides by themselves are insufficiently electrophilic to react with benzene. Aluminum chloride serves as a Lewis acid catalyst to enhance the electrophilicity of the alkylating agent. With tertiary and secondary alkyl halides, the addition of aluminum chloride can lead to the formation of carbocations, and these are the electrophilic species that attack the aromatic ring.

$$(CH_3)_3C-\ddot{\underset{\cdot\cdot}{Cl}}: + AlCl_3 \longrightarrow (CH_3)_3C-\overset{+}{\ddot{\underset{\cdot\cdot}{Cl}}}-\overset{-}{AlCl_3}$$

tert-Butyl chloride | Aluminum chloride | Lewis acid–Lewis base complex

$$(CH_3)_3C-\overset{+}{\ddot{\underset{\cdot\cdot}{Cl}}}-\overset{-}{AlCl_3} \longrightarrow (CH_3)_3C^+ + \overset{-}{AlCl_4}$$

tert-Butyl chloride–aluminum chloride complex | *tert*-Butyl cation | Tetrachloroaluminate anion

Once formed, the carbocation reacts with benzene in the same way as other electrophiles do.

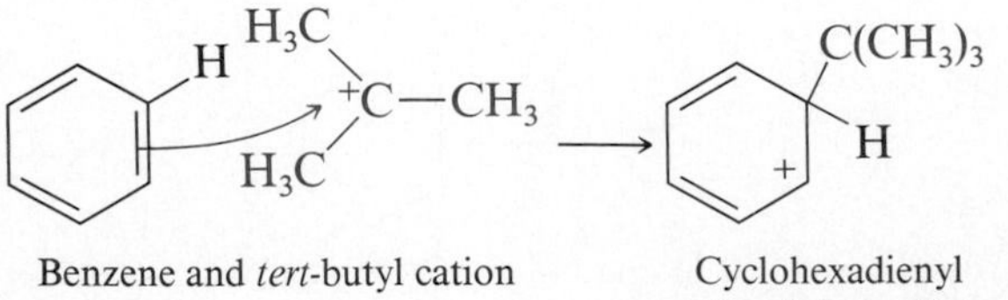

Benzene and *tert*-butyl cation | Cyclohexadienyl cation intermediate

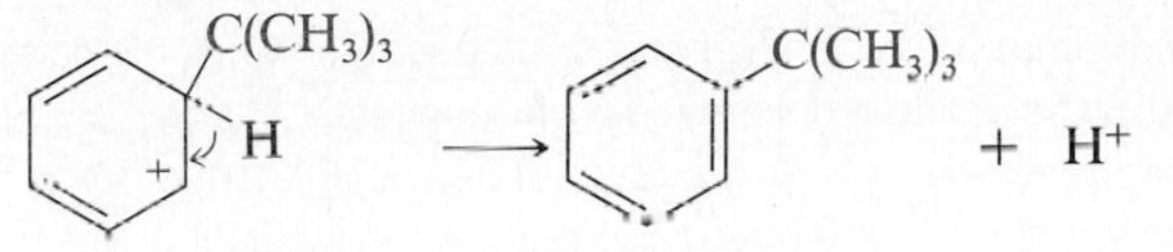

Cyclohexadienyl cation   *tert*-Butylbenzene   Proton
intermediate

The aluminum chloride catalyst is regenerated and goes on to catalyze another alkylation.

$$H^+ \;+\; \overline{Al}Cl_4 \longrightarrow HCl \;+\; AlCl_3$$

Proton   Tetrachloroaluminate   Hydrogen   Aluminum
ion   chloride   chloride

Methyl and ethyl halides do not form carbocations under these conditions. They do alkylate benzene, however. The aluminum chloride complexes of methyl halides and primary alkyl halides contain highly polarized carbon-halogen bonds and are the electrophilic species that react with benzene.

$$CH_3 - \overset{+}{\underset{\cdot\cdot}{\ddot{X}}} - \overline{Al}X_3 \qquad\qquad RCH_2 - \overset{+}{\underset{\cdot\cdot}{\ddot{X}}} - \overline{Al}X_3$$

Methyl halide–aluminum   Primary halide aluminum
halide complex   halide complex

One drawback to Friedel-Crafts alkylation with primary alkyl halides is that secondary or tertiary carbocations can be formed from primary alkyl halides by rearrangement under the reaction conditions. This leads to the isolation of alkylbenzenes which are different from those to be expected in the absence of such rearrangements. For example, a Friedel-Crafts alkylation using isobutyl chloride (a primary alkyl halide) yields only *tert*-butylbenzene.

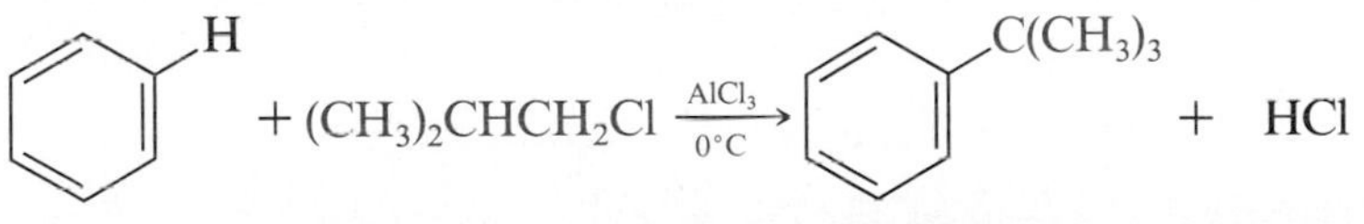

Benzene   Isobutyl chloride   *tert*-Butylbenzene   Hydrogen
(66%)   chloride

Here, the electrophile must be the *tert*-butyl cation formed by a hydride migration that accompanies ionization of the carbon-chlorine bond.

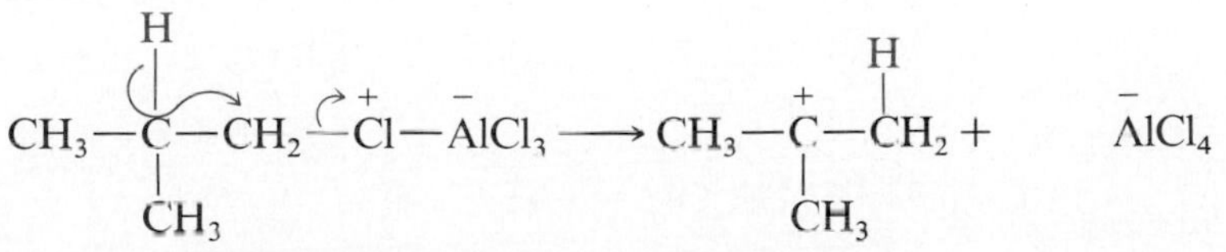

Isobutyl chloride –   *tert*-Butyl cation   Tetrachloroaluminate
aluminum chloride complex   ion

PROBLEM 13.4 In an attempt to prepare *n*-propylbenzene, a chemist alkylated benzene with *n*-propyl chloride and aluminum chloride. To our chemist's chagrin, two isomeric hydrocarbons were obtained in a ratio of 2 : 1, the desired *n*-propylbenzene being the minor component. What do you think is the major component? How did it arise?

Other Lewis acid catalysts used in Friedel-Crafts reactions include ferric chloride ($FeCl_3$), zinc chloride ($ZnCl_2$), and boron trifluoride (BF_3).

Since electrophilic attack on benzene is simply another reaction available to a carbocation, alternative carbocation precursors can be used in place of alkyl halides. For example, alkenes, which are converted to carbocations by protonation, can serve to alkylate benzene.

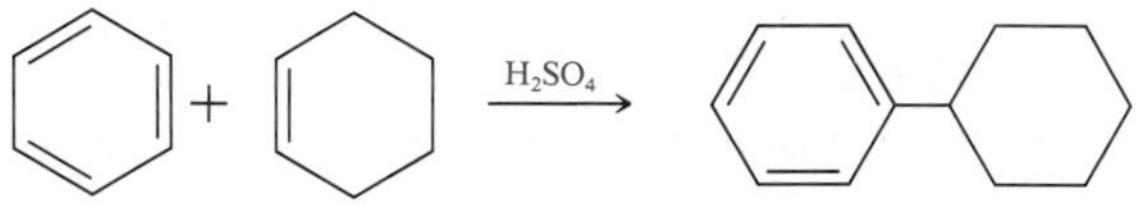

Benzene   Cyclohexene   Cyclohexylbenzene (65–68%)

Liquid hydrogen fluoride is sometimes used as both the proton donor and the solvent in these reactions.

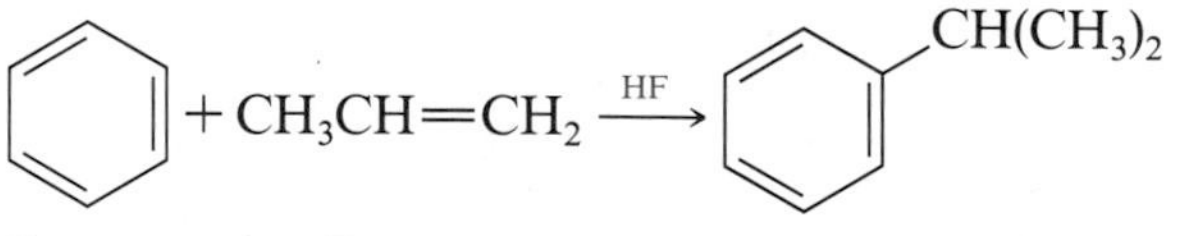

Benzene   Propene   Isopropylbenzene (75%)

PROBLEM 13.5 Write a reasonable mechanism for the formation of isopropylbenzene from benzene and propene in liquid hydrogen fluoride.

Alcohols undergo a comparable reaction in acidic media or when treated with strong Lewis acids, leading to alkylation of benzene.

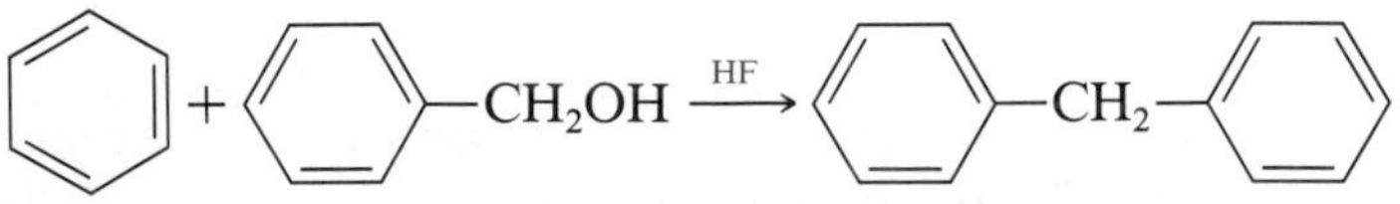

Benzene   Benzyl alcohol   Diphenylmethane (65%)

Rearrangement of carbon skeletons can occur in these procedures, just as they do in Friedel-Crafts reactions of alkyl halides.

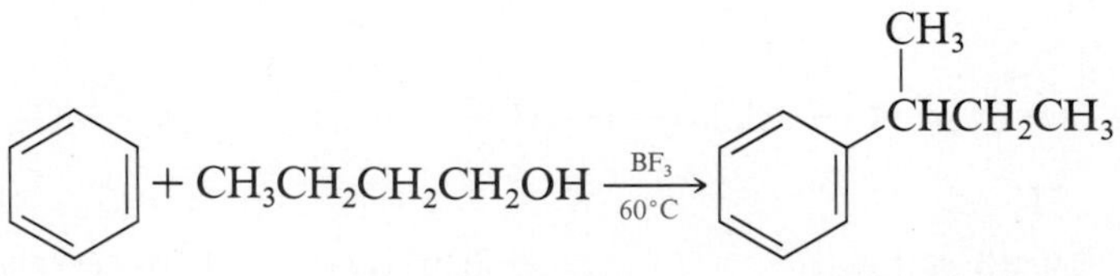

Benzene   1-Butanol   *sec*-Butylbenzene (80%)

13.7 FRIEDEL-CRAFTS ACYLATION OF BENZENE

Another version of the Friedel-Crafts reaction uses *acyl* halides and yields *acylbenz-*

enes. (An acyl group has the general formula RC with the structure $RC{=}O$ where R can be alkyl or aryl.)

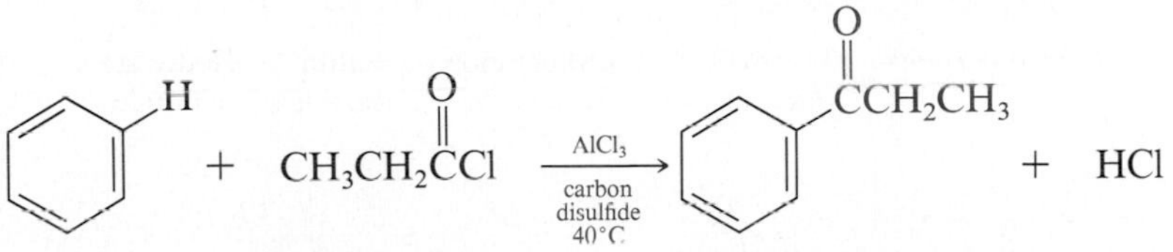

| Benzene | Propionyl chloride (propanoyl chloride) | Propiophenone (88%) (1-phenyl-1-propanone) | Hydrogen chloride |

The electrophile in a Friedel-Crafts acylation reaction is an *acyl cation* (also referred to as an *acylium ion*). Acyl cations are stabilized by resonance. The acyl cation derived from propionyl chloride is represented by the two resonance forms

$$CH_3CH_2\overset{+}{C}{=}\ddot{O}: \longleftrightarrow CH_3CH_2C{\equiv}\overset{+}{O}:$$

Most stable resonance form;
all atoms have octets of electrons

Acyl cations form by coordination of an acyl chloride with aluminum chloride, followed by cleavage of the carbon-chlorine bond.

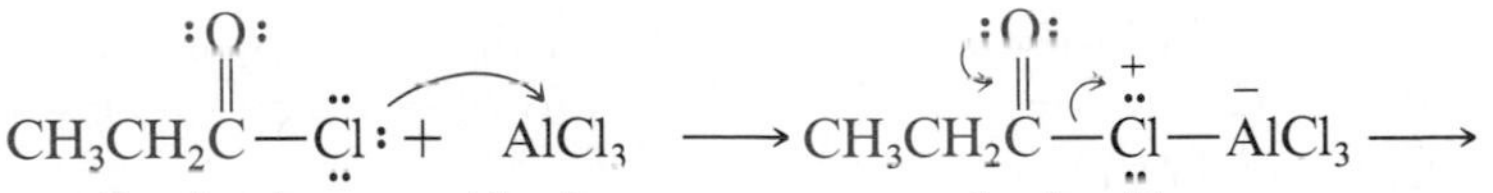

| Propionyl chloride | Aluminum chloride | Lewis acid– Lewis base complex |

| | Propionyl cation | Tetrachloro-aluminate ion |

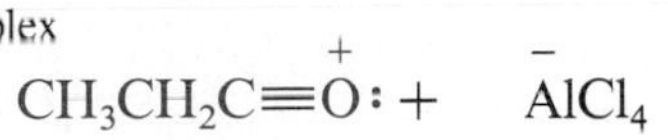

The electrophilic site of an acyl cation is its acyl carbon; it reacts with benzene in a manner analogous to that of other electrophilic reagents.

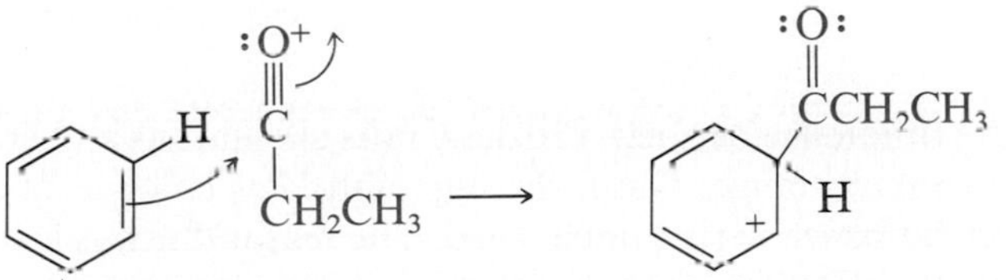

| Benzene and propionyl cation | Cyclohexadienyl cation intermediate |

| Cyclohexadienyl cation intermediate | Propiophenone | Proton |

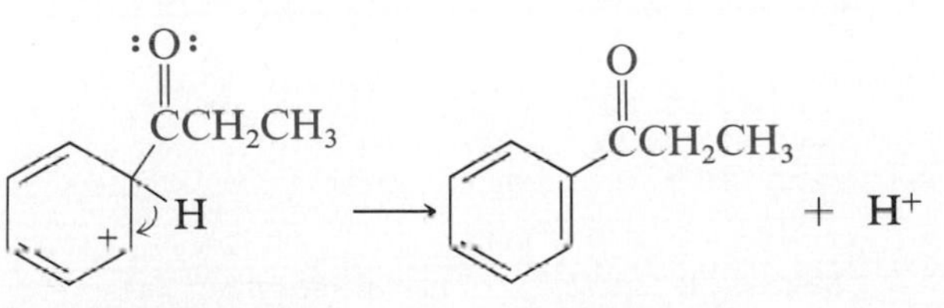

Acyl chlorides are readily available. They are prepared by the reactions of carboxylic acids with thionyl chloride.

$$\underset{\text{Carboxylic acid}}{\text{RCOH}} + \underset{\substack{\text{Thionyl} \\ \text{chloride}}}{\text{SOCl}_2} \longrightarrow \underset{\text{Acyl chloride}}{\text{RCCl}} + \underset{\substack{\text{Sulfur} \\ \text{dioxide}}}{\text{SO}_2} + \underset{\substack{\text{Hydrogen} \\ \text{chloride}}}{\text{HCl}}$$

Carboxylic acid anhydrides, compounds of the type RCOCR, can also serve as sources of acyl cations and, in the presence of aluminum chloride, can acylate benzene. One acyl unit of an acid anhydride becomes attached to the benzene ring, while the other becomes part of a carboxylic acid.

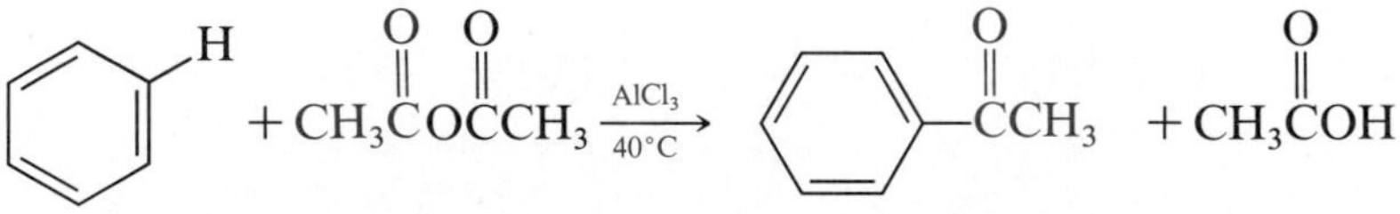

Benzene Acetic anhydride Acetophenone (76–83%) Acetic acid

Excess aluminum chloride must be used in Friedel-Crafts acylation reactions because the acylbenzene product forms a strong complex with it. Complexation of the aluminum chloride by the acylbenzene ties it up in a form in which it can no longer function as a Lewis acid catalyst.

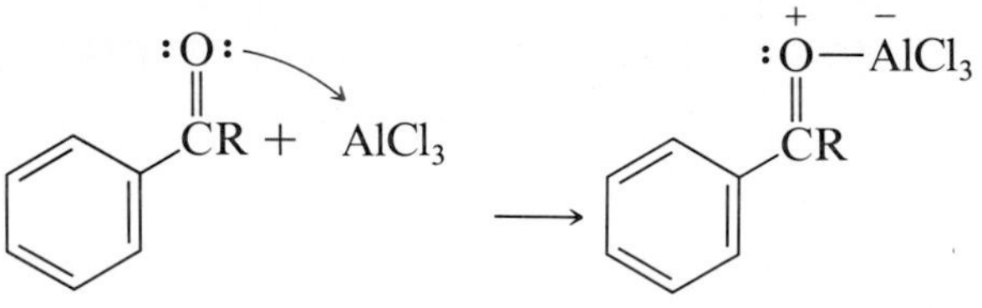

Acylbenzene Aluminum chloride Acylbenzene–aluminum chloride complex

An important difference between Friedel-Crafts alkylations and acylations is that acylium ions do not rearrange. The acyl group of the acyl chloride or acid anhydride is transferred to the benzene ring unchanged. The reason for this is that an acylium ion is so strongly stabilized by resonance that it is more stable than any ion that could conceivably arise from it by a hydride or alkyl group shift.

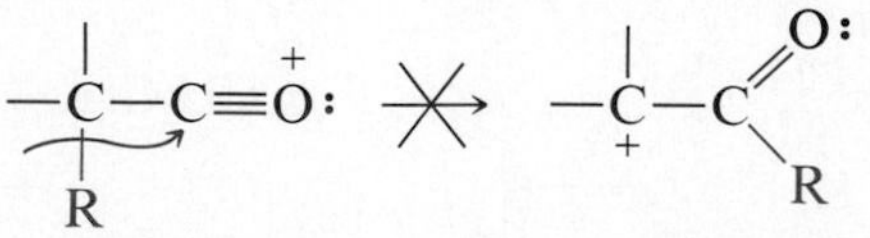

More stable cation; all atoms have octets of electrons   Less stable cation; six electrons at carbon

13.8 SYNTHESIS OF ALKYLBENZENES BY ACYLATION-REDUCTION

Because acylation of an aromatic ring can be accomplished without rearrangement of the carbon skeleton of the electrophile, it is frequently used as the first step in a procedure for the *alkylation* of aromatics by *acylation-reduction*. As we saw in Section 13.6, the direct alkylation of benzene with primary alkyl halides in the Friedel-Crafts reaction is normally not a practical route to compounds that bear a primary alkyl group as a substituent on the ring.

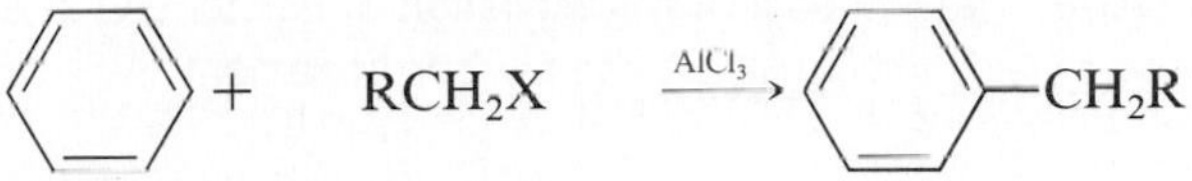

Benzene   Primary alkyl halide   Alkylbenzene

Primary alkyl halides (except ethyl and primary benzyl halides) yield products having rearranged alkyl groups as substituents. When a compound of the type $ArCH_2R$ is desired, a two-step transformation is used instead.

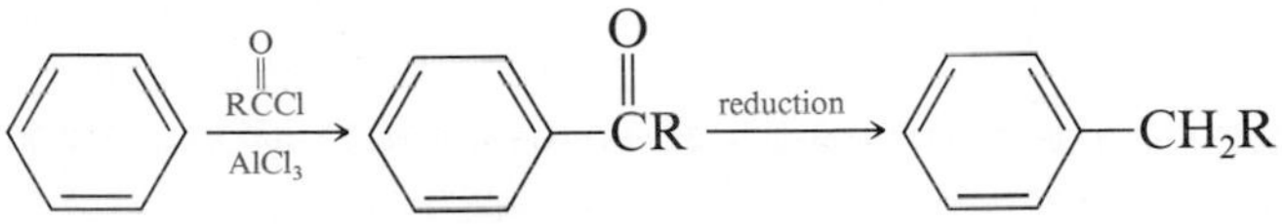

Benzene   Acylbenzene   Alkylbenzene

The first step is a Friedel-Crafts acylation using an acyl chloride or anhydride. The second step is a reduction of the carbonyl group ($\diagup\diagdown C=O$) to a methylene group (CH_2).

The most commonly used procedure for reducing an acylbenzene to an alkylbenzene involves treatment with a zinc-mercury amalgam in concentrated hydrochloric acid and is called the *Clemmensen reduction*.

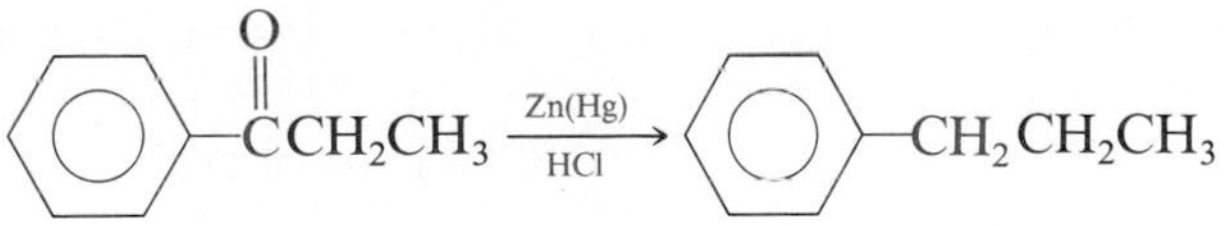

Propiophenone   Propylbenzene (88%)

The synthesis of butylbenzene illustrates the use of the acylation-reduction sequence.

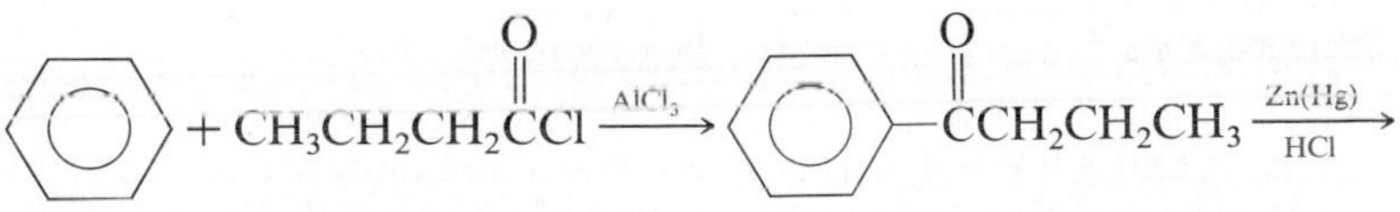

Benzene   Butanoyl chloride   1-Phenyl-1-butanone (86%)

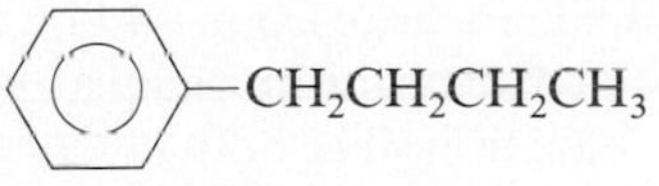

Butylbenzene (73%)

Direct alkylation of benzene using 1-chlorobutane and aluminum chloride would yield *sec*-butylbenzene by rearrangement and so could not be used.

PROBLEM 13.6 Using benzene and any necessary organic or inorganic reagents, suggest efficient syntheses of

(*a*) Isobutylbenzene, $C_6H_5CH_2CH(CH_3)_2$
(*b*) Neopentylbenzene, $C_6H_5CH_2C(CH_3)_3$

SAMPLE SOLUTION (*a*) Friedel-Crafts alkylation of benzene with isobutyl chloride is not suitable because it yields *tert*-butylbenzene by rearrangement.

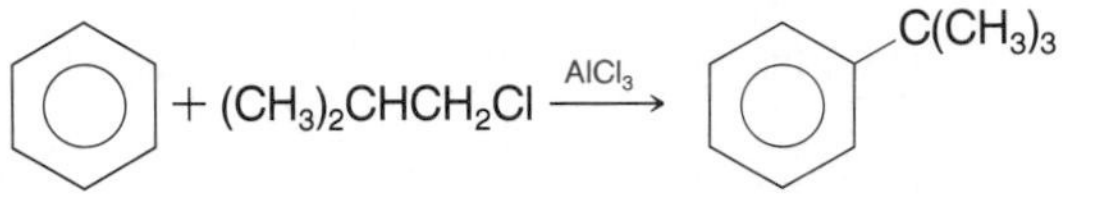

Benzene Isobutyl chloride *tert*-Butylbenzene (66%)

The two-step acylation-reduction sequence will be effective. Acylation of benzene with $(CH_3)_2CH\overset{O}{\overset{\|}{C}}Cl$ puts the side chain on the ring with the correct carbon skeleton. Clemmensen reduction converts the carbonyl group to a methylene group.

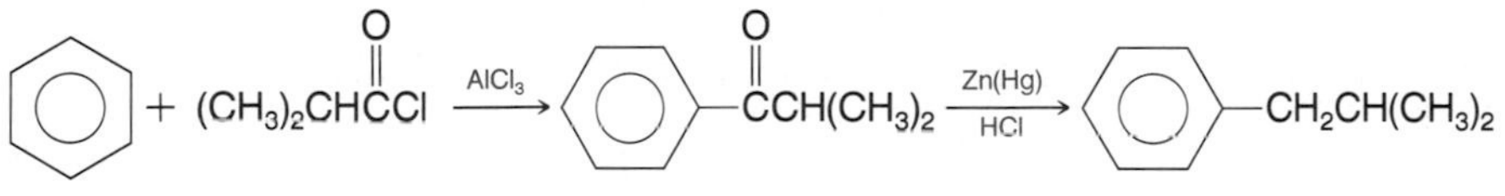

Benzene Isobutyryl chloride Isobutyrophenone (84%) Isobutylbenzene (80%)

An alternative method for the reduction of aldehyde and ketone carbonyl groups is the *Wolff-Kishner reduction.* Heating an aldehyde or ketone with hydrazine (H_2NNH_2) and sodium or potassium hydroxide in a high-boiling alcohol such as diethylene glycol ($HOCH_2CH_2OCH_2CH_2OH$, bp 245°C) or triethylene glycol ($HOCH_2CH_2OCH_2CH_2OCH_2CH_2OH$, bp 287°C) converts the carbonyl to a CH_2 group.

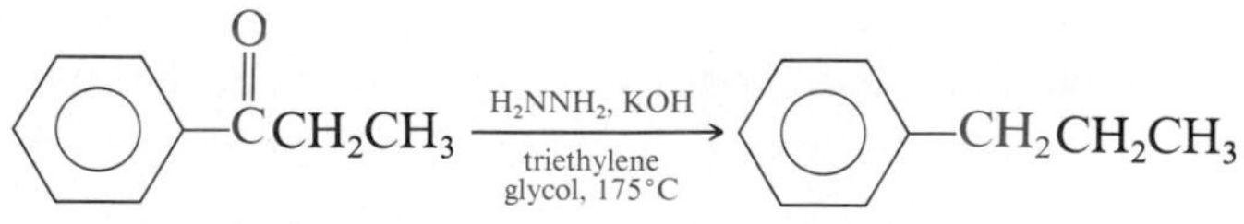

Propiophenone Propylbenzene (82%)

Both the Clemmensen and the Wolff-Kishner reductions are designed to carry out a specific functional group transformation, the reduction of an aldehyde or ketone carbonyl to a methylene group. Neither will reduce the carbonyl group of a carboxylic acid, nor are carbon-carbon double or triple bonds affected by these methods. We will not discuss the mechanism of either the Clemmensen reduction or the Wolff-Kishner reduction since both involve chemistry that is beyond the scope of that which we have covered to this point.

13.9 RATE AND ORIENTATION IN ELECTROPHILIC AROMATIC SUBSTITUTION

So far we have been concerned only with electrophilic substitution on benzene. Two important questions arise when we turn to analogous substitutions on arenes that already bear at least one substituent.

1. What is the effect of a substituent on the *rate* of electrophilic aromatic substitution?
2. What is the effect of a substituent on the *regioselectivity* (orientation) of electrophilic aromatic substitution?

To illustrate substituent effects on rate, consider the nitration of benzene, toluene, and (trifluoromethyl)benzene.

Toluene
(most reactive)   Benzene   (Trifluoromethyl)benzene
(least reactive)

Toluene is more reactive than benzene. It undergoes nitration some 20 to 25 times as fast as benzene. Because toluene is more reactive than benzene, we say that a methyl group *activates* the ring toward electrophilic aromatic substitution. (Trifluoromethyl)benzene is much less reactive than benzene. It undergoes nitration about 40,000 times more slowly than benzene. We say that a trifluoromethyl group *deactivates* the ring toward electrophilic aromatic substitution. Thus, the *rate* of electrophilic aromatic substitution depends markedly on a substituent already on the ring. Before we offer an explanation for this fact, let us examine the effect of a methyl group and of a trifluoromethyl group on the regioselectivity of substitution.

Three products are possible from nitration of toluene: *o*-nitrotoluene, *m*-nitrotoluene, and *p*-nitrotoluene. All are formed, but not in equal amounts. The meta isomer is formed to only a very small extent (3 percent). Together, the ortho- and para-substituted isomers comprise 97 percent of the product mixture.

Toluene   *o*-Nitrotoluene   *m*-Nitrotoluene   *p*-Nitrotoluene
                   (63%)   (3%)   (34%)

Because substitution in toluene occurs primarily at positions ortho and para to methyl, we say that *a methyl substituent is an ortho, para director*. Nitration of (trifluoromethyl)benzene, on the other hand, yields almost exclusively *m*-nitrotoluene (91 percent). The ortho- and para-substituted isomers are minor components of the reaction mixture.

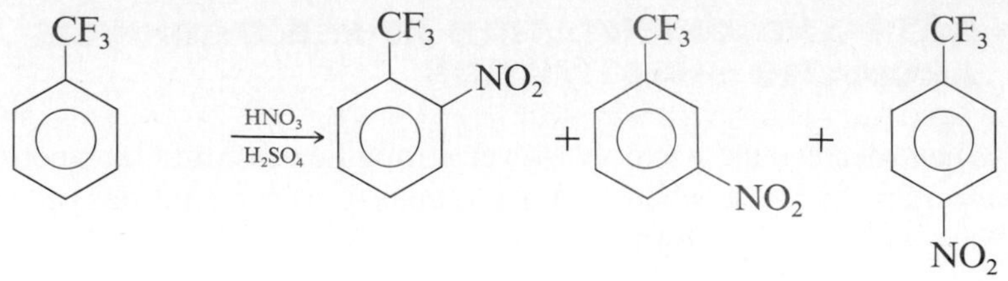

(Trifluoromethyl)benzene   o-Nitro(trifluoro-methyl)benzene (6%)   m-Nitro(trifluoro-methyl)benzene (91%)   p-Nitro(trifluoro-methyl)benzene (3%)

Because substitution in (trifluoromethyl)benzene occurs primarily at positions meta to the substituent, we say that *a trifluoromethyl group is a meta director.*

The *regioselectivity* of substitution, like the rate, is strongly affected by the substituent. In the following two sections we will examine the relationship between the structure of the substituent and its effect on rate and regioselectivity of electrophilic aromatic substitution.

13.10 ANALYSIS OF RATE AND ORIENTATION EFFECTS IN THE NITRATION OF TOLUENE

Why is there such a marked difference between methyl and trifluoromethyl substituents in their influence on the course of electrophilic aromatic substitution? Methyl is activating and ortho, para–directing; trifluoromethyl is deactivating and meta-directing. The first point to remember is that the regioselectivity of substitution is set once the cyclohexadienyl cation intermediate is formed. If we can explain why the intermediates leading to ortho and para nitration of toluene

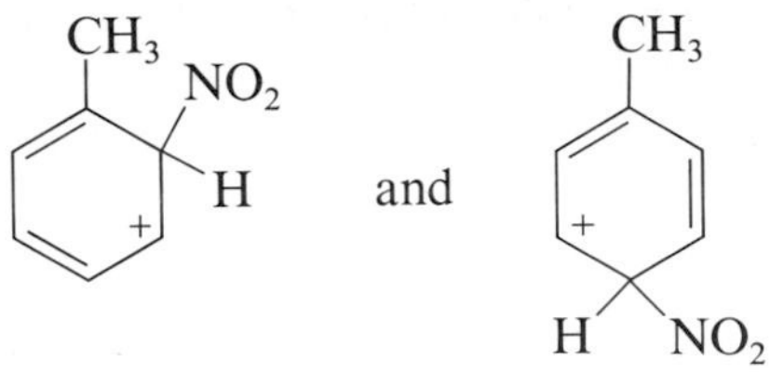

   and

are formed *faster* than the intermediate leading to meta nitration

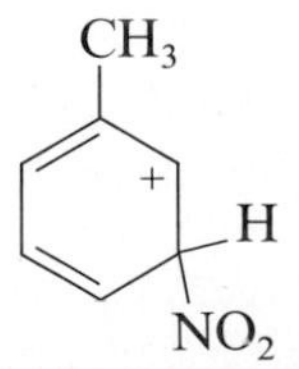

we will understand the reasons for the orientational preference. A principle we have invoked previously serves us well here: a more stable carbocation is formed faster than a less stable one. The most likely reason for the directing effect of methyl must be that the cyclohexadienyl cation precursors to o- and p-nitrotoluene are more stable than the one leading to m-nitrotoluene.

One way to assess the relative stabilities of these various intermediates is to examine electron delocalization in them using a resonance description. The cyclohexadienyl cations leading to *o*- and *p*-nitrotoluene have tertiary carbocation character. Each has one principal resonance form in which the positive charge is on the carbon bearing the methyl group.

Ortho attack:

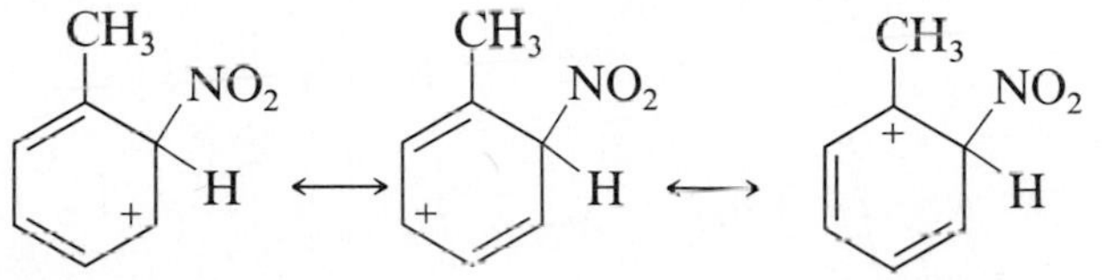

This resonance form
is a tertiary carbo-
cation

Para attack:

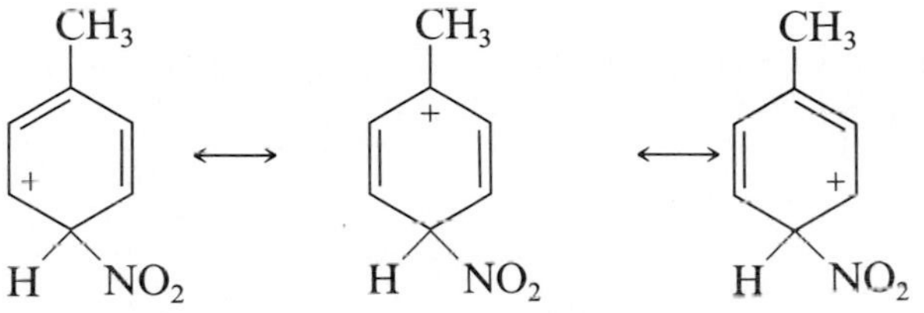

This resonance form
is a tertiary carbocation

The three resonance structures of the cyclohexadienyl cation intermediate leading to meta substitution are all secondary carbocations.

Meta attack:

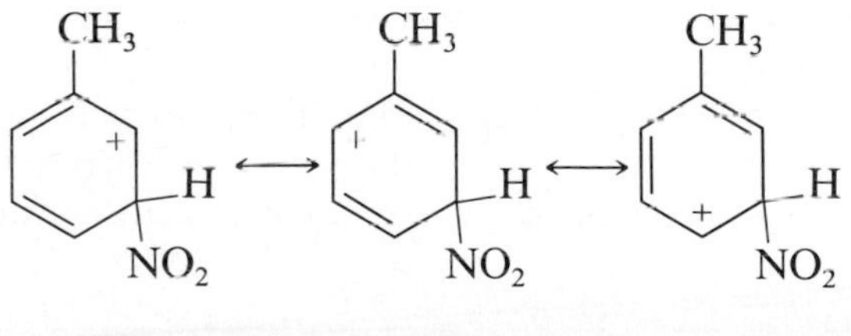

Because of their tertiary carbocation character the cyclohexadienyl cation intermediates leading to ortho and to para substitution are more stable and are formed faster than the secondary carbocation intermediate leading to meta substitution. They are also more stable than the secondary cyclohexadienyl cation intermediate formed during nitration of benzene. A methyl group is an activating substituent because it stabilizes the carbocation intermediate formed in the rate-determining step more than a hydrogen substituent does. It is ortho, para–directing because it stabilizes the carbocation formed by electrophilic attack at these positions more than it stabilizes the intermediate formed by attack at the meta position. Figure 13.2 compares the energies of activation for attack at the various positions of toluene.

All the available ring positions of toluene are activated toward electrophilic attack. The ortho and para positions are activated more than the meta position. The relative

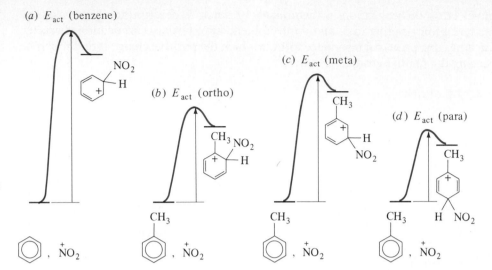

FIGURE 13.2   Comparative energy diagrams for nitronium ion attack on (*a*) benzene and at the (*b*) ortho, (*c*) meta, and (*d*) para positions of toluene. E_{act}(benzene) > E_{act}(meta) > E_{act}(ortho) > E_{act}(para).

rates of attack at the various positions in toluene compared with a single position in benzene are as shown (for nitration at 25°C):

These relative rate data per position are experimentally determined and are known as *partial rate factors.* They offer a convenient way to express substituent effects in electrophilic aromatic substitution reactions.

The major influence of the methyl group is *electronic.* The most important factor is relative carbocation stability. To a small extent the methyl group exerts a steric effect that slightly retards attack at the ortho positions. Thus, attack is slightly more likely at the para carbon than at a single ortho carbon. However, para substitution is at a statistical disadvantage since there are two equivalent ortho positions but only one para position.

PROBLEM 13.7   The rates of nitration relative to a single position of benzene at the various positions of *tert*-butylbenzene are as shown.

(*a*)   How reactive is *tert*-butylbenzene toward nitration compared with benzene?

(*b*)   How reactive is *tert*-butylbenzene toward nitration compared with toluene?

(*c*)   Predict the distribution among the various mononitration products of *tert*-butylbenzene.

SAMPLE SOLUTION (*a*) Benzene has six equivalent sites at which nitration can occur. Each is assigned a relative rate of 1.0 since benzene is our standard of comparison. Summing the individual relative rates of attack at each position in *tert*-butylbenzene and benzene, we obtain

$$\frac{tert\text{-Butylbenzene}}{\text{Benzene}} = \frac{2(4.5) + 2(3) + 75}{6(1)} = \frac{90}{6} = 15$$

tert-Butylbenzene undergoes nitration 15 times as fast as benzene.

All alkyl groups, not just methyl, are activating substituents and ortho, para directors. This is because any alkyl group, be it methyl, ethyl, isopropyl, *tert*-butyl, etc., stabilizes a carbocation site to which it is directly attached. When R = alkyl

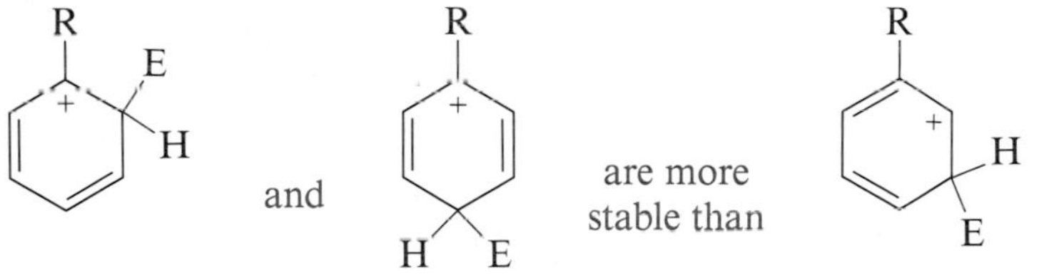

where E is any electrophile. All three structures are more stable for R = alkyl than for R = H and are formed at faster rates.

13.11 ANALYSIS OF RATE AND ORIENTATION EFFECTS IN THE NITRATION OF (TRIFLUOROMETHYL)BENZENE

Turning now to electrophilic aromatic substitution in (trifluoromethyl)benzene, we consider the electronic properties of a trifluoromethyl group. Because of their high electronegativity the three fluorine atoms polarize the electron distribution in their σ bonds to carbon, so that carbon bears a partial positive charge.

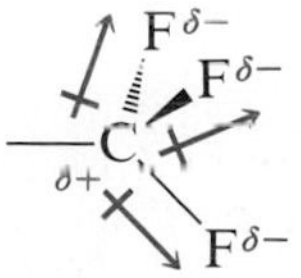

Unlike a methyl group, which is slightly electron-releasing, a trifluoromethyl group is a powerful electron-withdrawing substituent. Consequently, a CF_3 group destabilizes a carbocation site to which it is attached,

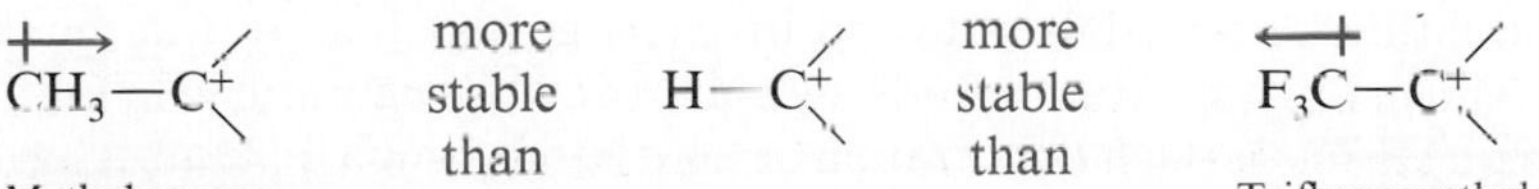

Methyl group
releases electrons,
stabilizes carbocation

Trifluoromethyl
group withdraws
electrons, destabilizes
carbocation

When we examine the cyclohexadienyl cation intermediates involved in the nitration of (trifluoromethyl)benzene, we find that those leading to ortho and para substitution are strongly destabilized.

Ortho attack:

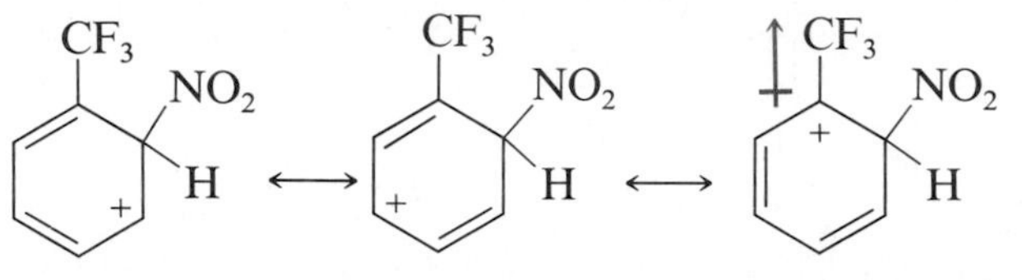

Positive charge on carbon bearing trifluoromethyl group; very unstable

Para attack:

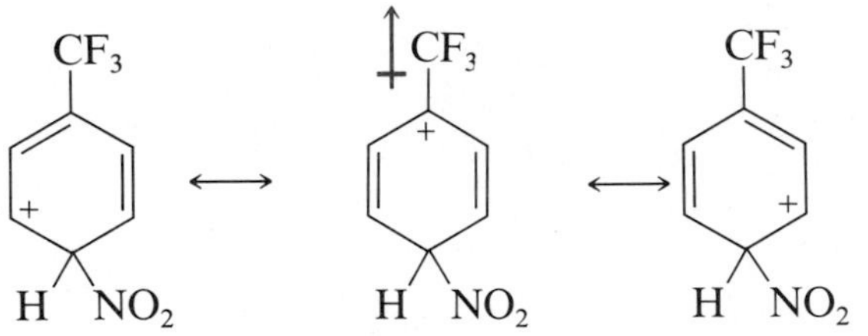

Positive charge on carbon bearing trifluoromethyl group; very unstable

None of the three major resonance forms of the intermediate formed during attack at the meta position has a positive charge on the carbon bearing the trifluoromethyl substituent.

Meta attack:

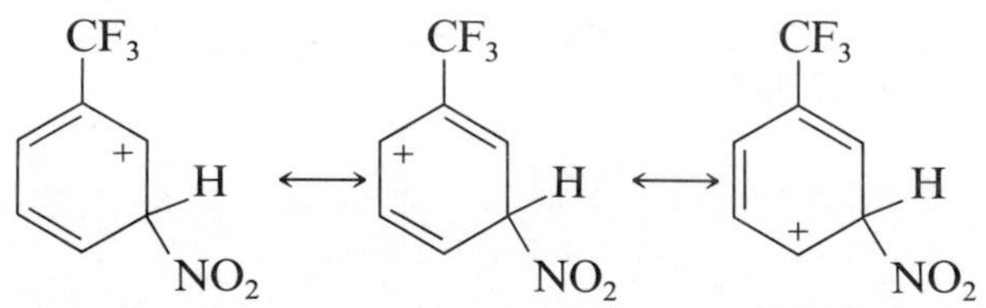

Attack at the meta position leads to a more stable intermediate than attack at either the ortho or para positions, so predominant meta substitution is observed in (trifluoromethyl)benzene. Even the intermediate corresponding to meta attack, however, is very unstable and is formed with difficulty. The trifluoromethyl group is only one bond farther removed from the positive charge here than it is in the ortho and para intermediates and so still exerts a significant, although somewhat diminished, destabilizing effect through polarization of the σ bonds which separate it from the site of positive charge.

All the ring positions of (trifluoromethyl)benzene are deactivated as compared with benzene. The meta position is simply deactivated less than the ortho and para

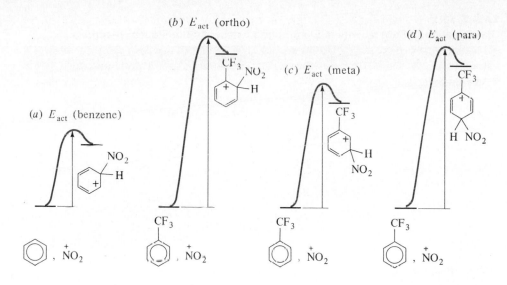

FIGURE 13.3 Comparative energy diagrams for nitronium ion attack at (a) benzene and the (b) ortho, (c) meta, and (d) para positions of (trifluoromethyl)benzene. E_{act}(ortho) ~ E_{act}(para) > E_{act}(meta) > E_{act}(benzene).

positions. We find that the relative rates of nitration of the various sites of (trifluoromethyl)benzene are

Figure 13.3 compares the energy profile for nitronium ion attack at benzene with those for attack at the ortho, meta, and para positions of (trifluoromethyl)benzene. The presence of the electron-withdrawing trifluoromethyl group raises the activation energy for attack at all the ring positions, but the increase is least for attack at the meta position.

13.12 SUBSTITUENT EFFECTS IN ELECTROPHILIC AROMATIC SUBSTITUTION. ACTIVATING SUBSTITUENTS

Our analysis of substituent effects has so far centered upon two groups, methyl and trifluoromethyl. We have seen that a methyl substituent is activating and ortho, para–directing. A trifluoromethyl group is strongly deactivating and meta-directing. What about other substituents and their effects on rate and regioselectivity in electrophilic aromatic substitution?

Table 13.1 summarizes orientation and rate effects in electrophilic aromatic substitution reactions for a variety of frequently encountered substituents. It is arranged in order of decreasing activating power: the most strongly activating substituents are at the top, the most strongly deactivating substituents are at the bottom. The main features of the table can be summarized as follows:

1. All activating substituents are ortho, para directors.
2. Halogen substituents are slightly deactivating but are ortho, para–directing.
3. Substituents more deactivating than halogen are meta directors.

TABLE 13.1

Classification of Substituents in Electrophilic Aromatic Substitution Reactions

Effect on rate	Substituent		Effect on orientation
Very strongly activating	$-\ddot{N}H_2$	(amino)	Ortho, para–directing
	$-\ddot{N}HR$	(alkylamino)	
	$-\ddot{N}R_2$	(dialkylamino)	
	$-\ddot{O}H$	(hydroxyl)	
Strongly activating	$-\ddot{N}HCR$ (with $\overset{\overset{\displaystyle O}{\|}}{}$)	(acylamino)	Ortho, para–directing
	$-\ddot{O}R$	(alkoxy)	
	$-\ddot{O}CR$ (with $\overset{\overset{\displaystyle O}{\|}}{}$)	(acyloxy)	
Activating	$-\ddot{R}$	(alkyl)	Ortho, para–directing
	$-Ar$	(aryl)	
	$-CH{=}CR_2$	(alkenyl)	
Standard of comparison	$-H$	(hydrogen)	
Deactivating	$-X$ $(X = F, Cl, Br, I)$	(halogen)	Ortho, para–directing
	$-CH_2X$	(halomethyl)	
Strongly deactivating	$-\overset{\overset{\displaystyle O}{\|}}{C}H$	(formyl)	Meta-directing
	$-\overset{\overset{\displaystyle O}{\|}}{C}R$	(acyl)	
	$-\overset{\overset{\displaystyle O}{\|}}{C}OH$	(carboxylic acid)	
	$-\overset{\overset{\displaystyle O}{\|}}{C}OR$	(ester)	
	$-\overset{\overset{\displaystyle O}{\|}}{C}Cl$	(acyl chloride)	
	$-C{\equiv}N$	(cyano)	
	$-SO_3H$	(sulfonic acid)	
Very strongly deactivating	$-CF_3$	(trifluoromethyl)	Meta-directing
	$-NO_2$	(nitro)	

Some of the most powerful activating substituents are those in which an oxygen atom is attached directly to the ring. These substituents include the hydroxyl group as well as alkoxy and acyloxy groups.

$$H\ddot{O}- \qquad R\ddot{O}- \qquad R\overset{\overset{\displaystyle O}{\|}}{C}\ddot{O}-$$
$$\text{Hydroxyl} \qquad \text{Alkoxy} \qquad \text{Acyloxy}$$

A hydroxyl group is such a strongly activating substituent that phenol undergoes nitration about 1000 times faster than benzene. A mixture of *o*-nitrophenol and *p*-nitrophenol is formed, with only a trace of the meta isomer.

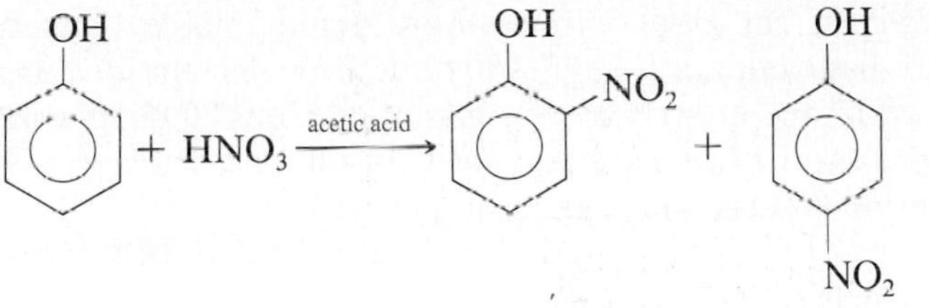

Phenol   o-Nitrophenol (44%)   p-Nitrophenol (56%)

Hydroxyl, alkoxy, and acyloxy groups activate the ring to such an extent that bromination occurs rapidly even in the absence of a ferric bromide catalyst.

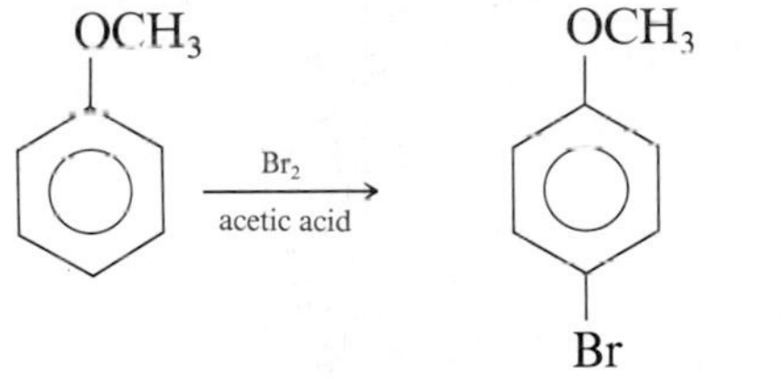

Anisole   p-Bromoanisole (90%)

It is the presence of the unshared electron pairs of the oxygen attached to the ring that is responsible for the activating and ortho, para–directing properties of these substituents. Attack at positions ortho and para to the one that bears the substituent yields cyclohexadienyl cations that are stabilized by donation of an unshared electron pair from oxygen.

Ortho attack:

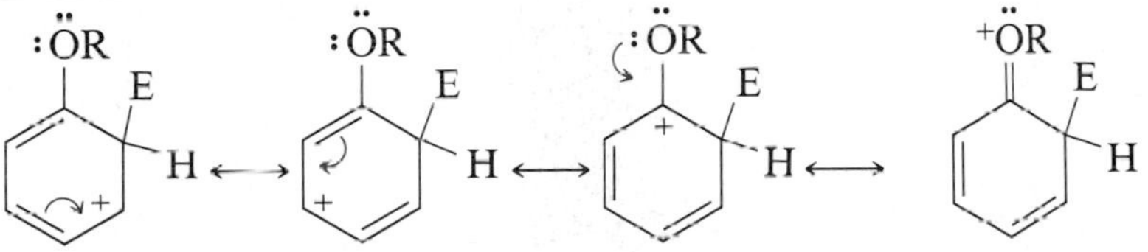

Most stable resonance form; oxygen and all carbons have octets of electrons

Para attack:

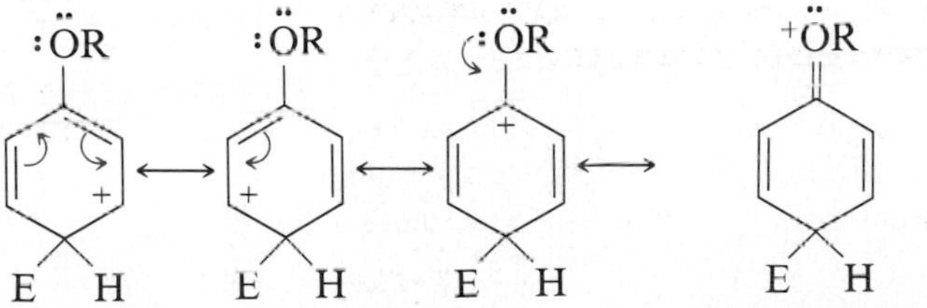

Most stable resonance form; oxygen and all carbons have octets of electrons

lone pair of e⁻ activating [handwritten annotation]

Oxygen-stabilized carbocations of this type are far more stable than tertiary carbocations. They are best represented by structures in which the positive charge is on oxygen because all the atoms have octets of electrons in such a structure. Their stability permits them to be formed rapidly, resulting in rates of electrophilic aromatic substitution that are much faster than that of benzene.

The lone pair on oxygen cannot be directly involved in carbocation stabilization when attack is meta to the substituent.

Meta attack:

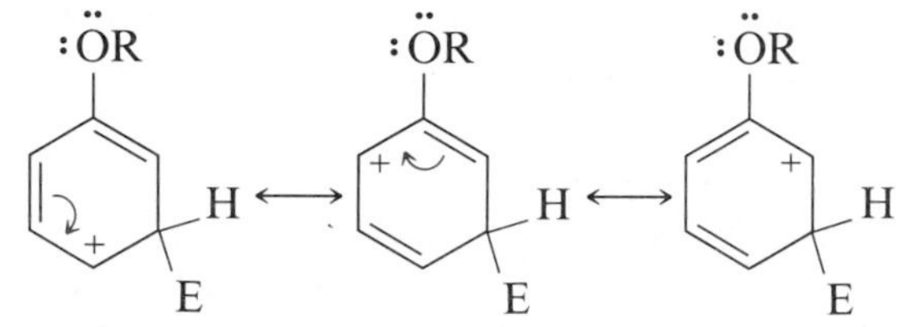

Oxygen lone pair cannot be used to stabilize positive charge
in any of these structures; all have six electrons around
positively charged carbon.

The greater stability of the carbocations arising from attack at the ortho and para positions compared with the carbocation formed by attack at the position meta to the oxygen substituent explains the ortho, para–directing property of hydroxyl, alkoxy, and acyloxy groups.

Nitrogen-containing substituents related to the amino group are even more strongly activating than the corresponding oxygen-containing substituents.

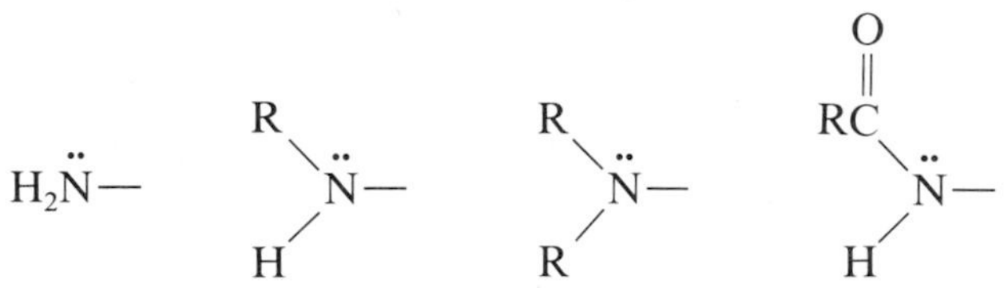

The nitrogen atom in each of these groups bears an electron pair that, like the unshared pair of an oxygen substituent, stabilizes a carbocation site to which it is attached. Since nitrogen is less electronegative than oxygen, it is a better electron pair donor and stabilizes the cyclohexadienyl cation intermediates in electrophilic aromatic substitution more than does the corresponding oxygen-containing substituent.

PROBLEM 13.8   Write structural formulas for the cyclohexadienyl cations formed from aniline ($C_6H_5NH_2$) during

(a)   Ortho bromination (four resonance structures)
(b)   Meta bromination (three resonance structures)
(c)   Para bromination (four resonance structures)

SAMPLE SOLUTION   (a)   There are the customary three resonance structures for the cyclohexadienyl cation plus a resonance structure (the most stable one) derived by delocalization of the nitrogen lone pair into the ring.

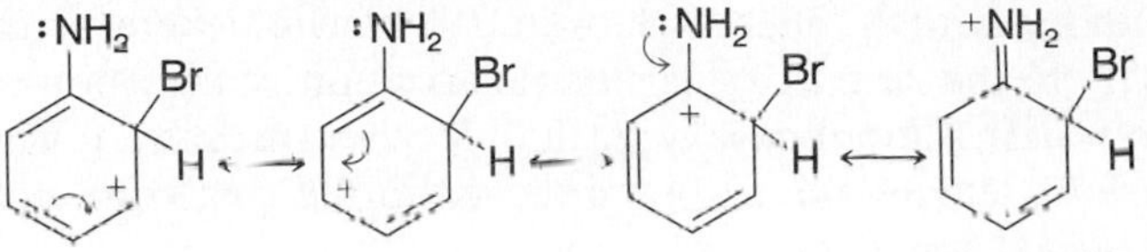

Most stable resonance structure

Alkyl groups are, as we saw when we discussed the nitration of toluene in Section 13.10, activating and ortho, para-directing substituents. Aryl and alkenyl substituents resemble alkyl groups in this respect; they too are activating and ortho, para-directing.

PROBLEM 13.9 Treatment of biphenyl (see Section 12.15 to remind yourself of its structure) with a mixture of nitric acid and sulfuric acid gave two principal products both having the molecular formula $C_{12}H_9NO_2$. What are these two products?

The next group of substituents in Table 13.1 consists of the halogens. They are slightly deactivating and ortho, para-directing. We shall defer discussing the reasons for this behavior until Section 13.14. Instead, we shall turn our present attention to the substituents at the end of the table, those that are strongly deactivating and meta-directing.

13.13 SUBSTITUENT EFFECTS IN ELECTROPHILIC AROMATIC SUBSTITUTION. DEACTIVATING SUBSTITUENTS

As Table 13.1 indicates, there are a variety of substituent types that are strongly deactivating and meta-directing. We have already discussed one of these, the trifluoromethyl group. Several of the others are characterized by the presence of a carbonyl group ($\gt C{=}O$) at their point of attachment to the aromatic ring.

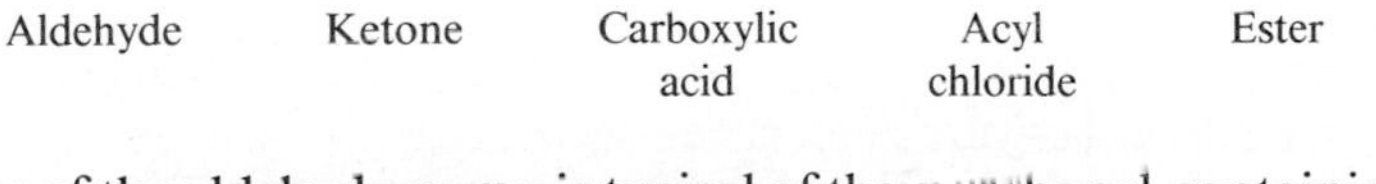

Aldehyde | Ketone | Carboxylic acid | Acyl chloride | Ester

C=O = deactivating meta dir.
C≡N "

The behavior of the aldehyde group is typical of these carbonyl-containing substituents. Nitration of benzaldehyde takes place several thousand times more slowly than that of benzene and yields *m*-nitrobenzaldehyde as the major product.

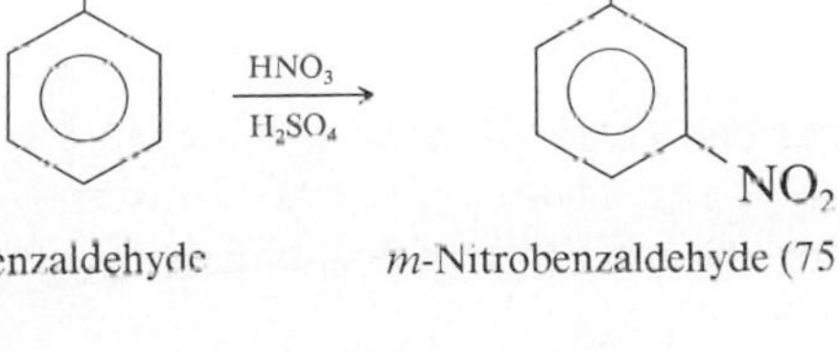

Benzaldehyde → *m*-Nitrobenzaldehyde (75–84%)

To help us understand the effect of substituents in which there is a carbonyl group attached directly to the ring, consider the polarization of a carbon-oxygen double bond. The electrons in the carbon-oxygen double bond are drawn toward oxygen and away from carbon, leaving the carbon attached to the ring with a partial positive charge. Using benzaldehyde as an example

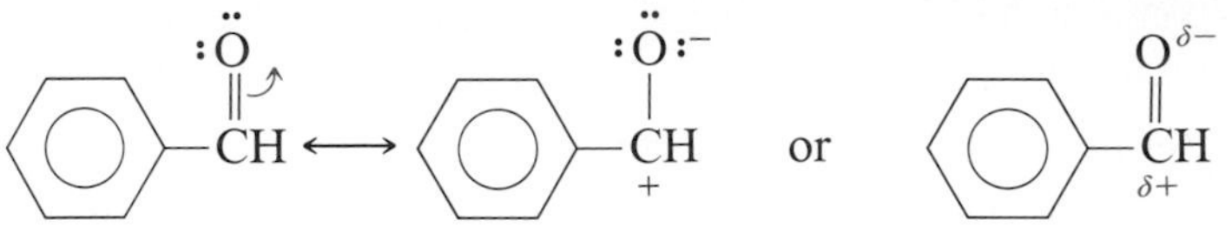

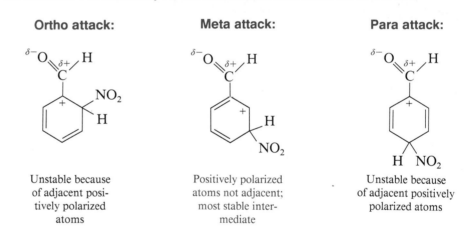

Because the carbon atom attached to the ring is positively polarized, a carbonyl group behaves in much the same way as a trifluoromethyl group and destabilizes all the cyclohexadienyl cation intermediates in electrophilic aromatic substitution reactions. Attack at any ring position in benzaldehyde is slower than attack in benzene. The intermediates that lead to ortho and para substitution are particularly unstable because each is characterized by a resonance structure in which there is a positive charge on the carbon that bears the electron-withdrawing substituent. The intermediate leading to meta substitution avoids this unfavorable juxtaposition of positive charge, is not as unstable, and gives rise to most of the product. For the specific case of the nitration of benzaldehyde, the relevant intermediates are as shown.

Ortho attack:	Meta attack:	Para attack:
Unstable because of adjacent positively polarized atoms	Positively polarized atoms not adjacent; most stable intermediate	Unstable because of adjacent positively polarized atoms

PROBLEM 13.10   Each of the following reactions has been reported in the chemical literature and the principal organic product has been isolated in good yield. Write a structural formula for the isolated product of each reaction.

(a)   Treatment of benzoyl chloride ($C_6H_5\overset{\displaystyle O}{\overset{\displaystyle \|}{C}}Cl$) with chlorine and ferric chloride

(b)   Treatment of methyl benzoate ($C_6H_5\overset{\displaystyle O}{\overset{\displaystyle \|}{C}}OCH_3$) with nitric acid and sulfuric acid

(c)   Nitration of propiophenone ($C_6H_5\overset{\displaystyle O}{\overset{\displaystyle \|}{C}}CH_2CH_3$)

SAMPLE SOLUTION   (a)   Benzoyl chloride has a carbonyl group attached directly to the ring. A $\overset{\overset{\displaystyle O}{\parallel}}{C}Cl$ substituent is meta-directing. The reaction conditions, namely, use of chlorine and ferric chloride, are those that introduce a chlorine onto the ring. The product is *m*-chlorobenzoyl chloride.

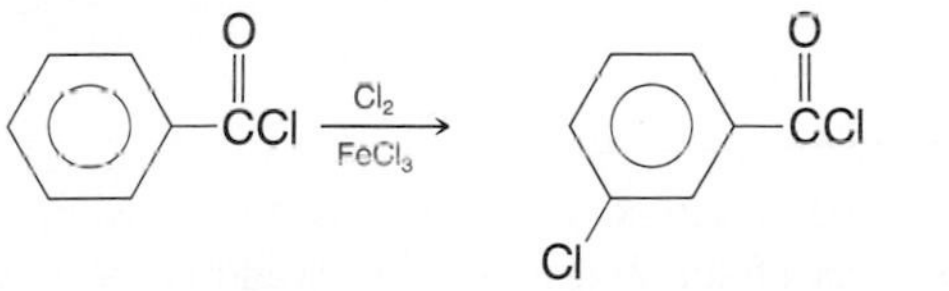

Benzoyl chloride   *m*-Chlorobenzoyl chloride
                          (isolated in 62% yield)

A cyano substituent is similar to a carbonyl group for analogous reasons involving resonance of the type

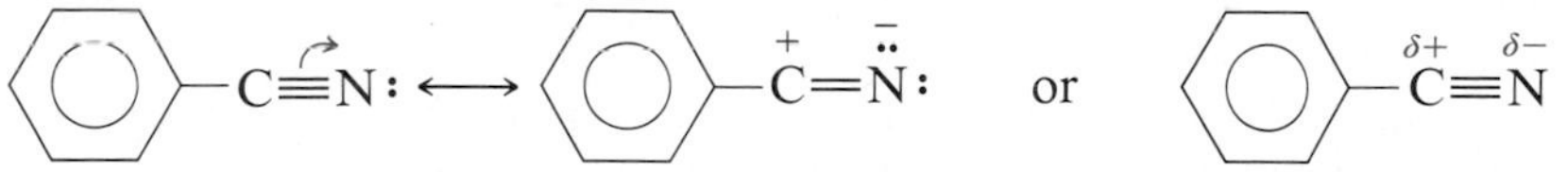

Cyano groups are electron-withdrawing, deactivating, and meta-directing.

Sulfonic acid groups are electron-withdrawing because sulfur has a formal positive charge in several of the principal resonance forms of benzenesulfonic acid.

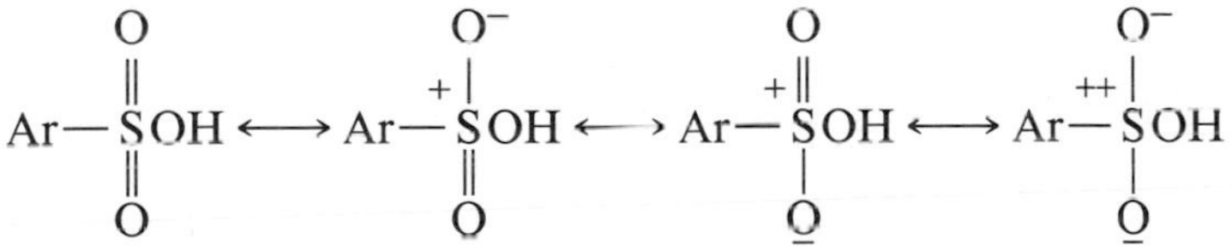

When benzene undergoes disulfonation, *m*-benzenedisulfonic acid is formed. The first sulfonic acid group to go on directs the second one to the meta position.

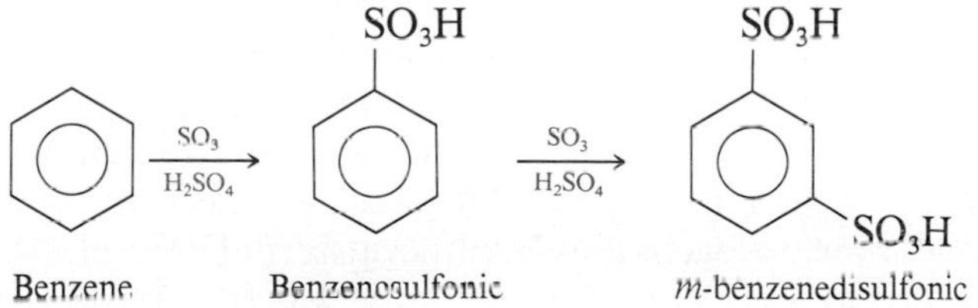

Benzene   Benzenesulfonic   *m*-benzenedisulfonic
                 acid   acid (90%)

The nitrogen atom of a nitro group bears a full positive charge in its most important contributing Lewis structures.

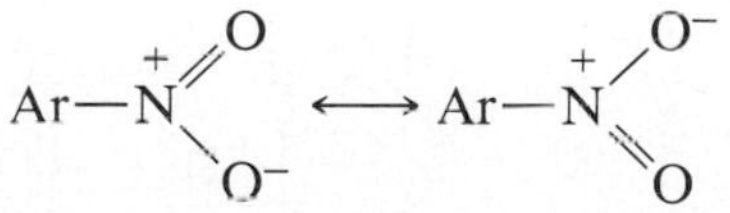

This makes it a powerful electron-withdrawing deactivating substituent and a meta director.

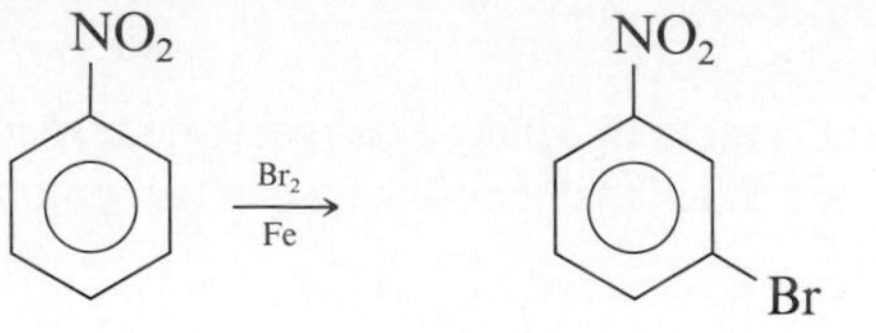

Nitrobenzene   *m*-Bromonitrobenzene (60–75%)

PROBLEM 13.11 Would you expect the substituent —N̈(CH₃)₃⁺ to more closely resemble —N̈(CH₃)₂ or —NO₂ in its effect on rate and orientation in electrophilic aromatic substitution? Why?

13.14 SUBSTITUENT EFFECTS IN ELECTROPHILIC AROMATIC SUBSTITUTION. HALOGEN SUBSTITUENTS

Returning to Table 13.1, notice that halogen substituents direct an incoming electrophile to the ortho and para positions but deactivate the ring toward substitution. Nitration of chlorobenzene is a typical example of electrophilic aromatic substitution in a halobenzene; it proceeds at a rate that is some 30 times slower than the corresponding nitration of benzene. The principal products are *o*-chloronitrobenzene and *p*-chloronitrobenzene.

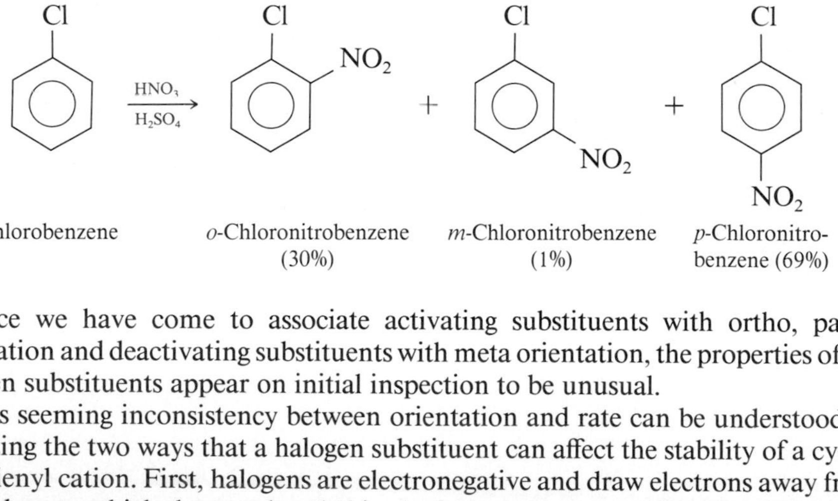

Chlorobenzene   *o*-Chloronitrobenzene (30%)   +   *m*-Chloronitrobenzene (1%)   +   *p*-Chloronitrobenzene (69%)

Since we have come to associate activating substituents with ortho, para–orientation and deactivating substituents with meta orientation, the properties of the halogen substituents appear on initial inspection to be unusual.

This seeming inconsistency between orientation and rate can be understood by analyzing the two ways that a halogen substituent can affect the stability of a cyclohexadienyl cation. First, halogens are electronegative and draw electrons away from the carbon to which they are bonded in the same way that a trifluoromethyl group does. Thus, all the cyclohexadienyl cation intermediates formed by electrophilic attack on a halobenzene are less stable than the corresponding cyclohexadienyl cation for benzene, and halobenzenes are less reactive than benzene.

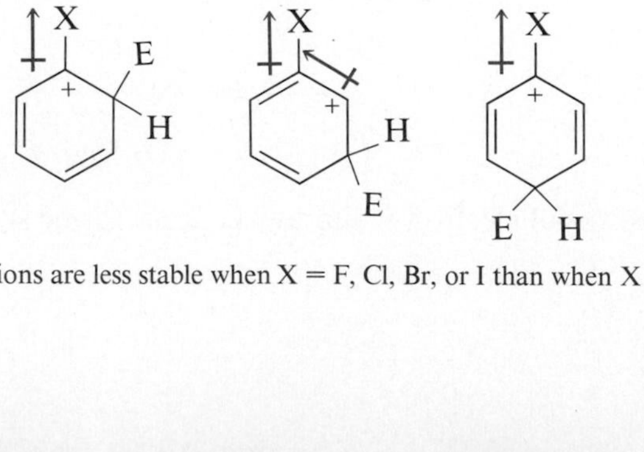

All these ions are less stable when X = F, Cl, Br, or I than when X = H

However, like hydroxyl groups and amino groups, halogen substituents possess unshared electron pairs that can be donated to a positively charged carbon. This electron donation stabilizes the intermediates derived from ortho and from para attack.

Resonance involving the halogen lone pair:

Ortho attack:

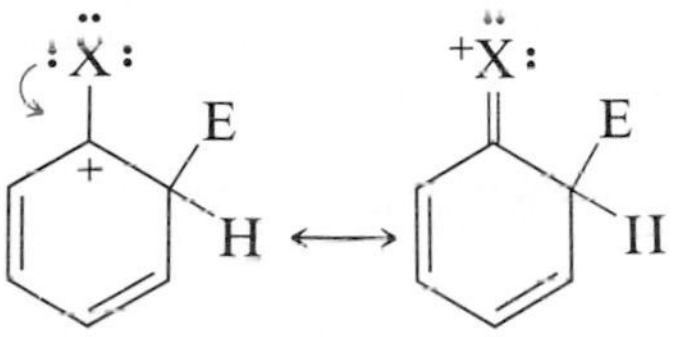

Para attack:

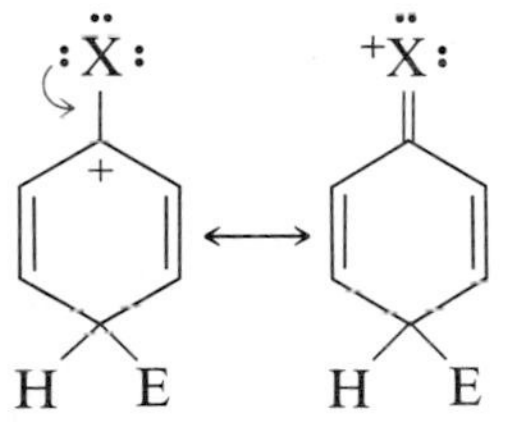

Comparable resonance stabilization of the intermediate leading to meta substitution is not possible. Thus, resonance involving halogen lone pairs causes electrophilic attack to be favored at the ortho and para positions but is not large enough to overcome the electron-withdrawing effect of the halogen, which deactivates all the ring positions. The partial rate factors for nitration of chlorobenzene illustrate the rate and orientation properties of a chlorine substituent.

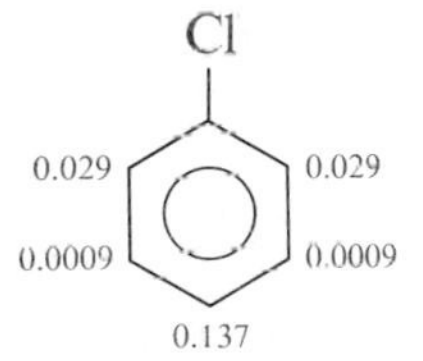

The other halogens resemble chlorine in their effects on rate and orientation in electrophilic aromatic substitution.

13.15 REGIOSELECTIVE SYNTHESIS OF DISUBSTITUTED AROMATIC COMPOUNDS

Since the position of electrophilic attack on an aromatic ring is controlled by the directing effects of substituents already present, the preparation of disubstituted aromatic compounds requires that careful thought be given to the order of introduction of the two groups.

Compare the independent preparations of *m*-bromoacetophenone and *p*-bromoacetophenone from benzene. Both syntheses require a Friedel-Crafts acylation step and a bromination step, but the major product is determined by the *order* in which the two steps are carried out. When the meta-directing acetyl group is introduced first, the final product is *m*-bromoacetophenone.

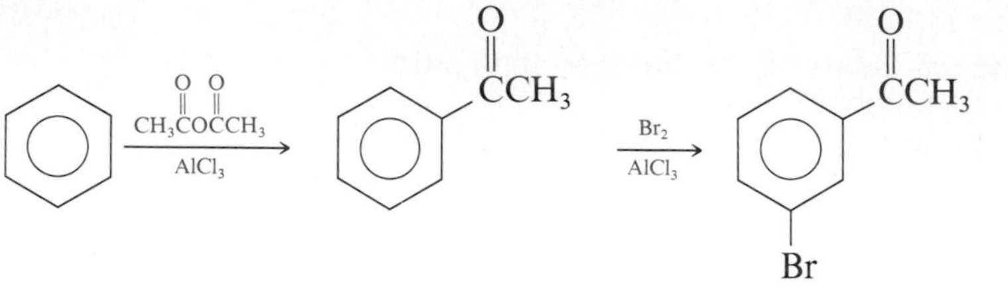

Benzene   Acetophenone (76–83%)   *m*-Bromoacetophenone
                                                            (59%)

When the ortho, para–directing bromo substituent is introduced first, the final product is *p*-bromoacetophenone (along with some of its ortho isomer, from which it is separated by distillation).

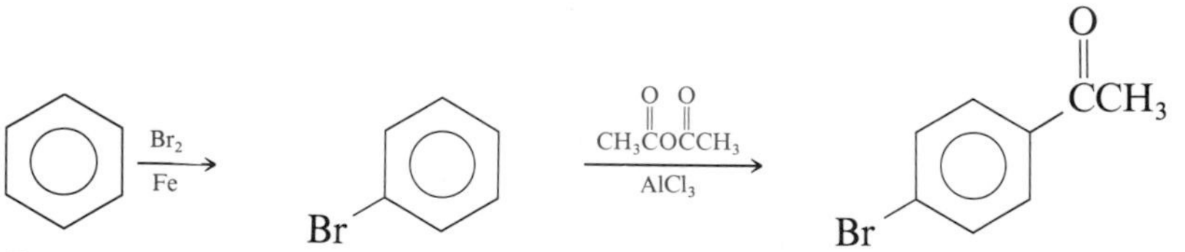

Benzene   Bromobenzene (65–75%)   *p*-Bromoacetophenone (69–79%)

PROBLEM 13.12   Write chemical equations showing how you could prepare *m*-bromonitrobenzene as the principal organic product, starting with benzene and using any necessary organic or inorganic reagents. How could you prepare *p*-bromonitrobenzene?

A less obvious example of a situation in which the success of a synthesis depends on the order of introduction of substituents is illustrated by the preparation of *m*-nitroacetophenone. Here, even though both substituents are meta-directing, the only practical synthesis is the one in which Friedel-Crafts acylation is carried out first.

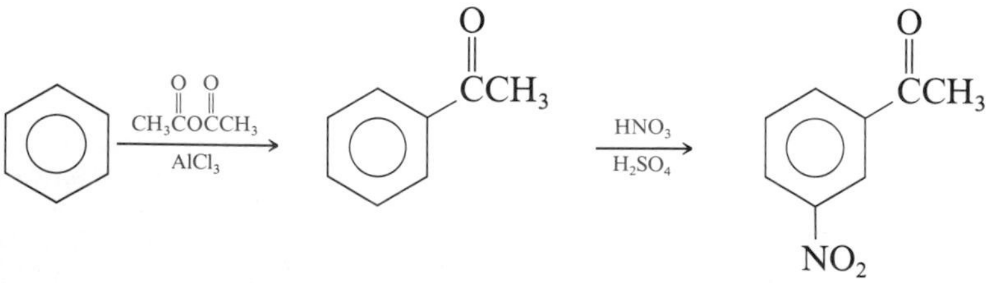

Benzene   Acetophenone (76–83%)   *m*-Nitroacetophenone (55%)

When the reverse order of steps is attempted, it is observed that the Friedel-Crafts acylation of nitrobenzene fails.

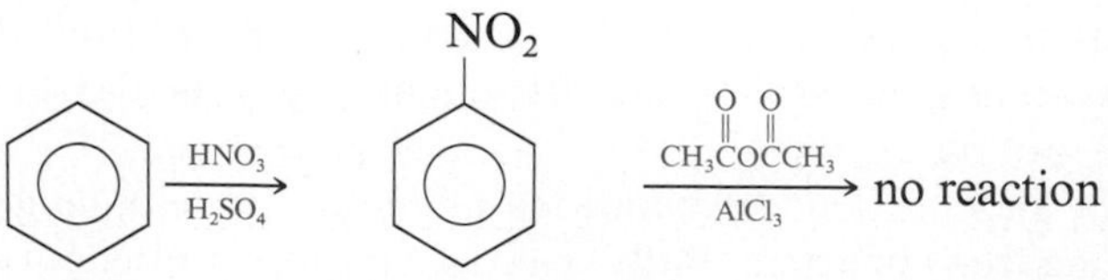

Benzene   Nitrobenzene (95%)

Neither Friedel-Crafts acylation nor alkylation reactions can be carried out on nitrobenzene. The presence of a strongly deactivating substituent such as a nitro group on an aromatic ring so depresses its reactivity that Friedel-Crafts reactions do not take place. The practical limit for Friedel-Crafts alkylation and acylation reactions is effectively a monohalobenzene. An aromatic ring more deactivated than a monohalobenzene cannot be alkylated or acylated under Friedel-Crafts conditions.

Sometimes the orientation of two substituents in an aromatic compound precludes its straightforward synthesis. *m*-Chloroethylbenzene, for example, has two ortho, para–directing groups in a meta relationship and so cannot be prepared either from chlorobenzene or from ethylbenzene. In cases such as this it is frequently possible to couple electrophilic aromatic substitution with functional group manipulation to produce the desired compound. The key here is to recognize that an ethyl substituent can be introduced by Friedel-Crafts acylation followed by a Clemmensen or Wolff-Kishner reduction step later in the synthesis. If the chlorine is introduced prior to reduction it will be directed meta to the acetyl group, giving the correct substitution pattern.

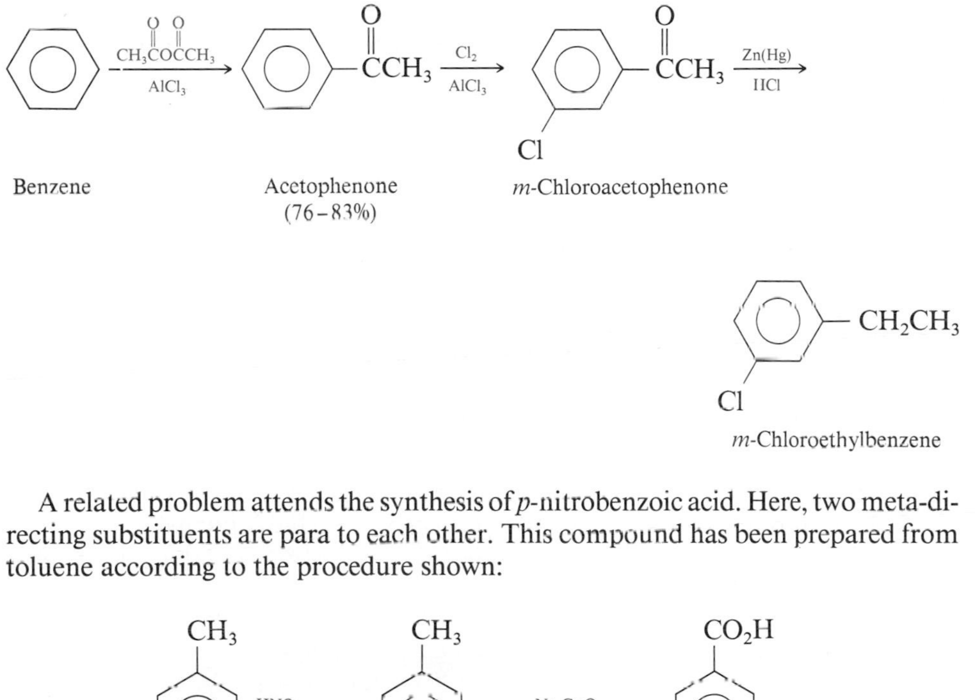

Benzene   Acetophenone   *m*-Chloroacetophenone
               (76–83%)

m-Chloroethylbenzene

A related problem attends the synthesis of *p*-nitrobenzoic acid. Here, two meta-directing substituents are para to each other. This compound has been prepared from toluene according to the procedure shown:

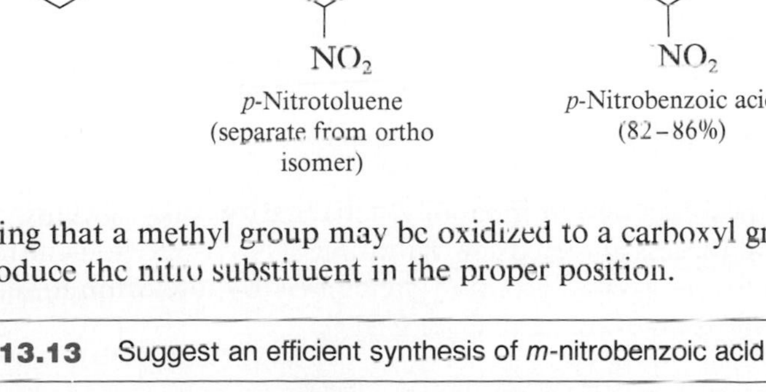

p-Nitrotoluene   *p*-Nitrobenzoic acid
(separate from ortho   (82–86%)
isomer)

By recognizing that a methyl group may be oxidized to a carboxyl group, it can be used to introduce the nitro substituent in the proper position.

PROBLEM 13.13   Suggest an efficient synthesis of *m*-nitrobenzoic acid from toluene.

13.16 MULTIPLE SUBSTITUENT EFFECTS

When a benzene ring bears two or more substituents, both its reactivity and the site of further substitution can in most cases be predicted from the cumulative effects of its substituents.

In the simplest cases all the available sites are equivalent and substitution at any one of them gives the same product.

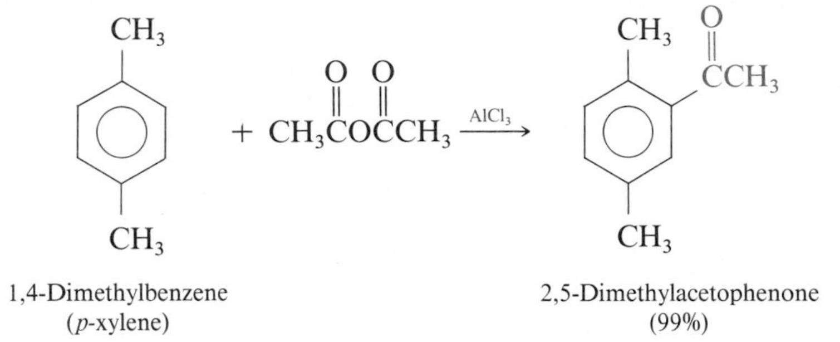

1,4-Dimethylbenzene
(*p*-xylene)

2,5-Dimethylacetophenone
(99%)

Often the directing effects of substituents reinforce each other. Bromination of *p*-nitrotoluene, for example, takes place at the position which is ortho to the ortho, para–directing methyl group and meta to the meta-directing nitro group.

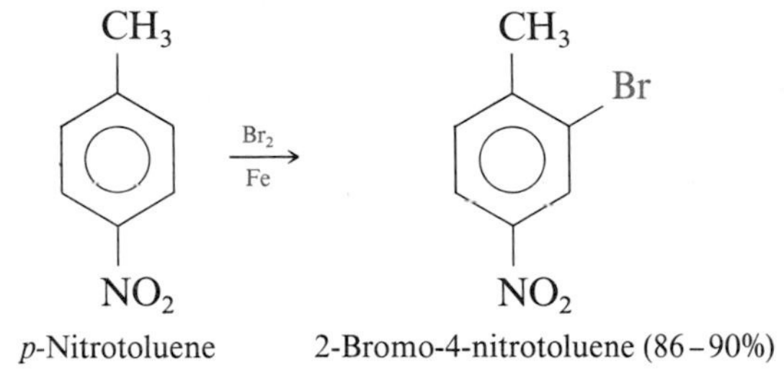

p-Nitrotoluene   2-Bromo-4-nitrotoluene (86–90%)

In cases in which the directing effects of individual substituents oppose each other, it is the more activating substituent that controls the regioselectivity of electrophilic aromatic substitution. Thus, bromination occurs ortho to the *N*-methylamino group in 4-chloro-*N*-methylaniline because this group is a very powerful activating substituent while the chlorine is weakly deactivating.

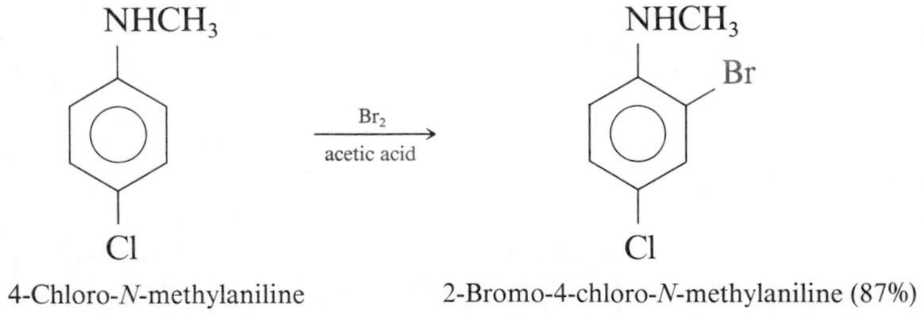

4-Chloro-*N*-methylaniline   2-Bromo-4-chloro-*N*-methylaniline (87%)

When two positions are comparably activated by alkyl groups, substitution usually occurs at the less hindered site. Nitration of *p-tert*-butyltoluene takes place at positions ortho to the methyl group in preference to those ortho to the larger *tert*-butyl group. This is an example of a *steric effect*.

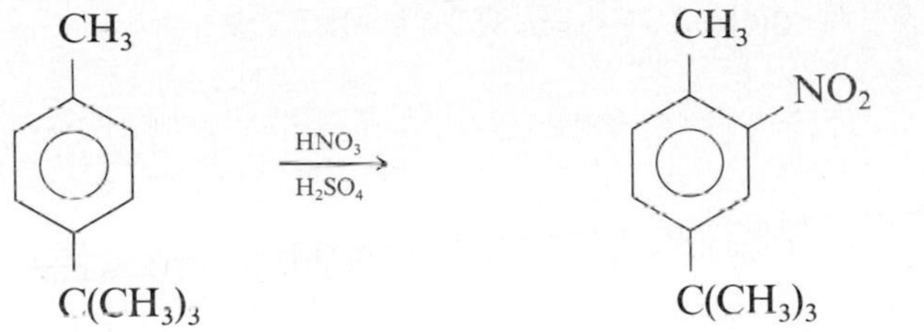

p-tert-Butyltoluene → 4-tert-Butyl-2-nitrotoluene (88%)

Nitration of *m*-xylene is directed ortho to one methyl group and para to the other.

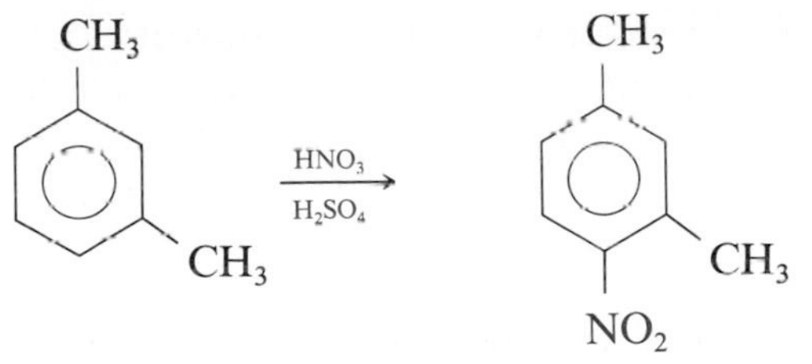

m-Xylene → 2,4-Dimethyl-1-nitrobenzene (98%)

The ortho position between the two methyl groups is less reactive because it is more sterically hindered.

When neither group of a disubstituted benzene is strongly activating, mixtures of regioisomers are usually formed. Friedel-Crafts acylation of *m*-chlorotoluene gives just as much product by attack para to chlorine as by attack para to methyl.

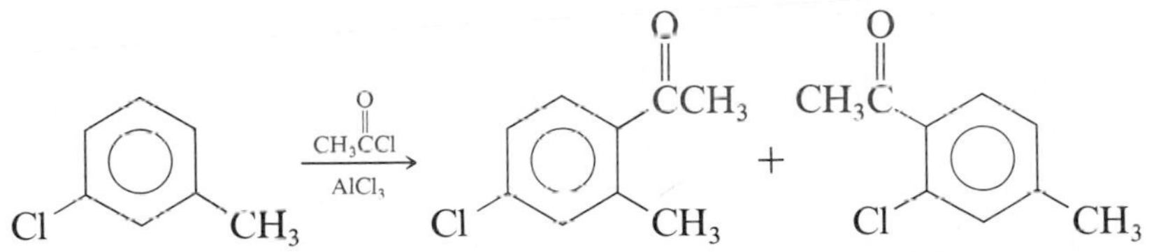

m-Chlorotoluene → 4-Chloro-2-methylaceto-phenone (~50%) + 2-Chloro-4-methylaceto-phenone (~50%)

PROBLEM 13.14   Write the structure of the principal organic product obtained on nitration of each of the following:

(a) p-Methylbenzoic acid
(b) m-Dichlorobenzene
(c) m-Dinitrobenzene
(d) p-Methoxyacetophenone
(e) p-Methylanisole
(f) 2,6-Dibromoanisole

SAMPLE SOLUTION   (a) Of the two substituents in p-methylbenzoic acid, the methyl group is more activating and so controls the regioselectivity of electrophilic aromatic substitution. The position para to the ortho, para–directing methyl group already bears a substituent (the carboxyl group), so substitution occurs ortho to the methyl group. This position is meta to the *m*-directing carboxyl group, so the orienting properties of the two substituents reinforce each other. The product is 4-methyl-3-nitrobenzoic acid.

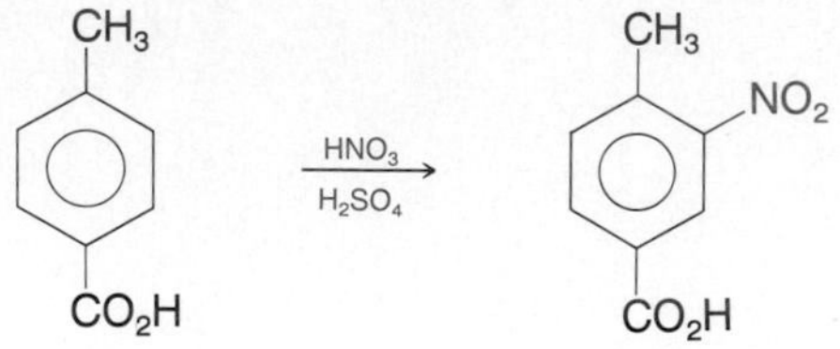

p-Methylbenzoic acid   4-Methyl-3-nitrobenzoic acid

13.17 SUBSTITUTION IN NAPHTHALENE

Polycyclic aromatic hydrocarbons undergo electrophilic aromatic substitution when treated with the same reagents that react with benzene. In general, polycyclic aromatics are more reactive than benzene. Since, however, most lack the symmetry of benzene, mixtures of products may be formed even on monosubstitution. Among polycyclic aromatic hydrocarbons we will discuss only naphthalene, and that only briefly.

Two sites are available for substitution in naphthalene, C-1 and C-2, C-1 being normally the preferred site of electrophilic attack.

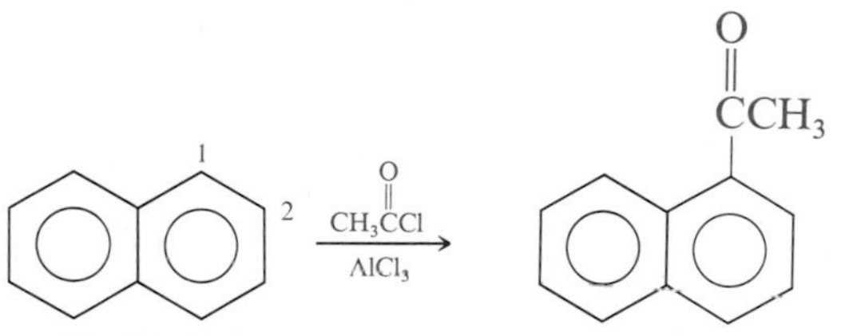

Naphthalene   1-Acetylnaphthalene (90%)

The C-1 position is the more reactive because the arenium ion formed by electrophilic attack there is a relatively stable one. Benzenoid character is retained in one ring while the positive charge is delocalized by allylic resonance.

Attack at C-1:

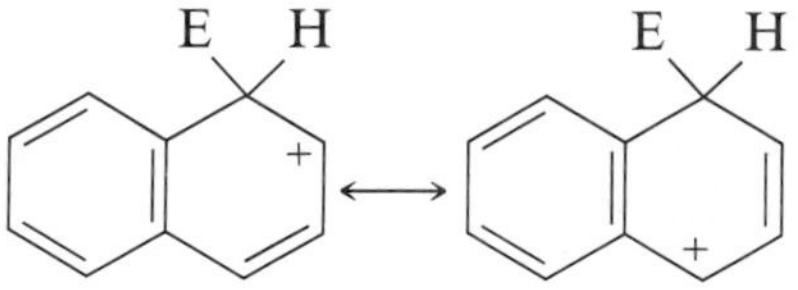

Attack at C-2:

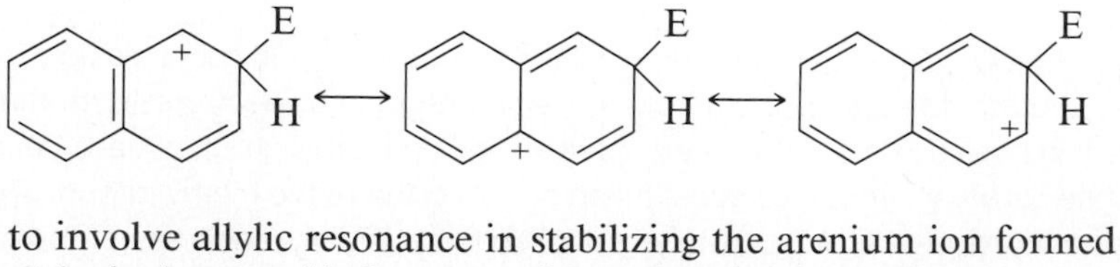

In order to involve allylic resonance in stabilizing the arenium ion formed during attack at C-2, the benzenoid character of the other ring is sacrificed.

13.18 SUMMARY

On reaction with electrophilic reagents, aromatic substances undergo substitution rather than elimination.

$$ArH + \quad E^+ \quad \longrightarrow \quad ArE \quad + \ H^+$$

Arene   Electrophile   Product of   Proton
electrophilic
aromatic substitution

Substitution occurs by attack of the electrophile on the π electrons of the aromatic ring in the rate-determining step, to form a cyclohexadienyl cation intermediate. Loss of a proton from this intermediate restores the aromaticity of the ring and yields the product of *electrophilic aromatic substitution.*

Benzene   Electrophile   Cyclohexadienyl   Product of   Proton
cation intermediate   electrophilic
aromatic
substitution

Table 13.2 presents some typical examples of electrophilic aromatic substitution reactions; it illustrates conditions for carrying out the nitration, sulfonation, halogenation, and Friedel-Crafts alkylation and acylation of aromatic substances.

Some of the aromatic starting materials in the table bear substituents on the ring, and these substituents influence both the *rate* at which reaction occurs and the *regioselectivity* of substitution. Substituents are classified as activating or deactivating according to whether they cause the ring to react more rapidly or less rapidly than benzene toward electrophilic aromatic substitution. Substituents are arranged into three major categories:

1. *Activating and ortho, para–directing:* These substituents stabilize the cyclohexadienyl cation formed in the rate-determining step. They include $-\overset{..}{N}R_2$, $-\overset{..}{\underset{..}{O}}R, -R, -Ar$, and related species. The most strongly activating members of this group are bonded to the ring by a nitrogen or oxygen atom that bears an unshared pair of electrons.

2. *Deactivating and ortho, para–directing:* The halogens are the most prominent members of this class. They withdraw electron density from all the ring positions by an inductive effect, making halobenzenes less reactive than benzene. Lone pair electron donation stabilizes the cyclohexadienyl cations corresponding to attack at the ortho and para positions more than those formed by attack at the meta positions, giving rise to the observed regioselectivity.

3. *Deactivating and meta-directing:* These substituents are strongly electron-withdrawing and destabilize carbocations. They include $-CF_3$, $-\overset{\displaystyle O}{\overset{\|}{C}}R$, $-C{\equiv}N, -NO_2$, and related species. All the ring positions are deactivated

TABLE 13.2

Representative Electrophilic Aromatic Substitution Reactions

Reaction (section) and comments	General equation and specific example
Nitration (Section 13.3) The active electrophile in the nitration of benzene and its derivatives is nitronium cation ($:\overset{..}{O}=\overset{+}{N}=\overset{..}{O}:$). It is generated by reaction of nitric acid and sulfuric acid. Very reactive arenes —those that bear strongly activating substituents — undergo nitration in nitric acid alone.	$$ArH \ + \ HNO_3 \xrightarrow{H_2SO_4} ArNO_2 \ + \ H_2O$$ Arene Nitric acid Nitroarene Water Fluorobenzene p-Fluoronitrobenzene (80%)
Sulfonation (Section 13.4) Sulfonic acids are formed when aromatic compounds are treated with sources of sulfur trioxide. These sources can be concentrated sulfuric acid (for very reactive arenes) or solutions, called *oleum,* of sulfur trioxide in sulfuric acid (for benzene and arenes less reactive than benzene).	$$ArH \ + \ SO_3 \longrightarrow ArSO_3H$$ Arene Sulfur trioxide Arenesulfonic acid 1,2,4,5-Tetramethylbenzene 2,3,5,6-Tetramethylbenzenesulfonic acid (94%)
Halogenation (Section 13.5) Chlorination and bromination of arenes is carried out by treatment with the appropriate halogen in the presence of a Lewis acid catalyst. Very reactive arenes undergo halogenation in the absence of a catalyst.	$$ArH \ + \ X_2 \xrightarrow{FeX_3} ArX \ + \ HX$$ Arene Halogen Aryl halide Hydrogen halide Phenol p-Bromophenol (80–84%)
Friedel-Crafts Alkylation (Section 13.6) Carbocations, usually generated from an alkyl halide and aluminum chloride, attack the aromatic ring to yield alkylbenzenes. The arene must be at least as reactive as a halobenzene. Carbocation rearrangements can occur, especially with primary alkyl halides.	$$ArH \ + \ RX \xrightarrow{AlCl_3} ArR \ + \ HX$$ Arene Alkyl halide Alkylarene Hydrogen halide Benzene Cyclopentyl bromide Cyclopentylbenzene (54%)
Friedel-Crafts acylation (Section 13.7) Acyl cations (acylium ions) generated by treating an acyl chloride or acid anhydride with aluminum chloride attack aromatic rings to yield ketones. The arene must be at least as reactive as a halobenzene. Acyl cations are relatively stable, and do not rearrange.	$$ArH \ + \ RCCl \xrightarrow{AlCl_3} ArCR \ + \ HCl$$ Arene Acyl chloride Ketone Hydrogen chloride or $$ArH \ + \ RCOCR \xrightarrow{AlCl_3} ArCR \ + \ RCOH$$ Arene Acid anhydride Ketone Carboxylic acid Anisole p-Methoxyacetophenone (90–94%)

but since the *meta* positions are deactivated less than the ortho and para, meta substitution is favored.

Friedel-Crafts reactions cannot be carried out effectively on arenes which are more deactivated than a monohalobenzene. Orientation in arenes which bear two or more substituents is generally controlled by the directing effect of the more powerful activating substituent.

When electrophilic aromatic substitution reactions are used in multistep syntheses, careful thought needs to be given to the order of introduction of various substituents, since they can affect the observed regiochemistry. Regiochemistry can sometimes be controlled by using functional group manipulations to alter the directing properties of substituents.

PROBLEMS

13.15 Give reagents suitable for effecting each of the following reactions and write the principal products. If an ortho-para mixture is expected, show both. If the meta isomer is the expected major product, write only that isomer.

(a) Nitration of benzene
(b) Nitration of the product of (a)
(c) Bromination of toluene
(d) Bromination of (trifluoromethyl)benzene
(e) Sulfonation of anisole

(f) Sulfonation of acetanilide ($C_6H_5NHCCH_3$, with $\overset{O}{\underset{\|}{}}$ on the C)
(g) Chlorination of bromobenzene
(h) Friedel-Crafts alkylation of anisole with benzyl chloride
(i) Friedel-Crafts acylation of benzene with benzoyl chloride
(j) Nitration of the product from (i)
(k) Clemmensen reduction of the product from (i)
(l) Wolff-Kishner reduction of the product from (i)

13.16 Write a structural formula for the most stable cyclohexadienyl cation intermediate formed in each of the following reactions. Is the cyclohexadienyl cation more or less stable than the corresponding intermediate formed by electrophilic attack on benzene?

(a) Bromination of p-xylene
(b) Chlorination of m-xylene
(c) Nitration of acetophenone

(d) Friedel-Crafts acylation of anisole with CH_3CCl (with $\overset{O}{\underset{\|}{}}$ on the C)
(e) Nitration of isopropylbenzene
(f) Bromination of nitrobenzene

13.17 In each of the following pairs of compounds choose which one will react faster with the indicated reagent and write a chemical equation for the faster reaction:

(a) Toluene or chlorobenzene with nitric acid and sulfuric acid
(b) Fluorobenzene or (trifluoromethyl)benzene with benzyl chloride and aluminum chloride

(c) Methyl benzoate ($C_6H_5\overset{\displaystyle O}{\overset{\|}{C}}OCH_3$) or phenyl acetate ($C_6H_5O\overset{\displaystyle O}{\overset{\|}{C}}CH_3$) with bromine in acetic acid

(d) Acetanilide ($C_6H_5NH\overset{\displaystyle O}{\overset{\|}{C}}CH_3$) or nitrobenzene with sulfur trioxide in sulfuric acid

(e) *p*-Dimethylbenzene (*p*-xylene) or *p*-di-*tert*-butylbenzene with acetyl chloride and aluminum chloride

(f) Benzophenone ($C_6H_5\overset{\displaystyle O}{\overset{\|}{C}}C_6H_5$) or biphenyl ($C_6H_5—C_6H_5$) with chlorine and ferric chloride

13.18 Arrange the following five compounds in order of decreasing rate of bromination: benzene, toluene, *o*-xylene, *m*-xylene, 1,3,5-trimethylbenzene (the relative rates are 2×10^7, 5×10^4, 5×10^2, 60, and 1).

13.19 Each of the following reactions has been carried out under conditions such that disubstitution or trisubstitution occurred. Identify the principal organic product in each case.

(a) Nitration of *p*-chlorobenzoic acid (disubstitution)
(b) Bromination of aniline (trisubstitution)
(c) Bromination of *o*-aminoacetophenone (disubstitution)
(d) Nitration of benzoic acid (disubstitution)
(e) Bromination of *p*-nitrophenol (disubstitution)
(f) Reaction of biphenyl with *tert*-butyl chloride and ferric chloride (disubstitution)
(g) Sulfonation of phenol (disubstitution)

13.20 Write equations showing how you could prepare each of the following from benzene or toluene and any necessary organic or inorganic reagents. If an ortho, para mixture is formed in any step of your synthesis, assume that you can separate the two isomers.

(a) Isopropylbenzene
(b) *p*-Isopropylbenzenesulfonic acid
(c) 2-Bromo-2-phenylpropane
(d) 4-*tert*-Butyl-2-nitrotoluene
(e) *m*-Chloroacetophenone
(f) *p*-Chloroacetophenone
(g) 3-Bromo-4-methylacetophenone
(h) 2-Bromo-4-ethyltoluene
(i) 1-Bromo-3-nitrobenzene
(j) 1-Bromo-2,4-dinitrobenzene
(k) 3-Bromo-5-nitrobenzoic acid
(l) 2-Bromo-4-nitrobenzoic acid
(m) Diphenylmethane
(n) 1-Phenyloctane
(o) 1-Phenyl-1-octene
(p) 1-Phenyl-1-octyne

13.21 Write equations showing how you could prepare each of the following from anisole and any necessary organic or inorganic reagents. If an ortho, para mixture is formed in any step of your synthesis, assume that you can separate the two isomers.

(a) *p*-Methoxybenzenesulfonic acid

(b)   2-Bromo-4-nitroanisole

(c)   4-Bromo-2-nitroanisole

(d)   *p*-Methoxystyrene

13.22 How many products are capable of being formed from toluene in each of the following reactions?

(a)   Mononitration (HNO_3, H_2SO_4, 40°C)

(b)   Dinitration (HNO_3, H_2SO_4, 80°C)

(c)   Trinitration (HNO_3, H_2SO_4, 110°C)

The explosive TNT (trinitrotoluene) is the major product obtained on trinitration of toluene. Which trinitrotoluene isomer is TNT?

13.23 Friedel-Crafts acylation of the individual isomers of xylene with acetyl chloride and aluminum chloride yields a single product, different for each xylene isomer, in high yield in each case. Write the structures of the products of acetylation of *o*-, *m*-, and *p*-xylene.

13.24 Each of the following reactions has been reported in the chemical literature and gives a predominance of a single product in synthetically acceptable yield. Write the structure of the product. Only monosubstitution is involved in each case unless otherwise indicated.

(a)

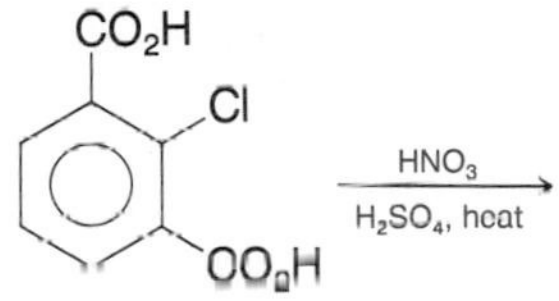

(b)

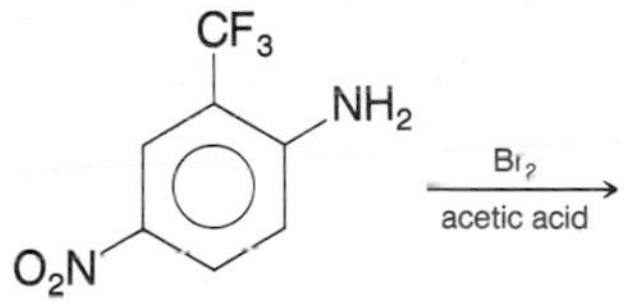

(c)

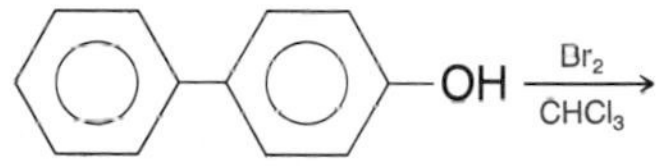

(d)

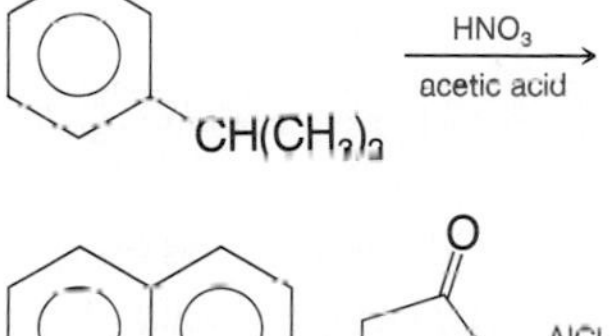

(e)

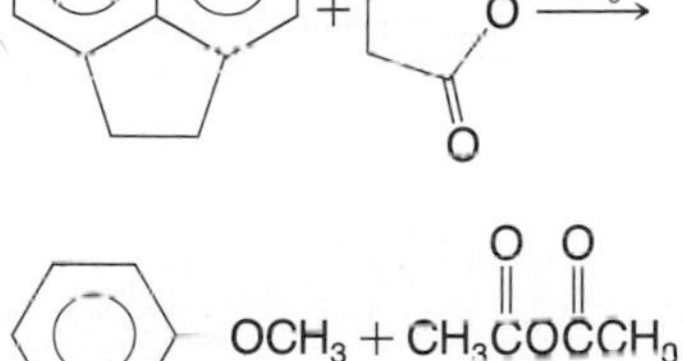

(f)

(g)

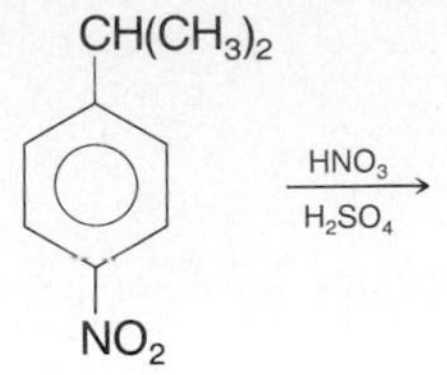

$\xrightarrow[\text{H}_2\text{SO}_4]{\text{HNO}_3}$

(h)

$+ \ (\text{CH}_3)_2\text{C}{=}\text{CH}_2 \ \xrightarrow{\text{H}_2\text{SO}_4}$

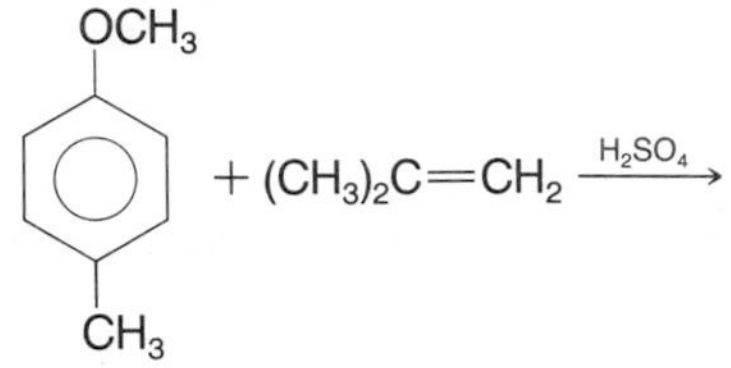

(i)

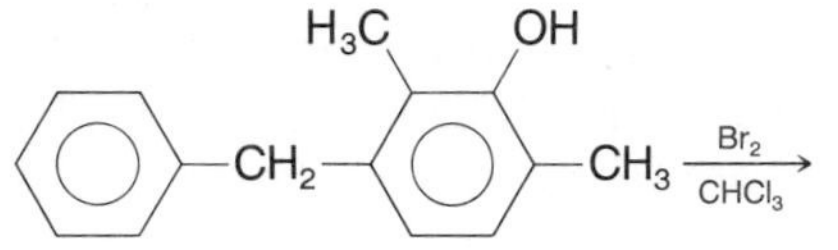

$\xrightarrow[\text{CHCl}_3]{\text{Br}_2}$

(j)

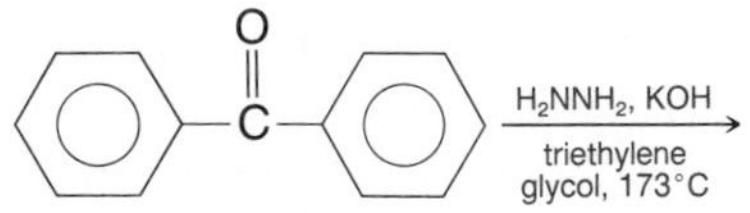

$\xrightarrow[\substack{\text{triethylene} \\ \text{glycol, 173°C}}]{\text{H}_2\text{NNH}_2, \ \text{KOH}}$

(k)

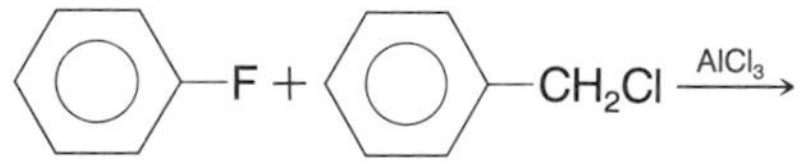

$\xrightarrow{\text{AlCl}_3}$

(l)

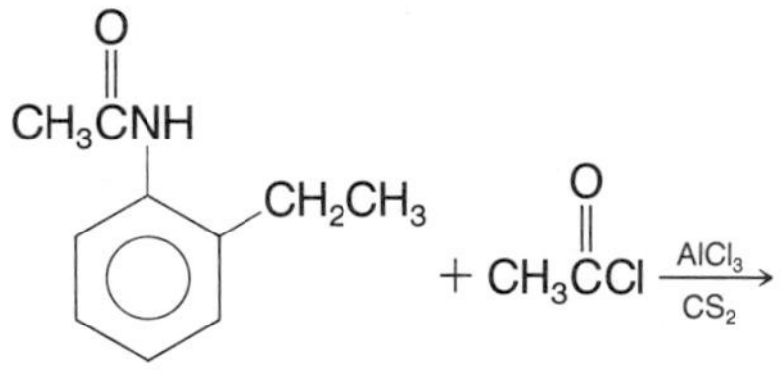

$+ \ \text{CH}_3\overset{\text{O}}{\overset{\|}{\text{C}}}\text{Cl} \ \xrightarrow[\text{CS}_2]{\text{AlCl}_3}$

(m)

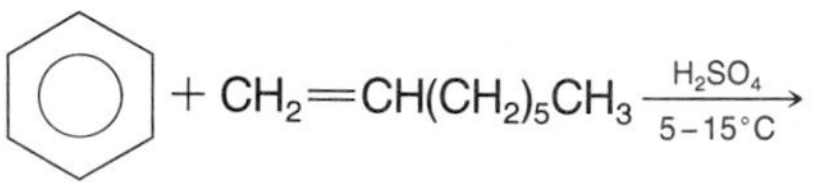

$+ \ \text{CH}_2{=}\text{CH(CH}_2)_5\text{CH}_3 \ \xrightarrow[5-15°\text{C}]{\text{H}_2\text{SO}_4}$

(n)

$\xrightarrow[\text{HCl}]{\text{Zn(Hg)}}$

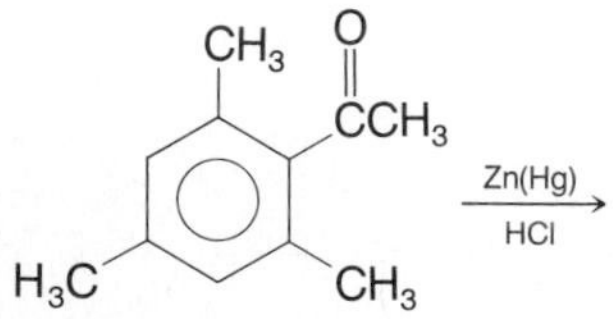

13.25 What combination of acyl chloride or acid anhydride and arene would you choose to prepare each of the following compounds by a Friedel-Crafts acylation reaction?

(a) $\text{C}_6\text{H}_5\overset{\text{O}}{\overset{\|}{\text{C}}}\text{CH}_2\text{C}_6\text{H}_5$

(b)

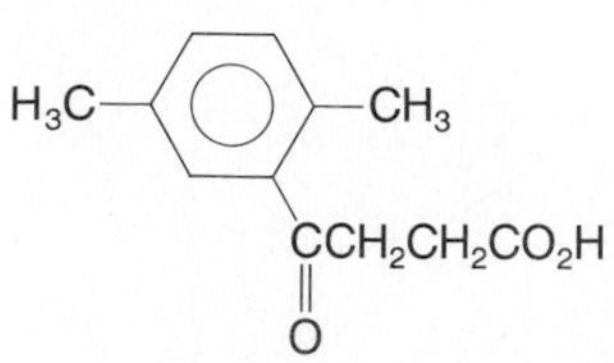

(c) O_2N-⬡$-\overset{\overset{O}{\|}}{C}-$⬡

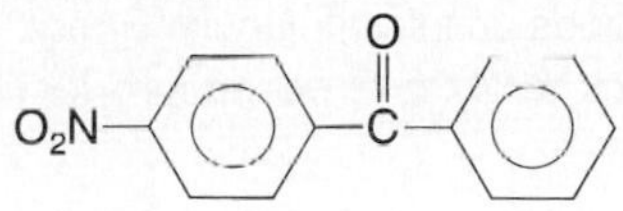

(e) 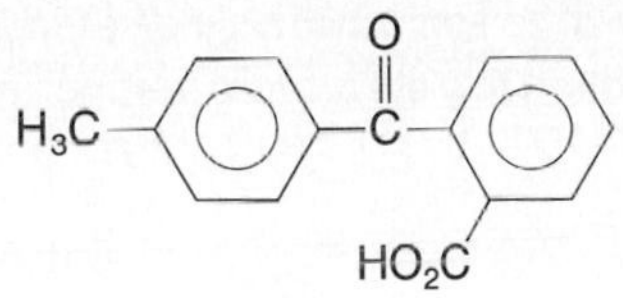 H_3C-⬡$-\overset{\overset{O}{\|}}{C}-$⬡ (with HO_2C substituent)

(d) 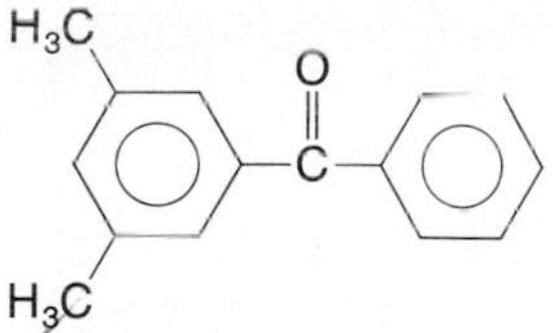   (ring with H_3C and H_3C substituents)$-\overset{\overset{O}{\|}}{C}-$⬡

13.26 Suggest a suitable series of reactions for carrying out each of the following synthetic transformations:

(a) 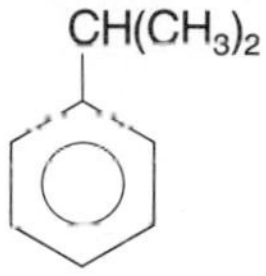 $CH(CH_3)_2$-substituted benzene   to   benzene with CO_2H and SO_3H

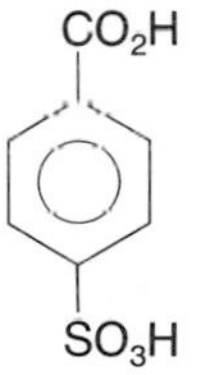

(b) 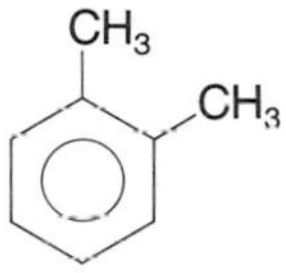   o-xylene (CH_3, CH_3)   to   benzene with CO_2H, CO_2H, $C(CH_3)_3$

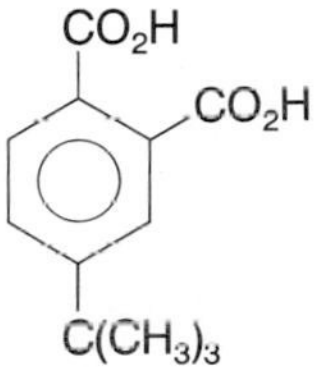

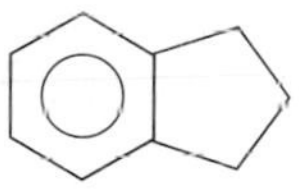

(c)   indane   to   indane with $\overset{\overset{O}{\|}}{C}CH_3$ and $(CH_3)_3C$ substituents

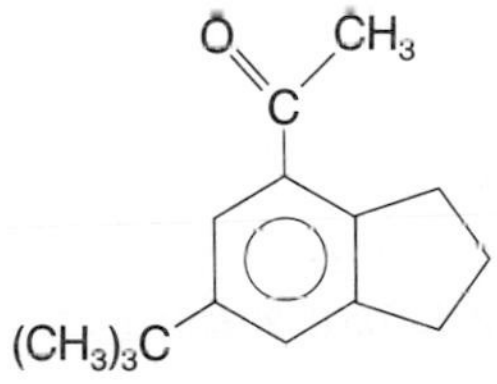

(d) 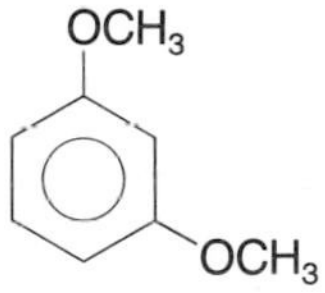   1,3-dimethoxybenzene (OCH_3, OCH_3)   to   benzene with OCH_3, O_2N, OCH_3, $C(CH_3)_3$

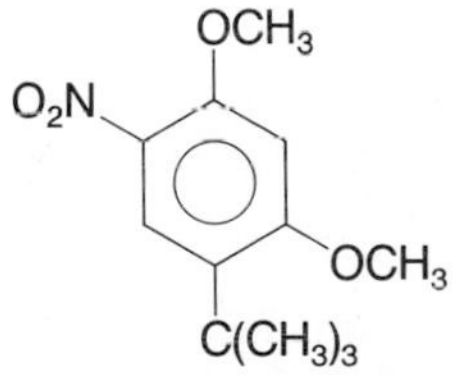

13.27 A standard synthetic sequence for building a six-membered cyclic ketone onto an existing aromatic ring is shown in outline below. Specify the reagents necessary for each step.

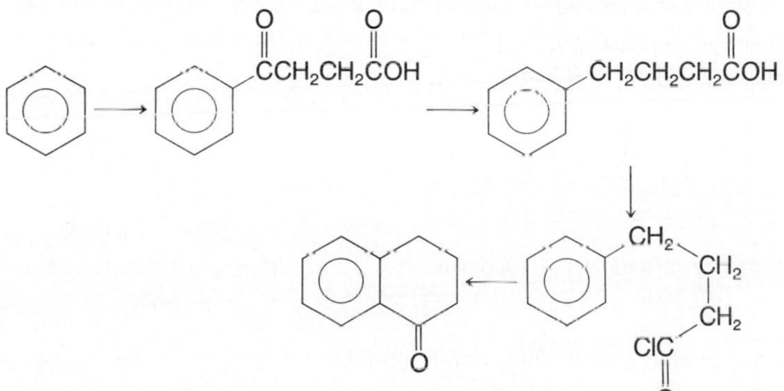

⬡ → ⬡$-\overset{\overset{O}{\|}}{C}CH_2CH_2\overset{\overset{O}{\|}}{C}OH$ → ⬡$-CH_2CH_2CH_2\overset{\overset{O}{\|}}{C}OH$

→ ⬡$-CH_2-CH_2-CH_2-\overset{\overset{O}{\|}}{C}Cl$ → (tetralone ketone)

13.28 Each of the compounds indicated below undergoes an intramolecular Friedel-Crafts acylation reaction to yield a cyclic ketone. Write the structure of the expected product in each case.

(a)

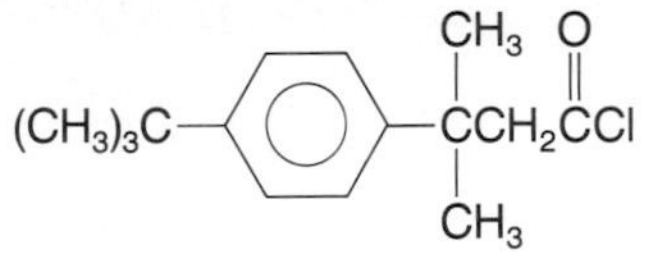

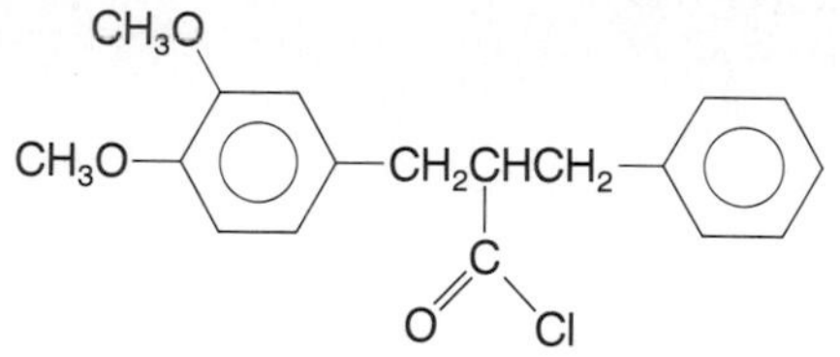

(b)

(c)

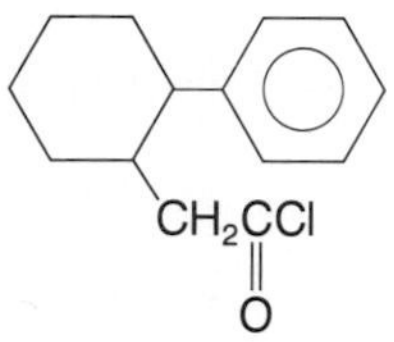

13.29 The relative rates of attack at the individual positions of biphenyl compared with that at a single position of benzene (taken as 1) in electrophilic chlorination are as shown.

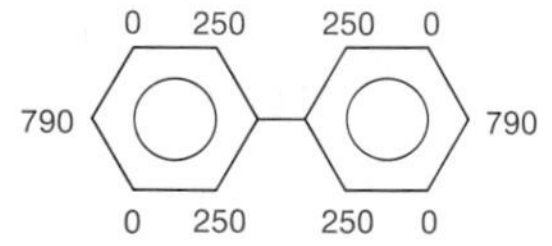

(a) What is the relative rate of chlorination of biphenyl compared with benzene?
(b) If, in a particular chlorination reaction, 10 g of *o*-chlorobiphenyl were formed, how much *p*-chlorobiphenyl would you expect to find?

13.30 When 2-isopropyl-1,3,5-trimethylbenzene is heated with aluminum chloride (trace of HCl present) at 50°C, the major material present after 4 h is 1-isopropyl-2,4,5-trimethylbenzene. Suggest a reasonable mechanism for this isomerization.

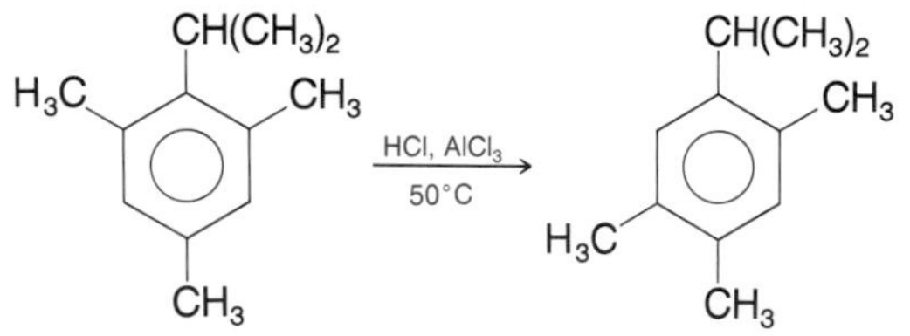

13.31 When a dilute solution of 6-phenylhexanoyl chloride in carbon disulfide was slowly added (over a period of 8 days!) to a suspension of aluminum chloride in the same solvent, it yielded a product A ($C_{12}H_{14}O$) in 67 percent yield. Heating A with potassium permanganate gave benzene-1,2-dicarboxylic acid.

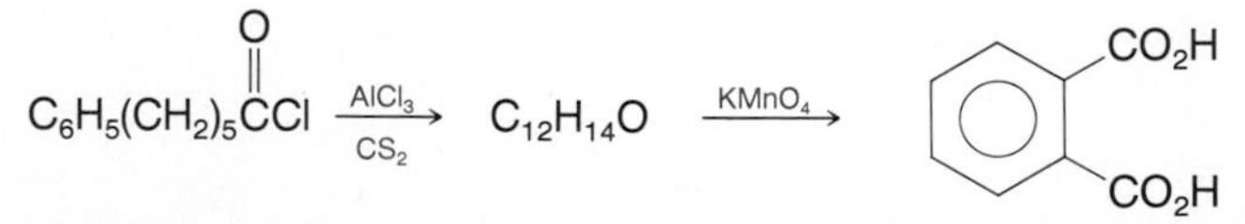

| 6-Phenylhexanoyl chloride | Compound A | Benzene-1,2-dicarboxylic acid |

Formulate a reasonable structure for compound A.

13.32 Reaction of hexamethylbenzene with methyl chloride and aluminum chloride gave a salt B, which, on being treated with aqueous sodium bicarbonate solution, yielded compound C. Suggest a mechanism for the conversion of hexamethylbenzene to C by correctly inferring the structure of B.

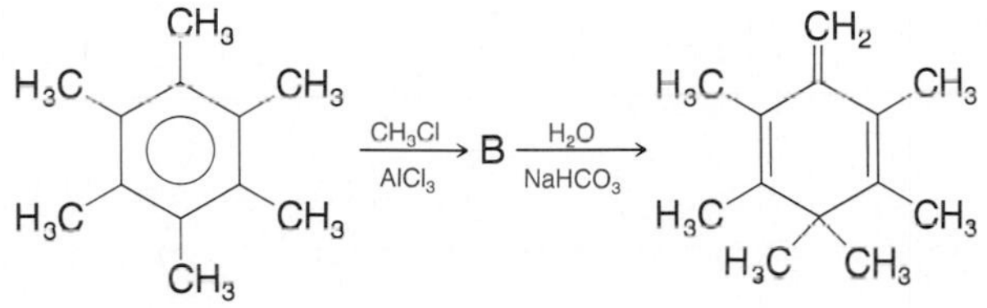

Hexamethylbenzene   Compound C

13.33 The synthesis of compound F was achieved by using compounds D and E as the sources of all carbon atoms. Suggest a synthetic sequence involving no more than three steps by which D and E may be converted to F.

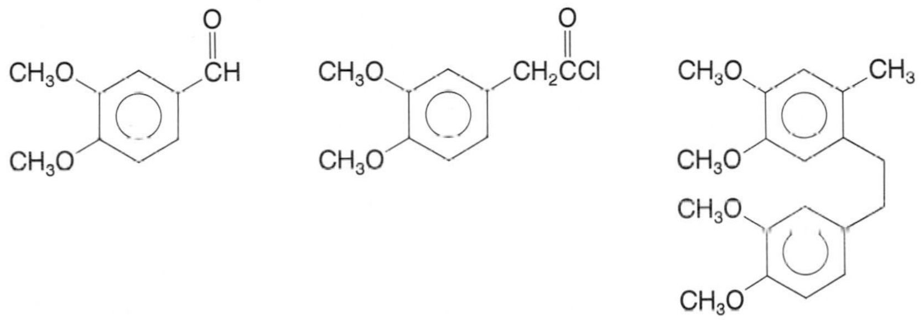

Compound D   Compound E   Compound F

13.34 When styrene is refluxed with aqueous sulfuric acid, two "styrene dimers" are formed as the major products. One of these styrene dimers is 1,3-diphenyl-1-butene; the other is 1-methyl-3-phenylindan. Suggest a reasonable mechanism for the formation of each of these compounds.

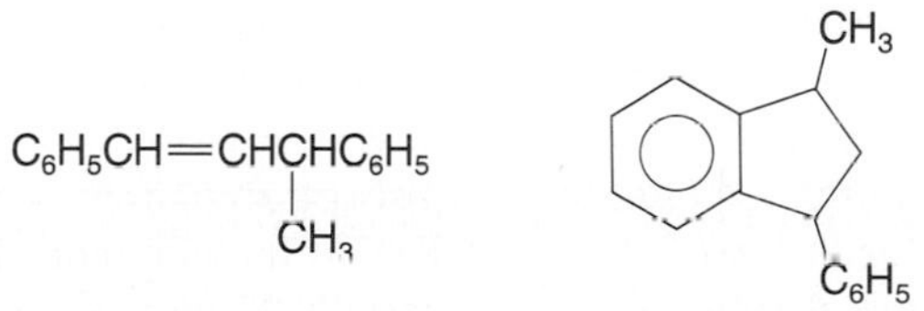

$C_6H_5CH{=}CHCHC_6H_5$
            |
           CH_3

1,3-Diphenyl-1-butene   1-Methyl-3-phenylindan

13.35 Treatment of the alcohol whose structure is given below with sulfuric acid gave as the major organic product a tricyclic hydrocarbon of molecular formula $C_{16}H_{16}$. Suggest a reasonable structure for this hydrocarbon.

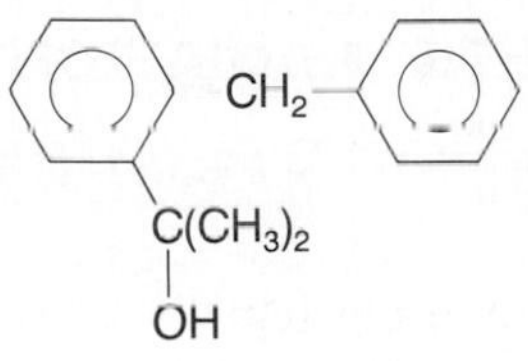

SPECTROSCOPY

$\mathbf{P}$rior to the second half of this century, the structure of a substance—a new natural product, for example—was determined using information obtained from chemical reactions. This information included the identification of functional groups by chemical tests, along with the results of degradation experiments in which the substance was cleaved into smaller, more readily identified fragments. Typical of this approach is the demonstration of the presence of a double bond in an alkene by catalytic hydrogenation and subsequent determination of its location by ozonolysis. After considering all the available chemical evidence, the chemist proposed a candidate structure (or structures) consistent with the observations. Proof of structure was provided either by converting the substance to some already known compound or by an independent synthesis.

Qualitative tests and chemical degradation as structural probes have been supplemented and to a large degree superseded in present-day organic chemistry by instrumental methods of structure determination. The most prominent of these techniques are *nuclear magnetic resonance (nmr) spectroscopy, infrared (ir) spectroscopy, ultraviolet-visible (uv-vis) spectroscopy, and mass spectrometry (ms).* As diverse as these techniques are, all of them are based on the absorption of energy by a molecule, and all examine how a molecule responds to that absorption of energy. In describing these techniques in the present chapter our emphasis will be on their application to the task of structure determination. We begin with a brief discussion of the nature of electromagnetic radiation, a topic that is fundamental to understanding the physical bases upon which molecular spectroscopy depends.

14.1 PRINCIPLES OF MOLECULAR SPECTROSCOPY. ELECTROMAGNETIC RADIATION

Electromagnetic radiation, of which visible light is but one example, is said to have a dual nature. It has the properties of both particles and waves. The particles are called *photons* and each possesses an amount of energy referred to as a *quantum*. In 1900 the

German physicist Max Planck related the energy of a photon to its frequency by the expression

$$E = h\nu$$

where E and ν are the energy and the frequency, respectively, of a photon. The units of frequency are reciprocal seconds (s^{-1}). These units are sometimes referred to as cycles per second, but the modern term is Hz (after the nineteenth century physicist Heinrich R. Hertz). The constant of proportionality h is called *Planck's constant* and has the value

$$h = 6.62 \times 10^{-27} \text{ erg} \cdot \text{sec}$$

The equation gives the energy of a photon in ergs; 1 erg per molecule is equivalent to 1.439×10^{13} kcal/mol.

Electromagnetic radiation is propagated at the speed of light. The speed of light (c) is 3.0×10^8 m/s and is equal to the product of the frequency ν and the wavelength λ.

$$c = \nu\lambda$$

The range of photon energies is called the electromagnetic spectrum and is shown in Figure 14.1. Visible light occupies a very small region of the electromagnetic

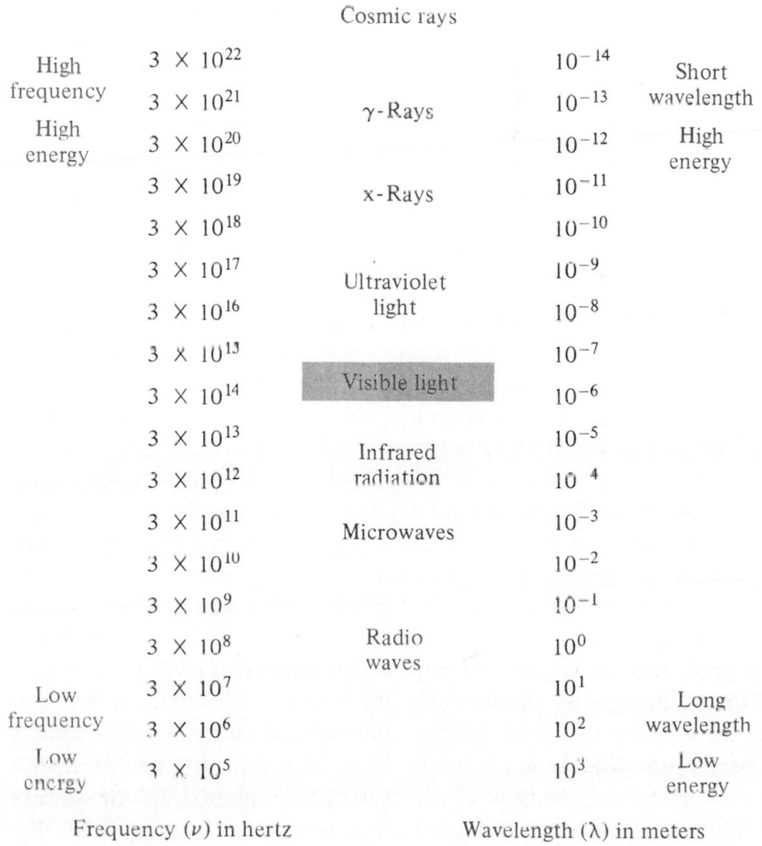

	Frequency (ν) in hertz		Wavelength (λ) in meters	
		Cosmic rays		
High frequency	3×10^{22}		10^{-14}	Short wavelength
	3×10^{21}	γ-Rays	10^{-13}	
High energy	3×10^{20}		10^{-12}	High energy
	3×10^{19}	x-Rays	10^{-11}	
	3×10^{18}		10^{-10}	
	3×10^{17}	Ultraviolet light	10^{-9}	
	3×10^{16}		10^{-8}	
	3×10^{15}		10^{-7}	
	3×10^{14}	Visible light	10^{-6}	
	3×10^{13}	Infrared radiation	10^{-5}	
	3×10^{12}		10^{-4}	
	3×10^{11}	Microwaves	10^{-3}	
	3×10^{10}		10^{-2}	
	3×10^{9}		10^{-1}	
	3×10^{8}	Radio waves	10^{0}	
Low frequency	3×10^{7}		10^{1}	Long wavelength
	3×10^{6}		10^{2}	
Low energy	3×10^{5}		10^{3}	Low energy

FIGURE 14.1 The electromagnetic spectrum.

spectrum. It is characterized by wavelengths of 4×10^{-7} m (violet) to 8×10^{-7} m (red). When examining Figure 14.1, it is helpful to keep the following two relationships in mind:

1. Frequency is inversely proportional to wavelength; the greater the frequency, the shorter the wavelength.
2. Energy is proportional to frequency; electromagnetic radiation of higher frequency possesses more energy than radiation of lower frequency.

Depending on its source, a photon can have a vast amount of energy; cosmic rays and x-rays are streams of very high energy photons. Radio waves are of relatively low energy. Ultraviolet radiation is of higher energy than the violet end of visible light. Infrared radiation is of lower energy than the red end of visible light. When a molecule is exposed to electromagnetic radiation, it may absorb a photon, thereby increasing its energy by an amount equal to the energy of the photon. Molecules are highly selective with respect to the frequencies of radiation that they absorb. Only photons of certain specific frequencies are absorbed by a molecule. The particular photon energies absorbed by a molecule depend on molecular structure and can be measured with instruments called *spectrometers*. The data obtained are very sensitive indicators of molecular structure and have revolutionized the practice of chemical analysis.

14.2 PRINCIPLES OF MOLECULAR SPECTROSCOPY. QUANTIZED ENERGY STATES

What determines whether a photon is absorbed by a molecule or not? The most important condition is that the energy of the photon must match the energy difference between two states of the molecule. Consider, for example, the two energy states designated E_1 and E_2 in Figure 14.2. The energy difference between them is $E_2 - E_1$, or ΔE. In nuclear magnetic resonance (nmr) spectroscopy these are two different spin states of an atomic nucleus; in infrared (ir) spectroscopy, they are two different vibrational energy states; in ultraviolet-visible (uv-vis) spectroscopy, they are two different electronic energy states. Unlike kinetic energy, which is continuous, meaning that all values of kinetic energy are available to a molecule, only certain energies are possible for electronic, vibrational, and nuclear spin states. These energy states are said to be *quantized*. More of the molecules exist in the lower energy state E_1 than in the higher energy state E_2. Excitation of a molecule from a lower state to a higher one requires the addition of an increment of energy equal to ΔE. Thus, when electromagnetic radiation is incident upon a molecule, only the frequency whose corresponding energy equals ΔE is absorbed. All other frequencies are transmitted.

Spectrometers are designed to measure the absorption of electromagnetic radiation by a sample. Basically, a spectrometer consists of a source of radiation, a compartment containing the sample through which the radiation passes, and a detector. The frequency of radiation is continuously varied, and its intensity at the detector is compared with that at the source. When the frequency is reached at which the sample absorbs radiation, the detector senses a decrease in intensity. The relation between frequency and absorption is plotted on a strip chart and is called a *spectrum*. A spectrum consists of a series of peaks at particular frequencies; its interpretation can provide structural information. Each type of spectroscopy developed independently of the others, so the format followed in presenting the data is different for each one.

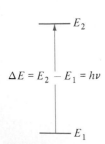

$$\Delta E = E_2 - E_1 = h\nu$$

FIGURE 14.2 Two energy states of a molecule. Absorption of energy equal to $E_2 - E_1$ excites a molecule from its lower energy state to the next higher state.

An nmr spectrum looks different from an ir spectrum, and both look different from a uv-vis spectrum.

With this general background, we will now discuss spectroscopic techniques individually. Nmr, ir, and uv-vis spectroscopy provide complementary information, and all are useful. Among them, nmr provides the information most directly related to molecular structure and is the one we shall examine first.

14.3   PROTON MAGNETIC RESONANCE (^{1}H NMR) SPECTROSCOPY

As noted in the preceding section, nmr spectroscopy is based on transitions between _nuclear spin states_. What do we mean by nuclear spin and how can it provide structural information?

A proton, like an electron, possesses the property of spin. Also, like an electron, a proton has two (nuclear) spin states with spin quantum numbers of $+\frac{1}{2}$ and $-\frac{1}{2}$. There is no difference in energy between these two nuclear spin states; a proton is just as likely to have a spin of $+\frac{1}{2}$ as $-\frac{1}{2}$. Since both spin states have the same energy, transitions between the two seem to offer no promise as the basis for a spectroscopic technique. There is a simple way, however, to render the energies of the two nuclear spin states unequal and that is by placing the proton in a strong magnetic field.

A proton is a spinning charge and has associated with it a magnetic moment coinciding with the axis of rotation (Figure 14.3). In the presence of an external magnetic field H_0, the two nuclear spin states no longer have the same energy. The state in which the nuclear magnetic moment is aligned with the external field is lower in energy than the state in which it opposes the applied field. As depicted in Figure 14.4, the difference in energy between the two states is directly proportional to the strength of the applied field. Even in very powerful magnetic fields, the energy difference is quite small. At a field strength of 14,100 gauss the energy difference between the two proton spin states is only 6×10^{-6} kcal/mol. At a field strength of 23,500 gauss the energy difference between the two states is 10^{-5} kcal/mol. Energy differences of this magnitude correspond to the radiofrequency region of the electromagnetic spectrum (see Figure 14.1).

Figure 14.5 illustrates the essential features of a nuclear magnetic resonance spectrometer. At its heart is a powerful magnet, either a permanent magnet or an electromagnet. Consider the situation in which it is a permanent magnet of field strength 14,100 gauss. The sample is placed between the pole faces of the magnet, which causes the nuclear spins to align themselves either with the field or against the field. There is a small excess of nuclei that have their spins aligned with the field as compared with those aligned against it. For a sample of 1 million protons at 25°C and 14,100 gauss, approximately 500,005 are in the lower-energy state compared with 499,995 in the higher state. The sample cavity is surrounded by a radiofrequency

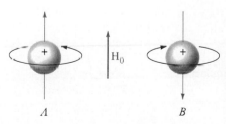

A B

FIGURE 14.3 Nuclear spin states of a proton. Spin state A, in which the nuclear magnetic moment is parallel to the applied field H_0, is of lower energy than spin state B, in which the nuclear magnetic moment is antiparallel to the applied field. *(From Pine, Hendrickson, Cram, and Hammond, Organic Chemistry, McGraw-Hill, New York, 1980.)*

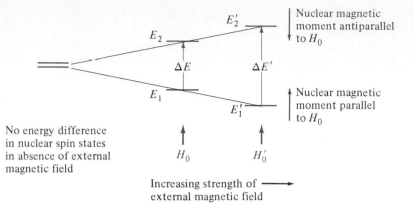

FIGURE 14.4 An external magnetic field causes the two nuclear spin states to have different energies. The difference in energy ΔE is proportional to the strength of the applied field.

source and the frequency is continuously varied. When the frequency of the source matches the energy difference between nuclear spin states, the two are said to be in *resonance* with each other. For protons at 14,100 gauss, this frequency is approximately 60×10^6 Hz, or 60 MHz. Energy is absorbed and nuclei undergo a *spin flip* from the lower-energy orientation to the higher one. The absorption of energy is detected in a radiofrequency receiver and shown as a peak on the nmr spectrum.

Nuclear magnetic resonance spectrometers that use electromagnets operate in a complementary manner. The frequency of the source is maintained at a constant value, say 60 MHz, and the magnetic field strength is varied until the energy gap between spin states matches that of the source. Both types of spectrometers are available, and the nmr spectra from permanent magnet instruments are identical to those from electromagnet spectrometers. Spectrometers that operate at higher field strengths, corresponding to 90 MHz, 100 MHz, and 220 MHz for protons, are also fairly common in research laboratories.

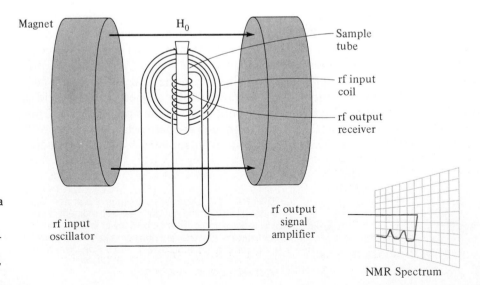

FIGURE 14.5 Diagram of a nuclear magnetic resonance spectrometer. *(From Pine, Hendrickson, Cram, and Hammond, Organic Chemistry, McGraw-Hill, New York, 1980, p. 136.)*

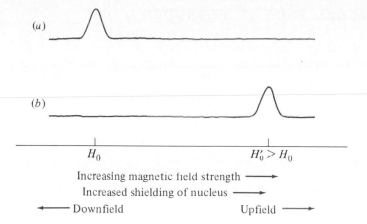

14.4 NUCLEAR SHIELDING AND CHEMICAL SHIFT

Our discussion to this point has centered on transitions between the two spin states of a "bare" proton, that is, an isolated hydrogen nucleus. In real molecules a proton is bonded to another atom by a two-electron covalent bond. These electrons, indeed all the electrons in a molecule, affect the magnetic environment of the proton. Since our concern is organic chemistry, let us examine the difference between a bare proton and a proton in some organic compound.

$$H^+$$
Bare proton

Proton in an organic molecule

Assuming that we carry out the nmr experiment under conditions such that the radiofrequency source emits at a constant frequency of 60 MHz and the magnetic field strength is slowly increased until the resonance condition is realized, we find that a bare proton undergoes its spin flip at some value of the applied field H_0 close to 14,100 gauss. Figure 14.6a is an idealized representation of the spectrum obtained. Under the same conditions, as shown in Figure 14.6b, we find that a field strength slightly greater than H_0 is required to flip the spin of the proton in an organic molecule. We say that the nmr signal of the bound proton appears at higher field than the signal of the bare proton and call this difference between the two a *chemical shift.* *A chemical shift is the change in the resonance position of a nucleus which is brought about by its molecular environment.* (A peak at higher field than another is said to be *upfield;* a peak at lower field than another is said to be *downfield.*)

The reason that higher field strengths are required to achieve the same separation of energy levels of bound protons compared with bare protons is that the molecule's electrons *shield* the bound proton from the external field. Under the influence of an external field there is an induced magnetic field associated with the electrons that opposes the applied field. This is illustrated in Figure 14.7. Thus, the magnetic field "felt" by a proton is less than the actual field strength H_0. In order to achieve a separation between levels equal to the energy of the radiofrequency radiation, the strength of the applied field must be increased by an amount equal to the strength of the opposing induced magnetic fields of the electrons. Increased shielding of a nucleus requires higher magnetic field strengths to achieve resonance.

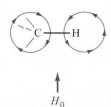

FIGURE 14.7 The induced magnetic field of the electrons in the carbon-hydrogen bond opposes the external magnetic field. The resultant magnetic field experienced by the proton is slightly less than H_0.

14.5 HOW CHEMICAL SHIFT IS MEASURED

Chemists compare the shielding of protons in organic molecules by specifying their chemical shifts relative to a standard substance. This substance is tetramethylsilane $(CH_3)_4Si$, abbreviated as TMS. The protons of TMS are more shielded than those of almost all organic compounds. In a solution containing TMS, all the relevant signals appear to the left of the TMS peak. The orientation of the spectrum on the chart is adjusted electronically so that the TMS peak coincides with the zero grid line. Peak positions are measured in frequency units (hertz) downfield from the TMS peak. (Frequency units are more convenient to use than units of magnetic field strength. The two sets of units are directly proportional to each other.)

Figure 14.8 is the 60-MHz nmr spectrum of chloroform $(CHCl_3)$ containing a few drops of TMS. The signal due to the proton in chloroform appears 437 Hz downfield from the TMS peak. By common agreement, chemical shifts (δ) are reported in parts per million (ppm) from the TMS peak.

$$\text{Chemical shift } (\delta) = \frac{\text{position of signal} - \text{position of TMS peak}}{\text{spectrometer frequency}} \times 10^6$$

Thus, the chemical shift for the proton in chloroform is:

$$\delta = \frac{437 \text{ Hz} - 0 \text{ Hz}}{60 \times 10^6 \text{ Hz}} \times 10^6 = 7.28 \text{ ppm}$$

Nuclear magnetic resonance spectra are recorded on chart paper that is calibrated in both parts per million (ppm) and hertz, and both are referred to the TMS peak as the zero point. When reporting chemical shifts in frequency units (hertz) the field strength of the instrument must be specified. A 60-MHz nmr spectrometer separates the energy of nuclear spin states only 60 percent as much as does a 100-MHz

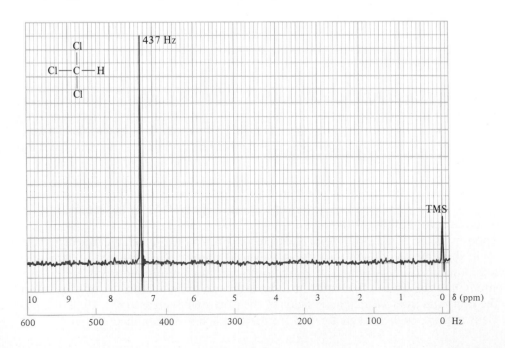

FIGURE 14.8 The proton magnetic resonance (¹H nmr) spectrum of chloroform $(CHCl_3)$.

spectrometer. Thus, a chemical shift of 437 Hz with a 60-MHz instrument appears at 728 Hz with a 100-MHz spectrometer. By reporting chemical shifts in parts per million, this effect of field strength is taken into account; thus irrespective of magnetic field strength, the signal due to the proton of chloroform appears at 7.28 ppm. All chemical shifts in this text are reported in parts per million.

PROBLEM 14.1   Calculate the chemical shift in parts per million for each of the following compounds, given their shifts in hertz as measured downfield from tetramethylsilane on a spectrometer of 60-MHz field strength.

  (a)   Bromoform (CHBr$_3$), 413 Hz
  (b)   Iodoform (CHI$_3$), 322 Hz
  (c)   Methyl chloride (CH$_3$Cl), 184 Hz

SAMPLE SOLUTION   (a)   The chemical shift in bromoform is calculated from the equation given above:

$$\delta = \frac{413 \text{ Hz (CHBr}_3) - 0 \text{ Hz (TMS)}}{60 \times 10^6 \text{ Hz}} \times 10^6 = 6.88 \text{ ppm}$$

Although chloroform is a good solvent for organic compounds, it is not widely used as a medium for measuring nmr spectra; its own nmr signal could potentially obscure a signal in the sample. Chloroform-d (CDCl$_3$) is used instead. Because the magnetic properties of deuterium, the mass 2 isotope of hydrogen (D = ^{2}H), are different from those of protium (^{1}H), chloroform-d does not give an nmr signal under the conditions employed for measuring proton magnetic resonance spectra. Chloroform-d exhibits no peaks in the spectrum that would obscure those of an organic molecule being examined.

14.6   CHEMICAL SHIFT AND MOLECULAR STRUCTURE

What makes nuclear magnetic resonance spectroscopy such a powerful tool for structure determination is that protons in different environments experience different degrees of shielding and have different chemical shifts. In compounds of the type CH$_3$X, for example, atoms or groups that we recognize as electronegative cause the methyl protons to be less shielded than those in which X is not an electron-attracting substituent. In the series of methyl halides, the protons of methyl iodide are the most shielded while those of methyl fluoride are the least shielded.

> Increased shielding of methyl protons
> Decreasing electronegativity of halogen

	CH$_3$F	CH$_3$Cl	CH$_3$Br	CH$_3$I
	Methyl fluoride	Methyl chloride	Methyl bromide	Methyl iodide
Chemical shift of methyl protons (δ), ppm:	4.3	3.1	2.7	2.2

Similarly, in a group of compounds in which a methyl group is bonded to elements in the same row of the periodic table, shielding increases as electronegativity decreases.

> Increased shielding of methyl protons
> Decreasing electronegativity of attached atom

	CH_3F	CH_3OCH_3	$(CH_3)_3N$	CH_3CH_3
	Methyl fluoride	Dimethyl ether	Trimethylamine	Ethane
Chemical shift of methyl protons (δ), ppm:	4.3	3.2	2.2	0.9

An electronegative substituent polarizes the electron distribution in its bond to carbon, decreases the electron density at the methyl carbon, and decreases the shielding of the methyl protons. The effects are cumulative, as the chemical shift data for the various chlorinated derivatives of methane indicate.

	$CHCl_3$	CH_2Cl_2	CH_3Cl
	Chloroform (trichloromethane)	Methylene chloride (dichloromethane)	Methyl chloride (chloromethane)
Chemical shift (δ), ppm:	7.3	5.3	3.1

Vinyl protons in alkenes and aryl protons in arenes are substantially less shielded than protons in alkanes.

	Benzene	Ethylene	Ethane
Chemical shift (δ), ppm:	7.3	5.3	0.9

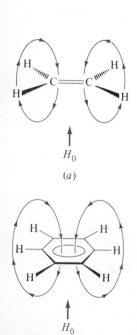

FIGURE 14.9 The induced magnetic field of the π electrons of (a) an alkene and (b) an arene reinforce the applied fields in the regions where vinyl and aryl protons are located.

One reason for the decreased shielding of vinyl and aryl protons is related to the directional properties of the induced magnetic field of the π electrons. As Figure 14.9 shows, the induced magnetic field due to the π electrons is just like that due to electrons in σ bonds; it opposes the applied magnetic field. However, all magnetic fields close upon themselves, and protons attached to a carbon-carbon double bond or an aromatic ring lie in a region where the induced field reinforces the applied field. This diminishes the shielding of vinyl and aryl protons.

A similar, although much smaller, effect of π-electron systems is seen in the chemical shifts of benzylic and allylic hydrogens. The methyl hydrogens in hexamethylbenzene and in 2,3-dimethyl-2-butene are less shielded than those in ethane.

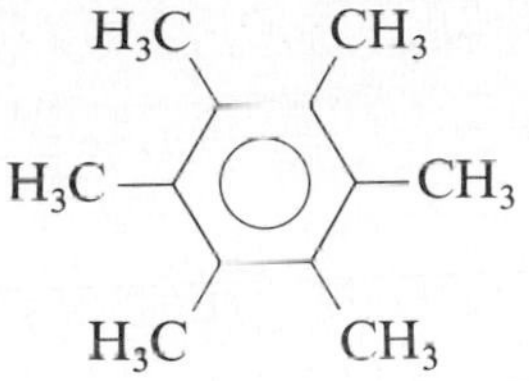

$$\text{Hexamethylbenzene} \qquad \text{2,3-Dimethyl-2-butene}$$

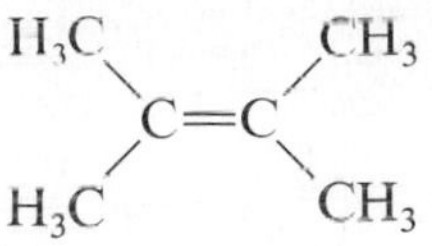

Chemical shift
(δ), ppm: 2.2   1.7

Table 14.1 collects chemical shift information for protons of various types. Within each type, methyl (CH_3) protons are more shielded than methylene (CH_2) protons and methylene protons are more shielded than methine (CH) protons. These differences are small—only about 0.7 ppm separates a methyl proton from a methine proton of the same type. Overall, proton chemical shifts among common organic compounds encompass a range of about 12 ppm. The protons in alkanes are the most shielded while O—H protons of carboxylic acids are the least shielded.

14.7 INTERPRETING PROTON NMR SPECTRA

Analyzing an nmr spectrum in terms of a unique molecular structure makes use of the information contained in Table 14.1. By knowing the chemical shifts characteris-

TABLE 14.1
Chemical Shifts of Representative Types of Protons

Type of proton	Chemical shift (δ), ppm*	Type of proton	Chemical shift (δ), ppm*
H—C—R	0.9–1.8	H—C—NR	2.2–2.9
H—C—C≡C	1.6–2.6	H—C—Cl	3.1–4.1
H—C—C(=O)—	2.1–2.5	H—C—Br	2.7–4.1
H—C≡C—	2.5	H—C—O	3.3–3.7
H—C Ar	2.3–2.8	H—NR	1–3†
H—C=C<	4.5–6.5	H—OR	0.5–5†
H—Ar	6.5–8.5	H—OAr	6–8†
H—C(=O)—	9–10	H—OC(=O)—	10–13†

* Approximate values relative to tetramethylsilane; other groups within the molecule can cause a proton signal to appear outside of the range cited.

† The chemical shifts of protons bonded to nitrogen and oxygen are temperature- and concentration-dependent.

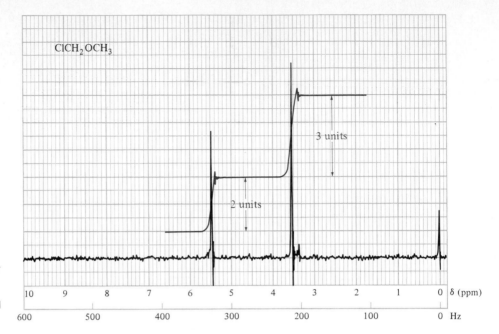

FIGURE 14.10 The ^1H nmr spectrum of chloromethyl methyl ether ($ClCH_2OCH_3$). Both the normal spectrum and its integral are shown.

tic of various proton environments, the presence of a particular structural unit in an unknown compound may be inferred. An nmr spectrum also provides other information that aids the task of structure determination. This additional information includes

1. The number of signals
2. The intensity of the signals, as measured by the area under each peak
3. The multiplicity or *splitting* of each signal

Protons that have different chemical shifts are said to be *chemical-shift nonequivalent* (or *chemically nonequivalent*). A separate nmr signal is given for each chemical-shift nonequivalent proton in a substance. The proton (^1H) nmr spectrum of chloromethyl methyl ether, $ClCH_2OCH_3$, shown in Figure 14.10, contains two peaks. One peak corresponds to the methylene protons, the other to the methyl protons. The methylene group bears two electronegative substituents, a chlorine and an oxygen, and so is less shielded than the methyl group, which bears only an oxygen. The signal that corresponds to the methylene protons appears at $\delta = 5.5$ ppm; the signal corresponding to the methyl protons is at $\delta = 3.5$ ppm.

Another way in which the nmr peaks may be assigned to the appropriate protons of chloromethyl methyl ether is based on their intensities. The three equivalent protons of the methyl group give rise to a more intense peak than the two equivalent protons of the methylene group. Intensities, as measured by peak areas, are proportional to the number of equivalent protons responsible for the signal. The area of the methyl signal in chloromethyl methyl ether is 50 percent greater than that of the methylene signal.

Peak areas are measured electronically. The spectrum is recorded in the usual manner and the nmr spectrometer is then switched from the normal mode to the integral mode. As the spectrum is scanned a second time, the instrument continuously adds the areas of all the peaks and superposes it as a series of steps on the original spectrum. The height of each step is proportional to the area under the peak and proportional to the number of protons responsible for the peak.

FIGURE 14.11 The ^1H nmr spectrum of *p*-xylene. Integrated areas are relative rather than absolute. The 6 : 4 ratio between methyl and aryl protons is reflected in a 1.5 : 1 ratio of peak areas.

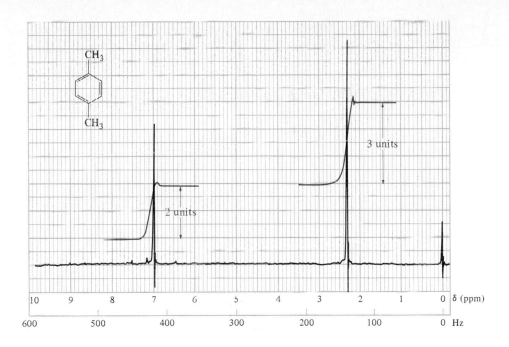

Figure 14.11 shows the ^1H nmr spectrum of *p*-xylene. Like the spectrum of chloromethyl methyl ether (Figure 14.10), it exhibits two peaks in a 3 : 2 ratio (integrated areas are relative; a 3 : 2 ratio of areas is just as consistent with a 6 : 4 ratio of protons as is a 3 : 2 ratio). The six methyl protons of *p*-xylene are equivalent to each other and give rise to a single peak. Similarly, the four aryl protons are equivalent to each other.

Protons are chemically equivalent to each other when they are in equivalent environments. To test for chemical equivalence, replace one proton by some test group. Then replace another by the same test group. If the two structures produced are the same, the two protons are chemically equivalent. To illustrate, replace one of the methyl protons of propane by chlorine to give 1-chloropropane. Replace a proton in the other methyl group by chlorine; the product is again 1-chloropropane. Thus, the six methyl protons of propane are all equivalent.

$CH_3CH_2CH_3$	$ClCH_2CH_2CH_3$	$CH_3CH_2CH_2Cl$
Propane	1-Chloropropane	1-Chloropropane

The two methylene protons of propane are equivalent to each other. Replacement of either one generates 2-chloropropane. Neither of the methylene protons is equivalent to any of the methyl protons. The ^1H nmr spectrum of propane contains two signals, one for the 6-equivalent methyl protons, the other for the 2-equivalent methylene protons.

PROBLEM 14.2 How many signals would you expect to find in the ^1H nmr spectrum of each of the following compounds?

(*a*) 1-Bromobutane
(*b*) 1-Butanol
(*c*) Butane
(*d*) 1,4-Dibromobutane
(*e*) 2,2-Dibromobutane
(*f*) 2,2,3,3-Tetrabromobutane
(*g*) 1,1,4-Tribromobutane
(*h*) 1,1,1-Tribromobutane

SAMPLE SOLUTION (*a*) To test for chemical-shift equivalency replace the protons at C-1, C-2, C-3, and C-4 of 1-bromobutane by some test group such as chlorine. Four constitutional isomers result.

$CH_3CH_2CH_2CHBr$	$CH_3CH_2CHCH_2Br$	$CH_3CHCH_2CH_2Br$	$ClCH_2CH_2CH_2CH_2Br$
$\quad\ \ \vert$	$\quad\ \ \vert$	$\quad\vert$	
Cl	Cl	Cl	
1-Bromo-1-Chlorobutane	1-Bromo-2-Chlorobutane	1-Bromo-3-Chlorobutane	1-Bromo-4-Chlorobutane

Thus, separate signals will be seen for the protons at C-1, C-2, C-3, and C-4. Barring any accidental overlap, we expect to find four signals in the nmr spectrum of 1-bromobutane.

Chemical-shift nonequivalence can occur when two environments are stereochemically different. The two vinyl protons of 2-bromopropene have different chemical shifts.

$$\begin{array}{ccc} Br & & H \quad \delta = 5.3 \text{ ppm} \\ & C{=}C & \\ H_3C & & H \quad \delta = 5.5 \text{ ppm} \end{array}$$

2-Bromopropene

One of the vinyl protons is cis to bromine; the other is cis to the methyl group. Replacing one of the vinyl protons by some test group, say chlorine, gives the *Z* isomer of 2-bromo-1-chloropropene; replacing the other gives the *E* stereoisomer. The *E* and *Z* forms of 2-bromo-1-chloropropene are stereoisomers that are not enantiomers; they are diastereomers. Protons that yield diastereomers on being replaced by some test group are described as *diastereotopic*. The vinyl protons of 2-bromopropene are diastereotopic. Diastereotopic protons can have different chemical shifts. Because their environments are similar, however, the difference in chemical shift between diastereotopic protons is usually slight and it sometimes happens that two diastereotopic protons accidentally have the same chemical shift.

PROBLEM 14.3 How many signals would you expect to find in the ^1H nmr spectrum of each of the following compounds?

(*a*) Vinyl bromide

(*b*) 1,1-Dibromoethene

(*c*) *cis*-1,2-Dibromoethene

(*d*) *trans*-1,2-Dibromoethene

(*e*) Allyl bromide

(*f*) 2-Methyl-2-butene

SAMPLE SOLUTION (*a*) Each of the protons of vinyl bromide is unique and has a chemical shift different from the other two. The least shielded proton is attached to the carbon that bears the bromine. The pair of protons at C-2 are diastereotopic with respect to each other; one is cis to bromine while the other is trans to bromine. There are three proton signals in the nmr spectrum of vinyl bromide. Their observed chemical shifts are as indicated.

$$\begin{array}{ccc} Br & & H \quad \delta = 5.7 \text{ ppm} \\ & C{=}C & \\ \delta = 6.4 \text{ ppm} \ \ H & & H \quad \delta = 5.8 \text{ ppm} \end{array}$$

When enantiomers are generated by replacing first one proton and then another by a test group, the pair of protons is said to be *enantiotopic*. Enantiotopic protons are chemical-shift equivalent.

At the beginning of this section we noted that an nmr spectrum provides structural information based on chemical shift, the number of peaks, the intensities of the peaks as measured by their integrated areas, and the multiplicity, or splitting, of the peaks. We have discussed the first three of these features of ^{1}H nmr spectroscopy. Let us now direct our attention to peak splitting and see what kind of information it offers.

14.8 SPIN-SPIN SPLITTING IN NMR SPECTROSCOPY

Each individual peak in the nmr spectra of chloroform (Figure 14.8), chloromethyl methyl ether (Figure 14.10), and *p*-xylene (Figure 14.11) consists of a single sharp absorption, referred to as a *singlet*. It is quite common to see nmr spectra, however, in which the signal due to a particular proton is not a singlet but instead appears as a collection of peaks. The signal may be split into two peaks (a *doublet*), three peaks (a *triplet*), four peaks (a *quartet*), etc. Figure 14.12 shows the ^{1}H nmr spectrum of 1,1-dichloroethane. The methyl protons appear as a doublet centered at $\delta =$ 2.0 ppm, and the signal for the methine proton is a quartet at $\delta = 5.9$ ppm.

There is a simple rule that allows us to predict splitting patterns in ^{1}H nmr spectroscopy. The number of peaks into which the signal for a particular proton is split, i.e., its *multiplicity*, is equal to one more than the number of protons that are vicinal to it. The three methyl protons of 1,1-dichloroethane are vicinal to the methine proton and split it into a quartet. The single methine proton, in turn, splits the methyl protons into a doublet.

This proton splits the signal for the methyl protons into a doublet

These three protons split the signal for the methine proton into a quartet

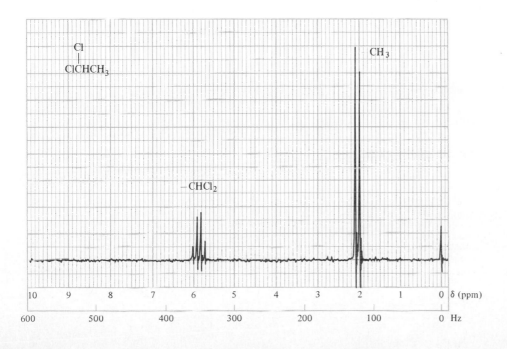

FIGURE 14.12 The ^{1}H nmr spectrum of 1,1-dichloroethane, showing the methine proton as a quartet and the methyl protons as a doublet.

The physical basis for peak splitting can be described by examining the methyl doublet in the nmr spectrum of 1,1-dichloroethane. Splitting of the methyl signal into a doublet occurs because the chemical shift of the methyl protons is influenced by the spin of the methine proton on the adjacent carbon. (All the methyl protons in 1,1-dichloroethane are equivalent and all have the same chemical shift. Whatever we say about one of these methyl protons applies equally to the other two.)

Figure 14.13a recalls the fundamentals of an nmr experiment carried out at a constant frequency. There is a certain field strength H_0 at which the energy required to flip the spin of the methyl protons (designated H_a in the figure) is exactly equal to the energy of the radiofrequency source. The magnetic moment of the methine proton (designated H_b) can be aligned parallel (Figure 14.13b) or antiparallel (Figure 14.13c) to the applied field. The net field experienced by the methyl protons is the aggregate of the applied field, the induced fields due to the electrons, *and* the tiny magnetic field of the methine proton. When the magnetic moment of the methine proton is parallel to the applied field, as in Figure 14.13b, it reinforces it and slightly deshields the methyl protons. Thus, the magnetic field strength required to achieve resonance is less than H_0, and the signal due to the methyl protons appears at lower field. When the magnetic moment of the methine proton is antiparallel to the applied field, it opposes it, increasing the shielding of the methyl protons and causing a shift in their signal to higher field. Since both spin states are almost equally likely, the methyl protons of half of the molecules undergo a shift to slightly higher field while those of the other half are deshielded by the methine proton and are shifted to slightly lower field. Instead of a single peak for the methyl protons, two peaks of equal intensity appear. The true chemical shift for the methyl protons is the midpoint of the doublet.

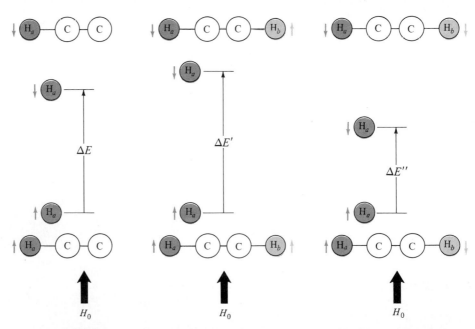

FIGURE 14.13 Diagram showing how two possible spin orientations of nucleus H_b split the signal of nucleus H_a into a doublet.

(a) Energy equal to ΔE is required to flip spin of nucleus H_a. This corresponds to an external magnetic field strength of H_0.

(b) Magnetic moment of H_b is parallel to direction of external magnetic field and deshields nucleus H_a. A weaker external field is sufficient to make $\Delta E' = \Delta E$.

(c) Magnetic moment of H_b is antiparallel to direction of external magnetic field and shields nucleus H_a. A stronger external field is required in order to make $\Delta E'' = \Delta E$.

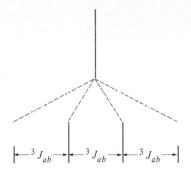

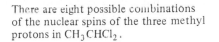

There are eight possible combinations of the nuclear spins of the three methyl protons in CH_3CHCl_2.

These eight combinations cause the signal of the $CHCl_2$ proton to be split into a quartet, in which the intensities of the peaks are in the ratio $1:3:3:1$.

FIGURE 14.14 The methyl protons of 1,1-dichloroethane split the signal of the methine proton into a quartet.

Turning now to the methine proton, its signal is split by the methyl protons into a quartet. The same kind of analysis applies here and is outlined in Figure 14.14. The methine proton "sees" eight different combinations of nuclear spins for the methyl protons. In one combination, the magnetic moments of all three methyl protons reinforce the applied field. At the other extreme, the magnetic moments of all three methyl protons oppose the applied field. There are three combinations in which the magnetic moments of two methyl protons reinforce the applied field while one opposes it. Finally, there are three combinations in which the magnetic moments of two methyl protons oppose the applied field and one reinforces it. These eight possible combinations give rise to four distinct peaks for the methine proton, with a ratio of intensities of $1:3:3:1$.

We describe the observed splitting of nmr signals as *spin-spin splitting* and the physical basis for it as *spin-spin coupling*. It has its origin in the communication of nuclear spin information between nuclei. This information is transmitted by way of the electrons in the bonds that intervene between the nuclei rather than through space. Its effect is greatest when the number of bonds is small. Vicinal protons are separated by three bonds and coupling between vicinal protons, as in 1,1-dichloroethane, is called *three-bond coupling* or *vicinal coupling*. Four-bond couplings are weaker and not normally observable at 60 MHz.

A very important characteristic of spin-spin splitting is that protons that have the same chemical shift do not split each other. Ethane, for example, shows only a single sharp peak in its nmr spectrum. Even though there is a vicinal relationship between the protons of one methyl group and those of the other, they do not split each other because they are equivalent.

PROBLEM 14.4   Describe the appearance of the ¹H nmr spectrum of each of the following compounds. How many signals would you expect to find, and into how many peaks will each signal be split?

(a) 1,2-Dichloroethane
(b) 1,1,1-Trichloroethane
(c) 1,1,2-Trichloroethane

(d) 1,2,2-Trichloropropane
(e) 1,1,1,2-Tetrachloropropane

SAMPLE SOLUTION   (a) All the protons of 1,2-dichloroethane ($ClCH_2CH_2Cl$) are chemically equivalent and have the same chemical shift. Protons that have the same chemical shift do not split each other, so the nmr spectrum of 1,2-dichloroethane consists of a single sharp peak.

Mutual coupling of nuclei requires that they split each other's signal equally. The separation in hertz between the two halves of the methyl doublet in 1,1-dichloroethane is equal to the separation between any two adjacent peaks of the methine quartet. The extent to which two nuclei are coupled is known as the *coupling constant J* and in simple cases is equal to the separation between adjacent lines of the signal of a particular proton. The three-bond coupling constant $^3J_{ab}$ in 1,1-dichloroethane has a value of 7 Hz. The size of the coupling constant is independent of the field strength; the separation between adjacent peaks in 1,1-dichloroethane is 7 Hz irrespective of whether the spectrum is recorded at 60 MHz, 100 MHz, or 360 MHz.

14.9 PATTERNS OF SPIN-SPIN SPLITTING. THE ETHYL GROUP

At first glance splitting may seem to complicate the interpretation of nmr spectra. In fact, it makes the task of structure determination easier because it is a source of additional information. It tells us how many protons are vicinal to a proton responsible for a particular signal. With practice, the eye becomes trained to pick out characteristic patterns of peaks, associating them with particular structural types. One of the most common of these patterns is that of the ethyl group, represented in the nmr spectrum of ethyl bromide in Figure 14.15.

In compounds of the type CH_3CH_2X, especially where X is an electronegative atom or group, such as bromine in ethyl bromide, the ethyl group appears as a triplet-quartet pattern. The methylene proton signal is split into a quartet by coupling with the methyl protons. The signal for the methyl protons is a triplet because of vicinal coupling to the two protons of the adjacent methylene group.

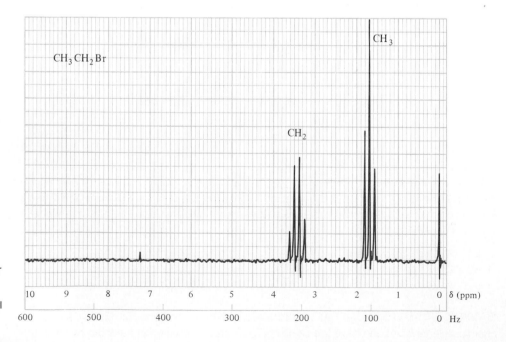

FIGURE 14.15 The ^{1}H nmr spectrum of ethyl bromide, showing the characteristic triplet-quartet pattern of an ethyl group.

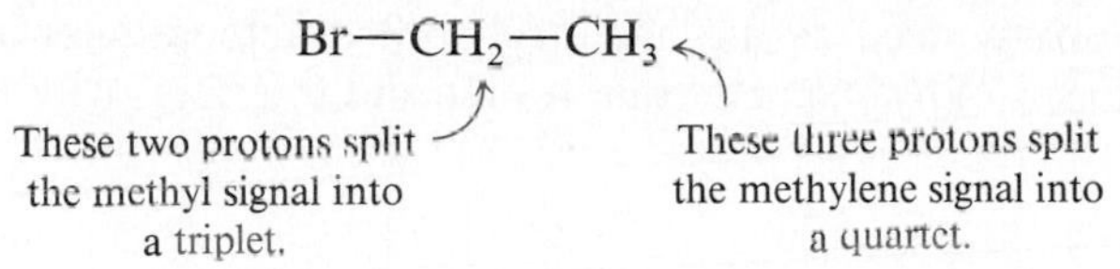

$$Br\text{—}CH_2\text{—}CH_3$$

These two protons split
the methyl signal into
a triplet.

These three protons split
the methylene signal into
a quartet.

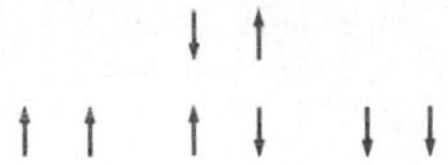

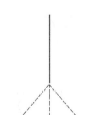

There are four possible combinations of the nuclear spins of the two methylene protons in CH_3CH_2Br.

We have discussed in the preceding section why methyl groups split the signals due to vicinal protons into a quartet. Splitting by a methylene group gives a triplet corresponding to the spin combinations shown in Figure 14.16 for ethyl bromide. The relative intensities of the peaks of this triplet are $1:2:1$.

PROBLEM 14.5 Describe the appearance of the ^1H nmr spectrum of each of the following compounds. How many signals would you expect to find, and into how many peaks will each signal be split?

(a) $ClCH_2OCH_2CH_3$
(b) $CH_3CH_2OCH_3$
(c) $CH_3CH_2OCH_2CH_3$
(d) *p*-Diethylbenzene
(e) $ClCH_2CH_2OCH_2CH_3$
(f) $Cl_2CHCH(OCH_2CH_3)_2$

SAMPLE SOLUTION (a) Along with the triplet-quartet pattern of the ethyl group, the nmr spectrum of this compound will contain a singlet for the two protons of the chloromethyl group.

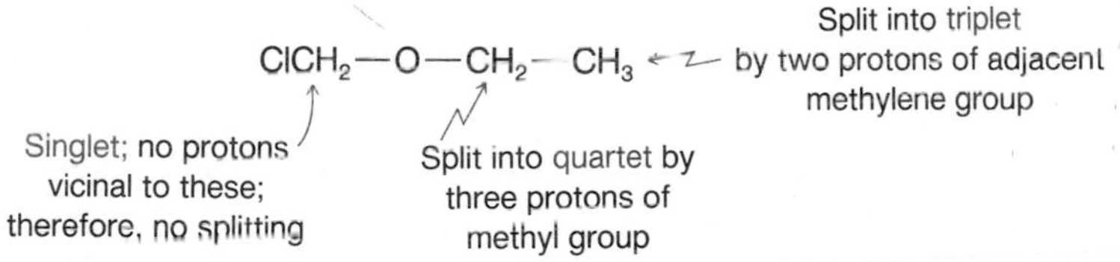

$$ClCH_2\text{—}O\text{—}CH_2\text{—}CH_3$$

Split into triplet by two protons of adjacent methylene group

Singlet; no protons vicinal to these; therefore, no splitting

Split into quartet by three protons of methyl group

These four combinations cause the signal of the CH_3 protons to be split into a triplet, in which the intensities of the peaks are in the ratio $1:2:1$.

FIGURE 14.16 The methylene protons of ethyl bromide split the signal of the methyl protons into a triplet.

Table 14.2 summarizes the splitting patterns and peak intensities expected for coupling to various numbers of protons. The intensities are equal to the coefficients of the binomial expansion.

The spectrum of ethyl bromide presented in Figure 14.15 illustrates a facet of splitting patterns that often occurs in nmr spectroscopy. The multiplets are not perfectly symmetrical but are skewed toward each other. The methylene quartet is not quite a $1:3:3:1$ pattern, and the methyl triplet is not quite $1:2:1$. The distortion

TABLE 14.2
Splitting patterns of common multiplets

Number of equivalent protons to which nucleus is coupled	Appearance of multiplet	Intensities of lines in multiplet
1	Doublet	$1:1$
2	Triplet	$1:2:1$
3	Quartet	$1:3:3:1$
4	Pentet	$1:4:6:4:1$
5	Hextet	$1:5:10:10:5:1$
6	Heptet	$1:6:15:20:15:6:1$

becomes more pronounced as the chemical-shift difference between the protons decreases, a property of nmr spectra that sometimes interferes with their interpretation.

14.10 PATTERNS OF SPIN-SPIN SPLITTING. THE ISOPROPYL GROUP

The nmr spectrum of isopropyl iodide, shown in Figure 14.17, illustrates the appearance of an isopropyl group. The six equivalent methyl protons are split by the C-2 methine proton into a doublet at $\delta = 1.9$ ppm. The methine proton is equally coupled to all six methyl protons, so its signal is split into a heptet. A doublet-heptet pattern is characteristic of an isopropyl group. What sometimes happens, however, and is evident in Figure 14.17 is that the outermost lines of the heptet are so weak that they are indistinguishable from the background *noise* present in all spectra. There should be seven peaks in the signal for the methine proton centered at $\delta = 4.3$ ppm, but only five of them are readily discernible.

14.11 PATTERNS OF SPIN-SPIN SPLITTING. PAIRS OF DOUBLETS

The last of the commonly encountered splitting patterns that we will examine involves only two protons. Our simple splitting rules tell us that the signal for each proton will be split into a doublet because of coupling to the other. A representative example of this splitting pattern can be found in the nmr spectrum of 2,3,4-trichloroanisole, shown in Figure 14.18.

The singlet at $\delta = 3.9$ ppm corresponds to the protons of the $-OCH_3$ group. The two aryl protons are nonequivalent and mutually coupled; they give rise to two

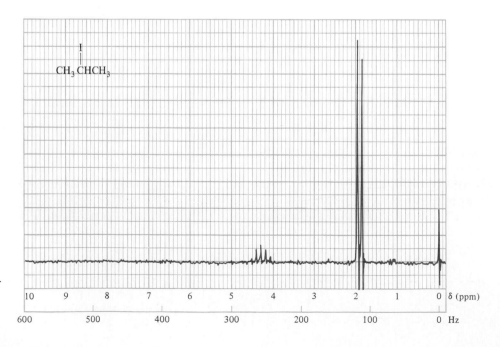

FIGURE 14.17 The ^1H nmr spectrum of isopropyl iodide, showing the doublet-heptet pattern of an isopropyl group.

FIGURE 14.18 The ^{1}H nmr spectrum of 2,3,4-trichloroanisole, showing the splitting of the ring protons into a pair of doublets.

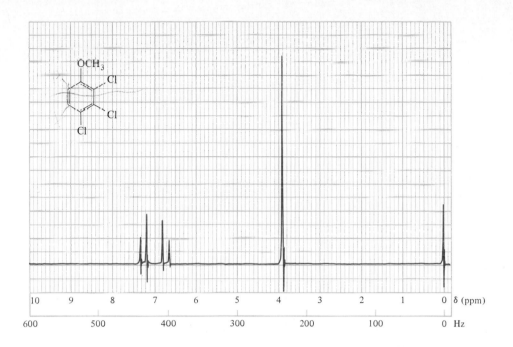

doublets at $\delta = 6.7$ ppm and $\delta = 7.25$ ppm. (The aryl proton that is ortho to the electron-releasing methoxyl group is slightly more shielded than the one meta to it.)

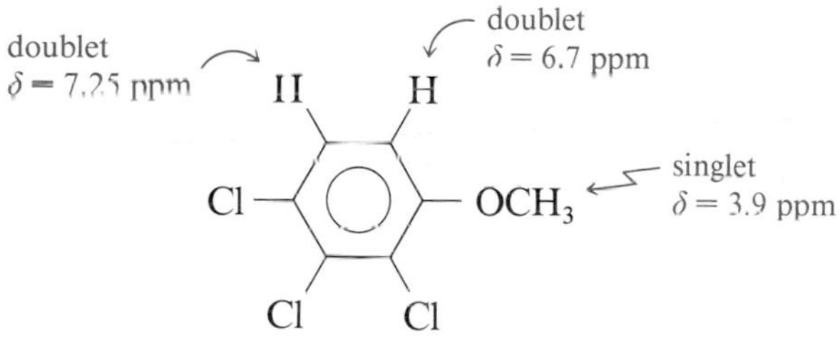

2,3,4-Trichloroanisole

In neither of the doublets is the intensity ratio of the peaks 1 : 1. Both doublets are skewed in the sense that the outermost two peaks are weaker than the innermost two. Skewing is minimal when the chemical-shift difference between two coupled protons is much larger than their coupling constant. Conversely, skewing is most pronounced when the chemical-shift difference is small.

PROBLEM 14.6 To which one of the following compounds does the nmr spectrum of Figure 14.19 correspond?

$ClCH_2C(Cl)(OCH_2CH_3)_2$ $Cl_2CHCH(OCH_2CH_3)_2$ $CH_3CH_2OCH(Cl)CH(Cl)OCH_2CH_3$

A similar pattern is seen in the mutual splitting of the signals of nonequivalent geminal protons. Two protons may be bonded to the same carbon and yet have

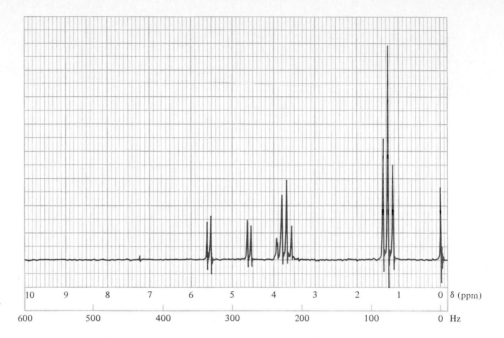

FIGURE 14.19 The ^{1}H nmr spectrum for Problem 14.6.

different chemical shifts because they are diastereotopic. Protons such as these are coupled through two bonds and split each other into a doublet by an amount equal to $^{2}J_{ab}$ when their chemical shifts are different. This is called *geminal coupling*.

14.12 PROTON NMR SPECTRA OF ALCOHOLS

The hydroxyl proton of a primary alcohol RCH_2OH is vicinal to two protons and would be expected to be split into a triplet. Under certain conditions splitting of alcohol protons is observed, but usually it is not. Figure 14.20 presents the nmr spectrum of benzyl alcohol, showing the methylene and hydroxyl protons as singlets at $\delta = 4.7$ and 1.8 ppm, respectively. (The aromatic protons also appear as a singlet, but that is because they all accidentally have the same chemical shift and so cannot split each other.)

The reason that splitting of the hydroxyl proton of an alcohol is not observed is that it is involved in rapid exchange reactions with other alcohol molecules. Transfer of the hydroxyl proton from an oxygen of one alcohol molecule to the oxygen of another is quite fast and effectively *decouples* it from other protons in the molecule. Factors that slow down this exchange of hydroxyl protons between molecules, such as lowering the temperature or increasing the crowding around the hydroxyl group, can cause splitting of hydroxyl resonances.

The chemical shift of the hydroxyl proton is variable, with a range of $\delta = 0.5$ to 5 ppm, depending on the solvent, the temperature at which the spectrum is recorded, and the concentration of the solution. The alcohol proton shifts to lower field in more concentrated solutions.

An easy way to verify that a particular signal belongs to a hydroxyl proton is to add D_2O. The hydroxyl proton is replaced by deuterium according to the equation:

$$RCH_2OH + D_2O \rightleftharpoons RCH_2OD + DOH$$

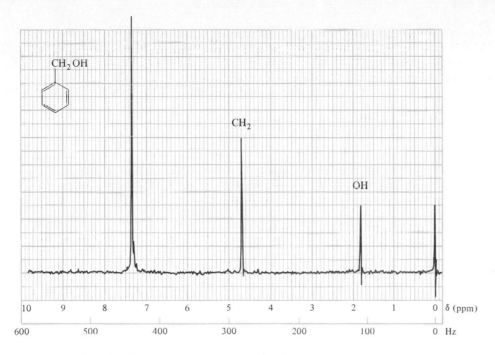

FIGURE 14.20 The ¹H nmr spectrum of benzyl alcohol. The hydroxyl proton and the methylene protons are vicinal but do not split each other because of the rapid intermolecular exchange of hydroxyl protons.

Deuterium does not give a signal under the conditions of proton nmr spectroscopy. Thus, exchange of a hydroxyl proton by deuterium leads to the disappearance of its nmr peak. Protons bonded to nitrogen and sulfur also undergo exchange with D_2O. Those bound to carbon normally do not, so this technique is useful for assigning the proton resonances of —OH, $>$NH, and —SH groups.

14.13 NMR AND CONFORMATIONS

We know from Chapter 3 that the protons in cyclohexane exist in two distinctly different environments, six being axial protons and six equatorial. The nmr spectrum of cyclohexane, however, shows only a single sharp peak at $\delta = 1.4$ ppm. All the protons of cyclohexane appear to be equivalent in the nmr spectrum. Why is this?

The answer is related to the rate of ring flipping in cyclohexane. Recall that the rate of conformational change is very fast.

One of the inherent properties of nmr spectroscopy is that it is too slow a technique to "see" the individual conformations of cyclohexane. What nmr sees is the *average* environment of the protons. Since chair-chair interconversion in cyclohexane converts each axial proton to an equatorial one and vice versa, the average environments of all the protons are the same. A single peak is observed which has a chemical shift midway between the true chemical shifts of the axial and the equatorial protons.

A similar effect occurs when measuring coupling constants. The magnitudes of vicinal coupling constants J depend on a variety of factors, including stereochemistry. Protons that are anti to each other have values of $^3J = 12$ to 15 Hz, while those that are gauche have $^3J = 2$ to 4 Hz. Most observed vicinal coupling constants, as measured from the splitting of peaks in the nmr spectrum, are on the order of 6 to 7 Hz, a value that reflects the weighted average of the coupling constants between two protons over all the conformations of the molecule.

14.14   CARBON-13 NUCLEAR MAGNETIC RESONANCE. THE SENSITIVITY PROBLEM

Magnetic resonance spectroscopy of nuclei other than protons is also possible. Two nuclei that have been extensively studied are ^{19}F and ^{31}P. Like ^{1}H, both ^{19}F and ^{31}P are the principal isotopes present in the "natural" state of each element, and both have nuclear spins of $\pm\frac{1}{2}$. What about carbon? When we use proton nmr as a tool for identifying organic compounds, we infer the structure of the carbon skeleton on the basis of the environments of the hydrogen substituents. Clearly, a spectroscopic method that examines the carbons directly would simplify the task of structure determination. The major isotopic form of carbon (^{12}C), however, has a nuclear spin of zero and is incapable of giving an nmr signal. The mass 13 isotope of carbon has a nuclear spin of $\pm\frac{1}{2}$ and so would be a suitable candidate for study by nmr were it not for a severe sensitivity problem. (By *sensitivity* we mean the ease with which signals may be detected under the conditions of measurement.) The nmr signal given by a ^{13}C nucleus is inherently much weaker than a proton signal, and in addition only about 1 percent of all carbon atoms in a sample are ^{13}C. Thus, even with an nmr spectrometer properly tuned for ^{13}C magnetic resonance (15 MHz at 14,100 gauss), the peaks are too weak to be detected and are lost in the background noise. This hampered the development of ^{13}C nmr (cmr) spectroscopy as a routine technique for organic structure determination until a new generation of nmr spectrometers incorporating special sensitivity-enhancing features became available in the 1970s.

The strategy behind sensitivity enhancement is based on the fact that background noise is random but the signals of a particular sample, even though they may be weak, always appear at the same chemical shift regardless of how many times the spectrum is scanned. By scanning the spectrum of a compound hundreds of times, storing the spectra in a computer, and then adding all the spectra together, the spectrometer accumulates the nmr signals more than the background noise. An increase in the signal-to-noise ratio results.

Within the solution to the sensitivity problem lies another problem. All the spectra that we have seen to this point required several minutes for the instrument to scan from low field to high field. To record several hundred spectra of the same sample under these conditions would take many hours or even days, and ^{13}C nmr spectroscopy would be a technique used only in special cases. Fortunately, there is another way to measure nmr spectra, a way in which a complete spectrum can be obtained in about 1 second and 1000 spectra can be obtained in 1000 seconds. This technique involves irradiating the sample with a brief but intense pulse of radiofrequency energy, exciting all the ^{13}C nuclei in the sample to their higher spin state. The excited nuclei then relax to their lower energy state, and it is this process that is monitored. By a mathematical technique known as a *Fourier transform,* the data are converted to a spectrum that looks very much like one obtained in the usual manner. Instruments that perform in this way are known as *Fourier transform (FT) nmr spectrometers*. As a technique for sensitivity enhancement, FT nmr is practically essential for ^{13}C nmr.

14.15 CARBON-13 NUCLEAR MAGNETIC RESONANCE. INTERPRETATION OF SPECTRA

Figure 14.21*a* and *b* shows, respectively, the ^{1}H and the ^{13}C nmr spectra of 1-chloro-pentane. Comparing the two will serve to orient us in the appearance of ^{13}C nmr spectra and the information that ^{13}C nmr provides.

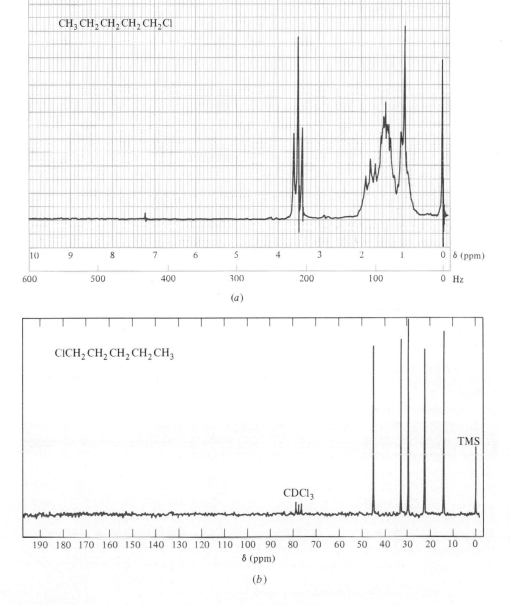

FIGURE 14.21 (*a*) The ^{1}H nmr spectrum of 1-chloropentane. (*b*) The ^{13}C nmr spectrum of 1-chloropentane. *(The ^{13}C spectrum is taken from Carbon-13 NMR Spectra: A Collection of Assigned, Coded, and Indexed Spectra, by LeRoy F. Johnson and William C. Jankowski, Wiley-Interscience, New York, 1972. Reprinted by permission of John Wiley & Sons, Inc.)*

First, and most important, ^{13}C nmr is much more able to discriminate between nonequivalent nuclei than is ^1H nmr. The only well-defined signal in the proton spectrum of 1-chloropentane in Figure 14.21a is the triplet of the $-CH_2Cl$ group at $\delta = 3.5$ ppm. There is a badly skewed triplet at $\delta = 0.9$ ppm for the methyl group, while all the other protons appear as a broad multiplet at $\delta = 1$ to 2 ppm. The ^{13}C nmr spectrum in Figure 14.21b, on the other hand, consists of five well-defined peaks, one for each of the nonequivalent carbons of 1-chloropentane. Notice, too, the widely separated chemical shifts of these five peaks; they cover a range of over 30 ppm compared with less than 3 ppm for the proton signals of the same compound. In general, proton signals span a range of about 12 ppm; ^{13}C signals span a range of over 200 ppm.

PROBLEM 14.7   How many signals would you expect to see in the ^{13}C nmr spectrum of each of the following compounds?

(a) Propylbenzene
(b) Isopropylbenzene
(c) 1,2,3-Trimethylbenzene

(d) 1,2,4-Trimethylbenzene
(e) 1,3,5-Trimethylbenzene

SAMPLE SOLUTION   (a)   The two ring carbons that are ortho to the propyl substituent are equivalent and so must have the same chemical shift. Similarly, the two ring carbons that are meta to the propyl group are equivalent to each other. The carbon atom para to the substituent is unique, as is the carbon that bears the substituent. Thus, there will be four signals for the ring carbons, designated, *w, x, y,* and *z* in the structural formula. These four signals for the ring carbons added to those for the three nonequivalent carbons of the propyl group yields a total of *seven* signals.

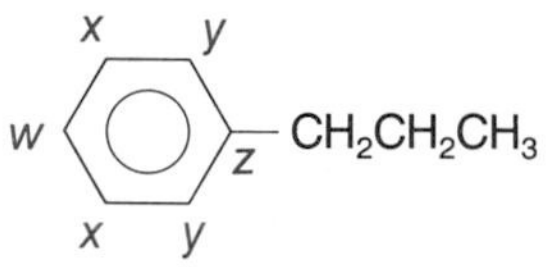

Propylbenzene

Chemical shifts in ^1H nmr are measured relative to the protons of tetramethylsilane; chemical shifts in ^{13}C nmr are measured relative to the carbons of tetramethylsilane. Table 14.3 lists typical chemical-shift ranges for some representative types of carbon atoms.

TABLE 14.3
Chemical Shifts of Representative Carbons

Type of carbon	Chemical shift (δ), ppm*	Type of carbon	Chemical shift (δ), ppm*
RCH_3	0–35	$\diagdown C = C \diagup$	100–150
R_2CH_2	15–40		
R_3CH	25–50		
RCH_2NH_2	35–50	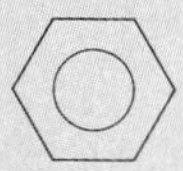 (benzene ring)	110–175
RCH_2OH	50–65		
$-C \equiv C-$	65–90	$\diagdown C = O$	190–220

* Approximate values relative to tetramethylsilane.

A second aspect of the ^{13}C nmr spectrum of Figure 14.21b is that all the peaks are singlets. The reason for the lack of $^{13}C-^{13}C$ coupling is that only 1 percent of all the carbons in any given sample are ^{13}C. The probability that two ^{13}C atoms are present in the same molecule is quite small, and the probability that they are separated by only a few bonds is smaller still. The lack of splitting due to $^{13}C-H$ coupling, however, reflects a deliberate decision to measure the spectrum under conditions that suppress such splitting. A ^{13}C nucleus can couple strongly to protons directly attached to it as well as to protons separated from it by two, three, or even more bonds. The myriad of $^{13}C-H$ couplings that would result would make the spectrum too complicated for ready interpretation. A technique known as *broad-band decoupling* removes all the proton-carbon couplings, producing the kind of spectrum shown in Figure 14.21b.

There is another technique, known as *off-resonance decoupling* that removes all the carbon-proton couplings except those involving nuclei that are directly bonded. Thus, a methine (CH) carbon appears as a doublet, a methylene (CH_2) carbon as a triplet, and a methyl (CH_3) carbon as a quartet. Carbons that bear no hydrogen substituents are called *quaternary* carbons and appear as a singlet. Typically, a carbon spectrum is recorded in the broad-band decoupled mode, signals are tentatively assigned on the basis of their chemical shifts, and then an off-resonance decoupled spectrum is taken to confirm the assignments. In the off-resonance decoupled spectrum of 1-chloropentane, all the signals are triplets except for the peak at $\delta = 13.9$ ppm, which is a quartet and must correspond to the methyl group.

PROBLEM 14.8 Into how many peaks would each of the signals be split in the off-resonance decoupled ^{13}C nmr spectrum of each of the following compounds?

(a) Propylbenzene

(b) Isopropylbenzene

(c) 1,2,3-Trimethylbenzene

(d) 1,2,4-Trimethylbenzene

(e) 1,3,5-Trimethylbenzene

SAMPLE SOLUTION (a) The number of peaks into which the signal for a particular carbon is split is one more than the number of protons bonded to it. The ring carbon that bears the propyl group has no hydrogen substituents and so appears as a singlet. All the other ring carbons bear one hydrogen and each is split into a doublet. The two methylene groups of the side chain are each split into a triplet by their two hydrogens, and the methyl group appears as a quartet because of splitting by three hydrogens.

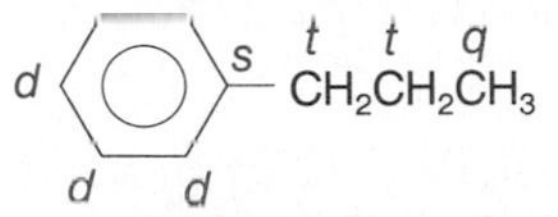

(s = singlet; d = doublet; t = triplet; q = quartet)

The last feature of ^{13}C nmr spectra to be pointed out concerns peak intensities. Because of differences in the rates of relaxation of nuclei, spectra measured by the pulse relaxation technique used in ^{13}C FT nmr spectroscopy are subject to distortion of their signal intensities. This distortion is quite pronounced for carbons that do not have hydrogen substituents, as the ^{13}C nmr spectrum of m-methylphenol (m-cresol) shown in Figure 14.22 illustrates. In addition to the signal for the methyl group at $\delta = 21$ ppm, there are six signals for the six aromatic carbons in the range $\delta = 112$ to

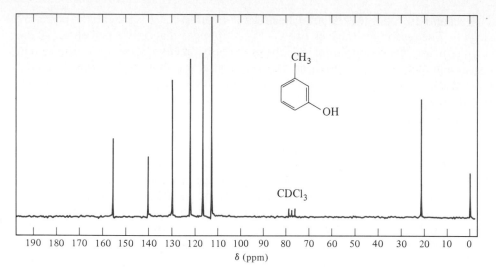

FIGURE 14.22 The ^{13}C nmr spectrum of *m*-cresol. *(The ^{13}C spectrum is taken from Carbon-13 NMR Spectra: A Collection of Assigned, Coded, and Indexed Spectra, by LeRoy F. Johnson and William C. Jankowski, Wiley-Interscience, New York, 1972. Reprinted by permission of John Wiley & Sons, Inc.)*

155 ppm. Two of these peaks are clearly smaller than the others and correspond to the ring carbons that bear the methyl group and the hydroxyl group.

PROBLEM 14.9   Which one of the compounds of Problems 14.7 and 14.8 corresponds to the broad-band decoupled ^{13}C nmr spectrum of Figure 14.23?

Both ^{13}C and ^1H nmr are powerful techniques for organic structure determination. They share the quality of being "nondestructive"; that is, the sample can be recovered in its entirety after the spectrum is taken. Thus, one never faces the problem of choosing whether to take a proton spectrum or a carbon spectrum — both can be recorded in successive experiments on the same sample. The two forms of magnetic resonance spectroscopy provide complementary information. In the next section we will examine another nondestructive analytical method, infrared spectroscopy, which is a useful adjunct to nmr.

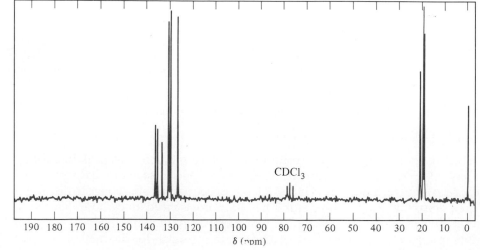

FIGURE 14.23   The ^{13}C nmr spectrum of the unknown compound of Problem 14.9. *(The ^{13}C spectrum is taken from Carbon-13 NMR Spectra: A Collection of Assigned, Coded, and Indexed Spectra, by LeRoy F. Johnson and William C. Jankowski, Wiley-Interscience, New York, 1972. Reprinted by permission of John Wiley & Sons, Inc.)*

14.16   INFRARED SPECTROSCOPY

Prior to the introduction of nmr spectroscopy, infrared (ir) spectroscopy was the instrumental method most often applied to structure determination of organic compounds. While nmr spectroscopy is, in general, more revealing of the structure of an unknown compound, ir still retains an important place in the chemist's inventory of spectroscopic methods because of its usefulness in identifying the presence of certain functional groups within a molecule.

Infrared radiation comprises the portion of the electromagnetic spectrum (Figure 14.1) in which the wavelengths range from approximately 10^{-4} to 10^{-6} m. It lies between microwaves and visible light. The fraction of the infrared region of most use for structure determination lies between 2.5×10^{-6} m and 16×10^{-6} m. Two derived units that are commonly employed in infrared spectroscopy are the *micrometer* and the *wave number*. One micrometer (μm) is 10^{-6} m, and infrared spectra record the region from 2.5 μ to 16 μ. Wave numbers are reciprocal centimeters (cm^{-1}), so the region 2.5 to 16 μ corresponds to 4000 to 625 cm^{-1}. An advantage to using wave numbers is that they are directly proportional to energy while wavelengths are inversely proportional to energy. Thus, 4000 cm^{-1} is the high-energy end of the scale and 625 cm^{-1} is the low-energy end.

Electromagnetic radiation in the 4000 to 625 cm^{-1} region corresponds to the separation between adjacent *vibrational energy states* in organic molecules. Absorption of a photon of infrared radiation excites a molecule from its lowest, or *ground,* vibrational state to a higher one. These vibrations include stretching and bending modes of the type illustrated for a methylene group in Figure 14.24. A single molecule can have a large number of distinct vibrations available to it, and infrared spectra of different molecules, like fingerprints and snowflakes, are different. Superposability of their infrared spectra is commonly offered as proof that two compounds are the same.

A typical infrared spectrum, such as that of hexane in Figure 14.25, appears as a series of absorption peaks of varying shape and intensity. Almost all organic compounds exhibit a peak or group of peaks near 3000 cm^{-1} because it is in this region

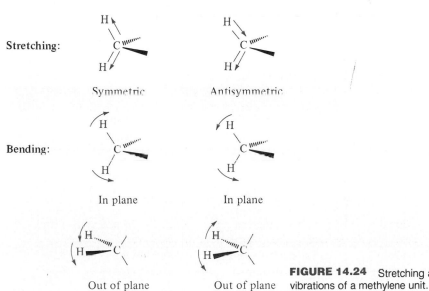

FIGURE 14.24 Stretching and bending vibrations of a methylene unit.

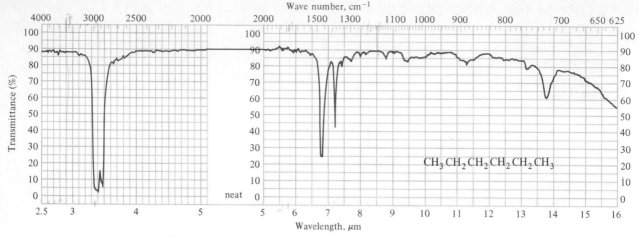

FIGURE 14.25 The infrared spectrum of hexane.

TABLE 14.4

Infrared Absorption Frequencies of Some Common Structural Units

Single bonds		Double bonds	
Structural unit	**Frequency, cm⁻¹**	**Structural unit**	**Frequency, cm⁻¹**
Stretching vibrations			
—O—H (alcohols)	3200–3600	C=C	
—O—H (carboxylic acids)	2500–3600	C=O	1620–1680
N—H	3350–3500	Aldehydes and ketones	1710–1750
sp C—H	3310–3320	Carboxylic acids	1700–1725
sp² C—H	3000–3100	Acid anhydrides	1800–1850 and 1740–1790
sp³ C—H	2850–2950	Acyl halides	1770–1815
		Esters	1730–1750
sp² C—O	1200	Amides	1680–1700
sp³ C—O	1025–1200		

Triple bonds	
—C≡C—	2100–2200
—C≡N	2240–2280

Bending vibrations of diagnostic value

Alkenes:		*Substituted derivatives of benzene:*	
Cis-disubstituted	665–730	Monosubstituted	730–770 and 690–710
Trans-disubstituted	960–980	Ortho-disubstituted	735–770
Trisubstituted	790–840	Meta-disubstituted	750–810 and 680–730
		Para-disubstituted	790–840

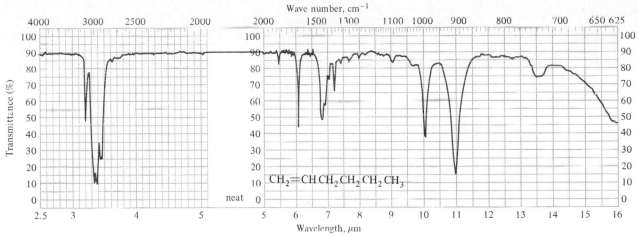

FIGURE 14.26 The infrared spectrum of 1-hexene.

that absorption due to carbon-hydrogen stretching vibrations occurs. The peaks at 1460, 1380, and 725 cm^{-1} are due to various bending vibrations.

Infrared spectra can be recorded on a sample regardless of its physical state — solid, liquid, gas, or dissolved in some solvent. The spectrum in Figure 14.25 was taken on the neat sample, meaning the pure liquid. A drop or two of hexane was placed between two sodium chloride disks to give a thin film of sample, through which the infrared beam is passed. Solids may be dissolved in a suitable solvent such as carbon tetrachloride or chloroform. More commonly, though, a solid sample is mixed with potassium bromide and the mixture pressed into a thin wafer, which is placed in the path of the infrared beam.

In using infrared spectroscopy for structure determination, peaks in the range 1600 to 4000 cm^{-1} are usually emphasized because this is the region in which the vibrations characteristic of particular functional groups are found. The region 1300 to 625 cm^{-1} is known as the *fingerprint region;* it is here that the pattern of peaks varies most from compound to compound. Table 14.4 lists the frequencies (in wave numbers) associated with a variety of groups commonly found in organic compounds.

To illustrate how structural features affect infrared spectra, compare the spectrum of hexane (Figure 14.25) with that of 1-hexene (Figure 14.26). The two are quite different. In the C—H stretching region of 1-hexene, there is a peak at 3095 cm^{-1}, while all the C—H stretching vibrations of hexane appear below 3000 cm^{-1}. A peak or peaks above 3000 cm^{-1} is characteristic of a hydrogen bonded to sp^2 hybridized carbon. The ir spectrum of 1-hexene also displays a peak at 1640 cm^{-1} corresponding to its C=C stretching vibration. The peaks near 1000 and 900 cm^{-1} in the spectrum of 1-hexene, absent in the spectrum of hexane, are bending vibrations involving the hydrogens of the doubly bonded carbons.

Carbon-hydrogen stretching vibrations with frequencies above 3000 cm^{-1} also are found in arenes such as *tert*-butylbenzene, as shown in Figure 14.27. This spectrum also contains two intense bands at 760 and 700 cm^{-1}, which are characteristic of monosubstituted benzene rings. Other substitution patterns, some of which are listed in Table 14.4, give different combinations of peaks.

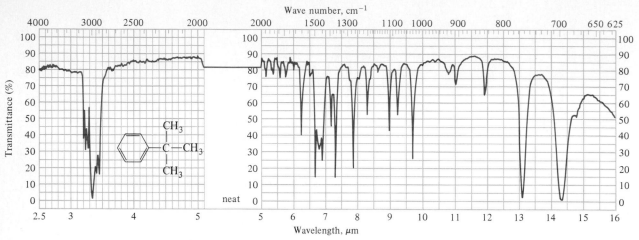

FIGURE 14.27 The infrared spectrum of *tert*-butylbenzene.

In addition to sp^2 C—H stretching modes, there are other stretching vibrations that appear at frequencies above 3000 cm^{-1}. The most important of these is the O—H stretch of alcohols. Figure 14.28 shows the ir spectrum of 2-hexanol. It contains a broad peak at 3300 cm^{-1} ascribable to O—H stretching of hydrogen-bonded alcohol groups. In dilute solution, where hydrogen bonding is less and individual alcohol molecules are present as well as hydrogen-bonded aggregates, an additional peak appears at approximately 3600 cm^{-1}.

Carbonyl groups rank among the structural units most readily revealed by ir spectroscopy. The carbon-oxygen double bond stretching mode gives rise to a very strong peak in the 1650 to 1800 cm^{-1} region. This peak is clearly evident in the spectrum of 2-hexanone, shown in Figure 14.29. The position of the carbonyl peak varies with the nature of the substituents on the carbonyl group. Thus, characteristic frequencies are associated with aldehydes and ketones, amides, esters, etc., as summarized in Table 14.4.

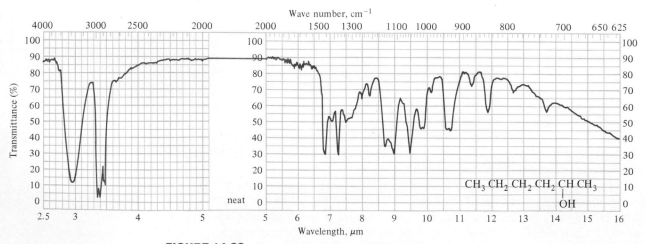

FIGURE 14.28 The infrared spectrum of 2-hexanol.

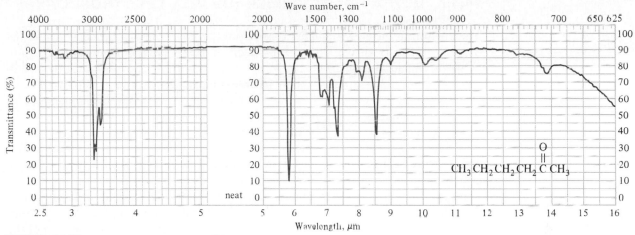

FIGURE 14.29 The infrared spectrum of 2-hexanone.

PROBLEM 14.10 Which one of the following compounds is most consistent with the infrared spectrum given in Figure 14.30? Explain your reasoning.

Phenol   Acetophenone   Benzoic acid   Benzyl alcohol

In later chapters when families of compounds are discussed in detail, the infrared frequencies associated with each type of functional group will be described.

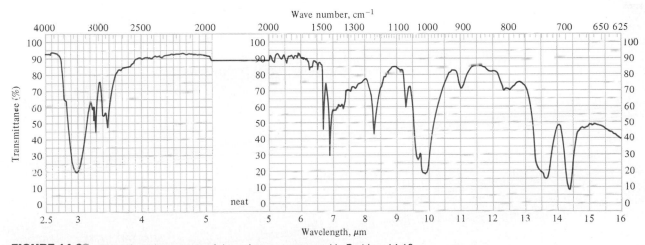

FIGURE 14.30 The infrared spectrum of the unknown compound in Problem 14.10.

14.17   ULTRAVIOLET-VISIBLE (UV-VIS) SPECTROSCOPY

The portion of the electromagnetic spectrum (Figure 14.1) that lies just beyond the infrared region is visible light. Visible light occupies the relatively narrow region between 12,500 and 25,000 cm^{-1}. Wave numbers are directly proportional to energy, so visible light is approximately 10 times more energetic than infrared radiation. Red light is the low-energy end of the visible region, violet light the high-energy end. Ultraviolet radiation lies beyond the violet end of visible light; it encompasses the region from 25,000 to 50,000 cm^{-1}. Positions of absorption in visible spectroscopy are customarily expressed in units of 10^{-9} m, or *nanometers* (nm). Thus, the visible region corresponds to 800 to 400 nm, and the ultraviolet region to 400 nm to 200 nm.

Figure 14.31 shows the ultraviolet (uv) spectrum of a conjugated diene, *cis, trans*-1,3-cyclooctadiene, measured in ethanol as solvent in the range 200 to 280 nm. There are no additional absorption maxima beyond 280 nm, so that portion of the spectrum has been omitted. As is typical of most uv spectra, the absorption peak is rather broad. The wavelength at which absorption is a maximum is referred to as the λ_{max} of the sample. In this case λ_{max} is 230 nm. The absorbance A of a sample is proportional to its concentration in solution and the path length through which the beam of ultraviolet radiation passes. To correct for concentration and path length, absorbance is converted to *molar absorptivity ε* by dividing it by the concentration c in moles per liter and the path length l in centimeters.

$$\varepsilon = \frac{A}{c \cdot l}$$

Molar absorptivity, when measured at λ_{max}, is cited as ε_{max}. It is normally expressed without units. Both λ_{max} and ε_{max} are affected by the solvent, so it is explicitly indicated when reporting uv-vis spectroscopic data. Thus, you might find a literature reference that includes these data expressed in the form

cis, trans-1,3-Cyclooctadiene
$\lambda_{max}^{ethanol}$ 230 nm
$\varepsilon_{max}^{ethanol}$ 2630

As might be expected from the physical appearance of a uv-vis spectrum or from the limited data contained in a literature citation, the structural information we derive from uv-vis spectroscopy is less than that derived from the other methods discussed to this point. Ultraviolet spectroscopy, however, does tell us some things

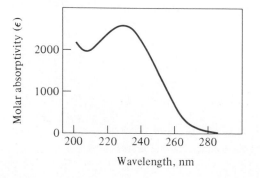

FIGURE 14.31 The ultraviolet spectrum of *cis,trans*-1,3-cyclooctadiene.

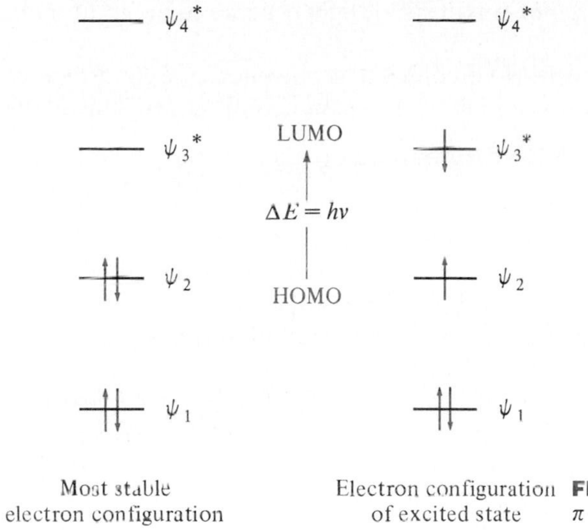

Most stable electron configuration

Electron configuration of excited state

FIGURE 14.32 The nature of the $\pi \to \pi^*$ transition in 1,3-butadiene.

that ir and nmr spectroscopy do not; it probes the electron distribution in a molecule and is particularly useful when conjugated π-electron systems are present.

The transitions involved in uv-vis spectroscopy are those between electron energy levels. In these transitions an electron is promoted from the highest occupied molecular orbital (HOMO) of a molecule to the lowest unoccupied molecular orbital (LUMO). In alkenes and polyenes the HOMO is the highest-energy π orbital and the LUMO is the lowest-energy π^* orbital. Excitation of a π electron from a bonding to an antibonding orbital is referred to as a $\pi \to \pi^*$ transition. Figure 14.32 depicts the π-electron configuration of 1,3-butadiene in its ground (most stable) electron configuration and in its excited state. The HOMO-LUMO energy gap is quite large (132 kcal/mol) and the $\pi \to \pi^*$ transition in 1,3-butadiene requires electromagnetic radiation of 217 nm.

Substituents, especially those that extend conjugation, decrease the HOMO-LUMO energy difference and cause a shift of λ_{max} to longer wavelengths. Thus, while λ_{max} is 217 nm for 1,3-butadiene, it is 227 nm for 2,3-dimethyl-1,3-butadiene (methyl substitution effect) and 258 nm for 1,3,5-hexatriene (extended conjugation effect). A striking example of the effect of conjugation on the absorption of light can be found in the case of a compound known as *lycopene*. Lycopene is a red substance that contributes to the color of ripe tomatoes; it has a conjugated system of 11 double bonds and absorbs visible light ($\lambda_{max} = 505$ nm). By absorbing the blue-green fraction of visible light, lycopene appears red.

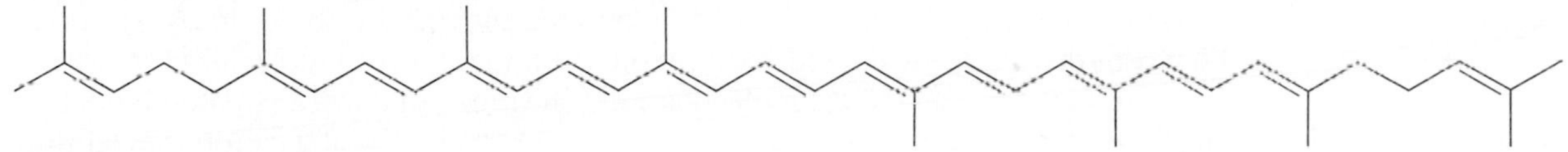

Lycopene

Many organic compounds such as lycopene are colored because their HOMO-LUMO energy gap is small enough so that λ_{max} appears in the visible range of the spectrum. All that is required for a compound to be colored, however, is that it possess some absorption in the visible range. It often happens that a compound will have its λ_{max} in the uv region, but that the peak is broad and extends into the visible.

Absorption of the blue to violet components of visible light occurs and the compound appears yellow.

In addition to the $\pi \rightarrow \pi^*$ transition of conjugated polyenes, a second type of absorption is important in uv-vis examination of organic compounds. This is the $n \rightarrow \pi^*$ transition of the carbonyl ($\diagdown$C$=$O) group. One of the electrons in a lone pair orbital of oxygen is excited to an antibonding orbital of the carbonyl group. The n in $n \rightarrow \pi^*$ identifies the electron as one of the nonbonded ones of oxygen. This transition gives rise to relatively weak absorption peaks ($\varepsilon_{max} < 100$) in the region 270 to 300 nm. Compounds that contain a carbon-carbon double bond in conjugation with a carbonyl group exhibit both an intense $\pi \rightarrow \pi^*$ absorption and a weak $n \rightarrow \pi^*$ absorption.

The structural unit associated with the electronic transition in uv-vis spectroscopy is called a *chromophore*. Chemists often refer to *model compounds* to help interpret uv-vis spectra. An appropriate model is a simple compound of known structure that incorporates the chromophore suspected of being present in the sample. Because remote substituents do not affect λ_{max} of the chromophore, a strong similarity between the spectrum of the model compound and that of the unknown can serve to identify the kind of π-electron system present in the sample. There is also a substantial body of data concerning the uv-vis spectral properties associated with a great many chromophores, as well as empirical correlations of substituent effects on λ_{max}. Such data are helpful in using uv-vis spectroscopy as a tool for structure determination.

14.18 MASS SPECTROMETRY

Mass spectrometry is the instrumental method of structure determination that is the most different from the others in the group discussed in this chapter. It does not depend on the selective absorption of particular frequencies of electromagnetic radiation but rather examines what happens to a molecule when it is bombarded with high-energy electrons. If an electron having an energy of about 10 electron-volts (10 eV = 230.5 kcal/mol) collides with an organic molecule, the energy transferred as a result of that collision is sufficient to dislodge one of the molecule's electrons.

$$\text{A:B} \; + \; e^- \; \longrightarrow \; \text{A}\overset{+}{\cdot}\text{B} \; + \; 2e^-$$

| Molecule | Electron | Cation radical | Two electrons |

We say the molecule AB has been ionized by *electron impact*. The species that results, called the *molecular ion,* is positively charged and has an odd number of electrons— it is a *cation radical.* The molecular ion has the same mass (less the negligible mass of a single electron) as the molecule from which it is formed.

While energies of about 10 eV are required, energies of about 70 eV are used. Electrons this energetic not only bring about the ionization of a molecule but impart a large amount of energy to the molecular ion. The molecular ion dissipates this excess energy by dissociating into smaller fragments. Dissociation of a cation radical produces a neutral fragment and a positively charged one. The molecular ion A $\overset{+}{\cdot}$ B can dissociate in several ways. Among them are the fragmentations

$$\text{A}\overset{+}{\underset{\curvearrowright}{}}\text{B} \; \longrightarrow \; \text{A}^+ \; + \; \text{B}\cdot$$

| Cation radical | | Cation | Radical |

$$\text{A}\overset{+}{\underset{\curvearrowright}{}}\text{B} \; \longrightarrow \; \text{A}\cdot \; + \; \text{B}^+$$

| Cation radical | | Radical | Cation |

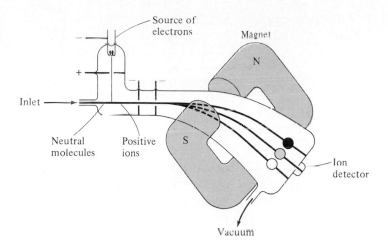

FIGURE 14.33 Schematic diagram of a mass spectrometer. Only positive ions are detected. The cation ● has the highest mass/charge ratio and its path is deflected least by the magnet. The cation ○ has the lowest mass/charge ratio and its path is deflected most. The mass/charge ratio of cation ◎ is such that its path carries it through a small slit to the detector. *(Adapted from John B. Russell, General Chemistry, McGraw-Hill, New York, 1980, p. 111.)*

Ionization and fragmentation produce a mixture of particles, some neutral and some positively charged. To understand what follows, we need to examine the design of an electron-impact mass spectrometer, shown in a schematic diagram in Figure 14.33. The sample is bombarded with 70-eV electrons and the resulting positively charged ions (the molecular ion as well as fragment ions) are directed into an analyzer tube surrounded by a magnet. This magnet deflects the ions from their original trajectory, causing them to adopt a circular path, the radius of which depends on their mass/charge ratio (m/z). Ions of small m/z are deflected more than those of larger m/z. By varying either the magnetic field strength or the degree to which the ions are accelerated on entering the analyzer, ions of a particular m/z can be selectively focused through a narrow slit onto a detector, where they are counted. Scanning all m/z values gives the distribution of positive ions, called a *mass spectrum,* characteristic of a particular compound.

Modern mass spectrometers are interfaced with computerized data handling systems capable of displaying the mass spectrum according to a number of different formats. Bar graphs on which relative intensity is plotted versus m/z are the most common. Figure 14.34 shows the mass spectrum of benzene in bar graph form.

The mass spectrum of benzene is relatively simple and illustrates some of the information that mass spectrometry provides. The most intense peak in the mass spectrum is called the *base peak* and is assigned a relative intensity of 100. Ion abundances are proportional to peak intensities and are reported as intensities relative to the base peak. The base peak in the mass spectrum of benzene corresponds to the molecular ion (M^+) at $m/z = 78$.

| Benzene | Electron | Molecular ion of benzene | Two electrons |

$$ + \quad e^- \longrightarrow \quad + \quad 2e^- $$

Benzene does not undergo extensive fragmentation; none of the fragment ions in its mass spectrum are as abundant as the molecular ion.

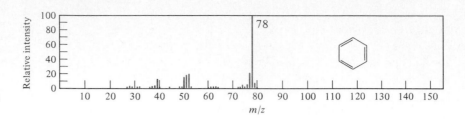

FIGURE 14.34   The mass spectrum of benzene.

There is a small peak one mass unit higher than M$^+$ in the mass spectrum of benzene. What is the origin of this peak? What we see in Figure 14.34 as a single mass spectrum is actually a superposition of the spectra of three isotopically distinct benzenes. Most of the benzene molecules contain only ^{12}C and ^{1}H and have a molecular mass of 78. Smaller proportions of benzene molecules contain ^{13}C in place of one of the ^{12}C atoms or ^{2}H in place of one of the protons. Both these species have a molecular mass of 79.

93.4%
(all carbons are ^{12}C)
gives M$^+$ 78

6.5%
(* = ^{13}C)
gives M$^+$ 79

0.1%
(all carbons are ^{12}C)
gives M$^+$ 79

Not only the molecular ion peak but all the peaks in the mass spectrum of benzene are accompanied by a smaller peak one mass unit higher. Indeed, since all organic compounds contain carbon and most contain hydrogen, similar *isotopic clusters* will appear in the mass spectra of all organic compounds.

Isotopic clusters are especially apparent when atoms such as bromine and chlorine are present in an organic compound. The natural ratios of isotopes in these elements are

$$\frac{^{35}\text{Cl}}{^{37}\text{Cl}} = \frac{100}{32.7} \qquad \frac{^{79}\text{Br}}{^{81}\text{Br}} = \frac{100}{97.5}$$

Figure 14.35 presents the mass spectrum of chlorobenzene. There are two prominent molecular ion peaks, one at m/z 112 for $C_6H_5{}^{35}$Cl and the other at m/z 114 for $C_6H_5{}^{37}$Cl. The peak at m/z 112 is three times as intense as the one at m/z 114.

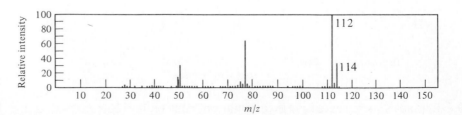

FIGURE 14.35   The mass spectrum of chlorobenzene.

PROBLEM 14.11   Knowing what to look for with respect to isotopic clusters can aid in interpreting mass spectra. How many peaks would you expect to see for the molecular ion in each of the following compounds? At what m/z values would these peaks appear? (Disregard the small peaks due to ^{13}C and ^{2}H.)

(a) *p*-Dichlorobenzene   (c) *p*-Dibromobenzene
(b) *o*-Dichlorobenzene   (d) *p*-Bromochlorobenzene

SAMPLE SOLUTION   (a)   The two isotopes of chlorine are ^{35}Cl and ^{37}Cl. There will be three isotopically different forms of *p*-dichlorobenzene present. They have the structures shown below. Each one will give an M^+ peak at a different value of m/z.

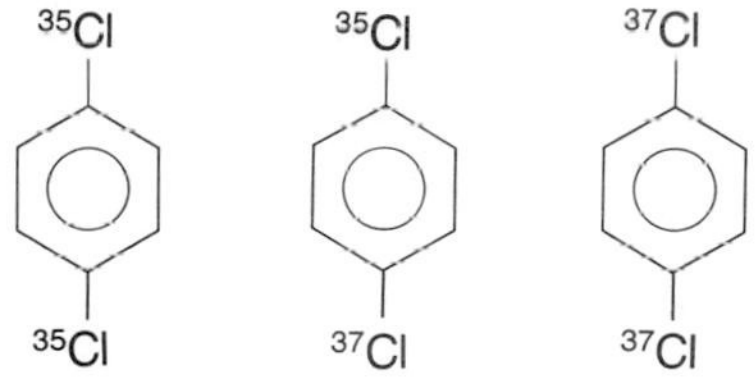

m/z 146 m/z 148 m/z 150

Unlike the case of benzene, in which ionization involves loss of a π electron from the ring, electron impact–induced ionization of chlorobenzene involves loss of an electron from an unshared pair of chlorine. The molecular ion then fragments by carbon-chlorine bond cleavage.

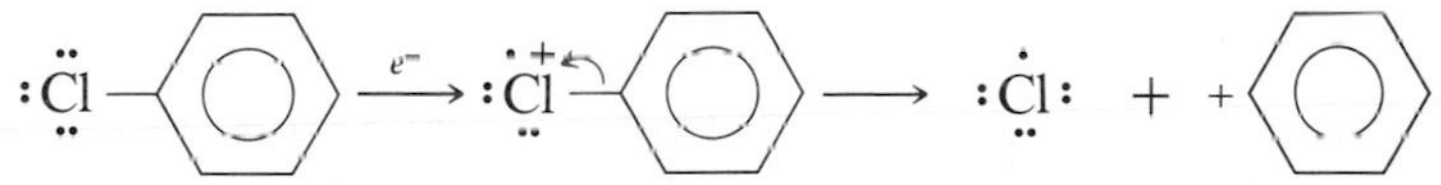

Chlorobenzene   Molecular ion   Chlorine   Phenyl cation
                     of chlorobenzene   atom m/z 77

The peak at m/z 77 in the mass spectrum of chlorobenzene in Figure 14.35 is attributed to the fragmentation shown above. Because there is no peak of significant intensity two atomic mass units higher, we know the cation responsible for the peak at m/z 77 cannot contain chlorine.

Some classes of compounds are so prone to fragmentation that the molecular ion peak is very weak. The base peak in most unbranched alkanes, for example, is m/z 43, which is followed by peaks of decreasing intensity at m/z values of 57, 71, 85, etc. These peaks correspond to cleavage of each possible carbon-carbon bond in the molecule. This pattern is evident in the mass spectrum of decane, depicted in Figure 14.36. The points of cleavage are indicated in the diagram below.

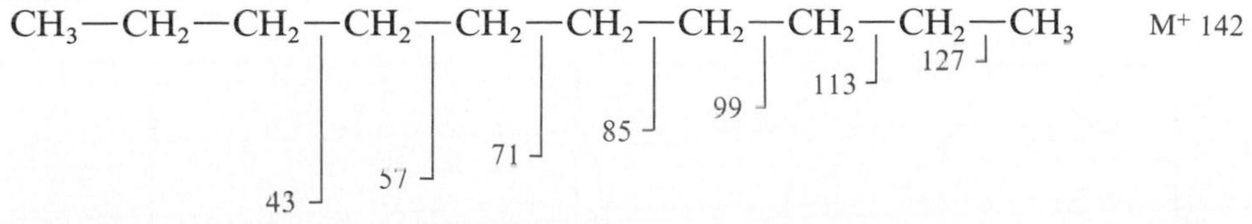

Many of the fragmentation processes in mass spectrometry proceed so as to form a stable carbocation, and the principles that we have developed regarding carbocation stability are applicable. Alkylbenzenes of the type $C_6H_5CH_2R$ undergo cleavage of

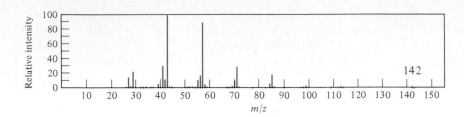

FIGURE 14.36 The mass spectrum of decane.

the bond to the benzylic carbon to give m/z 91 as the base peak. The mass spectrum in Figure 14.37 and the fragmentation diagram shown below illustrate this for propylbenzene.

$$\text{C}_6\text{H}_5-\text{CH}_2-\text{CH}_2-\text{CH}_3 \qquad \text{M}^+\ 120$$

While this cleavage is probably driven by the stability of benzyl cation, evidence has been obtained suggesting that tropylium cation, formed by rearrangement of benzyl cation, is actually the species responsible for the peak.

PROBLEM 14.12   The base peak appears at m/z 105 for one of the compounds shown below and at m/z 119 for the other two. Match the compounds with the appropriate m/z value for their base peaks.

CH₂CH₃ ... CH₃ ... CH₃ ... CH₂CH₂CH₃ ... CH₃ ... CH₃CHCH₃ ... CH₃

Understanding how molecules fragment upon electron impact permits a mass spectrum to be analyzed in sufficient detail to deduce the structure of an unknown compound. Thousands of compounds of known structure have been examined by mass spectrometry, and the fragmentation patterns that characterize different classes are well documented. As various groups are covered in subsequent chapters, aspects of their fragmentation behavior under conditions of electron impact will be described.

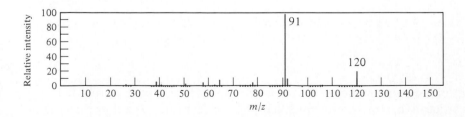

FIGURE 14.37 The mass spectrum of propylbenzene.

14.19 SUMMARY

The task of structure determination in modern-day organic chemistry relies heavily on instrumental methods. The most generally used methods are

1. *^{1}H Nuclear magnetic resonance spectroscopy:* In the presence of an external magnetic field, the $+\frac{1}{2}$ and $-\frac{1}{2}$ nuclear spin states of a proton have slightly different energies. The energy required to "flip" the spin of a proton depends on the extent to which a nucleus is shielded from the external magnetic field. Protons in different environments within a molecule have different chemical shifts. The *number of signals* in a ^{1}H nmr spectrum reveals the number of chemical shift nonequivalent protons in a molecule; the *integrated areas* tell us their relative ratios; their *chemical shifts* indicate the kind of environment surrounding the proton; and the *splitting pattern* is related to the number of protons on adjacent carbons.

2. *^{13}C Nuclear magnetic resonance spectroscopy:* The ^{13}C nucleus has a spin of $\pm\frac{1}{2}$ but is present to the extent of only 1.1 percent at natural abundance. By using special techniques for signal enhancement, high-quality ^{13}C nmr spectra may be obtained and these provide a useful complement to proton spectra. Carbon-13 chemical shifts cover a range of over 200 ppm. In many substances a separate signal is observed for each carbon atom, and the chemical shifts are characteristic of particular structural types. Carbon signals are normally presented as singlets, but through a technique known as off-resonance decoupling they appear as multiplets in which the number of peaks is one more than the number of directly bonded hydrogens.

3. *Infrared spectroscopy:* This method probes molecular structure by examining transitions between vibrational energy levels using electromagnetic radiation in the 625 to 4000 cm^{-1} range. It is useful for determining the presence of certain *functional groups* based on their characteristic absorption frequencies.

4. *Ultraviolet-visible spectroscopy:* Transitions between electronic energy levels involving electromagnetic radiation in the 200- to 800-nm range form the basis of uv-vis spectroscopy. The absorption peaks tend to be broad but are sometimes useful in indicating the presence of particular *conjugated π-electron* systems within a molecule.

5. *Mass spectrometry:* Mass spectrometry exploits the information obtained when a molecule is ionized by electron impact and then dissociates to smaller fragments. Positive ions are separated and detected according to their mass/charge ratio. By examining the fragments and by knowing how classes of molecules dissociate on electron impact, one can deduce the structure of a compound. Mass spectrometry is quite sensitive; as little as 10^{-9} g of compound is sufficient.

PROBLEMS

14.13 Each of the following compounds is characterized by a ^{1}H nmr spectrum that consists of only a single peak having the chemical shift indicated. Identify each compound.

(a) C_8H_{18} $\delta = 0.9$ ppm
(b) C_5H_{10} $\delta = 1.5$ ppm
(c) C_8H_8 $\delta = 5.8$ ppm
(d) C_4H_9Br $\delta = 1.8$ ppm

(e) $C_2H_4Cl_2$ $\delta = 3.7$ ppm
(f) $C_2H_3Cl_3$ $\delta = 2.7$ ppm
(g) $C_5H_8Cl_4$ $\delta = 3.7$ ppm

14.14   Deduce the structure of each of the following compounds on the basis of their ^1H nmr spectra and molecular formulas:

(a) C_8H_{10} $\delta = 1.2$ ppm (triplet, 3H)
            $\delta = 2.6$ ppm (quartet, 2H)
            $\delta = 7.1$ ppm (broad singlet, 5H)

(b) $C_{10}H_{14}$ $\delta = 1.3$ ppm (singlet, 9H)
            $\delta = 7.0$ to 7.5 ppm (multiplet, 5H)

(c) C_6H_{14} $\delta = 0.8$ ppm (doublet, 12H)
            $\delta = 1.4$ ppm (heptet, 2H)

(d) C_6H_{12} $\delta = 0.9$ ppm (triplet, 3H)
            $\delta = 1.6$ ppm (singlet, 3H)
            $\delta = 1.7$ ppm (singlet, 3H)
            $\delta = 2.0$ ppm (pentet, 2H)
            $\delta = 5.1$ ppm (triplet, 1H)

(e) $C_4H_6Cl_4$ $\delta = 3.9$ ppm (doublet, 4H)
            $\delta = 4.6$ ppm (triplet, 2H)

(f) $C_4H_6Cl_2$ $\delta = 2.2$ ppm (singlet, 3H)
            $\delta = 4.1$ ppm (doublet, 2H)
            $\delta = 5.7$ ppm (triplet, 1H)

(g) C_3H_7ClO $\delta = 2.0$ ppm (pentet, 2H)
            $\delta = 2.8$ ppm (singlet, 1H)
            $\delta = 3.7$ ppm (triplet, 2H)
            $\delta = 3.8$ ppm (triplet, 2H)

(h) $C_{14}H_{14}$ $\delta = 2.9$ ppm (singlet, 4H)
            $\delta = 7.1$ ppm (broad singlet, 10H)

14.15   From among the isomeric compounds of molecular formula C_4H_9Cl, choose the one having a ^1H nmr spectrum that

(a)   Contains only a single peak
(b)   Has several peaks including a doublet at $\delta = 3.4$ ppm
(c)   Has several peaks including a triplet at $\delta = 3.5$ ppm
(d)   Has several peaks including two distinct three-proton signals, one of them a triplet at $\delta = 1.0$ ppm and the other a doublet at $\delta = 1.5$ ppm

14.16   Identify the C_3H_5Br isomers on the basis of the following information:

(a)   Isomer A has the ^1H nmr spectrum shown in Figure 14.38.
(b)   Isomer B has three peaks in its ^{13}C nmr spectrum: $\delta = 32.6$ ppm (triplet); 118.8 ppm (triplet); and 134.2 ppm (doublet).
(c)   Isomer C has two peaks in its ^{13}C nmr spectrum: $\delta = 12.0$ ppm (triplet) and 16.8 ppm (doublet). The peak at lower field is only one-half as intense as the one at higher field.

14.17   Identify each of the $C_4H_{10}O$ isomers on the basis of their ^{13}C nmr spectra.

(a) $\delta = 18.9$ ppm (quartet) (two carbons)
      $\delta = 30.8$ ppm (doublet) (one carbon)
      $\delta = 69.4$ ppm (triplet) (one carbon)

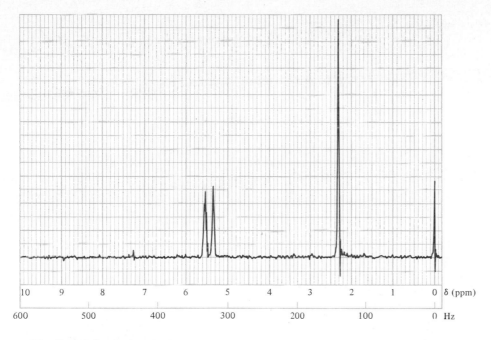

(*b*) $\delta = 10.0$ ppm (quartet)
     $\delta = 22.7$ ppm (quartet)
     $\delta = 32.0$ ppm (triplet)
     $\delta = 69.2$ ppm (doublet)

(*c*) $\delta = 31.2$ ppm (quartet) (three carbons)
     $\delta = 68.9$ ppm (singlet) (one carbon)

14.18   Identify the C_6H_{14} isomers on the basis of their ^{13}C nmr spectra:

(*a*) $\delta = 19.1$ ppm (quartet)
     $\delta = 33.9$ ppm (doublet)

(*b*) $\delta = 13.7$ ppm (quartet)
     $\delta = 22.8$ ppm (triplet)
     $\delta = 31.9$ ppm (triplet)

(*c*) $\delta = 11.1$ ppm (quartet)
     $\delta = 18.4$ ppm (quartet)
     $\delta = 29.1$ ppm (triplet)
     $\delta = 36.4$ ppm (doublet)

(*d*) $\delta = 8.5$ ppm (quartet)
     $\delta = 28.7$ ppm (quartet)
     $\delta = 30.2$ ppm (singlet)
     $\delta = 36.5$ ppm (triplet)

(*e*) $\delta = 14.0$ ppm (quartet)
     $\delta = 20.5$ ppm (triplet)
     $\delta = 22.4$ ppm (quartet)
     $\delta = 27.6$ ppm (doublet)
     $\delta = 41.6$ ppm (triplet)

14.19   Compound D (C_4H_6) has two signals of approximately equal intensity in its ^{13}C nmr spectrum; one is a triplet at $\delta = 30.2$ ppm, the other a doublet at $\delta = 136$ ppm. Identify compound D.

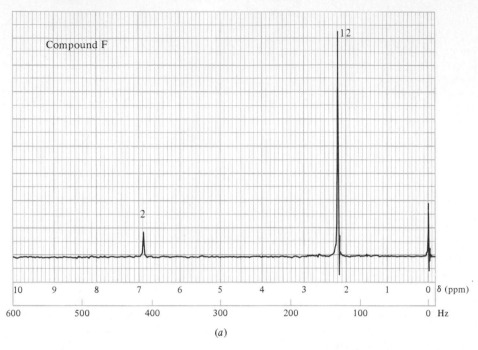

(a)

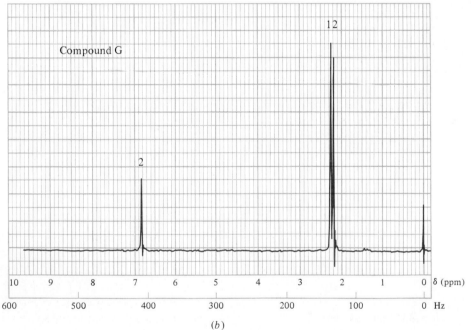

FIGURE 14.39 The ¹H nmr spectra of (a) compound F and (b) compound G ($C_{10}H_{14}$) (Problem 14.22).

(b)

14.20 Compound E ($C_3H_7ClO_2$) exhibited three peaks in its ¹³C nmr spectrum at $\delta = 46.8$ (triplet), 63.5 (triplet), and 72.0 ppm (doublet). What is the most reasonable structure for this compound?

14.21 From among the compounds chlorobenzene, o-dichlorobenzene, and p-dichlorobenzene, choose the one that:

(a) Gives the simplest ¹H nmr spectrum

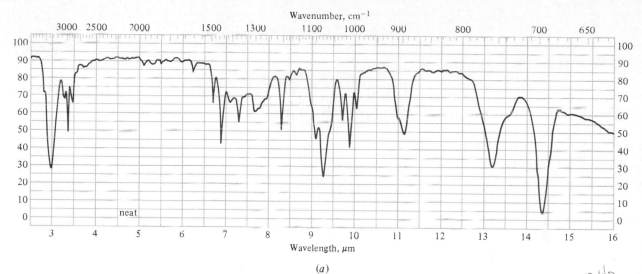

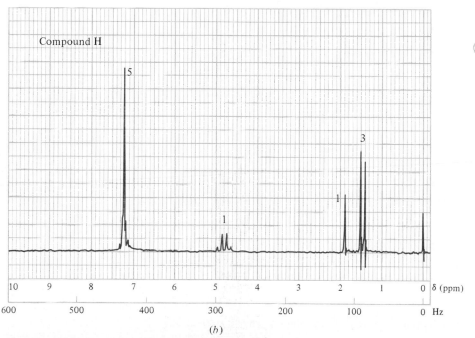

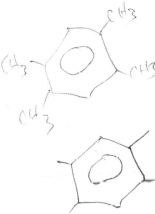

FIGURE 14.40 Infrared (a) and ¹H nmr (b) spectra of compound H ($C_9H_{10}O$) (Problem 14.23).

(b) Gives the simplest ¹³C nmr spectrum

(c) Has three peaks in its ¹³C nmr spectrum

(d) Has four peaks in its ¹³C nmr spectrum

14.22 Compounds F and G are isomers of molecular formula $C_{10}H_{14}$. Identify each one on the basis of the ¹H nmr spectra presented in Figure 14.39.

14.23 Compound H ($C_8H_{10}O$) has the infrared and ¹H nmr spectra presented in Figure 14.40. What is the structure of compound H?

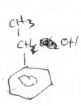

14.24 Deduce the structure of compound I on the basis of its mass spectrum and ¹H nmr spectrum presented in Figure 14.41.

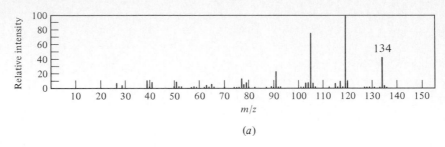

(a)

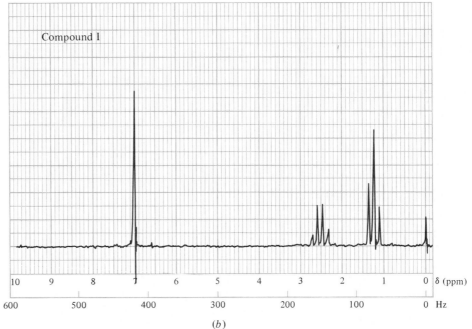

Compound I

FIGURE 14.41 Mass spectrum (a) and ^{1}H nmr spectrum (b) of compound I (Problem 14.24).

(b)

14.25 Figure 14.42 presents several types of spectroscopic data (ir, ^{1}H nmr, ^{13}C nmr, and mass spectra) for compound J. What is compound J?

14.26 [18]-Annulene exhibits a ^{1}H nmr spectrum that is unusual in that in addition to a peak at $\delta = 8.8$ ppm, it contains a second peak having a chemical shift δ of -1.9 ppm. A negative value for the chemical shift δ indicates that the protons are *more* shielded than those of tetramethylsilane. This peak is 1.9 ppm *upfield* from the TMS peak. The high-field peak has half the area of the low-field peak. Suggest an explanation for these observations.

[18]-Annulene

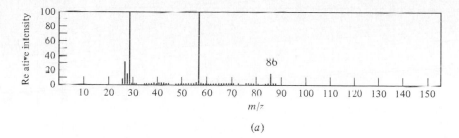

(a)

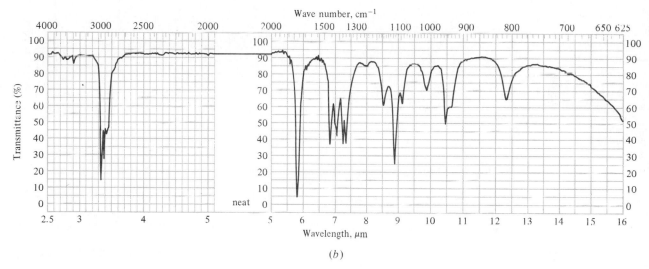

(b)

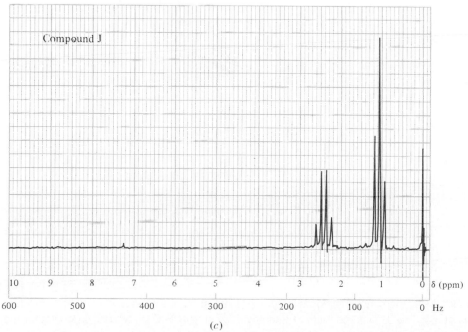

(c)

FIGURE 14.42 Mass (*a*), infrared (*b*), ¹H nmr (*c*), and (see following page) ¹³C nmr (*d*) spectra of compound J (Problem 14.25).

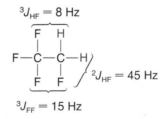

211.4

CDCl₃

δ (ppm)

190 180 170 160 150 140 130 120 110 100 90 80 70 60 50 40 30 20 10 0

(d)

FIGURE 14.42 *(Continued)* *(The ¹³C spectrum is taken from Carbon-13 NMR Spectra: A Collection of Assigned, Coded, and Indexed Spectra, by LeRoy F. Johnson and William C. Jankowski, Wiley-Interscience, New York, 1972. Reprinted by permission of John Wiley & Sons, Inc.)*

14.27 ¹⁹F is the only isotope of fluorine that occurs naturally and it has a nuclear spin of $\pm\frac{1}{2}$.

(a) Into how many peaks will the proton signal in the ¹H nmr spectrum of methyl fluoride be split?

(b) Into how many peaks will the fluorine signal in the ¹⁹F nmr spectrum of methyl fluoride be split?

(c) The chemical shift of the protons in methyl fluoride is $\delta = 4.3$ ppm. Given that the geminal ¹H−¹⁹F coupling constant is 45 Hz, specify the δ values at which peaks are observed in the proton spectrum of this compound at 60 MHz.

(d) Repeat (c) for the ¹H nmr spectrum run at 100 MHz.

14.28 Given the coupling constants of 1,1,1,2-tetrafluoroethane as shown in the diagram, answer the following questions about the ¹H and ¹⁹F nmr spectra of this compound:

$$^3J_{HF} = 8 \text{ Hz}$$

$$\overbrace{\quad\quad\quad}$$

$$\begin{array}{ccc} F & H \\ | & | \\ F-C-C-H \\ | & | \\ F & F \end{array} \quad ^2J_{HF} = 45 \text{ Hz}$$

$$\underbrace{\quad\quad\quad}$$

$$^3J_{FF} = 15 \text{ Hz}$$

(a) How many signals are there in the ¹H nmr spectrum of this compound? Into how many peaks are these signals split?

(b) How many signals are there in the ¹⁹F nmr spectrum of this compound? Into how many peaks are these signals split?

14.29 ³¹P is the only phosphorus isotope present at natural abundance and has a nuclear spin of $\pm\frac{1}{2}$. The ¹H nmr spectrum of trimethyl phosphite $(CH_3O)_3P$ exhibits a doublet for the methyl protons with a splitting of 12 Hz.

(a) Into how many peaks is the ³¹P signal split?

(b) What is the difference in chemical shift (in hertz) between the lowest- and highest-field peaks of the ³¹P multiplet?

ORGANOMETALLIC COMPOUNDS

CHAPTER

15

Organometallic compounds occupy the interface at which organic and inorganic chemistry meet; they are molecules that contain a carbon-to-metal bond. You probably are already familiar with several organometallic substances—mercurochrome and tetraethyllead, for example. You will meet numerous others in this chapter.

Organometallic compounds have rather different properties from those of other classes that we have encountered thus far. Most prominently, many organometallic compounds serve as powerful sources of nucleophilic carbon. It is this feature that makes them especially valuable to the synthetic organic chemist. Synthetic procedures employing organometallic reagents are the principal methods for carbon-carbon bond formation in organic chemistry. In this chapter you will learn how to prepare organic derivatives of lithium, magnesium, copper, and zinc, as well as some other metals, and how to apply them to the task of building carbon chains.

15.1 CARBON-METAL BONDS IN ORGANOMETALLIC COMPOUNDS

Situated as it is in group IV of the periodic table, carbon is neither strongly electropositive nor strongly electronegative. When carbon is bonded covalently to an atom more electronegative than itself, such as oxygen or chlorine, the electron distribution in the bond is polarized so that carbon is slightly positive and the more electronegative atom is slightly negative. Conversely, when carbon is bonded to an atom that is more electropositive, such as a metal, the electrons in the bond are more strongly attracted toward carbon.

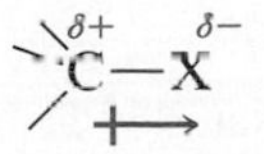

$$\overset{\delta+}{C}-\overset{\delta-}{X} \qquad \overset{\delta-}{C}-\overset{\delta+}{M}$$

X is more electronegative   M is more electropositive
than carbon   than carbon

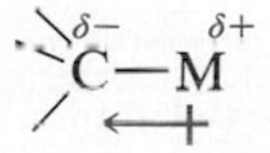

521

The electronegativity of carbon on the Pauling scale (Table 1.2) is 2.5; that of oxygen is 3.5 and chlorine is 3.0. The electronegativities of most metals are in the range 0.8 to 1.7. Hydrogen, with an electronegativity of 2.1, is only slightly more electropositive than carbon, and the carbon-hydrogen bond is not very polar.

A species containing a negatively charged carbon is referred to as a *carbanion.* Covalently bonded organometallic compounds are said to have *carbanionic charac-ter.* As the metal becomes more electropositive, the ionic character of the carbon-metal bond becomes more pronounced. Organosodium and organopotassium com-pounds have ionic carbon-metal bonds; organolithium and organomagnesium compounds tend to have covalent, but rather polar, carbon-metal bonds with signifi-cant carbanionic character. It is the carbanionic character of such compounds that is responsible for their usefulness as synthetic reagents.

15.2 ORGANOMETALLIC NOMENCLATURE

Organometallic compounds are named as substituted derivatives of metals. The metal is the base name and the attached alkyl groups are identified by the appropriate prefix.

$$CH_2{=}CHNa \qquad (CH_3CH_2)_2Mg$$

| Cyclopropyllithium | Vinylsodium | Diethylmagnesium |

When the metal bears a substituent other than carbon, the substituent is treated as if it were an anion and named separately.

$$CH_3MgI \qquad\qquad (CH_3CH_2)_2AlCl$$

| Methylmagnesium iodide | Diethylaluminum chloride |

PROBLEM 15.1   Each of the following organometallic reagents will be encountered later in this chapter. Suggest a suitable name for each.

(a) $(CH_3)_3CLi$

(c) $(CH_3CH_2)_2Cd$

(b) [cyclohexyl structure with H and MgCl]

(d) ICH_2ZnI

SAMPLE SOLUTION   (a)   The metal lithium provides the base name for this sub-stance. The alkyl group to which lithium is bonded is *tert*-butyl, so the name of this organo-metallic compound is *tert*-butyllithium. In systematic nomenclature it is 1,1-dimethylethyl-lithium.

An exception to this type of nomenclature is NaC≡CH, which is normally re-ferred to as sodium acetylide. Both sodium acetylide and ethynylsodium are accept-able IUPAC names.

15.3 PREPARATION OF ORGANOLITHIUM COMPOUNDS

Before we describe the applications of organometallic reagents to organic synthesis, let us examine the preparation of these compounds themselves. Organolithium compounds and other group I organometallics are prepared by the reaction of an alkyl halide with the appropriate metal.

$$RX + 2M \longrightarrow RM + M^+X^-$$

Alkyl halide	Group I metal	Group I organometallic	Metal halide

$$CH_3CH_2CH_2CH_2Br + 2Li \xrightarrow[-10°C]{\text{diethyl ether}} CH_3CH_2CH_2CH_2Li + LiBr$$

Butyl bromide · · · Lithium · · · Butyllithium (80-90%) · · · Lithium bromide

$$(CH_3)_3CCl + 2Li \xrightarrow[-30°C]{\text{diethyl ether}} (CH_3)_3CLi + LiCl$$

tert-Butyl chloride · · · Lithium · · · *tert*-Butyllithium (75%) · · · Lithium chloride

Unlike elimination and nucleophilic substitution reactions, formation of organolithium compounds does not require that the halogen be bonded to sp^3 hybridized carbon. Compounds such as vinyl halides and aryl halides, in which the halogen is bonded to sp^2 hybridized carbon, react in the same way as alkyl halides, but somewhat less readily.

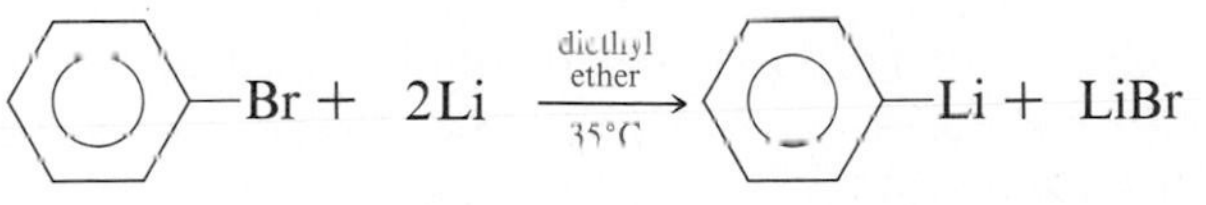

Bromobenzene · · · Lithium · · · Phenyllithium (95-99%) · · · Lithium bromide

---

PROBLEM 15.2 Write an equation showing the formation of each of the following from the appropriate alkyl, alkenyl, or aryl bromide:

(a) Isopropenyllithium   (c) Benzylsodium

(b) *sec*-Butyllithium   (d) *m*-Fluorophenylpotassium

SAMPLE SOLUTION (a) In the preparation of organolithium compounds from organic halides, lithium becomes bonded to the carbon which bore the halogen. Therefore, isopropenyllithium must arise from isopropenyl bromide.

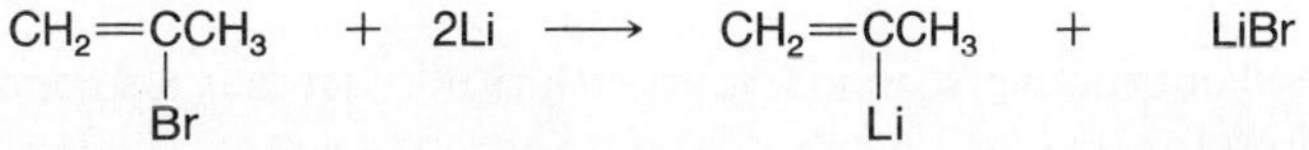

Isopropenyl bromide · · · Lithium · · · Isopropenyllithium · · · Lithium bromide

---

Reaction with an alkyl halide takes place at the metal surface. In the first step, an electron is transferred from the metal to the alkyl halide.

$$R:\ddot{\underset{\cdot\cdot}{X}}:\ +\ Li\cdot\ \longrightarrow\ [R:\ddot{\underset{\cdot\cdot}{X}}:]^{\bar{\cdot}}\ +\ \ \ Li^+$$

   Alkyl halide   Lithium   Anion radical   Lithium cation

Since the alkyl halide has gained one electron, it is negatively charged and has an odd number of electrons. It is an *anion radical.* The extra electron occupies an antibonding orbital. This anion radical fragments to an alkyl radical and a halide anion.

$$[R:\ddot{\underset{\cdot\cdot}{X}}:]^{\bar{\cdot}}\ \longrightarrow\ \ \ R\cdot\ \ +\ \ [:\ddot{\underset{\cdot\cdot}{X}}:]^-$$

   Anion radical   Alkyl radical   Halide anion

Following fragmentation, the alkyl radical rapidly combines with a metal atom to form the organometallic.

$$R\cdot\ \ +\ \ Li\cdot\ \longrightarrow\ \ R:Li$$

   Alkyl radical   Lithium   Alkyllithium

Among alkyl halides the order of reactivity is I > Br > Cl > F. Bromides combine high reactivity with ready availability and are used most often. Fluorides are relatively unreactive except under the most vigorous conditions.

Organolithium compounds are sometimes prepared in hydrocarbon solvents such as pentane and hexane, but normally diethyl ether is used. It is especially important that the solvent be anhydrous. Even trace amounts of water or alcohols will react with lithium to form insoluble hydroxide or alkoxide salts that coat the surface of the metal and prevent it from reacting with the alkyl halide. Furthermore, organolithium reagents are strong bases and react rapidly with even weak proton sources to form alkanes. We shall discuss this property of organolithium reagents in Section 15.5.

15.4 PREPARATION OF ORGANOMAGNESIUM COMPOUNDS. GRIGNARD REAGENTS

The most important organometallic reagents in organic chemistry are organomagnesium compounds. They are called *Grignard reagents* after the French chemist Victor Grignard. Grignard developed efficient methods for the preparation of organic derivatives of magnesium and demonstrated their application in the synthesis of alcohols. For these achievements he was awarded the 1912 Nobel prize in chemistry.

Grignard reagents are prepared directly from organic halides by reaction with magnesium, a group II metal.

$$RX\ \ +\ \ Mg\ \longrightarrow\ \ \ RMgX$$

   Organic halide   Magnesium   Organomagnesium halide

(R may be methyl or primary, secondary, or tertiary alkyl; it may also be a cycloalkyl, alkenyl, or aryl group.)

   Cyclohexyl chloride   Magnesium   Cyclohexylmagnesium chloride (96%)

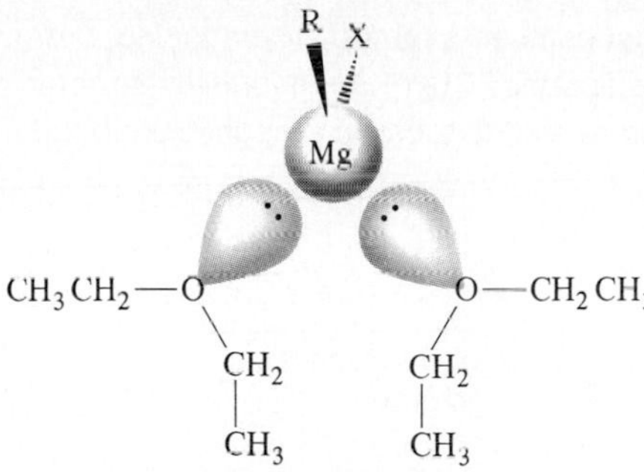

FIGURE 15.1 Solvation of an alkylmagnesium halide by diethyl ether. R, X, and the two diethyl ether molecules are arranged around magnesium in a tetrahedral geometry.

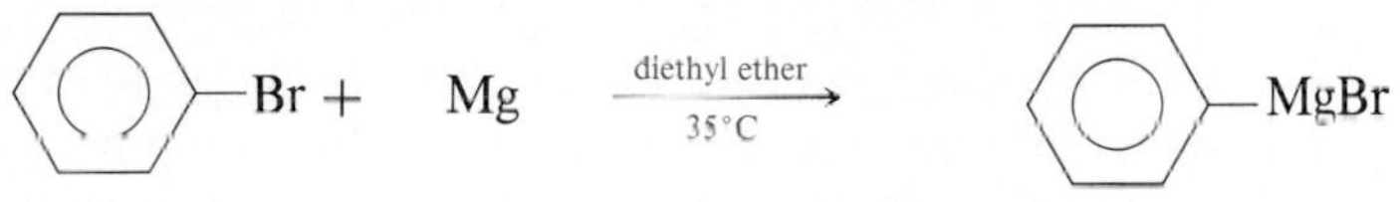

Bromobenzene   Magnesium   Phenylmagnesium bromide (95%)

Anhydrous diethyl ether is the customary solvent for the preparation of organomagnesium compounds. It stabilizes the Grignard reagent by forming a Lewis acid–Lewis base complex with it, as shown in Figure 15.1. Sometimes the reaction of an organic halide does not begin readily, but once started it is exothermic and maintains the temperature of the reaction mixture at the boiling point of diethyl ether (35°C).

The order of halide reactivity is I > Br > Cl > F, and alkyl halides are more reactive than aryl and vinyl halides. When more vigorous reaction conditions are required, such as when vinyl or aryl chlorides are used, tetrahydrofuran (THF) is chosen as the solvent instead of diethyl ether. Tetrahydrofuran's boiling point (65°C) is somewhat higher than that of diethyl ether and, more importantly, it forms a more stable complex with the Grignard reagent, which seems to increase the rate of its formation.

$$CH_2{=}CHCl \xrightarrow[\text{THF, 60°C}]{Mg} CH_2{=}CHMgCl$$

Vinyl chloride   Vinylmagnesium chloride (92%)

Grignard reagents exist in solution as an equilibrium mixture of an alkylmagnesium halide and a dialkylmagnesium.

$$2RMgX \underset{}{\overset{\text{fast}}{\rightleftharpoons}} R_2Mg \quad + \quad MgX_2$$

Alkylmagnesium   Dialkylmagnesium   Dihalide salt
halide   of magnesium

The alkylmagnesium halide is normally present in greater concentration than the dialkylmagnesium. For simplicity, organic chemists represent Grignard reagents as alkylmagnesium halides (RMgX).

PROBLEM 15.3 Write the structure of the alkylmagnesium halide or arylmagnesium halide formed from each of the following compounds on reaction with magnesium in diethyl ether:

(a) *p*-Bromofluorobenzene   (c) Iodocyclobutane

(b) Allyl chloride   (d) 1-Bromocyclohexene

SAMPLE SOLUTION (a) Of the two halogen substituents on the aromatic ring, bromine reacts much faster with magnesium than does fluorine. Therefore, fluorine is left intact on the ring while the carbon-bromine bond is converted to a carbon-magnesium bond.

$$F\text{—}\langle\bigcirc\rangle\text{—}Br \ + \ Mg \ \xrightarrow{\text{diethyl ether}} \ F\text{—}\langle\bigcirc\rangle\text{—}MgBr$$

p-Bromofluorobenzene   Magnesium p-Fluorophenylmagnesium bromide

---

The formation of a Grignard reagent can be viewed as taking place in a manner analogous to that of organolithium reagents except that each magnesium atom can participate in two separate one-electron transfer steps.

$$R\!:\!\ddot{\underset{\cdot\cdot}{X}}\!: \ + \ Mg\!\cdot \ \longrightarrow \ [R\!:\!\ddot{\underset{\cdot\cdot}{X}}\!:]^{\overline{\cdot}} \ + \ \overset{+}{Mg}\!\cdot$$

Alkyl halide   Magnesium   Anion radical

$$[R\!:\!\ddot{\underset{\cdot\cdot}{X}}\!:]^{\overline{\cdot}} \ + \ \overset{+}{Mg}\!\cdot \ \longrightarrow \ [R\!:\!Mg]^{+}\,[\!:\!\ddot{\underset{\cdot\cdot}{X}}\!:]^{-}$$

Anion radical   Alkylmagnesium halide

Organolithium and organomagnesium compounds find their most significant applications in the preparation of alcohols by reaction with aldehydes and ketones. Before discussing these reactions, let us first examine the reactions of these organometallic compounds with proton donors.

15.5 ORGANOLITHIUM AND ORGANOMAGNESIUM COMPOUNDS AS BRÖNSTED BASES

Organolithium and organomagnesium compounds are stable species when prepared in suitable solvents such as diethyl ether. The polar nature of their carbon-metal bonds, however, makes them strongly basic, and they react instantly with proton donors even as weakly acidic as water and alcohols. A proton is transferred from the hydroxyl group to the negatively polarized carbon of the organometallic to form an alkane.

$$\overset{\delta-}{R} \ \ \overset{\delta+}{H} \atop \underset{\delta+}{M} \ \ \underset{\delta-}{OR'} \ \longrightarrow \ RH + R'O^{-}M^{+}$$

$$CH_3CH_2CH_2CH_2Li + H_2O \longrightarrow CH_3CH_2CH_2CH_3 + \quad LiOH$$

Butyllithium   Water   Butane (100%)   Lithium hydroxide

$$\langle\bigcirc\rangle\text{—}MgBr + CH_3OH \longrightarrow \langle\bigcirc\rangle + \ CH_3OMgBr$$

Phenylmagnesium bromide   Methanol   Benzene (100%)   Methoxymagnesium bromide

Because of their basicity organolithium compounds and Grignard reagents cannot be prepared or used in the presence of any material that bears a hydroxyl group. Nor are these reagents compatible with —NH or —SH groups, which can also function as proton donors and convert an organolithium or organomagnesium compound to a hydrocarbon.

Organolithium and organomagnesium compounds possess a significant degree of carbanionic character in their carbon-metal bonds. Carbanions rank among the strongest bases that we will encounter in this text. Their conjugate acids are hydrocarbons — very weak acids indeed. The equilibrium constants K_a for ionization of hydrocarbons are much smaller than K_a for water and alcohols.

$$\underset{\text{Hydrocarbon}}{\overset{\diagup}{\underset{\diagdown}{C}}\!-\!H} \ \rightleftharpoons \ \underset{\text{Proton}}{H^+} \ + \ \underset{\text{Carbanion}}{\overset{\diagup}{\underset{\diagdown}{C}}\!:}$$

Table 15.1 presents some approximate data for the acid strengths of representative hydrocarbons.

Notice that the acidity of a proton depends strongly on the hybridization of the carbon to which it is bonded. Alkanes are weaker acids than alkenes, which are weaker than alkynes. This is because the electron pair of the carbanion is more stable when it is in an orbital with more s character. (Remember, s electrons are held more strongly than p electrons.) As the diagram in Figure 15.2 illustrates, the orbital occupied by the unshared pair is sp^3 hybridized in alkyl anions, sp^2 hybridized in alkenyl and aryl anions, and sp hybridized in alkynyl anions.

Acidity increases in progressing from the top of Table 15.1 to the bottom. An acid will transfer a proton to the conjugate base of any acid above it in the table. Organolithium compounds and Grignard reagents function as carbanion equivalents and will abstract a proton from any species more acidic than a hydrocarbon. Thus, N—H

TABLE 15.1

Approximate Acidities of Some Hydrocarbons and Reference Materials

Compound	Formula*	K_a	pK_a	Conjugate base
2-Methylpropane	$(CH_3)_3C\!-\!H$	10^{-71}	71	$(CH_3)_3\overset{\,\cdot}{C}\!:$
Ethane	$CH_3CH_2\!-\!H$	10^{-62}	62	$CH_3\overset{\,\cdot}{C}H_2$
Methane	$CH_3\!-\!H$	10^{-60}	60	$H_3\overset{\,\cdot}{C}\!:$
Ethylene	$CH_2\!=\!CH\!-\!H$	10^{-45}	45	$CH_2\!=\!\overset{\cdots}{C}H$
Benzene	$C_6H_5\!-\!H$	10^{-43}	43	$C_6H_5\!:$
Ammonia	$H_2N\!-\!H$	10^{-36}	36	$H_2\overset{\cdot\cdot}{N}\!:^-$
Acetylene	$HC\!\equiv\!CH$	10^{-26}	26	$HC\!\equiv\!\overset{\,\cdot}{C}\!:$
Ethanol	$CH_3CH_2O\!-\!H$	10^{-16}	16	$CH_3CH_2O^-$
Water	$HO\!-\!H$	1.8×10^{-16}	15.7	HO^-

* The acidic proton in each compound is in color.

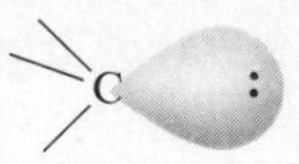

The electron pair of an alkyl anion occupies an orbital (sp^3) with 25 percent s character; alkyl anions are the least stable class of simple hydrocarbon anions.

(*a*)

   and

The electron pair of an alkenyl or aryl anion occupies an orbital (sp^2) with $33\frac{1}{3}$ percent s character.

(*b*)

The electron pair of alkynyl anion occupies an orbital (sp) with 50 percent s character; alkynyl anions are the most stable class of simple hydrocarbon anions.

(*c*)

FIGURE 15.2 Hybridization of the orbital that is occupied by the un-shared pair of electrons in (*a*) an alkyl anion; (*b*) an alkenyl and an aryl anion; and (*c*) an alkynyl anion.

groups and terminal acetylenes ($RC\equiv C-H$) are converted to their conjugate bases by proton transfer to organolithium and organomagnesium compounds.

$$CH_3Li \ + \ NH_3 \ \longrightarrow \ CH_4 \ + \ LiNH_2$$

Methyllithium	Ammonia	Methane	Lithium amide
(stronger base)	(stronger acid: $K_a = 10^{-36}$)	(weaker acid: $K_a \cong 10^{-60}$)	(weaker base)

$$CH_3CH_2MgBr + \ HC\equiv CH \ \longrightarrow \ CH_3CH_3 \ + \ HC\equiv CMgBr$$

Ethylmagnesium bromide	Acetylene	Ethane	Ethynylmagnesium bromide
(stronger base)	(stronger acid: $K_a \cong 10^{-26}$)	(weaker acid: $K_a \cong 10^{-62}$)	(weaker base)

PROBLEM 15.4 Butyllithium is commercially available and is frequently used by organic chemists as a strong base. Show how you could use butyllithium to prepare solutions containing

(*a*) Lithium diethylamide, $(CH_3CH_2)_2NLi$
(*b*) Lithium 1-hexanolate, $CH_3(CH_2)_4CH_2OLi$
(*c*) Lithium benzenethiolate, C_6H_5SLi

SAMPLE SOLUTION When butyllithium is used as a base it abstracts a proton, in this case a proton attached to nitrogen. The source of lithium diethylamide must be diethylamine.

$$(CH_3CH_2)_2NH + CH_3CH_2CH_2CH_2Li \longrightarrow (CH_3CH_2)_2NLi + CH_3CH_2CH_2CH_3$$

Diethylamine	Butyllithium	Lithium diethylamide	Butane
(stronger acid)	(stronger base)	(weaker base)	(weaker acid)

While diethylamine is not specifically listed in Table 15.1, its strength as an acid ($K_a \cong 10^{-36}$) is, as might be expected, similar to that of ammonia.

It is sometimes necessary in a synthesis to reduce an alkyl halide to a hydrocarbon. In these cases converting the halide to a Grignard reagent and then adding water or an alcohol as a proton source is a satisfactory procedure.

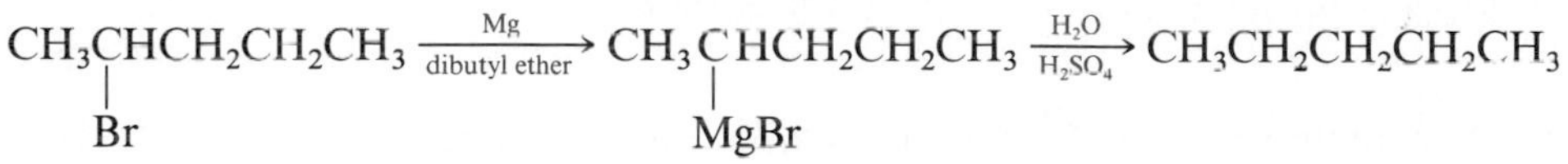

$$CH_3CHCH_2CH_2CH_3 \xrightarrow[\text{dibutyl ether}]{Mg} CH_3CHCH_2CH_2CH_3 \xrightarrow[H_2SO_4]{H_2O} CH_3CH_2CH_2CH_2CH_3$$

Br	MgBr	
2-Bromopentane	1-Methylbutylmagnesium bromide	Pentane (50–53%)

Dibutyl ether was used as the solvent in this example because its boiling point of 142°C is much higher than that of the product. Pentane boils at 36°C and can be distilled directly from the reaction mixture. Had diethyl ether (bp 35°C) been used as the solvent, isolation of the pentane product would have been very difficult.

Grignard reagents and organolithium reagents react with oxygen and so are usually prepared and used under an inert atmosphere such as nitrogen or argon. Nitro groups also react with these organometallics, and so the preparation will not succeed if a $-NO_2$ function is present in the alkyl halide from which the reagent is prepared, the solvent, or the substance with which it is designed to react.

15.6 SYNTHESIS OF ALCOHOLS USING GRIGNARD REAGENTS

The principal synthetic application of Grignard reagents is their reaction with certain carbonyl-containing compounds to produce alcohols. A new carbon-carbon bond is formed by the rapid exothermic reaction of a Grignard reagent with an aldehyde or ketone.

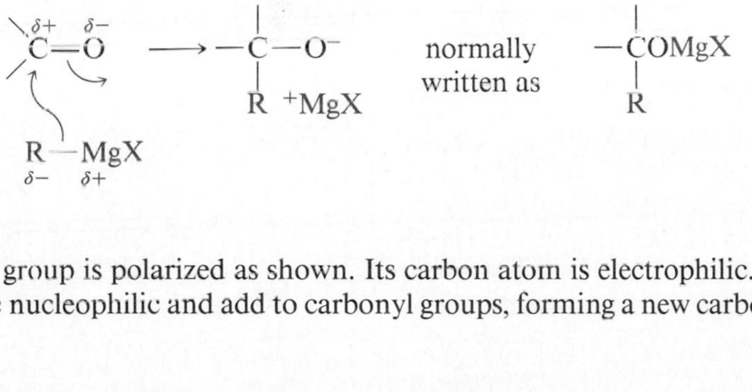

A carbonyl group is polarized as shown. Its carbon atom is electrophilic. Grignard reagents are nucleophilic and add to carbonyl groups, forming a new carbon-carbon bond. This addition step leads to formation of the alkoxymagnesium halide, which in the second stage of the synthesis is hydrolyzed to an alcohol.

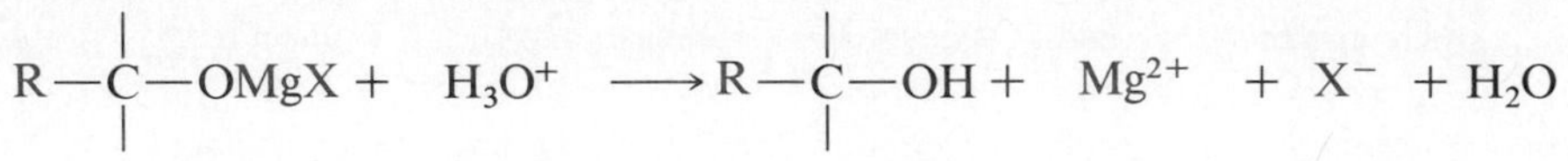

$$\underset{\substack{\text{Alkoxymagnesium} \\ \text{halide}}}{R-\overset{|}{\underset{|}{C}}-OMgX} + \underset{\substack{\text{Hydronium} \\ \text{ion}}}{H_3O^+} \longrightarrow \underset{\text{Alcohol}}{R-\overset{|}{\underset{|}{C}}-OH} + \underset{\substack{\text{Magnesium} \\ \text{ion}}}{Mg^{2+}} + \underset{\substack{\text{Halide} \\ \text{ion}}}{X^-} + \underset{\text{Water}}{H_2O}$$

The type of alcohol produced depends on the carbonyl-containing compound used. Substituents present on the carbonyl group in the reactant stay there — they become substituents on the carbon that bears the hydroxyl group in the product. Thus, formaldehyde reacts with Grignard reagents to yield primary alcohols, aldehydes yield secondary alcohols, and ketones yield tertiary alcohols.

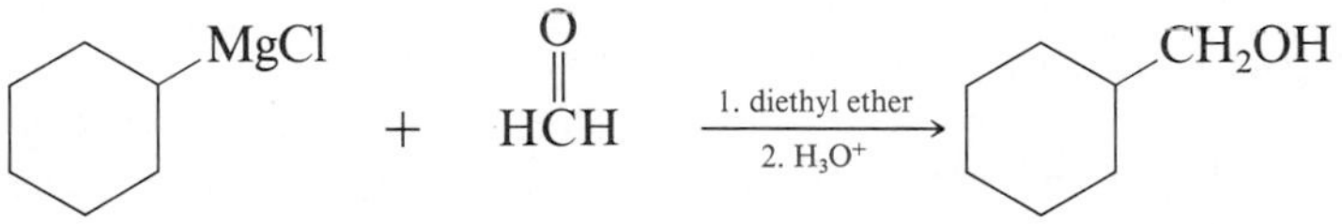

(a) Reaction of a Grignard reagent with formaldehyde

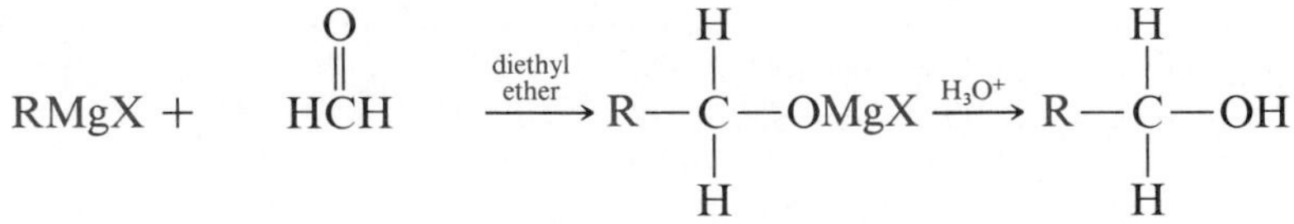

(b) Reaction of a Grignard reagent with an aldehyde

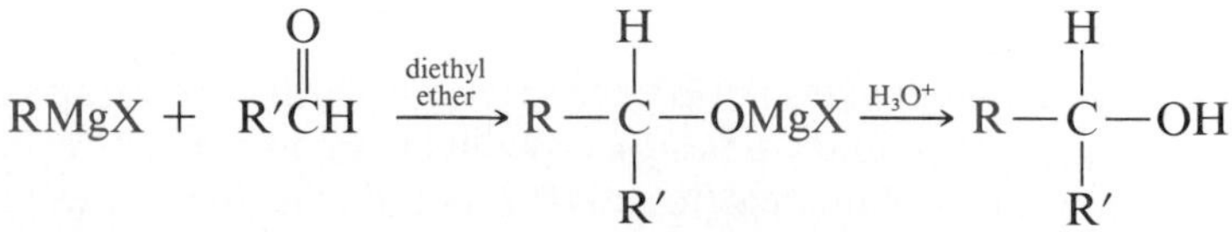

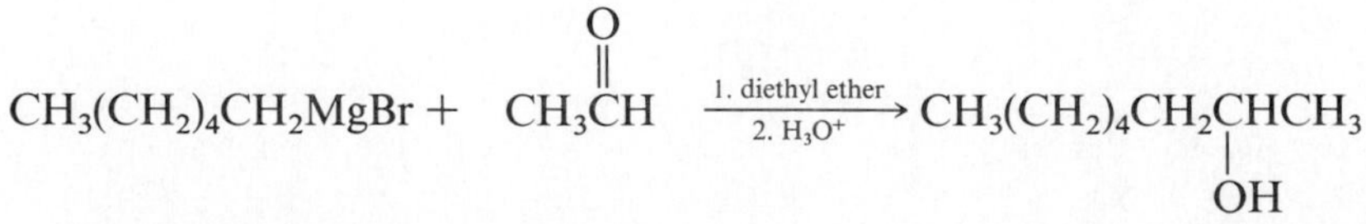

(c) Reaction of a Grignard reagent with a ketone

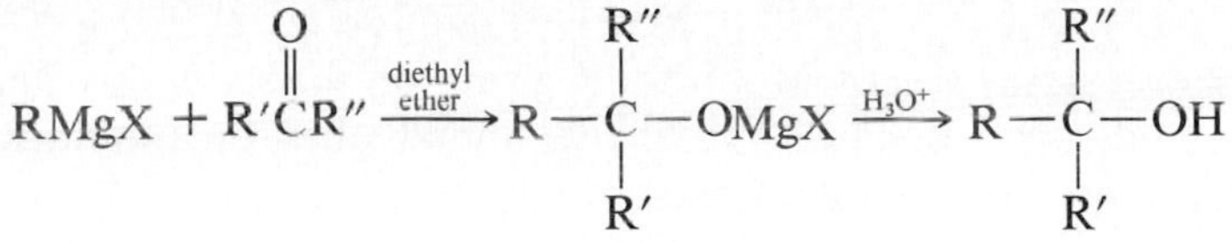

$$RMgX + R'\overset{\overset{\displaystyle O}{\|}}{C}R'' \xrightarrow[\text{ether}]{\text{diethyl}} R-\overset{\overset{\displaystyle R''}{|}}{\underset{\underset{\displaystyle R'}{|}}{C}}-OMgX \xrightarrow{H_3O^+} R-\overset{\overset{\displaystyle R''}{|}}{\underset{\underset{\displaystyle R'}{|}}{C}}-OH$$

| Grignard reagent | Ketone | Tertiary alkoxymagnesium halide | Tertiary alcohol |

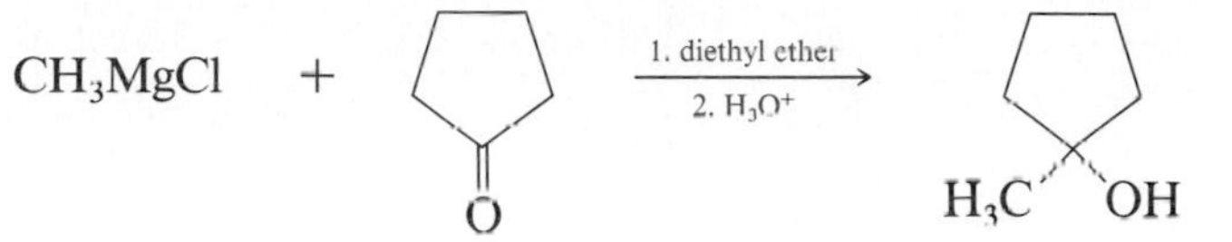

$$CH_3MgCl \quad + \quad \text{(cyclopentanone)} \xrightarrow[\text{2. } H_3O^+]{\text{1. diethyl ether}} \text{(1-methylcyclopentanol)}$$

| Methylmagnesium chloride | Cyclopentanone | 1-Methylcyclopentanol (62%) |

PROBLEM 15.5   Write the structure of the product of the reaction of propylmagnesium bromide with each of the following electrophiles. Assume the reactions are worked up by the addition of dilute aqueous acid in the usual manner.

(a)   Formaldehyde $H\overset{\overset{\displaystyle O}{\|}}{C}H$

(b)   Benzaldehyde $C_6H_5\overset{\overset{\displaystyle O}{\|}}{C}H$

(c)   Cyclohexanone   (cyclohexanone ring)$=O$

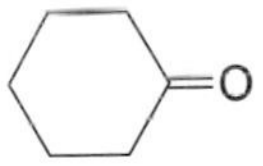

(d)   2-Butanone $CH_3\overset{\overset{\displaystyle O}{\|}}{C}CH_2CH_3$

SAMPLE SOLUTION   (a)   Grignard reagents react with formaldehyde to give primary alcohols having one more carbon atom than the alkyl halide from which the Grignard reagent was prepared. The product is 1-butanol.

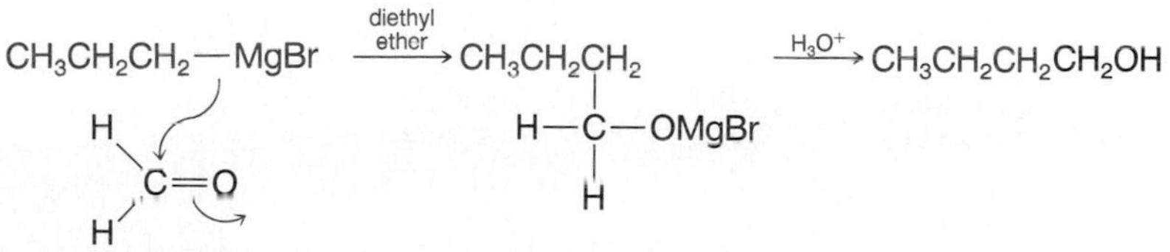

$$CH_3CH_2CH_2-MgBr \xrightarrow[\text{ether}]{\text{diethyl}} CH_3CH_2CH_2 \xrightarrow{H_3O^+} CH_3CH_2CH_2CH_2OH$$

Propylmagnesium bromide + formaldehyde

1-Butanol

The availability of efficient methods for the formation of carbon-carbon bonds is fundamental to organic synthesis. The addition of Grignard reagents to aldehydes and ketones is one of the most frequently used reactions in synthetic organic chemistry. Not only does it permit the extension of carbon chains, but since the product is an alcohol, a wide variety of subsequent functional group transformations is possible.

15.7 SYNTHESIS OF ALCOHOLS USING ORGANOLITHIUM REAGENTS

Organolithium reagents react with carbonyl groups in the same way that Grignard reagents do. In their reactions with aldehydes and ketones, organolithium reagents are somewhat more reactive than Grignard reagents.

$$RLi \ + \ \underset{\substack{\text{Aldehyde} \\ \text{or ketone}}}{\overset{}{\ce{C=O}}} \longrightarrow \underset{\text{Lithium alkoxide}}{R-\overset{|}{\underset{|}{C}}-OLi} \xrightarrow{H_3O^+} \underset{\text{Alcohol}}{R-\overset{|}{\underset{|}{C}}-OH}$$

Alkyl-
lithium
compound

$$CH_2\text{=}CHLi \ + \ \text{benzaldehyde} \xrightarrow[\text{2. } H_3O^+]{\text{1. diethyl ether}} \text{1-Phenyl-2-propen-1-ol (76\%)}$$

Vinyllithium   Benzaldehyde   1-Phenyl-2-propen-1-ol (76%)

15.8 SYNTHESIS OF ACETYLENIC ALCOHOLS

The preparation of acetylide anions was described in Sections 10.6 and 10.7, where their reaction with alkyl halides was shown to be an effective synthesis of higher alkynes. The nucleophilic reactivity of acetylide anions is also evident in their reactions with aldehydes and ketones to form alkynyl alcohols. (The reactions of sodium acetylides are normally carried out in liquid ammonia because the organometallic reagent is prepared in that solvent by the reaction of a terminal alkyne with sodium amide.)

$$\underset{\substack{\text{Terminal} \\ \text{alkyne}}}{RC\equiv CH} + \underset{\substack{\text{Sodium} \\ \text{amide}}}{NaNH_2} \xrightarrow[-33°C]{NH_3} \underset{\substack{\text{Sodium} \\ \text{alkynide}}}{RC\equiv CNa} + \underset{\text{Ammonia}}{NH_3}$$

The reaction of acetylide anions with aldehydes and ketones is entirely analogous to the reactions of Grignard reagents and organolithium reagents with carbonyl compounds.

$$\underset{\substack{\text{Sodium} \\ \text{alkynide}}}{RC\equiv CNa} + \underset{\substack{\text{Aldehyde} \\ \text{or ketone}}}{R'\overset{O}{\overset{||}{C}}R''} \xrightarrow{NH_3} \underset{\substack{\text{Sodium salt of an} \\ \text{alkynyl alcohol}}}{RC\equiv C-\overset{R''}{\underset{R'}{\overset{|}{C}}}-ONa} \xrightarrow{H_3O^+} \underset{\substack{\text{Alkynyl} \\ \text{alcohol}}}{RC\equiv C\overset{R''}{\underset{R'}{\overset{|}{C}}OH}}$$

HC≡CNa + (cyclohexanone) $\xrightarrow[\text{2. H}_3\text{O}^+]{\text{1. NH}_3}$ (1-ethynylcyclohexanol)

Sodium acetylide   Cyclohexanone   1-Ethynylcyclohexanol
                                                (65–75%)

Alkynyl Grignard reagents are customarily prepared by an acid-base reaction between a terminal alkyne and a Grignard reagent.

$$\text{CH}_3(\text{CH}_2)_3\text{C}\equiv\text{CH} + \text{CH}_3\text{CH}_2\text{MgBr} \xrightarrow{\text{diethyl ether}} \text{CH}_3(\text{CH}_2)_3\text{C}\equiv\text{CMgBr} + \text{CH}_3\text{CH}_3$$

1-Hexyne   Ethylmagnesium   1-Hexynylmagnesium   Ethane
                    bromide   bromide

$$\text{CH}_3(\text{CH}_2)_3\text{C}\equiv\text{CMgBr} + \overset{\text{O}}{\overset{\|}{\text{HCH}}} \xrightarrow[\text{2. H}_3\text{O}^+]{\text{1. diethyl ether}} \text{CH}_3(\text{CH}_2)_3\text{C}\equiv\text{CCH}_2\text{OH}$$

1-Hexynylmagnesium   Formaldehyde   2-Heptyn-1-ol (82%)
      bromide

15.9 RETROSYNTHETIC ANALYSIS

In earlier discussions of synthesis, the importance of reasoning backward from the target molecule to suitable starting materials has been stressed. A name for this process is *retrosynthetic analysis*. Organic chemists have employed this approach for many years but the term was invented and a formal statement of its principles was set forth only relatively recently by E. J. Corey at Harvard University. Beginning in the 1960s, Corey began studies aimed at making the strategy of organic synthesis sufficiently systematic so that the power of electronic computers could be applied to assist synthetic planning.

A symbol used to indicate a retrosynthetic step is an open arrow written from product to suitable precursors or fragments of those precursors.

Target molecule ⟹ precursors

Often the precursor is not defined completely, but rather its chemical nature is emphasized by writing it as a species to which it is equivalent for synthetic purposes. Thus, a Grignard reagent or organolithium reagent might be considered as synthetically equivalent to a carbanion.

RMgX or RLi   is synthetically equivalent to   R

To illustrate the application of retrosynthetic analysis as a guide to synthetic design, consider the preparation of alcohols by addition of Grignard reagents to aldehydes and ketones. To help decide what combination of carbonyl compound and Grignard reagent is required in order to prepare a target alcohol, direct your attention to the carbon atom that bears the hydroxyl group. Originally, this must have been the

Step 1:   Locate the hydroxyl-bearing carbon.

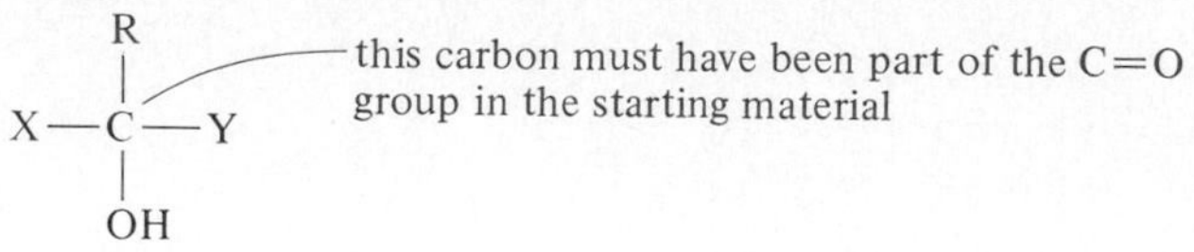

$$
\begin{array}{c}
\text{R} \\
| \\
\text{X} \!-\! \text{C} \!-\! \text{Y} \\
| \\
\text{OH}
\end{array}
$$

— this carbon must have been part of the C=O
group in the starting material

Step 2:   Disconnect one of the organic substituents attached to the carbon that
bears the hydroxyl group.

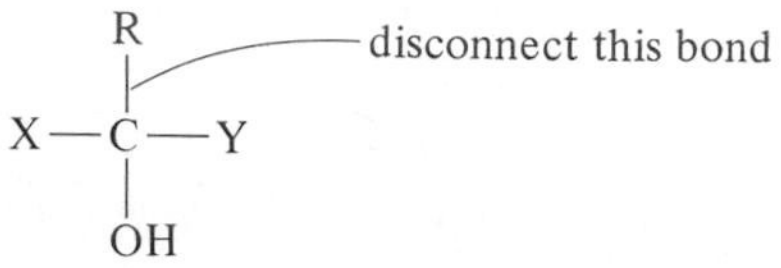

$$
\begin{array}{c}
\text{R} \\
| \\
\text{X} \!-\! \text{C} \!-\! \text{Y} \\
| \\
\text{OH}
\end{array}
$$

— disconnect this bond

Step 3:   Steps 1 and 2 reveal the carbonyl-containing substrate and the
carbanionic fragment.

$$
\begin{array}{c}
\text{R} \\
\sim\!\sim \\
\text{X} \!-\! \text{C} \!-\! \text{Y} \\
| \\
\text{OH}
\end{array}
\qquad \Longrightarrow \qquad
\begin{array}{c}
\text{R}^- \\
\text{X} \qquad \text{Y} \\
\diagdown\;\diagup \\
\text{C} \\
\| \\
\text{O}
\end{array}
$$

Step 4:   Since a Grignard reagent may be considered as synthetically
equivalent to a carbanion, this suggests the synthesis shown.

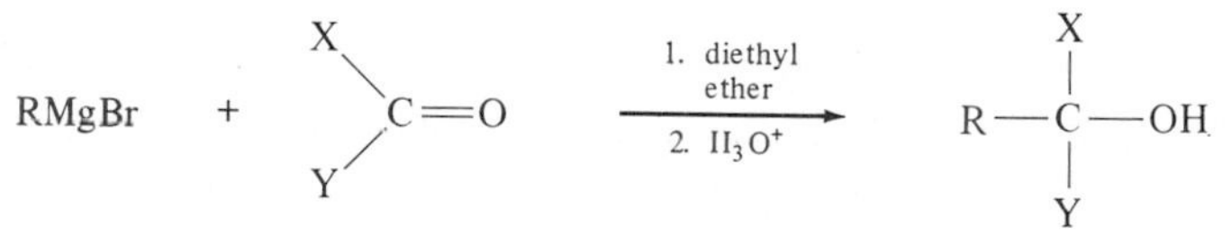

FIGURE 15.3   A retrosynthetic analysis of alcohol preparation by way of the addition
of a Grignard reagent to an aldehyde or ketone.

carbonyl carbon of the aldehyde or ketone (Figure 15.3, step 1). Next, as shown in
Figure 15.3, step 2, mentally disconnect a bond between that carbon and one of its
attached groups (other than hydrogen). The attached group is the group that is to be
transferred from the Grignard reagent. Once you recognize these two structural
fragments, the carbonyl partner and the carbanion that attacks it (Figure 15.3, step
3), you can readily determine the synthetic mode wherein a Grignard reagent is used
as the synthetic equivalent of a carbanion (Figure 15.3, step 4).

Primary alcohols, by this analysis, are seen to be the products of Grignard addition
to formaldehyde.

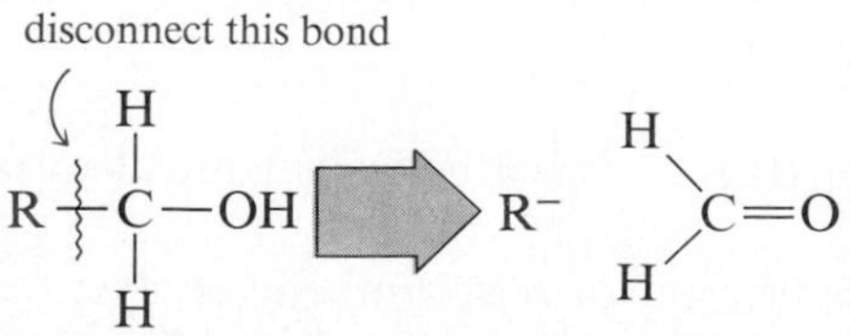

disconnect this bond

$$
\begin{array}{c}
\text{H} \\
| \\
\text{R} \!-\!\!\!\!\mid\!\!\!\!- \text{C} \!-\! \text{OH} \\
| \\
\text{H}
\end{array}
\quad \Longrightarrow \quad
\text{R}^-
\qquad
\begin{array}{c}
\text{H} \\
\diagdown \\
\text{C} \!=\! \text{O} \\
\diagup \\
\text{H}
\end{array}
$$

Secondary alcohols may be prepared by *two* different combinations of Grignard
reagent and aldehyde.

(a) Disconnect the R—C bond

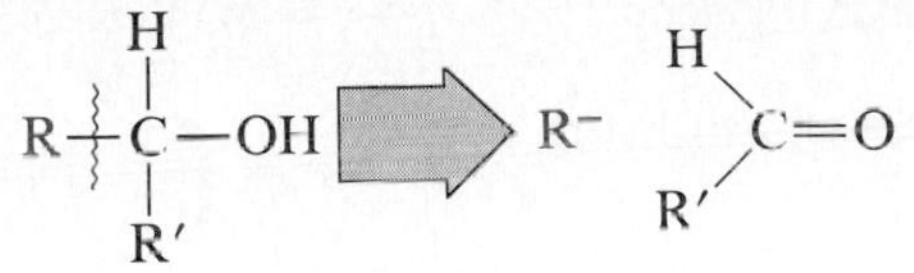

$$R \overset{H}{\underset{R'}{\vdash C}} -OH \Longrightarrow R^- \quad \overset{H}{\underset{R'}{\diagdown}}C=O$$

(b) Disconnect the R'—C bond

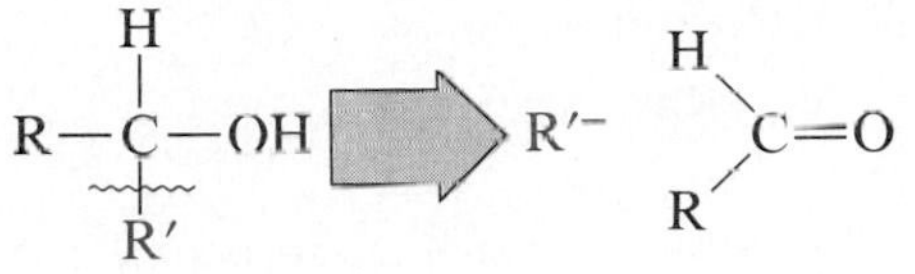

$$R -\overset{H}{\underset{R'}{\overset{|}{C}}} -OH \Longrightarrow R'^- \quad \overset{H}{\underset{R}{\diagdown}}C=O$$

There are three combinations of Grignard reagent and ketone that give rise to tertiary alcohols.

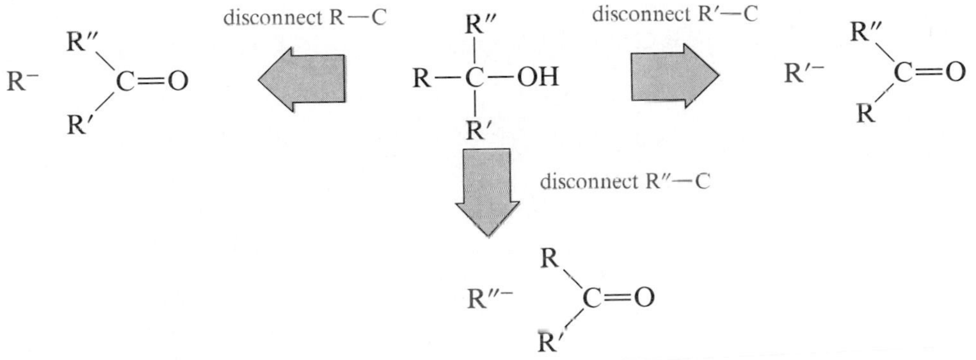

$$R^- \quad \overset{R''}{\underset{R'}{\diagdown}}C=O \xleftarrow{\text{disconnect R—C}} R -\overset{R''}{\underset{R'}{\overset{|}{C}}} -OH \xrightarrow{\text{disconnect R'—C}} R'^- \quad \overset{R''}{\underset{R}{\diagdown}}C=O$$

$$\downarrow \text{disconnect R''—C}$$

$$R''^- \quad \overset{R}{\underset{R'}{\diagdown}}C=O$$

Usually, there is little to choose among the various routes leading to a particular target alcohol. For example, all three of the routes shown below have been used to prepare the tertiary alcohol 2-phenyl-2-butanol.

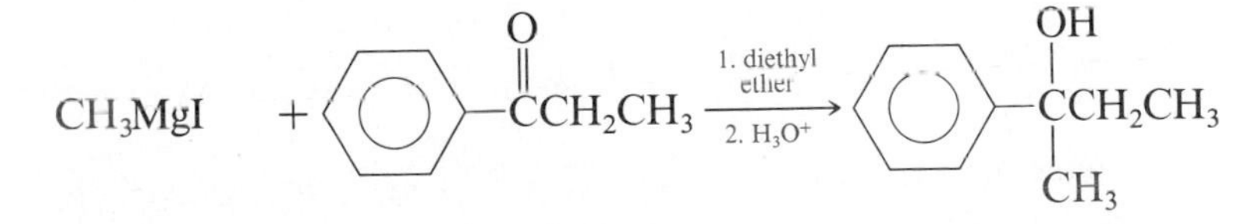

$$CH_3MgI \; + \; \langle \text{Ph} \rangle -\overset{O}{\overset{\|}{C}}CH_2CH_3 \xrightarrow[\text{2. } H_3O^+]{\text{1. diethyl ether}} \langle \text{Ph} \rangle -\overset{OH}{\underset{CH_3}{\overset{|}{C}}}CH_2CH_3$$

| Methylmagnesium iodide | Propiophenone | 2-Phenyl-2-butanol |

$$CH_3CH_2MgBr \; + \; \langle \text{Ph} \rangle -\overset{O}{\overset{\|}{C}}CH_3 \xrightarrow[\text{2. } H_3O^+]{\text{1. diethyl ether}} \langle \text{Ph} \rangle -\overset{OH}{\underset{CH_3}{\overset{|}{C}}}CH_2CH_3$$

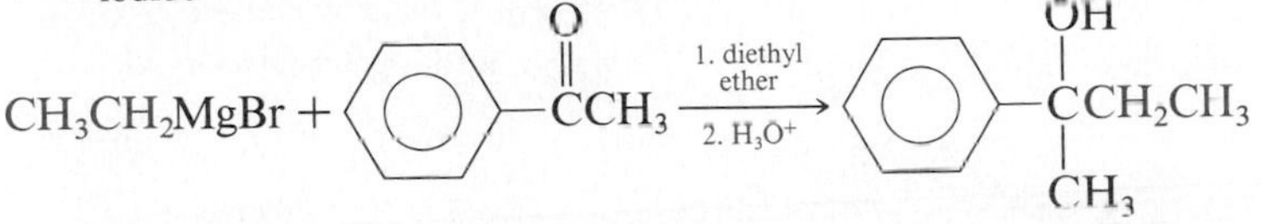

| Ethylmagnesium bromide | Acetophenone | 2-Phenyl-2-butanol |

$$\langle \text{Ph} \rangle -MgBr \; + \; CH_3\overset{O}{\overset{\|}{C}}CH_2CH_3 \xrightarrow[\text{2. } H_3O^+]{\text{1. diethyl ether}} \langle \text{Ph} \rangle -\overset{OH}{\underset{CH_3}{\overset{|}{C}}}CH_2CH_3$$

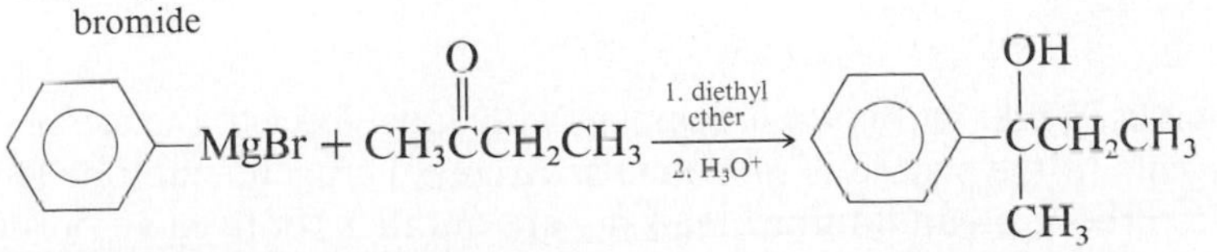

| Phenylmagnesium bromide | 2-Butanone | 2-Phenyl-2-butanol |

PROBLEM 15.6 Suggest two ways in which each of the following alcohols might be prepared by using a Grignard reagent:

(a) 2-Hexanol, $CH_3CHCH_2CH_2CH_2CH_3$
  |
  OH

(b) 1-Phenyl-1-propanol, $C_6H_5CHCH_2CH_3$
  |
  OH

(c) 2-Phenyl-2-propanol, $C_6H_5C(CH_3)_2$
  |
  OH

SAMPLE SOLUTION (a) Since 2-hexanol is a secondary alcohol, it arises by the reaction of a Grignard reagent with an aldehyde. Disconnection of bonds to the hydroxyl-bearing carbon generates two pairs of structural fragments:

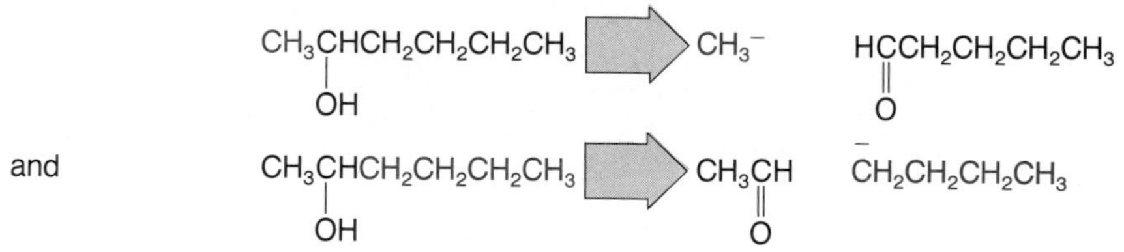

$CH_3CHCH_2CH_2CH_2CH_3$ ⟹ CH_3^- $HCCH_2CH_2CH_2CH_3$
  | ‖
  OH   O

and

$CH_3CHCH_2CH_2CH_2CH_3$ ⟹ CH_3CH $CH_2CH_2CH_2CH_3$
  | ‖
  OH   O

Therefore, one of the routes involves the addition of a methyl Grignard reagent to a five-carbon aldehyde

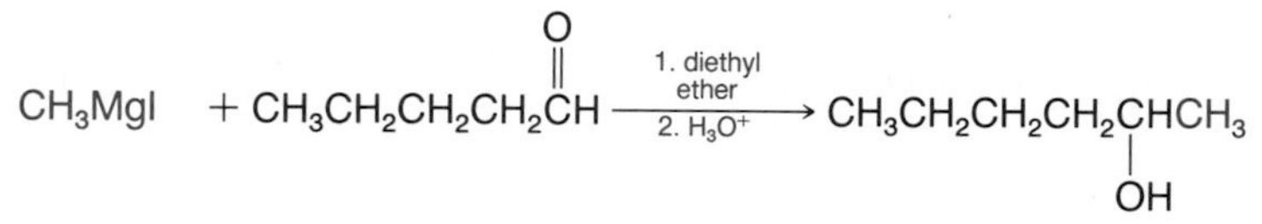

$$CH_3MgI + CH_3CH_2CH_2CH_2\overset{O}{\overset{\|}{C}}H \xrightarrow[\text{2. H}_3\text{O}^+]{\text{1. diethyl ether}} CH_3CH_2CH_2CH_2\underset{\underset{OH}{|}}{C}HCH_3$$

Methylmag-   Pentanal   2-Hexanol
nesium iodide

while the other route requires addition of a butylmagnesium halide to a two-carbon aldehyde.

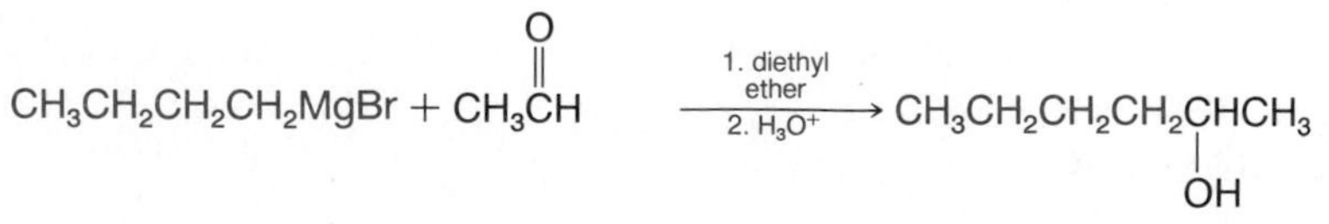

$$CH_3CH_2CH_2CH_2MgBr + CH_3\overset{O}{\overset{\|}{C}}H \xrightarrow[\text{2. H}_3\text{O}^+]{\text{1. diethyl ether}} CH_3CH_2CH_2CH_2\underset{\underset{OH}{|}}{C}HCH_3$$

Butylmagnesium   Acetaldehyde   2-Hexanol
bromide

All that has been said in this section applies with equal force to the use of organolithium reagents in the synthesis of alcohols. Grignard reagents are one source of nucleophilic carbon; organolithium reagents are another. Both have pronounced carbanionic character in their carbon-metal bonds and undergo the same kind of reaction with aldehydes and ketones.

15.10 PREPARATION OF TERTIARY ALCOHOLS FROM ESTERS AND GRIGNARD REAGENTS

Tertiary alcohols can also be prepared by a variation of the Grignard synthesis using an ester as the carbonyl component. The reaction of an ester — a methyl ester, for example — with a Grignard reagent follows the course

$$
\underset{\substack{\text{Grignard} \\ \text{reagent}}}{\text{RMgX}} + \underset{\substack{\text{Methyl} \\ \text{ester}}}{\text{R}'\overset{\overset{\displaystyle O}{\|}}{\text{C}}\text{OCH}_3} \xrightarrow{\text{diethyl ether}} \underset{\text{Ketone}}{\text{R}'\overset{\overset{\displaystyle O}{\|}}{\text{C}}\text{R}} + \underset{\substack{\text{Methoxymagnesium} \\ \text{halide}}}{\text{CH}_3\text{OMgX}}
$$

$$
\xrightarrow[\substack{\text{1. RMgX, diethyl ether} \\ \text{2. H}_3\text{O}^+}]{}
$$

$$
\underset{\text{Tertiary alcohol}}{\text{R}'\overset{\overset{\displaystyle OH}{|}}{\underset{\underset{\displaystyle R}{|}}{\text{C}}}\text{R}}
$$

Two moles of a Grignard reagent are required per mole of ester. Two of the groups attached to the carbon that bears the hydroxyl must therefore be identical, since both are derived from the Grignard reagent. A ketone is an intermediate in the reaction but reacts with the second equivalent of the Grignard reagent as fast as it is formed. Ketones are more reactive than esters toward Grignard reagents, so it is not possible to stop the reaction at the ketone stage even if only one equivalent of the Grignard reagent is used.

Methyl and ethyl esters are more readily available than most other types and are the ones most often used.

$$
\underset{\substack{\text{Methylmagnesium} \\ \text{bromide}}}{2\text{CH}_3\text{MgBr}} + \underset{\substack{\text{Methyl} \\ \text{2-methylpropanoate}}}{(\text{CH}_3)_2\text{CH}\overset{\overset{\displaystyle O}{\|}}{\text{C}}\text{OCH}_3} \xrightarrow[\text{2. H}_3\text{O}^+]{\text{1. diethyl ether}} \underset{\substack{\text{2,3-Dimethyl-} \\ \text{2-butanol (73\%)}}}{(\text{CH}_3)_2\text{CH}\overset{\overset{\displaystyle OH}{|}}{\underset{\underset{\displaystyle CH_3}{|}}{\text{C}}}\text{CH}_3} + \underset{\text{Methanol}}{\text{CH}_3\text{OH}}
$$

PROBLEM 15.7   What combination of ester and Grignard reagent could you use to prepare each of the following tertiary alcohols?

(a) 3-Methyl-3-pentanol
(b) 6-Methyl-6-undecanol
(c) $(\text{C}_6\text{H}_5)_2\text{COH}$

$\triangle$

(d) $(\text{C}_6\text{H}_5)_3\text{COH}$

SAMPLE SOLUTION   (a)   The carbon that bears the hydroxyl substituent has two ethyl groups and a methyl group attached to it.

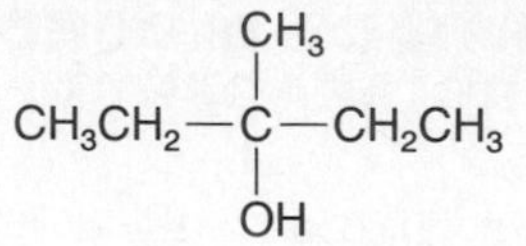

$$CH_3CH_2 - \underset{\underset{OH}{\overset{\overset{CH_3}{|}}{|}}}{C} - CH_2CH_3 \qquad \text{3-Methyl-3-pentanol}$$

The two groups that are the same, ethyl in this case, come from the Grignard reagent. Therefore, use an ethyl Grignard reagent and an acetate ester.

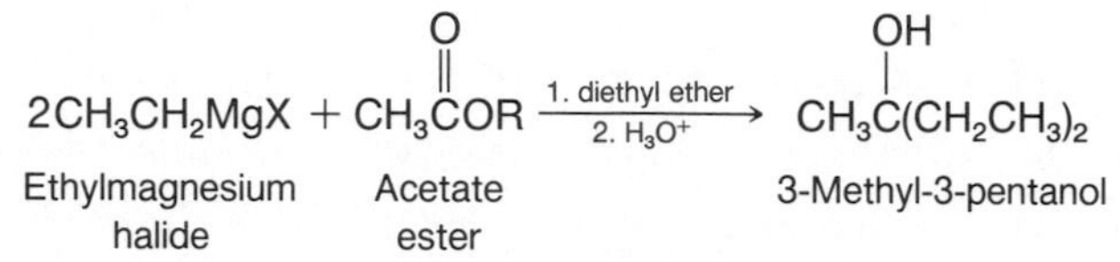

$$2CH_3CH_2MgX + CH_3\overset{\overset{O}{||}}{C}OR \xrightarrow[\text{2. } H_3O^+]{\text{1. diethyl ether}} CH_3\overset{\overset{OH}{|}}{C}(CH_2CH_3)_2$$

Ethylmagnesium   Acetate   3-Methyl-3-pentanol

halide   ester

Appropriate choices would be, for example, ethylmagnesium bromide and ethyl acetate.

15.11 ALKANE SYNTHESIS USING ORGANOCOPPER REAGENTS

Organometallic compounds of copper have been known for a long time, but their versatility as reagents in synthetic organic chemistry has only recently been recognized. The most useful organocopper reagents are the lithium dialkylcuprates, which result when a copper(I) (cuprous) halide reacts with two equivalents of an alkyllithium in diethyl ether or tetrahydrofuran.

$$2RLi + CuX \xrightarrow[\text{THF}]{\substack{\text{diethyl} \\ \text{ether or}}} R_2CuLi + LiX$$

Alkyllithium   Cu(I) halide   Lithium   Lithium

(X=Cl, Br, I)   dialkylcuprate   halide

In the first stage of the preparation, one molar equivalent of alkyllithium displaces halide from copper to give an alkylcopper(I) species.

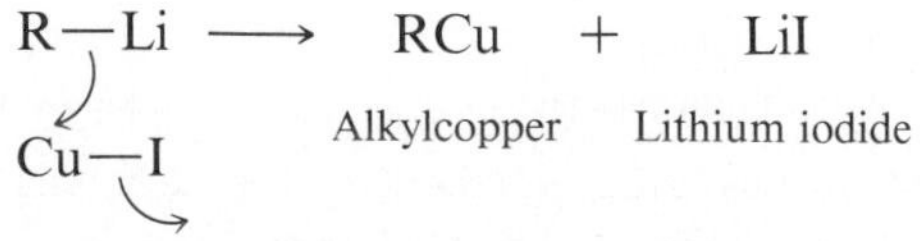

$$R - Li \longrightarrow RCu + LiI$$
$$Cu - I$$

Alkylcopper   Lithium iodide

The second molar equivalent of the alkyllithium adds to the alkylcopper to give a negatively charged dialkyl-substituted derivative of copper(I) called a *dialkylcuprate* anion. It is formed as its lithium salt, a lithium dialkylcuprate.

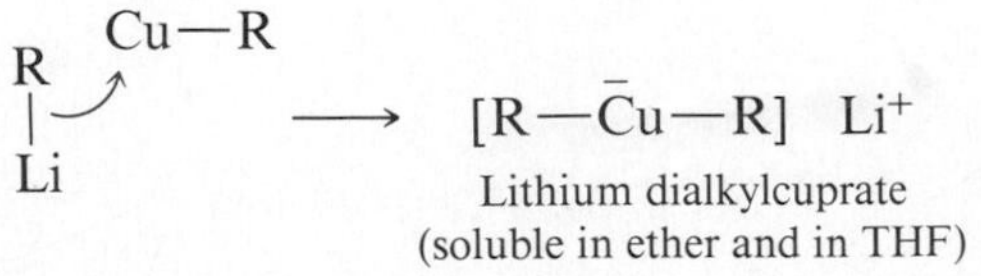

$$\underset{Li}{\overset{Cu - R}{\overset{|}{R}}} \longrightarrow [R - \bar{C}u - R] \; Li^+$$

Lithium dialkylcuprate

(soluble in ether and in THF)

Lithium dialkylcuprates react with alkyl halides to produce alkanes by carbon-

carbon bond formation between the alkyl group of the alkyl halide and the alkyl group of the dialkylcuprate.

$$R_2CuLi \ + \ R'X \ \longrightarrow R-R' + \ RCu \ + \ LiX$$

| Lithium dialkylcuprate | Alkyl halide | Alkane | Alkylcopper | Lithium halide |

Primary alkyl halides, especially iodides, are the best substrates. Elimination becomes a problem with secondary and tertiary alkyl halides.

$$(CH_3)_2CuLi \ + CH_3(CH_2)_8CH_2I \xrightarrow[0°C]{ether} CH_3(CH_2)_8CH_2CH_3$$

Lithium dimethyl-cuprate   1-Iododecane   Undecane (90%)

$$(CH_3CH_2CH_2CH_2)_2CuLi + CH_3(CH_2)_3CH_2Cl \xrightarrow[25°C]{THF} CH_3(CH_2)_7CH_3$$

Lithium dibutylcuprate   1-Chloropentane   Nonane (80%)

Lithium diarylcuprates are prepared in the same way as lithium dialkylcuprates and undergo comparable reactions with primary alkyl halides.

$$(C_6H_5)_2CuLi \ + ICH_2(CH_2)_6CH_3 \xrightarrow[ether]{diethyl} C_6H_5CH_2(CH_2)_6CH_3$$

Lithium diphenyl-cuprate   1-Iodooctane   1-Phenyloctane (99%)

The most frequently used organocuprates are those in which the alkyl group is primary. Steric hindrance seems to make organocuprates that bear secondary and tertiary alkyl groups less reactive toward alkyl halides. They are less stable as well and tend to decompose before they react with the substrate. The reaction of cuprate reagents with alkyl halides follows the usual S_N2 order: $CH_3 >$ primary $>$ secondary $>$ tertiary and $I > Br > Cl > F$. p-Toluenesulfonate esters are suitable substrates and are somewhat more reactive than halides. Because the alkyl halide and dialkylcuprate reagent should both be primary in order to produce satisfactory yields of coupled products, the reaction is limited to the formation of RCH_2-CH_2R' and RCH_2-CH_3 bonds in alkanes.

A key step in the reaction mechanism appears to be nucleophilic attack on the alkyl halide by the negatively charged copper atom. The intermediate thus formed is unstable and fragments to the observed products.

$$R_2\overset{-}{Cu} \ + R'-X \longrightarrow [R_2CuR' \ X^-] \longrightarrow RR' \ + \ RCu \ + \ X^-$$

Lithium dialkylcuprate   Alkyl halide   Alkane   Alkylcopper   Halide ion

However, the intimate details of the reaction mechanism are not well understood. Indeed, there is probably more than one mechanism by which cuprates react with organic halogen compounds. Vinyl halides and aryl halides are known to be very unreactive toward nucleophilic attack, yet react smoothly with lithium dialkylcuprates.

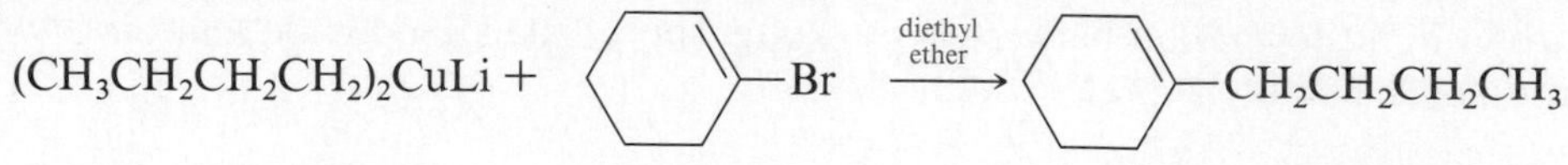

$$(CH_3CH_2CH_2CH_2)_2CuLi + \underset{\text{1-Bromocyclohexene}}{\text{⬡—Br}} \xrightarrow[\text{ether}]{\text{diethyl}} \underset{\text{1-Butylcyclohexene (80%)}}{\text{⬡—CH}_2CH_2CH_2CH_3}$$

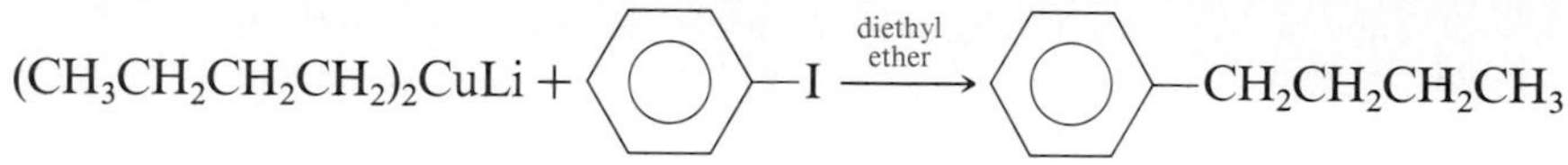

Lithium dibutylcuprate

$$(CH_3CH_2CH_2CH_2)_2CuLi + \underset{\text{Iodobenzene}}{\text{⬡—I}} \xrightarrow[\text{ether}]{\text{diethyl}} \underset{\text{Butylbenzene (75%)}}{\text{⬡—CH}_2CH_2CH_2CH_3}$$

Lithium dibutylcuprate

PROBLEM 15.8   Suggest a combination of organic halide and cuprate reagent appropriate for the preparation of each of the following compounds:

(a)   2-Methylbutane
(b)   1,3,3-Trimethylcyclopentene

SAMPLE SOLUTION   (a)   First inspect the target molecule to see which bonds are capable of being formed by reaction of an alkyl halide and a cuprate, bearing in mind that neither the alkyl halide nor the alkyl group of the lithium dialkylcuprate should be secondary or tertiary.

A bond between a methyl group and a methylene group can be formed ——→

$$CH_3—CH_2—\overset{\overset{\displaystyle CH_3}{|}}{\underset{|}{CH}}—CH_3$$

None of the bonds to the tertiary carbon can be formed efficiently

There are two combinations, both acceptable, that give rise to the $CH_3—CH_2$ bond.

$$\underset{\substack{\text{Lithium dimethyl-}\\\text{cuprate}}}{(CH_3)_2CuLi} + \underset{\substack{\text{1-Bromo-2-methyl-}\\\text{propane}\\\text{(isobutyl bromide)}}}{BrCH_2CH(CH_3)_2} \longrightarrow \underset{\text{2-Methylbutane}}{CH_3CH_2CH(CH_3)_2}$$

$$\underset{\substack{\text{Iodomethane}\\\text{(methyl iodide)}}}{CH_3I} + \underset{\text{Lithium diisobutylcuprate}}{LiCu[CH_2CH(CH_3)_2]_2} \longrightarrow \underset{\text{2-Methylbutane}}{CH_3CH_2CH(CH_3)_2}$$

15.12   ORGANOZINC INTERMEDIATES IN SYNTHESIS

Zinc reacts with alkyl halides in a manner similar to that of magnesium.

$$\underset{\text{Alkyl halide}}{RX} + \underset{\text{Zinc}}{Zn} \xrightarrow{\text{ether}} \underset{\text{Alkylzinc halide}}{RZnX}$$

Indeed, Victor Grignard was led to study organomagnesium compounds because of earlier work he performed in the area of organic derivatives of zinc. Organozinc reagents are not nearly as reactive toward aldehydes and ketones as Grignard reagents and organolithium compounds but are involved as intermediates in a number of common laboratory transformations.

In acidic media zinc reduces alkyl halides to the corresponding alkanes.

$$RX \xrightarrow[H^+]{Zn} RH$$

Alkyl halide   Alkane

$$CH_3(CH_2)_{14}CH_2I \xrightarrow[\substack{\text{acetic acid,} \\ \text{HCl}}]{Zn} CH_3(CH_2)_{14}CH_3$$

1-Iodohexadecane   Hexadecane (85%)

Presumably, an alkylzinc halide intermediate is formed, which then abstracts a proton from the acid.

$$RX \quad + Zn \longrightarrow \quad RZnX$$

Alkyl halide   Zinc   Alkylzinc halide

$$\overset{\delta-}{R}\!\!-\!\!\overset{\delta+}{Zn}X \ + HX \longrightarrow \ RH \ + \quad ZnX_2$$

Alkylzinc halide   Acid   Alkane   Zinc salt of acid

Vicinal dihalides undergo dehalogenation to form alkenes on treatment with zinc.

$$\begin{array}{c} R_2C-CR_2 \\ \; | \quad\;\; | \\ X \quad X \end{array} + Zn \xrightarrow[\text{or ethyl alcohol}]{\text{diethyl ether}} R_2C{=}CR_2 + \quad ZnX_2$$

Vicinal dihalide   Zinc   Alkene   Zinc dihalide

$$\begin{array}{c} \qquad\qquad\quad CH_3 \\ \qquad\qquad\quad | \\ C_6H_5CHCHCHC_6H_5 \\ \;\;\; | \;\; | \\ \;\;\; Br \; Br \end{array} \xrightarrow[\text{ethanol}]{Zn} \begin{array}{c} \;\; CH_3 \\ \;\; | \\ C_6H_5CH{=}CHCHC_6H_5 \end{array}$$

1,2-Dibromo-1,3-diphenylbutane   1,3-Diphenyl-1-butene (89%)

Mechanistically these eliminations can be understood as reactions that take place when there is a halide leaving group β to a carbon-zinc bond. This bond serves as a source of electrons to trigger the *β elimination* process.

$$\begin{array}{c} X \\ | \\ R_2C-CR_2 \\ | \\ X \end{array} + Zn \longrightarrow \begin{array}{c} X \\ | \\ R_2C-CR_2 \\ | \\ ZnX \end{array} \longrightarrow R_2C{=}CR_2 + ZnX_2$$

Vicinal dihalide   Zinc β-Haloalkylzinc   Alkene   Zinc
                          halide   dihalide

Dihalides in which halogens are in a 1,3 relationship are converted to cyclopropane derivatives by treatment with zinc in ethanol. This reaction may be classified as a γ (gamma) *elimination*.

$$R_2C \overset{\alpha}{\underset{X}{\overset{\beta}{\diagdown}}} \overset{C}{\underset{\gamma}{\diagup}} CR_2 + Zn \longrightarrow R_2C \overset{C}{\diagdown\diagup} CR_2 + ZnX_2$$

1,3-Dihalide   Zinc   Cyclopropane derivative   Zinc dihalide

$$\begin{array}{cc} CH_3CH_2 & CH_2CH_3 \\ & C \\ BrCH_2 & CH_2Br \end{array} \xrightarrow[\substack{CH_3CH_2OH, \\ H_2O}]{Zn} \begin{array}{cc} CH_3CH_2 & CH_2CH_3 \\ & \triangle \end{array}$$

3,3-Di(bromomethyl)pentane   1,1-Diethylcyclopropane (92%)

An organozinc intermediate is formed in these cyclization reactions, and the polarized carbon-zinc bond acts as an internal nucleophile to displace halide from a nearby carbon.

$$R_2C \overset{C}{\underset{X}{\diagdown}} \overset{C}{\underset{X}{\diagup}} CR_2 + Zn \longrightarrow R_2C \overset{C}{\underset{ZnX}{\diagdown}} CR_2{-}X \longrightarrow R_2C \overset{C}{\diagdown\diagup} CR_2 + ZnX_2$$

When halogens are farther removed from each other, the yields of ring-closed products are generally poor, so the zinc-promoted cyclization of dihalides is limited to the preparation of cyclopropanes.

PROBLEM 15.9   What two constitutionally isomeric compounds having the molecular formula $C_4H_8Br_2$ yield methylcyclopropane on treatment with zinc in aqueous ethanol?

Cyclopropanes may be synthesized by another reaction involving organozinc reagents. A zinc-copper couple [Zn(Cu)], that is, zinc that has had its surface activated with a little copper, reacts with diiodomethane in ether to give a solution containing iodomethylzinc iodide, ICH_2ZnI.

$$ICH_2I \quad + \quad Zn \xrightarrow[Cu]{\text{diethyl ether}} \quad ICH_2ZnI$$

Diiodomethane   Zinc   Iodomethylzinc iodide

Iodomethylzinc iodide is referred to as the *Simmons-Smith reagent* after Howard E. Simmons and Ronald D. Smith of E.I. DuPont de Nemours & Co., who first described its value as an effective agent for methylene transfer to alkenes. When an alkene is added to a solution of iodomethylzinc iodide in ether, a cyclopropane is formed.

$$CH_2{=}C \overset{CH_2CH_3}{\underset{CH_3}{\diagup}} \xrightarrow[\text{diethyl ether}]{CH_2I_2, \ Zn(Cu)} \quad \triangle \overset{CH_2CH_3}{\underset{CH_3}{\diagdown}}$$

2-Methyl-1-butene   1-Ethyl-1-methylcyclopropane (79%)

The reaction seems to proceed by a direct transfer of a methylene (CH_2) unit from the organometallic to the alkene.

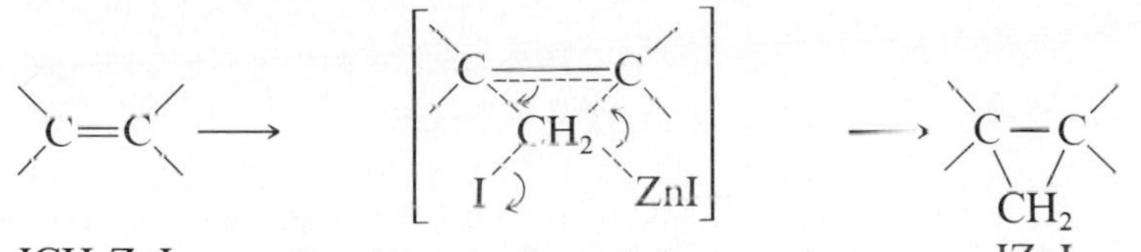

ICH₂ZnI — Transition state for methylene transfer — IZnI

PROBLEM 15.10 What alkenes would you choose as starting materials in order to prepare each of the following cyclopropane derivatives by reaction with iodomethylzinc iodide?

(a)

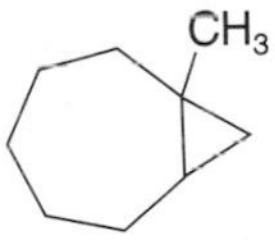

(c)

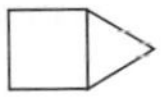

(b)

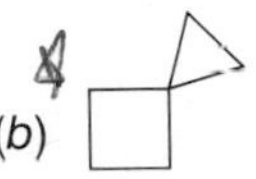

(d)

SAMPLE SOLUTION (a) In a cyclopropane synthesis using the Simmons-Smith reagent, you should remember that a CH_2 unit is transferred. Therefore, retrosynthetically disconnect the bonds to a CH_2 group of a three-membered ring to identify the starting alkene.

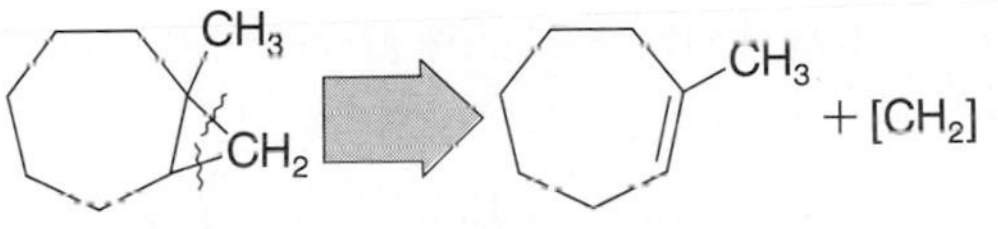

The complete synthesis is:

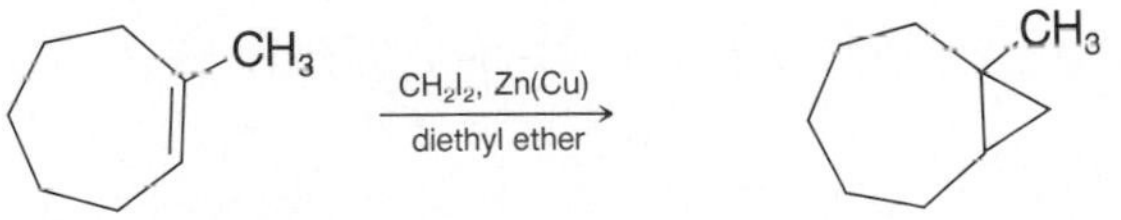

1-Methylcycloheptene → 1-Methylbicyclo[5.1.0]octane (55%)

Methylene transfer from iodomethylzinc iodide is *stereospecific*. Substituents that were cis in the alkene remain cis in the cyclopropane.

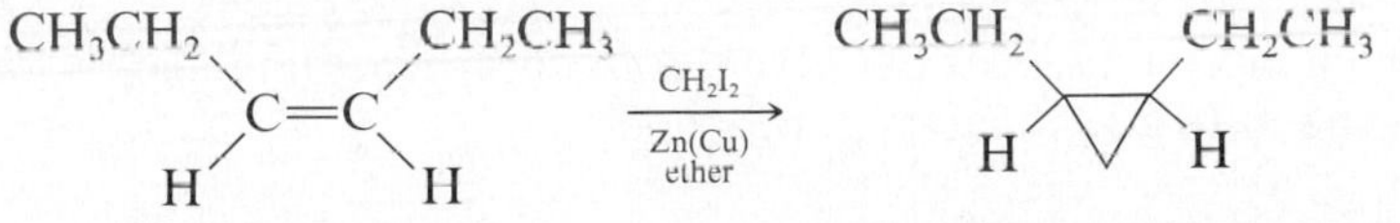

(Z)-3-Hexene → cis 1,2-Diethylcyclopropane (34%)

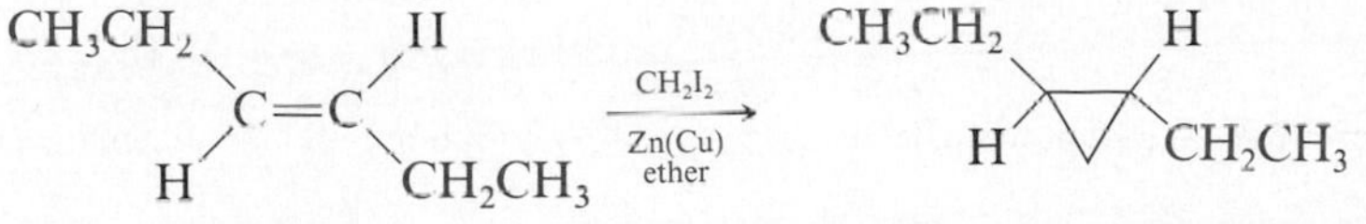

(E)-3-Hexene → trans-1,2-Diethylcyclopropane (15%)

Yields in Simmons-Smith reactions are sometimes low. Nevertheless, since it often provides the only feasible route to a particular cyclopropane derivative, it is a valuable addition to the organic chemist's store of synthetic methods.

15.13 CARBENES AND CARBENOIDS

Iodomethylzinc iodide is often referred to as a *carbenoid,* meaning that it resembles a *carbene* in its chemical reactions. Carbenes are neutral molecules that have no unpaired electrons and contain a *divalent* carbon atom. A divalent carbon is one with only two substituents and no multiple bonds. Iodomethylzinc iodide reacts as if it were a source of the carbene H—$\ddot{\text{C}}$—H.

It is clear that free $:$CH$_2$ is not involved in the Simmons-Smith reaction, but there is substantial evidence to indicate that carbenes are formed as intermediates in certain other reactions that convert alkenes to cyclopropanes. The most studied examples of these reactions involve dichlorocarbene and dibromocarbene.

$$\underset{\text{Dichlorocarbene}}{\overset{\ddot{\text{C}}}{\underset{\text{Cl}}{\diagup}}\underset{\text{Cl}}{\diagdown}} \qquad \underset{\text{Dibromocarbene}}{\overset{\ddot{\text{C}}}{\underset{\text{Br}}{\diagup}}\underset{\text{Br}}{\diagdown}}$$

Carbon is associated with only six valence electrons in a carbene, and carbenes are rather electrophilic. They are too reactive to be isolated and stored in the manner of stable organic compounds, but they have been trapped in frozen argon for spectroscopic study at very low temperatures.

Dihalocarbenes are formed when trihalomethanes are treated with a strong base, such as potassium *tert*-butoxide. The trihalomethyl anion produced on proton abstraction dissociates to a dihalocarbene and a halide anion.

$$\underset{\substack{\text{Tribromomethane}}}{\text{Br}_3\text{C}-\text{H}} \quad + \; {}^{-}\!\!:\!\underset{\substack{\textit{tert}\text{-Butoxide} \\ \text{ion}}}{\ddot{\text{O}}\text{C}(\text{CH}_3)_3} \longrightarrow \quad \underset{\substack{\text{Tribromomethide} \\ \text{ion}}}{\text{Br}_3\bar{\text{C}}\!:} \quad + \; \underset{\substack{\textit{tert}\text{-Butyl} \\ \text{alcohol}}}{\text{H}-\ddot{\text{O}}\text{C}(\text{CH}_3)_3}$$

$$\underset{\text{Tribromomethide ion}}{\overset{:\ddot{\text{B}}\text{r}}{\underset{\text{Br}}{\diagdown}}\overset{\text{Br}}{\underset{}{\diagup}}\text{C}\!:^{-}} \longrightarrow \underset{\text{Dibromocarbene}}{\overset{\text{Br}}{\underset{\text{Br}}{\diagdown}}\overset{}{\diagup}\text{C}\!:} \quad + \quad \underset{\text{Bromide ion}}{:\ddot{\text{B}}\text{r}\!:^{-}}$$

When generated in the presence of an alkene, dihalocarbenes undergo cycloaddition to the double bond to give dihalocyclopropanes.

$$\underset{\text{Cyclohexene}}{\bigcirc} + \underset{\text{Tribromomethane}}{\text{CHBr}_3} \xrightarrow[\text{(CH}_3)_3\text{COH}]{\text{KOC(CH}_3)_3} \underset{\text{7,7-Dibromobicyclo[4.1.0]heptane (75\%)}}{\overset{\text{Br}}{\underset{\text{Br}}{\diagup}}}$$

The process in which a dihalocarbene is formed from a trihalomethane corresponds to an elimination reaction in which a proton and a halide are lost from the same carbon. It is an α *elimination* reaction proceeding via the organometallic intermediate $K^+ [:CX_3]^-$. (Simple monohalo- and dihaloalkanes do not undergo α elimination under these conditions but react instead by the familiar concerted β elimination pathway described earlier for the preparation of alkenes and alkynes.)

15.14 ORGANIC DERIVATIVES OF CADMIUM AND MERCURY

Unlike the closely related group IIb element zinc, cadmium and mercury do not react readily with alkyl halides. Organocadmium compounds are normally prepared by adding a cadmium halide salt to a solution of a Grignard reagent.

$$2RMgX + CdX_2 \longrightarrow R_2Cd + 2MgX_2$$

| Grignard reagent | Cadmium halide | Dialkyl- cadmium | Magnesium halide |

$$CH_3CH_2MgBr \xrightarrow[\text{diethyl ether}]{CdCl_2} (CH_3CH_2)_2Cd$$

Ethylmagnesium bromide   Diethylcadmium

Exchange of cadmium for magnesium in the organometallic is driven by formation of the magnesium halide salt. Magnesium is more electropositive than cadmium and better able to exist as a dipositive ion in its dihalide salt. The major application of organocadmium compounds is in the synthesis of ketones from acyl chlorides. This reaction will be discussed in Chapter 18.

Organomercury compounds can be prepared in the same way.

$$2RMgX + HgX_2 \longrightarrow R_2Hg + 2MgX_2$$

| Grignard reagent | Mercuric halide | Dialkylmer- cury | Magnesium halide |

$$(CH_3)_2CHMgBr \xrightarrow[\text{diethyl ether}]{HgCl_2} (CH_3)_2CHHgCH(CH_3)_2$$

Isopropylmagnesium bromide   Diisopropylmercury (60%)

We have already encountered hydroxyalkylmercuric acetates—organometallic compounds formed by oxymercuration of alkenes—in Section 7.7. Carbon-mercury bonds are not very polar, and so carbon possesses negligible carbanionic properties in organomercury compounds.

Mercury derivatives find various uses as fungicides and antiseptic agents. One of the most familiar of these antiseptics is *mercurochrome*, an organometallic which contains mercury but no chromium (the "chrome" portion of its name derives from its bright red color in water).

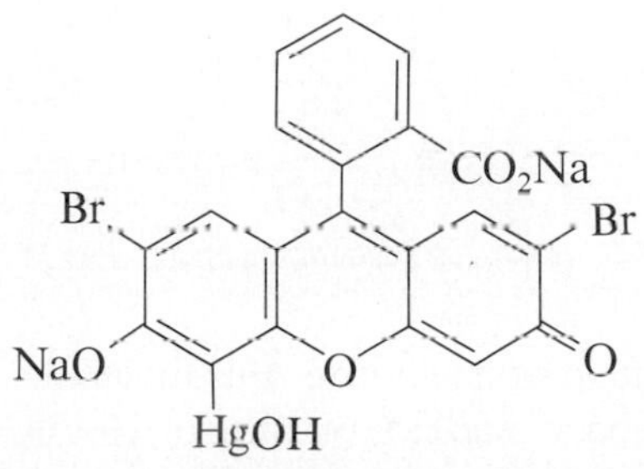

Mercurochrome

Cadmium and mercury compounds of all kinds are quite toxic to most organisms. For a long time wastewater containing mercuric salts from industrial operations was dumped in rivers and lakes in large amounts. There, inorganic mercury (Hg^{2+}) is converted by bacterial action to methylmercury (CH_3Hg^+) and dimethylmercury (CH_3HgCH_3), materials that accumulate in the tissues of fish. At even modest mercury levels fish populations are threatened. It is dangerous to eat fish caught in mercury-polluted waters, and hundreds of incidents of poisoning are known to have occurred under these circumstances. While stricter controls on waste disposal have helped, so much mercury remains in the environment that it will take a long time for it to be reduced to a reasonable level in some riverbeds.

15.15 TRANSITION METAL ORGANOMETALLIC COMPOUNDS

A large number of organometallic compounds are based on transition metals. Examples include organic derivatives of iron, nickel, chromium, platinum, and rhodium. Many important industrial processes are catalyzed by transition metals or their complexes.

A noteworthy feature of many organic derivatives of transition metals is that the organic group is bonded to the metal through its π system rather than by a σ bond. The compound (benzene)tricarbonylchromium, for example, has three carbon monoxide ligands and a *benzene* ring—not a phenyl group—attached to chromium.

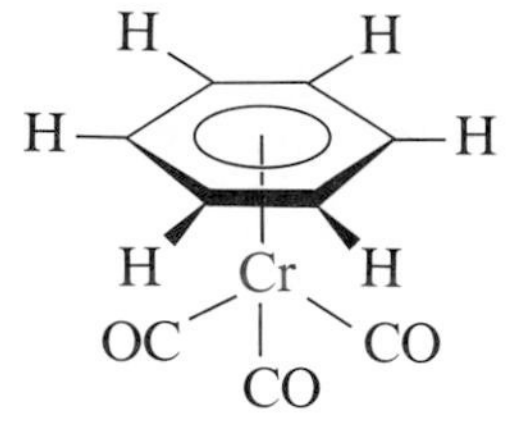

(Benzene)tricarbonylchromium

Strong bonding is exhibited between chromium and the π electrons of the benzene ring. (Benzene)tricarbonylchromium is a yellow solid, stable in air, and soluble in organic solvents such as benzene and ether.

It is not uncommon to find stable transition metal organometallics bearing as their organic ligand species that are known to be highly reactive in their free or uncoordinated state. One such organometallic is (cyclobutadiene)tricarbonyliron.

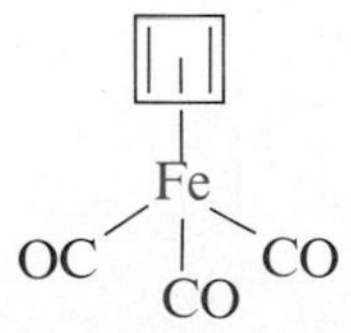

(Cyclobutadiene)tricarbonyliron

Bonding of cyclobutadiene to iron stabilizes this molecule. Many of the reactions of free cyclobutadiene have been studied by using (cyclobutadiene)tricarbonyliron as the source of cyclobutadiene. Oxidation causes dissociation of the complex, liberating free cyclobutadiene in solution.

One of the earliest and best examples of π-bonded organometallic compounds is *ferrocene*. It was expected that the σ-bonded species dicyclopentadienyliron could be prepared by adding iron(II) chloride (ferrous chloride) to cyclopentadienylsodium. (Cyclopentadienylsodium is ionic. Its anion is the cyclopentadienide ion, which contains six π electrons.)

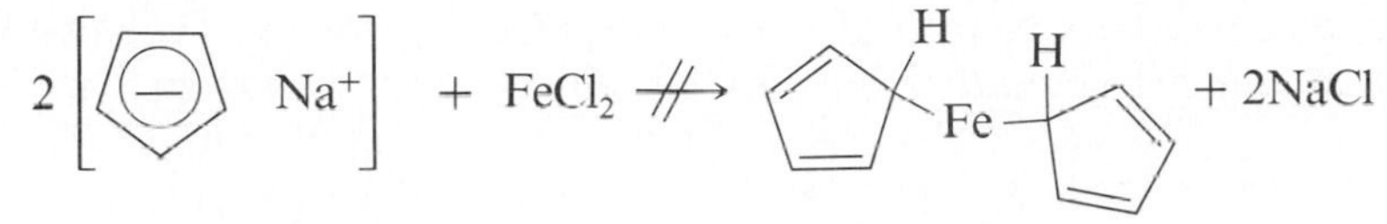

Cyclopentadienylsodium   Iron(II) chloride   (not formed)

However, the isolated product, ferrocene, clearly was structurally much different from that expected. It was subsequently demonstrated to have the doubly π-bonded *sandwich* structure shown.

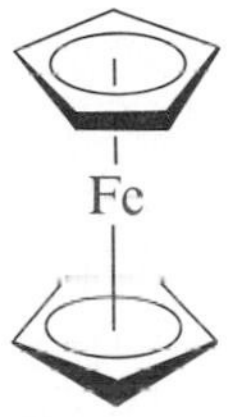

Ferrocene

Since then, numerous related molecules have been prepared—even some in which uranium is the metal atom! The precise nature of the bonding in such compounds served to foster development of bonding theories in both organic and inorganic

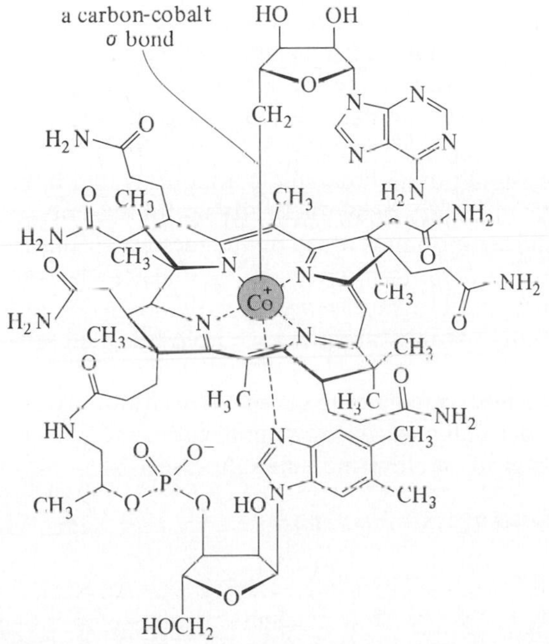

FIGURE 15.4   Structure of coenzyme B_{12}.

chemistry and stimulate research at the point where these two disciplines merge. Two of the leading figures in the growth of organometallic chemistry of the transition metals are E. O. Fischer of the Technical University in Munich and Sir Geoffrey Wilkinson of Imperial College, London. They shared the Nobel prize in chemistry in 1973.

Naturally occurring compounds with carbon-metal bonds are very rare. The best example of such an organometallic is coenzyme B_{12}, which has a carbon-cobalt σ bond (Figure 15.4). Pernicious anemia results from a coenzyme B_{12} deficiency and can be treated by adding sources of cobalt to the diet. One source of cobalt is vitamin B_{12}, a compound structurally related to, but not identical with, coenzyme B_{12}.

15.16 SUMMARY

Organometallic compounds contain a carbon-metal bond and this bond is polarized so that carbon bears a partial to complete negative charge and the metal bears a partial to complete positive charge.

| Methyllithium has a covalent carbon-lithium bond | Sodium acetylide has an ionic bond between carbon and sodium |

A species that has a negatively charged carbon is called a *carbanion.* Organometallic substances with a significant degree of carbanionic character are strong bases and good nucleophiles. Many organometallic compounds are used extensively in synthesis because their nucleophilicity can be exploited in reactions that lead to carbon-carbon bond formation.

The most useful organometallic reagents are derived from magnesium and are called *Grignard reagents.* They are prepared by the reaction of magnesium and an organic halide, usually in diethyl ether. Alkyllithium reagents are prepared in an analogous way and undergo chemical reactions that are similar to those of Grignard reagents. The preparation of Grignard reagents and alkyllithium reagents are summarized in Table 15.2. The table also illustrates the preparation of *lithium dialkylcuprates* and *iodomethylzinc iodide,* two other organometallic compounds that have found a place in the organic chemist's inventory of synthetically useful reagents.

Table 15.3 shows how organometallic compounds based on magnesium, lithium, copper, and zinc are used in *carbon-carbon bond-forming reactions.* In particular, the reactions of organolithium compounds and Grignard reagents with aldehydes and ketones to produce alcohols rank among the most frequently employed of all synthetic procedures.

Organometallic compounds are often intermediates in chemical reactions involving metals. Alkyl halides are reduced to alkanes by treatment with zinc in acid solution, presumably by proton transfer to an alkylzinc halide intermediate.

$$RX \ + \ Zn \ \longrightarrow \ [RZnX] \ \xrightarrow{\ H^+\ } \ RH$$

| Alkyl halide | Zinc | Alkylzinc halide | Alkane |

TABLE 15.2

Preparation of Organometallic Reagents Used in Synthesis

Type of organometallic reagent (section) and comments	General equation for preparation and specific example
Organolithium reagents (Section 15.3) Lithium metal reacts with organic halides to produce organolithium compounds. The organic halide may be alkyl, alkenyl, or aryl. Iodides react most and fluorides least readily; bromides are used most often. Suitable solvents include hexane, diethyl ether, and tetrahydrofuran.	$RX + 2Li \longrightarrow RLi + LiX$ Alkyl halide — Lithium — Alkyllithium — Lithium halide $CH_3CH_2CH_2Br \xrightarrow[\text{diethyl ether}]{Li} CH_3CH_2CH_2Li$ Propyl bromide — Propyllithium (78%)
Grignard reagents (Section 15.4) Grignard reagents are prepared in a manner similar to that used for organolithium compounds. Diethyl ether and tetrahydrofuran are appropriate solvents.	$RX + Mg \longrightarrow RMgX$ Alkyl halide — Magnesium — Alkylmagnesium halide (Grignard reagent) $C_6H_5CH_2Cl \xrightarrow[\text{diethyl ether}]{Mg} C_6H_5CH_2MgCl$ Benzyl chloride — Benzylmagnesium chloride (93%)
Lithium dialkylcuprates (Section 15.11) These reagents contain a negatively charged copper atom and are formed by the reaction of a cuprous salt with two equivalents of an organolithium reagent.	$2RLi + CuX \longrightarrow R_2CuLi + LiX$ Alkyllithium — Cuprous halide — Lithium dialkylcuprate — Lithium halide $2CH_3Li + CuI \xrightarrow[\text{ether}]{\text{diethyl}} (CH_3)_2CuLi + LiI$ Methyllithium — Cuprous iodide — Lithium dimethylcuprate — Lithium iodide
Iodomethylzinc iodide (Section 15.12) This is the Simmons-Smith reagent. It is prepared by the reaction of zinc (usually in the presence of copper) with diiodomethane.	$CH_2I_2 + Zn \xrightarrow[\text{ether}]{\text{diethyl}} ICH_2ZnI$ Diiodomethane — Zinc — Iodomethylzinc iodide

Vicinal dihalides are converted to alkenes by reaction with zinc:

Vicinal dihalide — β-Haloalkylzinc halide intermediate — Alkene

Cyclopropanes are formed when 1,3-dihalides react with zinc:

1,3-Dihaloalkane — Zinc — γ-Haloalkylzinc halide intermediate — Cyclopropane derivative

TABLE 15.3
Synthetic Applications of Organometallic Reagents

Reaction (section) and comments	General equation and specific example
Alcohol synthesis via the reaction of Grignard reagents with carbonyl compounds (Section 15.6) This is one of the most useful reactions in synthetic organic chemistry. Grignard reagents react with formaldehyde to yield primary alcohols, with aldehydes to give secondary alcohols, and with ketones to form tertiary alcohols.	$$RMgX + R'\overset{\overset{O}{\|}}{C}R'' \xrightarrow[\text{2. } H_3O^+]{\text{1. diethyl ether}} R\underset{\underset{R''}{\|}}{\overset{\overset{R'}{\|}}{C}}OH$$ Grignard Aldehyde Alcohol reagent or ketone $$CH_3MgI + CH_3CH_2CH_2\overset{\overset{O}{\|}}{C}H \xrightarrow[\text{2. } H_3O^+]{\text{1. diethyl ether}} CH_3CH_2CH_2\underset{\underset{OH}{\|}}{C}HCH_3$$ Methylmagnesium Butanal 2-Pentanol (82%) iodide
Reaction of Grignard reagents with esters (Section 15.10) Tertiary alcohols in which two of the substituents on the hydroxyl carbon are the same may be prepared by the reaction of an ester with two equivalents of a Grignard reagent.	$$2RMgX + R'\overset{\overset{O}{\|}}{C}OR'' \xrightarrow[\text{2. } H_3O^+]{\text{1. diethyl ether}} R\underset{\underset{R}{\|}}{\overset{\overset{R'}{\|}}{C}}OH$$ Grignard Ester Tertiary reagent alcohol $$C_6H_5MgBr + C_6H_5\overset{\overset{O}{\|}}{C}OCH_2CH_3 \xrightarrow[\text{2. } H_3O^+]{\text{1. diethyl ether}} (C_6H_5)_3COH$$ Phenylmagnesium Ethyl benzoate Triphenylmethanol bromide (89–93%)

Synthesis of alcohols using organolithium reagents (Section 15.7) Organolithium reagents react with aldehydes and ketones in a manner similar to that of Grignard reagents to produce alcohols.	$$RLi + R'\overset{\overset{O}{\|}}{C}R'' \xrightarrow[\text{2. } H_3O^+]{\text{1. diethyl ether}} R\underset{\underset{R''}{\|}}{\overset{\overset{R'}{\|}}{C}}OH$$ Alkyllithium Aldehyde Alcohol
	           or ketone $$\triangleright\!\!-Li + CH_3\overset{\overset{O}{\|}}{C}C(CH_3)_3 \xrightarrow[\text{2. } H_3O^+]{\text{1. diethyl ether}} \triangleright\!\!-\underset{\underset{CH_3}{\|}}{\overset{\overset{OH}{\|}}{C}}C(CH_3)_3$$ Cyclopropyllithium 3,3-Dimethyl-2-butanone 2-Cyclopropyl-
	                                                           3,3-dimethyl-
	                                                           2-butanol (71%)

Preparation of alkanes using lithium dialkylcuprates (Section 15.11) Two alkyl groups may be coupled together to form an alkane by the reaction of an alkyl halide with a lithium dialkylcuprate. Both alkyl groups must be primary (or methyl). Aryl and vinyl halides may be used in place of alkyl halides.	$$R_2CuLi + R'CH_2X \longrightarrow RCH_2R'$$ Lithium Primary Alkane dialkylcuprate alkyl halide $$(CH_3)_2CuLi + C_6H_5CH_2Cl \xrightarrow{\text{diethyl ether}} C_6H_5CH_2CH_3$$ Lithium Benzyl Ethylbenzene (80%) dimethylcuprate chloride

TABLE 15.3 (continued)

Reaction (section) and comments	General equation and specific example
The Simmons-Smith reaction (Section 15.12) Methylene transfer from iodo-methylzinc iodide converts alkenes to cyclopropanes. The reaction is a stereospecific syn addition of a CH_2 group to the double bond.	$R_2C{=}CR_2 \;+\;$ ICH_2ZnI $\xrightarrow{\text{diethyl ether}}$ cyclopropane derivative $+\; ZnI_2$ Alkene · Iodomethylzinc iodide · Cyclopropane derivative · Zinc Iodide Cyclopentene $\xrightarrow[\text{diethyl ether}]{CH_2I_2, \; Zn(Cu)}$ Bicyclo[3.1.0]hexane (53%)

A wide variety of organometallic compounds is known and novel structural types continue to be discovered. Organometallic chemistry is one of the most active areas of contemporary scientific research and promises to be a fertile field for continued development.

PROBLEMS

15.11 Write structural formulas for each of the following compounds. Specify which compounds qualify as organometallic compounds.

- (a) Cyclopentyllithium
- (b) Ethoxymagnesium chloride
- (c) 2-Phenylethylmagnesium iodide
- (d) Lithium divinylcuprate
- (e) Mercuric acetate
- (f) Benzylpotassium
- (g) Sodium *p*-toluenesulfonate

15.12 Suggest appropriate methods for the preparation of each of the compounds given below from the starting material of your choice:

- (a) $CH_3CH_2CH_2CH_2CH_2MgI$
- (b) $(CH_3CH_2CH_2CH_2CH_2)_3B$
- (c) $CH_3CH_2CH_2CH_2CH_2Li$
- (d) $(CH_3CH_2CH_2CH_2CH_2)_2CuLi$
- (e) $(CH_3CH_2CH_2CH_2CH_2)_2Cd$

15.13 Which compound in each of the following pairs would you expect to have the more polar carbon-metal bond? *electropositivity*

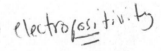

- (a) CH_3CH_2Li or $(CH_3CH_2)_3B$
- (b) $(CH_3)_2Be$ or $(CH_3)_2Mg$
- (c) $(CH_3)_4Si$ or $(CH_3)_2Mg$
- (d) CH_3Na or CH_3AlH_2

15.14 Write the structure of the principal organic product of each of the following reactions:

(a) 1-Bromopropane with lithium in diethyl ether
(b) 1-Bromopropane with magnesium in diethyl ether
(c) 2-Iodopropane with lithium in diethyl ether
(d) 2-Iodopropane with magnesium in diethyl ether
(e) 2-Iodopropane with zinc in acetic acid
(f) Product of (a) with copper(I) iodide
(g) Product of (f) with 1-bromobutane
(h) Product of (f) with iodobenzene
(i) Product of (b) with D_2O and DCl
(j) Product of (d) with D_2O and DCl
(k) Product of (a) with formaldehyde in ether, followed by dilute acid
(l) Product of (b) with benzaldehyde in ether, followed by dilute acid
(m) Product of (c) with cycloheptanone in ether, followed by dilute acid

(n) Product of (d) with $CH_3\overset{\overset{\displaystyle O}{\|}}{C}CH_2CH_3$ in ether, followed by dilute acid

(o) Product of (b) with $C_6H_5\overset{\overset{\displaystyle O}{\|}}{C}OCH_3$ in ether, followed by dilute acid
(p) 1-Octene with diiodomethane and zinc-copper couple in ether
(q) (E)-2-Decene with diiodomethane and zinc-copper couple in ether
(r) (Z)-3-Decene with diiodomethane and zinc-copper couple in ether
(s) 1,2-Diiodopropane with zinc in ethanol
(t) 1,3-Diiodopropane with zinc in ethanol

15.15 Using 1-bromobutane and any necessary organic or inorganic reagents, suggest efficient syntheses of each of the following alcohols:

(a) 1-Pentanol
(b) 2-Hexanol
(c) 1-Phenyl-1-pentanol
(d) 3-Methyl-3-heptanol
(e) 1-Butylcyclobutanol

15.16 Using bromobenzene and any necessary organic or inorganic reagents, suggest efficient syntheses of each of the following alcohols:

(a) Benzyl alcohol
(b) 1-Phenyl-1-hexanol
(c) 1,1-Diphenylmethanol
(d) 4-Phenyl-4-heptanol
(e) 1-Phenylcyclooctanol

15.17 Analyze the following structures so as to determine all the practical combinations of Grignard reagent and carbonyl compound that will give rise to each:

(a) $CH_3CH_2\underset{\underset{\displaystyle OH}{|}}{CH}CH_2CH(CH_3)_2$

(b)

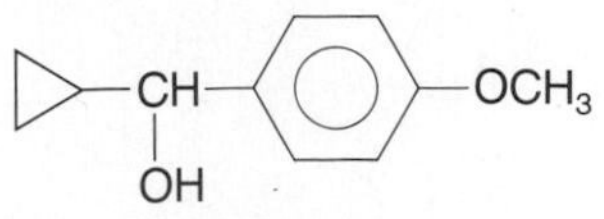

(c) $(CH_3)_3CCH_2OH$

(d) 6-Methyl-5-hepten-2-ol

(e)

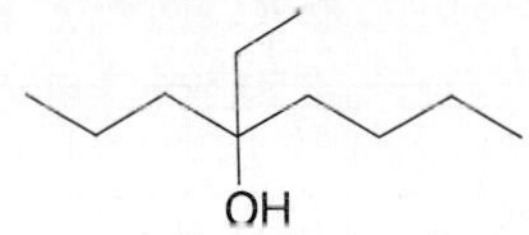

15.18 A number of drugs are prepared by reactions of the type described in this chapter. Indicate what you believe would be a reasonable last step in the synthesis of each of the following:

(a) 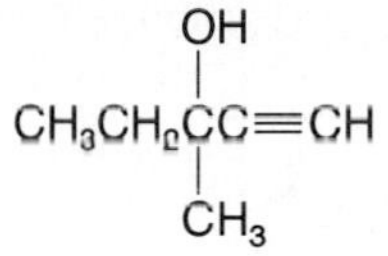
$$CH_3CH_2\underset{\underset{CH_3}{|}}{\overset{\overset{OH}{|}}{C}}C\equiv CH$$ *Meparfynol*, a mild hypnotic or sleep-inducing agent

(b) 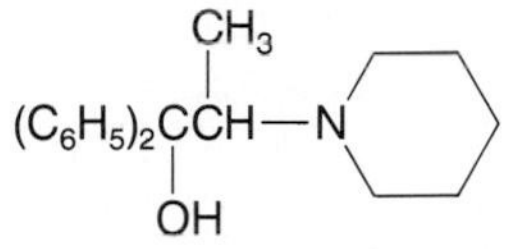
$$(C_6H_5)_2\underset{\underset{OH}{|}}{\overset{\overset{CH_3}{|}}{C}}CH-N\bigcirc$$ *Diphepanol*, an antitussive (cough suppressant)

(c)

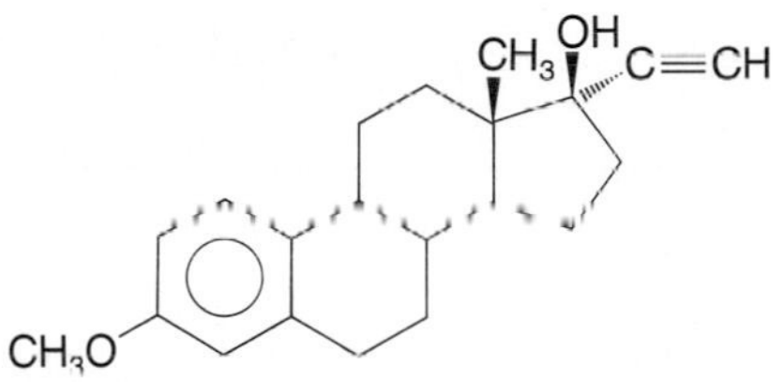

Mestranol, an estrogenic used as a component of oral contraceptive drugs

15.19 Predict the principal organic product of each of the following reactions:

(a) $+ NaC\equiv CH \xrightarrow[\text{2. } H_3O^+]{\text{1. liquid ammonia}}$

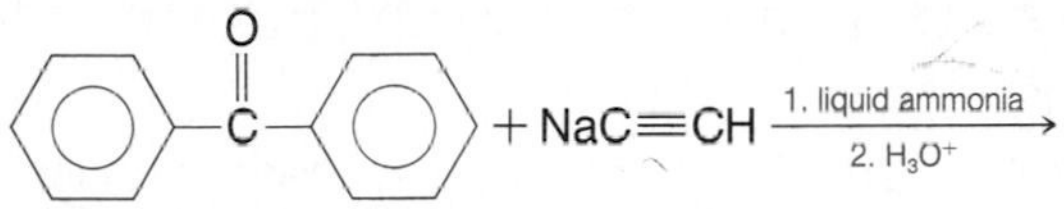

(b) $+ CH_3CH_2Li \xrightarrow[\text{2. } H_3O^+]{\text{1. diethyl ether}}$

(c) $-Br \xrightarrow[\substack{\text{2. O} \\ \overset{\|}{HCH} \\ \text{3. } H_3O^+}]{\text{1. Mg, THF}}$

(d) $\xrightarrow[\substack{Zn(Cu) \\ \text{diethyl ether}}]{CH_2I_2}$

(e) $\xrightarrow[\substack{Zn(Cu) \\ \text{ether}}]{CH_2I_2}$

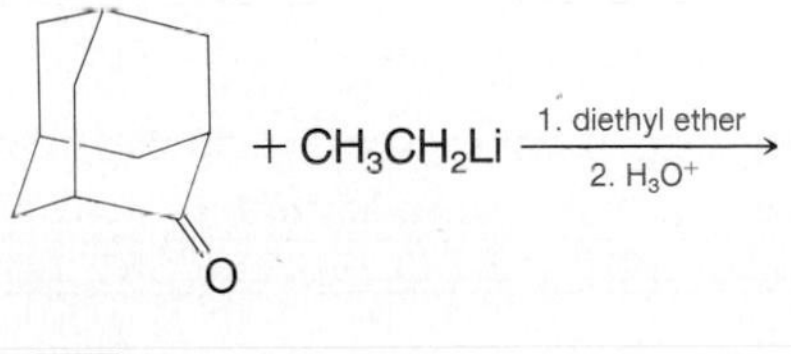

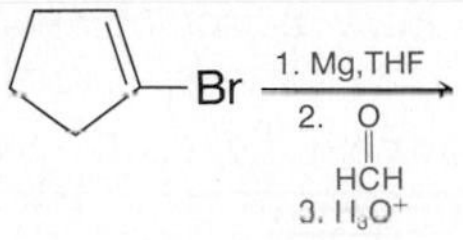

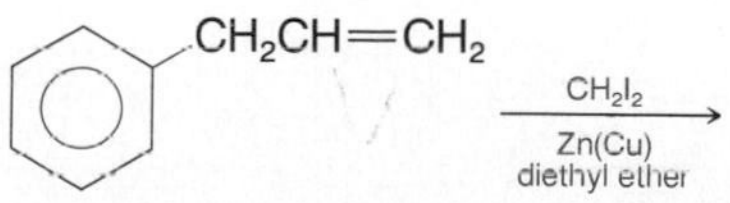

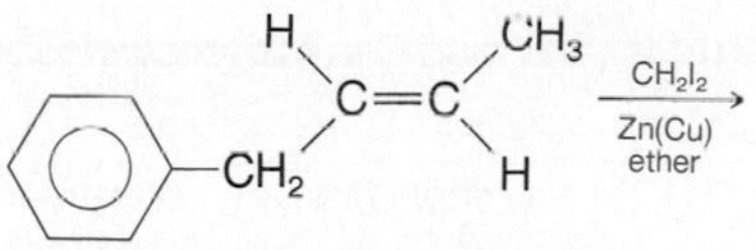

(f) + LiCu(CH₃)₂ $\longrightarrow$

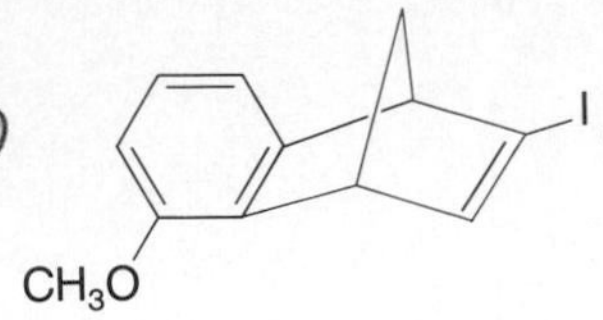

(g) $(R)-CH_3CH_2CH_2CH_2CHCH_2I$ $\xrightarrow[\text{acetic acid}]{\text{Zn}}$
   with CH_2CH_3 substituent

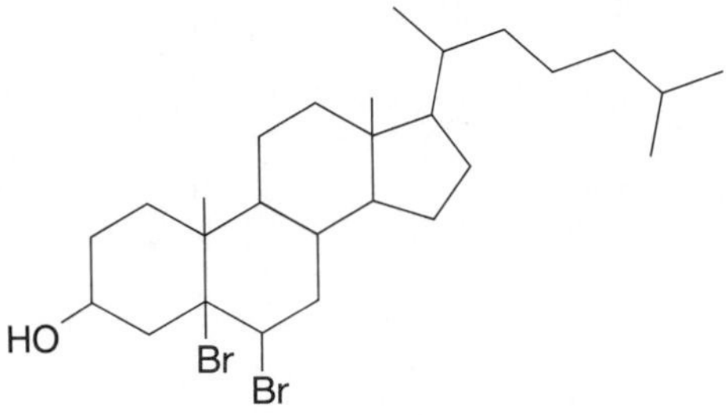

$\xrightarrow[\text{diethyl ether}]{\text{Zn}}$

(h)

(i) $CH_3CHCH_2CH_2Br$ $\xrightarrow[\text{ethanol}]{\text{Zn}}$
   with Br substituent

15.20 Addition of phenylmagnesium bromide to 4-*tert*-butylcyclohexanone gives two iso-meric tertiary alcohols as products. Both alcohols yield the same alkene when subjected to acid-catalyzed dehydration. Suggest reasonable structures for these two alcohols.

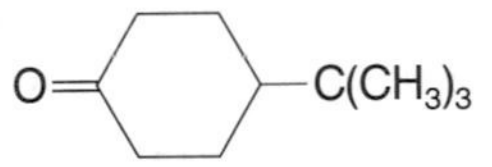

$O=$⬡$-C(CH_3)_3$

4-*tert*-Butylcyclohexanone

15.21 (a) Unlike other esters, which react with Grignard reagents to give

tertiary alcohols, ethyl formate $(H\overset{O}{\overset{||}{C}}OCH_2CH_3)$ yields a different class of alcohols on treatment with Grignard reagents. What kind of alcohol is formed in this case and why?

(b) Diethyl carbonate $(CH_3CH_2O\overset{O}{\overset{||}{C}}OCH_2CH_3)$ reacts with excess Grignard reagent to yield alcohols of a particular type. What is the structural feature that characterizes alcohols prepared in this way?

15.22 Reaction of lithium diphenylcuprate with optically active 2-bromobutane yields 2-phenylbutane, with high net inversion of configuration. When the 2-bromobutane used has the stereostructure shown below, will the 2-phenylbutane formed have the *R* or the *S* configu-ration?

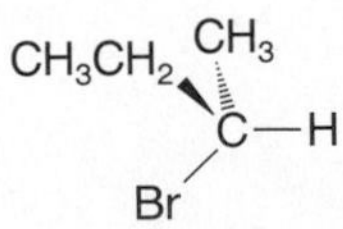

CH_3CH_2 CH_3
       $C-H$
Br

15.23 Suggest reasonable structures for compounds A, B, and C in the following reactions:

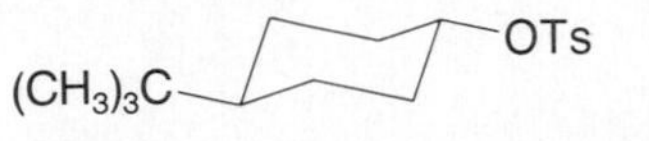

$(CH_3)_3C-$⬡$-OTs$ $\xrightarrow{\text{LiCu(CH}_3)_2}$ compound A + compound B

trans-4-*tert*-Butylcyclohexyl *p*-toluenesulfonate

                                                    $(C_{11}H_{22})$ $(C_{10}H_{18})$

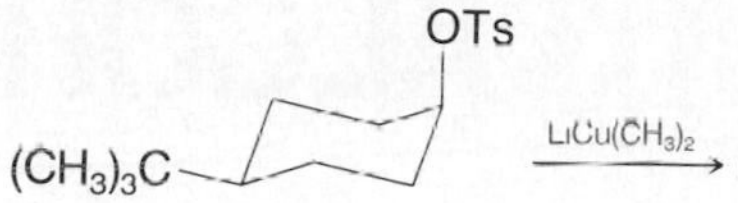

$(CH_3)_3C$ ⟶ [OTs cyclohexane structure] $\xrightarrow{\text{LiCu(CH}_3)_2}$ compound B + compound C
$(C_{11}H_{22})$

Compound C is thermodynamically more stable than compound A.

15.24 Sometimes the strongly basic properties of Grignard reagents can be turned to synthetic advantage. A chemist needed samples of butane specifically labeled with deuterium, the mass 2 isotope of hydrogen, as shown:

(a) $CH_3CH_2CH_2CH_2D$
(b) $CH_3CHDCH_2CH_3$

Suggest methods for the preparation of each of these using heavy water (D_2O) as the source of deuterium, butanols of your choice, and any necessary organic or inorganic reagents.

15.25 Reactions of dialkylmagnesium species (R_2Mg) have been studied. In the preparation of these compounds, a Grignard reagent is generated in the usual manner in diethyl ether solution and then dioxane is added. Based on the fact that magnesium halides are insoluble in dioxane, explain why this procedure allows dialkylmagnesium compounds to be studied in the absence of alkylmagnesium halides.

15.26 Diphenylmethane is significantly more acidic than benzene, and triphenylmethane is more acidic than either. Identify the most acidic proton in each compound and suggest a reason for the trend in acidity.

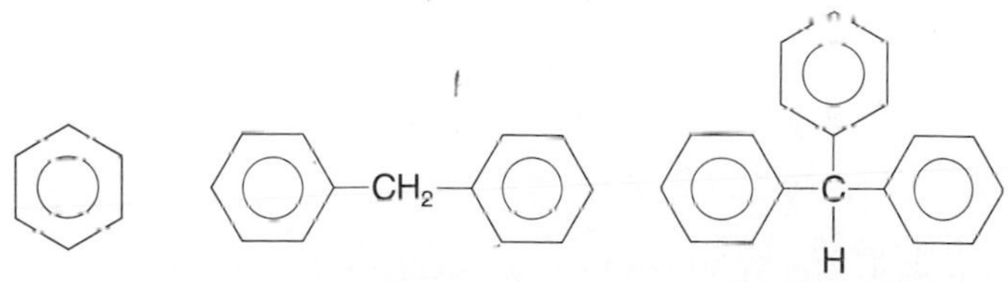

Benzene
$K_a \cong 10^{-45}$

Diphenylmethane
$K_a \cong 10^{-34}$

Triphenylmethane
$K_a \cong 10^{-32}$

15.27 Deduce whether the zinc-promoted debromination of vicinal dihalides is a syn or anti elimination on the basis of the observation that *meso*-2,3-dibromobutane yields only (E)-2-butene while the chiral diastereomer of 2,3-dibromobutane yields only (Z)-2-butene.

15.28 Identify compounds D, E, F, and G in the following reaction sequences:

(a) [cyclopentane]—$CH(CH_3)_2$ $\xrightarrow[hv]{Br_2 \text{ (excess)}}$ compound D $\xrightarrow[\substack{\text{acetic} \\ \text{acid}}]{Zn}$ compound E
$(C_8H_{14}Br_2)$ (C_8H_{14})

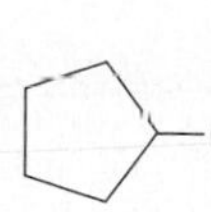

(b) $C_6H_5CH_2CH_2CH_2Br$ $\xrightarrow[\substack{\text{benzoyl} \\ \text{peroxide, heat}}]{N\text{-bromosuccinimide}}$ compound F $\xrightarrow{Zn}$ compound G
$(C_9H_{10}Br_2)$ (C_9H_{10})

The material in the next several chapters deals with the chemistry of various oxygen-containing functional groups. The interplay of these important classes of compounds—alcohols, ethers, aldehydes, ketones, carboxylic acids, and derivatives of carboxylic acids—is fundamental to the science of organic chemistry.

$$\text{ROH} \qquad \text{ROR}' \qquad \underset{\text{RCH}}{\overset{\overset{\textstyle O}{\|}}{}} \qquad \underset{\text{RCR}'}{\overset{\overset{\textstyle O}{\|}}{}} \qquad \underset{\text{RCOH}}{\overset{\overset{\textstyle O}{\|}}{}}$$

| Alcohol | Ether | Aldehyde | Ketone | Carboxylic acid |

We begin by discussing in more detail a class of compounds already familiar to us, *alcohols.* Alcohols were introduced in Chapter 4 and have appeared regularly since then. With this chapter we extend our knowledge of alcohols, particularly with respect to their relationship to carbonyl-containing compounds. This chapter is a transitional one. It ties together much of the material encountered earlier and sets the stage for our study of the other oxygen-containing functional groups in the chapters that follow.

16.1 SOURCES OF ALCOHOLS

The simple alcohols methanol, ethanol, and isopropyl alcohol are abundant and cheap. Methanol—7×10^9 lb/year of it in the United States—is produced by hydrogenation of carbon monoxide.

$$\text{CO} \quad + \quad 2\text{H}_2 \quad \xrightarrow[400°C]{\text{ZnO}-\text{Cr}_2\text{O}_3} \text{CH}_3\text{OH}$$

| Carbon monoxide | Hydrogen | | Methanol |

Ethanol and isopropyl alcohol are prepared by the hydration of ethylene and propene, respectively (Section 7.6).

$$CH_2{=}CH_2 \xrightarrow{H_2SO_4} CH_3CH_2OSO_2OH \xrightarrow{H_2O,\ heat} CH_3CH_2OH$$

Ethylene   Ethyl hydrogen sulfate   Ethanol

$$CH_3CH{=}CH_2 \xrightarrow{H_2SO_4} \underset{\underset{OSO_2OH}{|}}{CH_3CHCH_3} \xrightarrow{H_2O,\ heat} \underset{\underset{OH}{|}}{CH_3CHCH_3}$$

Propene   Isopropyl hydrogen sulfate   Isopropyl alcohol

There is a second source of ethanol, and that is natural products. When vegetable matter ferments, its carbohydrates are converted to ethanol and carbon dioxide. Fermentation of barley produces beer, grapes give wine. The maximum ethanol content that can be obtained under these conditions is on the order of 15 percent. Distillation of the fermentation broth gives a distillate enriched in ethanol. Whiskey is the aged distillate of fermented grain; it is almost 50 percent ethanol. Brandy and cognac are the aged distillates of fermented grapes and other fruits. The characteristic flavors, odors, and colors of alcoholic beverages depend on their origin and the way in which they are aged.

Methanol, ethanol, and isopropyl alcohol are included among the readily available starting materials commonly found in laboratories where organic synthesis is

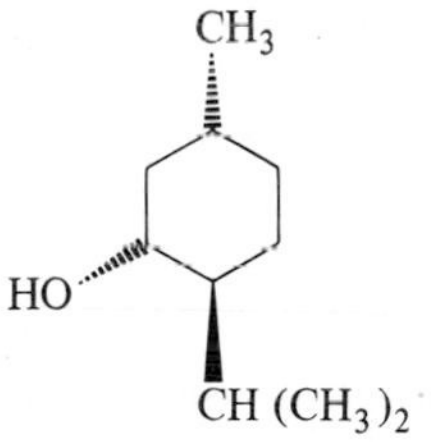

Menthol (obtained from oil of peppermint and used to flavor tobacco and food)

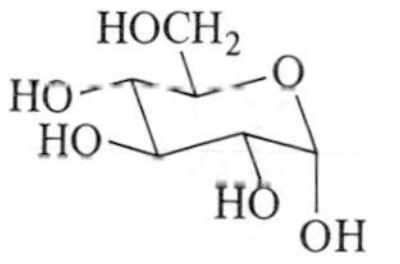

Glucose (a carbohydrate)

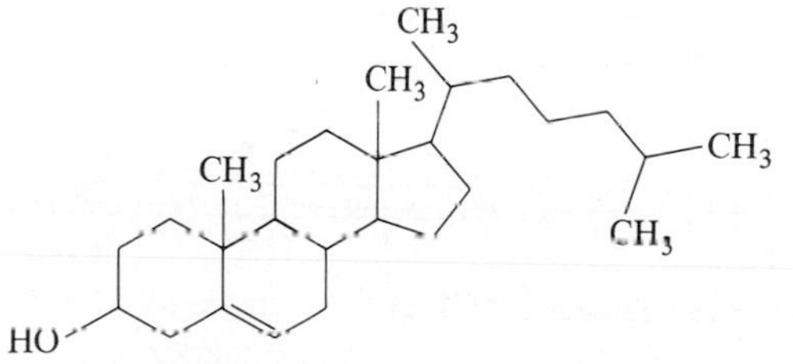

Cholesterol (principal constituent of gallstones and biosynthetic precursor of the steroid hormones)

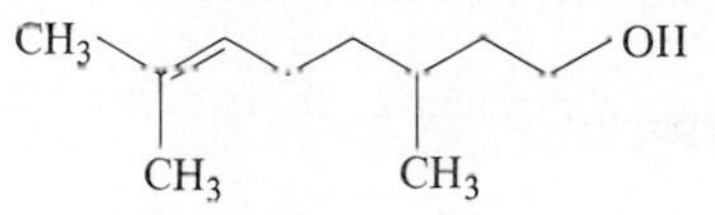

Citronellol (found in rose and geranium oil and used in perfumery)

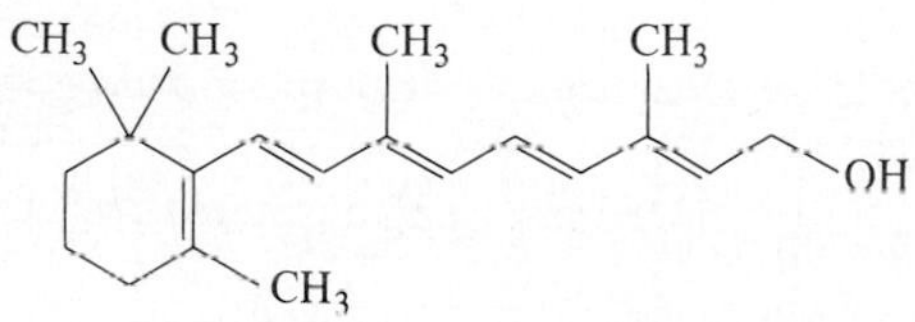

Retinol (vitamin A, an important substance in vision)

FIGURE 16.1 Some naturally occurring alcohols.

carried out. So, too, are many other alcohols. All alcohols of four carbons or less, as well as most of the five- and six-carbon alcohols and many higher alcohols, are commercially available at low cost. Some occur naturally; others are the products of efficient syntheses. Figure 16.1 presents the structures of a few naturally occurring alcohols. How alcohols are prepared in the laboratory is introduced in the following section.

16.2 REACTIONS THAT LEAD TO ALCOHOLS. A REVIEW AND A PREVIEW

A reaction that is characteristic of one functional group often serves as a synthesis for another. Alcohols can be prepared by a number of reactions described in earlier chapters. Table 16.1 summarizes those reactions. Several new methods for preparing alcohols will be described in succeeding sections of this chapter. Some of these methods involve *reduction* of carbonyl groups:

$$\underset{\text{Carbonyl compound}}{\overset{\displaystyle O}{\underset{\displaystyle \parallel}{C}}} \quad \xrightarrow{\text{reducing agent}} \quad \underset{\text{Alcohol}}{\overset{\displaystyle H \quad OH}{C}}$$

Before discussing specific reduction methods, let us review the sequence of oxidation states in organic compounds. Among one-carbon compounds, as summarized in Table 16.2, methane and carbon dioxide represent extremes in oxidation state. Methane is the most reduced form of carbon, carbon dioxide is the most oxidized form. In general, oxidation of a substance increases its oxygen content, reduction decreases it. Stated in other terms, oxidation at a carbon atom decreases the number of its bonds to hydrogen, and reduction increases the number of its bonds to hydrogen.

Compounds that are related by a hydration-dehydration process occupy the same oxidation state. Both carbon dioxide and carbonic acid, for example, contain the same number of carbon-oxygen bonds and are in equivalent oxidation states.

$$O=C=O \ + \ H_2O \longrightarrow \underset{\substack{\text{Carbonic acid} \\ \text{(four carbon-} \\ \text{oxygen bonds)}}}{\overset{\displaystyle O}{\underset{\displaystyle HO \quad OH}{\overset{\displaystyle \parallel}{C}}}}$$

Carbon dioxide   Water
(four carbon-
oxygen bonds)

(handwritten: Same # of C-O bonds)

PROBLEM 16.1   Carbon monoxide can be used to prepare several important industrial chemicals. Identify compounds A through C in the equations below, and specify whether carbon monoxide is oxidized, is reduced, or undergoes no change in oxidation state during the reaction.

(a) CO + compound A $\longrightarrow$ CH_3OH

(b) CO + compound B $\longrightarrow$ H_2 + CO_2

(c) CO + compound C $\longrightarrow$ $\overset{\displaystyle O}{\overset{\displaystyle \parallel}{HCOH}}$

TABLE 16.1

Summary of Reactions Discussed in Earlier Chapters That Yield Alcohols

Reaction (section) and comments	General equation and specific example
Acid-catalyzed hydration of alkenes (Section 7.6) The elements of water add to the double bond in accordance with Markovnikov's rule. This reaction is not used frequently for laboratory-scale synthesis of alcohols.	$R_2C=CR_2 + H_2O \xrightarrow{H^+} R_2CHCR_2$ $\qquad\qquad\qquad\qquad\qquad\quad \vert$ $\qquad\qquad\qquad\qquad\qquad\quad OH$ Alkene　　Water　　　Alcohol $(CH_3)_2C=CHCH_3 \xrightarrow[H_2SO_4]{H_2O} CH_3\underset{\underset{OH}{\vert}}{C}CH_2CH_3$ with CH_3 above 2-Methyl-2-butene　　　2-Methyl-2-butanol (90%)
Oxymercuration-demercuration of alkenes (Section 7.7) Markovnikov addition of the elements of water to the double bond occurs. This is a useful synthetic reaction. Rearrangements do not occur.	$R_2C=CR_2 \xrightarrow[2.\ NaBH_4]{1.\ Hg(O_2CCH_3)_2,\ THF\text{-}H_2O} R_2CHCR_2$ $\qquad\qquad\qquad\qquad\qquad\qquad\qquad\qquad \vert$ $\qquad\qquad\qquad\qquad\qquad\qquad\qquad\qquad OH$ Alkene　　　　　　　　　Alcohol 2-Phenylpropene　　　　2-Phenyl-2-propanol (95%)
Hydroboration-oxidation of alkenes (Section 7.8) The elements of water add to the double bond with regioselectivity opposite to that of Markovnikov's rule. This is a very good synthetic method; addition is syn and no rearrangements are observed.	$R_2C=CR_2 \xrightarrow[2.\ H_2O_2,\ HO^-]{1.\ B_2H_6} R_2CHCR_2$ $\qquad\qquad\qquad\qquad\qquad\qquad\quad \vert$ $\qquad\qquad\qquad\qquad\qquad\qquad\quad OH$ Alkene　　　　　　Alcohol $CH_3(CH_2)_7CH=CH_2 \xrightarrow[2.\ H_2O_2,\ HO^-]{1.\ B_2H_6,\ diglyme} CH_3(CH_2)_7CH_2CH_2OH$ 1-Decene　　　　　　　　　1-Decanol (93%)
Hydrolysis of alkyl halides (Section 9.1) A reaction useful only with substrates that do not undergo E2 elimination readily. It is rarely used for the synthesis of alcohols since alkyl halides are normally prepared from alcohols.	$RX + HO^- \longrightarrow ROH + X^-$ Alkyl　Hydroxide　　Alcohol　Halide halide　ion　　　　　　　　ion 2,4,6-Trimethylbenzyl chloride　　2,4,6-Trimethylbenzyl alcohol (78%)

TABLE 16.1 (continued)

Reaction (section) and comments	General equation and specific example
Reaction of Grignard reagents with aldehydes and ketones (Section 15.6) A method that allows for alcohol preparation with formation of new carbon-carbon bonds. Primary, secondary, and tertiary alcohols can all be prepared.	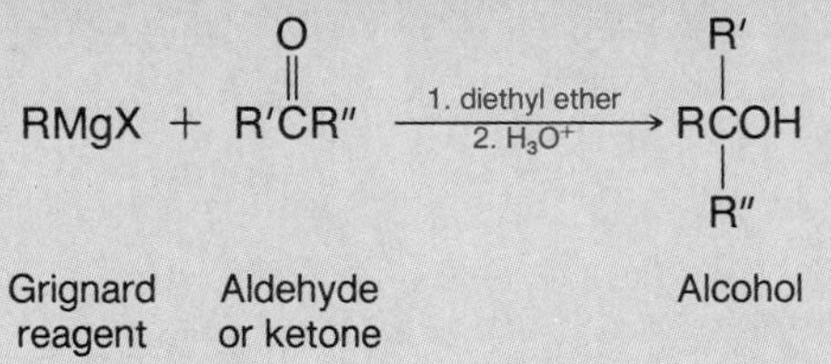 $RMgX + R'CR'' \xrightarrow[\text{2. H}_3O^+]{\text{1. diethyl ether}} RCOH$ with R', R'' substituents Grignard reagent / Aldehyde or ketone / Alcohol Cyclopentylmagnesium bromide + Formaldehyde (HCH) $\xrightarrow[\text{2. H}_3O^+]{\text{1. diethyl ether}}$ Cyclopentylmethanol (CH_2OH) (62–64%)
Reaction of organolithium reagents with aldehydes and ketones (Section 15.7) Organolithium reagents react with aldehydes and ketones in a manner similar to that of Grignard reagents to form alcohols.	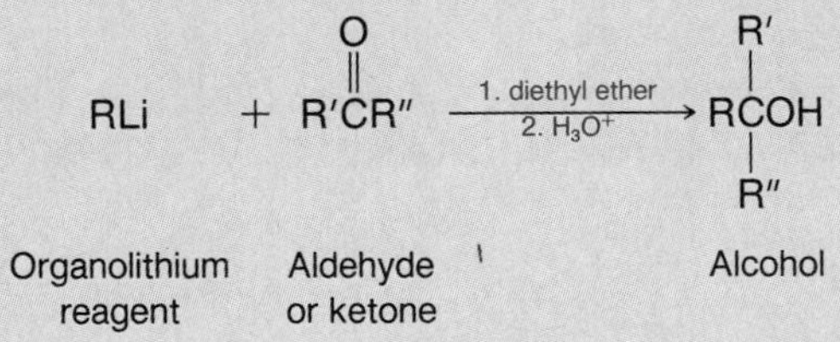 $RLi + R'CR'' \xrightarrow[\text{2. H}_3O^+]{\text{1. diethyl ether}} RCOH$ Organolithium reagent / Aldehyde or ketone / Alcohol $CH_3CH_2CH_2CH_2Li$ + Acetophenone ($C_6H_5CCH_3$) $\xrightarrow[\text{2. H}_3O^+]{\text{1. diethyl ether}}$ $CH_3CH_2CH_2CH_2-C(C_6H_5)(CH_3)-OH$ Butyllithium / Acetophenone / 2-Phenyl-2-hexanol (67%)
Reaction of Grignard reagents with esters (Section 15.10) Tertiary alcohols are produced in which two of the substituents on the hydroxyl-bearing carbon are derived from the Grignard reagent.	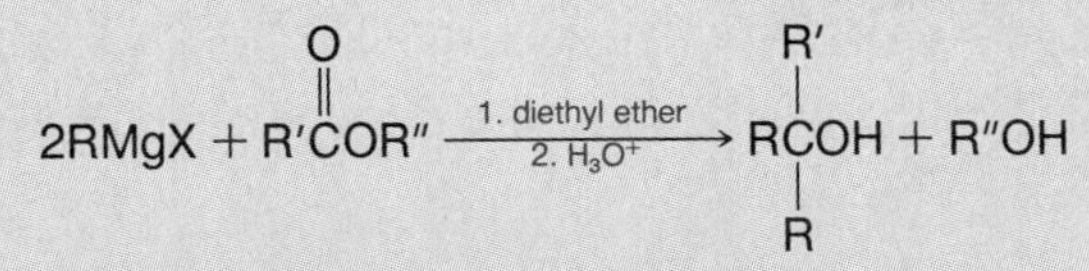 $2RMgX + R'COR'' \xrightarrow[\text{2. H}_3O^+]{\text{1. diethyl ether}} RCOH + R''OH$ $CH_3CH_2CH_2CH_2CH_2MgBr + CH_3COCH_2CH_3 \xrightarrow[\text{2. H}_3O^+]{\text{1. diethyl ether}}$ Pentylmagnesium bromide / Ethyl acetate $CH_3C(OH)(CH_2CH_2CH_2CH_2CH_3)CH_2CH_2CH_2CH_2CH_3$ 6-Methyl-6-undecanol (75%)

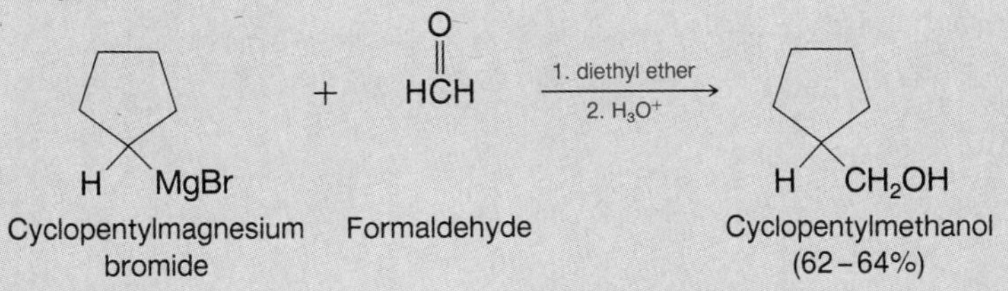

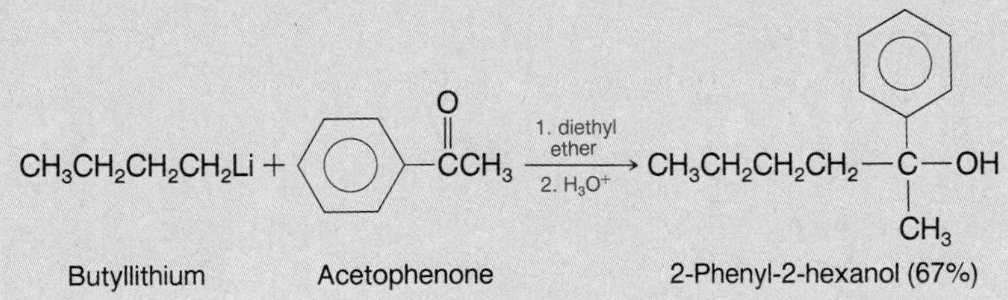

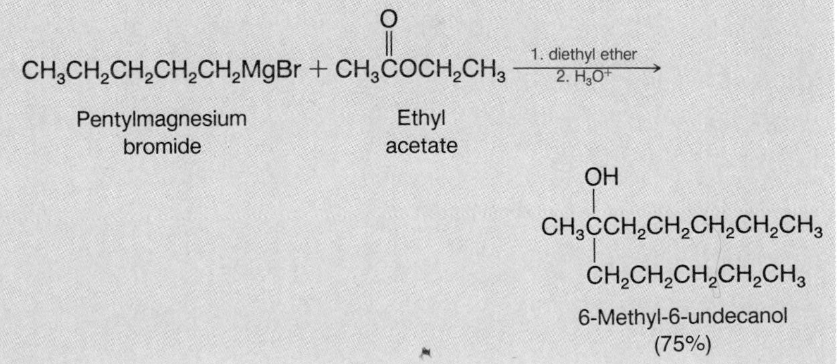

TABLE 16.2

Oxidation States of One-Carbon Compounds

			Number of carbon-oxygen bonds	Number of carbon-hydrogen bonds
Highest oxidation state	Carbon dioxide	O=C=O	4	0
	Formic acid	$\overset{\overset{\text{O}}{\|\|}}{\text{HCOH}}$	3	1
	Formaldehyde	$\overset{\overset{\text{O}}{\|\|}}{\text{HCH}}$	2	2
	Methanol	CH_3OH	1	3
Lowest oxidation state	Methane	CH_4	0	4

(arrow at left: increasing oxidation state)

SAMPLE SOLUTION (a) Converting starting material to product requires the addition of four hydrogen atoms to carbon monoxide. Since the hydrogen content of carbon increases, carbon monoxide is reduced. The reaction requires 2 mol hydrogen as the reducing agent per mole carbon monoxide.

$$CO \ + \ 2H_2 \ \longrightarrow \ CH_3OH$$

Carbon monoxide · · · · · · Hydrogen · · · · · · Methanol

The task of choosing the correct reagent for a particular functional group transformation is simplified if one keeps in mind the progression of oxidation states of common organic substances. The conversion of methanol to formaldehyde requires an oxidizing agent; the preparation of methanol from formic acid requires a reducing agent. The following section describes a number of synthetic methods for the preparation of alcohols by reduction reactions.

16.3 PREPARATION OF ALCOHOLS BY REDUCTION OF ALDEHYDES AND KETONES

The most obvious way to reduce an aldehyde or ketone to an alcohol is by hydrogenation of the carbon-oxygen double bond. Like the hydrogenation of alkenes, such a reaction is exothermic but exceedingly slow in the absence of a catalyst. Finely divided metals such as platinum, palladium, nickel, and ruthenium are effective catalysts for the hydrogenation of aldehydes and ketones. Aldehydes yield primary alcohols:

$$\overset{\overset{\text{O}}{\|\|}}{\text{RCH}} \ + \ H_2 \ \xrightarrow{\text{Pt, Pd, Ni, or Ru}} \ RCH_2OH$$

Aldehyde · · · · · hydrogen · · · · · · · · · · · · · · · · · Primary alcohol

$$CH_3O-\text{C}_6H_4-\overset{\overset{\text{O}}{\|\|}}{\text{CH}} \ \xrightarrow[\text{ethanol}]{H_2,\ Pt} \ CH_3O-\text{C}_6H_4-CH_2OH$$

p-Methoxybenzaldehyde · · · · · · · · · · · *p*-Methoxybenzyl alcohol (92%)

Ketones yield secondary alcohols:

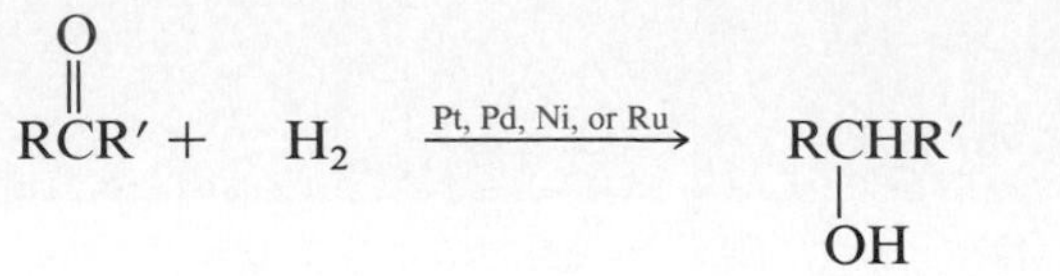

$$\underset{\substack{\text{Ketone}}}{\overset{\overset{\displaystyle O}{\|}}{R\overset{}{C}R'}} + \underset{\substack{\text{Hydrogen}}}{H_2} \xrightarrow{\text{Pt, Pd, Ni, or Ru}} \underset{\substack{\text{Secondary alcohol}}}{\overset{\displaystyle R\overset{}{C}HR'}{\underset{\displaystyle OH}{|}}}$$

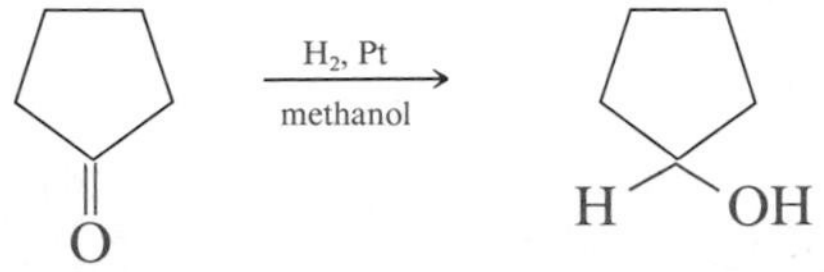

Cyclopentanone $\xrightarrow[\text{methanol}]{\text{H}_2,\ \text{Pt}}$ Cyclopentanol (93–95%)

PROBLEM 16.2 Which of the isomeric $C_4H_{10}O$ alcohols can be prepared by hydrogenation of aldehydes? Which can be prepared by hydrogenation of ketones? Which cannot be prepared by hydrogenation of a carbonyl compound?

For most laboratory-scale reductions of aldehydes and ketones, catalytic hydrogenation has been replaced by methods based on metal hydride reducing agents. The two reagents that are most commonly used are sodium borohydride and lithium aluminum hydride.

$$Na^+ \left[\begin{array}{c} H \\ | \\ H-B-H \\ | \\ H \end{array} \right] \qquad Li^+ \left[\begin{array}{c} H \\ | \\ H-Al-H \\ | \\ H \end{array} \right]$$

Sodium borohydride (NaBH₄)   Lithium aluminum hydride (LiAlH₄)

Sodium borohydride is particularly easy to use. It is soluble in water and in alcohols, and all that is required is to add it to an aqueous or alcoholic solution of an aldehyde or ketone.

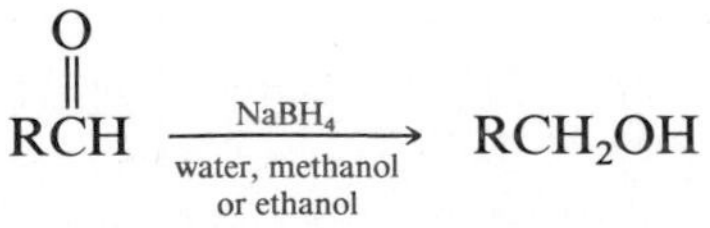

$$\underset{\substack{\text{Aldehyde}}}{\overset{\overset{\displaystyle O}{\|}}{R\overset{}{C}H}} \xrightarrow[\substack{\text{water, methanol} \\ \text{or ethanol}}]{\text{NaBH}_4} \underset{\substack{\text{Primary alcohol}}}{RCH_2OH}$$

m-Nitrobenzaldehyde $\xrightarrow[\text{methanol}]{\text{NaBH}_4}$ *m*-Nitrobenzyl alcohol (82%)

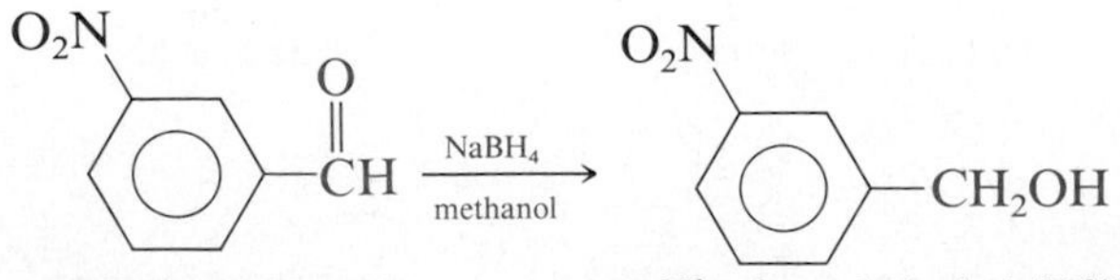

$$
\underset{\text{Ketone}}{\overset{\displaystyle\overset{O}{\|}}{R C R'}} \xrightarrow[\substack{\text{water, methanol} \\ \text{or ethanol}}]{NaBH_4} \underset{\text{Secondary alcohol}}{\overset{\displaystyle\underset{\displaystyle OH}{|}}{R CHR'}}
$$

$$
\underset{\text{4,4-Dimethyl-2-pentanone}}{\overset{\displaystyle\overset{O}{\|}}{CH_3C\,CH_2C(CH_3)_3}} \xrightarrow[\text{ethanol}]{NaBH_4} \underset{\text{4,4-Dimethyl-2-pentanol (85\%)}}{\overset{\displaystyle\underset{\displaystyle OH}{|}}{CH_3CH\,CH_2C(CH_3)_3}}
$$

Lithium aluminum hydride reacts with water and alcohols and so must be used in solvents such as anhydrous diethyl ether or tetrahydrofuran. Following reduction, a separate hydrolysis step is required to liberate the alcohol product.

$$
\underset{\text{Aldehyde}}{\overset{\displaystyle\overset{O}{\|}}{R CH}} \xrightarrow[\text{2. } H_2O]{\text{1. } LiAlH_4,\text{ diethyl ether}} \underset{\text{Primary alcohol}}{R CH_2OH}
$$

$$
\underset{\text{Heptanal}}{\overset{\displaystyle\overset{O}{\|}}{CH_3(CH_2)_5CH}} \xrightarrow[\text{2. } H_2O]{\text{1. } LiAlH_4,\text{ diethyl ether}} \underset{\text{1-Heptanol (86\%)}}{CH_3(CH_2)_5CH_2OH}
$$

$$
\underset{\text{Ketone}}{\overset{\displaystyle\overset{O}{\|}}{R CR'}} \xrightarrow[\text{2. } H_2O]{\text{1. } LiAlH_4,\text{ diethyl ether}} \underset{\text{Secondary alcohol}}{\overset{\displaystyle\underset{\displaystyle OH}{|}}{R CHR'}}
$$

$$
\underset{\text{1,1-Diphenyl-2-propanone}}{\overset{\displaystyle\overset{O}{\|}}{(C_6H_5)_2CHC\,CH_3}} \xrightarrow[\text{2. } H_2O]{\text{1. } LiAlH_4,\text{ diethyl ether}} \underset{\text{1,1-Diphenyl-2-propanol (84\%)}}{\overset{\displaystyle\underset{\displaystyle OH}{|}}{(C_6H_5)_2CHCHCH_3}}
$$

Sodium borohydride and lithium aluminum hydride react with carbonyl compounds in much the same way that Grignard reagents do, except that they function as *hydride donors* rather than as carbanion sources. Borohydride transfers a hydrogen with its pair of bonding electrons to the positively polarized carbon of a carbonyl group. The negatively polarized oxygen attacks boron. Ultimately, all four of the hydrogens of borohydride undergo transfer to carbonyl groups and a tetraalkoxyborate is formed.

$$
\underset{\underset{\delta+}{R_2C}\overset{}{=}\underset{\delta-}{O}}{\overset{H\,\rlap{\,\,\}}{}BH_3}{}} \longrightarrow \underset{R_2C-O}{\overset{H\quad \overline{B}H_3}{|\qquad}} \xrightarrow{3R_2C=O} \underset{\text{Tetraalkoxyborate}}{(R_2CHO)_4\overline{B}}
$$

Hydrolysis or alcoholysis converts the tetraalkoxyborate intermediate to the corresponding alcohol. The equation below illustrates the process for reactions carried out in water. An analogous process occurs in methanol or ethanol and yields the alcohol and $(CH_3O)_4B^-$ or $(CH_3CH_2O)_4B^-$.

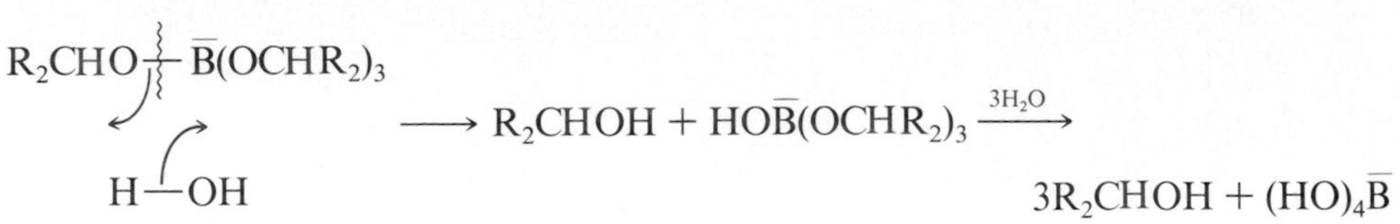

$$R_2CHO \overset{}{-} \bar{B}(OCHR_2)_3$$

$$\longrightarrow R_2CHOH + HO\bar{B}(OCHR_2)_3 \xrightarrow{3H_2O}$$

$$H{-}OH$$

$$3R_2CHOH + (HO)_4\bar{B}$$

A similar series of hydride transfer reactions occurs when aldehydes and ketones are treated with lithium aluminum hydride.

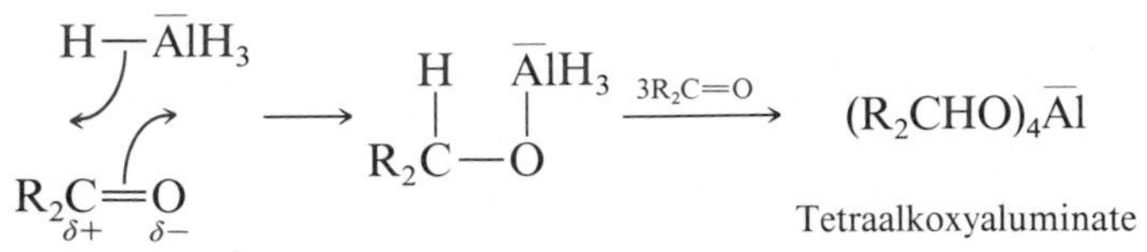

$$H{-}\bar{A}lH_3$$

$$R_2C\underset{\delta+\quad\delta-}{=}O$$

$$\longrightarrow \quad \underset{R_2C-O}{\overset{H\quad \bar{A}lH_3}{|\quad |}} \xrightarrow{3R_2C=O} (R_2CHO)_4\bar{A}l$$

Tetraalkoxyaluminate

Aqueous workup converts the tetraalkoxyaluminate to the desired alcohol.

$$(R_2CHO)_4\bar{A}l \quad + 4H_2O \longrightarrow 4R_2CHOH + \bar{A}l(OH)_4$$

Tetraalkoxyaluminate   Alcohol

PROBLEM 16.3 Sodium borodeuteride ($NaBD_4$) and lithium aluminum deuteride ($LiAlD_4$) are convenient reagents for introducing deuterium, the mass 2 isotope of hydrogen, into organic compounds. Write the structure of the organic product of the following reactions, clearly showing the position of all the deuterium atoms in each.

(a) Reduction of $CH_3\overset{O}{\overset{\|}{C}}H$ (acetaldehyde) with $NaBD_4$ in H_2O

(b) Reduction of $CH_3\overset{O}{\overset{\|}{C}}CH_3$ (acetone) with $NaBD_4$ in CH_3OD

(c) Reduction of $C_6H_5\overset{O}{\overset{\|}{C}}H$ (benzaldehyde) with $NaBD_4$ in CD_3OH

(d) Reduction of $H\overset{O}{\overset{\|}{C}}H$ (formaldehyde) with $LiAlD_4$ in diethyl ether, followed by addition of D_2O

SAMPLE SOLUTION (a) Sodium borodeuteride transfers deuterium to the carbonyl group of acetaldehyde, forming a C—D bond.

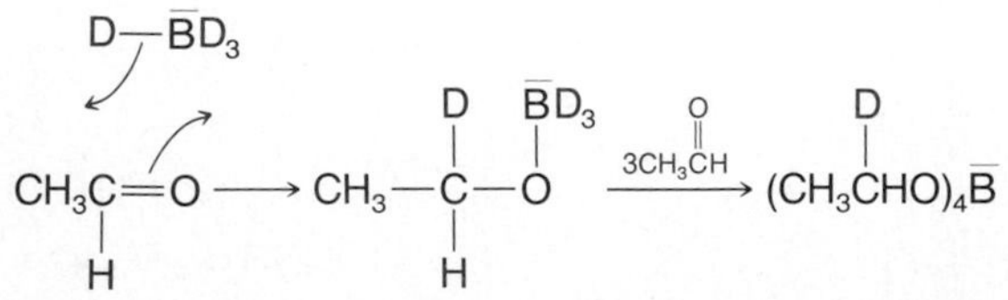

$$D{-}\bar{B}D_3$$

$$CH_3C{=}O \longrightarrow \underset{H}{\overset{D\quad \bar{B}D_3}{CH_3-\underset{|}{\overset{|}{C}}-O}} \xrightarrow{3CH_3\overset{O}{\overset{\|}{C}}H} (CH_3CHO)_4\bar{B}$$

Hydrolysis of $(CH_3CHDO)_4\bar{B}$ in H_2O leads to the formation of ethanol, with retention of the C—D bond formed in the above-mentioned step and formation of an O—H bond.

$$CH_3\underset{\overset{|}{D}}{CH}-O-\bar{B}(OCHDCH_3)_3 \longrightarrow CH_3\underset{\overset{|}{D}}{CH} + \bar{B}(OCHDCH_3)_3 \xrightarrow{3H_2O} CH_3\underset{\overset{|}{D}}{CH}OH + \bar{B}(OH)_4$$

H—OH   Ethanol-1-d

Neither sodium borohydride nor lithium aluminum hydride reduces isolated carbon-carbon double bonds. This makes possible the selective reduction of a carbonyl group in a molecule that contains both carbon-carbon and carbon-oxygen double bonds.

$$(CH_3)_2C=CHCH_2CH_2\overset{\overset{\displaystyle O}{\|}}{C}CH_3 \xrightarrow[\text{2. } H_2O]{\text{1. LiAlH}_4,\text{ diethyl ether}} (CH_3)_2C=CHCH_2CH_2\underset{\overset{|}{OH}}{CH}CH_3$$

6-Methyl-5-hepten-2-one   6-Methyl-5-hepten-2-ol (90%)

Note that catalytic hydrogenation would not be suitable for this transformation because hydrogen adds to carbon-carbon double bonds faster than it reduces carbonyl groups.

The oldest method for converting ketones to secondary alcohols is by reduction with sodium in ethanol.

$$CH_3\overset{\overset{\displaystyle O}{\|}}{C}(CH_2)_4CH_3 \xrightarrow[\text{ethanol}]{Na} CH_3\underset{\overset{|}{OH}}{CH}(CH_2)_4CH_3$$

2-Heptanone   2-Heptanol (62–65%)

When aldehydes are treated with sodium in ethanol, side reactions occur, limiting the effectiveness of this procedure as a route to primary alcohols.

16.4 PREPARATION OF ALCOHOLS BY REDUCTION OF CARBOXYLIC ACIDS

Carboxylic acids, compounds of the type $R\overset{\overset{\displaystyle O}{\|}}{C}OH$, are exceedingly difficult to reduce. Acetic acid, for example, is often used as a solvent in catalytic hydrogenations because it is inert under the reaction conditions. A very powerful reducing agent is required in order to convert a carboxylic acid to a primary alcohol. Lithium aluminum hydride is that reducing agent.

$$R\overset{\overset{\displaystyle O}{\|}}{C}OH \xrightarrow[\text{2. } H_2O]{\text{1. LiAlH}_4,\text{ diethyl ether}} RCH_2OH$$

Carboxylic acid   Primary alcohol

$$\text{▷—CO}_2\text{H} \xrightarrow[\text{2. H}_2\text{O}]{\text{1. LiAlH}_4, \text{ diethyl ether}} \text{▷—CH}_2\text{OH}$$

Cyclopropanecarboxylic   Cyclopropylmethanol (78%)
acid

Sodium borohydride is not nearly as potent a hydride donor as lithium aluminum hydride and does not reduce carboxylic acids.

16.5 PREPARATION OF ALCOHOLS BY REDUCTION OF ESTERS OF CARBOXYLIC ACIDS

Esters of carboxylic acids, compounds of the type $\overset{\text{O}}{\overset{\|}{\text{RCOR}'}}$, are much more easily reduced than are carboxylic acids themselves. Two alcohol molecules are formed from each ester molecule.

$$\overset{\text{O}}{\overset{\|}{\text{RCOR}'}} \longrightarrow \text{RCH}_2\text{OH} + \text{R}'\text{OH}$$

Ester   Primary   Alcohol
                      alcohol

Since hydrogenation is accompanied by cleavage *(lysis)*, the overall process is referred to as the *hydrogenolysis* of an ester.

Hydrogenolysis of esters is normally carried out over a combination of copper-chromium oxides called *copper chromite*. Reductions using this catalyst are performed at high temperature and pressure.

$$\bigcirc\!\!\!-\overset{\text{O}}{\overset{\|}{\text{C}}}\text{OCH}_2\text{CH}_3 \xrightarrow[\substack{\text{copper chromite}\\250°\text{C}}]{\text{H}_2(200 \text{ atm})} \bigcirc\!\!\!-\text{CH}_2\text{OH} + \text{CH}_3\text{CH}_2\text{OH}$$

Ethyl cyclohexanecarboxylate   Cyclohexylmethanol   Ethanol
                                              (97%)

Reduction occurs in two stages. In the first stage the ester is converted to an aldehyde, which then undergoes rapid reduction. The bond between the carbonyl group and the ester oxygen is cleaved on hydrogenolysis. The carbonyl group becomes the —CH$_2$OH group of a primary alcohol.

$$\overset{\text{O}}{\overset{\|}{\text{RC}}}\!-\!\text{OR}' \xrightarrow[\substack{\text{catalyst}\\(\text{slow})}]{\text{H}_2} \overset{\text{O}}{\overset{\|}{\text{RCH}}} + \text{R}'\text{OH}$$

                                        Aldehyde   Alcohol

The first hydrogenation
step cleaves this bond
of the ester

$$\xrightarrow[\substack{\text{catalyst}\\(\text{fast})}]{\text{H}_2} \text{RCH}_2\text{OH} \quad \text{(a primary alcohol)}$$

This carbon and this
oxygen are those of
the original C=O
group

PROBLEM 16.4   Which of the following esters will yield a mixture containing equimolar amounts of benzyl alcohol and isopropyl alcohol on hydrogenolysis?

$$C_6H_5CH_2COCH(CH_3)_2 \qquad C_6H_5COCH(CH_3)_2 \qquad (CH_3)_2CHCOCH_2C_6H_5$$

Lithium aluminum hydride smoothly converts esters to primary alcohols in high yield. Aldehydes are likely intermediates in these reactions.

$$\langle\!\!\langle\bigcirc\rangle\!\!\rangle\text{--COCH}_2\text{CH}_3 \xrightarrow[\text{2. H}_2\text{O}]{\text{1. LiAlH}_4,\ \text{diethyl ether}} \langle\!\!\langle\bigcirc\rangle\!\!\rangle\text{--CH}_2\text{OH} + \text{CH}_3\text{CH}_2\text{OH}$$

Ethyl benzoate   Benzyl alcohol (90%)   Ethanol

Sodium borohydride slowly reduces esters to alcohols but is rarely used for this purpose. Lithium aluminum hydride is preferred.

Esters are reduced by sodium in alcohol. This is a method of long standing in organic chemistry known as the *Bouveault-Blanc reduction.*

$$CH_3(CH_2)_{10}COCH_2CH_3 \xrightarrow[\text{ethanol}]{\text{Na}} CH_3(CH_2)_{10}CH_2OH + CH_3CH_2OH$$

Ethyl dodecanoate   1-Dodecanol (65–75%)   Ethanol

PROBLEM 16.5   Bouveault-Blanc reduction of butyl oleate yields the primary alcohol oleyl alcohol. The carbon-carbon double bond remains intact in this process.

$$CH_3(CH_2)_7CH=CH(CH_2)_7COCH_2CH_2CH_2CH_3 \xrightarrow[\text{1-butanol}]{\text{Na}}$$

Butyl oleate

$$CH_3(CH_2)_7CH=CH(CH_2)_7CH_2OH + CH_3CH_2CH_2CH_2OH$$

Oleyl alcohol (82–84%)   1-Butanol

Could hydrogenolysis over a metal catalyst be used to prepare oleyl alcohol from butyl oleate? Could lithium aluminum hydride be used?

16.6 PREPARATION OF ALCOHOLS FROM EPOXIDES

Although the chemical reactions of epoxides will not be covered in detail until the following chapter, we shall describe their use in the synthesis of alcohols here.

Grignard reagents react with ethylene oxide to yield primary alcohols containing two more carbon atoms than the alkyl halide from which the organometallic was prepared.

$$RMgX + H_2C\!\!-\!\!-\!\!-\!\!CH_2 \xrightarrow[\text{2. H}_3\text{O}^+]{\text{1. diethyl ether}} RCH_2CH_2OH$$

$$\underset{O}{\diagdown\diagup}$$

Grignard   Ethylene oxide   Primary alcohol
reagent   (oxirane)

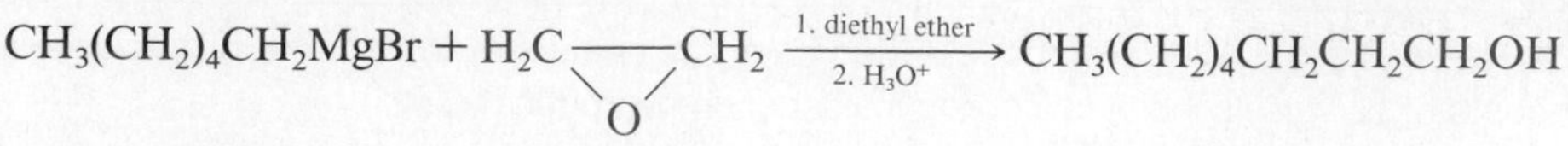

$$CH_3(CH_2)_4CH_2MgBr + H_2C{-}{-}CH_2 \xrightarrow[\text{2. H}_3\text{O}^+]{\text{1. diethyl ether}} CH_3(CH_2)_4CH_2CH_2CH_2OH$$

Hexylmagnesium bromide	Ethylene oxide (oxirane)	1-Octanol (71%)

PROBLEM 16.6   Each of the following alcohols has been prepared by reaction of a Grignard reagent with ethylene oxide. Select the appropriate Grignard reagent in each case.

(a)   [structure: o-methylphenyl—CH₂CH₂OH]

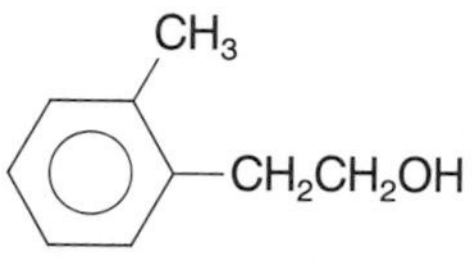

(c) $CH_3(CH_2)_7CH_2OH$

(b)   [cyclohexyl—CH₂CH₂OH structure]

(d)   [naphthalenyl—CH₂CH₂OH structure]

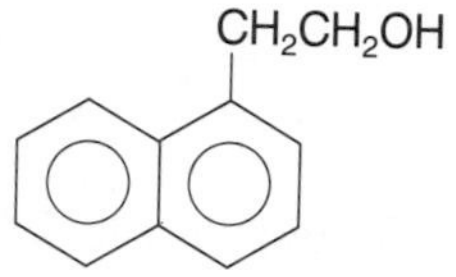

SAMPLE SOLUTION   (a)   Reaction with ethylene oxide results in the addition of a —CH₂CH₂OH unit to the Grignard reagent. The Grignard reagent derived from o-bromotoluene (or o-chlorotoluene or o-iodotoluene) is appropriate here.

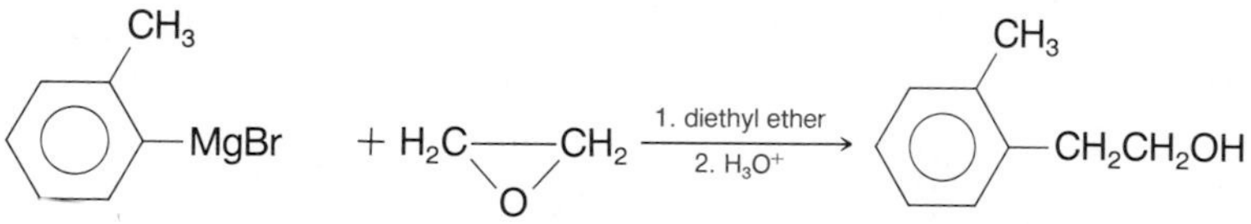

[structure: o-methylphenyl—MgBr] + H₂C——CH₂ (ethylene oxide) $\xrightarrow[\text{2. H}_3\text{O}^+]{\text{1. diethyl ether}}$ [structure: o-methylphenyl—CH₂CH₂OH]

o-Methylphenylmagnesium bromide	Ethylene oxide	2-(o-Methylphenyl)ethanol (66%)

Organolithium reagents react with epoxides in a manner similar to that of Grignard reagents. Epoxide rings are readily opened and undergo carbon-oxygen bond cleavage when attacked by nucleophiles. Grignard reagents and organolithium reagents react with ethylene oxide by serving as sources of nucleophilic carbon.

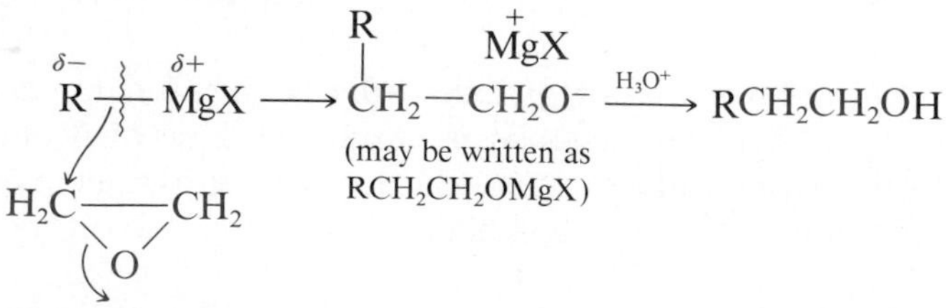

$$\overset{\delta-}{R}{\raisebox{0.2em}{\scriptsize\}}}\overset{\delta+}{MgX} \longrightarrow \overset{\overset{\displaystyle R}{|}}{CH_2}{-}CH_2O^- \xrightarrow{H_3O^+} RCH_2CH_2OH$$

$$H_2C{-}{-}CH_2$$
$$\underset{O}{}$$

$\overset{+}{MgX}$

(may be written as RCH_2CH_2OMgX)

This kind of chemical reactivity of epoxides is rather general. Nucleophiles other than Grignard reagents react with epoxides, and epoxides more elaborate than ethylene oxide may be used. All these features will be discussed in Sections 17.14 and 17.15.

16.7 PREPARATION OF DIOLS

Diols—compounds that bear two hydroxyl groups—are prepared by many of the same reactions employed for alcohols by using compounds having two carbonyl groups as starting materials. The examples shown below illustrate: (1) catalytic hydrogenation of a dialdehyde; (2) metal hydride reduction of a diketone; and (3) reduction of a diester by sodium in ethanol

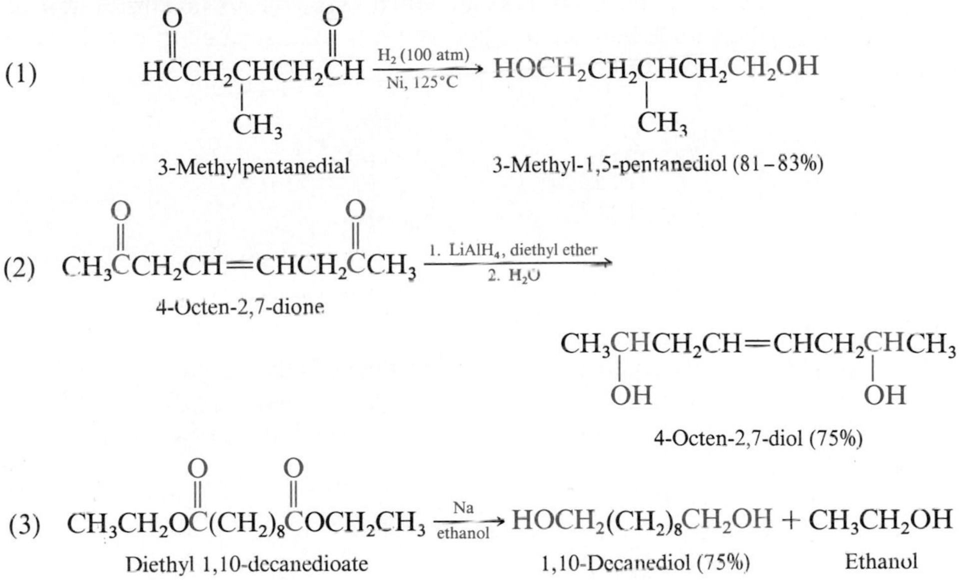

(1)
$$\underset{\text{3-Methylpentanedial}}{\overset{O}{\underset{|}{\overset{\parallel}{HCCH_2CHCH_2CH}}}} \underset{CH_3}{} \xrightarrow[\text{Ni, 125°C}]{H_2 \text{ (100 atm)}} \underset{\text{3-Methyl-1,5-pentanediol (81–83\%)}}{\underset{CH_3}{\overset{}{HOCH_2CH_2CHCH_2CH_2OH}}}$$

(2)
$$\underset{\text{4-Octen-2,7-dione}}{\overset{O}{\overset{\parallel}{CH_3CCH_2CH}}=CHCH_2\overset{O}{\overset{\parallel}{CCH_3}}} \xrightarrow[\text{2. } H_2O]{\text{1. LiAlH}_4\text{, diethyl ether}}$$

$$\underset{\text{4-Octen-2,7-diol (75\%)}}{\underset{OH}{\overset{}{CH_3CHCH_2CH}}=CHCH_2\underset{OH}{\overset{}{CHCH_3}}}$$

(3)
$$\underset{\text{Diethyl 1,10-decanedioate}}{\overset{O}{\overset{\parallel}{CH_3CH_2OC(CH_2)_8}}\overset{O}{\overset{\parallel}{COCH_2CH_3}}} \xrightarrow[\text{ethanol}]{Na} \underset{\text{1,10-Decanediol (75\%)}}{HOCH_2(CH_2)_8CH_2OH} + \underset{\text{Ethanol}}{CH_3CH_2OH}$$

Vicinal diols have hydroxyl groups on adjacent carbons. Two commonly encountered vicinal diols are 1,2-ethanediol and 1,2-propanediol.

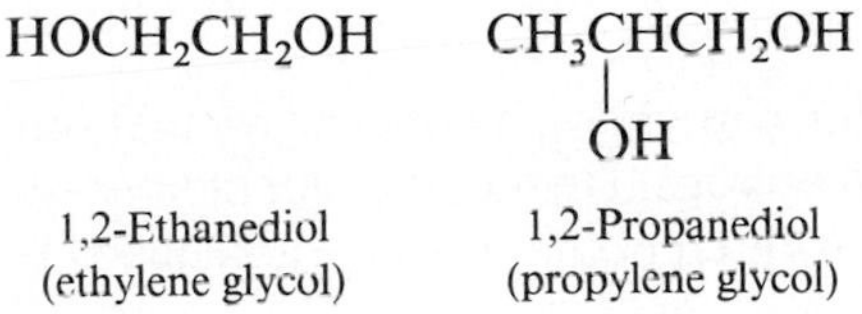

$$\underset{\substack{\text{1,2-Ethanediol} \\ \text{(ethylene glycol)}}}{HOCH_2CH_2OH} \qquad \underset{\substack{\text{1,2-Propanediol} \\ \text{(propylene glycol)}}}{\underset{OH}{\overset{}{CH_3CHCH_2OH}}}$$

Ethylene glycol and propylene glycol are common names for these two diols and are acceptable IUPAC names. Aside from these two compounds, the IUPAC system does not use the word "glycol" for naming diols.

Vicinal diols are intermediates in the cleavage of alkenes by potassium permanganate (Section 7.17).

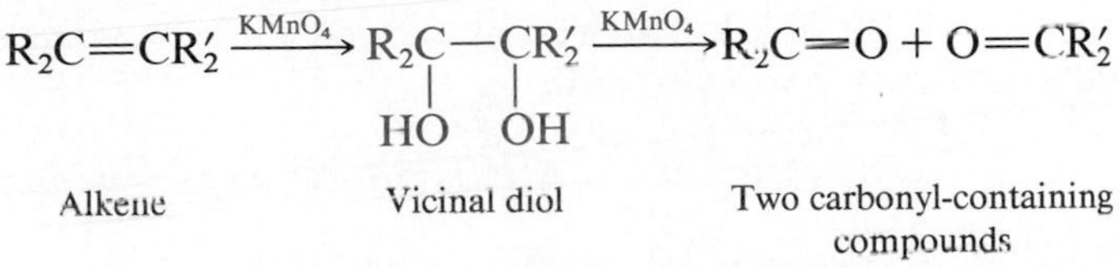

$$\underset{\text{Alkene}}{R_2C{=}CR_2'} \xrightarrow{KMnO_4} \underset{\substack{HO \quad OH \\ \text{Vicinal diol}}}{R_2C{-}CR_2'} \xrightarrow{KMnO_4} \underset{\substack{\text{Two carbonyl-containing} \\ \text{compounds}}}{R_2C{=}O + O{=}CR_2'}$$

The step in which the vicinal diol is formed is referred to as *hydroxylation* of an alkene. By carefully controlling the temperature and the duration of the reaction, it is possible to isolate the diol intermediate. This reaction has been used for the preparation of vicinal diols, although the yields are sometimes low.

$$CH_2{=}CHCH(CH_3)_2 \xrightarrow[\text{H}_2\text{O, cold}]{\text{KMnO}_4} HOCH_2\underset{\underset{OH}{|}}{CH}CH(CH_3)_2$$

3-Methyl-1-butene   3-Methyl-1,2-butanediol (50%)

An important feature of the reaction is that both hydroxyl groups are introduced at the same face of the double bond. Thus, potassium permanganate brings about the syn hydroxylation of alkenes.

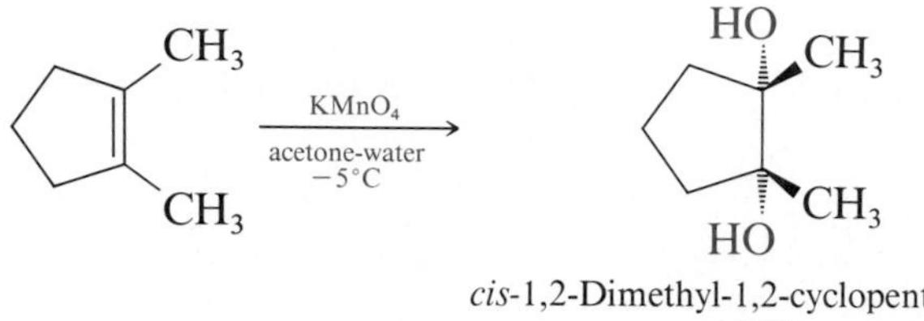

cis-1,2-Dimethyl-1,2-cyclopentanediol
(45%)

Both oxygens of diols obtained by hydroxylation of alkenes are derived from permanganate. A cyclic manganate ester is an intermediate. Hydrolysis of this manganate ester under the conditions of its formation yields the diol and manganese dioxide.

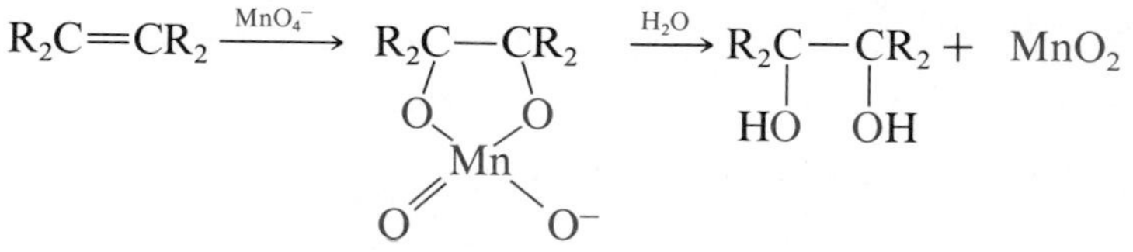

| Alkene | Cyclic manganate ester | Vicinal diol | Manganese dioxide |

A better procedure for converting alkenes to vicinal diols employs osmium tetraoxide as the hydroxylating agent in the presence of *tert*-butyl hydroperoxide. In the first stage osmium tetraoxide combines with the alkene to form a cyclic osmate ester:

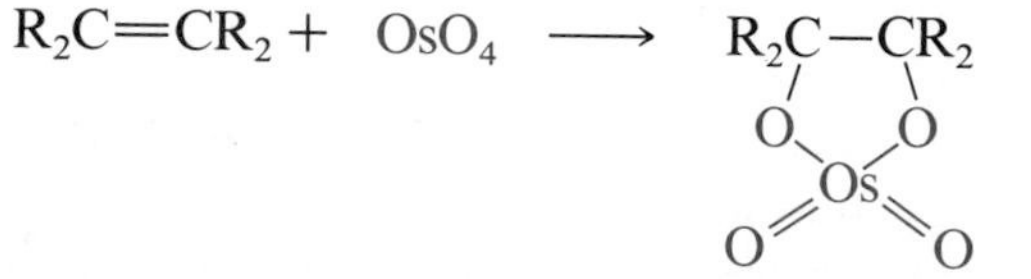

| Alkene | Osmium tetraoxide | Cyclic osmate ester |

Osmate esters are fairly stable but are readily cleaved in the presence of an oxidizing agent such as *tert*-butyl hydroperoxide.

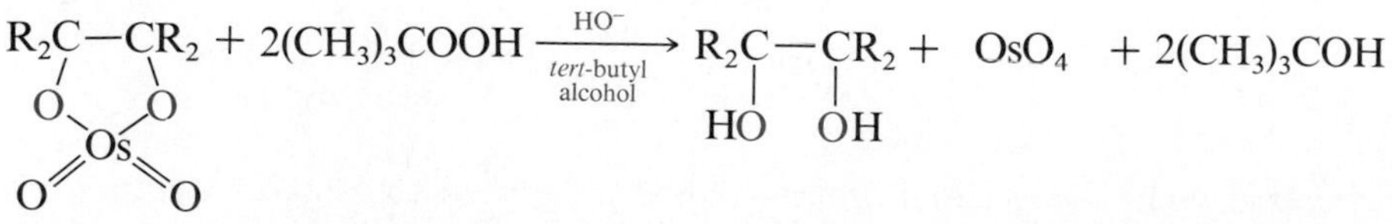

| *tert*-Butyl hydroperoxide | Vicinal diol | Osmium tetraoxide | *tert*-Butyl alcohol |

Since osmium tetraoxide is regenerated in the second step, the overall reaction can be carried out by using only catalytic amounts of osmium tetraoxide. This is fortunate because osmium tetraoxide is both toxic and expensive. The entire process is performed in a single operation by simply allowing a solution of the alkene and *tert*-butyl hydroperoxide in *tert*-butyl alcohol containing a small amount of osmium tetraoxide and base to stand for several hours.

$$CH_3(CH_2)_7CH{=}CH_2 \xrightarrow[\text{\textit{tert}-butyl alcohol, HO}^-]{\text{(CH}_3)_3\text{COOH, OsO}_4\text{(cat)}} CH_3(CH_2)_7\overset{\displaystyle |}{\underset{\displaystyle \underset{OH}{|}}{C}}HCH_2OH$$

1-Decene   1,2-Decanediol (73%)

As was also the case with permanganate hydroxylation, this method leads to stereospecific syn addition of hydroxyl groups.

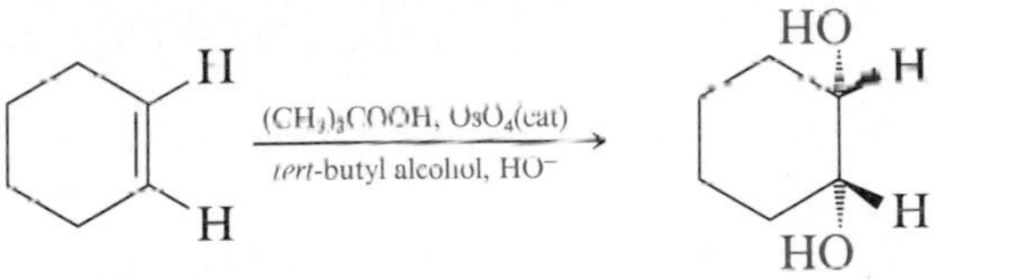

Cyclohexene   *cis*-1,2-Cyclohexanediol
                                           (62%)

A method for anti hydroxylation of alkenes by way of the hydrolysis of epoxides will be described in Section 17.15.

16.8 REACTIONS OF ALCOHOLS. A REVIEW AND A PREVIEW

Alcohols are among the most versatile starting materials for the preparation of a variety of organic functional groups. Several of the reactions of alcohols have already been seen in earlier chapters and are summarized in Table 16.3.

You should be aware of the fact that alcohols can undergo reactions involving various combinations of the bonds to carbon and oxygen. The ionization of alcohols when they act as weak acids and the formation of alkoxides when they react with metals are reactions that take place at the O—H bond.

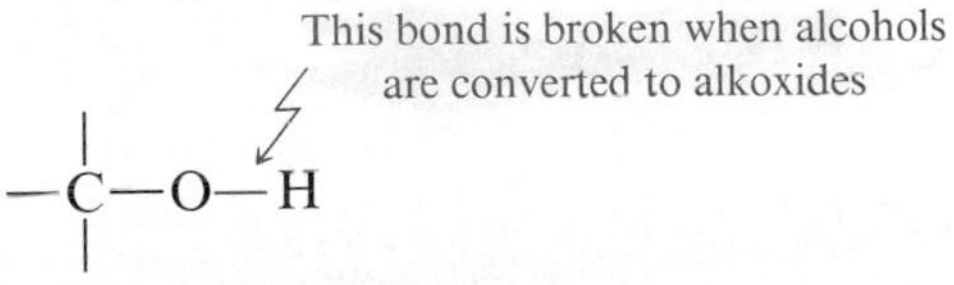

This bond is broken when alcohols
are converted to alkoxides

The carbon-oxygen bond of alcohols is cleaved when alcohols are converted to alkyl halides or undergo acid-catalyzed dehydration.

This bond is broken when alcohols are
subjected to acid-catalyzed dehydration
or converted to alkyl halides through
the use of HX, SOCl$_2$, or PBr$_3$

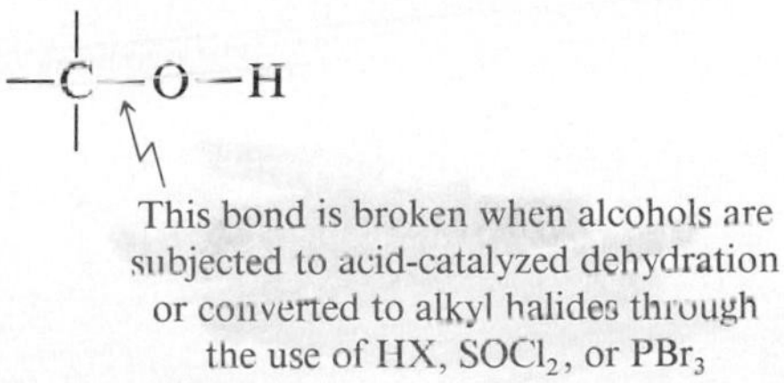

TABLE 16.3

Summary of Reactions of Alcohols Discussed in Earlier Chapters

Reaction (section) and comments	General equation and specific example
Conversion to alkoxides (Sections 4.6– 4.7) Alcohols are slightly less acidic than water; alkoxides are slightly more basic than hydroxide. Alcohols react readily with group I and group II metals to liberate hydrogen and form metal alkoxides.	$2ROH + 2M \longrightarrow 2RO^-M^+ + H_2$ Alcohol Metal Metal alkoxide Hydrogen $CH_3CHOH \xrightarrow{Na} CH_3CHO^-Na^+$ $\quad\ \mid CH_3 \qquad\qquad\qquad \mid CH_3$ 2-Propanol Sodium 2-propanolate (100%)
Reaction with hydrogen halides (Section 4.9) The order of alcohol reactivity parallels the order of carbocation stability: 3° > 2° > 1°. Benzylic alcohols react readily. The order of hydrogen halide reactivity parallels their acidities: HI > HBr > HCl ≫ HF.	$ROH + HX \longrightarrow RX + H_2O$ Alcohol Hydrogen Alkyl Water $\qquad\qquad$ halide $\quad$ halide CH_3O ⬡ $-CH_2OH \xrightarrow{HBr} CH_3O$ ⬡ $-CH_2Br$ m-Methoxybenzyl alcohol m-Methoxybenzyl bromide (98%)
Reaction with thionyl chloride (Section 4.17) Thionyl chloride converts alcohols to alkyl chlorides.	$ROH + SOCl_2 \longrightarrow RCl + SO_2 + HCl$ Alcohol Thionyl Alkyl Sulfur Hydrogen $\qquad\qquad$ chloride $\quad$ chloride dioxide chloride $(CH_3)_2C{=}CHCH_2CH_2CHCH_3 \xrightarrow[\text{diethyl ether}]{SOCl_2,\ pyridine} (CH_3)_2C{=}CHCH_2CH_2CHCH_3$ $\qquad\qquad\qquad\qquad \mid OH \qquad\qquad\qquad\qquad\qquad\qquad\qquad \mid Cl$ 6-Methyl-5-hepten-2-ol 6-Chloro-2-methyl-2-heptene (67%)
Reaction with phosphorus trihalides (Section 4.17) Phosphorus trihalides convert alcohols to alkyl halides.	$3ROH + PX_3 \longrightarrow 3RX + P(OH)_3$ Alcohol Phosphorus trihalide Alkyl halide Phosphorous acid ⬠$-CH_2OH \xrightarrow{PBr_3}$ ⬠$-CH_2Br$ Cyclopentylmethanol (Bromomethyl)cyclopentane (50%)
Acid-catalyzed dehydration (Section 6.2) This is a frequently used procedure for the preparation of alkenes. The order of alcohol reactivity parallels the order of carbocation stability: 3° > 2° > 1°. Benzylic alcohols react readily. Rearrangements are sometimes observed.	$R_2CCHR_2 \xrightarrow[\text{heat}]{H^+} R_2C{=}CR_2 + H_2O$ $\quad \mid OH \qquad\qquad\qquad$ Alkene $\quad$ Water Alcohol Br ⬡ $-CHCH_2CH_3 \xrightarrow[\text{heat}]{KHSO_4} Br$ ⬡ $-CH{=}CHCH_3$ $\qquad\qquad \mid OH$ 1-(m-Bromophenyl)propanol 1-(m-Bromophenyl)propene (71%)

TABLE 16.3 (continued)

Reaction (section) and comments	General equation and specific example
Conversion to *p*-toluenesulfonate esters (Section 9.15) Alcohols react with *p*-toluenesulfonyl chloride to give *p*-toluenesulfonate esters. Sulfonate esters are reactive substrates for nucleophilic substitution and elimination reactions.	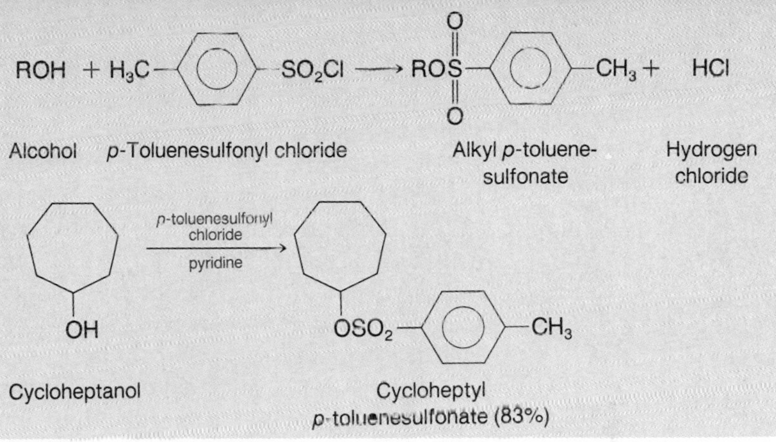 $ROH + H_3C-C_6H_4-SO_2Cl \longrightarrow ROS(O)_2-C_6H_4-CH_3 + HCl$ Alcohol *p*-Toluenesulfonyl chloride Alkyl *p*-toluenesulfonate Hydrogen chloride Cycloheptanol $\xrightarrow[\text{pyridine}]{\text{*p*-toluenesulfonyl chloride}}$ Cycloheptyl *p*-toluenesulfonate (83%)

Some additional reactions of alcohols that we will examine in this chapter occur by O—H bond breaking and some by C—O bond breaking. Additionally, primary and secondary alcohols can exhibit a third reaction type in which a carbon-oxygen double bond is formed by cleavage of both an O—H bond and a C—H bond.

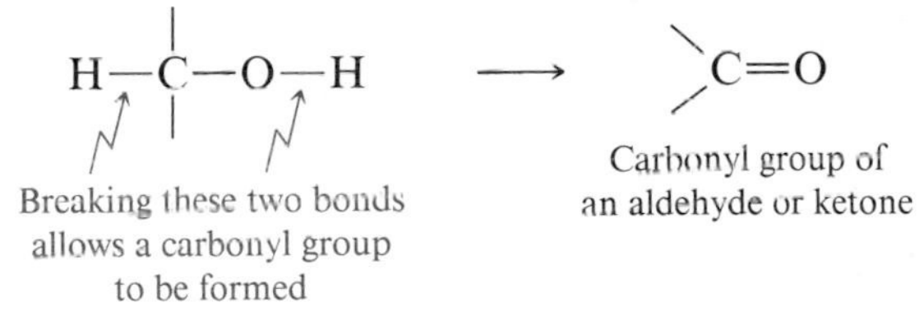

$$H-\overset{|}{\underset{|}{C}}-O-H \longrightarrow \;>\!C=O$$

Breaking these two bonds allows a carbonyl group to be formed   Carbonyl group of an aldehyde or ketone

Transformations such as this correspond to the loss of the elements of hydrogen from an alcohol and are sometimes termed *dehydrogenation* reactions. More normally, however, organic chemists classify the conversion of an alcohol to a carbonyl compound as an *oxidation* of the alcohol. These reactions are especially important in the practice of organic synthesis.

16.9 CONVERSION OF ALCOHOLS TO ETHERS

Primary alcohols are converted to ethers on heating in the presence of an acid catalyst, usually sulfuric acid.

$$2RCH_2OH \xrightarrow{\text{H}^+,\text{ heat}} RCH_2OCH_2R + H_2O$$

Primary alcohol   Dialkyl ether   Water

This kind of reaction is said to be a *condensation reaction*. In a condensation reaction two molecules combine to form a larger one while liberating a small molecule. In this case two alcohol molecules combine to give an ether and water.

$$2CH_3CH_2CH_2CH_2OH \xrightarrow[130°C]{H_2SO_4} CH_3CH_2CH_2CH_2OCH_2CH_2CH_2CH_3 + H_2O$$

1-Butanol   Dibutyl ether (60%)   Water

When applied to the synthesis of ethers, the reaction is effective only with primary alcohols. Elimination to form alkenes is the principal reaction that occurs with secondary and tertiary alcohols.

The mechanism by which two molecules of a primary alcohol condense to give a dialkyl ether and water is outlined for the specific case of ethanol in Figure 16.2. The individual steps of this mechanism are analogous to those seen earlier. Nucleophilic attack on a protonated alcohol was encountered in the reaction of primary alcohols with hydrogen halides (Section 4.17), and the nucleophilic properties of alcohols were discussed in the context of solvolysis reactions (Section 9.7). Both the first and last steps are proton transfer reactions between oxygens.

Diethyl ether is prepared on an industrial scale by heating ethanol with sulfuric acid at 140°C. At higher temperatures elimination predominates and ethylene is the major product.

Diols react intramolecularly to form cyclic ethers when a five-membered or six-membered ring can result.

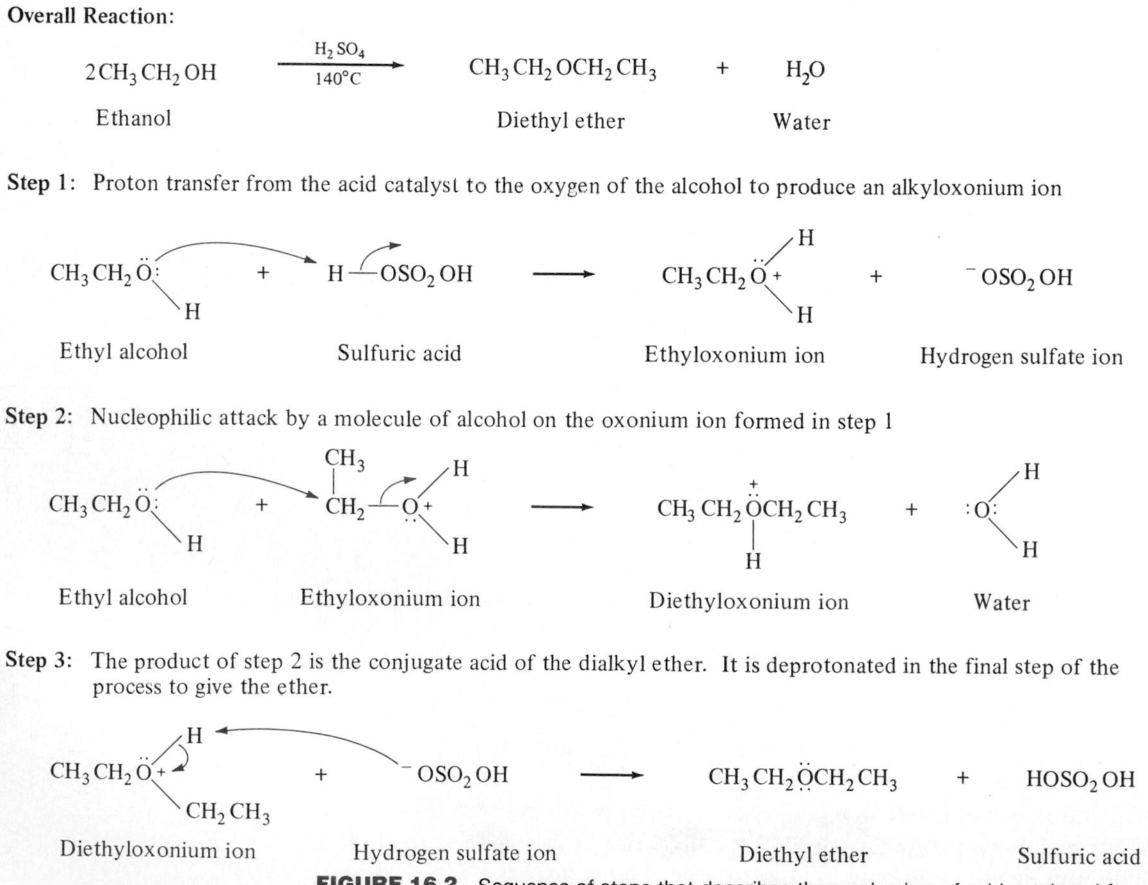

Overall Reaction:

$$2CH_3CH_2OH \xrightarrow[140°C]{H_2SO_4} CH_3CH_2OCH_2CH_3 + H_2O$$

Ethanol   Diethyl ether   Water

Step 1: Proton transfer from the acid catalyst to the oxygen of the alcohol to produce an alkyloxonium ion

Ethyl alcohol   Sulfuric acid   Ethyloxonium ion   Hydrogen sulfate ion

Step 2: Nucleophilic attack by a molecule of alcohol on the oxonium ion formed in step 1

Ethyl alcohol   Ethyloxonium ion   Diethyloxonium ion   Water

Step 3: The product of step 2 is the conjugate acid of the dialkyl ether. It is deprotonated in the final step of the process to give the ether.

Diethyloxonium ion   Hydrogen sulfate ion   Diethyl ether   Sulfuric acid

FIGURE 16.2 Sequence of steps that describes the mechanism of acid-catalyzed formation of diethyl ether from ethyl alcohol.

$$HOCH_2CH_2CH_2CH_2CH_2OH \xrightarrow[\text{heat}]{H_2SO_4} \qquad + H_2O$$

1,5-Pentanediol   Oxane (76%)   Water

In these intramolecular ether-forming reactions, the alcohol functions may be primary, secondary, or tertiary.

PROBLEM 16.7 On heating 1,2,4-butanetriol in the presence of an acid catalyst, a cyclic ether of molecular formula $C_4H_8O_2$ was obtained in 81 to 88 percent yield. Suggest a reasonable structure for this product.

16.10 ESTERIFICATION

Acid-catalyzed condensation of an alcohol and a carboxylic acid yields an ester and water.

$$ROH + \underset{\text{Carboxylic acid}}{R'\overset{O}{\overset{\|}{C}}OH} \underset{H^+}{\overset{}{\rightleftharpoons}} \underset{\text{Ester}}{R'\overset{O}{\overset{\|}{C}}OR} + \underset{\text{Water}}{H_2O}$$

Alcohol   Carboxylic acid   Ester   Water

$$CH_3CH_2OH + CH_3\overset{O}{\overset{\|}{C}}OH \underset{H^+}{\overset{}{\rightleftharpoons}} CH_3\overset{O}{\overset{\|}{C}}OCH_2CH_3 + H_2O$$

Ethanol   Acetic acid   Ethyl acetate   Water

The direct formation of an ester from an alcohol and a carboxylic acid is known as the *Fischer esterification reaction.* It is a reversible process, and the position of equilibrium lies slightly to the side of products when the reactants are simple alcohols and carboxylic acids. When the reaction is used for preparative purposes, the position of equilibrium can be made more favorable by using either the alcohol or the carboxylic acid in excess.

$$CH_3OH + \text{(benzoic acid)} \xrightarrow[\text{heat}]{H_2SO_4} \text{(methyl benzoate)} + H_2O$$

Methanol   Benzoic acid   Methyl benzoate   Water
(0.6 mol)   (0.1 mol)   (isolated in 70%
                                         yield based on
                                         benzoic acid)

$$\text{(cyclohexanol)} + CH_3\overset{O}{\overset{\|}{C}}OH \xrightarrow[\text{heat}]{HCl} \text{(cyclohexyl acetate)} + H_2O$$

Cyclohexanol   Acetic acid   Cyclohexyl acetate   Water
(1 mol)   (2.2 mol)   (isolated in 53%
                                       yield based on
                                       cyclohexanol)

Another way to shift the position of equilibrium to favor the formation of ester is to remove the other product (water) from the reaction mixture as it is formed. This is accomplished by adding benzene as a cosolvent and distilling the benzene-water azeotrope.

$$CH_3CHCH_2CH_3 + CH_3\overset{\overset{\displaystyle O}{\|}}{C}OH \xrightarrow[\text{benzene, heat}]{H^+} CH_3\overset{\overset{\displaystyle O}{\|}}{C}OCHCH_2CH_3 + H_2O$$

OH		CH$_3$	
sec-Butyl alcohol (0.20 mol)	Acetic acid (0.25 mol)	*sec*-Butyl acetate (isolated in 71% yield based on *sec*-butyl alcohol)	Water (codistills with benzene)

For steric reasons, the order of alcohol reactivity in the Fischer esterification is CH_3OH > primary > secondary > tertiary. Phenols are much less reactive than aliphatic alcohols.

PROBLEM 16.8 Write the structure of the ester formed in each of the following reactions:

(a) $CH_3CH_2CH_2CH_2OH + CH_3CH_2\overset{\overset{\displaystyle O}{\|}}{C}OH \xrightarrow[\text{heat}]{H_2SO_4}$

(b) $C_6H_5CH_2OH + (CH_3)_2CHCH_2\overset{\overset{\displaystyle O}{\|}}{C}OH \xrightarrow[\text{heat}]{H_2SO_4}$

(c) $2CH_3OH + H\overset{\overset{\displaystyle O}{\|}}{O}C\text{—}\bigcirc\text{—}\overset{\overset{\displaystyle O}{\|}}{C}OH \xrightarrow[\text{heat}]{H_2SO_4} (C_{10}H_{10}O_4)$

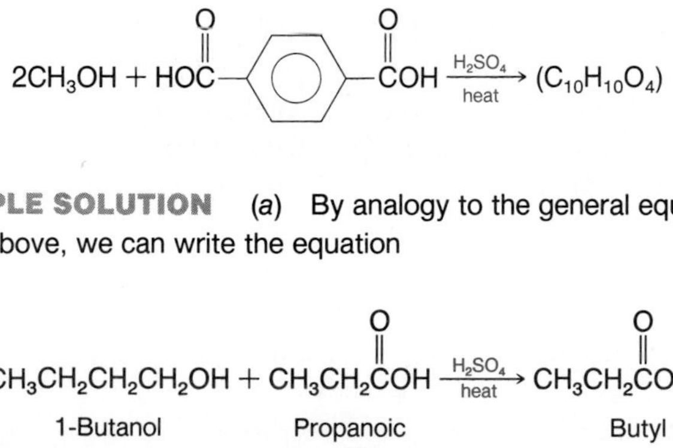

SAMPLE SOLUTION (a) By analogy to the general equation and to the examples cited above, we can write the equation

$$CH_3CH_2CH_2CH_2OH + CH_3CH_2\overset{\overset{\displaystyle O}{\|}}{C}OH \xrightarrow[\text{heat}]{H_2SO_4} CH_3CH_2\overset{\overset{\displaystyle O}{\|}}{C}OCH_2CH_2CH_2CH_3 + H_2O$$

1-Butanol	Propanoic acid	Butyl propanoate	Water

As actually carried out in the laboratory, 3 mol of propanoic acid was used per mole of 1-butanol and the desired ester was obtained in 78 percent yield.

Esters are also formed by the reaction of alcohols with acyl chlorides:

$$ROH + R'\overset{\overset{\displaystyle O}{\|}}{C}Cl \longrightarrow R'\overset{\overset{\displaystyle O}{\|}}{C}OR + HCl$$

Alcohol	Acyl chloride	Ester	Hydrogen chloride

The reaction of an alcohol with an acyl chloride is normally carried out in the presence of a weak base such as pyridine. Pyridine not only captures the hydrogen chloride that is formed but also exerts a catalytic effect.

$$(CH_3)_2CHCH_2OH + \underset{\substack{O_2N \\ \\ O_2N}}{\bigcirc}-\overset{\overset{\displaystyle O}{\|}}{C}Cl \xrightarrow{\text{pyridine}} \underset{\substack{O_2N \\ \\ O_2N}}{\bigcirc}-\overset{\overset{\displaystyle O}{\|}}{C}OCH_2CH(CH_3)_2$$

| Isobutyl alcohol | 3,5-Dinitrobenzoyl chloride | Isobutyl 3,5-dinitrobenzoate (86%) |

Carboxylic acid anhydrides react similarly to acyl chlorides.

$$\underset{\text{Alcohol}}{ROH} + \underset{\substack{\text{Carboxylic} \\ \text{acid anhydride}}}{R'\overset{\overset{\displaystyle O}{\|}}{C}O\overset{\overset{\displaystyle O}{\|}}{C}R'} \longrightarrow \underset{\text{Ester}}{R'\overset{\overset{\displaystyle O}{\|}}{C}OR} + \underset{\substack{\text{Carboxylic} \\ \text{acid}}}{R'\overset{\overset{\displaystyle O}{\|}}{C}OH}$$

$$\underset{\text{2-Phenylethanol}}{C_6H_5CH_2CH_2OH} + \underset{\substack{\text{Trifluoroacetic} \\ \text{anhydride}}}{CF_3\overset{\overset{\displaystyle O}{\|}}{C}O\overset{\overset{\displaystyle O}{\|}}{C}CF_3} \xrightarrow{\text{pyridine}} \underset{\substack{\text{2-Phenylethyl} \\ \text{trifluoroacetate} \\ (83\%)}}{C_6H_5CH_2CH_2O\overset{\overset{\displaystyle O}{\|}}{C}CF_3} + \underset{\substack{\text{Trifluoroacetic} \\ \text{acid}}}{CF_3\overset{\overset{\displaystyle O}{\|}}{C}OH}$$

The mechanism of the Fischer esterification reaction and the mechanism of the esterification of alcohols using acyl chlorides and anhydrides can be best understood after some fundamental principles of carbonyl group reactivity have been developed and will be discussed in detail in Chapter 20. For the present, it is sufficient to point out that most of the reactions that convert alcohols to esters proceed by cleavage of the O—H bond of the alcohol.

$$RO{-}H \longrightarrow R'\overset{\overset{\displaystyle O}{\|}}{C}{-}OR$$

| This bond is broken in conversion of an alcohol to an ester | This oxygen is the same oxygen that was attached to the group R in the starting alcohol ROH |

In these reactions the hydroxylic oxygen atom of the alcohol acts as a nucleophile. The acyl group of the carboxylic acid, acyl chloride, or anhydride is transferred to the oxygen of the alcohol. This is most clearly evident in the esterification of chiral alcohols. Since it is the O—H bond that is cleaved, *retention of configuration* at the chiral center is observed.

$$\underset{\substack{(R)\text{-}(+)\text{-2-Phenyl-} \\ \text{2-butanol} \\ (81\% \text{ optically pure})}}{\overset{\substack{CH_3CH_2 \quad CH_3}}{\underset{C_6H_5}{C}}{-}OH} + \underset{\substack{p\text{-Nitrobenzoyl} \\ \text{chloride}}}{O_2N{-}\bigcirc{-}\overset{\overset{\displaystyle O}{\|}}{C}Cl} \xrightarrow{\text{pyridine}} \underset{\substack{(R)\text{-}(-)\text{-2-Phenyl-2-butyl} \\ p\text{-nitrobenzoate (63\% yield)} \\ (81\% \text{ optically pure})}}{\overset{\substack{CH_3CH_2 \quad CH_3}}{\underset{C_6H_5}{C}}{-}O\overset{\overset{\displaystyle O}{\|}}{C}{-}\bigcirc{-}NO_2}$$

PROBLEM 16.9 A similar conclusion may be drawn by considering the individual re-actions of the cis and trans isomers of 4-*tert*-butylcyclohexanol with acetic anhydride. Based on the information just presented, predict the product in each case.

The reaction of alcohols with carboxylic acid chlorides is analogous to their reaction with *p*-toluenesulfonyl chloride described earlier (Section 9.15). In those reactions, a *p*-toluenesulfonate ester was formed by displacement of chloride from the sulfonyl group by the oxygen of the alcohol. Carboxylic esters arise by displacement of chloride from a carbonyl group by the alcohol oxygen.

16.11 ESTERS OF INORGANIC ACIDS

While the term *ester,* used without a modifier, is normally taken to mean an ester of a carboxylic acid, alcohols can react with inorganic acids in a process similar to the Fischer esterification. The products are esters of inorganic acids. For example, alkyl nitrates are esters formed by the reaction of alcohols with nitric acid.

$$ROH + HONO_2 \xrightarrow{\text{H}^+} RONO_2 + H_2O$$

Alcohol　　Nitric acid　　Alkyl nitrate　　Water

$$CH_3OH + HONO_2 \xrightarrow{\text{H}_2\text{SO}_4} CH_3ONO_2 + H_2O$$

Methanol　　Nitric acid　　Methyl nitrate (66–80%)　　Water

PROBLEM 16.10 Nitroglycerin, the explosive component of dynamite, is the trinitrate of glycerol (1,2,3-propanetriol). Write a structural formula for nitroglycerin.

Dialkyl sulfates are esters of sulfuric acid, trialkyl phosphites are esters of phosphorous acid (H_3PO_3), and trialkyl phosphates are esters of phosphoric acid (H_3PO_4).

$$\overset{\displaystyle O}{\underset{\displaystyle O}{\overset{\|}{\underset{\|}{CH_3OSOCH_3}}}} \qquad (CH_3O)_3P\!:\qquad (CH_3O)_3\overset{+}{P}\!-\!\ddot{\underset{..}{O}}\!:^{-}$$

Dimethyl sulfate　　Trimethyl phosphite　　Trimethyl phosphate

Some esters of inorganic acids, such as dimethyl sulfate, are used as reagents in synthetic organic chemistry. Certain naturally occurring alkyl phosphates play an important role in biological processes.

16.12 OXIDATION OF ALCOHOLS

As noted in Section 16.8, one type of reaction that alcohols undergo is oxidation to a carbonyl compound. Whether oxidation leads to an aldehyde, a ketone, or a carboxylic acid depends on the alcohol and on the oxidizing agent.

Primary alcohols may be oxidized either to an aldehyde or to a carboxylic acid.

$$RCH_2OH \xrightarrow{\text{oxidize}} \underset{\text{Aldehyde}}{R\overset{\displaystyle O}{\overset{\|}{C}}H} \xrightarrow{\text{oxidize}} \underset{\text{Carboxylic acid}}{R\overset{\displaystyle O}{\overset{\|}{C}}OH}$$

Primary alcohol

Vigorous oxidation leads to the formation of a carboxylic acid, but there are a number of methods that permit us to stop the oxidation at the intermediate aldehyde stage. The reagents that are most commonly used for oxidizing alcohols are based on high-oxidation-state transition metals, particularly manganese(VII) and chromium(VI).

Oxidation with potassium permanganate converts primary alcohols to carboxylic acids:

$$\underset{\underset{CH_3}{|}}{CH_3CH_2CH(CH_2)_4CH_2OH} \xrightarrow[\text{H}_2\text{SO}_4,\ \text{H}_2\text{O}]{\text{KMnO}_4} \underset{\underset{CH_3}{|}}{CH_3CH_2CH(CH_2)_4\overset{\displaystyle O}{\overset{\|}{C}}OH}$$

6-Methyl-1-octanol   6-Methyloctanoic acid (66%)

When the oxidation is carried out in basic solution, the carboxylic acid exists as a carboxylate salt, so that a separate acidification step is required in order to isolate the carboxylic acid.

$$\underset{\underset{CH_2CH_3}{|}}{CH_3CH_2CH_2CH_2CHCH_2OH} \xrightarrow[\text{2. H}^+]{\text{1. KMnO}_4,\ \text{HO}^-} \underset{\underset{CH_2CH_3}{|}}{CH_3CH_2CH_2CH_2CH\overset{\displaystyle O}{\overset{\|}{C}}OH}$$

2-Ethyl-1-hexanol   2-Ethylhexanoic acid (74%)

Chromic acid (H_2CrO_4) is a good oxidizing agent and is formed when solutions containing chromate (CrO_4^{2-}) or dichromate ($Cr_2O_7^{2-}$) are acidified. It is not as powerful an oxidant as permanganate. Sometimes it is possible to obtain aldehydes in satisfactory yield before they are further oxidized, but in most cases carboxylic acids are the major products isolated on treatment of primary alcohols with chromic acid.

$$FCH_2CH_2CH_2OH \xrightarrow[\text{H}_2\text{SO}_4,\ \text{H}_2\text{O}]{\text{K}_2\text{Cr}_2\text{O}_7} FCH_2CH_2\overset{\displaystyle O}{\overset{\|}{C}}OH$$

3-Fluoropropanol   3-Fluoropropanoic acid (74%)

Conditions that do permit the easy isolation of aldehydes in good yield by oxidation of primary alcohols utilize various Cr(VI) species as the oxidant in *anhydrous* media. One such combination is a chromium trioxide–pyridine complex having the formula $(C_5H_5N)_2CrO_3$ in dichloromethane as the reaction medium. This combination is referred to as *Collins' reagent.*

$$CH_3CH_2CH(CH_2)_4CH_2OH \xrightarrow[CH_2Cl_2]{(C_5H_5N)_2CrO_3} CH_3CH_2CH(CH_2)_4\overset{\overset{\textstyle O}{\|}}{C}H$$
$$\qquad\qquad | \qquad\qquad\qquad\qquad\qquad\qquad\qquad | $$
$$\qquad\qquad CH_3 \qquad\qquad\qquad\qquad\qquad\qquad CH_3$$

6-Methyl-1-octanol   6-Methyloctanal (69%)

Two related oxidants are pyridinium chlorochromate (PCC), $C_5H_5NH^+ \ ClCrO_3^-$, and pyridinium dichromate (PDC), $(C_5H_5NH)_2^{2+} \ Cr_2O_7^{2-}$. Like Collins' reagent, both PCC and PDC are sources of Cr(VI) and are used in dichloromethane.

$$CH_3(CH_2)_5CH_2OH \xrightarrow[CH_2Cl_2]{PCC} CH_3(CH_2)_5\overset{\overset{\textstyle O}{\|}}{C}H$$

1-Heptanol   Heptanal (78%)

$$(CH_3)_3C-\hspace{-0.5em}\bigcirc\hspace{-0.5em}-CH_2OH \xrightarrow[CH_2Cl_2]{PDC} (CH_3)_3C-\hspace{-0.5em}\bigcirc\hspace{-0.5em}-\overset{\overset{\textstyle O}{\|}}{C}H$$

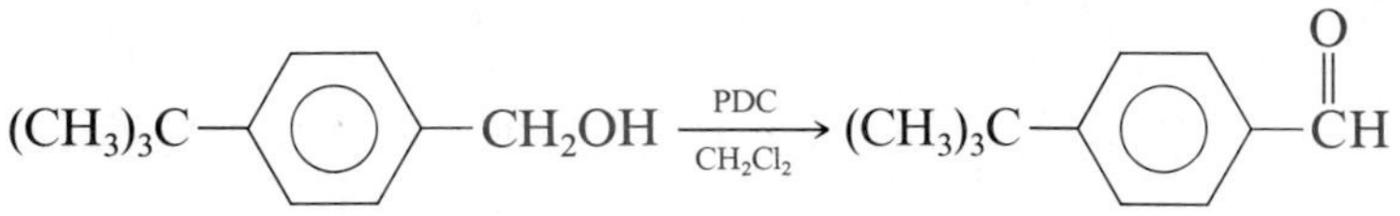

p-tert-Butylbenzyl alcohol   *p-tert*-Butylbenzaldehyde (94%)

Secondary alcohols are oxidized to ketones by the same reagents that oxidize primary alcohols.

$$\begin{array}{c} OH \\ | \\ R\overset{}{C}HR' \end{array} \xrightarrow{oxidize} \begin{array}{c} O \\ \| \\ R\overset{}{C}R' \end{array}$$

Secondary alcohol   Ketone

$$\bigcirc\hspace{-0.5em}-\overset{}{C}HCH_2CH_2CH_2CH_3 \xrightarrow[\substack{H_2O, \ acetic \\ acid}]{KMnO_4} \bigcirc\hspace{-0.5em}-\overset{\overset{\textstyle O}{\|}}{C}CH_2CH_2CH_2CH_3$$
$$\qquad\qquad | $$
$$\qquad\quad OH$$

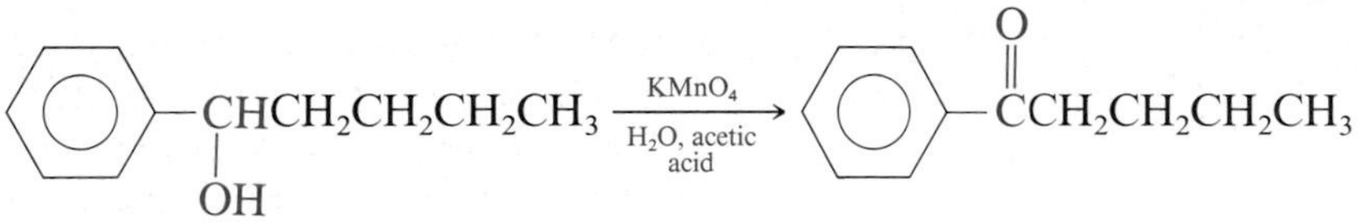

1-Phenyl-1-pentanol   1-Phenyl-1-pentanone (96%)

The chromium-based reagents are more commonly used than potassium permanganate for the oxidation of secondary alcohols because they are somewhat milder oxidants and less prone to degrade the ketone product by overoxidation.

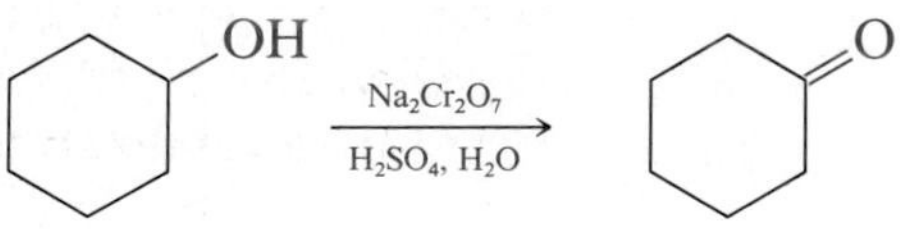

Cyclohexanol $\xrightarrow[H_2SO_4, \ H_2O]{Na_2Cr_2O_7}$ Cyclohexanone (85%)

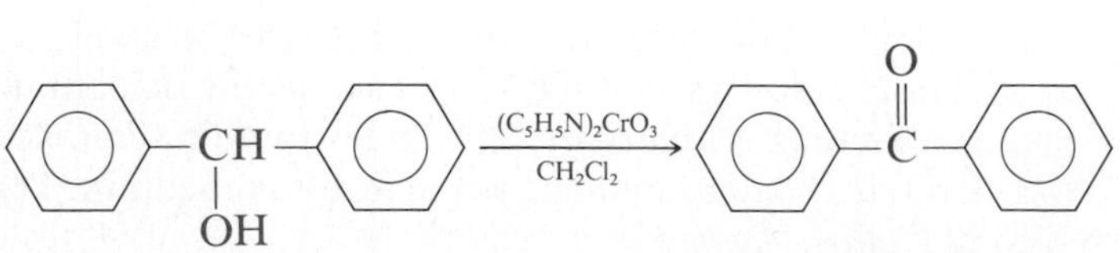

Diphenylmethanol $\xrightarrow[CH_2Cl_2]{(C_5H_5N)_2CrO_3}$ Benzophenone (96%)

Tertiary alcohols have no hydrogen on their hydroxyl-bearing carbon and do not undergo oxidation readily.

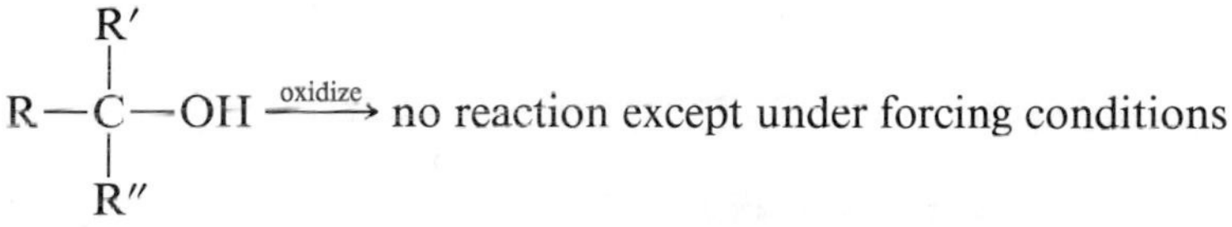

$$R-\underset{\underset{R''}{|}}{\overset{\overset{R'}{|}}{C}}-OH \xrightarrow{\text{oxidize}} \text{no reaction except under forcing conditions}$$

In the presence of strong oxidizing agents at elevated temperatures, oxidation of tertiary alcohols leads to cleavage of the various carbon-carbon bonds at the hydroxyl-bearing carbon atom, and a complex mixture of products is formed.

PROBLEM 16.11   Predict the principal organic product of each of the following reactions:

(a) $ClCH_2CH_2CH_2CH_2OH \xrightarrow[H_2SO_4, \, H_2O]{K_2Cr_2O_7}$

(b) $CH_3\underset{\underset{OH}{|}}{C}HCH_2CH_2CH_2CH_2CH_2CH_3 \xrightarrow[H_2SO_4, \, H_2O]{Na_2Cr_2O_7}$

(c) $CH_3CH_2CH_2CH_2CH_2CH_2CH_2OH \xrightarrow[CH_2Cl_2]{(C_5H_5N)_2CrO_3}$

(d)

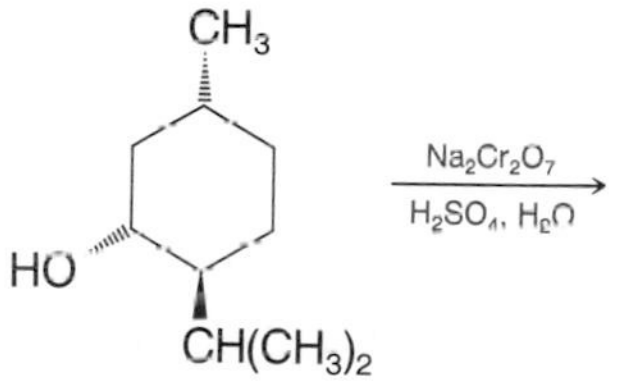

$\xrightarrow[H_2SO_4, \, H_2O]{Na_2Cr_2O_7}$

SAMPLE SOLUTION   (a)   The substrate is a primary alcohol so can be oxidized either to an aldehyde or to a carboxylic acid. Aldehydes are the major products only when the oxidation is carried out in anhydrous media. Carboxylic acids are formed when water is present. The reaction shown produced 4-chlorobutanoic acid in 56 percent yield.

$$ClCH_2CH_2CH_2CH_2OH \xrightarrow[H_2SO_4, \, H_2O]{K_2Cr_2O_7} ClCH_2CH_2CH_2\overset{\overset{O}{\|}}{C}OH$$

4-Chloro-1-butanol   4-Chlorobutanoic acid

The mechanisms by which transition metal oxidizing agents convert alcohols to aldehydes and ketones are rather complicated and will not be dealt with in detail. In broad outline, chromic acid oxidation involves initial formation of an alkyl chromate:

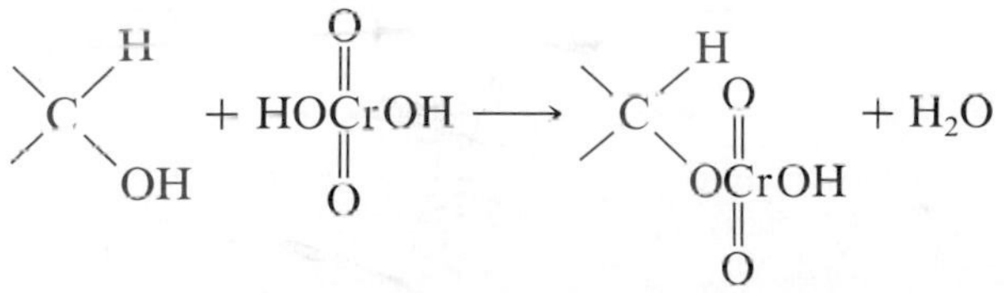

Alcohol   Chromic acid   Alkyl chromate

This alkyl chromate then undergoes an elimination reaction to form the carbon-oxygen double bond.

Alkyl chromate   Aldehyde
                        or ketone

In the elimination step, chromium is reduced from Cr(VI) to Cr(IV). Since the eventual product is Cr(III), further electron-transfer steps are also involved.

16.13 BIOLOGICAL OXIDATION OF ALCOHOLS

Many biological processes involve oxidation of alcohols to carbonyl compounds or the reverse process, reduction of carbonyl compounds to alcohols. Ethanol, for example, is metabolized in the liver to acetaldehyde. Such processes are catalyzed by enzymes; the enzyme that catalyzes the oxidation of ethanol is called *alcohol dehydrogenase.*

$$CH_3CH_2OH \xrightleftharpoons[]{alcohol\ dehydrogenase} \overset{\overset{\displaystyle O}{\|}}{CH_3CH}$$

Ethanol   Acetaldehyde

In addition to enzymes, biological oxidations require substances known as *coenzymes.* Coenzymes are organic molecules that, in concert with an enzyme, act upon a substrate to bring about chemical change. Most of the substances that we call vitamins are coenzymes. The coenzyme contains a functional group that is complementary to a functional group of the substrate; the enzyme catalyzes the interaction of these mutually complementary functional groups. Ethanol loses the elements of hydrogen when it is oxidized to acetaldehyde, so some substance must be present that can take up two protons plus two electrons. If ethanol is oxidized, some other substance must be reduced. This other substance is the oxidized form of the coenzyme *nicotinamide adenine dinucleotide* (NAD). Chemists and biochemists abbreviate the oxidized form of this coenzyme as NAD^+ and its reduced form as NADH. More completely, the chemical equation for the biological oxidation of ethanol may be written:

$$CH_3CH_2OH + \quad NAD^+ \xrightleftharpoons[]{alcohol\ dehydrogenase} \overset{\overset{\displaystyle O}{\|}}{CH_3CH} + \quad NADH + H^+$$

Ethanol   Oxidized form   Acetaldehyde   Reduced
              of NAD coenzyme   form of NAD
                                                                 coenzyme

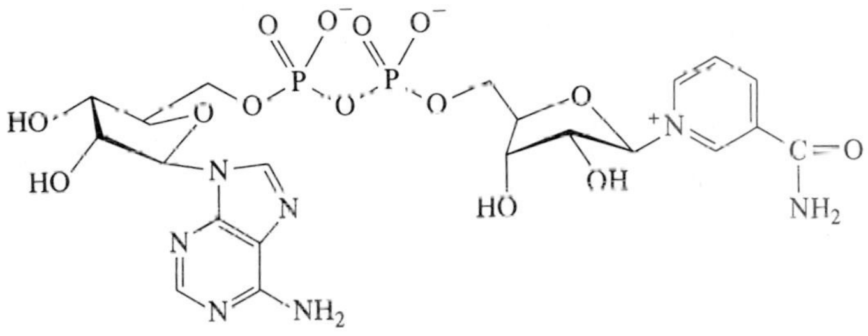

FIGURE 16.3 Structure of NAD$^+$, the oxidized form of the coenzyme nicotinamide adenine dinucleotide.

The structure of the oxidized form of nicotinamide adenine dinucleotide is shown in Figure 16.3. The only portion of the coenzyme that undergoes chemical change in an oxidation-reduction reaction is the substituted pyridine ring of the nicotinamide unit (shown in color in Figure 16.3). If the remainder of the coenzyme molecule is represented by R, its role as an oxidizing agent is shown in the equation:

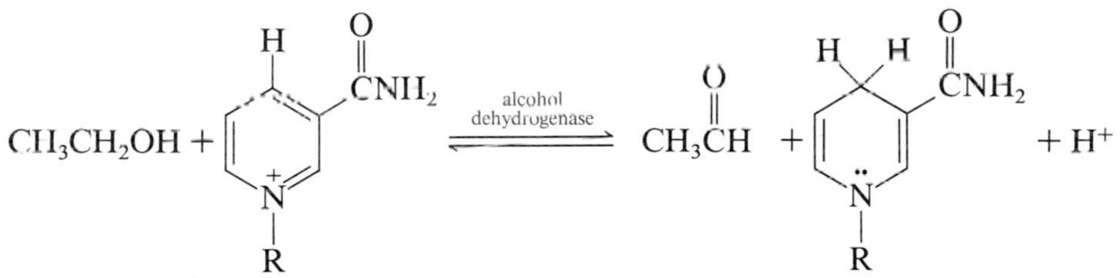

| Ethanol | NAD$^+$ | | Acetaldehyde | NADH |

According to one mechanistic interpretation, a hydrogen with a pair of electrons is transferred from ethanol to NAD$^+$, forming acetaldehyde and converting the positively charged pyridinium ring to a dihydropyridine.

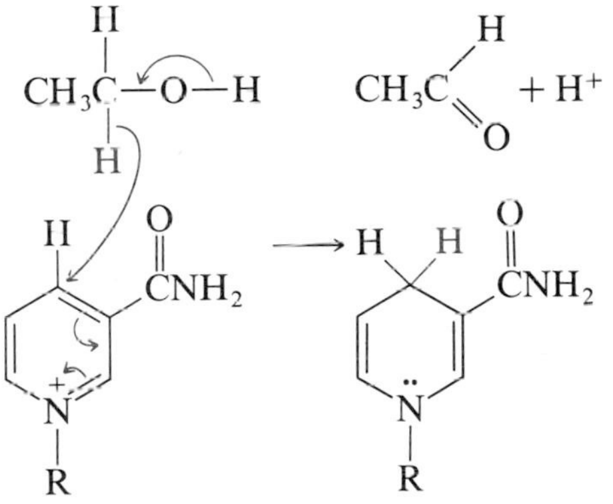

The pyridinium ring of NAD$^+$ serves as an acceptor of hydride ion in this picture of its role in biological oxidation.

PROBLEM 16.12 The mechanism of enzymatic oxidation has been studied by isotopic labeling with the aid of deuterated derivatives of ethanol. Specify the number of deuterium atoms that you would expect to find attached to the dihydropyridine ring of the reduced form of the nicotinamide adenine dinucleotide coenzyme following enzymatic oxidation of each of the alcohols given:

(a) CD_3CH_2OH (b) CH_3CD_2OH (c) CH_3CH_2OD

SAMPLE SOLUTION Examination of the proposed mechanism for biological oxidation of ethanol reveals that the hydrogen that is transferred to the coenzyme comes from C-1 of ethanol. Therefore, the dihydropyridine ring will bear no deuterium atoms when CD_3CH_2OH is oxidized because all the deuterium atoms of the substrate are bound to C-2.

$$CD_3CH_2OH \quad + \quad [\text{NAD}^+] \quad \underset{}{\overset{\text{alcohol}}{\underset{\text{dehydrogenase}}{\rightleftharpoons}}} \quad CD_3\overset{O}{\overset{\|}{C}}H \quad + \quad [\text{NADH}] \quad + \ H^+$$

2,2,2-Trideuterioethanol   NAD$^+$ 2,2,2-Trideuterioethanal   NADH

The reverse reaction also is observed to occur in living systems; NADH reduces acetaldehyde to ethanol in the presence of alcohol dehydrogenase. In this process, NADH serves as a hydride donor and is oxidized to NAD$^+$ while acetaldehyde is reduced.

The NAD$^+$-NADH coenzyme system is involved in a large number of biological oxidation-reduction reactions. Another reaction similar to the ethanol-acetaldehyde conversion is the oxidation of lactic acid to pyruvic acid by NAD$^+$ and the enzyme *lactic acid dehydrogenase.*

$$\underset{\underset{\displaystyle\text{OH}}{|}}{CH_3CHCOH} + NAD^+ \xrightarrow{\text{lactic acid dehydrogenase}} CH_3\overset{O\ O}{\overset{\|\ \|}{CC}}OH + NADH + H^+$$

         Lactic acid   Pyruvic acid

We shall encounter other biological processes in which the NAD$^+$ $\rightleftharpoons$ NADH interconversion plays a prominent role as we proceed through our coverage of organic chemistry.

16.14 OXIDATIVE CLEAVAGE OF VICINAL DIOLS

A reaction characteristic of vicinal diols is their oxidative cleavage on treatment with periodic acid (HIO_4). The carbon-carbon bond of the vicinal diol unit is broken and two carbonyl groups result. Periodic acid is reduced to iodic acid (HIO_3).

$$\underset{\underset{\displaystyle HO\ \ OH}{|\ \ \ |}}{\overset{\overset{\displaystyle R\ \ \ R''}{|\ \ \ |}}{R-C-C-R'}} + HIO_4 \longrightarrow \underset{R}{\overset{R}{>}}C{=}O + \underset{R'}{\overset{R'}{>}}C{=}O + HIO_3 + H_2O$$

Vicinal   Periodic   Aldehyde   Aldehyde   Iodic   Water
diol   acid   or ketone   or ketone   acid

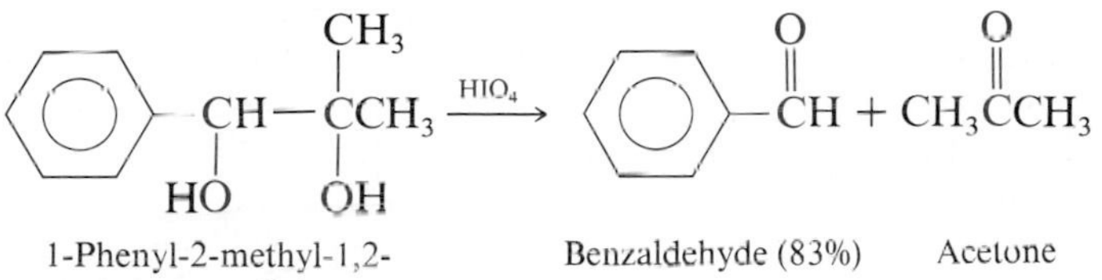

1-Phenyl-2-methyl-1,2-propanediol   Benzaldehyde (83%)   Acetone

This reaction occurs only when the hydroxyl groups are on adjacent carbons.

PROBLEM 16.13 Predict the products formed on oxidation of each of the following with periodic acid:

(a) $HOCH_2CH_2OH$

(b) $(CH_3)_2CHCH_2CHCHCH_2C_6H_5$
                  HO OH

(c)

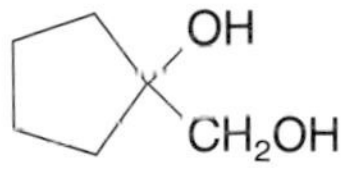

SAMPLE SOLUTION (a) The carbon-carbon bond of 1,2-ethanediol is cleaved by periodic acid to give two molecules of formaldehyde.

     1,2-Ethanediol   Formaldehyde

Cyclic diols react to give dicarbonyl compounds. The reactions are faster when the hydroxyl groups are cis than when they are trans, but both stereoisomers are oxidized by periodic acid.

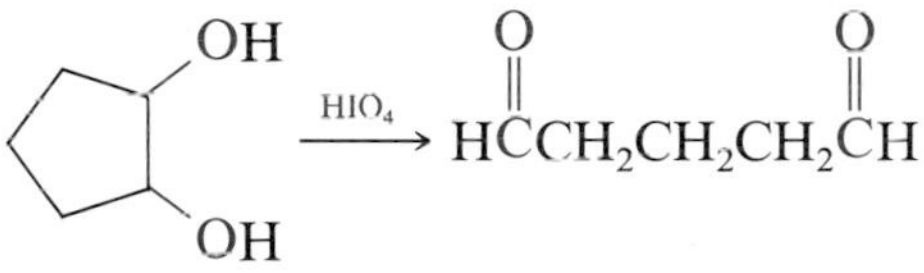

1,2-Cyclopentanediol   Pentanedial
(either stereoisomer)

Periodic acid cleavage of vicinal diols is often used for analytical purposes as an aid in structure determination. By identifying the carbonyl compounds produced, the constitution of the starting diol may be deduced. This technique has found its widest application with carbohydrates and will be discussed more fully in Chapter 27.

16.15 SPECTROSCOPIC ANALYSIS OF ALCOHOLS

Characteristic features of the infrared spectra of alcohols were discussed in Section 14.16. The O—H stretching mode is particularly easy to identify, appearing in the 3200 to 3650 cm^{-1} region. As the infrared spectrum of cyclohexanol, presented in Figure 16.4, demonstrates, this peak is seen as a broad absorption of moderate

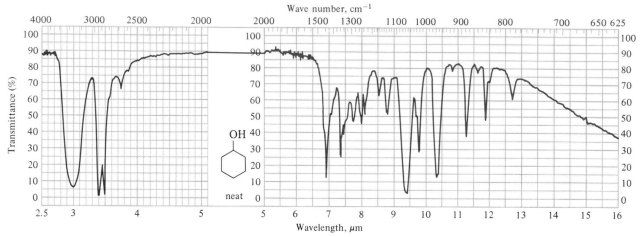

FIGURE 16.4   The ir spectrum of cyclohexanol.

intensity. The C—O bond stretching of alcohols gives rise to a moderate to strong absorbance between 1025 and 1200 cm^{-1}. It appears at 1070 cm^{-1} in cyclohexanol, a typical secondary alcohol, but is shifted to slightly higher energy in tertiary alcohols and slightly lower energy in primary alcohols.

Features that are helpful in identifying alcohols by nuclear magnetic resonance spectroscopy include the presence of a hydroxyl proton signal in the spectrum and the chemical shift of the proton of an H—C—O unit when the alcohol is primary or secondary.

$$-\overset{|}{\underset{|}{C}}-O-H \qquad H-\overset{|}{\underset{|}{C}}-O$$

$$\delta = 0.5-5 \text{ ppm} \qquad \delta = 3.3-4.0 \text{ ppm}$$

The chemical shift of the hydroxyl proton signal is variable, depending on solvent, temperature, and concentration. Its precise position is not particularly significant in structure determination. Because the signals due to hydroxyl protons are not usually split by other protons in the molecule and are often rather broad, they are often fairly easy to identify. To illustrate, Figure 16.5 shows the ^{1}H nmr spectrum of 2-phenylethanol, in which the hydroxyl proton signal appears as a singlet at $\delta = 1.6$ ppm. Of the two triplets in this spectrum, the one at lower field ($\delta = 3.8$ ppm) corresponds to the protons of the CH$_2$O unit. The higher-field triplet at $\delta = 2.8$ ppm arises from the benzylic CH$_2$ group. The assignment of a particular signal to the hydroxyl proton can be confirmed by adding D$_2$O. The hydroxyl proton is replaced by deuterium, and its ^{1}H nmr signal disappears.

In relation to the ^{13}C nmr spectrum of an alkane, the electronegative oxygen of an alcohol causes a pronounced downfield shift in the position of the carbon to which it is attached. This can be seen by comparing chemical shifts for the carbons of ethyl alcohol and propane.

17.9 ppm   57.3 ppm   15.6 ppm   16.1 ppm

CH$_3$CH$_2$OH   CH$_3$CH$_2$CH$_3$

Ethyl alcohol   Propane

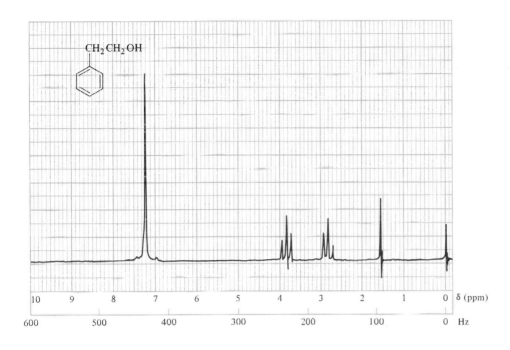

FIGURE 16.5 The ¹H nmr spectrum of 2-phenylethanol ($C_6H_5CH_2CH_2OH$).

16.16 MASS SPECTROMETRY OF ALCOHOLS

Peaks due to the molecular ion are normally quite small in the mass spectra of alcohols. A peak corresponding to loss of water from the molecular ion (m/z M − 18) is often evident. Alcohols also fragment readily by a pathway in which the molecular ion loses an alkyl group from the hydroxyl-bearing carbon to form a stable cation. Thus, the mass spectra of most primary alcohols exhibit a prominent peak at m/z 31.

$$RCH_2\ddot{O}H \longrightarrow R\overset{\frown}{-}CH_2\overset{\curvearrowleft}{=}\overset{\cdot+}{\ddot{O}}H \longrightarrow R\cdot + CH_2=\overset{+}{\ddot{O}}H$$

Primary alcohol   Molecular ion   Alkyl   Conjugate acid of
                                            radical   formaldehyde, m/z 31

PROBLEM 16.14   Three of the most intense peaks in the mass spectrum of 2-methyl-2-butanol appear at m/z 59, 70, and 73. Explain the origin of these peaks.

16.17 SUMMARY

Functional group interconversions involving alcohols either as reactants or as reaction products are the focus of this chapter. A number of reactions which produce alcohols had been encountered in earlier chapters and were summarized in Table 16.1 (Section 16.2). As we saw in Chapter 15 and as is evident in Table 16.1, Grignard reagents and organolithium reagents are among the most effective tools in alcohol synthesis. The present chapter described yet another way in which these organometallic reagents could be used. Treatment of a Grignard or organolithium reagent with ethylene oxide yields a primary alcohol (Section 16.6).

TABLE 16.4
Preparation of Alcohols by Reduction of Carbonyl Functional Groups

Carbonyl compound	Product of reduction of carbonyl compound by specified reducing agent			
	Lithium aluminum hydride (LiAlH₄)	Sodium borohydride (NaBH₄)	Hydrogen (in the presence of a catalyst)	Sodium (or potassium) in alcohol
Aldehyde RCH=O (Section 16.3)	Primary alcohol RCH_2OH	Primary alcohol RCH_2OH	Primary alcohol RCH_2OH	Reduction to primary alcohol occurs, but competing side reactions make reaction of little practical value
Ketone RCR'=O (Section 16.3)	Secondary alcohol RCHR'—OH	Secondary alcohol RCHR'—OH	Secondary alcohol RCHR'—OH	Secondary alcohol RCHR'—OH
Carboxylic acid RCOH=O (Section 16.4)	Primary alcohol RCH_2OH	Not reduced	Not reduced	Not reduced
Carboxylic ester RCOR'=O (Section 16.5)	Primary alcohol RCH_2OH plus R'OH	Reduced too slowly to be of practical value	Primary alcohol RCH_2OH plus R'OH	Primary alcohol RCH_2OH plus R'OH

$$RMgX \quad + H_2C \overset{\diagup\diagdown}{\underset{O}{\quad}} CH_2 \xrightarrow[\text{2. H}_3\text{O}^+]{\text{1. diethyl ether}} RCH_2CH_2OH$$

Grignard reagent   Ethylene oxide   Primary alcohol

$$CH_3CH_2CH_2CH_2MgBr + H_2C \overset{\diagup\diagdown}{\underset{O}{\quad}} CH_2 \xrightarrow[\text{2. H}_3\text{O}^+]{\text{1. diethyl ether}} CH_3CH_2CH_2CH_2CH_2CH_2OH$$

Butylmagnesium   Ethylene oxide   1-Hexanol (60–62%)
bromide

The oxidation-reduction relationship between alcohols and carbonyl compounds is evident when examining the preparation of alcohols and the reactions of alcohols:

1. Alcohols can be prepared from carbonyl compounds by reduction. Table 16.4 summarizes the types of alcohols formed from each of the various classes of carbonyl compounds—aldehydes, ketones, carboxylic acids and esters—and the reagents that are appropriate for each reduction.
2. Alcohols are converted to carbonyl compounds by *oxidation*. Table 16.5 summarizes these reactions.

Alcohols undergo a number of reactions useful in the synthesis of classes of compounds discussed in earlier chapters. These reactions were reviewed in Table 16.3. Table 16.6 lists the reactions discussed for the first time in the present chapter.

Compounds that contain two hydroxyl groups are called diols, those with three are called triols, etc. A reaction unique to the preparation of vicinal diols is the hydroxyl-

TABLE 16.5
Oxidation of Alcohols

Class of alcohol	Desired product	Suitable oxidizing agent(s)
Primary, RCH_2OH	Aldehyde $\overset{\text{O}}{\overset{\|}{R\text{C}\text{H}}}$	Collins' reagent* PCC† PDC‡
Primary, RCH_2OH	Carboxylic acid $\overset{\text{O}}{\overset{\|}{R\text{C}\text{OH}}}$	$KMnO_4$ $Na_2Cr_2O_7, H_2SO_4$ H_2CrO_4
Secondary, $\underset{\text{OH}}{RCHR'}$	Ketone $\overset{\text{O}}{\overset{\|}{R\text{C}R'}}$	Collins' reagent PCC PDC $KMnO_4$ $Na_2Cr_2O_7, H_2SO_4$ H_2CrO_4

* Collins' reagent is $(C_5H_5N)_2CrO_3$ in dichloromethane.
† PCC is pyridinium chlorochromate; it is used in dichloromethane.
‡ PDC is pyridinium dichromate; it is used in dichloromethane.

TABLE 16.6

Summary of Reactions of Alcohols Presented in This Chapter

Reaction (section) and comments	General equation and specific example

Conversion to dialkyl ethers (Section 16.9) On being heated in the presence of an acid catalyst, two molecules of a primary alcohol combine to form an ether and water. Diols can undergo an intramolecular condensation if a five-membered or six-membered cyclic ether results.

$$2RCH_2OH \xrightarrow[\text{heat}]{H^+} RCH_2OCH_2R + H_2O$$

Alcohol   Dialkyl ether   Water

$$2(CH_3)_2CHCH_2CH_2OH \xrightarrow[150^\circ C]{H_2SO_4} (CH_3)_2CHCH_2CH_2OCH_2CH_2CH(CH_3)_2$$

Isoamyl alcohol   Diisoamyl ether (27%)

Fischer esterification (Section 16.10) Alcohols and carboxylic acids yield an ester and water in the presence of an acid catalyst. The reaction is an equilibrium process that can be driven to completion by using either the alcohol or acid in excess or by removing the water as it is formed.

$$ROH + R'\overset{\overset{\displaystyle O}{\|}}{C}OH \xrightarrow{H^+} R'\overset{\overset{\displaystyle O}{\|}}{C}OR + H_2O$$

Alcohol   Carboxylic   Ester   Water
          acid

$$CH_3CH_2CH_2CH_2CH_2OH + CH_3\overset{\overset{\displaystyle O}{\|}}{C}OH \xrightarrow{H^+} CH_3\overset{\overset{\displaystyle O}{\|}}{C}OCH_2CH_2CH_2CH_2CH_3$$

1-Pentanol   Acetic acid   1-Pentyl acetate (71%)

Esterification with carboxylic acid chlorides (Section 16.10) Acyl chlorides react with alcohols to give esters. The reaction is usually carried out in the presence of pyridine.

$$ROH + R'\overset{\overset{\displaystyle O}{\|}}{C}Cl \longrightarrow R'\overset{\overset{\displaystyle O}{\|}}{C}OR + HCl$$

Alcohol   Acyl   Ester   Hydrogen
       chloride   chloride

$$(CH_3)_3COH + CH_3\overset{\overset{\displaystyle O}{\|}}{C}Cl \xrightarrow{\text{pyridine}} CH_3\overset{\overset{\displaystyle O}{\|}}{C}OC(CH_3)_3$$

tert-Butyl alcohol   Acetyl   *tert*-Butyl
          chloride   acetate (62%)

Esterification with carboxylic acid anhydrides (Section 16.10) Carboxylic acid anhydrides react with alcohols to form esters in the same way that acyl chlorides do.

$$ROH + R'\overset{\overset{\displaystyle O}{\|}}{C}O\overset{\overset{\displaystyle O}{\|}}{C}R' \longrightarrow R'\overset{\overset{\displaystyle O}{\|}}{C}OR + R'\overset{\overset{\displaystyle O}{\|}}{C}OH$$

Alcohol   Carboxylic   Ester   Carboxylic
      acid anhydride   acid

m-Methoxybenzyl alcohol + Acetic anhydride $\xrightarrow{\text{pyridine}}$ *m*-Methoxybenzyl acetate (99%)

Formation of esters of inorganic acids (Section 16.11) Alkyl nitrates, dialkyl sulfates, trialkyl phosphites, and trialkyl phosphates are examples of alkyl esters of inorganic acids. In some cases, these compounds are prepared by the direct reaction of an alcohol and the inorganic acid.

$$ROH + HONO_2 \xrightarrow{H^+} RONO_2 + H_2O$$

Alcohol   Nitric acid   Alkyl nitrate   Water

Cyclopentanol   Cyclopentyl nitrate (69%)

Oxidation of alcohols (Section 16.12) See Table 16.5.

ation of alkenes (Section 16.7). Reagents for alkene hydroxylation include potassium permanganate and osmium tetraoxide–*tert*-butyl hydroperoxide. Syn addition of hydroxyl groups to the double bond is observed.

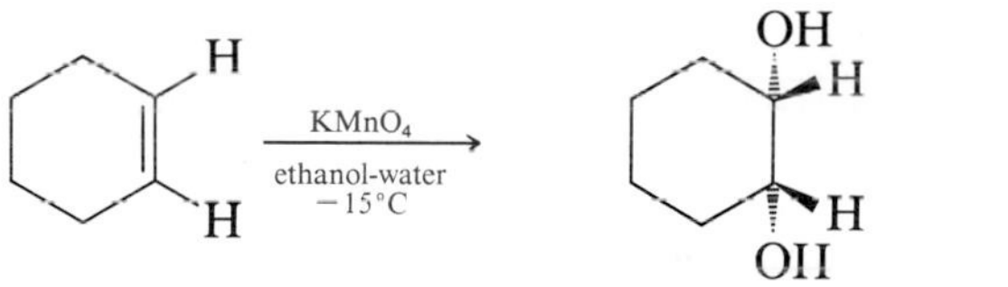

Cyclohexene | *cis*-1,2-Cyclohexanediol (33%)

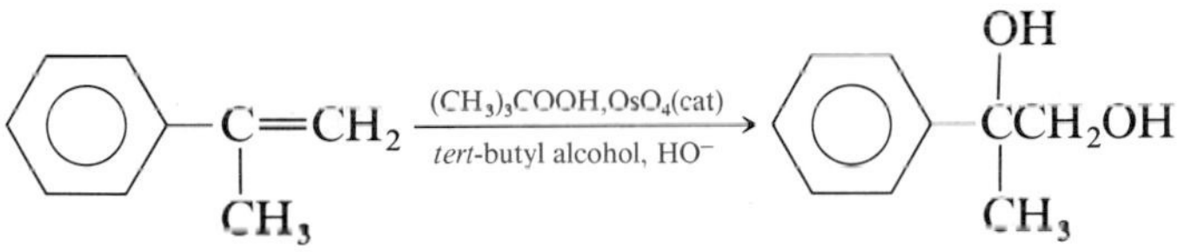

2-Phenylpropene | 2-Phenyl-1,2-propanediol (71%)

Diols undergo many of the same reactions that simple alcohols do. A reaction characteristic of vicinal diols is their cleavage to two carbonyl compounds on treatment with periodic acid (Section 16.14).

$$\underset{\text{Diol}}{R_2\underset{HO}{\underset{|}{C}}-\underset{OH}{\underset{|}{C}}R_2} \xrightarrow{HIO_4} \underset{\text{Two carbonyl-containing compounds}}{R_2C{=}O + O{=}CR_2}$$

$$\underset{\text{9,10-Dihydroxyoctadecanoic acid}}{CH_3(CH_2)_7\underset{HO}{\underset{|}{C}}H-\underset{OH}{\underset{|}{C}}H(CH_2)_7\overset{O}{\overset{\|}{C}}OH} \xrightarrow{HIO_4} \underset{\text{Nonanal (89\%)}}{CH_3(CH_2)_7\overset{O}{\overset{\|}{C}}H} + \underset{\text{9-Oxononanoic acid (76\%)}}{H\overset{O}{\overset{\|}{C}}(CH_2)_7\overset{O}{\overset{\|}{C}}OH}$$

PROBLEMS

16.15 Write chemical equations, showing all necessary reagents, for the preparation of 1-butanol by each of the following methods:

(*a*) Hydroboration-oxidation of an alkene
(*b*) Use of a Grignard reagent
(*c*) Use of a Grignard reagent in a way different from (*b*)
(*d*) Reduction of a carboxylic acid
(*e*) Hydrogenolysis of a methyl ester
(*f*) Reduction of an ethyl ester with sodium in ethanol
(*g*) Hydrogenation of an appropriate aldehyde or ketone
(*h*) Reduction with sodium borohydride

16.16 Write chemical equations, showing all necessary reagents, for the preparation of 2-butanol by each of the following methods:

(a) Hydroboration-oxidation of an alkene
(b) Oxymercuration-demercuration of an alkene
(c) Use of a Grignard reagent
(d) Use of a Grignard reagent different from that used in (c)
(e)–(h) Four methods for reducing a carbonyl-containing compound

16.17 Write chemical equations, showing all necessary reagents, for the preparation of 2-methyl-2-propanol by each of the following methods:

(a) Oxymercuration-demercuration of an alkene
(b) Use of a Grignard reagent
(c) Use of the same Grignard reagent used in (b) with a different carbonyl-containing compound

16.18 Which of the isomeric $C_5H_{12}O$ alcohols can be prepared by sodium borohydride reduction of a carbonyl compound?

16.19 Evaluate the feasibility of preparing

(a) 1-Butanol from butane
(b) 2-Methyl-2-propanol from 2-methylpropane
(c) Benzyl alcohol from toluene

$$RH \xrightarrow[\text{light or heat}]{Br_2} RBr \xrightarrow{KOH} ROH$$

Of the three alcohols given, which one could be prepared most efficiently by this process?

16.20 Sorbitol is a sweetener often substituted for cane sugar since it is better tolerated by diabetics. It is also an intermediate in the commercial synthesis of vitamin C. Sorbitol is prepared by high-pressure hydrogenation of glucose over a nickel catalyst. What is the structure (including stereochemistry) of sorbitol?

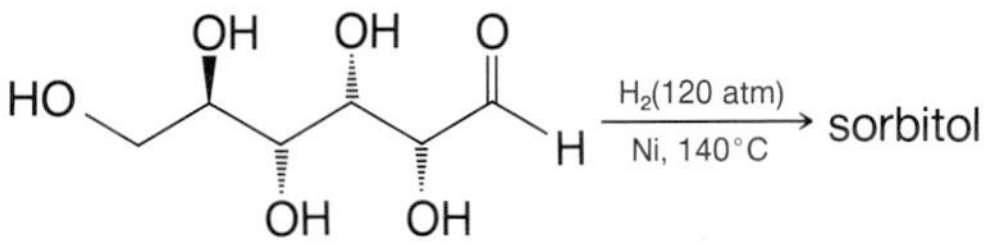

Glucose

16.21 Write equations showing how 1-phenylethanol ($C_6H_5CHCH_3$) could be prepared from each of the following starting materials:

(a) Bromobenzene
(b) Benzaldehyde
(c) Benzyl alcohol
(d) Styrene
(e) Acetophenone
(f) Benzene

16.22 Write equations showing how 2-phenylethanol ($C_6H_5CH_2CH_2OH$) could be prepared from each of the following starting materials:

(a) Bromobenzene
(b) Styrene
(c) Phenylacetylene
(d) 2-Phenylethanal ($C_6H_5CH_2CHO$)

(e) Ethyl 2-phenylethanoate ($C_6H_5CH_2CO_2CH_2CH_3$)
(f) 2-Phenylethanoic acid ($C_6H_5CH_2CO_2H$)

16.23 Outline practical syntheses of each of the following compounds from benzene and alcohols containing four or less carbon atoms, using any necessary organic or inorganic reagents:

(a) 2-Hexanol

(b) 2-Hexanone, $CH_3\overset{\overset{\displaystyle O}{\|}}{C}CH_2CH_2CH_2CH_3$
(c) Hexanoic acid, $CH_3(CH_2)_4CO_2H$

(d) Ethyl hexanoate, $CH_3(CH_2)_4\overset{\overset{\displaystyle O}{\|}}{C}OCH_2CH_3$
(e) 2-Methyl-1,2-propanediol

(f) 2,2-Dimethylpropanal, $(CH_3)_3\overset{\overset{\displaystyle O}{\|}}{C}CH$
(g) 1-Chloro-2-phenylethane

(h) 2-Methyl-1-phenyl-1-propanone, $C_6H_5\overset{\overset{\displaystyle O}{\|}}{C}CH(CH_3)_2$
(i) Isobutylbenzene, $C_6H_5CH_2CH(CH_3)_2$

16.24 Write the structure of the principal organic product formed in the reaction of 1-propanol with each of the following reagents:

(a) Sulfuric acid (catalytic amount), heat at 140°C
(b) Sulfuric acid (catalytic amount), heat at 200°C
(c) Chromium trioxide–pyridine complex [$(C_5H_5N)_2CrO_3$] in dichloromethane
(d) Potassium permanganate in dilute sodium hydroxide, followed by acidification of the reaction mixture
(e) Potassium dichromate ($K_2Cr_2O_7$) in sulfuric acid, heat
(f) Metallic sodium
(g) Sodium amide ($NaNH_2$)

(h) Sodium acetate ($Na\overset{\overset{\displaystyle O}{\|}}{O}CCH_3$)

(i) Acetic acid ($CH_3\overset{\overset{\displaystyle O}{\|}}{C}OH$) in the presence of dissolved hydrogen chloride

(j) 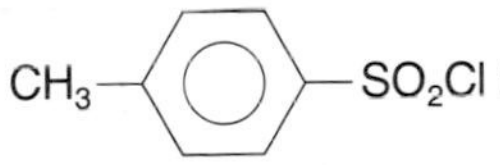 $CH_3\!-\!\langle\bigcirc\rangle\!-\!SO_2Cl$ in the presence of pyridine

(k) 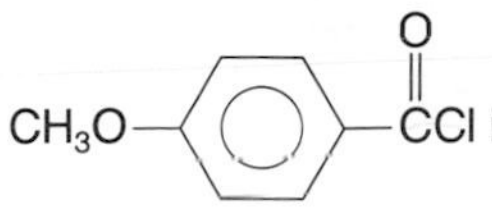 $CH_3O\!-\!\langle\bigcirc\rangle\!-\!\overset{\overset{\displaystyle O}{\|}}{C}Cl$ in the presence of pyridine

(l) $C_6H_5\overset{\overset{\displaystyle O}{\|}}{C}O\overset{\overset{\displaystyle O}{\|}}{C}C_6H_5$ in the presence of pyridine

(m) 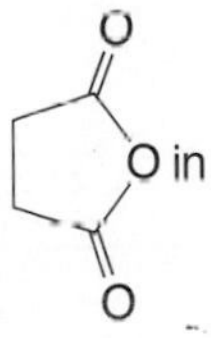 [succinic anhydride structure] O in the presence of pyridine

16.25 Repeat the preceding problem for 2-propanol.

16.26 Each of the following reactions has been reported in the chemical literature. Predict the product in each case, showing stereochemistry where appropriate.

(a) $\xrightarrow[\text{heat}]{\text{H}_2\text{SO}_4}$

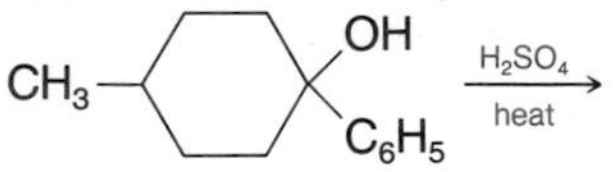

(b) $CH_2{=}CHCH{=}CHCH_2CH_2CH_2OH \xrightarrow[\text{CH}_2\text{Cl}_2]{(\text{C}_5\text{H}_5\text{N})_2\text{CrO}_3}$

(c) $(CH_3)_2C{=}C(CH_3)_2 \xrightarrow[(\text{CH}_3)_3\text{COH, HO}^-]{(\text{CH}_3)_3\text{COOH, OsO}_4(\text{cat})}$

(d) $\xrightarrow[\text{2. H}_2\text{O}_2, \text{HO}^-]{\text{1. B}_2\text{H}_6, \text{diglyme}}$

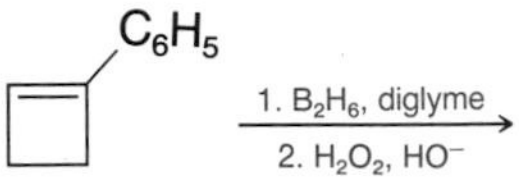

(e) $\xrightarrow[\text{ethanol}]{\text{Na}}$

(f) $CH_3CHC{\equiv}C(CH_2)_3CH_3 \xrightarrow[\text{H}_2\text{SO}_4, \text{H}_2\text{O}]{\text{H}_2\text{CrO}_4}$ acetone
   with OH below the first CH

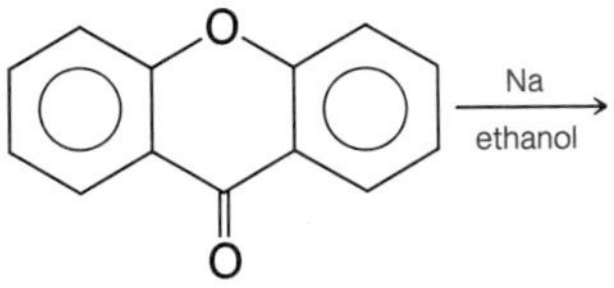

(g) $\xrightarrow[\text{cold}]{\text{KMnO}_4}$

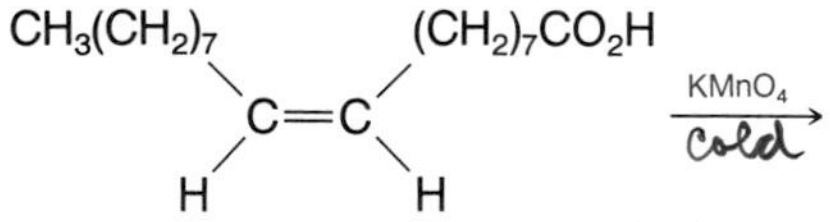

(h) $CH_3\overset{O}{\overset{\|}{C}}CH_2CH{=}CHCH_2\overset{O}{\overset{\|}{C}}CH_3 \xrightarrow[\text{2. H}_2\text{O}]{\text{1. LiAlH}_4, \text{diethyl ether}}$

(i) $+$ $\xrightarrow{\text{pyridine}}$

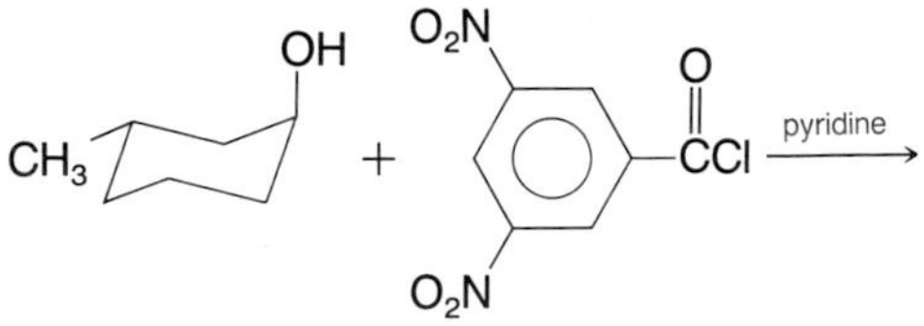

(j) $+ CH_3\overset{O}{\overset{\|}{C}}O\overset{O}{\overset{\|}{C}}CH_3 \longrightarrow$

(k) $\xrightarrow[\text{CH}_3\text{OH}]{\text{HIO}_4}$

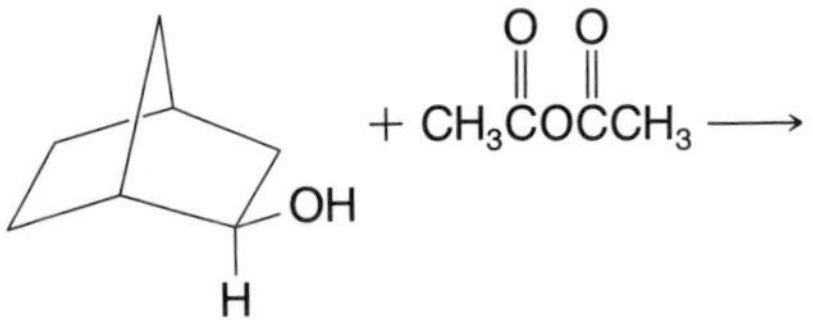

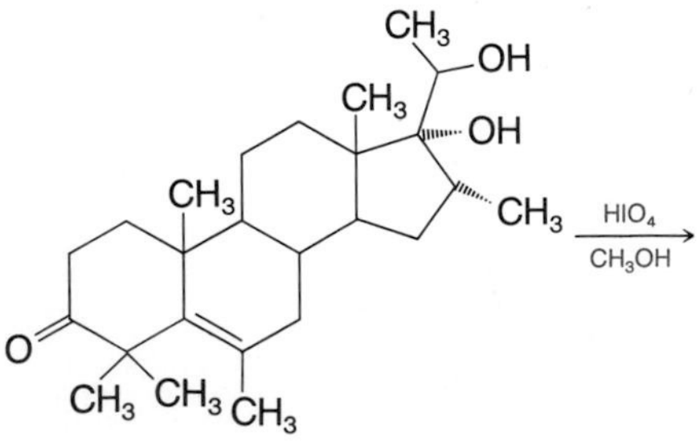

(*l*) $\xrightarrow[\text{H}_2\text{SO}_4]{\text{CH}_3\text{OH}}$

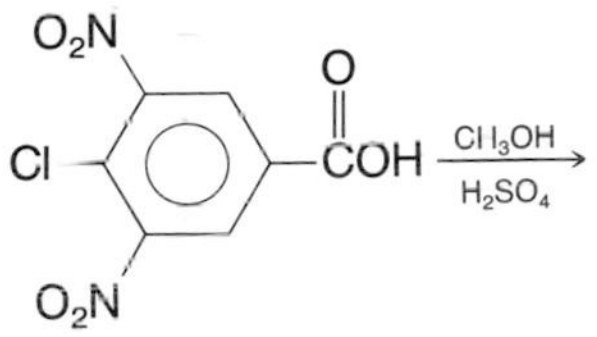

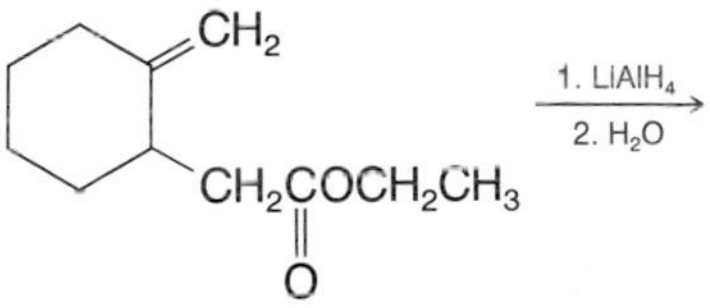

(*m*) $\xrightarrow[\text{2. H}_2\text{O}]{\text{1. LiAlH}_4}$

(*n*) $\xrightarrow[\text{2. H}_2\text{O}]{\text{1. LiAlH}_4}$

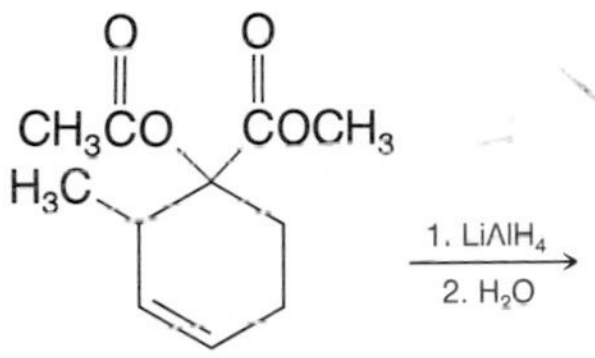

(*o*) Product of (*n*) $\xrightarrow[\text{CH}_3\text{OH, H}_2\text{O}]{\text{HIO}_4}$

16.27 Over 100 years ago, Kekulé studied the hydroxylation of maleic acid and of fumaric acid by potassium permanganate. One of these compounds gave only *meso*-tartaric acid on hydroxylation, the other gave only racemic tartaric acid.

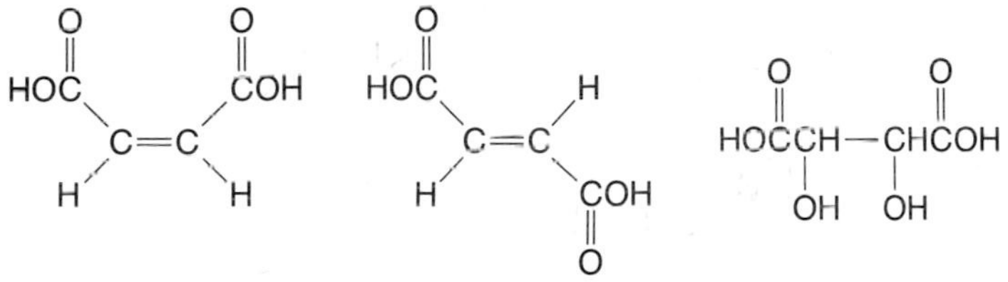

Maleic acid   Fumaric acid   Tartaric acid

Which starting material yielded which stereoisomeric form of tartaric acid? Why?

16.28 The *E* and *Z* stereoisomers of 4-octene yield diastereomeric 4,5-octanediols on treatment with the osmium tetraoxide–*tert*-butyl hydroperoxide reagent. Which alkene leads to the meso diol? Which alkene forms the chiral diol?

16.29 Suggest reaction sequences and reagents suitable for carrying out each of the following conversions. Two synthetic operations are required in each case.

(*a*)

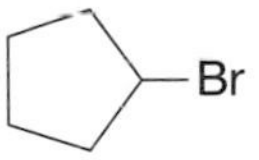

(*b*) (CH$_3$)$_3$C

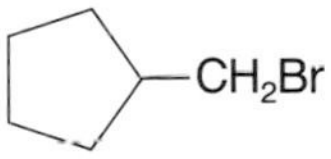

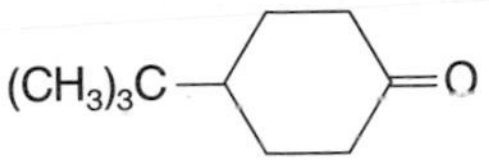

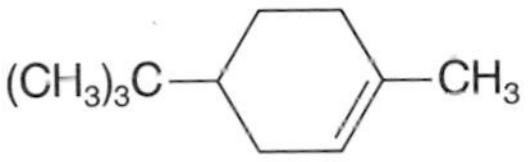

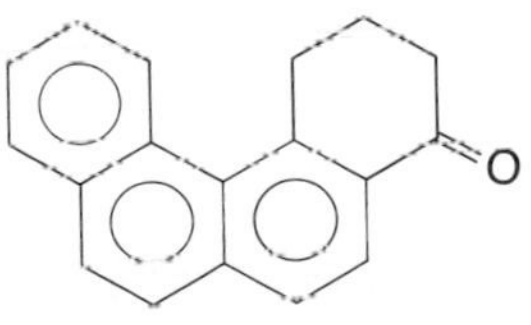

(*c*)

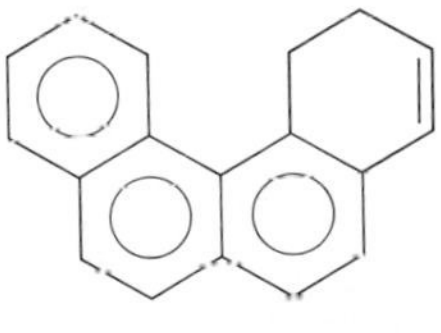

(d)

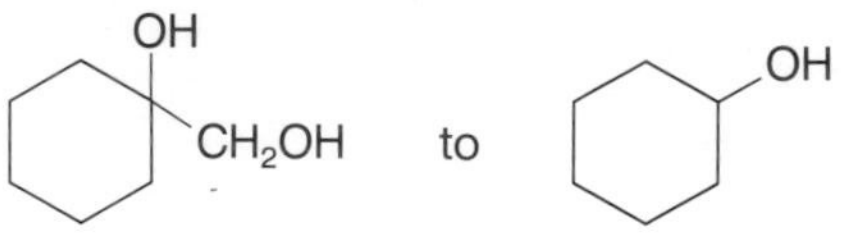

(e)

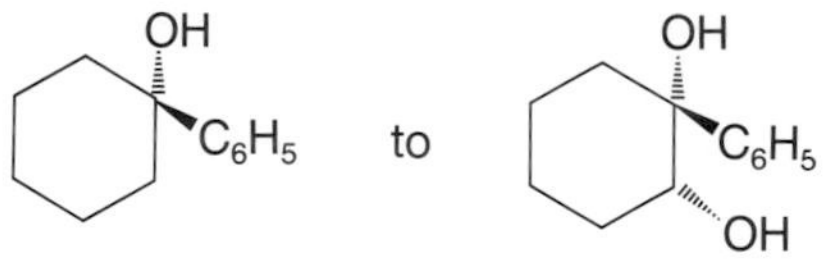

16.30 The fungus responsible for Dutch elm disease is spread by European bark beetles when they burrow into the tree. Other beetles congregate at the site, attracted by the scent of a mixture of chemicals, some emitted by other beetles and some coming from the tree. One of the compounds given off by female bark beetles is 4-methyl-3-heptanol. Suggest an efficient synthesis of this pheromone from alcohols of five or less carbon atoms.

16.31 Show by a series of equations how you could prepare 3-methylpentane from ethanol and any necessary inorganic reagents.

16.32 The cis isomer of 3-hexen-1-ol (CH_3CH_2CH=$CHCH_2CH_2OH$) has the characteristic odor of green leaves and grass. Suggest a synthesis for this compound from acetylene and any necessary organic or inorganic reagents.

16.33 R. B. Woodward was the leading organic chemist of the middle part of this century. Known primarily for his achievements in the synthesis of complex natural products, he was awarded the Nobel prize in chemistry in 1965. He entered Massachusetts Institute of Technology as a 16-year-old freshman in 1933 and four years later was awarded the Ph.D. While a student there he carried out a synthesis of *estrone,* a female sex hormone. The early stages of Woodward's estrone synthesis required the conversion of *m*-methoxybenzaldehyde to *m*-methoxybenzyl cyanide, which was accomplished in three steps.

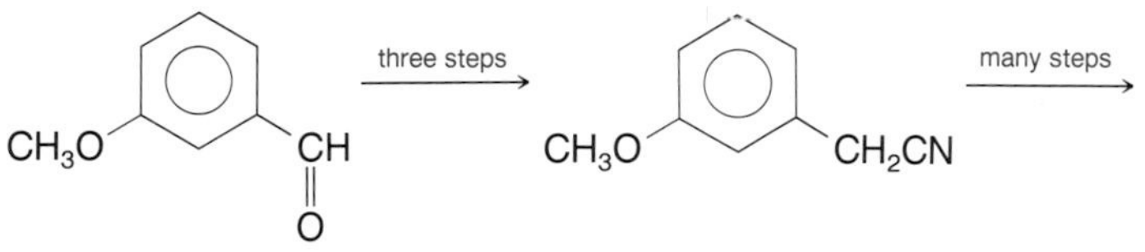

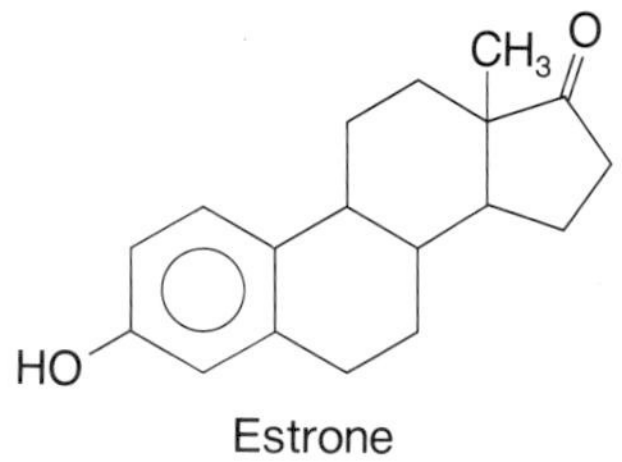

Estrone

Suggest a reasonable three-step sequence, showing all necessary reagents, for the preparation of *m*-methoxybenzyl cyanide from *m*-methoxybenzaldehyde.

16.34 Complete the following series of equations by writing structural formulas for compounds A through F:

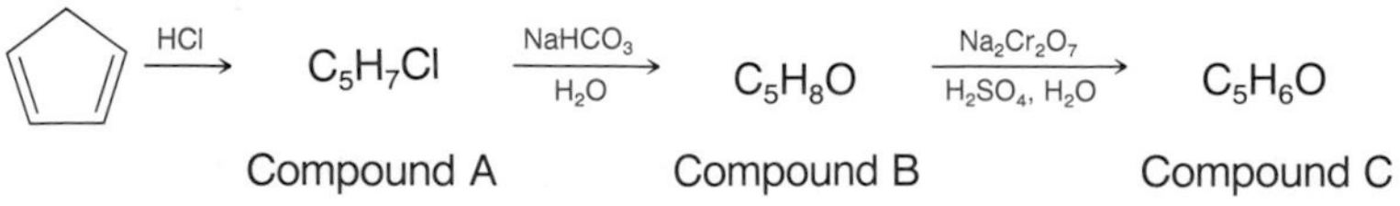

(b) CH_2=$CHCH_2CH_2CHCH_3$ $\xrightarrow{\text{SOCl}_2}_{\text{pyridine}}$ $C_6H_{11}Cl$ $\xrightarrow[\text{2. reductive workup}]{1.\ O_3}$
                    |
                    OH   Compound D

C_5H_9ClO $\xrightarrow{\text{NaBH}_4}$ $C_5H_{11}ClO$

Compound E   Compound F

16.35 Suggest a chemical test that would permit you to distinguish between the two glycerol monobenzyl ethers shown.

$C_6H_5CH_2OCH_2CHCH_2OH$ $HOCH_2CHCH_2OH$
                     | |
                     OH $OCH_2C_6H_5$

1-*O*-Benzylglycerol   2-*O*-Benzylglycerol

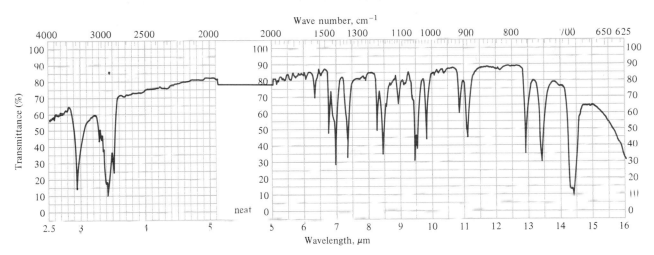

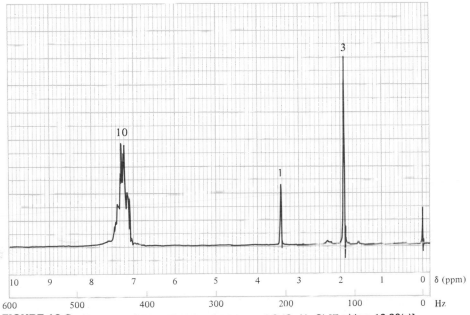

FIGURE 16.6   The ir and ¹H nmr spectra of compound G ($C_{14}H_{14}O$) [Problem 16.36(a)].

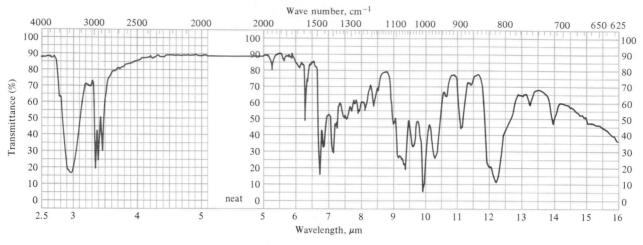

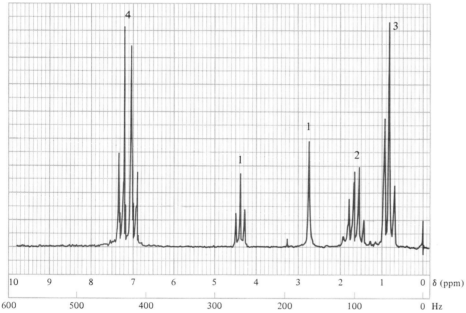

FIGURE 16.7 The ir and ^{1}H nmr spectra of compound H ($C_9H_{11}BrO$) [Problem 16.36(b)].

16.36 Identify the following compounds on the basis of their ir and ^{1}H nmr spectra:

(a) Compound G ($C_{14}H_{14}O$) (Figure 16.6)

(b) Compound H ($C_9H_{11}BrO$) (Figure 16.7)

16.37 Compound I ($C_8H_{18}O_2$) does not react with periodic acid. Its ir and ^{1}H nmr spectra are shown in Figure 16.8. What is compound I?

16.38 Identify each of the following $C_4H_{10}O$ isomers on the basis of their ^{13}C nmr spectra:

(a) 31.2 ppm: quartet
   68.9 ppm: singlet

(b) 10.0 ppm: quartet
   22.7 ppm: quartet
   32.0 ppm: triplet
   69.2 ppm: doublet

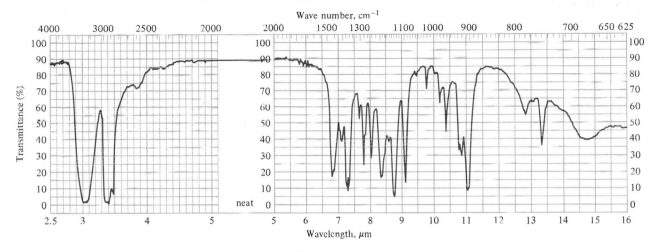

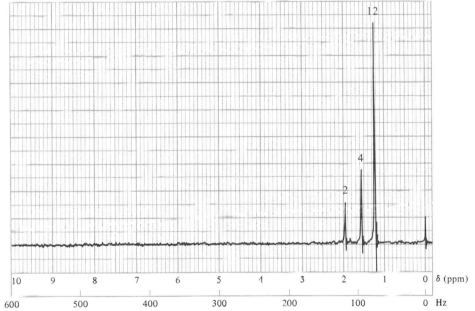

FIGURE 16.8 The ir and ¹H nmr spectra of compound I ($C_8H_{18}O_2$) (Problem 16.37).

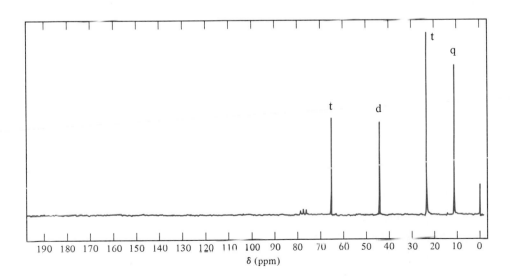

FIGURE 16.9 The ¹³C nmr spectrum of compound K ($C_6H_{14}O$) (Problem 16.40). *(The ¹³C spectrum is taken from "Carbon-13 NMR Spectra: A Collection of Assigned, Coded, and Indexed Spectra," by LeRoy F. Johnson and William C. Jankowski, Wiley-Interscience, New York, 1972. Reprinted by permission of John Wiley & Sons, Inc.)*

(c)   18.9 ppm: quartet, area 2
30.8 ppm: doublet, area 1
69.4 ppm: triplet, area 1

16.39   Compound J ($C_3H_7ClO_2$) exhibited three peaks in its ^{13}C nmr spectrum at 46.8 (triplet), 63.5 (triplet), and 72.0 ppm (doublet). What is the structure of compound J?

16.40   Compound K ($C_6H_{14}O$) has the ^{13}C nmr spectrum shown in Figure 16.9. Its mass spectrum has a prominent peak at m/z 31. Suggest a reasonable structure for compound K.

ETHERS AND EPOXIDES

In contrast to alcohols with their rich chemical reactivity, ethers—compounds containing a C—O—C unit—undergo relatively few chemical reactions. As you saw when we discussed Grignard reagents in Chapter 15 and lithium aluminum hydride reductions in Chapter 16, this lack of reactivity of ethers makes them valuable as solvents in a number of synthetically important transformations. In the present chapter you will learn of the conditions in which an ether linkage acts as a functional group, as well as the methods by which ethers are prepared. Both aspects —the preparation of ethers and the reactions that they undergo—illustrate the application of principles developed earlier to the chemistry of a new functional group.

Unlike most ethers, epoxides—compounds in which the C—O—C unit forms a three-membered ring—are very reactive substances. The principles of nucleophilic substitution reactions are important in understanding the preparation and properties of epoxides.

17.1 CLASSES OF ETHERS. NOMENCLATURE

Ethers are described as *symmetrical* or *unsymmetrical* depending on whether the two groups bonded to oxygen in ROR′ are the same or different. Unsymmetrical ethers are also called *mixed* ethers.

$$CH_3CH_2OCH_2CH_3 \qquad CH_3CH_2OCH_3$$

Diethyl ether   Ethyl methyl ether
(a symmetrical ether)   (an unsymmetrical ether)

Ethers are named by specifying the two alkyl groups in alphabetical order as separate words, then adding the word "ether" at the end. The prefix "di-" is used in symmetrical ethers, in which both alkyl groups are the same.

Cyclic ethers have their ether oxygen as part of a ring. Tetrahydrofuran and tetrahydropyran are cyclic ethers.

Tetrahydrofuran (oxolane)

Tetrahydropyran (oxane)

In general, the properties of cyclic ethers are very much like those of their noncyclic counterparts. An exception, as we shall see, is the case of epoxides. Epoxides are cyclic ethers in which the ether oxygen forms part of a three-membered ring. The nomenclature of epoxides was described in Section 7.15. Examples of epoxides include the following:

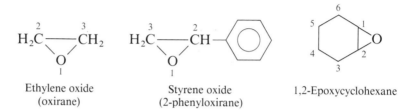

Ethylene oxide
(oxirane)

Styrene oxide
(2-phenyloxirane)

1,2-Epoxycyclohexane

PROBLEM 17.1   Each of the following ethers has been shown to be or is suspected to be a *mutagen,* which means it can induce mutations in test cells. Write the structures of each of these ethers.

(*a*)   Chloromethyl methyl ether
(*b*)   2-(Chloromethyl)oxirane (also known as epichlorohydrin)
(*c*)   3,4-Epoxy-1-butene (2-vinyloxirane)

SAMPLE SOLUTION   (*a*)   Chloromethyl methyl ether has a chloromethyl group ($ClCH_2$—) and a methyl group (CH_3—) attached to oxygen. Its structure is $ClCH_2OCH_3$.

It is not unusual for substances to have more than one ether linkage. Two such compounds, often used as solvents, are the *diethers* 1,2-dimethoxyethane and 1,4-dioxane. Diglyme, also a commonly used solvent, is a *triether.*

$CH_3OCH_2CH_2OCH_3$

1,2-Dimethoxyethane

1,4-Dioxane

$CH_3OCH_2CH_2OCH_2CH_2OCH_3$

Bis(2-methoxyethyl) ether
(diethylene glycol dimethyl ether;
diglyme)

Molecules that contain several ether functions are referred to as *polyethers.* Polyethers have received much recent attention and some examples of them will appear in Sections 17.4 and 17.5.

17.2   STRUCTURE AND BONDING IN ETHERS AND EPOXIDES

Bonding in ethers is readily understood by analogy to the bonding models developed for water and alcohols. In neutral molecules in which oxygen bears two substituents, each of them is attached by a two-electron σ bond, leaving oxygen with two unshared electron pairs. The four pairs of electrons—the two bonded pairs plus the two unshared pairs—are arranged in an approximately tetrahedral orientation around oxygen. The precise bond angles at oxygen vary among the compounds water, methanol, and dimethyl ether in a way expected on the basis of a combination of van der Waals and valence-shell electron pair repulsive forces.

Water Methanol Dimethyl ether

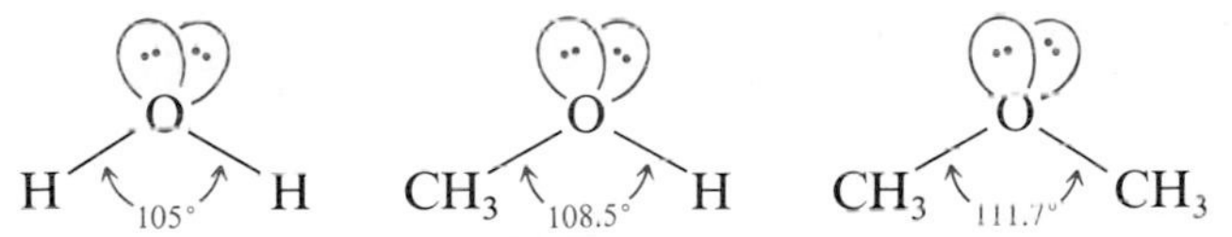

Opposing the repulsive interaction between the two nonbonded electron pairs on oxygen is the van der Waals repulsion between its two substituents. When these two substituents are small, such as the hydrogens in water, the bond angle at oxygen is smaller than the tetrahedral value of 109.5°. As the substituents become larger, the repulsive force between them increases and so does the bond angle.

The carbon-oxygen bond distance in ethers is about the same as it is in alcohols.

Methanol Dimethyl ether

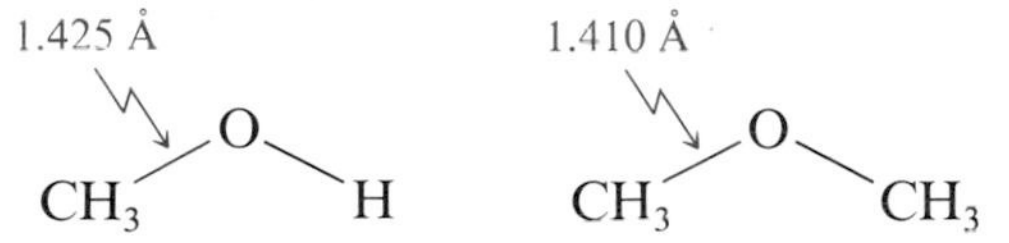

A single oxygen atom in a hydrocarbon chain affects the conformation of a molecule in much the same way that a CH_2 unit does. The most stable conformation of diethyl ether is the all-staggered anti conformation. Tetrahydropyran adopts the chair form in its most stable conformation.

Anti conformation of diethyl ether

Chair conformation of tetrahydropyran

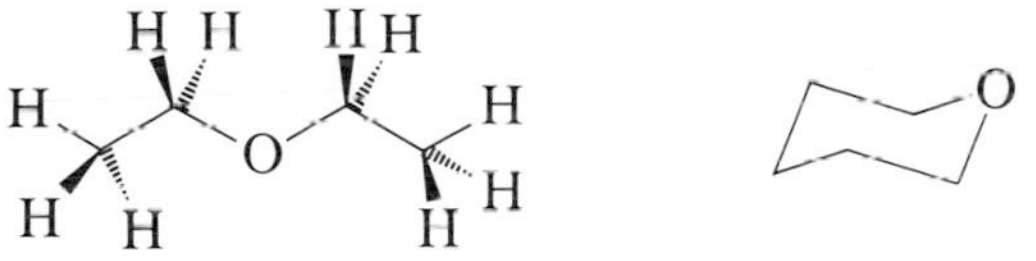

Incorporating an oxygen atom into a three-membered ring requires its bond angle to be seriously distorted from the normal tetrahedral value. In ethylene oxide, for example, the bond angle at oxygen is 61.5°.

C—O—C angle 61.5°

C—C—O angle 59.2°

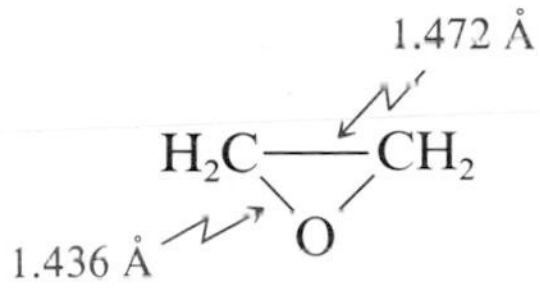

Thus epoxides, like cyclopropanes, are strained. They tend to undergo reactions that open the three-membered ring by cleavage of one of the carbon-oxygen bonds.

PROBLEM 17.2 The heats of combustion of 1,2-epoxybutane (2-ethyloxirane) and tetrahydrofuran have been measured: one is 597.8 kcal/mol, the other is 609.1 kcal/mol. Match the heats of combustion with the respective compounds.

The combination of bond angles that are close to tetrahedral at the ether oxygen with the polar character of their carbon-oxygen bonds causes most ethers to be polar molecules. Dimethyl ether, for example, has a dipole moment of 1.3 D. Cyclic ethers have larger dipole moments; ethylene oxide and tetrahydrofuran have dipole moments in the 1.7 to 1.8 D range—about the same as that of water.

17.3 PHYSICAL PROPERTIES OF ETHERS

It is instructive to compare ethers, alkanes, and alcohols with respect to certain of their physical properties. The boiling points of ethers (Table 17.1) are very similar to those of alkanes of the same chain length but are substantially lower than the boiling points of comparably constituted alcohols.

$$CH_3CH_2OCH_2CH_3 \qquad CH_3CH_2CH_2CH_2CH_3 \qquad CH_3CH_2CH_2CH_2OH$$

Diethyl ether	Pentane	1-Butanol
(bp 34.6°C)	(bp 36°C)	(bp 117°C)

The solubility properties of ethers, on the other hand, resemble those of alcohols more than they do those of alkanes. Diethyl ether and 1-butanol dissolve in water to almost exactly the same extent—7.5 and 9 g/100 mL, respectively—while pentane is essentially insoluble in water.

What structural features are responsible for these differences? In general, the boiling points of alcohols are unusually high because alcohols are extensively hydro-

TABLE 17.1
Physical Properties of Some Ethers

Ether	Condensed structural formula	Melting point, °C	Boiling point, °C (1 atm)	Solubility, g/100 mL H$_2$O
Dimethyl ether	CH$_3$OCH$_3$	−138.5	−24	Very soluble
Ethyl methyl ether	CH$_3$CH$_2$OCH$_3$		7.6	Soluble
Diethyl ether	CH$_3$CH$_2$OCH$_2$CH$_3$	−116.3	34.6	7.5
Dipropyl ether	CH$_3$CH$_2$CH$_2$OCH$_2$CH$_2$CH$_3$	−122	90.1	Slight
Diisopropyl ether	(CH$_3$)$_2$CHOCH(CH$_3$)$_2$	−60	68.5	0.2
Dibutyl ether	(CH$_3$CH$_2$CH$_2$CH$_2$)$_2$O	−97.9	142.4	Less than 0.05
1,2-Dimethoxyethane	CH$_3$OCH$_2$CH$_2$OCH$_3$		83	∞
Diethylene glycol dimethyl ether (diglyme)	CH$_3$OCH$_2$CH$_2$OCH$_2$CH$_2$OCH$_3$		161	∞
Tetrahydrofuran		−108.5	65	∞
1,4-Dioxane		11.8	101.4	∞
Ethylene oxide		−111.7	10.7	∞

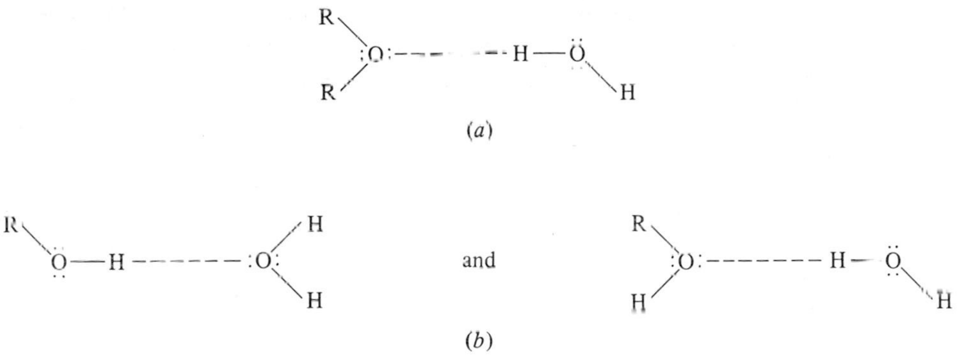

(a)

and

(b)

FIGURE 17.1 (a) Hydrogen bonding between an ether molecule and a water molecule. (b) Two ways in which an alcohol molecule and a water molecule can form a hydrogen bond.

gen-bonded in the liquid phase (Section 4.4). Ethers and alkanes, lacking hydroxyl groups, cannot form analogous hydrogen-bonded aggregates. The oxygen of an ether, however, can use one of its unshared electron pairs to accept a proton in a hydrogen bond with water, as shown in Figure 17.1. Thus, low-molecular-weight ethers dissolve in water because they can associate with water molecules in much the same way that alcohols do. Alkanes, lacking unshared electron pairs, cannot act as proton acceptors in hydrogen bonds and are practically insoluble in water.

PROBLEM 17.3 Ethers tend to dissolve in alcohols and vice versa. Represent the hydrogen bonding interaction between an alcohol molecule and an ether molecule.

17.4 CROWN ETHERS

The polar nature of carbon-oxygen bonds and the presence of unshared electron pairs at oxygen are responsible for another property of ethers, namely, their ability to form Lewis acid–Lewis base complexes with metal ions.

| Ether | Metal ion | Ether–metal ion |
| (Lewis base) | (Lewis acid) | complex |

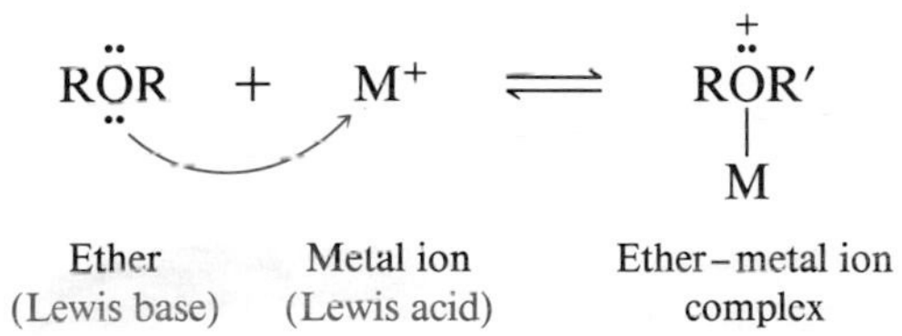

The strength of this bonding depends on the kind of ether. Simple ethers form relatively weak complexes with metal ions. A major advance in the area came in 1967 when Charles J. Pedersen of DuPont described the preparation and properties of a class of *polyethers* that form much more stable complexes with metal ions than do simple ethers.

Pedersen prepared a series of *macrocyclic polyethers*, cyclic compounds containing 4 or more ether linkages in a ring of 12 or more atoms. He called these compounds *crown ethers* because their molecular models resemble crowns. Systematic nomenclature of crown ethers is somewhat cumbersome, so Pedersen devised an appropriate shorthand description whereby the word "crown" is preceded by the total number of atoms in the ring and is followed by the number of oxygen atoms.

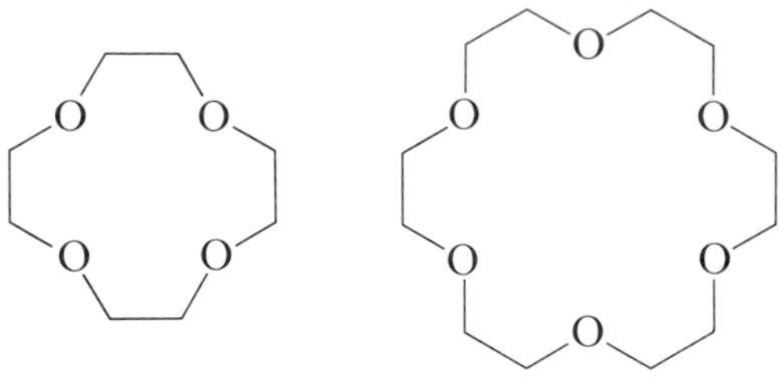

12-Crown-4   18-Crown-6

12-Crown-4 and 18-crown-6 are a cyclic tetramer and hexamer, respectively, of repeating —OCH_2CH_2— units; they are polyethers based on ethylene glycol $HOCH_2CH_2OH$ as the parent alcohol.

PROBLEM 17.4   An organic compound with which you are already familiar might be described as "6-crown-2" according to this terminology. What is this compound?

The metal ion–complexing properties of crown ethers can be illustrated by their effects on the solubility and reactivity of salts in nonpolar media. Potassium fluoride is ionic and practically insoluble in benzene, but 0.05 M solutions can be prepared when 18-crown-6 is present. This increased solubility of potassium fluoride in benzene is attributable to coordination of potassium cation by the crown ether.

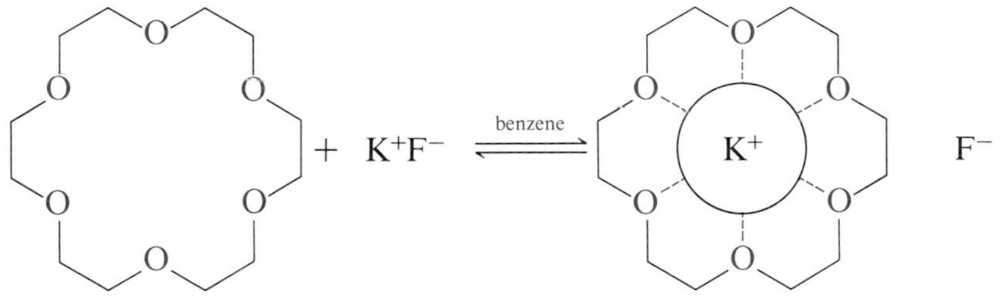

18-Crown-6   Potassium   18-Crown-6–potassium F^-
                    fluoride   fluoride complex
                    (solid)   (in solution)

The internal cavity of 18-crown-6 is about 2.6 to 3.2 Å in diameter. Potassium ion, with a diameter of 2.66 Å, fits comfortably within this cavity and is close enough to each of the six oxygens for electrostatic attraction to exist between them. Coordination of a metal ion disperses its positive charge and lowers the energy required for transferring it from the solid phase, where ionic bonding to its anionic partner is strong, to the dissolved phase. In order to maintain electrical neutrality, every potassium ion that is transferred from the solid to the solution is accompanied by fluoride, and a solution of strongly complexed potassium ions and relatively unsolvated fluoride ions is formed.

In media such as water and alcohols, fluoride ion is strongly solvated by hydrogen bonding and is neither very basic nor very nucleophilic. On the other hand, the poorly solvated or "naked" fluoride ions that are present when potassium fluoride dissolves in benzene in the presence of a crown ether are better able to express their anionic reactivity. Thus, alkyl halides react with potassium fluoride in benzene containing 18-crown-6, thereby providing a method for the preparation of otherwise difficulty accessible alkyl fluorides.

$$CH_3(CH_2)_6CH_2Br \xrightarrow[\text{18-crown-6}]{\text{KF, benzene, 90°C}} CH_3(CH_2)_6CH_2F$$

1-Bromooctane   1-Fluorooctane (92%)

No reaction is observed when the reaction is carried out under comparable conditions but with the crown ether omitted.

Catalysis by crown ethers has been demonstrated in numerous organic reactions that involve anions as reactants.

17.5 MONENSIN: A POLYETHER ANTIBIOTIC

Many ethers, especially cyclic ones, occur naturally. A group that bears mention includes the *polyether antibiotics,* of which monensin (Figure 17.2) is an example. Monensin binds sodium ions in a manner reminiscent of crown ether–metal ion binding. Four of monensin's ether oxygens and two of its hydroxyl groups surround a sodium ion. The alkyl groups are oriented toward the outside of the monensin-Na$^+$ complex while the polar oxygens and the metal ion are on the inside. The hydrocarbonlike surface of the complex permits it to carry its sodium ion through the hydrocarbonlike interior of a cell membrane. This disrupts the normal balance of sodium ions within the cell and interferes with important processes of cellular respiration. Small amounts of monensin are added to poultry feed in order to kill parasites that live in the intestines of chickens.

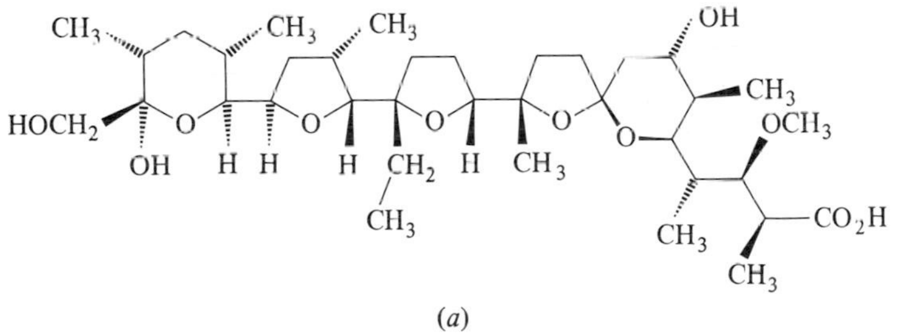

(a)

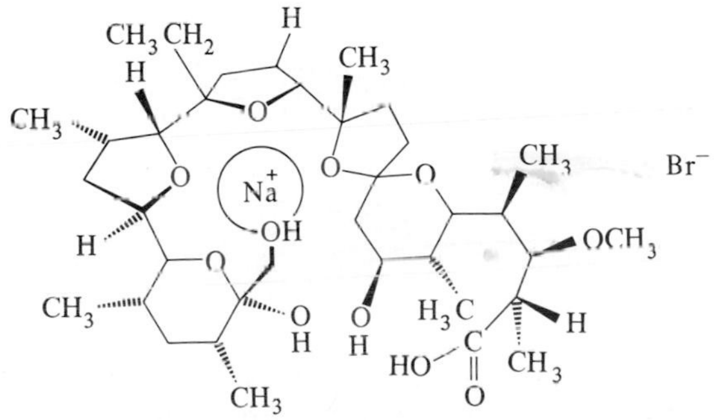

(b)

FIGURE 17.2 (a) The structure of monensin; (b) the structure of the monensin–sodium bromide complex showing coordination of sodium ion by oxygen atoms of monensin.

Compounds such as monensin and the crown ethers that affect metal ion transport are referred to as *ionophores* ("ion bearers").

17.6 PREPARATION OF ETHERS

Because they are widely used as solvents, many simple dialkyl ethers are commercially available. Diethyl ether and dibutyl ether, for example, are prepared by acid-catalyzed condensation of the corresponding alcohols, as described earlier in Section 16.9.

$$2CH_3CH_2CH_2CH_2OH \xrightarrow[130°C]{H_2SO_4} CH_3CH_2CH_2CH_2OCH_2CH_2CH_2CH_3 + H_2O$$

1-Butanol   Dibutyl ether (60%)   Water

In the following sections two additional methods, each more versatile than acid-catalyzed condensation of alcohols, will be presented. These methods are

1. *The Williamson ether synthesis* (Section 17.7), a reaction in which an ether is formed by nucleophilic substitution of an alkyl halide by an alkoxide:

$$RO^- + R'X \longrightarrow ROR' + X^-$$

Alkoxide   Alkyl   Ether   Halide ion
   ion   halide

2. *Solvomercuration-demercuration* (Section 17.8), a modification of the oxy-mercuration-demercuration procedure using alcohols instead of aqueous tetrahydrofuran as the reaction medium:

$$\begin{array}{c} \diagdown \\ C=C \\ \diagup \end{array} + ROH \xrightarrow[2.\ NaBH_4]{1.\ Hg(OAc)_2} -CH-C-OR$$

Alkene   Alcohol   Ether

Let us now examine these two methods for the preparation of ethers in more detail.

17.7 THE WILLIAMSON ETHER SYNTHESIS

A method of long standing for the preparation of ethers is the *Williamson ether synthesis.* Nucleophilic substitution of an alkyl halide by an alkoxide leads to carbon-oxygen bond formation between the substrate and the nucleophile.

$$RO^- \curvearrowright R'-X \longrightarrow ROR' + X^-$$

Alkoxide   Alkyl   Ether   Halide ion
   ion   halide

Preparation of ethers by the Williamson ether synthesis is most successful when the alkyl halide substrate is one which is reactive toward S_N2 substitution. Methyl halides and primary alkyl halides are the best substrates.

$$CH_3CH_2CH_2CH_2ONa + CH_3CH_2I \longrightarrow CH_3CH_2CH_2CH_2OCH_2CH_3 + NaI$$

| Sodium butoxide | Iodoethane | Butyl ethyl ether (71%) | Sodium iodide |

PROBLEM 17.5 Show how each of the following ethers could be prepared by the Williamson ether synthesis:

(a) $CH_3CH_2OCH_2CH_3$
(b) $CH_3OCH_2CH_2CH_3$ (two ways)
(c) $C_6H_5CH_2OCH_2CH_3$ (two ways)

SAMPLE SOLUTION (a) Ethers are characterized by a C—O—C unit. One of the carbon-oxygen bonds comes from the alkoxide ion used as a nucleophile, while the other carbon-oxygen bond arises by an S_N2 reaction of the alkoxide with an alkyl halide. The preparation of diethyl ether requires sodium ethoxide as the nucleophile and an ethyl halide as the substrate.

$$CH_3CH_2ONa + CH_3CH_2Br \longrightarrow CH_3CH_2OCH_2CH_3 + NaBr$$

| Sodium ethoxide | Ethyl bromide | Diethyl ether | Sodium bromide |

Secondary and tertiary alkyl halides are poor substrates because they tend to react with alkoxide bases by E2 elimination rather than by S_N2 substitution. Whether the alkoxide base is primary, secondary, or tertiary is much less important than the nature of the alkyl halide. Thus benzyl isopropyl ether is prepared in high yield from benzyl chloride, a primary chloride that is incapable of undergoing elimination, and sodium isopropoxide.

$$(CH_3)_2CHONa + \underset{\text{Benzyl chloride}}{\text{⬡—CH}_2\text{Cl}} \longrightarrow \underset{\text{Benzyl isopropyl ether}}{(CH_3)_2CHOCH_2\text{—⬡}} + NaCl$$

Sodium isopropoxide   Benzyl chloride   Benzyl isopropyl ether (84%)   Sodium chloride

The alternative synthetic route using the sodium salt of benzyl alcohol and an isopropyl halide would be much less effective because of increased competition from the elimination reaction.

$$\text{⬡—CH}_2\text{ONa} + (CH_3)_2CHX \longrightarrow \text{⬡—CH}_2\text{OH} + CH_3CH{=}CH_2$$

Sodium benzoxide   Isopropyl halide   Benzyl alcohol   Propene

PROBLEM 17.6 Only one combination of alkyl halide and alkoxide is appropriate for the preparation of each of the following ethers by the Williamson ether synthesis. What is the correct combination in each case?

(a) CH_3CH_2O—⬠

(c) $(CH_3)_3COCH_2C_6H_5$

(b) $CH_2{=}CHCH_2OCH(CH_3)_2$

SAMPLE SOLUTION (a) The ether linkage of cyclopentyl ethyl ether involves a primary carbon and a secondary one. Choose the alkyl halide corresponding to the primary alkyl group, leaving the secondary alkyl group to arise from the alkoxide nucleophile.

| Sodium cyclopentanolate | Ethyl bromide | | Cyclopentyl ethyl ether |

The alternative combination, cyclopentyl bromide and sodium ethoxide, is not appropriate since elimination will be the principal reaction.

| Sodium ethoxide | Bromocyclopentane (cyclopentyl bromide) | | Ethanol | Cyclopentene (major products) |

Both reactants in the Williamson ether synthesis usually originate in alcohol precursors. Sodium and potassium alkoxides are prepared by direct reaction of an alcohol with the appropriate metal, and alkyl halides are most commonly made from alcohols by reaction with a hydrogen halide (Section 4.9), thionyl chloride (Section 4.17), or phosphorus tribromide (Section 4.17). Alternatively, alkyl *p*-toluenesulfonates may be used in place of alkyl halides; alkyl *p*-toluenesulfonates are also prepared from alcohols as their immediate precursors (Section 9.15).

For the preparation of methyl ethers by the Williamson method, dimethyl sulfate is often used in place of the more expensive methyl halides. Nucleophiles, including alkoxides, attack dimethyl sulfate at carbon in an S_N2 process.

| Alkoxide | Dimethyl sulfate | Methyl ether | Methyl sulfate anion |

| 2-Butanol | | *sec*-Butyl methyl ether (56%) |

Extending the scope of ether synthesis to include mixed ethers in which, for example, both carbons of the C—O—C unit are secondary requires that some other method be available. Such a method is described in the following section and is an extension of some chemistry you have already learned.

17.8 PREPARATION OF ETHERS BY SOLVOMERCURATION-DEMERCURATION OF ALKENES

We have learned in Section 7.7 of a method for the Markovnikov hydration of alkenes known as oxymercuration-demercuration.

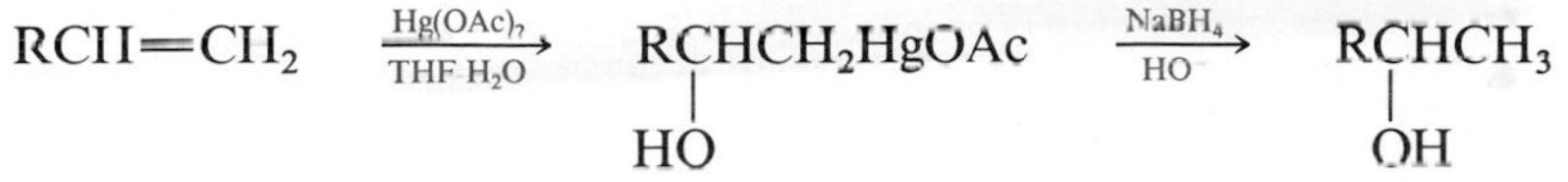

$$RCH=CH_2 \xrightarrow[\text{THF H}_2\text{O}]{\text{Hg(OAc)}_2} \underset{\underset{\text{HO}}{|}}{RCHCH_2HgOAc} \xrightarrow[\text{HO}^-]{\text{NaBH}_4} \underset{\underset{\text{OH}}{|}}{RCHCH_3}$$

Alkene	Oxymercuration step	Hydroxyalkyl mercuric acetate	Demercuration step	Alcohol

The hydroxyl group of the alcohol product is derived from a water molecule acting as a nucleophile toward a mercury-substituted carbocation.

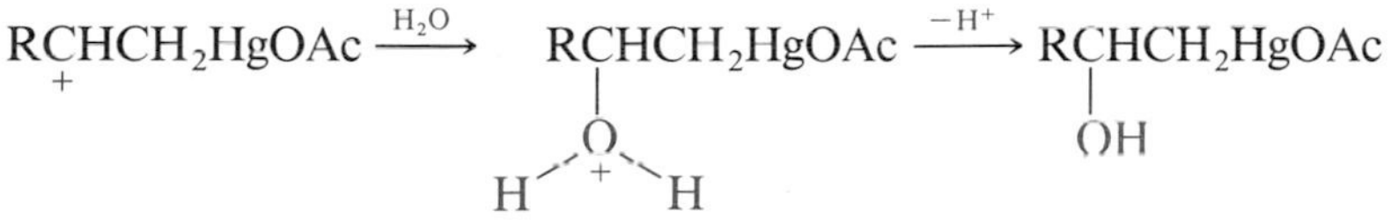

$$\underset{+}{RCHCH_2HgOAc} \xrightarrow{\text{H}_2\text{O}} \underset{\underset{\underset{H\overset{+}{\diagdown}H}{O}}{|}}{RCHCH_2HgOAc} \xrightarrow{-\text{H}^+} \underset{\underset{\text{OH}}{|}}{RCHCH_2HgOAc}$$

Mercury-substituted carbocation

Hydroxyalkylmercuric acetate

This reaction sequence can be modified to serve as an ether synthesis. If, instead of tetrahydrofuran-water an alcohol is employed as the solvent, then the mercury-substituted carbocation intermediate is captured by the alcohol to produce an alkoxyalkylmercuric acetate.

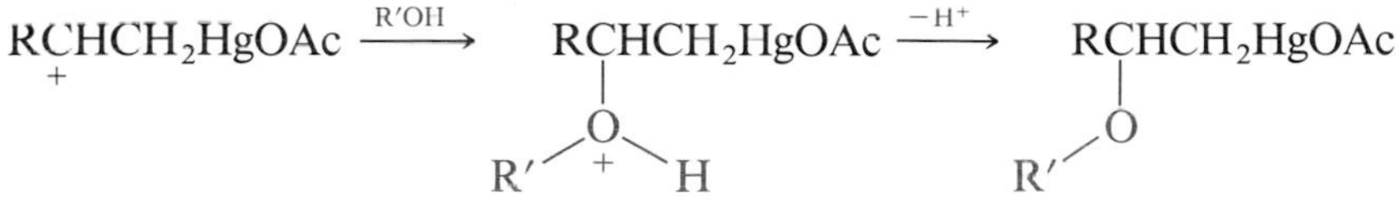

$$\underset{+}{RCHCH_2HgOAc} \xrightarrow{\text{R}'\text{OH}} \underset{\underset{\underset{R'\overset{+}{\diagdown}H}{O}}{|}}{RCHCH_2HgOAc} \xrightarrow{-\text{H}^+} \underset{\underset{\underset{R'\diagdown}{O}}{|}}{RCHCH_2HgOAc}$$

Mercury-substituted carbocation

Alkoxyalkylmercuric acetate

Demercuration of the alkoxyalkylmercuric acetate by sodium borohydride generates the product, in this case an ether.

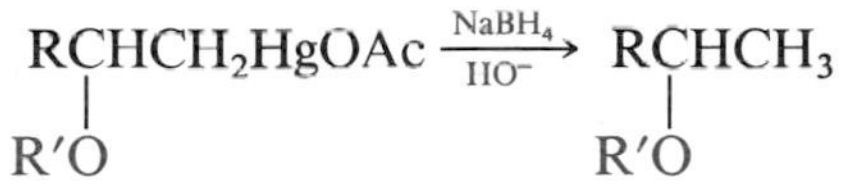

$$\underset{\underset{\text{R}'\text{O}}{|}}{RCHCH_2HgOAc} \xrightarrow[\text{HO}^-]{\text{NaBH}_4} \underset{\underset{\text{R}'\text{O}}{|}}{RCHCH_3}$$

Alkoxyalkylmercuric acetate

Ether

The term *solvomercuration-demercuration* has been applied to this general class of reactions. Ether synthesis by solvomercuration-demercuration, like alcohol synthesis by oxymercuration-demercuration, corresponds to addition to the alkene according to Markovnikov's rule.

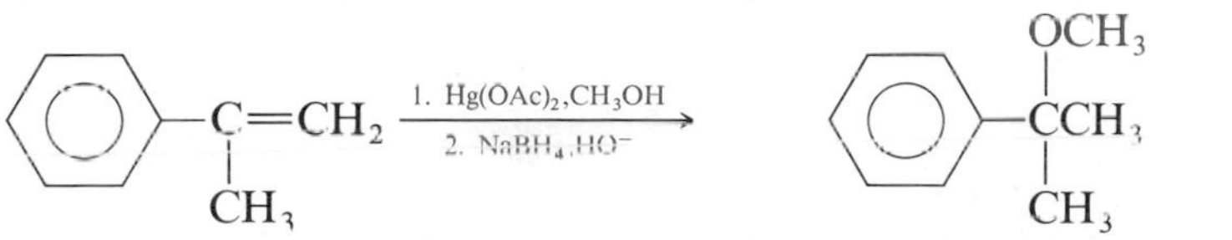

2-Phenylpropene

2-Methoxy-2-phenylpropane (100%)

Ethers in which both alkyl groups are secondary or tertiary, which are difficultly accessible by the Williamson method, are readily prepared by solvomercuration-demercuration.

$$CH_2{=}\underset{\underset{CH_3}{|}}{C}CH_2CH_3 \xrightarrow[\text{2. NaBH}_4,\ \text{HO}^-]{\text{1. Hg(OAc)}_2,\ (CH_3)_2CHOH} CH_3{-}\underset{\underset{CH_3}{|}}{\overset{\overset{OCH(CH_3)_2}{|}}{C}}CH_2CH_3$$

2-Methyl-1-butene   2-Isopropoxy-2-methylbutane
(81%)

Rearrangement of the carbon skeleton does not occur.

$$CH_3{-}\underset{\underset{CH_3}{|}}{\overset{\overset{CH_3}{|}}{C}}{-}CH{=}CH_2 \xrightarrow[\text{2. NaBH}_4,\ \text{HO}^-]{\text{1. Hg(OAc)}_2,\ CH_3OH} CH_3{-}\underset{\underset{CH_3OCH_3}{|}}{\overset{\overset{CH_3}{|}}{C}}{-}CHCH_3$$

3,3-Dimethyl-1-butene   2-Methoxy-3,3-dimethylbutane
(83%)

PROBLEM 17.7   Show how each of the following ethers could be efficiently prepared by solvomercuration-demercuration:

(a) $C_6H_5\underset{\underset{CH_3}{|}}{C}HOCH_2CH_2CH_3$

(b)   Cyclohexyl methyl ether

(c) $CH_3\underset{\underset{OCH_2CH_3}{|}}{C}HCH_2CH_2CH_2CH_3$

(d) $(CH_3)_2\underset{\underset{OCH_3}{|}}{C}CH_2CH_3$

SAMPLE SOLUTION   (a)   A decision must be made as to which one of the groups bonded to the ether oxygen is derived from an alkene and which is derived from the alcohol used as the solvent. The two groups are

$$C_6H_5\underset{\underset{CH_3}{|}}{C}H{-} \qquad \text{and} \qquad {-}CH_2CH_2CH_3$$

Since the overall reaction follows Markovnikov's rule, a propyl group can only originate in the alcohol. Markovnikov addition to propene yields an isopropyl group. Therefore, the correct alkene is styrene and the correct alcohol is 1-propanol.

$$C_6H_5CH{=}CH_2 \xrightarrow[\text{2. NaBH}_4,\ \text{HO}^-]{\text{1. Hg(OAc)}_2,\ CH_3CH_2CH_2OH} C_6H_5\underset{\underset{OCH_2CH_2CH_3}{|}}{C}HCH_3$$

Styrene   1-Phenylethyl propyl ether

When the solvent is a tertiary alcohol, its reaction with the mercury-substituted carbocation intermediate is relatively hindered. The carbocation is captured instead by acetate anion from mercuric acetate. Ethers of tertiary alcohols are therefore better prepared by using mercuric trifluoroacetate. Since trifluoroacetate anion is a much weaker nucleophile than acetate, capture of the mercury-substituted carbocation by the tertiary alcohol becomes more favorable.

Cyclohexene   1. $Hg(OCCT_3)_2,(CH_3)_3COH$ 2. $NaBH_4,HO^-$ tert-Butoxycyclohexane (90%)
(tert-butyl cyclohexyl ether)

(The yield of *tert*-butyl cyclohexyl ether was only 18 percent when mercuric acetate was used.)

17.9 REACTIONS OF ETHERS. A REVIEW AND A PREVIEW

Up to this point in the text, no reactions of dialkyl ethers have been presented. Indeed, ethers are one of the least reactive of the functional groups we shall study. It is this low level of reactivity, along with an ability to dissolve nonpolar substances, that makes ethers so often used as solvents when carrying out organic reactions. Nevertheless, most ethers are hazardous materials and precautions must be taken when using them. Diethyl ether is extremely flammable and because of its high volatility can form explosive mixtures in air relatively quickly. Open flames must never be present in laboratories where diethyl ether is being used. Other low-molecular-weight ethers must also be treated as fire hazards.

PROBLEM 17.8 Combustion in air is, of course, a chemical property of ethers that is shared by many other organic compounds. Write balanced chemical equations for the complete combustion (in air) of each of the following:

(a) Diethyl ether

(b) Dipentyl ether

(c) (tetrahydrofuran)

SAMPLE SOLUTION (a) When burned in sufficient oxygen, all organic compounds that contain carbon, hydrogen, and oxygen form carbon dioxide and water.

$$CH_3CH_2OCH_2CH_3 + 6O_2 \longrightarrow 4CO_2 + 5H_2O$$

    Diethyl ether   Oxygen   Carbon dioxide   Water

Another dangerous property of ethers is the ease with which they undergo oxidation in air to form explosive peroxides. Air oxidation of diethyl ether proceeds according to the equation

$$CH_3CH_2OCH_2CH_3 + O_2 \longrightarrow CH_3\underset{\underset{\displaystyle HOO}{|}}{C}HOCH_2CH_3$$

    Diethyl ether   Oxygen   1-Ethoxyethyl hydroperoxide

The reaction is a free-radical one and oxidation occurs at the carbon that bears the ether oxygen to form a hydroperoxide, a compound of the type ROOH. Hydroperoxides tend to be unstable and shock-sensitive. On standing, they form related peroxidic derivatives, which are also prone to violent decomposition. Air oxidation leads to peroxides within a few days if ethers are even briefly exposed to atmospheric oxygen.

For this reason, one should never use old bottles of dialkyl ethers, and extreme care must be exercised in their disposal.

17.10 ACID-CATALYZED CLEAVAGE OF ETHERS

Just as the carbon-oxygen bond of alcohols is cleaved on reaction with hydrogen halides (Section 4.9), so too is an ether linkage broken:

$$ROH \ + \quad HX \quad \longrightarrow \quad RX \ + H_2O$$

Alcohol   Hydrogen halide   Alkyl   Water
halide

$$ROR' \ + \quad HX \quad \longrightarrow \quad RX \ + R'OH$$

Ether   Hydrogen halide   Alkyl   Alcohol
halide

The cleavage of ethers is normally carried out under conditions (excess hydrogen halide, heat) such that the alcohol formed as one of the original products is subsequently converted to an alkyl halide. Thus, the reaction typically leads to two alkyl halide molecules:

$$ROR' + \ 2HX \ \xrightarrow{heat} \ RX + R'X \ + H_2O$$

Ether   Hydrogen   Two alkyl halides   Water
halide

$$CH_3CHCH_2CH_3 \ \xrightarrow[heat]{HBr} \ CH_3CHCH_2CH_3 \ + \quad CH_3Br$$
$$\quad\quad | \qquad\qquad\qquad\qquad\qquad | $$
$$\quad\quad OCH_3 \qquad\qquad\qquad\qquad Br$$

sec-Butyl methyl ether   2-Bromobutane (81%)   Bromomethane
(*sec*-butyl bromide)   (methyl bromide)

Cyclic ethers yield one molecule of a dihalide.

$$\xrightarrow[150°C]{HI} \ ICH_2CH_2CH_2CH_2I$$

Tetrahydrofuran   1,4-Diiodobutane (65%)

The order of hydrogen halide reactivity is HI > HBr ≫ HCl. Hydrogen fluoride is not effective in ether cleavage. A convenient alternative to hydrogen iodide is a solution of potassium iodide in phosphoric acid.

$$(CH_3)_2CHOCH(CH_3)_2 \ \xrightarrow[H_3PO_4, \ heat]{KI} \quad 2(CH_3)_2CHI$$

Diisopropyl ether   Isopropyl iodide (90%)
(2-iodopropane)

PROBLEM 17.9   A series of dialkyl ethers was allowed to react with excess hydrogen bromide with the following results. Identify the ether in each case.

(*a*) One ether gave a mixture of bromocyclopentane and 1-bromobutane.
(*b*) Another ether gave only benzyl bromide.
(*c*) A third ether gave one mole of 1,5-dibromopentane per mole of ether.

SAMPLE SOLUTION (*a*) In the reaction of dialkyl ethers with excess hydrogen bromide, each alkyl group of the ether function is cleaved and forms an alkyl bromide. Since bromocyclopentane and 1-bromobutane are the products, the starting ether must be butyl cyclopentyl ether.

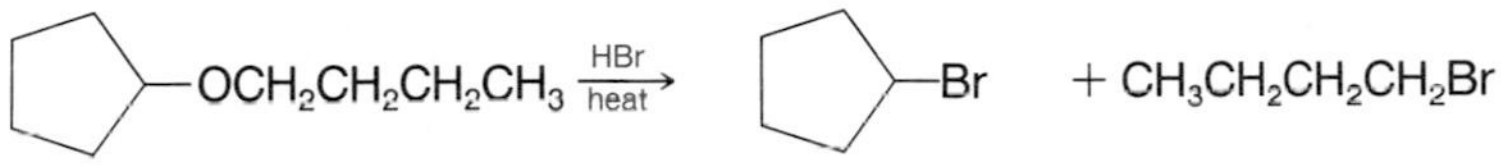

Butyl cyclopentyl ether — Bromocyclopentane — 1-Bromobutane

Taking diethyl ether as a simple example, the mechanism for its cleavage by hydrogen iodide begins with protonation of the ether oxygen.

$$CH_3CH_2\ddot{\underset{\cdot\cdot}{O}}CH_2CH_3 + H{-}I \overset{fast}{\rightleftharpoons} CH_3CH_2\overset{+}{\underset{|\atop H}{\ddot{O}}}CH_2CH_3 + I^-$$

Diethyl ether — Hydrogen iodide — Diethyloxonium ion — Iodide

Iodide ion is a good nucleophile and attacks this oxonium ion, displacing ethanol and forming a molecule of ethyl iodide.

$$I^- \quad \overset{CH_3}{\overset{|}{CH_2}}{-}\overset{+}{\underset{|\atop H}{\ddot{O}}}CH_2CH_3 \xrightarrow{slow} CH_3CH_2I + H\ddot{\underset{\cdot\cdot}{O}}CH_2CH_3$$

Iodide — Diethyloxonium ion — Ethyl iodide (iodoethane) — Ethanol

The ethanol produced in this step then reacts with hydrogen iodide in the way described earlier for the reaction of primary alcohols with hydrogen halides (Section 4.9):

$$CH_3CH_2\ddot{\underset{\cdot\cdot}{O}}H + HI \longrightarrow CH_3CH_2\overset{+}{\underset{\cdot\cdot}{O}}H_2 + I^-$$

Ethanol — Hydrogen iodide — Ethyloxonium ion — Iodide

$$I^- \quad \overset{CH_3}{\overset{|}{CH_2}}{-}\overset{+}{\underset{\cdot\cdot}{O}}H_2 \longrightarrow CH_3CH_2I + H_2\ddot{O}:$$

Iodide — Ethyloxonium ion — Ethyl iodide (iodoethane) — Water

---

PROBLEM 17.10 Di-*tert*-butyl ether is rapidly cleaved, even by hydrogen chloride at room temperature.

$$(CH_3)_3COC(CH_3)_3 \xrightarrow{\text{HCl}} 2(CH_3)_3CCl$$

Di-*tert*-butyl ether   *tert*-Butyl chloride (95%)
(2-chloro-2-methylpropane)

The oxonium ion formed by protonation of di-*tert*-butyl ether is much too crowded to be attacked directly by chloride ion. Suggest an alternative process by which *tert*-butyl chloride is formed from di-*tert*-butyloxonium ion.

With mixed ethers of the type ROR′, the question arises as to which carbon-oxygen bond is broken first. While some studies have been carried out on this point of mechanistic detail, it is not one that we need examine at our level of study.

17.11 PREPARATION OF EPOXIDES. A REVIEW AND A PREVIEW.

There are two principal laboratory methods for the preparation of epoxides:

1. Epoxidation of alkenes by reaction with peroxy acids
2. Base-promoted ring closure of vicinal halohydrins

Epoxidation of alkenes was discussed in Section 7.15 and is represented by the general equation

$$R_2C{=}CR_2 + R'\overset{\displaystyle O}{\overset{\displaystyle \|}{C}}OOH \longrightarrow R_2C\overset{\displaystyle }{\underset{\displaystyle O}{\diagdown\diagup}}CR_2 + R'\overset{\displaystyle O}{\overset{\displaystyle \|}{C}}OH$$

Alkene   Peroxy acid   Epoxide   Carboxylic acid

The reaction is easy to carry out and yields are usually high. Epoxidation is stereospecific in the sense that substituents that are cis to each other in the alkene remain cis in the epoxide.

(*E*)-1,2-Diphenylethene   Peroxyacetic acid   *trans*-2,3-Diphenyloxirane (78–83%)   Acetic acid

The base-promoted ring closure of vicinal halohydrins will be discussed in detail in Section 17.12. It proceeds according to the general equation

$$\underset{\substack{| \; | \\ HO \; X}}{R_2C{-}CR_2} + HO^- \longrightarrow R_2C\overset{\displaystyle }{\underset{\displaystyle O}{\diagdown\diagup}}CR_2 + X^- + H_2O$$

Halohydrin   Hydroxide ion   Epoxide   Halide ion   Water

Since halohydrins are ordinarily prepared by hypohalous acid addition to alkenes (Section 7.14), both methods are based on alkenes as starting materials for the synthesis of epoxides.

Both ethylene oxide and 1,2-epoxypropane are prepared in large quantities industrially. Much of the 1,2-epoxypropane is made by the halohydrin route. Ethylene oxide is made by direct oxidation of ethylene with air over a special silver catalyst.

17.12 CONVERSION OF VICINAL HALOHYDRINS TO EPOXIDES

The formation of vicinal halohydrins from alkenes was described in Section 7.14. Reaction of an alkene with chlorine or bromine in aqueous media leads to addition of the elements of hypochlorous (HOCl) or hypobromous (HOBr) acid across the double bond.

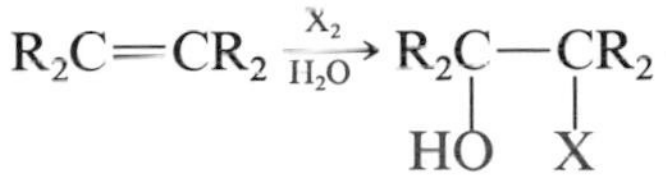

$$R_2C{=}CR_2 \xrightarrow[H_2O]{X_2} R_2C{-}CR_2$$
$$\qquad\qquad\qquad HO \quad X$$

Alkene Halohydrin

Halohydrins are readily converted to epoxides on treatment with base in what amounts to an *intramolecular* Williamson ether synthesis. Base treatment first brings the alcohol functional group into equilibrium with its corresponding alkoxide.

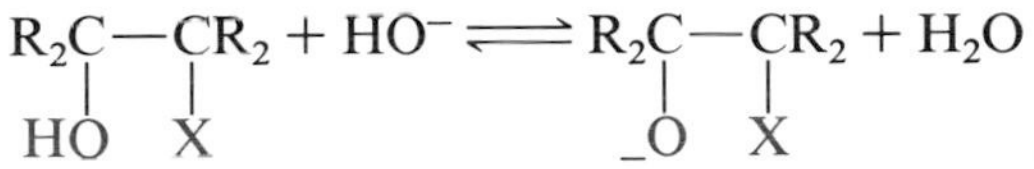

$$R_2C{-}CR_2 + HO^- \rightleftharpoons R_2C{-}CR_2 + H_2O$$
$$HO \quad X \qquad\qquad\quad {_-}O \quad X$$

Halohydrin

This alkoxide contains both a nucleophile (the alkoxide oxygen) and a leaving group (the halogen) in close proximity to each other. In a step faster than the reaction of the halohydrin with any external nucleophile, the alkoxide oxygen attacks the carbon that bears the leaving group, giving an epoxide. As in other nucleophilic displacement reactions, the nucleophile approaches carbon from the side opposite the leaving group.

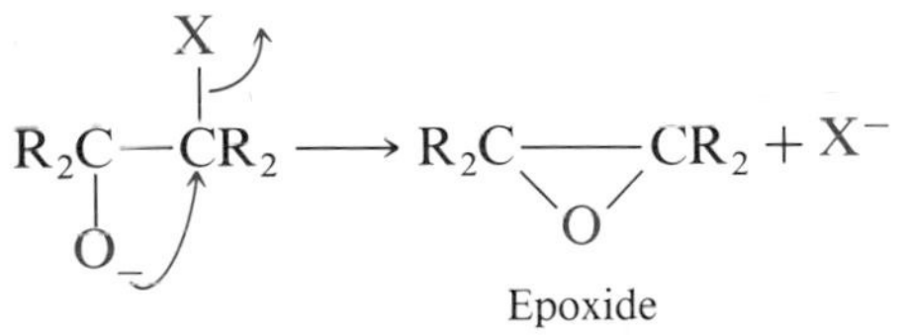

$$R_2C{-}CR_2 \longrightarrow R_2C{-}\!\!-\!\!{-}CR_2 + X^-$$

Epoxide

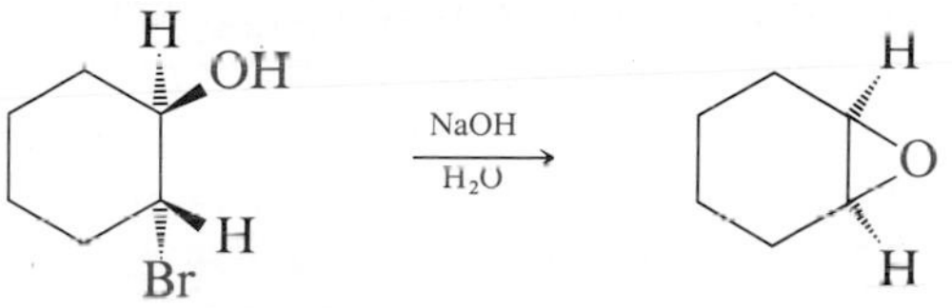

trans-2-Bromocyclohexanol   1,2-Epoxycyclohexane (81%)

Overall, the stereospecificity of this method for epoxide preparation is the same as that observed in peroxy acid oxidation of alkenes. Substituents that are cis to each other in the alkene remain cis in the epoxide. This is because the addition of hypobromous acid is an anti addition, and the ensuing intramolecular nucleophilic substitution reaction takes place with inversion of configuration at the carbon that bears the halogen.

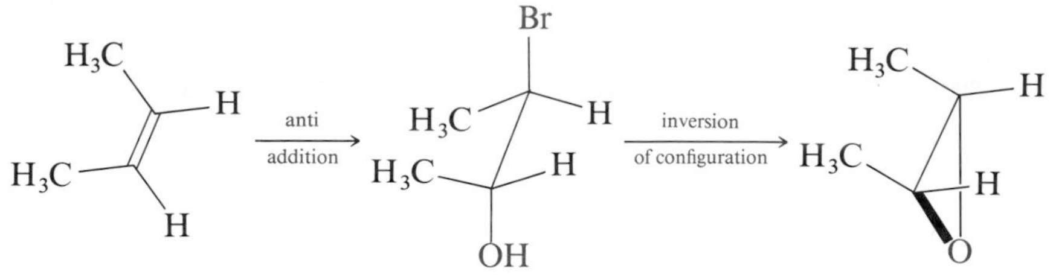

(Z)-2-Butene
(cis-2-butene)

cis-2,3-Epoxybutane

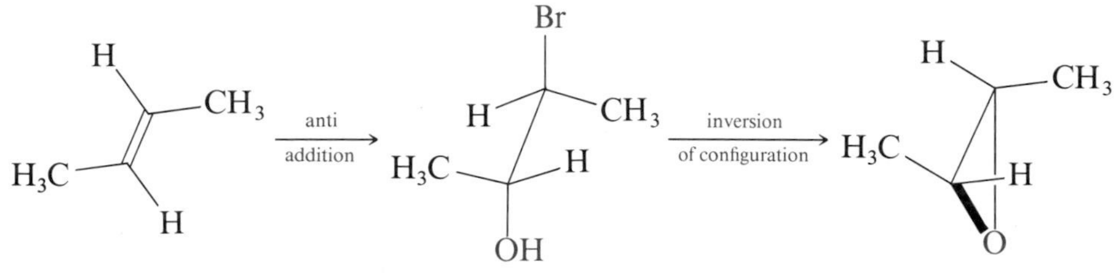

(E)-2-Butene
(trans-2-butene)

trans-2,3-Epoxybutane

PROBLEM 17.11 Is either of the epoxides formed in the above reactions chiral? Is either epoxide optically active when prepared from the alkene by this method?

About 2×10^9 lb/yr of 1,2-epoxypropane is produced in the United States as an intermediate in the preparation of various polymeric materials, including polyurethane plastics and foams and polyester resins. A large fraction of the 1,2-epoxypropane is made from propene through formation and base-promoted ring closure of the chlorohydrin.

17.13 REACTIONS OF EPOXIDES. A REVIEW AND A PREVIEW

The chemical property that most distinguishes epoxides is their far greater reactivity toward nucleophilic reagents compared with that exhibited by simple ethers. Epoxides react rapidly with nucleophiles under conditions in which other ethers are inert. This enhanced reactivity results from the ring strain of epoxides. Reactions that lead to ring opening relieve this strain and are energetically favored.

We saw an example of nucleophilic ring opening of epoxides in Section 16.6, where the reaction of Grignard reagents with ethylene oxide was described as a synthetic route to primary alcohols.

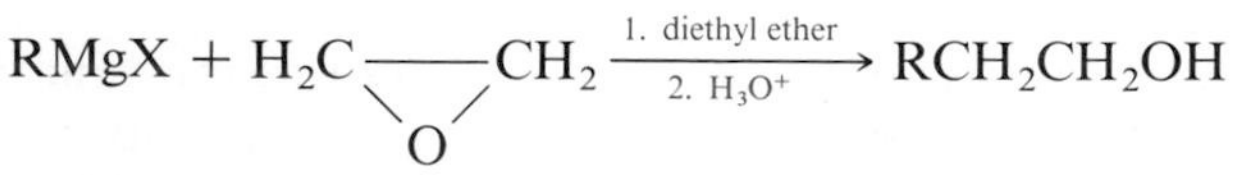

$$RMgX + H_2C \underset{O}{\overset{}{\diagdown\diagup}} CH_2 \xrightarrow[\text{2. } H_3O^+]{\text{1. diethyl ether}} RCH_2CH_2OH$$

Grignard   Ethylene oxide   Primary alcohol
reagent

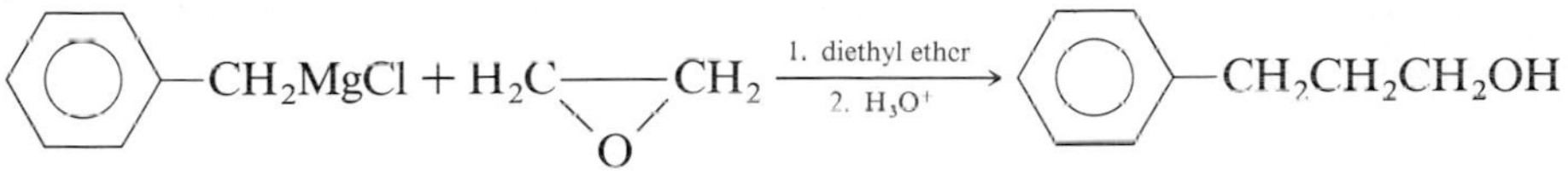

$$\text{Benzylmagnesium chloride} \quad \text{—CH}_2\text{MgCl} + \text{H}_2\text{C}\text{——}\text{CH}_2 \xrightarrow[\text{2. H}_3\text{O}^+]{\text{1. diethyl ether}} \quad \text{—CH}_2\text{CH}_2\text{CH}_2\text{OH}$$

Benzylmagnesium   Ethylene oxide   3-Phenylethanol (71%)
chloride

Nucleophiles other than Grignard reagents also lead to ring opening of epoxides. There are two fundamental ways in which these reactions are carried out. The first to be discussed (Section 17.14) involves anionic nucleophiles and leads to an alkoxide ion.

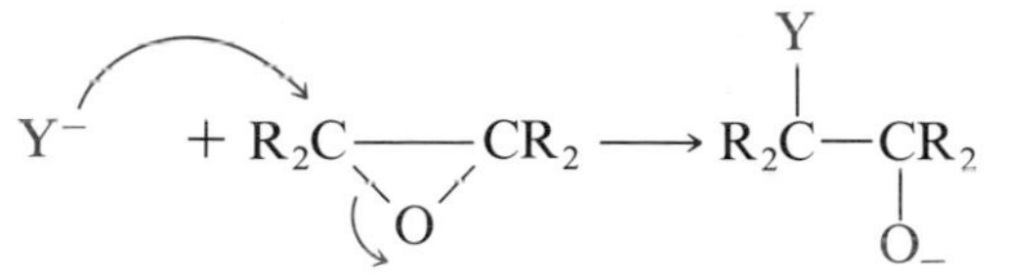

$$Y^- \quad + \quad R_2C\text{——}CR_2 \longrightarrow R_2\overset{\displaystyle Y}{\underset{\displaystyle O_-}{C}}\text{—}CR_2$$

Nucleophile   Epoxide   Alkoxide ion

These reactions are usually performed in water or alcohols as solvents, and the alkoxide ion produced by nucleophilic attack is rapidly transformed to an alcohol by proton transfer.

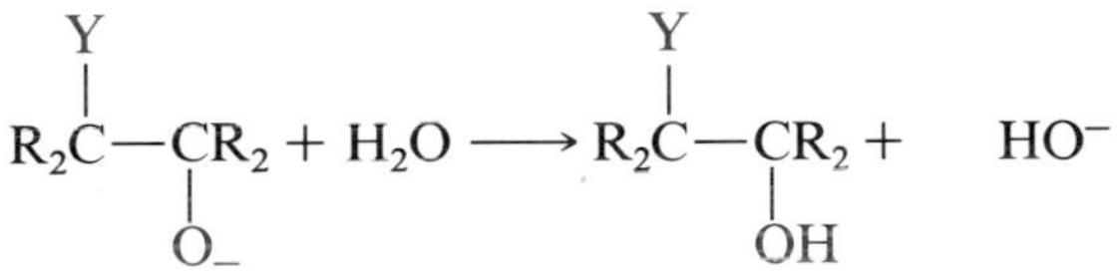

$$R_2\overset{\displaystyle Y}{\underset{\displaystyle O_-}{C}}\text{—}CR_2 + H_2O \longrightarrow R_2\overset{\displaystyle Y}{\underset{\displaystyle OH}{C}}\text{—}CR_2 + \quad HO^-$$

Alkoxide   Water   Alcohol   Hydroxide ion

Nucleophilic ring-opening reactions of epoxides are often carried out under conditions of acid catalysis. Here the nucleophile is not an anion but rather a solvent molecule.

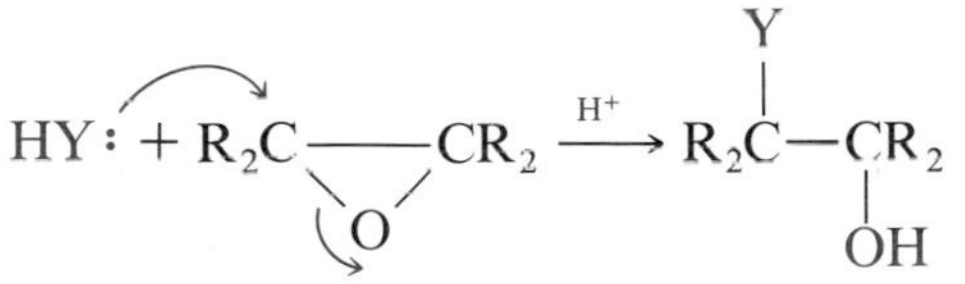

$$HY: + R_2C\text{——}CR_2 \xrightarrow{H^+} R_2\overset{\displaystyle Y}{\underset{\displaystyle OH}{C}}\text{—}CR_2$$

Epoxide   Alcohol

Acid-catalyzed ring opening of epoxides is discussed in Section 17.15.

There is an important difference in the regiochemistry of ring-opening reactions of epoxides depending on the reaction conditions. Unsymmetrically substituted epoxides tend to react with anionic nucleophiles at the less hindered carbon of the ring. Under conditions of acid catalysis, however, the more highly substituted carbon is attacked.

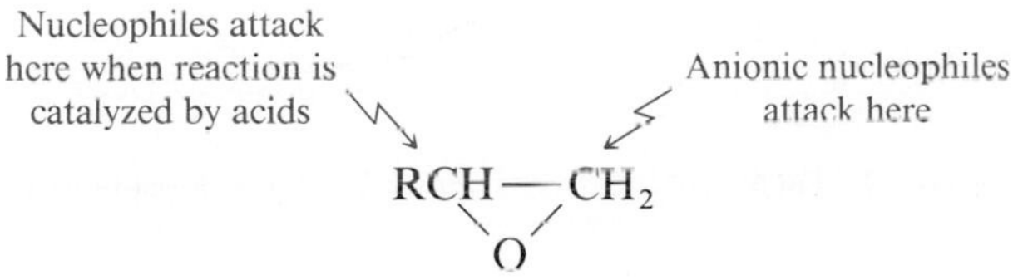

Nucleophiles attack
here when reaction is   Anionic nucleophiles
catalyzed by acids   attack here

$$RCH\text{——}CH_2$$
$$\underset{O}{}$$

The underlying reasons for this difference in site of nucleophilic attack will be explained in Section 17.15.

17.14   NUCLEOPHILIC RING-OPENING REACTIONS OF EPOXIDES

Ethylene oxide is a very reactive substance. It reacts readily, even exothermically, with anionic nucleophiles to yield 2-substituted derivatives of ethanol by cleaving the carbon-oxygen bond of the ring.

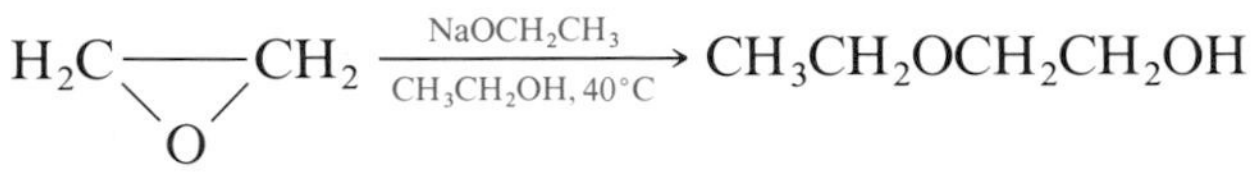

$$H_2C\overset{\displaystyle\diagdown}{\underset{O}{}}CH_2 \xrightarrow[CH_3CH_2OH,\,40°C]{NaOCH_2CH_3} CH_3CH_2OCH_2CH_2OH$$

Ethylene oxide   2-Ethoxyethanol (50%)
(oxirane)

$$H_2C\overset{\displaystyle\diagdown}{\underset{O}{}}CH_2 \xrightarrow[ethanol-water,\,0°C]{KSCH_2CH_2CH_2CH_3} CH_3CH_2CH_2CH_2SCH_2CH_2OH$$

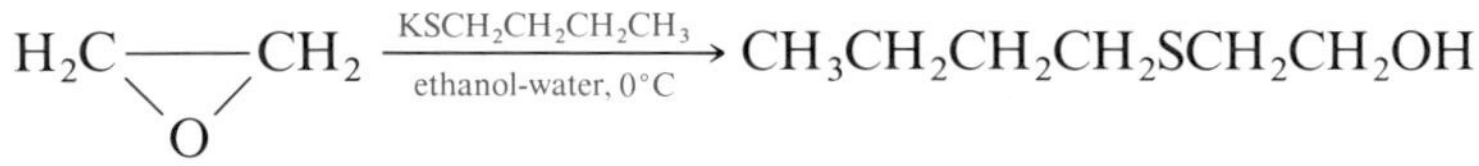

Ethylene oxide   2-(Thiobutyl)ethanol (99%)
(oxirane)

PROBLEM 17.12   What is the principal organic product formed in the reaction of ethylene oxide with each of the following?

(a)   Sodium cyanide (NaCN) in aqueous ethanol
(b)   Sodium azide (NaN$_3$) in aqueous ethanol
(c)   Sodium hydroxide (NaOH) in water
(d)   Phenyllithium (C$_6$H$_5$Li) in ether, followed by addition of dilute sulfuric acid
(e)   1-Butynylsodium (CH$_3$CH$_2$C≡CNa) in liquid ammonia

SAMPLE SOLUTION   (a)   Sodium cyanide is a source of the nucleophilic cyanide anion. Cyanide ion attacks ethylene oxide, opening the ring and forming 2-cyanoethanol.

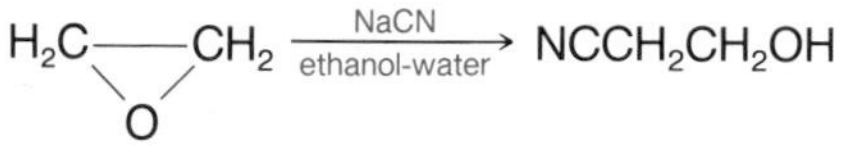

$$H_2C\overset{\displaystyle\diagdown}{\underset{O}{}}CH_2 \xrightarrow[ethanol-water]{NaCN} NCCH_2CH_2OH$$

Ethylene oxide   2-Cyanoethanol

Nucleophilic ring opening of epoxides has many of the features of an S$_N$2 reaction. Inversion of configuration is observed at the carbon at which substitution occurs.

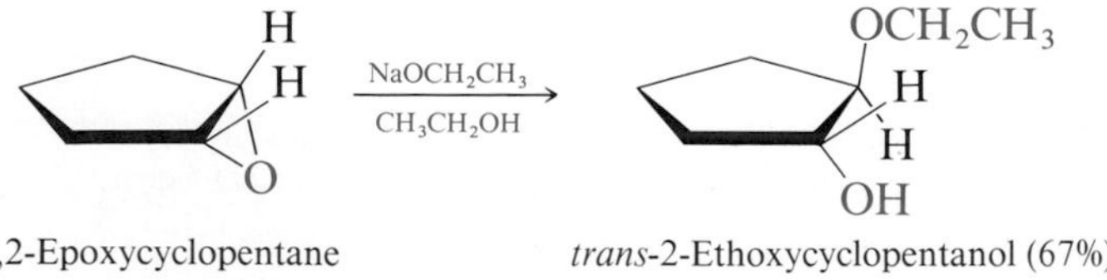

1,2-Epoxycyclopentane   *trans*-2-Ethoxycyclopentanol (67%)

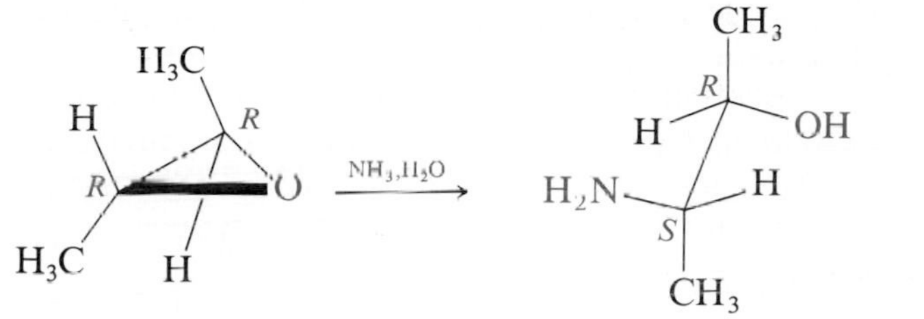

(2R,3R)-2,3-Epoxybutane → (2R,3S)-3-Amino-2-butanol (70%)

Unsymmetrical epoxides are preferentially attacked at the less substituted, less sterically hindered carbon of the ring.

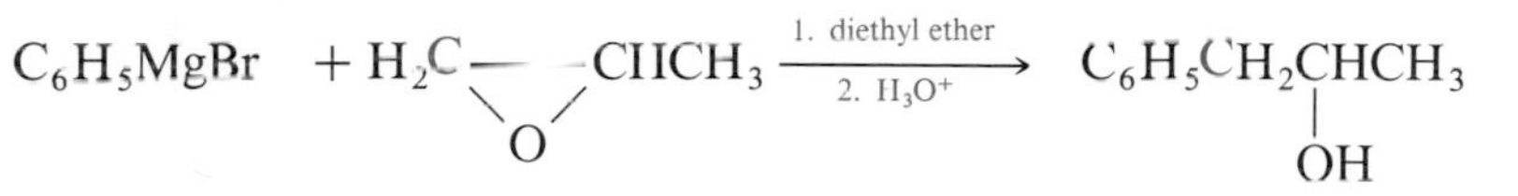

$$C_6H_5MgBr + H_2C\text{---}CHCH_3 \xrightarrow[\text{2. } H_3O^+]{\text{1. diethyl ether}} C_6H_5CH_2CHCH_3$$

Phenylmagnesium bromide   1,2-Epoxypropane   1-Phenyl-2-propanol (60%)

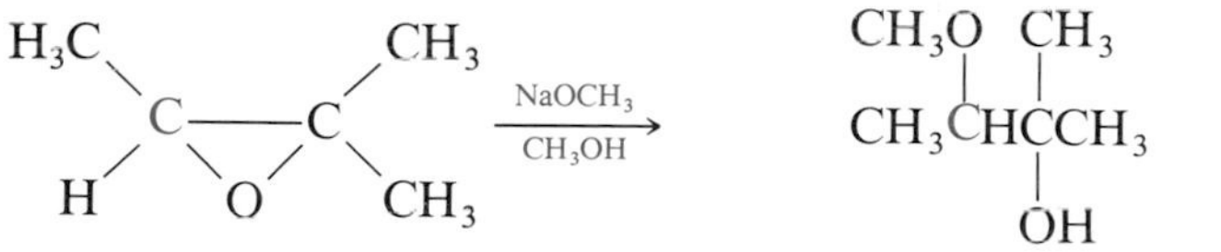

2,2,3-Trimethyloxirane $\xrightarrow[\text{CH}_3\text{OH}]{\text{NaOCH}_3}$ 3-Methoxy-2-methyl-2-butanol (53%)

PROBLEM 17.13 Given the starting material 1-methyl-1,2-epoxycyclopentane, of absolute configuration as shown, decide which one of the compounds A through C correctly represents the product of its reaction with sodium methoxide in methanol.

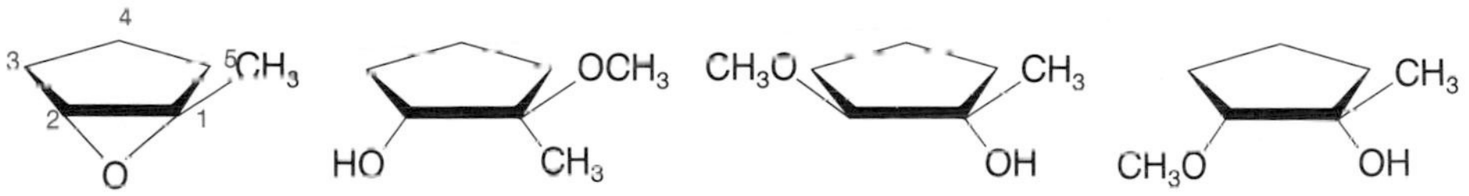

1,2-Epoxy-1-methylcyclopentane   Compound A   Compound B   Compound C

Application of the fundamental principles of nucleophilic substitution reactions leads to the mechanistic picture of epoxide ring opening shown in Figure 17.3. The

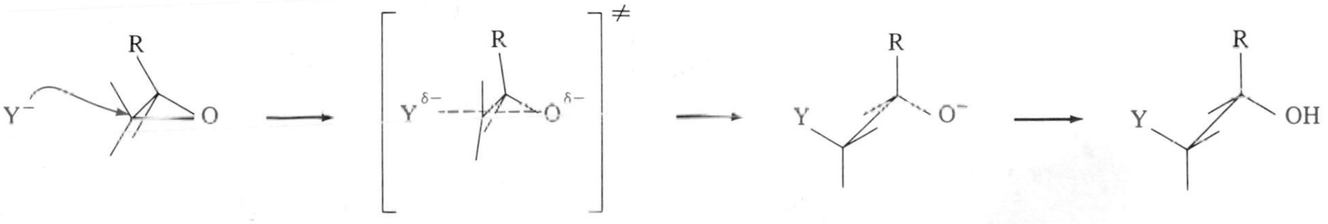

Nucleophile   Epoxide   Activated complex   Alkoxide ion   β-Substituted alcohol

FIGURE 17.3 Nucleophilic ring opening of an epoxide.

nucleophile attacks the less crowded carbon from the side opposite the carbon-oxygen bond. Bond formation with the nucleophile accompanies carbon-oxygen bond breaking, and a substantial portion of the strain in the three-membered ring is relieved as it begins to open in the activated complex. The initial product of nucleophilic substitution is an alkoxide anion, which rapidly abstracts a proton from the solvent to give a β-substituted alcohol as the isolated product.

17.15 ACID-CATALYZED RING-OPENING REACTIONS OF EPOXIDES

Epoxides react with nucleophiles under conditions of acid catalysis to yield ring-opened substitution products. One industrial preparation of ethylene glycol proceeds by hydrolysis of ethylene oxide in dilute sulfuric acid.

$$H_2C\underset{O}{\overset{\textstyle\diagdown\diagup}{\rule{1.2cm}{0.4pt}}}CH_2 + H_2O \xrightarrow[60°C]{H_2SO_4} HOCH_2CH_2OH$$

Ethylene oxide   Ethylene glycol
(epoxyethane)   (1,2-ethanediol)

Other nucleophiles react with ethylene oxide in an analogous manner to yield 2-substituted derivatives of ethanol.

$$H_2C\underset{O}{\overset{\textstyle\diagdown\diagup}{\rule{1.2cm}{0.4pt}}}CH_2 \xrightarrow[10°C]{HBr} BrCH_2CH_2OH$$

Ethylene oxide   2-Bromoethanol (87–92%)

$$H_2C\underset{O}{\overset{\textstyle\diagdown\diagup}{\rule{1.2cm}{0.4pt}}}CH_2 \xrightarrow[H_2SO_4, 25°C]{CH_3CH_2OH} CH_3CH_2OCH_2CH_2OH$$

Ethylene oxide   2-Ethoxyethanol (85%)

In acid-catalyzed nucleophilic ring opening of epoxides, the first step is protonation of the ring oxygen by the acid catalyst.

$$H_2C\underset{\ddot{O}\,\ddot{}}{\overset{\textstyle\diagdown\diagup}{\rule{1cm}{0.4pt}}}CH_2 + H^+ \underset{}{\overset{fast}{\rightleftharpoons}} H_2C\underset{\underset{H}{\overset{|}{O_+}}}{\overset{\textstyle\diagdown\diagup}{\rule{1cm}{0.4pt}}}CH_2$$

Ethylene oxide   Ethyleneoxonium ion

Attack by the nucleophile is rate-determining. When epoxides react with water or alcohols in the presence of an acid, the nucleophile is a weak one and a fair measure of positive character develops at the ring carbon in the activated complex. Breaking of the ring carbon-oxygen bond is more advanced than formation of the bond to the nucleophile in the transition state.

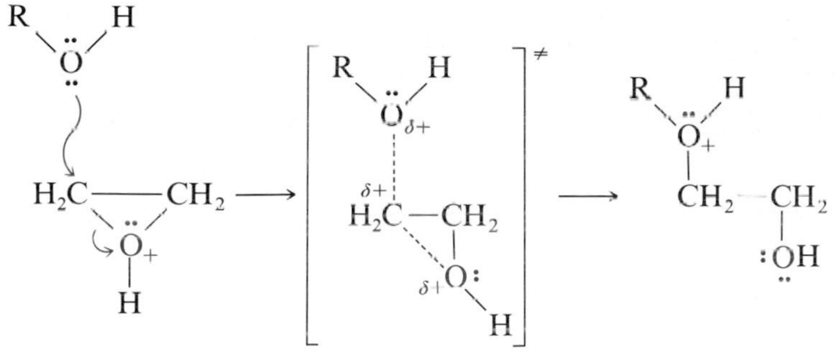

Activated complex

The acid catalyst is regenerated in the last step.

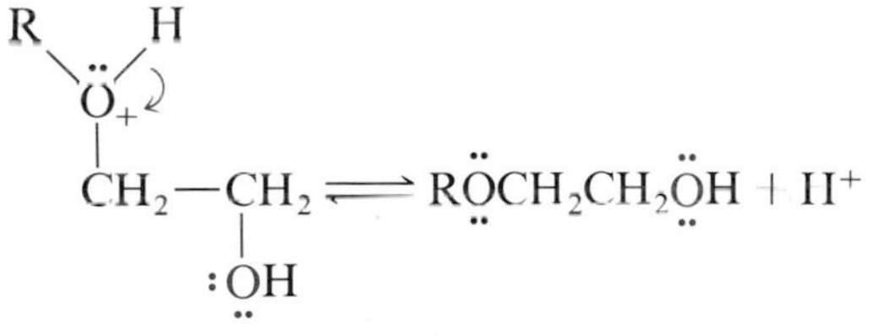

Because carbocation character develops at the carbon that has its bond to the ring oxygen broken in the rate-determining step, substitution is favored at the carbon that can better support a developing positive charge. Thus, in contrast to the reaction of epoxides with relatively basic nucleophiles, in which S_N2-like attack is faster at the less crowded carbon of the three-membered ring, acid catalysis promotes substitution at the position that can better stabilize a positive charge. Usually this is the ring carbon that bears the greater number of alkyl groups.

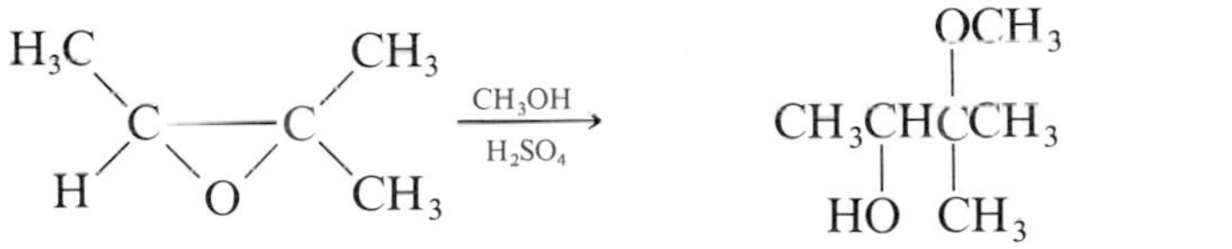

2,2,3-Trimethyloxirane   3-Methoxy-3-methyl-2-butanol (76%)

Aryl substituents stabilize carbocations effectively, making the benzylic position the one that is substituted when styrene oxide undergoes nucleophilic ring opening.

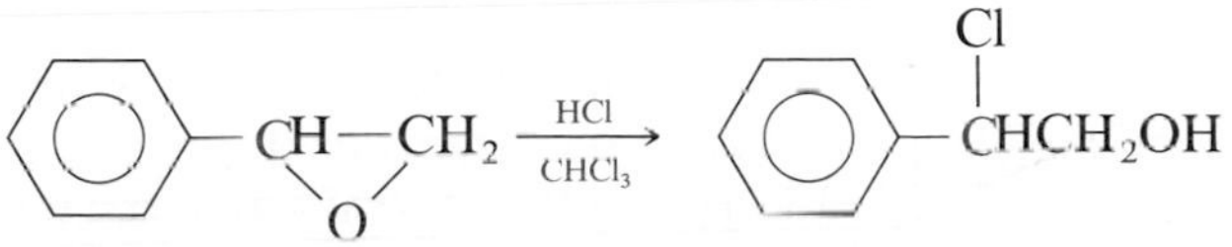

1,2-Epoxy-1-phenylethane   2-Chloro-2-phenylethanol (71%)
(styrene oxide)

While nucleophilic participation in the transition state is slight, it is enough to ensure that substitution proceeds with inversion of configuration.

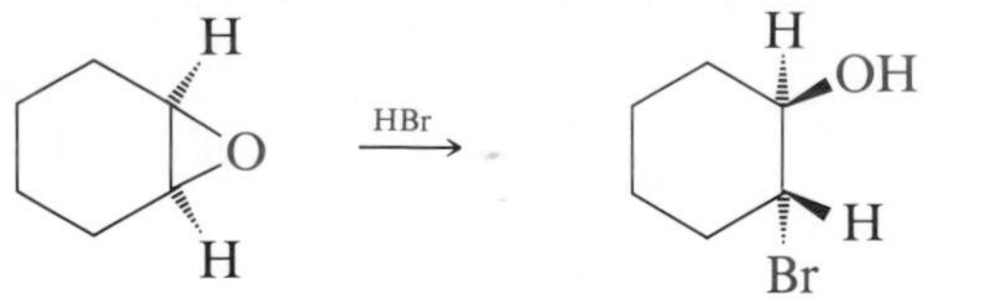

1,2-Epoxycyclohexane (cyclohexene oxide) — *trans*-2-Bromocyclohexanol (73%)

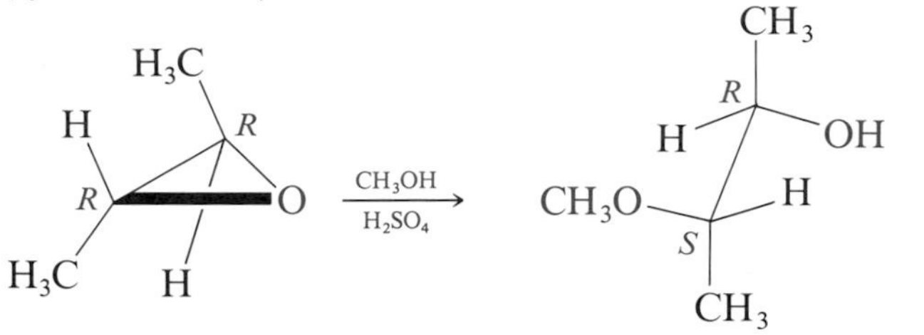

(2*R*,3*R*)-2,3-Epoxybutane — (2*R*,3*S*)-3-Methoxy-2-butanol (57%)

PROBLEM 17.14 Which product, compound A, B, or C, would you expect to be formed when 1-methyl-1,2-epoxycyclopentane of the absolute configuration shown is allowed to stand in methanol containing a few drops of sulfuric acid?

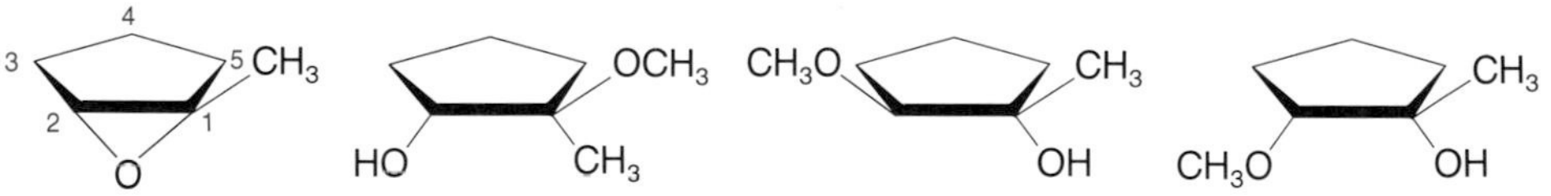

1,2-Epoxy-1-methylcyclopentane — Compound A — Compound B — Compound C

A method for achieving net anti hydroxylation of alkenes combines two stereospecific processes: epoxidation of the double bond and hydrolysis of the derived epoxide.

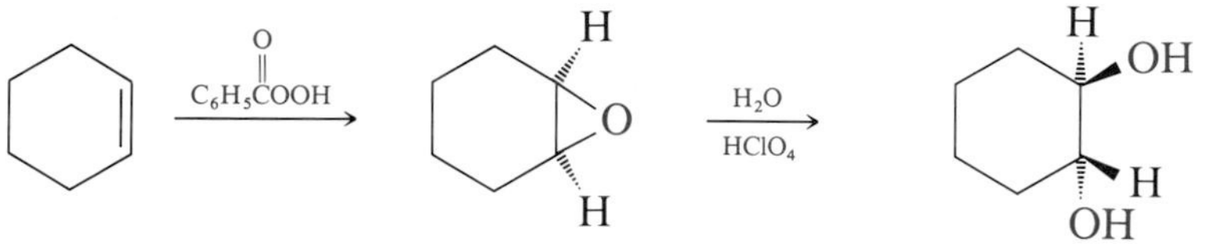

Cyclohexene — 1,2-Epoxycyclohexane — *trans*-1,2-Cyclohexanediol (80%)

17.16 EPOXIDES IN BIOLOGICAL PROCESSES

Many naturally occurring substances are epoxides. You have seen two examples of such compounds already in disparlure, the sex attractant of the gypsy moth (Section 7.15), and in the carcinogenic epoxydiol formed from benz[a]pyrene (Section 12.13). In most cases, epoxides are biosynthesized by the enzyme-catalyzed transfer of one of the oxygen atoms of an O_2 molecule to an alkene. Since only one of the atoms of O_2 is transferred to the substrate, the enzymes that catalyze such transfers are classified as *monooxygenases*. A biological reducing agent, usually NADH (Section 16.13), is required as well.

$$R_2C{=}CR_2 + O_2 + H^+ + NADH \xrightarrow{\text{enzyme}} R_2C\underset{O}{\diagdown\!\diagup}CR_2 + H_2O + NAD^+$$

A prominent example of such a reaction is the biological epoxidation of the polyene hydrocarbon squalene.

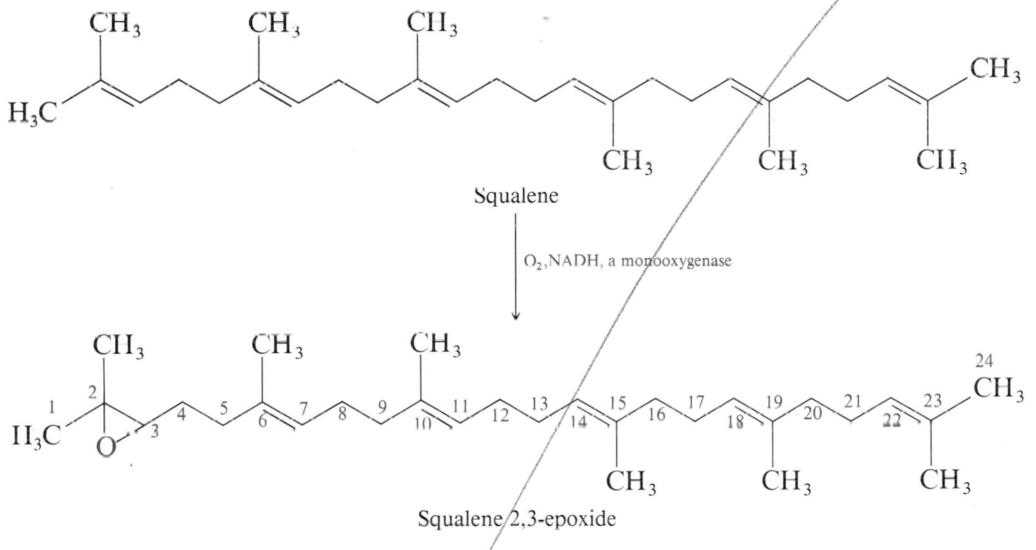

Squalene

O_2,NADH, a monooxygenase

Squalene 2,3-epoxide

The reactivity of epoxides toward nucleophilic ring opening is responsible for one of the biological roles they play. Squalene 2,3-epoxide, for example, is the biological precursor to cholesterol and the steroid hormones, including testosterone, progesterone, estrone, and cortisone. The pathway from squalene 2,3-epoxide to these compounds will be described in Chapter 28. It is related, in its early stages, to the biosynthesis of certain plant materials called *triterpenes* by cyclization of squalene 2,3-epoxide.

The cyclization of squalene 2,3-epoxide is triggered by epoxide ring opening, accompanied by nucleophilic attack by the π electrons of the C-6–C-7 double bond. Participation of the C-10–C-11, C-14–C-15, and C-18–C-19 double bonds leads to a carbocation having a tetracyclic carbon skeleton. The reaction is enzyme catalyzed.

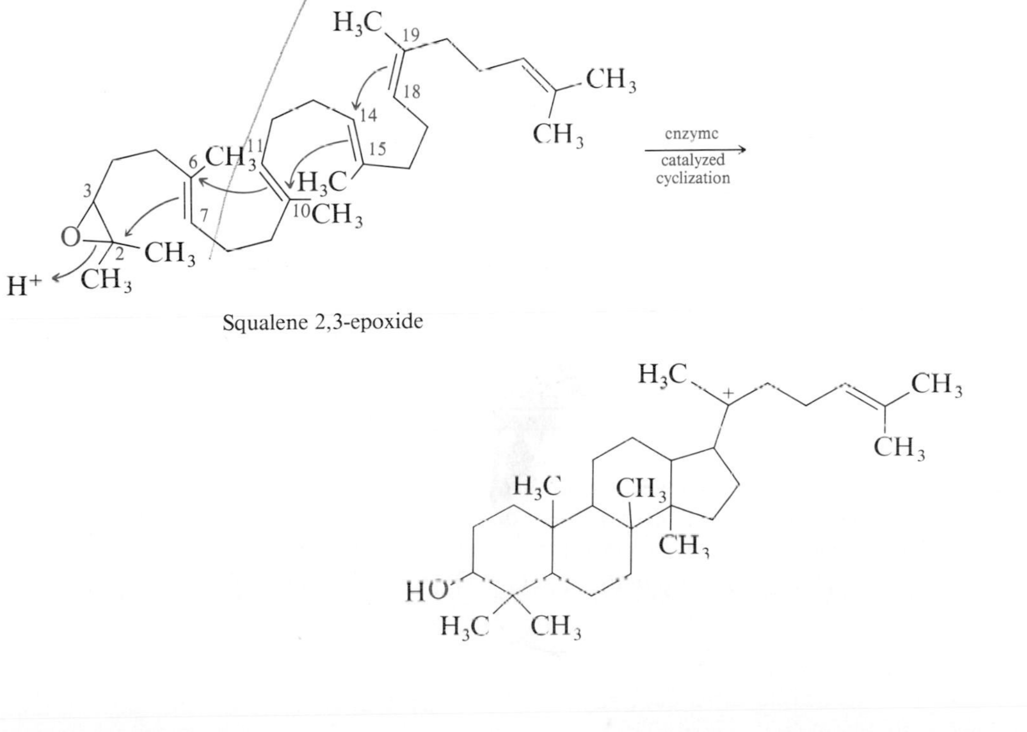

Squalene 2,3-epoxide

enzyme catalyzed cyclization

Loss of a proton from the cyclized carbocation produces a natural product known as dammaradienol, which is found in the resin of an East Indian tree.

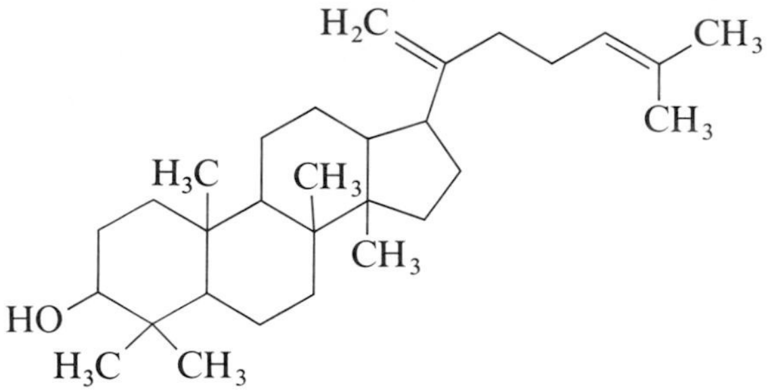

Dammaradienol

PROBLEM 17.15 The same trees that produce dammaradienol also yield the related substance dammarenediol. How might this compound arise biosynthetically?

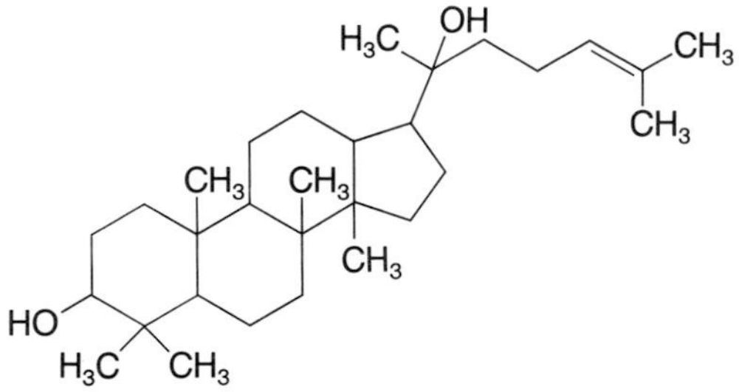

Dammarenediol

The cyclization process combines elements of epoxide ring opening (Sections 17.14–17.15) with the capture of carbocations by carbon-carbon double bonds (Section 7.18). When we discuss steroid biosynthesis in Chapter 28, you will see how carbocation rearrangements become involved and affect the structural outcome of the cyclization.

17.17 SPECTROSCOPIC ANALYSIS OF ETHERS

The infrared spectra of ethers are characterized by a strong, rather broad band due to C—O—C stretching between 1070 and 1150 cm^{-1}. Dialkyl ethers exhibit this band consistently at 1120 cm^{-1}, as illustrated in the infrared spectrum of dipropyl ether (Figure 17.4).

Infrared absorbances of epoxide rings, as shown for 1,2-epoxypropane in Figure 17.5, include three bands in the regions 750 to 840 cm^{-1}, 810 to 950 cm^{-1}, and 1250 cm^{-1}. Sometimes a weak band for the C—H stretch of an epoxide ring can be seen in the 3000 to 3050 cm^{-1} region.

The chemical shift of the proton in the H—C—O—C unit of an ether is very similar to that of the proton in the H—C—OH unit of an alcohol. A range $\delta = 3.3$ to

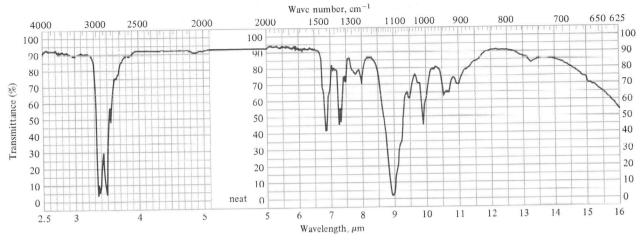

FIGURE 17.4 The infrared spectrum of dipropyl ether (CH₃CH₂CH₂OCH₂CH₂CH₃).

4.0 ppm is appropriate. In the ¹H nmr spectrum of dipropyl ether, shown in Figure 17.6, the assignment of signals to the various protons in the molecule is

$$\delta = 0.9 \text{ ppm} \longrightarrow \qquad \begin{array}{c} \delta = 1.5 \text{ ppm} \\ \end{array} \qquad \delta = 0.9 \text{ ppm}$$

$$\mathrm{CH_3CH_2CH_2OCH_2CH_2CH_3}$$

$$\delta = 3.4 \text{ ppm}$$

Protons attached directly to the three-membered ring of an epoxide are somewhat more shielded than those of a noncyclic HC—O—C unit. The chemical shift of epoxide ring protons is in the range $\delta = 2.2$ to 3.0 ppm.

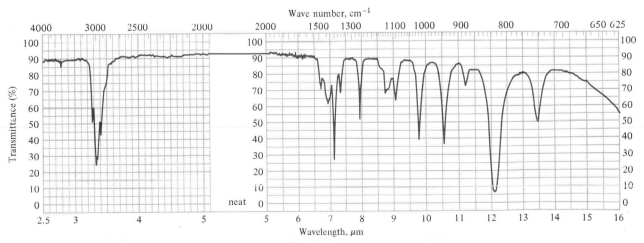

FIGURE 17.5 The infrared spectrum of 1,2-epoxypropane, CH₃CH—CH₂.

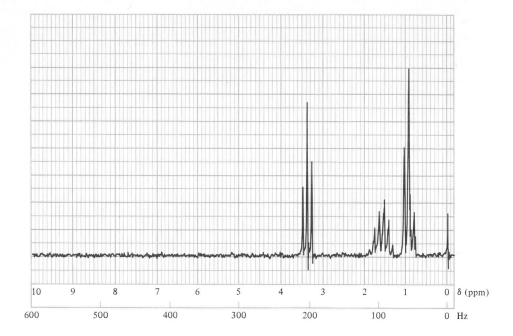

FIGURE 17.6 The ¹H nmr spectrum of dipropyl ether ($CH_3CH_2CH_2OCH_2CH_2CH_3$).

17.18 MASS SPECTROMETRY OF ETHERS

Ethers, like alcohols, lose an alkyl radical from their molecular ion to give an oxygen-stabilized cation. Thus, m/z 73 and m/z 87 are both more abundant than the molecular ion in the mass spectrum of *sec*-butyl ethyl ether.

$$\overset{\bullet +}{CH_3CH_2\overset{\bullet\bullet}{O}}-CHCH_2CH_3$$
$$|$$
$$CH_3$$

$$m/z\ 102$$

$$CH_3CH_2\overset{+}{\underset{\bullet\bullet}{O}}{=}CHCH_3 + \cdot CH_2CH_3 \qquad CH_3CH_2\overset{+}{\underset{\bullet\bullet}{O}}{=}CHCH_2CH_3 + \cdot CH_3$$
$$m/z\ 73 \qquad\qquad\qquad\qquad\qquad m/z\ 87$$

PROBLEM 17.16 There is another oxygen-stabilized cation of m/z 87 capable of being formed by fragmentation of the molecular ion in the mass spectrum of *sec*-butyl ethyl ether. Suggest a reasonable structure for this ion.

Rearrangement processes accompany fragmentation in the mass spectra of epoxides, making their mass spectrometric behavior more complicated than that of simple ethers.

17.19 SUMMARY

While ethers are commonplace substances in organic chemistry, there are only a few methods by which they are prepared, and the number of chemical reactions that they undergo is limited. Table 17.2 summarizes the preparation of ethers.

The only important reaction of ethers is their cleavage by hydrogen halides (Section 17.10).

$$ROR' + 2HX \longrightarrow RX + R'X + H_2O$$

| Ether | Hydrogen halide | Alkyl halide | Alkyl halide | Water |

The order of hydrogen halide reactivity is $HI > HBr > HCl \gg HF$.

$$\text{Benzyl ethyl ether} \quad CH_2OCH_2CH_3 \xrightarrow[\text{heat}]{HBr} \quad CH_2Br + CH_3CH_2Br$$

Benzyl ethyl ether   Benzyl bromide   Ethyl bromide

TABLE 17.2
Preparation of Ethers

Reaction (section) and comments	General equation and specific example
Acid-catalyzed condensation of alcohols (Sections 16.9 and 17.6) Two molecules of an alcohol condense in the presence of an acid catalyst to yield a dialkyl ether and water. The reaction is limited to the synthesis of symmetrical ethers from primary alcohols.	$2RCH_2OH \xrightarrow{H^+} RCH_2OCH_2R + H_2O$ Alcohol Ether Water $CH_3CH_2CH_2OH \xrightarrow[\text{heat}]{H_2SO_4} CH_3CH_2CH_2OCH_2CH_2CH_3$ Propyl alcohol Dipropyl ether
The Williamson ether synthesis (Section 17.7) An alkoxide ion displaces a halide or similar leaving group in an S_N2 reaction. The alkyl halide cannot be one that is prone to elimination, so this reaction is limited to primary alkyl halides. There is no limitation on the alkoxide ion that can be used.	$RO^- + R'CH_2X \longrightarrow ROCH_2R' + X^-$ Alkoxide ion Primary alkyl halide Ether Halide ion $(CH_3)_2CHCH_2ONa + CH_3CH_2Br \longrightarrow (CH_3)_2CHCH_2OCH_2CH_3 + NaBr$ Sodium isobutoxide Ethyl bromide Ethyl isobutyl ether (66%) Sodium bromide

Solvomercuration-demercuration (Section 17.8) Alkenes are converted to ethers by reaction with mercuric acetate in an alcohol solvent. The regioselectivity of addition follows Markovnikov's rule. Solvomercuration reactions involving tertiary alcohols are carried out in the presence of mercuric trifluoroacetate instead of mercuric acetate.	$R_2C{=}CHR \xrightarrow[\text{R'OH}]{Hg(O_2CCH_3)_2} R_2C\!\!-\!\!CHR \xrightarrow[\text{HO}^-]{NaBH_4} R_2C\!\!-\!\!CH_2R$
	                           R'O HgO_2CCH_3 R'O
	 Alkene Alkoxyalkylmercuric acetate Ether
	
	 $CH_3CH_2CH_2CH_2CH{=}CH_2 \xrightarrow[\substack{2.\ NaBH_4,\\ HO^-}]{\substack{1.\ Hg(OAc)_2,\\ CH_3CH_2OH}} CH_3CH_2CH_2CH_2CHCH_3$
	                                                 CH_3CH_2O
	 1-Hexene 2-Ethoxyhexane (98%)

TABLE 17.3
Preparation of Epoxides

Reaction (section) and comments	General equation and specific example
Peroxy acid oxidation of alkenes (Sections 7.15 and 17.11) Peroxy acids transfer oxygen to alkenes to yield epoxides. Stereospecific syn addition is observed.	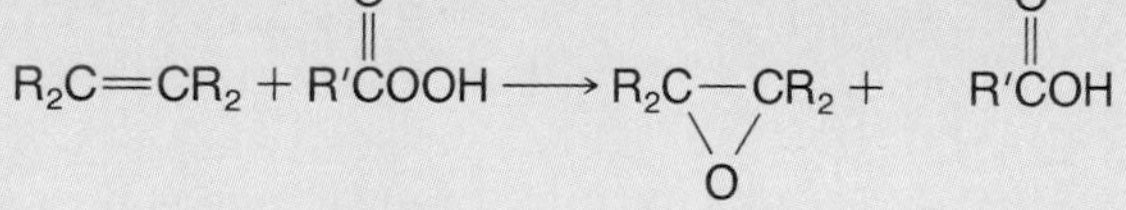 Alkene; Peroxy acid; Epoxide; Carboxylic acid 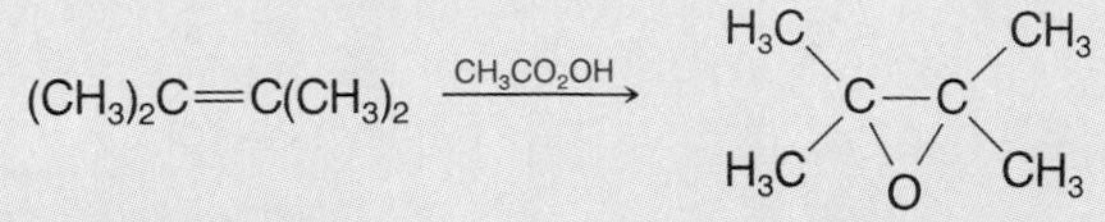 2,3-Dimethyl-2-butene; 2,2,3,3-Tetramethyloxirane (70–80%)
Base-promoted cyclization of vicinal halohydrins (Section 17.12) This reaction is an intramolecular version of the Williamson ether synthesis. The alcohol function of a vicinal halohydrin is converted to its conjugate base, which then displaces halide from the adjacent carbon to give an epoxide.	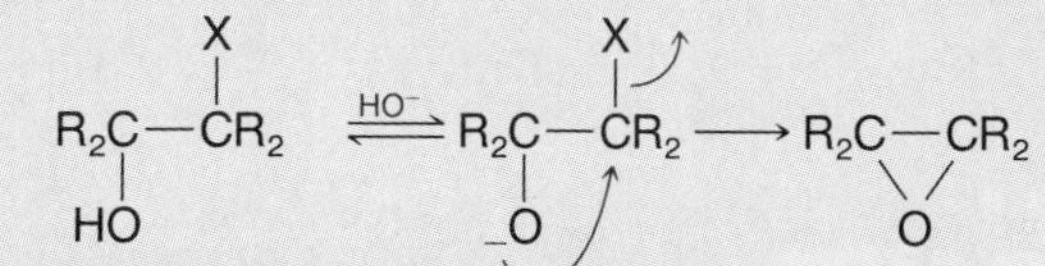 Vicinal halohydrin; Epoxide 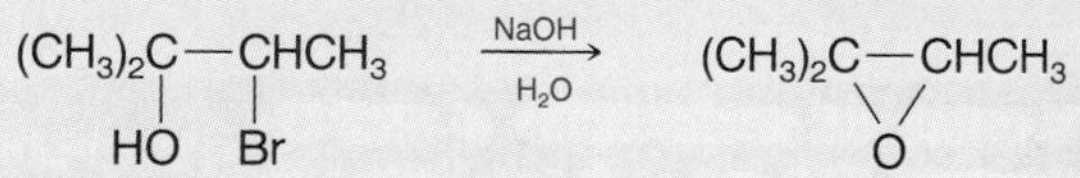 3-Bromo-2-methyl-2-butanol; 2,2,3-Trimethyloxirane (78%)

Epoxides are prepared by the methods listed in Table 17.3. Epoxides are much more reactive than ethers, especially in reactions that lead to cleavage of their three-membered ring. Relief of ring strain provides the driving force for these reactions. Anionic nucleophiles usually attack the less substituted carbon of the epoxide in an S_N2-like fashion.

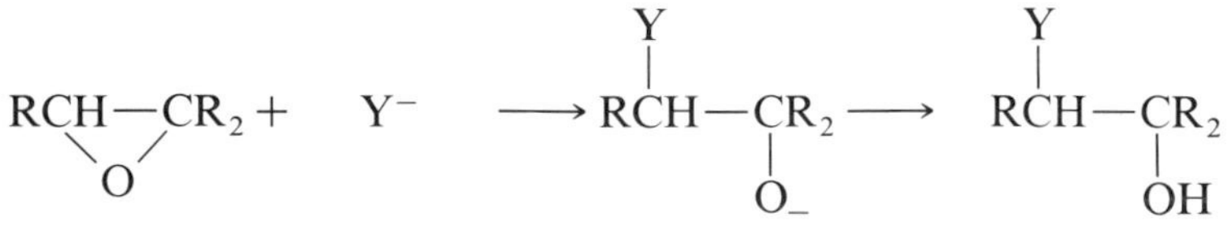

Epoxide Nucleophile *β*-Substituted alcohol

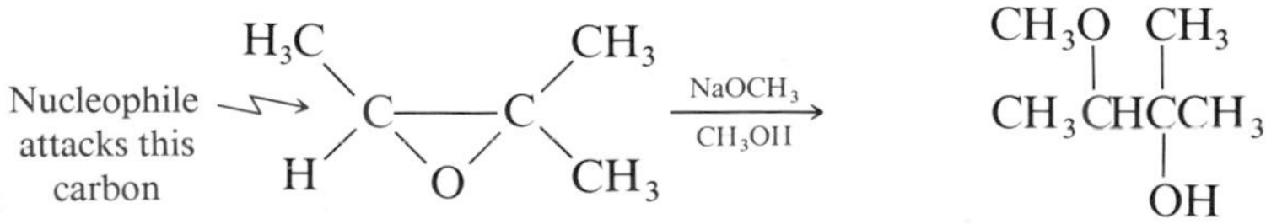

2,2,3-Trimethyloxirane 3-Methoxy-2-methyl-2-butanol (53%)

Under conditions of acid catalysis, nucleophiles attack the carbon that can better support a positive charge. Carbocation character is developed in the transition state.

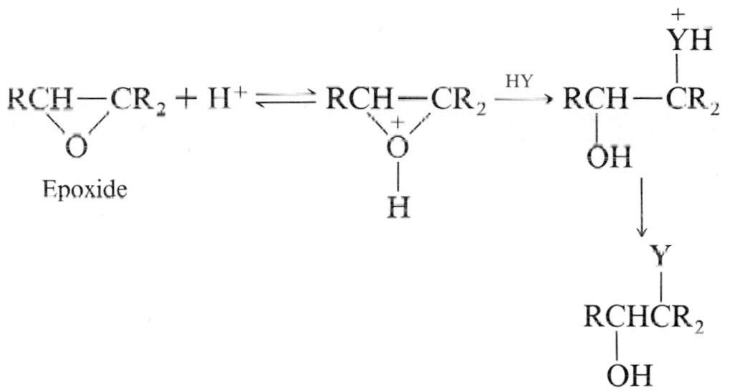

$$RCH-CR_2 + H^+ \rightleftharpoons RCH-CR_2 \xrightarrow{HY} RCH-CR_2$$

Epoxide

β-Substituted alcohol

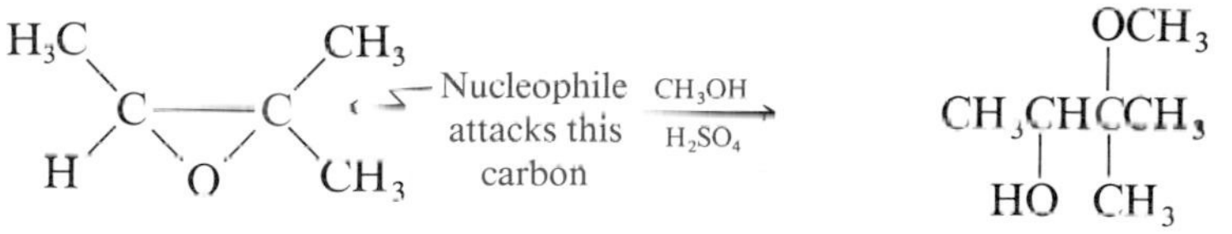

2,2,3-Trimethyloxirane

3-Methoxy-3-methyl-2-butanol (76%)

Inversion of configuration is observed at the carbon that undergoes nucleophilic attack, irrespective of whether the reaction occurs in acid or in base.

PROBLEMS

17.17 Write the structures of all the constitutionally isomeric ethers of molecular formula $C_5H_{12}O$ and give an acceptable name for each.

17.18 Many ethers, including diethyl ether, are effective as general anesthetics. Because simple ethers are quite flammable, their place in medical practice has been taken by highly halogenated nonflammable ethers. Two such general anesthetic agents are *isoflurane* and *enflurane*. These compounds are isomeric; isoflurane is 1-chloro-2,2,2-trifluoroethyl difluoromethyl ether while enflurane is 2-chloro-1,1,2-trifluoroethyl difluoromethyl ether. Write the structural formulas of isoflurane and enflurane.

17.19 While epoxides are always considered to have their oxygen atom as part of a three-membered ring, the prefix *epoxy* in the IUPAC system of nomenclature can be used to denote a cyclic ether of various sizes. Thus

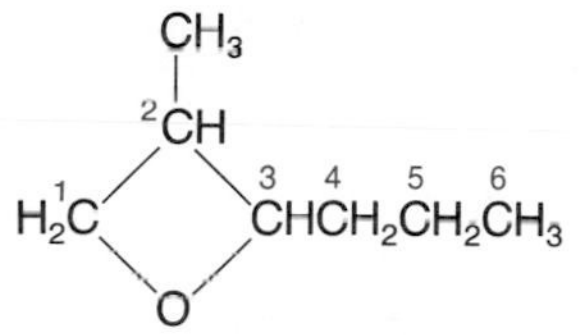

may be named 2-methyl-1,3-epoxyhexane. Using the epoxy prefix in this way, name each of the following compounds.

(a)

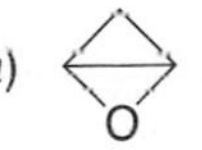

(b)

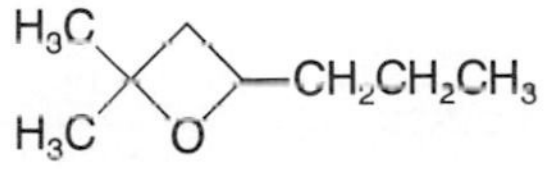

(c)

(d)

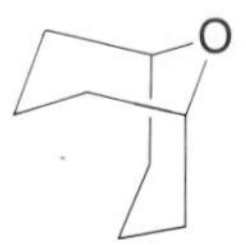

17.20 The IUPAC name for the parent six-membered oxygen-containing heterocycle is *oxane*. Oxane is a cyclic ether also known as tetrahydropyran. Numbering of the ring is sequential and begins with oxygen as atom 1. Multiple incorporation of oxygen in the ring is indicated by the prefixes di-, tri-, etc.

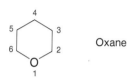

Oxane

(a) How many methyl-substituted oxanes are there? Which ones are chiral?
(b) Write structural formulas for 1,3-dioxane, 1,4-dioxane, and 1,3,5-trioxane.
(c) One dioxane isomer is not correctly classified as an ether. Which one? To what structural class does this compound belong?

17.21 The most stable conformation of 1,3-dioxan-5-ol is the chair form that has its hydroxyl group in an axial orientation. Suggest a reasonable explanation for this fact.

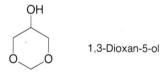

1,3-Dioxan-5-ol

17.22 Write chemical equations representing the preparation of 2-ethoxypentane by

(a) The Williamson ether synthesis
(b) Solvomercuration-demercuration

17.23 The octane rating of gasoline may be boosted by adding a small amount of *tert*-butyl methyl ether. Suggest a synthesis of this compound from methanol and 2-methylpropene.

17.24 Outline the steps in the preparation of each of the constitutionally isomeric ethers of molecular formula $C_4H_{10}O$, starting with the appropriate alcohols. Use the Williamson ether synthesis as your key reaction.

17.25 Predict the principal organic product of each of the following reactions. Specify stereochemistry where appropriate.

(a) $CH_3CH_2Br + CH_3CH_2CHCH_3 \longrightarrow$
$\qquad\qquad\qquad\qquad\qquad |$
$\qquad\qquad\qquad\qquad\quad ONa$

(b) 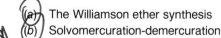 $-Br + CH_3CH_2CHCH_3 \longrightarrow$
$\qquad\qquad\qquad\qquad\qquad\qquad |$
$\qquad\qquad\qquad\qquad\qquad ONa$

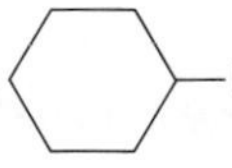

(c) $CH_3CH=CHCH_2Cl + (CH_3)_3COK \longrightarrow$

(d) $CH_3CH_2I +$
$\qquad\qquad\qquad\quad CH_3CH_2\underset{C_6H_5}{\overset{CH_3}{\underset{|}{\overset{|}{C}}}}-OK \longrightarrow$

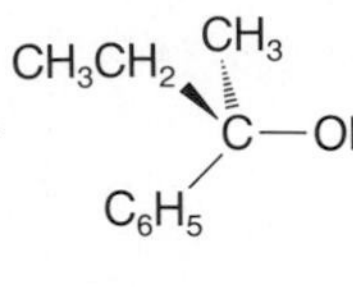

(e) 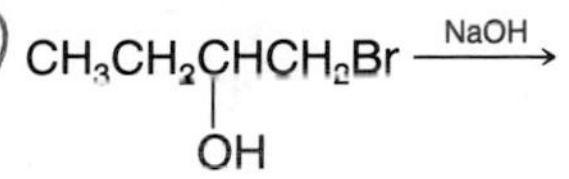 $CH_3CH_2CHCH_2Br$ $\xrightarrow{NaOH}$
            |
           OH

(f) 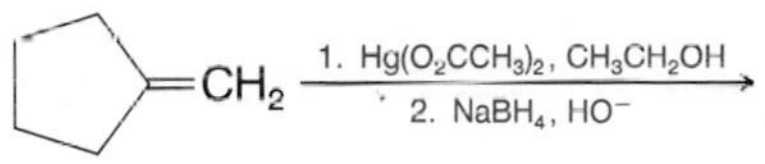 $=CH_2$ $\xrightarrow{\text{1. } Hg(O_2CCH_3)_2, CH_3CH_2OH}{\text{2. } NaBH_4, HO^-}$

(g) 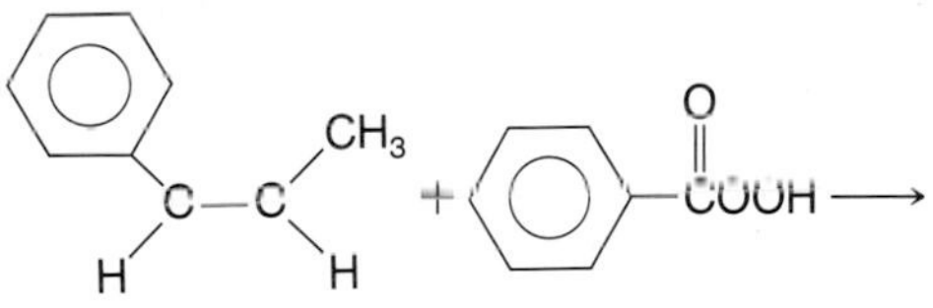 $C-C$ $+$ $-COOH \longrightarrow$

(h) 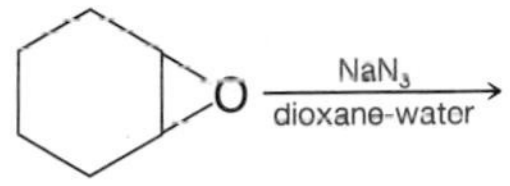 O $\xrightarrow{NaN_3}{\text{dioxane-water}}$

(i) 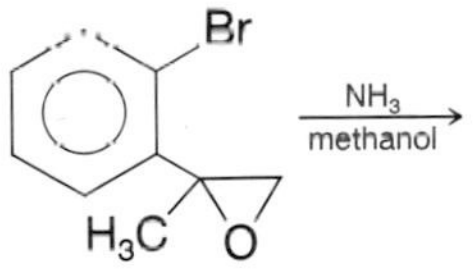 $\xrightarrow{NH_3}{\text{methanol}}$ (with Br, H_3C, O)

(j) 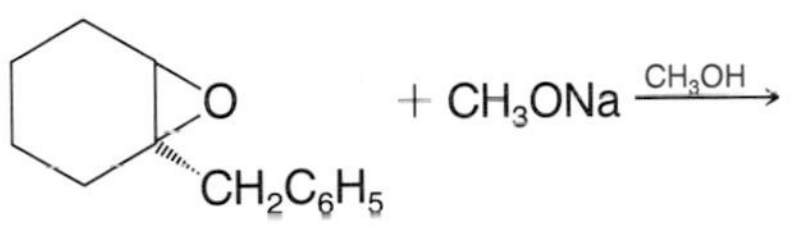 O $+ CH_3ONa$ $\xrightarrow{CH_3OH}$ ($CH_2C_6H_5$)

17.26 Select reaction conditions that would allow you to carry out each of the following stereospecific transformations:

(a) 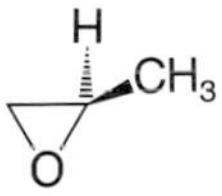   (epoxide with H, CH_3, O) $\longrightarrow$ (R)-1,2-propanediol

(b) 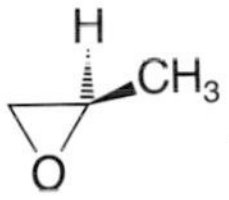   (epoxide with H, CH_3, O) $\longrightarrow$ (S)-1,2-propanediol

17.27 Suggest short, efficient reaction sequences suitable for preparing each of the following compounds from the given starting materials and any necessary organic or inorganic reagents:

(a) 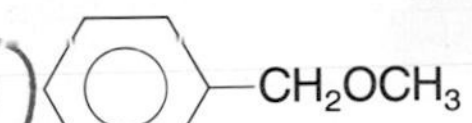 $-CH_2OCH_3$ from $-\overset{\overset{\displaystyle O}{\|}}{C}OCH_3$

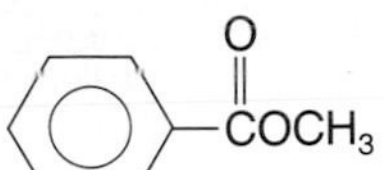

(b) 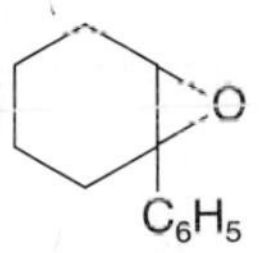   (cyclohexane epoxide with C_6H_5) from bromobenzene and cyclohexanol

(c) $C_6H_5CH_2CHCH_3$ from bromobenzene and isopropyl alcohol
              |
              Br

(d) $C_6H_5CH_2CH_2CH_2OCH_2CH_3$ from benzyl alcohol and ethanol

(e)  from 1,3-cyclohexadiene and ethanol

17.28 Among the ways in which 1,4-dioxane may be prepared are the routes expressed in the equations shown:

(a) 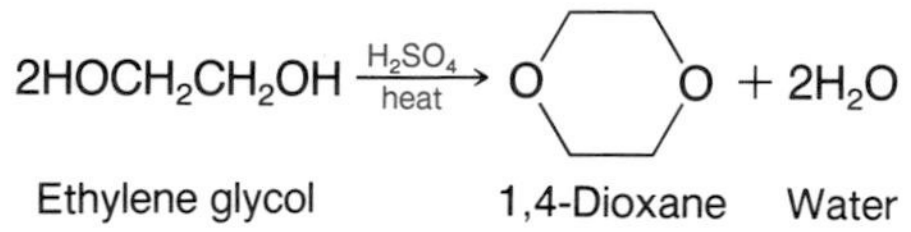 $2HOCH_2CH_2OH \xrightarrow[\text{heat}]{H_2SO_4}$ O⟨1,4-Dioxane⟩O $+ 2H_2O$

Ethylene glycol   1,4-Dioxane   Water

(b) 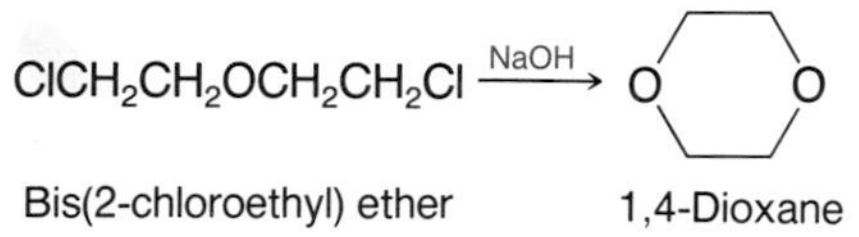 $ClCH_2CH_2OCH_2CH_2Cl \xrightarrow{NaOH}$ O⟨1,4-Dioxane⟩O

Bis(2-chloroethyl) ether   1,4-Dioxane

Suggest reasonable mechanisms for each of these reactions.

17.29 Deduce the identity of the missing compounds in the following reaction sequences:

(a) 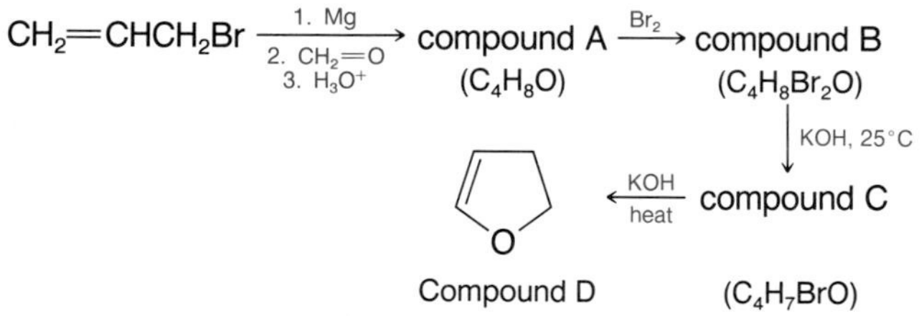 $CH_2{=}CHCH_2Br \xrightarrow[\substack{2.\ CH_2=O \\ 3.\ H_3O^+}]{1.\ Mg}$ compound A $\xrightarrow{Br_2}$ compound B
(C_4H_8O) $(C_4H_8Br_2O)$

$\Big\downarrow$ KOH, 25°C

Compound D $\xleftarrow[\text{heat}]{KOH}$ compound C

Compound D (C_4H_7BrO)

(b)

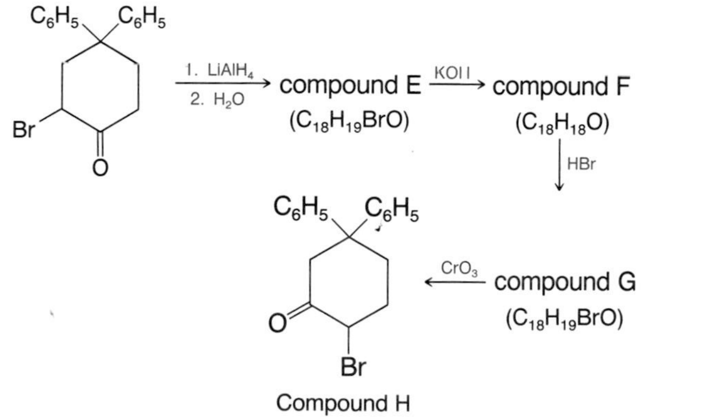

$\xrightarrow[\substack{2.\ H_2O}]{1.\ LiAlH_4}$ compound E $\xrightarrow{KOH}$ compound F
$(C_{18}H_{19}BrO)$ $(C_{18}H_{18}O)$

$\Big\downarrow$ HBr

compound G
$(C_{18}H_{19}BrO)$

$\xleftarrow{CrO_3}$

Compound H

(c)   Compound I $(C_7H_{12}) \xrightarrow[\text{acetone-water}]{KMnO_4}$ compound J $(C_7H_{14}O_2)$
                                                               (a liquid)

$\Big\downarrow C_6H_5CO_2OH$

Compound K $\xrightarrow[\text{H}_2\text{SO}_4]{H_2O}$ compound L $(C_7H_{14}O_2)$
                              (mp 99.5–101°C)

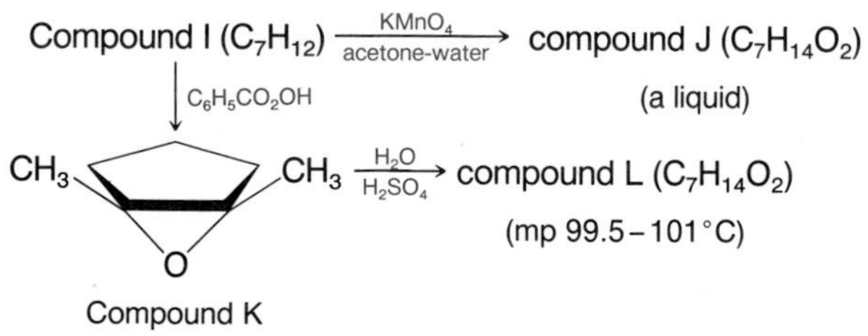

17.30 Cineole is the principal component of eucalyptus oil, has the molecular formula $C_{10}H_{18}O$, and contains no double or triple bonds. It reacts with hydrochloric acid to give the dichloride shown.

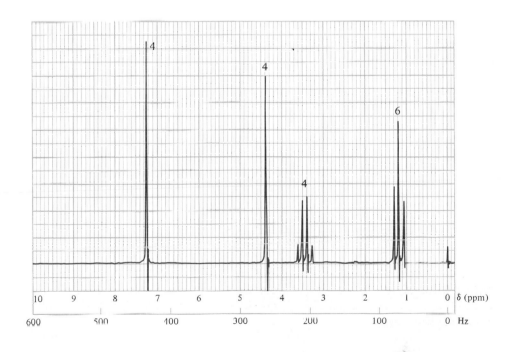

Deduce the structure of cineole.

17.31 All the following questions pertain to ^1H nmr spectra of isomeric ethers having the molecular formula $C_5H_{12}O$.

(a) Which one has only singlets in its ^1H nmr spectrum?

(b) Along with other signals, this ether has a coupled doublet-heptet pattern. None of the protons responsible for this pattern are coupled to protons anywhere else in the molecule. Identify this ether.

(c) In addition to other signals in its ^1H nmr spectrum, this ether exhibits two signals at relatively low field. One is a singlet; the other is a doublet. What is the structure of this ether?

(d) In addition to other signals in its ^1H nmr spectrum, this ether exhibits two signals at relatively low field. One is a triplet; the other is a quartet. Which ether is this?

17.32 The ^1H nmr spectrum of compound M (C_8H_8O) consists of two singlets of equal area at $\delta = 5.1$ (sharp) and 7.2 ppm (broad). On treatment with excess hydrogen bromide, compound M is converted to a single dibromide ($C_8H_8Br_2$). The ^1H nmr spectrum of the dibromide is similar to that of M in that it exhibits two singlets of equal area at $\delta = 4.7$ (sharp) and 7.3 ppm (broad). Suggest reasonable structures for compound M and the dibromide derived from it.

17.33 The ^1H nmr spectrum of compound N ($C_{12}H_{18}O_2$) is shown in Figure 17.7. Vigorous

FIGURE 17.7 The ^1H nmr spectrum of compound N, $C_{12}H_{18}O_2$ (Problem 17.33).

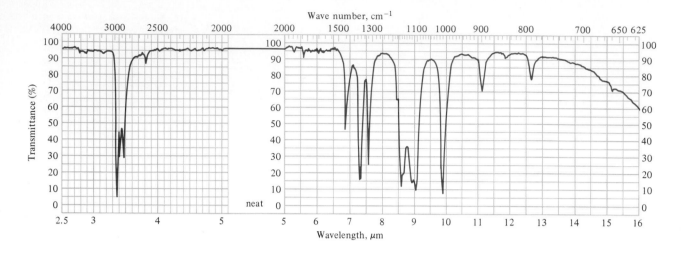

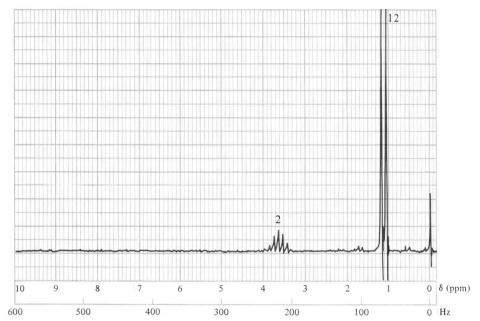

FIGURE 17.8   The infrared (top) and ^{1}H nmr (bottom) spectra of compound O, $C_6H_{14}O$ (Problem 17.34a).

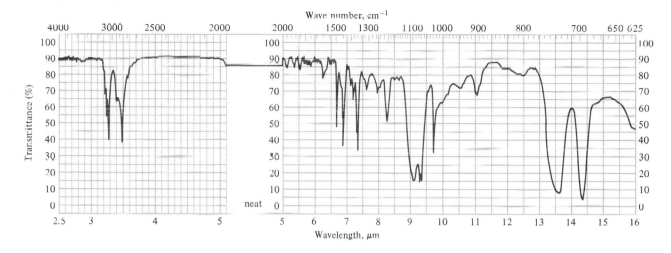

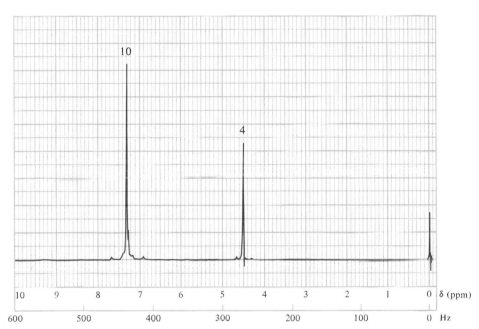

FIGURE 17.9 The infrared (top) and ^{1}H nmr (bottom) spectra of compound P, $C_{14}H_{14}O$ (Problem 17.34b).

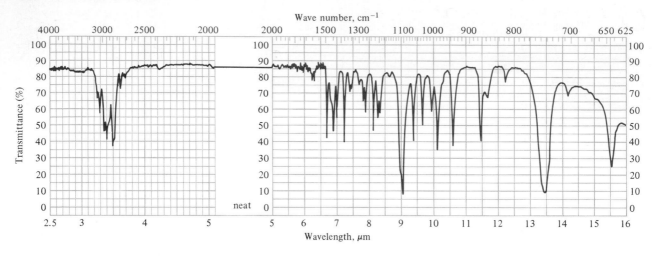

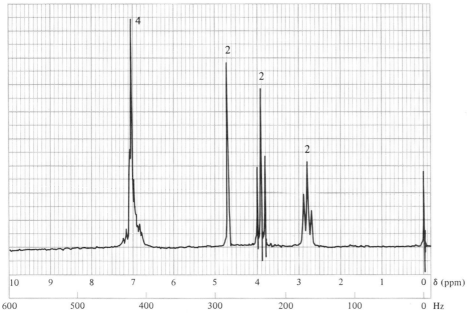

FIGURE 17.10   The infrared (top) and ^{1}H nmr (bottom) spectra of compound Q, $C_9H_{10}O$ (Problem 17.34c).

oxidation of compound N with potassium permanganate converts it to 1,4-benzenedicarboxylic acid.

$$\text{Compound N} \xrightarrow[\text{2. H}^+]{\text{1. KMnO}_4} HO_2C\!-\!\!\bigcirc\!\!-\!CO_2H$$

What is the structure of compound N?

17.34   Identify compounds O through Q on the basis of the information provided.

(a)   Compound O is $C_6H_{14}O$. Its infrared and ^{1}H nmr spectra are shown in Figure 17.8.

(b)   Compound P is $C_{14}H_{14}O$. Its infrared and ^{1}H nmr spectra are shown in Figure 17.9.

(c)   Compound Q is $C_9H_{10}O$. Its infrared and ^{1}H nmr spectra are shown in Figure 17.10.

ALDEHYDES AND KETONES. NUCLEOPHILIC ADDITION TO THE CARBONYL GROUP

CHAPTER

18

Aldehydes and ketones are characterized by the presence of an acyl group $RC\overset{\displaystyle O}{\overset{\|}{}}—$ bonded either to hydrogen or to another carbon.

$$\underset{\text{Aldehyde}}{\overset{\displaystyle O}{\overset{\|}{RCII}}} \qquad \underset{\text{Ketone}}{\overset{\displaystyle O}{\overset{\|}{RCR'}}}$$

While the present chapter includes the usual collection of topics that serve to acquaint us with a particular class of compounds, its central focus is on a fundamental reaction type, *nucleophilic addition to carbonyl groups.* The principles of nucleophilic addition to the carbonyl groups of aldehydes and ketones developed here will be seen to have broad applicability in later chapters, in which the transformations of various derivatives of carboxylic acids are discussed.

18.1 NOMENCLATURE

In naming aldehydes one identifies the longest continuous chain that contains the $—\overset{\displaystyle O}{\overset{\|}{C}}H$ group. This provides the base name. The *-e* ending of the corresponding alkane name is replaced by *-al* and substituents are specified in the usual way. It is not necessary to specify the location of the $—\overset{\displaystyle O}{\overset{\|}{C}}H$ group in the name since the chain must be numbered by starting with this group as C-1. The suffix *-dial* is added to the appropriate alkane name when the compound contains two aldehyde functions.

639

$$CH_3CH_2CH_2CH_2\overset{\overset{\displaystyle O}{\|}}{C}H$$

Pentanal

$$CH_3\overset{\overset{\displaystyle CH_3}{|}}{\underset{\underset{\displaystyle CH_3}{|}}{C}}CH_2CH_2\overset{\overset{\displaystyle O}{\|}}{C}H$$

4,4-Dimethylpentanal

$$CH_2{=}CHCH_2CH_2CH_2\overset{\overset{\displaystyle O}{\|}}{C}H$$

5-Hexenal

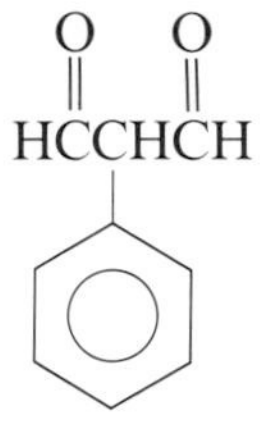

$$H\overset{\overset{\displaystyle O}{\|}}{C}CH\overset{\overset{\displaystyle O}{\|}}{C}H$$

2-Phenylpropanedial

When a formyl group ($-\overset{\overset{\displaystyle O}{\|}}{C}H$) is attached to a ring, the ring name is followed by the suffix -*carbaldehyde.*

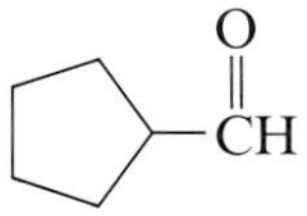

Cyclopentanecarbaldehyde

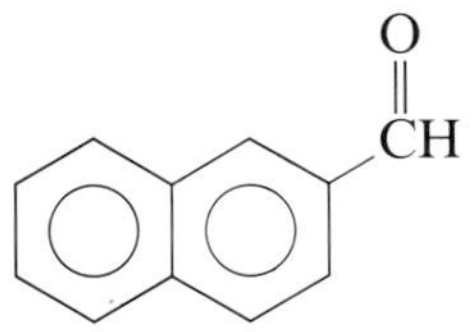

2-Naphthalenecarbaldehyde

(The suffix -*carboxaldehyde* is synonymous with -*carbaldehyde.*)

Certain common names of familiar aldehydes are acceptable as IUPAC names. A few examples include

$$H\overset{\overset{\displaystyle O}{\|}}{C}H$$

Formaldehyde
(methanal)

$$CH_3\overset{\overset{\displaystyle O}{\|}}{C}H$$

Acetaldehyde
(ethanal)

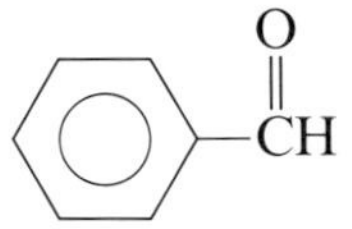

Benzaldehyde
(benzenecarbaldehyde)

These names are derived from the common names of the corresponding carboxylic acids by replacing the -*ic acid* or -*oic acid* suffix with -*aldehyde.*

PROBLEM 18.1   The common names and structural formulas of a few aldehydes are given below. Provide an alternative IUPAC name.

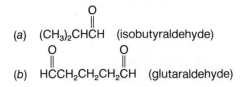

(*a*) $(CH_3)_2CH\overset{\overset{\displaystyle O}{\|}}{C}H$ (isobutyraldehyde)

(*b*) $H\overset{\overset{\displaystyle O}{\|}}{C}CH_2CH_2CH_2\overset{\overset{\displaystyle O}{\|}}{C}H$ (glutaraldehyde)

(c) $C_6H_5CH=CHCH$ (cinnamaldehyde)

with a carbonyl O above the last CH

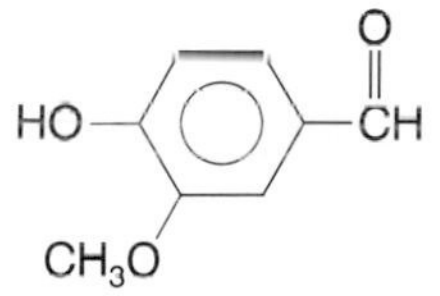

(d) HO—⟨ring⟩—CH (vanillin)

CH₃O

SAMPLE SOLUTION (a) Do not be fooled by the fact that the common name is iso-butyraldehyde. The longest continuous chain is three carbons, so the base name is propanal. There is a methyl group at C-2, so the compound is 2-methylpropanal.

$$\overset{3}{CH_3}\overset{2}{CH}\overset{1}{CH}$$ with O and CH₃

2-Methylpropanal
(isobutyraldehyde)

With ketones, the -*e* ending of an alkane is replaced by -*one* in the longest continuous chain containing the carbonyl group. The location of the carbonyl is specified by numbering the chain in the direction that provides the lower number for this group.

$CH_3CH_2CCH_2CH_2CH_3$ $CH_3CHCH_2CCH_3$ CH_3—⟨cyclohexane⟩=O

CH₃

3-Hexanone
(*not* 4-hexanone)

4-Methyl-2-pentanone (*not* 2-methyl-4-pentanone)

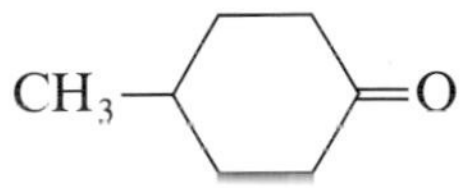

4-Methylcyclohexanone

Although systematic names are preferred, the IUPAC rules also permit ketones to be named by separately designating both the groups attached to the carbonyl followed by the word "ketone." The groups are listed alphabetically.

$CH_3CH_2CCH_2CH_2CH_3$ ⟨benzene⟩—$CH_2CCH_2CH_3$ $CH_2=CHCCH=CH_2$

Ethyl propyl
ketone

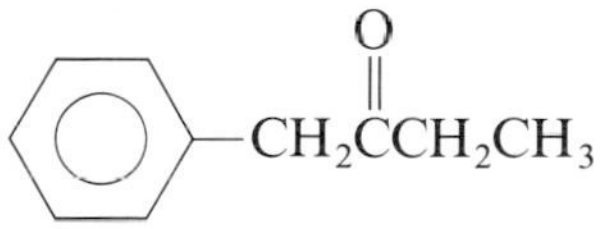

Benzyl ethyl ketone

Divinyl ketone

A few of the common names acceptable for ketones in the IUPAC system are

CH_3CCH_3 ⟨benzene⟩—CCH_3 ⟨benzene⟩—C—⟨benzene⟩

Acetone
(propanone)

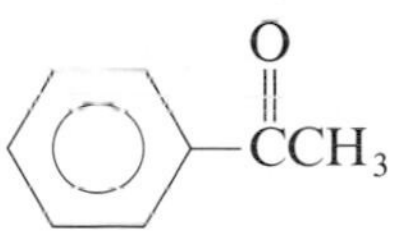

Acetophenone
(methyl phenyl ketone)

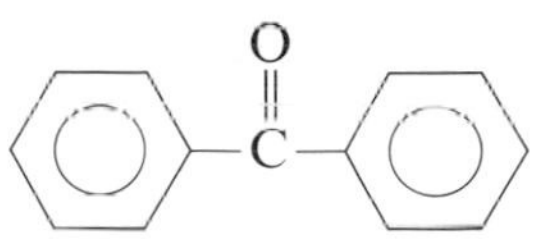

Benzophenone
(diphenyl ketone)

(The -*phenone* suffix indicates that the acyl group is attached to a benzene ring.)

PROBLEM 18.2 While the following are all permissible IUPAC names for ketones, each can be named in a different way based on the longest-continuous-chain method. Convert each name into one based on the latter system.

(a) Dibenzyl ketone

(b) Benzyl *tert*-butyl ketone

(c) Ethyl isopropyl ketone

(d) Methyl neopentyl ketone

(e) Allyl methyl ketone

SAMPLE SOLUTION (a) First write the structure corresponding to the name. Dibenzyl ketone has two benzyl groups attached to a carbonyl.

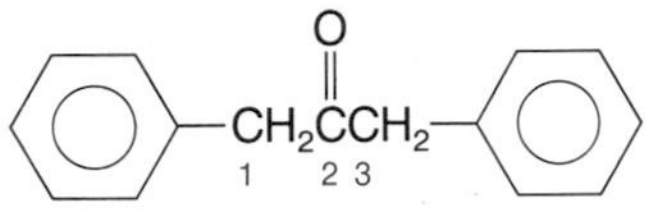

Dibenzyl ketone

The longest continuous chain contains three carbons, and C-2 is the carbon of the carbonyl group. A systematic IUPAC name for this ketone is *1,3-diphenyl-2-propanone*.

18.2 STRUCTURE AND BONDING. THE CARBONYL GROUP

Two of the more notable aspects of the carbonyl group are its geometry and its polarity. The carbonyl group and the atoms attached to it lie in the same plane. Formaldehyde, for example, is a planar molecule. The bond angles involving the carbonyl group are close to 120° and do not vary much among simple aldehydes and ketones.

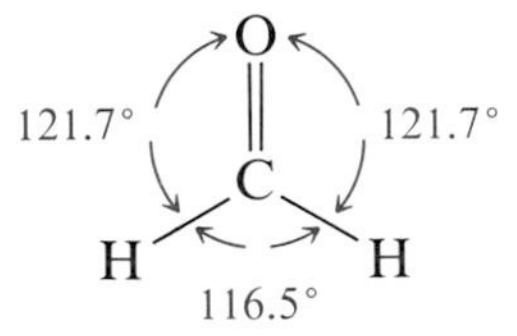

| Formaldehyde | Acetaldehyde | Acetone |

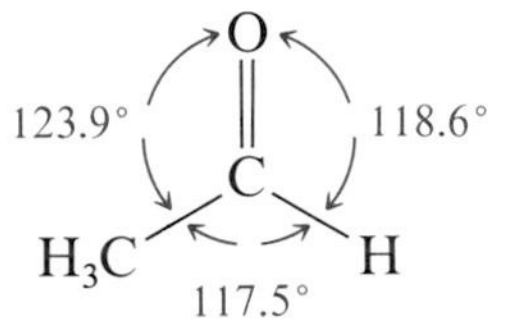

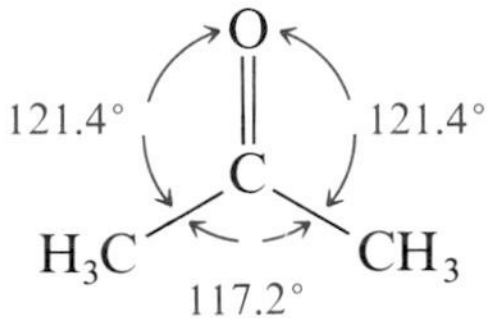

The average carbon-oxygen double bond distance of 1.22 Å in aldehydes and ketones is significantly shorter than the typical carbon-oxygen single bond distance of 1.41 Å in alcohols and ethers.

The presence of a carbonyl group makes aldehydes and ketones rather polar. Their molecular dipole moments, for example, are substantially larger than those of comparable compounds that contain carbon-carbon double bonds.

$$CH_3CH_2CH{=}CH_2 \qquad CH_3CH_2CH{=}O$$

1-Butene

dipole moment: 0.34 D

Propanal

dipole moment: 2.52 D

Bonding in formaldehyde is represented according to an sp^2 hybridization model for carbon, as shown in Figure 18.1. Using its three sp^2 hybridized orbitals, carbon forms σ bonds to two hydrogen atoms and an oxygen atom. An unhybridized p orbital on carbon participates in π bonding by overlapping with an oxygen $2p$ orbital.

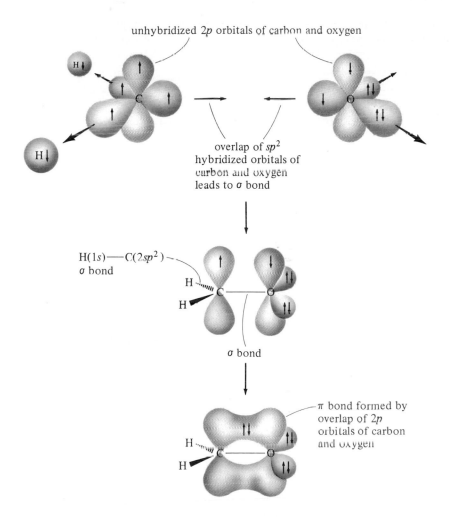

unhybridized 2p orbitals of carbon and oxygen

overlap of sp^2 hybridized orbitals of carbon and oxygen leads to σ bond

H(1s)—C($2sp^2$)—σ bond

σ bond

π bond formed by overlap of 2p orbitals of carbon and oxygen

FIGURE 18.1 A description of bonding in formaldehyde based on sp^2 hybridization of carbon and oxygen.

The short carbon-oxygen bond distance reflects the strong bonding that results from the combination of σ and π components.

Because of the high electronegativity of oxygen, the electron density in both the σ and π components of the carbon-oxygen double bond is displaced toward oxygen. The carbonyl group is polarized so that carbon is partially positive and oxygen is partially negative.

In resonance terms, electron delocalization in the carbonyl group is represented by contributions from two principal resonance forms:

Of these two forms, A, having one more covalent bond and avoiding the separation of positive and negative charges that characterizes B, is the one that better approximates the bonding in a carbonyl group. Nonetheless, contributions from the dipolar resonance form B are significant, and carbonyl groups are appreciably stabilized by resonance.

Alkyl substituents stabilize a carbonyl group in much the same way that they stabilize carbon-carbon double bonds, namely, by releasing electrons to sp^2 hybridized carbon. Thus, as measured by their heats of combustion, the ketone 2-butanone is more stable than its aldehyde isomer butanal.

$$\underset{\text{Butanal}}{CH_3CH_2CH_2\overset{\displaystyle O}{\overset{\|}{C}}H} + \underset{\text{Oxygen}}{\tfrac{11}{2}O_2} \longrightarrow \underset{\substack{\text{Carbon}\\\text{dioxide}}}{4CO_2} + \underset{\text{Water}}{4H_2O} \qquad \Delta H° = -592.1 \text{ kcal/mol}$$

$$\underset{\text{2-Butanone}}{CH_3CH_2\overset{\displaystyle O}{\overset{\|}{C}}CH_3} + \underset{\text{Oxygen}}{\tfrac{11}{2}O_2} \longrightarrow \underset{\substack{\text{Carbon}\\\text{dioxide}}}{4CO_2} + \underset{\text{Water}}{4H_2O} \qquad \Delta H° = -584.2 \text{ kcal/mol}$$

Less heat is given off when 2-butanone is burned because it has less potential energy. A ketone carbonyl has two electron-releasing alkyl groups that contribute to its stabilization, while an aldehyde has only one. Structural effects on carbonyl group *stability* are an important factor in the *relative reactivities* of aldehydes and ketones, as will become evident later in this chapter.

18.3 PHYSICAL PROPERTIES

Table 18.1 collects selected physical constants of some representative aldehydes and ketones. Aldehydes and ketones have higher boiling points than hydrocarbons of similar molecular weight because of intermolecular dipole-dipole attractive forces involving the carbonyl group. Hydrogen bonds between the carbonyl oxygen and water cause aldehydes and ketones to have a higher level of water solubility than hydrocarbons. Acetone, for example, is miscible with water in all proportions. The degree of solubility in water decreases with increasing molecular weight. Benzaldehyde dissolves in water to the extent of only 0.3 g/100 mL, solubility lower by a factor of 10 than the solubility of benzyl alcohol but 4 times as high as that of benzene.

18.4 SOURCES OF ALDEHYDES AND KETONES

Low-molecular-weight aldehydes and ketones are important industrial chemicals. While specialized procedures have been developed for making many of them, most can be prepared by oxidation (or dehydrogenation) of the corresponding alcohol. Formaldehyde, a starting material for a number of plastics, is prepared by oxidation of methanol over a silver or iron oxide – molybdenum oxide catalyst at elevated temperature.

$$\underset{\text{Methanol}}{CH_3OH} + \underset{\text{Oxygen}}{\tfrac{1}{2}O_2} \xrightarrow[500°C]{\text{catalyst}} \underset{\text{Formaldehyde}}{H\overset{\displaystyle O}{\overset{\|}{C}}H} + \underset{\text{Water}}{H_2O}$$

TABLE 18.1
Physical Properties of Some Aldehydes and Ketones

Compound	Condensed structural formula	Melting point, °C	Boiling point, °C (1 atm)	Solubility g/100 mL H_2O
Aldehydes				
Formaldehyde	$\overset{O}{\overset{\|}{HCH}}$	−92	−21	Very soluble
Acetaldehyde	$\overset{O}{\overset{\|}{CH_3CH}}$	−123.5	20.2	∞
Propanal	$\overset{O}{\overset{\|}{CH_3CH_2CH}}$	−81	49.5	20
Butanal	$\overset{O}{\overset{\|}{CH_3CH_2CH_2CH}}$	−99	75.7	4
Pentanal	$\overset{O}{\overset{\|}{CH_3CH_2CH_2CH_2CH}}$	−92	103.4	Slight
Benzaldehyde	$C_6H_5\overset{O}{\overset{\|}{CH}}$	−26	178	0.3
Ketones				
Acetone	$\overset{O}{\overset{\|}{CH_3CCH_3}}$	−94.8	56.2	∞
2-Butanone	$\overset{O}{\overset{\|}{CH_3CCH_2CH_3}}$	−86.9	79.6	37
2-Pentanone	$\overset{O}{\overset{\|}{CH_3CCH_2CH_2CH_3}}$	−77.8	102.4	Slight
3-Pentanone	$\overset{O}{\overset{\|}{CH_3CH_2CCH_2CH_3}}$	−39.9	102.0	4.7
Cyclopentanone	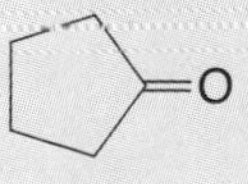 cyclopentanone $=O$	−51.3	130.7	43.3
Cyclohexanone	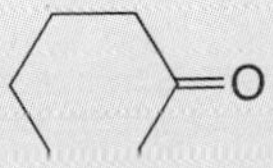 cyclohexanone $=O$	−45	155	—
Acetophenone	$\overset{O}{\overset{\|}{C_6H_5CCH_3}}$	21	202	Insoluble
Benzophenone	$\overset{O}{\overset{\|}{C_6H_5CC_6H_5}}$	48	306	Insoluble

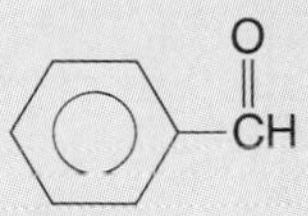

Similar processes are used to convert ethanol to acetaldehyde and isopropyl alcohol to acetone.

Acetaldehyde can be prepared by hydration of acetylene (Section 10.13), but a more economical procedure is air oxidation of ethylene in the presence of palladium chloride and cupric chloride as catalysts.

$$CH_2{=}CH_2 + \tfrac{1}{2}O_2 \xrightarrow[H_2O]{PdCl_2,\ CuCl_2} CH_3\overset{\overset{\displaystyle O}{\|}}{C}H$$

Ethylene Oxygen   Acetaldehyde

This is known as the *Wacker process.* A organopalladium compound is an intermediate.

Organocobalt compounds are intermediates in the *oxo process,* whereby alkenes are converted to aldehydes containing an additional carbon atom.

$$RCH{=}CH_2 + CO + H_2 \xrightarrow{Co_2(CO)_8} RCH_2CH_2\overset{\overset{\displaystyle O}{\|}}{C}H$$

Alkene Carbon Hydrogen   Aldehyde
     monoxide

Benzaldehyde is prepared industrially by hydrolysis of (dichloromethyl)benzene,

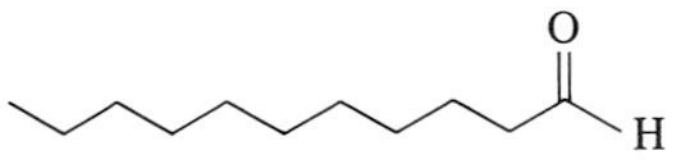

Undecanal
(sex pheromone of greater wax moth)

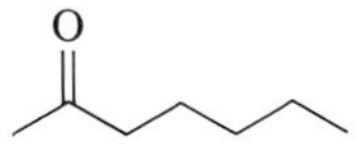

2-Heptanone
(component of alarm pheromone of bees)

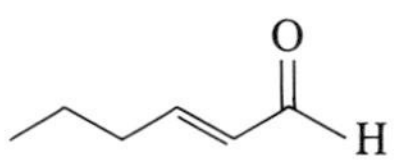

trans-2-Hexenal
(alarm pheromone of myrmicine ant)

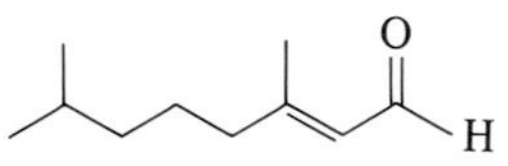

Citral
(present in lemon grass oil)

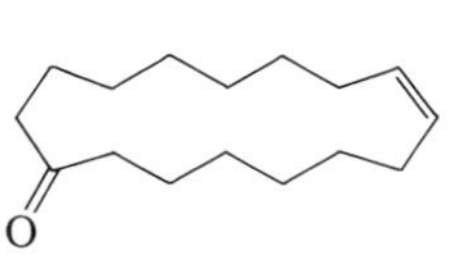

Civetone
(obtained from scent glands of
African civet cat)

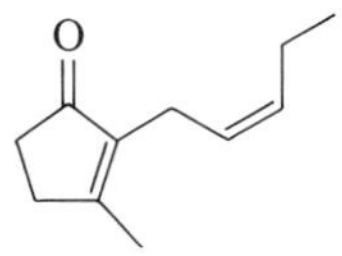

Jasmone
(found in oil of jasmine)

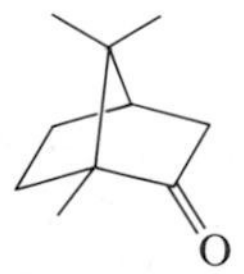

Camphor
(isolated from certain Indonesian
trees; has characteristic odor)

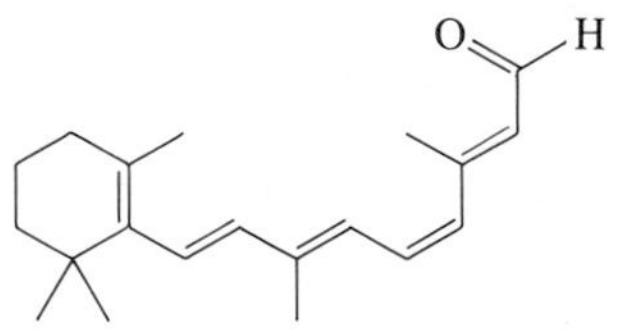

11-*cis*-Retinal
(critically important in chemistry
of vision)

FIGURE 18.2 Some naturally occurring aldehydes and ketones.

also known as benzal dichloride. Benzal dichloride is made by free-radical chlorination of toluene.

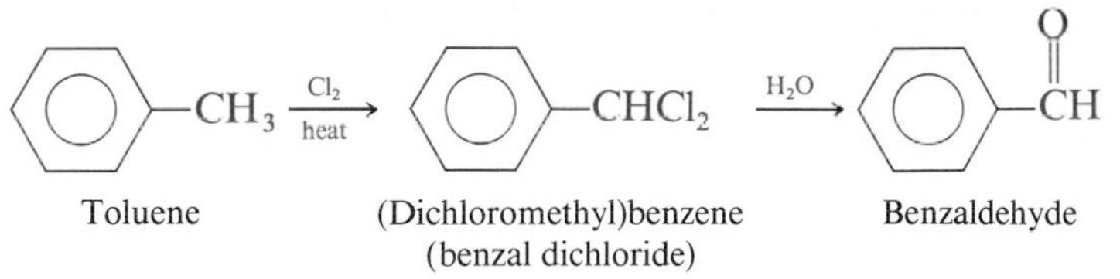

Toluene   (Dichloromethyl)benzene   Benzaldehyde
(benzal dichloride)

A number of aldehydes and ketones are prepared industrially and in the laboratory as well by a reaction known as the *aldol condensation*. This will be discussed in detail in Sections 19.9 through 19.12.

Many aldehydes and ketones occur naturally. Several of these are shown in Figure 18.2.

18.5 REACTIONS THAT LEAD TO ALDEHYDES AND KETONES. A REVIEW AND A PREVIEW

Reactions that yield aldehydes and ketones have appeared earlier among several functional group classes that we have already examined, namely, alkenes, alkynes, arenes, and alcohols. These reactions are summarized in Table 18.2.

This chapter introduces three procedures by which acyl chlorides may be converted to aldehydes or ketones. Acyl chlorides can be converted to aldehydes by reduction or to ketones by reaction with lithium diorganocuprates or diorganocadmium reagents.

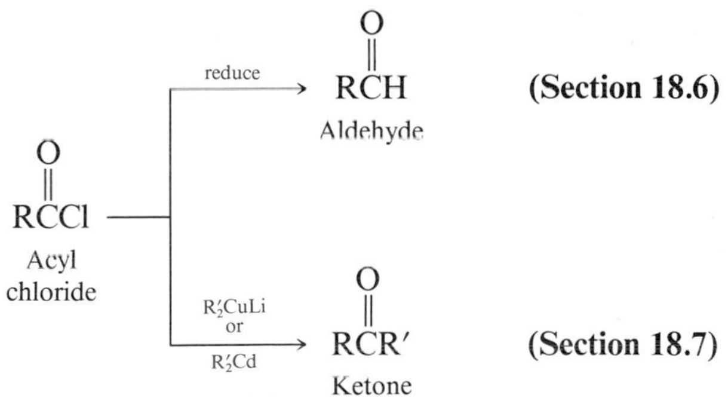

18.6 PREPARATION OF ALDEHYDES FROM CARBOXYLIC ACID DERIVATIVES

The number of one-step procedures for reducing carboxylic acids directly to aldehydes is quite limited, and organic chemists more routinely approach this transformation in an indirect way. The most widely used of these indirect methods is simply a combination of reactions with which you are already familiar. It is a two-stage procedure in which a carboxylic acid is reduced to a primary alcohol, which is then oxidized to an aldehyde.

TABLE 18.2

Summary of Reactions Discussed in Earlier Chapters that Yield Aldehydes and Ketones

Reaction (section) and comments	General equation and specific example
Ozonolysis of alkenes (Section 7.16) This cleavage reaction is more often seen in structural analysis than in synthesis. The substitution pattern around a double bond is revealed by identifying the carbonyl-containing compounds that comprise the product. Hydrolysis of the ozonide intermediate in the presence of zinc (reductive workup) permits aldehyde products to be isolated without further oxidation.	$$\underset{\text{Alkene}}{\overset{R}{\underset{R'}{>}}C=C\overset{H}{\underset{R''}{<}}} \xrightarrow[\text{2. }H_2O,\ Zn]{\text{1. }O_3} \underset{\text{Two carbonyl compounds}}{R\overset{O}{\overset{\|}{C}}R' + R''\overset{O}{\overset{\|}{C}}H}$$ 2,6-Dimethyl-2-octene $\xrightarrow[\text{2. }H_2O,\ Zn]{\text{1. }O_3}$ $CH_3\overset{O}{\overset{\|}{C}}CH_3$ Acetone + $H\overset{O}{\overset{\|}{C}}CH_2CH_2\overset{CH_3}{\underset{}{CH}}CH_2CH_3$ 4-Methylhexanal (91%)
Hydration of alkynes (Section 10.13) Reaction occurs by way of an enol intermediate formed by Markovnikov addition of water to the triple bond.	$$\underset{\text{Alkyne}}{RC\equiv CR'} + H_2O \xrightarrow[HgSO_4]{H_2SO_4} \underset{\text{Ketone}}{R\overset{O}{\overset{\|}{C}}CH_2R'}$$ $HC\equiv C(CH_2)_5CH_3 + H_2O \xrightarrow[HgSO_4]{H_2SO_4} CH_3\overset{O}{\overset{\|}{C}}(CH_2)_5CH_3$ 1-Octyne → 2-Octanone (91%)
Friedel-Crafts acylation of aromatic compounds (Section 13.7) Acyl chlorides and carboxylic acid anhydrides acylate aromatic rings in the presence of aluminum chloride. The reaction is one of electrophilic aromatic substitution in which acylium ions are generated and attack the ring.	$$ArH + R\overset{O}{\overset{\|}{C}}Cl \xrightarrow{AlCl_3} Ar\overset{O}{\overset{\|}{C}}R + HCl \quad \text{or}$$ $$ArH + R\overset{O}{\overset{\|}{C}}O\overset{O}{\overset{\|}{C}}R \xrightarrow{AlCl_3} Ar\overset{O}{\overset{\|}{C}}R + RCO_2H$$ $CH_3O-\bigcirc\!\!\!\!\bigcirc$ + $CH_3\overset{O}{\overset{\|}{C}}O\overset{O}{\overset{\|}{C}}CH_3$ $\xrightarrow{AlCl_3}$ $CH_3O-\bigcirc\!\!\!\!\bigcirc-\overset{O}{\overset{\|}{C}}CH_3$ Anisole + Acetic anhydride → p-Methoxyacetophenone (90–94%)
Oxidation of primary alcohols to aldehydes (Section 16.12) Use of the chromium trioxide–pyridine complex (Collins' reagent) in anhydrous media such as dichloromethane permits oxidation of primary alcohols to aldehydes while avoiding overoxidation to the corresponding carboxylic acids.	$$\underset{\text{Primary alcohol}}{RCH_2OH} \xrightarrow[CH_2Cl_2]{(C_5H_5N)_2CrO_3} \underset{\text{Aldehyde}}{R\overset{O}{\overset{\|}{C}}H}$$ $CH_3(CH_2)_8CH_2OH \xrightarrow[CH_2Cl_2]{(C_5H_5N)_2CrO_3} CH_3(CH_2)_8\overset{O}{\overset{\|}{C}}H$ 1-Decanol → Decanal (63–66%)
Oxidation of secondary alcohols to ketones (Section 16.12) Many oxidizing agents are available for converting secondary alcohols to ketones. Collins' reagent may be used, as well as other Cr(VI)-based reactions such as chromic acid or potassium dichromate and sulfuric acid. Potassium permanganate may also be used.	$$\underset{\text{Secondary alcohol}}{R\underset{\underset{OH}{\|}}{C}HR'} \xrightarrow{Cr(VI)} \underset{\text{Ketone}}{R\overset{O}{\overset{\|}{C}}R'}$$ $C_6H_5\underset{\underset{OH}{\|}}{C}HCH_2CH_2CH_2CH_3 \xrightarrow[\text{water}]{CrO_3 \atop \text{acetic acid–}} C_6H_5\overset{O}{\overset{\|}{C}}CH_2CH_2CH_2CH_3$ 1-Phenyl-1-pentanol → 1-Phenyl-1-pentanone (93%)

648

$$RCO_2H \xrightarrow{\text{reduce}} RCH_2OH \xrightarrow{\text{oxidize}} \overset{\displaystyle O}{\overset{\|}{R C H}}$$

Carboxylic acid   Primary alcohol   Aldehyde

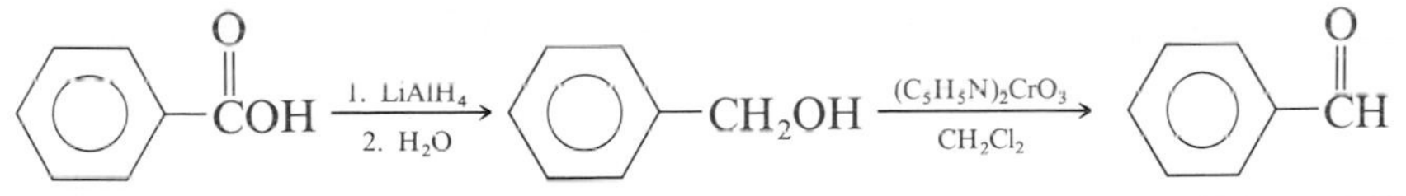

Benzoic acid $\xrightarrow[\text{2. } H_2O]{\text{1. } LiAlH_4}$ Benzyl alcohol (81%) $\xrightarrow[CH_2Cl_2]{(C_5H_5N)_2CrO_3}$ Benzaldehyde (89%)

PROBLEM 18.3   Can catalytic hydrogenation be used to reduce a carboxylic acid to a primary alcohol in the first step of this sequence?

PROBLEM 18.4   Esters can be converted to aldehydes by reduction to a primary alcohol followed by oxidation with Collins' reagent, pyridinium chlorochromate (PCC), or pyridinium dichromate (PDC). What reducing agents are appropriate for the production of primary alcohols from esters?

In another two-step transformation, an acyl chloride is prepared from the corresponding carboxylic acid and is then reduced by catalytic hydrogenation to the desired aldehyde. The hydrogenation step is called the *Rosenmund reduction* and requires a special catalyst, palladium metal supported on barium sulfate. This is a weak hydrogenation catalyst, good enough to bring about reduction of the highly reactive acyl chloride but not active enough to reduce the aldehyde to the primary alcohol.

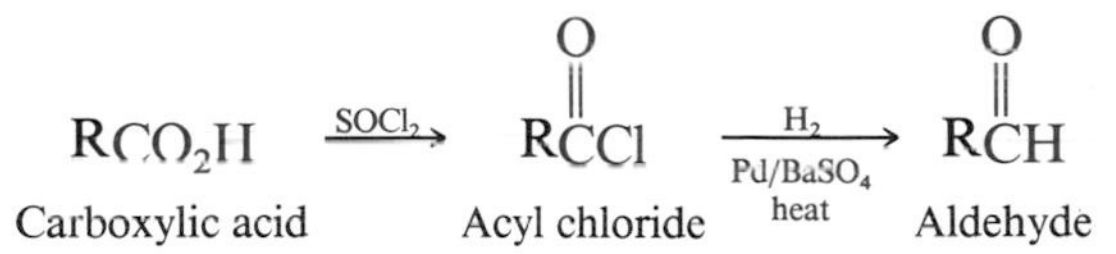

$$RCO_2H \xrightarrow{SOCl_2} \overset{\displaystyle O}{\overset{\|}{R C Cl}} \xrightarrow[\substack{Pd/BaSO_4 \\ \text{heat}}]{H_2} \overset{\displaystyle O}{\overset{\|}{R C H}}$$

Carboxylic acid   Acyl chloride   Aldehyde

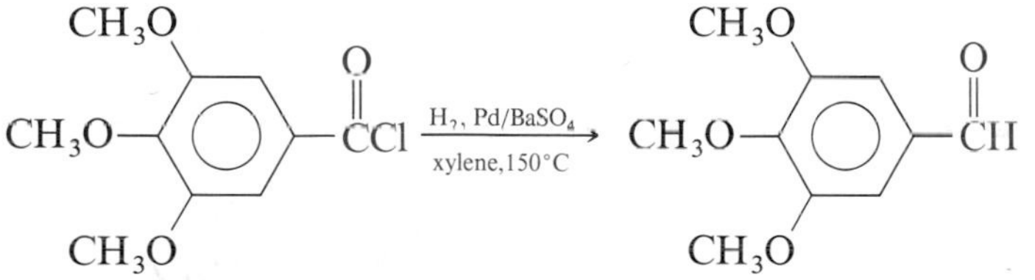

3,4,5-Trimethoxybenzoyl chloride $\xrightarrow[\text{xylene,150°C}]{H_2, Pd/BaSO_4}$ 3,4,5-Trimethoxybenzaldehyde (71%)

Sometimes the catalyst is further deactivated (poisoned) by adding certain sulfur-containing compounds.

18.7 PREPARATION OF KETONES FROM CARBOXYLIC ACID DERIVATIVES

Acyl chlorides react with a number of organometallic compounds to form ketones. Most of these organometallics, Grignard reagents for example, react further to convert the initially formed ketone to a tertiary alcohol.

$$\underset{\text{Acyl chloride}}{\overset{O}{\underset{\|}{R C Cl}}} \xrightarrow{R'M} \underset{\text{Ketone}}{\overset{O}{\underset{\|}{R C R'}}} \xrightarrow{R'M} \underset{\text{Tertiary alcohol}}{\overset{OH}{\underset{\underset{R'}{|}}{\underset{|}{R C R'}}}}$$

In order to have a synthetic method that is useful for preparing ketones, one must choose an organometallic reagent that reacts much faster with the starting acyl chloride than it does with the product ketone. Two types of organometallic compounds satisfy this requirement, organocuprate reagents and organocadmium reagents.

You have already learned something of the chemistry of organocuprates in their reactions with alkyl halides as a synthesis of alkanes (Section 15.11). Recall that lithium dialkyl- and diarylcuprates are prepared through the reaction of an alkyl- or aryllithium reagent with a cuprous salt.

$$\underset{\substack{\text{Alkyl- or}\\\text{aryllithium}}}{2 R Li} + \underset{\substack{\text{Cuprous}\\\text{halide}}}{CuX} \xrightarrow{\text{diethyl ether}} \underset{\substack{\text{Lithium dialkyl-}\\\text{or diarylcuprate}}}{LiCuR_2} + \underset{\substack{\text{Lithium}\\\text{halide}}}{LiX}$$

Adding an acyl chloride to the solution of the lithium diorganocuprate yields a ketone.

$$\underset{\substack{\text{Lithium}\\\text{dialkyl- or}\\\text{diarylcuprate}}}{LiCuR_2} + \underset{\substack{\text{Acyl}\\\text{chloride}}}{\overset{O}{\underset{\|}{R'CCl}}} \xrightarrow[-78°C]{\text{diethyl ether}} \underset{\text{Ketone}}{\overset{O}{\underset{\|}{R'CR}}} + \underset{\substack{\text{Alkyl}\\\text{or arylcopper}}}{RCu} + \underset{\substack{\text{Lithium}\\\text{chloride}}}{LiCl}$$

$$\underset{\substack{\text{Lithium}\\\text{dimethylcuprate}}}{LiCu(CH_3)_2} + \underset{\substack{\text{2,2-Dimethylpropanoyl}\\\text{chloride}}}{\overset{O}{\underset{\|}{(CH_3)_3CCCl}}} \xrightarrow[-78°C]{\text{diethyl ether}} \underset{\substack{\text{3,3-Dimethyl-2-butanone (84\%)}\\\text{(\textit{tert}-butyl methyl ketone)}}}{\overset{O}{\underset{\|}{(CH_3)_3CCCH_3}}}$$

$$\underset{\substack{\text{Lithium}\\\text{diphenylcuprate}}}{LiCu(C_6H_5)_2} + \underset{\text{Acetyl chloride}}{\overset{O}{\underset{\|}{CH_3CCl}}} \xrightarrow[-78°C]{\text{diethyl ether}} \underset{\text{Acetophenone (55\%)}}{\overset{O}{\underset{\|}{C_6H_5CCH_3}}}$$

As noted when the reactions of organocuprates with alkyl halides were discussed, secondary and tertiary alkylcuprates are less stable than primary alkyl- and arylcuprates and tend to decompose before they react. The same is true for their reactions with acyl halides. Ketone synthesis using organocuprate reagents is limited to methyl, primary, and arylcuprates.

PROBLEM 18.5 Using ethyl alcohol as the source of all the carbon atoms, outline a synthesis of 2-butanone via an organocuprate reagent.

Organocadmium reagents react with acyl chlorides in much the same way that organocuprates do. As described in Section 15.14, dialkylcadmium reagents are prepared by treating Grignard reagents with cadmium chloride.

$$2CH_3CH_2MgBr \xrightarrow[\text{diethyl ether}]{CdCl_2} (CH_3CH_2)_2Cd$$

Ethylmagnesium bromide   Diethylcadmium

$$2C_6H_5MgBr \xrightarrow[\text{diethyl ether}]{CdCl_2} (C_6H_5)_2Cd$$

Phenylmagnesium bromide   Diphenylcadmium

Addition of an acyl chloride to the solution of the organocadmium reagent yields the desired ketone. Suitable solvents include diethyl ether and benzene.

$$(CH_3CH_2)_2Cd + \underset{\text{Benzoyl chloride}}{C_6H_5\overset{\displaystyle O}{\overset{\|}{C}}Cl} \longrightarrow \underset{\substack{\text{1-Phenyl-1-propanone (84\%)}\\ \text{(propiophenone)}}}{C_6H_5\overset{\displaystyle O}{\overset{\|}{C}}CH_2CH_3}$$

Diethylcadmium

$$(C_6H_5)_2Cd + \underset{\text{Propanoyl chloride}}{CH_3CH_2\overset{\displaystyle O}{\overset{\|}{C}}Cl} \longrightarrow \underset{\substack{\text{1-Phenyl-1-propanone (81\%)}\\ \text{(propiophenone)}}}{C_6H_5\overset{\displaystyle O}{\overset{\|}{C}}CH_2CH_3}$$

Diphenylcadmium

As these examples illustrate, the reagent may be either a dialkyl- or diarylcadmium. Secondary and tertiary alkylcadmium reagents, like their organocopper counterparts, decompose easily and are not suitable for ketone synthesis.

PROBLEM 18.6   Using acetic acid as the source of all its carbon atoms, outline a synthesis of 2-butanone via an organocadmium reagent.

18.8 REACTIONS OF ALDEHYDES AND KETONES. A REVIEW AND A PREVIEW

A summary of our earlier encounters with the chemical reactions of aldehydes and ketones is presented in Table 18.3. All the transformations shown there are valuable tools of the synthetic organic chemist. Carbonyl groups provide access to hydrocarbons via Clemmensen or Wolff-Kishner reduction (Section 13.8), to alcohols of the same carbon skeleton by a variety of reduction methods (Section 16.3), and to alcohols of more complex structure by reaction with organolithium or Grignard reagents (Sections 15.6 and 15.7).

One of the characteristics of the carbonyl group is its tendency to undergo *nucleophilic addition* reactions of the type represented by the general equation

$$\overset{\delta+}{\diagdown}C=\overset{\delta-}{O} + \overset{\delta+}{X}-\overset{\delta-}{Y} \longrightarrow \overset{OX}{\underset{Y}{\diagdown C}}$$

Aldehyde   Product of
or ketone   nucleophilic addition

TABLE 18.3

Summary of Reactions of Aldehydes and Ketones Discussed in Earlier Chapters

Reaction (section) and comments	General equation and specific example
Reduction to hydrocarbons (Section 13.8) Two methods for converting carbonyl groups to methylene units are the Clemmensen reduction (zinc amalgam and concentrated hydrochloric acid) and the Wolff-Kishner reduction (heat with hydrazine and potassium hydroxide in a high-boiling alcohol).	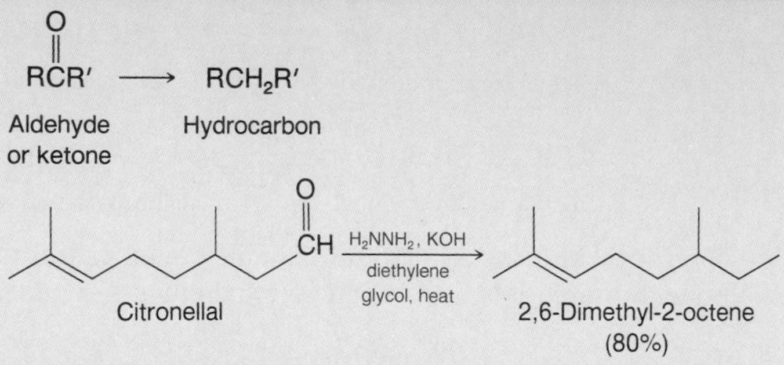
Reduction to alcohols (Section 16.3) Aldehydes are reduced to primary alcohols and ketones are reduced to secondary alcohols by a variety of reducing agents. Catalytic hydrogenation over a metal catalyst and reduction with sodium borohydride or lithium aluminum hydride are general methods. Sodium in ethanol is useful for reduction of ketones.	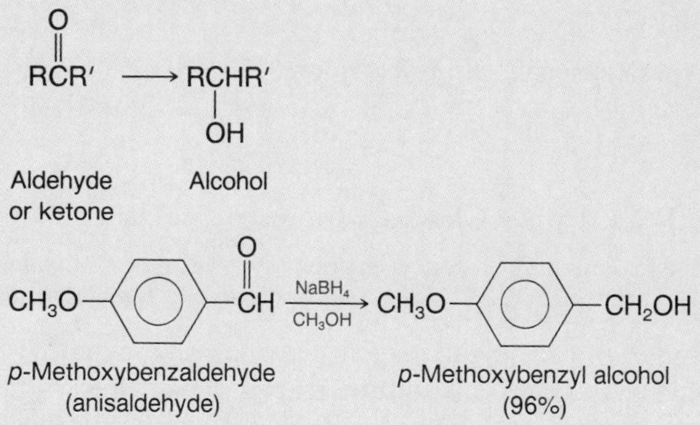
Addition of Grignard reagents and organolithium compounds (Sections 15.6–15.7) Aldehydes are converted to secondary alcohols and ketones to tertiary alcohols.	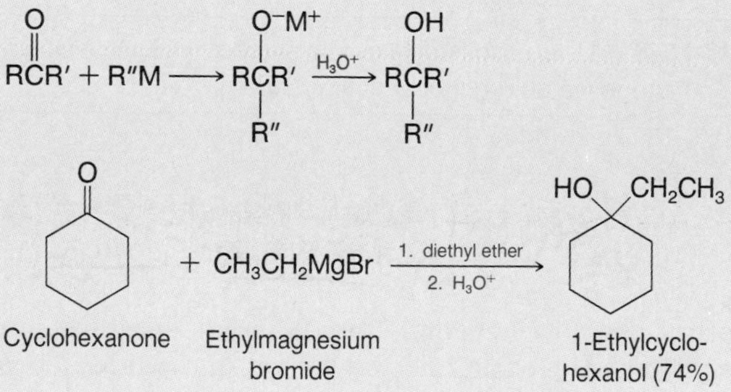

A negatively polarized atom or group is transferred to the positively polarized carbon of the carbonyl group in the rate-determining step of these reactions. Grignard reagents, organolithium reagents, lithium aluminum hydride, and sodium borohydride all react with carbonyl compounds by nucleophilic addition.

The next section presents the mechanistic features of nucleophilic addition to aldehydes and ketones through a discussion of their *hydration,* the addition of a water molecule to the carbonyl group. Once we develop the principles of nucleophilic addition, we will survey a number of nucleophilic addition reactions of aldehydes and ketones; some of these are of synthetic interest, others are of mechanistic importance, and a few possess both qualities.

18.9 PRINCIPLES OF NUCLEOPHILIC ADDITION TO CARBONYL GROUPS. HYDRATION OF ALDEHYDES AND KETONES

Aldehydes and ketones react with water in a rapidly reversible equilibrium process.

$$\underset{\substack{\text{Aldehyde} \\ \text{or ketone}}}{\overset{\displaystyle O \atop \displaystyle \| \atop \displaystyle RCR'}{}} + \underset{\text{Water}}{H_2O} \underset{}{\overset{\text{fast}}{\rightleftharpoons}} \underset{\substack{\text{Geminal diol} \\ \text{(hydrate)}}}{\overset{\displaystyle OH \atop \displaystyle | \atop \displaystyle RCR' \atop \displaystyle | \atop \displaystyle OH}{}} \qquad K_{\text{hydr}} = \frac{[\text{hydrate}]}{[\text{carbonyl compound}][\text{water}]}$$

Overall, the reaction is classified as an *addition reaction.* The elements of water add to the carbonyl group. Hydrogen becomes bonded to the negatively polarized carbonyl oxygen, hydroxyl to the positively polarized carbon.

Table 18.4 compares the equilibrium constants K_{hydr} for hydration of some simple aldehydes and ketones. The position of equilibrium depends strongly on the nature of the carbonyl group and is influenced by a combination of *electronic* and *steric* effects.

Consider first the electronic effect of substituents on the stabilization of the carbonyl group. As the starting material becomes more stable, the smaller will be its equilibrium constant for hydration. Formaldehyde has no alkyl substituents to stabilize its carbonyl group and is converted almost completely to its hydrate in aqueous solution. The carbonyl of acetaldehyde is stabilized by one alkyl substituent, the carbonyl of acetone by two. The proportion of hydrate present in an aqueous solution of a typical aldehyde is much less than that in an aqueous solution of formaldehyde, while ketones are converted to their hydrates to an even smaller extent.

TABLE 18.4
Equilibrium Constants (K_{hydr}) for Hydration of Aldehydes and Ketones

Carbonyl compound	Hydrate	K_{hydr}*	Percent conversion to hydrate†
$\overset{\displaystyle O \atop \displaystyle \|}{HCH}$	$CH_2(OH)_2$	41	99.96
$\overset{\displaystyle O \atop \displaystyle \|}{CH_3CH}$	$CH_3CH(OH)_2$	1.8×10^{-2}	50
$\overset{\displaystyle O \atop \displaystyle \|}{(CH_3)_3CCH}$	$(CH_3)_3CCH(OH)_2$	4.1×10^{-3}	19
$\overset{\displaystyle O \atop \displaystyle \|}{CH_3CCH_3}$	$(CH_3)_2C(OH)_2$	2.5×10^{-5}	0.14

* $K_{\text{hydr}} = \dfrac{[\text{hydrate}]}{[\text{carbonyl compound}][\text{water}]}$ Units of K_{hydr} are M^{-1}.

† Total concentration (hydrate plus carbonyl compound) assumed to be 1 *M*. Water concentration is 55.5 *M*.

A striking example of an electronic effect on carbonyl group stability and its relation to the equilibrium constant for hydration is seen in the case of hexafluoroacetone. In contrast to the almost negligible hydration of acetone, hexafluoroacetone is completely hydrated.

$$\underset{\substack{\text{Hexafluoroacetone}}}{CF_3\overset{\displaystyle O}{\overset{\|}{C}}CF_3} + \underset{\substack{\text{Water}}}{H_2O} \rightleftharpoons \underset{\substack{\text{1,1,1,3,3,3-Hexafluoro-}\\ \text{2,2-propanediol}}}{CF_3\overset{\displaystyle OH}{\underset{\displaystyle OH}{\overset{|}{\underset{|}{C}}}}CF_3} \qquad K_{hydr} = 22,000$$

Instead of stabilizing the carbonyl group by electron donation as alkyl substituents do, trifluoromethyl groups destabilize it by withdrawing electrons. A less stabilized carbonyl group is associated with a greater equilibrium constant for addition.

To understand the role played by steric effects, let us examine the geminal diol product (hydrate). The carbon that bears the two hydroxyl groups is sp^3 hybridized. Its substituents are more crowded than they are in the starting aldehyde or ketone. Increased crowding can be better tolerated when the substituents are hydrogen than when they are alkyl groups.

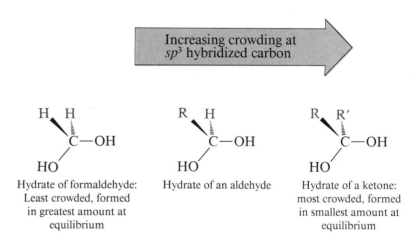

Hydrate of formaldehyde: Least crowded, formed in greatest amount at equilibrium

Hydrate of an aldehyde

Hydrate of a ketone: most crowded, formed in smallest amount at equilibrium

Electronic and steric effects operate in the same direction to make the equilibrium constants for hydration of aldehydes more favorable than those of ketones.

Let us turn now to structural effects and the effects of catalysts on the *rate* of hydration. While the equilibrium for the hydration of aldehydes and ketones is rapidly established under neutral conditions, it is markedly catalyzed by both acids and bases.

Figure 18.3 presents the two steps of the base-catalyzed mechanism. In the first step the nucleophile, a hydroxide ion, adds to the carbonyl group by forming a bond to the carbonyl carbon. The product of the nucleophilic addition step is an alkoxide anion. It abstracts a proton from water in the second step to yield the geminal diol product and regenerate hydroxide ion. The proton transfer step, like other proton transfers between oxygens that we have seen, is fast. The first step is rate-determining.

The role of the basic catalyst (HO^-) is to increase the rate of the nucleophilic addition step. Hydroxide ion, the nucleophile in the base-catalyzed reaction, is much more powerful than a water molecule, the nucleophile in neutral media.

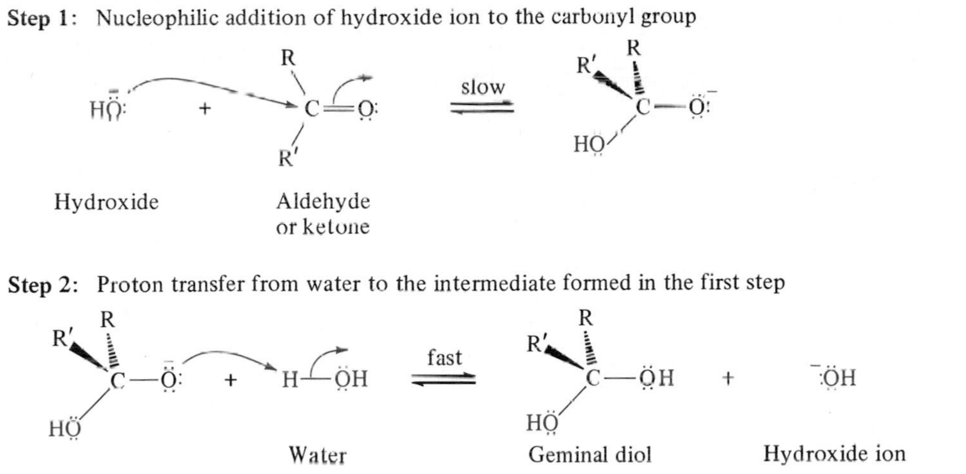

Step 1: Nucleophilic addition of hydroxide ion to the carbonyl group
slow
Hydroxide
Aldehyde or ketone
Step 2: Proton transfer from water to the intermediate formed in the first step
fast
Water
Geminal diol
Hydroxide ion

FIGURE 18.3 Sequence of steps that describes the mechanism of hydration of an aldehyde or ketone under base-catalyzed conditions.

Aldehydes react faster than ketones for almost the same reasons that their equilibrium constants for hydration are more favorable. The $sp^2 \rightarrow sp^3$ hybridization change that the carbonyl carbon undergoes during the hydration process is partially developed in the transition state for the rate-determining nucleophilic addition step (Figure 18.4). Alkyl groups at the reaction site increase the activation energy by

Energy
E_{act}
sp^2
sp^3
sp^3
Reaction coordinate

FIGURE 18.4 Potential energy diagram for base-catalyzed hydration of a carbonyl compound.

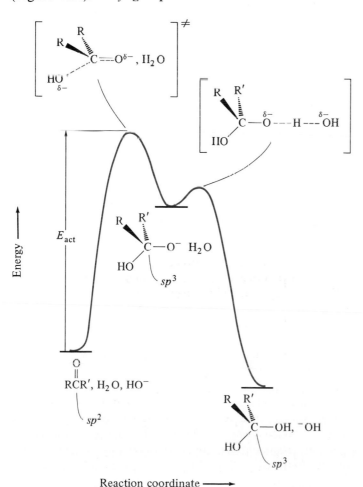

Step 1: Protonation of the carbonyl oxygen

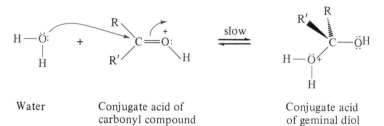

| Aldehyde or ketone | Hydronium ion | Conjugate acid of carbonyl compound | Water |

Step 2: Nucleophilic addition to the protonated aldehyde or ketone

Water   Conjugate acid of carbonyl compound   Conjugate acid of geminal diol

Step 3: Proton transfer from the conjugate acid of the geminal diol to a water molecule

FIGURE 18.5 Sequence of steps that describes the mechanism of hydration of an aldehyde or ketone under acid-catalyzed conditions.

Conjugate acid of geminal diol   Water   Geminal diol   Hydronium ion

simultaneously lowering the energy of the starting state (ketones have a more stabilized carbonyl group than aldehydes) and raising the energy of the transition state (a steric crowding effect).

Three steps are involved in the acid-catalyzed hydration reaction, as shown in Figure 18.5. The first and last are rapid proton transfer processes. The second is the nucleophilic addition step. The role of the acid catalyst is to activate the carbonyl group toward attack by a weakly nucleophilic water molecule. Protonation of oxygen makes the carbonyl carbon of an aldehyde or ketone much more electrophilic. Expressed in resonance terms, the protonated carbonyl has a greater degree of carbocation character than its neutral counterpart.

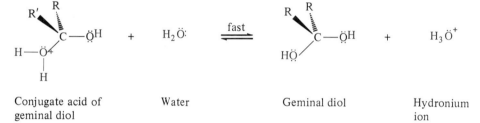

Electron delocalization in the neutral carbonyl involves a charge-separated dipolar resonance form. Electron delocalization in the protonated carbonyl is more pronounced because there is no separation of opposite charges in either resonance form.

Steric and electronic effects influence the rate of nucleophilic addition to a protonated carbonyl group in much the same way as they do in the case of a neutral one. The carbocation character of the protonated carbonyl is stabilized by electron-releasing alkyl groups, so the rate-determining addition of water to protonated ketones is slower than the corresponding addition to protonated aldehydes. The activated complex for nucleophilic addition is more crowded for protonated ketones than for protonated aldehydes and is formed more slowly.

With this as background, let us proceed to a related reaction, the addition of alcohols to aldehydes and ketones.

18.10 ACETAL FORMATION

Aldehydes react with alcohols under conditions of acid catalysis to yield geminal diethers known as *acetals*.

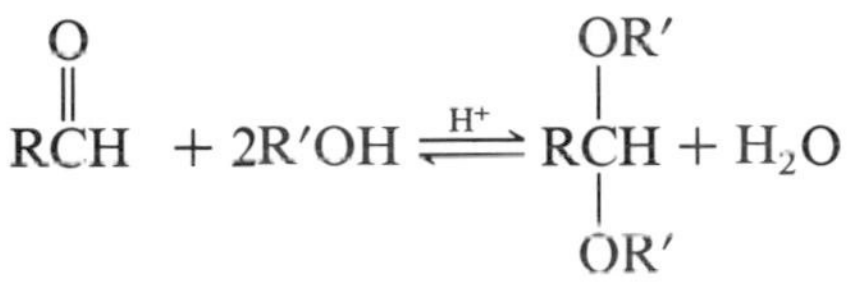

$$
\underset{\text{Aldehyde}}{\text{RCH}}^{\displaystyle \overset{O}{\parallel}} + \underset{\text{Alcohol}}{2R'OH} \underset{H^+}{\rightleftharpoons} \underset{\text{Acetal}}{\overset{OR'}{\underset{OR'}{\text{RCH}}}} + \underset{\text{Water}}{H_2O}
$$

Benzaldehyde + 2CH$_3$CH$_2$OH $\xrightarrow{\text{HCl}}$ CH(OCH$_2$CH$_3$)$_2$

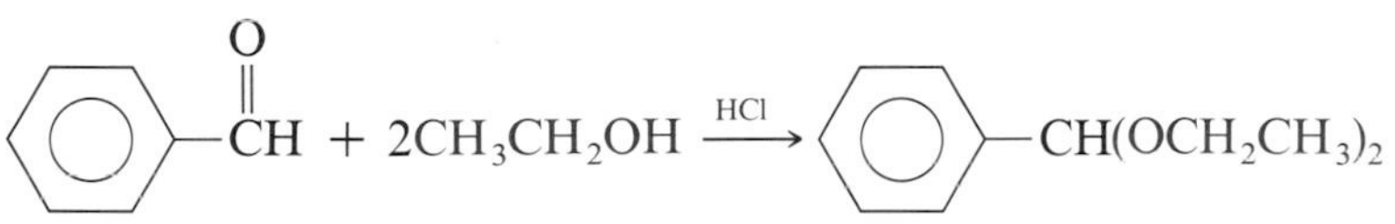

Benzaldehyde   Ethanol   Benzaldehyde diethyl acetal (66%)

Acetal formation is reversible. An equilibrium is established between the reactants, i.e., the carbonyl compound and the alcohol, and the acetal product. The position of equilibrium is favorable for acetal formation from most aldehydes, especially when excess alcohol is present as the reaction solvent. For most ketones the position of equilibrium is unfavorable and other methods must be used for the preparation of acetals from ketones.

At one time it was customary to designate the products of addition of alcohols to ketones as *ketals*. This term has been dropped from the IUPAC system of nomenclature and the term "acetal" is now applied to the adducts of both aldehydes and ketones.

The overall reaction proceeds in two stages. In the first stage one molecule of the alcohol undergoes acid-catalyzed nucleophilic addition to the aldehyde to yield a *hemiacetal*. The mechanism of this step is analogous to that of the acid-catalyzed hydration of an aldehyde.

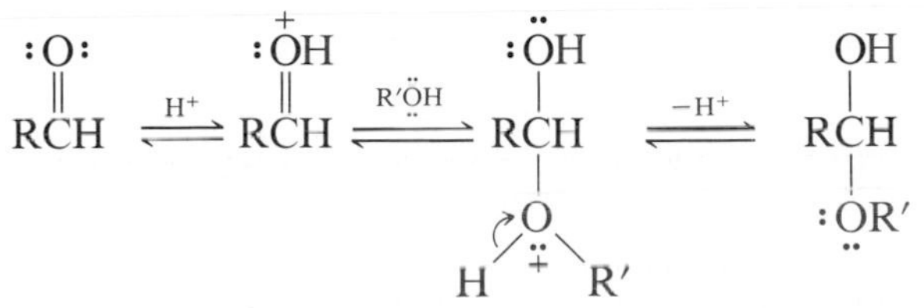

Under the conditions of its formation, the hemiacetal is converted to an acetal by way of a carbocation intermediate.

$$
\underset{\substack{\text{Hemiacetal}}}{\overset{\displaystyle :\!\overset{..}{\text{O}}\text{H}}{\underset{\displaystyle :\text{OR}}{\text{RCH}}}}
\;\xrightarrow[\text{H}^+,\ \text{fast}]{}\;
\underset{\displaystyle :\text{OR}'}{\overset{\displaystyle \overset{\text{H}\,\overset{+}{}\,\text{H}}{\overset{\displaystyle \overset{..}{\text{O}}}{\mid}}}{\text{RCH}}}
\;\xrightarrow[\text{slow}]{}\;
\underset{\substack{\text{Carbocation}}}{\overset{\displaystyle \text{R}\quad\text{H}}{\underset{\displaystyle :\text{OR}'}{\overset{+}{\text{C}}}}}
\;+\;\underset{\substack{\text{Water}}}{\text{H}_2\overset{..}{\text{O}}\!:}
$$

This carbocation is stabilized by electron release from its oxygen substituent.

$$
\underset{\substack{\text{A particularly stable}\\ \text{resonance form; both}\\ \text{carbon and oxygen have}\\ \text{octets of electrons}}}{
\overset{\displaystyle \text{R}\quad\text{H}}{\underset{\displaystyle :\overset{..}{\text{O}}\text{R}''}{\overset{+}{\text{C}}}}
\;\longleftrightarrow\;
\overset{\displaystyle \text{R}\quad\text{H}}{\underset{\displaystyle {}^{+}\overset{..}{\text{O}}\text{R}'}{\overset{\displaystyle \|}{\text{C}}}}
}
$$

Nucleophilic capture of the carbocation intermediate by an alcohol molecule leads to an acetal.

$$
\underset{\substack{\text{Alcohol}}}{\overset{\displaystyle \text{R}'}{\underset{\displaystyle \text{H}}{:\overset{..}{\text{O}}\!:}}}
\;+\;
\overset{\displaystyle \text{R}\quad\text{H}}{\underset{\displaystyle {}^{+}\overset{..}{\text{O}}\text{R}'}{\overset{\displaystyle \|}{\text{C}}}}
\;\xrightarrow[\text{fast}]{}\;
\underset{\displaystyle :\overset{..}{\text{O}}\text{R}'}{\overset{\displaystyle \overset{\text{H}\ \overset{+}{}\ \text{R}'}{\overset{..}{\text{O}}}}{\text{RCH}}}
\;\xrightarrow[-\text{H}^+,\ \text{fast}]{}\;
\underset{\substack{\text{Acetal}}}{\underset{\displaystyle :\overset{..}{\text{O}}\text{R}'}{\overset{\displaystyle :\overset{..}{\text{O}}\text{R}'}{\text{RCH}}}}
$$

PROBLEM 18.7   Consider acid-catalyzed acetal formation between acetaldehyde with methanol. Write structural formulas for

(a)   The hemiacetal intermediate
(b)   The carbocation intermediate (two resonance forms)
(c)   The acetal product

SAMPLE SOLUTION   (a)   The hemiacetal intermediate corresponds to addition of methanol to the carbonyl group.

$$
\underset{\substack{\text{Acetaldehyde}}}{\overset{\displaystyle \text{O}}{\overset{\displaystyle \|}{\text{CH}_3\text{CH}}}}
\;+\;
\underset{\substack{\text{Methanol}}}{\text{CH}_3\text{OH}}
\;\underset{}{\overset{\text{H}^+}{\rightleftharpoons}}\;
\underset{\substack{\text{Hemiacetal}\\\text{intermediate}}}{\overset{\displaystyle \text{OCH}_3}{\overset{\displaystyle \mid}{\text{CH}_3\text{CHOH}}}}
$$

(b)   This hemiacetal is converted to a carbocation by protonation followed by loss of a water molecule from the protonated form.

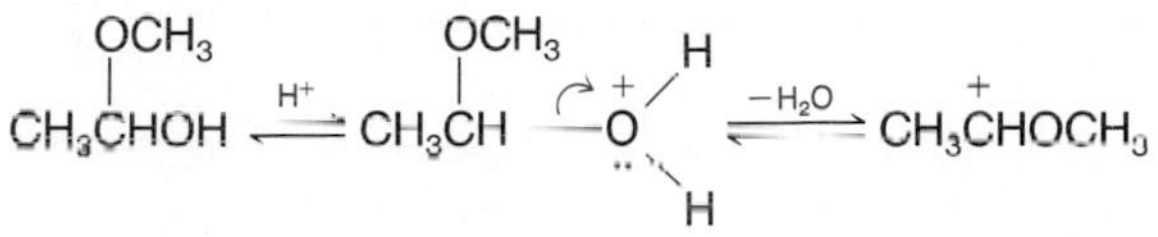

$$CH_3\overset{\underset{\displaystyle |}{OCH_3}}{C}HOH \underset{}{\overset{H^+}{\rightleftharpoons}} CH_3\overset{\underset{\displaystyle |}{OCH_3}}{C}H \overset{+}{\underset{\underset{\displaystyle H}{\displaystyle |}}{O}} \overset{\displaystyle H}{} \underset{}{\overset{-H_2O}{\rightleftharpoons}} CH_3\overset{+}{C}HOCH_3$$

Hemiacetal
intermediate

Carbocation
intermediate

The two principal forms of this carbocation are

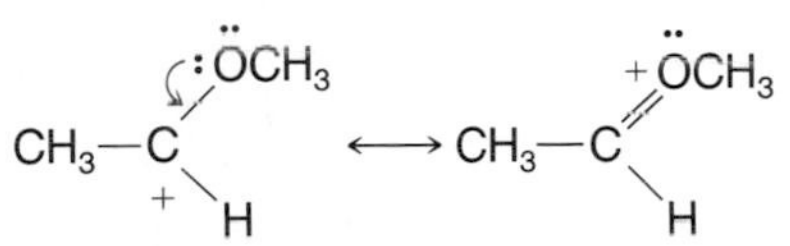

$$CH_3-\overset{\underset{\displaystyle +}{\displaystyle C}}{\underset{\displaystyle H}{}}\overset{:\ddot{O}CH_3}{} \longleftrightarrow CH_3-C\overset{+\ddot{O}CH_3}{\underset{\displaystyle H}{}}$$

(c) The dimethyl acetal of acetaldehyde is $CH_3CH(OCH_3)_2$.

PROBLEM 18.8 Repeat the preceding problem for the diethyl acetal of benzaldehyde.

Diols that bear two hydroxyl groups in a 1,2- or 1,3-relationship to each other yield *cyclic acetals* on reaction with either aldehydes or ketones. The five-membered cyclic acetals derived from ethylene glycol are the most commonly encountered examples. Often the position of equilibrium is made more favorable by removing the water formed in the reaction by azeotropic distillation with benzene or toluene.

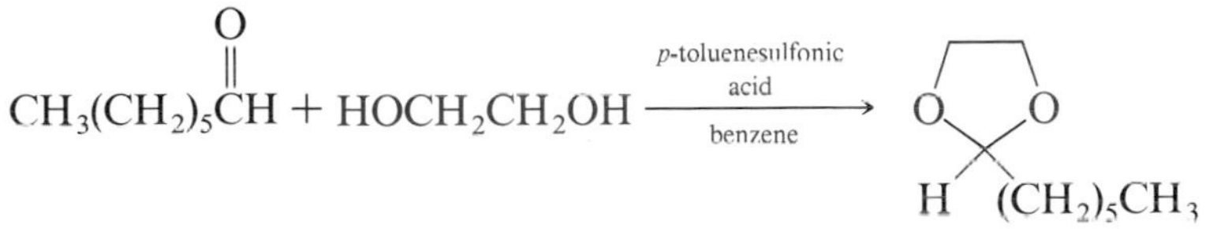

$$CH_3(CH_2)_5\overset{\underset{\displaystyle \|}{O}}{C}H + HOCH_2CH_2OH \xrightarrow[\text{benzene}]{\underset{\text{acid}}{p\text{-toluenesulfonic}}}$$

Heptanal

Ethylene glycol
(1,2-ethanediol)

2-Heptyl-1,3-dioxolane
(81%)

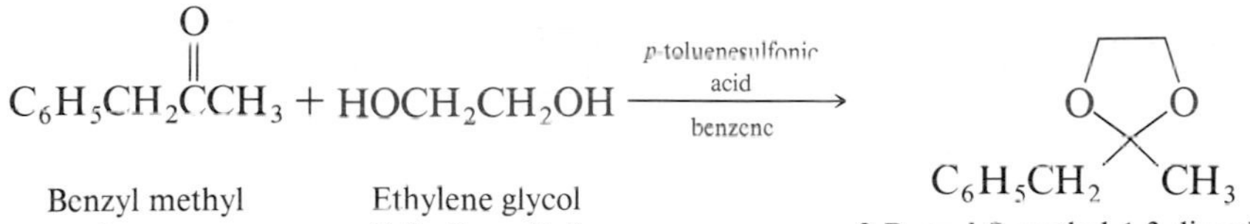

$$C_6H_5CH_2\overset{\underset{\displaystyle \|}{O}}{C}CH_3 + HOCH_2CH_2OH \xrightarrow[\text{benzene}]{\underset{\text{acid}}{p\text{-toluenesulfonic}}}$$

Benzyl methyl
ketone

Ethylene glycol
(1,2-ethanediol)

2-Benzyl-2-methyl-1,3-dioxolane
(78%)

PROBLEM 18.9 Write the structures of the cyclic acetals derived from:

(a) Cyclohexanone and ethylene glycol
(b) Benzaldehyde and 1,3-propanediol
(c) Isobutyl methyl ketone and ethylene glycol
(d) Isobutyl methyl ketone and 2,2-dimethyl-1,3-propanediol.

SAMPLE SOLUTION   (a) The cyclic acetals derived from ethylene glycol contain a five-membered 1,3-dioxolane ring.

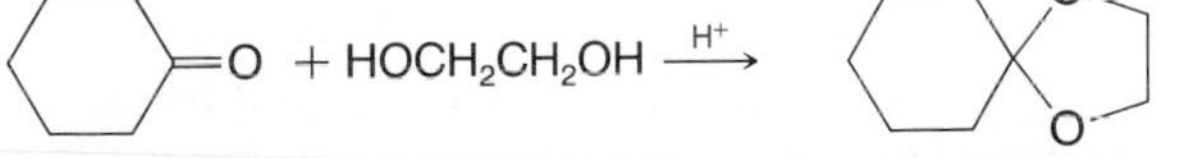

$$\text{Cyclohexanone} + HOCH_2CH_2OH \xrightarrow{H^+}$$

Cyclohexanone   Ethylene glycol

Acetal of cyclohexanone
and ethylene glycol

Acetals are susceptible to hydrolysis in aqueous acid.

$$\underset{\substack{\text{Acetal}}}{\overset{\displaystyle OR''}{\underset{\displaystyle OR''}{RCR'}}} + H_2O \;\underset{\xleftarrow{\hspace{1em}}}{\overset{H^+}{\rightleftharpoons}}\; \underset{\substack{\text{Aldehyde} \\ \text{or ketone}}}{\overset{\displaystyle O}{\overset{\|}{RCR'}}} + \underset{\substack{\text{Alcohol}}}{2R''OH}$$

This reaction is simply the reverse of the reaction by which acetals are formed — acetal formation is favored by excess alcohol, acetal hydrolysis by excess water. Acetal formation and acetal hydrolysis share the same mechanistic pathway but traverse that pathway in opposite directions. In the following section you will see how acetal formation and hydrolysis have been applied to synthetic organic chemistry as a means of carbonyl group protection.

Acetal formation from carbonyl compounds and alcohols requires acid catalysis to convert the hemiacetal to the essential carbocation intermediate. Hemiacetals are accessible under conditions of base catalysis but the reaction does not proceed further.

$$\underset{\substack{\text{Aldehyde} \\ \text{or ketone}}}{\overset{\displaystyle O}{\overset{\|}{RCR'}}} + \underset{\substack{\text{Alkoxide} \\ \text{ion}}}{R''O^-} \rightleftharpoons \overset{\displaystyle O^-}{\underset{\displaystyle OR''}{RCR'}} \rightleftharpoons \underset{\substack{\text{Hemiacetal}}}{\overset{\displaystyle OH}{\underset{\displaystyle OR''}{RCR'}}} \;\xcancel{\longrightarrow}\; \underset{\substack{\text{Acetal} \\ \text{(not formed} \\ \text{in base)}}}{\overset{\displaystyle OR''}{\underset{\displaystyle OR''}{RCR'}}}$$

Converting a hemiacetal to an acetal in basic media requires loss of hydroxide ion as a leaving group, a process not normally observed. Hemiacetals easily revert to reactants and cannot be isolated. Thus, while base-catalyzed nucleophilic addition of alcohols to aldehydes and ketones leads to hemiacetals, the reaction is unproductive in practice and of little consequence.

18.11   ACETALS AS PROTECTING GROUPS

In the practice of organic synthesis, it frequently happens that one of the reactants contains a functional group that is incompatible with the reaction conditions. Consider, for example, the conversion

$$\underset{\substack{\text{5-Hexyn-2-one}}}{\overset{\displaystyle O}{\overset{\|}{CH_3CCH_2CH_2C\equiv CH}}} \longrightarrow \underset{\substack{\text{5-Heptyn-2-one}}}{\overset{\displaystyle O}{\overset{\|}{CH_3CCH_2CH_2C\equiv CCH_3}}}$$

Seemingly straightforward enough, what is needed is to prepare the acetylenic anion, then alkylate it with methyl iodide (Section 10.7). There is a complication, however. The carbonyl group in the starting alkyne will neither tolerate the strongly basic conditions required for anion formation nor survive in a solution containing carbanions. Acetylide ions add to carbonyl groups (Section 15.8). Thus, the necessary anion

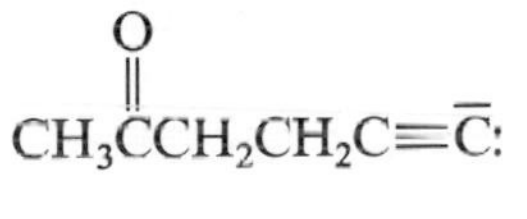

$$CH_3\overset{\overset{\displaystyle O}{\|}}{C}CH_2CH_2C\equiv\bar{C}:$$

is inaccessible.

The strategy that is routinely followed is to *protect* the carbonyl group during the reactions with which it is incompatible and then to *remove* the protecting group in a subsequent step. Acetals, especially those derived from ethylene glycol, are among the most useful groups for carbonyl protection because they can be introduced and removed readily. A fact of vital importance is that they are inert to many of the reagents, such as hydride reducing agents and organometallics, that react readily with carbonyl groups. The sequence shown is the one actually used to bring about the desired transformation.

(a) Protection of carbonyl group:

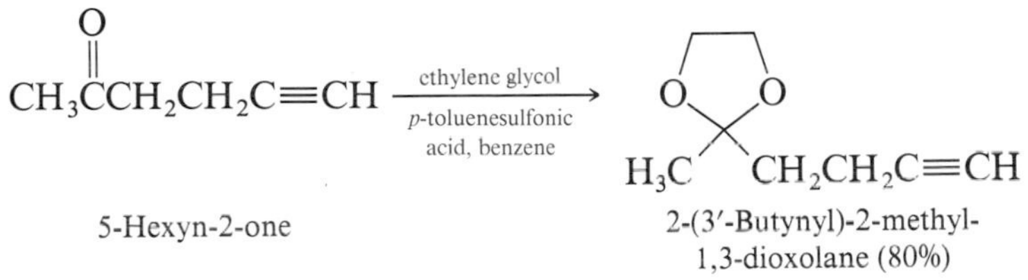

5-Hexyn-2-one

2-(3′-Butynyl)-2-methyl-
1,3-dioxolane (80%)

(b) Alkylation of alkyne:

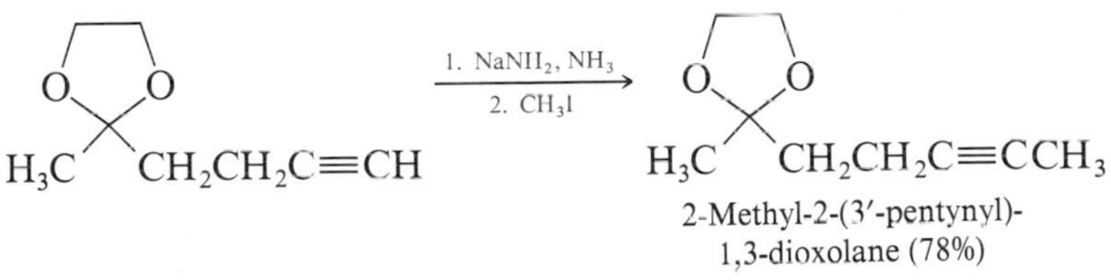

2-Methyl-2-(3′-pentynyl)-
1,3-dioxolane (78%)

(c) Unmasking of the carbonyl group by hydrolysis:

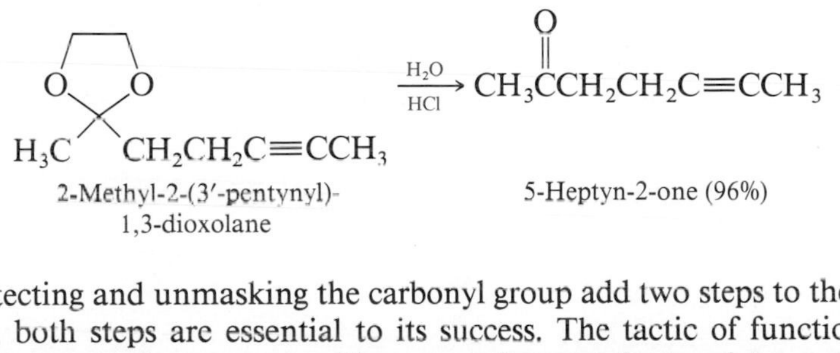

2-Methyl-2-(3′-pentynyl)-
1,3-dioxolane

5-Heptyn-2-one (96%)

While protecting and unmasking the carbonyl group add two steps to the synthetic procedure, both steps are essential to its success. The tactic of functional group protection is frequently encountered in preparative organic chemistry, and considerable attention has been paid to the design of effective protecting groups for a variety of functionalities.

18.12 CYANOHYDRIN FORMATION

Structurally a cyanohydrin corresponds to the addition of hydrogen cyanide to the carbonyl group of an aldehyde or ketone:

$$
\underset{\substack{\text{Aldehyde} \\ \text{or ketone}}}{\overset{\overset{\displaystyle O}{\|}}{R\!C\!R'}} + \underset{\substack{\text{Hydrogen} \\ \text{cyanide}}}{HC\!\equiv\!N} \longrightarrow \underset{\text{Cyanohydrin}}{\overset{\overset{\displaystyle OH}{|}}{\underset{\underset{\displaystyle C\equiv N}{|}}{R\!C\!R'}}}
$$

The mechanism of this reaction is analogous to that of base-catalyzed hydration. Cyanide ion is the nucleophile that attacks the carbonyl carbon in the first step, and hydrogen cyanide is the proton donor in the second.

Nucleophilic addition step:

$$
\underset{\text{Cyanide ion}}{:N\!\equiv\!C\overset{-}{:}} + \underset{\text{Aldehyde or ketone}}{\overset{R}{\underset{R'}{C}}\!=\!\overset{..}{O}\!:} \longrightarrow \underset{\text{Conjugate base of cyanohydrin}}{:N\!\equiv\!C\!-\!\overset{R}{\underset{R'}{C}}\!-\!\overset{..}{\underset{..}{O}}\!:^{-}}
$$

Proton transfer step:

$$
\underset{\substack{\text{Conjugate base of} \\ \text{cyanohydrin}}}{:N\!\equiv\!C\!-\!\overset{R}{\underset{R'}{C}}\!-\!\overset{..}{\underset{..}{O}}\!:^{-}} + \underset{\substack{\text{Hydrogen} \\ \text{cyanide}}}{H\!-\!C\!\equiv\!N\!:} \longrightarrow \underset{\text{Cyanohydrin}}{N\!\equiv\!C\!-\!\overset{R}{\underset{R'}{C}}\!-\!\overset{..}{O}H} + \underset{\text{Cyanide ion}}{^{-}\!:C\!\equiv\!N\!:}
$$

Overall, the reaction consumes hydrogen cyanide and is catalytic in cyanide ion. Adding an acid to a solution containing an aldehyde or ketone and sodium or potassium cyanide ensures that sufficient cyanide ion is always present to cause the reaction to proceed at a reasonable rate.

$$
\underset{\text{Acetone}}{\overset{\overset{\displaystyle O}{\|}}{CH_3CCH_3}} \xrightarrow[\text{then H}_2\text{SO}_4]{\text{NaCN, H}_2\text{O}} \underset{\substack{\text{2-Cyano-2-propanol (77–78\%)} \\ \text{(acetone cyanohydrin)}}}{\overset{\overset{\displaystyle OH}{|}}{\underset{\underset{\displaystyle C\equiv N}{|}}{CH_3CCH_3}}}
$$

The reaction is reversible and the position of equilibrium depends on the usual steric and electronic factors that govern nucleophilic addition to carbonyl groups. Aldehydes and unhindered ketones give good yields of cyanohydrins.

PROBLEM 18.10   Equilibrium constants for the dissociation (K_{diss}) of cyanohydrins according to the equation

$$
\underset{\text{Cyanohydrin}}{\overset{\overset{\displaystyle OH}{|}}{\underset{\underset{\displaystyle CN}{|}}{R\!-\!C\!-\!R'}}}
\quad\xrightarrow{\;K_{diss}\;}\quad
\underset{\substack{\text{Aldehyde}\\\text{or ketone}}}{\overset{\overset{\displaystyle O}{\|}}{R\!-\!C\!-\!R'}}
\;+\;
\underset{\substack{\text{Hydrogen}\\\text{cyanide}}}{HCN}
$$

have been measured for a number of cyanohydrins. Which cyanohydrin in each of the following pairs has the greater dissociation constant?

(a) $\underset{\displaystyle}{CH_3CH_2\overset{\overset{\displaystyle OH}{|}}{C}HCN}$ or $\underset{\displaystyle}{(CH_3)_2\overset{\overset{\displaystyle OH}{|}}{C}CN}$ /

(b) $C_6H_5\overset{\overset{\displaystyle OH}{|}}{C}HCN$ or $C_6H_5\underset{\underset{\displaystyle CH_3}{|}}{\overset{\overset{\displaystyle OH}{|}}{C}}CN$ √

(c)   [cyclohexane with HO and CN on C1]   or   [1,2,3,4-tetrahydronaphthalene with HO and CN on C1] √

SAMPLE SOLUTION   (a)   The reaction as written is the reverse of cyanohydrin formation, and the principles that govern equilibria in nucleophilic addition to carbonyl groups apply in reverse order to the dissociation of cyanohydrins to aldehydes and ketones. Cyanohydrins of ketones dissociate more at equilibrium than do cyanohydrins of aldehydes. More strain due to crowding is relieved when a ketone cyanohydrin dissociates and a more stabilized carbonyl group is formed. The equilibrium constant K_{diss} is larger for

$$
\underset{\text{Acetone cyanohydrin}}{\overset{\overset{\displaystyle OH}{|}}{\underset{\underset{\displaystyle CN}{|}}{CH_3\!-\!C\!-\!CH_3}}}
\quad\xrightarrow{\;K_{diss}\;}\quad
\underset{\text{Acetone}}{\overset{\overset{\displaystyle O}{\|}}{CH_3\!-\!C\!-\!CH_3}}
\;+\;
\underset{\text{Hydrogen cyanide}}{HCN}
$$

than it is for

$$
\underset{\text{Propanal cyanohydrin}}{\overset{\overset{\displaystyle OH}{|}}{CH_3CH_2\!-\!C\!-\!HCN}}
\quad\xrightarrow{\;K_{diss}\;}\quad
\underset{\text{Propanal}}{\overset{\overset{\displaystyle O}{\|}}{CH_3CH_2\!-\!C\!-\!H}}
\;+\;
\underset{\text{Hydrogen cyanide}}{HCN}
$$

Cyanohydrin formation is a reaction of synthetic value in that a new carbon-carbon bond is made by this process and a cyano group may be converted to a carboxylic acid function (by hydrolysis, to be discussed in Section 20.12) or to an amine (by reduction, to be discussed in Section 23.10).

A few cyanohydrins and ethers of cyanohydrins occur naturally. One species of millipede stores benzaldehyde cyanohydrin, along with an enzyme that catalyzes its cleavage to benzaldehyde and hydrogen cyanide, in separate compartments above its legs. When attacked, the insect ejects a mixture of the cyanohydrin and the enzyme, repelling the invader by spraying it with hydrogen cyanide.

18.13   REACTION WITH PRIMARY AMINES. NUCLEOPHILIC ADDITION-ELIMINATION

Primary amines are compounds of the type RNH_2 or $ArNH_2$. They react with aldehydes and ketones to form the corresponding N-alkyl- or N-aryl-substituted *imines*.

$$
\underset{\substack{\text{Aldehyde} \\ \text{or ketone}}}{R\overset{\displaystyle O}{\overset{\|}{C}}R'} \; + \; \underset{\substack{\text{Primary amine}}}{R''NH_2} \; \longrightarrow \; \underset{\substack{N\text{-substituted} \\ \text{imine}}}{R\overset{\displaystyle NR''}{\overset{\|}{C}}R'} \; + \; \underset{\substack{\text{Water}}}{H_2O}
$$

N-Substituted imines of the type formed in this reaction are known as *Schiff's bases*.

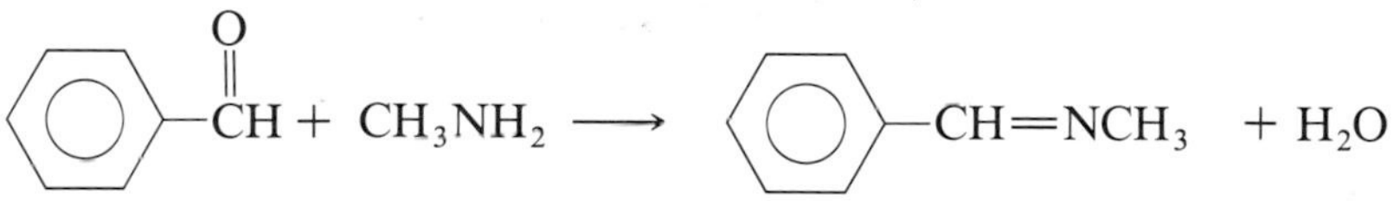

Benzaldehyde   Methylamine N-Benzylidenemethylamine (70%)   Water

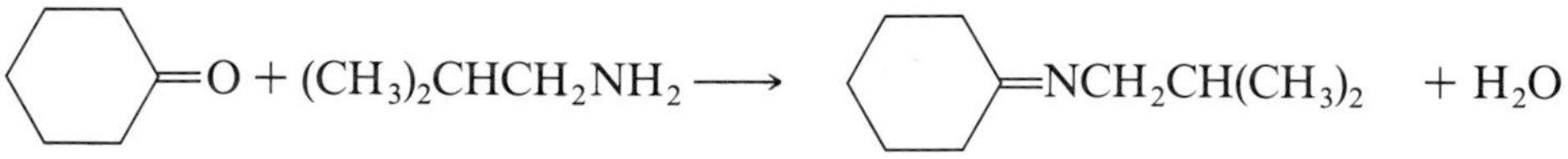

Cyclohexanone   Isobutylamine N-Cyclohexylideneisobutylamine (79%)   Water

Schiff's base formation occurs in two stages. The first stage is nucleophilic addition of the amine to the carbonyl compound to give an intermediate known as a *carbinolamine*.

Nucleophilic addition stage

$$
\underset{\substack{\text{Aldehyde} \\ \text{or ketone}}}{R\overset{\displaystyle O}{\overset{\|}{C}}R'} \; + \; \underset{\substack{\text{Primary} \\ \text{amine}}}{R''NH_2} \; \rightleftharpoons \; \underset{\substack{\text{Carbinolamine}}}{R\underset{\displaystyle HNR''}{\overset{\displaystyle OH}{\underset{|}{\overset{|}{C}}}}R'}
$$

Once formed, the carbinolamine undergoes elimination of water, yielding the imine.

Elimination stage

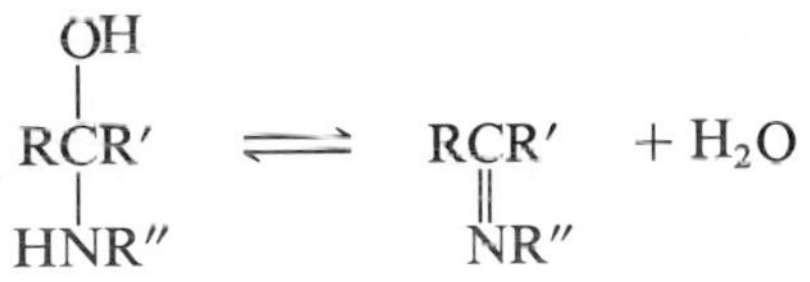

$$
\underset{\substack{\text{Carbinolamine}}}{\overset{\overset{\displaystyle OH}{\overset{\displaystyle |}{}}}{\underset{\underset{\displaystyle HNR''}{\underset{\displaystyle |}{}}}{RCR'}}} \;\rightleftharpoons\; \underset{\substack{N\text{-Substituted}\\ \text{imine}}}{\overset{\overset{\displaystyle }{}}{\underset{\underset{\displaystyle NR''}{\overset{\displaystyle \|}{}}}{RCR'}}} \;+\; \underset{\substack{\text{Water}}}{H_2O}
$$

Both the addition and the elimination phase of the reaction are sensitive to acid catalysis. Careful control of pH is essential, since sufficient acid must be present to give a reasonable equilibrium concentration of the protonated form of the aldehyde or ketone. However, too acidic a reaction medium converts the amine to its protonated form, a form that is not nucleophilic, and retards reaction.

PROBLEM 18.11   Write the structure of the carbinolamine intermediate and the imine product formed in the reaction of each of the following:

(a) Acetaldehyde and benzylamine, $C_6H_5CH_2NH_2$
(b) Benzaldehyde and butylamine, $CH_3CH_2CH_2CH_2NH_2$
(c) Cyclohexanone and *tert*-butylamine, $(CH_3)_3CNH_2$

(d) Acetophenone and cyclohexylamine, ⬡—NH_2

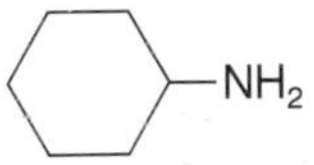

SAMPLE SOLUTION   (a)   The carbinolamine is formed by nucleophilic addition of the amine to the carbonyl group. Its dehydration gives the imine product.

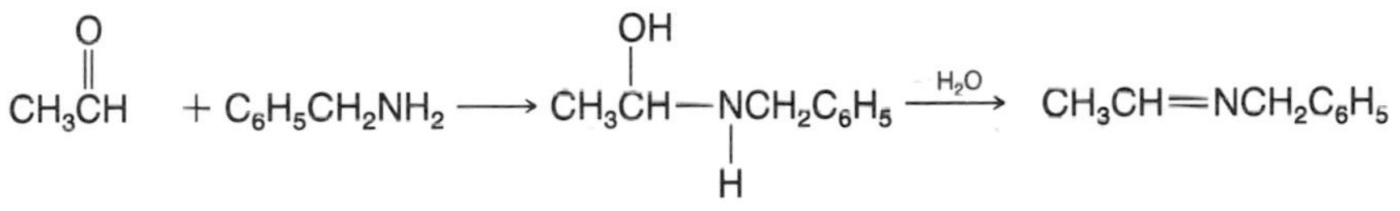

$$
\underset{\substack{\text{Acetaldehyde}}}{\overset{\overset{\displaystyle O}{\overset{\displaystyle \|}{}}}{CH_3CH}} \;+\; \underset{\substack{\text{Benzylamine}}}{C_6H_5CH_2NH_2} \longrightarrow \underset{\substack{\text{Carbinolamine}\\ \text{intermediate}}}{\overset{\overset{\displaystyle OH}{\overset{\displaystyle |}{}}}{\underset{\underset{\displaystyle H}{\underset{\displaystyle |}{}}}{CH_3CH-NCH_2C_6H_5}}} \;\xrightarrow{H_2O}\; \underset{\substack{\text{Imine product}\\ (N\text{-ethylidenebenzylamine})}}{CH_3CH{=}NCH_2C_6H_5}
$$

Imine formation is a reversible reaction. Both imine formation and imine hydrolysis have been extensively studied from a mechanistic perspective because of their relevance to biochemical processes. Many biological reactions involve initial binding of a carbonyl compound to an enzyme or coenzyme by way of Schiff's base formation.

18.14   REACTION WITH SECONDARY AMINES. ENAMINES

Secondary amines are compounds of the type R_2NH. They are nucleophilic and add to aldehydes and ketones to form carbinolamines, but their carbinolamine intermediates can dehydrate to a neutral product only in the direction that leads to a carbon-carbon double bond.

$$\underset{\substack{\text{Aldehyde}\\\text{or ketone}}}{RCH_2\overset{\displaystyle O}{\overset{\|}{C}}R'} + \underset{\substack{\text{Secondary}\\\text{amine}}}{R''_2\overset{..}{N}H} \ \rightleftharpoons\ \underset{\text{Carbinolamine}}{RCH_2\underset{\underset{\displaystyle R''\quad R''}{\diagdown N\diagup}}{\overset{\displaystyle OH}{\overset{|}{C}}}-R'} \ \overset{-H_2O}{\rightleftharpoons}\ \underset{\text{Enamine}}{RCH\!=\!\underset{\underset{\displaystyle R''\quad R''}{\diagdown N\diagup}}{CR'}}$$

The product of this dehydration is an alkenyl-substituted amine, or *enamine*.

Cyclopentanone   Pyrrolidine   *N*-(1-Cyclopentenyl)   Water
pyrrolidine (80–90%)

PROBLEM 18.12   Write the structure of the carbinolamine intermediate and the enamine product formed in the reaction of each of the following:

(a) Propanal and dimethylamine,
     CH₃NHCH₃

(b) 3-Pentanone and pyrrolidine

(c) Acetophenone and HN⬡

SAMPLE SOLUTION   (a) Nucleophilic addition of dimethylamine to the carbonyl group of propanal produces a carbinolamine.

$$\underset{\text{Propanal}}{CH_3CH_2\overset{\displaystyle O}{\overset{\|}{C}}H} + \underset{\substack{\text{Dimethylamine}}}{CH_3\underset{\underset{\displaystyle H}{|}}{N}CH_3} \ \longrightarrow\ \underset{\text{Carbinolamine intermediate}}{CH_3CH_2\underset{\underset{\displaystyle OH}{|}}{CH}-N\overset{\displaystyle CH_3}{\underset{\displaystyle CH_3}{\diagdown}}}$$

Dehydration of this carbinolamine yields the enamine.

$$\underset{\text{Carbinolamine intermediate}}{CH_3CH_2\underset{\underset{\displaystyle OH}{|}}{CH}-N\overset{\displaystyle CH_3}{\underset{\displaystyle CH_3}{\diagdown}}} \ \overset{-H_2O}{\longrightarrow}\ \underset{\text{\textit{N}-(1-Propenyl)dimethylamine}}{CH_3CH\!=\!CH-N\overset{\displaystyle CH_3}{\underset{\displaystyle CH_3}{\diagdown}}}$$

Enamines are used in synthetic organic chemistry in ways that will be described in Section 19.17.

18.15 REACTION WITH DERIVATIVES OF AMMONIA

A number of compounds of the type ZNH_2 react with aldehydes and ketones in a manner analogous to that of primary amines:

$$\underset{\substack{\text{Aldehyde} \\ \text{or ketone}}}{\overset{\overset{\textstyle O}{\|}}{RCR'}} + \underset{\substack{\text{Derivative} \\ \text{of ammonia}}}{ZNH_2} \longrightarrow \overset{\overset{\textstyle NZ}{\|}}{RCR'} + H_2O$$

The $\text{C}{=}\text{O}$ functional group is converted to $\text{C}{=}\text{NZ}$ and a molecule of water is formed. Among nucleophiles of the type ZNH_2 those most often encountered are hydroxylamine ($HONH_2$), hydrazine (H_2NNH_2), phenylhydrazine ($C_6H_5NHNH_2$),

and semicarbazide ($H_2N\overset{\overset{\textstyle O}{\|}}{C}NHNH_2$).

Hydroxylamine reacts with aldehydes and ketones to give *oximes:*

$$\underset{\text{Heptanal}}{\overset{\overset{\textstyle O}{\|}}{CH_3(CH_2)_5CH}} + \underset{\text{Hydroxylamine}}{HONH_2} \longrightarrow \underset{\substack{\text{Heptanal oxime} \\ (81-93\%)}}{\overset{\overset{\textstyle NOH}{\|}}{CH_3(CH_2)_5CH}} + \underset{\text{Water}}{H_2O}$$

The products obtained on reaction of aldehydes and ketones with hydrazine are called *hydrazones.*

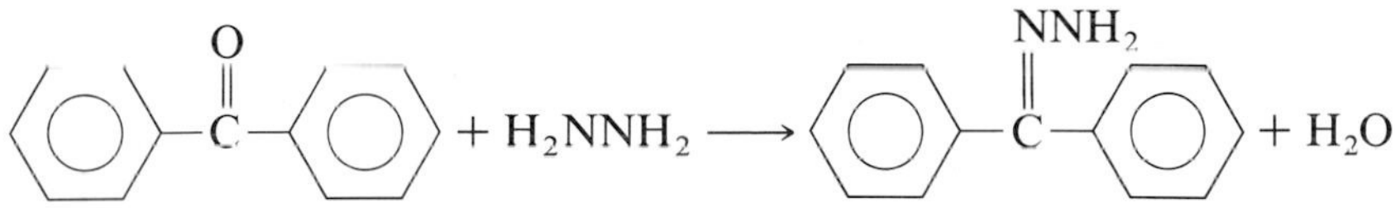

Benzophenone   Hydrazine   Benzophenone hydrazone (73%)   Water

Hydrazones are intermediates in the Wolff-Kishner reduction of aldehydes and ketones (Section 13.8).

Phenylhydrazones result from reaction of phenylhydrazine with aldehydes and ketones.

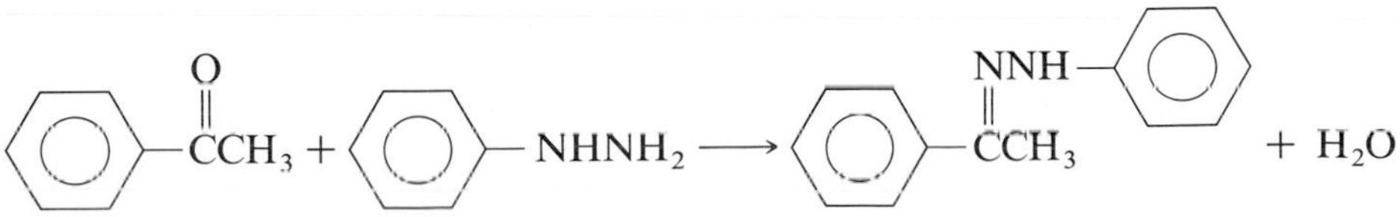

Acetophenone   Phenylhydrazine   Acetophenone phenylhydrazone (87-91%)   Water

Substituted derivatives of phenylhydrazine such as *p*-nitrophenylhydrazine and 2,4-dinitrophenylhydrazine react similarly.

Semicarbazide converts aldehydes and ketones to semicarbazones.

$$CH_3\overset{\displaystyle O}{\overset{\|}{C}}(CH_2)_9CH_3 + H_2N\overset{\displaystyle O}{\overset{\|}{N}}CNHNH_2 \xrightarrow[\substack{\text{ethanol-}\\\text{water}}]{\substack{\text{sodium}\\\text{acetate}}} CH_3\overset{\displaystyle \overset{\displaystyle O}{\overset{\|}{NNHCNH_2}}}{\overset{\|}{C}}(CH_2)_9CH_3 + H_2O$$

2-Dodecanone   Semicarbazide   2-Dodecanone   Water
                                                    semicarbazone (93%)

The mechanisms by which these various derivatives of ammonia react with aldehydes and ketones are all similar to the nucleophilic addition-elimination mechanism described for the reaction of aldehydes and ketones with primary amines.

18.16 THE WITTIG REACTION

A synthetic method of broad scope uses *phosphorus ylides* to convert aldehydes and ketones to alkenes:

$$\overset{R}{\underset{R'}{\diagdown}}C=O + (C_6H_5)_3\overset{+}{P}-\overset{..}{\overset{-}{C}}\overset{A}{\underset{B}{\diagup}} \longrightarrow \overset{R}{\underset{R'}{\diagdown}}C=C\overset{A}{\underset{B}{\diagup}} + (C_6H_5)_3\overset{+}{P}-O^-$$

Aldehyde or   Triphenylphosphonium   Alkene   Triphenylphosphine
ketone   ylide   oxide

The synthetic potential of this reaction was demonstrated by the German chemist Georg Wittig (Nobel prize in chemistry 1979). It is called the *Wittig reaction* and is a standard method for the preparation of alkenes.

Wittig reactions may be carried out in a number of different solvents; those chosen most often are tetrahydrofuran (THF) and dimethyl sulfoxide (DMSO).

$$\bigcirc\!\!=\!O + (C_6H_5)_3\overset{+}{P}-\overset{..}{\overset{-}{C}}H_2 \xrightarrow{\text{DMSO}} \bigcirc\!\!=\!CH_2 + (C_6H_5)_3\overset{+}{P}-O^-$$

Cyclohexanone   Methylenetri-   Methylene-   Triphenyl-
              phenylphosphorane   cyclohexane (86%)   phosphine oxide

One of the features of the Wittig reaction that makes it a particularly attractive synthetic procedure is its absolute regiospecificity. The position at which the double bond is introduced is never in doubt. The double bond is formed between the carbonyl carbon of the aldehyde or ketone and the negatively charged carbon of the ylide.

PROBLEM 18.13   Identify the alkene product in each of the following Wittig reactions:

(a)   Benzaldehyde + $(C_6H_5)_3\overset{+}{P}-\overset{-}{\diagdown}\!\!\bigcirc$

(b)   Benzaldehyde + $(C_6H_5)_3\overset{+}{P}-\overset{..}{\overset{-}{C}}HCH_3$

(c)   Formaldehyde + $(C_6H_5)_3\overset{+}{P}-\overset{..}{\overset{-}{C}}HC_6H_5$

(d)   Butanal + $(C_6H_5)_3\overset{+}{P}-\overset{..}{\overset{-}{C}}HCH=CH_2$

(e)   Cyclohexyl methyl ketone + $(C_6H_5)_3\overset{+}{P}-\overset{..}{\overset{-}{C}}H_2$

SAMPLE SOLUTION   (a)   In a Wittig reaction the negatively charged substituent attached to phosphorus is transferred to the aldehyde or ketone, replacing the carbonyl oxygen. The reaction shown has been used to prepare the indicated alkene in 65 percent yield.

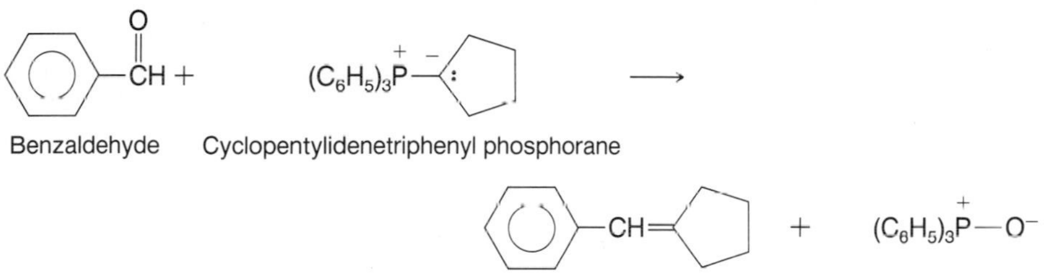

Benzaldehyde   Cyclopentylidenetriphenyl phosphorane

Benzylidenecyclopentane (65%)   Triphenylphosphine oxide

Before the mechanism of the Wittig reaction is described, a brief note about ylides is in order. Ylides are neutral molecules that have two oppositely charged atoms, each with an octet of electrons, directly bonded to each other. In the ylides used in the Wittig reaction, phosphorus has eight electrons and is positively charged; its attached carbon also has eight electrons and is negatively charged. The negatively charged carbon of an ylide has carbanionic character and can act as a nucleophile toward carbonyl groups.

The ability of ylides to act as sources of nucleophilic carbon permits them to add to the carbonyl group of an aldehyde or ketone. The product of this nucleophilic addition step is a dipolar intermediate called a *betaine*.

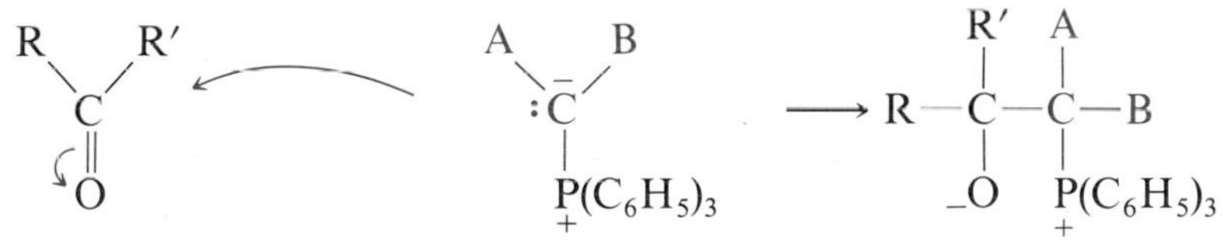

Aldehyde or   Triphenylphosphonium   Betaine
ketone   ylide

This betaine intermediate is unstable and rapidly fragments, probably by way of a second intermediate containing a four-membered ring, to an alkene and triphenylphosphine oxide.

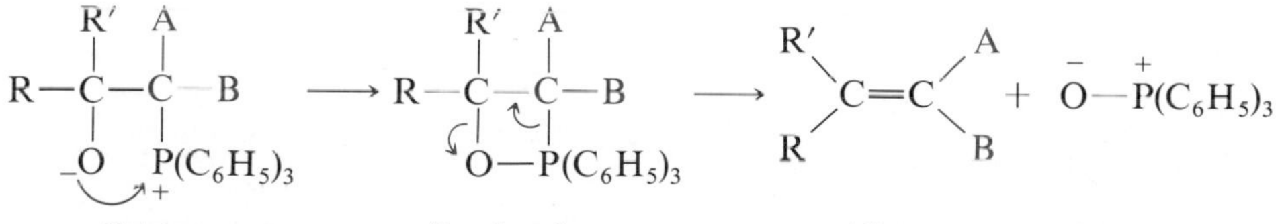

Betaine   Oxaphosphetane   Alkene   Triphenylphosphine
                                                                                              oxide

The driving force for the Wittig reaction is the formation of the very strong phosphorus-oxygen bond of triphenylphosphine oxide (bond energy ~130 kcal/mol).

18.17 PLANNING AN ALKENE SYNTHESIS VIA THE WITTIG REACTION

In order to identify the carbonyl compound and the ylide required to produce a given alkene, mentally disconnect the double bond so that one of its carbons is derived from a carbonyl group and the other is derived from an ylide. Taking styrene as a representative example, we see that two such disconnections are possible; either benzaldehyde or formaldehyde is an appropriate precursor.

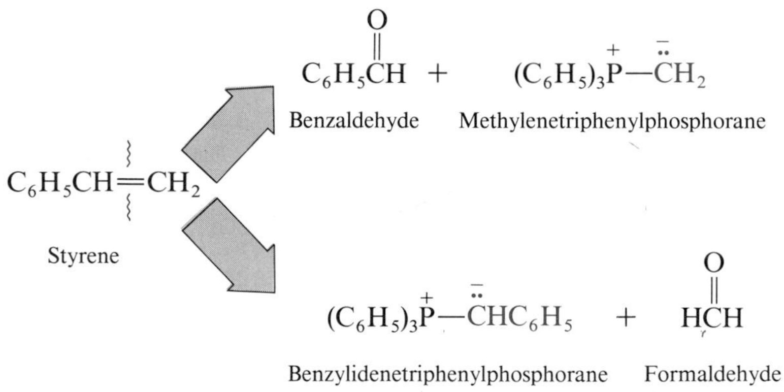

$$C_6H_5CH\overset{\}{=}CH_2$$

Styrene

$$\underset{\text{Benzaldehyde}}{C_6H_5\overset{O}{\overset{||}{C}}H} + \underset{\text{Methylenetriphenylphosphorane}}{(C_6H_5)_3\overset{+}{P}-\overset{..}{C}H_2}$$

$$\underset{\text{Benzylidenetriphenylphosphorane}}{(C_6H_5)_3\overset{+}{P}-\overset{..}{C}HC_6H_5} + \underset{\text{Formaldehyde}}{H\overset{O}{\overset{||}{C}}H}$$

Either route is a feasible one, and indeed styrene has been prepared from both combinations of reactants. Typically there will be two Wittig routes to an alkene and any choice between them is made on the basis of availability of the particular starting materials.

PROBLEM 18.14 What combinations of carbonyl compound and ylide could you use to prepare each of the following alkenes?

(a) $CH_3CH_2CH_2CH{=}CCH_2CH_3$
$\qquad\qquad\qquad\qquad\quad |$
$\qquad\qquad\qquad\qquad\ CH_3$

(b) $CH_3CH_2CH_2CH{=}CH_2$

(c) $C_6H_5CH_2CH{=}C(CH_2CH_3)_2$

(d)

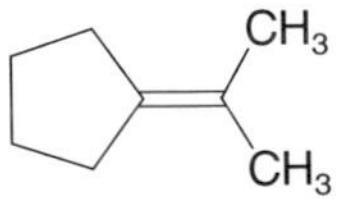

SAMPLE SOLUTION (a) There are two Wittig reaction routes that lead to the target molecule. One is represented by

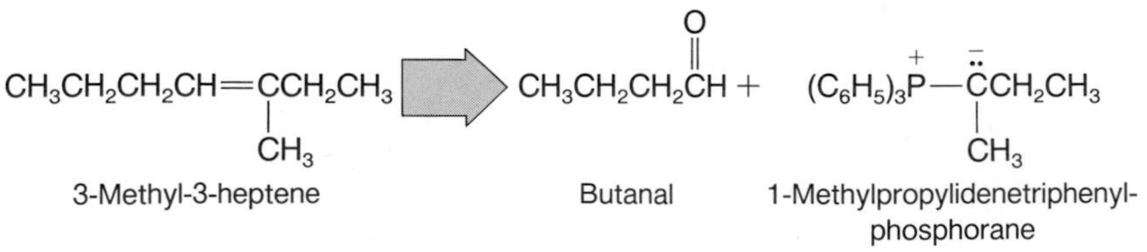

$$\underset{\text{3-Methyl-3-heptene}}{CH_3CH_2CH_2CH{=}CCH_2CH_3} \Longrightarrow \underset{\text{Butanal}}{CH_3CH_2CH_2\overset{O}{\overset{||}{C}}H} + \underset{\substack{\text{1-Methylpropylidenetriphenyl-}\\\text{phosphorane}}}{(C_6H_5)_3\overset{+}{P}-\overset{..}{C}CH_2CH_3}$$

The other is

$$CH_3CH_2CH_2CH=CCH_2CH_3 \quad \Rightarrow \quad CH_3CH_2CH_2\overset{..}{C}H-\overset{+}{P}(C_6H_5)_3 \; + \; Cl \; I_3\overset{\overset{O}{\|}}{C}CH_2CH_3$$
$$\underset{CH_3}{|}$$

| 3-Methyl-3-heptene | Butylidenetriphenylphosphorane | 2-Butanone |

Phosphorus ylides are prepared from alkyl halides according to a two-step sequence. The first step is a nucleophilic substitution of the S_N2 type by triphenylphosphine on an alkyl halide to give an alkyltriphenylphosphonium salt.

$$(C_6H_5)_3P\!: \quad + \quad \underset{B}{\overset{A}{\diagdown}}CH\!\!\overset{\curvearrowright}{-}\!X \longrightarrow (C_6H_5)_3\overset{+}{P}\!\!-\!\!\overset{\overset{A}{|}}{C}H\!-\!B \quad X^-$$

| Triphenylphosphine | Alkyl halide | Alkyltriphenylphosphonium halide |

Triphenylphosphine is a very powerful nucleophile, yet is not strongly basic. Consequently, methyl, primary, and secondary alkyl halides are all suitable substrates.

$$(C_6H_5)_3P\!: \quad + \quad CH_3Br \quad \xrightarrow{\text{benzene}} \quad (C_6H_5)_3\overset{+}{P}\!\!-\!CH_3 \; Br^-$$

| Triphenylphosphine | Bromomethane | Methyltriphenylphosphonium bromide (99%) |

The alkyltriphenylphosphonium salt products are ionic and crystallize in high yield from the nonpolar solvents in which they are prepared. After isolation of the alkyltriphenylphosphonium halide, it is converted to the desired ylide by deprotonation with a strong base.

$$(C_6H_5)_3\overset{+}{P}\!\!-\!\!\underset{\underset{H}{|}}{\overset{\overset{A}{|}}{C}}\!\!-\!B \quad + \; Y^- \longrightarrow \quad (C_6H_5)_3\overset{+}{P}\!\!-\!\overset{..}{C}\!\!\underset{B}{\overset{A}{\diagup}} \quad + \quad HY$$

| Alkyltriphenylphosphonium salt | Base | Triphenylphosphonium ylide | Conjugate acid of base used |

Suitable strong bases include the sodium salt of dimethyl sulfoxide (in dimethyl sulfoxide as the solvent) and organolithium reagents (in diethyl ether or tetrahydrofuran).

$$(C_6H_5)_3\overset{+}{P}\!\!-\!CH_3 \; Br^- + NaCH_2\overset{\overset{O}{\|}}{S}CH_3 \xrightarrow{\text{DMSO}} (C_6H_5)_3\overset{+}{P}\!\!-\!\overset{..}{C}H_2 + CH_3\overset{\overset{O}{\|}}{S}CH_3 + NaBr$$

| Methyltriphenylphosphonium bromide | Sodiomethyl methyl sulfoxide | Methylenetriphenylphosphorane | Dimethyl sulfoxide | Sodium bromide |

Normally the ylides are not isolated but are treated directly in the solution in which they were generated with the appropriate aldehyde or ketone.

18.18   STEREOSELECTIVE ADDITION TO CARBONYL GROUPS

Nucleophilic addition to carbonyl groups sometimes leads to a mixture of stereoisomeric products. The preferred direction of attack is frequently controlled by steric factors, with the nucleophile approaching the carbonyl group at its less hindered face. Sodium borohydride reduction of 7,7-dimethylbicyclo[2.2.1]heptan-2-one offers an instructive illustration of this point.

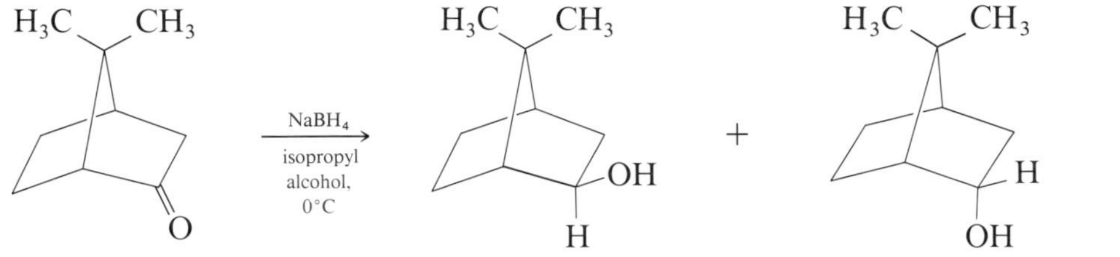

| 7,7-Dimethylbicyclo[2.2.1] heptan-2-one | *exo*-7,7-Dimethylbicyclo[2.2.1] heptan-2-ol (80%) | *endo*-7,7-Dimethylbicyclo[2.2.1] heptan-2-ol (20%) |

Approach of borohydride to the upper face of the carbonyl group is sterically hindered by one of the methyl groups. The bottom face of the carbonyl group is less congested and the major product is formed by hydride transfer from this direction.

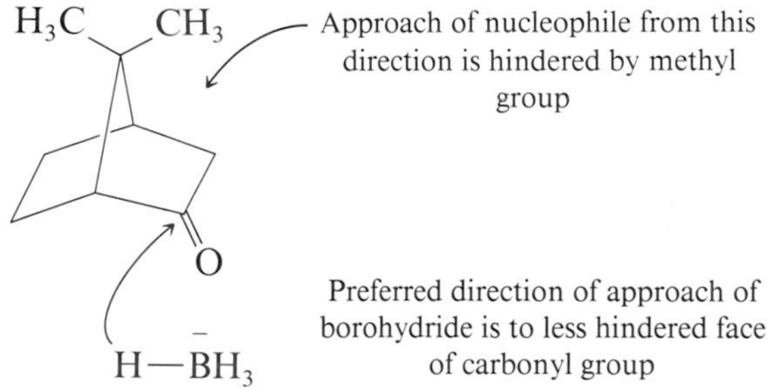

Approach of nucleophile from this direction is hindered by methyl group

Preferred direction of approach of borohydride is to less hindered face of carbonyl group

The reduction reaction is *stereoselective*. A single starting material has the potential of forming two stereoisomeric forms of product but yields one isomer preferentially.

It is possible to predict the preferred stereochemical path of nucleophilic addition if one face of a carbonyl group is significantly more hindered to the approach of the reagent than the other. When no clear distinction between the two faces is evident, other more subtle effects, which are still incompletely understood, come into play.

Enzyme-catalyzed reductions of carbonyl groups are, more often than not, completely stereoselective. Pyruvic acid is converted exclusively to (S)-$(+)$-lactic acid by the lactate dehydrogenase–NADH system (Section 16.13). The enantiomer (R)-$(-)$-lactic acid is not formed.

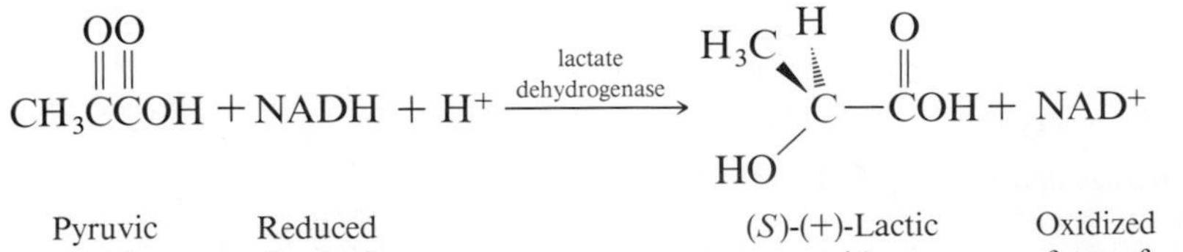

| Pyruvic acid | Reduced form of coenzyme | | (S)-$(+)$-Lactic acid | Oxidized form of coenzyme |

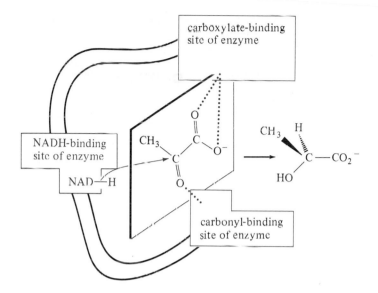

FIGURE 18.6 Schematic representation of enzyme-mediated reduction of pyruvate to (*S*)-(+)-lactate. A preferred orientation of binding of the substrate to the enzyme, coupled with a prescribed location of the reducing agent, the coenzyme NADH, leads to hydrogen transfer exclusively to a single face of the carbonyl group.

Here the enzyme, a chiral molecule, binds the coenzyme and substrate in such a way that hydrogen is transferred exclusively to the face of the carbonyl group that leads to (*S*)-(+)-lactic acid (Figure 18.6).

The stereochemical outcome of enzyme-mediated reactions depends heavily on the way the protein chain is folded. Aspects of protein conformation will be discussed in Chapter 29.

18.19 OXIDATION OF ALDEHYDES

Aldehydes are readily oxidized to carboxylic acids by a number of reagents, including those based on Cr(VI) and Mn(VII).

$$\underset{\text{Aldehyde}}{\overset{\displaystyle O}{\underset{\|}{R\text{C}H}}} \xrightarrow{\text{oxidize}} \underset{\text{Carboxylic acid}}{\overset{\displaystyle O}{\underset{\|}{R\text{C}OH}}}$$

$$\underset{\text{Heptanal}}{\overset{\displaystyle O}{\underset{\|}{CH_3(CH_2)_5CH}}} \xrightarrow[\substack{H_2SO_4,\ H_2O \\ 20°C}]{KMnO_4} \underset{\text{Heptanoic acid (76–78\%)}}{\overset{\displaystyle O}{\underset{\|}{CH_3(CH_2)_5COH}}}$$

Furfural $\xrightarrow[H_2SO_4,\ H_2O]{K_2Cr_2O_7}$ Furoic acid (75%)

Mechanistically these reactions probably proceed through the hydrate of the aldehyde and follow a course similar to that of alcohol oxidation.

$$
\underset{\substack{\text{Aldehyde}}}{\overset{\displaystyle O}{\overset{\|}{RCH}}} + H_2O \ \rightleftharpoons\ \underset{\substack{\text{Geminal diol} \\ \text{(hydrate)}}}{\overset{\displaystyle OH}{\underset{\displaystyle OH}{RCH}}} \xrightarrow{\text{oxidize}} \underset{\substack{\text{Carboxylic} \\ \text{acid}}}{\overset{\displaystyle O}{\overset{\|}{RCOH}}}
$$

Silver oxide is a very mild oxidizing agent. Only a few functional groups, including aldehydes, are affected by this reagent.

| Vanillin | | Vanillic acid (83–95%) |

Silver oxide is reduced to metallic silver in the process. This reaction forms the basis of the *Tollens test,* a qualitative test for aldehydes. The compound in question is treated with ammoniacal silver nitrate in a clean test tube. Formation of a shiny mirror of silver on the walls of the test tube is taken as a positive indication of the presence of an aldehyde or other easily oxidized functional group.

18.20 SPECTROSCOPIC ANALYSIS OF ALDEHYDES AND KETONES

Carbonyl groups are among the easiest functional groups to detect by infrared spectroscopy. The carbon-oxygen stretching mode of aldehydes and ketones gives rise to a strong absorption in the region 1710 to 1750 cm^{-1}. Conjugation of the carbonyl group with an aromatic ring or with a double bond lowers this value to 1670 to 1700 cm^{-1}.

Aldehydes exhibit weak absorptions due to C—H stretching of the formyl group near 2720 and 2820 cm^{-1}. The infrared spectrum of butanal is presented in Figure 18.7. Can you locate the peaks characteristic of the $\overset{\displaystyle O}{\overset{\|}{-CH}}$ group?

A very reliable way to identify aldehyde functions is by ^{1}H nmr spectroscopy. The chemical shift of the $\overset{\displaystyle O}{\overset{\|}{-CH}}$ proton occurs in a region ($\delta = 9$ to 10 ppm) where few other proton signals appear. As an illustration of this characteristic of the nmr spectra of aldehydes, examine that of 2-methylpropanal in Figure 18.8. The aldehyde proton appears as a doublet because it is coupled to the proton at C-2.

The presence of carbonyl groups can often be inferred by locating nmr absorptions corresponding to protons on adjacent carbons. Methyl ketones, for example, ethyl methyl ketone as shown in Figure 18.9, exhibit a sharp methyl singlet near $\delta = 2.0$ ppm. A methyl group bonded to a carbonyl is less shielded than the methyl of an

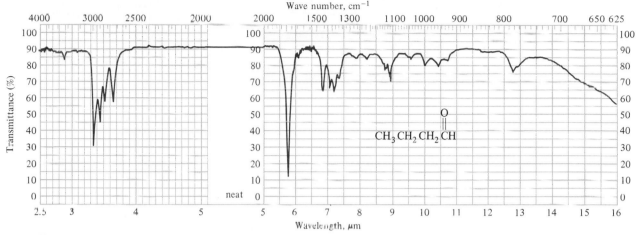

FIGURE 18.7   Infrared spectrum of butanal.

alkyl group. Similarly, the quartet centered at $\delta = 2.4$ ppm is due to the methylene protons of the ethyl group and appears at lower field than a methylene group in an alkane.

Aldehydes and ketones are also readily identified by their ^{13}C nmr spectra. Carbonyl carbons appear at very low field, some 190 to 220 ppm downfield from the tetramethylsilane internal standard. Figure 18.10 illustrates this for 3-heptanone, where the chemical shift of the carbonyl carbon is 211 ppm. All seven carbons of 3-heptanone are clearly evident in this spectrum. Note that the intensity of the peak due to the carbonyl carbon is significantly less than that of the others, even though each peak in the spectrum corresponds to a single carbon atom. This decreased intensity is a characteristic of carbons that bear no hydrogen substituents.

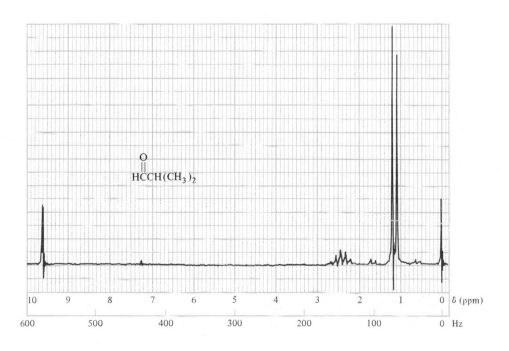

FIGURE 18.8   The ^1H nmr spectrum of 2-methylpropanal, showing the aldehyde proton as a doublet at low field.

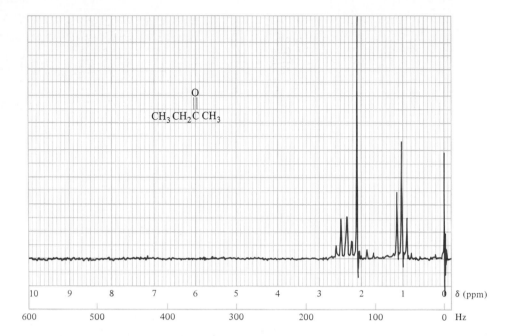

FIGURE 18.9 The ^{1}H nmr spectrum of ethyl methyl ketone.

Off-resonance decoupling of ^{13}C nmr spectra provides a ready means of distinguishing aldehydes from ketones. The carbonyl carbon of an aldehyde appears as a doublet because of coupling to its attached proton, whereas a ketone carbonyl bears only carbon substituents and its ^{13}C nmr signal is a singlet.

18.21 MASS SPECTROMETRY OF ALDEHYDES AND KETONES

Aldehydes and ketones typically give a prominent molecular ion peak in their mass spectra. Aldehydes also exhibit an M-1 peak. A major fragmentation pathway for both aldehydes and ketones leads to formation of acyl cations (acylium ions) by

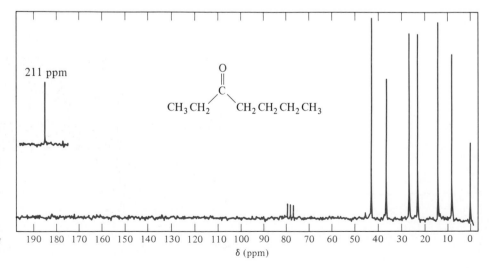

FIGURE 18.10 The ^{13}C nmr spectrum of 3-heptanone. *(Taken from "Carbon-13 NMR Spectra: A Collection of Assigned, Coded, and Indexed Spectra," by LeRoy F. Johnson and William C. Jankowski, Wiley-Interscience, New York, 1972. Reprinted by permission of John Wiley & Sons, Inc.)*

cleavage of an alkyl group from the carbonyl. The most intense peak in the mass spectrum of diethyl ketone, for example, is m/z 57, corresponding to loss of the ethyl radical from the molecular ion.

$$CH_3CH_2\overset{\overset{\displaystyle :O^{\;\dot+}}{\|}}{C}CH_2CH_3 \longrightarrow CH_3CH_2C{\equiv}\overset{..}{O}{}^+ + \cdot CH_2CH_3$$

m/z 86 $\qquad\qquad\qquad\qquad$ m/z 57

18.22 SUMMARY

There are a large number of synthetic routes to aldehydes and ketones. Those which appeared in earlier chapters are reviewed in Table 18.2. Several new methods that permit acyl chlorides to be converted to aldehydes and ketones have been described and are summarized in Table 18.5.

TABLE 18.5

Preparation of Aldehydes and Ketones from Acyl Chlorides

Product	Reagent or process	General equation and typical example
Aldehyde	Hydrogenation over a palladium–on–barium sulfate catalyst, the Rosenmund reduction (Section 10.6)	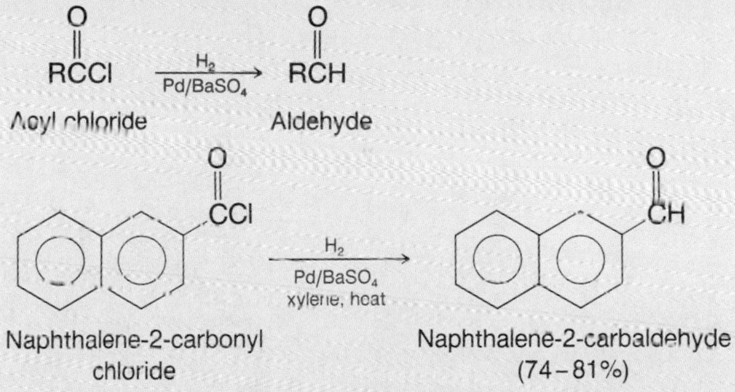
Ketone	Lithium diorganocuprate (Section 18.7)	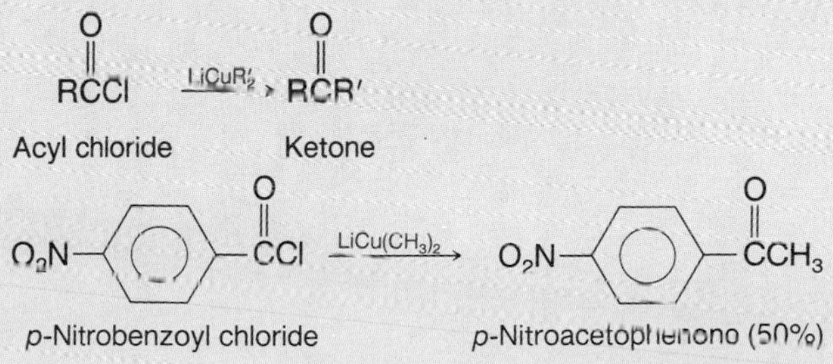
Ketone	Diorganocadmium reagent (Section 18.7)	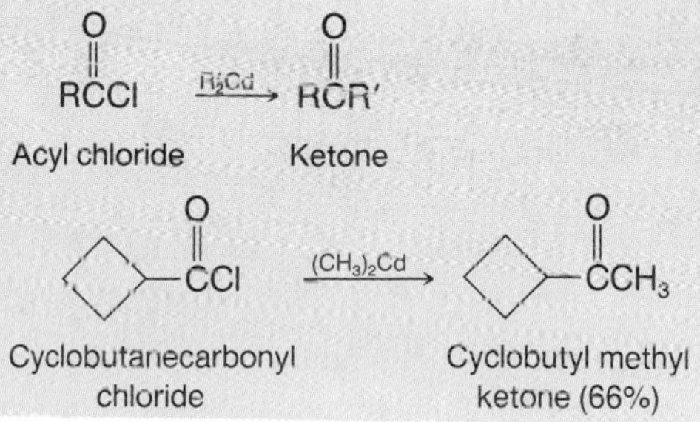

TABLE 18.6

Nucleophilic Addition to Aldehydes and Ketones

Reaction (section) and comments	General equation and typical example

Hydration (Section 18.9) Can be either acid- or base-catalyzed. Equilibrium constant is normally unfavorable for hydration of ketones unless R and/or R′ are strongly electron-withdrawing.

$$\underset{\substack{\text{Aldehyde}\\\text{or ketone}}}{\overset{\displaystyle O}{\overset{\|}{R C R'}}} + \underset{\text{Water}}{H_2O} \rightleftharpoons \underset{\text{Geminal diol}}{\overset{\displaystyle OH}{\underset{\displaystyle OH}{R C R'}}}$$

$$\underset{\substack{\text{Chloroacetone}\\\text{(90\% at equilibrium)}}}{\overset{\displaystyle O}{\overset{\|}{ClCH_2CCH_3}}} \underset{H_2O}{\rightleftharpoons} \underset{\substack{\text{Chloroacetone hydrate}\\\text{(10\% at equilibrium)}}}{\overset{\displaystyle OH}{\underset{\displaystyle OH}{ClCH_2CCH_3}}}$$

Acetal formation (Sections 18.10–18.11) Reaction is acid-catalyzed. Equilibrium constant normally favorable for aldehydes, unfavorable for ketones. Cyclic acetals from vicinal diols form readily.

$$\underset{\substack{\text{Aldehyde}\\\text{or ketone}}}{\overset{\displaystyle O}{\overset{\|}{R C R'}}} + \underset{\text{Alcohol}}{2R''OH} \overset{H^+}{\rightleftharpoons} \underset{\text{Acetal}}{\overset{\displaystyle OR''}{\underset{\displaystyle OR''}{R C R'}}} + \underset{\text{Water}}{H_2O}$$

m-Nitrobenzaldehyde   Methanol   *m*-Nitrobenzaldehyde dimethyl acetal (76–85%)

Cyanohydrin formation (Section 18.12) Reaction is base-catalyzed. Cyanohydrins are useful synthetic intermediates; cyano group can be hydrolyzed to —CO_2H or reduced to —CH_2NH_2.

$$\underset{\substack{\text{Aldehyde}\\\text{or ketone}}}{\overset{\displaystyle O}{\overset{\|}{R C R'}}} + \underset{\substack{\text{Hydrogen}\\\text{cyanide}}}{HCN} \rightleftharpoons \underset{\text{Cyanohydrin}}{\overset{\displaystyle OH}{\underset{\displaystyle CN}{R C R'}}}$$

$$\underset{\text{3-Pentanone}}{\overset{\displaystyle O}{\overset{\|}{CH_3CH_2CCH_2CH_3}}} \overset{KCN}{\underset{H^+}{\longrightarrow}} \underset{\text{3-Cyano-3-pentanol (75\%)}}{\overset{\displaystyle OH}{\underset{\displaystyle CN}{CH_3CH_2CCH_2CH_3}}}$$

Reaction with primary amines (Section 18.13) Isolated product is an imine (Schiff's base). A carbinolamine intermediate is formed, which undergoes dehydration to imine.

$$\underset{\substack{\text{Aldehyde}\\\text{or ketone}}}{\overset{\displaystyle O}{\overset{\|}{R C R'}}} + \underset{\substack{\text{Primary}\\\text{amine}}}{R''NH_2} \rightleftharpoons \underset{\text{Imine}}{\overset{\displaystyle NR''}{\overset{\|}{R C R''}}} + \underset{\text{Water}}{H_2O}$$

$$\underset{\text{2-Methylpropanal}}{\overset{\displaystyle O}{\overset{\|}{(CH_3)_2CHCH}}} + \underset{\textit{tert}\text{-Butylamine}}{(CH_3)_3CNH_2} \longrightarrow \underset{\substack{N\text{-(2-Methyl-1-propylidene)-}\\\textit{tert}\text{-butylamine (50\%)}}}{(CH_3)_2CHCH{=}NC(CH_3)_3}$$

TABLE 18.6 (continued)

Reaction (section) and comments	General equation and typical example
Reaction with secondary amines (Section 18.14) Isolated product is an enamine. Carbinolamine intermediate cannot dehydrate to a stable imine.	$$\underset{\substack{\text{Aldehyde}\\\text{or ketone}}}{RCCH_2R'} + \underset{\substack{\text{Secondary}\\\text{amine}}}{(R'')_2NH} \rightleftharpoons \underset{\text{Enamine}}{RC{=}CHR'} + \underset{\text{Water}}{H_2O}$$ Cyclohexanone + Morpholine $\xrightarrow[\text{heat}]{\text{benzene}}$ 1-Morpholinocyclohexene (85%)
Reaction with hydroxylamine (Section 18.15) An aldehyde or ketone is converted to an oxime by a nucleophilic addition-elimination reaction.	$$\underset{\substack{\text{Aldehyde}\\\text{or ketone}}}{RCR'} + \underset{\text{Hydroxylamine}}{H_2NOH} \longrightarrow \underset{\text{Oxime}}{\overset{NOH}{RCR'}} + \underset{\text{Water}}{H_2O}$$ $$\underset{\text{Propanal}}{CH_3CH_2CH} + \underset{\text{Hydroxylamine}}{H_2NOH} \xrightarrow{\text{water}} \underset{\text{Propanal oxime (77\%)}}{CH_3CH_2CH{=}NOH}$$
Reaction with hydrazine and derivatives of hydrazine (Section 18.15) Analogous to the reaction of other amines with aldehydes and ketones, a carbinolamine is formed by nucleophilic addition and then undergoes dehydration. These hydrazones are crystalline solids used as identification derivatives.	$$\underset{\substack{\text{Aldehyde}\\\text{or ketone}}}{RCR'} + \underset{\text{Hydrazine}}{H_2NNHR''} \longrightarrow \underset{\text{Hydrazone}}{\overset{NNHR''}{RCR'}} + \underset{\text{Water}}{H_2O}$$ 2-Dodecanone 2,4-Dinitrophenylhydrazine $\xrightarrow[\text{ethanol}]{H^+}$ 2-Dodecanone 2,4-Dinitrophenylhydrazone (94%)
The Wittig reaction (Sections 18.16–18.17) Reaction of a phosphorus ylide with aldehydes and ketones leads to the formation of an alkene. A very versatile method for the preparation of alkenes.	$$\underset{\substack{\text{Aldehyde}\\\text{or ketone}}}{RCR'} + \underset{\substack{\text{Wittig}\\\text{reagent (an ylide)}}}{(C_6H_5)_3\overset{+}{P}{-}\ddot{C}} \longrightarrow \underset{\text{Alkene}}{C{=}C} + \underset{\substack{\text{Triphenylphosphine}\\\text{oxide}}}{(C_6H_5)_3\overset{+}{P}{-}O^-}$$ $$\underset{\text{Acetone}}{CH_3CCH_3} + \underset{\substack{\text{1-Pentylidenetriphenyl-}\\\text{phosphorane}}}{(C_6H_5)_3\overset{+}{P}{-}\ddot{C}HCH_2CH_2CH_2CH_3} \xrightarrow{\text{DMSO}}$$ $$\underset{\substack{\text{2-Methyl-2 heptene}\\(56\%)}}{(CH_3)_2C{=}CHCH_2CH_2CH_2CH_3} + \underset{\substack{\text{Triphenyl-}\\\text{phosphino oxide}}}{(C_6H_5)_3\overset{+}{P}{-}O^-}$$

Nucleophilic addition to their rather polar carbon-oxygen double bonds is the characteristic chemical reaction of aldehydes and ketones and the principal topic of this chapter. Reagents of the type HY react according to the general equation

$$\overset{\delta+}{\underset{}{C}}\!\!=\!\!\overset{\delta-}{O} + \overset{\delta+}{H}\!\!-\!\!\overset{\delta-}{Y} \rightleftharpoons \quad Y\!-\!\overset{|}{\underset{|}{C}}\!-\!OH$$

| Aldehyde or ketone | Product of nucleophilic addition to carbonyl group |

Aldehydes undergo nucleophilic addition more readily and have more favorable equilibrium constants for addition than do ketones. A summary of the nucleophilic addition reactions to aldehydes and ketones introduced in this chapter is presented in Table 18.6.

The step in which the nucleophile attacks the carbonyl carbon is rate-determining in both base-catalyzed and acid-catalyzed nucleophilic addition. In the base-catalyzed mechanism this is the first step.

$$Y^- \,+\, \overset{}{\underset{}{C}}\!\!=\!\!O \xrightarrow{\text{slow}} Y\!-\!\overset{|}{\underset{|}{C}}\!-\!O^-$$

| Nucleophile | Aldehyde or ketone |

$$Y\!-\!\overset{|}{\underset{|}{C}}\!-\!O^- + H\!-\!Y \xrightarrow{\text{fast}} Y\!-\!\overset{|}{\underset{|}{C}}\!-\!OH \,+\, Y^-$$

Product of nucleophilic addition

Under conditions of acid catalysis, the nucleophilic addition step follows protonation of the carbonyl oxygen. Protonation increases the carbocation character of a carbonyl group and makes it more electrophilic.

$$\overset{}{\underset{}{C}}\!\!=\!\!\overset{..}{O}\!: \,+\, H\!-\!Y\!: \xrightarrow{\text{fast}} \overset{}{\underset{}{C}}\!\!=\!\!\overset{+}{\underset{..}{O}}H \longleftrightarrow \overset{+}{\underset{}{C}}\!-\!\overset{..}{O}H$$

| Aldehyde or ketone | Resonance forms of protonated aldehyde or ketone |

$$HY\!: \,+\, \overset{}{\underset{}{C}}\!\!=\!\!\overset{+}{\underset{..}{O}}H \xrightarrow{\text{slow}} \overset{+}{HY}\!-\!\overset{|}{\underset{|}{C}}\!-\!\overset{..}{O}H \xrightarrow{-H^+} Y\!-\!\overset{|}{\underset{|}{C}}\!-\!\overset{..}{O}H$$

Product of nucleophilic addition

Often the product of nucleophilic addition is not isolated but is an intermediate leading to the ultimate product. The last five entries in Table 18.6 exemplify this type of reaction.

Aldehydes are easily oxidized to carboxylic acids. Ketones are oxidized less readily and require carbon-carbon bond cleavage.

PROBLEMS

18.15 (a) Write structural formulas and provide IUPAC names for all the isomeric aldehydes and ketones that have the molecular formula $C_5H_{10}O$. Include stereoisomers.

(b) Which of the isomers in (a) yield chiral alcohols on reaction with sodium borohydride?

(c) Which of the isomers in (a) yield chiral alcohols on reaction with methylmagnesium iodide?

18.16 Each of the following aldehydes or ketones is known by a common name. Its IUPAC name is provided in parentheses. Write a structural formula for each one.

(a) Chloral (2,2,2-trichloroethanal)
(b) Pivaldehyde (2,2-dimethylpropanal)
(c) Acrolein (2-propenal)
(d) Crotonaldehyde [(E)-2-butenal]
(e) Citral [(E)-3,7-dimethyl-2,6-octadienal]
(f) Pinacolone (3,3-dimethyl-2-butanone)
(g) Deoxybenzoin (1,2-diphenylethanone)
(h) Diacetone alcohol (4-hydroxy-4-methyl-2-pentanone)
(i) Mesityl oxide (4-methyl-3-penten-2-one)
(j) Carvone (5-isopropenyl-2-methyl-2-cyclohexenone)
(k) Biacetyl (2,3-butanedione)
(l) Dimedone (5,5-dimethyl-1,3-cyclohexanedione)
(m) Dypnone (1,3-diphenyl-2-buten-1-one)

18.17 Predict the product of reaction of propanal with each of the following:

(a) Lithium aluminum hydride
(b) Sodium borohydride
(c) Hydrogen (nickel catalyst)
(d) Methylmagnesium iodide, followed by dilute acid
(e) Sodium acetylide, followed by dilute acid
(f) Phenyllithium, followed by dilute acid
(g) Methanol containing dissolved hydrogen chloride
(h) Ethylene glycol, p-toluenesulfonic acid, benzene
(i) Aniline ($C_6H_5NH_2$)
(j) Dimethylamine, p-toluenesulfonic acid, benzene
(k) Hydroxylamine
(l) Hydrazine
(m) Product of (l) heated in triethylene glycol with sodium hydroxide
(n) p-Nitrophenylhydrazine
(o) Semicarbazide
(p) Ethylidenetriphenylphosphorane [$(C_6H_5)_3P$—$\ddot{C}HCH_3$]
(q) Sodium cyanide with addition of sulfuric acid
(r) Silver oxide
(s) Chromic acid
(t) Potassium permanganate

18.18 Repeat the preceding problem for cyclopentanone.

18.19 Hydride reduction ($LiAlH_4$ or $NaBH_4$) of each of the following ketones has been reported in the chemical literature and gives a mixture of two diastereomeric alcohols in each case. Give the structures of both alcohol products for each ketone.

(a) (S)-3-Phenyl-2-butanone

(b) 4-tert-Butylcyclohexanone

(c)

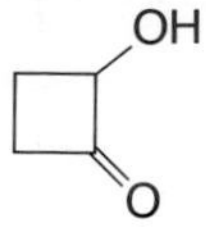

(d)

(e)

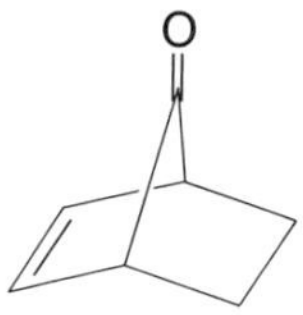

18.20 Choose which member in each of the following pairs reacts faster or has the more favorable equilibrium constant for reaction with the indicated reagent. Explain your reasoning.

(a) $C_6H_5\overset{O}{\overset{\|}{C}}H$ or $C_6H_5\overset{O}{\overset{\|}{C}}CH_3$ (rate of reduction with sodium borohydride)

(b) $C_6H_5\overset{O}{\overset{\|}{C}}H$ or $CH_3\overset{O}{\overset{\|}{C}}H$ (rate of reaction with $H_2NNH\overset{O}{\overset{\|}{C}}NH_2$)

(c) $Cl_3\overset{O}{\overset{\|}{C}}CH$ or $CH_3\overset{O}{\overset{\|}{C}}H$ (equilibrium constant for hydration)

(d) Cyclopropanone or cyclopentanone (equilibrium constant for hydration)

(e) Acetone or 3,3-dimethyl-2-butanone (equilibrium constant for cyanohydrin formation)

(f) Acetone or 3,3-dimethyl-2-butanone (rate of reduction with sodium borohydride)

(g) $CH_2(OCH_2CH_3)_2$ or $(CH_3)_2C(OCH_2CH_3)_2$ (rate of acid-catalyzed hydrolysis)

18.21 Each of the following reactions has been reported in the chemical literature and gives a single product in good yield. What is the principal organic product in each reaction?

(a)
1. KMnO₄, H₂O, heat
2. H⁺

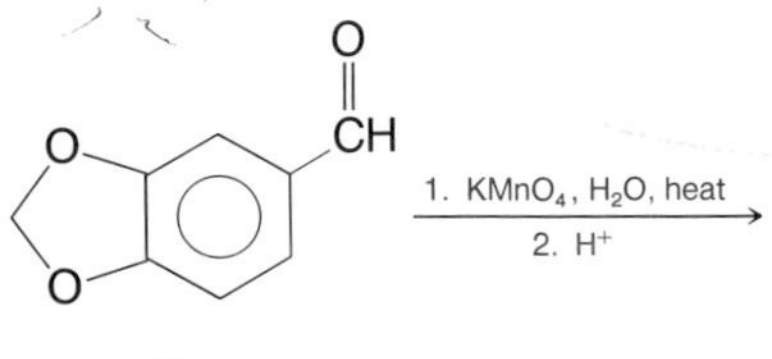

(b)
1. Ag₂O, NaOH, H₂O
2. H⁺

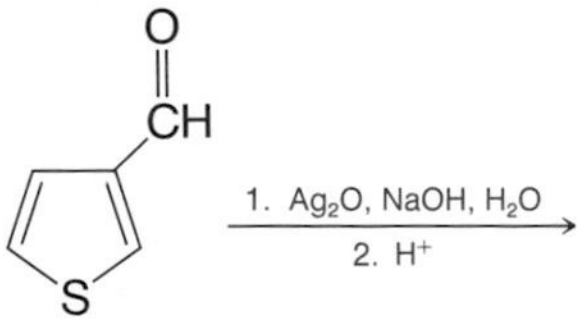

(c)
H₂NNH₂, KOH
ethylene glycol, heat

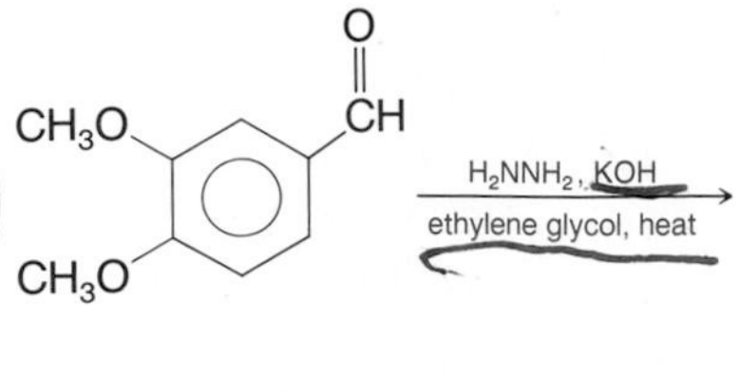

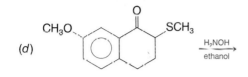

(d)   [structure: 7-methoxy-2-(methylthio)-1-tetralone] $\xrightarrow[\text{ethanol}]{H_2NOH}$

Good to know
Wolf Kishner

(e) $CH_3(CH_2)_{16}\overset{O}{\overset{\|}{C}}Cl + (C_6H_5CH_2CH_2)_2Cd \xrightarrow{\text{benzene}}$

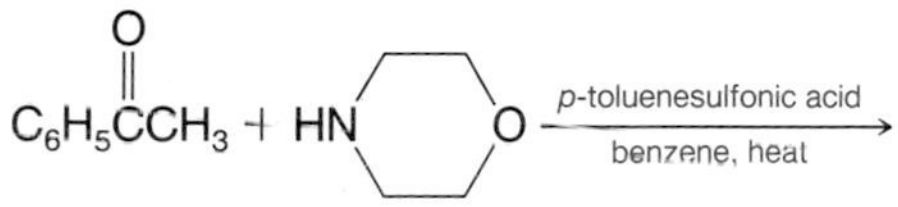

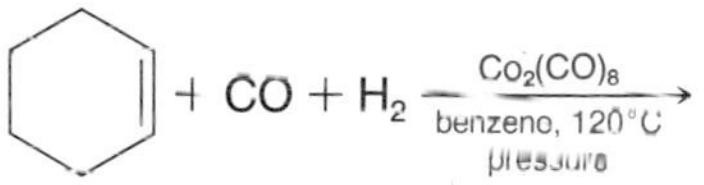

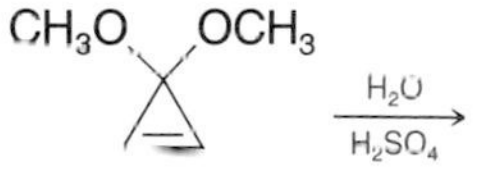

(f)   [bicyclic ketone structure] $\xrightarrow[\text{diethylene glycol, heat}]{H_2NNH_2, \text{ KOH}}$

(g) $C_6H_5\overset{O}{\overset{\|}{C}}CH_3 \xrightarrow[\text{HCl}]{\text{NaCN}}$

(h) $C_6H_5\overset{O}{\overset{\|}{C}}CH_3 + HN\!\!\overset{\frown}{\underset{\smile}{\quad}}\!\!O \xrightarrow[\text{benzene, heat}]{p\text{-toluenesulfonic acid}}$?

(i)   [cyclohexene structure] $+ CO + H_2 \xrightarrow[\substack{\text{benzene, 120°C} \\ \text{pressure}}]{Co_2(CO)_8}$

(j)   [cyclopropane with two OCH$_3$ groups] $\xrightarrow[H_2SO_4]{H_2O}$

(k)   [pyrrole-2-carbaldehyde structure] $\overset{O}{\overset{\|}{C}}H + H_2NNH\overset{O}{\overset{\|}{C}}NH_2 \xrightarrow[\text{ethanol-water}]{\text{sodium acetate}}$

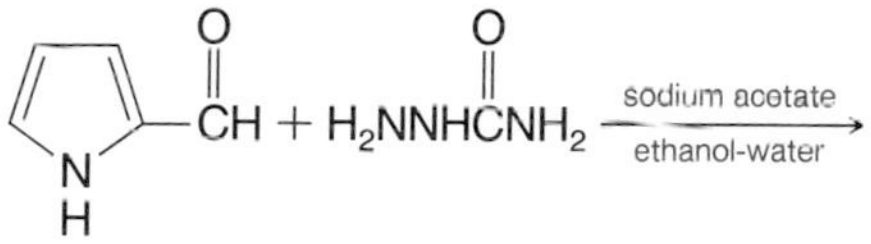

18.22 The triketone $C_6H_5\overset{OOO}{\overset{\|\|\|}{CCCC}}C_6H_5$ forms a stable hydrate $C_{15}H_{12}O_4$. What is the most reasonable structure for this hydrate?

18.23 On standing in ^{17}O-labeled water, both formaldehyde and its hydrate are found to have incorporated the ^{17}O isotope of oxygen. Suggest a reasonable explanation for this observation.

18.24 Reaction of benzaldehyde with 1,2-octanediol in benzene containing a small amount of *p*-toluenesulfonic acid yields almost equal quantities of two products in a combined yield of 94 percent. Both products have the molecular formula $C_{15}H_{22}O_2$. Suggest reasonable structures for these products.

1, 2 Alcohols
yield
Cyclic
Acetals

18.25 Compounds that contain both carbonyl and alcohol functional groups are often more stable as cyclic hemiacetals or cyclic acetals than as open-chain compounds. Examples of several of these are shown below. Deduce the structure of the open-chain form of each.

(a)   [tetrahydropyran-2-ol structure] OH

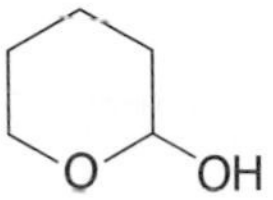

(b)

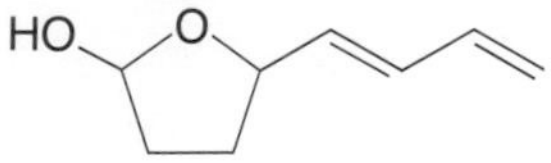

(c) 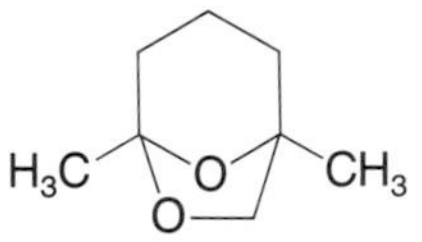 Frontalin (aggregating pheromone of the Southern pine bettle)

H_3C CH_3

(d) H_3C CH_3 CH_2CH_3 Multistriatin (aggregating pheromone of European elm bark beetle)

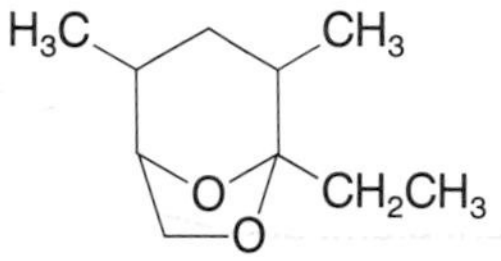

(e) Brevicomin (sex attractant of Western pine beetle)

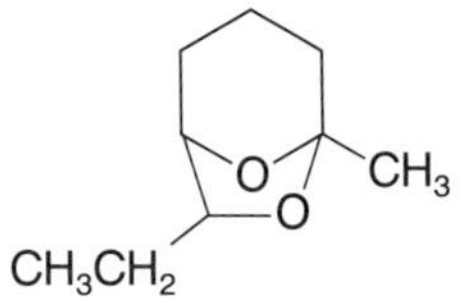

CH_3

CH_3CH_2

(f) CH_2CH_3 Talaromycin A (a toxic substance produced by a fungus that grows on poultry house litter)

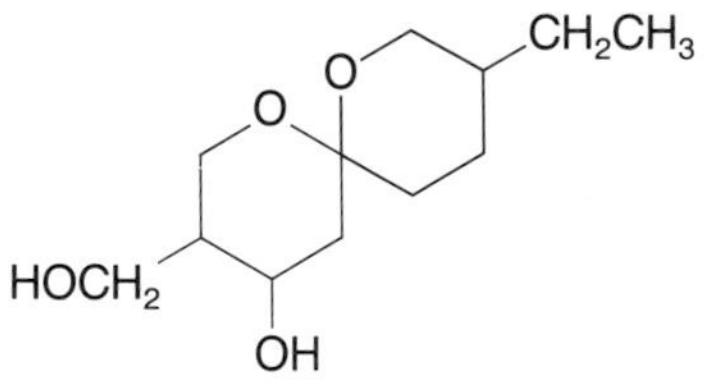

$HOCH_2$ OH

18.26 Compounds that contain a carbon-nitrogen double bond are capable of stereoisomerism much like that seen in alkenes. The structures

$$\underset{R'}{\overset{R}{>}}C=N\overset{..}{\underset{X}{<}} \quad \text{and} \quad \underset{R'}{\overset{R}{>}}C=N\overset{X}{\underset{..}{<}}$$

are stereoisomeric. Specifying stereochemistry in these systems is best done by using *E-Z* descriptors and considering the nitrogen lone pair to be the lowest-priority group. Write the structures, clearly showing stereochemistry, of the following:

(a) (Z)-$CH_3CH=NCH_3$
(b) (E)-Acetaldehyde oxime
(c) (Z)-2-Butanone hydrazone
(d) (E)-Acetophenone semicarbazone

18.27 Compounds known as *nitrones* are formed when *N*-substituted derivatives of hydroxylamine react with aldehydes and ketones:

$$\underset{\text{Butanal}}{CH_3CH_2CH_2\overset{\overset{\textstyle O}{\|}}{C}H} + \underset{\textit{N}\text{-Phenylhydroxylamine}}{C_6H_5NHOH} \longrightarrow \underset{\substack{\textit{N}\text{-(Butylidene)aniline} \\ \textit{N}\text{-oxide (a nitrone) (80\%)}}}{CH_3CH_2CH_2CH=\overset{\overset{\textstyle O^-}{|}}{\underset{+}{N}}C_6H_5} + \underset{\text{Water}}{H_2O}$$

Write a reasonable sequence of steps that describes the mechanism of this reaction.

18.28 Suggest reasonable mechanisms for each of the following reactions:

(a) $(CH_3)_3C\overset{O}{\underset{Cl}{C}}\text{—}CH_2 \xrightarrow[CH_3OH]{NaOCH_3} (CH_3)_3C\overset{O}{C}CH_2OCH_3$ (88%)

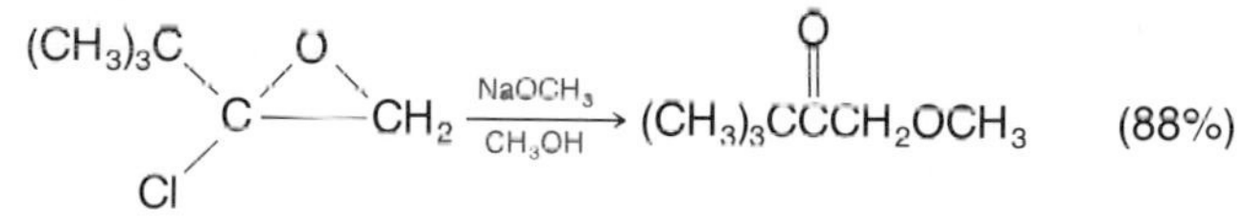

(b) $(CH_3)_3C\overset{O}{C}CH\underset{Cl}{C}H \xrightarrow[CH_3OH]{NaOCH_3} (CH_3)_3C\overset{}{C}CH\underset{OH}{C}H(OCH_3)_2$ (72%)

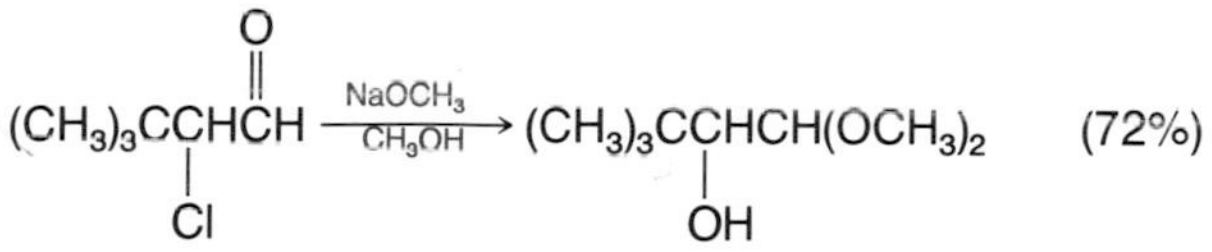

18.29 Hydrolysis of amygdalin, a substance present in peach, plum, and almond pits, liberates the R-enantiomer of mandelonitrile. (Mandelonitrile is the cyanohydrin of benzaldehyde.) Given the orientation shown below and with all the atoms of benzaldehyde lying in the plane of the page, determine whether cyanide ion must add to the top or to the bottom face of the carbonyl group in order to form (R)-mandelonitrile.

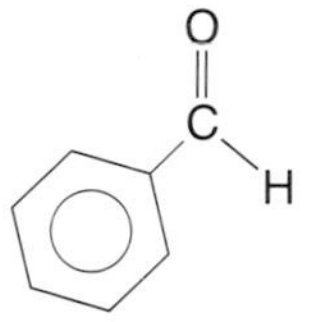

18.30 Using ethanol as the source of all the carbon atoms, describe efficient syntheses of each of the following, using any necessary organic or inorganic reagents:

(a) $CH_3CH(OCH_2CH_3)_2$

(b)

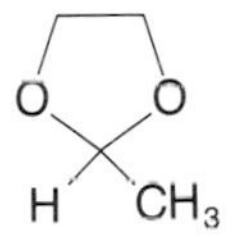

(c)

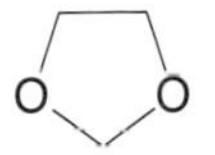

(d) $CH_3\underset{OH}{C}HC\equiv CH$

(e) $H\overset{O}{C}CH_2C\equiv CH$

18.31 Describe reasonable syntheses of benzophenone, $C_6H_5\overset{O}{C}C_6H_5$, from the starting materials given below and any necessary inorganic reagents.

(a) Benzoyl chloride and benzene
(b) Benzoyl chloride and bromobenzene (two ways)
(c) Benzyl alcohol and bromobenzene
(d) Diphenylmethyl bromide, $(C_6H_5)_2CHBr$
(e) Dimethoxydiphenylmethane, $(C_6H_5)_2C(OCH_3)_2$
(f) 1,1,2,2-Tetraphenylethene, $(C_6H_5)_2C=C(C_6H_5)_2$

18.32 The sex attractant of the female winter moth has been identified as $CH_3(CH_2)_8CH=CHCH_2CH=CHCH_2CH=CHCH=CH_2$. Devise a synthesis of this material from 3,6-hexadecadien-1-ol and allyl alcohol.

18.33 Suggest reasonable structures for compounds A and B.

$$(C_6H_5)_2CHCHCCl \xrightarrow{AlCl_3} \text{compound A}$$

(mp 100–101°C)

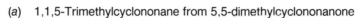

compound B

(mp 151–154°C)

Both A and B have the molecular formula $C_{21}H_{16}O$ and exhibit a strong absorption peak in the infrared around 1680 cm^{-1}.

18.34 Hydrolysis of compound C in dilute aqueous hydrochloric acid gave (along with methanol) compound D, mp 164–165°C. Compound D had the molecular formula $C_{16}H_{16}O_4$; it exhibited hydroxyl absorption in its infrared spectrum at 3550 cm^{-1} but had no peaks in the carbonyl region. What is a reasonable structure for compound D?

Compound C

18.35 Syntheses of each of the following compounds have been reported in the chemical literature. Using the indicated starting material and any necessary organic or inorganic reagents, describe short sequences of reactions that would be appropriate for each transformation.

(a) 1,1,5-Trimethylcyclononane from 5,5-dimethylcyclononanone

(b) from

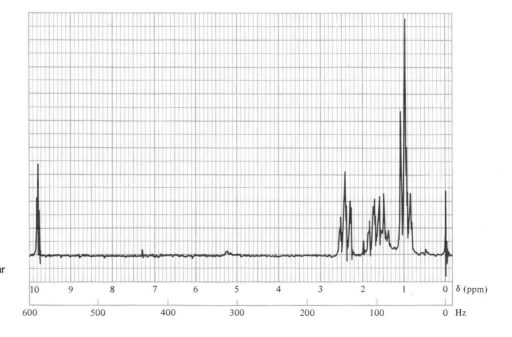

FIGURE 18.11 The ^1H nmr spectrum of compound E (C_4H_8O) (Problem 18.37).

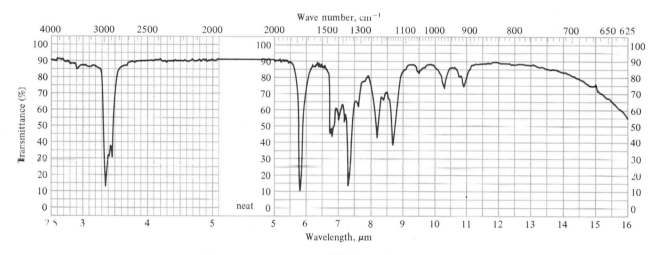

(c)

CH$_3$ CH$_2$

CCH$_2$CH$_2$CH$_2$CH=CH$_2$

from *o*-bromotoluene and 5-hexenal

(d) CH$_3$(CH$_2$)$_{18}$CH$_3$ from CH$_3$(CH$_2$)$_{16}$CCl
                                              ‖
                                              O

(e) CH$_3$CCH$_2$CH$_2$C(CH$_2$)$_5$CH$_3$ from HC≡CCH$_2$CH$_2$CH$_2$OH
        ‖   ‖
        O   O

(f)

Cl CH$_3$

CH$_2$OCH$_3$ from 3-chloro-2-methylbenzaldehyde

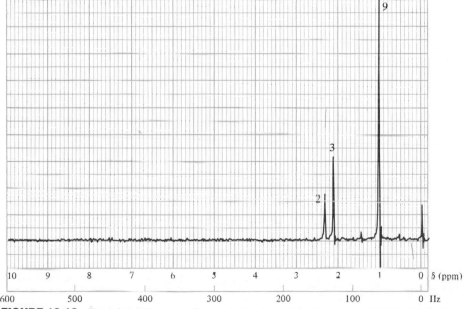

FIGURE 18.12 The infrared (top) and ¹H nmr (bottom) spectra of compound F (C$_7$H$_{14}$O) (Problem 18.38a).

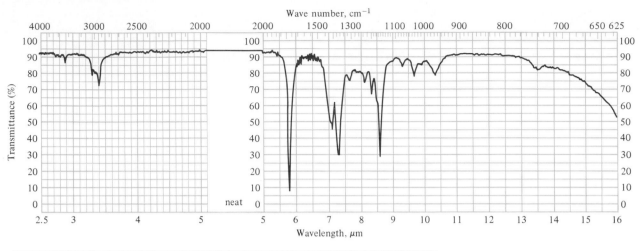

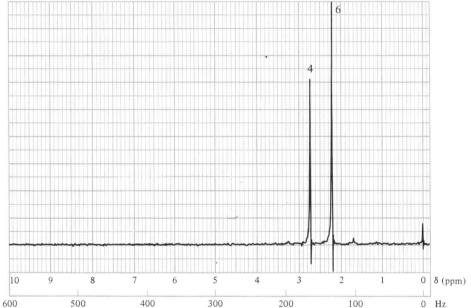

FIGURE 18.13 The infrared (top) and ^{1}H nmr (bottom) spectra of compound G ($C_6H_{10}O_2$) (Problem 18.38b).

18.36 Increased "single bond character" in a carbonyl group is associated with a decreased carbon-oxygen stretching frequency. Among the three compounds benzaldehyde, 2,4,6-trimethoxybenzaldehyde, and 2,4,6-trinitrobenzaldehyde, which one will have the lowest-frequency carbonyl absorption? Which one will have the highest?

18.37 Compound E has the molecular formula C_4H_8O and contains a carbonyl group. Identify compound E on the basis of its ^{1}H nmr spectrum shown in Figure 18.11.

18.38 Identify each of the following compounds on the basis of its infrared and ^{1}H nmr spectra:

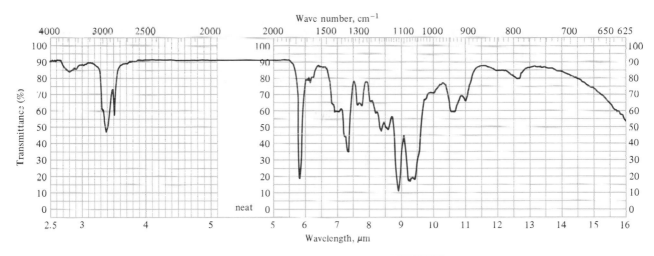

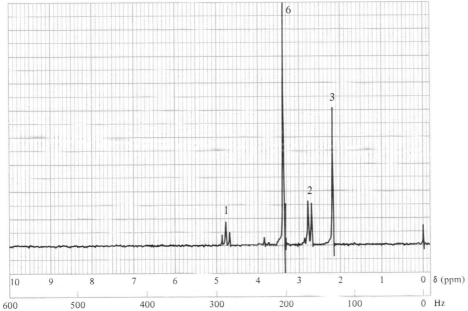

FIGURE 18.14 The infrared (top) and ¹H nmr (bottom) spectra of compound H ($C_6H_{12}O_3$) (Problem 18.38c).

(a) Compound F ($C_7H_{14}O$) (Figure 18.12).
(b) Compound G ($C_6H_{10}O_2$) (Figure 18.13).
(c) Compound H ($C_6H_{12}O_3$) (Figure 18.14).

18.39 Identify each of the following compounds on the basis of its infrared and ¹H nmr spectra:

(a) Compound I ($C_9H_{10}O$) (Figure 18.15).
(b) Compound J (C_9H_9ClO) (Figure 18.16).
(c) Compound K (C_9H_9BrO) (Figure 18.17).

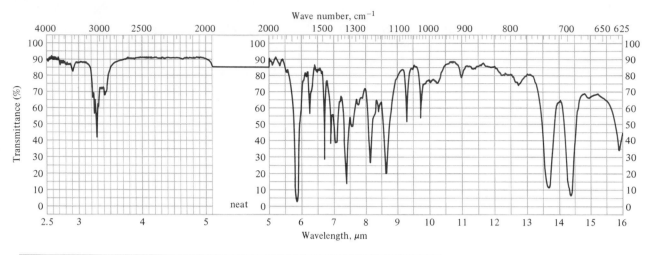

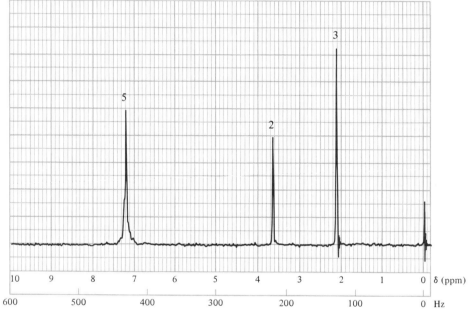

FIGURE 18.15 The infrared (top) and ^{1}H nmr (bottom) spectra of compound I ($C_9H_{10}O$) (Problem 18.39a).

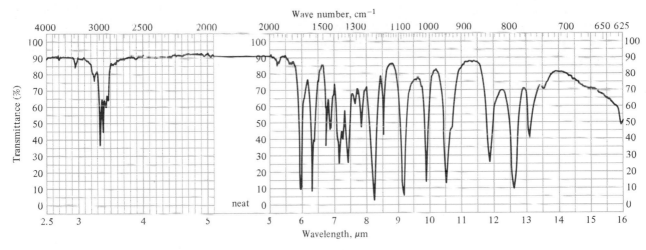

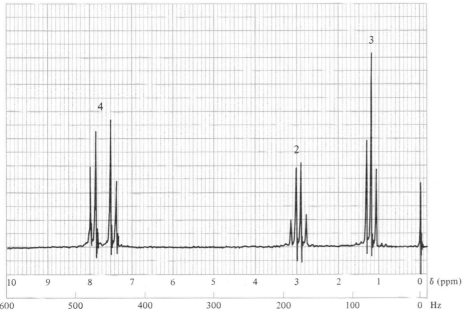

FIGURE 18.16 The infrared (top) and ^{1}H nmr (bottom) spectra of compound J (C_9H_9ClO) (Problem 18.39*b*).

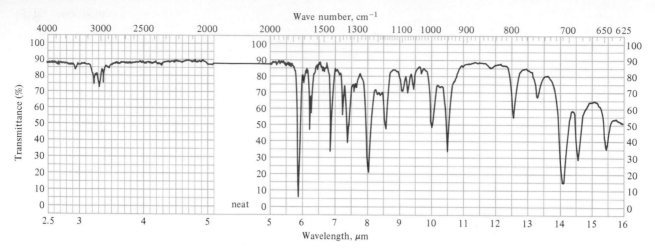

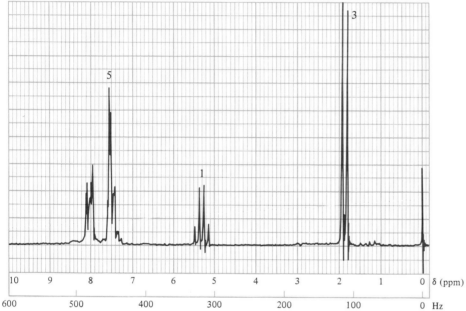

FIGURE 18.17 The infrared (top) and ^{1}H nmr (bottom) spectra of compound K (C_9H_9BrO) (Problem 18.39c).

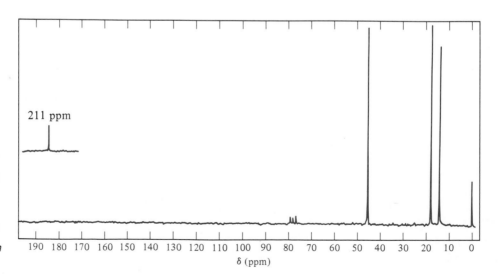

FIGURE 18.18 The ^{13}C nmr spectrum of compound L ($C_7H_{14}O$) (Problem 18.40). *(Taken from ''Carbon-13 NMR Spectra: A Collection of Assigned, Coded, and Indexed Spectra,'' by LeRoy F. Johnson and William C. Jankowski, Wiley-Interscience, New York, 1972. Reprinted by permission of John Wiley & Sons, Inc.)*

211 ppm

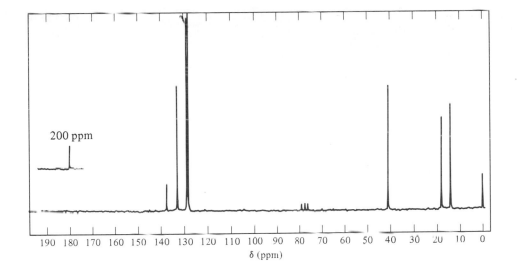

FIGURE 18.19 The ^{13}C nmr spectrum of compound M $(C_{10}H_{12}O)$ (Problem 18.41). *(Taken from "Carbon-13 NMR Spectra: A Collection of Assigned, Coded, and Indexed Spectra," by LeRoy F. Johnson and William C. Jankowski, Wiley-Interscience, New York, 1972. Reprinted by permission of John Wiley & Sons, Inc.)*

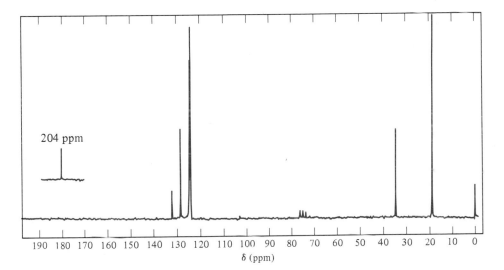

FIGURE 18.20 The ^{13}C nmr spectrum of compound N $(C_{10}H_{12}O)$ (Problem 18.41). *(Taken from "Carbon-13 NMR Spectra: A Collection of Assigned, Coded, and Indexed Spectra," by LeRoy F. Johnson and William C. Jankowski, Wiley-Interscience, New York, 1972. Reprinted by permission of John Wiley & Sons, Inc.)*

18.40 Compound L is a ketone of molecular formula $C_7H_{14}O$. Its ^{13}C nmr spectrum is shown in Figure 18.18. What is the structure of compound L?

18.41 Compound M and compound N are isomers having the molecular formula $C_{10}H_{12}O$. The mass spectrum of each compound contains an abundant peak at m/z 105. The ^{13}C nmr spectra of compound M (Figure 18.19) and compound N (Figure 18.20) are shown. Identify these two compounds.

ENOLS, ENOLATES, AND ENAMINES

In the preceding chapter you learned that nucleophilic addition to carbonyl groups is one of the fundamental reaction types of organic chemistry. Aside from its own intrinsic reactivity, a carbonyl group can affect the chemical properties of aldehydes and ketones in other ways. Aldehydes and ketones participate in an equilibrium with their *enol* isomers.

$$
\underset{\substack{\text{Aldehyde or}\\\text{ketone}}}{R_2CHCR'} \rightleftharpoons \underset{\text{Enol}}{R_2C=CR'}
$$

(with O double bond on left structure, OH on right structure)

In this chapter you will see a number of processes in which it is the enol form of an aldehyde or ketone rather than the carbonyl compound itself which is the reactive species.

There is also an important group of reactions in which the carbonyl group acts as a powerful electron-withdrawing substituent, increasing the acidity of protons on the carbons to which it is attached.

$$
\underset{\substack{|\\H}}{R_2CCR'}
$$

(O double bond above, H below) — This proton is far more acidic than protons attached to simple alkyl groups

As an electron-withdrawing group on a carbon-carbon double bond, a carbonyl group renders the π system of the double bond susceptible to nucleophilic attack.

$$
R_2C=CHCR'
$$

(O double bond above) — Normally, carbon-carbon double bonds are attacked by electrophilic reagents; a carbon-carbon double bond that is conjugated to a carbonyl group is attacked by nucleophilic reagents

Substituent effects arising from the presence of a carbonyl group in a molecule make possible a number of chemical reactions that are of great synthetic and mechanistic importance. This chapter is complementary to the preceding one; the two chapters taken together demonstrate the extraordinary range of chemical reactions available to aldehydes and ketones.

19.1 THE α CARBON ATOM AND ITS HYDROGENS

It is convenient to identify various carbon atoms in aldehydes and ketones by using the Greek letters α, β, γ, etc. to specify their location in relation to the carbonyl group. The carbon atom adjacent to the carbonyl group is the α carbon atom, the next one down the chain is the β carbon, and so on. Butanal for example, has an α carbon, a β carbon, and a γ carbon.

$$CH_3CH_2CH_2CH$$

Carbonyl group is reference point;
no Greek letter assigned to it

Hydrogen substituents are identified by the same Greek letter as the carbon atom to which they are attached. A hydrogen connected to the α carbon atom is an α hydrogen. Butanal has two α protons, two β protons and three γ protons. No Greek letter is assigned to the hydrogen attached directly to the carbonyl group of an aldehyde.

PROBLEM 19.1   How many α hydrogens are there in each of the following?

(a)   3,3-Dimethyl-2-butanone
(b)   2,2-Dimethylpropanal
(c)   Benzyl methyl ketone
(d)   Cyclohexanone

SAMPLE SOLUTION   (a)   This ketone has two different α carbons but only one of them has hydrogen substituents. There are three equivalent α hydrogens. The other nine hydrogens are attached to β carbon atoms.

3,3-Dimethyl-2-butanone

Other than nucleophilic addition to the carbonyl group, the most important reactions of aldehydes and ketones involve substitution of an α hydrogen. Activation of the α position toward substitution is a direct result of its location with respect to the carbonyl group. A particularly well studied example is halogenation of aldehydes and ketones. A careful examination of this reaction will lay the groundwork for understanding mechanistically related processes.

19.2 ACID-CATALYZED HALOGENATION OF ALDEHYDES AND KETONES

Aldehydes and ketones react with halogens by substitution of halogen for one of the α hydrogens.

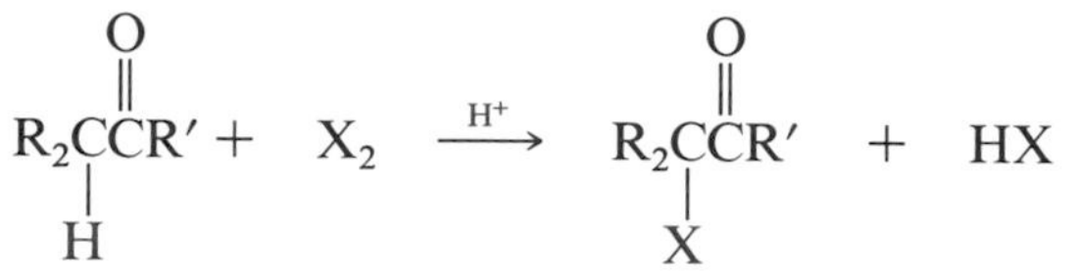

| Aldehyde or ketone | Halogen | α-Halo aldehyde or ketone | Hydrogen halide |

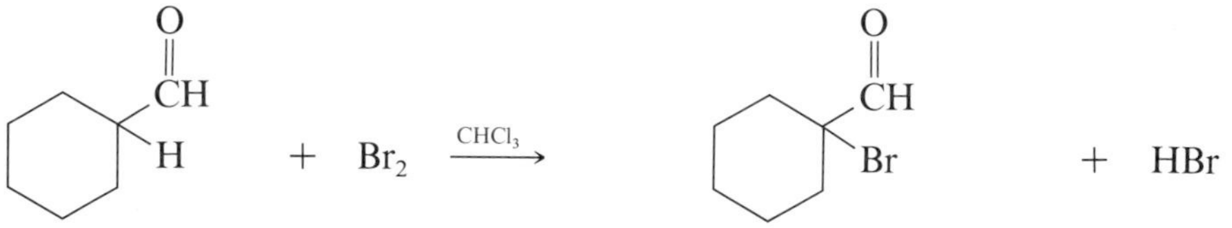

Cyclohexanecarbaldehyde — Bromine — 1-Bromocyclohexanecarbaldehyde (80%) — Hydrogen bromide

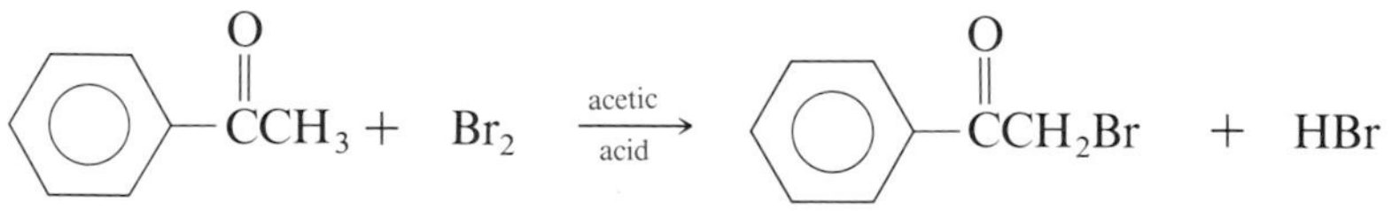

Acetophenone — Bromine — Bromomethyl phenyl ketone (phenacyl bromide) (77–92%) — Hydrogen bromide

---

PROBLEM 19.2 Chlorination of 2-butanone yields two isomeric products each having the molecular formula C_4H_7ClO. Identify these two compounds.

The reaction is catalyzed by acids. (We shall see later that it is also catalyzed by bases.) Since one of the reaction products, the hydrogen halide, is an acid and therefore a catalyst for the reaction, the process is said to be *autocatalytic*. Mechanistically, acid-catalyzed halogenation of aldehydes and ketones is much different from free-radical halogenation of alkanes. While both processes lead to the replacement of a hydrogen by a halogen substituent, they do so by pathways that are unrelated to one another.

19.3 MECHANISM OF α HALOGENATION OF ALDEHYDES AND KETONES

In one of the earliest systematic investigations of a reaction mechanism in organic chemistry, Arthur Lapworth discovered in 1904 that the rates of chlorination and bromination of acetone were the same. Later he found that iodination of acetone proceeded at the same rate as chlorination and bromination. Moreover, the rates of all three halogenation reactions, while first-order in acetone, are independent of the halogen concentration. Thus, the halogen does not participate in the reaction until *after* the rate-determining step. These kinetic observations, coupled with the fact that substitution occurs exclusively at the α carbon atom, led Lapworth to propose that the rate-determining step is the conversion of acetone to a more reactive form, its enol isomer.

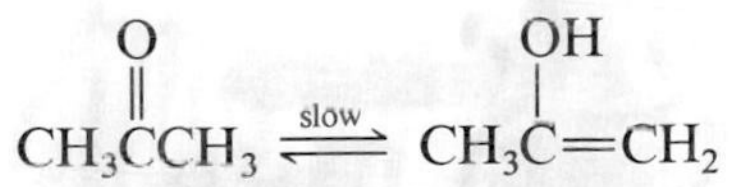

$$CH_3CCH_3 \xrightleftharpoons{\text{slow}} CH_3C{=}CH_2$$

O ‖	OH │
Acetone	Propen-2-ol (enol form of acetone)

Once formed, this enol reacts rapidly with the halogen to form an α-halo ketone.

$$CH_3C{=}CH_2 + X_2 \xrightarrow{\text{fast}} CH_3CCH_2X + HX$$

OH │		O ‖	
Propen-2-ol (enol form of acetone)	Halogen	α-Halo derivative of acetone	Hydrogen halide

---

PROBLEM 19.3 Write the structures of the enol forms of 2-butanone that react with chlorine to give 1-chloro-2-butanone and 3-chloro-2-butanone.

Both phases of the Lapworth mechanism, enol formation and enol halogenation, are new to us. Let us examine them in reverse order. We can understand enol halogenation by analogy to halogen addition to alkenes. An enol is a very reactive kind of alkene. Its carbon-carbon double bond bears an electron-releasing hydroxyl group, which activates it toward attack by electrophiles.

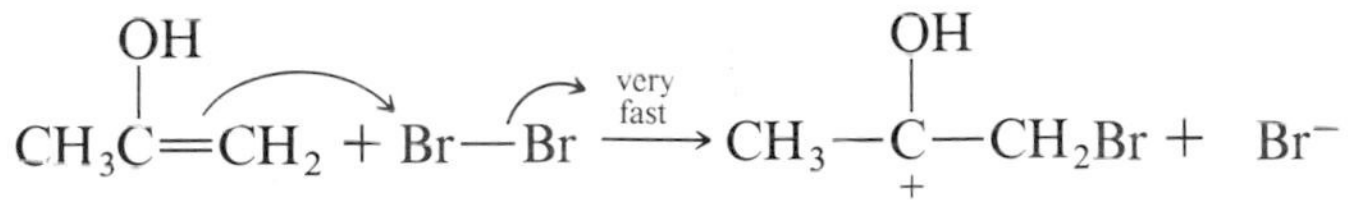

$$CH_3C{=}CH_2 + Br{-}Br \xrightarrow[\text{fast}]{\text{very}} CH_3{-}\overset{+}{C}{-}CH_2Br + Br^-$$

OH │		OH │	
Propen-2-ol (enol form of acetone)	Bromine	Stabilized cationic intermediate	Bromide ion

The hydroxyl group stabilizes the cationic intermediate by delocalization of one of the unshared electron pairs of oxygen.

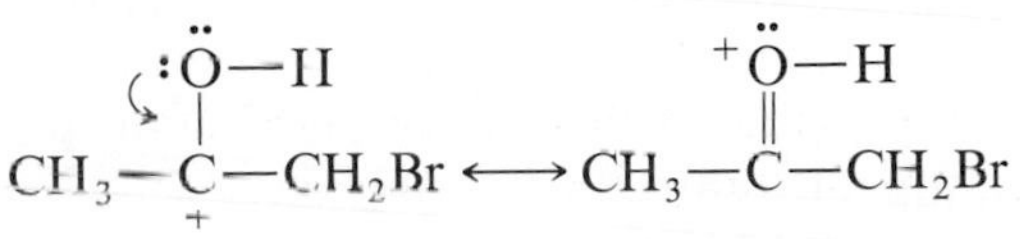

$$CH_3{-}\overset{+}{C}{-}CH_2Br \longleftrightarrow CH_3{-}C{-}CH_2Br$$

:Ö—H │		$\overset{+}{\text{O}}$—H ‖	
Less stable resonance form; six electrons on positively charged carbon	More stable resonance form; all atoms (except hydrogen) have octets of electrons

Participation by the enolic oxygen in the bromination step is responsible for the rapid attack on the carbon-carbon double bond of an enol. Organic chemists usually represent this participation explicitly:

$$\overset{\overset{\displaystyle \ddot{\ddot{O}}H}{|}}{CH_3C=CH_2} \quad \longrightarrow \quad \overset{\overset{\displaystyle +\ddot{\ddot{O}}H}{\|}}{CH_3-C-CH_2Br} + Br^-$$
$$\underset{Br-Br}{\searrow}$$

Representing the bromine addition step in this way both emphasizes the increased nucleophilicity of the enol double bond and identifies the source of that increased nucleophilicity as the enolic oxygen.

PROBLEM 19.4   Represent the reaction of chlorine with each of the enol forms of 2-butanone (see Problem 19.3) according to the curved arrow formalism just described.

The cationic intermediate is simply the protonated form (conjugate acid) of the α-halo ketone. Deprotonation of the cationic intermediate gives the products.

$$\overset{\overset{\displaystyle +\ddot{O}-H \quad {}^-Br}{\|}}{\underset{\text{Cationic intermediate}}{CH_3-C-CH_2Br}} \quad \longrightarrow \quad \underset{\text{Bromoacetone}}{\overset{\overset{\displaystyle :\ddot{O}:}{\|}}{CH_3CCH_2Br}} + \quad \underset{\text{Hydrogen bromide}}{HBr}$$

Having now seen how an enol, once formed, reacts with a halogen, let us consider the process of enolization itself.

19.4 ENOLIZATION

Enols are related to an aldehyde or ketone by a proton transfer equilibrium known as *keto-enol tautomerism.* (*Tautomerism* refers to an interconversion between two structures that differ by the placement of an atom or group.)

$$\underset{\text{Keto form}}{\overset{\overset{\displaystyle O}{\|}}{RCH_2CR'}} \quad \underset{\xrightarrow{\text{tautomerism}}}{\rightleftharpoons} \quad \underset{\text{Enol form}}{\overset{\overset{\displaystyle OH}{|}}{RCH=CR'}}$$

The mechanism of enolization involves solvent-mediated proton transfer steps rather than a direct intramolecular "proton jump" from carbon to oxygen. In aqueous acid, for example, the mechanism of enolization proceeds as shown in Figure 19.1. A hydronium ion transfers a proton to the carbonyl oxygen in step 1, and a water molecule acts as a Brönsted base to remove a proton from the α carbon atom in step 2. The second step is slower than the first. The first step involves proton transfer between oxygens, while the second is a proton transfer from carbon to oxygen.

You have had an earlier encounter with enols in the hydration of alkynes (Section 10.13). The mechanism of enolization of aldehydes and ketones is precisely the reverse of the mechanism by which an enol is converted to a carbonyl compound.

The amount of enol present at equilibrium, the *enol content,* is quite small for

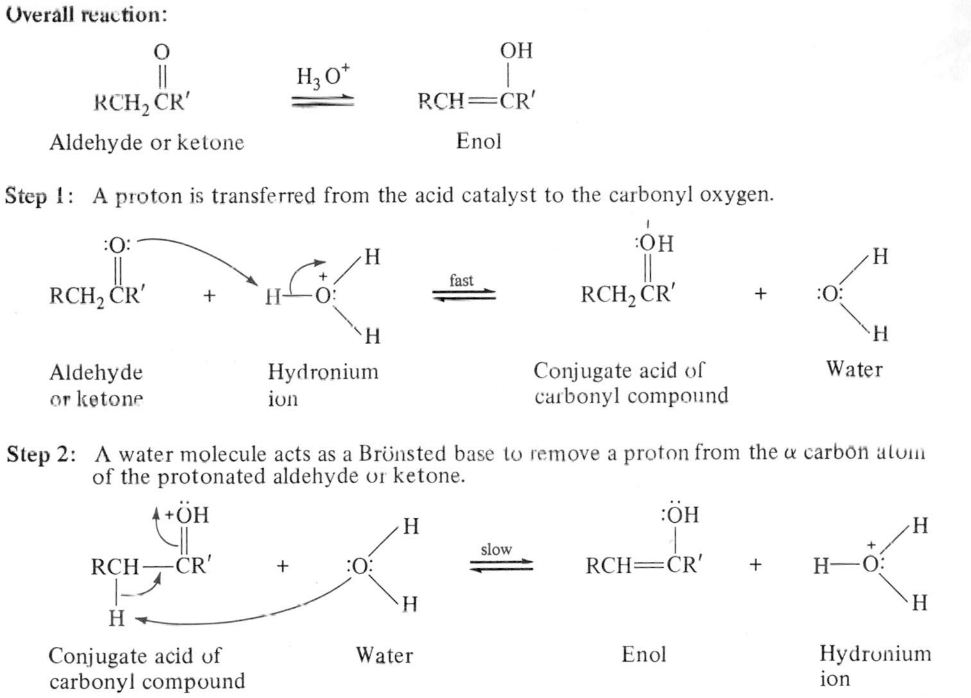
Overall reaction:
Aldehyde or ketone
Enol
Step 1: A proton is transferred from the acid catalyst to the carbonyl oxygen.
fast
Aldehyde or ketone
Hydronium ion
Conjugate acid of carbonyl compound
Water
Step 2: A water molecule acts as a Brönsted base to remove a proton from the α carbon atom of the protonated aldehyde or ketone.
slow
Conjugate acid of carbonyl compound
Water
Enol
Hydronium ion

FIGURE 19.1 Sequence of steps that describes the acid-catalyzed enolization of an aldehyde or ketone in aqueous solution.

simple aldehydes and ketones. The equilibrium constants for enolization, as shown by the following examples, are much less than unity.

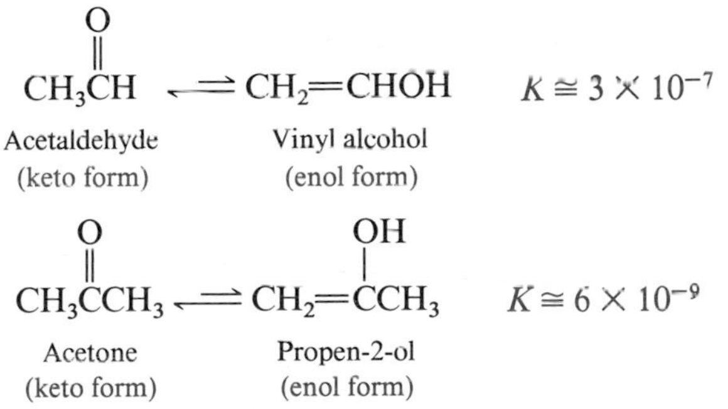
Acetaldehyde (keto form)
Vinyl alcohol (enol form)
K ≅ 3 × 10⁻⁷
Acetone (keto form)
Propen-2-ol (enol form)
K ≅ 6 × 10⁻⁹

In these and numerous other simple cases, the keto form is more stable by some 11 to 14 kcal/mol than the enol. The principal reason for this difference is the greater resonance energy of a carbonyl group as compared with a carbon-carbon double bond.

With unsymmetrical ketones, enolization may occur in either direction:

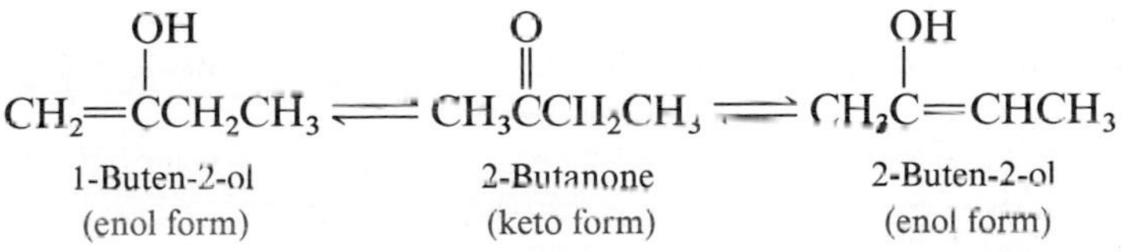
1-Buten-2-ol (enol form)
2-Butanone (keto form)
2-Buten-2-ol (enol form)

The ketone is by far the principal species present at equilibrium. Both enols are also present, but in very small concentrations. The enol with the more highly substituted double bond is the more stable of the two and is present in higher concentration than the other.

PROBLEM 19.5 Write structural formulas corresponding to

(a) The enol form of 3-pentanone
(b) The enol form of acetophenone
(c) The enol form of cyclohexanone
(d) The two enol forms of 2-methylcyclohexanone

SAMPLE SOLUTION   (a) Remember that it is the α carbon atom that is deprotonated in the enolization process. The ketone 3-pentanone gives a single enol since the two α carbons are equivalent.

$$\underset{\substack{\text{3-Pentanone} \\ \text{(keto form)}}}{CH_3CH_2\overset{\overset{\displaystyle O}{\|}}{C}CH_2CH_3} \rightleftharpoons \underset{\substack{\text{2-Penten-3-ol} \\ \text{(enol form)}}}{CH_3CH=\overset{\overset{\displaystyle OH}{|}}{C}CH_2CH_3}$$

It is important to recognize that an enol is a real substance, capable of independent existence. An enol is *not* a resonance form of a carbonyl compound. Interconversion of enol and keto forms requires a reorganization of atomic positions. A carbonyl compound and its enol form are constitutional isomers of each other.

19.5 STABILIZED ENOLS

Certain structural features can act to make the keto $\rightleftharpoons$ enol equilibrium more favorable by stabilizing the enol form. Enolization of 2,4-cyclohexadienone is one such example.

2,4-Cyclohexadienone   Phenol $K > 10$
(keto form, not   (enol form, aromatic)
aromatic)

The enol is *phenol* and the gain in resonance stabilization that attends formation of the aromatic ring is more than sufficient to overcome the normal preference for the keto form. The amount of the keto form present in phenol at equilibrium is far too small to measure.

A 1,3 arrangement of two carbonyl groups (these compounds are called β-diketones) leads to a situation in which the keto and enol forms are of comparable stability.

$$\underset{\substack{\text{2,4-Pentanedione (20\%)} \\ \text{(keto form)}}}{CH_3\overset{\overset{\displaystyle O}{\|}}{C}CH_2\overset{\overset{\displaystyle O}{\|}}{C}CH_3} \rightleftharpoons \underset{\substack{\text{4-Hydroxy-3-penten-2-one (80\%)} \\ \text{(enol form)}}}{CH_3\overset{\overset{\displaystyle OH}{|}}{C}=CH\overset{\overset{\displaystyle O}{\|}}{C}CH_3} \qquad K = 4$$

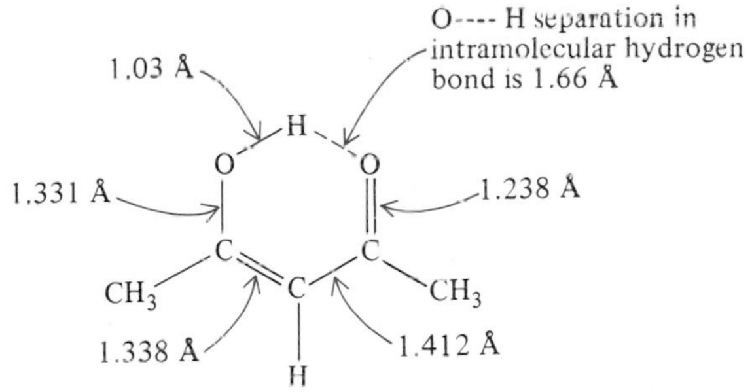

FIGURE 19.2 Structure of the enol form of 2,4-pentanedione.

The two principal structural features that stabilize the enol of a β-dicarbonyl compound are (1) conjugation of its double bond with the remaining carbonyl group; and (2) the presence of a strong intramolecular hydrogen bond between the enolic hydroxyl group and the carbonyl oxygen (Figure 19.2).

In β-diketones it is the methylene group flanked by the two carbonyls that is involved in enolization. The alternative enol

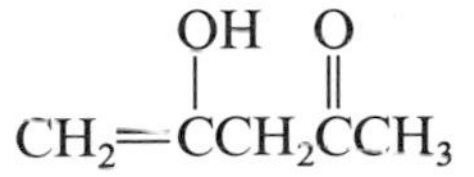

$$\underset{\text{4-hydroxy-4-penten-2-one}}{CH_2=CCH_2CCH_3}$$

does not have its carbon-carbon double bond conjugated with the carbonyl group, is not as stable, and is present in negligible amounts at equilibrium.

PROBLEM 19.6 Write structural formulas corresponding to

(a) The two most stable enol forms of CH_3CCH_2CH (with two C=O groups)

(b) The two most stable enol forms of 2,4-hexanedione

(c) The two most stable enol forms of 1-phenyl-1,3-butanedione

SAMPLE SOLUTION (a) Enolization of this 1,3-dicarbonyl compound can involve either of the two carbonyl groups:

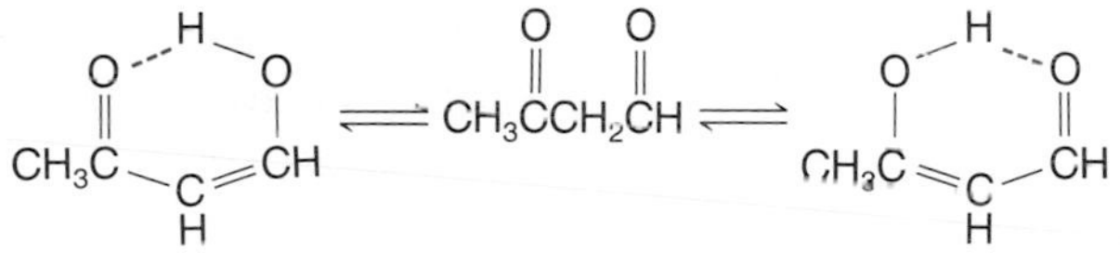

Both enols have their carbon-carbon double bonds conjugated to a carbonyl group and can form an intramolecular hydrogen bond. They are of comparable stability.

19.6 BASE-CATALYZED ENOLIZATION. ENOLATE ANIONS

The solvent-mediated proton transfer equilibrium that interconverts a carbonyl compound and its enol can be catalyzed by bases as well as by acids. Figure 19.3 illustrates the roles of hydroxide ion and water in a base-catalyzed enolization occurring in aqueous medium. As in acid-catalyzed enolization, protons are transferred

Overall reaction:

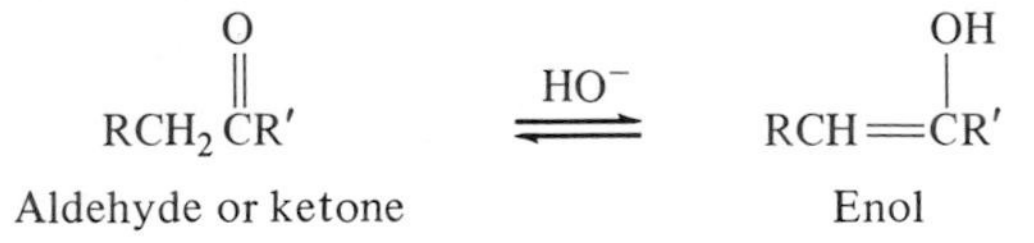

$$\underset{\substack{\text{Aldehyde or ketone}}}{RCH_2\overset{\displaystyle O}{\overset{\displaystyle \|}{C}}R'} \quad \underset{\displaystyle \rightleftharpoons}{\overset{\displaystyle HO^-}{}} \quad \underset{\substack{\text{Enol}}}{RCH=\overset{\displaystyle OH}{\overset{\displaystyle |}{C}}R'}$$

Step 1: A proton is abstracted by hydroxide ion from the α carbon atom of the carbonyl compound.

| Aldehyde or ketone | Hydroxide ion | | Conjugate base of carbonyl compound | Water |

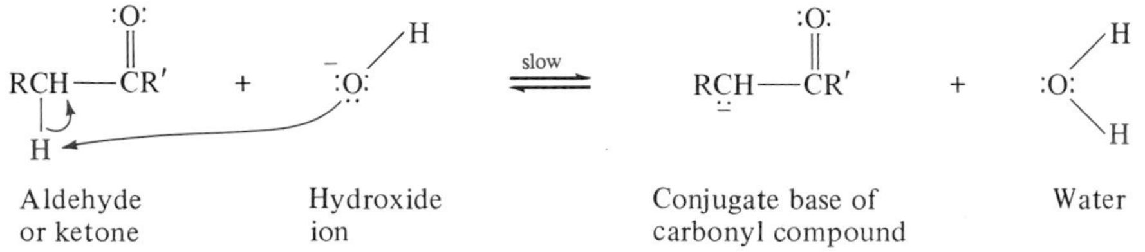

Step 2: A water molecule acts as a Brönsted acid to transfer a proton to the oxygen of the enolate ion.

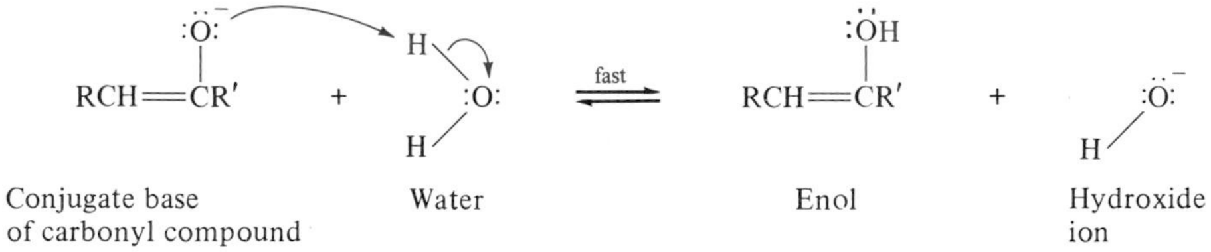

| Conjugate base of carbonyl compound | Water | | Enol | Hydroxide ion |

FIGURE 19.3 Sequence of steps that describes the base-catalyzed enolization of an aldehyde or ketone in aqueous solution.

sequentially rather than in a single step. First (step 1), the base abstracts a proton from the α carbon atom to yield an anion. This anion is a resonance-stabilized species. Its negative charge is shared by the α carbon atom and the carbonyl oxygen.

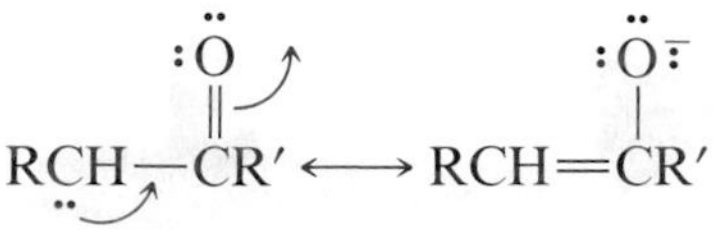

Resonance structures of
conjugate base

Protonation of this anion can occur either at the α carbon or at oxygen. Protonation of the α carbon simply returns the anion to starting aldehyde or ketone. Protonation at oxygen, as shown in step 2 of Figure 19.3, produces the enol.

The key intermediate in this process, the conjugate base of the carbonyl compound, is a *carbanion.* It contains a negatively charged carbon. More frequently, organic chemists refer to it as an *enolate ion* since it is at the same time the conjugate base of an enol. The term *enolate* is more descriptive of the electron distribution in this intermediate in that oxygen bears a greater share of the negative charge than does the α carbon atom.

The slow step in base-catalyzed enolization is formation of the enolate ion. The second step, proton transfer from water to the enolate oxygen, is very fast, as are almost all proton transfers from one oxygen atom to another.

Our experience to this point has been that C—H bonds are not very acidic. Compared with most hydrocarbons, however, aldehydes and ketones have relatively acidic protons on their α carbon atoms. Equilibrium constants for enolate formation from simple aldehydes and ketones are in the 10^{-16} to 10^{-20} range.

$$(CH_3)_2CHCH \rightleftharpoons H^+ + (CH_3)_2C = CH \qquad \begin{aligned} K_a &= 3 \times 10^{-16} \\ (pK_a &= 15.5) \end{aligned}$$

2-Methylpropanal

$$C_6H_5CCH_3 \rightleftharpoons H^+ + C_6H_5C = CH_2 \qquad \begin{aligned} K_a &= 1.6 \times 10^{-16} \\ (pK_a &= 15.8) \end{aligned}$$

Acetophenone

It is the delocalization of the negative charge onto the electronegative oxygen that is responsible for the enhanced acidity of aldehydes and ketones. With K_a's in the 10^{-16} to 10^{-20} range, aldehydes and ketones are about as acidic as water and alcohols. Thus, hydroxide ion and alkoxide ions are sufficiently strong bases to produce solutions containing significant concentrations of enolate ions at equilibrium.

β-Diketones, such as 2,4-pentanedione, are even more acidic:

$$CH_3CCH_2CCH_3 \rightleftharpoons H^+ + CH_3C = CHCCH_3 \qquad \begin{aligned} K_a &= 10^{-9} \\ (pK_a &= 9) \end{aligned}$$

In the presence of bases such as hydroxide, methoxide, ethoxide, etc., these β-diketones are converted completely to their enolate ions. Notice that it is the methylene group flanked by the two carbonyl groups that is deprotonated. Both carbonyl groups participate in stabilizing the enolate by delocalizing its negative charge.

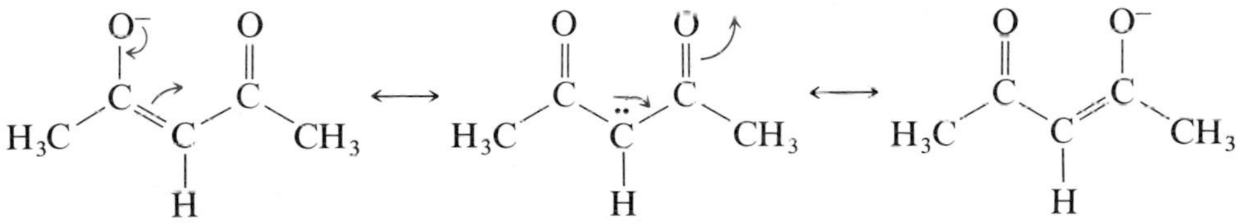

PROBLEM 19.7   Write the structure of the enolate ion derived from each of the following β-dicarbonyl compounds. Give the three principal resonance forms of each.

(a)   2-Methyl-1,3-cyclopentanedione   (c)
(b)   1-Phenyl-1,3-butanedione

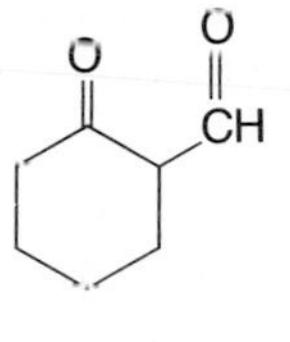

SAMPLE SOLUTION   (a)   First identify the proton that is removed by the base. It is on the carbon between the two carbonyl groups.

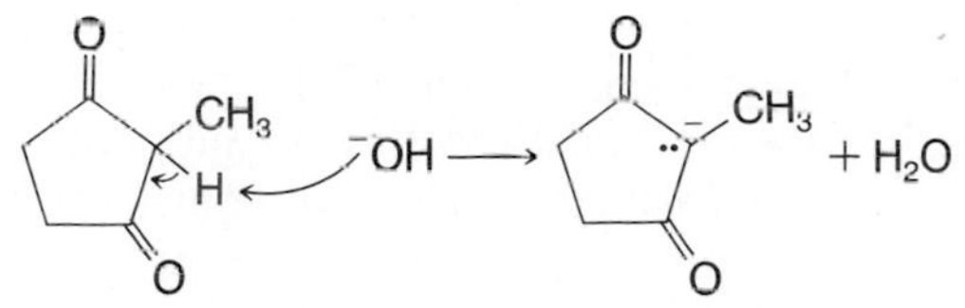

The principal resonance forms of this anion are

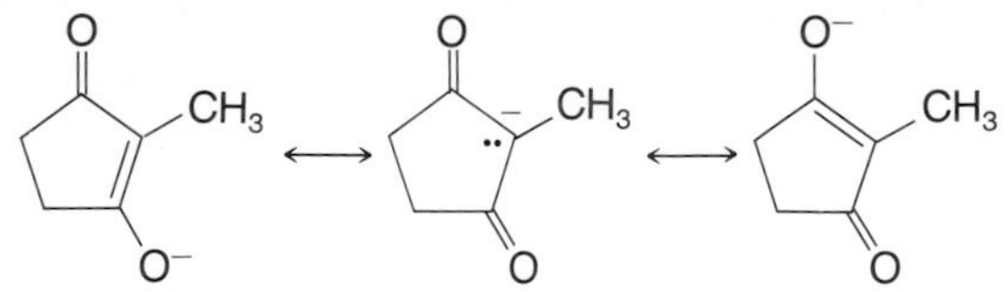

Enolate ions of β-dicarbonyl compounds are useful intermediates in organic synthesis. We shall see some examples of how they are employed in this way later in the chapter.

19.7 THE HALOFORM REACTION

Generation of an enolate ion in the presence of a halogen leads to rapid halogenation at the α carbon atom.

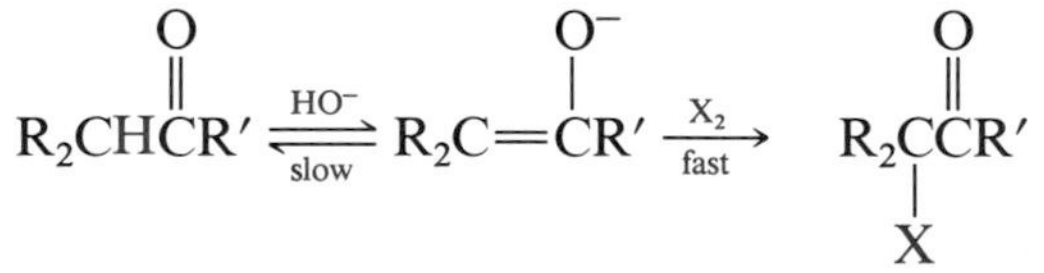

Aldehyde or ketone — Enolate — α-Halo aldehyde or ketonc

As in the acid-catalyzed halogenation of aldehydes and ketones, the reaction rate is independent of the concentration of the halogen, and chlorination, bromination, and iodination all occur at the same rate. The rate-determining step is formation of the reactive intermediate, the enolate ion. Once formed, the enolate ion reacts rapidly with the halogen.

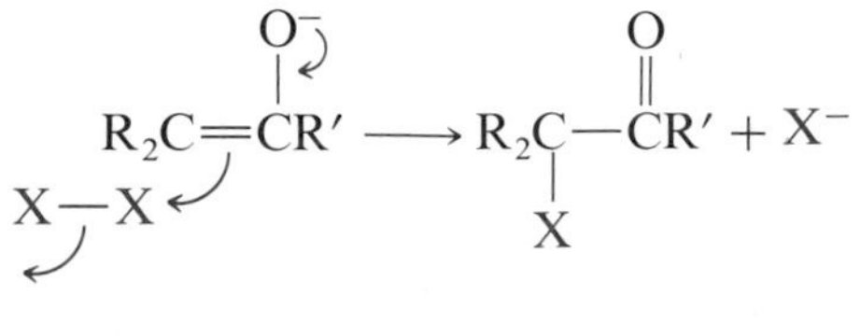

Unlike the acid-catalyzed process, however, this reaction normally cannot be limited to monohalogenation. Methyl ketones, for example, undergo a novel polyhalogenation and cleavage process on treatment with a halogen in aqueous base.

Methyl ketone — Carboxylate salt — Trihalomethane (haloform)

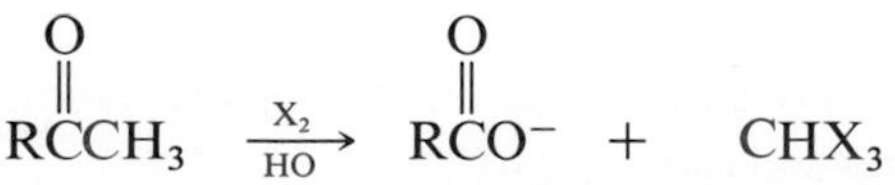

This reaction is called the *haloform reaction* because chloroform, bromoform, or iodoform is one of the products, depending, of course, on the particular halogen.

Acidification of the reaction mixture yields a carboxylic acid. The haloform reaction is sometimes used for the preparation of carboxylic acids from methyl ketones.

$$(CH_3)_3CCCH_3 \xrightarrow[\text{2. H}^+]{\text{1. Br}_2,\ \text{NaOH, H}_2\text{O}} (CH_3)_3CCO_2H \ + \ CHBr_3$$

3,3-Dimethyl-2-butanone   2,2-Dimethylpropanoic   Bromoform
                                     acid (71–74%)

Once one halogen is introduced at the α carbon atom, it increases the acidity of the remaining protons at that position because of its electron-withdrawing inductive effect. Thus, the rate of each successive halogenation step is faster than that of the one that precedes it.

$$RCCH_3 \xrightarrow[\text{HO}^-]{X_2} RCCH_2X \xrightarrow[\text{HO}^-]{X_2} RCCHX_2 \xrightarrow[\text{HO}^-]{X_2} RCCX_3$$

        (Slowest   (Fastest
       halogenation   halogenation
            step)   step)

Nucleophilic addition of hydroxide to the carbonyl group of the trihalomethyl ketone gives an anionic intermediate that undergoes carbon-carbon bond cleavage.

$$RCCX_3 \ + \ HO^- \rightleftharpoons RC\overset{\displaystyle O^-}{\underset{\displaystyle OH}{-}}CX_3 \ \longrightarrow \ RCOH \ + \ \ :\!CX_3$$

                                         $\downarrow$ HO$^-$ $\downarrow$ H$_2$O

                                          RCO_2^- HCX_3

Trihalomethyl   Carboxylate   Trihalomethane
   ketone   anion   (a haloform)

Cleavage occurs because of the stability of the resulting trihalomethide anion, $:\!\overline{C}X_3$. Its three electron-withdrawing halogen substituents stabilize the negative charge and permit its formation under the reaction conditions.

The haloform reaction is sometimes used as a qualitative test for methyl ketones. When iodine is used as the halogen, a positive test is indicated by the formation of iodoform, a yellow solid with a characteristic "hospital-like" odor (it is sometimes used as an antiseptic), melting at 120°C. In addition to methyl ketones, positive tests are given by secondary alcohols of the type $CH_3\overset{\displaystyle OH}{\underset{\displaystyle |}{C}}HR$ and by ethanol. In these cases, oxidation of the secondary alcohol to a methyl ketone and of ethanol to acetaldehyde under the reaction conditions generates the necessary $CH_3\overset{\displaystyle O}{\overset{\displaystyle \|}{C}}-$ unit. The iodoform test is less commonly employed than it was before the advent of spectroscopic methods for identifying functional groups.

19.8 SOME CHEMICAL AND STEREOCHEMICAL CONSEQUENCES OF ENOLIZATION

A number of novel reactions involving the α carbon atom of aldehydes and ketones are understandable on the basis of enol and enolate anion intermediates.

Substitution of deuterium for hydrogen at the α carbon atom of an aldehyde or ketone is a convenient way to introduce an isotopic label into a molecule and is readily carried out by treating the carbonyl compound with deuterium oxide (heavy water, D_2O) and base.

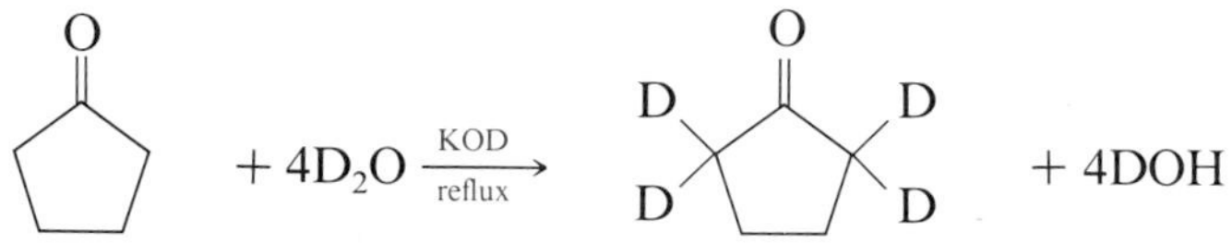

$$\text{Cyclopentanone} + 4D_2O \xrightarrow[\text{reflux}]{\text{KOD}} \text{Cyclopentanone-2,2,5,5-}d_4 + 4DOH$$

Only the α hydrogens are replaced by deuterium in this reaction. The key intermediate is the enolate ion formed by proton abstraction from the α carbon atom of cyclopentanone. Transfer of deuterium from the solvent D_2O to the enolate gives cyclopentanone containing a deuterium atom in place of one of the hydrogens at the α carbon.

Formation of the enolate:

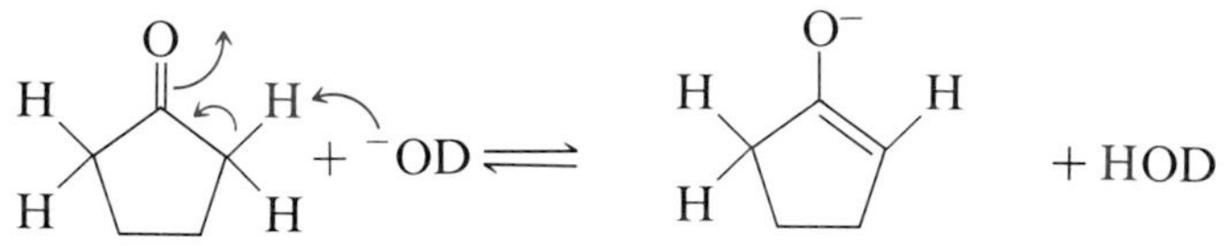

$$\text{Cyclopentanone} + {}^-OD \rightleftharpoons \text{Enolate of cyclopentanone} + HOD$$

Deuterium transfer to the enolate:

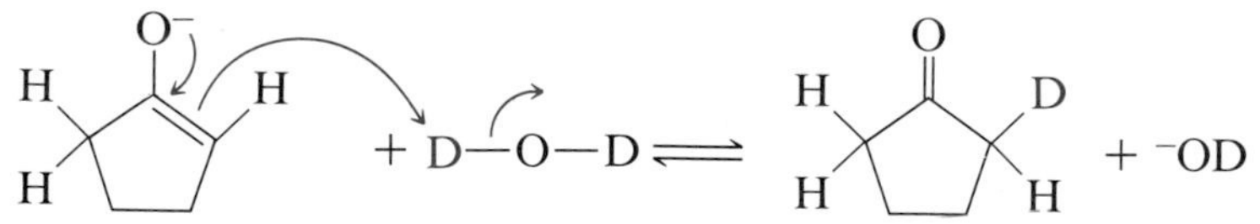

$$\text{Enolate of cyclopentanone} + D{-}O{-}D \rightleftharpoons \text{Cyclopentanone-2-}d_1 + {}^-OD$$

In excess D_2O the process continues until all four α protons are eventually replaced by deuterium.

If the α carbon atom of an aldehyde or ketone is a chiral center, its stereochemical integrity is lost on enolization. Enolization of optically active *sec*-butyl phenyl ketone leads to its racemization by way of the achiral enol form with which it is in equilibrium.

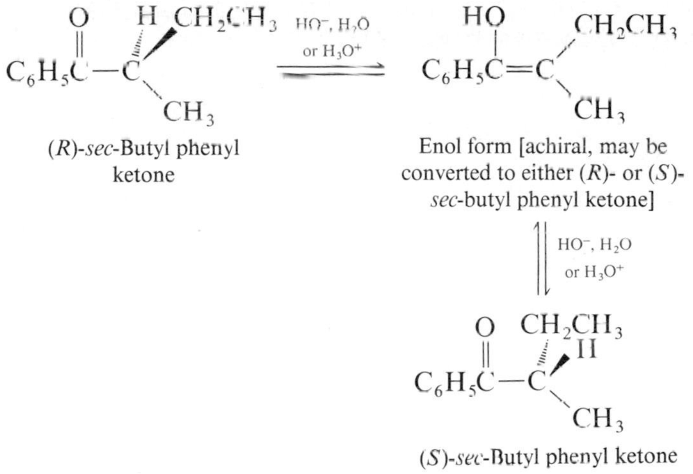

(R)-sec-Butyl phenyl ketone

Enol form [achiral, may be converted to either (R)- or (S)- sec-butyl phenyl ketone]

(S)-sec-Butyl phenyl ketone

Each act of proton abstraction from the α carbon atom converts a chiral molecule to an achiral enol or enolate anion. Careful kinetic studies have established that the rate of loss of optical activity of sec-butyl phenyl ketone is equal to its rate of hydrogen-deuterium exchange, its rate of bromination, and its rate of iodination. In each of the reactions the rate-determining step is conversion of the starting ketone to the enol or enolate anion.

19.9 THE ALDOL CONDENSATION. ALDEHYDES

As noted earlier, an aldehyde is partially converted to its enolate anion by bases such as hydroxide ion and alkoxide ions.

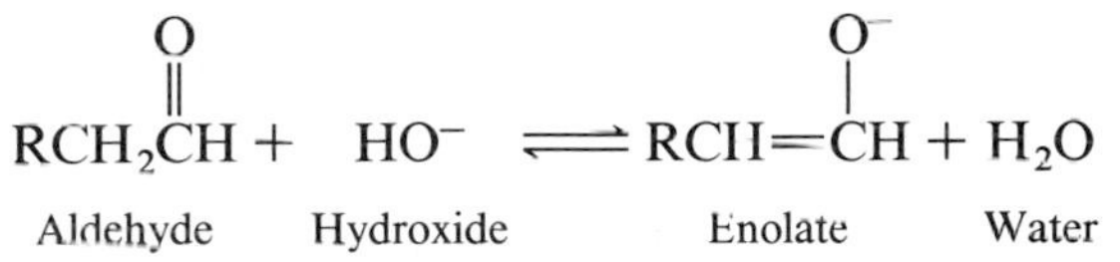

| Aldehyde | Hydroxide | Enolate | Water |

In a solution that contains both an aldehyde and its enolate ion, the enolate undergoes nucleophilic addition to the carbonyl group. This addition is exactly analogous to the addition reactions of other nucleophilic reagents to aldehydes and ketones described in Chapter 18.

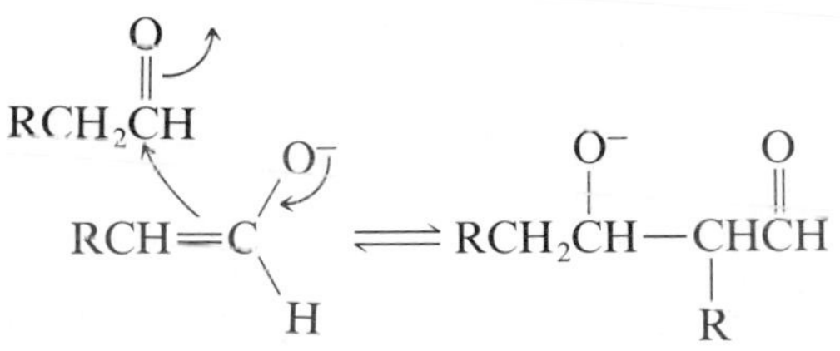

The alkoxide formed in this nucleophilic addition step then abstracts a proton from the solvent (usually water or ethanol) to yield the product of *aldol addition*. This product is known as an *aldol* because it contains both an aldehyde function and a hydroxyl group (*ald* + *ol* = *aldol*).

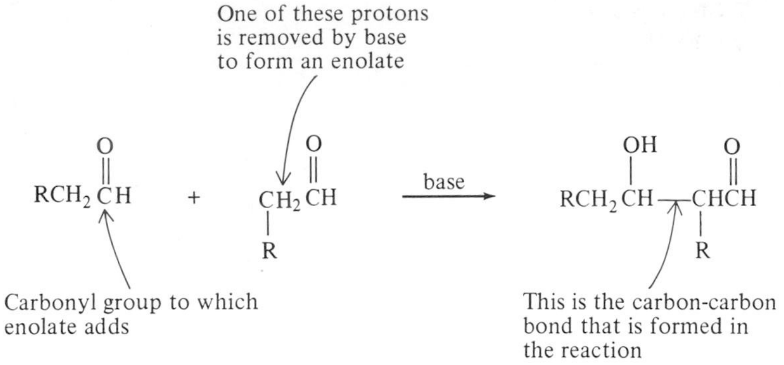

One of these protons
is removed by base
to form an enolate

$$RCH_2\overset{\overset{\displaystyle O}{\|}}{C}H \quad + \quad \overset{\overset{\displaystyle O}{\|}}{\underset{\underset{\displaystyle R}{|}}{C}H_2CH} \quad \xrightarrow{\text{base}} \quad RCH_2\overset{\overset{\displaystyle OH}{|}}{C}H \overset{}{-}\overset{\overset{\displaystyle O}{\|}}{\underset{\underset{\displaystyle R}{|}}{C}HCH}$$

FIGURE 19.4 The reactive sites in aldol addition are the carbonyl group of one aldehyde molecule and the α carbon atom of another.

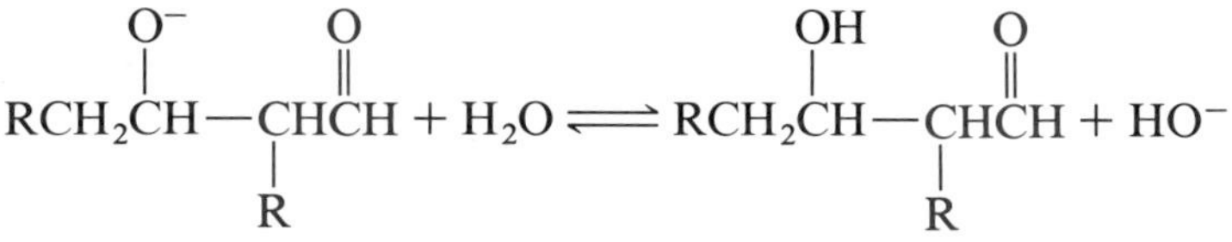

Carbonyl group to which enolate adds

This is the carbon-carbon bond that is formed in the reaction

$$RCH_2\overset{\overset{\displaystyle O^-}{|}}{C}H - \overset{\overset{\displaystyle O}{\|}}{\underset{\underset{\displaystyle R}{|}}{C}HCH} + H_2O \rightleftharpoons RCH_2\overset{\overset{\displaystyle OH}{|}}{C}H - \overset{\overset{\displaystyle O}{\|}}{\underset{\underset{\displaystyle R}{|}}{C}HCH} + HO^-$$

Product of aldol
addition

An important feature of the aldol addition process is that carbon-carbon bond formation occurs between the α carbon atom of one aldehyde and the carbonyl group of another. This is because carbanion (enolate) generation can *only* involve proton abstraction from the α carbon atom. The overall transformation can be represented schematically as shown in Figure 19.4.

Aldol addition occurs readily with aldehydes.

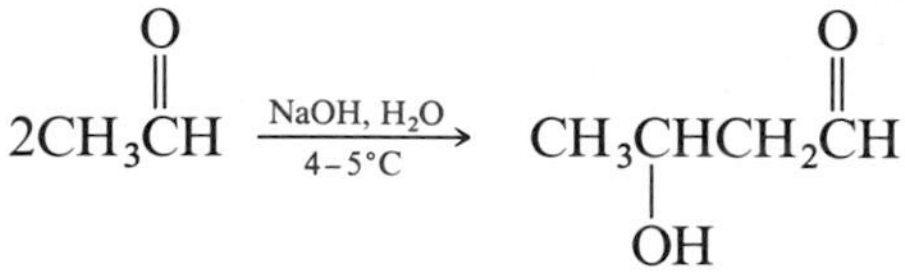

$$2CH_3\overset{\overset{\displaystyle O}{\|}}{C}H \xrightarrow[4-5°C]{\text{NaOH, H}_2\text{O}} CH_3\underset{\underset{\displaystyle OH}{|}}{C}HCH_2\overset{\overset{\displaystyle O}{\|}}{C}H$$

Acetaldehyde   3-Hydroxybutanal (50%)
                              (acetaldol)

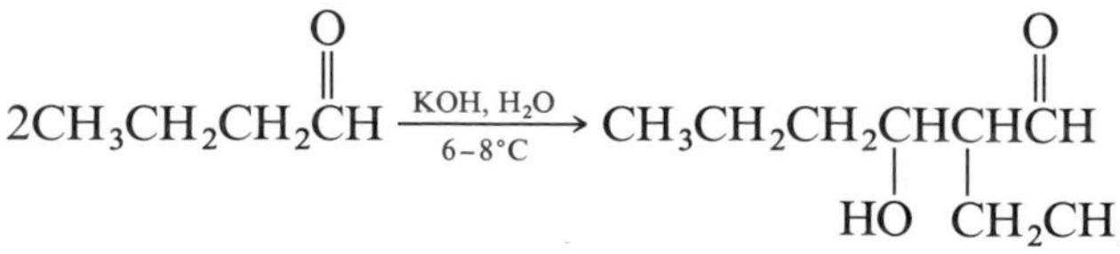

$$2CH_3CH_2CH_2\overset{\overset{\displaystyle O}{\|}}{C}H \xrightarrow[6-8°C]{\text{KOH, H}_2\text{O}} CH_3CH_2CH_2\underset{\underset{\displaystyle HO}{|}}{C}H\underset{\underset{\displaystyle CH_2CH_3}{|}}{C}H\overset{\overset{\displaystyle O}{\|}}{C}H$$

Butanal   2-Ethyl-3-hydroxyhexanal (75%)

PROBLEM 19.8   Write the structure of the aldol addition product of

(a)   Pentanal, $CH_3CH_2CH_2CH_2\overset{\overset{\displaystyle O}{\|}}{C}H$

(b)   2-Methylbutanal, $CH_3CH_2\underset{\underset{\displaystyle CH_3}{|}}{C}H\overset{\overset{\displaystyle O}{\|}}{C}H$

(c)   3-Methylbutanal, $(CH_3)_2CHCH_2CH$
$$\overset{O}{\underset{\|}{}}$$

SAMPLE SOLUTION   (a)   A good way to correctly identify the aldol addition product of any aldehyde is to work through the process mechanistically. Remember that the first step is enolate formation and that this *must* involve proton abstraction from the α carbon atom of a molecule of the starting carbonyl compound.

$$CH_3CH_2CH_2CH_2\overset{O}{\underset{\|}{C}}H + HO^- \rightleftharpoons CH_3CH_2CH_2\overset{\cdot\cdot}{C}H\overset{O}{\underset{\|}{C}}H \longleftrightarrow CH_3CH_2CH_2CH=\overset{O^-}{C}H$$

| Pentanal | Hydroxide | | Enolate of pentanal |

Now use the negatively charged carbon of the enolate to form a new carbon-carbon bond to the carbonyl group. Proton transfer from the solvent to the alkoxide completes the process.

$$CH_3CH_2CH_2CH_2\overset{O}{\underset{\|}{C}}H \quad :\overset{O}{\underset{\|}{C}}HCH$$

$$\underset{CH_2CH_2CH_3}{|}$$

| Pentanal | Enolate of pentanal |

$$\longrightarrow CH_3CH_2CH_2CH_2\overset{O^-}{\underset{|}{C}}H\overset{O}{\underset{\|}{C}}HCH \xrightarrow{H_2O}$$

$$\underset{CH_2CH_2CH_3}{|}$$

$$CH_3CH_2CH_2CH_2\overset{OH}{\underset{|}{C}}H\overset{O}{\underset{\|}{C}}HCH$$

$$\underset{CH_2CH_2CH_3}{|}$$

3-Hydroxy-2-propylheptanal
(aldol addition product
of pentanal)

The β-hydroxy aldehyde products of aldol addition undergo dehydration on heating, to yield α,β-unsaturated aldehydes.

$$RCH_2\overset{OH}{\underset{|}{C}}HCHCH \xrightarrow{heat} RCH_2CH=CCH + H_2O$$

$$\underset{R}{|} \qquad\qquad \underset{R}{|}$$

| β-Hydroxy aldehyde | α,β-Unsaturated aldehyde | Water |

Conjugation of the newly formed double bond with the carbonyl group stabilizes the α,β-unsaturated aldehyde, provides the driving force for the dehydration process, and controls its regioselectivity. Dehydration can be effected by heating the aldol with acid or base. Normally, if the α,β-unsaturated aldehyde is the desired product, all that is done is to carry out the base-catalyzed aldol addition reaction at elevated temperature. Under these conditions, once the aldol addition product is formed, it rapidly loses water to form the α,β-unsaturated aldehyde.

$$2CH_3CH_2CH_2\overset{\displaystyle O}{\overset{\displaystyle \|}{C}}H \xrightarrow[80-100°C]{NaOH,\ H_2O} CH_3CH_2CH_2CH{=}\overset{\displaystyle O}{\overset{\displaystyle \|}{C}}CH \quad via \quad \left[CH_3CH_2CH_2\overset{\displaystyle OH}{\overset{\displaystyle |}{C}}H\overset{\displaystyle O}{\overset{\displaystyle \|}{C}}HCH \right]$$

Butanal $\qquad\qquad$ 2-Ethyl-2-hexenal (86%) $\qquad\qquad$ 2-Ethyl-3-hydroxyhexanal
(not isolated; dehydrates
under reaction conditions)

The process by which two molecules of an aldehyde combine to form an α,β-unsaturated aldehyde and a molecule of water is called the *aldol condensation* reaction.

PROBLEM 19.9   Write the structure of the aldol condensation product of each of the aldehydes in Problem 19.8. One of these aldehydes can undergo aldol addition, but not aldol condensation. Which one? Why?

SAMPLE SOLUTION   *(a)*   Dehydration of the product of aldol addition of pentanal introduces the double bond between C-2 and C-3 to give an α,β-unsaturated aldehyde.

$$CH_3CH_2CH_2CH_2\overset{OH}{\overset{|}{C}}H\overset{O}{\overset{\|}{C}}HCH \xrightarrow{-H_2O} CH_3CH_2CH_2CH_2CH{=}\overset{O}{\overset{\|}{C}}CH$$

Product of aldol addition of
pentanal (3-hydroxy-2-propyl-
heptanal) $\qquad\qquad$ Product of aldol condensation
of pentanal (2-propyl-2-
heptenal)

The point has been made earlier (Section 6.4) that alcohols require acid catalysis in order to undergo dehydration to alkenes. Thus, it may seem strange that aldol addition products can be dehydrated in base. This is another example of the way in which the enhanced acidity of protons at the α carbon atom affects the reactions of carbonyl compounds. Elimination may take place in a concerted E2 fashion or it may be stepwise and proceed through an enolate ion.

$$RCH_2\overset{OH}{\overset{|}{C}}H\overset{O}{\overset{\|}{C}}HCH + HO^- \underset{fast}{\rightleftharpoons} RCH_2\overset{OH}{\overset{|}{C}}H\overset{\cdot\cdot}{C}{-}\overset{O}{\overset{\|}{C}}H + HOH$$

β-Hydroxy aldehyde $\qquad\qquad$ Enolate ion of
β-hydroxy aldehyde

$$RCH_2\overset{OH}{\overset{|}{C}}H{-}\overset{\cdot\cdot}{C}{-}\overset{O}{\overset{\|}{C}}H \xrightarrow{slow} RCH_2CH{=}\overset{O}{\overset{\|}{C}}CH + HO^-$$

Enolate of
β-hydroxy aldehyde $\qquad\qquad$ α,β-Unsaturated
aldehyde

19.10 THE ALDOL CONDENSATION. KETONES

As with other reversible nucleophilic addition reactions, the equilibria for aldol additions are less favorable for ketones than they are for aldehydes. For example, only 2 percent of the aldol addition product of acetone is present at equilibrium.

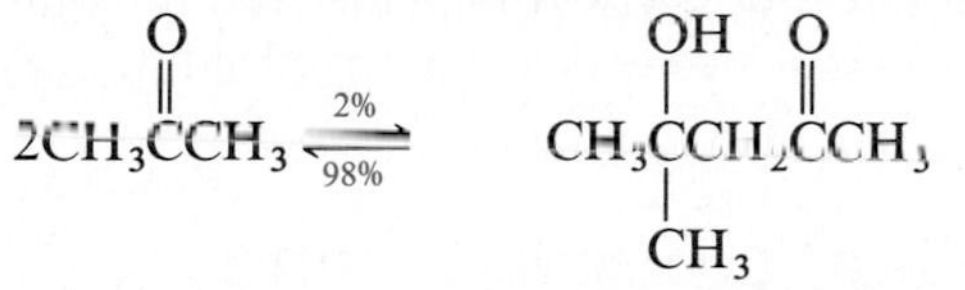

| Acetone | 4-Hydroxy-4-methyl-2-pentanone (diacetone alcohol) |

The situation is similar for aldol addition reactions of other ketones.

When conditions are chosen so as to favor dehydration of the aldol addition product, the position of equilibrium shifts to the right and α,β-unsaturated ketones result. One method uses aluminum tri-*tert*-butoxide as the base at elevated temperature.

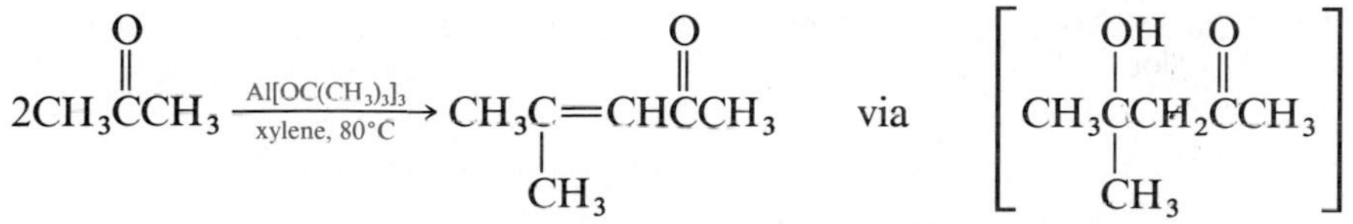

Acetone — 4-Methyl-3-penten-2-one (37%) (mesityl oxide) — via — 4-Hydroxy-4-methyl-2-pentanone (not isolated)

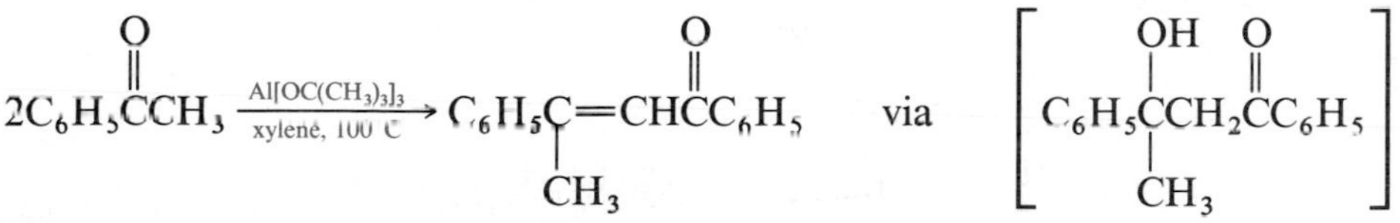

Acetophenone — 1,3-Diphenyl-2-buten-1-one (77%) — via — 3-Hydroxy-1,3-diphenyl-1-butanone (not isolated)

PROBLEM 19.10   Cyclohexanone undergoes efficient aldol condensation on being heated with aluminum tri-*tert*-butoxide in xylene. Write the structure of the aldol condensation product along with that of its β-hydroxy ketone precursor.

Dicarbonyl compounds, that is, compounds having two carbonyl groups within the same molecule, undergo intramolecular aldol condensation reactions. Even bases as weak as sodium carbonate are adequate catalysts in these cases.

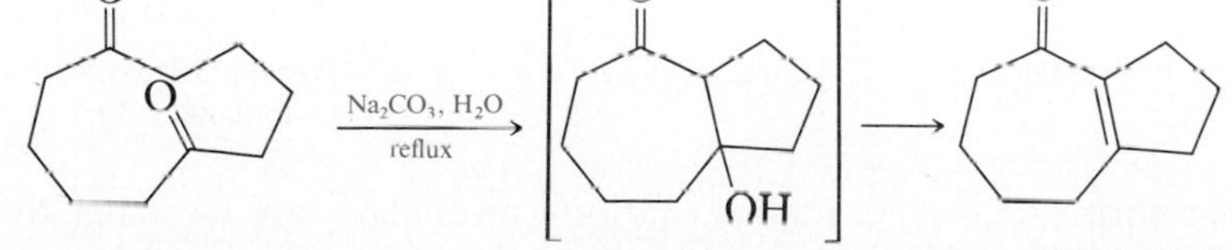

1,6-Cyclodecanedione — Not isolated; dehydrates under reaction conditions — Bicyclo[5.3.0]dec-1(7)-en-2-one (96%)

Intramolecular aldol condensations proceed best when five, six, or seven-membered rings result.

Since the aldol condensation leads to carbon-carbon bond formation, organic chemists have exploited it as a synthetic tool. Its usefulness has been extended beyond the *self-condensation* of aldehydes and ketones just described to include *mixed* (or *crossed*) *aldol condensations* between two different carbonyl compounds. Let us now examine the circumstances that lead to efficient mixed aldol condensation reactions.

19.11 MIXED ALDOL CONDENSATIONS

Mixed aldol condensations are effective only when the number of reaction possibilities is limited. It would not be a useful procedure, for example, to treat a solution of acetaldehyde and propanal with base. A mixture of four aldol addition products forms under these conditions. Two of the products are those of self-addition:

$$\underset{\substack{\text{3-Hydroxybutanal}\\ \text{(from addition of enolate}\\ \text{of acetaldehyde to}\\ \text{acetaldehyde)}}}{\text{CH}_3\text{CHCH}_2\overset{\displaystyle O}{\overset{\|}{\text{CH}}}} \qquad \underset{\substack{\text{3-Hydroxy-2-methylpentanal}\\ \text{(from addition of enolate}\\ \text{of propanal to propanal)}}}{\text{CH}_3\text{CH}_2\text{CHCHCH}}$$

(with OH substituent on the CH of 3-hydroxybutanal; HO and CH₃ substituents on 3-hydroxy-2-methylpentanal)

Two are the products of mixed addition:

$$\underset{\substack{\text{3-Hydroxy-2-methylbutanal}\\ \text{(from addition of enolate}\\ \text{of propanal to acetaldehyde)}}}{\text{CH}_3\text{CHCHCH}} \qquad \underset{\substack{\text{3-Hydroxypentanal}\\ \text{(from addition of enolate}\\ \text{of acetaldehyde to propanal)}}}{\text{CH}_3\text{CH}_2\text{CHCH}_2\text{CH}}$$

(3-Hydroxy-2-methylbutanal with OH and CH₃ substituents; 3-hydroxypentanal with OH substituent)

The mixed aldol condensations that are the most synthetically useful are those in which one of the reactants is an aldehyde that cannot form an enolate. Formaldehyde, for example, has often been used successfully.

$$\underset{\text{Formaldehyde}}{\overset{\displaystyle O}{\overset{\|}{\text{HCH}}}} + \underset{\text{3-Methylbutanal}}{(\text{CH}_3)_2\text{CHCH}_2\overset{\displaystyle O}{\overset{\|}{\text{CH}}}} \xrightarrow[\substack{\text{water-}\\ \text{ether}}]{\text{K}_2\text{CO}_3} \underset{\substack{\text{2-Hydroxymethyl-3-}\\ \text{methylbutanal (52\%)}}}{(\text{CH}_3)_2\text{CHCHCH}}$$

(product bearing CH₂OH substituent)

Not only is formaldehyde incapable of forming an enolate, but it is so reactive toward nucleophilic addition that it suppresses self-condensation of the other component by reacting rapidly with any enolate present.

Aromatic aldehydes cannot form enolates and a large number of mixed aldol condensation reactions have been carried out in which an aromatic aldehyde is used as the carbonyl component.

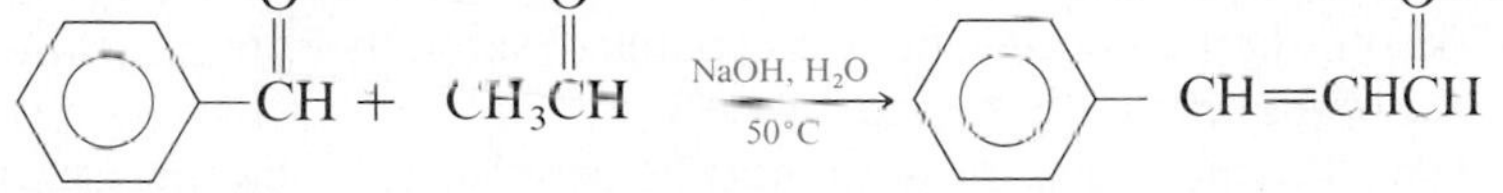

$$\underset{\text{Benzaldehyde}}{C_6H_5-CH} + \underset{\text{Acetaldehyde}}{CH_3CH} \xrightarrow[50°C]{\text{NaOH, H}_2\text{O}} \underset{\substack{\text{3-Phenylpropenal (90\%)}\\\text{(cinnamaldehyde)}}}{C_6H_5-CH=CHCH}$$

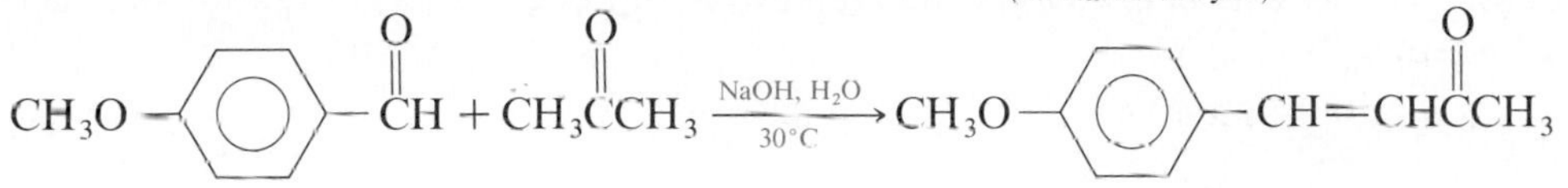

$$\underset{\text{Anisaldehyde}}{CH_3O-C_6H_4-CH} + \underset{\text{Acetone}}{CH_3CCH_3} \xrightarrow[30°C]{\text{NaOH, H}_2\text{O}} \underset{\substack{\text{4-p-Methoxyphenyl-3-}\\\text{buten-2-one (83\%)}}}{CH_3O-C_6H_4-CH=CHCCH_3}$$

Mixed aldol condensations using aromatic aldehydes are referred to as *Claisen-Schmidt condensations*. They always involve dehydration of the product of mixed addition and yield a product in which the double bond is conjugated to both the aromatic ring and the carbonyl group.

PROBLEM 19.11 Give the structure of the mixed aldol condensation product of benzaldehyde with

(a) Acetophenone, $C_6H_5\overset{O}{\overset{\|}{C}}CH_3$

(b) *tert*-Butyl methyl ketone, $(CH_3)_3\overset{O}{\overset{\|}{C}}CCH_3$

(c) Cyclohexanone

SAMPLE SOLUTION (a) The enolate of acetophenone reacts with benzaldehyde to yield the product of mixed addition. Dehydration of the intermediate occurs, giving the α,β-unsaturated ketone.

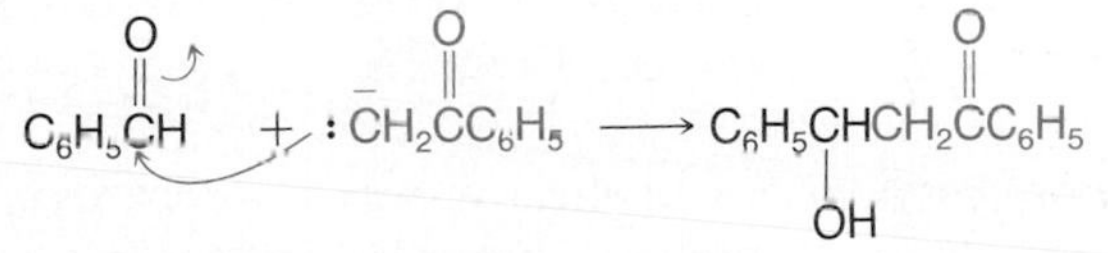

$$\underset{\text{Benzaldehyde}}{C_6H_5\overset{O}{\overset{\|}{C}}H} + \underset{\substack{\text{Enolate of}\\\text{acetophenone}}}{:\overset{-}{C}H_2\overset{O}{\overset{\|}{C}}C_6H_5} \longrightarrow C_6H_5\underset{\overset{|}{O}H}{C}HCH_2\overset{O}{\overset{\|}{C}}C_6H_5$$

$$\downarrow -\text{H}_2\text{O}$$

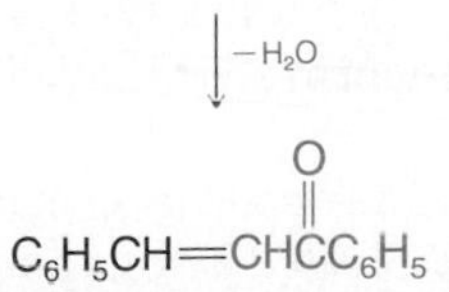

$$\underset{\text{1,3-Diphenyl-2-propen-1-one}}{C_6H_5CH=CH\overset{O}{\overset{\|}{C}}C_6H_5}$$

As actually carried out, the mixed aldol condensation product, 1,3-diphenyl-2-propen-1-one, has been isolated in 85 percent yield on treating benzaldehyde with acetophenone in an aqueous ethanol solution of sodium hydroxide at 15 to 30°C.

19.12 THE ALDOL CONDENSATION IN SYNTHETIC ORGANIC CHEMISTRY

Aldol condensations are one of the fundamental carbon-carbon bond-forming processes of synthetic organic chemistry. Furthermore, since the products of these aldol condensations contain functional groups capable of subsequent modification, access to a host of useful materials is gained.

To illustrate how aldol condensation may be coupled to functional group modification, consider the synthesis of 2-ethyl-1,3-hexanediol, a compound used as an insect repellent. This 1,3-diol is prepared by reduction of the aldol addition product of butanal.

$$
\underset{\text{Butanal}}{CH_3CH_2CH_2\overset{\displaystyle O}{\overset{\|}{C}}H} \xrightarrow[\text{addition}]{\text{aldol}} \underset{\text{2-Ethyl-3-hydroxyhexanal}}{CH_3CH_2CH_2\underset{\displaystyle CH_2CH_3}{\overset{\displaystyle OH}{\overset{|}{C}H}}\overset{\displaystyle O}{\overset{\|}{C}H}} \xrightarrow[\text{Ni}]{H_2} \underset{\text{2-Ethyl-1,3-hexanediol}}{CH_3CH_2CH_2\underset{\displaystyle CH_2CH_3}{\overset{\displaystyle OH}{\overset{|}{C}H}}CHCH_2OH}
$$

It is usually the carbon skeleton that provides a clue as to the appropriate aldol route to a particular compound. Take, for example, 4-methyl-2-pentanone (methyl isobutyl ketone, or MIBK, a common industrial solvent). It has the same carbon skeleton as the aldol condensation product of acetone.

$$
\underset{\substack{\text{4-Methyl-2-pentanone} \\ \text{(MIBK)}}}{CH_3\underset{\displaystyle CH_3}{\overset{|}{C}H}CH_2\overset{\displaystyle O}{\overset{\|}{C}}CH_3} \Longrightarrow \underset{\substack{\text{4-Methyl-3-penten-2-one} \\ \text{(mesityl oxide)}}}{CH_3\underset{\displaystyle CH_3}{\overset{|}{C}}=CH\overset{\displaystyle O}{\overset{\|}{C}}CH_3} \Longrightarrow \underset{\text{Acetone}}{2CH_3\overset{\displaystyle O}{\overset{\|}{C}}CH_3}
$$

Indeed, MIBK is prepared from acetone as its ultimate precursor by hydrogenation of the carbon-carbon double bond of the aldol condensation product of acetone.

19.13 EFFECTS OF CONJUGATION IN α,β-UNSATURATED ALDEHYDES AND KETONES

Aldol condensation reactions are an effective way to prepare α,β-unsaturated aldehydes and ketones. These compounds have some interesting properties that result from conjugation of the carbon-carbon double bond with the carbonyl group. As shown in Figure 19.5, the π systems of the carbon-carbon and carbon-oxygen double bonds overlap to form an extended π system that permits increased electron delocalization.

This electron delocalization stabilizes a conjugated system. The equilibrium between a β,γ-unsaturated ketone and an α,β-unsaturated analog favors the conjugated isomer.

$$
\underset{\substack{\text{4-Hexen-2-one (17\%)} \\ (\beta,\gamma\text{-unsaturated ketone})}}{CH_3CH=CHCH_2\overset{\displaystyle O}{\overset{\|}{C}}CH_3} \underset{25°C}{\overset{K=4.8}{\rightleftharpoons}} \underset{\substack{\text{3-Hexen-2-one (83\%)} \\ (\alpha,\beta\text{-unsaturated ketone})}}{CH_3CH_2CH=CH\overset{\displaystyle O}{\overset{\|}{C}}CH_3}
$$

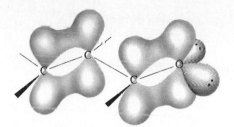

(a) Individual π orbitals of the carbon-carbon and carbon-oxygen double bonds of an α, β-unsaturated carbonyl compound.

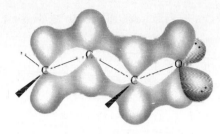

(b) Extended π system formed by overlap of the individual π systems.

FIGURE 19.5 Conjugation in α,β-unsaturated aldehydes and ketones leads to an extended π system and increases the delocalization of the π electrons.

PROBLEM 19.12 Commercial mesityl oxide $(CH_3)_2C\!=\!CHCCH_3$ (with a carbonyl O shown above) is often contaminated with about 10 percent of an isomer having the same carbon skeleton. What is a likely structure for this compound?

In resonance terms, electron delocalization in α,β-unsaturated carbonyl compounds is represented by contributions from three principal resonance structures:

A
Most stable structure

B

C

Electrophilic reagents, bromine and peroxy acids for example, react more slowly with the carbon-carbon double bond of α,β-unsaturated carbonyl compounds than with simple alkenes. Conjugation with the electron-withdrawing carbonyl group reduces the π electron density of the carbon-carbon double bond, making it less nucleophilic.

On the other hand, the polarization of electron density in α,β-unsaturated carbonyl compounds makes their β carbon atoms rather electrophilic. Some chemical consequences of this enhanced electrophilicity are described in the following section.

19.14 CONJUGATE NUCLEOPHILIC ADDITION TO α,β-UNSATURATED CARBONYL COMPOUNDS

α,β-Unsaturated carbonyl compounds contain two electrophilic sites — the carbonyl carbon and the carbon atom that is β to it. Nucleophiles such as organolithium and Grignard reagents and lithium aluminum hydride tend to react by nucleophilic addition to the carbonyl group. This is called *direct addition,* or *1,2 addition.* With certain other nucleophiles addition at the β carbon atom is observed. This is called *conjugate addition,* or *1,4 addition.* The nucleophile becomes attached to the β carbon, and the α carbon accepts a proton from the solvent or from the conjugate acid of the nucleophile.

Conjugate addition to an α,β-unsaturated carbonyl compound:

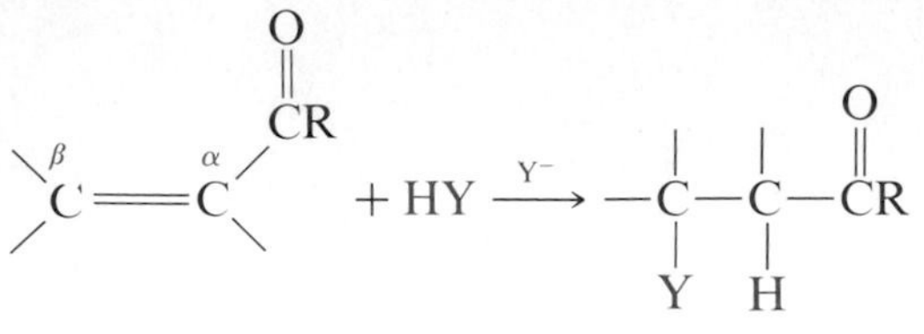

α,β-Unsaturated
carbonyl compound

Product of con-
jugate addition

In most cases, conjugate addition is observed under conditions of thermodynamic control. A carbon-oxygen double bond is stabilized by resonance to a greater extent than a carbon-carbon double bond. The product of conjugate addition retains the carbonyl group and is more stable than the product of direct addition, which retains the carbon-carbon double bond.

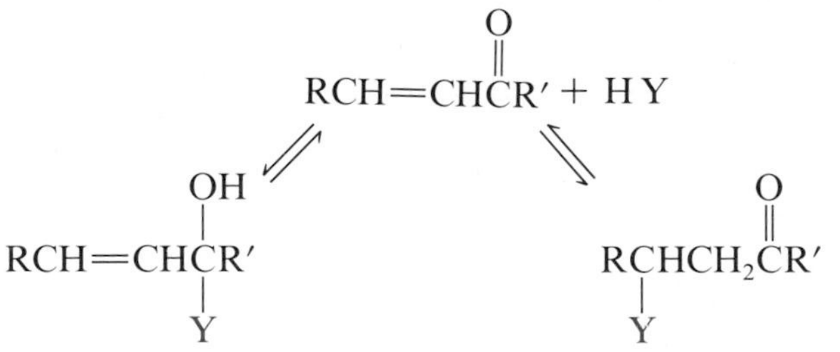

Product of direct
addition
(less stable)

Product of conjugate
addition
(more stable)

Conjugate addition is most often observed when the nucleophile is weakly basic.

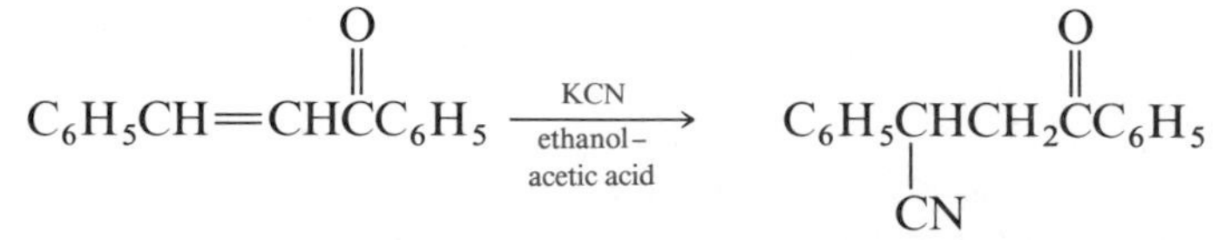

1,3-Diphenyl-2-propen-1-one

3-Cyano-1,3-diphenyl-1-propanone
(93–96%)

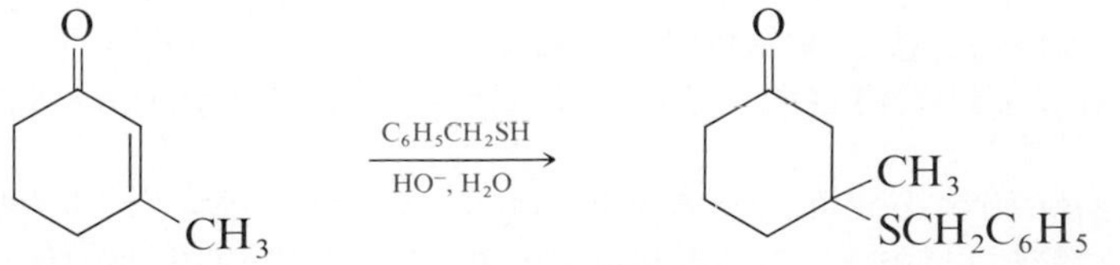

3-Methyl-2-cyclohexen-1-one

3-Benzylthio-3-methylcyclohexanone
(58%)

Ordinarily, nucleophilic addition to simple carbon-carbon double bonds is quite slow because a high-energy carbanion is formed in the rate-determining step.

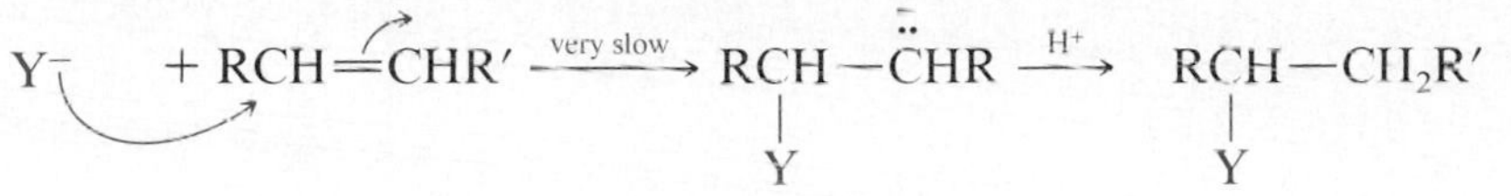

$$Y^- \;+\; RCH{=}CHR' \xrightarrow{\text{very slow}} \underset{\underset{Y}{|}}{RCH}{-}\overset{..}{\underset{}{C}}HR' \xrightarrow{H^+} \underset{\underset{Y}{|}}{RCH}{-}CH_2R'$$

| Nucleophile | Alkene | Carbanion | Product of nucleophilic addition to alkene; not a commonplace reaction |

Since carbonyl groups stabilize carbanions to which they are attached by enolate resonance, nucleophilic addition to the carbon-carbon double bond takes place much more rapidly in α,β-unsaturated carbonyl compounds than it does in simple alkenes.

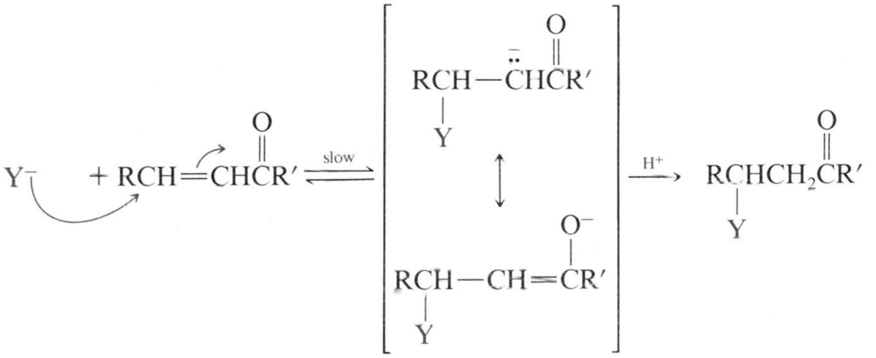

$$Y^- + RCH{=}CHCR' \;\underset{}{\overset{\text{slow}}{\rightleftharpoons}}\; \left[\begin{array}{c} \overset{\displaystyle O}{\overset{\|}{RCH{-}\overset{..}{C}HCR'}} \\ {\underset{Y}{|}} \\ \updownarrow \\ RCH{-}CH{=}\overset{\displaystyle O^-}{\underset{}{CR'}} \\ {\underset{Y}{|}} \end{array} \right] \xrightarrow{H^+} \underset{\underset{Y}{|}}{RCH}CH_2\overset{\displaystyle O}{\overset{\|}{C}}R'$$

| Nucleophile | α,β-Unsaturated carbonyl compound | Anionic intermediate is an enolate | Product of conjugate nucleophilic addition to α,β-unsaturated carbonyl compound |

Conjugate addition to α,β-unsaturated carbonyl compounds is another result of the stabilization of carbanions by the electron-withdrawing carbonyl group.

Enolate anions have been employed as nucleophiles in conjugate additions to α,β-unsaturated ketones. Reactions of this type are known as *Michael additions.* (Arthur Michael, an American chemist whose career spanned the period between the 1870s and 1930s, systematically studied these reactions.)

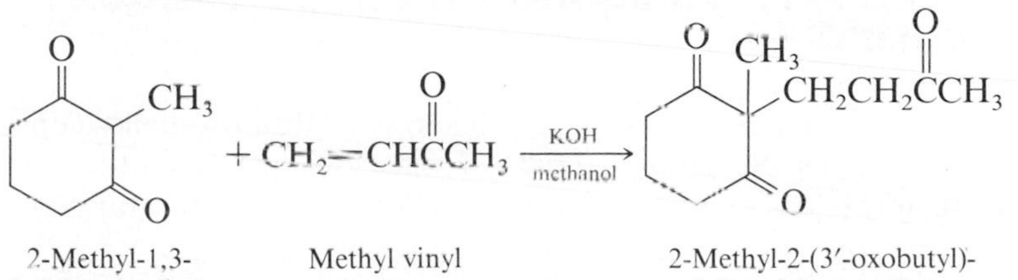

| 2-Methyl-1,3-cyclohexanedione | Methyl vinyl ketone | 2-Methyl-2-(3'-oxobutyl)-1,3-cyclohexanedione (85%) |

$$+ \; CH_2{=}CHCCH_3 \xrightarrow[\text{methanol}]{KOH}$$

The reaction shown above is the first step in a process of synthetic value in that the product of Michael addition has the necessary functionality to undergo an intramolecular aldol condensation.

2-Methyl-2-(3′-oxobutyl)-
1,3-cyclohexanedione

$\xrightarrow[\text{benzene, heat}]{\text{Al[OC(CH}_3)_3]_3}$

Intramolecular aldol addition
product; not isolated

$\xrightarrow{-\text{H}_2\text{O}}$

Δ^9-9-Methyloctalin-3,8-dione (40%)

The synthesis of cyclohexenone derivatives by Michael addition followed by intra-molecular aldol condensation is called the *Robinson annulation,* after Sir Robert Robinson, who popularized its use. By *annulation* we mean the building of a ring onto some starting molecule. (The alternative spelling "annelation" is also often used.)

PROBLEM 19.13 Both the conjugate addition step and the intramolecular aldol condensation step can be carried out in one synthetic operation without isolating any of the intermediates along the way. For example, consider the reaction

$C_6H_5CH_2CCH_2C_6H_5$ + CH_2=CHCCH$_3$ $\xrightarrow[\text{CH}_3\text{OH}]{\text{NaOCH}_3}$ CH$_3$ —C$_6$H$_5$

Dibenzyl ketone   Methyl vinyl
                        ketone

3-Methyl-2,6-diphenyl-2-
cyclohexen-1-one (55%)

Write structural formulas corresponding to the intermediates formed in the conjugate addition step and in the aldol addition step.

19.15 CONJUGATE ADDITION OF ORGANOCOPPER REAGENTS TO α,β-UNSATURATED CARBONYL COMPOUNDS

The preparation and some synthetic applications of lithium dialkylcuprates have been described earlier (Sections 15.11, 18.7). The most prominent feature of these reagents is their capacity to undergo conjugate addition to α,β-unsaturated aldehydes and ketones.

R_2C=CHCR′ + LiCuR″$_2$ $\xrightarrow[\text{2. H}_2\text{O}]{\text{1. ether}}$ $R_2CCH_2CR′$
                                                                                         |
                                                                                         R″

α,β-Unsatu-
rated aldehyde
or ketone

Lithium
dialkylcuprate

Aldehyde or ketone
alkylated at the
β position

O
+ LiCu(CH$_3$)$_2$ $\xrightarrow[\text{2. H}_2\text{O}]{\text{1. ether}}$

O
CH$_3$
CH$_3$

3-Methyl-2-
cyclohexen-1-one

Lithium
dimethylcuprate

3,3-Dimethylcyclohexanone
(98%)

PROBLEM 19.14   Outline two ways in which 4-methyl-2-octanone can be prepared by conjugate addition of an organocuprate reagent to an α,β-unsaturated ketone.

SAMPLE SOLUTION   Mentally disconnect one of the bonds to the β carbon so as to identify the group that comes from the lithium dialkylcuprate.

Disconnect this bond

O
CH$_3$CH$_2$CH$_2$CH$_2$ CHCH$_2$CCH$_3$
CH$_3$
4-Methyl-2-octanone

$\Longrightarrow$

O
CH$_3$CH$_2$CH$_2$$\bar{\text{C}}H_2$ + CH$_3$CH$=$CHCCH$_3$

According to this disconnection, the butyl group is derived from lithium dibutylcuprate. A suitable preparation is

O
CH$_3$CH$=$CHCCH$_3$ + LiCu(CH$_2$CH$_2$CH$_2$CH$_3$)$_2$ $\xrightarrow[\text{2. H}_2\text{O}]{\text{1. ether}}$ CH$_3$CH$_2$CH$_2$CH$_2$CHCH$_2$CCH$_3$
CH$_3$

O

3-Buten-2-one

Lithium dibutylcuprate

4-Methyl-2-octanone

Now see if you can identify the second possibility.

This reaction has been primarily developed by Professor Herbert House (now at the Georgia Institute of Technology) through systematic studies begun in the mid-1960s. Like other carbon-carbon bond-forming reactions, organocuprate addition to enones is a powerful tool in organic synthesis.

19.16   ALKYLATION OF ENOLATE ANIONS

Since enolate anions are sources of nucleophilic carbon, one potential use in organic synthesis is their reaction with alkyl halides to give α-alkyl derivatives of aldehydes and ketones.

O
R$_2$CHCR' $\xrightarrow{\text{base}}$ R$_2$C$=$CR' $\xrightarrow{\text{R''X}}$ R$_2$C$-$CR'
O$^-$
O
R''

Aldehyde
or ketone

Enolate anion

α-Alkyl derivative
of an aldehyde or ketone

In practice, this reaction is difficult to carry out with simple aldehydes and ketones. Aldol condensation competes with alkylation, and it is not always possible to limit the reaction to the introduction of a single alkyl group. The most successful alkylation procedures utilize β-diketones as starting materials. Because they are so acidic, β-diketones can be converted quantitatively to their enolate ions by relatively weak bases and do not self-condense. Ideally, the alkyl halide should be a methyl or primary alkyl halide.

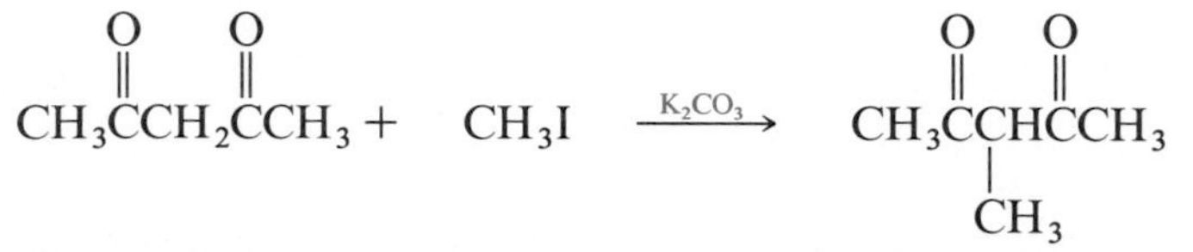

$$CH_3CCH_2CCH_3 + \quad CH_3I \xrightarrow{K_2CO_3} \quad CH_3CCHCCH_3$$

$$\overset{CH_3}{|}$$

| 2,5-Pentanedione | Iodomethane | 3-Methyl-2,5-pentanedione (75–77%) |

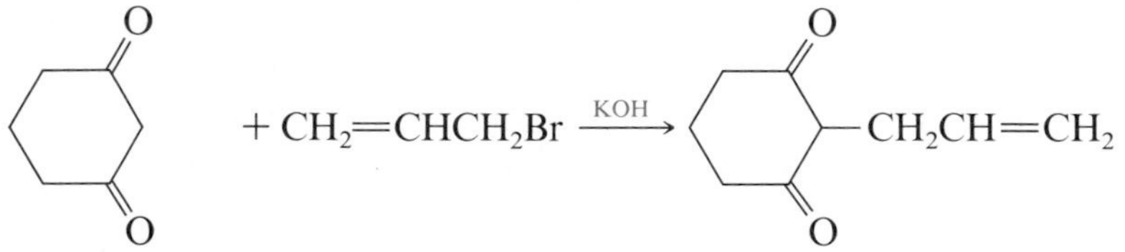

1,3-Cyclohexanedione + $CH_2{=}CHCH_2Br$ $\xrightarrow{\text{KOH}}$ 2-Allyl-1,3-cyclohexanedione (75%)

Alkylation of simple aldehydes and ketones can be achieved by an alternative method using an enamine as the source of nucleophilic carbon.

19.17 ALKYLATION OF ENAMINES

The preparation of enamines from aldehydes and ketones and secondary amines was described in Section 18.14. The primary motivation for preparing enamines lies in their use as reagents for carbon-carbon bond formation. The carbon-carbon double bond of an enamine is considered to be electron-rich. Electron release from nitrogen increases the electron density at carbon in an enamine.

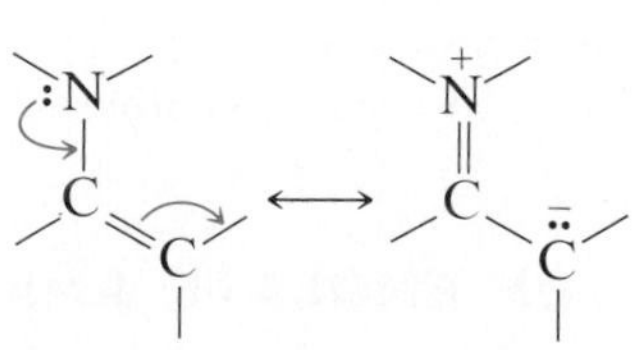

This is analogous to, but more pronounced than, the corresponding situation in which electron release from oxygen increases the electron richness of the double bond of an enol. Nitrogen is less electronegative than oxygen and is a better electron pair donor. Enamines are more nucleophilic than enols, and this heightened nucleophilicity is exploited in certain synthetic applications.

In particular, enamines react with certain alkyl halides while enols do not. The alkyl halides that react with enamines are those that are the most reactive in S_N2 processes, namely, methyl and primary alkyl halides, α-halo ethers, and α-halo carbonyl compounds.

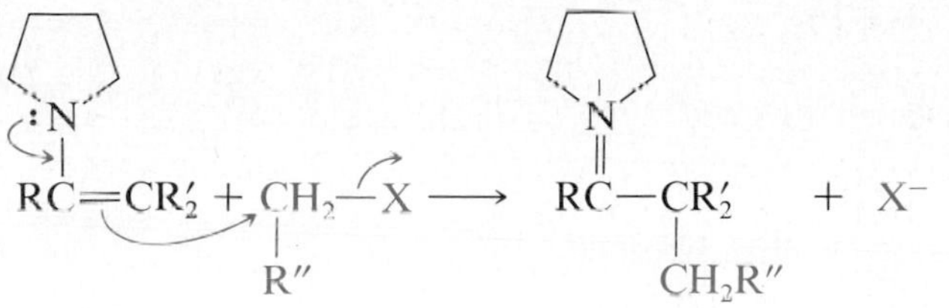

| Pyrrolidine enamine | Primary alkyl halide | Alkylated derivative of enamine | Halide ion |

These alkylated derivatives of enamines are readily hydrolyzed to aldehydes and ketones.

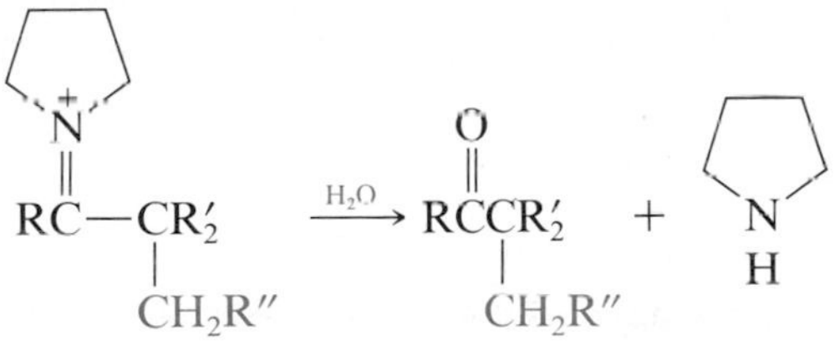

| Alkylated derivative of enamine | Ketone | Pyrrolidine |

Both reactions, alkylation of an enamine and hydrolysis of the derived product, can be carried out sequentially in a single reaction vessel to yield an aldehyde or ketone alkylated at its α carbon atom.

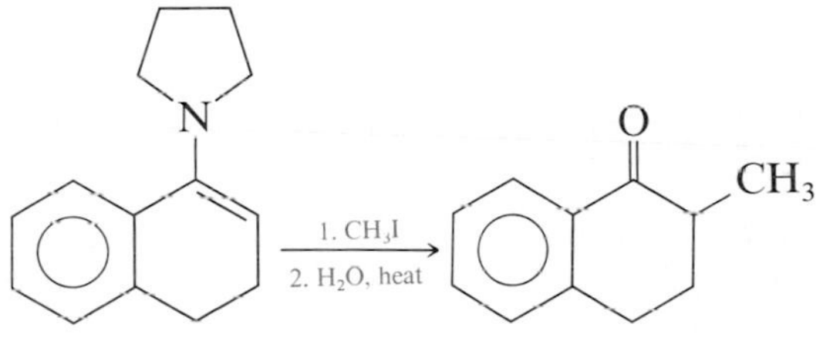

1-Methyl-2-tetralone (81%)

Thus, the transformation described as "difficult" in the preceding section, monoalkylation of an aldehyde or ketone at its α carbon, can be accomplished by the following three-step procedure:

1. Preparation of an enamine
2. Alkylation of an enamine
3. Hydrolysis of the alkylated enamine

PROBLEM 19.15 Show how each of the following conversions can be carried out by alkylation of a pyrrolidine enamine.

(a) Cyclohexanone to 2-methylcyclohexanone
(b) Cyclohexanone to 2-butylcyclohexanone
(c) 3-Pentanone to 2 methyl-1-phenyl-3-pentanone

SAMPLE SOLUTION (a) The first step is formation of the pyrrolidine enamine of cyclohexanone.

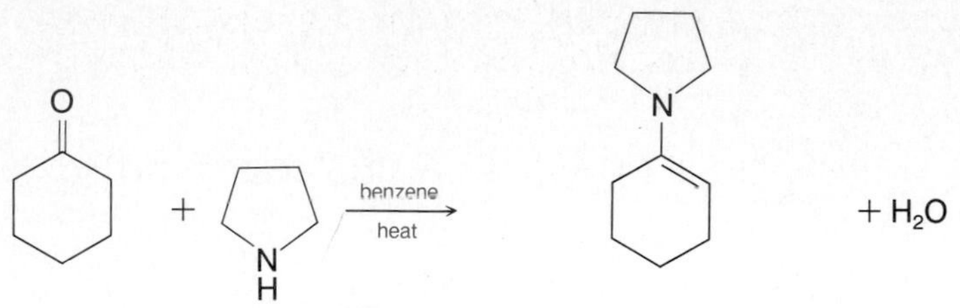

Cyclohexanone   Pyrrolidine   N-(1-Cyclohexenyl)pyrrolidine

Methylation of the enamine and subsequent hydrolysis yield the desired 2-methylcyclohexanone.

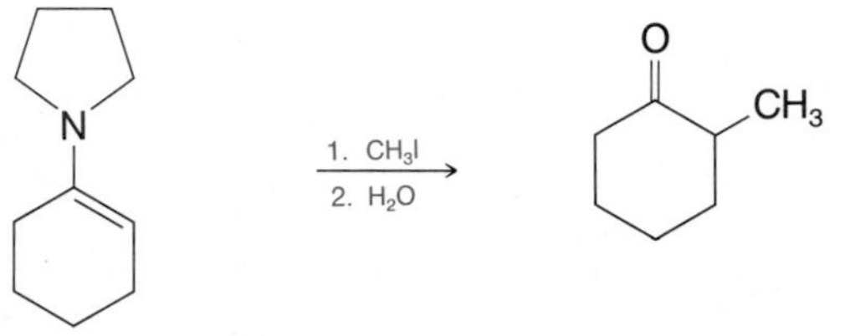

N-(1-Cyclohexenyl)pyrrolidine   2-Methylcyclohexanone

As actually carried out, the enamine has been formed as shown in 85 to 90 percent yield and converted to 2-methylcyclohexanone in 44 percent yield.

Enamines are also sufficiently nucleophilic to undergo conjugate addition to α,β-unsaturated carbonyl compounds.

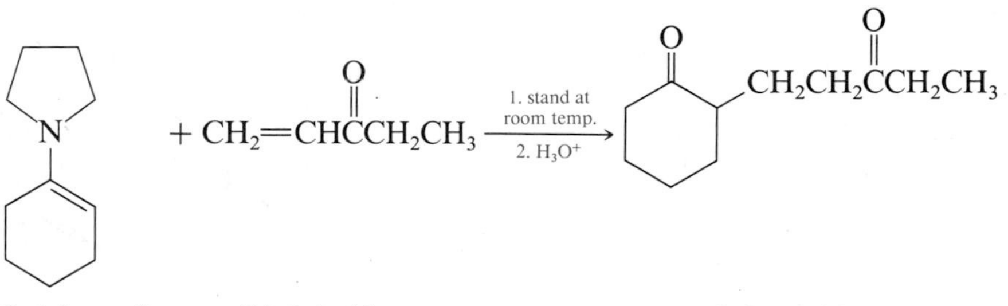

$+ \ CH_2\!=\!CHCCH_2CH_3 \ \xrightarrow[\text{2. }H_3O^+]{\text{1. stand at room temp.}}$

N-(1-Cyclohexenyl)-
pyrrolidine   Ethyl vinyl ketone   1-(2-Oxocyclohexyl)-3-
pentanone (65%)

Enamine addition, followed by intermolecular aldol condensation of the resulting diketone, is an alternative method for bringing about the Robinson annulation.

19.18 SUMMARY

Because aldehydes and ketones exist in a dynamic equilibrium with their corresponding enol forms, they can express a variety of different kinds of chemical reactivity.

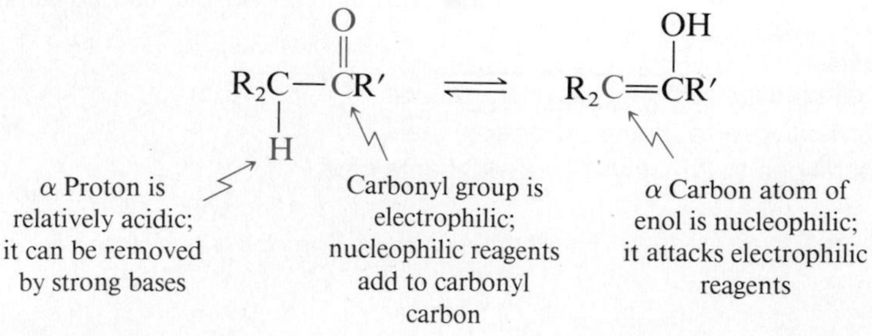

$$R_2C - \underset{H}{\overset{O}{\overset{\|}{C}}}R' \rightleftharpoons R_2C = \underset{}{\overset{OH}{C}}R'$$

α Proton is
relatively acidic;
it can be removed
by strong bases

Carbonyl group is
electrophilic;
nucleophilic reagents
add to carbonyl
carbon

α Carbon atom of
enol is nucleophilic;
it attacks electrophilic
reagents

TABLE 19.1

Reactions of Aldehydes and Ketones That Involve Enol or Enolate Ion Intermediates

Reaction (section) and comments	General equation and typical example
Enolization (Sections 19.4 through 19.6) Aldehydes and ketones exist in equilibrium with their enol forms. The rate at which equilibrium is achieved is increased by acidic or basic catalysts. The enol content of simple aldehydes and ketones is quite small; β-diketones, however, are extensively enolized.	$R_2CH-\overset{\displaystyle O}{\overset{\|}{C}}R' \rightleftharpoons R_2C=\overset{\displaystyle OH}{\overset{\|}{C}}R'$ Aldehyde or ketone $\qquad$ Enol $(CH_3)_2CH\overset{\displaystyle O}{\overset{\|}{C}}H \overset{K}{\rightleftharpoons} \underset{H_3C}{\overset{H_3C}{>}}C=C\underset{H}{\overset{OH}{<}} \qquad K = 1.3 \times 10^{-4}$ 2-Methylpropanal $\qquad$ 2-Methyl-1-propen-1-ol Cyclopentanone $\overset{K}{\rightleftharpoons}$ Cyclopenten-1-ol $\qquad K = 1 \times 10^{-8}$
α-Halogenation (Sections 19.2 and 19.3) Halogens react with aldehydes and ketones by substitution; an α hydrogen is replaced by halogen. Reaction occurs by electrophilic attack of the halogen on the carbon-carbon double bond of the enol form of the aldehyde or ketone.	$R_2CH\overset{\displaystyle O}{\overset{\|}{C}}R' + X_2 \longrightarrow R_2\overset{\displaystyle O}{\underset{X}{\overset{\|}{C}}}R' + HX$ Aldehyde $\quad$ Halogen $\qquad$ α-Halo aldehyde $\quad$ Hydrogen or ketone $\qquad\qquad\qquad$ or ketone $\qquad\qquad$ halide $Br-C_6H_4-\overset{\displaystyle O}{\overset{\|}{C}}CH_3 + Br_2 \xrightarrow{\text{acetic acid}} Br-C_6H_4-\overset{\displaystyle O}{\overset{\|}{C}}CH_2Br + HBr$ p-Bromoacetophenone $\quad$ Bromine $\qquad$ p-Bromophenacyl $\quad$ Hydrogen $\qquad\qquad\qquad\qquad\qquad\qquad$ bromide (69–72%) $\quad$ bromide
Enolate ion formation (Section 19.6) An α proton of an aldehyde or ketone is more acidic than most other protons bound to carbon. Aldehydes and ketones are weak acids, with K_a's in the range 10^{-16} to 10^{-20} (pK_a 16 to 20). Their enhanced acidity is due to the electron-withdrawing effect of the carbonyl group and the resonance stabilization of the enolate anion.	$R_2CH\overset{\displaystyle O}{\overset{\|}{C}}R' + HO^- \rightleftharpoons R_2\overset{\displaystyle O}{\overset{\|}{\ddot{C}}}R' + H_2O$ Aldehyde $\quad$ Hydroxide $\qquad$ Enolate $\quad$ Water or ketone $\qquad$ ion $\qquad\qquad$ anion $R_2\overset{\displaystyle O}{\overset{\|}{\ddot{C}}}-CR' \longleftrightarrow R_2C=\overset{\displaystyle O^-}{\overset{\|}{C}}R'$ (Principal resonance forms of enolate ion) $CH_3CH_2\overset{\displaystyle O}{\overset{\|}{C}}CH_2CH_3 + HO^- \rightleftharpoons CH_3CH=\overset{\displaystyle O^-}{\overset{\|}{C}}CH_2CH_3 + H_2O$ 3-Pentanone $\qquad$ Hydroxide ion $\qquad$ Enolate anion $\quad$ Water $\qquad\qquad\qquad\qquad\qquad\qquad$ of 3-pentanone
Haloform reaction (Section 19.7) Methyl ketones are cleaved on reactions with excess halogen in the presence of base. The products are a trihalomethane (haloform) and a carboxylate salt.	$R\overset{\displaystyle O}{\overset{\|}{C}}CH_3 + 3X_2 \xrightarrow{HO^-} R\overset{\displaystyle O}{\overset{\|}{C}}O^- + HCX_3$ Methyl $\quad$ Halogen $\qquad$ Carboxylate $\quad$ Trihalomethane ketone $\qquad\qquad\qquad$ ion $\qquad\qquad$ (haloform) $(CH_3)_3C\overset{\displaystyle O}{\overset{\|}{C}}CH_2CCH_3 \xrightarrow[\text{2. H}^+]{\text{1. Br}_2,\text{NaOH}} (CH_3)_3CCH_2CO_2H + CHBr_3$ 4,4-Dimethyl-2-pentanone $\qquad$ 3,3-Dimethylbutanoic $\quad$ Bromoform $\qquad\qquad\qquad\qquad\qquad\qquad\qquad$ acid (89%)

723

TABLE 19.1 (continued)

Reaction (section) and comments	General equation and typical example
Aldol condensation (Sections 19.9 through 19.12) A reaction of great synthetic value for carbon-carbon bond formation. Nucleophilic addition of an enolate ion to a carbonyl group, followed by dehydration of the β-hydroxy aldehyde or ketone, yields an α, β-unsaturated carbonyl compound.	$$2RCH_2CR' \xrightarrow{HO^-} RCH_2C{=}CCR' + H_2O$$ Aldehyde or ketone α,β-Unsaturated aldehyde or ketone Water $$CH_3(CH_2)_6CH \xrightarrow[CH_3CH_2OH]{NaOCH_2CH_3} CH_3(CH_2)_6CH{=}C(CH_2)_5CH_3$$ $HC{=}O$ Octanal 2-Hexyl-2-decenal (79%)
Claisen-Schmidt reaction (Section 19.11) A mixed aldol condensation in which an aromatic aldehyde reacts with an enolizable aldehyde or ketone.	$$ArCH + RCH_2CR' \xrightarrow{HO^-} ArCH{=}CCR' + H_2O$$ R Aromatic aldehyde Aldehyde or ketone α,β-Unsaturated carbonyl compound Water $$C_6H_5CH + (CH_3)_3CCCH_3 \xrightarrow[\substack{ethanol-\\water}]{NaOH} C_6H_5CH{=}CHCC(CH_3)_3$$ Benzaldehyde 3,3-Dimethyl-2-butanone 4,4-Dimethyl-1-phenyl-1-penten-3-one (88–93%)
Conjugate addition to α,β-unsaturated carbonyl compounds (Sections 19.14 and 19.15) The β carbon atom of an α,β-unsaturated carbonyl compound is electrophilic; nucleophiles, especially weakly basic ones, yield the products of conjugate addition to α,β-unsaturated aldehydes and ketones. Organocuprate reagents undergo efficient conjugate addition to α,β-unsaturated carbonyl compounds.	$$R_2C{=}CHCR' + HY\text{:} \longrightarrow R_2CCH_2CR'$$ Y α,β-Unsaturated aldehyde or ketone Nucleophile Product of conjugate addition $$(CH_3)_2C{=}CHCCH_3 \xrightarrow[H_2O]{NH_3} (CH_3)_2CCH_2CCH_3$$ NH_2 4-Methyl-3-penten-2-one (mesityl oxide) 4-Amino-4-methyl-2-pentanone (63–70%)
Robinson annulation (Section 19.14) A combination of conjugate addition of an enolate anion to an α,β-unsaturated ketone with subsequent intramolecular aldol condensation.	2-Methylcyclohexanone $+ CH_2{=}CHCCH_3$ Methyl vinyl ketone $\xrightarrow[\substack{2.\ KOH,\ heat}]{1.\ NaOCH_2CH_3,\ CH_3CH_2OH}$ 6-Methylbicyclo-[4.4.0]-1-decen-3-one (46%)

TABLE 19.1 (continued)

Reaction (section) and comments	General equation and typical example
α-Alkylation of aldehydes and ketones (Sections 19.16 and 19.17) Alkylation of simple aldehydes and ketones via their enolates is difficult. β-Diketones can be converted quantitatively to their enolate anions, which react efficiently with primary alkyl halides. An effective way to alkylate aldehydes and ketones is by way of an enamine.	$\underset{\text{$\beta$-Diketone}}{\overset{\displaystyle O \quad\quad O}{RCCH_2CR}} \xrightarrow{RCH_2X, HO^-} \underset{\text{α-Alkyl-β-diketone}}{\overset{\displaystyle O \quad\quad O}{RCCHCR}}$

2 Benzyl-1,3-cyclohexanedione $+ C_6H_5CH_2Cl \xrightarrow[\text{ethanol}]{KOCH_2CH_3}$ 2,2-Dibenzyl-1,3-cyclohexanedione (69%)

Benzyl chloride

Enolate anions, generated by proton abstraction from the α carbon atom of an aldehyde or ketone, are more nucleophilic than enols and are important intermediates in base-promoted reactions of carbonyl compounds.

$$\underset{\substack{\text{Aldehyde} \\ \text{or ketone}}}{\overset{\displaystyle O}{R_2C-CR'} \atop \overset{|}{H}} \xleftrightarrow[H_2O]{HO^-} \left[\underset{\text{Enolate ion}}{\overset{\displaystyle O \atop \| \atop R_2\overline{C}-CR'}{\updownarrow} \atop \overset{O^- \atop |}{R_2C=CR'}} \right] \xleftrightarrow[HO^-]{H_2O} \underset{\text{Enol}}{\overset{OH \atop |}{R_2C=CR'}}$$

A summary of the reactions described in this chapter that proceed by way of enol or enolate ion intermediates is presented in Table 19.1. Most of the reactions described in the table are of synthetic value. In particular, the aldol condensation and variations of it provide one of the most useful methods for creating carbon-carbon bonds.

PROBLEMS

19.16 (a) Write structural formulas for all the isomeric aldehydes and ketones of molecular formula C_4H_6O.
  (b) Are any of these compounds stereoisomeric?
  (c) Are any of these compounds chiral?
  (d) Which of these are α,β-unsaturated aldehydes or α,β-unsaturated ketones?
  (e) Which of these can be prepared by a simple (i.e., not mixed) aldol condensation?
  (f) Which one of these has four equivalent α hydrogens?

19.17 In each of the following pairs of compounds, choose the one that has the greater enol content and write the structure of its enol form.

(a) $(CH_3)_3CCH$ $\overset{\displaystyle O}{\|}$ or $(CH_3)_2CHCH$ $\overset{\displaystyle O}{\|}$

(b) $C_6H_5\overset{\overset{O}{\|}}{C}C_6H_5$ or $C_6H_5CH_2\overset{\overset{O}{\|}}{C}CH_2C_6H_5$

(c) $C_6H_5\overset{\overset{O}{\|}}{C}CH_2\overset{\overset{O}{\|}}{C}C_6H_5$ or $C_6H_5CH_2\overset{\overset{O}{\|}}{C}CH_2C_6H_5$

(d)   [cyclohexanone structure] =O   or   [cyclohexadienone structure] =O

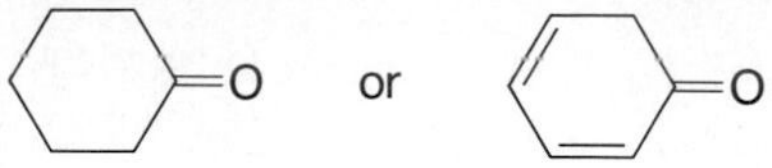

(e)   [cyclopentanone structure] =O   or   [cyclopentadienone structure] =O

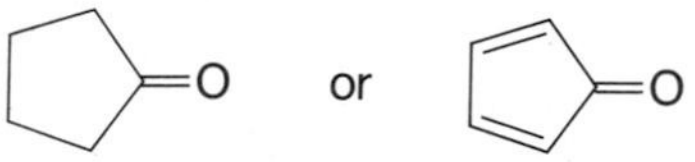

(f)   [1,3-cyclohexanedione structure]   or   [1,4-cyclohexanedione structure]

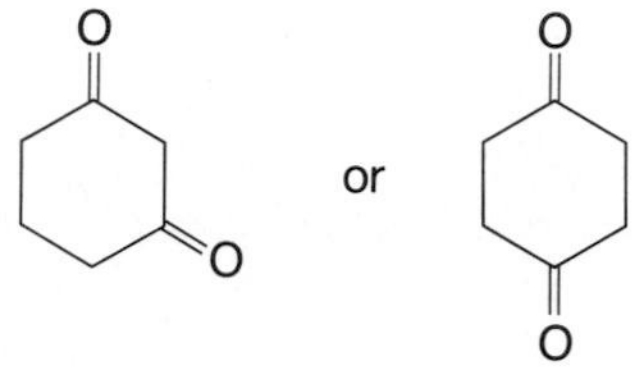

19.18   Give the structure of the expected organic product in the reaction of 3-phenylpropanal with each of the following:

(a)   Chlorine in acetic acid
(b)   Sodium hydroxide in ethanol, 10°C
(c)   Sodium hydroxide in ethanol, 70°C
(d)   Product of (c) with sodium borohydride in ethanol
(e)   Product of (c) with sodium cyanide in acidic ethanol
(f)   Product of (c) with $(C_6H_5)_3\overset{+}{P}—\overset{-}{\underset{..}{C}}H_2$

19.19   Each of the following reactions has been reported in the chemical literature. Write the structure of the product(s) formed in each case.

(a)   [ortho-chlorophenyl ketone] $\overset{\overset{O}{\|}}{C}CH_2CH_3$ $\xrightarrow[CH_2Cl_2]{Cl_2}$

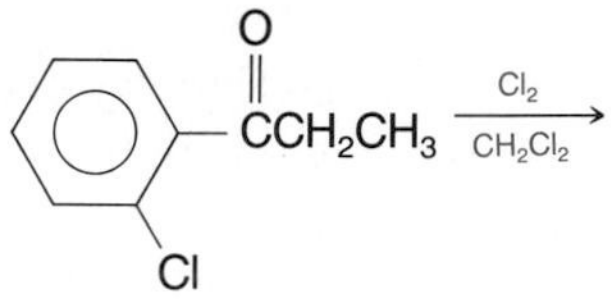

(b)   [(CH₃)₃C and H₃C, CH₃ substituted indane with CCH₃ ketone group, O] $\xrightarrow[\text{2. H}^+]{\text{1. Br}_2\text{ (excess), KOH}}$

(c)   [cyclopentanone with two C_6H_5 groups] $\xrightarrow[\text{diethyl ether}]{Br_2}$

(d)   Cl—[phenyl]—$\overset{\overset{O}{\|}}{C}H$ + [cyclohexanone with two C_6H_5 groups] $\xrightarrow[\text{ethanol}]{KOH}$

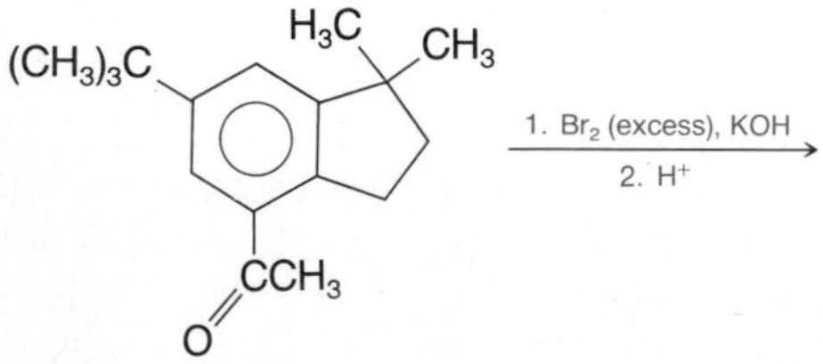

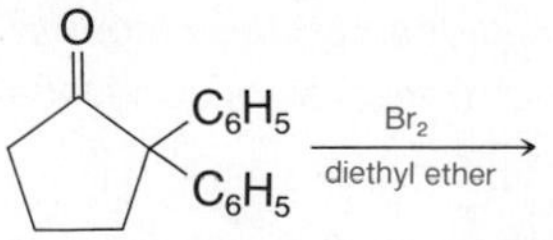

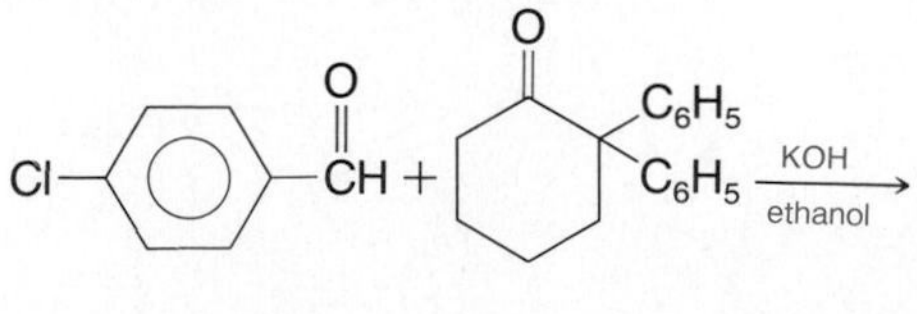

(e)

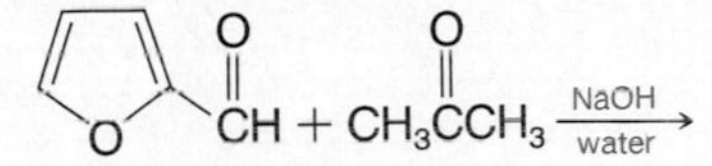

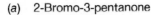

(f)

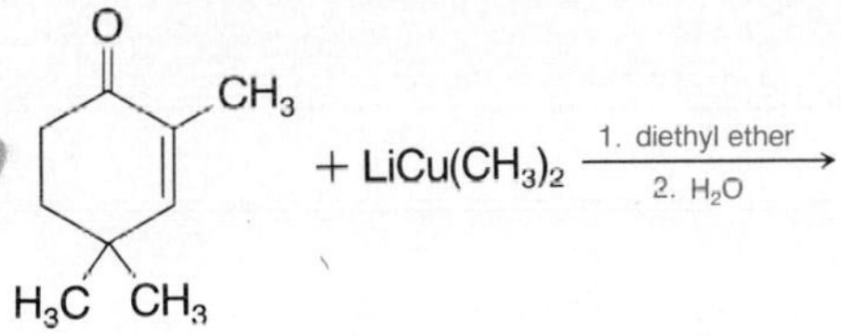

(g)

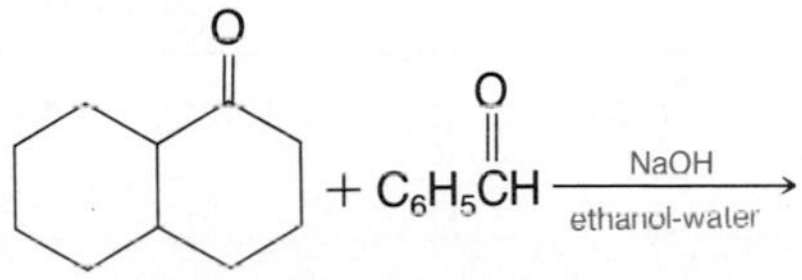

19.20 Show how each of the following compounds could be prepared from 3-pentanone. In most cases more than one synthetic transformation will be necessary.

(a) 2-Bromo-3-pentanone
(b) 1-Penten-3-one
(c) 1-Penten-3-ol
(d) 3-Hexanone
(e) 2-Methyl-3-pentanone
(f) 2-Methyl-1-phenyl-1-penten-3-one

19.21 Show how you could prepare each of the following compounds from cyclopentanone, D_2O, and any necessary organic or inorganic reagents.

(a)

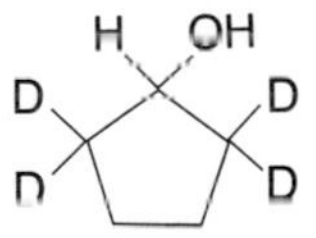

(c)

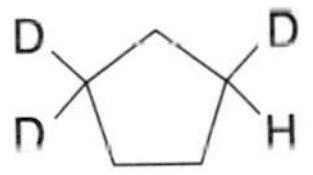

(b)

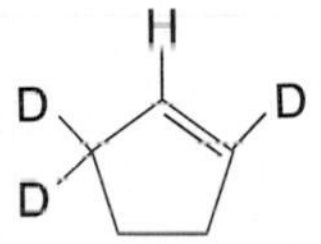

(d)

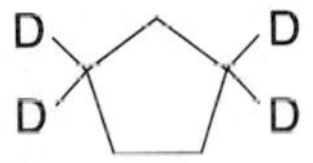

19.22 (a) At present, butanal is prepared industrially by the oxo process (Section 18.4). What starting materials are required? Write a chemical equation for this industrial synthesis.

(b) Prior to about 1970, the principal industrial preparation of butanal was from acetaldehyde. Outline a practical synthesis of butanal from acetaldehyde.

19.23 Identify the reagents appropriate for each step in the following syntheses:

(a)

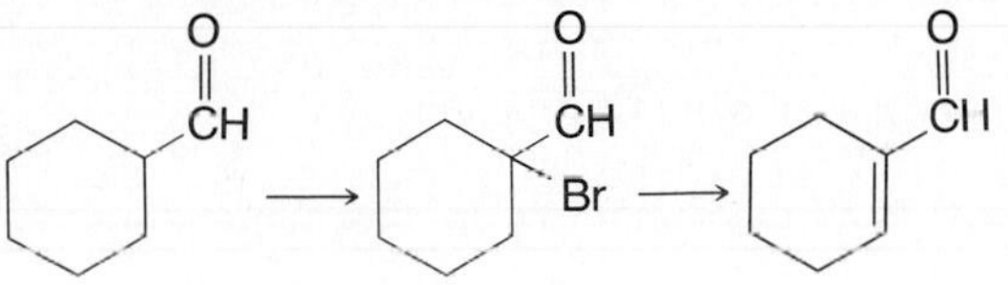

(b)

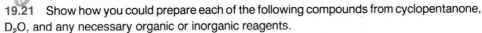

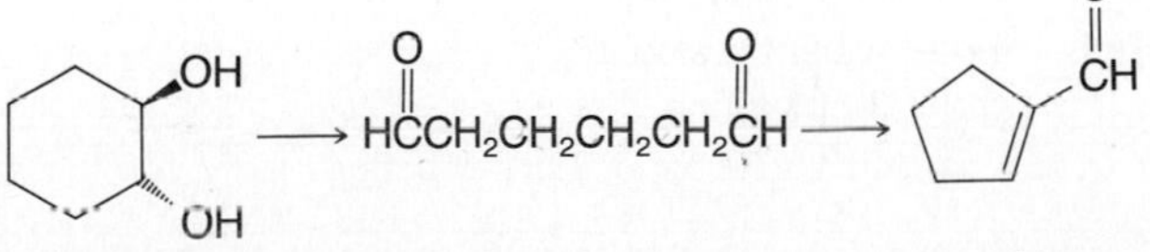

(c)

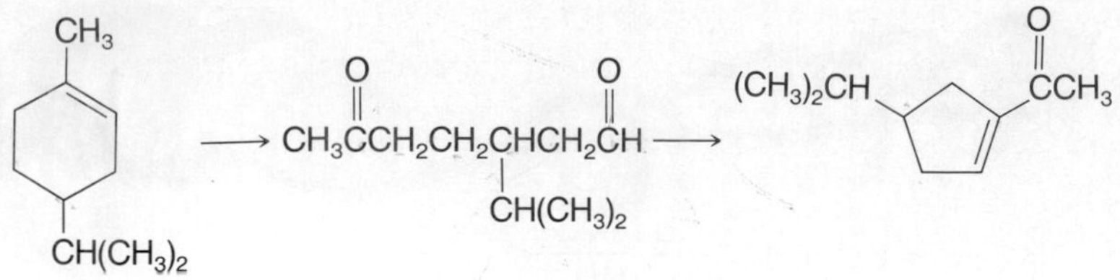

$$CH_3CCH_2CH_2CHCH_2CH \longrightarrow$$
(with two C=O groups, and $CH(CH_3)_2$ substituent)

(d) 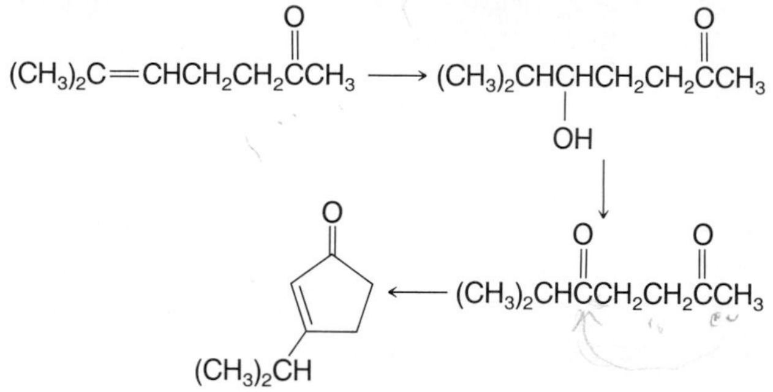 $(CH_3)_2C=CHCH_2CH_2CCH_3 \longrightarrow (CH_3)_2CHCHCH_2CH_2CCH_3$
with OH below

$\longrightarrow (CH_3)_2CHCCH_2CH_2CCH_3$

19.24 A multistep synthesis was used to prepare 3,3-dimethyl-6-*tert*-butylindanone from acetone, bromobenzene, and *tert*-butyl chloride. The sequence of steps shown below is composed entirely of reactions that you have learned about. Identify the reagents required for each transformation.

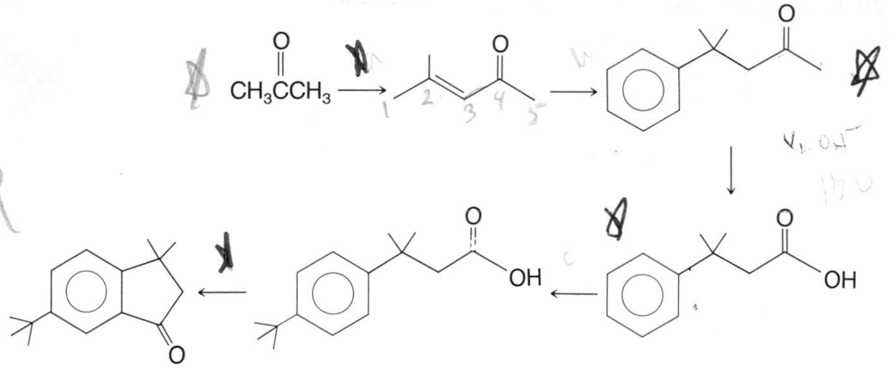

$$CH_3CCH_3 \longrightarrow$$

3,3-Dimethyl-6-*tert*-butylindanone

19.25 Prepare each of the following compounds from the starting materials given and any necessary organic or inorganic reagents.

(a) $(CH_3)_2CHCHCHCCH_2OH$ from $(CH_3)_2CHCH_2OH$
with CH_3 above, HO and CH_3 below

(b) $C_6H_5CH=CCH_2OH$ from benzyl alcohol and 1-propanol
with CH_3 below

(c) from acetophenone, 4-methylbenzyl alcohol, and 1,3-butadiene

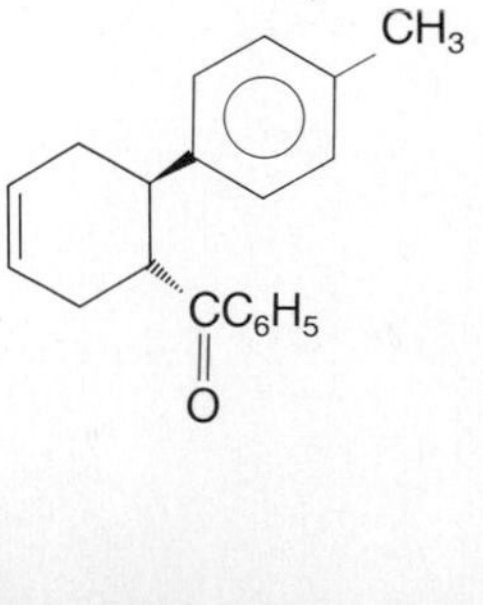

19.26 Terreic acid is a naturally occurring antibiotic substance. Its actual structure is an enol isomer of the structure shown. Write the two most stable enol forms of terreic acid and choose which of those two is more stable.

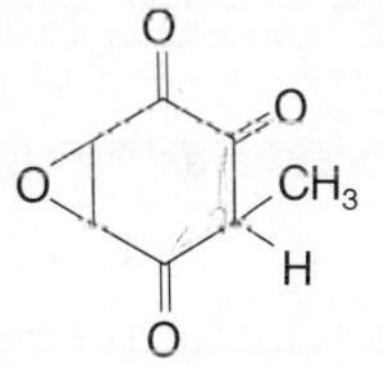

19.27 *(a)* For a long time attempts to prepare compound A were thwarted by its ready isomerization to compound B. The isomerization is efficiently catalyzed by traces of base. Write a reasonable mechanism for this isomerization.

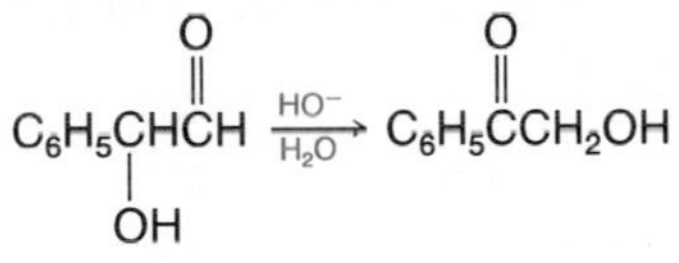

$$C_6H_5CHCH \xrightarrow[H_2O]{HO^-} C_6H_5CCH_2OH$$

Compound A	Compound B

(b) Another attempt to prepare compound A by hydrolysis of its dimethyl acetal gave only the 1,4-dioxane derivative C. How was C formed?

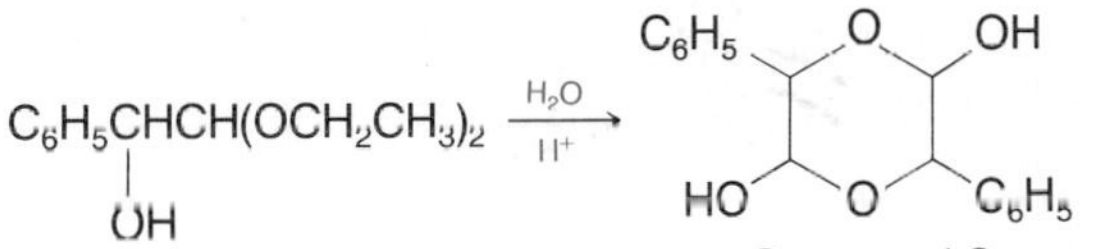

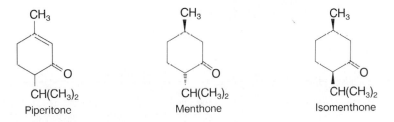

$$C_6H_5CHCH(OCH_2CH_3)_2 \xrightarrow[H^+]{H_2O}$$

Compound C

19.28 Consider the ketones piperitone, menthone, and isomenthone.

Piperitone	Menthone	Isomenthone

Suggest reasonable explanations for each of the following observations:

(a) Optically active piperitone $(\alpha_D - 32°)$ is converted to racemic piperitone on standing in a solution of sodium ethoxide in ethanol.

(b) Menthone is converted to a mixture of menthone and isomenthone on treatment with 90 percent sulfuric acid.

19.29 Many nitrogen-containing compounds engage in a proton transfer equilibrium that is analogous to keto-enol tautomerism.

$$HX-N=Z \rightleftharpoons X=N-ZH$$

Each of the compounds shown is the less stable partner of such a tautomeric pair. Write the structure of the more stable partner for each one.

(a) $CH_3CH_2N\!=\!O$

(b) $(CH_3)_2C\!=\!CHNHCH_3$

(c) $CH_3CH\!=\!\overset{+}{N}\!\!\begin{array}{c}\overset{O^-}{\diagup}\\[-2pt]\diagdown\\OH\end{array}$

(d)

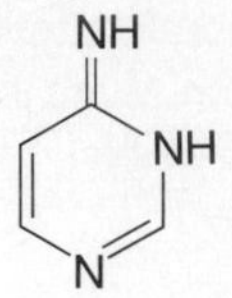

(e) $HN\!=\!C\!\!\begin{array}{c}\overset{OH}{\diagup}\\[-2pt]\diagdown\\NH_2\end{array}$

19.30 Outline reasonable mechanisms for each of the following reactions:

(a) $CH_3\overset{O}{\overset{\|}{C}}CH_2CH_2CH_2Cl \xrightarrow[\text{H}_2\text{O, heat}]{\text{NaOH}} CH_3\overset{O}{\overset{\|}{C}}\!\!\triangleleft$ (77–83%)

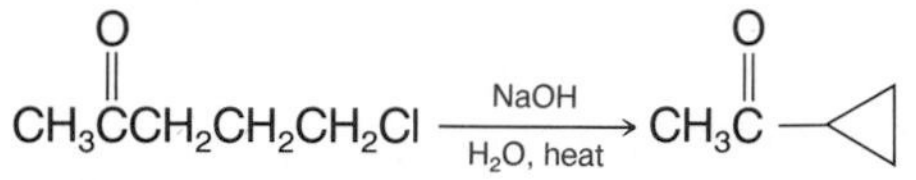

(b)

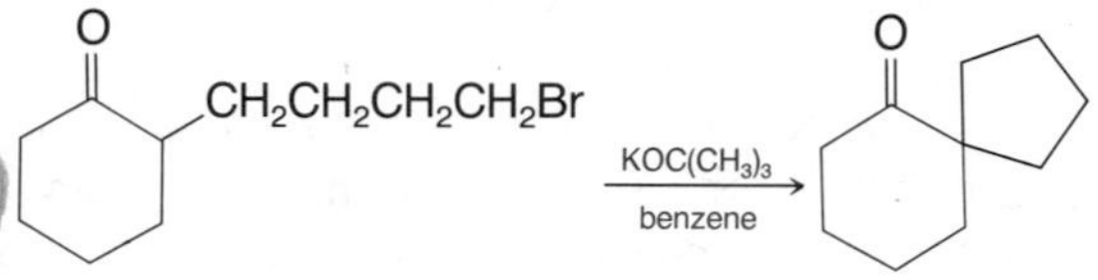

$\xrightarrow[\text{benzene}]{\text{KOC(CH}_3)_3}$ (76%)

(c) $(CH_3)_2C\!=\!CHCH_2CH_2\overset{}{C}\!=\!CHCH \;\;\underset{CH_3}{\big|}\;\; \xrightarrow[\text{heat}]{\text{HO}^-} (CH_3)_2C\!=\!CHCH_2CH_2\overset{O}{\overset{\|}{C}}CH_3 + CH_3\overset{O}{\overset{\|}{C}}H$

(96%)

(d) $H\overset{O}{\overset{\|}{C}}CH_2CH_2\underset{CH_3}{\overset{}{C}}H\overset{O}{\overset{\|}{C}}CH_3 \xrightarrow[\text{H}_2\text{O, CH}_3\text{OH}]{\text{KOH}}$

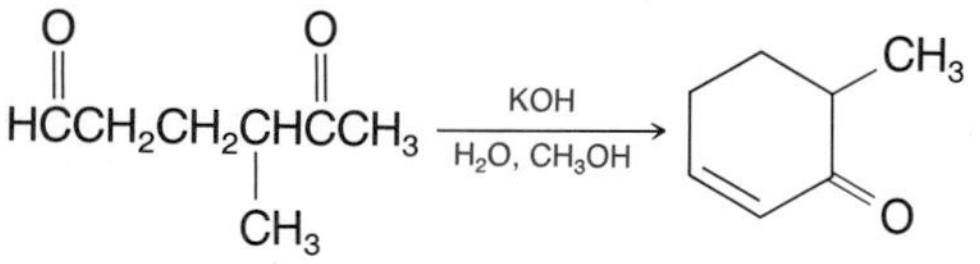

(40%)

(e)

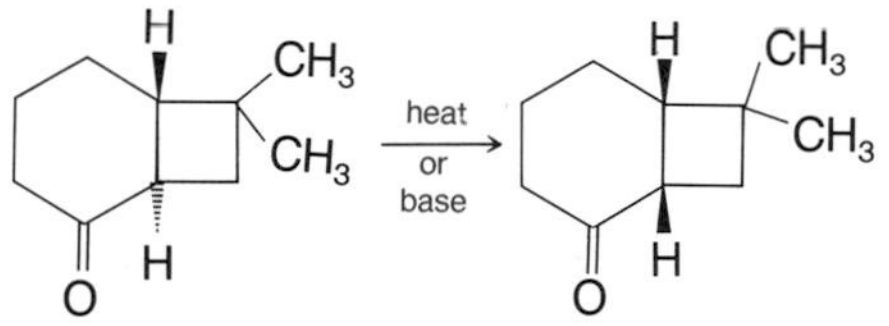

$\xrightarrow[\substack{\text{or}\\\text{base}}]{\text{heat}}$

(f) $C_6H_5\overset{O}{\overset{\|}{C}}CH_2CH_2CH_2\overset{O}{\overset{\|}{C}}C_6H_5 \xrightarrow[\text{CH}_3\text{OH}]{\text{NaOH, I}_2}$

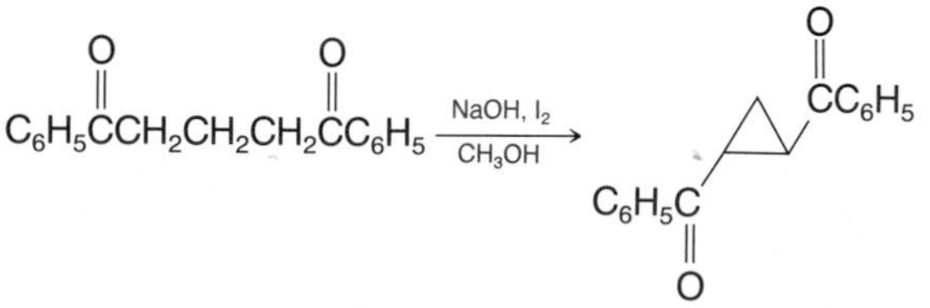

(66–72%)

(g) $C_6H_5\overset{O\;O}{\overset{\|\;\|}{CC}}C_6H_5 + C_6H_5CH_2\overset{O}{\overset{\|}{C}}CH_2C_6H_5 \xrightarrow[\text{ethanol}]{\text{KOH}}$

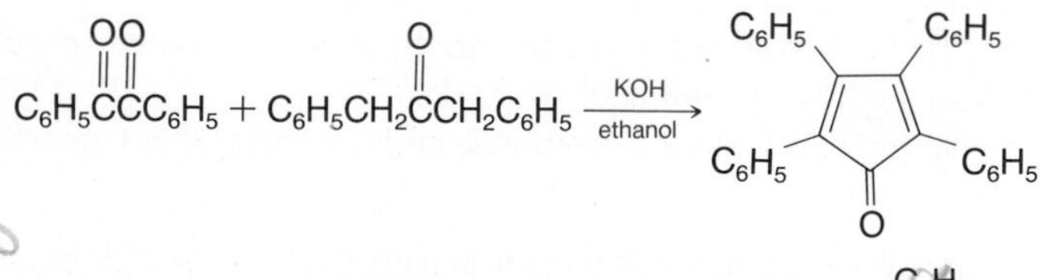

(91–96%)

(h) $C_6H_5CH_2\overset{O}{\overset{\|}{C}}CH_2CH_3 + CH_2\!=\!\overset{O}{\overset{\|}{C}}\underset{C_6H_5}{\overset{}{C}}C_6H_5 \xrightarrow[\text{CH}_3\text{OH}]{\text{NaOCH}_3}$

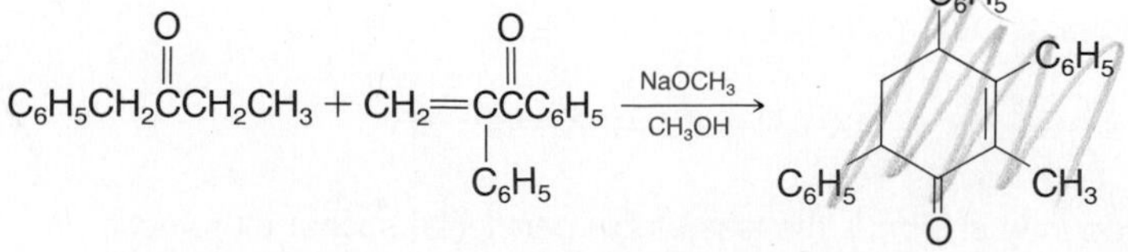

(51%)

19.31 Suggest reasonable explanations for each of the following observations:

(a) The C=O stretching frequency of α,β-unsaturated ketones (about 1675 cm^{-1}) is less than that of typical dialkyl ketones (1710 to 1750 cm^{-1}).

(b) The C=O stretching frequency of cyclopropenone (1640 cm^{-1}) is lower than that of typical α,β-unsaturated ketones (1675 cm^{-1}).

(c) The dipole moment of diphenylcyclopropenone ($\mu = 5.1$D) is substantially larger than that of benzophenone ($\mu = 3.0$D).

(d) The β carbon of an α,β-unsaturated ketone is less shielded than the corresponding carbon of an alkene. Typical ^{13}C nmr chemical shift values are

$$\underset{(\delta \cong 129 \text{ ppm})}{CH_2{=}CH\overset{\overset{\displaystyle O}{\|}}{C}R} \qquad \underset{(\delta \cong 114 \text{ ppm})}{CH_2{=}CHCH_2R}$$

19.32 Bromination of 3-methyl-2-butanone yielded two compounds, each having the molecular formula C_5H_9BrO, in a 95:5 ratio. The ^{1}H nmr spectrum of the major isomer D was characterized by a doublet at $\delta = 1.17$ ppm (six protons), a heptet at $\delta = 3.02$ ppm (one proton), and a singlet at $\delta = 4.10$ ppm (two protons). The ^{1}H nmr spectrum of the minor isomer E exhibited two singlets, one at $\delta = 1.89$ ppm and the other at $\delta = 2.46$ ppm. The lower-field singlet had one-half the area of the higher-field singlet. Suggest reasonable structures for these two compounds.

CARBOXYLIC ACIDS

Carboxylic acids, compounds of the type $\overset{\overset{O}{\parallel}}{R}COH$, comprise one of the most frequently encountered classes of organic compounds. Countless natural products are carboxylic acids. Some, such as acetic acid, have been known for centuries. Others, such as the prostaglandins, have been isolated only recently, and scientists continue to discover new roles that these substances play as regulators of biological processes.

Acetic acid

CH_3COH

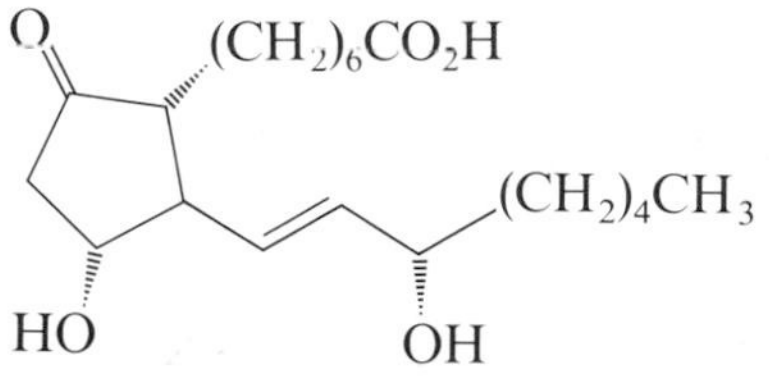

PGE$_1$ (a prostaglandin; a small amount of PGE$_1$ lowers blood pressure significantly)

$(CH_2)_6CO_2H$

$(CH_2)_4CH_3$

HO   OH

The chemistry of carboxylic acids is the central theme of this chapter. The significance of carboxylic acids is magnified when one realizes that they are the parent compounds of a large group of derivatives, which includes acyl chlorides, acid anhydrides, esters, and amides. Those classes of compounds will be discussed in the chapter following this one. Together, this chapter and the next tell the story of some of the most fundamental structural types and functional group transformations in organic and biological chemistry.

20.1 CARBOXYLIC ACID NOMENCLATURE

Nowhere in organic chemistry are common names as prevalent as they are among carboxylic acids. Many carboxylic acids are better known by common names than by

TABLE 20.1

Systematic and Common Names of Some Carboxylic Acids

Entry number	Structural formula	Systematic name	Common name
1.	HCO_2H	Methanoic acid	Formic acid
2.	CH_3CO_2H	Ethanoic acid	Acetic acid
3.	$CH_3(CH_2)_{16}CO_2H$	Octadecanoic acid	Stearic acid

4.	CH_3CHCO_2H
	     |
	    OH
		2-Hydroxypropanoic acid	Lactic acid

5.		2-Hydroxy-2-phenylethanoic acid	Mandelic acid
6.	$CH_2{=}CHCO_2H$	Propenoic acid	Acrylic acid
7.		(Z)-9-Octadecenoic acid	Oleic acid
8.		Benzenecarboxylic acid	Benzoic acid
9.		o-Hydroxybenzenecarboxylic acid	Salicylic acid
10.	$HO_2CCH_2CO_2H$	Propanedioic acid	Malonic acid
11.	$HO_2CCH_2CH_2CO_2H$	Butanedioic acid	Succinic acid
12.		1,2-Benzenedicarboxylic acid	Phthalic acid

their systematic names, and the framers of the IUPAC nomenclature rules have taken a liberal view toward accepting these common names as permissible alternatives to the systematic ones. Table 20.1 lists both the common and the systematic names of a number of important carboxylic acids.

Systematic names for carboxylic acids are derived by counting the number of carbons in the longest continuous chain that includes the carboxyl group and replacing the -*e* ending of the corresponding alkane by -*oic acid*. The first three acids in the table, methanoic (1 carbon), ethanoic (2 carbons), and octadecanoic acid (18 carbons) illustrate this point. When substituents are present, their locations are identified by number; numbering of the carbon chain always begins at the carboxyl group. This is illustrated in entries 4 and 5 in the table.

Double bonds in the main chain are signaled by the ending -*enoic acid* and their position is designated by a numerical prefix. Entries 6 and 7 are representative carboxylic acids that contain double bonds. Double bond stereochemistry is specified by using the *E-Z* notation.

In cases in which a carboxyl group is attached to a ring, the parent ring is named (retaining the final -*e*) and the suffix -*carboxylic acid* is added, as shown in entries 8 and 9.

Compounds with two carboxyl groups, such as those shown in entries 10 through 12, are distinguished by the suffix *-dioic acid* or *-dicarboxylic acid* as appropriate. The final *-e* in the base name of the alkane is retained.

PROBLEM 20.1 The list of carboxylic acids in Table 20.1 is by no means exhaustive insofar as common names are concerned. Many others are known by their common names, a few of which are listed below. Give a systematic IUPAC name for each.

(a) $CH_2{=}CCO_2H$ (methacrylic acid)
       |
       CH_3

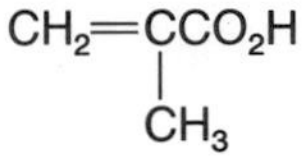

(b) $(CH_3)_3CCO_2H$ (pivalic acid)

(c) H_3C H (crotonic acid)
       $C{=}C$
     H CO_2H

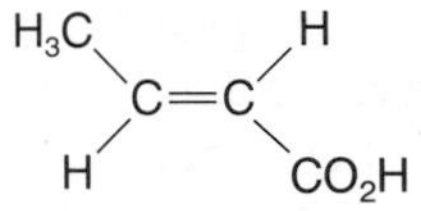

(d) HO_2CCO_2H (oxalic acid)

(e) H H (maleic acid)
       $C{=}C$
     HO_2C CO_2H

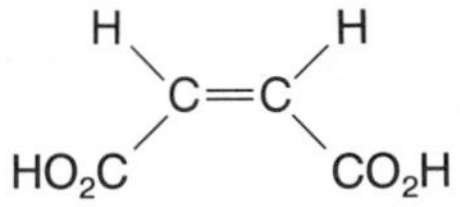

(f) CH_3—⬡—CO_2H (*p*-toluic acid)

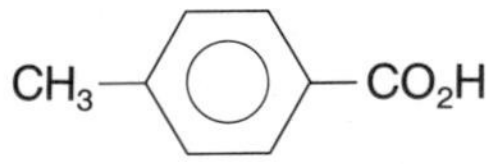

SAMPLE SOLUTION (a) Methacrylic acid is an industrial chemical used in the preparation of transparent plastics such as *Lucite* and *Plexiglas*. The carbon chain that includes both the carboxylic acid and the double bond is three carbon atoms in length. The compound is named as a derivative of *propenoic acid*. It is not necessary to locate the position of the double bond by number, as in "2-propenoic acid," because no other positions are structurally possible for it. The methyl group is at C-2, so the correct systematic name for methacrylic acid is *2-methylpropenoic acid*.

20.2 STRUCTURE AND BONDING

The essential structural features of carboxylic acids can be seen by referring to the simplest one, formic acid. Formic acid is a planar molecule, with one of its carbon-oxygen bonds significantly shorter than the other.

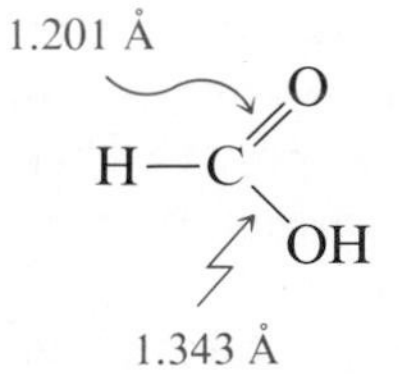

Bond distances in formic acid 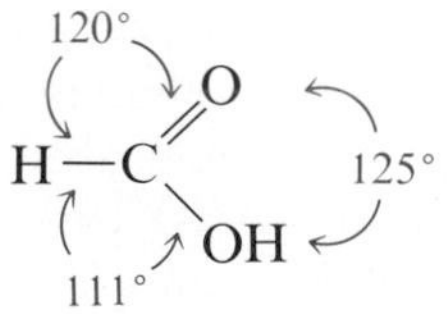 — Bond angles in formic acid

We have come to associate trigonal planar geometry at carbon with sp^2 hybridization and to picture shortened bond distances as arising from multiple bonding of the $\sigma + \pi$ type. By analogy to aldehydes and ketones, the orbital description of bonding in carboxylic acids may be portrayed as in Figure 20.1a. Notice that one of the electron pairs of the hydroxyl oxygen can be delocalized if its orbital overlaps with the π orbital of the carbonyl group to form an extended π system (Figure 20.1b).

In resonance terms, conjugation of the hydroxyl oxygen with the carbonyl group is represented as

$$H-C\overset{\displaystyle \overset{..}{O}:}{\underset{\displaystyle \underset{..}{O}H}{}} \longleftrightarrow H-\overset{+}{C}\overset{\displaystyle \overset{..}{O}:^-}{\underset{\displaystyle \underset{..}{O}H}{}} \longleftrightarrow H-C\overset{\displaystyle \overset{..}{O}:^-}{\underset{\displaystyle \underset{..}{O}H}{}}$$

Lone pair donation from the hydroxyl group stabilizes the carbonyl group and makes it less electrophilic than that of an aldehyde or ketone. Electron density is increased at the carbonyl oxygen. The bond between carbon and the hydroxyl group has a degree of "double bond character."

The separation of charge implied in the resonance formulation of the carboxyl group makes it fairly polar, and simple carboxylic acids such as acetic acid, propanoic acid, and benzoic acid have dipole moments in the range 1.7 to 1.9D.

20.3 PHYSICAL PROPERTIES

A summary of physical properties of some representative carboxylic acids is presented in Table 20.2. The melting points and boiling points of carboxylic acids are higher than those of alcohols of comparable molecular weight and indicate strong attractive forces between molecules.

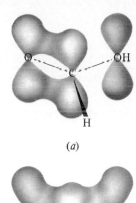

(a)

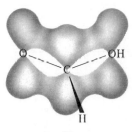

(b)

FIGURE 20.1 Orbital overlap in formic acid. (a) A p orbital of the hydroxyl oxygen of formic acid and the π system of the carbonyl group are shown in an orientation that permits orbital overlap. (b) Orbital overlap generates an extended π system, which includes the carbon and both oxygen atoms of formic acid.

TABLE 20.2

Physical Properties of Some Carboxylic Acids and Dicarboxylic Acids

Compound	Condensed structural formula	Melting point, °C	Boiling point, °C (1 atm)	Solubility, g/100 mL H$_2$O
Carboxylic acids				
Formic acid	HCO$_2$H	8.4	101	∞
Acetic acid	CH$_3$CO$_2$H	16.6	118	∞
Propanoic acid	CH$_3$CH$_2$CO$_2$H	−20.8	141	∞
Butanoic acid	CH$_3$CH$_2$CH$_2$CO$_2$H	−5.5	164	∞
Pentanoic acid	CH$_3$(CH$_2$)$_3$CO$_2$H	−34.5	186	3.3 (16°C)
Decanoic acid	CH$_3$(CH$_2$)$_8$CO$_2$H	31.4	269	0.003 (15°C)
Benzoic acid	C$_6$H$_5$CO$_2$H	122.4	250	0.21 (17°C)
Dicarboxylic acids				
Oxalic acid	HO$_2$CCO$_2$H	186	Sublimes	10 (20°C)
Malonic acid	HO$_2$CCH$_2$CO$_2$H	130–135	Decomp.	138 (16°C)
Succinic acid	HO$_2$CCH$_2$CH$_2$CO$_2$H	189	235	6.8 (20°C)
Glutaric acid	HO$_2$CCH$_2$CH$_2$CH$_2$CO$_2$H	97.5		63.9 (20°C)

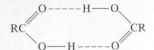

FIGURE 20.2 Intermolecular hydrogen bonding between two carboxylic acid molecules.

A unique hydrogen bonding arrangement, shown in Figure 20.2, contributes to these attractive forces. The hydroxyl group of one carboxylic acid molecule acts as a proton donor toward the carbonyl oxygen of a second. In a reciprocal fashion, the hydroxyl proton of the second carboxyl function interacts with the carbonyl oxygen of the first. This results in the two carboxylic acid molecules being held together by *two* hydrogen bonds. So efficient is this hydrogen bonding that some carboxylic acids exist as hydrogen-bonded dimers even in the gas phase. In the pure liquid a mixture of hydrogen-bonded dimers and higher aggregates is present.

In aqueous solution intermolecular association between carboxylic acid molecules is replaced by hydrogen bonding to water (Figure 20.3). The solubility properties of carboxylic acids are similar to those of alcohols. Carboxylic acids of four carbon atoms or less are miscible with water in all proportions.

20.4 ACIDITY OF CARBOXYLIC ACIDS

Their acidity is the most notable property of carboxylic acids, which are the most acidic class of compounds that contain only carbon, hydrogen, and oxygen. With ionization constants K_a on the order of 10^{-5} ($pK_a \sim 5$), they are much stronger acids than water and alcohols. The case should not be overstated, however. Carboxylic acids are weak acids and do not ionize completely. A $0.1M$ solution of acetic acid in water, for example, is only 1.3 percent ionized.

To understand the greater acidity of carboxylic acids compared with water and alcohols, compare the structural changes that accompany the ionization of a representative alcohol (ethanol) and a representative carboxylic acid (acetic acid). The equilibria that define K_a are

Ionization of ethanol

$$CH_3CH_2OH \rightleftharpoons H^+ + CH_3CH_2O^- \qquad K_a = \frac{[H^+][CH_3CH_2O^-]}{[CH_3CH_2OH]} = 10^{-16}$$

Ethanol   Ethoxide ion

Ionization of acetic acid

$$CH_3\overset{\displaystyle O}{\overset{\|}{C}}OH \rightleftharpoons H^+ + CH_3\overset{\displaystyle O}{\overset{\|}{C}}O^- \qquad K_a = \frac{[H^+][CH_3CO_2^-]}{[CH_3CO_2H]} = 1.8 \times 10^{-5}$$

Acetic acid   Acetate ion

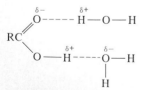

FIGURE 20.3 Hydrogen bonding interactions between a carboxylic acid and water.

Ionization of ethanol yields an alkoxide ion in which the negative charge is localized on oxygen. Solvation forces are the principal means by which ethoxide ion is stabilized. The small ionization constant K_a of 10^{-16} for ethanol translates to a substantial free energy requirement, 21.7 kcal/mol, for its ionization. An energy diagram portraying this relationship is presented in Figure 20.4. Since it is *equilibria,* not *rates* of ionization, that are being compared, the diagram shows only the initial and final states. It is not necessary to be concerned about the energy of activation of this process, since that does not affect the extent of ionization but only its rate.

In comparing the energies of ionization in Figure 20.4, notice that the energy of the

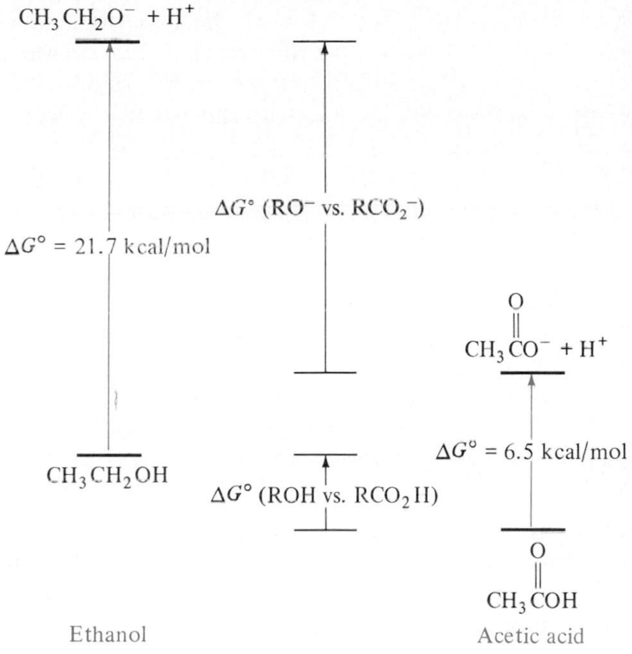

$CH_3CH_2O^- + H^+$

ΔG° (RO$^-$ vs. RCO$_2$$^-$)

$\Delta G^\circ = 21.7$ kcal/mol

$$CH_3\overset{\displaystyle O}{\overset{\displaystyle \|}{C}}O^- + H^+$$

CH_3CH_2OH

ΔG° (ROH vs. RCO$_2$II)

$\Delta G^\circ = 6.5$ kcal/mol

$$CH_3\overset{\displaystyle O}{\overset{\displaystyle \|}{C}}OH$$

Ethanol

Acetic acid

FIGURE 20.4 Diagram comparing the free energies of ionization of ethanol and acetic acid in water.

starting state of acetic acid is placed somewhat lower than that of ethanol. This is because acetic acid is stabilized to some degree by resonance of the type described in Section 20.2.

$$CH_3C\overset{O:}{\underset{OH}{\diagdown}} \longleftrightarrow CH_3C\overset{\ddot{O}:^-}{\underset{\overset{OH}{+}}{\diagdown}}$$

Since this delocalization is accompanied by separation of positive and negative charge, its effect is small. A very large stabilizing effect, however, accompanies ionization. Acetate ion is stabilized both by solvation forces and by electron delocalization, which permits the negative charge to be shared equally by both oxygens.

$$CH_3C\overset{O:}{\underset{\ddot{O}:^-}{\diagdown}} \longleftrightarrow CH_3C\overset{\ddot{O}:^-}{\underset{\ddot{O}:}{\diagdown}} \quad \text{or} \quad CH_3C\overset{O^{-\frac{1}{2}}}{\underset{O^{-\frac{1}{2}}}{\diagdown}}$$

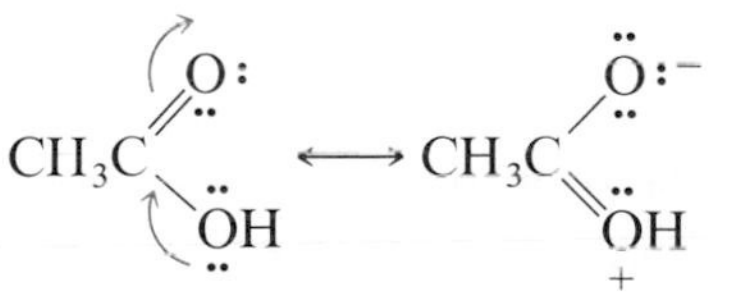

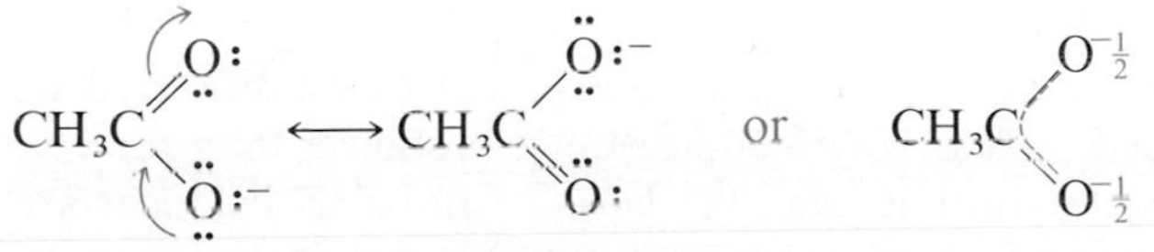

In acetate ion resonance stabilization is associated with dispersal of charge and is far more effective than in acetic acid, where electron delocalization leads to charge separation. Because acetate ion is strongly stabilized by electron delocalization, the free energy of ionization is only 6.5 kcal/mol, much lower than that of ethanol, and the extent of ionization is greater.

A description of electron delocalization in acetate ion is presented in Figure 20.5 showing the conjugated system through which the π electrons of the carboxylate group are delocalized.

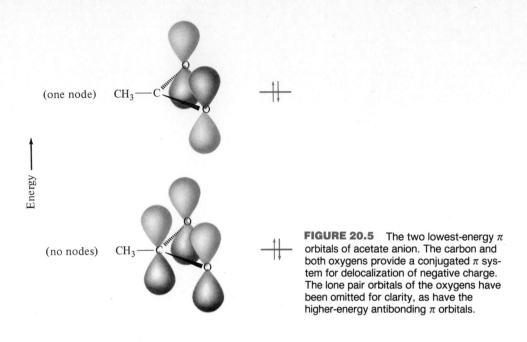

(one node) CH_3—C

(no nodes) CH_3—C

Energy

FIGURE 20.5 The two lowest-energy π orbitals of acetate anion. The carbon and both oxygens provide a conjugated π system for delocalization of negative charge. The lone pair orbitals of the oxygens have been omitted for clarity, as have the higher-energy antibonding π orbitals.

$$\overset{\displaystyle O}{\overset{\displaystyle \|}{}}$$

PROBLEM 20.2 Peroxyacetic acid (CH_3COOH) is a weaker acid than acetic acid; its K_a is 6.3×10^{-9} (pK_a 8.2) versus 1.8×10^{-5} for acetic acid (pK_a 4.7). Why are peroxy acids weaker than carboxylic acids?

Electron delocalization in acetate ion, as well as in carboxylate ions generally, is supported by studies that reveal significant differences in the pattern of carbon-oxygen bond distances of carboxylic acids and their carboxylate anions.

1.21 Å
$$CH_3C \overset{\displaystyle O}{\underset{\displaystyle OH}{}}$$
1.36 Å

1.25 Å
$$CH_3C \overset{\displaystyle O^{-\frac{1}{2}}}{\underset{\displaystyle O_{-\frac{1}{2}}}{}} \qquad \overset{+}{NH_4}$$
1.25 Å

As expected, the two carbon-oxygen bond distances in acetic acid are different from each other. The C=O double bond distance is shorter than the C—O single bond distance. In ammonium acetate, the two C—O bond distances are equal, as they should be according to both the resonance and molecular orbital pictures of bonding in carboxylate anions.

20.5 SALTS OF CARBOXYLIC ACIDS

In the presence of strong bases such as sodium hydroxide, carboxylic acids are neutralized instantly and quantitatively.

$$\underset{\substack{\text{Carboxylic} \\ \text{acid} \\ \text{(stronger} \\ \text{acid)}}}{\text{RCOH}} + \underset{\substack{\text{Hydroxide} \\ \text{ion} \\ \text{(stronger} \\ \text{base)}}}{\text{HO}^-} \xrightarrow{K \sim 10^{11}} \underset{\substack{\text{Water} \\ \ \\ \text{(weaker} \\ \text{acid)}}}{\text{H}_2\text{O}} + \underset{\substack{\text{Carboxylate} \\ \text{ion} \\ \text{(weaker base)}}}{\text{RCO}^-}$$

PROBLEM 20.3   Write an ionic equation for the reaction of acetic acid with each of the following and specify whether the equilibrium favors starting materials or products:

(a) Sodium ethoxide

(b) Potassium *tert*-butoxide

(c) Sodium bromide

(d) Sodium acetylide

(e) Potassium nitrate

(f) Lithium amide

SAMPLE SOLUTION   (a)   The reaction is an acid-base reaction; ethoxide ion is the base.

$$\underset{\substack{\text{Acetic acid} \\ \text{(stronger acid)}}}{\text{CH}_3\text{CO}_2\text{H}} + \underset{\substack{\text{Ethoxide ion} \\ \text{(stronger base)}}}{\text{CH}_3\text{CH}_2\text{O}^-} \ \rightarrow \ \underset{\substack{\text{Acetate ion} \\ \text{(weaker base)}}}{\text{CH}_3\text{CO}_2^-} + \underset{\substack{\text{Ethanol} \\ \text{(weaker acid)}}}{\text{CH}_3\text{CH}_2\text{OH}}$$

The position of equilibrium lies well to the right. Ethanol with a K_a of 10^{-16} (pK_a 16) is a much weaker acid than acetic acid.

The metal carboxylate salts formed on neutralization of carboxylic acids are named by first specifying the metal ion and then adding the name of the acid modified by replacing -*ic acid* by -*ate*. Monocarboxylate salts of diacids are designated by naming both the cation and hydrogen as substituents of carboxylate groups.

$$\underset{\substack{\text{Lithium} \\ \text{acetate}}}{\text{CH}_3\text{COLi}} \qquad \underset{\substack{\text{Sodium } p\text{-chlorobenzoate}}}{\text{Cl}\!-\!\!\bigcirc\!\!-\!\text{CONa}} \qquad \underset{\substack{\text{Potassium hydrogen} \\ \text{hexanedioate}}}{\text{HOC(CH}_2)_4\text{COK}}$$

Most metal carboxylate salts are ionic, and as long as the molecular weight is not too high, the sodium and potassium salts of carboxylic acids are soluble in water. Carboxylic acids therefore may be extracted from ether solutions into aqueous sodium or potassium hydroxide as their carboxylate salts.

The solubility behavior of salts of straight-chain carboxylic acids having 12 to 18 carbons is unusual and can be illustrated by considering sodium stearate.

Sodium stearate
(sodium octadecanoate)

Sodium stearate has a polar carboxylate group at one end of a long hydrocarbon chain. The carboxylate group is *hydrophilic* ("water-loving") and tends to confer

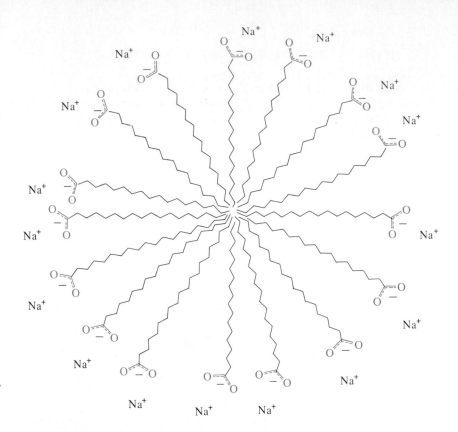

FIGURE 20.6 Idealized representation of the cross section of a sodium stearate micelle. A real micelle is likely to be more irregular and to contain voids and channels.

water solubility on the molecule. The hydrocarbon chain is *lipophilic* ("fat-loving") and tends to associate with other hydrocarbon chains. The compromise achieved by sodium stearate when it is placed in water is to form a colloidal dispersion of spherical aggregates called *micelles.* Each micelle is composed of 50 to 100 individual molecules. Micelles form spontaneously when the carboxylate concentration exceeds a certain minimum value called the *critical micelle concentration.* A representation of a micelle is shown in Figure 20.6.

Polar carboxylate groups dot the surface of the micelle. There they bind to water molecules and to sodium ions. The nonpolar hydrocarbon chains are directed toward the interior of the micelle, where individually weak but cumulatively significant dispersion forces (attractive van der Waals forces) bind them together. Micelles are approximately spherical because a sphere encloses the maximum volume of material for a given surface area and disrupts the water structure least. Because their surfaces are negatively charged, two micelles repel each other rather than clustering to form higher aggregates.

It is the formation of micelles and their properties that is responsible for the cleansing action of soaps. Water that contains sodium stearate removes grease by dissolving it in the hydrocarbon-like interior of the micelles. The grease is washed away with the water, not because it dissolves in it but because it dissolves in the micelles that are dispersed in the water. Sodium stearate is an example of a soap; sodium and potassium salts of other C_{12} through C_{18} straight-chain carboxylic acids possess similar properties.

Detergents are substances, including soaps, that cleanse by micellar action. A large number of synthetic detergents are known. One example is sodium lauryl sulfate.

Sodium lauryl sulfate has a long hydrocarbon chain terminating in a polar sulfate ion and forms soaplike micelles in water.

Sodium lauryl sulfate
(sodium 1-dodecyl sulfate)

Detergents are designed to be effective in hard water, i.e., water containing calcium salts that form insoluble calcium carboxylates with soaps. These precipitates rob the soap of its cleansing power and form an unpleasant scum. The calcium salts of synthetic detergents such as sodium lauryl sulfate, however, are soluble and retain their micelle-forming ability in water.

20.6 SUBSTITUENTS AND ACID STRENGTH

The acidity of a carboxylic acid is little affected by alkyl substituents. The ionization constants of all acids that have the general formula $C_nH_{2n+1}CO_2H$ are very similar to each other and equal approximately 10^{-5} (pK_a 5). Table 20.3 lists some representative examples.

TABLE 20.3

Effect of Substituents on Acidity of Carboxylic Acids

Name of acid	Structure	Ionization constant, K_a*	pK_a
Standard of comparison			
Acetic acid	CH_3CO_2H	1.8×10^{-5}	4.7
Alkyl substituents have a negligible effect on acidity			
Propanoic acid (propionic acid)	$CH_3CH_2CO_2H$	1.3×10^{-5}	4.9
2-Methylpropanoic acid (isobutyric acid)	$(CH_3)_2CHCO_2H$	1.6×10^{-5}	4.8
2,2-Dimethylpropanoic acid (pivalic acid)	$(CH_3)_3CCO_2H$	0.9×10^{-5}	5.1
Heptanoic acid	$CH_3(CH_2)_5CO_2H$	1.3×10^{-5}	4.9
α-Halogen substituents increase acidity			
Fluoroacetic acid	FCH_2CO_2H	2.5×10^{-3}	2.6
Chloroacetic acid	$ClCH_2CO_2H$	1.4×10^{-3}	2.9
Bromoacetic acid	$BrCH_2CO_2H$	1.4×10^{-3}	2.9
Dichloroacetic acid	Cl_2CHCO_2H	5.0×10^{-2}	1.3
Trichloroacetic acid	Cl_3CCO_2H	1.3×10^{-1}	0.9
Other electronegative groups also increase acidity			
Methoxyacetic acid	$CH_3OCH_2CO_2H$	2.7×10^{-4}	3.6
Cyanoacetic acid	$N{\equiv}CCH_2CO_2H$	3.4×10^{-3}	2.5
Nitroacetic acid	$O_2NCH_2CO_2H$	2.1×10^{-2}	1.7

* In water at 25°C.

Electronegative substituents, particularly when they are attached to the α carbon atom, significantly increase the acidity of a carboxylic acid. As the data in Table 20.3 show, all the monohaloacetic acids are about 100 times more acidic than acetic acid. Multiple halogen substitution increases the acidity even more; trichloroacetic acid is almost as strong as sulfuric acid.

Organic chemists have long attributed the acid-strengthening effect of electronegative atoms or groups to an *inductive effect* of the substituent transmitted through the σ bonds of the molecule. According to this model, the σ electrons in the carbon-chlorine bond of chloroacetate ion are drawn toward chlorine, leaving the α carbon atom with a slight positive charge. The α carbon, because of this positive character, attracts electrons from the negatively charged carboxylate, thus dispersing the charge and stabilizing the anion. The more stable the anion, the greater the equilibrium constant for its formation.

The carboxylate anion is stabilized by electron-withdrawing effect of chlorine; the equilibrium constant for its formation is greater than that for corresponding formation of acetate ion from acetic acid.

Inductive effects of this kind diminish rapidly as the number of σ bonds between the reaction site and the polar substituent increase. The acid-strengthening effect of a halogen decreases as it becomes more remote from the carboxyl group.

$ClCH_2CO_2H$	$ClCH_2CH_2CO_2H$	$ClCH_2CH_2CH_2CO_2H$
Chloroacetic acid	3-Chloropropanoic acid	4-Chlorobutanoic acid
$K_a = 1.4 \times 10^{-3}$	$K_a = 1.0 \times 10^{-4}$	$K_a = 3.0 \times 10^{-5}$
$pK_a = 2.9$	$pK_a = 4.0$	$pK_a = 4.5$

PROBLEM 20.4   Which is the stronger acid in each of the following pairs?

(a) $(CH_3)_3CCH_2CO_2H$ or $(CH_3)_3\overset{+}{N}CH_2CO_2H$
(b) $CH_3CH_2CO_2H$ or CH_3CHCO_2H with OH
(c) $(CH_3)_2CHCO_2H$ or $CH_2{=}CHCO_2H$
(d) CH_3CCO_2H (C=O)   or $CH_2{=}CHCO_2H$
(e) $CH_3CH_2CH_2CO_2H$ or $CH_3SCH_2CO_2H$ (S with two O)

SAMPLE SOLUTION   (a)   The two compounds are viewed as substituted derivatives of acetic acid. A *tert*-butyl group is slightly electron-releasing and should have a modest effect on acidity. The compound $(CH_3)_3CCH_2CO_2H$ is expected to have an acid strength similar to that of acetic acid. A trimethylammonium substituent, on the other hand, is positively charged and is a powerful electron-withdrawing substituent. The compound

$(CH_3)_3\overset{+}{N}CH_2CO_2H$ is expected to be a much stronger acid than $(CH_3)_3CCH_2CO_2H$. The measured ionization constants, shown below, confirm this prediction.

$$(CH_3)_3CCH_2CO_2H \qquad (CH_3)_3\overset{+}{N}CH_2CO_2H$$

Weaker acid	Stronger acid
$K_a = 5 \times 10^{-6}$	$K_a = 1.5 \times 10^{-2}$
($pK_a = 5.30$)	($pK_a = 1.83$)

Another proposal that has been advanced to explain the acid-strengthening effect of polar substituents holds that the electron-withdrawing effect is transmitted through the medium rather than through successive polarization of σ bonds. This is referred to as a *field effect*. Both field and inductive contributions to the polar effect tend to operate in the same direction, and it is believed that both are important.

It is a curious fact that substituents affect the entropy of ionization more than they do the enthalpy term in the expression

$$\Delta G° = \Delta H° - T\,\Delta S°$$

The enthalpy term $\Delta H°$ is close to zero for the ionization of most carboxylic acids, regardless of their strength. The free energy of ionization $\Delta G°$ is dominated by the $-T\,\Delta S°$ term. Ionization is accompanied by an increase in solvation forces, leading to a decrease in the entropy of the system; $\Delta S°$ is negative and $-T\,\Delta S°$ is positive. Anions that incorporate substituents capable of dispersing negative charge require less solvent reorganization for their stabilization, and less entropy is lost in their production than in the generation of anions that lack electron-withdrawing groups.

Solvation entropy effects are almost entirely responsible for the modest acid-weakening effect of alkyl substituents. Propanoic acid is a slightly weaker acid than acetic acid as measured by their ionization constants, which by definition apply to dilute aqueous solution. When their relative acidities are compared in the gas phase, however, propanoic acid is a slightly stronger acid than acetic.

20.7 IONIZATION OF SUBSTITUTED BENZOIC ACIDS

A considerable body of data is available on the acidity of substituted benzoic acids. Benzoic acid itself is a somewhat stronger acid than acetic acid. Its carboxyl group is attached to an sp^2 hybridized carbon and ionizes to a greater extent than one that is attached to an sp^3 hybridized carbon. Remember, carbon becomes more electron-withdrawing as its s character increases.

$$CH_3CO_2H \qquad CH_2{=}CHCO_2H \qquad \text{(benzene ring)}{-}CO_2H$$

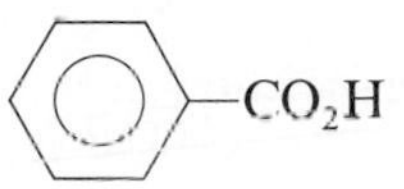

Acetic acid	Acrylic acid	Benzoic acid
$K_a\ 1.8 \times 10^{-5}$	$K_a\ 5.5 \times 10^{-5}$	$K_a\ 6.3 \times 10^{-5}$
($pK_a\ 4.8$)	($pK_a\ 4.3$)	($pK_a\ 4.2$)

PROBLEM 20.5 What is the most acidic neutral molecule characterized by the molecular formula $C_3H_xO_2$?

TABLE 20.4

Acidity of Some Substituted Benzoic Acids

Substituent in $XC_6H_4CO_2H$	K_a (pK_a)* for different positions of substituent X		
	Ortho	Meta	Para
1. H	6.3×10^{-5} (4.2)	6.3×10^{-5} (4.2)	6.3×10^{-5} (4.2)
2. CH_3	1.2×10^{-4} (3.9)	5.3×10^{-5} (4.3)	4.2×10^{-5} (4.2)
3. F	5.4×10^{-4} (3.3)	1.4×10^{-4} (3.9)	7.2×10^{-5} (4.1)
4. Cl	1.2×10^{-3} (2.9)	1.5×10^{-4} (3.8)	1.0×10^{-4} (4.0)
5. Br	1.4×10^{-3} (2.8)	1.5×10^{-4} (3.8)	1.1×10^{-4} (4.0)
6. I	1.4×10^{-3} (2.9)	1.4×10^{-4} (3.9)	9.2×10^{-5} (4.0)
7. CH_3O	8.1×10^{-5} (4.1)	8.2×10^{-5} (4.1)	3.4×10^{-5} (4.5)
8. O_2N	6.7×10^{-3} (2.2)	3.2×10^{-4} (3.5)	3.8×10^{-4} (3.4)

* In water at 25°C.

Substituents influence the acidity of substituted benzoic acids through a combination of inductive, field, resonance, and solvation effects. As is the case with aliphatic carboxylic acids, changes in the free energy of ionization arise primarily from changes in the entropy associated with solvation. Table 20.4 lists the ionization constants of some substituted benzoic acids. The most pronounced effects are observed when strongly electron-withdrawing substituents are present at positions ortho to the carboxyl group. An *o*-nitro substituent, for example, increases the acidity of benzoic acid 100-fold. Substituent effects are smaller at positions meta and para to the carboxyl group.

20.8 DICARBOXYLIC ACIDS

Dicarboxylic acids are characterized by separate ionization constants, designated K_1 and K_2, respectively, for each ionization step.

1,4-Benzenedicarboxylic acid

Hydrogen 1,4-benzenedicarboxylate (monoanion)

$K_1 = 3.1 \times 10^{-4}$
($pK_1 = 3.5$)

The first ionization constant of dicarboxylic acids is larger than K_a for monocarboxylic analogs. One reason is statistical. There are two potential sites for ionization rather than one. Further, one carboxyl group acts as an electron-withdrawing group to facilitate dissociation of the other. This is particularly noticeable when the two carboxyl groups are separated by only a few bonds. Oxalic and malonic acid, for example, are several orders of magnitude stronger than simple alkyl derivatives of acetic acid. Heptanedioic acid, in which the carboxyl groups are well separated from each other, is only slightly stronger than acetic acid.

HO_2CCO_2H
Oxalic acid
K_1 6.5×10^{-2}
(pK_1 1.2)

$HO_2CCH_2CO_2H$
Malonic acid
K_1 1.4×10^{-3}
(pK_1 2.8)

$HO_2C(CH_2)_5CO_2H$
Heptanedioic acid
K_1 3.1×10^{-5}
(pK_1 4.3)

20.9 CARBONIC ACID

Through an accident of history, the simplest dicarboxylic acid, carbonic acid

$$\overset{O}{\overset{\|}{HOCOH}},$$ is not even classified as an organic compound. Because many minerals are carbonate salts, nineteenth century chemists placed carbonates, bicarbonates, and carbon dioxide in the inorganic realm. Nevertheless, the essential features of carbonic acid and its salts are easily understood on the basis of our knowledge of carboxylic acids.

Carbonic acid is formed when carbon dioxide reacts with water. Hydration of carbon dioxide is far from complete, however. Almost all the carbon dioxide that is dissolved in water exists as carbon dioxide; only 0.3 percent of it is converted to carbonic acid. Carbonic acid is a weak acid and ionizes to a small extent to bicarbonate ion.

$$CO_2 + H_2O \rightleftharpoons \overset{O}{\overset{\|}{HOCOH}} \rightleftharpoons H^+ + \overset{O}{\overset{\|}{HOCO^-}}$$

Carbon   Water   Carbonic   Bicarbonate
dioxide   acid   ion

The equilibrium constant for the overall reaction

$$CO_2 + H_2O \overset{K}{\rightleftharpoons} H^+ + HCO_3^-$$

Carbon   Water   Bicarbonate
dioxide   ion

is related to an apparent equilibrium constant K_1 for carbonic acid ionization by the expression

$$K_1 = K[H_2O] = \frac{[H^+][HCO_3^-]}{[CO_2]} = 4.3 \times 10^{-7}$$

These equations tell us that the reverse process, proton transfer from acids to bicarbonate to form carbon dioxide, will be favorable when K_a of the acid exceeds 4.3×10^{-7} (p$K_a < 6.4$). Among compounds containing carbon, hydrogen, and oxygen, only carboxylic acids are acidic enough to meet this requirement. They dissolve in aqueous sodium bicarbonate with the evolution of carbon dioxide. This behavior is the basis of a qualitative test for carboxylic acids.

PROBLEM 20.6 The value cited for the "apparent K_1" of carbonic acid, 4.3×10^{-7}, is the one normally given in reference books. It is determined by measuring the pH of water to which a known amount of carbon dioxide has been added. When we recall that only 0.3 percent of carbon dioxide is converted to carbonic acid in water, what is the "true K_1" of carbonic acid?

Carbonic anhydrase is an enzyme that catalyzes the hydration of carbon dioxide to bicarbonate. The uncatalyzed hydration of carbon dioxide is too slow to be effective in transporting carbon dioxide from the tissues to the lungs, so mammals have developed catalysts to speed this process. The activity of carbonic anhydrase is remarkable; it has been estimated that one molecule of this enzyme can catalyze the hydration of 3.6×10^7 molecules of carbon dioxide per minute.

As with other dicarboxylic acids, the second ionization constant of carbonic acid is far smaller than the first.

$$\underset{\text{Bicarbonate ion}}{HO\overset{\overset{\displaystyle O}{\|}}{C}O^-} \underset{K_2}{\rightleftharpoons} H^+ + \underset{\text{Carbonate ion}}{{}^-O\overset{\overset{\displaystyle O}{\|}}{C}O^-}$$

The value of K_2 is 5.6×10^{-11} (pK_a 10.2). Bicarbonate is a weaker acid than carboxylic acids but a stronger acid than water and alcohols.

20.10 SOURCES OF CARBOXYLIC ACIDS

Many carboxylic acids were first isolated from natural sources and were given names indicative of their origin. Formic acid (Latin *formica*, "ant") was obtained by distilling ants. Since ancient times acetic acid (Latin *acetum*, "vinegar") has been known to be present in wine that has turned sour. Butyric acid (Latin *butyrum*, "butter") contributes to the odor of rancid butter, and lactic acid (Latin *lac*, "milk") has been isolated from sour milk.

While these humble origins make interesting historical notes, in most cases the large-scale preparation of carboxylic acids relies on chemical synthesis. Virtually none of the 3×10^9 lb of acetic acid produced in the United States each year is obtained from vinegar. Instead, several catalytic processes have been developed. Two of these involve oxidation of petroleum-derived starting materials:

$$\underset{\text{Butane}}{2CH_3CH_2CH_2CH_3} + \underset{\text{Oxygen}}{5O_2} \longrightarrow \underset{\text{Acetic acid}}{2CH_3CO_2H} + \underset{\text{Water}}{6H_2O}$$

$$\underset{\text{Acetaldehyde}}{2CH_3\overset{\overset{\displaystyle O}{\|}}{C}H} + \underset{\text{Oxygen}}{O_2} \xrightarrow[\substack{\text{manganese acetate,} \\ \text{cobalt acetate, or} \\ \text{copper acetate}}]{} \underset{\text{Acetic acid}}{2CH_3CO_2H}$$

The third route uses carbon monoxide and methanol, both of which are derived from coal.

$$\underset{\text{Methanol}}{CH_3OH} + \underset{\substack{\text{Carbon} \\ \text{monoxide}}}{CO} \xrightarrow[\substack{\text{rhodium catalyst} \\ \text{heat, pressure}}]{\text{cobalt or}} \underset{\text{Acetic acid}}{CH_3CO_2H}$$

The principal end use of acetic acid is in the production of vinyl acetate for paints and adhesives.

The carboxylic acid produced in the greatest amounts is 1,4-benzenedicarboxylic acid (terephthalic acid). About 5×10^9 lb/yr is produced in the United States as a starting material for the preparation of polyester fibers. One important process converts *p*-xylene to terephthalic acid by oxidation with nitric acid.

p-Xylene   1,4-Benzenedicarboxylic acid (terephthalic acid)

TABLE 20.5

Summary of Reactions Discussed in Earlier Chapters That Yield Carboxylic Acids

Reaction (section) and comments	General equation and specific example
Oxidative cleavage of alkenes (Section 7.17) Potassium permanganate cleaves alkenes to two carbonyl compounds. If one of the substituents at the double bond is hydrogen, the cleavage product is an aldehyde, which is rapidly oxidized to a carboxylic acid under the reaction conditions.	$RCH{=}CR'_2 \xrightarrow{KMnO_4} RCO_2H + R'_2C{=}O$ Alkene Carboxylic acid Carbonyl compound $CH_3CH(CH_2)_3CHCH{=}CH_2 \xrightarrow[H_2O]{KMnO_4} CH_3CH(CH_2)_3CHCO_2H + CO_2$ CH₃ CH₃ CH₃ CH₃ 3,7-Dimethyl-1-octene 2,6-Dimethylheptanoic acid (45%) Carbon dioxide
Side-chain oxidation of alkylbenzenes (Section 12.18) A primary or secondary alkyl side chain on an aromatic ring is degraded to a carboxyl group by reaction with a strong oxidizing agent such as potassium permanganate or chromic acid.	$ArCHR_2 \xrightarrow[K_2Cr_2O_7, H_2SO_4]{KMnO_4 \text{ or}} ArCO_2H$ Alkylbenzene Arenecarboxylic acid 3-Methoxy-4-nitrotoluene $\xrightarrow[\text{2. } H^+]{\text{1. } KMnO_4, HO^-}$ 3-Methoxy-4-nitrobenzoic acid (100%)
Oxidation of primary alcohols (Section 16.12) Potassium permanganate and chromic acid convert primary alcohols to carboxylic acids by way of the corresponding aldehyde.	$RCH_2OH \xrightarrow[K_2Cr_2O_7, H_2SO_4]{KMnO_4 \text{ or}} RCO_2H$ Primary alcohol Carboxylic acid $(CH_3)_3CCHC(CH_3)_3 \xrightarrow[H_2O, H_2SO_4]{H_2CrO_4} (CH_3)_3CCHC(CH_3)_3$ CH₂OH CO₂H 2-tert-Butyl-3,3-dimethyl-1-butanol 2-tert-Butyl-3,3-dimethylbutanoic acid (82%)
Oxidation of aldehydes (Section 18.19) Aldehydes are particularly sensitive to oxidation and are converted to carboxylic acids by a number of oxidizing agents, including potassium permanganate, chromic acid, and silver oxide.	$RCH{=}O \xrightarrow{\text{oxidizing agent}} RCO_2H$ Aldehyde Carboxylic acid Furan-2-carbaldehyde (furfural) $\xrightarrow[H_2SO_4, H_2O]{K_2Cr_2O_7}$ Furan-2-carboxylic acid (furoic acid) (75%)
Haloform cleavage of methyl ketones (Section 19.7) In the presence of aqueous base, halogens react with methyl ketones to yield a trihalomethane (a haloform) and a carboxylate salt. Acidification gives the corresponding acid.	$RCCH_3 \xrightarrow[HO^-]{X_2} CHX_3 + RCO_2^- \xrightarrow{H^+} RCO_2H$ Methyl ketone Trihalomethane Carboxylate salt Carboxylic acid $(CH_3)_2C{=}CHCCH_3 \xrightarrow[\text{2. } H^+]{\text{1. } Cl_2, HO^-} (CH_3)_2C{=}CHCO_2H + CHCl_3$ 4-Methyl-3-penten-2-one 3-Methyl-2-butenoic acid (49–53%) Chloroform

You will recognize the side-chain oxidation of *p*-xylene to terephthalic acid as a reaction type discussed previously (Section 12.18). Examples of other reactions encountered earlier that can be applied to the synthesis of carboxylic acids are collected in Table 20.5.

The reactions summarized in the table lead to carboxylic acids that have either the same number of or fewer carbon atoms than the starting material. The reactions to be described in the next two sections permit carboxylic acids to be prepared by extending a chain by one carbon atom and are of great value in laboratory syntheses of carboxylic acids.

20.11 SYNTHESIS OF CARBOXYLIC ACIDS BY THE CARBOXYLATION OF GRIGNARD REAGENTS

You have learned of the reaction of Grignard reagents with the carbonyl group of aldehydes, ketones, and esters and have seen their applications in organic synthesis. Grignard reagents react in much the same way with *carbon dioxide* to yield magnesium salts of carboxylic acids. Acidification converts these magnesium carboxylate salts to the desired carboxylic acids.

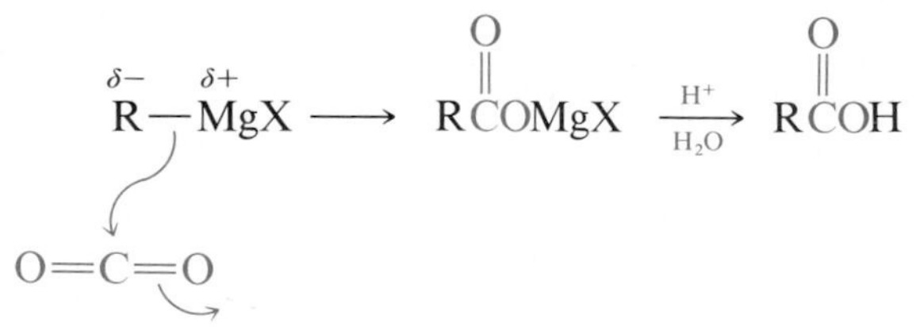

Grignard reagent acts as a nucleophile toward carbon dioxide

Halomagnesium carboxylate

Carboxylic acid

$$\underset{\text{Cl}}{\text{CH}_3\text{CHCH}_2\text{CH}_3} \xrightarrow[\substack{2.\ \text{CO}_2 \\ 3.\ \text{H}_3\text{O}^+}]{1.\ \text{Mg, diethyl ether}} \underset{\text{CO}_2\text{H}}{\text{CH}_3\text{CHCH}_2\text{CH}_3}$$

2-Chlorobutane → 2-Methylbutanoic acid (76–86%)

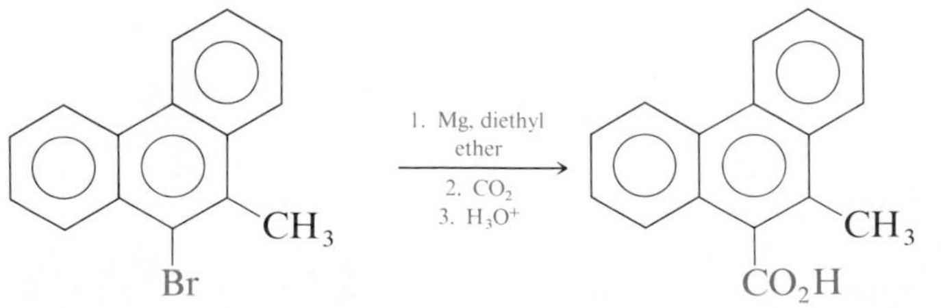

9-Bromo-10-methylphenanthrene → 10-Methylphenanthrene-9-carboxylic acid (82%)

Overall, the carboxylation of Grignard reagents transforms an alkyl or aryl halide to a carboxylic acid in which the carbon skeleton has been extended by one carbon atom.

$$RX \xrightarrow[\text{diethyl ether}]{Mg} RMgX \xrightarrow[\text{2. } H_3O^+]{1. CO_2} RCO_2H$$

Alkyl halide   Grignard reagent   Carboxylic acid

The only limitation to this procedure is that the alkyl or aryl halide must not bear substituents that are incompatible with Grignard reagents, such as OH, NH, SH, C=O, or NO_2.

20.12 SYNTHESIS OF CARBOXYLIC ACIDS BY THE PREPARATION AND HYDROLYSIS OF NITRILES

Primary and secondary alkyl halides may be converted to the next higher carboxylic acid by a two-step synthetic sequence involving the preparation and hydrolysis of *nitriles*. Nitriles, also known as alkyl cyanides, are prepared by a nucleophilic substitution reaction.

$$RX + :\overset{-}{C}\equiv N: \longrightarrow RC\equiv N + X^-$$

Primary or   Cyanide ion   Nitrile   Halide ion
secondary alkyl   (alkyl cyanide)
halide

The reaction is of the S_N2 type and is most effective with primary alkyl halides. Secondary alkyl halides react more slowly and give somewhat lower yields; elimination is the only reaction observed with tertiary alkyl halides. Aryl and vinyl halides do not react. Dimethyl sulfoxide is the preferred solvent for this reaction, but alcohols and water-alcohol mixtures have also been used.

Once the nitrile group has been introduced into the molecule, it is subjected to a separate hydrolysis step. Usually this is carried out in aqueous acid at reflux.

$$RC\equiv N + 2H_2O + H^+ \longrightarrow R\overset{\overset{\displaystyle O}{\|}}{C}OH + NH_4^+$$

Nitrile   Water   Carboxylic   Ammonium
                          acid   ion

The mechanism of nitrile hydrolysis will be presented in more detail in Section 21.19.

Benzyl chloride   Benzyl cyanide (92%)   Phenylacetic acid (77%)

Dicarboxylic acids have been prepared from dihalides by this method.

$$BrCH_2CH_2CH_2Br \xrightarrow[H_2O]{NaCN} NCCH_2CH_2CH_2CN \xrightarrow[\text{heat}]{H_2O,HCl} HO\overset{\overset{\displaystyle O}{\|}}{C}CH_2CH_2CH_2\overset{\overset{\displaystyle O}{\|}}{C}OH$$

1,3-Dibromopropane   1,3 Dicyanopropane   Glutaric acid
                              (77–86%)   (83–85%)

PROBLEM 20.7 Only one of the two procedures just described, preparation and carboxylation of a Grignard reagent or formation and hydrolysis of a nitrile, is appropriate to each of the following RX → RCO_2H conversions. Identify the correct procedure in each case, and specify why the other will fail.

(*a*) Bromobenzene → benzoic acid
(*b*) 2-Chloroethanol → 3-hydroxypropanoic acid
(*c*) *tert*-Butyl chloride → 2,2-dimethylpropanoic acid
(*d*) *p*-Nitrobenzyl bromide → *p*-nitrophenylacetic acid

SAMPLE SOLUTION (*a*) Bromobenzene is an aryl halide and is unreactive toward nucleophilic substitution by cyanide ion. The route $C_6H_5Br \rightarrow C_6H_5CN \rightarrow C_6H_5CO_2H$ fails because the first step fails. The route proceeding through the Grignard reagent is perfectly satisfactory and appears as an experiment in a number of introductory organic chemistry laboratory texts.

Bromobenzene Phenylmagnesium bromide Benzoic acid

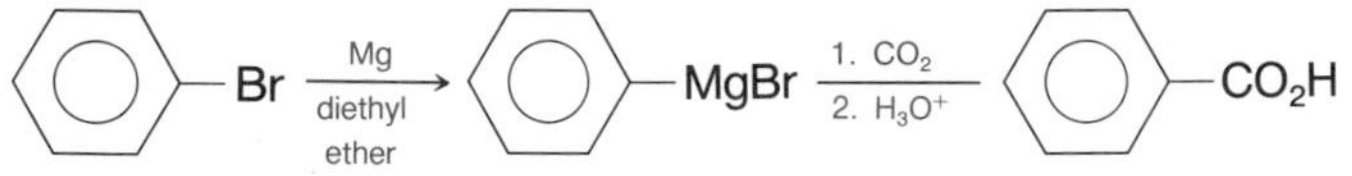

Nitrile functional groups in cyanohydrins (Section 18.12) are hydrolyzed under conditions similar to those of alkyl cyanides. Cyanohydrin formation followed by hydrolysis provides a route to the preparation of α-hydroxy carboxylic acids.

2-Pentanone 2-Cyano-2-pentanol 2-Hydroxy-2-methylpentanoic acid (60% from 2-pentanone)

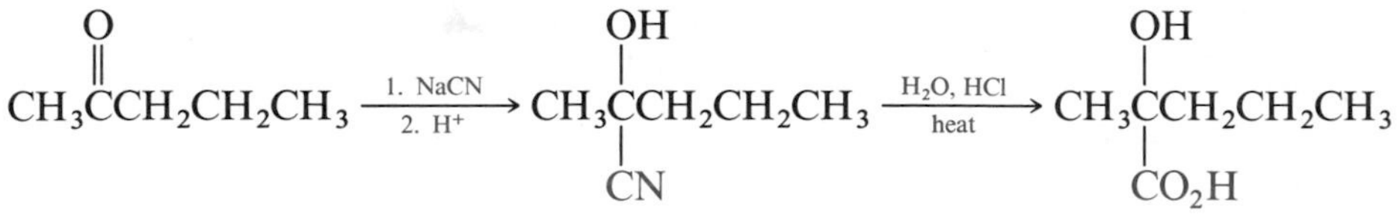

20.13 REACTIONS OF CARBOXYLIC ACIDS. A REVIEW AND A PREVIEW

The most apparent chemical property of carboxylic acids, their acidity, has already been examined in earlier sections of this chapter. Three reactions of carboxylic acids—conversion to acyl chlorides, reduction, and esterification—have been encountered in previous chapters and are reviewed in Table 20.6. Acid-catalyzed esterification of carboxylic acids is one of the fundamental reactions of organic chemistry, and this portion of the chapter begins with an examination of the mechanism by which it occurs. Later, in Sections 20.16 and 20.17, two new reactions of carboxylic acids that are of synthetic value will be described.

TABLE 20.6

Summary of Reactions of Carboxylic Acids Discussed in Earlier Chapters

Reaction (section) and comments	General equation and specific example
Formation of acyl chlorides (Section 13.7) Thionyl chloride reacts with carboxylic acids to yield acyl chlorides. Phosphorus pentachloride may be used instead.	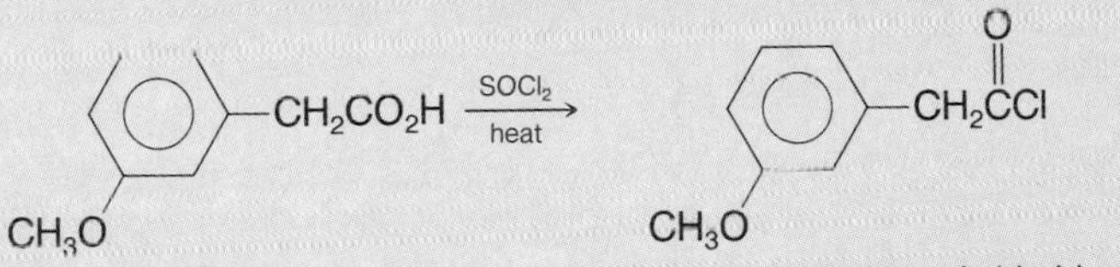 RCO_2H + $SOCl_2$ $\longrightarrow$ $\overset{\displaystyle O}{\overset{\displaystyle \|}{R C}}Cl$ + SO_2 + HCl Carboxylic acid Thionyl chloride Acyl chloride Sulfur dioxide Hydrogen chloride *m*-Methoxyphenylacetic acid → *m*-Methoxyphenylacetyl chloride (85%) $CH_3(CH_2)_{16}CO_2H \xrightarrow{PCl_5} CH_3(CH_2)_{16}\overset{\displaystyle O}{\overset{\displaystyle \|}{C}}Cl$ Octadecanoic acid Octadecanoyl chloride (70%)
Lithium aluminum hydride reduction (Section 16.4) Carboxylic acids are reduced to primary alcohols by the powerful reducing agent lithium aluminum hydride.	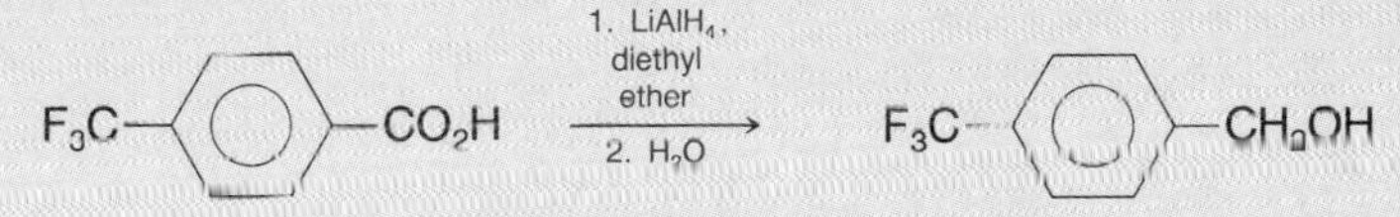 $RCO_2H \xrightarrow[\text{2. } H_2O]{\text{1. LiAlH}_4\text{, diethyl ether}} RCH_2OH$ Carboxylic acid Primary alcohol *p*-(Trifluoromethyl)benzoic acid → *p*-(Trifluoromethyl)benzyl alcohol (96%)
Esterification (Section 16.10) In the presence of an acid catalyst, carboxylic acids and alcohols react to form esters. The reaction is an equilibrium process but can be driven to favor the ester by removing the water that is formed.	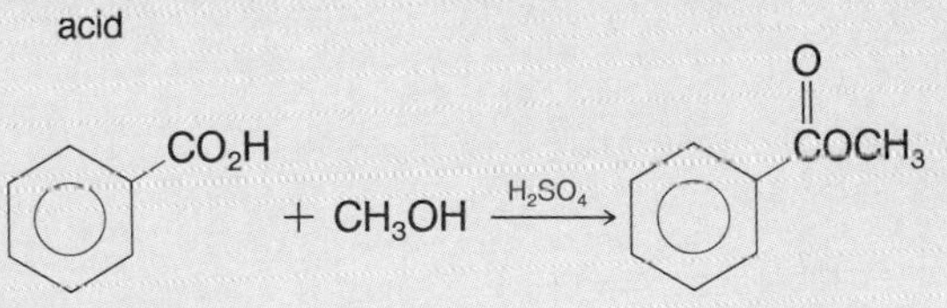 $RCO_2H + R'OH \underset{}{\overset{H^+}{\rightleftharpoons}} \overset{\displaystyle O}{\overset{\displaystyle \|}{R C}}OR' + H_2O$ Carboxylic acid Alcohol Ester Water Benzoic acid Methanol Methyl benzoate (70%)

20.14 MECHANISM OF ACID-CATALYZED ESTERIFICATION

An important question concerning the mechanism of acid-catalyzed esterification has to do with the origin of the alkoxy oxygen. For example, does the methoxy oxygen in methyl benzoate come from methanol or is it derived from benzoic acid?

$$\text{C}_6\text{H}_5-\overset{\displaystyle O}{\overset{\|}{\text{C}}}\text{O}\,\text{CH}_3$$

Is this the oxygen originally present in benzoic acid or is it the oxygen of methanol?

The answer to this question is critical to mechanistic understanding because it tells us whether it is the carbon-oxygen bond of the alcohol or a carbon-oxygen of the carboxylic acid that is broken during the process of ester formation.

A clear-cut answer was provided by Irving Roberts and Harold C. Urey of Columbia University in 1938. They prepared methanol that had been enriched in the mass 18 isotope of oxygen. When this sample of methanol was esterified with benzoic acid, the methyl benzoate product contained all the ^{18}O label that was originally present in the methanol.

$$\underset{\substack{\text{Benzoic acid}}}{\text{C}_6\text{H}_5\overset{\displaystyle O}{\overset{\|}{\text{C}}}\text{OH}} + \underset{\substack{^{18}\text{O-Enriched} \\ \text{methanol}}}{\text{CH}_3\bullet\text{H}} \xrightarrow{\text{H}^+} \underset{\substack{^{18}\text{O-Enriched} \\ \text{methyl benzoate}}}{\text{C}_6\text{H}_5\overset{\displaystyle O}{\overset{\|}{\text{C}}}\bullet\text{CH}_3} + \underset{\substack{\text{Water}}}{\text{H}_2\text{O}}$$

In the above equation $\bullet$ = oxygen enriched in its mass 18 isotope; analysis of isotopic enrichment was performed by mass spectrometry.

The results of the Roberts-Urey experiment require that any mechanism proposed for esterification occur in such a way that the carbon-oxygen bond of the alcohol is preserved. The oxygen that is lost (as a water molecule) must come from the carboxylic acid.

A mechanism that is consistent with these facts is one in which methanol acts as a nucleophile. In the first step, the acid catalyst protonates benzoic acid at its carbonyl oxygen.

$$\underset{\substack{\text{Benzoic acid}}}{\text{C}_6\text{H}_5\text{C}\overset{\displaystyle \ddot{\text{O}}:}{\underset{\ddot{\text{O}}\text{H}}{\diagup}}} \underset{\text{H}^+, \text{ fast}}{\rightleftharpoons} \underset{\substack{\text{Conjugate acid of benzoic acid}}}{\text{C}_6\text{H}_5\text{C}\overset{\displaystyle \overset{+}{\text{O}}\text{H}}{\underset{\ddot{\text{O}}\text{H}}{\diagup}}}$$

Why is it that the carbonyl oxygen is protonated rather than the hydroxyl oxygen? Examine the relative stability of the ions formed by protonation at either of the two possible sites. Protonation of the carbonyl oxygen yields a resonance-stabilized cation:

$$\text{C}_6\text{H}_5\text{C}\overset{\displaystyle \overset{+}{\text{O}}\text{H}}{\underset{\ddot{\text{O}}\text{H}}{\diagup}} \longleftrightarrow \text{C}_6\text{H}_5\text{C}\overset{\displaystyle \ddot{\text{O}}\text{H}}{\underset{\overset{\ddot{\text{O}}\text{H}}{+}}{\diagup}}$$

**Electron delocalization in carbonyl-
protonated benzoic acid**

Delocalization of an unshared electron pair of the hydroxyl group permits the positive charge to be shared equally between both oxygens. Protonation of the hydroxyl oxygen, on the other hand, yields a less stable cation:

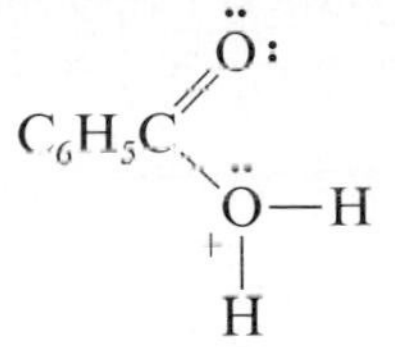

$$C_6H_5C \overset{\displaystyle \overset{\cdot\cdot}{O}:}{\underset{\overset{\cdot\cdot}{O}-H}{}} \quad \underset{\overset{+}{|}}{H}$$

Localized positive charge in hydroxyl-protonated benzoic acid

The positive charge in this cation cannot be shared by the two oxygens; it is localized on one of them. Since protonation of the carbonyl oxygen gives a more stable cation, it is that cation that is formed preferentially.

As was seen to be the case with aldehydes and ketones, protonation of a carbonyl group increases its electrophilicity. The protonated carboxylic acid is susceptible to attack by nucleophiles such as methanol.

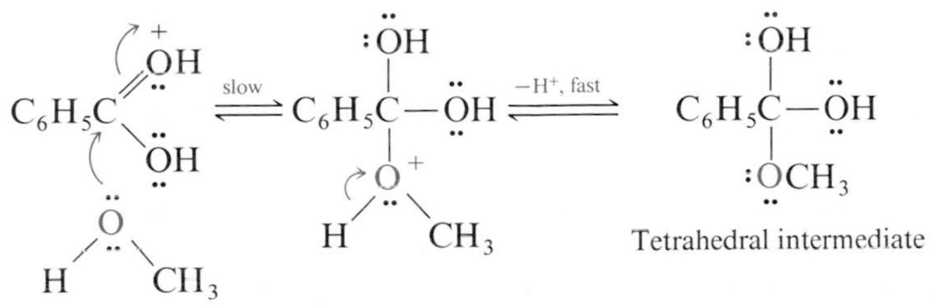

The product of nucleophilic addition of methanol to the carbonyl group followed by elimination of a proton is the key intermediate in the process. It is called the *tetrahedral intermediate* because the hybridization at carbon has changed from sp^2 in benzoic acid to sp^3 in the intermediate. Addition of an alcohol molecule to the carbonyl group, which is the first step in formation of the tetrahedral intermediate, is usually the rate-determining step in esterification of carboxylic acids.

Addition of methanol to benzoic acid to form the tetrahedral intermediate is exactly analogous to the addition of an alcohol to an aldehyde or ketone to form a hemiacetal (Section 18.10). The tetrahedral intermediate cannot be isolated. It is unstable under the acid-catalyzed conditions of its formation and undergoes dehydration to form methyl benzoate.

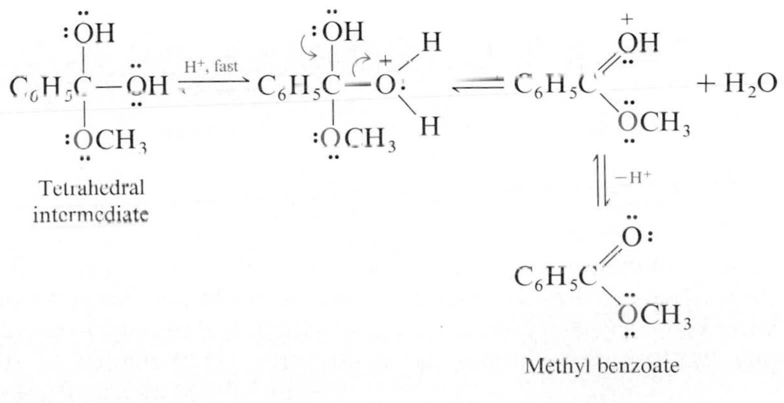

Methyl benzoate

Notice that the ^{18}O-label of methanol is incorporated into the ester according to this mechanism, as the observations of the Roberts-Urey experiment require it to be.

PROBLEM 20.8 When benzoic acid is allowed to stand in water that is enriched in ^{18}O, the isotopic label becomes incorporated into the benzoic acid. The reaction is catalyzed by acids. Suggest an explanation for this observation.

In the next chapter the three elements of the mechanism just described will be seen again as part of the general theme that unites the chemistry of carboxylic acid derivatives. These elements are

1. Activation of the carbonyl group by protonation of the carbonyl oxygen
2. Nucleophilic addition to the protonated carbonyl to form a tetrahedral intermediate
3. Elimination from the tetrahedral intermediate to restore the carbonyl group

This sequence is one of the fundamental mechanistic patterns of organic chemistry.

20.15 INTRAMOLECULAR ESTER FORMATION. LACTONES

Hydroxy acids, i.e., compounds that contain both a hydroxyl and a carboxylic acid function within the same molecule, have the capacity to form cyclic esters called *lactones*. This intramolecular esterification reaction takes place spontaneously and is particularly favored when the ring that is formed is five-membered or six-membered. Lactones that contain a five-membered cyclic ester are referred to as *γ-lactones* because they arise from *γ*-hydroxy carboxylic acids. Their six-membered analogs are known as *δ-lactones*.

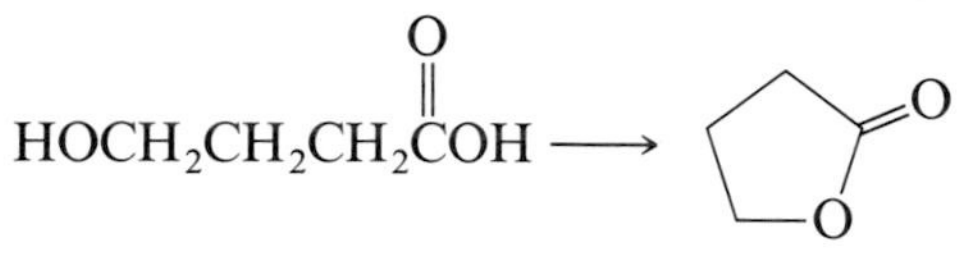

4-Hydroxybutanoic acid
(*γ*-hydroxybutyric acid)

4-Butanolide
(*γ*-butyrolactone)

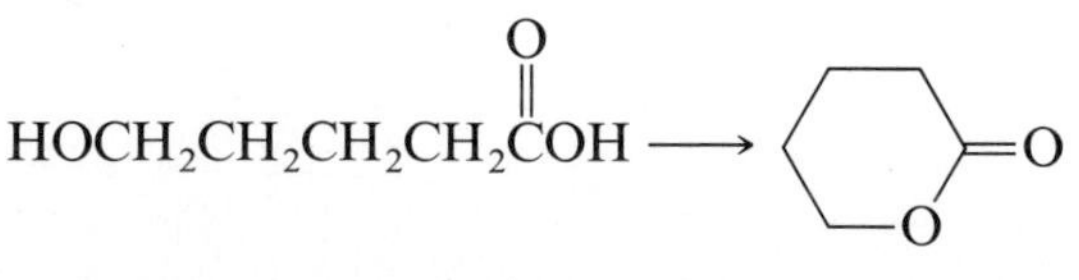

5-Hydroxypentanoic acid
(*δ*-hydroxyvaleric acid)

5-Pentanolide
(*δ*-valerolactone)

A lactone is named by replacing the *-oic acid* ending of the parent carboxylic acid by *-olide* and identifying its oxygenated carbon by number. This system is illustrated in the lactones shown in the preceding equations. Both 4-butanolide and 5-pentanolide are better known by their common names, *γ*-butyrolactone and *δ*-valerolactone, respectively, and these two common names are permitted by the IUPAC rules.

Reactions that are expected to produce hydroxy acids often yield the derived lactones instead if a five- or six-membered ring can be formed.

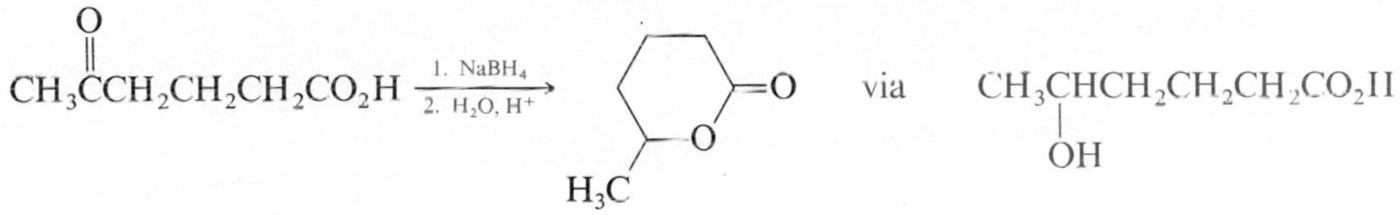

5-Oxohexanoic acid   5-Hexanolide (78%)   5-Hydroxyhexanoic acid

Many natural products are lactones and it is not unusual to find examples in which the ring size is rather large. A few naturally occurring lactones are shown in Figure 20.7. The *macrolide antibiotics,* of which erythromycin is one example, are macrocyclic (large-ring) lactones. The lactone ring of erythromycin is 14-membered.

PROBLEM 20.9   Write the structure of the hydroxy acid corresponding to each of the following lactones. The structure of each lactone is given in Figure 20.7.

(a) Mevalonolactone
(b) Pentadecanolide
(c) Vernolepin

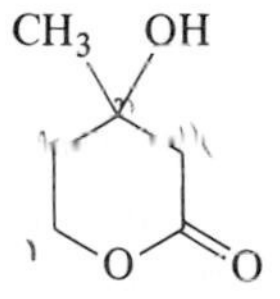

Mevalonolactone

(an intermediate in the
biosynthesis of terpenes
and steroids)

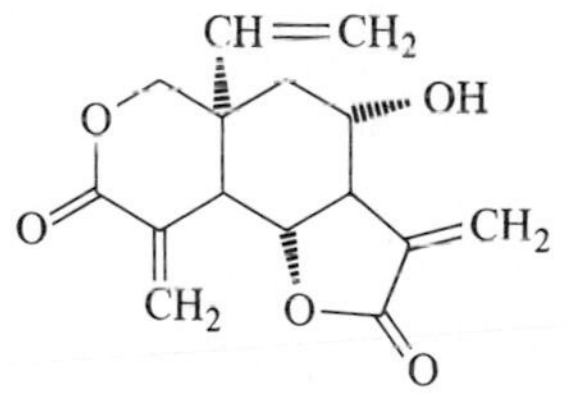

Vernolepin

(a tumor-inhibitory substance
that incorporates both a
γ lactone and a δ lactone
into its tricyclic framework)

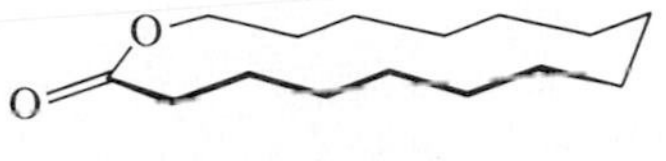

15-Pentadecanolide

(an odor-enhancing substance
used in perfume)

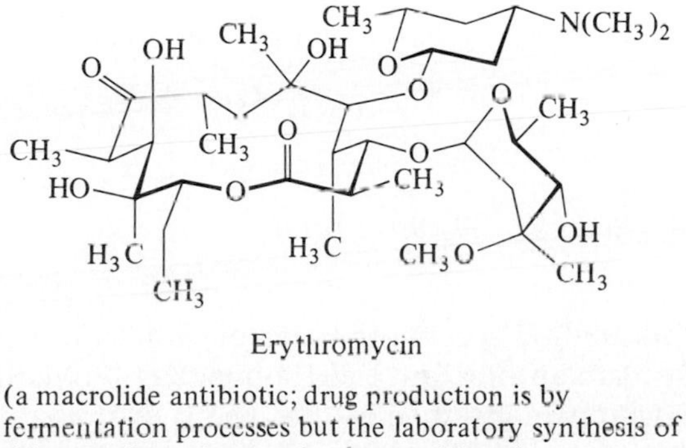

Erythromycin

(a macrolide antibiotic; drug production is by
fermentation processes but the laboratory synthesis of
this complex substance has been achieved)

FIGURE 20.7 Some naturally occurring lactones.

SAMPLE SOLUTION (a) The ring oxygen of the lactone is derived from the hydroxyl group of the hydroxy acid, while the carbonyl group corresponds to that of the carboxyl function. To identify the hydroxy acid, disconnect the O—C(O) bond of the ester.

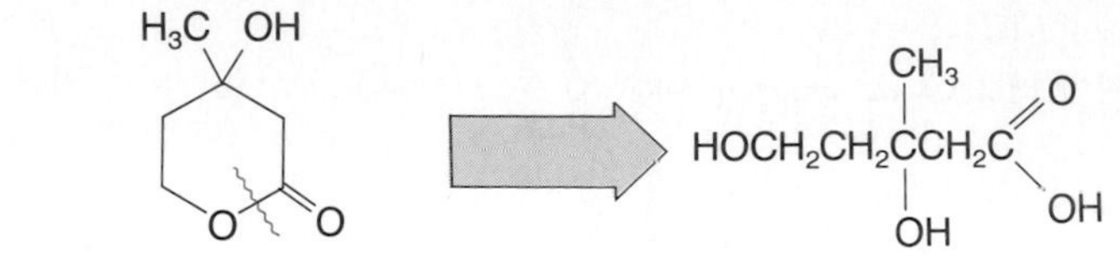

Mevalonolactone
(disconnect bond indicated)

Mevalonic acid

Lactones whose rings are three- or four-membered (α-lactones and β-lactones) are very reactive, making their isolation difficult. Special methods are normally required for the laboratory synthesis of small-ring lactones as well as those that contain rings larger than six-membered.

20.16 α-HALOGENATION OF CARBOXYLIC ACIDS. THE HELL-VOLHARD-ZELINSKY REACTION

Esterification of carboxylic acids involves nucleophilic addition to the carbonyl group as a key step. In this respect the carbonyl group of a carboxylic acid resembles that of an aldehyde or ketone. Do carboxylic acids resemble aldehydes and ketones in other ways? Do they, for example, form enols, and can they be halogenated at their α carbon atom via an enol in the way that aldehydes and ketones can?

The enol content of carboxylic acids is far less than that of aldehydes and ketones, so introduction of a halogen substituent at the α carbon atom requires a slightly different set of reaction conditions. Bromination is the reaction that is normally carried out and the usual procedure involves treatment of the carboxylic acid with bromine in the presence of a small amount of phosphorus trichloride as a catalyst.

| Carboxylic acid | Bromine | α-Bromo carboxylic acid | Hydrogen bromide |

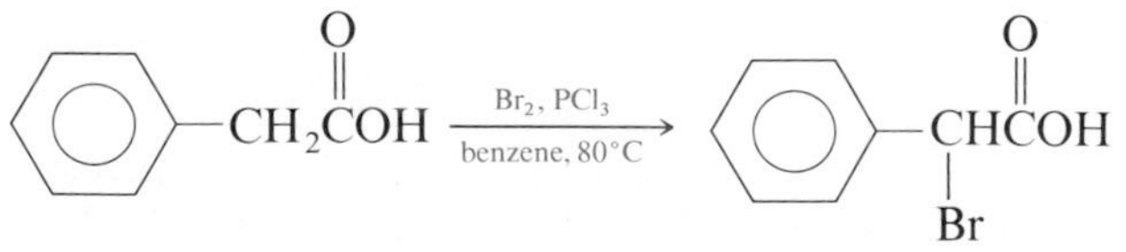

Phenylacetic acid

α-Bromophenylacetic acid
(60–62%)

This method of α bromination of carboxylic acids is called the *Hell-Volhard-Zelinsky* reaction after Carl Hell (Stuttgart), J. Volhard, and Nicolaus Zelinsky (Göttingen), who demonstrated its use. The Hell-Volhard-Zelinsky reaction is sometimes carried out by using a small amount of phosphorus instead of phosphorus trichloride as the catalyst. Phosphorus reacts with bromine to yield phosphorus tribromide as the active catalyst under these conditions.

Figure 20.8 presents the generally accepted mechanism for the Hell-Volhard-Zelinsky reaction. It is based on the observation that the enol content of an acyl chloride is substantially greater than that of its corresponding carboxylic acid. In step 1, the phosphorus trichloride catalyst converts a small amount of the carboxylic acid to an acyl chloride. Enolization of the acyl chloride in step 2 gives an intermediate capable of reacting rapidly with bromine in step 3. Only catalytic amounts of phosphorus trichloride are needed because the exchange reaction of step 4 not only converts the α-bromo acyl chloride to the desired α-bromo acid but produces a second molecule

Overall Reaction:

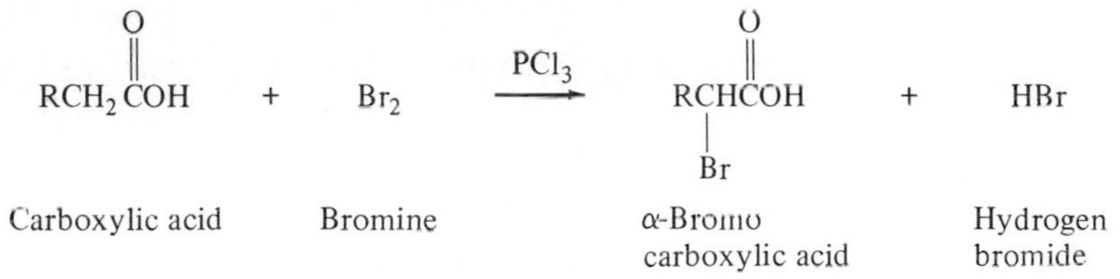

Carboxylic acid		Bromine			α-Bromo carboxylic acid		Hydrogen bromide

Step 1: The phosphorus trichloride catalyst converts a molecule of the carboxylic acid to the corresponding acyl chloride.

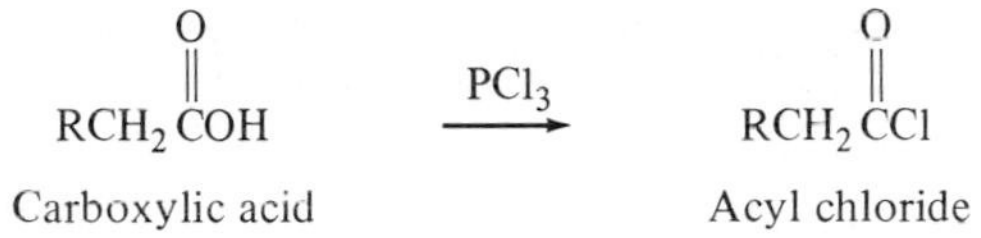

Carboxylic acid   Acyl chloride

Step 2: The acyl chloride undergoes enolization.

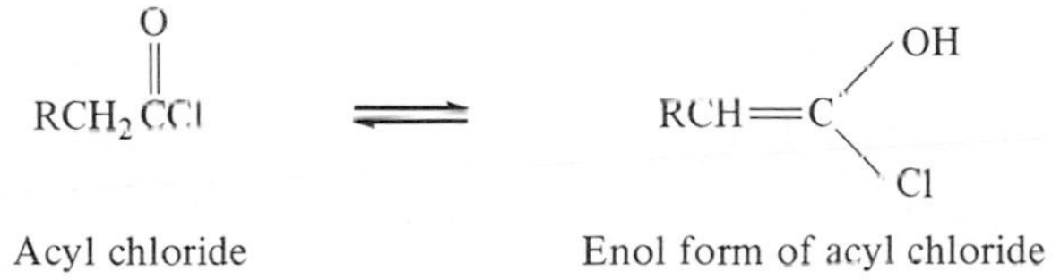

Acyl chloride   Enol form of acyl chloride

Step 3: The enol form of the acyl chloride reacts with bromine.

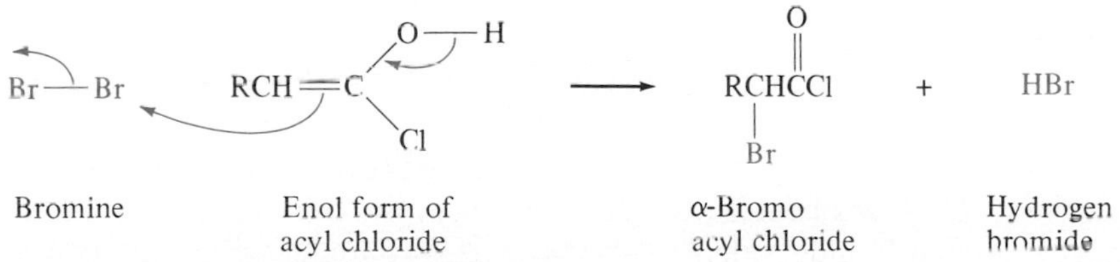

Bromine	Enol form of acyl chloride	α-Bromo acyl chloride	Hydrogen bromide

Step 4: Exchange of groups between α-bromo acyl chloride and carboxylic acid. This step produces the observed reaction product, the α bromo acid, and gives a molecule of acyl chloride that can reenter the sequence at step 2.

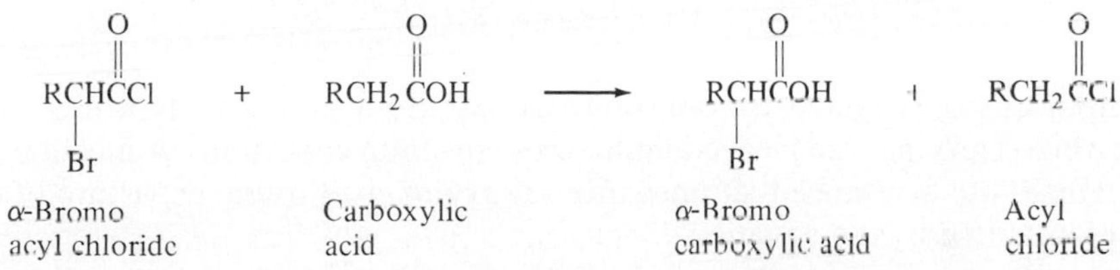

α-Bromo acyl chloride	Carboxylic acid	α-Bromo carboxylic acid	Acyl chloride

FIGURE 20.8 Sequence of steps that describes the mechanism of the α bromination of a carboxylic acid by the Hell-Volhard-Zelinsky reaction.

of acyl chloride, which then goes through the same sequence of steps. Eventually, all the carboxylic acid is transformed to its α-bromo derivative.

The Hell-Volhard-Zelinsky reaction is of synthetic value in that the α halogen can be displaced by nucleophilic substitution.

$$CH_3CH_2CH_2CO_2H \xrightarrow[P]{Br_2} CH_3CH_2\underset{\underset{Br}{|}}{C}HCO_2H \xrightarrow[H_2O,\ heat]{K_2CO_3} CH_3CH_2\underset{\underset{OH}{|}}{C}HCO_2H$$

| Butanoic acid | 2-Bromobutanoic acid (77%) | 2-Hydroxybutanoic acid (69%) |

A standard method for the preparation of an α-amino acid uses α-bromo carboxylic acids as the substrate and aqueous ammonia as the nucleophile.

$$(CH_3)_2CHCH_2CO_2H \xrightarrow[PCl_3]{Br_2} (CH_3)_2CH\underset{\underset{Br}{|}}{C}HCO_2H \xrightarrow[H_2O]{NH_3} (CH_3)_2CH\underset{\underset{NH_2}{|}}{C}HCO_2H$$

| 3-Methylbutanoic acid | 2-Bromo-3-methylbutanoic acid (88%) | 2-Amino-3-methylbutanoic acid (48%) |

PROBLEM 20.10 α-Iodo acids are not normally prepared by direct iodination of carboxylic acids under conditions of the Hell-Volhard-Zelinsky reaction. Show how you could convert octadecanoic acid to its 2-iodo derivative by an efficient sequence of reactions.

20.17 DECARBOXYLATION OF MALONIC ACID AND RELATED COMPOUNDS

The loss of a molecule of carbon dioxide from a carboxylic acid is known as a *decarboxylation* reaction.

$$RCO_2H \longrightarrow RH + CO_2$$

| Carboxylic acid | Alkane | Carbon dioxide |

Decarboxylation of simple carboxylic acids takes place with great difficulty and is rarely encountered.

Compounds that do undergo ready thermal decarboxylation include those related to malonic acid. On being heated above its melting point, malonic acid is converted to acetic acid and carbon dioxide.

$$HO_2CCH_2CO_2H \xrightarrow{150°C} CH_3CO_2H + CO_2$$

| Malonic acid (propanedioic acid) | Acetic acid (ethanoic acid) | Carbon dioxide |

It is important to recognize that only one carboxyl group is lost in this process. The second carboxyl group is not cleaved under these reaction conditions. A mechanism that recognizes the assistance that one carboxyl group gives to the departure of the other is represented by the equation

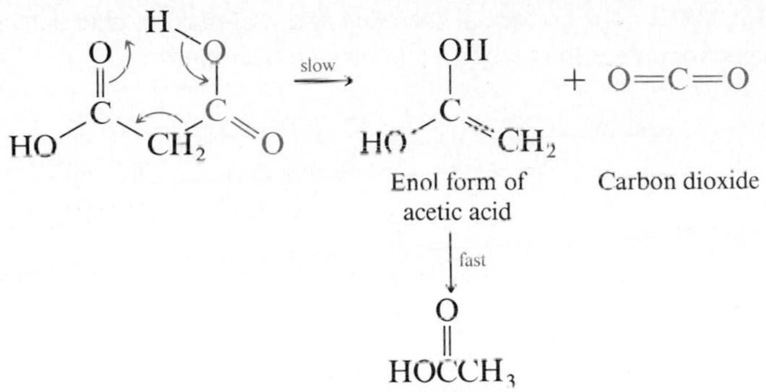

The activated complex involves the carbonyl oxygen of one carboxyl group—the one that stays behind—acting as a proton acceptor toward the hydroxyl group of the carboxyl that is lost. Carbon-carbon bond cleavage leads to the enol form of acetic acid, along with an equivalent amount of carbon dioxide.

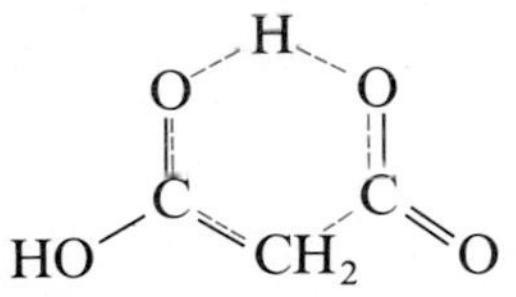

Representation of activated complex in thermal decarboxylation of malonic acid

The enol intermediate is subsequently transformed to acetic acid by proton transfer processes.

The protons attached to C-2 of malonic acid are not directly involved in the process and so may be replaced by other substituents without much effect on the ease of decarboxylation. Analogs of malonic acid substituted at C-2 undergo efficient thermal decarboxylation.

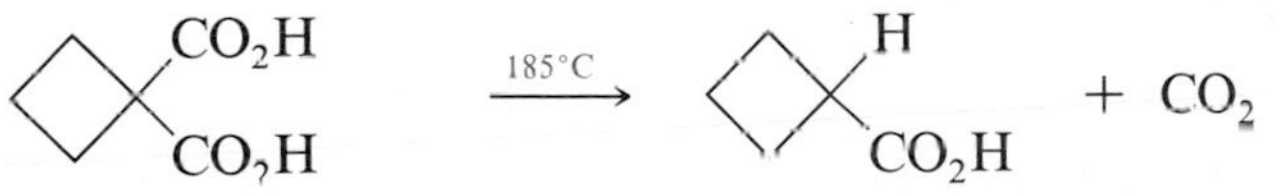

1,1-Cyclobutanedicarboxylic acid $\xrightarrow{185°C}$ Cyclobutanecarboxylic acid (74%) + Carbon dioxide

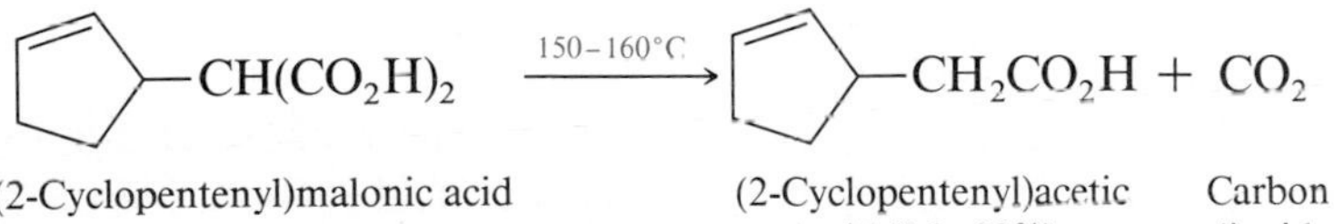

2-(2-Cyclopentenyl)malonic acid $\xrightarrow{150-160°C}$ (2-Cyclopentenyl)acetic acid (96–99%) + Carbon dioxide

---

PROBLEM 20.11 What will be the product isolated after thermal decarboxylation of each of the following? Using curved arrows in the manner shown above, represent the bond changes that take place at the transition state.

(a) $(CH_3)_2C(CO_2H)_2$

(b) $CH_3(CH_2)_6CHCO_2H$
       $\quad\quad\quad\quad\quad CO_2H$

(c) CO_2H
       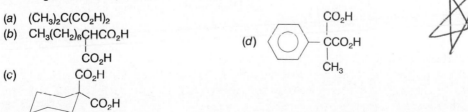

(d) CO_2H
       $\quad -CCO_2H$
       $\quad\quad CH_3$

SAMPLE SOLUTION (a) Thermal decarboxylation of malonic acid derivatives leads to the replacement of one of the carboxyl groups by a hydrogen.

$$(CH_3)_2C(CO_2H)_2 \xrightarrow{\text{heat}} (CH_3)_2CHCO_2H + CO_2$$

| 2,2-Dimethylmalonic acid | 2-Methylpropanoic acid | Carbon dioxide |

The activated complex incorporates a cyclic array of six atoms:

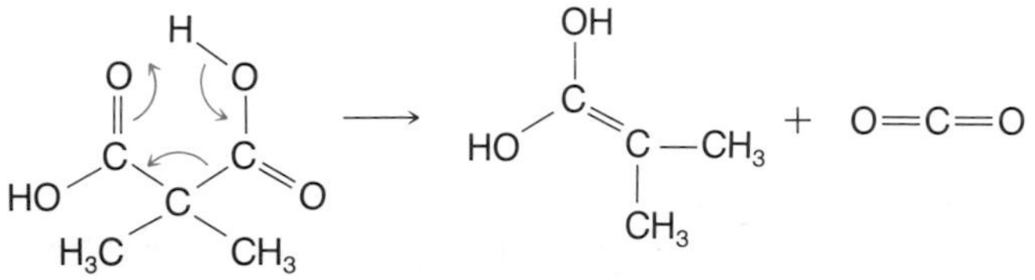

| 2,2-Dimethylmalonic acid | Enol form of 2-methylpropanoic acid | Carbon dioxide |

Tautomerization of the enol form to 2-methylpropanoic acid completes the process.

The thermal decarboxylation of malonic acid derivatives is the last step in a multistep synthesis of carboxylic acids known as the *malonic ester synthesis.* This synthetic method will be described in Section 22.7.

Notice that the carboxyl group that stays behind during the decarboxylation of malonic acid has a hydroxyl function that is not directly involved in the process. Compounds that have substituents other than hydroxyl groups at this position would be expected to undergo an analogous decarboxylation.

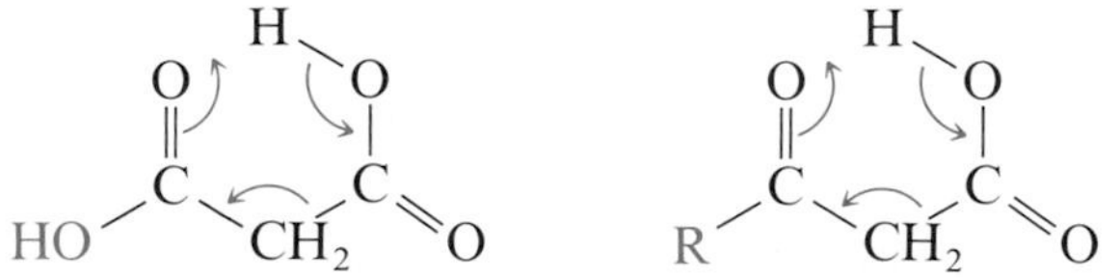

Bonding changes during decarboxylation of malonic acid

Bonding changes during decarboxylation of a β-keto acid

The compounds most frequently encountered in this reaction are β-keto acids, that is, carboxylic acids in which the β carbon is a carbonyl function. Decarboxylation of β-keto acids leads to ketones.

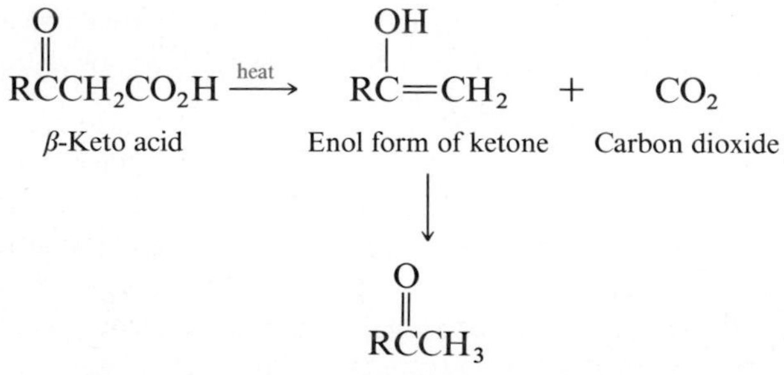

$$\underset{\text{β-Keto acid}}{R\overset{O}{\overset{\|}{C}}CH_2CO_2H} \xrightarrow{\text{heat}} \underset{\text{Enol form of ketone}}{R\overset{OH}{\overset{|}{C}}=CH_2} + \underset{\text{Carbon dioxide}}{CO_2}$$

$$\downarrow$$

$$\underset{\text{Ketone}}{R\overset{O}{\overset{\|}{C}}CH_3}$$

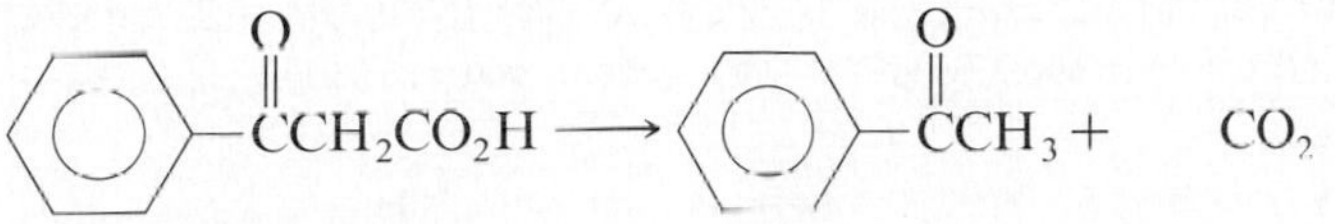

Benzoylacetic acid   Acetophenone   Carbon dioxide

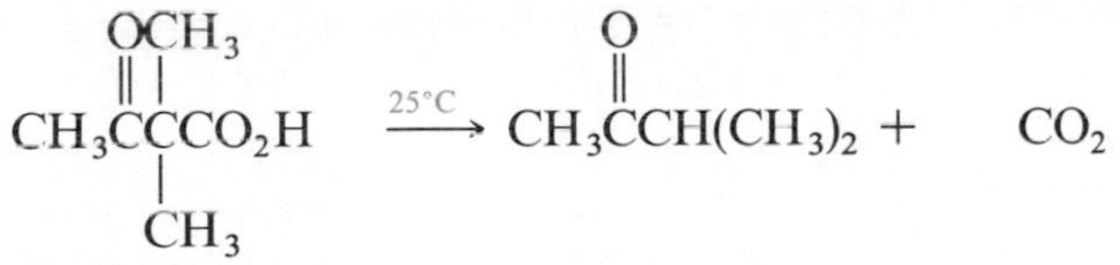

2,2-Dimethylacetoacetic   3-Methyl-2-butanone   Carbon dioxide
acid

PROBLEM 20.12   Show the bonding changes that occur and write the structure of the intermediate formed in the thermal decarboxylation of:

(a)   Benzoylacetic acid
(b)   2,2-Dimethylacetoacetic acid

SAMPLE SOLUTION   (a)   By analogy to the thermal decarboxylation of malonic acid, we represent the corresponding reaction of benzoylacetic acid as:

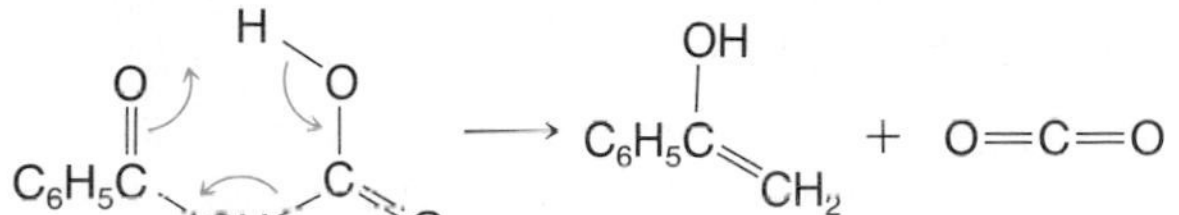

Benzoylacetic acid   Enol form of   Carbon dioxide
                                acetophenone

Acetophenone is the isolated product; it is formed from its enol by proton transfer reactions.

The thermal decarboxylation of β-keto acids is the last step in a ketone synthesis known as the *acetoacetic ester synthesis*. The acetoacetic ester synthesis is discussed in Section 22.6.

20.18 SPECTROSCOPIC PROPERTIES OF CARBOXYLIC ACIDS

The infrared and proton nuclear magnetic resonance spectra of a representative carboxylic acid, 2-(p-chlorophenyl)propanoic acid, are shown in Figure 20.9. The hydroxyl absorptions overlap with the C—H stretching frequencies to produce a broad absorption in the 3500 to 2500 cm^{-1} region of the infrared spectrum. The carbonyl group gives rise to a strong band at 1700 cm^{-1}. In general, the position of the carbonyl absorption of carboxylic acids is found in the range 1650 to 740 cm^{-1}.

The hydroxyl proton of a carboxyl group is normally the least shielded of all the protons of an nmr spectrum. Carboxyl protons appear 10 to 12 ppm downfield from tetramethylsilane (TMS), often as a broad peak. Since the customary sweep width of a ^{1}H nmr spectrum is 10 ppm, the spectrum must be offset as shown in Figure 20.9 in order to display this signal on the chart paper. The chemical shift of the carboxyl

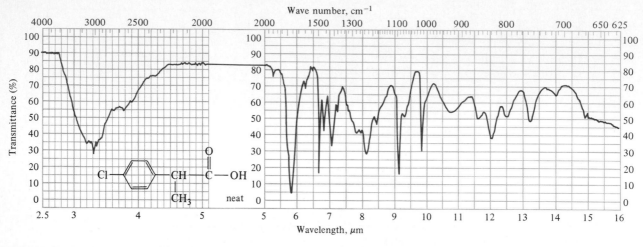

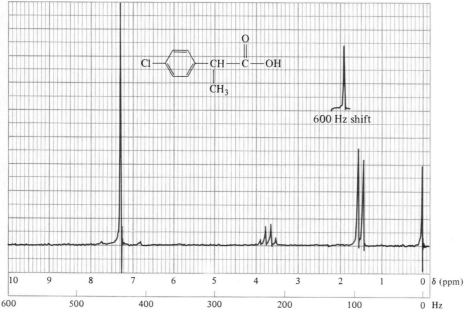

FIGURE 20.9 The infrared (top) and ¹H nmr (bottom) spectra of 2-(*p*-chlorophenyl)propanoic acid.

proton is 11.9 ppm and is equal to the sum of the offset (600 Hz = 10 ppm at 60 MHz) and the observed chart position (1.9 ppm). As with other acidic protons, the carboxyl proton can be identified by adding D_2O to the sample. Hydrogen-deuterium exchange converts $-CO_2H$ to $-CO_2D$, and the signal corresponding to the carboxyl proton disappears.

Carboxylic acids resemble aldehydes and ketones in their ¹³C nmr spectra. The chemical shift of the carbonyl carbon typically appears about 190 ppm downfield from TMS.

20.19 SUMMARY

Carboxylic acids are compounds of the type $R\overset{O}{\overset{\|}{C}}OH$, and it is the carboxyl group that is responsible for their characteristic properties. They are weak acids, which dissociate to a modest extent in water but are deprotonated quantitatively by bicarbonate. The dissociation constants K_a of most carboxylic acids are in the 10^{-4} to 10^{-5} range. The greater acidity of carboxylic acids compared with alcohols results from efficient electron delocalization in the carboxylate anion. The negative charge is shared by both oxygens of the anion.

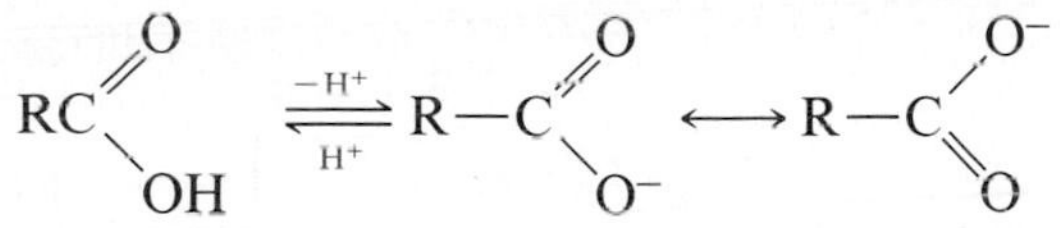

Carboxylic acid   Resonance description of electron delocalization in carboxylate anion

Electron-withdrawing substituents, especially when located close to the carboxyl group, increase the acidity of carboxylic acids. Thus, trifluoroacetic acid and 2,4,6-trinitrobenzoic acid are rather strong acids.

CF_3CO_2H

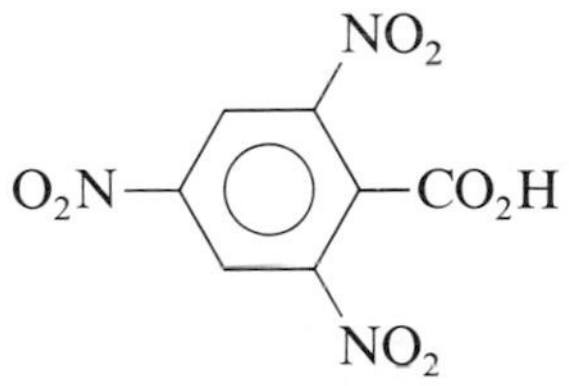

Trifluoroacetic acid
$K_a = 5.9 \times 10^{-1}$
$(pK_a = 0.2)$

2,4,6-Trinitrobenzoic acid
$K_a - 2.2 \times 10^{-1}$
$(pK_a = 0.6)$

Several reactions that have been encountered earlier lead to carboxylic acids and can be used for their synthesis. These are summarized in Table 20.5. Two new methods were introduced in this chapter and are particularly useful:

1. Carboxylation of Grignard reagents (Section 20.11)
2. Preparation and hydrolysis of nitriles (Section 20.12)

Grignard reagents are nucleophilic and add to carbon dioxide to yield carboxylic acids after acidification.

$$RX \xrightarrow[\text{diethyl ether}]{\text{Mg}} RMgX \xrightarrow{CO_2} RCOMgX \xrightarrow[\text{H}^+]{\text{H}_2\text{O}} RCO_2H$$

Alkyl halide   Grignard reagent   Halomagnesium carboxylate   Carboxylic acid

Primary and secondary alkyl halides undergo nucleophilic substitution by cyanide ion to form nitriles. Nitriles are converted to carboxylic acids by hydrolysis.

$$RX \xrightarrow{CN^-} RCN \xrightarrow[\text{H}^+]{\text{H}_2\text{O}} RCO_2H$$

Primary or secondary alkyl halide   Nitrile   Carboxylic acid

Carboxylic acids can be transformed to acyl chlorides with thionyl chloride, to primary alcohols by reduction with lithium aluminum hydride, and to esters by acid-catalyzed condensation with alcohols (Table 20.6). The mechanism of acid-catalyzed esterification involves some key elements that are important to understanding the chemistry of derivatives of carboxylic acids.

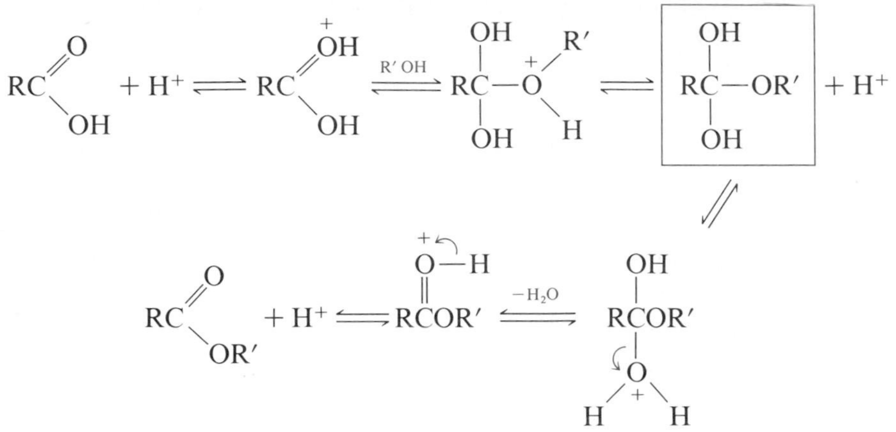

Protonation of the carbonyl oxygen activates the carbonyl group toward nucleophilic addition. Addition of methanol gives a tetrahedral intermediate (shown in the box in the above equation), which has the capacity to revert to starting materials or to undergo dehydration to yield an ester. It will be seen in the next chapter that esters can be hydrolyzed to carboxylic acids and alcohols under conditions of acid catalysis. The mechanism of that reaction is precisely the reverse of the one shown above.

An intramolecular analog of the esterification reaction can occur when a molecule contains both a hydroxyl and a carboxyl group. Cyclic esters are called *lactones* and are most stable when the ring is five- or six-membered (Section 20.15).

Halogenation at the α carbon atom of carboxylic acids can be accomplished by the *Hell-Volhard-Zelinsky reaction* (Section 20.16). An acid is treated with chlorine or bromine in the presence of a catalytic quantity of phosphorus or a phosphorus trihalide:

$$\underset{\text{Carboxylic acid}}{R_2CHCO_2H} + \underset{\text{Halogen}}{X_2} \xrightarrow{\text{P or PX}_3} \underset{\alpha\text{-Halo acid}}{R_2C(X)CO_2H} + HY$$

This reaction involves the enol form of the carboxylic acid chloride or bromide as an intermediate. It is of synthetic value in that α-halo acids are reactive substrates in nucleophilic substitution reactions.

1,1-Dicarboxylic acids and β-keto acids are subject to ready thermal decarboxylation by a mechanism in which a β carbonyl group assists the departure of carbon dioxide (Section 20.17).

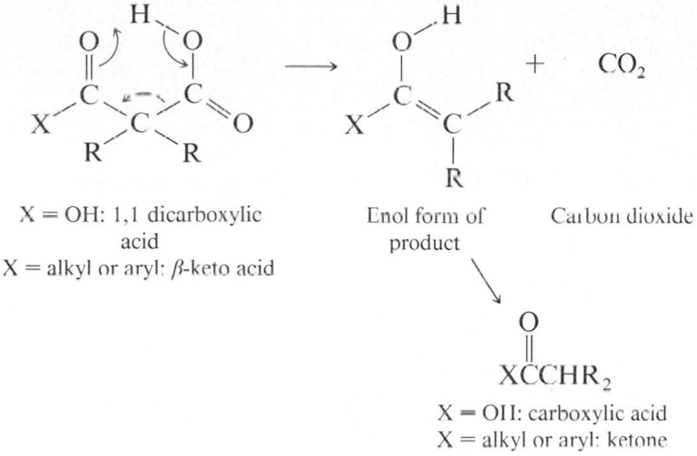

X = OH: 1,1 dicarboxylic acid
X = alkyl or aryl: β-keto acid

Enol form of product   Carbon dioxide

X = OH: carboxylic acid
X = alkyl or aryl: ketone

PROBLEMS

20.13 Many carboxylic acids are much better known by their common names than by their systematic names. Some of these are given below. Provide a structural formula for each one on the basis of its systematic name.

(a) 2-Hydroxypropanoic acid (better known as *lactic acid,* it is found in sour milk and is formed in the muscles as a waste product during exercise).

(b) 2-Hydroxy-2-phenylethanoic acid (also known as *mandelic acid* and obtained from plums, peaches, and other fruits).

(c) Tetradecanoic acid (also known as *myristic acid,* it can be obtained from a variety of fats).

(d) 10-Undecenoic acid (also called *undecylenic acid,* It Is used, In combination with its zinc salt, to treat fungal infections such as athlete's foot).

(e) 3,5-Dihydroxy-3-methylpentanoic acid (also called *mevalonic acid,* it is an important intermediate in the biosynthesis of terpenes and steroids).

(f) (E)-2-Methyl-2-butenoic acid (also known as *tiglic acid,* it is a constituent of various natural oils).

(g) 2-Hydroxybutanedioic acid (also known as *malic acid* and found in apples and other fruits).

(h) 2-Hydroxy-1,2,3-propanetricarboxylic acid (better known as *citric acid,* it contributes to the tart taste of citrus fruits).

(i) 2-(p-Isobutylphenyl)propanoic acid (an anti-inflammatory drug better known as *ibuprofen*).

(j) o-Hydroxybenzenecarboxylic acid (better known as *salicylic acid* and obtained from willow bark).

20.14 Give an acceptable systematic IUPAC name for each of the following:

(a) $CH_3(CH_2)_6CO_2H$

(b) $CH_3(CH_2)_6CO_2K$

(c) $CH_2{=}CH(CH_2)_5CO_2H$

(d)

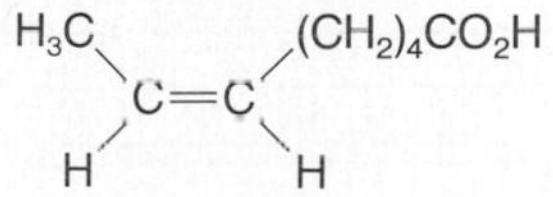

$$H_3C\diagdown\diagup(CH_2)_4CO_2H$$
$$C{=}C$$
$$H\diagup\diagdown H$$

(e) $HO_2C(CH_2)_6CO_2H$
(f) 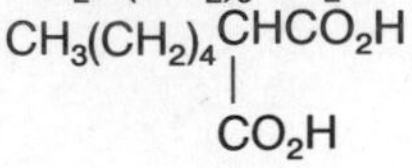 $CH_3(CH_2)_4\underset{\underset{CO_2H}{|}}{CH}CO_2H$

(g)

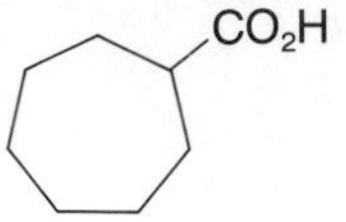

(h)

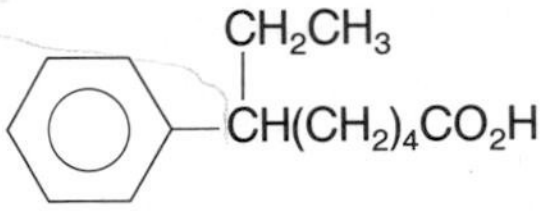

20.15 Rank the compounds in each group below in order of decreasing acidity:

(a) Acetic acid, ethane, ethanol
(b) Benzene, benzoic acid, benzyl alcohol
(c) Propanedial, 1,3-propanediol, propanedioic acid, propanoic acid
(d) Acetic acid, ethanol, trifluoroacetic acid, 2,2,2-trifluoroethanol, trifluoromethanesulfonic acid (CF_3SO_2OH)
(e) Cyclopentanecarboxylic acid, 2,4-pentanedione, cyclopentanone, cyclopentene.

20.16 In spite of the fact that entropy effects on the ionization of carboxylic acids are larger than enthalpy effects, reliable predictions concerning relative acidities can usually be made by considering how well substituents stabilize carboxylate anions by electron withdrawal. Using this procedure, identify the more acidic compound in each of the following pairs.

(a) $CF_3CH_2CO_2H$ or $CF_3CH_2CH_2CO_2H$
(b) $CH_3CH_2CH_2CO_2H$ or $CH_3C\equiv CCO_2H$

(c)

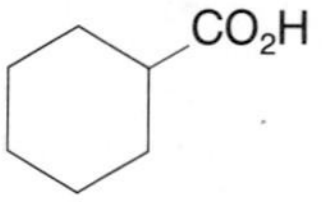

or

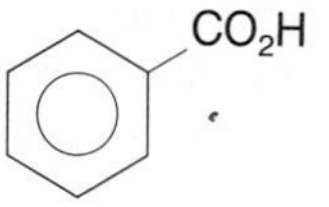

(d)

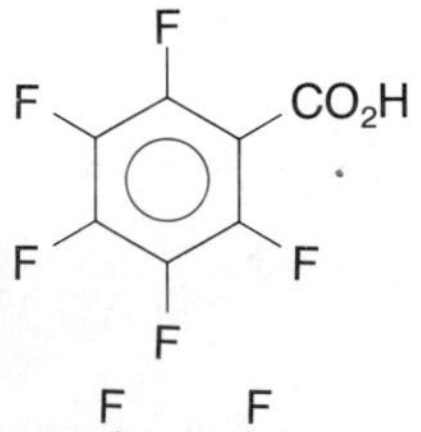

or

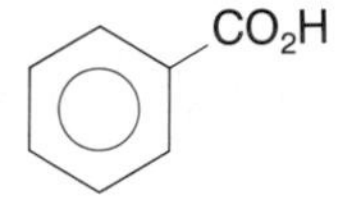

(e)

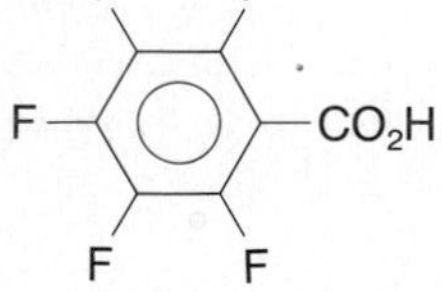

or

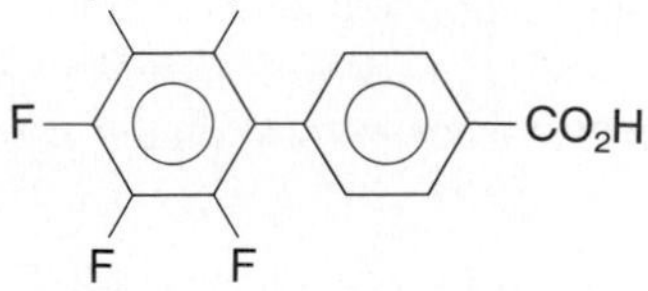

(f)

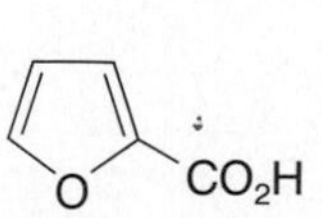

or

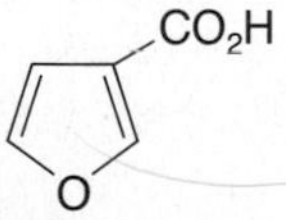

(g)

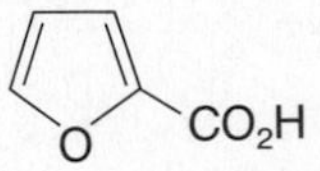

or

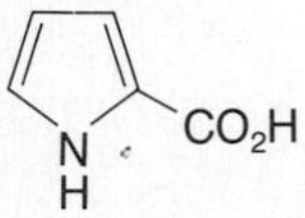

20.17 Propose methods for preparing butanoic acid from each of the following.

(a) 1-Butanol
(b) Butanal
(c) 1-Butene
(d) 1-Propanol
(e) 2-Propanol

(f) 4-Octene
(g) Acetaldehyde
(h) Ethylpropanedioic acid
      (ethylmalonic acid)

20.18 It is sometimes necessary to prepare isotopically labeled samples of organic substances for probing biological transformations and reaction mechanisms. Various sources of the radioactive mass 14 carbon isotope are available. Describe synthetic procedures by which benzoic acid, labeled with ^{14}C at its carbonyl carbon, could be prepared from benzene and the ^{14}C-labeled precursors listed below. You may use any necessary organic or inorganic reagents. (In the formulas shown, an asterisk indicates ^{14}C.)

Since most halogenated Alkanes are prepared by Alcohols   easier by Grignard

(a) $\overset{*}{C}H_3Cl$

(b) $H\overset{*}{C}H$ (with O double bond)
      $\overset{O}{\|}$

(c) $CH_3\overset{*}{C}O_2H$

(d) $\overset{*}{C}O_2$

20.19 Give the product of the reaction of pentanoic acid with each of the following reagents:

(a) Sodium hydroxide
(b) Sodium bicarbonate
(c) Thionyl chloride
(d) Phosphorus tribromide
(e) Benzyl alcohol, sulfuric acid (catalytic amount)
(f) Chlorine, phosphorus tribromide (catalytic amount)
(g) Bromine, phosphorus trichloride (catalytic amount)
(h) Product of (g) treated with sodium iodide in acetone
(i) Product of (g) treated with aqueous ammonia
(j) Lithium aluminum hydride, then hydrolysis
(k) Phenylmagnesium bromide

20.20 Show how butanoic acid may be converted to each of the following compounds:

(a) 1-Butanol
(b) Butanoyl chloride
(c) Butanal
(d) 1-Chlorobutane
(e) Phenyl propyl ketone (1-phenyl-1-butanone)
(f) 4-Octanone
(g) 2-Bromobutanoic acid
(h) 2-Butenoic acid

20.21 Show by a series of equations, using any necessary organic or inorganic reagents, how acetic acid can be converted to each of the following compounds:

(a) $H_2NCH_2CO_2H$
(b) $C_6H_5OCH_2CO_2H$
(c) $NCCH_2CO_2H$
(d) $HO_2CCH_2CO_2H$

(e) ICH_2CO_2H
(f) $BrCH_2CO_2CH_2CH_3$
(g) $(C_6H_5)_3\overset{+}{P}-\overset{..}{\overset{-}{C}}HCO_2CH_2CH_3$
(h) $C_6H_5CH=CHCO_2CH_2CH_3$

20.22 Each of the following reactions has been reported in the chemical literature and gives a single product in good yield. What is the product in each reaction?

(a)
$$H_3C, H$$
$$C=C$$
$$H, CH$$
$$\parallel$$
$$O$$
1. Ag_2O, water
2. H^+

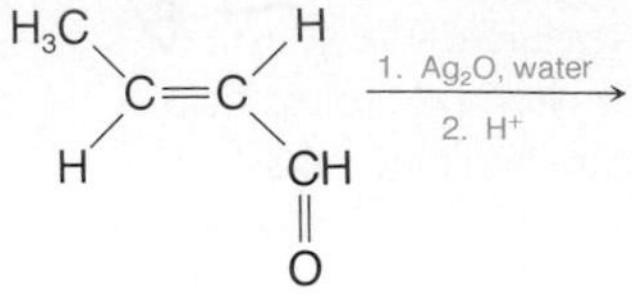

(b)
OCH₃
F
1. $KMnO_4$, H_2O
2. H^+
$$O=CCH_3$$

(c)
$$H_3C, CH_3$$
$$C=C$$
$$H, CO_2H$$
ethanol, H_2SO_4

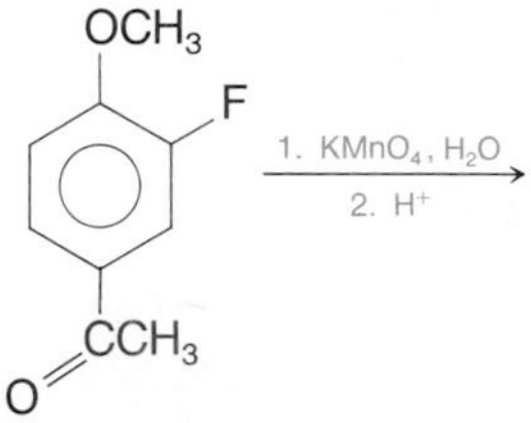

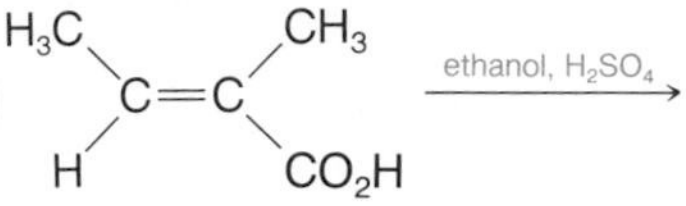

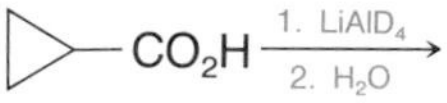

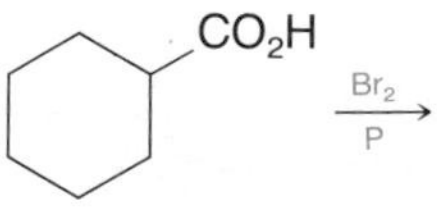

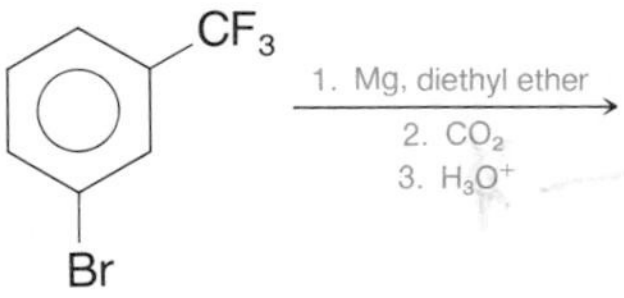

(d) ▷—CO_2H
1. $LiAlD_4$
2. H_2O

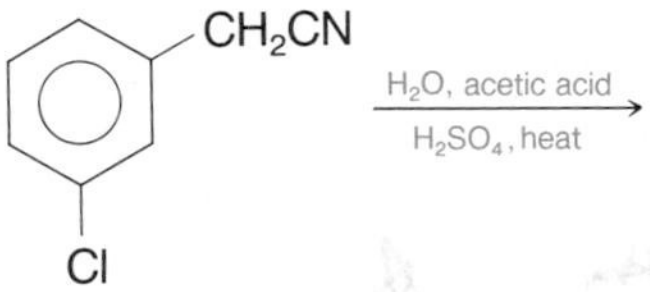

(e)
CO_2H (cyclohexane)
Br_2
P

(f)
CF_3
1. Mg, diethyl ether
2. CO_2
3. H_3O^+
Br

(g)
CH_2CN
H_2O, acetic acid
H_2SO_4, heat
Cl

(h)
Cl
$$O$$
$$\parallel$$
$$CCH_3$$
1. Cl_2, HO^-
2. H^+
Cl   Cl

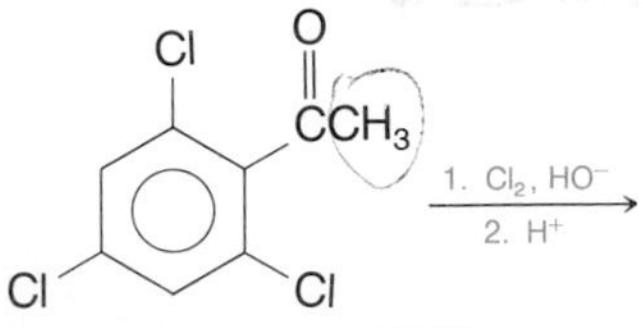

(i) $CH_3C\equiv CCO_2H$ $\xrightarrow{CH_3OH}{H_2SO_4}$

(j) $CH_2=CH(CH_2)_8CO_2H$ $\xrightarrow{HBr}{benzoyl\ peroxide}$

(k)
CO_2H
CO_2H
1. $KMnO_4$, HO^-
2. H^+

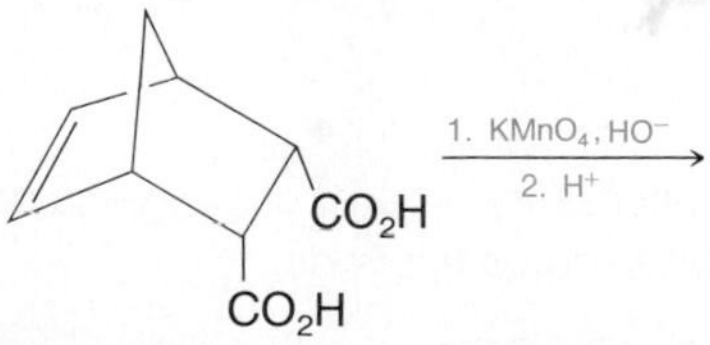

20.23 Show by a series of equations how you could synthesize each of the following compounds from the indicated starting material and any necessary organic or inorganic reagents:

(a) 2-Methylpropanoic acid from *tert*-butyl alcohol
(b) 3-Methylbutanoic acid from *tert*-butyl alcohol

(c) $HO_2C(CH_2)_5CO_2H$ from $HO_2C(CH_2)_3CO_2H$

(d) 3-Phenyl-1-butanol from CH_3CHCH_2CN
$\qquad\qquad\qquad\qquad\qquad$ |
$\qquad\qquad\qquad\qquad\qquad C_6H_5$

(e)

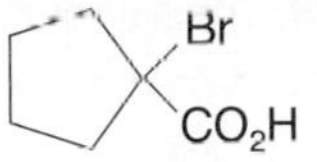

from cyclopentyl bromide

(f)

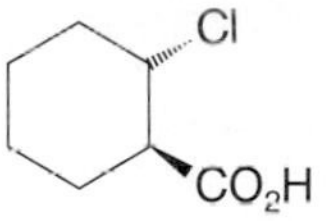

from (E)-$ClCH\!=\!CHCO_2H$

(g)

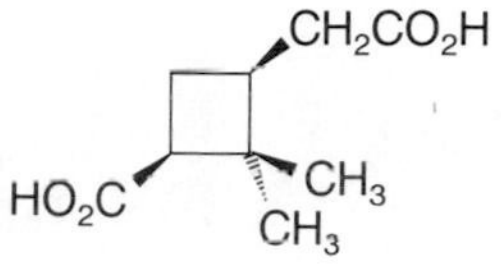

from

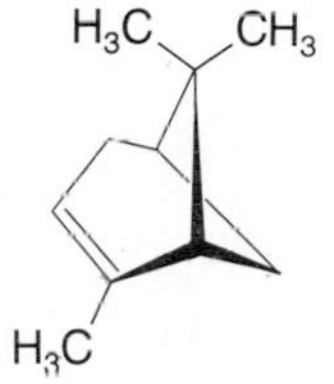

(h) 2,4-Dimethylbenzoic acid from m-xylene
(i) 4-Chloro-3-nitrobenzoic acid from p-chlorotoluene
(j) (Z)-$CH_3CH\!=\!CHCO_2H$ from propyne

(k) $(CH_3)_2C\!=\!CHCH_2OH$ from $(CH_3)_2C\!=\!CHCCH_3$ (with carbonyl $\overset{O}{\overset{\|}{C}}$)

20.24 Suggest reasonable explanations for each of the following observations:

(a) Both hydrogens are anti to each other in the most stable conformation of formic acid.
(b) Oxalic acid has a dipole moment of zero in the gas phase.
(c) The dissociation constant of o-hydroxybenzoic acid is greater (by a factor of 12) than that of o-methoxybenzoic acid.
(d) One of the diastereomers of 3-hydroxycyclohexanecarboxylic acid readily forms a lactone while the other does not.
(e) Ascorbic acid (vitamin C), while not a carboxylic acid, is sufficiently acidic to cause carbon dioxide liberation on being dissolved in aqueous sodium bicarbonate.

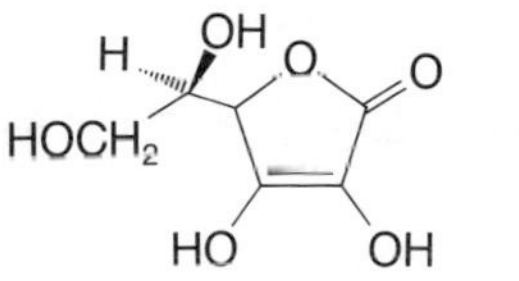

Ascorbic acid

20.25 When compound A is heated, two isomeric products are formed. What are these two products?

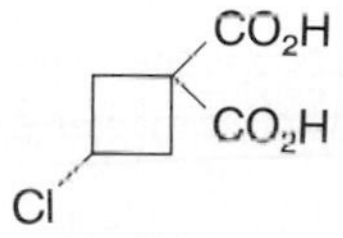

Compound A

20.26 A certain carboxylic acid ($C_{14}H_{26}O_2$), which can be isolated from whale blubber or sardine oil, yields nonanoic acid and pentanedioic acid on oxidation with potassium permanganate. What is the structure of this acid?

20.27 When levulinic acid ($CH_3\overset{\overset{\displaystyle O}{\|}}{C}CH_2CH_2CO_2H$) was hydrogenated at high pressure over a nickel catalyst at 220°C, a single product, compound B ($C_5H_8O_2$), was isolated in 94 percent yield. Compound B lacks hydroxyl absorption in its infrared spectrum and does not immediately liberate carbon dioxide on being shaken with sodium bicarbonate. What is a reasonable structure for compound B?

20.28 On standing in dilute aqueous acid, compound C is smoothly converted to mevalonolactone.

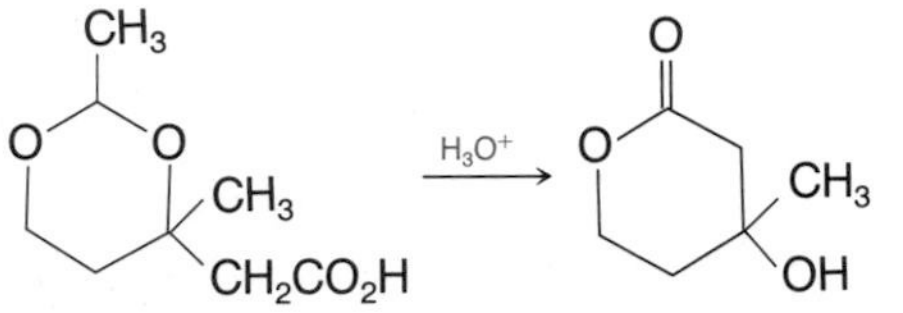

Compound C   Mevalonolactone

Suggest a reasonable mechanism for this reaction. What other organic product is also formed?

20.29 Suggest reaction conditions suitable for the preparation of compound D from 5-hydroxy-2-hexynoic acid.

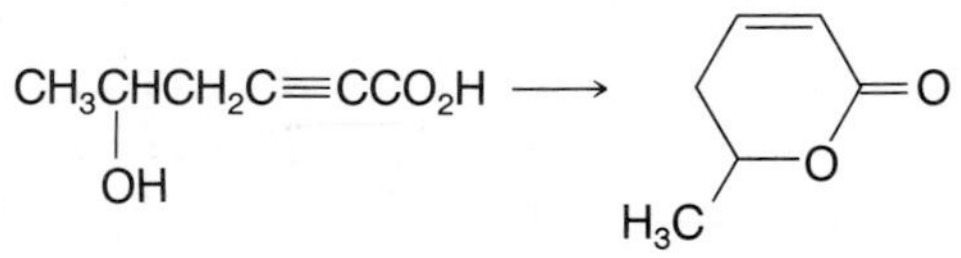

5-Hydroxy-2-hexynoic acid   Compound D

20.30 A method for converting one carboxylic acid to another that has one less carbon atom is known as the *Barbier-Wieland* degradation. It has been used to prepare pentadecanoic acid from hexadecanoic acid as shown below. Identify compounds E through H in this sequence.

$$CH_3(CH_2)_{14}CO_2H \xrightarrow[\text{H}^+]{CH_3CH_2OH} \text{compound E} \xrightarrow[\text{2. H}_3\text{O}^+]{\substack{\text{1. C}_6\text{H}_5\text{MgBr,}\\ \text{diethyl ether}}} \text{compound F}$$

$$(C_{18}H_{36}O_2) \qquad\qquad (C_{28}H_{42}O)$$

$$\Big\downarrow \text{H}^+\text{, heat}$$

$$CH_3(CH_2)_{13}CO_2H + \text{compound H} \xleftarrow[\text{2. H}^+]{\text{1. KMnO}_4} \text{compound G}$$

$$(C_{13}H_{10}O) \qquad\qquad (C_{28}H_{40})$$

20.31 In the presence of the enzyme *aconitase,* the double bond of aconitic acid undergoes hydration. The reaction is reversible and the following equilibrium is established:

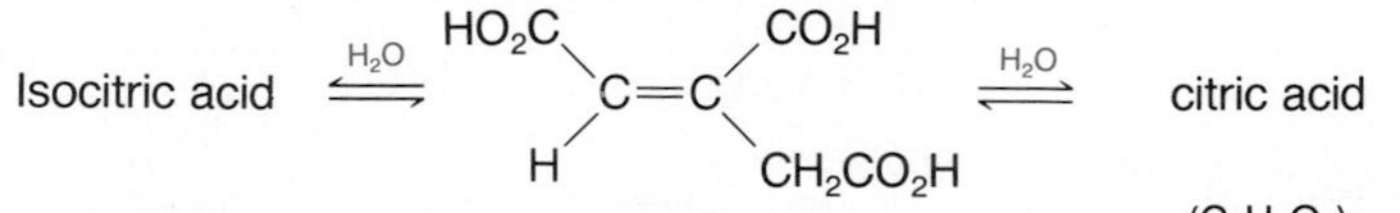

Isocitric acid   Aconitic acid   citric acid

$(C_6H_8O_7)$ $(C_6H_8O_7)$

(6% at equilibrium)   (4% at equilibrium)   (90% at equilibrium)

(a) The major tricarboxylic acid present is *citric acid*, the substance responsible for the tart taste of citrus fruits. Citric acid is achiral. What is its structure?

(b) What must be the constitution of isocitric acid? (Assume no rearrangements accompany hydration.) How many stereoisomers are possible for isocitric acid?

20.32 The ^1H nmr spectra of formic acid (HCO_2H), maleic acid (*cis*-$HO_2CCH{=}CHCO_2H$), and malonic acid ($HO_2CCH_2CO_2H$) are similar in that each is characterized by two singlets of equal intensity. Match these compounds with the designations I, J, and K on the basis of the appropriate ^1H nmr chemical shift data.

Compound I: signals at $\delta = 3.2$ and 12.1 ppm
Compound J: signals at $\delta = 6.3$ and 12.4 ppm
Compound K: signals at $\delta = 8.0$ and 11.4 ppm

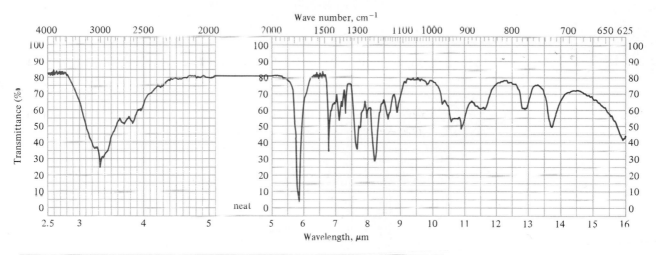

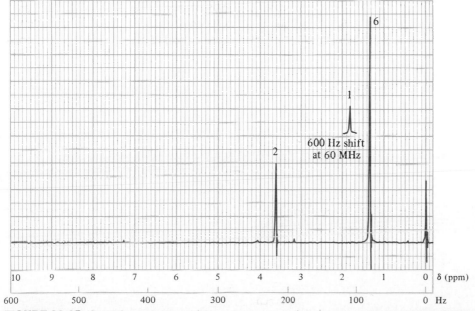

FIGURE 20.10 The infrared (top) and ^1H nmr (bottom) spectra of compound L ($C_5H_9ClO_2$) (Problem 20.33a).

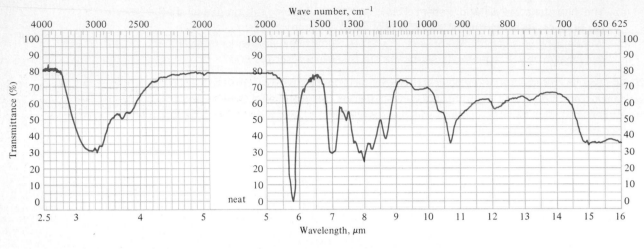

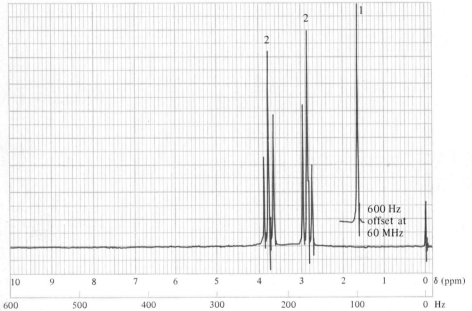

FIGURE 20.11 The infrared (top) and ^1H nmr (bottom) spectra of compound M ($C_3H_5ClO_2$) (Problem 20.33*b*).

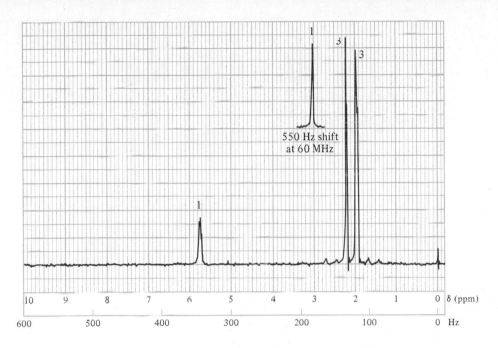

FIGURE 20.12 The ^{1}H nmr spectrum of compound N ($C_5H_8O_2$) (Problem 20.34a).

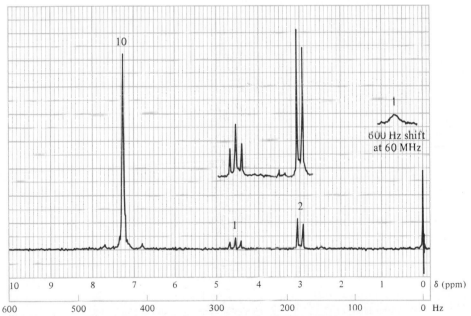

FIGURE 20.13 The ^{1}H nmr spectrum of compound O ($C_{15}H_{14}O_2$) (Problem 20.34b).

20.33 Identify compounds L and M on the basis of their infrared and ^{1}H nmr spectra.

(a) Compound L ($C_5H_9ClO_2$) (Figure 20.10)

(b) Compound M ($C_3H_5ClO_2$) (Figure 20.11)

20.34 Compounds N and O are carboxylic acids. Identify them on the basis of their ^{1}H nmr spectra.

(a) Compound N ($C_5H_8O_2$) (Figure 20.12)

(b) Compound O ($C_{15}H_{14}O_2$) (Figure 20.13). Compound O is achiral.

ACYL TRANSFER REACTIONS

This chapter differs from preceding ones in that it deals not with a single functional group but with several related classes of compounds. Included are:

$$\begin{matrix} & & O \\ & & \| \end{matrix}$$

1. Acyl chlorides, RCCl

$$\begin{matrix} & & O \quad O \\ & & \| \quad \| \end{matrix}$$

2. Carboxylic acid anhydrides, RCOCR

$$\begin{matrix} & & O \\ & & \| \end{matrix}$$

3. Esters of carboxylic acids, RCOR′

$$\begin{matrix} & & O \qquad\quad O \qquad\qquad\quad O \\ & & \| \qquad\quad \| \qquad\qquad\quad \| \end{matrix}$$

4. Carboxamides, RCNH$_2$, RCNHR′, and RCNR$_2'$

These classes of compounds are classified as *carboxylic acid derivatives,* since all may be converted to carboxylic acids by hydrolysis.

$$\begin{matrix} O & & & & O & & \\ \| & & & & \| & & \\ RCX & + H_2O & \longrightarrow & & RCOH & + & HX \end{matrix}$$

| Carboxylic acid derivative | Water | Carboxylic acid | Conjugate acid of leaving group |

The hydrolysis of a carboxylic acid derivative is an example of an *acyl transfer*

$$\begin{matrix} O \\ \| \end{matrix}$$

reaction. An acyl group RC is transferred from a carboxylic acid derivative to the oxygen of water. It is a *nucleophilic acyl substitution reaction.*

Acyl transfer reactions connect the various classes of carboxylic acid derivatives, with the preparation of one class often serving as a principal reaction of another. These reactions provide the basis for a large number of functional group transformations both in synthetic organic chemistry and in biological chemistry.

Also included in this chapter is a discussion of the chemistry of *nitriles*, compounds of the type RC≡N. Nitriles may be hydrolyzed to carboxylic acids or to amides and so are indirectly related to the other functional groups presented here.

21.1 NOMENCLATURE OF CARBOXYLIC ACID DERIVATIVES

Systematic names of acyl halides, acid anhydrides, esters, and amides are based on the name of the corresponding acid. *Acyl halides* are named by placing the name of the appropriate halide after that of the acyl group.

$$CH_3CCl \qquad CH_2{=}CHCH_2CCl \qquad F{-}C_6H_4{-}CBr$$

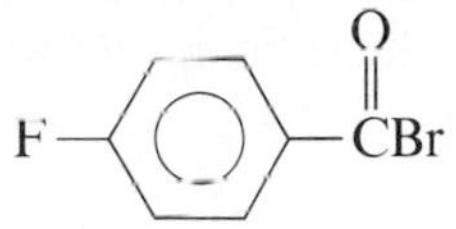

Acetyl chloride   3-Butenoyl   *p*-Fluorobenzoyl bromide
                               chloride

While acyl fluorides, bromides, and iodides are all known classes of organic compounds, they are encountered far less frequently than are acyl chlorides. Acyl chlorides will be the only acyl halides discussed in this chapter.

In naming *carboxylic acid anhydrides* in which both acyl groups are the same, one replaces the word "acid" by "anhydride." When the acyl groups are different, they are cited in alphabetical order.

$$CH_3COCCH_3 \qquad C_6H_5COCC_6H_5 \qquad C_6H_5COC(CH_2)_5CH_3$$

Acetic anhydride   Benzoic anhydride   Benzoic heptanoic anhydride

Cyclic anhydrides have their $-COC-$ unit as part of a ring. They are related to dicarboxylic acids and are named accordingly.

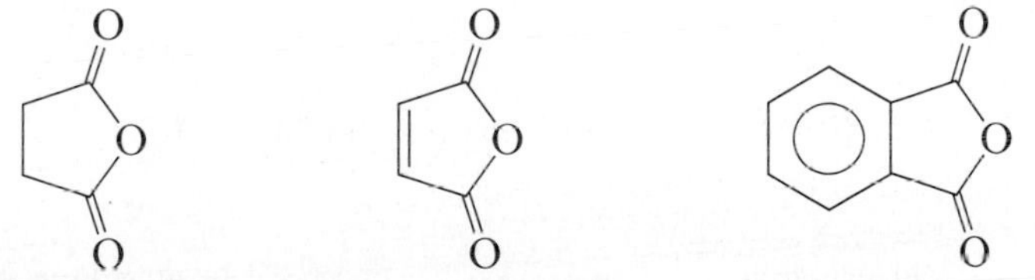

Succinic anhydride   Maleic anhydride   Phthalic anhydride

The alkyl group and the acyl group of an *ester* are specified independently. Esters are named as *alkyl alkanoates*. The alkyl group R′ of RCOR′ is cited first, followed by the acyl portion RC—. The acyl portion is named by substituting the suffix *-ate* for the *-ic* ending of the corresponding acid.

$$CH_3\overset{\overset{\displaystyle O}{\|}}{C}OCH_2CH_3 \qquad CH_3CH_2\overset{\overset{\displaystyle O}{\|}}{C}OCH_3 \qquad \text{⟨O⟩}-\overset{\overset{\displaystyle O}{\|}}{C}OCH_2CH_2Cl$$

Ethyl acetate   Methyl propanoate   2-Chloroethyl benzoate

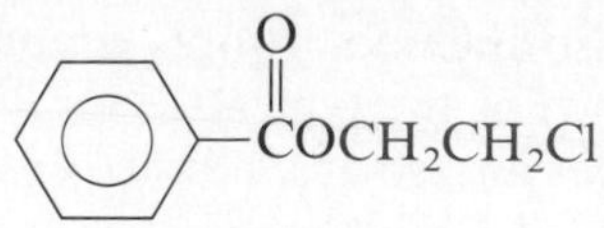

Aryl esters, that is, compounds of the type $R\overset{\overset{\displaystyle O}{\|}}{C}OAr$, are named in an analogous way.

The names of *amides* of the type $R\overset{\overset{\displaystyle O}{\|}}{C}NH_2$ are derived from carboxylic acids by replacing the suffixes *-oic acid* or *-ic acid* by *-amide.*

$$CH_3\overset{\overset{\displaystyle O}{\|}}{C}NH_2 \qquad C_6H_5\overset{\overset{\displaystyle O}{\|}}{C}NH_2 \qquad (CH_3)_2CHCH_2\overset{\overset{\displaystyle O}{\|}}{C}NH_2$$

Acetamide   Benzamide   3-Methylbutanamide

We name compounds of the type $R\overset{\overset{\displaystyle O}{\|}}{C}NHR'$ and $R\overset{\overset{\displaystyle O}{\|}}{C}NR'_2$ as *N*-alkyl and *N,N*-dialkyl-substituted derivatives of a parent amide.

$$CH_3\overset{\overset{\displaystyle O}{\|}}{C}NHCH_3 \qquad C_6H_5\overset{\overset{\displaystyle O}{\|}}{C}N(CH_2CH_3)_2 \qquad CH_3CH_2CH_2\overset{\overset{\displaystyle O}{\|}}{C}\underset{\underset{\displaystyle CH_3}{|}}{N}CH(CH_3)_2$$

N-Methylacetamide   *N,N*-Diethylbenzamide   *N*-Isopropyl-*N*-methyl butanamide

Modern systematic names for *nitriles* add the suffix *-nitrile* to the name of the parent hydrocarbon chain that includes the carbon of the cyano group. Nitriles may also be named by replacing the *-ic acid* or *-oic acid* ending of the corresponding carboxylic acid by *-onitrile.* Alternatively, they are sometimes given common names as alkyl cyanides.

$$CH_3C\equiv N \qquad C_6H_5C\equiv N \qquad CH_3\underset{\underset{\displaystyle C\equiv N}{|}}{C}HCH_3$$

Ethanenitrile   Benzonitrile   2-Methylpropanenitrile
(acetonitrile)   (isopropyl cyanide)

PROBLEM 21.1   Write a structural formula for each of the following compounds:

(a)   2-Phenylbutanoyl bromide
(b)   2-Phenylbutanoic anhydride
(c)   Butyl 2-phenylbutanoate
(d)   2-Phenylbutyl butanoate
(e)   2-Phenylbutanamide

(f)   N-Ethyl-2-phenylbutanamide

(g)   2-Phenylbutanenitrile

SAMPLE SOLUTION   (a)   A 2-phenylbutanoyl group is a four-carbon acyl unit that bears a phenyl substituent at C-2. When the name of an acyl group is followed by "halide," it designates an *acyl halide.*

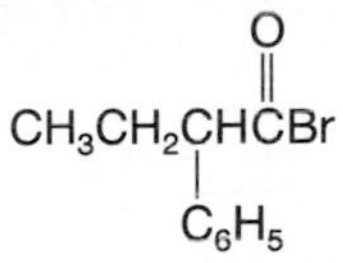

$$\underset{\underset{C_6H_5}{|}}{CH_3CH_2CHCBr} \quad \text{2-Phenylbutanoyl bromide}$$

21.2 STRUCTURE OF CARBOXYLIC ACID DERIVATIVES

In common with the other carbonyl-containing compounds that we have studied — aldehydes, ketones, and carboxylic acids — derivatives of carboxylic acids have a planar arrangement of bonds to their carbonyl group (Figure 21.1). An important structural feature of acyl chlorides, anhydrides, esters, and amides is that the atom attached to the acyl group bears an unshared pair of electrons that is capable of interacting with the carbonyl π system. Figure 21.2 portrays the arrangement of p atomic orbitals that gives rise to a delocalized π molecular orbital encompassing both the carbonyl group and its substituent.

Electron delocalization in carboxylic acids and its derivatives is represented in resonance terms by contributions from the structures:

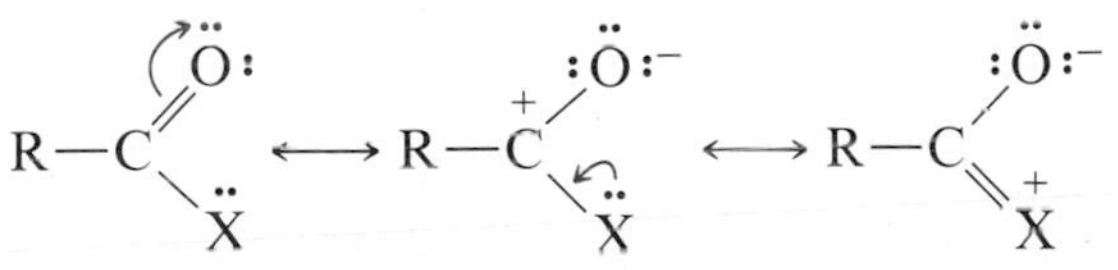

$$R-C\overset{\ddot{O}:}{\underset{\ddot{X}}{\diagdown}} \longleftrightarrow R-\overset{+}{C}\overset{:\ddot{O}:^-}{\underset{\ddot{X}}{\diagdown}} \longleftrightarrow R-C\overset{:\ddot{O}:^-}{\underset{\overset{+}{X}}{\diagup\!\!\diagdown}}$$

Electron release from the substituent stabilizes the carbonyl group and decreases its electrophilic character. The extent of this electron delocalization depends on the electron-donating properties of the substituent X. Generally, the less electronegative X is, the more it will donate electrons to the carbonyl group and the greater will be its stabilizing effect.

Resonance stabilization in acyl chlorides is not nearly as pronounced as in other derivatives of carboxylic acids.

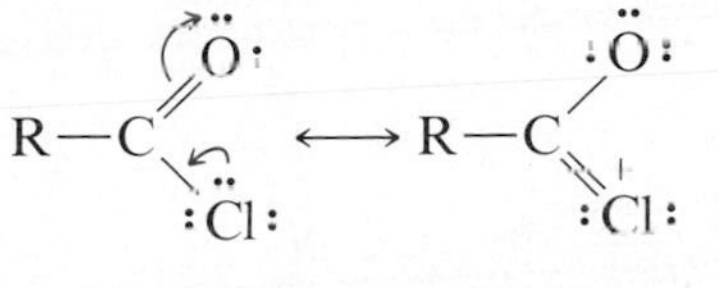

$$R-C\overset{\ddot{O}:}{\underset{:\ddot{C}l:}{\diagdown}} \longleftrightarrow R-C\overset{:\ddot{O}:}{\underset{:\overset{+}{C}l:}{\diagdown}} \qquad \textbf{weak resonance stabilization}$$

Because the carbon-chlorine bond is so long — typically on the order of 1.79 Å for acyl chlorides — the degree to which the electron pairs of chlorine are delocalized into the carbonyl π system is quite small. The carbonyl group of an acyl chloride feels the normal electron-withdrawing inductive effect of a chlorine substituent without a significant compensating electron-releasing effect due to resonance. This makes the

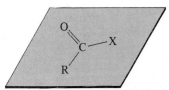

X = OH: carboxylic acid

X = Cl: acyl chloride

X = OCR: acid anhydride
            ‖
            O

X = OR: ester

X = NR$_2$: amide

FIGURE 21.1   All the bonds to the carbonyl carbon of a carboxylic acid derivative lie in the same plane.

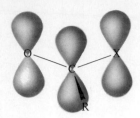

Individual *p* orbitals of carboxylic acid derivative

Extended π system of carboxylic acid derivative

FIGURE 21.2 The *p* orbitals of the carbonyl carbon, its oxygen, and the atom attached to the acyl group overlap to form an extended π system through which electrons are delocalized.

carbonyl carbon of an acyl chloride more susceptible to attack by nucleophiles than that of other carboxylic acid derivatives.

Acid anhydrides are better stabilized by electron delocalization than are acyl chlorides. The lone pair electrons of oxygen are delocalized more effectively into the carbonyl group. Resonance involves both carbonyl groups of an acid anhydride.

The carbonyl group of an ester is stabilized more than is that of an anhydride. Since both acyl groups of an anhydride compete for the oxygen lone pair, each carbonyl is stabilized less than the single carbonyl group of an ester.

Ester   is more effective than   Acid anhydride

Esters are stabilized by resonance to about the same extent as carboxylic acids but not as much as amides. Nitrogen is less electronegative than oxygen and so is better able to donate an electron pair to a carbonyl group.

Very effective resonance stabilization

Amide resonance is a powerful stabilizing force and gives rise to a number of structural effects. Unlike the pyramidal arrangement of bonds in ammonia and amines, the bonds to nitrogen in amides lie in the same plane. Formamide, for example, as shown in Figure 21.3, is a planar molecule. The carbon-nitrogen bond has considerable double bond character and, at 1.35 Å, is substantially shorter than the normal 1.47 Å carbon-nitrogen single bond distance observed in amines.

The barrier to rotation about the carbon-nitrogen bond in amides is 18 to 20 kcal/mol.

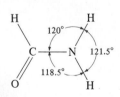

C-N bond distance = 1.352 Å

FIGURE 21.3 Structural features of formamide. The molecule is planar; the molecular plane corresponds to the plane of the paper.

$$E_{act} = 18 \text{ to } 20 \text{ kcal/mol}$$

This is an unusually high rotational energy barrier for a single bond and indicates that the carbon-nitrogen bond possesses a significant degree of double bond character, as the resonance picture suggests.

PROBLEM 21.2   The ^{1}H nuclear magnetic resonance spectrum of *N,N*-dimethylformamide shows a separate signal for each of the two methyl groups. Can you explain why?

Efficient electron release from nitrogen not only stabilizes the carbonyl group of amides but also decreases the rate at which nucleophiles attack the carbonyl carbon. Nucleophilic reagents attack electrophilic sites in a molecule; if electrons are donated to an electrophilic site in a molecule by a substituent, then the tendency of that molecule to react with external nucleophiles is moderated.

An extreme example of carbonyl group stabilization is seen in carboxylate anions.

The negatively charged oxygen substituent is a powerful electron donor to the carbonyl group. Resonance in carboxylate anions disperses the negative charge. It is more effective than resonance in carboxylic acids, acyl chlorides, anhydrides, esters, and amides, in all of which electron delocalization is accompanied by the separation of positive from negative charge.

Table 21.1 summarizes the stabilizing effects of substituents on carbonyl groups to which they are attached. In addition to a qualitative ranking, quantitative estimates of the relative rates of hydrolysis of the various classes of acyl derivatives are given. A weakly stabilized carboxylic acid derivative reacts with water faster than does a more stabilized one.

Most of the common methods for their preparation convert one class of carboxylic acid derivative to another, and the order of carbonyl group stabilization given in Table 21.1 bears directly on the means by which these transformations may be

TABLE 21.1
Relative Stability and Reactivity of Carboxylic Acid Derivatives

Carboxylic acid derivative		Stabilization	Relative rate of hydrolysis*
Acyl chloride	$RCCl$ (C=O)	Very small	10^{11}
Anhydride	$RCOCR$ (two C=O)	Small	10^7
Ester	$RCOR'$ (C=O)	Moderate	1.0
Amide	$RCNR'_2$ (C=O)	Large	$<10^{-2}$
Carboxylate anion	RCO^- (C=O)	Very large	

* Rates are approximate and are relative to ester as standard substrate at pH 7.

achieved. A reaction that converts a particular carboxylic acid derivative to one that lies below it in the table is practical; a reaction that converts it to one that lies above it in the table is not. This is another way of saying that one carboxylic acid derivative can be converted to another by acyl transfer if the reaction leads to a more stabilized carbonyl group. Numerous examples of reactions of this type will be presented in the sections that follow. We begin with reactions of acyl chlorides.

21.3 ACYL TRANSFER REACTIONS OF ACYL CHLORIDES

Acyl chlorides are among the most readily available and versatile acyl transfer agents. They are normally prepared from carboxylic acids by reaction with thionyl chloride (Section 13.7).

$$
\underset{\substack{\text{Carboxylic} \\ \text{acid}}}{\text{RCOH}} + \underset{\substack{\text{Thionyl} \\ \text{chloride}}}{\text{SOCl}_2} \longrightarrow \underset{\substack{\text{Acyl} \\ \text{chloride}}}{\text{RCCl}} + \underset{\substack{\text{Sulfur} \\ \text{dioxide}}}{\text{SO}_2} + \underset{\substack{\text{Hydrogen} \\ \text{chloride}}}{\text{HCl}}
$$

$$
\underset{\substack{\text{2-Methylpropanoic acid}}}{(\text{CH}_3)_2\text{CHCOH}} \xrightarrow[\text{heat}]{\text{SOCl}_2} \underset{\substack{\text{2-Methylpropanoyl chloride (90\%)}}}{(\text{CH}_3)_2\text{CHCCl}}
$$

Several synthetically useful reactions of acyl chlorides have been described in previous chapters. These are summarized in Table 21.2.

On treatment with the appropriate nucleophile, an acyl chloride may be converted to an acid anhydride, an ester, an amide, or a carboxylic acid. Examples illustrating these transformations of acyl chlorides are presented in Table 21.3.

PROBLEM 21.3   Apply the knowledge gained by studying Table 21.3 to help you predict the major organic product obtained by reaction of benzoyl chloride with each of the following:

(a) Acetic acid   (d) Methylamine, CH_3NH_2
(b) Benzoic acid   (e) Dimethylamine, $(CH_3)_2NH$
(c) Ethanol

SAMPLE SOLUTION   (a) As noted in Table 21.3, the reaction of an acyl chloride with a carboxylic acid yields an acid anhydride.

$$
\underset{\substack{\text{Benzoyl chloride}}}{\text{C}_6\text{H}_5\text{CCl}} + \underset{\substack{\text{Acetic acid}}}{\text{CH}_3\text{COH}} \longrightarrow \underset{\substack{\text{Acetic benzoic anhydride}}}{\text{C}_6\text{H}_5\text{COCCH}_3}
$$

The product is a mixed anhydride. Acetic acid acts as a nucleophile and substitutes for chloride on the benzoyl group.

TABLE 21.2

Summary of Reactions of Acyl Chlorides Discussed In Earlier Chapters

Reaction (section) and comments	General equation and specific example
Friedel-Crafts acylation (Section 13.7) A very effective method for attaching a carbon chain onto an aromatic ring. The reaction proceeds by electrophilic aromatic substitution; an acyl cation $RC{\equiv}\overset{..}{O}{}^{+}$ is an intermediate.	$$ArH + RCCl \xrightarrow{AlCl_3} ArCR + HCl$$ Arene / Acyl chloride / Acylbenzene / Hydrogen chloride 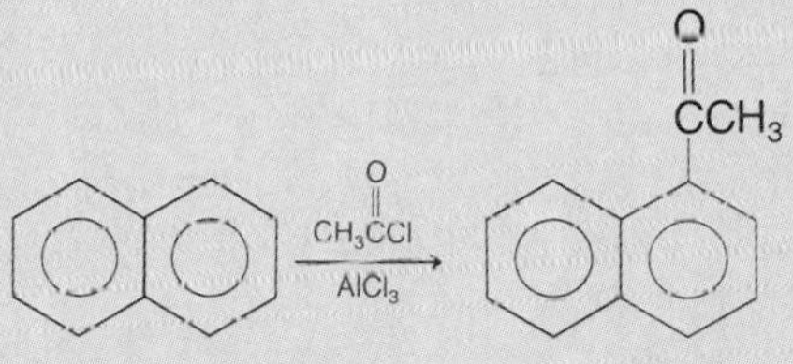 Naphthalene $\xrightarrow[AlCl_3]{CH_3CCl}$ 1-Acetylnaphthalene (90%)
Hydrogenation (Section 18.6) Acyl chlorides are reduced to aldehydes by hydrogenation over a palladium-on-barium sulfate catalyst. This procedure is known as the Rosenmund reduction.	$$RCCl + H_2 \xrightarrow[heat]{Pd/BaSO_4} RCH + HCl$$ Acyl chloride / Hydrogen / Aldehyde / Hydrogen chloride 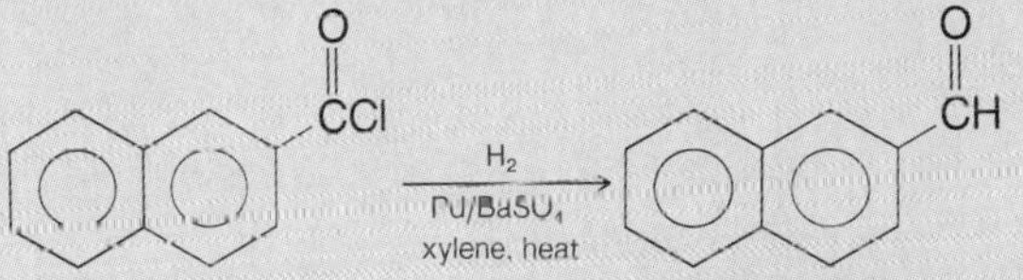 Naphthalene-2-carbonyl chloride $\xrightarrow[\substack{Pd/BaSO_4 \\ xylene, heat}]{H_2}$ Naphthalene-2-carbaldehyde (74–81%)
Reaction with lithium diorganocuprates (Section 18.7) Acyl chlorides react with lithium diorganocuprates to give ketones.	$$RCCl + LiCuR'_2 \longrightarrow RCR' + R'Cu + LiCl$$ Acyl chloride / Lithium diorganocuprate / Ketone / Organocopper / Lithium chloride 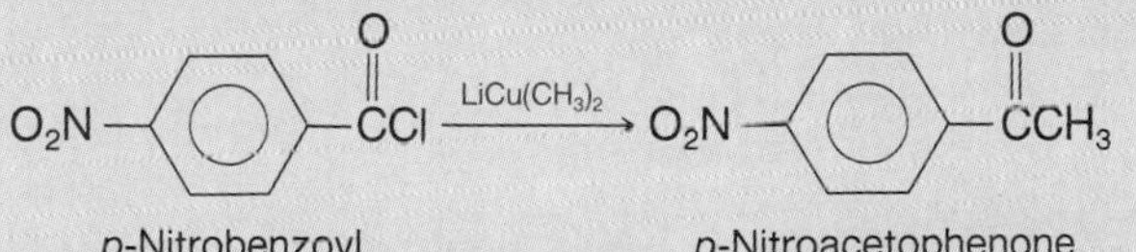 p-Nitrobenzoyl chloride $O_2N{-}\bigcirc{-}CCl \xrightarrow{LiCu(CH_3)_2} O_2N{-}\bigcirc{-}CCH_3$ p-Nitroacetophenone (50%)
Reaction with Diorganocadmium Reagents (Section 18.7) Acyl chlorides react with diorganocadmium reagents to give ketones.	$$RCCl + R'_2Cd \longrightarrow RCR' + R'CdCl$$ Acyl chloride / Diorganocadmium reagent / Ketone / Organocadmium chloride $CH_3(CH_2)_{16}CCl \xrightarrow{(CH_3CH_2)_2Cd} CH_3(CH_2)_{16}CCH_2CH_3$ Octadecanoyl chloride / 3-Icosanone (62%)

TABLE 21.3

Conversion of Acyl Chlorides to Other Carboxylic Acid Derivatives

Reaction (section) and comments	General equation and specific example
Reaction with carboxylic acids (Section 21.4) Acyl chlorides react with carboxylic acids to yield acid anhydrides. When this reaction is used for preparative purposes, a weak organic base such as pyridine is normally added. Pyridine is a catalyst for the reaction and also acts as a base to neutralize the hydrogen chloride that is formed.	$$\underset{\substack{\text{Acyl}\\\text{chloride}}}{RCCl} + \underset{\substack{\text{Carboxylic}\\\text{acid}}}{R'COH} \longrightarrow \underset{\substack{\text{Acid}\\\text{anhydride}}}{RCOCR'} + \underset{\substack{\text{Hydrogen}\\\text{chloride}}}{HCl}$$ $$\underset{\substack{\text{Heptanoyl}\\\text{chloride}}}{CH_3(CH_2)_5CCl} + \underset{\substack{\text{Heptanoic}\\\text{acid}}}{CH_3(CH_2)_5COH} \xrightarrow{\text{pyridine}} \underset{\substack{\text{Heptanoic anhydride}\\(78-83\%)}}{CH_3(CH_2)_5COC(CH_2)_5CH_3}$$
Reaction with alcohols (Section 16.10) Acyl chlorides react with alcohols to form esters. The reaction is typically carried out in the presence of pyridine.	$$\underset{\substack{\text{Acyl}\\\text{chloride}}}{RCCl} + \underset{\text{Alcohol}}{R'OH} \longrightarrow \underset{\text{Ester}}{RCOR'} + \underset{\substack{\text{Hydrogen}\\\text{chloride}}}{HCl}$$ $$\underset{\substack{\text{Benzoyl}\\\text{chloride}}}{C_6H_5CCl} + \underset{\substack{\textit{tert}\text{-Butyl}\\\text{alcohol}}}{(CH_3)_3COH} \xrightarrow{\text{pyridine}} \underset{\substack{\textit{tert}\text{-Butyl}\\\text{benzoate (80\%)}}}{C_6H_5COC(CH_3)_2}$$
Reaction with ammonia and amines (Section 21.3) Acyl chlorides react with ammonia and amines to form amides. A base such as sodium hydroxide is normally added to react with the hydrogen chloride produced.	$$\underset{\substack{\text{Acyl}\\\text{chloride}}}{RCCl} + \underset{\substack{\text{Ammonia}\\\text{or amine}}}{R_2'NH} + \underset{\text{Hydroxide}}{HO^-} \longrightarrow \underset{\text{Amide}}{RCNR_2'} + \underset{\text{Water}}{H_2O} + \underset{\substack{\text{Chloride}\\\text{ion}}}{Cl^-}$$ $$\underset{\substack{\text{Benzoyl}\\\text{chloride}}}{C_6H_5CCl} + \underset{\text{Piperidine}}{HN} \xrightarrow[\text{H}_2\text{O}]{\text{NaOH}} \underset{\substack{\textit{N}\text{-Benzoylpiperidine}\\(87-91\%)}}{C_6H_5C-N}$$
Hydrolysis (Section 21.3) Acyl chlorides react with water to yield carboxylic acids. In base, the acid is converted to its carboxylate salt. The reaction has little preparative value because the acyl chloride is nearly always prepared from the carboxylic acid rather than vice versa.	$$\underset{\substack{\text{Acyl}\\\text{chloride}}}{RCCl} + \underset{\text{Water}}{H_2O} \longrightarrow \underset{\substack{\text{Carboxylic}\\\text{acid}}}{RCOH} + \underset{\substack{\text{Hydrogen}\\\text{chloride}}}{HCl}$$ $$\underset{\substack{\text{Phenylacetyl}\\\text{chloride}}}{C_6H_5CH_2CCl} + \underset{\text{Water}}{H_2O} \longrightarrow \underset{\substack{\text{Phenylacetic}\\\text{acid}}}{C_6H_5CH_2COH} + \underset{\substack{\text{Hydrogen}\\\text{chloride}}}{HCl}$$

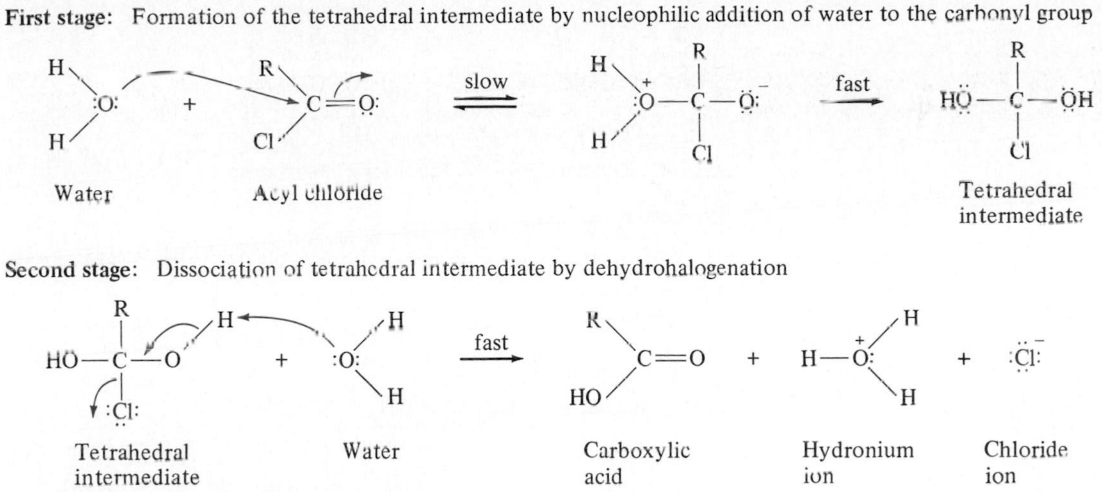

First stage: Formation of the tetrahedral intermediate by nucleophilic addition of water to the carbonyl group

Water   Acyl chloride   Tetrahedral intermediate

Second stage: Dissociation of tetrahedral intermediate by dehydrohalogenation

Tetrahedral intermediate   Water   Carboxylic acid   Hydronium ion   Chloride ion

FIGURE 21.4 Hydrolysis of an acyl chloride proceeds by way of a tetrahedral intermediate. Formation of the tetrahedral intermediate is rate-determining.

The mechanisms of all the reactions cited in Table 21.3 are similar to the mechanism of hydrolysis of an acyl chloride.

$$\underset{\substack{\text{Acyl}\\\text{chloride}}}{\text{RCCl}} + \underset{\text{Water}}{\text{H}_2\text{O}} \longrightarrow \underset{\substack{\text{Carboxylic}\\\text{acid}}}{\text{RCOH}} + \underset{\substack{\text{Hydrogen}\\\text{chloride}}}{\text{HCl}}$$

They differ with respect to the nucleophile that attacks the carbonyl group, which in hydrolysis is a water molecule. The accepted mechanism for hydrolysis of an acyl chloride is outlined in Figure 21.4.

In the first stage of the mechanism, water undergoes nucleophilic addition to the carbonyl group to form a tetrahedral intermediate. This stage of the process is analogous to the hydration of aldehydes and ketones discussed in Section 18.9.

The tetrahedral intermediate has three potential leaving groups on carbon, two hydroxyl groups and a chlorine atom. In the second stage of the reaction, the tetrahedral intermediate dissociates. Loss of chloride from the tetrahedral intermediate is faster than loss of hydroxide; chloride is less basic than hydroxide and is a better leaving group. The tetrahedral intermediate dissociates because this dissociation restores the resonance-stabilized carbonyl group.

PROBLEM 21.4 Write the structure of the tetrahedral intermediate formed in each of the reactions given in Problem 21.3. Using curved arrows, show how each tetrahedral intermediate dissociates to the appropriate products.

SAMPLE SOLUTION (a) The tetrahedral intermediate arises by nucleophilic addition of acetic acid to benzoyl chloride.

$$\underset{\substack{\text{Benzoyl}\\\text{chloride}}}{C_6H_5\overset{\displaystyle O}{\overset{\|}{C}}Cl} + \underset{\text{Acetic acid}}{CH_3\overset{\displaystyle O}{\overset{\|}{C}}OH} \longrightarrow \underset{\text{Tetrahedral intermediate}}{\left[C_6H_5\underset{\displaystyle Cl}{\overset{\substack{\displaystyle HO \quad O \\ \|}}{C}}OCCH_3 \right]}$$

Loss of a proton and chloride ion from the tetrahedral intermediate yields the mixed anhydride.

$$\underset{\substack{\text{Tetrahedral}\\\text{intermediate}}}{C_6H_5\underset{\displaystyle Cl}{\overset{\substack{H \\ \displaystyle O \quad O \\ \|}}{C}}OCCH_3} \longrightarrow \underset{\substack{\text{Acetic benzoic}\\\text{anhydride}}}{C_6H_5\overset{\substack{\displaystyle O \quad O \\ \| \quad \|}}{C}OCCH_3} + \underset{\substack{\text{Hydrogen}\\\text{chloride}}}{HCl}$$

Nucleophilic substitution in *acyl* chlorides occurs much more readily than nucleophilic substitution in *alkyl* chlorides. Benzoyl chloride ($C_6H_5\overset{\displaystyle O}{\overset{\|}{C}}Cl$), for example, is approximately 1000 times as reactive as benzyl chloride ($C_6H_5CH_2Cl$) toward solvolysis (80% ethanol : 20% water, 25°C). Nucleophilic acyl substitution does not involve carbocation intermediates such as those formed in S_N1 reactions of alkyl halides, nor does it proceed by way of the crowded pentacoordinate activated complexes that characterize the S_N2 mechanism of nucleophilic alkyl substitution. The transition state for nucleophilic acyl substitution leads directly to an intermediate that has a normal arrangement of bonds. This transition state can be achieved with a relatively modest expenditure of energy, especially when the carbonyl carbon is fairly electrophilic, as it is in an acyl halide.

21.4 PREPARATION OF CARBOXYLIC ACID ANHYDRIDES

After acyl halides, the next most reactive acyl transfer agents are carboxylic acid anhydrides. The most readily available acid anhydride is acetic anhydride. Millions of tons of acetic anhydride are prepared anually by the reaction of acetic acid with a substance known as *ketene*.

$$\underset{\text{Ketene}}{CH_2{=}C{=}O} + \underset{\text{Acetic acid}}{CH_3\overset{\displaystyle O}{\overset{\|}{C}}OH} \longrightarrow \underset{\text{Acetic anhydride}}{CH_3\overset{\substack{\displaystyle O \quad O \\ \| \quad \|}}{C}OCCH_3}$$

Ketene is a highly reactive substance. It is prepared by the dehydration of acetic acid at high temperature. In one industrial process, triethyl phosphate is used as a catalyst.

$$\underset{\substack{\displaystyle \| \\[-2pt] \text{CH}_3\text{COH}}}{\overset{\displaystyle \text{O}}{}} \xrightarrow[700°C]{(CH_3CH_2O)_3PO} \text{CH}_2{=}\text{C}{=}\text{O} + \text{H}_2\text{O}$$

Acetic acid   Ketene   Water

Acetic anhydride has several commercial applications including the synthesis of aspirin (Section 26.10) and the preparation of cellulose acetate for use in plastics and fibers. Two cyclic anhydrides, phthalic anhydride and maleic anhydride, are industrial chemicals.

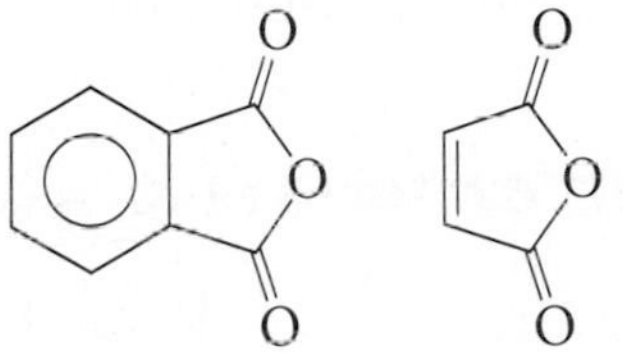

Phthalic anhydride   Maleic anhydride

The customary method for the laboratory synthesis of acid anhydrides is the reaction of acyl chlorides with carboxylic acids (Table 21.3).

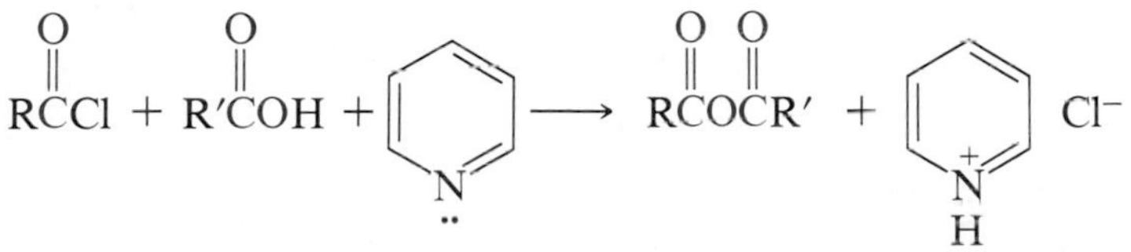

| Acyl chloride | Carboxylic acid | Pyridine | | Carboxylic acid anhydride | Pyridinium chloride |

This procedure is applicable to the preparation of both symmetrical anhydrides (R and R′ the same) and mixed anhydrides (R and R′ different).

PROBLEM 21.5 Benzoic anhydride has been prepared in excellent yield by adding one molar equivalent of water to two molar equivalents of benzoyl chloride. How do you suppose this reaction takes place?

Cyclic anhydrides in which the ring is five-membered or six-membered are sometimes prepared by heating the corresponding dicarboxylic acids in an inert solvent,

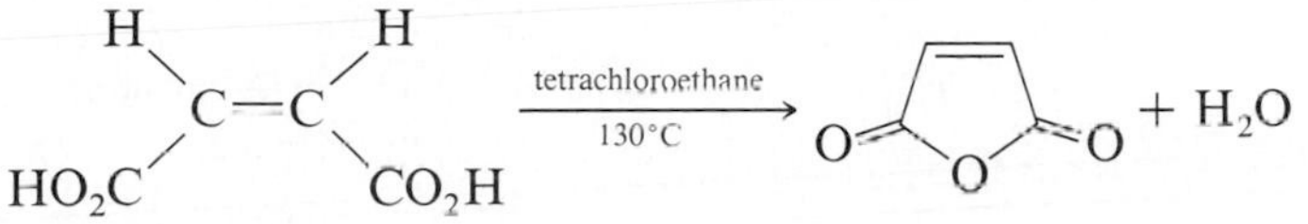

Maleic acid   Maleic anhydride   Water
                                     (89%)

More often, however, cyclization is achieved by heating the diacid in the presence of acetic anhydride.

$$\underset{\substack{\text{Succinic acid} \\ \text{(bp 235°C)}}}{\text{HOCCH}_2\text{CH}_2\text{COH}} + \underset{\substack{\text{Acetic} \\ \text{anhydride} \\ \text{(bp 139°C)}}}{\text{CH}_3\text{COCCH}_3} \xrightarrow{\text{heat}} \underset{\substack{\text{Succinic} \\ \text{anhydride (72\%)} \\ \text{(bp 261°C)}}}{} + \underset{\substack{\text{Acetic} \\ \text{acid} \\ \text{(bp 118°C)}}}{2\text{CH}_3\text{COH}}$$

The reaction is an equilibrium process. Acetic acid is the lowest-boiling component of the reaction mixture and is removed by distillation. The position of equilibrium shifts to favor formation of succinic anhydride.

21.5 REACTIONS OF CARBOXYLIC ACID ANHYDRIDES

The most important reactions of acid anhydrides involve cleavage of a bond between oxygen and one of the carbonyl groups. One acyl group is transferred to an attacking nucleophile; the other retains its single bond to oxygen and becomes the acyl group of a carboxylic acid.

$$\underset{\substack{\text{Bond cleavage} \\ \text{occurs here in} \\ \text{an acid anhy-} \\ \text{dride}}}{\text{RC}+\text{OCR}} + \underset{\text{Nucleophile}}{\text{HY:}} \longrightarrow \underset{\substack{\text{Product of} \\ \text{acyl transfer}}}{\text{RC}-\overset{..}{\text{Y}}} + \underset{\substack{\text{Carboxylic} \\ \text{acid}}}{\text{HOCR}}$$

One reaction of this type is familiar to you, namely, the Friedel-Crafts acylation reaction (Section 13.7).

$$\underset{\substack{\text{Acid} \\ \text{anhydride}}}{\text{RCOCR}} + \underset{\text{Arene}}{\text{ArH}} \xrightarrow{\text{AlCl}_3} \underset{\text{Ketone}}{\text{RCAr}} + \underset{\substack{\text{Carboxylic} \\ \text{acid}}}{\text{RCOH}}$$

$$\underset{\substack{\text{Acetic} \\ \text{anhydride}}}{\text{CH}_3\text{COCCH}_3} + \underset{o\text{-Fluoroanisole}}{\text{F}} \xrightarrow{\text{AlCl}_3} \underset{\substack{\text{3-Fluoro-4-methoxyacetophenone} \\ \text{(70–80\%)}}}{\text{CH}_3\text{C}} + \underset{\substack{\text{Acetic} \\ \text{acid}}}{\text{CH}_3\text{CO}_2\text{H}}$$

An acyl cation is an intermediate in Friedel-Crafts acylation reactions.

PROBLEM 21.6 Write a structural formula for the acyl cation intermediate in the preceding reaction.

TABLE 21.4
Conversion of Acid Anhydrides to Other Carboxylic Acid Derivatives

Reaction (section) and comments	General equation and specific example
Reaction with alcohols (Section 16.10) Acid anhydrides react with alcohols to form esters. The reaction may be carried out in the presence of pyridine or it may be catalyzed by acids. In the example shown, only one acetyl group of acetic anhydride becomes incorporated into the ester; the other becomes the acetyl group of an acetic acid molecule.	$R\overset{O}{\overset{\|}{C}}O\overset{O}{\overset{\|}{C}}R + R'OH \longrightarrow R\overset{O}{\overset{\|}{C}}OR' + R\overset{O}{\overset{\|}{C}}OH$ Acid anhydride; Alcohol; Ester; Carboxylic acid $CH_3\overset{O}{\overset{\|}{C}}O\overset{O}{\overset{\|}{C}}CH_3 + HO\underset{CH_3}{\underset{\|}{C}H}CH_2CH_3 \xrightarrow{H_2SO_4} CH_3\overset{O}{\overset{\|}{C}}O\underset{CH_3}{\underset{\|}{C}H}CH_2CH_3$ Acetic anhydride; *sec*-Butyl alcohol; *sec*-Butyl acetate (60%)
Reaction with ammonia and amines (Section 21.13) Acid anhydrides react with ammonia and amines to form amides. Two molar equivalents of amine are required. In the example shown, only one acetyl group of acetic anhydride becomes incorporated into the amide; the other becomes the acetyl group of the amine salt of acetic acid.	$R\overset{O}{\overset{\|}{C}}O\overset{O}{\overset{\|}{C}}R + 2R'_2NH \longrightarrow R\overset{O}{\overset{\|}{C}}NR'_2 + R\overset{O}{\overset{\|}{C}}O^- \; H_2\overset{+}{N}R'_2$ Acid anhydride; Amine; Amide; Ammonium carboxylate salt 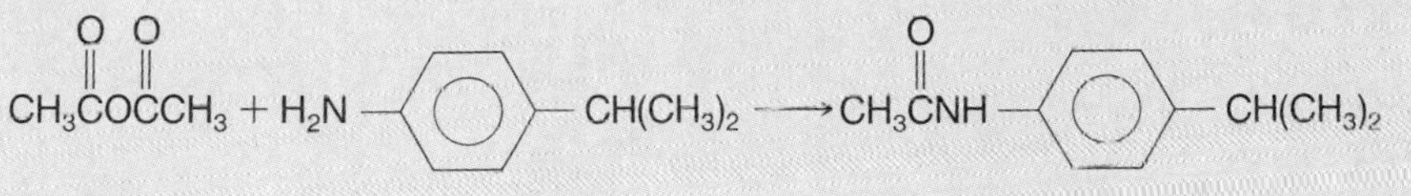 Acetic anhydride; *p*-Isopropylaniline; *p*-Isopropylacetanilide (98%)
Hydrolysis (Section 21.5) Acid anhydrides react with water to yield two carboxylic acid functions. Cyclic anhydrides yield dicarboxylic acids.	$R\overset{O}{\overset{\|}{C}}O\overset{O}{\overset{\|}{C}}R' + H_2O \longrightarrow 2R\overset{O}{\overset{\|}{C}}OH$ Acid anhydride; Water; Carboxylic acid 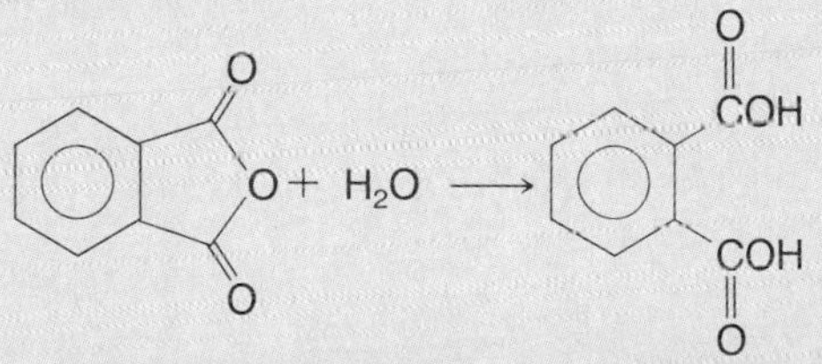 Phthalic anhydride; Water; Phthalic acid

Conversions of acid anhydrides to other carboxylic acid derivatives are illustrated in Table 21.4. Since a more highly stabilized carbonyl group must result in order for acyl transfer to be effective, acid anhydrides are readily converted to carboxylic acids, esters, and amides but not to acyl chlorides.

PROBLEM 21.7 Apply the knowledge gained by studying Table 21.4 to help you predict the major organic product of each of the following reactions.

Mechanism 21.4

(a) Benzoic anhydride + methanol $\xrightarrow{H^+}$

(b) Acetic anhydride + ammonia (2 mol) $\longrightarrow$

(c) Phthalic anhydride + $(CH_3)_2NH$ (2 mol) $\longrightarrow$

(d) Phthalic anhydride + sodium hydroxide (2 mol) $\longrightarrow$

SAMPLE SOLUTION (a) Nucleophilic acyl substitution by an alcohol on an acid anhydride yields an ester.

$$C_6H_5\overset{\overset{O}{\|}}{C}O\overset{\overset{O}{\|}}{C}C_6H_5 + CH_3OH \xrightarrow{H^+} C_6H_5\overset{\overset{O}{\|}}{C}OCH_3 + C_6H_5\overset{\overset{O}{\|}}{C}OH$$

| Benzoic anhydride | Methanol | Methyl benzoate | Benzoic acid |

The first example cited in Table 21.4 introduces a new aspect of acyl transfer processes that is relevant not only to the reactions of acid anhydrides but to those of acyl chlorides, esters, and amides as well. Acyl transfer reactions are subject to catalysis by acids.

We can describe how an acid catalyst can increase the rate of nucleophilic acyl substitution by considering the hydrolysis of an acid anhydride. Formation of the tetrahedral intermediate is rate-determining, and it is this step that is accelerated by the catalyst. The acid anhydride is activated toward nucleophilic addition by protonation of one of its carbonyl groups.

$$R\overset{\overset{:\ddot{O}}{\|}}{C}O\overset{\overset{\ddot{O}:}{\|}}{C}R + H^+ \xrightarrow{\text{fast}} R\overset{\overset{+}{\overset{H\ddot{O}}{\|}}}{C}O\overset{\overset{\ddot{O}:}{\|}}{C}R$$

| Acid anhydride | Proton | Protonated form of acid anhydride |

The protonated form of the acid anhydride is present to only a very small extent, but it is quite electrophilic. Water (and other nucleophiles) add to a protonated carbonyl group much faster than they do to a neutral one. Thus, the rate-determining nucleophilic addition of water to form a tetrahedral intermediate takes place more rapidly in the presence of an acid than in its absence.

| Water | Protonated form of an acid anhydride | | Tetrahedral intermediate |

An acid catalyst also facilitates the dissociation of the tetrahedral intermediate. Protonation of its carbonyl permits the leaving group to depart as a neutral carboxylic acid molecule, which is a less basic leaving group than a carboxylate anion.

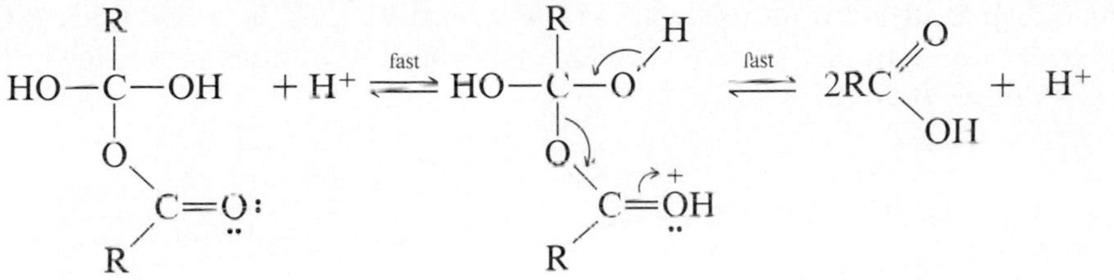

Tetrahedral intermediate → Two carboxylic acid molecules + Proton

This pattern of increased reactivity resulting from carbonyl group protonation has been seen before in nucleophilic additions to aldehydes and ketones (Section 18.9) and in the mechanism of the acid-catalyzed esterification of carboxylic acids (Section 20.14).

PROBLEM 21.8 Write the structure of the tetrahedral intermediate formed in each of the reactions given in Problem 21.7. Using curved arrows, show how each tetrahedral intermediate dissociates to the appropriate products.

SAMPLE SOLUTION (a) The reaction given is the acid-catalyzed esterification of methanol by benzoic anhydride. The first step is the activation of the anhydride toward nucleophilic addition by protonation.

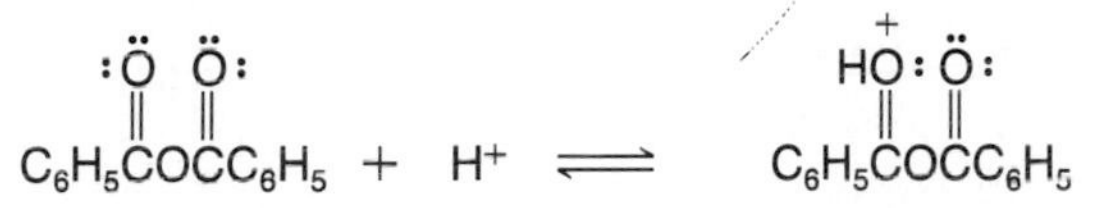

$C_6H_5COCC_6H_5$ (Benzoic anhydride) + H^+ ⇌ Protonated form of benzoic anhydride

The tetrahedral intermediate is formed by nucleophilic addition of methanol to the protonated carbonyl group.

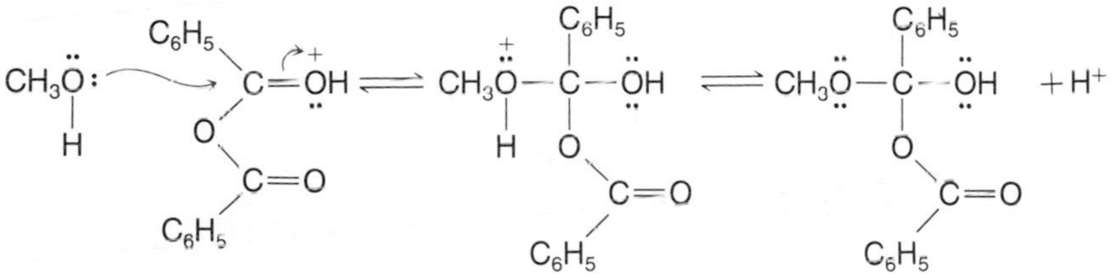

Methanol   Protonated form of benzoic anhydride → Tetrahedral intermediate + H^+

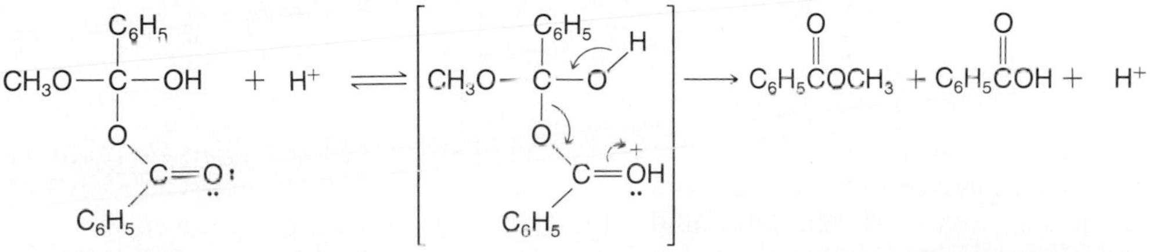

Tetrahedral intermediate + Proton ⇌ [ ] → Methyl benzoate + Benzoic acid + Proton

Acid anhydrides are more stable and less reactive than acyl chlorides. Acetic anhydride, for example, undergoes hydrolysis about 10^5 times more slowly than acetyl chloride at 25°C.

21.6 SOURCES OF ESTERS

Many esters are naturally occurring substances. Those of low molecular weight are fairly volatile, and many have pleasing odors. Esters often comprise a significant fraction of the fragrant oil of fruits and flowers.

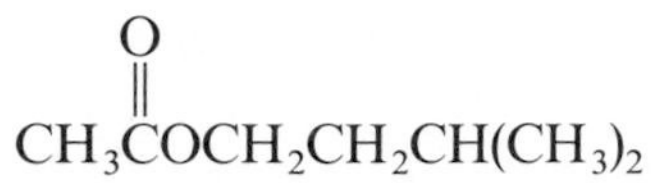

Isopentyl acetate
(contributes to characteristic
odor of bananas)

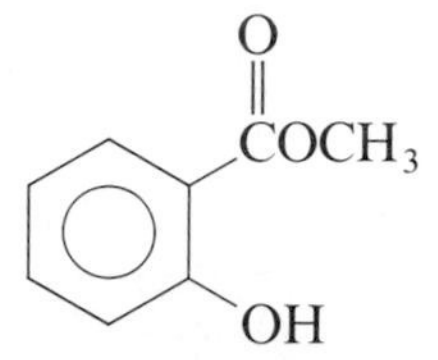

Methyl salicylate
(principal component of oil
of wintergreen)

Among the chemicals used by insects to communicate with one another, esters occur frequently.

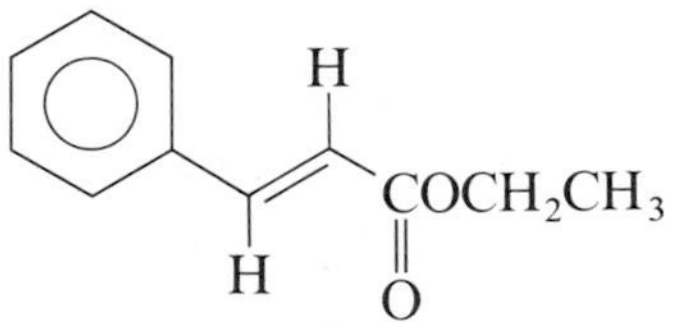

Ethyl cinnamate
(one of the constituents of
the sex pheromone of the
male Oriental fruit moth)

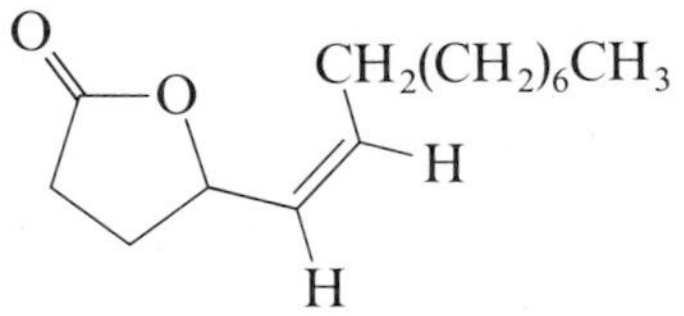

(*Z*)-5-Tetradecen-4-olide
(sex pheromone of female
Japanese bettle)

Esters of glycerol, called *glycerol triesters, triacylglycerols,* or *triglycerides,* are abundant natural products. The most important group of glycerol triesters are those in which each acyl group is unbranched and has 14 or more carbon atoms.

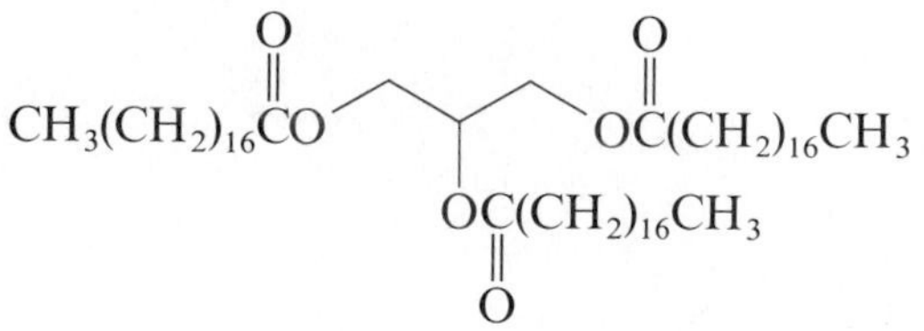

Tristearin, a trioctadecanoyl ester
of glycerol found in many animal and vegetable fats
(the three carbons and three
oxygens of glycerol are shown
in color)

Fats and *oils* are naturally occurring mixtures of glycerol triesters. Fats are mixtures that are solids at room temperature, whereas oils are liquids. The long-chain carboxylic acids obtained from fats and oils by hydrolysis are known as *fatty acids.*

The three principal methods used to prepare esters in the laboratory have all been described earlier, and all proceed by acylation of alcohols. These methods are summarized in Table 21.5.

TABLE 21.5
Preparation of Esters from Alcohols

Reaction (section) and comments	General equation and specific example
From carboxylic acids (Sections 16.10 and 20.14) In the presence of an acid catalyst, alcohols and carboxylic acids react to form an ester and water. This is the Fischer esterification reaction.	$$RCOH + R'OH \underset{}{\overset{H^+}{\rightleftharpoons}} RCOR' + H_2O$$ Carboxylic acid Alcohol Ester Water $$CH_3CH_2COH + CH_3CH_2CH_2CH_2OH \xrightarrow{H_2SO_4} CH_3CH_2COCH_2CH_2CH_2CH_3 + H_2O$$ Propanoic acid 1-Butanol Butyl propanoate (85%) Water
From acyl chlorides (Sections 16.10 and 21.3) Alcohols react with acyl chlorides by nucleophilic acyl substitution to yield esters. These reactions are typically performed in the presence of a weak base such as pyridine.	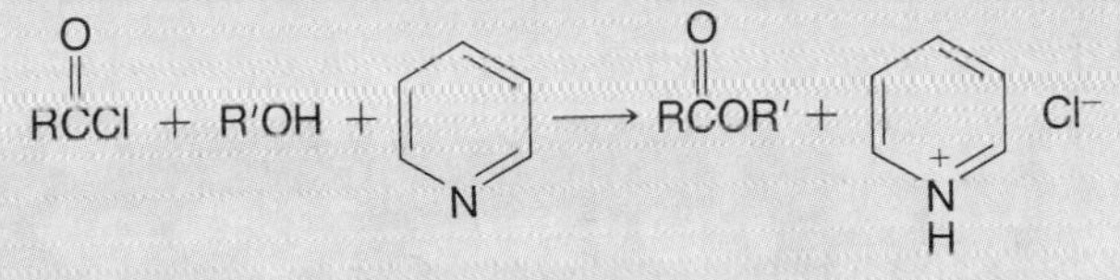 $$RCCl + R'OH + \text{(pyridine)} \longrightarrow RCOR' + \text{(pyridinium chloride)} + Cl^-$$ Acyl chloride Alcohol Pyridine Ester Pyridinium chloride 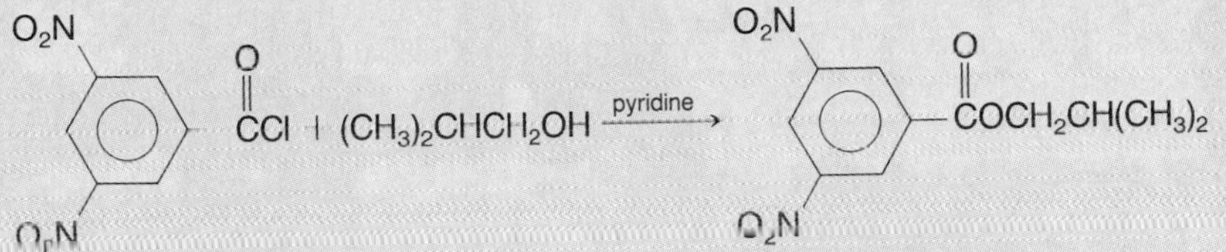 3,5-Dinitrobenzoyl chloride + $(CH_3)_2CHCH_2OH \xrightarrow{pyridine}$ Isobutyl 3,5-dinitrobenzoate (85%) Isobutyl alcohol
From carboxylic acid anhydrides (Sections 16.10 and 21.5) Acyl transfer from an acid anhydride to an alcohol is a standard method for the preparation of esters. The reaction is subject to catalysis by either acids (H₂SO₄) or bases (pyridine).	$$RCOCR + R'OH \longrightarrow RCOR' + RCOH$$ Acid anhydride Alcohol Ester Carboxylic acid 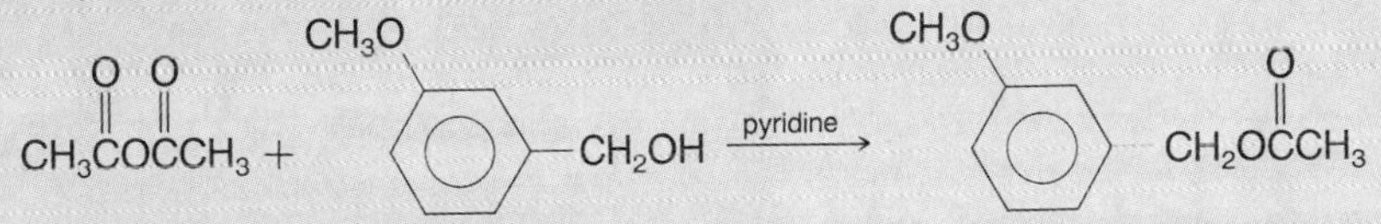 CH_3COCCH_3 + m-Methoxybenzyl alcohol $\xrightarrow{pyridine}$ m-Methoxybenzyl acetate (99%) Acetic anhydride

21.7 PHYSICAL PROPERTIES OF ESTERS

Esters are moderately polar, with dipole moments in the 1.5 to 2.0 D range. Dipole-dipole interactions contribute to the intermolecular attractive forces that cause esters to have higher boiling points than hydrocarbons of similar shape and molecular weight. Because they lack hydroxyl groups, ester molecules cannot form hydrogen bonds to each other; consequently, esters have lower boiling points than alcohols of comparable molecular weight.

$$\underset{\substack{\text{2-Methylbutane:} \\ \text{mol wt 72, bp 28°C}}}{\overset{\overset{\displaystyle CH_3}{|}}{CH_3CHCH_2CH_3}} \qquad \underset{\substack{\text{Methyl acetate:} \\ \text{mol wt 74, bp 57°C}}}{\overset{\overset{\displaystyle O}{\parallel}}{CH_3COCH_3}} \qquad \underset{\substack{\text{2-Butanol:} \\ \text{mol wt 74, bp 99°C}}}{\overset{\overset{\displaystyle OH}{|}}{CH_3CHCH_2CH_3}}$$

Using the unshared electron pairs of its oxygen atoms, an ester can participate in hydrogen bonds with substances that contain hydroxyl groups (water, alcohols, carboxylic acids). This confers some measure of water solubility on low-molecular-weight esters; methyl acetate, for example, dissolves in water to the extent of

TABLE 21.6
Summary of Reactions of Esters Discussed in Earlier Chapters

Reaction (section) and comments	General equation and specific example
Reaction with Grignard reagents (Section 15.10) Esters react with two equivalents of a Grignard reagent to produce tertiary alcohols. Two of the groups bonded to the carbon that bears the hydroxyl group in the tertiary alcohol are derived from the Grignard reagent.	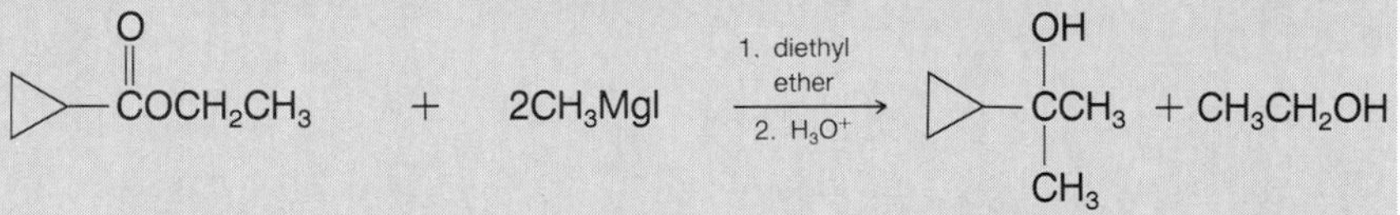
Hydrogenation (Section 16.5) Esters are cleaved when subjected to hydrogenation over a copper chromite catalyst. The reaction is carried out at high temperature and pressure (250°C, 200 atm). The acyl group is reduced to a primary alcohol. Nickel catalysts are sometimes used instead of copper chromite.	
Reduction with lithium aluminum hydride (Section 16.5) Lithium aluminum hydride cleaves esters by reduction to give the same products as are obtained by catalytic hydrogenation.	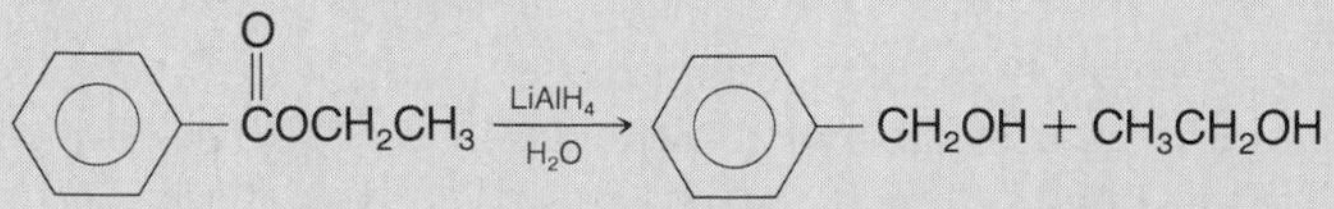

33 g/100 mL. Water solubility decreases as carbon content of the ester increases. Fats and oils, the glycerol esters of long-chain carboxylic acids, are practically insoluble in water.

21.8 REACTIONS OF ESTERS. A REVIEW AND A PREVIEW

Certain reactions of esters, useful in the preparation of alcohols, have been described in earlier chapters and are reviewed in Table 21.6.

Acyl transfer reactions of esters are summarized in Table 21.7. As acylating agents, esters are less reactive than acyl chlorides and acid anhydrides. Nucleophilic acyl substitution in esters, especially ester hydrolysis, has been extensively investigated from a mechanistic perspective. Indeed, much of what we know concerning the general topic of acyl group transfer derives from studies carried out on esters. The following sections describe those mechanistic studies.

21.9 ACID-CATALYZED ESTER HYDROLYSIS

Ester hydrolysis is the most studied and best understood of all acyl transfer reactions. Esters are fairly stable in neutral aqueous media but are cleaved when heated with water in the presence of strong acids or bases. The hydrolysis of esters in dilute aqueous acid is the reverse of the Fischer esterification reaction (Sections 16.10 and 20.14).

TABLE 21.7
Conversion of Esters to Other Carboxylic Acid Derivatives

Reaction (section) and comments	General equation and specific example
Reaction with ammonia and amines (Section 21.13) Esters react with ammonia and amines to form amides. Methyl and ethyl esters are the most reactive.	$\underset{\text{Ester}}{\text{RCOR}'} + \underset{\text{Amine}}{\text{R}''_2\text{NH}} \longrightarrow \underset{\text{Amide}}{\text{RCNR}''_2} + \underset{\text{Alcohol}}{\text{R}'\text{OH}}$ $\underset{\substack{\text{Ethyl}\\\text{fluoroacetate}}}{\text{FCH}_2\text{COCH}_2\text{CH}_3} + \underset{\text{Ammonia}}{\text{NH}_3} \xrightarrow{\text{H}_2\text{O}} \underset{\substack{\text{Fluoroacetamide}\\(90\%)}}{\text{FCH}_2\text{CNH}_2} + \underset{\text{Ethanol}}{\text{CH}_3\text{CH}_2\text{OH}}$
Hydrolysis (Sections 21.9 and 21.10) Ester hydrolysis may be catalyzed either by acids or by bases. Acid-catalyzed hydrolysis is an equilibrium-controlled process, the reverse of the Fischer esterification. Hydrolysis in base is irreversible and is the method usually chosen for preparative purposes.	$\underset{\text{Ester}}{\text{RCOR}'} + \underset{\text{Water}}{\text{H}_2\text{O}} \longrightarrow \underset{\substack{\text{Carboxylic}\\\text{acid}}}{\text{RCOH}} + \underset{\text{Alcohol}}{\text{R}'\text{OH}}$ Methyl m-nitrobenzoate $\xrightarrow[\text{2. H}^+]{\text{1. H}_2\text{O, NaOH}}$ m-Nitrobenzoic acid (90–96%) + Methanol

$$\underset{\text{Ester}}{\text{RCOR}'} + \underset{\text{Water}}{\text{H}_2\text{O}} \overset{\text{H}^+}{\rightleftharpoons} \underset{\substack{\text{Carboxylic} \\ \text{acid}}}{\text{RCOH}} + \underset{\text{Alcohol}}{\text{R}'\text{OH}}$$

When esterification is the objective, water is removed from the reaction mixture to encourage ester formation. When ester hydrolysis is the objective, the reaction is carried out in the presence of a generous excess of water.

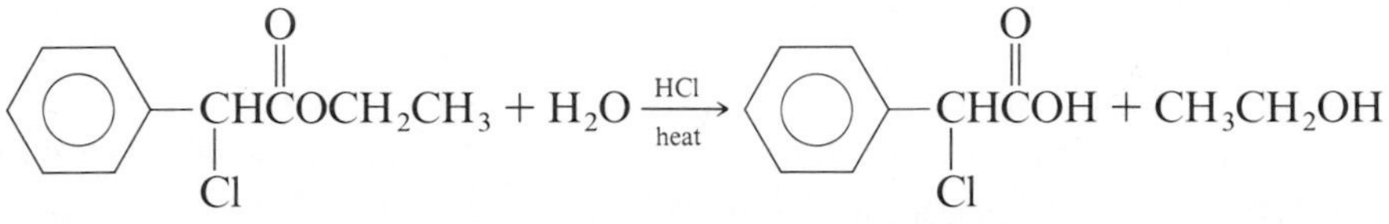

| Ethyl 2-chloro-2-phenylacetate | Water | 2-Chloro-2-phenylacetic acid (80–82%) | Ethyl alcohol |

PROBLEM 21.9 The compound having the structure shown was heated with dilute sulfuric acid to give a product having the molecular formula $C_5H_{12}O_3$ in 63 to 71 percent yield. Propose a reasonable structure for this product. What other organic compound is formed in this reaction?

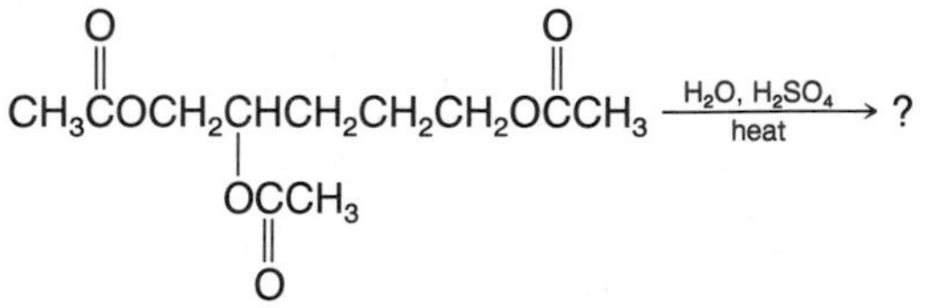

The mechanism of acid-catalyzed ester hydrolysis is presented in Figure 21.5. It is precisely the reverse of that for acid-catalyzed ester formation. Like other acyl transfer reactions, it proceeds in two stages. A tetrahedral intermediate is formed in the first stage, and this tetrahedral intermediate dissociates to products in the second stage.

A key feature of the first stage is the site at which the starting ester is protonated. Protonation of the carbonyl oxygen, as shown in step 1 of Figure 21.5, gives a cation that is stabilized by resonance.

Protonation of carbonyl oxygen

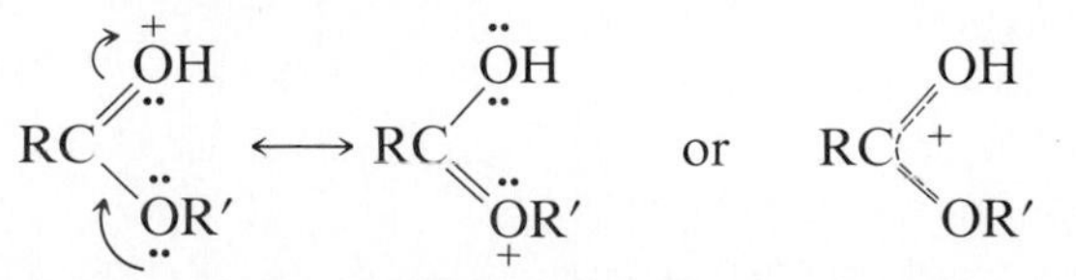

Positive charge is delocalized

The alternative site of protonation, the alkoxy oxygen, gives rise to a much less stable cation.

Step 1: Protonation of the carbonyl oxygen of the ester

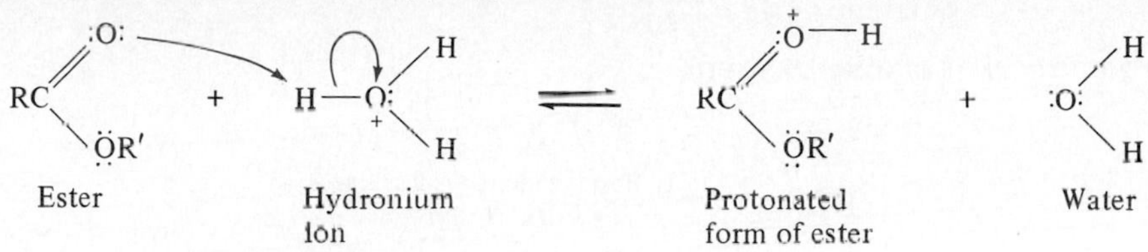

Ester Hydronium ion Protonated form of ester Water

Step 2: Nucleophilic addition of water to protonated form of ester

Water Protonated form of ester Oxonium ion

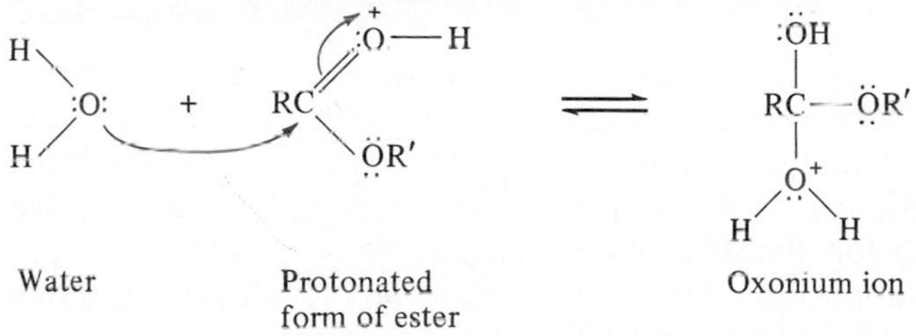

Step 3: Deprotonation of the oxonium ion to give the neutral form of the tetrahedral intermediate

Oxonium ion Water Tetrahedral intermediate Hydronium ion

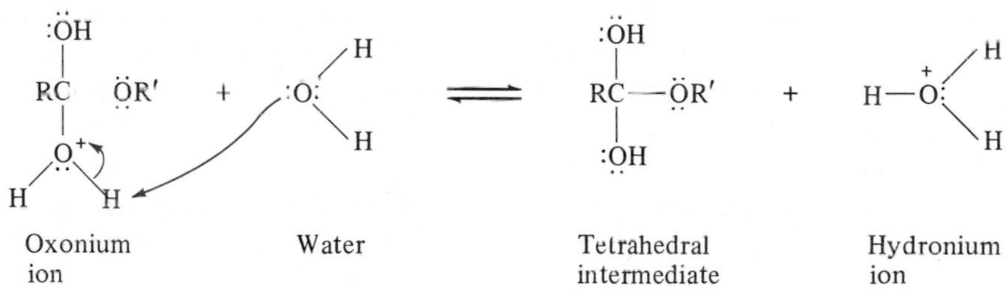

Step 4: Protonation of the tetrahedral intermediate at its alkoxy oxygen

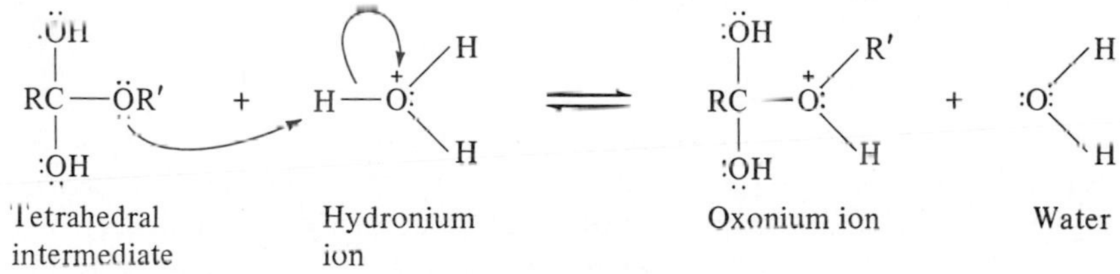

Tetrahedral intermediate Hydronium ion Oxonium ion Water

Step 5: Dissociation of the protonated form of the tetrahedral intermediate to an alcohol and the protonated form of the carboxylic acid

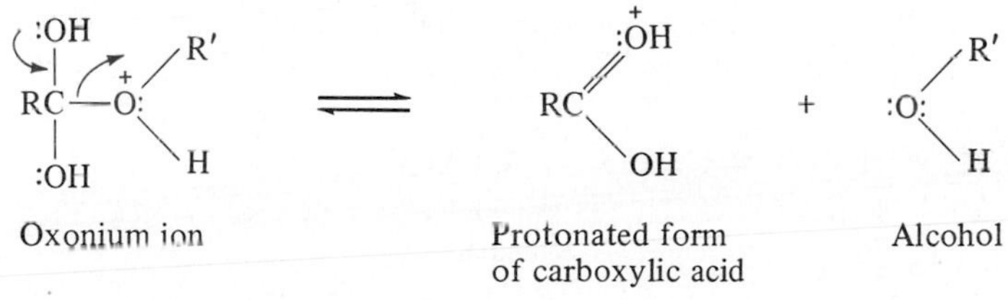

Oxonium ion Protonated form of carboxylic acid Alcohol

Step 6: Deprotonation of the protonated carboxylic acid

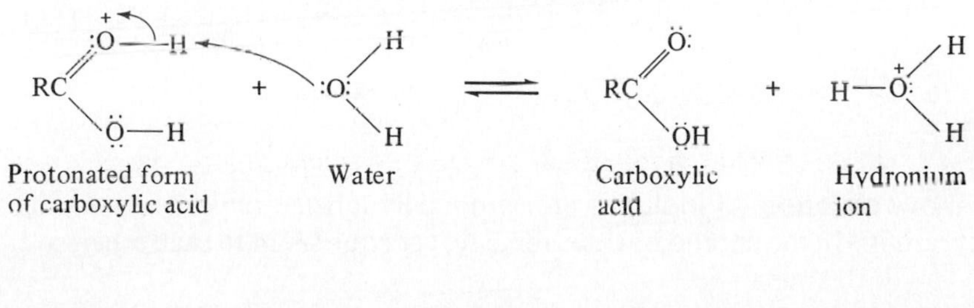

Protonated form of carboxylic acid Water Carboxylic acid Hydronium ion

FIGURE 21.5 Sequence of steps that describes the mechanism of acid-catalyzed ester hydrolysis. The formation of a tetrahedral intermediate is shown in steps 1 through 3. The dissociation of the tetrahedral intermediate is shown in steps 4 through 6.

Protonation of alkoxy oxygen

$$
RC \overset{\displaystyle O:}{\underset{\displaystyle \underset{\displaystyle H}{+\ddot{O}R'}}{\diagdown}}
$$

Positive charge is localized
on a single oxygen

Protonation of the carbonyl oxygen, as noted earlier in the reactions of aldehydes and ketones, makes the carbonyl group more susceptible to nucleophilic attack. A water molecule adds to the carbonyl group of the protonated ester in step 2. Loss of a proton from the resulting oxonium ion gives the neutral form of the tetrahedral intermediate in step 3 and completes the first stage of the mechanism.

Once formed, the tetrahedral intermediate can revert to starting materials by merely reversing the reactions that formed it, or it can continue onward to products. In the second stage of the reaction, the tetrahedral intermediate dissociates to an alcohol and a carboxylic acid. In step 4 of Figure 21.5, protonation of the tetrahedral intermediate at its alkoxy oxygen gives a new oxonium ion, which loses a molecule of alcohol in step 5. Along with the alcohol, the protonated form of the carboxylic acid arises by dissociation of the tetrahedral intermediate. Its deprotonation in step 6 completes the process.

PROBLEM 21.10   Based on the general mechanism for acid-catalyzed ester hydrolysis shown in Figure 21.5, write an analogous sequence of steps for the specific case of ethyl benzoate hydrolysis.

The most important facet of the mechanism outlined in Figure 21.5 is the tetrahedral intermediate. Evidence in support of the existence of such a species was developed by Professor Myron Bender on the basis of isotopic labeling experiments he carried out at the University of Chicago. Bender prepared ethyl benzoate, labeled with the mass 18 isotope of oxygen at the carbonyl oxygen, and then subjected it to acid-catalyzed hydrolysis in ordinary (unlabeled) water. He found that ethyl benzoate, recovered from the reaction before hydrolysis was complete, had lost a portion of its isotopic label. This observation is consistent only with the reversible formation of a tetrahedral intermediate under the reaction conditions.

In the above equation, O indicates an isotopically labeled oxygen atom. The two hydroxyl groups in the tetrahedral intermediate are equivalent to each other, so either the labeled or the unlabeled one can be lost when the tetrahedral intermediate reverts to ethyl benzoate. Both are retained when the tetrahedral intermediate goes on to form benzoic acid.

PROBLEM 21.11 In a similar experiment unlabeled γ-butyrolactone was allowed to stand in an acidic solution in which the water had been labeled with ^{18}O. When the lactone was extracted from the solution after 4 days, it was found to contain ^{18}O. Which oxygen of the lactone do you think became isotopically labeled?

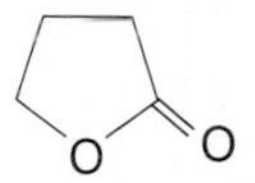

γ-Butyrolactone

21.10 BASE-PROMOTED ESTER HYDROLYSIS. SAPONIFICATION

Unlike its acid-catalyzed counterpart, ester hydrolysis in aqueous base is irreversible.

$$\underset{\text{Ester}}{\overset{\displaystyle O}{\overset{\|}{R\text{C}}}OR'} + \underset{\text{Hydroxide ion}}{HO^-} \longrightarrow \underset{\text{Carboxylate ion}}{\overset{\displaystyle O}{\overset{\|}{R\text{C}}}O^-} + \underset{\text{Alcohol}}{R'OH}$$

This is because carboxylic acids are converted to their corresponding carboxylate anions under these conditions, and these anions are incapable of acyl transfer to alcohols. Base is consumed in the reaction, so we speak of it as *base-promoted* rather than base-catalyzed.

In cases in which ester hydrolysis is carried out for preparative purposes, basic conditions are normally chosen. Base-promoted ester hydrolysis is both irreversible and usually occurs faster than hydrolysis in acid.

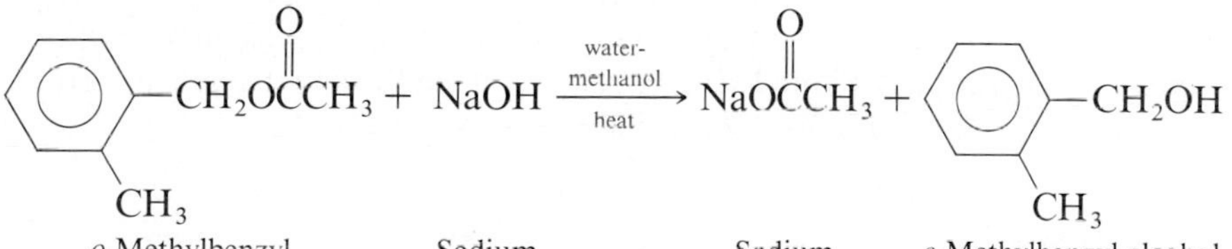

o-Methylbenzyl acetate + Sodium hydroxide → Sodium acetate + *o*-Methylbenzyl alcohol (95–97%)

When one wishes to isolate the carboxylic acid product, a separate acidification step following hydrolysis is necessary. Acidification converts the carboxylate salt to the free acid.

$$\underset{\substack{\text{Methyl 2-methylpropenoate}\\\text{(methyl methacrylate)}}}{CH_2{=}\overset{\displaystyle O}{\overset{\|}{\underset{\underset{\displaystyle CH_3}{|}}{C}C}}OCH_3} \xrightarrow[\text{2. } H_2SO_4]{\text{1. NaOH, } H_2O\text{, heat}} \underset{\substack{\text{2-Methylpropenoic}\\\text{acid (87\%)}\\\text{(methacrylic acid)}}}{CH_2{=}\overset{\displaystyle O}{\overset{\|}{\underset{\underset{\displaystyle CH_3}{|}}{C}C}}OH} + \underset{\text{Methyl alcohol}}{CH_3OH}$$

Base-promoted ester hydrolysis is called *saponification,* which means "soap-making." Over 2000 years ago, the Phoenicians made soap by heating animal fat with wood ashes. Animal fat is rich in glycerol triesters, and wood ashes are a source of potassium carbonate. Base-promoted cleavage of the fats produced a mixture of long-chain carboxylic acids as their potassium salts.

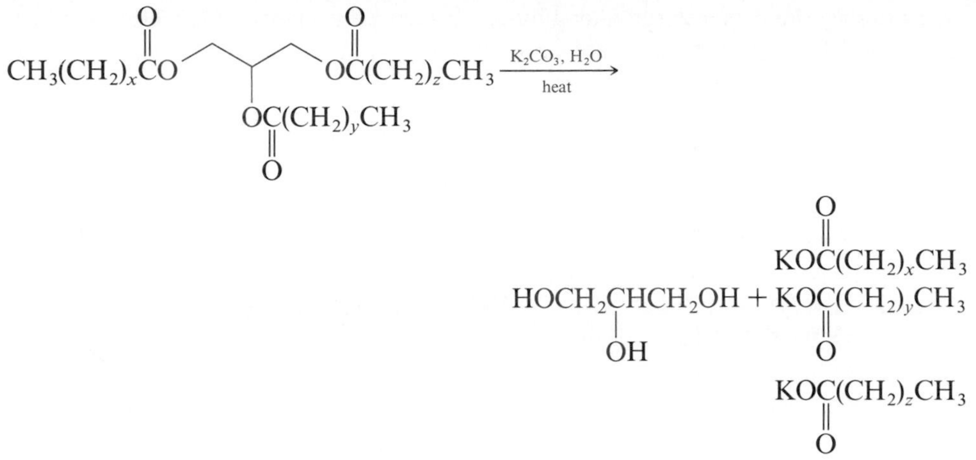

$$CH_3(CH_2)_x\overset{O}{\underset{}{C}}O\text{—}\overset{\displaystyle OC(CH_2)_yCH_3}{\underset{}{}}\text{—}OC(CH_2)_zCH_3 \xrightarrow[\text{heat}]{K_2CO_3,\ H_2O}$$

$$HOCH_2CHCH_2OH \underset{\displaystyle OH}{} + \begin{array}{l} KOC(CH_2)_xCH_3 \\ KOC(CH_2)_yCH_3 \\ KOC(CH_2)_zCH_3 \end{array}$$

Glycerol Potassium carboxylate salts

Potassium and sodium salts of long-chain carboxylic acids form micelles that dissolve grease (Section 20.5) and have cleansing properties. The carboxylic acids obtained by saponification of fats are called *fatty acids.*

PROBLEM 21.12 *Trimyristin* is obtained from coconut oil and has the molecular formula $C_{45}H_{86}O_6$. On being heated with aqueous sodium hydroxide followed by acidification, trimyristin was converted to glycerol and tetradecanoic acid as the only products. What is the structure of trimyristin?

In one the earliest kinetic studies of an organic reaction, carried out over a century ago, the rate of hydrolysis of ethyl acetate in aqueous sodium hydroxide was shown to be first-order in ester and first-order in base.

$$CH_3\overset{O}{\underset{}{C}}OCH_2CH_3 + NaOH \longrightarrow CH_3\overset{O}{\underset{}{C}}ONa + CH_3CH_2OH$$

Ethyl acetate Sodium hydroxide Sodium acetate Ethanol

$$Rate = k[CH_3\overset{O}{\underset{}{C}}OCH_2CH_3][NaOH]$$

Overall, the reaction exhibits second-order kinetics. Both the ester and the base are involved in the rate-determining step or in a rapid step that precedes it.

Two processes that are consistent with second-order kinetics both involve hydroxide ion as a nucleophile but differ in the site of nucleophilic attack. One of these processes is simply an S_N2 reaction in which hydroxide displaces carboxylate from the alkyl group of the ester. We say that this pathway involves *alkyl-oxygen cleavage* because it is the bond between oxygen and the alkyl group of the ester that breaks. The other process involves *acyl-oxygen cleavage,* with hydroxide attacking the carbonyl group.

Alkyl-oxygen cleavage

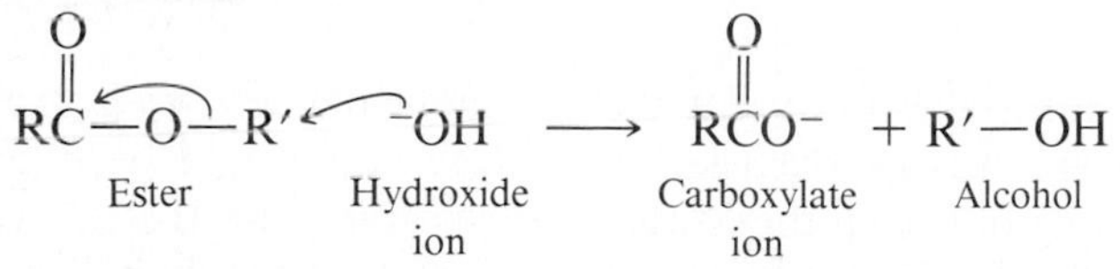

Ester	Hydroxide ion		Carboxylate ion	Alcohol

Acyl-oxygen cleavage

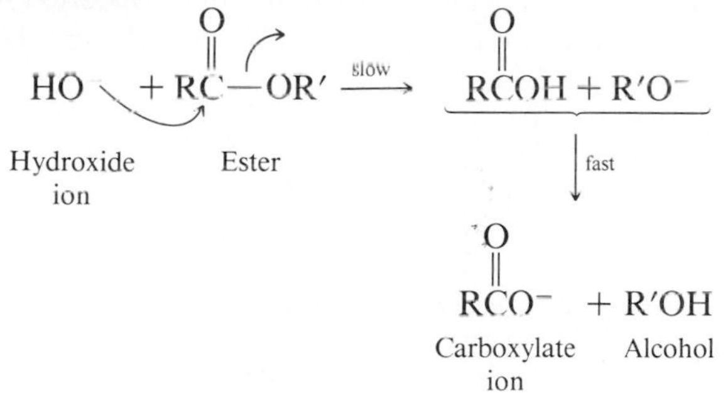

Convincing evidence that base-promoted ester hydrolysis proceeds by the second of these two paths, namely acyl-oxygen cleavage, has been obtained from several sources. In one experiment, ethyl propanoate labeled with ^{18}O in the ethoxy group was hydrolyzed in base. On isolating the products, all the ^{18}O was found in the ethyl alcohol; there was no ^{18}O enrichment in the sodium propanoate (as shown in the following equation, in which O indicates ^{18}O).

^{18}O-Labeled ethyl propanoate	Sodium hydroxide	Sodium propanoate	^{18}O-Labeled ethyl alcohol

Therefore, the carbon-oxygen bond broken in the process is the one between oxygen and the propanoyl group. The bond between oxygen and the ethyl group remains intact.

PROBLEM 21.13 In a similar experiment, pentyl acetate was subjected to saponification with ^{18}O-labeled hydroxide in ^{18}O-labeled water. What product do you think became isotopically labeled here, acetate ion or 1-pentanol?

Identical conclusions in support of acyl-oxygen cleavage have been obtained from stereochemical studies. Saponification of esters of optically active alcohols proceeds with *retention of configuration.*

$$CH_3\overset{\displaystyle O}{\overset{\displaystyle \|}{C}}-O-\overset{\displaystyle H}{\underset{\displaystyle CH_3}{C}}C_6H_5 \xrightarrow[\text{ethanol-water}]{KOH} CH_3\overset{\displaystyle O}{\overset{\displaystyle \|}{C}}OH + HO-\overset{\displaystyle H}{\underset{\displaystyle CH_3}{C}}C_6H_5$$

(R)-$(+)$-1-Phenylethyl acetate   Acetic acid (R)-$(+)$-1-Phenylethyl alcohol (80% yield; same optical purity as ester)

None of the bonds to the chiral center are broken when acyl-oxygen cleavage occurs. Had alkyl-oxygen cleavage occurred instead, it would have been accompanied by inversion of configuration at the chiral center to give (S)-$(-)$-1-phenylethyl alcohol.

Once it was established that hydroxide ion attacks the carbonyl group in base-promoted ester hydrolysis, the next question to be addressed concerned whether the reaction is concerted or involves an intermediate. In a concerted reaction acyl-oxygen cleavage occurs at the same time that hydroxide ion attacks the carbonyl group.

$$HO^- + R\overset{\displaystyle O}{\overset{\displaystyle \|}{C}}OR' \longrightarrow \left[HO\overset{\delta-}{\cdots}\underset{\displaystyle \overset{\displaystyle \|}{O}}{\overset{\displaystyle R}{C}}\overset{\delta-}{\cdots}OR' \right]^{\neq} \longrightarrow R\overset{\displaystyle O}{\overset{\displaystyle \|}{C}}OH + R'O^-$$

Hydroxide ion   Ester   Representation of activated complex for concerted displacement   Carboxylic acid   Alkoxide ion

In an extension of the work described in the preceding section, Bender showed that base-promoted ester hydrolysis was *not* concerted and, like acid hydrolysis, took place by way of a tetrahedral intermediate. The nature of the experiment was the same, and the results were similar to those observed in the acid-catalyzed reaction. Ethyl benzoate enriched in ^{18}O at the carbonyl oxygen was subjected to base-promoted hydrolysis and samples were isolated before saponification was complete. The recovered ethyl benzoate had lost a portion of its isotopic label, an observation consistent with the formation of a tetrahedral intermediate.

$$\underset{\substack{\text{Ethyl benzoate}\\ \text{(labeled with } ^{18}O)}}{C_6H_5\overset{\displaystyle O}{\overset{\displaystyle \|}{\underset{\displaystyle}{C}}}OCH_2CH_3} + \underset{\text{Water}}{H_2O} \rightleftharpoons \underset{\substack{\text{Tetrahedral}\\ \text{intermediate}}}{C_6H_5\overset{\displaystyle HO\ \ OH}{\underset{\displaystyle}{C}}OCH_2CH_3} \xrightarrow{HO^-}$$

$$C_6H_5\overset{\displaystyle O}{\overset{\displaystyle \|}{\underset{\displaystyle}{C}}}OCH_2CH_3 + H_2O$$

Ethyl benzoate   Water (labeled with ^{18}O)

All these facts—the observation of second-order kinetics, acyl-oxygen cleavage, and the involvement of a tetrahedral intermediate—are accommodated by the reaction mechanism shown in Figure 21.6. Like the acid-catalyzed mechanism, there are

Step 1: Nucleophilic addition of hydroxide ion to the carbonyl group

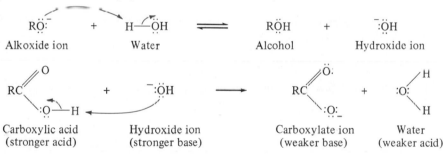

Hydroxide ion   Ester   Anionic form of tetrahedral intermediate

Step 2: Proton transfer to anionic form of tetrahedral intermediate

Anionic form of tetrahedral intermediate   Water   Tetrahedral intermediate   Hydroxide ion

Step 3: Dissociation of tetrahedral intermediate

Hydroxide ion   Tetrahedral intermediate   Water   Carboxylic acid   Alkoxide ion

Step 4: Proton transfer steps yield an alcohol and a carboxylate anion

Alkoxide ion   Water   Alcohol   Hydroxide ion

Carboxylic acid (stronger acid)   Hydroxide ion (stronger base)   Carboxylate ion (weaker base)   Water (weaker acid)

FIGURE 21.6 Sequence of steps that describes the mechanism of base-promoted ester hydrolysis.

two distinct stages, namely, formation of the tetrahedral intermediate followed by its dissociation. All the steps are reversible except the last one. The equilibrium constant for proton abstraction from the carboxylic acid by hydroxide is so large that step 4 is, for all intents and purposes, irreversible, and this makes the overall reaction irreversible.

Steps 2 and 4 are proton transfer reactions and are very fast. Nucleophilic addition to the carbonyl group has a higher activation energy than dissociation of the tetrahedral intermediate; step 1 is rate-determining.

PROBLEM 21.14 Based on the general mechanism for base-promoted ester hydrolysis shown in Figure 21.6, write an analogous sequence of steps for the saponification of ethyl benzoate.

21.11 REACTION OF ESTERS WITH AMMONIA AND AMINES

Esters function as acyl transfer agents to the nitrogen of ammonia to form amides.

$$
\underset{\text{Ester}}{\text{RCOR}'} + \underset{\text{Ammonia}}{\text{NH}_3} \longrightarrow \underset{\text{Amide}}{\text{RCNH}_2} + \underset{\text{Alcohol}}{\text{R}'\text{OH}}
$$

Ammonia is more nucleophilic than water and thus this reaction can be carried out in aqueous solution.

$$
\underset{\substack{\text{Methyl 2-methylpropenoate}}}{\text{CH}_2{=}\overset{\text{O}}{\overset{\|}{\text{C}}}\text{COCH}_3} \; + \; \underset{\text{Ammonia}}{\text{NH}_3} \xrightarrow{\text{H}_2\text{O}} \underset{\substack{\text{2-Methylpropenamide}\\(75\%)}}{\text{CH}_2{=}\overset{\text{O}}{\overset{\|}{\text{C}}}\text{CNH}_2} \; + \; \underset{\text{Methyl alcohol}}{\text{CH}_3\text{OH}}
$$

(with CH₃ substituent on each propene carbon)

Amines, which are substituted derivatives of ammonia, react similarly.

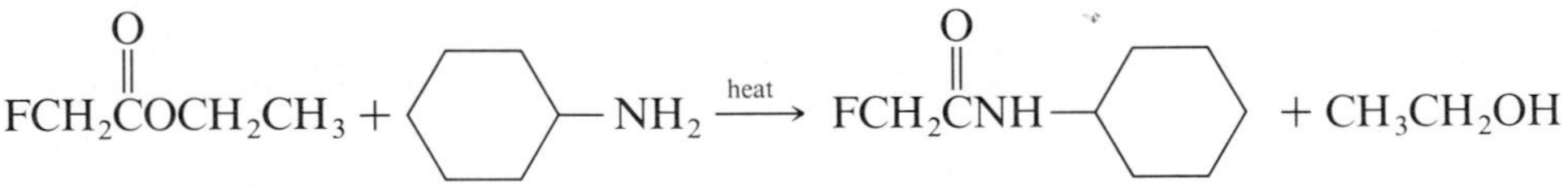

Ethyl fluoroacetate $\text{FCH}_2\text{COCH}_2\text{CH}_3$ + Cyclohexylamine $\langle\text{C}_6\text{H}_{11}\rangle\text{—NH}_2$ $\xrightarrow{\text{heat}}$ N-Cyclohexylfluoroacetamide (61%) $\text{FCH}_2\text{CNH—}\langle\text{C}_6\text{H}_{11}\rangle$ + Ethyl alcohol $\text{CH}_3\text{CH}_2\text{OH}$

The amine must be primary (RNH_2) or secondary (R_2NH). Tertiary amines (R_3N) cannot form amides because they have no proton on nitrogen that can be replaced by an acyl group.

PROBLEM 21.15 Give the structure of the expected product of the following reaction:

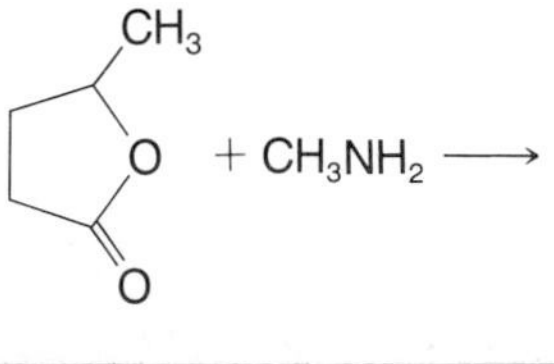

$$\text{+ CH}_3\text{NH}_2 \longrightarrow$$

The reaction of ammonia and amines with esters follows the same general mechanistic course as other acyl transfer reactions. A tetrahedral intermediate is formed in the first stage of the process and dissociates in the second stage.

Formation of tetrahedral intermediate

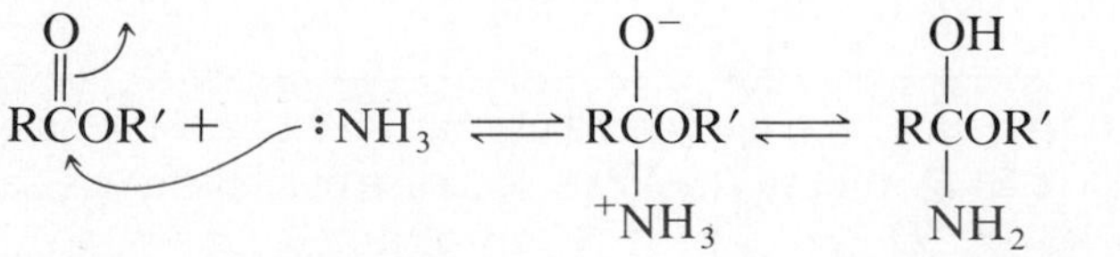

$$
\underset{\text{Ester}}{\text{RCOR}'} + \underset{\text{Ammonia}}{:\text{NH}_3} \rightleftharpoons \underset{\overset{|}{^+\text{NH}_3}}{\text{R}\overset{\text{O}^-}{\underset{|}{\text{C}}}\text{OR}'} \rightleftharpoons \underset{\substack{\text{Tetrahedral}\\\text{intermediate}}}{\text{R}\overset{\text{OH}}{\underset{\underset{\text{NH}_2}{|}}{\text{C}}}\text{OR}'}
$$

Dissociation of tetrahedral intermediate

$$RC\underset{NH_2}{\overset{O-H}{\mid}}OR' \rightleftharpoons RC\underset{NH_2}{\overset{O}{\diagup}} + R'OH$$

Tetrahedral   Amide   Alcohol
intermediate

While both stages are written as equilibrium processes, the position of equilibrium of the overall reaction lies far to the right because the amide carbonyl is stabilized to a much greater extent than the ester carbonyl.

21.12 THIOESTERS

Thioesters, compounds of the type $RCSR'$ (with $\overset{O}{\overset{\parallel}{}}$), undergo the same kinds of acyl transfer reactions as esters and by similar mechanisms. Nucleophilic acyl substitution of a thioester gives a *thiol* along with the product of acyl transfer. For example:

$$CH_3CSCH_2CH_2OC_6H_5 + CH_3OH \xrightarrow{HCl} CH_3COCH_3 + HSCH_2CH_2OC_6H_5$$

S-2-Phenoxyethyl   Methanol   Methyl   2-Phenoxyethanethiol
ethanethioate   acetate   (90%)

In this reaction an acetyl group is transferred from sulfur to the oxygen of methanol.

PROBLEM 21.16   Write the structure of the tetrahedral intermediate formed in the reaction just described.

The carbon-sulfur bond of a thioester is rather long—typically on the order of 1.8 Å—and delocalization of the sulfur lone pair electrons into the π orbital of the carbonyl group is not as pronounced as in esters. Acyl transfer reactions of thioesters are characterized by a more negative $\Delta G°$ than those of simple esters. A number of important biological processes involve thioesters; several of these are described in Chapter 28.

21.13 PREPARATION OF AMIDES

Amides are readily prepared by acylation of ammonia and amines with acyl chlorides, anhydrides, or esters.

Acylation of *ammonia* (NH_3) yields an amide ($R'CNH_2$, with $\overset{O}{\overset{\parallel}{}}$).

Primary amines (RNH_2) yield *N*-substituted amides ($R'CNHR$, with $\overset{O}{\overset{\parallel}{}}$).

Secondary amines (R_2NH) yield *N,N*-disubstituted amides ($R'CNR_2$, with $\overset{O}{\overset{\parallel}{}}$).

Examples illustrating these reactions may be found in Tables 21.3, 21.4, and 21.7.

Two molar equivalents of amine are required in the reaction with acyl chlorides and acid anhydrides; one molecule of amine acts as a nucleophile, the second as a Brönsted base.

$$2R_2NH + \underset{\text{Acyl chloride}}{R'\overset{\displaystyle O}{\overset{\|}{C}}Cl} \longrightarrow \underset{\text{Amide}}{R'\overset{\displaystyle O}{\overset{\|}{C}}NR_2} + \underset{\substack{\text{Hydrochloride salt} \\ \text{of amine}}}{R_2\overset{+}{N}H_2 \ Cl^-}$$

$$2R_2NH + \underset{\text{Acid anhydride}}{R'\overset{\displaystyle O}{\overset{\|}{C}}O\overset{\displaystyle O}{\overset{\|}{C}}R'} \longrightarrow \underset{\text{Amide}}{R'\overset{\displaystyle O}{\overset{\|}{C}}NR_2} + \underset{\substack{\text{Carboxylate salt} \\ \text{of amine}}}{R_2\overset{+}{N}H_2 \ ^-O\overset{\displaystyle O}{\overset{\|}{C}}R'}$$

It is possible to use only one molar equivalent of amine in these reactions provided some other base, such as sodium hydroxide, is present in the reaction mixture to react with the hydrogen chloride or carboxylic acid that is formed. This is a useful procedure in those cases in which the amine is a valuable one or is available only in small quantities.

Esters and amines react in a 1:1 molar ratio to give amides. No acidic product is formed from the acylating agent, so no additional base is required.

$$\underset{\text{Amine}}{R_2NH} + \underset{\text{Methyl ester}}{R'\overset{\displaystyle O}{\overset{\|}{C}}OCH_3} \longrightarrow \underset{\text{Amide}}{R'\overset{\displaystyle O}{\overset{\|}{C}}NR_2} + \underset{\text{Methanol}}{CH_3OH}$$

PROBLEM 21.17 Write an equation showing the preparation of the following amides from the indicated carboxylic acid derivative:

(a) $(CH_3)_2CH\overset{\displaystyle O}{\overset{\|}{C}}NH_2$ from an acyl chloride

(b) $CH_3\overset{\displaystyle O}{\overset{\|}{C}}NHCH_3$ from an acid anhydride

(c) $H\overset{\displaystyle O}{\overset{\|}{C}}N(CH_3)_2$ from a methyl ester

SAMPLE SOLUTION (a) Amides of the type $R\overset{\displaystyle O}{\overset{\|}{C}}NH_2$ are derived by acylation of ammonia.

$$\underset{\substack{\text{2-Methylpropanoyl} \\ \text{chloride}}}{(CH_3)_2CH\overset{\displaystyle O}{\overset{\|}{C}}Cl} + \underset{\text{Ammonia}}{2NH_3} \longrightarrow \underset{\text{2-Methylpropanamide}}{(CH_3)_2CH\overset{\displaystyle O}{\overset{\|}{C}}NH_2} + \underset{\text{Ammonium chloride}}{NH_4Cl}$$

Two molecules of ammonia are needed because its acylation produces, in addition to the desired amide, a molecule of hydrogen chloride. Hydrogen chloride (an acid) reacts with ammonia (a base) to give ammonium chloride.

All these reactions proceed by nucleophilic addition of the amine to the carbonyl group. Dissociation of the tetrahedral intermediate proceeds in the direction that leads to an amide.

$$\underset{\substack{\text{Acylating}\\\text{agent}}}{RCX} + \underset{\text{Amine}}{R_2'NH} \rightleftharpoons \underset{\substack{\text{Tetrahedral}\\\text{intermediate}}}{RC(O-H)(NR_2')X} \rightleftharpoons \underset{\text{Amide}}{RCNR_2'} + \underset{\substack{\text{Conjugate acid}\\\text{of leaving group}}}{HX}$$

The carbonyl group of an amide is stabilized to a greater extent than that of an acyl chloride, anhydride, or ester; amides are formed rapidly and in high yield from each of these carboxylic acid derivatives.

Amides are sometimes prepared directly from carboxylic acids and amines by a two-step process. The first step is an acid-base reaction in which the acid and the amine combine to form an ammonium carboxylate salt. On heating, the ammonium carboxylate salt loses water to form an amide.

$$\underset{\substack{\text{Carboxylic}\\\text{acid}}}{RCOH} + \underset{\text{Amine}}{R_2'NH} \longrightarrow \underset{\substack{\text{Ammonium}\\\text{carboxylate salt}}}{RCO^- \; R_2'NH_2^+} \xrightarrow{heat} \underset{\text{Amide}}{RCNR_2'} + \underset{\text{Water}}{H_2O}$$

In practice, both steps may be combined in a single operation by simply heating a carboxylic acid and an amine together.

$$\underset{\text{Benzoic acid}}{C_6H_5COH} + \underset{\text{Aniline}}{C_6H_5NH_2} \xrightarrow{225°C} \underset{\substack{N\text{-Phenylbenzamide}\\\text{(benzanilide) (80–84\%)}}}{C_6H_5CNHC_6H_5} + \underset{\text{Water}}{H_2O}$$

A similar reaction in which ammonia and carbon dioxide are heated under pressure is the basis of the industrial synthesis of *urea*. Here, the reactants first combine, yielding a salt called *ammonium carbamate:*

$$\underset{\text{Ammonia}}{H_3N:} + \underset{\text{Carbon dioxide}}{O=C=O} \longrightarrow H_3N^+-C(O)(O^-) \xrightarrow{NH_3} \underset{\text{Ammonium carbamate}}{H_2N-C(O)(O^-\,NH_4^+)}$$

On being heated, ammonium carbamate undergoes dehydration to form urea:

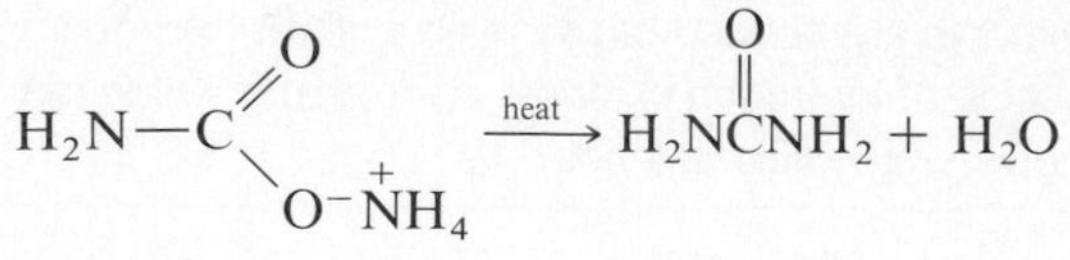

Ammonium carbamate   Urea   Water

Over 10^{10} lb of urea—most of it used as fertilizer—is produced annually in the United States by this method.

These thermal methods for preparing amides are limited in their generality. Most often amides are prepared in the laboratory from acyl chlorides, acid anhydrides, or esters, and these are the methods that you should apply to solving synthetic problems.

21.14 LACTAMS

Lactams are cyclic amides and are analogous to lactones, which are cyclic esters. Most lactams are known by their common names, as the examples shown illustrate.

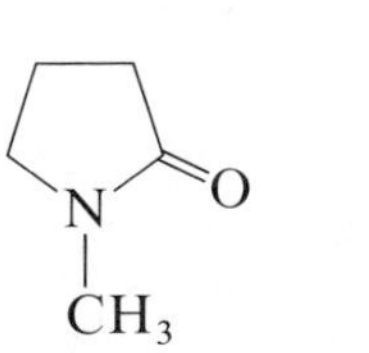

N-Methylpyrrolidone
(a polar aprotic
solvent)

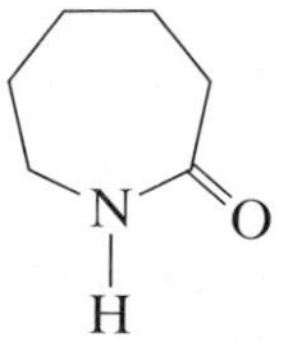

ϵ-Caprolactam
(industrial chemical
used to prepare nylon)

In contrast to their lactone counterparts, β-lactams, i.e., four-membered cyclic amides, are relatively stable. The β-lactam antibiotics, most specifically the penicillins and cephalosporins, are useful in treating bacterial infections.

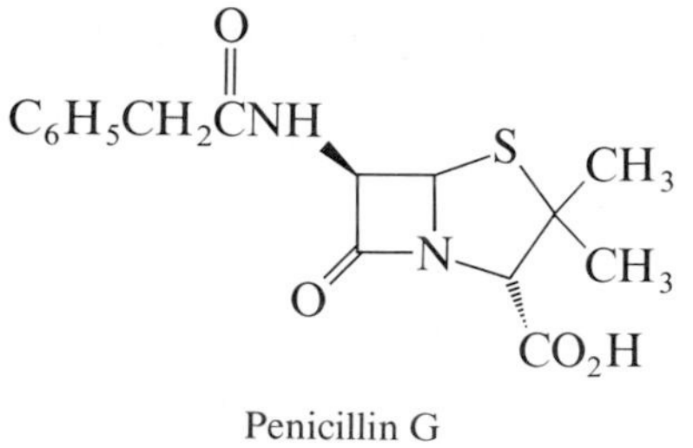

Penicillin G 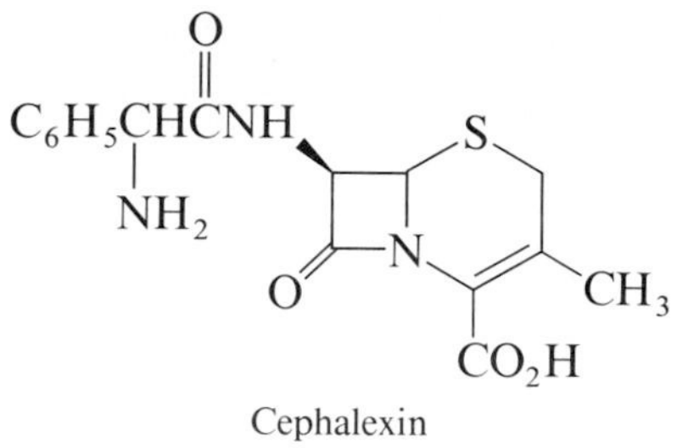   Cephalexin

21.15 IMIDES

Compounds that have two acyl groups bonded to a single nitrogen are known as *imides*. The most common imides are cyclic ones.

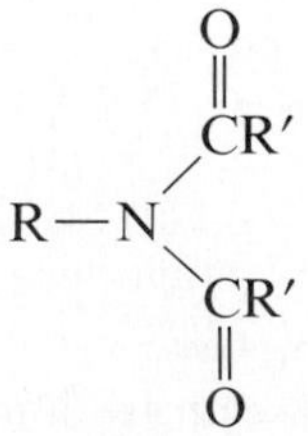

Imide 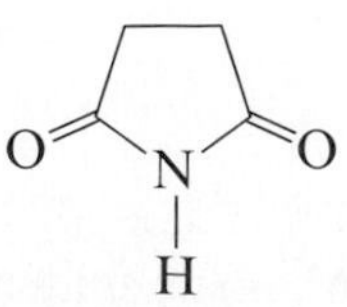   Succinimide 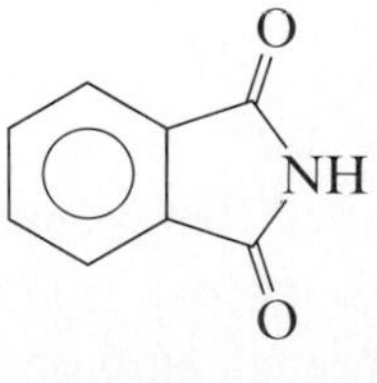   Phthalimide

Cyclic imides can be prepared by heating the ammonium salts of dicarboxylic acids.

$$\underset{\text{Succinic acid}}{\text{HOCCH}_2\text{CH}_2\text{COH}} + \underset{\text{Ammonia}}{2\text{NH}_3} \longrightarrow \underset{\text{Ammonium succinate}}{\text{NH}_4{}^+{}^-\text{OCCH}_2\text{CH}_2\text{CO}^-\text{NH}_4{}^+} \xrightarrow{\text{heat}} \underset{\text{Succinimide (82–83\%)}}{\text{NH}}$$

PROBLEM 21.18   Phthalimide has been prepared in 95 percent yield by heating the compound formed on reaction of phthalic anhydride (Section 21.4) with excess ammonia. This compound has the molecular formula $C_8H_{10}N_2O_3$. What is its structure?

21.16   HYDROLYSIS OF AMIDES

The only acyl transfer reaction that amides undergo is hydrolysis. Amides are fairly stable in water, but the amide bond is cleaved on heating in the presence of strong acids or bases. Nominally, this cleavage produces an amine and a carboxylic acid.

$$\underset{\text{Amide}}{\text{RCN}\overset{R'}{\underset{R'}{\Big\langle}}} + \underset{\text{Water}}{\text{H}_2\text{O}} \longrightarrow \underset{\substack{\text{Carboxylic} \\ \text{acid}}}{\text{RCOH}} + \underset{\text{Amine}}{\text{H}-\text{N}\overset{R'}{\underset{R'}{\Big\langle}}}$$

In acid, however, the amine is protonated, giving an ammonium ion, $R_2'\overset{+}{\text{NH}}_2$.

$$\underset{\text{Amide}}{\text{RCNR}_2'} + \underset{\text{Hydronium ion}}{\text{H}_3\text{O}^+} \longrightarrow \underset{\substack{\text{Carboxylic} \\ \text{acid}}}{\text{RCOH}} + \underset{\text{Ammonium ion}}{\text{R}'-\overset{\overset{H}{\overset{+}{|}}}{\underset{\underset{H}{|}}{\text{N}}}-\text{R}'}$$

In base the carboxylic acid is deprotonated, giving a carboxylate ion.

$$\underset{\text{Amide}}{\text{RCNR}_2'} + \underset{\text{Hydroxide ion}}{\text{HO}^-} \longrightarrow \underset{\text{Carboxylate ion}}{\text{RCO}^-} + \underset{\text{Amine}}{\text{R}'-\text{N}\overset{R'}{\underset{H}{\Big\langle}}}$$

The acid-base reactions that occur after the amide bond is broken render the overall hydrolysis irreversible in both cases. The amine product is removed by protonation in acid; the carboxylic acid is deprotonated in base.

$$CH_3CH_2\overset{\displaystyle O}{\overset{\|}{C}}CHCNH_2 \xrightarrow[\text{heat}]{H_2O,\ H_2SO_4} CH_3CH_2CHCOH + \overset{+}{N}H_4 \quad HSO_4^-$$

2-Phenylbutanamide   2-Phenylbutanoic   Ammonium hydrogen
           acid   sulfate
           (88–90%)

$$CH_3\overset{\displaystyle O}{\overset{\|}{C}}NH-\!\!\!\bigcirc\!\!\!-Br \xrightarrow[\text{water, heat}]{\text{KOH} \atop \text{ethanol-}} CH_3CO^-K^+ + H_2N-\!\!\!\bigcirc\!\!\!-Br$$

N-(4-Bromophenyl)acetamide   Potassium   *p*-Bromoaniline (95%)
  (*p*-bromoacetanilide)   acetate

Mechanistically amide hydrolysis is similar to the hydrolysis of other carboxylic acid derivatives. The mechanism of the acid-promoted hydrolysis is presented in Figure 21.7. It proceeds in two stages; a tetrahedral intermediate is formed in the first stage and dissociates in the second.

The amide is activated toward nucleophilic attack by protonation of its carbonyl oxygen. The cation produced in this step is stabilized by resonance involving the nitrogen lone pair. Electron delocalization permits the positive charge to be dispersed.

$$RC\overset{\overset{+}{O}H}{\underset{NH_2}{\diagup}} \longleftrightarrow RC\overset{\ddot{O}H}{\underset{\overset{+}{N}H_2}{\diagup}}$$

Most stable resonance forms of
an *O*-protonated amide

Proton transfer to the amide nitrogen gives a less stable cation.

$$RC\overset{\ddot{O}:}{\underset{\overset{+}{N}-H}{\diagup}}$$
      H   H

An acylammonium ion; the positive charge is localized on nitrogen

Protonation of nitrogen destroys the amide-type resonance that stabilizes the carbonyl group; protonation of oxygen enhances the electron delocalization involved in amide resonance.

Once formed, the *O*-protonated intermediate is attacked by a water molecule in step 2. The intermediate formed in this step loses a proton in step 3 to give the neutral form of the tetrahedral intermediate. The tetrahedral intermediate has its amino group ($-NH_2$) attached to sp^3 hybridized carbon, so the nitrogen lone pair is not delocalized by resonance. Therefore, the amino nitrogen is the most basic atom in the tetrahedral intermediate and is the site at which protonation occurs in step 4. Cleavage of the carbon-nitrogen bond in step 5 yields the protonated form of the carboxylic acid, along with a molecule of ammonia. In acid solution ammonia is immediately

Step 1: Protonation of the carbonyl oxygen of the amide

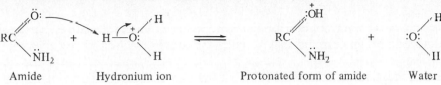

Amide   Hydronium ion   Protonated form of amide   Water

Step 2: Nucleophilic addition of water to the protonated form of the amide

Water   Protonated form of amide   Oxonium ion

Step 3: Deprotonation of the oxonium ion to give the neutral form of the tetrahedral intermediate

Oxonium ion   Water   Tetrahedral intermediate   Hydronium ion

Step 4: Protonation of the tetrahedral intermediate at its amino nitrogen

Tetrahedral intermediate   Hydronium ion   Ammonium ion   Water

Step 5: Dissociation of the *N*-protonated form of the tetrahedral intermediate to ammonia and the protonated form of the carboxylic acid

Ammonium ion   Protonated form of carboxylic acid   Ammonia

Step 6: Proton transfer processes yielding ammonium ion and the carboxylic acid

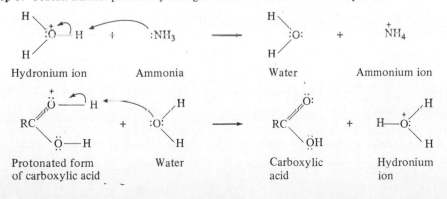

Hydronium ion   Ammonia   Water   Ammonium ion

Protonated form of carboxylic acid   Water   Carboxylic acid   Hydronium ion

FIGURE 21.7 Sequence of steps that describes the mechanism of acid-promoted amide hydrolysis. The formation of a tetrahedral intermediate is shown in steps 1 through 3. The dissociation of the tetrahedral intermediate is shown in steps 4 through 6.

protonated to give ammonium ion, as shown in step 6. This protonation step has such a large equilibrium constant that it makes the overall reaction irreversible.

PROBLEM 21.19 On the basis of the general mechanism for acid-promoted amide hydrolysis shown in Figure 21.7, write an analogous sequence of steps for the hydrolysis of

acetanilide, $CH_3CNHC_6H_5$.

In base the tetrahedral intermediate is formed in a manner analogous to that proposed for ester saponification. Steps 1 and 2 in Figure 21.8 show the formation of

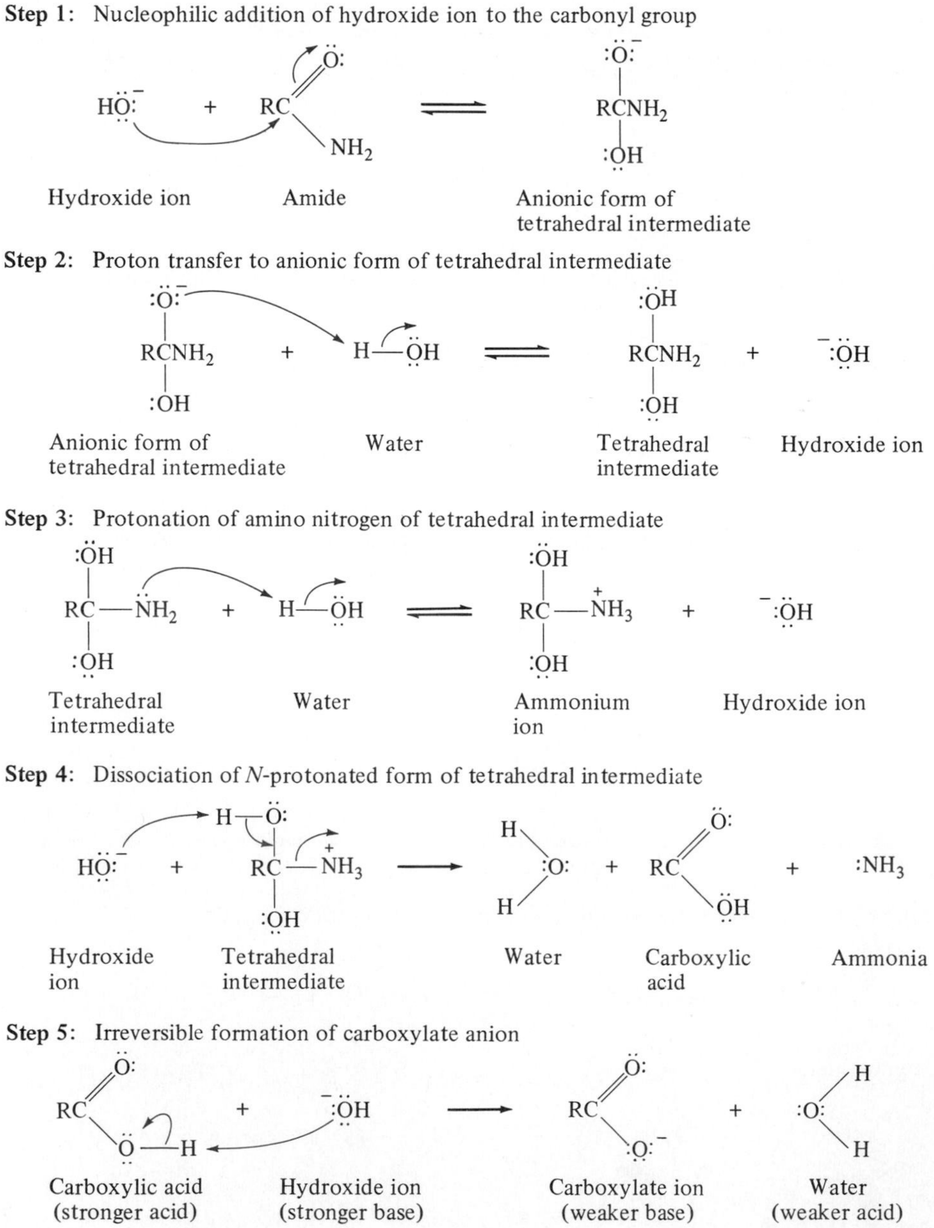

Step 1: Nucleophilic addition of hydroxide ion to the carbonyl group

Hydroxide ion   Amide   Anionic form of tetrahedral intermediate

Step 2: Proton transfer to anionic form of tetrahedral intermediate

Anionic form of tetrahedral intermediate   Water   Tetrahedral intermediate   Hydroxide ion

Step 3: Protonation of amino nitrogen of tetrahedral intermediate

Tetrahedral intermediate   Water   Ammonium ion   Hydroxide ion

Step 4: Dissociation of N-protonated form of tetrahedral intermediate

Hydroxide ion   Tetrahedral intermediate   Water   Carboxylic acid   Ammonia

Step 5: Irreversible formation of carboxylate anion

FIGURE 21.8 Sequence of steps that describes the mechanism of base-promoted amide hydrolysis.

Carboxylic acid (stronger acid)   Hydroxide ion (stronger base)   Carboxylate ion (weaker base)   Water (weaker acid)

the tetrahedral intermediate in the base-promoted hydrolysis of amides. In step 3 the basic amino group of the tetrahedral intermediate abstracts a proton from water, and in step 4 the derived ammonium ion undergoes base-promoted dissociation. Conversion of the carboxylic acid to its corresponding carboxylate anion in step 5 completes the process and renders the overall reaction irreversible.

PROBLEM 21.20   On the basis of the general mechanism for base-promoted hydrolysis shown in Figure 21.8, write an analogous sequence for the hydrolysis of *N,N*-dimethyl-

$$\overset{O}{\overset{\|}{H C}} N(CH_3)_2 .$$

formamide, $HCN(CH_3)_2$.

21.17 THE HOFMANN REARRANGEMENT OF *N*-BROMO AMIDES

On treatment with bromine in basic solution, amides of the type $R\overset{O}{\overset{\|}{C}}NH_2$ undergo an interesting reaction that leads to amines. This reaction was discovered by the German chemist August W. Hofmann over 100 years ago and is called the *Hofmann rearrangement*.

$$R\overset{O}{\overset{\|}{C}}NH_2 + Br_2 + 4HO^- \longrightarrow RNH_2 + 2Br^- + CO_3^{2-} + 2H_2O$$

Amide	Bromine	Hydroxide ion	Amine	Bromide ion	Carbonate ion	Water

The group R attached to the carboxamide function may be alkyl or aryl.

The relationship of the amine product to the amide reactant is rather remarkable. The overall reaction appears as if the carbonyl group has been plucked out of the amide, leaving behind a primary amine having one less carbon atom than the amide.

$$(CH_3)_3CCH_2\overset{O}{\overset{\|}{C}}NH_2 \xrightarrow[H_2O]{Br_2,\ NaOH} (CH_3)_3CCH_2NH_2$$

3,3-Dimethylbutanamide   2,2-Dimethylpropanamine
                               (neopentylamine) (94%)

m-Bromobenzamide $\xrightarrow[H_2O]{Br_2,\ KOH}$ *m*-Bromoaniline (87%)

PROBLEM 21.21   Outline an efficient synthesis of propanamine ($CH_3CH_2CH_2NH_2$) from butanoic acid.

The mechanism of the Hofmann rearrangement is presented in Figure 21.9. It involves three stages:

1. Formation of an *N*-bromo amide intermediate (steps 1 and 2)
2. Rearrangement of the *N*-bromo amide to an isocyanate (steps 3 and 4)
3. Hydrolysis of the isocyanate (steps 5 and 6)

Formation of the *N*-bromo amide intermediate is relatively straightforward. The base converts the amide to its corresponding anion (step 1), which acts as a nucleophile toward bromine (step 2).

Conversion of the *N*-bromo amide to its conjugate base in step 3 is also easy to understand. It is an acid-base reaction exactly analogous to that of step 1. The anion produced in step 3 is a key intermediate, which rearranges in step 4 by migration of the alkyl (or aryl) group from carbon to nitrogen, with loss of bromide from nitrogen. The product of this rearrangement is an isocyanate. The isocyanate formed in the

Overall Reaction

$$\underset{\text{Amide}}{\overset{\overset{\displaystyle O}{\|}}{\text{RCNH}_2}} \;+\; \underset{\text{Bromine}}{\text{Br}_2} \;+\; \underset{\text{Hydroxide ion}}{4\,\text{HO}^-} \;\longrightarrow\; \underset{\text{Amine}}{\text{RNH}_2} \;+\; \underset{\text{Bromide ion}}{2\,\text{Br}^-} \;+\; \underset{\text{Carbonate ion}}{\text{CO}_3^{2-}} \;+\; \underset{\text{Water}}{2\,\text{H}_2\text{O}}$$

Step 1: Deprotonation of the amide. Amides of the type $\overset{\overset{\displaystyle O}{\|}}{\text{RCNH}_2}$ are about as acidic as water, so appreciable quantities of the conjugate base are present at equilibrium in aqueous base. The conjugate base of an amide is stabilized by electron delocalization in much the same way that an enolate anion is.

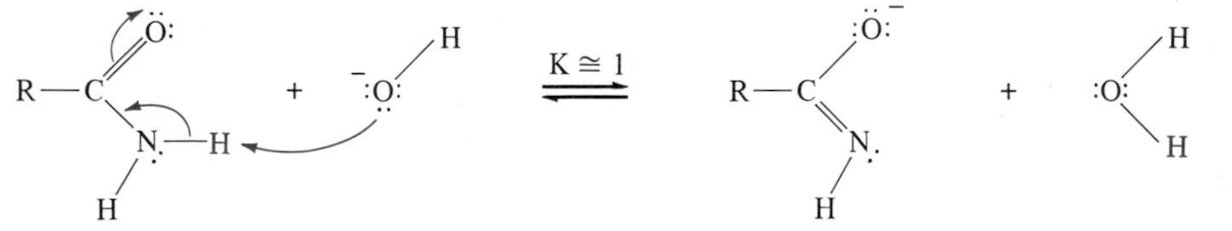

| Amide | Hydroxide ion | Conjugate base of amide | Water |

Step 2: Reaction of the conjugate base of the amide with bromine. The product of this step is an *N*-bromo amide.

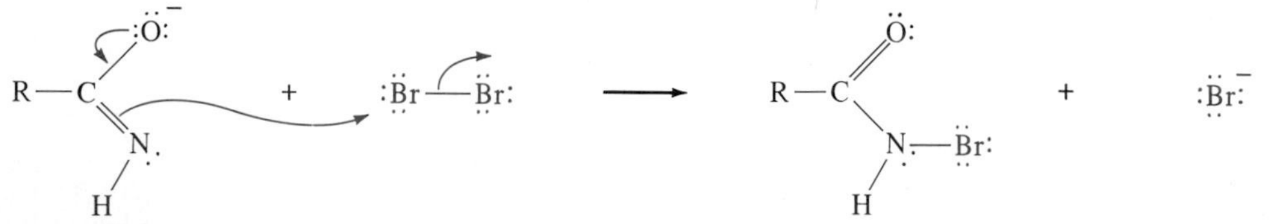

Conjugate base of amide   Bromine   *N*-Bromo amide   Bromide ion

Step 3: Deprotonation of the *N*-bromo amide. The electron-withdrawing effect of the bromine substituent reinforces that of the carbonyl group and makes the *N*-bromo amide even more acidic than the starting amide.

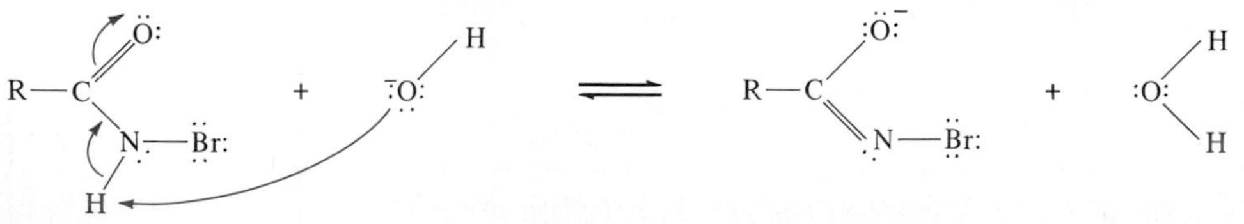

N-bromo amide   Hydroxide ion   Conjugate base of *N*-bromo amide   Water

FIGURE 21.9 Sequence of steps that describes the mechanism of the Hofmann rearrangement.

Step 4: Rearrangement of the conjugate base of the *N*-bromo amide. The group R migrates from carbon to nitrogen, and bromide ion is lost as a leaving group from nitrogen. The product of this rearrangement is an *N*-alkyl isocyanate.

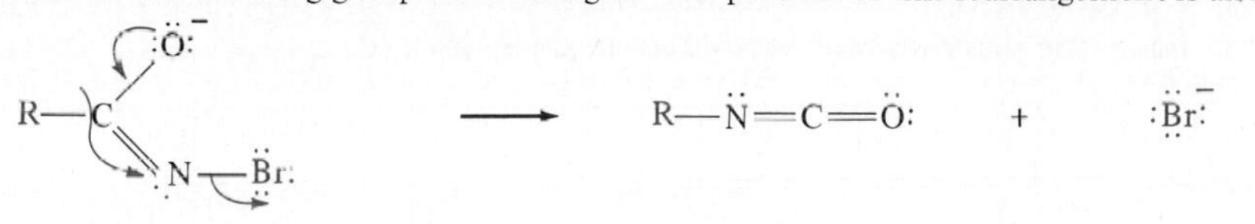

Conjugate base of *N*-bromo amide   *N*-Alkyl isocyanate   Bromide ion

Step 5: Hydrolysis of the isocyanate begins by base-catalyzed addition of water to form an *N*-alkylcarbamic acid.

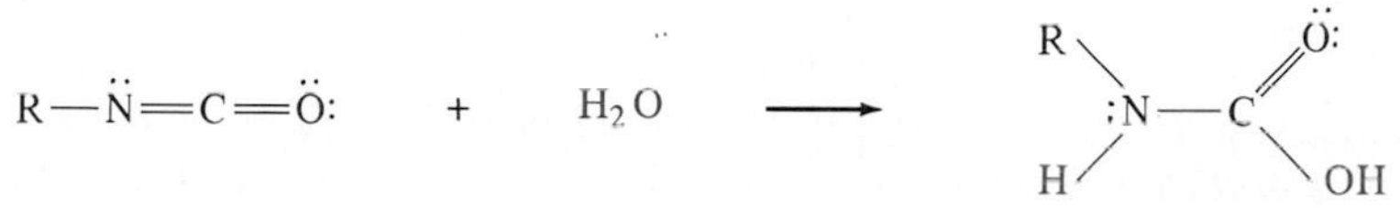

N-Alkyl isocyanate   *N*-Alkylcarbamic acid

Step 6: The *N*-alkylcarbamic acid is unstable and dissociates to an amine and carbon dioxide. Carbon dioxide is converted to carbonate ion in base. (Several steps are actually involved; in the interests of brevity, they are summarized as shown.)

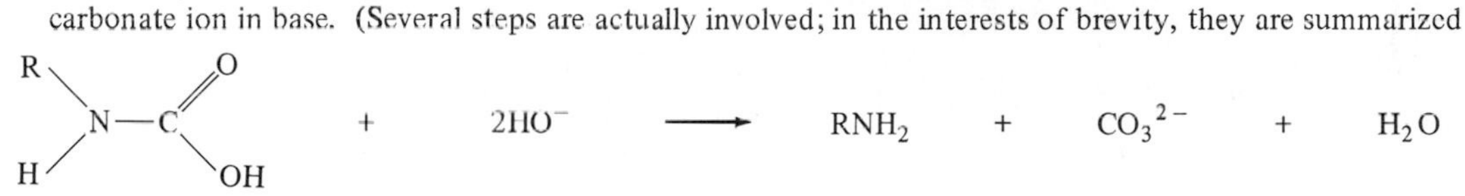

N-Alkylcarbamic acid   Hydroxide ion   Amine   Carbonate ion   Water

FIGURE 21.9 (continued)

rearrangement step then undergoes base-promoted hydrolysis in steps 5 and 6 to give the observed amine.

Among the experimental observations that contributed to elaboration of the mechanism shown in Figure 21.9 are the following:

1. Only amides of the type $\overset{\displaystyle O}{\overset{\displaystyle \|}{R C}} NH_2$ undergo the Hofmann rearrangement. The amide nitrogen must have *two* protons attached to it, of which one is replaced by bromine to give the *N*-bromo amide, while abstraction of the second by base is necessary to trigger the rearrangement. Amides of the type $\overset{\displaystyle O}{\overset{\displaystyle \|}{R C}} NHR'$ form *N*-bromo amides under the reaction conditions, but these *N*-bromo amides do not rearrange.

2. Rearrangement proceeds with *retention of configuration* at the migrating group.

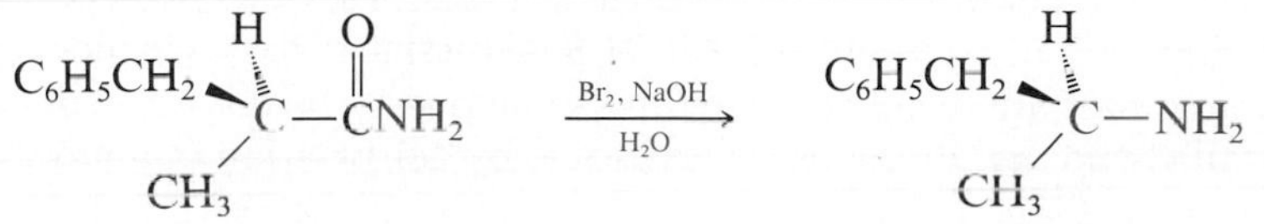

(*S*)-(+)-2-Methyl-3-phenylpropanamide   (*S*)-(+)-3-Phenyl-2-propanamine

The new carbon-nitrogen bond is formed at the same face of the migrating carbon as the bond that is broken. The rearrangement step depicted in Figure 21.9 satisfies this requirement. Presumably, carbon-nitrogen bond formation is concerted with carbon-carbon bond cleavage.

3. Isocyanates are intermediates. When the reaction of an amide with bromine is carried out in methanol containing sodium methoxide instead of in aqueous base, the product that is isolated is a *carbamate*.

$$CH_3(CH_2)_{14}\overset{O}{\overset{\|}{C}}NH_2 \quad \xrightarrow[CH_3OH]{Br_2,\ NaOCH_3} \quad CH_3(CH_2)_{14}NH\overset{O}{\overset{\|}{C}}OCH_3$$

Hexadecanamide   Methyl *N*-pentadecylcarbamate (84–94%)

Carbamates are esters of *carbamic acid* ($H_2N\overset{O}{\overset{\|}{C}}OH$). Carbamates are also known as *urethans*. They are relatively stable and are formed by addition of alcohols to isocyanates.

$$RN{=}C{=}O + CH_3OH \longrightarrow RNH\overset{O}{\overset{\|}{C}}OCH_3$$

Isocyanate   Methanol   Methyl *N*-alkylcarbamate

Carbamic acid itself ($H_2N\overset{O}{\overset{\|}{C}}OH$) and *N*-substituted derivatives of carbamic acid are unstable; they decompose spontaneously to carbon dioxide and ammonia or an amine. Thus in aqueous solution, an isocyanate intermediate yields an amine via the corresponding carbamic acid; in methanol, an isocyanate is converted to an isolable methyl carbamate. If desired, the carbamate can be isolated, purified, and converted to an amine in a separate hydrolysis operation.

While the Hofmann rearrangement is complicated with respect to mechanism, it is easy to carry out and gives amines that are sometimes difficult to prepare by other methods.

21.18 PREPARATION OF NITRILES

Nitriles are organic compounds that contain the $-C{\equiv}N$ functional group. We have already discussed the two principal reactions by which alkyl cyanides are prepared, namely the nucleophilic substitution of alkyl halides by cyanide and the conversion of aldehydes and ketones to cyanohydrins. Table 21.8 reviews aspects of these reactions. Neither of the reactions in Table 21.8 is suitable for aryl cyanides ($ArC{\equiv}N$); these compounds are readily prepared by a reaction to be discussed in Chapter 24.

Both alkyl and aryl cyanides are accessible by dehydration of amides.

$$R\overset{O}{\overset{\|}{C}}NH_2 \longrightarrow RC{\equiv}N + H_2O$$

Amide   Nitrile   Water
(R may be alkyl   (R may be alkyl
or aryl)   or aryl)

TABLE 21.8
Preparation of Nitriles

Reaction (section) and comments	General equation and specific example
Nucleophilic substitution by cyanide ion (Sections 9.1, 9.13) Cyanide ion is a good nucleophile and reacts with alkyl halides to give alkyl cyanides. The reaction is of the S_N2 type and is limited to primary and secondary alkyl halides. Tertiary alkyl halides undergo elimination; aryl and vinyl halides do not react.	$:N{\equiv}\bar{C}:\ +\ R{-}X \longrightarrow RC{\equiv}N\ +\ X^-$ Cyanide ion Alkyl halide Nitrile Halide ion $CH_3(CH_2)_8CH_2Cl \xrightarrow[\text{water}]{\text{KCN} \atop \text{ethanol-}} CH_3(CH_2)_8CH_2CN$ 1-Chlorodecane Undecanenitrile (95%)
Cyanohydrin formation (Section 18.12) Hydrogen cyanide adds to the carbonyl group of aldehydes and ketones.	$\underset{\text{Aldehyde or ketone}}{RCR'(=O)}\ +\ \underset{\text{Hydrogen cyanide}}{HCN}\ \longrightarrow\ \underset{\text{Cyanohydrin}}{RCR'(OH)(C{\equiv}N)}$ $CH_3CH_2CCH_2CH_3(=O) \xrightarrow[\text{H}^+]{\text{KCN}} CH_3CH_2CCH_2CH_3(OH)(CN)$ 3-Pentanone 2-Cyano-3-pentanol (75%)

Among the reagents used to effect the dehydration of amides is the compound P_4O_{10}. This compound is known by the common name "phosphorus pentoxide" because it was once thought to have the molecular formula P_2O_5. Indeed, one often finds it written that way in many chemical equations. Phosphorus pentoxide is the anhydride of phosphoric acid and is used in a number of reactions that require a dehydrating agent.

$$(CH_3)_2CHCNH_2(=O) \xrightarrow[200°C]{P_4O_{10}} (CH_3)_2CHC{\equiv}N$$

2-Methylpropanamide   2-Methylpropanenitrile (69–86%)

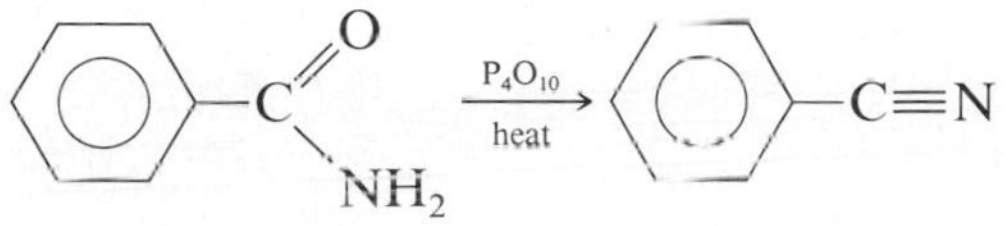

$$C_6H_5C(=O)NH_2 \xrightarrow[\text{heat}]{P_4O_{10}} C_6H_5{-}C{\equiv}N$$

Benzamide   Benzonitrile (74%)

PROBLEM 21.22 Show how ethyl alcohol could be used to prepare (a) CH_3CN and (b) CH_3CH_2CN. Along with ethyl alcohol you may use any necessary inorganic reagents.

An important nitrile is *acrylonitrile,* CH_2=CHCN. It is prepared industrially from propene, ammonia, and oxygen in the presence of a special catalyst. Polymers of acrylonitrile have many applications, the most prominent being their use in the preparation of acrylic fibers.

21.19 HYDROLYSIS OF NITRILES

Nitriles are classified as carboxylic acid derivatives because they are converted to carboxylic acids on hydrolysis. The conditions required are similar to those for the analogous conversion of amides, namely, heating in aqueous acid or base for several hours. Like the hydrolysis of amides, nitrile hydrolysis is irreversible in the presence of acids or bases. Acid hydrolysis yields ammonium ion and a carboxylic acid.

$$RC{\equiv}N + H_2O + \ \underset{\substack{\text{Hydronium}\\\text{ion}}}{H_3O^+} \longrightarrow \underset{\substack{\text{Carboxylic}\\\text{acid}}}{RCOH} + \underset{\substack{\text{Ammonium}\\\text{ion}}}{\overset{+}{N}H_4}$$

$$\underset{\substack{\text{Nitrile}\qquad\text{Water}}}{}$$

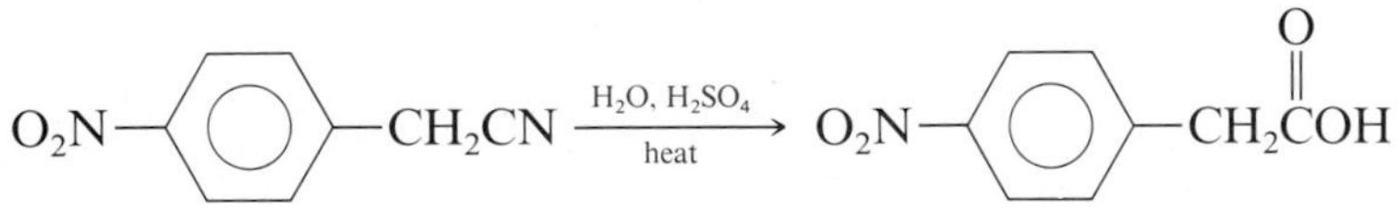

p-Nitrobenzyl cyanide — *p*-Nitrophenylacetic acid (92–95%)

In aqueous base hydroxide ion abstracts a proton from the carboxylic acid. In order to isolate the acid a subsequent acidification step is required.

$$RC{\equiv}N + H_2O + \ \underset{\substack{\text{Hydroxide}\\\text{ion}}}{HO^-} \longrightarrow \underset{\substack{\text{Carboxylate}\\\text{ion}}}{RCO^-} + \underset{\text{Ammonia}}{NH_3}$$

$$\underset{\substack{\text{Nitrile}\qquad\text{Water}}}{}$$

$$CH_3(CH_2)_9CN \xrightarrow[\text{2. } H^+]{\text{1. KOH,}H_2O\text{,heat}} CH_3(CH_2)_9COH$$

Undecanenitrile — Undecanoic acid (80%)

Nitriles are susceptible to nucleophilic addition. In their hydrolysis water adds across the carbon-nitrogen triple bond. In a series of proton transfer steps, an amide is produced.

Nitrile — Water — Imino acid — Amide

We have already discussed both the acid and base-promoted hydrolysis of amides (Section 21.16). All that remains to complete the mechanistic picture of nitrile hydrolysis is to examine the conversion of the nitrile to the corresponding amide.

Nucleophilic addition to the nitrile may be either acid- or base-catalyzed. In aqueous base hydroxide adds to the carbon-nitrogen triple bond.

$$HO:^- + RC{\equiv}N: \rightleftharpoons RC\begin{smallmatrix}\ddot{O}H\\N:^-\end{smallmatrix} \xrightarrow[OH^-]{H_2O} RC\begin{smallmatrix}\ddot{O}H\\NH\end{smallmatrix}$$

Hydroxide Nitrile   Imino acid
ion

The imino acid is transformed to the amide by the sequence

$$RC\begin{smallmatrix}O{-}H\\\ddot{N}H\end{smallmatrix} + :\ddot{O}H \rightleftharpoons RC\begin{smallmatrix}O\\\ddot{N}H\end{smallmatrix} + H{-}\ddot{O}H \rightleftharpoons RC\begin{smallmatrix}O\\NH_2\end{smallmatrix} + :\ddot{O}H$$

Imino Hydroxide Amide Water Amide Hydroxide
acid   ion   anion   ion

In acid, the nitrile is protonated on nitrogen, and water acts as the nucleophile.

$$H_2\ddot{O}: + RC{\equiv}\overset{+}{N}{-}H \rightleftharpoons RC\begin{smallmatrix}\overset{+}{O}H_2\\NH\end{smallmatrix} \underset{H_3O^+}{\overset{H_2O}{\rightleftharpoons}} RC\begin{smallmatrix}\ddot{O}H\\NH\end{smallmatrix}$$

Water Protonated   Imino acid
   form of nitrile

A series of proton transfers converts the imino acid to an amide.

$$RC\begin{smallmatrix}OH\\NH\end{smallmatrix} + H{-}\overset{+}{O}\begin{smallmatrix}H\\H\end{smallmatrix} \rightleftharpoons RC\begin{smallmatrix}O{-}H\\\overset{+}{NH_2}\end{smallmatrix} + :O:\begin{smallmatrix}H\\H\end{smallmatrix} \rightleftharpoons RC\begin{smallmatrix}O\\NH_2\end{smallmatrix} + H{-}\overset{+}{O}\begin{smallmatrix}H\\H\end{smallmatrix}$$

Imino Hydronium Conjugate acid Water Amide Hydronium
acid   ion   of amide   ion

Nucleophiles other than water can also add to the carbon-nitrogen triple bond of nitriles. In the following section we will see a synthetic application of such a nucleophilic addition.

21.20 ADDITION OF GRIGNARD REAGENTS TO NITRILES

The carbon-nitrogen triple bond of nitriles is much less reactive toward nucleophilic addition than is the carbon-oxygen double bond of aldehydes and ketones. Strongly basic nucleophiles such as Grignard reagents, however, do react with nitriles in a reaction that is of synthetic value.

$$RC{\equiv}N + R'MgX \xrightarrow[\text{2. }H_2O]{\substack{\text{1. diethyl}\\\text{ether}}} RC\overset{\overset{\displaystyle NH}{\|}}{}R' \xrightarrow[\text{heat}]{H_2O,H^+} RC\overset{\overset{\displaystyle O}{\|}}{}R'$$

Nitrile Grignard   Imine   Ketone
   reagent

The product of nucleophilic addition of the Grignard reagent to the nitrile is normally not isolated and purified but is hydrolyzed directly to a ketone. The overall sequence is used as a means of preparing ketones.

$$CH_3CH_2CN + C_6H_5MgBr \xrightarrow[\text{2. } H_2O,H^+,\text{heat}]{\text{1. diethyl ether}} C_6H_5\overset{\displaystyle O}{\overset{\|}{C}}CH_2CH_3$$

| Propanenitrile | Phenylmagnesium bromide | Ethyl phenyl ketone (91%) |

$$\underset{F_3C}{\text{(arene)}}—CN \quad + \quad CH_3MgI \xrightarrow[\substack{\text{2. } H_2O, H^+, \\ \text{heat}}]{\substack{\text{1. diethyl} \\ \text{ether}}} \underset{F_3C}{\text{(arene)}}—\overset{\displaystyle O}{\overset{\|}{C}}CH_3$$

m-(Trifluoromethyl)benzonitrile   Methylmagnesium iodide   m-(Trifluoromethyl)acetophenone (79%)

Organolithium reagents react in the same way and are often used instead of Grignard reagents.

21.21 SPECTROSCOPIC ANALYSIS OF CARBOXYLIC ACID DERIVATIVES

Infrared spectroscopy has been quite useful as an aid in determining structures of carboxylic acid derivatives. The carbonyl stretching vibration in particular is quite strong, and its position is sensitive to the nature of the carbonyl group. In general, electron donation from the substituent decreases the double bond character of the bond between carbon and oxygen and decreases the stretching frequency. Two distinct absorptions are observed for the symmetric and antisymmetric stretching vibrations of the anhydride function.

$$\underset{\substack{\text{Acetyl} \\ \text{chloride}}}{CH_3\overset{\displaystyle O}{\overset{\|}{C}}Cl} \qquad \underset{\substack{\text{Acetic} \\ \text{anhydride}}}{CH_3\overset{\displaystyle O}{\overset{\|}{C}}O\overset{\displaystyle O}{\overset{\|}{C}}CH_3} \qquad \underset{\substack{\text{Methyl} \\ \text{acetate}}}{CH_3\overset{\displaystyle O}{\overset{\|}{C}}OCH_3} \qquad \underset{\substack{\text{Acetamide}}}{CH_3\overset{\displaystyle O}{\overset{\|}{C}}NH_2}$$

$\nu_{C=O} = 1822$ cm^{-1} $\nu_{C=O} = 1748$ cm^{-1} $\nu_{C=O} = 1736$ cm^{-1} $\nu_{C=O} = 1694$ cm^{-1}
and 1815 cm^{-1}

Nitriles are readily identified by absorption due to —C≡N stretching in the 2210 to 2260 cm^{-1} region.

Chemical shift data in ^{1}H nuclear magnetic resonance spectroscopy permit assignment of structure in esters. Consider the two isomeric esters ethyl acetate and methyl propanoate. As Figure 21.10 shows, the number of signals and their multiplicities are the same for both esters. Both have a methyl singlet and a triplet-quartet pattern for their ethyl group.

singlet
$\delta = 2.0$ ppm
quartet
$\delta = 4.1$ ppm

$$CH_3\overset{\displaystyle O}{\overset{\|}{C}}OCH_2CH_3$$
triplet
$\delta = 1.3$ ppm

Ethyl acetate

singlet
$\delta = 3.6$ ppm
quartet
$\delta = 2.3$ ppm

$$CH_3O\overset{\displaystyle O}{\overset{\|}{C}}CH_2CH_3$$
triplet
$\delta = 1.2$ ppm

Methyl propanoate

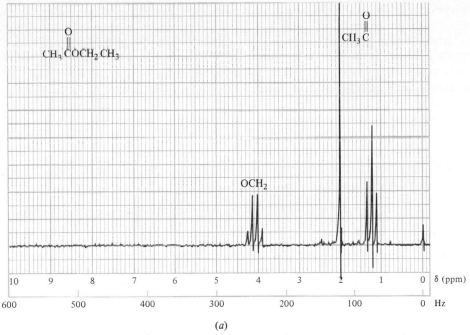

(a)

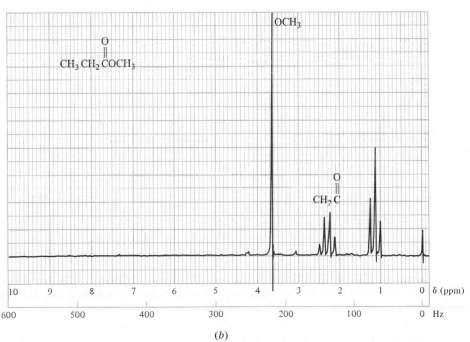

(b)

FIGURE 21.10 Proton nmr spectra of (a) ethyl acetate and (b) methyl propanoate.

Notice, however, that there is a significant difference in the chemical shifts of the corresponding signals in each spectrum. The methyl singlet is more shielded ($\delta =$ 2.0 ppm) when it is bonded to the carbonyl group of ethyl acetate than when it is bonded to the oxygen of methyl propanoate ($\delta =$ 3.6 ppm). The methylene quartet is more shielded ($\delta -$ 2.3 ppm) when it is bonded to the carbonyl group of methyl propanoate than when it is bonded to the oxygen of ethyl acetate ($\delta =$ 4.1 ppm). Analysis of the number of peaks and their splitting patterns will not provide an unambiguous answer to structure assignment in esters; chemical shift data, based on the observation that a carbonyl group is more shielding than oxygen, must also be used.

The chemical shift of the N—H proton of amides appears in the range $\delta = 5$ to 8 ppm. It is often a very broad peak; sometimes it is so broad that it does not rise much over the base line and can be lost in the background noise.

The ^{13}C nmr spectra of carboxylic acid derivatives, like the spectra of carboxylic acids themselves, are characterized by a low-field resonance for the carbonyl carbon in the range $\delta = 160$ to 180 ppm.

21.22 MASS SPECTROMETRY OF CARBOXYLIC ACID DERIVATIVES

A prominent peak in the mass spectra of most carboxylic acid derivatives is due to the corresponding acylium ion derived by cleavage of the bond to the carbonyl group.

$$R-C{\overset{\overset{+}{\ddot{O}}}{\underset{\ddot{X}\colon}{}}} \longrightarrow R-C\equiv\overset{+}{O}\colon + \cdot X\colon$$

Amides, however, tend to cleave in the opposite direction to produce a nitrogen-stabilized acylium ion.

$$R-C{\overset{\overset{+}{\ddot{O}}\colon}{\underset{\ddot{N}R_2'}{}}} \longrightarrow R\cdot + [\colon\overset{+}{O}\equiv C-\ddot{N}R_2' \longleftrightarrow \colon\ddot{O}=C=\overset{+}{N}R_2']$$

21.23 SUMMARY

The characteristic reaction of acyl chlorides, carboxylic acid anhydrides, esters, and amides is acyl transfer, or nucleophilic acyl substitution. Addition to the carbonyl group occurs to form a tetrahedral intermediate when these compounds react with nucleophilic reagents of the type HY :. The tetrahedral intermediate does not accumulate but dissociates to the product of nucleophilic acyl substitution.

$$\underset{\substack{\text{Carboxylic} \\ \text{acid derivative}}}{\overset{O}{\overset{\|}{R\overset{}{C}}}-X} + \underset{\text{Nucleophile}}{HY\colon} \rightleftharpoons \underset{\substack{\text{Tetrahedral} \\ \text{intermediate}}}{\overset{OH}{\underset{Y}{R\overset{|}{\underset{|}{C}}-X}}} \rightleftharpoons \underset{\substack{\text{Product of} \\ \text{nucleophilic} \\ \text{acyl substitution}}}{\overset{O}{\overset{\|}{R\overset{}{C}}-Y}} + \underset{\substack{\text{Conjugate acid} \\ \text{of leaving} \\ \text{group}}}{HX\colon}$$

Acyl chlorides are powerful acylating agents. Their carbonyl carbon is only weakly stabilized by the chlorine substituent, and chloride is a good leaving group, since it is not very basic. When an acyl chloride is converted to an anhydride, an ester, or an amide, the product has a more highly stabilized carbonyl group than that of the acyl chloride.

$$\underset{\substack{\text{Acyl} \\ \text{chloride}}}{\text{R}\overset{\text{O}}{\overset{\|}{\text{C}}}\text{Cl}} + \underset{\substack{\text{Carboxylic} \\ \text{acid}}}{\text{R}'\overset{\text{O}}{\overset{\|}{\text{C}}}\text{OH}} \longrightarrow \underset{\substack{\text{Acid} \\ \text{anhydride}}}{\text{R}\overset{\text{O}}{\overset{\|}{\text{C}}}\text{O}\overset{\text{O}}{\overset{\|}{\text{C}}}\text{R}'} + \underset{\substack{\text{Hydrogen} \\ \text{chloride}}}{\text{HCl}}$$

$$\underset{\substack{\text{Acyl chloride}}}{\text{R}\overset{\text{O}}{\overset{\|}{\text{C}}}\text{Cl}} + \underset{\text{Alcohol}}{\text{R}'\text{OH}} \longrightarrow \underset{\text{Ester}}{\text{R}\overset{\text{O}}{\overset{\|}{\text{C}}}\text{OR}'} + \underset{\substack{\text{Hydrogen} \\ \text{chloride}}}{\text{HCl}}$$

$$\underset{\substack{\text{Acyl} \\ \text{chloride}}}{\text{R}\overset{\text{O}}{\overset{\|}{\text{C}}}\text{Cl}} + \underset{\text{Amine}}{2\text{R}_2'\text{NH}} \longrightarrow \underset{\text{Amide}}{\text{R}\overset{\text{O}}{\overset{\|}{\text{C}}}\text{NR}_2'} + \underset{\substack{\text{Ammonium} \\ \text{chloride salt}}}{\overset{+}{\text{R}_2'\text{NH}_2}\ \ \text{Cl}^-}$$

Examples of each of these reactions may be found in Table 21.3.

Acid anhydrides are less reactive acylating agents than are acyl chlorides but are useful reagents for the preparation of esters and amides.

$$\underset{\substack{\text{Acid} \\ \text{anhydride}}}{\text{R}\overset{\text{O}}{\overset{\|}{\text{C}}}\text{O}\overset{\text{O}}{\overset{\|}{\text{C}}}\text{R}} + \underset{\text{Alcohol}}{\text{R}'\text{OH}} \longrightarrow \underset{\text{Ester}}{\text{R}\overset{\text{O}}{\overset{\|}{\text{C}}}\text{OR}'} + \underset{\substack{\text{Carboxylic} \\ \text{acid}}}{\text{R}\overset{\text{O}}{\overset{\|}{\text{C}}}\text{OH}}$$

$$\underset{\substack{\text{Acid} \\ \text{anhydride}}}{\text{R}\overset{\text{O}}{\overset{\|}{\text{C}}}\text{O}\overset{\text{O}}{\overset{\|}{\text{C}}}\text{R}} + \underset{\text{Amine}}{2\text{R}_2'\text{NH}} \longrightarrow \underset{\text{Amide}}{\text{R}\overset{\text{O}}{\overset{\|}{\text{C}}}\text{NR}_2'} + \underset{\substack{\text{Ammonium} \\ \text{carboxylate salt}}}{\overset{+}{\text{R}_2'\text{NH}_2}\ \ {}^-\text{O}\overset{\text{O}}{\overset{\|}{\text{C}}}\text{R}}$$

Table 21.4 presents examples of these reactions.

Esters react with amines to give amides (Table 21.7).

$$\underset{\text{Ester}}{\text{R}\overset{\text{O}}{\overset{\|}{\text{C}}}\text{OR}'} + \underset{\text{Amine}}{\text{R}_2''\text{NH}} \longrightarrow \underset{\text{Amide}}{\text{R}\overset{\text{O}}{\overset{\|}{\text{C}}}\text{NR}_2''} + \underset{\text{Alcohol}}{\text{R}'\text{OH}}$$

All these compounds—acyl chlorides, anhydrides, esters, and amides—may be converted to carboxylic acids by hydrolysis.

$$\underset{\substack{\text{Carboxylic} \\ \text{acid derivative}}}{\text{R}\overset{\text{O}}{\overset{\|}{\text{C}}}\text{X}:} + \underset{\text{Water}}{\text{H}_2\text{O}} \longrightarrow \underset{\substack{\text{Carboxylic} \\ \text{acid}}}{\text{R}\overset{\text{O}}{\overset{\|}{\text{C}}}\text{OH}} + \underset{\substack{\text{Conjugate acid} \\ \text{of leaving group}}}{\text{HX}:}$$

Hydrolysis is irreversible in base because the carboxylic acid is converted to the corresponding carboxylate anion under these conditions. Amide hydrolysis is irreversible in acid as well because of protonation of the amine product.

Nitriles, compounds of the type RC≡N, are useful materials. They may be hydrolyzed to carboxylic acids or converted to ketones by reaction with Grignard reagents.

$$\underset{\text{Nitrile}}{RC\equiv N} \xrightarrow[\substack{\text{or}\\ \text{1. } H_2O,\ HO^-\\ \text{2. } H^+}]{H_2O,\ H^+} \underset{\text{Carboxylic acid}}{R\overset{\displaystyle O}{\overset{\|}{C}}OH}$$

$$\underset{\text{Nitrile}}{RC\equiv N} + \underset{\text{Grignard reagent}}{R'MgX} \xrightarrow[\text{2. } H_2O,\ H^+]{\text{1. diethyl ether}} \underset{\text{Ketone}}{R\overset{\displaystyle O}{\overset{\|}{C}}R'}$$

PROBLEMS

21.23 Write a structural formula for each of the following compounds:

(a) *m*-Chlorobenzoyl bromide
(b) Trifluoroacetic anhydride
(c) *cis*-1,2-Cyclopropanedicarboxylic anhydride
(d) Ethyl cycloheptanecarboxylate
(e) 1-Phenylethyl acetate
(f) 2-Phenylethyl acetate
(g) *p*-Ethylbenzamide
(h) *N*-Ethylbenzamide
(i) 2-Methylhexanenitrile

21.24 Give an acceptable IUPAC name for each of the following compounds:

(a) $\underset{\substack{|\\ Cl}}{CH_3CHCH_2}\overset{\displaystyle O}{\overset{\|}{C}}Br$

(b) $CH_3\overset{\displaystyle O}{\overset{\|}{C}}OCH_2 -$ ⬡

(c) $CH_3O\overset{\displaystyle O}{\overset{\|}{C}}CH_2 -$ ⬡

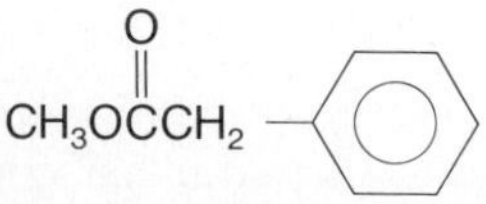

(d) $ClCH_2CH_2\overset{\displaystyle O}{\overset{\|}{C}}O\overset{\displaystyle O}{\overset{\|}{C}}CH_2CH_2Cl$

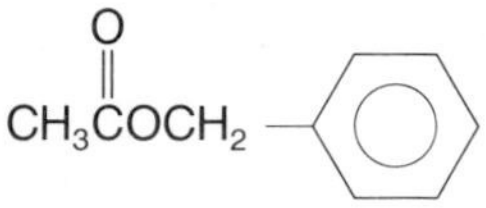

(e) (structure with H₃C and H₃C groups on ring with two C=O and O)

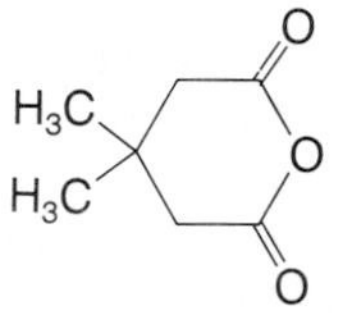

(f) $(CH_3)_2CHCH_2CH_2C\equiv N$

(g) $(CH_3)_2CHCH_2CH_2\overset{\displaystyle O}{\overset{\|}{C}}NH_2$

(h) $(CH_3)_2CHCH_2CH_2\overset{\displaystyle O}{\overset{\|}{C}}NHCH_3$

(i) $(CH_3)_2CHCH_2CH_2\overset{\displaystyle O}{\overset{\|}{C}}N(CH_3)_2$

21.25 Write a structural formula for the principal organic product or products of each of the following reactions:

(a) Acetyl chloride and bromobenzene, $AlCl_3$
(b) Benzoyl chloride and lithium dimethylcuprate
(c) Propanoyl chloride and sodium propanoate
(d) Butanoyl chloride and benzyl alcohol
(e) *p*-Chlorobenzoyl chloride and ammonia

(f) 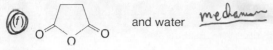 and water

 Also

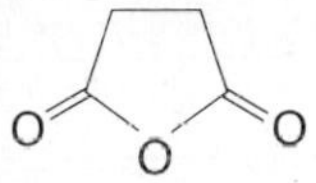

(g) 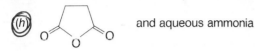 and aqueous sodium hydroxide

(h) 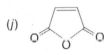 and aqueous ammonia

(i)  and benzene, AlCl₃

(j) and 1,3-pentadiene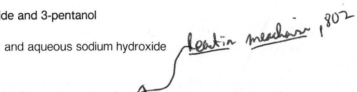

(k) Acetic anhydride and 3-pentanol

(l) 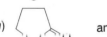 and aqueous sodium hydroxide  reaction mechanism, 802

(m) and aqueous ammonia

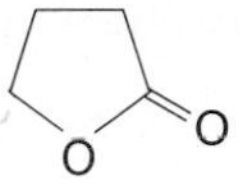

(n) and lithium aluminum hydride, then H₂O

(o) 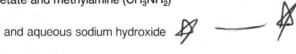 and excess methylmagnesium bromide, then H₃O⁺

(p) Ethyl phenylacetate and methylamine (CH₃NH₂)

(q) and aqueous sodium hydroxide

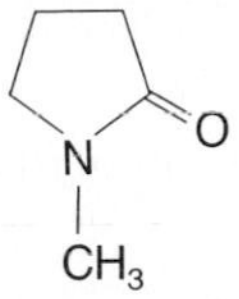

(r) and aqueous hydrochloric acid, heat — Mechanism

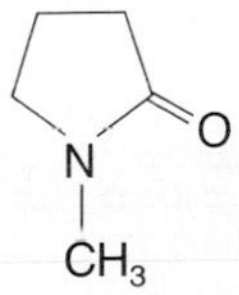

(s) and aqueous sodium hydroxide

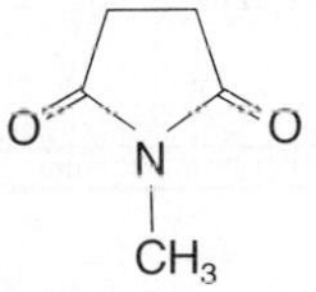

(t) and aqueous hydrochloric acid, heat — mechanism

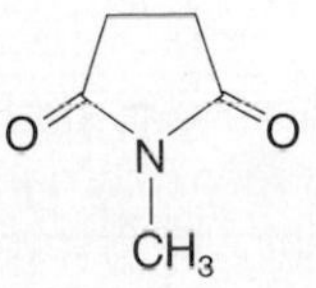

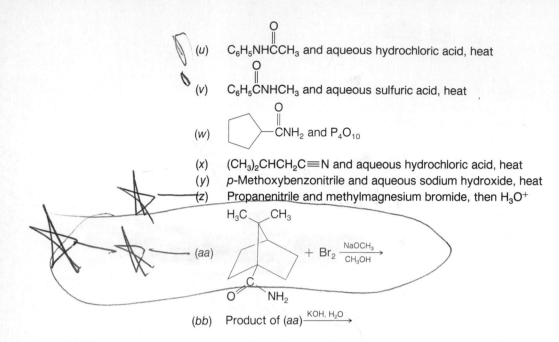

(u) C$_6$H$_5$NHCCH$_3$ and aqueous hydrochloric acid, heat

(v) C$_6$H$_5$CNHCH$_3$ and aqueous sulfuric acid, heat

(w) [cyclopentane]—CNH$_2$ and P$_4$O$_{10}$

(x) (CH$_3$)$_2$CHCH$_2$C≡N and aqueous hydrochloric acid, heat

(y) p-Methoxybenzonitrile and aqueous sodium hydroxide, heat

(z) Propanenitrile and methylmagnesium bromide, then H$_3$O$^+$

(aa) [bicyclic structure with H$_3$C, CH$_3$ and C(=O)NH$_2$] + Br$_2$ $\xrightarrow[\text{CH}_3\text{OH}]{\text{NaOCH}_3}$

(bb) Product of (aa) $\xrightarrow{\text{KOH, H}_2\text{O}}$

21.26 Using ethanol as the ultimate source of all the carbon atoms, along with any necessary inorganic reagents, show how you could prepare each of the following:

(a) Acetyl chloride
(b) Acetic anhydride
(c) Ethyl acetate
(d) Ethyl bromoacetate
(e) 2-Bromoethyl acetate
(f) Ethyl cyanoacetate
(g) Acetamide
(h) Methylamine (CH$_3$NH$_2$)
(i) 2-Hydroxypropanoic acid

21.27 Using toluene as the ultimate source of all the carbon atoms, along with any necessary inorganic reagents, show how you could prepare each of the following:

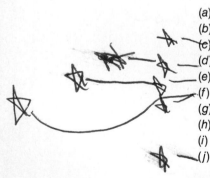

(a) Benzoyl chloride
(b) Benzoic anhydride
(c) Benzyl benzoate
(d) Benzamide
(e) Benzonitrile
(f) Benzyl cyanide
(g) Phenylacetic acid
(h) p-Nitrobenzoyl chloride
(i) m-Nitrobenzoyl chloride
(j) Aniline

21.28 The saponification of ^{18}O-labeled ethyl propanoate has been described in Section 21.10 as one of the significant experiments that demonstrated acyl-oxygen cleavage in ester hydrolysis. The ^{18}O-labeled ethyl propanoate used in this experiment was prepared from ^{18}O-labeled ethyl alcohol, which in turn was obtained from acetaldehyde and ^{18}O-enriched water. Write a series of equations showing the preparation of CH$_3$CH$_2$COCH$_2$CH$_3$ (where O = ^{18}O) from these starting materials.

21.29 Suggest a reasonable explanation for each of the following observations:

(a) The second-order rate constant k for saponification of ethyl trifluoroacetate is over 1 million times as great as that for ethyl acetate (25°C).

(b) The second-order rate constant for saponification of ethyl trimethylacetate (CH$_3$)$_3$CCO$_2$CH$_2$CH$_3$ is almost 100 times as small as that for ethyl acetate (30°C).

(c) The second-order rate constant k for saponification of methyl acetate is 100 times as great as that for *tert*-butyl acetate (25°C).

(d) The second-order rate constant k for saponification of methyl *m*-nitrobenzoate is 40 times as great as that for methyl benzoate (25°C).

(e) The second-order rate constant k for saponification of 5-pentanolide is over 20 times as great as that for 4-butanolide (25°C).

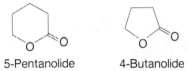

5-Pentanolide   4-Butanolide

(f) The second-order rate constant k for saponification of ethyl *trans*-4-*tert*-butylcyclo-hexanecarboxylate is 20 times as great as that for its cis diastereomer (25°C).

CO$_2$CH$_2$CH$_3$

~CO$_2$CH$_2$CH$_3$

Ethyl *trans*-4-*tert*-butyl-
cyclohexanecarboxylate

Ethyl *cis*-4-*tert*-butylcyclo-
hexanecarboxylate

21.30 The preparation of *cis*-4-*tert*-butylcyclohexanol from its trans stereoisomer was carried out by the sequence of steps described below. Write structural formulas, including stereochemistry, for compounds A and B.

Step 1:   ─OH + CH$_3$─⟨○⟩─ SO$_2$Cl $\xrightarrow{\text{pyridine}}$ compound A
(C$_{17}$H$_{26}$O$_3$S)

Step 2:   Compound A + ⟨○⟩─$\overset{\overset{\text{O}}{\|}}{C}$ONa $\xrightarrow[\text{heat}]{\text{dimethyl-formamide}}$ compound B
(C$_{17}$H$_{24}$O$_2$)

Step 3:   Compound B $\xrightarrow[\text{H}_2\text{O}]{\text{NaOH}}$ ─OH

21.31 Ambrettolide is obtained from hibiscus and has a musklike odor. Its preparation from methyl 9,10,16-trihydroxyhexadecanoate is outlined below. Write structural formulas, ignoring stereochemistry, for compounds C through K in this synthesis.

HOCH$_2$(CH$_2$)$_5$CH─CH(CH$_2$)$_7$CO$_2$CH$_3$ ⟶
                          | |
                          HO   OH

Methyl 9,10,16-trihydroxy-
hexadecanoate

Ambrettolide

Step	Reactant	Reagents	Product
1.	Methyl 9,10,16-trihydroxyhexadecanoate	Acetone, H_2SO_4	Compound C ($C_{20}H_{38}O_5$)
2.	Compound C	$KMnO_4$, then H^+	Compound D ($C_{20}H_{36}O_6$)
3.	Compound D	Sodium-ethanol, then H^+	Compound E ($C_{19}H_{36}O_5$)
4.	Compound E	H_2O, H^+, heat	Compound F ($C_{16}H_{32}O_5$)
5.	Compound F	HBr	Compound G ($C_{16}H_{29}Br_3O_2$)
6.	Compound G	Ethanol, H_2SO_4	Compound H ($C_{18}H_{33}Br_3O_2$)
7.	Compound H	Zinc, ethanol	Compound I ($C_{18}H_{33}BrO_2$)
8.	Compound I	Sodium acetate, acetic acid	Compound J ($C_{20}H_{36}O_4$)
9.	Compound J	KOH, ethanol, then H^+	Compound K ($C_{16}H_{30}O_3$)
10.	Compound K	Heat	Ambrettolide ($C_{16}H_{28}O_2$)

21.32 The preparation of the sex pheromone of the boll worm moth, (*E*)-9,11-dodecadien-1-yl acetate, from compound L has been described. Suggest suitable reagents for each step in this sequence.

(a) $HOCH_2CH{=}CH(CH_2)_7CO_2CH_3 \longrightarrow \overset{\overset{\displaystyle O}{\|}}{H}CCH{=}CH(CH_2)_7CO_2CH_3$

   Compound L   Compound M

(b) Compound M $\longrightarrow CH_2{=}CHCH{=}CH(CH_2)_7CO_2CH_3$

   Compound N

(c) Compound N $\longrightarrow CH_2{=}CHCH{=}CH(CH_2)_7CH_2OH$

   Compound O

(d) Compound O $\longrightarrow CH_2{=}CHCH{=}CH(CH_2)_7CH_2O\overset{\overset{\displaystyle O}{\|}}{C}CH_3$

   (*E*)-9,11-Dodecadien-1-yl acetate

21.33 Identify compounds P through R in the following equations:

(a)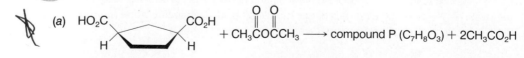

$HO_2C \cdots CO_2H + CH_3\overset{O}{\overset{\|}{C}}O\overset{O}{\overset{\|}{C}}CH_3 \longrightarrow$ compound P ($C_7H_8O_3$) + $2CH_3CO_2H$

(b) $CH_3\overset{O}{\overset{\|}{C}}CH_2CH_2\overset{O}{\overset{\|}{C}}OCH_2CH_3 \xrightarrow[\text{2. } H_3O^+]{\text{1. } CH_3MgI, \text{ diethyl ether}}$ compound Q ($C_6H_{10}O_2$)

(c)

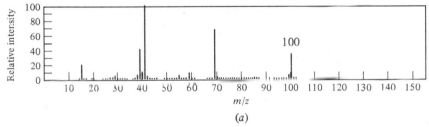

heat → compound R ($C_9H_4O_5$) + H_2O

21.34 When compounds of the type represented by S are allowed to stand in pentane, they are converted to a constitutional isomer.

$$RNHCH_2CH_2O\overset{O}{\overset{\|}{C}}-\bigcirc-NO_2 \longrightarrow compound\ T$$

Compound S

Hydrolysis of either S or T yields $RNHCH_2CH_2OH$ and p-nitrobenzoic acid.

(a) Suggest a reasonable structure for compound T.

(b) Demonstrate your understanding of the mechanism of this reaction by writing the structure of the key intermediate in the conversion of compound S to compound T.

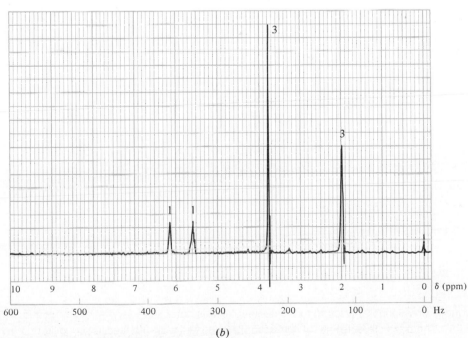

FIGURE 21.11 The (a) mass spectrum and (b) ^1H nmr spectrum of methyl methacrylate (Problem 21.37).

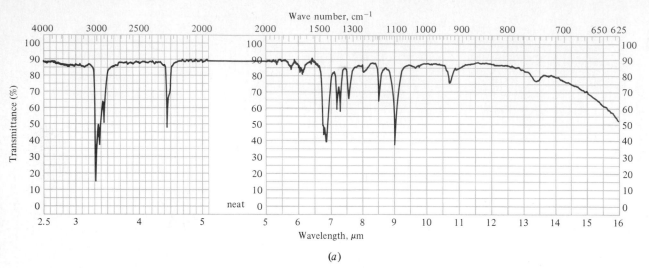

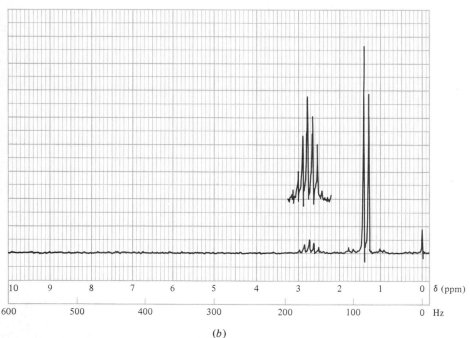

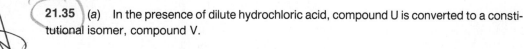

FIGURE 21.12 The (a) infrared and (b) ¹H nmr spectra of compound X (Problem 21.39).

21.35 (a) In the presence of dilute hydrochloric acid, compound U is converted to a constitutional isomer, compound V.

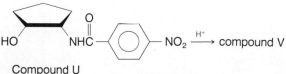

Compound U

Suggest a reasonable structure for compound V. (It is not a stereoisomer of compound U.)

(b) The trans stereoisomer of compound U is stable under the reaction conditions. Why does it not rearrange?

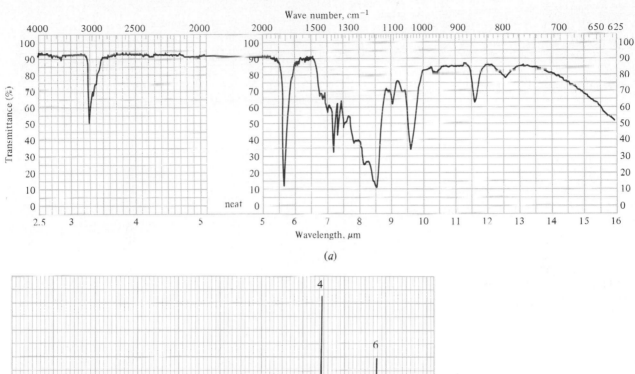

FIGURE 21.13 The (a) infrared and (b) ^1H nmr spectra of compound Y ($C_8H_{14}O_4$) (Problem 21.40).

21.36 Poly(vinyl alcohol) is a useful water-soluble polymer. It cannot be prepared directly from vinyl alcohol because of the rapidity with which vinyl alcohol (CH_2=CHOH) isomerizes to acetaldehyde. Vinyl acetate, however, does not rearrange and can be polymerized to poly(vinyl acetate). How could you make use of this fact to prepare poly(vinyl alcohol)?

$$\left(-CH_2CHCH_2CH-\right)_n$$
$$OH \quad OH$$

$$\left(-CH_2CHCH_2CH-\right)_n$$
$$CH_3CO \quad OCCH_3$$
$$\parallel \quad \parallel$$
$$O \quad O$$

Poly(vinyl alcohol)   Poly(vinyl acetate)

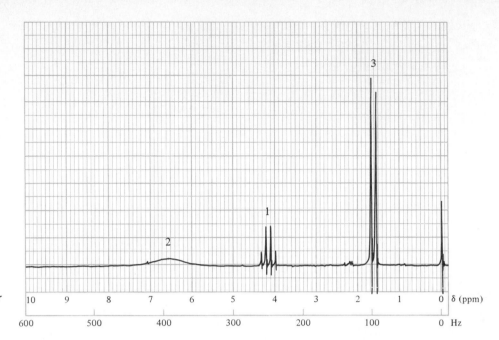

FIGURE 21.14 The ^{1}H nmr spectrum of compound Z (C_3H_6ClNO) (Problem 21.41).

21.37 *Lucite* is a polymer of methyl methacrylate, which is prepared from acetone by way of acetone cyanohydrin. The ^{1}H nmr and mass spectra of methyl methacrylate are shown in Figure 21.11. What is the structure of this important industrial chemical?

21.38 Compound W ($C_4H_6O_2$) has a strong band in the infrared at 1760 cm^{-1}. Its ^{13}C nmr spectrum exhibits signals at $\delta = 20.2$ (quartet), 96.8 (triplet), 141.8 (doublet), and 167.6 ppm (singlet). The ^{1}H nmr spectrum of compound W has a three-proton singlet at $\delta = 2.1$ ppm along with three other signals, each of which is a doublet of doublets, at $\delta = 4.7$, 4.9, and 7.3 ppm. What is the structure of compound W?

21.39 Compound X has a molecular weight of 69 and contains nitrogen. Identify compound X on the basis of the infrared and ^{1}H nmr spectra shown in Figure 21.12.

21.40 The infrared and ^{1}H nmr spectra of compound Y ($C_8H_{14}O_4$) are shown in Figure 21.13. What is the structure of compound Y?

21.41 The infrared spectrum of compound Z (C_3H_6ClNO) has an intense peak at 1680 cm^{-1}. Its ^{1}H nmr spectrum is shown in Figure 21.14. What is the structure of compound Z? How could you prepare compound Z from propanoic acid?

ESTER ENOLATES

You have already had considerable experience with carbanionic compounds and their applications in synthetic organic chemistry. The first carbanion you studied in detail was acetylide ion in Chapter 10. This was followed in Chapter 15 by an examination of organometallic compounds where you saw that species of this type —Grignard reagents, for example—act as sources of negatively polarized carbon. In Chapter 19 you learned that enolate ions—reactive intermediates generated from aldehydes and ketones—are nucleophilic and that this property can be used to advantage in organic synthesis as a method for carbon-carbon bond formation.

The present chapter extends our study of carbanions to the enolate ions derived from esters. *Ester enolates* are important reagents in synthetic organic chemistry. The stabilized enolates derived from *β-keto esters* are particularly useful.

$$R \overset{\displaystyle \overset{O}{\|}}{\underset{\beta}{C}} \overset{}{\underset{\alpha}{CH_2}} \overset{\displaystyle \overset{O}{\|}}{C} OR'$$

β-Keto ester: a ketone carbonyl is *β* to
the carbonyl group of the ester

A proton attached to the *α* carbon atom of a *β*-keto ester is relatively acidic. Typical acid dissociation constants K_a for *β*-keto esters are $\sim 10^{-11}$ (pK_a 11). The *α* carbon atom is flanked by two electron-withdrawing carbonyl groups, so a carbanion formed at this site is highly stabilized. The electron delocalization in the anion of a *β*-keto ester is represented by the resonance structures:

Principle resonance structures of the anion of a *β*-keto ester

This chapter begins by describing the preparation and properties of β-keto esters, proceeds to a discussion of their synthetic applications, continues to an examination of related species, and concludes by exploring some recent developments in the active field of synthetic carbanion chemistry.

22.1 BASE-PROMOTED ESTER CONDENSATION. THE CLAISEN CONDENSATION

Before describing how β-keto esters are used as reagents for organic synthesis, we need to see how these compounds themselves are prepared. The principal method for the preparation of β-keto esters is a reaction known as the *Claisen condensation.*

$$2RCH_2\overset{\displaystyle O}{\overset{\displaystyle \|}{C}}OR' \xrightarrow[\text{2. H}_3\text{O}^+]{\text{1. NaOR'}} RCH_2\overset{\displaystyle O}{\overset{\displaystyle \|}{C}}\underset{\displaystyle R}{\overset{\displaystyle O}{\overset{\displaystyle \|}{C}}}HCOR' + R'OH$$

$$\text{Ester} \qquad\qquad \text{β-Keto ester} \qquad \text{Alcohol}$$

On treatment with alkoxide bases, esters undergo self-condensation to give a β-keto ester and an alcohol. Ethyl acetate, for example, undergoes a Claisen condensation on treatment with sodium ethoxide to give a β-keto ester known by its common name *ethyl acetoacetate* (also called *acetoacetic ester*).

$$2CH_3\overset{\displaystyle O}{\overset{\displaystyle \|}{C}}OCH_2CH_3 \xrightarrow[\text{2. H}_3\text{O}^+]{\text{1. NaOCH}_2\text{CH}_3} CH_3\overset{\displaystyle O}{\overset{\displaystyle \|}{C}}CH_2\overset{\displaystyle O}{\overset{\displaystyle \|}{C}}OCH_2CH_3 + CH_3CH_2OH$$

$$\text{Ethyl acetate} \qquad\qquad \text{Ethyl acetoacetate (75\%)} \qquad \text{Ethanol}$$
$$\text{(acetoacetic ester)}$$

The systematic name of ethyl acetoacetate is *ethyl 3-oxobutanoate.* The presence of a ketone carbonyl group is indicated by the designation *"oxo"* along with the appropriate locant. Thus, there are four carbon atoms in the acyl group of ethyl 3-oxobutanoate, C-3 being the carbonyl carbon of the ketone function.

The mechanism of the Claisen condensation of ethyl acetate is presented in Figure 22.1. The first two steps of the mechanism are analogous to those of the aldol addition reaction of aldehydes (Section 19.9). An enolate ion is generated in step 1, which undergoes nucleophilic addition to the carbonyl group of a second ester molecule in step 2. The species formed in this step is a tetrahedral intermediate of the same type that we encountered in our discussion of acyl transfer reactions of esters in Section 21.10. It dissociates by expelling an ethoxide ion, as shown in step 3, which restores the carbonyl group to give the β-keto ester. Steps 1 to 3 show two different types of reactivity by esters: one molecule of the ester gives rise to an enolate; the second molecule acts as an acyl transfer agent.

Claisen condensation reactions are performed in two distinct experimental operations. The first stage concludes in step 4 of Figure 22.1, where the base removes a proton from C-2 of the β-keto ester. Because this proton is relatively acidic, the position of equilibrium for step 4 lies far to the right.

In general, the equilibrium represented by the sum of steps 1 to 3 is not favorable

Overall reaction:

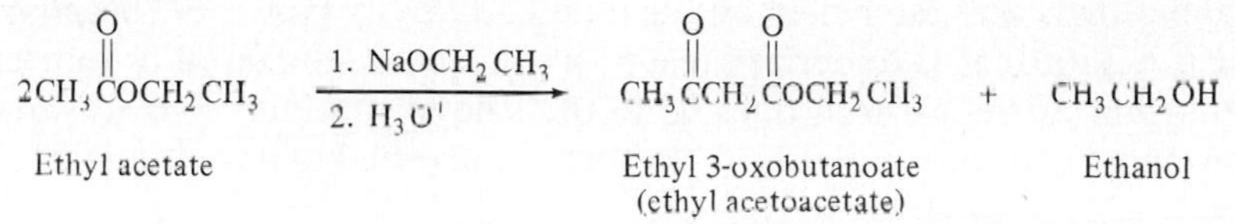

$$2CH_3COCH_2CH_3 \xrightarrow[\text{2. } H_3O^+]{\text{1. } NaOCH_2CH_3} CH_3CCH_2COCH_2CH_3 \quad + \quad CH_3CH_2OH$$

Ethyl acetate   Ethyl 3-oxobutanoate   Ethanol
                 (ethyl acetoacetate)

Step 1. Proton abstraction from the α carbon atom of ethyl acetate to give the corresponding enolate

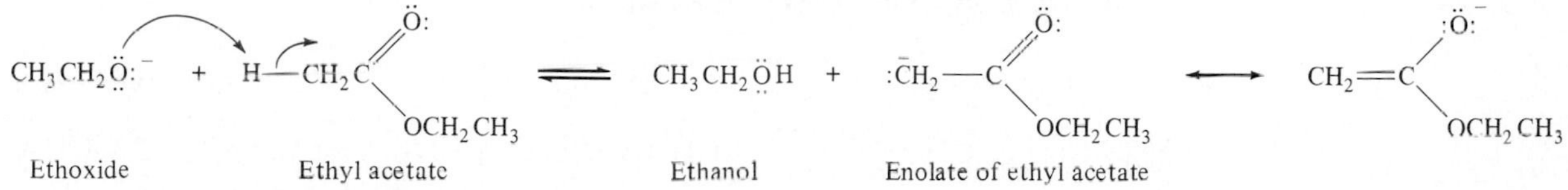

Ethoxide   Ethyl acetate   Ethanol   Enolate of ethyl acetate

Step 2: Nucleophilic addition of the ester enolate to the carbonyl group of the neutral ester. The product is the anionic form of the tetrahedral intermediate.

Ethyl acetate   Enolate of ethyl acetate   Anionic form of
                                        tetrahedral intermediate

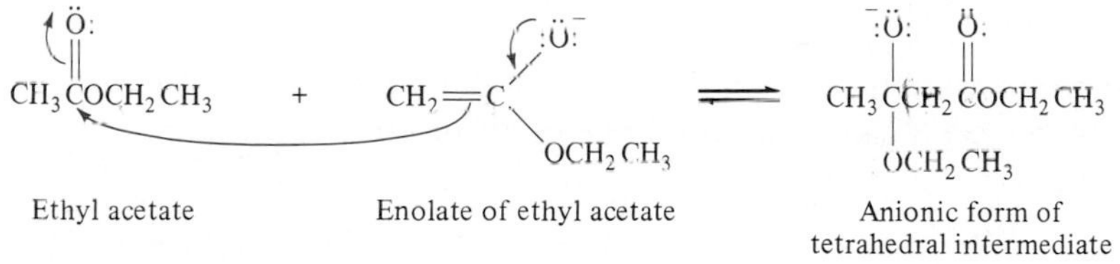

Step 3: Dissociation of the tetrahedral intermediate.

Anionic form of   Ethyl 3-oxobutanoate   Ethoxide ion
tetrahedral intermediate

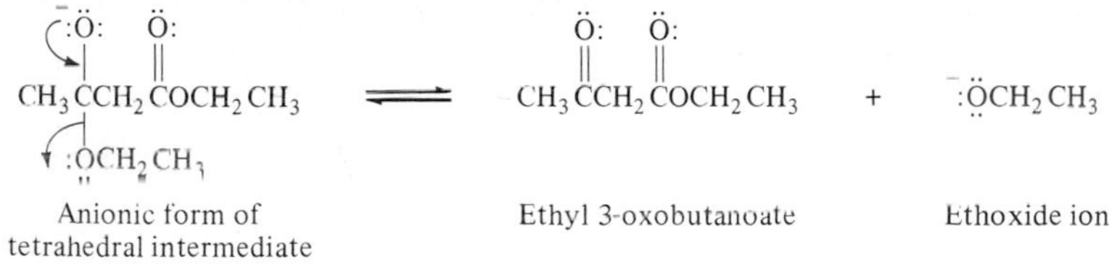

Step 4: Deprotonation of the β-keto ester product.

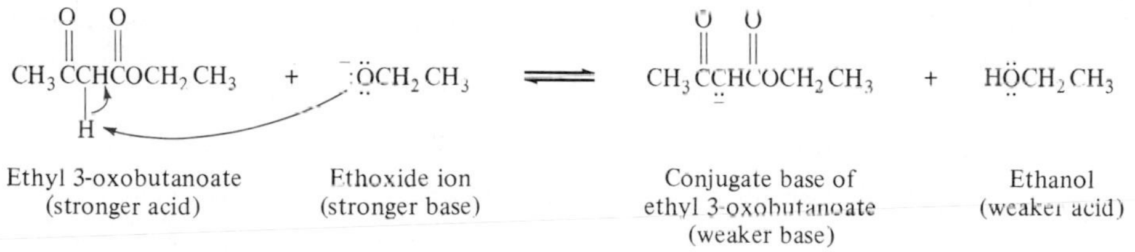

Ethyl 3-oxobutanoate   Ethoxide ion   Conjugate base of   Ethanol
(stronger acid)   (stronger base)   ethyl 3-oxobutanoate   (weaker acid)
                                     (weaker base)

Step 5: Acidification of the reaction mixture. This is performed in a separate synthetic operation to give the product in its neutral form for eventual isolation.

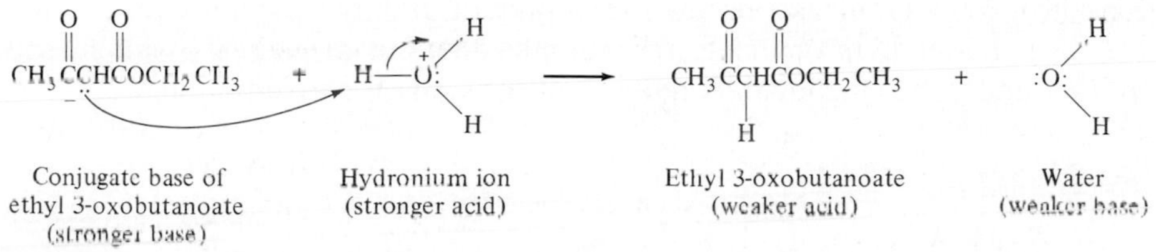

Conjugate base of   Hydronium ion   Ethyl 3-oxobutanoate   Water
ethyl 3-oxobutanoate   (stronger acid)   (weaker acid)   (weaker base)
(stronger base)

FIGURE 22.1 Sequence of steps that describes the mechanism of the Claisen condensation of ethyl acetate.

for condensation of two ester molecules to a β-keto ester. (Two ester carbonyl groups are more stable than one ester plus one ketone carbonyl.) However, because the β-keto ester is deprotonated under the reaction conditions, the equilibrium represented by the sum of steps 1 to 4 does lie to the side of products. On subsequent acidification (step 5), the anion of the β-keto ester is converted to its neutral form and isolated.

Organic chemists sometimes write equations for the Claisen condensation in a form that shows both stages explicitly.

$$2CH_3\overset{\overset{O}{\|}}{C}OCH_2CH_3 \xrightarrow{\text{NaOCH}_2\text{CH}_3} CH_3\overset{\overset{O}{\|}}{C}\overset{}{C}H\overset{\overset{O}{\|}}{C}OCH_2CH_3 \xrightarrow{\text{H}_3\text{O}^+} CH_3\overset{\overset{O}{\|}}{C}CH_2\overset{\overset{O}{\|}}{C}OCH_2CH_3$$
$$\overset{..}{N}a^+$$

| Ethyl acetate | Sodium salt of ethyl acetoacetate | Ethyl acetoacetate |

Like aldol condensation reactions, Claisen condensations always involve bond formation between the α carbon atom of one molecule and the carbonyl carbon of another.

$$2CH_3CH_2\overset{\overset{O}{\|}}{C}OCH_2CH_3 \xrightarrow[\text{2. H}_3\text{O}^+]{\text{1. NaOCH}_2\text{CH}_3} CH_3CH_2\overset{\overset{O}{\|}}{C}\overset{}{C}H\overset{\overset{O}{\|}}{C}OCH_2CH_3 + CH_3CH_2OH$$
$$\overset{}{\underset{CH_3}{|}}$$

| Ethyl propanoate | Ethyl 2-methyl-3-oxopentanoate (81%) | Ethanol |

PROBLEM 22.1 One of the following esters cannot undergo the Claisen condensation reaction. Which one? Write structural formulas for the Claisen condensation products of the other two.

$$CH_3CH_2CH_2CH_2CO_2CH_2CH_3 \qquad C_6H_5CO_2CH_2CH_3 \qquad C_6H_5CH_2CO_2CH_2CH_3$$

| Ethyl pentanoate | Ethyl benzoate | Ethyl phenylacetate |

Unless the β-keto ester product is capable of forming a stable anion by deprotonation, the Claisen condensation product is present in only trace amounts at equilibrium. Ethyl 2-methylpropanoate, for example, does not give any of its condensation product under the customary conditions of the Claisen condensation.

$$2(CH_3)_2CH\overset{\overset{O}{\|}}{C}OCH_2CH_3 \xrightleftharpoons{\text{NaOCH}_2\text{CH}_3} (CH_3)_2CH\overset{\overset{O}{\|}}{\underset{}{C}}\overset{}{\underset{\underset{H_3C\quad CH_3}{|}}{C}}\overset{\overset{O}{\|}}{C}OCH_2CH_3$$

| Ethyl 2-methylpropanoate | Ethyl 2,2,4-trimethyl-3-oxopentanoate (cannot form a stable anion; formed in no more than trace amounts) |

22.2 INTRAMOLECULAR CLAISEN CONDENSATION. THE DIECKMANN CONDENSATION

Esters of dicarboxylic acids undergo an intramolecular version of the Claisen condensation when a five- or six-membered ring can be formed.

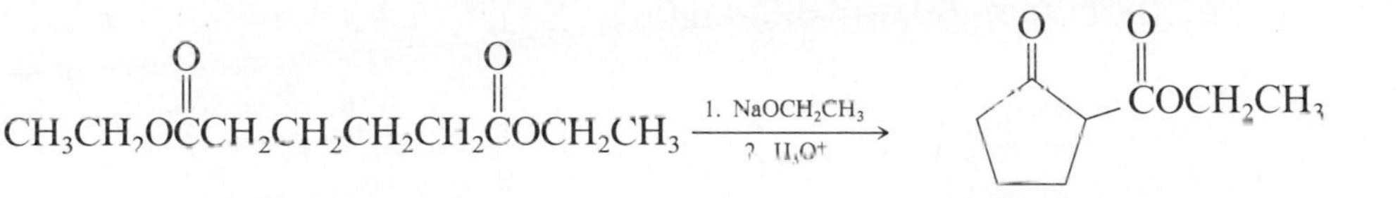

Diethyl hexanedioate
(diethyl adipate)

Ethyl (2-oxocyclopentane)carboxylate
(74–81%)

This cyclization reaction is an example of a *Dieckmann condensation.* The anion formed by proton abstraction at the carbon α to one carbonyl group attacks the other carbonyl to form a five-membered ring,

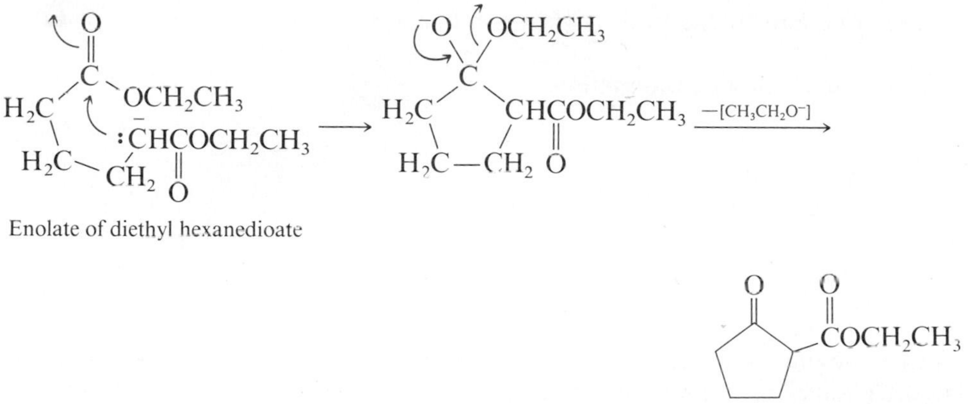

Enolate of diethyl hexanedioate

Ethyl (2-oxocyclopentane)-
carboxylate

PROBLEM 22.2   Write the structure of the Dieckmann condensation product formed on treatment of each of the following diesters with sodium ethoxide, followed by acidification.

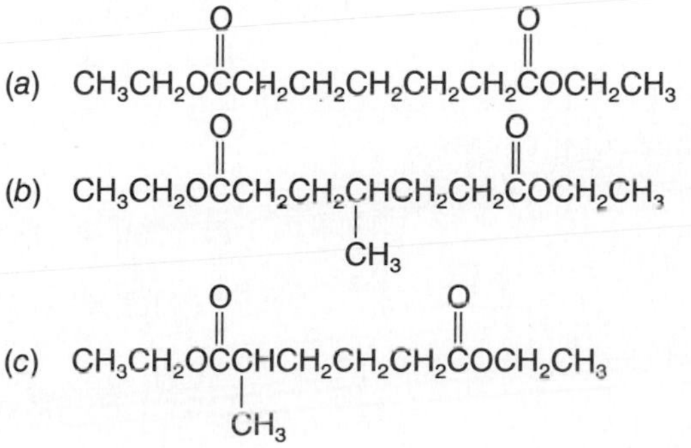

SAMPLE SOLUTION   (a)   Diethyl heptanedioate has one more methylene group in its chain than the diester cited in the example (diethyl hexanedioate). Its Dieckmann condensation product contains a six-membered ring instead of the five-membered ring formed from diethyl hexanedioate.

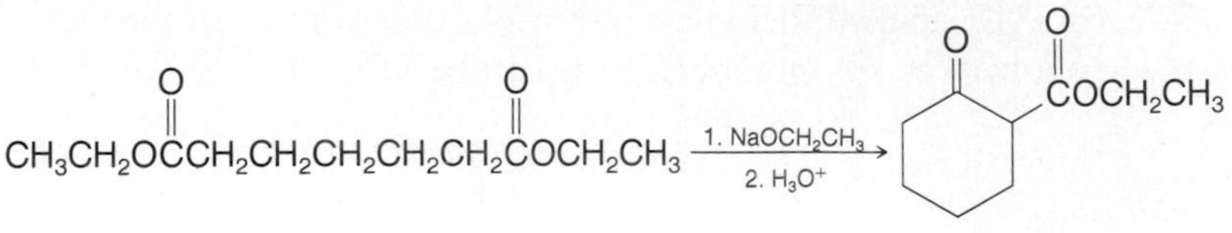

$$CH_3CH_2O\overset{\overset{\displaystyle O}{\|}}{C}CH_2CH_2CH_2CH_2CH_2\overset{\overset{\displaystyle O}{\|}}{C}OCH_2CH_3 \xrightarrow[\text{2. } H_3O^+]{\text{1. NaOCH}_2CH_3}$$

Diethyl heptanedioate
(diethyl pimelate)

Ethyl (2-oxocyclo-hexane)carboxylate

22.3 MIXED CLAISEN CONDENSATIONS

In analogy with the mixed aldol condensations discussed in Section 19.11, mixed Claisen condensations involve carbon-carbon bond formation between the α carbon atom of one ester and the carbonyl carbon of another.

A mixed Claisen condensation:

$$R\overset{\overset{\displaystyle O}{\|}}{C}OCH_2CH_3 + R'CH_2\overset{\overset{\displaystyle O}{\|}}{C}OCH_2CH_3 \xrightarrow[\text{2. } H_3O^+]{\text{1. NaOCH}_2CH_3} R\overset{\overset{\displaystyle O}{\|}}{C}\underset{\underset{\displaystyle R'}{|}}{C}H\overset{\overset{\displaystyle O}{\|}}{C}OCH_2CH_3$$

Ester   Another ester β-Keto ester

The best results are obtained in those mixed Claisen condensations in which one of the ester components, the acylating agent, is incapable of forming an enolate. Esters of this type include the following:

$$H\overset{\overset{\displaystyle O}{\|}}{C}OR \qquad RO\overset{\overset{\displaystyle O}{\|}}{C}OR \qquad RO\overset{\overset{\displaystyle O\ O}{\|\ \|}}{CC}OR \qquad \overset{\overset{\displaystyle O}{\|}}{C}OR$$

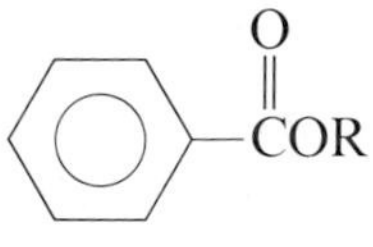

Formate esters   Carbonate esters   Oxalate esters   Benzoate esters

The equation shows an example of a mixed Claisen condensation in which a benzoate ester is used as the nonenolizable component.

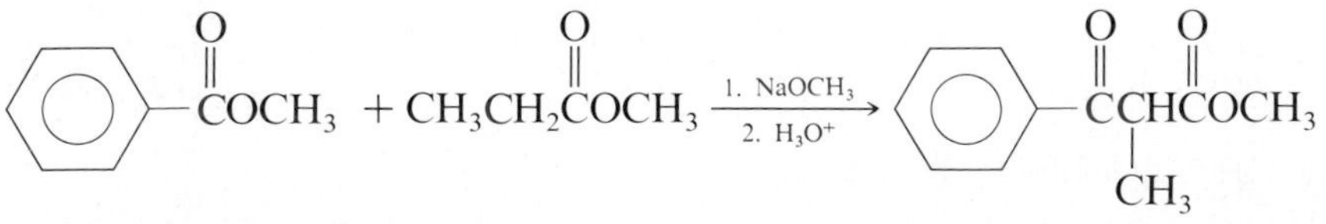

Methyl benzoate
(cannot form an enolate)   Methyl propanoate   Methyl 2-methyl-3-phenyl-3-oxopropanoate (60%)

Use of equimolar quantities of methyl benzoate and methyl propanoate ensures that the mixed condensation product will be present in far greater amounts at equilibrium than the self-condensation product of methyl propanoate.

PROBLEM 22.3 Give the structure of the product obtained when ethyl phenylacetate ($C_6H_5CH_2CO_2CH_2CH_3$) is treated with each of the following esters under conditions of the mixed Claisen condensation:

(*a*) Diethyl carbonate   (*b*) Diethyl oxalate   (*c*) Ethyl formate

SAMPLE SOLUTION (*a*) Diethyl carbonate cannot form an enolate, but ethyl phenylacetate can. Diethyl carbonate acts as an acyl transfer agent toward the enolate of ethyl phenylacetate.

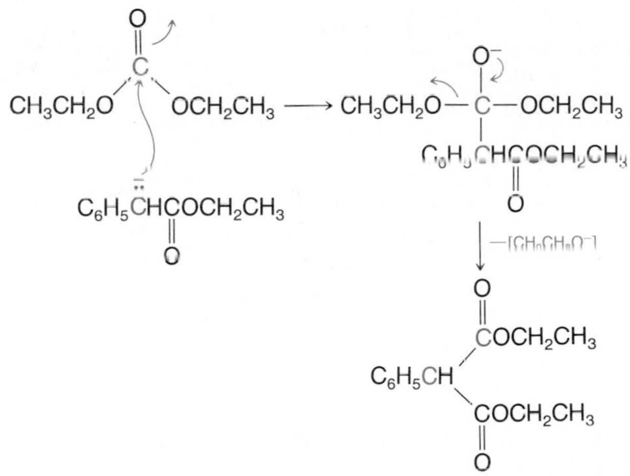

Diethyl 2-phenylpropanedioate
(diethyl phenylmalonate)

Note that when diethyl carbonate is used as the nonenolizable component in a mixed Claisen condensation, the product is a derivative of diethyl propanedioate (diethyl malonate). In this exercise, the product is diethyl phenylmalonate—it has been prepared in 86 percent yield by this method.

22.4 ACYLATION OF KETONES WITH ESTERS

In a reaction that is related to the mixed Claisen condensation, nonenolizable esters are used as acylating agents for ketone enolates. Ketones (via their enolates) are converted to β-keto esters by reaction with diethyl carbonate.

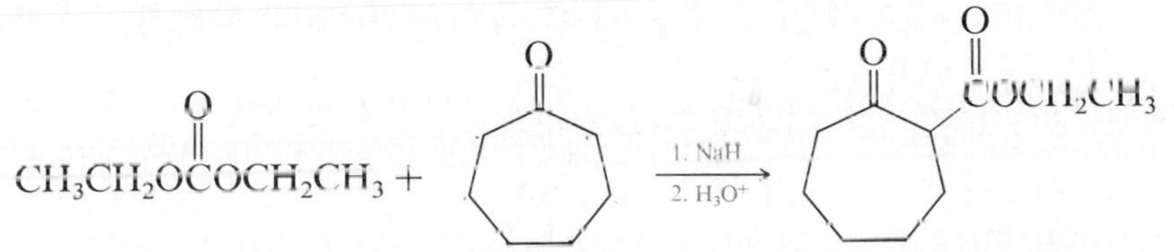

Diethyl carbonate   Cycloheptanone   Ethyl (2-oxocyclo-
heptane)carboxylate (91–94%)

Sodium hydride was used as the base in this example. It is a commonly encountered substitute for sodium ethoxide in these reactions.

Esters of nonenolizable monocarboxylic acids such as ethyl benzoate give β-diketones on reaction with ketone enolates.

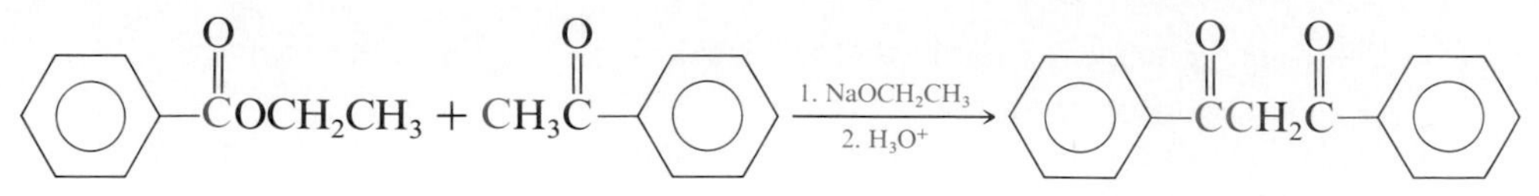

Ethyl benzoate   Acetophenone   1,3-Diphenyl-1,3-
                                                      propanedione (62–71%)

Intramolecular acylation of ketones yields cyclic β-diketones when the ring that is formed is five- or six-membered.

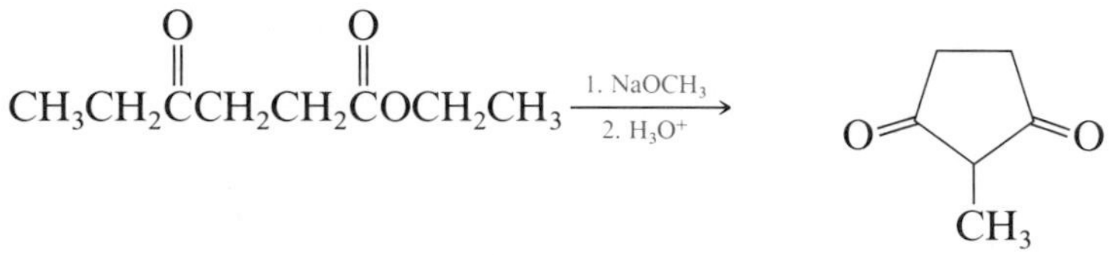

Ethyl 4-oxohexanoate   2-Methyl-1,3-cyclopentanedione
                                                (70–71%)

PROBLEM 22.4   Write an equation for the carbon-carbon bond-forming step in the cyclization reaction just cited. Show clearly the structure of the enolate ion and use curved arrows to represent its nucleophilic addition to the appropriate carbonyl group. Write a second equation showing dissociation of the tetrahedral intermediate formed in the carbon-carbon bond-forming step.

Even though ketones have the potential to react with themselves by aldol addition, recall that the position of equilibrium for such reactions lies to the side of the starting materials (Section 19.10). On the other hand, acylation of ketone enolates gives products (β-keto esters or β-diketones) that are converted to delocalized anions under the reaction conditions. Consequently, ketone acylation is observed to the exclusion of aldol addition when ketones are treated with base in the presence of esters.

22.5   APPLICATION OF THE CLAISEN AND DIECKMANN CONDENSATIONS TO THE SYNTHESIS OF KETONES

The carbon-carbon bond-forming potential inherent in the Claisen and Dieckmann condensations has been extensively exploited in organic synthesis. Subsequent transformations of the β-keto ester products permit the synthesis of other functional groups. One of these transformations converts β-keto esters to ketones; it is based on the fact that β-keto *acids* (not esters!) undergo decarboxylation readily (Section 20.17). Indeed, β-keto acids, and their corresponding carboxylate anions as well, lose carbon dioxide so easily that they tend to decarboxylate under the conditions of their formation.

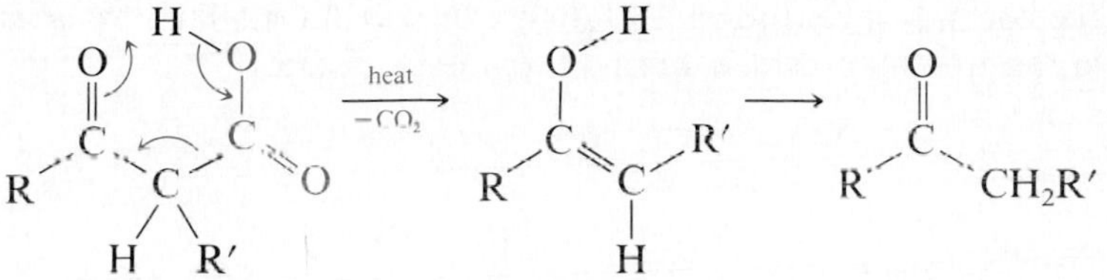
heat
$-CO_2$

β-Keto acid Enol form of ketone Ketone

Thus, 5-nonanone has been prepared from ethyl pentanoate by the sequence

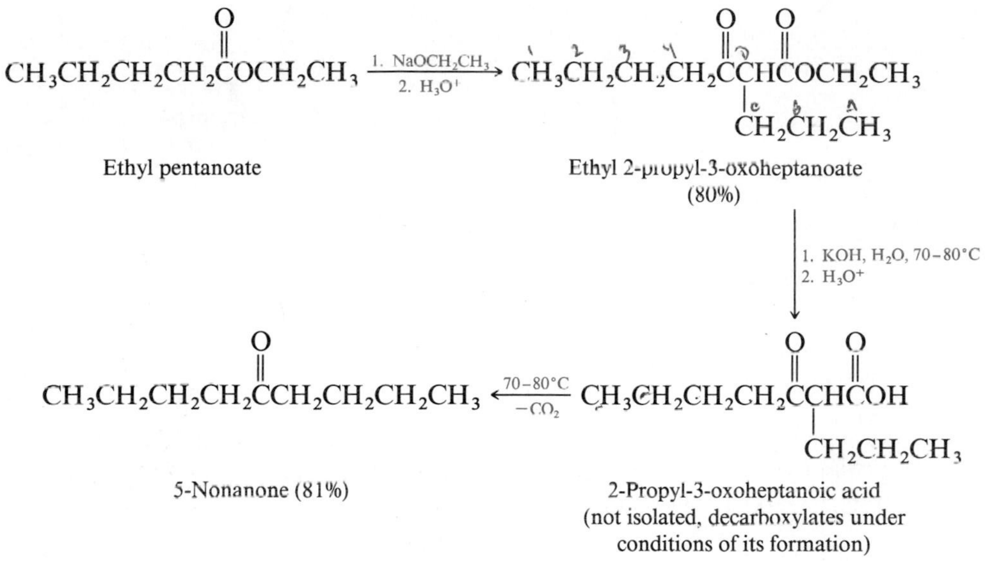
Ethyl pentanoate
1. $NaOCH_2CH_3$
2. H_3O^+
Ethyl 2-propyl-3-oxoheptanoate (80%)
1. KOH, H_2O, 70–80°C
2. H_3O^+
2-Propyl-3-oxoheptanoic acid (not isolated, decarboxylates under conditions of its formation)
70–80°C
$-CO_2$
5-Nonanone (81%)

The sequence begins with a Claisen condensation of ethyl pentanoate to give a β-keto ester. The ester is hydrolyzed, and the resulting β-keto acid decarboxylates to yield the desired ketone.

PROBLEM 22.5 Write appropriate chemical equations showing how you could prepare cyclopentanone from diethyl hexanedioate.

The most prominent application of β-keto esters to organic synthesis employs a similar pattern of ester saponification and decarboxylation as its final stage, as described in the following section.

22.6 THE ACETOACETIC ESTER SYNTHESIS

Ethyl acetoacetate (acetoacetic ester), available by the Claisen condensation of ethyl acetate, possesses properties that make it a useful starting material for the preparation of ketones. These properties are:

1. The acidity of the α-proton
2. The ease with which acetoacetic acid undergoes thermal decarboxylation

Ethyl acetoacetate is a stronger acid than ethanol and is quantitatively converted to its anion on treatment with sodium ethoxide in ethanol.

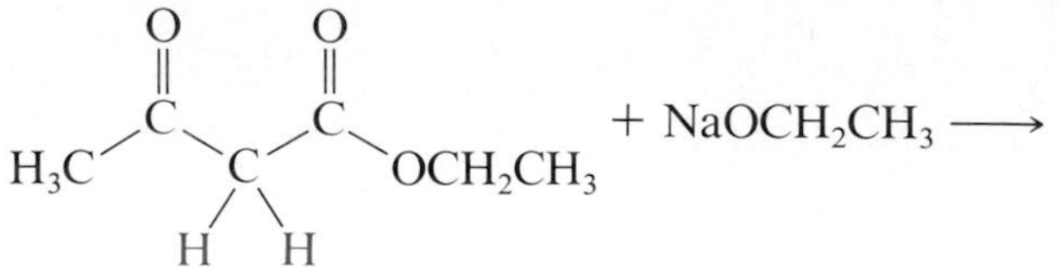

Ethyl acetoacetate
(stronger acid)
K_a 10^{-11}
(pK_a 11)

Sodium ethoxide
(stronger base)

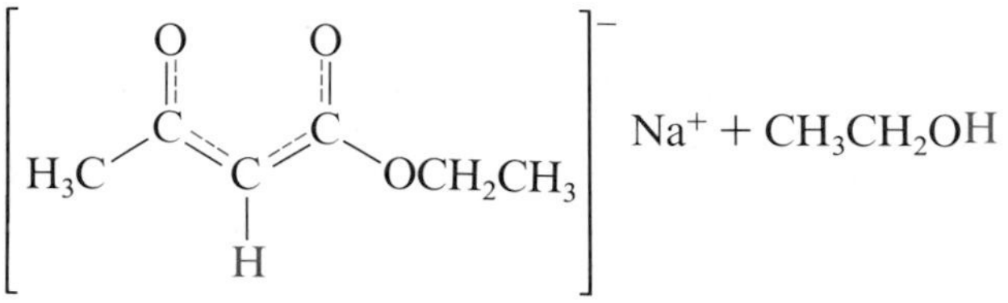

Sodium salt of ethyl
acetoacetate
(weaker base)

Ethanol
(weaker acid)
K_a 10^{-16}
(pK_a 16)

The anion produced by proton abstraction from ethyl acetoacetate is nucleophilic. Adding an alkyl halide to a solution of the sodium salt of ethyl acetoacetate leads to alkylation at its α carbon atom.

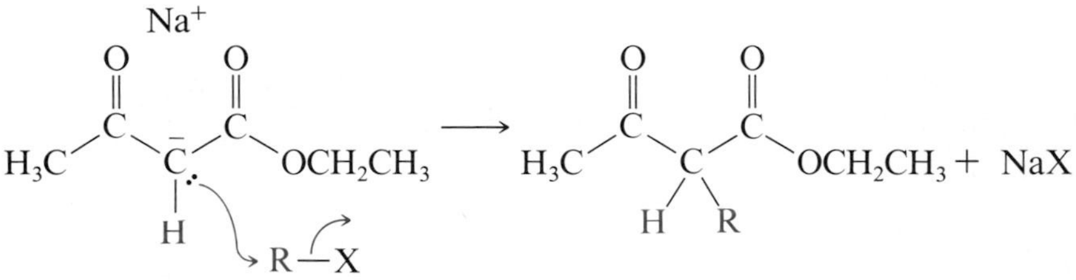

Sodium salt of ethyl acetoacetate;
alkyl halide

2-Alkyl derivative of
ethyl acetoacetate

Sodium
halide

The new carbon-carbon bond is formed by an S_N2-type reaction, so the alkyl halide must be one that is not sterically hindered. Methyl and primary alkyl halides work best, while secondary alkyl halides give lower yields. Tertiary alkyl halides react only by elimination.

Saponification and decarboxylation of the alkylated derivative of ethyl acetoacetate yields a ketone.

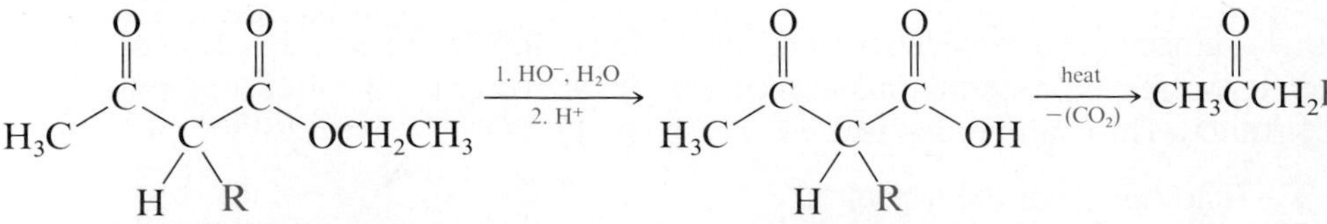

2-Alkyl derivative of
ethyl acetoacetate

2-Alkyl derivative of
acetoacetic acid

Ketone

This reaction sequence is called the *acetoacetic ester synthesis.* It is a standard procedure for the preparation of ketones from alkyl halides, as the conversion of 1-bromobutane to 2-heptanone illustrates.

$$CH_3\overset{O}{\overset{||}{C}}CH_2\overset{O}{\overset{||}{C}}OCH_2CH_3 \xrightarrow[\text{2. }CH_3CH_2CH_2CH_2Br]{\substack{\text{1. NaOCH}_2CH_3, \\ \text{ethanol}}} CH_3\overset{O}{\overset{||}{C}}\overset{}{C}H\overset{O}{\overset{||}{C}}OCH_2CH_3 \xrightarrow[\substack{\text{3. heat,} \\ (-CO_2)}]{\substack{\text{1. NaOH, H}_2O \\ \text{2. H}^+}}$$

$$\underset{\substack{CH_2CH_2CH_2CH_3}}{}$$

Ethyl
acetoacetate   Ethyl 2-(1-butyl)-3-
                                 oxobutanoate (70%)

$$CH_3\overset{O}{\overset{||}{C}}CH_2CH_2CH_2CH_2CH_3$$

2-Heptanone
(60%)

The acetoacetic ester synthesis accomplishes the overall transformation

$$R-X \longrightarrow R-CH_2\overset{O}{\overset{||}{C}}CH_3$$

Primary or secondary   α-Alkylated derivative
     alkyl halide   of acetone

Thus an alkyl halide is converted to an alkyl derivative of acetone.

We call a structural unit in a molecule that is related to a synthetic operation a *synthon.* The three-carbon unit $-CH_2\overset{O}{\overset{||}{C}}CH_3$ is a synthon that alerts us to the possibility that a particular molecule may be accessible by the acetoacetic ester synthesis.

PROBLEM 22.6   Show how you could prepare each of the following ketones from ethyl acetoacetate and any necessary organic or inorganic reagents:

(a)   1-Phenyl-1,4-pentanedione
(b)   4-Phenyl-2-butanone
(c)   5-Hexen-2-one

SAMPLE SOLUTION   (a)   Approach these syntheses in a retrosynthetic way. Identify the synthon $-CH_2\overset{O}{\overset{||}{C}}CH_3$ and mentally disconnect the bond to the α carbon atom. The $-CH_2\overset{O}{\overset{||}{C}}CH_3$ synthon is derived from ethyl acetoacetate; the remainder of the molecule originates in the alkyl halide.

disconnect here
1-Phenyl-1,4-pentanedione

Required alkyl halide

Derived from ethyl acetoacetate

Analyzing the target molecule in this way reveals that the required alkyl halide is an α-halo ketone. Thus, a suitable starting material would be 2-bromo-1-phenylethanone (phenacyl bromide).

Phenacyl bromide   Ethyl acetoacetate

1-Phenyl-1,4-pentanedione

It is reasonable to ask why one would prepare a ketone by the multistep process of the acetoacetic synthesis rather than by direct alkylation of the enolate of acetone. In practice, the monoalkylation of ketones via their enolates is a difficult reaction to carry out in good yield and requires careful control of reaction conditions. (Remember, however, *acylation* of ketone enolates can be achieved readily, as described in Section 22.4.) Because the delocalized enolate of ethyl acetoacetate is far less basic than the enolate of acetone, it gives a much higher ratio of substitution to elimination in its reaction with alkyl halides. This can be quite important in those syntheses in which the alkyl halide is expensive or difficult to obtain.

The anion of ethyl acetoacetate is said to be *synthetically equivalent* to the enolate of acetone. The use of synthetically equivalent groups is a common tactic in synthetic organic chemistry. One of the skills that characterizes the most creative practitioners of organic synthesis is an ability to recognize situations in which otherwise difficult transformations can be achieved through the use of synthetically equivalent reagents.

22.7 THE MALONIC ESTER SYNTHESIS

The *malonic ester synthesis* is a method for the preparation of carboxylic acids and is represented by the general equation

$$\text{RX} + \text{CH}_2(\text{COOCH}_2\text{CH}_3)_2 \xrightarrow[\text{ethanol}]{\text{NaOCH}_2\text{CH}_3} \text{RCH}(\text{COOCH}_2\text{CH}_3)_2 \xrightarrow[\substack{2.\ \text{H}^+ \\ 3.\ \text{heat}}]{1.\ \text{HO}^-,\ \text{H}_2\text{O}} \text{RCH}_2\text{COOH}$$

Alkyl halide

Diethyl malonate (malonic ester)

α-Alkylated derivative of diethyl malonate

Carboxylic acid

The malonic ester synthesis is conceptually analogous to the acetoacetic ester synthesis. The overall transformation is

Primary or secondary alkyl halide

α-Alkylated derivative of acetic acid

Diethyl malonate (also known as malonic ester) serves as a source of the synthon

$$-CH_2\overset{\displaystyle O}{\overset{\|}{C}}OH$$ in exactly the same way that ethyl acetoacetate serves as a source of the

synthon $$-CH_2\overset{\displaystyle O}{\overset{\|}{C}}CH_3.$$

The properties of diethyl malonate that make the malonic ester synthesis a useful procedure are the same as those responsible for the synthetic value of ethyl acetoacetate. The proton at C-2 of diethyl malonate is relatively acidic and is readily removed on treatment with sodium ethoxide.

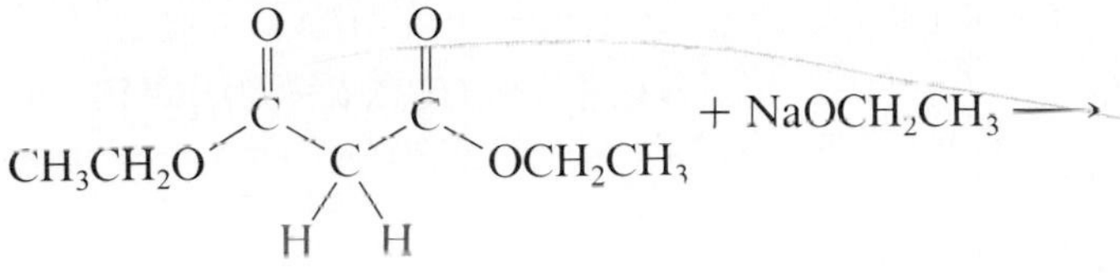

CH₃CH₂O—C(=O)—CH₂—C(=O)—OCH₂CH₃ + NaOCH₂CH₃ ⟶

Diethyl malonate
(stronger acid)
K_a 10^{-13}
(pK_a 13)

Sodium ethoxide
(stronger base)

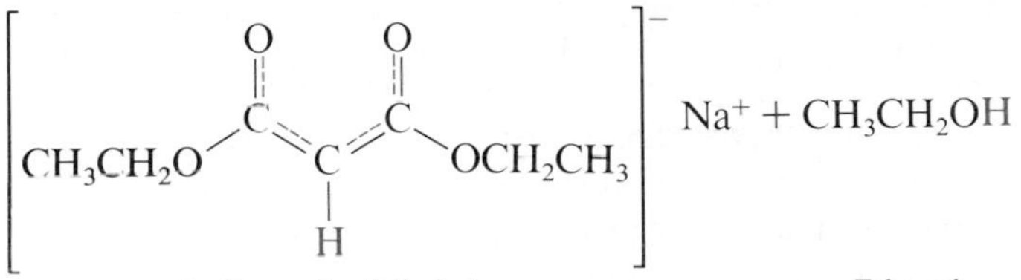

[CH₃CH₂O—C(=O)—C(H)=C—OCH₂CH₃]⁻ Na⁺ + CH₃CH₂OH

Sodium salt of diethyl
malonate
(weaker base)

Ethanol
(weaker acid)
K_a 10^{-16}
(pK_a 16)

Treatment of the anion of diethyl malonate with alkyl halides leads to alkylation at C-2.

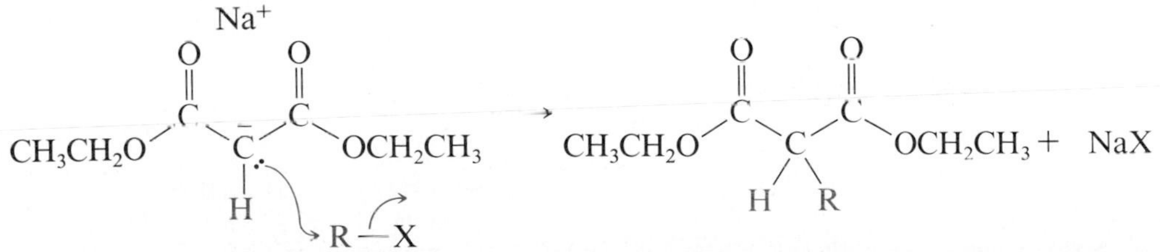

Na⁺

CH₃CH₂O—C(=O)—C⁻(H)—C(=O)—OCH₂CH₃   R—X ⟶ CH₃CH₂O—C(=O)—C(H)(R)—C(=O)—OCH₂CH₃ + NaX

Sodium salt of diethyl malonate;
alkyl halide

2-Alkyl derivative of
diethyl malonate

Sodium
halide salt

Conversion of the C-2 alkylated derivative to the corresponding malonic acid derivative by ester hydrolysis gives a compound susceptible to thermal decarboxylation. Temperatures of approximately 180°C are normally required.

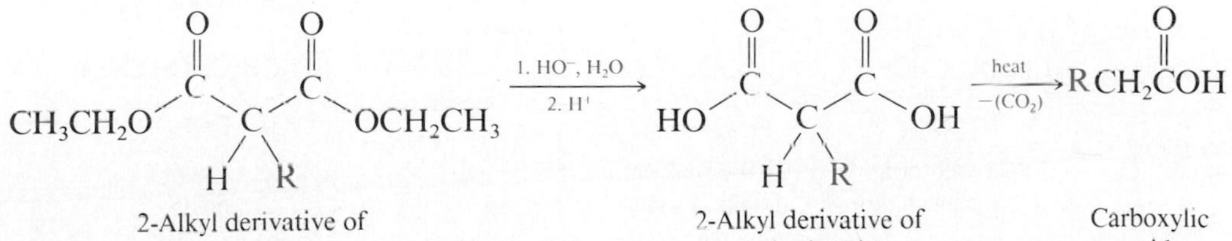

CH₃CH₂O—C(=O)—C(H)(R)—C(=O)—OCH₂CH₃ $\xrightarrow[\text{2. H}^+]{\text{1. HO}^-, \text{H}_2\text{O}}$ HO—C(=O)—C(H)(R)—C(=O)—OH $\xrightarrow[-(\text{CO}_2)]{\text{heat}}$ RCH₂COH

2-Alkyl derivative of
diethyl malonate

2-Alkyl derivative of
malonic acid

Carboxylic
acid

In a typical example of the malonic ester synthesis, 6-heptenoic acid has been prepared from 5-bromo-1-pentene.

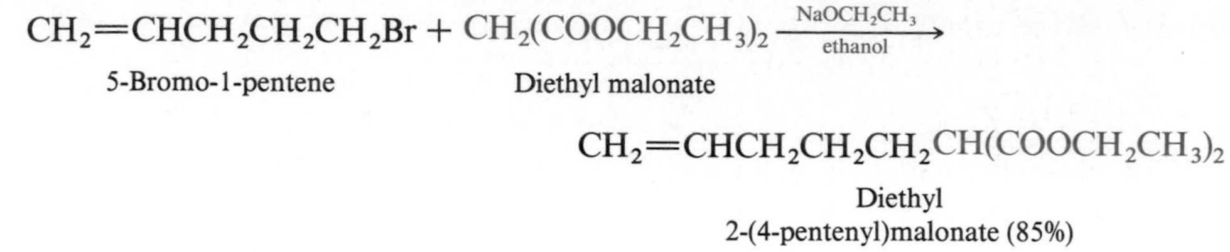

$$CH_2{=}CHCH_2CH_2CH_2Br + CH_2(COOCH_2CH_3)_2 \xrightarrow[\text{ethanol}]{NaOCH_2CH_3}$$

5-Bromo-1-pentene   Diethyl malonate

$$CH_2{=}CHCH_2CH_2CH_2CH(COOCH_2CH_3)_2$$

Diethyl
2-(4-pentenyl)malonate (85%)

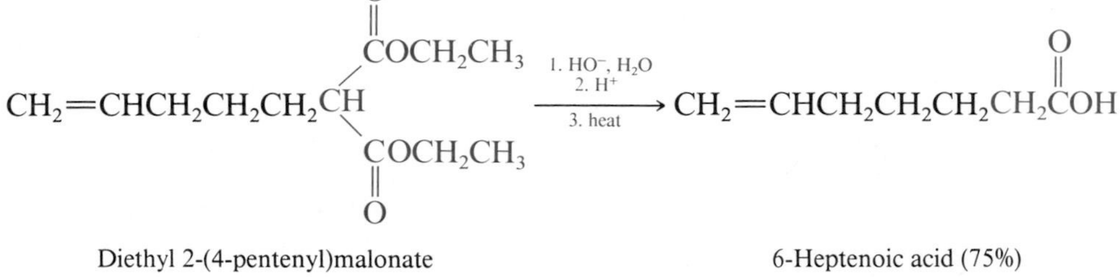

Diethyl 2-(4-pentenyl)malonate   6-Heptenoic acid (75%)

PROBLEM 22.7   Show how you could prepare each of the following carboxylic acids from diethyl malonate and any necessary organic or inorganic reagents:

(a) 3-Methylpentanoic acid
(b) Nonanoic acid
(c) 4-Methylhexanoic acid
(d) 3-Phenylpropanoic acid

SAMPLE SOLUTION   (a)   Analyze the target molecule retrosynthetically by mentally disconnecting a bond to the α carbon atom.

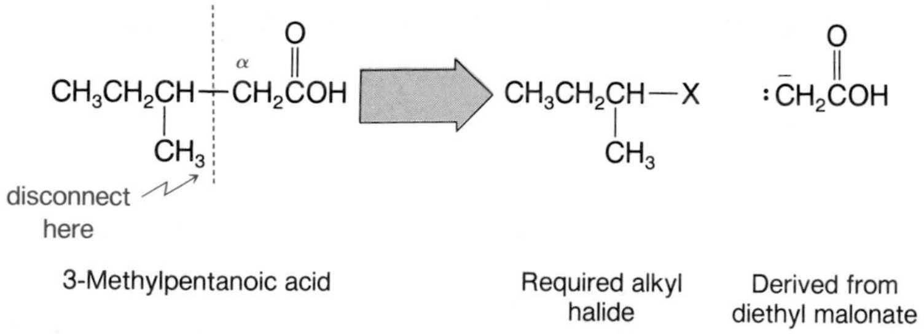

3-Methylpentanoic acid   Required alkyl halide   Derived from diethyl malonate

We see that a secondary alkyl halide is needed as the alkylating agent. The anion of diethyl malonate is a weaker base than ethoxide ion and reacts with secondary alkyl halides by substitution rather than elimination. Thus, the synthesis of 3-methylpentanoic acid begins with the alkylation of the anion of diethyl malonate by 2-bromobutane.

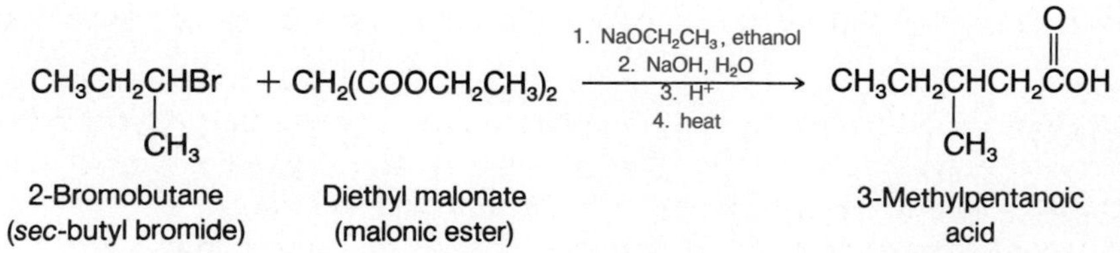

2-Bromobutane   Diethyl malonate   3-Methylpentanoic
(sec-butyl bromide)   (malonic ester)   acid

As actually carried out and reported in the chemical literature, diethyl malonate has been alkylated with 2-bromobutane in 83 to 84 percent yield and the product of that reaction converted to 3-methylpentanoic acid by saponification, acidification, and decarboxylation in 62 to 65 percent yield.

By performing two successive alkylation steps, the malonic ester synthesis can be applied to the synthesis of α,α-disubstituted derivatives of acetic acid.

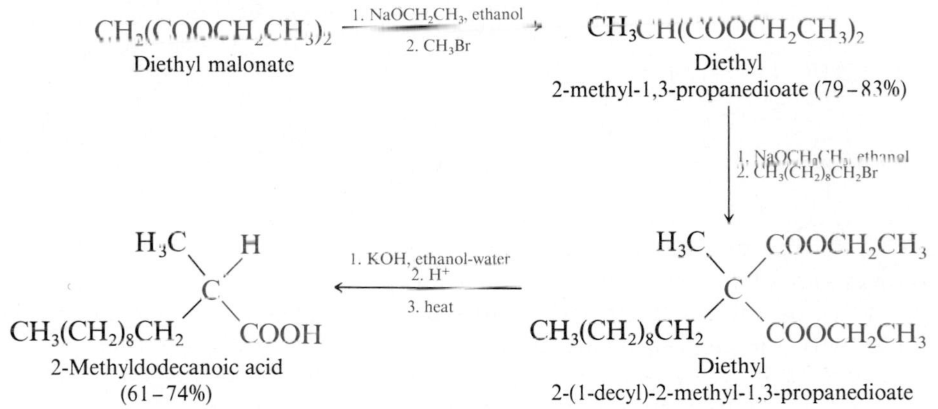

$$CH_2(COOCH_2CH_3)_2 \xrightarrow[\text{2. CH}_3\text{Br}]{\text{1. NaOCH}_2\text{CH}_3, \text{ ethanol}} CH_3CH(COOCH_2CH_3)_2$$

Diethyl malonate

Diethyl
2-methyl-1,3-propanedioate (79–83%)

1. NaOCH$_2$CH$_3$, ethanol
2. CH$_3$(CH$_2$)$_8$CH$_2$Br

H$_3$C, COOCH$_2$CH$_3$
C
CH$_3$(CH$_2$)$_8$CH$_2$, COOCH$_2$CH$_3$

Diethyl
2-(1-decyl)-2-methyl-1,3-propanedioate

1. KOH, ethanol-water
2. H$^+$
3. heat

H$_3$C, H
C
CH$_3$(CH$_2$)$_8$CH$_2$, COOH

2-Methyldodecanoic acid
(61–74%)

PROBLEM 22.8 Ethyl acetoacetate may also be subjected to double alkylation. Show how you could prepare 3-methyl-2-butanone by double alkylation of ethyl acetoacetate.

The malonic ester synthesis has been adapted to the preparation of cycloalkane-carboxylic acids from dihaloalkanes.

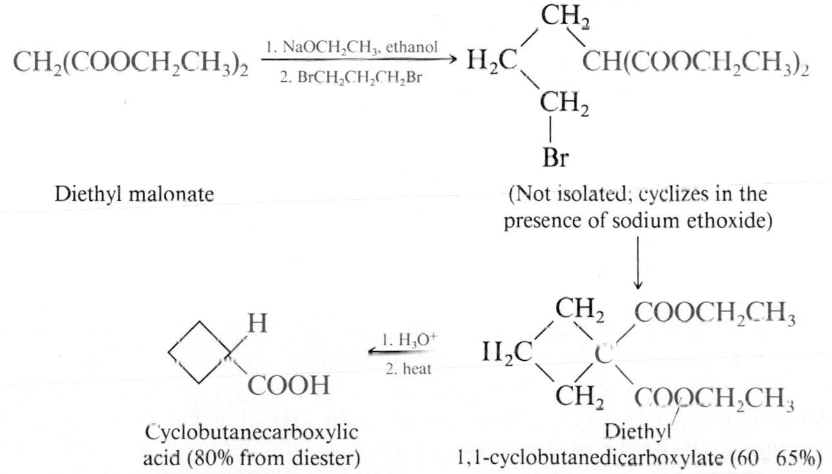

$$CH_2(COOCH_2CH_3)_2 \xrightarrow[\text{2. BrCH}_2\text{CH}_2\text{CH}_2\text{Br}]{\text{1. NaOCH}_2\text{CH}_3, \text{ ethanol}}$$

Diethyl malonate

(Not isolated; cyclizes in the presence of sodium ethoxide)

Diethyl
1,1-cyclobutanedicarboxylate (60–65%)

1. H$_3$O$^+$
2. heat

Cyclobutanecarboxylic
acid (80% from diester)

The cyclization step is limited to the formation of rings of seven or less carbons.

PROBLEM 22.9 Cyclopentyl methyl ketone has been prepared from 1,4-dibromobutane and ethyl acetoacetate. Outline the steps in this synthesis by writing a series of equations showing starting materials, reagents, and isolated intermediates.

22.8   BARBITURATES

Diethyl malonate has uses other than in the synthesis of carboxylic acids. One particularly valuable application lies in the preparation of *barbituric acid* by reaction with urea.

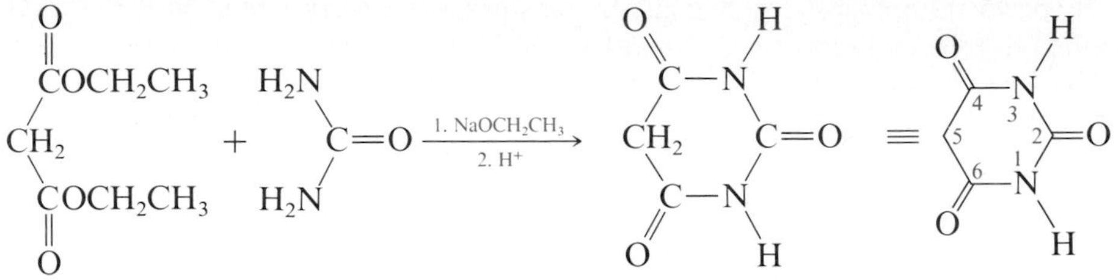

Diethyl malonate   Urea   Barbituric acid (72–78%)

Barbituric acid is the parent of a group of compounds known as *barbiturates.* The barbiturates are classified as *sedative-hypnotic agents,* meaning that they decrease the responsiveness of the central nervous system and promote sleep. Thousands of derivatives of the parent ring system of barbituric acid have been tested for sedative-hypnotic activity; the most useful are the 5,5-disubstituted derivatives.

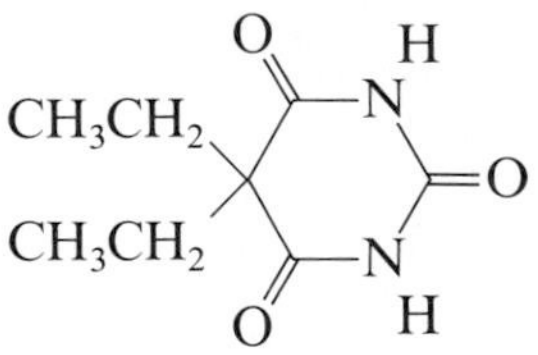

5,5-Diethylbarbituric acid   5-Ethyl-5-(1-methylbutyl)-   5-Allyl-5-(1-methylbutyl)-
(barbital; Veronal)   barbituric acid   barbituric acid
                                 (pentobarbital; Nembutal)   (secobarbital; Seconal)

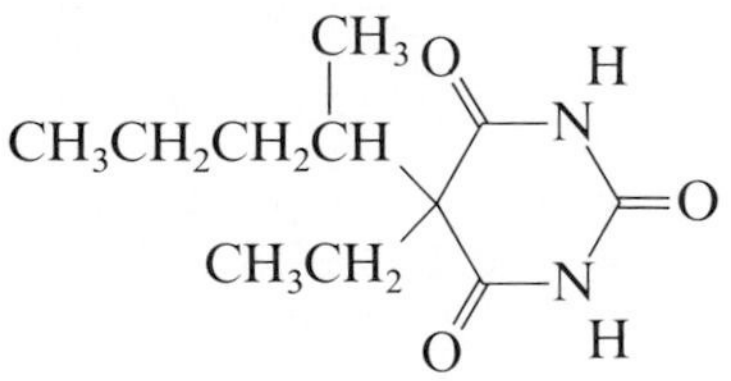

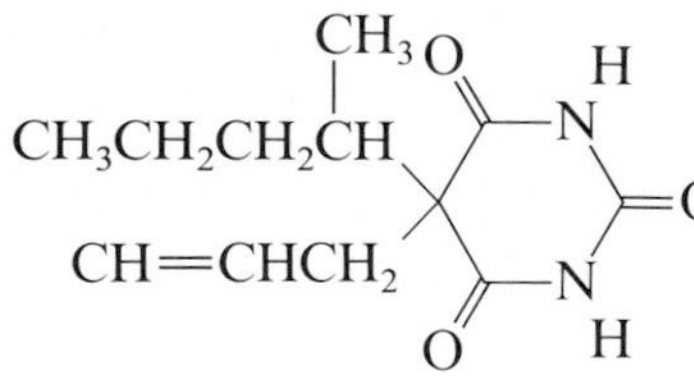

These compounds are prepared in a manner analogous to that of barbituric acid itself. Diethyl malonate is alkylated twice, then treated with urea.

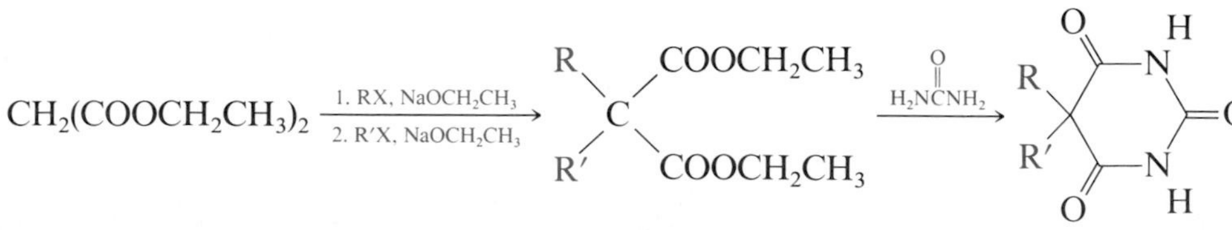

Diethyl malonate   Dialkylated   5,5-Disubstituted
                                 derivative of   derivative of
                                 diethyl malonate   barbituric acid

---

PROBLEM 22.10   Show, by writing a suitable sequence of reactions, how you could prepare pentobarbital from diethyl malonate. (The structure of pentobarbital is shown above.)

Barbituric acids, as their name implies, are weakly acidic and are converted to their sodium salts (sodium barbiturates) in aqueous sodium hydroxide. Sometimes the drug is dispensed in its neutral form, sometimes the sodium salt is used. The salt is designated by appending the word "sodium" to the name of the barbituric acid — *pentobarbital sodium,* for example.

PROBLEM 22.11 Thiourea ($H_2N\overset{\overset{\text{S}}{\|}}{C}NH_2$) reacts with diethyl malonate and its alkyl deriv-
atives in the same way that urea does. Give the structure of the product obtained when
thiourea is used instead of urea in the synthesis of pentobarbital. The anesthetic *thiopental
(Pentothal) sodium* is the sodium salt of this product. What is the structure of this compound?

PROBLEM 22.12 Aryl halides react too slowly to undergo substitution by the S_N2
mechanism with the sodium salt of diethyl malonate, so the phenyl substituent of *phenobarbital* cannot be introduced in the way that alkyl substituents can.

5-Ethyl-5-phenylbarbituric acid
(phenobarbital)

One synthesis of phenobarbital begins with ethyl phenylacetate and diethyl carbonate.
Using these starting materials and any necessary organic or inorganic reagents, devise a
synthesis of phenobarbital. (*Hint:* See the sample solution to Problem 22.3*a*.)

The various barbiturates differ in the time required for the onset of sleep and in the
duration of their effects. All the barbiturates must be used only in strict accordance
with instructions to avoid potentially lethal overdoses. Drug dependence in some
individuals is also a problem.

22.9 THE MICHAEL REACTION

Stabilized anions exhibit a pronounced tendency to undergo conjugate addition to
α,β-unsaturated carbonyl compounds. This reaction, called *the Michael reaction,* has
been described for anions derived from β-diketones in Section 19.14. The enolates of
ethyl acetoacetate and diethyl malonate also undergo Michael addition to the β
carbon atom of α,β-unsaturated aldehydes, ketones, and esters. For example

$$CH_3\overset{\overset{\text{O}}{\|}}{C}CH=CH_2 + CH_2(COOCH_2CH_3)_2 \xrightarrow[\text{ethanol}]{\text{KOH}} CH_3\overset{\overset{\text{O}}{\|}}{C}CH_2CH_2CH(COOCH_2CH_3)_2$$

Methyl vinyl ketone	Diethyl malonate	Ethyl 2-carboethoxy-5-oxohexanoate (83%)

In this reaction the enolate of diethyl malonate adds to the β carbon of methyl vinyl
ketone.

$$CH_3\overset{\overset{\text{O}}{\|}}{C}-CH=CH_2 + \quad :\overset{-}{C}H(COOCH_2CH_3)_2 \longrightarrow$$

$$CH_3\overset{\overset{-\text{O}}{|}}{C}=CH-CH_2-CH(COOCH_2CH_3)_2$$

The intermediate formed in the nucleophilic addition step abstracts a proton from the solvent to give the observed product.

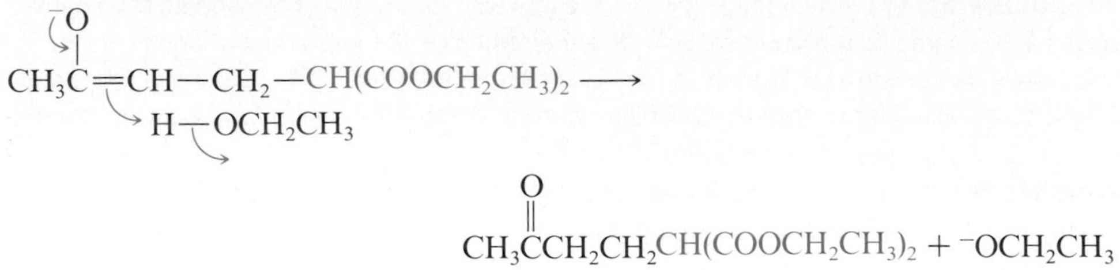

$$CH_3\overset{\overset{\displaystyle O}{|}}{C}=CH-CH_2-CH(COOCH_2CH_3)_2 \longrightarrow$$
$$H-OCH_2CH_3$$

$$CH_3\overset{\overset{\displaystyle O}{\|}}{C}CH_2CH_2CH(COOCH_2CH_3)_2 + {}^-OCH_2CH_3$$

PROBLEM 22.13   Give the structures of the products formed on treatment of each of the following compounds with diethyl malonate in the presence of sodium ethoxide:

(a)   Acrolein (CH_2=CHCHO)
(b)   Ethyl acrylate (CH_2=CHCOOCH$_2$CH$_3$)
(c)   Ethyl crotonate (CH_3CH=CHCOOCH$_2$CH$_3$)
(d)   Diethyl fumarate (*trans*-CH$_3$CH$_2$OOCCH=CHCOOCH$_2$CH$_3$)
(e)   2-Cyclopentenone

SAMPLE SOLUTION   (a)   Acrolein is an α,β-unsaturated aldehyde. By way of its anion, diethyl malonate adds to the carbon-carbon double bond of acrolein.

$$(CH_3CH_2OOC)_2CH_2 + CH_2\!\!=\!\!CH\overset{\overset{\displaystyle O}{\|}}{C}H \xrightarrow[\text{ethanol}]{\text{NaOCH}_2\text{CH}_3} (CH_3CH_2OOC)_2CH-CH_2CH_2\overset{\overset{\displaystyle O}{\|}}{C}H$$

Diethyl malonate   Acrolein

New carbon-carbon bond

After isolation, the Michael adducts of diethyl malonate and α,β-unsaturated carbonyl compounds may be subjected to ester hydrolysis and decarboxylation. When α,β-unsaturated ketones are carried through this sequence, the final products are 5-keto acids (δ-keto acids).

$$CH_3\overset{\overset{\displaystyle O}{\|}}{C}CH_2CH_2CH(COOCH_2CH_3)_2 \xrightarrow[\substack{\text{2. H}^+ \\ \text{3. heat}}]{\text{1. KOH, ethanol-water}} CH_3\overset{\overset{\displaystyle O}{\|}}{C}CH_2CH_2CH_2\overset{\overset{\displaystyle O}{\|}}{C}OH$$

Ethyl 2-carboethoxy-5-oxohexanoate
(from diethyl malonate and
methyl vinyl ketone)

5-Oxohexanoic acid
(42%)

Ethyl acetoacetate is similar to diethyl malonate in its reactivity toward α,β-unsaturated carbonyl compounds. The Michael addition of ethyl acetoacetate to α,β-unsaturated ketones has been used as the key carbon-carbon bond-forming step in the preparation of δ-diketones.

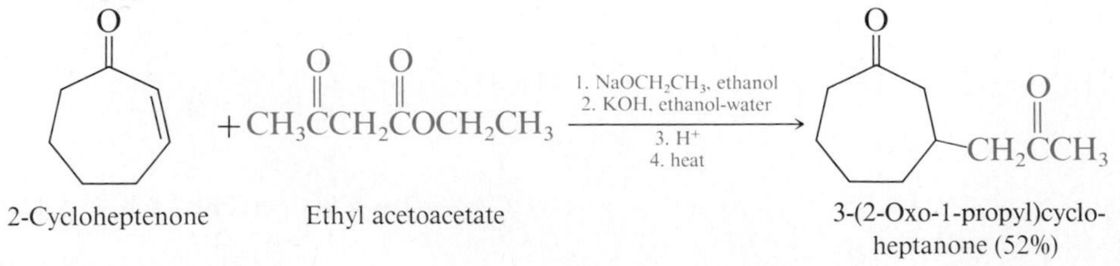

$$+ CH_3\overset{\overset{\displaystyle O}{\|}}{C}CH_2\overset{\overset{\displaystyle O}{\|}}{C}OCH_2CH_3 \xrightarrow[\substack{\text{3. H}^+ \\ \text{4. heat}}]{\substack{\text{1. NaOCH}_2\text{CH}_3, \text{ ethanol} \\ \text{2. KOH, ethanol-water}}}$$

2-Cycloheptenone   Ethyl acetoacetate

$$CH_2\overset{\overset{\displaystyle O}{\|}}{C}CH_3$$

3-(2-Oxo-1-propyl)cyclo-
heptanone (52%)

What about simple aldehydes and ketones? How do diethyl malonate and ethyl acetoacetate react with carbonyl groups when conjugate addition cannot take place?

22.10 THE KNOEVENAGEL CONDENSATION

Diethyl malonate reacts with aldehydes and ketones in a manner that is reminiscent of a mixed aldol condensation (Section 19.11). Bond formation between the α carbon atom of diethyl malonate and the carbonyl group of the aldehyde or ketone is followed by an elimination step to give a double bond between these two positions.

$$\underset{\substack{\text{Aldehyde} \\ \text{or ketone}}}{\overset{\overset{\displaystyle O}{\parallel}}{RCR'}} + \underset{\substack{\text{Diethyl} \\ \text{malonate}}}{CH_2(COOCH_2CH_3)_2} \xrightarrow{\text{amine}} \underset{\substack{\text{Diethyl} \\ \text{alkylidenemalonate}}}{\overset{R}{\underset{R'}{>}}C=C(COOCH_2CH_3)_2} + \underset{\text{Water}}{H_2O}$$

Amines such as piperidine are effective catalysts for this reaction, which is called the *Knoevenagel condensation.*

$$\underset{\substack{\text{Benzaldehyde}}}{\overset{\overset{\displaystyle O}{\parallel}}{C_6H_5CH}} + \underset{\text{Diethyl malonate}}{CH_2(COOCH_2CH_3)_2} \xrightarrow{\text{piperidine}} \underset{\substack{\text{Diethyl} \\ \text{benzylidenemalonate (74–86\%)}}}{C_6H_5CH=C(COOCH_2CH_3)_2}$$

Like diethyl malonate, ethyl acetoacetate undergoes Knoevenagel condensation reactions under these conditions.

$$\underset{\substack{\text{Butanal}}}{\overset{\overset{\displaystyle O}{\parallel}}{CH_3CH_2CH_2CH}} + \underset{\substack{\text{Ethyl acetoacetate}}}{\overset{\overset{\displaystyle O}{\parallel}\;\;\overset{\displaystyle O}{\parallel}}{CH_3CCH_2COCH_2CH_3}} \xrightarrow{\text{piperidine}} \underset{\substack{\text{Ethyl} \\ \text{butylideneacetoacetate (81\%)}}}{CH_3CH_2CH_2CH=C\overset{\displaystyle \overset{O}{\parallel}}{\underset{\displaystyle \underset{O}{\parallel}}{<}}\begin{array}{l}CCH_3 \\ COCH_2CH_3\end{array}}$$

22.11 ESTER ENOLATES BY REDUCTION OF α-BROMO ESTERS. THE REFORMATSKY REACTION

There is another aldol-type reaction of ester enolates called the *Reformatsky reaction,* in which the enolate is generated by reduction of an α-bromo ester with zinc.

$$\underset{\substack{\text{Aldehyde} \\ \text{or ketone}}}{\overset{\overset{\displaystyle O}{\parallel}}{RCR'}} + \underset{\substack{\alpha\text{-Bromo} \\ \text{ester}}}{\overset{\overset{\displaystyle O}{\parallel}}{\underset{\underset{\displaystyle Br}{|}}{R''CHCOCH_2CH_3}}} \xrightarrow[\text{2. } H_3O^+]{\text{1. Zn, benzene}} \underset{\substack{\text{Ethyl ester of a} \\ \beta\text{-hydroxy acid}}}{\overset{\overset{\displaystyle OH}{|}}{\underset{\underset{\displaystyle R'}{|}}{RC}}-\overset{\overset{\displaystyle O}{\parallel}}{\underset{\underset{\displaystyle R''}{|}}{CHCOCH_2CH_3}}}$$

Zinc acts as a reducing agent toward the α-bromo ester to give a zinc enolate. Except for the metal ion, the enolate formed by zinc reduction of an α-bromo ester is the same as the one that would be formed by proton abstraction from the α carbon atom of a simple ester.

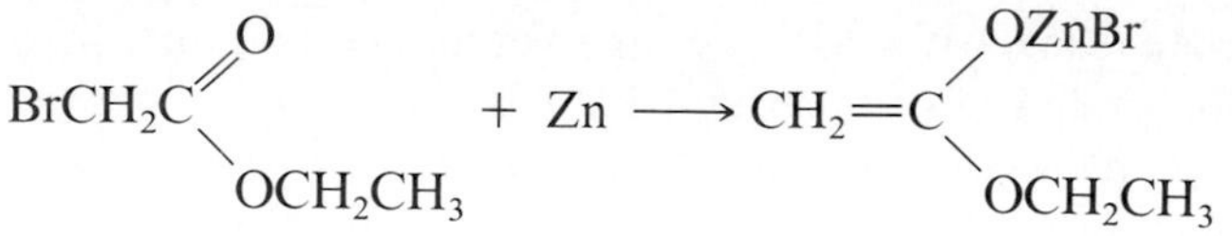

$$\text{BrCH}_2\text{C}\!\!\underset{\text{OCH}_2\text{CH}_3}{\overset{\text{O}}{\diagup}} \quad + \text{ Zn} \longrightarrow \text{CH}_2\!\!=\!\!\text{C}\!\!\underset{\text{OCH}_2\text{CH}_3}{\overset{\text{OZnBr}}{\diagup}}$$

Ethyl bromoacetate   Zinc   Zinc enolate

When generated in the presence of an aldehyde or ketone, the zinc enolate undergoes nucleophilic addition to the carbonyl group to give a β-hydroxy ester.

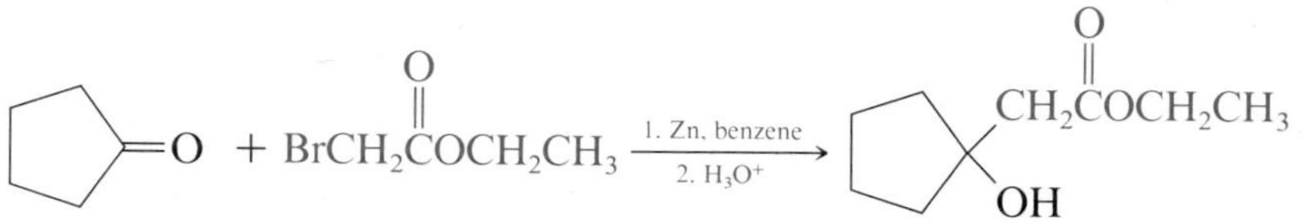

Cyclopentanone   Ethyl bromoacetate   Ethyl 2-(1-hydroxycyclopentyl)-acetate (95%)

PROBLEM 22.14   Describe syntheses of each of the following compounds by a sequence that includes a Reformatsky reaction as a key step:

(a)   2-Methyl-2-butenoic acid
(b)   3-Methyl-2-butenoic acid
(c)   3-Methyl-1,3-butanediol
(d)   2,3-Dimethyl-1,3-butanediol

SAMPLE SOLUTION   (a)   The β-hydroxy esters that are the products of the Reformatsky reaction can be converted to α,β-unsaturated esters by acid-catalyzed dehydration. Therefore, construct the carbon skeleton by a Reformatsky reaction between acetaldehyde and ethyl 2-bromopropanoate.

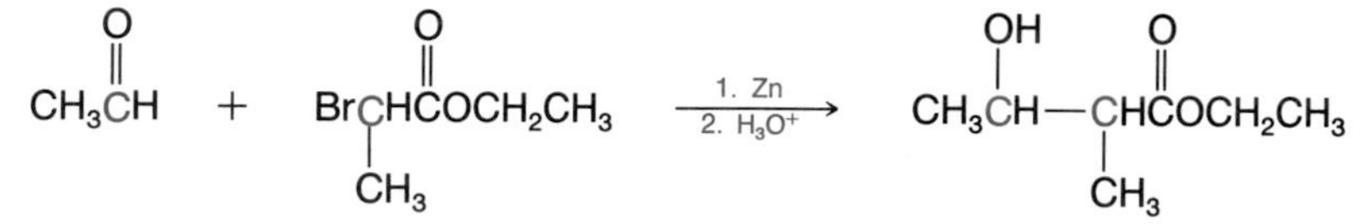

Acetaldehyde   Ethyl 2-bromopropanoate   Ethyl 3-hydroxy-2-methylbutanoate

Dehydration, followed by ester hydrolysis, yields the desired product.

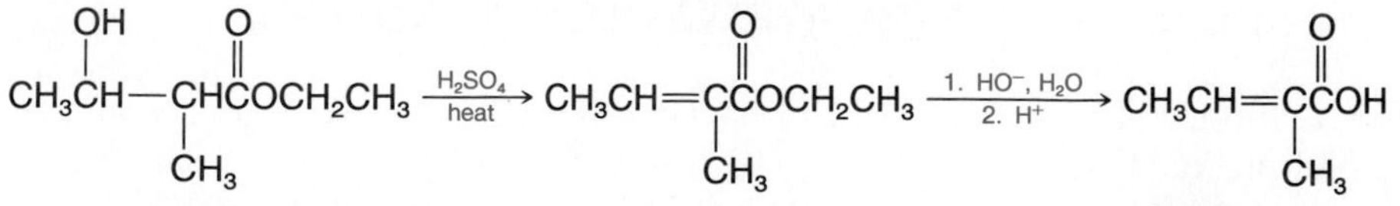

Ethyl 3-hydroxy-2-methylbutanoate   Ethyl 2-methyl-2-butenoate   2-Methyl-2-butenoic acid

---

The Reformatsky reaction was discovered in the nineteenth century but is still used extensively. It employs an α-bromo ester as the source of an ester enolate. In a relatively recent development, synthetic chemists have learned how to prepare enolates directly from esters by deprotonation with very strong bases. This approach is described in the following section.

22.12 DEPROTONATION OF ESTERS BY LITHIUM DIALKYLAMIDES

Consider the deprotonation of an ester as represented by the acid-base reaction

$$RCHCOR' + :B^- \rightleftharpoons RCH=C\overset{O^-}{\underset{OR'}{\diagdown}} + H-B$$

| Ester | Base | Ester enolate | Conjugate acid of base |

We already know what happens when esters of the type shown are treated with alkoxide bases—they undergo the Claisen condensation (Section 22.1). Simple esters have acid-dissociation constants K_a of approximately 10^{-22} (pK_a 22) and are incompletely converted to their enolates with alkoxide bases. The small amount of enolate that is formed reacts by nucleophilic addition to the carbonyl group of the neutral ester.

What about the situation in which the base is much stronger than an alkoxide ion? If the base is strong enough, it will convert the ester completely to its enolate. Under these conditions the Claisen condensation is suppressed because there is no neutral ester present for the enolate to add to. A very strong base is one that is derived from a very weak acid. Referring to the table of acidities (Table 4.3, p. 103), we see that ammonia is quite a weak acid; its K_a is 10^{-36} (pK_a 36). Therefore, amide ion ($H_2N:^-$) is a very strong base—more than strong enough to deprotonate an ester quantitatively. However, amide ion also tends to add to the carbonyl group of esters; to avoid this complication, highly hindered analogs of $H_2N:^-$ are used instead. The most frequently used base for ester enolate formation is *lithium diisopropylamide* (LDA).

$$\left[\begin{array}{c} H_3C \\ \diagdown \\ CH-\overset{..}{\underset{..}{N}}-CH \\ \diagup \\ H_3C \end{array} \begin{array}{c} CH_3 \\ \diagup \\ \diagdown \\ CH_3 \end{array} \right] Li^+ \qquad \text{Lithium diisopropylamide}$$

Lithium diisopropylamide is a strong enough base to abstract a proton from the α carbon atom of an ester, but because it is so sterically hindered, it does not add readily to the carbonyl group. To illustrate:

$$CH_3CH_2CH_2\overset{O}{\overset{||}{C}}OCH_3 + [(CH_3)_2CH]_2NLi \longrightarrow CH_3CH_2CH=C\overset{OLi}{\underset{OCH_3}{\diagdown}} + [(CH_3)_2CH]_2NH$$

| Methyl butanoate (stronger acid) K_a 10^{-22} (pK_a 22) | Lithium diisopropylamide (stronger base) | Lithium enolate of methyl butanoate (weaker base) | Diisopropylamine (weaker acid) K_a 10^{-36} (pK_a 36) |

Direct alkylation of esters has been achieved by formation of the ester enolate with LDA followed by addition of an alkyl halide. Tetrahydrofuran (THF) is the solvent most often used in these reactions.

$$CH_3CH_2CH_2\overset{\overset{\displaystyle O}{\|}}{C}OCH_3 \xrightarrow[\text{THF}]{\text{LDA}} CH_3CH_2CH{=}C\overset{\displaystyle OLi}{\underset{\displaystyle OCH_3}{\Big\backslash}} \xrightarrow{CH_3CH_2I} CH_3CH_2\overset{\overset{\displaystyle O}{\|}}{C}HCOCH_3$$
$$\underset{\displaystyle CH_3CH_2}{\Big|}$$

| Methyl butanoate | Lithium enolate of methyl butanoate | Methyl 2-ethylbut-anoate (92%) |

Ester enolates generated by proton abstraction with dialkylamide bases add to alde-hydes and ketones to give products identical to those formed by the Reformatsky method.

$$CH_3\overset{\overset{\displaystyle O}{\|}}{C}OCH_2CH_3 \xrightarrow[\text{THF}]{\text{LiNR}_2} CH_2{=}C\overset{\displaystyle OLi}{\underset{\displaystyle OCH_2CH_3}{\Big\backslash}} \xrightarrow[\text{2. H}_3O^+]{\text{1. (CH}_3)_2C{=}O} CH_3\overset{\overset{\displaystyle HO}{|}}{C}CH_2\overset{\overset{\displaystyle O}{\|}}{C}OCH_2CH_3$$
$$\underset{\displaystyle CH_3}{\Big|}$$

| Ethyl acetate | Lithium enolate of ethyl acetate | Ethyl 3-hydroxy-3-methylbutanoate (90%) |

Lithium dialkylamides are excellent bases for the preparation of ketone enolates as well. Ketone enolates generated in this way can be alkylated with alkyl halides or, as illustrated in the following equation, treated with an aldehyde or ketone to give the product of nucleophilic addition.

$$CH_3CH_2\overset{\overset{\displaystyle O}{\|}}{C}C(CH_3)_3 \xrightarrow[\text{THF}]{\text{LDA}} CH_3CH{=}C\overset{\displaystyle OLi}{\underset{\displaystyle C(CH_3)_3}{\Big\backslash}} \xrightarrow[\text{2. H}_3O^+]{\text{1. CH}_3CH_2CH} CH_3\overset{\overset{\displaystyle O}{\|}}{C}HCC(CH_3)_3$$
$$\underset{\displaystyle HOCHCH_2CH_3}{\Big|}$$

| 2,2-Dimethyl-3-pentanone | Lithium enolate of 2,2-dimethyl-3-pentanone | 5-Hydroxy-2,2,4-trimethyl-3-heptanone (81%) |

As the equation shows, mixed aldol additions can be achieved by the tactic of quantitative enolate formation using LDA followed by addition of a different alde-hyde or ketone.

PROBLEM 22.15 Outline efficient syntheses of each of the following compounds from readily available aldehydes, ketones, esters, and alkyl halides according to the methods described in this section.

(a) $(CH_3)_2CHCH\overset{\overset{\displaystyle O}{\|}}{C}OCH_2CH_3$
$\underset{\displaystyle CH_3}{\Big|}$

(b) $C_6H_5CH\overset{\overset{\displaystyle O}{\|}}{C}OCH_3$
$\underset{\displaystyle CH_3}{\Big|}$

(c) a cyclohexanone ring with $\overset{\displaystyle OH}{\overset{\displaystyle |}{CH}}C_6H_5$ substituent

(d) a cyclohexane ring with $\overset{\displaystyle OH}{|}$ and $CH_2\overset{\overset{\displaystyle O}{\|}}{C}OC(CH_3)_3$

SAMPLE SOLUTION   (*a*)   The α carbon atom of the ester has two different alkyl groups attached to it.

$$(CH_3)_2CH \overset{\textcircled{a}}{-} \underset{\underset{CH_3}{|\textcircled{b}}}{\overset{\overset{O}{\parallel}}{CHCOCH_2CH_3}}$$

disconnect bond ⓐ → $(CH_3)_2CHX + CH_3\overset{\overset{O}{\parallel}}{\overset{\cdot\cdot}{C}HCOCH_2CH_3}$

disconnect bond ⓑ → $CH_3X + (CH_3)_2CH\overset{\overset{O}{\parallel}}{\overset{\cdot\cdot}{C}HCOCH_2CH_3}$

The critical carbon-carbon bond-forming step requires nucleophilic substitution on an alkyl halide by an ester enolate. Methyl halides are more reactive than isopropyl halides in S_N2 reactions and cannot undergo elimination as a competing process. Therefore, choose the synthesis in which bond ⓑ is formed by alkylation.

$$(CH_3)_2CHCH_2\overset{\overset{O}{\parallel}}{C}OCH_2CH_3 \xrightarrow[\text{2. CH}_3\text{I}]{\text{1. LDA, THF}} (CH_3)_2CH\underset{\underset{CH_3}{|}}{\overset{\overset{O}{\parallel}}{C}HCOCH_2CH_3}$$

Ethyl 3-methylbutanoate   Ethyl 2,3-dimethylbutanoate

(This synthesis has been reported in the chemical literature and gives the desired product in 95 percent yield.)

22.13   SUMMARY

β-Keto esters are useful reagents for a number of carbon-carbon bond-forming reactions. They may be prepared by the self-condensation of esters (the Claisen and Dieckmann condensations) or by the acylation of ketone enolates with diethyl carbonate. These procedures are summarized in Table 22.1.

β-Keto esters are characterized by K_a's of about 10^{-11} (pK_a 11) and are quantitatively converted to their enolates on treatment with alkoxide bases.

$$\underset{R}{\overset{\overset{O}{\parallel}}{C}}\underset{CH_2}{\overset{\overset{O}{\parallel}}{C}}OR'$$

Most acidic proton of a β-keto ester

$R'O^-$

$$R\overset{O^-}{\underset{CH}{C}}\overset{O}{\underset{}{C}}OR' \longleftrightarrow R\overset{O}{\underset{\overset{\cdot\cdot}{C}H}{C}}\overset{O}{\underset{}{C}}OR' \longleftrightarrow R\overset{O}{\underset{CH}{C}}\overset{O^-}{\underset{}{C}}OR'$$

Resonance forms illustrating charge delocalization in enolate of a β-keto ester

TABLE 22.1

Preparation of β-Keto Esters

Reaction (section) and comments	General equation and specific example

Claisen condensation (Section 22.1) Esters of the type

$$RCH_2COR'$$ (with C=O)

are converted to β-keto esters on treatment with alkoxide bases. One molecule of an ester is converted to its enolate; a second molecule of ester acts as an acylating agent toward the enolate.

General equation:

$$2RCH_2COR' \xrightarrow[\text{2. } H^+]{\text{1. NaOR'}} RCH_2CCHCOR' + R'OH$$

Ester   β-Keto ester   Alcohol

(with R substituent below, and C=O groups)

Specific example:

$$2CH_3CH_2CH_2COCH_2CH_3 \xrightarrow[\text{2. } H^+]{\text{1. NaOCH}_2CH_3} CH_3CH_2CH_2CCHCOCH_2CH_3$$

with CH_2CH_3 substituent

Ethyl butanoate → Ethyl 2-ethyl-3-oxohexanoate (76%)

Dieckmann condensation (Section 22.2) An intramolecular analog of the Claisen condensation. Cyclic β-keto esters in which the ring is five- to seven-membered may be formed by using this reaction.

(Structure: benzene ring with two $CH_2COCH_2CH_3$ groups)

$$\xrightarrow[\text{2. } H^+]{\text{1. NaOCH}_2CH_3}$$

(Structure: indanone with $COCH_2CH_3$ group)

Diethyl 1,2-benzenediacetate → Ethyl indan-2-one-1-carboxylate (70%)

Mixed Claisen condensations (Section 22.3) Diethyl carbonate, diethyl oxalate, ethyl formate, and benzoate esters cannot form ester enolates but can act as acylating agents toward other ester enolates.

General equation:

$$RCOCH_2CH_3 + R'CH_2COCH_2CH_3 \xrightarrow[\text{2. } H^+]{\text{1. NaOCH}_2CH_3} RCCHCOCH_2CH_3$$

with R' substituent

Ester   Another ester   β-Keto ester

Specific example:

$$CH_3CH_2COCH_2CH_3 + CH_3CH_2OCCOCH_2CH_3 \xrightarrow[\text{2. } H^+]{\text{1. NaOCH}_2CH_3} CH_3CHCOCH_2CH_3$$

with $C{-}COCH_2CH_3$ (two C=O)

Ethyl propanoate   Diethyl oxalate   Diethyl 3-methyl-2-oxobutanedioate (60–70%)

Acylation of ketones (Section 22.4) Diethyl carbonate and diethyl oxalate can be used to acylate ketone enolates to give β-keto esters.

General equation:

$$RCH_2CR' + CH_3CH_2OCOCH_2CH_3 \xrightarrow[\text{2. } H^+]{\text{1. NaOCH}_2CH_3} RCHCR'$$

with $COCH_2CH_3$ substituent

Ketone   Diethyl carbonate   β-Keto ester

Specific example:

$$(CH_3)_3CCH_2CCH_3 + CH_3CH_2OCOCH_2CH_3 \xrightarrow[\text{2. } H^+]{\text{1. NaOCH}_2CH_3} (CH_3)_3CCH_2CCH_2COCH_2CH_3$$

4,4-Dimethyl-2-pentanone   Diethyl carbonate   Ethyl 5,5-dimethyl-3-oxohexanoate (66%)

The anion of a β-keto ester is alkylated at carbon on treatment with an alkyl halide.

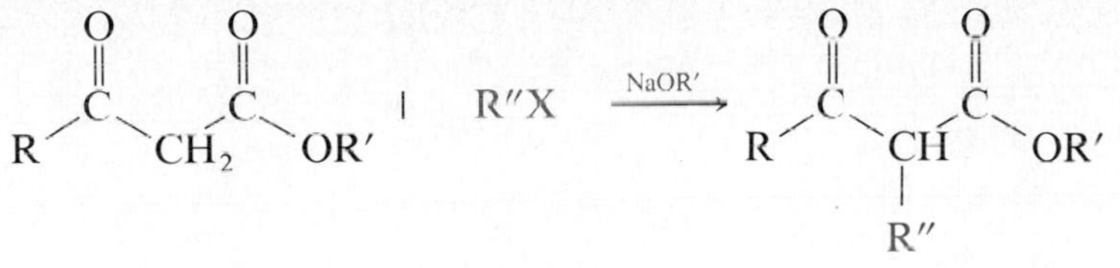

| β-Keto ester | Alkyl halide | Alkylated β-keto ester |

Saponification and decarboxylation of an alkylated β-keto ester converts it to a ketone.

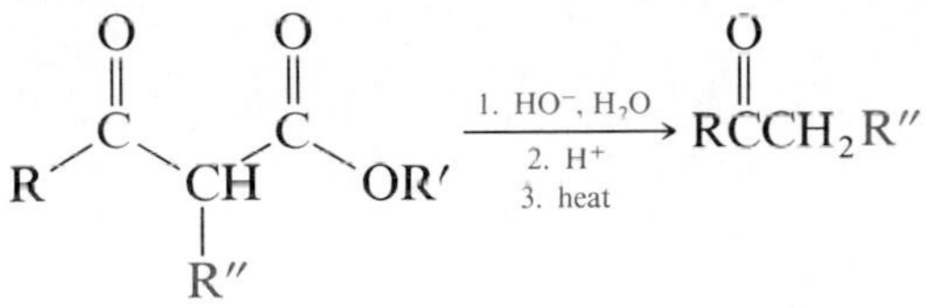

Alkylated β-keto ester   Ketone

The procedure whereby alkyl halides (RX) are converted to ketones of the type

$$RCH_2\overset{O}{\overset{\|}{C}}CH_3$$ with ethyl acetoacetate is called the *acetoacetic ester synthesis.*

The acetoacetic ester synthesis

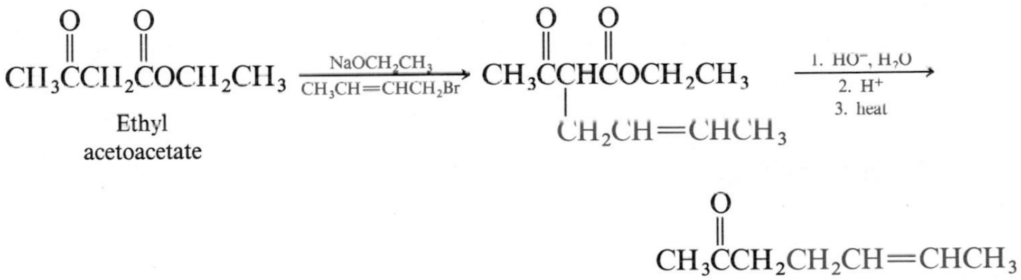

$$CH_3\overset{O}{\overset{\|}{C}}CH_2\overset{O}{\overset{\|}{C}}OCH_2CH_3 \xrightarrow[CH_3CH=CHCH_2Br]{NaOCH_2CH_3} CH_3\overset{O}{\overset{\|}{C}}\underset{\underset{CH_2CH=CHCH_3}{|}}{CH}\overset{O}{\overset{\|}{C}}OCH_2CH_3 \xrightarrow[\substack{2.\ H^+ \\ 3.\ heat}]{1.\ HO^-,\ H_2O}$$

Ethyl acetoacetate

$$CH_3\overset{O}{\overset{\|}{C}}CH_2CH_2CH=CHCH_3$$

5-Hepten-2-one
(81%)

The *malonic ester synthesis* is related to the acetoacetic ester synthesis. Alkyl halides (RX) are converted to carboxylic acids of the type RCH_2COOH by reaction with the enolate ion derived from diethyl malonate, followed by saponification and decarboxylation.

The malonic ester synthesis

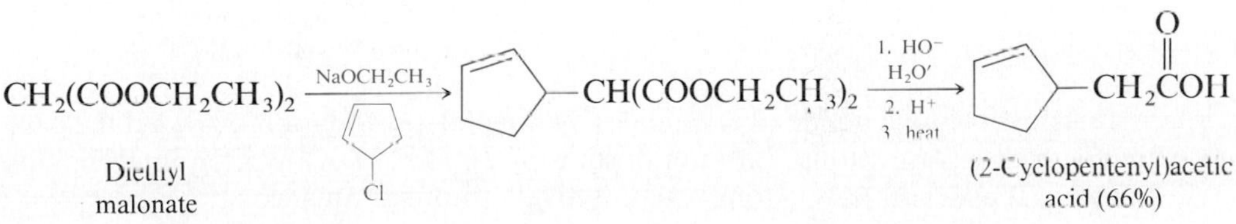

Diethyl malonate   (2-Cyclopentenyl)acetic acid (66%)

Michael addition of the enolate ions derived from ethyl acetoacetate and diethyl malonate provides an alternative method for their alkylation.

Michael addition

$$CH_2(COOCH_2CH_3)_2 + CH_3CH{=}CHCOCH_2CH_3 \xrightarrow[CH_3CH_2OH]{NaOCH_2CH_3} CH_3CHCH_2COCH_2CH_3$$

$$\underset{\text{CH(COOCH}_2\text{CH}_3)_2}{\quad}$$

Diethyl malonate	Ethyl 2-butenoate	Triethyl 2-methylpropane-1,1,3-tricarboxylate (95%)

Diethyl malonate and β-keto esters react with aldehydes and ketones in the presence of amines such as piperidine to give alkenes. This reaction is called the *Knoevenagel condensation*.

The Knoevenagel condensation

$$CH_2(COOCH_2CH_3)_2 + (CH_3)_2CHCH{=}O \xrightarrow[\text{acetic acid}]{\text{piperidine}} (CH_3)_2CHCH{=}C(COOCH_2CH_3)_2$$

Diethyl malonate	2-Methylpropanal	Diethyl isobutylidenemalonate (90–92%)

Ester enolates are intermediates in the *Reformatsky reaction*. Treatment of an α-halo ester with zinc in the presence of an aldehyde or ketone gives a β-hydroxy ester. A zinc enolate is an intermediate.

The Reformatsky reaction

$$CH_3CHCOCH_2CH_3 + CH_3CH_2CH_2CH_2CHCH{=}O \xrightarrow[2.\ H_3O^+]{1.\ Zn}$$

$$\underset{Br}{\quad} \qquad \underset{CH_3CH_2}{\quad}$$

Ethyl 2-bromopropanoate	2-Ethylhexanal

$$CH_3CH_2CH_2CH_2CHCHCHCOCH_2CH_3$$

with substituents OH, O, CH_3CH_2, CH_3

Ethyl 4-ethyl-3-hydroxy-2-methyloctanoate (87%)

It is possible to generate ester enolates by direct deprotonation provided the base that is used is very strong. Lithium diisopropylamide (LDA) is often used for this purpose. It also converts ketones quantitatively to their enolates.

$$CH_3CH_2\overset{\overset{\displaystyle O}{\|}}{C}C(CH_3)_3 \xrightarrow[\text{THF}]{\text{LDA}} CH_3CH=\overset{\overset{\displaystyle OLi}{|}}{C}C(CH_3)_3 \xrightarrow[\text{2. H}_3\text{O}^+]{\text{1. C}_6\text{H}_5\text{CH}} C_6H_5\overset{\overset{\displaystyle OH}{|}}{C}H\overset{}{C}H\overset{\overset{\displaystyle O}{\|}}{C}C(CH_3)_3$$

$$\underset{\text{CH}_3}{\quad}$$

2,2-Dimethyl-3-pentanone

5 Hydroxy-2,2,4-
trimethyl-5-phenyl-
3-pentanone (78%)

PROBLEMS

22.16 The following questions pertain to the esters shown and their behavior under conditions of the Claisen condensation.

$$CH_3CH_2CH_2CH_2\overset{\overset{\displaystyle O}{\|}}{C}OCH_2CH_3 \quad CH_3CH_2\overset{\overset{\displaystyle O}{\|}}{C}H\overset{}{C}OCH_2CH_3 \quad CH_3CHCH_2\overset{\overset{\displaystyle O}{\|}}{C}OCH_2CH_3 \quad (CH_3)_3C\overset{\overset{\displaystyle O}{\|}}{C}OCH_2CH_3$$

	CH₃	CH₃	
Ethyl	Ethyl	Ethyl	Ethyl
pentanoate	2-methylbutanoate	3-methylbutanoate	2,2-dimethylpropanoate

(a) Two of these esters are converted to β-keto esters in good yield on treatment with sodium ethoxide and subsequent acidification of the reaction mixture. Which two are these? Write the structure of the Claisen condensation product of each one.

(b) One ester is capable of being converted to a β-keto ester on treatment with sodium ethoxide, but the amount of β-keto ester that can be isolated after acidification of the reaction mixture is quite small. Which ester is this?

(c) One ester is incapable of reaction under conditions of the Claisen condensation. Which one? Why?

22.17 (a) Give the structure of the Claisen condensation product of ethyl phenylacetate ($C_6H_5CH_2COOCH_2CH_3$).

(b) What ketone would you isolate after saponification and decarboxylation of this Claisen condensation product?

(c) What ketone would you isolate after treatment of the Claisen condensation product of ethyl phenylacetate with sodium ethoxide and allyl bromide, followed by saponification and decarboxylation?

(d) Give the structure of the mixed Claisen condensation product of ethyl phenylacetate and ethyl benzoate.

(e) What ketone would you isolate after saponification and decarboxylation of the product in (d)?

(f) What ketone would you isolate after treatment of the product in (d) with sodium ethoxide and allyl bromide, followed by saponification and decarboxylation?

22.18 All the following questions concern ethyl 2-oxo-1-cyclohexanecarboxylate.

Ethyl 2-oxo-1-cyclohexanecarboxylate

(a) Write a chemical equation showing how you could prepare ethyl 2-oxo-1-cyclohexanecarboxylate by a Dieckmann condensation.

(b) Write a chemical equation showing how you could prepare ethyl 2-oxo-1-cyclohexanecarboxylate by acylation of a ketone.

(c) Write structural formulas for the two most stable enol forms of ethyl 2-oxo-1-cyclohexanecarboxylate.

(d) Write the three most stable resonance forms for the most stable enolate derived from ethyl 2-oxo-1-cyclohexanecarboxylate.

(e) Show how you could use ethyl 2-oxo-1-cyclohexanecarboxylate to prepare 2-methylcyclohexanone.

(f) Give the structure of the product formed on treatment of ethyl 2-oxo-1-cyclohexanecarboxylate with acrolein ($CH_2{=}CHCH$)

$$\overset{O}{\underset{\|}{}}$$

in ethanol in the presence of sodium ethoxide.

22.19 Give the structure of the product formed on reaction of ethyl acetoacetate with each of the following:

(a) 1-Bromopentane and sodium ethoxide

(b) Saponification and decarboxylation of the product in (a)

(c) Methyl iodide and the product in (a) treated with sodium ethoxide

(d) Saponification and decarboxylation of the product in (c)

(e) 1-Bromo-3-chloropropane and one equivalent of sodium ethoxide

(f) Product in (e) treated with a second equivalent of sodium ethoxide

(g) Saponification and decarboxylation of the product in (f)

(h) Phenyl vinyl ketone and sodium ethoxide

(i) Saponification and decarboxylation of the product in (h)

(j) 3-Pentanone in the presence of piperidine

22.20 Repeat the preceding problem for diethyl malonate.

22.21 (a) Only a small amount (less than 0.01 percent) of the enol form of diethyl malonate is present at equilibrium. Write a structural formula for this enol.

(b) Enol forms are present to the extent of about 8 percent in ethyl acetoacetate. There are three constitutionally isomeric enols possible. Write structural formulas for these three enols. Which one do you think is the most stable? The least stable? Why?

(c) Bromine reacts rapidly with both diethyl malonate and ethyl acetoacetate. The reaction is acid-catalyzed and liberates hydrogen bromide. What is the product formed in each reaction?

22.22 (a) On addition of one equivalent of methylmagnesium iodide to ethyl acetoacetate, the Grignard reagent is consumed but the only organic product obtained after working up the reaction mixture is ethyl acetoacetate. Why? What happened to the Grignard reagent?

(b) On repeating the reaction but using D_2O and DCl to work up the reaction mixture, it is found that the recovered ethyl acetoacetate contains deuterium. Where is this deuterium located?

22.23 Give the structure of the principal organic product of each of the following reactions:

(a) Ethyl octanoate $\xrightarrow[\text{2. H}^+]{\text{1. NaOCH}_2\text{CH}_3}$

(b) Product of (a) $\xrightarrow[\substack{\text{2. H}^+ \\ \text{3. heat}}]{\text{1. NaOH, H}_2\text{O}}$

(c) Ethyl acetoacetate + 1-bromobutane $\xrightarrow{\text{NaOCH}_2\text{CH}_3,\ \text{ethanol}}$

(d) Product of (c) $\xrightarrow[\substack{2.\ H^+ \\ 3.\ heat}]{1.\ NaOH,\ H_2O}$

(e) Product of (c) + 1-iodobutane $\xrightarrow{NaOCH_2CH_3,\ ethanol}$

(f) Product of (e) $\xrightarrow[\substack{2.\ H^+ \\ 3.\ heat}]{1.\ NaOH,\ H_2O}$

(g) Acetophenone + diethyl carbonate $\xrightarrow[2.\ H^+]{1.\ NaOCH_2CH_3}$

(h) Acetone + diethyl oxalate $\xrightarrow[2.\ H^+]{1.\ NaOCH_2CH_3}$

(i) Diethyl malonate + 1-bromo-2-methylbutane $\xrightarrow{NaOCH_2CH_3,\ ethanol}$

(j) Product of (i) $\xrightarrow[\substack{2.\ H^+ \\ 3.\ heat}]{1.\ NaOH,\ H_2O}$

(k) Diethyl malonate + 6-methyl-2-cyclohexenone $\xrightarrow{NaOCH_2CH_3,\ ethanol}$

(l) Product of (k) $\xrightarrow{H_2O,\ HCl,\ heat}$

(m) Ethyl acetoacetate + 2,3-dimethoxybenzaldehyde $\xrightarrow[acetic\ acid]{piperidine}$

(n) *tert*-Butyl acetate $\xrightarrow[\substack{2.\ benzaldehyde \\ 3.\ H^+}]{1.\ [(CH_3)_2CH]_2NLi,\ THF}$

22.24 Give the structure of the principal organic product of each of the following reactions:

(a) $\xrightarrow[heat]{H_2O,\ H_2SO_4}$ $C_7H_{12}O$

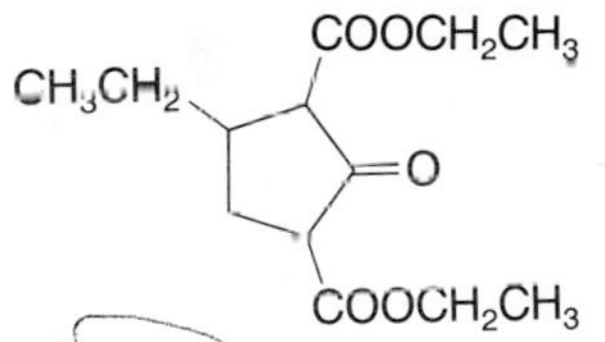

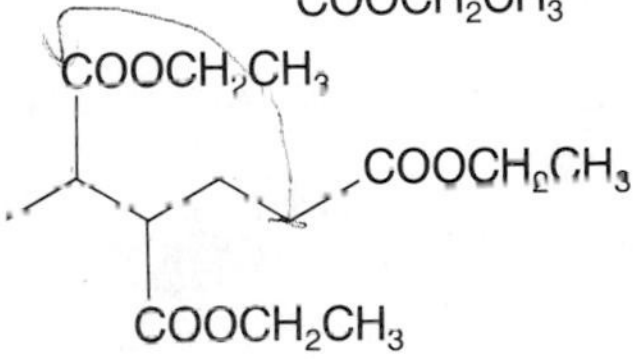

(h) $\xrightarrow[2.\ H^+]{1.\ NaOCl\ I_2CH_3}$ $C_{12}H_{18}O_5$

(c) Product of (b) $\xrightarrow[heat]{H_2O,\ H^+}$ $C_7H_{10}O_3$

(d) $\xrightarrow[2.\ H^+]{1.\ NaOCH_2CH_3}$ $C_9H_{12}O_3$

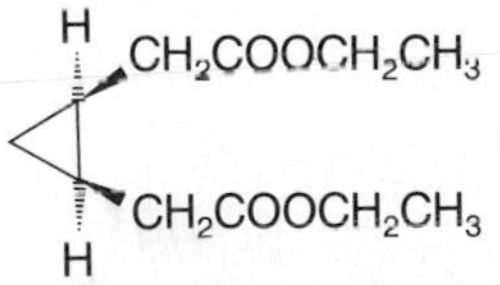

(e) Product of (d) $\xrightarrow[\substack{2.\ H^+ \\ 3.\ heat}]{1.\ HO^-,\ H_2O}$ C_6H_9O

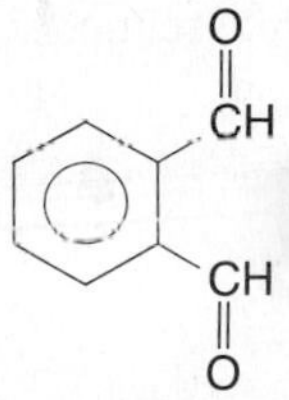

(f) + $CH_3CH_2OCCH_2CCH_2COCH_2CH_3$ $\xrightarrow[acetic\ acid]{piperidine}$ $C_{17}H_{16}O_5$

22.25 Show how you could prepare each of the following compounds. Use the starting material indicated along with ethyl acetoacetate or diethyl malonate and any necessary inorganic reagents. Assume also that the customary organic solvents are freely available.

(a) 4-Phenyl-2-butanone from benzyl alcohol
(b) 3-Phenylpropanoic acid from benzyl alcohol
(c) Diethyl benzylidenemalonate [$C_6H_5CH{=}C(COOCH_2CH_3)_2$] from benzyl alcohol
(d) 2-Allyl-1,3-propanediol from propene
(e) 4-Penten-1-ol from propene
(f) 5-Hexen-2-ol from propene
(g) Cyclopropanecarboxylic acid from 1,2-dibromoethane

(h) 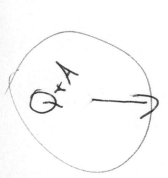 from 1,2-dibromoethane

(i) $HO_2C(CH_2)_{10}CO_2H$ from $HO_2C(CH_2)_6CO_2H$

22.26 *Diphenadione* inhibits the clotting of blood, i.e., it is an *anticoagulant.* It is used to control vampire bat populations in South America by a "Trojan horse" strategy. A few bats are trapped, smeared with diphenadione, and then released back into their normal environment. Other bats, in the normal course of grooming these diphenadione-coated bats, ingest the anticoagulant and bleed to death, either internally or through accidental bites and scratches.

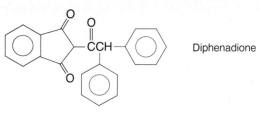

Diphenadione

Suggest a synthesis of diphenadione from 1,1-diphenylacetone and dimethyl 1,2-benzenedicarboxylate.

22.27 *Phenylbutazone* is a frequently prescribed anti-inflammatory drug. It is prepared by the reaction shown.

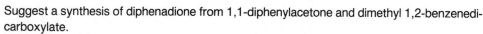

$CH_3CH_2CH_2CH_2CH(COOCH_2CH_3)_2$ + $C_6H_5NHNHC_6H_5$ $\longrightarrow$ $C_{19}H_{20}N_2O_2$

Diethyl butylmalonate   1,2-Diphenylhydrazine   Phenylbutazone

What is the structure of phenylbutazone?

22.28 The reaction of the sodium salt of ethyl acetoacetate with chloroacetone ($ClCH_2\overset{O}{\overset{\|}{C}}CH_3$) yields compound A ($C_9H_{14}O_4$). Treatment of compound A with sulfuric acid gives ethyl 2,5-dimethylfuran-3-carboxylate (compound B).

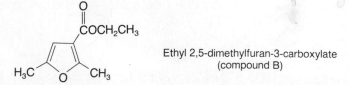

Ethyl 2,5-dimethylfuran-3-carboxylate
(compound B)

What is the structure of compound A? Suggest a reasonable mechanism for the conversion of compound A to compound B.

22.29   The use of epoxides as alkylating agents for diethyl malonate provides a useful route to γ-lactones. Write equations illustrating such a sequence for styrene oxide as the starting epoxide. Is the lactone formed by this reaction 3-phenylbutanolide or is it 4-phenylbutanolide?

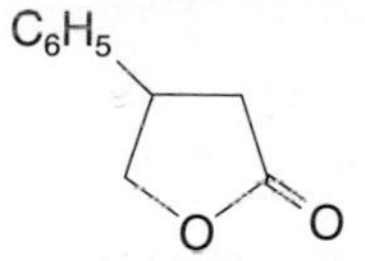

3-Phenylbutanolide 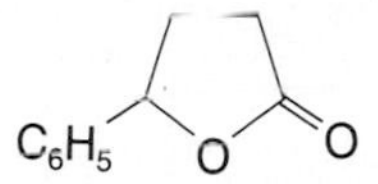   4-Phenylbutanolide

22.30   Diethyl malonate is prepared commercially by hydrolysis and esterification of ethyl cyanoacetate.

$$N\equiv CCH_2\overset{\displaystyle O}{\overset{\displaystyle \|}{C}}OCH_2CH_3 \quad \text{Ethyl cyanoacetate}$$

The preparation of ethyl cyanoacetate proceeds via ethyl chloroacetate and begins with acetic acid. Write a sequence of reactions describing this synthesis.

22.31   Ethyl cyanoacetate undergoes many of the same kinds of reactions as diethyl malonate and has a number of useful synthetic applications. It is, for example, the starting material used in the preparation of the anticonvulsant drug *phensuximide.*

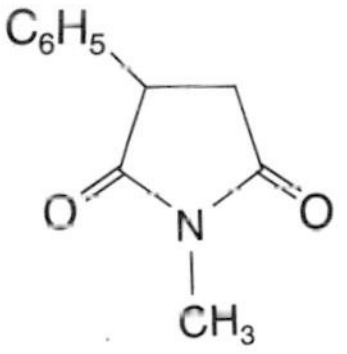

Phensuximide

(a)   The first step in the preparation of phensuximide is a Knoevenagel condensation between ethyl cyanoacetate and benzaldehyde. What is the product formed in this reaction?

(b)   In the second step the product of the Knoevenagel condensation is treated with sodium cyanide. A Michael addition takes place. What is the product of this reaction?

(c)   Hydrolysis of the Michael adduct in aqueous hydrochloric acid is accompanied by decarboxylation to give a diacid. When the diacid is heated with methylamine (CH$_3$NH$_2$), phensuximide is produced. What is the structure of the diacid intermediate?

22.32   The tranquilizing drug *meprobamate* has the structure shown.

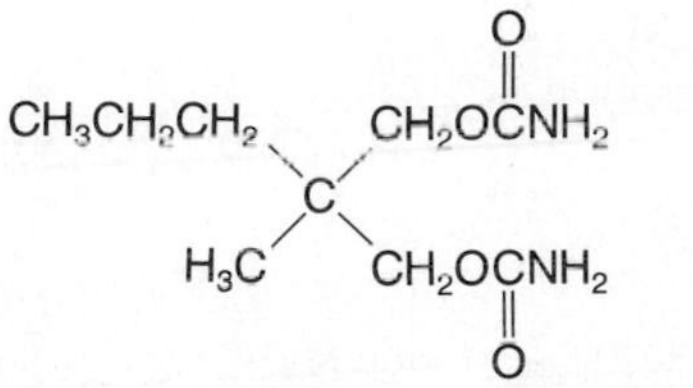

Meprobamate

Devise a synthesis of meprobamate from diethyl malonate and any necessary organic or inorganic reagents. *Hint: Carbamate esters,* i.e., compounds of the type $ROCNH_2$, are prepared from alcohols by the sequence of reactions

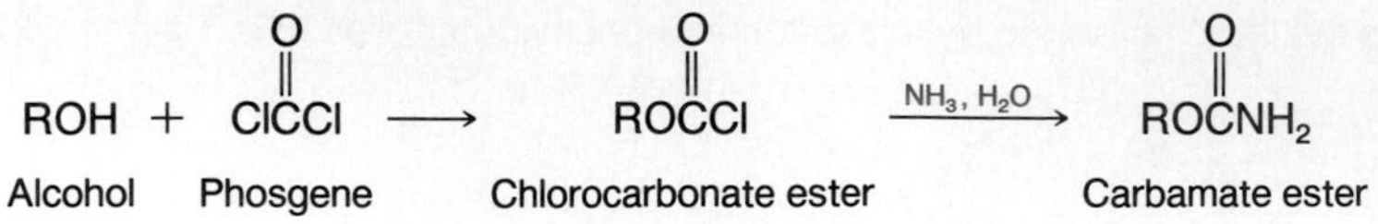

$$
\underset{\text{Alcohol}}{ROH} + \underset{\text{Phosgene}}{ClCCl} \longrightarrow \underset{\text{Chlorocarbonate ester}}{ROCCl} \xrightarrow{\text{NH}_3,\ \text{H}_2\text{O}} \underset{\text{Carbamate ester}}{ROCNH_2}
$$

ALKYLAMINES

Nitrogen-containing compounds are essential to life and are ultimately derived from atmospheric nitrogen. By a process known as *nitrogen fixation,* atmospheric nitrogen is reduced to ammonia, then converted to organic nitrogen compounds. This chapter describes the chemistry of *alkylamines,* organic derivatives of ammonia in which nitrogen is bonded to sp^3 hybridized carbon. *Arylamines,* in which the nitrogen is bonded directly to an sp^2 hybridized carbon of an aromatic ring, will be discussed separately in the following chapter.

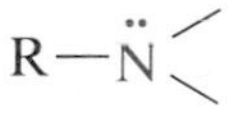

$$R—\ddot{N}\diagdown \qquad\qquad Ar—\ddot{N}\diagdown$$

R = alkyl group: Ar = aryl group:
alkylamine   arylamine

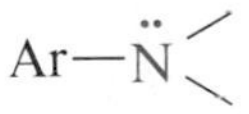

Examples of alkylamines include methylamine and epinephrine.

$$CH_3NH_2$$

Methylamine: Epinephrine (adrenaline): a
an industrial organic   hormone secreted by the adrenal
chemical   gland; prepares organism for
                "flight or fight"

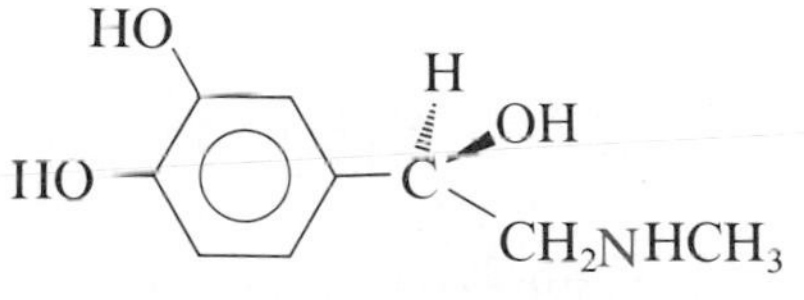

Methylamine is the simplest alkylamine. Epinephrine contains an aromatic ring in its structure, but is classified as an alkylamine because nitrogen is not bonded directly to the ring; both carbon-nitrogen bonds of epinephrine are to sp^3 hybridized carbon atoms.

Alkylamines, like ammonia, are weak bases. They are, however, the strongest bases that are found in significant quantities under physiological conditions. Amines are usually the bases involved in biological acid-base reactions; they are often the nucleophiles in biological nucleophilic substitution reactions.

Our word "vitamin" was coined in 1912 in the belief that the substances present in the diet that prevented scurvy, pellagra, beriberi, rickets, and other diseases were "vital amines." In many cases that belief was confirmed; certain vitamins did prove to be amines. In many other cases, however, vitamins were not amines. Nevertheless, the name "vitamin" entered our language and stands as a reminder that early chemists were acutely aware of the crucial place occupied by amines in biological processes.

23.1 NOMENCLATURE

Unlike alcohols and alkyl halides, which are classified as primary, secondary, or tertiary according to the degree of substitution at the carbon that bears the functional group, amines are classified according to their *degree of substitution at the nitrogen atom.* An amine with one carbon substituent on nitrogen is a *primary amine,* an amine with two is a *secondary amine,* and an amine in which all three of the bonds from nitrogen are to carbon is a *tertiary amine.*

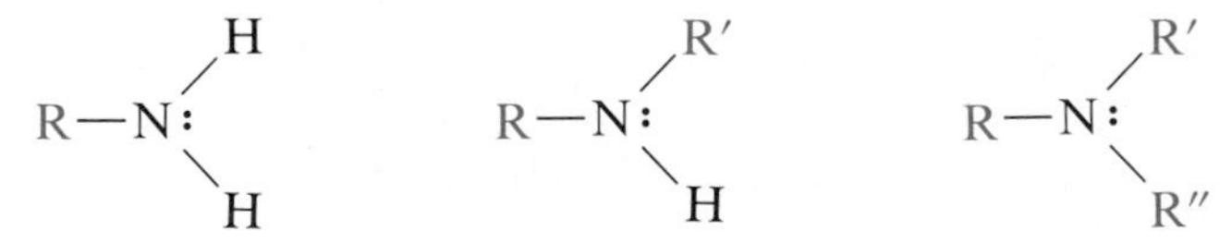

|Primary amine|Secondary amine|Tertiary amine|

Amines may be named in two principal ways, either as *alkylamines* or as *alkanamines.* When primary amines are named as alkylamines, the ending *-amine* is added to the name of the alkyl group that bears the nitrogen. When named as alkanamines, the alkyl group is named as an alkane and the *-e* ending replaced by *-amine.*

$CH_3CH_2NH_2$

Ethylamine
(ethanamine)

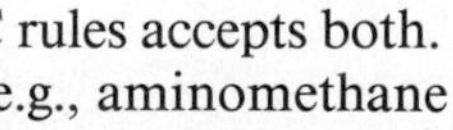

Cyclohexylamine
(cyclohexanamine)

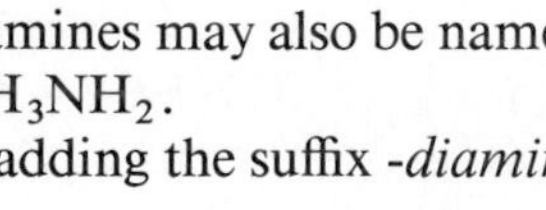

$CH_3CHCH_2CH_2CH_3$ with NH_2

1-Methylbutylamine
(2-pentanamine)

The alkanamine naming system was introduced by *Chemical Abstracts* and is more versatile and easier to use than the older IUPAC system of alkylamine names. The latest revision of the IUPAC rules accepts both. Alkylamines may also be named as amino-substituted alkanes, e.g., aminomethane for CH_3NH_2.

Compounds with two amino groups are named by adding the suffix *-diamine* to the full name of the corresponding hydrocarbon.

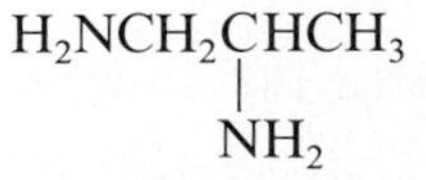

$H_2NCH_2CHCH_3$ with NH_2

1,2-Propanediamine

$H_2NCH_2CH_2CH_2CH_2CH_2CH_2NH_2$

1,6-Hexanediamine

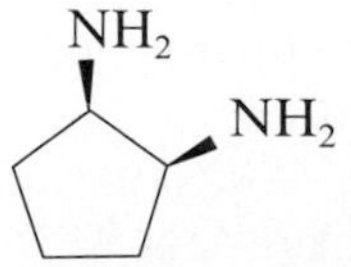

cis-1,2-Cyclopentane-diamine

PROBLEM 23.1   Give an acceptable alkylamine or alkanamine name for each of the following primary amines:

(a) $C_6H_5CH_2CH_2NH_2$
(b) $C_6H_5CHCH_3$
            |
           NH_2
(c) $(CH_3)_3CCH_2NH_2$

(d)

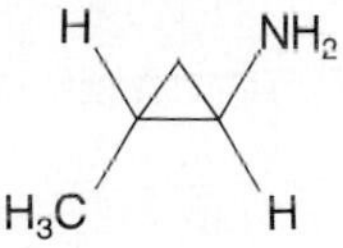

(e) $CH_2{=}CHCH_2NH_2$

SAMPLE SOLUTION   (a) The amino substituent is bonded to an ethyl group that bears a phenyl substituent at C-2. The compound $C_6H_5CH_2CH_2NH_2$ may be named as either 2-phenylethylamine or 2-phenylethanamine.

Secondary and tertiary amines are named as *N*-substituted derivatives of primary amines. The parent primary amine is taken to be the one with the longest carbon chain. The prefix *N*- is added to identify substituents on the amino nitrogen as needed.

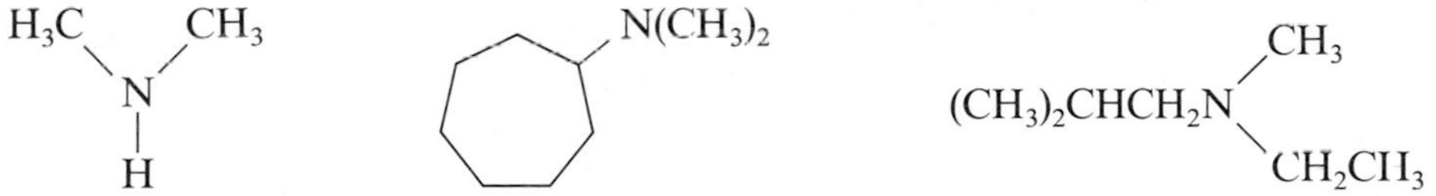

Dimethylamine
(a secondary
amine)

N,N-Dimethylcycloheptyl-
amine
(a tertiary amine)

N-Ethyl-*N*-methyliso-
butylamine
(a tertiary amine)

PROBLEM 20.2   Name each of the amines shown above as an *N*-substituted alkanamine.

SAMPLE SOLUTION   The first compound, CH_3NHCH_3, is an *N*-substituted derivative of methanamine; it is *N*-methylmethanamine.

Amino groups rank rather low in seniority when the parent compound is identified for naming purposes. Hydroxyl groups, as well as the higher oxidation states of carbon, outrank amino groups. In these cases the amino group is named as a substituent.

$$HOCH_2CH_2NH_2 \qquad (CH_3)_2NCH_2CO_2H$$

2-Aminoethanol
(not 2-hydroxyethylamine)

N,N-Dimethylaminoacetic acid

A nitrogen that bears four substituents is positively charged and is referred to as an ammonium ion. The anion that is associated with it is also identified in the name.

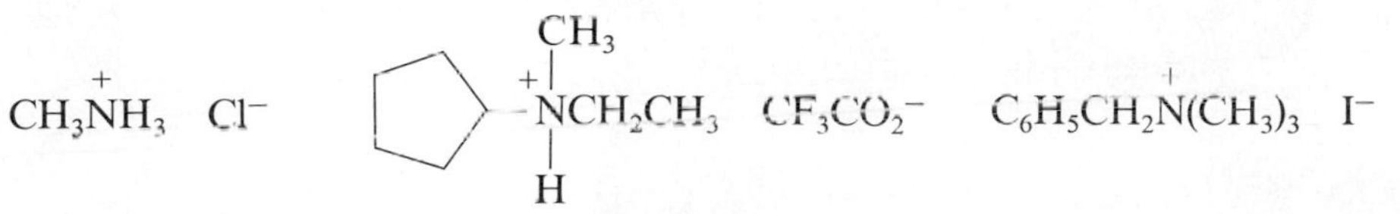

Methylammonium
chloride

N-Ethyl-*N*-methylcyclopentyl-
ammonium trifluoroacetate

Benzyltrimethyl-
ammonium iodide
(a quaternary ammonium
salt)

Ammonium salts that have four alkyl groups bonded to nitrogen are called *quaternary ammonium salts.*

23.2 STRUCTURE AND BONDING

The structure of alkylamines can be illustrated by reference to methylamine. Methylamine, like ammonia, has a pyramidal arrangement of substituents at nitrogen.

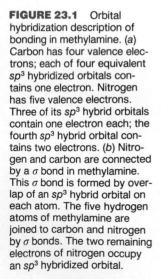

Typical H—N—H angles in amines are slightly smaller than tetrahedral, while C—N—H and C—N—C angles are slightly larger. Carbon-nitrogen bond distances range between 1.45 and 1.48 Å; C—N bonds are shorter than C—C single bonds, but longer than C—O bonds.

An orbital hybridization description of bonding in methylamine is shown in Figure 23.1. Nitrogen and carbon are both sp^3 hybridized and are joined by a σ bond. The unshared electron pair on nitrogen occupies an sp^3 hybridized orbital. It is this lone pair that is involved in reactions in which amines act as bases or nucleophiles.

Amines, like ammonia, undergo pyramidal inversion.

This process is exceedingly rapid. Its energy of activation is only 6 to 7 kcal/mol, a little greater than the barrier to rotation about the C—C bond in ethane but less than that for conformational ring inversion in cyclohexane. This means that while amines

FIGURE 23.1 Orbital hybridization description of bonding in methylamine. (*a*) Carbon has four valence electrons; each of four equivalent sp^3 hybridized orbitals contains one electron. Nitrogen has five valence electrons. Three of its sp^3 hybrid orbitals contain one electron each; the fourth sp^3 hybrid orbital contains two electrons. (*b*) Nitrogen and carbon are connected by a σ bond in methylamine. This σ bond is formed by overlap of an sp^3 hybrid orbital on each atom. The five hydrogen atoms of methylamine are joined to carbon and nitrogen by σ bonds. The two remaining electrons of nitrogen occupy an sp^3 hybridized orbital.

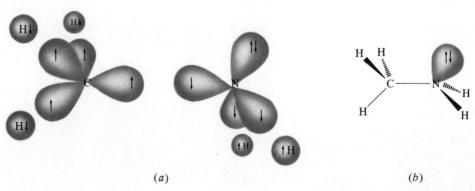

(*a*)   (*b*)

bearing three different alkyl groups are chiral, interconversion of the two enantio-meric forms is so fast that they cannot be resolved at room temperature.

Quaternary ammonium salts have a tetrahedral geometry at nitrogen and are configurationally stable. A number of chiral quaternary ammonium salts have been resolved into their enantiomeric forms. Their absolute configuration is specified as *R* or *S* in exactly the same way as the absolute configuration of compounds in which the chiral center is tetrahedral carbon.

PROBLEM 23.3 What is the configuration (*R* or *S*) of the quaternary ammonium salt whose structure is given below?

$$C_6H_5CH_2 \quad CH_3$$
$$N^+—CH_2CH_2CH_3 \quad I^-$$
$$C_6H_5$$

23.3 PHYSICAL PROPERTIES

A collection of physical constants of some representative amines is presented in Table 23.1. Most commonly encountered amines are liquids with unpleasant, "fishlike" odors.

Amines are more polar than hydrocarbons but less polar than alcohols and ethers.

$$CH_3CH_2CH_3 \qquad CH_3CH_2NH_2 \qquad CH_3CH_2OH$$

Propane	Ethylamine	Ethanol
$\mu = 0$ D	$\mu = 1.2$ D	$\mu = 1.7$ D
bp $-42°$C	bp $17°$C	bp $78°$C

Dipole-dipole interactions and hydrogen bonding are stronger in amines than in hydrocarbons of comparable molecular weight, so amines have higher boiling points than similarly constituted alkanes. The less polar nature of amines as compared with alcohols, however, makes these intermolecular attractive forces in amines weaker than in alcohols, and so amines have lower boiling points than comparable alcohols. Among isomeric amines, primary amines have the highest boiling points and tertiary amines the lowest.

$$CH_3CH_2CH_2NH_2 \qquad CH_3CH_2NHCH_3 \qquad (CH_3)_3N$$

Propylamine	*N*-Methylethylamine	Trimethylamine
(a primary amine)	(a secondary amine)	(a tertiary amine)
bp $50°$C	bp $34°$C	bp $3°$C

Primary and secondary amines can participate in intermolecular hydrogen bonding while tertiary amines cannot.

Amines having fewer than six or seven carbon atoms are soluble in water. All amines, even tertiary amines, can act as proton acceptors in hydrogen bonding to water molecules. The basicity of the unshared electron pair on nitrogen enhances the

TABLE 23.1
Physical Properties of Some Amines

Compound	Condensed structural formula	Melting point, °C	Boiling point, °C	Solubility, g/100 mL H_2O
Primary amines				
Methylamine	CH_3NH_2	−92.5	−6.7	Very high
Ethylamine	$CH_3CH_2NH_2$	−80.6	16.6	∞
Propylamine	$CH_3CH_2CH_2NH_2$	−83	50	∞
Isopropylamine	$(CH_3)_2CHNH_2$	−101	33	∞
Butylamine	$CH_3CH_2CH_2CH_2NH_2$	−50	77.8	∞
Isobutylamine	$(CH_3)_2CHCH_2NH_2$	−85	68	∞
sec-Butylamine	$CH_3CH_2CHNH_2$ $\quad\quad\quad\;CH_3$	−104	66	∞
tert-Butylamine	$(CH_3)_3CNH_2$	−67.5	45.2	
Pentylamine	$CH_3(CH_2)_4NH_2$	−55	103	Soluble
Hexylamine	$CH_3(CH_2)_5NH_2$	−19	129	Slightly soluble
Cyclohexylamine	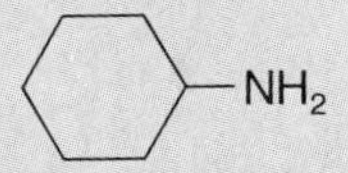 ⬡—NH_2	−18	134.5	∞
Benzylamine	$C_6H_5CH_2NH_2$	10	184.5	∞
Secondary amines				
Dimethylamine	$(CH_3)_2NH$	−92.2	6.9	Very soluble
Diethylamine	$(CH_3CH_2)_2NH$	−50	55.5	Very soluble
N-Methylpropylamine	$CH_3NHCH_2CH_2CH_3$		62.4	Soluble
Pyrrolidine	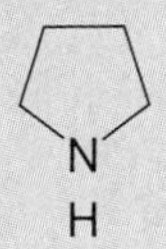 (ring) N–H	2	90	Very soluble
Piperidine	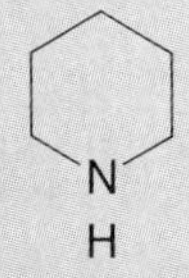 (ring) N–H	−10.5	106.4	∞
Dipropylamine	$(CH_3CH_2CH_2)_2NH$	−39.6	110	Soluble
Tertiary amines				
Trimethylamine	$(CH_3)_3N$	−117.1	2.9	41
Triethylamine	$(CH_3CH_2)_3N$	−114.7	89.4	∞
N-Methylpiperidine	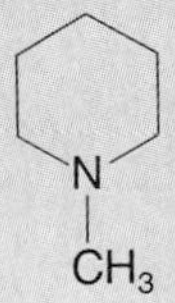 (ring) N–CH_3	3	107	

solvation of amines by water, alcohols, and other proton donors. Amines tend to be somewhat more soluble in water than are the corresponding alcohols.

23.4 BASICITY OF ALKYLAMINES

As a class amines are the strongest bases of all neutral molecules. They react with strong acids by proton transfer to nitrogen to form ammonium salts.

$$R_3N: \quad + \quad H-X \quad \longrightarrow \quad R_3\overset{+}{N}-H \quad X^-$$
$$\text{Amine} \qquad\qquad \text{Acid} \qquad\qquad \text{Ammonium salt}$$

These ammonium salts are ionic and are more soluble in polar solvents than in nonpolar ones. Amines can be extracted from an ether solution into water by shaking with dilute hydrochloric acid. Conversely, adding sodium hydroxide to an aqueous solution of an ammonium salt converts it to the free amine, which can be removed from the aqueous phase by extraction with ether.

$$R_3\overset{+}{N}-H \quad + \quad {}^-OH \quad \longrightarrow \quad R_3N: \quad + \quad H_2O$$

Ammonium ion	Hydroxide ion	Amine	Water
(stronger acid)	(stronger base)	(weaker base)	(weaker acid)

By convention, the base strength of an amine is related to the equilibrium constant for the reaction:

$$R_3N: \quad + \quad H-OH \quad \overset{K}{\rightleftharpoons} \quad R_3\overset{|}{N}H \quad + \quad {}^-OH$$

Amine	Water	Ammonium ion	Hydroxide ion
(weaker base)	(weaker acid)	(stronger acid)	(stronger base)

The basicity constant K_b for an amine is related to K, the equilibrium constant for the reaction, according to the expression:

$$K_b = K[H_2O] = \frac{[R_3\overset{+}{N}H][HO^-]}{[R_3N]}$$

and

$$pK_b = -\log K_b$$

Alkylamines are weak bases and have basicity constants in the range of 10^{-3} to 10^{-5}. They are weaker bases than hydroxide ion. Basicity constants K_b for some representative alkylamines are presented in Table 23.2.

In general, the order of amine basicity is

$$NH_3 \quad < \quad RNH_2 \quad \sim \quad R_3N \quad < \quad R_2NH$$

Ammonia	Primary amine	Tertiary amine	Secondary amine
(least basic)			(most basic)

Diethylamine, for example, is more basic than either ethylamine or triethylamine and all these compounds are more basic than ammonia, as measured by the value of their basicity constants.

TABLE 23.2
Basicity Constants of Alkylamines

Compound	Structure	Basicity constant, K_b*	pK_b
Ammonia	NH_3	1.8×10^{-5}	4.7
Primary amines			
Methylamine	CH_3NH_2	4.4×10^{-4}	3.4
Ethylamine	$CH_3CH_2NH_2$	5.6×10^{-4}	3.2
Propylamine	$CH_3CH_2CH_2NH_2$	3.9×10^{-4}	3.4
Isopropylamine	$(CH_3)_2CHNH_2$	4.3×10^{-4}	3.4
tert-Butylamine	$(CH_3)_3CNH_2$	2.8×10^{-4}	3.6
Cyclohexylamine	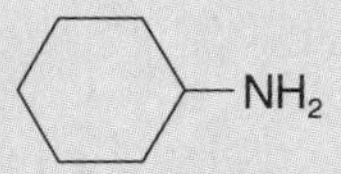 (cyclohexyl)—NH_2	4.4×10^{-4}	3.4
Benzylamine	$C_6H_5CH_2NH_2$	2.0×10^{-5}	4.7
Secondary amines			
Dimethylamine	$(CH_3)_2NH$	5.1×10^{-4}	3.3
Diethylamine	$(CH_3CH_2)_2NH$	1.3×10^{-3}	2.9
Dipropylamine	$(CH_3CH_2CH_2)_2NH$	8.2×10^{-4}	3.1
Piperidine	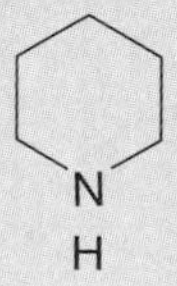 (piperidine ring, N-H)	1.6×10^{-3}	2.8
Tertiary amines			
Trimethylamine	$(CH_3)_3N$	5.3×10^{-5}	4.3
Triethylamine	$(CH_3CH_2)_3N$	5.6×10^{-4}	3.2
Tripropylamine	$(CH_3CH_2CH_2)_3N$	5×10^{-4}	3.3

* In water at 25°C

Basicity of amines in aqueous solution

$$NH_3 \quad < CH_3CH_2NH_2 \sim (CH_3CH_2)_3N < (CH_3CH_2)_2NH$$

Ammonia	Ethylamine	Triethylamine	Diethylamine
K_b 1.8×10^{-5}	K_b 5.6×10^{-4}	K_b 5.6×10^{-4}	K_b 1.3×10^{-3}
$(pK_b\ 4.7)$	$(pK_b\ 3.2)$	$(pK_b\ 3.2)$	$(pK_b\ 2.9)$

The discontinuity in basicity among the various classes of amines suggests that there are at least two substituent effects involved and that they operate in opposite directions.

An alkyl group can increase the base strength of an amine by releasing electrons through a field effect or an inductive effect. The positive charge of an ammonium ion is dispersed better by having alkyl groups instead of hydrogen substituents on nitrogen.

Ethyl group
releases electrons
to nitrogen;
positive charge is
dispersed and ion
is stabilized

$$CH_3CH_2\overset{+}{N}\overset{H}{\underset{H}{\diagdown}}H \qquad \text{more stable} \qquad H-\overset{+}{N}\overset{H}{\underset{II}{\diagdown}}H$$

Ethylammonium   more stable   Ammonium
ion   than   ion

By stabilizing the positive charge, alkyl groups increase the equilibrium constant for amine protonation. Ethylamine is more basic than ammonia because $CH_3CH_2\overset{+}{N}H_3$ is stabilized better than NH_4^+.

Were this the only effect of alkyl groups, the basicity of amines would increase with increasing alkyl substitution. Indeed, this is precisely what is observed for proton transfer to amines in the *gas phase*.

Gas-phase basicity of amines

$$NH_3 \quad < CH_3CH_2NH_2 < (CH_3CH_2)_2NH < (CH_3CH_2)_3N$$

Ammonia   Ethylamine   Diethylamine   Triethylamine
(least basic)   (most basic)

Electron release from alkyl groups provides the principal mechanism by which the conjugate acid of an amine is stabilized in the gas phase. The more alkyl groups that are attached to the positively charged nitrogen, the more stable the alkylammonium ion becomes.

Basicity as measured by K_b, however, refers to equilibrium measurements made in dilute aqueous solution. The altered order of amine basicities in solution, as compared with those in the gas phase, must arise from *solvation effects*.

While alkyl substituents increase the ability of an ammonium ion to disperse its positive charge, they decrease its ability to form hydrogen bonds to water molecules. Dialkylammonium ions, formed by protonation of secondary amines, have two hydrogen substituents on nitrogen that can participate in hydrogen bonding. Trialkylammonium ions have only one and are therefore less stabilized by solvation than are their dialkyl counterparts.

Diethylammonium ion
(two protons on nitrogen
available for hydrogen
bonding)

Triethylammonium ion
(only one proton on nitrogen
available for hydrogen
bonding)

Dialkylamines are more basic than either primary or tertiary amines because their conjugate acids possess the best combination of alkyl and hydrogen substituents to permit stabilization both by electron release from alkyl groups and by solvation due to hydrogen bonding.

PROBLEM 23.4 Identify the stronger base in each of the following pairs. Explain your reasoning.

(a) $FCH_2CH_2NH_2$ or $CH_3CH_2NH_2$
(b) $F_2CHCH_2NH_2$ or $F_3CCH_2NH_2$
(c) $CH_3OCH_2CH_2NH_2$ or $CH_3CH_2CH_2CH_2NH_2$

SAMPLE SOLUTION (a) The strongly electronegative fluorine substituent diminishes the ability of the substituent to stabilize the protonated form of the amine.

$$\overset{\longleftarrow}{FCH_2CH_2\overset{+}{N}H_3} \text{ is less stabilized than } \overset{\longrightarrow}{CH_3CH_2\overset{+}{N}H_3}$$

As measured by their respective K_b's, ethylamine is 90 times more basic than its 2-fluoro derivative.

$CH_3CH_2NH_2$	$FCH_2CH_2NH_2$
Ethylamine	2-Fluoroethylamine
stronger base: K_b 5.6×10^{-4}	weaker base: K_b 6.3×10^{-6}
(pK_b 3.2)	(pK_b 5.2)

23.5 AMINES AS NATURAL PRODUCTS

The ease with which amines are extracted into aqueous acid, combined with their regeneration on treatment with base, makes it a simple matter to separate amines from other plant materials, and nitrogen-containing natural products were among the earliest organic compounds to be studied. Their basic properties led them to be called *alkaloids*. A few of the more than 5000 naturally occurring alkaloids are shown in Figure 23.2. Some are relatively simple, others are complex. All the alkaloids in Figure 23.2 are characterized by a high level of biological activity; it is their biological potency that has spurred research on the isolation, structure determination, and synthesis of this class of compounds.

Among the more important amine derivatives found in the body are a group of compounds known as *polyamines,* which contain two to four nitrogen atoms separated by several methylene units.

Putrescine

Spermidine

Spermine

These compounds are present in almost all mammalian cells, where they are believed to be involved in cell differentiation and proliferation. Because each nitrogen of a

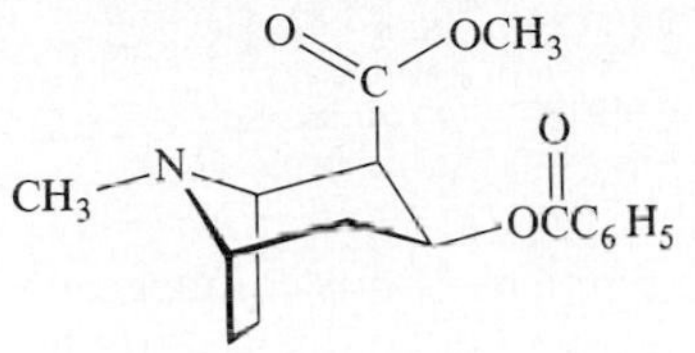

Cocaine

(a central nervous system stimulant obtained from the leaves of the coca plant)

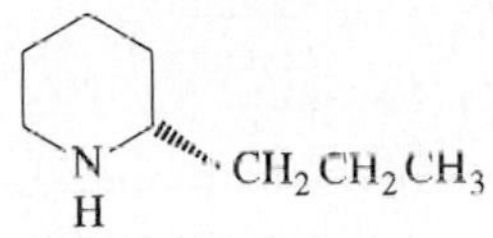

Coniine

(present along with other alkaloids in the hemlock extract used to poison Socrates)

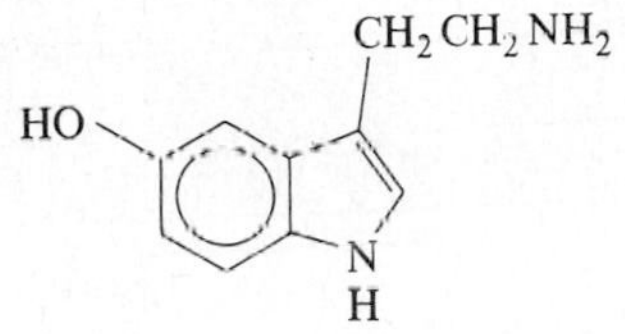

Serotonin

(not strictly an alkaloid because it does not occur in plants—serotonin is a hormone synthesized in the pineal gland. Certain mental disorders are believed to be related to serotonin levels in the brain.)

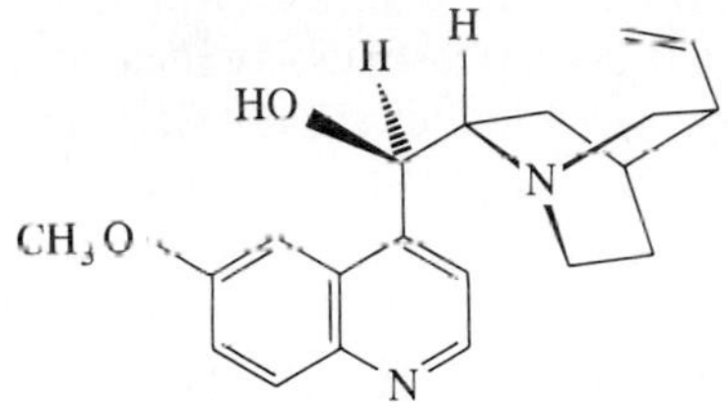

Quinine

(alkaloid of cinchona bark used to treat malaria)

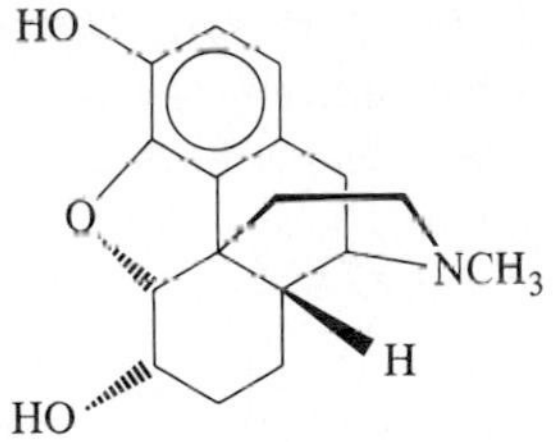

Morphine

(an opium alkaloid. While an excellent analgesic, its use is restricted because of potential for addiction. Heroin is the diacetate ester of morphine.)

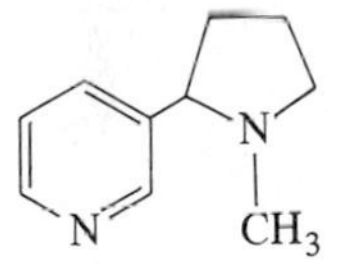

Nicotine

(an alkaloid present in tobacco; a very toxic compound sometimes used as an insecticide)

FIGURE 23.2 Some naturally occurring alkaloids and related compounds.

polyamine is protonated at physiological pH (7.4), putrescine, spermidine, and spermine exist as cations with a charge of +2, +3, and +4, respectively, in body fluids. Structural studies suggest that these polyammonium ions affect the conformation of biological macromolecules by electrostatic binding to specific anionic sites—the negatively charged phosphate groups of DNA, for example.

23.6 TETRAALKYLAMMONIUM SALTS AS PHASE TRANSFER CATALYSTS

In spite of their ionic nature, many quaternary ammonium salts are soluble in nonpolar media. The four alkyl groups attached to nitrogen shield its positive charge and impart *lipophilic* character to the tetraalkylammonium ion. The two quaternary ammonium salts shown below, for example, are soluble in solvents of low polarity such as benzene, decane, and halogenated hydrocarbons.

$$CH_3\overset{+}{N}(CH_2CH_2CH_2CH_2CH_2CH_2CH_2CH_3)_3 \quad Cl^-$$

Methyltrioctylammonium chloride

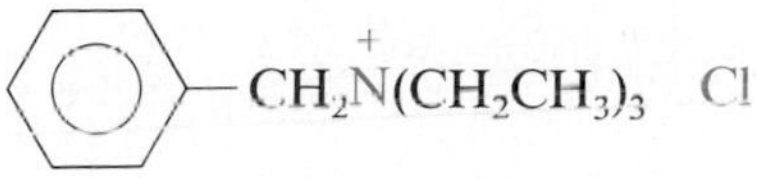

$$C_6H_5CH_2\overset{+}{N}(CH_2CH_3)_3 \quad Cl^-$$

Benzyltriethylammonium chloride

This property of quaternary ammonium salts is used to advantage in an experimental technique known as *phase transfer catalysis.* Imagine that you wish to carry out the reaction

$$CH_3CH_2CH_2CH_2Br + NaCN \longrightarrow CH_3CH_2CH_2CH_2CN + NaBr$$

Butyl bromide   Sodium   Pentanenitrile   Sodium
                 cyanide   bromide

Sodium cyanide does not dissolve in butyl bromide. The two reactants contact each other only at the surface of the solid sodium cyanide and the rate of reaction under these conditions is too slow to be of synthetic value. Dissolving the sodium cyanide in water is of little help, since butyl bromide is not soluble in water and reaction can occur only at the interface between the two phases. Adding a small amount of benzyltrimethylammonium chloride, however, causes pentanenitrile to form rapidly even at room temperature. The quaternary ammonium salt is acting as a *catalyst;* it increases the rate of the nucleophilic substitution reaction. How does it do this?

Quaternary ammonium salts catalyze the reaction between an anion and an organic substrate by transferring the anion from the aqueous phase, where it cannot contact the substrate, to the organic phase. In the example just cited, the first step occurs in the aqueous phase and is an exchange of the anionic partner of the quaternary ammonium salt for cyanide ion.

$$C_6H_5CH_2\overset{+}{N}(CH_3)_3 \ Cl^- + \ CN^- \ \underset{}{\overset{fast}{\rightleftharpoons}} \ C_6H_5CH_2\overset{+}{N}(CH_3)_3 \ CN^- + \ Cl^-$$

Benzyltrimethylammonium   Cyanide   Benzyltrimethylammonium   Chloride
       chloride   ion   cyanide   ion
        (aqueous)   (aqueous)   (aqueous)   (aqueous)

The benzyltrimethylammonium ion migrates to the butyl bromide phase, carrying a cyanide ion along with it.

$$C_6H_5CH_2\overset{+}{N}(CH_3)_3 \ CN^- \ \underset{}{\overset{fast}{\rightleftharpoons}} \ C_6H_5CH_2\overset{+}{N}(CH_3)_3 \ CN^-$$

Benzyltrimethylammonium   Benzyltrimethylammonium cyanide
       cyanide   (in butyl bromide)
        (aqueous)

Once in the organic phase, cyanide ion is only weakly solvated and is far more reactive than it is in water or ethanol, where it is strongly solvated by hydrogen bonding. Nucleophilic substitution takes place rapidly.

$$CH_3CH_2CH_2CH_2Br + C_6H_5CH_2\overset{+}{N}(CH_3)_3 \ CN^- \ \underset{}{\overset{fast}{\rightleftharpoons}}$$

Butyl bromide   Benzyltrimethylammonium
                                  cyanide
                           (in butyl bromide)

$$CH_3CH_2CH_2CH_2CN + C_6H_5CH_2\overset{+}{N}(CH_3)_3 \ Br^-$$

Pentanenitrile   Benzyltrimethylammonium
(in butyl bromide)   bromide
                        (in butyl bromide)

The benzyltrimethylammonium bromide formed in this step returns to the aqueous phase, where it can repeat the cycle.

Phase transfer catalysis succeeds for two reasons. First, it provides a mechanism for introducing an anion into the medium that contains the reactive substrate. More important, the anion is introduced in a weakly solvated, highly reactive state. You have already encountered phase transfer catalysis in another form in Section 17.4, where the metal-complexing properties of crown ethers were described. Crown ethers permit metal salts to dissolve in nonpolar solvents by surrounding the cation with a lipophilic cloak, leaving the anion free to react without the encumbrance of strong solvation forces.

Many nucleophilic substitution reactions can be accelerated by phase transfer catalysts. Numerous examples of other kinds of reactions, involving anions such as permanganate, borohydride, etc., have also been carried out under phase transfer conditions.

23.7 REACTIONS THAT LEAD TO ALKYLAMINES. A REVIEW AND A PREVIEW

Methods for the preparation of amines address either or both of the following questions:

1. How is the required carbon-nitrogen bond to be formed?
2. Given a nitrogen-containing organic compound such as an amide, nitrile, nitro compound, etc., how is the correct oxidation state of the desired amine to be achieved?

A number of reactions that lead to carbon-nitrogen bond formation have been presented in earlier chapters and are summarized in Table 23.3. In three of these reactions (the first, fourth, and fifth entries in the table) the product is not an amine but is a compound of a type that can be converted to an amine by reduction. Methods for reducing azides, amides, and imines to amines will be described in this chapter. All the reactions in Table 23.3 except for the Hofmann rearrangement (the sixth entry) are based on nitrogen acting as a nucleophile.

As methods for the preparation of amines are presented in the following sections, you will see that they are largely applications of principles that you have already learned. You will encounter some new reagents and some new uses for familiar reagents, but very little in the way of new reaction types is involved.

23.8 PREPARATION OF AMINES BY ALKYLATION OF AMMONIA

As alkyl derivatives of ammonia, amines are, in principle, capable of being prepared by nucleophilic substitution reactions of alkyl halides with ammonia.

$$RX + 2NH_3 \longrightarrow RNH_2 + \overset{+}{N}H_4 X$$

Alkyl   Ammonia   Primary   Ammonium
halide   amine   halide salt

As is the case with other nucleophilic substitution reactions of the S_N2 type, primary alkyl halides are the preferred substrates. Tertiary alkyl halides undergo elimination reactions under these conditions. Secondary alkyl halides, especially α-halo acids and esters, as illustrated in the third entry of Table 23.3, can be used. As a general method, however, the reaction is of limited value because the yields of primary amines obtained are typically rather low. The reason for this is that the primary

TABLE 23.3

Methods for Carbon-Nitrogen Bond Formation Discussed in Earlier Chapters

Reaction (section) and comments	General equation and specific example
Nucleophilic substitution by azide ion on an alkyl halide (Sections 9.1, 9.14) Azide ion is a very good nucleophile and reacts with primary and secondary alkyl halides to give alkyl azides. Phase transfer catalysts accelerate the rate of reaction.	$\ddot{N}{=}\overset{+}{N}{=}\ddot{N}{:}\ + \ R{-}X \longrightarrow \ddot{N}{=}\overset{+}{N}{=}\ddot{N}{-}R\ +\ X^-$ Azide ion — Alkyl halide — Alkyl azide — Halide ion $CH_3CH_2CH_2CH_2CH_2Br \xrightarrow[\text{Phase transfer catalyst}]{NaN_3} CH_3CH_2CH_2CH_2CH_2N_3$ Pentyl bromide (1-bromopentane) — Pentyl azide (89%) (1-azidopentane)
Nucleophilic ring opening of epoxides by ammonia (Section 17.14) The strained ring of an epoxide is opened on nucleophilic attack by ammonia and amines to give β-amino alcohols. Azide ion also reacts with epoxides; the products are β-azido alcohols.	$H_3N{:}\ +\ R_2C{-}CR_2 \ \longrightarrow \ H_2N{-}\overset{\displaystyle R\ \ R}{\underset{\displaystyle R\ \ R}{C{-}C}}{-}OH$ (O bridging the two carbons) Ammonia — Epoxide — β-Amino alcohol (2R, 3R)-2,3-Epoxybutane $\xrightarrow[\text{water}]{NH_3}$ (2R, 3S)-3-Amino-2-butanol (70%)
Nucleophilic substitution by ammonia on α-halo acids (Section 20.16) The α-halo acids obtained by halogenation of carboxylic acids under conditions of the Hell-Volhard-Zelinsky reaction are reactive substrates in nucleophilic substitution processes. A standard method for the preparation of α-amino acids is displacement of halide from α-halo acids by nucleophilic substitution using excess aqueous ammonia.	$H_3N{:}\ +\ \underset{\displaystyle X}{RCHCO_2H} \longrightarrow \underset{\displaystyle NH_2}{RCHCO_2H}\ +\ NH_4X$ Ammonia (excess) — α-Halo carboxylic acid — α-Amino acid — Ammonium halide $\underset{\displaystyle Br}{(CH_3)_2CHCHCO_2H} \xrightarrow[\text{H}_2\text{O}]{NH_3} \underset{\displaystyle NH_2}{(CH_3)_2CHCHCO_2H}$ 2-Bromo-3-methylbutanoic acid — 2-Amino-3-methylbutanoic acid (47–48%)
Nucleophilic acyl substitution (Section 21.11) Acylation of ammonia and amines by acyl transfer is an exceptionally effective method for the formation of carbon-nitrogen bonds.	$R_2\ddot{N}H\ +\ R'\overset{\displaystyle O}{\underset{\displaystyle X}{C}} \longrightarrow R_2N\overset{\displaystyle O}{C}R'\ +\ HX$ Primary or secondary amine, or ammonia — Acyl transfer agent — Amide $2\ \text{(pyrrolidine)} + CH_3CCl \longrightarrow \text{(N-acetylpyrrolidine)} + \text{(pyrrolidine hydrochloride, Cl}^-)$ Pyrrolidine (excess) — Acetyl chloride — N-Acetylpyrrolidine (79%) — Pyrrolidine hydrochloride

TABLE 23.3 (continued)

Reaction (section) and comments	General equation and specific example
Nucleophilic addition of amines to aldehydes and ketones (Sections 18.13, 18.14) Primary amines undergo nucleophilic addition to the carbonyl group of aldehydes and ketones to form carbinolamines. These carbinolamines dehydrate under the conditions of their formation to give N-substituted imines. Secondary amines yield enamines.	RNH_2 + $R'CR''$ $\longrightarrow$ $R'CR''$ + H_2O (Primary amine) (Aldehyde or ketone) (Imine) (Water) CH_3NH_2 + C_6H_5CH $\longrightarrow$ $C_6H_5CH{=}NCH_3$ Methylamine Benzaldehyde N-Benzylidenemethylamine (70%)
The Hofmann rearrangement (Section 21.17) Amides are converted to amines by reaction with bromine in basic media. An N-bromo amide is an intermediate; it rearranges to an isocyanate. Hydrolysis of the isocyanate yields an amine.	$RCNH_2$ $\xrightarrow[H_2O]{Br_2, \ HO^-}$ RNH_2 Amide Amine $(CH_3)_3CCNH_2$ $\xrightarrow[H_2O]{Br_2, \ HO^-}$ $(CH_3)_3CNH_2$ 2,2-Dimethylpropanamide _tert_-Butylamine (64%)

amine product is itself a nucleophile and competes with ammonia for the alkyl halide.

$$RX + RNH_2 + NH_3 \longrightarrow R_2NH + \overset{+}{N}H_4 \ X^-$$

Alkyl   Primary   Ammonia   Secondary   Ammonium
halide   amine   amine   halide salt

When 1-bromooctane, for example, is allowed to react with ammonia, both the primary amine and the secondary amine are isolated in comparable amounts.

$$CH_3(CH_2)_6CH_2Br \xrightarrow{NH_3(2 \ mol)} CH_3(CH_2)_6CH_2NH_2 + [CH_3(CH_2)_6CH_2]_2NH$$

1-Bromooctane   Octylamine   N,N-Dioctylamine
(1 mol)   (45%)   (43%)

In a similar manner, competitive alkylation may continue, resulting in formation of a trialkylamine.

$$RX + R_2NH + NH_3 \longrightarrow R_3N + \overset{+}{N}H_4 \ X^-$$

Alkyl   Secondary   Ammonia   Tertiary   Ammonium
halide   amine   amine   halide salt

Even the tertiary amine competes with ammonia for the alkylating agent. The product is a quaternary ammonium salt.

$$RX + R_3N \longrightarrow R_4\overset{+}{N} \ X^-$$

Alkyl halide   Tertiary   Quaternary
                amine   ammonium salt

Because alkylation of ammonia can lead to a complex mixture of products, it is used for the preparation of primary amines only when the starting alkyl halide is not particularly expensive and the desired amine can be easily separated from the other components of the reaction mixture.

PROBLEM 23.5   Alkylation of ammonia is sometimes employed in industrial processes and the resulting mixture of amines is separated by distillation. The ultimate starting materials for the industrial preparation of allylamine are propene, chlorine, and ammonia. Write a series of equations showing the industrial preparation of allylamine from these starting materials. (Allylamine has a number of uses, including the preparation of the diuretic drugs *meralluride* and *mercaptomerin*.)

23.9   THE GABRIEL SYNTHESIS

A method that achieves the same end result as that desired by alkylation of ammonia but which avoids the formation of secondary and tertiary amines as by-products is the *Gabriel synthesis.* Alkyl halides are converted to primary alkylamines without contamination by secondary or tertiary amines. The key reagent is the potassium salt of phthalimide, prepared by the reaction

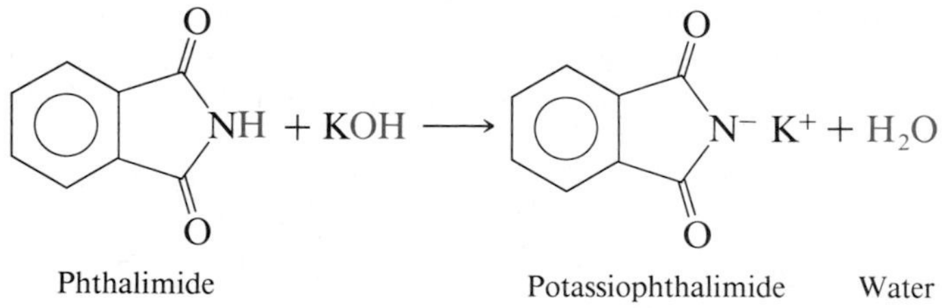

Phthalimide   Potassiophthalimide   Water

Phthalimide, with a K_a of 5×10^{-9} (pK_a 8.3), can be quantitatively converted to its potassium salt by an acid-base reaction with potassium hydroxide. The potassium salt of phthalimide has a negatively charged nitrogen atom, which acts as a nucleophile toward primary alkyl halides in a bimolecular nucleophilic substitution (S_N2) process.

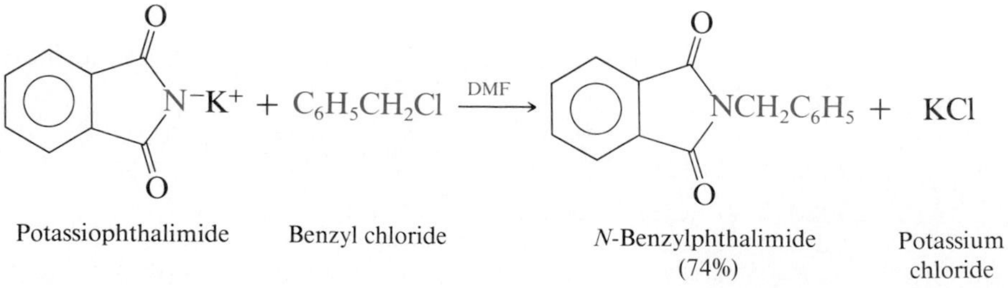

Potassiophthalimide   Benzyl chloride   *N*-Benzylphthalimide   Potassium
                                                     (74%)   chloride

DMF is an abbreviation for *N,N*-dimethylformamide, $\overset{\text{O}}{\overset{\|}{\text{HC}}}\text{N(CH}_3)_2$; it is a polar aprotic solvent of the kind described in Section 9.13, and is an excellent medium for S_N2-type reactions.

The product of this reaction is an imide (Section 21.15), a diacyl derivative of an amine. Either aqueous acid or base can be used to hydrolyze its two amide bonds and liberate the desired primary amine. A more effective method of cleaving the two amide bonds is by acyl transfer to hydrazine.

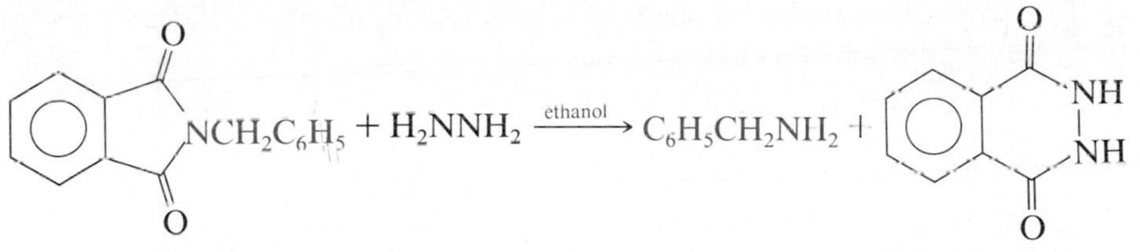

N-Benzylphthalimide	Hydrazine	Benzylamine	Phthalhydrazide
		(97%)	

PROBLEM 23.6 Which of the following amines can be prepared by the Gabriel synthesis? Which ones cannot? Write equations showing the successful applications of this method.

(a) Butylamine
(b) Isobutylamine
(c) *tert*-Butylamine

(d) 2-Phenylethylamine
(e) N-Methylbenzylamine
(f) Aniline

SAMPLE SOLUTION (a) The Gabriel synthesis is limited to preparation of amines of the type RCH$_2$NH$_2$, that is, primary alkylamines in which the amino group is bonded to a primary carbon. Butylamine may be prepared from butyl bromide by this method.

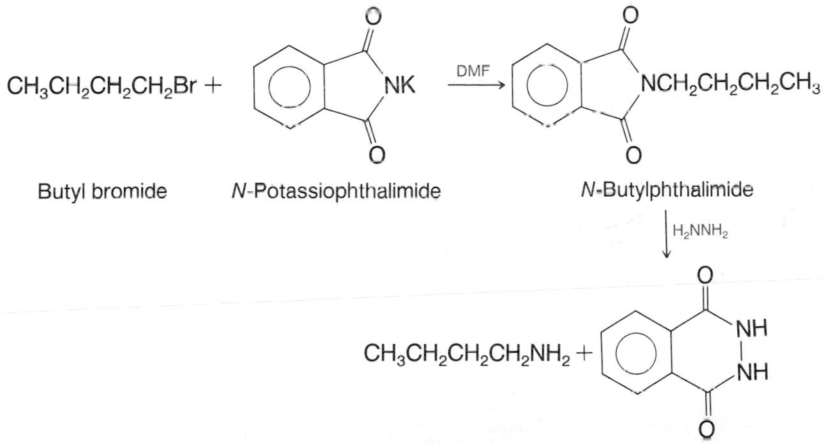

Butyl bromide	N-Potassiophthalimide	N-Butylphthalimide

Butylamine	Phthalhydrazide

Among compounds other than simple alkyl halides, α-halo ketones and α-halo esters have been employed as substrates in the Gabriel synthesis. Alkyl *p*-toluenesulfonate esters have also been used. Because phthalimide can undergo only a single alkylation, the formation of secondary and tertiary amines does not occur and the Gabriel synthesis is a valuable procedure for the laboratory preparation of primary amines.

23.10 PREPARATION OF ALKYLAMINES BY REDUCTION

Almost any nitrogen-containing organic compound can be reduced to an amine. The synthesis of amines then becomes a question of the availability of suitable precursors and the choice of an appropriate reducing agent.

Alkyl *azides,* prepared by nucleophilic substitution of alkyl halides by sodium azide, as shown in the first entry of Table 23.3, are reduced to amines by a variety of reagents, including lithium aluminum hydride.

$$R—\ddot{N}=\overset{+}{N}=\ddot{N}:^{-} \xrightarrow{\text{reduce}} R\ddot{N}H_2$$

Alkyl azide   Primary amine

$$C_6H_5CH_2CH_2N_3 \xrightarrow[\substack{\text{diethyl ether}\\ \text{2. } H_2O}]{\text{1. LiAlH}_4,} C_6H_5CH_2CH_2NH_2$$

2-Phenylethyl azide   2-Phenylethylamine (89%)

Catalytic hydrogenation is also effective.

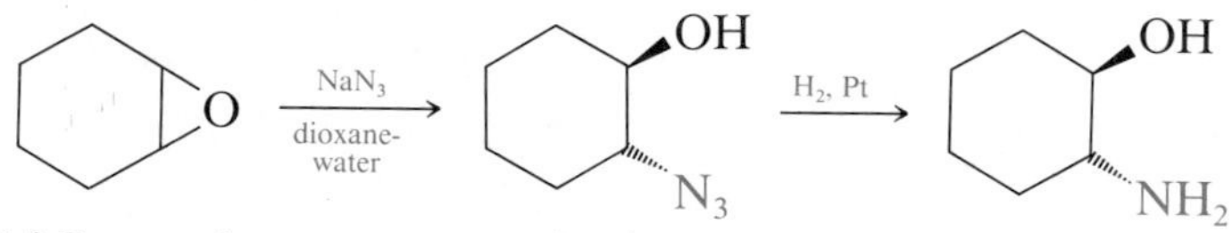

1,2-Epoxycyclo-hexane   *trans*-2-Azidocyclo-hexanol (61%)   *trans*-2-Aminocyclo-hexanol (81%)

In its overall design, this procedure is similar to the Gabriel synthesis; a nitrogen nucleophile is used in a carbon-nitrogen bond-forming operation and then converted to an amino group in a subsequent transformation.

The same reduction methods may be applied to the conversion of *nitriles* to primary amines.

$$RC≡N \xrightarrow[H_2, \text{ catalyst}]{\text{LiAlH}_4 \text{ or}} RCH_2NH_2$$

Nitrile   Primary amine

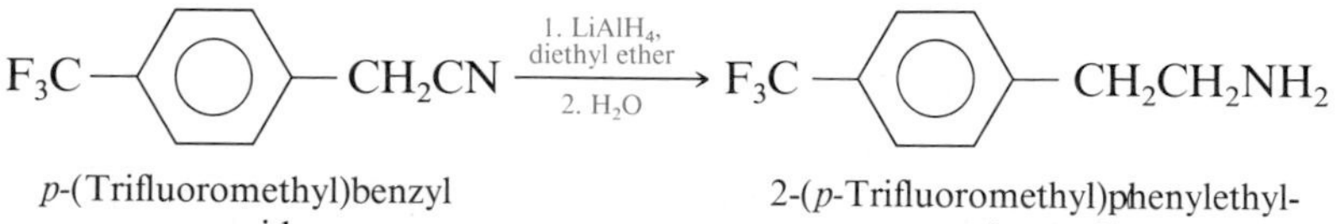

p-(Trifluoromethyl)benzyl cyanide   2-(*p*-Trifluoromethyl)phenylethyl-amine (53%)

$$CH_3CH_2CH_2CH_2CN \xrightarrow[\text{diethyl ether}]{H_2(100 \text{ atm}), \text{ Ni}} CH_3CH_2CH_2CH_2CH_2NH_2$$

Pentanenitrile   1-Pentanamine (56%)

Since nitriles can be prepared from alkyl halides by a nucleophilic substitution reaction with cyanide ion, the overall process $RX \rightarrow RC≡N \rightarrow RCH_2NH_2$ leads to primary amines that have one more carbon atom than the starting alkyl halide.

Oximes, prepared from aldehydes and ketones by reaction with hydroxylamine (Section 18.15), are converted to primary amines by reduction with sodium in ethanol.

$$\underset{\substack{\text{Oxime}}}{\overset{\displaystyle \text{NOH}}{\overset{\displaystyle \|}{RCR'}}} \xrightarrow[\text{ethanol}]{\text{Na}} \underset{\substack{\text{Primary amine}}}{\overset{\displaystyle \text{NH}_2}{\overset{\displaystyle |}{RCHR'}}}$$

$$\underset{\substack{\text{2-Octanone oxime}}}{\overset{\displaystyle \text{NOH}}{\overset{\displaystyle \|}{CH_3(CH_2)_5CCH_3}}} \xrightarrow[\text{ethanol}]{\text{Na}} \underset{\substack{\text{2-Octanamine (62–69\%)}}}{CH_3(CH_2)_5\underset{\displaystyle \underset{\displaystyle NH_2}{|}}{CHCH_3}}$$

Nitro groups are readily reduced to primary amine functions by a variety of methods, including catalytic hydrogenation. This is especially useful in the preparation of arylamines (Section 24.5) since the corresponding nitro-substituted arenes are available by nitration.

Reduction of an azide, a nitrile, an oxime, or a nitro compound furnishes a primary amine. A method that can provide access to primary, secondary, or tertiary amines is reduction of the carbonyl group of an amide by lithium aluminum hydride.

$$\underset{\substack{\text{Amide}}}{\overset{\displaystyle \text{O}}{\overset{\displaystyle \|}{RCNR_2'}}} \xrightarrow[\text{2. H}_2\text{O}]{\text{1. LiAlH}_4} \underset{\substack{\text{Amine}}}{RCH_2NR_2'}$$

Reduction of amides by lithium aluminum hydride is probably the most frequently used method for the laboratory preparation of amines. The nature of the amine product is determined by the number of carbon substituents bonded to nitrogen in the starting amide. Primary amides yield primary amines.

$$\underset{\substack{\text{3-Phenylbutanamide}}}{C_6H_5\underset{\displaystyle \underset{\displaystyle CH_3}{|}}{CH}CH_2\overset{\displaystyle \overset{\displaystyle O}{\|}}{C}NH_2} \xrightarrow[\text{2. H}_2\text{O}]{\substack{\text{1. LiAlH}_4, \\ \text{diethyl ether}}} \underset{\substack{\text{3-Phenyl-1-butanamine (59\%)}}}{C_6H_5\underset{\displaystyle \underset{\displaystyle CH_3}{|}}{CH}CH_2CH_2NH_2}$$

N-Alkyl amides yield secondary amines.

$$\underset{\substack{\text{N-Butylpentanamide}}}{CH_3CH_2CH_2CH_2\overset{\displaystyle \overset{\displaystyle O}{\|}}{C}NHCH_2CH_2CH_2CH_3} \xrightarrow[\text{2. H}_2\text{O}]{\substack{\text{1. LiAlH}_4, \\ \text{diethyl ether}}}$$

$$\underset{\substack{\text{N-Butyl-1-pentanamine (88\%)}}}{CH_3CH_2CH_2CH_2CH_2NHCH_2CH_2CH_2CH_3}$$

N,N-Dialkyl amides yield tertiary amines.

N,N-Dimethylcyclohexane carboxamide → N,N-Dimethyl(cyclohexylmethyl)-amine (88%)

Because amides are so easy to prepare, this is a versatile and general method for the preparation of amines.

PROBLEM 23.7   You will find it helpful at this point to review the origins of a number of the starting materials that were reduced in the reactions just described. Show how you could prepare each of the following from the compound given.

(a) 2-Phenylethyl azide from styrene
(b) Pentanenitrile from 1-butanol
(c) *p*-(Trifluoromethyl)benzyl cyanide from *p*-(trifluoromethyl)benzoic acid
(d) 2-Octanone oxime from 2-octanol
(e) *N*-Butylpentanamide from 1-butanamine and pentanoic acid

SAMPLE SOLUTION   (a)   Alkyl azides are prepared from alkyl halides by nucleophilic substitution with azide ion.

$$C_6H_5CH_2CH_2Br \xrightarrow{NaN_3} C_6H_5CH_2CH_2N_3$$

2-Phenylethyl bromide   2-Phenylethyl azide

Styrene is converted to the primary bromide by free-radical addition of hydrogen bromide. Addition is regioselective and takes place contrary to Markovnikov's rule.

$$C_6H_5CH{=\!=}CH_2 \xrightarrow[peroxides]{HBr} C_6H_5CH_2CH_2Br$$

Styrene   2-Phenylethyl bromide

The preparation of amines by the methods just described involves the prior synthesis and isolation of some reducible material that has a carbon-nitrogen bond — an azide, a nitrile, an oxime, or an amide. A method known as *reductive amination* that combines the two steps of carbon-nitrogen bond formation and reduction into a single operation is described in the following section. Like amide reduction, it offers the possibility of preparing primary, secondary, or tertiary amines by proper choice of appropriate starting materials.

23.11 REDUCTIVE AMINATION

A class of nitrogen-containing compounds that was omitted from the section just discussed includes *imines* and their derivatives. Imines are formed by the reaction of aldehydes and ketones with ammonia and can be reduced to primary amines by catalytic hydrogenation.

$$\underset{\substack{\text{Aldehyde} \\ \text{or ketone}}}{\overset{\overset{\displaystyle O}{\|}}{RCR'}} + \underset{\text{Ammonia}}{NH_3} \longrightarrow \underset{\text{Imine}}{\overset{\overset{\displaystyle NH}{\|}}{RCR'}} \xrightarrow[\text{catalyst}]{H_2} \underset{\text{Primary amine}}{\overset{\overset{\displaystyle NH_2}{|}}{RCHR'}}$$

The overall reaction is one which converts a carbonyl compound to an amine by carbon-nitrogen bond formation and reduction; it is commonly known as *reductive amination*. What makes it a particularly valuable synthetic procedure is that it can be carried out in a single operation by hydrogenation of a solution containing both ammonia and the carbonyl compound along with a hydrogenation catalyst. The intermediate imine is not isolated but undergoes reduction under the conditions of its formation. Also, the reaction is broader in scope than implied by the equation shown above. All classes of amines— primary, secondary, and tertiary—may be prepared by reductive amination.

When primary amines are desired, the reaction is carried out as just described.

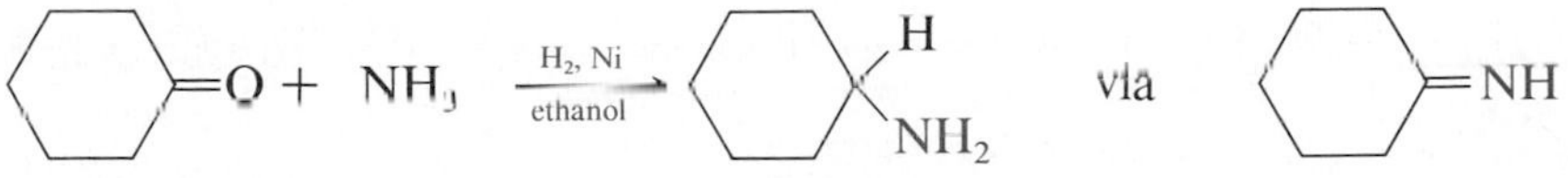

| Cyclohexanone | Ammonia | | Cyclohexylamine (80%) | via | |

Secondary amines are prepared by hydrogenation of a carbonyl compound in the presence of a primary amine. An *N*-substituted imine, or *Schiff's base,* is an intermediate.

$$(CH_3)_2CHCH + CH_3CH_2CH_2CH_2NH_2 \xrightarrow[\text{ethanol}]{H_2, \text{Pt}} (CH_3)_2CHCH_2NHCH_2CH_2CH_2CH_3$$

2-Methylpropanal Butylamine *N*-Isobutylbutylamine (92%)

$$\text{via} \quad (CH_3)_2CHCH{=}NCH_2CH_2CH_2CH_3$$

Reductive amination has been successfully applied to the preparation of tertiary amines from carbonyl compounds and secondary amines even though a neutral imine is not possible in this case.

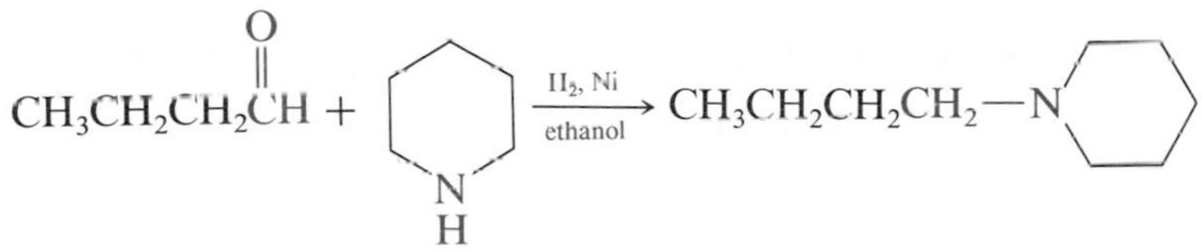

Butanal   Piperidine   *N*-Butylpiperidine (93%)

Presumably, the species that undergoes reduction here is a carbinolamine or an iminium ion derived from it.

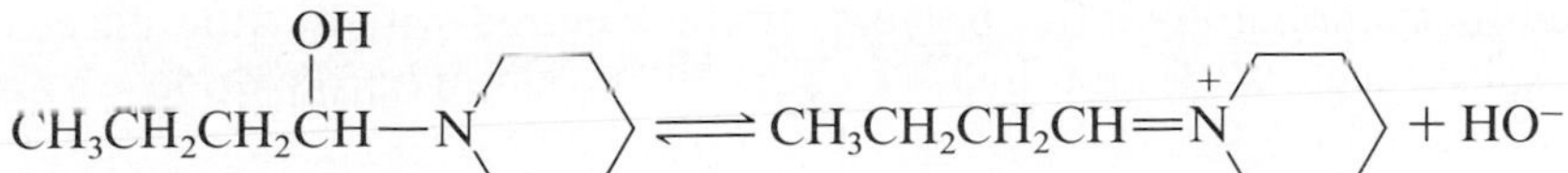

Carbinolamine   Iminium ion

PROBLEM 23.8 Show how you could prepare each of the following amines from benzaldehyde by reductive amination:

(a) Benzylamine
(b) Dibenzylamine

(c) *N,N*-Dimethylbenzylamine
(d) *N*-Benzylpiperidine

SAMPLE SOLUTION (a) Since benzylamine is a primary amine, it is derived from ammonia and benzaldehyde.

$$C_6H_5\overset{\overset{\textstyle O}{\|}}{C}H \ + \ NH_3 \ + \ H_2 \ \xrightarrow{\text{Ni}} \ C_6H_5CH_2NH_2 + H_2O$$

Benzaldehyde	Ammonia	Hydrogen	Benzylamine	Water
			(89%)	

The reaction proceeds by initial formation of the imine $C_6H_5CH{=}NH$, followed by its hydrogenation.

A variation of the classical reductive amination procedure that uses sodium cyanoborohydride ($NaBH_3CN$) instead of hydrogen as the reducing agent is better suited to amine syntheses in which only a few grams of material are needed. All that is done is to add sodium cyanoborohydride to an alcohol solution of the carbonyl compound and an amine. When primary amines are desired, ammonium acetate serves as a convenient source of ammonia.

$$C_6H_5\overset{\overset{\textstyle O}{\|}}{C}CH_3 + CH_3CO_2^- \ \overset{+}{N}H_4 \xrightarrow[\text{methanol}]{NaBH_3CN} \ C_6H_5\underset{\underset{\textstyle NH_2}{|}}{C}HCH_3$$

Acetophenone   Ammonium acetate   1-Phenylethylamine (77%)

Secondary and tertiary amines are also available by this method.

$$C_6H_5\overset{\overset{\textstyle O}{\|}}{C}H \ + CH_3CH_2NH_2 \xrightarrow[\text{methanol}]{NaBH_3CN} C_6H_5CH_2NHCH_2CH_3$$

Benzaldehyde   Ethylamine   N-Ethylbenzylamine (91%)

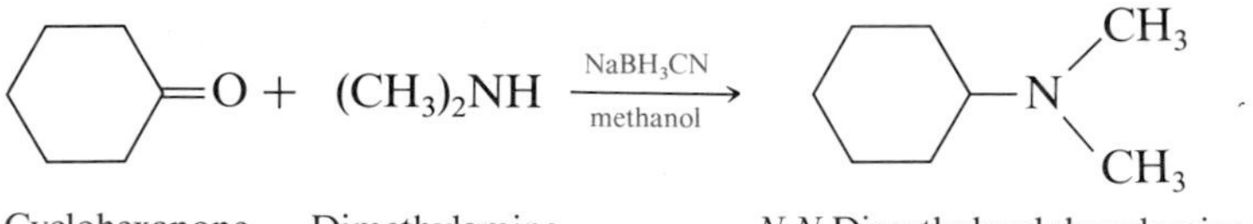

Cyclohexanone   Dimethylamine   N,N-Dimethylcyclohexylamine (52–54%)

Sodium cyanoborohydride, because of the electron-withdrawing effect of its cyano substituent, is a poorer hydride donor than sodium borohydride.

In order to act as a hydride donor, hydrogen must be transferred along with both electrons of its bond to boron.

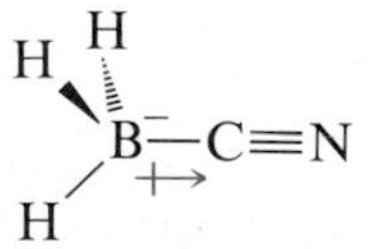

Cyano group withdraws electrons and makes it more difficult for hydride to be transferred from boron.

Consequently, sodium cyanohydride reduces aldehydes and ketones rather slowly, more slowly than these carbonyl compounds react with ammonia to form imines.

$$R_2C=O + NH_3 \xrightarrow{\text{fast}} R_2C=NH + H_2O$$

Aldehyde   Ammonia   Imine   Water
or ketone

Imines are basic and are in rapid equilibrium with their protonated form in the pH region (pH 6 to 8) at which these reactions are carried out. The protonated form of an imine, an *iminium ion,* is very electrophilic because of its positive charge and is reduced readily by cyanoborohydride.

$$R_2C=NH \xrightarrow{\text{H}^+, \text{fast}} R_2C=\overset{+}{N}H_2 \xrightarrow{\text{NaBH}_3\text{CN}} R_2CH-NH_2$$

Imine   Iminium ion   Amine

Thus, sodium cyanoborohydride is too weak a reducing agent to reduce aldehydes and ketones at an appreciable rate but does reduce the iminium ions that are present in a rapidly reversible equilibrium with the parent carbonyl compound.

23.12 REACTIONS OF AMINES. A REVIEW AND A PREVIEW

The noteworthy properties of amines are their *basicity* and their *nucleophilicity*. The basicity of amines has been discussed in Section 23.4. Among the most important reactions of amines as nucleophiles are several that have already been encountered in earlier chapters. These are summarized in Table 23.4.

Both the basicity and the nucleophilicity of amines originate in the unshared electron pair of nitrogen. When amines act as bases, they utilize this electron pair to abstract a proton from a suitable donor. When amines undergo the reactions summarized in Table 23.4, the first step in each case is the attack of nitrogen, using its unshared electron pair, on the positively polarized carbon of a carbonyl group.

Amine acting as a base   Amine acting as a nucleophile

The sections that follow extend the scope of nucleophilic reactions of amines to include processes in which nitrogen attacks

1. An alkyl halide

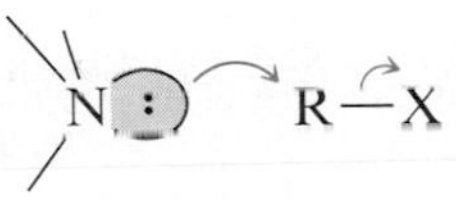

2. Hydrogen peroxide

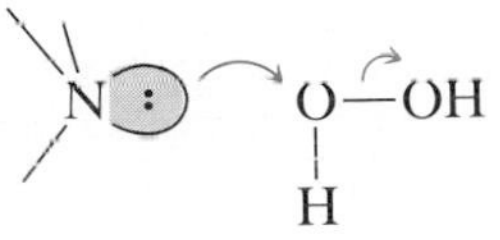

3. Nitrous acid

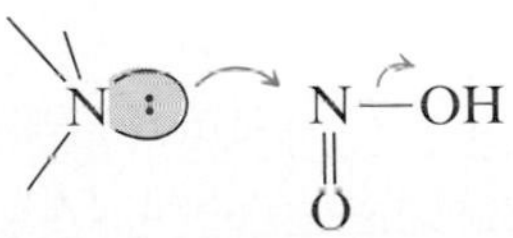

TABLE 23.4

Reactions of Amines Discussed in Previous Chapters

Reaction (section) and comments	General equation and specific example
Reaction of primary amines with aldehydes and ketones (Section 18.13) Imines are formed by nucleophilic addition of a primary amine to the carbonyl group of an aldehyde or ketone. The key step is formation of a carbinolamine intermediate, which then dehydrates to the imine.	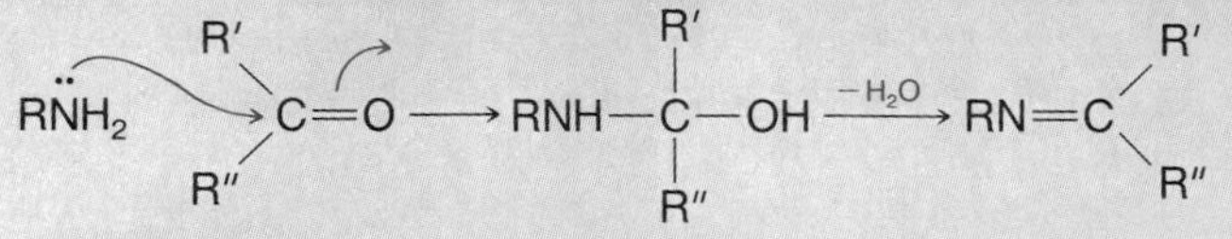 RNH_2 + $C=O$ (R′, R″) $\longrightarrow$ $RNH-C-OH$ (R′, R″) $\xrightarrow{-H_2O}$ $RN=C$ (R′, R″) Primary amine / Aldehyde or ketone / Carbinolamine / Imine CH_3NH_2 + C_6H_5CH (O) $\longrightarrow$ $C_6H_5CH=NCH_3$ + H_2O Methylamine / Benzaldehyde / N-Benzylidenemethylamine (70%) / Water
Reaction of secondary amines with aldehydes and ketones (Section 18.14) Enamines are formed in the corresponding reaction of secondary amines with aldehydes and ketones.	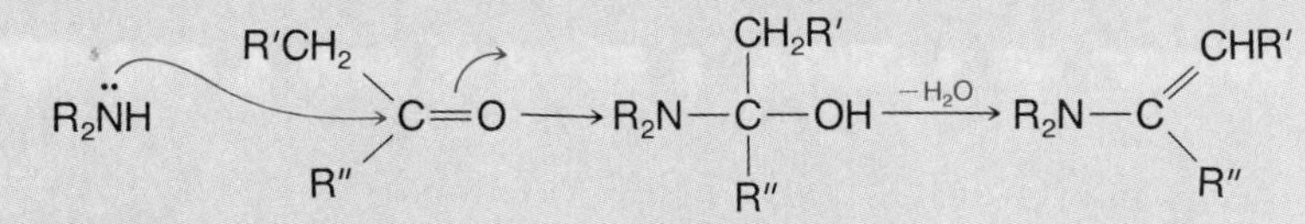 R_2NH + $C=O$ (R′CH₂, R″) $\longrightarrow$ $R_2N-C-OH$ (CH₂R′, R″) $\xrightarrow{-H_2O}$ R_2N-C (CHR′, R″) Secondary amine / Aldehyde or ketone / Carbinolamine / Enamine 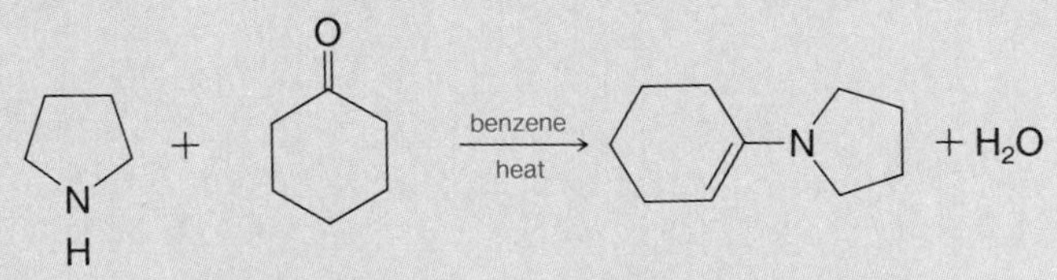 Pyrrolidine + Cyclohexanone $\xrightarrow[\text{heat}]{\text{benzene}}$ N-(1-cyclohexenyl)-pyrrolidine (85–90%) + H_2O
Reaction of amines with carboxylic acid chlorides (Section 21.3) Amines are converted to amides on reaction with acyl chlorides. Other acylating agents, such as carboxylic acid anhydrides and esters, may also be used but are less reactive.	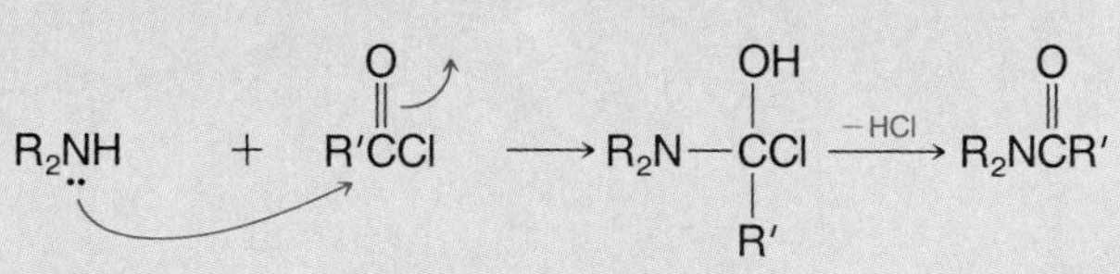 R_2NH + $R'CCl$ (O) $\longrightarrow$ R_2N-CCl (OH, R′) $\xrightarrow{-HCl}$ R_2NCR' (O) Primary or secondary amine / Acyl chloride / Tetrahedral intermediate / Amide $CH_3CH_2CH_2CH_2NH_2$ + $CH_3CH_2CH_2CH_2CCl$ (O) $\longrightarrow$ $CH_3CH_2CH_2CH_2CNHCH_2CH_2CH_2CH_3$ (O) Butylamine / Pentanoyl chloride / N-Butylpentanamide (81%)

The products formed in each of these processes possess certain unique properties that make them suitable topics for further examination.

23.13 REACTION OF AMINES WITH ALKYL HALIDES

Nucleophilic substitution results when primary alkyl halides are treated with amines.

$$\overset{\cdot\cdot}{R}NH_2 + R'CH_2X \longrightarrow \underset{\underset{H}{|}}{\overset{\overset{H}{|}^+}{RN}}{-}CH_2R' \ X^- \longrightarrow \underset{\underset{H}{|}}{\overset{\cdot\cdot}{RN}}{-}CH_2R' + HX$$

| Primary amine | Primary alkyl halide | Ammonium halide salt | Secondary amine | Hydrogen halide |

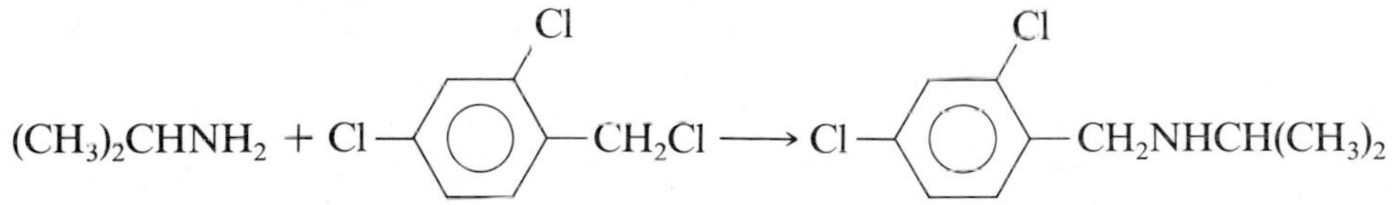

$$(CH_3)_2CHNH_2 + Cl\text{—}\langle\bigcirc\rangle\text{—}CH_2Cl \longrightarrow Cl\text{—}\langle\bigcirc\rangle\text{—}CH_2NHCH(CH_3)_2$$

| Isopropyl- amine | 2,4-Dichlorobenzyl chloride | N-Isopropyl-2,4-dichlorobenzylamine (71%) |

If the alkyl halide and the amine are not sterically hindered, a second alkylation may follow, converting the secondary amine to a tertiary amine. Alkylation need not stop there; the tertiary amine may itself be alkylated, giving a quaternary ammonium salt.

$$\overset{\cdot\cdot}{R}NH_2 \xrightarrow{R'CH_2X} \overset{\cdot\cdot}{R}NHCH_2R' \xrightarrow{R'CH_2X} \overset{\cdot\cdot}{R}N(CH_2R')_2 \xrightarrow{R'CH_2X} \overset{+}{R}N(CH_2R')_3 \ X^-$$

| Primary amine | Secondary amine | Tertiary amine | Quaternary ammonium salt |

Because of its high reactivity toward nucleophilic substitution, methyl iodide is the alkyl halide most frequently encountered in amine alkylations designed to proceed to the quaternary ammonium salt stage.

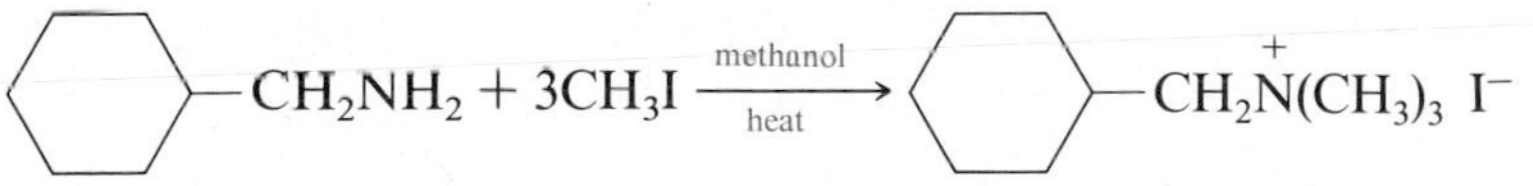

$$\langle\bigcirc\rangle\text{—}CH_2NH_2 + 3CH_3I \xrightarrow[\text{heat}]{\text{methanol}} \langle\bigcirc\rangle\text{—}CH_2\overset{+}{N}(CH_3)_3 \ I^-$$

| (Cyclohexylmethyl)- amine | Methyl iodide | (Cyclohexylmethyl)trimethyl- ammonium iodide (99%) |

| Cycloheptylamine | | Cycloheptyltrimethylammonium iodide (95%) |

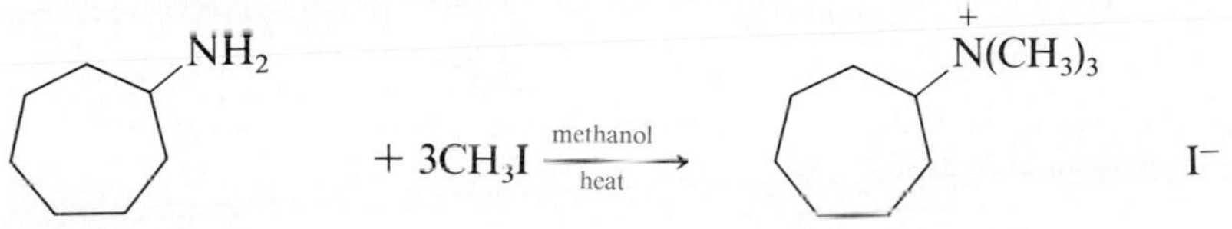

PROBLEM 23.9 *Choline* is a key substance involved in the chemical transmission of nerve impulses. It can be prepared in the laboratory by the reaction of trimethylamine with ethylene oxide in water. Choline is a hydroxide salt represented by the formula $(C_5H_{14}NO)^+$ $(OH)^-$. What is its structure?

Quaternary ammonium salts, as we have seen, are useful in synthetic organic chemistry as phase transfer catalysts. In another, more direct application, quaternary ammonium *hydroxides* are used as substrates in an elimination reaction to form alkenes.

23.14 THE HOFMANN ELIMINATION

The halide anion of quaternary ammonium iodides may be replaced by hydroxide by treatment with an aqueous slurry of silver oxide. Silver iodide precipitates and a solution of the quaternary ammonium hydroxide is formed.

$$2(R_4\overset{+}{N}\ I^-) \ + Ag_2O + H_2O \longrightarrow 2(R_4\overset{+}{N}\ \overline{O}H) + 2AgI$$

Quaternary ammonium iodide	Silver oxide	Water	Quaternary ammonium hydroxide	Silver iodide

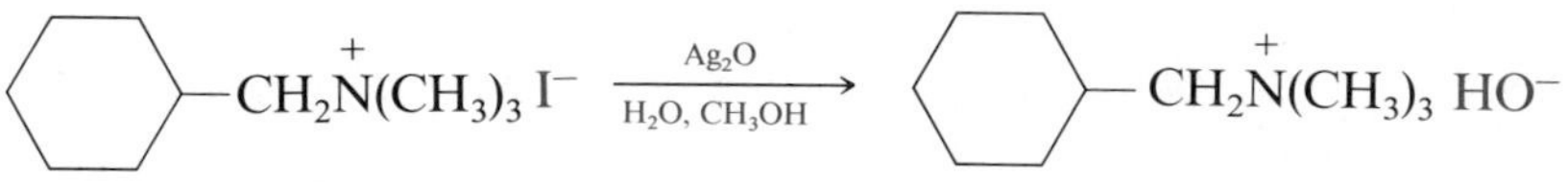

(Cyclohexylmethyl)trimethyl-
ammonium iodide

(Cyclohexylmethyl)trimethylammonium
hydroxide

An alternative method involves treating the quaternary ammonium halide with an ion-exchange resin, a solid polymeric material that replaces halide ions by hydroxide ions.

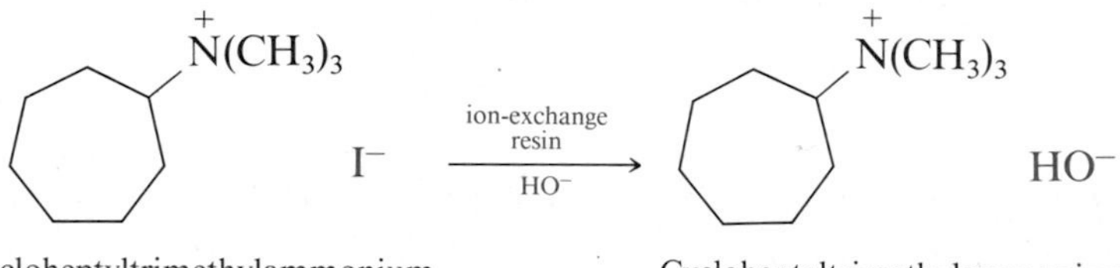

Cycloheptyltrimethylammonium
iodide

Cycloheptyltrimethylammonium
hydroxide

When quaternary ammonium hydroxides are heated, they undergo a β-elimination reaction to form an alkene and an amine.

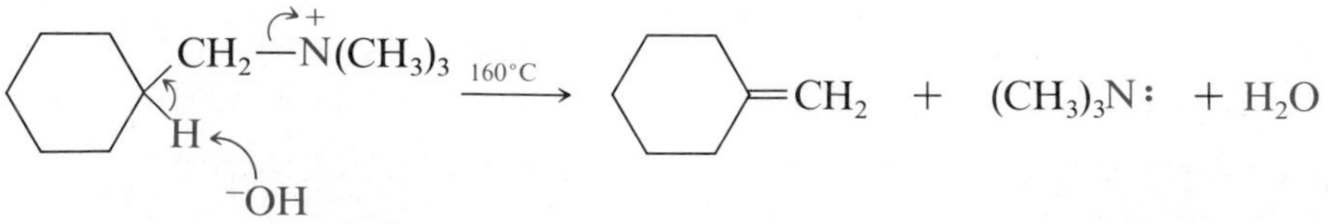

(Cyclohexylmethyl)trimethyl-ammonium hydroxide	Methylenecyclohexane (69%)	Trimethylamine	Water

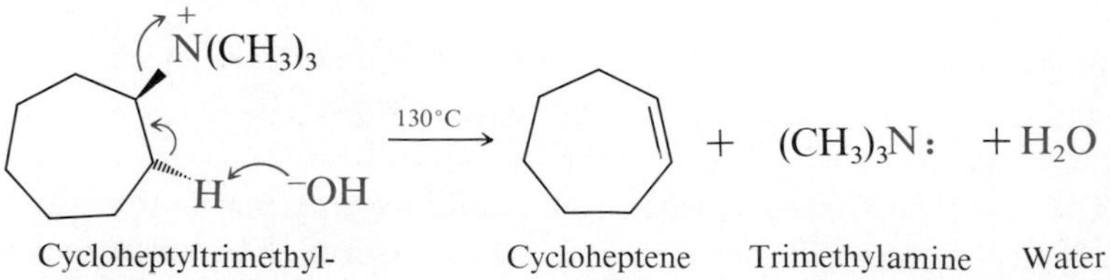

Cycloheptyltrimethyl-ammonium hydroxide	Cycloheptene (87%)	Trimethylamine	Water

This reaction is known as the *Hofmann elimination;* it was developed by August W. Hofmann in the mid-nineteenth century and is used both as a synthetic method to prepare alkenes and as a degradative tool for structure determination.

Hofmann was the first to use his reaction as an analytical method. Piperidine, the structure of which was unknown at the time (1881), had been isolated from piperine, an alkaloid present in pepper. An elimination sequence starting with piperidine yielded a nitrogen-containing alkene.

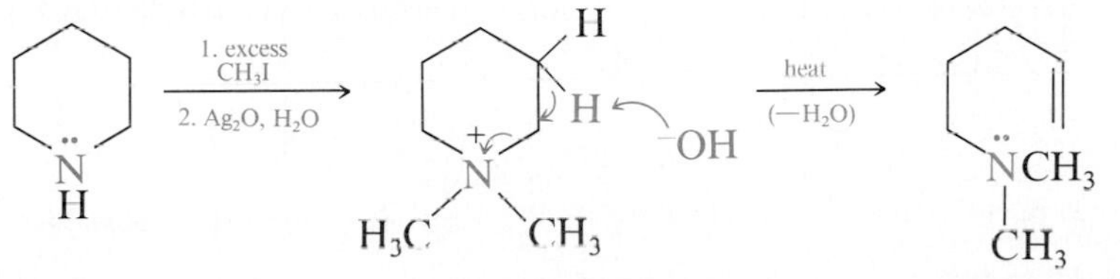

| Piperidine | *N,N*-Dimethylpiperidinium hydroxide | *N,N*-Dimethyl-4-pentenylamine |

This nitrogen-containing alkene was then subjected to a second elimination sequence and gave 1,4-pentadiene and trimethylamine.

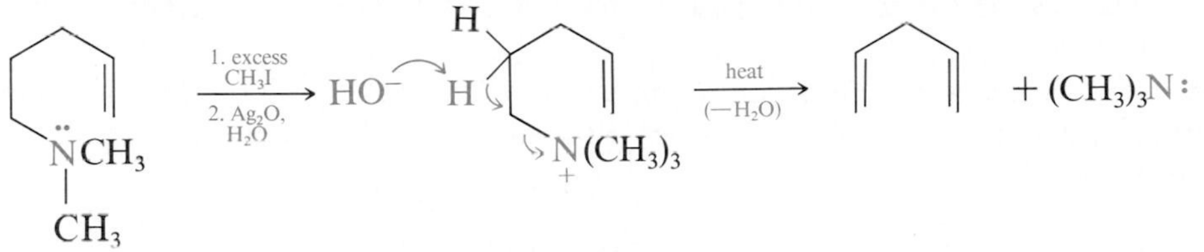

| *N,N*-Dimethyl-4-pentenylamine | *N,N,N*-Trimethyl-4-pentenyl-ammonium hydroxide | 1,4-Pentadiene | Trimethyl-amine |

Since two elimination sequences were required to completely remove the amine nitrogen from piperidine, Hofmann correctly concluded that the amine function must be part of a ring. This evidence, in conjunction with other chemical data, permitted the structure of piperidine to be established.

Prior to the advent of spectroscopic methods, Hofmann elimination reactions were routinely employed in the identification of alkaloid structures. The number of Hofmann eliminations required to remove nitrogen from the molecule gives the number of points of attachment to the amino function, and the identity of the alkene produced reveals the carbon skeleton.

PROBLEM 23.10 Some of the principal structural types found in alkaloids are shown below. How many Hofmann elimination sequences are required to completely remove nitrogen from each one?

(*a*) β-Phenylethylamine type

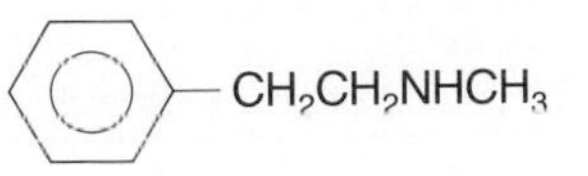

$CH_2CH_2NHCH_3$

(*c*) Benzylisoquinoline type

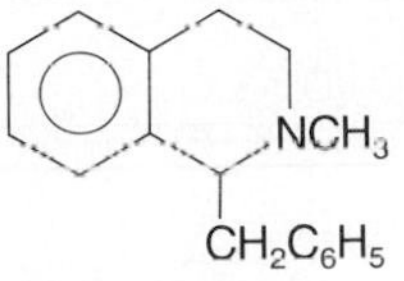

NCH_3

$CH_2C_6H_5$

(*b*) Tropane type

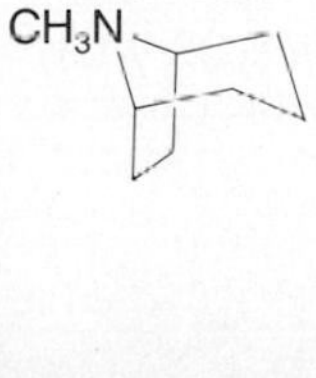

CH_3N

(*d*) Quinolizidine type

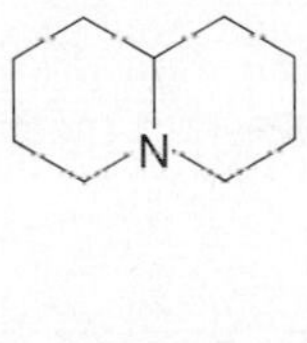

SAMPLE SOLUTION (a) There is only one point of attachment of nitrogen to the main portion of the molecule, so a single Hofmann elimination sequence suffices to remove nitrogen as trimethylamine.

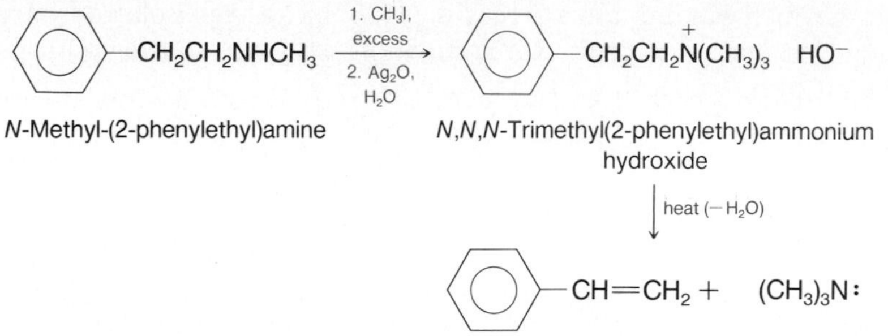

N-Methyl-(2-phenylethyl)amine

N,N,N-Trimethyl(2-phenylethyl)ammonium hydroxide

heat (−H₂O)

Styrene   Trimethylamine

23.15 REGIOSELECTIVITY OF THE HOFMANN ELIMINATION

A novel aspect of the Hofmann elimination is its regioselectivity. Elimination in alkyltrimethylammonium hydroxides proceeds in the direction that gives the less substituted alkene.

$$CH_3CHCH_2CH_3 \ HO^- \xrightarrow[\substack{-H_2O \\ -(CH_3)_3N}]{heat} CH_2{=}CHCH_2CH_3 + CH_3CH{=}CHCH_3$$

with $^+N(CH_3)_3$ below the first carbon.

sec-Butyltrimethylammonium hydroxide   1-Butene (95%)   2-Butene (5%) (cis and trans)

It is the less sterically hindered β hydrogen that is removed by the base in Hofmann elimination reactions. Methyl groups are deprotonated in preference to methylene groups, and methylene groups are deprotonated in preference to methines. The regioselectivity of Hofmann elimination is opposite to that predicted by the Zaitsev rule (Section 6.3). Base-promoted elimination reactions of alkyltrimethylammonium salts are said to obey the *Hofmann rule;* they yield the less substituted alkene.

PROBLEM 23.11 Give the structure of the alkene formed in major amounts when the hydroxide salts of each of the following quaternary ammonium ions are heated.

(a) (cyclopentane with CH₃ and $^+N(CH_3)_3$ substituents)

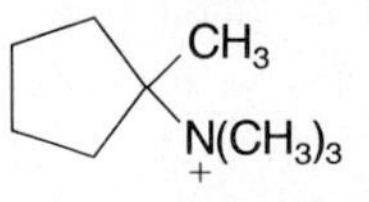

(c) $CH_3CH_2\overset{\overset{\displaystyle CH_3}{\overset{+}{|}}}{\underset{\underset{\displaystyle CH_3}{|}}{N}}CH_2CH_2CH_2CH_3$

(b) $(CH_3)_3CCH_2\overset{\underset{\displaystyle ^+N(CH_3)_3}{|}}{C}(CH_3)_2$

(d) $CH_3CH_2CH_2\overset{\overset{\displaystyle CH_3}{\overset{+}{|}}}{\underset{\underset{\displaystyle CH_3}{|}}{N}}CH_2CH(CH_3)_2$

SAMPLE SOLUTION (a) Two alkenes are capable of being formed by β elimination, methylenecyclopentane and 1-methylcyclopentene.

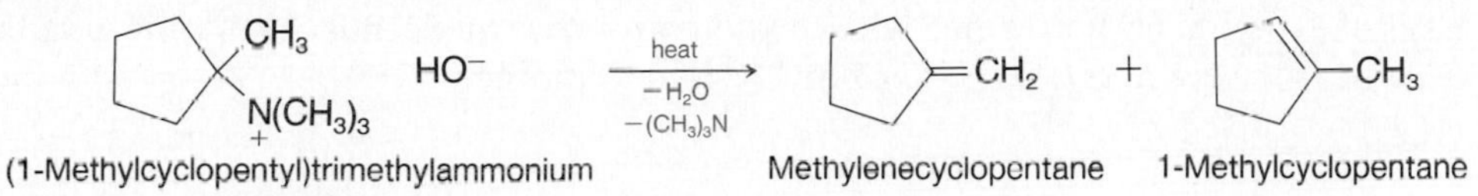

(1-Methylcyclopentyl)trimethylammonium hydroxide $\xrightarrow[\substack{-H_2O \\ -(CH_3)_3N}]{\text{heat}}$ Methylenecyclopentane + 1-Methylcyclopentane

Methylenecyclopentane has the less substituted double bond and is the major product. The reported isomer distribution is 91 percent methylenecyclopentane and 9 percent 1-methylcyclopentene.

Differences between the Zaitsev and Hofmann modes of elimination can be discerned by comparing the activated complexes for E2 elimination of an alkyl halide and an alkyltrimethylammonium salt. In the base-promoted elimination of an alkyl halide, the activated complex is believed to have significant double bond character. Substituents such as alkyl groups that stabilize double bonds also stabilize the transition state leading to their formation.

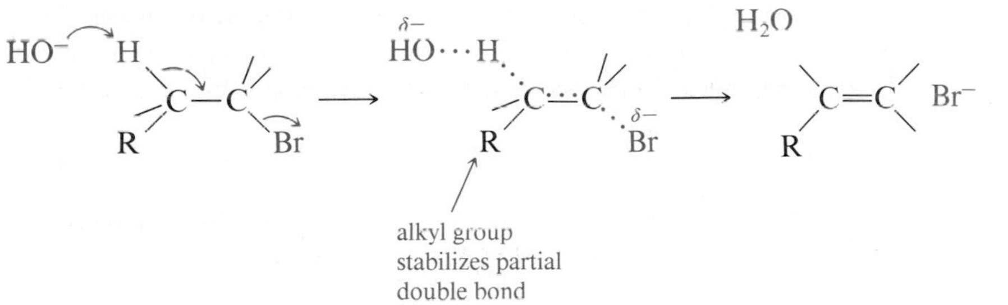

alkyl group stabilizes partial double bond

Because the nitrogen of a tetraalkylammonium salt is positively charged, it is strongly electron withdrawing and increases the acidity of protons on the α and β carbons. The increased acidity of the β hydrogens is believed to lead to an activated complex characterized by a significant degree of carbon-hydrogen bond cleavage.

extensive carbon-hydrogen bond cleavage

alkyl group sterically hinders attack by base and destabilizes carbanionic character

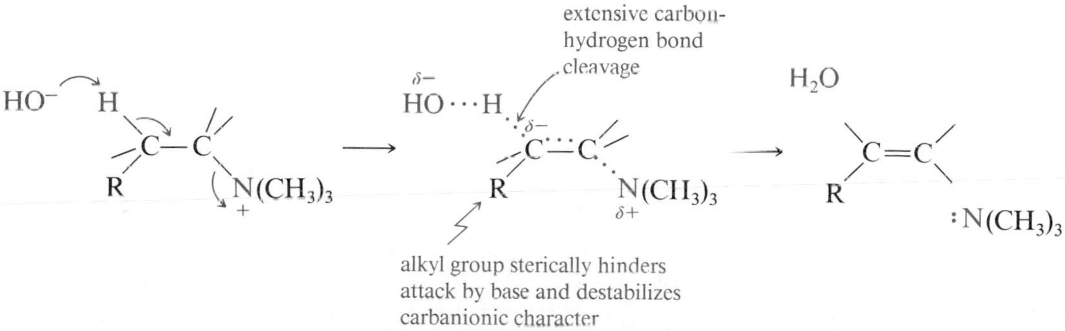

An alkyl substituent both sterically hinders proton removal by the attacking base and destabilizes the partial negative charge on the β carbon. These two effects combine to favor proton removal from the less hindered β carbon to give the less substituted alkene.

23.16 STEREOSPECIFICITY OF THE HOFMANN ELIMINATION

The preferred orientation of the atoms in the H—C—C—N unit at the transition state for a Hofmann elimination reaction has been demonstrated to be anti on the basis of experiments performed with the stereoisomeric quaternary ammonium salts derived from *cis*- and *trans*-4-*tert*-butylcyclohexylamine.

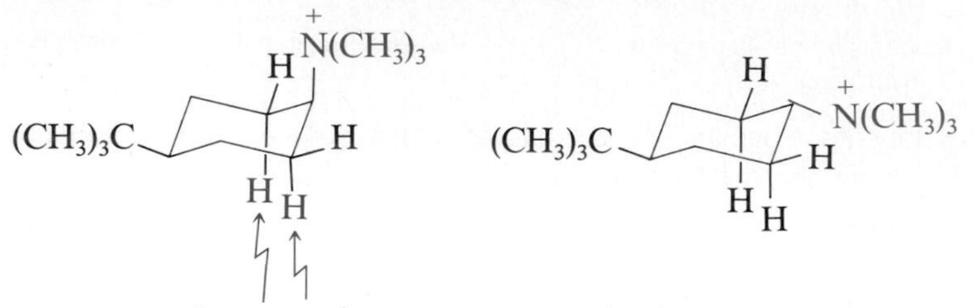

(protons anti to leaving group)

cis-(4-*tert*-Butylcyclohexyl)-trimethylammonium ion

(no β protons anti to leaving group)

trans-(4-*tert*-Butylcyclohexyl)-trimethylammonium ion

Only the cis isomer undergoes the Hofmann elimination readily. It alone has β hydrogens that are anti to the leaving group in a stable chair conformation. The trans isomer does not undergo elimination; its β hydrogens are gauche with respect to the leaving group. The activated complex for elimination from the trans isomer has a significantly higher energy than the cis complex.

23.17 ELIMINATION REACTIONS OF TRIALKYLAMINE *N*-OXIDES. THE COPE ELIMINATION

The Hofmann elimination provides a way to convert alkylamines to alkenes. An alternative method that brings about the same kind of transformation involves preparation of and thermal elimination from trialkylamine *N*-oxides.

Treatment of trialkylamines with hydrogen peroxide or with peroxy acids leads to oxidation of nitrogen and formation of a trialkylamine *N*-oxide.

Trialkyl amine   Hydrogen peroxide   Trialkylamine *N*-oxide   Water

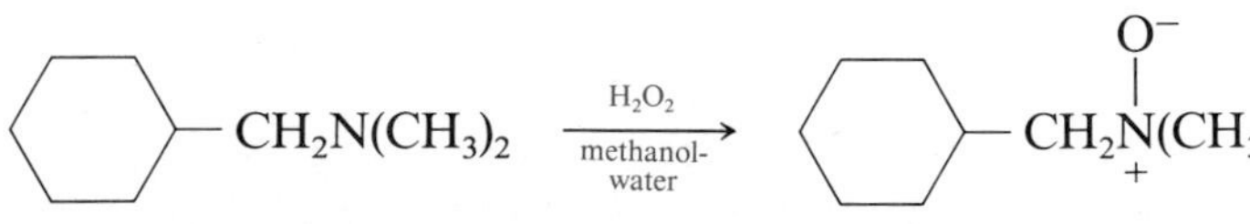

N,N-Dimethyl(cyclohexylmethyl)-amine

N,N-Dimethyl(cyclohexylmethyl)-amine *N*-oxide

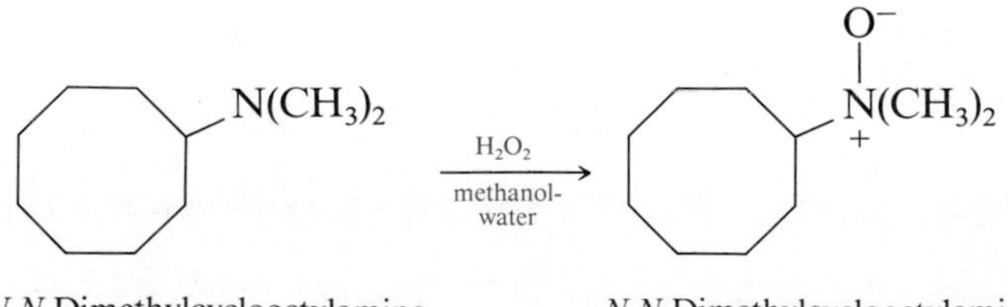

N,N-Dimethylcyclooctylamine

N,N-Dimethylcyclooctylamine *N*-oxide

Heating the above amine oxides causes them to lose *N,N*-dimethylhydroxylamine to form an alkene.

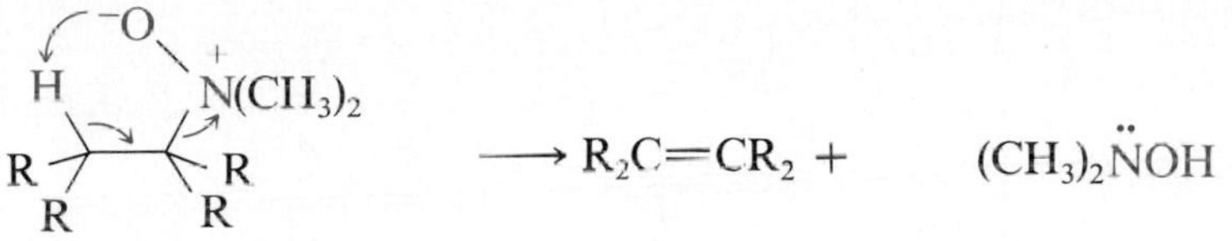

N,N-Dimethylalkylamine *N*-oxide $\longrightarrow$ $R_2C{=}CR_2$ + $(CH_3)_2\ddot{N}OH$

N,N-Dimethylalkylamine *N*-oxide   Alkene   *N,N*-Dimethylhydroxylamine

The base that removes the β proton is the negatively charged oxygen of the amine oxide.

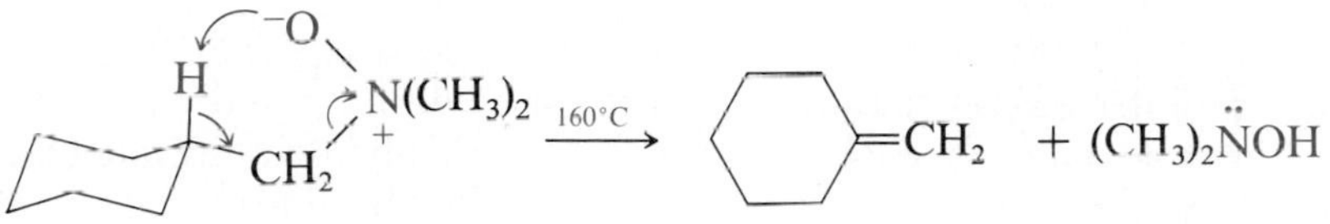

N,N-Dimethyl(cyclohexyl-methyl)amine *N*-oxide   Methylenecyclohexane (79–88%)   *N,N*-Dimethylhydroxylamine (90–96%)

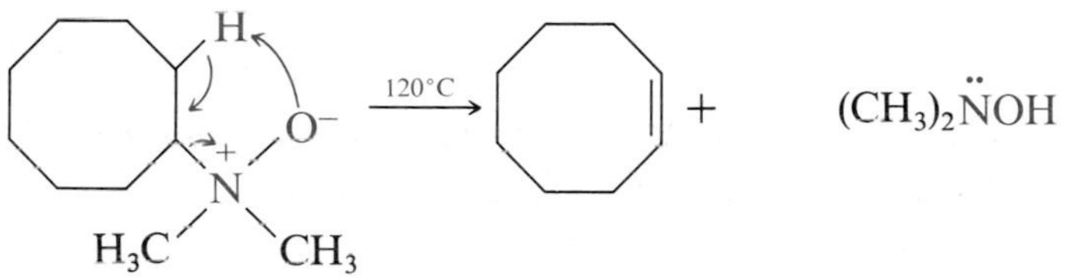

N,N-Dimethylcyclooctyl-amine *N*-oxide   Cyclooctene (90%)   *N,N*-Dimethylhydroxylamine (95%)

In cases in which elimination can occur in more than one direction, a mixture of regioisomeric alkenes is normally obtained.

The preparation of alkenes by thermolysis of amine oxides was developed by Arthur C. Cope of the Massachusetts Institute of Technology and is known as the *Cope elimination*.

23.18 STEREOSPECIFICITY OF THE COPE ELIMINATION

A number of stereochemical studies have been performed that provide information about the structure of the activated complex in Cope elimination reactions. In one such study using deuterium-labeled derivatives of *N,N*-dimethylcyclooctylamine, the stereoisomer with deuterium trans to nitrogen was found to give cyclooctene that retained *all* the deuterium originally present in the amine oxide.

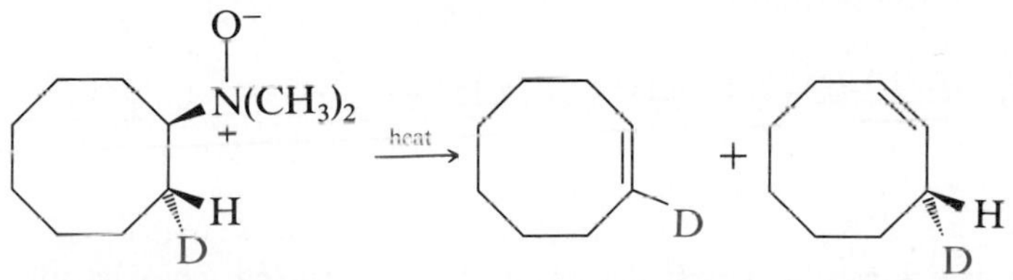

N,N-Dimethyl-(*trans*-2-deuteriocyclooctyl)amine *N*-oxide   Cyclooctene-1-*d*   Cyclooctene-3-*d*

When deuterium was cis to the leaving group, the cyclooctene product lost a portion of the deuterium that was originally present in the starting material.

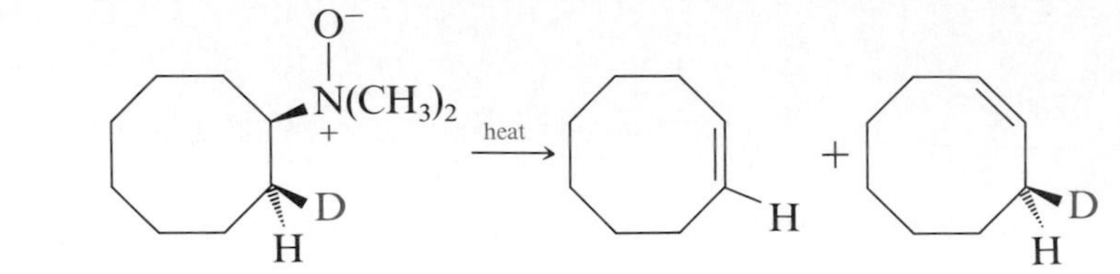

N,N-Dimethyl-(*cis*-2-
deuteriocyclooctyl)amine
N-oxide

Cyclooctene

Cyclooctene-3-*d*

These experiments show that the hydrogen (or deuterium) and the nitrogen are lost from the *same face* of the developing double bond and demonstrate that Cope elimination reactions are syn eliminations. The activated complex is believed to be as shown:

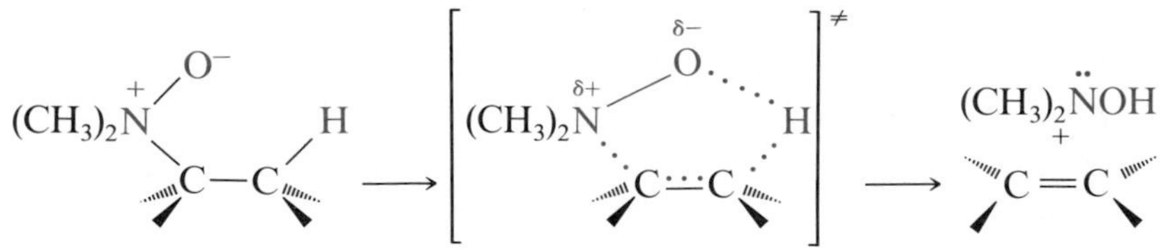

Activated complex

The activated complex is characterized by a cyclic arrangement of five atoms — the oxygen that functions as an internal base, the β proton, the two carbons involved in double bond formation, and the nitrogen. Each of these five atoms undergoes a bonding change during the process.

23.19 NITROSATION OF AMINES

When solutions of sodium nitrite are acidified, a number of species are formed that act as *nitrosating agents.* That is, they react with nucleophiles as if they were sources of nitrosyl cation $:\overset{+}{N}=\overset{..}{O}:$. In order to facilitate discussion, organic chemists simply group all these species together and speak of the chemistry of one of them, *nitrous acid,* as applying to nitrosating agents in general. For all but the most precise and detailed studies of nitrosation reactions, this is a satisfactory approximation and is the practice we will follow here.

Nitrite ion
(from sodium nitrite NaNO$_2$)

Nitrous acid

Nitrosation of amines is best illustrated by examining what happens when the amine is secondary. The amine acts as a nucleophile, attacking the nitrogen of the nitrosating agent.

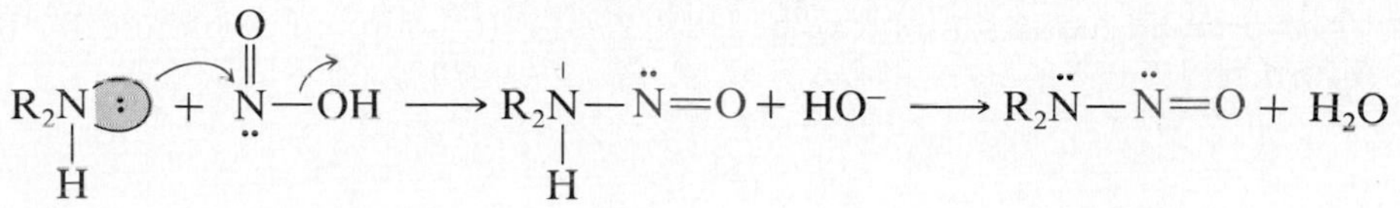

$$R_2\overset{\cdot\cdot}{N} \overset{\cdot\cdot}{:} + \overset{\overset{O}{\|}}{\underset{H}{N}}-OH \longrightarrow R_2\overset{+}{N}-\overset{\cdot\cdot}{N}=O + HO^- \longrightarrow R_2\overset{\cdot\cdot}{N}-\overset{\cdot\cdot}{N}=O + H_2O$$
$$\underset{H}{}$$

Secondary   Nitrous acid   N-Nitroso   Water
alkylamine   amine

The intermediate that is formed in the nucleophilic substitution step loses a proton to give an *N*-nitroso amine as the isolated product.

$$(CH_3)_2NH \xrightarrow[\text{H}_2\text{O}]{\text{NaNO}_2,\ \text{HCl}} (CH_3)_2N-N=O$$

Dimethylamine   *N*-Nitrosodimethylamine
                                    (88–90%)

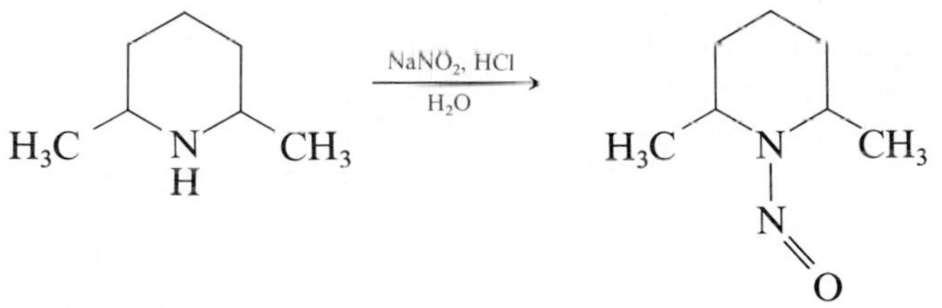

2,6-Dimethylpiperidine   *N*-Nitroso-2,6-dimethylpiperidine
                                                     (72%)

PROBLEM 23.12   *N*-Nitroso amines are believed to be stabilized by resonance. Write the two most stable resonance forms of *N*-nitrosodimethylamine $(CH_3)_2NNO$.

N-Nitroso amines are more often called *nitrosamines* and because many of them are potent carcinogens, they have been the object of much recent investigation. We encounter nitrosamines in the environment on a daily basis. A few of these, all of which are known carcinogens, are shown below.

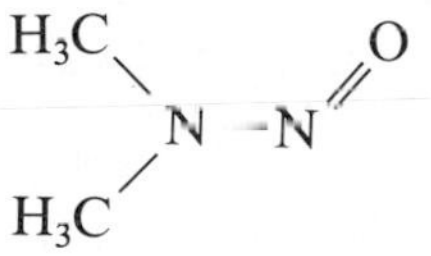

N-Nitrosodimethylamine   *N*-Nitrosopyrrolidine   *N*-Nitrosonornicotine
(formed during   (formed when bacon   (present in tobacco
tanning of leather;   that has been cured   smoke)
also found in beer   with sodium nitrite
and herbicides)   is fried)

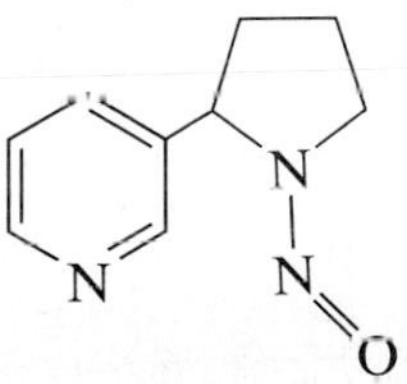

Nitrosamines are formed whenever nitrosating agents come in contact with secondary amines. Indeed, more nitrosamines are probably synthesized within our body than enter it by environmental contamination. Enzyme-catalyzed reduction of nitrate (NO_3^-) produces nitrite (NO_2^-), which combines with amines present in the body to form *N*-nitroso amines.

When primary amines are nitrosated, the *N*-nitroso compounds produced can react further.

$$RNH_2 \xrightarrow{\text{HONO}} R\ddot{N}\diagdown\overset{H}{\underset{N=O:}{}} \xrightarrow{H^+} R-\ddot{N}\diagdown\overset{H}{\underset{N=OH}{}}$$

Primary   (Not isolable)   (Not isolable)
alkylamine

$$\xrightarrow{-H^+}$$

$$R\overset{+}{N}\equiv N: \xleftarrow{-H_2O} R\ddot{N}=N-\overset{+}{O}H_2 \xleftarrow{H^+} R\ddot{N}=N-\ddot{O}H$$

Alkyldiazonium   (Not isolable)   (Not isolable)
ion

The product of this sequence of steps is an alkyldiazonium ion. The amine, in being converted to a diazonium ion, is said to have been *diazotized*. Alkyldiazonium ions are not very stable, usually decomposing rapidly under the conditions of their formation. Molecular nitrogen is a leaving group par excellence, and the reaction products arise by solvolysis of the diazonium ion. Usually, a carbocation intermediate is involved.

$$\begin{array}{ccc}
\overset{CH_3}{\underset{NH_2}{CH_3CH_2\overset{|}{\underset{|}{C}}CH_3}} & \xrightarrow{\text{HONO}} & \overset{CH_3}{\underset{\overset{N^+}{\underset{\overset{|||}{N}}{}}}{CH_3CH_2\overset{|}{C}CH_3}} \longrightarrow \overset{CH_3}{\underset{+}{CH_3CH_2\overset{|}{C}CH_3}} + :N\equiv N:
\end{array}$$

tert-Pentyl-   *tert*-Pentyldia-   *tert*-Pentyl   Nitrogen
amine   zonium ion   cation

$$\xleftarrow{-H^+} \qquad\qquad\qquad \xrightarrow{H_2O}$$

$$CH_3CH=C(CH_3)_2 + \underset{\underset{CH_3}{|}}{CH_3CH_2C}=CH_2 + \underset{\underset{OH}{|}}{CH_3CH_2\overset{\overset{CH_3}{|}}{C}CH_3}$$

2-Methyl-2-butene   2-Methyl-1-butene   2-Methyl-2-butanol
(2%)   (3%)   (80%)

Since nitrogen-free products result from the formation and decomposition of diazonium ions, these reactions are often referred to as *deamination reactions*. Alkyldiazonium ions are rarely used in synthetic work but have been studied extensively to probe the behavior of carbocations generated under conditions such that the leaving group is lost rapidly and irreversibly.

PROBLEM 23.13   Nitrous acid deamination of neopentylamine $(CH_3)_3CCH_2NH_2$ gives the same products as were indicated as being formed from *tert*-pentylamine in the preceding example. Suggest a mechanistic explanation for the formation of these compounds from neopentylamine.

Aryldiazonium ions, prepared by nitrous acid diazotization of primary arylamines, are substantially more stable than alkyldiazonium ions and are of enormous synthetic value. Their utility in the synthesis of substituted aromatic compounds will be described in Chapter 24.

The nitrosation of tertiary alkylamines is rather complicated and no generally useful chemistry is associated with reactions of this type.

23.20 SPECTROSCOPIC ANALYSIS OF AMINES

The absorptions of interest in the infrared spectra of amines are those associated with N—H and N—C vibrations. Primary amines exhibit two peaks in the range 3000 to 3500 cm^{-1}, which are due to symmetric and antisymmetric N—H stretching modes.

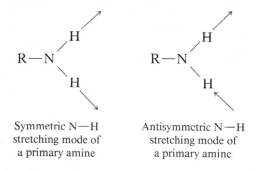

Symmetric N—H
stretching mode of
a primary amine

Antisymmetric N—H
stretching mode of
a primary amine

These two vibrations are clearly visible at 3270 and 3380 cm^{-1} in the infrared spectrum of butylamine, shown in Figure 23.3a. Secondary amines such as diethylamine, shown in Figure 23.3b, exhibit only one peak, which is due to N—H stretching, at 3280 cm^{-1}. Tertiary amines, of course, are transparent in this region since they have no N—H bonds. Stretching vibrations of N—C bonds are often difficult to assign and do not ordinarily provide much in the way of structural information.

Characteristics of the nuclear magnetic resonance spectra of amines may be illustrated by comparing p-methylbenzylamine (Figure 23.4a) with p-methylbenzyl alco-

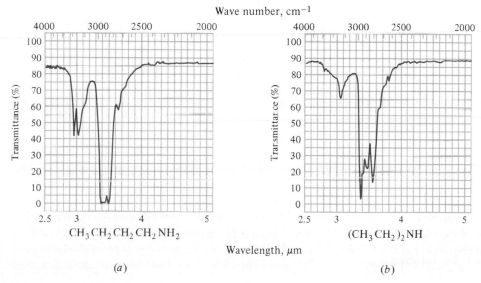

FIGURE 23.3 Portions of the infrared spectrum of (a) butylamine and (b) diethylamine. Primary amines exhibit two peaks due to N—H stretching in the 3300 to 3350 cm^{-1} region while secondary amines show only one.

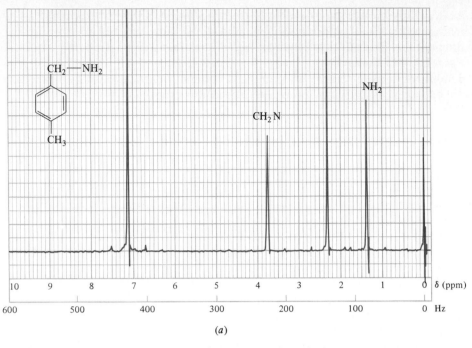

(a)

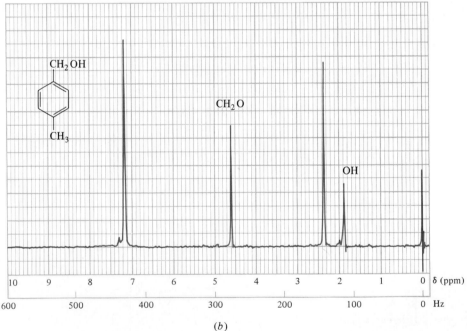

(b)

FIGURE 23.4 The ¹H nmr spectra of (a) p-methylbenzyl-amine and (b) p-methylbenzyl alcohol. The singlet corresponding to CH₂N in (a) is more shielded than that of CH₂O in (b).

hol (Figure 23.4b). Nitrogen is less electronegative than oxygen and so shields neighboring nuclei to a greater extent. The benzylic methylene group attached to nitrogen in p-methylbenzylamine appears at higher field (δ 3.7 ppm) than the benzylic methylene of p-methylbenzyl alcohol (δ 4.6 ppm). The N—H protons are somewhat more shielded than the O—H protons of an alcohol. In p-methylbenzylamine the protons of the amino group correspond to the signal at δ 1.4 ppm while the hydroxyl proton signal of 4-methylbenzyl alcohol is found at δ 1.9 ppm. The chemical shifts of amino group protons, like those of hydroxyl protons, are variable and are sensitive to solvent, concentration, and temperature.

Similarly, carbons that are bonded to nitrogen are more shielded than those bonded to oxygen, as revealed by comparing the ^{13}C chemical shifts of methylamine and methanol.

26.9 ppm   48.0 ppm

CH_3NH_2 CH_3OH

Methylamine   Methanol

23.21 MASS SPECTROMETRY OF AMINES

A number of features combine to make amines easily identifiable by mass spectrometry.

First, the peak for the molecular ion M^+ for all compounds that contain only carbon, hydrogen, and oxygen has an m/z value that is an even number. The presence of a nitrogen atom in the molecule requires that the m/z value for the molecular ion be odd. An odd number of nitrogens corresponds to an odd value of the molecular weight; an even number of nitrogens corresponds to an even molecular weight.

Second, nitrogen is exceptionally good at stabilizing adjacent carbocation sites. The fragmentation pattern seen in the mass spectra of amines is dominated by cleavage of groups from the carbon atom attached to the nitrogen, as the data for the pair of constitutionally isomeric amines given below illustrate.

$$(CH_3)_2\ddot{N}CH_2CH_2CH_2CH_3 \xrightarrow{e^-} (CH_3)_2\overset{+}{\ddot{N}}\!-\!CH_2\!-\!CH_2CH_2CH_3 \longrightarrow$$

N,N-Dimethylbutanamine M^+ (m/z 101)

$$(CH_3)_2\overset{+}{N}\!=\!CH_2 + \cdot CH_2CH_2CH_3$$

(m/z 58)
(most intense peak)

$$CH_3\ddot{N}HCH_2CH_2CH(CH_3)_2 \xrightarrow{e^-} CH_3\overset{+}{\ddot{N}}H\!-\!CH_2\!-\!CH_2CH(CH_3)_2 \longrightarrow$$

N,3-Dimethylbutanamine M^+ (m/z 101)

$$CH_3\overset{+}{N}H\!=\!CH_2 + \cdot CH_2CH(CH_3)_2$$

(m/z 44)
(most intense peak)

23.22 SUMMARY

Alkylamines are compounds of the type shown, where R, R′, and R″ are alkyl groups.

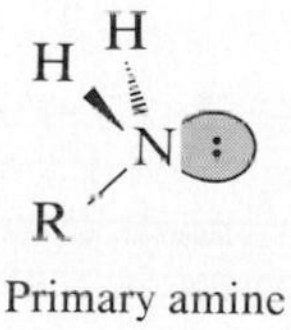

Primary amine 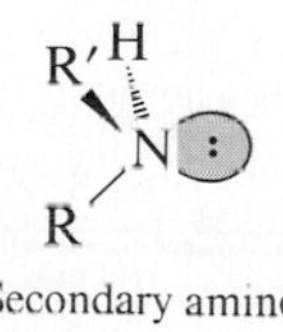   Secondary amine 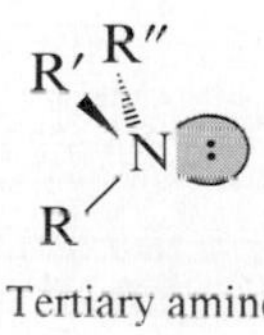   Tertiary amine

TABLE 23.5

Preparation of Amines

Reaction (section) and comments	General equation and specific example

Alkylation methods

Alkylation of ammonia (Section 23.8) Ammonia can act as a nucleophile toward primary and some secondary alkyl halides to give primary amines. Yields tend to be modest because the primary amine is itself a nucleophile and undergoes alkylation. Alkylation of ammonia can lead to a mixture containing a primary amine, a secondary amine, a tertiary amine, and a quaternary ammonium salt.

$$RX + 2NH_3 \longrightarrow RNH_2 + NH_4X$$

Alkyl halide   Ammonia   Alkylamine   Ammonium halide

$$C_6H_5CH_2Cl \xrightarrow{NH_3(8\ mol)} C_6H_5CH_2NH_2 + (C_6H_5CH_2)_2NH$$

Benzyl chloride   Benzylamine   Dibenzylamine
(1 mol)   (53%)   (39%)

Alkylation of phthalimide. The Gabriel synthesis (Section 23.9) The potassium salt of phthalimide reacts with alkyl halides to give *N*-alkylphthalimide derivatives. Hydrolysis or hydrazinolysis of this derivative yields a primary amine.

Alkyl halide   Potassiophthalimide   *N*-Alkylphthalimide

N-Alkylphthalimide   Hydrazine   Primary amine   Phthalhydrazide

$$CH_3CH{=}CHCH_2Cl \xrightarrow[\text{2. }H_2NNH_2,\ \text{ethanol}]{\text{1. }N\text{-potassiophthalimide, DMF}} CH_3CH{=}CHCH_2NH_2$$

1-Chloro-2-butene   2-Buten-1-amine (95%)

Reduction methods

Reduction of alkyl azides (Section 23.10) Alkyl azides, prepared by nucleophilic substitution by azide ion in primary or secondary alkyl halides, are reduced to primary amines by lithium aluminum hydride and by catalytic hydrogenation.

$$R\ddot{N}{=}\overset{+}{N}{=}\ddot{N} \xrightarrow{\text{reduce}} R\ddot{N}H_2$$

Alkyl azide   Primary amine

$$CF_3CH_2\underset{\displaystyle N_3}{CH}CO_2CH_2CH_3 \xrightarrow{H_2,\ Pd} CF_3CH_2\underset{\displaystyle NH_2}{CH}CO_2CH_2CH_3$$

Ethyl 2-azido-4,4,4-trifluorobutanoate   Ethyl 2-amino-4,4,4-trifluorobutanoate (96%)

Reduction of nitriles (Section 23.10) Nitriles are reduced to primary amines by lithium aluminum hydride and by catalytic hydrogenation.

$$RC{\equiv}N \xrightarrow{\text{reduce}} RCH_2NH_2$$

Nitrile   Primary amine

Cyclopropyl cyanide   Cyclopropylmethanamine (75%)

TABLE 23.5 (continued)

Reaction (section) and comments	General equation and specific example
Reduction of oximes (Section 23.10) Oximes are readily prepared by nucleophilic addition of hydroxylamine to aldehydes and ketones. Sodium in ethanol reduces oximes to primary amines. Catalytic hydrogenation may also be used.	$R_2C\!\!=\!\!NOH \xrightarrow{\text{reduce}} R_2CHNH_2$ Oxime $\qquad$ Primary amine $CH_3(CH_2)_5CH\!\!=\!\!NOH \xrightarrow[\text{ethanol}]{Na} CH_3(CH_2)_5CH_2NH_2$ Heptanal oxime $\qquad$ Heptanamine (60–73%)
Reduction of amides (Section 23.10) Lithium aluminum hydride reduces the carbonyl group of an amide to a methylene group. Primary, secondary, or tertiary amines may be prepared by proper choice of the starting amide.	$\overset{O}{\overset{\|}{R}\!C\!NR'_2} \xrightarrow{\text{reduce}} RCH_2NR'_2$ Amide $\qquad\qquad$ Amine $\overset{O}{\overset{\|}{CH_3}\!C\!NHC(CH_3)_3} \xrightarrow[\text{2. } H_2O]{\text{1. LiAlH}_4} CH_3CH_2NHC(CH_3)_3$ N-tert-Butylacetamide $\qquad$ N-Ethyl-tert-butylamine (60%)
Reductive amination (Section 23.11) Reaction of ammonia or an amine with an aldehyde or ketone in the presence of a reducing agent is an effective method for the preparation of primary, secondary, or tertiary amines. The reducing agent may be either hydrogen in the presence of a metal catalyst or sodium cyanoborohydride.	$\overset{O}{\overset{\|}{R}\!C\!R'} + R''_2NH \xrightarrow{\text{reducing agent}} \overset{NR''_2}{\underset{H}{R\!-\!C\!-\!R'}}$ Aldehyde $\quad$ Ammonia or $\qquad$ Amine or ketone $\quad$ an amine $\overset{O}{\overset{\|}{CH_3}\!C\!CH_3} +$ (cyclohexylamine) $\xrightarrow{H_2,\ Pt}$ (N-Isopropylcyclohexylamine) Acetone $\quad$ Cyclohexylamine $\qquad$ N-Isopropylcyclohexylamine (79%)

Amines have a pyramidal arrangement of bonds to sp^3 hybridized nitrogen. The unshared pair of electrons on nitrogen is of paramount importance in understanding the properties of amines.

Alkylamines are weak bases; they can use their unshared electron pair to abstract a proton from a suitable donor. Basicity constants K_b of alkylamines are in the range 10^{-3} to 10^{-5} (pK_b 3 to 5).

$$R_3N\!:\ +\ H_2O \rightleftharpoons R_3\overset{+}{N}H + HO^- \qquad K_b = \frac{[R_3\overset{+}{N}H][HO^-]}{[R_3N]}$$

The order of amine basicities in the gas phase reflects the fact that alkyl groups attached to nitrogen assist dispersal of the positive charge in the ammonium ion. Increasing alkyl substitution increases basicity. Tertiary amines are the most basic amines when basicity is measured in the gas phase. Replacement of a hydrogen substituent on nitrogen by an alkyl group, however, decreases its ability to stabilize the positive charge by solvation. The basicity of amines in aqueous solution reflects a compromise between the competing forces of stabilization of the ammonium ion by its substituents and by solvation. Secondary amines are the most basic class of amines in aqueous solution.

Methods for the preparation of amines were described in Sections 23.7 through 23.11 and are summarized in Table 23.5.

In their reactions, amines function as nucleophiles. Reactions of amines with carbonyl compounds were summarized in Table 23.4. Reactions with other electrophilic reagents include:

1. Reactions with alkyl halides (Section 23.13)
2. Oxidation (Section 23.17)
3. Nitrosation (Section 23.19)

Amines are alkylated by primary alkyl halides. The product may be a secondary or tertiary amine or a tetraalkylammonium salt.

$$RNH_2 \xrightarrow{R'CH_2X} RNHCH_2R' \xrightarrow{R'CH_2X} RN(CH_2R')_2 \xrightarrow{R'CH_2X} R\overset{+}{N}(CH_2R')_3 \ X^-$$

| Primary amine | Secondary amine | Tertiary amine | Quaternary ammonium salt |

Alkyltrimethylammonium iodides are readily prepared by alkylation of amines with excess methyl iodide.

$$RNH_2 \xrightarrow[\text{excess}]{CH_3I} R\overset{+}{N}(CH_3)_3 \ I^-$$

Amine   Alkyltrimethylammonium iodide

The iodide counterion of an alkyltrimethylammonium iodide may be exchanged for hydroxide by treatment with moist silver oxide or with an ion-exchange resin.

$$R\overset{+}{N}(CH_3)_3 \ I^- \xrightarrow[\text{or} \atop HO^-, \text{resin}]{Ag_2O, \ H_2O} R\overset{+}{N}(CH_3)_3 \ \bar{O}H$$

Alkyltrimethylammonium iodide   Alkyltrimethylammonium hydroxide

When alkyltrimethylammonium hydroxides that have β hydrogens are heated, they undergo an anti elimination, forming an alkene and trimethylamine (Sections 23.14 through 23.16).

$$HO^- \curvearrowright \overset{H}{\underset{\beta}{C}} - \overset{}{\underset{\alpha}{C}} \overset{}{\underset{\overset{+}{N}(CH_3)_3}{}} \xrightarrow{\text{heat}} H_2O + \overset{}{\underset{}{C}} = \overset{}{\underset{}{C}} + \ :N(CH_3)_3$$

| Alkyltrimethylammonium hydroxide | Water | Alkene | Trimethylamine |

This reaction, known as the *Hofmann elimination,* is useful both as a method for the preparation of alkenes and as an analytical tool for determining the structure of alkaloids. Hydroxide ion tends to abstract a proton from the less substituted β carbon to produce the less substituted alkene.

Tertiary amines are oxidized by hydrogen peroxide or peroxy acids to give trialkylamine *N*-oxides.

$$R_3N \ + \ H_2O_2 \ \longrightarrow \ R_3\overset{+}{N}\!\!-\!\!O^- + H_2O$$

Tertiary   Hydrogen   Trialkylamine   Water
amine   peroxide   *N*-oxide

When these trialkylamine *N*-oxides are heated, they undergo a syn β elimination to give an alkene and a hydroxylamine derivative.

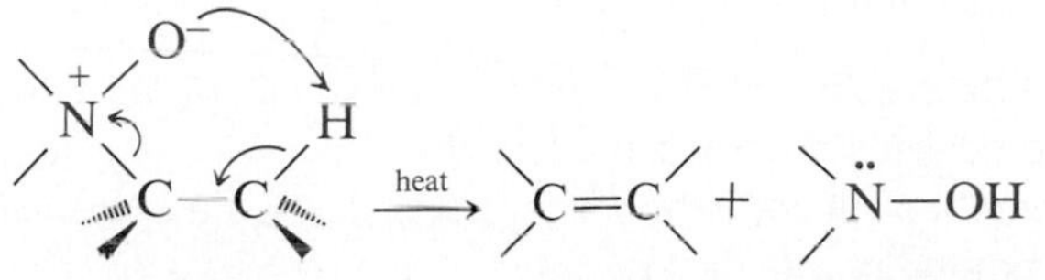

Trialkylamine   Alkene   Hydroxylamine
N-oxide   derivative

This reaction is known as the *Cope elimination* (Sections 23.17 and 23.18).

Nitrosation of amines occurs on acidification of a solution containing sodium nitrite in the presence of an amine. While several nitrosating agents may be present, organic chemists simply represent the active species as nitrous acid (HONO). Secondary amines react with nitrosating agents to give *N*-nitroso amines.

$$R_2NH \ \xrightarrow{\text{NaNO}_2,\ \text{HCl}} \ R_2N\!\!-\!\!N\!\!\overset{\displaystyle O}{\diagdown}$$

Secondary amine   *N*-Nitroso amine

Primary amines yield alkyldiazonium salts.

$$RNH_2 \ \xrightarrow{\text{NaNO}_2,\ \text{HCl}} \ R\overset{+}{N}\!\!\equiv\!\!N$$

Primary amine   Diazonium ion

Alkyldiazonium ions are very unstable, rapidly undergoing solvolysis under the conditions of their formation.

$$R\!\!-\!\!\overset{+}{N}\!\!\equiv\!\!N \ \xrightarrow{-N_2} \ \text{products of nucleophilic substitution and elimination}$$

PROBLEMS

23.14 Write structural formulas for all the amines of molecular formula $C_4H_{11}N$. Give an acceptable name for each one and classify it as a primary, secondary, or tertiary amine.

23.15 Provide a structural formula for each of the following compounds.

(a) Neopentylamine
(b) 2-Ethyl-1-butanamine
(c) *N*-Ethyl-1-butanamine
(d) Dibenzylamine
(e) Tribenzylamine
(f) Tetraethylammonium hydroxide
(g) *N*-Allylcyclohexylamine

(h) *N*-Allylpiperidine
(i) Benzyl 2-aminopropanoate
(j) 4-(*N*,*N*-Dimethylamino)cyclohexanone
(k) 2,2-Dimethyl-1,3-propanediamine

23.16 Many naturally occurring nitrogen compounds and many nitrogen-containing drugs are better known by common names than by their systematic names. A few of these are given below. Write a structural formula for each one.

(a) 4-Aminobutanoic acid, better known as *γ-aminobutyric acid* or *GABA:* involved in metabolic processes occurring in the brain
(b) 2-(3,4,5-Trimethoxyphenyl)ethylamine, better known as *mescaline:* a hallucinogen obtained from the peyote cactus
(c) 1,4-Butanediamine, better known as *putrescine:* formed by bacterial-induced decay of protein-containing organic matter
(d) *trans*-2-Phenylcyclopropylamine, better known as *tranylcypromine:* an antidepressant drug
(e) *N*-Benzyl-*N*-methyl-2-propynylamine, better known as *pargyline:* a drug used to treat high blood pressure
(f) 1-Phenyl-2-propanamine, better known as *amphetamine:* a stimulant
(g) *N*-Methyl-1-phenyl-2-propanamine, better known as *methamphetamine:* a stimulant
(h) 1-(*m*-Hydroxyphenyl)-2-(methylamino)ethanol: better known as *phenylephrine:* a nasal decongestant

23.17 Arrange the following compounds and/or anions in each group in order of decreasing basicity:

(a) H_3C^-, H_2N^-, HO^-, F^-
(b) H_2O, NH_3, HO^-, H_2N^-
(c) HO^-, H_2N^-, $:\bar{C}\equiv N:$, NO_3^-
(d)

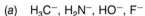

23.18 Which amine in each of the following pairs is the stronger base? Explain your reasoning.

(a) $Cl_3CCH_2NH_2$ or $Cl_3CCH_2CH_2NH_2$
(b) $(CH_3CH_2)_3N$ or $(ClCH_2CH_2)_3N$
(c) $CH_3CH_2CH_2NH_2$ or $HC\equiv CCH_2NH_2$
(d)

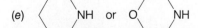

(e)

NH   or   O   NH

(f) CH_3CHCH_3 or $CH_3CHCO_2CH_2CH_3$
     | |
     NH_2 NH_2

23.19 Describe procedures for preparing each of the following compounds, using ethanol as the source of all their carbon atoms. Once you prepare a compound, you need not repeat its synthesis in a succeeding problem.

(a) Ethylamine
(b) N-Ethylacetamide
(c) Diethylamine
(d) N,N-Diethylacetamide
(e) Triethylamine
(f) Tetraethylammonium bromide
(g) Triethylamine N-oxide
(h) N,N-Diethylhydroxylamine

23.20 Show by writing the appropriate sequence of equations how you could carry out each of the following transformations:

(a) 1-Butanol to 1-pentanamine
(b) tert-Butyl chloride to neopentylamine
(c) Cyclohexanol to N-methylcyclohexylamine
(d) Isopropyl alcohol to 1-amino-2-methyl-2-propanol
(e) Isopropyl alcohol to 1-amino-2-propanol
(f) Isopropyl alcohol to 1-(N,N-dimethylamino)-2-propanol
(g)

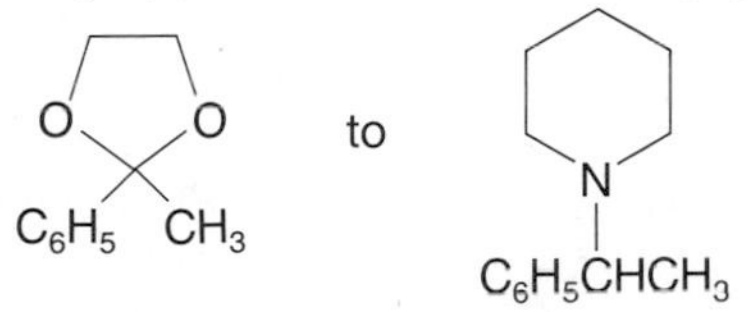

 to

23.21 Each of the following dihaloalkanes gives an N-(haloalkyl)phthalimide on reaction with one equivalent of the potassium salt of phthalimide. Write the structure of the phthalimide derivative formed in each case and explain the basis for your answer.

(a) FCH_2CH_2Br
(b) $BrCH_2CH_2CH_2CHCH_3$
$\qquad\qquad\qquad\qquad |$
$\qquad\qquad\qquad\quad Br$

(c) $BrCH_2\overset{\displaystyle CH_3}{\underset{\displaystyle CH_3}{\overset{|}{\underset{|}{C}}}}CH_2CH_2Br$

23.22 Give the structure of the expected product formed when benzylamine reacts with each of the following reagents:

(a) Hydrogen bromide
(b) Sulfuric acid
(c) Acetic acid
(d) Acetyl chloride
(e) Acetic anhydride
(f) Acetone
(g) Acetone and sodium cyanoborohydride
(h) Ethylene oxide
(i) 1,2-Epoxypropane
(j) Excess methyl iodide
(k) Sodium nitrite in dilute hydrochloric acid

23.23 Identify the principal organic products of each of the following reactions:

(a) Cyclohexanone + cyclohexylamine $\xrightarrow{H_2, Ni}$

(b) 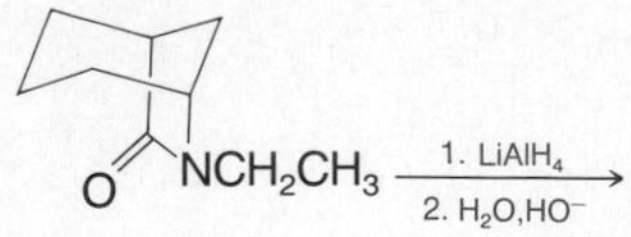 $\xrightarrow[\text{2. H}_2\text{O, HO}^-]{\text{1. LiAlH}_4}$

(c) $C_6H_5CH_2CH_2CH_2OH$ $\xrightarrow[\text{2. (CH}_3)_2\text{NH (excess)}]{\substack{\text{1. } p\text{-toluenesulfonyl chloride,} \\ \text{pyridine}}}$

(d) Product of (c) $\xrightarrow[\text{2. 160°C}]{\text{1. CH}_3\text{CO}_2\text{OH}}$

(e) 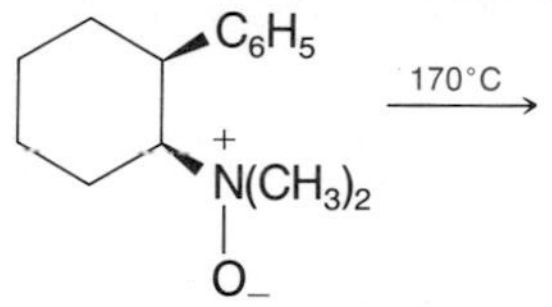 $\xrightarrow{170°\text{C}}$

(f) $(CH_3)_2CHNH_2 +$ $\longrightarrow$

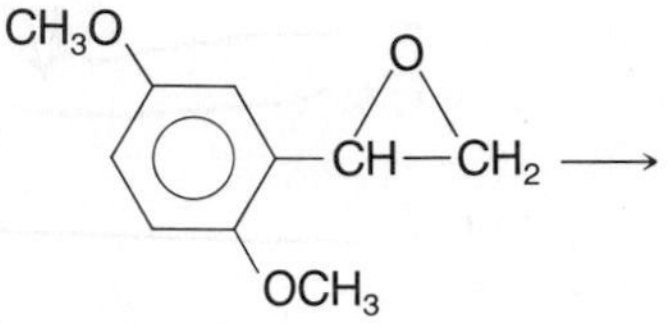

(g) $(C_6H_5CH_2)_2NH + CH_3\overset{\text{O}}{\overset{\|}{C}}CH_2Cl$ $\xrightarrow[\text{THF}]{\text{triethylamine}}$

(h) 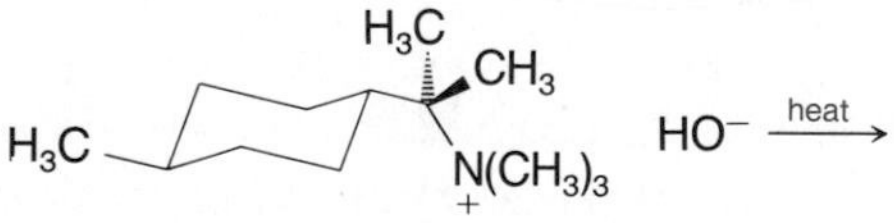 $HO^- \xrightarrow{\text{heat}}$

(i) $(CH_3)_2CHNHCH(CH_3)_2$ $\xrightarrow[\text{HCl, H}_2\text{O}]{\text{NaNO}_2}$

23.24 Provide a reasonable explanation for each of the following observations:

(a) 4-Methylpiperidine has a higher boiling point than *N*-methylpiperidine.

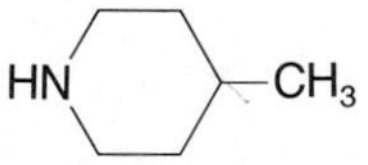

4-Methylpiperidine
(bp 129°C)

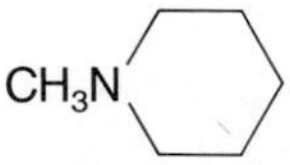

N-Methylpiperidine
(bp 106°C)

(b) Two isomeric quaternary ammonium salts are formed in comparable amounts when 4-*tert*-butyl-*N*-methylpiperidine is treated with benzyl chloride.

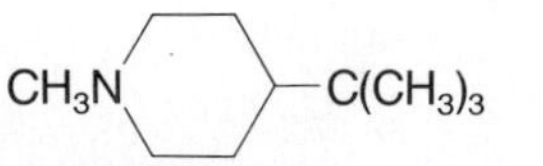

4-*tert*-Butyl-*N*-methylpiperidine

(c) Aziridines, compounds that contain nitrogen in a three-membered ring, undergo pyramidal inversion much more slowly than do other alkylamines.

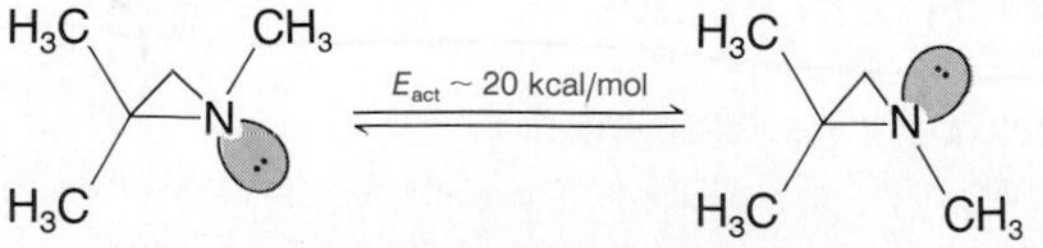

(d) The rate constant for formation of tetraethylammonium iodide from triethylamine and ethyl iodide is over 800 times greater (at 100°C) when the solvent is acetone than when it is hexane. The respective dielectric constants are 21 for acetone and 1.9 for hexane.

(e) When tetramethylammonium hydroxide is heated at 130°C, trimethylamine and methanol are formed.

(f) The major product formed on treatment of 1-propanamine with sodium nitrite in dilute hydrochloric acid is 2-propanol.

(g) Addition of perchloric acid to compound A gives a stable salt $(C_8H_{16}NO)^+ClO_4^-$. This salt gives no evidence of a carbonyl group in its infrared spectrum. (A molecular model of compound A will prove helpful in solving this problem.)

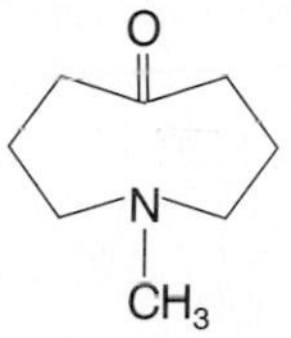

Compound A

23.25 Give the structures, including stereochemistry, of compounds B through D.

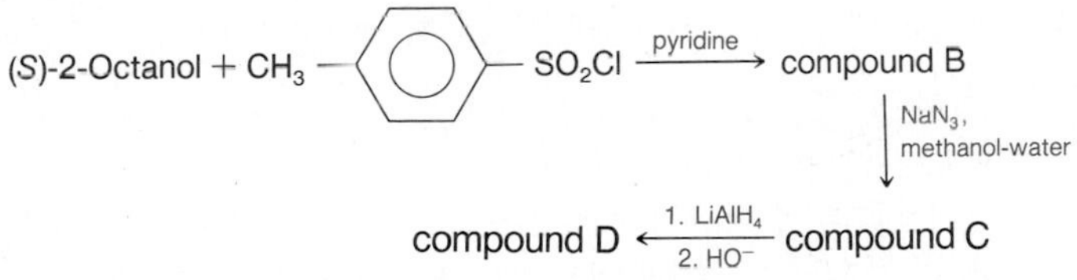

23.26 The bicyclic amine quinuclidine has been prepared from compound E by treatment with sodium hydroxide. Quinuclidine has also been prepared from compound F. Devise a procedure (more than one step is required) for the conversion of compound F to quinuclidine.

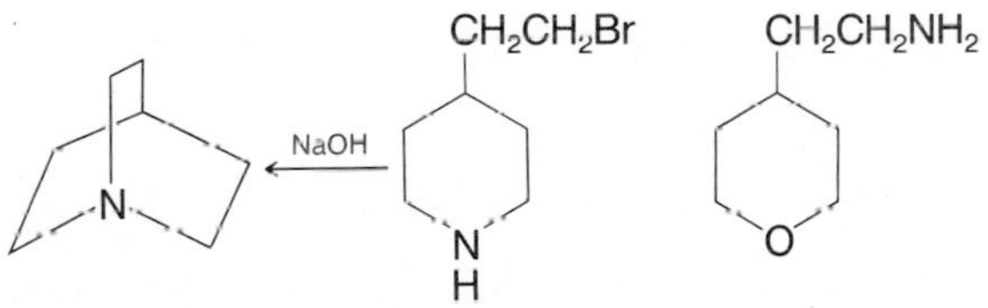

23.27 Two isomeric alcohols each having the molecular formula $C_6H_{14}O$ were obtained in comparable amounts when 3,3-dimethyl-1-butanamine was treated with sodium nitrite in dilute perchloric acid. Neither isomer is a tertiary alcohol. Identify these two alcohols.

23.28 Suggest a reasonable mechanism for the reaction:

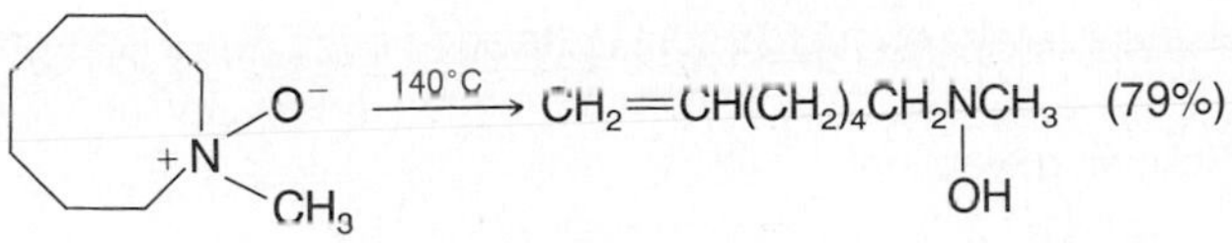

23.29 Under conditions in which ethyl aminoacetate is nitrosated, its diazonium salt is converted to the neutral compound ethyl diazoacetate, which can be isolated by extraction into dichloromethane.

$$H_2NCH_2CO_2CH_2CH_3 \xrightarrow[H_3O^+]{NaNO_2} \ :N\!\!\equiv\!\!\overset{+}{N}CH_2CO_2CH_2CH_3 \longrightarrow \ :\overset{..}{N}\!\!=\!\!\overset{+}{N}\!\!=\!\!CHCO_2CH_2CH_3 + H^+$$

Ethyl aminoacetate   Diazonium ion   Ethyl diazoacetate

Simple alkanamines do not undergo an analogous reaction. What are the structural features of the diazonium ion of ethyl aminoacetate that permit its conversion to ethyl diazoacetate under these conditions?

23.30 Devise efficient syntheses of each of the following compounds from the designated starting materials. You may also use any necessary organic or inorganic reagents.

(a) 3,3-Dimethyl-1-butanamine from neopentyl bromide

(b) $CH_2\!=\!CH(CH_2)_8CH_2\!-\!N$⟨pyrrolidine⟩ from 10-undecenoic acid and pyrrolidine

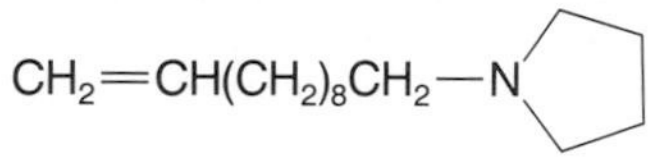

(c) [cyclopentane with OC₆H₅ and NH₂ substituents] from [cyclopentane with OC₆H₅ and OH substituents]

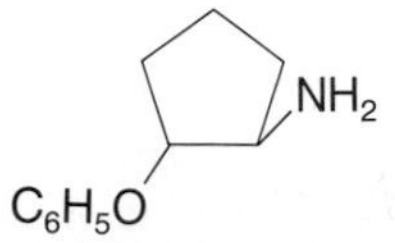

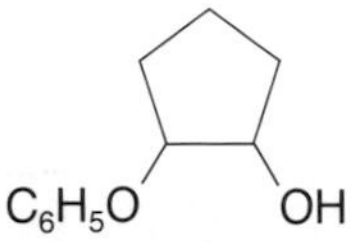

(d) $C_6H_5CH_2NCH_2CH_2CH_2CH_2NH_2$ from $C_6H_5CH_2NHCH_3$ and $BrCH_2CH_2CH_2CN$
 $\quad\quad\quad\quad |$
 $\quad\quad\quad CH_3$

(e) $(CH_3CH_2CH_2CH_2CH_2)_2NOH$ from $(CH_3CH_2CH_2CH_2CH_2)_3N$

(f) $NC\!-\!\langle\bigcirc\rangle\!-\!CH_2N(CH_3)_2$ from $NC\!-\!\langle\bigcirc\rangle\!-\!CH_3$

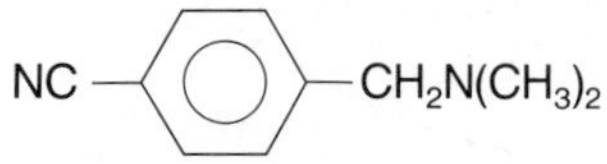

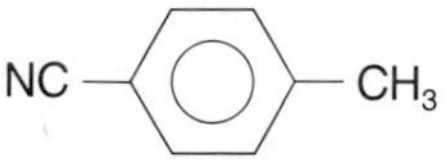

23.31 A synthesis of 3-phenyl-1-butene has been described in the chemical literature and is outlined below. Specify the reagents required for each step in this synthesis.

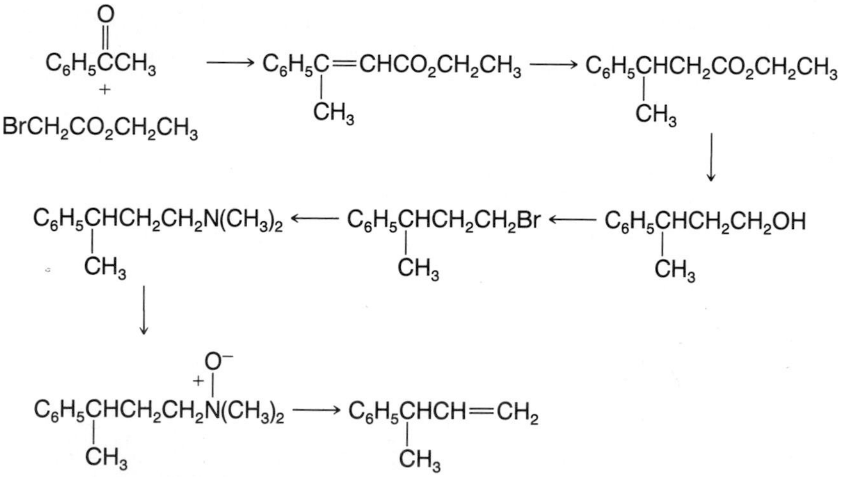

$$C_6H_5\overset{O}{\overset{||}{C}}CH_3 \;+\; BrCH_2CO_2CH_2CH_3 \longrightarrow C_6H_5\underset{CH_3}{C}\!=\!CHCO_2CH_2CH_3 \longrightarrow C_6H_5\underset{CH_3}{CH}CH_2CO_2CH_2CH_3$$

$$C_6H_5\underset{CH_3}{CH}CH_2CH_2N(CH_3)_2 \longleftarrow C_6H_5\underset{CH_3}{CH}CH_2CH_2Br \longleftarrow C_6H_5\underset{CH_3}{CH}CH_2CH_2OH$$

$$C_6H_5\underset{CH_3}{CH}CH_2CH_2\overset{O^-}{\overset{+}{N}}(CH_3)_2 \longrightarrow C_6H_5\underset{CH_3}{CH}CH\!=\!CH_2$$

23.32 Ammonia and amines undergo conjugate addition to α,β-unsaturated carbonyl compounds (Section 19.14). On the basis of this information, predict the principal organic product of each of the following reactions.

(a) $(CH_3)_2C\!=\!CH\overset{O}{\overset{||}{C}}CH_3 + NH_3 \longrightarrow$

(b) [cyclohexenone] $=O + HN$⟨piperidine⟩ $\longrightarrow$

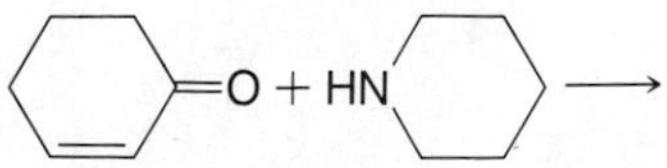

(c) $C_6H_5\overset{O}{\overset{||}{C}}CH\!=\!CH\overset{O}{\overset{||}{C}}C_6H_5 + (CH_3CH_2)_2NH \longrightarrow$

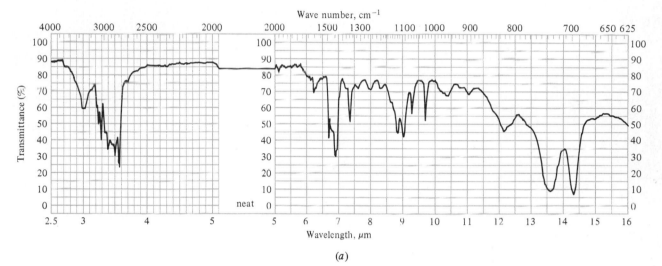

(d) $C_6H_5\overset{\displaystyle O}{\overset{\displaystyle \|}{C}}CH=CHC_6H_5 + HN\underset{}{\overset{}{\bigcirc}}O \longrightarrow$

(e) [structure with cyclohexenone ring bearing $(CH_2)_3CH(CH_2)_4CH_3$ and NH_2 substituents] $\xrightarrow{\text{spontaneous}}$

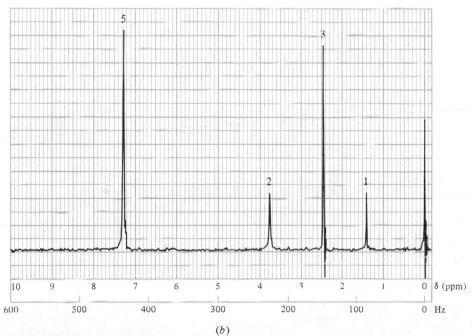

FIGURE 23.5 Infrared (a) and ^{1}H nmr (b) spectra of compound H (Problem 23.34).

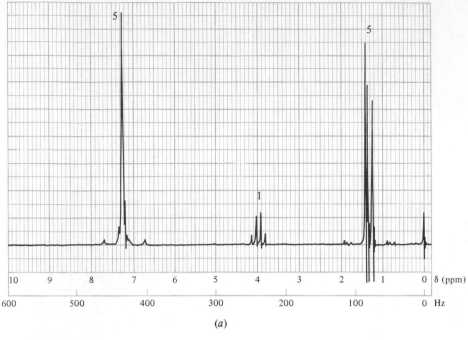

(a)

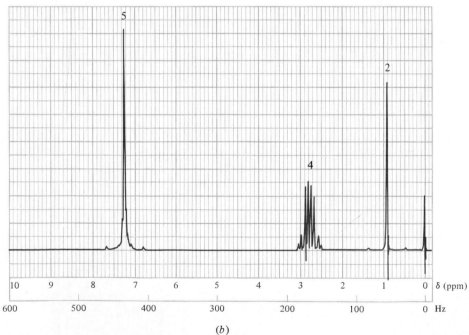

(b)

FIGURE 23.6 The ¹H nmr spectra of compound I (top) and compound J (bottom) (Problem 23.35).

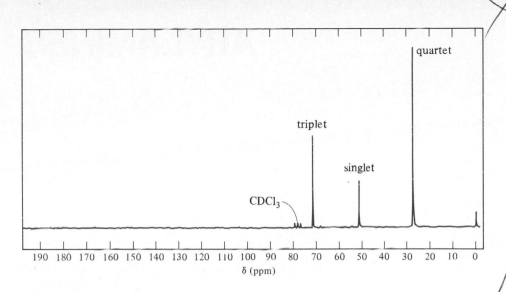

FIGURE 23.7 The ^{13}C nmr spectrum of the compound described in Problem 23.36. *(Taken from "Carbon-13 NMR Spectra: A Collection of Assigned, Coded, and Indexed Spectra," by LeRoy F. Johnson and William C. Jankowski, Wiley-Interscience, New York, 1972. Reprinted by permission of John Wiley & Sons, Inc.)*

23.33 A number of compounds of the type represented by G were prepared for evaluation as potential analgesic drugs. Their preparation is described in a retrosynthetic format as shown.

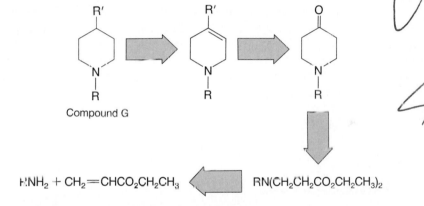

Based on this retrosynthetic analysis, design a synthesis of *N*-methyl-4-phenylpiperidine (compound G, where R = CH$_3$, R' = C$_6$H$_5$). Present your answer as a series of equations, showing all necessary reagents and isolated intermediates.

23.34 Identify compound H (C$_8$H$_{11}$N) on the basis of its infrared and ^{1}H nmr spectra (Figure 23.5).

23.35 Compounds I and J are isomers of compound H. Their ^{1}H nmr spectra are presented in Figure 23.6. Identify these two compounds.

23.36 Does the ^{13}C nmr spectrum shown in Figure 23.7 correspond to that of 1-amino-2-methyl-2-propanol or to 2-amino-2-methyl-1-propanol? Could this compound be prepared by reaction of an epoxide with ammonia?

ARYLAMINES

*A**rylamines,* amines in which nitrogen is bonded to a benzene or benzenoid ring, are *difunctional compounds.* The reactivity of the amine group is affected by its aryl substituent and the reactivity of the aromatic ring is affected by its amine substituent. In many ways arylamines resemble alkylamines and this chapter can be considered to be a continuation of the preceding one. The presence of an aryl substituent on nitrogen, however, gives rise to some unusual properties, and by being aware of how these two structural units interact you will gain a better understanding of both amines and aromatic compounds.

24.1 NOMENCLATURE

Aniline (rather than phenylamine) is the parent IUPAC name for amino-substituted derivatives of benzene. The locations of substituents are specified relative to the amine group as the point of reference.

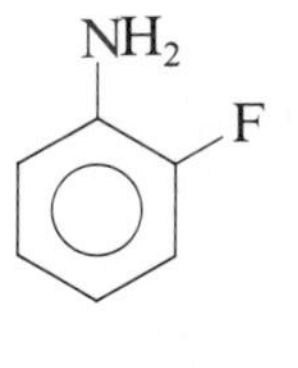

o-Fluoroaniline 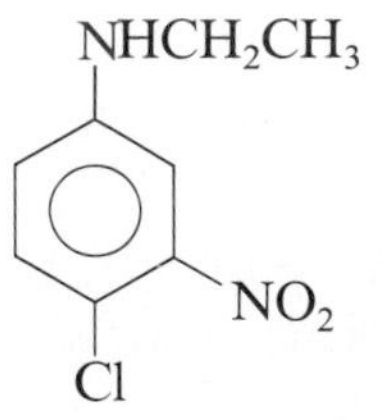   4-Chloro-*N*-ethyl-3-nitroaniline 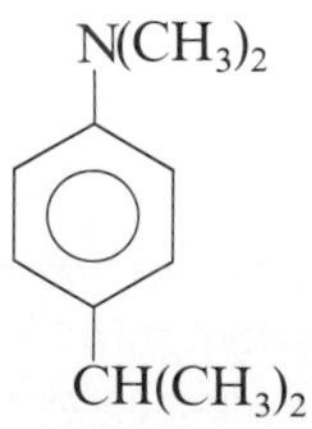   *p*-Isopropyl-*N,N*-dimethylaniline

Hydroxyl groups and oxygen-bearing carbon substituents outrank amine functions in determining the base name. In these cases the amino group is named as a substituent.

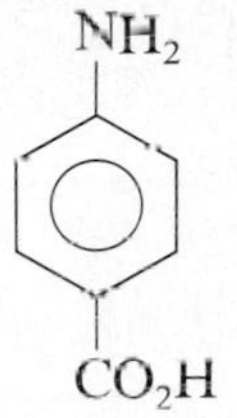

p-Aminobenzoic acid
(p-aminobenzenecarboxylic acid)

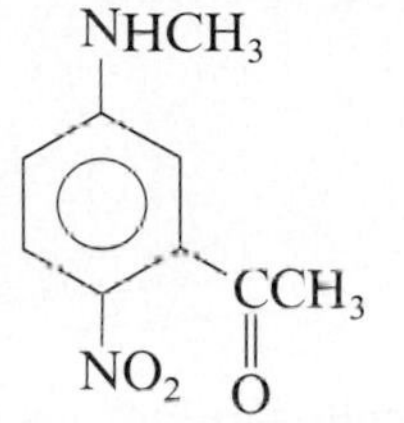

5-(N-Methylamino)-2-nitroacetophenone

PROBLEM 24.1 Give an acceptable IUPAC name for each of the following amines.

(a)

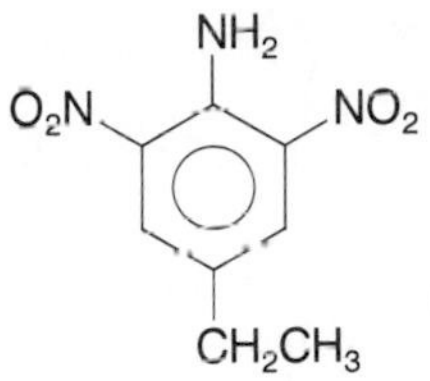

(b)

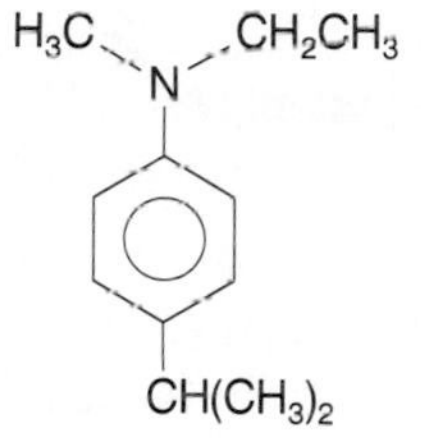

(c)

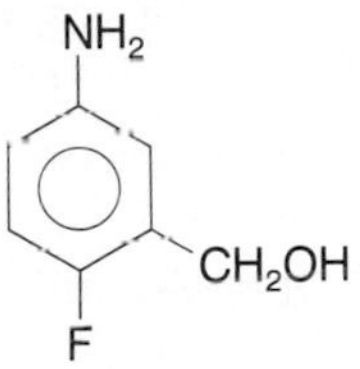

SAMPLE SOLUTION (a) Naming this compound as a derivative of aniline, there are, in alphabetical order, an ethyl substituent at C-4 and nitro groups at C-2 and C-6. The compound is 4-ethyl-2,6-dinitroaniline.

Certain derivatives of aniline are known by common names that have been incorporated into the IUPAC system.

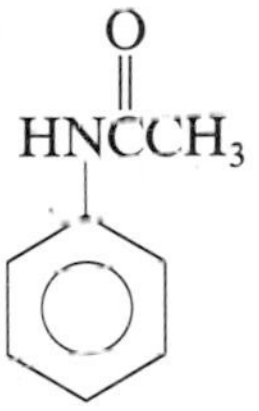

Acetanilide

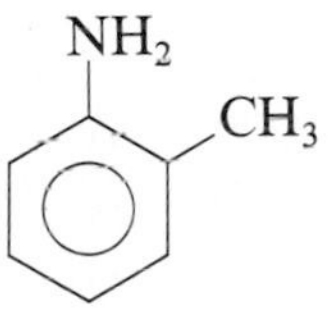

o-Toluidine
(similarly, m- and
p-toluidine)

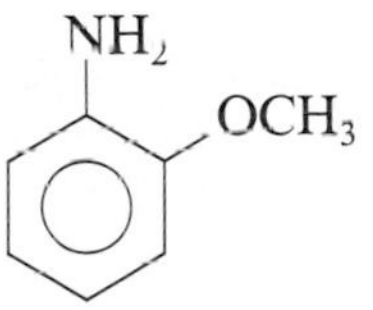

o-Anisidine
(similarly, m- and
p-anisidine)

Arylamines are named as "arenamines" in the *Chemical Abstracts* nomenclature system. The final -*e* of the corresponding arene is replaced by -*amine*. Thus, *benzenamine* is the *Chemical Abstracts* name for aniline.

24.2 STRUCTURE AND BONDING

As has been noted on earlier occasions, bonds to sp^2 hybridized carbon are shorter than those to sp^3 hybridized carbon. This property is evident when comparing the structures of arylamines and alkylamines. The carbon-nitrogen bond of aniline (1.402 Å) is shorter than that of methylamine (1.474 Å).

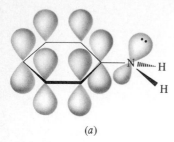

(a)

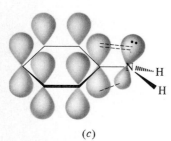

(b)

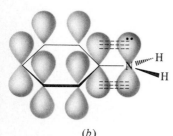

(c)

FIGURE 24.1 Orbital hybridization and lone pair bonding in aniline. (a) The lone pair is strongly bound by nitrogen when it occupies an sp^3 hybridized orbital. (b) The lone pair is delocalized best into the ring when it occupies a p orbital that is aligned for overlap with the aromatic π system. (c) The actual structure combines features of both (a) and (b); nitrogen adopts an orbital hybridization that is between sp^3 and sp^2.

1.474 Å

$CH_3—NH_2$

Methylamine

1.402 Å

NH_2

Aniline

Aniline, like methylamine, has a pyramidal arrangement of bonds around nitrogen, but its pyramid is somewhat shallower. One measure of the extent of this flattening is given by the angle between the carbon-nitrogen bond and the bisector of the H—N—H angle in the H—N—H plane.

CH_3 —N— H, ~125°, H

Methylamine

N— H, 142.5°, H

Aniline

H C N H, O 180° H

Formamide

For sp^3 hybridized nitrogen, this angle (not the same as the C—N—H bond angle!) is 125° and the measured angles in simple alkylamines are close to that. The corresponding angle for sp^2 hybridization at nitrogen with a planar arrangement of bonds, as in amides, for example, is 180°. The measured value for this angle in aniline is 142.5°, indicative of a hybridization somewhat closer to sp^3 than to sp^2.

The structure of aniline reflects a compromise between two modes of binding the nitrogen lone pair. These electrons are bound both by the attractive force exerted by the nitrogen nucleus and by their delocalization into the aromatic ring (Figure 24.1). The electrons are more strongly attracted to nitrogen when they are in an orbital with some s character—an sp^3 hybridized orbital, for example—than when they are in a p orbital. Increasing the amount of s character in an orbital brings the electrons closer to the nucleus. On the other hand, delocalization of these electrons into the aromatic π system is better achieved if they occupy a p orbital. A p orbital of nitrogen is better aligned for overlap with the p orbitals of the benzene ring to form an extended π system than is an sp^3 hybridized orbital. As a result of these two opposing forces, nitrogen adopts an orbital hybridization that is between sp^3 and sp^2. The orbital that contains the unshared electron pair has more p character than an sp^3 hybridized orbital, which permits the electrons to be partially delocalized into the aromatic π system while yet retaining a strong interaction with the nitrogen nucleus.

Delocalization of the nitrogen lone pair electrons into the aromatic π system strengthens the carbon-nitrogen bond of aniline, shortens it, and gives it "partial double bond character." In resonance terms, this is expressed as indicating significant contributions from dipolar resonance forms.

:NH$_2$ — H, H, H, H

Most stable Lewis structure for aniline

$^+$NH$_2$ — H, H, H, H

$^+$NH$_2$ — H, H, H, H

$^+$NH$_2$ — H, H, H, H

Dipolar resonance forms of aniline

PROBLEM 24.2 The carbon-nitrogen bond distance in *p*-aminoacetophenone (1.376 Å) is shorter than that in aniline (1.402 Å). Offer a resonance explanation for this observation.

The orbital and resonance models for bonding in arylamines are simply alternative ways of describing the same phenomenon. Delocalization of the nitrogen lone pair decreases the electron density at nitrogen while increasing it in the π system of the aromatic ring. We have already seen one chemical consequence of this in the high level of reactivity of aniline in electrophilic aromatic substitution reactions (Section 13.12). Other ways in which electron delocalization affects the physical and chemical properties of arylamines are described in the following sections of this chapter.

24.3 PHYSICAL PROPERTIES

Arylamines resemble alkylamines in their physical properties. Hydrogen bonding affects their melting points, boiling points, and solubility in water in the same way that it affects alkylamines; Table 24.1 lists these properties for some representative arylamines.

As seen in Table 24.1, *o*-nitroaniline has a lower melting point and a lower boiling

TABLE 24.1

Physical Properties of Some Arylamines

Compound name	Melting point, °C	Boiling point, °C
Primary amines		
Aniline	−6.3	184
o-Toluidine	−14.7	200
m-Toluidine	−30.4	203
p-Toluidine	44	200
o-Chloroaniline	−14	209
m-Chloroaniline	−10	230
p-Chloroaniline	72.5	232
o-Nitroaniline	71.5	284
m-Nitroaniline	114	306
p-Nitroaniline	148	332
Secondary amines		
N-Methylaniline	−57	196
N-Ethylaniline	−63	205
N-Methyl-4-nitroaniline	68	
Diphenylamine	54	302
Tertiary amines		
N,*N*-Dimethylaniline	2.4	194
N-Phenylpiperidine		257
Triphenylamine	127	365

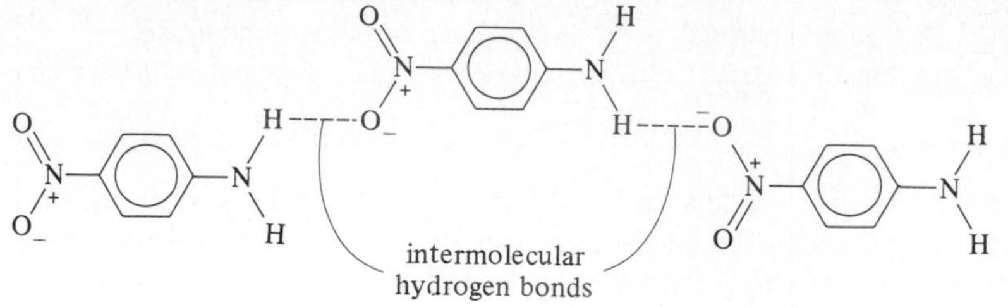

FIGURE 24.2 Intermolecular hydrogen bonding in *p*-nitroaniline.

intermolecular hydrogen bonds

point than either its meta or para isomer. The principal reason for this difference is the intramolecular hydrogen bond that exists in *o*-nitroaniline.

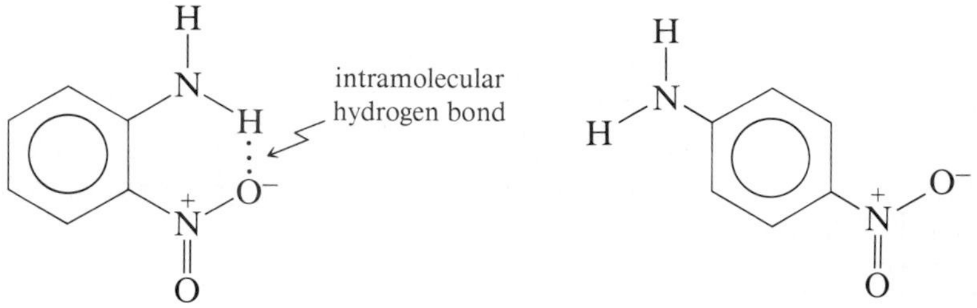

o-Nitroaniline
(individual molecules stabilized by intramolecular hydrogen bond: mp 71.5°C, bp 284°C)

p-Nitroaniline
(no intramolecular hydrogen bond possible: mp 148°C, bp 332°C)

The crystalline and liquid states of all the nitroaniline isomers are characterized by a network of hydrogen bonds between molecules, as shown for *p*-nitroaniline in Figure 24.2. Some of these hydrogen bonds are disrupted when the compounds melt; all of them are broken when the molecule passes into the gas phase. These phase changes require less energy for *o*-nitroaniline than for its meta and para isomers because intramolecular hydrogen bonds between its adjacent amino and nitro groups replace the intermolecular hydrogen bonds that are broken. Only *o*-nitroaniline has its amino and nitro groups close enough to each other to permit an internal hydrogen bond to form.

Similar behavior is observed for other aniline derivatives bearing substituents that can engage in hydrogen bonding to amino groups.

PROBLEM 24.3 The boiling point of *o*-aminoacetophenone (250°C) is substantially lower than that of its para isomer (293°C). Explain.

The dipole moments of arylamines are similar to those of alkylamines, but the directions of their moments are reversed. Nitrogen is the negative end of the dipole in alkylamines but is polarized positively in arylamines.

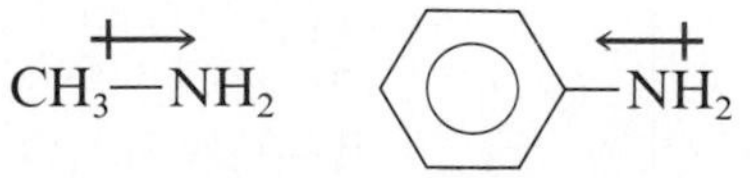

Methylamine
(μ 1.3 D)

Aniline
(μ 1.3 D)

The direction of polarization is not revealed directly by dipole moment measurements but can be deduced by examining the effects of substituents. The dipole moment of *p*-(trifluoromethyl)aniline, for example, is approximately equal to the sum of the separate dipole moments of aniline and (trifluoromethyl)benzene.

Aniline	(Trifluoromethyl)benzene	*p*-(Trifluoromethyl)aniline
μ 1.3 D	μ 2.9 D	μ 4.3 D

The separate effects of the amino group and the trifluoromethyl group must reinforce, rather than oppose, each other. Since the trifluoromethyl group attracts electrons, the amino group must release them.

PROBLEM 24.4   Which would you expect to have the greater dipole moment, *p*-dinitrobenzene or *p*-nitroaniline? Why?

Because of its electronegativity nitrogen tends to withdraw electrons from carbon by polarization of the electron distribution in σ bonds. Because nitrogen has an unshared pair of electrons, it can donate them to adjacent π systems. Dipole moment data reveal that the π donor effect of an amino substituent on an aromatic ring substantially exceeds its electron-withdrawing effect on σ bonds.

24.4 BASICITY OF ARYLAMINES

Aromatic amines are several orders of magnitude less basic than alkylamines; while K_b for most alkylamines is on the order of 10^{-5} (pK_b 5), arylamines have K_b's in the 10^{-10} range. The sharply decreased basicity of arylamines arises because the stabilizing effect of lone pair electron delocalization is sacrificed on protonation.

Amine is stabilized by delocalization of lone pair into π system of ring, decreasing the electron density at nitrogen	Lone pair electrons transformed to N-H bonded pair

The aromatic ring does very little to disperse the positive charge in the ammonium ion. Indeed, since the ring carbon attached to nitrogen is sp^2 hybridized, it is electron withdrawing and destabilizes the ammonium ion. Stabilization of the amine and destabilization of the ammonium ion combine to make the equilibrium constant for amine protonation smaller for arylamines than for alkylamines. This relationship is depicted in Figure 24.3, where the free energies of protonation of cyclohexylamine

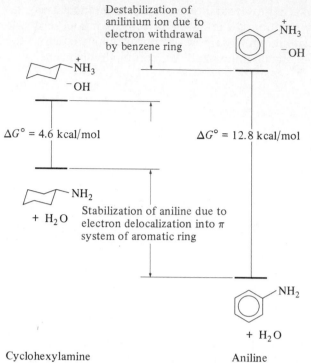

Destabilization of anilinium ion due to electron withdrawal by benzene ring

$\Delta G° = 4.6$ kcal/mol

$\Delta G° = 12.8$ kcal/mol

Stabilization of aniline due to electron delocalization into π system of aromatic ring

Cyclohexylamine

Aniline

FIGURE 24.3 Free energy changes accompanying protonation of aniline and cyclohexylamine by water.

and aniline are compared. As measured by their respective K_b's, cyclohexylamine is almost 1 million times more basic than aniline.

Aniline   Water   Anilinium ion   Hydroxide ion   (K_b 3.8×10^{-10}; pK_b 9.4)

Cyclohexylamine   Water   Cyclohexylammonium ion   Hydroxide ion   (K_b 4.4×10^{-4}; pK_b 3.4)

When the proton donor is a strong acid, arylamines can be completely protonated. Aniline is extracted from an ether solution into 1 N hydrochloric acid because it is converted to a water-soluble anilinium ion salt under these conditions.

PROBLEM 24.5 The two amines shown differ by a factor of 40,000 in their K_b values. Which is the stronger base? Why?

Tetrahydroquinoline   Tetrahydroisoquinoline

Conjugation of the amino group with a second aromatic ring, then a third, reduces its basicity even further. Diphenylamine is 6300 times less basic than aniline, while triphenylamine is scarcely a base at all, being estimated as 10^8 times less basic than aniline and 10^{14} times less basic than ammonia.

$C_6H_5NH_2$ $(C_6H_5)_2NH$ $(C_6H_5)_3N$

Aniline   Diphenylamine   Triphenylamine

(K_b 3.8 $\times$ 10^{-10};   (K_b 6 $\times$ 10^{-14};   ($K_b \sim$ 10^{-19};

pK_b 9.4)   pK_b 13.2)   p$K_b \sim$ 19)

The effects of some representative aryl substituents on the basicity of arylamines are summarized in Table 24.2. In general, electron-donating groups increase the basicity of aniline slightly while electron-withdrawing groups decrease it, in some cases dramatically. Thus, the basicity constant K_b of p-toluidine is 5 to 6 times greater than that of aniline, but aniline is 220 times more basic than its p-trifluoromethyl derivative.

NH$_2$ NH$_2$ NH$_2$

CH$_3$ CF$_3$

p-Toluidine   Aniline   p-(Trifluoromethyl)aniline

(K_b 2 $\times$ 10^{-9};   (K_b 3.8 $\times$ 10^{-10};   (K_b 1.7 $\times$ 10^{-12};

pK_b 8.7)   pK_b 9.4)   pK_b 11.5)

Most basic: methyl group   Least basic: trifluoromethyl

donates electrons   group withdraws electrons

TABLE 24.2

Basicities of Some Arylamines

Substituent in H$_2$NC$_6$H$_4$X	Position of substituent X and K_b (pK_b)*		
	Ortho	Meta	Para
Standard of comparison is aniline			
H	3.8 $\times$ 10^{-10} (9.4)	3.8 $\times$ 10^{-10} (9.4)	3.8 $\times$ 10^{-10} (9.4)
Electron-releasing substituents increase basicity slightly			
OCH$_3$	3.8 $\times$ 10^{-10} (9.4)	1.6 $\times$ 10^{-10} (9.8)	2.2 $\times$ 10^{-9} (8.7)
CH$_3$	3.3 $\times$ 10^{-10} (9.5)	5.5 $\times$ 10^{-10} (9.3)	2 $\times$ 10^{-9} (8.7)
Electron-withdrawing substituents decrease basicity			
Cl	4.5 $\times$ 10^{-12} (11.3)	3.8 $\times$ 10^{-11} (10.4)	7.2 $\times$ 10^{-11} (10.2)
$\overset{O}{\overset{\|}{C}}CH_3$	2.5 $\times$ 10^{-12} (11.6)	4 $\times$ 10^{-11} (10.4)	5 $\times$ 10^{-12} (11.3)
CN	8.9 $\times$ 10^{-14} (13.1)	5.6 $\times$ 10^{-12} (11.2)	5.5 $\times$ 10^{-13} (12.3)
NO$_2$	5.5 $\times$ 10^{-15} (14.3)	2.9 $\times$ 10^{-12} (11.5)	1 $\times$ 10^{-13} (13.0)

* In water at 25°C.

The methyl group in *p*-toluidine donates electrons to the ring and helps disperse the positive charge in the derived anilinium ion. A trifluoromethyl substituent enhances the tendency of the aryl group to attract electrons from nitrogen, thereby lowering its basicity.

Electron-withdrawing groups that are conjugated to the amine nitrogen have a very large base-weakening effect; K_b for *p*-nitroaniline, for example, is 3800 times smaller than K_b for aniline. A resonance interaction of the type shown leads to extensive delocalization of the unshared electron pair of the amine group and stabilizes the free arylamine.

Electron delocalization in *p*-nitroaniline

When the amine group is protonated, this type of resonance stabilization is no longer possible because the unshared electron pair of the amine nitrogen has been converted to a bonded pair of an ammonium ion. More resonance stabilization is lost when *p*-nitroaniline is protonated than when aniline is, which makes K_b smaller for *p*-nitroaniline.

PROBLEM 24.6   Each of the following is a much weaker base than aniline. Present a resonance argument to explain the effect of the substituent in each case.

(*a*) *o*-Cyanoaniline
(*b*) *p*-Cyanoaniline

(*c*) $C_6H_5NHCCH_3$ (with O double-bonded to C)

(*d*) *p*-Aminoacetophenone

SAMPLE SOLUTION   (*a*)   A cyano substituent is strongly electron-withdrawing. When present at a position ortho to an amino group on an aromatic ring, a cyano substituent increases the delocalization of the amine lone pair electrons by a direct resonance interaction.

This resonance stabilization is lost when the amine group becomes protonated, and *o*-cyanoaniline is therefore a weaker base than aniline.

Multiple substitution by strongly electron-withdrawing groups diminishes the basicity of arylamines still more. As just noted, aniline is 3800 times as strong a base as *p*-nitroaniline; however, it is 10^9 times more basic than 2,4-dinitroaniline. A practical consequence of this is that arylamines that bear two or more strongly electron-withdrawing groups are often not capable of being extracted from ether solution into dilute aqueous acid.

24.5 PREPARATION OF ARYLAMINES

Aniline was first isolated in 1826 as a degradation product of indigo, a dark blue dye obtained from the West Indian plant *Indigofera anil*. Subsequently, a more direct synthesis from benzene was developed, which is still used today for both industrial and laboratory-scale preparations.

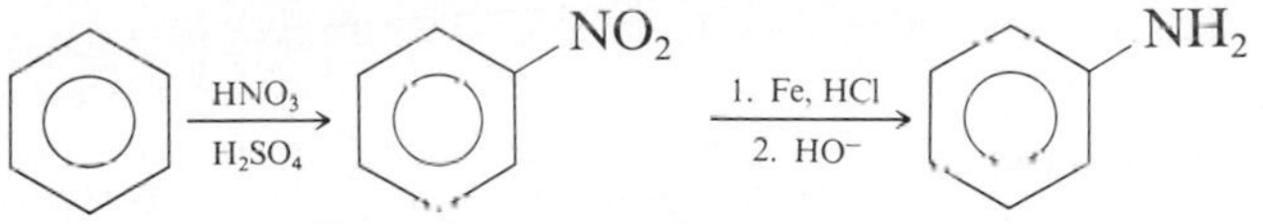

Benzene   Nitrobenzene (95%)   Aniline (97%)

The sequence $ArH \rightarrow ArNO_2 \rightarrow ArNH_2$ is the most general method for the preparation of arylamines. A nitric acid–sulfuric acid mixture is normally employed in the nitration step (Section 13.3). A variety of reducing agents is available to efficiently convert nitroarenes to arylamines. Catalytic hydrogenation over platinum, palladium, or nickel is often used, as is chemical reduction using either iron or tin in hydrochloric acid.

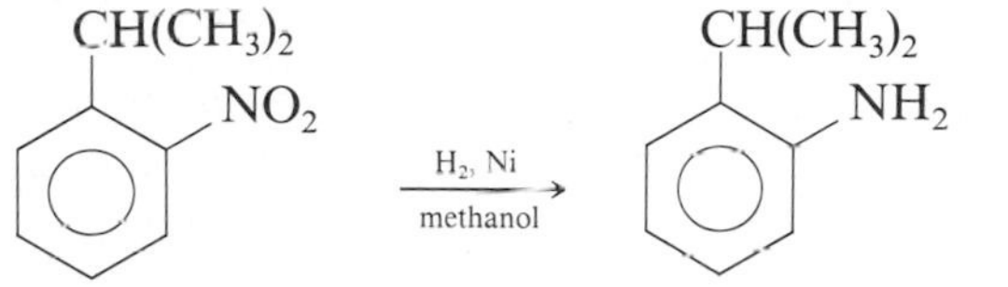

o-Isopropylnitrobenzene   *o*-Isopropylaniline (92%)

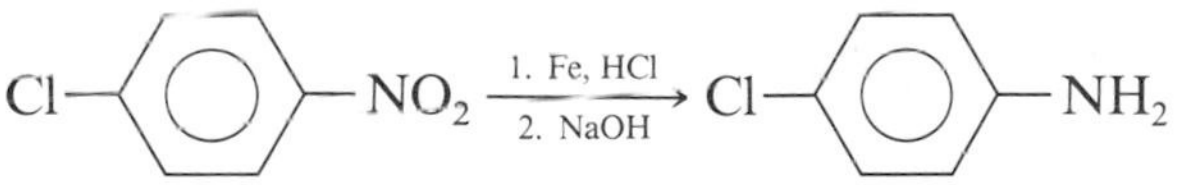

p-Chloronitrobenzene   *p*-Chloroaniline (95%)

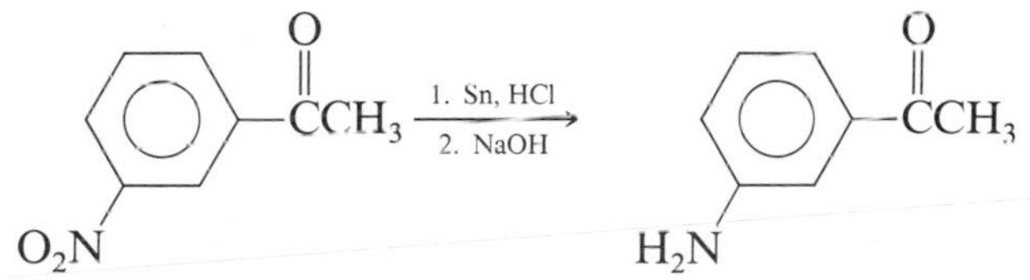

m-Nitroacetophenone   *m*-Aminoacetophenone (82%)

PROBLEM 24.7   Outline syntheses of each of the following arylamines from benzene:

(a) *o*-Isopropylaniline
(b) *p*-Isopropylaniline
(c) 4-Isopropyl-1,3-benzenediamine
(d) *p*-Chloroaniline
(e) *m*-Aminoacetophenone

SAMPLE SOLUTION   (a) The last step in the synthesis of *o*-isopropylaniline, the reduction of the corresponding nitro compound by catalytic hydrogenation, is given as one of the examples shown above. The necessary nitroarene is obtained by fractional distillation of the ortho-para mixture formed during nitration of isopropylbenzene.

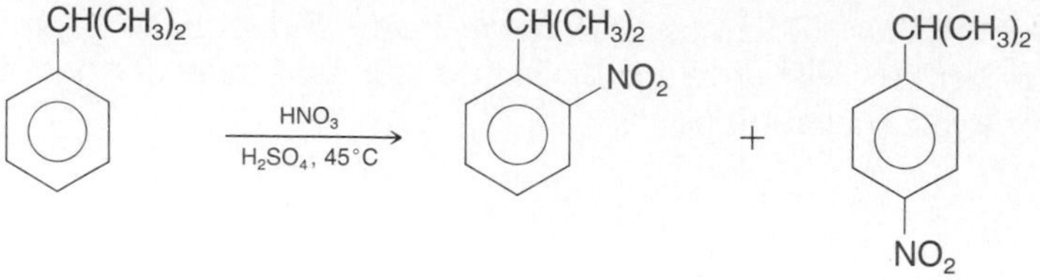

Isopropylbenzene → (HNO₃, H₂SO₄, 45°C) → o-Isopropylnitrobenzene (bp 110°C) + p-Isopropylnitrobenzene (bp 131°C)

As actually performed, a 62 percent yield of a mixture of ortho and para nitration products has been obtained with an ortho-para ratio of about 1:3.

Isopropylbenzene is prepared by the Friedel-Crafts alkylation of benzene using isopropyl chloride and aluminum chloride or propene in the presence of an acid catalyst (Section 13.6).

The reliability of each step has made the nitration-reduction sequence one of the mainstays in the organic chemist's inventory of synthetic procedures for over a century.

As noted in Section 21.17, the Hofmann rearrangement of substituted benzamides has been applied to the synthesis of arylamines. In the example cited there, *m*-bromobenzamide was converted to *m*-bromoaniline in 87 percent yield on treatment with bromine in aqueous base.

N-Alkyl derivatives are prepared from *N*-arylamines by reactions analogous to the preparation of secondary and tertiary alkylamines. Lithium aluminum hydride reduction of *N*-acyl derivatives of arylamines has been employed.

$$C_6H_5NHCCH_3 \xrightarrow[\text{2. H}_2\text{O,HO}^-]{\text{1. LiAlH}_4\text{,diethyl ether}} C_6H_5NHCH_2CH_3$$

Acetanilide   *N*-Ethylaniline (92%)

Reductive amination reactions have also been adapted to the preparation of *N*-alkylarylamines.

$$C_6H_5NH_2 + (CH_3)_2CHCH \xrightarrow[\text{methanol}]{\text{NaBH}_3\text{CN}} C_6H_5NHCH_2CH(CH_3)_2$$

Aniline   2-Methylpropanal   *N*-Isobutylaniline (78%)

24.6 REACTIONS OF ARYLAMINES. A REVIEW AND A PREVIEW

As noted in the introduction to this chapter, arylamines are difunctional compounds; they undergo reactions characteristic both of amine functions and of benzenoid rings. A selection of reactions of the amino group of arylamines that are analogous to those of alkylamines is presented in Table 24.3. In general, arylamines are less basic

TABLE 24.3

Summary of Amine Reactions Encountered in Earlier Chapters Applied to Arylamines

Reaction (section) and comments	General equation and specific example

Basicity (Section 24.4) Arylamines are weak bases with K_b's of $\sim 10^{-10}$. Their K_b values are approximately 10^5 times smaller than those of alkylamines.

$$ArNR_2 + H_2O \rightleftharpoons Ar\overset{H}{\underset{+}{N}}R_2 + HO^-$$

Arylamine   Water   Arylammonium ion   Hydroxide ion

N-Methylaniline   Water   N-Methylanilinium ion   Hydroxide ion

$$C_6H_5\text{NHCH}_3 + H_2O \rightleftharpoons C_6H_5\overset{H}{\underset{H}{\overset{+}{N}}}CH_3 + HO^-$$

Acylation (Section 21.13) The nitrogen atom of arylamines is readily acylated by acyl chlorides and anhydrides.

$$ArNHR + R'\overset{O}{\overset{\|}{C}}Cl \xrightarrow{HO^-} Ar\overset{}{\underset{R}{N}}\overset{O}{\overset{\|}{C}}R'$$

Arylamine   Acyl chloride   Amide

2-Fluoro-4-methylaniline   Acetic anhydride   2-Fluoro-4-methylacetanilide (100%)

$$+ CH_3\overset{O}{\overset{\|}{C}}O\overset{O}{\overset{\|}{C}}CH_3 \longrightarrow$$

Alkylation (Section 23.13) Arylamines are weakly nucleophilic and react with primary alkyl halides to give N-alkyl derivatives. By using excess arylamine, monoalkylation has been achieved.

$$2ArNHR + R'CH_2X \longrightarrow Ar\overset{R}{\underset{CH_2R'}{N}} + Ar\overset{+}{N}H_2R \ \ X^-$$

Arylamine   Primary alkyl halide   Alkylated arylamine   Arylammonium halide

$$C_6H_5NH_2 + C_6H_5CH_2Cl \xrightarrow[90°C]{NaHCO_3} C_6H_5NHCH_2C_6H_5$$

Aniline (4 mol)   Benzyl chloride (1 mol)   N-Benzylaniline (85–87%)

Reaction with aldehydes and ketones. Schiff's base formation (Section 18.13) Primary arylamines condense with aldehydes and ketones to produce imines. The reaction is directly analogous to that observed with primary alkylamines.

$$ArNH_2 + R\overset{O}{\overset{\|}{C}}R' \longrightarrow ArN=\overset{R}{\underset{R'}{C}} + H_2O$$

Primary arylamine   Aldehyde or ketone   Imine (Schiff's base)   Water

$$C_6H_5NH_2 + C_6H_5\overset{O}{\overset{\|}{C}}H \longrightarrow C_6H_5N=CHC_6H_5 + H_2O$$

Aniline   Benzaldehyde   N-Benzylideneaniline (84–87%)   Water and less nucleophilic than alkylamines because of the delocalization of the unshared electron pair of the amine nitrogen into the aromatic ring.

The same electron delocalization that reduces electron density at nitrogen increases it in the aryl group, and arylamines are extremely reactive substrates in electrophilic aromatic substitution reactions. Electrophilic aromatic substitution in arylamines has been mentioned briefly in Section 13.12, where it is pointed out that the $-\overset{..}{N}H_2$, $-\overset{..}{N}HR$, and $-\overset{..}{N}R_2$ groups are very powerful activating and ortho-para directing substituents. This facet of the chemical reactivity of arylamines is considered in more detail in the following section.

One of the most important reactions involving arylamines involves substitution in their derived diazonium salts. Diazotization of primary arylamines opens the door to a large number of synthetically useful transformations and comprises the most significant body of new chemistry in this chapter, which is presented in Sections 24.9 and 24.10.

24.7 ELECTROPHILIC AROMATIC SUBSTITUTION IN ARYLAMINES

Only rarely are electrophilic aromatic substitution reactions performed directly on arylamines. They are so reactive toward halogenation, for example, that it is difficult to limit the reaction to monosubstitution. Generally, halogenation proceeds rapidly until all available ring positions that are ortho or para to the amino group have been substituted.

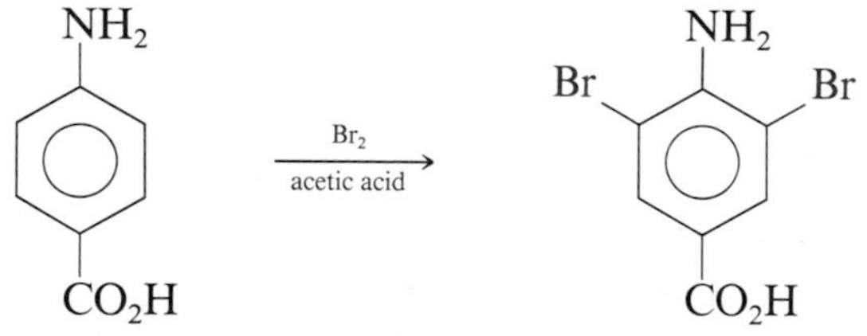

p-Aminobenzoic acid

4-Amino-3,5-dibromobenzoic acid
(82%)

Direct nitration of aniline and other arylamines is difficult to carry out, as it is accompanied by oxidation and leads to the formation of dark-colored "tars." As a solution to this problem it has become standard practice to first protect the amino group by acylation with either acetyl chloride or acetic anhydride.

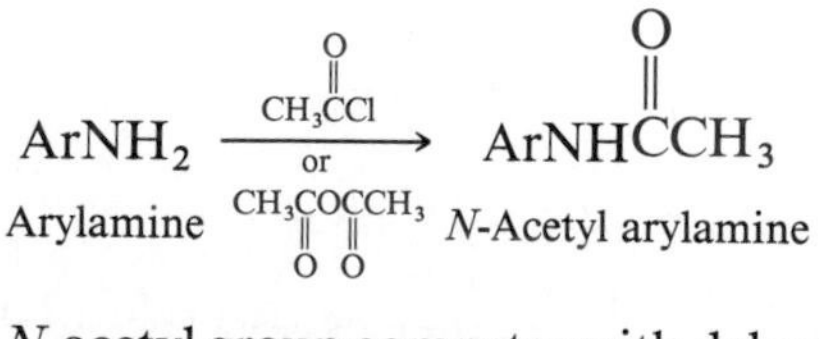

Amide resonance in the N-acetyl group competes with delocalization of the nitrogen lone pair into the ring.

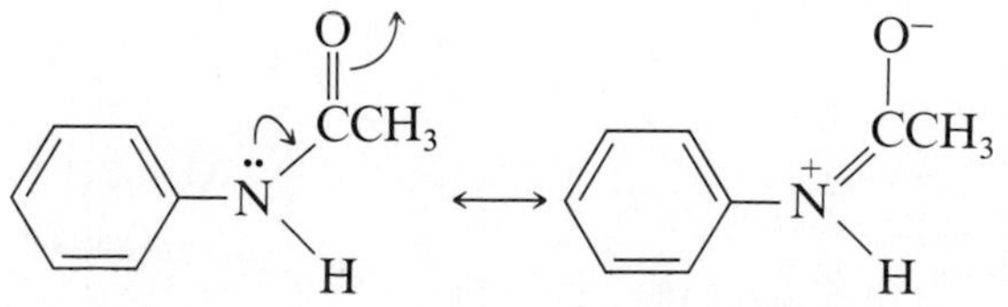

Amide resonance in acetanilide

Protecting the amino group of an arylamine in this way moderates its reactivity and permits nitration of the ring to be achieved. The acetamido group is activating toward electrophilic aromatic substitution and is ortho-para directing.

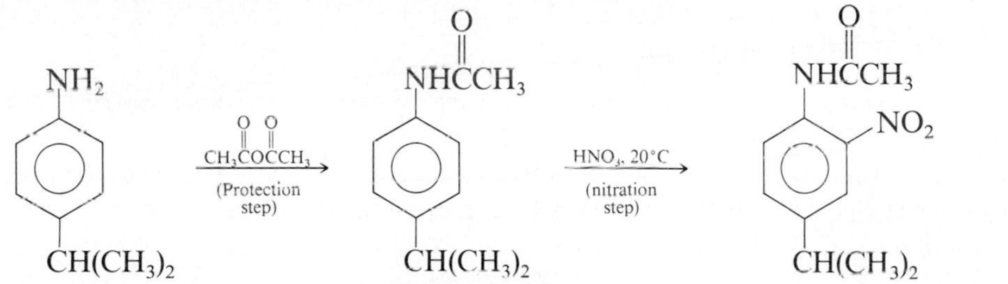

p-Isopropylaniline → p-Isopropylacetanilide (98%) → 4-Isopropyl-2-Nitroacetanilide (94%)

Another feature of the N-acetyl protecting group is that after it has served its purpose, it may be removed by hydrolysis, liberating a free amino group.

$$ ArNHCCH_3 \xrightarrow[\substack{or \\ 1.\ H_3O^+ \\ 2.\ HO^-}]{H_2O,\ HO^-} ArNH_2 $$

N-Acetylarylamine → Arylamine

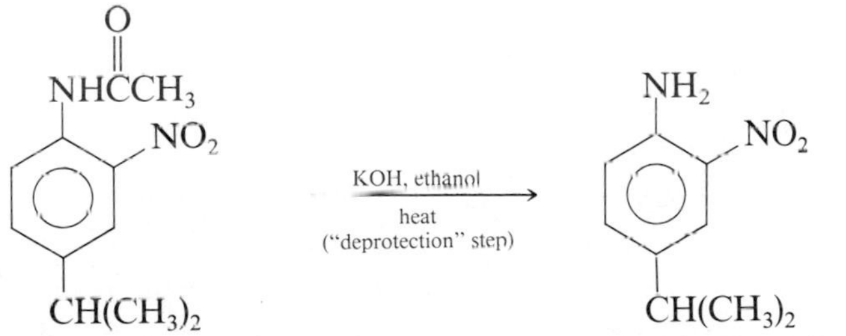

4-Isopropyl-2-nitroacetanilide → KOH, ethanol, heat ("deprotection" step) → 4-Isopropyl-2-nitroaniline (100%)

The net effect of the sequence *protect – nitrate – deprotect* is the same as if the substrate had been successfully nitrated directly. Because direct nitration is impossible, however, the indirect route is the method of choice.

PROBLEM 24.8 Outline syntheses of each of the following from aniline and any necessary organic or inorganic reagents:

(a) p-Nitroaniline
(b) 2,4-Dinitroaniline
(c) p-Aminoacetanilide

SAMPLE SOLUTION (a) It has already been stated that direct nitration of aniline is not a practical reaction. The amino group must first be protected as its N-acetyl derivative.

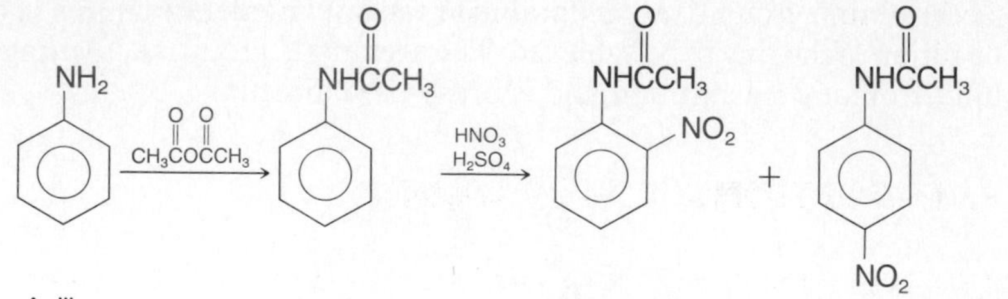

Aniline   Acetanilide   o-Nitroacetanilide   p-Nitroacetanilide

Nitration of acetanilide yields a mixture of ortho and para substitution products. The para isomer is separated, then subjected to hydrolysis to give p-nitroaniline.

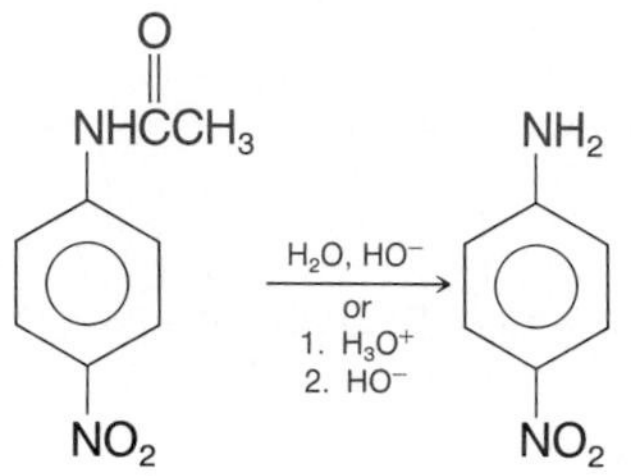

By reducing the electron-donating ability of an amino group through acetylation, it is possible to limit halogenation to monosubstitution.

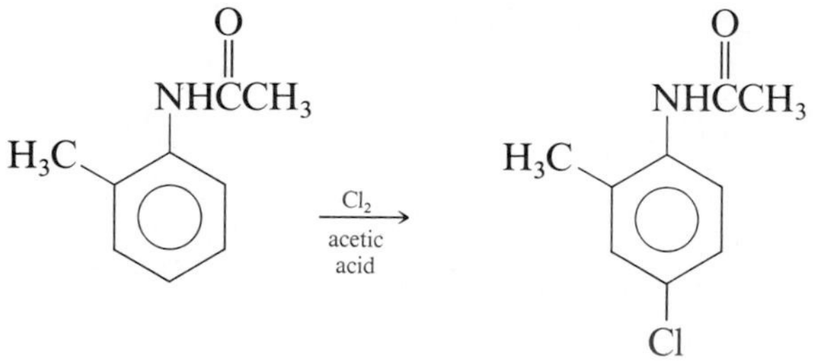

o-Methylacetanilide   4-Chloro-2-methylacetanilide (74%)

Friedel-Crafts alkylations and acylations of N-protected arylamines can be accomplished readily.

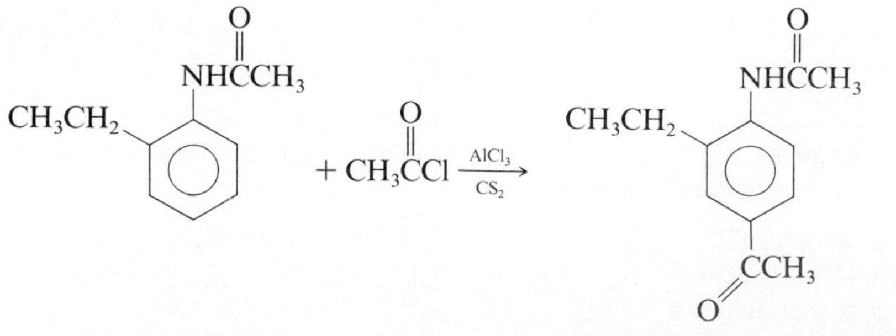

o-Ethylacetanilide   Acetyl chloride   4-Acetamido-3-ethylacetophenone (57%)

Friedel-Crafts reactions are normally not successful when attempted on the free (unprotected) arylamine.

24.8 NITROSATION OF ARYLAMINES

We learned in the preceding chapter (Section 23.19) that different reactions are observed when the various classes of alkylamines — primary, secondary, and tertiary — react with nitrosating agents. While no useful chemistry attends the nitrosation of tertiary alkylamines, electrophilic aromatic substitution by nitrosonium cation $(:N\overset{+}{\equiv}\overset{..}{O}:)$ takes place with *N,N*-dialkylarylamines.

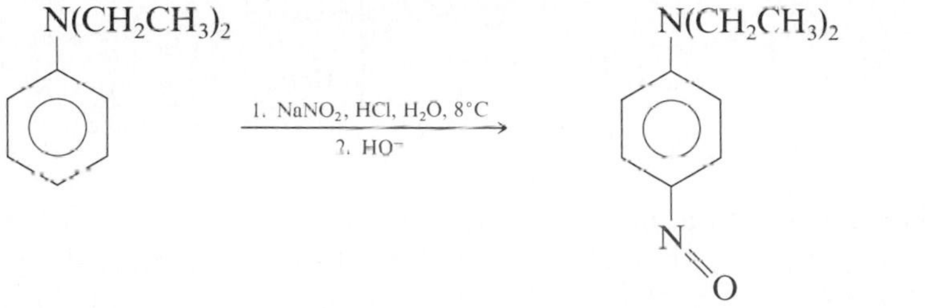

N,N-Diethylaniline $\qquad\qquad\qquad$ N,N-Diethyl-p-nitrosoaniline (95%)

Nitrosonium cation is a relatively weak electrophile and only attacks very strongly activated aromatic rings.

N-Alkylarylamines resemble secondary alkylamines in that they form *N*-nitroso compounds on reaction with nitrous acid.

$$C_6H_5NHCH_3 \xrightarrow[\text{H}_2\text{O, 10°C}]{\text{NaNO}_2,\ \text{HCl}} C_6H_5N-N=O$$
$$\qquad\qquad\qquad\qquad\qquad\qquad | $$
$$\qquad\qquad\qquad\qquad\qquad CH_3$$

N-Methylaniline $\qquad\qquad$ N-Methyl-N-nitrosoaniline (87–93%)

Primary arylamines, like primary alkylamines, form diazonium ion salts on nitrosation. Aryl diazonium ions are considerably more stable than their alkyl counterparts. Whereas alkyl diazonium ions decompose under the conditions of their formation, aryl diazonium salts are stable enough to be stored in aqueous solution at 0 to 5°C for reasonable periods of time. Loss of nitrogen from an aryl diazonium ion generates an unstable aryl cation and is much slower than loss of nitrogen from an alkyl diazonium ion.

$$C_6H_5NH_2 \xrightarrow[\text{H}_2\text{O, 0–5°C}]{\text{NaNO}_2,\ \text{HCl}} C_6H_5\overset{+}{N}\equiv N: \ Cl^-$$

Aniline $\qquad\qquad$ Benzenediazonium chloride

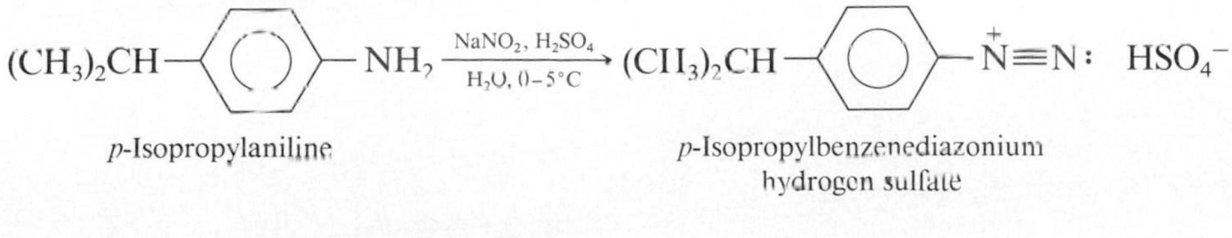

p-Isopropylaniline $\qquad\qquad$ p-Isopropylbenzenediazonium hydrogen sulfate

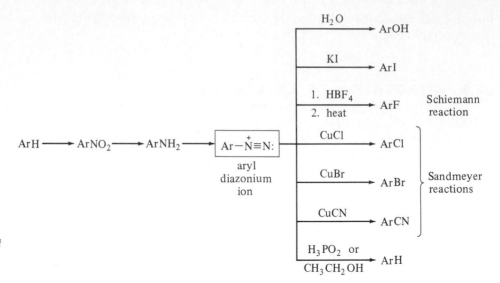

FIGURE 24.4 Flow chart showing the synthetic origin of aryl diazonium ions and their most useful transformations.

Aryl diazonium ions undergo a variety of reactions that make them versatile intermediates for the preparation of a host of ring-substituted aromatic compounds. In these reactions, summarized in Figure 24.4 and discussed individually in the following section, molecular nitrogen acts as a leaving group and is replaced by another atom or group. All the reactions are regiospecific; the entering group becomes bonded to precisely the ring position from which nitrogen departs.

24.9 SYNTHETIC TRANSFORMATIONS OF ARYL DIAZONIUM SALTS

An important reaction of aryl diazonium ions is their conversion to *phenols* by hydrolysis.

$$\overset{+}{Ar}N\equiv N\text{:} \quad + H_2O \longrightarrow ArOH + H^+ + \text{:}N\equiv N\text{:}$$

Aryl diazonium ion   Water   A phenol   Nitrogen

This is the most general method for the preparation of phenols. It is easily performed; the aqueous acidic solution in which the diazonium salt is prepared is heated and gives the phenol directly. An aryl cation is probably generated, which is then captured by water acting as a nucleophile.

$$(CH_3)_2CH-\underset{\text{}}{\bigcirc}-NH_2 \xrightarrow[\text{2. } H_2O,\text{ heat}]{\text{1. NaNO}_2,\text{ H}_2SO_4,\text{ H}_2O} (CH_3)_2CH-\underset{\text{}}{\bigcirc}-OH$$

p-Isopropylaniline   *p*-Isopropylphenol (73%)

m-Bromoaniline, NH$_2$ ring with Br $\xrightarrow[\text{2. } H_2O,\text{ heat}]{\text{1. NaNO}_2,\text{ H}_2SO_4,\text{ H}_2O,\text{ 0–5°C}}$ OH ring with Br

m-Bromoaniline   *m*-Bromophenol (66%)

Sulfuric acid is normally used instead of hydrochloric acid in the diazotization step so as to minimize the competition with water for capture of the cationic intermediate. Hydrogen sulfate anion (HSO_4^-) is less nucleophilic than chloride.

PROBLEM 24.9 Design a synthesis of *m*-bromophenol from benzene.

The reaction of an aryl diazonium salt with potassium iodide is the standard method for the preparation of *aryl iodides*. The diazonium salt is prepared from a primary aromatic amine in the usual way, a solution of potassium iodide is then added and the reaction mixture brought to room temperature or heated to accelerate the reaction.

$$Ar-\overset{+}{N}\equiv N: + \ I^- \longrightarrow \ ArI \ + :N\equiv N:$$

| Aryl diazonium ion | Iodide ion | Aryl iodide | Nitrogen |

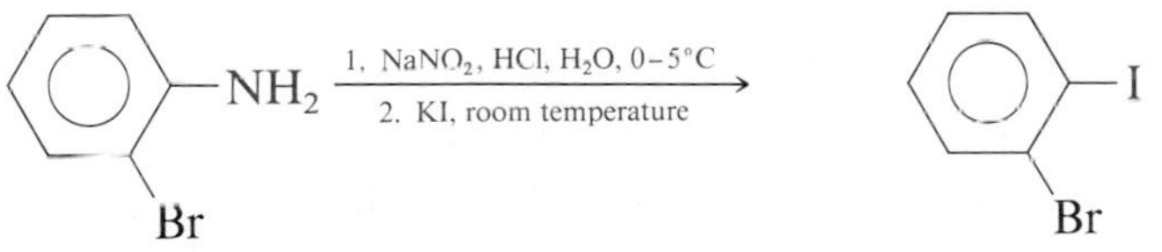

o-Bromoaniline   →   *o*-Bromoiodobenzene (72–83%)

1. $NaNO_2$, HCl, H_2O, 0–5°C
2. KI, room temperature

PROBLEM 24.10 Show by a series of equations how you could prepare *m*-bromoiodobenzene from benzene.

Diazonium salt chemistry provides the principal synthetic method for the preparation of *aryl fluorides* through a process known as the *Schiemann reaction*. In this procedure the aryl diazonium ion is isolated as its fluoroborate salt, which then yields the desired aryl fluoride on being heated.

$$Ar-\overset{+}{N}\equiv N: \ \bar{B}F_4 \xrightarrow{\text{heat}} ArF + \ BF_3 \ + :N\equiv N:$$

| Aryldiazonium fluoroborate | Aryl fluoride | Boron trifluoride | Nitrogen |

A standard way to form the aryl diazonium fluoroborate salt is to add fluoroboric acid (HBF_4) or a fluoroborate salt to the diazotization medium.

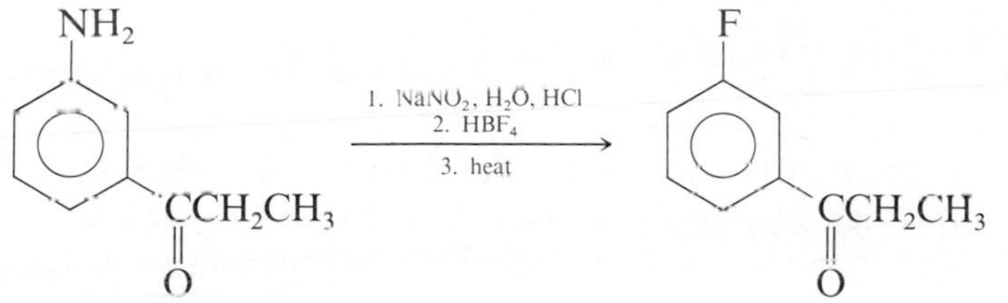

m-Aminopropiophenone   →   *m*-Fluoropropiophenone (68%)

1. $NaNO_2$, H_2O, HCl
2. HBF_4
3. heat

PROBLEM 24.11 Show the proper sequence of synthetic transformations in the conversion of benzene to *m*-fluoropropiophenone.

While *aryl chlorides* and *aryl bromides* are capable of being prepared by electrophilic aromatic substitution, it is often necessary to prepare these compounds from an aromatic amine. The amine is converted to the corresponding diazonium salt and then treated with cuprous chloride or cuprous bromide as appropriate.

$$Ar-\overset{+}{N}\equiv N\colon \xrightarrow{\text{CuX}} ArX \quad +\colon N\equiv N\colon$$

Aryl diazonium ion   Aryl chloride or bromide   Nitrogen

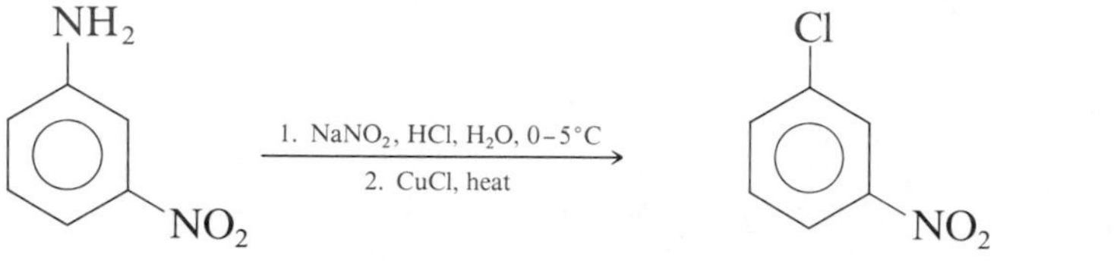

m-Nitroaniline

1. NaNO$_2$, HCl, H$_2$O, 0–5°C
2. CuCl, heat

m-Chloronitrobenzene (68–71%)

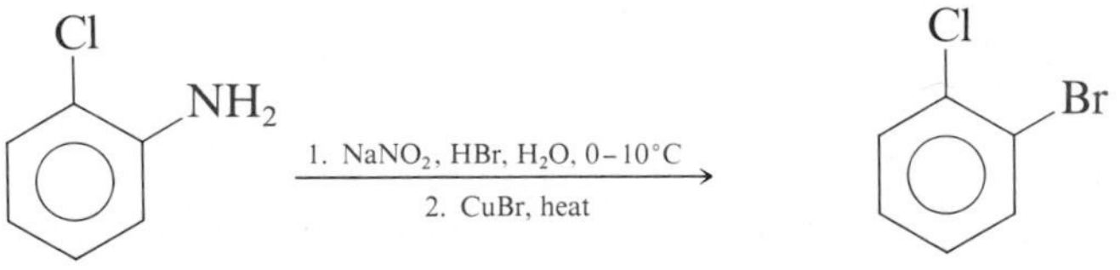

o-Chloroaniline

1. NaNO$_2$, HBr, H$_2$O, 0–10°C
2. CuBr, heat

o-Bromochlorobenzene (89–95%)

Reactions that employ cuprous salts as reagents for replacement of nitrogen in diazonium salts are called *Sandmeyer reactions.* The Sandmeyer reaction using cuprous cyanide is a good method for the preparation of aromatic *nitriles* (aryl cyanides).

$$Ar-\overset{+}{N}\equiv N\colon \xrightarrow{\text{CuCN}} ArCN +\colon N\equiv N\colon$$

Aryl diazonium ion   Aryl cyanide   Nitrogen

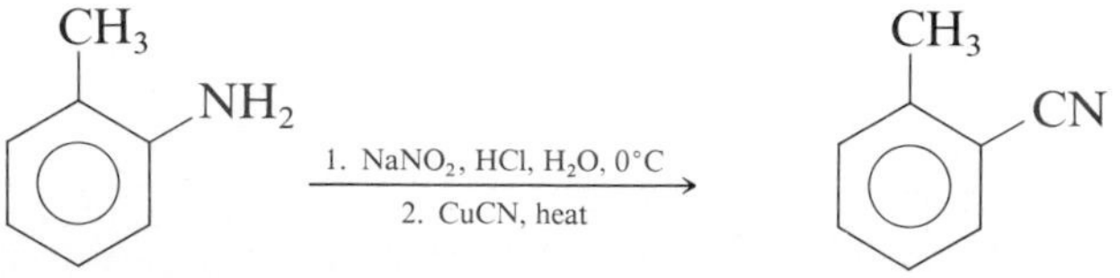

o-Toluidine

1. NaNO$_2$, HCl, H$_2$O, 0°C
2. CuCN, heat

o-Methylbenzonitrile (64–70%)

Since cyano groups may be hydrolyzed to carboxylic acids, the Sandmeyer preparation of aryl nitriles is a key step in the conversion of arylamines to substituted benzoic acids. In the example just cited, the *o*-methylbenzonitrile that was formed was subsequently subjected to acid-catalyzed hydrolysis and gave *o*-toluic acid (*o*-methylbenzoic acid) in 80 to 89 percent yield.

The preparation of aryl chlorides, bromides, and cyanides by the Sandmeyer reaction is mechanistically complicated and may involve aryl copper intermediates.

It is possible to replace amino substituents on an aromatic nucleus by hydrogen by reducing a diazonium salt with hypophosphorous acid H_2POH (H_3PO_2) or with ethanol. These reductions are free-radical reactions in which ethanol or hypophosphorous acid acts as a hydrogen atom donor.

$$Ar\overset{+}{-}N\equiv N: \xrightarrow[\text{CH}_3\text{CH}_2\text{OH}]{\text{H}_3\text{PO}_2 \text{ or}} ArH + :N\equiv N:$$

Aryl diazonium   Arene   Nitrogen
ion

Reactions of this type are known as *reductive deaminations.*

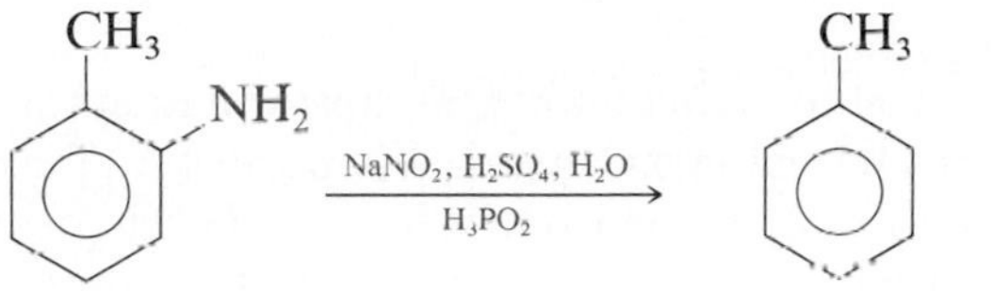

o-Toluidine   Toluene (70–75%)

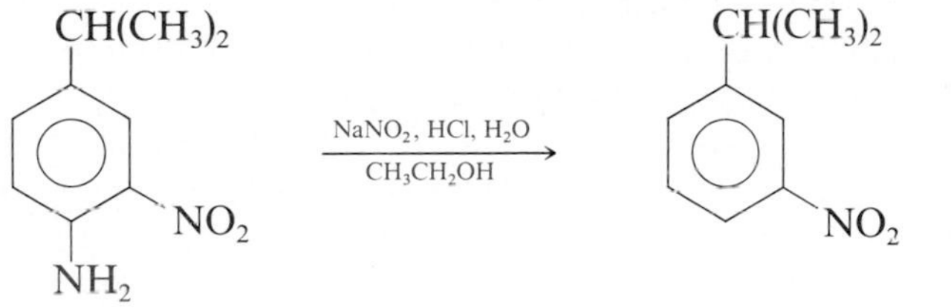

4-Isopropyl-2-nitroaniline   *m*-Isopropylnitrobenzene (59%)

Sodium borohydride has also been used to reduce aryl diazonium salts in reductive deamination reactions.

PROBLEM 24.12 Cumene (isopropylbenzene) is a relatively inexpensive commercially available starting material. Show how you could prepare *m*-isopropylnitrobenzene from cumene.

The value of diazonium salts in synthetic organic chemistry rests on two principal points. Through the use of diazonium salt chemistry:

1. Substituents that are otherwise difficultly accessible, such as fluoro, iodo, and hydroxyl, may be introduced onto a benzene ring.
2. Compounds that have substitution patterns not directly available by electrophilic aromatic substitution can be prepared.

The first of these two features is readily apparent and is illustrated by Problems 24.9 to 24.11. If you have not done these problems yet, you are strongly encouraged to attempt them now.

The second point is somewhat less obvious but is readily illustrated by the synthesis of 1,3,5-tribromobenzene. This particular substitution pattern cannot be obtained by direct bromination of benzene because bromine is an ortho-para director. Instead, advantage is taken of the powerful activating and ortho-para directing effects of the amino group in aniline. Bromination of aniline yields 2,4,6-tribromoaniline in quantitative yield. Diazotization of the resulting 2,4,6-tribromoaniline and reduction of the diazonium salt gives the desired 1,3,5-tribromobenzene.

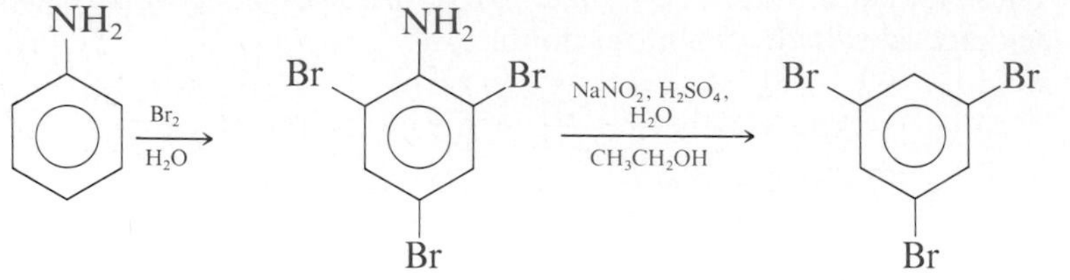

Aniline   2,4,6-Tribromoaniline (100%)   1,3,5-Tribromobenzene
                                                           (74–77%)

In order to exploit the versatility inherent in the transformations available to aryl diazonium salts, be prepared to reason backward. When you see a fluorine substituent in a synthetic target, for example, realize that it probably will have to be introduced by a Schiemann reaction of an arylamine; realize that the required arylamine is derived from a nitroarene, and that the nitro group is introduced by nitration. Be aware that an unsubstituted position of an aromatic ring need not have always been that way. It might have borne an amino group that was used to control the orientation of electrophilic aromatic substitution reactions prior to being removed by reductive deamination. The strategy of synthesis is intellectually demanding and a considerable sharpening of your reasoning power can be gained by attacking the synthesis problems at the end of each chapter. Remember, plan your sequence of accessible intermediates by reasoning backward from the target, then fill in the details on how each transformation is to be carried out.

24.10 AZO COUPLING

A reaction of aryl diazonium salts that does not involve loss of nitrogen takes place when they react with phenols and arylamines. Aryl diazonium ions are relatively weak electrophiles but have sufficient reactivity to attack strongly activated aromatic rings. The reaction is known as *azo coupling;* two aryl groups are joined together by an azo ($-N{=}N-$) function.

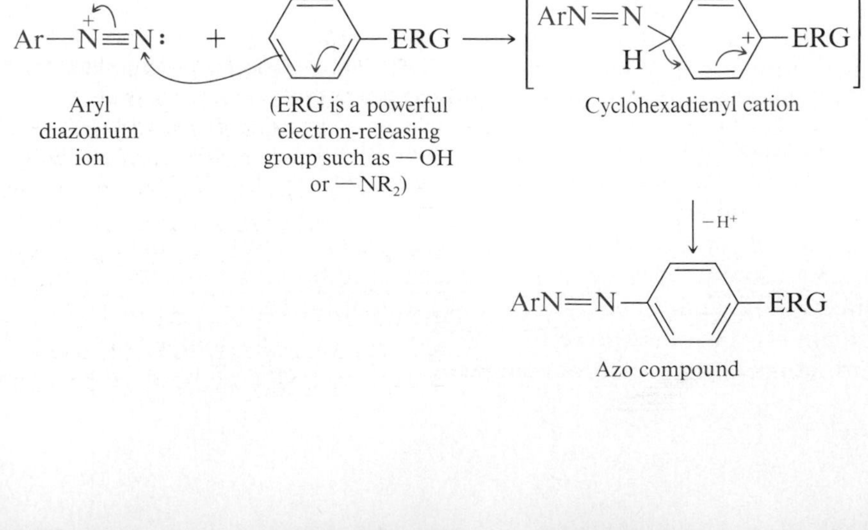

Aryl   (ERG is a powerful   Cyclohexadienyl cation
diazonium   electron-releasing
ion   group such as —OH
                  or —NR₂)

Azo compound

Azo compounds are often highly colored and many of them are used as dyes.

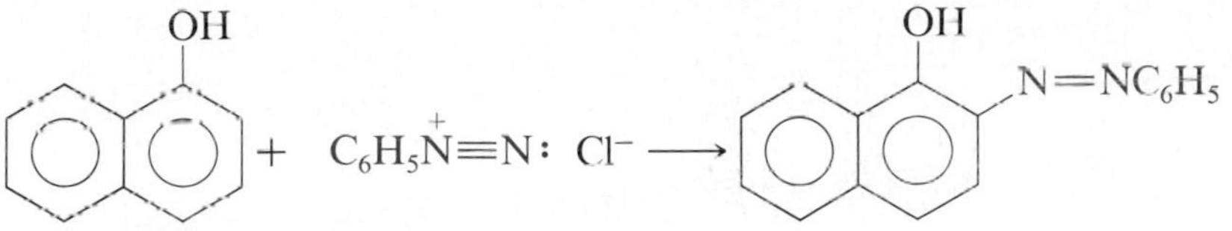

1-Naphthol   Benzenediazonium   2-(Phenylazo)-1-naphthol
                  chloride

The colors of azo compounds vary with the nature of the aryl group, with its substituents, and with pH. Substituents also affect the water solubility of azo dyes and how well they bind to a particular fabric. Countless combinations of diazonium salts and aromatic substrates have been examined with a view toward obtaining azo dyes suitable for a particular application.

24.11 SULFA DRUGS

In 1932 it was revealed that an azo dye, *Prontosil,* possesses antibacterial activity. Subsequent research established that Prontosil exerts its antibacterial effect by first undergoing an in vivo degradation to sulfanilamide. Sulfanilamide is the drug actually responsible for the observed biological activity.

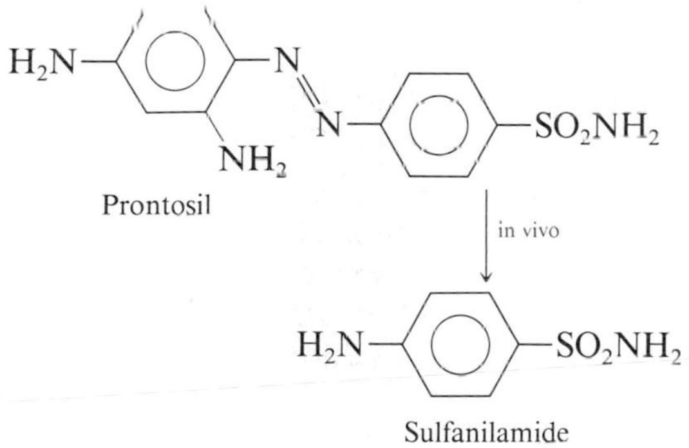

Prontosil

in vivo

Sulfanilamide

Bacteria require *p*-aminobenzoic acid in order to biosynthesize folic acid, a growth factor. Structurally, sulfanilamide resembles *p*-aminobenzoic acid and is mistaken for it by the bacteria, so that folic acid biosynthesis is inhibited and bacterial growth is halted. Since animals do not biosynthesize folic acid but obtain it in their food, sulfanilamide halts the growth of bacteria without harm to the host.

Identification of the means by which Prontosil combats bacterial infections was an early triumph of *pharmacology,* a branch of science at the interface of physiology and biochemistry that studies the mechanism of drug action.

The preparation of sulfanilamide as the most fundamental of the *sulfa drugs* is included in a number of introductory organic chemistry laboratory texts. The requisite carbon-sulfur bond is introduced by treating acetanilide with chlorosulfonic acid.

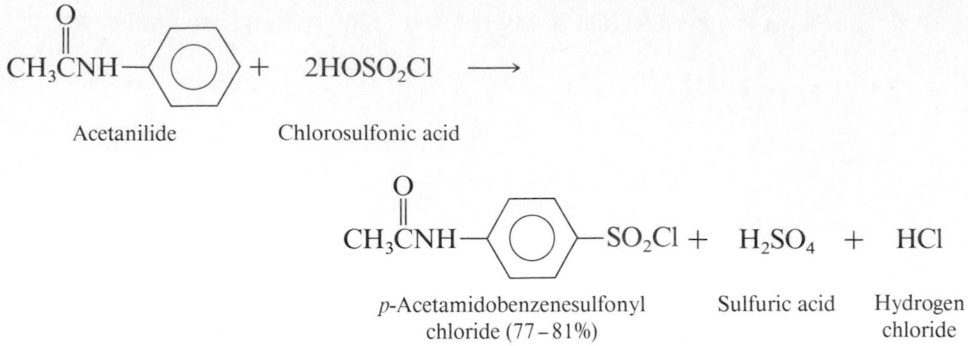

Acetanilide · Chlorosulfonic acid

p-Acetamidobenzenesulfonyl chloride (77–81%) · Sulfuric acid · Hydrogen chloride

Reaction of the arenesulfonyl chloride with ammonia converts it to the derived sulfonamide. Hydrolysis removes the N-acetyl protecting group, yielding sulfanilamide.

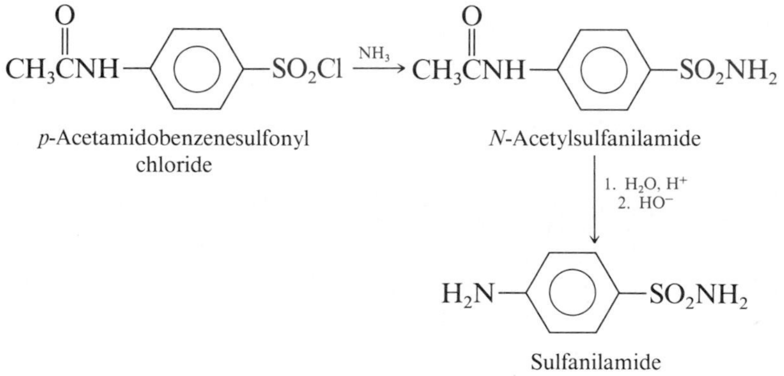

p-Acetamidobenzenesulfonyl chloride

N-Acetylsulfanilamide

1. H_2O, H^+
2. HO^-

Sulfanilamide

Thousands of compounds related to sulfanilamide have been synthesized and tested as agents against bacterial infections. Some of these sulfa drugs include

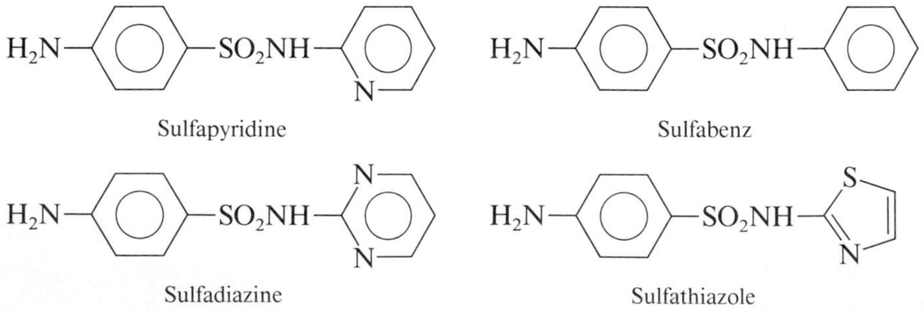

Sulfapyridine

Sulfabenz

Sulfadiazine

Sulfathiazole

PROBLEM 24.13 Suggest how you could adapt the sulfanilamide synthesis just shown so as to prepare

(a) Sulfapyridine   (c) Sulfadiazine

(b) Sulfabenz   (d) Sulfathiazole

SAMPLE SOLUTION (a) The sulfa drug we wish to prepare differs from sulfanilamide with respect to the substituent at sulfur. Recall that this substituent in sulfanilamide arises from reaction of a sulfonyl chloride intermediate with ammonia. Modify the sulfanila-mide synthesis so as to use a nucleophile other than ammonia in this reaction. In the case of sulfapyridine, the appropriate nucleophile is revealed by the retrosynthesis shown.

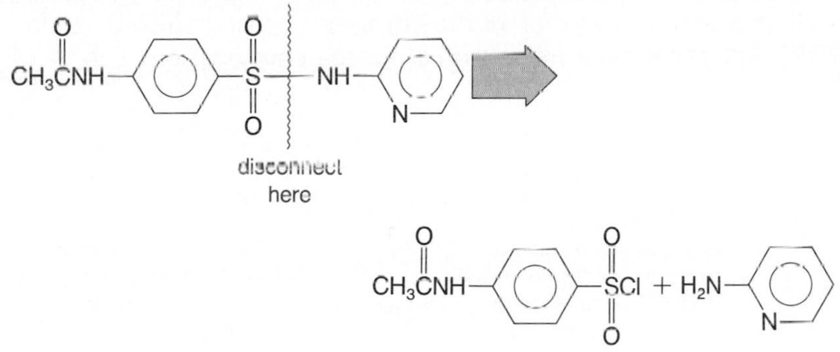

disconnect
here

p-Acetamidobenzenesulfonyl   2-Aminopyridine
chloride

Thus, preparation of p-acetamidobenzenesulfonyl chloride exactly as described above in connection with sulfanilimide synthesis is followed by reaction with 2-aminopyridine to give the N-acetyl derivative of sulfapyridine. Hydrolysis of the N-acetyl protecting group yields the desired sulfa drug.

Most present-day applications of sulfa drugs are in the area of veterinary medicine. While sulfa drugs were extensively used to treat bacterial infections in humans during the 1930s and 1940s, they have been superseded by more effective antibiotics, such as the penicillins and the tetracyclines.

24.12 SUMMARY

Arylamines are most often prepared by reduction of nitro-substituted arenes (Section 24.5).

$$ArNO_2 \xrightarrow{\text{reduce}} ArNH_2$$

Nitroarene   Primary arylamine

Typical reducing agents include iron in hydrochloric acid, tin in hydrochloric acid, and stannous chloride. Catalytic hydrogenation using a platinum, palladium, or Raney nickel catalyst is also an excellent method.

N-Alkylation of arylamines may be achieved by reaction with an alkyl halide, by reductive amination, or by lithium aluminum hydride reduction of an N-acylaryla-mine.

$$ArNH_2 \longrightarrow ArNHR \longrightarrow ArNR_2$$

Primary arylamine   N-Alkylarylamine   N,N-Dialkylarylamine

The two most prominent ways in which arylamines differ from alkylamines are

1. Arylamines are thousands to millions of times less basic than alkylamines (Section 24.4).

2. The diazonium ions derived from primary arylamines are much more stable than those of primary alkylamines and rank among the most useful intermediates in all of synthetic organic chemistry (Sections 24.8 to 24.10).

Delocalization of the nitrogen lone pair into the π system of the aromatic ring stabilizes arylamines. Protonation of the nitrogen is accompanied by a loss of this electron delocalization and is more exothermic than protonation of an alkylamine.

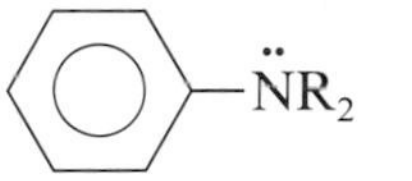

An arylamine is stabilized both by benzene-type resonance and by π donation of the lone pair to the ring.

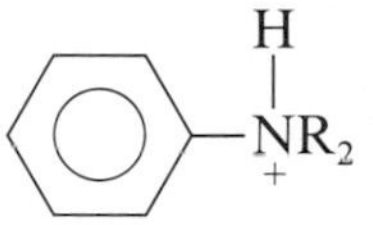

The protonated form of an arylamine has lost the increment of stabilization that results from lone pair delocalization.

The benzene ring of arylamines is very reactive toward electrophilic aromatic substitution. It is customary to protect arylamines as their *N*-acyl derivatives prior to carrying out ring nitration, chlorination, bromination, sulfonation, or Friedel-Crafts reactions (Section 24.7).

Nitrosation of arylamines occurs when they are treated with sodium nitrite in acid solution. The product of nitrosation varies with the class of arylamine. Primary arylamines yield aryl diazonium salts.

$$ArNH_2 \xrightarrow[H_2O]{NaNO_2,\ H^+} Ar\overset{+}{N}\equiv N\colon$$

Primary arylamine   Aryl diazonium salt

Secondary arylamines yield *N*-nitroso compounds.

$$ArNHR \xrightarrow[H_2O]{NaNO_2,\ H^+} ArN-N=O$$
$$\hspace{6cm} |$$
$$\hspace{6cm} R$$

Secondary arylamine   *N*-Alkyl-*N*-nitrosoarylamine

Tertiary arylamines undergo nitrosation of the ring.

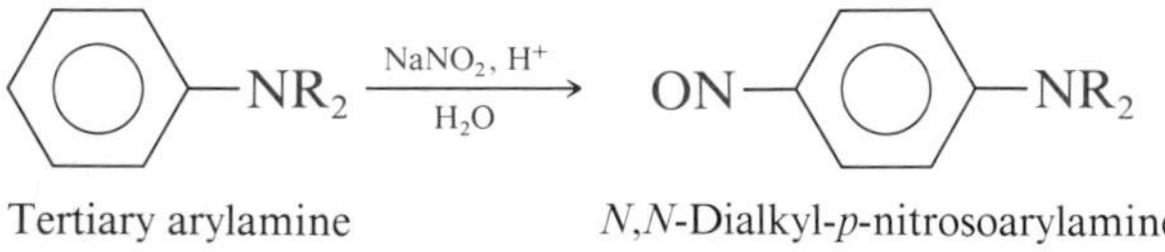

Tertiary arylamine   *N,N*-Dialkyl-*p*-nitrosoarylamine

Aryl diazonium salts are versatile intermediates. They serve as synthetic precursors to a number of classes of aryl derivatives, as summarized in Table 24.4 and Figure 24.4.

Aryl diazonium salts react with strongly activated aromatic rings to form azo compounds.

$$Ar\overset{+}{N}\equiv N\colon + \ Ar'H \longrightarrow ArN=NAr' + H^+$$

Aryldiazonium   Arylamine   Azo
ion   or phenol   compound

Most of these azo compounds are highly colored and many are used as dyes.

TABLE 24.4

Synthetically Useful Transformations Involving Aryl Diazonium Ions

Reaction and comments	General equation and specific example
Preparation of aryl diazonium salts The first stage is the conversion of a primary arylamine to an aryl diazonium salt by nitrosation. Usually this is accomplished by treating the arylamine with sodium nitrite and acid in aqueous media. Aryl diazonium salts have modest stability at low temperature.	$ArNH_2 \xrightarrow[H_2O, 0-5°C]{NaNO_2, H^+} Ar\overset{+}{N}{\equiv}N\colon$ Primary arylamine → Aryl diazonium ion 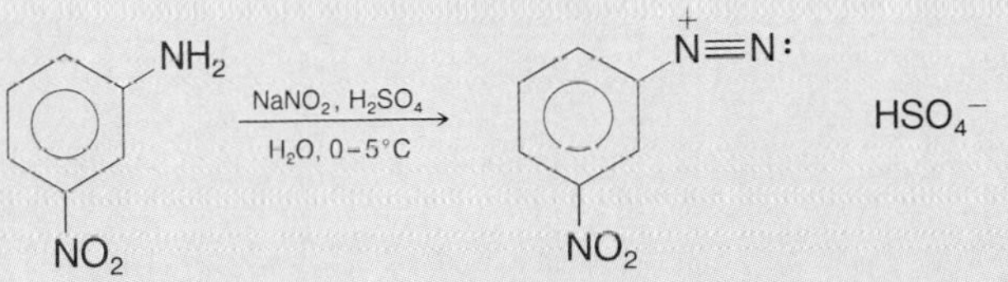 *m*-Nitroaniline $\xrightarrow[H_2O, 0-5°C]{NaNO_2, H_2SO_4}$ *m*-Nitrobenzenediazonium hydrogen sulfate (HSO_4^-)
Preparation of phenols Heating its aqueous acidic solution converts a diazonium salt to a phenol. This is the most general method for the synthesis of phenols.	$ArNH_2 \xrightarrow[\text{2. } H_2O, \text{ heat}]{\text{1. } NaNO_2, H_2SO_4, H_2O} ArOH$ Primary arylamine → Phenol 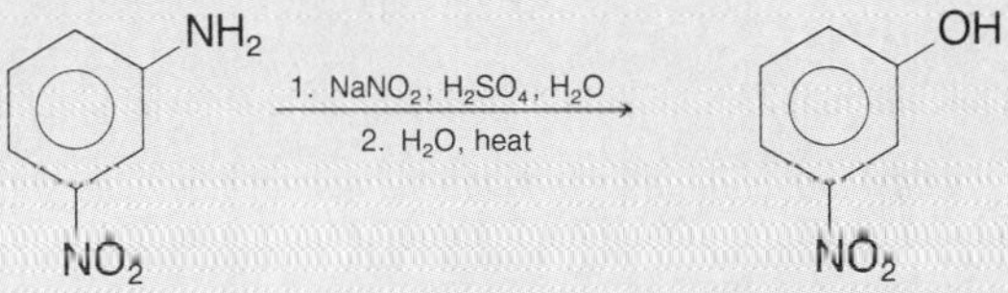 *m*-Nitroaniline $\xrightarrow[\text{2. } H_2O, \text{ heat}]{\text{1. } NaNO_2, H_2SO_4, H_2O}$ *m*-Nitrophenol (81–86%)
Preparation of aryl fluorides Addition of fluoroboric acid to a solution of a diazonium salt causes the precipitation of an aryl diazonium fluoroborate. When the dry aryl diazonium fluoroborate is heated, an aryl fluoride results. This is the Schiemann reaction; it is the most general method for the preparation of aryl fluorides.	$ArNH_2 \xrightarrow[\text{2. } HBF_4]{\text{1. } NaNO_2, H^+, H_2O} Ar\overset{+}{N}{\equiv}N\colon\ BF_4^- \xrightarrow{\text{heat}} ArF$ Primary arylamine → Aryl diazonium fluoroborate → Aryl fluoride 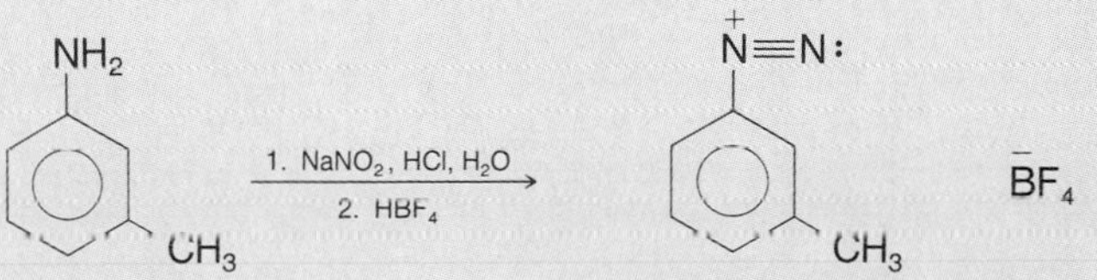 *m*-Toluidine $\xrightarrow[\text{2. } HBF_4]{\text{1. } NaNO_2, HCl, H_2O}$ *m*-Methylbenzenediazonium fluoroborate (76–84%) 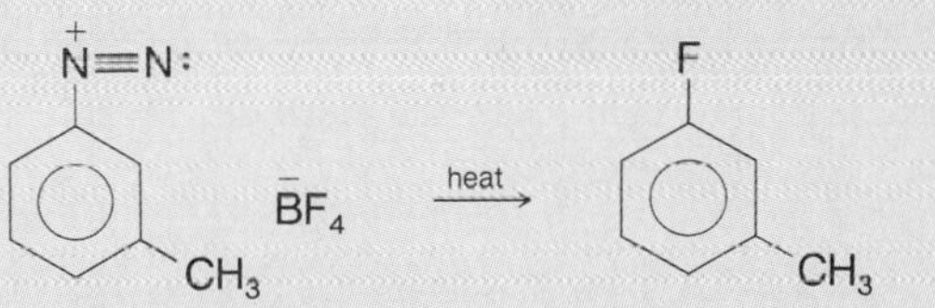 *m*-Methylbenzenediazonium fluoroborate $\xrightarrow{\text{heat}}$ *m*-Fluorotoluene (89%)

TABLE 24.4 (continued)

Reaction and comments	General equation and specific example

Preparation of aryl iodides Aryl diazonium salts react with sodium or potassium iodide to form aryl iodides. This is the most general method for the synthesis of aryl iodides.

$$ArNH_2 \xrightarrow[\text{2. NaI or KI}]{\text{1. NaNO}_2\text{, H}^+\text{, H}_2\text{O}} ArI$$

Primary arylamine → Aryl iodide

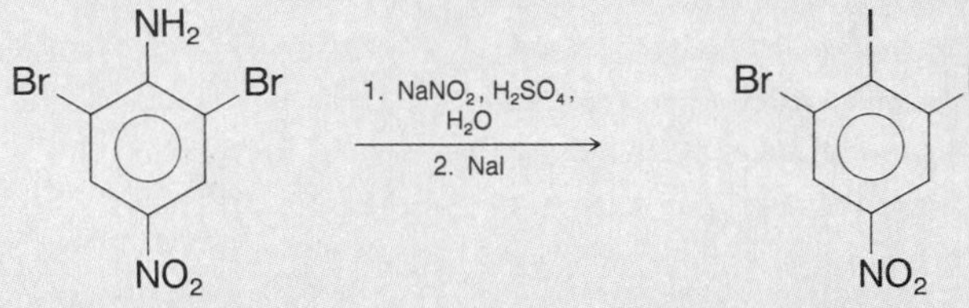

2,6-Dibromo-4-nitroaniline

1. NaNO$_2$, H$_2$SO$_4$, H$_2$O
2. NaI

1,3-Dibromo-2-iodo-5-nitrobenzene (84–88%)

Preparation of aryl chlorides In the Sandmeyer reaction a solution containing an aryl diazonium salt is treated with cuprous chloride to give an aryl chloride.

$$ArNH_2 \xrightarrow[\text{2. CuCl}]{\text{1. NaNO}_2\text{, HCl, H}_2\text{O}} ArCl$$

Primary arylamine → Aryl chloride

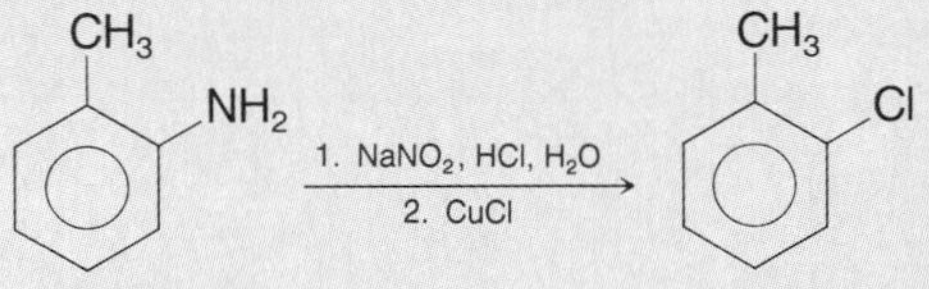

o-Toluidine

1. NaNO$_2$, HCl, H$_2$O
2. CuCl

o-Chlorotoluene (74–79%)

Preparation of aryl bromides The Sandmeyer reaction using cuprous bromide is applicable to the conversion of primary arylamines to aryl bromides.

$$ArNH_2 \xrightarrow[\text{2. CuBr}]{\text{1. NaNO}_2\text{, HBr, H}_2\text{O}} ArBr$$

Primary arylamine → Aryl bromide

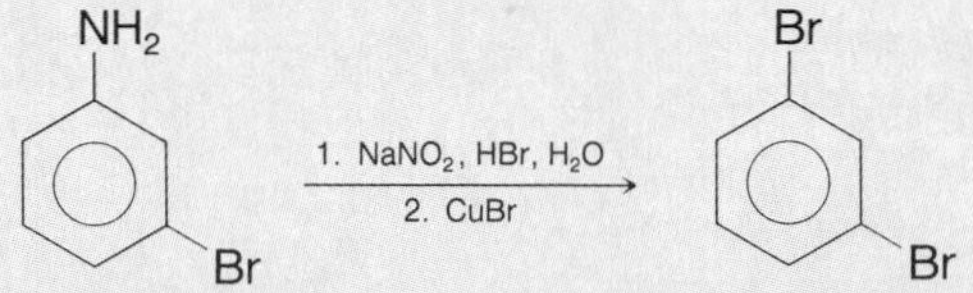

m-Bromoaniline

1. NaNO$_2$, HBr, H$_2$O
2. CuBr

m-Dibromobenzene (80–87%)

Preparation of aryl cyanides Cuprous cyanide converts aryl diazonium salts to aryl cyanides.

$$ArNH_2 \xrightarrow[\text{2. CuCN}]{\text{1. NaNO}_2\text{, H}_2\text{O}} ArCN$$

Primary arylamine → Aryl cyanide

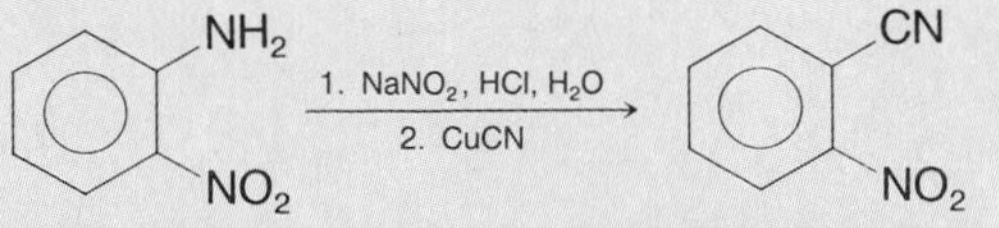

o-Nitroaniline

1. NaNO$_2$, HCl, H$_2$O
2. CuCN

o-Nitrobenzonitrile (87%)

TABLE 24.4 (continued)

Reaction and comments	General equation and specific example
Reductive deamination of primary arylamines The amino substituent of an arylamine can be replaced by hydrogen by treatment of its derived diazonium salt with ethanol or with hypophosphorous acid.	$ArNH_2 \xrightarrow[\text{2. } CH_3CH_2OH \text{ or } H_3PO_2]{\text{1. } NaNO_2, H^+, H_2O} ArH$ 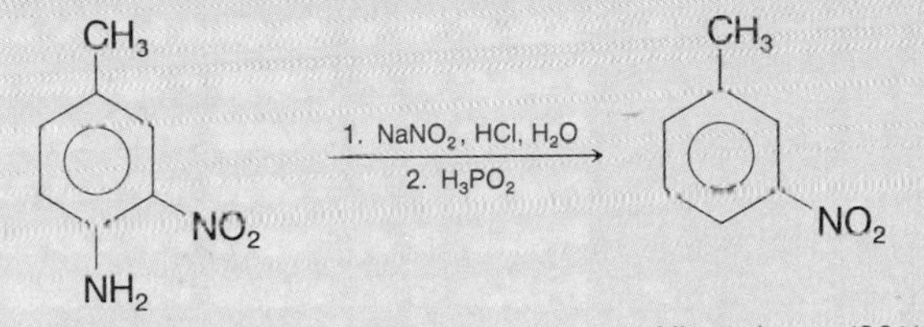

4-Methyl-2-nitroaniline

$\xrightarrow[\text{2. } H_3PO_2]{\text{1. } NaNO_2, HCl, H_2O}$

m-Nitrotoluene (80%)

PROBLEMS

24.14 (a) Give the structures and an acceptable name for all the isomers of molecular formula C_7H_9N that contain a benzene ring.

(b) Which one of these isomers is the strongest base?

(c) Which, if any, of these isomers yield an *N*-nitroso amine on treatment with sodium nitrite and hydrochloric acid?

(d) Which, if any, of these isomers undergo nitrosation of their benzene ring on treatment with sodium nitrite and hydrochloric acid?

24.15 Arrange the members of each group in order of decreasing basicity:

(a) Ammonia, aniline, methylamine
(b) Acetanilide, aniline, *N*-methylaniline
(c) 2,4-Dichloroaniline, 2,4-dimethylaniline, 2,4-dinitroaniline
(d) 3,4-Dichloroaniline, 4-chloro-2-nitroaniline, 4-chloro-3-nitroaniline
(e) Dimethylamine, diphenylamine, *N*-methylaniline
(f) *p*-Aminoacetophenone, *p*-anisidine, 1,4-benzenediamine

24.16 *Physostigmine* is an alkaloid obtained from a West African plant; it is used in the treatment of glaucoma. Treatment of physostigmine with methyl iodide gives a quaternary ammonium salt. What is the structure of this salt?

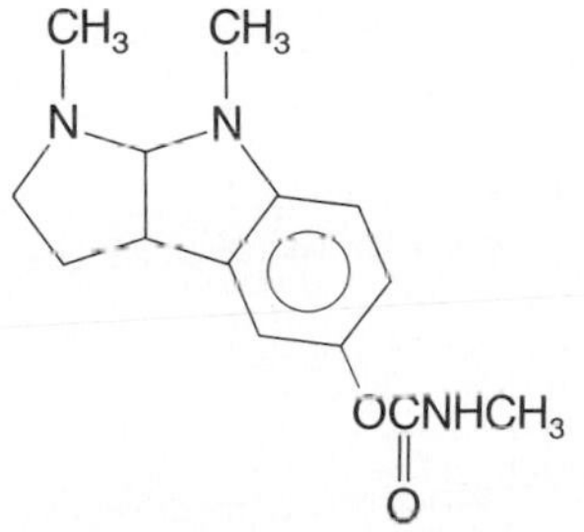

Physostigmine

24.17 Write the structure of the product formed on reaction of aniline with each of the following:

(a) Hydrogen bromide
(b) Excess methyl iodide

(0-p-) metu Q+A

(c)   Acetaldehyde
(d)   Acetaldehyde, H_2, Raney nickel
(e)   Acetic anhydride
(f)   Benzoyl chloride
(g)   Benzenesulfonyl chloride
(h)   Sodium nitrite, aqueous sulfuric acid, 0–5°C
(i)   Product of (h), heated in aqueous acid
(j)   Product of (h), treated with cuprous chloride
(k)   Product of (h), treated with cuprous bromide
(l)   Product of (h), treated with cuprous cyanide
(m)   Product of (h), treated with hypophosphorous acid
(n)   Product of (h), treated with potassium iodide
(o)   Product of (h), treated with fluoroboric acid, then heated
(p)   Product of (h), treated with phenol
(q)   Product of (h), treated with N,N-dimethylaniline

24.18 Write the structure of the product formed on reaction of acetanilide with each of the following:

(a)   Lithium aluminum hydride
(b)   Nitric acid and sulfuric acid
(c)   Sulfur trioxide and sulfuric acid
(d)   Chlorosulfonic acid
(e)   Product of (d), treated with dimethylamine
(f)   Bromine in acetic acid
(g)   tert-Butyl chloride, aluminum chloride
(h)   Acetyl chloride, aluminum chloride
(i)   6M hydrochloric acid, reflux
(j)   Aqueous sodium hydroxide, reflux

24.19 Each of the following reactions has been reported in the chemical literature and proceeds in good yield. Identify the principal organic product of each reaction.

(a)   1,2-Diethyl-4-nitrobenzene $\xrightarrow[\text{ethanol}]{H_2,\ Pt}$

(b)   1,3-Dimethyl-2-nitrobenzene $\xrightarrow[\text{2. HO}^-]{\text{1. SnCl}_2,\ \text{HCl}}$

(c)   Product of (b) + $\overset{\overset{\displaystyle O}{\|}}{\text{ClCH}_2\text{CCl}} \longrightarrow$

(d)   Product of (c) + $(CH_3CH_2)_2NH \longrightarrow$

(e)   Product of (d) + HCl $\longrightarrow$

(f) $C_6H_5NH\overset{\overset{\displaystyle O}{\|}}{C}CH_2CH_2CH_3 \xrightarrow[\text{2. HO}^-]{\text{1. LiAlH}_4}$

(g)   Aniline + heptanal $\xrightarrow{H_2,\ Ni}$

(h)   Acetanilide + $\overset{\overset{\displaystyle O}{\|}}{\text{ClCH}_2\text{CCl}} \xrightarrow{\text{AlCl}_3}$

(i) 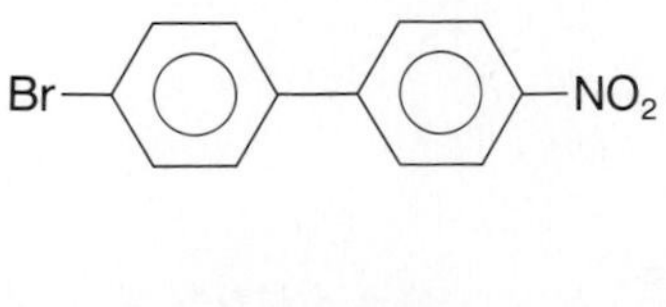   Br—⬡—⬡—NO_2 $\xrightarrow[\text{2. HO}^-]{\text{1. Fe, HCl}}$

(*j*)   Product of (*i*) $\xrightarrow[\text{2. H}_2\text{O, heat}]{\text{1. NaNO}_2,\text{ H}_2\text{SO}_4,\text{ H}_2\text{O}}$

(*k*)   2,6-Dinitroaniline $\xrightarrow[\text{2. CuCl}]{\text{1. NaNO}_2,\text{ H}_2\text{SO}_4,\text{ H}_2\text{O}}$

(*l*)   *m*-Bromoaniline $\xrightarrow[\text{2. CuBr}]{\text{1. NaNO}_2,\text{ HBr, H}_2\text{O}}$

(*m*)   *o*-Nitroaniline $\xrightarrow[\text{2. CuCN}]{\text{1. NaNO}_2,\text{ HCl, H}_2\text{O}}$

(*n*)   2,6-Diiodo-4-nitroaniline $\xrightarrow[\text{2. KI}]{\text{1. NaNO}_2,\text{ H}_2\text{SO}_4,\text{ H}_2\text{O}}$

(*o*) $:N\equiv\overset{+}{N}$—⟨O⟩—⟨O⟩—$\overset{+}{N}\equiv N:$ $2\overset{-}{B}F_4$ $\xrightarrow{\text{heat}}$

(*p*)   2,4,6-Trinitroaniline $\xrightarrow[\text{H}_2\text{O, H}_3\text{PO}_2]{\text{NaNO}_2,\text{ H}_2\text{SO}_4}$

(*q*)   2-Amino-5-iodobenzoic acid $\xrightarrow[\text{2. CH}_3\text{CH}_2\text{OH}]{\text{1. NaNO}_2,\text{ HCl, H}_2\text{O}}$

(*r*)   Aniline $\xrightarrow[\text{2. 2,3,6-trimethylphenol}]{\text{1. NaNO}_2,\text{ H}_2\text{SO}_4,\text{ H}_2\text{O}}$

(*s*) $(CH_3)_2N$—⟨O⟩ $\xrightarrow[\text{2. HO}^-]{\text{1. NaNO}_2,\text{ HCl, H}_2\text{O}}$
          CH$_3$

(*t*)   *o*-Aminobenzoic acid + *p*-toluenesulfonyl chloride $\longrightarrow$

24.20 Each of the following compounds has been prepared from *p*-nitroaniline. Outline a reasonable series of steps leading to each one.

(*a*)   *p*-Nitrobenzonitrile
(*b*)   3,4,5-Trichloroaniline
(*c*)   1,3-Dibromo-5-nitrobenzene
(*d*)   3,5-Dibromoaniline
(*e*)   *p*-Acetamidophenol (*acetaminophen*)

24.21 Each of the following compounds has been prepared from *o*-anisidine (*o*-methoxyaniline). Outline a series of steps leading to each one.

(*a*)   *o*-Bromoanisole
(*b*)   *o*-Fluoroanisole
(*c*)   3-Fluoro-4-methoxyacetophenone
(*d*)   3-Fluoro-4-methoxybenzonitrile
(*e*)   3-Fluoro-4-methoxyphenol

24.22 Design syntheses of each of the following compounds from the indicated starting material and any necessary organic or inorganic reagents:

(*a*)   *p*-Aminobenzoic acid from *p*-toluidine

(*b*)   *p*-FC$_6$H$_4$$\overset{\overset{\text{O}}{\|}}{\text{C}}CH_2CH_3$ from benzene
(*c*)   1-Bromo-2-fluoro-3,5-dimethylbenzene from *m*-xylene

(d) 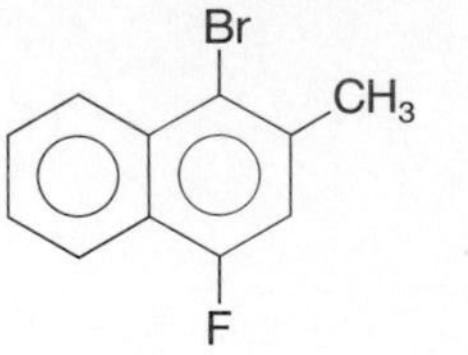 [structure: 1-bromo-2-methyl-4-fluoronaphthalene] from [structure: 1-amino-2-methyl-4-nitronaphthalene]

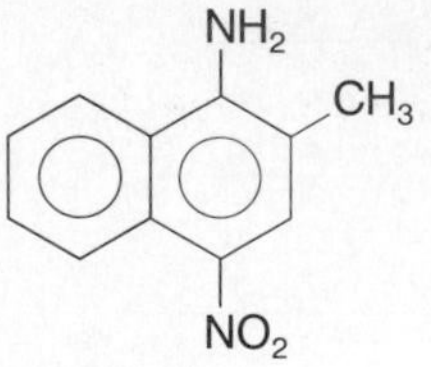

(e) o-BrC₆H₄C(CH₃)₃ from p-O₂NC₆H₄C(CH₃)₃

(f) m-ClC₆H₄C(CH₃)₃ from p-O₂NC₆H₄C(CH₃)₃

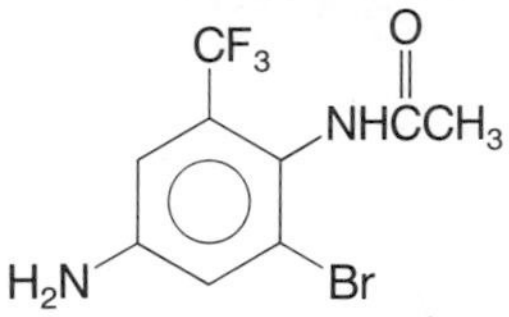

(g) 1-Bromo-3,5-diethylbenzene from m-diethylbenzene

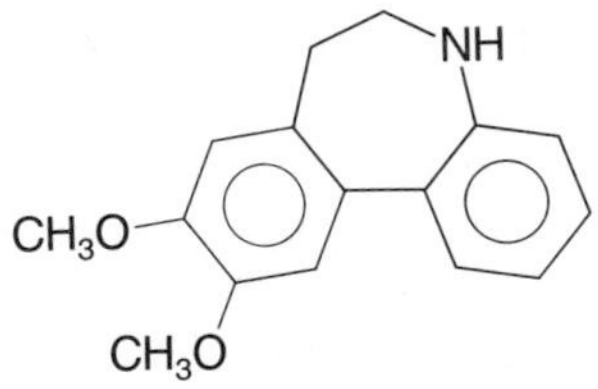

(h) 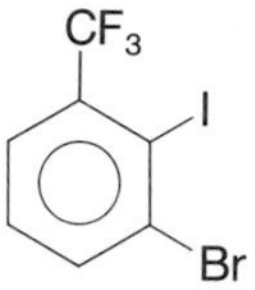 [structure: 1-CF₃, 2-I, 3-Br benzene] from [structure: benzene with CF₃, NHCCH₃ (O), Br, H₂N]

(i) [structure: tricyclic amine with two CH₃O groups, NH] from [structure: biphenyl with CH₂COCH₃ (O), CH₃O, CH₃O, O₂N]

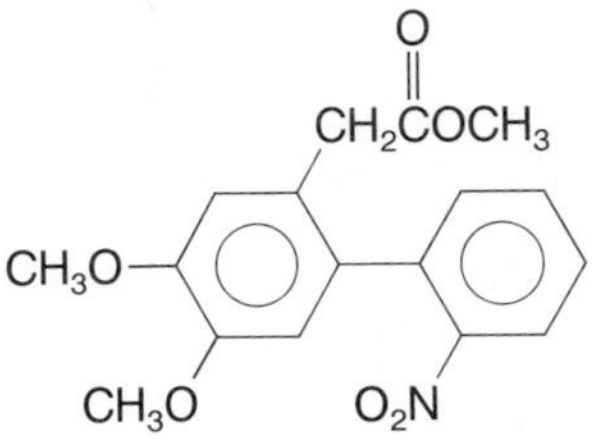

(24.23) The basicity constants of N,N-dimethylaniline and pyridine are almost the same, while 4-(N,N-dimethylamino)pyridine is considerably more basic than either.

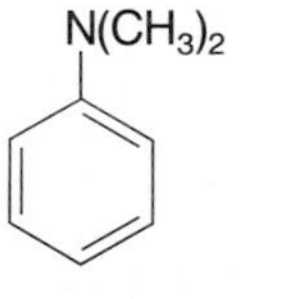

[structures shown below]

N,N-Dimethylaniline
K_b 1.3 × 10⁻⁹
pK_b 8.9

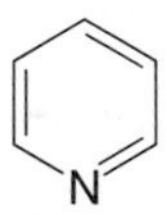

Pyridine
K_b 2 × 10⁻⁹
pK_b 8.7

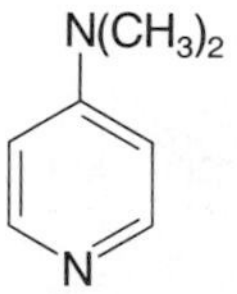

4-(N,N-Dimethylamino)pyridine
K_b = 5 × 10⁻⁵
pK_b 4.3

Identify the more basic of the two nitrogens of 4-(N,N-dimethylamino)pyridine and suggest an explanation for its enhanced basicity as compared with pyridine and N,N-dimethylaniline.

(24.24) Nathan Kornblum and his students at Purdue University discovered that while primary arylamines are routinely converted to their diazonium salts by nitrosation at pH 1 and below, primary alkylamines are inert below pH 3. This observation permitted them to carry out the selective reduction of an arylamine in the presence of an alkylamine.

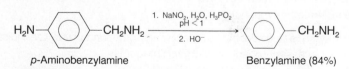

p-Aminobenzylamine   Benzylamine (84%)

Suggest an explanation for the selectivity observed in this reaction.

ARYL HALIDES

T he value of *alkyl halides* as starting materials for the preparation of a variety of organic functional groups has been stressed many times. In those earlier discussions, it was noted that *aryl halides* are normally much less reactive than alkyl halides in reactions that involve carbon-halogen bond cleavage. In the present chapter you will see that aryl halides can exhibit their own patterns of chemical reactivity, and that these reactions are novel, useful, and mechanistically interesting.

25.1 BONDING IN ARYL HALIDES

Aryl halides are compounds in which a halogen substituent is attached directly to an aromatic ring. Representative aryl halides include

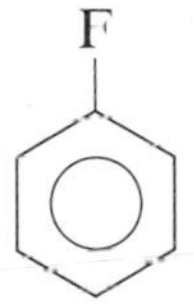

Fluorobenzene 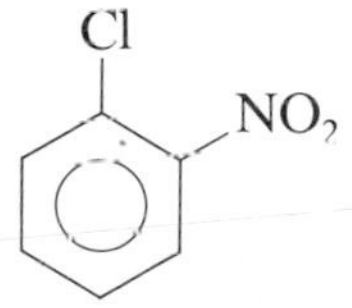   *o*-Chloronitrobenzene 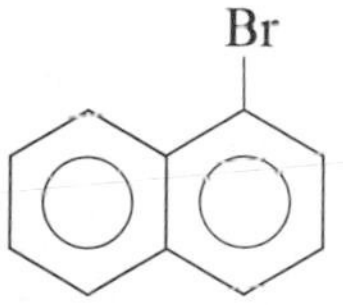   1-Bromonaphthalene 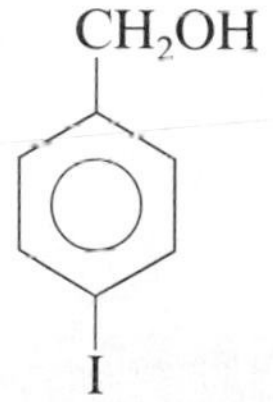   *p*-Iodobenzyl alcohol

Halogen-containing organic compounds in which the halogen substituent is not directly bonded to an aromatic ring, even though an aromatic ring may be present, are not aryl halides. Benzyl chloride ($C_6H_5CH_2Cl$), for example, is not an aryl halide.

As the data in Table 25.1 illustrate, the carbon-chlorine bond of an aryl chloride is both shorter and stronger than the carbon-chlorine bond of an alkyl chloride. In this respect aryl halides resemble vinyl halides more than they resemble alkyl halides. Since carbon-hydrogen and carbon-halogen bond distances and bond dissociation energies exhibit a similar dependence on structure, a hybridization effect seems to be

TABLE 25.1

Bond Distances and Bond Dissociation Energies of Selected Compounds

Compound	Hybridization of carbon to which X is attached	X = H		X = Cl	
		Bond distance, Å	Bond energy, kcal/mol	Bond distance, Å	Bond energy, kcal/mol
CH_3CH_2X	sp^3	1.11	98	1.79	81
$CH_2{=}CHX$	sp^2	1.10	108	1.72	88
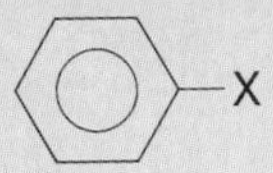 —X	sp^2	1.08	112	1.74	97

responsible. An increase in s character from 25 percent (sp^3 hybridization) to 33.3 percent (sp^2 hybridization) increases the tendency of carbon to attract electrons and to bind substituents more strongly.

PROBLEM 25.1 Consider all the isomers of C_7H_7Cl containing a benzene ring and write the structure of the one that has the weakest carbon-chlorine bond as measured by its bond dissociation energy.

Resonance contributions to the increased carbon-halogen bond strengths appear to be relatively unimportant except in aryl fluorides, where electron donation from fluorine into the ring imparts double bond character to the carbon-fluorine bond.

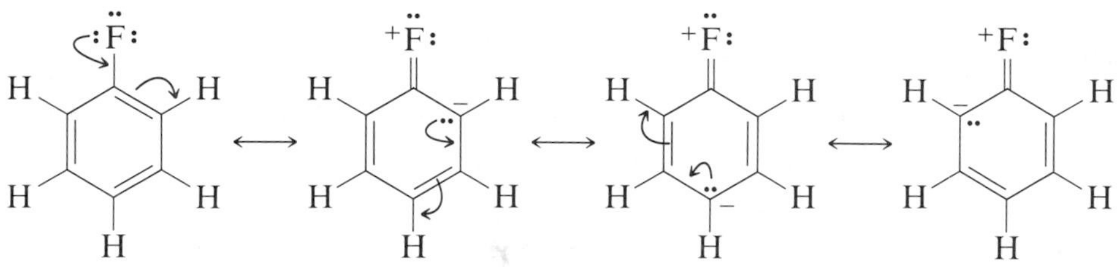

(Most stable Lewis structure)   Dipolar resonance forms of fluorobenzene

---

PROBLEM 25.2 For each of the following compounds, write a resonance structure that contains both a carbon-fluorine double bond and a carbon-nitrogen double bond:

(a) *p*-Fluoronitrobenzene
(b) *o*-Fluoronitrobenzene
(c) 4-Fluoro-1,2-dinitrobenzene

SAMPLE SOLUTION (a) Fluorine donates an electron pair to the aromatic ring by resonance; nitro groups withdraw an electron pair. Conjugation of an unshared electron pair of fluorine with a para nitro group generates a resonance form that has both a carbon-fluorine double bond and a carbon-nitrogen double bond.

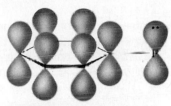

FIGURE 25.1 Overlap of a fluorine 2p orbital with a neighboring π system permits delocalization of an unshared electron pair of the halogen in an aryl fluoride.

Alternatively, the partial double bond character implied in the resonance description of fluorobenzene may be represented in orbital overlap terms, as shown in Figure 25.1. Overlap of a fluorine 2p orbital with the π system of the aromatic ring permits an unshared electron pair of the halogen to be delocalized into the aromatic ring. This adds a π bonding component to the carbon-halogen bond and strengthens it. The carbon-halogen bond distances of the other halogens are so long that overlap of a 2p orbital of chlorine, bromine, or iodine with the π system of the ring is not as effective and the degree of stabilization due to electon delocalization is less.

A situation comparable with that in halobenzenes exists in haloalkenes. Halogen is attached to the sp^2 hybridized carbon by a bond that is both shorter and stronger than that of an alkyl halide.

The strength of their carbon-halogen bonds causes aryl halides to react very slowly in those nucleophilic substitution reactions in which carbon-halogen bond cleavage is rate-determining. Later in this chapter we will see examples of nucleophilic substitution reactions of aryl halides that do take place at reasonable rates but proceed by mechanisms distinctly different from the classical S_N1 and S_N2 pathways.

25.2 SOURCES OF ARYL HALIDES

The two principal methods for the preparation of aryl halides—direct halogenation of arenes by electrophilic aromatic substitution and preparation by way of aryl diazonium salts—have been described earlier and are reviewed in Table 25.2. A number of aryl halides occur naturally, some of which are shown in Figure 25.2.

25.3 PHYSICAL PROPERTIES OF ARYL HALIDES

Aryl halides resemble alkyl halides in many of their physical properties. All are practically insoluble in water and are denser than water. Physical constants are listed for some representative examples in Table 25.3. All the monohalobenzenes are liquids. Among the various dihalobenzenes, the para isomers have the highest melting points because their high symmetry allows for the most efficient packing in the crystal. The boiling points of dihalobenzenes, however, are relatively insensitive to the substitution pattern.

Aryl halides are polar molecules but are less polar than alkyl halides.

Chlorocyclohexane
μ 2.2 D

Chlorobenzene
μ 1.7 D

TABLE 25.2

Summary of Reactions Discussed in Earlier Chapters That Yield Aryl Halides

Reaction (section) and comments	General equation and specific example
Halogenation of arenes (Section 13.5) Aryl chlorides and bromides are conveniently prepared by electrophilic aromatic substitution. The reaction is limited to chlorination and bromination. Fluorination is difficult to control; iodination is too slow to be useful.	ArH + X_2 $\xrightarrow[\text{or FeX}_3]{\text{Fe}}$ ArX + HX Arene · Halogen · Aryl halide · Hydrogen halide O_2N—⬡ + Br_2 $\xrightarrow{\text{Fe}}$ O_2N—⬡(Br) Nitrobenzene · Bromine · m-Bromonitrobenzene (85%)
The Sandmeyer reaction (Section 24.9) Diazotization of a primary arylamine followed by treatment of the diazonium salt with cuprous bromide or cuprous chloride yields the corresponding aryl bromide or aryl chloride.	$ArNH_2$ $\xrightarrow[\text{2. CuX}]{\text{1. NaNO}_2\text{, H}_3\text{O}^+}$ ArX Primary arylamine · Aryl halide 1-Amino-8-chloronaphthalene $\xrightarrow[\text{2. CuBr}]{\text{1. NaNO}_2\text{, HBr}}$ 1-Bromo-8-chloronaphthalene (62%)
The Schiemann reaction (Section 24.9) Diazotization of an arylamine followed by treatment with fluoroboric acid gives an aryl diazonium fluoroborate salt. Heating this salt converts it to an aryl fluoride.	$ArNH_2$ $\xrightarrow[\text{2. HBF}_4]{\text{1. NaNO}_2\text{, H}_3\text{O}^+}$ $Ar\overset{+}{N}\!\equiv\!N\!:\ BF_4^-$ $\xrightarrow{\text{heat}}$ ArF Primary arylamine · Aryl diazonium fluoroborate · Aryl fluoride $C_6H_5NH_2$ $\xrightarrow[\text{2. HBF}_4 \text{ 3. heat}]{\text{1. NaNO}_2\text{, H}_2\text{O, HCl}}$ C_6H_5F Aniline · Fluorobenzene (51–57%)
Reaction of aryl diazonium salts with iodide ion (Section 24.9) Adding potassium iodide to a solution of an aryl diazonium ion leads to the formation of an aryl iodide.	$ArNH_2$ $\xrightarrow[\text{2. KI}]{\text{1. NaNO}_2\text{, H}_3\text{O}^+}$ ArI Primary arylamine · Aryl iodide $C_6H_5NH_2$ $\xrightarrow[\text{2. KI}]{\text{1. NaNO}_2\text{, HCl, H}_2\text{O}}$ C_6H_5I Aniline · Iodobenzene (74–76%)

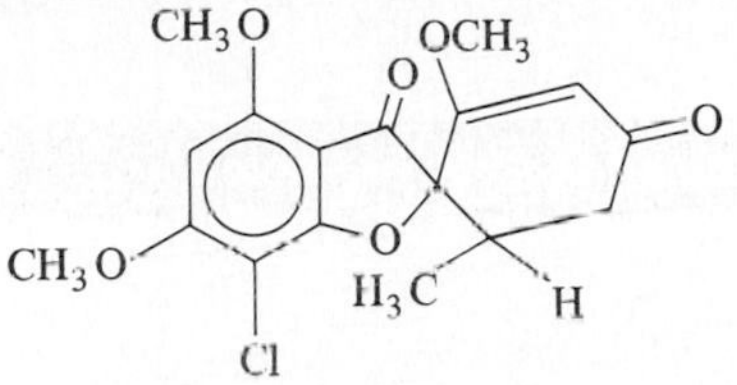

Griseofulvin: biosynthetic product of a particular microorganism, used as an orally administered antifungal agent.

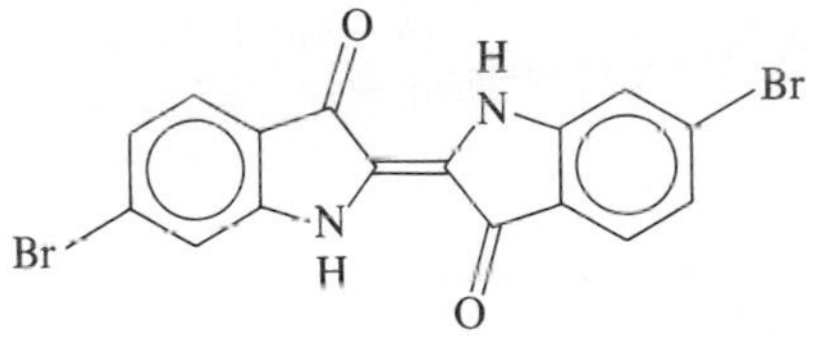

Dibromoindigo: principal constituent of a dye known as Tyrian purple, which is isolated from a species of Mediterranean sea snail and was much prized by the ancients for its vivid color

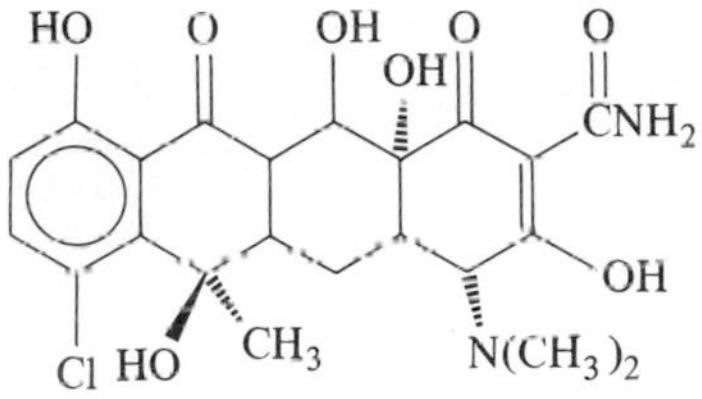

Chlortetracycline: an antibiotic

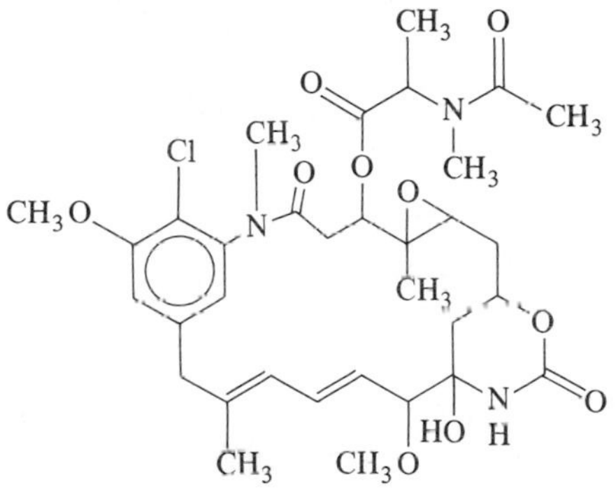

Maytansine: a potent antitumor agent isolated from a bush native to Kenya; 10 tons of plant yielded 6 g maytansine.

FIGURE 25.2 Some naturally occurring aryl halides.

TABLE 25.3

Boiling Points and Melting Points of Some Aryl Halides*

| | Halogen substituent (X) | | | | | | | |
| | Fluorine | | Chlorine | | Bromine | | Iodine | |
Compound	mp	bp	mp	bp	mp	bp	mp	bp
C_6H_5X	−41	85	−45	132	−31	156	−31	188
$o\text{-}C_6H_4X_2$	−34	91	−17	180	7	225	27	286
$m\text{-}C_6H_4X_2$	−59	83	−25	173	−7	218	35	285
$p\text{-}C_6H_4X_2$	−13	89	53	174	87	218	129	285
$1,3,5\text{-}C_6H_3X_3$	−5	76	63	208	121	271	184	
C_6X_6	5	80	230	322	327		350	

* All boiling points and melting points cited are in °C.

TABLE 25.4

Summary of Reactions of Aryl Halides Discussed in Earlier Chapters

Reaction (section) and comments	General equation and specific example
Electrophilic aromatic substitution (Section 13.14) Halogen substituents are slightly deactivating and ortho-para directing.	Br—C₆H₄ $\xrightarrow[\text{AlCl}_3]{\text{CH}_3\text{COCCH}_3}$ Br—C₆H₄—CCH₃ Bromobenzene → p-Bromoacetophenone (69–79%)
Formation of aryl Grignard reagents (Section 15.4) Aryl halides react with magnesium to form the corresponding arylmagnesium halide. Aryl iodides are the most reactive, aryl fluorides the least. A similar reaction occurs with lithium to give aryllithium reagents (Section 15.3).	ArX + Mg $\xrightarrow{\text{diethyl ether}}$ ArMgX Aryl halide · Magnesium · Arylmagnesium halide C₆H₅—Br + Mg $\xrightarrow{\text{diethyl ether}}$ C₆H₅—MgBr Bromobenzene · Magnesium · Phenylmagnesium bromide (95%)

Since carbon is sp^2 hybridized in chlorobenzene, it is more electronegative than the sp^3 hybridized carbon of chlorocyclohexane. Consequently, the distortion of electron density away from carbon toward chlorine is less pronounced in aryl halides than it is in alkyl halides and the molecular dipole moment is smaller.

25.4 REACTIONS OF ARYL HALIDES. A REVIEW AND A PREVIEW

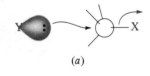

(a)

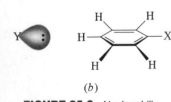

(b)

FIGURE 25.3 Nucleophilic substitution with inversion of configuration is blocked by the benzene ring of an aryl halide. (a) Alkyl halide: The new bond is formed by attack of the nucleophile at carbon from the side opposite the bond to the leaving group. Inversion of configuration is observed. (b) Aryl halide: The aromatic ring blocks approach of the nucleophile to carbon at side opposite bond to the leaving group. Inversion of configuration is impossible.

Table 25.4 summarizes the reactions of aryl halides that we have encountered to this point. Noticeably absent from the table are nucleophilic substitution reactions. Chlorobenzene, for example, is essentially inert to aqueous sodium hydroxide at room temperature. Reaction temperatures over 300°C are required for nucleophilic substitution to proceed at a reasonable rate.

C₆H₅—Cl $\xrightarrow[\text{2. H}^+]{\text{1. NaOH, H}_2\text{O, 370°C}}$ C₆H₅—OH

Chlorobenzene → Phenol (97%)

Aryl halides are much less reactive than alkyl halides in nucleophilic substitution reactions. The carbon-halogen bonds of aryl halides are too strong, and aryl cations are too high in energy, to permit aryl halides to ionize readily in S_N1-type processes. Further, as Figure 25.3 depicts, the optimal transition-state geometry required for S_N2 processes cannot be achieved. Nucleophilic attack from the side opposite the carbon-halogen bond is blocked by the aromatic ring.

Nevertheless, nucleophilic aromatic substitution reactions can take place when either one of the following special circumstances exists:

1. The aromatic ring bears a strongly electron-withdrawing substituent such as a nitro group in a position ortho or para to a halogen leaving group

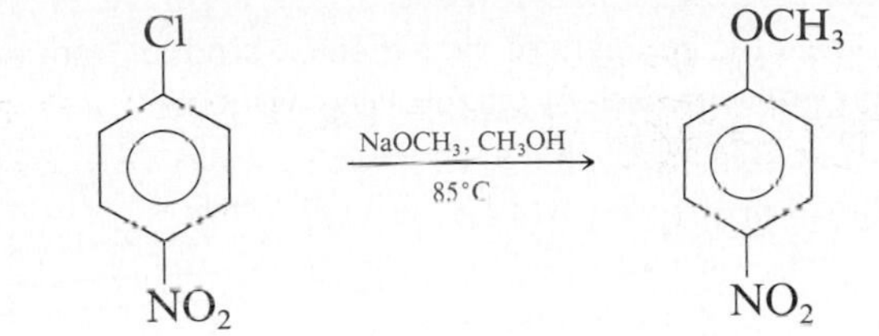

p-Chloronitrobenzene p-Nitroanisole (92%)

2. The nucleophile is an exceptionally strong base such as sodium amide or potassium amide

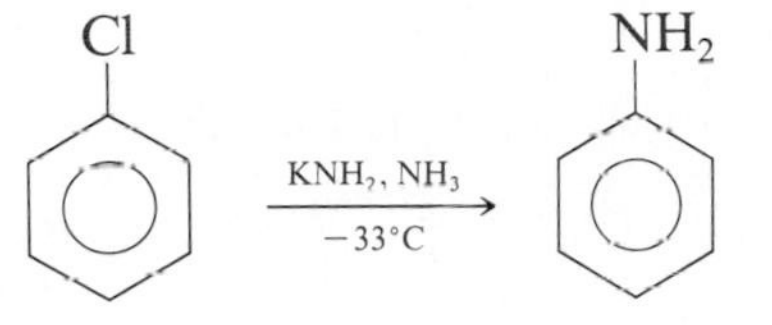

Chlorobenzene   Aniline (52%)

These reactions take place by mechanisms that are different from any we have discussed so far. The nucleophilic aromatic substitution reaction shown in (1) proceeds by an *addition-elimination mechanism;* the one shown in (2) proceeds by an *elimination-addition mechanism.* Let us examine each of these mechanisms in detail.

25.5 THE ADDITION-ELIMINATION MECHANISM OF NUCLEOPHILIC AROMATIC SUBSTITUTION

Although chlorobenzene reacts readily with concentrated sodium hydroxide only at temperatures over 300°C, 1-chloro-2,4-dinitrobenzene is converted to 2,4-dinitrophenol merely by heating with the weak hydroxide ion source, aqueous sodium carbonate, on a steam bath.

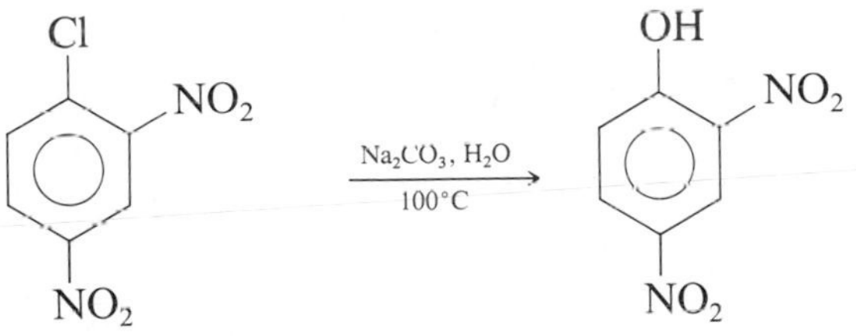

1-Chloro-2,4-dinitrobenzene   2,4-Dinitrophenol (90%)

Other nucleophiles such as ammonia react similarly, displacing chloride from the aromatic ring.

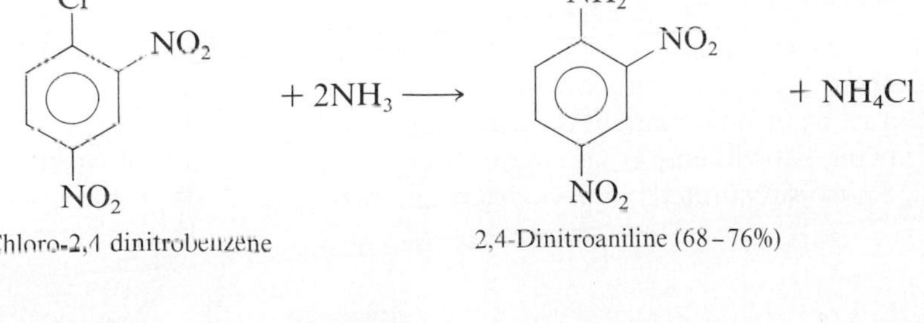

1-Chloro-2,4 dinitrobenzene   2,4-Dinitroaniline (68–76%)

PROBLEM 25.3   Write the structure of the expected product from the reaction of 1-chloro-2,4-dinitrobenzene with each of the following reagents:

(a) CH_3CH_2ONa

(c) CH_3NH_2

(b) $C_6H_5CH_2SNa$

SAMPLE SOLUTION   (a) Sodium ethoxide is a source of the nucleophile $CH_3CH_2O^-$, which displaces chloride from 1-chloro-2,4-dinitrobenzene.

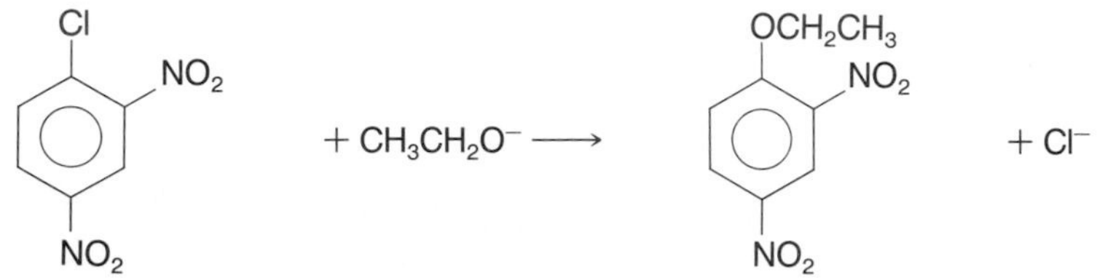

| 1-Chloro-2,4-dinitrobenzene | Ethoxide anion | 1-Ethoxy-2,4-dinitrobenzene |

In contrast to nucleophilic aliphatic substitution, fluoride ion can serve as a leaving group in nucleophilic aromatic substitution.

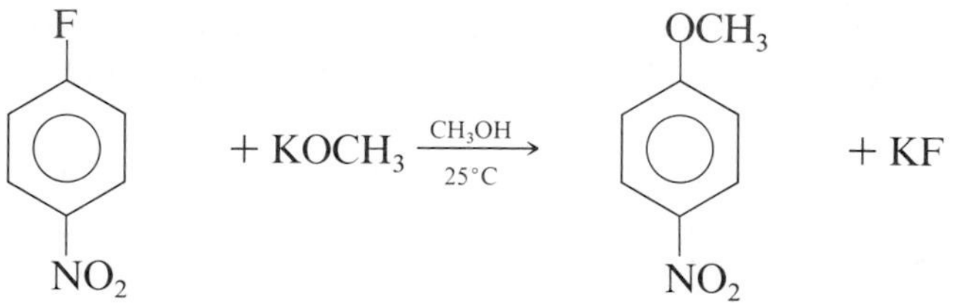

p-Fluoronitrobenzene   *p*-Nitroanisole (93%)

Among the experimental facts that bear on mechanism are the results of kinetic studies. The rate of the reaction depicted in the preceding equation is directly proportional to the concentration of the aryl halide and to the concentration of the nucleophile.

$$\text{Rate} = k\,[p\text{-fluoronitrobenzene}][\text{potassium methoxide}]$$

Since the reaction is characterized by second-order kinetics, it is likely that its rate-determining step is bimolecular and involves both the aryl halide and the nucleophile.

Substituent effects also serve as guides to mechanism. Here, there is a very obvious one; a strongly electron-withdrawing substituent must be present on the aryl halide and it must be ortho or para to the leaving group. The cumulative effect of electron-withdrawing substituents is apparent on comparing the rates of substitution in a series of nitro-substituted chlorobenzene derivatives.

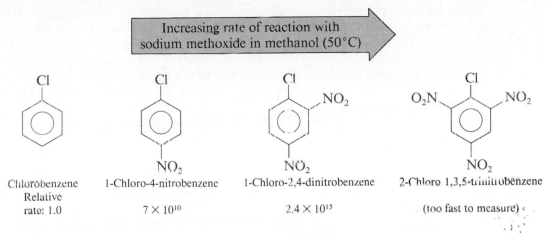

Increasing rate of reaction with
sodium methoxide in methanol (50°C)

Chlorobenzene	1-Chloro-4-nitrobenzene	1-Chloro-2,4-dinitrobenzene	2-Chloro 1,3,5-trinitrobenzene
Relative rate: 1.0	7×10^{10}	2.4×10^{15}	(too fast to measure)

Nitro groups strongly *activate* positions ortho and para to themselves toward nucleophilic aromatic substitution; they strongly *deactivate* these positions toward electrophilic aromatic substitution. The same property of a nitro group, its ability to attract electrons, is responsible for both effects.

In electrophilic aromatic substitution an electron pair is transferred from the aromatic π system to the attacking electrophile. A positively charged intermediate, a cyclohexadienyl cation, is formed in the rate-determining step.

Nitrobenzene and an electrophile → Cyclohexadienyl cation intermediate; nitro group is destabilizing → Product of electrophilic aromatic substitution

A nitro substituent markedly decreases the ability of an aromatic ring to transfer electrons to the electrophile. It destabilizes the cyclohexadienyl cation intermediate and raises the energy of the transition state leading to that intermediate.

In nucleophilic aromatic substitution the species that attacks the aromatic ring is an electron pair donor. The aryl halide is an electron pair acceptor. *Addition* of a nucleophile to the carbon that bears the leaving group generates a cyclohexadienyl anion intermediate. *Elimination* of a halide ion from this cyclohexadienyl anion intermediate gives the product of nucleophilic aromatic substitution.

o-Halonitrobenzene (X = F, Cl, Br, or I) and a nucleophile → Cyclohexadienyl anion intermediate; nitro group is stabilizing → Product of nucleophilic aromatic substitution

The aromatic character of a benzene ring is lost when it is transformed into a cyclohexadienyl anion, so addition of a nucleophile is normally quite endothermic and occurs very slowly. The activation energy is diminished if substituents that are effective at delocalizing negative charge, such as nitro groups, are present on the ring. Electron-withdrawing groups stabilize the cyclohexadienyl anion and lower the activation energy for its formation. This charge delocalization can be represented by the resonance structures shown for the case of an *ortho*-nitro substituent.

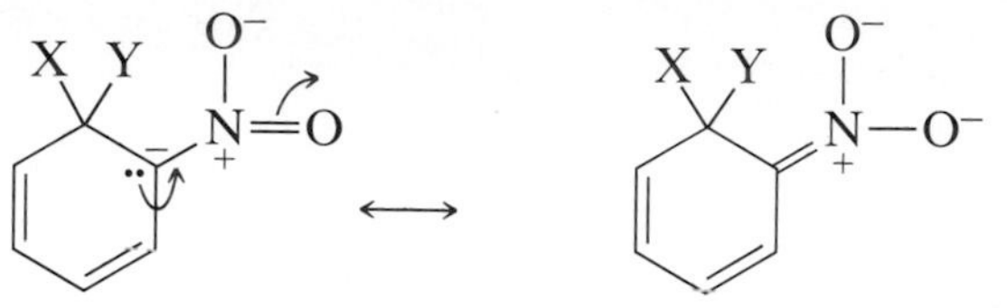

(Most stable resonance structure; negative charge is carried by most electronegative element, oxygen)

Resonance structures for a cyclohexadienyl anion bearing a *para*-nitro group include the following:

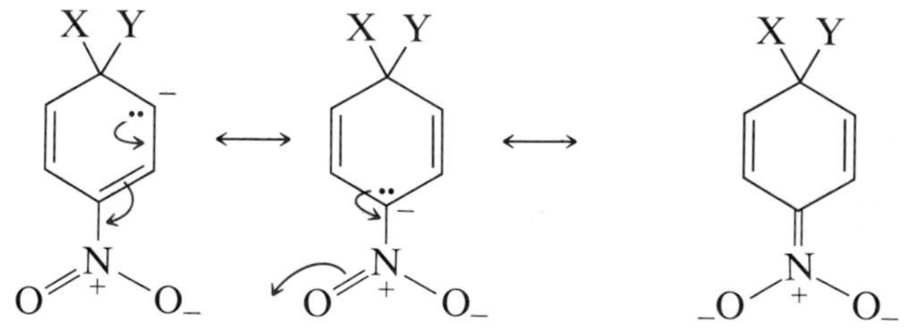

(Most stable resonance structure; negative charge is on oxygen)

Direct conjugation of the negatively charged carbon and the nitro group is not possible in the cyclohexadienyl anion intermediate derived from a *meta*-halonitrobenzene. Substrates of this type react slowly in nucleophilic aromatic substitution reactions.

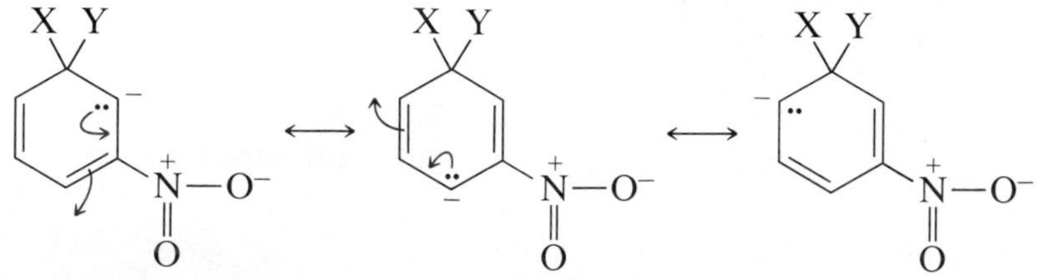

(Negative charge is restricted to carbon in all resonance forms of cyclohexadienyl anion derived from a *meta*-halonitrobenzene.)

Once formed, the cyclohexadienyl anion intermediate is converted to the product of nucleophilic aromatic substitution by expelling the anion of the leaving group. Halide is lost from an sp^3 hybridized carbon. This step restores the aromaticity of the ring.

Figure 25.4 presents an energy diagram depicting the addition-elimination mechanism for the reaction of *p*-fluoronitrobenzene with methoxide ion.

p-Fluoronitrobenzene + CH$_3$O$^-$ $\xrightarrow[\text{slow}]{\text{addition}}$ Cyclohexadienyl anion intermediate $\xrightarrow[\text{last}]{\text{elimination}}$ p-Nitroanisole + F$^-$

p-Fluoronitrobenzene	
Methoxide ion	
Cyclohexadienyl anion intermediate	
p-Nitroanisole	Fluoride ion

That the addition step, rather than the elimination step, is rate-determining is suggested by the observed order of reactivity in a series of p-nitro substituted aryl halides, which is ArF > ArCl > ArBr > ArI.

Relative reactivity toward sodium methoxide in methanol (50°C):

X = F	312
X = Cl	1.0
X = Br	0.8
X = I	0.4

If loss of halide were rate-determining, the aryl halide with the weakest carbon-halogen bond, the aryl iodide, would be the most reactive while the one with the strongest

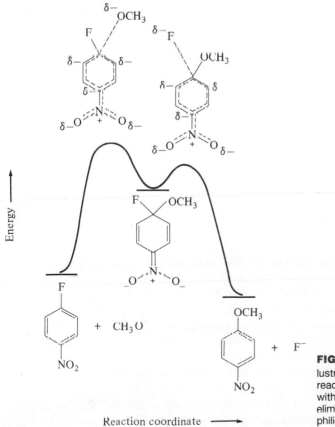

FIGURE 25.4 Energy diagram illustrating energy changes during reaction of p-fluoronitrobenzene with methoxide ion by the addition-elimination mechanism of nucleophilic aromatic substitution.

carbon-halogen bond, the aryl fluoride, would be the least reactive. Exactly the opposite substituent dependence was observed, so carbon-halogen bond cleavage cannot be involved in any significant way in the rate-determining step. Therefore, it must be the addition step that is rate-determining. Aryl fluorides are the most reactive of the aryl halides because the strongly electronegative fluorine substituent best stabilizes the cyclohexadienyl anion intermediate.

is more stable than

Fluorine stabilizes
cyclohexadienyl anion
by withdrawing electrons

Chlorine is less electronegative
than fluorine and does not
stabilize cyclohexadienyl
anion to as great an extent

Cyclohexadienyl anion intermediates derived by nucleophilic addition to aryl fluorides, being more stable than those derived from other aryl halides, are formed faster, and the rate of their formation determines the rate of the overall reaction.

PROBLEM 25.4 Reaction of 1,2,3-tribromo-5-nitrobenzene with sodium ethoxide in ethanol gave a single product, $C_8H_7Br_2NO_3$, in quantitative yield. Suggest a reasonable structure for this compound.

PROBLEM 25.5 Of the two compounds shown, one reacts with sodium methoxide in methanol 800 times faster than the other at 50°C. Which compound is more reactive? Why?

o-Chloronitrobenzene

1-Chloro-2-nitro-4-
(trifluoromethyl)benzene

The addition-elimination mechanism for nucleophilic aromatic substitution is important not only because it provides a reasonable explanation for the high reactivity of nitro-substituted aryl halides but also because it illustrates a principle you should remember. The words "activating" and "deactivating" as applied to substituent effects in organic chemistry are without meaning when they stand alone. When we say that a group is activating or deactivating, we need to specify the reaction type that is being considered. A nitro group is a strongly *deactivating* substituent in electrophilic aromatic substitution but is a strongly *activating* substituent in nucleophilic aromatic substitution. A nitro group *destabilizes a cyclohexadienyl cation,* the key intermediate in electrophilic aromatic substitution, while it *stabilizes a cyclohexadienyl anion,* the key intermediate in nucleophilic aromatic substitution. By being aware of the connection between reactivity and substituent effects, you will sharpen your perception of how chemical reactions occur.

25.6 THE ELIMINATION-ADDITION MECHANISM OF NUCLEOPHILIC AROMATIC SUBSTITUTION. BENZYNE

Very strong bases such as sodium or potassium amide react readily with aryl halides, even those without electron-withdrawing substituents, to give products corresponding to nucleophilic substitution of halide by the base.

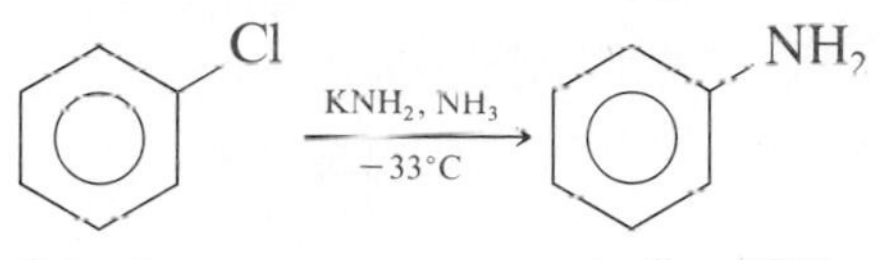

Chlorobenzene $\xrightarrow[-33^\circ C]{KNH_2,\ NH_3}$ Aniline (52%)

For a long time observations concerning the regiochemistry of these reactions presented organic chemists with a puzzle. With some substrates amide ion simply replaced the halogen on the same carbon atom.

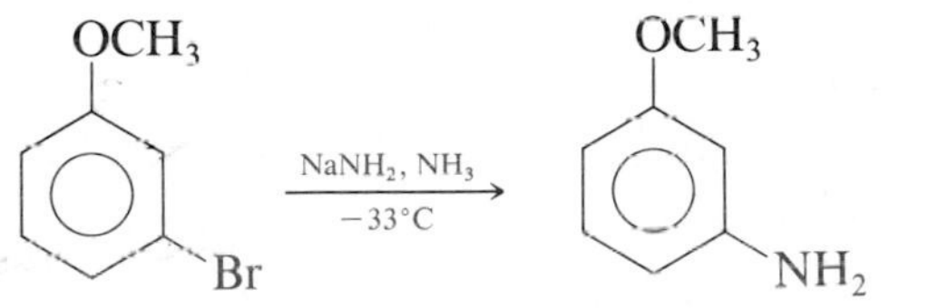

m-Bromoanisole $\xrightarrow[-33^\circ C]{NaNH_2,\ NH_3}$ m-Methoxyaniline (59%)

Other substrates gave products in which the amino group was attached to a carbon atom adjacent to the one bearing the leaving group.

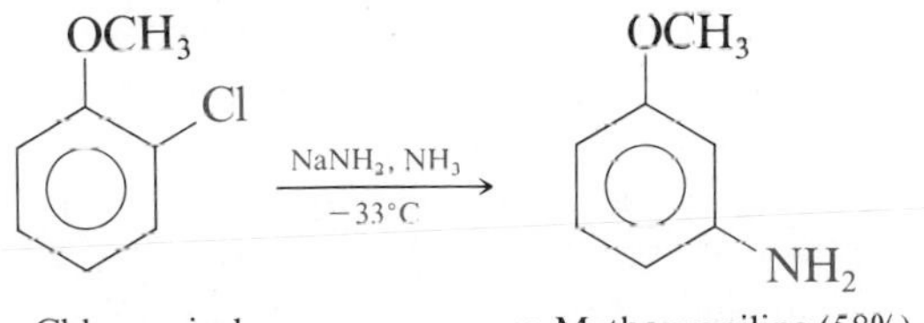

o-Chloroanisole $\xrightarrow[-33^\circ C]{NaNH_2,\ NH_3}$ m-Methoxyaniline (58%)

Some substrates yielded mixtures corresponding to substitution both at the position that bore the leaving group and at positions adjacent to it.

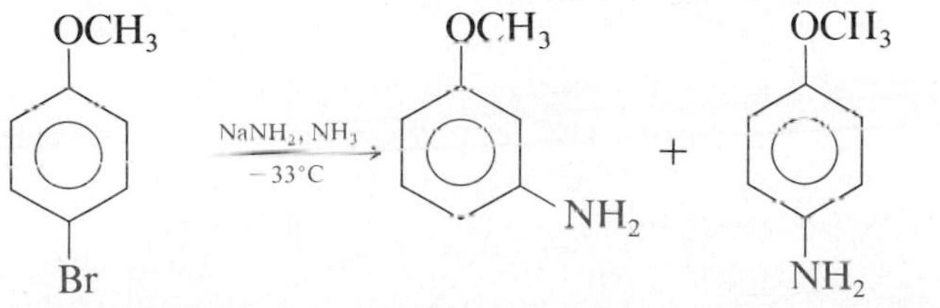

p-Bromoanisole $\xrightarrow[-33^\circ C]{NaNH_2,\ NH_3}$ m-Methoxyaniline + p-Methoxyaniline
(31% yield: equal amounts of m- and p-methoxyaniline)

These results rule out nucleophilic aromatic substitution by the addition-elimination mechanism, since that mechanism requires the attacking nucleophile to attach itself to the carbon from which the leaving group departs.

A solution to the question of the mechanism of these reactions was provided by John D. Roberts in 1953 on the basis of an imaginative experiment. Roberts prepared a sample of chlorobenzene in which one of the carbons, the one bearing the chlorine, was the radioactive mass-14 isotope of carbon. Reaction with potassium amide in liquid ammonia yielded aniline containing almost exactly one-half of its ^{14}C label at C-1 and one-half at C-2.

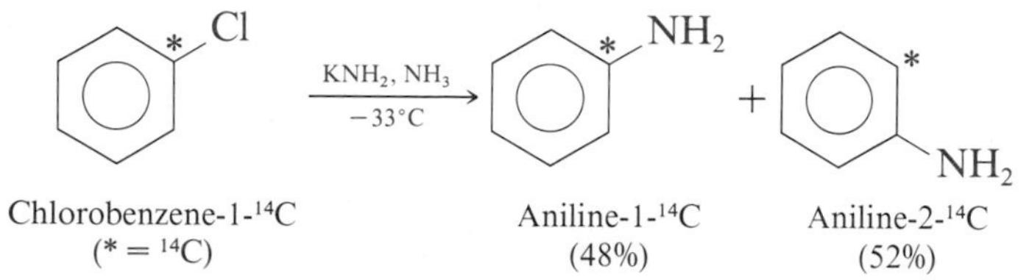

Chlorobenzene-1-^{14}C
(* = ^{14}C)

$\xrightarrow[-33°C]{KNH_2, NH_3}$

Aniline-1-^{14}C
(48%)

+

Aniline-2-^{14}C
(52%)

The mechanism most consistent with the observations of this isotopic labeling experiment is an *elimination-addition* mechanism. The first stage in this mechanism is a base-promoted dehydrohalogenation of chlorobenzene.

Elimination stage:

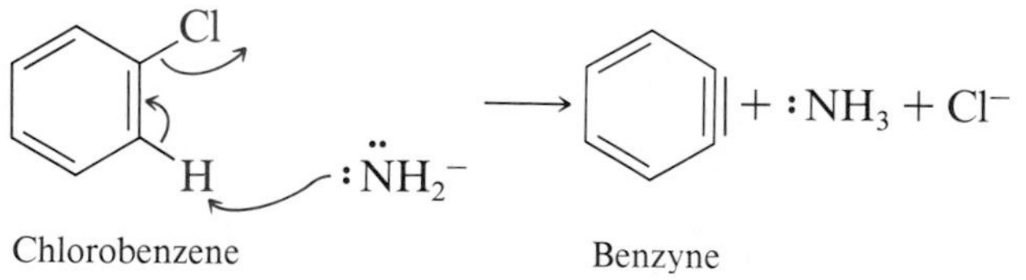

Chlorobenzene $+ :NH_3 + Cl^-$ Benzyne

The intermediate formed in this step contains a triple bond in an aromatic ring and is called *benzyne* or *dehydrobenzene*. Aromatic compounds related to benzyne are known as *arynes*. The triple bond in benzyne is somewhat different from the usual triple bond of an alkyne, however. In benzyne one of the π components of the triple bond is part of the delocalized π system of the aromatic ring. The second π component results from overlapping sp^2 hybridized orbitals (*not p-p overlap*), lies in the plane of the ring, and does not interact with the aromatic π system. This π bond is relatively weak since, as illustrated in Figure 25.5, its contributing sp^2 orbitals are not oriented properly for effective overlap.

Because the ring prevents linearity of the C—C≡C—C unit and because the π bonding in that unit is weak, benzyne is strained and highly reactive. This enhanced reactivity is evident in the second stage of the elimination-addition mechanism.

The degree of overlap of these orbitals is smaller than in the triple bond of an alkyne.

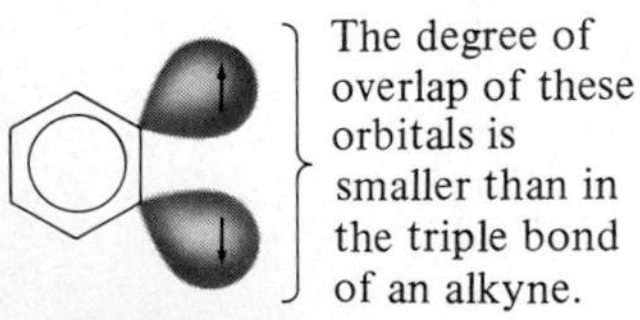

FIGURE 25.5 The sp^2 orbitals in the plane of the ring in benzyne are not properly aligned for good overlap and π bonding is weak.

Addition stage:

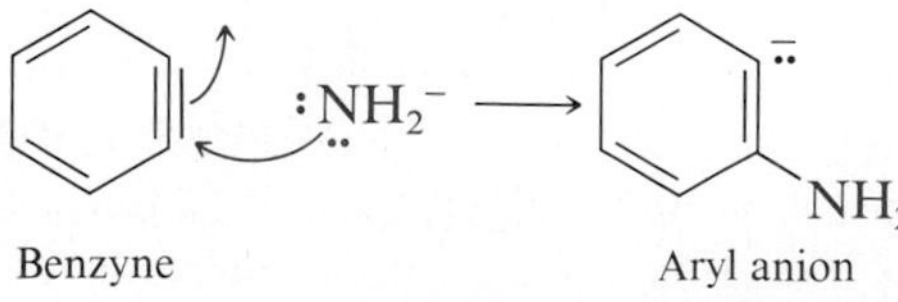

Benzyne $:NH_2^-$ Aryl anion

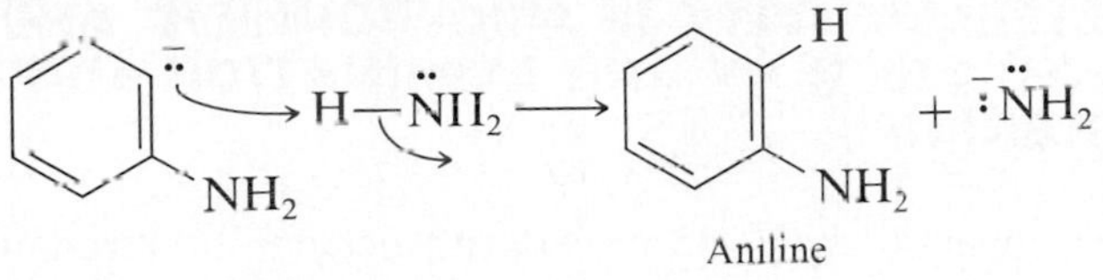

Aniline

In this stage the base acts as a nucleophile and adds to the strained bond of benzyne to form a carbanion, which then abstracts a proton from ammonia to yield the observed product.

The carbon that bears the leaving group and a carbon ortho to it become equivalent in the benzyne intermediate. Thus when chlorobenzene-1-^{14}C is the substrate, the amino group may be introduced with equal likelihood at either of two positions.

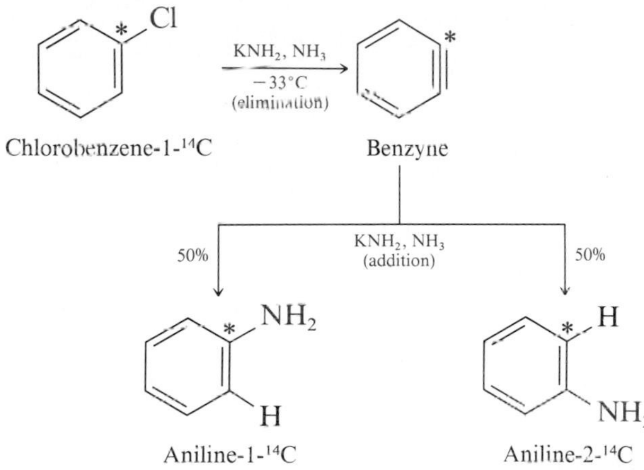

Chlorobenzene-1-^{14}C

$\xrightarrow[\substack{-33°C \\ (elimination)}]{KNH_2, NH_3}$

Benzyne

$\xrightarrow[(addition)]{KNH_2, NH_3}$

50%   50%

Aniline-1-^{14}C   Aniline-2-^{14}C

PROBLEM 25.6 2-Bromo-1,3-dimethylbenzene is inert to nucleophilic aromatic substitution on treatment with sodium amide in liquid ammonia. It is recovered unchanged even after extended contact with the reagent. Suggest an explanation for this lack of reactivity.

While nucleophilic aromatic substitution by the elimination-addition mechanism is most commonly seen with very strong amide bases, it also occurs with bases such as hydroxide ion at high temperatures. A ^{14}C-labeling study revealed that hydrolysis of chlorobenzene proceeds by way of a benzyne intermediate.

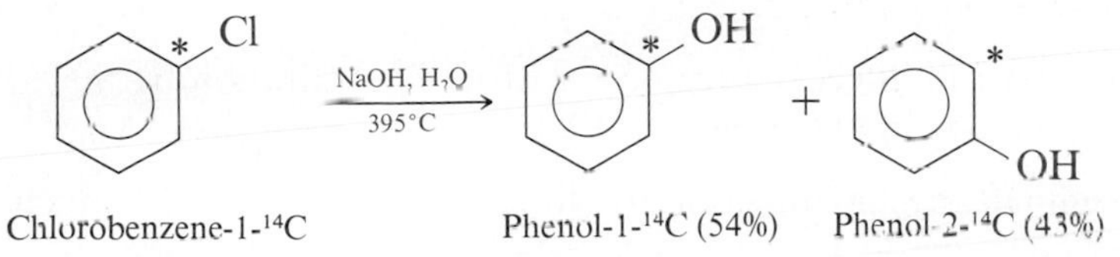

Chlorobenzene-1-^{14}C

$\xrightarrow[395°C]{NaOH, H_2O}$

Phenol-1-^{14}C (54%)   Phenol-2-^{14}C (43%)

PROBLEM 25.7 Two isomeric phenols are obtained in comparable amounts on hydrolysis of p-iodotoluene with 1 M sodium hydroxide at 300°C. Suggest reasonable structures for these two products.

25.7 REGIOSELECTIVITY IN NUCLEOPHILIC AROMATIC SUBSTITUTION BY THE ELIMINATION-ADDITION MECHANISM

Once the intermediacy of benzyne in certain nucleophilic aromatic substitution reactions was established, the influence of substituents on the regioselectivity of these reactions became more readily understood.

Substituent effects can be illustrated by examining the various trifluoromethyl-substituted chlorobenzenes. Both *o*- and *m*-chloro(trifluoromethyl)benzene yield only *m*-(trifluoromethyl)aniline on treatment with potassium amide in liquid ammonia.

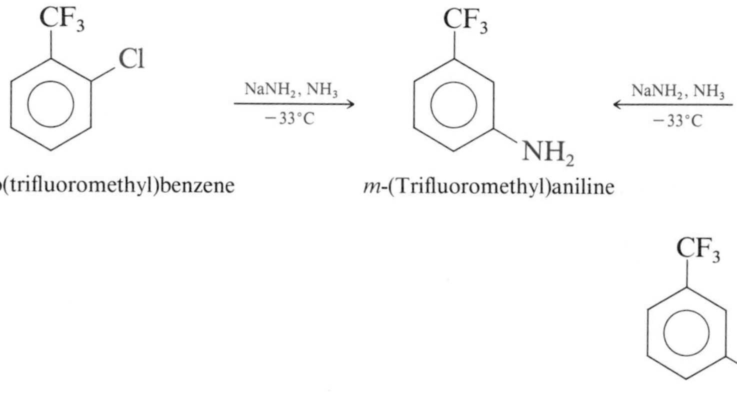

o-Chloro(trifluoromethyl)benzene   *m*-(Trifluoromethyl)aniline

m-Chloro(trifluoromethyl)benzene

p-Chloro(trifluoromethyl)benzene yields equal amounts of *m*- and *p*-(trifluoro-methyl)aniline under the same conditions.

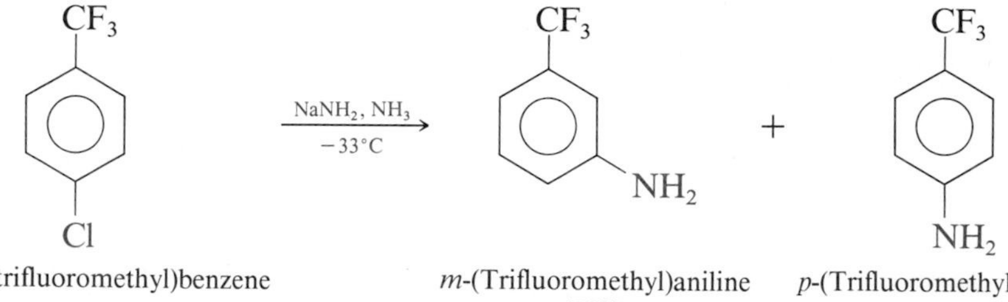

p-Chloro(trifluoromethyl)benzene   *m*-(Trifluoromethyl)aniline (50%)   *p*-(Trifluoromethyl)aniline (50%)

A substituent can affect the regiochemistry of substitution by the elimination-addition mechanism in two ways:

1. By influencing the direction of elimination in the step in which the aryne intermediate is formed
2. By influencing the orientation of nucleophilic addition to the aryne intermediate

Only one direction of elimination is possible for *o*-chloro(trifluoromethyl)benzene; the elimination step must lead to 3-(trifluoromethyl)benzyne.

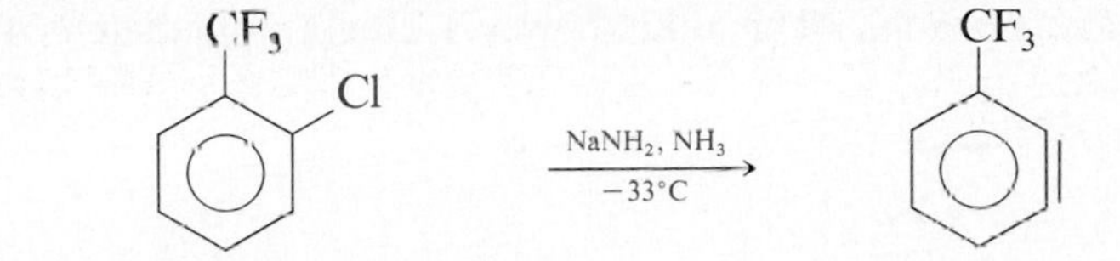

o-Chloro(trifluoromethyl)benzene   3-(Trifluoromethyl)benzyne

Amide ion may add to this intermediate at a position either ortho or meta to the trifluoromethyl group.

Ortho addition of amide ion:

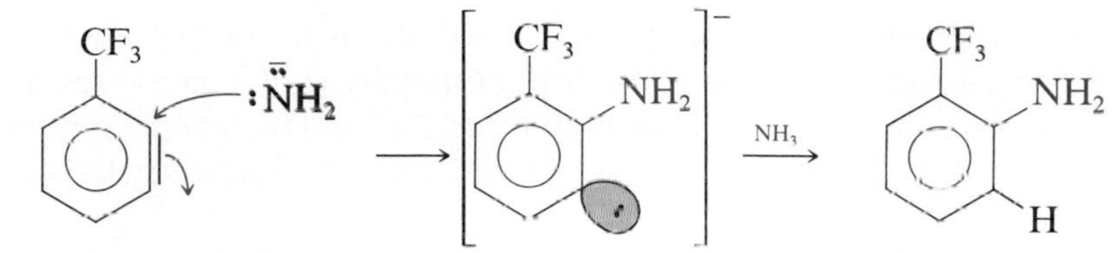

3-(Trifluoromethyl)benzyne   Aryl anion   o-(Trifluoromethyl)aniline

Meta addition of amide ion:

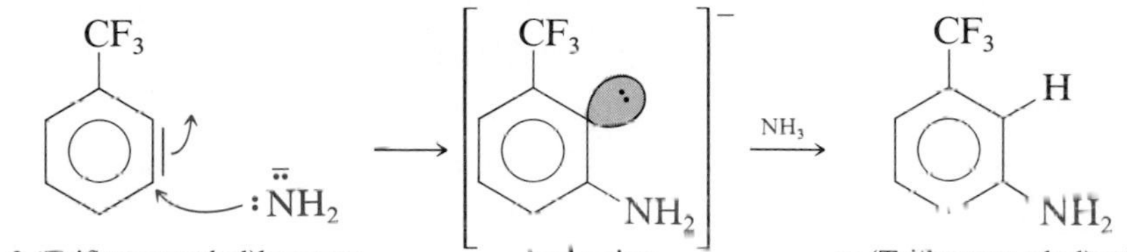

3-(Trifluoromethyl)benzyne   Aryl anion   m-(Trifluoromethyl)aniline

The preferred orientation of addition is the one that provides the more stable carbanion. Of the two aryl anions the one derived from meta addition of amide is the more stable. The effectiveness of the trifluoromethyl group as a carbanion-stabilizing substituent depends on how close it is to the negatively charged carbon. The fewer the number of σ bonds that separate the negatively charged carbon from the electron-withdrawing trifluoromethyl group, the greater the degree of stabilization.

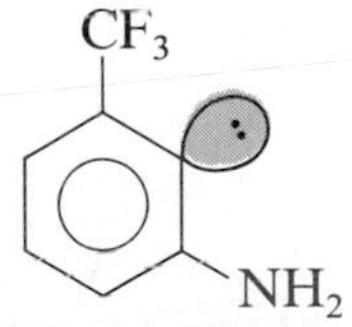

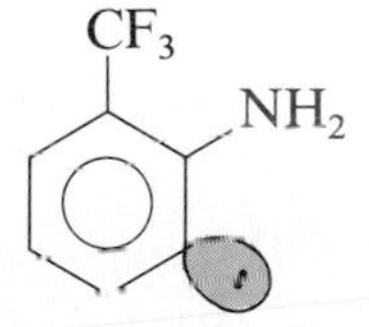

Intermediate corresponding to meta addition of amide: trifluoromethyl group ortho to negatively charged carbon

more stable than

Intermediate corresponding to ortho addition of amide: trifluoromethyl group meta to negatively charged carbon

Since it is more stable, the carbanion intermediate leading to m-(trifluoromethyl)aniline is formed faster than the one leading to the ortho isomer, and the product of meta substitution is the only one formed from o-chloro(trifluoromethyl)benzene.

When p-chloro(trifluoromethyl)benzene is the substrate, again only one benzyne intermediate is possible. This time, however, the intermediate is 4-(trifluoromethyl)benzyne.

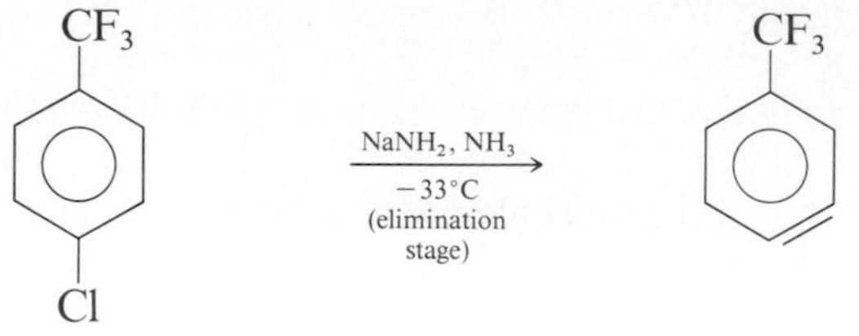

p-Chloro(trifluoromethyl)benzene   4-(Trifluoromethyl)benzyne

There is no regioselective preference for addition to this benzyne intermediate; approximately equal amounts of *m*- and *p*-(trifluoromethyl)aniline are formed. The trifluoromethyl group is remote enough from the triple bond that there is practically no difference in its ability to stabilize negative charge at the meta versus the para carbon.

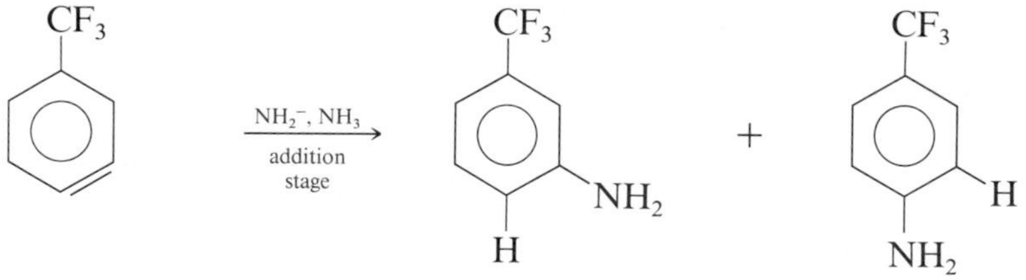

4-(Trifluoromethyl)benzyne   *m*-(Trifluoromethyl)aniline   *p*-(Trifluoromethyl)aniline
                                      (50%)   (50%)

When *m*-chloro(trifluoromethyl)benzene reacts with sodium amide in liquid ammonia, only *m*-(trifluoromethyl)aniline is formed. Therefore, 4-(trifluoromethyl)benzyne cannot be an intermediate. As described in the preceding paragraph, 4-(trifluoromethyl)benzyne *must* give both *m*- and *p*-(trifluoromethyl)aniline in the nucleophilic addition step. The only intermediate formed from *m*-chloro(trifluoromethyl)benzene is 3-(trifluoromethyl)benzyne, and this reacts with amide anion to give *m*-(trifluoromethyl)aniline.

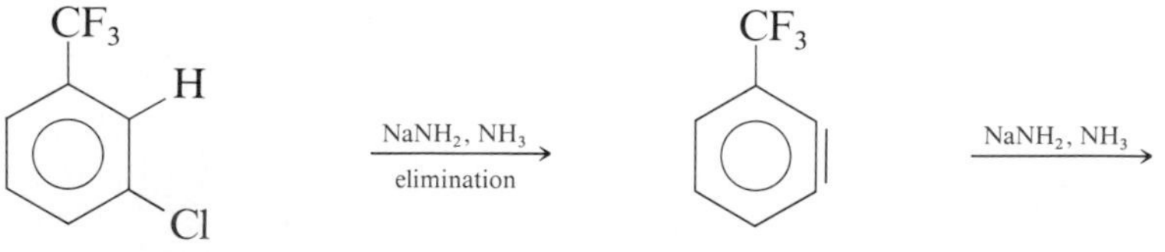

m-Chloro(trifluoromethyl)benzene   3-(Trifluoromethyl)benzyne

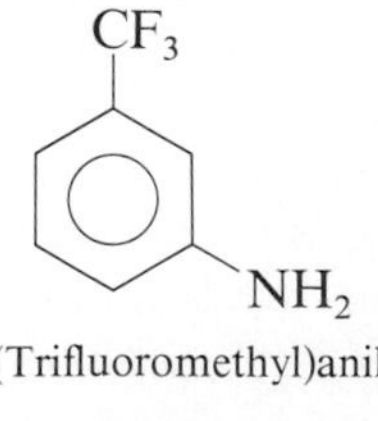

m-(Trifluoromethyl)aniline

The electron-withdrawing effect of the trifluoromethyl group makes a hydrogen ortho to it somewhat more acidic than protons para to it. Base-promoted elimination to 3-(trifluoromethyl)benzyne involves loss of a more acidic proton than elimination to 4-(trifluoromethyl)benzyne and takes place faster.

A methoxy group affects the regioselectivity of substitution by the elimination-addition mechanism in the same way that a trifluoromethyl group does, namely, by acting as a carbanion-stabilizing substituent. While methoxy groups are very powerful π electron donors, resonance effects of this type are relatively unimportant in reactions that proceed by the elimination-addition mechanism of nucleophilic aromatic substitution. What is important is the ability of oxygen to stabilize an electron pair in an sp^2 orbital of a carbon ortho to itself by withdrawing electron density from the σ framework of the ring. An electron pair in the sp^2 orbital of an aryl anion does not interact with the aromatic π system and is not directly conjugated to a substituent on the ring.

The effects of alkyl groups in these reactions are small. Mixtures of isomeric substitution products are formed from reaction of o-, m-, and p-chlorotoluene with sodium amide in liquid ammonia.

PROBLEM 25.8   Predict the products expected to be formed in the reaction of each of the following with sodium amide in liquid ammonia:

(a) o-Chlorotoluene
(b) m-Chlorotoluene
(c) p-Chlorotoluene

SAMPLE SOLUTION   (a)   With o-chlorotoluene only one benzyne intermediate, 3-methylbenzyne, is formed. Because the methyl group exerts only a small effect on the regiochemistry of addition, both o- and m-methylaniline are expected to be produced.

| o-Chlorotoluene | 3-Methylbenzyne | o-Methylaniline | m-Methylaniline |

Indeed, the product has been observed to be a 45:55 mixture of the o- and m-substituted isomers.

25.8 SUMMARY

Aryl halides are relatively unreactive toward nucleophilic substitution under conditions in which alkyl halides react readily. When conditions are chosen, however, to favor substitution by an addition-elimination or by an elimination-addition mechanism, these processes can occur rapidly.

The addition-elimination mechanism proceeds by way of a cyclohexadienyl anion intermediate and is favored by the presence of a strongly electon-withdrawing substituent on the aromatic ring. Most often this substituent is a nitro group in a position ortho or para to the leaving group.

Addition-elimination mechanism for nucleophilic aromatic substitution:

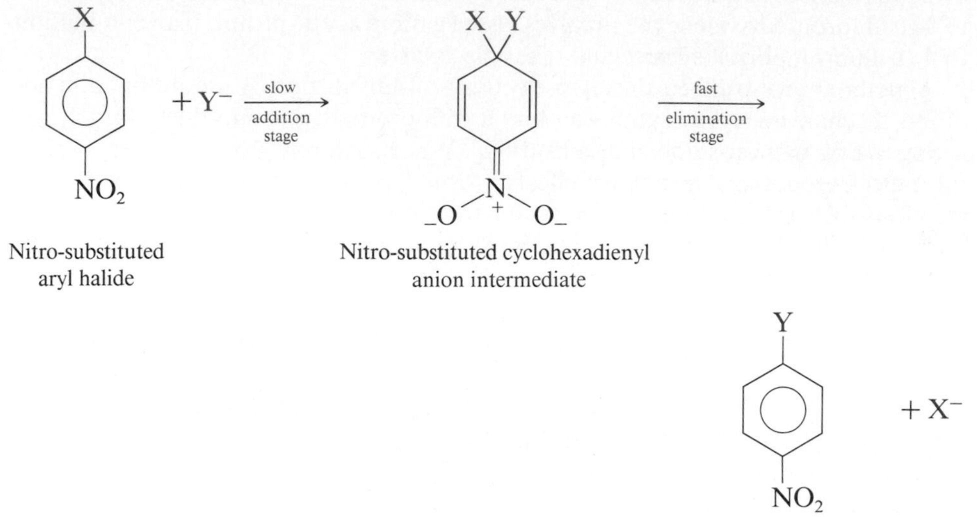

Nitro-substituted
aryl halide

Nitro-substituted cyclohexadienyl
anion intermediate

Product of nucleophilic
aromatic substitution

The elimination-addition mechanism is favored when the nucleophile is strongly basic; a very basic nucleophile can bring about elimination of an aryl halide to an aryne intermediate. The triple bond of the aryne is strained and highly reactive; it undergoes nucleophilic addition to yield the observed product.

Elimination-addition mechanism for nucleophilic aromatic substitution:

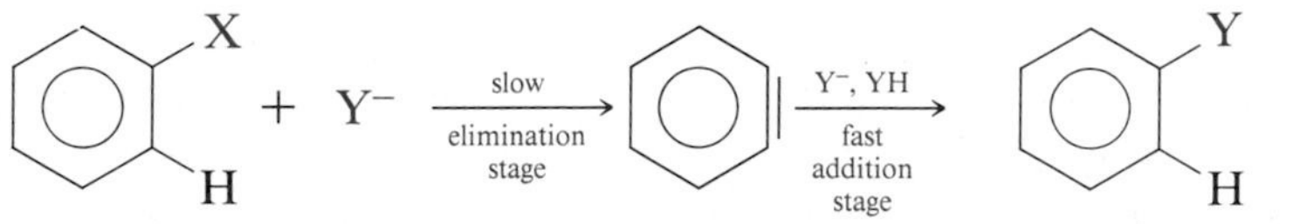

Aryl halide   Strong
base

Benzyne

Product of
nucleophilic
aromatic substitution

Substitution by the elimination-addition mechanism is most common with amide bases such as sodium amide and potassium amide.

Nucleophilic aromatic substitution by the addition-elimination mechanism always leads to replacement of the leaving group by the nucleophile on the same carbon atom of the ring. Substitution by the elimination-addition mechanism, however, can lead to substitution at the same carbon or at a carbon ortho to the one that bears the leaving group. The regioselectivity of these reactions depends on how substituents affect the regioselectivity of the elimination step and of the addition step.

PROBLEMS

25.9 Write a structural formula for each of the following:

(a) *m*-Chlorotoluene
(b) 2,6-Dibromoanisole

(c) p-Fluorostyrene

(d) 4,4'-Diiodobiphenyl

(e) 2-Bromo-1-chloro-4-nitrobenzene

(f) 1-Chloro-1-phenylethane

(g) p-Bromobenzyl chloride

(h) 2-Chloronaphthalene

(i) 1,8-Dichloronaphthalene

(j) 9-Fluorophenanthrene

25.10 Identify the principal organic product of each of the following reactions. If two regio-isomers are formed in appreciable amounts, show them both.

(a) Chlorobenzene + acetyl chloride $\xrightarrow{\text{AlCl}_3}$

(b) Bromobenzene + magnesium $\xrightarrow{\text{diethyl ether}}$

(c) Product of (b) + dilute hydrochloric acid $\longrightarrow$

(d) Iodobenzene + lithium dimethylcuprate $\xrightarrow{\text{diethyl ether}}$

(e) Bromobenzene + sodium amide $\xrightarrow{\text{liquid ammonia, }-33°\text{C}}$

(f) p-Bromotoluene + sodium amide $\xrightarrow{\text{liquid ammonia, }-33°\text{C}}$

(g) 1-Bromo-4-nitrobenzene + ammonia $\longrightarrow$

(h) p-Bromobenzyl bromide + sodium cyanide $\longrightarrow$

(i) p-Chlorobenzenediazonium chloride + N,N-dimethylaniline $\longrightarrow$

25.11 Potassium *tert*-butoxide reacts with halobenzenes on heating in dimethyl sulfoxide to give *tert*-butyl phenyl ether.

(a) o-Fluorotoluene yields *tert*-butyl o-methylphenyl ether almost exclusively under these conditions. By which mechanism (addition-elimination or elimination-addition) do aryl fluorides react with potassium *tert*-butoxide in dimethyl sulfoxide?

(b) At 100°C, bromobenzene reacts over 20 times as fast as fluorobenzene. By which mechanism do aryl bromides react?

25.12 Highly fluorinated derivatives of benzene undergo reactions that do not normally proceed readily with simpler compounds such as benzene or fluorobenzene. Two of these reactions are described below.

(a) Treatment of m-difluorobenzene with butyllithium in tetrahydrofuran gives a solution containing a compound that affords 2,6-difluorobenzoic acid in 87 percent yield on reaction with carbon dioxide followed by acidification. What is the species that reacts with carbon dioxide? Why does it form so readily under these conditions?

(b) 1,2,3,4,5,6-Hexafluorobenzene reacts with sodium methoxide in methanol at 65°C to give 2,3,4,5,6-pentafluoroanisole in 72 percent yield. By what mechanism do you think this reaction takes place? Why do you think it occurs so readily?

25.13 Choose the compound in each of the following pairs that reacts faster with sodium methoxide in methanol at 50°C:

(a) Chlorobenzene or o-chloronitrobenzene

(b) o-Chloronitrobenzene or m-chloronitrobenzene

(c) 4-Chloro-3-nitroacetophenone or 4-chloro-3-nitrotoluene

(d) 2-Fluoro-1,3-dinitrobenzene or 1-fluoro-3,5-dinitrobenzene

(e) 1,4-Dibromo-2-nitrobenzene or 1-bromo-2,4-dinitrobenzene

25.14 In each of the following reactions, an amine or a lithium amide derivative reacts with an aryl halide. Give the structure of the expected product and specify the mechanism by which it is formed.

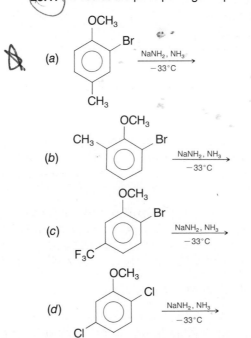

(a)

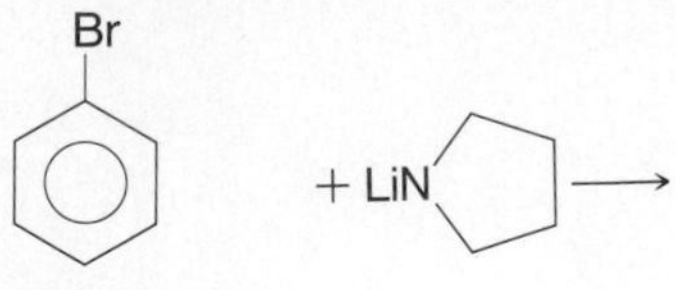

Br + LiN⟨pyrrolidine⟩ →

(b)

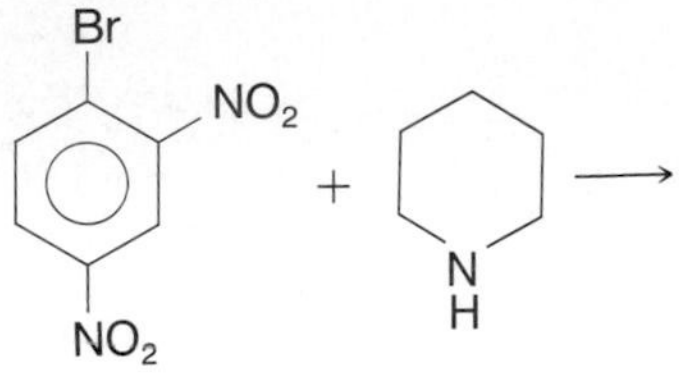

Br, NO$_2$, NO$_2$ + piperidine (NH) →

(c)

Br, NO$_2$, Br + piperidine (NH) →

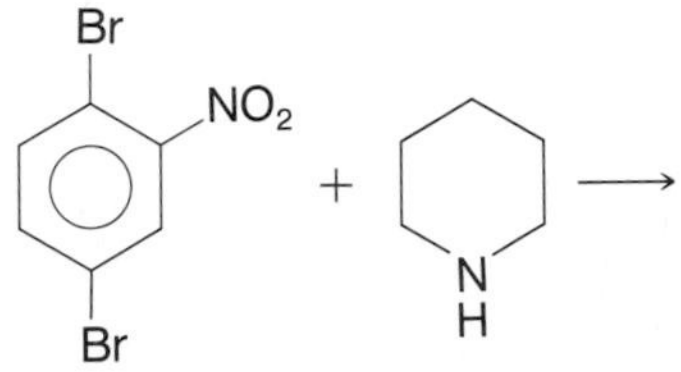

25.15 Piperidine, the amine reactant in parts (b) and (c) of the preceding problem, reacts with 1-bromonaphthalene on heating at 230°C to give a single product, compound A (C$_{15}$H$_{17}$N), as a noncrystallizable liquid. The same reaction using 2-bromonaphthalene yielded an isomeric product, compound B, a solid melting at 50 to 53°C. Mixtures of A and B were formed when either 1- or 2-bromonaphthalene was allowed to react with sodium piperidide in piperidine. Suggest reasonable structures for compounds A and B and offer an explanation for their formation under each set of reaction conditions.

25.16 1,2,3,4,5-Pentafluoro-6-nitrobenzene reacts readily with sodium methoxide in methanol at room temperature to yield two major products, each having the molecular formula C$_7$H$_3$F$_4$NO$_3$. Suggest reasonable structures for these two compounds.

25.17 Predict the principal organic product in each of the following reactions.

(a)

OCH$_3$, Br, CH$_3$ $\xrightarrow[-33°C]{NaNH_2, \ NH_3}$

(b)

CH$_3$, OCH$_3$, Br $\xrightarrow[-33°C]{NaNH_2, \ NH_3}$

(c)

OCH$_3$, Br, F$_3$C $\xrightarrow[-33°C]{NaNH_2, \ NH_3}$

(d)

OCH$_3$, Cl, Cl $\xrightarrow[-33°C]{NaNH_2, \ NH_3}$

(e)

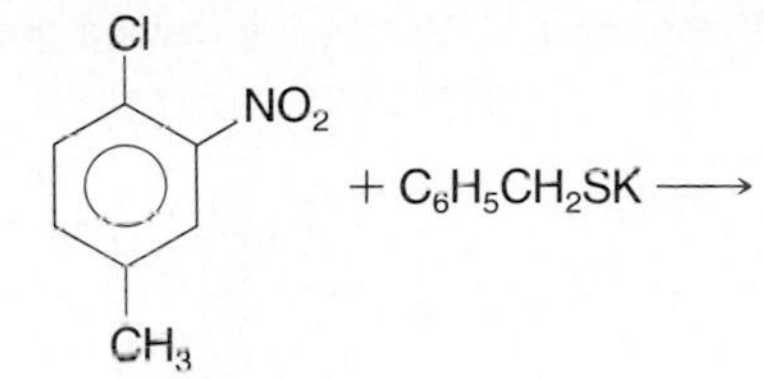

$+ C_6H_5CH_2SK \longrightarrow$

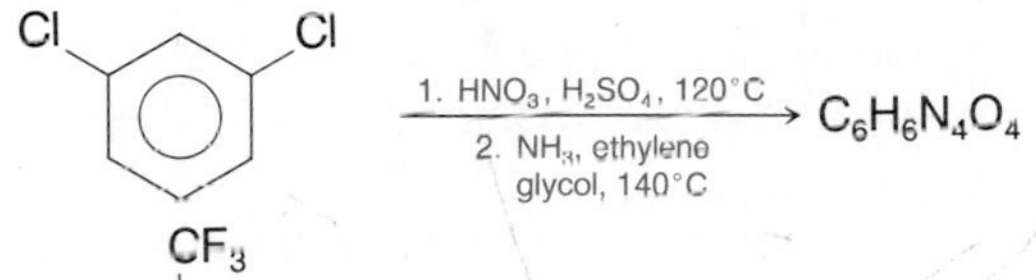

(f)

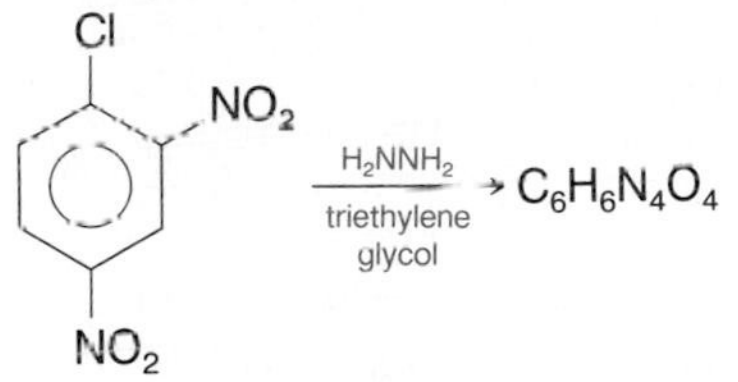

$\xrightarrow[\text{triethylene glycol}]{H_2NNH_2}$ $C_6H_6N_4O_4$

(g)

$\xrightarrow[\text{2. NH}_3, \text{ ethylene glycol, 140°C}]{\text{1. HNO}_3, \text{ H}_2\text{SO}_4, 120°C}$ $C_6H_6N_4O_4$

(h)

$\xrightarrow[\text{2. NaOCH}_3, \text{CH}_3\text{OH}]{\text{1. HNO}_3, \text{ H}_2\text{SO}_4}$ $C_8H_6F_3NO_3$

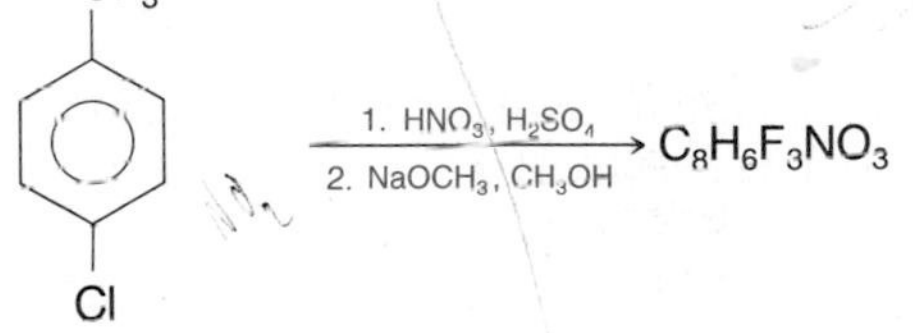

25.18 Hydrolysis of *p*-bromotoluene with aqueous sodium hydroxide at 300°C yields *m*-methylphenol and *p*-methylphenol in a 5 : 4 ratio. What is the meta-para ratio for the same reaction carried out on *p*-chlorotoluene?

25.19 After benzyne was shown to be an intermediate in the reaction of chlorobenzene with potassium amide, chemists explored alternative routes to its generation as a reactive intermediate. Two such procedures are described below.

(a) In one method the diazonium salt derived from *o*-aminobenzoic acid (benzenediazonium-2-carboxylate) is heated, yielding benzyne. Using curved arrows, show how this reaction occurs. In addition to benzyne, two other compounds, both inorganic, are formed. What are these two compounds?

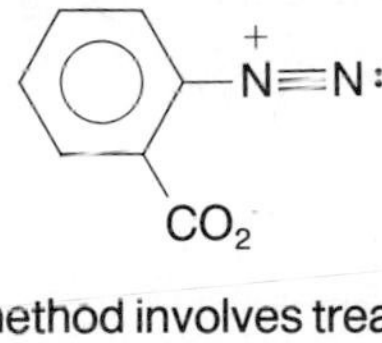

Benzenediazonium-2-carboxylate

(b) A second method involves treating *o*-bromofluorobenzene with lithium. An organometallic compound is formed, which rapidly dissociates to benzyne. What is this organometallic compound? Using curved arrows, show how it forms benzyne on dissociation.

25.20 Benzyne is a very reactive dienophile in Diels-Alder reactions. In many of the reactions that produce benzyne, it is generated in the presence of a diene and trapped as a Diels-Alder adduct. An example of such a reaction is shown in which benzyne, prepared as described in part (a) of the previous problem, reacts with furan. Write a structural formula for the product.

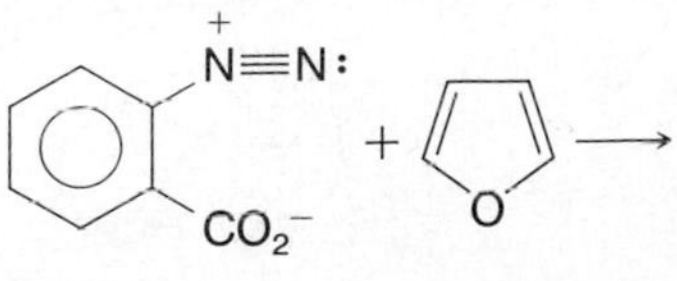

25.21 Nitro-substituted aromatic compounds that do not bear halide leaving groups react with nucleophiles according to the reaction:

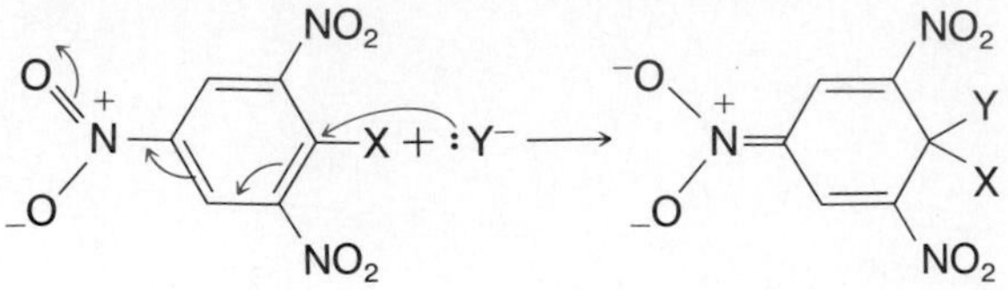

The product of this reaction, as its sodium salt, is called a *Meisenheimer complex* after the German chemist Jacob Meisenheimer, who reported on their formation and reactions in 1902. A Meisenheimer complex corresponds to the product of the nucleophilic addition stage in the addition-elimination mechanism for nucleophilic aromatic substitution.

(a) Give the structure of the Meisenheimer complex formed by addition of sodium ethoxide to 2,4,6-trinitroanisole.

(b) What other combination of reactants yields the same Meisenheimer complex as that of (a)?

(c) Write the structure of a suitable starting material that gives the Meisenheimer complex shown below on treatment with sodium hydroxide.

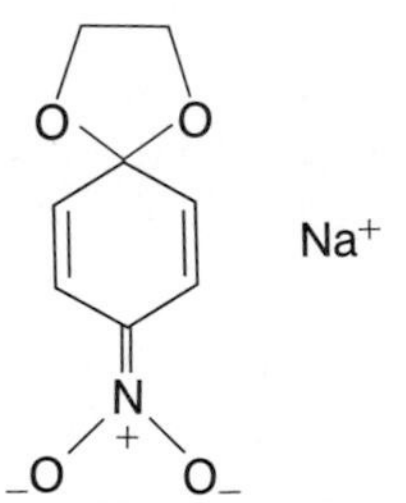

25.22 A careful study of the reaction of 2,4,6-trinitroanisole with sodium methoxide revealed that two different Meisenheimer complexes were present. Suggest reasonable structures for these two complexes.

25.23 Would you expect nucleophilic aromatic substitution by the addition-elimination mechanism to take place more readily or less readily in 2-chloropyridine than in chlorobenzene? Explain your reasoning.

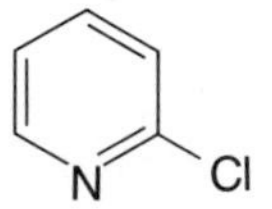

2-Chloropyridine

25.24 Suggest a reasonable mechanism for each of the following reactions:

(a) $C_6H_5Br + CH_2(COOCH_2CH_3)_2 \xrightarrow[\text{2. } H_3O^+]{\text{1. excess NaNH}_2, \text{ NH}_3} C_6H_5CH(COOCH_2CH_3)_2$ (51%)

(b) (54%)

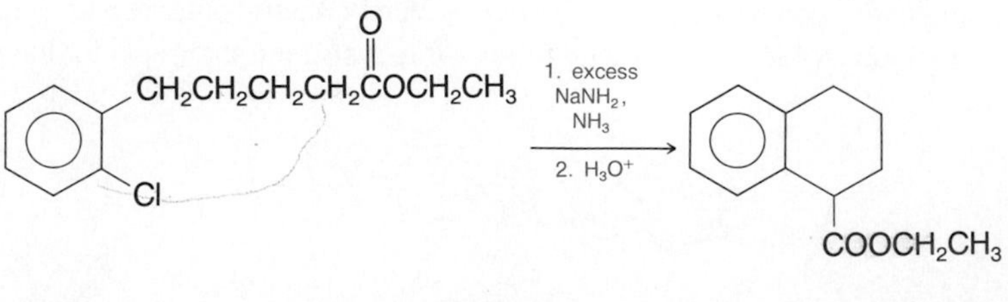

(c)

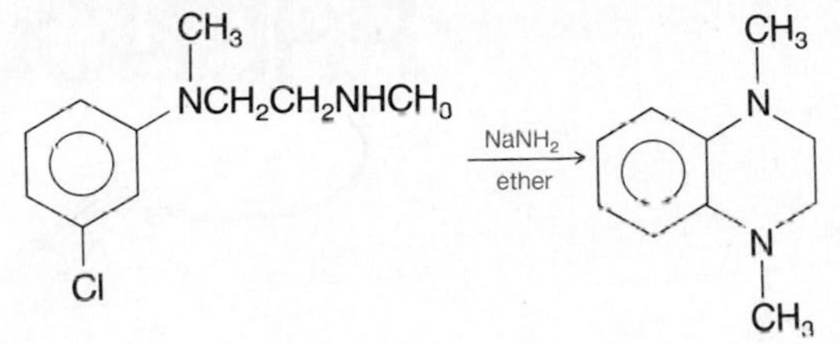

(64%)

25.25 Mixtures of chlorinated derivatives of biphenyl, called *polychlorinated biphenyls* or *PCBs,* were once prepared industrially on a large scale as insulating materials in electrical equipment. As equipment containing PCBs was discarded, the PCBs entered the environment at a rate that reached an estimated 25,000 lb/year. PCBs are very stable and accumulate in the fatty tissue of fish, birds, and mammals. They have been shown to be *teratogenic,* meaning they induce mutations in the offspring of affected individuals. Some countries have banned the use of PCBs. A large number of chlorinated biphenyls are possible, and the commercially produced material is a mixture of many compounds.

(a) How many monochloro derivatives of biphenyl are possible?

(b) How many dichloro derivatives are possible?

(c) How many octachloro derivatives are possible?

(d) How many nonachloro derivatives are possible?

25.26 DDT resistant insects have the ability to convert DDT to a less toxic substance called DDE. The mass spectrum of DDE shows a cluster of peaks for the molecular ion at m/z 316, 318, 320, 322, and 324. Suggest a reasonable structure for DDE.

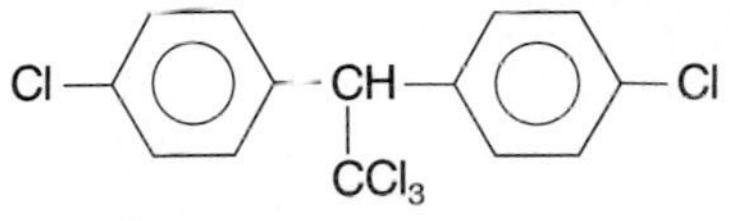

DDT (*d*ichloro*d*iphenyl*t*richloroethane)

*P*henols are compounds that have a hydroxyl group bonded directly to a benzene or benzenoid ring. The parent compound of this group, C_6H_5OH, called simply *phenol,* is an important industrial chemical. Many of the properties of phenols are analogous to those of alcohols, but this similarity is something of an oversimplification. Like arylamines, phenols are difunctional compounds; the hydroxyl group and the aromatic ring interact strongly, affecting each other's reactivity. This interaction leads to some novel and useful properties of phenols. A key step in the synthesis of aspirin, for example, is without parallel in the reactions of either alcohols or arenes. With periodic reminders of the ways in which phenols resemble alcohols and arenes, this chapter emphasizes the ways in which phenols are unique.

26.1 NOMENCLATURE

An old name for benzene was *phene,* and its hydroxyl derivative came to be called *phenol.* This, like many other entrenched common names, is an acceptable IUPAC name. Likewise, *o-*, *m-*, and *p*-cresol are acceptable names for the various ring-substituted hydroxyl derivatives of toluene. More highly substituted compounds are named as derivatives of phenol. Numbering of the ring begins at the hydroxyl-substituted carbon and proceeds in the direction that gives the lower number to the next carbon atom bearing a substituent. Substituents are cited in alphabetical order.

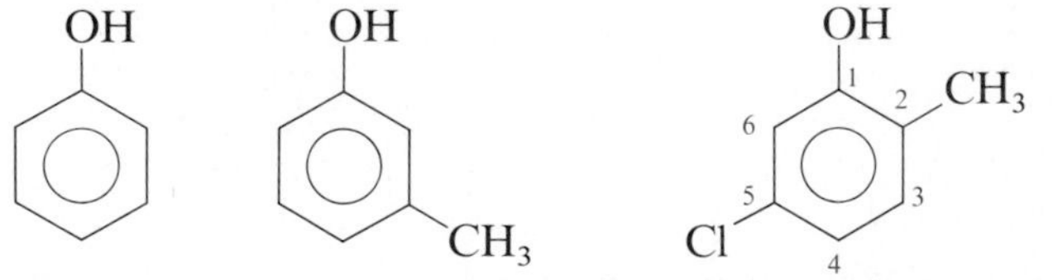

| Phenol | *m*-Cresol | 5-Chloro-2-methylphenol |

The three dihydroxy derivatives of benzene may be named as 1,2-, 1,3-, and 1,4-benzenediol, respectively, but each is more familiarly known by the common name indicated in parenthesis below. These common names are permissible IUPAC names.

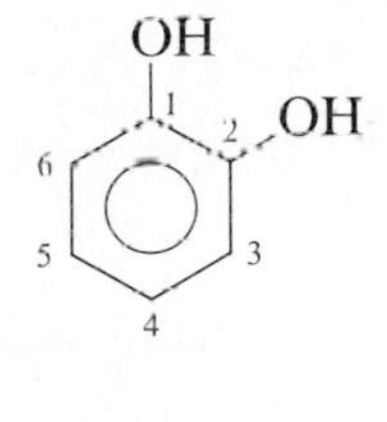

| 1,2-Benzenediol (pyrocatechol) | 1,3-Benzenediol (resorcinol) | 1,4-Benzenediol (hydroquinone) |

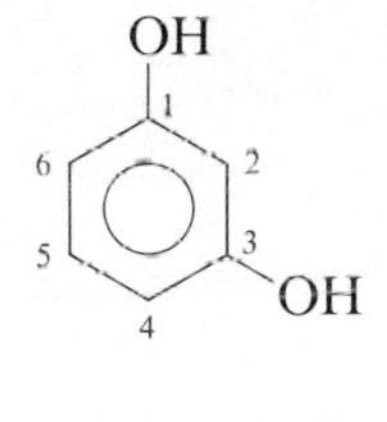

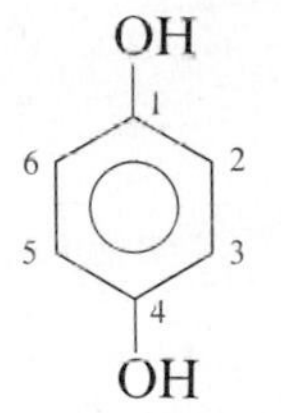

The common names for the two hydroxy derivatives of naphthalene are 1-naphthol and 2-naphthol. These are also acceptable IUPAC names.

PROBLEM 26.1   Write structural formulas for each of the following compounds.

(a) Pyrogallol (1,2,3-benzenetriol)
(b) o-Benzylphenol
(c) 3-Nitro-1-naphthol
(d) 4-Chlororesorcinol

SAMPLE SOLUTION   (a) Like the dihydroxybenzenes, the isomeric trihydroxybenzenes have unique names. Pyrogallol, used as a developer of photographic film, is 1,2,3-benzenetriol. The three hydroxyl groups occupy adjacent positions on a benzene ring.

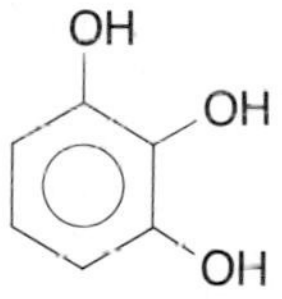

Pyrogallol
(1,2,3-benzenetriol)

Carboxyl and acyl groups take precedence over the phenolic hydroxyl in determining the base name. The hydroxyl is treated as a substituent in these cases.

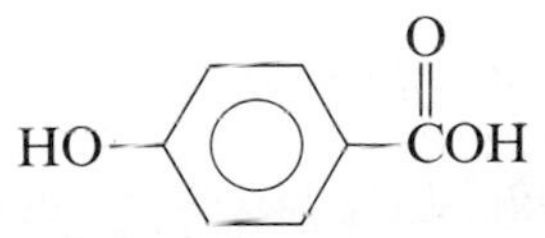

| p-Hydroxybenzoic acid | 2-Hydroxy-4-methylacetophenone |

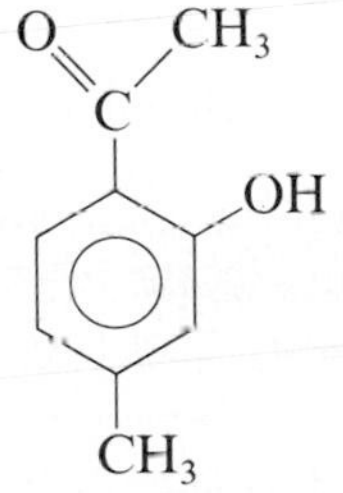

26.2 STRUCTURE AND BONDING

Phenol is planar with a C—O—H angle of 109°, almost the same as the tetrahedral angle and not much different from the 108.5° C—O—H angle of methanol.

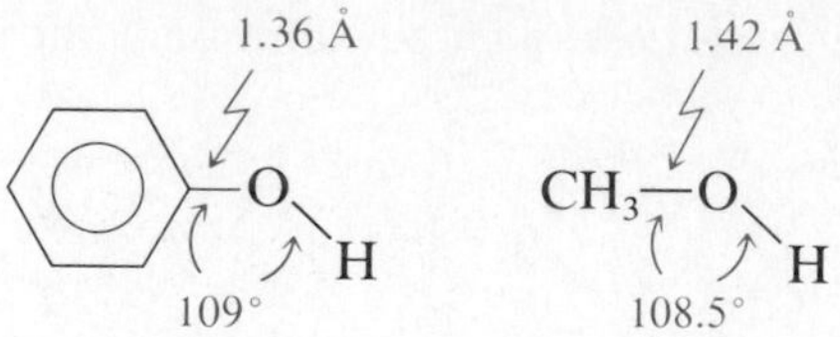

As we have noted on a number of occasions, bonds to sp^2 hybridized carbon are shorter than those to sp^3 hybridized carbon, and the case of phenols is no exception. The carbon-oxygen bond distance in phenol is slightly less than that in methanol.

In resonance terms, the shorter carbon-oxygen bond distance in phenol is attributed to the partial double bond character that results from conjugation of the unshared electron pair of oxygen with the aromatic ring.

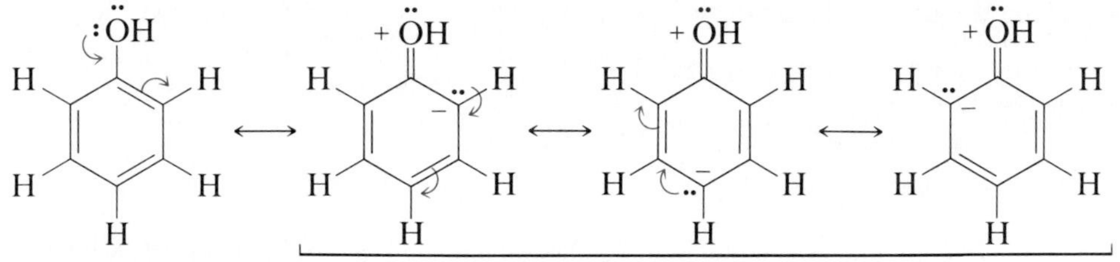

Most stable Lewis structure for phenol

Dipolar resonance forms of phenol

The hydroxyl substituent is an electron donor with respect to the ring. While, for example, the dipole moment of phenol has almost the same magnitude as that of methanol, it is opposite in its direction. In methanol oxygen withdraws electrons from carbon; in phenol oxygen donates electrons to the ring.

Methanol
μ 1.7 D

Phenol
μ 1.6 D

Phenol resembles aniline (Section 24.2) with respect to the direction of its dipole moment. As with aniline, the direction of electron polarization is deduced by examining the effects of substituents on the magnitude of the dipole moment.

PROBLEM 26.2 The dipole moment of *p*-nitrophenol (5.0 D) is greater than that of either phenol (1.6 D) or nitrobenzene (4.0 D). Explain.

Many of the differences between alcohols and phenols reflect the contrasting electron polarization in the two classes of compounds. The hydroxyl oxygen is more basic in alcohols; the hydroxyl proton is more acidic in phenols. How much more acidic phenols are than alcohols is discussed in Section 26.4.

TABLE 26.1
Physical Properties of Some Phenols

Compound name	Melting point, °C	Boiling point, °C	Solubility, g/100 mL H₂O
Phenol	43	182	8.2
o-Cresol	31	191	2.5
m-Cresol	12	203	0.5
p-Cresol	35	202	1.8
o-Chlorophenol	7	175	2.8
m-Chlorophenol	32	214	2.6
p-Chlorophenol	42	217	2.7
o-Nitrophenol	45	217	0.2
m-Nitrophenol	96		1.3
p-Nitrophenol	114	279	1.6
1-Naphthol	96	279	slight
2-Naphthol	122	285	0.1
Pyrocatechol	105	246	45.1
Resorcinol	110	276	147.3
Hydroquinone	170	285	6

26.3 PHYSICAL PROPERTIES

Table 26.1 lists selected physical properties of some representative phenols.

As shown in Figure 26.1a, the hydroxyl groups of phenol can participate in intermolecular hydrogen bonds to other phenol molecules. These hydrogen bonds stabilize the solid and liquid states and cause phenols to have higher melting points and boiling points than arenes and aryl halides of similar molecular weight. Hydrogen bonds between phenols and water molecules, represented in Figure 26.1b, contribute to the comparatively high — relative to arenes and aryl halides — water solubility of phenols. Table 26.2 compares phenol, toluene, and fluorobenzene with regard to the physical properties that are most influenced by intermolecular hydrogen bonding.

Some ortho-substituted phenols, such as o-nitrophenol, have boiling points that are significantly lower than those of their meta and para isomers. This is because an *intramolecular* hydrogen bond between the hydroxyl group and the substituent

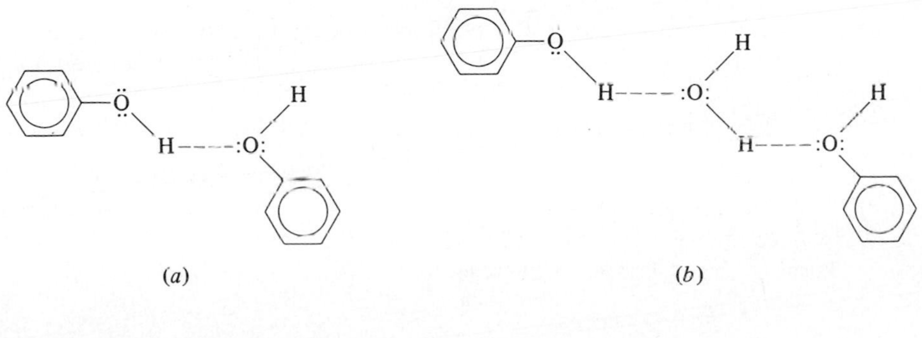

(a)

(b)

FIGURE 26.1 (a) A hydrogen bond between two phenol molecules; (b) hydrogen bonds between water and phenol molecules.

TABLE 26.2

Comparison of Physical Properties of an Arene, a Phenol, and an Aryl Halide

	Compound		
Physical property	Toluene, $C_6H_5CH_3$	Phenol, C_6H_5OH	Fluorobenzene, C_6H_5F
Molecular weight	92	94	96
Melting point	$-95°C$	$43°C$	$-41°C$
Boiling point (1 atm):	$111°C$	$132°C$	$85°C$
Solubility in water (25°C)	0.05 g/100 mL	8.2 g/100 mL	0.2 g/100 mL

forms at the expense of *intermolecular* hydrogen bonds and decreases the degree of association between molecules.

Intramolecular hydrogen bond in *o*-nitrophenol

PROBLEM 26.3 One of the hydroxybenzoic acids is known by the common name *salicylic acid*. Its methyl ester, methyl salicylate, occurs in oil of wintergreen. Methyl salicylate boils over 50°C lower than either of the other two methyl hydroxybenzoates. What is the structure of methyl salicylate? Why is its boiling point so much lower than that of either of its regioisomers?

26.4 ACIDITY OF PHENOLS

The most characteristic property of phenols is their acidity. Phenols are more acidic than alcohols but less acidic than carboxylic acids. Recall that carboxylic acids have ionization constants K_a of approximately 10^{-5} (pK_a 5), while the K_a's of alcohols are in the 10^{-16} to 10^{-20} range (pK_a 16 to 20). The K_a for most phenols is about 10^{-10} (pK_a 10).

> Because of its acidity, phenol was known as *carbolic acid* when Joseph Lister introduced it as an antiseptic in 1865 to prevent postoperative bacterial infections that were then a life-threatening hazard to even minor surgical procedures.

To understand the factors that cause phenols to be more acidic than alcohols, compare the ionization equilibria for phenol and ethanol. In particular, consider the differences in charge delocalization in ethoxide ion and in phenoxide ion. The negative charge in ethoxide ion is localized on oxygen and is stabilized only by solvation forces.

$$CH_3CH_2\ddot{O}H \rightleftharpoons H^+ + CH_3CH_2\ddot{O}{:}^- \qquad K_a = 10^{-16}\ (pK_a = 16)$$

Ethanol Proton Ethoxide ion

The negative charge in phenoxide ion is stabilized both by solvation and by electron delocalization into the ring.

$$\text{Phenol} \rightleftharpoons H^+ + \text{Phenoxide ion} \qquad K_a = 10^{-10} \quad (pK_a = 10)$$

Electron delocalization in phenoxide is represented by resonance between the structures:

The negative charge in phenoxide ion is shared by the oxygen and the carbons that are ortho and para to it. Delocalization of its negative charge strongly stabilizes phenoxide ion.

An energy diagram comparing the ionization of phenol with that of ethanol is shown in Figure 26.2. The energy difference between the two product states is quite large because of electron delocalization in phenoxide ion. The energy difference between the two starting states is smaller because electron delocalization in phenol, as described in Section 26.2, is accompanied by charge separation. Overall, the free

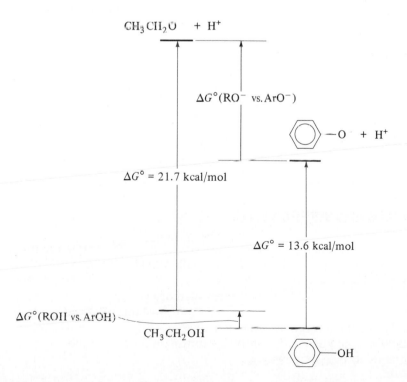

FIGURE 26.2 Free energies of ionization of ethanol and phenol in water. Most of the difference between the two is due to the large difference in stabilization energy of the phenoxide anion relative to the ethoxide anion.

energy change for ionization of phenol is considerably less than for ionization of ethanol and the equilibrium constant is larger.

To place the acidity of phenol in perspective, note that while phenol is more than 1 million times more acidic than ethanol, it is over 100,000 times weaker than acetic acid. Thus, phenols may be separated from alcohols because they are more acidic and from carboxylic acids because they are less acidic. On shaking an ether solution of an alcohol and a phenol with dilute sodium hydroxide, the phenol is converted quantitatively to its sodium salt, which is extracted into the aqueous phase. The alcohol remains in the ether phase.

Phenol	Hydroxide ion	Phenoxide ion	Water
(stronger acid)	(stronger base)	(weaker base)	(weaker acid)

On shaking an ether solution of a phenol and a carboxylic acid with dilute sodium bicarbonate, the carboxylic acid is converted quantitatively to its sodium salt and extracted into the aqueous phase. The phenol remains in the ether phase.

Phenol	Bicarbonate ion	Phenoxide ion	Carbonic acid
(weaker acid)	(weaker base)	(stronger base)	(stronger acid)

It is necessary to keep the acidity of phenols in mind when their preparation and reactions are discussed. Reactions that produce phenols, when carried out in basic solution, require an acidification step in order to convert the phenoxide ion to the neutral form of the phenol.

Phenoxide ion	Hydronium ion	Phenol	Water
(stronger base)	(stronger acid)	(weaker acid)	(weaker base)

Many synthetic reactions involving phenols as nucleophiles are carried out in the presence of sodium or potassium hydroxide. Under these conditions the phenol is converted to the phenoxide ion, which is a far better nucleophile.

26.5 SUBSTITUENT EFFECTS ON THE ACIDITY OF PHENOLS

As Table 26.3 shows, most phenols have ionization constants that are similar to that of phenol itself. Substituent effects, in general, are small. Alkyl substitution produces negligible changes in acidities, as do weakly electronegative groups attached to the ring. Only when the substituent is strongly electron-withdrawing, as is a nitro group, is a substantial change in acidity noted. The ionization constants of *o*- and *p*-nitrophenol are several hundred times greater than that of phenol. An ortho- or para-nitro group greatly stabilizes the phenoxide ion by permitting a portion of the negative charge to be borne by its own oxygens.

TABLE 26.3
Acidities of Some Phenols

Compound name	Ionization constant, K_a	pK_a
Monosubstituted phenols		
Phenol	1.0×10^{-10}	10.0
o-Cresol	4.7×10^{-11}	10.3
m-Cresol	8.0×10^{-11}	10.1
p-Cresol	5.2×10^{-11}	10.3
o-Chlorophenol	2.7×10^{-9}	8.6
m-Chlorophenol	7.6×10^{-11}	9.1
p-Chlorophenol	3.9×10^{-11}	9.4
o-Methoxyphenol	1.0×10^{-10}	10.0
m-Methoxyphenol	2.2×10^{-10}	9.6
p-Methoxyphenol	6.3×10^{-11}	10.2
o-Nitrophenol	5.9×10^{-8}	7.2
m-Nitrophenol	4.4×10^{-9}	8.4
p-Nitrophenol	6.9×10^{-8}	7.2
Di- and trinitrophenols		
2,4-Dinitrophenol	1.1×10^{-4}	4.0
3,5-Dinitrophenol	2.0×10^{-7}	6.7
2,4,6-Trinitrophenol	4.2×10^{-1}	0.4
Naphthols		
1-Naphthol	5.9×10^{-10}	9.2
2-Naphthol	3.5×10^{-10}	9.5

Electron delocalization in o-nitrophenoxide ion:

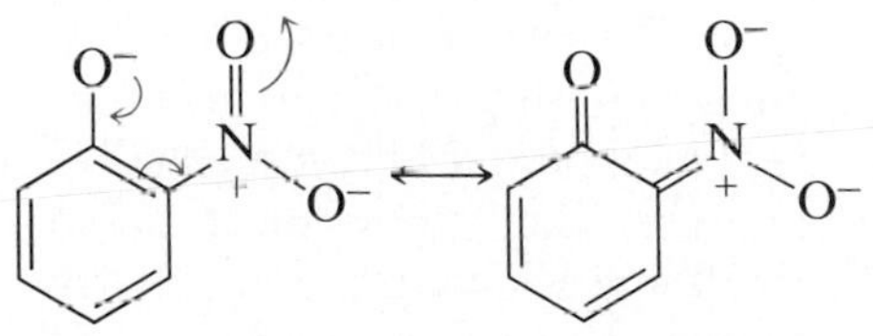

Electron delocalization in p-nitrophenoxide ion:

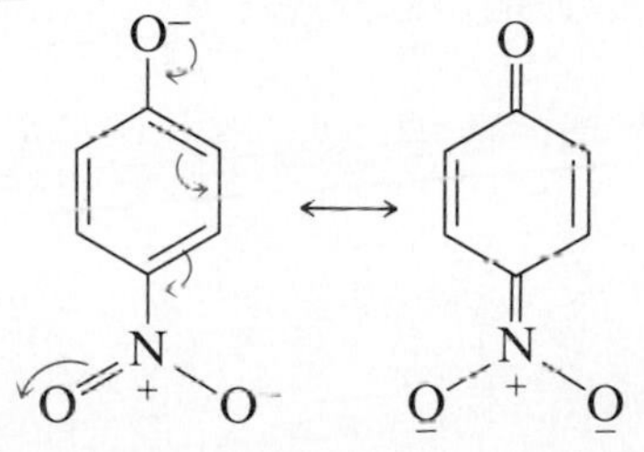

A meta-nitro group is not directly conjugated to the phenoxide oxygen and thus stabilizes a phenoxide ion to a smaller extent. *m*-Nitrophenol is more acidic than phenol but less acidic than either *o*- or *p*-nitrophenol.

PROBLEM 26.4   Which is the stronger acid in each of the following pairs? Explain your reasoning.

  (*a*)   Phenol or *p*-hydroxybenzaldehyde
  (*b*)   *m*-Cyanophenol or *p*-cyanophenol
  (*c*)   *o*-Fluorophenol or *p*-fluorophenol

SAMPLE SOLUTION   (*a*) The best approach when comparing acidities of phenols is to assess opportunities for stabilization of negative charge in the derived anions. Electron delocalization in the anion of *p*-hydroxybenzaldehyde is very effective because of conjugation with the formyl group.

A formyl substituent, like a nitro group, is strongly electron-withdrawing and increases the acidity of phenol, especially when ortho or para to the hydroxyl group. *p*-Hydroxybenzaldehyde, with a K_a of 2.4×10^{-8}, is a stronger acid than phenol.

Multiple substitution by strongly electron-withdrawing groups markedly increases the acidity of phenols, as the K_a values for 2,4-dinitrophenol (K_a 1.1×10^{-4}) and 2,4,6-trinitrophenol (K_a 4.2×10^{-1}) in Table 26.3 attest.

26.6   SOURCES OF PHENOLS

Phenol was first isolated in the early nineteenth century from coal tar, and a small portion of the more than 2 billion lb of phenol produced in the United States each year comes from this source. While significant quantities of phenol are used to prepare aspirin and dyes, most of it is converted to phenolic resins used in adhesives and plastics. Almost all the phenol produced commercially is synthetic, with several different processes in current use. These are summarized in Table 26.4.

The reaction of benzenesulfonic acid with sodium hydroxide (first entry in Table 26.4) proceeds by the addition-elimination mechanism of nucleophilic aromatic substitution (Section 25.5). Hydroxide replaces sulfite ion (SO_3^{2-}) at the carbon atom that bears the leaving group. Thus, *p*-toluenesulfonic acid is converted exclusively to *p*-cresol by an analogous reaction.

p-Toluenesulfonic acid   *p*-Cresol (63–72%)

TABLE 26.4

Industrial Syntheses of Phenol

Reaction and comments	Chemical equation
Reaction of benzenesulfonic acid with sodium hydroxide This is the oldest method for the preparation of phenol. Benzene is sulfonated and the benzenesulfonic acid heated with molten sodium hydroxide. Acidification of the reaction mixture gives phenol.	SO₃H →(1. NaOH 300–350 °C 2. H⁺)→ OH Benzenesulfonic acid · · · · Phenol
Hydrolysis of chlorobenzene Heating chlorobenzene with aqueous sodium hydroxide at high pressure gives phenol after acidification.	Cl →(1. NaOH, H₂O 370 °C 2. H⁺)→ OH Chlorobenzene · · · · Phenol
From cumene Oxidation of cumene takes place at the benzylic position to give a hydroperoxide. On treatment with dilute sulfuric acid, this hydroperoxide is converted to phenol and acetone.	—CH(CH₃)₂ →(O₂)→ —C(CH₃)₂ with OOH Isopropylbenzene (cumene) · · · · 1-Methyl-1-phenylethyl hydroperoxide

(second row of chemical equations)

—C(CH₃)₂ with OOH →(H₂O / H₂SO₄)→ —OH + (CH₃)₂C=O

1-Methyl-1-phenylethyl hydroperoxide · · · · Phenol · · · · Acetone

---

PROBLEM 26.5 Write a stepwise mechanism for the conversion of *p*-toluenesulfonic acid to *p*-cresol under the conditions shown in the preceding equation.

On the other hand, ¹⁴C-labeling studies have shown that the base-promoted hydrolysis of chlorobenzene (second entry in Table 26.4) proceeds by the elimination-addition mechanism and involves benzyne as an intermediate.

PROBLEM 26.6 Write a stepwise mechanism for the hydrolysis of chlorobenzene under the conditions shown in Table 26.4.

The most commonly used industrial synthesis of phenol is based on isopropylbenzene (cumene) as the starting material and is shown in the third entry of Table 26.4. The economically attractive features of this process are its use of cheap reagents (oxygen and sulfuric acid) and the fact that it yields two high-volume industrial chemicals, phenol and acetone. The mechanism of this novel synthesis forms the basis of Problem 26.32 at the end of this chapter.

The most important synthesis of phenols in the laboratory is from amines by hydrolysis of their corresponding diazonium salts, as described in Section 24.9.

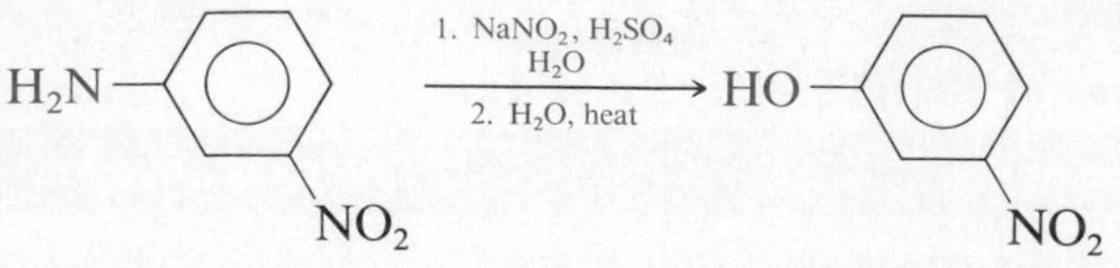

m-Nitroaniline

m-Nitrophenol (81–86%)

26.7 NATURALLY OCCURRING PHENOLS

Phenolic compounds are commonplace natural products. Figure 26.3 presents a sampling of some naturally occurring phenols. Phenolic natural products can arise by a number of different biosynthetic pathways. In mammals aromatic rings are hydroxylated by way of arene oxide intermediates formed by the enzyme-catalyzed reaction between an aromatic ring and molecular oxygen.

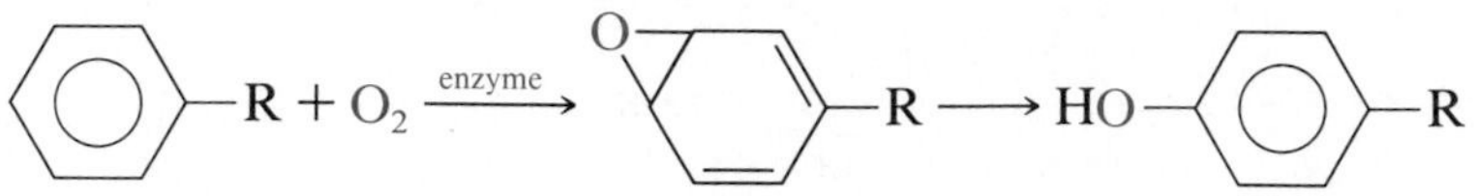

Arene   Arene oxide   Phenol

In plants phenol biosynthesis proceeds by building the aromatic ring from carbohydrate precursors that already contain the required hydroxyl group.

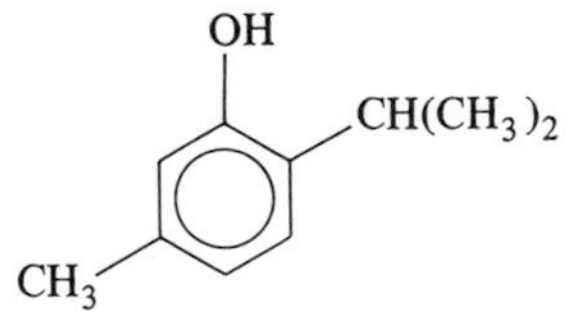

Thymol
(major constituent of oil of thyme)

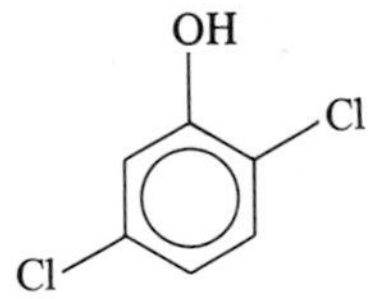

2,5-Dichlorophenol
(isolated from defensive secretion
of a species of grasshopper)

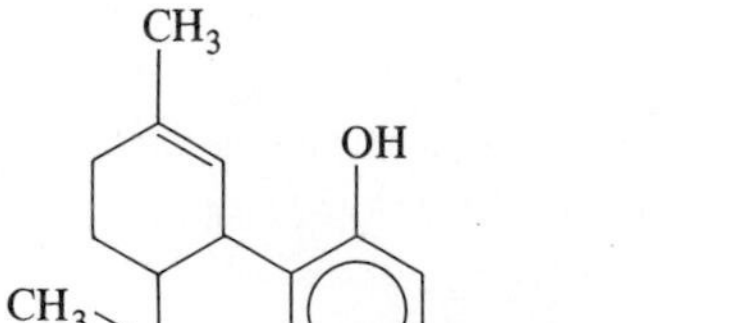

Δ^9-Tetrahydrocannabinol
(active component of marijuana)

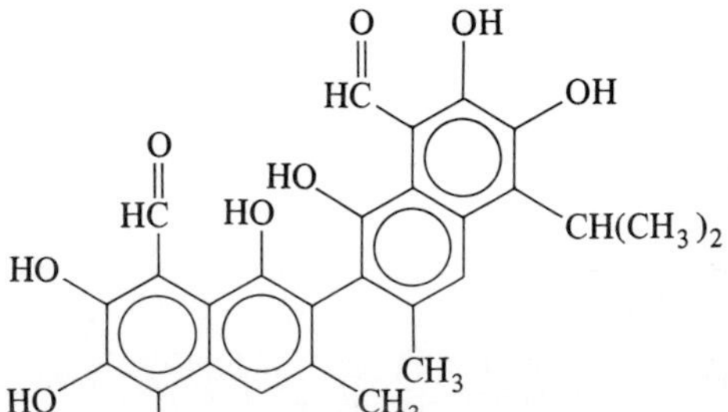

Gossypol
(About 10^9 lb of this material
are obtained each year in the United States
as a by-product of cotton-oil production; it
has been undergoing wide-scale testing as a
male contraceptive in the People's Republic
of China with encouraging results.)

FIGURE 26.3 Some naturally occurring phenols.

26.8 REACTIONS OF PHENOLS. ELECTROPHILIC AROMATIC SUBSTITUTION

In most of their reactions phenols behave as nucleophiles and the reagents that act upon them are electrophiles. Either the hydroxyl oxygen or the aromatic ring may be the site of nucleophilic reactivity in a phenol. Reactions that take place on the ring lead to electrophilic aromatic substitution; Table 26.5 summarizes the behavior of phenols in reactions of this type.

A hydroxyl group is a very powerful activating substituent and electrophilic aromatic substitution in phenols occurs far faster, and under milder conditions, than in benzene. The first entry in Table 26.5, for example, depicts the monobromination of phenol in high yield at low temperature and in the absence of any catalyst. In this case, the reaction was carried out in the nonpolar solvent 1,2-dichloroethane. In polar solvents such as water it is difficult to limit the bromination of phenols to monosubstitution. In the example shown below, all three positions that are ortho or para to the hydroxyl undergo rapid substitution.

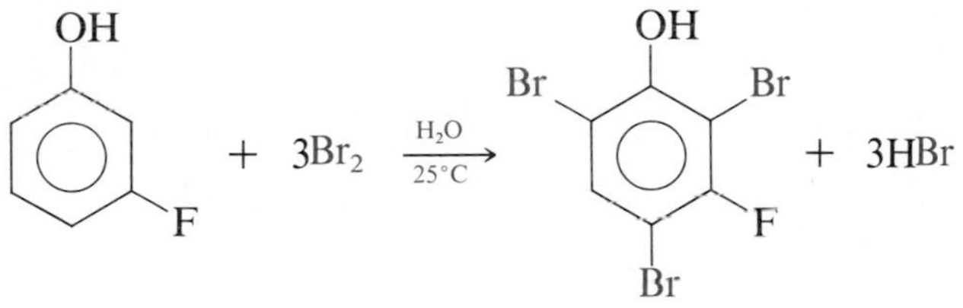

| *m*-Fluorophenol | Bromine | 2,4,6-Tribromo-3-fluorophenol (95%) | Hydrogen bromide |

Other typical electrophilic aromatic substitution reactions—nitration (second entry), sulfonation (fourth entry), and Friedel-Crafts alkylation and acylation (fifth and sixth entries)—take place readily and are synthetically useful. Phenols also undergo electrophilic substitution reactions that are limited to only the most active aromatic compounds; these include nitrosation (second entry) and coupling with diazonium salts (seventh entry).

PROBLEM 26.7 Each of the following reactions has been reported in the chemical literature and gives a single organic product in high yield. Identify the principal product in each case.

(a) 3-Benzyl-2,6-dimethylphenol treated with bromine in chloroform
(b) 4-Bromo-2-methylphenol treated with 2-methylpropene and sulfuric acid
(c) 2-Isopropyl-5-methylphenol (thymol) treated with sodium nitrite and dilute hydrochloric acid
(d) *p*-Cresol treated with propanoyl chloride and aluminum chloride

SAMPLE SOLUTION (a) The ring that bears the hydroxyl group is much more reactive than the other. In electrophilic aromatic substitution reactions of rings that bear several

TABLE 26.5

Electrophilic Aromatic Substitution Reactions of Phenols

Reaction and comments	Specific example
Halogenation Bromination and chlorination of phenols occur readily even in the absence of a catalyst. Substitution occurs primarily at the position para to the hydroxyl group. When the para position is blocked, ortho substitution is observed.	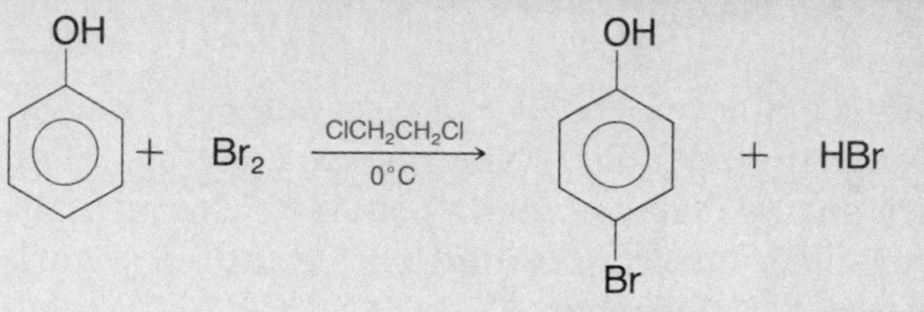 Phenol Bromine p-Bromophenol Hydrogen (93%) bromide
Nitration Phenols are nitrated on treatment with a dilute solution of nitric acid in either water or acetic acid. It is not necessary to use mixtures of nitric and sulfuric acids because of the high reactivity of phenols.	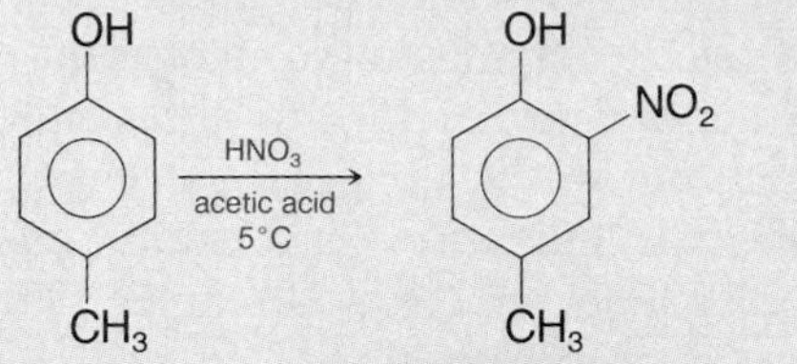 p-Cresol 4-Methyl-2-nitrophenol (73–77%)
Nitrosation On acidification of aqueous solutions of sodium nitrite, the nitrosonium ion ($:\overset{+}{N}\equiv O:$) is formed, which is a weak electrophile and attacks the strongly activated ring of a phenol. The product is a nitroso phenol.	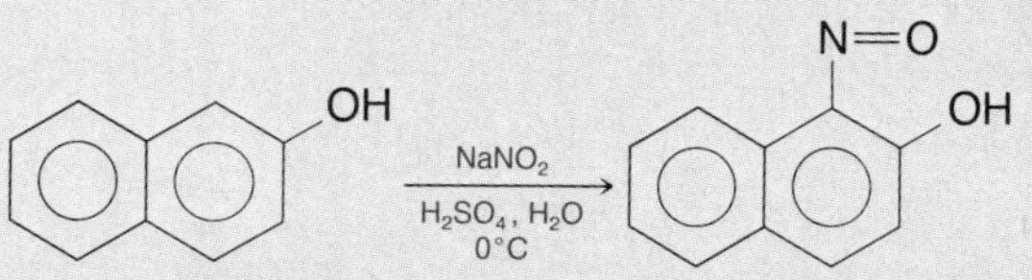 2-Naphthol 1-Nitroso-2-naphthol (99%)

Sulfonation Heating a phenol with concentrated sulfuric acid causes sulfonation of the ring.	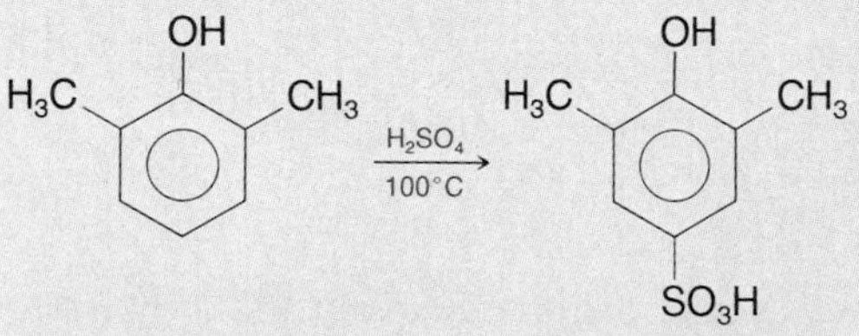
	
	2,6-Dimethylphenol 4-Hydroxyl-3,5-
	                      dimethylbenzenesulfonic
	                      acid (69%)

Friedel-Crafts alkylation Alcohols in combination with acids serve as sources of carbocations. Attack of a carbocation on the electron-rich ring of a phenol brings about its alkylation.	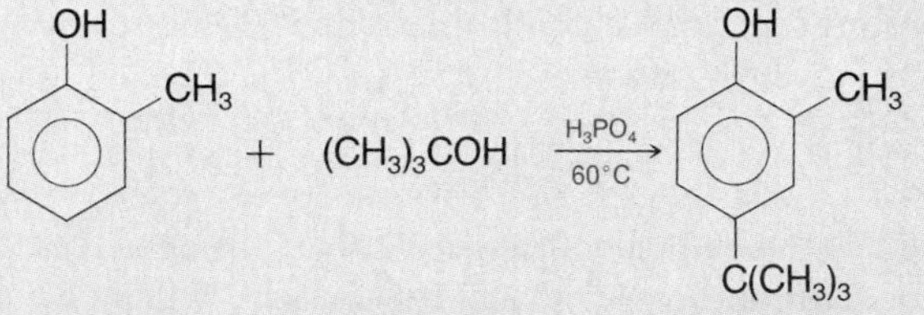
	
	o-Cresol tert-Butyl alcohol 4-tert-Butyl-
	                          2-methylphenol
	                          (63%)

Friedel-Crafts acylation In the presence of aluminum chloride, acyl chlorides and carboxylic acid anhydrides acylate the aromatic ring of phenols.	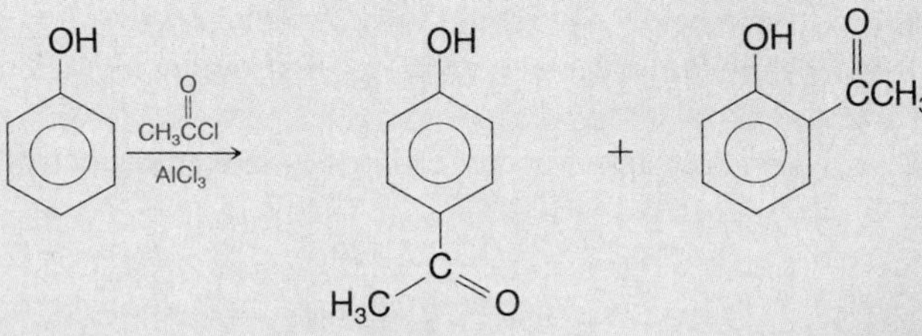 Phenol p-Hydroxyacetophenone o-Hydroxyacetophenone (74%) (16%)

TABLE 26.5 (continued)

Reaction and comments	Specific example
Reaction with arenediazonium salts Addition of a phenol to a solution of a diazonium salt formed from a primary aromatic amine leads to formation of an azo compound. The reaction is carried out at a pH such that a significant portion of the phenol is present as its phenoxide ion. The diazonium ion acts as an electrophile toward the strongly activated ring of the phenoxide ion.	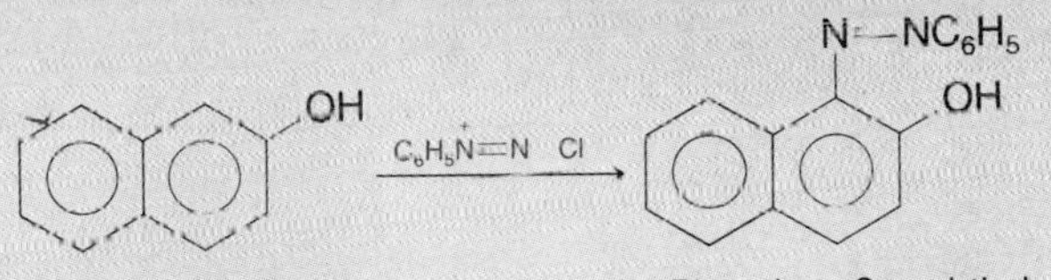

substituents, it is the most activating substituent that controls the orientation. Bromination occurs para to the hydroxyl group.

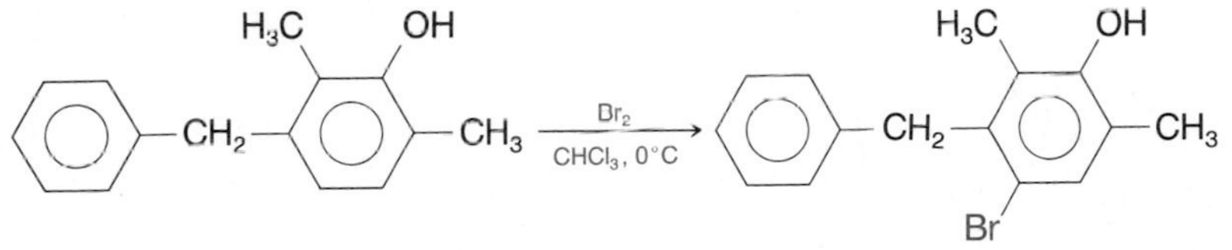

The aromatic ring of a phenol, like that of an arylamine, is regarded as an electron-rich functional unit and is capable of a variety of reactions. In some cases, however, it is the hydroxyl oxygen that reacts instead. An example of this kind of chemical reactivity is described in the following section.

26.9 ESTERIFICATION OF PHENOLS

Acylating agents, such as acyl chlorides and carboxylic acid anhydrides, can react with phenols either at the aromatic ring (*C*-acylation) or at the hydroxyl oxygen (*O*-acylation).

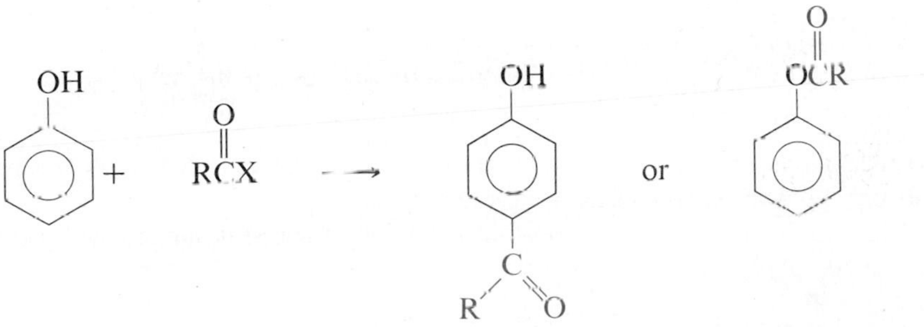

As shown in the sixth entry of Table 26.5 in the preceding section, *C*-acylation of phenols is observed under the customary conditions of the Friedel-Crafts reaction (treatment with an acyl chloride or acid anhydride in the presence of aluminum chloride). In the absence of aluminum chloride, however, *O*-acylation occurs instead.

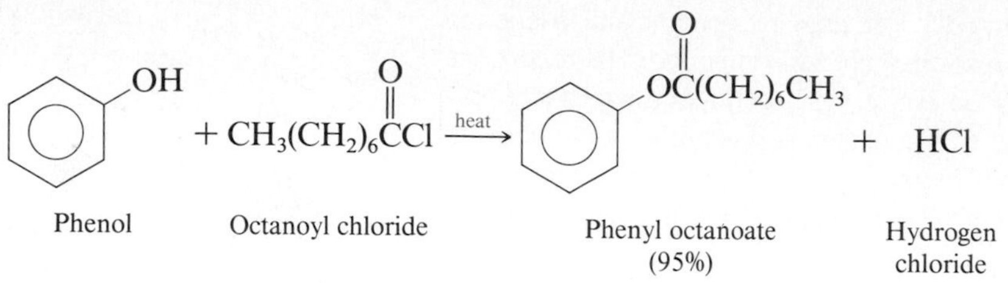

$$\text{Phenol} + CH_3(CH_2)_6\overset{O}{\overset{\|}{C}}Cl \xrightarrow{\text{heat}} \text{Phenyl octanoate (95\%)} + HCl$$

| Phenol | Octanoyl chloride | Phenyl octanoate (95%) | Hydrogen chloride |

The *O*-acylation of phenols with carboxylic acid anhydrides can be conveniently catalyzed in either of two ways. One method involves converting the acid anhydride to a more powerful acyl transfer agent by protonation of one of its carbonyl oxygens. Addition of a few drops of sulfuric acid is usually sufficient.

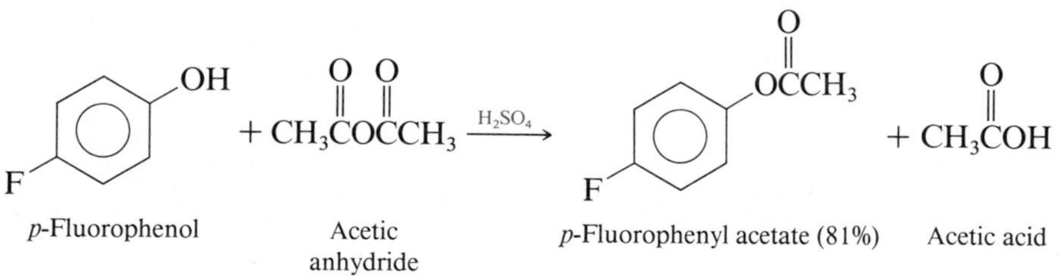

$$p\text{-Fluorophenol} + CH_3\overset{O}{\overset{\|}{C}}O\overset{O}{\overset{\|}{C}}CH_3 \xrightarrow{H_2SO_4} p\text{-Fluorophenyl acetate (81\%)} + CH_3\overset{O}{\overset{\|}{C}}OH$$

| *p*-Fluorophenol | Acetic anhydride | *p*-Fluorophenyl acetate (81%) | Acetic acid |

An alternative approach is to increase the nucleophilicity of the phenol by converting it to its phenoxide anion in basic solution.

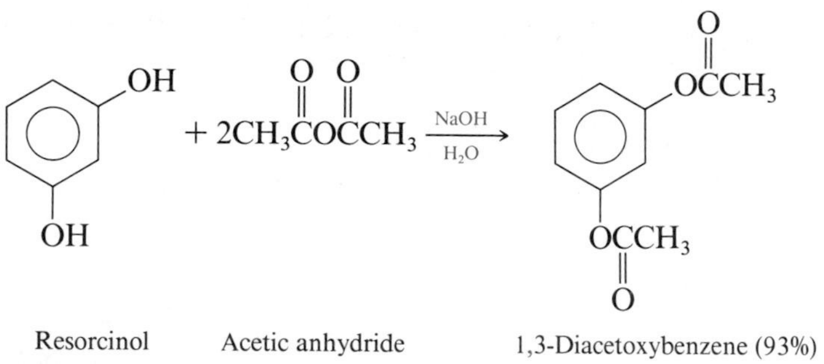

$$\text{Resorcinol} + 2CH_3\overset{O}{\overset{\|}{C}}O\overset{O}{\overset{\|}{C}}CH_3 \xrightarrow[H_2O]{NaOH} \text{1,3-Diacetoxybenzene (93\%)}$$

| Resorcinol | Acetic anhydride | 1,3-Diacetoxybenzene (93%) |

PROBLEM 26.8 Write chemical equations expressing each of the following transformations:

(a) Preparation of *o*-nitrophenyl acetate by sulfuric acid catalysis of the reaction between a phenol and a carboxylic acid anhydride

(b) Esterification of 2-naphthol with acetic anhydride in aqueous sodium hydroxide

(c) Reaction of phenol with benzoyl chloride

SAMPLE SOLUTION (a) The problem specifies that an acid anhydride be used. Therefore, use acetic anhydride to prepare the acetate ester of *o*-nitrophenol.

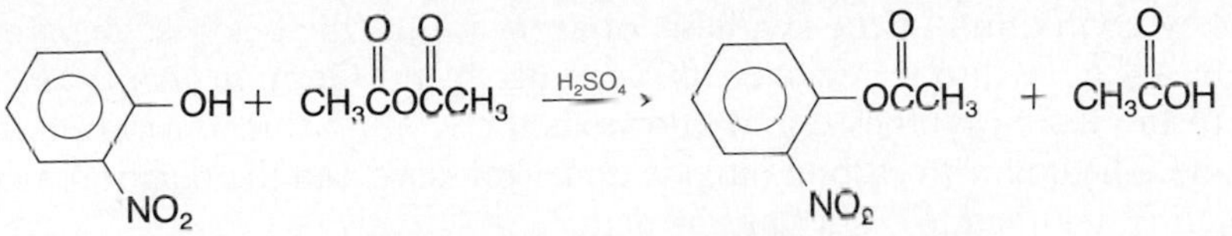

o-Nitrophenol   Acetic anhydride   o-Nitrophenyl acetate   Acetic acid
(isolated in 93% yield by
this method)

The preference for *O*-acylation of phenols arises because these reactions are *kinetically controlled*. *O*-Acylation is faster than *C*-acylation. The *C*-acyl isomers are more stable, however, and it is known that aluminum chloride is a very effective catalyst for the conversion of aryl esters to aryl ketones. (This isomerization is called the *Fries rearrangement*.)

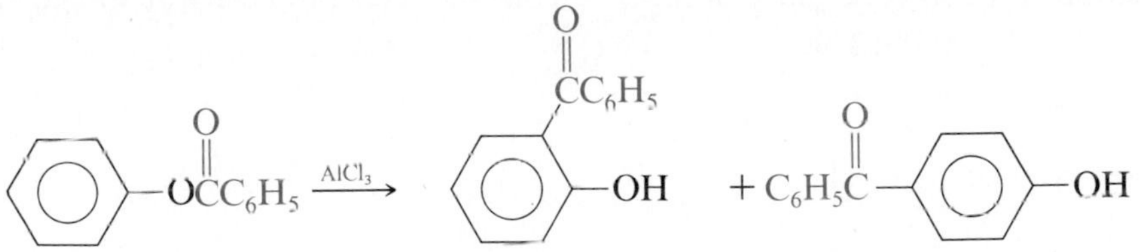

Phenyl benzoate   o-Hydroxybenzophenone   p-Hydroxybenzophenone
(9%)   (64%)

Thus, ring acylation of phenols is observed under Friedel-Crafts conditions because the presence of aluminum chloride causes that reaction to be subject to *thermodynamic (equilibrium) control* (Section 11.10).

Fischer esterification, in which a phenol and a carboxylic acid condense in the presence of an acid catalyst, is not often used for the preparation of aryl esters.

26.10 CARBOXYLATION OF PHENOLS. ASPIRIN AND THE KOLBE-SCHMITT REACTION

The best-known aryl ester is *O*-acetylsalicylic acid, better known as *aspirin*. It is prepared by acetylation of the phenolic hydroxyl group of salicylic acid.

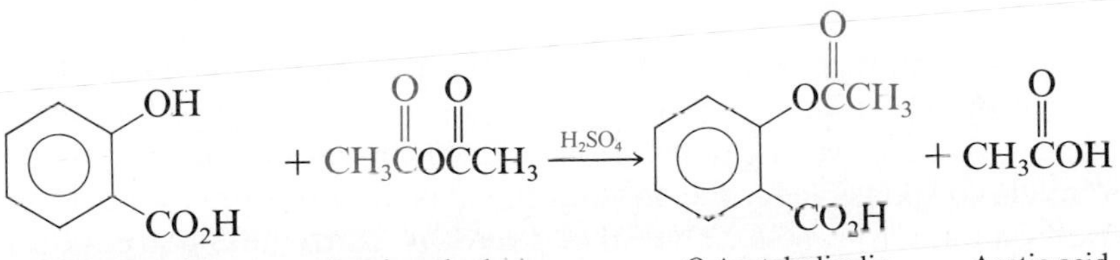

Salicylic acid   Acetic anhydride   O-Acetylsalicylic   Acetic acid
(o-hydroxybenzoic acid)   acid (aspirin)

Aspirin possesses a number of properties that make it the most often recommended drug. It is an analgesic, effective in relieving headache pain. It is also an anti-inflammatory agent, providing some relief from the swelling associated with arthritis and minor injuries. Aspirin is an antipyretic compound, that is, it reduces fever. Each year, more than 40 million lb of aspirin is produced in the United States, a rate that translates to 300 tablets per year for every man, woman, and child.

The key compound in the synthesis of aspirin, salicylic acid, is prepared from phenol by a process discovered over 100 years ago by the German chemist Hermann Kolbe. In the Kolbe synthesis, also known as the Kolbe-Schmitt reaction, sodium phenoxide is heated with carbon dioxide under pressure and the reaction mixture is subsequently acidified to yield salicylic acid.

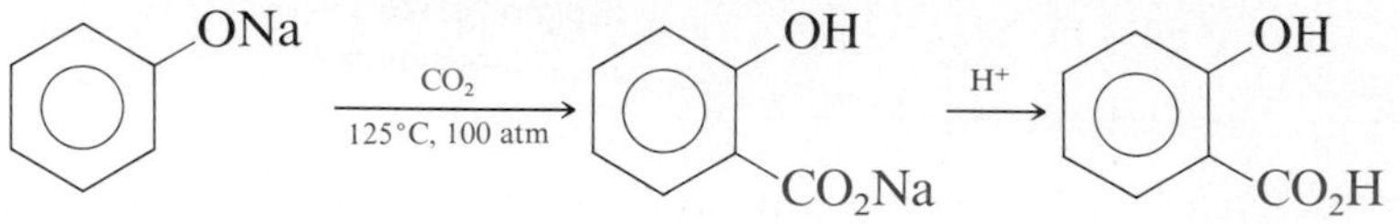

| Sodium phenoxide | Sodium salicylate | Salicylic acid (79%) |

While a hydroxyl group strongly activates an aromatic ring toward electrophilic attack, an oxyanion substituent is an even more powerful activator. Electron delocalization in phenoxide anion leads to increased electron density at the positions ortho and para to oxygen.

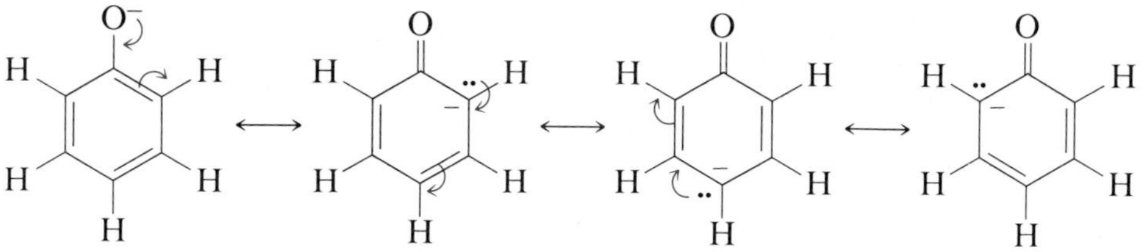

The increased nucleophilicity of the ring permits it to react with carbon dioxide. An intermediate is formed which is simply the keto form of salicylate anion.

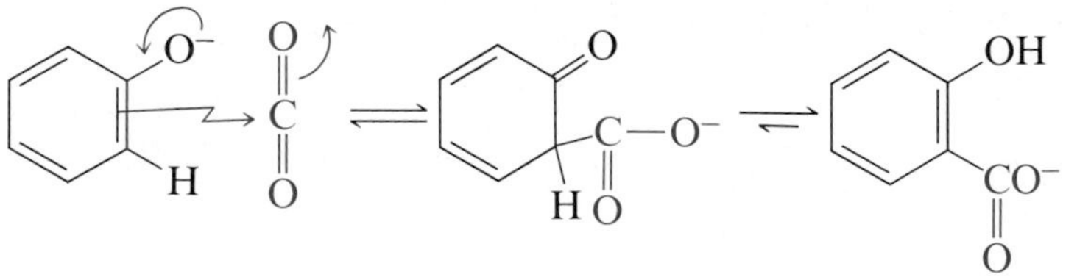

| Phenoxide anion (stronger base) | Carbon dioxide | Cyclohexadienone intermediate | Salicylate anion (weaker base) |

The Kolbe-Schmitt reaction is an equilibrium process governed by thermodynamic control. The position of equilibrium favors formation of the weaker base (salicylate ion) at the expense of the stronger one (phenoxide) ion. Thermodynamic control is also responsible for the pronounced bias toward ortho over para substitution. Salicylate anion is a weaker base than p-hydroxybenzoate and so is the predominant species when equilibrium is established between the two.

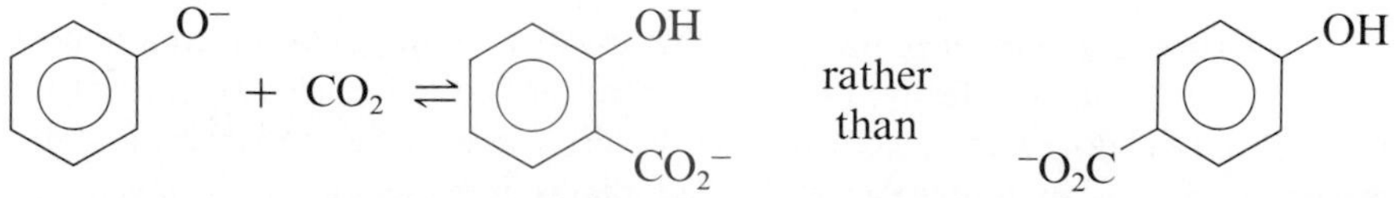

rather than

Phenoxide ion (strongest base; K_a of conjugate acid, 10^{-10})

Carbon dioxide

Salicylate anion (weakest base; K_a of conjugate acid, 1.06×10^{-3})

p-Hydroxybenzoate anion (K_a of conjugate acid, 3.3×10^{-5})

Salicylate anion is a weaker base than p-hydroxybenzoate because it is stabilized by intramolecular hydrogen bonding.

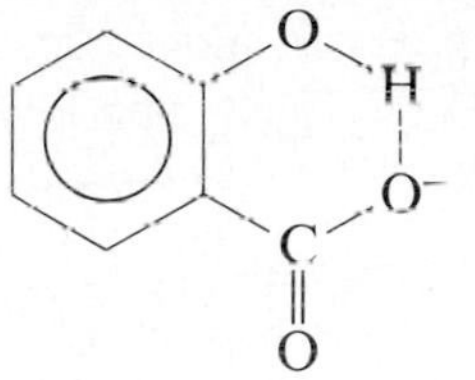

Intramolecular hydrogen bonding in salicylate anion disperses its charge

The Kolbe-Schmitt reaction has been applied to the preparation of other o-hydroxybenzoic acids. Alkyl derivatives of phenol behave very much like phenol itself.

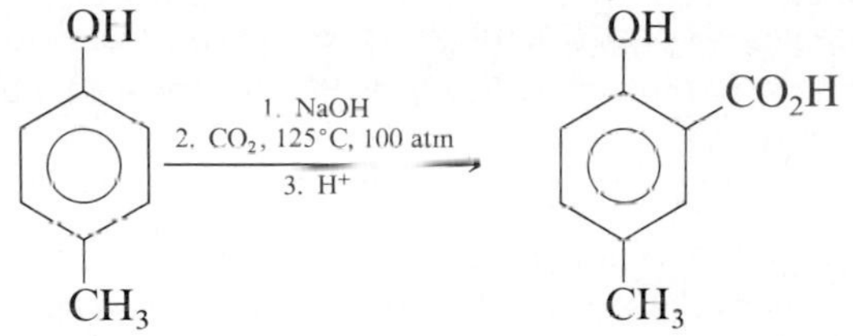

OH
p-Cresol

1. NaOH
2. CO_2, 125°C, 100 atm
3. H^+

OH CO_2H
CH_3
2-Hydroxy-5-methylbenzoic acid (78%)

Phenols that bear strongly electron-withdrawing substituents usually give low yields of carboxylated products; their derived phenoxide anions are less basic and the equilibrium constants for their carboxylation are smaller.

26.11 PREPARATION OF ARYL ETHERS

Aryl ethers are best prepared by the Williamson method (Section 17.7). Alkylation of the hydroxyl oxygen of a phenol takes place readily on reaction of a phenoxide anion with an alkyl halide.

$$ArO^- \ + \ R\!-\!X \longrightarrow ArOR \ + \ X^-$$

Phenoxide anion   Alkyl halide   Alkyl aryl ether   Halide anion

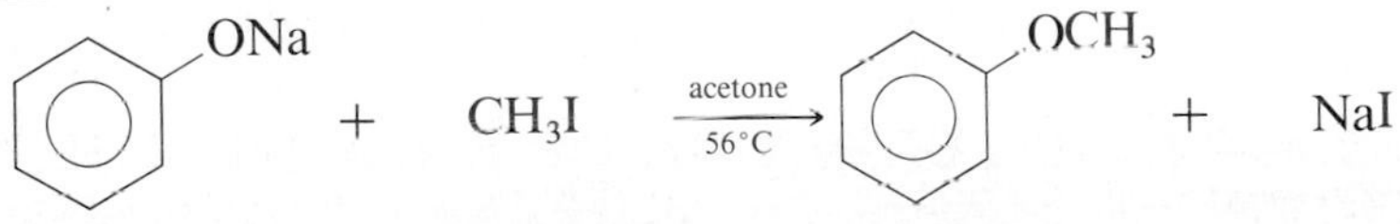

ONa
Sodium phenoxide

$+$ CH_3I
Iodomethane (methyl iodide)

$\xrightarrow[56°C]{acetone}$

OCH_3
Anisole (95%)

$+$ NaI
Sodium iodide

As the synthesis is normally performed, a solution of the appropriate phenol and alkyl halide is simply heated in the presence of sodium hydroxide or some other base.

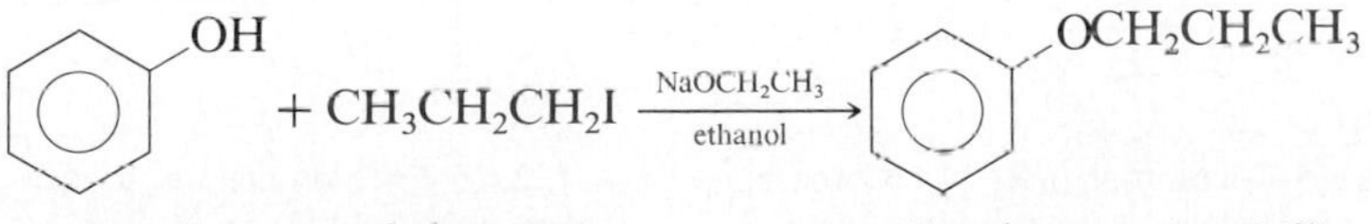

OH
Phenol

$+ CH_3CH_2CH_2I$
1-Iodopropane (propyl iodide)

$\xrightarrow[ethanol]{NaOCH_2CH_3}$

$OCH_2CH_2CH_3$
Phenyl propyl ether (74%)

The alkyl halide must be one that reacts readily in an S_N2 process. Thus, methyl and primary alkyl halides are the most effective alkylating agents. An economical alternative to a methyl halide is dimethyl sulfate.

$$\text{2-Naphthol} \quad + \quad (CH_3O)_2SO_2 \xrightarrow[\text{H}_2\text{O}]{\text{NaOH}} \quad \text{2-Methoxynaphthalene (65–73\%)}$$

2-Naphthol   Dimethyl sulfate   2-Methoxynaphthalene (65–73%)
(methyl 2-naphthyl ether)

Elimination becomes competitive with substitution when secondary alkyl halides are used and is the only reaction observed with tertiary alkyl halides.

PROBLEM 26.9 Reaction of phenol with 1,2-epoxypropane in aqueous sodium hydroxide at 150°C gives a single product, $C_9H_{12}O_2$, in 90 percent yield. Suggest a reasonable structure for this compound.

The reaction between an alkoxide ion and an aryl halide is a feasible approach to the preparation of alkyl aryl ethers only when the aryl halide is one that reacts rapidly by the addition-elimination mechanism of nucleophilic aromatic substitution (Section 25.5).

$$\text{p-Fluoronitrobenzene} \xrightarrow[\text{CH}_3\text{OH, 25°C}]{\text{KOCH}_3} \text{p-Nitroanisole (93\%)}$$

p-Fluoronitrobenzene   *p*-Nitroanisole (93%)

PROBLEM 26.10 Which of the two combinations of reactants listed below is more appropriate for the preparation of *p*-nitrophenyl phenyl ether?

(a) Fluorobenzene and *p*-nitrophenol
(b) *p*-Fluoronitrobenzene and phenol

26.12 ACCIDENTAL FORMATION OF A DIARYL ETHER. DIOXIN

A widely used herbicide is 2,4,5-trichlorophenoxyacetic acid (2,4,5-T). It is prepared by reaction of 2,4,5-trichlorophenol with chloroacetic acid.

$$\text{2,4,5-Trichlorophenol} + ClCH_2CO_2H \longrightarrow \text{2,4,5-Trichlorophenoxyacetic acid (2,4,5-T)}$$

2,4,5-Trichlorophenol   Chloroacetic   2,4,5-Trichlorophenoxyacetic
acid   acid (2,4,5-T)

The starting material for this process, 2,4,5-trichlorophenol, is made by treating 1,2,4,5-tetrachlorobenzene with aqueous base. Nucleophilic aromatic substitution of one of the chlorines by an addition-elimination mechanism yields 2,4,5-trichlorophenol.

1,2,4,5-Tetrachlorobenzene   2,4,5-Trichlorophenol

In the course of making 2,4,5-trichlorophenol, it almost always becomes contaminated with small amounts of 2,3,7,8-tetrachlorodibenzo-*p*-dioxin, better known simply as *dioxin*.

2,3,7,8-Tetrachlorodibenzo-*p*-dioxin
(dioxin)

PROBLEM 26.11   Can you propose a mechanism for the formation of dioxin from 1,2,4,5-tetrachlorobenzene and 2,4,5-trichlorophenol in the presence of sodium hydroxide?

Dioxin is carried along when 2,4,5-trichlorophenol is converted to 2,4,5-T, and enters the environment when 2,4,5-T is sprayed on vegetation. Typically, the amount of dioxin present in 2,4,5-T is very small. Agent orange, a 2,4,5-T-based defoliant used on a large scale in Vietnam, contained about 2 ppm of dioxin.

Tests with animals have revealed that dioxin is one of the most toxic substances known. Toward mice it is about 2000 times more toxic than strychnine and about 150,000 times more toxic than sodium cyanide. Fortunately, however, available evidence indicates that humans are far more resistant to dioxin than are test animals, and so far there have been no human fatalities directly attributable to dioxin. The most prominent symptom seen so far has been a severe skin disorder known as *chloracne*. Yet to be determined is the answer to the question of long-term effects. A connection between exposure to dioxin and the occurrence of a rare type of cancer has been claimed on the basis of a statistical study. Since 1979 the use of 2,4,5-T has been regulated in the United States.

26.13 CLEAVAGE OF ARYL ETHERS BY HYDROGEN HALIDES

The cleavage of dialkyl ethers by hydrogen halides has been discussed in Section 17.10, where it was noted that the same pair of alkyl halides results irrespective of the order in which the carbon-oxygen bonds of the ether were broken.

$$ROR' \quad + \quad 2HX \quad \longrightarrow RX + R'X + H_2O$$

Dialkyl ether   Hydrogen halide   Two alkyl halides   Water

Cleavage of alkyl aryl ethers by hydrogen halides always proceeds so that the alkyl-oxygen bond is broken and yields an alkyl halide and a phenol.

$$\text{ArOR} + \text{HX} \longrightarrow \text{ArOH} + \text{RX}$$

| Alkyl aryl ether | Hydrogen halide | Phenol | Alkyl halide |

Since phenols are not converted to aryl halides by reaction with hydrogen halides, reaction proceeds no further.

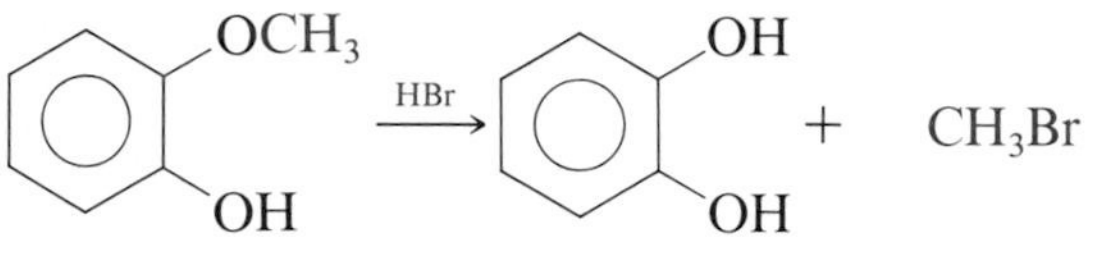

Guaiacol

Pyrocatechol (85–87%)

Methyl bromide (57–72%)

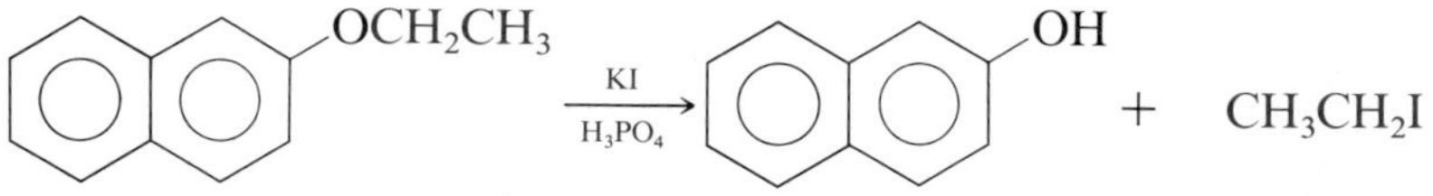

Ethyl 2-Naphthyl ether

2-Naphthol (95%)

Ethyl iodide (78%)

The first step in the reaction of an alkyl aryl ether with a hydrogen halide is protonation of oxygen to form an alkylaryloxonium ion.

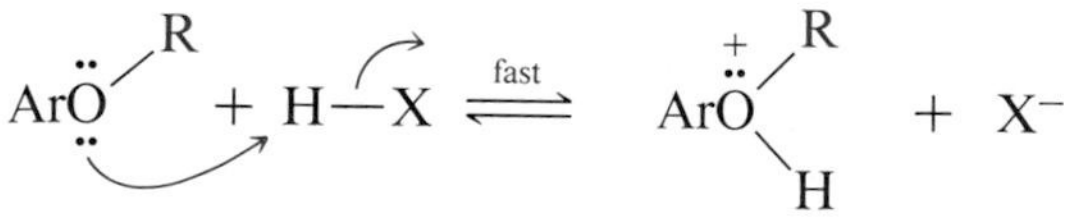

Alkyl aryl ether

Hydrogen halide

Alkylaryloxonium ion

Halide ion

This is followed by a nucleophilic substitution step.

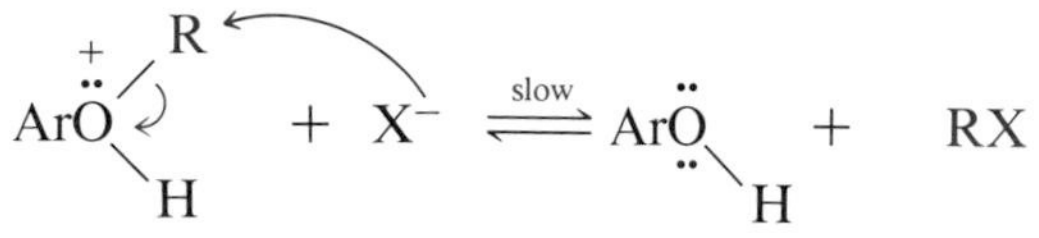

Alkylaryloxonium ion

Halide ion

Phenol

Alkyl halide

Attack by the halide nucleophile at the sp^3 hybridized carbon of the alkyl group is analogous to what takes place in the cleavage of dialkyl ethers. Attack at the sp^2 hybridized carbon of the aromatic ring is much slower. Indeed, it does not occur at all under these conditions.

26.14 OXIDATION OF PHENOLS. QUINONES

Phenols are more easily oxidized than alcohols, and a large number of inorganic oxidizing agents have been used for this purpose. The phenol oxidations that are of the most use to the organic chemist are those involving derivatives of 1,2-benzenediol (pyrocatechol) and 1,4-benzenediol (hydroquinone). Oxidation of compounds of this type with silver oxide or with chromic acid yields conjugated dicarbonyl compounds called *quinones*.

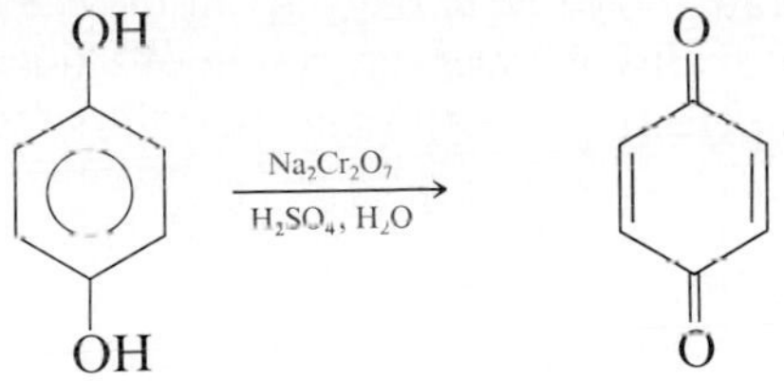

$$\text{Hydroquinone} \xrightarrow[\text{H}_2\text{SO}_4,\ \text{H}_2\text{O}]{\text{Na}_2\text{Cr}_2\text{O}_7} \text{p-Benzoquinone (76–81\%)}$$

Hydroquinone   p-Benzoquinone (76–81%)

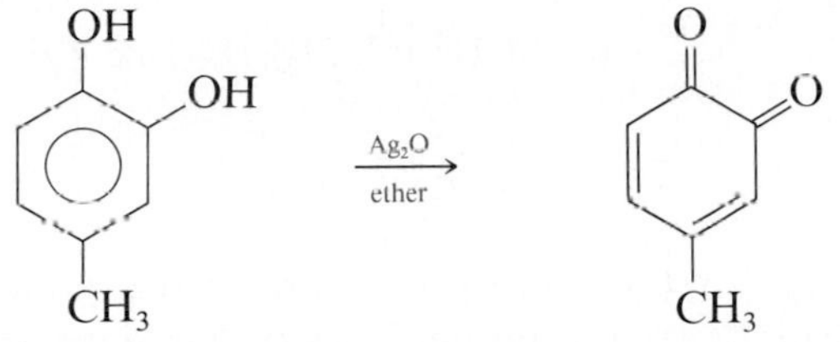

$$\text{4-Methylpyrocatechol} \xrightarrow[\text{ether}]{\text{Ag}_2\text{O}} \text{4-Methyl-1,2-benzoquinone (68\%)}$$

4-Methylpyrocatechol
(4-methyl-1,2-benzenediol)   4-Methyl-1,2-benzoquinone (68%)

Quinones are colored; *p*-benzoquinone, for example, is yellow. Many occur naturally and have been used as dyes. Alizarin is a red pigment extracted from the roots of the madder plant. Its preparation from anthracene, a coal tar derivative, in 1868 was a significant step in the development of the synthetic dyestuff industry.

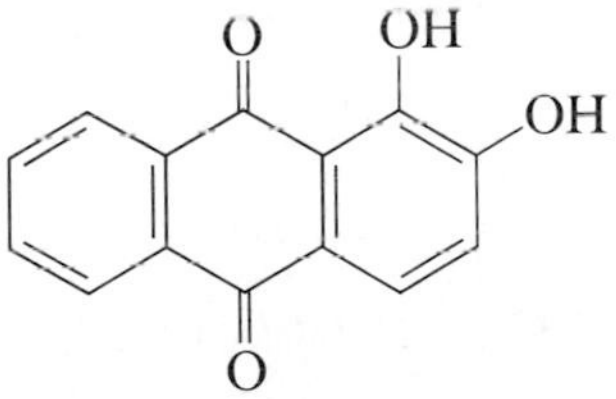

Alizarin

The oxidation-reduction process that connects hydroquinone and benzoquinone involves two 1-electron transfer reactions.

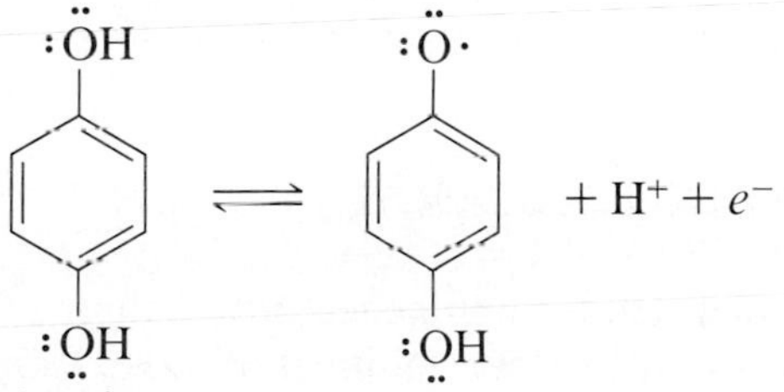

$$\text{Hydroquinone} \rightleftharpoons \qquad + \text{H}^+ + e^-$$

Hydroquinone

$$\rightleftharpoons \qquad + \text{H}^+ + e^-$$

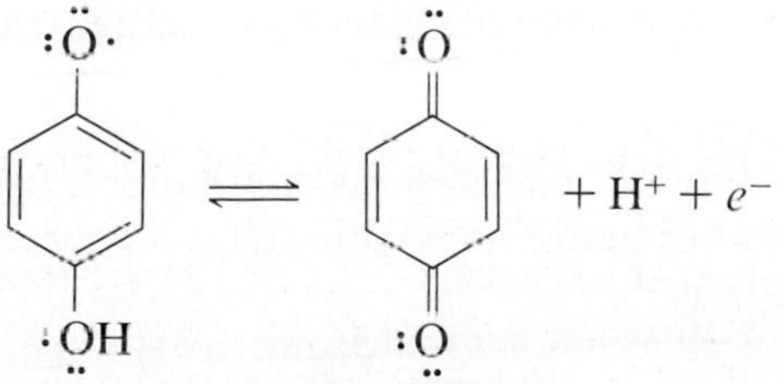

Benzoquinone

The ready reversibility of this reaction is essential to the role that quinones play in cellular respiration, the process by which an organism utilizes molecular oxygen to convert its food to carbon dioxide, water, and energy. Electrons are not transferred directly from the substrate molecule to oxygen but instead are transferred by way of an *electron transport chain* involving a succession of oxidation-reduction reactions. A key component of this electron transport chain is the substance known as *ubiquinone,* or coenzyme Q.

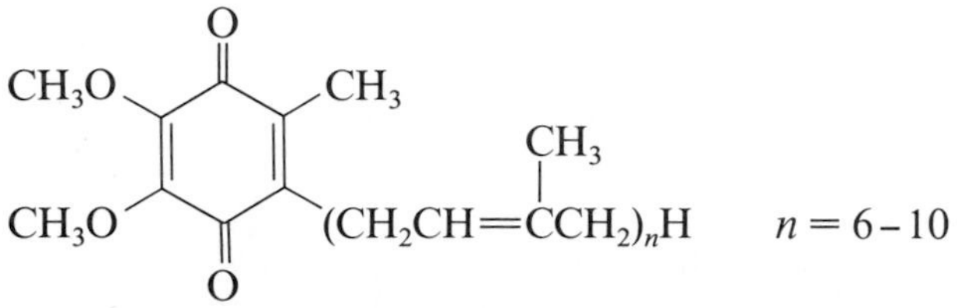

Ubiquinone (coenzyme Q)

The name "ubiquinone" is a shortened form of "ubiquitous quinone," a term coined to describe the observation that this substance can be found in all cells. The length of its side chain varies among different organisms; the most common form in vertebrates has $n = 10$, while ubiquinones in which $n = 6$ to 9 are found in yeasts and plants.

Another physiologically important quinone is vitamin K. Here "K" stands for *Koagulation* (Danish), since this substance was first identified as essential for the normal clotting of blood.

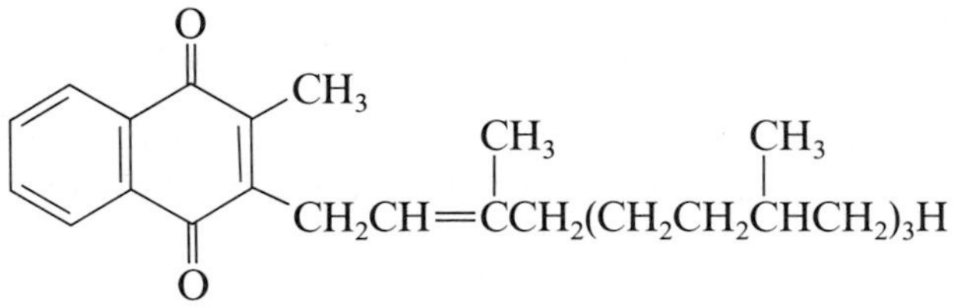

Vitamin K

Some vitamin K is provided in the normal diet, but a large proportion of that required by humans is produced by their intestinal flora.

26.15 SPECTROSCOPIC ANALYSIS OF PHENOLS

The infrared spectra of phenols combine features of those of alcohols and aromatic compounds. Hydroxyl absorbances resulting from O—H stretching are found in the 3600 cm^{-1} region and the peak due to C—O stretching appears around 1200 to 1250 cm^{-1}. These features can be seen in the infrared spectrum of *p*-cresol, shown in Figure 26.4.

The ^{1}H nmr signals for the hydroxyl protons of phenols generally appear at lower field than those of alcohols. Chemical shifts of protons in phenolic hydroxyl groups are in the range $\delta = 4$ to 12 ppm. Often the signal is very broad. Figure 26.5 is the ^{1}H nmr spectrum of *p*-cresol.

A peak for the molecular ion is usually quite prominent in the mass spectra of phenols. It is, for example, the most intense peak in phenol.

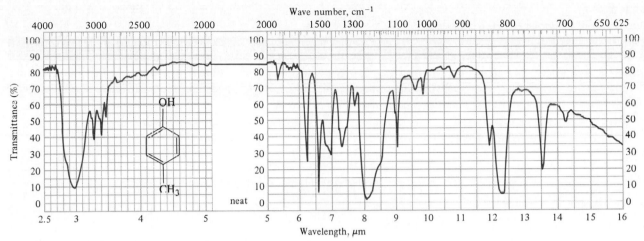

FIGURE 26.4   The infrared spectrum of *p*-cresol.

26.16   SUMMARY

Phenol is both an important industrial chemical and the parent of a large class of compounds that occur in wide distribution as natural products. Significant derivatives of phenols include aryl ethers and quinones. Quinones occupy a prominent place in biological chemistry.

The reactions of phenols fall into two main classes:

1.   Substitution on the aromatic ring
2.   Reactions that involve the phenolic hydroxyl group

The most common reaction of the aromatic ring of phenols is electrophilic aromatic substitution. Typical examples of these reactions are summarized in Table 26.5

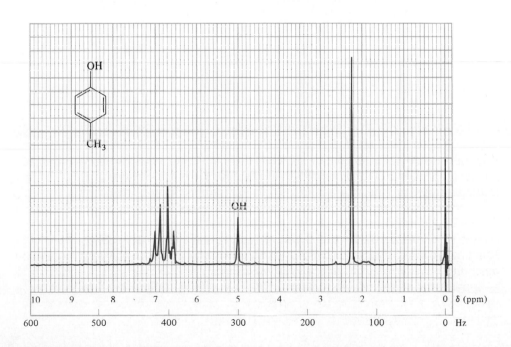

FIGURE 26.5   The ^{1}H nmr spectrum of *p*-cresol.

TABLE 26.6
Reactions of Phenolic Hydroxyl Groups

Reaction (section) and comments	General equation and specific example
Acidity (Sections 26.4–26.5) Phenols are weak acids with K_a's of about 10^{-10} (pK_a 10). Strongly electron-withdrawing substituents increase the acidity. Quantitative conversion to phenoxide anions is achieved in the presence of hydroxide.	$ArOH \rightleftharpoons H^+ \quad ArO^-$ Phenol · Proton · Phenoxide anion 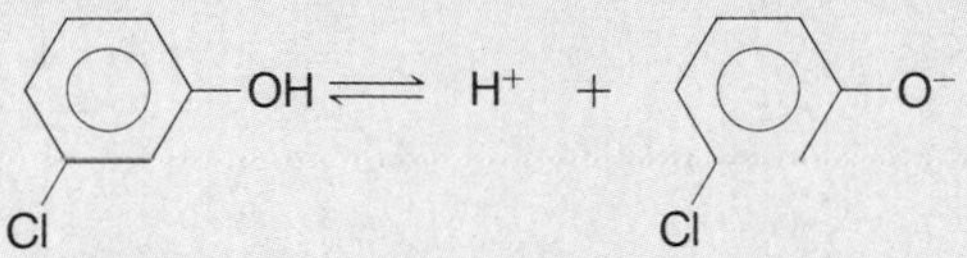 m-Chlorophenol (K_a 7.6 × 10⁻¹⁰) · Proton · m-Chlorophenoxide ion
Esterification (Section 26.9) Phenols are converted to aryl esters by acyl chlorides and acid anhydrides. The reaction may be catalyzed by either acids or bases.	$ArOH + RCX \longrightarrow ArOCR + HX$ Phenol · Acylating agent · Aryl ester Hydroquinone · 1,4-Diacetoxybenzene (96–98%)
Alkylation (Section 26.11) Phenoxide ions, prepared from phenols in base, react with alkyl halides to give alkyl aryl ethers.	$ArO^- + RX \longrightarrow ArOR + X^-$ Phenoxide anion · Alkyl halide · Alkyl aryl ether · Halide anion 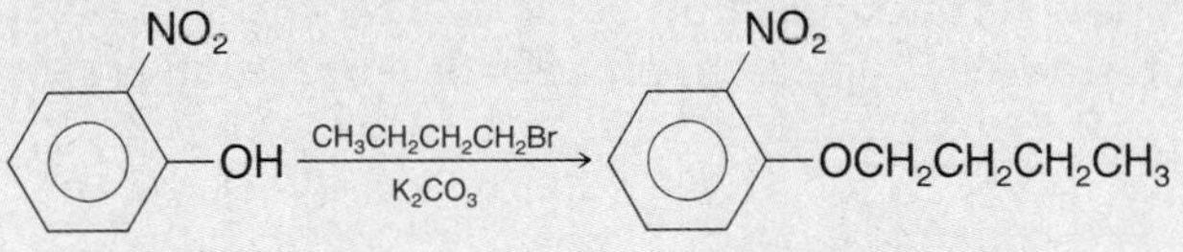 o-Nitrophenol · Butyl o-nitrophenyl ether (75–80%)

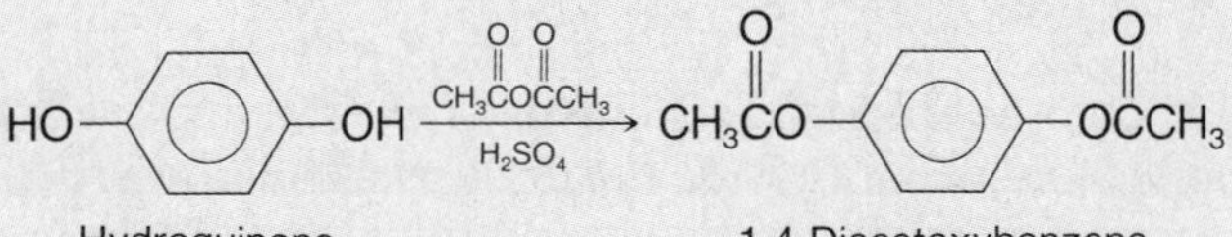

in Section 26.8. The hydroxyl group of a phenol is a very powerful activating substituent, and electrophilic aromatic substitution reactions occur with great ease in phenols. The incoming electrophile is directed primarily to positions ortho and para to the hydroxyl group, with para substitution predominating.

Reactions involving the hydroxyl group of phenols are summarized in Table 26.6.

The Kolbe-Schmitt synthesis of salicylic acid is a vital step in the preparation of aspirin (Section 26.10). It combines two modes of phenol reactivity—proton abstraction from the hydroxyl group to give a phenoxide anion and reaction of a ring position of the phenoxide ion with carbon dioxide.

Oxidation of 1,2- and 1,4-benzenediols gives colored compounds known as quinones.

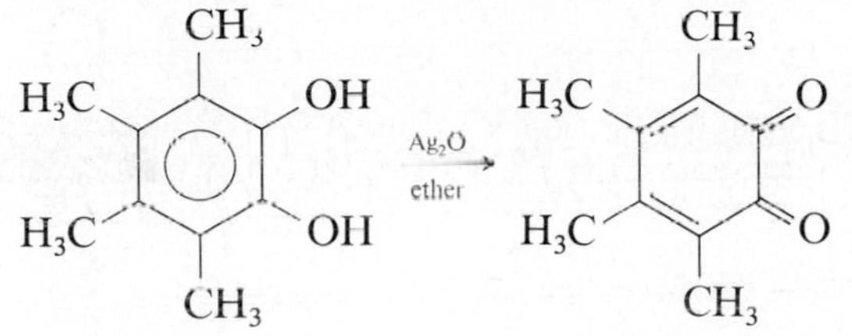

3,4,5,6-Tetramethyl-1,2-benzenediol → 3,4,5,6-Tetramethyl-1,2-benzoquinone (81%)

(Ag₂O, ether)

PROBLEMS

26.12 The IUPAC rules permit the use of common names for a number of familiar phenols and aryl ethers. These common names are listed below along with their systematic names. Write the structure of each compound.

(a) *Vanillin* (4-hydroxy-3-methoxybenzaldehyde): a component of vanilla bean oil, which contributes to its characteristic flavor

(b) *Thymol* (2-isopropyl-5-methylphenol): obtained from oil of thyme

(c) *Carvacrol* (5-isopropyl-2-methylphenol): present in oil of thyme and marjoram

(d) *Eugenol* (4-allyl-2-methoxyphenol): obtained from oil of cloves

(e) *Gallic acid* (3,4,5-trihydroxybenzoic acid): prepared by hydrolysis of tannins derived from plants

(f) *Salicyl alcohol* (*o*-hydroxybenzyl alcohol): obtained from bark of poplar and willow trees

26.13 Name each of the following compounds:

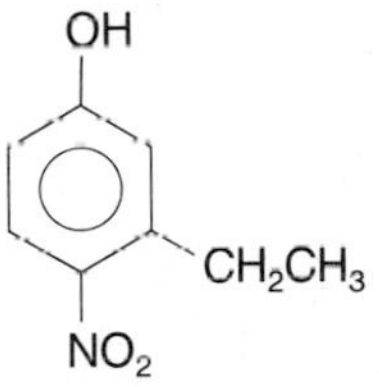

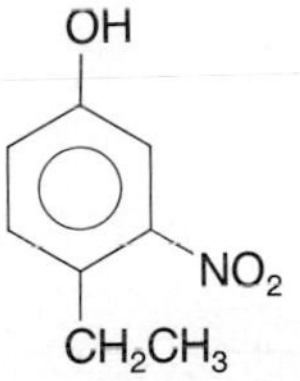

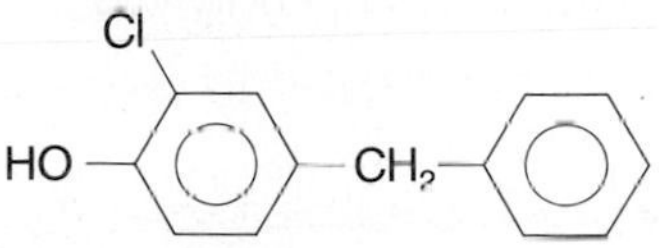

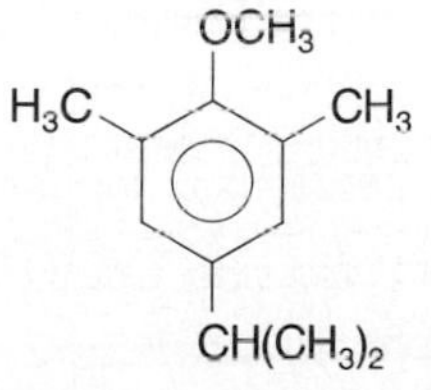

(e)

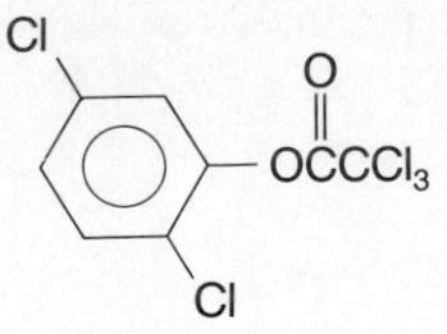

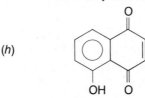

(f)

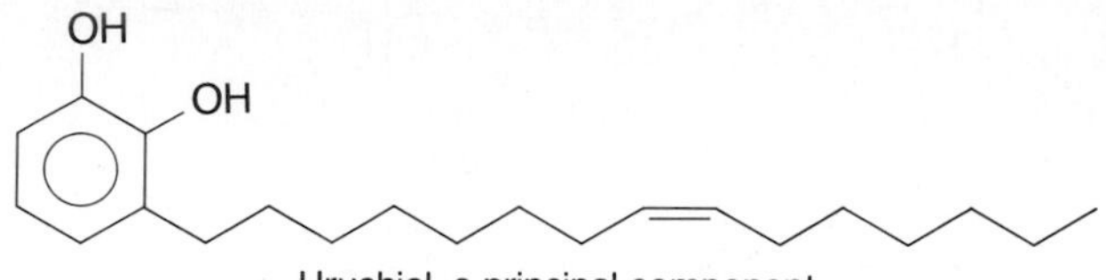

Urushiol, a principal component
of the allergenic oil of poison ivy

(g)

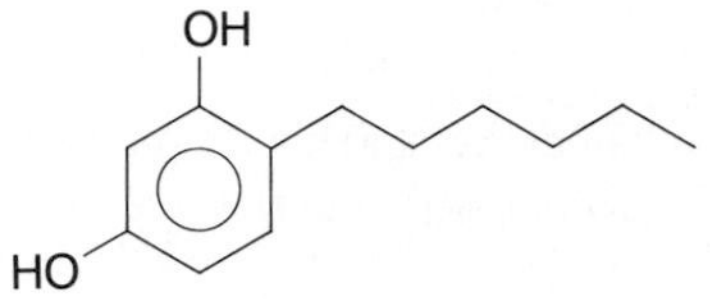

A commonly used antiseptic

(h)

Juglone, obtained from walnut hulls

26.14 Write a balanced chemical equation for each of the following reactions:

(a) Phenol + sodium hydroxide
(b) Product of (a) + ethyl bromide
(c) Product of (a) + dimethyl sulfate
(d) Product of (a) + butyl *p*-toluenesulfonate
(e) Product of (a) + acetic anhydride
(f) *o*-Cresol + benzoyl chloride
(g) *m*-Cresol + ethylene oxide
(h) 2,6-Dichlorophenol + bromine
(i) *p*-Cresol + excess aqueous bromine
(j) Isopropyl phenyl ether + excess hydrogen bromide + heat

26.15 Outline four different syntheses of phenol, beginning with benzene and using any necessary organic or inorganic reagents.

26.16 Which phenol in each of the following pairs is more acidic? Justify your choice.

(a) 2,4,6-Trimethylphenol or 2,4,6-trinitrophenol
(b) 2,6-Dichlorophenol or 3,5-dichlorophenol
(c) 3-Nitrophenol or 4-nitrophenol
(d) Phenol or 4-cyanophenol
(e) 2,5-Dinitrophenol or 2,6-dinitrophenol

26.17 Choose the reaction in each of the following pairs that proceeds at the faster rate. Explain your reasoning.

(a) Base-promoted hydrolysis of phenyl acetate or *m*-nitrophenyl acetate
(b) Base-promoted hydrolysis of *m*-nitrophenyl acetate or *p*-nitrophenyl acetate

(c)   Reaction of ethyl bromide with phenol or with the sodium salt of phenol

(d)   Reaction of ethylene oxide with the sodium salt of phenol or with the sodium salt of p-nitrophenol

(e)   Bromination of phenol or phenyl acetate

26.18   Pentafluorophenol is readily prepared by heating hexafluorobenzene with potassium hydroxide in *tert*-butyl alcohol.

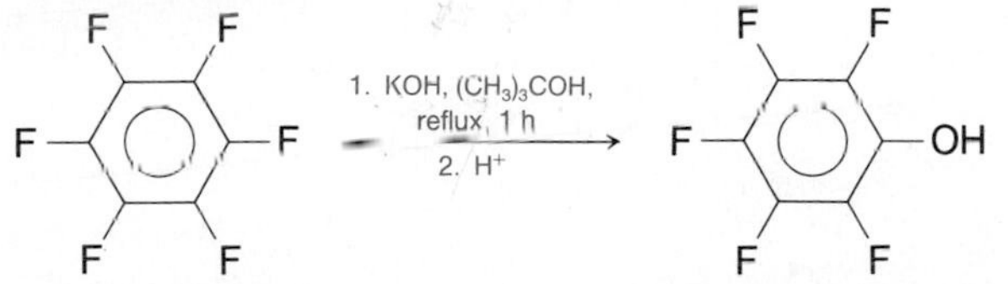

Hexafluorobenzene → [1. KOH, (CH₃)₃COH, reflux, 1 h 2. H⁺] → Pentafluorophenol (71%)

What is the most reasonable mechanism for this reaction? Comment on the comparative ease with which this conversion occurs.

26.19   Each of the following reactions has been reported in the chemical literature and proceeds cleanly in good yield. Identify the principal organic product in each case.

(a)   [2-methoxyphenol] + CH₂=CHCH₂Br $\xrightarrow[\text{acetone}]{\text{K}_2\text{CO}_3}$

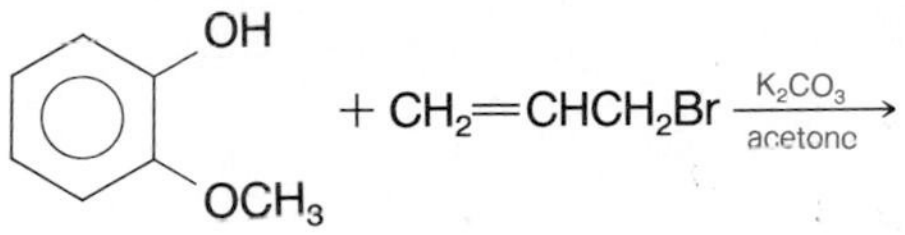

(b)   [2-acetyl-1,4-dihydroxybenzene] + (CH₃O)₂SO₂ (excess) $\xrightarrow[\text{H}_2\text{O}]{\text{NaOH}}$

(c)   [sodium phenoxide] + ClCH₂CHCH₂OH (with OH) $\longrightarrow$

(d)   [4-hydroxy-3-methoxybenzaldehyde] $\xrightarrow[\substack{\text{acetic acid,} \\ \text{heat}}]{\text{HNO}_3}$

(e)   [3-methoxyphenyl]—CH₂CO₂H + HI $\xrightarrow{\text{heat}}$

(f)   [2-ethoxy-4-nitrophenol] + Br₂ $\xrightarrow{\text{acetic acid}}$

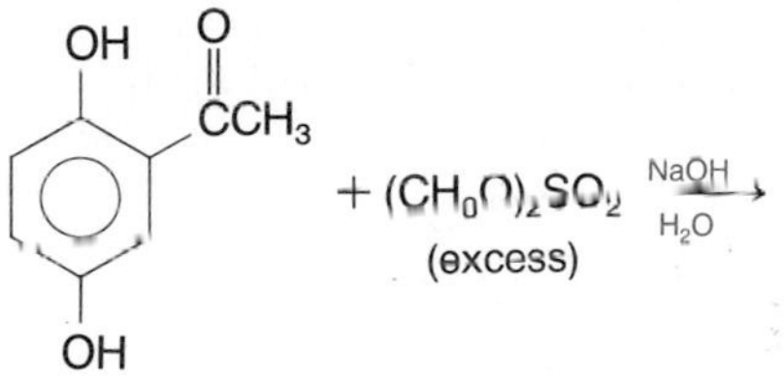

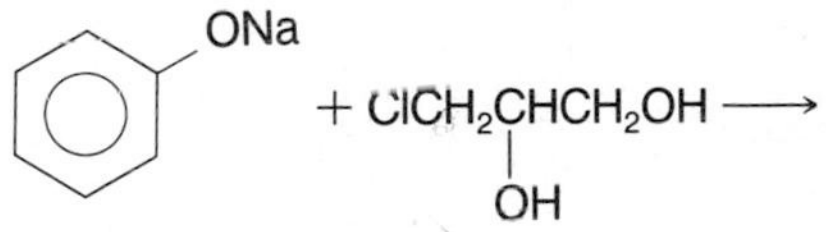

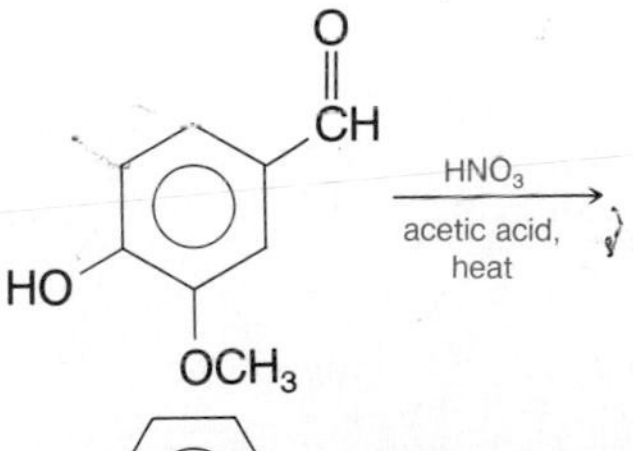

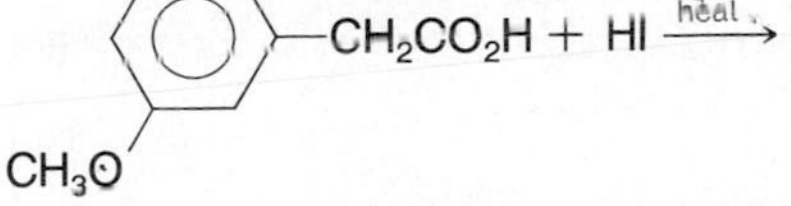

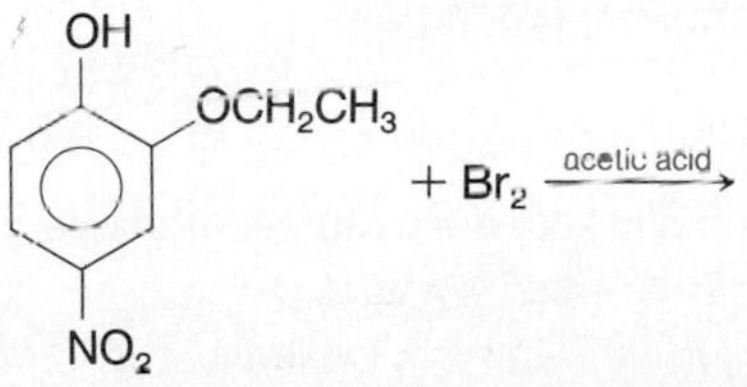

(g)

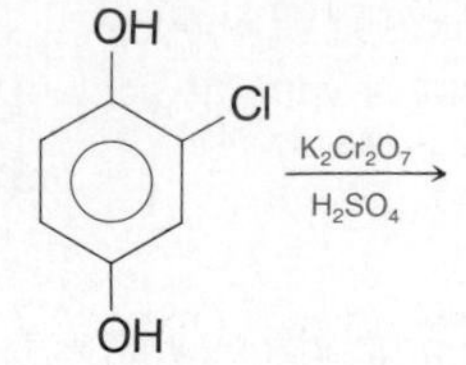

OH
Cl
OH

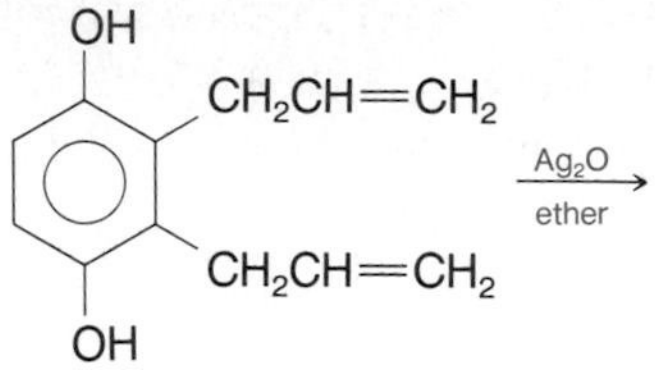

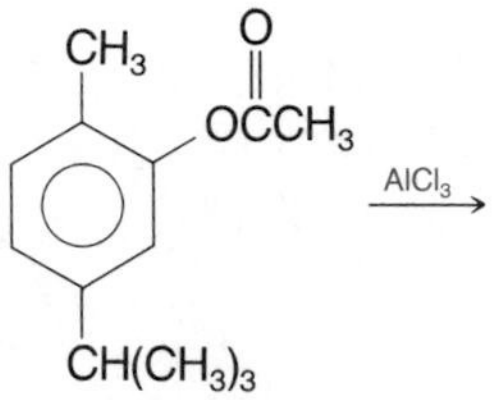

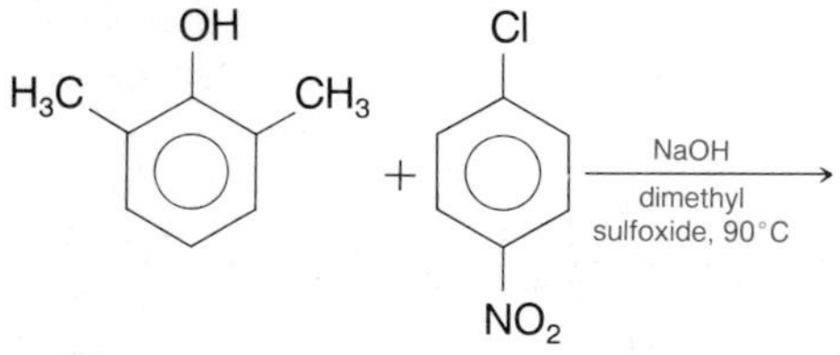

$\xrightarrow[\text{H}_2\text{SO}_4]{\text{K}_2\text{Cr}_2\text{O}_7}$

(h)

OH
$\text{CH}_2\text{CH}{=}\text{CH}_2$
$\text{CH}_2\text{CH}{=}\text{CH}_2$
OH

$\xrightarrow[\text{ether}]{\text{Ag}_2\text{O}}$

(i)

CH_3
$\text{O} \text{CCH}_3$ (with O double bond)
$\text{CH(CH}_3)_3$

$\xrightarrow{\text{AlCl}_3}$

(j)

OH
H_3C CH_3

+

Cl

NO_2

$\xrightarrow[\substack{\text{dimethyl} \\ \text{sulfoxide, 90}^\circ\text{C}}]{\text{NaOH}}$

(k)

Cl
OH
Cl

$+ 2\text{Cl}_2 \xrightarrow{\text{acetic acid}}$

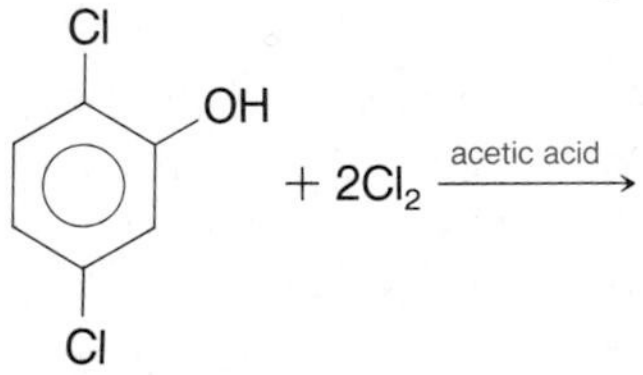

(l)

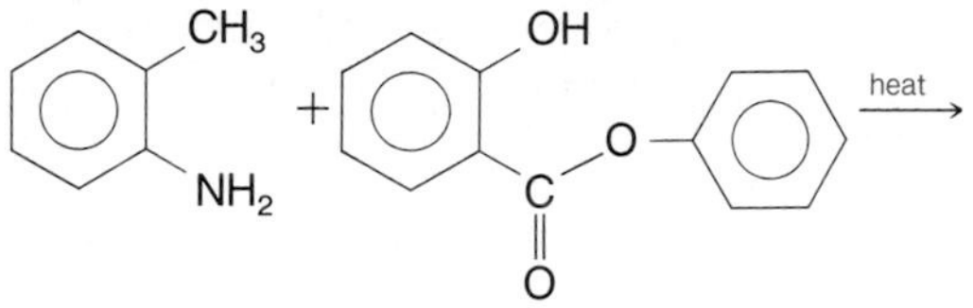

CH_3
NH_2

+

OH
C
O (ester to phenyl)
O

$\xrightarrow{\text{heat}}$

(m)

OH
Cl
Cl
Cl

$+ \text{C}_6\text{H}_5\overset{+}{\text{N}}{\equiv}\text{N} \ \ \text{Cl}^- \longrightarrow$

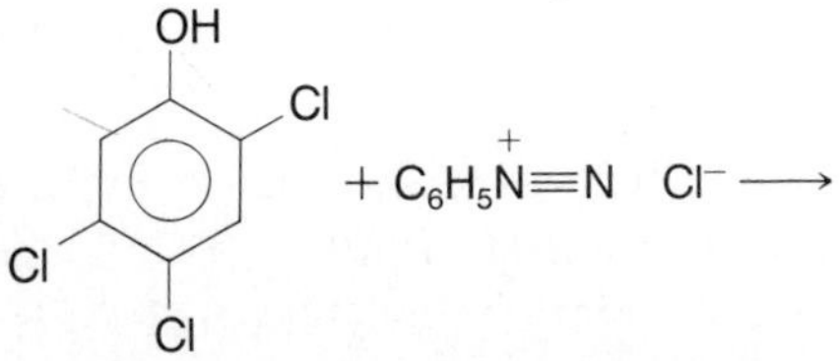

26.20 The synthesis of the analgesic substance *phenacetin* is outlined in the equation shown below. What is the structure of phenacetin?

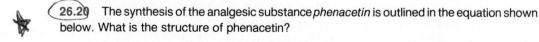

p-Nitrophenol $\xrightarrow[\substack{\text{2. Fe, HCl; then HO}^- }]{\text{1. (CH}_3\text{CH}_2\text{O})_2\text{SO}_2\text{, NaOH}}$ phenacetin

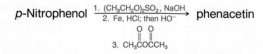

3. $\text{CH}_3\overset{\text{O}}{\text{C}}\text{O}\overset{\text{O}}{\text{C}}\text{CH}_3$

26.21 Identify compounds **A** through **C** in the synthetic sequence represented by equations (*a*) through (*c*).

(a) Phenol + H_2SO_4 $\xrightarrow{\text{heat}}$ compound A ($C_6H_6O_7S_2$)

(b) Compound A + Br_2 $\xrightarrow[\text{2. } H^+]{\text{1. } HO^-}$ compound B ($C_6H_5BrO_7S_2$)

(c) Compound B + H_2O $\xrightarrow[\text{heat}]{H^+}$ compound C (C_6H_5BrO)

26.22 Treatment of 3,5-dimethylphenol with dilute nitric acid, followed by steam distillation of the reaction mixture, gave compound D ($C_8H_9NO_3$, mp 66°C) in 36 percent yield. The nonvolatile residue from the steam distillation gave compound E ($C_8H_9NO_3$, mp 108°C) in 25 percent yield on extraction with chloroform. Identify compounds D and E.

26.23 Outline a reasonable synthesis of 4-nitrophenyl phenyl ether from chlorobenzene and phenol.

26.24 As an allergen for testing purposes, synthetic 3-pentadecylcatechol is more useful than natural poison ivy extracts (of which it is one component). A stable crystalline solid, it is efficiently prepared in pure form from readily available starting materials. Outline a reasonable synthesis of this compound from 2,3-dimethoxybenzaldehyde and any necessary organic or inorganic reagents.

OH
OH
$(CH_2)_{14}CH_3$
3-Pentadecylcatechol

26.25 In a general reaction known as the *cyclohexadienone-phenol rearrangement*, cyclohexadienones are converted to phenols under conditions of acid catalysis. An example is

O

$\xrightarrow{H^+}$

OH (100%)

Write a reasonable mechanism for this reaction.

26.26 Treatment of *p*-hydroxybenzoic acid with aqueous bromine leads to the evolution of carbon dioxide and the formation of 2,4,6-tribromophenol. Explain.

26.27 Treatment of phenol with excess aqueous bromine is actually more complicated than expected. A white precipitate forms rapidly, which on closer examination is not 2,4,6-tribromophenol but is instead 2,4,4,6-tetrabromocyclohexadienone. Explain the formation of this product.

26.28 Treatment of 2,4,6-*tri-tert*-butylphenol with bromine in cold acetic acid gives compound F ($C_{18}H_{29}BrO$) in quantitative yield. The infrared spectrum of compound F contains absorptions at 1630 and 1655 cm^{-1}. Its ^1H nmr spectrum shows only three peaks (all singlets), at $\delta = 1.19$, 1.26, and 6.90 ppm, in the ratio 9:18:2. What is a reasonable structure for compound F?

26.29 Compound G undergoes hydrolysis of its acetal function in dilute sulfuric acid to yield 1,2-ethanediol and compound H ($C_6H_6O_2$), mp 54°C. Compound H exhibits a carbonyl stretching band in the infrared at 1690 cm^{-1} and has two singlets in its ^{1}H nmr spectrum, at $\delta = 2.88$ and 6.70 ppm, in the ratio $2:1$. In standing in water or ethanol, compound H is converted cleanly to an isomeric substance, compound I, mp 172 to 173°C. Compound I has no peaks attributable to carbonyl groups in its infrared spectrum. Identify compounds H and I.

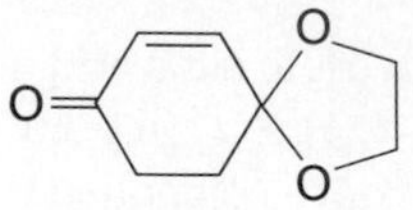

Compound G

26.30 Fomecin A ($C_8H_8O_5$) is an antibacterial fungal metabolite. It is soluble in dilute sodium hydroxide solution, forms a red 2,4-dinitrophenylhydrazone, and rapidly reduces ammoniacal silver nitrate. Its ^{1}H nmr spectrum includes singlets at $\delta = 4.58$ ppm (2 H), 6.53 ppm (1 H), and 10.2 ppm (1 H) as well as broad signals for four O—H protons. Treatment of fomecin A with acetic anhydride in pyridine gave compound J. Suggest a reasonable structure for fomecin A.

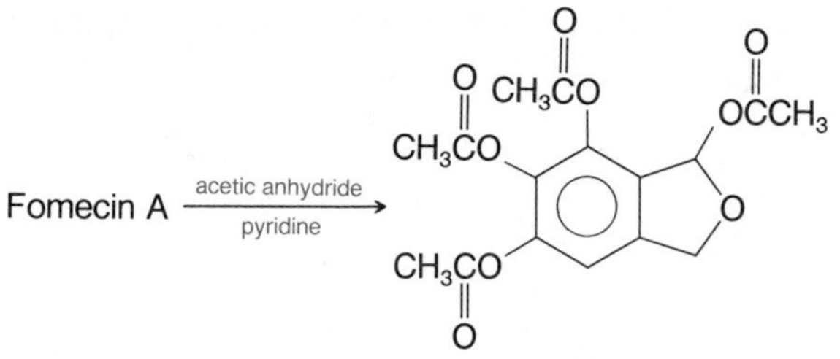

$$\text{Fomecin A} \xrightarrow[\text{pyridine}]{\text{acetic anhydride}}$$

Compound J

26.31 The equilibrium constant for reduction of 1,4-benzoquinone according to the equation

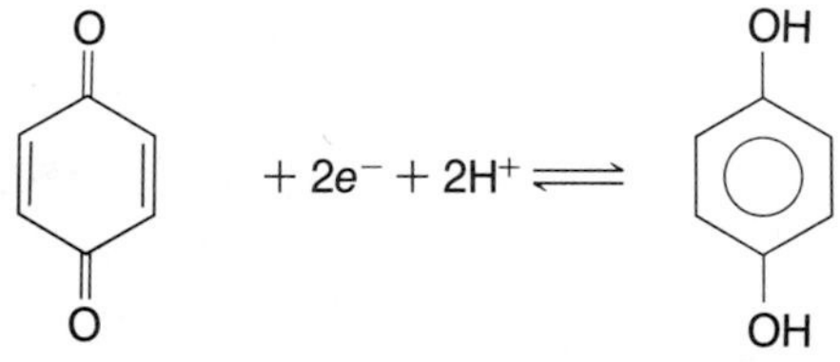

$$+ \ 2e^- + 2H^+ \rightleftharpoons$$

1,4-Benzoquinone   Hydroquinone

is more favorable than that for reduction of 1,4-naphthoquinone. 1,4-Naphthoquinone reduces more easily than 9,10-anthraquinone. What structural effect do you suppose is responsible for this order?

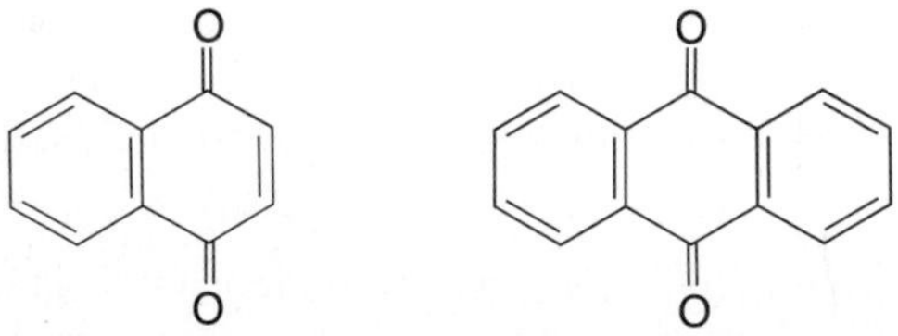

1,4-Naphthoquinone   9,10-Anthraquinone

26.32 One of the industrial processes for the preparation of phenol, discussed in Section 26.6, includes an acid-catalyzed rearrangement of cumene hydroperoxide as a key step. This reaction proceeds by way of an intermediate hemiacetal.

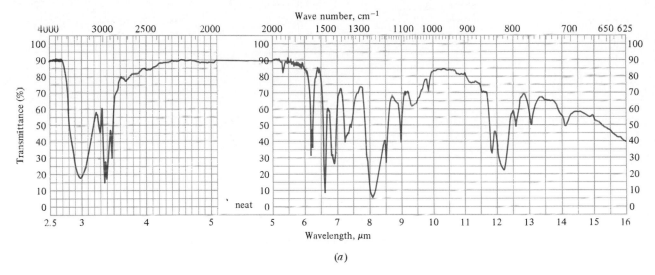

Cumene hydroperoxide   Hemiacetal   Phenol   Acetone

You have learned in Section 18.10 of the relationship between hemiacetals, ketones, and alcohols so the formation of phenol and acetone is simply an example of hemiacetal decomposition. The formation of the hemiacetal intermediate is a key step in the synthetic procedure; it is the step in which the aryl-oxygen bond is generated. Can you suggest a reasonable mechanism for this step?

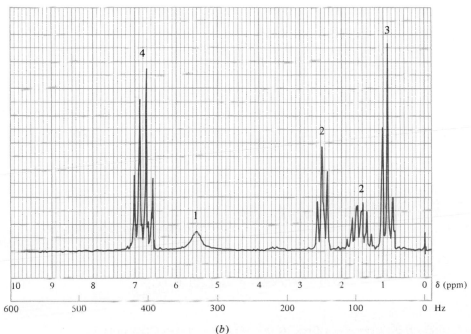

(a)

(b)

FIGURE 26.6 Infrared (a) and ¹H nmr (b) spectra of compound K ($C_9H_{12}O$) (Problem 26.34a).

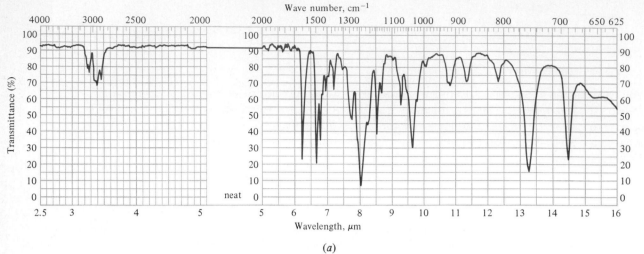

(a)

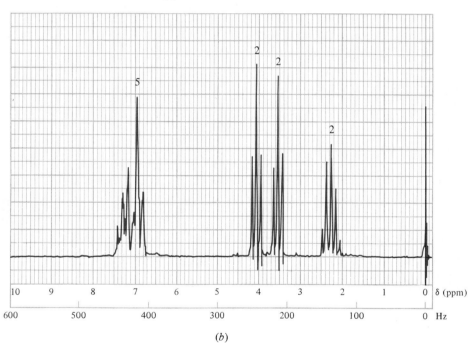

(b)

FIGURE 26.7 Infrared (a) and ^{1}H nmr (b) spectra of compound L ($C_9H_{11}BrO$) (Problem 26.34b).

26.33 A reaction related to that of Problem 26.32 is the *Baeyer-Villiger* oxidation of aryl ketones. Here, treatment with a peroxy acid converts an aryl ketone to an aryl ester.

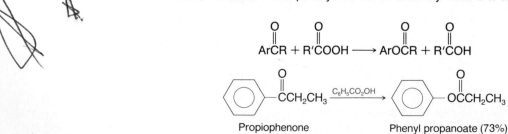

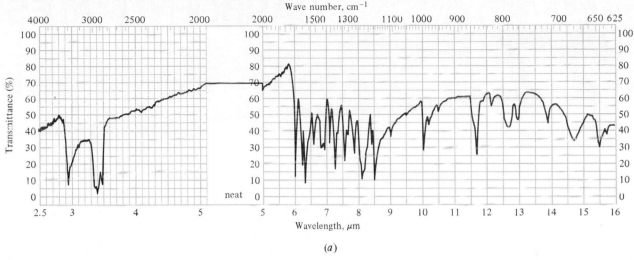

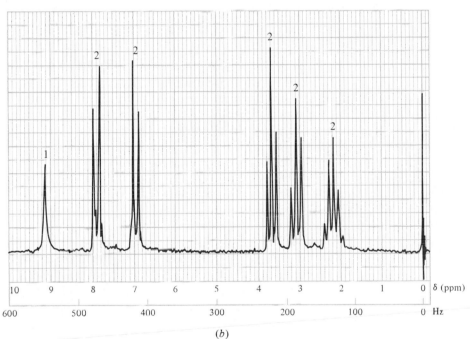

FIGURE 26.8 Infrared (a) and ¹H nmr b) spectra of compound M ($C_{10}H_{11}ClO_2$) (Problem 26.34c).

Given that the first step is the nucleophilic addition of peroxybenzoic acid to the ketone, show how phenyl propanoate is formed by Baeyer-Villiger oxidation of propiophenone.

26.34 Identify compounds **K** through **M** on the basis of the information provided.

(a) Compound K is $C_9H_{12}O$. Its infrared and ¹H nmr spectra are shown in Figure 26.6.

(b) Compound L is $C_9H_{11}BrO$. Its infrared and ¹H nmr spectra are shown in Figure 26.7.

(c) Compound M is $C_{10}H_{11}ClO_2$. Its infrared and ¹H nmr spectra are shown in Figure 26.8.

CHAPTER
27

CARBOHYDRATES

The major classes of substances common to living systems are lipids, proteins, nucleic acids, and carbohydrates. Carbohydrates are very familiar to us—we call many of them "sugars." They comprise a substantial portion of the food we eat and provide most of the energy that keeps the human engine running. Carbohydrates are structural components of the walls of plant cells and the wood of trees. Genetic information is stored and transferred by way of nucleic acids, specialized derivatives of carbohydrates, which will be examined in more detail in Chapter 29.

Historically, carbohydrates were once considered to be "hydrates of carbon" because their molecular formulas in many (but not all) cases correspond to $C_n(H_2O)_m$. It is more realistic to define a carbohydrate as a *polyhydroxy aldehyde or polyhydroxy ketone,* a point of view closer to structural reality and more suggestive of chemical reactivity.

This chapter is divided into two parts. The first and major portion is devoted to carbohydrate structure. You will see how the principles of stereochemistry and conformational analysis combine to aid our understanding of this complex subject. The remainder of the chapter describes chemical reactions of carbohydrates, most of which are extensions of what you have already learned concerning alcohols, aldehydes, ketones, and acetals.

27.1 CLASSIFICATION OF CARBOHYDRATES

The Latin word for "sugar" is *saccharum,* and the derived term *saccharide* is the basis of a system of carbohydrate classification. A *monosaccharide* is a simple carbohydrate, one that on attempted hydrolysis is not cleaved to smaller carbohydrates. Glucose ($C_6H_{12}O_6$), for example, is a monosaccharide. A *disaccharide* on hydrolysis is cleaved to two monosaccharide units, which may be the same or may be different. Sucrose—common table sugar—is a disaccharide that yields one molecule of glucose and one of fructose on hydrolysis.

$$\text{Sucrose } (C_{12}H_{22}O_{11}) + H_2O \longrightarrow \text{glucose } (C_6H_{12}O_6) + \text{fructose } (C_6H_{12}O_6)$$

TABLE 27.1
Some Classes of Monosaccharides

Number of carbon atoms	Aldose	Ketose
Four	Tetrose	Tetrulose
Five	Pentose	Pentulose
Six	Hexose	Hexulose
Seven	Heptose	Heptulose
Eight	Octose	Octulose

An *oligosaccharide* (*oligos* is a Greek word meaning "few") yields 3 to 10 monosaccharide units on hydrolysis. *Polysaccharides* are hydrolyzed to more than 10 monosaccharide units. Cellulose is a polysaccharide molecule that gives thousands of glucose molecules when completely hydrolyzed.

Over 200 different monosaccharides are known. They can be grouped according to the number of carbon atoms they contain and whether they are polyhydroxy aldehydes or polyhydroxy ketones. Monosaccharides that are polyhydroxy aldehydes are called *aldoses;* those that are polyhydroxy ketones are *ketoses.* Aldoses are grouped by appending the suffix *-ose* to the customary Greek stems that define the number of carbon atoms in the main chain; ketoses are grouped in the same way with *-ulose* as the suffix. Table 27.1 lists the names derived in this way for monosaccharides having four to eight carbon atoms.

27.2 FISCHER PROJECTION FORMULAS AND THE D,L SYSTEM OF STEREOCHEMICAL NOTATION

Stereochemistry is the key to understanding carbohydrate structure, a fact that was clearly appreciated by the German chemist Emil Fischer. Fischer was the foremost organic chemist of the late nineteenth century and is responsible for some of the most important fundamental discoveries in the areas of carbohydrate and protein chemistry. The Nobel prize in chemistry was awarded to Fischer in 1902, in part for the combination of intellectual brilliance and experimental achievement he brought to bear on the question of the structure of glucose. An artistic device used by Fischer to represent stereochemistry in chiral molecules is particularly well suited to carbohydrates and will be used in this chapter.

Consider a chiral molecule of the type C (abcd), oriented as shown in Figure 27.1. A projection of that three-dimensional molecule onto the plane of the page forms a cross, with the chiral carbon at its center. Bonds described by horizontal lines are

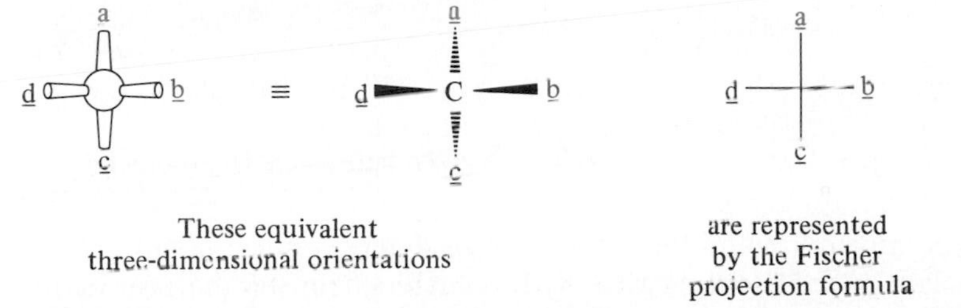

These equivalent three-dimensional orientations

are represented by the Fischer projection formula

FIGURE 27.1 The Fischer projection formula of a chiral molecule is derived from a particular orientation of the bonds to the chiral center. The horizontal bonds in a Fischer projection are directed outward toward the viewer; the vertical bonds are directed away from the viewer.

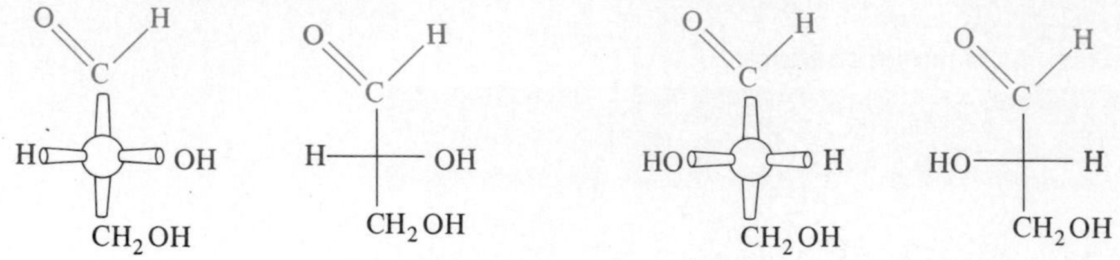

FIGURE 27.2 Three-dimensional representations and Fischer projection formulas of the enantiomers of glyceraldehyde.

(R)-$(+)$-Glyceraldehyde (S)-$(-)$-Glyceraldehyde

understood to be directed from the chiral center toward the viewer. Vertical bonds from the chiral center are directed away from the viewer.

Stereochemical representations of the type shown in Figure 27.1 are known as *Fischer projection formulas.* Figure 27.2 illustrates their application to the enantiomers of glyceraldehyde, a fundamental molecule in carbohydrate stereochemistry. When the Fischer projection is oriented as shown in the figure, with the carbon chain vertical and its aldehyde carbon at the top, the C-2 hydroxyl group points to the right in $(+)$-glyceraldehyde and to the left in $(-)$-glyceraldehyde.

Techniques for determining the absolute configuration of chiral molecules were not developed until the 1950s, so it was not possible for Fischer and his contemporaries to relate the sign of rotation of any substance to its absolute configuration. A system came to be adopted based on the arbitrary assumption, later shown to be correct, that the enantiomers of glyceraldehyde have the signs of rotation and absolute configurations shown in Figure 27.2. Two stereochemical descriptors were defined, D and L. The absolute configuration of $(+)$-glyceraldehyde, as depicted in the figure, was said to be D, and the spatial arrangement of atoms in its enantiomer, $(-)$-glyceraldehyde, was defined as the L configuration. Compounds that had a spatial arrangement of substituents analogous to those of D- $(+)$ and L-$(-)$-glyceraldehyde were said to have the D and L configurations, respectively.

PROBLEM 27.1 Identify each of the following Fischer projections as either D- or L-glyceraldehyde.

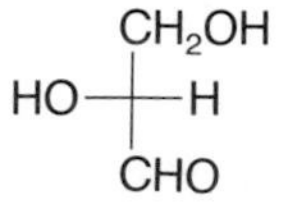

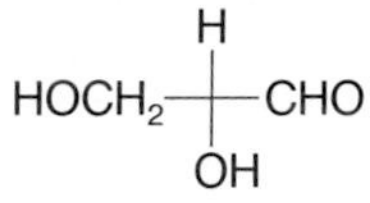

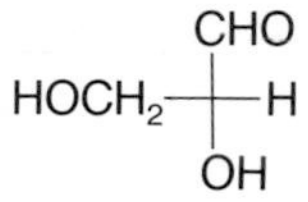

SAMPLE SOLUTION (*a*) Redraw the Fischer projection so as to more clearly show the true spatial orientation of the groups. Next, reorient the molecule so that its relationship to the glyceraldehyde enantiomers in Figure 27.2 is apparent.

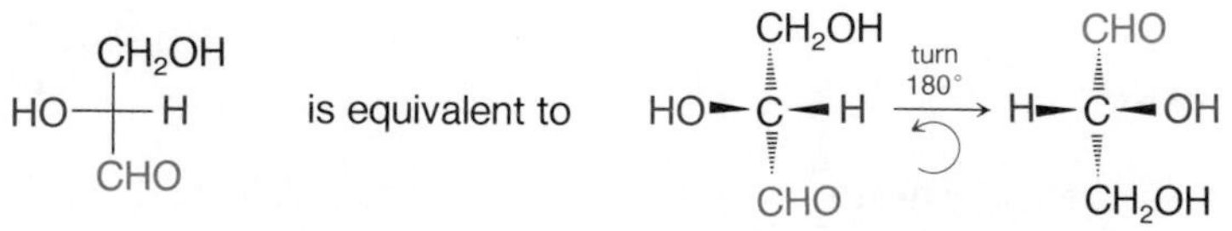

The structure is the same as that of $(+)$-glyceraldehyde in the figure. It is D-glyceraldehyde.

Fischer projection formulas and the D,L system of stereochemical notation both proved to be so helpful in representing carbohydrate stereochemistry that the chemical and biochemical literature is replete with their use. To read that literature you need to be acquainted with both of these devices, in addition to the more modern Cahn-Ingold-Prelog system.

27.3 THE ALDOTETROSES

Glyceraldehyde is considered to be the simplest chiral carbohydrate. It is an *aldotriose* and, since it contains one chiral center, exists in two stereoisomeric forms, the D and L enantiomers. Moving up the scale in complexity, next come the *aldotetroses*. Examination of their structures illustrates the application of the Fischer system to compounds that contain more than one chiral center.

The aldotetroses are the four stereoisomers of 2,3,4-trihydroxybutanal. Fischer projection formulas are constructed by orienting the molecule in an eclipsed conformation with the aldehyde group at the top. The four carbon atoms define the main chain of the Fischer projection and are arranged vertically. Horizontal bonds are directed outward, vertical bonds back.

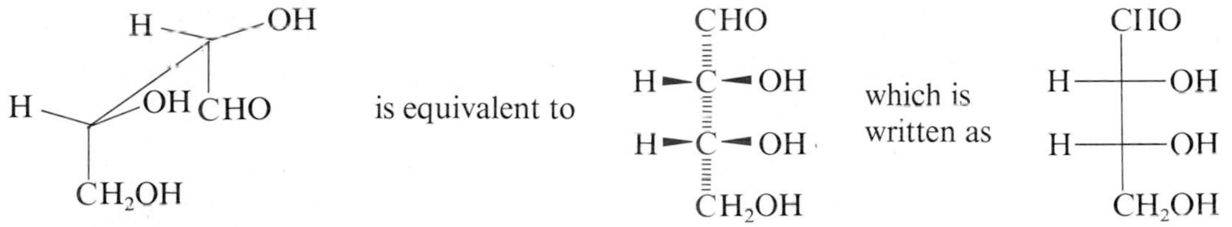

Eclipsed conformation of an aldotetrose ... is equivalent to ... which is written as ... Fischer projection formula of an aldotetrose

The particular aldotetrose shown above is called D-erythrose. The configurational descriptor D specifies that the configuration at the highest-numbered chiral center is analogous to that of D-(+)-glyceraldehyde. Its mirror image is L-erythrose.

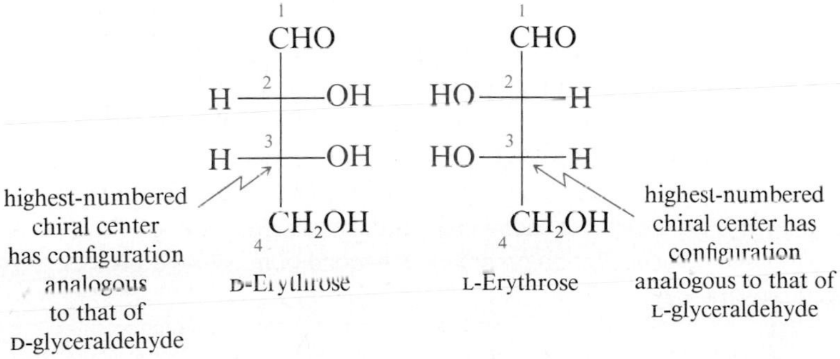

D-Erythrose — highest-numbered chiral center has configuration analogous to that of D-glyceraldehyde. L-Erythrose — highest-numbered chiral center has configuration analogous to that of L-glyceraldehyde.

Both hydroxyl groups are on the same side in Fischer projections of the erythrose enantiomers. The remaining two stereoisomers have their hydroxyl groups on opposite sides in Fischer projection formulas. They are diastereomers of D- and L-erythrose and are called D- and L-threose. The D and L descriptors again specify the configuration of the highest-numbered chiral center. D-Threose and L-threose are enantiomers of each other.

$$
\begin{array}{ccc}
& \overset{1}{\text{CHO}} \\
\text{HO}\!-\!\!\overset{2}{\rule{0pt}{1em}}\!\!-\!\text{H} \\
\text{H}\!-\!\!\overset{3}{\rule{0pt}{1em}}\!\!-\!\text{OH} \\
& \underset{4}{\text{CH}_2\text{OH}}
\end{array}
\qquad
\begin{array}{ccc}
& \overset{1}{\text{CHO}} \\
\text{H}\!-\!\!\overset{2}{\rule{0pt}{1em}}\!\!-\!\text{OH} \\
\text{HO}\!-\!\!\overset{3}{\rule{0pt}{1em}}\!\!-\!\text{H} \\
& \underset{4}{\text{CH}_2\text{OH}}
\end{array}
$$

highest-numbered chiral center has configuration analogous to that of D-glyceraldehyde   D-Threose   L-Threose   highest-numbered chiral center has configuration analogous to that of L-glyceraldehyde

---

PROBLEM 27.2   Which of the four aldotetroses just discussed is the following?

As shown for the aldotetroses, an aldose belongs to the D or the L series according to the configuration of the chiral center farthest removed from the aldehyde function. Individual names, such as erythrose and threose, specify the particular arrangement of chiral centers within the molecule relative to each other. Optical activities cannot be determined directly from configurational descriptors. As it turns out, both D-erythrose and D-threose are levorotatory, while D-glyceraldehyde is dextrorotatory.

27.4   ALDOPENTOSES AND ALDOHEXOSES

Aldopentoses have three chiral centers. The eight stereoisomers are divided into a set of four D aldopentoses and an enantiomeric set of four L aldopentoses. The aldopentoses are named ribose, arabinose, xylose, and lyxose. Fischer projection formulas of the D stereoisomers of the aldopentoses are given in Figure 27.3. Notice that all these diastereomers have the same configuration at C-4 and that this configuration is analogous to that of D-(+)-glyceraldehyde.

PROBLEM 27.3   L-(+)-Arabinose is a naturally occurring L sugar. It is obtained by acid hydrolysis of the polysaccharide present in mesquite gum. Write a Fischer projection formula for L-(+)-arabinose.

Among the aldopentoses, D-ribose is a component of many biologically important substances, most notably the ribonucleic acids, while D-xylose is very abundant and is isolated by hydrolysis of the polysaccharides present in corn cobs and the wood of trees.

The aldohexoses include some of the most familiar of the monosaccharides, as well as the most abundant organic compound on earth, D-(+)-glucose. With four chiral centers, 16 stereoisomeric aldohexoses are possible; 8 belong to the D series and

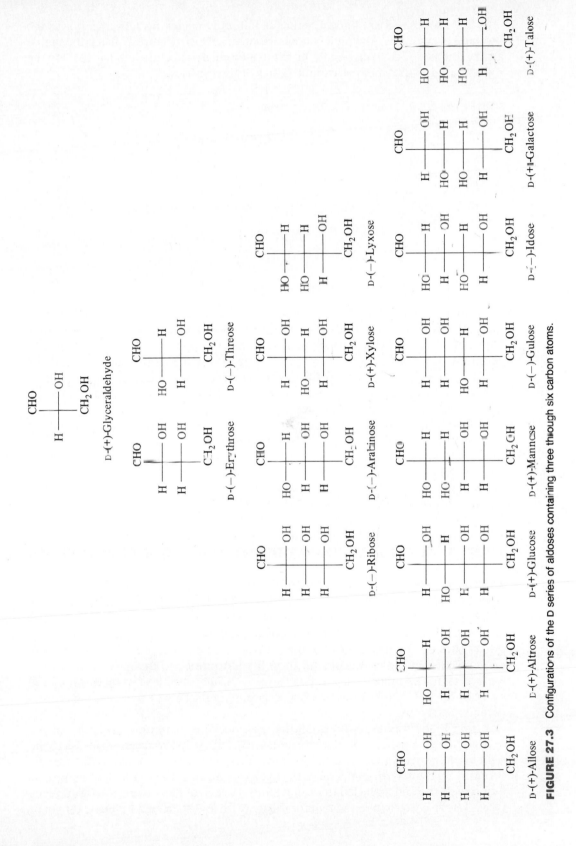

FIGURE 27.3 Configurations of the D series of aldoses containing three through six carbon atoms.

8 to the L series. All are known, either as naturally occurring substances or as the products of synthesis. The eight D aldohexoses are given in Figure 27.3; it is the spatial arrangement at C-5, hydrogen to the left in a Fischer projection and hydroxyl to the right, that identifies them as carbohydrates of the D-series.

PROBLEM 27.4   Name the following sugar.

```
            CHO
    H ———————— OH
    H ———————— OH
    H ———————— OH
   HO ———————— H
           CH₂OH
```

Of all the monosaccharides, D-(+)-glucose is the best known, most important, and most abundant. Its formation from carbon dioxide, water, and sunlight is the central theme of photosynthesis. Carbohydrate formation by photosynthesis is estimated to be on the order of 10^{11} tons per year, a source of stored energy utilized, directly or indirectly, by all higher forms of life on the planet. Glucose was isolated from raisins in 1747 and by hydrolysis of starch in 1811. Its structure was determined, in work culminating in 1900, by Emil Fischer.

D-(+)-Galactose is a constituent of numerous polysaccharides. It is best obtained by acid hydrolysis of lactose (milk sugar), a disaccharide of D-glucose and D-galactose. L-(−)-Galactose also occurs naturally and can be prepared by hydrolysis of flaxseed gum and agar. The principal source of D-(+)-mannose is hydrolysis of the polysaccharide of the ivory nut, a large nutlike seed obtained from a South American palm.

27.5   A MNEMONIC FOR CARBOHYDRATE CONFIGURATIONS

The task of relating carbohydrate configurations to names requires either a world-class memory or an easily recalled mnemonic. The mnemonic device that serves us well here was popularized by the husband-wife team of Louis F. and Mary Fieser of Harvard University in their 1956 textbook, *Organic Chemistry*. As with many mnemonic devices, it is not clear who actually invented it and references to this particular one appeared in the chemical education literature prior to the publication of the Fiesers' text. The mnemonic has two features: (1) a system for setting down all the stereoisomeric D-aldohexoses in a logical order; and (2) a way to assign the correct name to each one.

A systematic way to set down all the D aldohexoses is to draw skeletons of the necessary eight Fischer projection formulas, placing the hydroxyl group at C-5 to the right in each so as to guarantee that each member of the group belongs to the D series. Working up the carbon chain, place the hydroxyl group at C-4 to the right in the first four structures, and to the left in the next four. In each of these two sets of four, place the C-3 hydroxyl group to the right in the first two and to the left in the next two; in each of the resulting four sets of two, place the C-2 hydroxyl group to the right in the first one and to the left in the second.

Once the eight Fischer projections have been written, they are named in order with the aid of the sentence: All altruists gladly make gum in gallon tanks. The words of the sentence stand for *allose, altrose, glucose, mannose, gulose, idose, galactose, talose.*

An analogous pattern of configurations can be seen in the aldopentoses when they are arranged in the order *ribose, arabinose, xylose, lyxose.* (RAXL is an easily remembered nonsense word that gives the correct sequence). This pattern is discernible even in the aldotetroses erythrose and threose.

27.6 CYCLIC FORMS OF CARBOHYDRATES: FURANOSE FORMS

Aldoses have two types of functional group, an aldehyde group and the hydroxyl groups, which are capable of reacting with each other. Nucleophilic addition of an alcohol to an aldehyde leads to a hemiacetal, and when the hydroxyl and aldehyde functions are part of the same molecule, a *cyclic hemiacetal* is formed. The relationship of a hydroxy aldehyde to a cyclic hemiacetal is shown in Figure 27.4.

Cyclic hemiacetal formation is most favorable when the ring that results is five-membered or six-membered. Five-membered cyclic hemiacetals of carbohydrates are called *furanose* forms; six-membered ones are called *pyranose* forms. The ring carbon that is derived from the carbonyl group, the one that bears two oxygen substituents, is called the *anomeric* carbon.

Aldoses exist almost exclusively as their cyclic hemiacetals; very little of the open-chain form is present at equilibrium. In order to understand their structures and chemical reactions, we need to be able to translate Fischer projection formulas of carbohydrates into their cyclic hemiacetal forms. Consider first cyclic hemiacetal

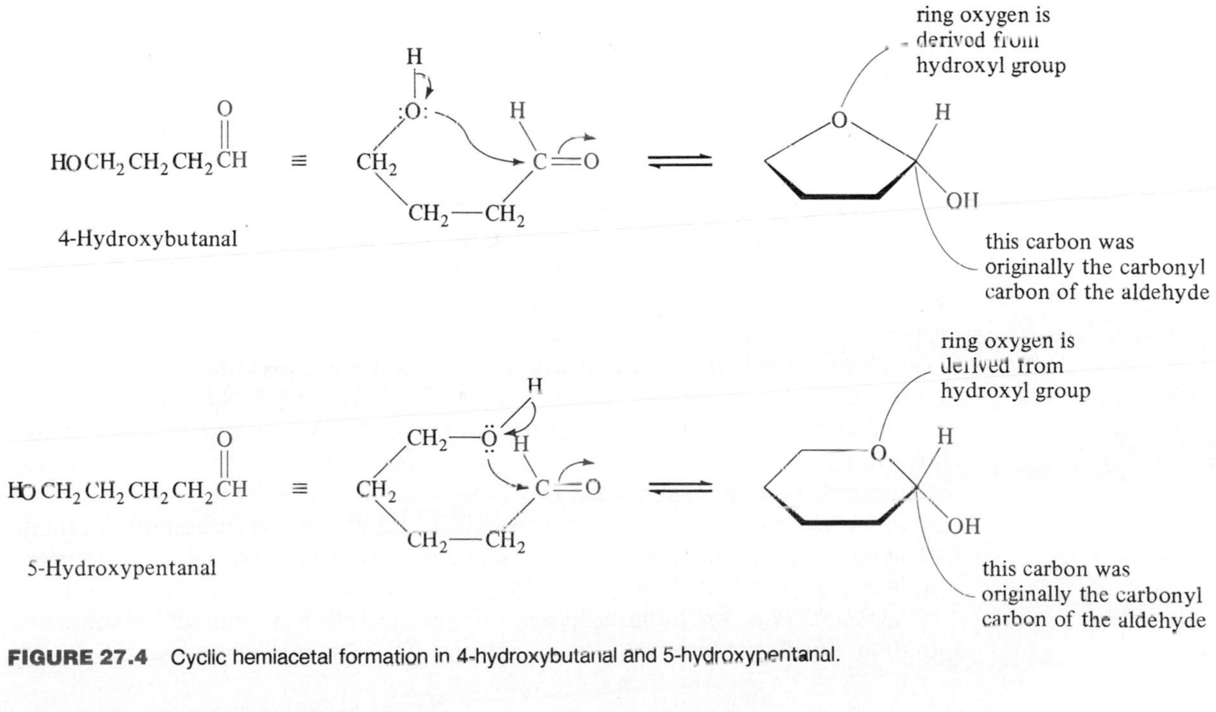

FIGURE 27.4 Cyclic hemiacetal formation in 4-hydroxybutanal and 5-hydroxypentanal.

formation in D-erythrose. So as to visualize furanose ring formation more clearly, redraw the Fischer projection in a form more suited to cyclization, being careful to maintain the stereochemistry at each chiral center.

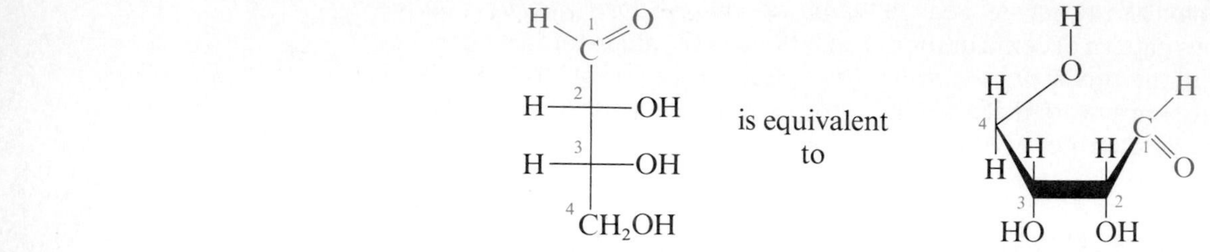

D-Erythrose

is equivalent to

Reoriented "coiled" form of D-erythrose showing C-4 hydroxyl group in position to add to carbonyl group

Hemiacetal formation between the carbonyl group and the terminal hydroxyl yields the five-membered furanose ring form. The anomeric carbon becomes a new chiral center; its hydroxyl group can be either cis or trans to the other hydroxyl groups of the molecule.

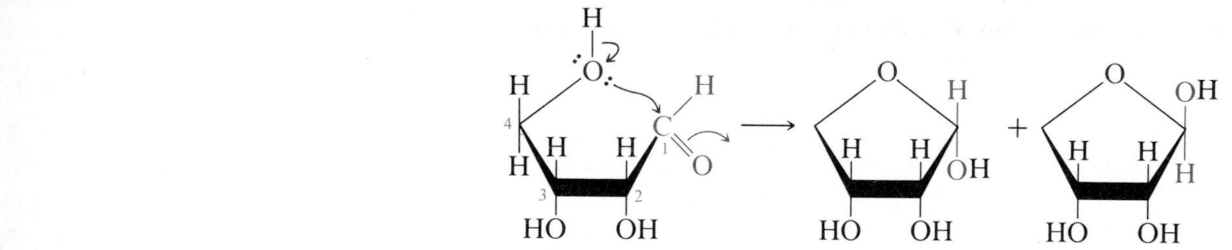

Coiled form of D-erythrose

α-D-Erythrofuranose
(hydroxyl group at anomeric carbon is down)

β-D-Erythrofuranose
(hydroxyl group at anomeric carbon is up)

Representations of carbohydrates constructed in this manner are called *Haworth formulas,* after the British carbohydrate chemist Sir Norman Haworth (St. Andrew's University and the University of Birmingham). Early in his career Haworth contributed to the discovery that sugars exist as cyclic hemiacetals rather than in open-chain forms. Later he collaborated on an efficient synthesis of vitamin C from carbohydrate precursors. This was the first chemical synthesis of a vitamin and provided an inexpensive route to its preparation on a commercial scale. Haworth was a corecipient of the Nobel prize for chemistry in 1937.

The two stereoisomeric furanose forms of D-erythrose are named α-D-erythrofuranose and β-D-erythrofuranose. The prefixes α and β designate the configuration at the anomeric carbon. When the Haworth formula of a carbohydrate in the D-series has the hydroxyl group at the anomeric carbon "down," it is α. When the anomeric hydroxyl is "up" in the D-series, it is β. These descriptors are exactly reversed in the L-series.

Substituents that are to the right in a Fischer projection are "down" in the corresponding Haworth formula.

Generating Haworth formulas for furanose forms of higher aldoses is slightly more complicated stereochemically in that they require an additional operation. Furanose forms of D-ribose are frequently encountered building blocks in biologically important organic molecules and involve hemiacetal formation between the aldehyde group at the end of the D-ribose chain and the hydroxyl at C-4.

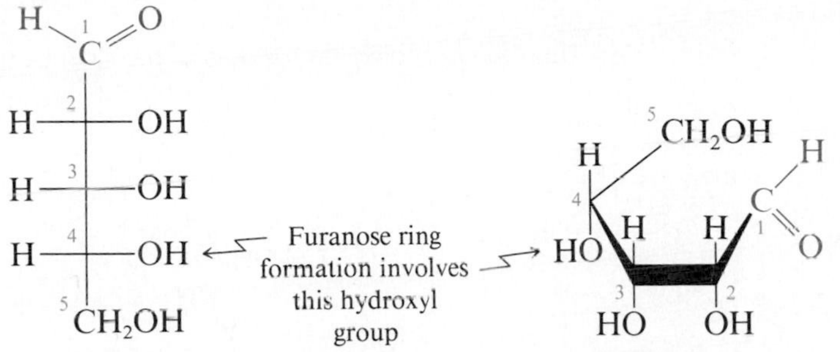

D-Ribose

Furanose ring formation involves this hydroxyl group

Coiled form of D-ribose

Observe that the "coiled" form of D-ribose derived directly from the Fischer projection does not have its C-4 hydroxyl group properly oriented for furanose ring formation. We must redraw the coiled form in a conformation that permits the five-membered cyclic hemiacetal to form. This is accomplished by rotation about the C(3)—C(4) bond, taking care that the configuration at C-4 is not changed.

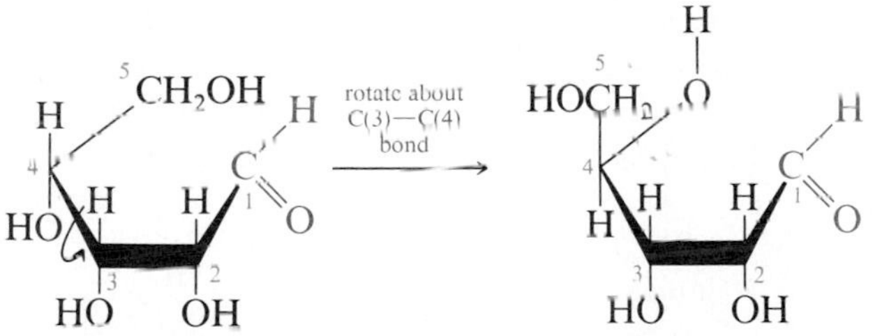

rotate about C(3)—C(4) bond

Original coiled form of D-ribose

Conformation of coiled form suitable for furanose ring formation

As viewed in the drawing, a 120° anticlockwise rotation of C-4 places its hydroxyl group in the proper position. At the same time, this rotation moves the CH_2OH group to a position such that it will become a substituent that is "up" on the five-membered ring. The hydrogen at C-4 then will be "down" in the furanose form.

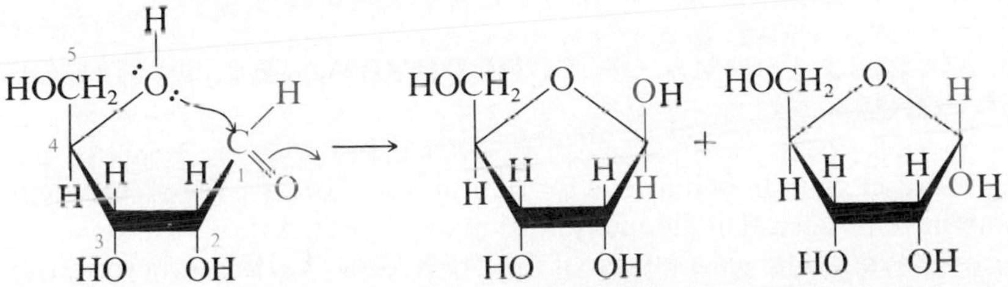

β-D-Ribofuranose   α-D-Ribofuranose

(handwritten margin notes)

furanose

CH₂OH R-up

CH₂OH L-Down

PROBLEM 27.5   Write Haworth formulas corresponding to the furanose forms of each of the following carbohydrates.

(a) D-Xylose

(b) D-Arabinose

(c) L-Arabinose

(d) D-Threose

SAMPLE SOLUTION   (a)   The Fischer projection formula of D-xylose is given in Figure 27.3.

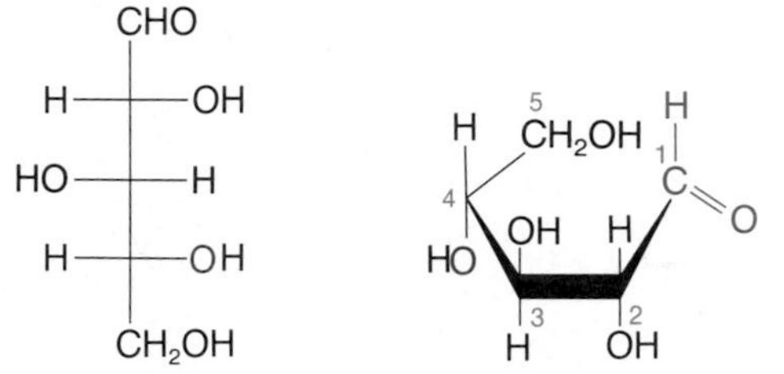

D-Xylose

Coiled form of D-xylose

Carbon-4 of the coiled form of D-xylose must be rotated in an anticlockwise sense in order to bring its hydroxyl group into the proper orientation for furanose ring formation.

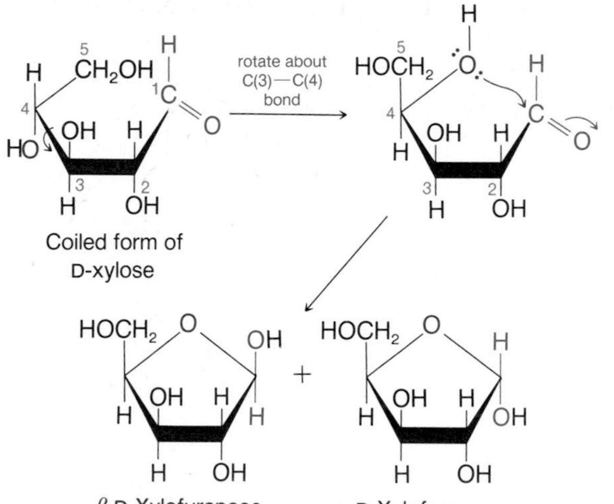

Coiled form of
D-xylose

rotate about
C(3)—C(4)
bond

β-D-Xylofuranose   α-D-Xylofuranose

27.7   CYCLIC FORMS OF CARBOHYDRATES: PYRANOSE FORMS

During the discussion of hemiacetal formation in D-ribose in the preceding section, you may have noticed that aldopentoses have the potential of forming a six-membered cyclic hemiacetal via addition of the C-5 hydroxyl to the carbonyl group. This mode of ring closure leads to α- and β-*pyranose* forms.

β-D-Ribopyranose + α-D-Ribopyranose

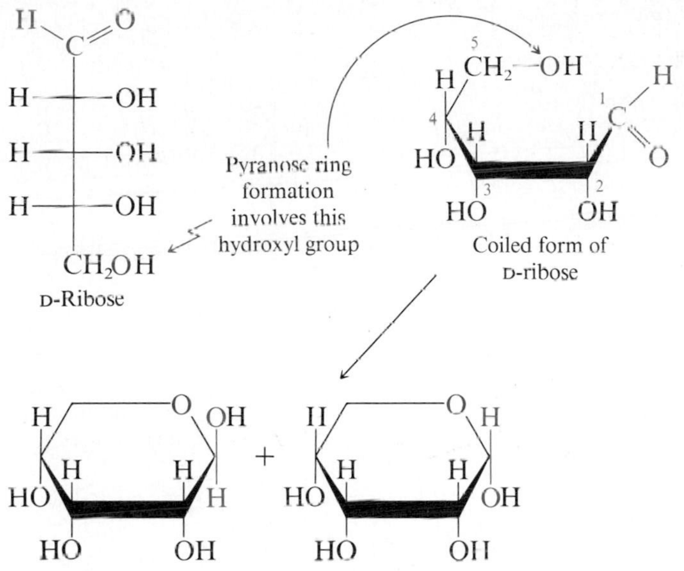

Aldohexoses such as D-glucose are capable of forming two furanose forms (α and β) and two pyranose forms (α and β). The Haworth representations of the pyranose forms of D-glucose are constructed as shown in Figure 27.5; each has a CH_2OH group as a substituent on the six-membered ring. In the α isomer the anomeric hydroxyl is trans to the CH_2OH group, whereas in the β isomer the anomeric hydroxyl and CH_2OH side chain are cis to each other.

Haworth formulas are satisfactory for representing *configurational* relationships in pyranose forms but are uninformative as to carbohydrate *conformations.* X-ray crystallographic studies of a large number of carbohydrates reveal that the six-membered pyranose ring of D-glucose adopts a chair conformation.

β-D-Glucopyranose

α-D-Glucopyranose

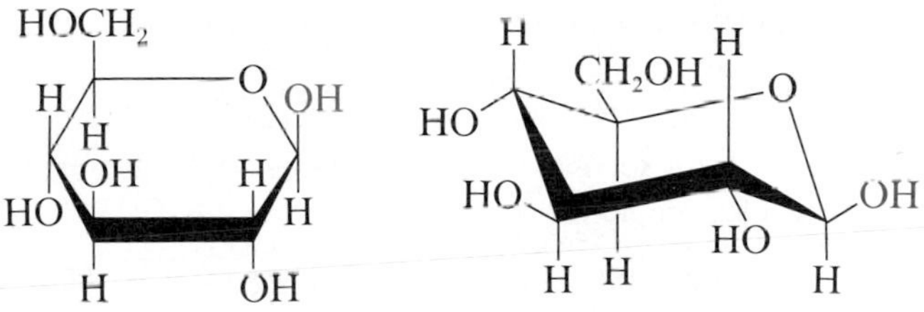

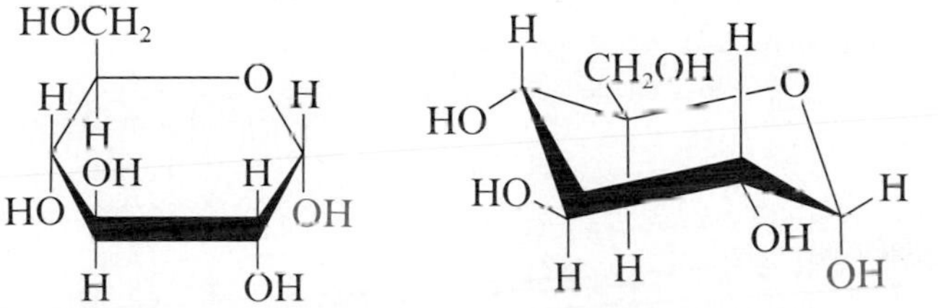

All the ring substituents other than hydrogen in β-D-glucopyranose are equatorial in the most stable chair conformation. Only the anomeric hydroxyl group is axial in the α isomer; all the other substituents are equatorial.

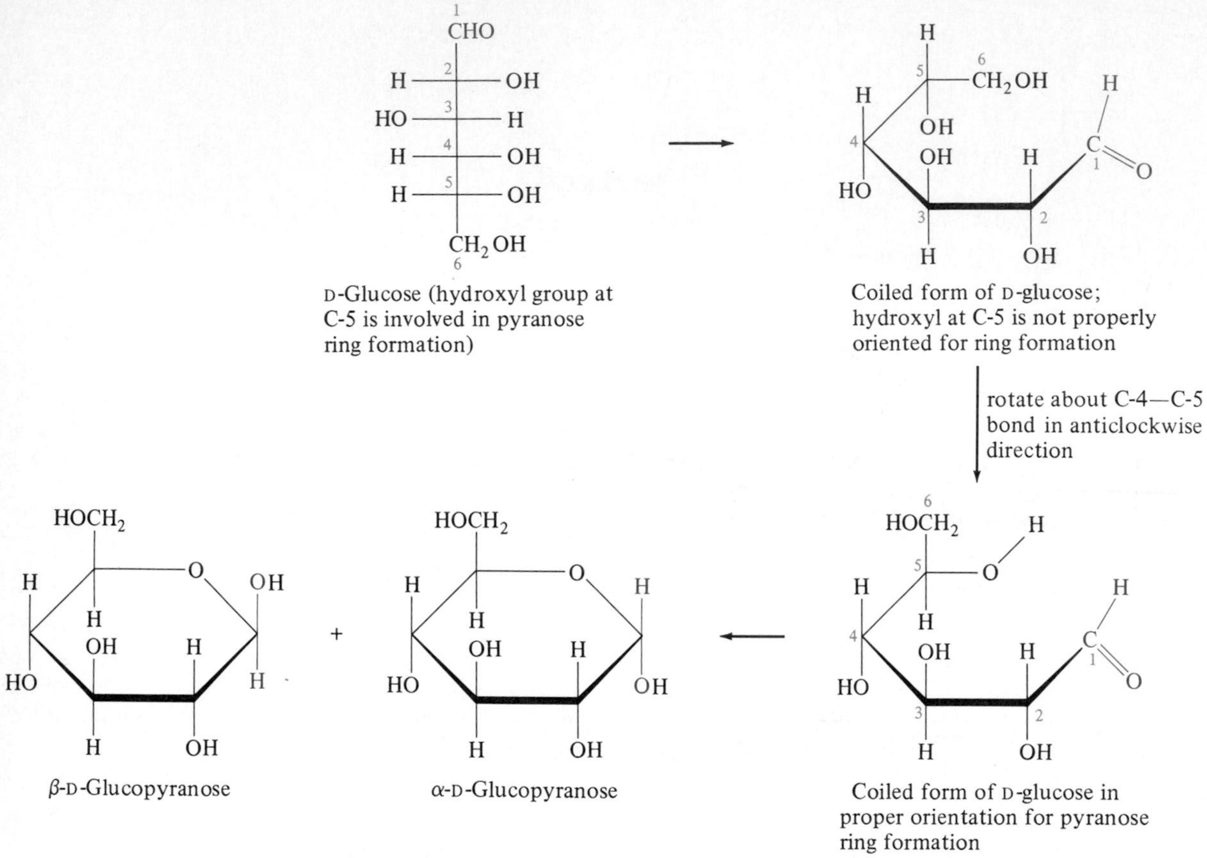

FIGURE 27.5 Haworth formulas for α- and β-pyranose forms of D-glucose.

Other aldohexoses behave similarly in adopting chair conformations that permit the CH_2OH substituent to occupy an equatorial orientation. Normally the CH_2OH group is the bulkiest, most conformationally demanding substituent in the pyranose form of an aldohexose.

PROBLEM 27.6 Clearly represent the most stable conformation of the β-pyranose form of each of the following sugars:

(a) D-Galactose   (b) D-Mannose   (c) L-Mannose   (d) L-Ribose

SAMPLE SOLUTION (a) In analogy with the procedure outlined for D-glucose in Figure 27.5, first generate a Haworth formula for β-D-galactopyranose.

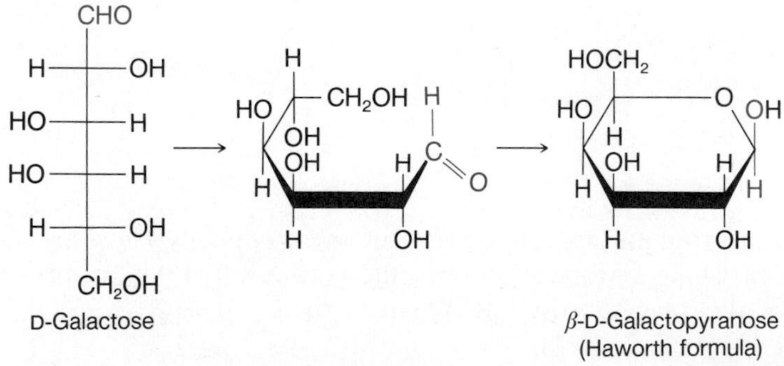

Next, redraw the planar Haworth formula more realistically as a chair conformation, choosing the one that has the CH_2OH group equatorial.

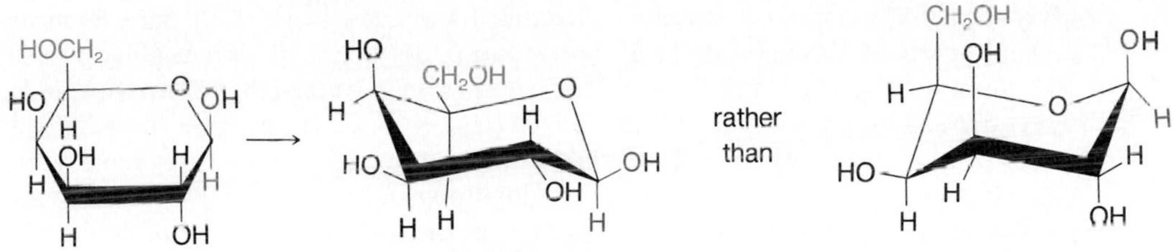

Most stable chair
conformation of
β-D-galactopyranose

rather
than

Less stable chair;
CH_2OH group is axial

Galactose differs from glucose in configuration at C-4. The C-4 hydroxyl is axial in β-D-galactopyranose while it is equatorial in β-D-glucopyranose.

Since six-membered rings are normally less strained than five-membered ones, pyranose forms are usually present in greater amounts than furanose forms at equilibrium and the concentration of the open-chain form is quite small. The distribution of carbohydrates among their various hemiacetal forms has been examined by using ^1H and ^{13}C nmr spectroscopy. In aqueous solution, for example, D-ribose is found to contain the various α and β furanose and pyranose forms in the amounts shown.

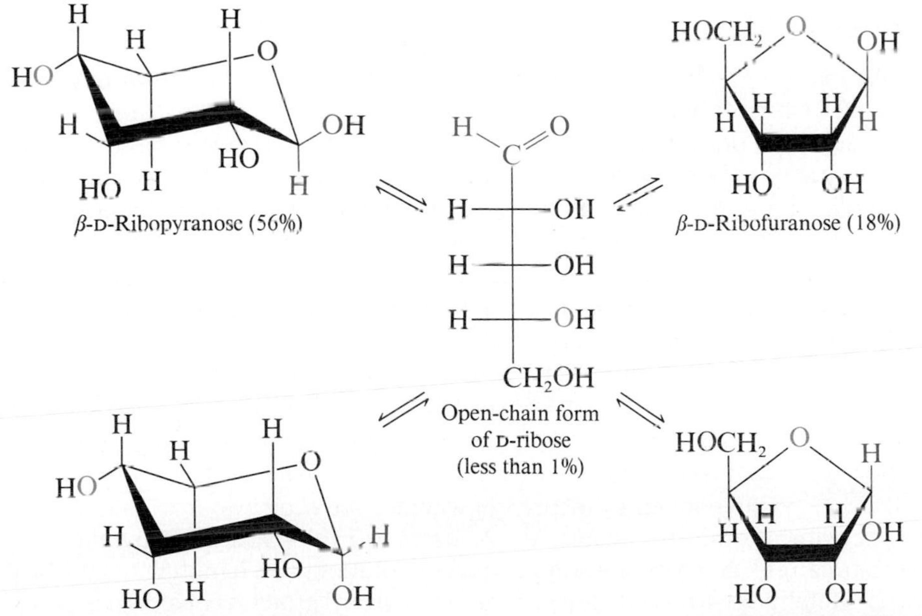

β-D-Ribopyranose (56%)

Open-chain form
of D-ribose
(less than 1%)

β-D-Ribofuranose (18%)

α-D-Ribopyranose (20%)

α-D-Ribofuranose (6%)

The concentration of the open-chain form at equilibrium is too small to measure directly, but it nevertheless occupies a central position in that interconversions of α and β anomers and furanose and pyranose forms take place by way of the open-chain form as an intermediate. As will be seen later, certain chemical reactions also proceed by way of the open-chain form.

27.8 MUTAROTATION

In spite of their easy interconversion in solution, α and β forms of carbohydrates are capable of independent existence and many have been isolated in pure form as crystalline solids. When crystallized from ethanol, D-glucose yields α-D-glucopyranose, mp 146°C, $[\alpha]_D$ + 112.2°. Crystallization from a water-ethanol mixture produces β-D-glucopyranose, mp 148 to 155°C, $[\alpha]_D$ + 18.7°. In the solid state the two forms do not interconvert and are stable indefinitely. Their structures have been unambiguously confirmed by x-ray crystallography.

The optical rotations just cited for each isomer are those measured immediately after each one is dissolved in water. On standing, the rotation of the solution containing the α isomer decreases from + 112.2° to + 52.5°; the rotation of the solution of the β isomer increases from + 18.7° to the same value of + 52.5°. This phenomenon is called *mutarotation*. What is happening is that each solution, initially containing only one anomeric form, undergoes equilibration to the same mixture of α and β pyranose forms. The open-chain form is an intermediate in the process.

α-D-Glucopyranose
(mp 146°C; $[\alpha]_D$
+ 112.2°)

Open-chain form
of D-glucose

β-D-Glucopyranose
(mp 148–155°C;
$[\alpha]_D$ + 18.7°)

The distribution between the α and β anomeric forms at equilibrium is readily calculated from the optical rotations of the pure isomers and the final optical rotation of the solution and is determined to be 36 percent α to 64 percent β. Independent measurements have established that only the pyranose forms of D-glucose are present in significant quantities at equilibrium. Furanose forms are effectively absent, and the open-chain form is present only to the extent of about 0.02 percent.

PROBLEM 27.7 The specific optical rotations of pure α- and β-D-mannopyranose are +29.3° and −17.0°, respectively. When either form is dissolved in water, mutarotation occurs and the observed rotation of the solution changes until a final rotation of +14.2° is observed. Assuming that only α- and β-pyranose forms are present, calculate the percent of each isomer at equilibrium.

It is not possible to tell by inspection whether the α or β anomer of a particular pyranose form is the more stable one. As described above, β-D-glucopyranose is the predominant form at equilibrium in aqueous solution. It is tempting to ascribe this greater stability to the fact that the anomeric hydroxyl group is equatorial in β-D-glucopyranose while it is axial in the α isomer. On the other hand (see the solution to Problem 27.7), α-D-mannopyranose, with its anomeric hydroxyl group in an axial orientation, is more stable than its β isomer. The question of relative stability of anomeric forms is a complicated one involving interactions of the anomeric hydroxyl group not only with other substituents but with solvent molecules (hydrogen bonding to water, for example) as well. Furthermore, an electronic interaction of the anomeric hydroxyl group with the ring oxygen, called the *anomeric effect*, stabilizes the α form relative to the β. We will not attempt to explain the physical basis underlying the anomeric effect but will only point out that because of it an electronegative substituent on the anomeric carbon is better stabilized by interaction with the lone pairs of the ring oxygen when it is axial than when it is equatorial.

27.9 KETOSES

Up to this point all our attention has been directed toward aldoses, carbohydrates having an aldehyde function in their open-chain form. Aldoses are more common than ketoses and their role in biological processes has been more thoroughly studied. Nevertheless, a large number of ketoses are known and several of them are pivotal intermediates in carbohydrate biosynthesis and metabolism. Examples of some ketoses include D-ribulose, L-xylulose, and D-fructose.

$$
\begin{array}{ccc}
\text{CH}_2\text{OH} & \text{CH}_2\text{OH} & \text{CH}_2\text{OH} \\
\text{C}{=}\text{O} & \text{C}{=}\text{O} & \text{C}{=}\text{O} \\
\text{H}\text{—OH} & \text{H}\text{—OH} & \text{HO}\text{—H} \\
\text{H}\text{—OH} & \text{HO}\text{—H} & \text{H}\text{—OH} \\
\text{CH}_2\text{OH} & \text{CH}_2\text{OH} & \text{H}\text{—OH} \\
& & \text{CH}_2\text{OH}
\end{array}
$$

D-Ribulose	L-Xylulose	D-Fructose
(a 2-pentulose that is a key compound in photosynthesis as well as many other aspects of carbohydrate biochemistry)	(a 2-pentulose excreted in excessive amounts in the urine of persons afflicted with the mild genetic disorder pentosuria)	(a 2-hexulose also known as "levulose", it is found in honey and is significantly sweeter than table sugar)

In these three examples the carbonyl group is located at C-2, which is the most common location for the carbonyl function in naturally occurring ketoses.

PROBLEM 27.8 How many tetruloses are possible? Write Fischer projection formulas for each.

Free ketoses, like aldoses, exist mainly as cyclic hemiacetals. In the case of D-ribulose, furanose forms result from addition of the C-5 hydroxyl to the carbonyl group.

Coiled form of D-ribulose	β-D-Ribulofuranose	α-D-Ribulofuranose

The anomeric carbon of a furanose or pyranose form of a ketose bears both a hydroxyl group and a carbon substituent. In the case of 2-ketoses, this substituent is a CH_2OH group. As with aldoses, the anomeric carbon of a cyclic hemiacetal is readily identifiable because it is bonded to two oxygens.

27.10 DEOXY SUGARS

A commonplace variation on the general pattern seen in carbohydrate structure is the replacement of one or more of the hydroxyl substituents by some other atom or group. In *deoxy sugars* the hydroxyl group is replaced by hydrogen. Two examples of deoxy sugars are 2-deoxy-D-ribose and L-rhamnose.

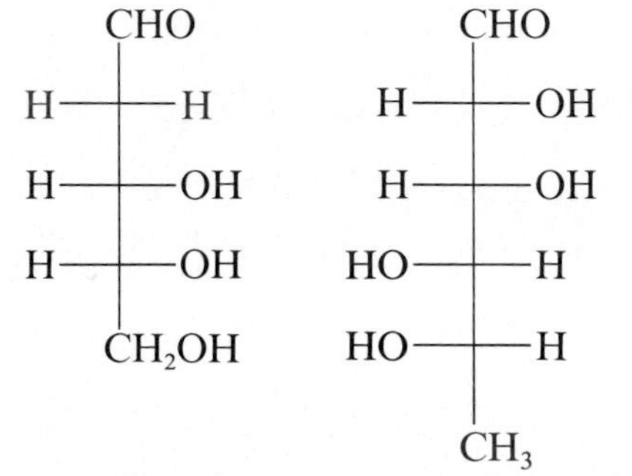

2-Deoxy-D-ribose

L-Rhamnose
(6-deoxy-L-mannose)

The hydroxyl at C-2 in D-ribose is absent in 2-deoxy-D-ribose. In Chapter 29 we shall see how derivatives of 2-deoxy-D-ribose, called *deoxyribonucleotides,* are the fundamental building blocks of deoxyribonucleic acid (DNA), the material responsible for storing genetic information. L-Rhamnose is a compound isolated from a number of plants. Its carbon chain terminates in a methyl rather than a CH_2OH group.

PROBLEM 27.9   Write Fischer projection formulas for

(a) Cordycepose (3-deoxy-D-ribose): a deoxy sugar isolated by hydrolysis of the antibiotic substance *cordycepin*

(b) L-Fucose (6-deoxy-L-galactose): obtained from seaweed

SAMPLE SOLUTION   (a)   The hydroxyl group at C-3 in D-ribose is replaced by hydrogen in 3-deoxy-D-ribose.

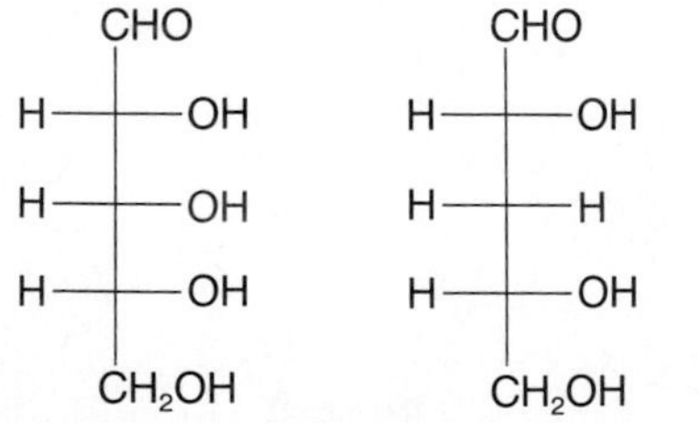

D-Ribose
(from Figure 27.3)

3-Deoxy-D-ribose
(cordycepose)

27.11 AMINO SUGARS

Another structural variation is the replacement of a hydroxyl group in a carbohydrate by an amine group. Such compounds are called *amino sugars*. More than 60 amino sugars are presently known, many of them having been isolated and identified only recently as components of antibiotic substances.

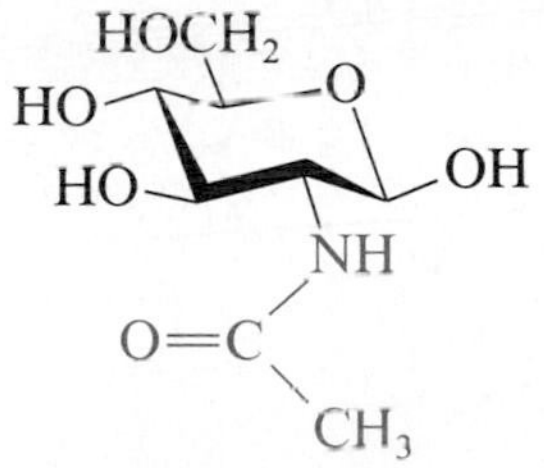

N-Acetyl-D-glucosamine
(shown in its β-pyranose
form; principal component
of the polysaccharide that
is major constituent of
chitin, the substance that
makes up the tough outer
skeleton of arthropods and
insects)

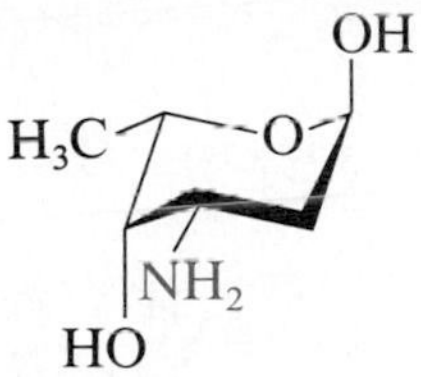

L-Daunosamine (shown in its
α-pyranose form; obtained
from daunomycin and adriamycin,
two powerful drugs used in
cancer chemotherapy)

27.12 BRANCHED-CHAIN CARBOHYDRATES

Carbohydrates that have a carbon substituent attached to the main chain are said to have a *branched chain.* D-Apiose and L-vancosamine are representative branched-chain carbohydrates.

Branching
group

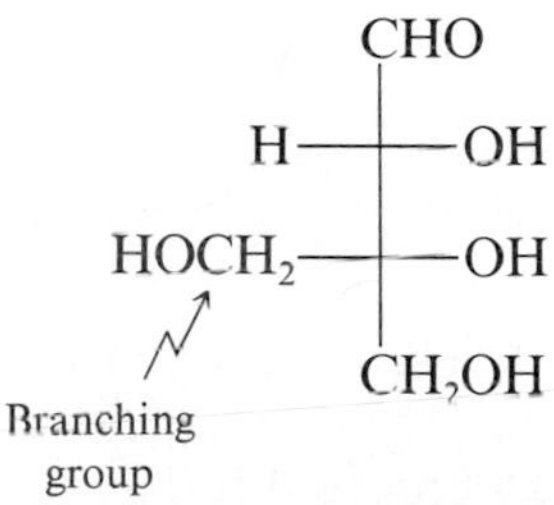

D-Apiose (shown in its
open-chain form; isolated
from parsley and
from the cell wall
polysaccharide of various
marine plants)

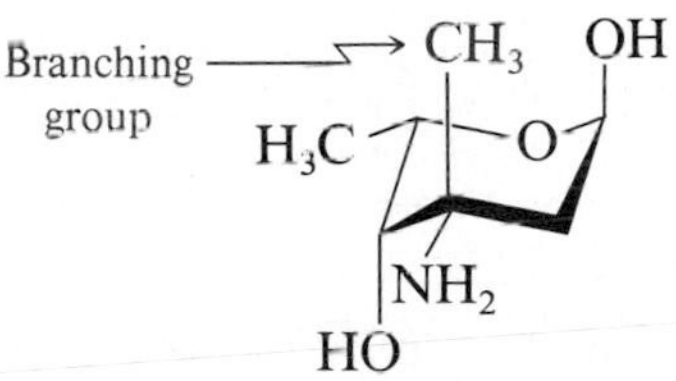

L-Vancosamine (shown in its
α-pyranose form; a component
of the antibiotic compound
vancomycin)

27.13 GLYCOSIDES

Glycosides are a large and very important class of carbohydrate derivatives characterized by the replacement of the anomeric hydroxyl group by some other substituent.

Glycosides are termed O-glycosides, N-glycosides, S-glycosides, etc. according to the atom attached to the anomeric carbon.

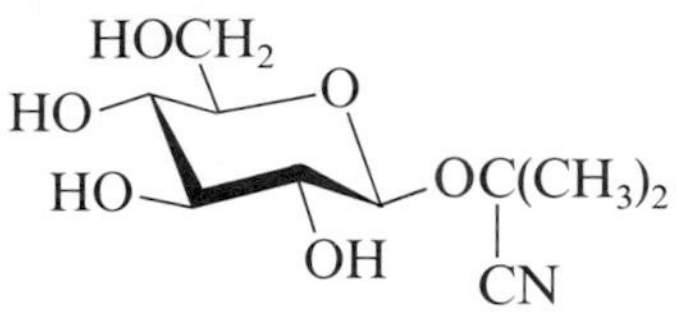

Linamarin
(an O-glycoside:
obtained from manioc,
a type of yam widely
distributed in
Southeast Asia)

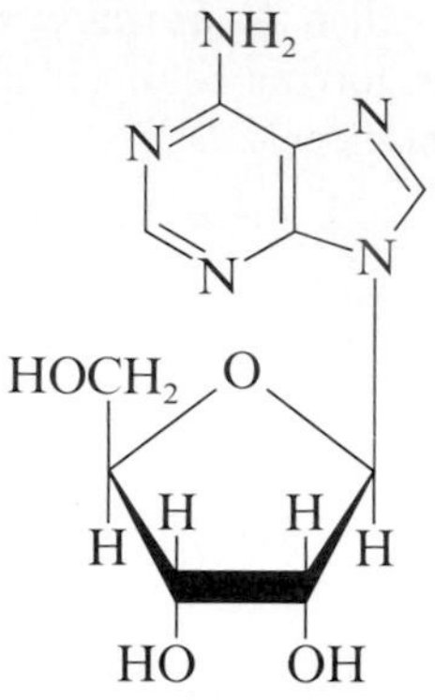

Adenosine
(an N-glycoside: also
known as a nucleoside;
adenosine is one of the
fundamental molecules
of biochemistry)

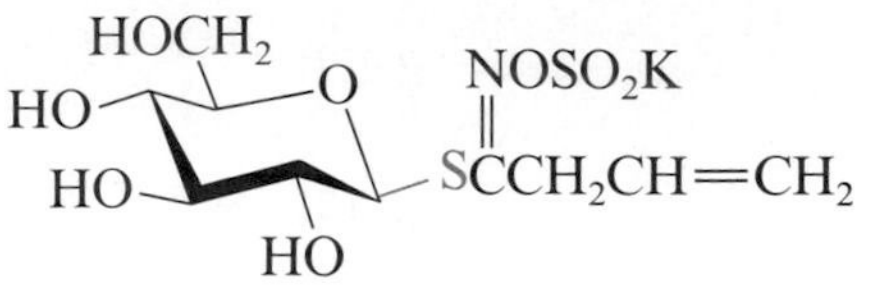

Sinigrin
(an S-glycoside:
contributes to the
characteristic flavor
of mustard and
horseradish)

Usually, the term "glycoside" without a prefix is taken to mean an O-glycoside and will be used that way in this chapter. Glycosides are classified as α or β in the customary way, according to the configuration at the anomeric carbon. All the glycosides shown above are β-glycosides.

Structurally, O-glycosides are mixed acetals; they are ethers that involve the anomeric position of furanose and pyranose forms of carbohydrates. Recall the sequence of intermediates in acetal formation (Section 18.10).

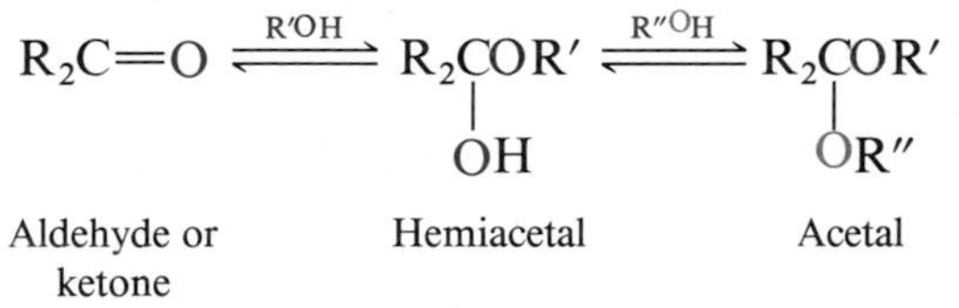

When this sequence is applied to carbohydrates, the first step takes place intramolecularly and spontaneously to yield a cyclic hemiacetal. The second step is *intermolecular,* requires an alcohol R″OH as a reactant, and proceeds readily only in the presence of an acid catalyst. An oxygen-stabilized carbocation is an intermediate.

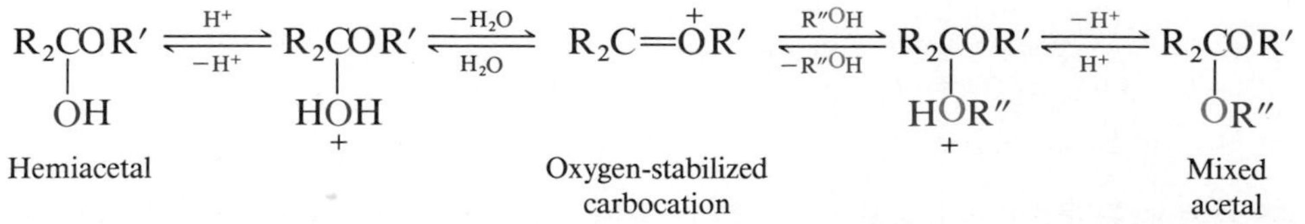

The preparation of glycosides in the laboratory is carried out by simply allowing a carbohydrate to react with an alcohol in the presence of an acid catalyst.

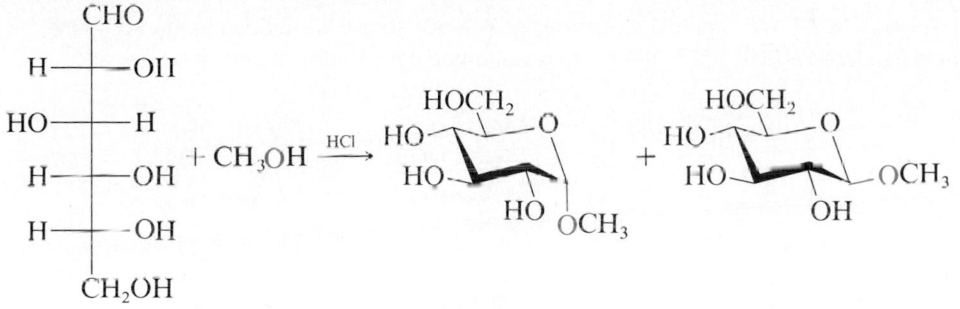

D-Glucose   Methanol

Methyl
α-D-glucopyranoside
(major product; isolated
in 49% yield)

Methyl
β-D-glucopyranoside
(minor product)

A point to be emphasized about glycoside formation is that, despite the presence of a number of other hydroxyl groups in the carbohydrate, it is *only the anomeric hydroxyl group that is replaced.* This is because a carbocation at the anomeric position is stabilized by the ring oxygen and is the only one capable of being formed under the reaction conditions.

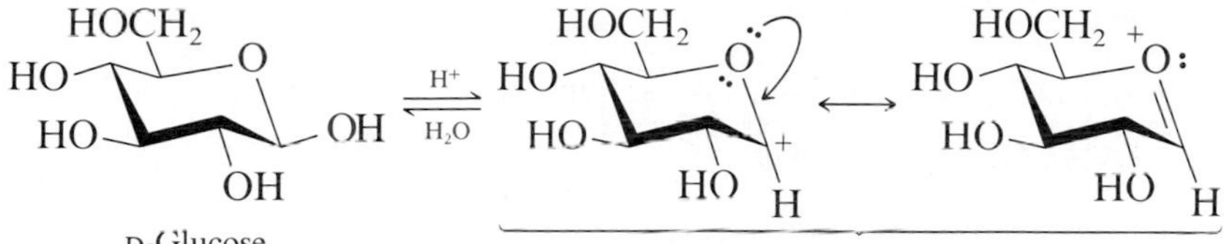

D-Glucose
(shown in β-pyranose
form)

Electron pair on ring oxygen
can stabilize carbocation at
anomeric position only

Once the carbocation is formed (slow step), it is captured by the alcohol acting as a nucleophile. Attack can occur at either the β face or the α face of the carbocation.

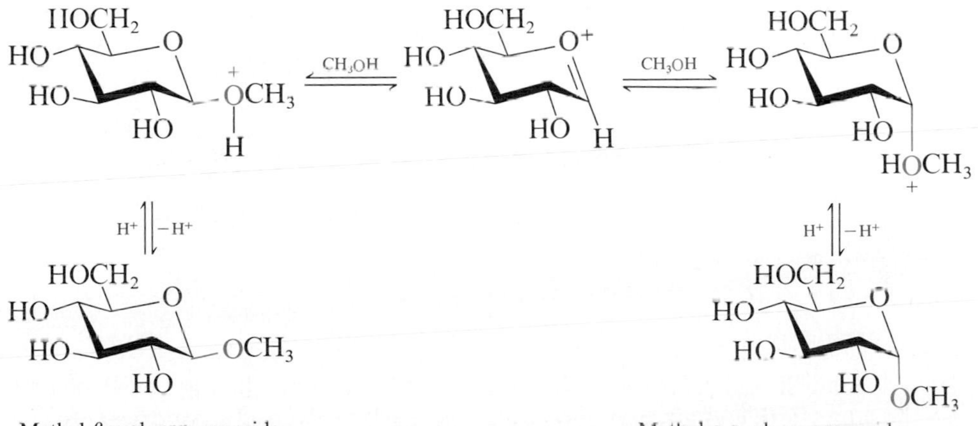

Methyl β-D-glucopyranoside   Methyl α-D-glucopyranoside

Equilibration of the glycosides usually takes place under the conditions of this reaction. It is an example of a reaction that is subject to *thermodynamic* (or *equilibrium*) *control.* Glucopyranosides are more stable than glucofuranosides and predominate at equilibrium.

PROBLEM 27.10 Methyl glycosides of 2-deoxy sugars have been prepared by the acid catalyzed addition of methanol to unsaturated carbohydrates known as *glycals*.

D-Galactal → Methyl 2-deoxy-α-D-lyxohexopyranoside (38%) + Methyl 2-deoxy-β-D-lyxohexopyranoside (36%)

Suggest a reasonable mechanism for the acid-catalyzed formation of glycosides from glycals.

Under neutral or basic conditions glycosides are configurationally stable; unlike the free sugars from which they are derived, they do not exhibit mutarotation. Converting the anomeric hydroxyl group to an ether function (hemiacetal → acetal) prevents its reversion to the open-chain form in neutral or basic media. In aqueous acid, acetal formation can be reversed and the glycoside hydrolyzed to an alcohol and the free sugar.

27.14 DISACCHARIDES

Disaccharides are carbohydrates that yield two monosaccharide molecules on hydrolysis. Structurally, disaccharides are *glycosides* in which the alkoxy group attached to the anomeric carbon is derived from a second sugar molecule.

Maltose and cellobiose are two isomeric disaccharides. Maltose is obtained by the hydrolysis of starch, cellobiose by the hydrolysis of cellulose.

Maltose

Cellobiose

How do maltose and cellobiose differ in structure? First, note the ways in which they are the same. Both contain two glucose units joined by a glycosidic bond between C-1 of one glucose molecule and C-4 of a second glucose. The configuration at C-4 is the same in both cellobiose and maltose, as it must be in order for the carbohydrate to be glucose. The configuration at C-1, however, is α in maltose, and β in cellobiose, i.e., maltose is an α-glucoside and cellobiose is a β-glucoside. The two compounds are diastereomers; they differ in configuration at the anomeric carbon involved in the glycosidic bond.

The stereochemistry and points of connection of glycosidic bonds are commonly designated by symbols such as $\alpha(1,4)$ for maltose and $\beta(1,4)$ for cellobiose; α and β designate the stereochemistry at the anomeric position while the numerals specify the ring carbons involved.

Both maltose and cellobiose have a free anomeric hydroxyl group that is not involved in a glycoside bond. While drawn in the β orientation in the structures shown, the configuration at the free anomeric center is variable and may be either α or β. Indeed, two stereoisomeric forms of maltose have been isolated; one has its anomeric hydroxyl group in an equatorial orientation, while the other has an axial anomeric hydroxyl.

PROBLEM 27.11 The two stereoisomeric forms of maltose just mentioned undergo mutarotation when dissolved in water. What is the structure of the key intermediate in this process?

Because maltose and cellobiose differ in structure, they differ in properties. One notable difference is their behavior toward enzyme-catalyzed hydrolysis. An enzyme known as *maltase* catalyzes the hydrolytic cleavage of the α-glycosidic bond of maltose but is without effect in promoting the hydrolysis of the β-glycosidic bond of cellobiose. A different enzyme, *emulsin*, produces the opposite result: emulsin catalyzes the hydrolysis of cellobiose but not of maltose. The behavior of each enzyme is general for glucosides. Maltase catalyzes the hydrolysis of α-glucosides and is also known as α-*glucosidase*, whereas emulsin catalyzes the hydrolysis of β-glucosides and is known as β-*glucosidase*. The specificity of these enzymes offers a useful tool for structure determination in that it allows the stereochemistry of glycosidic linkages to be assigned.

Lactose is found to the extent of 2 to 6 percent in milk and is known as "milk sugar." It is a disaccharide in which C-1 of D-galactose and C-4 of D-glucose are linked by a β-glycosidic bond.

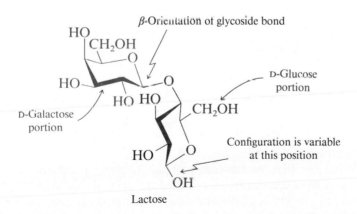

Lactose

Digestion of lactose is facilitated by the β-glycosidase *lactase*. A deficiency of this enzyme makes it difficult to digest lactose and causes abdominal discomfort. Lactose intolerance is a genetic disorder; it is treated by restricting the amount of milk in the diet.

The best known of all the carbohydrates is probably *sucrose*—common table sugar. Sucrose is a disaccharide in which D-glucose and D-fructose are joined at their anomeric carbons by a glycosidic bond. Its chemical composition is the same irrespective of its source; sucrose from cane and sucrose from sugar beets are chemically identical.

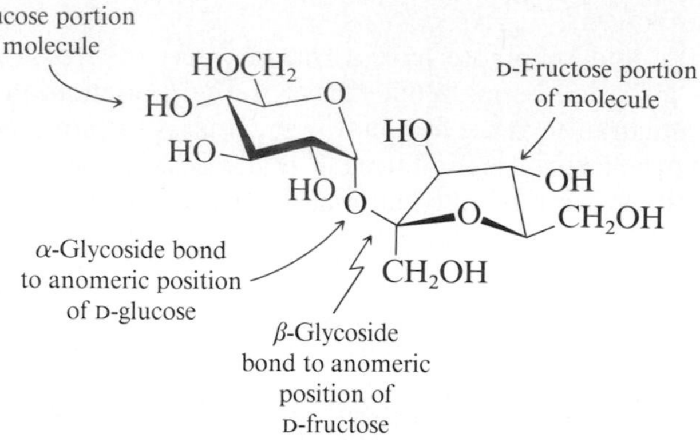

D-Glucose portion of molecule

D-Fructose portion of molecule

α-Glycoside bond to anomeric position of D-glucose

β-Glycoside bond to anomeric position of D-fructose

Sucrose

Since sucrose does not have a free anomeric hydroxyl group, it does not undergo mutarotation.

27.15 POLYSACCHARIDES

Cellulose is the principal structural component of vegetable matter. Wood is 30 to 40 percent cellulose, cotton over 90 percent. Photosynthesis in plants is responsible for the formation of 10^9 tons per year of cellulose. Structurally, cellulose is a polysaccharide composed of several thousand D-glucose units joined by $\beta(1,4)$ glycosidic linkages.

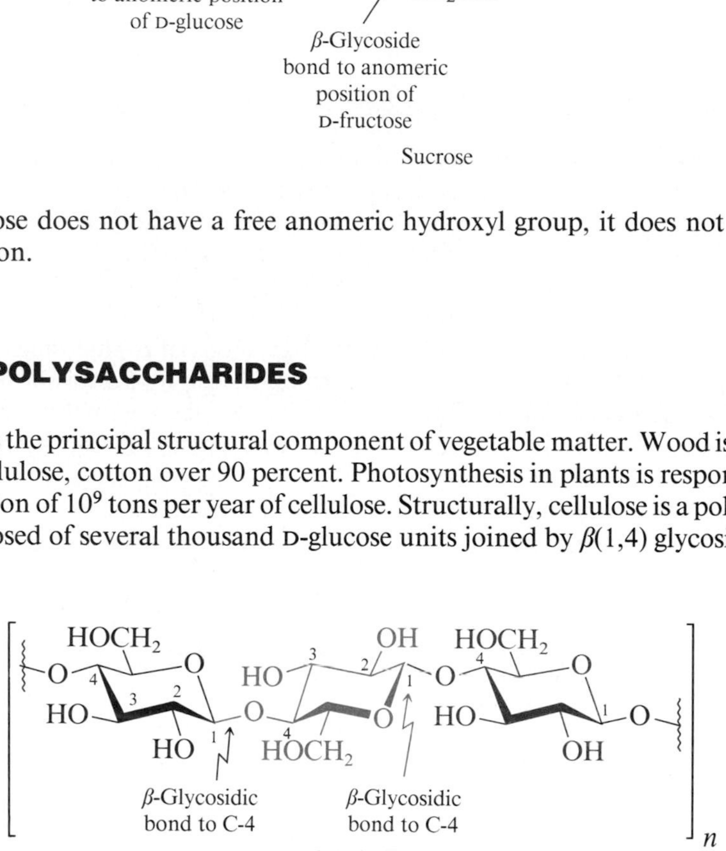

β-Glycosidic bond to C-4

β-Glycosidic bond to C-4

Cellulose

Complete hydrolysis of all the glycosidic bonds of cellulose yields D-glucose. The disaccharide fraction that results from partial hydrolysis is cellobiose.

Animals do not possess the enzymes necessary to catalyze the hydrolysis of cellulose and so cannot digest it. Cattle and other ruminants can use cellulose as a food source in an indirect way. Colonies of microorganisms that live in the digestive tract of these animals consume cellulose and in the process convert it to other substances that are digestible by the host.

A more direct source of energy for animals is provided by the starches found in many foods. Starch is a mixture of a water-dispersible fraction called *amylose* and a second component, *amylopectin.* Amylose is a polysaccharide made up of about 100 to several thousand D-glucose units joined by $\alpha(1,4)$ glycosidic bonds.

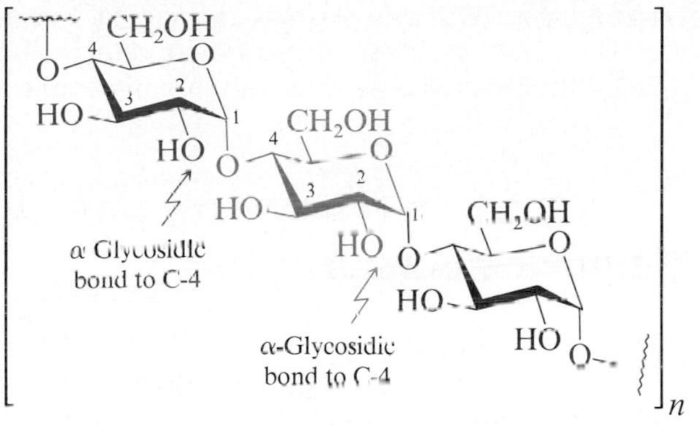

Amylose

PROBLEM 27.12 Complete the syllogism: Cellulose is to cellobiose as amylose is to _____.

Like amylose, amylopectin is a polysaccharide of $\alpha(1,4)$-linked D-glucose units. Instead of being a continuous length of $\alpha(1,4)$ units, however, amylopectin is branched. Attached to C-6 at various points on the main chain are short polysaccharide branches of 24 to 30 glucose units joined by $\alpha(1,4)$ glycosidic bonds. A portion of an amylopectin molecule is represented in Figure 27.6.

Starch is a plant's way of storing glucose to meet its energy needs. Animals can tap that source by eating starchy foods and, with the aid of their α-glycosidase enzymes,

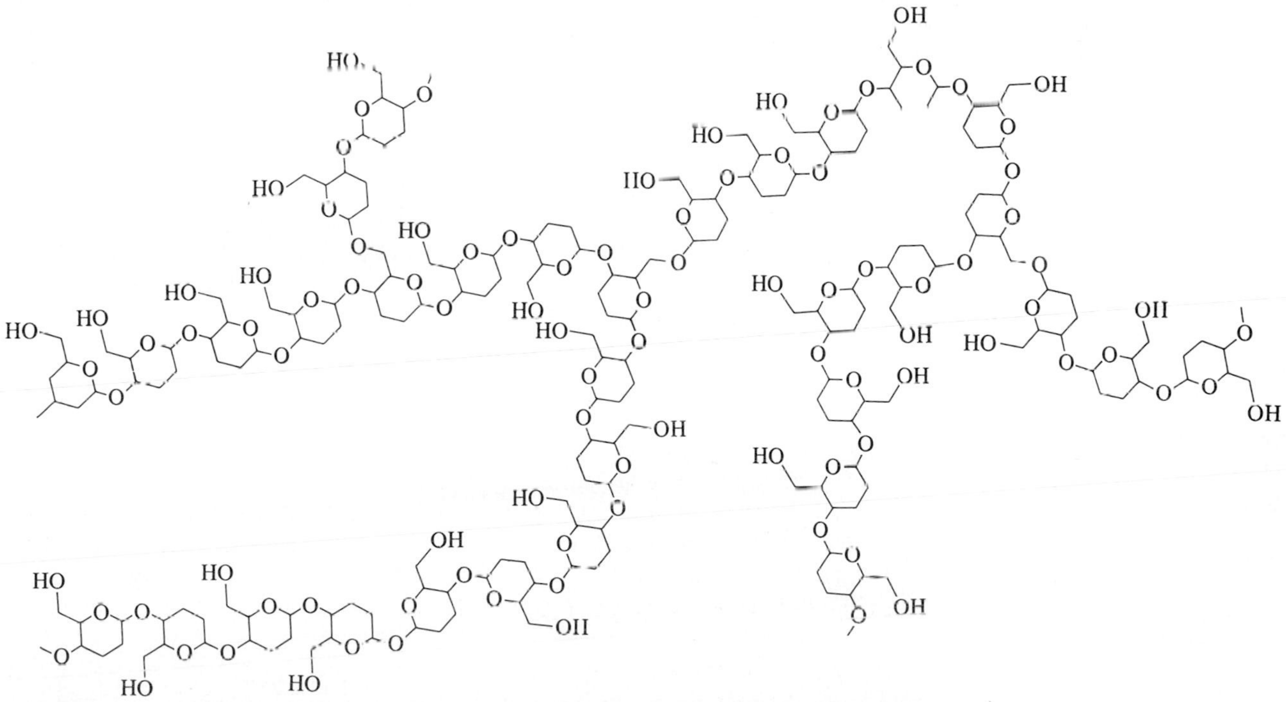

FIGURE 27.6 A portion of the amylopectin structure. α-Glycosidic bonds link C-1 and C-4 of adjacent glucose units. Glycoside bonds from C-6 to the polysaccharide side chains are designated by red oxygen atoms. Hydroxyl groups attached to the individual rings have been omitted for clarity.

hydrolyze the starch to glucose. When more glucose is available than is needed as fuel, animals store it as glycogen. *Glycogen* is similar to amylopectin in that it is a branched polysaccharide of $\alpha(1,4)$-linked D-glucose units with subunits connected to C-6 of the main chain.

27.16 CELL-SURFACE GLYCOPROTEINS

That carbohydrates play an informational role in biological interactions is a recent revelation of great importance. *Glycoproteins*, protein molecules covalently bound to carbohydrates, are often the principal species involved. When a cell is attacked by a virus or bacterium or when it interacts with another cell, the drama begins when the foreign particle attaches itself to the surface of the host cell. The invader recognizes the host by the glycoproteins on the cell surface. More specifically, it recognizes particular carbohydrate sequences at the end of the glycoprotein. For example, the receptor on the cell surface to which an influenza virus attaches itself has been identified as a glycoprotein terminating in a disaccharide of *N*-acetylgalactosamine and *N*-acetylneuraminic acid (Figure 27.7). Since attachment of the invader to the surface of the host cell is the first step in infection, one approach to disease prevention is to selectively inhibit this "host-guest" interaction. Identifying the precise nature of the interaction is the first step in the rational design of drugs that prevent it.

Human blood group substances offer another example of the informational role played by carbohydrates. The structure of the glycoproteins attached to the surface of blood cells determines whether blood is type A, B, AB, or O. Differences between the carbohydrate components of the various glycoproteins have been identified and are shown in Figure 27.8. Compatibility of blood types is dictated by *antigen-antibody* interactions. The cell-surface glycoproteins are *antigens*. *Antibodies* present in certain blood types can cause the blood cells of certain other types to clump together and sets practical limitations on transfusion procedures. The antibodies "recognize" the antigens they act upon by their terminal saccharide units.

Antigen-antibody interactions are the fundamental basis by which the immune system functions. These interactions are chemical in nature and often involve associations between glycoproteins of an antigen and complementary glycoproteins of the antibody. The precise chemical nature of antigen-antibody association is an area of active investigation, with significant implications for chemistry, biochemistry, and physiology.

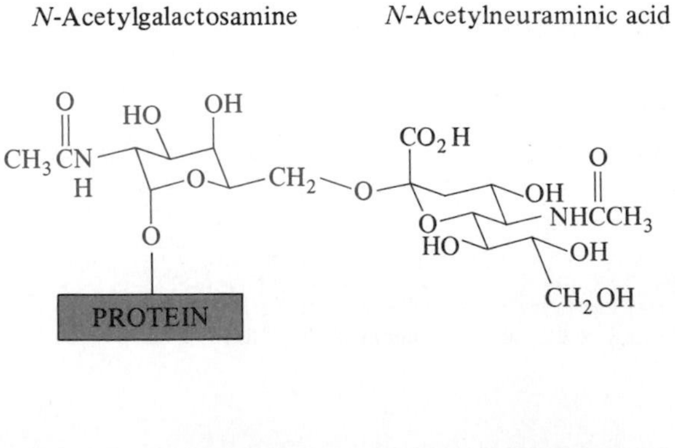

FIGURE 27.7 Schematic diagram of cell-surface glycoprotein, showing disaccharide unit that is recognized by invading influenza virus.

FIGURE 27.8 Terminal carbohydrate units of human blood group glycoproteins. The group R' is a polymer of *N*-acetylgalactosamine units attached to a protein. The structural difference between the type A, type B, and type O glycoproteins lies in the group designated R.

In notes

27.17 CARBOHYDRATE STRUCTURE DETERMINATION

Present-day techniques for structure determination in carbohydrate chemistry are substantially the same as those for any other type of compound. The full range of modern instrumental methods, including mass spectrometry and infrared and nuclear magnetic resonance spectroscopy, is brought to bear on the problem. If the unknown substance is crystalline, x ray diffraction can provide precise structural information which in the best cases is equivalent to taking a photograph of the compound.

Prior to the widespread availability of instrumental methods, the major approach to structure determination relied on a battery of chemical reactions and tests. The response of an unknown substance to various reagents and procedures provided a body of data from which the structure could be deduced. Some of these procedures are still used to supplement the information obtained by instrumental methods. To better understand the scope and limitations of these tests, a brief survey of the chemical reactions of carbohydrates is in order. In many cases these reactions are simply applications of chemistry you have already learned. Certain of the transformations, however, are virtually unique to carbohydrates.

27.18 REDUCTION OF CARBOHYDRATES

While carbohydrates exist almost entirely as cyclic hemiacetals in aqueous solution, they are in rapid equilibrium with their open-chain forms, and most of the reagents that react with simple aldehydes and ketones react in an analogous way with the carbonyl functional groups of carbohydrates.

The carbonyl group of carbohydrates can be reduced to an alcohol function. Typical procedures include catalytic hydrogenation and sodium borohydride reduction. Lithium aluminum hydride is not suitable because it is not compatible with the solvents (water, alcohols) that are required to dissolve carbohydrates. The products of carbohydrate reduction are called *alditols*. Since these alditols lack a carbonyl group, they are, of course, incapable of forming cyclic hemiacetals and exist exclusively in noncyclic forms.

α-D-Mannofuranose, or
β-D-Mannofuranose, or
α-D-Mannopyranose, or
β-D-Mannopyranose

$\rightleftharpoons$

CHO
HO——H
HO——H
H——OH
H——OH
CH$_2$OH
D-Mannose

$\xrightarrow[\text{ethanol–}\text{water}]{\text{H}_2,\ \text{Ni}}$

CH$_2$OH
HO——H
HO——H
H——OH
H——OH
CH$_2$OH
D-Mannitol (79%)

α-D-Galactofuranose, or
β-D-Galactofuranose, or
α-D-Galactopyranose, or
β-D-Galactopyranose

$\rightleftharpoons$

CHO
H——OH
HO——H
HO——H
H——OH
CH$_2$OH
D-Galactose

$\xrightarrow[\text{H}_2\text{O}]{\text{NaBH}_4}$

CH$_2$OH
H——OH
HO——H
HO——H
H——OH
CH$_2$OH
D-Galactitol (90%)

PROBLEM 27.13   Does sodium borohydride reduction of D-ribose yield an optically active product? Explain.

Another name for glucitol, obtained by reduction of D-glucose, is sorbitol; it is used as a sweetner, especially in special diets required to be low in sugar. Reduction of D-fructose yields a mixture of glucitol and mannitol, corresponding to the two possible configurations at the newly generated chiral center at C-2.

27.19   OXIDATION OF CARBOHYDRATES

A characteristic property of an aldehyde function is its sensitivity to oxidation. A solution of copper (II) sulfate as its citrate complex (*Benedict's reagent*) is capable of oxidizing aliphatic aldehydes to the corresponding carboxylic acid.

$$
\underset{\substack{\text{Aldehyde}}}{\overset{\overset{\displaystyle O}{\|}}{\text{RCH}}} + \underset{\substack{\text{Cupric ion}\\\text{[from copper(II)}\\\text{sulfate]}}}{2\text{Cu}^{2+}} + \underset{\substack{\text{Hydroxide}\\\text{ion}}}{5\text{HO}^-} \longrightarrow \underset{\substack{\text{Carboxy-}\\\text{late anion}}}{\overset{\overset{\displaystyle O}{\|}}{\text{RCO}^-}} + \underset{\substack{\text{Copper(I)}\\\text{oxide}}}{\text{Cu}_2\text{O}} + \underset{\substack{\text{Water}}}{3\text{H}_2\text{O}}
$$

The formation of a red precipitate of copper(I) oxide by reduction of Cu(II) is taken as a positive test for an aldehyde. Carbohydrates that give positive tests with Benedict's reagent are termed *reducing sugars.*

Aldoses are, of course, reducing sugars since they possess an aldehyde function in their open-chain form. Ketoses are also reducing sugars. Under the conditions of the test, ketoses equilibrate with aldoses by way of enediol intermediates and the aldoses are oxidized by the reagent.

$$
\begin{array}{ccc}
\text{CH}_2\text{OH} & \text{CHOH} & \text{H}\quad\text{O} \\
| & \| & \diagdown\!\!\diagup \\
\text{C}=\text{O} \rightleftharpoons & \text{C}-\text{OH} \rightleftharpoons & \text{C} \\
| & | & | \\
\text{R} & \text{R} & \text{CHOH} \\
& & | \\
& & \text{R}
\end{array}
\xrightarrow{\text{Cu}^{2+}}
\begin{array}{l}
\text{positive test} \\
(\text{Cu}_2\text{O formed})
\end{array}
$$

Ketose   Enediol   Aldose

The same kind of equilibrium is available to α-hydroxy ketones generally; such compounds give a positive test with Benedict's reagent. Any carbohydrate that contains a free hemiacetal function is a reducing sugar. The free hemiacetal is in equilibrium with the open-chain form and through it is susceptible to oxidation. Maltose, for example, gives a positive test with Benedict's reagent.

Maltose $\rightleftharpoons$ Open-chain form of maltose $\xrightarrow{\text{Cu}^{2+}}$ positive test (Cu₂O formed)

Glycosides, in which the anomeric carbon is part of an acetal function, do not give a positive test and are not reducing sugars.

Methyl α-D-glucopyranoside: not a reducing sugar   Sucrose: not a reducing sugar

---

PROBLEM 27.14 Which of the following would be expected to give a positive test with Benedict's reagent? Why?

(a) D-Mannitol

(b) L-Arabinose

(c) 1,3-Dihydroxyacetone

(d) D-Fructose

(e) Lactose

(f) Amylose

SAMPLE SOLUTION (a) Recall from the preceding section that D-mannitol has the structure shown. Lacking an aldehyde, α-hydroxy ketone, or hemiacetal function, it cannot be oxidized by Cu^{2+} and will not give a positive test with Benedict's reagent.

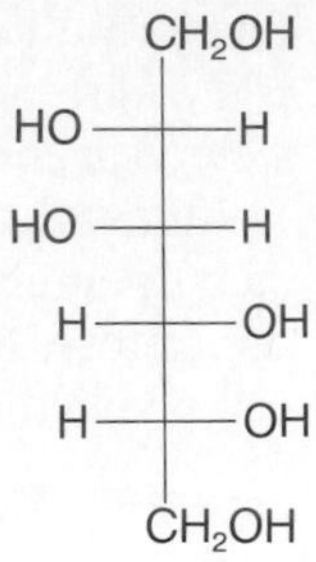

D-Mannitol

Know how these work

Benedict's reagent is the basis of a test kit available in drug stores that permits individuals to monitor the glucose levels in their urine.

Fehling's solution, a tartrate complex of copper(II) sulfate, and *Tollens' reagent* (Section 18.19) have also been used as tests for reducing sugars.

Derivatives of aldoses in which the terminal aldehyde function is oxidized to a carboxylic acid are called *aldonic acids.* Aldonic acids are named by replacing the "-*ose*" ending of the aldose by "-*onic acid.*" Oxidation with Benedict's reagent, Fehling's solution, or Tollens' reagent is accompanied by undesired side reactions and is not a satisfactory preparative procedure for converting aldoses to aldonic acids. Instead, oxidation of aldoses with bromine is the most commonly used method for the preparation of aldonic acids. The reaction involves oxidation of the anomeric carbon of the furanose or pyranose form of a carbohydrate to a lactone.

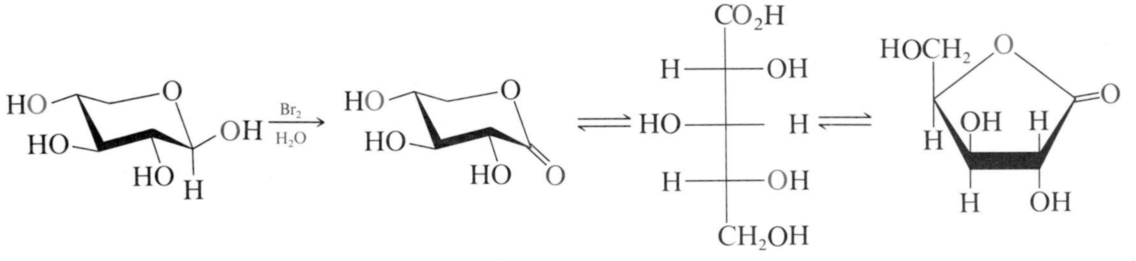

β-D-Xylopyranose

D-Xylonic acid
(90%)

Aldonic acids normally exist in equilibrium with their five- or six-membered cyclic lactones. They can be isolated as carboxylate salts of their open-chain forms on treatment with base.

The reaction of aldoses with nitric acid leads to the formation of *aldaric acids* by oxidation of both the aldehyde and the terminal primary alcohol function to carboxylic acid groups. Aldaric acids are also known as *saccharic acids* and are named by substituting -*aric acid* for the -*ose* ending of the corresponding carbohydrate.

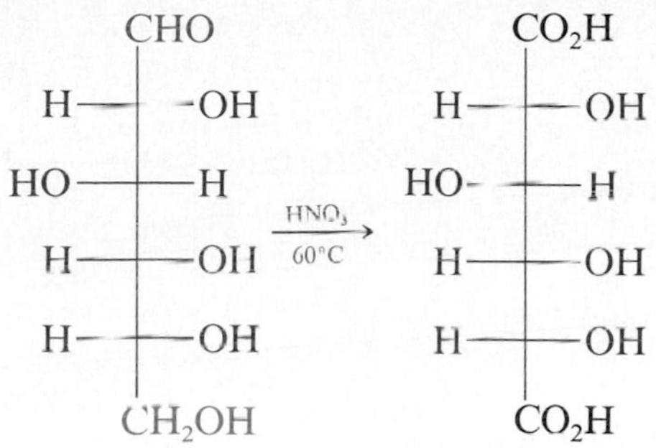

CHO
H——OH
HO——H
H——OH
H——OH
CH₂OH

$\xrightarrow[60^\circ C]{HNO_3}$

CO₂H
H——OH
HO——H
H——OH
H——OH
CO₂H

D-Glucose   D-Glucaric acid (41%)

Like aldonic acids, aldaric acids exist mainly as cyclic lactones.

PROBLEM 27.15   Another aldohexose gives the same aldaric acid on oxidation as does D-glucose. Which one?

Uronic acids represent an intermediate oxidation state between aldonic and aldaric acids. They have an aldehyde function at one end of their carbon chain and a carboxylic acid group at the other.

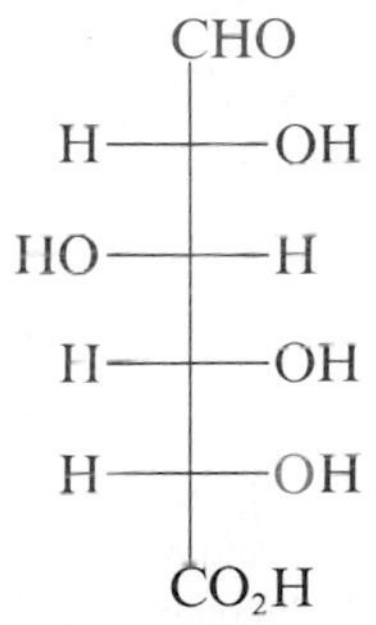

CHO
H——OH
HO——H
H——OH
H——OH
CO₂H

Fischer projection formula of D-glucuronic acid

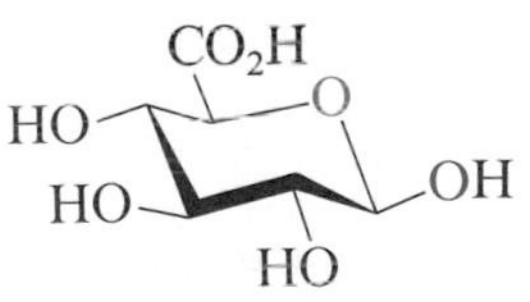

β-Pyranose form of D-glucuronic acid

Do A mechanism

Uronic acids are biosynthetic intermediates in various metabolic processes; ascorbic acid (vitamin C), for example, is biosynthesized by way of glucuronic acid. Many metabolic waste products are excreted in the urine as their glucuronate salts.

27.20 CYANOHYDRIN FORMATION AND CARBOHYDRATE CHAIN EXTENSION

The presence of an aldehyde function in their open-chain forms makes aldoses reactive toward addition of hydrogen cyanide. Addition yields a mixture of diastereomeric cyanohydrins.

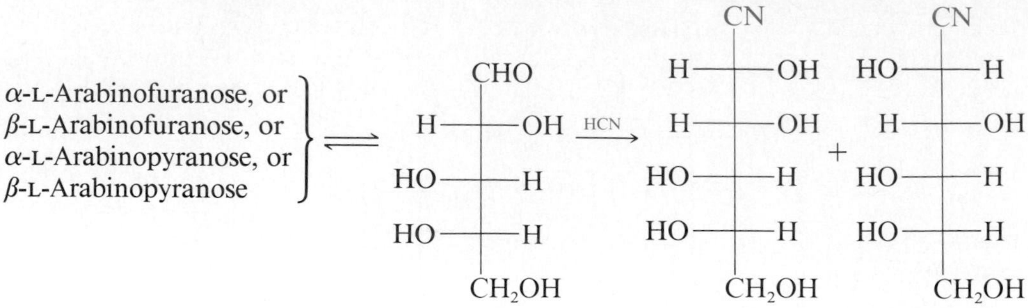

α-L-Arabinofuranose, or
β-L-Arabinofuranose, or
α-L-Arabinopyranose, or
β-L-Arabinopyranose
} ⇌

L-Arabinose   L-Mannononitrile   L-Gluconitrile

The reaction is used for the chain extension of aldoses in the synthesis of new or unusual sugars. In this case, L-arabinose is an abundant natural product and possesses the correct configurations at its three chiral centers for elaboration to the relatively rare L-enantiomers of glucose and mannose. After cyanohydrin formation, the cyano groups are converted to aldehyde functions by hydrogenation in aqueous solution. Under these conditions, —C≡N is reduced to —CH=NH and hydrolyzes rapidly to —CH=O. Use of a poisoned palladium-on-barium sulfate catalyst prevents further reduction to the alditols.

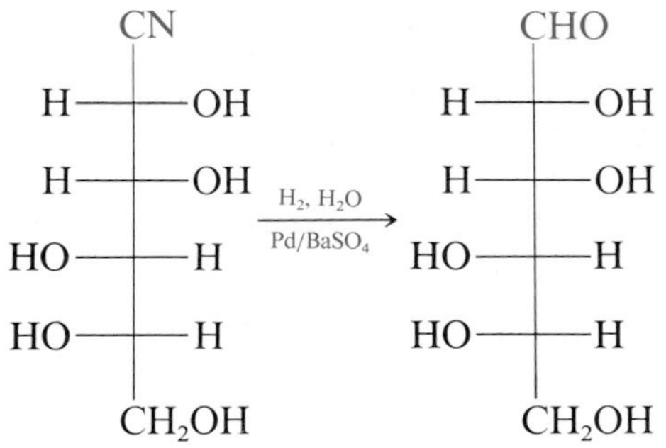

L-Mannononitrile   L-Mannose
                          (56% yield from
                             L-arabinose)

(Similarly, L-gluconitrile has been reduced to L-glucose; its yield was 26 percent from L-arabinose.)

An older version of this sequence is called the *Kiliani-Fischer* synthesis. It, too, proceeds through a cyanohydrin, but it uses a less efficient method for converting the cyano group to the required aldehyde.

27.21 OSAZONE FORMATION

Prior to the introduction of chromatographic methods, research in carbohydrate chemistry was hampered by the physical nature of many sugars. Their water solubility, coupled with a tendency to form noncrystallizable syrups, made isolation and purification difficult. An early approach to the problem was to convert the carbohydrate to a crystalline derivative with the customary reagents that react with carbonyl groups. In 1884 Emil Fischer discovered that while one equivalent of phenylhydrazine reacts with aldoses and ketoses to yield the expected phenylhydrazone, excess phenylhydrazine converts aldoses and ketoses to compounds containing *two* adjacent phenylhydrazone units, which he called *osazones*. Not only is the aldehyde function converted to a phenylhydrazone, but so also is the alcohol function on the adjacent carbon.

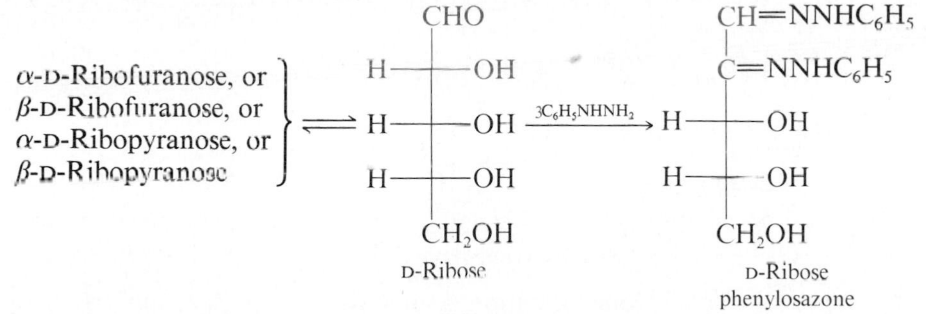

$$\alpha\text{-}D\text{-Ribofuranose, or}$$
$$\beta\text{-}D\text{-Ribofuranose, or}$$
$$\alpha\text{-}D\text{-Ribopyranose, or}$$
$$\beta\text{-}D\text{-Ribopyranose}$$

D-Ribose

D-Ribose
phenylosazone

Three equivalents of phenylhydrazine are required; two are involved in phenylhydrazone formation while the third undergoes reduction to a molecule of aniline and one of ammonia. The reaction is general for both aldoses and ketoses. It is always limited to adjacent positions on the chain and proceeds no further than the osazone no matter how much excess phenylhydrazine is present.

Phenylosazones are yellow crystalline solids and are insoluble in water. Carbohydrates are often characterized on the basis of the melting point of their osazone derivative. Osazone formation proved very helpful to Fischer as a tool for interrelating configurations of various sugars. For example, his observation that $(+)$-glucose, $(+)$-mannose, and $(-)$-fructose all gave the same osazone established that all three had the same configuration at C-3, C-4, and C-5.

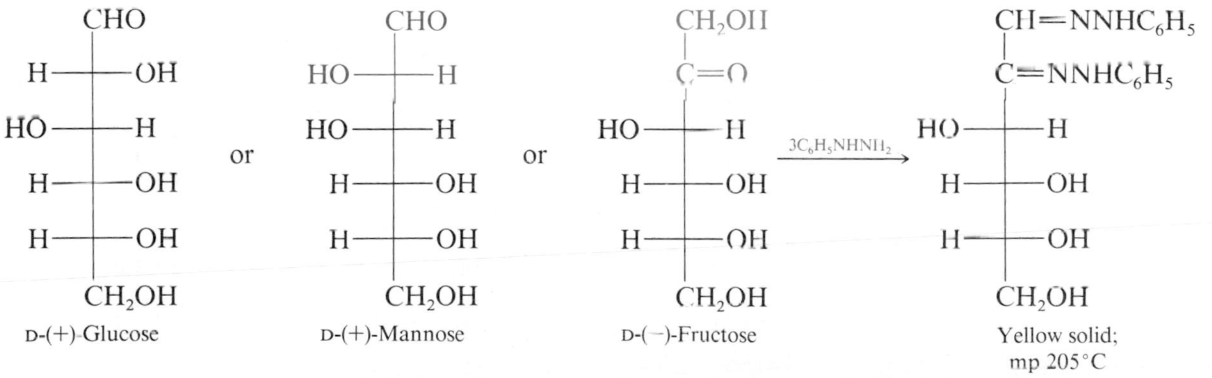

D-(+)-Glucose   D-(+)-Mannose   D-(−)-Fructose   Yellow solid;
mp 205°C

Aldoses that differ in configuration only at C-2 yield the same osazone. Ketoses having their carbonyl function at C-2 yield the same osazone as the corresponding aldose.

PROBLEM 27.16 Name another aldose that reacts with phenylhydrazine to give the same osazone as

(a) D-Ribose

(b) L-Threose

(c) D-Allose

(d) L-Fucose (6-deoxy-L-galactose)

SAMPLE SOLUTION (a) Osazone formation in aldoses leads to phenylhydrazone functions at C-1 and C-2. Thus, the aldose that gives the same phenylosazone as D-ribose is the one that has the opposite configuration at C-2, namely, D-arabinose.

CHO		
H——OH		CH=NNHC₆H₅

(structures)

CHO
H——OH
H——OH
H——OH
CH₂OH
D-Ribose

or

CHO
HO——H
H——OH
H——OH
CH₂OH
D-Arabinose

$\xrightarrow{3C_6H_5NHNH_2}$

CH=NNHC₆H₅
C=NNHC₆H₅
H——OH
H——OH
CH₂OH

The mechanism of osazone formation is somewhat complicated and its elaboration is not appropriate to an introductory treatment of carbohydrate chemistry.

27.22 EPIMERIZATION, ISOMERIZATION, AND RETRO-ALDOL CLEAVAGE REACTIONS OF CARBOHYDRATES

Carbohydrates undergo a number of isomerization and degradation reactions under both laboratory and physiological conditions. For example, a mixture of glucose, fructose, and mannose results when any one of the individual components is treated with aqueous base. The nature of this reaction can be understood by examining the consequences of base-catalyzed enolization of glucose.

CHO
H—C—OH
HO——H
H——OH
H——OH
CH₂OH
D-Glucose
(*R* configuration at C-2)

$\underset{\text{HO}^-,\ \text{H}_2\text{O}}{\rightleftharpoons}$

CHOH
‖
C—OH
HO——H
H——OH
H——OH
CH₂OH
Enediol
(C-2 not a chiral center)

$\underset{\text{HO}^-,\ \text{H}_2\text{O}}{\rightleftharpoons}$

CHO
HO—C—H
HO——H
H——OH
H——OH
CH₂OH
D-Mannose
(*S* configuration at C-2)

Because the configuration at C-2 is lost on enolization, the enediol intermediate can revert either to D-glucose or to D-mannose. Two stereoisomers that have multiple chiral centers but differ in configuration at only one of them are referred to as *epimers*. Glucose and mannose are epimeric at C-2. Under these conditions epimerization occurs only at C-2 because it alone is α to the carbonyl group.

There is another reaction available to the enediol intermediate. Proton transfer from water to C-1 converts the enediol not to an aldose but to the ketose D-fructose.

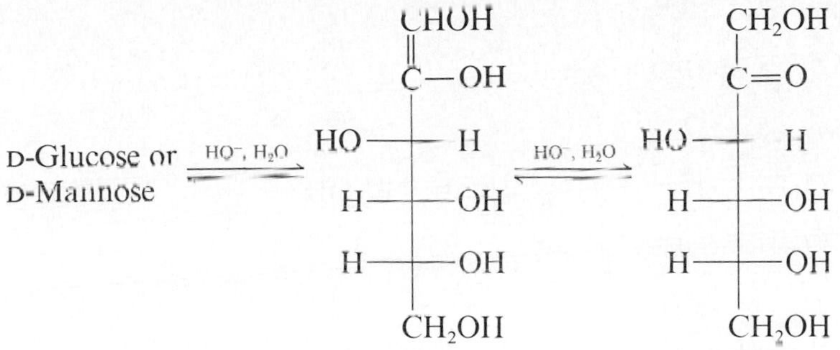

D-Glucose or D-Mannose $\xrightarrow{HO^-,\ H_2O}$ (Enediol) $\xrightleftharpoons{HO^-,\ H_2O}$ D-Fructose

The isomerization of D-glucose to D-fructose by way of an enediol intermediate is an important step in *glycolysis,* a complex process (11 steps!) by which an organism converts glucose to chemical energy in the absence of oxygen. The substrate is not glucose itself but its 6-phosphate ester. The enzyme that catalyzes the isomerization is called *phosphoglucose isomerase.*

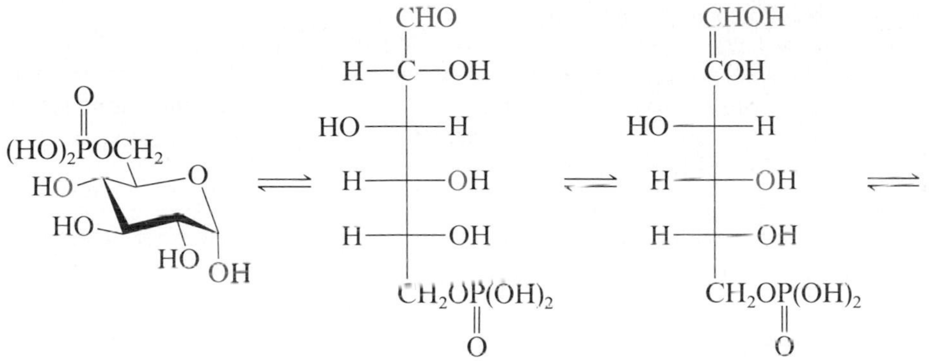

D-Glucose 6-phosphate ⇌ Open-chain form of D-glucose 6-phosphate ⇌ Enediol ⇌

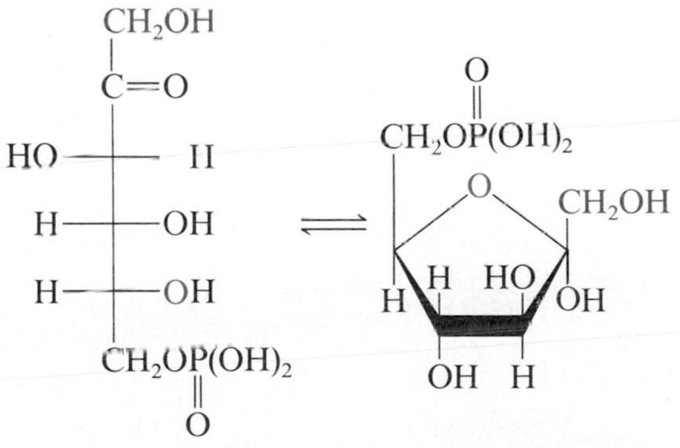

Open-chain form of D-fructose 6-phosphate ⇌ D-Fructose 6-phosphate

Following its formation, D-fructose 6-phosphate is converted to its corresponding 1,6-phosphate diester, which is then cleaved to two 3-carbon fragments under the influence of the enzyme *aldolase.*

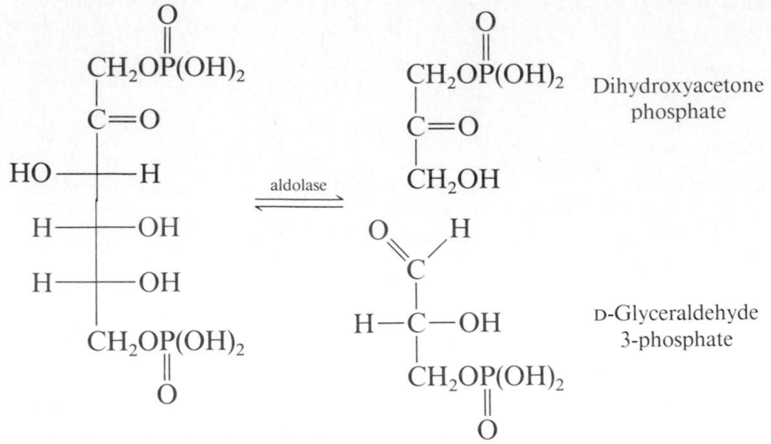

This cleavage is a *retro-aldol* reaction. It is the reverse of the process by which D-fructose 1,6-diphosphate would be formed by addition of the enolate of dihydroxyacetone phosphate to D-glyceraldehyde 3-phosphate. The enzyme aldolase catalyzes both the aldol condensation of the two components and, in glycolysis, the retro-aldol cleavage of D-fructose 1,6-diphosphate.

Further steps in glycolysis utilize the D-glyceraldehyde 3-phosphate formed in the aldolase-catalyzed cleavage reaction as a substrate. Its coproduct, dihydroxyacetone phosphate, is not wasted, however. The enzyme *triose phosphate isomerase* converts dihydroxyacetone phosphate to D-glyceraldehyde 3-phosphate, which enters the glycolysis pathway for further transformations.

PROBLEM 27.17   Suggest a reasonable structure for the intermediate in the conversion of dihydroxyacetone phosphate to D-glyceraldehyde 3-phosphate.

Cleavage reactions of carbohydrates also occur on treatment with aqueous base for prolonged periods as a consequence of base-catalyzed retro-aldol reactions. As pointed out in Section 19.9, aldol addition is a reversible process and β-hydroxy carbonyl compounds can be cleaved to an enolate and an aldehyde or ketone.

27.23   ACYLATION AND ALKYLATION OF HYDROXYL GROUPS IN CARBOHYDRATES

The alcohol groups of carbohydrates undergo chemical reactions typical of hydroxyl functions. They are converted to esters by reaction with acyl chlorides and carboxylic acid anhydrides.

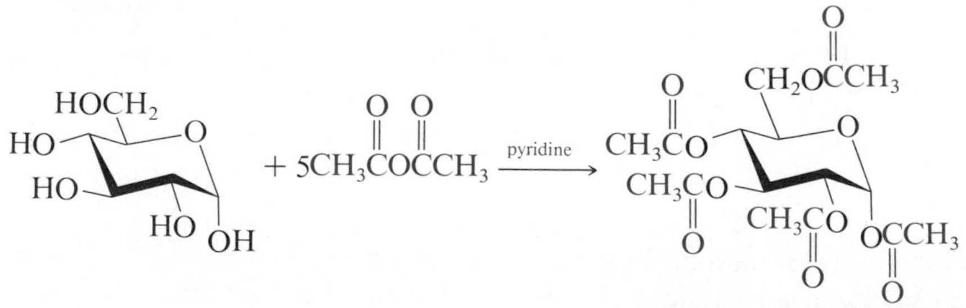

α-D-Glucopyranose   Acetic anhydride   1,2,3,4,6-Penta-*O*-acetyl-α-D-glucopyranose (88%)

Ethers are formed under conditions of the Williamson ether synthesis. Methyl ethers of carbohydrates are efficiently prepared by alkylation with methyl iodide in the presence of silver oxide.

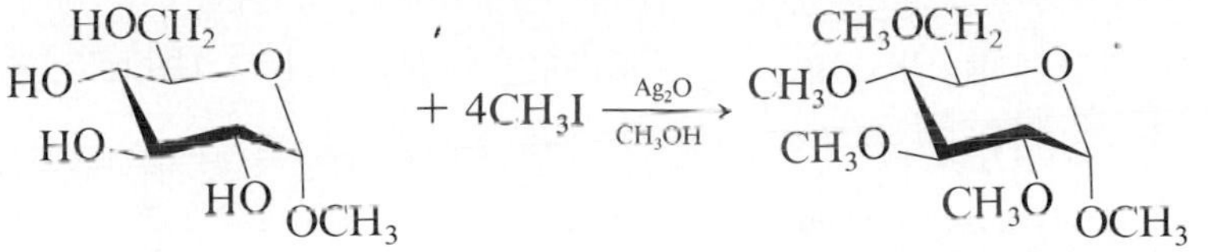

Methyl α-D-
glucopyranoside

Methyl
iodide

Methyl 2,3,4,6-tetra-*O*-methyl-
α-D-glucopyranoside (97%)

This reaction has been used in an imaginative way in determining the ring size of glycosides. Once all the free hydroxyl groups of a glycoside have been methylated, the glycoside is subjected to acid-catalyzed hydrolysis. Only the anomeric methoxy group is hydrolyzed under these conditions—another example of the ease of carbocation formation at the anomeric position.

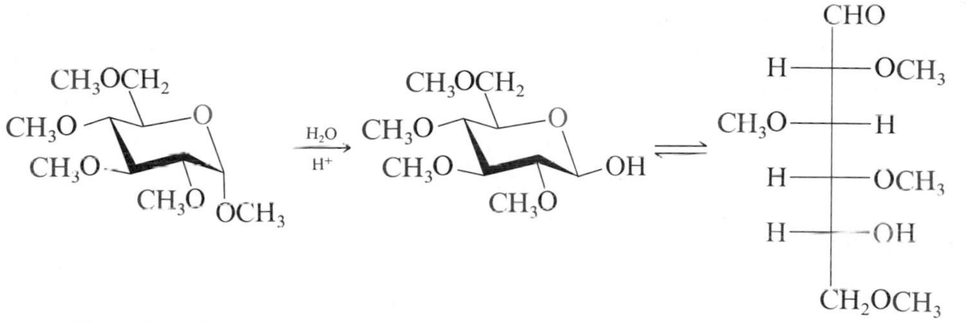

Methyl 2,3,4,6-
tetra-*O*-methyl-α-D-
glucopyranoside

2,3,4,6-Tetra-*O*-methyl-D-
glucose

Notice that all the hydroxyl groups in the free sugar except C-5 are methylated. Carbon-5 is not methylated because it was originally the site of the ring oxygen in the methyl glycoside. Once the position of the hydroxyl group in the free sugar has been determined, either by spectroscopy or by converting the sugar to a known compound, the ring size stands revealed.

27.24 PERIODIC ACID OXIDATION OF CARBOHYDRATES

Periodic acid oxidation (Section 16.12) finds extensive use as an analytical method in carbohydrate chemistry. Structural information is obtained by measuring the number of equivalents of periodic acid that react with a given compound and by identifying the reaction products. A vicinal diol consumes one equivalent of periodate and is cleaved to two carbonyl compounds.

$$R_2C-CR_2' + HIO_4 \longrightarrow R_2C=O + R_2'C=O + HIO_3 + H_2O$$
$$\ \ \ \ |\ \ \ \ |$$
$$\ \ HO\ \ OH$$

Vicinal
diol

Periodic
acid

Two carbonyl compounds

Iodic
acid

Water

α-Hydroxy carbonyl compounds are cleaved to a carboxylic acid and a carbonyl compound.

$$\underset{\substack{\text{OH}}}{\overset{\overset{\displaystyle O}{\|}}{R\overset{}{C}\underset{}{C}R_2'}} + HIO_4 \longrightarrow \overset{\overset{\displaystyle O}{\|}}{R\overset{}{C}OH} + R_2'C{=}O + HIO_3$$

| α-Hydroxy carbonyl compound | Periodic acid | Carboxylic acid | Carbonyl compound | Iodic acid |

When three contiguous carbons bear hydroxyl groups, two moles of periodate are consumed per mole of carbohydrate and the central carbon is oxidized to a molecule of formic acid.

$$\underset{\substack{\text{HO}\quad\text{OH}\quad\text{OH}}}{R_2C{-}CH{-}CR_2'} + 2HIO_4 \longrightarrow R_2C{=}O + \overset{\overset{\displaystyle O}{\|}}{H\overset{}{C}OH} + R_2'C{=}O + 2HIO_3$$

| Points at which cleavage occurs | Periodic acid | Carbonyl compound | Formic acid | Carbonyl compound | Iodic acid |

Ether and acetal functions are not affected by the reagent.

An example of the use of periodic acid oxidation in structure determination can be found in a case in which a previously unknown methyl glycoside was obtained by the reaction of D-arabinose with methanol and hydrogen chloride. The size of the ring was identified as five-membered because only one mole of periodic acid was consumed per mole of glycoside and no formic acid was produced. Were the ring six-membered, two moles of periodic acid would be required per mole of glycoside and one mole of formic acid would be produced.

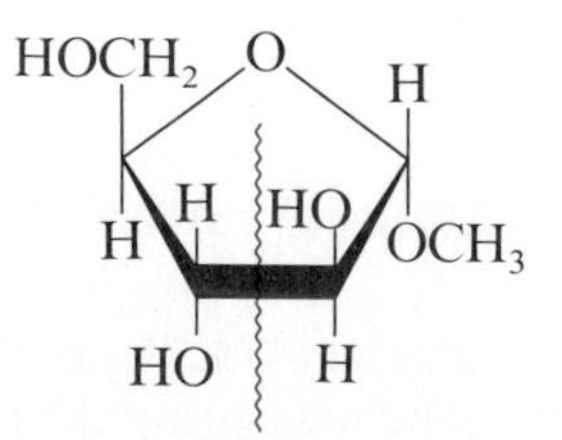

Only one site for periodic acid cleavage in methyl α-D-arabinofuranoside

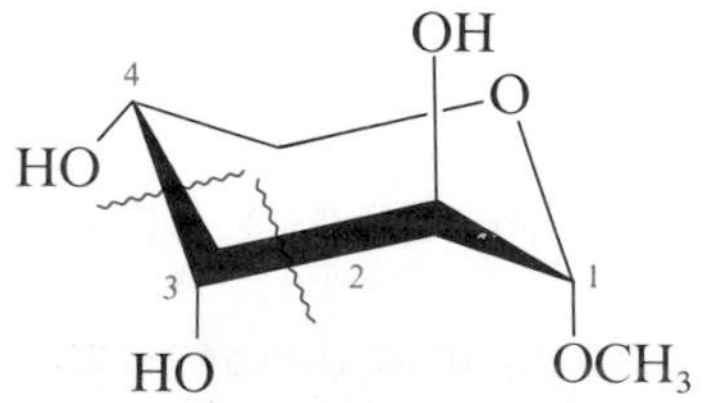

Two sites of periodic acid cleavage in methyl α-D-arabinopyranoside, C-3 lost as formic acid

Isolation of the product of periodate oxidation identified the new glycoside as methyl α-D-arabinofuranoside rather than its β-methyl isomer.

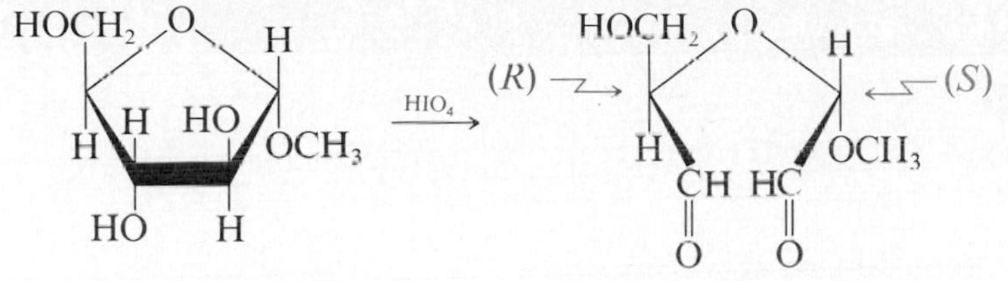

Methyl α-D-arabinofuranoside

Dialdehyde product; R configuration
at one chiral center, S
configuration at the other

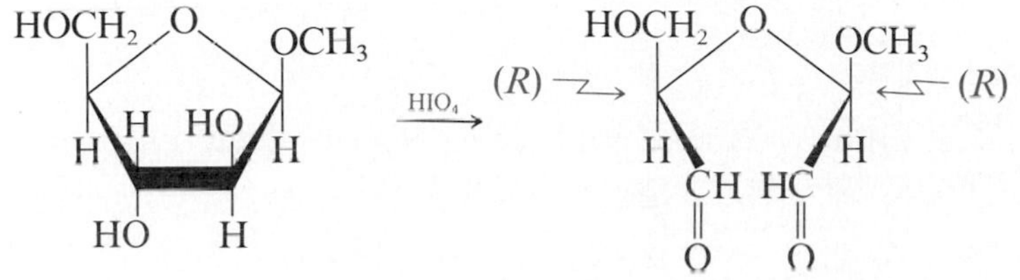

Methyl β-D-arabinofuranoside

Dialdehyde product; R configuration
at both chiral centers

The two possible dialdehyde products are diastereomers and both are known compounds. Comparing the periodate oxidation product with authentic dialdehyde of known configuration established the configuration at the anomeric center as α.

PROBLEM 27.18 Give the products of periodic acid oxidation of each of the following. How many moles of reagent will be consumed per mole of substrate in each case?

(a) D-Arabinose

(b) D-Ribose

(c) Methyl β-D-glucopyranoside

(d)

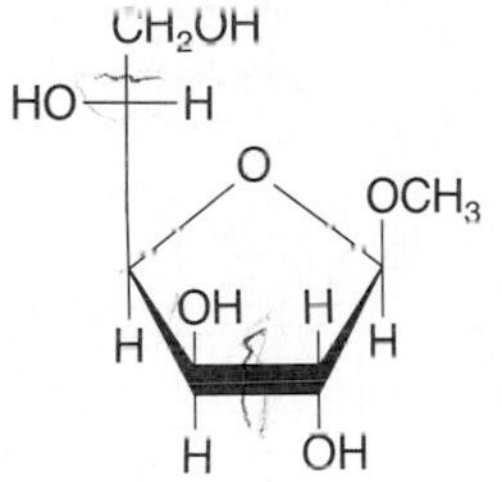

SAMPLE SOLUTION (a) The α-hydroxy aldehyde unit at the end of the sugar chain is cleaved, as well as all the vicinal diol functions. Four moles of periodic acid are required per mole of D-arabinose. Four moles of formic acid and one mole of formaldehyde are produced.

D-Arabinose, showing
points of cleavage
by periodic acid;
each cleavage
requires one
equivalent of HIO_4.

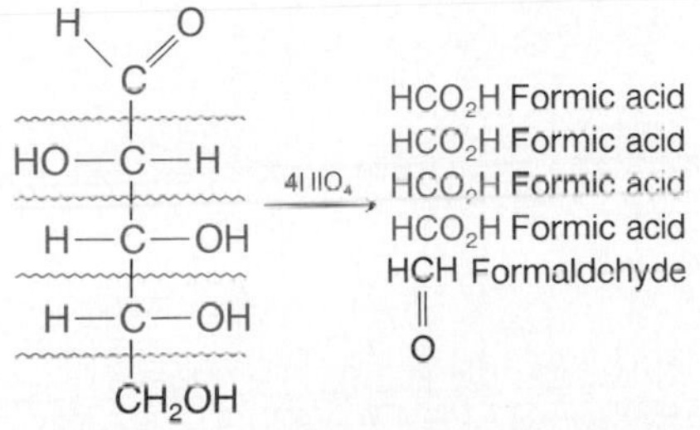

HCO_2H Formic acid
HCO_2H Formic acid
HCO_2H Formic acid
HCO_2H Formic acid
HCH Formaldehyde

TABLE 27.2
Summary of Reactions of Carbohydrates

Reaction (section) and comments	Example

Transformations of the carbonyl group

Reduction (Section 27.18) The carbonyl group of aldoses and ketoses is reduced by sodium borohydride or by catalytic hydrogenation. The products are called *alditols*.

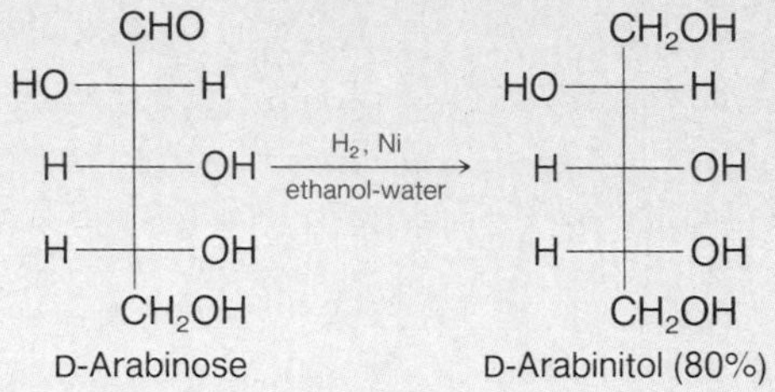

D-Arabinose → D-Arabinitol (80%)
(H_2, Ni, ethanol-water)

Oxidation with Benedict's reagent (Section 27.19) Sugars that contain a free hemiacetal function are called reducing sugars. They react with cupric sulfate in a sodium citrate–sodium carbonate buffer (Benedict's reagent) to form a red precipitate of cuprous oxide. Used as a qualitative test for reducing sugars.

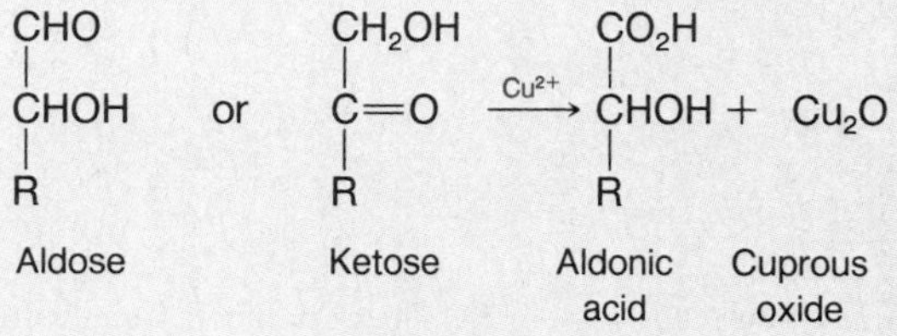

Aldose or Ketose $\xrightarrow{Cu^{2+}}$ Aldonic acid + Cuprous oxide (Cu_2O)

Oxidation with bromine (Section 27.19) When a preparative method for an aldonic acid is required, bromine oxidation is used. The aldonic acid is formed as its lactone. More properly described as a reaction of the anomeric hydroxyl group than of a free aldehyde.

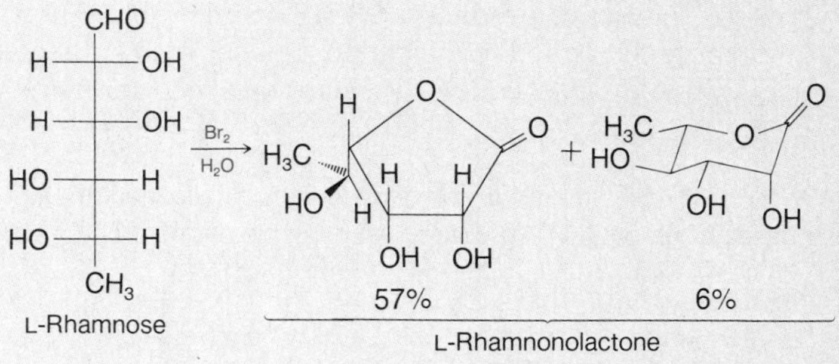

L-Rhamnose $\xrightarrow{Br_2, H_2O}$ L-Rhamnonolactone (57% + 6%)

Chain extension by way of cyanohydrin formation (Section 27.20) The Kiliani-Fischer synthesis proceeds by nucleophilic addition of HCN to an aldose, followed by conversion of the cyano group to an aldehyde. A mixture of stereoisomers results; the two aldoses are epimeric at C-2.

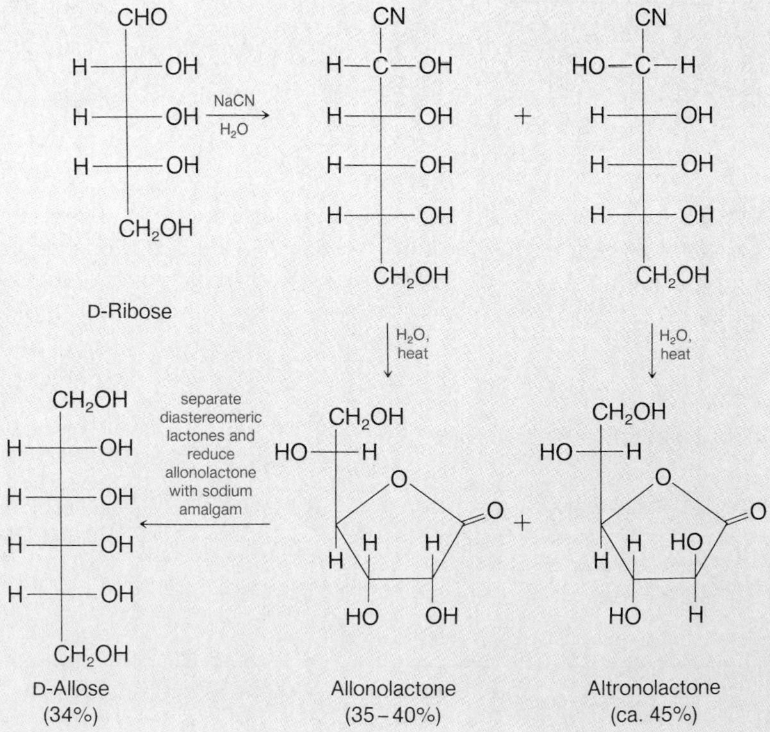

D-Ribose $\xrightarrow{NaCN, H_2O}$

separate diastereomeric lactones and reduce allonolactone with sodium amalgam

D-Allose (34%)   Allonolactone (35–40%) + Altronolactone (ca. 45%)
(H_2O, heat)

1040

TABLE 27.2 (continued)

Reaction (section) and comments	Example
Osazone formation (Section 27.21) Aldoses and ketoses react, by way of their open-chain forms, with three equivalents of phenylhydrazine to yield osazones. Most osazones crystallize readily and are used to characterize carbohydrates.	α-D-Ribofuranose, or β-D-Ribofuranose, or α-D-Ribopyranose, or β-D-Ribopyranose $\xrightarrow{3C_6H_5NHNH_2}$ Phenylosazone of D-ribose
Enediol formation (Section 27.22) Enolization of an aldose or ketose gives an enediol. Enediols can revert to aldoses or ketoses with loss of stereochemical integrity at the α-carbon atom.	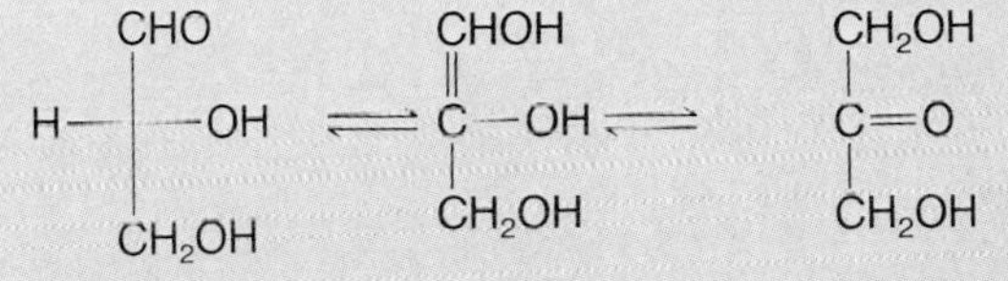 D-Glyceraldehyde Enediol 1,3-Dihydroxyacetone

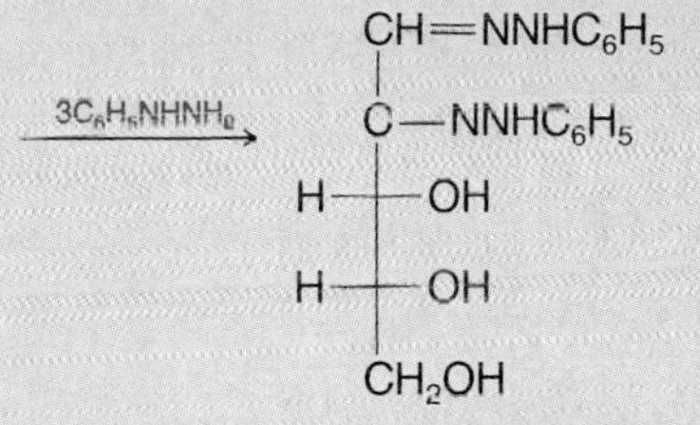

Reactions of the hydroxyl group

Reaction (section) and comments	Example
Acylation (Section 27.23) Esterification of the available hydroxyl groups occurs when carbohydrates are treated with acylating agents.	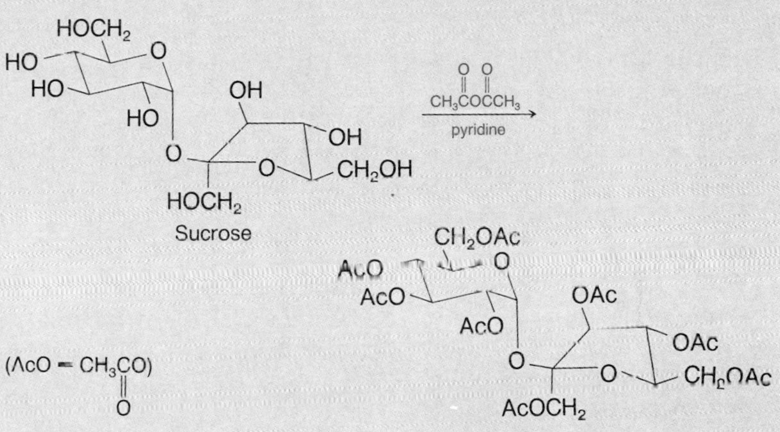 Sucrose $\xrightarrow[\text{pyridine}]{CH_3COCCH_3}$ $(AcO = CH_3CO)$ Sucrose octaacetate (66%)
Alkylation (Section 27.23) Alkyl halides react with carbohydrates to form ethers at the available hydroxyl groups. An application of the Williamson ether synthesis to carbohydrates.	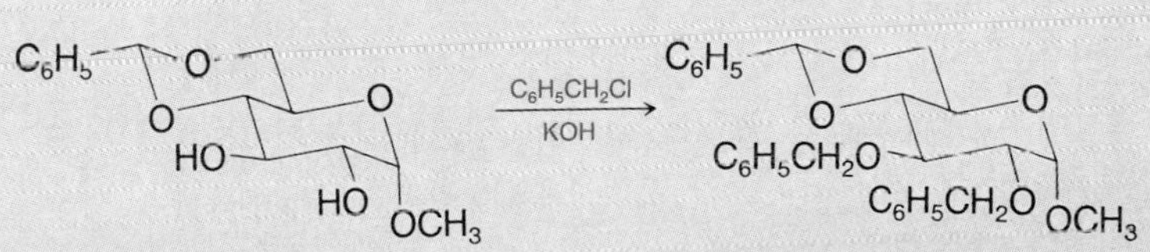 Methyl 4,6-O-benzylidene-α-D-glucopyranoside $\xrightarrow[KOH]{C_6H_5CH_2Cl}$ Methyl 2,3-di-O-benzyl-4,6-O-benzylidene-α-D-glucopyranoside (92%)
Periodic acid oxidation (Section 27.24) Vicinal diol and α-hydroxy carbonyl functions in carbohydrates are cleaved by periodic acid. Used analytically as a tool for structure determination.	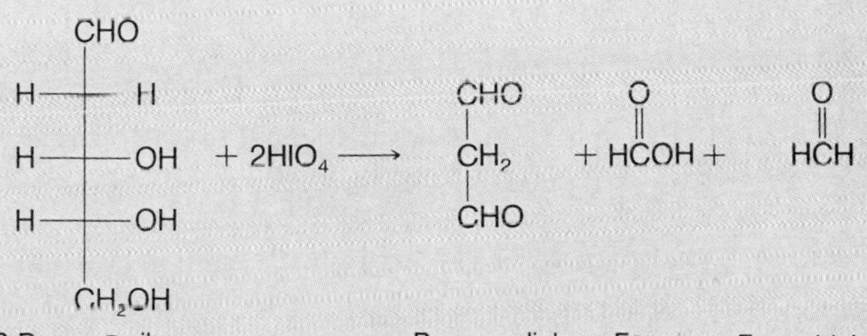 2-Deoxy-D-ribose $+ 2HIO_4 \longrightarrow$ Propanedial $+$ HCOH $+$ HCH Formic acid Formaldehyde

27.25 SUMMARY

Carbohydrates are marvelous molecules! In most of them, every carbon bears a functional group and the nature of these functional groups changes as the molecule interconverts between open-chain and cyclic hemiacetal forms. Any approach to understanding carbohydrates must begin with structure.

Fischer projection formulas are a convenient way to represent configurational relationships in molecules with multiple chiral centers. It is important to remember that vertical bonds are directed away from you, horizontal bonds point toward you.

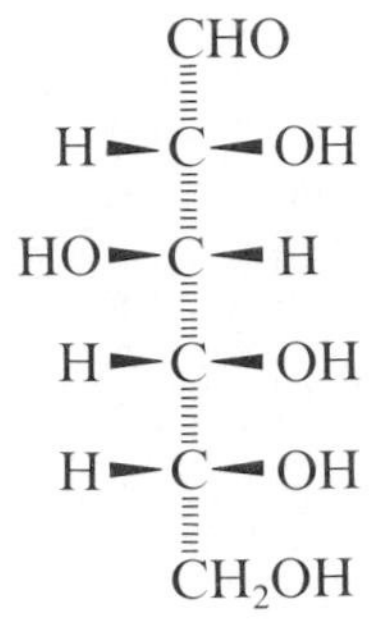

When the hydroxyl group at the highest-numbered chiral center is to the right, the sugar belongs to the D series; when this hydroxyl is to the left, the compound is an L sugar.

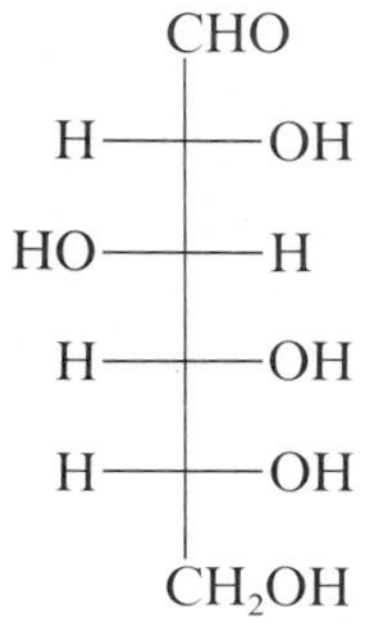

Carbohydrates that terminate in carbonyl groups are *aldoses;* those that contain a ketone carbonyl in the main carbon chain are *ketoses.* Both aldoses and ketoses exist as cyclic hemiacetals. Six-membered cyclic hemiacetals are called *pyranose* forms while five-membered ones are called *furanose* forms.

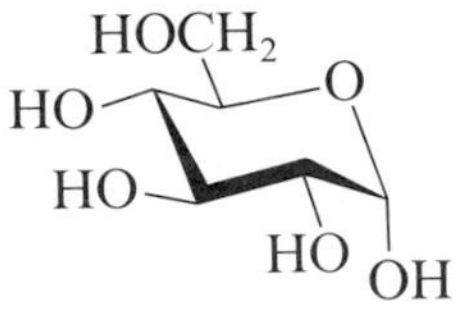

α-Pyranose form of the aldohexose D-glucose

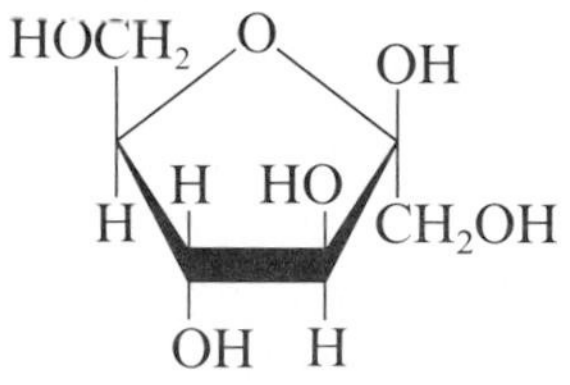

β-Furanose form of the ketohexose D-fructose

The symbols α and β refer to the configuration at the anomeric carbon. When the direction of the hydroxyl group at this carbon is "down" in the D series, this configuration is α; when the hydroxyl group direction is "up" in the D series, the configuration is β. The symbols are reversed in the L series.

Glycosides are acetals, compounds in which the anomeric hydroxyl group has been replaced by an alkoxy group. Glycosides are easily prepared in the laboratory by allowing a carbohydrate and an alcohol to stand in the presence of an acid catalyst. Glycosides are hydrolyzed to an alcohol and a carbohydrate either in aqueous acid or in the presence of enzymes called glycosidases.

D-Glucose + ROH $\underset{H^+ \text{ or enzyme}}{\overset{H^+ \text{ or enzyme}}{\rightleftharpoons}}$ 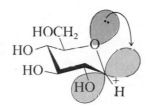 OR + H$_2$O

Glycoside

Both reactions, glycoside formation and glycoside hydrolysis, proceed through a carbocation intermediate. Carbocation formation at the anomeric position is facilitated by electron release from the ring oxygen.

An unshared electron pair on the ring oxygen stabilizes a carbocation at the anomeric position.

Disaccharides are molecules in which two monosaccharides are joined by a glycoside bond. *Polysaccharides* have many monosaccharide units connected through glycosidic linkages. Complete hydrolysis of disaccharides and polysaccharides cleaves the glycoside bonds, yielding the free monosaccharide components.

Carbohydrates undergo chemical reactions characteristic of aldehydes and ketones, alcohols, diols, etc., depending on their structure. A review of the reactions described in this chapter is presented in Table 27.2. While some of the reactions have synthetic value, many of them are used in analysis and structure determination.

PROBLEMS

27.19 Refer to the Fischer projection formula of D-(+)-xylose in Figure 27.3 (Section 27.4) and give structural formulas for

(a) (−)-Xylose (Fischer projection)
(b) D-Xylitol
(c) β-D-Xylopyranose
(d) α-L-Xylofuranose
(e) Methyl α-L-xylofuranoside
(f) D-Xylose phenylhydrazone
(g) D-Xylose phenylosazone
(h) D-Xylonic acid (open-chain Fischer projection)
(i) δ-Lactone of D-xylonic acid
(j) γ-Lactone of D-xylonic acid
(k) D-Xylaric acid (open-chain Fischer projection)

27.20 From among the carbohydrates shown in Figure 27.3, choose the D-aldohexoses that yield

(a) An optically inactive product on reduction with sodium borohydride
(b) An optically inactive product on oxidation with bromine
(c) An optically inactive product on oxidation with nitric acid

(d) The same enediol
(e) The same phenylosazone

 27.21 Write the Fischer projection formula of the open-chain form of each of the following:

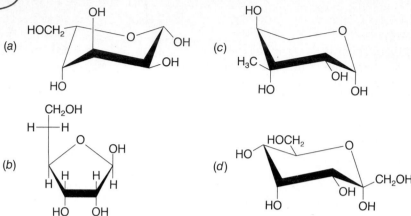

(a)

(b)

(c)

(d)

27.22 From among the carbohydrates shown in Problem 27.21 choose the one(s) that

(a) Belong to the L-series
(b) Are deoxy sugars
(c) Are branched-chain sugars
(d) Are ketoses
(e) Are furanose forms
(f) Have the α configuration at their anomeric carbon

27.23 How many pentuloses are possible? Write their Fischer projection formulas.

 27.24 The Fischer projection formula of the branched-chain carbohydrate D-apiose has been presented in Section 27.12.

(a) How many chiral centers are there in the open-chain form of D-apiose?
 (b) Does D-apiose form an optically active osazone?
(c) Does D-apiose form an optically active alditol on reduction?
(d) How many chiral centers are there in the furanose forms of D-apiose?
(e) How many stereoisomeric furanose forms of D-apiose are possible? Write their Haworth formulas.

27.25 Treatment of D-mannose with methanol in the presence of an acid catalyst yields four isomeric products having the molecular formula $C_7H_{14}O_6$. What are these four products?

27.26 Maltose and cellobiose (Section 27.14) are examples of disaccharides derived from D-glucopyranosyl units.

(a) How many other disaccharides are possible that meet this structural requirement?
(b) How many of these are reducing sugars?

 **27.27** Gentiobiose has the molecular formula $C_{12}H_{22}O_{11}$ and has been isolated from gentian root and by hydrolysis of amygdalin. Gentiobiose exists in two different forms, one melting at 86°C and the other at 190°C. The lower-melting form is dextrorotatory ($[\alpha]_D{}^{22} +16°$), while the higher-melting one is levorotatory ($[\alpha]_D{}^{22} -6°$). The rotation of an aqueous solution of either form, however, gradually changes until a final value of $[\alpha]_D{}^{22} +9.6°$ is observed. Hydrolysis of gentiobiose is efficiently catalyzed by emulsin and produces two moles of D-glucose

per mole of gentiobiose. Gentiobiose forms an octamethyl ether, which on hydrolysis in dilute acid yields 2,3,4,6-tetra-O-methyl-D-glucose and 2,3,4-tri-O-methyl-D-glucose. What is the structure of gentiobiose?

27.28 *Cyanogenic glycosides* are potentially toxic because they liberate hydrogen cyanide on enzyme-catalyzed or acidic hydrolysis. Give a mechanistic explanation for this behavior for the specific cases of

(a)

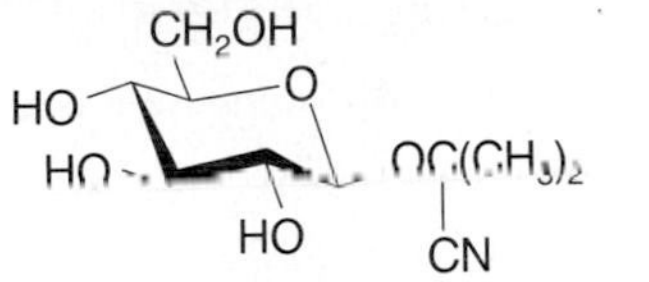

CH_2OH

Linamarin

(b)

CO_2H

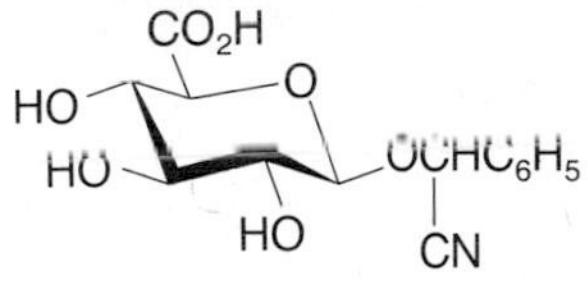

Laetrile

27.29 The more stable anomers of the pyranose forms of D-glucose, D-mannose, and D-galactose are shown below.

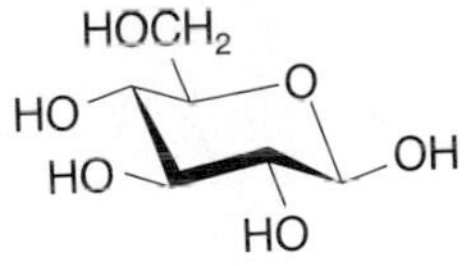

β-D-Glucopyranose
(64% at equilibrium)

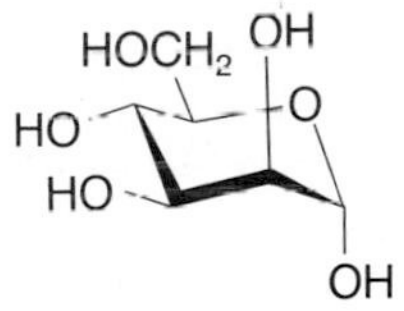

α-D-Mannopyranose
(68% at equilibrium)

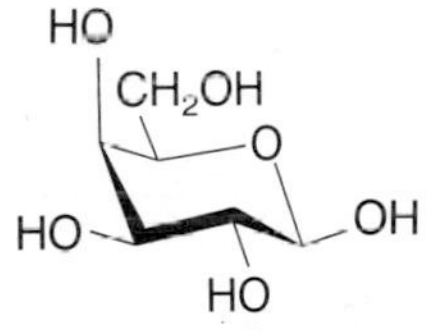

β-D-Galactopyranose
(64% at equilibrium)

Based on these empirical observations and your own knowledge of steric effects in six-membered rings, predict the preferred form (α- or β-pyranose) at equilibrium in aqueous solution for each of the following:

(a) D-Gulose

(b) D-Talose

(c) D-Xylose

(d) D-Lyxose

27.30 Basing your answers on the general mechanism for the first stage of acid-catalyzed acetal hydrolysis

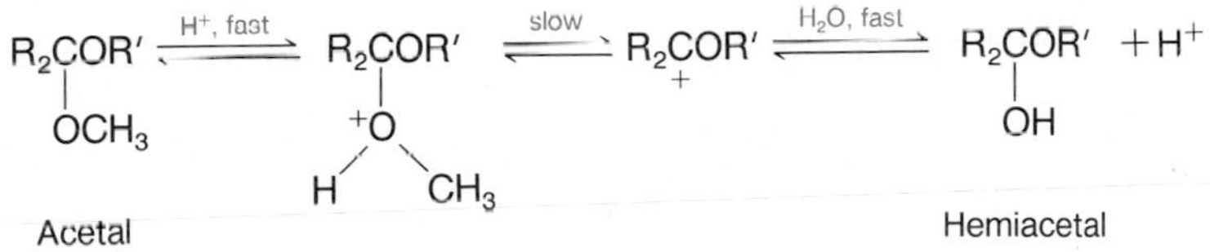

$$R_2COR' \underset{H^+, \text{ fast}}{\rightleftharpoons} R_2COR' \underset{\text{slow}}{\rightleftharpoons} R_2COR' \underset{H_2O, \text{ fast}}{\rightleftharpoons} R_2COR' + H^+$$

Acetal   Hemiacetal

suggest reasonable explanations for the following observations:

(a) Methyl α-D-fructofuranoside (A) undergoes acid-catalyzed hydrolysis some 10^5 times faster than methyl α-D-glucofuranoside (D).

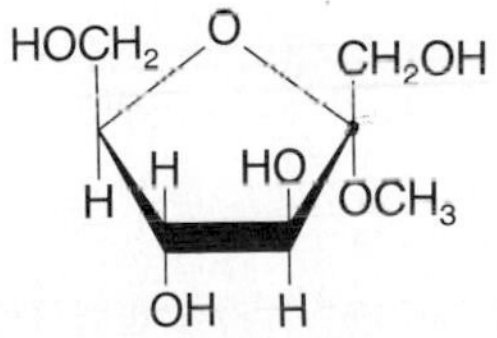

Compound A

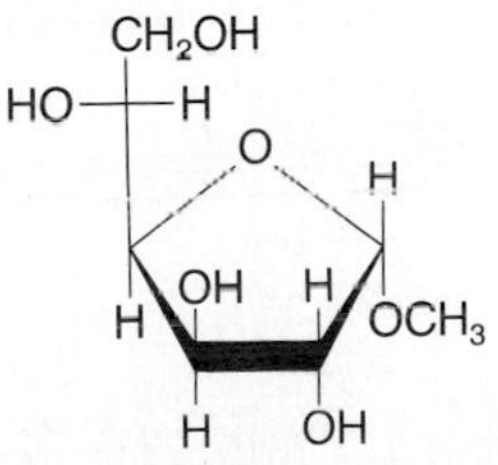

Compound B

(b) The β-methyl glucopyranoside of 2-deoxy-D-glucose (C) undergoes hydrolysis several thousand times faster than that of D-glucose (D).

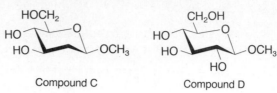

Compound C   Compound D

27.31 D-Altrosan is converted to D-altrose by dilute aqueous acid. Suggest a reasonable mechanism for this reaction.

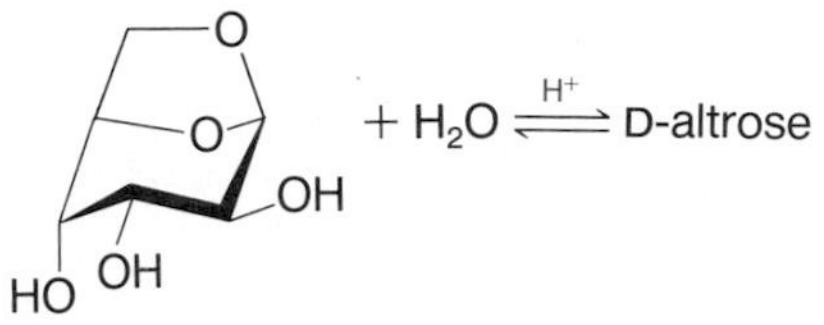

$+ H_2O \xrightleftharpoons[\text{}]{H^+}$ D-altrose

D-Altrosan

27.32 When D-galactose is heated at 165°C, a small amount of compound E was isolated.

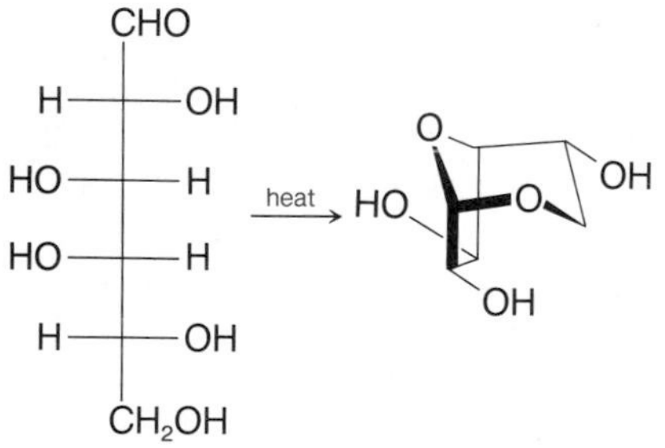

CHO
H——OH
HO——H
HO——H
H——OH
CH$_2$OH

$\xrightarrow{\text{heat}}$

D-Galactose   Compound E

The structure of compound E was established, in part, by converting it to known compounds. Treatment of E with excess methyl iodide in the presence of silver oxide, followed by hydrolysis with dilute hydrochloric acid, gave a trimethyl ether of D-galactose. Comparing this trimethyl ether with known trimethyl ethers of D-galactose allowed the structure of compound E to be deduced.

How many trimethyl ethers of D-galactose are there? Which one is the same as the product derived from E?

27.33 Phlorizin is obtained from the root bark of apple, pear, cherry, and plum trees. It has the molecular formula $C_{21}H_{24}O_{10}$ and yields compound F and D-glucose on hydrolysis in the presence of emulsin. When phlorizin was treated with excess methyl iodide in the presence of potassium carbonate, then subjected to acid-catalyzed hydrolysis, compound G was obtained. Deduce the structure of phlorizin from this information.

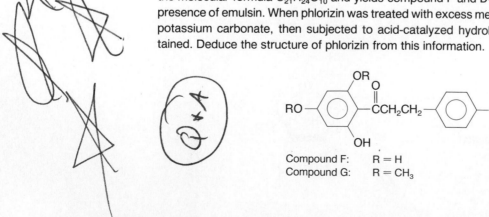

Compound F: R = H
Compound G: R = CH$_3$

HARD

27.34 Emil Fischer's determination of the structure of glucose was carried out as the nine-teenth century ended and the twentieth began. The structure of no other sugar was known at that time, and none of the spectroscopic techniques that aid organic analysis were then available. All Fischer had was information from chemical transformations, polarimetry, and his own intellect. Fischer realized that (+)-glucose could be represented by 16 possible stereo-structures. By arbitrarily assigning a particular configuration to the chiral center at C-5, the configurations of C-2, C-3, and C-4 could be determined relative to it, reducing the number of structural possibilities to eight. Thus, he started with a structural representation shown as below, where C-5 of (+)-glucose has what is now known as the D-configuration.

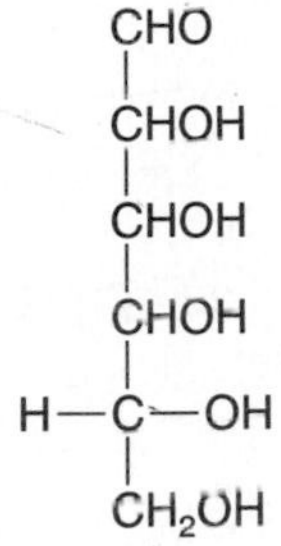

```
            CHO
             |
            CHOH
             |
            CHOH
             |
            CHOH
             |
       H — C — OH
             |
            CH₂OH
```

Eventually, Fischer's arbitrary assumption proved to be correct, and the structure he pro-posed for (+)-glucose is correct in an absolute as well as a relative sense. The following exercise utilizes information available to Fischer and leads you through a reasoning process similar to that employed in his determination of the structure of (+)-glucose. See if you can work out the configuration of (+)-glucose from the information provided, assuming the config-uration of C-5 as shown above.

1. Chain extension of the aldopentose (−)-arabinose by way of the derived cyanohydrin gave a mixture of (+)-glucose and (+)-mannose.
2. Oxidation of (−)-arabinose with warm nitric acid gave an optically active aldaric acid.
3. Both (+)-glucose and (+)-mannose were oxidized to optically active aldaric acids with nitric acid.
4. There is another sugar, (+)-gulose, that gives the same aldaric acid on oxidation as does (+)-glucose.

DO TONIGHT

CHAPTER

28

ACETATE-DERIVED NATURAL PRODUCTS

Photochemical energy is stored as chemical energy by the process of *photosynthesis* (Section 27.4). Carbon dioxide and water combine in the presence of visible light and chlorophyll to form carbohydrates. The energy that is stored in the carbohydrate is released during carbohydrate *metabolism*.

$$C_6H_{12}O_6 + 6O_2 \longrightarrow 6CO_2 + H_2O + energy$$

Glucose   Oxygen   Carbon dioxide   Water

During one stage of carbohydrate metabolism, called *glycolysis*, glucose is converted to lactic acid. Pyruvic acid is an intermediate.

$$C_6H_{12}O_6 \longrightarrow CH_3\overset{\displaystyle O}{\overset{\|}{C}}CO_2H \longrightarrow CH_3\underset{\displaystyle OH}{CH}CO_2H$$

Glucose   Pyruvic acid   Lactic acid
              (2-oxopropanoic acid)   (2-hydroxypropanoic acid)

In most biochemical reactions the pH of the medium is close to 7, a condition under which carboxylic acids are nearly completely converted to their conjugate bases. Thus, it is common practice in biological chemistry to specify the derived carboxylate anion rather than the carboxylic acid itself. For example, we say that glycolysis leads to lactate by way of pyruvate.

Pyruvate is utilized by living systems in a number of different ways. One pathway, the one leading to lactate and beyond, is concerned with energy storage and production. This is not the only pathway available to pyruvate, however. A significant fraction of it is converted to acetate for use as a starting material in the *biosynthesis* of more complex substances. These *acetate-derived natural products* play essential roles in the chemistry of life. Their diverse structural types and the biosynthetic thread that connects all of them to acetate are the principal concerns of this chapter.

We begin with a brief description of the biosynthetic origin of acetate.

28.1 ACETYL COENZYME A

The form in which acetate is utilized in most of its important biochemical reactions is acetyl coenzyme A (Figure 28.1a). Acetyl coenzyme A is a *thioester* (Section 21.12). Its formation from pyruvate involves several steps and is summarized in the overall equation:

$$CH_3CCOH + CoASH + NAD^+ \longrightarrow CH_3CSCoA + NADH + CO_2 + H^+$$

Pyruvic acid	Coenzyme A	Oxidized form of nicotinamide adenine dinucleotide	Acetyl coenzyme A	Reduced form of nicotinamide adenine dinucleotide	Carbon dioxide	Proton

All the individual steps are catalyzed by enzymes. NAD^+ (Section 16.13) is required as an oxidizing agent, and coenzyme A (Figure 28.1b) is the acetyl group acceptor. Coenzyme A is a *thiol;* its chain terminates in a *sulfhydryl* (-SH) group. Acetylation of the sulfhydryl group of coenzyme A gives acetyl coenzyme A.

Because sulfur does not donate electrons to an attached carbonyl group as well as oxygen does, compounds of the type $RCSR'$ are better acyl transfer agents than is $RCOR'$. They also contain a greater proportion of enol at equilibrium. Both properties are apparent in the properties of acetyl coenzyme A. In some of its reactions acetyl coenzyme A acts as an acetyl transfer agent, whereas in others the α carbon atom of the acetyl group is the reactive site.

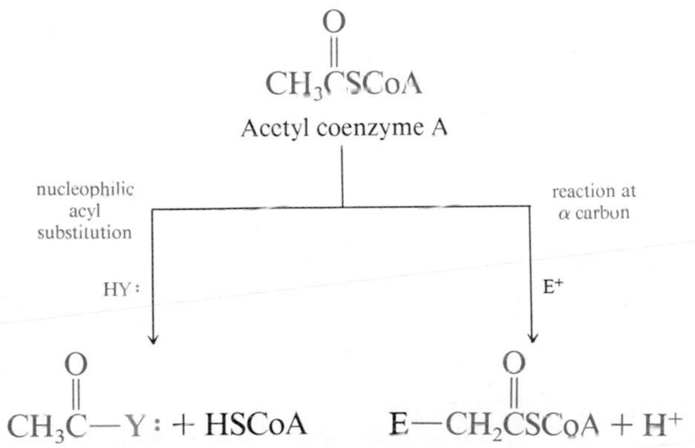

$$CH_3CSCoA$$
Acetyl coenzyme A

nucleophilic acyl substitution   reaction at α carbon

HY: E^+

$$CH_3C-Y: + HSCoA \qquad E-CH_2CSCoA + H^+$$

You will see numerous examples of both reaction types in the following sections. Keep in mind that in vivo reactions (reactions in living systems) are enzyme-catalyzed and occur at rates that are far greater than when the same transformations are carried out in vitro ("in glass") in the absence of enzymes. In spite of the rapidity with which enzyme-catalyzed reactions take place, the nature of these transformations is essentially the same as the fundamental processes of organic chemistry described throughout this text.

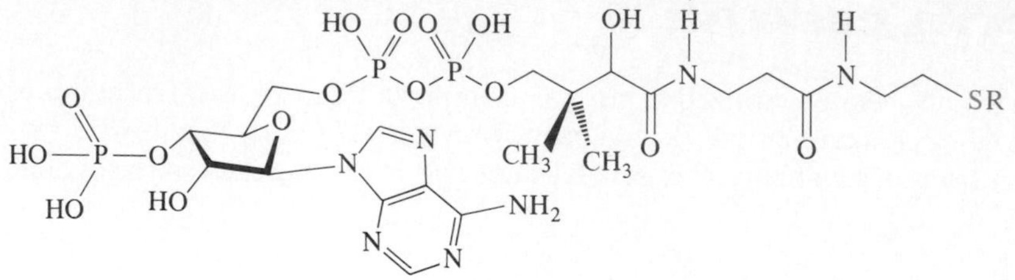

FIGURE 28.1 Structures of (a) acetyl coenzyme A and (b) coenzyme A.

(a) R = $\overset{O}{\overset{\|}{C}}CH_3$ Acetyl coenzyme A (abbreviation: $CH_3\overset{O}{\overset{\|}{C}}SCoA$)

(b) R = H   Coenzyme A (abbreviation: CoASH)

Among the natural products that are biosynthesized from acetate are a number of members of the *lipid* class. Lipids are naturally occurring substances that are soluble in nonpolar solvents. This is an operational, rather than a structural, distinction. Material from some living organism is shaken with a polar solvent (water or an alcohol-water mixture) and a nonpolar one (diethyl ether, hexane, or dichloromethane). Carbohydrates, proteins, nucleic acids, and related compounds are polar and do not dissolve in the nonpolar solvent—they either dissolve in the aqueous phase or remain behind undissolved. The portion of the natural material that dissolves in the nonpolar solvent is the lipid fraction.

Fats are one type of lipid. They have a number of functions in living systems, including that of energy storage. While carbohydrates serve as a source of readily available energy, an equal mass of fat delivers over twice the energy. It is more efficient for an organism to store energy in the form of fat because it requires less mass than storing the same amount of energy in carbohydrate molecules.

How living systems convert acetate to fats is an exceedingly complex story, one that is well understood in broad outline and becoming increasingly clear in detail as well. We will examine several aspects of this topic in the next few sections, focusing primarily on its structural and chemical features.

28.2 FATTY ACIDS

Fats and oils are naturally occurring mixtures of triacylglycerols. They differ in that fats are solids at room temperature while oils are liquids. Scientists generally ignore this distinction and refer to both groups of compounds as fats. Two typical triacylglycerols are shown in Figure 28.2. All three acyl groups in a triacylglycerol may be the same, all three may be different, or one may be different from the other two. The triacylglycerol shown in Figure 28.2a, 2-oleyl-1,3-distearoylglycerol, occurs naturally and comprises about 20 percent of the triacylglycerol mixture of cocoa butter. The presence of a cis double bond in the oleyl side chain interferes with the ability of neighboring triacylglycerol molecules to pack together into a crystal lattice and this substance is a relatively low-melting solid (mp 43°C). Hydrogenation of the double bond converts 2-oleyl-1,3-distearoylglycerol to *tristearin* (Figure 28.2b). Tristearin has a higher melting point (mp 72°C) than 2-oleyl-1,3-distearoylglycerol because all three of its acyl groups adopt the extended zigzag conformation that permits efficient crystal packing. Catalytic hydrogenation is commonly used in the food industry to transform readily available liquid vegetable oils to solid "shortenings."

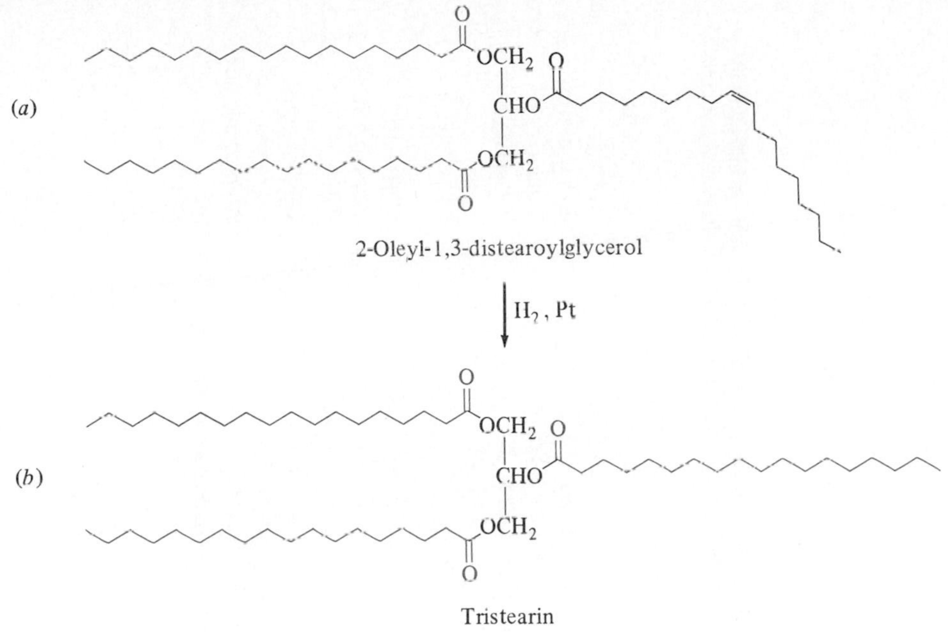

2-Oleyl-1,3-distearoylglycerol

$\downarrow$ H₂, Pt

Tristearin

FIGURE 28.2 The structures of two typical triacylglycerols. (a) 2-Oleyl-1,3-distearoylglycerol is a naturally occurring triacylglycerol found in cocoa butter. The cis double bond of its 2-acyl group gives the molecule a shape which interferes with efficient crystal packing. (b) Catalytic hydrogenation converts 2-oleyl-1,3-distearoylglycerol to tristearin. Tristearin has a higher melting point than 2-oleyl-1,3-distearoylglycerol.

Hydrolysis of fats yields glycerol and long-chain *fatty acids*. Thus, tristearin gives glycerol and three molecules of stearic acid (octadecanoic acid) on hydrolysis. Table 28.1 lists a few representative fatty acids. As the examples in Table 28.1 indicate, most naturally occurring fatty acids possess an even number of carbon atoms and an unbranched carbon chain. The carbon chain may be completely saturated or may incorporate one or more multiple bonds. When double bonds are present, they almost always have the cis (or Z) configuration. Acyl groups containing 14 to 20 carbon atoms are the most abundant in triacylglycerols.

Strictly speaking, the term "fatty acid" is restricted to those carboxylic acids that occur naturally in triacylglycerols. Many chemists and biochemists, however, refer to all unbranched carboxylic acids, irrespective of their origin and chain length, as fatty acids.

PROBLEM 28.1 What fatty acids are produced on hydrolysis of 2-oleyl-1,3-distearoylglycerol? What other triacylglycerol gives the same fatty acids and in the same proportions as 2-oleyl-1,3-distearoylglycerol?

In the form of triacylglycerols, fatty acids are found in both plants and animals, where they are biosynthesized from acetate by way of acetyl coenzyme A. The following section outlines the mechanism of fatty acid biosynthesis.

28.3 FATTY ACID BIOSYNTHESIS

The major elements of fatty acid biosynthesis may be described by considering the formation of butanoic acid from two molecules of acetyl coenzyme A. The "machinery" responsible for accomplishing this conversion is a complex of enzymes known as *fatty acid synthetase*. Certain portions of this complex, referred to as *acyl carrier protein* (ACP), bear a side chain that is structurally similar to coenzyme A. An

TABLE 28.1

Some Representative Fatty Acids

Structural formula	Systematic name	Common name
Saturated fatty acids		
$CH_3(CH_2)_{10}COOH$	Dodecanoic acid	Lauric acid
$CH_3(CH_2)_{12}COOH$	Tetradecanoic acid	Myristic acid
$CH_3(CH_2)_{14}COOH$	Hexadecanoic acid	Palmitic acid
$CH_3(CH_2)_{16}COOH$	Octadecanoic acid	Stearic acid
$CH_3(CH_2)_{18}COOH$	Icosanoic acid	Arachidic acid
Unsaturated fatty acids		
$CH_3(CH_2)_7CH{=}CH(CH_2)_7COOH$	(Z)-9-Octadecenoic acid	Oleic acid
$CH_3(CH_2)_4CH{=}CHCH_2CH{=}CH(CH_2)_7COOH$	(9Z,12Z)-9,12-Octadecadienoic acid	Linoleic acid
$CH_3CH_2CH{=}CHCH_2CH{=}CHCH_2CH{=}CH(CH_2)_7COOH$	(9Z,12Z,15Z)-9,12,15-Octadecatrienoic acid	Linolenic acid
$CH_3(CH_2)_4CH{=}CHCH_2CH{=}CHCH_2CH{=}CHCH_2CH{=}CH(CH_2)_3COOH$	(5Z,8Z,11Z,14Z)-5,8,11,14-Icosatetraenoic acid	Arachidonic acid

important early step in fatty acid biosynthesis is the transfer of the acetyl group from a molecule of acetyl coenzyme A to the sulfhydryl group of acyl carrier protein.

$$
\underset{\substack{\text{Acetyl} \\ \text{coenzyme A}}}{CH_3\overset{\displaystyle O}{\overset{\|}{C}}SCoA} + \underset{\substack{\text{Acyl carrier} \\ \text{protein}}}{HS-ACP} \longrightarrow \underset{\substack{S\text{-Acetyl acyl} \\ \text{carrier protein}}}{CH_3\overset{\displaystyle O}{\overset{\|}{C}}S-ACP} + \underset{\substack{\text{Coenzyme A}}}{HSCoA}
$$

PROBLEM 28.2 Using HSCoA and HS—ACP as abbreviations for coenzyme A and acyl carrier protein, respectively, write a structural formula for the tetrahedral intermediate in the above reaction.

A second molecule of acetyl coenzyme A reacts with carbon dioxide (actually bicarbonate ion at biological pH) to give malonyl coenzyme A.

$$
\underset{\substack{\text{Acetyl} \\ \text{coenzyme A}}}{CH_3\overset{\displaystyle O}{\overset{\|}{C}}SCoA} + \underset{\substack{\text{Bicarbonate}}}{HCO_3^-} \longrightarrow \underset{\substack{\text{Malonyl} \\ \text{coenzyme A}}}{{}^-O\overset{\displaystyle O}{\overset{\|}{C}}CH_2\overset{\displaystyle O}{\overset{\|}{C}}SCoA} + \underset{\substack{\text{Water}}}{H_2O}
$$

Formation of malonyl coenzyme A is followed by an acyl transfer reaction, which binds the malonyl group to the acyl carrier protein as a thioester.

$$
\underset{\substack{\text{Malonyl} \\ \text{coenzyme A}}}{{}^-O\overset{\displaystyle O}{\overset{\|}{C}}CH_2\overset{\displaystyle O}{\overset{\|}{C}}SCoA} + \underset{\substack{\text{Acyl carrier} \\ \text{protein}}}{HS-ACP} \longrightarrow \underset{\substack{S\text{-Malonyl acyl} \\ \text{carrier protein}}}{{}^-O\overset{\displaystyle O}{\overset{\|}{C}}CH_2\overset{\displaystyle O}{\overset{\|}{C}}S-ACP} + \underset{\substack{\text{Coenzyme A}}}{HSCoA}
$$

When both building block units are in place on the acyl carrier protein, carbon-carbon bond formation occurs between the α carbon atom of the malonyl group and the carbonyl carbon of the acetyl group. This is shown in step 1 of Figure 28.3. Carbon-carbon bond formation is accompanied by decarboxylation and produces a four-carbon acetoacetyl (3-oxobutanoyl) group bound to acyl carrier protein.

The acetoacetyl group is then transformed to a butanoyl group by the reaction sequence illustrated in steps 2 to 4 of Figure 28.3.

The four carbon atoms of the butanoyl group are derived from two molecules of acetyl coenzyme A. Carbon dioxide assists the reaction but is not incorporated into the product. The same carbon dioxide that is used to convert one molecule of acetyl coenzyme A to malonyl coenzyme A is regenerated in the decarboxylation step that accompanies carbon-carbon bond formation.

Successive repetitions of the steps shown in Figure 28.3 give unbranched acyl groups having 6, 8, 10, 12, 14, and 16 carbon atoms. In each case, chain extension occurs by reaction with a malonyl group bound to the acyl carrier protein. Thus, the biosynthesis of the 16-carbon acyl group of hexadecanoic (palmitic) acid can be represented by the equation:

$$
\underset{\substack{S\text{-Acetyl} \\ \text{acyl carrier} \\ \text{protein}}}{CH_3\overset{\displaystyle O}{\overset{\|}{C}}S-ACP} + 7\underset{\substack{S\text{-Malonyl} \\ \text{acyl carrier} \\ \text{protein}}}{HO\overset{\displaystyle O}{\overset{\|}{C}}CH_2\overset{\displaystyle O}{\overset{\|}{C}}S-ACP} \longrightarrow \underset{\substack{S\text{-Hexadecanoyl} \\ \text{acyl carrier} \\ \text{protein}}}{CH_3(CH_2)_{14}\overset{\displaystyle O}{\overset{\|}{C}}S-ACP} + 7\underset{\substack{\text{Carbon} \\ \text{dioxide}}}{CO_2} + 7\underset{\substack{\text{Acyl} \\ \text{carrier} \\ \text{protein}}}{HS-ACP}
$$

Step 1: An acetyl group is transferred to the α carbon atom of the malonyl group with evolution of carbon dioxide. Presumably, decarboxylation gives an enol, which attacks the acetyl group.

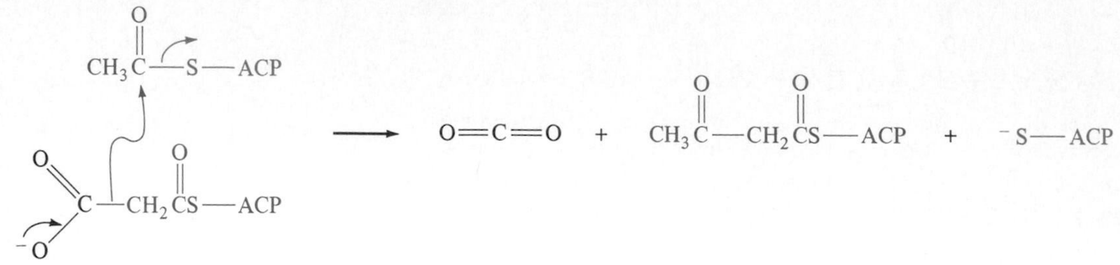

| Acetyl and malonyl groups bound to acyl carrier protein | Carbon dioxide | S-Acetoacetyl acyl carrier protein | Acyl carrier protein (anionic form) |

Step 2: The ketone carbonyl of the acetoacetyl group is reduced to an alcohol function. This reduction requires NADPH as a coenzyme. (NADPH is the phosphate ester of NADH and reacts similarly to it.)

$$CH_3\overset{O}{\overset{\|}{C}}CH_2\overset{O}{\overset{\|}{C}}S\text{---}ACP \;+\; NADPH \;+\; H_3O^+ \longrightarrow CH_3\underset{OH}{\overset{}{C}}HCH_2\overset{O}{\overset{\|}{C}}S\text{---}ACP \;+\; NADP^+ \;+\; H_2O$$

| S-Acetoacetyl acyl carrier protein | Reduced form of coenzyme | Hydronium ion | S-3-Hydroxybutanoyl acyl carrier protein | Oxidized form of coenzyme | Water |

Step 3: Dehydration of the β-hydroxy acyl group.

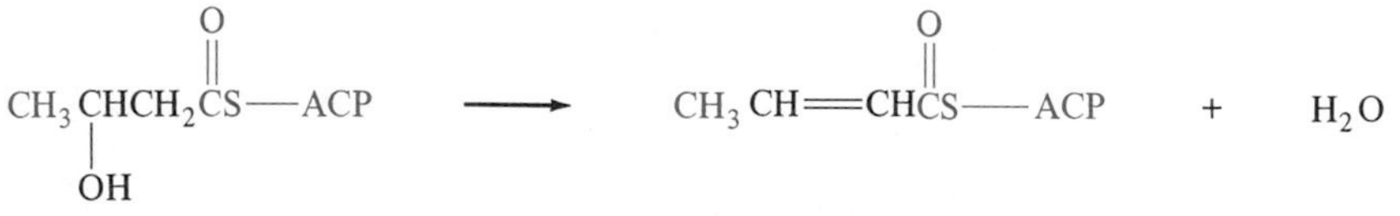

$$CH_3\underset{OH}{\overset{}{C}}HCH_2\overset{O}{\overset{\|}{C}}S\text{---}ACP \longrightarrow CH_3CH{=}CH\overset{O}{\overset{\|}{C}}S\text{---}ACP \;+\; H_2O$$

| S-3-Hydroxybutanoyl acyl carrier protein | S-2-Butenoyl acyl carrier protein | Water |

Step 4: Reduction of the double bond of the α, β-unsaturated acyl group. This step requires NADPH as a coenzyme.

$$CH_3CH{=}CH\overset{O}{\overset{\|}{C}}S\text{---}ACP \;+\; NADPH \;+\; H_3O^+ \longrightarrow CH_3CH_2CH_2\overset{O}{\overset{\|}{C}}S\text{---}ACP \;+\; NADP^+ \;+\; H_2O$$

| S-2-Butenoyl acyl carrier protein | Reduced form of coenzyme | Hydronium ion | S-Butanoyl acyl carrier protein | Oxidized form of coenzyme | Water |

FIGURE 28.3 Sequence of steps that describes the formation of a butanoyl group from acetyl and malonyl building blocks bound to acyl carrier protein.

PROBLEM 28.3 By analogy to the intermediates given in steps 1 to 4 of Figure 28.3, write the sequence of acyl groups that are attached to the acyl carrier protein in the con-version of $CH_3(CH_2)_{12}\overset{O}{\overset{\|}{C}}S\text{---}ACP$ to $CH_3(CH_2)_{14}\overset{O}{\overset{\|}{C}}S\text{---}ACP$.

This phase of fatty acid biosynthesis concludes with the transfer of the acyl group from acyl carrier protein to coenzyme A. The resulting acyl coenzyme A molecules can then undergo a number of subsequent biological transformations. One such transformation is chain extension, leading to acyl groups with more than 16 carbons. Another is the introduction of one or more carbon-carbon double bonds. A third is acyl transfer from sulfur to oxygen to form esters such as triacylglycerols. The process by which acyl coenzyme A molecules are converted to triacylglycerols involves a type of intermediate called a *phospholipid* and is discussed in the following section.

28.4 PHOSPHOLIPIDS

Triacylglycerols arise, not by acyl transfer to glycerol itself, but by a sequence of steps in which the first stage is acyl transfer to L-glycerol 3-phosphate (from reduction of dihydroxyacetone 3-phosphate, formed as described in Section 27.22). The product of this stage is called a *phosphatidic acid*.

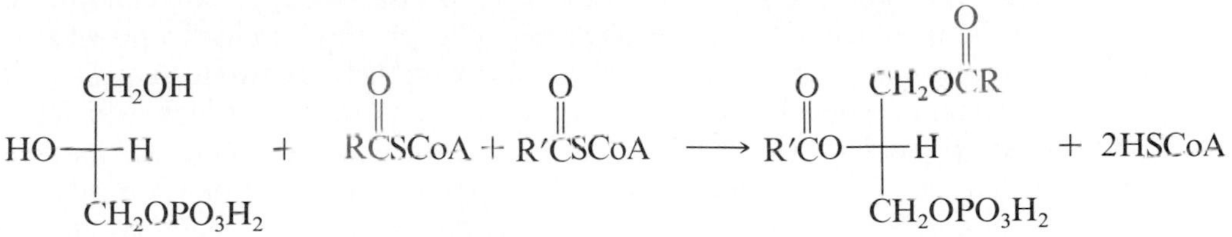

L-Glycerol 3-phosphate + Two acyl coenzyme A molecules (R and R' may be the same or they may be different) → Phosphatidic acid + Coenzyme A

PROBLEM 28.4 What is the absolute configuration (*R* or *S*) of L-glycerol 3-phosphate? What must be the absolute configuration of the naturally occurring phosphatidic acids biosynthesized from it?

Hydrolysis of the phosphate ester function of the phosphatidic acid gives a diacylglycerol, which then reacts with a third acyl coenzyme A molecule to produce a triacylglycerol.

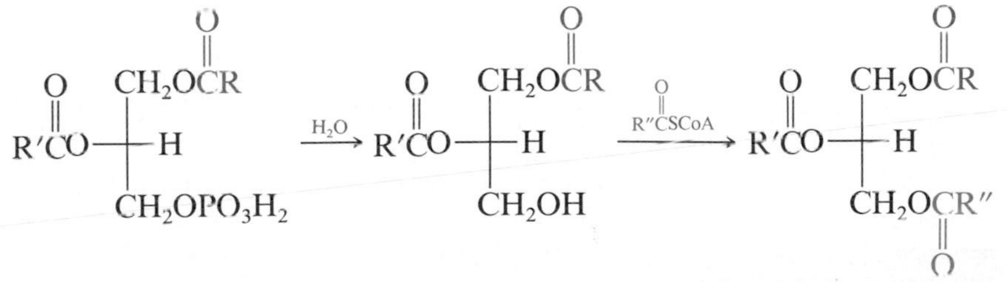

Phosphatidic acid → Diacylglycerol → Triacylglycerol

Phosphatidic acids not only are intermediates in the biosynthesis of triacylglycerols but also are biosynthetic precursors of other members of a group of compounds called *phosphoglycerides* or *glycerol phosphatides*. Phosphorus-containing derivatives of lipids are known as *phospholipids*, and phosphoglycerides are one type of phospholipid.

One important phospholipid is *phosphatidylcholine*, also called *lecithin*. Phosphatidylcholine is a mixture of diesters of phosphoric acid. One ester function is derived from a diacylglycerol whereas the other is a choline $[OCH_2CH_2\overset{+}{N}(CH_3)_3]$ unit.

$$
\begin{array}{c}
\underset{\underset{\displaystyle R'C}{\overset{\displaystyle O}{\parallel}}}{}\!\!-\!\!O\!\!-\!\!\begin{array}{c} CH_2OCR \\ | \\ H \\ | \\ CH_2OPO_2^- \end{array}\!\!-\!\! \\
\underset{OCH_2CH_2\overset{+}{N}(CH_3)_3}{}
\end{array}
$$

Phosphatidylcholine
(R and R′ are usually different)

We have seen in Section 20.5 that anions of fatty acids spontaneously associate in water to form spherical micelles. The polar carboxylate groups are located on the surface of the micelle, where they are solvated by water molecules. Water is excluded from the interior of the micelle, where attractive van der Waals forces cause the long-chain alkyl groups to associate with one another. Phosphatidylcholine possesses a polar "head group" (the positively charged choline and negatively charged phosphate units) and two nonpolar "tails" (the acyl groups). Under certain conditions, such as at the interface of two aqueous phases, phosphatidylcholine forms what is called a *lipid bilayer,* as shown in Figure 28.4. Because there are two long-chain acyl groups in each molecule, the most stable assembly is an extended sheet in which the polar groups are solvated by water molecules at the top and bottom surfaces and the lipophilic acyl groups are stacked in interlocking pairs in the interior of the bilayer.

Phosphatidylcholine is one of the principal components of cell membranes. It is believed that these membranes are composed of lipid bilayers analogous to those of Figure 28.4. Nonpolar materials can diffuse through the bilayer from one side to the other relatively easily; polar materials, particularly metal ions such as Na^+, K^+, and Ca^{2+}, cannot. The transport of metal ions through a membrane is usually assisted by certain proteins present in the lipid bilayer, which contain a metal ion binding site surrounded by a lipophilic exterior. The metal ion is picked up at one side of the lipid bilayer and delivered at the other, surrounded at all times by a polar environment on its passage through the hydrocarbonlike interior of the membrane. Ionophore antibiotics such as monensin (Section 17.5) disrupt the normal functioning of cells by facilitating metal ion transport across cell membranes.

28.5 WAXES

Waxes are solid materials that make up part of the protective coatings of a number of things, including the leaves of plants, the fur of animals, and the feathers of birds. They are usually mixtures of esters in which both the alkyl and acyl group are unbranched and contain a dozen or more carbon atoms. Beeswax, for example, contains the ester triacontyl hexadecanoate as one component of a complex mixture of hydrocarbons, alcohols, and esters.

$$
CH_3(CH_2)_{14}\overset{\overset{\displaystyle O}{\parallel}}{C}OCH_2(CH_2)_{28}CH_3
$$

Triacontyl hexadecanoate

PROBLEM 28.5 Spermaceti is a wax obtained from the sperm whale. It contains, among other materials, an ester known as *cetyl palmitate,* which is used as an emollient in

FIGURE 28.4 Schematic drawing of a phospholipid bilayer.

Water

$^+N(CH_3)_3$

Hydrophilic "head" groups

Lipophilic "tails"

Hydrophilic "head" groups

$(CH_3)_3\overset{+}{N}$

Water

a number of soaps and cosmetics. The systematic name for cetyl palmitate is *hexadecyl hexadecanoate*. Write a structural formula for this substance.

Fatty acids normally occur naturally as esters; fats, oils, phospholipids, and waxes all are unique types of fatty acid esters. There is, however, an important class of fatty acid derivatives that exists and carries out its biological role in the form of the free acid. This class of fatty acid derivatives is described in the following section.

28.6 PROSTAGLANDINS

Research in physiology carried out in the 1930s established that the lipid fraction of semen contains small amounts of substances that exert powerful effects on smooth muscle. Sheep prostate glands proved to be a convenient source of this material and yielded a mixture of structurally related substances referred to collectively as *prostaglandins*. We now know that prostaglandins are present in almost all animal tissues, where they carry out a variety of regulatory functions.

Prostaglandins are extremely potent substances and exert their physiological effects at very small concentrations. Because of this their isolation was difficult, and it was not until 1960 that the first members of this class, designated PGE_1 and $PGF_{1\alpha}$ (Figure 28.5), were obtained as pure compounds. More than a dozen structurally related prostaglandins have since been isolated and identified. All the prostaglandins are 20-carbon carboxylic acids and contain a cyclopentane ring. All have hydroxyl groups at C-11 and C-15 (for the numbering of the positions in prostaglandins, see Figure 28.5). Prostaglandins belonging to the F series have an additional hydroxyl group at C-9, while a carbonyl function is present at this position in the various PGE's. The subscript numerals in their abbreviated names indicate the number of double bonds.

Prostaglandins are believed to arise from unsaturated C_{20}-carboxylic acids by the process illustrated in Figure 28.6 for the biosynthesis of PGE_2 from arachidonic acid. Studies with isotopically enriched oxygen (^{18}O—^{18}O) have shown that all three oxygens of PGE_2 arise from O_2. The first stage of the biosynthesis is the formation of an *endoperoxide* (PGG_2) by what is believed to be a free-radical–induced carbon-carbon bond-forming reaction (steps 1 through 4). This stage is catalyzed by a *cyclooxygenase* enzyme. The hydroperoxide function of PGG_2 is reduced in step 5 to the endoperoxide alcohol (PGH_2). In the second stage of the process, the oxygen-oxygen bond of PGH_2 is cleaved, with formation of a ketone carbonyl at C-9 and a secondary hydroxyl at C-11. The product is PGE_2. Reductive cleavage of the endoperoxide gives the C-9, C-11 diol function of a prostaglandin known as $PGF_{2\alpha}$.

Mammals cannot biosynthesize arachidonic acid directly. They obtain linoleic acid (Table 28.1) from vegetable oils in their diet and extend the carbon chain of

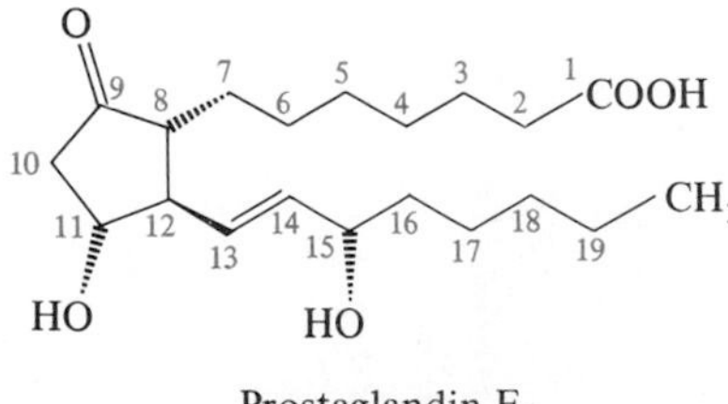

FIGURE 28.5 Structures of two representative prostaglandins. The numbering scheme is illustrated in the structure of PGE_1.

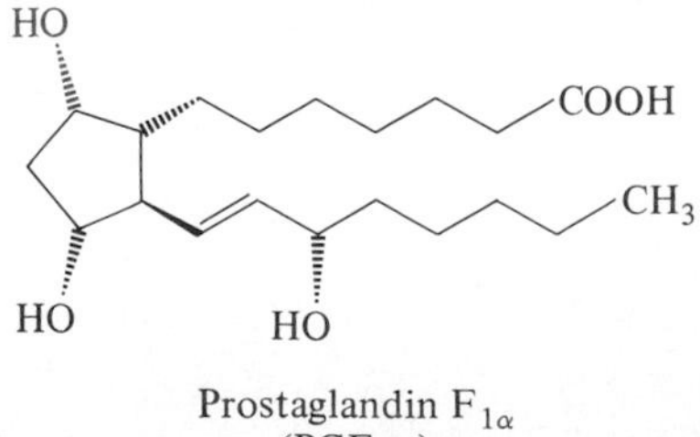

Prostaglandin E_1
(PGE_1)

Prostaglandin $F_{1\alpha}$
($PGF_{1\alpha}$)

Step 1: Oxygen abstracts a hydrogen atom from C-13 of arachidonic acid to give a free radical. This free radical is stabilized by allylic resonance involving the double bonds at both C-11 and C-14.

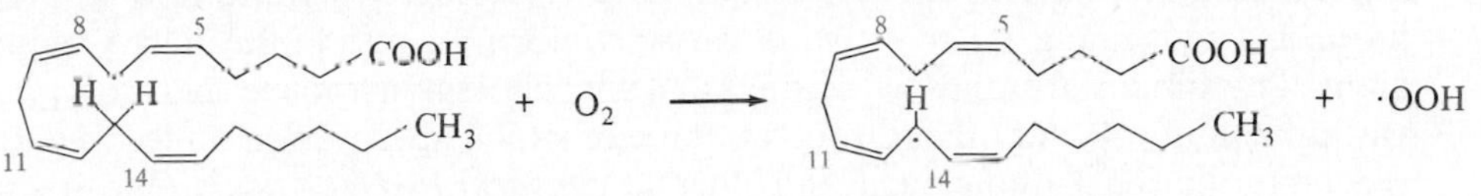

Arachidonic acid + Oxygen ⟶ "Doubly allylic" radical + Hydroperoxy radical

Step 2: A second molecule of oxygen adds to the free radical by bond formation at C-11. The product is an alkylperoxy radical.

Step 3: The alkylperoxy radical formed in step 2 cyclizes by attack of its peroxy oxygen on C-9 while another molecule of oxygen attacks C-15. Carbon-carbon bond formation occurs between C-8 and C-12.

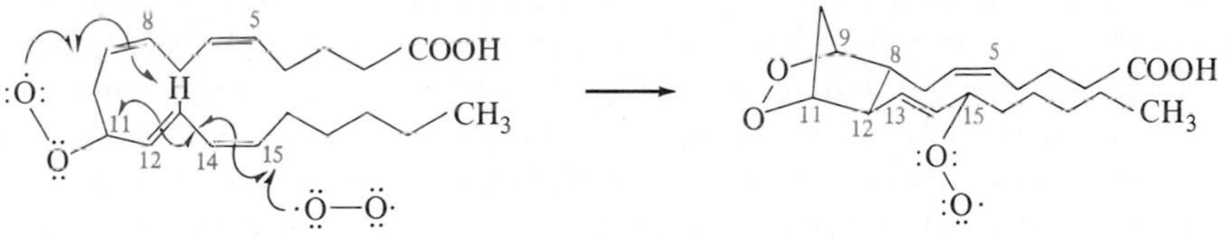

Step 4: The alkylperoxy radical formed in step 3 abstracts a hydrogen atom from some suitable donor present in the medium to give a substance (PGG$_2$) containing both an endoperoxide unit and a hydroperoxide function.

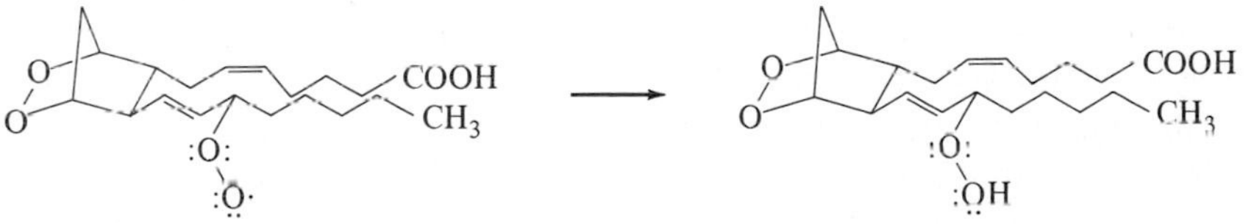

PGG$_2$

Step 5: The hydroperoxide is reduced to an alcohol. The endoperoxide unit is left intact.

PGG$_2$ PGH$_2$

Step 6: Dissociation of the endoperoxide unit occurs by cleavage of the oxygen-oxygen bond. Subsequent hydrogen atom transfer processes convert one alkoxy radical to a ketone carbonyl and the other to an alcohol in PGE$_2$.

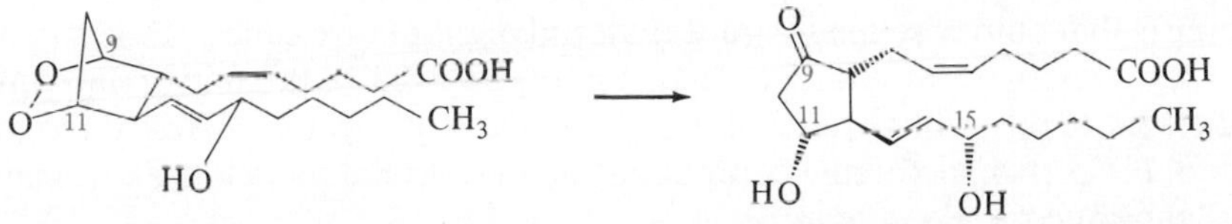

PGH$_2$ PGE$_2$

FIGURE 28.6 Sequence of steps illustrating the intermediates in the enzyme-catalyzed conversion of arachidonic acid to PGE$_2$.

linoleic acid from 18 to 20 carbons while introducing two more double bonds. Linoleic acid is said to be an *essential fatty acid,* forming part of the dietary requirement of mammals. Animals fed on diets that are deficient in linoleic acid grow poorly and suffer a number of other disorders, some of which are reversed on feeding them vegetable oils rich in linoleic acid and other *polyunsaturated fatty acids.* One function of these substances is to provide the raw materials for prostaglandin biosynthesis.

PROBLEM 28.6 PGE_1 (Figure 28.5) is biosynthesized by a pathway analogous to that of PGE_2. What fatty acid is the biosynthetic precursor to PGE_1? Modify the biosynthetic scheme of Figure 28.6 so that it illustrates the biosynthesis of PGE_1.

Physiological responses to prostaglandin levels in tissues encompass a variety of effects. Some prostaglandins relax bronchial muscle, others contract it. Some stimulate uterine contractions and have been used to induce therapeutic abortions. PGE_1 dilates blood vessels and lowers blood pressure; it inhibits the aggregation of platelets and offers promise as a drug to reduce the formation of blood clots.

The long-standing question of the mode of action of aspirin has been addressed in terms of its effects on prostaglandin biosynthesis. Prostaglandin biosynthesis is the body's response to tissue damage and is manifested by pain and inflammation at the affected site. Aspirin has been shown to inhibit the activity of the cyclooxygenase enzyme required for the first stage of prostaglandin formation. Aspirin reduces pain and inflammation—and probably fever as well—by reducing prostaglandin levels in the body.

Much of the fundamental work on prostaglandins and related compounds was carried out by Sune Bergstrom and Bengt Samuelsson of the Karolinska Institute (Sweden) and by John Vane of the Wellcome Foundation (Great Britain). These three shared the Nobel prize for physiology or medicine in 1982. Bergstrom began his research on prostaglandins because he was interested in the oxidation of fatty acids. That research led to the identification of a whole new class of biochemical mediators. Prostaglandin research has now revealed that other derivatives of oxidized polyunsaturated fatty acids, structurally distinct from the prostaglandins, are also physiologically important. These fatty acid derivatives include, for example, a group of substances known as the *leukotrienes,* which have been implicated as mediators in immunological processes.

28.7 TERPENES. THE ISOPRENE RULE

The word "essential" as applied to naturally occurring organic substances can have two different meanings. For example, as used in the previous section with respect to fatty acids, "essential" means "necessary." Linoleic acid is an "essential" fatty acid; it must be present in the diet in order for animals to grow properly because they lack the ability to biosynthesize it directly.

"Essential" is also used as the adjective form of the noun "essence." The mixtures of substances that comprise the fragrant material of plants are called "essential oils" because they contain the essence, i.e., the odor, of the plant. The study of the composition of essential oils ranks as one of the oldest areas of organic chemical research. Very often, the principal volatile component of an essential oil belongs to a class of chemical substances called the *terpenes.*

Myrcene, a hydrocarbon isolated from bayberry oil, is a typical terpene.

$$CH_2$$
$$\|$$
$$(CH_3)_2C=CHCH_2CH_2CCH=CH_2 \quad \equiv$$

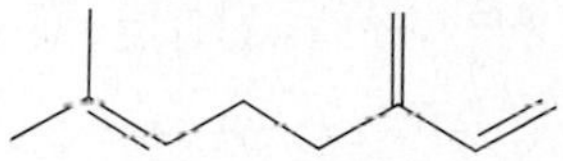

Myrcene

The structural feature that distinguishes terpenes from other natural products is the *isoprene unit*. The carbon skeleton of myrcene (exclusive of its double bonds) corresponds to the head-to-tail union of two isoprene units.

$$CH_3$$
$$|$$
$$CH_2=C-CH=CH_2 \quad \equiv$$

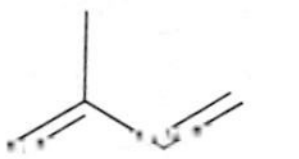

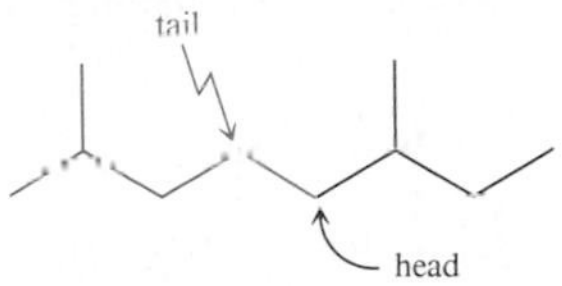

Isoprene
(2-methyl-1,3-butadiene)

Two isoprene units
linked head to tail

Terpenes are often referred to as *isoprenoid* compounds. They are classified according to the number of carbon atoms they contain, as summarized in Table 28.2.

While the term "terpene" once referred only to hydrocarbons, current usage includes functionally substituted derivatives as well. Figure 28.7 presents the structural formulas for a number of representative terpenes. The isoprene units in some of these are relatively easy to identify. The three isoprene units in the sesquiterpene farnesol, for example, are indicated in color below. They are joined in a head-to-tail fashion.

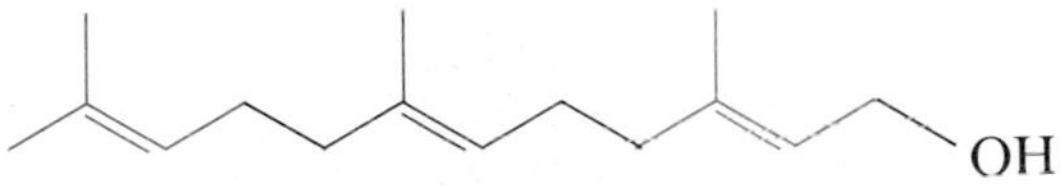

OH

Isoprene units in farnesol

Many terpenes contain one or more rings, but these also can be viewed as collections of isoprene units. An example is α-selinene. Like farnesol, it is made up of three isoprene units linked head-to tail.

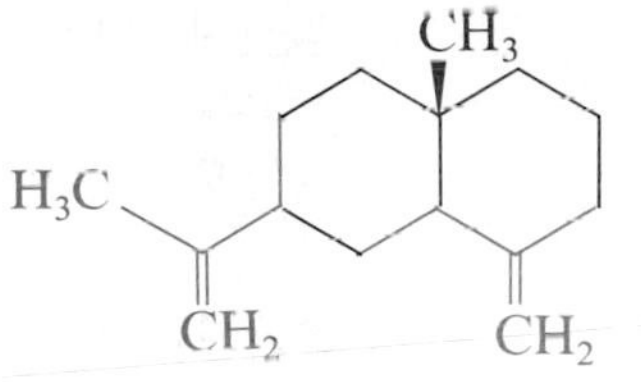

Isoprene units in α-selinene

TABLE 28.2

Classification of Terpenes

Class	Number of carbon atoms
Monoterpene	10
Sesquiterpene	15
Diterpene	20
Sesterpene	25
Triterpene	30
Tetraterpene	40

Monoterpenes

α-Phellandrene
(eucalyptus)

Menthol
(peppermint)

Citral
(lemon grass)

Sesquiterpenes

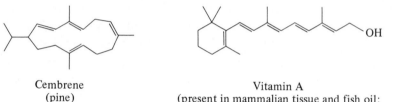

α-Selinene
(celery)

Farnesol
(ambrette)

Abscicic acid
(a plant hormone)

Diterpenes

Cembrene
(pine)

Vitamin A
(present in mammalian tissue and fish oil;
important substance in the chemistry of vision)

Triterpenes

Squalene
(shark liver oil)

Tetraterpenes

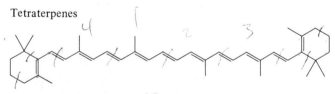

β-Carotene
(present in carrots and other vegetables;
enzymes in the body cleave β-carotene to vitamin A)

FIGURE 28.7 Some representative terpenes and related natural products. Structures are customarily depicted as carbon skeleton formulas when describing compounds of isoprenoid origin.

PROBLEM 28.7 Locate the isoprene units in each of the monoterpenes, sesquiterpenes, and diterpenes shown in Figure 28.7. (In some cases there are two equally correct arrangements.)

Tail-to-tail linkages of isoprene units sometimes occur, especially in the higher terpenes. The C(12)—C(13) bond of squalene unites two C_{15} units in a tail-to-tail manner. (Notice, however, that isoprene units are joined head-to-tail within each C_{15} unit of squalene.)

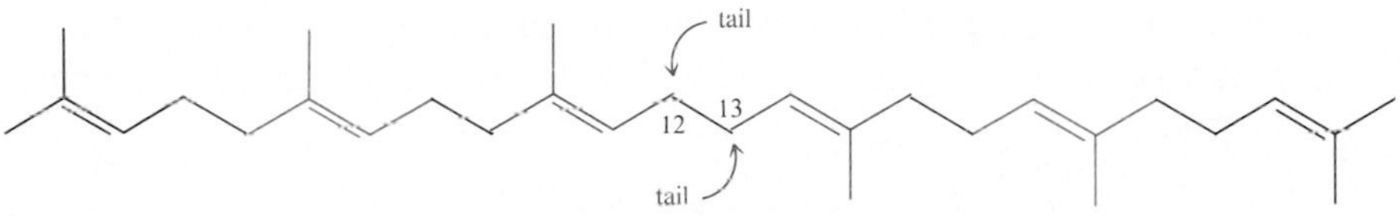

Isoprene units in squalene

PROBLEM 28.8 Identify the isoprene units in β-carotene (Figure 28.7). Which carbons are joined by a tail-to-tail link between isoprene units?

The work of the German chemist Otto Wallach (Nobel prize in chemistry, 1910) was fundamental in establishing the structures of many monoterpenes, and he is credited with recognizing that they can be viewed as collections of isoprene units. Leopold Ruzicka of the Swiss Federal Institute of Technology (Zürich), in his studies of sesquiterpenes and higher terpenes, extended and refined what we now know as the *isoprene rule.* He was a corecipient of the Nobel prize in chemistry in 1939. While exceptions to it are known, the isoprene rule is a useful guide to terpene structures and has stimulated research in the biosynthetic origin of these compounds. Terpenes contain isoprene units but isoprene does not occur naturally. What is the *biological isoprene unit,* how is it biosynthesized, and how do individual isoprene units combine to give terpenes?

28.8 ISOPENTENYL PYROPHOSPHATE. THE BIOLOGICAL ISOPRENE UNIT

Isoprenoid compounds are biosynthesized from acetate by a process that involves several stages. The first stage is the formation of *mevalonic acid* from three molecules of acetic acid.

$$3\ CH_3\overset{\displaystyle O}{\overset{\|}{C}}OH \xrightarrow[\text{steps}]{\text{several}} HO\overset{\displaystyle O}{\overset{\|}{C}}CH_2\overset{\displaystyle CH_3}{\overset{|}{\underset{\displaystyle OH}{\underset{|}{C}}}}CH_2CH_2OH$$

Acetic acid — Mevalonic acid

In the second stage mevalonic acid is converted to *3-methyl-3-butenyl pyrophosphate (isopentenyl pyrophosphate).*

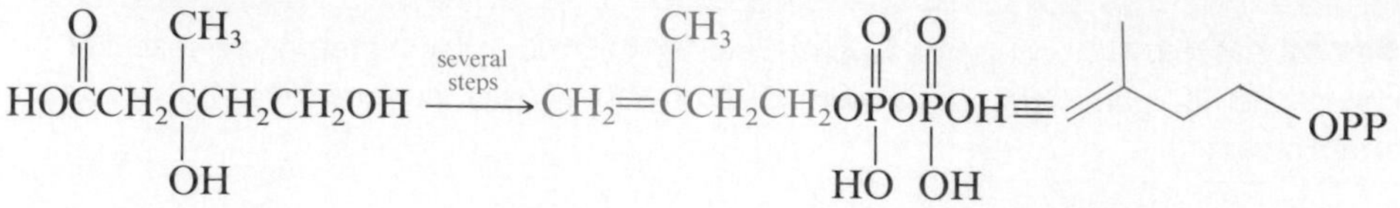

Mevalonic acid   Isopentenyl pyrophosphate

Isopentenyl pyrophosphate is the biological isoprene unit; it contains five carbon atoms connected in the same order as in isoprene. (It is convenient to use the symbol -OPP to represent the pyrophosphate group, as shown in the equation.)

Isopentenyl pyrophosphate undergoes an enzyme-catalyzed reaction that converts it, in an equilibrium process, to *3-methyl-2-butenyl pyrophosphate (dimethylallyl pyrophosphate)*.

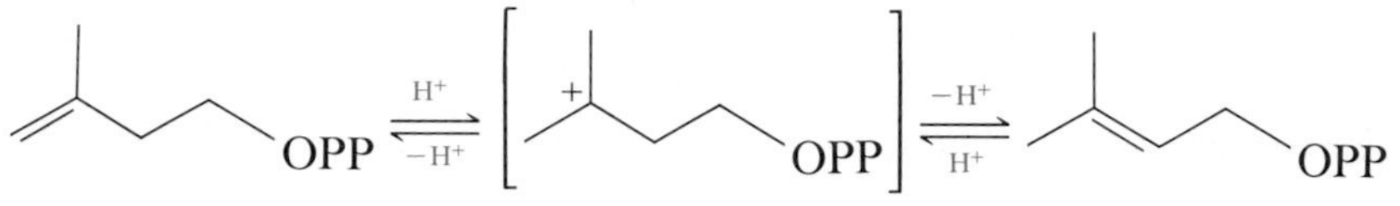

Isopentenyl   Carbocation intermediate   Dimethylallyl
pyrophosphate   pyrophosphate

Isopentenyl pyrophosphate and dimethylallyl pyrophosphate are structurally similar — both contain a double bond and a pyrophosphate ester unit — but the chemical reactivity expressed by each is different. The principal site of reaction in dimethylallyl pyrophosphate is the carbon that bears the pyrophosphate group. Pyrophosphate is a reasonably good leaving group in nucleophilic substitution reactions, especially when, as in dimethylallyl pyrophosphate, it is located at an allylic carbon. Isopentenyl pyrophosphate, on the other hand, does not have its leaving group attached to an allylic carbon and is far less reactive than dimethylallyl pyrophosphate toward nucleophilic reagents. The principal site of reaction in isopentenyl pyrophosphate is the carbon-carbon double bond, which, like the double bonds of simple alkenes, is reactive toward electrophiles.

28.9 CARBON-CARBON BOND-FORMING PROCESSES IN TERPENE BIOSYNTHESIS

The chemical properties of isopentenyl pyrophosphate and dimethylallyl pyrophosphate are complementary in a way that permits them to react with each other to form a carbon-carbon bond that unites two isoprene units. Using the π electrons of its double bond, isopentenyl pyrophosphate acts as a nucleophile and displaces pyrophosphate from dimethylallyl pyrophosphate.

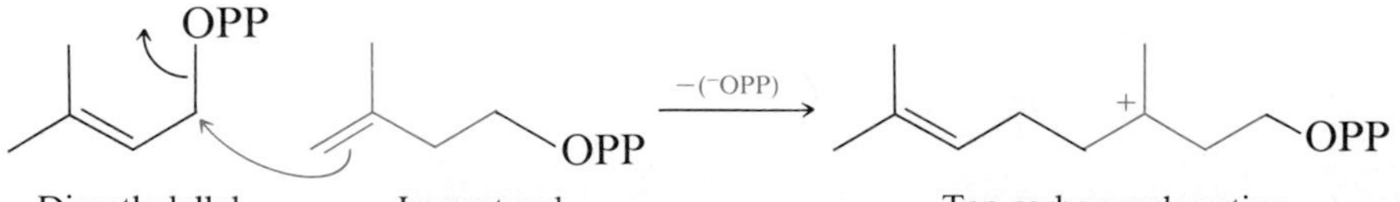

Dimethylallyl   Isopentenyl   Ten-carbon carbocation
pyrophosphate   pyrophosphate

The tertiary carbocation formed in this step can react according to any of the various reaction pathways available to carbocations. One of these is loss of a proton to give an alkene.

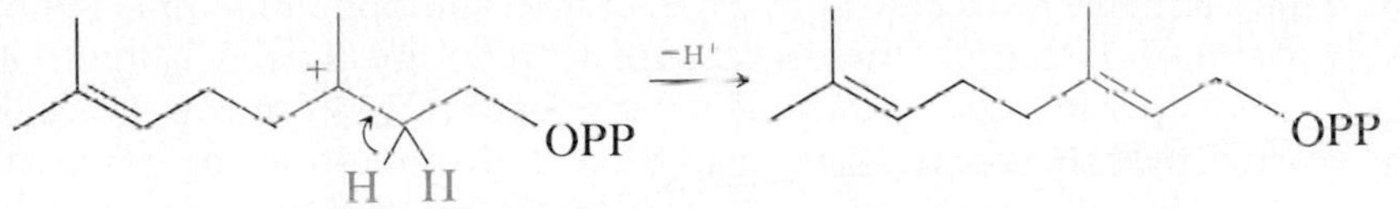

Geranyl pyrophosphate

The product of this reaction is geranyl pyrophosphate. Hydrolysis of the pyrophosphate ester group gives geraniol, a naturally occurring monoterpene found in rose oil.

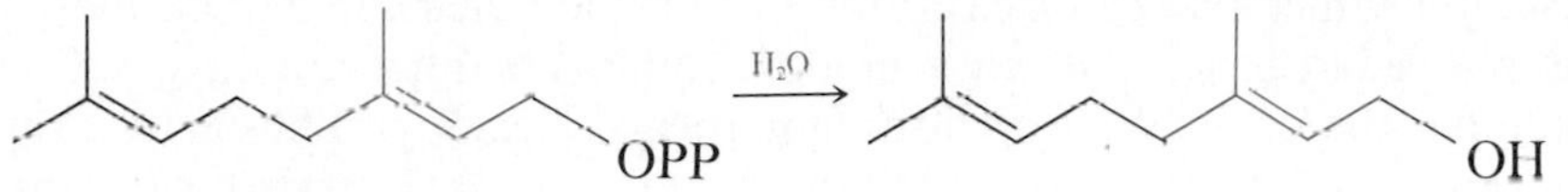

Geranyl pyrophosphate   Geraniol

Geranyl pyrophosphate is an allylic pyrophosphate and, like dimethylallyl pyrophosphate, can act as an alkylating agent toward a molecule of isopentenyl pyrophosphate. A 15-carbon carbocation is formed, which, on deprotonation, gives farnesyl pyrophosphate.

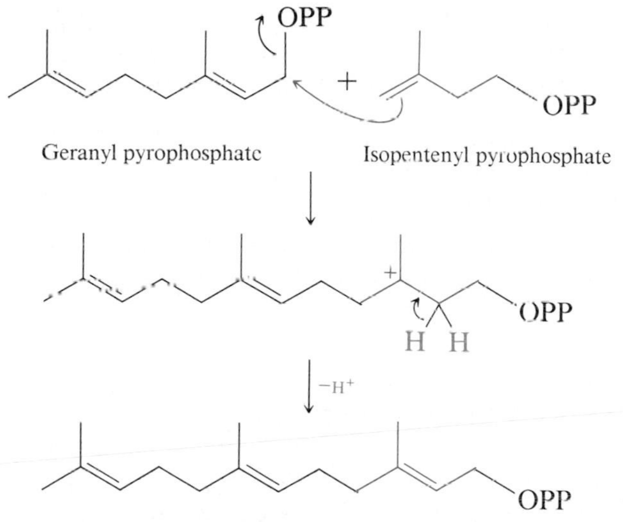

Geranyl pyrophosphate   Isopentenyl pyrophosphate

Farnesyl pyrophosphate

Hydrolysis of the pyrophosphate ester group converts farnesyl pyrophosphate to the corresponding alcohol farnesol (see Figure 28.7 for the structure of farnesol).

A repetition of the process just shown produces the diterpene geranylgeraniol from farnesyl pyrophosphate.

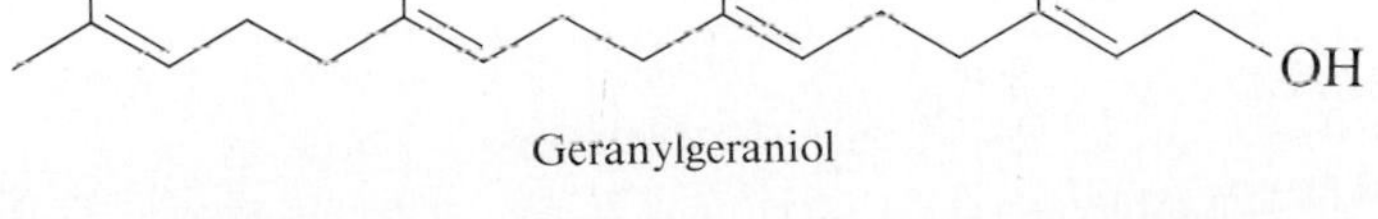

Geranylgeraniol

PROBLEM 28.9 Write a sequence of reactions that describes the formation of geranylgeraniol from farnesyl pyrophosphate.

The higher terpenes are formed not by successive addition of C_5 units but by the coupling of simpler terpenes. Thus, the triterpenes (C_{30}) are derived from two molecules of farnesyl pyrophosphate and the tetraterpenes (C_{40}) from two molecules of geranylgeranyl pyrophosphate. These carbon-carbon bond-forming processes involve tail-to-tail couplings and proceed by a more complicated mechanism than that just described.

The enzyme-catalyzed reactions that lead to geraniol and farnesol (as their pyrophosphate esters) are mechanistically related to the acid-catalyzed dimerization of alkenes discussed in Section 7.18. The reaction of an allylic pyrophosphate or a carbocation with a source of π electrons is a recurring theme in terpene biosynthesis and is invoked to explain the origin of more complicated structural types. Consider, for example, the formation of cyclic monoterpenes. Neryl pyrophosphate, formed by an enzyme-catalyzed isomerization of the E double bond in geranyl pyrophosphate, has the proper geometry to form a six-membered ring via intramolecular attack of the double bond on the allylic pyrophosphate unit.

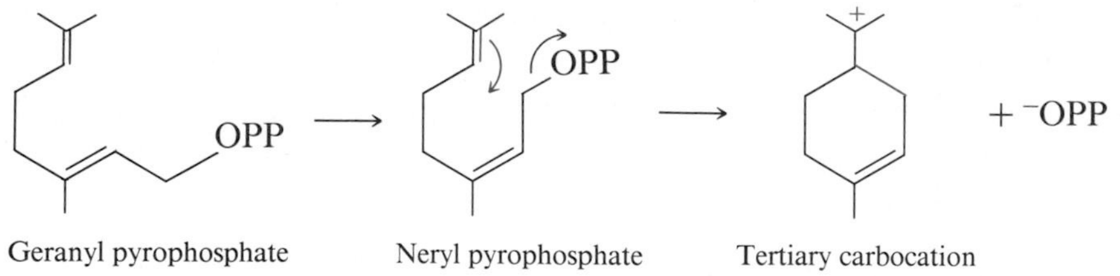

| Geranyl pyrophosphate | Neryl pyrophosphate | Tertiary carbocation |

Loss of a proton from the tertiary carbocation formed in this step gives limonene, an abundant natural product found in many citrus fruits. Capture of the carbocation by water gives α-terpineol, also a known natural product.

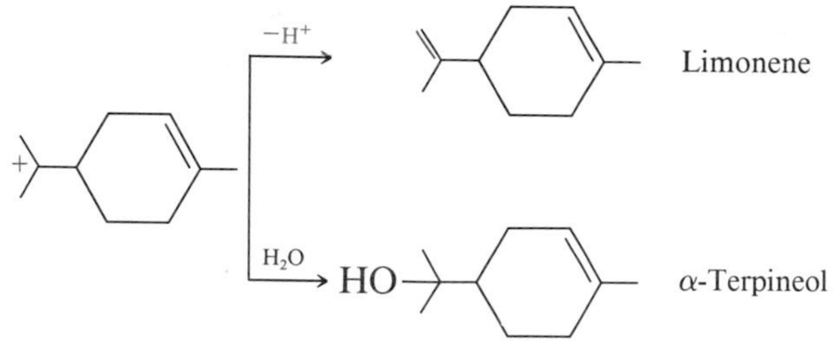

The same tertiary carbocation serves as the precursor to numerous bicyclic monoterpenes. A carbocation having a bicyclic skeleton is formed by intramolecular attack of the π electrons of the double bond on the positively charged carbon.

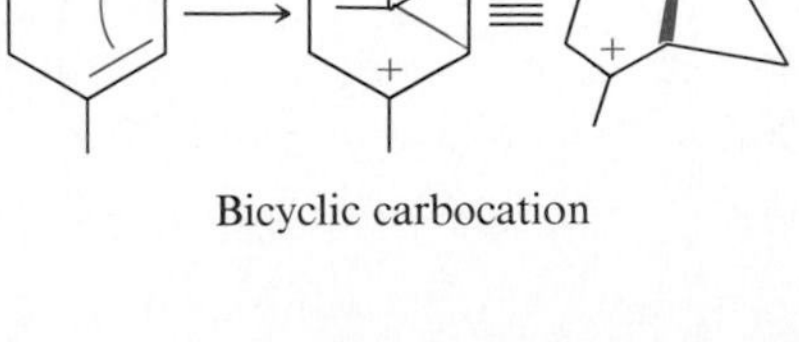

Bicyclic carbocation

A. Loss of a proton from the bicyclic carbocation yields α-pinene and
β-pinene. The pinenes are the most abundant of the monoterpenes.
They are the principal constituents of turpentine.

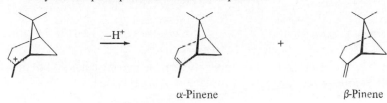

α-Pinene   β-Pinene

B. Capture of the carbocation by water, accompanied by rearrangement of
the bicyclo[3.1.1] carbon skeleton to a bicyclo[2.2.1] unit, yields borneol.
Borneol is found in the essential oil of certain trees that grow in Indonesia.

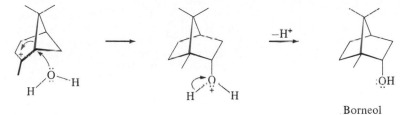

Borneol

FIGURE 28.8 Two of the
reaction pathways available to
the C_{10} bicyclic carbocation
formed from neryl pyrophosphate. The same carbocation
can lead to monoterpenes
based on either the bicy-
clo[3.1.1] or the bicyclo[2.2.1]
carbon skeleton.

This bicyclic carbocation then undergoes many of the reactions typical of carbocation intermediates to provide a variety of bicyclic monoterpenes, as outlined in
Figure 28.8.

PROBLEM 28.10   The structure of the bicyclic monoterpene borneol is shown in Figure 28.8. Isoborneol, a stereoisomer of borneol, can be prepared in the laboratory by a
two-step sequence. In the first step, borneol is oxidized to camphor by treatment with
chromic acid. In the second step, camphor is reduced with sodium borohydride to a mixture of 85 percent isoborneol and 15 percent borneol. On the basis of these transformations, deduce structural formulas for isoborneol and camphor.

Analogous processes involving cyclizations and rearrangements of carbocations
derived from farnesyl pyrophosphate produce a rich variety of structural types in the
sesquiterpene series. We will have more to say about the chemistry of higher terpenes,
especially the triterpenes, later in this chapter. For the moment, however, let us
return to smaller molecules in order to complete the picture of how isoprenoid
compounds arise from acetate.

28.10 THE PATHWAY FROM ACETATE TO ISOPENTENYL PYROPHOSPHATE

The introduction to Section 28.8 pointed out that mevalonic acid is the biosynthetic
precursor of isopentenyl pyrophosphate. The early steps in the biosynthesis of mevalonate from three molecules of acetic acid are analogous to those in fatty acid biosynthesis (Section 28.3) except that they do not involve acyl carrier protein. Thus, the
reaction of acetyl coenzyme A with malonyl coenzyme A yields a molecule of acetoacetyl coenzyme A.

$$\underset{\substack{\text{Acetyl} \\ \text{coenzyme A}}}{\text{CH}_3\overset{\displaystyle O}{\overset{\displaystyle \|}{\text{C}}}\text{SCoA}} + \underset{\substack{\text{Malonyl} \\ \text{coenzyme A}}}{{}^-\text{O}_2\text{CCH}_2\overset{\displaystyle O}{\overset{\displaystyle \|}{\text{C}}}\text{SCoA}} \longrightarrow \underset{\substack{\text{Acetoacetyl} \\ \text{coenzyme A}}}{\text{CH}_3\overset{\displaystyle O}{\overset{\displaystyle \|}{\text{C}}}\text{CH}_2\overset{\displaystyle O}{\overset{\displaystyle \|}{\text{C}}}\text{SCoA}} + \underset{\substack{\text{Carbon} \\ \text{dioxide}}}{\text{CO}_2}$$

Carbon-carbon bond formation then occurs between the ketone carbonyl of acetoacetyl coenzyme A and the α carbon of a molecule of acetyl coenzyme A.

$$\underset{\substack{\text{Acetoacetyl} \\ \text{coenzyme A}}}{\text{CH}_3\overset{O}{\overset{\|}{\text{C}}}\text{CH}_2\overset{O}{\overset{\|}{\text{C}}}\text{SCoA}} + \underset{\substack{\text{Acetyl} \\ \text{coenzyme A}}}{\text{CH}_3\overset{O}{\overset{\|}{\text{C}}}\text{SCoA}} \longrightarrow \underset{\substack{\beta\text{-Hydroxy-}\beta\text{-methylglutaryl} \\ \text{coenzyme A (HMG CoA)}}}{\text{CH}_3\overset{\text{HO}}{\underset{\underset{\overset{\displaystyle \|}{\text{O}}}{\text{CH}_2\text{COH}}}{\text{C}}}\text{CH}_2\overset{O}{\overset{\|}{\text{C}}}\text{SCoA}} + \underset{\text{Coenzyme A}}{\text{CoASH}}$$

The product of this reaction, known by the common name β-hydroxy-β-methylglutaryl coenzyme A (HMG CoA), has the carbon skeleton of mevalonic acid and is converted to it by enzymatic reduction.

$$\underset{\substack{\beta\text{-Hydroxy-}\beta\text{-methylglutaryl} \\ \text{coenzyme A (HMG CoA)}}}{\text{CH}_3\overset{\text{HO}}{\underset{\underset{\overset{\|}{\text{O}}}{\text{CH}_2\text{COH}}}{\text{C}}}\text{CH}_2\overset{O}{\overset{\|}{\text{C}}}\text{SCoA}} \longrightarrow \underset{\text{Mevalonic acid}}{\text{CH}_3\overset{\text{HO}}{\underset{\underset{\overset{\|}{\text{O}}}{\text{CH}_2\text{COH}}}{\text{C}}}\text{CH}_2\text{CH}_2\text{OH}}$$

In keeping with its biogenetic origin in three molecules of acetic acid, mevalonic acid has six carbon atoms. The conversion of mevalonate to isopentenyl pyrophosphate involves loss of the "extra" carbon as carbon dioxide. First, the alcohol hydroxyl groups of mevalonate are converted to phosphate ester functions—they are enzymatically *phosphorylated,* with introduction of a simple phosphate at the tertiary site and a pyrophosphate at the primary site. Decarboxylation, in concert with loss of the tertiary phosphate, introduces a carbon-carbon double bond and gives isopentenyl pyrophosphate, the fundamental building block for formation of isoprenoid natural products.

Mevalonate → (Unstable; undergoes rapid decarboxylation with loss of phosphate) $\xrightarrow[-\text{CO}_2]{-\text{PO}_4{}^{3-}}$ Isopentenyl pyrophosphate

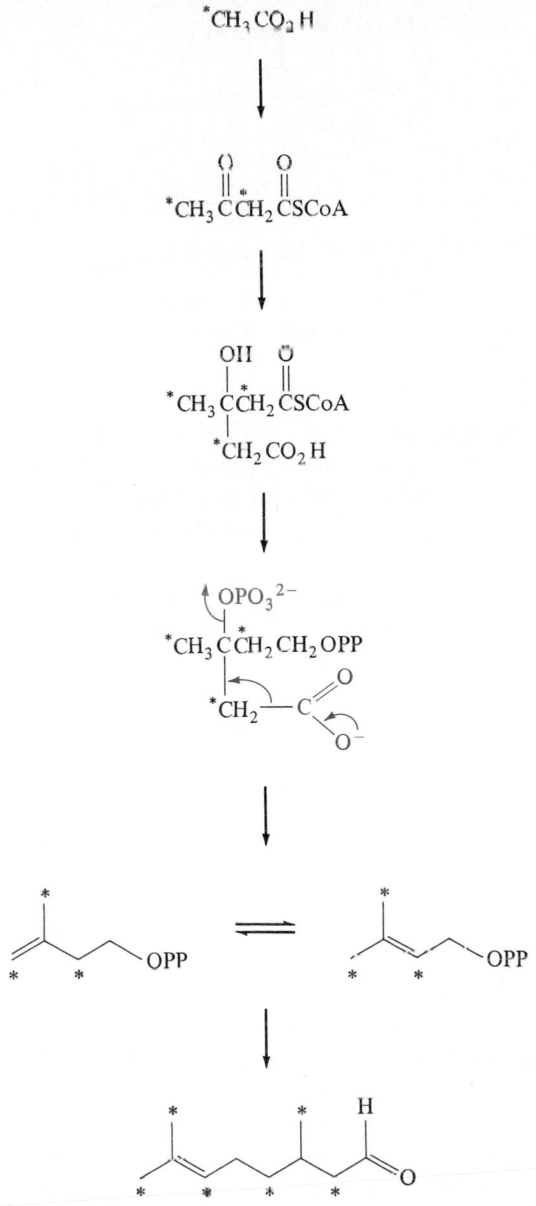

*CH$_3$CO$_2$H

↓

O O
‖ ‖
*CH$_3$CCH$_2$CSCoA

↓

OH O
| ‖
*CH$_3$CCH$_2$CSCoA
|
*CH$_2$CO$_2$H

↓

OPO$_3$$^{2-}$
|
*CH$_3$CCH$_2$CH$_2$OPP
|
*CH$_2$—C
O
O$^-$

↓

⇌

↓

FIGURE 28.9 Diagram showing the distribution of the ^{14}C label in citronellal biosynthesized from acetate in which the methyl carbon was isotopically enriched with ^{14}C.

Much of what we know concerning the pathway from acetate to mevalonate to isopentenyl pyrophosphate to terpenes has emerged as a result of "feeding" experiments, in which plants are grown in the presence of radioactively labeled organic substances and the distribution of the radioactive label is determined in the products of biosynthesis. To illustrate, eucalyptus plants were allowed to grow in a medium containing acetic acid enriched with ^{14}C in its methyl group. Citronellal was isolated from the mixture of monoterpenes produced by the plants and shown, by a series of chemical degradations, to contain the radioactive ^{14}C label at carbons 2,4,6, and 8, as well as at the carbons of both branching methyl groups.

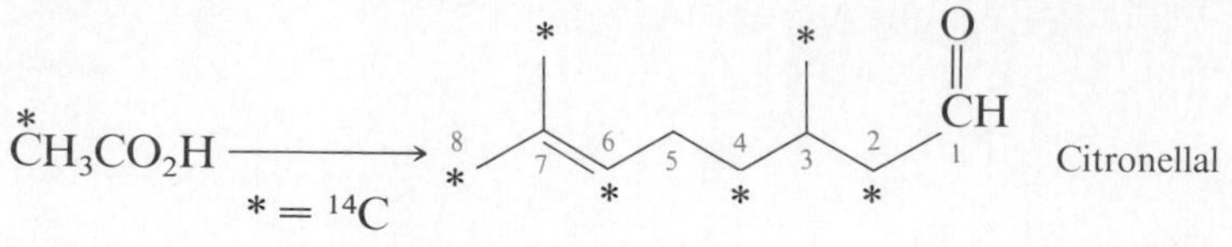

Figure 28.9 traces the ^{14}C label from its origin in acetic acid to its experimentally determined distribution in citronellal.

PROBLEM 28.11 How many carbon atoms of citronellal would be radioactively labeled if the acetic acid used in the experiment were enriched with ^{14}C at C-1 instead of at C-2? Identify these carbon atoms.

A more recent experimental technique employs ^{13}C as the isotopic label. Instead of locating the position of the label by a laborious degradation procedure, the ^{13}C nmr spectrum of the natural product is recorded. The signals for the carbons that are enriched in ^{13}C are far more intense than those corresponding to carbons in which ^{13}C is present only at the natural abundance level.

Isotope incorporation experiments have demonstrated the essential correctness of the scheme presented in this and preceding sections for terpene biosynthesis. Considerable effort has been expended toward its detailed elaboration because of the common biosynthetic origin of terpenes and another class of acetate-derived natural products, the steroids.

28.11 STEROIDS. CHOLESTEROL

Cholesterol is the central compound in any discussion of steroids. Its name is a combination of the Greek words for "bile" *(chole)* and "solid" *(stereos)* preceding the characteristic alcohol suffix *-ol*. It is the most abundant steroid present in humans and the most important one as well since all other steroids arise from it. An average adult has over 200 g of cholesterol; it is found in almost all body tissues with relatively large amounts present in the brain and spinal cord and in gallstones. Cholesterol is the principal constituent of the plaque that builds up on the walls of arteries and restricts the flow of blood in the circulatory disorder known as *atherosclerosis*.

Cholesterol was isolated in the eighteenth century but its structure is so complex that its correct constitution was not determined until 1932 and its stereochemistry not verified until 1955. Steroids are characterized by the tetracyclic ring system shown in Figure 28.10a. As shown in Figure 28.10b, cholesterol contains this tetracyclic skeleton modified to include an alcohol function at C-3, a double bond at C-5, methyl groups at C-10 and C-13, and a C_8H_{17} side chain at C-17. Isoprene units may

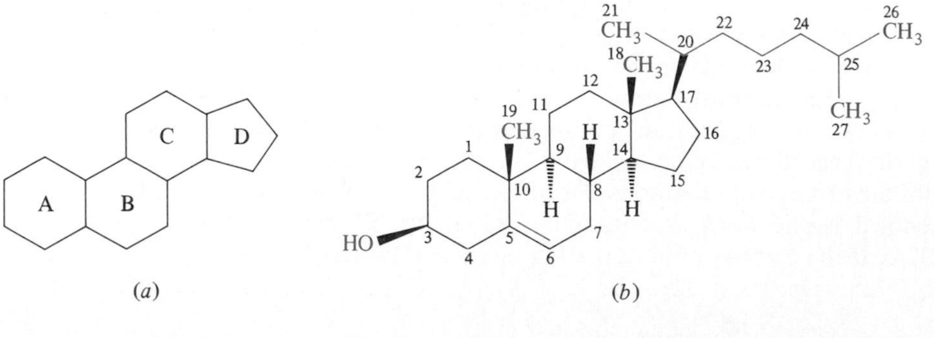

FIGURE 28.10 (*a*) The tetracyclic ring system characteristic of steroids. The rings are designated A, B, C, and D as shown. (*b*) The structure of cholesterol. A unique numbering system is used for steroids and is indicated in the structural formula.

be discerned in various portions of the cholesterol molecule but the overall correspondence with the isoprene rule is far from perfect. Indeed, cholesterol has only 27 carbon atoms, three too few for it to be classed as a triterpene.

Cholesterol occurs in both plants and animals. We accumulate cholesterol by eating meats and vegetables that contain it and are also able to biosynthesize it from acetate. The pioneering work that identified the key intermediates in the complicated pathway of cholesterol biosynthesis was carried out by Konrad Bloch (Harvard) and Feodor Lynen (Munich), corecipients of the 1962 Nobel prize for physiology and medicine. An important discovery was that the triterpene squalene (Figure 28.7) is an intermediate in the formation of cholesterol from acetate. Thus, the early stages of cholesterol biosynthesis are the same as those of terpene biosynthesis described in Sections 28.8 through 28.10. (In fact, a significant fraction of our knowledge of terpene biosynthesis is a direct result of experiments carried out in the area of steroid biosynthesis.)

How does the tetracyclic steroid cholesterol arise from the acyclic triterpene squalene? Figure 28.11 outlines the stages involved. It has been shown that the first step is oxidation of squalene to the corresponding 2,3-epoxide. Enzyme-catalyzed ring opening of this epoxide in step 2 is accompanied by a cyclization reaction, in which the electrons of four of the five double bonds of squalene 2,3-epoxide are used to close the A, B, C, and D rings of the potential steroid skeleton. (A related example of this process in the formation of a triterpene has been described earlier in Section 17.16.) The carbocation that results from the cyclization reaction of step 2 is then converted to a compound known as *lanosterol* by the rearrangement shown in step 3. Lanosterol is a triterpene obtained from lanolin, a mixture of substances which coats the wool fibers of sheep. Step 4 of Figure 28.11 simply indicates the structural changes that remain to be accomplished in the transformation of lanosterol to cholesterol.

PROBLEM 28.12   The biosynthesis of cholesterol as outlined in Figure 28.11 is admittedly quite complicated. It will aid your understanding of the process if you consider the following questions.

(a)   Which carbon atoms of squalene 2,3-epoxide correspond to the doubly bonded carbons of cholesterol?

(b)   Which two hydrogen atoms of squalene 2,3-epoxide are the ones that migrate in step 3?

(c)   Which methyl group of squalene 2,3-epoxide becomes the methyl group at the C, D ring junction of cholesterol?

(d)   What three methyl groups of squalene 2,3-epoxide are lost during the conversion of lanosterol to cholesterol?

SAMPLE SOLUTION   (a)   As the structural formula in step 4 of Figure 28.11 indicates, the double bond of cholesterol unites C-5 and C-6 (steroid numbering). The corresponding carbons in the cyclization reaction of step 2 in the figure may be identified as C-7 and C-8 of squalene 2,3-epoxide (systematic IUPAC numbering).

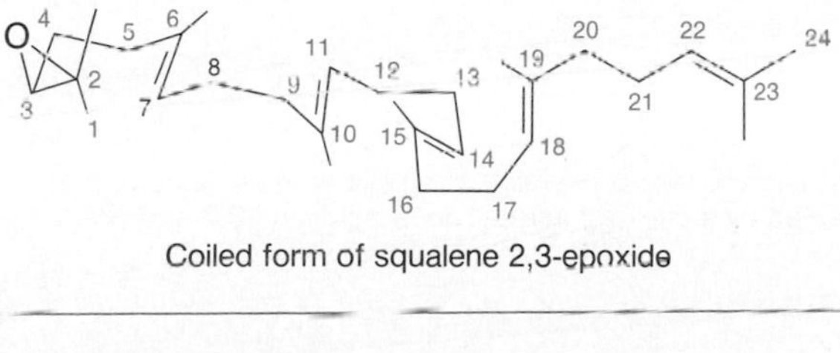

Coiled form of squalene 2,3-epoxide

Step 1: Squalene undergoes enzymic oxidation to the 2.3-epoxide. This reaction has been described earlier, in Section 17.16.

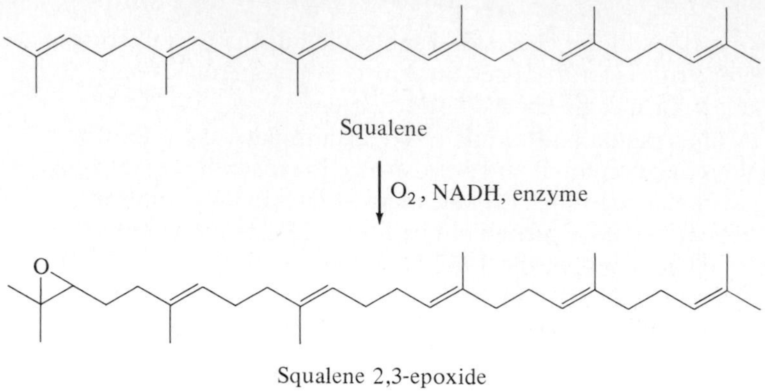

Squalene

O_2, NADH, enzyme

Squalene 2,3-epoxide

Step 2: Cyclization of squalene 2,3-epoxide, shown in its coiled form, is triggered by ring opening of the epoxide. Cleavage of the carbon-oxygen bond is assisted by protonation of oxygen and by nucleophilic participation of the π electrons of the neighboring double bond. A series of ring closures leads to the tetracyclic carbocation shown.

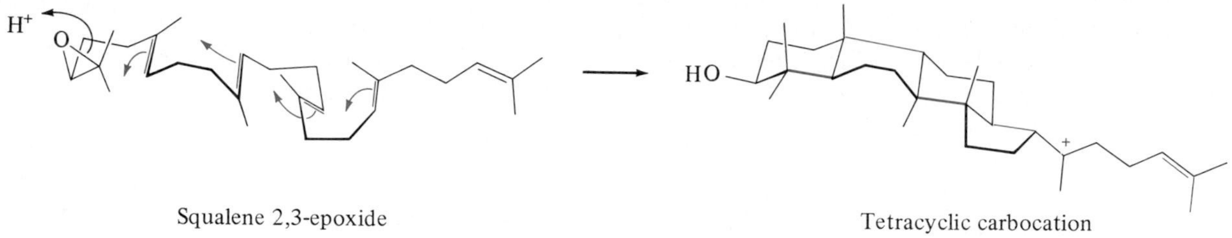

Squalene 2,3-epoxide   Tetracyclic carbocation

Step 3: Rearrangement of the tertiary carbocation formed by cyclization produces lanosterol. Two hydride shifts, from C-17 to C-20 and from C-13 to C-17, are accompanied by methyl shifts from C-14 to C-13 and from C-8 to C-14. A double bond is formed at C-8 by loss of the proton at C-9.

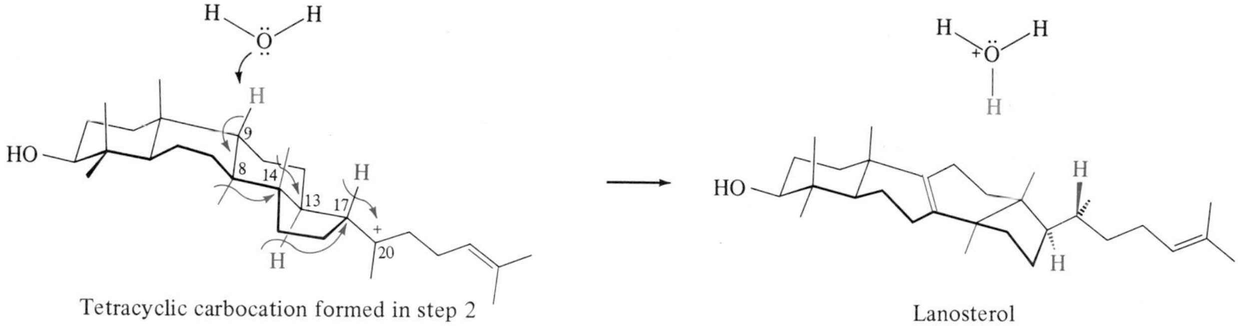

Tetracyclic carbocation formed in step 2   Lanosterol

Step 4: A series of enzyme-catalyzed reactions converts lanosterol to cholesterol. The three methyl groups shown in blue in the structural formula of lanosterol are lost via separate multistep operations, the C-8 and C-24 double bonds are reduced, and a new double bond is introduced at C-5.

Lanosterol   many steps   Cholesterol

FIGURE 28.11 The biosynthetic conversion of squalene to cholesterol proceeds through lanosterol. Lanosterol is formed by a cyclization reaction of squalene 2,3-epoxide.

PROBLEM 28.13   The biosynthetic pathway shown in Figure 28.11 was developed with the aid of isotopic labeling experiments. Which carbon atoms of cholesterol would you expect to be labeled when acetate enriched with ^{14}C in its methyl group ($^{14}CH_3COOH$) is used as the carbon source?

Once formed in the body, cholesterol can undergo a number of transformations. A very common one is esterification of its C-3 hydroxyl group by acyl transfer from coenzyme A derivatives of fatty acids. Other processes convert cholesterol to the biologically important steroids described in the following sections.

28.12 VITAMIN D

A steroid very closely related structurally to cholesterol is its 7-dehydro derivative. 7-Dehydrocholesterol is formed by enzymic oxidation of cholesterol and has a conjugated diene unit in its B ring. 7-Dehydrocholesterol is present in the tissues of the skin, where it is transformed to vitamin D_3 by a sunlight-induced photochemical reaction.

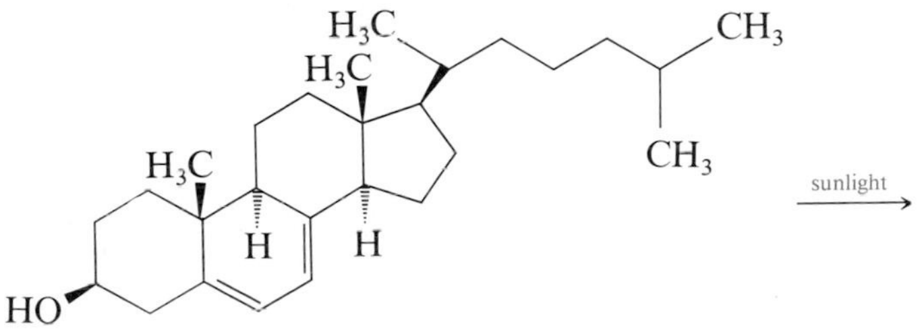

7-Dehydrocholesterol

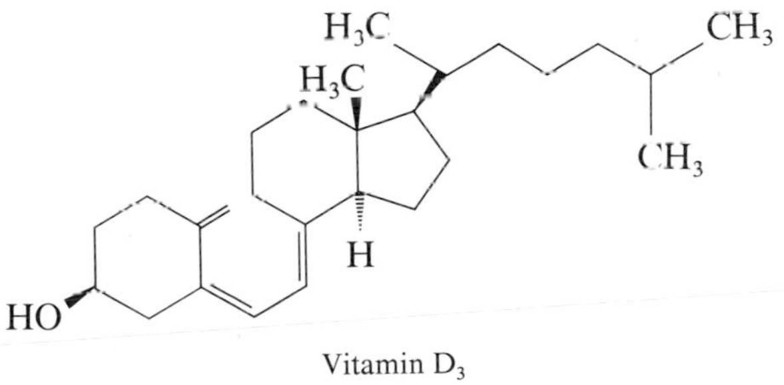

Vitamin D_3

Vitamin D_3 is a key compound in the process by which Ca^{2+} is absorbed from the intestine. Low levels of vitamin D_3 lead to Ca^{2+} concentrations in the body that are insufficient to support proper bone growth, resulting in the disease called *rickets*.

Rickets was once a serious health problem. It was thought to be a dietary deficiency disease because it could be prevented in children by feeding them fish liver oil. Actually, rickets is an environmental disease brought about by a deficiency of sunlight. Where the winter sun is weak, children may not be exposed to enough of its light to convert the 7-dehydrocholesterol in their skin to vitamin D_3 at levels sufficient to promote the growth of strong bones. Fish have adapted to an environment that screens them from sunlight, so they are not directly dependent on photochemistry for their vitamin D_3 and accumulate it by a different process. While fish liver oil is a good source of vitamin D_3, it is not very palatable. The major dietary source of D-type vitamins at present is ergosterol, a steroid found in yeast. Ergosterol is structurally similar to 7-dehydrocholesterol and, on irradiation with sunlight or artificial light, is converted to vitamin D_2, a substance analogous to vitamin D_3 and comparable with it in antirachitic activity. Irradiated ergosterol is added to milk and other foods to ensure that children receive enough vitamin D for their bones to develop properly.

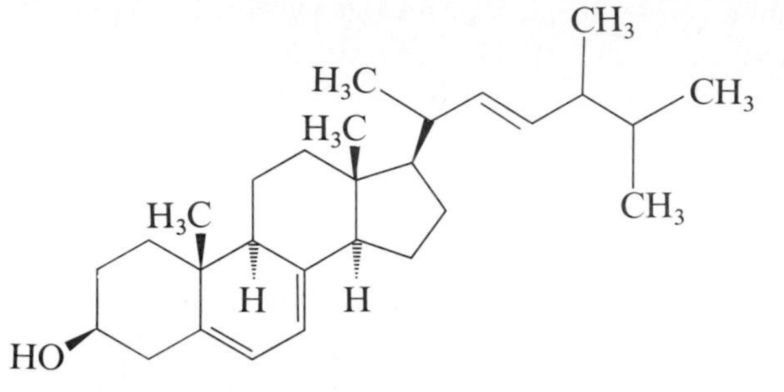

Ergosterol

PROBLEM 28.14   Suggest a reasonable structure for vitamin D_2.

28.13   BILE ACIDS

A significant fraction of the body's cholesterol is used to form *bile acids.* Oxidation in the liver removes a portion of the C_8H_{17} side chain, and additional hydroxyl groups are introduced at various positions on the steroid nucleus. Cholic acid is the most abundant of the bile acids. In the form of certain amide derivatives called *bile salts,* of which sodium taurocholate is one example, bile acids act as emulsifying agents to aid the digestion of fats. Bile salts have detergent properties similar to those of salts of long-chain fatty acids and promote the transport of lipids through aqueous media.

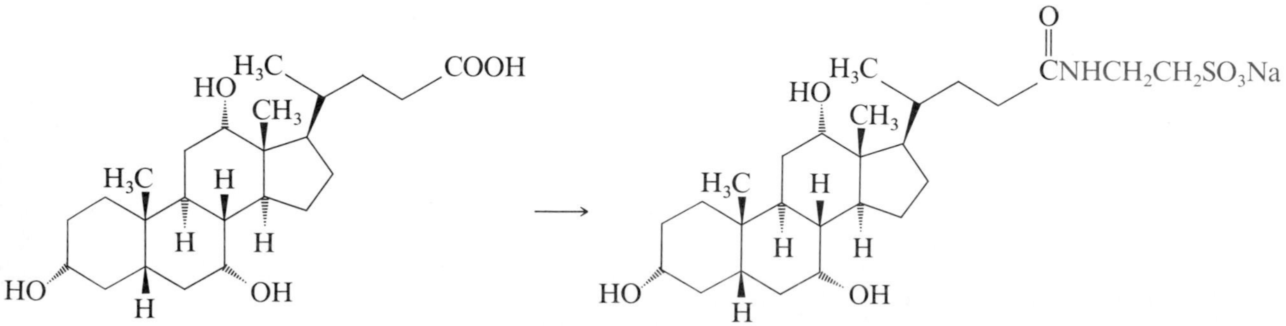

Cholic acid
(a bile acid)

Sodium taurocholate
(a bile salt)

28.14   CORTICOSTEROIDS

The outer layer, or *cortex,* of the adrenal gland is the source of a large group of substances known as *corticosteroids.* Like the bile acids, they are derived from cholesterol by oxidation, with cleavage of a portion of the alkyl substituent on the D ring.

Cortisol is the most abundant of the corticosteroids while cortisone is probably the best known. Cortisone is commonly prescribed as an anti-inflammatory drug, especially in the treatment of rheumatoid arthritis.

Cortisol   Cortisone

Corticosteroids exhibit a wide range of physiological effects. One important function is to assist in maintaining the proper electrolyte balance in body fluids. They also play a vital regulatory role in the metabolism of carbohydrates and in mediating the allergic response.

28.15 SEX HORMONES

Hormones are the chemical messengers of the body; they are secreted by the endocrine glands and regulate biological processes. Corticosteroids, described in the preceding section, are hormones produced by the adrenal glands. The sex glands — testes in males, ovaries in females — secrete a number of hormones which are involved in sexual development and reproduction. Testosterone is the principal male sex hormone; it is an *androgen*. Testosterone promotes muscle growth, deepening of the voice, the growth of body hair, and other male secondary sex characteristics. Testosterone is formed from cholesterol and is the biosynthetic precursor of estradiol, the principal female sex hormone, or *estrogen*. Estradiol is a key substance in the regulation of the menstrual cycle and the reproductive process. It is the hormone most responsible for the development of female secondary sex characteristics.

Testosterone   Estradiol

Testosterone and estradiol are present in the body in only minute amounts and their isolation and identification required heroic efforts. In order to obtain 0.012 g of estradiol for study, for example, 4 tons of sow ovaries had to be extracted!

A separate biosynthetic pathway leads from cholesterol to progesterone, a female sex hormone. One function of progesterone is to suppress ovulation at certain stages of the menstrual cycle and during pregnancy. Synthetic substances, such as *norethindrone*, have been developed that are superior to progesterone when taken orally to "turn off" ovulation. By inducing temporary infertility, they form the basis of "the pill," i.e., the oral contraceptive agents that have had a profound effect on population growth since their introduction in the 1960s.

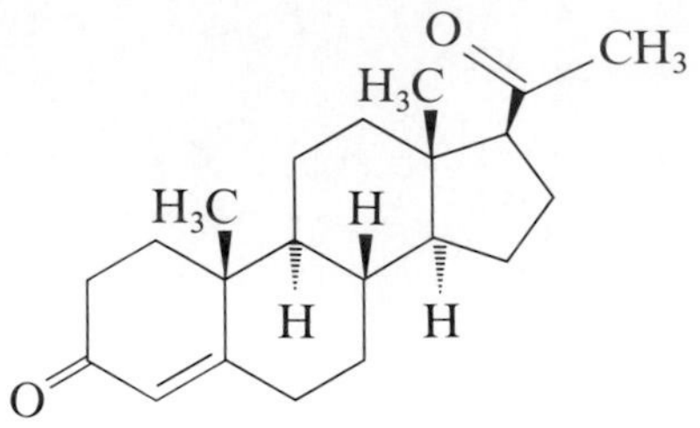

Progesterone

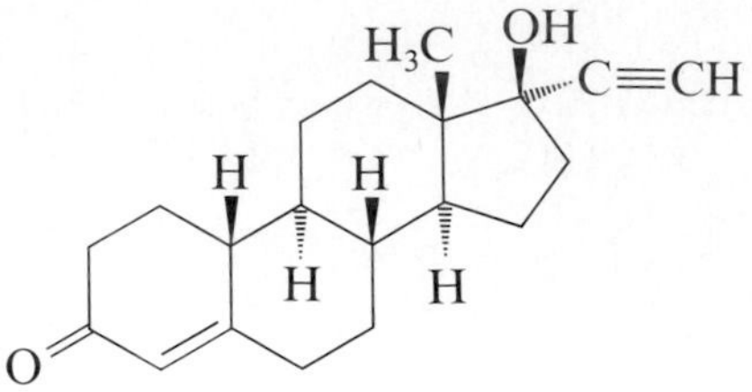

Norethindrone

28.16 CAROTENOIDS

Carotenoids are natural pigments characterized by a tail-to-tail linkage between two C_{20} units and an extended conjugated system of double bonds. They are the most widely distributed of the substances that give color to our world and occur in flowers, fruits, plants, insects, and animals. It has been estimated that biosynthesis from acetate produces approximately 100 million tons of carotenoids per year. The most familiar carotenoids are lycopene and β-carotene, pigments found in numerous plants and easily isolable from ripe tomatoes and carrots, respectively.

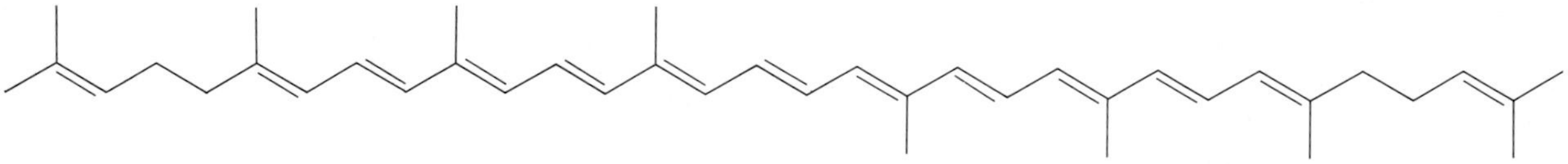

Lycopene

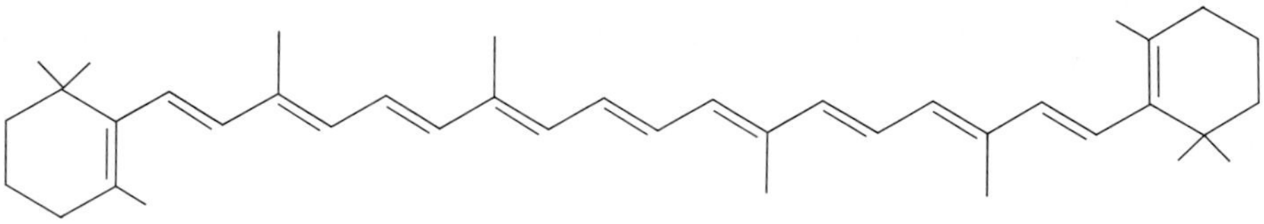

β-Carotene

Carotenoids absorb visible light (Section 14.17) and dissipate its energy as heat, thereby protecting the organism from any potentially harmful effects associated with sunlight-induced photochemistry. They are also indirectly involved in the chemistry of vision, owing to the fact that β-carotene is the biosynthetic precursor of vitamin A, also known as retinol, a key substance in the visual process.

There are two types of photoreceptor cells in the eye, rods and cones. Cones are sensitive to color but function only in bright light; rod vision operates even in dim light but only in black, white, and shades of gray. Because rods are larger and easier to work with in the laboratory, most of what we know of the visual process concerns rod vision. The complete process is quite complex, but can be illustrated in broad outline as shown in Figure 28.12. Oxidation of vitamin A in the liver followed by isomerization of one of its double bonds gives 11-*cis*-retinal. In the eye the aldehyde function of

β-Carotene obtained from the diet is cleaved at its central carbon-carbon bond to give vitamin A (retinol)

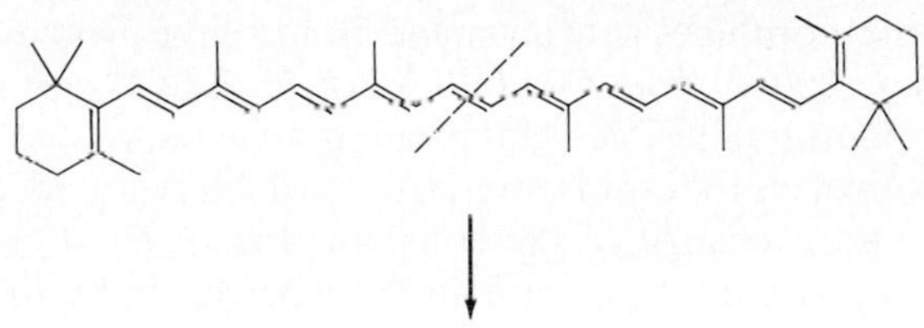

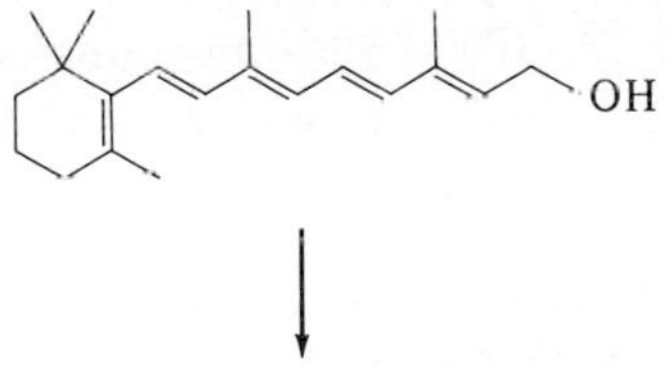

Oxidation of retinol converts it to the corresponding aldehyde retinal.

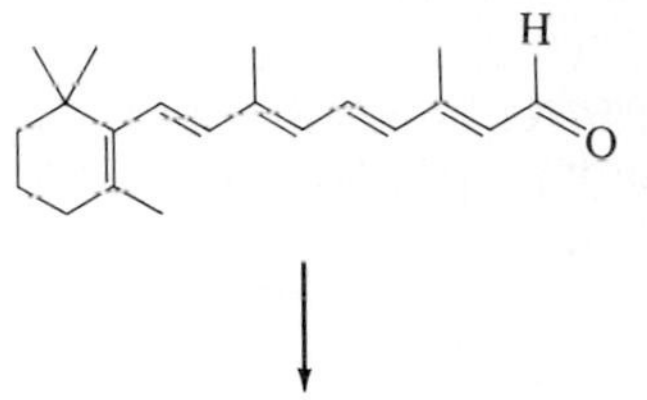

The double bond at C-11 is isomerized from the trans to the cis configuration.

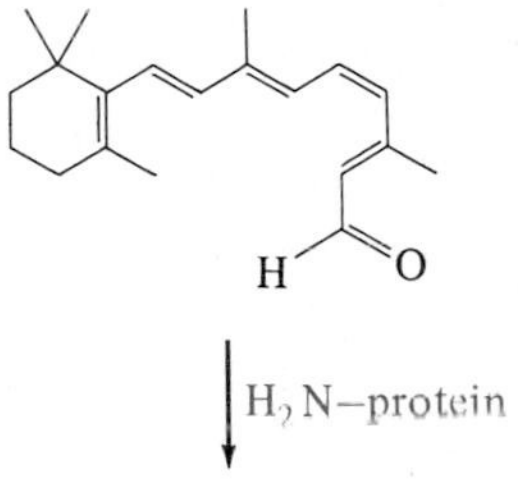

11-*cis*-Retinal is the biologically active stereoisomer and reacts with the protein opsin to form an imine. The covalently bound complex between 11-*cis*-retinal and opsin is called rhodopsin

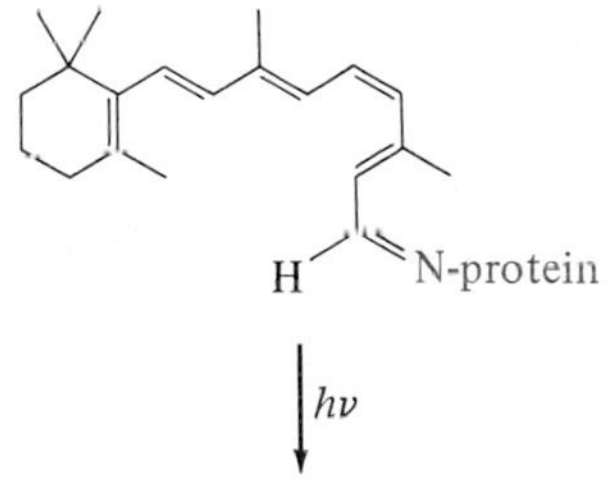

H_2N—protein

Rhodopsin absorbs a photon of light, causing the cis double at C-11 to undergo a photochemical transformation to trans, which triggers a nerve impulse detected by the brain as a visual image.

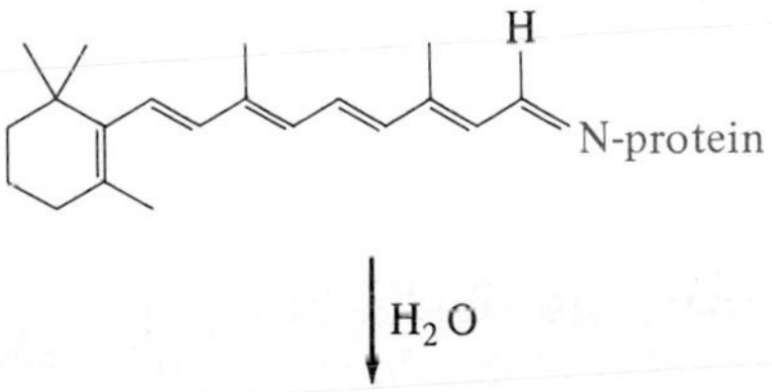

$h\nu$

Hydrolysis of the isomerized (inactive) form of rhodopsin liberates opsin and the all-trans isomer of retinal.

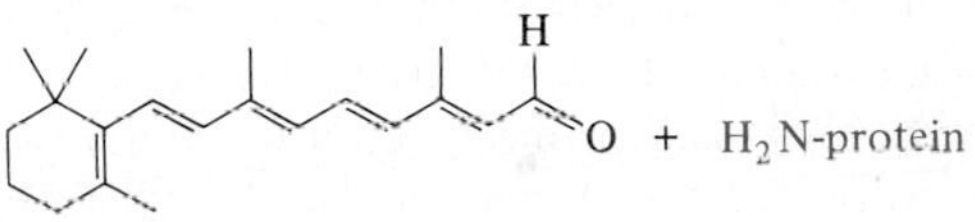

H_2O

$+ \ H_2N$-protein

FIGURE 28.12 Summary diagram illustrating the role of vitamin A in the chemistry of vision.

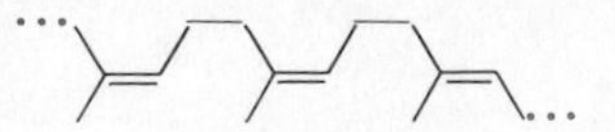

(a) Natural rubber

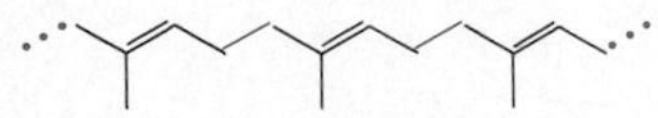

(b) Gutta percha

FIGURE 28.13 Portions of the carbon skeletons of (a) natural rubber and (b) gutta percha. Both substances are polyisoprenes but the double bonds are cis (Z) in natural rubber and trans (E) in gutta percha. Each substance typically contains several thousand isoprene units.

11-*cis*-retinal combines with an amino group of the protein *opsin* to form an imine or Schiff's base called *rhodopsin* (also referred to as *visual purple*). When rhodopsin absorbs a photon of visible light, the cis double bond of the retinal unit undergoes a photochemical cis to trans isomerization, which is attended by a dramatic change in its shape and a change in the conformation of rhodopsin. This conformational change is translated into a nerve impulse perceived by the brain as a visual image. Enzyme-promoted hydrolysis of the photochemically isomerized rhodopsin regenerates opsin and a molecule of all-*trans*-retinal. Once all-*trans*-retinal has been enzymatically converted to its 11-cis isomer, it and opsin reenter the cycle shown in Figure 28.12.

28.17 POLYISOPRENES. NATURAL RUBBER

On one of his voyages to the New World, Christopher Columbus observed some of the native population playing with balls made from the latex of trees indigenous to the region. Later, Joseph Priestley called this material "rubber" to describe its ability to remove pencil marks by rubbing.

Rubber is a naturally occurring *polyisoprene* or polymer of isoprene. Its structural formula may be represented as in Figure 28.13a. A noteworthy feature of its structure is the cis (or Z) configuration of its double bonds.

There is another naturally occurring polyisoprene called *gutta percha,* a tough, hornlike substance once used in materials such as golf ball covers. As illustrated in Figure 28.13b, gutta percha has the trans (or E) configuration of its double bonds.

The trans stereochemistry of the double bonds in gutta percha permits neighboring molecules to pack together efficiently, producing a three-dimensional network held together by attractive van der Waals forces. The corresponding forces are weaker in natural rubber, where the cis double bonds preclude close contact between neighboring chains. This allows adjacent molecules to slide past each other and makes it an easy matter to stretch natural rubber. The ability of natural rubber to return to its original shape, its *elasticity,* is not very great but is substantially enhanced by *vulcanization,* a process discovered by Charles Goodyear in 1839. When natural rubber is heated with sulfur, a chemical reaction occurs in which neighboring polyisoprene chains are connected by sulfur bridges. While these bridges permit movement of one chain with respect to another, their presence ensures that the rubber will return to its original shape once the distorting force is removed.

28.18 SUMMARY

Chemists and biochemists find it convenient to divide the principal organic substances present in cells into four main groups: carbohydrates, proteins, nucleic acids, and lipids. Structural differences separate carbohydrates from proteins, and both of these are structurally distinct from nucleic acids. Lipids, on the other hand, are characterized by a physical property, their solubility in nonpolar solvents, rather than by their structure. In this chapter we have examined a number of types of lipid molecules, those which are biosynthetically derived from acetate. The range of substances examined, while encompassing neither all classes of lipids nor all classes of acetate-derived natural products, does include the most important examples of both.

Acetyl coenzyme A (Figure 28.1, Section 28.1) is a key substance in biological chemistry. It is the biosynthetic building block that leads to the *fatty acids* (Section 28.3) and the *terpenes* (Section 28.10). Fatty acids and terpenes are not only of substantial interest in their own right, but they are also the biological precursors of other classes of important natural products.

Fatty acids most often occur naturally as esters. *Fats and oils* are glycerol esters of long-chain carboxylic acids. Typically, the carbon chains of fatty acids are unbranched and contain even numbers of carbon atoms.

Triacylglycerol
(R, R′, and R″ may be the same or different)

Phospholipids (Section 28.4) are intermediates in the biosynthesis of triacylglycerols from fatty acids and are the principal constituents of cell membranes. *Waxes* (Section 28.5) are mixtures of substances that usually contain esters of fatty acids and long-chain alcohols. A group of compounds called *prostaglandins* (Section 28.6) have hormonelike activity and are potent regulators of biochemical processes. The prostaglandins are biosynthesized from C_{20} fatty acids by a combination of reactions that lead to oxygen incorporation and five-membered ring formation. Terpenes and related *isoprenoid* compounds are biosynthesized from acetate by way of mevalonate and isopentenyl pyrophosphate. The processes that form carbon-carbon bonds between isoprene units can be understood on the basis of nucleophilic attack of the π electrons of a double bond on a carbocation or carbocation precursor (Section 28.9). The pathway from acetate to *steroids* proceeds by way of the triterpene squalene. Enzymatic cyclization of 2,3-epoxysqualene gives lanosterol, which in turn is converted to cholesterol (Section 28.11). Cholesterol is an abundant natural product and is the precursor of other physiologically important steroids and steroidal hormones. *Carotenoids* are tetraterpenes (Section 28.16), which most commonly occur as plant pigments. The diterpene vitamin A is derived by cleavage of β-carotene and is a key substance in the chemistry of vision.

PROBLEMS

28.15 Identify the carbon atoms expected to be labeled with ^{14}C when each of the following substances is biosynthesized from acetate enriched with ^{14}C in its methyl group:

(a) $CH_3(CH_2)_{14}CO_2H$ Palmitic acid

(b)

PGE$_2$

(c)

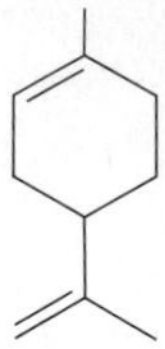

Limonene

(d)

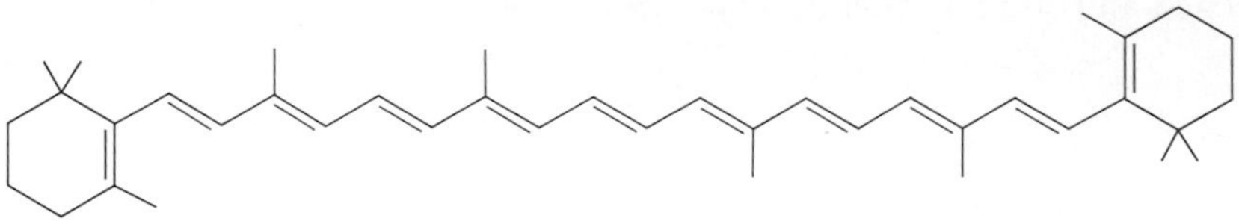

β-Carotene

28.16 The biosynthetic pathway to prostaglandins described in Figure 28.6 (Section 28.6) proceeds via endoperoxide intermediates of the PGG and PGH type. These intermediates are also precursors of a class of physiologically potent substances known as *prostacyclins*. Which carbon atoms of the prostacyclin shown below would you expect to be enriched in ^{14}C if it were biosynthesized from acetate labeled with ^{14}C in its methyl group?

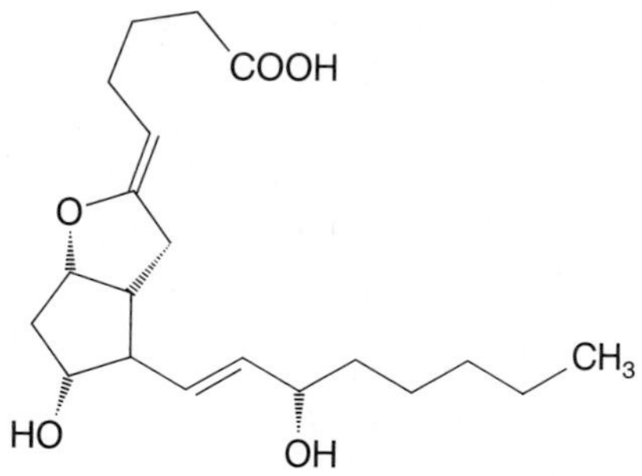

28.17 Identify the isoprene units in each of the following naturally occurring substances:

(a) *Ascaridole:* A naturally occurring peroxide present in chenopodium oil.

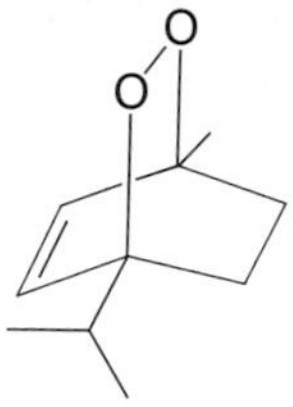

(b) *Dendrolasin:* A constituent of the defense secretion of a species of ant.

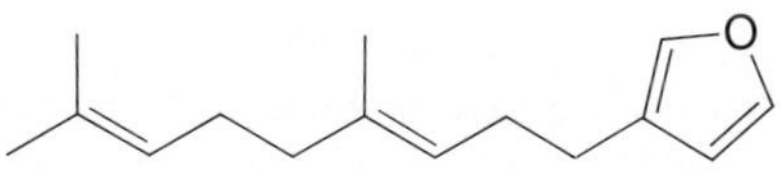

(c) *γ-Bisabolene:* A sesquiterpene found in the essential oils of a large number of plants.

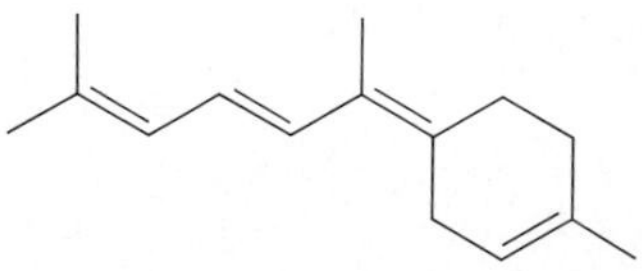

(d) *α-Santonin:* An anthelmintic substance isolated from artemisia flowers.

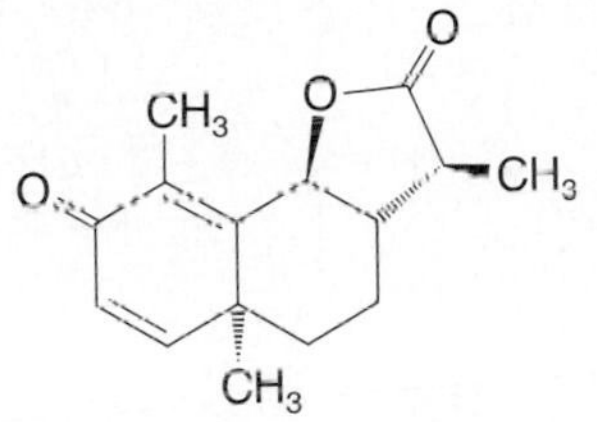

(e) *Tetrahymanol:* A pentacyclic triterpene isolated from a species of protozoa.

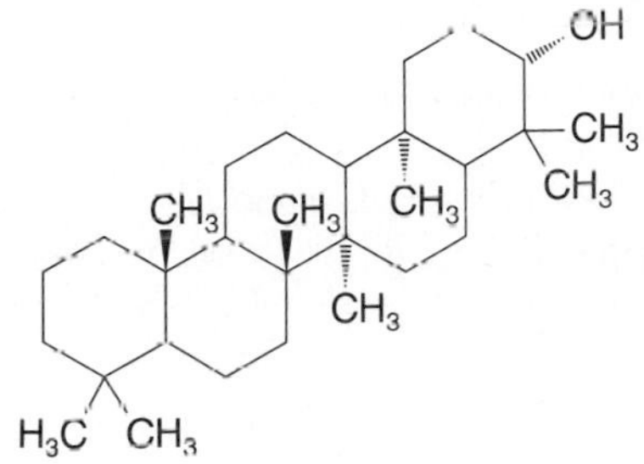

28.18 Cubitene is a diterpene present in the defense secretion of a species of African termite. What unusual feature characterizes the joining of isoprene units in cubitene?

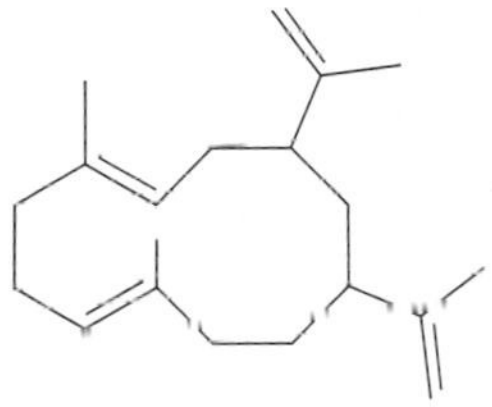

28.19 *Pyrethrins* are a group of naturally occurring insecticidal substances found in the flowers of various plants of the chrysanthemum family. The structure of a typical pyrethrin, *cinerin I,* is shown below (exclusive of stereochemistry).

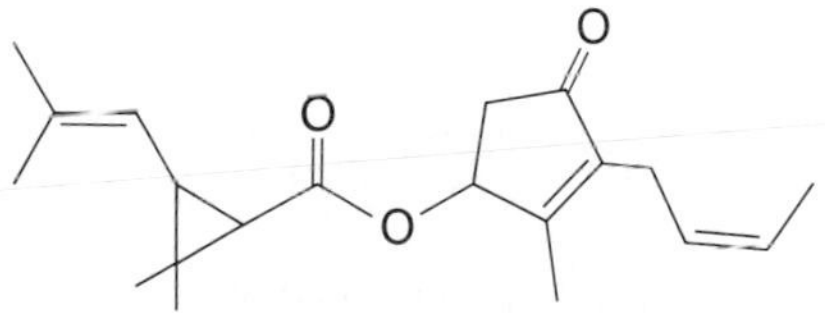

(a) Locate any isoprene units present in cinerin I.

(b) Hydrolysis of cinerin I gives an optically active carboxylic acid, (+)-chrysanthemic acid. Ozonolysis of (+)-chrysanthemic acid, followed by oxidation, gives acetone and an optically active dicarboxylic acid, (−)-caronic acid ($C_7H_{10}O_4$). What is the structure of (−)-caronic acid? Are the two carboxyl groups cis or trans to each other? What does this information tell you about the structure of (+)-chrysanthemic acid?

28.20 *Cerebrosides* are found in the brain and in the myelin sheath of nerve tissue. The structure of the cerebroside *phrenosine* is shown on page 1082.

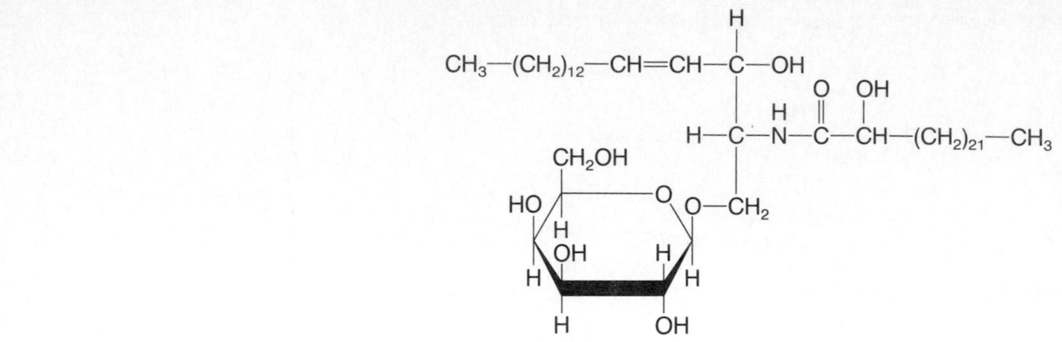

(a) What hexose is formed on hydrolysis of the glycoside bond of phrenosine? Is phreno-sine an α- or a β-glycoside?

(b) Hydrolysis of phrenosine gives, in addition to the hexose in (a), a fatty acid called *cerebronic acid* along with a third substance called *sphingosine*. Write structural formulas for both cerebronic acid and sphingosine.

28.21 Each of the following reactions has been reported in the chemical literature and proceeds in good yield. What are the principal organic products of each reaction? (In some of the exercises more than one diastereomer may be theoretically possible, but in such instances one diastereomer is either the major product or the only product. For those reactions in which one diastereomer is formed preferentially, indicate its expected stereochemistry.)

(a) $CH_3(CH_2)_7C\equiv C(CH_2)_7COOH + H_2 \xrightarrow{\text{Lindlar Pd}}$

(b) $CH_3(CH_2)_7C\equiv C(CH_2)_7COOH \xrightarrow[\text{2. H}^+]{\text{1. Li, NH}_3}$

(c) $(Z)-CH_3(CH_2)_7CH=CH(CH_2)_7\overset{\displaystyle O}{\overset{\displaystyle \|}{C}}OCH_2CH_3 + H_2 \xrightarrow{\text{Pt}}$

(d) $(Z)-CH_3(CH_2)_5\underset{\underset{\displaystyle OH}{|}}{C}HCH_2CH=CH(CH_2)_7\overset{\displaystyle O}{\overset{\displaystyle \|}{C}}OCH_3 \xrightarrow[\text{2. H}_2\text{O}]{\text{1. LiAlH}_4}$

(e) $(Z)-CH_3(CH_2)_7CH=CH(CH_2)_7COOH + C_6H_5CO_2OH \longrightarrow$

(f) Product of (e) + $H_3O^+ \longrightarrow$

(g) $(Z)-CH_3(CH_2)_7CH=CH(CH_2)_7COOH \xrightarrow[\text{2. H}^+]{\text{1. OsO}_4, \text{(CH}_3)_3\text{COOH, HO}^-}$

(h) $CH_3(CH_2)_{14}\overset{\displaystyle O}{\overset{\displaystyle \|}{C}}OCH_2(CH_2)_{14}CH_3 \xrightarrow[\text{copper chromite}]{\text{H}_2 \text{ (high pressure, 250°C}}$

(i)

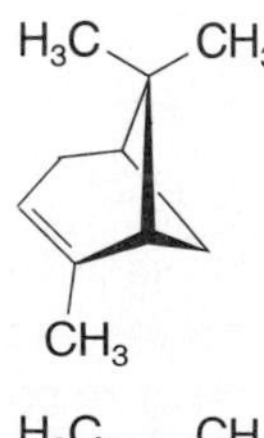

$\xrightarrow[\text{2. H}_2\text{O}_2, \text{HO}^-]{\text{1. B}_2\text{H}_6, \text{diglyme}}$

(j)

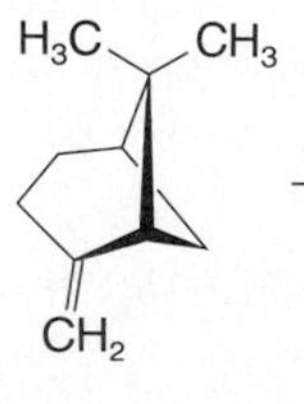

$\xrightarrow[\text{2. H}_2\text{O}_2, \text{HO}^-]{\text{1. B}_2\text{H}_6, \text{diglyme}}$

(k)

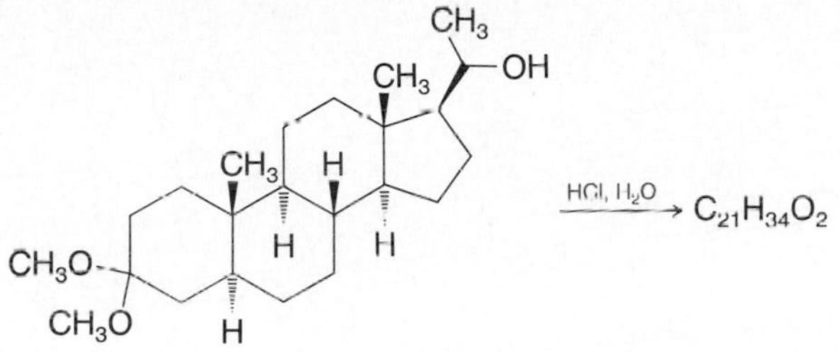

$$\xrightarrow{\text{HCl, H}_2\text{O}} C_{21}H_{34}O_2$$

(l)

$$\xrightarrow[\text{Zn}]{\text{BrCH}_2\text{COOCH}_2\text{CH}_3}$$

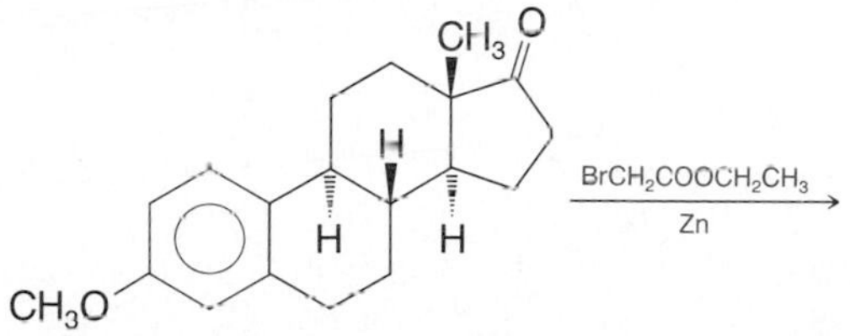

28.22 Describe an efficient synthesis of each of the following compounds from octadecanoic (stearic) acid using any necessary organic or inorganic reagents:

(a)	Octadecane	(e)	Icosanoic acid
(b)	1-Phenyloctadecane	(f)	1-Heptadecanamine
(c)	1-Phenylicosane	(g)	1-Octadecanamine
(d)	3-Ethylicosane	(h)	1-Nonadecanamine

28.23 A synthesis of triacylglycerols has been described that begins with the substance shown.

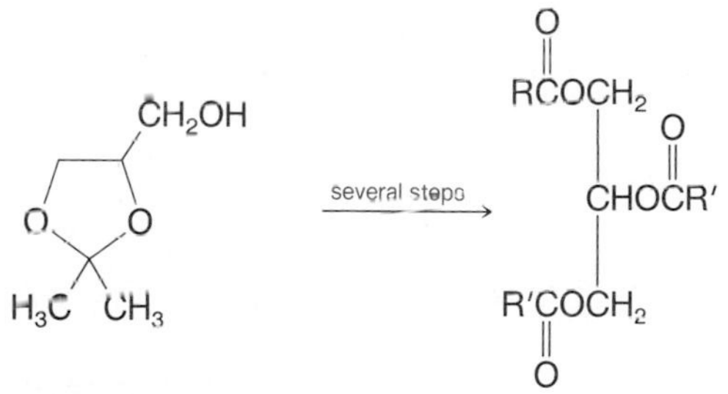

4-(Hydroxymethyl)-
2,2-dimethyl-1,3-dioxolane

Triacylglycerol

Outline a series of reactions suitable for the preparation of a triacylglycerol of the type illustrated in the equation, where R and R′ are different.

28.24 The isoprenoid compound shown is a scent marker present in the urine of the red fox. Suggest a reasonable synthesis for this substance from 3-methyl-3-buten-1-ol and any necessary organic or inorganic reagents.

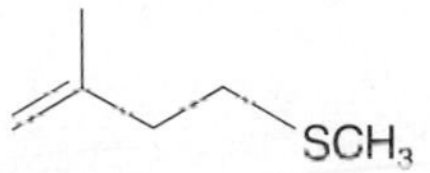

28.25 Sabinene is a monoterpene found in the oil of citrus fruits and plants. It has been synthesized from 6-methyl-2,5-heptanedione by the sequence shown on page 1084. Suggest reagents suitable for carrying out each of the indicated transformations.

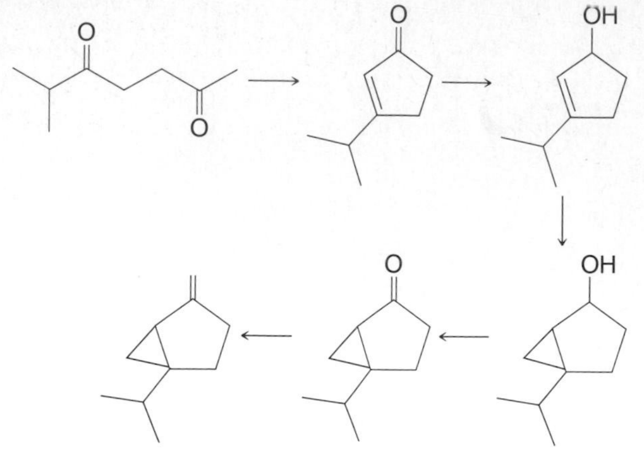

Sabinene

28.26 Isoprene has sometimes been used as a starting material in the laboratory synthesis of terpenes. In one such synthesis, the first step is the electrophilic addition of two moles of hydrogen bromide to isoprene to give 1,3-dibromo-3-methylbutane.

$$CH_2{=}\overset{\underset{\displaystyle CH_3}{|}}{C}CH{=}CH_2 \quad + \quad 2HBr \quad \longrightarrow \quad (CH_3)_2\overset{\underset{\displaystyle Br}{|}}{C}CH_2CH_2Br$$

| 2-Methyl-1,3-butadiene (isoprene) | Hydrogen bromide | 1,3-Dibromo-3-methylbutane |

Write a series of equations describing the mechanism of this reaction.

28.27 The ionones are fragrant substances present in the scent of iris and are used in perfume. A mixture of α- and β-ionone can be prepared by treatment of pseudoionone with sulfuric acid.

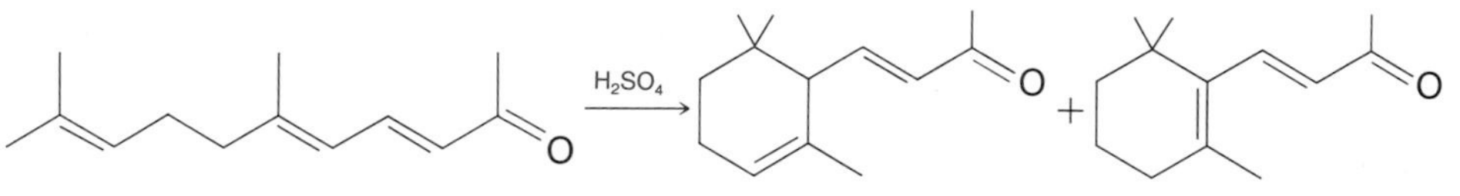

Pseudoionone $\xrightarrow{H_2SO_4}$ α-Ionone   + β-Ionone

Write a stepwise mechanism for this reaction.

28.28 β,γ-Unsaturated steroidal ketones represented by the partial structure shown below are readily converted in acid to their α,β-unsaturated isomers. Write a stepwise mechanism for this reaction.

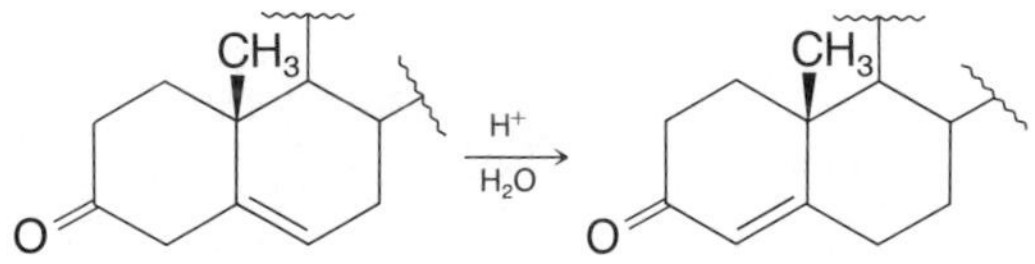

AMINO ACIDS, PEPTIDES, AND PROTEINS. NUCLEIC ACIDS

The relationship between structure and function achieves its ultimate expression in the chemistry of amino acids, peptides, and proteins.

Amino acids are carboxylic acids that contain an amine function. Under certain conditions the amine group of one molecule and the carboxyl group of a second can react, uniting the two amino acids by an amide bond.

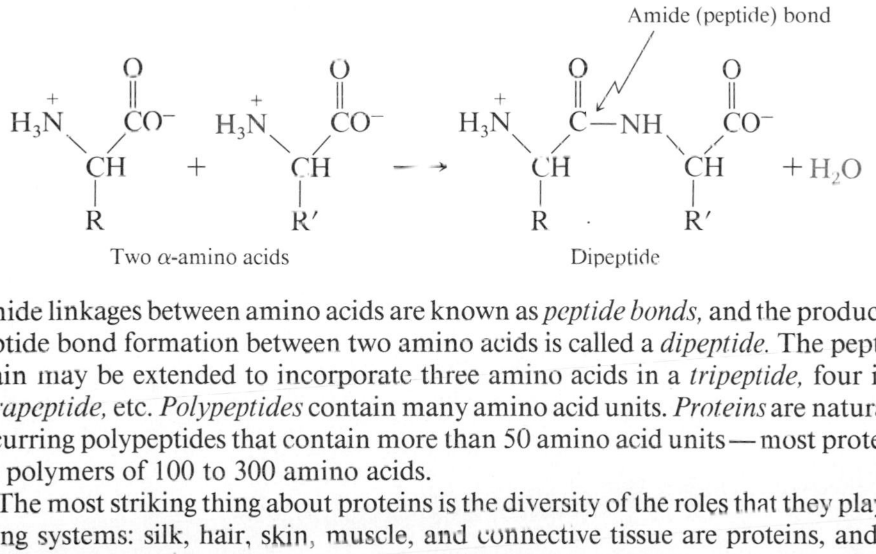

Amide (peptide) bond

Two α-amino acids → Dipeptide + H_2O

Amide linkages between amino acids are known as *peptide bonds,* and the product of peptide bond formation between two amino acids is called a *dipeptide.* The peptide chain may be extended to incorporate three amino acids in a *tripeptide,* four in a *tetrapeptide,* etc. *Polypeptides* contain many amino acid units. *Proteins* are naturally occurring polypeptides that contain more than 50 amino acid units — most proteins are polymers of 100 to 300 amino acids.

The most striking thing about proteins is the diversity of the roles that they play in living systems: silk, hair, skin, muscle, and connective tissue are proteins, and all enzymes are proteins. As in most aspects of chemistry and biochemistry, structure is the key to function. We will explore the facets of protein structure by first concentrating on their fundamental building block units, the α-amino acids. Then, after developing the principles of peptide structure and conformation, you will see how the insights gained from these smaller molecules aid our understanding of proteins.

The chapter concludes with a discussion of the *nucleic acids,* which are the genetic material of living systems and direct the biosynthesis of proteins. These two types of biopolymers, nucleic acids and proteins, are the organic chemicals of life.

TABLE 29.1

α-Amino Acids Found in Proteins

Name	Abbreviation	Structural formula†
Amino acids with nonpolar side chains		
Glycine	Gly (G)	$\overset{\overset{+}{N}H_3}{H-CHCO_2^-}$
Alanine	Ala (A)	$\overset{\overset{+}{N}H_3}{CH_3-CHCO_2^-}$
Valine*	Val (V)	$\overset{\overset{+}{N}H_3}{(CH_3)_2CH-CHCO_2^-}$
Leucine*	Leu (L)	$\overset{\overset{+}{N}H_3}{(CH_3)_2CHCH_2-CHCO_2^-}$
Isoleucine*	Ile (I)	$\overset{CH_3 \quad \overset{+}{N}H_3}{CH_3CH_2CH-CHCO_2^-}$
Methionine*	Met (M)	$\overset{\overset{+}{N}H_3}{CH_3SCH_2CH_2-CHCO_2^-}$
Proline	Pro (P)	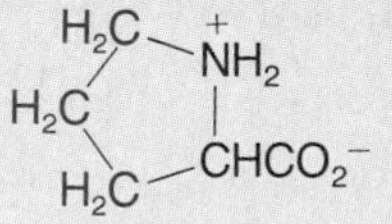 ![proline structure] $H_2C, H_2C, \overset{+}{N}H_2, H_2C-CHCO_2^-$
Phenylalanine*	Phe (F)	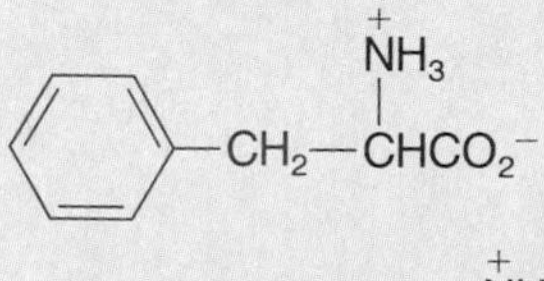 $C_6H_5-CH_2-\overset{\overset{+}{N}H_3}{CHCO_2^-}$
Tryptophan*	Trp (W)	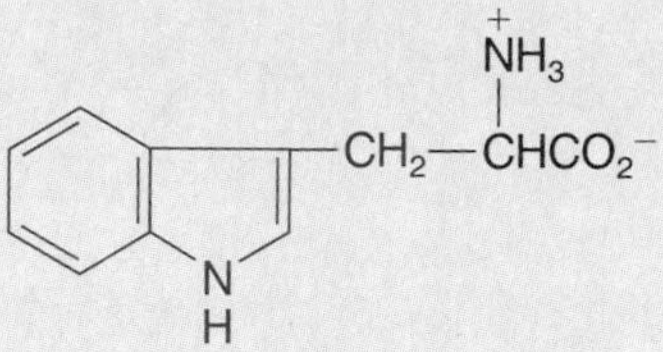 indole$-CH_2-\overset{\overset{+}{N}H_3}{CHCO_2^-}$
Amino acids with polar, but nonionized side chains		
Asparagine	Asn (N)	$\overset{O}{H_2NCCH_2}-\overset{\overset{+}{N}H_3}{CHCO_2^-}$
Glutamine	Gln (Q)	$\overset{O}{H_2NCCH_2CH_2}-\overset{\overset{+}{N}H_3}{CHCO_2^-}$
Serine	Ser (S)	$\overset{\overset{+}{N}H_3}{HOCH_2-CHCO_2^-}$
Threonine*	Thr (T)	$\overset{OH \quad \overset{+}{N}H_3}{CH_3CH-CHCO_2^-}$

TABLE 29.1 (continued)

Name	Abbreviation	Structural formula†
Amino acids with acidic side chains		
Aspartic acid	Asp (D)	$^-OCCH_2-CHCO_2^-$ with O and $\overset{+}{N}H_3$
Glutamic acid	Glu (E)	$^-OCCH_2CH_2-CHCO_2$ with O and $\overset{+}{N}H_3$
Tyrosine	Tyr (Y)	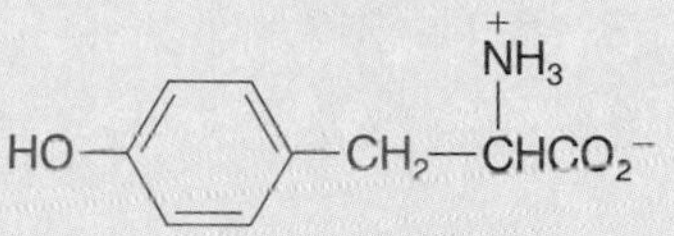 $HO-\!\!\left\langle\!\!\bigcirc\!\!\right\rangle\!\!-CH_2-CHCO_2^-$ with $\overset{+}{N}H_3$
Cysteine	Cys (C)	$HSCH_2-CHCO_2^-$ with $\overset{+}{N}H_3$
Amino acids with basic side chains		
Lysine*	Lys (K)	$H_3\overset{+}{N}CH_2CH_2CH_2CH_2-CHCO_2^-$ with $\overset{+}{N}H_3$
Arginine*	Arg (R)	$H_2\overset{+}{N}CNHCH_2CH_2CH_2-CHCO_2^-$ with $\overset{+}{N}H_2$ and $\overset{+}{N}H_3$
Histidine*	His (H)	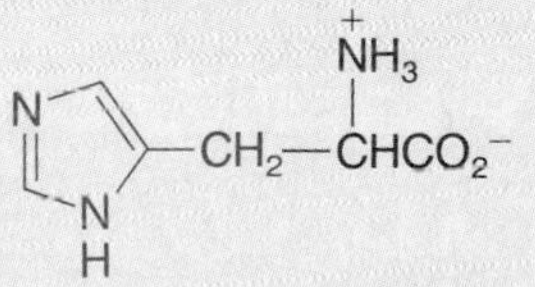 imidazole$-CH_2-CHCO_2^-$ with $\overset{+}{N}H_3$

An asterisk (*) designates an essential amino acid, which must be present in the diet of mammals to ensure normal growth.

† All amino acids are shown in the form present in greatest concentration at pH 7.

29.1 CLASSIFICATION OF AMINO ACIDS

Amino acids are classified as α, β, γ, etc. according to the location of the amine group on the carbon chain that contains the carboxylic acid function.

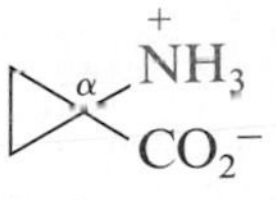

$\overset{+}{N}H_3$ / CO_2^- (1-aminocyclopropanecarboxylic acid structure)

1-Aminocyclopropanecarboxylic acid: an α-amino acid that is the biological precursor to ethylene in plants

$H_3\overset{+}{N}CH_2CH_2CO_2^-$
    β   α

3-Aminopropanoic acid: known as β-alanine, it is a β-amino acid that comprises one of the structural units of coenzyme A

$H_3\overset{+}{N}CH_2CH_2CH_2CO_2^-$
    γ   β   α

4-Aminobutanoic acid: known as γ-aminobutyric acid (GABA), it is a γ-amino acid and is involved in the transmission of nerve impulses

While several hundred different amino acids are known to occur naturally, chemists and biochemists recognize a group of 20 of them as deserving of special attention. These are the amino acids that are normally present in proteins and are listed in Table 29.1. All the amino acids from which proteins are derived are α-amino acids, and all but one of these contain a primary amino function and conform to the general structure

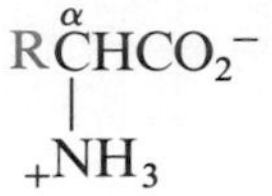

$$\overset{\alpha}{R\overset{|}{C}HCO_2^-}$$
$$\underset{+NH_3}{|}$$

Proline is a secondary amine in which the amino nitrogen and the substituent are involved in ring formation.

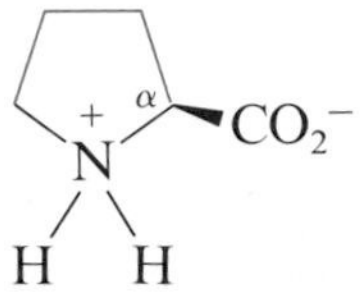

Proline

Table 29.1 includes the customary three-letter abbreviations for the common amino acids as well as their newer one-letter abbreviations.

While humans possess the capacity to biosynthesize some of the amino acids shown in the table, they must obtain certain of the others from their diet. Those that must be included in our dietary requirements are called *essential amino acids* and are identified as such in the table.

29.2 STEREOCHEMISTRY OF AMINO ACIDS

Glycine (aminoacetic acid) is the simplest of the amino acids in Table 29.1 and is the only one that is achiral. The α carbon atom is a chiral center in all the others. Configurations in amino acids are normally specified by the D, L notational system. All the chiral amino acids obtained from proteins have the L configuration at their α carbon atom.

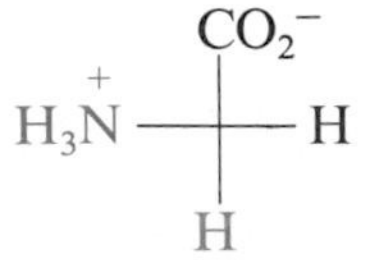

Glycine
(achiral)

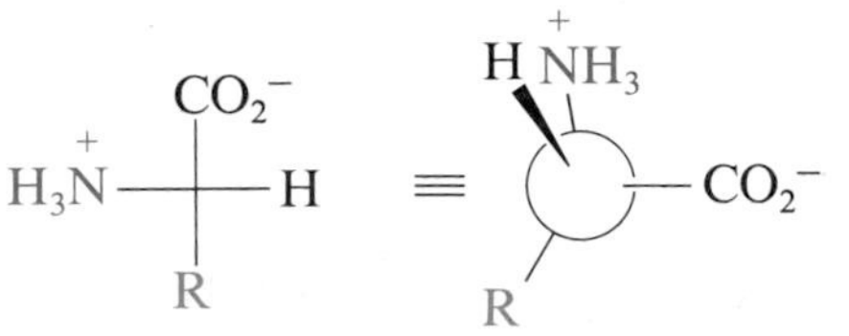

Fischer projection formula
of an L amino acid

PROBLEM 29.1   What is the absolute configuration (R or S) at the α carbon atom in each of the following L amino acids?

(a)   L-Serine   (b)   L-Cysteine   (c)   L-Methionine

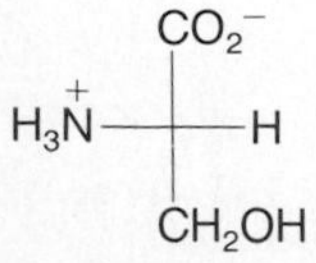

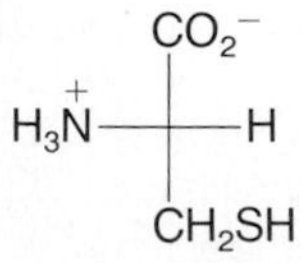

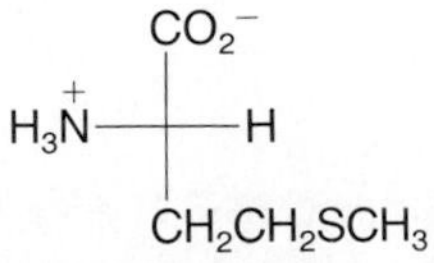

SAMPLE SOLUTION (a) First identify the four substituents attached directly to the chiral center and rank them in order of decreasing sequence rule precedence. For L-serine the substituents are:

$$H_3\overset{+}{N}- \quad > -CO_2^- > -CH_2OH > \quad H$$

Highest ranked   Lowest ranked

Next, translate the Fischer projection formula of L-serine to a three-dimensional represen-tation and orient it so the lowest-ranked substituent at the chiral center is directed away from you.

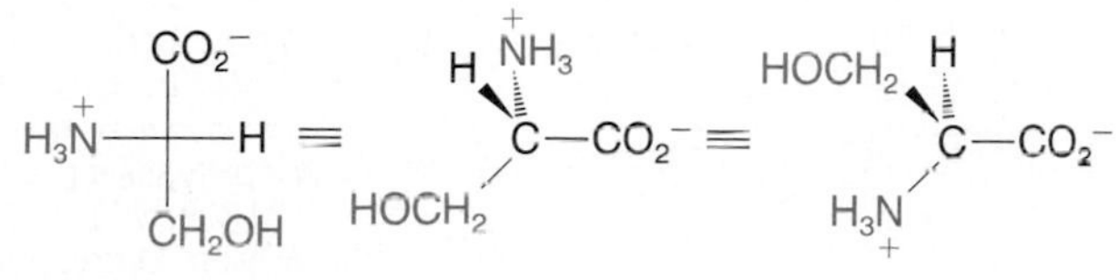

The order of decreasing sequence rule precedence of the three highest-ranked substitu-ents traces an anticlockwise path.

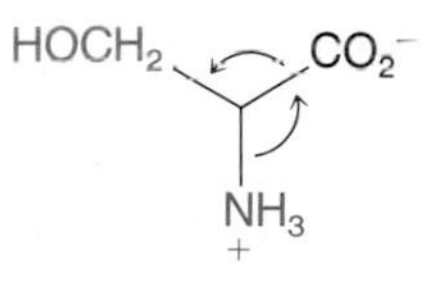

The absolute configuration of L-serine is *S*.

PROBLEM 29.2 Which of the amino acids in Table 29.1 have more than one chiral center?

While all the chiral amino acids obtained from proteins have the L configuration at their α carbon, that should not be taken to mean that D amino acids are unknown. There are, in fact, quite a number of naturally occurring D amino acids. D-Alanine, for example, is a constituent of bacterial cell walls. The significant point is that D amino acids are not constituents of proteins.

29.3 ACID-BASE BEHAVIOR OF AMINO ACIDS

A typical amino acid such as glycine has physical properties that suggest it is a very polar substance, much more polar than would be expected on the basis of its formu-lation as $H_2NCH_2CO_2H$. Glycine is a crystalline solid; it does not melt, but on being heated it eventually decomposes at 233°C. It is very soluble in water but practically insoluble in nonpolar organic solvents. These properties are attributed to the fact that glycine exists as a *zwitterion* or *inner salt*.

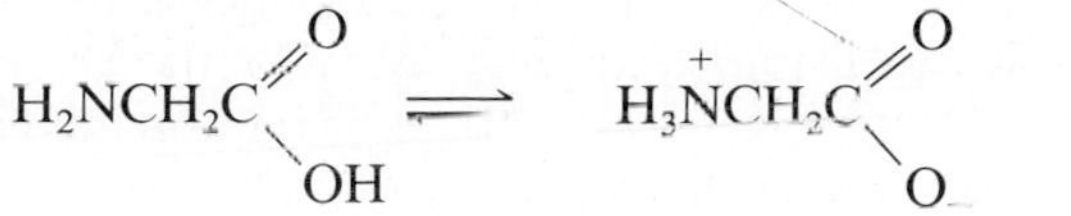

Zwitterionic form of glycine

The equilibrium expressed by the equation lies overwhelmingly to the side of the zwitterion.

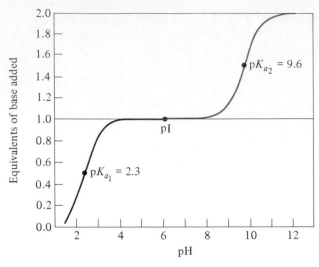

FIGURE 29.1 Representation of the titration curve of glycine. At pH values less than pK_{a_1}, $\overset{+}{H_3N}CH_2CO_2H$ is the major species present. At pH values between pK_{a_1} and pK_{a_2}, the principal species is the zwitterion $\overset{+}{H_3N}CH_2CO_2^-$. The concentration of the zwitterion is a maximum at the isoelectric point pI. At pH values greater than pK_{a_2}, $H_2NCH_2CO_2^-$ is the species present in greatest concentration.

Glycine, as well as other amino acids, is *amphoteric,* containing an acidic functional group and a basic functional group. The acidic functional group is the ammonium ion $\overset{+}{H_3N}-$; the basic functional group is the carboxylate ion $-CO_2^-$. How do we know this? Aside from the physical properties cited in the preceding paragraph, the acid-base properties of glycine require such a conclusion. Consider the titration behavior of glycine in proceeding from a strongly acidic medium to one that is strongly basic, as illustrated in Figure 29.1. In a strongly acidic medium the species present is $\overset{+}{H_3N}CH_2CO_2H$. As the pH is raised, a proton is removed from this species. Is the proton removed from the positively charged nitrogen or from the carboxyl group? We know what to expect for the relative acid strengths of $\overset{+}{R N H_3}$ and RCO_2H. A typical ammonium ion has $pK_a \cong 9$, and a typical carboxylic acid has $pK_a \cong 5$. The measured pK_a for the most acidic site in the conjugate acid of glycine is 2.35, a value much more in accord with that expected for deprotonation of the carboxyl group. As the pH is raised, a second deprotonation step, corresponding to removal of a proton from nitrogen of the zwitterion, is observed. The pK_a associated with this step is 9.78, much like that of typical alkylammonium ions.

$$\overset{+}{H_3N}CH_2C\!\!\begin{array}{c}O\\ \diagdown\\ OH\end{array} \underset{+H^+}{\overset{-H^+}{\rightleftharpoons}} \overset{+}{H_3N}CH_2C\!\!\begin{array}{c}O\\ \diagdown\\ O^-\end{array} \underset{+H^+}{\overset{-H^+}{\rightleftharpoons}} H_2NCH_2C\!\!\begin{array}{c}O\\ \diagdown\\ O^-\end{array}$$

Species present in strong acid | Zwitterion; predominant species in solutions near neutrality | Species present in strong base

Thus, glycine is characterized by two pK_a values: the one corresponding to the more acidic site is designated pK_{a_1}, the one corresponding to the less acidic site is designated pK_{a_2}. Table 29.2 lists pK_{a_1} and pK_{a_2} values for the α-amino acids that have neutral side chains, which are the first two groups of amino acids given in Table 29.1. In all cases their pK_a values are similar to those of glycine and indicate an acid-base behavior that parallels that of glycine.

Table 29.2 includes a column labeled pI, which gives *isoelectric point* values. The isoelectric point is the pH at which the amino acid bears no net charge; it corresponds

TABLE 29.2
Acid-Base Properties of Amino Acids with Neutral Side Chains

Amino acid	pK_{a_1}*	pK_{a_2}*	pI
Glycine	2.34	9.60	5.97
Alanine	2.34	9.69	6.00
Valine	2.32	9.62	5.96
Leucine	2.36	9.60	5.98
Isoleucine	2.36	9.60	6.02
Methionine	2.28	9.21	5.74
Proline	1.99	10.60	6.30
Phenylalanine	1.83	9.13	5.48
Tryptophan	2.83	9.39	5.89
Asparagine	2.02	8.80	5.41
Glutamine	2.17	9.13	5.65
Serine	2.21	9.15	5.68
Threonine	2.09	9.10	5.60

* In all cases pK_{a_1} corresponds to ionization of the carboxyl group; pK_{a_2} corresponds to ionization of the ammonium ion.

to the pH at which the concentration of the zwitterion is a maximum. For the amino acids in Table 29.2 this is the average of pK_{a_1} and pK_{a_2}. It lies slightly to the acid side of neutrality because amino acids are somewhat stronger acids than bases.

Some amino acids, including those listed in the last two sections of Table 29.1 have side chains that bear acidic or basic groups. As Table 29.3 indicates, these amino acids are characterized by three pK_a values. The "extra" pK_a value (it can be either pK_{a_2} or pK_{a_3}) reflects the nature of the function present in the side chain. The isoelectric points of the amino acids in Table 29.3 are midway between the pK_a values of the monocation and monoanion and are well removed from neutrality when the side chain bears a carboxyl function (aspartic acid, for example) or a basic amine function (lysine, for example).

TABLE 29.3
Acid-Base Properties of Amino Acids with Ionizable Side Chains

Amino acid	pK_{a_1}	pK_{a_2}*	pK_{a_3}	pI
Aspartic acid	1.88	3.65	9.60	2.77
Glutamic acid	2.19	4.25	9.67	3.22
Tyrosine	2.20	9.11	10.07	5.66
Cysteine	1.96	8.18	10.28	5.07
Lysine	2.18	8.95	10.53	9.74
Arginine	2.17	9.04	12.48	10.76
Histidine	1.82	6.00	9.17	7.59

In all cases pK_{a_1} corresponds to ionization of the carboxyl group of RCHCO$_2$H.
$$\underset{\overset{|}{\underset{+}{NH_3}}}{}$$

PROBLEM 29.3   Write the most stable structural formula for tyrosine:

(a) In its cationic form   (c) As a monoanion
(b) In its zwitterionic form   (d) As a dianion

SAMPLE SOLUTION   (a) The cationic form of tyrosine is the one present at low pH. The positive charge is on nitrogen and the species present is an ammonium ion.

$$HO-\bigcirc-CH_2CHCO_2H$$
$$\overset{\underset{|}{\overset{|}{NH_3}}}{} \atop +$$

PROBLEM 29.4   Write structural formulas for the principal species present when the pH of a solution containing lysine is raised from 1 to 9 to 13.

The acid-base properties of their side chains are one way in which individual amino acids differ. This is important in peptides and proteins, where the properties of the substance depend on its amino acid constituents, especially on the nature of the side chains. It is also important in analyses in which a complex mixture of amino acids is separated into its components by taking advantage of the differences in their proton-donating and proton-accepting abilities.

29.4 SYNTHESIS OF AMINO ACIDS

There are a variety of methods available for the laboratory synthesis of amino acids. One of the oldest dates back over 100 years and is simply a nucleophilic substitution in which ammonia reacts with an α-halo carboxylic acid.

$$\underset{\underset{\text{2-Bromopropanoic acid}}{Br}}{CH_3CHCO_2H} + \underset{\text{Ammonia}}{2NH_3} \xrightarrow{H_2O} \underset{\underset{\underset{+}{NH_3}}{\text{Alanine (65–70\%)}}}{CH_3CHCO_2^-} + \underset{\text{Ammonium bromide}}{NH_4Br}$$

The α-halo acid is normally prepared by the Hell-Volhard-Zelinsky reaction (Section 20.16).

PROBLEM 29.5   Outline the steps in a synthesis of valine from 3-methylbutanoic acid.

In the *Strecker synthesis* an aldehyde is converted to an α-amino acid with one more carbon atom by a two-stage procedure in which an α-amino nitrile is an intermediate. The α-amino nitrile is formed by reaction of the aldehyde with ammonia or an ammonium salt and a source of cyanide ion. Hydrolysis of the nitrile group to a carboxylic acid function completes the synthesis.

$$\underset{\text{Acetaldehyde}}{\overset{\overset{O}{\|}}{CH_3CH}} \xrightarrow[\text{NaCN}]{NH_4Cl} \underset{\underset{\text{2-Aminopropanenitrile}}{NH_2}}{CH_3CHC\equiv N} \xrightarrow[\text{2. HO}^-]{\text{1. H}_2\text{O, HCl, heat}} \underset{\underset{\underset{+}{NH_3}}{\text{Alanine (52–60\%)}}}{CH_3CHCO_2^-}$$

PROBLEM 29.6   Outline the steps in the preparation of valine by the Strecker synthesis.

The most widely used method for the laboratory synthesis of α-amino acids is a modification of the malonic ester synthesis (Section 22.7). The key reagent is *diethyl acetamidomalonate,* a derivative of malonic ester that already has the critical nitrogen substituent in place at the α carbon atom. The side chain is introduced by alkylating the anion of diethyl acetamidomalonate with an alkyl halide in exactly the same way as diethyl malonate itself is alkylated.

$$CH_3\overset{O}{\overset{\|}{C}}NHCH(CO_2CH_2CH_3)_2 \xrightarrow[CH_3CH_2OH]{NaOCH_2CH_3} CH_3\overset{O}{\overset{\|}{C}}NH\overset{-}{C}(CO_2CH_2CH_3)_2 \xrightarrow{C_6H_5CH_2Cl}$$
$$\overset{..}{Na^+}$$

Diethyl
acetamidomalonate   Sodium salt of
diethyl acetamidomalonate

$$CH_3\overset{O}{\overset{\|}{C}}NHC(CO_2CH_2CH_3)_2$$
$$|$$
$$CH_2C_6H_5$$

Diethyl
acetamidobenzylmalonate
(90%)

Hydrolysis of the alkylated derivative of diethyl acetamidomalonate removes the acetyl group from nitrogen and converts the two ester functions to carboxyl groups. Decarboxylation of the diacid gives the desired product.

$$CH_3\overset{O}{\overset{\|}{C}}NHC(CO_2CH_2CH_3)_2 \xrightarrow[H_2O,\ heat]{HBr} H_3\overset{+}{N}C(CO_2H)_2 \xrightarrow[-CO_2]{heat} C_6H_5CH_2CHCO_2^-$$
$$|\qquad\qquad\qquad\qquad\qquad |\qquad\qquad\qquad\qquad\qquad |$$
$$CH_2C_6H_5 \qquad\qquad\qquad CH_2C_6H_5 \qquad\qquad\qquad\qquad NH_3$$
$$\qquad\qquad\qquad\qquad\qquad\qquad\qquad\qquad\qquad\qquad\qquad\qquad +$$

Diethyl
acetamidobenzylmalonate   (not isolated)   Phenylalanine
(65%)

PROBLEM 29.7   Outline the steps in the synthesis of valine from diethyl acetamidomalonate. The overall yield of valine by this method is reported to be rather low (31 percent). Can you think of a reason why this synthesis is not very efficient?

Unless a resolution step is introduced into the reaction scheme, the α-amino acids prepared by the synthetic methods just described are racemic. Optically active amino acids, when desired, may be obtained by resolving a racemic mixture or by *enantioselective synthesis.* A synthesis is described as enantioselective if it produces one enantiomer of a chiral compound in an amount greater than its mirror image. Recall from Section 8.8 that optically inactive reactants cannot give optically active products. Therefore enantioselective syntheses of amino acids require an enantiomerically enriched chiral reagent or catalyst at some point in the process. If the chiral reagent or catalyst is enantiomerically homogeneous and if the reaction sequence is completely enantioselective, an optically pure amino acid is obtained. Chemists have succeeded in preparing α-amino acids by techniques that are more than 95 percent enantioselective. While this is an impressive feat, we must not lose sight of the fact that the reactions that produce amino acids in living systems do so with 100 percent enantioselectivity.

29.5 REACTIONS OF AMINO ACIDS

Amino acids undergo reactions characteristic of both amine and carboxylic acid functional groups. Acylation is a typical reaction of the amino group.

$$H_3\overset{+}{N}CH_2CO_2^- + CH_3\overset{O}{\underset{\|}{C}}O\overset{O}{\underset{\|}{C}}CH_3 \longrightarrow CH_3\overset{O}{\underset{\|}{C}}NHCH_2CO_2H + CH_3CO_2H$$

| Glycine | Acetic anhydride | N-Acetylglycine (89–92%) | Acetic acid |

Ester formation is a typical reaction of the carboxyl group.

$$CH_3\underset{\underset{+}{\underset{NH_3}{|}}}{CH}CO_2^- + CH_3CH_2OH \xrightarrow{\text{HCl}} CH_3\underset{\underset{+}{\underset{NH_3}{|}}}{CH}\overset{O}{\underset{\|}{C}}OCH_2CH_3 \ Cl^-$$

| Alanine | Ethanol | Hydrochloride salt of alanine ethyl ester (90–95%) |

A reaction of amino acids that is used to detect their presence is the formation of a purple color on treatment with *ninhydrin*. The same compound responsible for the purple color is formed from all amino acids in which the α-amino group is primary.

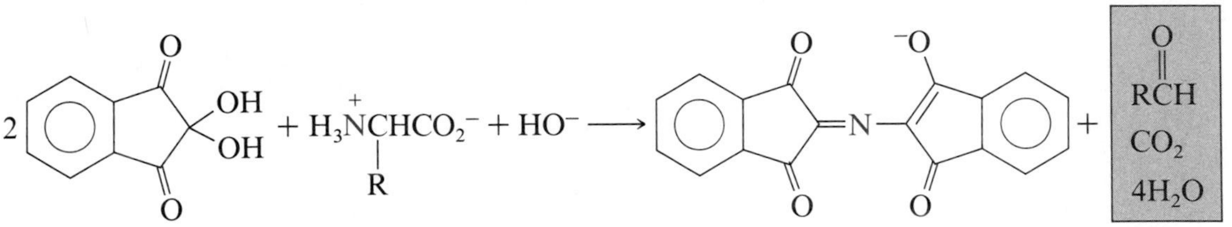

$$2 \text{ (Ninhydrin)} + H_3\overset{+}{N}\underset{\underset{R}{|}}{CH}CO_2^- + HO^- \longrightarrow \text{(Violet dye)} + \begin{array}{c} O \\ \| \\ RCH \\ CO_2 \\ 4H_2O \end{array}$$

| Ninhydrin | | Violet dye | (formed, but not normally isolated) |

Proline, in which the α-amino group is secondary, gives an orange compound on reaction with ninhydrin.

PROBLEM 29.8 Suggest a reasonable mechanism for the reaction of an α-amino acid with ninhydrin.

29.6 SOME BIOCHEMICAL REACTIONS INVOLVING AMINO ACIDS

The 20 amino acids listed in Table 29.1 are biosynthesized by a variety of diverse pathways and we will touch on only a few of them in an introductory way. We will examine the biosynthesis of glutamic acid first, since it illustrates a biochemical process analogous to a reaction we have discussed earlier in the context of amine synthesis, *reductive amination* (Section 23.11).

Glutamic acid is formed in most organisms from ammonia and α-ketoglutaric acid. α-Ketoglutaric acid is one of the intermediates in the *tricarboxylic acid cycle* (also called the *Krebs cycle*) and arises via metabolic breakdown of food sources—carbohydrates, fats, and proteins.

$$\underset{\alpha\text{-Ketoglutaric acid}}{HO_2CCH_2CH_2\overset{O}{\overset{\|}{C}}CO_2H} + \underset{\text{Ammonia}}{NH_3} \xrightarrow[\text{reducing agents}]{\text{enzymes}} \underset{\text{L-Glutamic acid}}{HO_2CCH_2CH_2\underset{\underset{+}{NH_3}}{\overset{|}{C}HCO_2^-}}$$

Ammonia reacts with the ketone carbonyl group to give an imine ($>C=NH$), and this imine is reduced to the amine function of the α-amino acid. Both imine formation and imine reduction are enzyme-catalyzed reactions. The reduced form of nicotinamide diphosphonucleotide (NADPH) is a coenzyme and acts as a reducing agent. The step in which the imine is reduced is the one in which the chiral center is introduced and gives only L-glutamic acid.

L-Glutamic acid is not an essential amino acid. It need not be present in the diet, since mammals possess the ability to biosynthesize it from sources of α-ketoglutaric acid. It is, however, a key intermediate in the biosynthesis of other amino acids by a process known as *transamination*. L-Alanine, for example, is formed from pyruvic acid by transamination from L-glutamic acid.

$$\underset{\text{Pyruvic acid}}{CH_3\overset{O}{\overset{\|}{C}}CO_2H} + \underset{\text{L-Glutamic acid}}{HO_2CCH_2CH_2\underset{\underset{+}{NH_3}}{\overset{|}{C}HCO_2^-}} \xrightarrow{\text{enzymes}} \underset{\text{L-Alanine}}{CH_3\underset{\underset{+}{NH_3}}{\overset{|}{C}HCO_2^-}} + \underset{\alpha\text{-Ketoglutaric acid}}{HO_2CCH_2CH_2\overset{O}{\overset{\|}{C}}CO_2H}$$

In transamination an amine group is transferred from L-glutamic acid to pyruvic acid. An outline of the mechanism of transamination is presented in Figure 29.2.

One amino acid often serves as the biological precursor to another. L-Phenylalanine is classified as an essential amino acid, whereas its *p*-hydroxy derivative, L-tyrosine, is not. This is because mammals have the ability to convert L-phenylalanine to L-tyrosine by hydroxylation of the aromatic ring. An *arene oxide* (Section 26.7) is an intermediate.

L-Phenylalanine   Arene oxide intermediate

L-Tyrosine

Step 1: The amine function of L-glutamate reacts with the ketone carbonyl of pyruvate to form an imine.

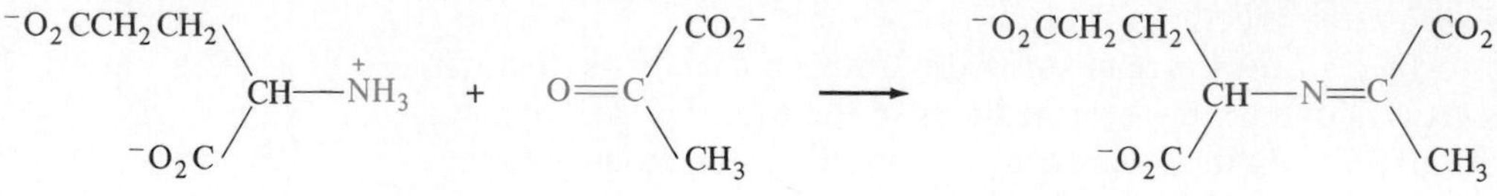

L-Glutamate   Pyruvate   Imine

Step 2: Enzyme-catalyzed proton transfer steps cause migration of the double bond, converting the imine formed in step 1 to an isomeric imine.

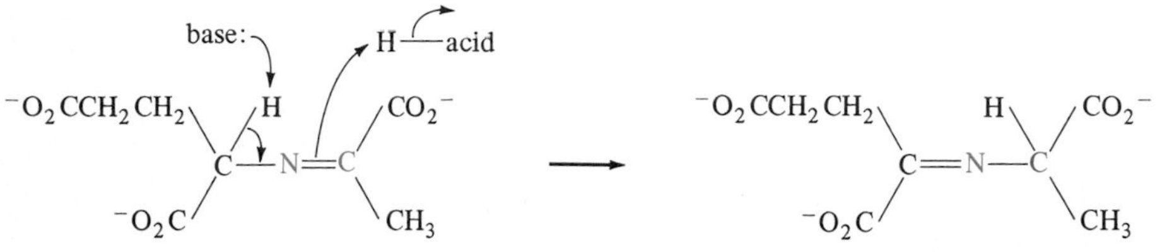

Imine from step (1)   Rearranged imine

Step 3: Hydrolysis of the rearranged imine gives L-alanine and α-ketoglutarate.

Rearranged imine   Water   α-Ketoglutarate   L-Alanine

FIGURE 29.2 Sequence of steps that describes the mechanism of transamination. All the steps are enzyme-catalyzed.

Some individuals lack sufficient quantities of the enzymes necessary to convert L-phenylalanine to L-tyrosine. The L-phenylalanine that they obtain from their diet is therefore diverted along an alternative metabolic pathway that leads to formation of phenylpyruvic acid.

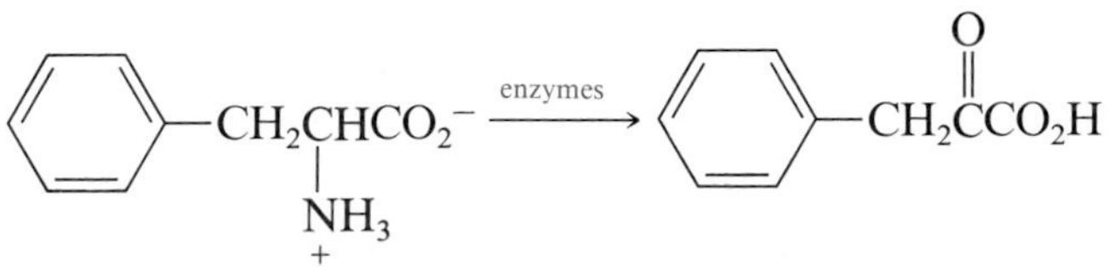

L-Phenylalanine   Phenylpyruvic acid

Phenylpyruvic acid is produced in sufficient quantities to cause mental retardation in infants who are deficient in the enzymes necessary to convert L-phenylalanine to L-tyrosine. They are said to suffer from *phenylketonuria,* or *PKU disease.* PKU disease can be detected by a simple test routinely administered to infants shortly after birth. It is an inborn error of metabolism and while not presently correctable, can be controlled by restricting the dietary intake of L-phenylalanine. In practice this means avoiding foods such as meat that are rich in L-phenylalanine.

Among the biochemical reactions that amino acids undergo is *decarboxylation* to amines. Decarboxylation of histidine, for example, gives histamine, a powerful vaso-

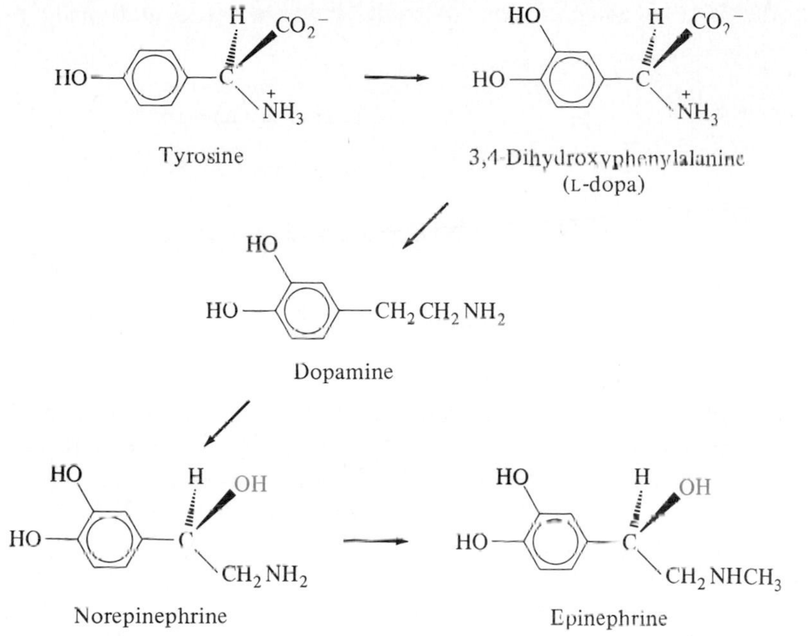

Tyrosine

3,4-Dihydroxyphenylalanine
(L-dopa)

Dopamine

Norepinephrine

Epinephrine

FIGURE 29.3 Tyrosine is the biosynthetic precursor to a number of neurotransmitters. Each transformation is enzyme-catalyzed. Hydroxylation of the aromatic ring of tyrosine converts it to 3,4-dihydroxyphenylalanine (L-dopa), decarboxylation of which gives dopamine. Hydroxylation of the benzylic carbon of dopamine converts it to norepinephrine (noradrenaline), and methylation of the amino group of norepinephrine yields epinephrine (adrenaline).

dilator normally present in tissue and formed in excessive amounts under conditions of traumatic shock.

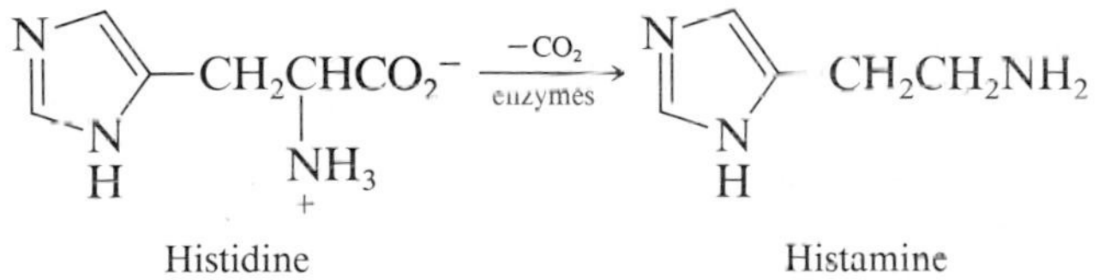

Histidine

Histamine

Histamine is responsible for many of the symptoms associated with hay fever and the common cold—runny nose, congestion, etc. An antihistamine relieves these symptoms by blocking the action of histamine.

PROBLEM 29.9 One of the amino acids in Table 29.1 is the biological precursor to γ-aminobutyric acid (4-aminobutanoic acid), which it forms by a decarboxylation reaction. Which amino acid is this?

The chemistry of the brain and central nervous system is affected by a group of substances called *neurotransmitters*. Several of these neurotransmitters arise from L-tyrosine by structural modification and decarboxylation, as outlined in Figure 29.3.

29.7 PEPTIDES

A principal biochemical reaction of amino acids is their conversion to peptides, polypeptides, and proteins. In all these substances amino acids are linked together by amide bonds. The amide bond between the amino group of one amino acid and the carboxyl of another is called a *peptide bond.* Alanylglycine is a representative dipeptide.

N-terminal amino acid   C-terminal amino acid

$$H_3\overset{+}{N}CHC - NHCH_2CO_2^-$$

$$\underset{CH_3}{|}$$

Alanylglycine
(Ala-Gly)

By agreement, peptide structures are written so that the amino group (as $H_3\overset{+}{N}$— or H_2N—) is at the left and the carboxyl group (as CO_2H or CO_2^-) is at the right. The left and right ends of the peptide are referred to as the *N terminus* (or amino terminus) and the *C terminus* (or carboxyl terminus), respectively. Alanine is the N-terminal amino acid in alanylglycine while glycine is the C-terminal amino acid. A dipeptide is named as an acyl derivative of the C-terminal amino acid. We call the precise order of bonding in peptides its amino acid *sequence.* The amino acid sequence is conveniently specified by using the three-letter amino acid abbreviations for each amino acid and connecting them by a hyphen. Individual amino acid components of peptides are often referred to as amino acid *residues.*

It is understood that α-amino acids occur as their L stereoisomers unless otherwise indicated. The D notation is explicitly shown when a D amino acid is present, and a racemic amino acid is identified by the prefix DL.

PROBLEM 29.10   Write structural formulas showing the constitution of each of the following dipeptides:

(*a*) Gly-Ala   (*d*) Gly-Glu
(*b*) Ala-Phe   (*e*) Lys-Gly
(*c*) Phe-Ala   (*f*) D-Ala-D-Ala

SAMPLE SOLUTION   (*a*) Gly-Ala is a constitutional isomer of Ala-Gly. Glycine is the N-terminal amino acid in Gly-Ala; alanine is the C-terminal amino acid.

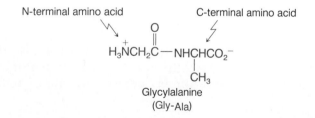

N-terminal amino acid   C-terminal amino acid

$$H_3\overset{+}{N}CH_2C - NHCHCO_2^-$$

$$\underset{CH_3}{|}$$

Glycylalanine
(Gly-Ala)

Figure 29.4 shows the structure of Ala-Gly as determined by x-ray crystallography. An important feature is the planar geometry associated with the peptide bond, and the most stable conformation with respect to this bond has the two α carbon atoms anti to each other. Rotation about the amide linkage is slow because delocalization of the unshared electron pair of nitrogen into the carbonyl group gives partial double-bond character to the carbon-nitrogen bond.

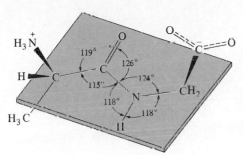

FIGURE 29.4 The structure of the dipeptide L-alanylglycine as revealed by x-ray crystallography. The six atoms shown in red all lie in the same plane and the α carbon atoms of the two amino acids are anti to one another.

PROBLEM 29.11 Expand your answer to Problem 29.10 by showing the structural formula for each dipeptide in a manner that reveals the stereochemistry at the α carbon atom.

SAMPLE SOLUTION (a) Glycine is achiral, so Gly-Ala has only one chiral center, the α carbon atom of the L-alanine residue. When the carbon chain is drawn in an extended zigzag fashion and L-alanine is the C terminus, its structure is as shown:

Glycyl-L-alanine (Gly-Ala)

Higher peptides are treated in an analogous fashion. Figure 29.5 gives the structural formula and amino acid sequence of a representative pentapeptide. This particular pentapeptide occurs naturally and is known as *leucine enkephalin*. Enkephalins are pentapeptide components of *endorphins,* polypeptides present in the brain that act as the body's own pain-killers. A second substance, known as *methionine enkephalin,* is also present in endorphins. Methionine enkephalin differs from leucine enkephalin only in having methionine instead of leucine as its C-terminal amino acid.

PROBLEM 29.12 What is the amino acid sequence (using three-letter abbreviations) of methionine enkephalin?

Peptides that have structures slightly different from those described to this point are known. One such variation is seen in the structure of the nonapeptide *oxytocin,*

FIGURE 29.5 The structure of the pentapeptide leucine enkephalin. Its amino acid sequence is Tyr-Gly-Gly-Phe-Leu, tyrosine being the N-terminal and leucine the C-terminal amino acid.

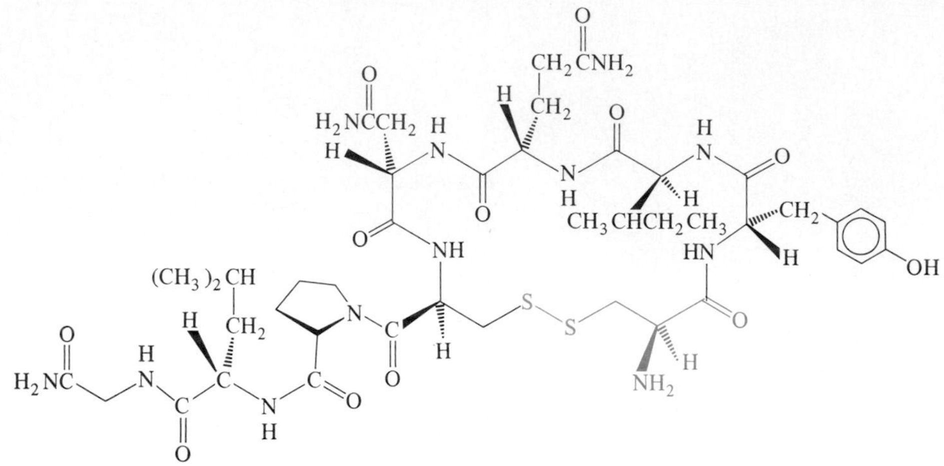

FIGURE 29.6 The structure of oxytocin, a nonapeptide containing a disulfide bond between two cysteine residues. One of these cysteines is the N-terminal amino acid; it is shown in blue. The C-terminal amino acid is the amide of glycine and is shown in red. There are no free carboxyl groups in the molecule; all exist in the form of carboxamides.

shown in Figure 29.6. Oxytocin is a hormone secreted by the pituitary gland that stimulates uterine contractions during childbirth. Rather than terminating in a carboxyl group, the terminal glycine residue in oxytocin has been modified so that it exists as the corresponding amide. Two cysteine units, one of them the N-terminal amino acid, are joined by the sulfur-sulfur bond of a large-ring cyclic disulfide unit. This is a common structural modification in polypeptides and proteins that contain cysteine residues. It provides a covalent bond between regions of peptide chains that may be many amino acid residues removed from each other.

29.8 PEPTIDE STRUCTURE DETERMINATION. AMINO ACID ANALYSIS

Chemists and biochemists distinguish among several levels of peptide structure. The *primary structure* of a peptide is its constitution and is in most cases equivalent to its amino acid sequence plus any disulfide links. The determination of the primary structure can be a truly formidable task, as the 20 amino acids of Table 29.1 provide a number of molecular building blocks large enough to produce a staggering number of dipeptides, tripeptides, tetrapeptides, etc. Given a peptide of unknown structure, how does one go about determining its amino acid sequence?

The first step is to determine which amino acids are present and the relative amounts of each one. The unknown peptide is subjected to acid-catalyzed hydrolysis by heating it in 6 N hydrochloric acid for about 24 h. Under these conditions the amide bonds are cleaved and a solution is obtained that contains all the amino acids originally present in the peptide. This mixture is then separated by a technique known as *ion-exchange chromatography.* Ion-exchange chromatography separates amino acids according to their acid-base properties and, to a lesser extent, depends on differences in the lipophilic properties of their side chains. After adsorption of the mixture of amino acids on a polymeric ion-exchange resin, the resin is washed with aqueous buffers of increasing pH. Individual amino acids pass through the column at different rates and appear in different fractions of the aqueous effluent. The exit of amino acids from the column is determined by mixing the effluent with ninhydrin. The intensity of the ninhydrin color is monitored electronically and the appearance of an amino acid recorded as a peak on a strip chart. A typical amino acid analysis is

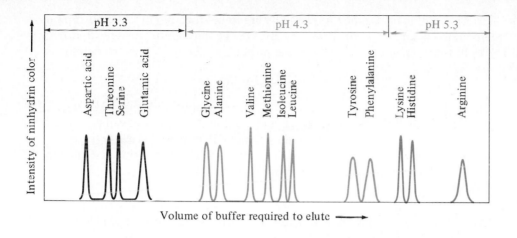

FIGURE 29.7 Analysis of a representative mixture of amino acids. Sodium citrate buffers were used to elute the amino acids from the column. The peaks shown in red correspond to amino acids eluted with a buffer of pH 3.3; those shown in yellow were eluted at pH 4.3. Because amino acids with basic side chains are strongly held by the adsorbent, a shorter column was used for their analysis. These were eluted with a buffer of pH 5.3 and are shown in blue at the right. The horizontal scale is different for these three amino acids than for the others.

illustrated in Figure 29.7. The volume of buffer of specified pH required to remove the various amino acids from the ion-exchange column is compared with that required for authentic samples of known amino acids and serves to identify each component in the mixture. Since peak areas are proportional to the amount of amino acid present, the molar ratios of amino acids are also determined. The entire operation is carried out automatically on an apparatus called an *amino acid analyzer.* The analytical procedure is extremely sensitive and can be performed routinely with 10^{-5} to 10^{-7} g of peptide.

An amino acid analysis identifies the component amino acids in a peptide and their ratios. Additional information typically includes molecular weight, which for small peptides can be determined exactly by mass spectrometry. Molecular weights of large peptides and proteins are estimated by a variety of specialized physical methods designed for this purpose.

29.9 PEPTIDE STRUCTURE DETERMINATION. PRINCIPLES OF SEQUENCE ANALYSIS

The principles of peptide sequencing may be illustrated by considering a representative example. Suppose we have a tetrapeptide of unknown sequence but which, on amino acid analysis, was shown to contain Ala, Gly, Phe, and Val in equimolar ratios. What additional information do we need in order to determine the sequence of this tetrapeptide?

While there are 24 possible tetrapeptides that may be derived from Ala, Gly, Phe, and Val, a relatively modest amount of information is required to distinguish among them. The three experimentally determined facts enumerated below are sufficient to define the structure of this tetrapeptide.

1. Valine is the N-terminal amino acid.
2. Hydrolysis of the tetrapeptide yielded a number of fragments, one of which was a tripeptide composed of Gly, Phe, and Val.
3. Also present in the hydrolysis mixture was a dipeptide containing Ala and Gly.

We interpret these data as follows. Since Val is the N-terminal amino acid and the tripeptide obtained on partial hydrolysis contains Val but not Ala, Ala must be the

C-terminal amino acid. Since Ala and Gly are found together in the same dipeptide fragment from a partial hydrolysis, Gly must be the third amino acid residue, counting from the N terminus. Therefore, the primary structure of the tetrapeptide is:

Val-Phe-Gly-Ala

Partial hydrolysis of the tetrapeptide can give dipeptide and tripeptide fragments other than those cited above, so more information is normally available than the three facts that yielded the solution to the problem. In practice, as many fragments as possible are isolated in pure form and their amino acid composition is determined. Thus, the internal consistency of the entire pattern of results serves as a check on the proposed sequence.

PROBLEM 29.13   List all the substances that might be present in a hydrolysis mixture obtained from Val-Phe-Gly-Ala.

Amino acid sequencing of higher peptides is a challenging task, one that requires experimental skill and a systematic analysis of the data. While the approach just outlined is, in principle, capable of being applied to a peptide of any length, peptide sequencing has been simplified by the introduction of a number of techniques that provide additional information. Some of these techniques are described in the following sections. We will begin with *end group analysis.* A very important piece of information in our example of peptide sequencing was the identification of valine as the N-terminal amino acid. How was the identity of this amino acid determined?

29.10   END GROUP ANALYSIS. THE N TERMINUS

Identification of the N-terminal amino acid rests on the fact that the α-amino groups of all the amino acids in a peptide chain, except the N-terminal amino acid, are incorporated into amide bonds. The α-amino group of the N-terminal amino acid is free and capable of acting as a nucleophile. One method for N-terminal amino acid identification involves treatment of a peptide with 1-fluoro-2,4-dinitrobenzene, a very reactive compound in reactions that proceed by the addition-elimination mechanism of nucleophilic aromatic substitution (Section 25.5).

1-Fluoro-2,4-dinitrobenzene

The amino group of the N-terminal amino acid displaces fluoride from 1-fluoro-2,4-dinitrobenzene and gives a peptide in which the N-terminal nitrogen is labeled with a 2,4-dinitrophenyl (DNP) group. This is shown for the case of Val-Phe-Gly-Ala in Figure 29.8. The 2,4-dinitrophenyl-labeled peptide DNP-Val-Phe-Gly-Ala is isolated and subjected to hydrolysis, after which the 2,4-dinitrophenyl derivative of the N-terminal amino acid is isolated and identified as DNP-Val by comparing its chromatographic behavior with that of standard samples of 2,4-dinitrophenyl-labeled amino acids. None of the other amino acid residues bear a 2,4-dinitrophenyl group; they appear in the hydrolysis product as the free amino acids.

The reaction is carried out by mixing the peptide and 1-fluoro-2,4-dinitrobenzene in the presence of a weak base such as sodium carbonate. In the first step the base abstracts a proton from the terminal $H_3\overset{+}{N}$ group to give a free amino function. The nucleophilic amino group attacks 1-fluoro-2,4-dinitrobenzene, displacing fluoride.

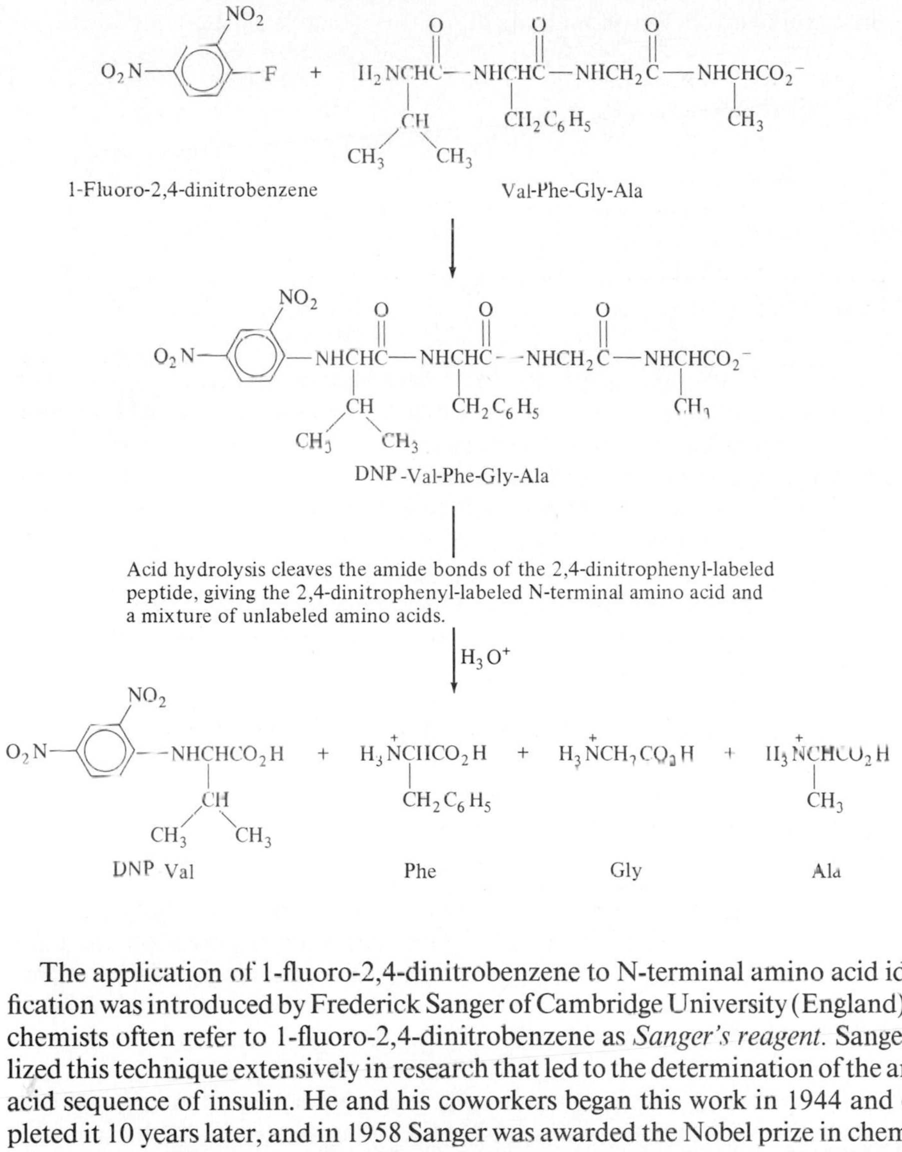

1-Fluoro-2,4-dinitrobenzene   Val-Phe-Gly-Ala

DNP-Val-Phe-Gly-Ala

Acid hydrolysis cleaves the amide bonds of the 2,4-dinitrophenyl-labeled peptide, giving the 2,4-dinitrophenyl-labeled N-terminal amino acid and a mixture of unlabeled amino acids.

H_3O^+

DNP Val   Phe   Gly   Ala

FIGURE 29.8 Use of 1-fluoro-2,4-dinitrobenzene as a reagent to identify the N-terminal amino acid of a peptide.

The application of 1-fluoro-2,4-dinitrobenzene to N-terminal amino acid identification was introduced by Frederick Sanger of Cambridge University (England), and chemists often refer to 1-fluoro-2,4-dinitrobenzene as *Sanger's reagent*. Sanger utilized this technique extensively in research that led to the determination of the amino acid sequence of insulin. He and his coworkers began this work in 1944 and completed it 10 years later, and in 1958 Sanger was awarded the Nobel prize in chemistry for this pioneering achievement. (Sanger was a corecipient of a second Nobel prize in chemistry in 1978 for studies in nucleic acid chemistry.)

A related method employs 5-(dimethylamino)napthalene-1-sulfonyl chloride as the reagent that reacts with the amino group of the N-terminal amino acid.

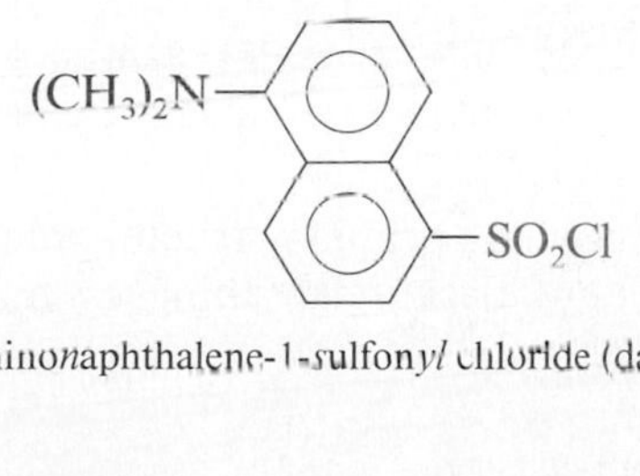

5-Dimethylaminonaphthalene-1-sulfonyl chloride (dansyl chloride)

This compound is known as *dansyl* chloride (dansyl is a combination of the letters indicated in italics in its systematic name) and the procedure is called *dansylation.* Identification of the *N* terminus by dansylation is carried out in exactly the same way as in Sanger's method. Dansyl chloride labels the peptide at the α-amino group of the N-terminal amino acid as a sulfonamide derivative which is isolated and identified after hydrolysis.

Val-Phe-Gly-Ala $\xrightarrow[\text{2. } 6 \text{ } N \text{ HCl, heat}]{\text{1. dansyl chloride, HO}^-}$

(CH$_3$)$_2$N— [naphthalene ring structure] —S—NHCHCO$_2$H, with O double bonds on S, and CH(CH$_3$)$_2$ branch

Dansyl derivative of valine

The amount of peptide required for dansylation is very small. The (dimethylamino)naphthyl group imparts a strong fluorescence to dansyl derivatives, and they can be located on paper or thin-layer chromatography plates in minute amounts, much smaller than those required to detect 2,4-dinitrophenyl derivatives.

The *Edman degradation,* developed by Pehr Edman (University of Lund, Sweden) is the most frequently used method for determining the N-terminal amino acid. It has an advantage over other methods in that it permits successive identification of amino acids, one by one, beginning at the N terminus. The Edman degradation is based on the chemistry shown in Figure 29.9. An amino acid reacts with phenyl isothiocyanate to give a *phenylthiocarbamoyl* (PTC) derivative, as shown in the first step. This PTC derivative is then treated with an acid in an *anhydrous* medium (Edman used nitromethane saturated with hydrogen chloride) to cleave the amide bond between the N-terminal amino acid and the remainder of the peptide. No other peptide bonds are cleaved in this step. Amide bond hydrolysis requires water. When the PTC derivative is treated with acid in an anhydrous medium, the sulfur atom of the $\diagdown$C$=$S acts as an internal nucleophile, and the only amide bond cleaved under these conditions is the one to the N-terminal amino acid. The product of this cleavage, called a *thiazolone,* is unstable under the conditions of its formation and rearranges to a *phenylthiohydantoin* (PTH), which is isolated and identified by comparing it with standard samples of PTH derivatives of known amino acids. This is normally done by chromatographic methods but mass spectrometry has also been used.

Notice that only the N-terminal amide bond is broken in the Edman degradation; the rest of the peptide chain remains intact. It can be isolated and subjected to a second Edman procedure to determine its new N terminus. Thus one can proceed along a peptide chain by beginning with the N terminus and determining each amino acid in order. The sequence is given directly by the structure of the PTH derivative formed in each successive degradation.

PROBLEM 29.14   Give the structure of the PTH derivative isolated in the second Edman cycle of our representative tetrapeptide Val-Phe-Gly-Ala.

Ideally, one could determine the primary structure of even the largest protein by repeating the Edman procedure. Because anything less than 100 percent conversion

Step 1: A peptide is treated with phenyl isothiocyanate to give a phenylthiocarbamoyl (PTC) derivative.

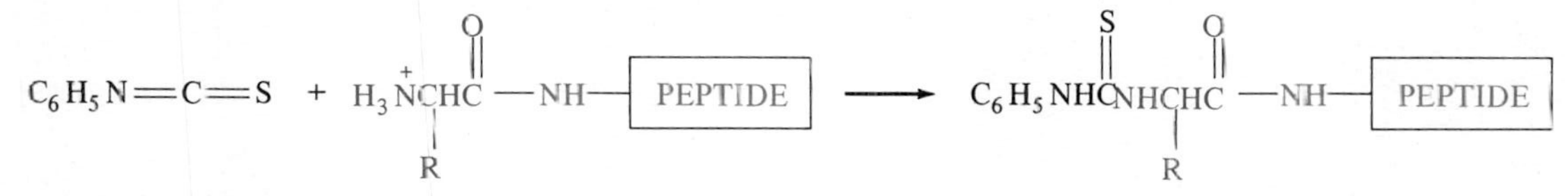

$$C_6H_5N=C=S \ + \ H_3\overset{+}{N}CHC-NH-\boxed{PEPTIDE} \longrightarrow C_6H_5NHCNHCHC-NH-\boxed{PEPTIDE}$$

Phenyl isothiocyanate

PTC derivative

Step 2: On reaction with hydrogen chloride in an anhydrous solvent, the thiocarbonyl sulfur of the PTC derivative attacks the carbonyl carbon of the N-terminal amino acid. The N-terminal amino acid is cleaved as a thiazolone derivative from the remainder of the peptide.

PTC derivative $\xrightarrow{\text{HCl}}$ Thiazolone $+$ $H_3\overset{+}{N}$—$\boxed{PEPTIDE}$

Remainder of peptide

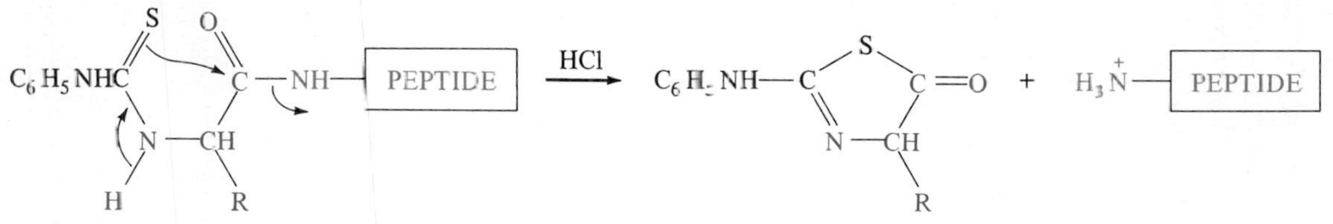

Step 3: Once formed, the thiazolone derivative isomerizes to a more stable phenylthiohydantoin (PTH) derivative, which is isolated and characterized, thereby providing identification of the N-terminal amino acid. The remainder of the peptide (formed in step 2) can be isolated and subjected to a second Edman degradation.

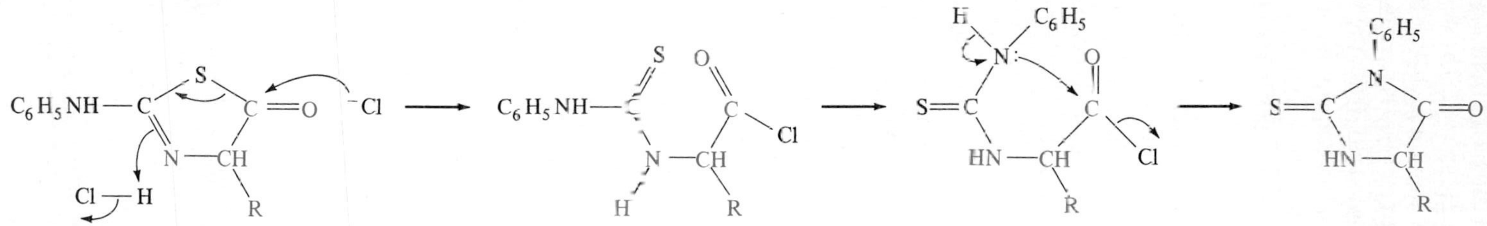

Thiazolone $\longrightarrow$ $\longrightarrow$ PTH derivative

FIGURE 29.9 The chemical reactions involved in the identification of the N-terminal amino acid of a peptide by Edman degradation.

in any single Edman degradation gives a mixture containing some of the original peptide along with the degraded one, two different PTH derivatives are formed in the following Edman cycle, and the ideal is not realized in practice. Nevertheless, some impressive results have been achieved. It is a fairly routine matter to sequence the first 20 amino acids from the N terminus by repetitive Edman cycles, and even 60 residues have been determined on a single sample of the protein myoglobin. The entire procedure has been automated and incorporated into a device called an *Edman sequenator*, which carries out all the operations under computer control.

29.11 END GROUP ANALYSIS. THE C TERMINUS

While there are a few purely chemical techniques for determining the amino acid at the carboxyl, or C, terminus of a peptide, the most widely used method is one based on enzymatic hydrolysis. Enzymes that catalyze the hydrolysis of peptide bonds are called *peptidases, proteases,* or *proteolytic enzymes.* Many are quite selective with respect to the type of peptide bonds that they cleave. One group of pancreatic enzymes, known as *carboxypeptidases,* catalyzes only the hydrolysis of the peptide bond involving the C-terminal amino acid (Figure 29.10). Once the C-terminal amino acid has been cleaved, yielding a peptide shorter by one amino acid, the carboxypeptidase then catalyzes the cleavage of the peptide bond to the new C-terminal amino acid. This property is used to advantage in sequence analysis. A peptide is incubated with a carboxypeptidase, and the concentration of free amino acids is monitored as a function of time. The order of their decreasing concentration indicates the position of each amino acid with respect to the C terminus of the original peptide. Normally, the sequence of only a few amino acids from the C terminus can be determined in this way.

29.12 SELECTIVE HYDROLYSIS OF PEPTIDES

Carboxypeptidase-catalyzed hydrolysis is an example of one way in which biochemical degradation of a peptide can be used in sequence determination. Living organisms contain a number of different peptidases, several of which find use in peptide sequencing because they permit the controlled hydrolysis of peptides by cleavage of certain specific amide bonds.

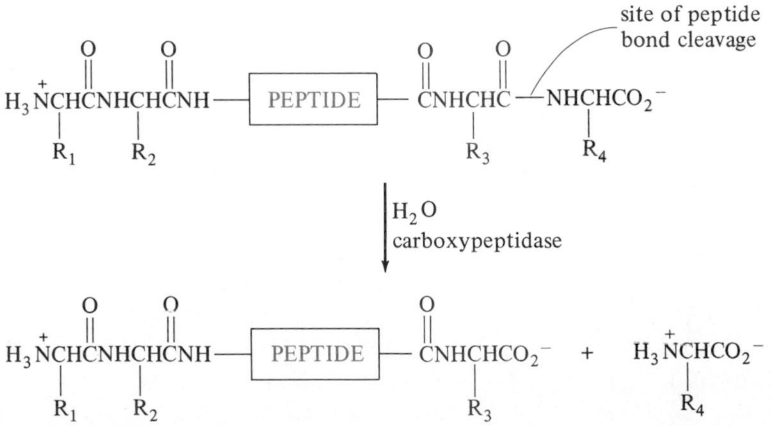

FIGURE 29.10 Diagram showing the site of carboxypeptidase-catalyzed hydrolysis. Carboxypeptidase is selective for cleavage of the peptide bond to the C-terminal amino acid. Once this peptide bond has been cleaved, carboxypeptidase proceeds to catalyze the hydrolytic cleavage of the new C-terminal amino acid.

```
1   2   3   4   5   6   7   8   9   10  11
Phe-Val-Asn-Gln-His-Leu-Cys-Gly-Ser-His-Leu
```

Ser-His-Leu-Val

Leu-Val-Glu-Ala

```
        12  13  14  15
        Val-Glu-Ala-Leu
```

Ala-Leu-Tyr

```
        16  17
        Tyr-Leu-Val-Cys
```

```
            18  19  20  21  22  23  24
            Val-Cys-Gly-Glu-Arg-Gly-Phe
```

```
                        25
                        Gly-Phe-Phe-Tyr-Thr-Pro-Lys
```

```
                            26  27  28  29  30
                            Tyr-Thr-Pro-Lys-Ala
```

```
1            5            10           15           20           25           30
Phe-Val-Asn-Gln-His-Leu-Cys-Gly-Ser-His-Leu-Val-Glu-Ala-Leu-Tyr-Leu-Val-Cys-Gly-Glu-Arg-Gly-Phe-Phe-Tyr-Thr-Pro-Lys-Ala
```

FIGURE 29.11 Diagram showing how the amino acid sequence of the B chain of bovine insulin can be determined by overlap of peptide fragments. Pepsin-catalyzed hydrolysis produced the fragments shown in blue, trypsin produced the one shown in yellow, and acid-catalyzed hydrolysis gave many fragments, including the four shown in red.

Trypsin, a digestive enzyme present in the intestine, is one such peptidase; it catalyzes only the hydrolysis of peptide bonds involving the carboxyl group of a lysine or arginine residue. *Chymotrypsin,* another digestive enzyme, is selective for the hydrolysis of peptide bonds involving the carboxyl group of amino acids with aromatic groups in their side chains, namely, phenylalanine, tyrosine, and tryptophan. The catalytic activity of *pepsin* is much like that of chymotrypsin, but pepsin also catalyzes the hydrolysis of peptide bonds to the carbonyl groups of methionine and leucine.

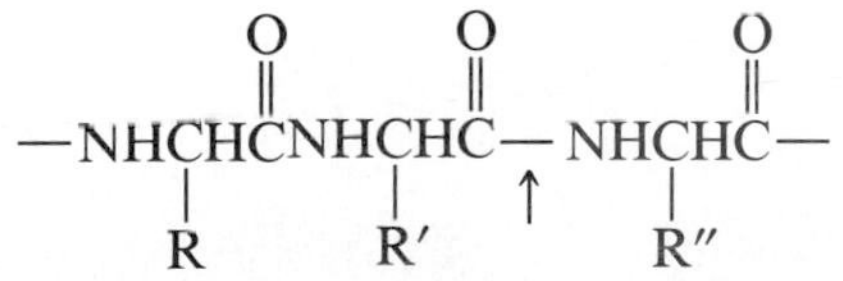

Site of trypsin-catalyzed
hydrolysis when red amino
acid residue is lysine
or arginine

Unlike hydrolysis in 6 *N* hydrochloric acid, which gives a complicated mixture of peptide fragments and free amino acids, selective enzyme-catalyzed hydrolysis produces only a limited number of fragments. After these fragments have been isolated and sequenced, their primary structures are examined for points of overlap in order to deduce the structure of the peptide that produced them. An example of this is shown in Figure 29.11, which depicts some of the work that led to Sanger's determination of the structure of insulin. Insulin, a protein containing 51 amino acids, has two principal structural units, a 21–amino acid peptide (the A chain) and a 30–amino acid peptide (the B chain). The A chain and the B chain are joined by disulfide bonds between two pairs of cysteine residues (CysS-SCys). Reaction with 1-fluoro-2,4-dinitrobenzene established that phenylalanine is the N terminus of the B chain. The four peptides shown in blue in Figure 29.11 were isolated by pepsin-catalyzed hydrolysis of the B chain, and the amino acid sequence of each was determined independently. Together these four peptides contain 28 of the 30 amino acids of the insulin B chain, but there are no points of overlap between them. Since, however, the sequences of two tetrapeptides isolated by acid hydrolysis (shown in red) overlap with both the N-terminal fragment and a second peptide from the pepsin-catalyzed hydrolysis, it is possible to establish the first 15 residues beginning with the N terminus. Similarly, overlaps with two other fragments from the acid hydrolysis (red) reveals that the sequence Tyr-Leu joins the first 15 residues to the third peptide obtained from pepsin-catalyzed hydrolysis. Thus, the first 24 residues are identified. The peptide shown in yellow in Figure 29.11 was isolated by trypsin-catalyzed hydrolysis; its amino acid sequence overlaps with two of the peptides from pepsin-catalyzed hydrolysis, providing the necessary information to complete the entire sequence of 30 amino acids.

Sanger also determined the sequence of the A chain and identified the cysteine residues involved in disulfide bonds between the A and B chains as well as in the disulfide linkage within the B chain. The complete insulin structure is shown in Figure 29.12. The structure shown is that of bovine insulin (from cattle). The A chains of human insulin and bovine insulin differ in only two amino acid residues; their B chains are identical except for the amino acid at the C terminus.

Once a peptide has been isolated and purified, a whole battery of techniques is brought to bear on the question of its amino acid sequence; these include Edman degradation and carboxypeptidase determination of the C terminus, along with partial hydrolysis and sequence analysis of smaller fragments. Since Sanger's successful determination of the primary structure of insulin in 1954, the amino acid sequences of hundreds of proteins, some with more than 300 amino acids in their chains, have been determined. In a recent innovation it has become possible to determine the nucleotide sequence of the gene that directs the synthesis of a peptide and, on the basis of the genetic code (Section 29.28), to deduce the sequence of the amino acids in the peptide.

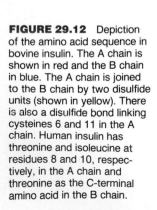

FIGURE 29.12 Depiction of the amino acid sequence in bovine insulin. The A chain is shown in red and the B chain in blue. The A chain is joined to the B chain by two disulfide units (shown in yellow). There is also a disulfide bond linking cysteines 6 and 11 in the A chain. Human insulin has threonine and isoleucine at residues 8 and 10, respectively, in the A chain and threonine as the C-terminal amino acid in the B chain.

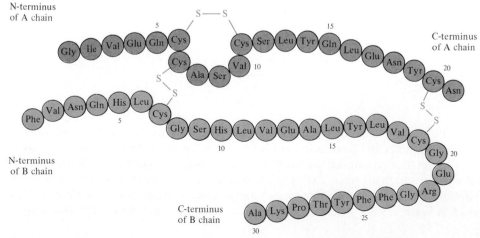

29.13 THE STRATEGY OF PEPTIDE SYNTHESIS

One way to confirm the structure proposed for a peptide is to synthesize a peptide having a specific sequence of amino acids and compare the two. This was done, for example, in the case of *bradykinin,* a peptide present in blood that acts to lower blood pressure. Excess bradykinin, formed as a response to the sting of wasps and other insects containing substances in their venom that stimulate bradykinin release, causes severe local pain. Bradykinin was originally believed to be an octapeptide containing two proline residues. However, a nonapeptide containing three prolines in the sequence shown below was synthesized and determined to be identical to natural bradykinin in every respect, including biological activity.

Arg-Pro-Pro-Gly-Phe-Ser-Pro-Phe-Arg   Bradykinin

A reevaluation of the original sequence data established that natural bradykinin was indeed the nonapeptide shown. Here the synthesis of a peptide did more than confirm structure; synthesis was instrumental in determining structure.

Chemists and biochemists also synthesize peptides in order to promote understanding of the mechanism of their action. By systematically altering the sequence it is sometimes possible to determine which amino acids are intimately involved in the reactions that involve a particular peptide. Many synthetic peptides have been prepared in searching for new drugs.

PROBLEM 29.15 The C-terminal methyl ester of the dipeptide Asp-Phe was prepared in connection with research on digestive enzymes known as *gastrins.* The chemist who synthesized it noted that it had a sweet taste. This synthetic dipeptide ester is now known as *aspartame;* it is 200 times sweeter than sucrose and is widely used as an artificial sweetener. Write a structural formula for aspartame.

The objective in peptide synthesis may be simply stated: to connect amino acids together in a prescribed sequence by amide bond formation between them. A number of very effective methods and reagents have been designed for peptide bond formation, so the joining together of amino acids by amide linkages is not difficult. (Some of these methods are described in Section 29.16.) The real difficulty lies in ensuring that the correct sequence is obtained. This can be illustrated by considering the synthesis of a representative dipeptide, Phe-Gly, for example. Random peptide bond formation in a mixture containing phenylalanine and glycine would be expected to lead to four dipeptides:

$$H_3\overset{+}{N}CHCO_2^- + H_3\overset{+}{N}CH_2CO_2^- \longrightarrow \text{Phe-Gly} + \text{Phe-Phe} + \text{Gly-Phe} + \text{Gly-Gly}$$
$$|$$
$$CH_2C_6H_5$$

Phenylalanine   Glycine

The amino groups of both phenylalanine and glycine can be involved in peptide bond formation. Similarly, the carboxyl groups of both amino acids can be involved.

In order to direct the synthesis so that only Phe-Gly is formed, the amino group of phenylalanine and the carboxyl group of glycine must be protected so that they cannot react under the conditions of peptide bond formation. We can represent the peptide bond formation step by the equation below, where X and Y are amine- and carboxyl-protecting groups, respectively.

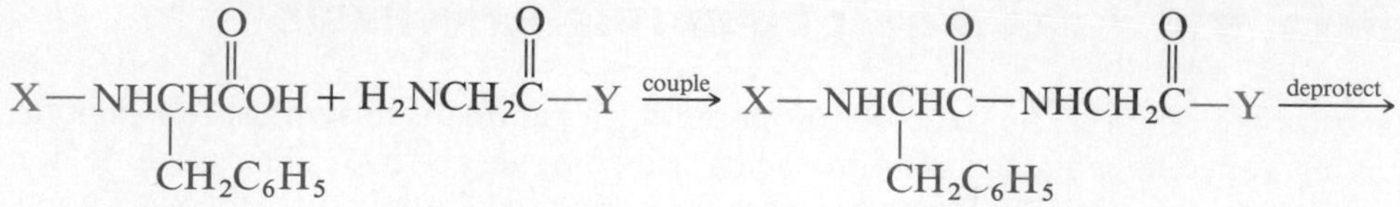

$$X-NHCHCOH + H_2NCH_2C-Y \xrightarrow{\text{couple}} X-NHCHC-NHCH_2C-Y \xrightarrow{\text{deprotect}}$$
$$\underset{CH_2C_6H_5}{|} \qquad\qquad\qquad\qquad \underset{CH_2C_6H_5}{|}$$

N-Protected   C-Protected   Protected Phe-Gly
phenylalanine   glycine

$$H_3\overset{+}{N}CHC-NHCH_2CO^-$$
$$\underset{CH_2C_6H_5}{|}$$

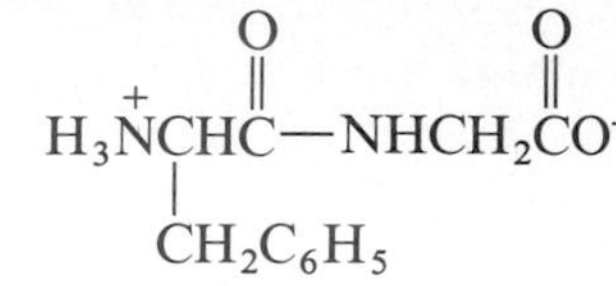

Phe-Gly

Thus, the synthesis of a dipeptide of prescribed sequence requires at least three operations:

1. *Protect* the amino group of the N-terminal amino acid and the carboxyl group of the C-terminal amino acid.
2. *Couple* the two protected amino acids by amide bond formation between them.
3. *Deprotect* the amino group at the N terminus and the carboxyl group at the C terminus.

Higher peptides are prepared in an analogous way by a direct extension of the logic just outlined for the synthesis of dipeptides.

Sections 29.14 through 29.17 elaborate on these features of peptide synthesis by describing the chemistry associated with the protection and deprotection of amino and carboxyl functions, along with methods for peptide bond formation.

29.14 AMINO GROUP PROTECTION

The reactivity of an amino group is suppressed by converting it to an amide function, and amino groups are most often protected by acylation. The benzyloxycarbonyl group ($C_6H_5CH_2O\overset{O}{\overset{||}{C}}-$) is one of the most often used amino-protecting groups. It is attached by acylation of an amino acid with benzyloxycarbonyl chloride.

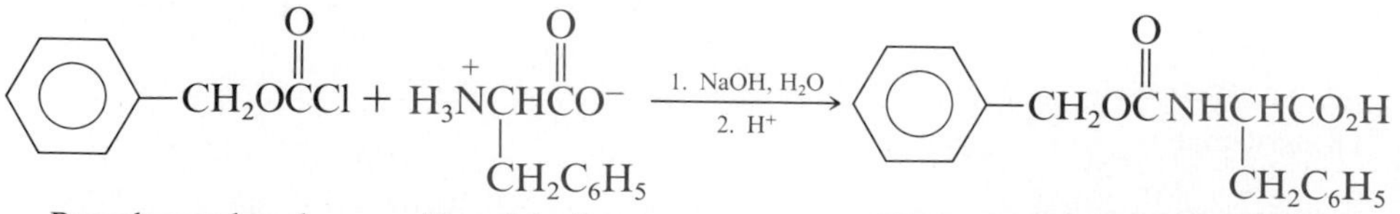

Benzyloxycarbonyl   Phenylalanine   N-Benzyloxycarbonylphenylalanine
chloride   (82–87%)

PROBLEM 29.16 Lysine reacts with two equivalents of benzyloxycarbonyl chloride to give a derivative containing two benzyloxycarbonyl groups. What is the structure of this compound?

Just as it is customary to identify individual amino acids by abbreviations, so too with protected amino acids. The approved abbreviation for a benzyloxycarbonyl group is the letter Z. Thus, *N*-benzyloxycarbonylphenylalanine is represented as

$$\underset{\underset{\displaystyle CH_2C_6H_5}{|}}{ZNHCHCO_2H} \qquad \text{or more simply as Z-Phe}$$

Another name for the benzyloxycarbonyl group is *carbobenzoxy,* abbreviated Cbz, and one finds this term used frequently in the older literature. However, this name is not as descriptive of structure as is benzyloxycarbonyl and is no longer an acceptable name.

The value of the benzyloxycarbonyl protecting group is that it is easily removed by reactions other than hydrolysis. In peptide synthesis amide bonds are formed. We protect the *N* terminus as an amide but need to remove the protecting group under conditions that do not cleave the very amide bonds we labored so hard to construct. Removing the protecting group by hydrolysis would surely bring about cleavage of peptide bonds as well. One advantage that the benzyloxycarbonyl protecting group enjoys over more familiar acyl groups such as acetyl is that it is can be removed by *hydrogenolysis* in the presence of palladium. The equation shown below illustrates this for the removal of the benzyloxycarbonyl protecting group from the ethyl ester of Z-Phe-Gly.

$$C_6H_5CH_2\overset{\overset{\displaystyle O}{\|}}{O}C\underset{\underset{\displaystyle CH_2C_6H_5}{|}}{N}HCH\overset{\overset{\displaystyle O}{\|}}{C}NHCH_2CO_2CH_2CH_3 \xrightarrow[\text{Pd}]{H_2} C_6H_5CH_3 + CO_2$$

N-Benzyloxycarbonylphenylalanylglycine ethyl ester   Toluene   Carbon dioxide

$$+ \; H_2NCH\overset{\overset{\displaystyle O}{\|}}{C}NHCH_2CO_2CH_2CH_3$$
$$\underset{\displaystyle CH_2C_6H_5}{|}$$

Phenylalanylglycine ethyl ester (100%)

The products derived from the protecting group, toluene and carbon dioxide, are easily removed, which facilitates purification. The resulting dipeptide has its carboxyl group protected and its amino group free. It can be coupled with an *N*-protected amino acid to give a tripeptide or subjected to ester hydrolysis to yield the dipeptide Phe-Gly.

Hydrogenolysis of the benzyloxycarbonyl protecting group involves initial cleavage of the benzyl-oxygen bond. The products of this cleavage are toluene and a carbamic acid (Section 21.17). The carbamic acid spontaneously loses carbon dioxide.

$$C_6H_5CH_2-\overset{\overset{\displaystyle O}{\|}}{O}C-NHR + H_2 \xrightarrow{Pd} C_6H_5CH_3 + [H\overset{\overset{\displaystyle O}{\|}}{O}C-NHR] \longrightarrow CO_2 + H_2NR$$

Z-Protected peptide   Hydrogen   Toluene   Carbamic acid (not isolated)   Carbon dioxide   Peptide

Alternatively, the benzyloxycarbonyl protecting group may be removed by treatment with hydrogen bromide in acetic acid.

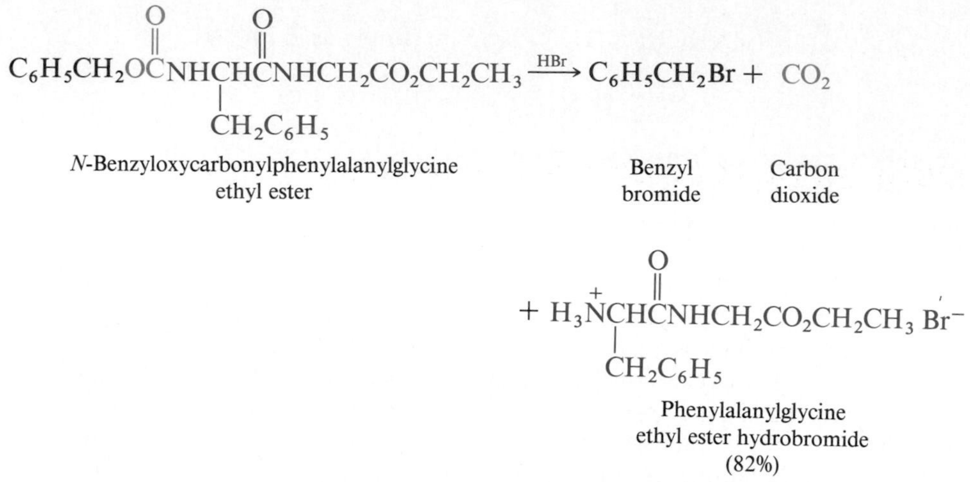

$$C_6H_5CH_2OCNHCHCNHCH_2CO_2CH_2CH_3 \xrightarrow{\text{HBr}} C_6H_5CH_2Br + CO_2$$

with the $\overset{O}{\overset{||}{}}$ carbonyls and the $CH_2C_6H_5$ side chain.

N-Benzyloxycarbonylphenylalanylglycine   Benzyl   Carbon
ethyl ester   bromide   dioxide

$$+ \ H_3\overset{+}{N}CHCNHCH_2CO_2CH_2CH_3 \ Br^-$$
$$\qquad\qquad CH_2C_6H_5$$

Phenylalanylglycine
ethyl ester hydrobromide
(82%)

Deprotection by this method is based on the ease with which benzyl esters are cleaved by nucleophilic attack at the benzylic carbon in the presence of strong acids. Bromide ion is the nucleophile and a carbamic acid is an intermediate.

A related N-terminal–protecting group is *tert*-butoxycarbonyl, abbreviated Boc.

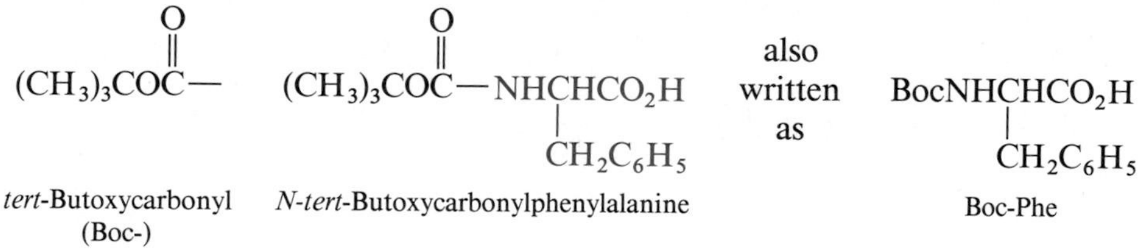

$$(CH_3)_3COC— \qquad (CH_3)_3COC—NHCHCO_2H \qquad\qquad BocNHCHCO_2H$$
$$\qquad\qquad\qquad\qquad\qquad CH_2C_6H_5 \qquad\qquad\qquad\qquad\qquad CH_2C_6H_5$$

also
written
as

tert-Butoxycarbonyl   *N-tert*-Butoxycarbonylphenylalanine   Boc-Phe
(Boc-)

Like the benzyloxycarbonyl-protecting group, the Boc group may be removed by treatment with hydrogen bromide. (It is stable to hydrogenolysis, however.)

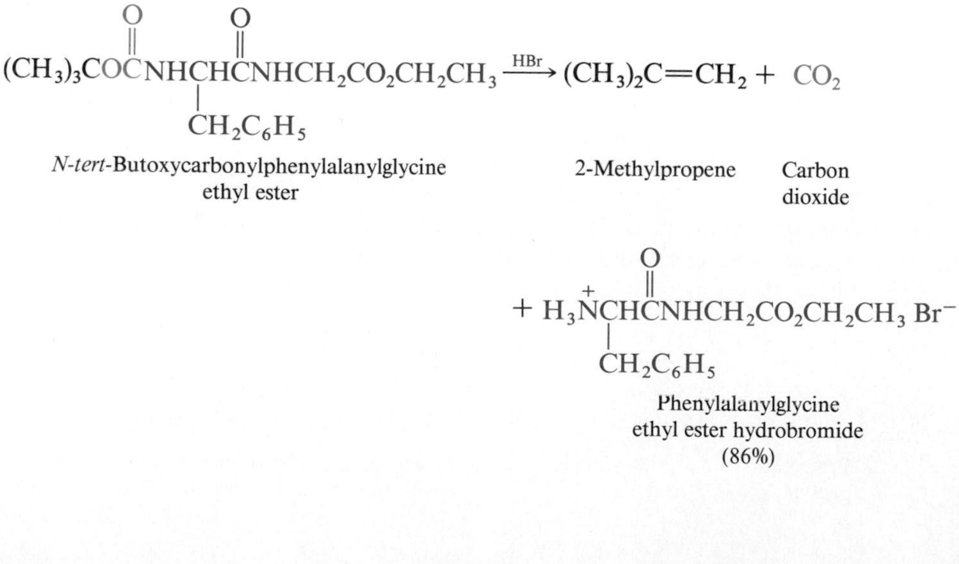

$$(CH_3)_3COCNHCHCNHCH_2CO_2CH_2CH_3 \xrightarrow{\text{HBr}} (CH_3)_2C{=}CH_2 + CO_2$$
$$\qquad\qquad CH_2C_6H_5$$

N-tert-Butoxycarbonylphenylalanylglycine   2-Methylpropene   Carbon
ethyl ester   dioxide

$$+ \ H_3\overset{+}{N}CHCNHCH_2CO_2CH_2CH_3 \ Br^-$$
$$\qquad\qquad CH_2C_6H_5$$

Phenylalanylglycine
ethyl ester hydrobromide
(86%)

The *tert*-butyl group is cleaved as the corresponding carbocation, leaving a carbamic acid, which loses carbon dioxide to give the unprotected amino group. Loss of a proton from *tert*-butyl cation converts it to 2-methylpropene. Because of the ease with which a *tert*-butyl group is cleaved as a carbocation, other acidic reagents, such as trifluoroacetic acid, may also be used.

29.15 CARBOXYL GROUP PROTECTION

Carboxyl groups of amino acids and peptides are normally protected as esters. Methyl and ethyl esters are prepared by Fischer esterification, as the example cited in Section 29.5 illustrates. Deprotection of methyl and ethyl esters is accomplished by hydrolysis in dilute base. Benzyl esters are a popular choice because they can be removed by hydrogenolysis. Thus a synthetic peptide, protected at both its N terminus with a Z group and at its C terminus as a benzyl ester, can be completely deprotected in a single operation.

$$C_6H_5CH_2OCNHCHCNHCH_2CO_2CH_2C_6H_5 \xrightarrow[Pd]{H_2} H_3\overset{+}{N}CHCNHCH_2CO_2^- + 2C_6H_5CH_3 + CO_2$$

with $\underset{CH_2C_6H_5}{}$ substituents shown on the respective structures

N-Benzyloxycarbonylphenylalanylglycine benzyl ester | Phenylalanylglycine (87%) | Toluene | Carbon dioxide

Several of the amino acids listed in Table 29.1 bear side-chain functional groups, which must also be protected during peptide synthesis. In most cases protecting groups are available that can be removed by hydrogenolysis.

29.16 PEPTIDE BOND FORMATION

In order to form a peptide bond between two suitably protected amino acids, the free carboxyl group of one of them must be *activated* so that it is a reactive acylating agent. The most familiar acylating agents are acyl chlorides and they were once extensively used to couple amino acids. Certain drawbacks to this approach became evident, however, stimulating chemists to seek alternative methods for the formation of amide bonds.

In one method treatment with N,N'-dicyclohexylcarbodiimide (DCCI) of a solution containing the N-protected and the C-protected amino acids leads directly to peptide bond formation.

$$ZNHCHCOH + H_2NCH_2COCH_2CH_3 \xrightarrow[chloroform]{DCCI} ZNHCHC-NHCH_2COCH_2CH_3$$

with $CH_2C_6H_5$ substituents shown on the respective structures

Z-Protected phenylalanine | Glycine ethyl ester | Z-Protected Phe-Gly ethyl ester (83%)

Overall reaction:

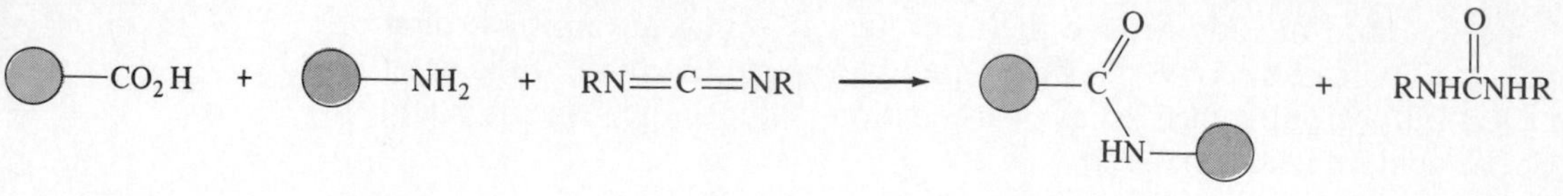

Carboxylic acid   Amine   DCCI   Amide N,N'-Dicyclohexylurea

DCCI = N,N'-dicyclohexylcarbodiimide; R = cyclohexyl

Mechanism:

Step 1: In the first stage of the reaction, the carboxylic acid adds to one of the double bonds of DCCI to give an O-acylisourea.

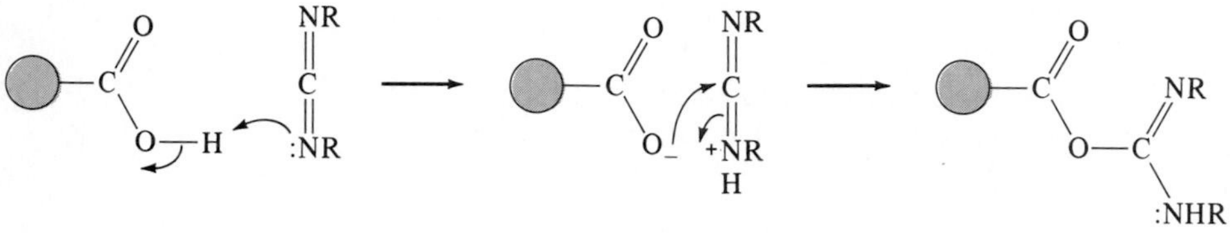

Carboxylic acid   DCCI O-Acylisourea

Step 2: Structurally, O-acylisoureas resemble carboxylic acid anhydrides and are powerful acylating agents. In the reaction's second stage the amine adds to the carbonyl group of the O-acylisourea to give a tetrahedral intermediate.

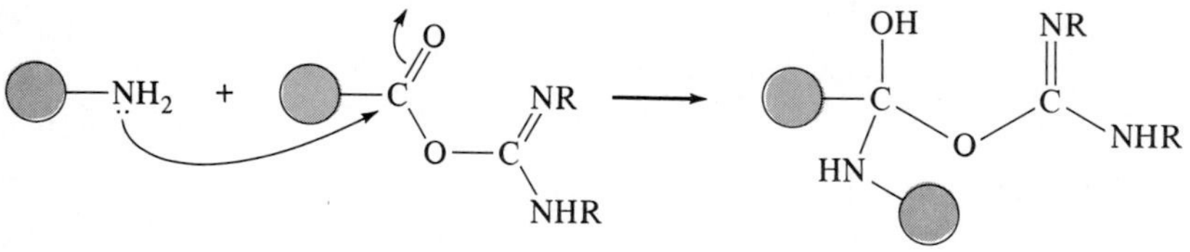

Amine O-Acylisourea   Tetrahedral intermediate

Step 3: The tetrahedral intermediate dissociates to an amide and N,N'-dicyclohexylurea.

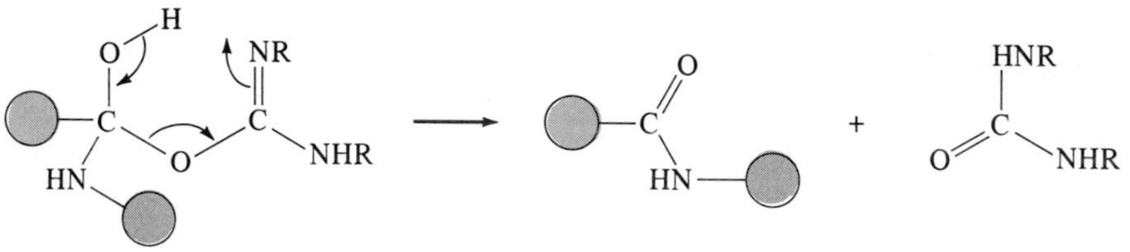

Tetrahedral intermediate   Amide N,N'-Dicyclohexylurea

FIGURE 29.13 Sequence of steps that describes the mechanism of amide bond formation by *N,N'*-dicyclohexylcarbodiimide-promoted condensation of a carboxylic acid and an amine.

N,N'-Dicyclohexylcarbodiimide has the structure shown.

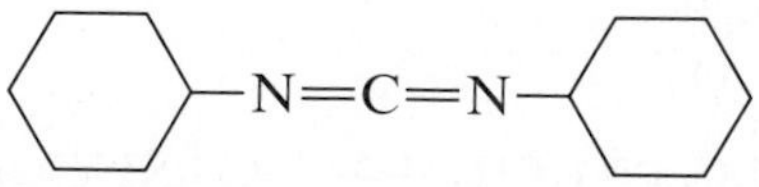

N,N'-Dicyclohexylcarbodiimide (DCCI)

The mechanism by which DCCI promotes the condensation of an amine and a carboxylic acid to give an amide is outlined in Figure 29.13.

PROBLEM 29.17 Show the steps involved in the synthesis of Ala-Leu from alanine and leucine using benzyloxycarbonyl and benzyl ester protecting groups and DCCI-promoted peptide bond formation.

In the second principal method of peptide synthesis the carboxyl group is activated toward acyl transfer by converting it to an *active ester*, usually a *p*-nitrophenyl ester. Recall from Section 21.11 that esters react with ammonia and amines to give amides. *p*-Nitrophenyl esters are much more reactive than *O*-alkyl esters in these reactions because *p*-nitrophenoxide is a better (less basic) leaving group than typical alkoxide leaving groups such as methoxide and ethoxide. Simply allowing the active ester and a *C*-protected amino acid to stand in a suitable solvent is sufficient to bring about peptide bond formation.

7-Protected phenylalanine *p*-nitrophenyl ester

Glycine ethyl ester

Z-Protected Phe-Gly ethyl ester (78%)

p-Nitrophenol

The *p*-nitrophenol formed as a by-product in this reaction is easily removed by extraction with dilute aqueous base. Unlike free amino acids and peptides, protected peptides are not zwitterionic and are more soluble in organic solvents than in water.

PROBLEM 29.18 *p*-Nitrophenyl esters are made from Z-protected amino acids by reaction with *p*-nitrophenol in the presence of *N,N'*-dicyclohexylcarbodiimide. Suggest a reasonable mechanism for this reaction.

PROBLEM 29.19 Show how you could convert the ethyl ester of Z-Phe-Gly to Leu-Phe-Gly (as its ethyl ester) by the active ester method.

Higher peptides are prepared either by stepwise extension of peptide chains, one amino acid residue at a time, or by coupling of fragments containing several residues (the *fragment condensation* approach). Human pituitary adrenocorticotropic hormone (ACTH), for example, has 39 amino acids and was synthesized by coupling of smaller peptides containing residues 1 through 10, 11 through 16, 17 through 24, and 25 through 39. An attractive feature of this approach is that the various protected peptide fragments may be individually purified, thereby simplifying the purification of the final product. Included among the substances that have been synthesized by fragment condensation are insulin (51 amino acids) and the protein ribonuclease A

(124 amino acids). In the stepwise extension approach, the starting peptide in a particular step differs from the coupling product by only one amino acid residue and the properties of the two peptides may be so similar as to preclude purification by conventional techniques. The following section describes a method by which many of the difficulties involved in the purification of intermediates have been surmounted.

29.17 SOLID-PHASE PEPTIDE SYNTHESIS. THE MERRIFIELD METHOD

In 1962 R. Bruce Merrifield of Rockefeller University reported the synthesis of the nonapeptide bradykinin (see Section 29.13) by a novel method. In Merrifield's method peptide coupling and deblocking are carried out not in homogeneous solution but at the surface of an insoluble polymer or *solid support*. Beads of a copolymer prepared from styrene containing about 2 percent divinylbenzene are treated with chloromethyl methyl ether and tin (IV) chloride to give a resin in which about 10 percent of the aromatic rings bear —CH_2Cl groups (Figure 29.14). The growing peptide is anchored to this polymer and excess reagents, impurities, and by-products are removed by thorough washing after each operation. This greatly simplifies the purification of intermediates.

The actual process of solid-phase peptide synthesis, outlined in Figure 29.15, begins with the attachment of the C-terminal amino acid to the chloromethylated polymer in step 1. Nucleophilic substitution by the carboxylate anion of an *N*-Boc-protected C-terminal amino acid displaces chloride from the chloromethyl group of the polymer to form an ester, protecting the C terminus while anchoring it to a solid support. Next, the Boc group is removed by treatment with acid (step 2) and the polymer containing the unmasked N terminus is washed with a series of organic solvents. By-products are removed, and only the polymer and its attached C-terminal amino acid residue remain. Next (step 3), a peptide bond to an *N*-Boc-protected amino acid is formed by condensation in the presence of *N,N'*-dicyclohexylcarbodiimide. Again, the polymer is washed thoroughly. The Boc protecting group is then removed by acid treatment (step 4) and, after washing, the polymer is now ready for the addition of another amino acid residue by a repetition of the cycle. When all the amino acids have been added, the synthetic peptide is removed from the polymeric support by treatment with hydrogen bromide in trifluoroacetic acid.

By successively adding amino acid residues to the C-terminal amino acid, it took Merrifield only 8 days to synthesize bradykinin in 68 percent yield. The biological activity of synthetic bradykinin was identical to that of natural material.

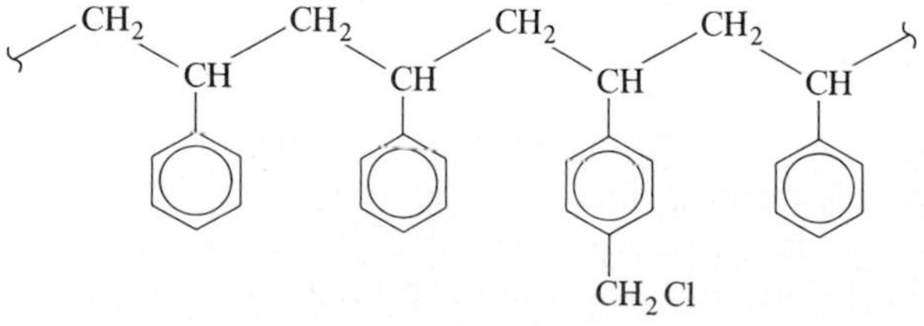

FIGURE 29.14 A section of polystyrene showing one of the benzene rings modified by chloromethylation. Individual polystyrene chains in the resin used in solid-phase peptide synthesis are connected to each other at various points (cross-linked) by adding a small amount of *p*-divinylbenzene to the styrene monomer. The chloromethylation step is carried out under conditions such that only about 10 percent of the benzene rings bear —CH_2Cl groups.

Step 1: The Boc-protected amino acid is anchored to the resin. Nucleophilic substitution of the benzylic chloride by the carboxylate anion gives an ester.

Step 2: The Boc protecting group is removed by treatment with hydrochloric acid in dilute acetic acid. After the resin has been washed, the C-terminal amino acid is ready for coupling.

Step 3: The resin-bound C-terminal amino acid is coupled to an N-protected amino acid by using N,N'-dicyclohexylcarbodiimide. Excess reagent and N,N'-dicyclohexylurea are washed away from the resin after coupling is complete.

Step 4: The Boc protecting group is removed as in step 2. If desired, steps 3 and 4 may be repeated to introduce as many amino acid residues as desired.

Step n: When the peptide is completely assembled, it is removed from the resin by treatment with hydrogen bromide in trifluoroacetic acid.

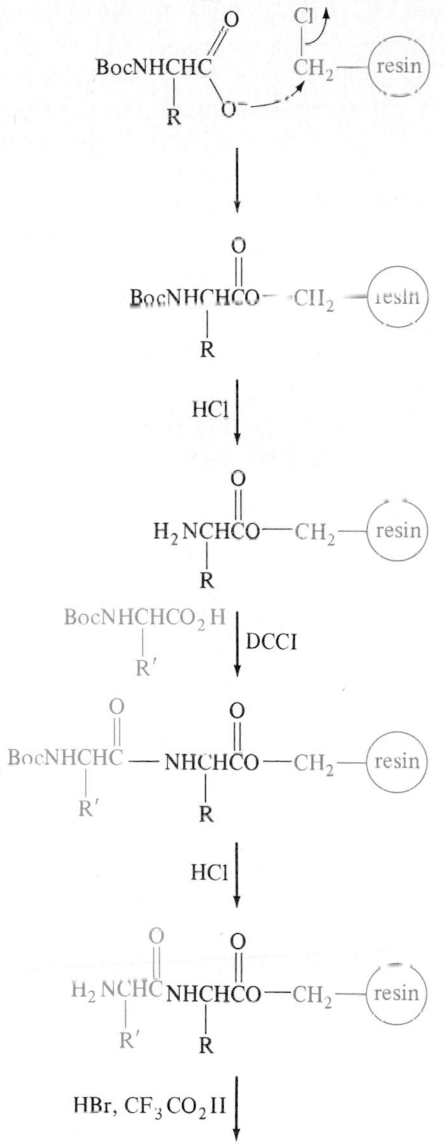

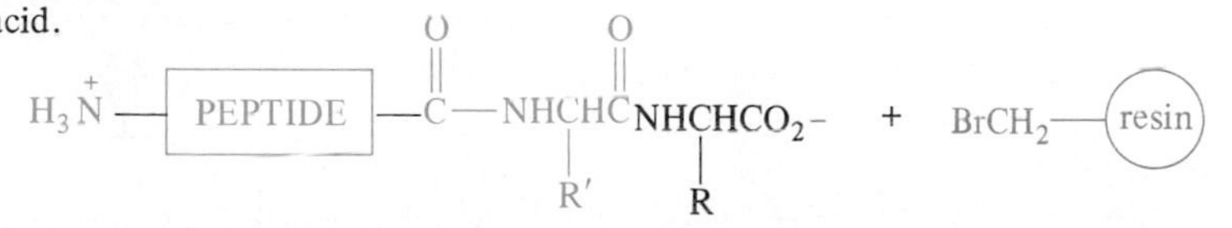

FIGURE 29.15 Diagram illustrating peptide synthesis by the solid-phase method of Merrifield. Amino acid residues are attached sequentially beginning at the C terminus.

PROBLEM 29.20 Starting with phenylalanine and glycine, outline the steps in the preparation of Phe-Gly by the Merrifield method.

Merrifield successfully automated all the steps in solid-phase peptide synthesis, and computer-controlled equipment is now commerically available. Using an early version of his "peptide synthesizer" and in collaboration with coworker Bernd Gutte, Merrifield reported the synthesis of the enzyme ribonuclease in 1969. It took them only 6 weeks to perform the 369 reactions and 11,391 steps necessary to assemble the sequence of 124 amino acids of ribonuclease.

Solid-phase peptide synthesis does not solve all purification problems, however. Even if every coupling step in the ribonuclease synthesis proceeded in 99 percent yield, the product would be contaminated with many different peptides containing 123 amino acids, 122 amino acids, etc. Thus, Merrifield and Gutte's 6 weeks of synthesis was followed by 4 months spent in purifying the final product. The technique has since been refined to the point that yields at the 99 percent level and greater are achieved with current instrumentation, and thousands of peptides and peptide analogs have been prepared by the solid-phase method. Merrifield was awarded the Nobel prize in chemistry in 1984.

29.18 SECONDARY STRUCTURES OF PEPTIDES AND PROTEINS

The primary structure of a peptide is its constitution, and its constitution is dictated by the amino acid sequence. We also speak of the *secondary structure* of a peptide, that is, the conformational relationship of nearest neighbor amino acids with respect to each other. We apply unique names to certain arrangements of peptide units, just as we identify certain conformations of small molecules by descriptive names such as gauche, anti, chair, boat, etc. On the basis of x-ray crystallographic studies and careful examination of molecular models, Linus Pauling and Robert B. Corey of the California Institute of Technology showed that certain peptide conformations were more stable than others. Two arrangements, the α *helix* and the β-*pleated sheet,* stood out as secondary structural units that are both particularly stable and commonly encountered. Both of these incorporate two features that appear to be especially important determinants of peptide structure:

1. The geometry of the peptide bond is planar and the main chain is arranged in an anti conformation (Section 29.7).

FIGURE 29.16 Diagram of the β-sheet secondary structure of a protein, showing hydrogen bonds between amide protons (blue) and carbonyl oxygens (red) of adjacent chains. Van der Waals repulsions between substituents (yellow) at the α carbon atoms are relieved by a conformational change within each chain, which has the effect of introducing vertical creases in the sheet. This motion places the substituents almost perpendicular to the plane of the page, alternating above and below the plane as one proceeds along the chain. The secondary structure that results is known as the β-pleated sheet.

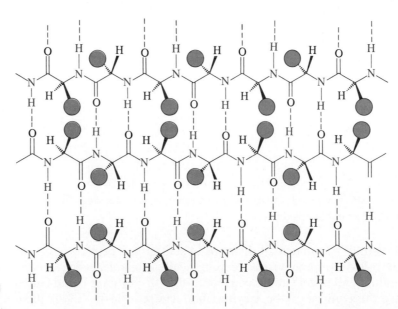

2. Hydrogen bonding can occur when the N—H group of one amino acid unit and the C=O group of another are close in space; conformations that maximize the number of these hydrogen bonds are stabilized by them.

Figure 29.16 depicts one type of secondary structure, a flat sheet characterized by C=O $\cdots$ H—N hydrogen bonds between two peptide chains. This is not a particularly stable structure because of van der Waals repulsions between the substituents on the α carbon atom of one chain and the corresponding substituents on the neighboring chain. Rotation about the single bond between the α carbon and nitrogen permits substituents on adjacent chains to move away from each other. Within each chain, then, the carbon atoms alternate with respect to their orientation—one is above the mean plane of the sheet, the next is below this mean plane, and so on. The peptide chains are not flat but are rippled and give the three-dimensional surface the appearance of a *pleated sheet*. The β-pleated sheet is an important secondary structure,

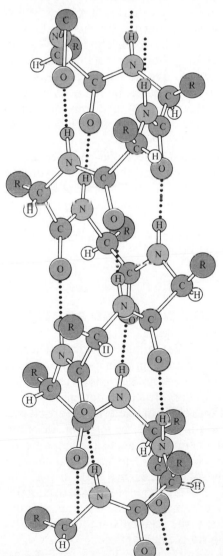

FIGURE 29.17 A protein α helix. The α helix is stabilized by hydrogen bonds within the chain between amide protons (blue) and carbonyl oxygens (red). The substituents at the α carbon are shown in yellow. When viewed along the helical axis, the chain turns in a clockwise direction and is designated as a right-handed helix

especially in proteins that are rich in amino acids with small substituents on their α carbon atoms. *Fibroin,* the major protein of most silk fibers, is almost entirely β-pleated sheet and over 80 percent of fibroin is a repeating sequence of the six-residue unit

-Gly-Ser-Gly-Ala-Gly-Ala-

The β-pleated sheet is flexible, but since the peptide chains are nearly in an extended conformation, it resists stretching.

Unlike the β-pleated sheet, in which hydrogen bonds are formed between two chains, the α *helix* is stabilized by hydrogen bonds within a single chain. Figure 29.17 illustrates a section of peptide α helix constructed from L amino acids. A right-handed helical conformation with about 3.6 amino acids per turn permits each carbonyl oxygen to be hydrogen-bonded to an amide proton and vice versa. The α helix is found in many proteins; the principal protein components of muscle (*myosin*) and wool (α-keratin), for example, contain high percentages of α helix. When wool fibers are stretched, these helical regions are elongated by the breaking of hydrogen bonds. Disulfide bonds between cysteine residues of neighboring α-keratin chains are too strong to be broken during stretching, however, and they limit the extent of distortion. After the stretching force is removed, the hydrogen bonds reform spontaneously and the wool fiber returns to its original shape. Wool has properties that are different from those of silk because the secondary structures of the two fibers are different, and their secondary structures are different because their primary structures are different.

Proline is the only amino acid in Table 29.1 that is a secondary amine, and its presence in a peptide chain introduces an amide nitrogen that has no hydrogen substituent available for hydrogen bonding. This disrupts the network of hydrogen bonds and divides the peptide into two separate regions of α helix. The presence of proline is often associated with a bend in the peptide chain.

Proteins, or sections of proteins, sometimes exist as *random coils,* an arrangement that lacks the regularity of the α helix or β-pleated sheet.

29.19 TERTIARY STRUCTURE OF PEPTIDES AND PROTEINS

The *tertiary structure* of a peptide or protein refers to the folding of the chain. The way the chain is folded affects both the physical properties of a protein and its biological function. Structural proteins, such as those present in skin, hair, tendons, wool, and silk, may have either helical or pleated-sheet secondary structures but in general are elongated in shape, with a chain length many times chain diameter. They are classed as *fibrous* proteins and, as befits their structural role, tend to be insoluble in water. Many other proteins, including most enzymes, operate in aqueous media; some are soluble but most are dispersed as colloids. Proteins of this type are called *globular* proteins. Globular proteins are approximately spherical. Figure 29.18 depicts the peptide backbone of carboxypeptidase A (Section 29.11), a globular protein containing 307 amino acids. A typical protein such as carboxypeptidase A incorporates elements of a number of secondary structures; some segments are helical, others pleated sheet, and still others correspond to no simple description.

The shape of a large protein is influenced by many factors including, of course, its

FIGURE 29.18 The peptide backbone of carboxypeptidase A. The substituents at the α carbon atom of each of the amino acids have been omitted so that the tertiary structure of the enzyme may be more easily seen. Amino acids are numbered beginning at the N terminus (bottom center). The C-terminal amino acid is residue 307 at the left. A disulfide bridge between two cysteines is shown in yellow and Zn^{2+} is shown in red. The amino acids that are coordinated to Zn^{2+} (His-69, His-196, and Glu-72) are indicated in blue.

primary and secondary structure. The disulfide bond shown in yellow in Figure 29.18 links Cys-138 of carboxypeptidase A to Cys-161 and is a contributor to the tertiary structure. Carboxypeptidase A contains a Zn^{2+} ion (shown in red in Figure 29.18); it is a *metalloenzyme*. The Zn^{2+} ion is essential to the catalytic activity of the enzyme, and its presence influences the tertiary structure. The Zn^{2+} ion lies near the center of the enzyme, where it is coordinated to the imidazole nitrogens of two histidine residues (His-69, His-196) and to the carboxylate side chain of Glu-72. (These ligands are shown in blue in the figure.)

Protein tertiary structure is also influenced by its environment. In water a globular protein usually adopts a shape that places its lipophilic groups toward the interior, with its polar groups on the surface, where they are solvated by water molecules. About 65 percent of the mass of most cells is water, and the proteins present in cells are said to be in their *native state*—the tertiary structure in which they express their biological activity. When the tertiary structure of a protein is disrupted by adding substances that cause the protein chain to unfold, the protein becomes *denatured* and loses most, if not all, of its activity. Evidence that supports the view that the tertiary structure is dictated by the primary structure includes experiments in which proteins are denatured and allowed to stand, whereupon they are observed to spontaneously readopt their native-state conformation with full recovery of biological activity.

Tertiary structures of proteins are determined by x-ray crystallography. Myoglobin, the oxygen storage protein of muscle, was the first protein to have its tertiary structure revealed (1957). For their work on myoglobin and hemoglobin, respectively, John C. Kendrew and Max F. Perutz were awarded the 1962 Nobel prize in chemistry. A knowledge of how the protein chain is folded is an indispensable step toward understanding the mechanism of enzyme action.

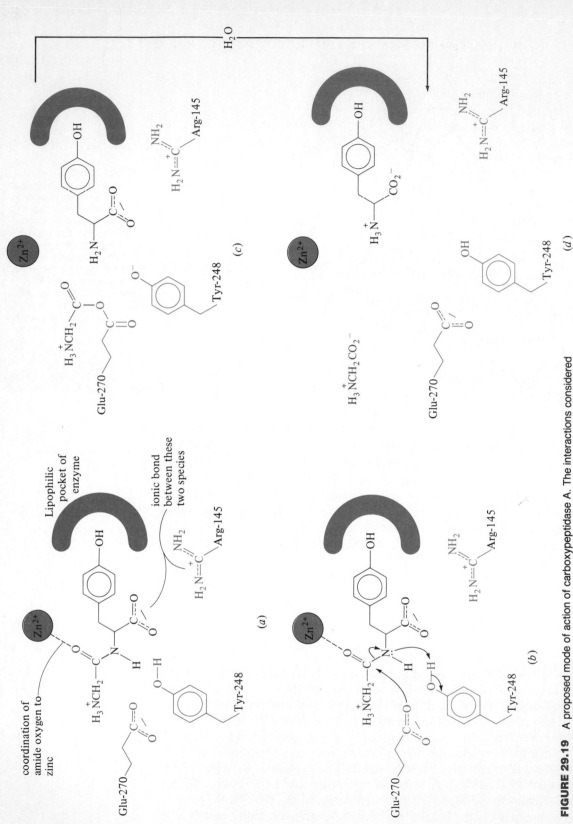

FIGURE 29.19 A proposed mode of action of carboxypeptidase A. The interactions considered significant in binding the substrate to the active site are show in (a). In (b) nucleophilic attack by the carboxylate group of Glu-270 cleaves the peptide bond and gives tyrosine and a mixed anhydride (c). Hydrolysis of the mixed anhydride gives glycine and returns Glu-270 to its catalytically active form (d).

29.20 HOW AN ENZYME WORKS. CARBOXYPEPTIDASE A

You saw in Section 29.11 that carboxypeptidase A catalyzes the hydrolysis of the peptide bond to the C-terminal amino acid. What are the features of this enzyme that cause it to accelerate the rate of this reaction, and why is it selective for the C terminus? A reasonable proposal for the mechanism of carboxypeptidase A action has been proposed by William N. Lipscomb of Harvard University. Lipscomb compared the x-ray crystal structure of carboxypeptidase A with that of carboxypeptidase A in which a molecule of glycyltyrosine (Gly-Tyr) had been trapped. He found that the tertiary structure of the enzyme adjusts to the presence of the dipeptide in ways that are chemically reasonable and biochemically significant.

On the basis of these structural studies and what was already known about the enzyme, Lipscomb suggested the mechanism shown in Figure 29.19 for the mode of action of carboxypeptidase A. First (Figure 29.19a), the peptide substrate is bound at or near the *active site,* the region of the enzyme's interior where the catalytically significant functional groups are found. In the case of carboxypeptidase A, it is that portion of the enzyme near Zn^{2+}. There is a lipophilic pocket close to the active site into which the aryl side chain of Gly-Tyr can fit. The C-terminal specificity of carboxypeptidase may be related to the presence of a nearby arginine residue (Arg-145), which can bind the carboxylate group of the C terminus of the peptide as an ammonium salt. The Zn^{2+} ion at the active site coordinates with the carbonyl oxygen of the peptide bond, increasing its susceptibility toward nucleophilic attack.

The nucleophile that attacks the peptide bond is believed to be the carboxylate group of a glutamate residue near the active site (Glu-270), as shown in Figure 29.19b. Cleavage of the peptide bond is assisted both by coordination of the carbonyl oxygen with Zn^{2+} and proton transfer from Tyr-248 to the nitrogen of the leaving group. Lipscomb found that Tyr-248 was substantially closer to the active site in the enzyme-substrate complex than in the pure enzyme alone. The C-terminal amino acid (tyrosine) is released in this step and the remainder of the peptide (glycine) is attached to Glu-270 as a mixed anhydride (Figure 29.19c). Rapid hydrolysis of this anhydride frees the glycine residue and restores Glu-270 to its catalytically active state (Figure 29.19d).

Living systems contain thousands of different enzymes. As we have seen, all are structurally quite complex, and there are no sweeping generalizations that can be made to include all aspects of enzymic catalysis. The case of carboxypeptidase A illustrates one mode of enzyme action, the bringing together of reactants in the presence of catalytically active functions at the active site.

29.21 COENZYMES

The molecules that undergo permanent chemical change in carboxypeptidase-catalyzed hydrolyses are the peptide and a water molecule. The various functional groups in the amino acid side chains play a catalytic role and are rapidly restored to their original form, ready to catalyze the cleavage of another C-terminal peptide bond. The number of chemical processes that protein side chains can engage in is rather limited. Most prominent among them are proton donation, proton abstraction, and nucleophilic addition to carbonyl groups. In many biological processes a richer variety of reactivity is required, and proteins often act in combination with nonprotein organic molecules to bring about the necessary chemistry. These "helper molecules," re-

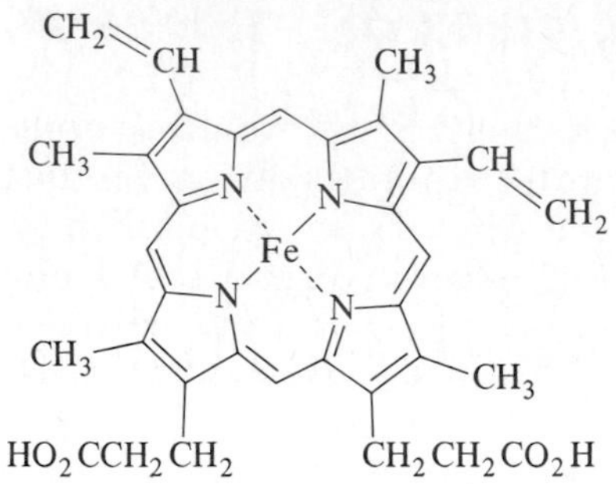

CH$_2$=CH

CH$_3$

CH$_3$

CH

CH$_2$

CH$_3$

CH$_3$

CH$_3$

HO$_2$CCH$_2$CH$_2$ CH$_2$CH$_2$CO$_2$H

FIGURE 29.20   The structure of heme.

ferred to as *coenzymes, cofactors,* or *prosthetic groups,* interact with both the enzyme and the substrate to produce the necessary chemical change. Acting alone, for example, proteins lack the necessary functionality to be effective oxidizing or reducing agents. They can catalyze biological oxidations and reductions, however, in the presence of a suitable coenzyme. In earlier sections we have seen numerous examples of these reactions in which the coenzyme NAD$^+$ acted as an oxidizing agent and others in which NADH acted as a reducing agent.

Heme (Figure 29.20) is an important prosthetic group in which ferrous iron, Fe(II), is coordinated with the four nitrogen atoms of a type of tetracyclic aromatic substance known as a *porphyrin.* The oxygen-storing protein of muscle *myoglobin,* represented schematically in Figure 29.21, consists of a heme group surrounded by a protein of 153 amino acids. Four of the six available coordination sites of Fe^{2+} are taken up by the nitrogens of the porphyrin, one by a histidine residue of the protein, and the last by a water molecule. Myoglobin stores oxygen obtained from the blood

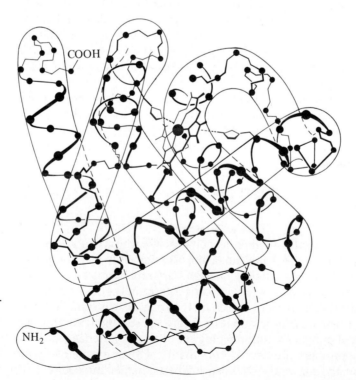

FIGURE 29.21   Diagram showing the folding of the protein chain of myoglobin. The heme portion is shown in red. The blue rings represent imidazole rings of histidine ligands available for coordination to iron.

by formation of an Fe-O_2 complex. The oxygen displaces water as the sixth ligand on iron and is held there until needed. The protein serves as a container for the heme and prevents oxidation of Fe^{2+} to Fe^{3+}, an oxidation state in which iron lacks the ability to bind oxygen. Separately, neither heme nor the protein binds oxygen in aqueous solution; together, they do it very well.

29.22 PROTEIN QUATERNARY STRUCTURE. HEMOGLOBIN

Rather than existing as a single polypeptide chain, some proteins are assemblies of two or more chains. The manner in which these subunits are organized is called the *quaternary structure* of the protein.

Hemoglobin is the oxygen-carrying protein of blood. It binds oxygen at the lungs and transports it to the muscles, where it is stored by myoglobin. Hemoglobin binds oxygen in very much the same way as myoglobin, using heme as the prosthetic group. Hemoglobin is much larger than myoglobin, however, having a molecular weight of 64,500, whereas that of myoglobin is 17,500; hemoglobin contains four heme units, myoglobin only one. Hemoglobin is an assembly of four hemes and four protein chains, including two identical chains called the *alpha chains* and two identical chains called the *beta chains*.

Substances such as CO and CN^- form strong bonds to the iron of heme, strong enough to displace O_2 from it. Carbon monoxide binds 30 to 50 times more effectively than oxygen to myoglobin and hundreds of times better than oxygen to hemoglobin. Strong binding of CO at the active site interferes with the ability of heme to perform its biological task of transporting and storing oxygen, with potentially lethal results.

How function depends on structure can be seen in the case of the genetic disorder *sickle cell anemia*. This is a debilitating, sometimes fatal disease in which red blood cells become distorted ("sickle-shaped") and interfere with the flow of blood through the capillaries. This condition results from the presence of an abnormal hemoglobin in affected individuals. The primary structures of the beta chain of normal and sickle cell hemoglobin differ by a single amino acid out of 149; sickle cell hemoglobin has valine in place of glutamic acid as the sixth residue from the *N* terminus. A tiny change in amino acid sequence can produce a life-threatening result! This modification is genetically controlled and probably became established in the gene pool because bearers of the trait have an increased resistance to malaria.

29.23 PYRIMIDINES AND PURINES

One of the major scientific achievements of the twentieth century has been the identification, at the molecular level, of the chemical interactions that are involved in the transfer of genetic information and the control of protein biosynthesis. The substances involved are biological macromolecules called *nucleic acids*. Nucleic acids were isolated over 100 years ago and, as their name implies, they are acidic substances present in the nuclei of cells. There are two major kinds of nucleic acids, ribonucleic acid (RNA) and deoxyribonucleic acid (DNA). In order to understand the complex structure of nucleic acids, we need to first examine some simpler substances, nitrogen-containing aromatic heterocycles called *pyrimidines* and *purines*. The parent substance of each class and the numbering system used are shown.

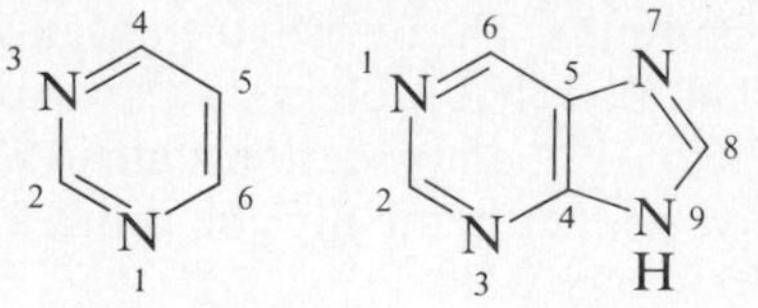

Pyrimidine

Purine

The pyrimidines that occur in DNA are cytosine and thymine. Cytosine is also a principal structural unit in RNA, which, however, contains uracil instead of thymine. Other pyrimidine derivatives are sometimes present but in small amounts.

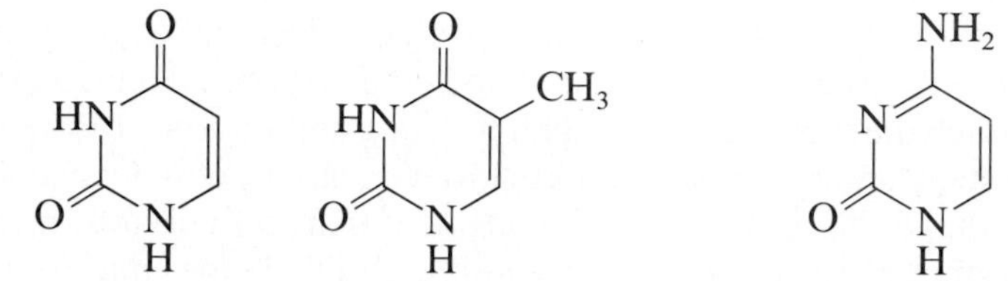

Uracil
(occurs in RNA)

Thymine
(occurs in DNA)

Cytosine
(occurs in both RNA and DNA)

PROBLEM 29.21   5-Fluorouracil is a drug used in cancer chemotherapy. What is its structure?

Adenine and guanine are the principal purines of both DNA and RNA.

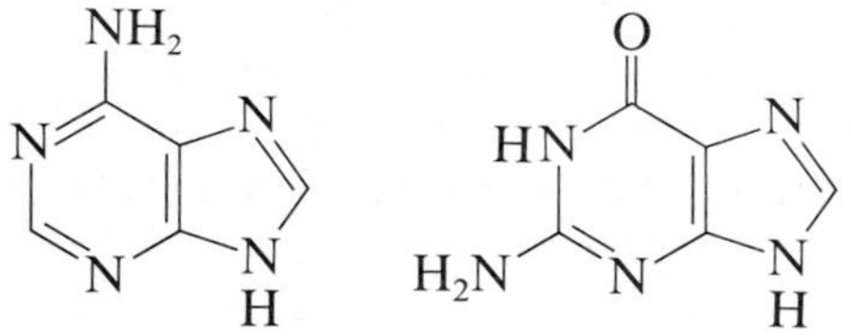

Adenine

Guanine

The rings of purines and pyrimidines are aromatic and planar. You will see that this flat shape is significant when we consider the structure of nucleic acids.

Pyrimidines and purines occur naturally in substances other than nucleic acids. Coffee, for example, is a familiar source of caffeine. Tea contains both caffeine and theobromine.

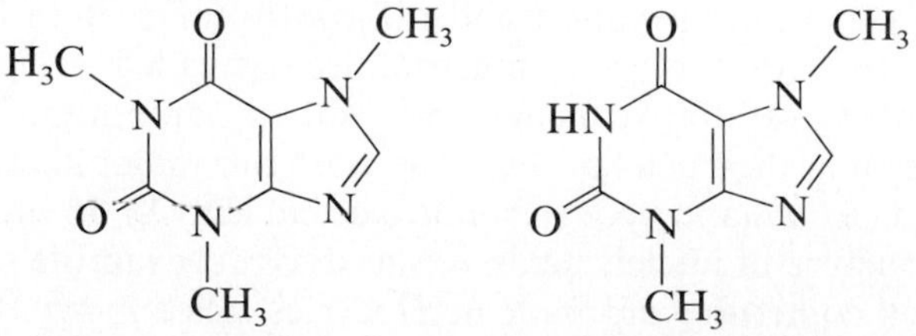

Caffeine

Theobromine

29.24 NUCLEOSIDES

The term *nucleoside* was once restricted to pyrimidine and purine *N*-glycosides of D-ribofuranose and 2-deoxy-D-ribofuranose because these are the substances present in nucleic acids. The term is used more liberally now with respect to the carbohydrate portion, but is still usually limited to pyrimidine and purine substituents at the anomeric carbon. Uridine is a representative pyrimidine nucleoside; it bears a D-ribofuranose group at N-1. Adenosine is a representative purine nucleoside; its carbohydrate unit is attached at N-9.

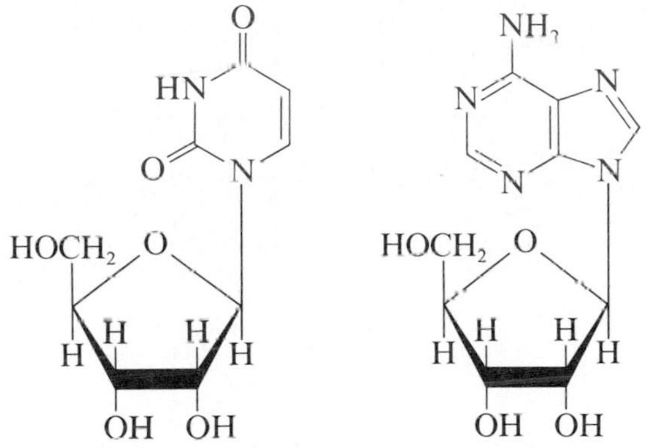

Uridine
(1-β-D-ribofuranosyluracil)

Adenosine
(9-β-D-ribofuranosyladenine)

It is customary to refer to the noncarbohydrate portion of a nucleoside as a purine or pyrimidine *base*.

PROBLEM 29.22 The names of the principal nucleosides obtained from RNA and DNA are given below. Write a structural formula for each one.

(a) Thymidine (thymine-derived nucleoside in DNA)
(b) Cytidine (cytosine-derived nucleoside in RNA)
(c) Guanosine (guanine-derived nucleoside in RNA)

SAMPLE SOLUTION (a) Thymine is a pyrimidine base present in DNA; its carbohydrate substituent is 2-deoxyribofuranose, which is attached to N-1 of thymine.

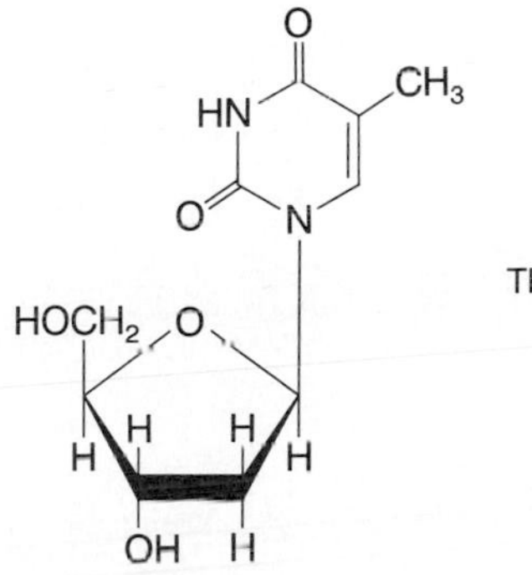

Thymidine

Nucleosides of 2-deoxyribose are named in the same way. Carbons in the carbohydrate portion of the molecule are identified as 1′, 2′, 3′, 4′, and 5′ to distinguish them from atoms in the purine or pyrimidine base. Thus, the adenine nucleoside of 2-deoxyribose is called 2′-deoxyadenosine or 9-β-2′-deoxyribofuranosyladenine.

29.25 NUCLEOTIDES

Nucleotides are phosphoric acid esters of nucleosides. The 5′-monophosphate of adenosine is called *5′-adenylic acid* or *adenosine 5′-monophosphate* (AMP).

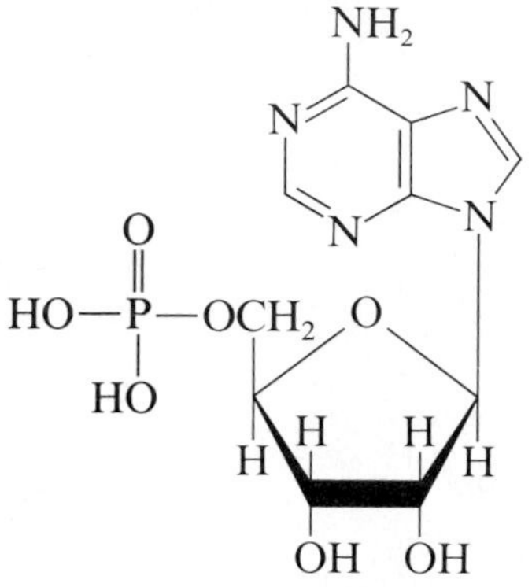

5′-Adenylic acid (AMP)

As its name implies, 5′-adenylic acid is an acidic substance; it is a dibasic acid with pK_a's for ionization of 3.8 and 6.2, respectively. In aqueous solution at pH 7, the major form is the dianion. (Crystalline 5′-adenylic acid exists as a zwitterion in which N-7 is protonated and one of the phosphate oxygens is deprotonated.)

The analogous D-ribonucleotides of the other purines and pyrimidines are uridylic acid, guanylic acid, and cytidylic acid. Thymidylic acid is the 5′-monophosphate of thymidine (the carbohydrate is 2-deoxyribose in this case).

Other important 5′-nucleotides of adenosine include adenosine diphosphate (ADP) and adenosine triphosphate (ATP).

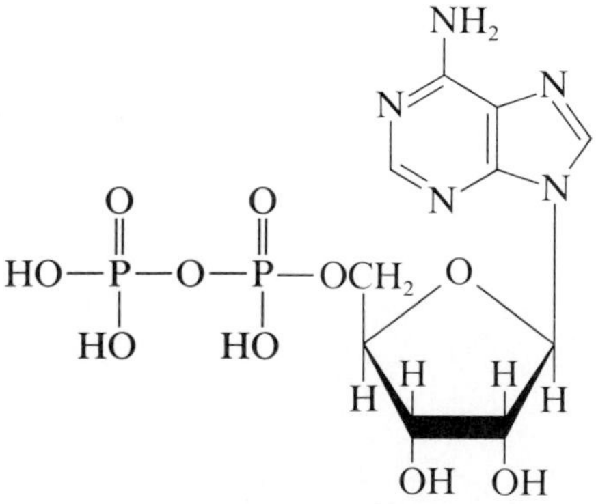

Adenosine diphosphate (ADP)

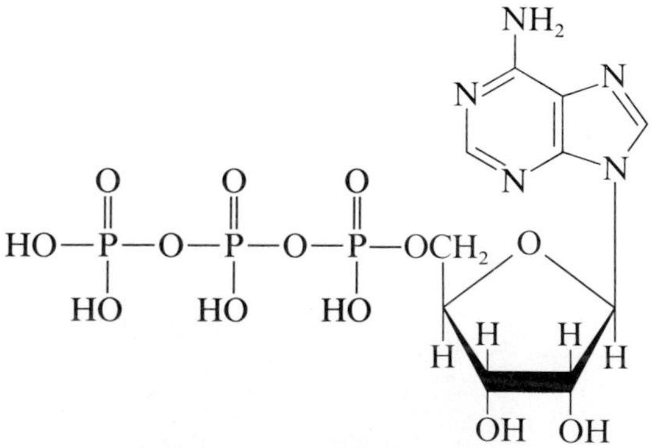

Adenosine triphosphate (ATP)

Each phosphorylation step in the sequence shown is endothermic.

$$\text{Adenosine} \xrightarrow[\text{enzymes}]{PO_4^{3-}} \text{AMP} \xrightarrow[\text{enzymes}]{PO_4^{3-}} \text{ADP} \xrightarrow[\text{enzymes}]{PO_4^{3-}} \text{ATP}$$

The energy to drive each step comes from carbohydrates by the process of glycolysis (Section 27.22). It is convenient to view ATP as the storage vessel for the energy released during conversion of carbohydrates to carbon dioxide and water. We describe AMP, ADP, and ATP as "high-energy compounds" and associate their high energy with the phosphate ester linkages. That energy becomes available to the cells when ATP undergoes phosphate ester hydrolysis; the hydrolysis of ATP to ADP and phosphate has a $\Delta G°$ value of -8.4 kcal/mol.

Adenosine 3'-5'-cyclic monophosphate *(cyclic AMP)* is an important regulator of a large number of biological processes. It is a cyclic ester of phosphoric acid and adenosine involving the hydroxyl groups at C-3' and C-5'.

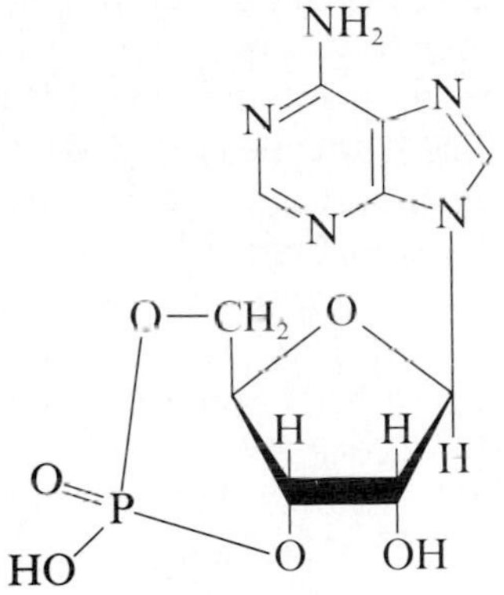

Adenosine 3'-5'-cyclic monophosphate *(cyclic AMP)*

29.26 NUCLEIC ACIDS

Nucleic acids are *polynucleotides* in which a phosphate ester unit links the 5' oxygen of one nucleotide to the 3' oxygen of another. Figure 29.22 is a generalized depiction of the structure of a nucleic acid. Nucleic acids are classified as ribonucleic acids (RNA) or deoxyribonucleic acids (DNA) depending on the carbohydrate present.

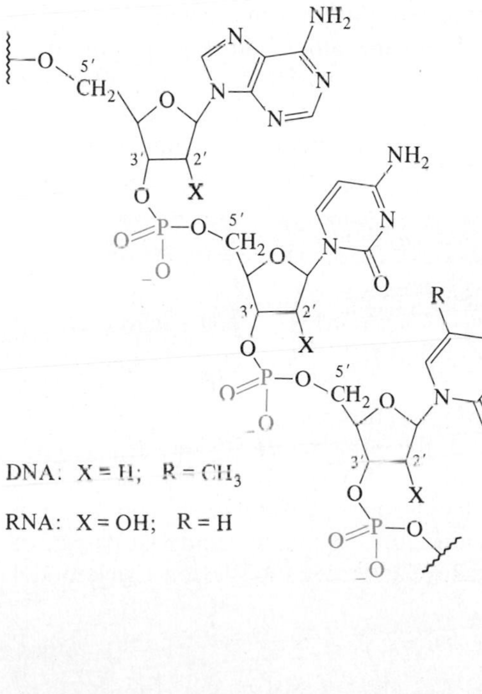

DNA: X = H; R = CH₃

RNA: X = OH; R = H

FIGURE 29.22 A portion of a polynucleotide chain.

Research on nucleic acids progressed slowly until it became evident during the 1940s that they played a role in the transfer of genetic information. It was known that the genetic information of an organism resides in the chromosomes present in each of its cells and that individual chromsomes are made up of smaller units called genes. When it became apparent that genes are DNA, interest in nucleic acids intensified. There was a feeling that once the structure of DNA was established, the precise way in which it carried out its designated role would become more evident. In some respects the problems are similar to those of protein chemistry. Knowing that DNA is a polynucleotide is comparable with knowing that proteins are polyamides. What is the nucleotide sequence (primary structure)? What is the precise shape of the polynucleotide chain (secondary and tertiary structure)? Is the genetic material a single strand of DNA or is it an assembly of two or more strands? The complexity of the problem can be indicated by noting that a typical strand of human DNA contains approximately 10^8 nucleotides; if uncoiled it would be several centimeters long, yet it and many others like it reside in cells too small to see with the naked eye.

In 1953 James D. Watson and Francis H. C. Crick pulled together data from biology, biochemistry, chemistry, and x-ray crystallography, along with the insight they gained from molecular models, to propose a structure for DNA and a mechanism for its replication. Their two brief papers paved the way for an explosive growth in our understanding of life processes at the molecular level, the field we now call *molecular biology.* Along with Maurice Wilkins, who was responsible for the x-ray crystallographic work, Watson and Crick shared the 1962 Nobel prize in physiology or medicine.

29.27 STRUCTURE AND REPLICATION OF DNA. THE DOUBLE HELIX

Watson and Crick were aided in their search for the structure of DNA by a discovery made by Erwin Chargaff (Columbia University). Chargaff found that there was a consistent pattern in the composition of DNA's from various sources. While there was a wide variation in the distribution of the bases among species, one-half of the bases in all samples of DNA were purines and the other half were pyrimidines. Furthermore, the ratio of the purine adenine (A) to the pyrimidine thymine (T) was always close to 1 : 1. Likewise, the ratio of the purine guanine (G) to the pyrimidine cytosine (C) was also close to 1 : 1. Analysis of human DNA, for example, revealed it to have the following composition:

Purine	Pyrimidine	Base ratio
Adenine (A) 30.3%	Thymine (T) 30.3%	A/T = 1.00
Guanine (G) 19.5%	Cytosine (C) 19.9%	G/C = 0.98
Total purines 49.8%	Total pyrimidines 50.1%	

Feeling that the constancy in the A/T and G/C ratios was no accident, Watson and Crick proposed that it resulted from a structural complementarity between A and T and between G and C. Consideration of various hydrogen bonding arrangements revealed that A and T could form the hydrogen-bonded *base pair* shown in Figure 29.23a and that G and C could associate as in Figure 29.23b. Specific base pairing of A to T and of G to C by hydrogen bonds is a key element in the Watson-Crick model

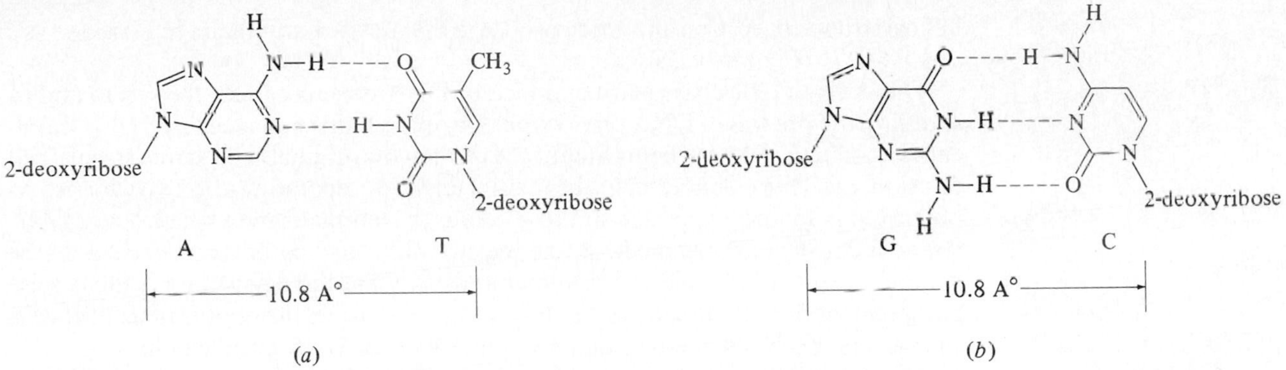

A   T

|—————10.8 A°—————|

(a)

G   C

|—————10.8 A°—————|

(b)

FIGURE 29.23 Base pairing between (a) adenine and thymine and (b) guanine and cytosine.

for the structure of DNA. We shall see that it is also a key element in the replication of DNA.

Because each hydrogen-bonded base pair contains one purine and one pyrimidine, A $\cdots$ T and G $\cdots$ C are approximately the same size. Thus, two nucleic acid chains may be aligned side by side with their bases in the middle, as illustrated in Figure 29.24. The two chains are joined by the network of hydrogen bonds between the paired bases A $\cdots$ T and G $\cdots$ C. Since x-ray crystallographic data indicated a

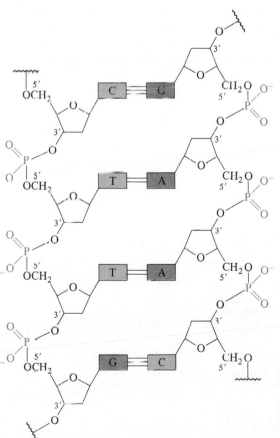

FIGURE 29.24 Hydrogen bonds between complementary bases (A and T and G and C) permit pairing of two DNA strands. The strands are antiparallel; the 5' end of the left strand is at the top while that of the right strand is at the bottom.

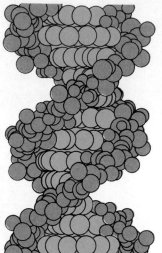

FIGURE 29.25 A space-filling model of a portion of a DNA double helix. The red atoms on the outside of the double helix belong to the carbohydrate-phosphate "backbone" of the molecule. The yellow atoms belong to the purine and pyrimidine bases and are stacked as steps on a spiral staircase on the inside.

helical structure, Watson and Crick proposed that the two strands are intertwined as a *double helix* (Figure 29.25).

The Watson-Crick base pairing model for DNA structure holds the key to understanding the process of DNA *replication.* During cell division a cell's DNA is duplicated, that in the new cell being identical to that in the original cell. At one stage of cell division the DNA double helix begins to unwind, separating the two chains. As portrayed in Figure 29.26 each strand serves as the template upon which a new DNA strand is constructed. Each new strand is exactly like the original partner because the A $\cdots$ T, G $\cdots$ C base-pairing requirement ensures that the new strand is the precise complement of the template, just as the old strand was. As the double helix unravels, each strand becomes one-half of a new and identical DNA double helix.

The structural requirements for the pairing of nucleic acid bases are also critical for utilizing genetic information, and in living systems this means protein biosynthesis.

29.28   DNA-DIRECTED PROTEIN BIOSYNTHESIS

Protein biosynthesis is directed by DNA through the agency of several types of ribonucleic acid called *messenger RNA (mRNA), transfer RNA (tRNA),* and *ribosomal RNA (rRNA).* There are two main stages in protein biosynthesis, *transcription* and *translation.*

In the transcription stage a molecule of mRNA having a nucleotide sequence complementary to one of the strands of a DNA double helix is constructed. A diagram illustrating transcription is presented in Figure 29.27. Transcription begins at the 5′ end of the DNA molecule and ribonucleotides with bases complementary to the DNA bases are polymerized with the aid of the enzyme *RNA polymerase.* Thymine does not occur in RNA; the base that pairs with adenine in RNA is uracil. Unlike DNA, RNA is single-stranded.

In the translation stage, the nucleotide sequence of the mRNA is decoded and "read" as an amino acid sequence to be constructed. Since there are only four different bases in mRNA and 20 amino acids to be coded for, codes using either one

FIGURE 29.26 During DNA replication the double helix unwinds and each of the original strands serves as a template for the synthesis of its complementary strand.

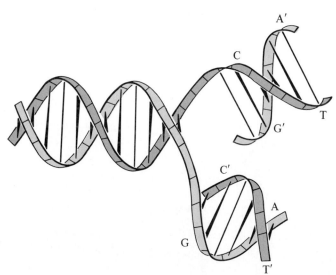

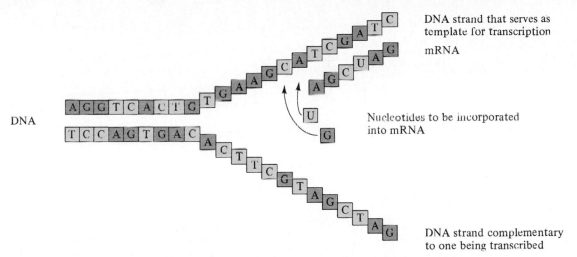

FIGURE 29.27 During transcription a molecule of mRNA is assembled by using DNA as a template.

nucleotide to one amino acid or two nucleotides to one amino acid are inadequate. If nucleotides are read in sets of three, however, the four mRNA bases (A, U, C, G) generate 64 possible "words," more than sufficient to code for 20 amino acids. It has been established that the *genetic code* is indeed made up of triplets of adjacent nucleotides called *codons*. The amino acids corresponding to each of the 64 possible codons of mRNA have been determined (Table 29.4).

PROBLEM 29.23 It was pointed out in Section 29.22 that sickle cell hemoglobin has valine in place of glutamic acid at one point in its protein chain. Compare the codons for valine and glutamic acid. How do they differ?

TABLE 29.4
The Genetic Code (Messenger RNA Codons)*

Alanine	Arginine	Asparagine	Aspartic acid	Cysteine
GCU GCA	CGU CGA AGA	AAU	GAU	UGU
GCC GCG	CGC CGG AGG	AAC	GAC	UGC
Glutamic acid	Glutamine	Glycine	Histidine	Isoleucine
GAA	CAA	GGU GGA	CAU	AUU AUA
GAG	CAG	GGC GGG	CAC	AUC
Leucine	Lysine	Methionine	Phenylalanine	Proline
UUA CUU CUA	AAA	AUG	UUU	CCU CCA
UUG CUC CUG	AAG		UUC	CCC CCG
Serine	Threonine	Tryptophan	Tyrosine	Valine
UCU UCA AGU	ACU ACA	UGG	UAU	GUU GUA
UCC UCG AGC	ACC ACG		UAC	GUC GUG

* The first letter of each triplet corresponds to the nucleotide nearer the 5′ terminus, the last letter to the nucleotide nearer the 3′ terminus. UAA, UGA, and UAG are not included in the table; they are chain-terminating codons.

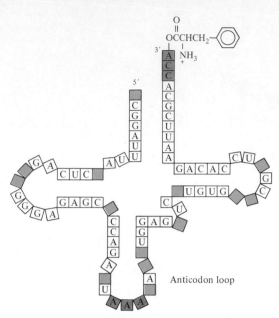

FIGURE 29.28 Phenylalanine tRNA. Transfer RNA's usually contain modified bases, slightly different from those in other RNA's, which are indicated in yellow. The anticodon for phenylalanine is shown in red, and the CCA triplet which bears the phenylalanine is in blue.

The mechanism of translation makes use of the same complementary base pairing principle used in replication and transcription. Each amino acid is associated with a particular tRNA. tRNA is much smaller than DNA and mRNA. It is single-stranded and contains 70 to 90 ribonucleotides arranged in a "cloverleaf" pattern (Figure 29.28). Its characteristic shape results from the presence of paired bases in some regions and their absence in others. All tRNA's have a CCA triplet at their 3′ terminus, to which is attached, by an ester linkage, an amino acid unique to that particular tRNA. At one of the loops of tRNA there is a nucleotide triplet called the *anticodon,* which is complementary to a codon of mRNA. The codons of mRNA are read by the anticodons of tRNA, and the proper amino acids are transferred in sequence to the growing protein.

29.29 SUMMARY

Organic chemistry began with the study of natural products, and all natural products are formed by enzyme-catalyzed reactions. All enzymes are proteins, high-molecular weight substances that are polymers of α-amino acids. A group of 20 amino acids, listed in Table 29.1, regularly appears as the hydrolysis products of proteins. With the exception of glycine, all are chiral and have the L configuration. The most stable structure of a neutral amino acid is a *zwitterion.* The pH of an aqueous solution at which the concentration of the zwitterion is a maximum is called the *isoelectric point* (pI).

Fischer projection formula of
L-valine in its zwitterionic form

Amino acids are synthesized in the laboratory from:

1. α-Halo acids by reaction with ammonia
2. Aldehydes by reaction with ammonia and cyanide ion (the Strecker synthesis)
3. Alkyl halides by reaction with the enolate anion derived from diethyl acetamidomalonate

The amino acids prepared by these methods are formed as racemic mixtures and are optically inactive.

Amino acids undergo reactions characteristic of the amino group and the carboxyl group.

An amide linkage between two α-amino acids is called a *peptide bond*. The *primary structure* of a peptide is given by its amino acid sequence plus any disulfide bonds between two cysteine residues. Determination of the amino acid sequence is facilitated by terminal residue analysis. The N terminus may be identified by reaction of the peptide with:

1. 1-Fluoro-2,4-dinitrobenzene (Sanger's reagent)
2. Phenyl isothiocyanate (Edman's reagent)
3. 5-Dimethylaminonaphthalene-1-sulfonyl chloride (dansyl chloride)

The C terminus is identified by hydrolysis with carboxypeptidase, an enzyme that selectively catalyzes the hydrolysis of the peptide bond to the C-terminal amino acid. Enzymes such as trypsin, chymotrypsin, and pepsin are also used in peptide sequencing. They promote selective hydrolysis of the peptide into a limited number of smaller fragments, each of which may be individually sequenced to provide data on the overall sequence.

Peptide synthesis requires that the number of possible reactions be limited by the judicious use of protecting groups. Amino-protecting groups include benzyloxycarbonyl (Z) and *tert*-butoxycarbonyl (Boc). Carboxyl groups are protected as esters. Peptide bond formation between a protected amino acid having a free carboxyl group and a protected amino acid having a free amino group can be accomplished with the aid of N,N'-dicyclohexylcarbodiimide (DCCI). Polypeptides are synthesized by two methods, the *fragment-condensation* approach and *solid-phase synthesis*.

Two *secondary structures* of proteins are particularly prominent. The *β-pleated sheet* is stabilized by hydrogen bonds between N—H and C=O groups of adjacent chains. The *α helix* is stabilized by hydrogen bonds within a single polypeptide chain.

The folding of a peptide chain is its *tertiary structure*. Many proteins consist of two or more chains, and the way in which the various units are assembled into the protein in its native state is its *quaternary structure*.

Enzymes are catalysts. They accelerate the rates of chemical reactions in biological systems, but the kinds of reactions that take place are the basic reactions of organic chemistry. One way in which enzymes accelerate these reactions is by bringing reactive functions together in the presence of catalytically active functions of the protein. Often those catalytically active functions are nothing more than proton donors and proton acceptors. In many cases a protein acts in cooperation with a *coenzyme*, a small molecule having the proper functionality to carry out a chemical change not otherwise available to the protein itself.

Protein biosynthesis is directed by nucleic acids. *Nucleic acids* are polymeric nucleotides; *nucleotides* are phosphate esters of nucleosides; and *nucleosides* are purine and pyrimidine N-glycosides of D-ribose and 2-deoxy-D-ribose. Deoxyribonucleic acid exists as a double helix, in which hydrogen bonds are responsible for complementary base pairing between adenine and thymine and between guanine and cytosine. During cell division the strands unwind and are duplicated. Each strand acts as a template upon which its complement is constructed. In the *transcription* stage of protein biosynthesis, a molecule of *messenger RNA* (mRNA) having a nucleotide sequence complementary to that of DNA is assembled. Transcription is followed by *translation,* in which triplets of nucleotides of mRNA called *codons* are recognized by the *transfer RNA* (tRNA) for a particular amino acid, and that amino acid is added to the growing peptide chain.

PROBLEMS

29.24 The imidazole ring of the histidine side chain acts as a proton acceptor in certain enzyme-catalyzed reactions. Which is the more stable protonated form of the histidine residue, A or B? Why?

A   B

29.25 Acrylonitrile (CH$_2$=CHC≡N) readily undergoes conjugate addition when treated with nucleophilic reagents. Describe a synthesis of β-alanine (H$_3$NCH$_2$CH$_2$CO$_2^-$) that takes advantage of this fact.

29.26 (*a*) DL-Isoleucine has been prepared by the following sequence of reactions. Give the structure of compounds C through F isolated as intermediates in this synthesis.

$$\text{CH}_3\text{CH}_2\text{CHCH}_3 \xrightarrow[\text{sodium ethoxide}]{\text{diethyl malonate}} \text{C} \xrightarrow[\text{2. HCl}]{\text{1. KOH}} \text{D (C}_7\text{H}_{12}\text{O}_4)$$
$$\overset{|}{\underset{\text{Br}}{}}$$

$$\text{D} \xrightarrow{\text{Br}_2} \text{E (C}_7\text{H}_{11}\text{BrO}_4) \xrightarrow{\text{heat}} \text{F} \xrightarrow[\text{H}_2\text{O}]{\text{NH}_3} \text{DL-isoleucine}$$

(*b*) An analogous procedure has been used to prepare DL-phenylalanine. What alkyl halide would you choose as the starting material for this synthesis?

29.27 Hydrolysis of compound G in concentrated hydrochloric acid for several hours at 100°C gives one of the amino acids in Table 29.1. Which one? Is it optically active?

Compound G

29.28 If you synthesized the tripeptide Leu-Phe-Ser from amino acids prepared by the Strecker synthesis, how many stereoisomers would you expect to be formed?

29.29 How many peaks would you expect to see on the strip chart after amino acid analysis of bradykinin?

Arg-Pro-Pro-Gly-Phe-Ser-Pro-Phe-Arg   Bradykinin

29.30 Automated amino acid analysis of peptides containing asparagine (Asn) and gluta-mine (Gln) residues gives a peak corresponding to ammonia. Why?

29.31 What are the products of each of the following reactions? Your answer should account for all the amino acid residues in the starting peptides.

(a) Reaction of Leu-Gly-Ser with 1-fluoro-2,4-dinitrobenzene
(b) Hydrolysis of the compound in (a) in concentrated hydrochloric acid (100°C)
(c) Treatment of Met-Val-Pro with dansyl chloride, followed by hydrolysis in concentrated hydrochloric acid (100°C)
(d) Treatment of Ile-Glu-Phe with C_6H_5N=C=S, followed by hydrogen bromide in nitro-methane
(e) Reaction of Asn-Ser-Ala with benzyloxycarbonyl chloride
(f) Reaction of the product of (e) with p-nitrophenol and N,N'-dicyclohexylcarbodiimide
(g) Reaction of the product of (f) with the ethyl ester of valine
(h) Hydrogenolysis of the product of (g) over palladium

29.32 Hydrazine cleaves amide bonds to form *acylhydrazides* according to the general mechanism of nucleophilic acyl substitution discussed in Chapter 21.

$$\underset{\text{Amide}}{\overset{\displaystyle O}{\overset{\|}{R\overset{}{C}NHR'}}} + \underset{\text{Hydrazine}}{H_2NNH_2} \longrightarrow \underset{\text{Acylhydrazide}}{\overset{\displaystyle O}{\overset{\|}{R\overset{}{C}NHNH_2}}} + \underset{\text{Amine}}{R'NH_2}$$

This reaction forms the basis of one method of terminal residue analysis. A peptide is treated with excess hydrazine in order to cleave all the peptide linkages. One of the terminal amino acids is cleaved as the free amino acid and identified, while all the other amino acid residues are converted to acylhydrazides. Which amino acid is identified by *hydrazinolysis*, the N terminus or the C terminus?

29.33 *Somatostatin* is a tetradecapeptide of the hypothalamus that inhibits the release of pituitary growth hormone. Its amino acid sequence has been determined by a combination of Edman degradations and enzymic hydrolysis experiments. On the basis of the following data, deduce the primary structure of somatostatin.

I.   Edman degradation gave PTH-Ala.
II. Selective hydrolysis gave peptides having the following indicated sequences:
     Phe-Trp
     Thr-Ser-Cys
     Lys-Thr-Phe
     Thr-Phe-Thr-Ser-Cys
     Asn-Phe-Phe-Trp-Lys
     Ala-Gly-Cys-Lys-Asn-Phe
III. Somatostatin has a disulfide bridge.

29.34 What protected amino acid would you anchor to the solid support in the first step of a synthesis of oxytocin (Figure 29.6) by the Merrifield method?

29.35 *Nebularine* is a toxic nucleoside isolated from a species of mushroom. Its systematic name is 9-β-D-ribofuranosylpurine. Write a structural formula for nebularine.

29.36 The nucleoside *vidarabine* (ara-A) shows promise as an antiviral agent. Its structure is identical to that of adenosine (Section 29.24) except that D-arabinose replaces D-ribose as the carbohydrate component. Write a structural formula for this substance.

29.37 In one of the early experiments designed to elucidate the genetic code, Marshall Nirenberg of the U.S. National Institutes of Health (Nobel prize in physiology or medicine, 1968) prepared a synthetic mRNA in which all the bases were uracil. He added this poly-U to a cell-free system containing all the necessary materials for protein biosynthesis. A polymer of a single amino acid was obtained. What amino acid was polymerized?

THE "SYSTEMS" OF SYSTEMATIC IUPAC NOMENCLATURE

There is no single "IUPAC name" for most compounds. The IUPAC rules permit the use of several approaches to systematic nomenclature.

1. *Substitutive nomenclature* identifies the carbon chain that bears the substituent and specifies the substituent through the use of a characteristic prefix or suffix. Examples of substitutive nomenclature include:

$$CH_3CHCH_2CH_3 \qquad CH_3CHCH_2CH_3 \qquad CH_3CHCH_2CO_2H$$
$$| \qquad\qquad\qquad | \qquad\qquad\qquad\quad |$$
$$Cl \qquad\qquad\qquad OH \qquad\qquad\qquad OH$$

$$\text{2-Chlorobutane} \qquad \text{2-Butanol} \qquad \text{3-Hydroxybutanoic acid}$$

In general, the names resulting from substitutive nomenclature are the preferred ones.

2. *Radicofunctional nomenclature* identifies the functional group as a separate word, which is preceded by the name of the group or groups attached to it.

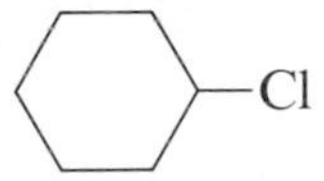

$$\text{Cyclohexyl—Cl} \qquad CH_3CH_2OH \qquad CH_3OCH_2CH_3$$

$$\text{Cyclohexyl chloride} \qquad \text{Ethyl alcohol} \qquad \text{Ethyl methyl ether}$$

Radicofunctional nomenclature sometimes has an advantage over substitutive nomenclature in being more readily recognizable in terms of structure. Most people, after some exposure to organic chemistry, can picture the structure of *tert*-butyl chloride more quickly from that name than they can when the same substance is called 2-chloro-2-methylpropane.

3. *Additive nomenclature* designates atoms that have been added to a particular structure by a particular prefix or suffix.

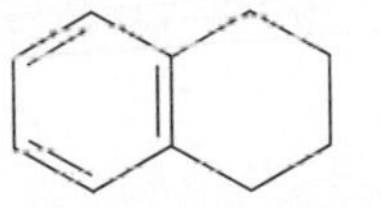

$$\text{1,2,3,4-Tetrahydronaphthalene} \qquad —CH_2—CH_2$$
$$\qquad\qquad\qquad\qquad\qquad\qquad\qquad O$$

$$\text{1,2,3,4-Tetrahydronaphthalene} \qquad\qquad \text{Styrene oxide}$$

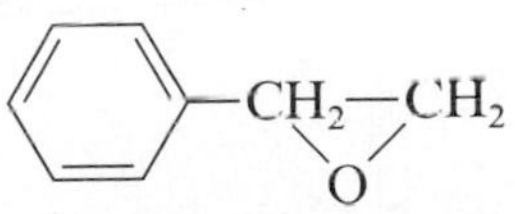

4. *Subtractive nomenclature* uses certain prefixes to express the removal of atoms or groups from some parent compound.

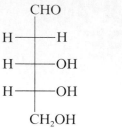

2-Deoxy-D-ribose
("deoxy" designates replacement
of a hydroxyl group by hydrogen)

Norcamphor
("nor" designates replacement
of all methyl groups by hydrogen)

5. *Conjunctive nomenclature* is useful when functionally substituted side chains are attached to a ring. Both units are named independently and the names joined together.

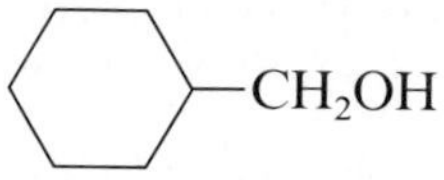

Cyclohexanemethanol

1,4-Benzenediacetic acid

6. *Replacement nomenclature* uses prefixes such as oxa-, aza-, and thia- to designate the replacement of a carbon by oxygen, nitrogen, or sulfur, respectively.

$CH_3CH_2OCH_2CH_2OCH_2CH_2Cl$ $CH_3CH_2N=NCH_2CH_2CH_3$

1-Chloro-3,6-dioxaoctane

3,4-Diaza-3-heptene

1,2-dithiacyclobutane

ANSWERS TO IN-TEXT PROBLEMS

Problems are of two types, in-text problems that appear within the body of each chapter and end-of-chapter problems. This appendix gives brief answers to all the in-text problems. More detailed discussions of in-text problems as well as detailed solutions to all the end-of-chapter problems are provided in a separate *Study Guide.* Answers to part (*a*) of those in-text problems with multiple parts have been provided in the form of a sample solution within each chapter and are not repeated here.

CHAPTER 1

1.1 All the third row elements have a neon core containing 10 electrons ($1s^2 2s^2 2p^6$). The elements in the third row, their atomic numbers Z, and their electron configurations beyond the neon core are: Na ($Z = 11$) $3s^1$; Mg ($Z = 12$) $3s^2$; Al ($Z = 13$) $3s^2 3p_x^1$; Si ($Z = 14$) $3s^2 3p_x^1 3p_y^1$; P ($Z = 15$) $3s^2 3p_x^1 3p_y^1 3p_z^1$; S ($Z = 16$) $3s^2 3p_x^2 3p_y^1 3p_z^1$; Cl ($Z = 17$) $3s^2 3p_x^2 3p_y^2 3p_z^1$; Ar ($Z = 18$) $3s^2 3p_x^2 3p_y^2 3p_z^2$.

1.2 Electron configuration of C^+: $1s^2 2s^2 2p^1$.

1.3 Those ions that possess a rare gas configuration are: (*a*) Na^+; (*c*) H^-; (*e*) Cl^-; and (*f*) Ca^{2+}.

1.4 Ionization of sodium is endothermic. Electron capture by chlorine is exothermic.

1.5 (*b*) H–N̈–H (with H below); (*c*) :N̈:F̈:; (*d*) :P̈:Cl̈:; (*e*) H:C̈:Cl̈:; (*f*) H:C̈:C̈:H

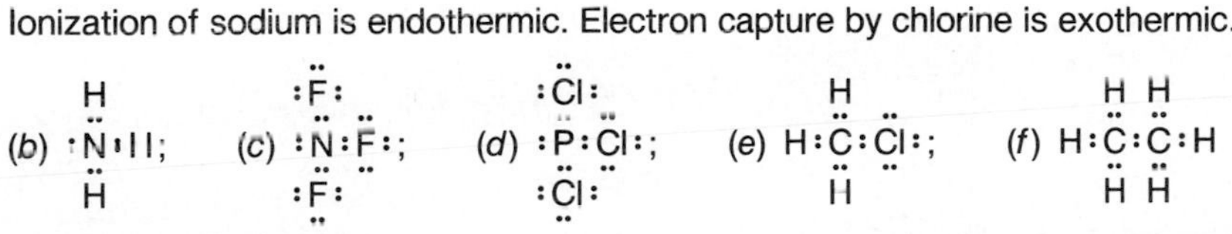

1.6 Fluorine is more electronegative than chlorine, so fluorine is the negative end of dipole in FCl. Iodine is less electronegative than chlorine, so chlorine is the negative end of dipole in ICl.

1.7 The formal charge of hydrogen in each species is 0. The formal charge of carbon is: (*b*) -1; (*c*) 0; (*d*) $+1$; (*e*) 0. In each species the net charge is equal to the formal charge of carbon.

1.8 (b) H—C—H (with $:O:$ double-bonded to C above); (c) $:\ddot{O}=C=\ddot{O}:$; (d) cyclopropene-type structure with $C=C$ (each bearing H) bonded to a CH_2 carbon;

(e) $:N≡C—C≡C—H$

1.9 (b) H—C bonded to $\ddot{O}:$ (double bond) and to $\ddot{O}$—H; (c) $CH_2=CH_2$; (d) $CH_2=CH—CH=CH_2$;

(e) $CH_2=CH—CH=\ddot{O}:$

1.10 (b) $^-:\ddot{O}—C$ (double-bonded O, and O—H) $\longleftrightarrow$ $:O=C$ (with $\ddot{O}^{\,\overline{}}$ and O—H)

(c) $^-:\ddot{O}—C$ (double-bonded O, with $\ddot{O}^{\,\overline{}}$) $\longleftrightarrow$ $:O=C$ (with $\ddot{O}^{\,\overline{}}$ and $\ddot{O}^{\,\overline{}}$)

(d) $^-:\ddot{O}—B$ (double-bonded O, with $\ddot{O}^{\,\overline{}}$) $\longleftrightarrow$ $:O=B$ (with $\ddot{O}^{\,\overline{}}$ and $\ddot{O}:$)

and $^-:\ddot{O}—B$ (with $\ddot{O}^{\,\overline{}}$ and $\ddot{O}:$) $\longleftrightarrow$ $^-:\ddot{O}—B$ (with $\ddot{O}:$ and double-bonded O)

1.11 (b) Tetrahedral; (c) trigonal pyramidal; (d) tetrahedral; (e) trigonal planar; (f) linear

1.12 (b) Oxygen is negative end of dipole moment directed along bisector of H—O—H angle; (c) no dipole moment; (d) dipole moment directed along axis of C—Cl bond, with chlorine at negative end, and carbon and hydrogens partially positive; (e) dipole moment directed along bisector of H—C—H angle, with oxygen at negative end; (f) dipole moment aligned with axis of linear molecule, with nitrogen at negative end

1.13 (b) Oxygen has two unshared pairs; (c) nitrogen has one unshared pair; (d) no unshared pairs

1.14 (b) $ClCH_2CH_2Cl$; (c) $(CH_3)_3CH$; (d) $CH_3CH_2OCH_2CH_3$ or $(CH_3CH_2)_2O$

1.15 (b) $(CH_3)_2CHCH(CH_3)_2$; (c) $HOCH_2CHCH(CH_3)_2$ with a CH_3 substituent on the middle CH; (d) a cyclohexane ring with a $CH—C(CH_3)_3$ group

1.16 (b) $CH_3CH_2CH_2CH_3$ and $(CH_3)_3CH$; (c) $CH_3CH_2CH_2CH_2CH_3$, $(CH_3)_2CHCH_2CH_3$, and $(CH_3)_4C$; (d) CH_3CHCl_2 and $ClCH_2CH_2Cl$; (e) $CH_3CH_2CH_2CH_2Br$, $(CH_3)_2CHCH_2Br$, $(CH_3)_3CBr$, and $CH_3CHCH_2CH_3$ with a Br substituent; (f) $CH_3CH_2CH_2OH$, $(CH_3)_2CHOH$, and $CH_3OCH_2CH_3$

CHAPTER 2

2.1   Endothermic

2.2   (*b*) 7 orbitals, 4 occupied;   (*c*) 8 orbitals, 7 occupied;   (*d*) 8 orbitals, 5 occupied;   (*e*) 12 orbitals, 8 occupied;   (*f*) 14 orbitals, 7 occupied

2.3   Each carbon is sp^3 hybridized. There are 10 σ bonds, of which 8 are $C(sp^3)$-$H(1s)$ and 2 are $C(sp^3)$-$C(sp^3)$.

2.4

2.5   (*b*) $CH_3(CH_2)_{26}CH_3$

2.6

2.7   (*b*) $CH_3CH_2CH_2CH_2CH_3$ (pentane), $(CH_3)_2CHCH_2CH_3$ (2-methylbutane), $(CH_3)_4C$ (2,2-dimethylpropane);   (*c*) 2,2,4-trimethylpentane;   (*d*) 2,2,3,3-tetramethylbutane

2.8 $CH_3CH_2CH_2CH_2CH_2$— (primary); $CH_3CH_2CH_2\overset{|}{C}HCH_3$ (secondary); $CH_3CH_2\overset{|}{C}HCH_2CH_3$ (secondary); $(CH_3)_2CHCH_2CH_2$— (primary);   —$CH_2CH(CH_3)CH_2CH_3$ (primary); $(CH_3)_2\overset{|}{C}CH_2CH_3$ (tertiary); $(CH_3)_2CH\overset{|}{C}HCH_3$ (secondary)

2.9   (*b*) 4-Ethyl-2-methylhexane;   (*c*) 8-Ethyl-4-isopropyl-2,6-dimethyldecane

2.10   (*b*) cyclodecane;   (*c*) 4-isopropyl-1,1-dimethylcyclodecane;   (*d*) 1,1-dicyclopropylbutane;   (*e*) 1,2-dicyclopropylbutane;   (*f*) cyclohexylcyclohexane

2.11   Hydrogen bonds in liquid HF cause it to have a higher boiling point (bp 19°C) than Ne (bp −246°C).

2.12   (*b*) 2245 kcal/mole;   (*c*) 3182 kcal/mol

2.13   Hexane > pentane > isopentane > neopentane

CHAPTER 3

3.1   (*b*) Butane;   (*c*) 2-methylbutane;   (*d*) 3-methylhexane

3.2   Shape of potential energy diagram is identical to that for ethane (Figure 3.3). Activation energy for internal rotation is higher than that of ethane, lower than that of butane.

3.3   150°

3.4   (*b*)   (*c*)   (*d*)

3.5 (*b*) Less stable (*c*) Methyl is equatorial and down

3.6

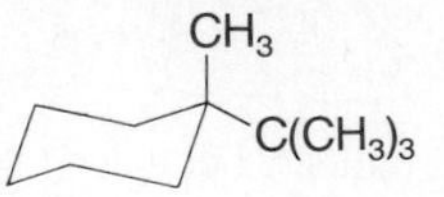

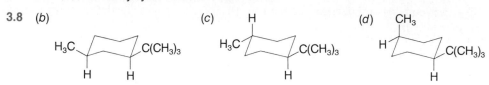

3.7 *cis*-1,3,5-Trimethylcyclohexane is more stable.

3.8 (*b*)   (*c*)   (*d*)

3.9 Van der Waals repulsions between methyl groups make *cis*-1,2-dimethylcyclohexane less stable than the trans isomer.

3.10 (*b*)   (*c*) 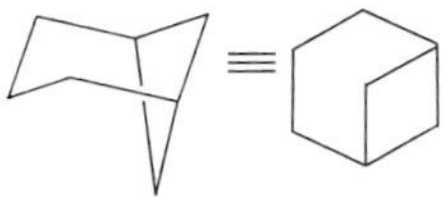 ≡   (*d*)

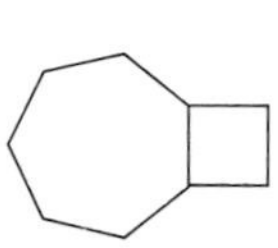

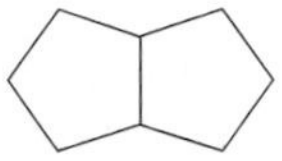

CHAPTER 4

4.1 $CH_3CH_2CH_2CH_2Cl$ (*n*-butyl chloride); $(CH_3)_2CHCH_2Cl$ (isobutyl chloride); $CH_3CHCH_2CH_3$ (*sec*-butyl chloride); $(CH_3)_3CCl$ (*tert*-butyl chloride)
|
Cl

4.2 $CH_3CH_2CH_2CH_2OH$ (1-butanol); $(CH_3)_2CHCH_2OH$ (2-methyl-1-propanol); $CH_3CHCH_2CH_3$ (2-butanol); $(CH_3)_3COH$ (2-methyl-2-propanol)
|
OH

4.3 $CH_3CH_2CH_2CH_2Cl$ (1-chlorobutane); $(CH_3)_2CHCH_2Cl$ (1-chloro-2-methylpropane); $CH_3CHCH_2CH_3$ (2-chlorobutane); $(CH_3)_3CCl$ (2-chloro-2-methylpropane)
|
Cl

4.4 $CH_3CH_2CH_2CH_2OH$ (primary); $(CH_3)_2CHCH_2OH$ (primary); $CH_3CHCH_2CH_3$ (secondary);
|
OH

$(CH_3)_3COH$ (tertiary)

4.5 The carbon-bromine bond is longer than the carbon-chlorine bond; therefore while the charge e in the dipole moment expression $\mu = e \cdot d$ is smaller for the bromine than for the chlorine compound, the distance d is larger.

4.6

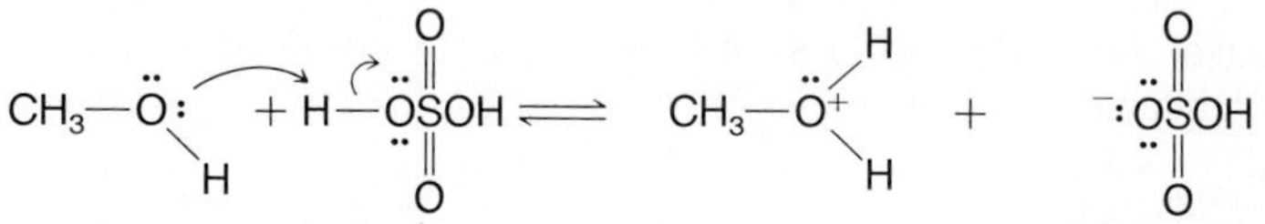

Methyl alcohol	Sulfuric acid	Methyloxonium ion	Hydrogen sulfate ion
(base)	(acid)	(conjugate acid)	(conjugate base)

4.7 (*b*) 6.7×10^{-5}; (*c*) 1.8×10^{-4}; (*d*) 6.5×10^{-2}. Order of decreasing acidity is oxalic acid > aspirin > formic acid > vitamin C.

4.8

$$H_2N\overset{\cdot\cdot}{} \; + \; H-\overset{\cdot\cdot}{\underset{\cdot\cdot}{O}}CH_3 \; \rightleftharpoons \; H_3N\cdot \; + \; \overset{\cdot\cdot}{}OCH_3$$

Stronger base   Stronger acid   Weaker acid   Weaker base

The position of equilibrium lies to the right.

4.9 (b) $(CH_3CH_2)_3COH + HCl \longrightarrow (CH_3CH_2)_3CCl + H_2O$

(c) $CH_3(CH_2)_{12}CH_2OH + HBr \longrightarrow CH_3(CH_2)_{12}CH_2Br + H_2O$

4.10 $CH_3CH_2CH_2CH_2^+$ (*n*-butyl cation or 1-butyl cation—primary); $CH_3\overset{+}{C}HCH_2CH_3$ (*sec*-butyl cation or 1-methylpropyl cation—secondary); $(CH_3)_2CHCH_2^+$ (isobutyl cation or 2-methylpropyl cation—primary); $(CH_3)_3C^+$ (*tert*-butyl cation or 1,1-dimethylethyl cation—tertiary)

4.11 $(CH_3)_2\overset{+}{C}CH_2CH_3$

4.12 **1-Butanol:**

1. $CH_3CH_2CH_2CH_2\overset{\cdot\cdot}{O}\overset{H}{\underset{}{}} \; + \; H-\overset{\cdot\cdot}{\underset{\cdot\cdot}{Br}} \; \longrightarrow \; CH_3CH_2CH_2CH_2\overset{H}{\underset{H}{\overset{+}{O}}} \; + \; \overset{\cdot\cdot}{\underset{\cdot\cdot}{Br}}$

2. $\overset{\cdot\cdot}{\underset{\cdot\cdot}{Br}} \cdots \overset{CH_3CH_2CH_2}{\underset{H}{\overset{|}{CH_2}}}-\overset{H}{\underset{H}{\overset{+}{O}}} \; \longrightarrow \; CH_3CH_2CH_2CH_2Br + \overset{H}{\underset{H}{\overset{\cdot\cdot}{O}}}$

2-Butanol:

1. $CH_3CH_2\overset{|}{\underset{\overset{|}{\overset{\cdot\cdot}{O}:}}{C}}HCH_3 + H-\overset{\cdot\cdot}{\underset{\cdot\cdot}{Br}} \; \longrightarrow \; CH_3CH_2\overset{|}{\underset{\overset{H}{\overset{|}{\overset{+}{O}}}\; H}{C}}HCH_3 + \overset{\cdot\cdot}{\underset{\cdot\cdot}{Br}}$

2. $CH_3CH_2\overset{|}{\underset{\overset{|}{\overset{+}{O}}}{C}}HCH_3 \; \longrightarrow \; CH_3CH_2\overset{+}{C}HCH_3 + \overset{\cdot\cdot}{\underset{H\;\;H}{O}}$

3. $\overset{\cdot\cdot}{\underset{\cdot\cdot}{Br}} \; + \; \overset{CH_3CH_2}{\underset{+}{\overset{|}{C}}}HCH_3 \; \longrightarrow \; CH_3CH_2\overset{|}{\underset{Br}{C}}HCH_3$

4.13 (b) The carbon-carbon bond dissociation energy is lower for 2-methylpropane because it yields a more stable (secondary) radical; propane yields a primary radical. (c) The carbon-carbon bond dissociation energy is lower for 2,2-dimethylpropane because it yields a still more stable tertiary radical.

4.14 CH_3CHCl_2 and $ClCH_2CH_2Cl$

4.15 1-Chloropropane (43%); 2-chloropropane (57%)

CHAPTER 5

5.1 (b) 3,3-Dimethyl-1-butene; (c) 2-methyl-2-hexene; (d) 2,4,4-trimethyl-2-pentene; (e) 3-isopropyl-2,6-dimethyl-3-heptene

5.2

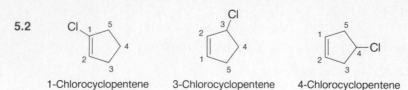

1-Chlorocyclopentene 3-Chlorocyclopentene 4-Chlorocyclopentene

5.3 $CH_2—CH_2$ (longest) > $CH_2—CH=$ > $CH=CH$ (shortest)

5.4

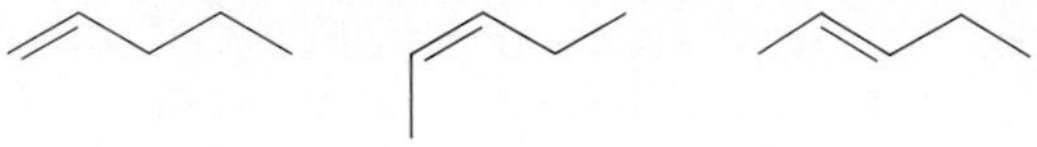

1-Pentene *cis*-2-Pentene *trans*-2-Pentene

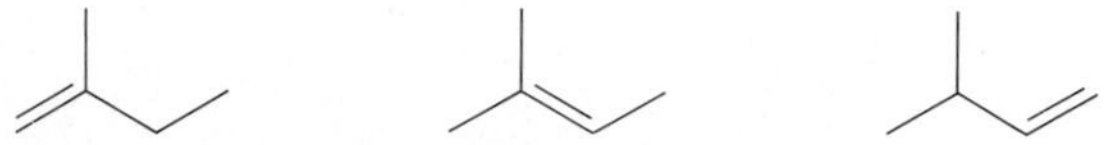

2-Methyl-1-butene 2-Methyl-2-butene 3-Methyl-1-butene

5.5 *cis*-1,2-Dichloroethene has a dipole moment; *trans*-1,2-dichloroethene does not.

5.6
CH3(CH2)7
(CH2)12CH3
C=C
H
H

5.7 (*b*) *Z*; (*c*) *E*; (*d*) *E*

5.8
O
H
CCH2CH3
C=C
CH3CH2CH
CH3
CH3

5.9
CH3CH2
CH3
C=C
H3C
H
H3C
CH3
C=C
CH3CH2
H
H3C
CH2CH3
C=C
H3C
H

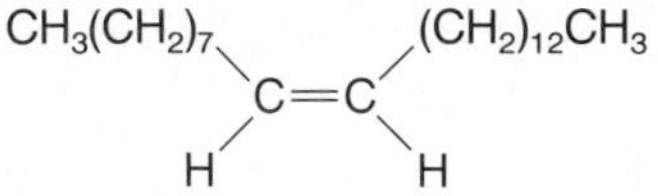

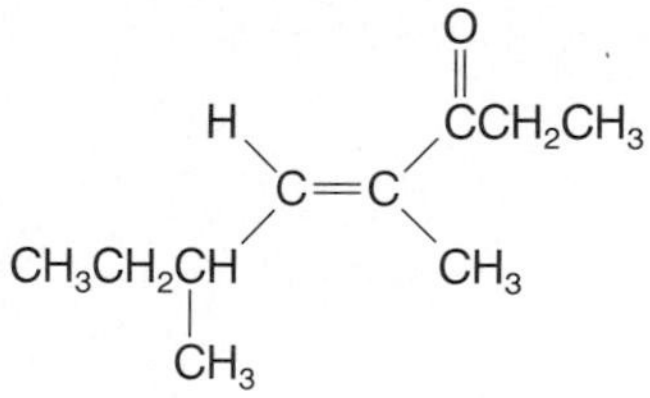

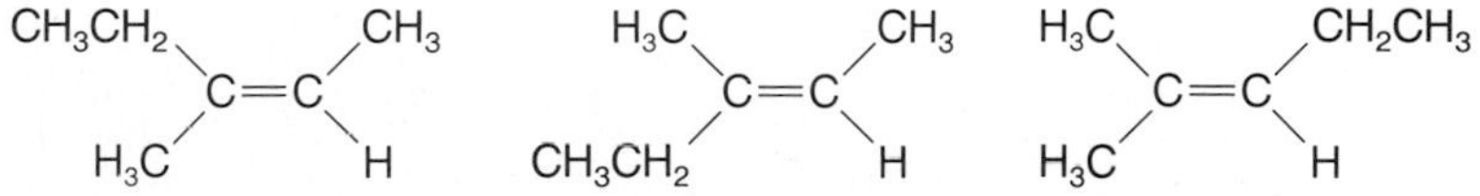

(*Z*)-3-Methyl-2-pentene (*E*)-3-Methyl-2-pentene 2-Methyl-2-pentene

5.10 2-Methyl-2-butene (most stable) > (*E*)-2-pentene > (*Z*)-2-pentene > 1-pentene (least stable)

5.11 2-Methyl-1-butene, 2-methyl-2-butene, and 3-methyl-1-butene

5.12 (*c*)

(*d*)

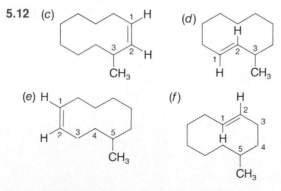

(*e*)

(*f*)

5.13 (*b*) 3; (*c*) 2; (*d*) 3; (*e*) 2; (*f*) 2

CHAPTER 6

6.1 (*b*) Propene; (*c*) propene; (*d*) 2,3,3-trimethyl-1-butene

6.2 (*b*)

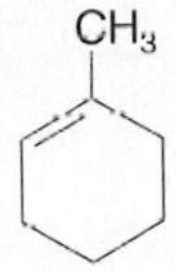

Major   and   Minor

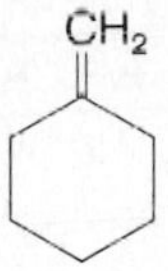

(*c*)

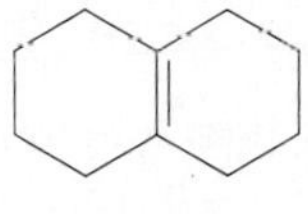

Major   and   Minor

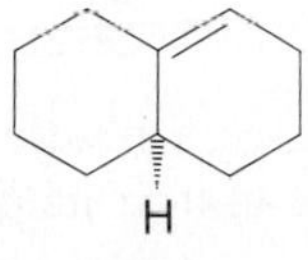

6.3 1-Pentene, *cis*-2-pentene, and *trans*-2-pentene

6.4 (*b*)

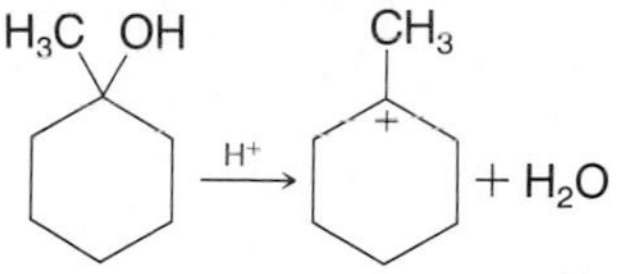

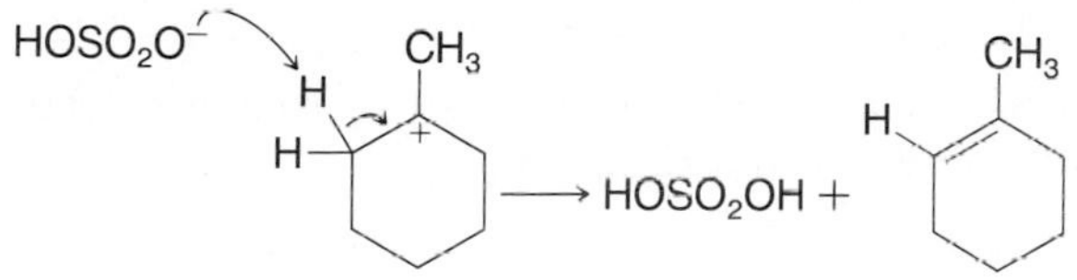

and

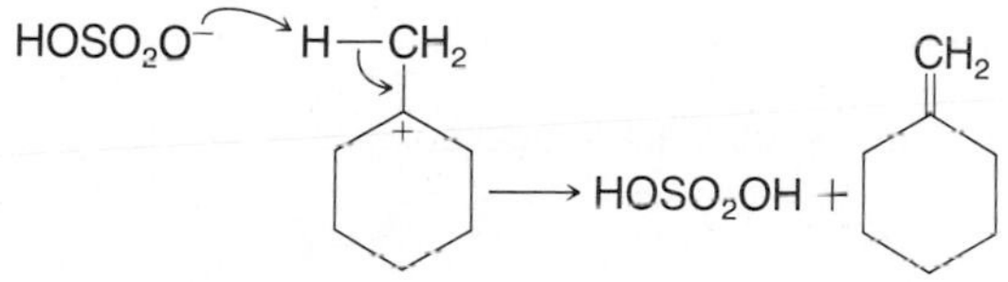

(*c*)

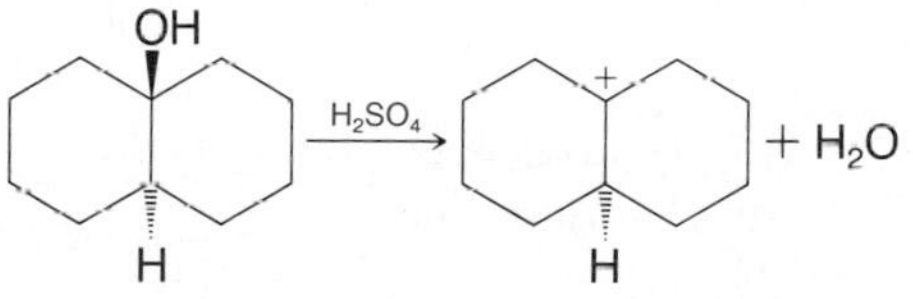

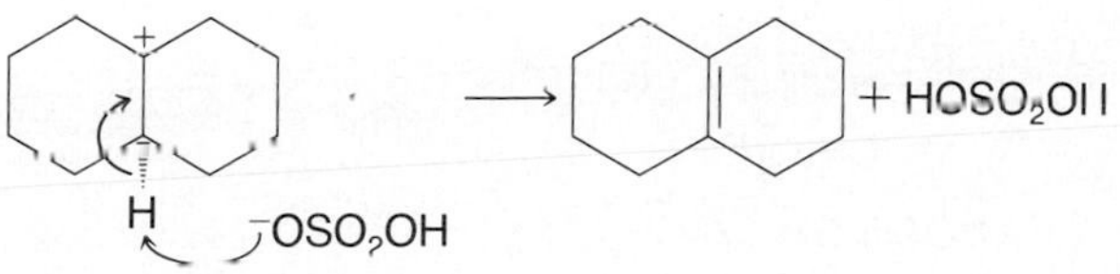

and

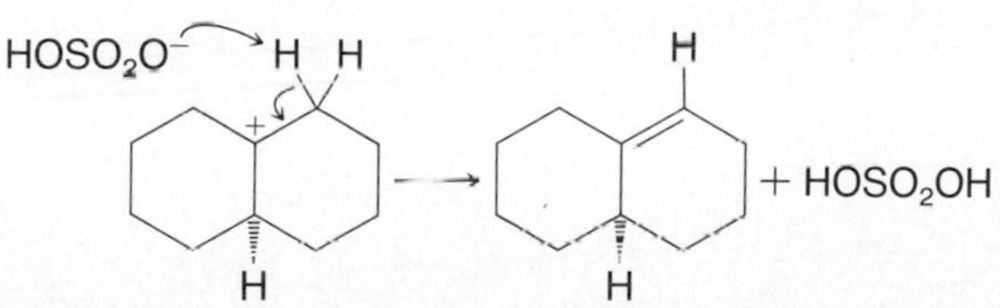

6.5

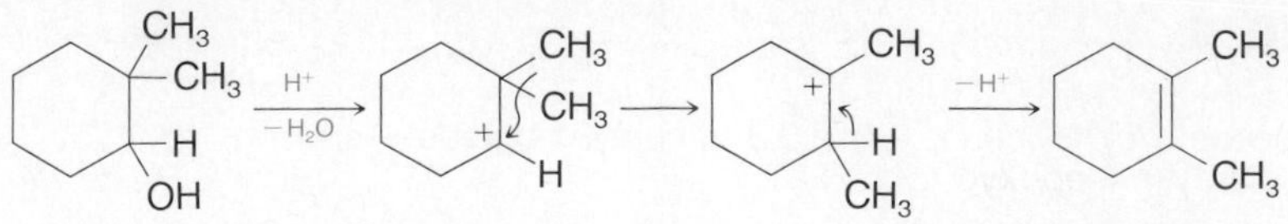

6.6 (*b*) $(CH_3)_2C\!=\!CH_2$; (*c*) $CH_3CH\!=\!C(CH_2CH_3)_2$; (*d*) $CH_3CH\!=\!C(CH_3)_2$ (major) and $CH_2\!=\!CHCH(CH_3)_2$ (minor); (*e*) $CH_2\!=\!CHCH(CH_3)_2$; 1-methylcyclohexene (major) and methylenecyclohexane (minor)

6.7 (*b*) $CH_2\!=\!CHCH_2CH_2CH_2CH_3$, *cis*-$CH_3CH\!=\!CHCH_2CH_2CH_3$, and *trans*-$CH_3CH\!=\!CHCH_2CH_2CH_3$; (*c*) $CH_2\!=\!CHCH_2CH(CH_3)_2$, *cis*-$CH_3CH\!=\!CHCH(CH_3)_2$, and *trans*-$CH_3CH\!=\!CHCH(CH_3)_2$; (*d*) *cis*-$(CH_3)_3CCH\!=\!CHCH_3$ and *trans*-$(CH_3)_3CCH\!=\!CHCH_3$

6.8 Neomenthyl chloride reacts faster because it can more easily adopt the anti arrangement of leaving group and proton required at the transition state.

CHAPTER 7

7.1 (*b*) $(CH_3)_2\overset{|}{\underset{Cl}{C}}CH_2CH_3$; (*c*) $CH_3\overset{|}{\underset{Cl}{C}}HCH_2CH_3$; (*d*) CH_3CH_2

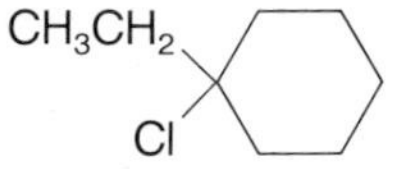

7.2 (*b*) $(CH_3)_2\overset{+}{C}CH_2CH_3$; (*c*) $CH_3\overset{+}{C}HCH_2CH_3$; (*d*) CH_3CH_2

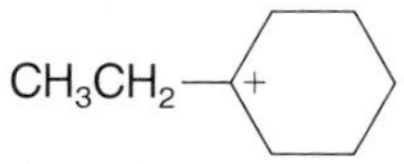

7.3 Addition in accordance with Markovnikov's rule gives 1,2-dibromopropane. Addition contrary to Markovnikov's rule gives 1,3-dibromopropane.

7.4 Absence of peroxides: (*b*) 2-bromo-2-methylbutane; (*c*) 2-bromobutane; (*d*) 1-bromo-1-ethylcyclohexane. Presence of peroxides: (*b*) 1-bromo-2-methylbutane; (*c*) 2-bromobutane; (*d*) 1-bromo-1-cyclohexylethane.

7.5 The concentration of hydroxide ion is too small in acid solution to be chemically significant.

7.6 E1

7.7 2-Pentanol and 3-pentanol

7.8 (*b*) 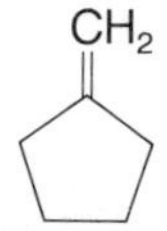 or 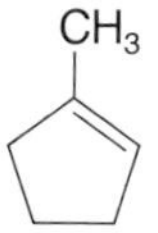   (*c*) *cis*-3-Hexene and *trans*-3-hexene

7.9 (*b*) $CH_3\overset{|}{\underset{OH}{C}}HCH_2CH_3$ (*c*) 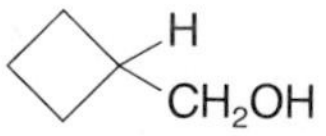 (*d*)

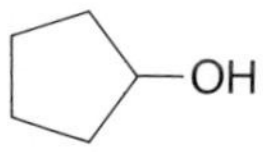

(*e*) $CH_3\overset{|}{\underset{OH}{C}}HCH(CH_2CH_3)_2$ (*f*) $HOCH_2CH_2CH(CH_2CH_3)_2$

7.10

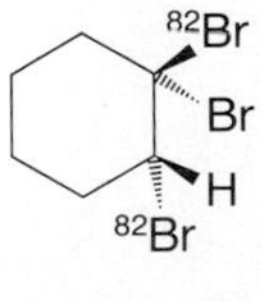

7.11   2-Methyl-2-butene (most reactive) > 2-methyl-1-butene > 3-methyl-1-butene (least reactive)

7.12   *cis*-2-Methyl-7,8-epoxyoctadecane

7.13   *cis*-$(CH_3)_2CHCH_2CH_2CH_2CH_2CH$=$CH(CH_2)_9CH_3$

7.14   2,4,4-Trimethyl-1-pentene

7.15

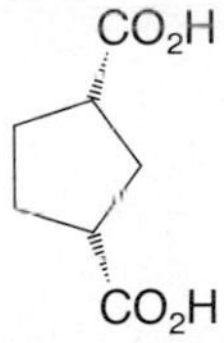

7.16   Hydrogenation over a metal catalyst such as platinum, palladium, or nickel.

CHAPTER 8

8.1   (*b*) Achiral;   (*c*) chiral;   (*d*) achiral

8.2   (*b*) (*Z*)-1,2-Dichloroethene is achiral. The plane of the molecule is a plane of symmetry. A second plane of symmetry is perpendicular to the plane of the molecule and bisects the carbon-carbon bond.

   (*c*) *cis*-1,2-Dichlorocyclopropane is achiral. It has a plane of symmetry that bisects the C-1–C-2 bond and passes through C-3.

   (*d*) *trans*-1,2-Dichlorocyclopropane is chiral. It has neither a plane of symmetry nor a center of symmetry.

8.3   (*c*) C-2 is a chiral center;   (*d*) no chiral centers

8.4   (*c*) C-2 is a chiral center;   (*d*) no chiral centers

8.5 $[\alpha]_D -39°$

8.6   Two-thirds (66.7 percent)

8.7   (*b*) *R*;   (*c*) *S*;   (*d*) *S*

8.8   (*b*)

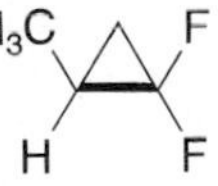

8.9   The product is chiral and should be optically active. No bonds are made or broken at the chiral center.

8.10   2*S*,3*R*

8.11   The midpoint of the C-2–C-3 bond is a center of symmetry.

8.12   *cis*-1,3-Dimethylcyclohexane

8.13   *RRR RRS RSR SRR SSS SSR SRS RSS*

8.14   Eight

8.15   (*Z*)-2-Butene

8.16

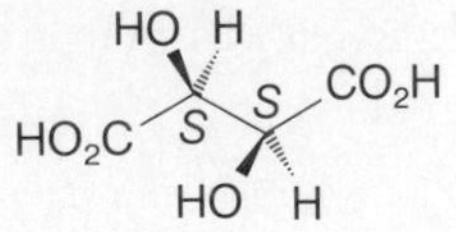

and

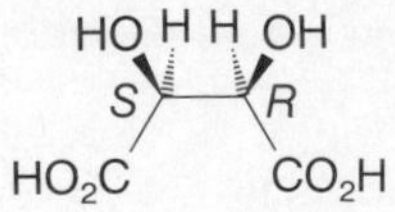

8.17   No

8.18   (S)-1-Phenylethylammonium (S)-malate

CHAPTER 9

9.1   (b) $(CH_3)_3COCH_3$;   (c) $CH_3N{=}\overset{+}{N}{=}\overset{-}{\underset{\cdot\cdot}{N}}{:}$;   (d) $CH_3C{\equiv}N$; (e) CH_3SH

9.2 $ClCH_2CH_2CH_2C{\equiv}N$

9.3   No

9.4   Hydrolysis of (R)-(−)-2-bromooctane by the S_N2 mechanism yields optically active (S)-(+)-2-octanol. The 2-octanol obtained by hydrolysis of racemic 2-bromooctane is not optically active.

9.5   (b) 1-Bromopentane;   (c) 2-chloropentane;   (d) 2-bromo-5-methylhexane;   (e) 1-bromodecane

9.6   (b) 1-Methylcyclopentyl iodide;   (c) cyclopentyl bromide;   (d) *tert*-butyl iodide

9.7   Both *cis*- and *trans*-1,4-dimethylcyclohexanol are formed in the hydrolysis of either *cis*- or *trans*-1,4-dimethylcyclohexyl bromide.

9.8   A hydride shift produces a tertiary carbocation; a methyl shift produces a secondary carbocation.

9.9   (b) ;   (c) $CH_3\underset{\underset{\displaystyle OCH_3}{|}}{C}HCH_2CH_3$;   (d) *cis*- and *trans*-

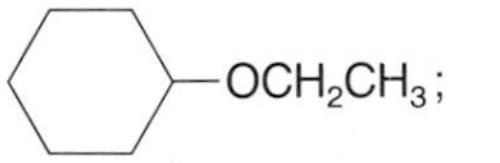

$CH_3CH{=}CHCH_3$ and $CH_2{=}CHCH_2CH_3$

9.10 $CH_3(CH_2)_{16}CH_2OH + $

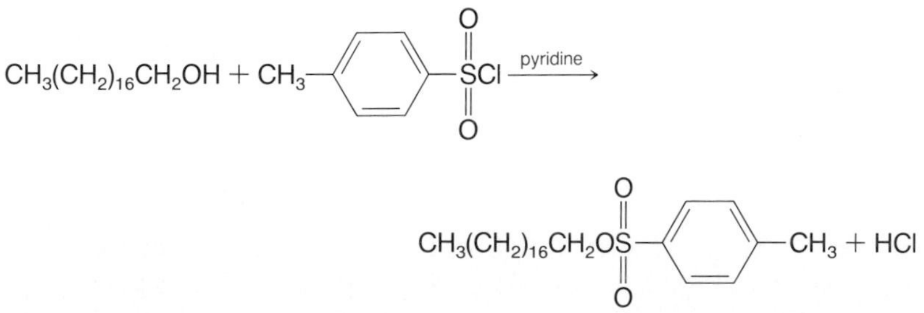

9.11   (b) $CH_3(CH_2)_{16}CH_2I$;   (c) $CH_3(CH_2)_{16}CH_2C{\equiv}N$;   (d) $CH_3(CH_2)_{16}CH_2SH$;
(e) $CH_3(CH_2)_{16}CH_2SCH_2CH_2CH_2CH_3$

9.12   The product has the R configuration and a specific rotation $[\alpha]_D$ of $-9.9°$.

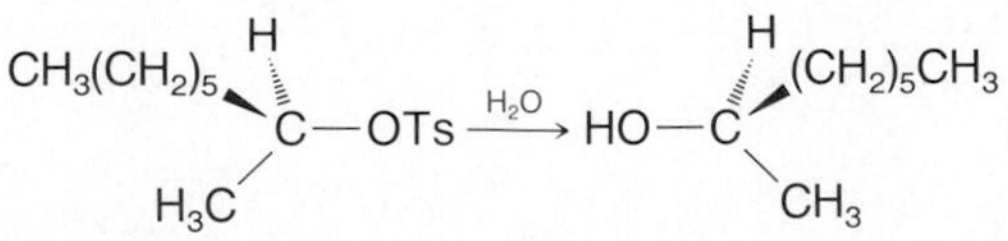

CHAPTER 10

10.1 $CH_3CH_2CH_2C\equiv CH$ (1-pentyne), $CH_3CH_2C\equiv CCH_3$ (2-pentyne), $(CH_3)_2CHC\equiv CH$ (3-methyl-1-butyne)

10.2 The bond from the methyl group in 1-butyne is to an sp^3 hybridized carbon and so is longer than the bond from the methyl group in 2-butyne, which is to an sp hybridized carbon.

10.3 *(b)* $HC\equiv C-H$ + $:CH_2CH_3$ $\overset{K\gg 1}{\rightleftharpoons}$ $HC\equiv C:$ + CH_3CH_3
  Acetylene   Ethyl anion   Acetylide ion   Ethane
  (stronger acid) (stronger base) (weaker base)   (weaker acid)

(c) $CH_2=CH-H$ + $:NH_2$ $\overset{K\ll 1}{\rightleftharpoons}$ $CH_2=CH$ + $:NH_3$
  Ethylene   Amide ion   Vinyl anion   Ammonia
  (weaker acid)   (weaker base)   (stronger base)   (stronger acid)

(d) $CH_3C\equiv CCH_2O-H$ + $:NH_2$ $\overset{K\gg 1}{\rightleftharpoons}$ $CH_3C\equiv CCH_2O:^-$ + $:NH_3$
  2-Butyn-1-ol   Amide ion   2-Butyn-1-oxide   Ammonia
                                       anion
  (stronger acid) (stronger base)   (weaker base)   (weaker acid)

10.4 *(b)* $HC\equiv CH \xrightarrow[\text{2. }CH_3Br]{\text{1. }NaNH_2,\ NH_3} CH_3C\equiv CH \xrightarrow[\text{2. }CH_3CH_2CH_2Br]{\text{1. }NaNH_2,\ NH_3} CH_3C\equiv CCH_2CH_2CH_3$
(or reverse order of steps)

(c) $HC\equiv CH \xrightarrow[\text{2. }CH_3CH_2CH_2Br]{\text{1. }NaNH_2,\ NH_3} CH_3CH_2CH_2C\equiv CH \xrightarrow[\text{2. }CH_3CH_2Br]{\text{1. }NaNH_2,\ NH_3} CH_3CH_2CH_2C\equiv CCH_2CH_3$
  (or reverse order of steps)

10.5 Both $CH_3CH_2CH_2CH_2C\equiv CH$ and $CH_3CH_2C\equiv CCH_3$ can be prepared by alkylation of acetylene. The alkyne $(CH_3)_2CHC\equiv CH$ cannot be prepared by alkylation of acetylene because the required alkyl halide, $(CH_3)_2CHBr$, is secondary and will react with the strongly basic acetylide ion by elimination.

10.6 $(CH_3)_3CCCH_3$ (with Br above and Br below on central C) or $(CH_3)_3CCH_2CHBr_2$ or $(CH_3)_3CCHCH_2Br$ (with Br below)

10.7 *(b)* $CH_3CH_2CH_2OH \xrightarrow[\text{heat}]{H_2SO_4} CH_3CH=CH_2 \xrightarrow{Br_2} CH_3CHCH_2Br \xrightarrow[\text{2. }H^+]{\text{1. }NaNH_2} CH_3C\equiv CH$ (Br above middle C)

(c) $(CH_3)_2CHBr \xrightarrow{NaOCH_2CH_3} CH_3CH=CH_2$; then proceed as in *(a)* and *(b)*.

(d) $CH_3CHCl_2 \xrightarrow[\text{2. }H_2O]{\text{1. }NaNH_2} HC\equiv CH \xrightarrow[\text{2. }CH_3Br]{\text{1. }NaNH_2} CH_3C\equiv CH$

(e) $CH_3CH_2OH \xrightarrow[\text{heat}]{H_2SO_4} CH_2=CH_2 \xrightarrow{Br_2} BrCH_2CH_2Br \xrightarrow{NaNH_2} HC\equiv CH$; then proceed as in *(d)*.

10.8 Oleic acid is *cis*-$CH_3(CH_2)_7CH=CH(CH_2)_7CO_2H$. Stearic acid is $CH_3(CH_2)_{16}CO_2H$.

10.9 Elaidic acid is *trans*-$CH_3(CH_2)_7CH=CH(CH_2)_7CO_2H$.

10.10

$CH_3C\equiv CH \xrightarrow[\text{2. }CH_3CH_2CH_2CH_2Br]{\text{1. }NaNH_2,\ NH_3} CH_3C\equiv CCH_2CH_2CH_2CH_3$

10.11 $CH_3CH_2Br \xrightarrow[\text{DMSO, heat}]{KOC(CH_3)_3} CH_2{=}CH_2 \xrightarrow{Br_2} BrCH_2CH_2Br \xrightarrow{NaOCH_2CH_3} CH_2{=}CHBr$

10.12 (b) $CH_2{=}CHCl \xrightarrow{HCl} CH_3CHCl_2$

(c) $CH_2{=}CHBr \xrightarrow[\text{2. H}_2\text{O}]{\text{1. NaNH}_2\text{, NH}_3} HC{\equiv}CH \xrightarrow{2HCl} CH_3CHCl_2$

(d) $CH_3CHBr_2 \xrightarrow[\text{2. H}_2\text{O}]{\text{1. NaNH}_2\text{, NH}_3} HC{\equiv}CH \xrightarrow{2HCl} CH_3CHCl_2$

10.13 $CH_3C{\equiv}CCH_3 \xrightarrow[\text{H}_2\text{SO}_4]{\text{H}_2\text{O, Hg}^{2+}} \left[\underset{\underset{OH}{|}}{CH_3C}{=}CHCH_3 \right] \longrightarrow CH_3\overset{\overset{O}{\|}}{C}CH_2CH_3$

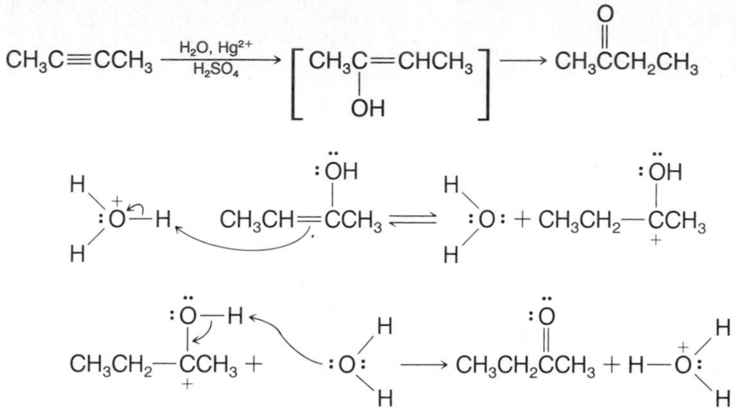

10.14 2-Pentyne is the most reactive of the C_5H_8 alkynes because it has the most highly substituted triple bond.

10.15 $CH_3(CH_2)_4C{\equiv}CCH_2CH_2C{\equiv}C(CH_2)_4CH_3$

10.16 $(CH_3)_2CHCH_2C{\equiv}CH$

CHAPTER 11

11.1 (b) $CH_2{=}\underset{\underset{CH_3}{|}}{C}{-}\overset{+}{C}H_2 \longleftrightarrow \overset{+}{C}H_2{-}\underset{\underset{CH_3}{|}}{C}{=}CH_2$

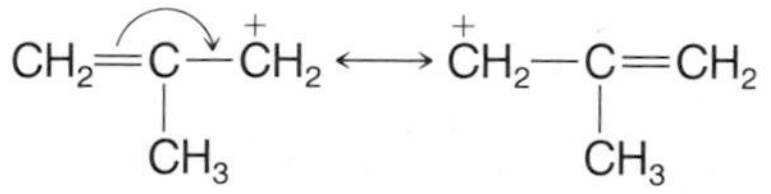

(c)

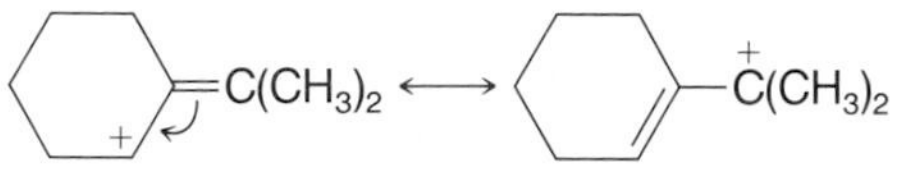

11.2 3-Bromo-1-methylcyclohexene and 3-chloro-3-methylcyclohexene

11.3 (b) (c)

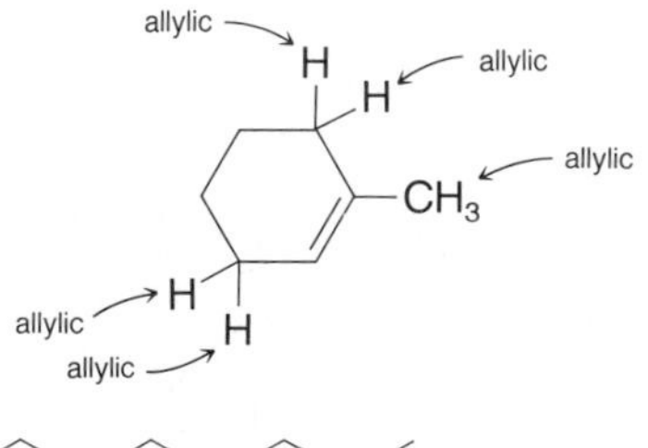

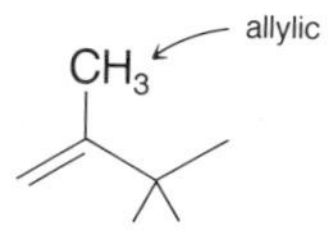

(d)

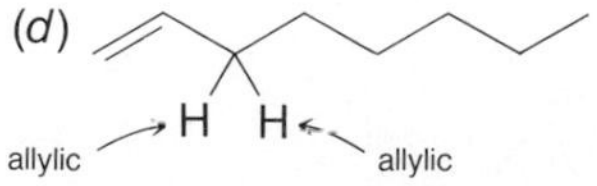

11.4 2,3,3-Trimethyl-1-butene gives only $(CH_3)_3CC{=}CH_2$.
$\underset{\underset{CH_2Br}{|}}{}$

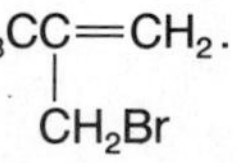

1-Octene gives a mixture of $CH_2{=}CH\underset{\substack{|\\ Br}}{C}H(CH_2)_4CH_3$ as well as the cis and trans stereoisomers of $BrCH_2CH{=}CH(CH_2)_4CH_3$.

11.5 (*b*) All the double bonds in humulene are isolated. (*c*) Two of the double bonds in cembrene are conjugated to one another but isolated from the remaining double bonds in the molecule. (*d*) The CH=C=CH unit is a cumulated double bond; it is conjugated to the double bond at C-2.

11.6 *cis*-Alloocimene

11.7 3,4-Dibromo-3-methyl-1-butene, 3,4-dibromo-2-methyl-1-butene, and 1,4-dibromo-2-methyl-2-butene

11.8

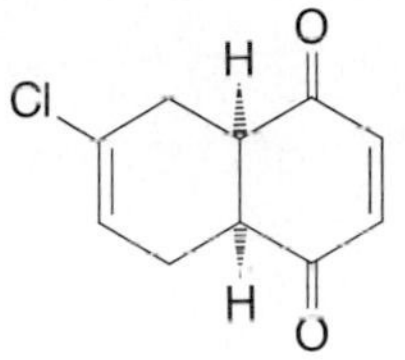

11.9 (*b*) $CH_2{=}CHCH{=}CH_2$ + *cis*-$N{\equiv}CCH{=}CHC{\equiv}N$;

(*c*)

$CH_3CH{=}CHCH{=}CH_2$ + 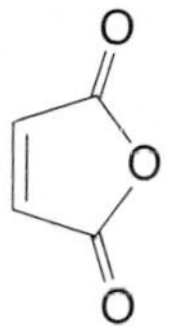

(*d*) 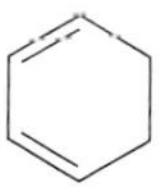 $+\ CH_3O\overset{O}{\overset{\|}{C}}C{\equiv}C\overset{O}{\overset{\|}{C}}OCH_3$

CHAPTER 12

12.1 (*b*) Yes; (*c*) yes; (*d*) no (three monosubstitution products)

12.2 (*a*)

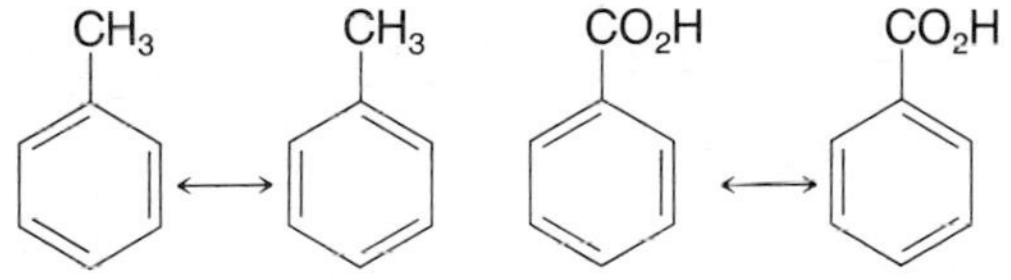

(*b*)

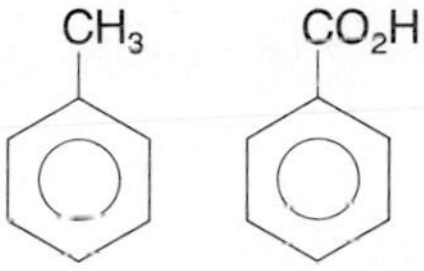

12.3 Styrene, 1050 kcal/mol; cyclooctatetraene, 1086 kcal/mol

12.4 Diels-Alder reaction

12.5 (*b*) Five doubly occupied bonding orbitals plus two half-filled nonbonding orbitals plus five vacant antibonding orbitals

12.6 (b) Cyclononatetraenide anion is aromatic.

12.7 (b) Two; (c) five

12.8 (b) CH=CH$_2$ (c) NH$_2$

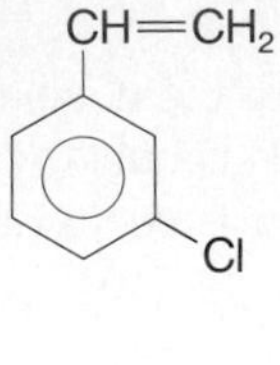

(structures: a benzene ring with CH=CH$_2$ and Cl substituents; a benzene ring with NH$_2$ and NO$_2$ substituents)

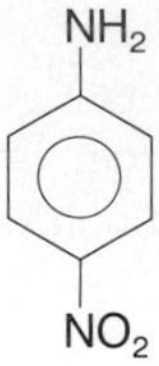

12.9 1,2-Dihydronaphthalene, 24.1 kcal/mol; 1,4-dihydronaphthalene, 27.1 kcal/mol

12.10

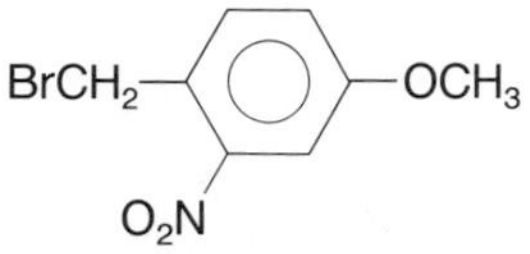

(structure: benzene ring with BrCH$_2$—, —OCH$_3$, and O$_2$N substituents)

12.11

(structure: benzene ring with (CH$_3$)$_3$C—, —CO$_2$H, and CO$_2$H substituents)

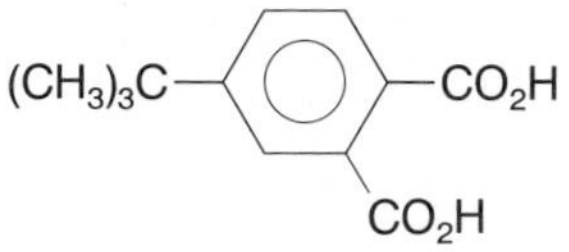

12.12 (b) C$_6$H$_5$CH$_2$OC(CH$_3$)$_3$; (c) C$_6$H$_5$CH$_2$ $\overset{+}{N}$=N=$\overset{..}{\underset{..}{N}}$$^-$; (d) C$_6H_5CH_2$SH; (e) C$_6H_5CH_2$I

12.13 (b) C$_6$H$_5$C(CH$_3$)$_2$; (c) C$_6$H$_5$CHCH$_2$OH; (d) C$_6$H$_5$CHCH$_2$Br;
     | | |
     OH   CH$_3$ OH

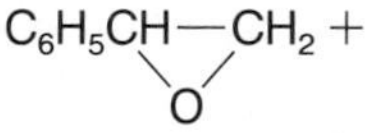

(e) C$_6$H$_5$CH—CH$_2$ + C$_6$H$_5$CO$_2$H
        \\ /
         O

CHAPTER 13

13.1

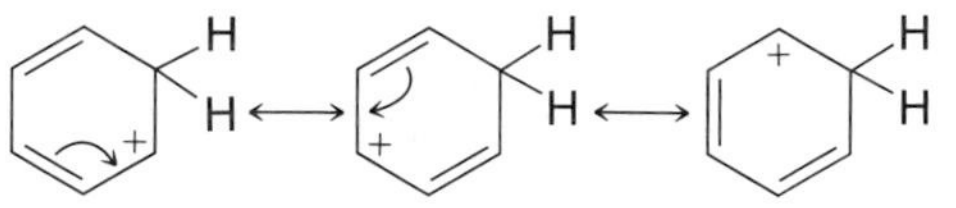

13.2

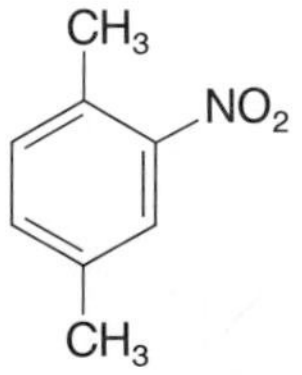

(structure: benzene ring with CH$_3$, NO$_2$, and CH$_3$ substituents)

13.3

(structure: benzene ring with SO$_3$H, H$_3$C, CH$_3$, H$_3$C, CH$_3$ substituents)

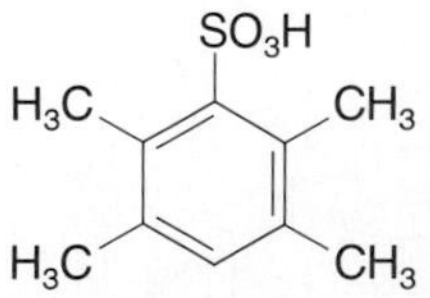

13.4 The major product is isopropylbenzene. Ionization of 1-chloropropane is accompanied by a hydride shift to give CH$_3\overset{+}{C}$HCH$_3$, which then attacks benzene.

13.5 $CH_3CH{=}CH_2 + H{-}F \longrightarrow CH_3\overset{+}{C}HCH_3 + F^-$

then

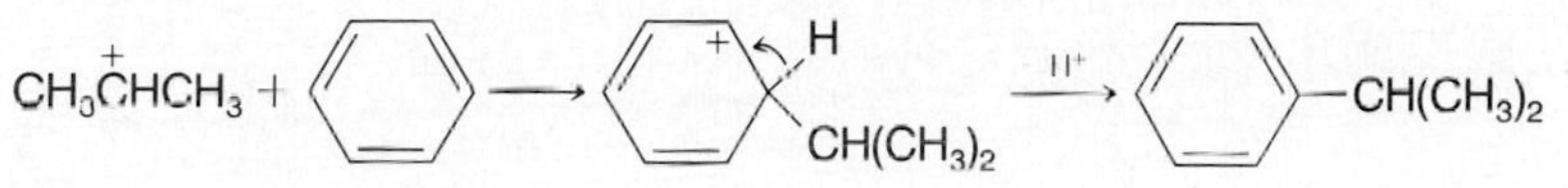

$CH_3\overset{+}{C}HCH_3 + $

13.6 (b) Friedel-Crafts acylation of benzene with $(CH_3)_3CCCl$, followed by reduction with Zn(Hg) and hydrochloric acid

13.7 (b) Toluene is 1.7 times more reactive than *tert*-butylbenzene.
  (c) Ortho (10%), meta (6.7%), para (83.3%)

13.8 (b)

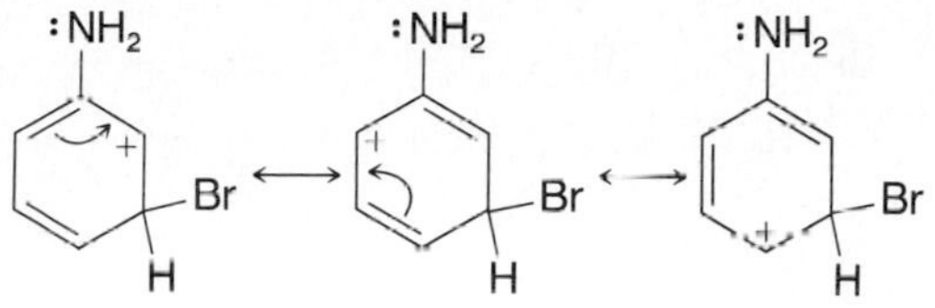

(c)

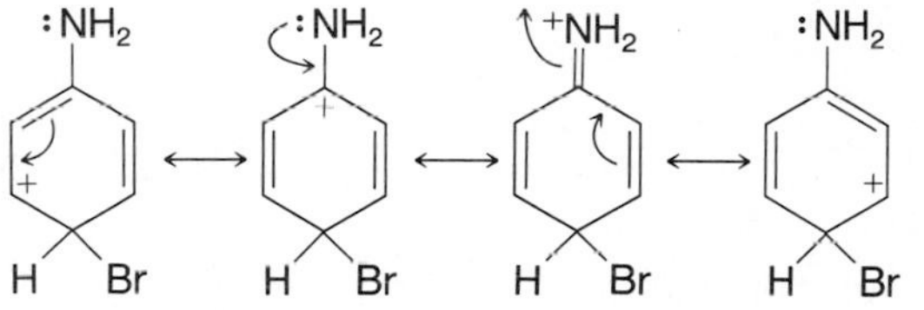

13.9

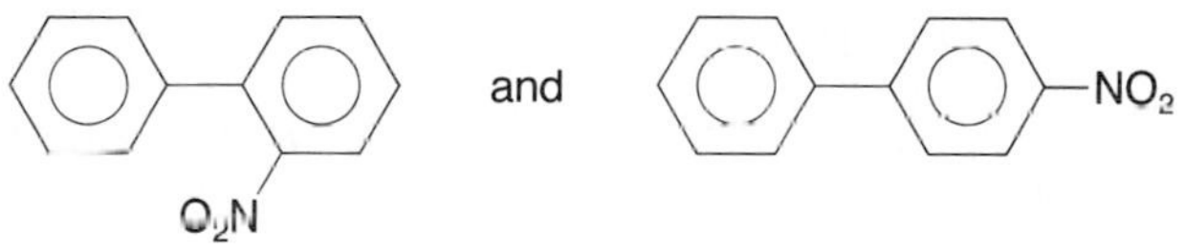

and

13.10 (b)

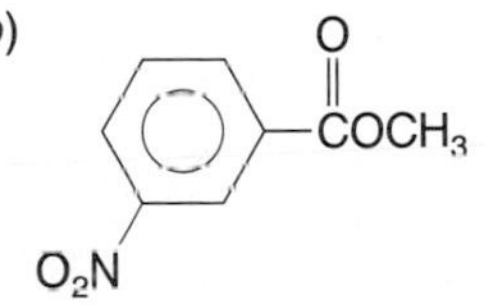

(c)

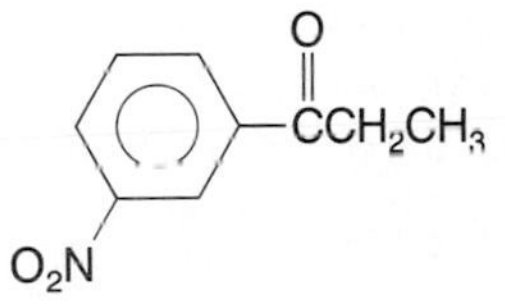

13.11 The group $-\overset{+}{N}(CH_3)_3$ is strongly deactivating and meta-directing. Its positively charged nitrogen makes it a powerful electron-withdrawing substituent. It resembles a nitro group.

13.12 *m*-Bromonitrobenzene:

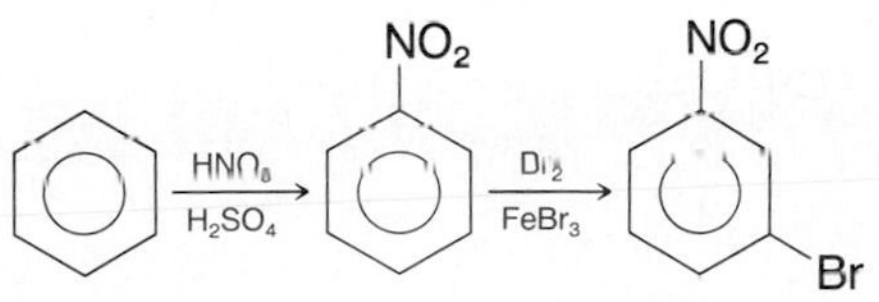

p-Bromonitrobenzene:

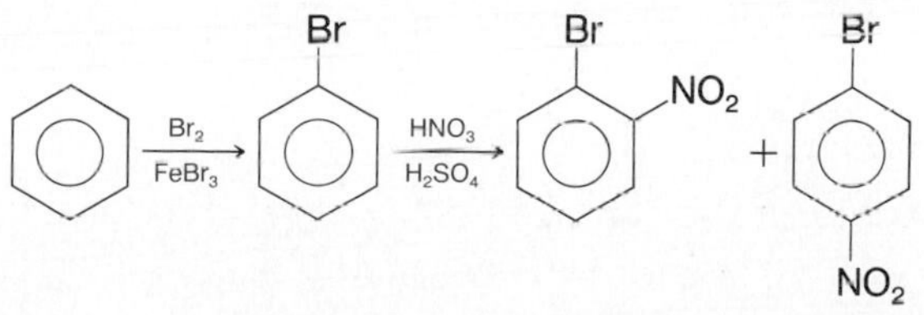

13.13

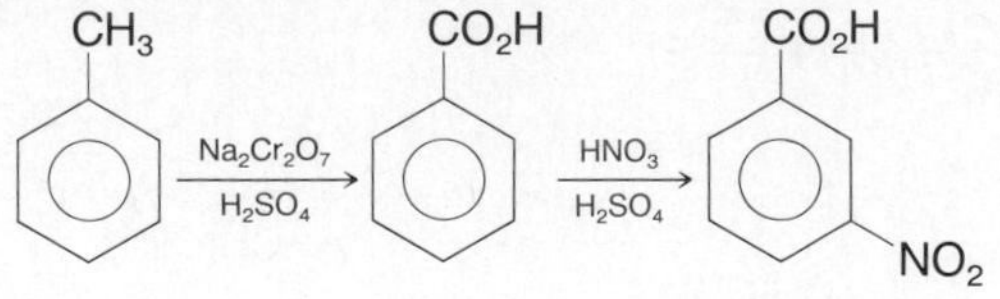

CH₃ → (Na₂Cr₂O₇ / H₂SO₄) → CO₂H → (HNO₃ / H₂SO₄) → CO₂H with NO₂

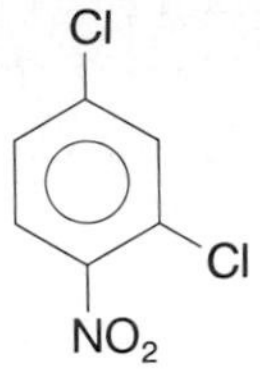

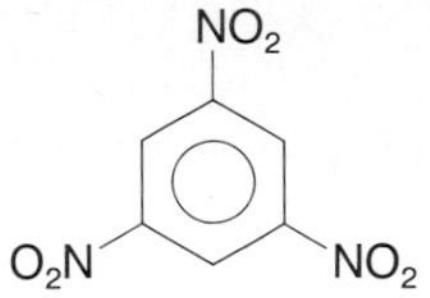

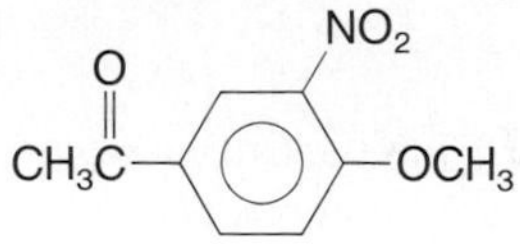

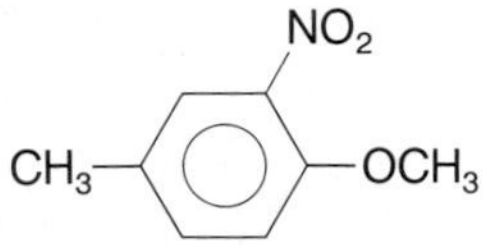

13.14 (b) Cl, Cl, NO₂ substituted benzene

(c) NO₂, O₂N, NO₂ substituted benzene

(d) CH₃CO, NO₂, OCH₃ substituted benzene (with C=O)

(e) NO₂, CH₃, OCH₃ substituted benzene

(f) OCH₃, Br, Br, NO₂ substituted benzene

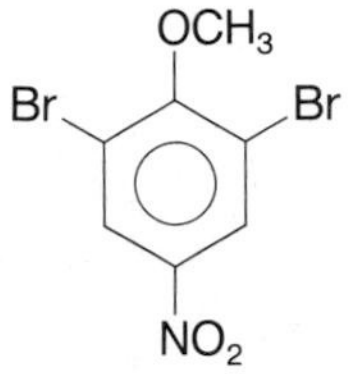

CHAPTER 14

14.1 (b) 5.37 ppm; (c) 3.07 ppm

14.2 (b) Five; (c) two; (d) two; (e) three; (f) one; (g) four; (h) three

14.3 (b) One; (c) one; (d) one; (e) four; (f) four

14.4 (b) One signal (singlet); (c) two signals (doublet and triplet); (d) two signals (both singlets); (e) two signals (doublet and quartet)

14.5 (b) Three signals (singlet, triplet, and quartet); (c) two signals (triplet and quartet); (d) three signals (singlet, triplet, and quartet); (e) four signals (three triplets and quartet); (f) four signals (two doublets, triplet, and quartet)

14.6 Cl₂CHCH(OCH₂CH₃)₂

14.7 (b) Six; (c) six; (d) nine; (e) three

14.8 (b)   (c)   (d) 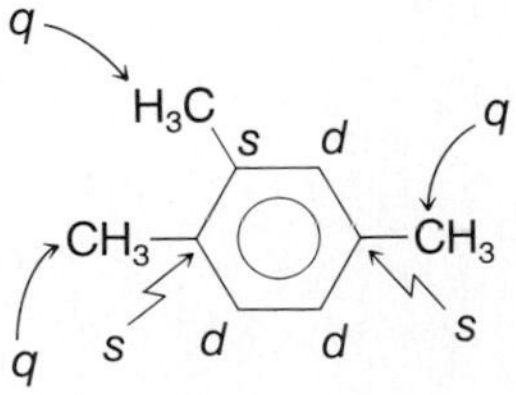   (e)

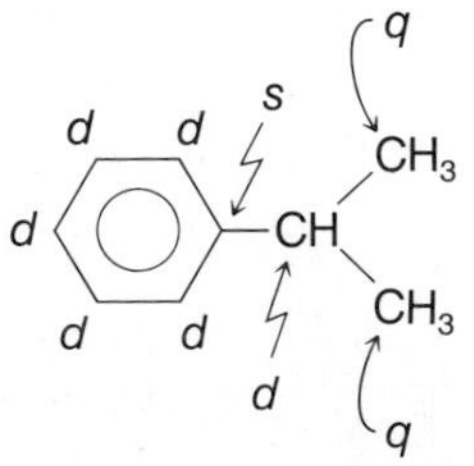

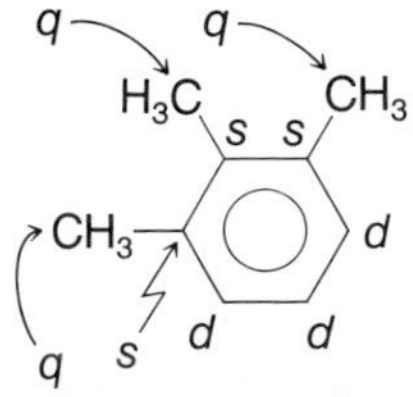

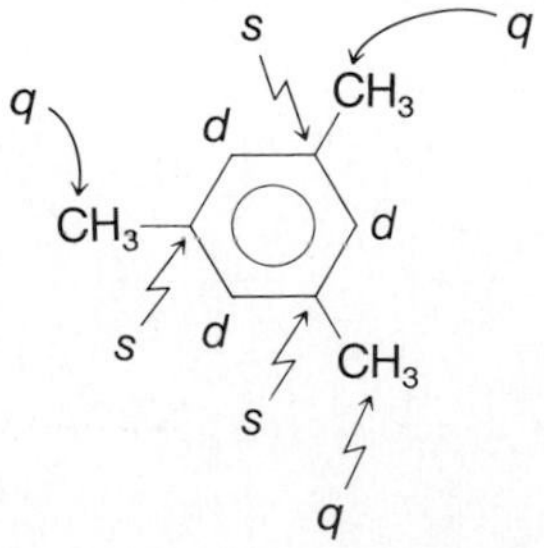

14.9 1,2,4-Trimethylbenzene

14.10 Benzyl alcohol

14.11 (*b*) Three peaks (*m/z* 146, 148, and 150); (*c*) three peaks (*m/z* 234, 236, and 238); (*d*) three peaks (*m/z* 190, 192, and 194)

14.12

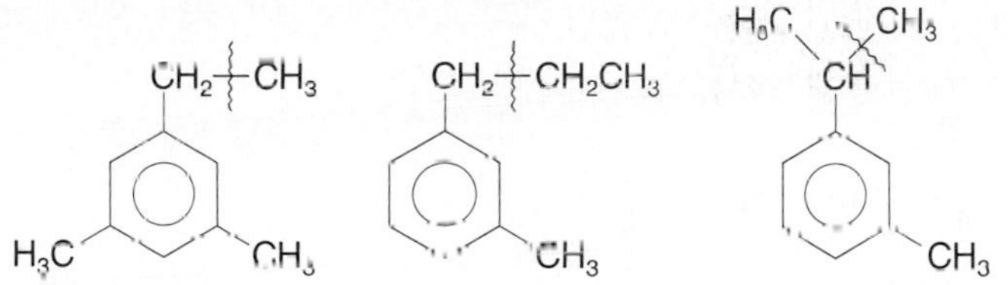

Base peak $C_9H_{11}^+$ (*m/z* 119) | Base peak $C_8H_9^+$ (*m/z* 105) | Base peak $C_9H_{11}^+$ (*m/z* 119)

CHAPTER 15

15.1 (*b*) Cyclohexylmagnesium chloride; (*c*) diethylcadmium; (*d*) iodomethylzinc iodide

15.2 (*b*) $CH_3\underset{Br}{\underset{|}{C}}HCH_2CH_3 + 2Li \longrightarrow CH_3\underset{Li}{\underset{|}{C}}HCH_2CH_3 + LiBr$

(*c*) $C_6H_5CH_2Br + 2Na \longrightarrow C_6H_5CH_2Na + NaBr$

(*d*)

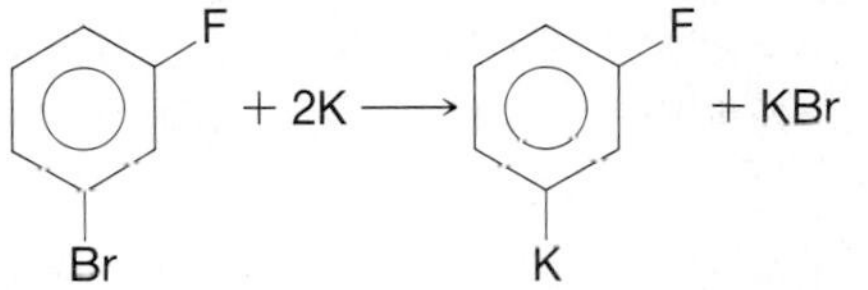

15.3 (*b*) $CH_2{=}CHCH_2MgCl$; (*c*) 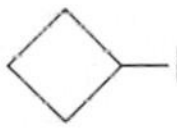—MgI; (*d*) 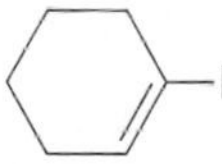—MgBr

15.4 (*b*) $CH_3(CH_2)_4CH_2OH + CH_3CH_2CH_2CH_2Li \longrightarrow CH_3CH_2CH_2CH_3 + CH_3(CH_2)_4CH_2OLi$ (*c*) $C_6H_5SH + CH_3CH_2CH_2CH_2Li \longrightarrow CH_3CH_2CH_2CH_3 + C_6H_5SLi$

15.5 (*b*) $C_6H_5\underset{OH}{\underset{|}{C}}HCH_2CH_2CH_3$; (*c*) 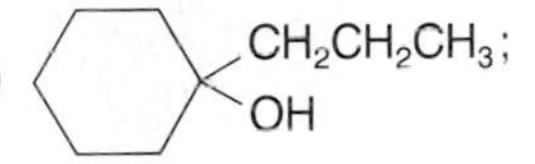; (*d*) $CH_3CH_2CH_2\underset{CH_3CH_2}{\underset{|}{\overset{\overset{CH_3}{|}}{C}}}OH$

15.6 (*b*) $C_6H_5MgBr + CH_3CH_2\overset{\overset{O}{\|}}{C}H \xrightarrow[\text{2. } H_3O^+]{\text{1. diethyl ether}} C_6H_5\underset{OH}{\underset{|}{C}}HCH_2CH_3$

and $CH_3CH_2MgBr + C_6H_5\overset{\overset{O}{\|}}{C}H \xrightarrow[\text{2. } H_3O^+]{\text{1. diethyl ether}} C_6H_5\underset{OH}{\underset{|}{C}}HCH_2CH_3$

(*c*) $CH_3MgI + C_6H_5\overset{\overset{O}{\|}}{C}CH_3 \xrightarrow[\text{2. } H_3O^+]{\text{1. diethyl ether}} C_6H_5\underset{OH}{\underset{|}{\overset{\overset{CH_3}{|}}{C}}}CH_3$

and $C_6H_5MgBr + CH_3\overset{\overset{O}{\|}}{C}CH_3 \xrightarrow[\text{2. } H_3O^+]{\text{1. diethyl ether}} C_6H_5\underset{OH}{\underset{|}{\overset{\overset{CH_3}{|}}{C}}}CH_3$

15.7 (*b*) $2CH_3CH_2CH_2CH_2CH_2MgBr + CH_3\overset{\overset{\text{O}}{\|}}{C}OCH_2CH_3$; (*c*) $2C_6H_5MgBr + {\triangleright}\!-\!\overset{\overset{\text{O}}{\|}}{C}OCH_2CH_3$;

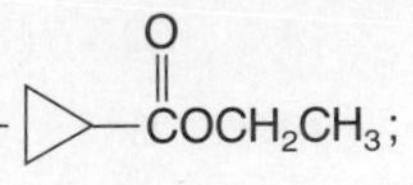

(*d*) $2C_6H_5MgBr + C_6H_5\overset{\overset{\text{O}}{\|}}{C}OCH_2CH_3$

15.8 (*b*) $LiCu(CH_3)_2 +$

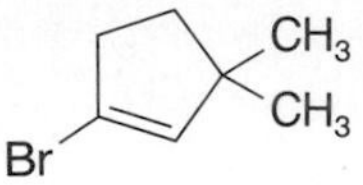

15.9 $BrCH_2CH_2\underset{\underset{\text{Br}}{|}}{C}HCH_3$ and $BrCH_2\underset{\underset{\text{CH}_3}{|}}{C}HCH_2Br$

15.10 (*b*) $=CH_2$; (*c*) 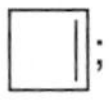 ; (*d*)

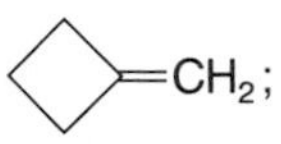

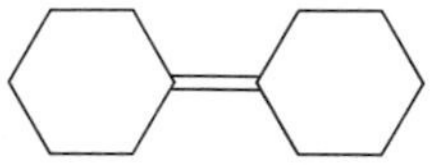

CHAPTER 16

16.1 (*b*) Carbon monoxide is oxidized. Compound B is H_2O. (*c*) Oxidation state remains the same. Compound C is H_2O.

16.2 The primary alcohols $CH_3CH_2CH_2CH_2OH$ and $(CH_3)_2CHCH_2OH$ can each be prepared by hydrogenation of an aldehyde. The secondary alcohol $CH_3\underset{\underset{\text{OH}}{|}}{C}HCH_2CH_3$ can be prepared by hydrogenation of a ketone. The tertiary alcohol $(CH_3)_3COH$ cannot be prepared by hydrogenation.

16.3 (*b*) $CH_3\overset{\overset{\text{D}}{|}}{\underset{\underset{\text{OD}}{|}}{C}}CH_3$; (*c*) $C_6H_5\overset{\overset{\text{D}}{|}}{\underset{\underset{\text{H}}{|}}{C}}OH$; (*d*) DCH_2OD

16.4 $C_6H_5\overset{\overset{\text{O}}{\|}}{C}OCH(CH_3)_2$

16.5 Hydrogenolysis over a metal catalyst could not be used because the carbon-carbon double bond would be hydrogenated. Lithium aluminum hydride could be used.

16.6 (*b*) MgBr (*c*) $CH_3(CH_2)_5CH_2MgBr$ (*d*)

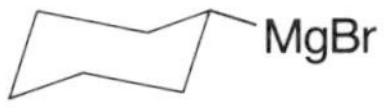

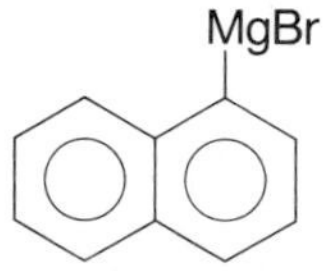

16.7 HO

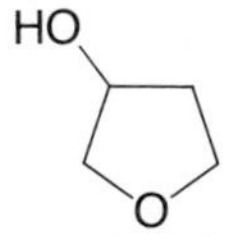

16.8 (*b*) $(CH_3)_2CHCH_2\overset{\overset{\text{O}}{\|}}{C}OCH_2C_6H_5$; (*c*) $CH_3O\overset{\overset{\text{O}}{\|}}{C}\!-\!\!\langle\bigcirc\rangle\!\!-\!\overset{\overset{\text{O}}{\|}}{C}OCH_3$

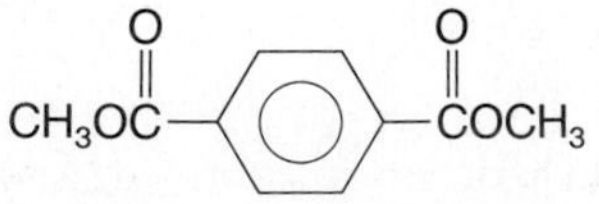

16.9

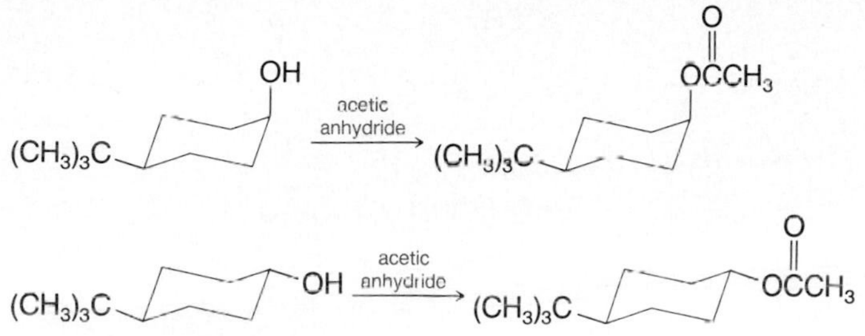

16.10 $O_2NOCH_2CHCH_2ONO_2$
$\quad\quad\quad\quad\quad\quad |$
$\quad\quad\quad\quad\quad ONO_2$

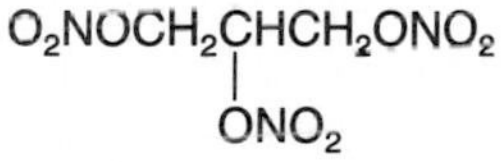

16.11 (b) $CH_3\overset{O}{\overset{||}{C}}(CH_2)_5CH_3$; (c) $CH_3(CH_2)_6\overset{O}{\overset{||}{C}}H$; (d) $(CH_3)_2CH\!-\!$

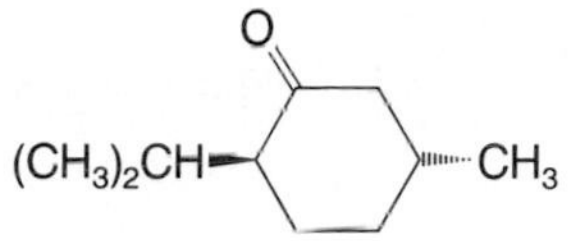

16.12 (b) One; (c) none

16.13 (b) $(CH_3)_2CHCH_2\overset{O}{\overset{||}{C}}H + C_6H_5CH_2\overset{O}{\overset{||}{C}}H$; (c)

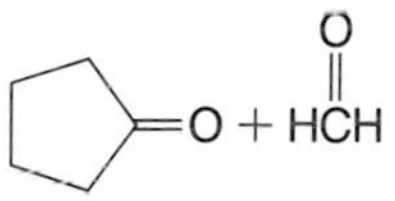

16.14 The peak at m/z 70 corresponds to loss of water from the molecular ion. The peaks at m/z 59 and 73 correspond to the cleavages indicated:

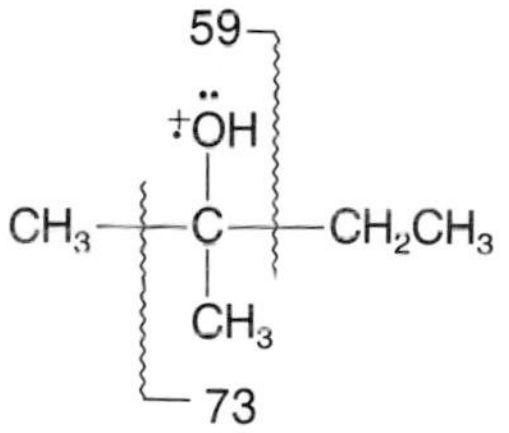

CHAPTER 17

17.1 (b) $CH_2\!\!-\!\!CHCH_2Cl$; (c) $CH_2\!\!-\!\!CHCH\!\!=\!\!CH_2$
$\quad\quad\quad\quad \underset{O}{\diagup}\quad\quad\quad\quad\quad\quad \underset{O}{\diagup}$

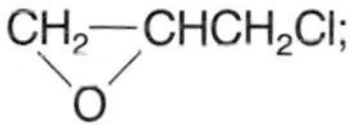

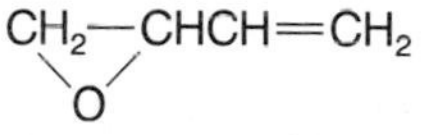

17.2 1,2-Epoxybutane (609.1 kcal/mol); tetrahydrofuran (597.8 kcal/mol)

17.3

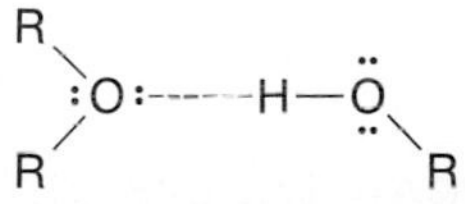

17.4 1,4-Dioxane

17.5 (b) $CH_3ONa + CH_3CH_2CH_2Br \longrightarrow CH_3OCH_2CH_2CH_3 + NaBr$
and $CH_3CH_2CH_2ONa + CH_3Br \longrightarrow CH_3OCH_2CH_2CH_3 + NaBr$
  (c) $C_6H_5CH_2ONa + CH_3CH_2Br \longrightarrow C_6H_5CH_2OCH_2CH_3 + NaBr$
and $CH_3CH_2ONa + C_6H_5CH_2Br \longrightarrow C_6H_5CH_2OCH_2CH_3 + NaBr$

17.6 (b) $(CH_3)_2CHONa + CH_2\!\!=\!\!CHCH_2Br \longrightarrow CH_2\!\!=\!\!CHCH_2OCH(CH_3)_2 + NaBr$
  (c) $(CH_3)_3COK + C_6H_5CH_2Br \longrightarrow (CH_3)_3COCH_2C_6H_5 + NaBr$

17.7 (b)

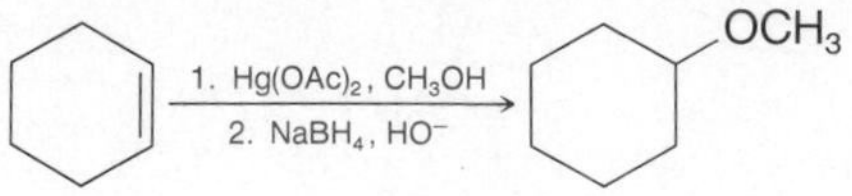

(c) $CH_2{=}CH_2$ $\xrightarrow[\text{2. NaBH}_4,\text{HO}^-]{\text{1. Hg(OAc)}_2,\text{CH}_3\text{CHCH}_2\text{CH}_2\text{CH}_2\text{CH}_3}$ $CH_3CHCH_2CH_2CH_2CH_3$
(with OH on the reagent and OCH_2CH_3 on the product)

and $CH_2{=}CHCH_2CH_2CH_2CH_3$ $\xrightarrow[\text{2. NaBH}_4,\text{HO}^-]{\text{1. Hg(OAc)}_2,\text{CH}_3\text{CH}_2\text{OH}}$ $CH_3CHCH_2CH_2CH_2CH_3$
(with OCH_2CH_3)

(d) $CH_2{=}CCH_2CH_3$ (with CH_3) $\xrightarrow[\text{2. NaBH}_4,\text{HO}^-]{\text{1. Hg(OAc)}_2,\text{CH}_3\text{OH}}$ $(CH_3)_2CCH_2CH_3$ (with OCH_3)

and $(CH_3)_2C{=}CHCH_3$ $\xrightarrow[\text{2. NaBH}_4,\text{HO}^-]{\text{1. Hg(OAc)}_2,\text{CH}_3\text{OH}}$ $(CH_3)_2CCH_2CH_3$ (with OCH_3)

17.8 (b) $CH_3(CH_2)_3CH_2OCH_2(CH_2)_3CH_3 + 15O_2 \longrightarrow 10CO_2 + 11H_2O$

(c) $+ \frac{11}{2}O_2 \longrightarrow 4CO_2 + 4H_2O$

17.9 (b) $C_6H_5CH_2OCH_2C_6H_5$; (c)

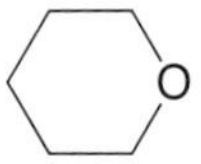

17.10 $(CH_3)_3COC(CH_3)_3$ $\xrightarrow{\text{HCl}}$ $(CH_3)_3\overset{+}{C}OC(CH_3)_3$ (with H)

$(CH_3)_3C{-}\overset{+}{O}C(CH_3)_3$ (with H) $\longrightarrow$ $(CH_3)_3C^+ + HOC(CH_3)_3$

$(CH_3)_3C^+ + Cl^- \longrightarrow (CH_3)_3CCl$

$(CH_3)_3COH + HCl \longrightarrow (CH_3)_3CCl + H_2O$

17.11 Only the trans epoxide is chiral. As formed in this reaction, neither product is optically active.

17.12 (b) $N_3CH_2CH_2OH$; (c) $HOCH_2CH_2OH$; (d) $C_6H_5CH_2CH_2OH$;
(e) $CH_3CH_2C{\equiv}CCH_2CH_2OH$

17.13 Compound B

17.14 Compound A

17.15 The cyclized carbocation is captured by a water molecule.

17.16 $CH_2{=}\overset{+}{O}CHCH_2CH_3$ (with CH_3)

CHAPTER 18

18.1 (b) Pentanedial; (c) 3-phenyl-2-propenal; (d) 4-hydroxy-3-methoxybenzaldehyde

18.2 (b) 3,3-Dimethyl-1-phenyl-2-butanone; (c) 2-methyl-3-pentanone; (d) 4,4-dimethyl-2-pentanone; (e) 4-penten-2-one

18.3 No. Carboxylic acids are inert to catalytic hydrogenation.

18.4 Lithium aluminum hydride, sodium in alcohol, and hydrogenation over copper chromite

18.5 $CH_3CH_2OH \xrightarrow[\text{or HBr}]{PBr_3} CH_3CH_2Br \xrightarrow{Li} CH_3CH_2Li \xrightarrow{CuI} (CH_3CH_2)_2CuLi$

$CH_3CH_2OH \xrightarrow[\text{2. H}^+]{\text{1. KMnO}_4} CH_3CO_2H \xrightarrow{SOCl_2} CH_3\overset{O}{\overset{\|}{C}}Cl$

$CH_3\overset{O}{\overset{\|}{C}}Cl + (CH_3CH_2)_2CuLi \xrightarrow[-78°C]{\text{diethyl ether}} CH_3\overset{O}{\overset{\|}{C}}CH_2CH_3$

18.6 $CH_3CO_2H \xrightarrow[\text{2. H}_2O]{\text{1. LiAlH}_4} CH_3CH_2OH \xrightarrow[\text{or HBr}]{PBr_3} CH_3CH_2Br \xrightarrow{Mg} CH_3CH_2MgBr \xrightarrow{CdCl_2} (CH_3CH_2)_2Cd$

$CH_3CO_2H \xrightarrow{SOCl_2} CH_3\overset{O}{\overset{\|}{C}}Cl$

$CH_3\overset{O}{\overset{\|}{C}}Cl + (CH_3CH_2)_2Cd \longrightarrow CH_3\overset{O}{\overset{\|}{C}}CH_2CH_3$

18.7 (solution in text)

18.8 (a) $C_6H_5\overset{OCH_2CH_3}{\underset{}{C}}HOH$; (b) $C_6H_5\overset{+}{C}HOCH_2CH_3$; (c) $C_6H_5CH(OCH_2CH_3)_2$

18.9 (b) [structure: 1,3-dioxane ring with C_6H_5 and H] (c) [structure: 1,3-dioxolane with $(CH_3)_2CHCH_2$ and CH_3] (d) [structure: dioxane ring with H_3C CH_3 and $(CH_3)_2CHCH_2$ CH_3]

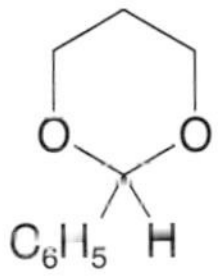

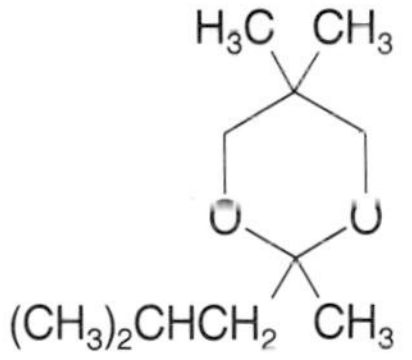

18.10 (b) $C_6H_5\overset{OH}{\underset{CH_3}{C}}CN$; (c) [structure: tetralin ring with HO CN]

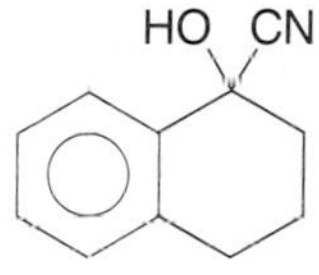

18.11 (b) $C_6H_5\overset{OH}{\underset{}{C}}HNHCH_2CH_2CH_2CH_3 \longrightarrow C_6H_5CH{=}NCH_2CH_2CH_2CH_3$

(c) [cyclohexane ring with OH, $NHC(CH_3)_3$] $\longrightarrow$ [cyclohexane ring with $=NC(CH_3)_3$]

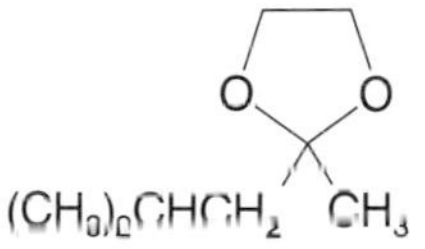

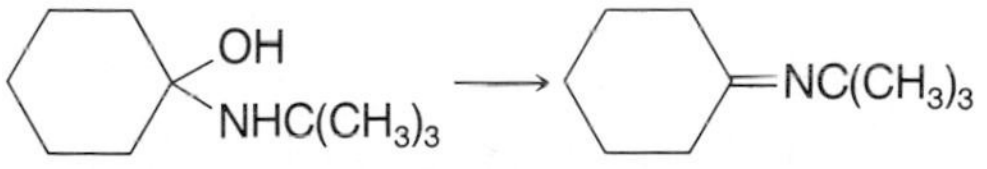

(d) $C_6H_5\overset{OH}{\underset{CH_3}{C}}{-}NH{-}$[cyclohexane] $\longrightarrow C_6H_5\overset{}{\underset{CH_3}{C}}{=}N{-}$[cyclohexane]

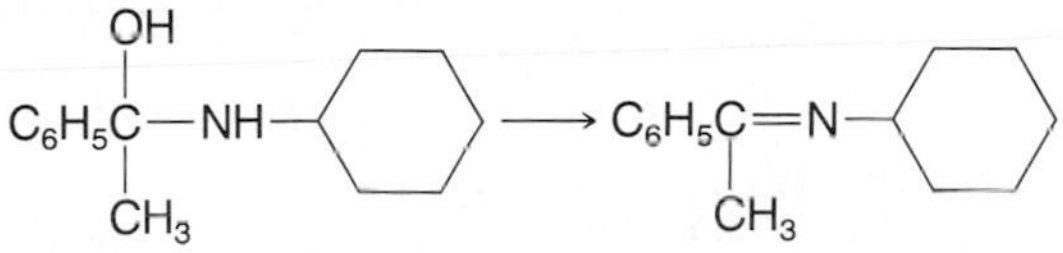

18.12 (b) $CH_3CH_2\overset{\text{[pyrrolidine]}}{\underset{OH}{C}}CH_2CH_3 \longrightarrow CH_3CH{=}\overset{\text{[pyrrolidine]}}{C}CH_2CH_3$

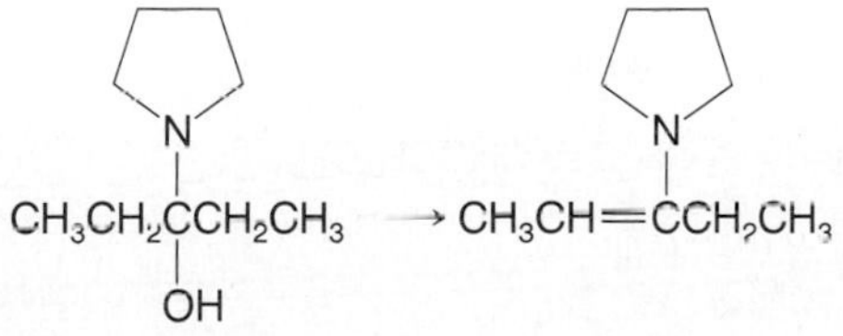

(c) $C_6H_5\overset{\overset{\displaystyle N}{|}}{\underset{\underset{\displaystyle OH}{|}}{C}}CH_3 \longrightarrow C_6H_5\overset{\overset{\displaystyle N}{|}}{C}=CH_2$

18.13 (b) $(E) + (Z)$-$C_6H_5CH=CHCH_3$; (c) $C_6H_5CH=CH_2$;

(d) $CH_3CH_2CH_2CH=CHCH=CH_2$ (e) [cyclohexyl]$-\overset{\overset{\displaystyle CH_2}{\|}}{C}CH_3$

18.14 (b) $CH_3CH_2CH_2\overset{\overset{\displaystyle O}{\|}}{C}H + (C_6H_5)_3\overset{+}{P}-\overset{..}{\overset{..}{C}}H_2$ or $H\overset{\overset{\displaystyle O}{\|}}{C}H + CH_3CH_2CH_2\overset{..}{\overset{..}{C}}H-\overset{+}{P}(C_6H_5)_3$

(c) $C_6H_5CH_2\overset{\overset{\displaystyle O}{\|}}{C}H + (C_6H_5)_3\overset{+}{P}-\overset{..}{\overset{..}{C}}(CH_2CH_3)_2$ or

$C_6H_5CH_2\overset{..}{\overset{..}{C}}H-\overset{+}{P}(C_6H_5)_3 + CH_3CH_2\overset{\overset{\displaystyle O}{\|}}{C}CH_2CH_3$

(d) [cyclopentyl]$=O + (C_6H_5)_3\overset{+}{P}-\overset{..}{\overset{..}{C}}(CH_3)_2$ or [cyclopentyl]$:-\overset{+}{P}(C_6H_5)_3 + CH_3\overset{\overset{\displaystyle O}{\|}}{C}CH_3$

CHAPTER 19

19.1 (b) Zero; (c) five; (d) four

19.2 $ClCH_2\overset{\overset{\displaystyle O}{\|}}{C}CH_2CH_3$ and $CH_3\overset{\overset{\displaystyle O}{\|}}{C}\overset{\underset{\underset{\displaystyle Cl}{|}}{}}{C}HCH_3$

19.3 $CH_2=\overset{\overset{\displaystyle OH}{|}}{C}CH_2CH_3 \xrightarrow{Cl_2} ClCH_2\overset{\overset{\displaystyle O}{\|}}{C}CH_2CH_3$ $CH_2=\overset{\overset{\displaystyle OH}{|}}{C}CHCH_3 \xrightarrow{Cl_2} CH_3\overset{\overset{\displaystyle O}{\|}}{C}\overset{\underset{\underset{\displaystyle Cl}{|}}{}}{C}HCH_3$

19.4 $CH_2=\overset{\overset{\displaystyle :\overset{..}{O}H}{|}}{C}CH_2CH_3 \longrightarrow ClCH_2\overset{\overset{\displaystyle +\overset{..}{O}H}{|}}{C}CH_2CH_3$ $CH_3\overset{\overset{\displaystyle :\overset{..}{O}H}{|}}{C}=CHCH_3 \longrightarrow CH_3\overset{\overset{\displaystyle +\overset{..}{O}H}{|}}{C}\overset{\underset{\underset{\displaystyle Cl}{|}}{}}{C}HCH_3$

$Cl-Cl$ Cl^- $Cl-Cl$ Cl^-

19.5 (b) $C_6H_5\overset{\overset{\displaystyle OH}{|}}{C}=CH_2$; (c) [cyclohexene]$-OH$; (d) [cyclohexene with CH_3]$-OH$ and [cyclohexene with CH_3]$-OH$

19.6 (b) $CH_3\overset{\overset{\displaystyle O}{\|}}{C}CH=\overset{\overset{\displaystyle HO}{|}}{C}CH_2CH_3$ and $CH_3\overset{\overset{\displaystyle OH}{|}}{C}=CHCCH_2CH_3$;

(c) $C_6H_5\overset{\overset{\displaystyle O}{\|}}{C}CH=\overset{\overset{\displaystyle HO}{|}}{C}CH_3$ and $C_6H_5\overset{\overset{\displaystyle OH}{|}}{C}=CHCCH_3$

19.7 (b)

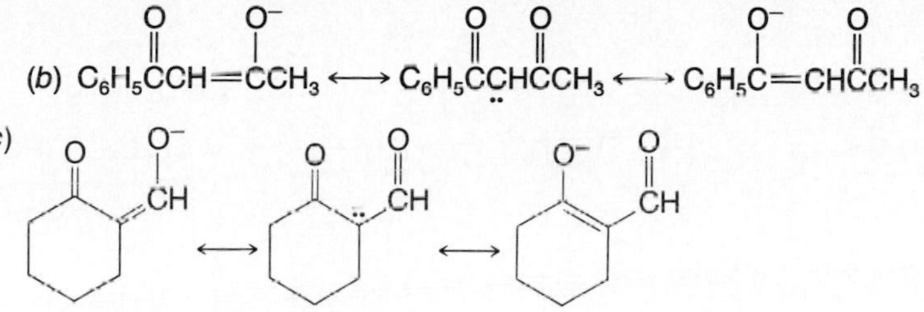

(c)

19.8 (b) (c)

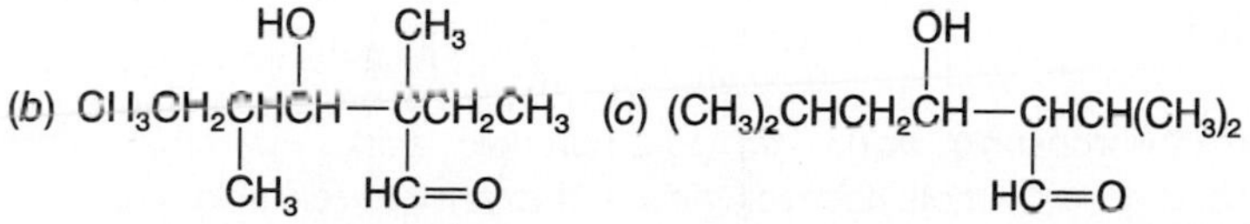

19.9 (b) (c)

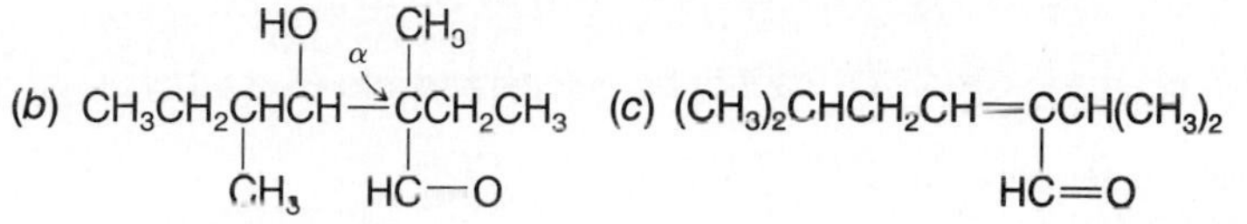

Cannot dehydrate; no protons
on α carbon atom

19.10

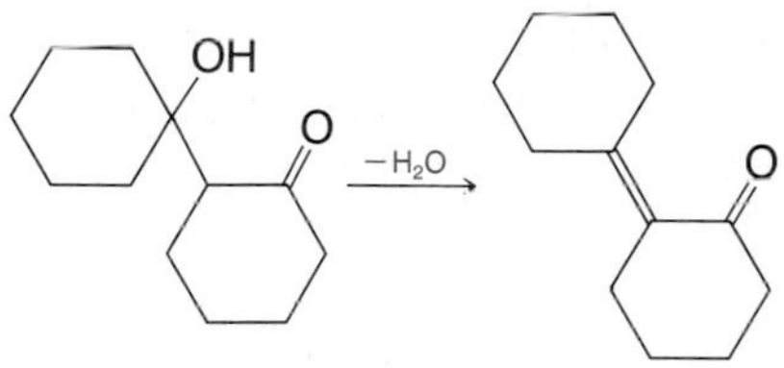

19.11 (b) $C_6H_5CH{=}CHCC(CH_3)_3$; (c)

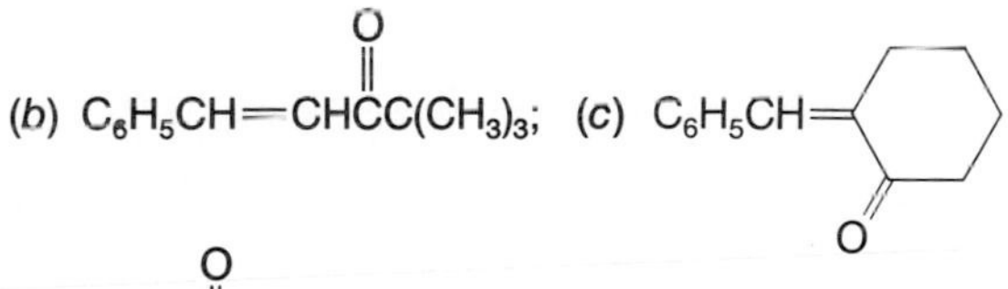

19.12

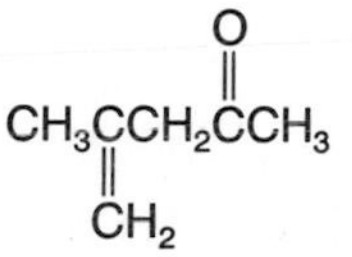

19.13   and

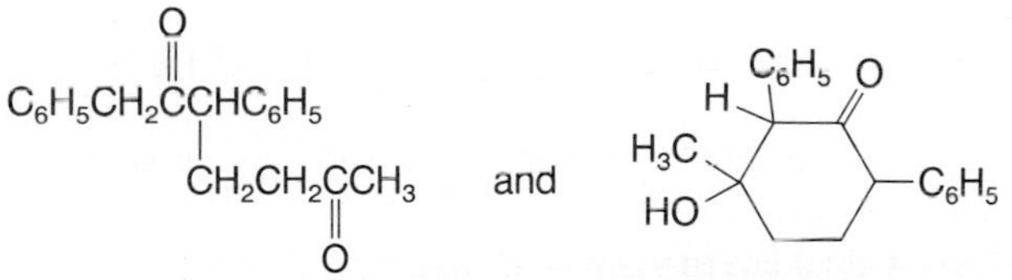

19.14

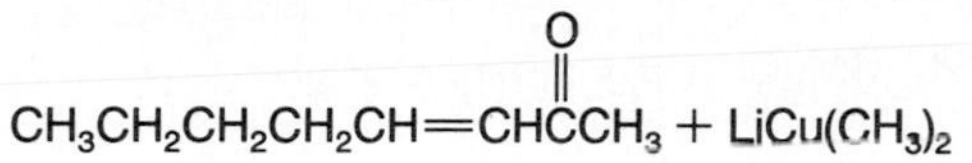

19.15 (b)

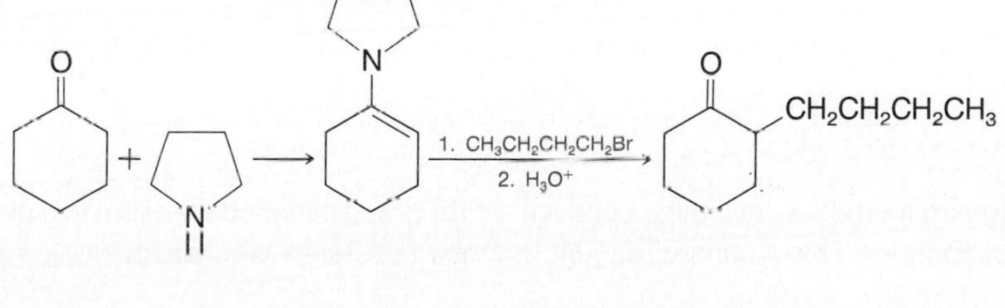

(c) $CH_3CH_2\overset{O}{\overset{\|}{C}}CH_2CH_3 +$ [pyrrolidine] $\longrightarrow CH_3CH_2C{=}CHCH_3 \xrightarrow[\text{2. } H_3O^+]{\text{1. } C_6H_5CH_2Br} CH_3CH_2\overset{O}{\overset{\|}{C}}CHCH_3$
$\underset{CH_2C_6H_5}{|}$

CHAPTER 20

20.1 (b) 2,2-Dimethylpropanoic acid; (c) (E)-2-butenoic acid; (d) ethanedioic acid; (e) (Z)-butenedioic acid; (f) p-methylbenzoic acid or 4-methylbenzoic acid.

20.2 The negative charge in $CH_3\overset{O}{\overset{\|}{C}}OO^-$ cannot be delocalized into the carbonyl group.

20.3 (b) $CH_3CO_2H + (CH_3)_3CO^- \rightleftharpoons CH_3CO_2^- + (CH_3)_3COH$
(The position of equilibrium lies to the right.)

(c) $CH_3CO_2H + Br^- \rightleftharpoons CH_3CO_2^- + HBr$
(The position of equilibrium lies to the left.)

(d) $CH_3CO_2H + HC{\equiv}C:^- \rightleftharpoons CH_3CO_2^- + HC{\equiv}CH$
(The position of equilibrium lies to the right.)

(e) $CH_3CO_2H + NO_3^- \rightleftharpoons CH_3CO_2^- + HNO_3$
(The position of equilibrium lies to the left.)

(f) $CH_3CO_2H + H_2N^- \rightleftharpoons CH_3CO_2^- + NH_3$
(The position of equilibrium lies to the right.)

20.4 (b) $CH_3\underset{OH}{\overset{|}{C}}HCO_2H$; (c) $CH_2{=}CHCO_2H$; (d) $CH_3\overset{O}{\overset{\|}{C}}CO_2H$; (e) $CH_3\overset{O}{\overset{\|}{\underset{\underset{O}{\|}}{S}}}CH_2CO_2H$

20.5 $HC{\equiv}CCO_2H$

20.6 The "true K_1" for carbonic acid is 1.4×10^{-4}.

20.7 (b) The conversion proceeding by way of the nitrile is satisfactory.

$$HOCH_2CH_2Cl \xrightarrow{NaCN} HOCH_2CH_2CN \xrightarrow{hydrolysis} HOCH_2CH_2CO_2H$$

Since 2-chloroethanol has a proton bonded to oxygen, it is not an appropriate substrate for conversion to a stable Grignard reagent.

(c) The procedure involving a Grignard reagent is satisfactory.

$$(CH_3)_3CCl \xrightarrow{Mg} (CH_3)_3CMgCl \xrightarrow[\text{2. } H_3O^+]{\text{1. } CO_2} (CH_3)_3CCO_2H$$

The reaction of tert-butyl chloride with cyanide ion proceeds by elimination rather than substitution.

(d) Formation and hydrolysis of the corresponding nitrile is satisfactory.

$O_2N{-}$[benzene ring]${-}CH_2Br \xrightarrow{NaCN} O_2N{-}$[benzene ring]${-}CH_2CN \xrightarrow{hydrolysis} O_2N{-}$[benzene ring]${-}CH_2CO_2H$

The Grignard method is unsuitable because a nitro group cannot be present in either the alkyl halide used to form the Grignard reagent or in the substance with which it is to react.

20.8 Water labeled with ^{18}O adds to benzoic acid to give the intermediate shown. This intermediate can lose unlabeled H_2O to give benzoic acid containing ^{18}O.

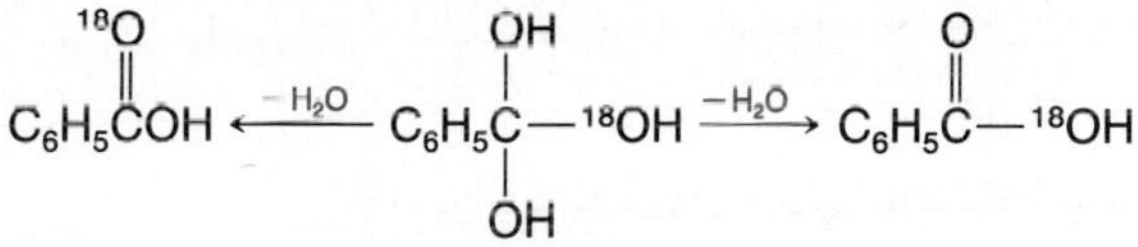

$$C_6H_5\overset{\overset{18O}{\|}}{C}OH \xleftarrow{-H_2O} C_6H_5\overset{\overset{OH}{|}}{\underset{\underset{OH}{|}}{C}}-{}^{18}OH \xrightarrow{-H_2O} C_6H_5\overset{\overset{O}{\|}}{C}-{}^{18}OH$$

20.9 (b) $HOCH_2(CH_2)_{13}CO_2H$; (c)

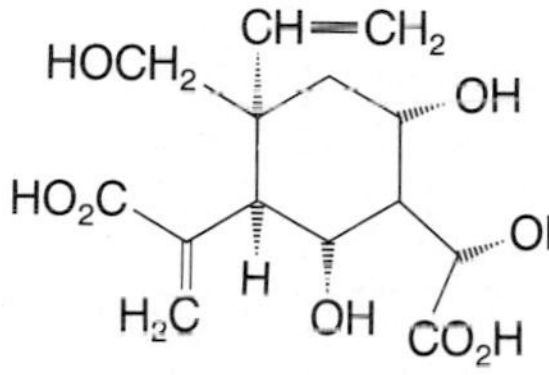

20.10 $CH_3(CH_2)_{15}CH_2CO_2H \xrightarrow[PCl_3]{Br_2} CH_3(CH_2)_{15}\underset{\underset{Br}{|}}{C}HCO_2H \xrightarrow[acetone]{NaI} CH_3(CH_2)_{15}\underset{\underset{I}{|}}{C}HCO_2H$

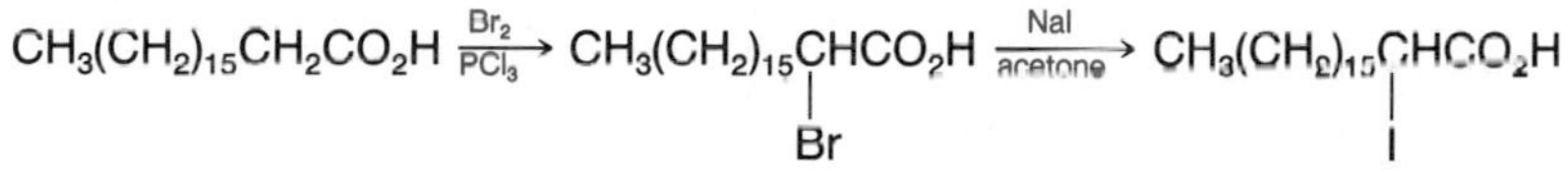

20.11 (b) $CH_3(CH_2)_6CH_2CO_2H$ via

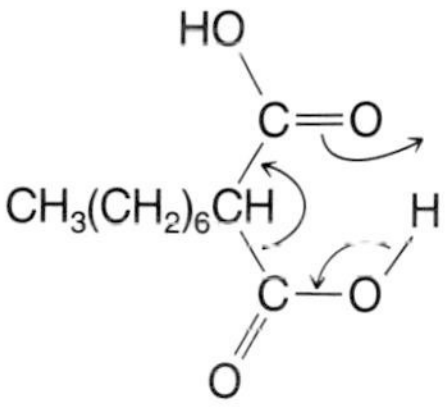

(c)

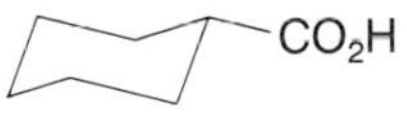

via

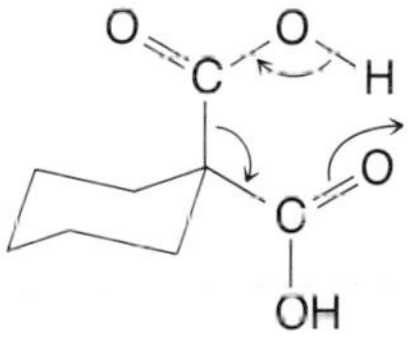

(d) $C_6H_5\underset{\underset{CH_3}{|}}{C}HCO_2H$ via

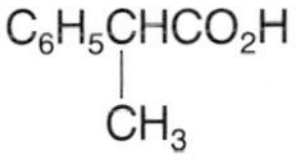

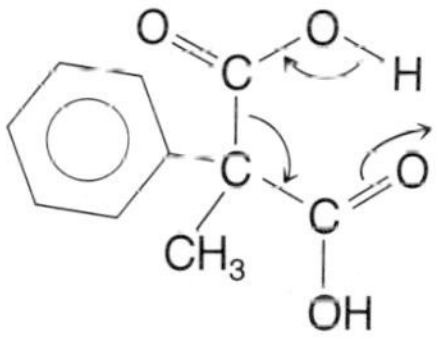

20.12 (b)

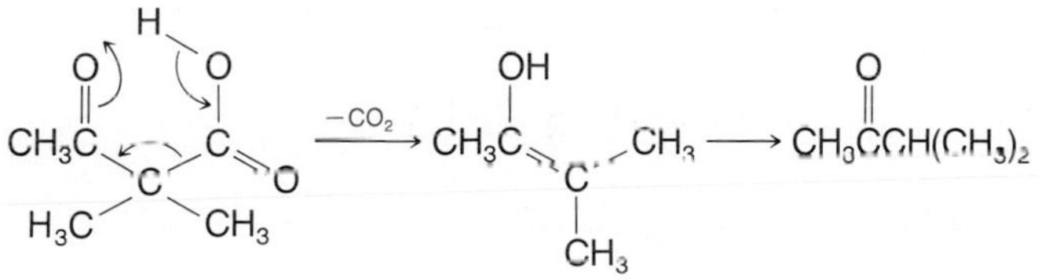

CHAPTER 21

21.1 (b) $CH_3CH_2\underset{\underset{C_6H_5}{|}}{C}H\overset{\overset{O}{\|}}{C}O\overset{\overset{O}{\|}}{C}\underset{\underset{C_6H_5}{|}}{C}HCH_2CH_3$; (c) $CH_3CH_2\underset{\underset{C_6H_5}{|}}{C}H\overset{\overset{O}{\|}}{C}OCH_2CH_2CH_2CH_3$;

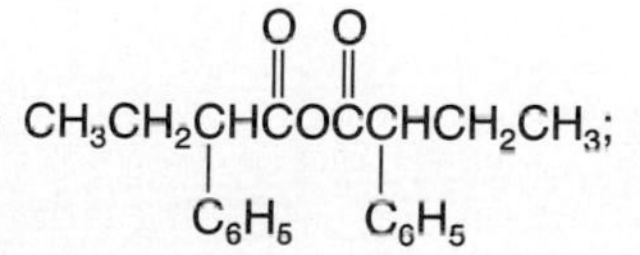

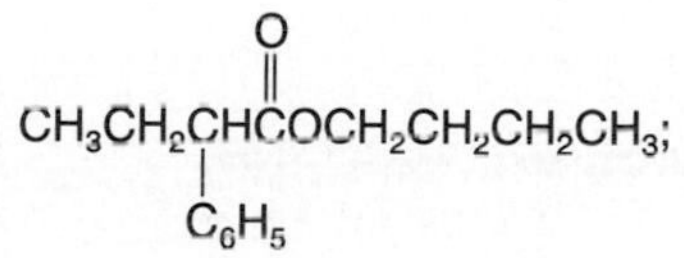

(d) $\underset{\text{O}}{\overset{\text{O}}{\parallel}}$ CH$_3$CH$_2$CH$_2$COCH$_2$CHCH$_2$CH$_3$; (e) CH$_3$CH$_2$CHCNH$_2$;
                                C$_6$H$_5$ C$_6$H$_5$

(f) CH$_3$CH$_2$CHCNHCH$_2$CH$_3$; (g) CH$_3$CH$_2$CHC≡N
             C$_6$H$_5$ C$_6$H$_5$

21.2 Rotation about the carbon-nitrogen bond is slow in amides. The methyl groups of *N,N*-dimethylformamide are nonequivalent because one is cis to oxygen, the other cis to hydrogen.

21.3 (b) C$_6$H$_5$COCC$_6$H$_5$; (c) C$_6$H$_5$COCH$_2$CH$_3$; (d) C$_6$H$_5$CNHCH$_3$; (e) C$_6$H$_5$CN(CH$_3$)$_2$

21.4 (b) C$_6$H$_5$COCC$_6$H$_5$ ⟶ C$_6$H$_5$COCC$_6$H$_5$ + HCl

(c) C$_6$H$_5$COCH$_2$CH$_3$ ⟶ C$_6$H$_5$COCH$_2$CH$_3$ + HCl

(d) C$_6$H$_5$CNHCH$_3$ ⟶ C$_6$H$_5$CNHCH$_3$ + CH$_3$$\overset{+}{\text{N}}H_3$ Cl$^-$

(e) C$_6$H$_5$CN(CH$_3$)$_2$ ⟶ C$_6$H$_5$CN(CH$_3$)$_2$ + (CH$_3$)$_2$$\overset{+}{\text{N}}H_2$ Cl$^-$

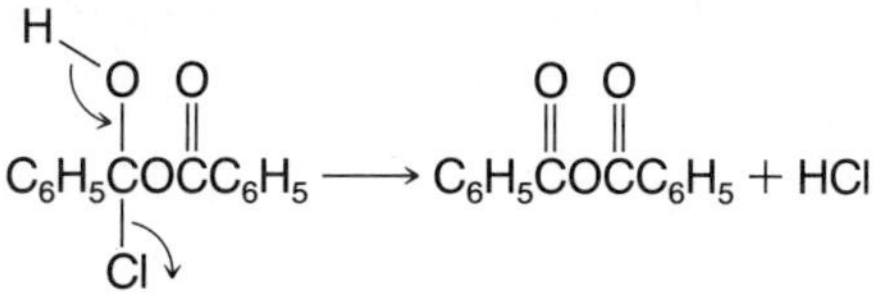

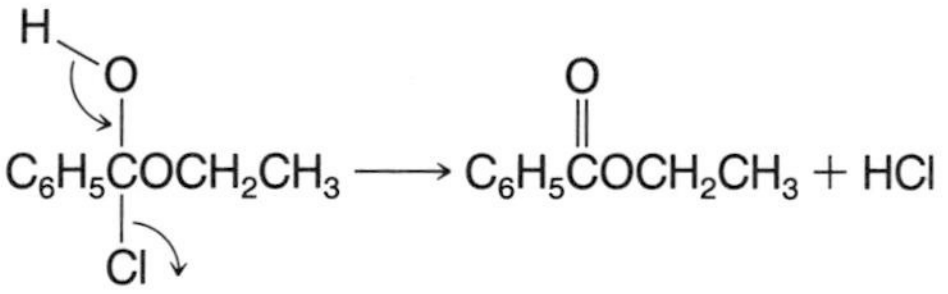

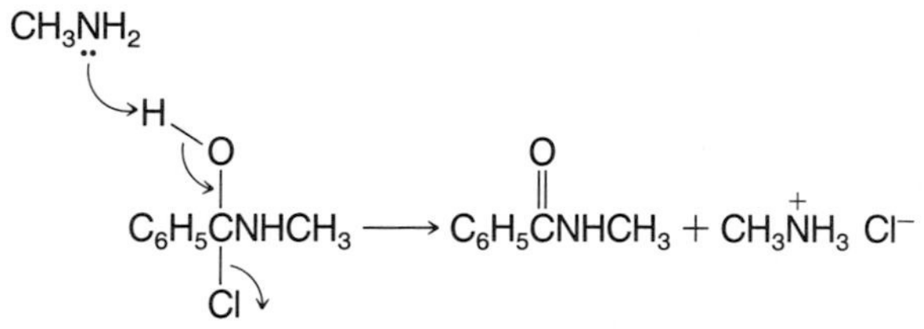

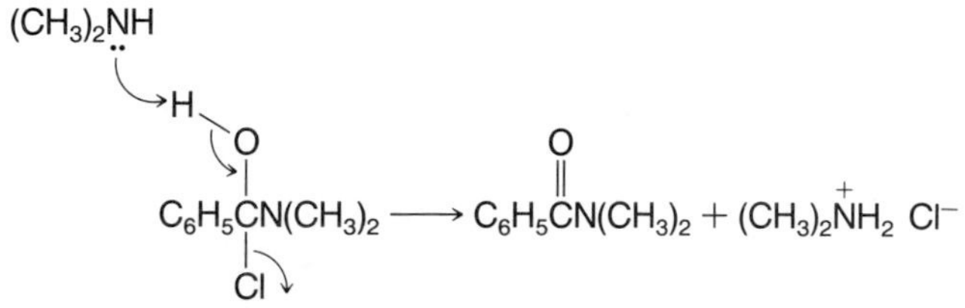

21.5 C$_6$H$_5$CCl + H$_2$O ⟶ C$_6$H$_5$COH + HCl

C$_6$H$_5$CCl + C$_6$H$_5$COH ⟶ C$_6$H$_5$COCC$_6$H$_5$ + HCl

21.6 CH$_3$$\overset{+}{\text{C}}$=Ö: ⟷ CH$_3$C≡$\overset{+}{\text{O}}$:

21.7 (b) CH$_3$CNH$_2$ and CH$_3$CO$_2^-$$\overset{+}{\text{N}}H_4$;

(c)

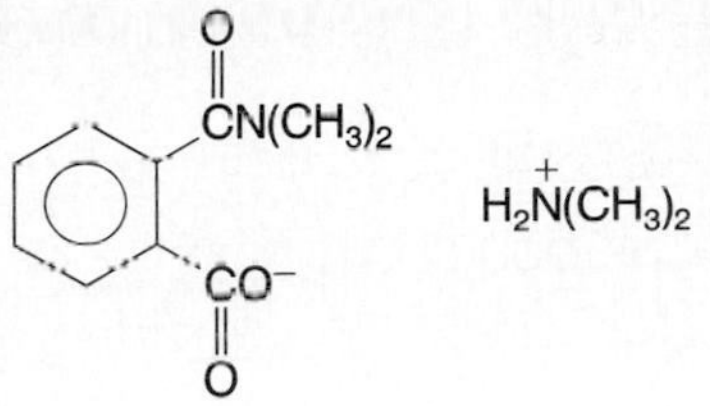

$H_2\overset{+}{N}(CH_3)_2$

(d)

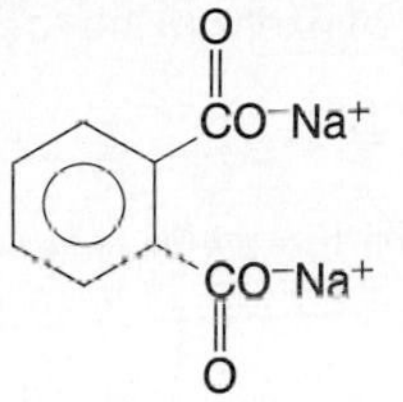

21.8 (b)

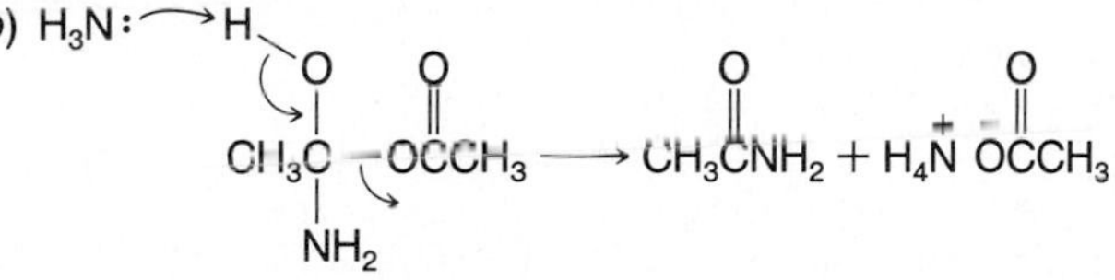

(c) $(CH_3)_2\overset{..}{N}H$

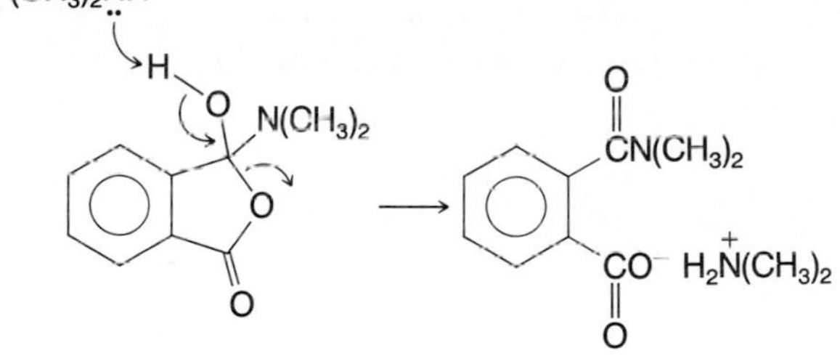

(d) HO^-

$+ H_2O$

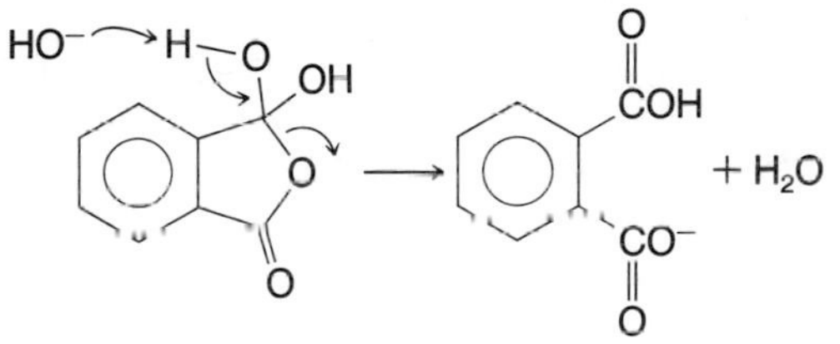

21.9 $HOCH_2CHCH_2CH_2CH_2OH$ $(C_5H_{12}O_3)$ and CH_3CO_2H
                    |
                   OH

21.10

Step 1: Protonation of the carbonyl oxygen

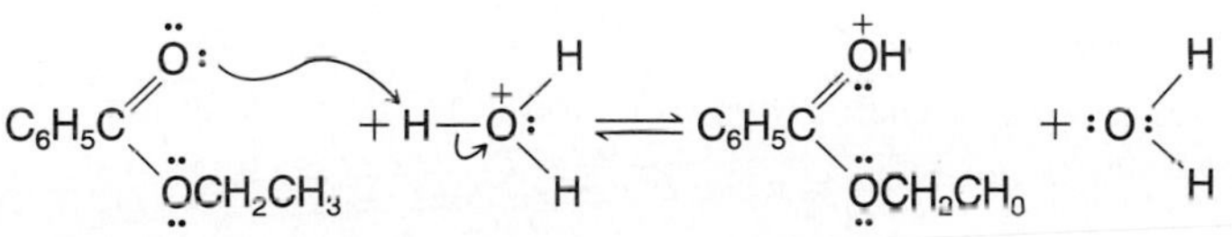

Step 2: Nucleophilic addition of water

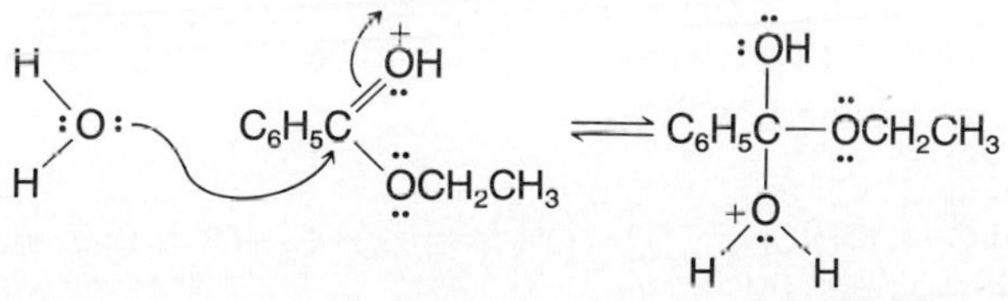

Step 3: Deprotonation of oxonium ion to give neutral form of tetrahedral intermediate

$$C_6H_5\overset{\ddot{O}H}{\underset{\overset{+}{\underset{H}{O}H}}{C}}-\ddot{O}CH_2CH_3 + :\overset{H}{\underset{H}{O}} \rightleftharpoons C_6H_5\overset{:OH}{\underset{\underset{HO:}{}}{C}}-\ddot{O}CH_2CH_3 + H-\overset{+}{\underset{H}{O}}\overset{H}{}$$

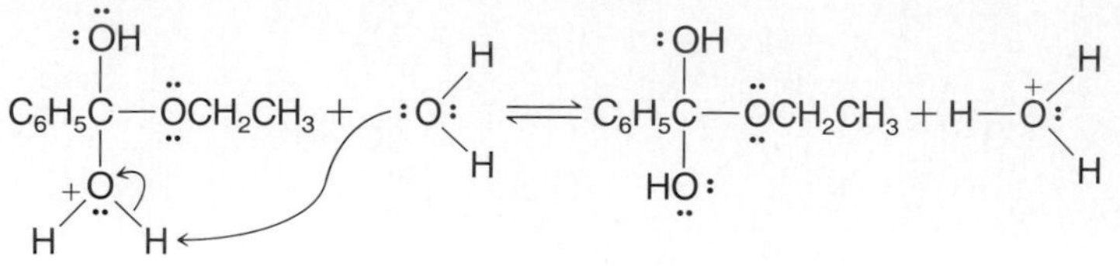

Step 4: Protonation of ethoxy oxygen

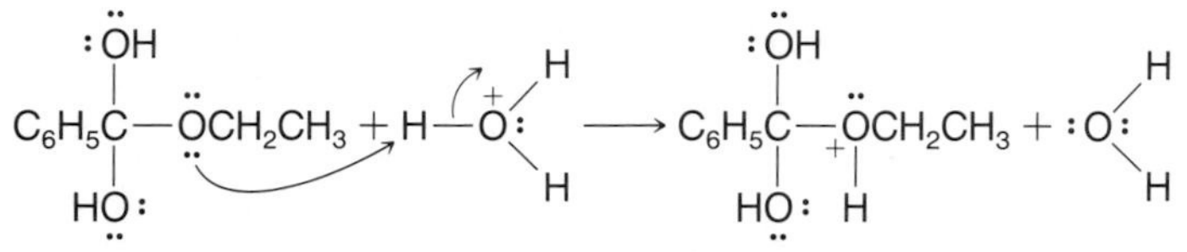

$$C_6H_5\overset{:\ddot{O}H}{\underset{\underset{HO:}{}}{C}}-\ddot{O}CH_2CH_3 + H-\overset{+}{\underset{H}{O}:} \longrightarrow C_6H_5\overset{:\ddot{O}H}{\underset{\underset{HO:\;H}{}}{C}}-\overset{+}{\underset{}{O}}CH_2CH_3 + :\overset{H}{\underset{H}{O}}:$$

Step 5: Dissociation of protonated form of tetrahedral intermediate

$$C_6H_5\overset{:\ddot{O}H}{\underset{\underset{:OH\;H}{}}{C}}-\overset{+}{\underset{}{O}}CH_2CH_3 \rightleftharpoons C_6H_5C\overset{\overset{+}{O}H}{\underset{:OH}{}} + H\ddot{O}CH_2CH_3$$

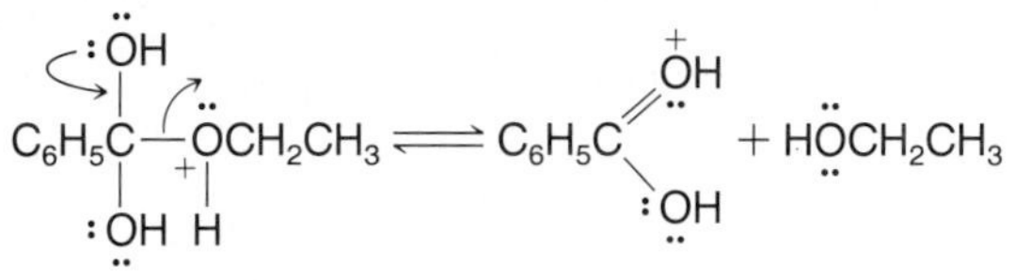

Step 6: Deprotonation of protonated form of benzoic acid

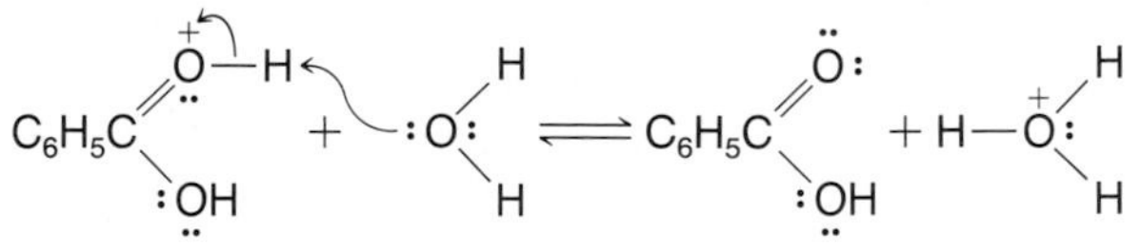

$$C_6H_5C\overset{\overset{+}{O}-H}{\underset{:OH}{}} + :\overset{H}{\underset{H}{O}}: \rightleftharpoons C_6H_5C\overset{\ddot{O}:}{\underset{:OH}{}} + H-\overset{+}{\underset{H}{O}}\overset{H}{}$$

21.11 The carbonyl oxygen of the lactone became labeled with ^{18}O.

21.12

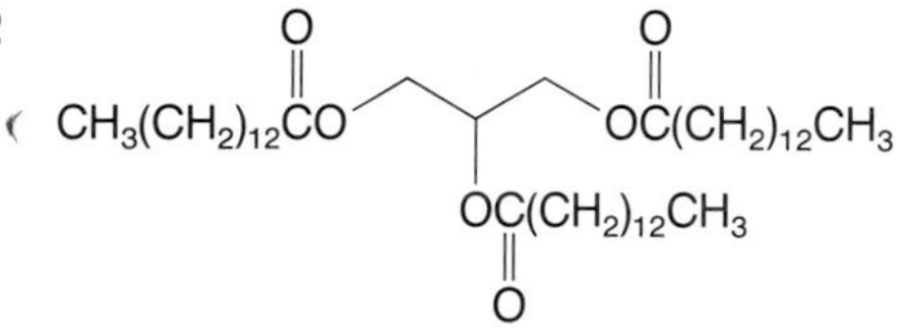

$$CH_3(CH_2)_{12}\overset{O}{\overset{\|}{C}}O-\underset{\underset{O\overset{O}{\underset{\|}{C}}(CH_2)_{12}CH_3}{}}{}-O\overset{O}{\overset{\|}{C}}(CH_2)_{12}CH_3$$

21.13 The isotopic label appeared in the acetate ion.

21.14

Step 1: Nucleophilic addition of hydroxide ion to the carbonyl group.

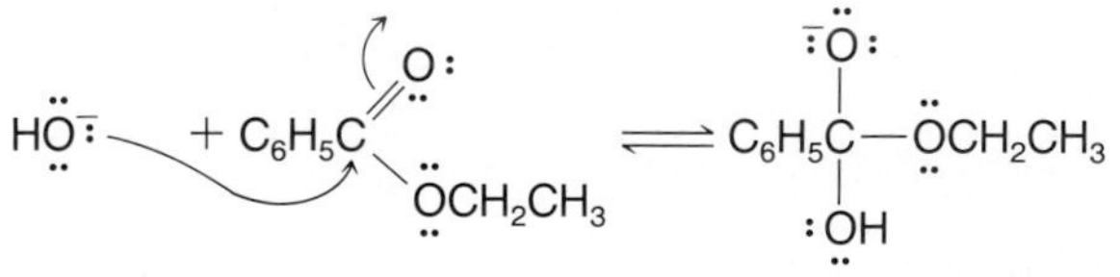

$$H\ddot{O}:^- + C_6H_5C\overset{O:}{\underset{\ddot{O}CH_2CH_3}{}} \rightleftharpoons C_6H_5\overset{:\overset{-}{\ddot{O}}:}{\underset{:OH}{}}C-\ddot{O}CH_2CH_3$$

Step 2: Proton transfer from water to give neutral form of tetrahedral intermediate.

$$C_6H_5\overset{:\ddot{O}:^-}{\underset{:OH}{}}C-\ddot{O}CH_2CH_3 + H-\ddot{O}H \rightleftharpoons C_6H_5\overset{:\ddot{O}H}{\underset{:OH}{}}C-\ddot{O}CH_2CH_3 + {}^-\ddot{O}H$$

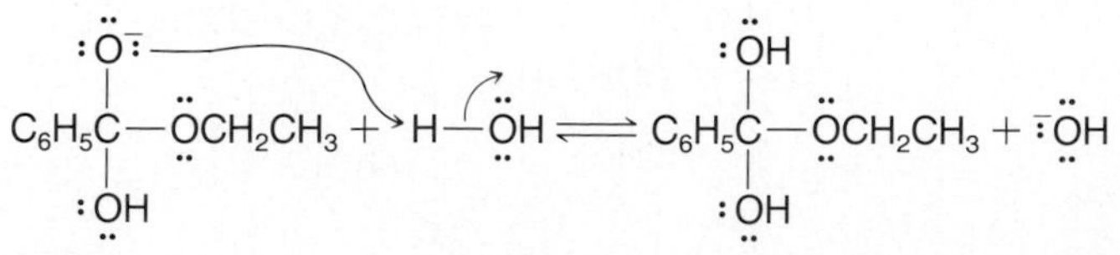

Step 3: Hydroxide ion–promoted dissociation of tetrahedral intermediate.

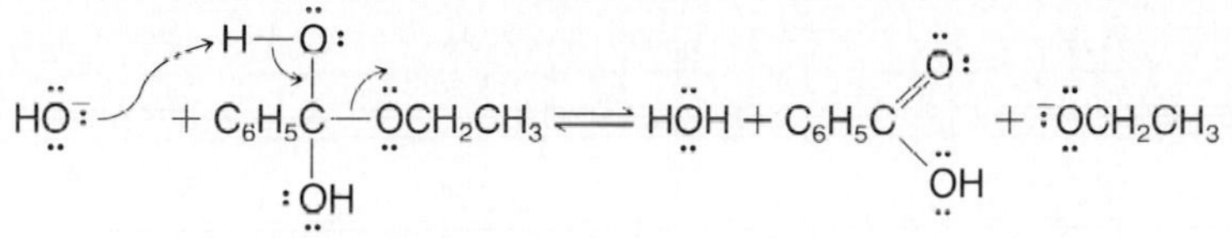

Step 4: Proton abstraction from benzoic acid.

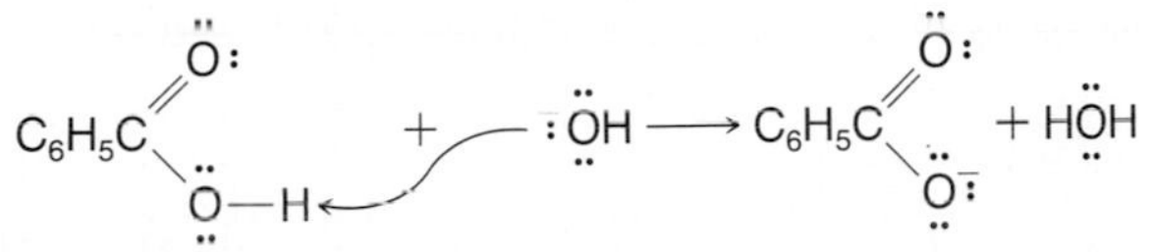

21.15

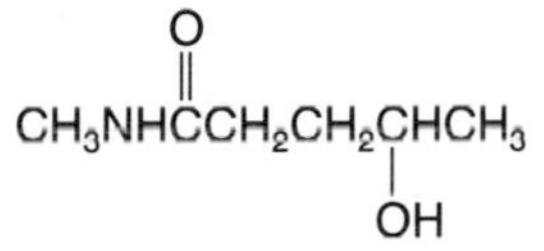

21.16

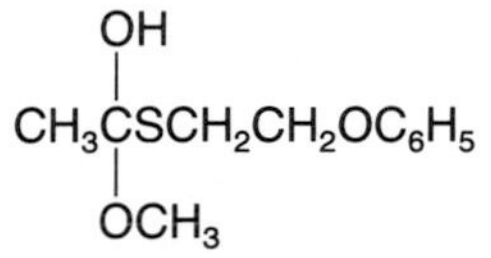

21.17 (b) $CH_3COCCH_3 + 2CH_3NH_2 \longrightarrow CH_3CNHCH_3 + CH_3CO^- \ CH_3\overset{+}{N}H_3$

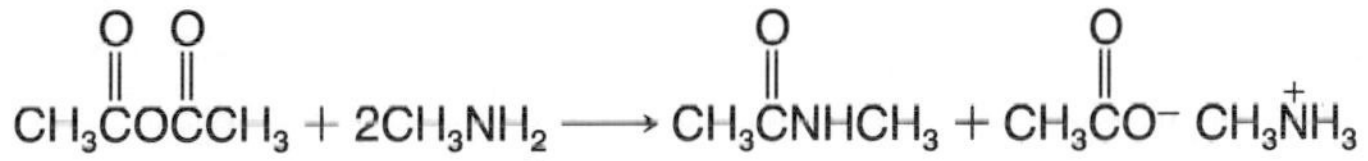

(c) $HCOCH_3 + HN(CH_3)_2 \longrightarrow HCN(CH_3)_2 + CH_3OH$

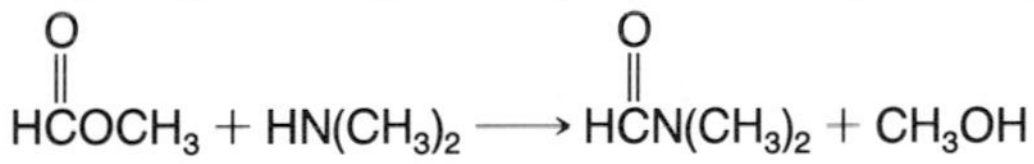

21.18

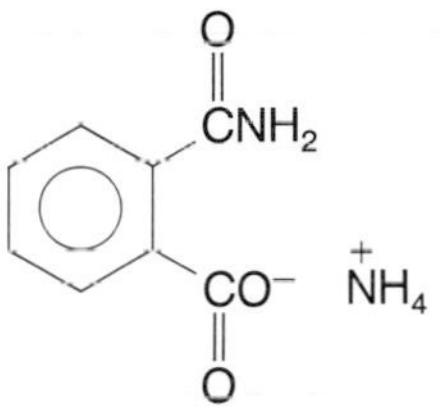

21.19

Step 1: Protonation of the carbonyl oxygen.

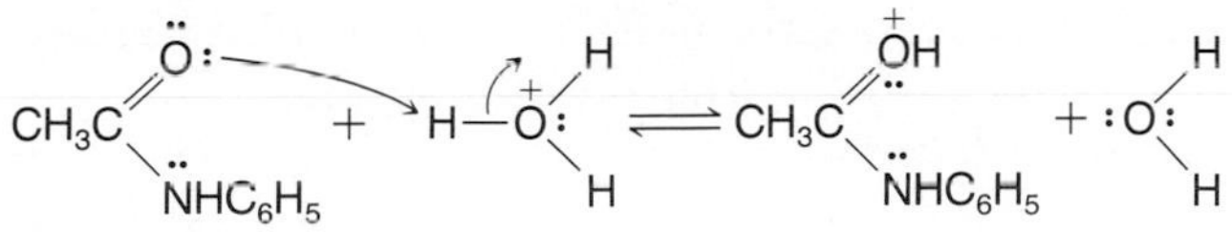

Step 2: Nucleophilic addition of water.

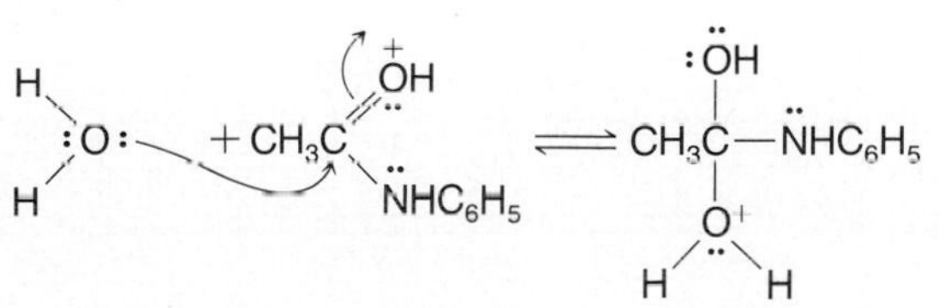

Step 3: Deprotonation of oxonium ion to give neutral form of tetrahedral intermediate.

$$CH_3\overset{\overset{\textstyle :\ddot{O}H}{|}}{\underset{\underset{\textstyle H}{\overset{\textstyle +}{\ddot{O}}}}{C}}{-}\ddot{N}HC_6H_5 + \ \ :\ddot{O}: \quad \rightleftharpoons \quad CH_3\overset{\overset{\textstyle :\ddot{O}H}{|}}{\underset{\underset{\textstyle :\ddot{O}H}{|}}{C}}{-}\ddot{N}HC_6H_5 + \ H{-}\overset{\overset{\textstyle H}{|}}{\underset{\underset{\textstyle H}{}}{\overset{+}{O}}}:$$

Step 4: Protonation of amino group of tetrahedral intermediate.

$$CH_3\overset{\overset{\textstyle :\ddot{O}H}{|}}{\underset{\underset{\textstyle :\ddot{O}H}{|}}{C}}{-}\ddot{N}HC_6H_5 + H{-}\overset{\overset{\textstyle H}{|}}{\underset{\underset{\textstyle H}{}}{\overset{+}{O}}}: \quad \rightleftharpoons \quad CH_3\overset{\overset{\textstyle :\ddot{O}H}{|}}{\underset{\underset{\textstyle :\ddot{O}H}{|}}{C}}{-}\overset{\overset{\textstyle H}{|}}{\underset{\underset{\textstyle H}{|}}{\overset{+}{N}}}C_6H_5 + \ :\overset{\overset{\textstyle H}{}}{\underset{\underset{\textstyle H}{}}{\ddot{O}}}:$$

Step 5: Dissociation of *N*-protonated form of tetrahedral intermediate.

$$CH_3\overset{\overset{\textstyle :\ddot{O}H}{|}}{\underset{\underset{\textstyle :\ddot{O}H}{|}}{C}}{-}\overset{\overset{\textstyle H}{|}}{\underset{\underset{\textstyle H}{|}}{\overset{+}{N}}}C_6H_5 \quad \rightleftharpoons \quad CH_3\overset{\overset{\textstyle +}{\overset{\textstyle OH}{}}}{\underset{\underset{\textstyle \ddot{O}H}{}}{C}} + \ H_2\ddot{N}C_6H_5$$

Step 6: Proton transfer processes.

$$\overset{\textstyle H}{\underset{\textstyle H}{\searrow}}\overset{+}{\ddot{O}}{-}H \ \longleftarrow \ + \ H_2\ddot{N}C_6H_5 \quad \rightleftharpoons \quad \overset{\textstyle H}{\underset{\textstyle H}{\searrow}}\ddot{O}: \ + \ H_3\overset{+}{N}C_6H_5$$

$$CH_3\overset{\overset{\textstyle +}{\overset{\textstyle O}{\|}}{-}H}{\underset{\underset{\textstyle \ddot{O}H}{}}{C}} \ + \ \ :\overset{\overset{\textstyle H}{}}{\underset{\underset{\textstyle H}{}}{\ddot{O}}}: \quad \rightleftharpoons \quad CH_3\overset{\overset{\textstyle :O:}{\|}}{\underset{\underset{\textstyle \ddot{O}H}{}}{C}} \ + \ H{-}\overset{\overset{\textstyle H}{}}{\underset{\underset{\textstyle H}{}}{\overset{+}{\ddot{O}}}}:$$

21.20

Step 1: Nucleophilic addition of hydroxide ion to the carbonyl group.

$$HO\overset{-}{:} \quad + \ \ H\overset{\overset{\textstyle :O:}{\|}}{C}N(CH_3)_2 \ \longrightarrow \ H\overset{\overset{\textstyle :\overset{-}{O}:}{|}}{\underset{\underset{\textstyle :OH}{|}}{C}}{-}\ddot{N}(CH_3)_2$$

Step 2: Proton transfer to give neutral form of tetrahedral intermediate.

$$H\overset{\overset{\textstyle :\overset{-}{O}:}{|}}{\underset{\underset{\textstyle :OH}{|}}{C}}{-}\ddot{N}(CH_3)_2 + \ \ H{-}\ddot{\underset{\textstyle }{O}}H \ \longrightarrow \ H\overset{\overset{\textstyle :\ddot{O}H}{|}}{\underset{\underset{\textstyle :OH}{|}}{C}}{-}\ddot{N}(CH_3)_2 + \ \overset{-}{:}\ddot{O}H$$

Step 3:   Proton transfer from water to nitrogen of tetrahedral intermediate.

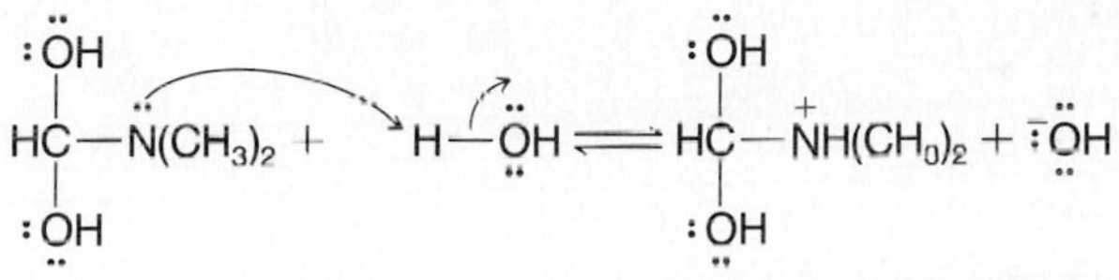

Step 4:   Dissociation of *N*-protonated form of tetrahedral intermediate.

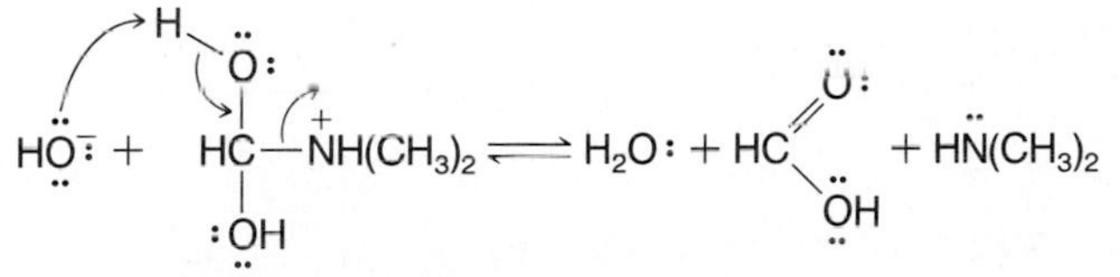

Step 5:   Irreversible formation of formate ion.

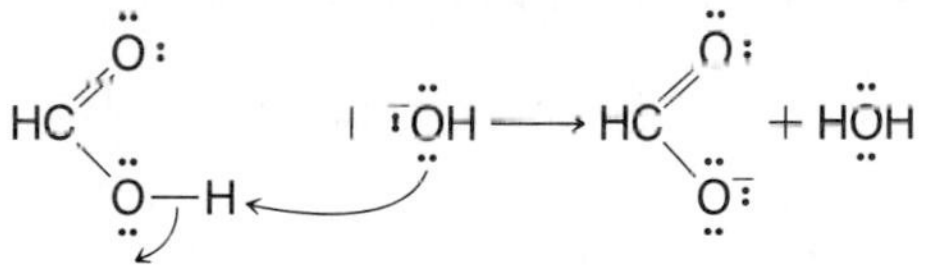

21.21 $CH_3CH_2CH_2CO_2H \xrightarrow[\text{2. NH}_3]{\text{1. SOCl}_2} CH_3CH_2CH_2\overset{\overset{\displaystyle O}{\|}}{C}NH_2 \xrightarrow[\text{H}_2\text{O, NaOH}]{\text{Br}_2} CH_3CH_2CH_2NH_2$

21.22   (*a*) $CH_3CH_2OH \xrightarrow{(C_5H_5N)_2CrO_3} CH_3\overset{\overset{\displaystyle O}{\|}}{C}H \xrightarrow{H_2NOH} CH_3CH{=}NOH \xrightarrow{P_4O_{10}} CH_3CN$

(*b*) $CH_3CH_2OH \xrightarrow[\text{or HBr}]{\text{PBr}_3} CH_3CH_2Br \xrightarrow{\text{NaCN}} CH_3CH_2CN$

CHAPTER 22

22.1   Ethyl benzoate cannot undergo the Claisen condensation.

Claisen condensation product of ethyl pentanoate:

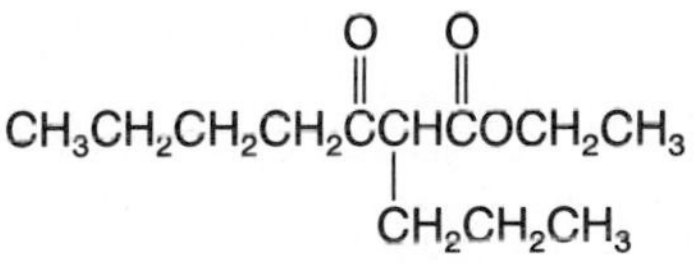

Claisen condensation product of ethyl phenylacetate:

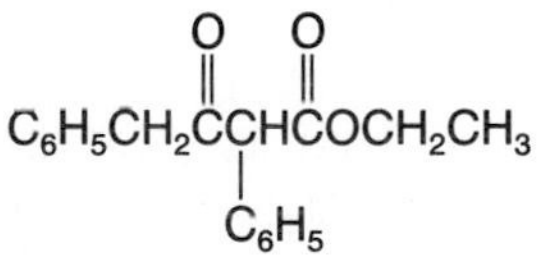

22.2   (*b*)

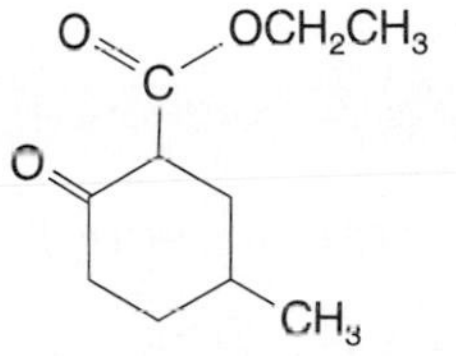

(*c*)

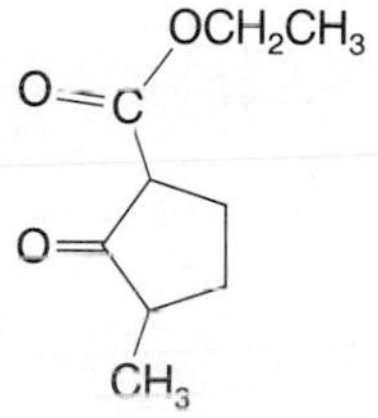

22.3   (*b*)

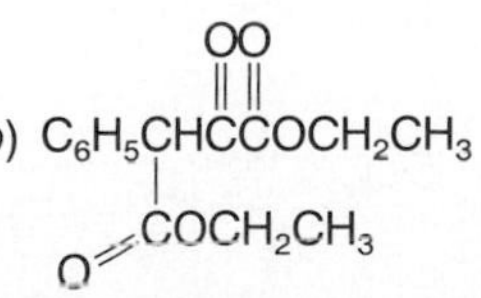

(*c*)

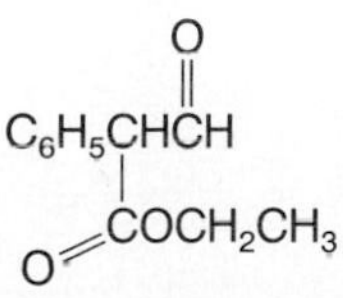

22.4

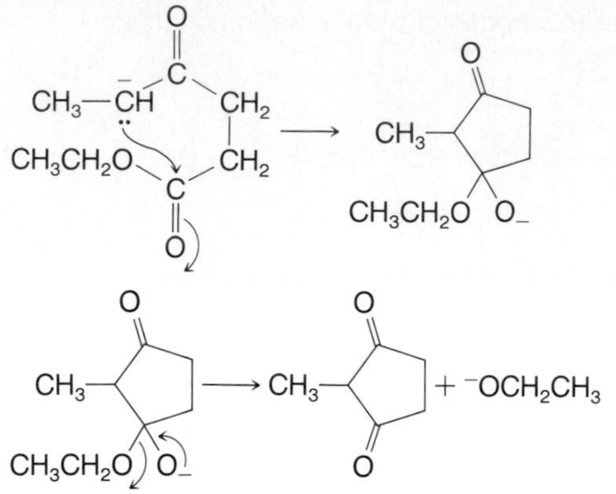

22.5 $CH_3CH_2OCCH_2CH_2CH_2CH_2COCH_2CH_3$ $\xrightarrow[\text{2. H}^+]{\text{1. NaOCH}_2\text{CH}_3}$

$\xrightarrow[\substack{\text{2. H}^+ \\ \text{3. heat}}]{\text{1. HO}^-, \text{H}_2\text{O}}$

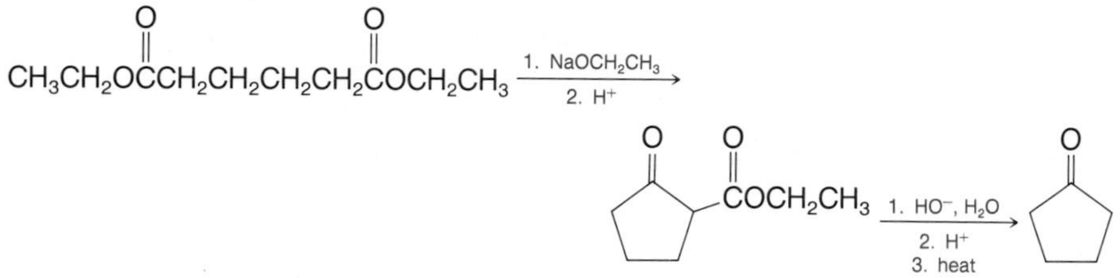

22.6 (b) $C_6H_5CH_2Br + CH_3CCH_2COCH_2CH_3$ $\xrightarrow[\substack{\text{2. HO}^-, \text{H}_2\text{O} \\ \text{3. H}^+ \\ \text{4. heat}}]{\text{1. NaOCH}_2\text{CH}_3}$ $C_6H_5CH_2CH_2CCH_3$

(c) $CH_2{=}CHCH_2Br + CH_3CCH_2COCH_2CH_3$ $\xrightarrow[\substack{\text{2. HO}^-, \text{H}_2\text{O} \\ \text{3. H}^+ \\ \text{4. heat}}]{\text{1. NaOCH}_2\text{CH}_3}$ $CH_2{=}CHCH_2CH_2CCH_3$

22.7 (b) $CH_3(CH_2)_5CH_2Br + CH_2(COOCH_2CH_3)_2$ $\xrightarrow[\text{ethanol}]{\text{NaOCH}_2\text{CH}_3}$

$CH_3(CH_2)_5CH_2CH(COOCH_2CH_3)_2$

$\downarrow \substack{\text{1. HO}^-, \text{H}_2\text{O} \\ \text{2. H}^+ \\ \text{3. heat}}$

$CH_3(CH_2)_5CH_2CH_2COH$

(c) $CH_3CH_2\underset{\underset{CH_3}{|}}{C}HCH_2Br + CH_2(COOCH_2CH_3)_2$ $\xrightarrow[\text{ethanol}]{\text{NaOCH}_2\text{CH}_3}$ $CH_3CH_2\underset{\underset{CH_3}{|}}{C}HCH_2CH(COOCH_2CH_3)_2$

$\downarrow \substack{\text{1. HO}^-, \text{H}_2\text{O} \\ \text{2. H}^+ \\ \text{3. heat}}$

$CH_3CH_2\underset{\underset{CH_3}{|}}{C}HCH_2CH_2COH$

(d) $C_6H_5CH_2Br + CH_2(COOCH_2CH_3)_2$ $\xrightarrow[\text{ethanol}]{\text{NaOCH}_2\text{CH}_3}$ $C_6H_5CH_2CH(COOCH_2CH_3)_2$

$\downarrow \substack{\text{1. HO}^-, \text{H}_2\text{O} \\ \text{2. H}^+ \\ \text{3. heat}}$

$C_6H_5CH_2CH_2COH$

22.8 $CH_3\overset{O}{\overset{\|}{C}}CH_2\overset{O}{\overset{\|}{C}}OCH_2CH_3 \xrightarrow[CH_3Br]{NaOCH_2CH_3} CH_3\overset{O}{\overset{\|}{C}}\underset{CH_3}{\overset{}{C}H}\overset{O}{\overset{\|}{C}}OCH_2CH_3$

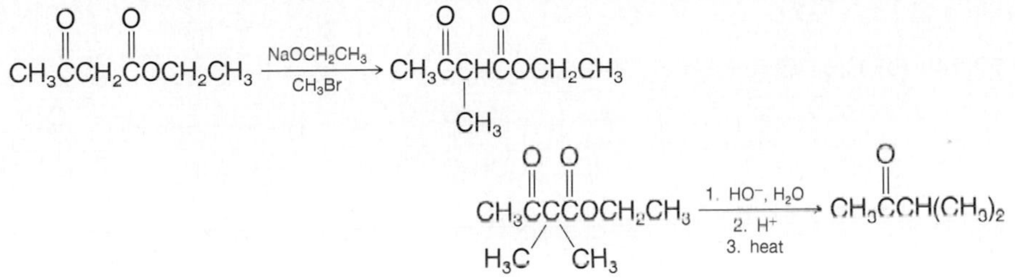

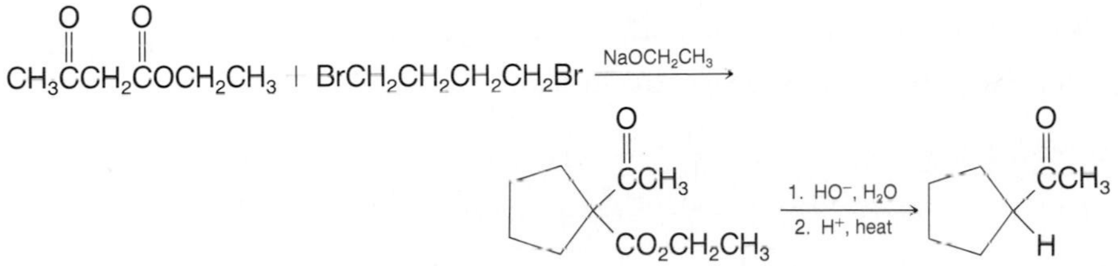

$CH_3\overset{O}{\overset{\|}{C}}\underset{H_3C\quad CH_3}{\overset{}{C}}\overset{O}{\overset{\|}{C}}OCH_2CH_3 \xrightarrow[\substack{2.\ H^+ \\ 3.\ heat}]{1.\ HO^-,\ H_2O} CH_3\overset{O}{\overset{\|}{C}}CH(CH_3)_2$

22.9 $CH_3\overset{O}{\overset{\|}{C}}CH_2\overset{O}{\overset{\|}{C}}OCH_2CH_3 + BrCH_2CH_2CH_2CH_2Br \xrightarrow{NaOCH_2CH_3}$

$\xrightarrow[\substack{2.\ H^+,\ heat}]{1.\ HO^-,\ H_2O}$

22.10 $CH_2(COOCH_2CH_3)_2 \xrightarrow[\substack{2.\ NaOCH_2CH_3,\ CH_3CH_2Br}]{\substack{1.\ NaOCH_2CH_3,\ CH_3CH_2CH_2CHCH_3 \\ Br}}$

$+ H_2N\overset{O}{\overset{\|}{C}}NH_2 \longrightarrow$

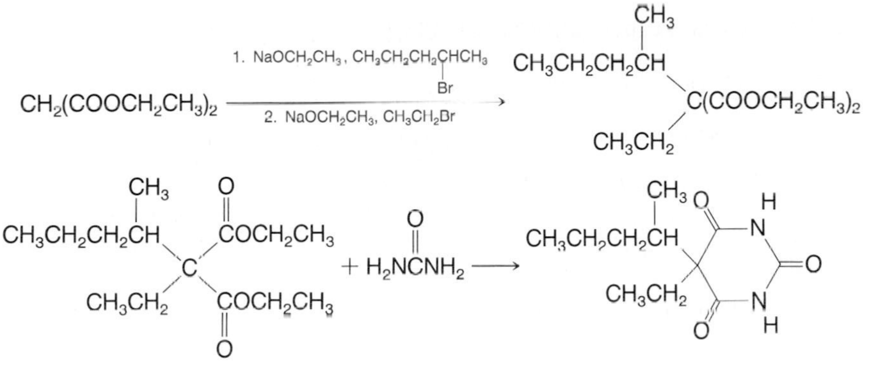

22.11

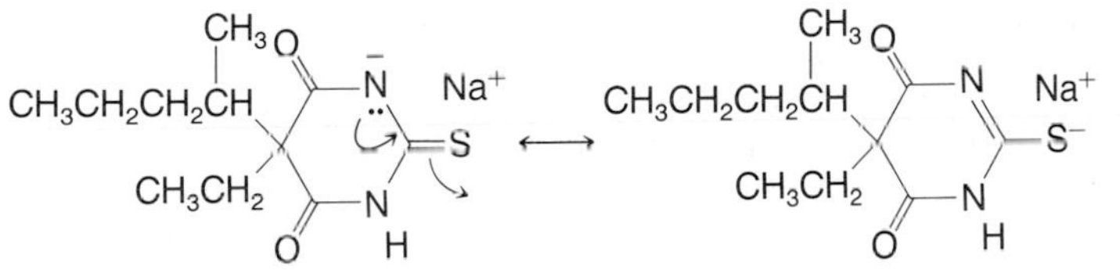

22.12 $C_6H_5CH_2\overset{O}{\overset{\|}{C}}CH_2CH_3 + CH_3CH_2O\overset{O}{\overset{\|}{C}}OCH_2CH_3 \xrightarrow{NaOCH_2CH_3} C_6H_5CH(COOCH_2CH_3)_2$

$C_6H_5CH(COOCH_2CH_3)_2 \xrightarrow[CH_3CH_2Br]{NaOCH_2CH_3} C_6H_5\underset{CH_2CH_2}{\overset{}{C}}(COOCH_2CH_3)_2 \xrightarrow{H_2NCNH_2}$

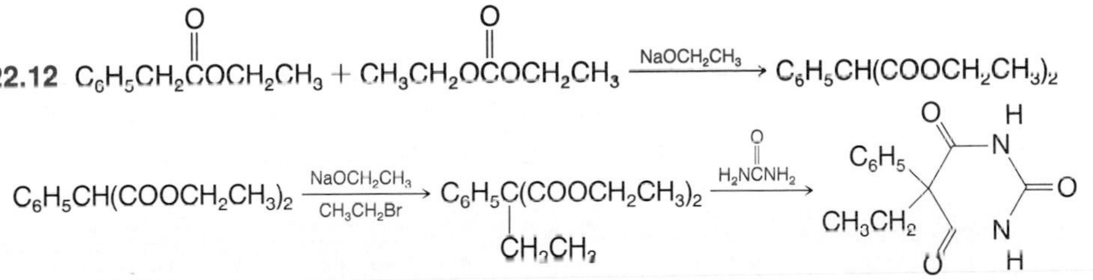

22.13 (b) $(CH_3CH_2OOC)_2CHCH_2CH_2\overset{O}{\overset{\|}{C}}OCH_2CH_3$ (c) $(CH_3CH_2OOC)_2CHCHCH_2\overset{O}{\overset{\|}{C}}OCH_2CH_3$
$\underset{CH_3}{}$

(d) $(CH_3CH_2OOC)_2CHCHCH_2COOCH_2CH_3$ (e)
$\overset{COOCH_2CH_3}{|}$

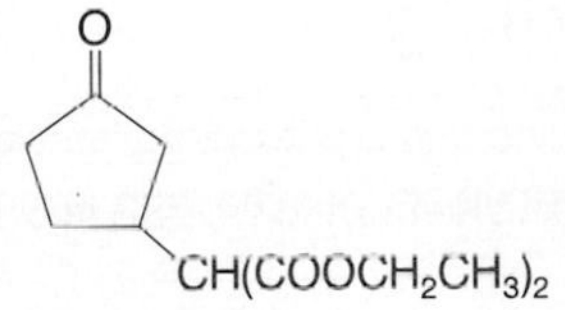

22.14 (b) $CH_3\overset{O}{\overset{\|}{C}}CH_3 + BrCH_2\overset{O}{\overset{\|}{C}}OCH_2CH_3 \xrightarrow[\text{2. H}_2\text{O}]{\text{1. Zn}}$

$$(CH_3)_2\underset{\underset{OH}{|}}{C}CH_2\overset{O}{\overset{\|}{C}}OCH_2CH_3 \xrightarrow[\text{heat}]{\text{H}_2\text{SO}_4} (CH_3)_2C=CH\overset{O}{\overset{\|}{C}}OCH_2CH_3$$

Then $(CH_3)_2C=CH\overset{O}{\overset{\|}{C}}OCH_2CH_3 \xrightarrow[\text{2. H}^+]{\text{1. HO}^-,\ \text{H}_2\text{O}} (CH_3)_2C=CHCO_2H$

(c) Reduce the hydroxy ester formed in (b).

$$(CH_3)_2\underset{\underset{OH}{|}}{C}CH_2\overset{O}{\overset{\|}{C}}OCH_2CH_3 \xrightarrow[\text{2. H}_2\text{O}]{\text{1. LiAlH}_4} (CH_3)_2\underset{\underset{OH}{|}}{C}CH_2CH_2OH$$

(d) Same as (c) except $CH_3\underset{\underset{Br}{|}}{C}HCOOCH_2CH_3$ is used in (b) instead of ethyl bromoacetate.

22.15 (b) $C_6H_5CH_2CO_2CH_3 \xrightarrow[\text{2. CH}_3\text{I}]{\text{1. LDA, THF}} C_6H_5\underset{\underset{CH_3}{|}}{C}HCO_2CH_3$

(c)

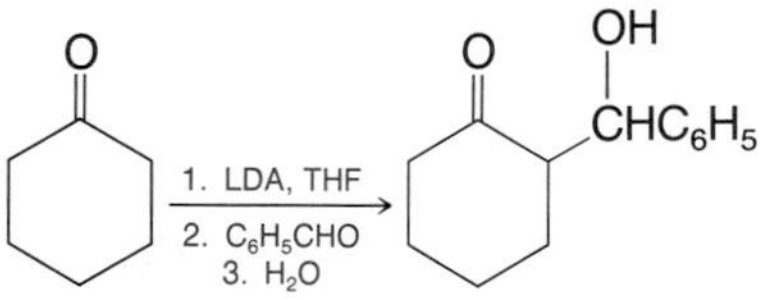

(d) $CH_3CO_2C(CH_3)_3 \xrightarrow[\substack{\text{2. cyclohexanone}\\\text{3. H}_2\text{O}}]{\text{1. LDA, THF}}$

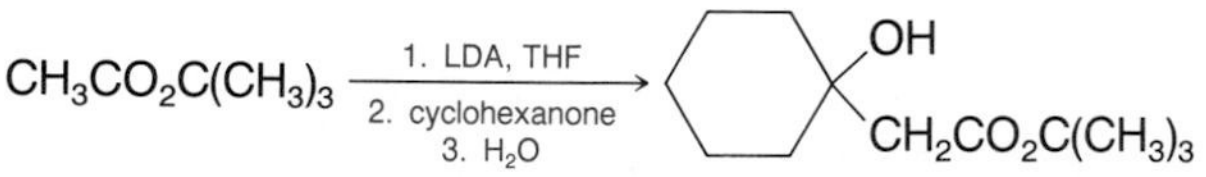

CHAPTER 23

23.1 (b) 1-Phenylethanamine or 1-phenylethylamine; (c) 2,2-dimethylpropanamine or neopentylamine; (d) *trans*-2-methylcyclopropanamine or *trans*-2-methylcyclopropylamine; (e) 2-propen-1-amine or allylamine

23.2 (b) *N,N*-Dimethylcycloheptanamine; (c) *N*-ethyl-*N*-methyl-2-methyl-1-propanamine

23.3 *R*

23.4 (b) $F_2CHCH_2NH_2$; (c) $CH_3CH_2CH_2CH_2NH_2$

23.5 $CH_2=CHCH_3 \xrightarrow[400°C]{Cl_2} CH_2=CHCH_2Cl \xrightarrow{NH_3} CH_2=CHCH_2NH_2$

23.6 Isobutylamine and 2-phenylethylamine can be prepared by the Gabriel synthesis. *tert*-Butylamine, *N*-methylbenzylamine, and aniline cannot.

(b) $(CH_3)_2CHCH_2Br$ + [phthalimide potassium salt] $\longrightarrow$ [N-alkyl phthalimide, $NCH_2CH(CH_3)_2$]

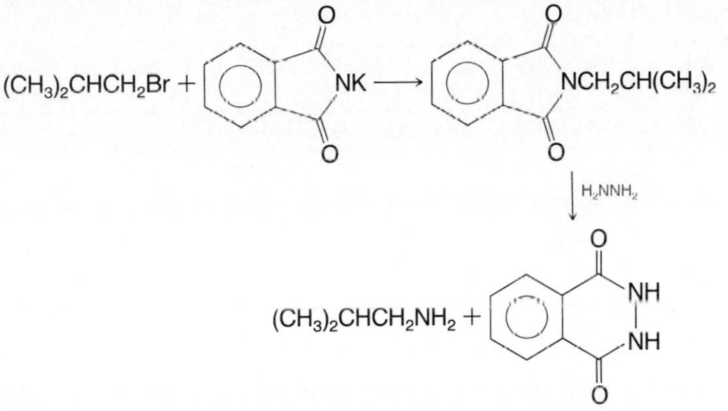

$\xrightarrow{H_2NNH_2}$

$(CH_3)_2CHCH_2NH_2$ + [phthalazine-1,4-dione]

(d) $C_6H_5CH_2CH_2Br$ + [phthalimide potassium salt] $\longrightarrow$ [N-alkyl phthalimide, $NCH_2CH_2C_6H_5$]

$\xrightarrow{H_2NNH_2}$

$C_6H_5CH_2CH_2NH_2$ + [phthalazine-1,4-dione]

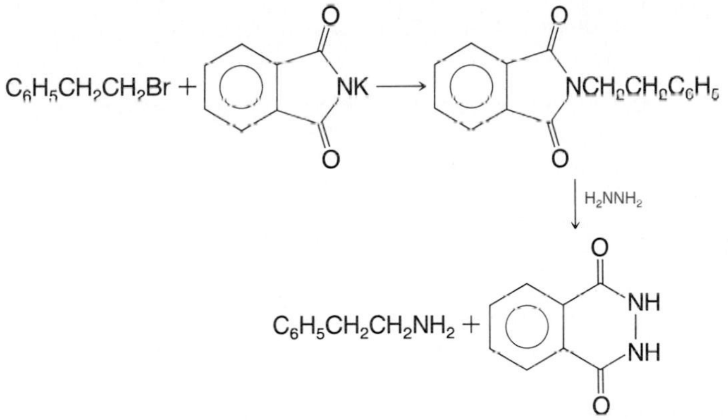

23.7 (b) $CH_3CH_2CH_2CH_2OH \xrightarrow[\text{or } PBr_3]{HBr} CH_3CH_2CH_2CH_2Br \xrightarrow{NaCN} CH_3CH_2CH_2CH_2CN$

(c) $F_3C-\!\!\!\bigcirc\!\!\!-CO_2H \xrightarrow[\text{2. } H_2O]{\text{1. LiAlH}_4} F_3C-\!\!\!\bigcirc\!\!\!-CH_2OH \xrightarrow{HBr}$

$F_3C-\!\!\!\bigcirc\!\!\!-CH_2Br \xrightarrow{NaCN} F_3C-\!\!\!\bigcirc\!\!\!-CH_2CN$

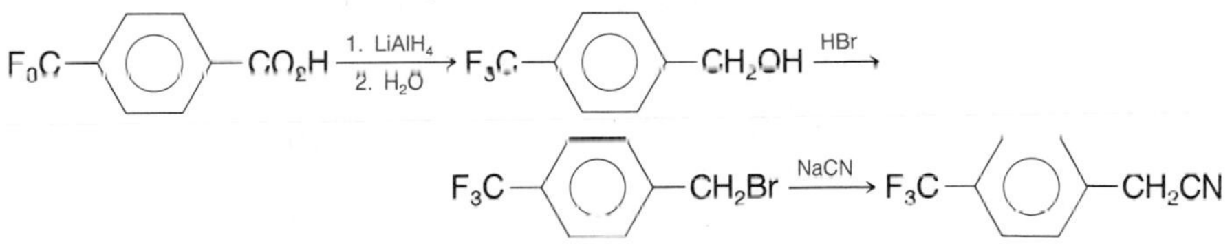

(d) $CH_3(CH_2)_5\underset{\underset{OH}{|}}{C}HCH_3 \xrightarrow[H_2SO_4]{Na_2Cr_2O_7} CH_3(CH_2)_5\overset{O}{\overset{||}{C}}CH_3 \xrightarrow{H_2NOH} CH_3(CH_2)_5\overset{NOH}{\overset{||}{C}}CH_3$

(e) $CH_3CH_2CH_2CH_2CO_2H \xrightarrow{SOCl_2} CH_3CH_2CH_2CH_2\overset{O}{\overset{||}{C}}Cl \xrightarrow{CH_3CH_2CH_2CH_2NH_2}$

$CH_3CH_2CH_2CH_2\overset{O}{\overset{||}{C}}NHCH_2CH_2CH_2CH_3$

23.8 (b) $C_6H_5\overset{O}{\overset{||}{C}}H + C_6H_5CH_2NH_2 \xrightarrow{H_2,\,Ni} C_6H_5CH_2NHCH_2C_6H_5$

(c) $C_6H_5\overset{O}{\overset{||}{C}}H + (CH_3)_2NH \xrightarrow{H_2,\,Ni} C_6H_5CH_2N(CH_3)_2$

(d) $C_6H_5\overset{O}{\overset{||}{C}}H + HN\!\!\bigcirc \xrightarrow{H_2,\,Ni} C_6H_5CH_2-N\!\!\bigcirc$

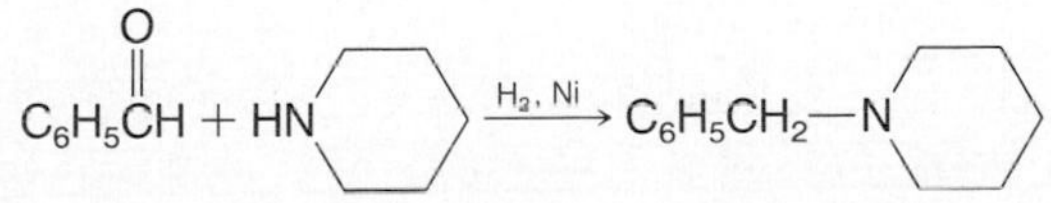

23.9 $(CH_3)_3\overset{+}{N}CH_2CH_2OH$ HO^-

23.10 (b) Two; (c) two; (d) three

23.11 (b) $(CH_3)_3CCH_2C=CH_2$; (c) $CH_2=CH_2$; (d) $CH_3CH=CH_2$
  |
  CH_3

23.12

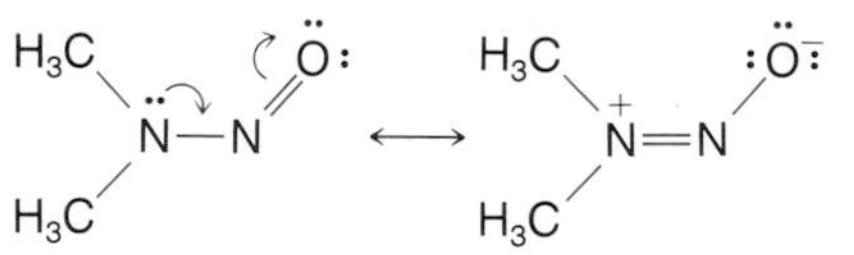

23.13 The diazonium ion from neopentylamine rearranges via a methyl shift on loss of nitrogen to give *tert*-pentyl cation.

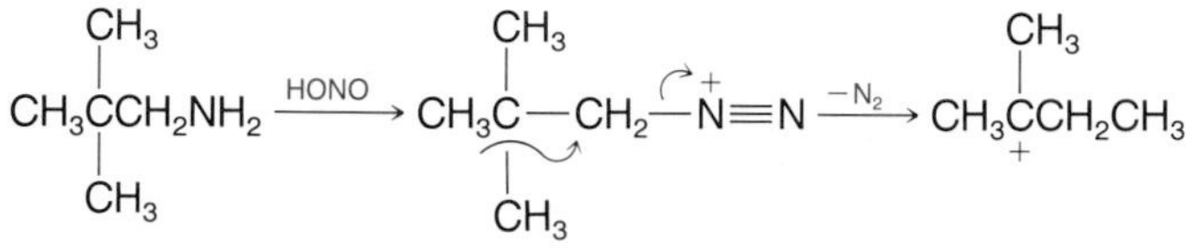

CHAPTER 24

24.1 (b) *N*-Ethyl-4-isopropyl-*N*-methylaniline; (c) 5-amino-2-fluorobenzyl alcohol

24.2 Electron release from the amino group into the ring gives partial double bond character to the carbon-nitrogen bond.

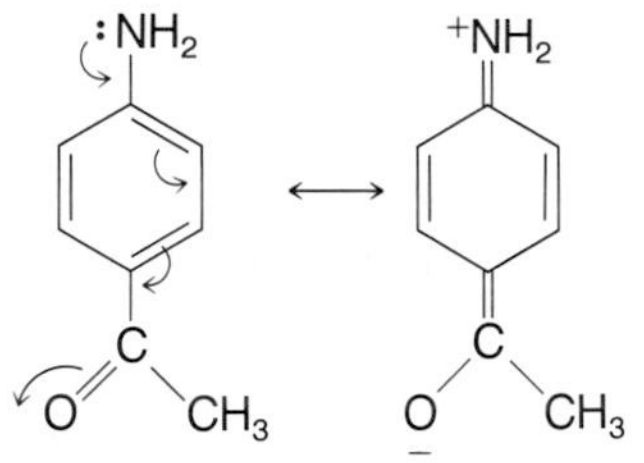

24.3 Hydrogen bonds between the amino group and the carbonyl oxygen are intramolecular in *o*-aminoacetophenone. Intermolecular hydrogen bonds in *p*-aminoacetophenone cause molecules to associate in the liquid phase, and these attractive forces must be broken in order for molecules to escape into the vapor phase.

24.4 *p*-Dinitrobenzene has no dipole moment. *p*-Nitroaniline has a substantial dipole moment; conjugation between the amino group and the nitro group leads to charge separation within the molecule.

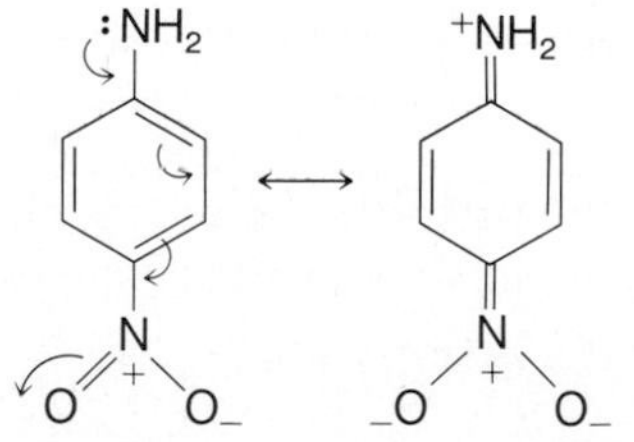

24.5 Tetrahydroisoquinoline is a stronger base than tetrahydroquinoline. The unshared electron pair of tetrahydroquinoline is delocalized into the aromatic ring, and this substance resembles aniline in its basicity, whereas tetrahydroisoquinoline resembles an alkylamine.

24.6 (b) The amino group is conjugated to the cyano group through the aromatic ring.

(c) The lone pair of nitrogen is delocalized into the carbonyl group by amide resonance.

(d) The amino group is conjugated to the carbonyl group through the aromatic ring.

24.7 (b) Prepare p-isopropylnitrobenzene as in (a), then reduce with H_2, Ni (or Fe + HCl or Sn + HCl, followed by base). (c) Prepare isopropylbenzene as in (a), then dinitrate with $HNO_3 + H_2SO_4$, then reduce both nitro groups; (d) Chlorinate benzene with $Cl_2 + FeCl_3$, then nitrate (HNO_3, H_2SO_4), separate the desired para isomer from the unwanted ortho isomer, and reduce. (e) Acetylate benzene by a Friedel-Crafts reaction (acetyl chloride + $AlCl_3$), then nitrate (HNO_3, H_2SO_4), then reduce the nitro group.

24.8 (b) Prepare acetanilide as in (a), dinitrate (HNO_3, H_2SO_4), then hydrolyze the amide in either acid or base. (c) Prepare p-nitroacetanilide as in (a), then reduce the nitro group with H_2 (or Fe + HCl or Sn + HCl, followed by base).

24.9 Intermediates: benzene to nitrobenzene to m-bromonitrobenzene to m-bromoaniline to m-bromophenol. Reagents: HNO_3, H_2SO_4; Br_2, $FeBr_3$; Fe, HCl then HO^-; $NaNO_2$, H_2SO_4, H_2O then heat in H_2O.

24.10 Prepare m-bromoaniline as in Problem 24.9; then $NaNO_2$, HCl, H_2O followed by KI.

24.11 Intermediates: benzene to propiophenone to m-nitropropiophenone to m-aminopropiophenone to m-fluoropropiophenone. Reagents: propionyl chloride, $AlCl_3$; HNO_3, H_2SO_4; Fe, HCl then HO^-; $NaNO_2$, H_2O, HCl then HBF_4 then heat.

24.12 Intermediates: isopropylbenzene to p-isopropylnitrobenzene to p-isopropylaniline to p-isopropylacetanilide to 4-isopropyl-2-nitroacetanilide to 4-isopropyl-2-nitroaniline to m-isopropylnitrobenzene. Reagents: HNO_3, H_2SO_4; Fe, HCl then HO^-; acetyl chloride, $AlCl_3$; HNO_3, H_2SO_4; acid or base hydrolysis; $NaNO_2$, HCl, H_2O and CH_3CH_2OH or H_3PO_2.

24.13 Treat *p*-acetamidobenzenesulfonyl chloride with (*b*) aniline;

(*c*)

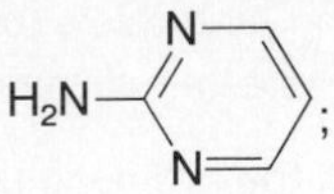

; (*d*)

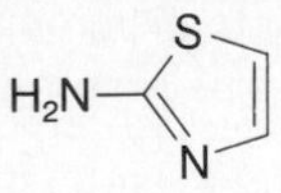

Complete the synthesis by hydrolysis of the acetyl group.

CHAPTER 25

25.1 $C_6H_5CH_2Cl$

25.2 (*b*)

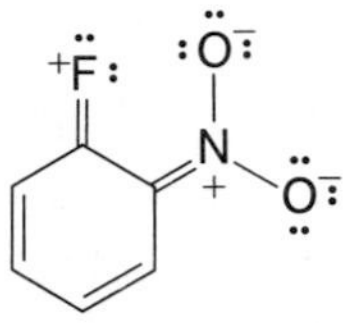

(*c*)

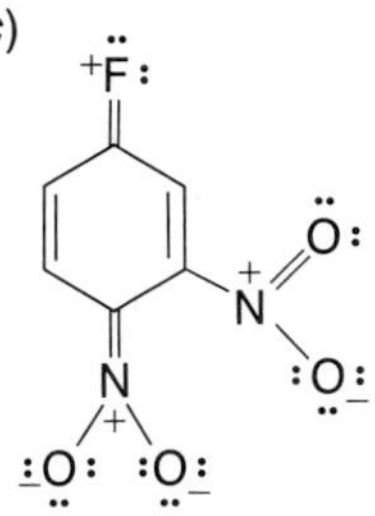

25.3 (*b*)

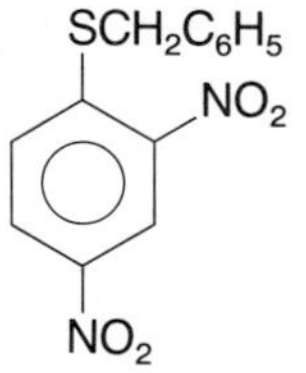

$SCH_2C_6H_5$

(*c*)

$NHCH_3$

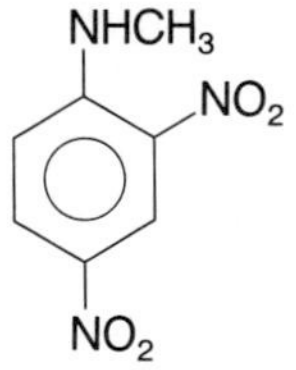

25.4

OCH_2CH_3

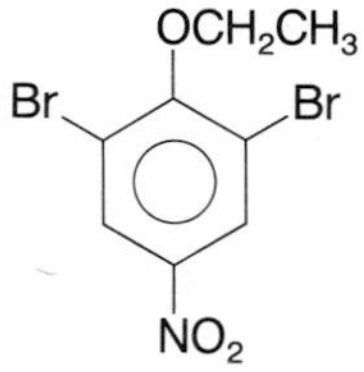

25.5 Its additional electron-withdrawing group makes 1-chloro-2-nitro-4-(trifluoromethyl)-benzene more reactive.

25.6 A benzyne intermediate is impossible because neither of the carbons ortho to the intended leaving group bears a proton.

25.7 3-Methylphenol and 4-methylphenol (*m*-cresol and *p*-cresol)

25.8 (*b*) Mixture of *o*-, *m*-, and *p*-methylaniline; (*c*) Mixture of *m*- and *p*-methylaniline

CHAPTER 26

26.1 (*b*)

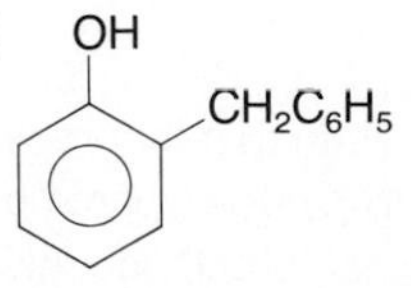

OH
$CH_2C_6H_5$

(*c*)

OH
NO_2

(*d*)

OH
OH
Cl

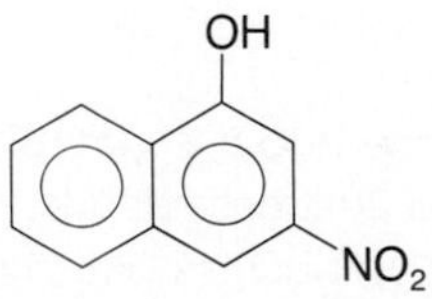

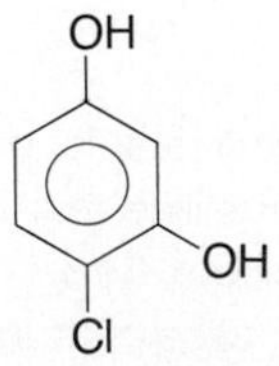

26.2 The hydroxyl group and the nitro group are directly conjugated.

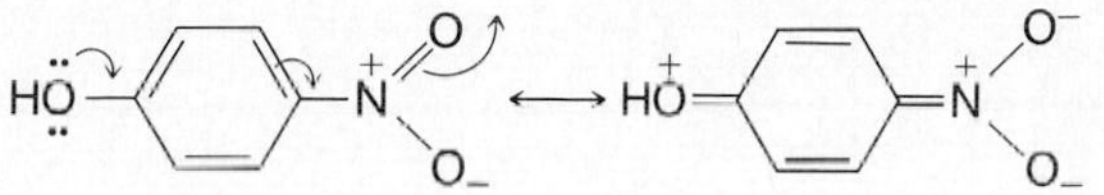

26.3 Methyl salicylate is the methyl ester of *o*-hydroxybenzoic acid. Intramolecular (rather than intermolecular) hydrogen bonding is responsible for its relatively low boiling point.

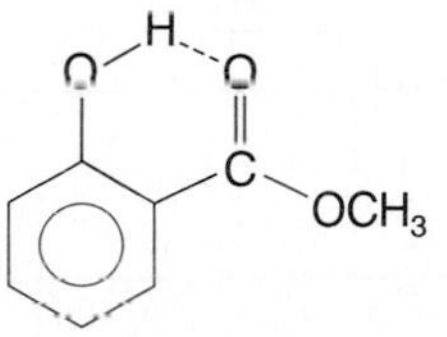

26.4 (*b*) *p*-Cyanophenol is stronger acid because of conjugation of cyano group with phenoxide oxygen. (*c*) *o*-Fluorophenol is stronger acid because electronegative fluorine substituent can stabilize negative charge better when fewer bonds intervene between it and the phenoxide oxygen.

26.5

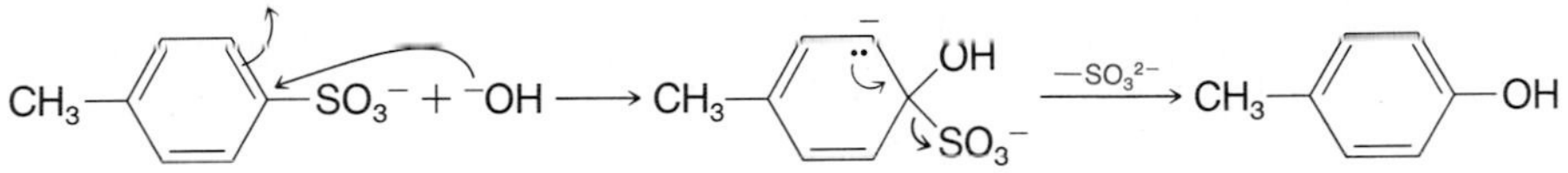

26.6

then

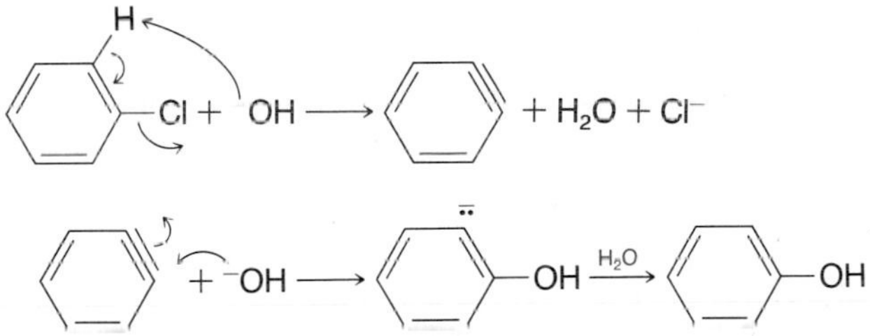

26.7 (*b*)

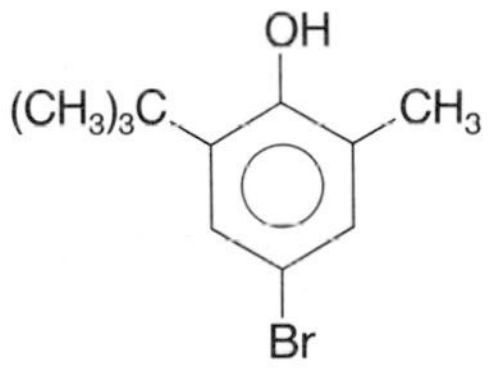

(*c*)

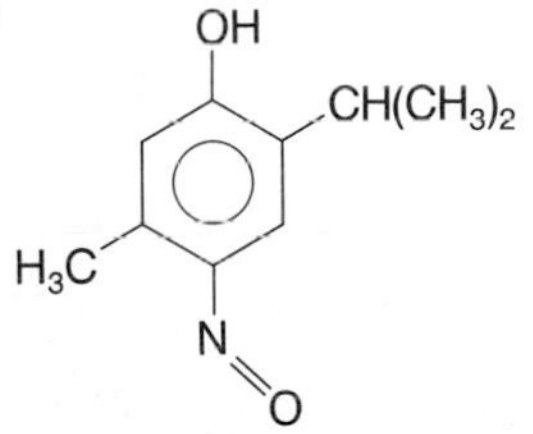

(*d*)

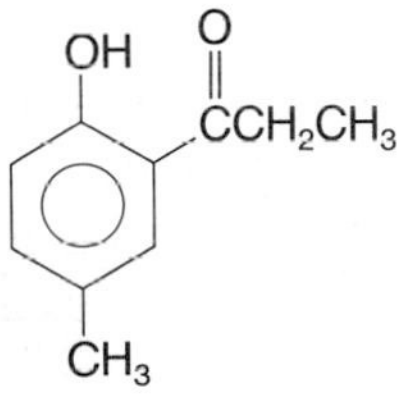

26.8 (*b*)

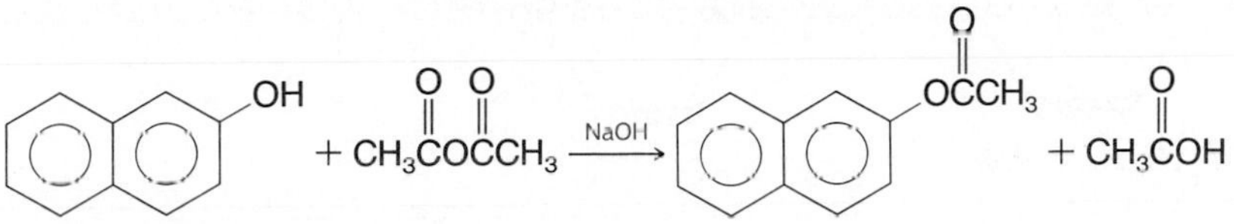

(*c*) $C_6H_5OH + C_6H_5\overset{O}{\overset{\|}{C}}Cl \longrightarrow C_6H_5O\overset{O}{\overset{\|}{C}}C_6H_5 + HCl$

26.9 $C_6H_5OCH_2\underset{\underset{OH}{|}}{C}HCH_3$

26.10 *p*-Fluoronitrobenzene and phenol (as its sodium or potassium salt)

26.11

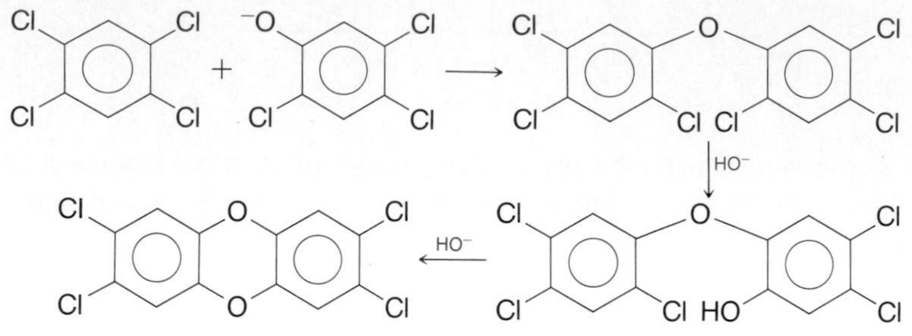

CHAPTER 27

27.1 (*b*) L-Glyceraldehyde; (*c*) D-glyceraldehyde

27.2 L-Erythrose

27.3

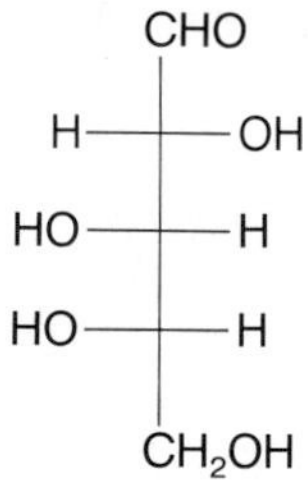

27.4 L-Talose

27.5 (*b*)

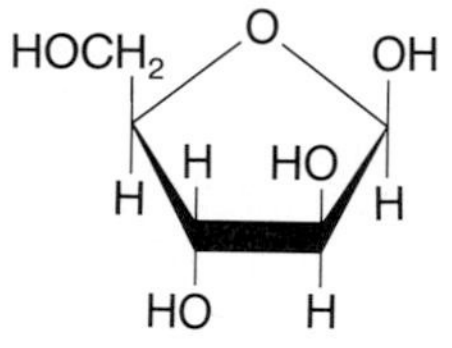

and

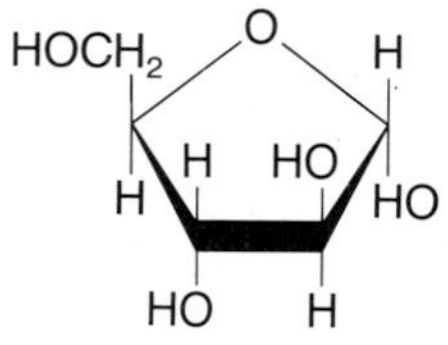

(*c*)

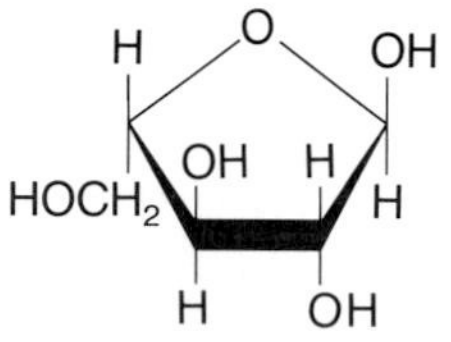

and

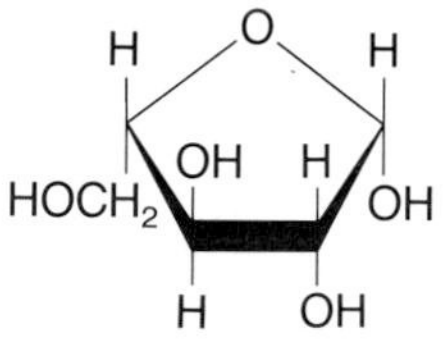

(*d*)

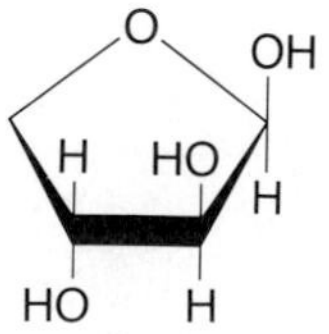

and

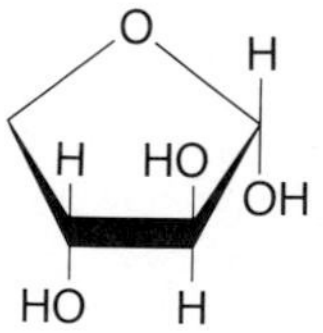

27.6 (*b*)

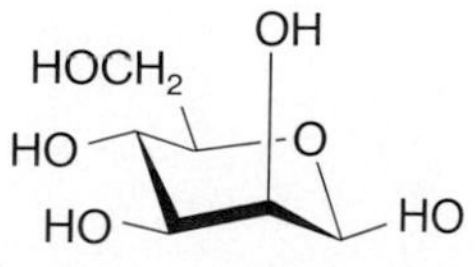

(*c*)

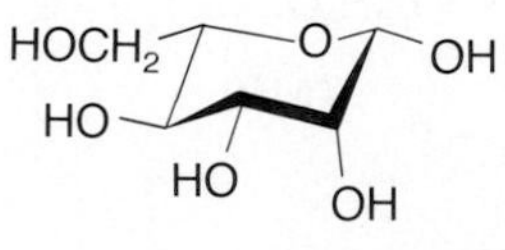

(*d*)

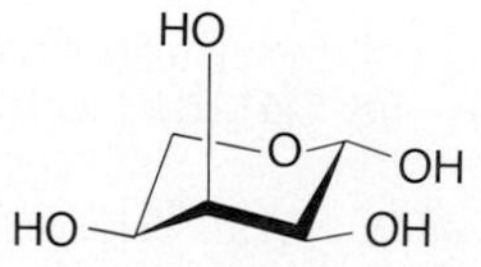

27.7 67% α, 33% β

27.8

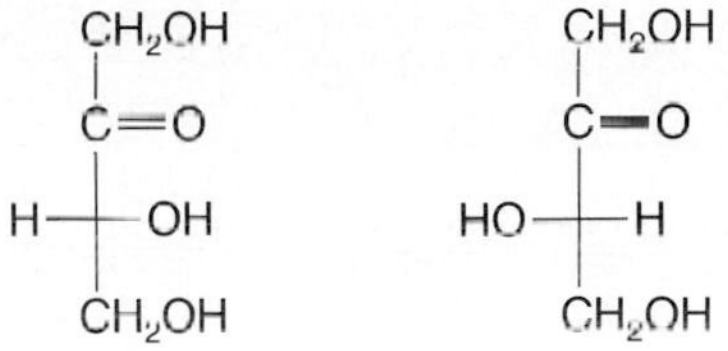

CH₂OH
|
C=O
|
H——OH
|
CH₂OH

CH₂OH
|
C=O
|
HO——H
|
CH₂OH

27.9

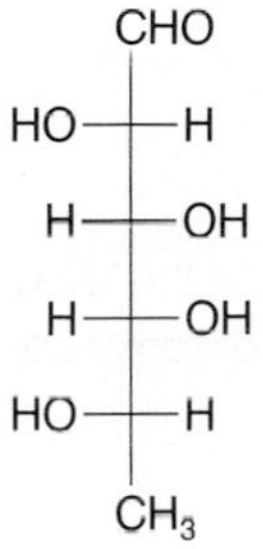

CHO
|
HO——H
|
H——OH
|
H——OH
|
HO——H
|
CH₃

27.10

$$\xrightarrow{} \xrightarrow{CH_3OH} \;+\; \beta$$

H—Cl

α

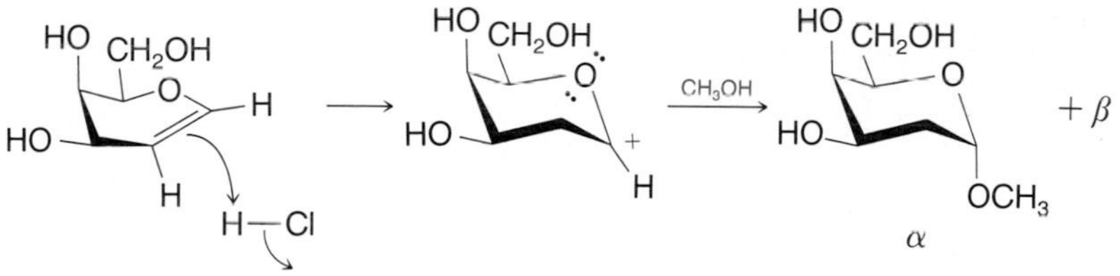

27.11

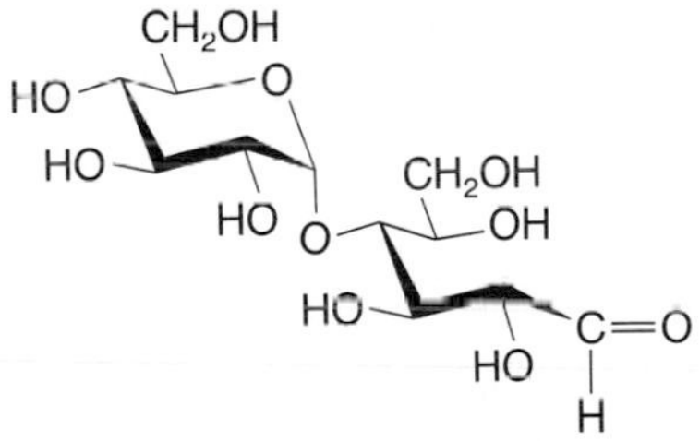

27.12 Maltose

27.13 No. The product is a meso form.

27.14 All (*b*) through (*f*) will give positive tests.

27.15 L-Gulose

27.16 (*b*) L-Erythrose; (*c*) D-Altrose; (*d*) 6-Deoxy-L-talose

27.17 The intermediate is an enol, $HOCH=CCH_2OP(OH)_2$
                                                   |
                                                  OH

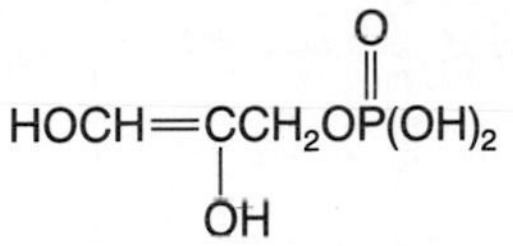

27.18 (*b*) Four equivalents of periodic acid are required. One molecule of formaldehyde and four molecules of formic acid are formed from each molecule of D-ribose.

(*c*) Two equivalents

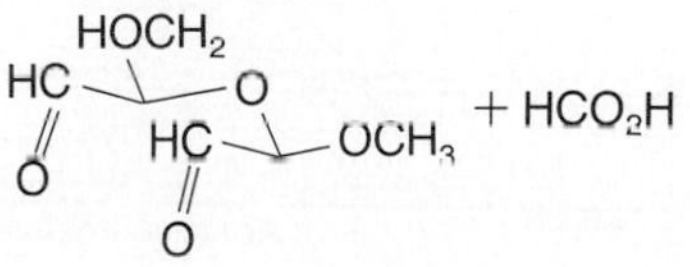

$+\ HCO_2H$

(*d*) Two equivalents

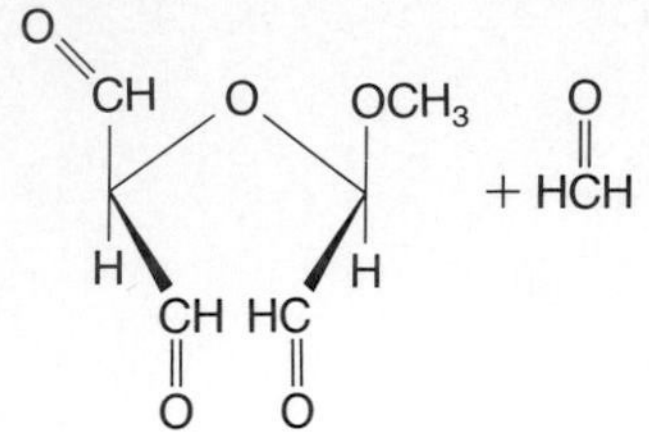

CHAPTER 28

28.1 Hydrolysis gives $CH_3(CH_2)_{16}CO_2H$ (two moles) and (Z)—$CH_3(CH_2)_7$-$CH{=}CH(CH_2)_7CO_2H$ (one mole). The same mixture of products is formed from 1-oleyl-2,3-distearoylglycerol.

28.2 $CH_3\overset{OH}{\underset{S-ACP}{C}}{-}SCoA$

28.3 $CH_3(CH_2)_{12}\overset{O}{\overset{\|}{C}}S{-}ACP$ $\quad CH_3(CH_2)_{12}\overset{O}{\overset{\|}{C}}CH_2\overset{O}{\overset{\|}{C}}S{-}ACP$

$CH_3(CH_2)_{12}\underset{OH}{CH}CH_2\overset{O}{\overset{\|}{C}}S{-}ACP$ $\quad CH_3(CH_2)_{12}CH{=}CH\overset{O}{\overset{\|}{C}}S{-}ACP$

$CH_3(CH_2)_{12}CH_2CH_2\overset{O}{\overset{\|}{C}}S{-}ACP$

28.4 *R* in both cases

28.5 $CH_3(CH_2)_{14}\overset{O}{\overset{\|}{C}}O(CH_2)_{15}CH_3$

28.6 The biosynthetic precursor is 8(Z),11(Z),14(Z)—$CH_3(CH_2)_4CH{=}CHCH_2$-$CH{=}CHCH_2CH{=}CH_2(CH_2)_6CO_2H$. The biosynthetic scheme is the same as for PGE_2 except all the substances lack the 5, 6 double bond.

28.7

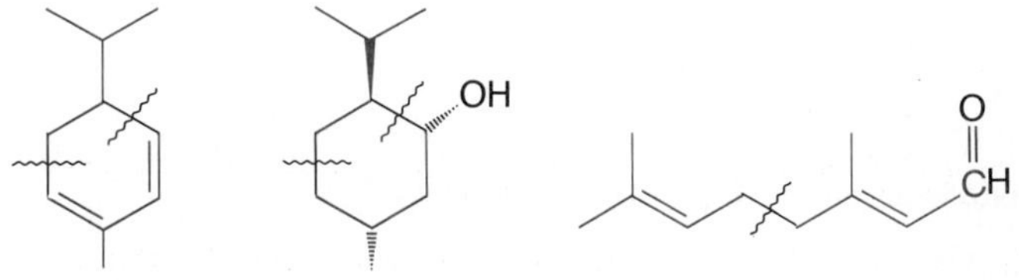

α-Phellandrene   Menthol   Citral

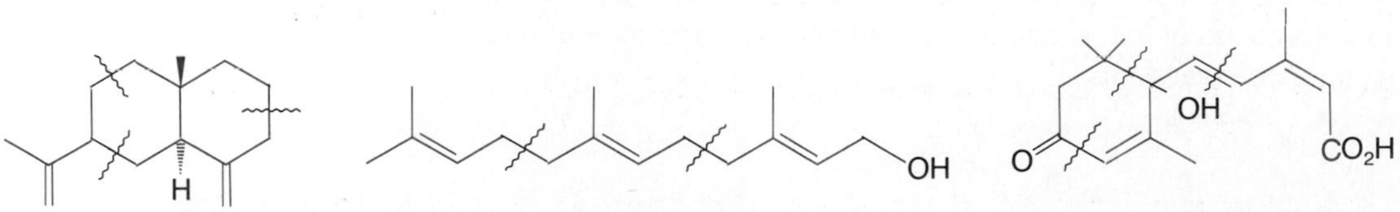

α-Selinene   Farnesol   Abscisic acid

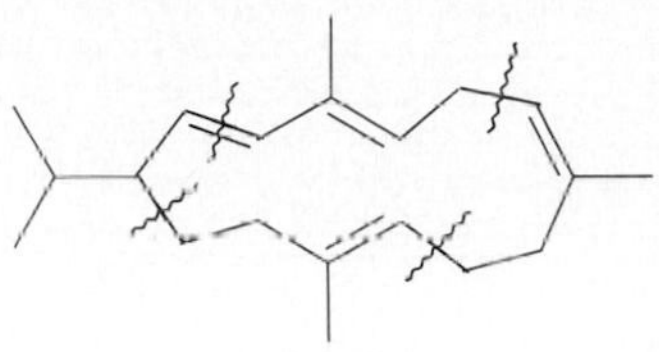

Cembrene

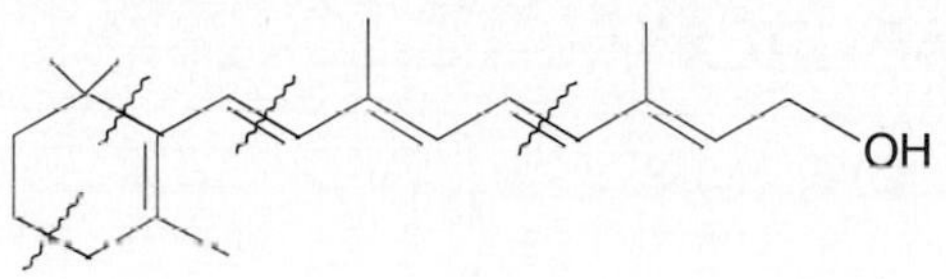

Vitamin A

28.8

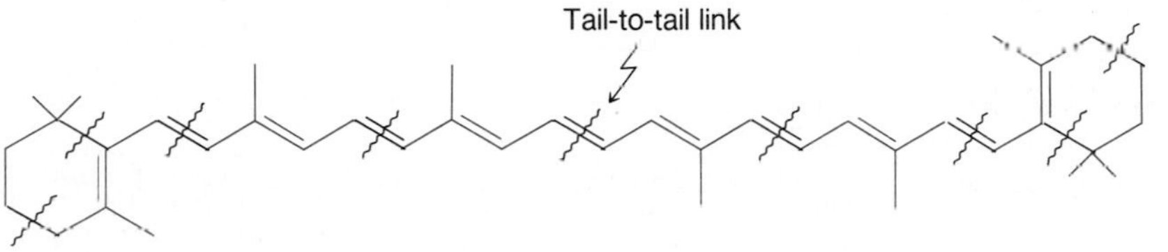

Tail-to-tail link

28.9

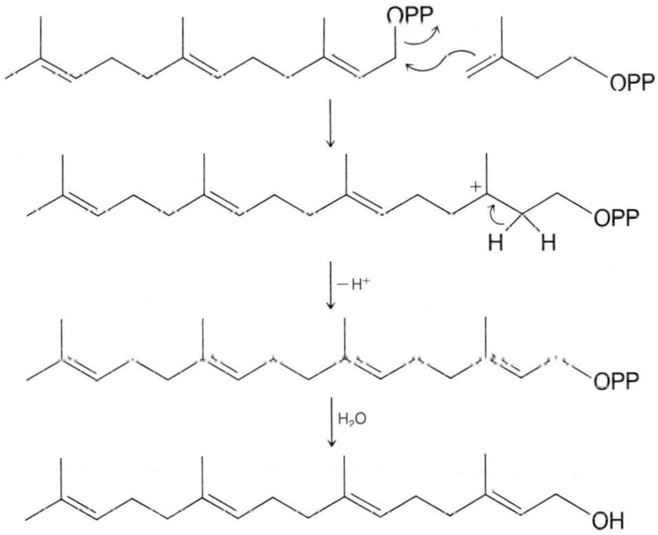

28.10

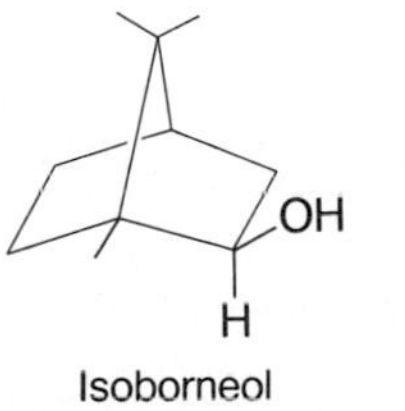

Isoborneol 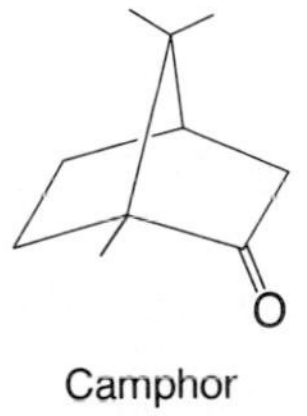   Camphor

28.11   Four carbons would be labeled with ^{14}C, they are C-1, C-3, C-5, and C-7.

28.12   (*b*) Hydrogens that migrate are those originally attached to C-13 and C-17 (steroid numbering).   (*c*) The methyl group attached to C-15 of squalene 2,3-epoxide.   (*d*) The methyl groups at C-2 and C-10 plus the terminal methyl group of squalene 2,3-epoxide.

28.13   All the methyl groups are labeled, plus C-1, C-3, C-5, C-7, C-9, C-13, C-15, C-17, C-20, and C-24 (steroid numbering).

28.14   The structure of vitamin D_2 is the same as that of vitamin D_3 except that vitamin D_2 has a double bond between C-22 and C-23 and a methyl substituent at C-24.

CHAPTER 29

29.1 (b) R; (c) S

29.2 Isoleucine and threonine

29.3 (b) HO—⟨benzene ring⟩—CH₂CHCO₂⁻ with +NH₃ substituent   (c) HO—⟨benzene ring⟩—CH₂CHCO₂⁻ with NH₂ substituent

or

(d) ⁻O—⟨benzene ring⟩—CH₂CHCO₂⁻ with NH₂ substituent

⁻O—⟨benzene ring⟩—CH₂CHCO₂⁻ with +NH₃ substituent

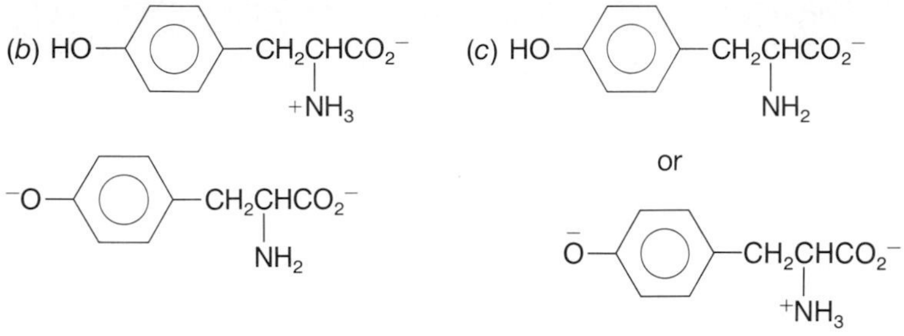

29.4 At pH 1

H₃⁺NCH₂CH₂CH₂CH₂CHCO₂H with +NH₃ substituent

at pH 9

H₃⁺NCH₂CH₂CH₂CH₂CHCO₂⁻ with NH₂ substituent

at pH 13

H₂NCH₂CH₂CH₂CH₂CHCO₂⁻ with NH₂ substituent

29.5 $(CH_3)_2CHCH_2CO_2H \xrightarrow[P]{Br_2} (CH_3)_2CHCHCO_2H \xrightarrow{NH_3} (CH_3)_2CHCHCO_2^-$

(with Br substituent) (with +NH₃ substituent)

29.6 $(CH_3)_2CHCH \xrightarrow[NaCN]{NH_4Cl} (CH_3)_2CHCHCN \xrightarrow[2.\ HO^-]{1.\ H_2O,\ HCl,\ heat} (CH_3)_2CHCHCO_2^-$

(aldehyde with =O) (with NH₂ substituent) (with +NH₃ substituent)

29.7 Treat the sodium salt of diethyl acetamidomalonate with isopropyl bromide. Remove the amide and ester functions by hydrolysis in aqueous acid, then heat to cause $(CH_3)_2CHC(CO_2H)_2$ (with +NH₃ substituent) to decarboxylate to give valine. The yield is low because isopropyl bromide is a secondary alkyl halide, because it is sterically hindered to nucleophilic attack, and because elimination competes with substitution.

29.8

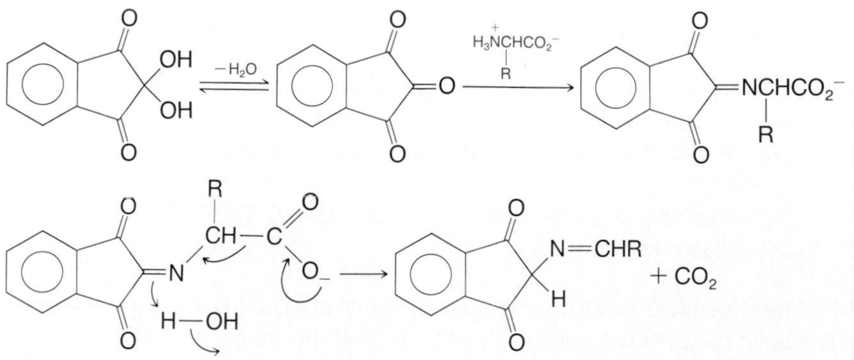

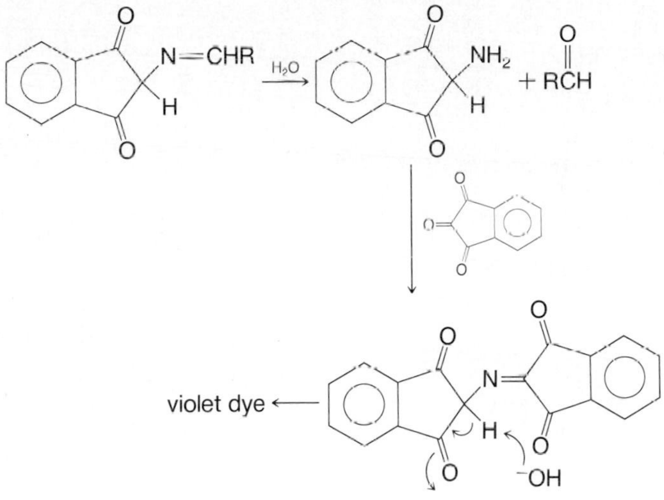

29.9 Glutamic acid

29.10 (b) $H_3\overset{+}{N}CHCNHCHCO_2^-$ (c) $H_3\overset{+}{N}CHCNHCHCO_2^-$ (d) $H_3\overset{+}{N}CH_2CNHCHCO_2^-$

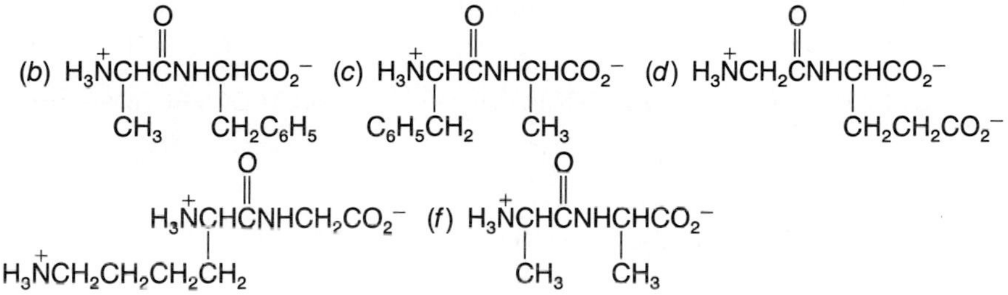

CH_3 $CH_2C_6H_5$ $C_6H_5CH_2$ CH_3 $CH_2CH_2CO_2^-$

(e) $H_3\overset{+}{N}CHCNHCH_2CO_2^-$ (f) $H_3\overset{+}{N}CHCNHCHCO_2^-$

$H_3\overset{+}{N}CH_2CH_2CH_2CH_2$ CH_3 CH_3

29.11 (b)

(c)

(d)

(e)

(f)

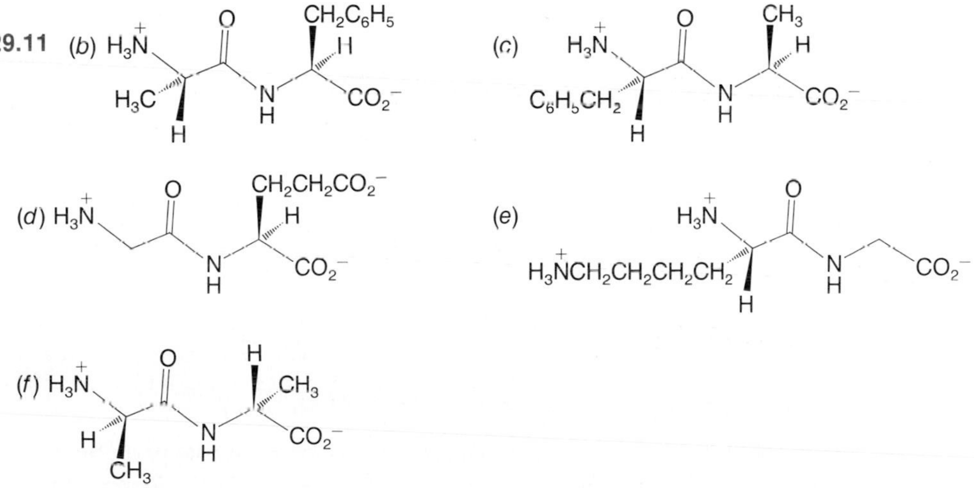

29.12 Tyr-Gly-Gly-Phe-Met

29.13 Val; Phe; Gly; Ala; Val-Phe; Phe-Gly; Gly-Ala; Val-Phe-Gly; Phe-Gly-Ala

29.14

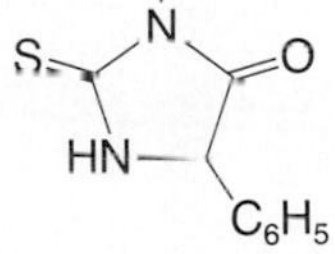

29.15 $H_3\overset{+}{N}CHCNHCHCOCH_3$
with O above both C=O groups, $^-O_2CCH_2$ below first CH, $CH_2C_6H_5$ below second CH

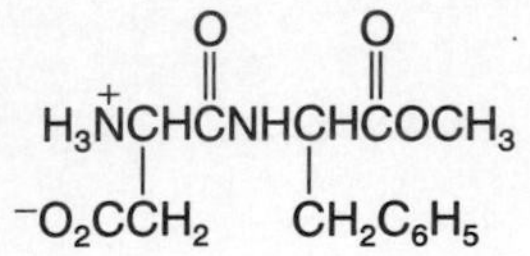

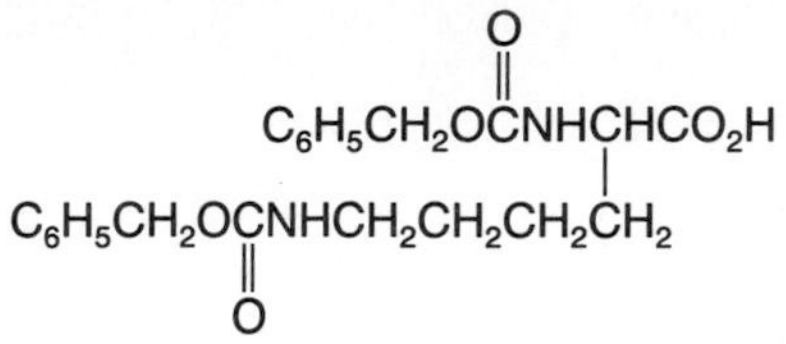

29.16 $C_6H_5CH_2O\overset{O}{C}NHCHCO_2H$

$C_6H_5CH_2O\overset{O}{C}NHCH_2CH_2CH_2CH_2$

29.17 $H_3\overset{+}{N}CHCO_2^- + C_6H_5CH_2O\overset{O}{C}Cl \longrightarrow C_6H_5CH_2O\overset{O}{C}NHCHCO_2H$
with CH_3 below

$H_3\overset{+}{N}CHCO_2^- + C_6H_5CH_2OH \xrightarrow[\text{2. HO}^-]{\text{1. H}^+,\text{ heat}} H_2NCHCO_2CH_2C_6H_5$
with $(CH_3)_2CHCH_2$ below both

$C_6H_5CH_2O\overset{O}{C}NHCHCO_2H + H_2N\overset{O}{C}HCOCH_2C_6H_5 \xrightarrow{\text{DCCI}} C_6H_5CH_2O\overset{O}{C}NHCHC\overset{O}{N}HCHCOCH_2C_6H_5$
with CH_3 $(CH_3)_2CHCH_2$ below left; CH_3 $CH_2CH(CH_3)_2$ below right

$C_6H_5CH_2O\overset{O}{C}NHCHC\overset{O}{N}HCHCOCH_2C_6H_5 \xrightarrow{\text{H}_2 \atop \text{Pd}} \text{Ala-Leu}$
with CH_3 $CH_2CH(CH_3)_2$ below

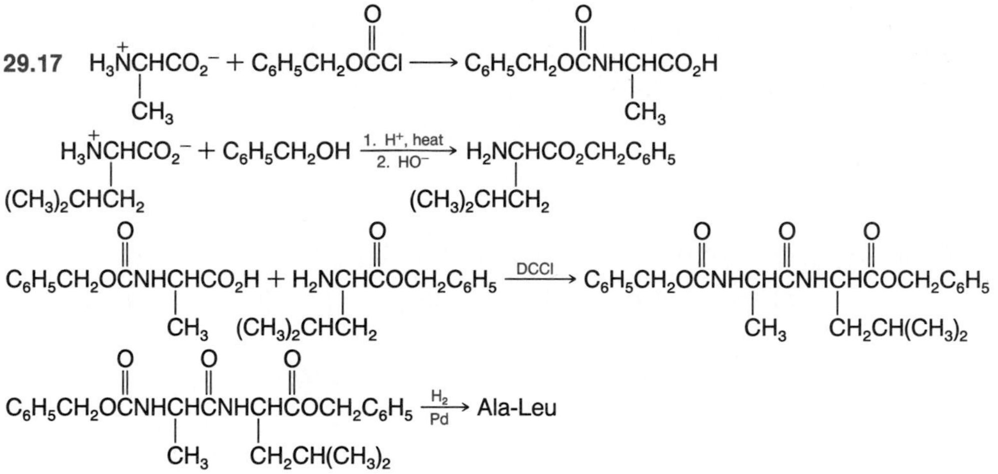

29.18   An *O*-acylisourea is formed by addition of the Z-protected amino acid to *N,N'*-dicyclohexylcarbodiimide, as shown in Figure 29.13. This *O*-acylisourea is attacked by *p*-nitrophenol.

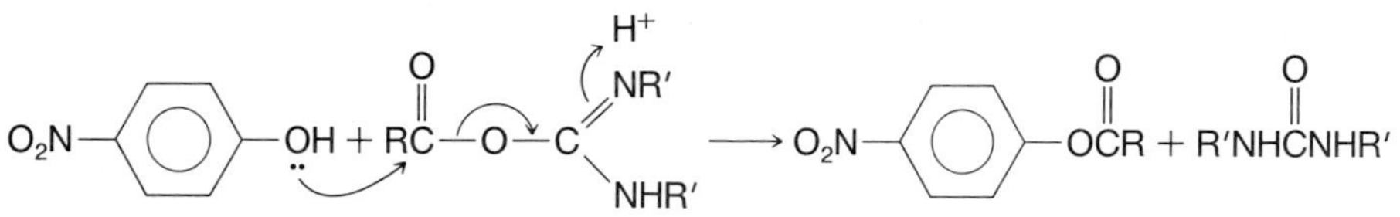

$O_2N\text{—}\langle\text{ring}\rangle\text{—OH} + R\overset{O}{C}\text{—O—}C\overset{NR'}{\underset{NHR'}{}} \longrightarrow O_2N\text{—}\langle\text{ring}\rangle\text{—O}\overset{O}{C}R + R'NH\overset{O}{C}NHR'$

29.19   Remove the Z protecting group from the ethyl ester of Z-Phe-Gly by hydrogenolysis. Couple with the *p*-nitrophenyl ester of Z-Leu, then remove the Z group of the ethyl ester of Z-Leu-Phe-Gly.

29.20   Protect glycine as its Boc derivative and anchor this to the solid support. Remove the protecting group and treat with Boc-protected phenylalanine and DCCI. Remove the Boc group with HCl, then treat with HBr in trifluoroacetic acid to cleave Phe-Gly from the solid support.

29.21

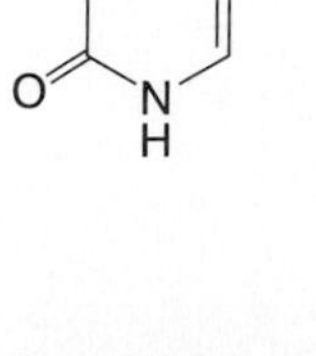

5-fluorouracil structure (pyrimidine ring with two C=O, HN and NH, and F substituent)

29.22 (*b*) Cytidine   (*c*) Guanosine

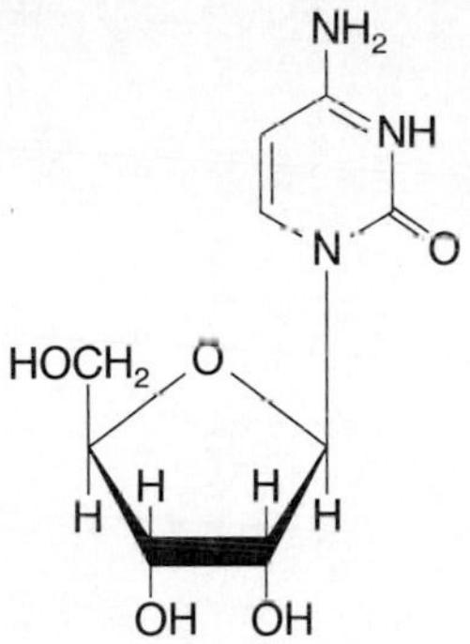

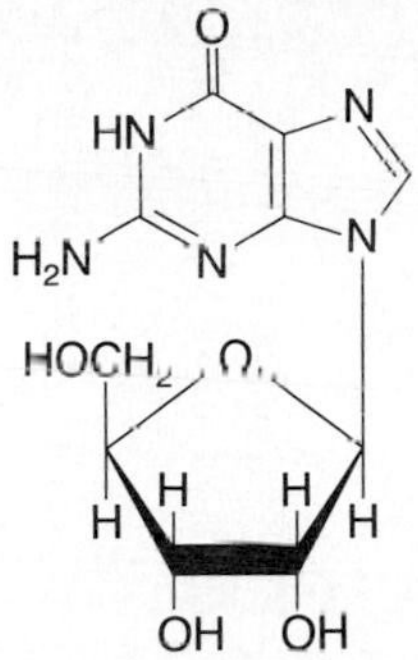

29.23 The codons for glutamic acid (GAA and GAG) differ by only one base from two of the codons for valine (GUA and GUG).

INDEX